Advanced Engineering Mathematics

Advanced Engineering Mathematics

Robert J. Lopez

Rose-Hulman Institute of Technology

Boston San Francisco New York
London Toronto Sydney Tokyo Singapore Madrid
Mexico City Munich Paris Cape Town Hong Kong Montreal

Sponsoring Editors: Carolyn Lee-Davis, Laurie Rosatone
Assistant Editor: RoseAnne Johnson
Project Editor: Rachel S. Reeve
Managing Editor: Karen Guardino
Marketing Manager: Michael Boezi
Editorial and Production Services: Techsetters, Inc.
Technical Art Specialist: Joseph K. Vetere
Senior Media Producer: Marlene Thom
CD Programmer: Greg Faron, Integre Technical Publishing Co., Inc.
Prepress Supervisor: Caroline Fell
Manufacturing Buyer: Evelyn Beaton
Senior Designer: Barbara T. Atkinson
Interior Design: Sandra Rigney
Cover Design: Night & Day Design
Cover Photograph: VCG/FPG International LLC

The fonts euex 10 and msam 10 are used in the Maple® worksheets
with permission from Richard J. Kinch, TrueTeX Typesetting Software
(http://www.truetex.com).

Maple is a registered trademark of Waterloo Software, Inc.

Library of Congress Cataloging-in-Publication Data

Lopez, Robert J., 1941-
    Advanced engineering mathematics / Robert J. Lopez.
    p. cm
    Includes bibliographical references and index.
    ISBN 0-201-38073-0
    1. Engineering mathematics. I. Title.

TA330.L67 2000
515′.02462–dc21

00-032752

Contents

The following optional unit is available on the CD-ROM enclosed with this book; on the book's Web site; and can be ordered through Addison-Wesley (ISBN 0-201-72204-6).

Preface

This is a book of applied and engineering mathematics, written in traditional notation and language, for students of science, engineering, and applied mathematics. It contains examples drawn from a wide spectrum of physical and mathematical disciplines, and provides a fairly complete curriculum in undergraduate applied mathematics.

In this text, results are typically stated "up front," either informally, or formally as a theorem, then illustrated with examples before being proved or verified. A conscious effort has been made to ensure that students will understand what a theorem is saying, *before* they are subjected to its proof. While this is not the standard ordering one finds in math texts in general, it is the ordering often found in classrooms where applications rule.

Nearly 7000 exercises are available for student practice and enrichment. Nearly all are completely solved in the *Instructor's Manual*.

This text is the outgrowth of more than ten years of using Maple in the classroom to teach science and engineering students courses in calculus, differential equations, linear algebra, boundary value problems, advanced calculus, vector calculus, complex variables, and statistics. The materials were conceived in laboratory/classrooms where each student sat at a desktop machine, and nurtured in an environment where each student carries a laptop computer the way we old-timers carried our slide rules. Over the years, the materials were presented at various stages in journal articles, conference talks, workshops, and seminars.

Throughout, the theme guiding their development has been the realization that modern computer algebra systems and software pose a new paradigm for teaching, learning, and doing applied mathematics. Indeed, the phrase "new apprenticeship" echoes in the writings and talks leading up to this present volume. It is no longer enough to acknowledge the power of such software while still adhering to the paradigm of pencil-and-paper, and the chalkboard.

Paralleling the text, therefore, is a collection of Maple worksheets which implement all the calculations and derivations found described in this book. In fact, each of the book's 273 sections is mirrored in a worksheet that includes both the prose and the mathematics, the mathematics being "live" in the worksheet. Students reading the text can have the parallel discussion on their computer screens, and can execute in Maple, the calculations the text is describing.

Yet, it is entirely possible to "lecture" from this text. There is ample opportunity for an instructor to reproduce calculations and derivations summarized in these pages. While such lectures are being delivered, it is hoped that students will interact with the material by using a computer algebra system to interpret the mathematics, and to work the exercises.

Two-thirds of the exercise sets are divided into problems of types A and B. Problems of type A are more conceptual, and less demanding computationally. The A-problems would be the ones a student might work by hand if they found that to be an effective way to learn. The B-problems are generally more computationally intensive. It is anticipated that technological tools of some sort will be used freely when working these exercises.

Not all in the math, science, and engineering communities have embraced the use of technology as the operational instrument for meeting, and mastering, mathematics. Many times, both on my own campus and on others, I have had to articulate the case for the active use of technology as the learning agent in math courses. Typically, I would try to show by examples how technology has improved pedagogy. Sometimes, I would say something like "The course of instruction at a school for operators of earth-moving machinery should not end in a test of dexterity with a shovel, nor should admission be limited to those capable of digging a ditch with one." But my favorite analogy is that of the Magic Skates, born of my experiences at the ice-hockey rinks in Canada where I lived for twelve years.

If you can't skate, you can't play hockey. Only youngsters who master the art of skating can experience the game of hockey. But suppose a poor skater acquires a pair of magic skates which transform the wearer into an adept skater capable of experiencing the thrill of the game of hockey. Is it viable to argue that the magic skates invalidate the player's ensuing encounter with the game?

This text embraces the magic skates of computer algebra systems. Every volunteer hockey coach I knew back in Canada would have paid for pairs of magic skates from their own pockets, just to see their teams play a better game of hockey. It was the game that mattered, the play, the experience, the participation in a really exciting sport. And if we can't make our students feel the same way about applied and engineering mathematics, our programs will retain only the dwindling handful willing to make the 5:30 AM practice before school.

Distinctive Features

1) New Paradigm   Access to computer algebra tools can be assumed throughout, and the pedagogy can be predicated on its availability and use. Although the text is written in traditional notation, its structure reflects the author's experience in using a computer as an active partner in teaching, learning, and doing applied and engineering mathematics.

2) New Apprenticeship   Insights into the deep results of classical applied mathematics are extracted from examples, as much by calculations and graphics as by subtle reasoning. This text shows how to use modern software tools to learn, do, and interpret applied and engineering mathematics.

3) Flexibility   A computer algebra system allows the instructor the option to bypass certain drills in skill-building, to concentrate on key ideas. Therefore, topics can be reordered more easily whenever supporting computations can be relegated to the computer.

4) "Big Picture" First   Reflecting the author's own learning style, most presentations begin with the "big picture," with computations and supporting graphics given first. Then, when the goal is clear, the supporting calculations and relationships are developed.

5) Parallel Worksheets   The 273 sections of the text are paralleled by a matching number of Maple worksheets containing, not only the calculations and graphics of the section, but also the text and explanations. The student using the Maple worksheet sees more than just a computer dialog. The complete text is included in the worksheets, with detailed explanations of both the mathematics and the Maple commands required to obtain it.

6) *Pervasive Access to Mathematical Tools*   Relying on a computer algebra system allows mathematical tools to be used before they are developed formally in the text. For example, numerical evaluation of integrals occurs well before the formal treatment of numeric integration in Unit Eight. Eigenvalues are computed numerically in Unit Three, before the chapters on numerical methods.

7) *Complete Integration of Numeric and Symbolic Results*   Numeric results are interwoven with symbolic calculations throughout the text. Numeric solutions for differential equations appear early enough to be used throughout the study of models based on differential equations. The perturbation techniques of Poincare, Lindstedt, and Krylov-Bogoliubov are in the same unit as the second-order IVP. Collocation, Rayleigh-Ritz, and Galerkin techniques for solving BVPs are contiguous with analytic techniques, and with finite-difference, finite-element, and shooting techniques. Later, in Unit Five, numeric methods for solving PDEs also appear in conjunction with the more classical symbolic results.

8) *Early Appearance of the Laplace Transform*   The Laplace transform as a tool for solving IVPs for ordinary differential equations appears in Chapter 6. This makes it available in Unit Three where systems of ODEs are studied.

9) *Integration of Matrix Algebra with Systems of ODEs*   Systems of first-order linear ODEs motivate and drive vector and matrix manipulations. Chemical mixing tanks provide the model, the Laplace transform is used to obtain solutions, and the vector-matrix structure in the model and its solution is deduced. This motivates a study of the eigenvalue problem, and leads to the fundamental matrix, first via the Laplace transform, then as the exponential of a matrix. Necessary matrix algebra is developed in the context of linear systems of ODEs.

10) *Two Types of Exercises, Part A and Part B*   The exercises in approximately two-thirds of the sections are divided into two categories. The A-exercises are generally more conceptual, and can be done without a suite of computer tools. The B-exercises generally presuppose access to appropriate computer tools, and provide both practice for the section and generalizations beyond the text.

11) *A Unit on Series*   A unit discussing power series, Fourier series, and asymptotic series sits between the two units on ordinary differential equations. Solutions represented in these forms then appear in Unit Three, the second unit on ODEs.

12) *A Unit on the Calculus of Variations*   A unit on the Calculus of Variations (Unit Nine) is available as a supplement.

13) *Socratic Chapter Reviews*   The many questions (rather than new exercises) in a Chapter Review aid the student in organizing the chapter's material.

Supplements

Unit 9: Calculus of Variations (0-201-72204-6)

This unit includes the chapters "Basic Formalisms," "Constrained Optimization" and "Variational Mechanics."

Instructor's Technology Resource & Solutions Manual (0-201-71001-3)

This manual includes:

- Introduction to & Tips for Maple®
- Introduction to & Tips for Mathematica®
- Solutions to A Exercises
- A CD-ROM in the back of the manual includes:
  - Fully worked solutions to B exercises in Maple® Worksheets
  - Fully worked solutions to B exercises in Mathematica® Notebooks
  - Free Mathematica® Reader

Student's Technology Resource & Solutions Manual (0-201-71004-8)

This manual includes:

- Introduction to & Tips for Maple®
- Introduction to & Tips for Mathematica®
- Solutions to Selected A Answers
- A CD-ROM in the back of the manual includes:
  - Fully worked selected solutions to B exercises in Maple® Worksheets
  - Fully worked selected solutions to B exercises in Mathematica® Notebooks
  - Free Mathematica® Reader

Web Site (http://www.awl.com/lopez)

- Unit Nine in PDF files
- Unit Nine in Maple® Files
- Chapter 1 of the Student's Technology Resource & Solutions Manual in PDF files
- Additional resources for instructors and students

Acknowledgments

I would like to express my thanks and appreciation to the reviewers who provided their time and advice:

John E. Beard	Michigan Technological University
Hongwei Chen	Christopher Newport University
Steve Dunbar	University of Nebraska-Lincoln
Amitabha Ghosh	Rochester Institute of Technology
Constant J. Goutziers	SUNY at Oneonta

Ronald B. Guenther	Oregon State University
Mark Haugan	Purdue University
Vernon Howe	La Sierra University
Jim Powell	Utah State University
Joseph Mahaffy	San Diego State University
Douglas B. Meade	University of South Carolina
Amnon J. Meir	Auburn Unversity
Jeffrey J. Morgan	Texas A&M University
William F. Moss	Clemson University
Vikas Prakash	Case Western Reserve University
C. Roth	McGill University
Thomas Stanford	University of South Carolina
Ross Taylor	Clarkson University
Stanley Zietz	Drexel University

The following colleagues checked the text for accuracy and I appreciate their care in detecting errors in the earlier drafts.

Gregory T. Adams	Bucknell University
D. Chris Arney	United States Military Academy
Neil Berger	University of Illinois at Chicago
Warren Burch	Brevard Community College-Cocoa Campus
Philip S. Crooke	Vanderbilt University
Elias Deeba	University of Houston-Downtown
William Siegmann	Rensselaer Polytechnic Institute
Jeff A. Suzuki	Dorchester, MA
James Thomas	Colorado State University
Marie M. Vanisko	Carroll College
Stephen Whalen	University of Minnesota-Twin Cities

An explicit acknowledgment is appropriate for Addison Wesley Longman, the company, and its staff who believed in this project and supported it from the very beginning. Without that support, this book would still be just an idea.

Douglas B. Meade, Department of Mathematics, University of South Carolina, Columbia, deserves thanks for updating the original Maple V Release 4 worksheets for this text, and for proof reading the exercises and the text's answer section.

Dale Doty, Donna Farrior, and Shirley Pomeranz, all at the University of Tulsa, also deserve thanks for providing Mathematica solutions for the B exercises.

It is a special satisfaction to acknowledge the work of Constant J. Goutziers, Department of Mathematical Sciences, SUNY College at Oneonta, who developed the answer section in the text, and in so doing, helped improve many of the exercises. I am pleased to call him colleague and friend.

Unit One

Ordinary Differential Equations—Part One

The typical calculus student is familiar with separable differential equations arising from exponential growth and decay models, the logistic model, and Newton's law of cooling. The techniques for solving these equations belong to the arena of the calculus. After a brief reminder of these ideas, including a review of the technique of separation of variables, we continue with a study of analytic methods for solving first-order differential equations. The variable-volume mixing tank is modeled with a first-order linear equation whose solution we examine from the perspectives of integrating factors and variation of parameters.

Many differential equations encountered in the applications cannot be solved with analytic techniques. Hence, we study a variety of numerical methods for solving first-order differential equations. Not so long ago a numeric method had to be programmed by each user. Now, numeric solutions of differential equations are available in preprogrammed software packages and on hand-held calculators. We rely on access to these newer technologies and slant our discussion on the conceptual bases for the numeric techniques presented.

We particularly emphasize the meaning of order of the numeric method and consider empirical tests for determining or verifying the order of a method. For example, the simple Euler method is first order, whereas the slightly more complex rk4 method is fourth order. We are more interested in the difference between a first-order and a fourth-order method than we are in the actual computer coding of the algorithms.

A second concept that is important for numerical techniques is that of error control. We explain how a numeric solver for a differential equation can be made to monitor the truncation error, and we show how this information can be used to return a solution of a guaranteed precision.

Linear, second-order, constant-coefficient differential equations occupy a preeminent position in engineering and applied science because they describe the oscillations of a spring-mass system that has both frictional losses and applied forces. They equally well describe the vibrations in an automobile shock-absorber and the oscillation of current in an electric circuit containing resistance, inductance, and capacitance. These equations, then, occupy a preeminent position in this text, and their discussion comprises the remaining two chapters in this unit.

In addition to the chapter devoted specifically to the linear, second-order, constant-coefficient differential equation, we also devote a chapter to the Laplace transform, a technique for converting linear differential equations into algebraic equations. This technique for solving differential equations is particularly effective in solving spring-mass and electric-circuit problems, especially when there are driving forces or voltages that are piecewise defined. The Laplace transform is but one example of an integral transform and provides a basis for mastering the Fourier transform later on in Unit Five. In addition, the Laplace transform is a very important component in the design of feedback control systems in all forms of engineering practice.

Chapter **1**

First-Order Differential Equations

INTRODUCTION   In this first chapter we will learn what ordinary differential equations are and how they are classified. We will see examples of their use in modeling physical systems. We will learn how to verify that a given function is a solution of a differential equation and what the general solution of a differential equation is. We will also introduce the initial value problem and the boundary value problem.

We will see the direction field determined by a first-order differential equation, and we will learn one theorem that indicates when an initial value problem has at least one, or exactly one, solution. We will also see Picard's iterative technique by which a solution of the first-order equation can, in theory, be constructed.

1.1 Introduction

Why Differential Equations?

Three of the nine units in this text are devoted to *differential equations;* and of the three, two are devoted to *ordinary* differential equations. The astute reader immediately asks the following two questions.

1.   Why do differential equations occur in engineering and science?
2.   Why bother with differential equations? What do they say about physical systems?

Answers

QUESTION 1   Knowledge of the physical universe derives from observation of changes, not of absolute quantities. Forces are determined not directly, but rather by measuring displacements, or changes, in lengths of springs; the dynamics of a system are inferred from observations of changes in position; distances are determined by measuring changes in angles, lengths, and light intensities.

Newton's Second Law of Motion, $\mathbf{F} = m\mathbf{a}$, contains the acceleration, which is the rate of change of velocity, itself a rate of change of position. Rates of change are derivatives, velocity being a first derivative of displacement, and acceleration being the second. Derivatives are inherent in a Newtonian formulation of the laws of motion.

As seen in elementary calculus, a rate of change is an instantaneous measure of a change per unit time. This is the derivative, the limiting value of the ratio of the change in a quantity divided by the length of the time interval over which the change has taken place. It is important to avoid confusing the notion of the *amount* of change with the *rate* at which the change took place.

If the rate of change is known, that is, if the derivative is known, then an approximation of the amount of change in a given interval of time is the *differential*. For example, if along a fixed axis, $x(t)$ measures the displacement of an object from a fixed origin O, then x' is the *velocity*, and the differential $dx = x' \, dt$ approximates the *displacement* that occurs during the interval dt. The *change in position* is approximately the differential dx, but the *rate of change* in position is exactly the derivative $\frac{dx}{dt}$.

Derivatives also appear in engineering and science, courtesy of conservation laws, or laws of balance. In ordinary circumstances, the amount of matter in a closed system remains a constant. If the amount of matter in the system changes, it must be that matter has entered or left the system. In fact, the change in the content of the system must be accounted for by the passage of content through the boundary of the system. Application of a conservation law usually involves derivatives, or rates of change. The *rate* of change of the conserved quantity is balanced against the *rate* of passage through the boundary of the system. Often, this boundary passage is determined by the rate of *inflow* less the rate of *outflow*.

Examples of the formulation of differential equations are given in this chapter, in Chapter 2, and in Unit Three. Such differential equations are said to be *mathematical models* of a physical system, and the act of formulating the appropriate equation is called *modeling*.

QUESTION 2   The examples below show not only how a model is framed by differential equations, but also what kinds of information can be extracted from the model. The purpose of the differential equation is not the solution, but rather the information contained in the solution. As we shall see, it is sometimes easier to extract information directly from the differential equation than from the solution. And every solution need not be an exact formula. Sometimes a numeric solution suffices, sometimes an approximate analytic expression.

The differential equation is written to model a system in the real world. If the model is valid, it contains information about the physical system, and the mathematical techniques for analyzing differential equations become the tools by which that information is extracted from the model. Proficiency in these techniques should be prompted by the desire to understand, design, and control physical systems.

EXAMPLE 1.1   **Mechanical work**   In elementary physics, the mechanical work done by a constant force acting along a line is defined by $W = fx$, the product of the force f and x, the distance through which it acts. Using calculus, this definition is generalized to the case of the variable force $f(x)$ via the definition $W = \int_{x_1}^{x_2} f(x) \, dx$. This expression is obtained by arguing that over a small interval dx, the force is essentially the constant value $f(x)$, so it does the differential amount of work $dW = f(x) \, dx$. If all such differential amounts of work are summed, the integral formula results. However, the integral formula for work is really the solution of the differential equation

$$\frac{dW}{dx} = f(x) \tag{1.1}$$

in which the derivative $\frac{dW}{dx}$ becomes the differential dW by multiplying through by the increment dx.

The force a spring exerts on an object is proportional to the stretch of the spring, resulting in the relationship $f(x) = -kx$. The minus sign indicates that the force opposes the stretch. Solving (1.1) tells us that the positive work done (on the system) when stretching the spring from its equilibrium length to x is $W = \int_0^x ky \, dy = \frac{1}{2}kx^2$.   ❖

EXAMPLE 1.2

Harmonic oscillator   A spring is hung from a support, and a weight is attached and allowed to sag to equilibrium. Displacements of the weight upward from equilibrium are measured as positive. The weight is raised one unit and released, thereafter undergoing periodic motion. The weight is then raised two units and again released, again undergoing periodic motion.

   The differential equation governing the displacement of the weight is

$$y'' + y = 0$$

with $y(0) = A$, $y'(0) = 0$ representing the starting conditions for the weight. Of course, A assumes the values 1 or 2. In Figure 1.1, graphs of the two solutions show the weight experiences periodic motion. In each case, the amplitude of the oscillation is the same as the initial displacement, but the frequencies of the motions are the same. The frequency is *independent* of the starting amplitude.

   This is not the case for the differential equation

$$y'' + y + y^3 = 0$$

In Figure 1.2, graphs of two motions starting, respectively, with initial displacements of 1 and 2, show that while the amplitudes remain 1 and 2, the frequencies are now different. The frequency is *dependent* on the starting amplitude.   ❖

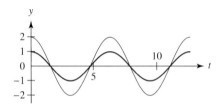

FIGURE 1.1   Solutions of $y'' + y = 0$, different amplitudes, same frequency

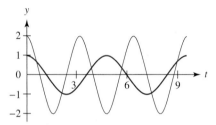

FIGURE 1.2   Solutions of $y'' + y + y^3 = 0$, different amplitudes, different frequencies

   We have demonstrated a fundamental difference between a linear and a nonlinear oscillator. The linear oscillator is the subject of Chapter 5, whereas the nonlinear oscillator appears in Sections 14.4–6. Definitions of *linear* and *nonlinear* appear in Section 1.2, although the appearance of the term y^3 in the second equation signals the meaning of *nonlinear*.

EXAMPLE 1.3

Law of mass action   In chemistry, a *mole* of a substance is that quantity which contains Avogadro's number (6.023×10^{23}) of molecules. The amount in a mole is related to the atomic number. For example, a mole of hydrogen atoms (H) has a mass of 1 gram while a mole of oxygen (O) has a mass of 16 grams. A mole of water (H_2O) has a mass of 18 grams.

   Using a catalyst and proper temperature control, the conversion of hydrogen and oxygen to water can take place at a rate considerably slower than explosive. Suppose 40 grams of oxygen (O) and 8 grams of hydrogen (H) are combined to form water. If the amount of water present at time t is given by $x(t)$ and it is assumed (the law of mass-action) that the rate of change of the amount of water is proportional to the product of the amounts of hydrogen and oxygen remaining, the differential equation

$$\frac{dx}{dt} = k\left(40 - \frac{16}{18}x\right)\left(8 - \frac{2}{18}x\right) \tag{1.2}$$

governs the reaction. (If at a particular moment there are 18 grams of water, that is, $x = 18$, then 16 of those grams came from oxygen and 2 came from hydrogen.)

   Since there is no water initially, $x(0) = 0$. she solution of the differential equation that satisfies this condition is

$$x(t) = 360\frac{e^{-8kt/3} - 1}{5e^{-8kt/3} - 8}$$

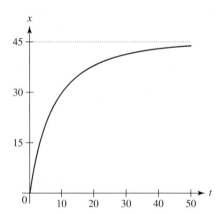

FIGURE 1.3   Formation of water from hydrogen and oxygen

(The differential equation is actually separable and can be solved with the techniques of elementary calculus discussed in Section 3.1.) If an observation provides the data that at time $t = 1$ there were 6 grams of water, then the constant of proportionality $k = \frac{3}{8} \ln \frac{55}{52}$ is determined, so

$$x(t) = 360 \frac{\lambda^t - 1}{5\lambda^t - 8} \qquad \lambda = \frac{52}{55}$$

The limiting value of the amount of water formed is $\frac{360}{8} = 45$ grams, or $\frac{5}{2}$ moles. Since the 40 grams of oxygen are also $\frac{5}{2}$ moles, all the oxygen is consumed. The 5 grams of hydrogen consumed are 5 moles of H but $\frac{5}{2}$ moles of binary hydrogen, H_2. Figure 1.3 shows the asymptote $x = 45$ and the dynamic history of the reaction. Thus, from (1.2) and an observation of $x(t)$, the rate constant k for this reaction can be determined. ❖

EXAMPLE 1.4   **Models of growth and decay**   If $y(t)$ measures the size of a population and it is assumed that the rate of change of the size is proportional to the size, then the differential equation

$$\frac{dy}{dt} = ry$$

models the population. For growth, the constant r is positive and is called the *growth-rate* constant. The solution of the equation is $y(t) = y_0 e^{rt}$, the law of exponential growth that predicts the population will grow without bound.

In an effort to make the model more realistic, the growth-rate constant r is modified to become a function of the population. Thus, the growth rate is allowed to vary, becoming zero as the population grows. This self-limitation is observed in closed biological systems when over-crowding or resource-depletion occurs.

The simplest model incorporating self-limiting growth is the *logistic model* in which the growth rate is taken as $r = k(a - y)$. When $y(t)$ is small, the growth-rate factor is nearly constant at ka. As the population approaches the value a, the growth rate approaches zero. The resulting model is then

$$\frac{dy}{dt} = k(a - y)y \qquad (1.3)$$

For example, taking $k = a = 1$ gives the differential equation $y' = (1 - y)y$, for which solutions satisfying the initial conditions of the form $y(0) = \alpha, 0 < \alpha < \frac{3}{2}$, are displayed in Figure 1.4. The solutions $y(t) = 0$ and $y(t) = 1$ are called *equilibrium solutions* because for them $y' = 0$ and $y(t)$ remains constant. Neighboring solutions tend to the *stable* solution $y = 1$ and away from the *unstable* solution $y = 0$.

The model expressed by (1.3) is more a statement of a theory than it is a device for predicting a population size. If experimental data shows (1.3) is a valid description of the dynamics of a population, the theorist then has confirmation that the appropriate features of the biological system have been identified. However, (1.3) does not account for the mechanisms that actually inhibit continued growth. Whether the population runs out of food or whether physical crowding actually inhibits reproduction remains unknown. The model only indicates that making the rate constant a function of population size can account for observed self-limitation in population sizes.

As studied in Section 2.2, we next consider the effect of harvesting the self-limited population. Harvesting at a constant rate h modifies the differential equation to

$$\frac{dy}{dt} = (1 - y)y - h$$

For each value of the harvesting rate h, there is an equilibrium population y determined by the algebraic equation $y' = 0$, that is, by the quadratic equation $y^2 - y + h = 0$. Figure 1.5

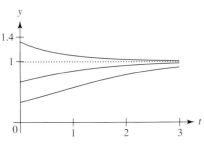

FIGURE 1.4   Solutions of $y' = (1 - y)y$

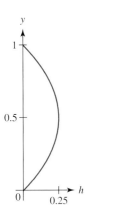

FIGURE 1.5   Equilibrium solutions for the harvested logistic model

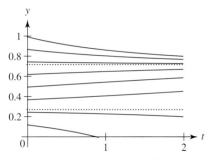

FIGURE 1.6   Solutions of the logistic model with $h < h_{\max}$

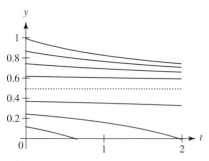

FIGURE 1.7   Solutions of the logistic model with $h = h_{\max}$

contains a graph of y against h and shows that for each value of $h < \frac{1}{4}$ there are two possible equilibrium populations. Thus, if the population starts at the equilibrium solution $y = 1$ with no harvesting and over a period of years is subjected to gradually increasing harvesting rates, it can remain at slowly decreasing equilibria. The maximum harvest rate at which an equilibrium can exist is $h = \frac{1}{4}$, corresponding to $y = \frac{1}{2}$.

It is extremely risky to attempt harvesting at this rate because if for some reason the population falls slightly below $y = \frac{1}{2}$, then it will rapidly become extinct. Indeed, consider the harvesting rate $h = \frac{1}{5}$ for which the corresponding equilibrium populations are $\frac{5 \pm \sqrt{5}}{10}$ or, approximately, 0.72 and 0.28. Figure 1.6 shows these equilibrium solutions amongst others generated with this harvesting rate. The two equilibrium solutions are the horizontal lines. The larger equilibrium solution is stable, since nearby solutions tend toward it as time increases. However, the smaller equilibrium solution is unstable, since nearby solutions move away from it as time increases. Solutions slightly above the smaller equilibrium solution tend toward the larger equilibrium, while solutions below the smaller equilibrium solution tend toward extinction.

In Figure 1.7 the horizontal line at $y = \frac{1}{2}$, the only equilibrium solution, corresponds to $h = \frac{1}{4}$, the maximum sustainable harvest rate. Populations that start out larger than $\frac{1}{2}$ decrease toward $\frac{1}{2}$, but populations that start out smaller become extinct. Harvesting at the maximum rate is perilous. If the population drops slightly below the equilibrium level, it becomes extinct.

Of course, if the harvesting rate exceeds $\frac{1}{4}$, the population becomes extinct no matter what the starting level might have been, as shown in Figure 1.8 where $h = \frac{1}{3}$. Models such as these are used by resource managers who are charged with regulating national and international fisheries.   ❖

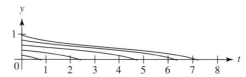

FIGURE 1.8   Solutions of the logistic model with $h > h_{\max}$

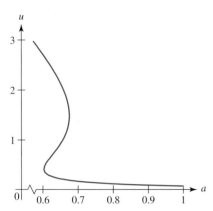

FIGURE 1.9   Equilibrium solutions for the equation of Example 1.5

EXAMPLE 1.5   **Temperatures in a reaction vessel**   In the differential equation

$$\frac{du}{dt} = \frac{9 - u}{a} e^{-5/(1+u)} - u = f(u)$$

from [55], the quantity $u + 1$ is the dimensionless temperature in a reaction vessel through which flows a stream of reactant with constant concentration and temperature. The analysis of the behavior of this system is similar to that of the harvested logistic in Example 1.4. The equilibrium solutions determined by $u' = 0$ are the roots of the equation $f(u) = 0$. For each value of the parameter a, proportional to the flow rate through the reaction chamber, there is an equilibrium solution u_a. The points (a, u_a) lie on the curve shown in Figure 1.9. At the two turning points (see Table 1.1) on the graph of u_a there are two equilibrium solutions. For values of a between the turning points, there are three equilibrium solutions. Elsewhere, there is one equilibrium solution.

By adjusting the flow rate, the parameter a can be varied. Hence, it is possible to operate the reaction vessel with different equilibrium temperatures. Table 1.2 lists five equilibrium solutions corresponding to values of $a = 0.5, 0.65, 0.7$, respectively, to the left of the turning points, between the turning points, and to the right of the turning points.

u_a	$\frac{3}{7}$	$\frac{3}{2}$
a	$20e^{-7/2} \doteq 0.6039$	$5e^{-2} \doteq 0.6767$

TABLE 1.1   Turning points for u_a in Example 1.5

a	0.5	0.65	0.7
u	$u_1 = 3.65$	$u_2 = 2.17$ $u_3 = 0.92$ $u_4 = 0.24$	$u_5 = 0.185$

TABLE 1.2   Equilibrium solutions in Example 1.5

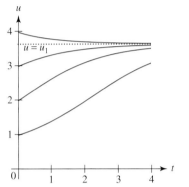

FIGURE 1.10   Solutions corresponding to $a = 0.5$ in Example 1.5

Figure 1.10 shows solutions corresponding to $a = 0.5 < 0.6039$. The horizontal line is the stable equilibrium solution $u = u_1$ that "attracts" solutions starting both above and below it. Figure 1.11 shows solutions corresponding to $a = 0.65 \in (0.6039, 0.6767)$. Both u_2 and u_4 are stable equilibria since solutions starting on either side tend toward them. The equilibrium solution $u = u_3$ is an unstable equilibrium because solutions starting on either side tend to move away. Figure 1.12 shows solutions corresponding to $a = 0.7 > 0.6767$. The one equilibrium solution is called *stable* since solutions on either side tend toward it with increasing time.   ❖

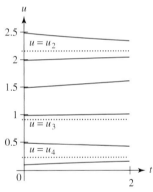

FIGURE 1.11   Solutions corresponding to $a = 0.65$ in Example 1.5

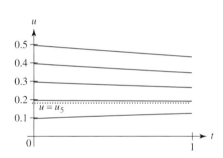

FIGURE 1.12   Solutions corresponding to $a = 0.7$ in Example 1.5

The harvested logistic model is certainly simpler to analyze than this model of the reaction chamber. However, the techniques developed for analyzing the first model guide and enlighten the analysis of the second.

EXAMPLE 1.6   **An epidemic model**   As an epidemic courses through a population, it is discovered that the population divides into two groups: the infected individuals and the rest. The infected individuals are *carriers* and can infect others, while the uninfected individuals are all *susceptible*.

If $i(t)$ is the proportion of infected individuals and $s(t)$ the proportion of susceptible, then $i + s = 1$. Assuming that the infected individuals move freely and uniformly through the population and that the rate of infection is proportional to the product $i(t)s(t)$ as a measure of the number of contacts between infected and susceptible individuals, we have the differential equation

$$\frac{di}{dt} = ki(t)s(t) = ki(t)(1 - i(t))$$

which is nothing more than a logistic model! The equilibrium solution $i(t) = 0$ is unstable, but the equilibrium solution $i(t) = 1$ is stable. Hence, the entire population will contract the disease.   ❖

Using similar techniques, Daniel Bernoulli modeled smallpox epidemics in 1760 and concluded that innoculation would extend the average life expectancy approximately 3 years on an existing expectancy of 27 years 7 months.

EXERCISES 1.1

1. If $x(t) = 5t^3 - 7t^2 - 4t + 3$ is the distance of an object from a fixed origin O as the object moves along a line, find the velocity and acceleration of the object.

2. If $x(t)$ is the distance of an object from a fixed origin O as the object moves along a line and $x''(t) = 4t^2 - 3t + 5$ is its acceleration, find $x(t)$ when $x(0) = 12$ and $x'(0) = -1$.

3. A spring whose natural length is 6 in is stretched to a length of 13 in by a force of 2 lb. Find the spring constant k and the work done (in foot-pounds) during the stretch.

4. The differential equation $x'(t) = (25 - \frac{5}{7}x)(12 - \frac{2}{7}x)$ is obtained by applying the law of mass-action to two substances A and B that react to form C, the amount of which is measured by $x(t)$. Describe the reaction. At equilibrium, $x' = 0$. Determine the amount of C this reaction generates.

5. Interpret the differential equation $y'(t) = y - 2$ as exponential growth with harvesting at a constant rate. Find the equilibrium solution and determine the fate of solutions starting with $y(0) = \frac{3}{2}$ and $\frac{5}{2}$.

6. In the epidemic model of Example 1.6, assume that infected individuals are removed from the population according to the rule $\frac{di}{dt} = -\frac{i}{10}$ and that susceptible individuals are infected according to the rule $\frac{ds}{dt} = -\frac{is}{5}$. If $i(0) = \frac{1}{20}$ and $s(0) = \frac{19}{20}$, obtain $i(t)$ and, hence, $s(t)$. What is the limiting value of $s(t)$?

7. Write a differential equation governing a function $y(x)$ for which the function times its derivative equals the sum of its first and second derivatives.

8. Write a differential equation governing a function $y(x)$ for which the angle made by the x-axis and a line tangent to its graph is twice the distance from the origin to the point of contact.

1.2 Terminology

Differential Equations

An algebraic equation such as $3x + 5 = 7$ is an *open sentence,* the truth or falsity of which is determined for each individual value ascribed to x, the placeholder or "opening" in the equation. This placeholder is usually called the *variable* in the equation, and the process of finding values for x that make the sentence true is called *solving the equation.* Another name for the variable x is the *unknown* in the equation.

Similarly, an equation such as

$$3\frac{dy}{dx} + y(x) = 7$$

containing one or more derivatives of an unknown function is called a *differential equation* (DE). *Ordinary* differential equations (ODEs) contain derivatives of functions of one variable, while *partial* differential equations (PDEs) contain derivatives of functions of two or more variables. Units One and Three are devoted to ordinary differential equations, whereas Unit Five treats the important partial differential equations of engineering and mathematical physics.

The notation $y = f(x)$ indicates that the values of the variable y are determined by the function f. The notation y' or $f'(x)$ would indicate the first derivative of $f(x)$. For emphasis, this derivative could be written as $\frac{dy}{dx}$, again indicating the distinction between the dependent variable y and independent variable x. An extremely handy notation, attributed to Isaac Newton, is the dot-notation in which $\dot{y}$ stands for $y'(t)$ and $\ddot{y}$ stands for $y''(t)$, where the independent variable t is time. However, on occasion, it is even useful to use $\dot{y}$ for $y'(x)$, especially when the derivative appears squared, as in $\sqrt{1 + \dot{y}^2}$, the integrand in the arc length integral of elementary calculus.

For the function $z = f(x, y)$ the variable z is dependent on the two independent variables x and y. Hence, partial derivatives of the form $f_x = \frac{\partial f}{\partial x}$ and $f_y = \frac{\partial f}{\partial y}$, or $f_{xx} = \frac{\partial^2 f}{\partial x^2}$ and $f_{yy} = \frac{\partial^2 f}{\partial y^2}$, could be found in a partial differential equation.

Just the way the algebraic equation poses the question "what values of the variable, if any, make the equation true?" so also does the differential equation pose the question "what functions, if any, make the differential equation true?" However, in the algebraic equation, the variable, say x, appears in only one form. In the differential equation, the unknown function, say $y(x)$, can appear, and so can one or more of its derivatives. So, the algebraic equation $3x + 5 = \frac{7}{x}$ has its variable appearing in two places, but it is the *same* variable in each place. The differential equation

$$y'' + 3y' + 5y = \sin x$$

has the single unknown $y(x)$, but in addition to $y(x)$ both its first and second derivatives appear. Other examples of differential equations are given later in this section.

Vocabulary

The following definitions are used to classify ordinary differential equations.

A *first-order* equation is one in which the highest ordered derivative would be a first derivative. A *second-order* equation would be one in which the highest ordered derivative is a second derivative.

A *first-degree* equation is one in which the highest ordered derivative appears to the first power. A *second-degree* equation would have that highest ordered derivative raised to the second power.

A *linear* equation is one in which the unknown function y and its derivatives $y^{(k)}$ appear as $\sum_{k=1}^{n} c_k y^{(k)}$, with coefficients c_k no worse than functions of the independent variable. A *nonlinear* equation is one that is not linear in the unknown function and its derivatives.

A linear equation is *homogeneous* if, when the unknown function and all its derivatives are placed on the left side of the equation, the right side is then zero. A linear equation is *nonhomogeneous* (inhomogeneous) otherwise. Thus, the terms homogeneous and nonhomogeneous rightly apply to linear equations.

These definitions are illustrated by the examples in Table 1.3.

Examples of Differential Equations

1. Equation (1) is a linear, nonhomogeneous, first-order, first-degree equation.
2. Equation (2) is also a linear, nonhomogeneous, first-order, first-degree equation. The coefficient x^2 multiplying $y(x)$ does not violate the condition of linearity.
3. Equation (3) is a nonlinear, first-order, first-degree equation. The product $y'(x)y(x)$ makes the equation nonlinear.

(1) $\frac{dy}{dx} = \sin x$

(2) $\frac{dy}{dx} + x^2 y(x) = 5$

(3) $y'(x)y(x) + 2y(x) = e^x$

(4) $[y'(x)]^2 + y(x) = \ln x$

(5) $f(t, y(t), y'(t)) = 0$

(5') $y'(t) = g(t, y(t))$

(6) $2y'' + 3y' + 4y = r(t)$

(7) $F(t, y(t), y'(t), y''(t)) = 0$

(8) $x'(t) = 3x + 4y$
     $y'(t) = 2x - 5y$

(9) $\frac{\partial^2 f}{\partial x^2} + \frac{\partial^2 f}{\partial y^2} + \alpha f(x, y) = 0$

(10) $\dot{x} = \sin t$

TABLE 1.3   Examples of differential equations

4. Equation (4) is again a nonlinear first-order equation, but now it is of second-degree because of the square on the derivative.

5. Equation (5) represents the most general first-order ordinary differential equation. Typically, it is required that the first derivative can be isolated to yield an equation of the form (5′). Not every equation (5) can be so simply represented as (5′). For example, if equation (5) contains the term $[y'(x)]^2$, then it takes *two* equations of the form (5′), namely, $y'(t) = \pm g(t, y(t))$, to express (5). Thus, solving for y' in the equation $(y')^2 - t^2 = 0$ gives *two* equations of the form (5), namely, $y' - t = 0$ and $y' + t = 0$.

6. Equation (6) is a second-order, first-degree, linear equation with constant coefficients. It is *homogeneous* if $r(t) = 0$ and *nonhomogeneous* otherwise. Also, notice how we infer the independent variable of differentiation from the explicit indication given in $r(t)$. This is usually a safe guess, but absolute precision would require that y'' and y' be replaced with symbols such as $y''(t)$ or $\frac{dy}{dt}$.

7. Equation (7) is the most general form of a second-order ordinary differential equation. As with equation (5), we would expect that we could solve for the highest ordered derivative to obtain an equation of the form $y'' = h(t, y(t), y'(t))$.

8. Equation (8) is a first-order, linear, homogeneous system. Its two equations are said to be *coupled* since no single equation contains just one variable and its derivatives.

   So far, all the differential equations presented have been ordinary differential equations because the derivatives appearing in them are *ordinary* derivatives.

9. Equation (9) is a second-order, linear, partial differential equation because the derivatives appearing in it are *partial* derivatives. In Unit Five, we will discover this equation to be a form of the *wave equation* governing the propagation of waves.

10. Equation (10) is a first-order, linear, ODE written in notation useful when discussing velocity and acceleration in physics.

Solution of a First-Order ODE

The algebraic equation $2x^2 + 3x - 5 = 0$ is quadratic in the unknown x; and therefore, its solutions, $x = -\frac{5}{2}, 1$, can be found by a specific algorithm called the *quadratic formula.* On the other hand, the two numbers that purport to be solutions can be verified by substitution. If substitution of a number into the equation results in an identity, then the numbers are solutions of the equation.

In a similar manner, on any interval I, the function $y = f(x) = cx^2 - \frac{1}{2}$ can be verified as the solution of the differential equation

$$xy' - 2y = 1 \qquad (1.4)$$

The function $f(x)$ has a first derivative and, if substituted into the differential equation, yields the identity $1 = 1$. Hence, it is a *solution,* by which we mean a *classical* or *pointwise* solution. At each point in a given interval I, the derivative exists, and under substitution, the differential equation is satisfied point-by-point. Other terms for such a solution are *strict* and *strong.* A slightly more formal definition is given at the end of the section.

The solution of (1.4) is found by the technique of *separation of variables,* studied in elementary calculus and reviewed in Section 3.1. The equation is written in the form

$$\frac{dy}{1 + 2y} = \frac{dx}{x}$$

and an antiderivative of both sides leads to

$$\tfrac{1}{2}\ln(1+2y) = \ln x + a \tag{1.5a}$$

$$1 + 2y = e^{2a}x^2 \tag{1.5b}$$

$$y = cx^2 - \tfrac{1}{2} \tag{1.5c}$$

(If constants of integration c_L and c_R are introduced on the left and right, respectively, then $a = c_R - c_L$, so only a single constant needs to be written on just one side of (1.5a).) For simplicity, we call the constant $c = e^{2a}$ in (1.5c) the *arbitrary constant of integration* introduced by antidifferentiation.

Geometrically, (1.5c) represents a *one-parameter family* of curves. For each value of the arbitrary constant c, (1.5c) defines a distinct curve in the xy-plane. Any solution of the differential equation not containing an arbitrary constant is called a *specific solution*. Clearly, if the arbitrary constant c in the one-parameter family (1.5c) is given a value, a specific solution results. If a one-parameter family of solutions for a first-order differential equation is known to contain every possible specific solution, then the one-parameter family is called the *general solution*.

There are first-order differential equations for which a one-parameter family of solutions does not contain all specific solutions. For example, the differential equation $(y')^2 + xy' - y = 0$ is satisfied by every member of the one-parameter family of linear functions $y = cx + c^2$, but the specific solution $y = -\frac{x^2}{4}$, a quadratic, is not a member of this family.

Sometimes, a specific solution is a member of a family of solutions if the family is expressed in a certain way. The differential equation $y' = y^2$ is satisfied by every member of the family $y = (c - x)^{-1}$ and by $y = 0$. However, for no finite value of c will $y = 0$ be a member of this family. Yet, if the family is expressed by $y = C(1 - Cx)^{-1}$, then $y = 0$ is the member corresponding to $C = 0$. Whether or not a solution is essentially an outlier is thus difficult to determine.

The first-order differential equation $(y' + y)(y' + 2y) = 0$ is satisfied by the members of the two one-parameter families of solutions $y = ae^{-x}$ and $y = be^{-2x}$. Hence, it is not even clear that first-order differential equations are always satisfied by a *single* one-parameter family of solutions.

Fortunately, an n-parameter family of solutions constituting the general solution can be found for linear ODEs of order n. The theory for first-order linear equations is developed in Sections 3.4 and 3.5, and the theory for linear equations of order greater than one is developed in Chapter 5.

Initial and Boundary Value Problems

An ordinary differential equation of order n can rightly be posed "alone," with no additional conditions placed on its solution. In that event, a general solution depending on an appropriate number of arbitrary constants would be the ideal solution sought. Failing that, an n-parameter family of solutions might be found. If additional conditions are included with the differential equation, we can have either an *initial value problem* (IVP) or a *boundary value problem* (BVP). Hence, there are three distinct senses in which the term "solution" might be applied.

For the differential equation itself, a *specific solution,* the *general solution* (if possible), or an n-parameter family of solutions could be found. Depending on the type of additional conditions included with the differential equation, either an IVP or a BVP might be formed.

In either of these two cases, we no longer speak of "solving the differential equation" but instead speak of solving the IVP or BVP.

IVPs   If the values of the dependent variable and its derivatives are specified at a single point, the resulting problem consisting of these conditions and the differential equation is called an initial value problem. Examples include

$$\frac{dy}{dx} + x^2 y(x) = 5 \qquad 2y'' + 3y' + 4y = \cos t$$
$$\text{and} \qquad y(0) = 1$$
$$y(1) = 2 \qquad y'(0) = -1$$

The first-order equation supports but one condition because its solution will generally contain but one "constant of integration." Applying any such condition results in an initial value problem. However, the second-order equation supports two conditions because its solution will generally contain two "constants of integration." Specifying both conditions at the same value of the independent variable results in the initial value problem. Incidentally, the second-order example, to be studied in detail later in Chapter 5, governs the forced motion of a damped linear oscillator whose initial position and initial velocity are prescribed.

BVPs   Since a second-order ordinary differential equation can support two side-conditions, it is possible to impose these conditions at two different values of the independent variable, thereby forming the *two-point boundary value problem*. In such a problem, we seek a function satisfying not only the given ODE but also the prescribed end-point conditions. The conditions on the left in Table 1.4 could arise in a heat-transfer problem where the temperatures at the left ($x = 0$) and right ($x = 1$) ends of a conducting rod were prescribed to be α and β, respectively. On the other hand, if the heat-transfer problem involved temperatures in a conducting ring, periodic conditions on the right in Table 1.4 would express smoothness of the temperatures across the left and right faces of the rod that were joined to make the ring.

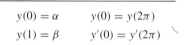

| $y(0) = \alpha$ | $y(0) = y(2\pi)$ |
| $y(1) = \beta$ | $y'(0) = y'(2\pi)$ |

TABLE 1.4   Common two-point boundary conditions

The One-Parameter Family of Curves

As c varies through all real numbers, the set of curves represented by $F = x^2 + c$ comprise a *one-parameter family*. Through each point in the plane, one member of the family passes. Each curve is an upward-opening parabola; and as c varies, the vertex of the parabola moves along the y-axis. Figure 1.13 shows several members of this family. When $c = 0$ the vertex of the parabola is at the origin. For $c < 0$, the vertex of the parabola is below the x-axis; but for $c > 0$, the vertex is above the x-axis.

If the function $F = x^2 + c$ is thought of as a function of two variables, that is, as $F(x, c)$, and graphed as a surface over the xc-plane, the plane sections $c = constant$ would be the individual members of the family. Figure 1.14 shows a portion of this surface and includes, as a thick line, the intersection of the surface with the cutting plane $c = 0$. The space curve determined by this intersection is the parabola corresponding to $c = 0$, namely, $y = x^2$.

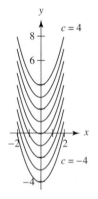

FIGURE 1.13   Members of the one-parameter family $F = x^2 + c$

The General Solution as a One-Parameter Family of Curves

The general solution of the first-order ODE $y' = f(x, y)$ has a geometric representation as a one-parameter family of curves. Generally, one curve passes through each point in the xy-plane. However, there can be exceptional points through which more than one curve passes. This occurs, for example, with $y = cx^2 - \frac{1}{2}$, the general solution of the differential

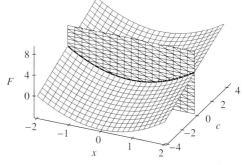

FIGURE 1.14 Plane sections of $F(x, c) = x^2 + c$ are the plane curves in Figure 1.13

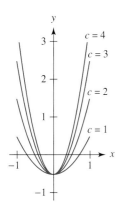

FIGURE 1.15 Members of the one-parameter family $y = cx^2 - \frac{1}{2}$

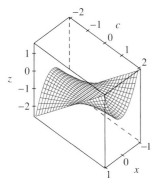

FIGURE 1.16 Plane sections of $F(x, c) = cx^2 - \frac{1}{2}$ are the plane curves in Figure 1.15

equation $xy' - 2y = 1$, several members of which are shown in Figure 1.15. Except for the point $(0, -\frac{1}{2})$, there is only one member passing through a given point. The characteristics of the differential equation responsible for this behavior will be examined more carefully in Section 14.1.

We conclude this look at the one-parameter family of curves by providing one alternate view of the family. We plot, therefore, $y(x)$ as a function of the two variables x and c, yielding the surface seen in Figure 1.16. The plane sections $c = constant$ include the individual curves seen in Figure 1.15.

Solution of a Second-Order ODE

As we will learn in Chapter 5, the general solution of the second-order differential equation $\frac{d^2 y}{dt^2} + 2\frac{dy}{dt} + 10y(t) = 0$ is $y(t) = e^{-t}(c_1 \cos 3t + c_2 \sin 3t)$ and now depends on two arbitrary constants, namely, c_1 and c_2. Just as for the first-order equation, the solution is validated by the substitution

$$(c_1 e^{-t} \cos 3t + c_2 e^{-t} \sin 3t)'' + 2(c_1 e^{-t} \cos 3t + c_2 e^{-t} \sin 3t)'$$
$$+ 10 e^{-t}(c_1 \cos 3t + c_2 \sin 3t) = 0$$

resulting in the identity $0 = 0$.

Since the general solution depends on two arbitrary constants, it forms a two-parameter family of curves. Unfortunately, the geometry does not lend itself to visualization as readily as it did in the case of the one-parameter family arising as the general solution of the first-order equation.

The Initial Value Problem

We give two examples of initial value problems for ordinary differential equations, a first-order equation, and a second-order equation.

EXAMPLE 1.7 The initial value problem associated with the first-order differential equation $x\frac{dy}{dx} - 2y(x) = 1$ consists of this equation and a condition of the form $y(x_0) = y_0$. As an example, take this condition to be $y(1) = 2$ and seek the single solution curve that passes though the point $(1, 2)$ in the xy-plane. This solution curve can be extracted from the one-parameter family of solutions (1.5c). Imposition of the initial condition on this solution leads to the equation $2 = c(1)^2 - \frac{1}{2}$, from which we obtain $c = \frac{5}{2}$ and $y(x) = \frac{5}{2}x^2 - \frac{1}{2}$ as the solution of the initial value problem. Its graph resembles one of the parabolas in Figure 1.15. ❖

EXAMPLE 1.8   The initial value problem for the second-order differential equation $y'' + 2y' + 10y(t) = 0$ consists of the differential equation and two conditions of the form $y(t_0) = y_0$ and $y'(t_0) = y_0'$. Initial values of the function and its first derivative are specified at a common value of the independent variable t. In the event $y(t)$ represents a position, then the derivative represents a velocity, so the IVP represents the motion of an object governed by the differential equation and set into motion at the location and with the velocity given by the initial conditions.

We have already verified that $y(t) = c_1 e^{-t} \cos 3t + c_2 e^{-t} \sin 3t$ is the general solution of the differential equation. If we impose the initial conditions $y(0) = 1$, $y'(0) = 2$, we obtain the algebraic equations $1 = c_1$, $2 = 3c_2 - c_1$ whose solution is $c_1 = 1$, $c_2 = 1$; so the solution of the IVP is $y(t) = e^{-t}(\cos 3t + \sin 3t)$. Its graph is the solution curve shown in Figure 1.17.   ❖

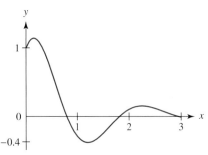

FIGURE 1.17   Solution of the second-order IVP in Example 1.8

Formal Definition of a Solution

On the interval $a < t < b$, the function $\phi(t)$ is a *classical solution* of the differential equation $y' = f(t, y)$ if at every point in the interval, the derivative ϕ' exists, and substitution of $\phi(t)$ for $y(t)$ in the differential equation yields the identity $\phi' = f(t, \phi(t))$.

The requirement that the derivative exists means ϕ is continuous, and hence defined, throughout the interval.

Other terms used for this notion of solution are *strict*, *strong*, and *pointwise*.

The literature provides a spectrum of precision in the definitions for "solution" of a differential equation. For example, consult the definitions in [11], [12], or [13].

EXERCISES 1.2–Part A

A1. Show that $y = \sqrt{1 - x^2}$ satisfies the equation $x^2 + y^2 = 1$.

For each of Exercises A2–13:

(a) Identify the differential equation as linear or nonlinear.

(b) If linear, determine whether the equation is homogeneous or nonhomogeneous.

(c) Determine the order and degree of the equation.

(d) Identify the dependent and independent variables.

A2. $y' = e^t$ **A3.** $\dfrac{dy}{dx} - 2xy = \cos x$

A4. $y'' + 3xy' + 5y = e^{-x} \sin 3x$ **A5.** $y'y = xy + 3$

A6. $\dfrac{d^2 y}{dx^2} - 3xyy' + 4y \sin 2x = x^2$ **A7.** $y' + \sin y = 1$

A8. $(y')^2 + 3y = \dfrac{5}{1 + y^2}$ **A9.** $y' + 3y = \dfrac{5}{1 + y^2}$

A10. $(y')^2 + 3y = 5$ **A11.** $y' + xy = \dfrac{5}{y}$ **A12.** $y' + \dfrac{x}{y} = \dfrac{5}{x}$

A13. $y''' + 2yy'' - 7xy' + 9y = \sin y$

For each of Exercises A14–19:

(a) Determine whether the system is an initial value problem or a boundary value problem.

(b) Determine whether there are the right number, too many, or too few conditions.

(c) State whether the differential equation is linear or nonlinear.

(d) Give the order of the equation.

A14. $y' + xy^2 = \ln x$, $y(1) = 1$

A15. $y'' + 3y' + 5y = \cos t$, $y(0) = 1$, $y'(0) = 0$

A16. $y''' - 2y'' + 5y' - 7y = 0$, $y(0) = 0$, $y'(1) = 1$

A17. $y'' + xy' - y^2 = 1$, $y(0) + y'(1) = 1$, $y'(0) - y(1) = 0$

A18. $y'' + y = 0$, $y(0) = 0$, $y'(\pi) = 0$

A19. $x^2 y'' + xy' + y = 0$, $y(1) = 1$, $y'(1) = 0$, $y''(1) = 2$

A20. Show that $y(x) = cx^3 - \dfrac{x}{2}$ satisfies the differential equation $xy' = x + 3y$. Find the value of c for which $y(1) = 1$, and graph the resulting solution.

EXERCISES 1.2–Part B

B1. Show that $y(x) = 9 + 5x + \frac{\sqrt{273+218x-23x^2}}{8}$ is a solution of $3x^2 - 5xy + 4y^2 - 8x - 9y = 12$.

B2. Show that $y(x) = \sin x + 2x^{-1}\cos x + (c - 2\sin x)x^{-2}$ satisfies $y' + \frac{2y}{x} = \cos x$. Find the value of c for which $y(\frac{\pi}{2}) = 1$, and graph the resulting solution.

B3. Show that $x^2y^3 - \frac{1}{4}y^4 = c$ implicitly defines $y(x)$ as a solution of $y' = \frac{2xy}{y-3x^2}$. Find the value of c for which $y(0) = 1$, and graph the resulting solution.

B4. Show that $y(x) = \frac{2}{c-x^2}$ satisfies $y' = xy^2$. Find the value of c for which $y(1) = 2$, and graph the resulting solution.

B5. Show that $y(x) = \frac{x(5x^2-7c)}{c-x^2}$ satisfies $x^2y' + 35x^2 + 11xy + y^2 = 0$. Find the value of c for which $y(1) = 1$, and graph the resulting solution.

B6. Show that $y(x) = x + \sqrt{2x^2+1}$ is a solution of the IVP $y' = \frac{y+x}{y-x}$, $y(0) = 1$. Graph this solution.

B7. Show that $y(x) = \frac{1}{3}(13e^{3x^2/2} - 1)$ is a solution of the IVP $y' = x(1 + 3y)$, $y(0) = 4$. Graph this solution.

B8. Graph representative curves from the family defined in

  (a) Exercise B2.   **(b)** Exercise B3.

  (c) Exercise B4.   **(d)** Exercise B5.

B9. Show that $y(x) = ae^{-2x} + be^{-x}$ satisfies $y'' + 3y' + 2y = 0$. Find values of a and b for which $y(0) = 2$, $y'(0) = -1$. Graph the resulting solution.

B10. Show that $y(x) = ae^{-4x} + be^{-x}$ satisfies $y'' + 5y' + 4y = 0$. Find values of a and b for which $y(0) = -1$, $y'(0) = 3$. Graph the resulting solution.

B11. Show that $y(x) = (a\cos x + b\sin x)e^{-x}$ satisfies $y'' + 2y' + 2y = 0$. Find values of a and b for which $y(0) = 4$, $y'(0) = -3$. Graph the resulting solution.

B12. Show that $y(x) = ae^{5x} + be^{-3x}$ satisfies $y'' - 2y' - 15y = 0$. Find values of a and b for which $y(0) = 2$, $y'(0) = -1$. Graph the resulting solution.

B13. Show that $y(x) = ax^{-1} + bx^{-5}$ satisfies $x^2y'' + 7xy' + 5y = 0$. Find values of a and b for which $y(1) = 1$, $y'(1) = 2$. Graph the resulting solution.

B14. Show that $y(x) = ax + b\ln x$ satisfies $x^2(1 - \ln x)y'' + xy' - y = 0$. Find values of a and b for which $y(1) = 2$, $y'(1) = -2$. Graph the resulting solution.

B15. Show that $y(x) = ax^2 + \sin x$ satisfies $(x^2\cos x - 2x\sin x)y'' + (x^2 + 2)\sin(x)y' - 2(x\sin x + \cos x)y = 0$.

B16. Show that $y(x) = ax + be^{2x}$ satisfies $(2x - 1)y'' - 4xy' + 4y = 0$. Find values of a and b for which $y(0) = 2$, $y'(0) = 1$. Graph the resulting solution.

B17. Show that $y(x) = axe^{2x} + be^x$ satisfies $(x + 1)y'' - (3x+4)y' + (2x+3)y = 0$. Find values of a and b for which $y(0) = -3$, $y'(0) = 5$. Graph the resulting solution.

1.3 The Direction Field

A Slope at Every Point

A first-order differential equation in the form $y'(x) = f(x, y(x))$ has an interesting geometric interpretation. At each point (x, y) in the xy-plane, the number $f(x, y)$ is the value of y', so it is the value of a slope on some solution curve. The totality of such slopes in a region of the plane constitutes a *direction field*. The typical graph used to represent such a field of slopes or tangents is a collection of arrows or line segments, each of whose slope corresponds to the slope defined at the point where the arrow or line segment is drawn. The graph itself is often called the direction field.

At every point on some solution curve $\phi(x)$, a member of the direction field is tangent. If at each point (x, y) a line segment having its slope determined by $f(x, y)$ is drawn, then the solution curve $\phi(x)$ will have one of these line segments tangent to it at each of its points. So, given the graph of the direction field, it is sometimes possible to sketch solution curves by following the flow of the arrows. We illustrate these ideas with an example.

EXAMPLE 1.9 Consider the differential equation $y'(x) = f(x, y(x)) = x + y^2$ and the region $0 \le x$, $y \le 1$. On this region establish a grid whose nodes are the points $(x_i, y_j) = (\frac{i}{4}, \frac{j}{4})$, $0 \le i$,

$j \leq 4$. If the function $f(x, y)$ is evaluated at each nodal point, we have 25 numbers which exist at the nodes of a rectangular grid, and which are listed in Table 1.5.

	$j = 0$	$j = 1$	$j = 2$	$j = 3$	$j = 4$
$i = 4$	0	$\frac{1}{16}$	$\frac{1}{4}$	$\frac{9}{16}$	1
$i = 3$	$\frac{1}{4}$	$\frac{5}{16}$	$\frac{1}{2}$	$\frac{13}{16}$	$\frac{5}{4}$
$i = 2$	$\frac{1}{2}$	$\frac{9}{16}$	$\frac{3}{4}$	$\frac{17}{16}$	$\frac{3}{2}$
$i = 1$	$\frac{3}{4}$	$\frac{13}{16}$	1	$\frac{21}{16}$	$\frac{7}{4}$
$i = 0$	1	$\frac{17}{16}$	$\frac{5}{4}$	$\frac{25}{16}$	2

TABLE 1.5   Slope field evaluated on a rectangular grid

Each number in Table 1.5 represents the value of a slope. One way to visualize these different slope values is by a graph in which heights are a measure of the slope, as shown in Figure 1.18. Even though the graph in Figure 1.18 is a mathematically correct way of representing the values of the slopes generated by $f(x, y)$, it is neither typical nor particularly revealing. These slopes are usually displayed as arrows such as seen in Figure 1.19. Solution curves emananting from the initial points $(0, 0.2)$, $(0, 0.4)$, $(0, 0.6)$, and $(0, 0.8)$ are superimposed on the direction field in Figure 1.20. ❖

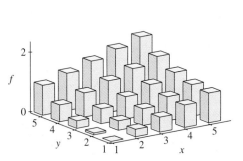

FIGURE 1.18   Block diagram for the slopes generated by $y' = x + y^2$

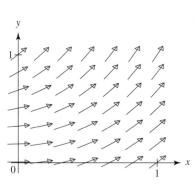

FIGURE 1.19   Direction field for $y' = x + y^2$

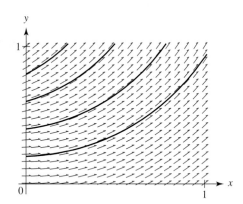

FIGURE 1.20   Direction field and solutions for $y' = x + y^2$

Notice how the arrows in the direction field are approximately tangent to each of the four solution curves. (It's hard to know where to draw tangents to a solution if the solution isn't known.) Consequently, a direction field can, at best, suggest the shape of solution curves. Modern numeric solvers can obtain the solution curves as easily as the direction field itself.

Isoclines

An *isocline* is a curve along which the slopes in a direction field are constant. Since these slope-values are determined by the function $f(x, y)$, an isocline is just a *level curve* of $f(x, y)$. (A level curve, or contour, of $f(x, y)$ is a curve along which $f(x, y)$ remains constant.) Several such contours are graphed on the surface $z = f(x, y) = x + y^2$ in

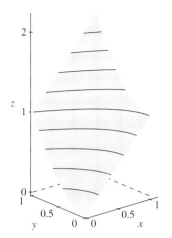

FIGURE 1.21 Surface and level curves for $f(x, y) = x + y^2$

Figure 1.21. Figure 1.22 shows the contours superimposed on the direction field and several solution curves.

Also in the literature is the term *nullcline,* indicating the particular isocline through points in the *xy*-plane where the direction field has slope zero. This particular level curve, determined by the equation $f(x, y) = 0$ to be $x = -y^2$, is shown in Figure 1.23.

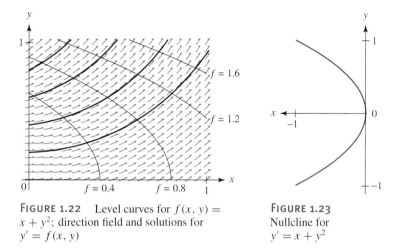

FIGURE 1.22 Level curves for $f(x, y) = x + y^2$; direction field and solutions for $y' = f(x, y)$

FIGURE 1.23 Nullcline for $y' = x + y^2$

With today's technological tools, it is usually as easy to obtain solution curves as it is to obtain the direction field. However, [59] uses the direction field to anaylze the qualitative behavior of the equation $v\frac{dv}{dx} = a(V - v)^2 - bx^2$, which governs x, the length of the gouge a grounding iceberg makes on the ocean floor as the berg drifts with velocity v in a current of velocity V. So, while the direction field isn't normally an efficient way to obtain solutions, it can occasionally provide insight into their behavior.

EXERCISES 1.3-Part A

A1. Sketch several solution curves on the direction field in Figure 1.24, computed for the differential equation $y' = |x| - 2|y|$. Sketch the nullclines. Sketch the isoclines along which y' assumes the values 1, 2, and 3.

A2. Sketch several solution curves on the direction field in Figure 1.25, computed from the differential equation $y' = \sqrt{x^2 + y^2}$. Sketch the nullclines. Sketch the isoclines along which y' assumes the values 1, 2, and 3.

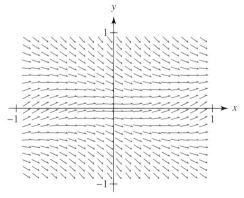

FIGURE 1.24 Direction field for Exercise A1

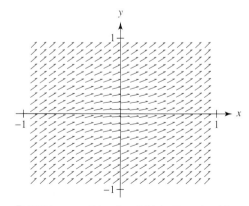

FIGURE 1.25 Direction field for Exercise A2

EXERCISES 1.3–Part B

B1. Construct a direction field for the differential equation $y' = x + 3y$, and on it show the isoclines where y' assumes the values 1, 2, and 3. In addition, sketch the solution emanating from the initial point $(\frac{1}{2}, \frac{1}{2})$.

B2. Construct a direction field for the differential equation $y' = x^2 + 3y$, and on it show the isoclines where y' assumes the values 1, 2, and 3. In addition, sketch the solution emanating from the initial point $(\frac{1}{2}, \frac{1}{2})$.

B3. Construct a direction field for the differential equation $y' = x + 3y^2$, and on it show the isoclines where y' assumes the values 1, 2, and 3. In addition, sketch the solution emanating from the initial point $(\frac{1}{2}, \frac{1}{2})$.

B4. Construct a direction field for the differential equation $y' = x^2 + 3y^2$, and on it show the isoclines where y' assumes the values $\frac{1}{2}$, 1, 2, and 3. In addition, sketch the solution emanating from the initial point $(\frac{1}{2}, \frac{1}{2})$.

B5. Construct a direction field for the differential equation $y' = x \sin y$, and on it show the isoclines where y' assumes the values $\frac{k}{10}$,

$k = 1, 3, 5, 7$. In addition, sketch the solution emanating from the initial point $(0, 1)$.

B6. Construct a direction field for the differential equation $y' = \frac{x}{y}$, and on it show the isoclines where y' assumes the values $0.1, 0.3, 0.5$, and 0.7. In addition, sketch the solution emanating from the initial point $(-0.3, 0.5)$.

B7. Construct a direction field for the differential equation $y' = xy^2$, and on it show the isoclines where y' assumes the values $0.1, 0.3$, and 0.5. In addition, sketch the solution emanating from the initial point $(-0.3, 0.5)$.

B8. Construct a first-quadrant direction field for the differential equation $yy' = (1 - y)^2 - x^2$, and on it show the nullcline, the isocline where y' assumes the value 0. In addition, sketch the solution emanating from the initial point $(0, 1)$. Use the direction field to argue that c, the x-intercept of this solution, satisfies $c > 1$.

B9. Determine the equilibrium solutions for the differential equation $y' = y^3 - 3y^2 + 2y$, $y \geq 0$. For $x \geq 0$, sketch the direction field and use it to determine the stability of the equilibrium solutions.

1.4 Picard Iteration

Picard Iteration

Given the initial value problem defined by the equation $y'(x) = f(x, y(x))$ and the initial condition $y(x_0) = y_0$, the formal integration $y(x) - y_0 = \int_{y_0}^{y(x)} du = \int_{x_0}^{x} f(t, y(t)) \, dt$ leads to the iterative scheme

$$\phi_1(x) = \phi_0(x) + \int_{x_0}^{x} f(t, \phi_0(t)) \, dt$$

$$\phi_2(x) = \phi_0(x) + \int_{x_0}^{x} f(t, \phi_1(t)) \, dt$$

$$\vdots$$

$$\phi_{n+1}(x) = \phi_0(x) + \int_{x_0}^{x} f(t, \phi_n(t)) \, dt$$

for generating $\phi_1(x), \phi_2(x), \ldots$, a sequence of approximations that Emile Picard (1856–1941) proved converges to a solution of the IVP. Known as Picard's theorem, this result, explored in Section 1.5, guarantees both the existence and uniqueness of a solution. The iteration scheme itself is called "Picard iteration."

EXAMPLE 1.10 If $\phi_0 = 1$, the IVP $y' = y$, $y(0) = 1$, with exact solution $y(x) = e^x$, generates Picard iterates that are nothing more than the partial sums of the Maclaurin expansion of e^x

$$\phi_1 = 1 + \int_0^x 1 \, dt = 1 + x \qquad \phi_2 = 1 + \int_0^x (1 + t) \, dt = 1 + x + \frac{x^2}{2}$$

$$\phi_3 = 1 + \int_0^x \left(1 + t + \frac{t^2}{2}\right) dt = 1 + x + \frac{x^2}{2} + \frac{x^3}{3!}$$

❖

EXAMPLE 1.11 As an example of a more complex computation, consider the differential equation $y'(x) = f(x, y(x)) = x + y^2$ and the initial condition $y(0) = 0$. The solution of this IVP, given in terms of the Bessel functions $J_\alpha(h(x))$ and $Y_\alpha(h(x))$, where $\alpha = \frac{1}{3}$ and $-\frac{2}{3}$ and $h(x) = \frac{2}{3}x^{3/2}$, is

$$y(x) = -\sqrt{x}\,\frac{J_{-2/3}(h(x)) - \sqrt{3}\,Y_{-2/3}(h(x))}{J_{1/3}(h(x)) - \sqrt{3}\,Y_{1/3}(h(x))}$$

Of course, Bessel functions of any kind are beyond the typical reader this early in a study of differential equations since Bessel functions are solutions of special second-order differential equations. (See Section 16.2.) However, a computer-generated series expansion of this solution is found to be $y(x) = \frac{1}{2}x^2 + \frac{1}{20}x^5 + \frac{1}{160}x^8 + \frac{7}{8800}x^{11} + \cdots$, and a graph of the solution is the solid curve in Figure 1.26. (See the companion Maple worksheet for details of the computation.)

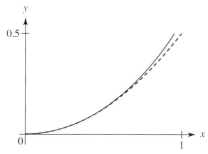

FIGURE 1.26   Exact solution (solid) and $\phi_1(x)$ (dashed) for the IVP of Example 1.11

The Picard iteration scheme, implemented with $\phi_0 = y_0 = 0$, yields the following as the next four iterates:

$$\phi_1(x) = \tfrac{1}{2}x^2 \qquad \phi_2(x) = \tfrac{1}{2}x^2 + \tfrac{1}{20}x^5 \qquad \phi_3(x) = \tfrac{1}{2}x^2 + \tfrac{1}{20}x^5 + \tfrac{1}{160}x^8 + \tfrac{1}{4400}x^{11}$$

$$\phi_4(x) = \tfrac{1}{2}x^2 + \tfrac{1}{20}x^5 + \tfrac{1}{160}x^8 + \tfrac{7}{8800}x^{11} + \tfrac{3}{49,280}x^{14} + \tfrac{87}{23,936,000}x^{17}$$

$$+ \tfrac{1}{7,040,000}x^{20} + \tfrac{1}{445,280,000}x^{23}$$

Visual inspection of the terms in each approximation confirms the claim about convergence to the solution. Figure 1.26 shows the exact solution (solid) and $\phi_1(x)$ (dashed). Subsequent iterates are so close to the exact solution as to be indistinguishable. Hence, for this example, the convergence is rapid.   ❖

EXERCISES 1.4–Part A

A1. Solve the IVP $y' = x$, $y(1) = 1$, exactly, and then apply Picard iteration starting with $\phi_0 = 1$.

A2. The exact solution of the IVP $y' = x + y$, $y(0) = 1$ is $y(x) = 2e^x - x - 1$. Set $\phi_0 = 1$, and show that the next two Picard iterates are the terms of the Maclaurin expansion of the exact solution.

A3. Obtain the exact solution of the IVP $y' = xy$, $y(0) = 1$. Set $\phi_0 = 1$, and show that the next three Picard iterates are the terms of the Maclaurin expansion of the exact solution.

A4. Obtain the exact solution of the IVP $y' = \frac{y}{x}$, $y(1) = 1$. Set $\phi_0 = 1$, and obtain the next three Picard iterates. Generalize, and show that the Picard iterates converge to the exact solution.

A5. Obtain the exact solution of the IVP $y' = y^2$, $y(0) = 1$. Set $\phi_0 = 1$, and obtain the next three Picard iterates. Compare the Maclaurin expansion of the exact solution to the series generated by Picard iteration.

EXERCISES 1.4–Part B

B1. Show that $y(x) = \frac{1}{4}(5e^{2x} - 1 - 2x - 2x^2)$ is a solution of the initial value problem $y' = x^2 + 2y$, $y(0) = 1$. For this solution, create a Taylor polynomial at the point $x = 0$; then use Picard approximation with $\phi_0 = 1$ and compare the result with the Taylor polynomial.

B2. Repeat the Picard iteration of Exercise B1, this time using $\phi_0 = x$ and $\phi_{k+1} = y(0) + \int_0^x f(t, \phi_k(t))\,dt$, $k > 0$. Observe that $\phi_0(0) \ne 1 = y(0)$.

B3. Show that $y(x) = \frac{1}{25}(76e^{5x} - 1 - 5x)$ is a solution of the initial value problem $y' = x + 5y$, $y(0) = 3$. For this solution, create a

Taylor polynomial at the point $x = 0$; then use Picard approximation with $\phi_0 = 3$ and compare the result with the Taylor polynomial.

B4. Repeat the Picard iteration of Exercise B3, this time using $\phi_0 = x$ and ϕ_{k+1} as given in Exercise B2. Observe that $\phi_0(0) \ne 3$.

B5. Show that $y(x) = \frac{1}{5}[2\sin x + \cos x + e^{-x}(4\cos x - 8\sin x)]$ is a solution of the initial value problem $y'' + 2y' + 2y = \cos x$, $y(0) = 1$, $y'(0) = -2$. For this solution, create a Taylor polynomial at the point $x = 0$, then, with $\phi_0 = y(0)$; use the following extension of Picard iteration, comparing the result with the Taylor polynomial.

$$y_{n+1}(x) = y(0) + \int_0^x y'(0)\,dt + \int_0^x \int_0^s f(t, y_n(t), y_n'(t))\,dt\,ds$$

B6. Repeat Exercise B5 for $y(x) = \frac{1}{78}[9\sin 2x - 6\cos 2x + e^{-x}(22\sin 3x - 150\cos 3x)]$ and the initial value problem $y'' + 2y' + 10y = \sin 2x$, $y(0) = -2$, $y'(0) = 3$.

1.5 Existence and Uniqueness for the Initial Value Problem

Existence and Uniqueness: Algebraic Equations

A system of linear algebraic equations such as

$$a_{11}x + a_{12}y = c_1 \qquad a_{21}x + a_{22}y = c_2$$

can have no solutions, exactly one solution, or an infinite number of solutions, depending on the values of the constant coefficients. For example, the system on the left in Table 1.6 is inconsistent and has no solution. On the other hand, the system in the center of Table 1.6 has the unique solution $x = \frac{1}{3}$, $y = \frac{10}{9}$. Finally, the system on the right in Table 1.6 has an infinite number of solutions, here represented as $x = 2 - \frac{3}{2}t$, $y = t$, for $-\infty < t < \infty$. Sections 12.4–12.7 show that linear systems such as these have a unique solution if the system determinant, the determinant of the coefficient matrix, $\begin{bmatrix} a_{11} & a_{12} \\ a_{21} & a_{22} \end{bmatrix}$, is nonzero and will have either no solution or an infinite number of solutions if this determinant is zero. For the examples in Table 1.6, we find the determinants to be $0, -9, 0$, respectively. Only the second example has a unique solution. The first example has no solution; the third, an infinite number.

We seek criteria for determining when an initial value problem for a first-order ordinary differential equation has a solution, when it has a unique solution, and when it has no solution.

$$2x + 3y = 4$$
$$2x + 3y = 5$$

$$2x + 3y = 4$$
$$5x + 3y = 5$$

$$2x + 3y = 4$$
$$4x + 6y = 8$$

TABLE 1.6   Linear algebraic equations with no solution, one solution, and infinitely many solutions, respectively

Initial Value Problems

Table 1.7 lists three examples of initial value problems of the form $f(x, y(x), y'(xd)) = 0$, $y(x_0) = y_0$, one of which has no solution, one that has exactly one solution, and one with an infinite number of solutions.

The initial value problem on the left in Table 1.7 has no solution. For real numbers, $a^2 + b^2 = 0$ implies $a = b = 0$, so the solution of the differential equation itself is $y(x) = 0$. Then, the initial condition cannot be satisfied, and there is no solution to the given IVP. The initial value problem in the center in Table 1.7 has the unique solution $y(x) = x - 1 + 2e^{-x}$. Verification that this is a solution is left to Exercise 19. That it is the *only* solution requires Picard's theorem. Finally, the initial value problem on the right in Table 1.7 has an infinite number of solutions. In fact, these solutions are given by $y(x) = cx^2 - \frac{1}{2}$. For any value of the constant c we have $y(0) = -\frac{1}{2}$. This family was studied in Section 1.2.

$$[y'(x)]^2 + [y(x)]^2 = 0$$
$$y(0) = 1$$

$$y' + y(x) = x$$
$$y(0) = 1$$

$$xy' - 2y(x) = 1$$
$$y(0) = -\frac{1}{2}$$

TABLE 1.7   IVPs with no solution, one solution, and infinitely many solutions

Picard's Theorem

The following existence and uniqueness theorem is attributed to Picard, who showed that the iteration scheme studied in Section 1.4 converges, under suitable conditions, to a unique solution of the initial value problem

$$y' = f(x, y) \qquad y(x_0) = y_0 \tag{1.6}$$

A proof of this theorem requires more mathematical tools than we have yet mastered; it can be found, for example, in [80].

THEOREM 1.1

1. R is the closed rectangle defined by $|x - x_0| \le a$ and $|y - y_0| \le b$, with a and b positive.

2. On R, $f(x, y)$ and $f_y(x, y)$ are both continuous.

3. On R, $|f(x, y)| \le M$ and $|f_y(x, y)| \le L$.

4. h is the smaller of a and $\frac{b}{M}$.

$\Longrightarrow$

The Picard iterates converge, on $|x - x_0| \le h$, to the unique solution of the IVP.

Hypothesis 2, which requires $f_y(x, y)$ to be continuous demands more than is needed and can be replaced with the condition that $f(x, y)$ obey a Lipschitz (Rudolf Lipschitz, 1832–1903) condition in y. The Lipschitz condition requires all possible "secant lines" to have bounded slopes, that is, $\left| \frac{f(x, y_2) - f(x, y_1)}{y_2 - y_1} \right| \le N$. Hypothesis 3 is actually a consequence of Hypothesis 2, since a fundamental theorem of analysis declares that a continuous function on a closed set is bounded.

This is a *local* theorem since it guarantees a solution exists in the bounded interval $|x - x_0| \le h$. A global existence and uniqueness theorem from [95] replaces the "local" Lipschitz condition with the uniform Lipschitz condition in which the same constant N works as a bound for all x and y. In this case, existence and uniqueness is guaranteed for all x.

EXAMPLE 1.12   The IVP (1.6) where $f(x, y) = 1 + y^2$ and $y(0) = 0$ has solution $y(x) = \tan x$, which cannot be extended outside the interval $(-\frac{\pi}{2}, \frac{\pi}{2})$ because the function $f(x, y) = 1 + y^2$ obeys only a local Lipschitz condition. In fact, $\left| \frac{y_2^2 - y_1^2}{y_2 - y_1} \right|$ simplifies to $|y_2 + y_1|$, which means the Lipschitz constant N depends on y and is, hence, local.   ❖

EXAMPLE 1.13   The IVP (1.6) with $y(0) = 0$ and $f(x, y) = \cos x$ has solution $y(x) = \sin x$, which is the unique solution for all values of x. This function f satisfies a uniform Lipschitz condition since

$$\frac{f(x, y_2) - f(x, y_1)}{y_2 - y_1} = \frac{\cos x - \cos x}{y_2 - y_1} = 0$$

so the Lipschitz constant N, which can be taken as zero, applies for all x and y.   ❖

EXAMPLE 1.14   This final example explores the relationship between the bounding constant M, which dominates the function $f(x, y)$, and the size of the interval $|x - x_0| \le h$ for which Picard's theorem guarantees existence and uniqueness. The theorem concludes that h is the smaller of a and $\frac{b}{M}$, where a and b are the dimensions of the rectangle R over which the bound M dominates $f(x, y)$.

Consider the initial value problem $y' = 2x$, $y(0) = 0$, which clearly has solution $y(x) = x^2$. The function $f(x, y)$ is $2x$ and we take the rectangle R to be centered at the origin. If the upper right corner of R rests at $(a, b) = (3, 5)$, then $M = 6$ bounds $f(x, y)$

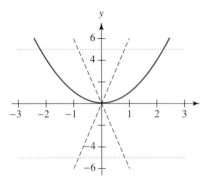

on R. Thus, h is the smaller of 3 and $\frac{5}{6}$ and $y(x) = x^2$ escapes the rectangle R before x gets to 3. In Figure 1.27, the lines $y = Mx$ and $y = -Mx$ are dashed and the lines $y = b$ and $y = -b$ are dotted. The intersection of the dashed and dotted lines determines the value of $h = \frac{b}{M}$. If the dashed line goes through the top of rectangle R, then $h < a$, otherwise $h = a$. Of course, the solution can exist and be unique outside the interval $|x - x_0| \le h$, but the theorem only guarantees existence and uniqueness within this interval. ❖

FIGURE 1.27   The rectangle R for the IVP in Example 1.14

EXERCISES 1.5

1. Show that if the homogeneous system $a_{11}x + a_{12}y = 0$, $a_{21}x + a_{22}y = 0$ has only the trivial solution $x = y = 0$, then the nonhomogeneous system $a_{11}x + a_{12}y = c_1, a_{21}x + a_{22}y = c_2$ has a unique solution. *Hint:* Assume two solutions, subtract, and reduce the problem to the homogeneous case.

For each of Exercises 2–11:

   (a) Determine if the system has a unique solution, no solution, or infinitely many solutions.

   (b) Exhibit any solutions that exist and give a reason why no solution exists if that is the case. Valid arguments could be based on a graph, computation of a determinant, or on some explicit solution technique.

2. $3x - 5y = 7, -6x + 10y = -14$
3. $3x - 5y = 7, -6x + 10y = -21$
4. $3x - 5y = 7, 6x + 10y = -14$
5. $3x - 5y = 0, -6x + 10y = 0$
6. $3x - 5y = 0, 6x + 10y = 0$
7. $3x - 2y + 5z = 1, 7y - z = 1, x - 31y + 6z = -4$
8. $3x - 2y + 5z = 1, 7y - z = 1, x - 31y + 6z = 4$
9. $3x - 2y + 5z = 1, 7y - z = 1, x + 31y + 6z = -4$
10. $3x - 2y + 5z = 0, 7y - z = 0, x - 31y + 6z = 0$

11. $3x - 2y + 5z = 0, 7y - z = 0, x + 31y + 6z = 0$
12. Show that both $y = 0$ and $y = \left(\frac{3}{4}x\right)^{4/3}$ are solutions of the initial value problem $y' = y^{1/4}, y(0) = 0$. Explain why this does, or does not, contradict Picard's theorem.
13. Repeat Exercise 12 for $y = 0$ and $y = x^4$ and the initial value problem $xy' - 4y = 0, y(0) = 0$.
14. What does Picard's theorem say about the existence and uniqueness of a solution to the initial value problem $y' = 1 + y^2, y(0) = 0$? Is the function $\tan(x)$ significant in this setting?
15. What does Picard's theorem say about the existence and uniqueness of a solution to the initial value problem $yy' = 1, y(0) = 0$? Are the functions $\pm\sqrt{2x}$ significant in this setting?

In each of Exercises 16–18:

   (a) What does Picard's theorem say about the existence and uniqueness of a solution?

   (b) Using the technique of separation of variables learned in calculus, obtain one or more solutions for each problem.

16. $y' = \sqrt{xy}, y(0) = 0$ 17. $y' = y^{2/3}, y(0) = 0$
18. $y' = \sqrt{y^2 - 1}, y(0) = 1$
19. Verify that $y(x) = x - 1 + 2e^{-x}$ is a solution of the middle IVP in Table 1.7.

Chapter Review

1. What is a differential equation?
2. What is a solution of a differential equation? How is a function shown to be a solution of a differential equation?
3. What is the difference between a differential equation of first order and one of first degree?
4. Give an example of a differential equation of first order and second degree.
5. Give an example of an ordinary differential equation of second order and first degree.
6. Give an example of a first-order, linear, differential equation.

7. Give an example of a first-order, nonlinear, differential equation.

8. Can a first-order, second-degree differential equation be linear?

9. Can a nonlinear differential equation ever by homogeneous?

10. Give an example of an IVP posed for a first-order ODE.

11. Give an example of an IVP posed for a second-order linear ODE.

12. Give an example of a BVP posed for a second-order linear ODE.

13. Does it make sense to pose a BVP for a first-order ODE? Why?

14. What is a one-parameter family of curves?

15. What is a general solution of a first-order ODE?

16. Give an explanation of the term *direction field*. What is its relevance to the study of differential equations?

17. What is an isocline?   **18.** What is a nullcline?

19. What does Picard iteration accomplish?

20. Give an example of the process of Picard iteration.

21. State a theorem governing the existence and uniqueness of solutions of a class of differential equations.

22. Explain the value of an existence-uniqueness theorem for differential equations.

Chapter 2

Models Containing ODEs

INTRODUCTION   This chapter examines four simple physical structures that are easily modeled by a first-order ordinary differential equation embedded in an initial value problem. The purpose here is to show the reader that differential equations are a natural vehicle for capturing a wide spectrum of physical processes.

At a deeper level, the modeling process is typical of the way science settles on theories. After observations are made, hypotheses are formulated, and these hypotheses are often crafted into a differential equation as a model. To test the validity of the hypotheses, the predictions of the model are determined by extracting information from the differential equation and its solution. Then, new experiments and observations are made to validate the predictions. If the data do not contradict the predictions, the model is upheld or, at least, not invalidated.

To the extent that new observations suggest a need for further refinement of the model, the process continues until a better model is obtained. However, the reader should be aware that models built from a differential equation help quantize phenomena but do not necessarily provide mechanistic explanations. For example, Schroedinger's differential equation of quantum mechanics describes atomic events but does not "explain" them beyond the primitive concepts upon which the differential equation itself is based.

2.1   Exponential Growth and Decay

Exponential Growth

Growth for which the rate of change of the increasing substance is proportional to the amount of substance already present is called exponential growth. This type of growth is captured in the first-order differential equation $\frac{dP}{dt} = kP(t), k > 0$, where $P(t)$ is the amount of substance present at time t. For bacterial colonies, human populations, and even for money compounding at interest, it is not unreasonable to contend that the rate of growth is indeed proportional to the existing size or amount.

If an initial amount $P(0) = P_0$ grows under this law, then the population at any later time will be given by $P(t) = P_0 e^{kt}$, the solution of the differential equation. The initial amount grows exponentially, a characteristic of a growth law where the rate of change is directly proportional to the present amount. The rate constant k determines how quickly growth takes place. Figure 2.1 shows exponential growth curves for $k = \frac{1}{2}$ and $k = 1$.

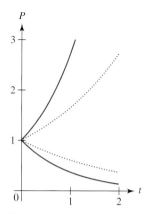

FIGURE 2.1 Growth curves (increasing) and decay curves (decreasing) for $k = \frac{1}{2}$ (dotted) and $k = 1$ (solid)

DOUBLING TIME The time it takes for a population to double is called the *doubling time*. It can be determined by solving the equation $2P_0 = P_0 e^{k\tau}$ for the doubling time $\tau = \frac{\ln 2}{k}$. A population for which $k = 3$ would have a doubling time $\tau = \frac{\ln 2}{3} \overset{\circ}{=} 0.23$, where the units would be determined by those of k.

When dealing with the growth of investments, the doubling time is a convenient quantization of the return. An investment earning 6% will double in $\tau = \frac{\ln 2}{0.06} \overset{\circ}{=} 11.55$ years. To simplify the arithmetic, $100 \ln 2 = 69.31471806$ is replaced by 72 since 72 has more factors than 69. Then, the doubling time for the 6% investment would be approximately $\frac{72}{6} = 12$. In financial circles, this approximation technique for the doubling time is called the *Rule of 72*.

Radioactive Decay

The radioactive decay of elements like uranium can be modeled with the first-order differential equation $\frac{dP}{dt} = -kP(t)$, $k > 0$. Unfortunately, this differential equation is often explained by saying that *the rate of decay of the amount of material $P(t)$ is proportional to the amount of material present*. However, this explanation anthropomorphizes the radioactive substance. It gives an intelligence to the material by claiming an individual atom adjusts its tendency to decay according to whether it is part of a large sample or a small sample. There can be no communication mechanism in uranium by which an individual atom knows the size of the sample in which it resides!

The differential equation should be written in the equivalent form

$$\frac{\frac{dP}{dt}}{P(t)} = -k$$

which supports the explanation that *the rate of decay per unit substance is constant*. This view of inert matter gives to Mother Nature a reasonable uniformity. (We could also have made the same observation about population growth, stating that *the rate of growth per individual is constant*.)

An initial amount P_0 will decay to an amount $P(t)$ according to the law determined by the solution of the governing differential equation, namely, $P(t) = P_0 e^{-kt}$. Figure 2.1 shows curves of exponential decay for $k = \frac{1}{2}$ and $k = 1$.

HALF-LIFE The length of time it takes for half the initial amount to decay is called the half-life of the radioactive substance. Since $k > 0$, we solve the equation $\frac{P_0}{2} = P_0 e^{-k\tau}$ for the *half-life* $\tau = \frac{\ln 2}{k}$. Growth and decay models differ only by the sign of the proportionality constant in the differential equation, so it is not surprising that the doubling time and the half-life are governed by the same expression.

EXERCISES 2.1

1. A population grows exponentially from P_0 to $13P_0$ in 8 years. What constant annual growth rate accounts for this?

2. A population growing exponentially doubles in 7 years. What is the annual growth rate? What is the population after 10 years?

3. A population growing exponentially triples in 11 years. What is the doubling time?

4. After an hour, a population is measured at 15,000, and an hour later, at 27,000. Find the initial population and the doubling time if the growth is exponential.

5. A radioactive element has a half-life of 5000 years. How long will it take for all but 1% of the initial amount to decay?

6. A radioactive element has a half-life of 100 minutes. How long will it take for all but 5% of the initial amount to decay?

7. The National Association of Investment Clubs (NAIC) recommends buying and holding stocks that have the potential to double in value within 5 years. If the value grows exponentially, what annual growth rate must a stock sustain to meet this target?

8. Write a differential equation modeling a growth rate inversely proportional to the square root of the population size.

9. Many college students invest $100,000 in their educations. If at age 18 a student invested that amount of money at 10%, what would it be worth upon retirement at age 65? Suppose retirement is postponed until age 67. How much more is the investment worth?

2.2 Logistic Models

Self-Limiting Growth

The exponential growth model of Section 2.1 predicts that a population will grow exponentially, forever, with no bound. Clearly, in a finite world, such infinite growth cannot occur. Hence, the exponential growth model is often modified to take into account the tendency of a population to self-limit its growth. Overcrowding and the depletion of resources combine to limit growth in most observable closed colonies. Hence, the logistic model

$$\frac{\frac{dP}{dt}}{P(t)} = k(a - P(t))$$

introduces $k(a - P(t))$, a population-dependent growth rate that is nearly ka when the population is small but tends to zero as the population increases to the value a. This behavior is called "self-limited" growth. The techniques of integral calculus yield

$$P(t) = \frac{aP_0}{P_0 + (a - P_0)e^{-akt}}$$

as the solution of the IVP containing the condition $P(0) = P_0$. Details of the solution process itself are deferred to Section 3.1. However, with $a > 0$ and $P_0 \neq 0$, the limiting value of the population is $\lim_{t \to \infty} P(t) = \frac{aP_0}{P_0} = a$, so the constant a is called the *carrying capacity* of the population. Moreover, if the differential equation is written in the form $P'(t) = kP(t)(a - P(t))$, we see $P(t) = a$ is a solution of the equation $P'(t) = 0$. The solution $P(t) = a$ is therefore an *equilibrium solution* since a population starting out at the carrying capacity will remain at that size indefinitely.

For the specific model where $k = 1$ and $a = 1$, solutions with initial populations $P_0 = \frac{1}{10}$ and $P_0 = \frac{3}{2}$ along with the equilibrium solution $P(t) = 1$ are plotted in Figure 2.2.

The equilibrium solution is the horizontal line. A population that starts out at the carrying capacity remains at that size indefinitely. The solid curve is a classic example of a logistic curve with its characteristic "S" shape. It shows the behavior of a population starting out below the carrying capacity, growing quickly at first, then experiencing a diminution of its growth rate. The population is asymptotic to the carrying capacity. Finally, the dotted curve shows the decay to carrying capacity for a population that starts out greater than the carrying capacity.

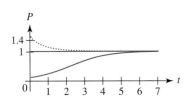

FIGURE 2.2 Solutions of the logistic equation

Logistic Growth with Harvesting

Section 1.1 contains an example of logistic growth with constant harvesting. Suppose, in that model, harvesting is not constant but is a periodic perturbation of a constant rate. In something like the fishing industry where harvesting represents catching fish, the rate of harvesting would probably not be readily maintained at a constant level. So, assuming it

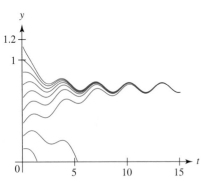

FIGURE 2.3 Solutions of the logistic equation with variable harvesting rate

varies periodically about a constant level, we write the differential equation

$$\frac{dP}{dt} = P(t)(1 - P(t)) - \frac{1}{5} + \frac{1}{10}\cos 2t \tag{2.1}$$

Unfortunately, (2.1) cannot be analyzed for large values of t by setting $P' = 0$ because the resulting expression for P is not independent of t. Instead, Figure 2.3 shows numeric solutions of (2.1) with different values of $P(0)$. For low enough values of the initial population, extinction occurs. For higher values, the population is asymptotic to what appears to be a periodically varying "steady state."

EXERCISES 2.2–Part A

A1. The logistic equation $P' = kP(a - P)$ can be solved by separation of variables. Relying on skills from elementary calculus, obtain $\frac{dP}{P(a-P)} = k\,dt$, integrate both sides, and solve for $P = y(t)$. Apply the initial condition $P(0) = A$ to determine the constant of integration.

A2. For $P' = kP(a - P) - h$, the general logistic equation with constant harvesting h, obtain an expression for $P_\infty(h)$, the limiting population corresponding to $P' = 0$. From $P_\infty(h)$, determine h_{max}, the maximum sustainable rate of harvesting. At this rate of harvesting, what is the new eqilibrium level of the population?

EXERCISES 2.2–Part B

B1. The growth of a population is modeled with a logistic equation. Determine the three parameters P_0, a, and k from the observations $10, 57, 94$ made at times $0, \frac{1}{2}, 1$, respectively. Exhibit the logistic curve, graph it, and use it to determine the population at time $t = \frac{3}{4}$. *Hints:* Clearly, P_0 is known, so there will be just two equations in two unknowns. Equally spaced time intervals yield $(e^{-akt_1})^2 = e^{-akt_2}$, from which an algebraic solution can be extracted "by hand." (Alternatively, use a computer algebra system.)

B2. The growth of a population is modeled with a logistic equation. Determine the three parameters P_0, a, and k from the observations $57, 94, 98$ made at times $\frac{1}{2}, 1, \frac{6}{5}$, respectively. Exhibit the logistic curve, graph it, and use it to determine the population at time $t = \frac{1}{3}$. *Hints:* An analytic solution is now impossible. A set of three equations in three unknowns must be solved numerically, best done with appropriate technology.

B3. Repeat Exercise B1 for the data points $(5, 497)$, $(10, 923)$, $(15, 993)$. Notice that now the initial population is not known, but

the hint still applies. Use the solution to determine the population at time $t = 12$.

B4. Repeat Exercise B2 for the data points $(1, 118)$, $(5, 497)$, $(10, 923)$. Again, a numeric solution of three equations in three unknowns must be computed. Use the solution to determine the population at time $t = 8$.

B5. Solve the IVP $y' = y(1 - y)$, $y(0) = A$, obtaining a solution $y(t)$ dependent on the parameter A. For what values of A does the graph of the solution have an inflection point? Where is that inflection point located?

B6. A population of size $P(t)$ experiences growth for which the growth rate $r = r(P)$ is quadratic in P. Formulate a differential equation modeling this growth. What can be said about the population if $r(P)$ has no real zeros?

2.3 Mixing Tank Problems—Constant and Variable Volumes

The Mixing Tank

Brine containing c pounds of salt per gallon that flows into a tank at a fixed rate of r_1 gallons per second is instantly mixed with the existing contents of the tank, and the fluid in the tank is drained at a fixed rate of r_2 gallons per second. The tank initially contained V gallons of fluid in which h pounds of salt were dissolved. We seek a differential equation and an initial value problem describing the amount of salt in the tank at any time t.

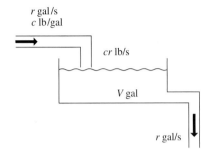

FIGURE 2.4   Schematic for the constant-volume mixing tank

The assumption of instantaneous mixing allows us to model the fluid in the tank as a homogeneous mixture at all times. If the rates of flow into and out of the tank are equal, namely, if $r_1 = r_2 = r$, the volume of fluid in the tank remains constant. If $r_1 > r_2$, the tank fills until it overflows. If $r_1 < r_2$, the tank drains until it is empty. A useful diagram for a tank problem with constant volume appears in Figure 2.4.

Let $x(t)$ be the amount of salt in the tank at any time t. If we account for changes in the salt content of the tank by recording the *rate* at which salt enters and leaves the tank, we obtain the initial value problem

$$\frac{dx}{dt} = rc - \frac{r}{V}x \qquad x(0) = h$$

The derivative $\frac{dx}{dt}$ is measured in pounds per second. The rate at which salt enters the tank is rc, also measured in pounds per second. Each second, r gallons of fluid leave the tank, taking with them some of the dissolved salt. In fact, since these r gallons represent the fraction $\frac{r}{V}$ of the volume of fluid in the tank, they take with them an equal fraction of the salt then in the tank, namely, they take the fraction $\frac{r}{V}$ of $x(t)$ pounds of salt—hence, the differential equation.

Constant Volume

EXAMPLE 2.1   Suppose brine containing 3 lb of salt per gallon flows at a rate of 4 gal/s into a tank that initially held 1000 gal of pure water. If the fluid in the tank mixes instantly and exits at a rate of 4 gal/s, model the salt content in the tank up to the moment the tank is full.

If $x(t)$ represents the amount of salt in the tank at any time t, then a differential equation governing that amount is $\frac{dx}{dt} = 3 \times 4 - \frac{4}{1000}x(t)$. With the initial condition $x(0) = 0$, a solution for the initial value problem associated with this tank would be $x(t) = 3000(1 - e^{-t/250})$.

Before performing any analysis on this solution, a moment's thought reveals significant information about the model. As t becomes infinite, the tank will contain 1000 gal of brine, each gallon of which will contain 3 lb of salt per gallon. Since the tank initially contained water, a graph of the solution $x(t)$ should be that of an increasing function that approaches the asymptote $x = 3000$ from below. Noting $x_\infty = 3000$ and observing Figure 2.5 with its plot of the solution $x(t)$ and the asymptote $x = 3000$, confirm this analysis. ❖

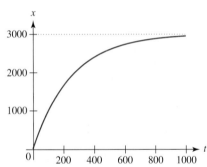

FIGURE 2.5   Salt content in a constant-volume mixing tank. Dotted line is the asymptote $x_\infty = 3000$

Varying Volume

EXAMPLE 2.2   **The filling tank**   Suppose brine containing 3 lb of salt per gallon flows at a rate of 4 gal/s into a 1000-gal tank that initially held 200 gal of pure water. If the fluid in the tank mixes instantly and exits at a rate of 2 gal/s, model the salt content in the tank.

The volume of fluid in the tank increases at a rate of 2 gal/s. Therefore, the volume of fluid in the tank is given by $V(t) = 200 + 2t$. After 400 s, the tank will overflow and any differential equation describing the salt content in the tank will cease to be valid.

If $x(t)$ represents the amount of salt in the tank at any time t, then a differential equation governing that amount is

$$\frac{dx}{dt} = 3 \times 4 - \frac{2}{V(t)}x(t) = 12 - \frac{2}{200 + 2t}x(t)$$

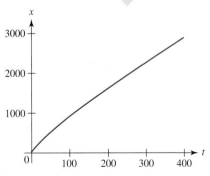

FIGURE 2.6   Salt content in a mixing tank with increasing volume

Along with the initial condition $x(0) = 0$, this differential equation forms an initial value problem whose solution, $x(t) = \frac{1200t + 6t^2}{100 + t}$, is plotted in Figure 2.6. We observe an increasing amount of salt in the tank, with a maximal amount of 2880 occurring at time $t = 400$. ❖

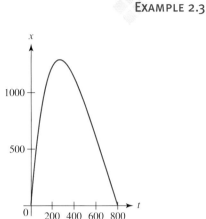

FIGURE 2.7   Salt content in a mixing tank with decreasing volume

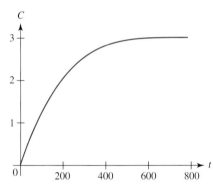

FIGURE 2.8   Concentration of salt in the decreasing-volume mixing tank

EXAMPLE 2.3   **The emptying tank**   Suppose brine containing 3 lb of salt per gallon flows at a rate of 4 gal/s into a 1000-gal tank that initially held 800 gal of pure water. If the fluid in the tank mixes instantly and exits at a rate of 5 gal/s, model the salt content in the tank.

The volume of fluid in the tank decreases at a rate of 1 gal/s. Hence, this varying volume is given by $V(t) = 800 - t$ and the tank will be empty in $t = 800$ s. After this time, the differential equation we now build is no longer valid.

If $x(t)$ represents the amount of salt in the tank at any time t, then a differential equation governing that amount is

$$\frac{dx}{dt} = 3 \times 4 - \frac{5}{V(t)} x(t) = 12 - \frac{5}{800 - t} x(t)$$

Since the initial condition is still $x(0) = 0$, the solution of the associated initial value problem is

$$x(t) = \frac{3}{409,600,000,000} t^5 - \frac{3}{102,400,000} t^4 + \frac{3}{64,000} t^3 - \frac{3}{80} t^2 + 12t$$

the graph of which appears in Figure 2.7.

We see that the amount of salt in the tank initially increases to a maximum, then, as the actual volume of fluid in the tank decreases, the amount of salt decreases. When the tank is empty, there is neither fluid nor salt present, so $x(800) = 0$. We can, however, ask for the limiting value of the concentration of salt in the tank at the moment the tank empties. An expression for the concentration at any time t is the ratio of the amount of salt to the volume of fluid. A graph of the concentration $C = \frac{x(t)}{V(t)}$ appears in Figure 2.8. The concentration at $t = 800$ is 3 lb of salt per gallon. The last few drops of fluid leaving the tank have this concentration, which is the concentration of the influx.   ❖

EXERCISES 2.3–Part A

A1. Write the differential equation governing $x(t)$, the amount of salt in a constant-volume mixing tank if the constant volume is V gallons, the flow rate is r gallons per second, and the concentration of salt in the influx is c pounds per gallon.

A2. Obtain the limiting amount of salt in the tank of Exercise A1. *Hint:* Set $x' = 0$ and solve for x. Obtain the limiting *concentration* of salt in the tank. Does this make sense physically? Explain.

A3. If the initial amount of salt in the tank in Exercise A1 is x_0, show that $x(t) = cV + (x_0 - cV)e^{-rt/V}$. Is this solution consistent with the results of Exercise A2?

A4. Show that the units for $x(t)$ in Exercise A3 are consistent.

A5. Sketch a representative curve determined by $x(t)$ in Exercise A3.

EXERCISES 2.3–Part B

B1. Brine containing $c = 2$ lb of salt per gallon that flows into a tank at a fixed rate of $r_1 = 3$ gal/s is instantly mixed with the existing contents of the tank, and the fluid in the tank is drained at the same rate of $r_2 = 3$ gal/s. The tank initially contained $V = 200$ gal of water in which $\lambda = 20$ lb of salt were dissolved. Formulate and solve an initial value problem for the amount of salt in the tank at any time t. Plot a graph of the solution, and determine the limiting amount of salt in the tank. *Hint:* The differential equation for this problem is variable separable and can be solved with methods learned in calculus. Alternatively, a computer algebra system can be used.

B2. Repeat Exercise B1 if the parameters are $c = 5, r_1 = 2, r_2 = 2,$ $V = 500$ and the tank initially held pure water.

B3. Repeat Exercise B1 if the parameters are $c = 6, r_1 = 7, r_2 = 7,$ $V = 50$, and $\lambda = 25$.

B4. Repeat Exercise B1 if the parameters are $c = 3, r_1 = 4, r_2 = 4,$ $V = 300$, and $4\lambda = 15$.

In Exercises B5–9, use a computer algebra system or other technology to obtain and plot either an analytic solution or a numeric solution.

B5. Brine containing $c = 2$ lb of salt per gallon that flows into a $C = 300$-gal tank at a fixed rate of $r_1 = 3$ gal/s is instantly mixed with the existing contents of the tank, and the fluid in the tank is drained at the rate of $r_2 = 2$ gal/s. The tank initially contained $V = 100$ gal of water in which $\lambda = 10$ lb were dissolved. Formulate an initial value problem for the amount of salt in the tank at any time t. Determine the amount of salt in the tank at the moment it overflows.

B6. Repeat Exercise B5 if the parameters are $c = 7, r_1 = 5, r_2 = 3,$ $V = 50$ and the $C = 300$-gal tank initially contained pure water.

B7. Repeat Exercise B5 if the parameters are $c = 2, r_1 = 6, r_2 = 4,$ $V = 400, \lambda = 200$, and $C = 1000$.

B8. Brine containing $c = \frac{1}{2}$ lb of salt per gallon that flows into a $C = 300$-gal tank at a fixed rate of $r_1 = 3$ gal/s is instantly mixed with the existing contents of the tank, and the fluid in the tank is drained at the rate of $r_2 = 5$ gal/s. The tank is initially filled with water in which 30 lb of salt are dissolved. Formulate an initial value problem for the amount of salt in the tank at any time t. Determine the limiting concentration of the fluid in the tank at the moment it empties.

B9. Repeat Exercise B8 if the parameters are $c = \frac{3}{4}, r_1 = 2, r_2 = 7,$ $C = 600$ and the tank is initially filled with pure water. In addition to a graph of the amount of salt in the tank, obtain a graph of the concentration of salt in the tank.

B10. At the time-dependent rate of $r = 3 + \frac{1}{2}\cos t$ gal/min, brine containing 2 lb of salt per gallon flows into a 500-gal tank which initially held 200 gal of pure water. If the instantly mixed fluid in the tank is drained at a rate of 1 gal/min, formulate a differential equation for the amount of salt in the tank up to the moment the tank overflows.

2.4 Newton's Law of Cooling

Newtonian Model of Heat Flow

To quantize the change in temperature observed when a body of one temperature is placed in an environment of a different temperature, we assume the rate of change of temperature is proportional to the difference in temperature between the body and the environment. This assumption includes the supposition that the instant a temperature gradient (spatial difference of temperature) exists, heat flows, and flows everywhere. Heat flow exhibits no inertia: as soon as a difference in temperature is established, heat instantly flows. The rate of change of temperature is proportional to the difference in temperature between the body and the surroundings.

The model of heat transfer so postulated and captured in the following differential equation is attributed to Isaac Newton. Letting $U(t)$ represent the temperature of the object in question and U_s represent the temperature of the surrounding environment, we also assume that U_s remains constant during the process of heat transfer to or from the body. Such an environment is called a *heat sink,* a reservoir of heat energy capable of absorbing or releasing heat without having its own temperature appreciably changed. The ocean, for example, is a heat sink for a thermal event on a passing ship.

Thus, we have the differential equation $\frac{dU}{dt} = k(U(t) - U_s)$, whose solution is $U(t) = U_s + Ae^{kt}$. Consequently, the model contains the three parameters U_s, k, and A. (The constant A arises from integration following a separation of variables. See Section 3.1 and the exercises.) It will take three pieces of data to specify all three constants. An initial condition of the form $U(0) = U_0$ gives an initial value problem whose solution is then $U(t) = U_s + e^{kt}(U_0 - U_s)$, leaving just U_s and k to be determined to specify the parameters of the model.

Notice carefully that no assumptions were made about the body cooling or warming. Some texts make this distinction; take k positive; adjust the sign in the exponential or in the difference $U - U_s$; and generate two versions of the model, one for heating and one for cooling. This is an unnecessary distinction: use one form of the differential equation and let the data determine the value of k.

EXAMPLE 2.4 At 1:27 P.M. a medical examiner (M.E.) notes the temperature of the body of a deceased person is 80°F, and the environment in which the body has been located is at 71°F. Being careful not to alter the surrounding temperature, the examiner waits 20 minutes and again checks the body's temperature, finding it to be 78°F. Show how the examiner estimates the time of death for the deceased.

Observation yields two data points $(t, U) = (0, 80)$ and $(20, 78)$ on the curve that is a solution to Newton's law of cooling. In addition, U_s, the surrounding temperature, is noted as 71°F. Hence, all three constants in the model can be determined, and the time at which the body's temperature was 98.6°F can be calculated.

Starting with the solution in the form $U(t) = 71 + Ae^{kt}$, the M.E. forms the two algebraic equations

$$71 + A = 80 \quad \text{and} \quad 71 + Ae^{20k} = 78$$

whose solution is $A = 9, k = \frac{1}{20}\ln\frac{7}{9}$. Hence, the equation for the model is $U(t) = 71 + 9e^{(t/20)\ln(7/9)} = 71 + 9\left(\frac{7}{9}\right)^{t/20}$. The M.E. then determines the time at which the body's temperature was 98.6 by solving the equation $U(t) = 98.6$ for $t = -89.17842108$ and concludes that 89.2 minutes prior to 1:27 P.M. the deceased expired. ❖

EXAMPLE 2.5 From [56], we pose the following problem:

At the spring picnic a cup of coffee is brewed; milk is added, bringing the coffee's temperature to 180°F; and the cup is set on the picnic table because a volleyball game just started. Seven minutes later the temperature of the coffee is 150°F; seven minutes after that, 127°F. What is the temperature that spring day?

Three pieces of data have been given, namely, the data points $(0, 180)$, $(7, 150)$, and $(14, 127)$. This is enough information to convert $U(t) = U_s + Ae^{kt}$ to the three equations

$$U_s + A = 180 \quad U_s + Ae^{7k} = 150 \quad U_s + Ae^{14k} = 127$$

in the three unknowns A, k, and U_s. The solution is $U_s = \frac{360}{7}, k = \frac{1}{7}\ln\frac{23}{30}, A = \frac{900}{7}$; so the ambient temperature, assumed constant during this experiment, is $\frac{360}{7} \stackrel{\circ}{=} 51.4°$F—a bit chilly for a spring picnic. ❖

EXAMPLE 2.6 Also from [56], we pose this final problem:

At the fall picnic, on a day when the temperature was a balmy 74°F, a cold drink was taken from the cooler just as the volleyball game was starting. The drink was set on the picnic table, and seven minutes later its temperature was 50°F. Seven minutes after that, its temperature was 60°F. How cold is the inside of the cooler?

Again, three pieces of data are given; two pieces are the data points $(7, 50)$ and $(14, 60)$, and one piece of data is the ambient temperature $U_s = 74°$F. We can form two equations in the two unknowns A and k, thereby determining the model, from which we then calculate $U(0)$. On the other hand, we can incorporate the value of $U(0) = x$ in a third equation, and solve for all three unknowns at once. Therefore, solve the equations

$$74 + A = x \quad 74 + Ae^{7k} = 50 \quad 74 + Ae^{14k} = 60$$

obtaining $A = -\frac{288}{7}, k = \frac{1}{7}\ln\frac{7}{12}, x = \frac{230}{7}$. The temperature in the cooler is a frosty $\frac{230}{7} \stackrel{\circ}{=} 32.9°$F! ❖

EXERCISES 2.4–Part A

A1. Using the techniques of elementary calculus, separate variables in Newton's law of cooling to obtain $dU/(U - U_s) = k\,dt$. Antidifferentiate (integrate) both sides, then solve for $U(t) = U_s + Ae^{kt}$.

A2. To $U(t) = U_s + Ae^{kt}$ apply the initial condition $U(0) = U_0$, thereby determining the constant A.

A3. Show that if the three data points used to fit Newton's law of cooling to observations are taken at equally spaced time intervals, then the system of equations reduces to a quadratic in e^{kt_1}. In particular, let the times of the observations be t_1, $t_2 = 2t_1$, $t_3 = 3t_1$

and form the three equations e_1, e_2, e_3. Then, let $e_4 = e_1 - e_2$ and $e_5 = e_2 - e_3$ and divide the corresponding sides of e_4 and e_5. This will lead to an equation linear in e^{kt_1}. Solving this equation for e^{kt_1} gives k via a logarithm, from which A and U_s can then be found.

A4. Write a differential equation modeling heat flow for which the rate of change of a body's temperature is directly proportional to the square of the difference in temperatures between the body and its surroundings.

EXERCISES 2.4–Part B

B1. A cool object of unknown temperature is placed in a warm environment, also of unknown temperature. Three observations of the object's temperature are then taken. If the observed temperatures $U(t)$ are $U(2) = 8$, $U(5) = 15$, $U(11) = 23$, find the initial temperature of the object and the temperature of the warm environment. Graph the object's temperature as a function of time. *Hint:* Form three equations in the three unknown parameters in the general solution of Newton's law of cooling and solve using Maple's **fsolve** command or the equivalent. Even though the object is warming, the constant k will be negative!

B2. A warm object of unknown temperature is placed in a cool environment, also of unknown temperature. Three observations of the object's temperature are then taken. If the observed temperatures $U(t)$ are $U(1) = 100$, $U(5) = 75$, $U(12) = 47$, find the initial temperature of the object and the temperature of the cool environment. Graph the object's temperature as a function of time.

Hint: Form three equations in the three unknown parameters in the general solution of Newton's law of cooling and solve using Maple's **fsolve** command or the equivalent. The constant k will again be negative!

B3. If, in Exercise B2, $U(12) = 30$, the solution of the three equations in three unknowns yields an ambient temperature higher than the initial temperature. Compute the solution for these three data points and conclude that although the Newton's law of cooling curve can be fit to this data, the data could not have been generated from any real physical system cooling according to that law.

B4. Example 2.4 is predicated on the assumption that the normal body temperature for a human being is 98.6°F. This happens to be a myth. Normal body temperatures can range as low as 96°F and as high as 99°F. If this variation in initial temperature is taken into account, how much difference in predicted time of death could there be in the conclusion of Example 2.4?

Chapter Review

1. In words, state the hypothesis that leads to the law of exponential growth.

2. A population obeys the exponential law $P(t) = 10e^{rt}$. Use the following examples to show that no assumptions need be made about the sign of r, the growth-rate constant, since that information follows from observational data.
   (a) If $P(1) = 20$, find r.   **(b)** If $P(1) = 5$, find r.

3. Explain the "Rule of 72."

4. What is the half-life of an exponentially decaying quantity? How can it be estimated from the Rule of 72?

5. What is the doubling time for an exponentially growing quantity? How can it be estimated from the Rule of 72?

6. Explain the term *self-limiting growth*.

7. What is the *carrying capacity* in the logistic model?

8. What is an equilibrium solution?

9. Formulate a mixing-tank problem in which the volume of fluid in the tank remains constant. Write a differential equation modeling the physical process.

10. Why make the assumption of "instantaneous mixing" for the fluid in the tank?

11. Formulate a mixing-tank problem in which the volume of fluid in the tank decreases. Write a differential equation modeling the physical process. For how long is the differential equation valid?

12. Formulate a mixing-tank problem in which the volume of the fluid in the tank increases. Write a differential equation modeling the physical process. For how long is the differential equation valid?

13. In words, state Newton's law for heat flow between an object and a surrounding heat-sink?

14. What is the differential equation which expresses the content of Newton's law for heat flow?

15. The solution to a model of exponential growth contains the exponential e^{rt} in which r, the growth-rate constant, is positive. For exponential decay, r is negative. It is tempting, but wrong, to impose this same type of conclusion on the solution of Newton's heat-flow problem. Recall that for the *warming* soft drink in Example 2.6, $k = \frac{1}{7}\ln\frac{7}{12} < 0$. To see why this difference exists,

let $U_s = 0$ in Newton's model.

(a) Show $U(t) = U_0 e^{kt}$ for any U_0 in Newton's model.

(b) If $U_0 = 10$, show that cooling must occur and that k is negative.

(c) If $U_0 = -10$, show that warming must occur, but k is again negative.

(d) What is the difference between the population model and the heat-transfer model that accounts for this anomaly?

Chapter *3*

Methods for Solving
First-Order ODEs

I N T R O D U C T I O N   Practitioners of every trade, art, or science must learn the fundamental techniques of their disciplines. The sculptor must know how to carve, the carpenter how to cut and join wood, and the lab technician how to acquire and interpret data. A study of differential equations likewise includes a component of methods and techniques.

In this chapter we learn some of the analytic techniques for obtaining solutions of first-order differential equations. Some, like separation of variables, might be familiar from elementary calculus. Others, like variation of parameters for a linear equation, will more likely be new. However, we hope the reader does not lose sight of the forest because of the trees. These techniques for solving differential equations are not the end-all and be-all of the discipline.

Techniques are just tools for determining how solutions of differential equations behave. It is the behavior of the solution, reflecting the behavior of a modeled physical system, that is the forest. Methods of solution are just the trees.

Computer software, and even calculators, can provide analytic solutions, as well as numeric ones. But, like using a calculator to learn arithmetic, technology should not replace understanding. While we heartily endorse use of technology as a labor-saving strategy, we also counsel that basic techniques not be ignored. These techniques can be learned in conjunction with technology, much like arithmetic, and even algebra, can be learned in conjunction with a calculator.

3.1 Separation of Variables

Variable-Separable Equations

A differential equation that can be put into the form $f(y)y'(x) = g(x)$ is said to be a *separable* ordinary differential equation. The equations for exponential growth and decay, the logistic model, the mixing tank with *constant* volume, and Newton's law of cooling are all separable differential equations. Any such equation can be solved by multiplication by the differential dx, yielding $f(y)y'\,dx = g(x)\,dx$. Since dy, the differential of $y(x)$ is *defined* to be $y'\,dx$, the equation really reads $f(y)\,dy = g(x)\,dx$. Integrating both sides, we find

$$\int f(y)\,dy = \int g(x)\,dx + C \tag{3.1}$$

where C is an arbitrary constant of integration. Solving the differential equation has been reduced to evaluating the integrals in (3.1) and perhaps applying some algebra that puts the solution into the explicit form $y = y(x)$. Finally, if the separable differential equation arises in the context of an initial value problem, then, as verified in Exercise A1, the initial condition $y(x_0) = y_0$ can be incorporated into the integrals via

$$\int_{y_0}^{y} f(t)\, dt = \int_{x_0}^{x} g(t)\, dt \tag{3.2}$$

EXAMPLE 3.1   **Exponential growth**   The equation for exponential growth is a prototypical separable equation. Its solution should be mastered by every student of differential equations. In detail, the calculus and algebra for solving this equation begin with writing the equation $P'(t) = kP(t)$ in the separated form $\frac{P'}{P} = k$. Multiplication by the differential dt yields $\frac{P'\,dt}{P} = k\,dt$, from which $\frac{dP}{P} = k\,dt$ follows by the definition of the differential. Integrating both sides, namely, $\int \frac{dP}{P} = \int k\,dt + C$, gives $\ln|P| = kt + C$. The solution of the differential equation has been effected. Isolating $P(t)$ is now effected by *algebra*.

Exponentiate both sides to obtain $e^{\ln|P|} = e^{kt+C} = e^{kt}e^{C}$, and hence $|P| = e^{C}e^{kt} = ae^{kt}$, where e^{C}, never zero, has been replaced with the different arbitrary constant, a. Since P, a population, is nonnegative, drop the absolute value. (Were P a negative quantity, the solution could still be written as $P(t) = be^{kt}$, where now $b < 0$.) Applying the initial condition $P(0) = P_0$ finally yields $P(t) = P_0 e^{kt}$.   ❖

EXAMPLE 3.2   **Logistic equation**   The logistic equation $\frac{dP}{dt} = kP(t)(a - P(t))$ is also variable separable, the separated form being $\frac{P'}{P(a-P)} = k$. Multiplication by the differential dt yields $\frac{dP}{P(a-P)} = k\,dt$, an integrable form leading to

$$\frac{1}{a}(\ln|P| - \ln|a - P|) = kt + C$$

Exponentiating is simplified if the equation is multiplied through by a and the two logarithms on the left are combined, producing

$$\ln\left|\frac{P}{a - P}\right| = akt + c$$

where $c = aC$. Exponentiating now yields $|\frac{P}{a-P}| = e^{akt}e^{c}$. Replacing e^{c} with b yields $|\frac{P}{a-P}| = be^{akt}$ and dropping the absolute value on the left, justified by renaming b to A, a constant that might contain any minus sign arising from dropping the absolute value, leads to $\frac{P}{a-P} = Ae^{akt}$. Solving for $P(t) = aAe^{akt}/(1 + Ae^{akt})$ completes the solution, except perhaps for applying the initial condition $P(0) = P_0 \neq 0$, in which case $P_0 = \frac{aA}{1+A}$ so $A = \frac{P_0}{a-P_0}$, and $P(t) = aP_0 e^{akt}/(a - P_0 + P_0 e^{akt})$. Some texts, however, multiply top and bottom by e^{-akt} to put the solution into the form

$$P(t) = \frac{aP_0}{P_0 + (a - P_0)e^{-akt}}$$

Evaluation of the integral $\int \frac{dx}{x(a-x)}$ can be obtained by a partial fraction decomposition of the integrand. Thus,

$$\frac{1}{x(a - x)} = \frac{\frac{1}{a}}{x} + \frac{\frac{1}{a}}{a - x}$$

and the integration to logarithms is immediate. A much more complicated solution can be based on completion of the square in the denominator, and integration to an inverse hyperbolic tangent, a function convertible to logarithms. This virtuosity in computation is left to the exercises. ❖

EXAMPLE 3.3 **A routine textbook exercise** Do not believe that the essence of differential equations is simply mastery of *solution* techniques. The goal is insight into the behavior of solutions, and this can sometimes require more than just rote application of a mantra such as "separate variables and integrate both sides."

Taken from a recent calculus text, the differential equation $\frac{dy}{dx} = \frac{x^2}{1+3y^2}$ can be solved mentally by separation of variables, leading to the implicit solution $y + y^3 = \frac{1}{3}x^3 + c$. (See [56] or [57].) Graphs of $y(x)$, obtained as the level curves of $f(x, y) = y + y^3 - \frac{1}{3}x^3$, and seen in Figure 3.1, appear to approach an oblique asymptote. Validation of this observation requires skills far beyond the routine steps of solving the differential equation itself.

For example, to obtain $y(x)$ explicitly, a cubic equation must be solved, leading to $y(x) = \frac{A}{6} - \frac{2}{A}$, where

$$A = \sqrt[3]{36x^3 + 108c + 12\sqrt{12 + 9x^6 + 54x^3c + 81c^2}}$$

$$= x\sqrt[3]{36 + \frac{108c}{x^3} + 12\sqrt{\frac{12}{x^6} + 9 + \frac{54c}{x^3} + \frac{81c^2}{x^6}}} \qquad x \neq 0$$

For large x, $A \to x\sqrt[3]{72}$, so $y \to Lx - \frac{L^2}{x}$, where $L = \frac{\sqrt[3]{72}}{6}$. Thus, $y' \to L$, a constant, and indeed the solutions of the differential equation tend toward the oblique asymptote $y = Lx$. ❖

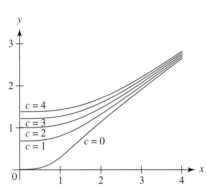

FIGURE 3.1 Level curves of $f = y^3 + y - \frac{1}{3}x^3$ are solutions of $y' = \frac{x^2}{1+3y^2}$.

EXERCISES 3.1–Part A

A1. Show that the solution $y(x)$ defined by (3.2) passes through the point (x_0, y_0).

A2. Verify the partial fraction decomposition $\frac{1}{x(a-x)} = \frac{1/a}{x} + \frac{1/a}{a-x}$ used to perform the integration in Example 3.2.

A3. Find a method for solving a cubic equation and use it to obtain $y(x) = \frac{A}{6} - \frac{2}{A}$ in Example 3.3.

A4. Obtain the general solution of $3y^2 = (x^2 + 1)y' + 4$.

A5. The law of mass action in Section 1.1 leads to separable differential equations of the form $x' = k(a - x)(b - x)$. Obtain the solution.

EXERCISES 3.1–Part B

B1. Evaluate the integral $\int \frac{dx}{x(a-x)}$ by completing the square in the denominator and integrating to an inverse hyperbolic tangent. Convert this function to its logarithmic equivalent, and show the result agrees with integrating by partial fractions.

For each of Exercises B2–7:

(a) Separate variables and obtain a general solution.

(b) Plot representative members of the family of solutions defined by the general solution.

(c) Apply the data to the general solution, thereby obtaining the solution to the given initial value problem.

(d) Plot the solution of the initial value problem.

(e) Solve the initial value problem by including the initial data in the integrations used in the separation of variables. Thus, $y'f(y) = g(x)$, $y(x_0) = y_0$, has the solution $\int_{y_0}^{y} f(t)\, dt = \int_{x_0}^{x} g(t)\, dt$.

B2. $xy' = y^2$, $y(1) = 1$ **B3.** $(3 - x)y' = y^2$, $y(1) = 1$

B4. $(y + 2) = 3xyy'$, $y(3) = 5$ **B5.** $x^3yy' = e^y$, $y(2) = 3$

B6. $(e^x + 3)y' = y^2$, $y(0) = 1$ **B7.** $yy' = e^{y-2x}$, $y(0) = 1$

Equations with Homogeneous Coefficients

Polynomials Homogeneous of Degree k

If m and n are integers, the total degree of the product $x^m y^n$ is $k = m + n$. Polynomials in x and y having all terms of the same total degree k are said to be *homogeneous of degree k*. For example, the polynomial $p(x, y) = 3x^4 y^2 - 5xy^5 + 7x^3 y^3$ has each of its terms of total degree 6 and is therefore said to be homogeneous of degree 6. As a consequence, we clearly have $p(\lambda x, \lambda y) = \lambda^6 p(x, y)$.

Functions Homogeneous of Degree k

As a generalization of the polynomial case, we *define* an arbitrary function $f(x, y)$ to be homogeneous of degree k if it satisfies the equation

$$f(\lambda x, \lambda y) = \lambda^k f(x, y)$$

For example, the function $f(x, y) = \frac{y^3}{3x^2 - 5xy + 4y^2}$ is homogeneous of degree 1 because

$$f(\lambda x, \lambda y) = \frac{(\lambda y)^3}{3(\lambda x)^2 - 5(\lambda x)(\lambda y) + 4(\lambda y)^2} = \lambda \frac{y^3}{3x^2 - 5xy + 4y^2} = \lambda f(x, y)$$

whereas the function $g(x, y) = \frac{3x+4y}{5x-7y}$ is homogeneous of degree zero because

$$g(\lambda x, \lambda y) = \frac{3(\lambda x) + 4(\lambda y)}{5(\lambda x) - 7(\lambda y)} = \frac{3x + 4y}{5x - 7y} = \lambda^0 g(x, y) = g(x, y) \qquad (3.3)$$

Differential Equations with Coefficients Homogeneous of Degree Zero

A first-order differential equation of the form $y'(x) = F(x, y)$ in which the function $F(x, y)$ is homogeneous of degree zero will become variable-separable under the change of variables $y(x) = xv(x)$. This is so because

$$F(\lambda x, \lambda y) = \lambda^0 F(x, y) = F(x, y) \quad \Rightarrow \quad F(x, xv) = x^0 F(1, v) = h(v)$$

where, on the right, x plays the role of λ. In effect, the degree-zero homogeneous function depends on the single variable $v = \frac{y}{x}$. For example, if the function $g(x, y)$ in (3.3) were to have its numerator and denominator both divided by x, then every appearance of the variable y would be in the form $\frac{y}{x}$ as shown by

$$\frac{3x + 4y}{5x - 7y} = \frac{3 + 4\frac{y}{x}}{5 - 7\frac{y}{x}} = \frac{3 + 4v}{5 - 7v}$$

The change of variables $y(x) = xv(x)$ converts the form $y'(x) = F(x, y)$ to $xv'(x) + v(x) = F(1, v(x))$, which has the separated form $\frac{v'}{F(1,v)-v} = \frac{1}{x}$. For example, the differential equation $y'(x) = g(x, y)$, under the change of variables $y(x) = xv(x)$, becomes $v(x) + xv'(x) = \frac{3+4v(x)}{5-7v(x)}$, a separable differential equation in x and v. Separation of variables is effected by moving $v(x)$ from the left to the right side of the equation, adding fractions on the right, and making the appropriate divisions. Thus,

$$xv'(x) = \frac{3 + 4v(x)}{5 - 7v(x)} - v(x) \qquad xv' = \frac{7v^2 - v + 3}{5 - 7v} \qquad \frac{(5 - 7v)\,dv}{7v^2 - v + 3} = \frac{dx}{x}$$

Integration of both sides is now possible, yielding

$$\frac{9}{\sqrt{83}} \arctan\left(\frac{14v-1}{\sqrt{83}}\right) - \frac{1}{2}\ln(7v^2 - v + 3) = \ln x + c$$

or

$$\frac{9}{\sqrt{83}} \arctan\left(\frac{14\frac{y}{x}-1}{\sqrt{83}}\right) - \frac{1}{2}\ln\left(7\left(\frac{y}{x}\right)^2 - \frac{y}{x} + 3\right) = \ln x + c \qquad (3.4)$$

upon replacing $v(x)$ with $\frac{y}{x}$.

Differential Equations with Coefficients Homogeneous of Degree $k \neq 0$

A first-order differential equation of the form $f(x, y)y'(x) + g(x, y) = 0$ in which both $f(x, y)$ and $g(x, y)$ are homogeneous of the *same degree* is also variable-separable under the same change of variables $y(x) = xv(x)$. This is so because in the then-equivalent differential equation $y'(x) = F(x, y) = -\frac{f(x,y)}{g(x,y)}$, the function $F(x, y)$ would be homogeneous of degree zero. Hence, this differential equation is equivalent to the first case discussed previously.

Any elation evinced by the certainty that a differential equation with homogeneous coefficients becomes variable-separable under the change of variables $y(x) = xv(x)$ is tempered by the realization that the resulting integrals may be difficult, or even impossible, to evaluate in closed form. Even when successful, the integrations often leave in their wake an implicitly defined solution such as the one in (3.4).

EXERCISES 3.2–Part A

A1. Show that for any $f(z)$, the function $F(x, y) = x^k f(\frac{y}{x})$ is homogeneous of degree k.

A2. Show that if $F(x, y)$ is homogeneous of degree k in x and y, then F can be written as $F = x^k f(\frac{y}{x})$. (Some texts take this as the *definition* of homogeneity.)

A3. Using Exercise A2, show that if $F(x, y)$ is homogeneous of degree k in x and y, then $x\frac{\partial F}{\partial x} + y\frac{\partial F}{\partial y} = kF$, a result known as Euler's theorem for homogeneous functions.

A4. Give an example of a polynomial $p(x, y)$ that is homogeneous of degree 2. Then, use it to illustrate Euler's theorem as stated in Exercise A3.

A5. Use the function $f(x, y) = \frac{y^3}{3x^2 - 5xy + 4y^2}$ to illustrate Euler's theorem as stated in Exercise A3.

EXERCISES 3.2–Part B

B1. If $F(x, y)$ is homogeneous in x and y, show that the differential equation $y''f(x) + F(x, y)(y - xy') = 0$ is separable under the change of variables $y(x) = xv(x)$.

In Exercises B2–5:

(a) Obtain a general solution via Maple's **dsolve** command (or the equivalent).

(b) Plot representative members of the family of curves contained in the general solution.

(c) Apply the initial condition to the general solution to obtain the solution of the given IVP; and graph this solution.

(d) Make the change of variables $y(x) = xv(x)$ to reduce the differential equation to a separable form, and solve. Show that the resulting general solution is equivalent to the one found in part (a).

B2. $(5x + 2y)y' = 3y - x,\ y(1) = 1$

B3. $(5y^2 - x^2)y' = xy,\ y(1) = 1$

B4. $(2x^2 + 3y^2)y' = 5xy,\ y(2) = 1$

B5. $(5x^2 + 3y^2)xy' + (x^2 + y^2)y = 0,\ y(1) = 1$

Equations of the form $\frac{dy}{dx} = \frac{ax+by+c}{px+qy+r}$ can be reduced to homogeneous equations provided the equations $ax + by + c = 0$, $px + qy + r = 0$

have a unique solution $(x, y) = (h, k)$. The change of variables $x = t + h, y = w + k$ then yields the new differential equation $\frac{dw}{dt} = \frac{at+bw}{pt+qw}$, which is homogeneous. (In particular, the second transformation equation defines $w = w(y(x))$, while the first defines $x = x(t)$ so that together we have $w = w(y(x(t)))$, from which it follows that $\frac{dw}{dt} = \frac{dy}{dx}$. The algebra on the right-hand side of the differential equation is obvious.) Use this technique to solve the equations in Exercises B6–8.

B6. $y' = \dfrac{3x + 2y - 5}{7x - 6y + 4}$ **B7.** $y' = \dfrac{8x + 3y + 1}{x - y}$

B8. $y' = \dfrac{x + 2y}{3x - 4y + 9}$

3.3 Exact Equations

Level Curves

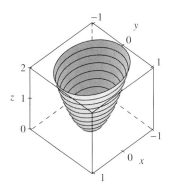

FIGURE 3.2 Surface corresponding to the function $f(x, y) = 2x^2 + 3y^2$

If $z = f(x, y)$ defines a surface, then its *level curves* (or contours) are curves $y(x)$ that satisfy $f(x, y(x)) = c$, where c is a constant. In effect, the level curves are obtained by solving the equation $f(x, y) = c$ for $y = y(x)$.

For example, the surface defined by the function $f(x, y) = 2x^2 + 3y^2$ is shown in Figure 3.2, where the level curves have been drawn right on the surface itself. The level curves themselves, which appear in Figure 3.3, can be obtained by solving the equation $2x^2 + 3y^2 = c$ for $y(x) = \pm\frac{1}{3}\sqrt{-6x^2 + 3c}$, which are the two branches, upper and lower, for the ellipses shown in the contour plot.

Differential Equation for Level Curves

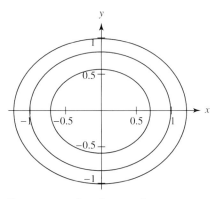

FIGURE 3.3 Level curves for $f(x, y) = 2x^2 + 3y^2$

Suppose the function $z = f(x, y)$ has level curves given by $y = y(x)$ and that these level curves have been found by solving the equation $f(x, y) = c$ for $y(x)$. Then, if $y = y(x)$ is substituted back into the equation, the identity $f(x, y(x)) = c$ results. If this identity is differentiated with respect to x, we get

$$f_x + f_y y' = 0 \tag{3.5}$$

as a differential equation satisfied by the level curves. Thus, starting with $f(x, y)$, we have level curves $y(x)$ that are solutions of the differential equation (3.5).

For the function $f(x, y) = 2x^2 + 3y^2$, the differential equation defining its level curves is $4x + 6yy' = 0$ and its solution is given implicitly by $f(x, y) = 2x^2 + 3y^2 = c$.

Exact Equations and First Integrals

A differential equation of the form (3.5) that arises as the equation for the level curves of the function $f(x, y)$ is called an *exact differential equation*, and the function $f(x, y)$ is called a *first integral* for the differential equation. Thus, the level curves of the first integral are the solution curves for the corresponding exact equation.

For example, the differential equation $4x + 6yy' = 0$ is an exact equation, and the function $f(x, y) = 2x^2 + 3y^2$ is a first integral for it.

In general, the partial derivatives f_x and f_y are functions of x and y. Hence, it is possible to represent the exact equation in either of the forms

$$M(x, y) + N(x, y)y' = 0 \quad \text{or} \quad M(x, y)\,dx + N(x, y)\,dy = 0 \tag{3.6}$$

where, clearly,

$$f_x = M(x, y) \quad \text{and} \quad f_y = N(x, y) \tag{3.7}$$

For the exact equation $4x + 6yy' = 0$, we then have $M(x, y) = 4x$ and $N(x, y) = 6y$.

Condition for Exactness

Not every differential equation of the form (3.6) is exact. According to the definition, the equation is exact only if $M(x, y) = f_x$ and $N(x, y) = f_y$ for some function $f(x, y)$. If $M(x, y)$ and $N(x, y)$ were picked at random, the resulting differential equation would not necessarily be exact since these two functions need not be the partial derivatives of the *same* function $f(x, y)$. Hence, we seek a condition that determines whether or not a given differential equation is exact.

If we *start* with the function $f(x, y)$ and *generate* the exact equation $M + N \frac{dy}{dx} = 0$, then we know for sure that (3.7) is valid. If the second partial derivatives of $f(x, y)$ are continuous, we are guaranteed that the mixed second-order partial derivatives (the "cross-partials") are equal. Thus, we are guaranteed

$$\frac{\partial^2 f}{\partial y \partial x} = f_{xy} = f_{yx} = \frac{\partial^2 f}{\partial x \partial y}$$

Consequently, we would also have $M_y = N_x$, which is precisely the same relationship between the mixed second partial derivatives f_{xy} and f_{yx}.

It turns out that if M, N, M_y, and N_x are continuous in a simply connected region, then the condition $M_y = N_x$ completely characterizes the exact equation and, therefore, guarantees the existence of a first integral $f(x, y)$, whose level curves are the solution of the equation.

Loosely speaking, a *simply connected region* is one with "no holes." A circle or rectangle is simply connected, but an annulus is not. Every closed curve in a simply connected region can be continuously shrunk to a point. In an annulus, a circle surrounding the "hole" cannot be shrunk to a point.

Of the following two differential equations, the first one is not exact but the second is.

EXAMPLE 3.4   Given the functions

$$M = y \arctan \frac{y}{x} + \frac{xy^2}{x^2 + y^2} \quad \text{and} \quad N = x \arctan \frac{y}{x} + \frac{x^2 y}{x^2 + y^2}$$

the differential equation $M\,dx + N\,dy = 0$ can be tested for exactness by computing the partial derivatives

$$M_y = \arctan \frac{y}{x} + \frac{3x^3 y + xy^3}{(x^2 + y^2)^2} \quad \text{and} \quad N_x = \arctan \frac{y}{x} - \frac{x^3 y - xy^3}{(x^2 + y^2)^2}$$

Since $M_y - N_x = \frac{4x^3 y}{(x^2 + y^2)^2} \neq 0$, it is clear that $M_y \neq N_x$ and the differential equation $M\,dx + N\,dy = 0$ is not exact.   ❖

EXAMPLE 3.5   Given the functions

$$M = \frac{x^2 - 2xy - 2y^2}{(x - y)^2} \quad \text{and} \quad N = \frac{2x^2 + 2xy - y^2}{(x - y)^2}$$

the differential equation $M\,dx + N\,dy = 0$ can be tested for exactness by computing the partial derivatives

$$M_y = -\frac{6xy}{(x - y)^3} \quad \text{and} \quad N_x = -\frac{6xy}{(x - y)^3}$$

That $M_y = N_x$ is obvious. Hence, the differential equation $M\,dx + N\,dy = 0$ is an exact differential equation.   ❖

Constructing the First Integral

Once it's clear that the equation $M\,dx + N\,dy = 0$ is exact, the obvious question is "how can the first integral $f(x, y)$ be found?" Exactness means (3.7) holds for some function $f(x, y)$. Thus, integrate the first equation with respect to x, holding y constant, to find an expression for $f(x, y)$. When adding the constant of integration, make it a function of y, since y was held constant in the integration. Thus,

$$f(x, y) = \int M(x, y)\,dx + g(y)$$

If this really is $f(x, y)$, then differentiation with respect to y should reproduce $N(x, y)$. This yields a condition on $g'(y)$, from which $g(y)$ is then determined. The process is demonstrated in the following example.

EXAMPLE 3.6   For the exact differential equation in Example 3.5, integrate $M(x, y)$ to obtain

$$f(x, y) = \int \frac{x^2 - 2xy - 2y^2}{(x-y)^2}\,dx = x + \frac{3y^2}{x-y} + g(y)$$

Since $f_y = N$, we obtain

$$\frac{6xy - 3y^2}{(x-y)^2} + g'(y) = \frac{2x^2 + 2xy - y^2}{(x-y)^2}$$

from which $g'(y) = 2$ and $g(y) = 2y$ follow. Consequently, we have $f(x, y) = x + \frac{3y^2}{x-y} + 2y = \frac{x^2 + xy + y^2}{x-y}$. ❖

EXERCISES 3.3–Part A

A1. Verify that $f(x, y)$, the first integral constructed in Example 3.6, satisfies $f_x = M$ and $f_y = N$.

A2. Rephrase the procedure for finding a first integral of an exact equation, showing how to start with an integral of $N(x, y)$. Apply your algorithm to the exact equation solved in Example 3.6.

A3. Show that $2xy^3 + 3x^2y^2y' = 0$ is exact, and find a first integral. From the first integral, explicitly obtain the solution curves $y = y(x)$.

A4. Sketch the surface $z = \sqrt{x^2 + y^2}$; sketch its level curves, and write them explicitly as $y = y(x)$. What exact differential equation do these level curves satisfy?

A5. Put $y' = \frac{y}{x}$ into the form $M\,dx + N\,dy = 0$ and test for exactness. If it is exact, obtain a first integral. If it is not exact, solve it by some other means, write the solution in the form $f(x, y) = c$, and compare $f_x\,dx + f_y\,dy = 0$ to $M\,dx + N\,dy = 0$.

EXERCISES 3.3–Part B

B1. Put the differential equation $y' = \frac{p(x,y)}{q(x,y)}$ into the form $M\,dx + N\,dy = 0$. What condition must p and q satisfy for the equation to be exact?

In each of Exercises B2–13 show that the given differential equation is exact, and obtain the general solution in the form $f(x, y) = c$, where $f(x, y)$ is a first integral. Plot several level curves for this first integral.

B2. $y' = \dfrac{3y(3 + 8x^2)}{27y^2 - 9x - 8x^3}$

B3. $y' = -\dfrac{5y^3 - 18xy^2 - 9x^2}{15xy^2 - 18x^2y - 5}$

B4. $y' = \dfrac{7y^2 + 30x^2 - 15yx^2}{12y - 14xy + 5x^3}$

B5. $y' = -\dfrac{y - 4x + 8xy^3}{x(1 + 12xy^2)}$

B6. $y' = \dfrac{16x - 3y^4 - 3}{12xy^3}$

B7. $y' = \dfrac{4xy^3 - 9y^4 + 5}{6xy^2(6y - x)}$

B8. $y' = -\dfrac{4x^3y}{9y^2 + x^4}$

B9. $y' = \dfrac{y(8xy^2 + 5y^2 - 3x^2 - 20x^3y^2)}{x(15x^3y^2 - 12xy^2 - 15y^2 + x^2)}$

B10. $y' = \dfrac{10x + 9x^2y^3 - 2y}{3 - 9x^3y^2 + 2x}$

B11. $y' = -\dfrac{y(5 + 12x + 12xy)}{5x + 6x^2 + 12yx^2 + 8y}$

B12. $y' = -\dfrac{2(6xy + 4 + 9xy^3)}{3x^2(2 + 9y^2)}$

B13. $y' = \dfrac{10 - 3y^2 - 14x + 27x^2}{6xy}$

3.4   Integrating Factors and the First-Order Linear Equation

EXAMPLE 3.7   Newton's law for heat transfer between a body and its surroundings has the rate of change of temperature of the body proportional to the difference of temperature between the body and its surroundings. From Section 2.4, this law is expressed by the differential equation $\frac{dU}{dt} = k(U(t) - U_s)$. As we have seen, if U_s is constant, the equation is variable-separable.

Suppose, however, that the surrounding temperature varies periodically about a constant value. Then, we would have a differential equation such as

$$U' = -\tfrac{1}{2}[U(t) - (1 + \cos t)] \qquad (3.8)$$

where we have taken $k = -\tfrac{1}{2}$ and $U_s(t) = 1 - \tfrac{1}{10}\cos t$. This equation is neither separable nor exact. However, some insight into the behavior of this system can be obtained from the computer-generated numerical solutions in Figure 3.4. After an initial transitory phase, the solutions seem to settle into periodically varying temperatures. To determine the nature of these "steady-state" temperatures, we need an explicit solution.   ❖

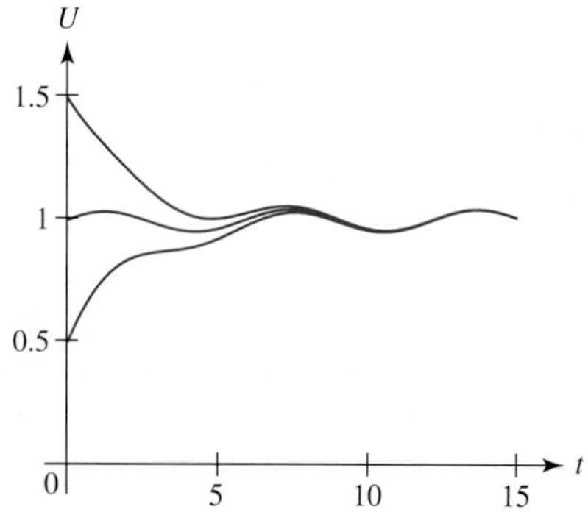

EXAMPLE 3.8   If a constant-volume mixing tank contains 100 gal of brine, has a flow-through rate of 10 gal/min, and is fed with incoming brine containing $\tfrac{1}{5}$ lb of salt per gallon, the techniques of Section 2.3 lead to the separable differential equation $\frac{dx}{dt} = 2 - \frac{10}{100}x(t)$.

Suppose, however, that the content of the inflow varied periodically so that it was $\tfrac{1}{5} + \tfrac{1}{10}\cos t$. Then, the differential equation would be

$$x'(t) = 2 + \cos t - \tfrac{1}{10}x(t) \qquad (3.9)$$

which is neither separable nor exact. As of yet, we do not have a solution technique for such an equation, but numeric solutions for different initial amounts of salt in the tank are shown in Figure 3.5. Instead of an exponential rise to the steady-state value of $x_\infty = 20$, the amount of salt in the tank varies at what appears to be a periodic rate. The exact nature of this variation would require an exact solution to discover.

Also in Section 2.3, we saw variable-volume mixing tanks governed by the differential equations

$$\frac{dx}{dt} = 12 - \frac{2}{200 + 2t}x(t) \quad \text{and} \quad \frac{dx}{dt} = 12 - \frac{5}{800 - t}x(t) \qquad (3.10)$$

In the first, the volume of brine in the tank was increasing, while in the second, decreasing. Neither equation is separable, neither equation is exact.   ❖

We need a technique for solving equations like those in Examples 3.7 and 3.8.

The First-Order Linear Equation

TERMINOLOGY   Table 3.1 contains the equations from Examples 3.7 and 3.8, rewritten in the form

$$y'(x) + p(x)y(x) = r(x) \qquad (3.11)$$

the *standard form* of the first-order linear equation. As in Section 1.2, if $r(x) = 0$, the linear equation is called *homogeneous*. If $r(x) \neq 0$, the linear equation is nonhomogeneous (inhomogeneous in some texts). The homogeneous equation is variable-separable. The nonhomogeneous equation requires new solution techniques.

FIGURE 3.4   Numeric solutions of Newton's law of cooling with variable ambient temperature

FIGURE 3.5   Numeric solutions for a constant-volume mixing tank into which salt flows at a variable rate

$$U' + \tfrac{1}{2}U = \tfrac{1}{2} + \tfrac{1}{20}\cos t$$

$$x' + \tfrac{1}{10}x = 2 + \cos t$$

$$x' - \frac{1}{100+t}x = 12$$

$$x' + \frac{5}{800-t}x = 12$$

TABLE 3.1   Equations of Examples 3.7 and 3.8 rewritten in standard form of first-order linear equation

THE HOMOGENEOUS EQUATION   The first-order linear homogeneous equation is of the form $y'(x) + p(x)y(x) = 0$. Consequently, its solution, by separation of variables, is $y(x) = ce^{-\int p(x)\,dx}$.

THE NONHOMOGENEOUS EQUATION   The solution of the nonhomogeneous first-order linear equation (3.11) is

$$y(x) = e^{-\int p(x)\,dx}\left(\int e^{\int p(x)\,dx}r(x)\,dx + c\right) \tag{3.12}$$

a result obtained by the following steps.

1.   Obtain the *integrating factor* $e^{\int p(x)\,dx}$.
2.   Multiply (3.11) by the integrating factor, obtaining $(y' + py)e^{\int p\,dx} = re^{\int p\,dx}$.
3.   Write this as $\frac{d}{dx}(ye^{\int p\,dx}) = re^{\int p\,dx}$, a step justified in the derivation that follows Examples 3.9 and 3.10.
4.   Integrate both sides, obtaining $ye^{\int p\,dx} = \int re^{\int p\,dx}dx + c$.
5.   Solve for y, obtaining (3.12).

This algorithm is motivated and derived right after the following two examples.

EXAMPLE 3.9   The differential equation $xy'(x) + 2y(x) = e^x$ is a first-order linear equation whose standard form would be

$$y' + \frac{2}{x}y = \frac{e^x}{x}$$

from which we recognize $p(x) = \frac{2}{x}$ and $r(x) = e^x/x$. Consequently, the integrating factor is

$$e^{\int 2/x\,dx} = e^{2\ln x} = e^{\ln x^2} = x^2$$

If the standard form of the equation is multiplied by the integrating factor, the left side is then the derivative of the product $y(x)x^2$. Hence,

$$\frac{d}{dx}(y(x)x^2) = x^2\frac{e^x}{x} = xe^x$$

Integrating both sides yields $y(x)x^2 = \int xe^x\,dx + c$, from which we obtain $y(x)x^2 = xe^x - e^x + c$. Isolating $y(x)$ gives the solution $y(x) = (xe^x - e^x + c)/x^2$.   ❖

EXAMPLE 3.10   Slightly varying the equation in Example 3.9, we have $y'(x) + 2xy(x) = \sin x$, which leads to integrals that cannot be evaluated in closed form. This should be expected as the norm rather than the exception. With $p(x) = 2x$ and $r(x) = \sin x$, the integrating factor is $e^{\int 2x\,dx} = e^{x^2}$ and the solution is $y(x) = e^{-x^2}(\int e^{x^2}\sin x\,dx + c)$, for which an antiderivative expressed in terms of elementary functions is not known.   ❖

THE FIRST-ORDER LINEAR NONHOMOGENEOUS EQUATION IS NOT SEPARABLE   The nonhomogeneous equation, in general, is not separable. Putting (3.11) into the form

$$\frac{dy}{r(x) - p(x)y(x)} = dx$$

does not achieve a separation of variables—there are x's on the left, along with the y's.

THE FIRST-ORDER LINEAR NONHOMOGENEOUS EQUATION IS NOT EXACT   Equation (3.11) is not exact, which we see by putting it into the form $M(x, y) \, dx + N(x, y) \, dy = 0$, and obtaining

$$(p(x)y - r(x)) \, dx + dy = 0$$

With $M = p(x)y - r(x)$ and $N = 1$, so that $M_y = p(x)$ and $N_x = 0$, it is unlikely that the equation is exact, since rarely will $p(x) = 0$.

Integrating Factors

DEFINITION 3.1

A function $u(x)$ with the property that

$$u(x)(M(x, y) \, dx + N(x, y) \, dy) = 0$$

is exact is called an *integrating factor* for the differential equation $M(x, y) \, dx + N(x, y) \, dy = 0$. We want, therefore, $(u(x)M(x, y))_y = (u(x)N(x, y))_x$ to hold.

INTEGRATING FACTOR FOR THE FIRST-ORDER LINEAR EQUATION   The differential form of (3.11) has $M = py - r$ and $N = 1$, so $u(x)$, an integrating factor for the first-order linear equation, must satisfy $u(x)p(x) = u'(x)$, a separable differential equation whose solution is $u(x) = e^{\int p(x) \, dx}$. The integrating factor is the reciprocal of the solution of the homogeneous equation!

We next observe that multiplication of (3.11) by the integrating factor $u(x) = e^{\int p(x) \, dx}$ converts the left side to the exact derivative of the product $y(x)e^{\int p(x) \, dx}$, that is,

$$\frac{d}{dx}\left(y(x)e^{\int p(x) \, dx}\right) = y'(x)e^{\int p(x) \, dx} + y(x)e^{\int p(x) \, dx}p(x)$$

Therefore the differential equation becomes

$$\frac{d}{dx}\left(y(x)e^{\int p(x) \, dx}\right) = e^{\int p(x) \, dx}r(x)$$

An integration of both sides then yields

$$y(x)e^{\int p(x) \, dx} = \left(\int e^{\int p(x) \, dx}r(x) \, dx + c\right)$$

Isolating $y(x)$ by multiplying through by $e^{-\int p(x) \, dx}$ yields

$$y(x) = e^{-\int p(x) \, dx}\left(\int e^{\int p(x) \, dx}r(x) \, dx + c\right) = e^{-\int p(x) \, dx}\int e^{\int p(x) \, dx}r(x) \, dx + ce^{-\int p(x) \, dx}$$

$$(3.13)$$

EXERCISES 3.4–Part A

A1. Obtain the general solution of (3.8). Does the exact solution tend toward a periodic function? If so, does this limiting periodic function have the same period as U_s?

A2. Obtain the general solution of (3.9). Does the exact solution tend toward a periodic function? If so, does this limiting periodic function have the same period as the perturbation in the inflow?

A3. Obtain the general solution of the leftmost equation in (3.10).

A4. Obtain the general solution of the rightmost equation in (3.10).

A5. Use the condition $(uM)_y = (uN)_x$ to determine $u(x)$, an integrating factor for the equation $y\,dx - x\,dy = 0$. Use this

integrating factor to convert the differential equation to an exact equation, and obtain its general solution by the techniques of Section 3.3.

EXERCISES 3.4–Part B

B1. What condition would the integrating factor $u(x, y)$ have to satisfy to render the equation $(2x + 3y)\,dx + (5x + 7y)\,dy = 0$ exact?

In Exercises B2–13, put the given equation into the standard form (3.11) and solve as a first-order linear equation. Graph representative members of the family of curves contained in the general solutions so found.

B2. $x^4 + 11x^3 + (y' + 10)x^2 + 2xy = 0$

B3. $x^2y' - (7 + 2y)x - 3 + 10x^{-1} = 0$

B4. $2y - 8x^5 - 8x^4 + 7x^3 + 6x^2 + xy' = 0$

B5. $(y' + 12)x - 2y + 8x^{-1} - 4x^{-2} = 0$

B6. $\frac{1}{2}y - 8x^{7/2} + 16x^{5/2} - 2x^{3/2} + xy' + \sqrt{x} = 0$

B7. $2y - 5x^5 - 8x^3 + 9x^2 + xy' = 0$

B8. $10x^3 - 6x^2 + (y' - 1)x + \frac{1}{2}y = 0$

B9. $10x^2 - 4x^3 + (y' + 1)x + 3 - \frac{1}{2}y = 0$

B10. $24x\cos x + y'\cos x - y\sin x + 5\cos x = 0$

B11. $24x^2 - 4y' + y'x^2 - 2xy - 6x^4 = 0$

B12. $y - 12x^2 - 4x^3 - 18x + 2y' + 2y'x - 10 = 0$

B13. $10x^2 + 5x + 2y' - y'x + \frac{1}{2}y - 6 = 0$

B14. Brine containing $c = 5$ lb of salt per gallon that flows into a 700-gal tank at a fixed rate of $r_1 = 3$ gal/s is instantly mixed with the existing contents of the tank, and the fluid in the tank is

drained at the rate of $r_2 = 1$ gal/s. The tank initially contained $V = 100$ gal in which 20 lb of salt were dissolved. Formulate and solve an initial value problem for the amount of salt in the tank at any time t. Plot the solution, and determine the amount of salt in the tank at the moment it overflows. Check the answer using Maple's **dsolve** command (or the equivalent).

B15. Repeat Exercise B14 if the parameters are $c = 2, r_1 = 8, r_2 = 5$, $V = 200$ and the 700-gal tank initially contained pure water.

B16. Repeat Exercise B14 if the parameters are $c = \frac{1}{2}, r_1 = 6, r_2 = 2$, and $V = 100$ for a 1000-gal tank in which 150 lb of salt were initially dissolved.

B17. Brine containing $c = \frac{3}{2}$ lb of salt per gallon that flows at a fixed rate of $r_1 = 3$ gal/s into an initially full 500-gal tank in which 75 lb of salt were initially dissolved is instantly mixed with the existing contents of the tank, and the fluid in the tank is drained at the rate of $r_2 = 4$ gal/s. Formulate and solve an initial value problem for the amount of salt in the tank at any time t. Plot the solution and determine the limiting concentration of the fluid in the tank at the moment it empties. Check the answer using Maple's **dsolve** command (or the equivalent).

B18. Repeat Exercise B17 if the parameters are $c = \frac{5}{4}, r_1 = 1$, and $r_2 = 5$ for a 900-gal tank that is initially filled with pure water.

3.5 Variation of Parameters and the First-Order Linear Equation

Solution of the First-Order Linear ODE

In Section 3.4, the general solution of the first-order, linear, nonhomogeneous differential equation (3.11) was found to be (3.13). Denoting the general solution by y_g, we next write (3.13) in the form $y_g = y_h + y_p$, where

$$y_h = ce^{-\int p(x)\,dx} \tag{3.14}$$

is the *homogeneous solution,* the solution of the homogeneous equation $y' + py = 0$, and

$$y_p = e^{-\int p(x)\,dx}\int e^{\int p(x)\,dx} r(x)\,dx$$

is the *particular solution,* one solution of the full nonhomogeneous equation (3.11). Thus, we have

$$y_g = e^{-\int p(x)\,dx} \left(\int e^{\int p(x)\,dx} r(x)\,dx + c \right)$$

$$= ce^{-\int p(x)\,dx} + e^{-\int p(x)\,dx} \int e^{\int p(x)\,dx} r(x)\,dx = y_h + y_p \qquad (3.15)$$

The resolution of the general solution into the two components y_h and y_p is a characteristic of linear equations, both algebraic and differential. In Chapter 5 the second-order linear differential equation will be shown to have a solution with the very same structure.

The Homogeneous Solution

The homogeneous solution, namely y_h, is the solution of the homogeneous equation formed by setting $r(x)$, the right-hand side of the differential equation, equal to zero. Thus, the homogeneous equation is $y'(x) + p(x)y(x) = 0$, a variable-separable equation whose solution is (3.14). The homogeneous solution always houses the arbitrary constants of integration. For a first-order equation, there is just one such constant. The homogeneous solution for an nth-order linear equation would have n arbitrary constants.

The Particular Solution—Variation of Parameters

The particular solution is any one solution of the nonhomogeneous equation. We will find a particular solution by the technique called *variation of parameters,* a technique we will use again in Chapter 5 when we find particular solutions of second-order linear equations.

Look for a particular solution in which c, the arbitrary constant, is taken as a function, say, $u(x)$. The parameter c is therefore taken as a variable—hence, the name, *variation of parameters*. Thus, seek a particular solution of the form

$$y_p = u(x)y_h(x)$$

and determine $u(x)$ by "brute force," setting $y(x) = y_p$ in the nonhomogeneous differential equation. This substitution into the differential equation yields

$$u'y_h + uy_h' + uy_h p(x) = r(x)$$

which we write as

$$u(y_h' + py_h) + u'y_h = r$$

Since $y_h' + py_h = 0$ for the homogeneous solution y_h, we are left with $u'y_h = r$, from which we determine $u(x) = \int r(x)/y_h(x)\,dx$. Consequently, we obtain $y_p = y_h \int r(x)/y_h(x)\,dx$. Recalling (3.14) and noting that $1/y_h = e^{\int p(x)\,dx}$, we recover (3.15), the general solution to the first-order linear equation.

EXAMPLE 3.11   In Section 2.3 the differential equations

$$x'(t) = 12 - \frac{2}{200 + 2t}x(t) \quad \text{and} \quad x'(t) = 12 - \frac{5}{800 - t}x(t)$$

were used to model the amount of salt in mixing tanks with increasing and decreasing volumes, respectively. The solutions to initial value problems containing these equations and the initial condition $x(0) = 0$ were $x(t) = \frac{1200t + 6t^2}{100 + t}$ and

$$x(t) = \frac{3}{409,600,000,000}t^5 - \frac{3}{102,400,000}t^4 + \frac{3}{64,000}t^3 - \frac{3}{80}t^2 + 12t \qquad (3.16)$$

respectively. Each of these equations is a first-order linear, nonhomogeneous equation amenable to the analysis of Section 3.4 as well as to the techniques of this present section.

Writing the first equation in the form $x'(t) + p(t)x(t) = r(t)$, we find $p(t) = \frac{1}{100+t}$ and $r(t) = 12$, giving the integrating factor $e^{\int 1/(100+t)\,dt} = 100 + t$. Thus,

$$x(t)(100 + t) = \int (100 + t)12\,dt = 1200t + 6t^2 + c$$

leads to $x(t) = \frac{1200t+6t^2+c}{100+t}$. Application of the condition $x(0) = 0$ leads to $c = 0$ and the solution on the left in (3.16).

Alternatively, the homogeneous solution is $x_h = ce^{-\int 1/(100+t)\,dt} = \frac{c}{100+t}$, while the particular solution, computed by the recipe

$$x_p = x_h \int \frac{r(t)}{x_h}\,dt = \frac{c}{100+t} \int \frac{12(100+t)}{c}\,dt$$

again is just $\frac{1200t+6t^2}{100+t}$ so the solution techniques produce the same results since the general solution is the sum $x_h + x_p = \frac{c}{100+t} + \frac{1200t+6t^2}{100+t}$.   ❖

EXERCISES 3.5–Part A

A1. By separation of variables, the general solution of the homogeneous linear equation $y' + py = 0$ is $y_h = ce^{-\int p(x)\,dx}$. Suppose an antiderivative of $p(x)$ is $P(x)$. Show that the general solution is unaffected by using the most general antiderivative, $P(x) + \alpha$.

A2. Show that $y_h = ce^{-\int p(x)\,dx}$ is equivalent to $y_h = Ae^{-\int_{x_0}^{x} p(\lambda)\,d\lambda}$, where c and A are both arbitrary constants. *Hint:* Let $P(x)$ be an antiderivative of $p(x)$.

A3. Show that $y(x) = 0$ is the *only* solution of the completely homogeneous IVP $y' + py = 0$, $y(x_0) = 0$. *Hint:* Show that it is a solution, then invoke the uniqueness theorem from Section 1.4.

A4. Suppose $u(x)$ and $v(x)$ are both solutions of the IVP $y' + py = r$, $y(x_0) = y_0$. Show that $w = u - v$ satisfies the completely homogeneous IVP $w' + pw = 0$, $w(x_0) = 0$ and, hence, by Exercise A3, $w = 0$, so $u = v$.

A5. Use variation of parameters to obtain a particular solution to $xy' + 2y = x^3 \sin x$.

A6. Let y_h be the homogeneous solution of the first-order linear ODE (3.11), and let $\tilde{y}$ be one specific solution contained in the family denoted by y_h. If y_p is a particular solution, show that $Y = y_p + \tilde{y}$ is also a particular solution. Hence, the particular solution of a linear ODE can contain an additive term that is properly a member of the family forming the homogeneous solution.

EXERCISES 3.5–Part B

B1. An object initially at $3°F$ is placed in an environment whose temperature is given by $U_s = 1 + e^{-t}$. After one minute, the temperature of the object is $2°F$. Use Newton's law of cooling to write a differential equation governing the cooling process. Obtain and graph its solution.

B2. An object initially at $0°F$ is placed in an environment whose temperature is given by $U_s = 1 - \frac{1}{5}e^{-t}$. After one minute, the temperature of the object is $\frac{1}{2}°F$. Use Newton's law of cooling to write a differential equation governing this thermal process. Obtain and graph its solution.

B3. A mixing tank contains 100 gal of water in which 50 lb of salt have been dissolved. Brine enters and leaves at a rate of 5 gal/min. The incoming brine contains $2 + te^{-t}$ lb of salt per gallon. Formulate and solve an IVP governing the amount of salt in the tank at any time t. Graph the solution and determine the limiting amount of salt in the tank. Compare this solution to that of a

system in which the incoming brine contains 2 lb of salt per gallon. Explain these results.

In Exercises B4–13, obtain the general solution by constructing the homogeneous and particular solutions and adding appropriately. Use the method of variation of parameters to construct the particular solutions. Compare the resulting general solution to that produced by the integrating-factor method. Check the results using a computer algebra system. Plot representative members of the family of curves defined by the general solution.

B4. $7x^3 - 6x^2 + xy' + 2y = 0$

B5. $x^2y' - 2xy + 14 - 3x^{-1} + 11x^{-2} = 0$

B6. $2y - 3x^4 + x^3 + 2x^2 + xy' = 0$

B7. $xy' - 16 - 2y - 2x^{-1} + 15x^{-2} = 0$

B8. $9x^{5/2} + 7x^{3/2} + xy' + \frac{1}{2}y = 0$

B9. $2y \cos 2x - 16x \sin 2x + y' \sin 2x - 3 \sin 2x = 0$

B10. $5x^2 \cos x - 4x \cos x + y' \cos x - y \sin x - 19 \cos x = 0$

B11. $9x^2 + 9x + y' - 2y - 14 = 0$

B12. $21x^3 - 4x^2 + xy' + 2y = 0$

B13. $4x^2 - 5x^3 + xy' + x + \frac{1}{2}y = 0$

Using a technique that includes variation of parameters, obtain an analytic solution for Exercises B14–18.

B14. Exercise 5B, Section 2.3   **B15.** Exercise 6B, Section 2.3

B16. Exercise 7B, Section 2.3   **B17.** Exercise 8B, Section 2.3

B18. Exercise 9B, Section 2.3

In Exercises B19–28, use variation of parameters as part of the technique for finding the general solution of the given equation.

B19. $xy' + 3x - \frac{1}{2}y + 11 = 0$

B20. $17x^{7/2} + 7x^{5/2} + 8x^{3/2} + xy' + y = 0$

B21. $3\sqrt{x} - 7x^{3/2} + xy' - y = 0$

B22. $3x^2 \cos x + 6x \cos x + y' \cos x - y \sin x - 4 \cos x = 0$

B23. $6x^2 \cosh x - x \cosh x + y' \cosh x + y \sinh x = 0$

B24. $y' + x^2 y' - 2xy - x - x^3 + 6x^2 + 6x^4 = 0$

B25. $x^2 y' - y' - 2xy - x + x^3 - x^2 + x^4 = 0$

B26. $x^2 y' - 4y' - 2xy + 20 - 21x^2 + 28x - 7x^3 + 4x^4 = 0$

B27. $12x^2 - 2x^3 + 4x + 2y' + 2xy' + y - 10 = 0$

B28. $\frac{1}{2}y - 28x^2 + 2y' - xy' - 14 = 0$

3.6   The Bernoulli Equation

EXAMPLE 3.12   Consider the following specific logistic model for population growth, namely, $y'(t) = 3y(2 - y)$, a separable differential equation with solution $y(t) = 2/(1 + 2ce^{-6t})$. Writing the equation in the form $y' - 6y = -3y^2$ suggests that while this is not a linear equation, if the exponent on the right-hand side could be changed to either 0 or 1, the equation would indeed become a linear equation since $y^0 = 1$ and $y^1 = y$. In this form, the separable equation is called a *Bernoulli equation* for which the change of variables $y(t) = z(t)^k$, for some k to be determined, leads to a first-order linear equation in $z(t)$.

In fact, carrying out this change of variables leads to

$$kz^{k-1}z' - 6z^k = -3z^{2k}$$

and, after division by kz^{k-1}, to

$$z' - \frac{6}{k}z = -\frac{3}{k}z^{2k-(k-1)} = -\frac{3}{k}z^{k+1}$$

from which we see that $k + 1 = 0$ or $k + 1 = 1$. The choice $k + 1 = 1$ means $k = 0$, in which case we would have made the meaningless change of variables $y(t) = z(t)^0 = 1$. Hence, we choose $k + 1 = 0$ and $k = -1$. The resulting equation, $z' + 6z = 3$, is now linear in $z(t)$, with $z(t) = ce^{-6t} + \frac{1}{2}$ and $y(t) = \frac{1}{z(t)} = 2/(1 + 2ce^{-6t})$.   ❖

The Bernoulli Equation

The *general Bernoulli equation*

$$y'(t) + p(t)y(t) = r(t)y(t)^s$$

under the change of variables $y(t) = z(t)^k$, becomes a first-order linear equation in $z(t)$ if $k = \frac{1}{1-s}$. Indeed, this substitution yields

$$kz^{k-1}z' + pz^k = rz^{ks}$$

and division by kz^{k-1} gives

$$z' + \frac{p}{k}z = \frac{r}{k}z^{ks-(k-1)} = \frac{r}{k}z^{k(s-1)+1}$$

Setting $k(s-1) + 1 = 0$ to make the equation in $z(t)$ linear requires, as claimed, $k = \frac{1}{1-s}$. After solving for $z(t)$, we get $y(t) = z(t)^{1/(1-s)}$.

EXAMPLE 3.13 In the Bernoulli equation $y'(t) + 5y(t) = y(t)^3$, the substitution $y(t) = z(t)^{1/(1-s)} = z(t)^{1/(1-3)} = z(t)^{-1/2}$ leads to $-\frac{1}{2}z^{-3/2}z' + 5z^{-1/2} = z^{-3/2}$. Divide through by $-\frac{1}{2}z^{-3/2}$ to get $z' - 10z = -2$, the solution of which is $z(t) = ce^{10t} + \frac{1}{5}$. Since $y(t) = z(t)^{-1/2}$, we finally obtain $y(t) = 1/\sqrt{ce^{10t} + \frac{1}{5}}$. ❖

EXERCISES 3.6–Part A

A1. Let $z(t)$ satisfy the first-order linear equation $z' + p(1-s)z = r(1-s)$, and set $y(t) = z^{1/(1-s)}$. Show that $y(t)$ satisfies the Bernoulli equation $y'(t) + p(t)y(t) = r(t)y(t)^s$.

In Exercises A2–6, put the given differential equation into the form of a Bernoulli equation $y' + p(x)y = r(x)y^s$. Set $y = z^k$ and, without resorting to a formula, determine the value of k leading to an equation linear in $z(t)$.

A2. $(y' + 9y^2)x^2 - 2xy - 16x^4y^2 = 0$

A3. $64x^2 - xy'y^3 + 12x - 8y^4 = 0$

A4. $x^3y' + 4x^2y + 10xy^3 + 18y^3 = 0$

A5. $x^3y' + x^2y - y^{3/2}x^2 - 5y^{3/2} = 0$

A6. $36x^{5/2}y^3 - 4x^{3/2}y^3 + xy' - 2\sqrt{x}y^3 - y = 0$

EXERCISES 3.6–Part B

B1. Solve the logistic equation $y' = y(1-y)$ as a separable equation and as a Bernoulli equation.

B2. Show that $y' = y(1-y) - h$, the logistic equation with constant harvesting, can be solved as a separable equation but not as a Bernoulli equation.

In Exercises B3–12, show that the given differential equation is a Bernoulli equation, transform it to a linear equation, and obtain the general solution by the method of either Section 3.4 or 3.5. Confirm the solution with a computer algebra system, and plot representative members of the one-parameter family of curves represented by the general solution.

B3. $2x^3y^2 + 5x^2y^2 + xy' - 14xy^2 - \frac{1}{2}y = 0$

B4. $10x^2y^{3/2} + 4xy' - 44xy^{3/2} + y + 4y^{3/2} = 0$

B5. $36x^{5/2} - 6x^{3/2} + xy'y^2 + 3y^3 - 27x^{7/2} = 0$

B6. $x^3y^{2/3} + 3x^2y' - xy = 0$

B7. $4x^2y^2\cos x + 13xy^2\cos x + y'\cos x + y\sin x - 3y^2\cos x = 0$

B8. $y^2y'\sinh x + \frac{1}{3}y^3\cosh x - \sinh x = 0$

B9. $y' + x^2y' + \frac{2}{3}xy + \frac{1}{3}y^4 + \frac{1}{3}x^2y^4 = 0$

B10. $2xy' - y + 2xy^3 = 0$ **B11.** $x^2y' + y' + 2xy + 2xy^2 = 0$

B12. $y' + 3y - 6xy^{1/3} = 0$

Chapter Review

1. Give an example of a variable-separable ODE. Obtain its solution.

2. Give an example of an ODE with homogeneous coefficients. Obtain its solution.

3. Give an example of an exact ODE. Obtain its solution.

4. Give an example of an ODE that is *not* exact. Why is it not exact?

5. What is an integrating factor? Describe it in words, then give an example.

6. Give an example of an ODE that is not exact but for which an integrating factor can be found to make it exact. Show how the integrating factor leads to a solution of the equation.

7. Give an example of a first-order linear ODE, nonhomogeneous and with constant coefficients. Obtain an integrating factor. Show that the integrating factor makes the equation exact. Use the equation's exactness to obtain a solution.

8. What is the homogeneous solution for the example in Question 7? What is the particular solution? How do the homogeneous and particular solutions lead to the general solution?

9. Give an example of a Bernoulli equation. Obtain its solution.

10. Is every Bernoulli equation linear? Is every first-order linear equation a Bernoulli equation?

Chapter *4*

Numeric Methods for Solving First-Order ODEs

I N T R O D U C T I O N   This chapter does two things. It presents a collection of numeric methods for solving first-order differential equations, and it discusses how to interpret these methods and the solutions they generate.

Problems in modern engineering and science are sufficiently complex to yield only to numeric computation. Analytic techniques provide a basic understanding of how solutions to differential equations behave, but the ability to compute solutions to otherwise intractable equations is an essential skill. That is why we make numeric methods available early in our study of differential equations.

Our discussion includes Euler's method, methods based on Taylor series, Runge–Kutta, multistep, and predictor-corrector methods. Each of these is an interesting study showing how a central idea about computing a solution numerically is expressed in a reliable algorithm. However, we will rely on technology and preprogrammed packages to implement these algorithms. It is rare that the beginner will write code as effective as that provided in a professionally written package, and we encourage readers to use the most convenient and effective technology they can find.

However, reliance on such hardware and software should not be an excuse for failing to learn the essential ideas associated with these numeric methods. The concept of a method's order and the difference between per-step and global truncation errors are brought out in the discussion of Euler's method and the fourth-order Runge–Kutta techniques. The ideas of stability, error monitoring, and adaptivity are brought out in the other sections of the chapter.

4.1   Fixed-Step Methods—Order and Error

Fixed-Step Methods

The initial value problem $\frac{dy}{dt} = f(t, y(t))$, $y(t_0) = y_0$ is solved numerically by *fixed-step methods* that compute values y_k, approximations to $y(t_k)$, at uniformly spaced grid points $t_k = t_0 + kh$. The stepsize h remains fixed during the computation. *Euler's* method is studied because it is simple; however, it is not adequate for many problems in science and engineering. (Accuracy can sometimes be achieved only at the expense of very small stepsizes, leading to an excessive amount of computation.) Members of the family of Runge–Kutta methods supercede Euler's method. In particular, the *fourth-order Runge–Kutta* (rk4) method is one of the more widely used fixed-step methods.

FORMULAE: EULER'S METHOD   The recipe for computing a numeric solution by Euler's method can be expressed through the following equations where the initial point (t_0, y_0) is the "starting" point:

$$t_k = t_0 + kh \qquad y_{k+1} = y_k + hf(t_k, y_k) \qquad k = 0, 1, \ldots$$

FORMULAE: FOURTH-ORDER RUNGE–KUTTA METHOD   The computation of a numerical solution by the rk4 method is a bit more involved, requiring, as it does, four function evaluations for each step. The benefit of this increased computational effort is greater accuracy.

$$F_1 = hf(t_k, y_k) \qquad\qquad F_2 = hf\left(t_k + \frac{h}{2}, y_k + \frac{1}{2}F_1\right)$$

$$F_3 = hf\left(t_k + \frac{h}{2}, y_k + \frac{1}{2}F_2\right) \qquad F_4 = hf(t_{k+1}, y_k + F_3)$$

$$y_{k+1} = y_k + \tfrac{1}{6}(F_1 + 2F_2 + 2F_3 + F_4) \qquad k = 0, 1, \ldots$$

Both Euler's method and the rk4 algorithm will be demonstrated later in this section.

Packaged Solvers

A generation ago, obtaining a numeric solution of a differential equation required programming in a language like FORTRAN, execution in batch mode via punch-cards, and no graphics. Technology has progressed so rapidly that numeric solvers are readily available in software packages for computers and even on hand-held calculators. These professional numeric solvers are efficiently programmed and implemented, often specifically tailored to the hardware on which they run. We call such systems *packaged solvers* and presume that the typical reader uses such tools instead of writing computer code from "scratch."

MODEL IVP   To illustrate the behavior of a typical packaged solver for numeric solutions to first-order differential equations, we take

$$y' = f(t, y) = t + y \qquad y(0) = 0 \tag{4.1}$$

with exact solution $y(t) = e^t - t - 1$, as a model IVP for testing the Euler and rk4 algorithms. The graph of this solution appears as the solid curve in Figure 4.1.

EULER'S METHOD   In Figure 4.1 the solid and dotted curves are, respectively, the exact and Euler ($h = 0.1$) solutions for the test IVP. The computations from which the dotted curve was drawn are

$$y_1 = y_0 + hf(t_0, y_0) = 0 + 0.1(0 + 0) = 0$$
$$y_2 = y_1 + hf(t_1, y_1) = 0 + 0.1(0.1 + 0) = 0.01$$
$$y_3 = y_2 + hf(t_2, y_2) = 0.01 + 0.1(0.2 + 0.01) = 0.031$$

These, and other values calculated by the Euler algorithm ($h = 0.1$), are listed in the second column of Table 4.1, in which the fourth column contains $y(t_k)$, the exact values obtained from $y(t)$, the exact solution. The growing inaccuracy of the Euler values can be seen by comparing the second and fourth columns.

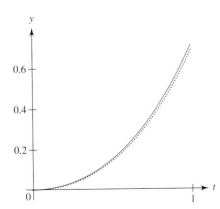

FIGURE 4.1   Exact solution (solid) and Euler solution (dotted), $h = 0.1$, for the IVP $y' = t + y$, $y(0) = 0$

t_k	y_k **(Euler)**	y_k **(rk4)**	$y(t_k)$ **(Exact)**
0.0	0	0	0
0.1	0	0.005170833	0.005170918
0.2	0.01	0.021402571	0.021402758
0.3	0.031	0.049858497	0.049858808
0.4	0.0641	0.091824240	0.091824698
0.5	0.11051	0.148720639	0.148721271
0.6	0.171561	0.222117962	0.222118800
0.7	0.2487171	0.313751627	0.313752707
0.8	0.34358881	0.425539563	0.425540928
0.9	0.457947691	0.559601414	0.559603111
1.0	0.5937424601	0.718279744	0.718281828

TABLE 4.1   Euler, rk4, and exact values for the IVP $y' = t + y$, $y(0) = 0$

FOURTH-ORDER RUNGE–KUTTA   Computing function values with the rk4 algorithm is considerably more tedious. For example, when $k = 0$ and $h = 0.1$, rk4 determines y_1 by the following arithmetic

$$F_1 = hf(t_0, y_0) = \frac{1}{10}(0 + 0) = 0$$

$$F_2 = hf\left(t_0 + \frac{h}{2}, y_0 + \frac{F_1}{2}\right) = \frac{1}{10}\left(\frac{1}{20} + 0\right) = \frac{1}{200}$$

$$F_3 = hf\left(t_0 + \frac{h}{2}, y_0 + \frac{F_2}{2}\right) = \frac{1}{10}\left(\frac{1}{20} + \frac{1}{400}\right) = \frac{21}{4000}$$

$$F_4 = hf(t_1, y_0 + F_3) = \frac{1}{10}\left(\frac{1}{10} + \frac{21}{4000}\right) = \frac{421}{40000}$$

$$y_1 = y_0 + \frac{1}{6}(F_1 + 2F_2 + 2F_3 + F_4) = \frac{1}{6}\left(0 + \frac{2}{200} + \frac{2 \times 21}{4000} + \frac{421}{40000}\right) = \frac{1241}{240000}$$

The third column of Table 4.1 contains this and other y_k values computed by the fourth-order Runge–Kutta algorithm, again with $h = 0.1$. Even a cursory glance at this table reveals the greater accuracy of this method. In fact, these values are accurate enough that graphs of the approximate and exact solutions are indistinguishable.

Order of a Method

DEFINITION 4.1

If a numerical procedure uses the fixed stepsize h to compute y_{approx} as an approximation to y_{exact}, the exact value, in such a way as to obey a relationship of the form $y_{exact} = y_{approx} + Ch^k$, we say the procedure is of *order k*.

In essence, this definition says the *truncation error* Ch^k is proportional to the kth power of the stepsize. Truncation error is inherent in the structure of the algorithm and would be present even if its calculations were done in exact arithmetic. Truncation error is therefore independent of *roundoff error*, the error that occurs because calculations are done on a computing device carrying only a finite number of digits. (See Section 37.1 for a more extensive discussion of roundoff error.)

Methods for solving differential equations numerically have two types of truncation error: the *per-step error* and the *global error* (which is an accumulation of error after n steps). Respectively, the global errors for the Euler and rk4 methods are proportional to h and h^4, so

$$y_{\text{exact}} = y_{\text{Euler}} + Mh \quad \text{and} \quad y_{\text{exact}} = y_{\text{rk4}} + Nh^4$$

We illustrate the validity of these statements in the next two examples.

ORDER: EULER METHOD We will use the model IVP (4.1) for a test of the order of the Euler method. Using appropriate computer software, we evaluate the exact solution at $t = 1$ and find $y_{\text{exact}}(1) = 0.71828182845904523560$, correct to 20 decimal places.

We then compute $y_{\text{Euler}}(1)$ numerically, with $h = 0.1/2^k$, $k = 0, 1, \ldots$, and check the constancy of $M = (y_{\text{exact}}(1) - y_{\text{Euler}}(1))/h$, displaying the results in the second row of Table 4.2. The values of M are reasonably constant, indicating that the Euler method has a global truncation error proportional to the stepsize h, making the method a first-order method.

h	$\frac{0.1}{1}$	$\frac{0.1}{2}$	$\frac{0.1}{4}$	$\frac{0.1}{8}$	$\frac{0.1}{16}$	$\frac{0.1}{32}$	$\frac{0.1}{64}$
M	1.245	1.300	1.329	1.344	1.351	1.355	1.357
N	0.0208	0.0217	0.0222	0.0224	0.0225	0.0226	0.0226

TABLE 4.2 Constancy of M (Euler's method) and N (rk4)

ORDER: FOURTH-ORDER RUNGE–KUTTA METHOD To show the rk4 method is fourth order, we have to show the constancy of $N = (y_{\text{exact}}(1) - y_{\text{rk4}}(1))/h^4$. Thus, we compute, with decreasing stepsizes, values for $y_{\text{rk4}}(1)$ and record results in the third row of Table 4.2. The values for N are reasonably constant, thereby supporting the claim that rk4 is a fourth-order method.

Error Estimation

If the truncation error of a numerical method is known to obey a rule of the form $y_{\text{exact}} = y_{\text{approx}} + Ch^k$ and two values of y_{approx} are computed for stepsizes h and $\frac{h}{2}$, $y_{\text{approx}}(h)$ and $y_{\text{approx}}(\frac{h}{2})$ are, respectively, a less and a more accurate approximation of y_{exact}, the "correct" value. From these two approximations alone, an estimate of the accuracy of the better approximation can be deduced. This simple observation is at the heart of *adaptive procedures* for both numeric integration and numeric solving of differential equations. In an adaptive procedure, the stepsize is halved and a test is made of the accuracy of the more accurate approximation. If it is within prescribed error bounds, the value is accepted. Otherwise, it is rejected and recomputed with a still smaller stepsize.

The adaptive relationship

$$y_{\text{exact}} - y_{\text{approx}}\left(\frac{h}{2}\right) = \frac{y_{\text{approx}}\left(\frac{h}{2}\right) - y_{\text{approx}}(h)}{2^k - 1} \tag{4.2}$$

says that the error in the better approximation is the difference between the better and worse approximations divided by a factor dependent on the known order of the numeric method. The expression follows from the order statements

$$y_{\text{exact}} = y_{\text{approx}}(h) + Ch^k \quad \text{and} \quad y_{\text{exact}} = y_{\text{approx}}\left(\frac{h}{2}\right) + C\left(\frac{h}{2}\right)^k \tag{4.3}$$

by eliminating Ch^k to get

$$y_{\text{exact}} = \left(\frac{1}{2^k - 1}\right)\left(2^k y_{\text{approx}}\left(\frac{h}{2}\right) - y_{\text{approx}}(h)\right) \tag{4.4}$$

and then subtracting $y_{\text{approx}}(\frac{h}{2})$ from both sides.

The result in (4.4) is the formula for *Richardson extrapolation*, which computes a weighted average of the better and worse approximations and, if the equations in (4.3) are exact, returns y_{exact}. Logically, the weighting favors the more exact value. This extrapolation scheme is used when differentiating numerically, is found in the Romberg method of numeric integration, and is also used in some numeric differential equation solvers. See Sections 42.2 and 43.2 for details.

MONITORING ERROR: EULER METHOD   We apply (4.2), the fundamental result on error monitoring, to the Euler method, using the values of $y_{\text{Euler}}(1)$ computed earlier with $h = 0.1/2^j$, $j = 0, 1, \ldots$. Since each was computed with half the stepsize of its predecessor, we can test (4.2) with $k = 1$ by evaluating both sides and listing the results in Table 4.3. Note that the denominator on the right is 1. The increasing match of the two columns in the table supports the validity of the adaptive error estimate (4.2).

$y_{\text{exact}} - y_{\text{approx}}\left(\dfrac{h}{2}\right)$	$\dfrac{y_{\text{approx}}\left(\frac{h}{2}\right) - y_{\text{approx}}(h)}{2^k - 1}$
0.06498412331462510143	0.05955524504442013397
0.03321799006907250388	0.03176613324555259755
0.01679688770570813701	0.01642110236336436687
0.00844625215126835539	0.00835063555443978162
0.00423518475136823752	0.00421106739990011787
0.00212062051103751173	0.00211456424033072579

TABLE 4.3   Adaptive error estimate for Euler's method

$y_{\text{exact}} - y_{\text{approx}}\left(\dfrac{h}{2}\right)$	$\dfrac{y_{\text{approx}}\left(\frac{h}{2}\right) - y_{\text{approx}}(h)}{2^k - 1}$
$0.13580271 \times 10^{-6}$	$0.129901411 \times 10^{-6}$
$0.86661892 \times 10^{-8}$	$0.847576814 \times 10^{-8}$
$0.54730581 \times 10^{-9}$	$0.541258890 \times 10^{-9}$
$0.34385198 \times 10^{-10}$	$0.341947077 \times 10^{-10}$
$0.21546779 \times 10^{-11}$	$0.214870130 \times 10^{-11}$
$0.13484276 \times 10^{-12}$	$0.134655680 \times 10^{-12}$

TABLE 4.4   Adaptive error estimate for rk4

MONITORING ERROR: FOURTH-ORDER RUNGE–KUTTA METHOD   We apply (4.2), the fundamental result on error monitoring, to the fourth-order Runge–Kutta method, using the values of $y_{\text{rk4}}(1)$ computed earlier with $h = 0.1/2^j$, $j = 0, 1, \ldots$. Since each was computed with half the stepsize of its predecessor, we can test (4.2) with $k = 4$ by evaluating both sides and listing the results in Table 4.4. Note that the denominator on the right is 15. The increasing match of the two columns in the table supports the validity of (4.2).

The error-monitoring formula (4.2) is essential when the exact solution is not known. It provides a means for determining the accuracy of a numeric result without comparing it to the unknown exact answer. Within an integration routine, it allows for an automatic return of an answer correct to within a prescribed tolerance. In an algorithm for solving differential equations numerically, it can be used to increase stepsize and, hence, the overall speed of the computation, by taking larger steps where possible and then automatically reducing the stepsize where needed.

EXERCISES 4.1–Part A

A1. Use Euler's algorithm with $h = 0.1$ to obtain y_1 and y_2 for the IVP $y' = t^2 + y^2$, $y(1) = 1$.

A2. Use the rk4 algorithm with $h = 0.1$ to obtain y_1 for the IVP of Exercise A1.

A3. Use (4.3) to obtain (4.2).

A4. Use (4.2) to obtain (4.4).

A5. Using $h = \frac{1}{10}$, apply Euler's method to $y' = -20y$, $y(0) = 1$, computing y_k, $k = 1, \ldots, 5$, in exact arithmetic. The exact solution to this IVP is $y(t) = e^{-20t}$. Can you find a stepsize h for which Euler's method gives reasonable values for this problem?

EXERCISES 4.1–Part B

B1. The IVP $y' = f(t, y)$, $y(0) = 0$, is to be solved, with $y(1)$ being the goal of the calculation.

   (a) If rk4 is used with $h = 0.1$, how many function evaluations will take place?

   (b) What accuracy can be expected for the answer computed as per part (a)?

   (c) How many function evaluations would it take Euler's method to compute $y(1)$ with the same accuracy?

   (d) What accuracy can be expected for $y(1)$ if it is computed with Euler's method using the same number of function evaluations as in part (a)?

In each of Exercises B2–18:

   (a) Use a computer algebra system or one of the methods of Chapter 3 to obtain the indicated function value. If using a computer algebra system, it might be necessary to obtain a general solution to which the initial condition is then applied.

   (b) Use Euler's method with $h = 0.1$ to approximate the indicated function value.

   (c) Test the order of the global truncation error by computing, at the indicated point, the (supposedly constant) ratio $(y_{\text{exact}} - y_{\text{approx}})/h^k$, $k = 1$, for $h = \left(\frac{1}{2}\right)^s$, $s = 1, 2, \ldots$. (For some IVPs, the global truncation error in the Euler method can be $O(h^k)$, $k > 1$.)

   (d) With $k = 1$, test the error estimation formula (4.2).

   (e) With $k = 1$, test the validity of (4.4), Richardson's extrapolation formula.

   (f) Use rk4 to obtain the indicated function value.

   (g) Test the order of the global truncation error by computing, at the indicated point, the (supposedly constant) ratio $(y_{\text{exact}} - y_{\text{approx}})/h^k$, $k = 4$, for $h = \left(\frac{1}{2}\right)^s$, $s = 1, 2, \ldots$.

   (h) With $k = 4$, test the error estimation formula (4.2).

   (i) With $k = 4$, test the validity of (4.4), Richardson's extrapolation formula.

B2. $36t^3y + 10ty + (9t^4 + 5t^2 - 5)y' = 0$, $y(-7) = 4$; $y(-6)$

B3. $32t^3 - (28y^3 + 6y)y' = 0$, $y(6) = -7$; $y(7)$

B4. $12t^2y^3 - 4t^3 + (12t^3y^2 + 27y^2)y' = 0$, $y(-4) = -4$; $y(-3)$

B5. $9y - 24t^3y^3 + (18t - 24t^4y^2)y' = 0$, $y(8) = -6$; $y(9)$

B6. $t + (1 + y)y' = 0$, $y(1) = 2$, $y(2)$

B7. $8 + 18t^2 - 27y^2y' = 0$, $y(8) = 2$; $y(9)$

B8. $3ty - (7y^2 + 6t^2)y' = 0$, $y(1) = 1$; $y(2)$

B9. $4t + (10y - 9)y' = 0$, $y(1) = 2$; $y(1.5)$

B10. $9 + (2t - 5y)y' = 0$, $y(2) = 1$; $y(3)$

B11. $5t - 7t^2 + (y + 1)y' = 0$, $y(1) = 2$; $y(2)$

B12. $7t^3 - 6t^2 + ty' + 2y = 0$, $y(1) = 2$; $y(2)$ See Exercise B4, Section 3.5.

B13. $t^2y' - 2ty + 14 - 3t^{-1} - 11t^{-2} = 0$, $y(1) = 1$; $y(2)$ See Exercise B5, Section 3.5.

B14. $2y - 3t^4 + t^3 + 2t^2 + ty' = 0$, $y(-2) = 4$; $y(-1)$ See Exercise B6, Section 3.5.

B15. $ty' - 16 - 2y - 2t^{-1} + 15t^{-2} = 0$, $y(1) = 3$; $y(2)$ See Exercise B7, Section 3.5.

B16. $9t^{5/2} + 7t^{3/2} + ty' + \frac{1}{2}y = 0$, $y(1) = 1$; $y(2)$ See Exercise B8, Section 3.5.

B17. $64t^2 - ty'y^3 + 12t - 8y^4 = 0$, $y(1) = 1$; $y(2)$

B18. $t^3y' + 4t^2y + 10ty^3 + 18y^3 = 0$, $y(2) = 1$; $y(3)$

4.2   The Euler Method

Rationale

The basis for the calculations in Euler's method is the tangent line approximation of a function $g(t)$. Given a function $g(t)$, a linear approximation, that is, an approximation along the tangent line constructed at $(t_0, g(t_0))$, is given by $y(t) = g(t_0) + g'(t_0)(t - t_0)$. Letting $h = t - t_0$ so $t = t_0 + h$, we can write this expansion as $y(t_0 + h) = g(t_0) + g'(t_0)h$. Generalizing from t_0 to t_k and writing y_k for $g(t_0 + kh)$, we have

$$t_k = t_0 + kh \qquad y_{k+1} = y_k + hg'(t_k)$$

Given the initial value problem $y' = f(t, y(t))$, $y(t_0) = y_0$, this tangent line approximation becomes the Euler algorithm

$$t_k = t_0 + kh \qquad y_{k+1} = y_k + hf(t_k, y_k) \qquad k = 0, 1, \ldots$$

Figure 4.2 illustrates this tangent line approximation for the curve defined by the function $y(t)$. The height $y(t_0 + h)$ is approximated by the height $y_0 + hy'(t_0)$, which is on the tangent line (constructed at (t_0, y_0)) above $t = t_0 + h$.

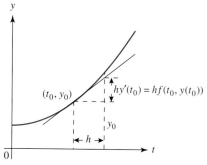

FIGURE 4.2   Euler's method gives the tangent-line approximation

EXAMPLE 4.1   To apply the Euler algorithm to the initial value problem $y' = t + y$, $y(0) = 0$, determine the t-values at which the solution $y(t)$ is to be approximated numerically. Taking $h = \frac{1}{10}$, we get $t_k = 0 + k(\frac{1}{10}) = \frac{k}{10}$. The very first y-value is given as initial data, namely, $y_0 = 0$. The remaining y-values must be computed by the Euler algorithm. Thus, with $f(t, y) = t + y$, we get

$$y_1 = y_0 + hf(t_0, y_0) = 0 + \tfrac{1}{10}(0 + 0) = 0$$

$$y_2 = y_1 + hf(t_1, y_1) = 0 + \tfrac{1}{10}\left(\tfrac{1}{10} + 0\right) = \tfrac{1}{100}$$

$$y_3 = y_2 + hf(t_2, y_2) = \tfrac{1}{100} + \tfrac{1}{10}\left(\tfrac{2}{10} + \tfrac{1}{100}\right) = \tfrac{31}{1000}$$

$$y_4 = y_3 + hf(t_3, y_3) = \tfrac{31}{1000} + \tfrac{1}{10}\left(\tfrac{3}{10} + \tfrac{31}{1000}\right) = \tfrac{641}{10000}$$

Clearly, continuing to work in exact arithmetic will become increasingly more difficult, and continuing to work "by hand" will soon become impossible. A computer implementation of these calculations is found, therefore, in the accompanying Maple worksheet.   ❖

EXAMPLE 4.2   Let us apply Euler's method to the IVP $y' = -10y$, $y(0) = 1$, taking $h = \frac{1}{10}$. Working in exact arithmetic, we get

$$y_1 = y_0 + hf(t_0, y_0) = 1 + \tfrac{1}{10}(-10 \times 1) = 0$$

$$y_2 = y_1 + hf(t_1, y_1) = 0 + \tfrac{1}{10}(-10 \times 0) = 0$$

$$y_3 = y_2 + hf(t_2, y_2) = 0 + \tfrac{1}{10}(-10 \times 0) = 0$$

that is, we get $y(t) = 0$. Since the actual solution of this IVP is $y(t) = e^{-10t}$, we can see there are difficulties with Euler's method. Both this IVP and the one in Exercise A5, Section 4.1 can be explained by the following analysis.   ❖

Consider the IVP $y' = \lambda y$, $y(0) = 1$, for which Euler's method gives

$$y_{k+1} = y_k + h(\lambda y_k) = (1 + h\lambda)y_k$$

Since $y_1 = (1 + h\lambda)y_0 = (1 + h\lambda)$ and $y_2 = (1 + h\lambda)y_1 = (1 + h\lambda)(1 + h\lambda) = (1 + h\lambda)^2$, we generalize and declare that $y_k = (1 + h\lambda)^k$. Thus, when $\lambda = -10$ and $h = \frac{1}{10}$, we will compute $y_k = 0$ for every k. If $\lambda = -20$ and $h = \frac{1}{10}$, then $y_k = (1 - 2)^k = (-1)^k$. The restriction this analysis places on the stepsize h is detailed in the exercises. Equations such as these for which the stepsize is controlled by a term that ceases to contribute to the solution are called *stiff* and require special numeric routines. An additional discussion of *stiffness* appears in Section 4.4.

EXERCISES 4.2–Part A

A1. Apply Euler's method to the initial value problem $y' = \lambda y$, $y(0) = y_0$, showing that with positive stepsize h the general step amounts to computing $y_{k+1} = y_k(1 + \lambda h)$, $k = 0, 1, \ldots$.

A2. Show that the difference equation $y_{k+1} = y_k(1 + \lambda h)$, $k = 0, 1, \ldots$, has solution $y_k = y_0(1 + \lambda h)^k$. This can be done by writing the y_k for $k = 0, 1, 2$, and then generalizing.

A3. If $\lambda > 0$, then the exponential solution in Exercise A1 will grow in magnitude, and so will the solution $y_k = y_0(1 + \lambda h)^k$, although the latter solution might grow more slowly. If $\lambda < 0$, then the exponential solution tends to 0 but y_k only tends to zero if $|1 + \lambda h| < 1$. Show that the solution of this inequality is $0 < h < \frac{2}{|\lambda|}$. (Euler's method is said to be *conditionally stable* [42] because of this restriction on stepsize.)

A4. Use Exercises A1–3 to explain the outcome in Exercise A5, Section 4.1.

A5. Is the result in Example 4.2 inconsistent with Exercises A1–3? Explain.

EXERCISES 4.2–Part B

B1. Using $h = 0.15$ and then $h = 0.25$, obtain the Euler solution of the IVP $y' = -10y$, $y(0) = 1$. For $t \le 1.5$, compare with the exact solution $y = e^{-10t}$. Explain these results in light of Exercises A1–3.

B2. For the IVP in Exercise B1, determine analytically or empirically a stepsize h for which Euler's method gives, on $[0, 1]$, an error no worse than 0.01.

B3. Use Euler's method to solve the initial value problem $y' + 15y = 30t^2 + 49t + 18$, $y(0) = 2$. First, use a stepsize of $h = 0.1$ and solve on the interval $0 \le t \le 10$; then use a stepsize of $h = 0.2$. Solve the equation exactly, and identify the homogeneous solution and the particular solution. Explain the behavior of the solution in light of these two components. Explain the numeric results in light of the condition articulated in Exercise A3. (The differential equation in this problem is stiff.)

B4. Repeat Exercise B3 for the initial value problem $y' + 18y = 2\frac{9 + 9t^2 - t}{(1 + t^2)^2}$, $y(0) = 3$.

B5. Repeat Exercise B3 for the initial value problem $y' + 50y = 100t + 101$, $y(0) = 1$, but use $h = 0.039$ and $h = 0.041$.

B6. Repeat Exercise B3 for the initial value problem $y' + 100y = t$, $y(0) = 5$, but use $h = 0.019$ and $h = 0.021$.

B7. In the differential equation $y' = f(t, y)$, the role of the constant λ in Exercise A1 is played by $\frac{\partial f}{\partial y}$. Using stepsizes $h = 0.008$ and $h = 0.009$, apply Euler's method to the initial value problem $y' + 3t^2y = t^2$, $y(0) = 2$ on the interval $0 \le t \le 10$. If $|1 + \lambda h| < 1$ must hold on the interval of integration and the role of λ is played by $\frac{\partial f}{\partial y}$, estimate the largest stepsize for which Euler's method remains stable for this problem.

B8. In light of Exercise B7, estimate the largest stepsize that can be used in Euler's method when solving the initial value problem $y' + e^{t/2}y = 1$, $y(0) = 1$ on the interval $0 \le t \le 10$. Test your deduction empirically by applying Euler's method with several different stepsizes.

Use Euler's method to obtain a numeric solution to the IVP in each of Exercises B9–15. In B9–11, $h = 1$, $0 \le t \le 200$. In B12–14, $h = 0.1$, $1 \le x \le 2$. In B15, $h = 0.1$, $0 \le x \le 1$. Sketch the numeric solution and compare to the exact solution obtained earlier.

B9. Exercise B1, Section 2.3   **B10.** Exercise B5, Section 2.3

B11. Exercise B8, Section 2.3   **B12.** Exercise B2, Section 3.2

B13. Exercise B3, Section 3.3, and use the initial condition $y(1) = 1$.

B14. Exercise B3, Section 3.4, and use the initial condition $y(1) = 1$.

B15. Exercise B1, Section 3.6, and use the initial condition $y(0) = 2$.

4.3   Taylor Series Methods

Taylor Series from the ODE

There are two ways in which a Taylor series can be used to construct an approximation to the solution of a differential equation. One can either use the differential equation to generate Taylor polynomials that approximate the solution, or one can use a Taylor polynomial as part of a numeric integration scheme for solving an associated initial value problem. We will illustrate both techniques, first examining a process whereby the differential equation $y' = f(t, y(t))$ is used recursively to generate Taylor polynomials. For simplicity, we will include an initial condition, thus illustrating how to solve an initial value problem.

Since the coefficients of the Taylor series for $y(t)$ contain derivatives of $y(t)$, construction of a Taylor polynomial for $y(t)$ requires computing $y''(a)$, $y'''(a)$, etc., where $(a, y(a))$ is a known initial point. Using the chain rule (or implicit differentiation), these derivatives can be computed from the differential equation itself. In general, for the differential equation $y' = f(t, y(t))$, the second and third derivatives would be

$$y'' = f' \equiv \frac{df}{dt} = \frac{\partial f}{\partial t} + \frac{\partial f}{\partial y}\frac{dy}{dt} = f_t + f_y f$$

$$y''' = f'' = \frac{d}{dt}(f_t + f_y f) = f_{tt} + f_{ty}f + f_y(f_t + f_y f) + f(f_{yt} + f_{yy}f)$$

For example, if $f(t, y) = -ty^2$, these second and third derivatives would be

$$y'' = -y^2 + 2t^2 y^3 \qquad y''' = 6ty^3 - 6t^3 y^4$$

In principle, given the initial condition $y(a) = y_a$, all derivatives $y^{(k)}$, if they exist, could be computed in this fashion, and $y(t)$ represented by the Taylor series

$$y(t) = \sum_{k=0}^{\infty} \frac{y^{(k)}(a)}{k!}(t-a)^k$$

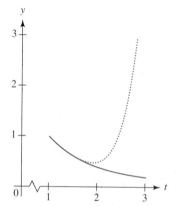

FIGURE 4.3   Exact solution (solid) and Taylor approximation (dotted) for the IVP $y' = -ty^2$, $y(1) = 1$

Unfortunately, except for the most trivial examples, the required differentiations quickly become too tedious to carry out by hand. Computer algebra systems can provide symbolic differentiations, making it possible to examine this series expansion process with a significant number of terms.

EXAMPLE 4.3   The exact solution of the initial value problem $y' = -ty^2$, $y(1) = 1$ is $y(t) = \frac{2}{t^2+1} = 1 - (t-1) + \frac{1}{2}(t-1)^2 - \frac{1}{4}(t-1)^4 + \frac{1}{4}(t-1)^5 + \cdots$. Figure 4.3, showing graphs of the exact solution (solid line) and of the fifth-degree Taylor approximation (dotted line), reveals the shortcoming of the method. The series solution, being a Taylor polynomial, will typically agree well with the exact solution in a neighborhood of the initial point but then diverge for values of the independent variable that are outside this neighborhood. This observation inspires a different use of the Taylor polynomial, one in which the neighborhood about the initial point is deliberately kept small but the solution is extended numerically by iterating the series-building process.   ❖

Numeric Taylor Methods

The IVP of Example 4.3 can be solved with the following second-order fixed-step method. Truncate the series in that example, keeping just terms through second order but use this result to compute just $y(a + h)$. The point $(a + h, y(a + h))$ is then taken as a new initial point and the second-order Taylor polynomial again calculated, this time at $t = a + h$. This process is repeated at each point of the grid $t_k = a + kh$, $k = 1, 2, \ldots, N$.

Thus, the numeric Taylor method with global truncation error of order two is given by

$$y_{k+1} = y_k + hf(t_k, y_k) + \frac{h^2}{2!} f'(t_k, y_k) \qquad y_0 = y(t_0)$$

As seen in the previous section, the derivative f' gives y'' and is computed either by the chain rule or by implicit differentiation.

EXAMPLE 4.4   For the initial value problem $y'(t) = -ty^2$, $y(1) = 1$, we take $t_k = 1 + kh$ and $f'(t, y) = y^2(2t^2y - 1)$ to obtain

$$y_{k+1} = y_k - h\left(1 + \left(k + \tfrac{1}{2}\right)h\right) y_k^2 + h^2(1 + kh)^2 y_k^3$$

The computation of y_1 and y_2, for $h = 0.1$, are

$$y_1 = y_0 - \tfrac{1}{10}\left(1 + \tfrac{1}{20}\right) y_0^2 + \tfrac{1}{100} y_0^3 = y_0 - \tfrac{21}{200} y_0^2 + \tfrac{1}{100} y_0^3 = \tfrac{181}{200} = 0.905$$

$$y_2 = y_1 - \tfrac{1}{10}\left(1 + \tfrac{3}{20}\right) y_1^2 + \tfrac{1}{100} \left(\tfrac{11}{10}\right)^2 y_1^3 = y_1 - \tfrac{23}{200} y_1^2 + \tfrac{121}{10000} y_1^3 = 0.81978$$

On the interval $[1, 3]$, the numeric solution computed with stepsize $h = 0.1$ differs from the exact solution by less than 0.0004. Hence, if the numeric solution were included in Figure 4.3, it would be indistinguishable from the exact solution's solid line. Thus, the numeric solution that uses the Taylor series expression for a stepsize h and then repeats the calculation of the Taylor expansion is far more accurate than the Taylor polynomial alone. The Taylor polynomial approximates $y(t)$ well only near $t = 1$; as t exceeds 1.5, the approximation begins to fail.

The maximum global error of -0.0003601022 occurs at $t = 3$ since $y(3) = \tfrac{1}{5} = 0.2$ and $y_{20} = 0.2003601022$. Recomputing with the stepsize cut in half, that is, with $h = \tfrac{1}{20}$, we find $y_{40} = 0.2000852413$ and a global error of -0.0000852413. Dividing the first error by $\left(\tfrac{1}{10}\right)^2$ and the second by $\left(\tfrac{1}{20}\right)^2$, we get -0.03601022 and -0.034096520, respectively. The approximate equality of these two numbers supports the claim that the errors obey

$$y_{\text{exact}} = y_{\text{approx}} + Ch^2$$

and, hence, that we have computed with a second-order method.

EXERCISES 4.3–Part A

A1. Using the differential equation $y' = -y$, generate a Taylor series solution satisfying $y(0) = \tfrac{1}{2}$.

A2. Using the differential equation $y' = xy$, generate a Taylor series solution satisfying $y(0) = 1$.

A3. Using the differential equation $y'' + xy' + y = e^x$, generate a Taylor series solution satisfying $y(0) = 1$, $y'(0) = -1$.

A4. For the differential equation $xy'' + y' + y = e^x$, show that a Taylor series cannot be generated at $x = 0$, no matter what the initial conditions. The point $x = 0$ is a *singular point* for the given DE.

A5. Legendre's equation $(1 - x^2)y'' - 2xy' + 2y = 0$ has $x = \pm1$ as singular points. Show that $y(x) = x$ is a solution that has a Taylor series expansion at $x = 1$. Does this contradict Exercise A4?

EXERCISES 4.3–Part B

B1. The IVP $y'' + 4y' + 13y = 0$, $y(0) = 1$, $y'(0) = 0$, models the motion of a car's shock absorber displaced one unit and released.

(a) Generate a fifth-degree Taylor series solution at $t = 0$ and sketch its graph.

(b) Expand the exact solution $y(t) = e^{-2t}(\cos 3t + \tfrac{2}{3} \sin 3t)$ in a Maclaurin series and compare it to part (a).

(c) Graph the series solution and the exact solution.

B2. Brine containing 2 lb of salt per gallon flows into a 200-gal tank at a rate of 5 gal/min, mixes instantly, and flows out at a rate of 8 gal/min. If the tank initially contained 50 gal of water:

(a) Formulate an IVP governing the amount of salt in the tank.

(b) Obtain an exact solution to the IVP.

(c) Expand the exact solution in a series about $t = 0$.

(d) Generate a Maclaurin series solution directly from the differential equation.

(e) Compare the series and exact solutions.

In Exercises B3–9:

(a) Obtain an exact solution and, from it, an accurate value at the indicated terminal point. If using a computer algebra system, applying the initial condition to an exact solution might be appropriate for some IVPs.

(b) At the initial point, expand the exact solution in a Taylor polynomial of degree p.

(c) Build a Taylor series numeric method of degree p and use it to calculate the value at the terminal point, using a stepsize of $h = 0.1$. Determine the error at the terminal point.

(d) Using $h = 0.05$, recalculate the value at the terminal point and again determine the error.

(e) Test the order of the method by examining the constancy of $(y_{\text{exact}} - y_{\text{approx}})/h^p$ for $h = 0.1/2^k$, $k = 0, 1, \ldots$.

(f) On the same set of axes, plot the exact solution, the Taylor polynomial, and the two numeric solutions in parts (c) and (d).

B3. $y' = \dfrac{(36t^3 + 10t)y}{5 - 5t^2 - 9t^4}$, $y(-7) = 4$; terminal point $(-6, y(-6))$, and degree $p = 3$

B4. $y' = \dfrac{16t^3}{14y^3 + 3y}$, $y(6) = -7$; terminal point $(7, y(7))$, and degree $p = 2$

B5. $y' = \dfrac{8 + 18t^2}{27y^2}$, $y(8) = 2$; terminal point $(9, y(9))$, and degree $p = 2$

B6. $y' = \dfrac{3ty - 10t^2}{7y^2 + 6t^2}$, $y(1) = 1$; terminal point $(2, y(2))$, and degree $p = 2$

B7. $y' = \dfrac{ty}{y^2 - t^2}$, $y(2) = 3$; terminal point $(3, y(3))$, and degree $p = 1$

B8. $y' = \dfrac{7t^2 - 5ty}{y^2 + t^2}$, $y(1) = 2$; terminal point $(2, y(2))$, and degree $p = 2$

B9. $y' = \dfrac{36t^3y + 10ty}{5 - 5t^2 - 9t^4}$, $y(1) = 2$; terminal point $(2, y(2))$, and degree $p = 2$

4.4   Runge–Kutta Methods

Generating Runge–Kutta Methods

Methods patterned after the work of the German mathematicians Carl Runge (1856–1927) and M. W. Kutta (1867–1944) are called Runge–Kutta methods. For each given global truncation order, there exists a family of Runge–Kutta methods. The family of second-order methods for the differential equation $y'(t) = f(t, y(t))$ obtains $y(t + h)$ from $y(t)$ by adding to it a weighted average of two estimates of the actual increment. Thus, we write

$$y(t + h) = y(t) + a_1 hf + a_2 hf(t + ph, y + qhf) \tag{4.5}$$

and the Taylor expansion

$$y(t + h) = y(t) + hy'(t) + \frac{h^2}{2}y''(t) + \cdots$$

Writing $y'(t)$ as f and $y''(t)$ as $\frac{d}{dt}f(t, y(t)) = \frac{\partial f}{\partial t} + \frac{\partial f}{\partial y}\frac{dy}{dt} = f_t + f_y f$, the Taylor expansion becomes

$$y(t + h) = y(t) + hf(t, y) + \frac{h^2}{2}f_t(t, y) + \frac{h^2}{2}f(t, y)f_y(t, y) + \cdots$$

The term $hf(t, y)$ is written as $\frac{h}{2}f + \frac{h}{2}f$, so we get

$$y(t + h) = y(t) + \frac{h}{2}f + \frac{h}{2}[f + hf_t + hff_y] + \cdots \tag{4.6}$$

The terms in the brackets are the start of the Taylor expansion

$$f(t + h, y + hf) = f + hf_t + hff_y + \cdots$$

so we replace that term with

$$f(t + ph, y + qhf) = f + phf_t + qhff_y + \cdots$$

and write (4.5) as

$$y(t + h) = y(t) + a_1 hf + a_2 h[f + phf_t + qhff_y] \qquad (4.7)$$

thereby obtaining two forms of $y(t + h)$, namely, (4.6) and (4.7).

In these two equations, equating coefficients of f yields the equation $a_1 + a_2 = 1$, equating coefficients of f_t yields the equation $pa_2 = \frac{1}{2}$, and equating coefficients of f_y yields the equation $qa_2 = \frac{1}{2}$. Expressing p, q, and a_1 in terms of a_2, we get $a_1 = 1 - a_2$ and $p = q = \frac{1}{2a_2}$. Various choices of a_2 therefore generate different second-order methods. In fact, in order to keep t_k and $t_k + ph$ in the interval $[t_k, t_{k+1}]$, we must have $p \leq 1$, requiring, therefore, $a_2 \geq \frac{1}{2}$.

A Spectrum of Runge–Kutta Methods

Figure 4.4 is a first step in analyzing the variety of methods just created. In it, the dashed line and solid curve are graphs of a_1 and $p = q$, respectively, as functions of a_2, consistent with the restriction $a_2 \geq \frac{1}{2}$. Specifying a_2 determines p, q, and a_1, and thereby determines a second-order Runge–Kutta method. For example, if a vertical line is drawn at $a_2 = 1$ in Figure 4.4, its intersection with the dashed line $a_1 = 1 - a_2$ determines $a_1 = 0$ and its intersection with the solid curve $p = q = \frac{1}{2a_2}$ determines $p = q = \frac{1}{2}$. This gives method B in Table 4.5.

Using the notation $f_k = f(t_k, y_k)$, Table 4.5 lists methods A, B, and C, three of the more common members of the family of second-order Runge–Kutta methods, along with methods D and E, two methods not of this family. Method D is a *multipoint* method and requires both y_0 and y_1 to start. Method E is *implicit* since y_{k+1} appears on both sides of the equation. The reasons for including them in Table 4.5 becomes clear upon inspecting Table 4.6.

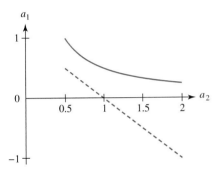

FIGURE 4.4 Second-order Runge–Kutta methods are parametrized by $a_2 \geq \frac{1}{2}$. Solid curve is $p = q = \frac{1}{2a_2}$, while dashed line is $a_1 = 1 - a_2$.

Method	a_2	Computational Formula
A	$\frac{1}{2}$	$y_{k+1} = y_k + \dfrac{h}{2}[f_k + f(t_{k+1}, y_k + hf_k)]$
B	1	$y_{k+1} = y_k + hf\left(t_k + \dfrac{h}{2}, y_k + \dfrac{h}{2}f_k\right)$
C	$\frac{3}{4}$	$y_{k+1} = y_k + \dfrac{h}{4}\left[f_k + 3f\left(t_k + \frac{2}{3}h, y_k + \frac{2}{3}hf_k\right)\right]$
D		$y_{k+2} = y_k + 2hf_{k+1}$
E		$y_{k+1} = y_k + \dfrac{h}{2}[f_k + f_{k+1}]$

TABLE 4.5 Second-order methods, including three from Runge–Kutta family

Method	Midpoint Rule	Improved Euler	Modified Euler	Heun
A		[12, p. 414][1] [51, p. 1037][2] [80, p. 74] [82, p. 566][3]	[16, p. 280] [38, p. 404]	[11, p. 46] [12, p. 414][1] [49, p. 582] [51, p. 1037][2] [60, p. 437] [82, p. 566][3]
B	[16, p. 279]		[12, p. 417] [49, p. 582] [60, p. 437] [66, p. 206]	
C				[16, p. 280]
D			[13, p. 368]	
E		[13, p. 372]		

TABLE 4.6   Nomenclature for the five methods in Table 4.5

The nomenclature for these methods is quite tangled in the literature. A scan of nearly a dozen texts yields the names Midpoint, Improved Euler, Modified Euler, and Heun, listed in Table 4.6. There is clearly no uniformity in naming the common second-order Runge–Kutta methods. Reference [16] designates method C as Heun's method, but no other text even mentions method C. Reference [13] gives an implicit method as the Improved Euler method, and [16] calls method B the Midpoint Rule, while [12], [49], [60], and [66] all call the same method the Modified Euler method. But [16] and [38] both call method A the Modified Euler method! Great care must therefore be taken when consulting different texts.

Method A

The exact solution of the IVP

$$y' = f(t, y) = -ty^2 \qquad y(1) = 1 \tag{4.8}$$

is $y_e = \frac{2}{t^2+1}$. To use method A, take $h = \frac{1}{10}$ and $t_k = 1 + kh$ so that $t_1 = 1 + h = \frac{11}{10}$, then find y_1 to be

$$y_1 = y_0 + \frac{h}{2}[f(t_0, y_0) + f(t_0 + h, y_0 + hf(t_0, y_0))]$$

$$= 1 + \frac{1}{20}\left[-1 - \left(\frac{11}{10}\left(1 + \frac{1}{10}(-1)\right)^2\right)\right] = \frac{18,109}{20,000} \doteq 0.90545$$

For the exact solution we find $y_e(1 + h) = \frac{200}{221} \doteq 0.904977$.

Method B

Again taking $h = \frac{1}{10}$ and applying method B to (4.8), we find y_1 to be

$$y_1 = y_0 + hf\left(t_0 + \frac{h}{2}, y_0 + \frac{h}{2}f(t_0, y_0)\right)$$

$$= 1 + \frac{1}{10}\left[-\left(1 + \frac{1}{20}\right)\left(1 + \frac{1}{20}(-1)\right)^2\right] = \frac{72,419}{80,000} \doteq 0.9052375$$

Method C

Once again taking $h = \frac{1}{10}$ and applying method C to (4.8), we find y_1 to be

$$y_1 = y_0 + \frac{h}{4}\left[f(t_0, y_0) + 3f\left(t_0 + \frac{2}{3}h, y_0 + \frac{2}{3}hf(t_0, y_0)\right)\right]$$

$$= 1 + \frac{1}{40}\left[-1 + 3(-1)\left(1 + \frac{2}{30}\right)\left(1 + \frac{2}{30}(-1)\right)^2\right] = \frac{40,739}{45,000} \doteq 0.905311$$

Method D

Method D is not self-starting since it requires *both* y_k and y_{k+1} to compute y_{k+2}. Hence, it is a *multipoint method* and requires some other method for determining y_1. In practice, a single-step method is used; but for simplicity, if we take y_1 as the exact value $\frac{200}{221}$, we find y_2 for the IVP (4.8) to be

$$y_2 = y_0 + 2hf(t_1, y_1) = 1 + \frac{2}{10}\left(-\left(1 + \frac{1}{10}\right)\left(\frac{200}{221}\right)^2\right) = \frac{40,041}{48,841} \doteq 0.819824$$

Method E

To apply method E to the initial value problem (4.8), note that the algorithm is

$$y_{k+1} = y_k + \frac{h}{2}[f(t_k, y_k) + f(t_{k+1}, y_{k+1})]$$

Since the unknown y_{k+1} appears on both sides of an algebraic equation, it must be computed as the solution of that equation at each step. For example, to compute y_1 with $h = \frac{1}{10}$, we would have to solve the equation $y_1 = \frac{19}{20} - \frac{11}{200}y_1^2$ for y_1. This is a quadratic equation and admits the exact solution $y_1 = \frac{-100 \pm \sqrt{12,090}}{11}$. The complication we now face is deciding which solution is appropriate in this context. Moreover, rarely will the differential equation be so amenable as to admit exact solutions. Hence, we solve the equation numerically to obtain $y_1 = 0.9049578236$ by restricting the solver to solutions in the interval $(0, 2)$, in keeping with the initial condition $y(1) = 1$.

Iterating this process for ten steps gives $y_{10} = 0.3997464021$. The exact solution is $y_e(2) = \frac{2}{5} = 0.4$, so the global truncation error at this point is -0.0002535979. If we repeat these same calculations with the stepsize cut in half, we find $y_{20} = 0.3999367065$, so the global truncation error is now -0.0000632935. Dividing the first error by $\left(\frac{1}{10}\right)^2$ and the second by $\left(\frac{1}{20}\right)^2$, we get -0.02535979 and -0.02531740000, respectively. The near-equality of these two numbers suggests that, indeed, method E is second order.

Stiff Equations and the Backward Euler Method

The astute reader will wonder, "Why use an implicit method if it introduces the additional complication of solving an algebraic equation at each step?" Our answer will be to remind the reader of the difficulties posed for the Euler algorithm by the stiff equations in Section 4.2. Surprisingly, a "backward" form of the Euler algorithm, namely,

$$y_{k+1} = y_k + hf_{k+1} \qquad k = 0, 1, \ldots, N - 1$$

another example of an implicit method, fares much better with the stiff equation on the interval $t_0 \le t \le t_0 + kN$.

We illustrate with the initial value problem $y' = f(t, y) = t^2 - 100y$, $y(0) = y_0 = \frac{1}{100}$. Recall from Section 4.2 that for a stiff equation, the stepsize h in Euler's method is restricted

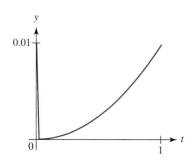

FIGURE 4.5   Forward Euler solution of IVP $y' = t^2 - 100y$, $y(0) = \frac{1}{100}$, using $h = 0.01$

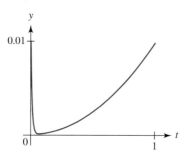

FIGURE 4.6   Exact solution of IVP $y' = t^2 - 100y$, $y(0) = \frac{1}{100}$

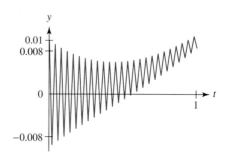

FIGURE 4.7   Forward Euler solution of IVP $y' = t^2 - 100y$, $y(0) = \frac{1}{100}$, with $h = 0.02$

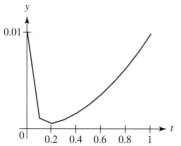

FIGURE 4.8   Backward Euler solution of IVP $y' = t^2 - 100y$, $y(0) = \frac{1}{100}$

by $0 < h < 2/|f_y| = \frac{2}{100} = 0.02$. A forward Euler solution with stepsize $h = 0.01$ is graphed in Figure 4.5; and the exact solution, $y(t) = \frac{t^2}{100} - \frac{t}{5000} + \frac{1}{500,000}(1 + 4999e^{-100t})$, is graphed in Figure 4.6. For a stepsize of $h = 0.01$, Euler's method is well behaved. However, for the stepsize $h = 0.02$, the situation is quite the contrary, as seen in Figure 4.7.

To demonstrate the behavior of the backward Euler method, and hence justify the introduction of implicit methods, we compute the solution on the interval $[0, 1]$ using a stepsize of $h = \frac{1}{10}$. The equation to be solved at each step of the backward Euler algorithm is

$$y_{k+1} = y_k + hf((k+1)h, y_{k+1}) = y_k + h[(k+1)^2 h^2 - 100 y_{k+1}]$$

which when $k = 0$ becomes $y_1 = \frac{11}{1000} - 10y_1$. In this instance, the equation is linear, so there is but the one solution, $y_1 = \frac{1}{1000}$. Continuing with similar calculations, we obtain Figure 4.8, which indicates the stable nature of the backward Euler method and reveals some of the attractions of implicit methods.

EXERCISES 4.4–Part A

A1. Use method A and $h = \frac{1}{10}$ to obtain y_1 and y_2 for the IVP $y' = t + y$, $y(1) = 2$.

A2. Use method B and $h = \frac{1}{10}$ to obtain y_1 and y_2 for the IVP $y' = t^2 - y^2$, $y(1) = 1$.

A3. Use method C and $h = \frac{1}{10}$ to obtain y_1 and y_2 for the IVP $y' = ty$, $y(1) = 3$.

A4. Use method D and $h = \frac{1}{10}$ to obtain y_2 and y_3 for the IVP $y' = \frac{y}{t}$, $y(1) = 2$, if $y(1.1) = 2.2$.

A5. Use method E and $h = \frac{1}{10}$ to obtain y_1 and y_2 for the IVP $y' = t - y^2$, $y(0) = 1$.

A6. Use the backward Euler method and $h = \frac{1}{100}$ to obtain y_1 and y_2 for the IVP $y' = t - 50y$, $y(1) = 1$.

EXERCISES 4.4–Part B

B1. Apply method E to the initial value problem $y' = ay$, $y(0) = y_0$, showing that with positive stepsize h the general step amounts to computing $y_{k+1} = \frac{2+ah}{2-ah} y_k$, $k = 0, 1, \ldots$.

B2. Show that the *difference equation* $y_{k+1} = \frac{2+ah}{2-ah} y_k$, $k = 0, 1, \ldots$, has solution $y_k = ((2+ah)/(2-ah))^k y_0$.

B3. Show that the exact solution to the initial value problem in Exercise B1 is $y(t) = y_0 e^{at}$.

B4. If $a > 0$, then the exponential solution in Exercise A3 will grow in magnitude (assuming forward integration with $h > 0$), and so

will the solution $y_k = (2 + ah)/(2 - ah)^k y_0$, although the latter solution might grow more slowly. If $a < 0$, then the exponential solution tends to zero but y_k only tends to zero if $\frac{2+ah}{2-ah} < 1$. Show that this inequality is then satisfied for any $a < 0$.

B5. Use method E to solve, on the interval $0 \le t \le 10$, the initial value problem $y' + 15y = 30t^2 + 49t + 18$, $y(0) = 2$. By experimenting with the stepsize h, determine if it is indeed true that this method is stable for any reasonable value of h. (See Exercise B3, Section 4.2.)

B6. Repeat Exercise B5 for the initial value problem $y' + 18y = 2\frac{9+9t^2-t}{(1+t^2)^2}$, $y(0) = 3$. (See Exercise B4, Section 4.2.)

B7. Repeat Exercise B5 for the initial value problem $y' + 50y = 100t + 101$, $y(0) = 1$. (See Exercise B5, Section 4.2.)

B8. Repeat Exercise B5 for the initial value problem $y' + 100y = t$, $y(0) = 5$. (See Exercise B6, Section 4.2.)

B9. Apply the backward Euler method to the initial value problem $y' = ay$, $y(0) = y_0$, showing that with positive stepsize h the general step amounts to computing $y_{k+1} = y_k/(1 - ah)$, $k = 0, 1, \ldots$.

B10. Show that the difference equation $y_{k+1} = y_k/(1 - ah)$, $k = 0, 1, \ldots$, has solution $y_k = (1/(1 - ah))^k y_0$.

B11. Show that the exact solution to the initial value problem in Exercise B9 is $y(t) = y_0 e^{at}$.

B12. If $a < 0$, then the exponential solution in Exercise B11 will decrease to zero (assuming forward integration with $h > 0$), and so will the solution $y_k = (1/(1 - ah))^k y_0$. The inequality $\left|\frac{1}{1-ah}\right| < 1$ is satisfied for any $a < 0$, so there is no restriction

placed on the stepsize h. Verify this empirically for the initial value problem of Exercise B6.

B13. Repeat Exercise B12 for the initial value problem of Exercise B7.

B14. Repeat Exercise B12 for the initial value problem of Exercise B8.

B15. Use the IVP $y' = t^2 + 2y$, $y(1) = 2$, to verify empirically that method A is second order. Use $h = 0.1/2^k$, $k = 0, 1, \ldots$, and compare values at $t = 2$.

B16. Use the IVP $y' = 3t^2 y$, $y(1) = 1$, to verify empirically that method B is second order. Use $h = 0.1/2^k$, $k = 0, 1, \ldots$, and compare values at $t = 2$.

B17. Use the IVP $y' = 2t + 3y$, $y(2) = -1$, to verify empirically that method C is second order. Use $h = 0.1/2^k$, $k = 0, 1, \ldots$, and compare values at $t = 3$.

B18. Use the IVP $y' = ty$, $y(1) = -1$, to verify empirically that method D is second order. Use $h = 0.1/2^k$, $k = 0, 1, \ldots$, and compare values at $t = 2$. Use the exact solution to obtain y_1.

B19. Use the IVP $y' = 3ty^2$, $y(1) = -1$, to verify empirically that method E is second order. Use $h = 0.1/2^k$, $k = 0, 1, \ldots$, and compare values at $t = 2$.

4.5 Adams–Bashforth Multistep Methods

Multistep Methods

The Euler, Taylor, and Runge–Kutta methods of the previous lessons are all single-step methods because they compute y_{k+1} using only the value of y_k. Since y_1 can thereby be computed from the initial value y_0, such methods are called "self-starting."

Methods that compute y_{k+1} from p previous values of y, namely, $y_k, y_{k-1}, \ldots, y_{k-p+1}$, are called *multistep* methods. In particular, a multistep method using p previous values is called a p-step method. The expectation is that using several previously calculated values will be more accurate than using just one. However, p-step methods are not self-starting and require the use of some other method to generate the $p - 1$ starting values in addition to y_0. (Writing method D in Table 4.5 as $y_{k+1} = y_{k-1} + 2hf(t_k, y_k)$ shows it to be a multistep method with $p = 2$.)

A class of multistep methods based on the integration of interpolating polynomials is called the class of *Adams–Bashforth* methods. This lesson demonstrates the construction of two such methods: the third- and fourth-order Adams–Bashforth algorithms.

Third-Order Adams–Bashforth Algorithm

The Adams–Bashforth method, globally of third order, is a three-step method that computes y_{n+1} from the three previous values y_n, y_{n-1}, y_{n-2}. In essence, the slope field $f(t, y)$ in the differential equation $y' = f(t, y)$ is approximated by $p_2(t)$, a quadratic polynomial that *interpolates,* or passes through, the three points $(t_n, f_n), (t_n - h, f_{n-1}), (t_n - 2h, f_{n-2})$. This polynomial is then integrated between t_n and $t_{n+1} = t_n + h$, the inspiration deriving from

$$y_{n+1} - y_n = \int_{y_n}^{y_{n+1}} dy = \int_{t_n}^{t_{n+1}} f(t, y(t))\, dt$$

and resulting in the (global) third-order Adams–Bashforth algorithm

$$y_{n+1} = y_n + \int_{t_n}^{t_n+h} p_2(\tau)\, d\tau = y_n + \frac{h}{12}(5f_{n-2} - 16f_{n-1} + 23f_n) \qquad (4.9)$$

The interpolating polynomial is $p_2(t) = A_2 t^2 + B_2 t + C_2$, where the coefficients are defined by

$$A_2 = \frac{1}{2h^2}(f_{n-2} - 2f_{n-1} + f_n)$$

$$B_2 = \frac{1}{2h^2}((h - 2t_n)f_{n-2} + 4(t_n - h)f_{n-1} + (3h - 2t_n)f_n)$$

$$C_2 = \frac{1}{2h^2}(t_n(t_n - h)f_{n-2} + 2t_n(2h - t_n)f_{n-1} + h(2h - 3t_n)f_n)$$

Empirical Error Analysis

The Adams–Bashforth three-step method, derived via a quadratic interpolating polynomial, has local truncation error of order four and global truncation order of three. We test these claims empirically with the initial value problem $y' = f(t, y) = -ty^2$, $y(1) = 1$, whose exact solution is $y(t) = \frac{2}{t^2+1}$. Set $h = 0.2$ and note that it takes three starting points, namely, (t_0, y_0), (t_1, y_1), and (t_2, y_2). Only the first point is known, so we generate the second and third from the exact solution. (In practice, a self-starting method such as a Runge–Kutta method would be used.) From these three points, compute $f_0 = f(t_0, y_0)$, $f_1 = (t_1, y_1)$, $f_2 = (t_2, y_2)$; and from the Adams–Bashforth formula, obtain y_3 as an approximation to $y(1.6)$.

We are really interested in the h-dependence of the ratio $C_k = (y(t_k) - y_k)/h^3$, the constancy of which signals a global truncation error of order three. Hence, after completing 98 steps of the calculation (thereby arriving at $t_{100} = 21$ where we have computed y_{100}), we obtain $\frac{1/221 - 0.004523449607}{(0.2)^3} = 0.00018$. Set $h = \frac{h}{2} = 0.1$ and recompute, after 198 steps, y_{200} as an approximation to $y(21)$, obtaining the ratio $\frac{1/221 - 0.00452472649}{(0.1)^3} = 0.00016$. The relative equality of the two values of C_k supports the claim that we created an Adams–Bashforth method of global truncation order three.

Fourth-Order Adams–Bashforth

An Adams–Bashforth method of global truncation order four is obtained by interpolating the four points (t_n, f_n), $(t_n - h, f_{n-1})$, $(t_n - 2h, f_{n-2})$, and $(t_n - 3h, f_{n-3})$, with $p_3(t)$, a cubic polynomial that is integrated between t_n and $t_{n+1} = t_n + h$, yielding

$$y_{n+1} = y_n + \int_{t_n}^{t_n+h} p_3(\tau)\, d\tau = y_n + \frac{h}{24}(55f_n - 59f_{n-1} + 37f_{n-2} - 9f_{n-3}) \qquad (4.10)$$

The cubic interpolating polynomial is $p_3(t) = A_3 t^3 + B_3 t^2 + C_3 t + D_3$, where the coefficients are given by

$$A_3 = \frac{1}{6h^2}(f_n - 3f_{n-1} + 3f_{n-2} - f_{n-3})$$

$$B_3 = -\frac{1}{2h^3}[(t_n - 2h)f_n - (3t_n - 5h)f_{n-1} + (3t_n - 4h)f_{n-2} - (t_n - h)f_{n-3}]$$

$$C_3 = \frac{(3t_n^2 - 12ht_n + 11h^2)f_n}{6h^3} - \frac{(3t_n^2 - 10ht_n + 6h^2)f_{n-1}}{2h^3}$$

$$+ \frac{(3t_n^2 - 8ht_n + 3h^2)f_{n-2}}{2h^3} - \frac{(3t_n^2 - 6ht_n + 2h^2)f_{n-3}}{6h^3}$$

$$D_3 = -\frac{(t_n^3 - 6ht_n^2 + 11h^2t_n - 6h^3)f_n}{6h^3} + \frac{t_n(t_n^2 - 5ht_n + 6h^2)f_{n-1}}{2h^3}$$

$$- \frac{t_n(t_n^2 - 4ht_n + 3h^2)f_{n-2}}{2h^3} + \frac{t_n(t_n^2 - 3ht_n + 2h^2)f_{n-3}}{6h^3}$$

TESTING THE ORDER   An empirical test of the order of the method just created is summarized in Table 4.7 where, for the IVP used previously, fifteen-digit approximates for $y(21)$ are computed with stepsizes $h = 0.1/2^k$, $k = 0, 1, \ldots, 7$. The relative constancy of the ratios in the fourth column of Table 4.7 supports the claim that the Adams–Bashforth method is of order four.

k	$\dfrac{0.1}{2^k}$	y_n	$\dfrac{\frac{1}{221} - y_n}{\left(\frac{0.1}{2^k}\right)^4}$	k	$\dfrac{0.1}{2^k}$	y_n	$\dfrac{\frac{1}{221} - y_n}{\left(\frac{0.1}{2^k}\right)^4}$
0	$\frac{0.1}{1}$	0.00452489721025165	0.0001033	4	$\frac{0.1}{16}$	0.00452488687793556	0.0000705
1	$\frac{0.1}{2}$	0.00452488740964568	0.0000851	5	$\frac{0.1}{32}$	0.00452488687783468	0.0000695
2	$\frac{0.1}{4}$	0.00452488690773257	0.0000766	6	$\frac{0.1}{64}$	0.00452488687782847	0.0000690
3	$\frac{0.1}{8}$	0.00452488687959708	0.0000725	7	$\frac{0.1}{128}$	0.00452488687782808	0.0000687

TABLE 4.7   Fourth-order Adams–Bashforth

EXERCISES 4.5–Part A

A1. The second-order Adams–Bashforth multistep method is given by $y_{n+1} = y_n + \frac{h}{2}(3f_n - f_{n-1})$. Use this method with $h = \frac{1}{10}$ to obtain y_2 for the IVP $y' = t^2 y^2$, $y(1) = \frac{1}{3}$. Use the exact solution to obtain the starting value, y_1.

A2. Use the third-order Adams–Bashforth method (4.9) with $h = \frac{1}{10}$ to obtain y_3 for the IVP $y' = ty^2$, $y(1) = \frac{1}{4}$. Since this is not a self-starting method, obtain y_1 and y_2 from the exact solution.

A3. Use the fourth-order Adams–Bashforth method (4.10) with $h = \frac{1}{10}$ to obtain y_4 for the IVP $y' = y^2$, $y(1) = \frac{1}{2}$. Use the exact solution to obtain the starting values y_1, y_2, and y_3.

A4. Interpolate the quadratic polynomial $p = ax^2 + bx + x$ through the three points $(0, 3)$, $(1, 1)$, $(2, 5)$. *Hint:* By substitution, form three equations in a, b, c and then solve.

A5. Let $p_1(t)$ linearly interpolate the two points (t_n, f_n) and $(t_n - h, f_{n-1})$. Integrate p_1 between t_n and $t_{n+1} = t_n + h$, thereby obtaining the second-order Adams–Bashforth multistep method $y_{n+1} = y_n + \frac{h}{2}(3f_n - f_{n-1})$.

EXERCISES 4.5–Part B

B1. In the programming language of your choice, implement the second-order Adams–Bashforth algorithm given in Exercise A1. Use the initial value problem $ty' + y = t^2$, $y(1) = 1$ to validate your code. Use the exact solution to obtain the starting value y_1, and take $h = 0.02$.

B2. Obtain empirical evidence that the Adams–Bashforth method explored in Exercise A1 is indeed of second order. *Hint:* Use the IVP in Exercise B1 and compute $y(2)$, with several decreasing stepsizes.

B3. In the programming language of your choice, implement the fifth-order Adams–Bashforth method $y_{n+1} = y_n + \frac{h}{720}(1901 f_n - 2774 f_{n-1} + 2616 f_{n-2} - 1274 f_{n-3} + 251 f_{n-4})$. Test your implementation on the IVP of Exercise B1, again using the exact solution to generate starting values.

B4. Obtain empirical evidence that the Adams–Bashforth method programmed in Exercise B3 is indeed of fifth order. *Hint:* Use the IVP in Exercise B1 and compute $y(2)$, with several decreasing stepsizes.

B5. Use a computer algebra system to derive the fifth-order Adams–Bashforth method by constructing and integrating the appropriate interpolating polynomial.

Exercises B6–9 anticipate a discussion of stability that appears in Section 4.7.

B6. The stability of a multistep method is determined by applying the method to the test equation $y' = ay$, $y(0) = y_0$. Show that for the second-order Adams–Bashforth method, this leads to the constant-coefficient second-order *difference equation* $y_{n+1} - (1 + \frac{3}{2}ah)y_n + \frac{ah}{2}y_{n-1} = 0$ or, equivalently, $y_{n+2} - (1 + \frac{3}{2}ah)y_{n-1} + \frac{ah}{2}y_n = 0$.

B7. A constant-coefficient second-order difference equation of the form $y_{n+2} + p y_{n+1} + q y_n = 0$ has a solution of the form $y_k = c_1 r_1^k + c_2 r_2^k$, where r_1 and r_2 are solutions of the *characteristic equation* $r^2 + pr + q = 0$. Show that this characteristic equation arises from the substitution $y_k = r^k$.

B8. For the difference equation derived in Exercise B6, set $ah = A$; then find the characteristic equation and its two roots. Plot the two roots as a function of A and include on the graph the horizontal lines ± 1.

B9. Analyze the graph in Exercise B8 as follows. If $a < 0$, then the solution of the test equation in Exercise B6 is $y(t) = y_0 e^{at}$, which remains bounded. Consequently, the roots r_1 and r_2 must satisfy $|r_j| \leqslant 1$, $j = 1, 2$, if the solution of the difference equation is to track the solution of the test equation. Show that these inequalities force the constraint $h \leqslant \frac{1}{|a|}$ when $a < 0$.

Exercise B9 says that an equation that appears stiff to the Euler method will appear stiff to the second-order Adams–Bashforth method constructed in Exercise A1. In Exercises B10–13, test this assertion on the given stiff equations from Section 4.4. Determine the largest value of h allowed by the restriction derived in Exercise B9, and use an h less than that value and an h greater than that value.

B10. Exercise B5, Section 4.4   **B11.** Exercise B6, Section 4.4
B12. Exercise B7, Section 4.4   **B13.** Exercise B8, Section 4.4

4.6 Adams–Moulton Predictor-Corrector Methods

Predictor-Corrector Methods

Multistep methods in which two formulas are used in tandem, first to "predict" y_{n+1} and then to "correct" the prediction, are called *predictor-corrector* methods. These methods are also generated by interpolation-and-extrapolation calculations similar to those used to generate the Adams–Bashforth methods. However, multistep predictor-corrector methods so generated are called Adams–Moulton or Adams–Bashforth–Moulton methods.

Since these are multistep methods, they are not self-starting. A self-starting method of comparable order must be used to generate the additional starting values needed. Although it would seem to be a bother to use a self-starting method such as a Runge–Kutta method to start a calculation and then switch to some other algorithm, there are two characteristics of the Adams–Moulton methods that recommend them to practitioners.

First, from an adroit use of the predicted and corrected values one can deduce a measure of the per-step error. Instead of computing with stepsize h and $\frac{h}{2}$ to estimate errors, the predicted and corrected values yield an error estimate.

Second, a desired error tolerance can be maintained with fewer function evaluations than with a Runge–Kutta algorithm of comparable order. Hence, in spite of the added complexity, Adams–Moulton predictor-corrector methods are useful.

Later in the section we derive the fourth-order Adams–Moulton predictor-corrector method, whose formulae are

$$y_{n+1,p} = y_n + \frac{h}{24}(55 f_n - 59 f_{n-1} + 37 f_{n-2} - 9 f_{n-3}) \tag{4.11a}$$

$$y_{n+1,c} = y_n + \frac{h}{24}(9 f_{n+1} + 19 f_n - 5 f_{n-1} + f_{n-2}) \tag{4.11b}$$

with local error terms obeying

$$y_{n+1} = y_{n+1,p} + C_p h^5 y^{(5)}(c_1) \qquad y_{n+1} = y_{n+1,c} + C_c h^5 y^{(5)}(c_2)$$

where $y_{n+1,p}$ is the predicted value, $y_{n+1,c}$ is the corrected value, $C_p = \frac{251}{720}$, and $C_c = -\frac{19}{720}$. The terms involving the fifth derivatives are the per-step (local) truncation errors, so the method is globally of fourth order.

MONITORING THE ERROR   The reasonable assumption that $y^{(5)}(c_1) = y^{(5)}(c_2) = \alpha$ leads to

$$y_{n+1,p} - y_{n+1,c} = (C_c - C_p)h^5\alpha = \left(-\frac{251}{720} - \frac{19}{720}\right)\left(\frac{C_c h^5 \alpha}{C_c}\right) = \frac{-\frac{270}{720}}{-\frac{19}{720}}[\text{error in } y_{n+1,c}]$$

and finally to

$$\text{error in } y_{n+1,c} = \tfrac{19}{270}[y_{n+1,p} - y_{n+1,c}] = \tfrac{1}{14.2}[y_{n+1,p} - y_{n+1,c}] \qquad (4.12)$$

From the predicted and corrected values of y_{n+1} we can estimate the error in the corrected value. Herein lies the attraction of the Adams–Moulton predictor-corrector schemes.

EXAMPLE 4.5   The initial value problem $y' = f(t, y) = -ty^2$, $y(1) = 1$, has exact solution $y(t) = \frac{2}{t^2+1}$. Designate $(1, 1)$, the initial point, by (t_0, y_0), set $h = 0.1$, and index the next three starting points by 1, 2, and 3. We then get for y_k and $f_k = f(t_k, y_k)$, $k = 0, \ldots, 3$, the values shown in Table 4.8.

Then, from the Adams–Moulton predictor formula we get

$$y_{4,p} = y_3 + \tfrac{0.1}{24}(55f_3 - 59f_2 + 37f_1 - 9f_0) = 0.6756242332$$

The predicted value is now used to determine $f_4 = f(t_4, y_4) = -0.6390553463$. The corrector formula can now be used to determine

$$y_{4,c} = y_3 + \tfrac{0.1}{24}(9f_4 + 19f_3 - 5f_2 + f_1) = 0.6756820491$$

The exact value of $y(1.4)$ is 0.6756756756, so the error in $y_{4,c}$ is -0.63735×10^{-5}, whereas $y_{4,p} - y_{4,c} = -0.0000578159$. The ratio (error in $y_{4,c}$)$/(y_{4,p} - y_{4,c}) = \frac{1}{9.071}$ is slightly larger than the expected $\frac{1}{14.2}$. This means the error in the corrected value is a bit worse than expected but still within the bounds of the error-monitoring rule. In fact, if we calculate $\frac{1}{14.2}(y_{4,p} - y_{4,c})$ we find $-0.4071542253 \times 10^{-5}$, comparing favorably with the computed value of the error in $y_{4,c}$. ❖

k	y_k	f_k
0	1	-1
1	0.9049773756	-0.9008824553
2	0.8196721311	-0.8062348830
3	0.7434944238	-0.7186191457

TABLE 4.8   Fourth-order Adams–Moulton solution to IVP $y' = -ty^2$, $y(1) = 1$

Derivation of Fourth-Order Adams–Moulton Method

The fourth-order Adams–Bashforth formula for y_{n+1}, derived in Section 4.5, is the predictor, $y_{n+1,p}$. To obtain the corrector formula, interpolate the four points $(t_n + kh, f_{n+k})$, $k = -2, \ldots, 1$, with a cubic polynomial $p_2(t)$ and integrate it between t_n and $t_n + h$ to obtain

$$y_{n+1,c} = y_n + \int_{t_n}^{t_n+h} p_2(t)\,dt = y_n + \frac{h}{24}(9f_{n+1} + 19f_n - 5f_{n-1} + f_{n-2})$$

Since the region of integration is within the span of the interpolated points, it is not an extrapolation, in contrast to the integration leading to the predictor (4.11a). The interpolating polynomial has the form $p_2(t) = \frac{1}{6h^3}\sum_{k=0}^{3} c_k t^k$, where each coefficient c_k is given by a sum $\sum_{k=-2}^{1} a_{n+k} f_{n+k}$, with the a's listed in Table 4.9. For example, the table gives $c_3 = f_{n+1} - 3f_n + 3f_{n-1} - f_{n-2}$.

	a_{n+1}	a_n	a_{n-1}	a_{n-2}
c_3	1	-3	3	-1
c_2	$3(h-t_n)$	$-6h+9t_n$	$3(h-3t_n)$	$3t_n$
c_1	$2h^2-6ht_n+3t_n^2$	$3(h^2+4ht_n-3t_n^2)$	$-6h^2-6ht_n+9t_n^2$	$h^2-3t_n^2$
c_0	$-2h^2t_n+3ht_n^2-t_n^3$	$6h^3-3h^2t_n-6ht_n^2+3t_n^3$	$6h^2t_n+3ht_n^2-3t_n^3$	$-h^2t_n+t_n^3$

TABLE 4.9   Coefficients for interpolating polynomial $p_2(t)$

Obtaining the Error Expressions

If the polynomial $p(t)$ that interpolates the points $(t_k, f(t_k)), k = 0, 1, 2, \ldots, N$, is integrated between $t = a$ and $t = b$ to obtain

$$\int_a^b p(t)\,dt = \sum_{k=0}^N w_k f(t_k)$$

then the error of the approximation is of the form $Cf^{(N+1)}(c)$, for some c in the smallest interval containing a, b, and all the sampled nodes t_k, and

$$C = \frac{1}{(N+1)!}\left[\frac{b^{N+2}-a^{N+2}}{N+2} - \sum_{k=0}^N w_k t_k^{N+1}\right]$$

provided $\prod_{k=0}^N (t - t_k)$ does not change sign in the open interval (a, b). (See [60, p. 374].)

In the derivation of both the aforementioned predictor and the corrector formulas, an interpolating polynomial was integrated between $a = t_n$ and $b = t_{n+1}$ for $N = 3$. Clearly, the proviso on the sign of $\prod_{k=0}^3 (t - t_k)$ is satisfied between t_n and t_{n+1}, so the error estimate involving C is valid. We therefore show that this estimate yields the errors stated in the first section. Note, however, that now $f = y'$, so $f^{(4)} = y^{(5)}$.

Table 4.10 contains the weights $w_k, k = 0, \ldots, 3$, for both the predictor and corrector formulas. Taking $a = t_n$ implies $t_{n-1} = a-h$, $t_{n-2} = a-2h$, $t_{n-3} = a-3h$ and calculation then gives $C_p = \frac{251}{720}h^5$ and $C_c = -\frac{19}{720}h^5$.

	w_0	w_1	w_2	w_3
Predictor weights	$-\frac{3}{8}h$	$\frac{37}{24}h$	$-\frac{59}{24}h$	$\frac{55}{24}h$
Corrector weights	$\frac{1}{24}h$	$-\frac{5}{25}h$	$\frac{19}{24}h$	$\frac{3}{8}h$

TABLE 4.10   Predictor and corrector weights

EXERCISES 4.6–Part A

A1. Use (4.11a) and (4.11b) to obtain y_4 for the IVP $y' = t^2 y^2$, $y(1) = \frac{1}{3}$. Generate starting values with the exact solution, and take $h = \frac{1}{10}$.

A2. For the calculation in Exercise A1, determine the error in y_4 and use it to test the validity of the error estimate (4.12).

A3. The second-order Adams–Bashforth–Moulton predictor-corrector algorithm is given by

$$y_{k+1,p} = y_k + \frac{h}{2}(3f_k - f_{k-1})$$

$$y_{k+1,c} = y_k + \frac{h}{2}(f(t_{k+1}, y_{k+1,p}) + f_k)$$

with per-step error estimates $\frac{5}{12}y'''(c_p)h^3$ and $-\frac{1}{12}y'''(c_c)h^3$ for the predictor and corrector steps, respectively. Use this algorithm to obtain y_2 for the IVP of Exercise A1. Generate starting values with the exact solution, and take $h = \frac{1}{10}$.

A4. For the calculation in Exercise A3, determine the error in y_2 and use it to test the validity of $(C_c/(C_c - C_p))(y_{k+1,p} - y_{k+1,c})$ as a measure of the per-step error.

A5. If (4.11b) is used without (4.11a), the resulting method is implicit. For the IVP $y' = -ty^2$, $y(1) = 1$, demonstrate how to calculate y_4 via this technique. Use $h = \frac{1}{10}$, and generate starting values from the exact solution.

EXERCISES 4.6–Part B

B1. Use Table 4.9 to write the interpolating polynomial $p_2(t)$ when $t_n = 0$ and $h = 1$. Show that it does indeed interpolate the points $(t_n + kh, f_{n+k}) = (k, f_{n+k})$, $k = -2, \ldots, 1$.

B2. By forming the appropriate interpolating polynomials and integrating, obtain the second-order Adams–Bashforth–Moulton predictor-corrector algorithm given in Exercise A3.

B3. In the language of your choice, implement the algorithm given in Exercise A3. Test the code on the initial value problem $y' = -ty^2$, $y(1) = 1$. Use the exact solution to compute starting values.

B4. Derive the per-step error estimates $\frac{5}{12}y'''(c_p)h^3$ and $-\frac{1}{12}y'''(c_c)h^3$ for the predictor and corrector steps, respectively, in the algorithm of Exercise A3.

B5. Using the IVP $y' = ty^2$, $y(1) = \frac{1}{4}$, empirically test, at $k = 1$, $(C_c/(C_c - C_p))(y_{k+1,p} - y_{k+1,c})$ as a measure of the per-step error in the algorithm of Exercise A3.

B6. Using the initial value problem in Exercise B5, show empirically that the method of Exercise A3 is globally a second-order method.

B7. By forming the appropriate interpolating polynomials and integrating, obtain the third-order Adams–Bashforth–Moulton predictor-corrector algorithm

$$y_{k+1,p} = y_k + \frac{h}{12}(23f_k - 16f_{k-1} + 5f_{k-2})$$

$$y_{k+1,c} = y_k + \frac{h}{12}(5f(t_{k+1}, y_{k+1,p}) + 8f_k - f_{k-1})$$

B8. Repeat Exercise B5 for the algorithm in Exercise B7 and the index $k = 2$.

B9. Derive the per-step error estimates $\frac{3}{8}y^{(4)}(c_p)h^4$ and $-\frac{1}{24}y^{(4)}(c_c)h^4$ for the predictor and corrector steps, respectively, in the algorithm of Exercise B7.

B10. Use the method of Exercise B7 to solve the IVP $y' = t^2 + y^3$, $y(0) = 1$.

B11. Using the IVP in Exercise B5, show empirically that the method of Exercise B7 is globally a third-order method.

B12. Implement as an implicit method, the third-order corrector formula from Exercise B7 and test it on the initial value problem from Exercise B5. Use the exact solution to obtain any required starting values.

B13. Show that the implicit method in Exercise B12 is *conditionally stable* by applying it to the test equation $y' = ay$, $y(0) = y_0$ as was explained in Exercises B6–9, Section 4.5. The difference equation $(1 - \frac{5ah}{12})y_{k+1} - (1 + \frac{2ah}{3})y_k + \frac{ah}{12}y_{k-1} = 0$ should result, as should the characteristic equation $(1 - \frac{5ah}{12})r^2 - (1 + \frac{2ah}{3})r + \frac{ah}{12} = 0$ when $y_k = r^k$ is tried for a solution. Solve this characteristic equation for the roots r_1 and r_2 and, along with horizontal lines of heights ± 1, plot these roots as a function of ah. From this graph, determine that *both* roots satisfy $|r| < 1$ if $-6 < ah < 0$.

B14. Devise and execute an empirical test to show that for the method in Exercise B12, $h < \frac{6}{|a|}$ is required for numeric stability. (Figure 4.7 shows the result of a computation that is numerically unstable.)

4.7 Milne's Method

Milne's Algorithm

Not every multistep method built by interpolation and integration is satisfactory. Milne's method, like the Adams–Bashforth and Adams–Moulton methods, approximates $f(t, y)$ with interpolating polynomials and, like the Adams–Moulton, is a predictor-corrector method. However, it will prove to be *conditionally stable,* in the sense that it can give unbounded solutions to equations with bounded solutions [42].

The Milne algorithm is defined by the formulas

$$y_{n+1,p} = y_{n-3} + \frac{4h}{3}(2f_n - f_{n-1} + 2f_{n-2}) \tag{4.13a}$$

$$y_{n+1,c} = y_{n-1} + \frac{h}{3}(f_{n+1} + 4f_n + f_{n-1}) \tag{4.13b}$$

with local error terms obeying

$$y_{n+1} = y_{n+1,p} + C_p h^5 y^{(5)}(c_1) \qquad y_{n+1} = y_{n+1,c} + C_c h^5 y^{(5)}(c_2)$$

where $y_{n+1,p}$ is the predicted value, $y_{n+1,c}$ is the corrected value, $C_p = \frac{28}{90}$, and $C_c = -\frac{1}{90}$. The terms involving the fifth derivatives are the per-step (local) truncation errors, so the method is globally of fourth order.

The reasonable assumption that $y^{(5)}(c_1) = y^{(5)}(c_2) = \alpha$ leads to

$$y_{n+1,p} - y_{n+1,c} = (C_c - C_p)h^5\alpha = \left(-\frac{1}{90} - \frac{28}{90}\right)\left(\frac{C_c h^5 \alpha}{C_c}\right) = \frac{-\frac{29}{90}}{-\frac{1}{90}}[\text{error in } y_{n+1,c}]$$

and finally to

$$\text{error in } y_{n+1,c} = \frac{1}{29}[y_{n+1,p} - y_{n+1,c}] \tag{4.14}$$

From the predicted and corrected values of y_{n+1} we can estimate the error in the corrected value as one twenty-ninth of the difference between the predicted and corrected values. Herein lies the seductive attraction of the Milne scheme. However, we will show this method, unlike the Adams–Moulton method, to be unstable for some stepsizes.

EXAMPLE 4.6   We illustrate Milne's algorithm for solving the familiar initial value problem $y' = f(t, y) = -ty^2$, $y(1) = 1$, whose exact solution, $y(t) = \frac{2}{t^2+1}$, is used to generate y_k and $f_k = f(t_k, y_k)$, $k = 1, \dots, 3$, since Milne's method, a multistep method, is not self-starting. Thus, with a stepsize of $h = 0.1$, we formulate Table 4.11, from which follows

$$y_{4,p} = y_0 + \tfrac{4}{3}(0.1)(2f_3 - f_2 + 2f_1) = 0.6756308910$$

Using $y_{4,p}$ for y_4, we estimate f_4 as $f(t_4, y_{4,p})$, obtaining $f_4 = -0.6390679413$. Applying the corrector formula, we obtain the corrected value of y_4 as

$$y_{4,c} = y_2 + \tfrac{0.1}{3}(f_4 + 4f_3 + f_2) = 0.6756794842$$

k	y_k	f_k
0	1	
1	0.9049773755442163	−0.9008824552
2	0.8196721310063224	−0.8062348828
3	0.7434944235687798	−0.7186191453

TABLE 4.11   Starting values for Milne's method applied to IVP $y' = -ty^2$, $y(1) = 1$

As with the Adams–Moulton method, we test the claim that the error in the corrected value is a known multiple of the difference in the values of the predicted and corrected y's.

The error in the corrected value is $y(1.4) - y_{4,c} = \frac{25}{37} - 0.6756794842 = -0.38088 \times 10^{-5}$, whereas the difference between the predicted and corrected values is $0.6756308910 - 0.6756794842 = -4.85932 \times 10^{-5}$. Hence, we expect the ratio $\frac{-4.85932}{-0.38088} = 12.76$ to be 29, but get half that. The error is therefore about twice what was expected but still within the spirit of the error-monitoring philosophy. ❖

Stability of Milne's Method

As explored in the exercises for Section 4.5, the stability or instability for a numerical method of solving an ODE is usually established by examining its behavior when applied to the test-equation $y' = f(t, y) = ay$. Milne's corrector formula (4.13b) applied to the test-equation yields

$$\left(1 - \frac{ah}{3}\right) y_{n+1} - \frac{4ah}{3} y_n - \left(1 + \frac{ah}{3}\right) y_{n-1} = 0$$

after bringing all terms to the left and grouping. Replacing $\frac{ah}{3}$ with the constant r, we then have

$$(1 - r)y_{n+1} - 4ry_n - (1 + r)y_{n-1} = 0 \qquad \text{(4.15)}$$

a linear, constant-coefficient, second-order difference equation whose solution is a sequence of numbers, $y_k, k = 0, 1, \ldots$.

We look for a solution of (4.15) in the form $y_n = \lambda^n$. Substitution into (4.15) and division by λ^{n-1} gives the quadratic $(1 - r)\lambda^2 - 4r\lambda - (1 + r) = 0$, whose solutions are

$$R_1 = \frac{2r - \sqrt{3r^2 + 1}}{1 - r} \quad \text{and} \quad R_2 = \frac{2r + \sqrt{3r^2 + 1}}{1 - r}$$

The most general solution for y_n in Milne's method is then a sequence whose members are of the form

$$y_n = c_1 R_1^n + c_2 R_2^n$$

for constants c_1 and c_2. Therefore, the behavior of this sequence is determined by the roots R_1 and R_2, which are graphed in Figure 4.9: R_1 as the solid line and R_2 as the dashed line.

Where $|R_1| < 1$, we see $|R_2| > 1$, and vice versa. Powers of one root will grow while powers of the other root will decay. If the solution of the DE being solved by Milne's corrector really grows, then the corrector will track it. If the solution of the DE decays, the corrector will always generate a solution with a growing component. Hence, Milne's corrector is called *weakly* stable. Milne's method, consisting of the predictor and corrector, turns out to be *conditionally* stable, that is, stable only for restricted stepsizes.

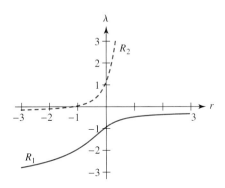

FIGURE 4.9 Characteristic roots R_1 (solid) and R_2 (dashed) for Milne's method applied to $y' = ay$

Derivation

The quadratic polynomial $p_1(t)$ that interpolates the three points $(t_n + kh, f_{n+k}), k = -2, -1, 0$, is integrated on the interval $t_n - 3h \leq t \leq t_n + h$, thereby extrapolating back one step and forward one step. In fact, integrating $y' = f(t, y)$ to obtain $y_{n+1} - y_{n-3} = \int_{t_n-3h}^{t_n+h} f(t, y(t)) \, dt \approx \int_{t_n-3h}^{t_n+h} p_1(t) \, dt$ leads to (4.13a).

The quadratic polynomial $p_2(t)$ that interpolates the three points $(t_n + kh, f_{n+k}), k = -1, 0, 1$, is integrated, without extrapolation, on the interval $t_n - h \leq t \leq t_n + h$. In fact, integrating $y' = f(t, y)$ to obtain $y_{n+1} - y_{n-1} = \int_{t_n-h}^{t_n+h} f(t, y(t)) \, dt \approx \int_{t_n-h}^{t_n+h} p_2(t) \, dt$ leads to (4.13b).

The coefficients of the interpolating polynomials $p_i(t) \frac{1}{2h^2}(A_i t^2 + B_i t + C_i), i = 1, 2$, are listed in Table 4.12.

$A_1 = f_n - 2f_{n-1} + f_{n-2}$

$B_1 = (3h - 2t_n)f_n - 4(h - t_n)f_{n-1} + (h - 2t_n)f_{n-2}$

$C_1 = (2h^2 - 3ht_n + t_n^2)f_n + (4ht_n - 2t_n^2)f_{n-1} - (ht_n - t_n^2)f_{n-2}$

$A_2 = f_{n+1} - 2f_n + f_{n-1}$

$B_2 = (h - 2t_n)f_{n+1} + 4t_nf_n - (h + 2t_n)f_{n-1}$

$C_2 = -t_n(h - t_n)f_{n+1} + 2(h^2 - t_n^2)f_n + (ht_n + t_n^2)f_{n-1}$

TABLE 4.12   Coefficients for quadratic interpolating polynomials $p_1(t)$ and $p_2(t)$

EXERCISES 4.7–Part A

A1. Use Milne's method with $h = \frac{1}{10}$ to obtain y_4 for the IVP $y' = t^2 y^2$, $y(1) = \frac{1}{3}$. Use the exact solution to obtain any starting values needed.

A2. Obtain y_4 exactly, and use this value to test the validity of (4.14).

A3. Show that $x_k = (-2)^k$ is the solution of the difference equation $x_{k+1} + 2x_k = 0$ and the initial condition $x_0 = 1$. Write out the first few terms of the sequence x_k.

A4. Solve $2x_{k+1} - 3x_k = 0$, $x_0 = \frac{1}{2}$, by requiring $x_k = Ar^k$ to satisfy the equation for r, constant. Write out the first few terms of the sequence x_k that solves this problem.

A5. Solve the problem of Exercise A4 by writing the difference equation as $x_{k+1} = \frac{3}{2}x_k$, then computing x_1 from x_0, etc.

A6. Solve $x_{k+2} + x_{k+1} - 6x_k = 0$, $x_0 = 2$, $x_1 = 3$, by requiring $x_k = Ar^k$ to satisfy the equation for r, constant. This leads to a quadratic in r, with two roots r_1 and r_2. The general solution is then $x_k = ar_1^k + br_2^k$. Determine a and b so that x_0 and x_1 assume the appropriate initial values. Write out the first few terms of the sequence x_k that solves this problem.

A7. Solve the problem of Exercise A6 by solving the difference equation for x_{k+2}, then computing x_2 from x_0 and x_1, etc.

EXERCISES 4.7–Part B

B1. In the language of your choice, code Milne's method and test it on the initial value problem $y' = -10y$, $y(1) = 1$. Show that even with a stepsize of $h = 0.1$, the method diverges for t large enough.

B2. Using $h = 0.1$, apply Milne's method to the IVP in Exercise B3, Section 4.1. For the case $n = 3$, test the validity of the error-tracking formula (4.14).

B3. Investigate the stability of Milne's method when applied to any of the initial value problems in Exercises B2–18, Section 4.1. Thus, use $h = 0.1$ and see if for $t \le 10$ there is any evidence that the calculation becomes unstable.

In each of Exercises B4–8:

(a) Use a computer algebra system to solve the given initial value problem for the difference equation.

(b) List and plot the first ten members of the solution sequence, expressing each as a sequence of real numbers.

(c) Use the difference equation itself to calculate, in order, $x_2, \ldots, x_{10}$.

(d) Construct the solution by assuming $x_k = Ar^k$, obtaining the characteristic equation for r and applying the initial conditions to the general solution $x_k = ar_1^k + br_2^k$, where r_i, $i = 1, 2$, are the roots of the characteristic equation.

B4. $x_{n+2} - 5x_{n+1} + 6x_n = 0$, $x_0 = 1$, $x_1 = 1$

B5. $6x_{n+2} + 5x_{n+1} + x_n = 0$, $x_0 = 1$, $x_1 = 1$

B6. $x_{n+2} + x_{n+1} - 6x_n = 0$, $x_0 = 1$, $x_1 = 1$

B7. $x_{n+2} - 4x_{n+1} + 13x_n = 0$, $x_0 = 1$, $x_1 = 1$

B8. $x_{n+2} - x_{n+1} + \frac{13}{36}x_n = 0$, $x_0 = 1$, $x_1 = 1$

4.8 rkf45, the Runge–Kutta–Fehlberg Method

Adaptive Methods

Throughout the previous discussions of numeric methods for solving initial value problems, we have stressed error estimation and demonstrated strategies for determining how accurate a computed answer is, even when the exact answer is not known. One strategy for this computes y_{n+1} from y_n using stepsizes h and $\frac{h}{2}$. Doing this with the fourth-order Runge–Kutta algorithm requires four function evaluations with stepsize h and seven with stepsize $\frac{h}{2}$. That two points coincide when recomputing y_{n+1} suggests there may be members of the Runge–Kutta family of methods whereby the 11 function evaluations required for recomputing with reduced stepsize can be reduced by an adroit placement of the evaluation points.

The Runge–Kutta–Fehlbergmethod combines fourth-order and fifth-order Runge–Kutta techniques to effect the step reduction strategy with only two additional function evaluations. Thus, rkf45 is a (global) fifth-order method requiring six evaluations of the function $f(t, y)$ in the differential equation $y' = f(t, y)$. It provides a fifth-order approximation of the solution and an estimate of the error at each step. It is the default numeric method in Maple.

Sophisticated procedures for solving differential equations both monitor error at each step and adjust the stepsize if the per-step error is greater than some prescribed limit. Numeric methods that adjust stepsize are called *adaptive* methods and are used for solving differential equations and evaluating integrals. Not only is the error monitored at each step, but stepsize automatically changes if the error is not small and increases if it appears the error tolerance is being easily maintained. Incidentally, Maple's default numeric integration scheme is also adaptive. (See Chapter 43.)

The rkf45 Formulae

The rkf45 formalism consists of the following nine formulas. Six function evaluations are involved in the computation of $F_1, F_2, \ldots, F_6$. From these values, $y_{n+1,4}$, with global truncation error of order four, is computed as an approximation to $y(t_{n+1})$. It is not directly used for computing $y_{n+1,5}$ which is taken as $y(t_{n+1})$; $y_{n+1,5}$ has a global truncation error of order five. The error expression is merely the difference $y_{n+1,5} - y_{n+1,4}$ and is used to determine the accuracy of $y_{n+1,5}$.

$$F_1 = hf(t_n, y_n)$$

$$F_2 = hf\left(t_n + \frac{h}{4}, y_n + \frac{F_1}{4}\right)$$

$$F_3 = hf\left(t_n + \frac{3h}{8}, y_n + \frac{3F_1}{32} + \frac{9F_2}{32}\right)$$

$$F_4 = hf\left(t_n + \frac{12h}{13}, y_n + \frac{1932F_1}{2197} - \frac{7200F_2}{2197} + \frac{7296F_3}{2197}\right)$$

$$F_5 = hf\left(t_n + h, y_n + \frac{439F_1}{216} - 8F_2 + \frac{3680F_3}{513} - \frac{845F_4}{4104}\right)$$

$$F_6 = hf\left(t_n + \frac{h}{2}, y_n - \frac{8F_1}{27} + 2F_2 - \frac{3544F_3}{2565} + \frac{1859F_4}{4104} - \frac{11F_5}{40}\right)$$

$$y_{n+1,4} = y_n + \frac{25F_1}{216} + \frac{1408F_3}{2565} + \frac{2197F_4}{4104} - \frac{F_5}{5}$$

$$y_{n+1,5} = y_n + \frac{16F_1}{135} + \frac{6656F_3}{12825} + \frac{28561F_4}{56430} - \frac{9F_5}{50} + \frac{2F_6}{55}$$

$$Error = \frac{F_1}{360} - \frac{128F_3}{4275} - \frac{2197F_4}{75240} + \frac{F_5}{50} + \frac{2F_6}{55}$$

EXAMPLE 4.7 Taking $h = 0.1$, we apply the rkf45 algorithm to the familiar initial value problem $y' = f(t, y) = -ty^2$, $y(1) = 1$, whose exact solution is $y(t) = \frac{2}{t^2+1}$, and then summarize the results in Table 4.13. The actual error appears to be about an order of magnitude smaller than the estimated error. Hence, the method's error estimator is conservative, claiming greater error than there really is.   ❖

t_k	$y(t_k)$	$y_{k,5}$	$\|y_{k,5} - y(t_k)\|$	*Error*
1.1	0.9049773756	0.9049773731	2.5×10^{-9}	9.2×10^{-8}
1.2	0.8196721311	0.8196721268	4.3×10^{-9}	8.7×10^{-8}
1.3	0.7434944238	0.7434944184	5.4×10^{-9}	7.8×10^{-8}
1.4	0.6756756757	0.6756756698	5.9×10^{-9}	6.6×10^{-8}
1.5	0.6153846154	0.6153846094	6.0×10^{-9}	5.5×10^{-8}
1.6	0.5617977528	0.5617977468	6.0×10^{-9}	4.5×10^{-8}

TABLE 4.13   rkf45 applied to IVP $y' = -ty^2$, $y(1) = 1$

EXERCISES 4.8

1. In the programming language of your choice, implement a nonadaptive version of the rkf45 algorithm.

2. For any of the initial value problems in Exercises B2–18, Section 4.1, use your code from Exercise 1 to test the validity of the error-tracking expression $y_{n+1,5} - y_{n+1,4}$ by comparing its value with the error in the numeric solution. Use the exact solution to establish the actual error made in the numeric solution.

Chapter Review

1. What does it mean to say a numeric method for solving a differential equation is a "fixed-step" method?

2. What is the truncation error for a fixed-step method?

3. What is the order of the truncation error of a fixed-step method?

4. What is the global order of Euler's method? What is the global order of the rk4 method? What does this say about the two methods?

5. State the formula by which error in a fixed-step method can be monitored.

6. State the formulae for Euler's method.

7. Describe in words what Euler's method actually does.

8. Give an example of an IVP, and implement the first three steps of Euler's method for this example.

9. Give an example where, from the differential equation and the initial condition, a Taylor series solution of an IVP is generated termwise by repeated differentiation.

10. Give an example of a numeric Taylor method, and implement it for an IVP of your choice. Compute the first two function values after the initial value.

11. What is an implicit method, and how does it differ from an "explicit" method like Euler's?

12. Using the backward Euler method, obtain the first two functions values in an IVP of your choice.

13. What is a multistep method?

14. What characterizes the class of multistep methods known as Adams–Bashforth methods?

15. What is a predictor-corrector method? Why are such methods used?

16. What does it mean to say that a method for numerically solving an IVP is *stable?* Are all the methods of this chapter stable?

17. Describe the concept of adaptivity for numeric solvers of IVPs. How is adaptivity achieved? How is it implemented?

Chapter 5

Second-Order Differential Equations

I N T R O D U C T I O N   The second-order linear differential equation with constant coefficients is the mathematical expression of a significant spectrum of physical processes. For example, it can be applied to both mechanical and electrical systems, readily modeling the vibrations of a damped spring-mass system or the oscillations of the electric current in a circuit containing resistance, inductance, and capacitance.

The chapter begins by modeling mechanical and electrical systems with the second-order, constant-coefficient, linear differential equation and continues with the physical interpretation of *initial conditions* typically imposed on spring-mass mechanical systems. A summary of physical outcomes and their mathematical counterparts is then provided.

The solution process for the homogeneous case is studied and culminates in an example of oscillatory motion experienced by a floating cylinder. The nonhomogeneous equation modeling forced or driven motion is studied next, opening the door to a study of *resonance* experienced by certain forced systems.

The Euler equation is a useful example of a linear equation with nonconstant coefficients and an exact solution that can be written in terms of the elementary functions.

The use of the *Green's function* for solving the initial value problem appears in an optional final section.

5.1   Springs 'n' Things

Mechanical Models

The vibrations of an object attached to a spring, and the flow of current in an RLC electric circuit containing resistance, inductance, and capacitance, are modeled by second-order linear differential equations with the same form. Only the physical meaning of the equations' constant coefficients distinguishes the two models. The *mathematical* properties of the solutions of both ODEs are the same. Hence, understanding the behavior of one model gives insight into the behavior of the other.

Since the mechanical system is easier to visualize (electrons are not visible!) it is reasonable to study spring-mass systems first. The mechanical system we study consists of a spring, hung from the ceiling, and an attached object of mass m. The object is assumed to move only in the vertical direction. When the particle comes to rest, its position is denoted as the equilibrium position.

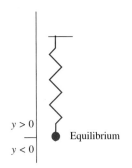

FIGURE 5.1   Coordinate systems for spring-mass system

A coordinate system in which $y(t)$ denotes vertical displacement from equilibrium is then used, the convention being that displacements upward are positive and displacements downward are negative. Upward displacements ($y(t) > 0$) compress the spring, while downward displacements ($y(t) < 0$) stretch the spring.

When resting at the equilibrium position, the object experiences no net force. The downward pull of gravity is balanced by the residual stretch of the spring as it sagged under the object's weight while the object sank to equilibrium just after being attached to the spring. See Figure 5.1.

THE FORCES   The differential equation for the motion of a particle is obtained by applying Newton's Second Law ($\mathbf{F} = m\mathbf{a}$) to the system. In Newton's law, the symbol $\mathbf{F}$ represents external forces acting on the particle. These forces include:

- Hooke's law restorative force applied to the particle by the spring.
- Viscous (frictional) forces.
- External driving forces like magnets, air blasts, vibrations of the ceiling support, etc.

Hooke's Law   An elastic force is said to be governed by Hooke's law when the stretch generated by the force is directly proportional to the force. Mathematically, this is the simplest relation between the applied force and the displacement. It is a linear relationship and can be thought of as an assumption or as a linearization of a more complicated relationship. However, experimental evidence shows that many elastic bodies support Hooke's law, at least in restricted ranges of displacement.

The force exerted on an object by the spring is such an elastic force. The spring in our discussion exerts on the particle a force that is directly proportional to the amount of stretch. If 1 pound of force stretches the spring 3 inches, then 2 pounds of force will stretch the spring 6 inches. Thus, including a sign to record the direction of displacement and consequently, the force, we have

$$\text{Force of spring } on \text{ object} = -ky(t)$$

The constant k is called the *spring constant* and has units of force per unit length. (In English units, it would be pounds per inch or an equivalent.) The minus sign says that the pull exerted by the spring opposes the displacement. If the object is moved upward 2 inches (so $y(t) = 2$) the spring pushes downward on the object and the force of the spring on the object is $-2k$ pounds. The minus sign, in our coordinate system, shows the force to be downward.

Viscous Forces   Viscosity refers to the "stickiness" of fluids. Molasses has a higher viscosity, is "stickier," and exerts more retarding force than water. Viscosity of a fluid is quantized by measuring the amount of time it takes for a standard ball-bearing to fall a fixed distance through the fluid.

An object moving in air has its motion restricted by collisions with the gas molecules composing the air. The resistive force so developed is called a *damping force*. In the spring-mass model we are developing, air-damping is generally included with any internal friction within the spring itself and a single term in the model is used to account for the frictional forces in the system.

One way to model the force of viscous damping is to assume it proportional to the velocity and directed so as to oppose the motion. (Perhaps the reader has experienced the force of air pressing against an open palm extended out the window of a moving automobile. The faster the auto goes, the more force is felt pushing back against the hand. What more natural way to model this force than to assume it is proportional to the speed of the auto?) Thus,

$$\text{Force of viscous damping} = -by'(t)$$

The minus sign indicates that the resistive force opposes the motion. If the object is moving upward ($y'(t) > 0$) the air resistance generated by the air molecules pushes downward against the object. Hence, the viscous force is negative and directed downward. The damping coefficient b has units *force/velocity* or, in the English system of units, something like *pounds/inches-per-second*.

This linear model of viscous damping is not the only possibility. In the design of aircraft and ship hulls, the drag forces of air and water are often taken as quadratic in the velocity. Thus, for certain flow regimes the fluid's drag is taken as proportional to $-|y'(t)|y'(t)$. The added complexity of determining the direction of the motion if $(y'(t))^2$ is used makes the absolute value preferable but still much harder to deal with than our simpler linear model. (Spring-mass systems with nonlinear damping cannot be solved analytically. Hence, they are treated in Unit Three.)

Driving Forces   Forces applied to the mass from sources other than the spring are called driving forces. Such forces would include the action of an electromagnet sitting under the mass (assumed to be magnetic), air blasts shot at the mass, or the periodic oscillation of the support of the spring. Such forces will be designated as $f(t)$ and must be included in the sum of all external forces acting on the object.

NEWTON'S SECOND LAW   Newton's Second Law of motion states that when a force is applied to a particle, the particle experiences an acceleration directly proportional to the force. The constant of proportionality is the mass m. The more massive the particle, the less effective the force is in changing the particle's acceleration. This is the content of the statement $F = ma$, the one-dimensional form of the second law. The acceleration is denoted by a and is given by $a = y''(t)$. The applied forces are all included in F. Hence, $F = ma$ becomes $F = my''(t)$ or $my''(t) = -ky(t) - by'(t) + f(t)$. Thus, we will study the differential equation

$$my''(t) + by'(t) + ky(t) = f(t) \qquad m, b, k \geq 0$$

This fundamental differential equation of science and engineering is a linear second-order nonhomogeneous differential equation. If we assume that the mass, damping coefficient, and spring constant all remain constant during the motion, we then have an equation with constant coefficients. The properties of such an equation, and its solutions, are the typical starting point for many more exotic studies of systems wherein frictional heating changes the damping and/or the stiffness of the spring. Equations with variable coefficients share characteristics with their constant-coefficient cousins, but the greater sophistication requires equally more sophisticated solution techniques.

If $f(t) \neq 0$ we say the motion is *forced* or *driven;* otherwise, we say the motion is free. If $b \neq 0$ we say the motion is *damped;* otherwise we say the motion is *undamped.* The relative amount of damping, that is, the relative size of b, determines three types of damped motion, namely, *overdamped, critically damped,* and *underdamped.* These distinctions, and their consequences, are the subject of the rest of this unit.

Electrical Models

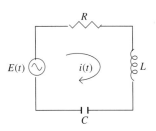

FIGURE 5.2   The basic RLC electric circuit

The electric circuit in Figure 5.2—in which a voltage source generating an "electromotive force" $E(t)$, a resistor with value R, an inductor with value L, and a capacitor with value C, are connected in a loop—supports a current whose value is $i(t)$. The connections between the value of the parameters R, L, and C and the current $i(t)$ resulting from the voltage source $E(t)$ are quantized in the differential equation

$$Li'' + Ri' + \frac{i}{C} = E'$$

This is again a linear, second-order, constant-coefficient ordinary differential equation. The analogy between this electric circuit and the damped mechanical oscillator is inescapable. The inductance L plays the role of the mass m; the resistance R plays the role of the damping coefficient b; and the reciprocal of the capacitance plays the role of the spring constant k.

EXERCISES 5.1–Part A

A1. Formulate a differential equation governing the motion of an object weighing 1 lb that stretches a vertical spring $3\frac{3}{4}$ in when attached and experiences a resistive force whose magnitude is one-sixteenth the object's speed.

A2. Formulate a differential equation for the object in Exercise A1 if the resistive force is three times the square of the speed.

A3. Verify by substitution that the differential equation $y'' + 2y' + 10y = 0$ has general solution $y(t) = e^{-t}(A \cos 3t + B \sin 3t)$, where A and B are arbitrary constants.

A4. Show that the solution of the initial value problem consisting of the ODE in Exercise A1, along with the initial conditions $y(0) = -3$, $y'(0) = 0$ is given by $y(t) = -e^{-t}(3 \cos 3t + \sin 3t)$. Thus, check that the solution satisfies both the differential equation and the initial conditions.

A5. Plot the solution given in Exercise A4. Estimate the first and second times at which $y(t) = 0$.

A6. Is the motion of the spring-mass system modeled by the IVP in Exercise A4 underdamped, overdamped, or critically damped? Provide a qualitative description of the motion of the object.

A7. Determine α and ϕ so that $3 \cos 3t + \sin 3t = \alpha \cos(3t - \phi)$ is an identity in t. *Hint:* Expand the trigonometric term on the right, and match coefficients of $\cos 3t$ and of $\sin 3t$.

A8. What does Exercise A7 say about the system modeled in Exercise A4?

A9. Determine the values of a and b so that $3 \cos(2t - \arctan \frac{1}{4}) = a \cos 2t + b \sin 2t$ is an identity in t. *Hint:* Simply expand the trigonometric term on the left side.

A10. Demonstrate how to use the initial conditions in Exercise A4 to determine the constants A and B in the general solution given in Exercise A3.

EXERCISES 5.1–Part B

B1. Use a computer algebra system to obtain the general solution of $y'' + 6y' + 13y = 0$.

B2. Use the solution found in Exercise B1 to obtain the solution of the IVP $y'' + 6y' + 13y = 0$, $y(0) = 1$, $y'(0) = -1$. Verify the calculation by using the computer algebra system to solve the IVP directly.

B3. Plot the solution found in Exercise B2, and calculate the first value of $t > 0$ for which $y(t) = 0$.

B4. Use a computer algebra system to obtain the general solution of the equation formulated in Exercise A1.

B5. Apply the initial conditions $y(0) = 1$, $y'(0) = -1$ to the solution obtained in Exercise B4. Plot the resulting solution and determine the first time beyond which the magnitude of the solution remains below 0.2.

The Initial Value Problem

The Differential Equation

A model for the mechanical system consisting of a damped spring, attached to a mass m, meeting linear viscous damping, and acted on by an external force $f(t)$ is the differential equation

$$my''(t) + by'(t) + ky(t) = f(t) \tag{5.1}$$

a linear, second-order, constant-coefficient, nonhomogeneous ordinary differential equation. If the mass is initially displaced by an amount α and tossed with a velocity β, then initial conditions of the form

$$y(0) = \alpha \qquad y'(0) = \beta$$

are included with the differential equation to form the initial value problem that is at the heart of the Newtonian view of the universe, the universe as the great clockwork. If the initial state of the system is known and the laws governing its progression in time are correct, then the solution of the initial value problem reveals the fate of the universe. The indeterminacy of quantum mechanics unleashed on physics at the turn of the 20th century effectively squelched this world view of Newtonian physics. However, the initial value problem has not lost its importance in the science and engineering of the macroscopic world.

Each modifier used to describe the differential equation reveals specific information about the system, the equation, and the nature of the solution. This unit develops an intuition for the implications of each term. Although the terms *order, linear,* and *homogeneous* have been met earlier in the context of the equation $f(t, y, y') = 0$, we repeat them here in the context of the equation $f(t, y, y', t'') = 0$.

SECOND ORDER When the highest derivative appearing in the differential equation is y'', the equation is said to be of *second order.* Equations (2), (3), (5), and (6) in Table 5.1 are all second order. Equations (1) and (4) are third order.

LINEAR VS. NONLINEAR Linear equations have y and its derivatives appearing in a "linear pattern." In Table 5.1, equations (1) and (2) are linear. The other four equations are nonlinear. Equation (1) has constant coefficients, whereas equation (2) has variable coefficients. Equation (3) is nonlinear because of the term $(y')^2$, equation (4) because of the product yy'', equation (5) because of the term $(y'')^2$, and equation (6) because of the term $\cos y''$. The physics captured in a linear equation is distinctly different from that captured in a nonlinear equation.

HOMOGENEOUS VS. NONHOMOGENEOUS A linear ODE is said to be *homogeneous* if, when y and all its derivatives are isolated on the left side of the equation, the right side is just zero. Equation (1) in Table 5.1 is homogeneous. Although equations (4) and (6) seem to have zero as a right-hand side, they are not called homogeneous because they are not linear equations.

A linear ODE that is not homogeneous is *nonhomogeneous* (or *inhomogeneous*). Equation (2) in Table 5.1 is nonhomogeneous. Although equations (3) and (5) seem to have nonzero right sides, they are not called nonhomogeneous because they are not linear equations.

---

(1) $3y''' - 2y'' + 5y' + 7y = 0$

(2) $t^2 y'' + y' \sin t - 4e^t y = t$

(3) $y'' + 2(y')^2 + y = t \cos 2t$

(4) $y''' + \frac{1}{2} yy'' = 0$

(5) $(y'')^2 + y' + y = e^{-t}$

(6) $\cos y'' + y' + y = 0$

TABLE 5.1 Examples of higher order ODEs

The Spring Constant

A spring specifically designed as a coil of steel can come with a range of stiffnesses. Given a spring, a simple test can be used to determine its spring constant k. Thus, if a weight of 2 lb stretches a spring 3 in,

$$k = \frac{2 \text{ lb}}{3 \text{ in}} = \frac{2 \text{ lb}}{3 \text{ in}} = \frac{2}{\frac{3}{12}} = 8 \text{ lb/ft}$$

If this test were actually done with an unstretched spring and the spring were allowed to come to rest with the weight attached, the equilibrium position attained would be designated as $y = 0$ and subsequent motions measured from this point. The gravitational force acting on the object is thus balanced by the initial stretch of the spring, and the two forces are in equilibrium and do not show up again in the spring-mass equation since they would only cancel out again.

Not all "springs" are wound coils of steel wire. The girders of a steel-framed building, the wings on an airplane, or the framework of the Statue of Liberty can all be modeled as spring-mass systems. The "spring constant" for such systems is a measure of the stiffness of the structure and can be a scalar or even a matrix. Chapter 12 treats such linear systems of differential equations.

Driving Forces

If an external driving force $f(t)$ such as a magnetic force, wind, earthquake, and vibrations of the ceiling support, act on the system, then Newton's Second Law ($\mathbf{F} = m\mathbf{a}$) becomes $my''(t) = -by'(t) - ky(t) + f(t)$ and, hence, (5.1). Such systems are said to be *forced* or *driven*. The force $f(t)$ is said to be the *forcing term* or the *driving term*. The motion governed by a nonhomogeneous differential equation is said to be *forced* (or *driven*) motion.

The ODE governing a spring-mass system with no forcing function is necessarily homogeneous, and the resulting motion is described as *free motion*.

Initial Conditions

For the coordinate system in which motion upward is positive, Table 5.2 matches the physical description of the initial conditions with the corresponding mathematical formalism. A positive initial velocity means the object was launched (or "tossed") upward or was in motion upward at the instant the clock for the observations was started. The word "released" typically means that the initial velocity was zero, that is, the object was simply displaced and allowed to go into motion under the action of just the spring and external force. No other agent added to its motion at $t = 0$.

	$y(0) > 0$	$y(0) < 0$	$y(0) = 0$
$y'(0) > 0$	Pull up, toss up	Pull down, toss up	From equilibrium toss up
$y'(0) < 0$	Pull up, toss down	Pull down, toss down	From equilibrium toss down
$y'(0) = 0$	Pull up, just release	Pull down, just release	From equilibrium just release

TABLE 5.2   Initial conditions for second-order ODEs

Engineering at a Glance

A great deal of mechanical, civil, and even electrical engineering, in addition to a great deal of physics, lurks in the initial value problem posed in this section. An understanding of the underlying mathematics supports an understanding of the science in this wide variety of fields. Hence, Table 5.3 outlines the mathematics of this chapter and the corresponding concepts in science and engineering to which this material relates.

Free motion (homogeneous ODE)		Forced motion (monhomogeneous ODE)	
Undamped	$(b = 0)$	Transient state	(TV set warming up)
Damped	$(b \neq 0)$	Steady state	(TV set warmed up)
Overdamped	$(b^2 > 4mk)$	Not resonating	$(b^2 \geq 2mk)$
Critically damped	$(b^2 = 4mk)$	Resonating	$(b^2 < 2mk)$
Underdamped	$(b^2 < 4mk$	Unreal resonance	$(b = 0)$
		Real resonance	$(b \neq 0)$

TABLE 5.3   Overview of spring-mass systems

During the course of the chapter, the concepts of *overdamping, critical damping,* and *underdamping* will be developed. The notion of *resonance,* and the restrictions on the parameters $b, m,$ and k that result in these different motions, will also be developed.

Free Overdamped Motion

We conclude this lesson with an example of *free overdamped motion.* Thus, we consider a spring-mass system with no driving term and enough damping to warrant the name *overdamped.*

EXAMPLE 5.1   A weight of 16 lb stretches a spring 8 in. A viscous force with damping coefficient 7 lb/ft/s acts so as to oppose the motion. We will solve for the motion of the object if it is initially pulled down 3 in from equilibrium and released.

The spring constant is determined first as

$$k = \frac{16\ \text{lb}}{8\ \text{in}} = \frac{16\ \text{lb}}{\frac{8}{12}\ \text{ft}} = 24\ \text{lb/ft}$$

The spring constant was computed in lb/ft because we next use $g = 32$ ft/s^2 in determining the mass. In English units, weight is a force, and mass is given by $m = \frac{16}{g} = \frac{16}{32} = \frac{1}{2}$ slug. The use of $g = 32$ ft/s^2 determines the units for the problem. Distances now have to be in feet and time must be in seconds. The resulting IVP is

$$\tfrac{1}{2}y'' + 7y' + 24y = 0 \quad \text{or} \quad y'' + 14y' + 48y = 0$$

and the associated initial conditions are $y(0) = \frac{3}{12} = \frac{1}{4}$ and $y'(0) = 0$. The solution of this initial value problem, $y(t) = e^{-6t} - \frac{3}{4}e^{-8t}$, is graphed in Figure 5.3, revealing that the initial displacement gradually diminishes with no oscillations ever taking place. This motion is therefore described as "overdamped" since there is sufficient damping for no vibrations to

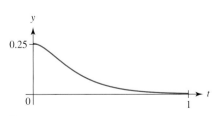

FIGURE 5.3   Overdamped motion

occur. This behavior is consistent with the negative exponentials comprising the solution. The reason for a solution with this structure is revealed later in the chapter where solution techniques are examined. ❖

EXERCISES 5.2–Part A

A1. Which of the following equations are linear and which are nonlinear?

(a) $y''' + e^t y'' - 5y' + t^2 y = \sin t$

(b) $y'' + 5yy'' - 5y' + ty' + 8y = 0$

(c) $\dfrac{d^4 y}{dt^4} + y''' \sin t - 2y'' + y^2 = 3$

A2. Which of the equations in Exercise A1 are homogeneous and which are nonhomogeneous?

A3. What is the order of each of the equations in Exercise A1?

A4. A weight of 3 lb stretches a spring 11 in. What is the spring constant in (a) pounds per inch; (b) in pounds per foot?

A5. Mechanical work is defined by the integral $\int_a^b f\, dx$, where f is a force. Calculate the work done (in foot-pounds) when the spring in Exercise A4 is stretched from its natural length to 9 in beyond its natural (unstretched) length.

A6. A mass of $\frac{1}{5}$ slugs stretches a 5-in spring to 13 in. Calculate the spring constant in pounds per inch.

EXERCISES 5.2–Part B

B1. A spring hangs from the ceiling at point A. A 4-lb weight stretches the spring to point B in such a way that the length AB is $\frac{2}{3}$ ft greater than the natural length of the spring. If the weight is further drawn to a position 5 in below B and released, find its velocity (if we neglect air resistance) as it passes the position B. (Formulate the initial value problem, solve using a computer algebra system, plot the solution, and answer the question posed.)

B2. For the system in Exercise B1, what is the acceleration of the weight when it reaches its maximum height above B?

B3. On the same set of axes, plot the position, velocity, and acceleration as functions of time for two cycles of the oscillation. Label each curve and the axes.

5.3 Overview of the Solution Process

Introduction

The initial value problem $y'' + 14y' + 48y = 0$, $y(0) = \frac{1}{4}$, $y'(0) = 0$, posed in Section 5.2, has solution $y(t) = e^{-6t} - \frac{3}{4}e^{-8t}$, a sum of exponential terms. The form of this solution suggests a technique of "guessing at an exponential" as the solution. We will refer to this educated guess as the *exponential guess* technique, formalized by assuming, for the second-order, linear, homogeneous differential equation with constant coefficients, a solution of the form $y(t) = e^{\lambda t}$. That such a guess is justified leads to the only possible solution and solves all such homogeneous linear equations with constant coefficients is the subject of this section and the next.

First-Order Motivation for Exponential Guess

The exponential guess is not a wild guess. There is good reason for suspecting this might be the form of the solution, as our experience with the first-order linear differential equation shows. Consider, for example, the equation $2y' + 3y = 0$ whose solution is $y(t) = Ce^{-3t/2}$, an obvious exponential, obtainable via the technique of separation of variables.

The Characteristic Equation

To carry out the exponential guess, substitute $y(t) = e^{\lambda t}$ into the differential equation $y'' + 14y' + 48y = 0$, obtaining

$$\lambda^2 e^{\lambda t} + 14\lambda e^{\lambda t} + 48 e^{\lambda t} = e^{\lambda t}(\lambda^2 + 14\lambda + 48) = 0$$

Since $e^{\lambda t} \neq 0$, we obtain the *characteristic equation*

$$\lambda^2 + 14\lambda + 48 = (\lambda + 6)(\lambda + 8) = 0$$

Thus, the exponential guess leads to the characteristic equation, a polynomial whose coefficients turn out to match those of the original differential equation.

The Characteristic Roots

The roots of the characteristic equation are called the *characteristic roots*. Here, these roots are $\lambda = -8, -6$. For each characteristic root we have a possible solution to the differential equation. The second-order differential equation gives rise to a quadratic characteristic equation having two characteristic roots. Hence, there are two possible solutions to the differential equation; and these two possible solutions, as a pair, are said to form a *fundamental set* of solutions.

The Fundamental Set

The set of solutions arising from the exponential guess is called a *fundamental set*. For this example, the set containing the two solutions $y_1 = e^{-8t}$ and $y_2 = e^{-6t}$, namely, $\{e^{-8t}, e^{-6t}\}$, is the fundamental set. Typically, the fundamental set for an nth-order linear differential equation contains n distinct solutions. (We will have more to say about this notion of *distinctness* later.)

General Solution

Having obtained a fundamental set of solutions, form

$$y_g = Ae^{-8t} + Be^{-6t}$$

the most general *linear combination* (linear sum) of its members. This sum is called the *general solution* because it contains all possible specific solutions within the two-parameter family it represents. For each specification of the constants A and B, the general solution reduces to a specific solution.

In general, the nth-order linear ODE with constant coefficients will have an nth-order polynomial equation as its characteristic equation and will therefore have n characteristic roots. A fundamental set for such an equation will therefore have n members, and the general solution will consist of a linear combination of these n solutions. Thus, the general solution of the nth-order linear ODE with constant coefficients has n arbitrary constants.

Application of Initial Data

To find that specific solution of the ODE that satisfies the initial conditions, apply the initial data. This then generates two equations in the two unknowns A and B, namely,

$$A + B = \tfrac{1}{4} \quad \text{and} \quad -8A - 6B = 0$$

whose solution is $A = -\tfrac{3}{4}$, $B = 1$. Hence, the solution of the IVP is $y(t) = e^{-6t} - \tfrac{3}{4}e^{-8t}$.

Interpretation of the Solution

The graph of this solution appeared in Figure 5.3 of the previous section where the motion described by the IVP was called *overdamped*. There are no oscillations for this damped spring-mass system. The mass moves in a resistive enough medium that the initial displacement is damped out without any oscillations taking place.

As we reflect on the nature of this solution and its relation to the physical system that it models, we are led to ask: "Why can we guess at a solution and not look for other possible solutions?" The answer lies in the following existence and uniqueness theorem.

Existence and Uniqueness Theorem

We will rely on the following existence and uniqueness theorem to justify the exponential guess. If there is but one solution to the initial value problem, then it makes little difference how we were led to it. No matter how we obtained it, it is the only one, and all "other" solutions must be equivalent to the one we found.

THEOREM 5.1

The solution to the Initial Value Problem (IVP)

$$y''(t) + p(t)y'(t) + q(t)y(t) = f(t) \qquad y(a) = \alpha \qquad y'(a) = \beta$$

exists and is unique in a closed interval containing $t = a$ provided $p(t)$, $q(t)$, and $f(t)$ are continuous in that interval.

For a proof of this theorem, see [80, p. 725] or [14, p. 409]. See also Exercise B8 for a proof of uniqueness.

EXAMPLE 5.2   The differential equation

$$t^2 y'' - 3ty' + 3y = 12 \tag{5.2}$$

along with the initial conditions

$$y(0) = 4 \quad \text{and} \quad y'(0) = 5$$

forms an IVP for which the solution is $y(t) = 4 + 5t + Ct^3$, where C is any constant. The solution is not unique since both the differential equation and the initial conditions are satisfied for any value of the constant C. Thus, there are multiple solutions possible for this IVP. To see why the solution is not unique, scrutinize the hypotheses of Theorem 5.1.

The standard form of the differential equation hypothesized in the existence and uniqueness theorem has leading coefficient 1. If (5.2) is divided by t^2, we have

$$y'' - \frac{3}{t}y' + \frac{3}{t^2}y = \frac{12}{t^2}$$

from which we see that none of $p(t) = -\frac{3}{t}$, $q(t) = 3t^{-2}$, $f(t) = 12t^{-2}$ are continuous in any neighborhood of $t = 0$. The differential equation does not satisfy the hypotheses of the theorem, and the solution is not unique. Moreover, if the initial condition were changed to $y(0) = a \neq 4$, there would be *no* solution, since at $t = 0$, the general solution of the differential equation, namely, $y(t) = 4 + ct + Ct^3$, can satisfy only the initial condition $y(0) = 4$. Consequently, without the guarantee provided by the existence and uniqueness theorem, an initial value problem can have many solutions or no solution at all. ❖

Additional Comments on Linear Combinations of "Distinct" Solutions

Given the set of functions $\{e^t, \sin(t), t^3\}$, a linear combination is the linear sum

$$ae^t + b\sin(t) + ct^3$$

The word *linear* is a mathematical term denoting the two properties hiding in $2x - 3y = 0$, the equation of a straight line through the origin. In the equation of the line, the variables are multiplied by constants and joined by addition or subtraction. That structure typifies the linear *form*. Writing that line as $y = \frac{3}{2}x$, we are led to examine the function $f(t) = \frac{3}{2}t$, which then has the algebraic property

$$f(ax + by) = \frac{3}{2}(ax + by) = a\frac{3}{2}x + b\frac{3}{2}y = af(x) + bf(y)$$

A function describing a line through the origin then has the algebraic property of *linearity,* namely,

$$f(ax + by) = af(x) + bf(y)$$

The function "goes past constants" and "acts separately on summands."

As a consequence, the derivative operator and the integral operator of elementary calculus are called *linear operators,* since they both treat constants and summands just the way the linear function does. The action of the linear function "goes past" the addition in $ax + by$ and "passes through" the scalar multiplications by a and b. Differentiation and integration treat the expression $au(x) + bv(x)$ the same way, yielding $au'(x) + bv'(x)$ and $a \int u\, dx + b \int v\, dx$, respectively.

Consequently, a sum of the functions $u(t)$ and $v(t)$ in the form $f = au(t) + bv(t)$ inherits the name *linear* combination because of the way constants and additions are employed.

The notion of the "distinctness" of the solutions found by the exponential guess is a "linear distinctness." We want the members of the fundamental set to be "different and distinct" but need to give a meaning to "differentness." The meaning given is "distinct in a linear sort of way." By this it is meant that no linear combination of the members of the fundamental set can add to zero for all t in an interval I (except if all the coefficients were taken as zero). If such a (nonzero) linear combination existed, then one member would be expressible, in a linear way, in terms of the others, and the members of the set would not be "linearly distinct."

For example, if

$$aU(t) + bV(t) + cW(t) = 0$$

has, for all t in an interval I, a solution in which the three constants a, b, c are not all zero, then one of the functions $U(t), V(t), W(t)$ can be isolated in terms of the other two. Suppose, for example, that at least $a \neq 0$, so that

$$U(t) = -\frac{b}{a}V(t) - \frac{c}{a}W(t)$$

Since $U(t)$ is thereby expressed in "a linear way" in terms of $V(t)$ and $W(t)$, we would judge these three functions as not being linearly distinct on the interval I.

Alternatively, consider the linear combination $a\cos x + b\sin x$, which can only vanish identically in x if $a = b = 0$. We then declare $\cos x$ and $\sin x$ to be *linearly* independent for all x, even though $\cos x$ and $\sin x$ are *functionally* related by $\cos(x - \frac{\pi}{2}) = \sin x$.

Additional discussion of the notion of "linear distinctness" or "linear independence" can be found in Section 5.4.

Superposition Statements

A major engineering principle is that *linear systems allow superposition*. Linear systems, in a sense, behave like linear operators, namely, they "map across sums and differences, and through scalar multiplication." There are two mathematical reflections of this idea: one for homogeneous differential equations, stated next, and one for nonhomogeneous differential equations, appearing in Section 5.10. The statement for homogeneous equations is important mathematically, while the statement for nonhomogeneous equations is important scientifically.

SUPERPOSITION FOR HOMOGENEOUS EQUATIONS   Superposition for the homogeneous differential equation can be expressed by the statement

> *a linear combination of solutions of a homogeneous linear differential equation is again itself a solution.*

For the differential equation $y'' + 14y' + 48y = 0$ solved previously, we found two different solutions, $y_1 = e^{-8t}$ and $y_2 = e^{-6t}$. Then any linear combination of the form $z(t) = ay_1 + by_2 = ae^{-8t} + be^{-6t}$ is itself again a solution of the differential equation. In fact, this most general linear combination is called the *general solution* y_g. Thus, for the homogeneous linear differential equation, any linear combination of solutions is itself a solution.

EXERCISES 5.3–Part A

A1. Use separation of variables to solve the first-order linear ODE $7y' + 5y = 0$. An exponential solution should result.

A2. Apply the exponential guess to the ODE in Exercise A1. Thus, substitute $y(t) = e^{\lambda t}$ into the DE and show that λ must be the very same number found by separation of variables.

A3. Show that the ODE $y'' + 10y' + 29y = 0$ is satisfied by both $y_1 = e^{-5t} \sin 2t$ and $y_2 = e^{-5t} \cos 2t$. Then show that $u = ay_1 + by_2 = e^{-5t}(a \sin 2t + b \cos 2t)$ is also a solution of the ODE. This demonstrates the superposition principle, whereby the sum of solutions is again a solution for linear ODEs.

A4. Repeat Exercise A3 for the ODE $t^2 y'' - 9ty' + 21y = 0$, which has a fundamental set consisting of the two solutions t^3 and t^7.

A5. Suppose $u(t)$ and $v(t)$ are solutions of the homogeneous DE $y'' + p(t)y' + q(t)y = 0$. Show that the linear combination $w = au(t) + bv(t)$ is also a solution. Thus, a general verification of superposition for homogeneous equations is given.

A6. Suppose $u(t)$ and $v(t)$ are solutions of $y'' + p(t)y' + q(t)y = f(t)$. Show that the linear combination

$w = au(t) + bv(t)$, $a + b \neq 1$, is *not* a solution of the nonhomogeneous DE. Thus, care must be taken with the term *superposition*. In the form stated in this section, it properly applies only to *homogeneous* DEs.

For each of Exercises A7–10:

(a) Apply the exponential guess to obtain the characteristic equation.

(b) Solve the characteristic equation to obtain the characteristic roots.

(c) Form a fundamental set.

(d) Form the most general linear combination of members of the fundamental set, thereby obtaining the general solution of the given differential equation.

A7. $y'' + 9y' + 20y = 0$ **A8.** $y'' + 4y' + 3y = 0$

A9. $y'' - 5y' + 6y = 0$ **A10.** $15y'' + 31y' + 14y = 0$

EXERCISES 5.3–Part B

B1. Use the exponential guess to obtain the characteristic equation and roots, a fundamental set, and the general solution for $70y''' - 79y'' - 63y' + 36y = 0$. Solving the characteristic equation, which is a cubic, should be done with appropriate technology.

B2. From first principles (the exponential guess), solve the IVP $y'' + 7y' + 12y = 0$, $y(0) = -2$, $y'(0) = 1$. Include the following steps in your solution.

(a) Obtain the characteristic equation and roots, a fundamental set, and the general solution as a linear combination of the members of the fundamental set.

(b) Apply the initial data to the general solution to obtain the solution of the given IVP.

(c) Solve the IVP with a computer algebra system.

(d) Plot the solution and calculate the positive time at which $y(t)$ has the value -1.

B3. Repeat steps (a)–(c) of Exercise B2 for the IVP $y'' + 11y' + 30y = 0$, $y(0) = 0$, $y'(0) = 2$. In part (d), calculate the maximum value attained by $y(t)$ and the time at which this maximum is attained.

B4. A spring-mass system is governed by the IVP $3y'' + by' + 5y = 0$, $y(0) = 1$, $y'(0) = 0$. Either analytically or empirically, determine that value of the damping coefficient $b > 0$ above which the motion is overdamped. *Hint:* Obtain the characteristic roots in terms of b. What is the smallest value of b for which the characteristic roots are real?

B5. A spring-mass system is governed by the IVP $3y'' + 5y' + ky = 0$, $y(0) = 1$, $y'(0) = 0$. Either analytically or empirically, determine that value of the spring constant $k > 0$ below which the motion is overdamped.

B6. A spring-mass system is governed by the IVP $my'' + 5y' + 7y = 0$, $y(0) = 1$, $y'(0) = 0$. Either analytically or empirically, determine that value of the mass $m > 0$ below which the motion is overdamped.

B7. Solve the IVP $y'' + 3y' + 2y = 0$, $y(0) = 1$, $y'(0) = 3$, and plot its solution. The solution, a sum of exponentials with negative powers, represents the motion of an overdamped system. Is it true that such motions always decay to zero monotonically?

B8. The IVP in Theorem 5.1 is called *completely homogeneous* if $f(t) = \alpha = \beta = 0$. Assume that the only solution of the completely homogeneous IVP is $y(t) = 0$. If $u(t)$ and $v(t)$ are two solutions of the nonhomogeneous IVP, show that $u - v$ is a solution of the completely homogeneous IVP and, therefore, the zero solution. Hence, conclude the solution of the nonhomogeneous IVP is unique.

5.4 **Linear Dependence and Independence**

Linear Dependence

As we saw in Section 5.3, the set of functions $\{u(t), v(t), w(t)\}$ is linearly dependent on an interval I if one is a linear combination (sum) of the others over that interval. Linearly dependent functions are not distinct in a linear sense. One function can be expressed in terms of the others using a linear expression.

The set $\{\sin t, \ln t, \ln t^2\}$ is linearly dependent for $t > 0$. For $t > 0$, the function $\ln t^2$ is just $2 \ln t$, so one of the functions can be expressed in terms of the others. As a consequence, there is a nontrivial linear combination of the three functions that adds up to zero. In fact, that linear combination has coefficients, not all zero, $a = 0, b = 1, c = -\frac{1}{2}$ and

$$a \sin t + b \ln t + c \ln t^2 = 0 \sin t + \ln t - \tfrac{1}{2} \ln t^2 = \ln t - \ln t = 0$$

The Wronskian

The *Wronskian* of the functions $\{u(t), v(t), w(t)\}$ is the determinant (see Section 12.4) of the matrix

$$\begin{bmatrix} u & v & w \\ u' & v' & w' \\ u'' & v'' & w'' \end{bmatrix}$$

and its utility lies in the theorem

$$\text{Linear Dependence on } I \Rightarrow \text{Wronskian} = 0 \text{ for all } t \text{ in } I \qquad (5.3)$$

Thus, functions that are linearly dependent on I will have a zero Wronskian on I, a conclusion easily reached by the following reasoning.

If the three functions $u(t), v(t), w(t)$ are linearly dependent on I, then there is a linear combination that is identically zero as an identity in t. Thus, that linear combination, and its first two derivatives, will all be zero on the interval I; and we have three equations in three unknowns to use for determining the coefficients

$$au(t) + bv(t) + cw(t) = 0$$
$$au'(t) + bv'(t) + cw'(t) = 0$$
$$au''(t) + bv''(t) + cw''(t) = 0$$

This is a homogeneous set of three equations in the three unknowns a, b, and c; in particular, we know that at least one of these constants is nonzero. In matrix terms, we have the homogeneous equations $A\mathbf{x} = \mathbf{0}$

$$\begin{bmatrix} u & v & w \\ u' & v' & w' \\ u'' & v'' & w'' \end{bmatrix} \begin{pmatrix} a \\ b \\ c \end{pmatrix} = \begin{pmatrix} 0 \\ 0 \\ 0 \end{pmatrix}$$

Since $\mathbf{x} \neq \mathbf{0}$, the determinant of A must be zero. (See Exercise A7, or Theorem 12.1, Section 12.6.) But the determinant of A is just the Wronskian.

Linear Independence

The negation of *linear dependence* is *linear independence*. Functions that are linearly independent are *distinct* in a *linear* sense. No linear combination adds up to zero except by taking all the constants zero. The set $\{t, t^2\}$ is linearly independent on any interval I because the only linear combination that equals zero for all t in I is the one for which all the coefficients are zero. Thus, $at + bt^2 = 0$, as an identity in t, can only be satisfied if $a = b = 0$.

The set $\{\sin t, \cos t\}$ is linearly independent on any interval I because there is no nontrivial linear combination that equals zero. As an identity in t, $a \sin t + b \cos t = 0$ can only be satisfied if $a = b = 0$. However, for $t > 0$, the set $\{\ln t, \ln t^2\}$ is linearly dependent because a nontrivial linear combination does equal zero. In fact, there are many solutions satisfying the identity $a \ln t + b \ln t^2 = 0$, namely, $a = -2b$ because $\ln t^2 = 2 \ln t$ for $t > 0$.

The *contrapositive* of the statement "$A \Rightarrow B$" is the equally valid statement "not $B \Rightarrow$ not A." For example, the statement "If a ball is to be used in a pro baseball game, then it must be white" is equivalent to "A ball that is not white cannot be used in a pro baseball game."

The contrapositive of (5.3) is the equivalent statement "Wronskian $\neq 0$ for at least one t in an interval $I \Rightarrow$ Linear Independence on the interval I."

EXAMPLE 5.3   Prior calculations convinced us that $\cos t$ and $\sin t$ are linearly independent on any interval I. Their Wronskian is the determinant of

$$\begin{bmatrix} \cos t & \sin t \\ -\sin t & \cos t \end{bmatrix}$$

which is 1. Since the Wronskian is not zero, the functions are not linearly dependent, that is, they are linearly independent. ❖

EXAMPLE 5.4   We have already convinced ourselves that $\{t, t^2\}$ is a linearly independent set on any interval I. The Wronskian is the determinant of

$$\begin{bmatrix} t & t^2 \\ 1 & 2t \end{bmatrix}$$

which is t^2. In any interval, even one that includes the origin, there is at least one t for which the Wronskian is not zero. Hence, the set is linearly independent on any interval I. ❖

CAUTION The two statements "Linear Dependence on $I \Rightarrow$ Wronskian = 0 on I" and "Wronskian $\neq 0$ for at least one t in $I \Rightarrow$ Linear Independence on I" are equivalent. The *converse* of the first statement, namely, "Wronskian = 0 on $I \Rightarrow$ Linear Dependence on I" is neither logically equivalent to either of the given (true) statements, nor mathematically true. It's a false statement.

EXAMPLE 5.5 The functions

$$f_1 = t^2 \quad \text{and} \quad f_2 = \begin{cases} t^2 & \text{if } t \geq 0 \\ -t^2 & \text{if } t < 0 \end{cases}$$

are linearly independent on any (open) interval containing zero, but their Wronskian is zero in every interval. Figure 5.4 showing f_1 as the solid curve and f_2 as the dotted, suggests that indeed the functions are linearly independent. The derivative

$$f_2' = \begin{cases} 2t & \text{if } t \geq 0 \\ -2t & \text{if } t < 0 \end{cases}$$

lets us calculate the Wronskian $f_1 f_2' - f_1' f_2$ as zero. The Wronskian is zero on any interval but the functions are linearly independent on an interval containing zero in its interior. ❖

In summary, the theorem (5.3) is equivalent to the contrapositive, "If the Wronskian is not zero for at least one t in the interval I, then the functions are linearly independent on the interval I." The converse, "If the functions are linearly independent on I, then the Wronskian is not zero on I," is a false statement.

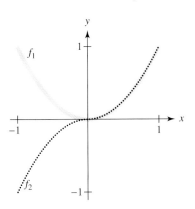

FIGURE 5.4 Linearly independent functions whose Wronskian is zero

EXERCISES 5.4–Part A

A1. Verify that $\{t^{-2}, t^{-3}\}$, $t > 0$, is a fundamental set (i.e., each member is a solution, and the set is linearly independent) for the differential equation $t^2 y'' + 6t y' + 6y = 0$.

A2. Verify that $\{2t^{-2} - 5t^{-3}, 5t^{-2} + 2t^{-3}\}$ is also a fundamental set for the differential equation in Exercise A1.

A3. Verify that $\{e^{-t}, e^{-2t}\}$ is a fundamental set for the differential equation $y'' + 3y' + 2y = 0$.

A4. Verify that $\{5e^{-t} - 7e^{-2t}, 4e^{-t} + 3e^{-2t}\}$ is also a fundamental set for the differential equation in Exercise A3.

A5. Write a second-order linear, homogeneous differential equation with constant coefficients for which $\{e^{-5t}, e^{3t}\}$ is a fundamental set.

A6. Write a third-order linear, homogeneous differential equation with constant coefficients for which $\{e^{2t}, e^{-3t}, e^{4t}\}$ is a fundamental set.

A7. Let $A = \begin{bmatrix} a_{11} & a_{12} \\ a_{21} & a_{22} \end{bmatrix}$, and $\mathbf{x} = \begin{pmatrix} x \\ 0 \end{pmatrix}$, $x \neq 0$. If $A\mathbf{x} = \mathbf{0}$, show $a_{11} = a_{21} = 0$, and hence $\det(A) = 0$.

EXERCISES 5.4–Part B

B1. Answer true or false, and give a reason: The fundamental set for a linear ODE is unique.

For the functions given in Exercises B2–4, establish linear independence on the interval (0, 1) by

(a) Showing that a general linear combination of the functions, set equal to zero, forces the coefficients to be zero.

(b) Showing the Wronskian (simplified) is not zero.

B2. $\{te^t \sin t, te^t \cos t, t^2 e^t\}$ **B3.** $\{e^{3t}, e^{5t}\}$ **B4.** $\{e^{-6t}, e^{-7t}, e^{5t}\}$

B5. Solve the IVP $y'' + 3y' + 2y = 0$, $y(0) = -3$, $y'(0) = -1$. Follow steps (a)–(d) spelled out in Exercise B2, Section 5.3. For part (d), determine the minimum value attained by y and the time at which this minimum is reached.

B6. Solve the IVP $y'' + 4y = 0$, $y(0) = 1$, $y'(0) = 1$, by seeking a solution of the form $y(t) = A \cos(\omega t - \phi)$. The constant $A > 0$ is the *amplitude* of the motion, and the constant ϕ, $-\pi < \phi \leq \pi$, is the *phase angle*. These two parameters are determined by the initial conditions. The *angular frequency* $\omega > 0$ is determined by making the trial solution fit the ODE. Plot the solution and on it indicate which feature represents the amplitude A, which feature represents the angular frequency ω, and which feature represents the phase angle ϕ.

B7. If $\{u(t), v(t)\}$ is known to be a fundamental set for a second-order linear ODE, what condition must a, b, c, d satisfy if $\{au + bv, cu + dv\}$ is also to be a fundamental set for the same equation?

B8. Let $u(x)$ and $v(x)$ be differentiable on the interval I, and let W be their Wronskian. Show that $\frac{dW}{dx} = uv'' - vu''$.

B9. Let $p(x)$ and $q(x)$ be continuous on the interval I, and let $u(x)$ and $v(x)$ be solutions of $y'' + p(x)y' + q(x)y = 0$ on I. If α is the DE satisfied by u and β is the DE satisfied by v, obtain $u\beta - v\alpha$ and show, using Exercise B8, that it is equivalent to $\frac{dW}{dx} + p(x)W = 0$, where W is the Wronskian of u and v. Show that $W = ce^{-\int p(x)\,dx}$ (called Abel's formula), where c is a constant. Hence, the Wronskian for solutions of the second-order linear ODE $y'' + p(x)y' + q(x)y = 0$ is either zero or nonzero. If it is zero at a single point in I, it is identically zero on I. If it is nonzero at a single point in I, it is nonzero on I.

B10. Use the results in Exercise B9 to make a statement about the linear dependence or independence of two solutions of the second-order linear ODE in Exercise B9.

Use of the Wronskian for showing the linear independence of functions requires that the functions be sufficiently differentiable for the Wronskian to exist. Alternatively, the continuous functions $f_k(x), k = 1, \ldots, n$,

are linearly dependent on $I = [a, b]$ if the *Grammian*, the determinant of the matrix $a_{ij} = \int_a^b f_i(x)f_j(x)\,dx$, is zero, and linearly independent if it is nonzero. In Exercises B11–13, use the Grammian to establish, on the interval $I = [0, 1]$, the linear dependence or independence of the given functions.

B11. The functions in Exercise B2

B12. The functions in Exercise B3

B13. The functions in Exercise B4

B14. Use a computer algebra system and brute force computation to show that Abel's formula in Exercise B9 generalizes to the linear third-order ODE.

B15. Let $\psi(x)$ be the determinant $\begin{vmatrix} a_1(x) & a_2(x) & a_3(x) \\ b_1(x) & b_2(x) & b_3(x) \\ c_1(x) & c_2(x) & c_3(x) \end{vmatrix}$ and show that

$$\frac{d\psi}{dx} = \begin{vmatrix} a_1' & a_2' & a_3' \\ b_1 & b_2 & b_3 \\ c_1 & c_2 & c_3 \end{vmatrix} + \begin{vmatrix} a_1 & a_2 & a_3 \\ b_1' & b_2' & b_3' \\ c_1 & c_2 & c_3 \end{vmatrix} + \begin{vmatrix} a_1 & a_2 & a_3 \\ b_1 & b_2 & b_3 \\ c_1' & c_2' & c_3' \end{vmatrix}.$$ *Hint:* Expand ψ by the first row, then differentiate.

B16. Let W be the Wronskian of the differentiable functions $u(x), v(x), w(x)$. Using the result in Exercise B15, show that

$$\frac{dW}{dx} = \begin{vmatrix} u & v & w \\ u' & v' & w' \\ u''' & v''' & w''' \end{vmatrix}.$$

B17. Suppose the functions $u(x), v(x), w(x)$ in Exercise B16 are solutions of $y''' + py'' + qy' + ry = 0$. Use the result in Exercise B16 to establish that $\frac{dW}{dx} = -pW$. Hence, it is possible to establish Abel's formula for the linear third-order ODE without the computational intensity experienced in Exercise B14.

5.5 ## Free Undamped Motion

The Harmonic Oscillator

Consider the spring-mass system whose motion is governed by the IVP $y'' + 16y = 0$, $y(0) = -1, y'(0) = 1$. This is a system in which viscous damping has been neglected. Initially, the mass is pulled down, tossed upward, and subsequently moves according to the solution $y(t) = \frac{1}{4} \sin 4t - \cos 4t$, graphed in Figure 5.5. Observe the trigonometric functions comprising the solution and the periodic motion shown in Figure 5.5. The solution is purely oscillatory, a motion described in physics as harmonic motion. Indeed, our system is an example of a harmonic oscillator, a spring-mass system without damping. But what happened to the exponential guess and uniqueness? How can exponentials yield the oscillatory behavior this solution exhibits?

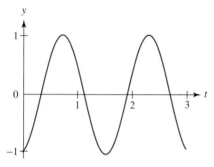

FIGURE 5.5   Solutions of the IVP
$y'' + 16y = 0, y(0) = -1, y'(0) = 1$

The Scientists' Solution

Because this harmonic oscillator does not seem to follow the pattern of the exponential guess, scientists often make the alternative guess

$$y(t) = A \cos \omega t + B \sin \omega t$$

one involving just trigonometric functions. Substitution into the differential equation gives

$$16(A \cos \omega t + B \sin \omega t) - \omega^2(A \cos \omega t + B \sin \omega t) = (16 - \omega^2)y(t) = 0$$

Ruling out the trivial solution $y(t) = 0$, we have no other choice but to take $\omega = \pm 4$. Because $\cos t$ is an even function, that is, $\cos(-t) = \cos t$, and $\sin t$ is an odd function, that is, $\sin(-t) = -\sin t$, we need only make use of one of these values of ω. (The functions $\{\cos(\pm \omega t), \sin(\pm \omega t)\}$ are linearly dependent, whereas the functions $\{\cos \omega t, \sin \omega t\}$ are linearly independent.) The constants A and B in the resulting general solution $y_g = A \cos 4t + B \sin 4t$ are determined by the initial conditions to be $A = -1$ and $B = -\frac{1}{4}$, thus re-creating the solution stated previously. However, we still do not know why no exponential solution appeared, but we do know that $\{\cos 4t, \sin 4t\}$ must be a fundamental set.

Solution by the Exponential Guess

Both curious and determined, we plunge ahead with the exponential guess, $y(t) = e^{\lambda t}$, obtaining $\lambda^2 e^{\lambda t} + 16e^{\lambda t} = 0$. The characteristic equation is $\lambda^2 + 16 = 0$, so the characteristic roots are $\lambda = \pm 4i$, and a fundamental set consists of the two imaginary exponentials $\{e^{4it}, e^{-4it}\}$.

Not only do we have to deal with multiple fundamental sets, we also have to deal with imaginary exponentials that are not the most obvious objects for students with just a background in Calculus. There must be some connection between these imaginary exponentials and the trigonometric functions that appeared in the earlier solutions, and there must be some connection between the two fundamental sets if the uniqueness theorem is to be believed.

Euler's Formulas

The link between imaginary exponentials and trigonometric functions is expressed via Euler's formulas, the two identities in the leftmost column of Table 5.4. A proof, based on Taylor expansions, is deferred to the exercises. However, we observe that one-half the sum, and $\frac{1}{2i}$ times the difference, isolates $\cos \theta$ and $\sin \theta$, respectively, as listed in the middle column of Table 5.4. This middle column therefore *defines* the trigonometric functions $\cos \theta$ and $\sin \theta$ in terms of imaginary exponentials. The rightmost column of Table 5.4 lists the definitions of the hyperbolic functions, which are given in terms of *real* exponentials.

Euler's Formulas	Trigonometric Functions	Hyperbolic Functions
$e^{i\theta} = \cos \theta + i \sin \theta$	$\cos \theta = \dfrac{e^{i\theta} + e^{-i\theta}}{2}$	$\cosh x = \dfrac{e^x + e^{-x}}{2}$
$e^{-i\theta} = \cos \theta - i \sin \theta$	$\sin \theta = \dfrac{e^{i\theta} - e^{-i\theta}}{2i}$	$\sinh x = \dfrac{e^x - e^{-x}}{2}$

TABLE 5.4   Euler's formulas, trigonometric and hyperbolic functions

New Fundamental Sets from Old

Given the fundamental set $\{e^{4it}, e^{-4it}\}$ form a new and equivalent fundamental set by taking two different linear combinations of its members. The justification for taking linear combinations of solutions is the Superposition Principle for homogeneous equations from Section

5.3, where we saw that linear combinations of solutions of the homogeneous equation are again solutions of the homogeneous equation.

The following linear combinations of the imaginary exponentials, by virtue of Euler's formulas, are "real" trigonometric functions

$$\tfrac{1}{2}e^{4it} + \tfrac{1}{2}e^{-4it} = \tfrac{1}{2}(e^{4it} + e^{-4it}) = \cos 4t$$

$$\frac{1}{2i}e^{4it} - \frac{1}{2i}e^{-4it} = \frac{1}{2i}(e^{4it} - e^{-4it}) = \sin 4t$$

But more important, they are guaranteed to be solutions by virtue of the Superposition Principle. So, instead of using the fundamental set $\{e^{4it}, e^{-4it}\}$, use the equivalent fundamental set $\{\cos 4t, \sin 4t\}$ and write the general solution as $y_g = A\cos 4t + B\sin 4t$.

Yet Another Form

It is common in engineering to cast the solution $y(t) = -\cos 4t + \tfrac{1}{4}\sin 4t$ into the form $y(t) = a\cos(4t - \phi)$, where a is the *amplitude* of the motion and ϕ is the *phase angle*. The amplitude is the maximum height reached by the oscillator, and the phase angle is the amount by which the graph of $\cos 4t$ is translated to the right. The equivalence is established by writing $A\cos 4t + B\sin 4t$ as

$$\sqrt{A^2 + B^2}\left(\cos(4t)\frac{A}{\sqrt{A^2 + B^2}} + \sin(4t)\frac{B}{\sqrt{A^2 + B^2}}\right) = a\cos(4t - \phi)$$

based on the trigonometric formula

$$\cos(x)\cos(y) + \sin(x)\sin(y) = \cos(x - y)$$

and the identifications

$$a = \sqrt{A^2 + B^2} \qquad \cos(\phi) = \frac{A}{\sqrt{A^2 + B^2}} \qquad \sin(\phi) = \frac{B}{\sqrt{A^2 + B^2}}$$

It is tempting to declare $\phi = \arctan\frac{B}{A}$, but since the range of the arctangent function is $(-\frac{\pi}{2}, \frac{\pi}{2})$, the temptation must be avoided. The phase angle ϕ lies in the interval $(-\pi, \pi]$, so we must consider the signs of both $\cos\phi$ and $\sin\phi$.

For the problem at hand, we find $a = \frac{\sqrt{17}}{4}$ and $\phi = \pi - \arctan\frac{1}{4}$.

Terminology for a $\cos(\omega t - \phi)$

Putting the harmonic oscillator's solution into the form $y(t) = a\cos(4t - \phi)$ allows an easy detection of the engineering parameters listed in Table 5.5. However, the reader is cautioned that some texts use the term *frequency* where they should use *angular frequency*.

Amplitude:	a
Period:	$T = \dfrac{2\pi}{\omega}$
Frequency:	$\dfrac{1}{T} = \dfrac{\omega}{2\pi}$
Angular (circular) frequency:	ω

TABLE 5.5   Engineering parameters for harmonic oscillator

EXERCISES 5.5–Part A

A1. Using Euler's formulas and taking linear combinations, convert the fundamental set $\{e^{3it}, e^{-3it}\}$ to the equivalent fundamental set $\{\cos 3t, \sin 3t\}$. Show clearly the linear combinations required to convert these complex exponentials to trigonometric functions. Compute the Wronskian of the new fundamental set and show it is not zero.

A2. Verify that both members of $\{e^{3it}, e^{-3it}\}$ are solutions of $y'' + 9y = 0$.

A3. Verify that both members of $\{\cos 3t, \sin 3t\}$ are solutions of $y'' + 9y = 0$.

A4. Evaluate e^{3it} and $\cos 3t$ at $t = \frac{\pi}{8}$. Are they the same number? Is it true that $e^{3it} = \cos 3t$?

A5. Evaluate e^{-it} at $t = \frac{k}{10}, k = 0, \ldots, 10$, obtaining floating-point complex numbers in the form $a + bi$.

A6. State the amplitude, period, frequency, and angular frequency of $3\cos 2t - 5\sin 2t$. *Hint:* Write it as $a\cos(2t - \phi)$.

A7. The sum and difference of the complex numbers $a + bi$ and $c + di$ are defined by $(a \pm c) + i(b \pm d)$, that is, by separately adding and subtracting the real and imaginary parts. Evaluate in floating-point form, each side of $\frac{1}{2}e^{2i} + \frac{1}{2}e^{-2i} = \cos 2$, verifying its validity.

A8. The *conjugate* of the complex number $z = a + bi$ is the complex number $\bar{z} = a - bi$. For each of the following, obtain both $z + \bar{z}$ and $z - \bar{z}$.

(a) $z = 2 + 3i$ (b) $z = 3 - 2i$

(c) $z = -5 + 7i$ (d) $z = -4 - 3i$

A9. The product of the complex numbers $a + bi$ and $c + di$ is defined as $(ac - bd) + i(ad + bc)$. This is obtained by multiplying the two numbers as a product of binomials and noting that $i^2 = -1$. (Multiplying two complex numbers requires forming all four possible products with the real and imaginary parts, just like multiplying the two real binomials $(x + \alpha)(y + \beta)$.) Obtain the product of each of the following pairs of complex numbers.

(a) $z_1 = 3 - 2i$, $z_2 = 4 + 5i$

(b) $z_1 = -\sqrt{2} + i\sqrt{3}$, $z_2 = 1 + 2i$

(c) $z_1 = e^{-2+3i}$, $z_2 = \frac{1}{2} - \frac{1}{3}i$ (d) $z_1 = 3 - 2i$, $z_2 = 3 + 2i$

(e) $z_1 = a + bi$, $z_2 = a - bi$

A10. The reciprocal of a complex number is obtained by writing

$$\frac{1}{a + bi} = \frac{1}{a + bi}\frac{a - bi}{a - bi} = \frac{a - bi}{a^2 + b^2}$$

which removes the complex number from the denominator. Obtain the reciprocal of each of the following complex numbers.

(a) $4 - 5i$ (b) i (c) $3 - 7i$

A11. Division by a complex number becomes multiplication by its reciprocal, as, for example, in

$$\frac{1 + i}{2 + i} = \frac{1 + i}{2 + i}\frac{2 - i}{2 - i} = \frac{3 + i}{5}$$

Express each of the following quotients in the form $a + bi$.

(a) $\dfrac{1 + 2i}{2 - i}$ (b) $\dfrac{3 + 2i}{i}$ (c) $\dfrac{4 - 5i}{3 - 2i}$

A12. For each of the following, evaluate both the left and right sides in floating-point form to verify the equality.

(a) $\dfrac{1}{2i}e^{2i} - \dfrac{1}{2i}e^{-2i} = \sin 2$ (b) $\dfrac{1}{2i}e^{i\sqrt{2}} - e^{-i\sqrt{2}} = \sin\sqrt{2}$

EXERCISES 5.5–Part B

B1. A typical verification of Euler's formulas uses series expansions of e^{it}, $\cos t$, $\sin t$. The following steps illustrate the essence of the verification.

(a) Use a computer algebra system to obtain a 10-term Maclaurin expansion of e^{it}.

(b) Separate the result in part (a) into its real and imaginary parts.

(c) Obtain Maclaurin expansions of $\cos t$ and $\sin t$.

(d) Show that the expansion in part (c) match the expansions representing the real and imaginary parts of e^{it}.

B2. Solve the IVP $y'' + 4y = 0$, $y(0) = -2$, $y'(0) = -1$, by performing the following steps:

(a) Use a computer algebra system to obtain the solution.

(b) Starting with an exponential guess, obtain the fundamental set in terms of imaginary exponentials.

(c) Show the specific linear combinations of the imaginary exponentials that yield an equivalent fundamental set containing trigonometric functions.

(d) Obtain the general solution in terms of the new fundamental set of trigonometric functions.

(e) Apply the initial conditions to the general solution. This process should result in the same solution found in part (a).

(f) Express your solution in the form $A\cos(\omega t - \phi)$.

(g) Plot the solution to this IVP and indicate on the graph which features correspond to the amplitude A, the angular frequency ω, and the phase angle ϕ.

5.6 Free Damped Motion

Overview of Damping

The linear, second-order, constant-coefficient equation governing the motion of the undriven (free) damped oscillator for which damping is assumed proportional to the velocity is $my'' + by' + ky = 0$. The characteristic equation arising from the exponential guess $y(t) = e^{\lambda t}$ is $m\lambda^2 + b\lambda + k = 0$ and the characteristic roots are $\lambda = \frac{-b \pm \sqrt{b^2 - 4mk}}{2m}$.

The nature of the roots, and hence of the solution, is determined by the *discriminant,* $b^2 - 4mk$. Consequently, there are three possible behaviors for a damped oscillator, behaviors corresponding to the mathematical possibilities for the characteristic roots. These outcomes are collected in Table 5.6.

Discriminant	Motion	Characteristic Roots
$b^2 > 4mk$	Overdamped	Real, distinct
$b^2 = 4mk$	Critically damped	Real, repeated
$b^2 < 4mk$	Underdamped	Complex conjugates

TABLE 5.6   Characteristic roots and the discriminant

EXAMPLES   This section develops the examples in Table 5.7, which illustrates a system exhibiting three distinct behaviors characterized by the amount of damping. The critically damped case is the boundary between underdamping and overdamping. However, some engineers claim that in the real world, the critically damped case does not occur since the parameters of a physical system cannot be established precisely enough to attain exact equality in the discriminant.

Motion	Equation	Characteristic Roots
Overdamped	$y'' + 10y' + 16y = 0$	$\lambda = -2, -8$
Critically damped	$y'' + 8y' + 16y = 0$	$\lambda = -4, -4$
Underdamped	$y'' + 4y' + 16y = 0$	$\lambda = -2 \pm 2\sqrt{3}i$

TABLE 5.7   Examples of damped systems

VARIATION WITH DAMPING   The differential equation for a spring-mass system with mass $m = 1$, spring constant $k = 16$, and unspecified damping b is $y'' + by' + 16y = 0$. Displacing the mass one unit upward and releasing it from rest so that the initial conditions are $y(0) = 1$, $y'(0) = 0$, yields the solution

$$y_b(t) = \frac{(b + \sqrt{b^2 - 64})e^{((-b+\sqrt{b^2-64})/2)t}}{2\sqrt{b^2 - 64}} - \frac{(b - \sqrt{b^2 - 64})e^{((-b-\sqrt{b^2-64})/2)t}}{2\sqrt{b^2 - 64}} \qquad b \neq 8$$

The surface shown in Figure 5.6 is a three-dimensional plot of the solution surface insofar as it depends on both t and the damping coefficient b. Plane sections $b = constant$ represent solutions corresponding to fixed values of b. Superimposed on the surface are the four solution curves $y_b(t)$, $b = 0, 4, 8, 10$. (Actually, $y_8(t) = \lim_{b \to 8} y_b(t)$.) In order, the curves represent undamped, underdamped, critically damped, and overdamped motions.

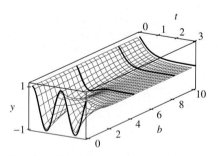

FIGURE 5.6   Solution surface $y_b(t)$ and solution curves corresponding to $b = 0, 4, 8, 10$

Overdamping

The initial value problem $y'' + 10y' + 16y = 0$, $y(0) = 1$, $y'(0) = 0$ models an overdamped spring-mass system in which the mass is displaced one unit upward and released from rest. The solution,

$$y(t) = \tfrac{1}{3}(4e^{-2t} - e^{-8t})$$

whose graph is the rightmost curve in the surface shown in Figure 5.6, does not oscillate, the damping coefficient $b = 10$ being too large to permit them. Mathematically, this is because the solution is a linear combination of the two distinct exponential terms e^{-2t} and e^{-8t}, arising from the characteristic equation $\lambda^2 + 10\lambda + 16 = 0$ whose roots are $\lambda = -2, -8$.

Critical Damping

The case of critical damping in which the roots of the characteristic equation are repeated divides the cases of overdamped and underdamped motions. Critical damping, characterized by $b^2 = 4mk$, falls at the interface of the two regions wherein b^2 is either greater or less than $4mk$. Critical damping, representing the minimum damping that will just prevent oscillation from occurring, is "unstable." Small changes in the system parameters can result in either overdamped or underdamped motions, which are very different.

To examine the implications of critical damping, consider the initial value problem $y'' + 8y' + 16y = 0$, $y(0) = 1$, $y'(0) = 0$ for which the exponential guess gives $\lambda^2 + 8\lambda + 16 = 0$ as the characteristic equation and $\lambda = -4, -4$ as the characteristic roots. If we use the same root twice, we will not have a fundamental set with two distinct solutions. We will see in Section 5.7 that an appropriate fundamental set is $\{e^{-4t}, te^{-4t}\}$, so the solution of the initial value problem is $y(t) = e^{-4t} + 4te^{-4t}$.

Figure 5.7 compares this solution (solid curve) with the overdamped solution $\frac{1}{3}(4e^{-2t} - e^{-8t})$ (dotted curve) for which the damping parameter had the value $b = 10$. The dashed upper curve is the curve for overdamping, while the solid lower curve is the curve for critical damping. The solid curve shows the mass closer to equilibrium at any instant t. Truly, overdamping leads to motion that is both devoid of oscillations and also "stickier" than motion subjected to critical damping.

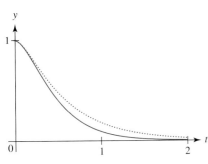

FIGURE 5.7 Critically damped (solid) and overdamped (dotted) solutions

Underdamping

The initial value problem $y'' + 4y' + 16y = 0$, $y(0) = 1$, $y'(0) = 0$ models an underdamped spring-mass system in which the mass is displaced one unit upward and released from rest. The solution,

$$y(t) = e^{-2t}\left(\cos 2\sqrt{3}t + \tfrac{1}{\sqrt{3}}\sin 2\sqrt{3}t\right)$$

whose graph is the solid curve in Figure 5.8, exhibits decaying oscillations, the damping coefficient $b = 4$ not being large enough to suppress them completely. This solution is a linear combination of the members of the fundamental set $\{e^{-2t}\cos 2\sqrt{3}t, e^{-2t}\sin 2\sqrt{3}t\}$ arising from the characteristic equation $\lambda^2 + 4\lambda + 16 = 0$ whose roots are $\lambda = -2 \pm 2\sqrt{3}i$. (The details appear in the following paragraph.) Figure 5.8 also shows as dotted curves the *exponential envelopes* $\pm\frac{2}{\sqrt{3}}e^{-2t}$ that bound the oscillations. In general, the underdamped solution is a sinusoid (sine or cosine function) modulated by an exponential envelope. Figure 5.9, showing graphs of $e^{-t/5}\cos 2t$ and the envelopes $\pm e^{-t/5}$, demonstrates this more dramatically.

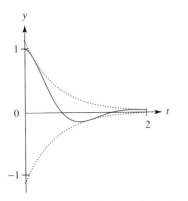

FIGURE 5.8 Underdamped solution (solid) and its exponential envelopes (dotted)

SOLUTION BY EXPONENTIAL GUESS The exponential guess for the underdamped equation in Table 5.7 is implemented by substituting $y = e^{\lambda t}$ into the differential equation, obtaining the characteristic equation $\lambda^2 + 4\lambda + 16 = 0$, the characteristic roots $\lambda = -2 \pm 2\sqrt{3}i$, and the exponential solutions

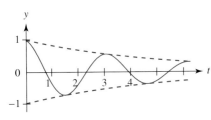

FIGURE 5.9 Exponential envelopes $\pm e^{-t/5}$ for $\cos 2t$

$$y_1 = e^{(-2+2\sqrt{3}i)t} = e^{-2t}e^{2\sqrt{3}it} \quad \text{and} \quad y_2 = e^{(-2-2\sqrt{3}i)t} = e^{-2t}e^{-2\sqrt{3}it}$$

In fact, the essence of the underdamped case is the appearance of complex conjugate characteristic roots and the ensuing imaginary exponentials. But, remembering Euler's formulas, we take the linear combinations $\frac{y_1+y_2}{2} = e^{-2t}\cos 2\sqrt{3}t$ and $\frac{y_1-y_2}{2i} = e^{-2t}\sin 2\sqrt{3}t$ to obtain real solutions whose Wronskian, $2\sqrt{3}e^{-4t}$, is the determinant of the matrix

$$\begin{bmatrix} e^{-2t}\cos 2\sqrt{3}t & e^{-2t}\sin 2\sqrt{3}t \\ -2e^{-2t}(\cos 2\sqrt{3}t + \sqrt{3}\sin 2\sqrt{3}t) & -2e^{-2t}(\sin 2\sqrt{3}t - \sqrt{3}\cos 2\sqrt{3}t) \end{bmatrix}$$

The nonzero value of the Wronskian guarantees these two solutions are linearly independent, so we can construct the general solution

$$y_g = e^{-2t}(A\cos 2\sqrt{3}t + B\sin 2\sqrt{3}t)$$

Applying the initial data to this solution leads to the two equations $A = 1$ and $-2A + 2\sqrt{3}B = 0$, from which we get $B = \frac{1}{\sqrt{3}}$ and

$$y(t) = e^{-2t}\left(\cos 2\sqrt{3}t + \frac{1}{\sqrt{3}}\sin 2\sqrt{3}t\right)$$

Conversion to the Form a cos(ωt − φ)   We wish to put the solution into the form $a\cos(\omega t - \phi)$ where the angular frequency is $\omega = 2\sqrt{3}$. From Section 5.5 we determine

$$a = \sqrt{A^2 + B^2} = \sqrt{1^2 + \left(\frac{1}{\sqrt{3}}\right)^2} = \frac{2}{\sqrt{3}} \quad \text{and} \quad \phi = \arctan\left(\frac{1}{\sqrt{3}}\right) = \frac{\pi}{6}$$

so $y(t) = \frac{2}{\sqrt{3}}e^{-2t}\cos(2\sqrt{3}t - \frac{\pi}{6})$. From this form of the solution, the engineering parameters of period $(T = \frac{2\pi}{2\sqrt{3}})$, frequency $(\frac{1}{T} = \frac{2\sqrt{3}}{2\pi})$, angular frequency $(2\sqrt{3})$, and phase angle $(\phi = \frac{\pi}{6})$ are more easily discerned.

EXERCISES 5.6–Part A

A1. Assume each given ODE governs a damped oscillator. Compute the characteristic roots and determine from them the nature of the motion (underdamped, overdamped, critically damped) of the oscillator. State the period, frequency, and angular frequency for each underdamped system.

 (a) $3y'' + 5y' + 2y = 0$ (b) $3y'' + 5y' + 4y = 0$

 (c) $3y'' + 6y' + 3y = 0$

A2. Each of the following differential equations governs an underdamped oscillator. For each, show how to proceed from exponential guess to a fundamental set with complex exponentials, to a fundamental set with real functions, to the general solution.

 (a) $y'' + 2y' + 5y = 0$ (b) $36y'' + 36y' + 13y = 0$

 (c) $2y'' + 2y' + y = 0$ (d) $9y'' + 6y' + 10y = 0$

 (e) $16y'' + 8y' + 17y = 0$

A3. If a spring-mass system has mass M and spring constant K, can the damping be adjusted so the system is exactly critically

damped? Answer *yes* or *no,* and give evidence! If the answer is "yes" give the value of b that results in critical damping.

A4. A 16-lb weight suspended from a spring vibrates freely, the resistance being numerically equal to twice the speed (in feet per seconds) at any instant. If the period of the motion is 6 s, what is the spring constant k (in pounds per foot)?

A5. A damped oscillator must be built so that its motion is governed by the solution $y = e^{-3t}(\cos 4t + \sin 4t)$. From this solution you can see that the motion obeys $y(0) = 1$, $y'(0) = 1$. Does this solution uniquely determine the system; that is, can m, b, and k be uniquely calculated just from the solution? *Hint: $my'' + by' + ky = 0 \Rightarrow m\lambda^2 + b\lambda + k = 0 \Rightarrow \lambda = \frac{-b\pm\sqrt{b^2-4mk}}{2m}$.* The fundamental set then contains $e^{((-b\pm\sqrt{b^2-4mk})/2m)t}$. Now, if you understand how the exponentials lead to the solution $y(t)$, you should be able to trace a path from the -3 and the 4 in $y(t)$ to the m, b, and k in the members of the fundamental set.

EXERCISES 5.6–Part B

B1. A mass of m slugs is suspended from a spring whose constant is 8 lb/ft. The motion of the mass is subject to a resistance (pounds) numerically equal to four times the speed (in feet per seconds). If the motion is to have a period of 4 s, what must the mass be? It turns out there are two values of m that work. Find both as floating point numbers. Their symbolic representation is messy. If you compute only one positive value for the mass, you need to re-examine your treatment of any square roots appearing in your solution.

The IVPs in Exercises B2–6 govern underdamped oscillators. For each:

    (a) Start with the exponential guess, and find the characteristic equation and its roots.

    (b) Write the fundamental solutions as complex exponentials and show how to convert these to equivalent solutions given in terms of trig functions.

    (c) Write the general solution from the new fundamental set.

    (d) Apply the initial data by forming and solving two equations in two unknowns.

    (e) Exhibit your solution and graph it.

    (f) Solve the IVP using a computer algebra system, making sure the two solutions agree.

    (g) Put the solution to this IVP into the form $Ae^{rt}\cos(\omega t - \phi)$, stating clearly the values of the period, frequency, angular frequency, and phase angle.

    (h) On the same set of axes plot the solution and the enveloping curves $Ae^{rt}, -Ae^{rt}$.

    (i) Find the time that must elapse for Ae^{rt}, the amplitude of the displacement $y(t)$, to remain less than 0.1.

B2. $y'' + 2y' + 5y = 0,\ y(0) = 2,\ y'(0) = 3$

B3. $y'' + 8y' + 25 = 0,\ y(0) = -3,\ y'(0) = 1$

B4. $y'' + 8y' + 41y = 0,\ y(0) = 5,\ y'(0) = -4$

B5. $y'' + 10y' + 41y = 0,\ y(0) = -2,\ y'(0) = 5$

B6. $9y'' + 12y + 85y = 0,\ y(0) = 3,\ y'(0) = -5$

5.7 Reduction of Order and Higher Order Equations

Reduction of Order

Reduction of order is a general technique for finding a second solution of a linear ODE when a first solution is known. We will use it to determine why, in the repeated root case of critical damping, the second solution is $te^{\lambda t}$. Note, however, that the method is used to find a second solution from a first solution in settings other than critical damping. In fact, the essence of the technique was seen in Section 3.5 where the *variation of parameters* concept was first explored for linear, first-order equations. Moreover, this same idea will shortly surface as the formal Variation of Parameters method in Section 5.9.

We use the following example to illustrate the reduction of order technique.

 EXAMPLE 5.6   Obtain the general solution to the differential equation $y'' + 8y' + 16y = 0$, the differential equation of critical damping from Section 5.6. The characteristic equation $\lambda^2 + 8\lambda + 16 = 0$ has the repeated roots $\lambda = -4, -4$, giving us the exponential solution $y_1 = e^{-4t}$. To find a second distinct solution we try

$$y_2 = u(t)y_1(t) = u(t)e^{-4t}$$

a product of the known solution with the unknown function $u(t)$ as a coefficient. This is the essence of the technique, looking for a second solution as the product of an unknown function with a known solution. Substitution of y_2 into the differential equation leads to

$$(ue^{-4t})'' + 8(ue^{-4t})' + 16ue^{-4t} = 0$$

a differential equation for the unknown function $u(t)$. Simplifying this differential equation to $u''e^{-4t} = 0$ gives $u(t) = a + bt$, since the exponential factor is not zero.

Constructing the general solution as $y_g = Ae^{-4t} + B(a + bt)e^{-4t}$ is worth a moment's thought. In the presence of Ae^{-4t}, the term Bae^{-4t} is redundant, so $u(t)$ need only have been taken as $u(t) = te^{-4t}$. The general solution is the more efficient

$$y_g = Ae^{-4t} + Bte^{-4t} = (A + Bt)e^{-4t}$$

and the fundamental set is best simplified to $\{e^{-4t}, te^{-4t}\}$.   ❖

EXAMPLE 5.7   For the general second-order linear equation

$$y'' + p(t)y' + q(t)y = 0 \qquad (5.4)$$

with one solution $y_1(t)$, a second trial solution of the form $y_2 = u(t)y_1$ leads to

$$u(y_1'' + py_1' + qy_1) + u''y_1 + u'(2y_1' + py_1) = 0 \qquad (5.5)$$

Since y_1 is a solution of (5.4), the first term on the left in (5.5) is zero, leaving

$$u''y_1 + u'(2y_1' + py_1) = 0$$

to determine $u(t)$. Setting $u' = v$, this equation becomes

$$v'y_1 + v(2y_1' + py_1) = 0 \qquad (5.6)$$

an equation of order one less than that of the original equation. Hence, the name, *reduction of order*. (Note that (5.6) is a first-order linear equation that can be solved by the methods of Sections 3.4 and 3.5.)   ❖

Higher Order Equations

Reduction of order applies to higher order equations, and the following five examples show both the technique and the solution of higher order linear equations.

EXAMPLE 5.8   Solve the initial value problem $y''' - 4y'' + y' + 6y = 0$, $y(0) = 1$, $y'(0) = -2$, $y''(0) = 8$. (Notice that for the *third*-order differential equation there are *three* initial conditions.)
    The exponential guess $y = e^{\lambda t}$ leads to the characteristic equation

$$\lambda^3 - 4\lambda^2 + \lambda + 6 = (\lambda - 2)(\lambda - 3)(\lambda + 1) = 0$$

whose roots are, clearly, $\lambda = 2, 3, -1$. The fundamental set is therefore $\{e^{2t}, e^{3t}, e^{-t}\}$ and its Wronskian, $12e^{4t}$, is the determinant of the matrix

$$\begin{bmatrix} e^{2t} & e^{3t} & e^{-t} \\ 2e^{2t} & 3e^{3t} & -e^{-t} \\ 4e^{2t} & 9e^{3t} & e^{-t} \end{bmatrix}$$

Since the Wronskian is nonzero, the three exponential solutions are linearly distinct (independent), so we can build the general solution as the linear combination $y_g = Ae^{2t} + Be^{3t} + Ce^{-t}$. Application of the initial data to this general solution leads to the three equations

$$A + B + C = 1 \qquad 2A + 3B - C = -2 \qquad 4A + 9B + C = 8$$

whose solution—$A = -3$, $B = C = 2$—yields $y(t) = -3e^{2t} + 2e^{3t} + 2e^{-t}$ as the solution of the initial value problem.   ❖

EXAMPLE 5.9   Obtain the general solution of the fourth-order differential equation

$$y^{(4)} - 7y^{(3)} + 18y'' - 20y' + 8y = 0 \qquad (5.7)$$

The exponential guess $y = e^{\lambda t}$ leads to the characteristic equation

$$\lambda^4 - 7\lambda^3 + 18\lambda^2 - 20\lambda + 8 = (\lambda - 1)(\lambda - 2)^3 = 0$$

so the characteristic roots are then $\lambda = 1, 2, 2, 2$. The repeated root $\lambda = 2$ occurs three times. Hence, the fundamental set will be $\{e^t, e^{2t}, te^{2t}, t^2e^{2t}\}$. Each repetition of the root $\lambda = 2$ must be accompanied by multiplication of a successively higher power of the independent variable t. Indeed, to see this, substitute $y_2 = u(t)e^{2t}$ into (5.7), obtaining $e^{2t}(u^{(4)} + u^{(3)}) = 0$, so $u = a + bt + ct^2 + de^{-t}$. (See Exercise A1.) The fundamental set follows from reasoning seen in Example 5.6.

The Wronskian, $2e^{7t}$, the determinant of the matrix

$$\begin{bmatrix} e^t & e^{2t} & te^{2t} & t^2e^{2t} \\ e^t & 2e^{2t} & (t+2)e^{2t} & (2t^2+2t)e^{2t} \\ e^t & 4e^{2t} & (4t+4)e^{2t} & (4t^2+8t+2)e^{2t} \\ e^t & 8e^{2t} & (8t+12)e^{2t} & (8t^2+24t+12)e^{2t} \end{bmatrix}$$

is nonzero, so the members of this fundamental set are linearly independent and the general solution can be written as $y_g = ae^t + (b + ct + dt^2)e^{2t}$.   ❖

EXAMPLE 5.10   Obtain the general solution to the differential equation $y^{(4)} + 2y^{(3)} + y'' = 0$.

By inspection we can see the characteristic equation is

$$\lambda^4 + 2\lambda^3 + \lambda^2 = \lambda^2(\lambda^2 + 2\lambda + 1) = \lambda^2(\lambda + 1)^2 = 0$$

and, consequently, the characteristic roots are $\lambda = 0, 0, -1, -1$. The fundamental set contains $e^{0t} = 1$ and $te^{0t} = t$ as well as e^{-t} and te^{-t}. The Wronskian, e^{-2t}, the determinant of the matrix

$$\begin{bmatrix} 1 & t & e^{-t} & te^{-t} \\ 0 & 1 & -e^{-t} & (1-t)e^{-t} \\ 0 & 0 & e^{-t} & (t-2)e^{-t} \\ 0 & 0 & -e^{-t} & (3-t)e^{-t} \end{bmatrix}$$

is nonzero, so the members of the fundamental set are linearly independent. Hence, the general solution is $y_g = a + bt + (c + dt)e^{-t}$.   ❖

EXAMPLE 5.11   Obtain the general solution to the differential equation $y''' - 3y'' + 9y' + 13y = 0$.

The exponential guess $y = e^{\lambda t}$ produces the characteristic equation $\lambda^3 - 3\lambda^2 + 9\lambda + 13 = 0$, whose roots are $\lambda = -1, 2 \pm 3i$. The fundamental set will undoubtedly contain the solution e^{-t} because of the characteristic root $\lambda = -1$. In addition, the imaginary exponentials $y_1 = e^{(2+3i)t} = e^{2t}e^{3it}$ and $y_2 = e^{(2-3i)t} = e^{2t}e^{-3it}$, through the linear combinations $\frac{1}{2}y_1 + \frac{1}{2}y_2 = e^{2t}\cos 3t$ and $\frac{1}{2i}y_1 - \frac{1}{2i}y_2 = e^{2t}\sin 3t$, complete the fundamental set to $\{e^{-t}, e^{2t}\cos 3t, e^{2t}\sin 3t\}$. The Wronskian, the determinant of the matrix

$$\begin{bmatrix} e^{-t} & e^{2t}\cos 3t & e^{2t}\sin 3t \\ -e^{-t} & e^{2t}(2\cos 3t - 3\sin 3t) & e^{2t}(3\cos 3t + 2\sin 3t) \\ e^{-t} & -e^{2t}(5\cos 3t + 12\sin 3t) & e^{2t}(12\cos 3t - 5\sin 3t) \end{bmatrix}$$

is the nonzero $54e^{3t}$, so the members of the fundamental set are linearly independent. The general solution is therefore $y_g = Ae^{-t} + Be^{2t}\cos 3t + ce^{2t}\sin 3t$.   ❖

EXAMPLE 5.12   Obtain the general solution of the differential equation $y^{(4)} + 8y'' + 16y = 0$.

The exponential guess $y = e^{\lambda t}$ leads to the characteristic equation $\lambda^4 + 8\lambda^2 + 16 = (\lambda^2 + 4)^2 = 0$, whose roots are $\lambda = \pm 2i, \pm 2i$. The first time the complex conjugate pair of roots appears, the fundamental set contains $\cos 2t$ and $\sin 2t$. The next time this pair of roots appears, the solutions associated have to be multiplied by t. (See Exercise A3.) Hence, the next two members of the fundamental set are $t\cos 2t$ and $t\sin 2t$. The complete fundamental set is then $\{\cos 2t, \sin 2t, t \cos 2t, t \sin 2t\}$. The Wronskian is the determinant of the matrix

$$\begin{bmatrix} \cos 2t & \sin 2t & t \cos 2t & t \sin 2t \\ -2 \sin 2t & 2 \cos 2t & -2t \sin 2t + \cos 2t & 2t \cos 2t + \sin 2t \\ -4 \cos 2t & -4 \sin 2t & -4t \cos 2t - 4 \sin 2t & -4t \sin 2t + 4 \cos 2t \\ 8 \sin 2t & -8 \cos 2t & 8t \sin 2t - 12 \cos 2t & -8t \cos 2t - 12 \sin 2t \end{bmatrix}$$

and has value $64 \neq 0$, so the four members of the fundamental set and the general solution can be constructed as the linear combination $y_g = a_1 \cos 2t + a_2 \sin 2t + t(a_3 \cos 2t + a_4 \sin 2t)$.   ❖

EXERCISES 5.7–Part A

A1. Show that the differential equation $u^{(4)} + u^{(3)} = 0$ can be integrated three times, resulting in a first-order linear equation in u, whose solution is then $u = a + bt + ct^2 + de^{-t}$ as claimed in Example 5.9.

A2. Show that u as given in Exercise A1 leads to the fundamental set given in Example 5.9.

A3. Apply the reduction of order technique to the equation of Example 5.12, setting $y_2 = e^{2it}u(t)$. Show this leads to an equation in $u(t)$ whose characteristic equation is $\lambda^2(\lambda^2 + 8i\lambda - 16) = 0$ and, hence, to $u = a + bt + ce^{-4it} + dte^{-4it}$. Show from this that the fundamental set is as given in the example.

A4. A linear constant-coefficient ODE has characteristic roots $0, 0, 1, 1, 1$. What is a fundamental set for this equation? What is the general solution?

A5. A linear constant-coefficient ODE has characteristic roots $-1 \pm i$, $-1 \pm i$. For this equation, write a fundamental set that contains only real functions. Write the general solution.

A6. A linear constant-coefficient ODE has characteristic roots $-1, -1$, $-1 \pm 2i$. Write the general solution in terms of real functions.

A7. A linear constant-coefficient ODE has characteristic roots $0, 0$, $0 \pm 3i$. Write the general solution in terms of real functions.

EXERCISES 5.7–Part B

B1. The ODE $y'' + 6y' + 9y = 0$ has characteristic roots $\lambda = -3, -3$. Hence, the exponential guess yields only the one solution $y_1 = e^{-3t}$. Use the method of reduction of order to find the second distinct solution $y_2 = te^{-3t}$. Then, compute the Wronskian and show it is not zero.

B2. Show that the ODE $x^2 y'' - 5xy' + 9y = 0$ has $y = x^3$ as one solution. (Do this by substitution back into the ODE.) Use reduction of order to find the second distinct solution, $x^3 \ln x$. Do this by trying $y_2 = x^3 u(x)$ and substituting back into the ODE. The function $u(x)$ satisfies by a differential equation that can be solved with a computer algebra system or by one of the methods of Chapter 3. Finally, compute the Wronskian of $\{y_1, y_2\}$ and show it is nonzero.

B3. Consider the IVP $y^{(4)} + 6y^{(3)} + 15y'' + 18y' + 10y = 0$, $y(0) = 3$, $y'(0) = -3$, $y''(0) = 2$, $y'''(0) = 1$.

(a) Use a computer algebra system to solve the equation.

(b) Plot the solution.

(c) Use the exponential guess to obtain the characteristic equation.

(d) Obtain the characteristic roots.

(e) Write the fundamental set (with only real functions) and show the Wronskian is nonzero.

(f) From your fundamental set, construct the general solution.

(g) Apply the initial data to the general solution; obtain and solve the appropriate four equations in four unknowns. Be sure the resulting solution agrees with the solution found in part (a).

B4. Repeat Exercise B3 for the initial data of Exercise B3 and the ODE $y^{(4)} + 6y^{(3)} + 18y'' + 30y' + 25y = 0$.

B5. Repeat Exercise B3 for the initial data of Exercise B3 and the ODE $y^{(4)} + 4y^{(3)} + 11y'' + 14y' + 10y = 0$.

5.8   The Bobbing Cylinder

Introduction

Undamped free motion for a spring-mass system is modeled by the differential equation $my'' + ky = 0$. Such a system is called a *harmonic oscillator* because the solution can be put into the form $y(t) = a\cos(\omega t - \phi)$, where $\omega = \sqrt{\frac{k}{m}}$ is the angular frequency. But nature provides more instances of harmonic oscillators than just spring-mass systems. In this section we examine the motion of a floating cylinder that bobs under the influence of buoyancy, a system governed by an equation of the form $y'' + \omega^2 y = 0$, and therefore is a harmonic oscillator.

Problem Statement

A cylinder of height H, radius R, and weight f floats with its axis vertical in a fluid whose weight per unit volume is ρ_w. If h is the height, at equilibrium, of the submerged part of the cylinder, describe the motion of the cylinder subsequent to its being pushed down into the fluid and released.

Figure 5.10 shows the water line as Plane 1, the base of the cylinder as Plane 2, the height H, the depth of submersion h, and the radius R.

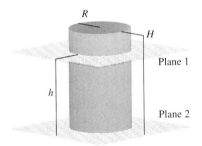

FIGURE 5.10   Cylinder floating at equilibrium

Analysis

This mechanical problem yields to Newton's Second Law, $\mathbf{F} = m\mathbf{a}$. The secret is realizing that the weight of the cylinder is a force balanced by the buoyant force exerted by the fluid. And this buoyant force is governed by Archimedes Principle, namely,

Buoyant force on cylinder = Weight of fluid displaced

$$F_b = w_f$$

The equilibrium and the nonequilibrium configurations must be considered. At equilibrium, where the amount of the cylinder beneath the surface of the fluid is measured by h, the volume of the submerged part is given by $\pi R^2 h$. The weight of the fluid displaced by this submerged volume is

$$w_f = \pi R^2 h \rho_w$$

Consequently, the buoyant force at equilibrium is given by

$$F_b = \pi R^2 h \rho_w$$

By Archimedes Principle, at equilibrium, the buoyant force $F_b = w_f$ balances f, the weight of the cylinder. Hence

$$\pi R^2 h \rho_w = f$$

This key equality links f, the weight of the cylinder, with h, the amount of the cylinder submerged at equilibrium.

Next, consider the nonequilibrium configuration. For this, paint a line around the cylinder at the equilibrium "water" line. Let $y(t)$ denote the amount by which this line on the cylinder is displaced above or below the surface of the fluid. Take $y(t)$ positive if the cylinder rises (the ring is visible) and negative if the cylinder descends (only the fish see the ring). Figure 5.11 depicts this configuration.

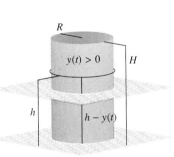

FIGURE 5.11   Coordinate system for bobbing cylinder

Now, $F_b(t)$, the time-dependent buoyant force on the cylinder is proportional to the length $h - y(t)$. Thus,

$$F_b(t) = \pi R^2 [h - y(t)] \rho_w$$

We are ready to invoke $F = ma = my''(t)$, the one-dimensional form of Newton's Second Law, where m, the mass of the cylinder, is given by $m = \frac{f}{g}$, with g being the acceleration of gravity. Thus, summing forces on the cylinder, we have

$$-f + F_b(t) = \frac{f}{g} y''(t)$$

which becomes

$$-f + \pi R^2 [h - y(t)] \rho_w = \frac{f}{g} y''(t)$$

But from Archimedes Law at equilibrium we found that $f = \pi R^2 \rho_w h$. Making this substitution and expanding the brackets, we find

$$-\pi R^2 \rho_w h + \pi R^2 \rho_w h - \pi R^2 \rho_w y(t) = \frac{f}{g} y''(t)$$

which collapses to

$$-\pi R^2 \rho_w y(t) = \frac{f}{g} y''(t)$$

Hence, we have the equation $y'' + \pi R^2 \rho_w g y / f = 0$, which we write as $y'' + \omega^2 y = 0$ by defining

$$\omega = R \sqrt{\frac{\pi \rho_w g}{f}}$$

We therefore have simple harmonic motion governed by the initial value problem for un-damped free motion

$$y'' + \omega^2 y = 0 \qquad y(0) = y_0 \qquad y'(0) = v_0$$

Solution

The solution to this initial value problem is $y(t) = y_0 \cos \omega t + \frac{v_0}{\omega} \sin \omega t$, which says the cylinder bobs with periodic motion whose angular frequency is ω and whose period is $T = \frac{2\pi}{\omega}$. The solution can also be written as $y(t) = \frac{\sqrt{v_0^2 + \omega^2 y_0^2}}{\omega} \cos(\omega t - \phi)$, where ϕ is determined by

$$\sin \phi = \frac{v_0}{\sqrt{v_0^2 + \omega^2 y_0^2}} \quad \text{and} \quad \cos \phi = \frac{\omega y_0}{\sqrt{v_0^2 + \omega^2 y_0^2}}$$

The reader is again cautioned not to define ϕ as $\arctan \frac{v_0}{\omega y_0}$ since the range of the arctangent function is $(-\frac{\pi}{2}, \frac{\pi}{2})$, not $(-\pi, \pi]$. A negative value for the initial displacement y_0 results in a second-quadrant phase angle; while a negative for v_0, the initial velocity, results in a third-quadrant angle. These angles would not be correctly obtained by a naive use of the arctangent function.

EXAMPLE 5.13   A cylinder of radius $R = 2$ in bobs, with a period of $\frac{1}{2}$ s, in water whose weight per cubic foot is $\rho_w = 62.5$ lb. Using the gravitational constant $g = 32$ ft/s^2, find f, the weight of the cylinder, and find the equilibrium height h.

This problem is realistic in that from an observable period, estimated by careful counting in the presence of a wristwatch, information about the cylinder can be deduced. The period is $T = \frac{2\pi}{\omega} = \frac{1}{2}$, so

$$\omega = 4\pi = R\sqrt{\frac{\pi\rho_w g}{f}} = \frac{2}{12}\sqrt{\frac{\pi(62.5)(32)}{f}}$$

from which we determine $f = 1.105242660$ lb as the weight of the cylinder. To find the equilibrium height h, use the equilibrium relationship $f = \pi R^2 \rho_w h$ or $1.105242660 = \pi\left(\frac{2}{12}\right)^2 62.5h$, from which we obtain $h = 0.2026423672$ ft or 2.43 in.   ❖

EXERCISES 5.8–Part A

A1. A cylinder weighing 22 lb/ft³ floats, with axis vertical, in water weighing 62.5 lb/ft³. The height of the cylinder is $H = 5$ ft. Compute the period and the amplitude of the bobbing if initially the cylinder is first raised so that the bottom is level with the surface of the water and then released.

A2. A cylinder with its axis vertical bobs in a liquid with weight per unit volume ρ_0. What is ρ_0 if the period of the motion is four-fifths the period for the same cylinder bobbing in water for which $\rho_w = 62.5$ lb/ft³?

A3. A cylinder whose cross-section is an isosceles triangle with sides 5 in, 5 in, and 6 in has height $H = 20$ in and weight 5 lb. After being pushed down so that its top is level with the surface of the water and released, the cylinder bobs, with axis vertical, in water whose weight per unit volume is 62.5 lb/ft³. Solve for the motion $y(t)$,

where $y(t)$ is the position relative to equilibrium. Plot the solution. Determine, and state clearly, the period and amplitude of the motion.

A4. A cylinder with rectangular cross-section of dimension $a \times b$ (feet) and with height H (feet) and weight w (pounds) bobs in a fluid whose weight per unit volume is $\tilde{\rho}$ lb/ft³. Develop a model for the motion.

A5. A harmonic oscillator is set into motion with the initial conditions $y(0) = \alpha$, $y'(0) = \beta$, and its solution is written in the form $y = a\cos(\omega t - \phi)$, with $a > 0$. Determine the quadrant in which the phase angle ϕ lies if

(a) $\alpha > 0$ but $\beta < 0$ **(b)** $\alpha < 0$ and $\beta < 0$

(c) $\alpha < 0$ but $\beta > 0$

EXERCISES 5.8–Part B

B1. A circular cylinder tall enough to bob has a radius of 6 in and a weight of 18 lb. It floats, with its axis vertical, in water whose weight is 62.5 lb/ft³. It is raised 3 in above equilibrium and pushed downward with an initial velocity of 4 in/s.

(a) Solve for the motion $y(t)$, where $y(t)$ is the displacement relative to the equilibrium position. (Be sure to use correct units.)

(b) Plot the solution.

(c) Determine, and clearly state, the period and the amplitude of the motion.

(d) Add linear damping to the system so that the governing differential equation contains the term $by'(t)$. Obtain the solution in terms of b, the damping coefficient which is to be small enough that the system is underdamped.

(e) Find, and plot, the angular frequency as a function of b.

(f) Find B, the value of b at which critical damping occurs.

(g) How sensitive is T, the period, to the damping coefficient? Compute the period for $b = \frac{1}{2}B$ and compare it to the period found in part (c) where $b = 0$.

(h) Plot a graph of the period as a function of the damping coefficient b.

(i) How sensitive is ω, the angular frequency, to the damping coefficient? Compute the angular frequency for $b = 0$ and $b = \frac{1}{2}B$.

B2. Develop a model for a bobbing cylinder with elliptic cross-section.

B3. Solve the IVP $y'' + 10y = 0$, $y(0) = 2$, $y'(0) = -3$. Write the solution in the form $y = a\cos(\omega t - \phi)$, and determine the amplitude, period, angular frequency, frequency, and phase angle. Graph the solution and, on the graph, indicate the amplitude, period, and phase angle.

B4. The "harmonic oscillator" $y = 2\cos(\omega t - \frac{\pi}{4})$ has the slowly decreasing angular frequency $\omega = 1 + 2e^{-t/10}$. Graph the motion and describe the behavior of the system.

B5. The "harmonic oscillator" $y = 3\cos(\omega t - \frac{\pi}{8})$ has the periodically varying angular frequency $\omega = 2 + \frac{1}{10}\cos\frac{t}{3}$. Graph the motion and describe the behavior of the system.

5.9   Forced Motion and Variation of Parameters

Forced Motion: Overview

- When a spring-mass system is "driven" or "forced" by an external forcing function $f(t)$ such as a magnet, an air-blast, or the shaking of an off-balance rotating shaft, the governing differential equation is the *nonhomogeneous equation*

$$my'' + by' + ky = f(t)$$

- The general solution of this nonhomogeneous differential equation splits into two parts (see Sections 3.4 and 3.5)

$$y_g = y_h + y_p$$

- y_h is the *homogeneous solution*. Containing two arbitrary constants, it is the general solution of the *homogeneous* equation $my'' + by' + ky = 0$.

- y_p is one *particular solution* to the full nonhomogeneous equation $my'' + by' + ky = f(t)$. It contains no arbitrary constants.

- There are two methods for finding the particular solution y_p, namely,

   1.   Variation of Parameters.

   2.   Undetermined Coefficients.

- Variation of Parameters always applies but can be computationally intractable.

- Undetermined Coefficients always works for the cases to which it applies.

- We call the method of Variation of Parameters the "Atomic Cannon 21-12" because, in principle, it always applies. No matter what the problem, it is valid. In addition, we designate it "21-12" because those numbers appear as subscripts in the Variation of Parameters formula. Unfortunately, this formula can result in difficult, if not impossible, integrations.

- We call the method of Undetermined Coefficients "Guess 'n' Test" because in it, we guess at the form of a solution and then use brute force to determine the coefficients in our guess. The algebra in this method is far simpler than the calculus in the alternative. Unfortunately, this method applies to a restricted class of driving forces $f(t)$. However, this class includes many functions that arise in engineering and scientific applications, making the method of Undetermined Coefficients worth learning.

Variation of Parameters

The Variation of Parameters formula for a particular solution is stated as a theorem.

> **THEOREM 5.2   VARIATION OF PARAMETERS**
>
> 1. $y'' + p(t)y' + q(t)y = r(t)$
> 2. $\{y_1(t), y_2(t)\}$ is a fundamental set for this equation.
> 3. $W(t)$ is the Wronskian for this fundamental set.
>
> $$\implies y_p = y_2 \int \frac{y_1 r(t)}{W(t)}\, dt - y_1 \int \frac{y_2 r(t)}{W(t)}\, dt$$

This formula for the particular solution y_p contains the subscripts 2, 1, 1, 2, in that order, separated in the middle by a minus sign. What else to call the method but 21-12 (twenty-one twelve)?

Students invariably fail to notice that the differential equation to which this method applies starts with y''. It does not start with $a(t)$ **times** y'', it starts with y''. To apply this formula, care must be exerted to ensure that $r(t)$ is what results in the given differential equation *after* division through by any lead coefficient y'' might have.

EXAMPLE 5.14   Obtain a particular solution for the inhomogeneous differential equation $y'' + y = \sec t$.

First, two independent solutions of the homogeneous differential equation, $y'' + y = 0$, must be found. The exponential guess will work for this constant-coefficient linear equation, so the characteristic equation is $\lambda^2 + 1 = 0$, the characteristic roots are $\lambda = \pm i$, and the desired solutions are $y_1 = \cos t$ and $y_2 = \sin t$. The Wronskian of these two functions is the determinant of the matrix

$$\begin{bmatrix} \cos t & \sin t \\ -\sin t & \cos t \end{bmatrix}$$

which evaluates to $W = 1$. The right-hand side of the nonhomogeneous differential equation is $r(t) = \sec t$. Hence, the Variation of Parameters formula dictates

$$y_p = \sin t \int \frac{\cos t \sec t}{1}\, dt - \cos t \int \frac{\sin t \sec t}{1}\, dt$$

$$= \sin t \int dt - \cos t \int \tan t\, dt = t \sin t - \cos t \ln|\cos t| \qquad \clubsuit$$

EXAMPLE 5.15   Obtain the general solution of the differential equation $y'' + y = \sec t \tan t$.

Since the homogeneous equation $y'' + y = 0$ is the same as in Example 5.14, we reuse $y_1 = \cos t$, $y_2 = \sin t$, and $W = 1$. However, $r(t) = \sec t \tan t$, so the Variation of Parameters formula gives

$$y_p = \sin t \int \frac{\cos t \sec t \tan t}{1}\, dt - \cos t \int \frac{\sin t \sec t \tan t}{1}\, dt$$

$$= \sin t \int \tan t\, dt - \cos t \int \tan^2 t\, dt$$

$$= -\sin t \ln|\cos t| - \cos t\,(\tan t - \arctan(\tan t))$$

$$= -\sin t \ln|\cos t| - \sin t + t \cos t$$

Simplifying $\arctan(\tan t)$ to t is valid only on the principal branch of the arctangent function. Recall, for example, that $\arctan(\tan 10) = 10 - 3\pi \neq 10$. Here, however, the correct simplification to $t - k\pi$ would generate $t \cos t$ and a constant multiple of $\cos t$ already a member of the homogeneous solution. Any such constant multiple of $\cos t$ and the term $-\sin t$ are already included in the homogeneous solution, so they need not be included in the particular solution. Consequently, we may take $y_p = -\sin t \ln|\cos t| + t \cos t$ as the particular solution. (See Exercise A1.)   ❖

EXAMPLE 5.16   Use Variation of Parameters to obtain a particular solution for the differential equation $y'' + 4y' + 16y = 6 \cos 2t$.

To implement the method of Variation of Parameters, we need two independent solutions of the homogeneous equation. Thus, we have the exponential guess $y = e^{\lambda t}$, the characteristic equation $\lambda^2 + 4\lambda + 16 = 0$, the characteristic roots $\lambda = -2 \pm 2\sqrt{3}i$, and the solutions $y_1 = e^{-2t} \cos 2\sqrt{3}t$ and $y_2 = e^{-2t} \sin 2\sqrt{3}t$. The Wronskian $W = 2\sqrt{3}e^{-4t}$ was computed in the example of underdamped motion in Section 5.6. Hence, the Variation of Parameters formula becomes

$$y_p = e^{-2t} \sin 2\sqrt{3}t \int \frac{e^{-2t} \cos(2\sqrt{3}t)\, 6 \cos 2t}{2\sqrt{3}e^{-4t}}\, dt$$

$$-e^{-2t} \cos 2\sqrt{3}t \int \frac{e^{-2t} \sin(2\sqrt{3}t)\, 6 \cos 2t}{2\sqrt{3}e^{-4t}}\, dt \tag{5.8}$$

The two integrals in (5.8) are extremely tedious to calculate and, after significant applications of algebra, trigonometry, and calculus (see the accompanying Maple worksheet), give the particular solution as

$$y_p = \tfrac{9}{26} \cos 2t + \tfrac{3}{13} \sin 2t$$

The driving term for this nonhomogeneous differential equation is $6 \cos 2t$. The particular solution has the simple form $a \cos 2t + b \sin 2t$. So, although Variation of Parameters led to difficult integrals in this example, there seems to be a connection between the driving term and the resulting particular solution that might be exploited to our advantage. Indeed, in Section 5.10, this connection between the form of the driving term and the form of the particular solution will lead to the computationally simpler method of Undetermined Coefficients. For the moment, we should note that Variation of Parameters can easily lead to integrals that are difficult, if not impossible, to evaluate.   ❖

Derivation of Variation of Parameters Formula

Given the equation $y'' + p(t)y' + q(t)y = r(t)$, let $y_1(t)$ and $y_2(t)$ be solutions of the homogeneous equation $y'' + p(t)y' + q(t)y = 0$. Thus, both

$$y_1'' + p(t)y_1' + q(t)y_1 \equiv 0 \quad \text{and} \quad y_2'' + p(t)y_2' + q(t)y_2 \equiv 0$$

hold identically. We then seek a particular solution in the form

$$y_p = u(t)y_1(t) + v(t)y_2(t)$$

reminiscent of the technique for first-order equations in Section 3.5 and for its cousin, reduction of order, in Section 5.7. Substitution into the differential equation will require computing both y_p' and y_p''. In fact,

$$y_p' = u'y_1 + uy_1' + v'y_2 + vy_2'$$

contains derivatives of the unknown functions $u(t)$ and $v(t)$. Since y_p'' will contain the second derivatives u'' and v'', we are interested in any device that would eliminate these two terms from the calculations. We therefore constrain these functions with the condition

$$u'y_1 + v'y_2 = 0 \tag{5.9}$$

so that

$$y_p' = uy_1' + vy_2' \qquad y_p'' = uy_1'' + u'y_1' + v'y_2' + vy_2''$$

knowing that if this approach succeeds, it is ultimately validated by Theorem 5.1, the uniqueness theorem in Section 5.3.

Marshalling expressions, we have

$$uy_1'' + u'y_1' + v'y_2' + vy_2'' + p(t)(uy_1' + vy_2') + q(t)(uy_1 + vy_2) = r(t)$$
$$u(y_1'' + py_1' + qy_1) + v(y_2'' + py_2' + qy_2) + u'y_1' + v'y_2' = r(t)$$
$$u\{0\} + v\{0\} + u'y_1' + v'y_2' = r(t)$$
$$u'y_1' + v'y_2' = r(t) \tag{5.10}$$

The third equation is a consequence of y_1 and y_2 being solutions of the homogeneous equation. The last equation and (5.9), the seemingly arbitrary condition imposed on y_p, give two equations for the unknown derivatives u' and v'. Writing W for the Wronskian $uv' - u'v$, we find

$$u' = -\frac{y_2 r}{W} \quad \text{and} \quad v' = \frac{y_1 r}{W} \tag{5.11}$$

and hence

$$u = -\int \frac{y_2 r}{W}\, dt \quad \text{and} \quad v = \int \frac{y_1 r}{W}\, dt$$

so that $y_p = uy_1 + vy_2 = [-\int \frac{y_2 r}{W}\, dt]y_1 + [\int \frac{y_1 r}{W}\, dt]y_2 = y_2 \int \frac{y_1 r(t)}{W(t)}\, dt - y_1 \int \frac{y_2 r(t)}{W(t)}\, dt.$

EXERCISES 5.9–Part A

A1. Verify that $y_p = -\sin t \ln|\cos t| + t\cos t$ is a particular solution of $y'' + y = \sec t \tan t$. To make the manipulations simpler, assume, for example, that t is restricted to an interval such as $(-\frac{\pi}{2}, \frac{\pi}{2})$, allowing the absolute value to be dropped from the cosine term.

A2. Show that the solution of (5.9) and (5.10) is given by (5.11).

A3. Use Variation of Parameters to obtain a particular solution for $y'' + y = 2$.

A4. Use Variation of Parameters to obtain a particular solution for $y'' + 4y = t$.

A5. Use Variation of Parameters to obtain a particular solution for $y'' + 3y' + 2y = r(t)$, if
   (a) $r(t) = 3$ **(b)** $r(t) = t$ **(c)** $r(t) = e^{-t}$

A6. Explain clearly why the Variation of Parameters formula can be written $y_p(t) = y_2 \int_a^t \frac{y_1 r}{W}\, dx - y_1 \int_a^t \frac{y_2 r}{W}\, dx$ for any fixed a in an appropriate interval.

EXERCISES 5.9–Part B

B1. Use Variation of Parameters to obtain a particular solution to each of the following nonhomogeneous Euler equations.
   (a) $x^2 y'' + 7xy' + 8y = x$ **(b)** $x^2 y'' + 6xy' + 4y = x^2$
   (c) $x^2 y'' + 8xy' + 10y = x^3$ **(d)** $x^2 y'' - 5xy' + 8y = x^5 \cos x$
   (e) $x^2 y'' - 6xy' + 6y = x^7 e^x$ **(f)** $x^2 y'' - 6xy' + 10y = x^6 \sin x$

In Exercises B2–7:
   (a) Use a computer algebra system to generate a general solution.
   (b) State y_h and y_p as contained in the general solution found in part (a).

(c) Extract the members of the fundamental set used to construct y_h in part (a).

(d) Compute the Wronskian for the fundamental set in part (c).

(e) Apply the Variation of Parameters formula, exhibiting and evaluating the required integrals.

(f) Resolve any discrepancies between the two versions of y_p. Note any terms that are redundant for the general solution.

B2. $y'' + 4y = \tan^2 2t$ **B3.** $y'' + 4y' + 8y = \cos 2t$

B4. $y'' + 6y' + 13y = \sin t$ **B5.** $y'' + 2y' + 10y = e^{-t}\sin 3t$

B6. $x^2y'' + 10xy' + 20y = x^{-4}$ **B7.** $x^2y'' - 5xy' + 9y = x$

B8. Consider the IVP
$$4y'' + 6y' + 9y = 3\cos 2t, \; y(0) = -2, \; y'(0) = 3.$$

(a) Solve with a computer algebra system and plot the resulting solution.

(b) Solve by completing all the steps in a traditional solution. Thus, start with the exponential guess, obtain the characteristic equation, characteristic roots, fundamental set, homogeneous solution, Wronskian, particular solution by Variation of Parameters, and general solution to which the data is to be applied. Resolve any discrepancies between your solution and the solution found in part (a).

B9. Given the IVP $5y'' + 4y' + 4y = 3\sin 2t, \; y(0) = -3, \; y'(0) = 2$:

(a) Use a computer algebra system to obtain the solution, and plot it.

(b) From the solution in part (a), extract the particular solution.

B10. Construct, from first principles, the solution to the IVP in Exercise B9. Thus:

(a) Obtain, from the characteristic equation, the fundamental set and the homogeneous solution.

(b) Obtain the particular solution via Variation of Parameters. Be sure to reconcile your particular solution with that found in Exercise B9, part (b).

(c) Write the general solution.

(d) Apply the data to the general solution, constructing and solving two equations in two unknowns. Exhibit the resulting solution for the IVP. Reconcile your solution with that found in Exercise B9.

B11. Apply Variation of Parameters to the third-order ODE $x^3y''' - 11x^2y'' + 46xy' - 70y = x$ via the following steps.

(a) Use a computer algebra system to obtain a general solution. Extract the particular solution and the fundamental set $\{y_1, y_2, y_3\}$.

(b) Write $y_p = u(x)y_1 + v(x)y_2 + w(x)y_3$, and impose the conditions $C_1 : u'y_1 + v'y_2 + w'y_3 = 0$ and $C_2 : u'y_1' + v'y_2' + w'y_3' = 0$ that prevent u''', v''', w''' from appearing in y_p'''. Substitute y_p into the differential equation, simplifying y_p' and y_p'' with C_1 and C_2 prior to obtaining y_p'''. The differential equation itself then gives a third condition C_3.

(c) Solve the three equations $C_k, k = 1, 2, 3$, for u', v', and w'.

(d) Integrate to find $u(x), v(x), w(x)$.

(e) Substitute these values into y_p and show you have the same particular solution found in part (a).

B12. Were the recipe in Exercise B11 to be carried out symbolically for the general third-order inhomogeneous ODE $y''' + a(x)y'' + b(x)y' + c(x)y = r(x)$, the resulting Variation of Parameters formula would be

$$y_p = y_1 \int \begin{vmatrix} y_2 & y_3 \\ y_2' & y_3' \end{vmatrix} \frac{r}{W} dx - y_2 \int \begin{vmatrix} y_1 & y_3 \\ y_1' & y_3' \end{vmatrix} \frac{r}{W} dx$$
$$+ y_3 \int \begin{vmatrix} y_1 & y_2 \\ y_1' & y_2' \end{vmatrix} \frac{r}{W} dx$$

where the determinant $\begin{vmatrix} y_i & y_j \\ y_i' & y_j' \end{vmatrix}$ is actually the Wronskian $W(y_i, y_j)$ and W is the Wronskian of the fundamental set $\{y_1, y_2, y_3\}$. Apply this formula to the ODE in Exercise B11 and show it results in the same particular solution already found.

5.10 Forced Motion and Undetermined Coefficients

EXAMPLE 5.17   Consider the differential equation

$$y'' + 4y' + 16y = 6\cos 2t \tag{5.12}$$

whose particular solution we found by Variation of Parameters in Example 5.16, Section 5.9. The characteristic equation $\lambda^2 + 4\lambda + 16 = 0$, a consequence of making an exponential guess, is written by inspection, and the characteristic roots are $\lambda = -2 \pm 2\sqrt{3}i$. A fundamental set consisting of real functions is then $\{e^{-2t}\cos 2\sqrt{3}t, e^{-2t}\sin 2\sqrt{3}t\}$.

We next observe that $6\cos 2t$ is a solution of the homogeneous equation $y'' + 4y = 0$. This inspires us to differentiate (5.12) twice and add four times (5.12) itself. This calculation can be expressed as

$$\left(\frac{d^2}{dt^2}+4\right)(y''+4y'+16y)=\left(\frac{d^2}{dt^2}+4\right)6\cos 2t$$

and it gives

$$y^{(4)}+4y'''+(4+16)y''+16y'+64y=0 \qquad \textbf{(5.13)}$$

the characteristic equation for which is

$$\lambda^4+4\lambda^3+20\lambda^2+16\lambda+64=(\lambda^2+4)(\lambda^2+4\lambda+16)=0$$

The characteristic roots for (5.13) are therefore $\pm 2i$ and $-2\pm 2\sqrt{3}i$. The first set of roots are associated with the driving term $6\cos 2t$ on the right in (5.12), while the second set are the characteristic roots for the homogeneous version of (5.12). The general solution for (5.13) would be built from the fundamental set for (5.12) and from $\{\cos 2t,\ \sin 2t\}$, which arises from the nonhomogeneous term in (5.12). Since these two "fundamental sets" are distinct, we do not have a repeated root case for (5.13) and its general solution does not contain any terms multiplied by powers of t.

The particular solution of (5.12) will contain that part of the general solution of (5.13) that is not part of the homogeneous solution of (5.12). Hence, for (5.12) we "guess" at a particular solution of the form $y_p = a\cos 2t + b\sin 2t$. Substituting this y_p into (5.12) gives

$$(12a+8b)\cos 2t + (12b-8a)\sin 2t = 6\cos 2t$$

which must be an identity in t. For this to be so, the coefficients of the trig terms on each side of the equation must match exactly. This matching condition yields the two equations

$$12a+8b=6 \quad \text{and} \quad 12b-8a=0$$

whose solution is $a=\frac{9}{26}$, $b=\frac{3}{13}$. The particular solution is then $y_p=\frac{9}{26}\cos 2t+\frac{3}{13}\sin 2t$, found with considerably less effort than invested in the Variation of Parameters calculation.

Having found the particular solution, we can write the general solution as the sum of the homogeneous solution and the particular solution, obtaining

$$y_g = Ae^{-2t}\cos 2\sqrt{3}t + Be^{-2t}\sin 2\sqrt{3}t + \frac{9}{26}\cos 2t + \frac{3}{13}\sin 2t$$

Initial data is applied to the general solution. For example, if $y(0)=1$, $y'(0)=1$, we would form the two equations

$$A+\frac{9}{26}=1 \quad \text{and} \quad -2A+\frac{6}{13}+2\sqrt{3}B=1$$

whose solution is $A=\frac{17}{26}$, $B=\frac{4\sqrt{3}}{13}$ so that the solution to the initial value problem would be

$$y(t)=\frac{e^{-2t}}{26}(17\cos 2\sqrt{3}t + 8\sqrt{3}\sin 2\sqrt{3}t) + \frac{9}{26}\cos 2t + \frac{3}{13}\sin 2t$$

Figure 5.12 contains a graph of this solution (solid) and a graph of y_p (dotted), revealing the rapid onset of the steady-state motion that is the system's response to the driving force $6\cos 2t$. Initially, there is a *transient phase* where the system is reacting to the initial conditions and, eventually, a *steady-state* phase where it is reacting just to the driving term. Since the homogeneous solution contains the factor e^{-2t}, the homogeneous solution constitutes the transient portion of the solution. The particular solution yields the steady-state portion. ❖

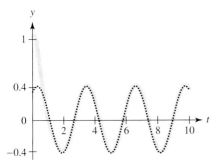

FIGURE 5.12   In Example 5.17, solution (solid) rapidly tends towards y_p (dotted)

For many of the initial value problems of science and engineering, the method of Undetermined Coefficients is computationally simpler that the method of Variation of Parameters, even in the context of a computer algebra system. The algebra of Undetermined Coefficients might be tedious, but the calculus of Variation of Parameters might be impossible.

Applicability of Undetermined Coefficients

Variation of Parameters always applies but is never certain to escape the confines of difficult integrations. Undetermined Coefficients, however, will always succeed. But Undetermined Coefficients only applies under the conditions

1. the ODE is linear with constant coefficients.

2. $r(t)$, the driving term, is a linear combination of terms that are themselves pieces of solutions of linear ODEs with constant coefficients. Thus, the terms in $r(t)$ must be obtainable from the exponential guess $e^{\lambda t}$ and can include terms that arise from repeated roots.

Table 5.8 lists particular examples of terms $r(t)$ can contain if Undetermined Coefficients is to work. The characteristic root $\lambda = 0$ puts $e^{0t} = 1$ into the fundamental set. Hence, a viable $r(t)$ can contain a constant multiple of 1. Moreover, if the root $\lambda = 0$ is repeated, then $te^{0t} = t$ would be in the fundamental set, so $r(t)$ can contain a constant multiple of t. Likewise, $r(t)$ can contain a linear combination of powers of t, and hence, a polynomial in t is a viable part of $r(t)$.

λ	**Multiplicity 1**	**Multiplicity 2**	**Multiplicity 3**
0	$e^{0t} = 1$	$te^{0t} = t$	$t^2 e^{0t} = t^2$
3	e^{3t}	te^{3t}	$t^2 e^{3t}$
$\pm 3i$	$e^{\pm 3it} \Rightarrow \begin{cases} \cos(3t) \\ \sin(3t) \end{cases}$	$te^{\pm 3it} \Rightarrow \begin{cases} t\cos(3t) \\ t\sin(3t) \end{cases}$	$t^2 e^{\pm 3it} \Rightarrow \begin{cases} t^2\cos(3t) \\ t^2\sin(3t) \end{cases}$
$2 \pm 3i$	$e^{(2\pm 3i)t} \Rightarrow \begin{cases} e^{2t}\cos(3t) \\ e^{2t}\sin(3t) \end{cases}$	$te^{(2\pm 3i)t} \Rightarrow \begin{cases} te^{2t}\cos(3t) \\ te^{2t}\sin(3t) \end{cases}$	$t^2 e^{(2\pm 3i)t} \Rightarrow \begin{cases} t^2 e^{2t}\cos(3t) \\ t^2 e^{2t}\sin(3t) \end{cases}$

TABLE 5.8   Sample nonhomogeneous terms for which Undetermined Coefficients will work

The real characteristic root $\lambda = 3$ puts e^{3t} into the fundamental set. Hence, a viable $r(t)$ can contain a constant multiple of a real exponential of the form e^{at}, where a is a real number. Likewise, if the root $\lambda = 3$ is repeated, then te^{3t} will be in the fundamental set, so $r(t)$ could therefore contain a constant multiple of te^{3t}. In fact, with additional repetition of the root $\lambda = 3$, $r(t)$ could contain a linear combination of $\{e^{3t}, te^{3t}, t^2 e^{3t}\}$, which could appear as the fragment $(2 - 5t + 7t^2)e^{3t}$.

The pure imaginary pair of conjugate roots $\lambda = \pm 3i$ would put both $\cos 3t$ and $\sin 3t$ into the fundamental set. Hence, $r(t)$ can contain a linear combination of $\cos 3t$ and $\sin 3t$. For example, $r(t)$ could contain $2\cos 3t - 5\sin 3t$ or $2\cos 3t + 0\sin 3t = 2\cos 3t$ or $0\cos 3t + 5\sin 3t = 5\sin 3t$. If the roots $\lambda = \pm 3i$ are repeated, then $t\cos 3t$ and $t\sin 3t$ as well as $\cos 3t$ and $\sin 3t$ can appear in the fundamental set. Hence, $r(t)$ can contain trig terms multiplied by powers of t.

Finally, the complex conjugate pair of roots $\lambda = 2 \pm 3i$ puts $e^{2t}\cos 3t$ and $e^{2t}\sin 3t$ into the fundamental set, so $r(t)$ can contain a linear combination such as $e^{2t}(3\cos 3t - 5\sin 3t)$ or $e^{2t}(3\cos 3t + 0\sin 3t) = 3e^{2t}\cos 3t$ or $e^{2t}(0\cos 3t - 5\sin 3t) = -5e^{2t}\sin 3t$. Of course, if the roots $\lambda = 2 \pm 3i$ are repeated, then $r(t)$ can contain linear combinations of $te^{2t} \cos 3t$ and $te^{2t} \sin 3t$.

In Table 5.9, the "characteristic roots" of the general solution of which $r(t)$ is a fragment are called *latent roots*. Success with the method of Undetermined Coefficients hinges on being able to reverse-engineer the full list of all possible latent roots from the evidence of the fragment of the solution, which remains as $r(t)$. Motivation for this is given by the relation between (5.12) and (5.13) in Example 5.17.

$r(t)$	Latent Roots	y_p
(1) $2e^{3t} + \cos 5t$	$3, \pm 5i$	$ae^{3t} + b \cos 5t + c \sin 5t$
(2) $-3 - 4 \sin 5t$	$0, \pm 5i$	$a + b \cos 5t + c \sin 5t$
(3) $e^{-t} - 2 \sin t + 3 \cos t$	$-1, \pm i$	$ae^{-t} + b \cos t + c \sin t$
(4) $e^{-t} + e^{-3t} \cos t$	$-1, -3 \pm i$	$ae^{-t} + e^{-3t}(b \cos t + c \sin t)$
(5) $t^2 + te^{2t}$	$0, 0, 0, 2, 2$	$a + bt + ct^2 + e^{2t}(\alpha + \beta t)$
(6) $t \cos 2t$	$\pm 2i, \pm 2i$	$a \cos 2t + b \sin 2t + t(c \cos 2t + d \sin 2t)$
(7) $te^{-3t} \cos 2t$	$-3 \pm 2i, -3 \pm 2i$	$e^{-3t}(a \cos 2t + b \sin 2t) + te^{-3t}(c \cos 2t + d \sin 2t)$
(8) $t + t^2$	$0, 0, 0$	$a + bt + ct^2$
(9) $e^{-2t} - 3t^2 e^{-2t}$	$-2, -2, -2$	$(a + bt + ct^2)e^{-2t}$
(10) $\sin t - 2t \cos t$	$\pm i, \pm i$	$a \cos t + b \sin t + t(c \cos t + d \sin t)$

TABLE 5.9   Examples of nonhomogeneous terms and the form of y_p they require

Thus, the left column in Table 5.9 gives examples of possible expressions for $r(t)$ that are viable candidates for Undetermined Coefficients. The middle column in the table gives the latent roots, those characteristic roots that would appear if the nonhomogeneous DE were manipulated to become homogeneous, as we did in Example 5.17. The right column of the table, the full form for y_p determined by the given $r(t)$, is the most general solution possible from the latent roots in the middle column. That is why success with Undetermined Coefficients hinges on being able to "think backward" and visualize the characteristic roots responsible for a general solution of which $r(t)$ is but a fragment.

In the first and second examples where just a single trig term appears, the list of latent roots contains the conjugate *pair* listed. Thus, the particular solution must contain both a sine and a cosine. In each of these two examples, there is an additional real latent root, so that in addition to the trig terms in y_p, there are real exponential terms. In the first case, y_p contains a constant multiple of e^{3t}; while in the second, a multiple of $e^{0t} = 1$.

In the third example, $r(t)$ contains both $\sin t$ and $\cos t$. The latent roots include the *single* conjugate pair $\pm i$, and y_p contains the *single* linear combination $b \cos t + c \sin t$.

In the fourth example, $r(t)$ contains $e^{-3t} \cos t$, but the latent roots contain the complex conjugate *pair* $-3 \pm i$ and y_p contains the linear combination $e^{-3t}(b \cos t + c \sin t)$.

In the fifth example, the t^2 term could only have appeared as a solution to a constant-coefficient, linear, homogeneous differential equation if the characteristic roots included $0, 0, 0$. Hence, these three roots appear as latent roots, and y_p must contain a linear combination of $1, t, t^2$. In addition, $r(t)$ contains te^{2t}, which could only have arisen from the *repeated* root $\lambda = 2, 2$. Hence, y_p must contain a linear combination of e^{2t} and te^{2t}.

In the sixth example, the product of t with $\cos 2t$ signals that the pair of conjugate roots $\pm 2i$ had to occur *twice*. Hence, y_p must contain a linear combination of $\cos 2t$ and $\sin 2t$ as well as a linear combination of $t \cos 2t$ and $t \sin 2t$.

The seventh example is like the sixth, whereby the factor t signals the complex conjugate pair $-3 \pm 2i$ must occur *twice*. Hence, y_p must contain a linear combination of $e^{-3t} \cos 2t$ and $e^{-3t} \sin 2t$ as well as a linear combination of $te^{-3t} \cos 2t$ and $te^{-3t} \sin 2t$.

In the eighth example, where both t and t^2 appear in $r(t)$, the list of latent roots need only include enough 0s to account for t^2. It would be overkill to attribute $0, 0$ to the t and then another $0, 0, 0$ to the t^2. Hence, y_p contains a linear combination of $1, t, t^2$ and not of the five terms $1, t, t^2, t^3, t^4$.

The ninth example is like the eighth, whereby e^{-2t} and $t^2 e^{-2t}$ appearing in $r(t)$ generate the latent roots $-2, -2, -2$, just enough to account for the term $t^2 e^{-2t}$ and not one -2 for e^{-2t} and *another* three for $t^2 e^{-2t}$. Hence, y_p contains a linear combination of just the three terms $e^{-2t}, te^{-2t}, t^2 e^{-2t}$.

Finally, the tenth example is again like the eighth and ninth, whereby *all* of $r(t)$ is accounted for by the latent roots $\pm i, \pm i$. It would be a mistake to look at the term $\sin t$ and count $\pm i$, then look at the term $t \cos t$, and count $\pm i, \pm i$ because of the factor t. The terms in $r(t)$ must be looked at collectively, not in isolation.

The Tricky Case

There is one subtlety in determining the form of the particular solution for Undetermined Coefficients. The case arises when a latent root hiding in $r(t)$ is also a characteristic root. In fact, suppose the root ρ is a characteristic root 2 times and is also a latent root 3 times. The guess for the particular solution would then be $y_p = e^{\rho t}(a + bt + ct^2)t^2$ instead of just $y_p = e^{\rho t}(a + bt + ct^2)$. The special rule is

increase, by the multiplicity of the characteristic root, the powers on the t's that the latent roots dictate.

EXAMPLE 5.18   Determine the form of the particular solution for

$$y'' = t + e^t + t \sin t$$

wherein the characteristic equation is $\lambda^2 = 0$, the characteristic roots are $\lambda = 0, 0$, and the latent roots of $r(t)$ are $0, 0, 1, \pm i, \pm i$. The guess for y_p is

$$y_p = (c_1 + c_2 t)t^2 + c_3 e^t + (c_4 \cos(t) + c_5 \sin(t)) + t(c_6 \cos(t) + c_7 \sin(t))$$

Because the characteristic roots already include 0 and 0, the latent roots 0 and 0, dicated by the term t, must be counted as the third and fourth appearance of $\lambda = 0$. Hence, corresponding to t in $r(t)$, we need not 1 and t in y_p but t^2 and t^3. The repeated pair of complex conjugate roots in the list of latent roots means there will be, in y_p, a sine and cosine term, and t times a sine and cosine pair.   ❖

EXAMPLE 5.19   Determine the form of the particular solution for

$$y'' - 2y' + y = t^2 e^t + \sin t$$

wherein the characteristic equation is $\lambda^2 - 2\lambda + 1 = (\lambda - 1)^2 = 0$, the characteristic roots are $\lambda = 1, 1$, and the latent roots of $r(t)$ are $1, 1, 1, \pm i$. The guess for y_p is

$$y_p = e^t(a + bt + ct^2)t^2 + \alpha \cos(t) + \beta \sin(t)$$

Because the latent roots $1, 1, 1$ are repetitions of the characteristic roots $1, 1$, the particular solution must contain, not just $e^{0t} = 1$, $te^{0t} = t$, and $t^2 e^{0t} = t^2$, but each of these terms with their exponents on t two powers higher. In effect, when counting the latent roots, count their repetitions amongst the characteristic roots *first* and continue the count for the latent roots. The guess for the particular solution is built from this higher number of repetitions.

EXAMPLE 5.20

Solve the initial value problem $y''' - y' = 4e^{-t} + 3e^{2t}$, $y(0) = 0$, $y'(0) = -1$, $y''(0) = 2$.

The characteristic equation is $\lambda^3 - \lambda = \lambda(\lambda^2 - 1) = 0$, the characteristic roots are $\lambda = 0, 1, -1$, and the homogeneous solution is $y_h = A + Be^t + Ce^{-t}$. The latent roots for $r(t)$ are -1 and 2. Hence, -1 is a "repeated" root, so the form of the particular solution will be $y_p = ae^{2t} + bte^{-t}$ and not $ae^{2t} + be^{-t}$. Substitution of y_p into the differential equation yields

$$6ae^{2t} + 2be^{-t} = 4e^{-t} + 3e^{2t}$$

from which matching of coefficients of like exponentials results in the two equations

$$2b = 4 \quad \text{and} \quad 6a = 3$$

and the solution $a = \frac{1}{2}$, $b = 2$. That gives the particular solution $y_p = \frac{1}{2}e^{2t} + 2te^{-t}$ and the general solution

$$y_g = A + Be^t + Ce^{-t} + \frac{1}{2}e^{2t} + 2te^{-t}$$

Finally, applying the initial data to the general solution generates the three equations

$$A + B + C + \frac{1}{2} = 0 \quad B - C + 3 = -1 \quad B + C - 2 = 2$$

whose solution is $A = -\frac{9}{2}$, $B = 0$, $C = 4$; hence, the solution of the initial value problem is $y(t) = -\frac{9}{2} + 4e^{-t} + \frac{1}{2}e^{2t} + 2te^{-t}$.

SUPERPOSITION FOR NONHOMOGENEOUS EQUATIONS In Section 5.3 we saw the Superposition Principle that for homogeneous linear equations, the sum of two solutions was again a solution. For nonhomogeneous linear DEs, there is a different superposition principle that we can now make because we know how to solve nonhomogeneous equations.

The solution of a linear equation driven by a forcing function that is a sum of inputs is the sum of responses to the individual and isolated inputs.

EXAMPLE 5.21

Table 5.10 shows the solutions to the three initial value problems $y'' + 14y' + 48y = f_k(t)$, $k = 1, 2, 3$, $y(0) = y'(0) = 0$. If the system is driven by $f_1(t) = f_2(t) + f_3(t)$, the response, $y_1(t)$, is the sum of the solutions $y_2(t)$ and $y_3(t)$, themselves responses to the driving terms $f_2(t)$ and $f_3(t)$, respectively. The driving term $f_1(t)$ can be decomposed into its individual summands and the response to each summand computed. The response to $f_1(t)$ is then the sum of the responses to the individual components in $f_1(t)$.

Driving Term	Response
$f_1(t) = 5\sin t + 3\cos t$	$y_1(t) = \frac{71}{2405}\cos t + \frac{277}{2405}\sin t - \frac{13}{74}e^{-6t} + \frac{19}{130}e^{-8t}$
$f_2(t) = 5\sin t$	$y_2(t) = -\frac{70}{2405}\cos t + \frac{235}{2405}\sin t + \frac{5}{74}e^{-6t} - \frac{5}{130}e^{-8t}$
$f_3(t) = 3\cos t$	$y_3(t) = \frac{141}{2405}\cos t + \frac{42}{2405}\sin t - \frac{18}{74}e^{-6t} + \frac{24}{130}e^{-8t}$

TABLE 5.10 Superposition for nonhomogeneous equations

EXERCISES 5.10–Part A

A1. Given $r(t)$ and the characteristics roots, provide a correct form for y_p.

(a) $r(t) = 2e^{3t} + \cos 5t - \sin 5t, \lambda = 3, -1 \pm 5i$

(b) $r(t) = -3 - 4 \cos 2t, \lambda = -3, 4$

(c) $r(t) = e^{-t} - 3 \sin t, \lambda = -1, 1$

(d) $r(t) = e^t + e^{-3t} \sin t, \lambda = -3, \pm i$

(e) $r(t) = t^2 + 3te^{2t}, \lambda = 0, 0, 2$

(f) $r(t) = t \sin 3t, \lambda = 0, 3, 3$

(g) $r(t) = te^{4t} \sin t, \lambda = 4, 4$

(h) $r(t) = 3t + 4t^2, \lambda = 0, 1, 2$

(i) $r(t) = e^{2t} - 3t^2 e^{2t}, \lambda = 0, 0, -2, -2$

(j) $r(t) = \cos t - 2t \sin t, \lambda = 0, \pm 2i$

A2. Let $y_k, k = 1, 2, 3$, be the respective solutions of $y'' + 3y' + 2y = f_k(t), y(0) = y'(0) = 0$, where $f_1(t) = t + \sin t, f_2(t) = t$, and $f_3(t) = \sin t$. Verify superposition for nonhomogeneous equations by showing $y_1(t) = y_2(t) + y_3(t)$ since $f_1(t) = f_2(t) + f_3(t)$.

A3. Does superposition hold if the initial conditions in Exercise A2 are replaced with $y(0) = 1, y'(0) = 0$?

A4. For $y'' + 2y' + 2y = \cos 2t$, obtain y_p; then put it into the form $A \cos(2t - \phi)$.

A5. Obtain y_p for $y'' + 2y' + 2y = \cos \omega t$ by starting with $y_p = A \cos(\omega t - \phi)$.

EXERCISES 5.10–Part B

B1. Consider the nonhomogeneous equation $y'' + 2y' + 2y = e^{-t} + \sin t$.

(a) Obtain the latent roots for $r(t)$.

(b) Construct a characteristic polynomial in λ corresponding to the latent roots.

(c) Replace each λ in part (b) with the differentiation operator d/dt, that is, with λ^n becomes d^n/dt^n, and apply the resulting operator to the given ODE. The resulting equation should now be homogeneous.

(d) Obtain the characteristic equation and characteristic roots for the homogeneous equation in part (c). Show that the roots are the union of the characteristic roots and the latent roots for the original second-order equation.

(e) Write the general solution corresponding to the roots in part (d), and show that it contains the y_p that would have been written from the latent roots found in part (a).

For the IVPs in Exercises B2–11:

(a) Use a computer algebra system to obtain and plot the solution.

(b) Obtain the characteristic roots and the latent roots hiding in $r(t)$. Distinguish carefully between both sets of roots.

(c) Write the homogeneous solution and the correct guess for the particular solution.

(d) Obtain the correct particular solution by completing the calculations of the method of Undetermined Coefficients.

(e) Form the general solution and apply the initial data to produce the correct solution to the IVP. Resolve any discrepancies with the solution found in part (a).

(f) From the latent roots, construct a polynomial in the differentiation operator d/dt, that is, with λ^n becomes d^n/dt^n.

Apply this operator to the given differential equation, obtaining a new, homogeneous DE.

(g) Obtain the characteristic roots for the DE formed in part (f), and construct the general solution corresponding to these roots. Show that y_p guessed at in part (c) is a part of this general solution.

B2. $y'' + 4y' + 4y = 5 \sin 3t, y(0) = -3, y'(0) = 1$

B3. $4y'' + 4y' + y = 3 \sin 2t - \cos 2t, y(0) = -2, y'(0) = 3$

B4. $y'' + 2y' + 2y = 3 \sin 5t, y(0) = 0, y'(0) = 3$

B5. $y'' + 2y' + 2y = e^{3t}, y(0) = -5, y'(0) = -4$

B6. $y^{(4)} - 5y'' + 4y = 3, y(0) = 0, y'(0) = 0, y''(0) = 0, y'''(0) = 0$

B7. $y'' + 3y' + 2y = 1 + 2te^t, y(0) = -1, y'(0) = 1$

B8. $y'' + 25y = 3t \sin 5t, y(0) = 2, y'(0) = -1$

B9. $y'' + 6y' + 13y = 5te^{-3t} \sin 2t, y(0) = 0, y'(0) = 1$

B10. $y^{(5)} + y^{(3)} = \sin t - 4t^2 + 2t - 5, y(0) = 1, y'(0) = -\frac{1}{2}, y''(0) = \frac{3}{4}, y'''(0) = 0, y^{(4)}(0) = -1$

B11. $y'' + 2y' + 2y = 5t^2 - 7t + 3, y(0) = -2, y'(0) = 3$

B12. A capacitor with capacitance $C = \frac{1}{500}$ farad, an inductor with coefficient of inductance $L = \frac{1}{20}$ henry, and a resistor with resistance $R = 1$ ohm are connected in series. If at $t = 0$, the current i equals zero, and the charge on the capacitor is 1 coulomb, find the current and the charge in the circuit due to the discharge of the capacitor. *Hint:* From Section 5.1, the current obeys $Li'' + Ri' + \frac{1}{C}i = E'(t)$. Here, $E = 0, i(0) = 0$, and (from [11]), $i'(0) = \frac{E(0) - i(0)R - q(0)}{c}/L$, where $q(t)$ is the charge, obtained from $i(t)$ as $q(t) = q(0) + \int_0^t i(s)\, ds$.

B13. For large t, find the limiting current in the RLC circuit for which $L = \frac{1}{10}, R = 20, C = 20$ and $E(t) = 5 \cos(10t)$.

5.11 Resonance

Introduction

Resonance can be characterized as that physical phenomenon whereby Mother Nature seems to overreact to a lesser, but periodic, stimulus. This overreaction to a small stimulus is called a temper tantrum in a child, "having a fit" in an adult, and "resonance" in Mother Nature. The response is simply out of proportion to the stimulus.

Of course, we are just creating imagery to focus attention. However, anyone who has seen the four-minute film clip of the Tacoma Narrows Bridge, Tacoma, Washington, twisting in the wind and collapsing in 1940 would need no additional metaphors for resonance, the phenomenon responsible for such a ruinous end to a magnificent work of engineering.

Resonance is quantized within the driven damped oscillator. A linear, damped oscillator is driven with a sinusoidal input having a variable angular frequency. The response of the system is observed as a function of the driving input frequency. The frequency for which the response is greatest is called the *resonant frequency,* a frequency at which the magnitude of the system's response can be far larger than the magnitude of the input driving force.

We distinguish two cases of resonance. If damping is assumed to be zero, the system is a fiction, since even at supercooled temperatures, friction does not disappear entirely. The resonance model in which no damping term appears has characteristics that just cannot be found in nature since a frictionless system is just not found in nature. For this reason, this model is called "unreal resonance."

The more realistic model for resonance includes frictional damping and is called "real resonance."

Unreal Resonance

The undamped oscillator with mass $m = 1$ and spring constant $k = 16$, driven by a periodic force of amplitude 1 and angular frequency $\omega = 4$, is modeled by the differential equation $y'' + 16y = \cos 4t$. If this system goes into motion with the inert initial conditions $y(0) = 0$, $y'(0) = 0$, we can describe the resulting motions by the solution $y(t) = \frac{t}{8} \sin 4t$, graphed as the solid curve in Figure 5.13. The oscillations of the system grow without bound, getting ever larger and larger. Of course, this is not what happens in reality where some frictional damping always exists. In this model with no damping, the envelope of the oscillations is linear, as shown by the dotted lines in Figure 5.13.

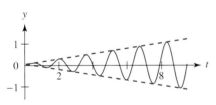

FIGURE 5.13 Resonant solution (solid) and envelopes $\pm\frac{t}{8}$ for zero damping

Fortunately, this unreal resonance is never seen in nature since nothing in nature takes place without frictional losses. But to see why the model predicts these unbounded motions, obtain the characteristic roots $\lambda = \pm 4i$ from the characteristic equation $\lambda^2 + 16 = 0$. The latent roots from the driving term $r(t)$ are themselves $\pm 4i$, so the particular solution will contain $t(a \cos 4t + b \sin 4t)$. Driving the system at its *natural* (angular) frequency $\omega = 4$ means the trig terms in the particular solution are multiplied by t. Physically, this correspondence means every imposed push lines up exactly with the natural peak in the motion, and the oscillations increase without bound.

The natural frequency of the undamped system is determined by the homogeneous equation $my'' + ky = 0$. The characteristic equation is $m\lambda^2 + k = 0$, so the characteristic roots are $\lambda = \pm\sqrt{\frac{k}{m}}i$. These roots are complex, so the fundamental set will be $\{\cos\sqrt{\frac{k}{m}}t, \sin\sqrt{\frac{k}{m}}t\}$. Clearly, $\omega_N = \sqrt{\frac{k}{m}}$ becomes the natural (angular) frequency since that is the frequency this system must necessarily exhibit if no external driving forces act on it. Driving the undamped system at its natural (angular) frequency then causes it to resonate, a resonance that predicts unbounded motions.

However, in real life, there is always damping. Hence, we next consider the same system ($m = 1, k = 16$) but add some damping to explore the case of real resonance. We will drive the system with a cosine term for which the frequency is not specified and seek to study the response of the system as a function of the driving frequency.

Real Resonance

Consider the damped oscillator for which the mass, damping coefficient, and spring constant are $m = 1, b = 4, k = 16$, respectively, and for which the driving term is $\cos \omega t$. A model for this system is the differential equation

$$y'' + 4y' + 16y = \cos \omega t$$

with characteristic equation $\lambda^2 + 4\lambda + 16 = 0$ and characteristic roots $\lambda = -2 \pm 2\sqrt{3}i$. The natural (angular) frequency for the system is $\omega_N = 2\sqrt{3}$, the fundamental set is $\{e^{-2t} \cos \omega_N t, e^{-2t} \sin \omega_N t\}$, and the homogeneous solution is $y_h = e^{-2t}(c_1 \cos \omega_N t + c_2 \sin \omega_N t)$.

At steady state when the term e^{-2t} has done its work and effectively becomes zero, the homogeneous solution hardly contributes to the general solution. The steady-state solution is essentially the particular solution, which we write in the form $y_p = A \cos(\omega t - \phi)$. Using the differential equation to determine A and ϕ, we have

$$-A\omega^2 \cos(\omega t - \phi) - 4A\omega \sin(\omega t - \phi) + 16A \cos(\omega t - \phi) = \cos \omega t$$

or, expanding the trig terms, we have

$$\{A(16 - \omega^2) \cos \phi + 4A\omega \sin \phi\} \cos \omega t$$
$$+ \{A(16 - \omega^2) \sin \phi - 4A\omega \cos \phi\} \cos \omega t = \cos \omega t \qquad \textbf{(5.14)}$$

By matching coefficients of $\cos \omega t$ and $\sin \omega t$ on both sides of (5.14), we find

$$A = \frac{1}{\sqrt{\omega^4 - 16\omega^2 + 256}}$$

$$\sin \phi = \frac{4\omega}{\sqrt{\omega^4 - 16\omega^2 + 256}}$$

$$\cos \phi = \frac{16 - \omega^2}{\sqrt{\omega^4 - 16\omega^2 + 256}}$$

The phase angle ϕ is determined jointly by the second and third equations. However, we are really only interested in the amplitude of the motion, graphed as a function of the driving frequency ω in Figure 5.14. For some value of the driving frequency ω, the magnitude of the steady-state's amplitude peaks. The driving frequency for which this amplitude is a maximum is called the *resonant* frequency, which can be found exactly by solving the equation

$$\frac{dA}{d\omega} = \frac{2\omega(\omega^2 - 8)}{(\omega^4 - 16\omega^2 + 256)^{3/2}} = 0$$

for the resonant frequency $\omega_R = \sqrt{8} = 2\sqrt{2}$.

The natural frequency is $2\sqrt{3}$, slightly larger than the resonant frequency $2\sqrt{2}$. For unreal (undamped) resonance, the natural frequency and the resonant frequency are the same. For real (damped) resonance, the resonant frequency is slightly smaller than the natural frequency.

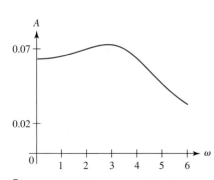

FIGURE 5.14   Response amplitude vs. frequency of the driving function

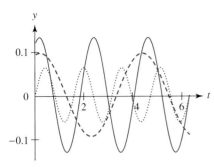

FIGURE 5.15 Graphs of $y_{\text{p}}(t)$ for $\omega = \frac{3}{2}, 2\sqrt{2}, 4$, as the dashed, solid, and dotted curves, respectively

As a function of t and ω, the particular solution is

$$y_{\text{p}} = \frac{\cos\left(\omega t - \arctan\frac{4\omega}{16 - \omega^2}\right)}{\sqrt{\omega^4 - 16\omega^2 + 256}}$$

Figure 5.15 shows $y_{\text{p}}(t)$ for the three values $\omega = \frac{3}{2}, 2\sqrt{2}, 4$, as the dashed, solid, and dotted curves, respectively. The accompanying Maple worksheet contains a complete animation of showing how $y_{\text{p}}(t)$ varies continuously with ω.

DEPENDENCE OF RESONANCE ON DAMPING   It is important to quantify the resonant frequency's dependence on damping. Hence, let the preceding damped oscillator now have variable damping b so that the system is modeled by the differential equation $y'' + by' + 16y = \cos \omega t$. As before, we want the system's steady-state response to the driving term, from which we will extract the value of the driving frequency ω maximizing the amplitude of the response. Again taking the particular solution as $y_{\text{p}} = A\cos(\omega t - \phi)$, the differential equation yields

$$-A\omega^2 \cos(\omega t - \phi) - bA\omega \sin(\omega t - \phi) + 16A\cos(\omega t - \phi) = \cos(\omega t)$$

from which, after expanding the trig terms and matching coefficients of $\cos \omega t$ and $\sin \omega t$, we find

$$A = \frac{1}{\sqrt{\omega^4 + (b^2 - 32)\omega^2 + 256}}$$

$$\sin \phi = \frac{4\omega}{\sqrt{\omega^4 + (b^2 - 32)\omega^2 + 256}}$$

$$\cos \phi = \frac{16 - \omega^2}{\sqrt{\omega^4 + (b^2 - 32)\omega^2 + 256}}$$

Again interested in how A, the amplitude of the steady-state response, varies with driving frequency ω, we give the damping coefficient b the values $1, 2, \ldots, 5$ and plot the resulting resonance curves in Figure 5.16. The more the damping, the smaller the maximum peak at resonance; and the less the damping, the greater the resonant peak. Also, if the damping is great enough there is no resonance peak at all. Hence, sufficient damping precludes resonance.

Figure 5.17 is another way of visualizing the dependence of the response amplitude $A(b, \omega)$ on both damping and input driving frequency. The function $A(b, \omega)$ is plotted as a surface over the $b\omega$-plane. The plane sections $b = $ constant are the curves seen in Figure 5.16. Both Figures 5.16 and 5.17 suggest that as the driving frequency ω is increased well past the resonant frequency, the amplitude of the steady-state response goes to zero.

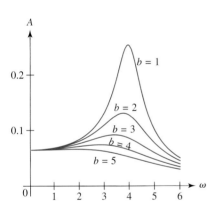

FIGURE 5.16   Amplitude vs. driving frequency for five values of damping parameter b

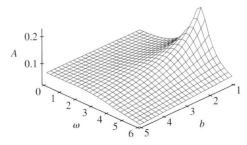

FIGURE 5.17   Amplitude as a function of driving frequency and damping

Analytically, showing that $\lim_{\omega \to \infty} A = 0$ is a routine exercise in elementary calculus. Physically, this means that for high enough driving frequency, the system cannot react to the rapidity of the swings being imposed on it and the system becomes "paralyzed." So if you have a motor shaking itself to pieces because of resonance, either add damping or increase the driving speed!

Finally, we obtain an expression for the resonant frequency as a function of the damping parameter b by again solving the equation

$$\frac{dA}{d\omega} = -\frac{\omega(2\omega^2 + b^2 - 32)}{(\omega^4 + (b^2 - 32)\omega^2 + 256)^{3/2}} = 0 \quad \text{for} \quad \omega_R = \tfrac{1}{2}\sqrt{64 - 2b^2}$$

Formulae for the dependence of the resonant frequency on all three parameters of mass, damping coefficient, and spring constant are obtained next.

DEPENDENCE OF RESONANCE ON m, b, AND k Complete formulae for the dependence of the resonant frequency on mass, damping, and spring constant can be obtained by finding the steady-state (particular) solution of the differential equation $my'' + by' + ky = \cos \omega t$. Assuming the particular solution $y_p = A \cos(\omega t - \phi)$, the differential equation yields

$$-mA\omega^2 \cos(\omega t - \phi) - bA\omega \sin(\omega t - \phi) + kA \cos(\omega t - \phi) = \cos(\omega t)$$

from which, by expanding the trig terms and matching coefficients of $\cos \omega t$ and $\sin \omega t$, we get

$$A = \frac{1}{\sqrt{m^2\omega^4 + (b^2 - 2mk)\omega^2 + k^2}}$$

The resonant frequency, the solution of

$$\frac{dA}{d\omega} = -\frac{\omega(2m^2\omega^2 + b^2 - 2mk)}{(m^2\omega^4 + (b^2 - 2mk)\omega^2 + k^2)^{3/2}} = 0$$

is

$$\omega_R = \sqrt{\frac{k}{m} - \frac{b^2}{2m^2}} = \sqrt{\frac{2mk - b^2}{2m^2}}$$

The first form for ω_R is more prevalent in texts, but the second makes it easier to see that $b^2 < 2mk$ is necessary for resonance.

Resonance Wrap-Up

Here are some questions about resonance and oscillations in damped spring-mass systems. Answering them will help clarify the connections, formulas, and insights generated throughout this unit. The answers are given later in this section.

Questions

- Can overdamped and critically damped systems be made to resonate?
- Can every underdamped system be made to resonate?
- What is the threshold of damping above which there can be no resonance?
- What is the threshold of damping above which there are no transient oscillations?

DATA   To answer these questions we recall formulae for the natural frequency and the resonant frequency. The quadratic formula for the characteristic roots of the characteristic equation gives the natural frequency

$$\lambda = \frac{-b \pm \sqrt{b^2 - 4mk}}{2m} \Rightarrow \omega_N = \sqrt{\frac{4mk - b^2}{4m^2}}$$

From $b^2 - 4mk$, the discriminant of the characteristic equation, we can classify the motion as a function of damping

$$\begin{cases} b^2 = 4mk & \text{critically damped} \\ b^2 > 4mk & \text{overdamped} \\ b^2 < 4mk & \text{underdamped} \end{cases}$$

The resonant frequency is $\omega_R = \frac{\sqrt{2mk - b^2}}{2m^2}$, from which we draw the two inferences

$$0 < b^2 < 4mk \Rightarrow \text{underdamped}$$
$$0 < b^2 < 2mk \Rightarrow \text{resonance}$$

Figure 5.18 summarizes all this data on a b^2-number line.

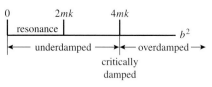

FIGURE 5.18   Which systems can be made to resonate?

Answers

- Overdamped, critically damped, and "half" the underdamped systems cannot be made to resonate. In particular,

$$0 < b^2 < 2mk \Rightarrow \text{underdamped, with resonance possible}$$
$$2mk \leq b^2 < 4mk \Rightarrow \text{underdamped, but resonance not possible}$$

  We quote the word "half" because systems for which $b^2 < 4mk$ are all underdamped but only those for which $b^2 < 2mk$ have a positive resonant frequency. In Figure 5.18 the underdamped systems for which resonance is possible appears to be "half" of the underdamped systems. Thus, the answer to the first question is "no, overdamped and critically damped systems cannot be made to resonate."

- The answer to the second question is "no, not every underdamped system can be made to resonate."

- The answer to the third question is "$b^2 = 2mk$ is the threshold above which a system cannot be made to resonate."

- And the answer to the last question is "$b^2 = 4mk$ is the threshold above which a system cannot be made to oscillate."

EXERCISES 5.11–Part A

A1. For the system governed by the ODE $3y'' + 7y = \cos \omega t$, find

    (a) ω_N, the natural frequency.   **(b)** ω_R, the resonant frequency.

    (c) $y(t)$ for $\omega = \omega_R$.

A2. For the system governed by $3y'' + 3y' + 7y = \cos \omega t$, find the natural and resonant frequencies and the magnitude of the amplitude at resonance.

A3. For the system governed by $3y'' + 6y' + 7y = \cos \omega t$, find the natural and resonant frequencies and the magnitude of the amplitude at resonance.

A4. For the damped oscillator driven by $\cos \omega t$, and with $m = 1$, $b = \frac{1}{11}$, $k = \frac{1}{200}$, find ω_N, ω_R, and the magnitude of the amplitude at resonance. Even in the presence of damping, this system's response at resonance is considerably larger than the magnitude of the driving force.

A5. From first principles, derive a formula for the resonant frequency of the system $my'' + by' + ky = \sin \omega t$. Compare to the formula derived for the damped oscillator driven by $\cos \omega t$.

EXERCISES 5.11–Part B

B1. Show that all three of the following expressions for the resonant frequency are equivalent:

$$\omega_R = \sqrt{\frac{2mk - b^2}{2m^2}} = \frac{\sqrt{4mk - 2b^2}}{2m} = \sqrt{\frac{k}{m} - \frac{b^2}{2m^2}}$$

For the systems governed by the ODEs in Exercises B2–4:

(a) Find the natural frequency.

(b) Find the resonant frequency by going through the steps of finding the magnitude of the steady-state response (the particular solution) and maximizing by differentiation.

(c) Plot a graph of the magnitude of the steady-state amplitude vs. driving frequency ω.

B2. $3y'' + 6y' + 10y = \cos \omega t$ **B3.** $8y'' + 13y' + 25y = \cos \omega t$

B4. $4y'' + 7y' + 11y = \cos \omega t$

B5. If $b^2 < 4mk$, obtain a formula for the natural frequency of the general system governed by the equation $my'' + by' + ky = \cos \omega t$.

B6. Obtain one of the expressions for ω_R given in Exercise B1 by applying the techniques of Exercises B2–4 to the equation in Exercise B5. (In Exercise B7, you will need $A = f(\omega)$, the expression for amplitude, so be sure to isolate this part of the computation and save it.)

B7. In the expression for $A = f(\omega)$ found in Exercise B6, set $m = 8$ and $k = 13$. For each value of $b = 1, 2, \ldots, 8$, obtain the corresponding function $f(\omega)$. Plot all eight functions on the same set of axes, being sure to label each curve with its value of b.

B8. For each of the eight resonance curves in Exercise B7 compute the resonant frequency ω_R and then plot a graph of these resonant frequencies vs. their corresponding values of damping b. This should be equivalent to taking the expression for resonant frequency found in Exercise B6, setting $m = 8$ and $k = 13$ and plotting the resulting expression as a function of b.

B9. In Exercise B6, you computed both $A = f(\omega)$, amplitude at steady state, and ω_R, the frequency at which $f(\omega)$ is maximized.

Obtain an expression for $f(\omega_R)$, the maximum amplitude the system attains at resonance.

B10. With $m = 8$ and $k = 13$ as in Exercise B7, plot, as a function of b for b in the interval $[1, 8]$, the expression for $f(\omega_R)$ found in Exercise B9. This graph shows how the "maximum amplitude at resonance" varies with damping (at fixed mass and spring constant).

B11. With m fixed at 8 and k fixed at 13, plot, as a function of b for b in the interval $[1, 8]$, the natural frequency ω_N. This graph shows how the natural frequency varies with damping (at fixed mass and spring constant).

B12. With m fixed at 8 and k fixed at 13, plot, as a function of b for b in the interval $[1, 8]$, ω_N/ω_R, the ratio of natural frequency to resonant frequency. From the graph, draw a valid engineering conclusion about the relationship between natural frequency and resonant frequency as damping varies at fixed mass and spring constant.

B13. With m fixed at 8 and k fixed at 13, plot the surface $f(\omega, b)$. This surface shows the response amplitude, at steady state, of a forced system with fixed mass and spring constant. The surface captures the dependence of this response amplitude on both damping and input driving frequency.

B14. Given the IVP $x^2 y'' + 3xy' + 6y = x^{-1}$, $y(1) = 2$, $y'(1) = -1$, use a computer algebra system to obtain and plot the solution.

B15. From first principles, provide a solution to the IVP in Exercise B14.

(a) From a fundamental set determined from Exercise 14 part (a), obtain y_h, the homogeneous solution. Verify that your choice for y_h actually works.

(b) Obtain the Wronskian.

(c) Obtain the particular solution by Variation of Parameters.

(d) Form the general solution.

(e) Apply the data. The solution should agree with that found in Exercise B14.

5.12 The Euler Equation

Euler's Differential Equation

Attributed to Euler, the differential equation $t^2 y'' + aty' + by = 0$, $t \neq 0$, is a useful example of a linear, nonconstant-coefficient differential equation with a known explicit solution, namely,

$$y(t) = c_1 t^{1/2(1-a+\sqrt{(a-1)^2-4b})} + c_2 t^{1/2(1-a-\sqrt{(a-1)^2-4b})}$$

In addition, it actually arises in the solution of Laplace's partial differential equation in polar and spherical coordinates (Sections 29.1 and 29.4, explicitly; Section 29.5, implicitly). The solution is a linear combination of two terms, each of which is t, the independent variable, raised to a constant power, as we will shortly verify.

Transformation to Constant-Coefficient Equivalent

For $t > 0$, the change of variables $t = e^x$, that is, $x = \ln t$, converts Euler's equation to a linear, constant-coefficient equation in the new variable $Y(x)$. In particular, this change of variables is implemented by carefully clarifying all variables appearing. Thus, $y(t) = y(t(x)) \equiv Y(x) = Y(x(t))$ or $y(t) = Y(x(t))$, to which we apply differentiation by the chain rule, obtaining

$$\frac{dy}{dt} = \frac{dY}{dx}\frac{dx}{dt} = \frac{dY}{dx}\frac{1}{t}$$

$$\frac{d^2y}{dt^2} = \frac{d}{dt}\left(\frac{dY}{dx}\right)\frac{1}{t} + \frac{dY}{dx}\frac{d}{dt}\left(\frac{1}{t}\right) = \frac{d^2Y}{dx^2}\frac{dx}{dt}\left(\frac{1}{t}\right) + \frac{dY}{dx}\left(-\frac{1}{t^2}\right) = \frac{d^2Y}{dx^2}\left(\frac{1}{t^2}\right) - \frac{1}{t^2}\frac{dY}{dx}$$

Substitution of these expressions for the derivatives transforms the Euler equation to

$$\frac{d^2Y}{dx^2} + (a - 1)\frac{dY}{dx} + bY = 0$$

an equation with constant coefficients. This connection to the constant-coefficient equation explains the solution technique for Euler's equation. Next, we describe the Euler-equation analog to the exponential guess.

Analog of Exponential Guess

If t of the Euler equation is related to x of the constant-coefficient cousin by $t = e^x$, then $e^{\lambda x} = t^\lambda$. Hence, the Euler equation should yield to the "power guess" much like the constant-coefficient equation yielded to the exponential guessof the characteristic equation is then $\lambda^2 + (a - 1)\lambda + b = 0$, obtained from $t^2[\lambda(\lambda - 1)t^{\lambda-2}] + at[\lambda t^{\lambda-1}] + bt^\lambda = t^\lambda[\lambda(\lambda - 1) + a\lambda + b] = 0$, and the analog of the characteristic root is

$$\lambda = \frac{1 - a \pm \sqrt{(a - 1)^2 - 4b}}{2}$$

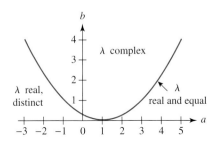

FIGURE 5.19   Discriminant for Euler's equation

Just as the constant-coefficient equation had three classes of solutions corresponding to the discriminant of the characteristic equation, so too can we categorize the solutions of Euler's equation by the discriminant $(a - 1)^2 - 4b$. If $b < \frac{1}{4}(a - 1)^2$, this discriminant is positive and λ is real. The boundary between regions where the discriminant is positive or negative is the parabola $b = b(a) = \frac{1}{4}(a - 1)^2$ in the ab-plane. On this boundary curve, the roots are equal; below it, real; and above it, complex. See Figure 5.19. Corresponding to the three possibilities for the discriminant, we consider the following three examples.

EXAMPLE 5.22   With $a = -4$, $b = 6$ the discriminant is 1, so the roots will be real and distinct. Euler's equation will be $t^2 y'' - 4t y' + 6y = 0$ and the "power guess" $y(t) = t^\lambda$ leads to $\lambda^2 - 5\lambda + 6 = 0$. Consequently, $\lambda = 2, 3$ and $y(t) = c_1 t^2 + c_2 t^3$.   ❖

EXAMPLE 5.23   With $a = -3$, $b = 13$ the discriminant is -36, so the roots will be the complex conjugate pair $\lambda = 2 \pm 3i$ and a fundamental set will contain $t^{2+3i} = t^2 t^{3i}$ and $t^{2-3i} = t^2 t^{-3i}$. Exponentiation of complex numbers, discussed at length in Section 35.6, is defined by

$$t^{i\beta} \equiv e^{i\beta \ln t} = \cos(\beta \ln t) + i\sin(\beta \ln t)$$

$$t^{-i\beta} \equiv e^{-i\beta \ln t} = \cos(\beta \ln t) - i\sin(\beta \ln t)$$

for $t > 0$, so the linear combinations

$$\frac{1}{2}t^{i\beta} + \frac{1}{2}t^{-i\beta} = \cos(\beta \ln t) \quad \text{and} \quad \frac{1}{2i}t^{i\beta} - \frac{1}{2i}t^{-i\beta} = \sin(\beta \ln t)$$

are now real-valued functions. Hence, the general solution to the Euler equation $t^2 y'' - 3ty' + 13y = 0$ can be expressed as $y(t) = c_1 t^2 \cos(3 \ln t) + c_2 t^2 \sin(3 \ln t)$. ❖

EXAMPLE 5.24 If $a = 3, b = 1$ the Euler equation is $t^2 y'' + 3ty' + y = 0$ and the roots $\lambda = -1, -1$ are determined by the "power guess" $y(t) = t^\lambda$. We are thus in the repeated-root case and initially have but the one solution $y_1 = \frac{1}{t}$. A second solution can be found by the reduction of order technique of Section 5.7. However, we should anticipate the result by analogy with the constant-coefficient equation where the solution $e^{\lambda x}$ generates the companion $xe^{\lambda x}$. Since $x = \ln t$, a second independent solution for the Euler equation should be $\ln(t)y_1 = \frac{\ln t}{t}$.

Reduction of order puts $y(t) = u(t)y_1 = \frac{u(t)}{t}$ into the Euler equation, resulting in $tu'' + u' = 0$ for the determination of $u(t)$. Setting $v(t) = u'(t)$ gives the equation $tv' + v = 0$, from which we get $v(t) = \frac{c_1}{t}$ and then $u(t) = c_1 \ln t + c_2$, as expected. ❖

EXERCISES 5.12–Part A

A1. Evaluate each of the following as a floating-point number in the form $a + bi$.

  (a) $4^{-i\sqrt{2}}$ **(b)** $7^{\pi i}$ **(c)** 2^{1-i} **(d)** 3^{-1-2i} **(e)** 5^{2+3i}

A2. For each of the following, write the general solution in terms of real-valued functions.

  (a) $2t^2 y'' - 4ty' + 5y = 0$ **(b)** $3t^2 y'' - 5ty' + 8y = 0$

  (c) $5t^2 y'' - 3ty' + 2y = 0$ **(d)** $4t^2 y'' + 5ty' + y = 0$
  (e) $6t^2 y'' + 2ty' + y = 0$

A3. Obtain a fundamental set for each of the following.

  (a) $t^2 y'' - ty' + y = 0$ **(b)** $t^2 y'' + 3ty' + y = 0$
  (c) $t^2 y'' - 5ty' + 9y = 0$ **(d)** $4t^2 y'' - 8ty' + 9y = 0$
  (e) $9t^2 y'' - 3ty' + 4y = 0$

EXERCISES 5.12–Part B

B1. Use the "power guess" $y(t) = t^\lambda$ to obtain the general solution of the third-order Euler's equation $t^3 y''' + 2t^2 y'' - 14ty' + 24y = 0$.

B2. Using the change of variables $t = e^x$, transform the equation in Exercise B1 to one with constant coefficients. Solve the constant coefficient equation and transform the solution back to $y(t)$, showing it agrees with the results of Exercise B1.

B3. Using each of the following, obtain the transform of the differential equation $t^2 y'' + aty' + by = 0$, where $t > 0$.

  (a) $t = cx, cx > 0$ **(b)** $t = x^2, x > 0$ **(c)** $t = \frac{1}{x}, x > 0$

  (d) $t = \frac{1}{x^2}, x > 0$ **(e)** $t = \cos x, 0 < x < \frac{\pi}{2}$

For the Euler differential equations in Exercises B4–26:

  (a) Obtain the general solution using a computer algebra system.

  (b) Use the "power guess" (and Variation of Parameters where necessary) to obtain the general solution in terms of real functions.

  (c) Make the change of variables $t = e^x$ to obtain a constant-coefficient equation in the variable x. Solve this new differential equation, and then restore the variable t in this general solution, being sure the transformed solution agrees with the solution found in parts (a) and (b).

B4. $t^2 y'' - 2ty' + 2y = 0$ **B5.** $t^2 y'' - ty' + y = 0$
B6. $t^2 y'' - ty' + 2y = 0$ **B7.** $t^2 y'' + 5ty' + 4y = 0$
B8. $t^2 y'' + 3ty' + 5y = 0$ **B9.** $t^2 y'' - 6y = 0$
B10. $6t^2 y'' + 5ty' - y = 0$ **B11.** $4t^2 y'' + y = 0$
B12. $t^2 y'' - 5ty' + 13y = 0$ **B13.** $9t^2 y'' + 3ty' + y = 0$
B14. $t^2 y'' - 3ty' + 8y = 0$ **B15.** $t^2 y'' - 9ty' + 21y = 0$
B16. $4t^2 y'' - 8ty' + 9y = 0$ **B17.** $36t^2 y'' + 13y = 0$
B18. $36t^2 y'' + 12ty' + 13y = 0$ **B19.** $6t^2 y'' + 7ty' - 12y = 0$
B20. $t^2 y'' - 7ty' + 41y = 0$ **B21.** $5t^2 y'' + 7ty' + 2y = 0$
B22. $9t^2 y'' + 39ty' + 25y = 0$ **B23.** $t^2 y'' + 4ty' + 2y = \ln t$
B24. $t^2 y'' + 9ty' + 7y = \ln^2 t$ **B25.** $t^2 y'' + 7ty' + 3y = t^{-2}$
B26. $t^2 y'' - 5ty' + 5y = 1 - t^2$

5.13   The Green's Function Technique for IVPS

Motivation

The solution of the initial value problem $y'(t) = r(t)$, $y(0) = 0$ is given by the integral $y(t) = \int_0^t r(x)\,dx$, which therefore provides a closed-form expression for a class of IVPs. In IVPs of this class, the differential equation is the simple $\frac{d}{dt}y(t) = r(t)$ wherein the differential operator $\frac{d}{dt}$ acts on $y(t)$ on the left of the equation. Writing the solution as an integral shows that it is possible to "invert" the differential operator by an integration, thereby obtaining an explicit representation of $y(t)$.

In this section, we seek to write the solution of a more complex initial value problem such as

$$y'' + 2y' + 10y = r(t) \qquad y(0) = y'(0) = 0 \qquad \textbf{(5.15)}$$

in the form of an integral $y(t) = \int_0^t G(t, x)r(x)\,dx$, where the function $G(t, x)$ is called the *Green's function* (after George Green, 1793–1841) for the given differential equation. The Green's function is the "kernel" of the integral operator that inverts the differential operator defining the left side of the differential equation.

In (5.15), the differential operator acting on $y(t)$ is more complicated than simply $\frac{d}{dt}$. It is the operator $A = \frac{d^2}{dt^2} + 2\frac{d}{dt} + 10$, for which the inverse is no longer just simple integration. The inverse of the operator A is an integral with the function $G(t, x)$ as the "kernel" or "core" of the integrand. Thus, differentiation is again inverted by an integration, but the more complicated differential operator in (5.15) requires a more complicated integration than that which suffices for $y' = r$.

The following example illustrates how to construct the Green's function for the IVP (5.15).

EXAMPLE 5.25   The solution of the IVP (5.15) is

$$y(t) = \frac{e^{-t}}{3}\left(\int_0^t \cos(3u)\sin(3t)e^u r(u)\,du - \int_0^t \cos(3t)\sin(3u)e^u r(u)\,du\right)$$

which is actually just y_p, the particular solution as determined by the Variation of Parameters formula. In fact, the characteristic equation is $\lambda^2 + 2\lambda + 10 = 0$, the characteristic roots are $\lambda = -1 \pm 3i$, the fundamental set is $\{e^{-t}\cos 3t, e^{-t}\sin 3t\}$, the homogeneous solution is $y_h = e^{-t}(A\cos 3t + B\sin 3t)$, the Wronskian is $W = 3e^{-2t}$, and by the Variation of Parameters formula the particular solution can be written as

$$y_p = e^{-t}\sin 3t \int \frac{e^{-t}\cos 3t}{3e^{-2t}} r(t)\,dt - e^{-t}\cos 3t \int \frac{e^{-t}\sin 3t}{3e^{-2t}} r(t)\,dt$$

If we replace the indefinite integrals with definite integrals, we introduce, at worst, multiples of y_1 and y_2 that we can lump in with the homogeneous solution. For example, if

$$\int \frac{y_1(t)}{W(t)} r(t)\,dt = \Phi(t)$$

then

$$y_2(t) \int_0^t \frac{y_1(x)}{W(x)} r(x)\,dx = y_2(t)[\Phi(t) - \Phi(0)] = y_2(t) \int \frac{y_1(t)}{W(t)} r(t)\,dt + Cy_2(t)$$

for some constant C. The term $Cy_2(t)$ can be moved into the homogeneous solution, which already contains a constant multiple of $y_2(t)$. Consequently, we are free to write the particular solution as

$$y_p(t) = y_2(t) \int_0^t \frac{y_1(x)}{W(x)} r(x)\,dx - y_1(t) \int_0^t \frac{y_2(x)}{W(x)} r(x)\,dx$$

and the general solution as

$$y_g = (Ay_1 + By_2) + y_2(t) \int_0^t \frac{y_1(x)}{W(x)} r(x)\,dx - y_1(t) \int_0^t \frac{y_2(x)}{W(x)} r(x)\,dx$$

Applying the initial conditions, we obtain the two equations

$$Ay_1(0) + By_2(0) = 0 \quad \text{and} \quad Ay_1'(0) + By_2'(0) = 0 \qquad (5.16)$$

whose solution is $A = B = 0$ if

$$y_1(0)y_2'(0) - y_1'(0)y_2(0) \neq 0 \qquad (5.17)$$

But the left side of (5.17) is the Wronskian evaluated at $t = 0$. Since y_1 and y_2 are independent, the Wronskian is nonzero, even at $t = 0$. Hence, $A = B = 0$, and the solution to the initial value problem is just the particular solution, which we now write as

$$y(t) = \int_0^t \frac{y_2(t)y_1(x) - y_1(t)y_2(x)}{W(x)} r(x)\,dx$$

This suggests defining the functions

$$g(t, x) = \frac{y_2(t)y_1(x) - y_1(t)y_2(x)}{W(x)} \quad \text{and} \quad G(t, x) = \begin{cases} 0 & \text{if } x > t \\ g(t, x) & \text{if } x < t \end{cases}$$

so we may write the solution to the initial value problem as

$$y(t) = \int_0^t G(t, x)r(x)\,dx$$

Since $x < t$ throughout the interval of integration, $G(t, x) = g(t, x)$ and the integral reproduces the variation of parameters solution for y_p. The function $G(t, x)$ is called the *Green's function* for the initial value problem.   ❖

Properties of the Green's Function

The Green's function for the example of the previous section requires us to form

$$g(t, x) = \tfrac{1}{3} \sin(3(t - x))e^{-(t-x)}$$

and hence

$$G(t, x) = \begin{cases} 0 & \text{if } x > t \\ \tfrac{1}{3} \sin(3(t - x))e^{-(t-x)} & \text{if } x < t \end{cases}$$

Figure 5.20(*a*), showing a graph of the Green's function as a surface over the *tx*-plane, suggests that along the line $t = x$ the function has a sharp edge. The function is continuous, but not differentiable, along this line. Shortly, we will see that for $t \neq x$, $G(t, x)$ is a solution of the differential equation, but along $t = x$ the derivative $\frac{dG}{dt}$ is discontinuous.

EVALUATING $y(t_0)$ To evaluate $y(t)$ at some $t = t_0$, the integral $y(t_0) = \int_0^{t_0} G(t_0, x)r(x)\,dx$ is evaluated by integrating along the line $t = t_0$ from $x = 0$ to $x = t_0$ on the 45°-line $x = t$. The line segment along which the integration takes place lies under the "rounded" portion of the surface shown in Figure 5.20(*a*).

THE DERIVATIVE $\frac{dG}{dt}$ Figure 5.20(*b*) contains a plot of the derivative $\frac{dG}{dt}$, graphed as a surface over the *tx*-plane. This derivative has a finite jump along the line $t = x$. The amount

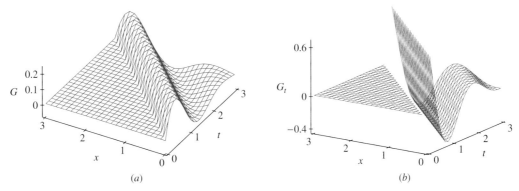

FIGURE 5.20   The Green's function in (*a*) is continuous along the line $t = x$. Its derivative $\frac{dG}{dt}$ along $t = x$ is discontinuous, as shown in (*b*).

of this jump, for this example, is 1. In fact, this jump in the derivative is actually given by the limit from the right less the limit from the left, that is, by

$$\lim_{t \to x^+} \frac{dG}{dt} - \lim_{t \to x^-} \frac{dG}{dt} = 1$$

In the final portion of this section we will generalize this result for an arbitrary second-order linear differential equation $a_0(t)y''(t) + a_1(t)y'(t) + a_2(t)y(t) = r(t)$ to

$$\lim_{t \to x^+} \frac{dG}{dt} - \lim_{t \to x^-} \frac{dG}{dt} = \frac{1}{a_0(x)} \qquad (5.18)$$

$G(t, x)$ **AS A SOLUTION OF THE HOMOGENEOUS DIFFERENTIAL EQUATION**   By direct calculation, it can be shown that over any region that does not include points of the line $x = t$, the Green's function $G(t, x)$ is a solution of the homogeneous differential equation from which it came. Because the derivative $\frac{dG}{dt}$ is discontinuous, $G(t, x)$ cannot be a classical solution of this equation on the rectangle $x > 0, t > 0$. However, the Green's function $G(t, x)$ is a solution of the homogeneous differential equation except along $t = x$, satisfies $G(0, x) = 0$, $\frac{dG}{dt}\big|_{t=0} = 0$, and by (5.18), $\frac{dG}{dt}\big|_{t=x} = \frac{1}{a_0(x)}$. The Green's function is therefore said to be a *generalized solution* of the IVP. (See Section 6.11 for a further discussion of the Green's function.)

Generalization

If $a_0(0) \neq 0$, the general second-order linear initial value problem

$$a_0(t)y''(t) + a_1(t)y'(t) + a_2(t)y(t) = r(t) \qquad y(0) = y'(0) = 0 \qquad (5.19)$$

is solved by

$$y(t) = y_2(t) \int_0^t \frac{y_1(x)}{W(x)a_0(x)} r(x)\,dx - y_1(t) \int_0^t \frac{y_2(x)}{W(x)a_0(x)} r(x)\,dx$$

$$= \int_0^t \frac{y_2(t)y_1(x) - y_1(t)y_2(x)}{W(x)a_0(x)} r(x)\,dx$$

The factor $\frac{1}{a_0(x)}$ enters each denominator when we divide the ODE in (5.19) by $a_0(x)$, making the right side become $\frac{r(t)}{a_0(t)}$. As we did earlier, defining the functions

$$g(t, x) = \frac{y_2(t)y_1(x) - y_1(t)y_2(x)}{W(x)a_0(x)} \quad \text{and} \quad G(t, x) = \begin{cases} 0 & \text{if } x > t \\ g(t, x) & \text{if } x < t \end{cases}$$

lets us write the solution to (5.19) as $y(t) = \int_0^t G(t, x) r(x)\, dx$. Therefore, $G(t, x)$ is the Green's function for (5.19). Computing

$$\lim_{t \to x^+} \frac{dG}{dt} - \lim_{t \to x^-} \frac{dG}{dt} = \frac{1}{a_0(x)}$$

for $G(t, x)$ establishes (5.18).

EXERCISES 5.13–Part A

A1. Given the IVP $y'' + 3y' + 2y = te^{-t}$, $y(0) = y'(0) = 0$:

    (a) Solve by finding $y_g = y_h + y_p$ and applying the initial conditions.

    (b) Solve by finding the Green's function $G(t, x)$ and computing $\int_0^t G(t, x) r(x)\, dx$.

    (c) Show that the solutions in parts (a) and (b) agree.

    (d) Verify (5.18).

A2. Repeat Exercise A1 for the IVP $t^2 y'' - 5ty' + 5y = 1 - t^2$, $y(1) = y'(1) = 0$. In part (b), the solution by Green's function would be $y(t) = \int_1^t G(t, x) \frac{r(x)}{a_0(x)}\, dx$.

EXERCISES 5.13–Part B

B1. If (5.17) holds, show that $A = B = 0$ is the only solution of (5.16).

For initial value problems in Exercises B2–12:

    (a) Use a computer algebra system to obtain the solution.

    (b) Obtain $G(t, x)$, the Green's function.

    (c) Obtain the solution to the IVP from the representation $y(t) = \int_{t_0}^t G(t, x) \frac{r(x)}{a_0(x)}\, dx$.

    (d) Plot the Green's function as a surface over the tx-plane.

    (e) Plot $\frac{dG}{dt}$ as a surface in the tx-plane.

    (f) Show that in accord with (5.18), $\frac{dG}{dt}$ jumps $\frac{1}{a_0(x)}$ across $t = x$.

B2. $y'' + 3y' + 2y = t \sin t$, $y(0) = 0$, $y'(0) = 0$

B3. $y'' + 5y' + 6y = \cos 2t$, $y(0) = 0$, $y'(0) = 0$

B4. $y'' + 4y' + 5y = 1 + t$, $y(0) = 0$, $y'(0) = 0$

B5. $y'' + 4y' + 5y = e^{-t} \cos t$, $y(0) = 0$, $y'(0) = 0$

B6. $y'' + 4y' + 4y = \sin 2t$, $y(0) = 0$, $y'(0) = 0$

B7. $6y'' + 17y' + 12y = 5 \cos 3t$, $y(0) = 0$, $y'(0) = 0$

B8. $15y'' + 13y' + 2y = 4te^{-t} \cos t$, $y(0) = 0$, $y'(0) = 0$

B9. $5t^2 y'' + 7ty' + 4y = 3t^5$, $y(1) = 0$, $y'(1) = 0$

B10. $6t^2 y'' + 5ty' - y = 2t^3$, $y(1) = 0$, $y'(1) = 0$

B11. $6t^2 y'' + 7ty' - 12y = 5t$, $y(1) = 0$, $y'(1) = 0$

B12. $t^2 y'' - 6ty' + 6y = t^4$, $y(1) = 0$, $y'(1) = 0$

Chapter Review

1. In the context of a spring-mass system, interpret each term in the ODE $3y'' + 5y' + 2y = 7 \cos t$.

2. A spring hangs vertically from a support. To its other end is attached a mass. In a coordinate system measuring displacement upward as positive, what initial conditions model the mass

    (a) being released from rest three units above equilibrium?

    (b) being tossed upward after being pulled two units below equilibrium?

    (c) being tossed downward from a position three units above equilibrium?

3. A spring hangs vertically from a support. A weight of 1 lb, attached to the spring, stretches it 3 in, at which point the system is at equilibrium. Formulate an IVP that models the motion of the weight if it is released from rest at a position 3 in above equilibrium.

4. The spring-mass system in Question 3, immersed in an oil bath, experiences a frictional force whose magnitude is one-half the velocity of the weight. Formulate an IVP that models the motion of the weight if it is tossed downward from a position 3 in below equilibrium.

5. What is the characteristic equation for the ODE in Question 1? What are the characteristic roots? What is the homogeneous solution for the equation?

6. Is $t^2 y'' + 3ty' + 4y = 0$ a linear ODE?

7. Form a linear combination of the functions $\cos t$ and $\sin t$. Is this the most general linear combination possible with these two functions?

8. Define linear dependence.

9. Define linear independence.

10. Show that $\{\cos^2 t, \sin^2 t\}$ are *linearly* independent but *functionally* dependent.

11. State Euler's formulas.

12. Express $2e^{3i}$ in the rectangular form $a + bi$.

13. Express $2e^{-3i}$ in rectangular form.

14. Explain how the fundamental set $\{e^{2it}, e^{-2it}\}$ can be replaced with the fundamental set $\{\cos 2t, \sin 2t\}$.

15. Demonstrate how the fundamental set $\{e^{(-2+3i)t}, e^{(-2-3i)t}\}$ can be replaced with the fundamental set $\{e^{-2t}\cos 3t, e^{-2t}\sin 3t\}$.

16. For each of the following, explain the term, give an example of a second-order linear ODE whose solution exhibits the given motion, and explain the connection between the motion and the characteristic roots.

    (a) free overdamped motion

    (b) free underdamped motion

    (c) free critically damped motion

17. The ODE for a spring-mass system has a characteristic root $-2 + 3i$. To the extent possible, find the mass, spring constant, and damping coefficient.

18. The homogeneous solution for a spring-mass system is $y = e^{-t}(\cos 2t - 3\sin 2t)$. To the extent possible, find the mass, spring constant, and damping coefficient.

19. A fourth-order linear ODE with constant coefficients has characteristic roots $-2, -2, -2, -2$. What is its homogeneous solution?

20. A fourth-order linear ODE with constant coefficients has characteristic roots $-1 \pm i, -1 \pm i$. What is its homogeneous solution?

21. An undamped harmonic oscillator is observed to vibrate with an amplitude of 5 and a period of 2. To what extent is the mass and spring constant for the system determined? To what extent is the initial position and initial velocity determined?

22. In each of the following, the solution of an IVP with a homogeneous ODE is given. Find a possible general solution.

    (a) $y = 2te^{-t} - 5\sin t$ (b) $y = 3t\cos 2t - 1$

    (c) $y = t^2 - 2t + 5$ (d) $y = 2\cos 5t + 3\sin 5t$

    (e) $y = 2\cos 5t + 5\sin 3t$ (f) $y = t\cos t + \sin t$

23. In the method of undetermined coefficients, what is the appropriate form for the particular solution in each of the following ODEs?

    (a) $y'' + 2y' + y = te^{-2t}$ (b) $y'' + 2y' + y = 1 + \sin t$

    (c) $y'' + 2y' + y = te^{-t}$ (d) $y'' + 5y' + 6y = \sin 2t + 3\cos 2t$

    (e) $y'' + 5y' + 6y = t^2 + t$ (f) $y'' + 5y' + 6y = 3e^{-t}$

    (g) $y'' + 5y' + 6y = t^2e^{-2t} + 3te^{-3t}$

    (h) $y'' + 2y' + 2y = e^{-2t}\cos t$

    (i) $y'' + 2y' + 2y = e^{-t}\sin 2t$

    (j) $y'' + 2y' + 2y = e^{-t}(3\cos t + 5\sin t)$

24. Give a verbal description of resonance.

25. Give a mathematical description of resonance.

26. Give the resonant and natural frequencies for the systems described by each of the following ODEs.

    (a) $2y'' + 3y' + 4y = f(t)$ (b) $3y'' + 2y' + y = f(t)$

    (c) $4y'' + 5y' + 4y = f(t)$ (d) $2y'' + 3y' + 3y = f(t)$

    (e) $y'' + 2y' + 2y = f(t)$

27. Write a pair of possible characteristic roots for

    (a) a critically damped oscillator.

    (b) an overdamped oscillator.

    (c) an underdamped oscillator.

    (d) an undamped harmonic oscillator.

    (e) an underdamped oscillator experiencing resonance.

    (f) an undamped harmonic oscillator experiencing resonance.

28. Give a verbal description of the term *steady state*. Define it mathematically. Give a physical example of a system in its steady state.

29. Give a verbal description of the term transient phase. Define it in mathematical terms. Give a physical example of a system in its transient phase.

30. Define the terms homogeneous solution, particular solution, and general solution. What is the connection between these three terms.

31. Answer true or false, and give a reason.

    (a) The homogeneous solution of a forced oscillator always decays to zero.

    (b) The homogeneous solution of a damped forced oscillator always decays to zero.

32. The homogeneous solution for $t^2y'' + 4ty' + 2y = \sqrt{t}$ is $y = at^{-1} + bt^{-2}$. Find a particular solution.

33. State the general second-order Euler equation. Transform it to an equation with constant coefficients.

34. Find the Green's function $G(t, x)$ for the IVP $y'' + 3y' + 2y = r(t)$, $y(0) = y'(0) = 0$.

35. Use the Green's function method to solve the IVP in Question 34 if $r(t) = t$.

36. Explain how the Green's function allows us to generalize the integration process which solves the IVP $y' = r(t)$, $y(0) = 0$.

Chapter 6

The Laplace Transform

I N T R O D U C T I O N   The *Laplace transform* is an example of an *integral transform* that will convert a differential equation into an algebraic equation. There are four *operational rules* that relate the transform of derivatives and integrals with multiplication and division. In addition, there are two *shifting laws* that relate multiplication by an exponential with translation. While not strictly necessary, a third shifting law sometimes appearing in the literature is also included.

The Laplace transform is an effective tool for solving constant-coefficient, linear differential equations. In particular, it readily handles equations in which the driving term (forcing function) is piecewise defined. It does this in conjunction with the Second Shifting Law and the Heaviside (unit-step) function, which represents the piecewise-defined function.

The trigonometric functions sine and cosine are common periodic forcing terms in engineering systems. Although we discover the Laplace transform of these two functions almost immediately, there is a special formula for obtaining the transform of a general periodic function. This formula is very useful when solving equations governing spring-mass systems driven by periodic functions for which there is no analytic formula.

Just as the derivative of a product is not the product of the derivatives, so too is the Laplace transform of the product of two functions not the product of the individual transforms. However, the Laplace transform of the *convolution* of two functions, a special type of multiplication for functions, is the product of the individual transforms. This convolution theorem is used to invert some Laplace transforms and to obtain convolutions without evaluating the convolution integral.

The last two sections of the chapter discuss $\delta(x)$, the Dirac delta "function," and its relation to the Green's function. Although a complete description of $\delta(x)$ is well beyond the scope of this text, the object itself proves to be so useful that it is hard to ignore it, even in a first course in differential equations.

The algebraic equation $x^2 + 1 = 0$ has no solution in the real numbers, but inventing the complex numbers allows this equation to have the solutions $x = \pm i$. Similarly, there is no pointwise-defined function $f(x)$ whose Laplace transform is 1. The fabrication of such an object, $\delta(x)$, takes us outside the realm of pointwise-defined functions, and into the domain of "distributions," where the solution of a BVP such as $y'' + 2y' + 10y = \delta(x - t)$, $y(0) = y(1) = 0$, is actually a Green's function.

6.1 Definition and Examples

Introduction

The Laplace transform $F(s)$ is given by the following.

DEFINITION 6.1

$$F(s) = L[f(t)] = \int_0^\infty f(t)e^{-st}\, dt \qquad (6.1)$$

It is an *integral transform* of $f(t)$ in which, by the defining integral, the variable t is "integrated out" so that only s remains. The input to this integral is a function $f(t)$; and the output is a function $F(s)$, called the *Laplace transform* of $f(t)$. We adopt the convention that if the input function is $f(t)$, then its Laplace transform, $L[f(t)]$, will be denoted by $F(s)$. The variable t is typically time, and we speak of $f(t)$ as being defined in the *time domain*. Similarly, we speak of $F(s)$ as being defined in the *transform domain*, sometimes called the *frequency domain*.

Because the upper limit of integration is infinite, the integral is improper and must be evaluated by the limiting process

$$\int_0^\infty f(t)e^{-st}\, dt = \lim_{T\to\infty} \int_0^T f(t)e^{-st}\, dt$$

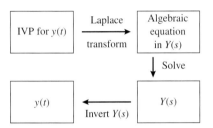

FIGURE 6.1 Schematic: Solution of IVP by Laplace transform

The principle use of the Laplace transform is the conversion of differential equations, into algebraic equations, which are then easier to solve. The schematic in Figure 6.1 diagrams the process for solving an initial value problem. The IVP is transformed to an algebraic equation in $Y(s)$, the Laplace transform of $y(t)$, the solution of the IVP. An algebraic equation is solved for $Y(s)$, and an inversion process maps $Y(s)$ back to $y(t)$. Before giving an example of this process in Section 6.2, we first calculate the following elementary transforms.

EXAMPLE 6.1   The Laplace transform of the function $f(t) = 1$ is an improper integral that converges if $s > 0$. Thus,

$$L[1] = F(s) = \int_0^\infty 1e^{-st}\, dt = \lim_{T\to\infty} \int_0^T e^{-st}\, dt = \frac{1}{s} - \lim_{T\to\infty} \frac{e^{-sT}}{s} = \frac{1}{s}$$

exists because $\lim_{T\to\infty} e^{-sT} = 0$ under the assumption on s. ❖

EXAMPLE 6.2   The Laplace transform of $f(t) = \sin\omega t$ is the function $F(s) = \frac{\omega}{s^2+\omega^2}$, again given by an improper integral that converges if $s > 0$. Thus,

$$L[\sin\omega t] = F(s) = \int_0^\infty \sin(\omega t)e^{-st}\, dt = \lim_{T\to\infty} \int_0^T \sin(\omega t)e^{-st}\, dt$$

$$= \frac{\omega}{s^2+\omega^2} - \lim_{T\to\infty} \frac{e^{-sT}}{s^2+\omega^2}(\omega\cos\omega T + s\sin\omega T) = \frac{\omega}{s^2+\omega^2}$$

exists because $\lim_{T\to\infty} e^{-sT} = 0$ under the assumption on s and $\omega\cos\omega T + s\sin\omega T$ is bounded as a function of T.

Similarly, for $f(t) = \cos\omega t$, we find $F(s) = \frac{s}{s^2+\omega^2}$, provided $s > 0$. Under the assumption on s, the improper integral in

$$L[\cos\omega t] = F(s) = \int_0^\infty \cos(\omega t)e^{-st}\, dt = \lim_{T\to\infty} \int_0^T \cos(\omega t)e^{-st}\, dt$$

$$= \frac{s}{s^2+\omega^2} + \lim_{T\to\infty} \frac{e^{-sT}}{s^2+\omega^2}(\omega\sin\omega T - s\cos\omega T) = \frac{s}{s^2+\omega^2}$$

converges, this time because $\omega \sin \omega T - s \cos \omega T$ is a bounded function of T, and $\lim_{T \to \infty} e^{-sT} = 0$.   ❖

EXAMPLE 6.3   Not every function can be expressed by simple analytic formulas. For example, the piecewise continuous function

$$f(t) = \begin{cases} 1 & \text{if } t \leq 1 \\ 0 & \text{if } t > 1 \end{cases}$$

graphed in Figure 6.2 requires two rules for its definition. Hence, its Laplace transform is given by

$$F(s) = L[f(t)] = \int_0^\infty f(t)e^{-st}\, dt = \int_0^1 1e^{-st}\, dt + \int_1^\infty 0e^{-st}\, dt = \frac{1 - e^{-s}}{s} \quad ❖$$

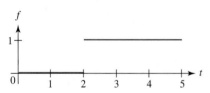

FIGURE 6.2   Piecewise continuous $f(t)$ in Example 6.3

In Section 6.5 we will express such piecewise-defined functions with the *Heaviside function* (after Oliver Heaviside, 1850–1925).

EXAMPLE 6.4   Figure 6.3 is a graph of the piecewise-defined function

$$f(t) = \begin{cases} 0 & \text{if } t \leq 2 \\ 1 & \text{if } t > 2 \end{cases}$$

whose Laplace transform is given by the improper integral

$$F(s) = L[f(t)] = \int_0^\infty f(t)e^{-st}\, dt = \int_0^2 0e^{-st}\, dt + \int_2^\infty 1e^{-st}\, dt$$

$$= \lim_{T \to \infty} \int_2^T e^{-st}\, dt = \frac{e^{-2s}}{s} - \lim_{T \to \infty} \frac{e^{-sT}}{s} = \frac{e^{-2s}}{s}$$

which converges, provided $s > 0$.   ❖

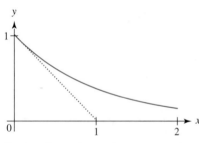

FIGURE 6.3   Piecewise continuous $f(t)$ in Example 6.4

Existence of Laplace Transforms

Not every function has a Laplace transform. Some functions grow too quickly for finite values of t, and some functions grow too quickly at infinity. In particular, the function

$$f_1 = \begin{cases} 0 & t = 0 \\ \dfrac{1}{t} & 0 < t < 1 \\ 0 & 1 \leq t \end{cases}$$

is too large for finite t, whereas the function $f_2 = e^{t^2}$ grows too quickly as t approaches infinity.

Indeed, to compute the Laplace transform of $f_1(t)$ the integral $\int_0^1 (e^{-st}/t)\, dt$ must be evaluated. However, its integrand is unbounded, and it does not exist as a Riemann integral. Figure 6.4 suggests that $e^{-x} \geq 1 - x$ on $x \geq 0$ (the minimum for $g(x) = e^{-x} - (1 - x)$ occurs at $x = 0$ since $g'(0) = 0$ and $g''(0) = 1$), so for $0 \leq t \leq 1$ and $s > 0$ we have $1 - st \leq e^{-st}$. Thus, for $t, s > 0$ we find

$$\int_0^1 \frac{e^{-st}}{t}\, dt > \int_0^1 \frac{1 - st}{t}\, dt = -\lim_{t \to 0^+} \ln(t) - s = \infty$$

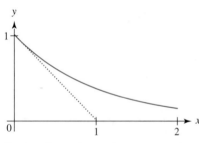

FIGURE 6.4   Comparison of e^{-x} (solid) and $1 - x$ (dotted)

To compute the Laplace transform of $f_2(t)$ the integral $\int_0^\infty e^{t^2} e^{-st}\, dt = \int_0^\infty e^{t(t-s)}\, dt$ must be evaluated. However, for $t > s + \varepsilon$, $\varepsilon > 0$, we find $e^{t(t-s)} > e^{\varepsilon t}$, so $\int_0^\infty e^{t(t-s)}\, dt >$

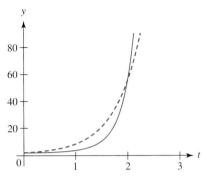

FIGURE 6.5   Comparison of e^{2t} (dashed) and e^{t^2} (solid)

$\int_0^\infty e^{\varepsilon t}\,dt = \infty$; so again, the improper integral, and hence $F_2(s)$, fail to exist. Figure 6.5 (where f_2 is graphed as a solid line and e^{2t} as the dashed line) suggests that eventually $e^{t^2} > e^{\alpha t}$ for any positive α because $e^{t^2} = e^{\alpha t}$ at $t = \alpha$. In fact, for any positive K and α, f_2 grows faster than $Ke^{\alpha t}$ because $\lim_{t\to\infty} f_2/Ke^{\alpha t} = \lim_{t\to\infty} e^{t(t-\alpha)}/K = \infty$.

Functions like f_1 fail to be piecewise continuous on $[0,\infty)$, that is, they fail to be continuous at all but a finite number of points that are either jump or removable discontinuities. Functions like f_2 fail to be of *exponential order;* that is, they fail to be bounded by an exponential of the form $Ke^{\alpha t}, \alpha > 0$. Ruling out these two shortcomings leaves us with those functions that have Laplace transforms. This distillation is formalized in the following existence theorem for Laplace transforms.

THEOREM 6.1

1. $f(t)$ is piecewise continuous on $[0,\infty)$.

2. $f(t)$ is of exponential order ($|f(t)| \le Ke^{\alpha t}$ for some K and $\alpha > 0$)

$$\implies F(s) = L[f(t)] = \int_0^\infty f(t)e^{-st}\,dt \text{ exists for } s > \alpha.$$

Proof

If $f(t)$ is of exponential order, then $|f(t)| \le Ke^{\alpha t}$ for $t > T$ and some $\alpha > 0$. Consequently,

$$\left| \int_0^\infty f(t)e^{-st}\,dt \right| \le \int_0^\infty |f(t)|e^{-st}\,dt = \int_0^T |f(t)|e^{-st}\,dt + \int_T^\infty |f(t)|e^{-st}\,dt$$

On the finite interval $[0, T]$, $f(t)$ is bounded by a constant M because of piecewise continuity. On the unbounded interval $[T, \infty)$, by the assumption of exponential order, it is bounded by $Ke^{\alpha t}$. Hence, we have

$$\left| \int_0^\infty f(t)e^{-st}\,dt \right| \le \int_0^\infty |f(t)|e^{-st}\,dt \le \int_0^T Me^{-st}\,dt + \int_T^\infty Ke^{\alpha t}e^{-st}\,dt$$

$$= M\frac{1 - e^{-sT}}{s} + K\frac{e^{T(\alpha - s)}}{s - \alpha}$$

where, for $s > \alpha$, each integral on the right is finite and therefore exists.

The conditions of piecewise continuity and exponential order are sufficient to guarantee the existence of the Laplace transform but not necessary. The function $f(t) = \frac{1}{\sqrt{t}}$ is not piecewise continuous on $[0, \infty)$ because $\lim_{t\to 0^+} f(t) = \infty$. Yet, its Laplace transform $F(s) = \frac{\sqrt{\pi}}{\sqrt{s}}$ exists.

EXERCISES 6.1–Part A

A1. Does $f(t) = t^{-2}, t > 0$, have a Laplace transform? Explain.

A2. If $L[1] = \frac{1}{s}$, what is $L[3]$? Explain.

A3. Without integrating, obtain $L[\cos(2t - 1)]$. *Hint:* Expand the trigonometric function.

A4. Without integrating, obtain $L[3\sin(5t + \frac{\pi}{4})]$.

A5. If $F(s) = \frac{1}{s^2+2}$, find $f(t)$.   **A6.** If $F(s) = \frac{3s}{s^2+2}$, find $f(t)$.

A7. If $F(s) = \frac{s+2}{s^2+5}$, find $f(t)$.

EXERCISES 6.1–Part B

B1. Use the definition (6.1) to obtain the Laplace transform of

    (a) $f(t) = t$ **(b)** $f(t) = e^{2t}$

For the functions given in Exercises B2–11, obtain the Laplace transforms

    (a) using the definition (6.1).

    (b) using a computer algebra system; be sure to reconcile the answer with the results in part (a).

B2. $f(t) = 6t - 13$ **B3.** $f(t) = 5t^2 + 3t - 4$

B4. $f(t) = 2te^{-3t}$ **B5.** $f(t) = 3t^2 \sin 3t$

B6. $f(t) = 4e^{-3t} \cos 5t$ **B7.** $f(t) = \sqrt{t}$

B8. $f(t) = 2te^{-3t} \cos \dfrac{t}{2}$ **B9.** $f(t) = \sin t \cos t$

B10. $f(t) = \cos^2 t$ **B11.** $f(t) = t + \sin t$

In Exercises B12–16, a function $f(t)$ is depicted by a graph in which the rightmost segment extends to infinity. For each:

    (a) Construct an analytic formula for the function.

    (b) Obtain the Laplace transform by applying the integral definition.

B12. See Figure 6.6.

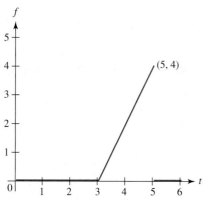

FIGURE 6.6   Exercise B12

B13. See Figure 6.7.

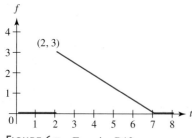

FIGURE 6.7   Exercise B13

B14. See Figure 6.8.

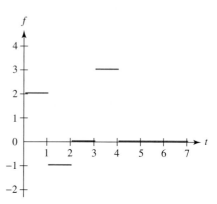

FIGURE 6.8   Exercise B14

B15. See Figure 6.9.

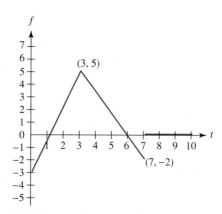

FIGURE 6.9   Exercise B15

B16. See Figure 6.10.

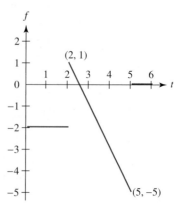

FIGURE 6.10   Exercise B16

6.2 Transform of Derivatives

Laplace Transform of Derivatives

As depicted in Figure 6.1, our primary interest in the Laplace transform is its use in solving initial value problems. Hence, we need to know how to take the transform of a derivative. Now in Section 6.1 we discovered that $L[\sin t] = \frac{1}{s^2+1}$ and $L[\cos t] = \frac{s}{s^2+1}$. The functions $\sin t$ and $\cos t$ are related by differentiation, and their transforms by a factor of s. Specifically, we note

$$L\left[\frac{d}{dt}\sin t\right] = L[\cos t] = \frac{s}{s^2+1} = s\left(\frac{1}{s^2+1}\right) = sL[\sin t]$$

$$L\left[\frac{d}{dt}\cos t\right] = L[-\sin t] = -\frac{1}{s^2+1} = s\left(\frac{s}{s^2+1}\right) - 1 = sL[\cos t] - \cos(0)$$

which are both manifestations of the *operational law*

$$L[y'(t)] = sL[y(t)] - y(0)$$

Thus, differentiation in the time domain is related to multiplication by s in the transform domain (also called the frequency domain). The Laplace transform of the derivative $y'(t)$ is essentially s times the transform of $y(t)$. Of course, there's a detail, namely, the subtraction of the value $y(0)$, but the basic connection is: differentiation of the function is related to multiplication of the function's transform by s.

The following display, in which $Y(s)$ is the transform of $y(t)$, articulates this rule for the first and higher derivatives. We make the observation that "the powers on s descend, while the order of the derivatives ascends" before moving to a derivation based on integration by parts

$$L[y'(t)] = sY(s) - y(0)$$
$$L[y''(t)] = s^2Y(s) - sy(0) - y'(0)$$
$$L[y'''(t)] = s^3Y(s) - s^2y(0) - sy'(0) - y''(0)$$
$$L[y^{(4)}(t)] = s^4Y(s) - s^3y(0) - s^2y'(0) - sy''(0) - y'''(0)$$

DERIVATION Integration by parts, applied to a definite integral, takes the form

$$\int_a^b u\,dv = uv]_a^b - \int_a^b v\,du \tag{6.2}$$

Taking $u = s^{-st}$ and applying (6.2) to the integral defining the Laplace transform of the derivative $y'(t)$, we get

$$L[y'(t)] = \int_0^\infty y'(t)e^{-st}\,dt = e^{-st}y(t)]_0^\infty - \int_0^\infty -sy(t)e^{-st}\,dt$$

$$= \left[\lim_{t\to\infty} se^{-st}y(t) - y(0)\right] + sY(s) = sY(s) - y(0)$$

Since the integral defining a Laplace transform is improper, we have evaluated the boundary term at infinity by taking the appropriate limit, which exists under the assumption that $y(t)$ is of exponential order. In particular, $|y(t)| < Ke^{\alpha t}$ for $t > T$ means $|e^{-st}y(t)| < Ke^{-t(s-\alpha)}$, so the limit in question is zero provided $s > \alpha$.

Transform Solution to an Initial Value Problem

As motivation for studying the Laplace transform, we solve the initial value problem $y'' + y = 1$, $y(0) = y'(0) = 1$ by Laplace transform techniques. We begin by taking the transform of both sides of the differential equation, obtaining, in light of the result on transforming derivatives,

$$s^2 Y(s) - sy(0) - y'(0) + Y(s) = \frac{1}{s} \quad \text{and hence} \quad (s^2 - 1)Y(s) - s - 1 = \frac{1}{s}$$

The unknown in this equation is now $Y(s) = L[y(t)]$, the Laplace transform of the unknown function $y(t)$. So, instead of having to solve a differential equation to determine $y(t)$, we need only solve an algebraic equation for $Y(s)$, getting

$$Y(s) = \frac{s^2 + s + 1}{s(s^2 + 1)} = \frac{1}{s} + \frac{1}{s^2 + 1} \tag{6.3}$$

where the fractions on the right are obtained by a partial fraction decomposition. (See the exercises for a review of this technique.) By inspection and memory, we notice that we have already met both of the transforms, the first being the transform of 1 and the second the transform of $\sin t$. Hence, (6.3) determines $y(t) = 1 + \sin t$.

That is the basic idea behind using the Laplace transform for solving initial value problems. Clearly, the more transforms we know, and the more we know about the behavior of the Laplace transform in general, the greater the range of initial value problems we can solve with this technique. So, it is to these objectives we now turn.

Transform of t^n

Recall that we have already obtained, by direct integration, $L[t^0] = L[1] = \frac{1}{s}$. Using the rule for the transform of a derivative, we can determine the transforms of $t, t^2, t^3, \ldots$, eventually generalizing to the transform of t^n. Table 6.1 shows how the transform of t^k, $k = 1, \ldots, 4$, can be recursively computed from the transform of t^{k-1}.

$$\frac{1}{s} = L[1] = L\left[\frac{d}{dt}(t)\right] = sL[t] \Rightarrow L[t] = \frac{1}{s^2}$$

$$\frac{1}{s^2} = L[t] = L\left[\frac{1}{2}\frac{d}{dt}(t^2)\right] = \frac{s}{2}L[t^2] \Rightarrow L[t^2] = \frac{2 \cdot 1}{s^3}$$

$$\frac{2 \cdot 1}{s^3} = L[t^2] = L\left[\frac{1}{3}\frac{d}{dt}(t^3)\right] = \frac{s}{3}L[t^3] \Rightarrow L[t^3] = \frac{3 \cdot 2 \cdot 1}{s^4}$$

$$\frac{3 \cdot 2 \cdot 1}{s^4} = L[t^3] = L\left[\frac{1}{4}\frac{d}{dt}(t^4)\right] = \frac{s}{4}L[t^4] \Rightarrow L[t^4] = \frac{4 \cdot 3 \cdot 2 \cdot 1}{s^5}$$

TABLE 6.1   Transforms of t^n, $n = 0, \ldots, 3$

The generalization for n, a nonnegative integer, is then inferred to be

$$L[t^n] = \frac{n!}{s^{n+1}}$$

a result that can be proved by induction. A further generalization to the case of $n = a \geq 0$

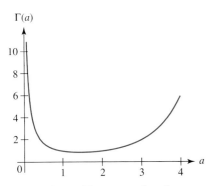

FIGURE 6.11   The gamma function $\Gamma(a), a > 0$

$f(t)$	$F(s)$
1	$\dfrac{1}{s}, s > 0$
t	$\dfrac{1}{s^2}, s > 0$
t^2	$\dfrac{2}{s^3}, s > 0$
t^n	$\dfrac{n!}{s^{n+1}}, s > 0$
$\sin \omega t$	$\dfrac{\omega}{s^2 + \omega^2}, s > 0$
$\cos \omega t$	$\dfrac{s}{s^2 + \omega^2}, s > 0$
e^{at}	$\dfrac{1}{s - a}, s > a$
$\sinh \omega t$	$\dfrac{\omega}{s^2 - \omega^2}, s > \omega$
$\cosh \omega t$	$\dfrac{s}{s^2 - \omega^2}, s > \omega$

TABLE 6.2   Table of Laplace transforms

would be

$$L = [t^a] = \frac{\Gamma(a+1)}{s^{a+1}} \tag{6.4}$$

where, for $a > 0$,

$$\Gamma(a) = \int_{0^+}^{\infty} e^{-z} z^{a-1}\, dz \tag{6.5}$$

is the *gamma function,* which, because $\Gamma(a+1) = a\Gamma(a)$, reduces to the factorial function when a is the positive integer n. Indeed, since (6.5) gives $\Gamma(1) = 1$, when n is a positive integer we have

$$\Gamma(n+1) = n\Gamma(n) = n(n-1)\Gamma(n-2) = \cdots = n(n-1)\cdots(1 \times \Gamma(1)) = n!$$

Figure 6.11 contains a graph of $\Gamma(a)$ for $a > 0$.

Finally, recall that in Section 6.1 we saw the transform $L[t^{-1/2}] = \frac{\sqrt{\pi}}{\sqrt{s}}$, which we can now obtain from (6.4) by taking $n = -\frac{1}{2}$ and using $\Gamma(\frac{1}{2}) = \sqrt{\pi}$. Thus, $L[t^{-1/2}] = \frac{\Gamma(-1/2+1)}{s^{-1/2+1}} = \frac{\Gamma(1/2)}{s^{1/2}} = \frac{\sqrt{\pi}}{\sqrt{s}}$.

Transform of e^{at}

The Laplace transform of $f(t) = e^{at}$ is determined from the defining integral

$$L[e^{at}] = \int_0^{\infty} e^{at} e^{-st}\, dt = \int_0^{\infty} e^{-(s-a)t}\, dt = \frac{1}{s-a}$$

either by fully evaluating the integral or by recognizing that with $s - a$ treated as s, it is the same as the integral for the transform of 1. Since $\int_0^{\infty} e^{-st}\, dt = \frac{1}{s}$, it is immediate that $\int_0^{\infty} e^{-(s-a)t}\, dt = \frac{1}{s-a}$.

We can immediately determine the transforms of $\sinh(\omega t)$ and $\cosh(\omega t)$ as follows:

$$L[\sinh(\omega t)] = L\left[\frac{e^{\omega t} - e^{-\omega t}}{2}\right] = \frac{1}{2}\left[\frac{1}{s-\omega} - \frac{1}{s+\omega}\right] = \frac{\omega}{s^2 - \omega^2}$$

$$L[\cosh(\omega t)] = L\left[\frac{e^{\omega t} + e^{-\omega t}}{2}\right] = \frac{1}{2}\left[\frac{1}{s-\omega} + \frac{1}{s+\omega}\right] = \frac{s}{s^2 - \omega^2}$$

Table of Transforms

Table 6.2 is a summary of the transforms we have learned.

EXERCISES 6.2–Part A

A1. To obtain the partial fraction decomposition in (6.3), write $\frac{s^2+s+1}{s(s^2+1)} = \frac{A}{s} + \frac{Bs+C}{s^2+1}$ and multiply through by $s(s^2 + 1)$ to obtain $s^2 + s + 1 = A(s^2 + 1) + s(Bs + C) = (A + B)s^2 + Cs + A$, which must be an identity in s. Matching coefficients of like powers of s gives the three equations $A + B = 1, C = 1, A = 1$, whose solution is $A = 1, B = 0, C = 1$. To obtain the partial fraction decomposition of each of the following fractions. (A repeated factor s^2 in the denominator requires $\frac{A}{s}$ as well as $\frac{B}{s^2}$.)

(a) $\dfrac{5s^2 - s - 9}{s(s^2 - 3)}$ (b) $\dfrac{6s^2 + 5s + 14}{s(s^2 + 7)}$

(c) $\dfrac{s^2 + 1}{s^2(s^2 + 4)}$ (d) $\dfrac{s^3 + 2s^2 + 4s + 2}{s^2(s^2 + 4)}$

A2. Use (6.5) to verify that $\Gamma(1) = 1$.

A3. Use (6.5) and integration by parts to verify $\Gamma(a + 1) = a\Gamma(a), a > 0$.

A4. Use the result in Exercises A2 and 3 to verify that $\Gamma(10) = 9!$.

A5. For each of the following IVPs, obtain $Y(s)$, the Laplace transform of the solution $y(t)$.

  (a) $3y'' + 4y' + 5y = t$, $y(0) = 1$, $y'(0) = -1$

  (b) $2y'' + 3y' + 6y = e^{-2t}$, $y(0) = 2$, $y'(0) = -3$

  (c) $5y'' + 8y' + 3y = \cos\sqrt{2}t$, $y(0) = -2$, $y'(0) = 1$

A6. By inspection, determine $f(t)$, the inverse Laplace transform for each of the following. *Hint:* where applicable, $\frac{a+b}{c+d} = \frac{a}{c+d} + \frac{b}{c+d}$.

  (a) $F(s) = \dfrac{2s+3}{s^2+5}$ (b) $F(s) = \dfrac{5-4s}{s^2-4}$ (c) $F(s) = \dfrac{3}{s+4}$

EXERCISES 6.2–Part B

B1. Using the well-known properties of the integral, show that
$L[af(t) + bg(t)] = aL[f(t)] + bL[g(t)]$.

In Exercises B2–6, obtain the Laplace transform $F(s)$ using

  (a) a computer algebra system's built-in command.

  (b) the linearity of the transform and the special properties of each function, not the definition (6.1).

  (c) the definition (6.1).

B2. $f(t) = \sin(3t - 5)$ *Hint:* Use a trigonometric formula for the sum/difference of two angles.

B3. $f(t) = 4\cos(2t + 7)$

B4. $f(t) = 3e^{4t+5}$ *Hint:* Use a special property of the exponential function.

B5. $f(t) = \sin^2(3t - 1)$ *Hint:* Use a half-angle formula.

B6. $f(t) = \cos^2(5t + 2)$

In Exercises B7–9, invert each Laplace transform $F(s)$ by

  (a) using a computer algebra system's built-in functionality.

  (b) applying pattern recognition to an algebraic rearrangement of $F(s)$.

B7. $F(s) = \dfrac{5s-7}{s^2+4}$ **B8.** $F(s) = \dfrac{3s+4}{s^2+9}$ **B9.** $F(s) = \dfrac{1-2s}{s^2-25}$

In Exercises B10–13, for the given function $f(t)$:

  (a) Obtain the Laplace transform $F(s)$.

  (b) Obtain the second derivative $f''(t)$.

  (c) Obtain the transform of $f''(t)$.

  (d) Demonstrate the validity of the formula $L[f''(t)] = s^2 F(s) - sf(0) - f'(0)$.

B10. $f(t) = te^{-3t}$ **B11.** $f(t) = t\sin 5t$

B12. $f(t) = te^{2t}\cos 3t$ **B13.** $f(t) = 4t^3 + 5t^2 - 7t - 9$

In Exercises B14–16, solve the given IVP by

  (a) obtaining the Laplace transform of the ODE, solving for the transform of the unknown solution, and inverting.

  (b) using a symbolic differential equation solver in a computer algebra system. Show that the two solutions agree.

B14. $y'' + 4y' + 4y = 2\sin 3t$, $y(0) = -2$, $y'(0) = 3$

B15. $y'' + 6y' + 9y = 3\cos 5t$, $y(0) = 1$, $y'(0) = -2$

B16. $y'' + 4y' + 13y = 5e^{-2t}$, $y(0) = 3$, $y'(0) = -1$

In Exercises B17–19, obtain $f(t)$ by

  (a) using a computer algebra system's built-in functionality.

  (b) applying pattern recognition to a partial fraction decomposition to the given $F(s)$.

B17. $F(s) = \dfrac{2s^4 + 15s + 45}{s^4(s+3)}$ **B18.** $F(s) = \dfrac{5s^2 - 6s + 57}{(s-2)(s^2+9)}$

B19. $F(s) = \dfrac{7s^4 + 6s^2 + 54}{s^3(s^2+9)}$

6.3 First Shifting Law

Operational Laws

The success of transform techniques in solving initial value problems hinges on their *operational* properties. Rules that govern how operations in the time domain translate to operations in the transform domain are called *operational laws*. We have already met one such law, namely, the rule for computing the transform of a derivative. The rule $L[y'(t)] = sL[y(t)] - y(0)$ is the first of *four* operational laws governing differentiating and integrating transforms and multiplying and dividing transforms by t. Section 6.4 is devoted to a study of all four of these operational laws.

However, before studying them, we examine the first of three *shifting* laws. It may be argued that the shifting laws are also operational rules, but we will reserve the phrase *operational laws* for the four laws linking multiplication and differentiation, division and integration.

In this section we examine the first of three shifting laws, the law that shows how to take the transform of a product in which one of the factors is an exponential. This **First Shifting Law,**

$$L[e^{at} f(t)] = F(s - a) \tag{6.6}$$

says that if you know $F(s)$ is the transform of $f(t)$, then you know the transform of $e^{at} f(t)$. It's just $F(s - a)$. Thus, multiplication by an exponential in the time domain is a shift in the transform domain.

EXAMPLE 6.5   The Laplace transform of $f(t) = \sin 3t$ is $F(s) = \frac{3}{s^2+9}$, so the transform of $e^{-2t} \sin 3t$ is $F(s - (-2)) = F(s + 2) = \frac{3}{(s+2)^2+9}$. ❖

EXAMPLE 6.6   The Laplace transform of $f(t) = \cosh 5t$ is $F(s) = \frac{s}{s^2-25}$, so the transform of $e^{-2t} \cosh 5t$ is $F(s + 2) = \frac{s+2}{(s+2)^2-25}$. ❖

EXAMPLE 6.7   Although we discovered the transform $L[e^{at}] = \frac{1}{s-a}$ by direct integration, it can also be obtained by applying (6.6) to the function $1 \times e^{at}$. With $f(t) = 1$ so $F(s) = \frac{1}{s}$, $L[1 \times e^{at}] = F(s - a) = \frac{1}{s-a}$. We see, then, there are sometimes several ways to obtain a given transform. ❖

EXAMPLE 6.8   This final example illustrates the use of the First Shifting Law "in reverse," that is, in service of *inverting* the transform $F(s) = \frac{s}{s^2+2s+3}$. We represent the inversion operation with notation $L^{-1}[F(s)] = f(t)$ and call $f(t)$ the *inverse Laplace transform* of $F(s)$. Although there is an integration process that produces this inversion (see Section 36.2), it is generally difficult to evaluate the integrals involved. Instead, we use a process of pattern recognition.

For example, no transform in Table 6.2 contains a complete quadratic in the denominator. The denominators in the transforms of the trigonometric and hyperbolic functions are of the form $s^2 \pm \omega^2$ and are not the complete quadratic form $s^2 + as + b$. Now, how can this complete quadratic form arise? Certainly if $s \to s - a$ in a denominator of the form $s^2 \pm \omega^2$, that is, if the First Shifting Law had been used to obtain the given transform. Thus, we are inspired to complete the square in the denominator of the given transform $F(s)$, hoping to see evidence of the shift of from s to $s - a$.

Thus, we write

$$F(s) = \frac{s}{(s+1)^2 + 2} = \frac{(s+1) - 1}{(s+1)^2 + 2} = \frac{s+1}{(s+1)^2 + 2} - \frac{1}{(s+1)^2 + 2}$$

where, after completing the square, we made sure *every* appearance of an s was in the form of $s + 1$ because we wanted $F(s)$ to have the form $\hat{F}(s + 1)$ for some new function $\hat{F}(S)$. Here, this was done by simply adding and subtracting 1 in a single location. We could also have achieved

$$\hat{F}(S) = \frac{S}{S^2 + 2} - \frac{1}{S^2 + 2}$$

by substituting $s \to S - 1$ in $F(s)$.

The function $\hat{F}(S)$ is inverted as a function of S, yielding $\hat{f}(t) = \cos\sqrt{2}t - \frac{1}{\sqrt{2}}\sin\sqrt{2}t$. This is the function that was multiplied by an exponential so that the First Shifting Law could be invoked in producing the original $F(s)$. Hence, the inverse of $F(s)$ is $f(t) = e^{-t}\hat{f}(t) = e^{-t}(\cos\sqrt{2}t - \frac{1}{\sqrt{2}}\sin\sqrt{2}t)$.

An alternative approach based on a partial fraction decomposition can also be effective. Here, the denominator factors over the complex field, a factorization favored in electrical and controls engineering. Thus, write

$$F(s) = \frac{\frac{1}{2} + \frac{i\sqrt{2}}{4}}{s + 1 - i\sqrt{2}} + \frac{\frac{1}{2} - \frac{i\sqrt{2}}{4}}{s + 1 + i\sqrt{2}}$$

and recognize the transform of an exponential in each fraction, obtaining

$$f(t) = \left(\frac{1}{2} + \frac{i\sqrt{2}}{4}\right)e^{(-1+i\sqrt{2})t} + \left(\frac{1}{2} - \frac{i\sqrt{2}}{4}\right)e^{(-1-i\sqrt{2})t}$$

$$= \frac{e^{-t}}{4}[(2+i\sqrt{2})(\cos\sqrt{2}t + i\sin\sqrt{2}t) + (2-i\sqrt{2})(\cos\sqrt{2}t - i\sin\sqrt{2}t)]$$

$$= e^{-t}\left(\cos\sqrt{2}t - \frac{1}{\sqrt{2}}\sin\sqrt{2}t\right) \qquad ❖$$

Justification

The validity of the First Shifting Law follows upon examining

$$L[e^{at}f(t)] = \int_0^\infty e^{at}f(t)e^{-st}\,dt = \int_0^\infty f(t)e^{-(s-a)t}\,dt = F(s-a)$$

where, in the second integral, the quantity $s - a$ becomes the argument of the transform of $f(t)$. Clearly, $s > a$ is required for this transform to exist.

EXERCISES 6.3–Part A

A1. For each of the following functions $f(t)$, use the First Shifting Law to obtain $F(s)$, the Laplace transform.

(a) $f(t) = t^2 e^{-3t}$ (b) $f(t) = e^{3-2t}\cos 4t$

(c) $f(t) = e^{-2+4t}\sin 3t$

A2. For each of the following functions $f(t)$, obtain the Laplace transform $F(s)$ using the First Shifting Law.

(a) $f(t) = e^{-3t}\cosh 4t$ (b) $f(t) = e^{2t}\sinh 5t$

A3. Obtain the Laplace transforms of the functions in Exercise A2 by first expressing each function in exponential form and then using the rule for the transform of an exponential. Be sure both solutions agree.

A4. Invert each of the following Laplace transforms using the First Shifting Law.

(a) $F(s) = \frac{1}{s^2 + 10s + 25}$ (b) $F(s) = \frac{s-3}{s^2 - 6s + 11}$

(c) $F(s) = \frac{e(s-2)}{s^2 - 4s + 1}$ (d) $F(s) = \frac{2e^{-1}}{s^2 + 2s - 3}$

(e) $F(s) = \frac{\pi e}{s^2 + 4s + 4 + \pi^2}$

In Exercises A5–7, obtain the inverse Laplace transform in each of two ways.

(a) Perform a partial fraction decomposition, then invert the individual fractions.

(b) Complete the square in the denominator, and use the First Shifting Law.

(c) Show that both methods give the same result.

A5. $F(s) = \frac{3s+4}{(s+3)(s+1)}$ **A6.** $F(s) = \frac{5s-9}{(s-1)(s-5)}$

A7. $F(s) = \frac{7s-8}{(s+6)(s-4)}$

EXERCISES 6.3–Part B

B1. Using the First Principles of exponential guess, characteristic equation, characteristic roots, fundamental set, homogeneous solution, Undetermined Coefficients, particular solution, and general solution, for example, solve the IVP $y'' + 2y' + 2y = te^{-3t}$, $y(0) = -1$, $y'(0) = 2$. Check your answer with a computer algebra system. Plot the solution.

B2. Solve the IVP in Exercise B1 using the Laplace transform. Use a computer algebra system as needed.

In Exercises B3–6, obtain $F(s)$, the Laplace transform of the given $f(t)$ by

   (a) applying the First Shifting Law.

   (b) using a computer algebra system's built-in functionality.

B3. $f(t) = e^{7t} \sinh 3t$ **B4.** $f(t) = e^{-3t+5} \cosh 2t$

B5. $f(t) = e^{-5t} \cos(7t - 3)$ **B6.** $f(t) = t^3 e^{5t}$

In Exercises B7–10, find $f(t)$, the inverse Laplace transform of the given $F(s)$, by

   (a) applying the First Shifting Law.

   (b) using a computer algebra system's built-in functionality.

B7. $F(s) = \dfrac{7}{s + 6}$ **B8.** $F(s) = \dfrac{s + 5}{s^2 + 6s - 2}$

B9. $F(s) = \dfrac{4s - 3}{s^2 + 8s - 5}$ **B10.** $F(s) = \dfrac{s}{(s - 5)(s + 2)}$

B11. Work Exercise B10 by splitting $F(s)$ into two fractions via a partial fraction decomposition. Invert each fraction. Show the answer is equivalent to the answer found using the First Shifting Law.

B12. Given the graph of the piecewise-defined function $f(t)$ shown in Figure 6.12, obtain its Laplace transform by writing a formula for $f(t)$ and then applying the integral definition of the Laplace transform. The rightmost segment of $f(t)$ extends to infinity.

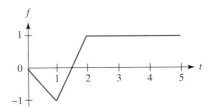

FIGURE 6.12   Exercise B12

6.4   Operational Laws

The Four Little (Operational) Laws

We have already seen the first of four operational laws governing the behavior of the Laplace transform. This first law, the rule for the transform of the derivative, showed that, to within a detail, taking a derivative in the time domain was equivalent to multiplication of the transform by s in the transform domain.

The First Shifting Law is another example of an operational law. We will discuss two other shifting laws, one in Section 6.5 and one in Section 6.6. Here, we concentrate on the paradigm of the derivative law: *differentiation* in the time domain becomes *multiplication* by s in the transform domain.

The four operations of differentiation, integration, multiplication, and division pair up in a curiously symmetric way, making up four operational rules that are best learned, and remembered, by the symmetry in the rules. This approach to the operational rules was presented to the author by Professor Syl Pagano of the Math Department of the University of Missouri-Rolla in 1965.

To within a *detail* in each case, differentiation and multiplication pair up in either direction, as do integration and division. The precise statement of the resulting four operational laws is given in Table 6.3. The detail in each instance can be seen in the statement of the rule. The rule for the transform of the derivative is stated for $y''(t)$, the *second* derivative, because the second derivative most often appears in the damped oscillator problems of science and engineering. The symbol $Y(s)$ is synonymous with $L[y(t)]$.

The detail for "multiply by $t \Rightarrow$ differentiate in s" is the minus sign needed in front of the derivative. The detail needed for "divide the transform by s" is "integrate $y(t)$ from

$$L[y''] = s^2 Y(s) - sy(0) - y'(0) \qquad L\left[\int_0^t y(x)\,dx\right] = \frac{1}{s}Y(s)$$

$$L[ty(t)] = -\frac{d}{ds}Y(s) \qquad L\left[\frac{y(t)}{t}\right] = \int_s^\infty Y(x)\,dx$$

TABLE 6.3   Formal definitions of the Four Operational Laws

0." The detail needed for "divide the function by t" is to integrate the transform from s to infinity.

A terse distillation of these operational rules would be to remember the pairings

$$(\text{differentiate} \Leftrightarrow \text{multiply}) \quad \text{and} \quad (\text{integrate} \Leftrightarrow \text{divide})$$

which are elucidated by the four lines in Table 6.4. If differentiation is applied in the time domain, then multiplication is applied in the transform domain. If multiplication is performed in the time domain, then differentiation is applied in the transform domain. Similarly, if integration is performed in the time domain, then division is performed in the transform domain. If division is performed in the time domain, then integration is performed in the transform domain.

Time Domain	Transform Domain
Transform a derivative	Multiply transform by s
Multiply function by t	Differentiate transform
Transform an integral	Divide transform by s
Divide function by t	Integrate transform

TABLE 6.4   Symmetries in the Four Operational Laws

EXAMPLE 6.9   We have already inferred the rule $L[t^n] = n!/s^{n+1}$ from the operational law for the transform of a derivative. We show here how this same result could be obtained by induction based on the operation law "multiplication by t becomes differentiation in s." A proof by induction begins with the validation of the claim for the first few cases, given in Table 6.5, and then shows the truth of the general case based on the truth of the preceding case.

$$L[t] = L[t \times 1] = -\frac{d}{ds}L[1] = -\frac{d}{ds}\frac{1}{s} = \frac{1}{s^2}$$

$$L[t^2] = L[t \times t] = -\frac{d}{ds}L[t] = -\frac{d}{ds}\frac{1}{s^2} = \frac{2}{s^3}$$

$$L[t^3] = L[t \times t^2] = -\frac{d}{ds}L[t^2] = -\frac{d}{ds}\frac{2}{s^3} = \frac{3!}{s^4}$$

TABLE 6.5   Obtaining $L[t^n] = \dfrac{n!}{s^{n+1}}$ by mathematical induction

An induction step would fit in here. To show the general case, namely, $L[t^n] = n!/s^{n+1}$, assume the truth of the previous case and show the general case follows from that previous case and the operational law. Thus, assume $L[t^{n-1}] = (n-1)!/s^n$ to be true, and bootstrap one more time via

$$L[t^n] = L[t \times t^{n-1}] = -\frac{d}{ds} L[t^{n-1}] = -\frac{d}{ds}\left[\frac{(n-1)!}{s^n}\right] = \frac{n!}{s^{n+1}} \qquad \diamond$$

EXAMPLE 6.10   The Laplace transform of the function $t \sin 3t$ is obtained by the operational law "multiplication by t becomes differentiation with respect to s." Thus,

$$L[t \sin 3t] = -\frac{d}{ds} L[\sin 3t] = -\frac{d}{ds}\frac{3}{s^2+9} = \frac{6s}{(s^2+9)^2} \qquad \diamond$$

EXAMPLE 6.11   The Laplace transform of the function $t \cos 3t$ is obtained by the operational law "multiplication by t becomes differentiation with respect to s." Thus,

$$L[t \cos 3t] = -\frac{d}{ds} L[\cos 3t] = -\frac{d}{ds}\frac{s}{s^2+9} = \frac{s^2-9}{(s^2+9)^2} \qquad \diamond$$

EXAMPLE 6.12   The Laplace transform of the function te^{-2t} is obtained by the operational law "multiplication by t becomes differentiation with respect to s." Thus,

$$L[te^{-2t}] = -\frac{d}{ds} L[e^{-2t}] = -\frac{d}{ds}\frac{1}{s-(-2)} = -\frac{d}{ds}\frac{1}{s+2} = \frac{1}{(s+2)^2}$$

We can equally well obtain this result by the First Shifting Law by finding the transform of just $f(t) = 1$ and then shifting the argument in the resulting $F(s)$. Thus,

$$L[te^{-2t}] = L[t]|_{s \to s+2} = \left.\frac{1}{s^2}\right|_{s \to s+2} = \frac{1}{(s+2)^2} \qquad \diamond$$

EXAMPLE 6.13   The Laplace transform of the function $\frac{\sin t}{t}$ is obtained by the operational law "division by t becomes integration with respect to s." Thus,

$$L\left[\frac{\sin t}{t}\right] = \int_s^\infty L[\sin t]\,dx = \int_s^\infty \frac{dx}{x^2+1} = \arctan x]_s^\infty = \frac{\pi}{2} - \arctan s$$

Notice how we anticipated the integration of the transform, an integration requiring the use of a "dummy" variable of integration. Hence, we wrote the transform as $F(x)$ instead of the usual $F(s)$. $\qquad \diamond$

EXAMPLE 6.14   Without performing the integration, obtain the Laplace transform of the function $f(t) = \int_0^t xe^{3x}\,dx$ using the operational law "integration of the function becomes division by s." Thus,

$$L\left[\int_0^t xe^{3x}\,dx\right] = \frac{1}{s}L[te^{3t}] = \frac{1}{s}[L[t]|_{s \to s-3}] = \frac{1}{s}\frac{1}{(s-3)^2}$$

where we recognized that the required transform could be obtained by dividing (by s) the transform of the integrand te^{3t}. The transform of te^{3t} is then available either using the rule "multiplication by t becomes differentiation with respect to s" or by the First Shifting Law. We chose to apply the First Shifting Law to the function t, whose transform is $\frac{1}{s^2}$, obtaining $\frac{1}{(s-3)^2}$. Additional experiments with this example appear in Exercises A3 and 4. $\qquad \diamond$

EXAMPLE 6.15   The Laplace transform of the function $f(t) = te^{2t}\sin 3t$ can be obtained by an adroit use of the operational laws. Thus,

$$L[te^{2t}\sin 3t] = L[t\sin 3t]|_{s\to s-2} = \left[-\frac{d}{ds}L[\sin 3t]\right]\Big|_{s\to s-2}$$

$$= \left[-\frac{d}{ds}\left[\frac{3}{s^2+9}\right]\right]\Big|_{s\to s-2} = \left[\frac{6s}{(s^2+9)^2}\right]\Big|_{s\to s-2} = \frac{6(s-2)}{((s-2)^2+9)^2}$$

in which both multiplication by t and the exponential e^{2t} suggest operational rules to apply. In fact, we have a choice of two starting points since we can think of $f(t)$ as either t times $e^{2t}\sin 3t$ (suggesting the rule "multiplication by t becomes differentiation with respect to s") or e^{2t} times $t\sin 3t$ (suggesting use of the First Shifting Law.)

We chose the second option because the "shifting" in the First Shifting Law is an easier transformation to make than differentiation in s. Thus, we reserve the simpler transformation for the very last step when the accumulation of intermediate results will be their most complicated. And this choice means we will differentiate a simpler transform than we would under the first option.

The First Shifting Law directs us to find the transform of that which the exponential multiplies, namely, of $f_1(t) = t\sin 3t$ whose transform, $F_1(s)$, is then shifted via $s \to s - 2$. To obtain $F_1(s)$ we have to transform $f_1(t) = t\sin 3t$, a function yielding to the rule "multiplication by t becomes differentiation with respect to s." Hence, we obtain the transform of just $\sin(3t)$, negate its derivative, and shift. ❖

EXAMPLE 6.16   Given the Laplace transform $Y(s) = \frac{3s-2}{s^2-s}$, the inverse Laplace transform, $y(t)$, is most easily found from a partial fraction decomposition. Thus, writing $Y(s) = \frac{2}{s} + \frac{1}{s-1}$, the inversion is immediate, yielding $y(t) = 2 + e^t$. The more complicated inversion based on completing the square in the denominator and the First Shifting Law is left to Exercise B1. ❖

EXAMPLE 6.17   Given the Laplace transform $Y(s) = \frac{s}{(s^2+1)^2}$, the inverse Laplace transform, $y(t)$, can be found if $Y(s)$ is observed to have the form $\frac{1}{2}\frac{du}{u^2}$ where $u = s^2 + 1$. This suggests the operational law "integration of the transform becomes division of the function by t." The integral of $Y(s)$ will be a new transform, this one of the function $\frac{y(t)}{t}$. Thus,

$$L\left[\frac{y(t)}{t}\right] = \int_s^\infty Y(x)\,dx = \int_s^\infty \frac{x}{(x^2+1)^2}\,dx = \frac{1}{2}\frac{1}{s^2+1}$$

which, by inspection, yields $\frac{y(t)}{t} = \frac{1}{2}\sin t$, and hence $y(t) = \frac{1}{2}t\sin t$. ❖

EXAMPLE 6.18   Given the Laplace transform $Y(s) = \ln\frac{s^2+1}{(s-1)^2}$ and the task of inverting it to find $y(t)$, we recall from calculus that differentiating a logarithm yields a rational function. There are no logarithms in Table 6.2, but there are rational functions. If this transform is to be inverted by pattern recognition, it seems like a good idea to do something to it that will convert it to a rational function. In fact, differentiation with respect to s becomes multiplication by t, or, more precisely, $-\frac{d}{ds}Y(s) = L[ty(t)]$. Hence, we compute

$$L[ty(t)] = -\frac{d}{ds}\left[\ln\frac{s^2+1}{(s-1)^2}\right] = \frac{2(s+1)}{(s^2+1)(s-1)} = \frac{2}{s-1} - \frac{2s}{s^2+1}$$

and invert by inspection to obtain $ty(t) = 2e^t - 2\cos t$, from which we finally get $y(t) = \frac{2}{t}(e^t - \cos t)$. ❖

Justifications

The twin conditions of piecewise continuity and exponential order that guaranteed the existence of the Laplace transform in Section 6.1 are also the basis for justifying the interchange of integration and differentiation in the following derivations. The technical details of *uniform* and *absolute* convergence of integrals needed here are beyond the scope of this text, so we simply point to [83], [21], and [43] for the finer details.

The Rule $L[tf(t)] = -\frac{d}{ds}F(s)$

Formal differentiation of $\int_0^\infty f(t)e^{-st}\,dt$, that is, passing the differential operator $-\frac{d}{ds}$ inside the integral, gives

$$-\frac{d}{ds}\int_0^\infty f(t)e^{-st}\,dt = -\int_0^\infty -tf(t)e^{-st}\,dt = L[tf(t)]$$

The Rule $L\left[\frac{f(t)}{t}\right] = \int_s^\infty F(\sigma)\,d\sigma$

If $F(s) = \int_0^\infty f(t)e^{-st}\,dt$ is the Laplace transform of $f(t)$, then, using σ as the dummy variable of integration, the integral of the transform can be written as $\int_s^\infty \int_0^\infty f(t)e^{-\sigma t}\,dt\,d\sigma$. Interchanging the order of integration yields the new inner integral $\int_s^\infty e^{-\sigma t}\,d\sigma = e^{-st}/t$, provided $t > 0$. Combining this result with the outer integral gives $\int_0^\infty \frac{f(t)}{t}e^{-st}\,dt = L[\frac{f(t)}{t}]$, the Laplace transform of $\frac{f(t)}{t}$.

The Rule $L\left[\int_0^t f(\tau)\,d\tau\right] = \frac{F(s)}{s}$

Define $g(t) = \int_0^t f(\tau)\,d\tau$ and recognize that we are trying to compute $G(s) = L[g(t)]$, the Laplace transform of $g(t)$. Consider, however, the Laplace transform of $g'(t) = f(t)$. Then, using the rule for the transform of a derivative, we find

$$L[g'(t)] = L[f(t)] = F(s) = sG(s) - g(0)$$

Noting that $g(0) = \int_0^0 f(t)\,dt = 0$, we solve for the unknown $G(s)$ and obtain $G(s) = \frac{F(s)}{s}$.

EXERCISES 6.4–Part A

A1. Let $F(s) = L[f(t)]$. Use the Four Operational Laws to obtain a formal representation of the following Laplace transforms.

(a) $L\left[e^{-3t}\int_0^t f(z)\,dz\right]$ **(b)** $L\left[\int_0^t e^{-3z}f(z)\,dz\right]$

(c) $L\left[t\int_0^t f(z)\,dz\right]$ **(d)** $L\left[\int_0^t zf(z)\,dz\right]$

(e) $L\left[te^{-3t}\int_0^t f(z)\,dz\right]$ **(f)** $L\left[\int_0^t ze^{-3z}f(z)\,dz\right]$

(g) $L\left[t\int_0^t e^{-3z}f(z)\,dz\right]$ **(h)** $L\left[e^{-3t}\int_0^t zf(z)\,dz\right]$

(i) $L[ty'(t)]$ **(j)** $L[e^{-3t}y'(t)]$ **(k)** $L[te^{-3t}y'(t)]$

A2. Use the appropriate Operational Laws to obtain the transforms of each of the following.

(a) $f(t) = 7t^2e^{-3t}$ **(b)** $f(t) = 3t\sin 4t$

(c) $f(t) = (3 - 2t)^2 e^{-5t}$ **(d)** $f(t) = te^{2t}\sin 3t$

(e) $f(t) = t\cos 3t$ **(f)** $f(t) = t^2 e^{t/2}\sin t$

A3. Verify the result in Example 6.14 by evaluating the integral and then transforming.

A4. Using the Operational Law for multiplication by t, obtain the transform of te^{3t} that appeared in Example 6.14.

A5. Invert $F(s) = \ln\frac{(s+5)^3}{\sqrt{s^2-4}}$.

EXERCISES 6.4–Part B

B1. Work Example 6.16 by completing the square in the denominator and using the First Shifting Law. Show that the results agree with the simpler solution in Example 6.16.

In Exercises B2–5, obtain the Laplace transform by

   (a) using a computer algebra's built-in functionality.

   (b) using the appropriate Operational Laws, obtaining an answer that agrees with part (a).

B2. $f(t) = \int_0^t xe^{-3x}\cos 2x\, dx$ **B3.** $f(t) = \int_0^t \dfrac{e^{-2x}\sin 5x}{x}\, dx$

B4. $f(t) = \int_0^t \dfrac{e^{-3x}-1}{x}\, dx$ **B5.** $f(t) = t^2 e^{-2t} y'(t)$

B6. Use $f(t) = \cos 2t$ to validate your answer in Exercise A1(a).

B7. Use $f(t) = \sin 3t$ to validate your answer in Exercise A1(b).

B8. Use $f(t) = t\cos t$ to validate your answer in Exercise A1(c).

B9. Use $f(t) = e^{-t}\sin t$ to validate your answer in Exercise A1(d).

B10. Use $f(t) = t\cos 2t$ to validate your answer in Exercise A1(e).

B11. Use $f(t) = \cos 5t$ to validate your answer in Exercise A1(f).

B12. Use $f(t) = t^2\sin t$ to validate your answer in Exercise A1(g).

B13. Use $f(t) = t\cos 4t$ to validate your answer in Exercise A1(h).

B14. Use $f(t) = e^t\sin t$ to validate your answer in Exercise A1(i).

B15. Use $f(t) = t^2\cos t$ to validate your answer in Exercise A1(j).

B16. Use $f(t) = \sin 2t$ to validate your answer in Exercise A1(k).

In Exercises B17–23, invert the given Laplace transform using

   (a) a computer algebra's built-in functionality.

   (b) an appropriate combination of partial fractions, Operational Laws, and inspection that reproduces the solution in part (a).

B17. $F(s) = \dfrac{3}{s(s+4)}$ **B18.** $F(s) = \dfrac{5}{s(s^2-4)}$

B19. $F(s) = \dfrac{2}{(s+3)^6}$ **B20.** $F(s) = \dfrac{1-s}{s(s^2-1)}$

B21. $F(s) = \dfrac{1}{s(s-4)}$ (Use partial fractions.)

B22. $F(s) = \ln\dfrac{s+5}{s-2}$ **B23.** $F(s) = \arctan(2s-3)$

B24. Work Exercise B21 using the Operational Law $L[\int_0^t y(x)\, dx] = \frac{1}{s}Y(s)$, not partial fractions.

B25. Solve the IVP $3y'' + 5y' + 4y = 7\sin 2t$, $y(0) = -3$, $y'(0) = 1$, using

   (a) a computer algebra system's built-in functionality.

   (b) the Laplace transform.

B26. Repeat Exercise B25 parts (a) and (b) for the IVP $y'' + 4y' + 13y = 5\cos 2t$, $y(0) = -1$, $y'(0) = 2$. In addition, as part (c), obtain a solution using First Principles (exponential guess, characteristic equation, characteristic roots, homogeneous solution, Undetermined Coefficients, particular solution, general solution, etc.).

6.5   Heaviside Functions and the Second Shifting Law

The Heaviside Function

The piecewise-defined function $f(t) = \begin{cases} 0 & \text{if } t < a \\ 1 & \text{if } t \geq a \end{cases}$ is a switch that is off until time $t = a$, then turns on, with value 1, at time $t = a$. For the specific value of $a = 2$, say, this function looks like the graph in Figure 6.13.

This type of function is important enough in applied and engineering mathematics that the Heaviside function $H(t)$ is defined as

$$H(t) = \begin{cases} 0 & \text{if } t < 0 \\ 1 & \text{if } t > 0 \end{cases}$$

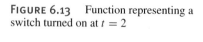

FIGURE 6.13   Function representing a switch turned on at $t = 2$

where we follow the convention of [51, p. 278], that at the point of discontinuity, the function is not defined. Figure 6.14 shows a plot of $H(t)$. Figure 6.15 shows a graph of a translation of the Heaviside function, namely, $H(t-3)$. Thus, the Heaviside function and its translations behave as unit cut-off functions.

Extinguishing

Used in conjunction with other functions, the Heaviside function and its translates "extinguish" functions prior to the switch-point. For example, consider the function $f(t) = $

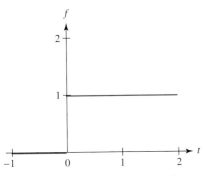

FIGURE 6.14   The Heaviside function H(t)

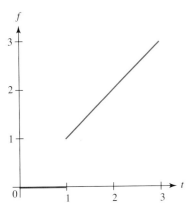

FIGURE 6.16   For $t < 3$, the function $f(t) = t$ is extinguished by $tH(t-3)$

$tH(t-1)$, graphed in Figure 6.16. Instead of getting a graph of $y(t) = t$, we get a graph where the portion corresponding to $t < 1$ is chopped, or cut off and **extinguished.** Multiplication by H($t-1$) is what both cut off *and* extinguished the graph in Figure 6.16. The Heaviside function is zero prior to being switched on, so the product of t with the translated Heaviside function shuts off, or extinguishes, $y(t) = t$ prior to $t = 1$.

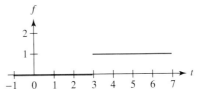

FIGURE 6.15   H($t - 3$), the Heaviside function H(t) translated three units to the right

Shifting and Extinguishing

At this point, it is imperative that the interactions between translation and chopping/extinguishing with a Heaviside function become completely clear. To this end, Table 6.6 lists the four possible alternatives. Examples 6.19 and 6.20 illustrate these four possibilities.

(a)	$f(t)$	Neither translated nor extinguished
(b)	$f(t)H(t-a)$	Just extinguished for $t < a$
(c)	$f(t-a)$	Just translated a units to right
(d)	$f(t-a)H(t-a)$	Right-translate by a, extinguish for $t < a$

TABLE 6.6   Alternatives for combining translation with extinguishing

EXAMPLE 6.19   For the function $f(t) = t^2$, Figure 6.17 shows, for $a = 1$, the four possibilities listed in Table 6.6. The function $f(t)$ itself is graphed in (a); the product $f(t)H(t-1)$ representing extinguishing is shown in (b); $f(t-1)$, the translation 1 unit to the right is shown in (c); and $f(t-1)H(t-1)$, the function translated to the right by 1, and extinguished for $t < 1$, is shown in (d).   ❖

EXAMPLE 6.20   For the function $f(t) = \sin t$, Figure 6.18 shows, for $a = 1$, the four possibilities listed in Table 6.6. The function $f(t)$ itself is graphed in (a); the product $f(t)H(t-1)$ representing extinguishing is shown in (b); $f(t-1)$, the translation 1 unit to the right is shown in (c); and $f(t-1)H(t-1)$, the function translated to the right by 1, and extinguished for $t < 1$, is shown in (d).   ❖

Second Shifting Law

Just as the four operational laws of Section 6.4 were best learned together and comparatively, we are now ready to discuss the Second Shifting Law in comparison to the First Shifting Law. Recall the First Shifting Law

$$L[e^{at} f(t)] = F(s - a)$$

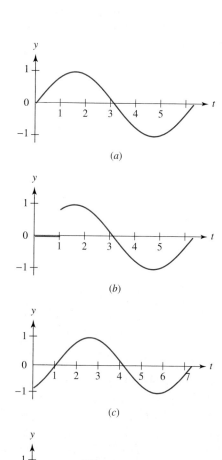

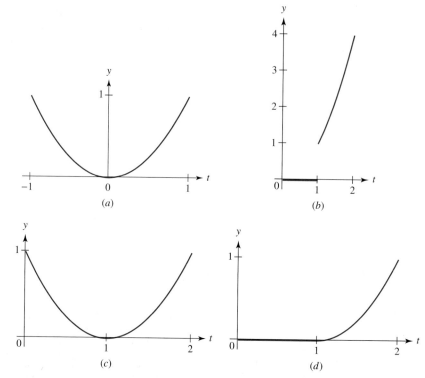

FIGURE 6.17   The function t^2 is (a) graphed; (b) extinguished for $t < 1$; (c) right-translated by 1; (d) right-translated by 1 and extinguished for $t < 1$

FIGURE 6.18   The function $\sin t$ is (a) graphed; (b) extinguished for $t < 1$; (c) right-translated by 1; (d) right-translated by 1 and extinguished for $t < 1$

which states that multiplication by an exponential in the time domain corresponds to shifting in the transform domain. It should not be surprising, then, that the Second Shifting Law states essentially that multiplication by an exponential in the transform domain corresponds to shifting in the time domain. The "detail" that accompanies this concept is the Heaviside function, which appears in the following statement of the Second Shifting Law

$$L[f(t-a)\mathrm{H}(t-a)] = e^{-as}F(s)$$

The translation, or shifting, that takes place in the time domain is in conjunction with chopping/extinguishing via a Heaviside function. The parallelism is remarkable, even beautiful. The delicacy of the translation in the time domain, however, can be problematic to the student who has not mastered the chopping/extinguishing action of the Heaviside function. For that reason, we rely on the following examples to clarify the use of the Second Shifting Law, both in a forward and in a reverse usage.

EXAMPLE 6.21   Obtain the Laplace transform of the properly translated, chopped, and extinguished function $g(t) = (t-2)\mathrm{H}(t-2)$ whose graph is shown in Figure 6.19. The required transform could be obtained by the direct computation

$$G(s) = L[g(t)] = \int_2^\infty (t-2)e^{-st}\, dt = \frac{e^{-2s}}{s^2}$$

The Second Shifting Law requires that we first obtain the transform of just $f(t) = t$, the function that is translated and then hit by the chopping/extinguishing action of the Heaviside function. Next, obtain $F(s) = \frac{1}{s^2}$, the transform of $f(t) = t$. According to the

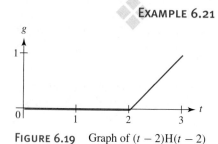

FIGURE 6.19   Graph of $(t-2)\mathrm{H}(t-2)$

Second Shifting Law, to get the transform of the translated, chopped, and extinguished version of $f(t) = t$, multiply the transform of t by e^{-2s}, again obtaining $G(s) = e^{-2s}/s^2$.

❖

EXAMPLE 6.22   Obtain the Laplace transform of the function $g(t) = \sin(t-2)\mathrm{H}(t-2)$, a properly translated, chopped, and extinguished function graphed in Figure 6.20. Again obtaining the required Laplace transform by directly integrating according to the definition, we compute

$$G(s) = L[g(t)] = \int_2^\infty \sin(t-2)e^{-st}\,dt = \frac{e^{-2s}}{s^2+1}$$

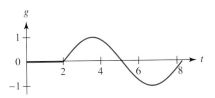

FIGURE 6.20   Graph of $\sin(t-2)\mathrm{H}(t-2)$

To apply the Second Shifting Law to obtain this same transform, first obtain the transform of just $f(t) = \sin t$, the function that is translated and then hit by the chopping/extinguishing action of the Heaviside function. Next, get $F(s) = \frac{1}{s^2+1}$, the Laplace transform of $f(t) = \sin t$. According to the Second Shifting Law, to get the transform of the translated, chopped, and extinguished version of $\sin t$ multiply the transform of $\sin t$ by e^{-2s}, again obtaining $G(s) = e^{-2s}/(s^2+1)$.

❖

EXAMPLE 6.23   The Second Shifting Law seems to give students most difficultly when it is used "in reverse," to invert a transform containing exponential factors. For example, consider the transform $G(s) = e^{-3s}/s^2$ containing e^{-3s}, the Second-Shift exponential "flag." Recognize the given transform as $G(s) = e^{-3s}F(s)$, where $F(s) = \frac{1}{s^2}$. Invert $F(s) = \frac{1}{s^2}$ to $f(t) = t$; then recall there was an exponential factor that indicates the need to translate $f(t) = t$ and to chop/extinguish this translation with an appropriate Heaviside function. Thus, the complete application of the Second Shifting Law yields

$$g(t) = f(t-3)\mathrm{H}(t-3) = (t-3)\mathrm{H}(t-3)$$

Again, the strategy is to recognize the exponential in the transform as a "flag," indicating the Second Shifting Law took place going forward. Cover the exponential term, invert what is visible, then translate, and multiply by the corresponding Heaviside function.   ❖

EXAMPLE 6.24   A second transform to invert via the Second Shifting Law is $G(s) = e^{-4s}/(s-3)$. Covering up e^{-4s}, the Second-Shift exponential flag, we are left with $F(s) = \frac{1}{s-3}$, which inverts to $f(t) = e^{3t}$. Translate $f(t) = e^{3t}$ to $e^{3(t-4)}$ and multiply by the corresponding Heaviside function to chop and extinguish, yielding

$$g(t) = f(t-4)\mathrm{H}(t-4) = e^{3(t-4)}\mathrm{H}(t-4)$$

❖

Justification

The validity of the Second Shifting Law is established as follows. Write the defining integral for the Laplace transform of $f(t-a)\mathrm{H}(t-a)$ and make the change of variables $z = t - a$, thereby obtaining

$$\int_0^\infty f(t-a)\mathrm{H}(t-a)e^{-st}\,dt = \int_{-a}^\infty f(z)\mathrm{H}(z)e^{-s(a+z)}\,dz$$

$$= e^{-as}\int_0^\infty f(z)e^{-sz}\,dz$$

$$= e^{-as}F(s)$$

The lower limit on the integral in the second line is zero because the Heaviside function is zero for $z < 0$. The Heaviside function disappears from that integral because its value is 1 for $z > 0$. After making these alterations and recognizing that now z is but a variable of integration, the integral in question becomes the defining integral for $F(s)$, the Laplace transform of $f(t)$.

EXERCISES 6.5–Part A

A1. The function $g(t) = tH(t - 1)$ is **not** in a form for which the Second Shifting Law applies. Every appearance of the variable t must occur, not as t, but as $t - 1$. For this function, the device $g(t) = (t - 1 + 1)H(t - 1) = (t - 1)H(t - 1) + H(t - 1)$ allows the Second Shifting Law to work. For each of the following, use an appropriate algebraic device that allows the Second Shifting Law to apply and then use that law to obtain the Laplace transform of the given function.

(a) $g(t) = t^2 H(t - 2)$ **(b)** $g(t) = \sin(t)H(t - 1)$

(c) $g(t) = e^t H(t - 2)$ **(d)** $g(t) = e^{-2t}H(t - 1)$

(e) $g(t) = \cos(t - 1)H(t - 2)$

A2. Use the Second Shifting Law to invert each of the following transforms.

(a) $G(s) = \dfrac{se^{-s}}{s^2 + 1}$ **(b)** $G(s) = \dfrac{e^{-s} + 2e^{-2s}}{s^2 + 4}$

(c) $G(s) = \dfrac{(1 + e^{-2s})^2}{s + 2}$ **(d)** $G(s) = \dfrac{4se^{-3s}}{(s^2 + 4)^2}$

(e) $G(s) = \dfrac{2e^{-4s}}{(3 + 3)^3}$

A3. For each of the following, sketch $f(t)$, $f(t)H(t - 2)$, $f(t - 2)$, and $f(t - 2)H(t - 2)$.

(a) $f(t) = 2t - 3$ **(b)** $f(t) = 1 + \cos t$ **(c)** $f(t) = \dfrac{1}{t^2 + 1}$

(d) $f(t) = 2 - 3\sin t$ **(e)** $f(t) = t^2 - 5t + 4$

A4. Sketch each of the following, then express each as piecewise-defined functions.

(a) $f(t) = 1 + H(t - 1)$ **(b)** $f(t) = 2H(t - 1) - 3H(t - 4)$

(c) $f(t) = (t - 1)H(t - 1) + (t - 2)^2 H(t - 2)$

A5. Use the Second Shifting Law to obtain the Laplace transform of $f(t) = \sin^2(t)H(t - 1)$. *Hint:* Use a half-angle formula, then consider Exercise A1.

EXERCISES 6.5–Part B

B1. Use a computer algebra system to obtain by the defining integral, the Laplace transforms in Exercise A1.

In Exercises B2 and 3:

   (a) Use a computer algebra system to obtain the Laplace transform.

   (b) Use the appropriate Operational Laws to reproduce the solution found in part (a).

B2. $f(t) = t^2 e^{-5t} \sin \omega t$ **B3.** $f(t) = \int_0^t x e^{3x} y'(x)\, dx$

B4. Invert the Laplace transform $F(s) = \dfrac{3s-5}{s^2(s^2+4)}$ using

   (a) a computer algebra system.

   (b) an appropriate combination of partial fractions, Operational Laws, and inspection, which reproduces the solution in part (a).

   (c) the Operational Law $L[\int_0^t y(x)\, dx] = \dfrac{Y(s)}{s}$ but not partial fractions.

In Exercises B5–10:

   (a) Plot the given function $f(t)$.

   (b) Obtain the Laplace transform by the integral definition of the transform.

   (c) Write the function in terms of the Heaviside function.

   (d) Obtain the Laplace transform by using a computer algebra system.

   (e) Apply the Second Shifting Law.

B5. $f(t) = \begin{cases} 0 & \text{if } t < 3 \\ 3 - t & \text{if } t \geq 3 \end{cases}$ **B6.** $f(t) = \begin{cases} 0 & \text{if } t < 1 \\ (t - 1)^2 & \text{if } t \geq 1 \end{cases}$

B7. $f(t) = \begin{cases} 0 & \text{if } t < \frac{1}{2} \\ e^{t+1} & \text{if } t \geq \frac{1}{2} \end{cases}$ **B8.** $f(t) = \begin{cases} 0 & \text{if } t \leq 2 \\ e^{2-t}(t - 2)^2 & \text{if } t > 2 \end{cases}$

B9. $f(t) = \begin{cases} 0 & \text{if } t \leq 1 \\ \sin(t - 1) & \text{if } t > 1 \end{cases}$ **B10.** $f(t) = \begin{cases} 0 & \text{if } t < \frac{1}{3} \\ e^{2t+1} & \text{if } t \geq \frac{1}{3} \end{cases}$

In Exercises B11–13, invert each Laplace transform

   (a) by use of a computer algebra system.

   (b) by implementing the Second Shifting Law stepwise.

   (c) by converting $f(t)$ (which is expressed in terms of the Heaviside function) to a piecewise-defined function.

   (d) by plotting the function $f(t)$ obtained as the inverse.

B11. $F(s) = \dfrac{5e^{-3s}}{s^2 - 16}$ **B12.** $F(s) = \dfrac{e^{-3s} - 4e^{-5s}}{s^2}$

B13. $F(s) = \dfrac{1 + 3e^{-5s}}{s}$

6.6 Pulses and the Third Shifting Law

The Unit Pulse

FIGURE 6.21 Unit pulse of duration $\tau = 2$, switched on at time $t_0 = 1$

Pressing a key on a computer keyboard sends through the computer's circuits a *pulse*, a current of constant height and finite duration. Figure 6.21 shows a graph of the piecewise-continuous function that is 1 for $1 < t < 3$, is undefined at $t = 1$ and 3, and is zero everywhere else. This is an example of a *unit pulse* of duration $\tau = 2$, switched on at time $t_0 = 1$. Clearly, the unit pulse is defined by giving either its starting time and duration or its on-off times.

An analytic representation of the graph in Figure 6.21 is given by

$$H(t-1) - H(t-3) = f(t) = \begin{cases} 1 & 1 < t < 3 \\ \text{undefined} & t = 1, 3 \\ 0 & \text{else} \end{cases} \qquad (6.7)$$

Because we have chosen $H(t)$ to be undefined at $t = 0$, the difference on the left in (6.7) is undefined at $t = 1$ and 3. Hence, the unit pulse can be specified either in the *echelon form* on the left in (6.7) or as the difference of Heaviside functions on the right. When writing the echelon form, it is often convenient to drop the explicit mention of the two points where the pulse is not defined.

Pulsed Functions

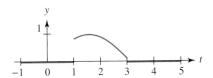

FIGURE 6.22 Product of $\sin t$ with unit pulse from Figure 6.21

Multiplication of the function $g(t)$ by a unit pulse will "filter out" that portion of $g(t)$ where the pulse is nonzero. Figure 6.22 shows the graph of the product of the unit pulse (6.7) with the function $g(t) = \sin t$. The unit pulse $f(t)$ has the effect of filtering out just that portion of $g(t)$ where the pulse has height 1.

The upper graph in Figure 6.23 shows, superimposed on a graph of $\sin t$, the graph of

$$p(t) = H(t-3) - H(t-5) \qquad (6.8)$$

a unit pulse of duration 2, starting at $t = 3$. Multiplying $\sin t$ by (6.8) filters out just that portion of $\sin t$ where t is in the interval $(3, 5)$. The lower graph in Figure 6.23 shows the filtering effect of this product.

Transform of Pulsed Functions

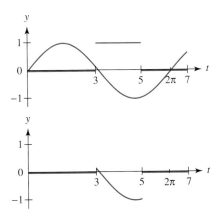

FIGURE 6.23 Product of $\sin t$ with unit pulse of duration $\tau = 2$, switched on at $t_0 = 3$

We next obtain the Laplace transform of the function $h(t) = p(t) \sin t$, where $p(t)$, given in (6.8), is a unit pulse of duration 2, starting at $t = 3$.

DIRECT INTEGRATION The Laplace transform of $h(t)$, obtained by direct integration, is

$$L[h(t)] = \int_3^5 \sin(t)e^{-st}\, dt = -\frac{e^{-5s}(\cos 5 + s \sin t)}{s^2 + 1} + \frac{e^{-3s}(\cos 3 + s \sin 3)}{s^2 + 1}$$

SECOND SHIFTING LAW Brute force integration is not elegant, and without a computer algebra system, it is at best, tedious. It is often preferable to use the Second Shifting Law, to which end we write

$$h(t) = \sin((t-3)+3)H(t-3) - \sin((t-5)+5)H(t-5)$$

Twice, the term $\sin t$ was arranged to appear shifted, with the amount of shift matching the shift in the corresponding Heaviside function, as required by the Second Shifting Law. We then apply the trigonometric expansion formula $\sin(A + B) = \sin A \cos B + \cos A \sin B$ to obtain

$$h(t) = \{\sin(t - 3) \cos 3 + \cos(t - 3) \sin 3\}H(t - 3)$$
$$-\{\sin(t - 5) \cos 5 + \cos(t - 5) \sin 5\}H(t - 5)$$

a form in which the Second Shifting Law can be applied. We therefore get

$$L[h(t)] = (e^{-3s} \cos 3 - e^{-5s} \cos 5)L[\sin t] + (e^{-3s} \sin 3 - e^{-5s} \sin 5)L[\cos t]$$
$$= \frac{e^{-3s} \cos 3}{s^2 + 1} + \frac{se^{-3s} \sin 3}{s^2 + 1} - \frac{e^{-5s} \cos 5}{s^2 + 1} - \frac{se^{-5s} \sin 5}{s^2 + 1}$$

THIRD SHIFTING LAW   Some texts (e.g., [11]) use the term *Third Shifting Law* for the following variant of the Second Shifting Law

$$L[H(t - a)g(t)] = e^{-as}L[g(t + a)]$$

The form on the left is precisely the type of problem faced in this ongoing example where we sought the transform of terms such as $H(t - 3) \sin t$. Direct use of the Second Shifting Law required each appearance of $g(t) = \sin t$ to have been in the form of functions with arguments shifted to $t - a$. However, the Third Shifting Law is "ice in the winter time" because the transform of $g(t + a)$ on the right embodies all the effort we exerted earlier with the trigonometric manipulations. It does not represent a radically new way to obtain the required transform.

Computing the transform of $\sin(t + 3)$ requires trigonometric manipulations equivalent to those performed when "rigging" the expression to use the Second Shifting Law directly. In fact, $\sin(t + 3) = \sin t \cos 3 + \cos t \sin 3$, so computing the transform of $\sin(t + 3)$ is equivalent to what we did when we used the Second Shifting Law.

EXAMPLE 6.25   Suppose the action of a shock absorber is modeled by the IVP $y'' + 3y' + 2y = f(t)$, $y(0) = 1$, $y'(0) = 1$, where $f(t)$ is the function whose graph is shown in Figure 6.24. The initial conditions indicate initial displacement and velocity of 1. Until time $t = 1$, the shock absorber is acting to damp out the induced vibrations. But between $t = 1$ and $t = 3$, a new force is introduced, a linearly decaying force that starts at magnitude 3 and decays to magnitude 1, at which point the force ceases to act. The driving force $f(t)$ is an example of a pulsed function, and solving the IVP by the Laplace transform requires we obtain first its representation and then its transform.

Taking the endpoints of the slanted line as $(1, 3)$ and $(3, 1)$, the equation of the slanted line segment would be $y = -(t - 3) + 1 = 4 - t$ so that $f(t) = (4 - t)(H(t - 1) - H(t - 3))$. The Third Shifting Law suggests defining $g(t) = 4 - t$, so we can write

$$L[f(t)] = e^{-s}L[g(t + 1)] - e^{-3s}L[g(t + 3)]$$
$$= e^{-s}L[4 - (t + 1)] - e^{-3s}L[4 - (t + 3)]$$
$$= e^{-s}\left(\frac{3}{s} - \frac{1}{s^2}\right) - e^{-3s}\left(\frac{1}{s} - \frac{1}{s^2}\right)$$

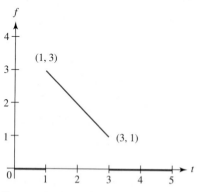

FIGURE 6.24   Driving term for shock absorber in Example 6.25

We can now take the Laplace transform of the given differential equation, obtaining

$$s^2Y - s - 1 + 3(sY - 1) + 2Y = e^{-s}\left(\frac{3}{s} - \frac{1}{s^2}\right) - e^{-3s}\left(\frac{1}{s} - \frac{1}{s^2}\right)$$

so that

$$Y = \frac{3}{s+1} - \frac{2}{s+2} + e^{-s}\left(\frac{9}{4s} - \frac{1}{2s^2} + \frac{7}{4(s+2)} - \frac{4}{s+1}\right)$$
$$+ e^{-3s}\left(\frac{1}{2s^2} - \frac{5}{4s} - \frac{3}{4(s+2)} + \frac{2}{s+1}\right)$$

whence

$$y(t) = 3e^{-t} - 2e^{-2t} + \left(\frac{9}{4} - \frac{t-1}{2} + \frac{7}{4}e^{-2(t-1)} - 4e^{-(t-1)}\right)H(t-1)$$
$$+ \left(2e^{-(t-3)} - \frac{3}{4}e^{-2(t-3)} - \frac{5}{4} + \frac{t-3}{2}\right)H(t-3)$$

The characteristic equation for the homogeneous differential equation is $\lambda^2 + 3\lambda + 2 = 0$, with characteristic roots $\lambda = -1, -2$, so the system is overdamped. Without the driving term $f(t)$ the solution would be $y_h(t) = 3e^{-t} - 2e^{-2t}$, graphed as the dotted curve in Figure 6.25. Without the driving term $f(t)$, the system would exhibit the effect of the initial displacement and velocity and then settle to rest asymptotically under the action of the damping. The pulsed driving term gives the system a "kick" of short duration. The system reacts to this kick and then resumes its determination to return to rest. The solid curve in Figure 6.25 is the driven solution, $y(t)$. The interval $1 < t < 3$ during which the driving term acts is marked as a thickened segment on the t-axis. ❖

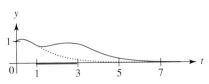

FIGURE 6.25   Solutions for driven (solid) and undriven (dotted) damped oscillator

EXERCISES 6.6–Part A

A1. Express in terms of Heaviside functions, the unit pulse that turns on at $t = 0$ and has duration 1.

A2. Sketch the product of $\cos t$ with the pulse in Exercise A1.

A3. To the pulse in Exercise A1, add a unit pulse that turns on at $t = \frac{1}{2}$ and has duration 2. Represent the sum and sketch its graph.

A4. Express the ramp function $f(t) = \begin{cases} t & 0 \le t \le 1 \\ 1 & t > 1 \end{cases}$ in terms of Heaviside functions (as the sum of two pulsed functions) and obtain the Laplace transform.

A5. If $f(t)$ is the function in Exercise A4, use Laplace transforms to solve the IVP $y'' + 2y' + 5y = f(t)$, $y(0) = y'(0) = 0$.

A6. For each of the following, express in terms of pulses and Heaviside functions and obtain the Laplace transforms using the Second Shifting Law.

(a) Exercise B12, Section 6.1   (b) Exercise B13, Section 6.1

(c) Exercise B14, Section 6.1   (d) Exercise B15, Section 6.1

(e) Exercise B16, Section 6.1   (f) Exercise B5, Section 6.5

EXERCISES 6.6–Part B

B1. Obtain the Laplace transform of $t^2 H(t-1)$ using

(a) the Second Shifting Law.   (b) the Third Shifting Law.

In Exercises B2–8, obtain the Laplace transform using

(a) a computer algebra system.

(b) the appropriate Operational Laws, for example, to reproduce the solution found in part (a).

B2. $f(t) = t^2 e^{-3t} \cos 4t$ **B3.** $f(t) = (t^2 e^{2t})^2$

B4. $f(t) = t \cos t H(t-1)$ **B5.** $f(t) = t^2 \sin 2t H(t-1)$

B6. $f(t) = te^{-t}H(t-5)$ **B7.** $f(t) = e^{-t} \cos t H(t-2)$

B8. $f(t) = e^{-2t} \cos 3t H(t-1)$

B9. Invert the Laplace transform $F(s) = \frac{2s-1}{s^2(s+5)}$ using

(a) a computer algebra system.

(b) an appropriate combination of partial fractions, Operational Laws, and inspection, which reproduces the solution found in part (a).

(c) the Operational Law $L[\int_0^t y(x)\,dx] = \frac{Y(s)}{s}$ but not partial fractions. (Integrate only once—write $F(s)$ as $\frac{G(s)}{s}$).

In Exercises B10–12:

(a) Plot $f(t)$.

(b) Use a computer algebra system to obtain the Laplace transform.

(c) Convert to a piecewise-defined function written in echelon form.

(d) Apply the Second Shifting Law, being sure the resulting transform agrees with part (b).

(e) Apply the Third Shifting Law to obtain the Laplace transform of $f(t)$.

(f) Obtain the Laplace transform by brute-force integration.

B10. $f(t) = (t - 5)H(t - 3)$ **B11.** $f(t) = \sin 2tH(t - 1)$

B12. $f(t) = te^{-3t}H(t - 2)$

For the IVPs in Exercises B13–21:

(a) Graph the driving force $r(t)$.

(b) Use a computer algebra system to obtain $R(s)$, the Laplace transform of $r(t)$.

(c) Express $r(t)$ as a piecewise-defined function in echelon form.

(d) Apply the Second Shifting Law to obtain the Laplace transform of $r(t)$.

(e) Apply the Third Shifting Law to obtain the Laplace transform of $r(t)$.

(f) Obtain $R(s)$ by brute-force integration.

(g) Solve for $y(t)$ using Laplace transforms.

(h) Express the solution $y(t)$ as a piecewise-defined function.

(i) Plot the solution $y(t)$ and explain how its behavior reflects the properties of the system.

B13. $y'' + 4y' + 13y = (t - 1)H(t - 2)$, $y(0) = 1$, $y'(0) = 0$

B14. $y'' + 2y' + 2y = 1 + t^2H(t - 2)$, $y(0) = 1$, $y'(0) = -1$

B15. $y'' + 6y' + 13y = t^2e^{-t} - tH(t - 1)$, $y(0) = -1$, $y'(0) = 1$

B16. $9y'' + 6y' + 2y = t(1 - H(t - 2))$, $y(0) = 2$, $y'(0) = 0$

B17. $2y'' + 2y' + y = \cos 2t(1 - H(t - 2\pi))$, $y(0) = -3$, $y'(0) = 1$

B18. $9y'' + 6y' + 5y = t(H(t - 1) - H(t - 2))$, $y(0) = -2$, $y'(0) = 0$

B19. $16y''' + 24y'' + 13y' + 5y = 1 - H(t - 5)$, $y(0) = 1$, $y'(0) = -1$, $y''(0) = 0$

B20. $25y''' + 45y'' + 33y' + 13y = te^{-t}(1 - H(t - 1))$, $y(0) = -1$, $y'(0) = 0$, $y''(0) = 1$

B21. $4y''' + 28y'' + 65y' + 51y = 1 - H(t - 2)$, $y(0) = 0$, $y'(0) = -2$, $y''(0) = 1$

6.7 Transforms of Periodic Functions

Laplace Transform of a Periodic Function

If $f(t)$ is a function periodic with period p, then its Laplace transform can be found by the formula

$$F(s) = \frac{1}{1 - e^{-ps}} \int_0^p f(t)e^{-st}\, dt \tag{6.9}$$

Prototypical examples of periodic functions would be the functions $\sin t$ and $\cos t$, each of which has period $p = 2\pi$. In fact, we use these functions in the first two of the three examples that follow.

EXAMPLE 6.26   The function $f(t) = \sin t$ has period $p = 2\pi$ and Laplace transform $F(s) = \frac{1}{s^2+1}$. The integral in the formula for periodic functions looks like the defining integral for the Laplace transform, except that the range of integration is just one period for the function. Hence, an alternate route to $F(s)$ must be

$$\frac{1}{1 - e^{-2\pi s}} \int_0^{2\pi} \sin(t)e^{-st}\, dt = \frac{1}{1 - e^{-2\pi s}} \left(\frac{1}{s^2 + 1} - \frac{e^{-2\pi s}}{s^2 + 1} \right) = \frac{1}{s^2 + 1} \quad \maltese$$

EXAMPLE 6.27   The function $f(t) = \cos \omega t$ has period $p = \frac{2\pi}{\omega}$ and Laplace transform $F(s) = \frac{s}{s^2+\omega^2}$. An alternate route to this transform must then be

$$\frac{1}{1 - e^{-2\pi s/\omega}} \int_0^{2\pi/\omega} \cos(\omega t)e^{-st}\, dt = -\frac{1}{1 - e^{-2\pi s/\omega}}s\frac{e^{-2\pi s/\omega} - 1}{s^2 + \omega^2} = \frac{s}{s^2 + \omega^2} \quad \maltese$$

EXAMPLE 6.28

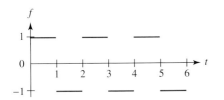

FIGURE 6.26   Square wave with amplitude 1 and period 2

This next transform is more complicated than those of $\sin t$ or $\cos \omega t$ because the formula for the function cannot be written in terms of a single elementary function. This example will show why (6.9) is so important. There really is no other reasonable way to obtain the transform of a periodic function whose rule cannot be easily written in terms of the elementary functions.

We will obtain the Laplace transform of $f(t)$, a square wave, which is the repeated replication of the function $g(t) = \begin{cases} 1 & \text{if } 0 \le t < 1 \\ -1 & \text{if } 1 \le t < 2 \end{cases}$ as a periodic function. Figure 6.26 shows three periods of the square wave; the leftmost two line segments would be the graph of $g(t)$. Clearly, the period is $p = 2$. By (6.9), the Laplace transform of $f(t)$ is

$$L[f(t)] = \frac{1}{1 - e^{-2s}} \left(\int_0^1 (1)e^{-st}\, dt + \int_1^2 (-1)e^{-st}\, dt \right) = \frac{1 - e^{-s}}{s(1 + e^{-s})}$$

To corroborate this result, represent $f(t)$ as the infinite sum of Heaviside functions

$$f(t) = H(t) + 2\sum_{k=1}^{\infty} H(t - 2k) - 2\sum_{k=0}^{\infty} H(t - (2k+1)) \qquad (6.10)$$

obtained by generalizing from Figure 6.26. The six line segments in that figure are represented, from left to right, in terms of Heaviside functions by

$$[H(t) - H(t - 1)] - [H(t - 1) - H(t - 2)] + [H(t - 2) - H(t - 3)]$$
$$- [H(t - 3) - H(t - 4)] + [H(t - 4) - H(t - 5)] - [H(t - 5) - H(t - 6)]$$
$$= H(t) - 2H(t - 1) + 2H(t - 2) - 2H(t - 3) + 2H(t - 4) - 2H(t - 5) + H(t - 6)$$

Transform (6.10) term-by-term and then sum, recognizing two instances of the geometric series $\sum_{k=0}^{\infty} x^k = \frac{1}{1-x}$. The independent corroboration we seek lies in

$$L[f(t)] = \frac{1}{s}\left[1 + 2\sum_{k=1}^{\infty} e^{-2ks} - 2\sum_{k=0}^{\infty} e^{-(2k+1)s} \right]$$
$$= \frac{1}{s}\left[1 + \frac{2e^{-2s}}{1 - e^{-2s}} - \frac{2e^{-s}}{1 - e^{-2s}} \right] = \frac{1}{s}\left[\frac{(1 - e^{-s})^2}{(1 - e^{-s})(1 + e^{-s})} \right] = \frac{1 - e^{-s}}{s(1 + e^{-s})} \qquad ❖$$

Derivation

To derive the formula for the transform of periodic functions, write the defining integral for the Laplace transform as a sum of integrals $\sum_{k=0}^{\infty} I_k$, where $I_k = \int_{kp}^{(k+1)p} f(x)e^{-sx}\, dx, k = 0, 1, \ldots$. In each integral I_k, $k = 0, 1, \ldots$, make the change of variables $t = x - kp$. Since each integral is over one period of the periodic function $f(t)$ for which $f(t + kp) = f(t)$, we obtain $I_k = \int_0^p f(t + kp)e^{-s(t+kp)}\, dt = e^{-kps} \int_0^p f(t)e^{-st}\, dt$.

$$L[f(t)] = \int_0^{\infty} f(t)e^{-st}\, dt$$

$$= \int_0^p f(x)e^{-sx}\, dx + \int_p^{2p} f(x)e^{-sx}\, dx + \int_{2p}^{3p} f(x)e^{-sx}\, dx + \cdots$$

$$= \int_0^p f(t)e^{-st}\, dt + e^{-sp} \int_0^p f(t)e^{-st}\, dt + e^{-2sp} \int_0^p f(t)e^{-st}\, dt + \cdots$$

$$= \left[\sum_{k=0}^{\infty} (e^{-ps})^k \right] \int_0^p f(t)e^{-st}\, dt = \frac{1}{1 - e^{-ps}} \int_0^p f(t)e^{-st}\, dt$$

where we have summed the geometric series $\sum_{k=0}^{\infty} x^k = \frac{1}{1-x}$ with $x = e^{-ps}$.

EXERCISES 6.7–Part A

A1. Sketch $f(t)$, the periodic extension of $g(t) = \sin t, 0 \le t \le \pi$. Use (6.9) to obtain the Laplace transform of $f(t)$, the full-wave rectification of the sine function.

A2. Sketch $f(t)$, the periodic extension of $g(t) = t^2, 0 \le t \le 1$. Use (6.9) to obtain the Laplace transform of $f(t)$.

A3. Sketch $f(t)$, the periodic extension of $g(t) = \begin{cases} t & 0 \le t \le 1 \\ 1 & 1 \le t < 2 \end{cases}$. Use (6.9) to obtain the Laplace transform of $f(t)$.

A4. Sketch $f(t)$, the periodic extension of $g(t) = \begin{cases} t & 0 \le t \le 1 \\ 2 - t & 1 \le t < 2 \end{cases}$. Use (6.9) to obtain the Laplace transform of $f(t)$.

A5. Sketch $f(t)$, the periodic extension of $g(t) = \begin{cases} \sin t & 0 \le t \le \pi \\ 0 & \pi \le t \le 2\pi \end{cases}$. Use (6.9) to obtain the Laplace transform of $f(t)$, the half-wave rectification of the sine function.

EXERCISES 6.7–Part B

B1. Each line segment in the graph of $f(t)$ in Figure 6.27 is a translation of $y(t) = t$.

    (a) State clearly, and graph, the function $g(t)$ that is being repeated periodically. *Hint:* Its domain is $[0, 2]$.

    (b) Use (6.9) to obtain the Laplace transform of the periodic function $f(t)$.

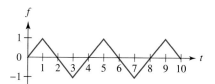

FIGURE 6.28   Exercise B2

the square wave given in Figure 6.26. Since $f(t)$ is both periodic and piecewise-defined, its transform is computed by (6.9). Hence, $Y(s)$ will be impossible to invert exactly, so expand $Y(s)$ in powers of e^{-s} by replacing e^{-s} with z, obtain a Taylor expansion (ten terms) about $z = 0$, and then replace z with its exponential equivalent. Invert the approximate $Y(s)$ and plot the solution.

B4. Repeat Exercise B3, this time taking $f(t)$ as the periodic function in Exercise B1. Expand $Y(s)$ in powers of e^{-2s}.

B5. Repeat Exercise B3, this time taking $f(t)$ as the periodic function in Exercise B2. Expand $Y(s)$ in powers of e^{-2s}.

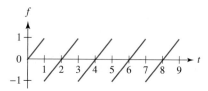

FIGURE 6.27   Exercise B1

B2. Repeat Exercise B1 for the function whose graph is shown in Figure 6.28. *Hint:* The domain of $g(t)$ is $[0, 4]$.

B3. Using the Laplace transform, solve the IVP $y'' + 2y' + 10y = f(t)$, $y(0) = 0$, $y'(0) = 0$, where $f(t)$ is the function whose graph is

6.8   Convolution and the Convolution Theorem

Introduction

The Laplace transform of a product is not the product of the transforms, since, for example, $L[t] = \frac{1}{s^2}$ but $L[t^2] = \frac{2}{s^3} \ne \left(\frac{1}{s^2}\right)^2$. Therefore, the product of two Laplace transforms does not invert back to the product of the inverses of the factors, so in general, $L^{-1}[F(s)G(s)] \ne f(t)g(t)$.

None of this should surprise the reader, since calculus has already provided at least two instances of results of this type. The derivative of a product is not the product of the derivatives. Instead, this is governed by the *product rule,* $(uv)' = uv' + vu'$. The integral of a product is not the product of the integrals. Instead, this is governed by the formula for *integration by parts*, $\int u\,dv = uv - \int v\,du$.

Likewise, the inverse of a product of two Laplace transforms is not the product of the inverses. Instead, it is the *convolution product,* the subject of this section.

The Convolution Product

The *convolution product* between two functions $f(t)$ and $g(t)$, denoted $f * g$, is a new function of t, defined, in the context of Laplace transforms, by either of the equivalent integrals in (6.11). The new *function* is $f * g$, and its *value* at t is $(f * g)(t)$

$$(f * g)(t) = \int_0^t f(t-x)g(x)\,dx = \int_0^t f(x)g(t-x)]\,dx \qquad \textbf{(6.11)}$$

That the integrals in (6.11) are indeed equivalent is demonstrated after Examples 6.29–6.31. Then, the *Convolution Theorem,* which relates the product of two Laplace transforms to the convolution product is given. First, however, we provide examples of this new object $f * g$, the "convolution product" of the two functions $f(t)$ and $g(t)$. In these and subsequent calculations, it might be helpful to articulate the integrands in the convolution product with language like "one function has the dummy argument, the other, t minus dummy," which always seems to amuse a classroom full of students but is nevertheless effective.

EXAMPLE 6.29   Find $t * e^t$, the convolution product of the two functions $f(t) = t$ and $g(t) = e^t$. According to (6.11), the convolution product $f * g$ is given by either of the integrals $\int_0^t (t - x)e^x\,dx$ or $\int_0^t x e^{t-x}\,dx$, both of which have the value $e^t - t - 1$, as calculation shows. Thus, $t * e^t$ is not the ordinary product te^t, but rather, it is $e^t - t - 1$. ❖

EXAMPLE 6.30   Find $1 * 1$, the convolution product of $f(t) = 1$ with $g(t) = 1$. Here, no matter which function has its argument shifted, the result is the same, namely, still 1. Hence, the convolution product $f * g = 1 * 1$ is given by the integral $\int_0^t 1\,dx = t$. ❖

EXAMPLE 6.31   The convolution product of $f(t) = e^{-2t}$ and $g(t) = \sin 3t$, namely, $e^{-2t} * \sin 3t$, is given by either of two integrals. Thus, we have

$$e^{-2t} * \sin 3t = \int_0^t e^{-2(t-x)} \sin 3x\,dx = \int_0^t e^{-2x} \sin(3(t - x))\,dx$$

$$= \tfrac{1}{13}(3e^{-2t} - 3\cos 3t + 2\sin 3t)$$

leading to Figure 6.29, where we see as a solid curve the graph of the ordinary product $e^{-2t} \sin 3t$ and as a dotted curve the graph of the convolution product $e^{-2t} * \sin 3t$, again showing how very different these two products really are. ❖

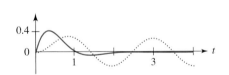

FIGURE 6.29   Ordinary product (solid) and convolution product (dotted) of e^{-2t} and $\sin 3t$

EQUIVALENCE OF CONVOLUTION PRODUCTS   To demonstrate in general that both convolution integrals are equivalent, make the change of variables $t - x = z$ in one integral and show it becomes the other. In particular, write

$$(f * g)(t) = \int_0^t f(x)g(t-x)\,dx = -\int_t^0 f(t-z)g(z)\,dz = \int_0^t f(t-z)g(z)\,dz$$

recognize that z in the last integral can be replaced by x, and conclude that both forms of the convolution integral are equivalent.

> **THEOREM 6.2   CONVOLUTION THEOREM**
>
> The Laplace transform of the convolution product $f * g$ is the product of the Laplace transforms, that is,
>
> $$L[f * g] = F(s)G(s)$$
>
> Equivalently, the product of two Laplace transforms inverts back to the convolution product of the individual inverses so that
>
> $$L^{-1}[F(s)G(s)] = f * g$$

By examples, we now illustrate the connection between $f * g$, the convolution product of $f(t)$ and $g(t)$, and $F(s)$ and $G(s)$, the Laplace transforms of $f(t)$ and $g(t)$, respectively.

EXAMPLE 6.32   The convolution $t * e^t = e^t - t - 1$ has already been computed. The Laplace transform of this convolution product is $L[t * e^t] = \frac{1}{s-1} - \frac{1}{s^2} - \frac{1}{s} = \frac{1}{s^2(s-1)}$ whereas the transforms of the factors $f(t) = t$ and $g(t) = e^t$ are $F(s) = \frac{1}{s^2}$ and $G(s) = \frac{1}{s-1}$, respectively. The ordinary product of these two transforms is $F(s)G(s) = \frac{1}{s^2(s-1)} = L[t * e^t]$. ❖

EXAMPLE 6.33   Inversion of the Laplace transform $L[y(t)] = \frac{1}{(s^2+1)^2}$ is a difficult project using only the four Operational Laws. The inversion is sketched here to show just how difficult it is using only the Operational Laws. To begin, observe that an integration with respect to s will lower the power of $(s^2 + 1)$ in the denominator from 2 to 1, leading to a transform recognizable as belonging to a trigonometric function. This integration is greatly assisted by having an s in the numerator so that the form $u^{-2}\,du$ appears as the integrand. To insert the s into the numerator, consider the transform of y', the derivative of y, related to the transform of y by multiplication by s according to the rule $L[y'] = sL[y] = \frac{s}{(s^2+1)^2}$. Integrating the transform of y' yields a new transform, the transform of $\frac{y'}{t}$. Hence,

$$L\left[\frac{y'}{t}\right] = \int_s^\infty \frac{x}{(x^2+1)^2}\,dx = \frac{1}{2(s^2+1)}$$

so $\frac{y'}{t} = \frac{1}{2}\sin t$, from which we determine y by the integration

$$y(t) = \int_0^t x \sin x \, dx = \frac{1}{2}(\sin t - t \cos t)$$

That is indeed a tedious calculation, and we have yet to justify not including the term $-y(0)$ in the very first step! But the point of the demonstration was to convince the reader of the difficulty in inverting this transform using only the four Operational Laws.

The convolution theorem leads to a much simpler solution. Writing the given transform as the product

$$L[y] = \left[\frac{1}{s^2+1}\right]\left[\frac{1}{s^2+1}\right]$$

we have the product of the transform of $\sin t$ with itself, explicitly stated as

$$L[y] = L[\sin t]L[\sin t]$$

The convolution theorem now tells us the function $y(t)$ is the convolution product of $\sin(t)$ with itself, namely,

$$y(t) = \sin(t) * \sin(t) = \int_0^t \sin x \sin(t - x)\,dx = \frac{1}{2}(\sin t - t \cos t)$$ ❖

Proof of the Convolution Theorem

The Laplace transform of the convolution product $f * g = \int_0^t f(t - x)g(x)\,dx$, namely,

$$L[f * g] = \int_0^\infty e^{-st} \int_0^t f(t - x)g(x)\,dx\,dt \tag{6.12}$$

can be interpreted as the iterated double integral

$$L[f * g] = \int_0^\infty \int_0^t e^{-st} f(t - x)g(x)\,dx\,dt \tag{6.13}$$

because (6.12) converges absolutely, and hence uniformly, for piecewise-continuous functions of exponential order. As we mentioned in Section 6.4, the notions of absolute and uniform convergence of integrals are beyond the scope of this text, so we are simply signaling the reader that the reversal of the order of integration, from $dx\,dt$ to $dt\,dx$ which we are about to do with (6.13), is mathematically justified.

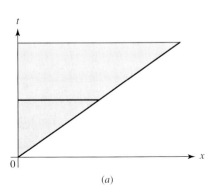

(a)

Figure 6.30(a) shows a portion of the xt-plane in which shading marks the region of integration for (6.13). In addition, an element of area for the order $dx\,dt$ is shown as the dark horizontal strip. Figure 6.30(b) shows an element of area for integration in the order $dt\,dx$.

Corresponding to (6.13) we therefore have the companion iterated integral

$$L[f * g] = \int_0^t \int_x^\infty e^{-st} f(t - x)g(x)\,dt\,dx = \int_0^t g(x)e^{-sx} \int_x^\infty e^{-s(t-x)} f(t - x)\,dt\,dx$$

in which the inner integral is subjected to the change of variables $t - x = y$, resulting in

$$L[f * g] = \int_0^\infty g(x)e^{-sx} \int_0^\infty e^{-sy} f(y)\,dy\,dx = \int_0^\infty g(x)e^{-sx}\,dx \int_0^\infty e^{-sy} f(y)\,dy$$

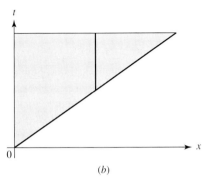

(b)

Figure 6.30   Changing the order of integration in the transform of a convolution

Notice how the change of variables on the inner integral changed the upper limit of integration for the outer integral. Also notice how, with a change of integration variables from x to t, and y to t, and an interpretation of the iterated integral as the product of two integrals, we have arrived at the desired result, namely, $L[f * g] = L[f(t)]L[g(t)] = F(s)G(s)$.

EXERCISES 6.8–Part A

A1. Compute the convolution product $1 * t$, and with it, verify the convolution theorem.

A2. Compute the convolution product $t * t^2$, and with it, verify the convolution theorem.

A3. Show that $t * \sin t = (\sin t) * t$.

A4. Obtain the convolution products $t * (\cos t * e^t)$ and $(e^t * t) * \cos t$. Comment on the outcome.

A5. Compute $e^t * \sin t$ by inverting the product of their Laplace transforms.

EXERCISES 6.8–Part B

B1. Sketch the region of integration for the iterated integral $\int_0^2 \int_{x^2}^4 f(x, y)\,dy\,dx$ and reverse the order of integration.

For the convolution products in Exercises B2–11:

(a) Exhibit both convolution integrals.

(b) Obtain the convolution product by evaluating both integrals in part (a), being sure results agree.

(c) Plot on the same set of axes both the ordinary product and the convolution product, indicating clearly which is which.

(d) Use a computer algebra system to obtain the transform of the convolution product in part (b).

(e) Obtain the product of the transforms of the factors in the given function. The convolution theorem says the results in parts (d) and (e) must be the same. Verify this is so.

(f) Use a computer algebra system to invert the product of the transforms of the factors. This should be the convolution product found in part (b).

B2. $t^2 * \cos 5t$ **B3.** $\cos 3t * \sin 2t$ **B4.** $\cos^2 3t * \sin 2t$

B5. $t^2 * e^{-3t}$ **B6.** $\sinh 2t * \cos 3t$ **B7.** $\cosh 3t * \sin 2t$

B8. $t * \sinh 2t$ **B9.** $3 * \cosh 5t$ **B10.** $e^{-2t} * e^{3t}$ **B11.** $\pi * e$

B12. Obtain the convolution product $t^2 * e^{3t} * \sin 5t$ by convolving the first two factors, then convolving that result with the third factor. Then obtain this same convolution product by inverting the product of the transforms of the individual factors.

B13. Repeat Exercise B12 for $t^2 * \sinh 2t * \cos t$.

For the Laplace transforms $F(s)$ given in Exercises B14–18:

   (a) Invert using a computer algebra system.

(b) Invert using the convolution theorem. Thus, write $F(s) = G(s)K(s)$ as suggested, invert G and K to $g(t)$ and $k(t)$, respectively, and compute the convolution $g * k$.

B14. $\dfrac{1}{(3s-2)^3} = \dfrac{1}{3s-2}\dfrac{1}{(3s-2)^2}$ **B15.** $\dfrac{1}{s^2(s+3)} = \dfrac{1}{s^2}\dfrac{1}{s+3}$

B16. $\dfrac{s}{(s+2)(s^2-9)} = \dfrac{1}{s+2}\dfrac{s}{s^2-9}$

B17. $\dfrac{s+1}{(s^2-1)^2} = \dfrac{s+1}{s^2-1}\dfrac{1}{s^2-1}$ **B18.** $\dfrac{2}{s(s^2+4)} = \dfrac{1}{s}\dfrac{2}{s^2+4}$

6.9   Convolution Products by the Convolution Theorem

Computing Convolutions by the Convolution Theorem

Section 6.8 presented the definition of the convolution product and demonstrated how to obtain this product by performing one of two alternate versions of the convolution integral. In addition, Section 6.8 revealed the convolution theorem whereby the product of two Laplace transforms is the transform of the convolution. In symbols we wrote both the "forward" version $L[f * g] = F(s)G(s)$ and the "backward" version $L^{-1}[F(s)G(s)] = f * g$.

Moreover, Section 6.8 showed that $f * g$, the convolution product of $f(t)$ and $g(t)$ could be obtained as the inverse Laplace transform of $F(s)G(s)$, the product of the Laplace transforms of $f(t)$ and $g(t)$. This present section will again demonstrate how to use the convolution theorem to compute the convolution product of two functions $f(t)$ and $g(t)$. In particular, we will be interested in cases where at least one of the two functions is a piecewise-defined function, especially one in which Heaviside functions appear.

When a factor in a convolution product is piecewise-defined, shifting the argument from t to $t - x$ in the convolution integral can be subtle because such functions don't have a single formula for their definition. The inequalities in the piecewise-defined function and the inequality $0 < x < t$, where x is the variable of integration, can be a challenge to coordinate.

The method explored here is the use of the convolution theorem, which takes both functions to the frequency domain, uses ordinary multiplication instead of convolution, and then inverts back out of the frequency domain to the time domain.

We illustrate with the following examples.

EXAMPLE 6.34   Compute the convolution of t^2 and $H(t-1)$ using (a) the convolution theorem and (b) the defining integral for convolution. The point of the example is to demonstrate how much easier the convolution theorem is to apply than direct integration, especially when one of the factors in the convolution is a piecewise-defined function like the Heaviside function.

❖

USING THE CONVOLUTION THEOREM   The convolution theorem says the Laplace transform of the convolution is the product of the transforms $L[t^2] = \frac{2}{s^3}$ and $L[H(t-1)] = e^{-s}/s$. Thus, $L[t^2 * H(t-1)] = 2e^{-s}/s^4$, so the desired convolution is

$$t^2 * H(t-1) = L^{-1}\left[\frac{2e^{-s}}{s^4}\right] = \frac{1}{3}(t-1)^3 H(t-1) \qquad (6.14)$$

as determined by the Second Shifting Law and the transform $L[t^3] = \frac{6}{s^4}$.

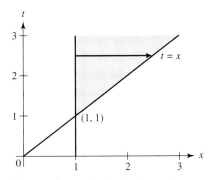

FIGURE 6.31 Region in the xt-plane where integrand for $t^2 * H(t-1)$ is nonzero

USING THE DEFINITION OF CONVOLUTION   Direct evaluation of the convolution integral

$$L[t^2 * H(t-1)] = \int_0^t H(x-1)(t-x)^2\, dx$$

requires a delicacy that starts with an analysis of the integrand. First, it is far easier to shift $t \to t-x$ in the function t^2 than it is in the Heaviside function. Second, because of the Heaviside function, the integrand is nonzero for $x-1 > 0$. Third, the very definition of integration requires $0 < x < t$ throughout the interval of integration. The region in the xt-plane satisfying both these constraints is shaded in Figure 6.31. A representative arrow shows the path of integration for fixed t.

Since the integration requires $0 < x < t$, if $t < 1$, then $x < 1$ and $H(x-1) = 0$. Hence, the integral yields the value 0 for $t < 1$. For $t > 1$, the integral is $\int_1^t (t-x)^2\, dx = \frac{1}{3}(t-1)^3$, so the convolution is again (6.14).

The hard part of this computation is sketching Figure 6.31, which shows the region of integration. The accompanying Maple worksheet demonstrates some built-in graphing tools that Maple makes available for sketching such regions.

 **EXAMPLE 6.35**   Obtain the convolution product $H(t-1) * H(t-2)$ using (a) the convolution theorem and (b) the defining integral for convolution.   ❖

BY THE CONVOLUTION THEOREM   To use the convolution theorem $L[f * g] = F(s)G(s)$, or actually, $f * g = L^{-1}[F(s)G(s)]$, obtain $F(s) = L[H(t-1)] = e^{-s}/s$ and $G(s) = L[H(t-2)] = e^{-2s}/s$. The resulting convolution

$$H(t-1) * H(t-2) = L^{-1}\left[\frac{e^{-3s}}{s^2}\right] = (t-3)H(t-3)$$

is computed by applying the Second Shifting Law in reverse to the transform $L[t] = \frac{1}{s^2}$.

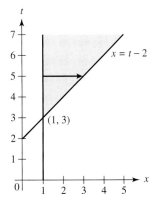

FIGURE 6.32 Region in the xt-plane where integrand for $H(t-1) * H(t-2)$ is nonzero

BY THE CONVOLUTION INTEGRAL   The integrand of the convolution integral $\int_0^t H(x-1) H(t-x-2)\, dx$ is nonzero for that region in the xt-plane in which the inequalities $x-1 > 0$ and $t-x-2 > 0$ are satisfied. This region is shaded in Figure 6.32. A representative arrow indicates a path of integration for a fixed value of t. The integrand is 0 if $t < 3$. Hence, the integral is 0 for $t < 3$. For $t > 3$ the integrand is 1 for $1 < x < t-2$ and the convolution integral is $\int_1^{t-2} dx = t-3$. Consequently, the convolution is again $(t-3)H(t-3)$.

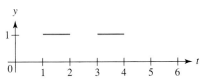 **EXAMPLE 6.36**   Obtain the convolution product of two unit pulses of duration $\tau = 1$, one starting at $t = 1$ and one starting at $t = 3$. These pulses, graphed in Figure 6.33, are represented in terms of Heaviside functions as $f(t) = H(t-1) - H(t-2)$ and $g(t) = H(t-3) - H(t-4)$.   ❖

FIGURE 6.33 Unit pulses of duration $\tau = 1$, starting at $t = 1$ and $t = 3$

BY THE CONVOLUTION THEOREM   To compute the convolution by Laplace transforms and the convolution theorem, obtain $L[f(t)] = (e^{-s} - e^{-2s})/s$ and $L[g(t)] = (e^{-3s} - e^{-4s})/s$ and invert the product of transforms $L[f(t)]L[g(t)] = (e^{-4s} - 2e^{-5s} + e^{-6s})/s^2$. The Second Shifting Law applied in reverse to the transform $L[t] = \frac{1}{s^2}$ gives

$$f * g = (t-4)H(t-4) - 2(t-5)H(t-5) + (t-6)H(t-6) = \begin{cases} 0 & 0 \le t < 4 \\ t-4 & 4 \le t < 5 \\ 6-t & 5 \le t < 6 \\ 0 & t \ge 6 \end{cases}$$

A graph of the convolution $f * g$ is found in Figure 6.34.

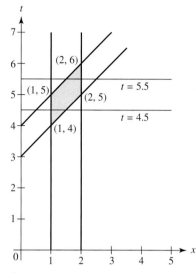

FIGURE 6.35   Region in the xt-plane where integrand for convolution of unit pulses in Figure 6.33 is nonzero

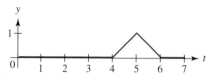

FIGURE 6.34   Convolution of unit pulses in Figure 6.33

BY THE CONVOLUTION INTEGRAL   Evaluating the convolution integral

$$f * g = \int_0^t [H(x - 1) - H(x - 2)][H(t - x - 3) - H(t - x - 4)]\, dx$$

requires knowing where, in the xt-plane, both $f(x)$ and $g(t-x)$ are simultaneously nonzero. Hence, the inequalities $x > 1$, $x < 2$, $t - x > 3$, $t - x < 4$ must all be satisfied. This region is shaded in Figure 6.35. Horizontal lines represent typical paths of integration along lines $t = $ constant. Any such line for which $t < 4$ or $t > 6$ yields a zero integrand and a value of zero for the convolution. Any such line between $t = 4$ and $t = 5$ yields an integrand of 1 and a value for $f * g$ that will be determined by the integral $\int_1^{t-3} dx = t - 4$. Any such line between $t = 5$ and $t = 6$ also yields an integrand of 1 and a value for $f * g$ that will be determined by the integral $\int_{t-4}^2 dx = 6 - t$. These values are consistent with the results from the convolution theorem.

EXERCISES 6.9–Part A

A1. Compute $H(t - 2) * \cos 3t$ by the convolution theorem.

A2. Compute the convolution of Exercise A1 by the defining integral after having sketched, in the xt-plane, the region where the integrand is nonzero.

A3. Compute $[H(t - 1) - H(t - 3)] * [H(t - 5) - H(t - 6)]$ by the convolution theorem.

A4. Compute the convolution of Exercise A3 by the defining integral after having sketched, in the xt-plane, the region where the integrand is nonzero.

A5. Compute $[H(t - 1) - H(t - 4)] * [H(t - 2) - H(t - 5)]$ by the convolution theorem.

A6. Compute the convolution of Exercise A5 by the defining integral after having sketched, in the xt-plane, the region where the integrand is nonzero.

EXERCISES 6.9–Part B

B1. Compute $[e^{-t} \sin t] * [t \cosh t]$ by the convolution theorem.

B2. Verify the convolution in Exercise B1 using the defining integral.

In Exercises B3–11, show graphically that the ordinary product and the convolution product are different after obtaining the indicated convolution product

    (a) using the Laplace transform and the convolution theorem.

    (b) by the defining convolution integral.

B3. $H(t - 3) * [t \cos t]$ **B4.** $H(t - 1) * [te^{-2t}]$

B5. $H\left(t - \dfrac{\pi}{3}\right) * \cos^2 2t$ **B6.** $H(t - 2) * [e^{-2t} \sin 5t]$

B7. $H(t - 1) * [t^2 \cosh 3t]$ **B8.** $H(t - \pi) * \sin^2 3t$

B9. $H\left(t - \dfrac{3}{2}\right) * \cosh^2 t$ **B10.** $H(t - 4) * [t \sinh 2t]$

B11. $H(t - 5) * \sqrt{t}$

In Exercises B12–21, let $P_{a,b}(t)$ represent a unit pulse between $t = a$ and $t = b$. Obtain the indicated convolution product

    (a) using the Laplace transform and the convolution theorem.

    (b) by the defining convolution integral.

B12. $P_{1,3}(t) * P_{2,4}(t)$ **B13.** $P_{2,5}(t) * P_{3,4}(t)$ **B14.** $e^{-3t} * P_{2,4}(t)$

B15. $t^2 * P_{2,5}(t)$ **B16.** $[t \sin t] * P_{0,\pi}(t)$

B17. $[e^{-2t} \cos 3t] * P_{\pi,2\pi}(t)$ **B18.** $H(t - 3) * P_{1,4}(t)$

B19. $[H(t - 3) \sin 2t] * H(t - 2)$ **B20.** $[H(t - 1) \cos t] * P_{2,4}(t)$

B21. $[t^2 P_{1,2}(t)] * P_{3,4}(t)$

6.10 The Dirac Delta Function

A Question

We now introduce a sophisticated object whose complete explanation is beyond the scope of this text. However, this object is so useful that it really needs to be available at every level. We meet this challenge of exposition by posing the following question:

Which function $f(t)$ has, as its Laplace transform, $F(s) = 1$?

The Answer

Surprisingly, **no** function $f(t)$ has $F(s) = 1$ as its Laplace transform. However, the symbol $\delta(t)$ has been invented as an object with the property that $L[\delta(t)] = 1$. Textbooks call this object the "delta function." Since the physicist Paul Dirac had a hand in promulgating its use, it is often called the "Dirac delta function."

Isn't this a contradiction? First we state that no *function* transforms to 1 and then we say that $L[\delta(t)] = 1$ for the *Dirac delta function* $\delta(t)$. This is not a contradiction. The object $\delta(t)$ is not a function. It's an invented "object" concocted to have Laplace transform 1.

The delta "function," $\delta(t)$, is a *distribution*, a member of a class of mathematical objects whose complete description would take considerably more time than we can afford here. Distributions are not point-valued functions but rather *functionals*, functions whose domain is a set of functions and whose range is the real numbers.

So, it is true that no point-valued function has 1 as its Laplace transform, but it is also true that defining an object whose Laplace transform is 1 is an extremely useful generalization.

The Dirac Delta as Impulse

For a first insight into the utility of the object $\delta(t)$, the Dirac delta function, consider the following two initial value problems

$$\text{IVP}_1: \ y'' + y = 0 \qquad y(0) = 0, y'(0) = 5$$
$$\text{IVP}_2: \ y'' + y = 5\delta(t) \qquad y(0) = 0, y'(0) = 0$$

The first problem is a standard initial value problem for a harmonic oscillator that is initially at the equilibrium position but moving with an initial velocity of 5. The solution of IVP_1 is therefore $y(t) = 5 \sin t$. The "solution" of IVP_2 is obtained by Laplace transforms. Hence,

$$s^2 Y + Y = 5$$

from which we get $Y(s) = \frac{5}{s^2+1}$ and therefore $y(t) = 5 \sin t$.

Each IVP delivered the same function, namely, $5 \sin t$. This is a classical solution for the first IVP but a "generalized" solution for the second. In fact, to get the second IVP to yield $5 \sin t$, the initial condition $y'(0) = 0$ was used, but $5 \sin t$ satisfies $y'(0) = 5$ not $y'(0) = 0$. The "solution" to the second IVP is actually a function with a discontinuity in its first derivative and is therefore not a classical solution.

In spite of the difficulty of dealing with the nonclassical solution generated by the Dirac delta function, it is still convenient to argue that an initial velocity of five units in the first problem, representing a physical "kick" or impulse imparted to the system, is replicated by driving the system with the delta function. A pragmatic interpretation of the Dirac delta is that it must be equivalent to delivering to the system an initial impulse or kick.

The Translated Dirac Delta: $\delta(t-a)$

If the kick is to be delivered at time $t = a$, then the Dirac delta is translated as $\delta(t-a)$ and its Laplace transform is defined by $L[\delta(t-a)] = e^{-as}$, consistent with the Second Shifting Law. The Second Shifting Law says that the shift $t = a$ in the time domain is equivalent to multiplication by e^{-as} in the transform domain. But the transform of the unshifted Dirac delta is $F(s) = 1$. Except for an unwritten Heaviside function, $L[\delta(t-a)] = e^{-as}$ is consistent with the Second Shifting Law. We are untroubled by the missing Heaviside function because the Dirac delta "function" is not a point-valued object. Heuristically, people say its value is zero except at $t = a$, but that is a myth. It just isn't a point-valued object, so it does not need a Heaviside function in the statement of the Second Shifting Law.

The Sifting Property of δ $(t-a)$

To add insult to injury, not only is the Dirac delta "function" not a function in the ordinary sense of the word, but we also have the audacity to put it inside an integral sign. Worse still, we then declare that the integral sign does not stand for integration. How can we say such things?

We stumbled upon the Dirac delta as an answer to the question, "Which function has Laplace transform $F(s) = 1$?" An alternate route to this same object is the question, "Which function $\delta(t)$ satisfies the equation

$$\int_{-\infty}^{\infty} \delta(t) f(t)\, dt = f(0)$$

for all input functions $f(t)$ continuous at $t = 0$?" Again, the answer is, "there is no such *function* $\delta(t)$." Thus, an object is invented so that the property holds, but the object cannot really be a function and the integral sign cannot really be an integral. However, the object, and its integral property, have proven so useful that both have been tolerated, and even encouraged, in the applied mathematics literature!

This integral rule generalizes and is called the "Sifting Property" of the Dirac delta function. In particular, under translation, we have

$$\int_{-\infty}^{\infty} \delta(t-a) f(t)\, dt = f(a) \tag{6.15}$$

for every function $f(t)$ continuous at $t = a$. Remember, we are not evaluating an actual integral. Neither we nor the underlying mathematics interprets these as real integrals. The "integral" in (6.15) is not evaluated by antidifferentiation but is interpreted as the Sifting Property.

Moreover, the Sifting Property will hold over any interval containing a. For example, with h taken as positive, the Sifting Property includes the following statements that should be interpreted as definitions:

$$\int_{a-h}^{a+h} \delta(t-a) f(t)\, dt = f(a)$$

$$\int_{a}^{a+h} \delta(t-a) f(t)\, dt = f(a)$$

$$\int_{a-h}^{a} \delta(t-a) f(t)\, dt = 0$$

Explaining $\delta(x)$ by "Concentrated Loads"

Because it is so difficult, at this level, to give a complete explanation of the logical structure in which the Dirac delta function properly resides, a considerably shorter but less rigorous argument typically prevails. This argument seeks to demonstrate the behavior of the Dirac delta function as the limiting behavior of a sequence of "real" functions.

At this point, the reader is warned about making a significant logical error. If a sequence of functions $f_n(x)$ all have a certain property and this sequence converges to a limit function $f(x)$, the limit function $f(x)$ need not have the same property shared by the members of the sequence! The typical example is the property of continuity. The sequence of polynomials $\{x^n\}_{n=0}^{\infty}$ are all continuous on the interval $[0, 1]$. The sequence converges to the discontinuous function

$$\begin{cases} 0 & 0 \le x < 1 \\ 1 & x = 1 \end{cases}$$

a result illustrated graphically in Figure 6.36.

Analytically, just remember that for x, any positive number less than 1, the value of $\lim_{n=\infty} x^n$ is 0. At $x = 1$ the value of 1^n is 1 for every n, so the limit function must be discontinuous. Each individual function in the sequence of functions is continuous, but the limiting function is discontinuous. The limit function need not inherit all the properties of the "nearby" functions.

It is imperative for the reader to keep this observation in mind during the following demonstration in which we set up a sequence of functions based on a unit pulse between $x = 0$ and $x = h$, with $h > 0$. Thus, define the unit pulse by $p(x) = H(x) - H(x - h)$ and then adjust the height (or "strength") of this pulse so that the area beneath the pulse is 1. Thus, define

$$y_h(x) = \frac{p(x)}{h} = \frac{1}{h}(H(x) - H(x - h))$$

where $H(x)$ is the Heaviside function. A plot of several members of the family of functions $y_h(x)$ is found in Figure 6.37. As $h \to 0$, the height of $y_h(x)$ increases. If $f(x)$ is integrable for $x \ge 0$ and continuous at $x = 0$, we have

$$\int_0^{\infty} y_h(x) f(x)\, dx = \int_0^h \frac{1}{h} f(x)\, dx = f(c) \tag{6.16}$$

for some c in $(0, h)$. By continuity, $f(c) \to f(0)$ as $h \to 0$, so it becomes tempting to claim that $\lim_{h \to 0} y_h(x) = \delta(x)$ and to say that, as the load becomes more and more concentrated, $y_h(x)$ approaches the delta function. It is thus appealing to explain the Dirac delta function in terms of the limit of a sequence of unit loads becoming more and more concentrated. Hence, the idea of the delta function as the "concentrated load" is born.

Alternatively, compute the Laplace transform

$$Y_h(s) = L[y_h(x)] = \frac{1 - e^{-hs}}{hs} \tag{6.17}$$

then "concentrate" the normalized pulse by letting $h \to 0$ calculating $\lim_{h \to 0} Y_h(s) = 1$. Again, the limit of an integral of $y_h(x)$ *behaves* like the delta function, but that does not mean the limit of $y_h(x)$ itself *is* the delta function. As we said at the beginning of the discussion, the limit function may not inherit all the properties of the converging nearby neighbors. The actions of the sequence $y_h(x)$ in (6.16) and (6.17) approach the action of the delta function, but the limit of the sequence $y_h(x)$ is not itself the delta function.

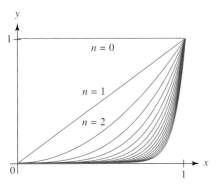

FIGURE 6.36   Continuous functions converging pointwise to a discontinuous function

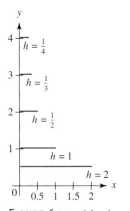

FIGURE 6.37   Members of the sequence $\frac{1}{n}(H(x) - H(x - n))$

EXERCISES 6.10–Part A

A1. Use the Laplace transform to obtain a formal solution to
$y'' + y = \delta(t - 1)$, $y(0) = y'(0) = 0$. Sketch the solution and its
derivative. Show that at $t = 1$ the solution has a jump of magnitude
1 in its derivative. Thus, the Laplace transform does not yield a
classical solution, but rather, a generalized one.

A2. Solve each of the following initial value problems by taking the
Laplace transform of the equation, solving for $Y(s)$, and inverting;
then plot the solution.

  (a) $y'' + 2y' + 5y = \delta(t - 1)$, $y(0) = 0$, $y'(0) = 0$
  (b) $3y'' + 4y' + 2y = \delta(t - 1)$, $y(0) = 0$, $y'(0) = 0$
  (c) $2y'' + 3y' + \frac{9}{4}y = \delta(t - 1)$, $y(0) = 0$, $y'(0) = 0$

A3. From Exercise A2, deduce the effect of having $\delta(t - 1)$ on the
right-hand side of the differential equation. Write complete and
grammatically correct English sentences to express your
observations and conclusions.

EXERCISES 6.10–Part B

B1. Let $f_a(t)$ be the piecewise linear function connecting the origin,
the point $(a, \frac{1}{a})$, and the point $(2a, 0)$. Its graph forms an isosceles
triangle with base $2a$, height $\frac{1}{a}$, and area 1. Show that $L[f_a(t)] \to 1$
as $a \to 0$, making it clear that there is more than one sequence
upon which to base the "concentrated load" explanation of the
Dirac delta function.

B2. Obtain, graph, and compare the solutions to

  (a) $y'' + 2y' + 5y = 0$, $y(0) = 0$, $y'(0) = 1$
  (b) $y'' + 2y' + 5y = \delta(t)$, $y(0) = 0$, $y'(0) = 0$

B3. Obtain, graph, and compare the solutions to

  (a) $3y'' + 4y' + 2y = 0$, $y(0) = 0$, $y'(0) = \frac{1}{3}$
  (b) $3y'' + 4y' + 2y = \delta(t)$, $y(0) = 0$, $y'(0) = 0$

6.11 Transfer Function, Fundamental Solution, and the Green's Function

Transfer Function and Fundamental Solution

The spring-mass system governed by the differential equation $y'' + 2y' + 2y = f(t)$ can be
viewed as a black box that takes an input $f(t)$ and delivers an output $y(t)$. In some sense,
the system is *transferring* the action of the input into an output. We follow up on this line
of thought.

Begin by finding out what the system does to a unit impulse $\delta(t)$. This corresponds to
giving the resting system a kick, or ping (like a sonar ping in the submarine movies). We
calculate the system's response to such a jolt.

Hence, solve the initial value problem $v'' + 2v' + 2v = \delta(t)$, $v(0) = v'(0) = 0$ for the
Laplace transform $V(s)$ and its inverse, $v(t)$. We will call $v(t)$ the **fundamental solution**
and $V(s)$ its Laplace transform, the **transfer function.** Hence, the Laplace transform of the
differential equation is

$$s^2 V(s) + 2s V(s) + 2V(s) = 1$$

from which we determine the *transfer function* to be

$$V(s) = \frac{1}{s^2 + 2s + 2}$$

and the *fundamental solution*, the inverse transform of the transfer function, to be

$$v(t) = e^{-t} \sin t$$

Table 6.7 summarizes the two definitions illustrated in this example.

| Fundamental solution | $v(t)$ | Inert system's response to unit impulse |
| Transfer function | $V(s)$ | Laplace transform of fundamental solution |

TABLE 6.7 Fundamental solution and transfer function

The Green's Function

Suppose we drive the system with forcing term $f(t) = \cos 2t$. To do this we solve, by transform techniques, the initial value problem $y'' + 2y' + 2y = \cos 2t$, $y(0) = y'(0) = 0$, obtaining

$$Y(s) = \frac{1}{s^2 + 2s + 2} \left[\frac{s}{s^2 + 4} \right] = V(s)F(s)$$

and the solution

$$y(t) = \frac{e^{-t}}{10} (\cos t - 3 \sin t) + \frac{1}{10} (2 \sin 2t - \cos 2t)$$

Presently, however, we are more interested in the transform $Y(s)$ that has been factored into the product of $V(s)$, the transfer function, and $F(s)$, the transform of the driving term $f(t)$. This is the product of two Laplace transforms, so the inverse, $y(t)$, is therefore the convolution of $v(t)$ and $f(t)$. This convolution is

$$y(t) = v * f = \int_0^t v(t - x) f(x) \, dx$$

and from Section 5.13, we are able to recognize $v(t - x)$ as $g(t, x)$ in the Green's function $G(t, x)$. In fact, $g(t, x)$ for this problem is

$$v(t - x) = e^{-(t-x)} \sin(t - x) = e^{-t}e^x \sin t \cos x - e^{-t}e^x \cos t \sin x = g(t, x)$$

Green's Function Revisited

We have just observed that the solution of the initial value problem $y'' + 2y' + 2y = \delta(t)$, $y(0) = y'(0) = 0$ is the fundamental solution $v(t)$. The $g(t, x)$ from the Green's function $G(t, x)$ is then supposedly $v(t - x)$. In fact, we will now show that the function $v(t - x)$ is actually the solution of the initial value problem $u'' + 2u' + 2u = \delta(t - x)$, $u(0) = u'(0) = 0$. Again, taking Laplace transforms, we get

$$s^2 U + 2sU + 2U = e^{-xs}$$

and hence,

$$U = \frac{e^{-xs}}{s^2 + 2s + 2}$$

and

$$u(t) = H(t - x)e^{-(t-x)} \sin(t - x)$$

where $H(x)$ is the Heaviside function. Except for the factor of $H(t - x)$ we have obtained $v(t - x)$. But in use, u would appear in an integral of the form

$$\int_0^t H(t - x)e^{-(t-x)} \sin(t - x) f(x) \, dx$$

and, since $x < t$ in the integral, $H(t - x) = 1$ there, so u is indeed the same as $v(t - x)$.

The diligent reader will not rest, however, until the process detailed in Section 5.13 is used to construct the very same Green's function as we have twice obtained in this section. Recall that recipe was $G(t, x) = \begin{cases} 0 & t < x \\ g(t, x) & t > x \end{cases}$, where $g(t, x) = \frac{y_2(t)y_1(x) - y_1(t)y_2(x)}{w(x)}, \{y_1, y_2\}$ is a fundamental set of solutions for the homogeneous differential equation, and $w(x)$ is their Wronskian. Since the characteristic equation is $\lambda^2 + 2\lambda + 2 = 0$, the characteristic roots are $\lambda = -1 \pm i$, yielding $y_1 = e^{-t} \cos t$, $y_2 = e^{-t} \sin t$, and $w(t) = e^{-2t}$. Consequently,

$$g(t, x) = \frac{e^{-t}\sin(t)e^{-x}\cos(x) - e^{-t}\cos(t)e^{-x}\sin(x)}{e^{-2x}} = e^{-(t-x)}\sin(t - x)$$

in agreement with $v(t - x)$.

A Final Interpretation

The response of $y'' + 2y' + 2y$ to the unit impulse $\delta(t)$ is the fundamental solution $v(t)$. If the kick were delivered at time $t = x$, the driving term for the system would be $\delta(t - x)$ and the response would be $v(t - x)$. If at each moment $t = x$ the driving function $f(t)$ is evaluated to $f(x)$ and that amount of impulsive kick delivered to the system, the driving term would be $\delta(t - x)f(x)$. The response to such a kick would now be $v(t - x)f(x)$.

Finally, if the action of $f(t)$ as a driving term for the system were decomposed into a collection of kicks $\delta(t - x)f(x)$ and the effects of all these individual kicks summed by integration over the corresponding responses $v(t - x)f(x)$, the system's total response would be

$$y(t) = \int_0^t v(t - x)f(x)\,dx$$

But this is just the convolution we computed earlier, and the Green's function appears naturally in the convolution integral.

EXERCISES 6.11–Part A

A1. The motion of a spring-mass system is governed by the IVP
$y'' + 2y' + 5y = 1 - H(t - 1)$, $y(0) = y'(0) = 0$.

(a) Use the Laplace transform to obtain the solution.

(b) Obtain the transfer function $V(s)$.

(c) Obtain the fundamental solution $v(t)$.

(d) Obtain the Green's function $v(t - x)$.

(e) The solution of the given IVP can be written as in integral with the Green's function. Obtain and evaluate this representation,

and then compare the result with the solution obtained in part (a).

(f) Use the Laplace transform to obtain a formal solution to $u'' + 2u' + 5u = \delta(t - x)$, $u(0) = u'(0) = 0$, and compare the solution to $v(t - x)$ obtained in part (d).

A2. Repeat Exercise A1 for the IVP $y'' + 3y' + 2y = t(1 - H(t - 2))$.

EXERCISES 6.11–Part B

B1. Use the Laplace transform to solve the IVP $y'' + 2y' + 10y = \delta(t) - \delta(t - 1) + \delta(t - 2) - \delta(t - 3) + \delta(t - 4) - \delta(t - 5)$, $y(0) = y'(0) = 0$, and then sketch the solution. Give a physical interpretation of the IVP and the resulting motion.

For the IVPs in Exercises B2–5, where the initial conditions are taken as $y(0) = y'(0) = 0$:

(a) Solve, using appropriate technology, and plot the solution.

(b) Obtain the transfer function $V(s)$.

(c) Find the fundamental solution $v(t)$.

(d) Write the solution as $y(t) = \int_0^t v(t-x)f(x)\,dx$, the convolution of $v(t)$ and $f(t)$, and show it is equivalent to that found in part (a).

(e) Find the Green's function $G(t,x) = \begin{cases} 0 & t < x \\ g(t,x) & x < t \end{cases}$ using the technique of Section 5.13.

(f) Show that $v(t-x)$ is $g(t,x)$.

(g) Find $v(t-x)$ as the solution to the IVP formed when the right-hand side is replaced by $\delta(t-x)$.

B2. $y'' + 2y' + 5y = 5\sin 3t$ **B3.** $y'' + 4y' + 13y = 10\cos 2t$

B4. $y'' + 6y' + 10y = f(t) = \begin{cases} t & 0 \le t < 1 \\ 1 & 1 \le t < 2 \\ 3-t & 2 \le t < 3 \\ 0 & t \ge 3 \end{cases}$

B5. $y'' + 5y' + 6y = t^2(1 - H(t-2))$

In Exercises B6–8, plot the solution after obtaining it using:

(a) An appropriate technology.

(b) The Laplace transform, via the following technique. Where the Laplace transform requires the initial values $y(0)$ and $y'(0)$, use the parameters a and b. Obtain $Y(s)$ in terms of these two parameters, and invert to obtain $y(t)$ containing a and b. Impose the given initial conditions on this solution, resulting in two equations in the two unknowns, a and b. Solve for a and b, and hence, obtain the solution to the given problem, which is called a *boundary value problem* and will be studied at length in Unit Three.

B6. $y'' + 2y' + 17y = \cos 2t$, $y\left(\frac{\pi}{3}\right) = -3$, $y'\left(\frac{2\pi}{3}\right) = 1$

B7. $6y'' + 5y' + y = t$, $y(0) = 1$, $y(1) = e$

B8. $4y'' + 7y' + 3y = t^2$, $y(0) = -1$, $y(1) = e$

Chapter Review

1. Give a verbal description of the Laplace transform.
2. State the mathematical definition of the Laplace transform.
3. Give an example of a function that does not have a Laplace transform.
4. Delineate a class of functions for which all members have a Laplace transform.
5. Give a verbal statement of each of the four operational rules for the Laplace transform. Highlight the symmetries in these rules.
6. In mathematical terms, state the four operational rules for Laplace transforms.
7. What are the Laplace transforms for t^k, $k = 0, 1, \ldots, 4$?
8. What are the Laplace transforms for $\cos \omega t$ and $\sin \omega t$?
9. What is the Laplace transform for e^{at}?
10. What are the Laplace transforms for $\cosh \omega t$ and $\sinh \omega t$?
11. Illustrate each of the four operational rules with an example.
12. Give a verbal statement of the First Shifting Law for the Laplace transform.
13. In mathematical terms, state the First Shifting Law. Give an example.
14. Define the Heaviside function. Draw its graph.
15. Use the Heaviside function to express a pulse of duration 3, starting at $t = 5$, and having height 7.
16. Use the Heaviside function to represent the function that is zero prior to $t = 2$, is $\sin t$ for $2 \le t \le 5$, and zero after $t = 5$.

17. Give a verbal statement of the Second Shifting Law for the Laplace transform. Highlight its symmetry with respect to the First Shifting Law.
18. In mathematical terms, state the Second Shifting Law.
19. Give an example of the use of the Second Shifting Law to obtain a Laplace transform.
20. Use the Second Shifting Law to obtain the Laplace transform of $t\text{Heaviside}(t-1)$. Check by writing and evaluating an appropriate integral.
21. Use the Second Shifting Law to invert the Laplace transform $F(s) = e^{-2s}/(s^2 + s)$.
22. Show how the Laplace transform is used to solve IVPs containing nonhomogeneous second-order linear ODEs with constant coefficients. Give an example.
23. Define the convolution product for two functions $f(t)$ and $g(t)$.
24. Obtain the convolution of $\sin t$ with $\cos t$.
25. State the convolution theorem for Laplace transforms.
26. Demonstrate how to use the convolution theorem to obtain the convolution of $\sin t$ with $\cos t$.
27. Use the convolution theorem to obtain the convolution of $\text{Heaviside}(t-1) - \text{Heaviside}(t-2)$ with $\text{Heaviside}(t-3) - \text{Heaviside}(t-5)$.
28. State the rule for finding the Laplace transform of a periodic function. Illustrate it with an example.
29. Define *fundamental solution* and *transfer function*. Give an example of each.

Unit Two

Infinite Series

The five chapters of Unit Two are really about representing functions, especially functions that are solutions of differential equations. For several important classes of differential equations that arise in science and engineering, Unit One showed how to find *exact* solutions in terms of the *elementary functions,* namely, polynomials, rational functions, trig functions and their inverses, exponentials, and logarithms.

However, many important differential equations appearing in practice cannot be solved in terms of these functions. Solutions can often only be expressed as infinite series, that is, infinite sums of simpler functions such as polynomials, or trig functions.

We must therefore give meaning to an infinite sum of constants, using this to give meaning to an infinite sum of functions. When the functions being added are the simple powers $(x - x_0)^k$, the sum is called a Taylor (power) series and if $x_0 = 0$ a Maclaurin series. When the functions are trig terms such as $\sin kx$ or $\cos kx$, the series might be a Fourier series, certain infinite sums of trig functions that can be made to represent arbitrary functions, even functions with discontinuities. This type of infinite series is also generalized to sums of other functions such as Legendre polynomials. Eventually, solutions of differential equations will be given in terms of infinite sums of Bessel functions, themselves infinite series.

When such infinite series are truncated, that is, terminated after a finite number of terms have been added, an approximate solution of the differential equation results. Although the availability of computing hardware and software has made *numeric* solutions of some of these equations viable, these additional symbolic techniques are sometimes imperative, and always useful, when the differential equation contains symbolic parameters.

But these are not the only ways of obtaining approximate analytic solutions of differential equations. Solutions of differential equations can also be given by asymptotic series, which can even be divergent. Surprisingly, such divergent asymptotic series can give useful approximations!

Since all these types of series are eventually going to be used to represent solutions of differential equations, it is imperative that we know when termwise operations such as differentiation and integration are valid. So, Unit Two contains some of the *tools* for approximating functions, whereas Unit Three will detail the *techniques* for obtaining series solutions and approximate solutions for differential equations. The notions of *convergence* and *divergence* developed in Unit Two are a useful, if not necessary, prelude to our continued work with differential equations.

Chapter 7

Sequences and Series of Numbers

INTRODUCTION   The key concept in Chapter 7 is *convergence,* both of sequences and series of constants. The infinite series is an infinite sum whose precise meaning is given by the behavior of its *sequence of partial sums*. Hence, we first study the convergence of sequences, in terms of which we define the convergence of infinite series. A little bit of rigor here is an innoculate against susceptibility to "proofs," based on a mishandling of infinite series, that $0 = 1$. For example, abusing the series $1 - 1 + 1 - 1 + \cdots$ leads to the contradictory outcome

$$0 = (1 - 1) + (1 - 1) + \cdots = 1 + (-1 + 1) + (-1 + 1) + \cdots = 1$$

Understanding convergence for a series of numbers paves the way for understanding, in Chapter 8, pointwise convergence for a series of functions. Ultimately, we wish to represent solutions of differential equations as infinite series of *functions,* so Chapters 7 and 8 are stepping stones along that path.

Some material in Chapter 7 may have been seen in an elementary calculus course that provided an introduction to sequences and series of numbers and the convergence thereof. For students already familiar with these ideas, this chapter is optional.

7.1 | Sequences

DEFINITION 7.1

A *sequence* is a function whose domain is the set of integers and whose range is the set of real or complex numbers. Typically, the domain is the set of counting numbers, namely, $0, 1, 2, \ldots$. Sometimes the domain can include negative integers, and sometimes the domain can start with an integer $n_0 > 0$. The values of the function, namely, $f(k)$, are the elements of the sequence, and these are generally written as $a_k = f(k)$.

In short, a sequence is an ordered list of numbers, as the following examples show.

EXAMPLE 7.1   The ordered list of numbers $1, \frac{1}{2}, \frac{1}{4}, \frac{1}{8}, \frac{1}{16}, \ldots$, which we label as $a_0, a_1, a_2, \ldots$, and form according to the rule $f : k \to 1/2^k$, $k = 0, 1, \ldots$, is an example of a *sequence* of numbers.

Figure 7.1, a graph of the points (k, a_k), suggests that if k were to increase without bound, the corresponding term a_k would approach zero. In fact, we know from elementary calculus that $\lim_{k\to\infty}(1/2^k) = 0$; so when the sequence has an analytic rule, the limiting value of a_k can often be found by treating the integer index as a continuous variable and applying the limit rules from elementary calculus. ❖

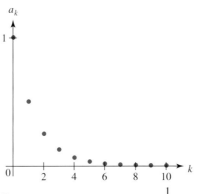

FIGURE 7.1   The sequence $a_k = \dfrac{1}{2^k}$

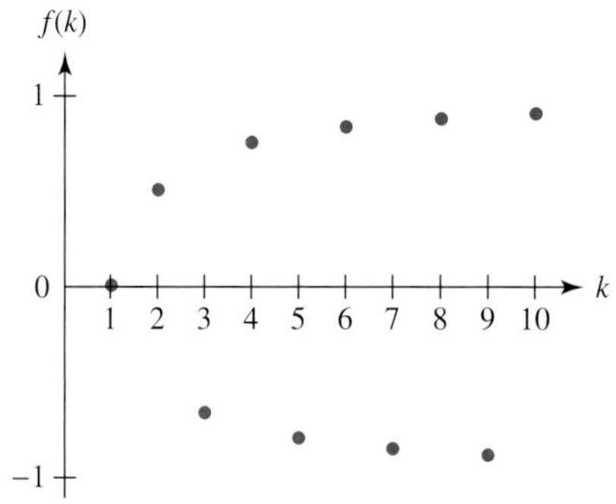

FIGURE 7.2   The sequence $a_k = (-1)^k \dfrac{k-1}{k}$

EXAMPLE 7.2   The ordered list of numbers $0, \frac{1}{2}, -\frac{2}{3}, \frac{3}{4}, -\frac{4}{5}, \frac{5}{6}, -\frac{6}{7}, \frac{7}{8}, -\frac{8}{9}, \frac{9}{10}, \dots$ formed according to the rule $f : k \to (-1)^k (k - 1)/k$, $k = 1, 2, \dots$, is graphed as $(k, f(k))$ in Figure 7.2. This graph shows the oscillatory effect of the factor $(-1)^k = -1, 1, -1, \dots$ and suggests there may not be a single limiting value for the members of the sequence. Every other member of the sequence tends to a different value, either 1 or -1, so the sequence does not have a unique limit. Finally, we note that $k = 0$ is not in the domain because $f(0)$ is undefined. ❖

Convergence and Divergence of Sequences

We give two definitions for the *convergence* of the sequence $\{a_k\}$ to the limit L. The first is intuitive and consistent with the formalism of elementary calculus. The second is precise and historically interesting but significantly more difficult to apply.

DEFINITION 7.2

The sequence $\{a_k\}$ converges to the limit L if $a_k = f(k)$ and $\lim_{k\to\infty} f(k) = L$.

DEFINITION 7.3

The sequence $\{a_k\}$ converges to the limit L if, for each positive number ε, there corresponds an index N beyond which every member of the sequence obeys the inequality $|a_k - L| < \varepsilon$.

A sequence that does not converge is said to *diverge*. A sequence can diverge because its members oscillate between two or more different values (as in Example 7.2) or because its members grow without bound and therefore $L = \infty$, which is not a finite number. In this latter event, the sequence is said to *diverge to infinity*.

EXAMPLE 7.3 The sequence whose general term is $a_k = 3^k/k^3$ diverges because $\lim_{k\to\infty}(3^k/k^3) = \infty$, a result supported by Figure 7.3.   ❖

EXAMPLE 7.4 The sequence which has $a_k = (5/k)^{1/k}$ for its general term converges to the limit $L = 1$ because $\lim_{k\to\infty}(5/k)^{1/k} = 1$, a result evident in Figure 7.4.   ❖

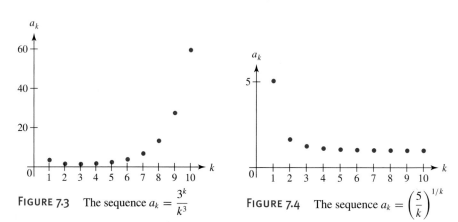

FIGURE 7.3   The sequence $a_k = \dfrac{3^k}{k^3}$ **FIGURE 7.4**   The sequence $a_k = \left(\dfrac{5}{k}\right)^{1/k}$

EXERCISES 7.1–Part A

A1. List the first five terms of the sequence $a_k = k^2/2^k$, $k = 0, 1, \ldots$, then analytically determine L, its limit.

A2. List the first five terms of the sequence $a_k = 3^k/k$, $k = 1, \ldots$, then analytically show it diverges (to infinity).

A3. For each of $n = 1, 3, 5$, find N, an integer for which $k > N$ guarantees $a_k > 10^n$, where a_k is defined as in Exercise A2.

A4. If a, b, c are constants, then $au_{k+2} + bu_{k+1} + cu_k = 0$ is a constant-coefficient, linear, homogeneous, second-order *difference*

equation with general solution $u_k = A\alpha^k + B\beta^k$, where α and β are roots of the quadratic equation $ar^2 + br + c = 0$. The arbitrary constants A and B can be determined by prescribing, for example, the values of u_1 and u_2. Show that the guess $u_k = r^k$ leads to this solution.

A5. Starting with $x_0 = 0$, obtain x_{k+1}, $k = 0, \ldots, 4$, if $x_{k+1} = g(x_k)$ and $g(x) = 1 + \frac{3}{16}x^2$. Find any fixed points of this iteration, that is, solutions of $x = g(x)$.

EXERCISES 7.1–Part B

B1. Show that the iteration in Exercise A5 diverges if $x_0 = 5$.

For the sequences given in Exercises B2–10:

   (a) List the first 10 members and plot the corresponding points (k, a_k).

   (b) Find L, the limit of the sequence.

   (c) For each $n = 1, 3, 5$, find N, an integer for which $k > N$ guarantees $|a_k - L| < 1/10^n$.

   (d) Find, if possible, an $N(\varepsilon)$ for which $k > N(\varepsilon)$ guarantees $|a_k - L| < \varepsilon$ for *any* small, positive ε.

B2. $\dfrac{k^2 + k}{2k^2 - 1}$, $k = 1, \ldots$ **B3.** $\dfrac{(-1)^k \sqrt{k}}{k + 1}$, $k = 1, \ldots$

B4. $\dfrac{\ln k}{k}$, $k = 1, \ldots$ **B5.** $\dfrac{\ln(\ln k)}{\sqrt{k}}$, $k = 2, \ldots$

B6. $\dfrac{k!}{k^k}$, $k = 1, \ldots$ **B7.** $\left(\dfrac{5k + 2}{5k + 1}\right)^k$, $k = 0, \ldots$

B8. $k - \sqrt{k^2 + k + 1}$, $k = 0, \ldots$ **B9.** $\left(1 + \dfrac{1}{k}\right)^{\ln k}$, $k = 1, \ldots$

B10. $\left(1 + \dfrac{1}{k!}\right)^{k^2}$, $k = 0, \ldots$

For each of the sequences in Exercises B11–17:

  (a) List the first 5 members and plot the corresponding (k, a_k).

  (b) Show that the sequence diverges (to infinity).

  (c) For each $n = 1, 3, 5$, find N, an integer for which $k > N$ guarantees $|a_k| > 10^n$.

B11. $\dfrac{(-1)^k 2^{k+1}}{k^2}, k = 1, \ldots$ **B12.** $\left(\dfrac{5k+2}{4k+1}\right)^k, k = 0, \ldots$

B13. $\dfrac{\cosh(\ln k)}{\sqrt{k}}, k = 1, \ldots$ **B14.** $k - \sqrt{2k^2 - k}, k = 1, \ldots$

B15. $\dfrac{7k-5}{3+2\sqrt{k}}, k = 0, \ldots$ **B16.** $\left(1 + \dfrac{1}{k}\right)^{k^2}, k = 0, \ldots$

B17. $\left(1 + \dfrac{1}{k!}\right)^{k^k}, k = 1, \ldots$

In Exercises B18–22, use the ideas of Exercise A4. For each:

  (a) Find an expression for the solution u_k.

  (b) Plot the solution.

  (c) Obtain $\lim_{k\to\infty} u_k$.

  (d) Show that u_3, u_4, and u_5 determined by the analytic solution are given by $u_{k+2} = -(bu_{k+1} + cu_k)/a, k = 1, 2, 3$.

B18. $u_{k+2} + 3u_{k+1} - u_k = 0, u_1 = 1, u_2 = -1$

B19. $2u_{k+2} - 5u_{k+1} - 3u_k = 0, u_1 = -2, u_2 = 3$

B20. $3u_{k+2} - u_{k+1} - 5u_k = 0, u_1 = 3, u_2 = 5$

B21. $u_{k+2} + 7u_{k+1} - 6u_k = 0, u_1 = -5, u_2 = -3$

B22. $-u_{k+2} + 3u_{k+1} + 2u_k = 0, u_1 = 0, u_2 = -2$

The ideas of Exercise A4 do not apply to difference equations that are nonlinear or do not have constant coefficients. In Exercises B23–27:

  (a) Determine if the equation is nonlinear or does not have constant coefficients (or both).

  (b) For $k = 1, \ldots, 5$, obtain u_{k+2} from u_k, u_{k+1} and the difference equation itself.

  (c) Plot the points $(k, u_k), k = 1, \ldots, 9$.

B23. $u_{k+2} + ku_{k+1} - u_k = 0, u_1 = 2, u_2 = -1$

B24. $u_{k+2} - k^2 u_{k+1} + \dfrac{1}{k} u_k = 0, u_1 = 1, u_2 = 1$

B25. $u_{k+2} = k\dfrac{u_{k+1}}{u_k}, u_1 = -1, u_2 = 2$

B26. $u_{k+2} = u_{k+1}^2 - 2u_k, u_1 = 1, u_2 = 2$

B27. $u_{k+2}u_{k+1} - 2u_k = k, u_1 = 2, u_2 = 3$

If the equation $f(x) = 0$ can be put into the form $x = g(x)$, then the sequence defined by $x_{k+1} = g(x_k)$ converges to a root r when x_0 is sufficiently near r and $|g'(r)| < 1$. For the functions $g(x)$ given in Exercises B28–34:

  (a) Use appropriate technology to determine r, a solution of $x - g(x) = 0$ for which $|g'(r)| < 1$.

  (b) Obtain $x_k, k = 1, \ldots, 5$, for an x_0 sufficiently near r to guarantee convergence.

B28. $g(x) = x^2 - \frac{13}{5}x + 3$ **B29.** $g(x) = 2x^3 - \frac{415}{36}x^2 + \frac{191}{12}x + 5$

B30. $g(x) = \cos x$ **B31.** $g(x) = \tan\left(\frac{225}{1606}x - \frac{83{,}492}{9721}\right)$

B32. $g(x) = \frac{7227}{264{,}985}x^2 + \sqrt{3 + \frac{27{,}529}{8030}x}$

B33. $g(x) = \dfrac{6050x^2 + 1369}{8000x + 6845}$

B34. $g(x) = -(4x^3 + 5x^2 + 2x + \frac{4}{5})$

B35. The continued fraction

$$a_1 + \cfrac{1}{a_2 + \cfrac{1}{a_3 + \cfrac{1}{a_4 + \cdots}}}$$

typically abbreviated to $a_1 + \frac{1}{a_2+} \frac{1}{a_3+} \frac{1}{a_4+} \cdots$, is defined to mean the limit, if it exists, of the sequence of *convergents*

$$a_1, a_1 + \frac{1}{a_2}, a_1 + \cfrac{1}{a_2 + \cfrac{1}{a_3}}, a_1 + \cfrac{1}{a_2 + \cfrac{1}{a_3 + \frac{1}{a_4}}}, \ldots$$

Given the continued fraction

$$y = 2 + \cfrac{1}{4 + \cfrac{1}{4 + \cfrac{1}{4 + \cdots}}}$$

  (a) write out the first six terms of the sequence of convergents.

  (b) find

$$y = x + 2 \ \text{if} \ x = \cfrac{1}{4 + \cfrac{1}{4 + \cfrac{1}{4 + \cdots}}}$$

so $x = \frac{1}{4+x}$, assuming, of course, that x exists.

The infinite product $\prod_{k=1}^{\infty}(1 + u_k)$ is defined as the limit of the sequence of *partial products* $P_n = \prod_{k=1}^{n}(1 + u_k) = (1 + u_1)(1 + u_2)\cdots(1 + u_n)$, provided $u_k \neq -1, k = 1, 2, \ldots$. For the infinite products in Exercises B36–40:

  (a) Write out the first six partial products.

  (b) Plot the points (n, P_n).

  (c) With a computer algebra system, try to determine the exact value of the infinite product.

  (d) Compare the partial products from part (a) with any analytic answers obtained in part (c).

B36. $\displaystyle\prod_{k=1}^{\infty}\left(1 + \frac{1}{k^2}\right)$ **B37.** $\displaystyle\prod_{k=2}^{\infty}\left(1 - \frac{1}{k^2}\right)$ **B38.** $\displaystyle\prod_{k=1}^{\infty}\left(1 + \frac{1}{k^3}\right)$

B39. $\displaystyle\prod_{k=2}^{\infty}\left(1 - \frac{1}{k^3}\right)$ **B40.** $\displaystyle\prod_{k=1}^{\infty}\left(1 + \frac{1}{k}\right)e^{-1/k}$

7.2 Infinite Series

Terminology

INFINITE SERIES   If $\{a_k\}$, $k = 0, 1, \ldots$, is a sequence of numbers, the ordered sum of all its terms, namely, $\sum_{k=0}^{\infty} a_k = a_0 + a_1 + a_2 + a_3 + \cdots$, is called an *infinite series*. The terms of the sequence are added in the order in which they appear in the sequence. Thus, for example, the sequence defined by $a : k \to 1/2^k$, for which the first few terms are $1, \frac{1}{2}, \frac{1}{4}, \frac{1}{8}, \frac{1}{16}, \frac{1}{32}$, gives rise to the series $\sum_{k=0}^{\infty} 1/2^k$, that is, to the sum $1 + \frac{1}{2} + \frac{1}{4} + \frac{1}{8} + \frac{1}{16} + \frac{1}{32} + \frac{1}{64} + \cdots$. We can give a meaning to this infinite chain of additions as soon as we make the following definition.

PARTIAL SUMS   Given the infinite series $\sum_{k=0}^{\infty} a_k$, the finite sum of terms up through a_n, namely, $\sum_{k=0}^{n} a_k$, is called a *partial sum*. For the series $\sum_{k=0}^{\infty} 1/2^k$, the partial sums up through $a_0, a_1, a_2, \ldots$ form a sequence of partial sums s_n, $n = 0, 1, \ldots$. In fact, we would have $s_0 = \sum_{j=0}^{0} 1/2^j$, $s_1 = \sum_{j=0}^{1} 1/2^j$, $s_2 = \sum_{j=0}^{2} 1/2^j$, $s_3 = \sum_{j=0}^{3} 1/2^j, \ldots$. Evaluating each such partial sum yields the sequence of partial sums $1, \frac{3}{2}, \frac{7}{4}, \frac{15}{8}, \ldots$.

SUM OF THE INFINITE SERIES   The *sum of an infinite series* is the limit of the sequence of its partial sums. Termwise addition is an infinite process and, if carried out literally, would be unending. Defining the sum of the series in terms of a limit is a prescription for obtaining, in finite time, the outcome of a task with an infinite number of steps.

In Figure 7.5 the graph of the sequence of partial sums for $\sum_{k=0}^{\infty} 1/2^k$ suggests this series sums to 2. In fact, Exercise A1 shows that $(\sum_{k=0}^{N} 1/2^k)(1 - \frac{1}{2}) = 1 - 1/2^{N+1}$; and, since $\lim_{N \to \infty} 1/2^{N+1} = 0$, we have $\frac{1}{2} \sum_{k=0}^{\infty} 1/2^k = 1$, and hence, $\sum_{k=0}^{\infty} 1/2^k = 2$.

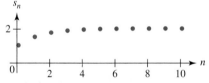

s_n

FIGURE 7.5   Sequence of partial sums for

$$\sum_{k=0}^{\infty} \frac{1}{2^k}$$

DIVERGENT SERIES   Series for which the sequence of partial sums diverges are called *divergent series*. Thus, an infinite series can diverge because its sum becomes infinite or because its partial sums oscillate. For example, the series $\sum_{k=0}^{\infty} 1$ diverges because its sequence of partial sums is $1, 2, 3, 4, 5, 6, 7, 8, 9, 10, 11, \ldots$, a sequence clearly diverging to infinity.

On the other hand, the series $1 - 1 + 1 - 1 + \cdots$, that is, the series $\sum_{k=0}^{\infty} (-1)^k$, also diverges, but this time because the sequence of partial sums $1, 0, 1, 0, 1, 0, 1, 0, 1, 0, 1, \ldots$ oscillates between 1 and 0.

GEOMETRIC SERIES   The series $\sum_{k=0}^{\infty} ar^k$, known as the *geometric series*, arises often and converges to the sum $S = \frac{a}{1-r}$, provided $|r| < 1$. If $1 \leq |r|$, the geometric series will diverge. This is established by letting S_N represent the partial sum up through $k = N$, forming the difference $S_N - rS_N$, and solving for $S_N = a(r^{N+1} - 1)/(r - 1)$. Provided $|r| < 1$, the term r^{N+1} will tend to zero as N becomes infinite. Hence, we obtain $S = \frac{a}{1-r}$.

For example, in the series $\sum_{k=0}^{N} 1/2^k$, $a = 1$ and $r = \frac{1}{2}$, so $S = \frac{1}{1-1/2} = 2$, as we saw earlier.

THE p-SERIES   A series of the form $\sum_{k=1}^{\infty} 1/k^p$ is called a *p-series*. The geometric series $\sum_{k=0}^{\infty} ar^k$ has a fixed constant r raised to integer powers, whereas the p-series has the integers k raised to a fixed power p. There is a single formula for the sum of the geometric series, but not one for the p-series. At this point, we don't have the tools for determining the behavior of the p-series, but Exercise B1, Section 7.3 shows the p-series $\sum_{k=1}^{\infty} \frac{1}{k^2}$ converges, while Exercise A1, Section 10.4 shows it can be summed to $\frac{\pi^2}{6}$.

THEOREM 7.1   THE nth-TERM RULE

1. $\displaystyle\sum_{n=0}^{\infty} a_n < \infty$

    $\Longrightarrow \displaystyle\lim_{n \to \infty} a_n = 0$

In words, this says that if the series $\sum_{n=0}^{\infty} a_n$ converges, the nth term must go to zero but not conversely. Thus, for a convergent series the condition $\lim_{n \to \infty} a_n = 0$ is *necessary* for convergence but not sufficient. If the series converges, the successive terms have to be getting smaller or the sum will not be bounded. But surprisingly, the rate at which the terms become smaller actually matters. If the rate of decay of the terms isn't fast enough, the series won't converge.

If the nth term goes to zero, the series need not converge, the classic counterexample being the *harmonic series* $\sum_{k=1}^{\infty} \frac{1}{k}$. This series diverges because, even though the general term goes to zero, it doesn't get small fast enough and the sum grows beyond all bounds. Below, the harmonic series is studied in detail.

The Harmonic Series

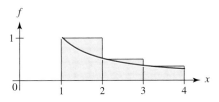

FIGURE 7.6   Divergence of the harmonic series

An intuitive feel for the truth of the nth term rule is important. Hence, we demonstrate the divergence of $\sum_{k=1}^{\infty} \frac{1}{k}$, the harmonic series, remarkable because the general term, namely $\frac{1}{k}$, goes to zero. Figure 7.6 shows a graph of $f(x) = \frac{1}{x}$, $1 \le x \le 4$, and a left-box rectangular approximation to the area under it. The interval is partitioned at the integers 1, 2, 3, and over each of the three subintervals $[k, k+1]$, $k = 1, 2, 3$, a rectangle of height $f(k)$ is constructed. Since the heights of the three rectangles in Figure 7.6 are determined at the left endpoint of the subinterval over which each sits.

The shaded area of the rectangles is clearly greater than the area under $f(x)$. Moreover, given the integer partition points for the rectangles, the areas of these three rectangles are $1, \frac{1}{2}$, and $\frac{1}{3}$, respectively. But those are the first three terms of the harmonic series. In fact, the first n terms of the harmonic series would be the same numbers as the areas of the first n rectangles of a similar integer-based rectangular approximation, between $x = 1$ and $x = n + 1$, of the area under $f(x) = \frac{1}{x}$. Generalizing from the figure, we conclude that the nth partial sum obeys the inequality

$$\int_1^{n+1} \frac{1}{x}\,dx = \ln(n+1) < \sum_{k=1}^{n} \frac{1}{k}$$

from which the divergence of the harmonic series follows because $\lim_{n \to \infty} \ln(n+1) = \infty$.

The sum of the first million terms is 14.39; and the values of the partial sums to 10^{10}, 10^{15}, and 10^{20} terms are, respectively, 23.60, 35.12, and 46.63. If one assumes the age of the universe to be 15 billion years, then there have been no more than 5.0×10^{17} seconds since time began. If one term of the harmonic series were added each second since time began, the sum would be less than 46. It takes a very large number of terms before the harmonic series becomes unboundedly large, but diverge, it does.

Termwise Arithmetic

If two series converge, then termwise *linear combinations* are valid, that is, if $\sum_{k=0}^{\infty} a_k = A$ and $\sum_{k=0}^{\infty} b_k = B$, then $\sum_{k=0}^{\infty}(\alpha a_k + \beta b_k) = \alpha A + \beta B$.

EXAMPLE 7.5 Consider the convergent series $\sum_{k=1}^{\infty} 1/2^k = 1$ and $\sum_{k=1}^{\infty} \frac{1}{k^2} = \frac{\pi^2}{6}$, the first being a geometric series and the second a *p*-series. A termwise linear combination of the two series would be $\sum_{k=1}^{\infty}(3/2^k + 7/k^2)$, with sum $3 + 7\frac{\pi^2}{6}$, the same linear combination of the sums of the original two series. ❖

EXERCISES 7.2–Part A

A1. Expand $(1 + r + r^2 + r^3)(1 - r)$ to obtain $1 - r^4$. Generalize this to $(\sum_{k=0}^{N} r^k)(1 - r) = 1 - r^{N+1}$, and thereby obtain the summation formula $\sum_{k=0}^{N} r^k = (1 - r^{N+1})/(1 - r)$. Test these calculations for $r = \frac{1}{2}$.

A2. Compute the number of seconds in 15 billion years.

A3. For the harmonic series:

   (a) Show $\frac{1}{3} + \frac{1}{4} > \frac{2}{4} = \frac{1}{2}$.

   (b) Show $\frac{1}{5} + \cdots + \frac{1}{8}$, the sum of the next 4 terms, exceeds $\frac{4}{8} = \frac{1}{2}$.

   (c) Show $\frac{1}{9} + \cdots + \frac{1}{16}$, the sum of the next 8 terms, exceeds $\frac{8}{16} = \frac{1}{2}$.

   (d) Show $\frac{1}{17} + \cdots + \frac{1}{32}$, the sum of the next 16 terms exceeds $\frac{16}{32} = \frac{1}{2}$.

   (e) Generalize these observations and construct a proof that the harmonic series diverges.

A4. Computationally, test the claim that $\sum_{k=1}^{\infty}(-1)^{k+1}\frac{1}{k} = \ln 2$. This series is called the *alternating harmonic series*.

A5. Computationally, test the claim that
$$\left| \sum_{k=1}^{N}(-1)^{k+1}\frac{1}{k} - \ln 2 \right| < \frac{1}{N+1}.$$

EXERCISES 7.2–Part B

B1. What sum would be attained by adding 10 million terms of the harmonic series each second since the universe began 15 billion years ago?

B2. On a fixed-precision computer, in a language such as BASIC, FORTRAN, Pascal, or C, sum the harmonic series termwise until it *appears* to have converged. Explain both the phenomenon of "apparent convergence" and the specific value generated by your program.

Exercises B3–12 give a_k, the general term of the series $\sum_{k=1}^{\infty} a_k$. For each:

   (a) Plot the points (k, a_k), $k = 1, \ldots, 10$.

   (b) Find $S_n = \sum_{k=1}^{n} a_k$, $n = 1, \ldots, 10$, the first ten partial sums.

   (c) Plot the points (n, S_n), $n = 1, \ldots, 10$.

   (d) Determine $\lim_{k \to \infty} a_k$.

   (e) Hazard a guess as to whether the series converges or diverges.

B3. $a_k = \dfrac{2}{k!}$ **B4.** $a_k = \dfrac{3}{k^2}$ **B5.** $a_k = \dfrac{(-1)^k}{k^3}$ **B6.** $a_k = \dfrac{\ln k}{k}$

B7. $a_k = \dfrac{k!}{k^k}$ **B8.** $a_k = \dfrac{(-1)^k k!}{2^k}$ **B9.** $a_k = \ln\left(\dfrac{k}{3k+2}\right)$

B10. $a_k = \left(\dfrac{1}{\sqrt{k}} - \dfrac{1}{\sqrt{k+1}}\right)$ **B11.** $a_k = \dfrac{k^2}{1+k}$

B12. $a_k = \dfrac{\sin\frac{k\pi}{5}}{k^3}$

Exercises B13–22 give a_k, the general term of the series $\sum_{k=1}^{\infty} a_k$. For each:

   (a) Find $S_n = \sum_{k=1}^{n} a_k$, $n = 1, \ldots, 10$, the first ten partial sums.

   (b) Plot the points (n, S_n), $n = 1, \ldots, 10$.

   (c) Sum the series.

   (d) Verify the *n*th-term rule.

B13. $a_k = \dfrac{3}{5^{k-1}}$ **B14.** $a_k = \dfrac{4}{3^{k-2}} + \dfrac{2}{7^{k+1}}$ **B15.** $a_k = e^{-3k}$

B16. $a_k = \dfrac{3^{k-2}}{5^{k+1}}$ **B17.** $a_k = (-1)^k 3^{1-k}$ **B18.** $a_k = (\sqrt{7})^{-k}$

B19. $a_k = \dfrac{1}{\pi^{2k}}$ **B20.** $a_k = \dfrac{4}{5^{3k+1}}$ **B21.** $a_k = \dfrac{6}{7^{2k-1}} - \dfrac{7}{8^{3k+2}}$

B22. $a_k = \left(\dfrac{-1}{2+\pi}\right)^k$

7.3   Series with Positive Terms

The Ratio and nth Root Tests

It is easier to determine whether or not a series converges than it is to determine the sum to which it might converge. There are few simple techniques for discovering the sum, but there are a number of "tests" for determining if convergence takes place. We remind the reader of two tests found in any calculus text.

THE RATIO TEST   For $\sum_{k=0}^{\infty} a_k$, a series with positive terms, let $\rho = \lim_{k \to \infty}(a_{k+1}/a_k)$. If $\rho < 1$, the series converges; if $1 < \rho$, the series diverges; if $\rho = 1$, the test is inconclusive since the series may diverge or converge.

Intuitively, the test compares the "next" term to the "previous" term. For a convergent series, the "next" term should be sufficiently smaller than the preceding term, else the series will grow too fast to converge. If this "continuing decrease" in the size of successive terms is maintained indefinitely, it is reasonable to expect the series to converge.

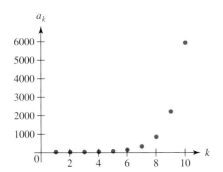

FIGURE 7.7   Divergence of the series in Example 7.7

THE *n*th Root Test   For $\sum_{n=0}^{\infty} a_n$, a series with nonnegative terms, let $r = \lim_{n \to \infty}(a_n)^{1/n}$. If $r < 1$, the series converges; if $r > 1$, the series diverges; if $r = 1$, the test is inconclusive since the series may diverge or converge.

It can be shown that if $\rho = \lim_{n \to \infty} a_{n+1}/a_n$ exists, then $r = \lim_{n \to \infty}(a_n)^{1/n}$ also exists and $\rho = r$. However, r may exist when ρ does not, so the nth root test is "more powerful." (See [65, p. 394].)

Typically, the ratio test is easier to apply, but there are series for which the nth root test can be convenient or even necessary.

EXAMPLE 7.6   To determine if the series $\sum_{k=0}^{\infty} k/2^k$ converges we use the ratio test. The ratio of two successive terms is $(k+1)2^k/k2^{k+1}$, the limit of which is $\rho = \frac{1}{2} < 1$. By the ratio test, the series converges. Alternatively, the nth root test requires us to calculate $r = \lim_{n \to \infty}(n/2^n)^{1/n} = \frac{1}{2}$, so both tests let us conclude the series converges.   ❖

EXAMPLE 7.7   To determine, via the ratio test, if the series $\sum_{k=1}^{\infty} 3^k/k$ converges or diverges, form the ratio between two successive terms, obtaining $k3^{k+1}/(k+1)3^k$, whose limit is $\rho = 3 > 1$. Hence, conclude the series diverges. In fact, Figure 7.7 shows how rapidly the terms are growing. Indeed, since $\lim_{k \to \infty}(3^k/k) = \infty$, the general term does not approach zero, a much simpler proof that the series diverges.

Alternatively, the nth root test requires us to compute $r = \lim_{n \to \infty}(3^n/n)^{1/n} = 3$, and both tests lead us to conclude the series diverges.   ❖

EXAMPLE 7.8   The series $\sum_{k=1}^{\infty} \frac{\ln k}{k}$ diverges because each term is larger than the terms in the harmonic series $\sum_{k=1}^{\infty} \frac{1}{k}$, and we have already seen that the harmonic series diverges. As with the harmonic series, the general term tends to zero, yet the series diverges. However, the ratio test is indeterminate since $\rho = \lim_{k \to \infty} \frac{k \ln(k+1)}{(k+1)\ln k} = 1$. Likewise, the nth root test is based on the number $r = \lim_{n \to \infty}\left(\frac{\ln n}{n}\right)^{1/n}$ whose value is, not surprisingly, 1. Thus, when the ratio test fails to be conclusive, so also will the nth root test fail to be conclusive.   ❖

EXAMPLE 7.9   The series $\sum_{n=1}^{\infty} \frac{1}{n^2}$ converges to $\frac{\pi^2}{6}$, but the ratio test again fails since $\rho = \lim_{k\to\infty} \frac{k^2}{(k+1)^2} = 1$. The nth root test gives $r = \lim_{n\to\infty} \left(\frac{1}{n^2}\right)^{1/n} = 1$, so we again see both the ratio and nth root tests failing to be conclusive. ❖

EXERCISES 7.3–Part A

A1. Apply both the ratio and nth root tests to the series $\sum_{k=1}^{\infty} k^k/2^{2k}$.

A2. Apply both the ratio and nth root tests to the series $\sum_{k=1}^{\infty} k^k/2^{k^2}$.

A3. If $\{a_k\}$ is the sequence defined by $a_1 = 3$, $a_{k+1} = a_k(2 - \cos k)/k$, determine if the series $\sum_{k=1}^{\infty} a_k$ converges or diverges.

A4. If
$$a_k = \begin{cases} \dfrac{1}{2^k} & k \text{ even} \\ \dfrac{k}{2^k} & k \text{ odd} \end{cases}$$
use the nth root test to show the series $\sum_{k=1}^{\infty} a_k$ converges.

A5. Test the series $\sum_{k=2}^{\infty} 1/(\ln k)^k$ for convergence or divergence.

EXERCISES 7.3–Part B

B1. The *integral test* for the series $\sum_{k=1}^{\infty} f(k)$, where $f(x)$ is a continuous, positive, decreasing function for $x \geq 1$, concludes that the series and the integral $\int_1^{\infty} f(x)\,dx$ converge or diverge together. Use this test to verify the convergence of the p-series determined by $f(k) = \frac{1}{k^2}$.

For the series in Exercises B2–15:

(a) Form and plot the first ten partial sums.

(b) Determine convergence or divergence by the ratio test, if possible.

(c) Determine convergence or divergence by the nth root test, if possible.

(d) If ρ from the ratio test and r from the nth root test both exist, are they the same number? They should be.

(e) If neither test is conclusive, try to establish convergence or divergence by comparison to more tractable series or make a conjecture based on empirical evidence.

B2. $\sum_{k=1}^{\infty} \dfrac{k!}{(3^k)^2}$ **B3.** $\sum_{k=1}^{\infty} \dfrac{5^k}{k^2 3^k}$ **B4.** $\sum_{k=2}^{\infty} \dfrac{1}{(\ln k)^2}$ **B5.** $\sum_{k=1}^{\infty} \dfrac{k!}{k^k}$

B6. $\sum_{k=1}^{\infty} \dfrac{k \ln k}{k!}$ **B7.** $\sum_{k=1}^{\infty} k^2 e^{-k}$ **B8.** $\sum_{k=1}^{\infty} \dfrac{k!}{5^k}$ **B9.** $\sum_{k=1}^{\infty} \dfrac{k^k}{10^k}$

B10. $\sum_{k=1}^{\infty} \dfrac{2\cdot 4\cdots(2k)}{1\cdot 3\cdots(2k-1)2^k}$ **B11.** $\sum_{k=1}^{\infty} \dfrac{k! \ln k}{(k+1)!\sqrt{k}}$

B12. $\sum_{k=1}^{\infty} \dfrac{\ln k}{k}$ **B13.** $\sum_{k=1}^{\infty} \dfrac{k^2 + 5k + 6}{k!}$

B14. $\sum_{k=1}^{\infty} ke^{-2k}$ **B15.** $\sum_{k=1}^{\infty} \dfrac{k^3}{3^k}$

B16. An exercise in [65] declares that if $a_k \geq 0$ and $\lim_{k\to\infty} ka_k$ exists and is positive, then $\sum_{k=1}^{\infty} a_k$ diverges. Which series, if any, in Exercises B2–15 can be declared to diverge on the basis of this proposition?

B17. An exercise in [65] declares that if $a_k \geq 0$ and there exists a number $p > 1$ for which $\lim_{k\to\infty} k^p a_k$ exists and is finite, then $\sum_{k=1}^{\infty} a_k$ converges. Which series, if any, in Exercises B2–15 can be declared to converge on the basis of this proposition?

B18. The *limit comparison test* for series of nonnegative terms states that (a) if $a_k \geq 0$, $b_k \geq 0$, and $\lim_{k\to\infty} a_k/b_k = \alpha$, where α is neither 0 nor ∞, then $\sum_{k=1}^{\infty} a_k$ and $\sum_{k=1}^{\infty} b_k$ either both converge or both diverge; (b) if $\alpha = 0$ and $\sum_{k=1}^{\infty} b_k$ converges, then $\sum_{k=1}^{\infty} a_k$ converges; (c) if $\alpha = \infty$ and $\sum_{k=1}^{\infty} b_k$ diverges, then $\sum_{k=1}^{\infty} a_k$ diverges. Which series, if any, in Exercises B2–15 can have their behavior decided on the basis of this test?

B19. If $a_1 = 1$, and $a_{k+1} = 1/(1 + a_k)$, the series $\sum_{k=1}^{\infty} a_k$ diverges. To see this,

(a) obtain and plot a_k, $k = 1, \ldots, 12$.

(b) show that $a_k \to \frac{\sqrt{5}-1}{2}$ is the Golden Ratio. (*Hint:* If $a_k \to A$, then $A = \frac{1}{1+A}$.) Since the terms of the series do not tend to zero, the series cannot converge.

B20. If
$$a_k = \begin{cases} \dfrac{1}{2^k} & k \text{ even} \\ \dfrac{k^2}{2^k} & k \text{ odd} \end{cases}$$
show that the series $\sum_{k=1}^{\infty} a_k$ converges. To this end,

(a) obtain and plot a_k, $k = 1, \ldots, 10$.

(b) obtain and plot the first ten partial sums $S_n = \sum_{k=1}^{n} a_k$, $n = 1, \ldots, 10$.

(c) show that ρ in the ratio test does not exist.

(d) show that r in the nth root test is $\frac{1}{2}$.

B21. Let $S_1, S_2, S_3, \ldots$ be the partial sums of a divergent series $\sum_{k=1}^{\infty} a_k$. If the sequence $\sigma_k = \frac{1}{k} \sum_{n=1}^{k} S_n$, the arithmetic means (or averages) of the first n terms of $S_1, S_2, S_3, \ldots$ converges to S, the series $\sum_{k=1}^{\infty} a_k$ is said to be summable to S by Cesàro's ([3], [65], and [97]; Césaro in [84]) method of arithmetic means of order 1, or $(C, 1)$ summable to S. It can be shown that if $\sum_{k=1}^{\infty} a_k$ itself converges, it converges to S, making Cesàro summability a *regular* method of summability.

(a) Show that σ_k generates the sequence $S_1, \frac{S_1+S_2}{2}, \frac{S_1+S_2+S_3}{3}, \ldots$.

(b) Show that $\sigma_n = \sum_{k=1}^{n} (1 - \frac{k-1}{n}) a_k$.

(c) Obtain the Cesàro sum for the convergent geometric series $\sum_{k=0}^{\infty} 1/2^k$.

(d) Obtain the Cesàro sum for the divergent series $\sum_{k=0}^{\infty} (-1)^k$.

(e) Show that $\frac{2}{3}$ is the $(C, 1$ sum of $1 + 0 - 1 + 1 + 0 - 1 + 1 + 0, 1 + \cdots$.

(f) Show that $\frac{3}{4}$ is the $(C, 1)$ sum of $1 + 0 + 0 - 1 + 1 + 0 + 0 - 1 + 1 + 0 + 0 - 1 + \cdots$.

(g) Show that $\sum_{k=1}^{\infty} (-1)^k k$ is not $(C, 1)$ summable.

(h) Sum the series in part (d) with your favorite computer algebra system, and compare its answer with the Cesàro sum.

B22. The *comparison test* of elementary calculus states that the series $\sum_{k=0}^{\infty} a_k$ converges if it has no negative terms and $\sum_{k=0}^{\infty} c_k$ is a convergent series for which $a_k \leq c_k$ for every $k > K$ where K is a fixed integer. Use this comparison test to show the convergence of each of the following.

(a) $\displaystyle\sum_{k=1}^{\infty} \frac{\sin^2 k}{2^k}$ (b) $\displaystyle\sum_{k=1}^{\infty} \frac{2 + \cos k}{k^2}$

7.4 Series with Both Negative and Positive Terms

Absolute and Conditional Convergence

The sequence of partial sums for a series with positive terms either converges or diverges to infinity. Each successive partial sum will be at least as large as the preceding one, so the partial sums form an increasing sequence that either has a finite limit or becomes infinite. Ruling out the latter case means the series must converge.

If there are positive terms *and* negative terms mixed in the series $\sum_{k=0}^{\infty} a_k$, convergence becomes a more delicate affair. Now, a partial sum can be either larger or smaller than its predecessor because the additional term added can be positive or negative. The sequence of partial sums can therefore have a finite limit or an infinite limit or can oscillate. Hence, there are now two ways for the series to diverge, so ruling out diverging to infinity does not imply convergence. Determining the convergence of a series with both positive and negative terms is more subtle than determining the convergence of a series of positive terms.

To deal with the added complexity of series with positive and negative terms, we distinguish between absolute and conditional convergence.

ABSOLUTE CONVERGENCE   A series with both positive and negative terms is said to converge *absolutely* if the series converges when all its negative terms are made positive. Thus, when $\sum_{k=0}^{\infty} |a_k|$ converges, the series $\sum_{k=0}^{\infty} a_k$ is said to converge *absolutely*. The convergence studied for series with positive terms was, in fact, absolute convergence.

The series

$$\sum_{k=0}^{\infty} \left(-\frac{1}{2}\right)^k = 1 - \frac{1}{2} + \frac{1}{4} - \frac{1}{8} + \frac{1}{16} - \frac{1}{32} + \cdots \tag{7.1}$$

converges to $\frac{1}{1-(-1/2)} = \frac{2}{3}$, whereas the series $\sum_{k=0}^{\infty} 1/2^k$, converges to 2. Thus, (7.1) converges to $\frac{2}{3}$ but converges absolutely to 2 and so is said to be absolutely convergent.

Ratio and Root Tests Extended   If in the ratio test and nth root test we compute $\rho = \lim_{k \to \infty} |a_{k+1}/a_k|$ and $r = \lim_{n \to \infty} |a_n|^{1/n}$, then the conclusions about convergence and divergence are as stated previously. Hence, these two tests can be used to determine the absolute convergence of a series.

CONDITIONAL CONVERGENCE   If the series $\sum_{k=0}^{\infty} a_k$ contains both positive and negative terms and converges when $\sum_{k=0}^{\infty} |a_k|$ diverges, we declare the convergence to be *conditional*. The convergence is contingent on the presence of the negative terms that serve to diminish the size of the partial sums. The classic example of this is the *alternating harmonic* series $\sum_{k=1}^{\infty} (-1)^k/k = 1 - \frac{1}{2} + \frac{1}{3} - \frac{1}{4} + \cdots = \ln 2$, which diverges when all the terms are made positive. (The value of the series is established in [54].)

THEOREM 7.2

An absolutely convergent series is convergent.

Intuitively, if a series of positive terms converges, the terms must be getting smaller. Making some of the terms negative would only make the partial sums smaller still, so a series that converges absolutely must converge in the sense of conditional convergence.

Note that the term "convergent" is used in three different contexts. A series $\sum_{k=0}^{\infty} a_k$ is said to *converge* if the sequence of its partial sums has a limit. If the series $\sum_{k=0}^{\infty} |a_k|$ converges, the original series is said to *converge absolutely*. Finally, if the original series converges but does not converge absolutely, then the series is said to *converge conditionally*. A series containing both positive and negative terms can therefore be said to converge, or it can be said to converge absolutely, or it can be said to converge conditionally. It can converge and converge absolutely, or it can converge and converge conditionally. It cannot converge absolutely and conditionally at the same time. To a certain extent, using the unmodified "converge" leaves the reader in the dark about whether the series converges absolutely or not.

Momentarily, we are going to see that *absolute* convergence is sufficient to guarantee that the terms of the series can be rearranged (reordered). Rearranging the terms of a *conditionally* convergent series leads to a new series that can diverge or can converge to an arbitrarily preassigned number (see [65]).

REGROUPING   If the series $\sum_{k=0}^{\infty} a_k$ converges (either conditionally or absolutely), then any series formed from it by *regrouping terms,* that is, by inserting parentheses while not changing the ordering of terms, converges to the same sum. For example, the alternating harmonic series converges (conditionally) to $\ln 2$, as do the regroupings

$$\left(1 - \tfrac{1}{2}\right) + \left(\tfrac{1}{3} - \tfrac{1}{4}\right) + \left(\tfrac{1}{5} - \tfrac{1}{6}\right) + \cdots = \tfrac{1}{2} + \tfrac{4-3}{(3)(4)} + \tfrac{6-5}{(5)(6)} + \cdots$$

$$= \tfrac{1}{(1)(2)} + \tfrac{1}{(3)(4)} + \tfrac{1}{(5)(6)} + \cdots$$

$$= \sum_{k=0}^{\infty} \frac{1}{(2k+1)(2k+2)}$$

and

$$1 + \left(-\tfrac{1}{2} + \tfrac{1}{3}\right) + \left(-\tfrac{1}{4} + \tfrac{1}{5}\right) + \left(-\tfrac{1}{6} + \tfrac{1}{7}\right) + \cdots = \tfrac{-3+2}{(2)(3)} + \tfrac{-5+4}{(4)(5)} + \tfrac{-7+6}{(6)(7)} + \cdots$$

$$= 1 - \tfrac{1}{(2)(3)} - \tfrac{1}{(4)(5)} - \tfrac{1}{(6)(7)} - \cdots$$

$$= 1 - \sum_{k=1}^{\infty} \frac{1}{2k(2k+1)}$$

REARRANGEMENTS   A series is said to have its terms *rearranged* if the order of its terms is changed. Dirichlet's theorem on rearranged series states that if the terms of an absolutely convergent series are rearranged, the new series converges absolutely to the same sum as the original. (See, e.g., [65, p. 404].) However, the terms of any conditionally convergent series can be rearranged to give either a divergent series or a conditionally convergent series converging to any desired real number.

Alternating Series

It is difficult to say much about the general series whose terms are both positive and negative. However, if the signs of the terms strictly alternate between positive and negative, the resulting series, $\sum_{k=0}^{\infty}(-1)^k a_k, a_k > 0$, is called an *alternating series,* for which we have the following convergence theorems.

ALTERNATING SERIES THEOREMS   The first alternating series theorem gives sufficient conditions that guarantee an alternating series does indeed converge. The second provides an estimate of the sum of an alternating series.

THEOREM 7.3   LEIBNIZ THEOREM

1. $\displaystyle\sum_{k=0}^{\infty}(-1)^k a_k$ with each a_k positive, is an alternating series.

2. For each k, $a_k \geq a_{k+1}$, that is, the sequence $\{a_k\}$ is monotonically decreasing.

3. $\displaystyle\lim_{k\to\infty} a_k = 0$

$\implies$ the series converges (at least conditionally).

In a word, the alternating series converges (at least conditionally) if its terms a_k monotonically decrease to zero.

EXAMPLE 7.10   For $\sum_{k=1}^{\infty}(-1)^k/k$, the alternating harmonic series, the terms $a_k = \frac{1}{k}$ form a monotone decreasing sequence whose limit is zero. Thus, Theorem 7.3 guarantees the series converges but does not determine what that limit might be.   ❖

THEOREM 7.4   ESTIMATION THEOREM

If $\sum_{k=0}^{\infty}(-1)^k a_k$ is an alternating series satisfying the hypotheses of Theorem 7.3, then the partial sum $S_N = \sum_{k=0}^{N}(-1)^k a_k$ approximates L, the sum of the series, with an error whose absolute value is less than a_{N+1}, the numerical value of the first unused term. A symbolic statement of this result would be

$$\left|\sum_{k=0}^{N}(-1)^k a_k - L\right| \leq a_{N+1}$$

EXAMPLE 7.11   Consider the alternating series $\sum_{k=1}^{\infty}(-1)^{(k+1)}/k^2$ that converges to the sum $\frac{\pi^2}{12}$ (obtained in Exercise A3, Section 10.4). Since the general term $a_k = \frac{1}{k^2}$ forms a sequence of decreasing positive terms whose limit is zero, we know by the theorem of Leibniz that the series converges. The series with all positive terms would be $\sum_{k=1}^{\infty}\frac{1}{k^2}$, a series whose sum is $\frac{\pi^2}{6}$ (obtained in Exercise A4, Section 10.4). Consequently, the original series converges absolutely.

  The partial sum after ten terms is $S_{10} = \frac{5194387}{6350400} = 0.8179621756$, which approximates the true sum $S = \frac{\pi^2}{12}$ to a tolerance of $S - S_{10} = 0.0045048580$. The approximation theorem claims this error is no worse than the absolute value of the first term neglected, namely, $\frac{1}{121} \overset{\circ}{=} 0.0083$, and indeed, this is the case.   ❖

EXERCISES 7.4–Part A

A1. Consider the alternating series $\sum_{k=2}^{\infty}(-1)^k/k + (-1)^k$.

  (a) List and plot the first 10 terms of the series.

  (b) Obtain and plot the first 10 partial sums of the series.

  (c) Show that the Leibniz convergence criterion does not apply.

  (d) Show that the series does not converge absolutely. *Hint:* An obvious rearrangement is nearly the harmonic series that diverges.

  (e) Show that a pairwise grouping of the terms, starting with the very first pair, leads to a series with partial sums $s_k = S_{2k+1} - 1$, where S_k are the partial sums of the alternating harmonic series.

  (f) Part (e) shows the given series converges to $\ln(2) - 1$. Obtain empirical evidence for this value of the series.

  (g) Show that the pairwise grouping described in part (e) leads to the series $\sum_{k=1}^{\infty} -\frac{1}{2k(2k+1)}$.

A2. Apply the integral test of Exercise B1, Section 7.3, to the general p-series $\sum_{k=1}^{\infty} 1/k^p$, concluding that the p-series converges for $p > 1$ and diverges otherwise.

A3. Consider the alternating series $\sum_{k=1}^{\infty}(-1)^{k+1}k/(k+1)$.

  (a) Show that the series is not absolutely convergent.

  (b) Show that the Leibniz convergence criterion does not apply.

  (c) Graph the first 10 partial sums, and from this, infer that the series is not conditionally convergent.

A4. Consider the alternating series

$$\frac{1}{5} - \frac{1}{3} + \frac{1}{25} - \frac{1}{9} + \frac{1}{125} - \frac{1}{27} + \cdots + \frac{1}{5^k} - \frac{1}{3^k} + \cdots$$

  (a) Plot the first 10 terms of the series.

  (b) Obtain and plot the first 15 partial sums, and from this, suspect that the series converges conditionally.

  (c) Show that the Leibniz convergence criterion does not apply.

  (d) Show that the series converges absolutely.

EXERCISES 7.4–Part B

B1. The series

$$\sum_{k=1}^{\infty}\frac{1}{k^2} = \frac{1}{1^2} + \frac{1}{2^2} + \frac{1}{3^2} + \cdots$$

converges (absolutely) to $\frac{\pi^2}{6}$ Hence any rearrangement must converge to the same sum. Show empirically that the rearrangement

$$\tfrac{1}{1^2} + \tfrac{1}{3^2} + \tfrac{1}{2^2} + \tfrac{1}{4^2} + \tfrac{1}{5^2} + \tfrac{1}{7^2} + \tfrac{1}{6^2} + \tfrac{1}{8^2} + \cdots$$

(two "odd" terms followed by two "even" terms) also converges to the sum $\frac{\pi^2}{6}$.

B2. The alternating harmonic series $\sum_{k=1}^{\infty}(-1)^{k+1}/k$ converges conditionally to $\ln 2$, so any rearrangement that converges need not necessarily converge to $\ln 2$. Show empirically that the rearrangement "two positive terms followed by one negative term," that is, $1 + \frac{1}{3} - \frac{1}{2} + \frac{1}{5} + \frac{1}{7} - \frac{1}{4} + \frac{1}{9} + \frac{1}{11} - \frac{1}{6} + \cdots$, converges to $\frac{3}{2}\ln 2$.

For the alternating series given in Exercises B3–14:

  (a) Plot the first 10 points $(k, |a_k|)$.

  (b) Apply the Leibniz convergence criterion to establish convergence.

  (c) Obtain and plot the first 10 members of the sequence of partial sums.

  (d) Use the estimation theorem to find an n for which the partial sum S_n is within 10^{-3} of the actual sum. (In B5–7, use 10^{-1}, 10^{-1}, 10^{-2}, respectively.)

  (e) Evaluate the partial sum S_n specified in part (d).

  (f) Use a computer algebra system to determine, where possible, the exact sum of the series, and compare the estimate to the exact value.

  (g) Test the series for absolute convergence.

B3. $\displaystyle\sum_{k=1}^{\infty} \frac{(-1)^{k+1}}{k^3}$ **B4.** $\displaystyle\sum_{k=1}^{\infty} \frac{(-1)^{k+1}}{k}$ **B5.** $\displaystyle\sum_{k=1}^{\infty} \frac{(-1)^{k+1}}{\sqrt{k}}$ **B11.** $\displaystyle\sum_{k=1}^{\infty} (-1)^{k+1}\frac{k}{e^{k^2}}$ **B12.** $\displaystyle\sum_{k=1}^{\infty} (-1)^{k+1}\frac{k+1}{k^2}$

B6. $\displaystyle\sum_{k=2}^{\infty} \frac{(-1)^k}{\ln k}$ **B7.** $\displaystyle\sum_{k=0}^{\infty} \frac{(-1)^k}{\sqrt{k+1}+\sqrt{k}}$ **B8.** $\displaystyle\sum_{k=1}^{\infty} (-1)^k\frac{\ln k}{k}$ **B13.** $\displaystyle\sum_{k=1}^{\infty} (-1)^{k+1}\frac{k^2}{2^k}$ **B14.** $\displaystyle\sum_{k=0}^{\infty} (-1)^k\frac{(k!)^2}{(2k)!}$

B9. $\displaystyle\sum_{k=1}^{\infty} (-1)^k\frac{k\ln k}{e^k}$ **B10.** $\displaystyle\sum_{k=1}^{\infty} (-1)^{k+1}\frac{k^3}{(k+1)!}$

Chapter Review

1. What does it mean for an infinite sequence of numbers to converge to a limit L?

2. What does it mean for an infinite sequence of numbers to diverge?

3. Give an example of an infinite sequence of numbers that converges to a limit.

4. Give an example of an infinite sequence of numbers that diverges. Explain why it diverges.

5. What is a partial sum of an infinite series?

6. What does it mean for an infinite series to converge?

7. Give an example of a convergent infinite series. What is its sum?

8. What does it mean for an infinite series to diverge? Give an example of a divergent series and explain why it diverges.

9. Answer true or false, and give a reason.

   (a) If a series converges, its nth term goes to zero.

   (b) If the nth term in an infinite series goes to zero, the series converges.

   (c) Two series can always be added or subtracted termwise.

10. Discuss the behavior of the harmonic series.

11. To what kinds of series does the ratio test apply? Give an example of its use.

12. To what kinds of series does the nth root test apply? Give an example of its use.

13. Describe the difference between conditional convergence and absolute convergence.

14. Give an example of a series that converges conditionally but not absolutely.

15. Answer true or false, and give an explanation.

   (a) A series that converges absolutely will converge conditionally.

   (b) A series that converges conditionally will converge absolutely.

16. What is an alternating series? Give an example. Is the series $1 + 2 - 3 - 4 + 5 + 6 - \cdots$ an alternating series?

17. State Leibniz' theorem on the convergence of alternating series. Give an example.

18. State the Estimation theorem for approximating the sum of an alternating series. Give an example.

Chapter *8*

Sequences and Series of Functions

INTRODUCTION   As in Chapter 7, convergence of an infinite series of *functions* is defined in terms of the convergence of the infinite sequence of its partial sums. However, there are several notions of convergence for a sequence of functions. We have already hinted that pointwise convergence simply means that at each fixed x_0, the series $\sum_{k=0}^{\infty} f_k(x_0)$ is just an infinite series of numbers.

In addition, there is the important idea of *uniform* convergence, a special case of pointwise convergence in which the rate of convergence is not dependent on x_0, the point at which convergence is being examined. Many termwise manipulations with infinite series can be justified in the presence of uniform convergence.

At the other extreme, we have *convergence in the mean,* a convergence that need not even be pointwise. A sequence of functions converges in the mean if certain *integrals* of these functions converge. But two functions which disagree at just one point are different, yet still have the same value for their definite integrals! As strange as this might seem, all of the theory of least-squares approximation, including the results on the convergence of Fourier series, is ultimately based on the notion of convergence in the mean.

8.1 Sequences of Functions

DEFINITION 8.1

The ordered list of functions $f_k(x), k = 0, 1, 2, \ldots,$ is called a *sequence of functions.*

Since a function is a rule (or formula) as well as a domain of admissible x's, it is typical for the common domain of the members of the sequence to be clearly stated. In some instances, we will find it convenient to call the first function f_0 as we did here; in other instances we will call the first function f_1.

Examples

Table 8.1 contains a list of 12 examples of sequences of functions, each starting with $f_1(x)$, except for the first sequence, which starts with $f_0(x)$. In subsequent discussions, these sequences will be referenced as SEQ-1–SEQ-12. Graphs of the first three members of each sequence appear in Table 8.2, where solid, dashed, and dotted curves are used, respectively. Members of the sequence SEQ-1 are also graphed in Figure 6.36.

SEQ	$f_k(x)$	Domain	SEQ	$f_k(x)$	Domain
1	x^k	$[0, 1]$	7	$\begin{cases} 4k^2 x & 0 \le x < \dfrac{1}{2k} \\[2mm] 4k(1-kx) & \dfrac{1}{2k} \le x < \dfrac{1}{k} \\[2mm] 0 & \dfrac{1}{k} \le x \le 1 \end{cases}$	$[0, 1]$
2	$\sin kx$	$[0, 2\pi]$	8	kxe^{-kx}	$[0, 1]$
3	$\dfrac{kx}{1 + k^2 x^2}$	$[0, \infty)$	9	$\begin{cases} \dfrac{1}{\sqrt{k}} & 0 \le x \le k \\[2mm] 0 & k < x < \infty \end{cases}$	$[0, \infty)$
4	$\dfrac{x}{k}$	$[-10, 10]$	10	$\begin{cases} 4k^{3/2} x & 0 \le x < \dfrac{1}{2k} \\[2mm] 4\sqrt{k}(1-kx) & \dfrac{1}{2k} \le x < \dfrac{1}{k} \\[2mm] 0 & \dfrac{1}{k} \le x \le 1 \end{cases}$	$[0, 1]$
5	xe^{-kx}	$[0, 1]$	11	$\dfrac{x}{1 + kx^2}$	$[-1, 1]$
6	$\dfrac{x^k}{k}$	$[0, 1]$	12	$\dfrac{1 - kx^2}{(1 + kx^2)^2}$	$[-1, 1]$

TABLE 8.1   Examples of sequences of functions

Although SEQ-7 and SEQ-10 look similar at first glance, the subtle changes in exponents on k in the rules defining their functions make for significant differences in behavior for the two sequences. This will be examined at length in the remaining sections of Chapter 8. Suffice it to say here that the areas under the functions in SEQ-10 form the decreasing sequence $\{\frac{1}{\sqrt{k}}\}$, whereas the areas under the functions in SEQ-7 remain a constant 1. (See Exercise A1.)

The functions in SEQ-5 and SEQ-8 differ only slightly, but in Section 8.3 their convergence properties will be seen to be significantly different. The functions in SEQ-12, the derivatives of the functions in SEQ-11, will also be useful in Section 8.3 where the convergence of differentiated sequences is discussed.

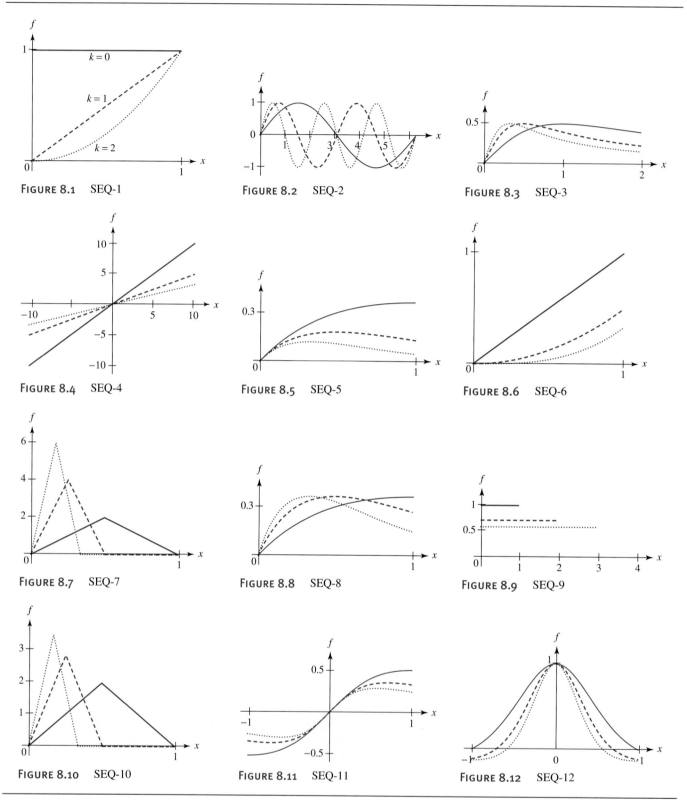

FIGURE 8.1   SEQ-1

FIGURE 8.2   SEQ-2

FIGURE 8.3   SEQ-3

FIGURE 8.4   SEQ-4

FIGURE 8.5   SEQ-5

FIGURE 8.6   SEQ-6

FIGURE 8.7   SEQ-7

FIGURE 8.8   SEQ-8

FIGURE 8.9   SEQ-9

FIGURE 8.10   SEQ-10

FIGURE 8.11   SEQ-11

FIGURE 8.12   SEQ-12

TABLE 8.2   Graphs of members of the sequences in Table 8.1. In each case, the first three members of the sequence are given by the solid, dashed, and dotted curves, respectively.

EXERCISES 8.1–Part A

A1. Show that $\int_0^1 f_k(x)\,dx = 1$ for the functions of SEQ-7 and $\frac{1}{\sqrt{k}}$ for the functions of SEQ-10.

A2. Sketch the first five members of the sequence $\{f_k'\}$ if $\{f_k\}$ is SEQ-1.

A3. Sketch the first five members of the sequence $\{f_k'\}$ if $\{f_k\}$ is SEQ-2.

A4. On the interval $[0, 2\pi]$, sketch the first five members of the sequence $\{\int_0^x f_k(t)\,dt\}$ if $\{f_k\}$ is SEQ-2.

A5. On the interval $[0, 1]$, sketch the first five members of the sequence $\{\int_0^x f_k(t)\,dt\}$ if $\{f_k\}$ is SEQ-5.

A6. On the interval $[-1, 1]$, sketch the first five members of the sequence $\{\int_{-1}^x f_k(t)\,dt\}$ if $\{f_k\}$ is SEQ-11.

EXERCISES 8.1–Part B

B1. In Section 6.10 the limiting function for SEQ-1 was said to be $f(x) = \begin{cases} 1 & x = 1 \\ 0 & 0 \le x < 1 \end{cases}$. Compare $\int_0^1 f(x)\,dx$ with $\lim_{k\to\infty} \int_0^1 f_k(x)\,dx$.

In Exercises B2–4, a sequence of functions $f_k(x)$, $k = 1, 2, \ldots$, is given. For each:

(a) Plot the first five members.

(b) Obtain the first five members of the sequence $f_k(1)$.

(c) Plot the first five members of the sequence of derivatives, $f_k'(x)$.

B2. $f_k(x) = \frac{2}{\pi} x \arctan(kx)$, $-1 \le x \le 1$

B3. $f_k(x) = \frac{1}{k} e^{-k^2 x^2}$, $-1 \le x \le 1$

B4. $f_k(x) = \frac{kx^k}{1 + kx^k}$, $0 \le x \le 2$

In Exercises B5–8, a sequence of functions $f_k(x)$, $k = 1, 2, \ldots$, is given. For each:

(a) Plot the first five members of the sequence.

(b) Obtain the first five members of the sequence $f_k(1)$.

(c) Plot the first five members of the sequence of derivatives, $f_k'(x)$.

(d) If a is the lower bound of the given domain, obtain the sequence $F_k(x) = \int_a^x f_k(t)\,dt$ and plot the first five members.

B5. $f_k(x) = \frac{kx^2}{1 + kx}$, $0 \le x \le 1$

B6. $f_k(x) = kxe^{-kx^2}$, $0 \le x \le 2$

B7. $f_k(x) = \frac{k^2 x}{1 + k^3 x^2}$, $-1 \le x \le 1$

B8. $f_k(x) = \frac{kx}{1 + kx}$, $0 \le x \le 1$

B9. For the sequence in Exercise B2:

(a) Show that at any fixed x, $\lim_{k\to\infty} f_k(x) = |x|$.

(b) Show that $f_k'(0)$ exists for each k, $k = 1, 2, \ldots$.

(c) Show that $|x|$, the limit function, does not have the differentiability property possessed by each member of the sequence.

B10. For the function $f(x) = e^x$, let $f_k(x)$ be the Taylor polynomial of degree k and $g_k(x) = f_{2k-1}(x)$ be the sequence of odd-degree polynomials. Each such polynomial has at least one real zero.

(a) Plot $g_k(x)$, $k = 1, \ldots, 5$, on an interval that shows the real zeros.

(b) Compute the real zero of each function graphed in part (a), and note that each member of the sequence $g_k(x)$ has a property that e^x, the limit function, does not have, namely, the members of the sequence each have a real zero but the limit function does not.

8.2 Pointwise Convergence

DEFINITION 8.2

Let $\{f_k(x)\}$, $k = 0, 1, \ldots$, defined on the interval $[a, b]$, be a sequence of functions. The function $f(x)$ defining the *pointwise limit* of this sequence is the function obtained by computing the limit of the sequence of values determined by this sequence at each fixed point x_0. Thus, $f(x_0) = \lim_{k\to\infty} f_k(x_0)$ defines the value of $f(x)$ at each point x_0. We then say that the sequence converges *pointwise* to the function $f(x)$.

More formally, we might also say the sequence $\{f_k(x)\}$, defined on $[a, b]$, converges pointwise to the function $f(x)$ whenever, given an $\varepsilon > 0$ and an x in $[a, b]$, there always exists an integer index N for which $n > N$ guarantees $|f_n(x) - f(x)| < \varepsilon$. In general, for pointwise convergence, the index N depends on x. At some points in the interval $[a, b]$ the value of N might have to be larger than at others.

Fortunately, the student can build a strong intuition about convergence of sequences of functions by studying their graphs, especially as was done in the examples of Section 8.1. In a course in mathematical analysis, the formalism of the ε-N definition is stressed. Here, we will take a more intuitive approach.

EXAMPLE 8.1   Obtain the pointwise limit of SEQ-1, Table 8.1, defined by $f_k(x) = x^k$, $k = 1, 2, \ldots$, on the interval $[0, 1]$. Figure 6.36 suggest the pointwise limit is the piecewise function $f(x) = \begin{cases} 0 & 0 \le x < 1 \\ 1 & x = 1 \end{cases}$. Analytically, the pointwise function can be found by using the limiting techniques learned in calculus, provided care is taken with domains. Clearly, at $x = 1$ each member of the sequence $\{f_k(1)\}$ is identically 1, so the sequence converges to 1 at this point. For any x in $[0, 1)$, we know that $\lim_{k \to \infty} x^k = 0$, so the limit function is 0 for $0 < x < 1$.

Before quitting this example, we make the following observation. Each member of the sequence $\{f_k(x)\}$ is a polynomial and is continuous. The limit function $f(x)$ is neither a polynomial nor a continuous function. Thus, in general, the pointwise limit of a sequence of functions can fail to have properties shared by each member of the sequence. This might be unintuitive to the novice but, nevertheless, signals important consequences for functions defined by limits of sequences.   ❖

EXAMPLE 8.2   Obtain the pointwise limit of SEQ-3, Table 8.1, defined by $f_k(x) = \frac{kx}{1+k^2x^2}$, $k = 1, 2, \ldots$, on the interval $[0, \infty)$. Figure 8.3 suggests the pointwise limit might be the function $f(x) = 0$.

For each fixed x, the sequence determined by $f_k(x)$ tends toward $\frac{kx}{k^2x^2} = \frac{1}{kx}$ and therefore has pointwise limit 0. More precisely, each function $f_k(x)$ has a maximum of $\frac{1}{2}$ at $x = \frac{1}{k}$. This highest point "moves" toward the origin. Eventually, at every positive x, the maximum value of $\frac{1}{2}$ passes to the left, and the sequence, at that x, is free to converge to zero. Hence, the convergence of SEQ-3 is more complicated than that of SEQ-1.

An explicit representation of N, the index beyond which $f_k(x)$ is within ε of $f(x) = 0$, is obtained by explicitly solving the equation $f_k(x) = \varepsilon$ for $k = \frac{1+\sqrt{1-4\varepsilon^2}}{2\varepsilon x}$, the larger root, rounded up to an integer. Thus, $N = N(x, \varepsilon)$, and as x nears zero, the index beyond which all sequence members are within ε of $f(x) = 0$ increases. For each fixed value of ε, the index k needed to guarantee $|f_k(x) - 0| < \varepsilon$ rises steeply as x nears the origin.   ❖

EXERCISES 8.2–Part A

A1. Find $f(x) = \lim_{k \to \infty} f_k(x)$ if $f_k(x) = \cos^k x$, $0 \le x \le \frac{\pi}{2}$.

A2. Find $f(x) = \lim_{k \to \infty} f_k(x)$ if $f_k(x) = \sqrt[k]{\cos x}$, $0 \le x \le \frac{\pi}{2}$.

A3. Find the pointwise limit of the sequence of functions defined by $f_k(x) = \frac{kx}{k+x}$, $0 \le x \le 1$.

A4. In Example 8.2, what is the smallest value of N that guarantees $|f_k(0.01)| < 0.001$ for all $k > N$?

A5. In [54], the sequence of functions

$$f_k(x) = \begin{cases} \dfrac{1 - (-x)^k}{1 + x} & 0 \le x < 1 \\ \dfrac{1}{2} & x = 1 \end{cases}$$

is used in the proof that the alternating harmonic series converges to $\ln 2$. What is $f(x)$, the pointwise limit of this sequence?

EXERCISES 8.2–Part B

B1. For $f_k(x)$ and $f(x)$ in Exercise A5, obtain $\lim_{k\to\infty}\int_0^1 f_k(x)\,dx$ and $\int_0^1 f(x)\,dx$.

In Exercises B2–10, for the indicated sequences taken from Table 8.1:

  (a) Determine $f(x)$, the pointwise limit of the sequence.

  (b) Determine $N(x,\varepsilon)$, an integer with the property that whenever the index k exceeds it, $|f(x)-f_k(x)|<\varepsilon$ for that x and that ε.

B2. SEQ-4 **B3.** SEQ-5 **B4.** SEQ-6 **B5.** SEQ-7 **B6.** SEQ-8
B7. SEQ 9 **B8.** SEQ-10 **B9.** SEQ-11 **B10.** SEQ-12

For the sequences in Exercises B11–14:

  (a) Determine $g(x)$, the pointwise limit of $\{f_k'(x)\}$, the sequence of derivatives.

  (b) Is $f'(x)=g(x)$?

  (c) Determine $F(x)$, the pointwise limit of the sequence of antiderivatives $F_k(x)=\int_a^x f_k(t)\,dt$, where a is the lower bound on the domain of the f_k.

  (d) Is $\int_a^x f(t)\,dt=F(x)$?

B11. $f_k(x)$ in Exercise B5 **B12.** $f_k(x)$ in Exercise B6
B13. $f_k(x)$ in Exercise B7 **B14.** $f_k(x)$ in Exercise B8

For the sequences in Exercises B15–17:

  (a) Determine $f(x)$, the pointwise limit of the sequence.

  (b) Determine $g(x)$, the pointwise limit of $\{f_k'(x)\}$, the sequence of derivatives.

  (c) Is $f'(x)=g(x)$?

B15. $f_k(x)$ in Exercise B2 **B16.** $f_k(x)$ in Exercise B3
B17. $f_k(x)$ in Exercise B4

B18. If $f_k(x)=\sum_{n=1}^{k}\frac{\sin nx}{n}, 0\le x\le 2\pi$, use graphical and numeric techniques to determine $f(x)=\lim_{k\to\infty}f_k(x)$. Be especially careful at the endpoints of the interval.

B19. If $f_k(x)=\sum_{n=1}^{k}\frac{1-\cos(n\pi/2)}{n}\sin nx, 0\le x\le\pi$, use graphical and numeric techniques to determine $f(x)=\lim_{k\to\infty}f_k(x)$. Be especially careful at $x=\frac{\pi}{2}$.

8.3 Uniform Convergence

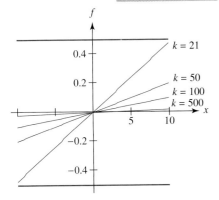

FIGURE 8.13 ε-band for sequence SEQ-4, Table 8.1

DEFINITION 8.3

The convergence of the sequence $\{f_k(x)\}$ to $f(x)$ is *uniform* on the interval $[a,b]$ whenever, given an $\varepsilon>0$, there always exists an integer index N that is independent of x and for which $n>N$ guarantees $|f_n(x)-f(x)|<\varepsilon$.

Thus, a sequence that converges uniformly converges pointwise, but in such a way that each given error tolerance ε can be attained with an index N that works all across the interval. Intuitively, if an ε-band is drawn about the limit function $f(x)$, then all sequence members with index $k>N$ lie completely within the confines of the band. Such an ε-band for the sequence described in Example 8.3 is seen in Figure 8.13.

EXAMPLE 8.3 The functions $f_k(x)=\frac{x}{k}, k=1,2,\ldots$, from SEQ-4, Table 8.1, converge uniformly to $f(x)=0$ on the interval $[-10,10]$. Elementary calculus and Figure 8.4 convince us that the pointwise limit of this sequence is $f(x)=0$. That the convergence is uniform follows from solving $|f_k(x)-0|=|\frac{x}{k}|<\varepsilon$ for $k>\frac{|x|}{\varepsilon}$, and taking $N=N(\varepsilon)=[\frac{10}{\varepsilon}]+1$, one more than the greatest integer in $\frac{10}{\varepsilon}$.

For example, remembering that $|x|\le 10$, choose N to be the first integer greater than or equal to $\frac{10}{\varepsilon}$, and every sequence member whose index is greater than N will lie in an ε-band of the limit function $f(x)=0$. If $\varepsilon=\frac{1}{2}, N=20$, so a band $\frac{1}{2}$-unit above and below $f(x)=0$ contains within it all sequence-members with indices greater than 20, as depicted in Figure 8.13. Every sequence-member indexed with an integer greater than 20 lies totally within the band of width 2ε. If ε is decreased to, say $\frac{1}{3}$, then $N = 30$. But all $f_k(x), k > 30$, will lie completely within this new ε-band. ❖

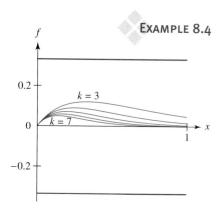

FIGURE 8.14 ε-band for sequence SEQ-5, Table 8.1

EXAMPLE 8.4   The functions $f_k(x) = xe^{-kx}, k = 1, 2, \ldots,$ from SEQ-5, Table 8.1, converge uniformly to $f(x) = 0$ on the interval $[0, 1]$. Elementary calculus and Figure 8.5 lead to a limiting function $f(x) = 0$. Uniform convergence is established as follows. Each function $f_k(x)$ has a maximum at the point $(\frac{1}{k}, 1/e^k)$. Consequently, $|f_k(x) - 0| \le |f_k(\frac{1}{k})| = 1/e^k < \frac{1}{k}$; to get $|f_k(x) - 0| < \varepsilon$, choose k so that $\frac{1}{k} < \varepsilon$ or $k > \frac{1}{\varepsilon}$. We therefore take $N(\varepsilon)$ as the first integer greater than $\frac{1}{\varepsilon}$; this is independent of x, and the convergence is uniform. Choose N to be the first integer greater than $\frac{1}{\varepsilon}$, and every sequence member whose index is greater than N will lie in an ε-band of the limit function $f(x) = 0$.

For example, if $\varepsilon = \frac{1}{3}$, $N = 3$, a band $\frac{1}{3}$-unit above and below $f(x) = 0$ will contain all sequence-members with indices greater than 3, as shown in Figure 8.14. Every sequence-member indexed with an integer greater than 3 lies totally within the band of width 2ε. If ε is decreased to, say $\frac{1}{10}$, then $N = 10$. But all $f_k(x), k > 10$, will lie completely within this new ε-band. ❖

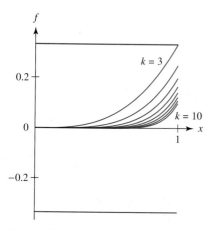

FIGURE 8.15 ε-band for sequence SEQ-6, Table 8.1

EXAMPLE 8.5   The functions $f_k(x) = x^k/k, k = 1, 2, \ldots,$ from SEQ-6, Table 8.1, converge uniformly to $f(x) = 0$ on the interval $[0, 1]$. Elementary calculus and Figure 8.6 lead to a limiting function $f(x) = 0$. To establish uniform convergence, note that on the interval $[0, 1]$, $f_k(x)$ satisfies the inequality $x^k/k \le \frac{1}{k}$ so that, as in Example 8.4, $N(\varepsilon)$ can be taken as the first integer greater than $\frac{1}{\varepsilon}$. Choose N to be the first integer greater than $\frac{1}{\varepsilon}$, and every sequence member whose index is greater than N will lie in an ε-band of the limit function $f(x) = 0$.

For example, if $\varepsilon = \frac{1}{3}$, $N = 3$, a band $\frac{1}{3}$-unit above and below $f(x) = 0$ will contain all sequence-members with indices greater than 3, as shown in Figure 8.15. Every sequence-member indexed with an integer greater than 3 lies totally within the band of width 2ε. If ε is decreased to, say $\frac{1}{10}$, then $N = 10$. But all $f_k(x), k > 10$, will lie completely within this new ε-band. ❖

Negation of Uniform Convergence

To say that the sequence $\{f_k(x)\}$ does *not* converge uniformly to $f(x)$ is to say that for *some* $\varepsilon > 0$, no matter what N is declared, there will always be at least one x for which $|f_k(x) - f(x)| > \varepsilon$. This negation is equivalent to the statement that there exists one $\varepsilon > 0$ and an x_k for each f_k for which $|f_k(x_k) - f(x_k)| > \varepsilon$.

Alternatively, uniform convergence is negated by claiming it is not true that given *any* $\varepsilon > 0$ there exists an N that works for every x and for which $k > N$ guarantees $|f_k(x) - f(x)| < \varepsilon$. This negates the avowal that for every $\varepsilon > 0$ something happens, because, for some one ε at least, it is impossible to find an N for which $k > N$ guarantees $|f_k(x) - f(x)| < \varepsilon$ for *every* x.

EXAMPLE 8.6   The functions $f_k(x)$ from SEQ-7, Table 8.1, converge pointwise, on the closed interval $[0, 1]$, to the limit function $f(x) = 0$. The convergence is pointwise but not uniform. Figure 8.7 shows that each function $f_k(x)$ is a "tent," and a bit of algebra shows each has a maximum height of $2k$ at $x = \frac{1}{2k}$. Since the maximum of each tent increases, it is impossible to obtain $|f_k(x)| < \varepsilon$ for all x's in $[0, 1]$ and all $k > N$. In particular, it would have to be true that $|f_k(\frac{1}{2k})| < \varepsilon$ for all $k > N$, but because $f_k(\frac{1}{2k}) = 2k > \varepsilon$ for $k > \frac{\varepsilon}{2}$, N cannot be independent of x and the convergence cannot be uniform.

Figure 8.16 shows the first five members of the sequence and an ε-band corresponding to $\varepsilon = \frac{1}{2}$. No matter what ε-band is drawn and no matter how far out in the sequence we begin looking, we will always find members of the sequence beyond that place in the sequence where peaks are outside the ε-band.   ❖

EXAMPLE 8.7   The functions $f_k(x) = \frac{kx}{1+k^2x^2}$ from SEQ-3, Table 8.1, were shown, in Section 8.2 to converge pointwise to $f(x) = 0$ throughout $[0, \infty)$. Because $f_k(\frac{1}{k}) = \frac{1}{2}$, the convergence cannot be uniform on any interval containing $x = 0$. In particular, consider the interval $[0, 1]$, and, for $\varepsilon < \frac{1}{2}$, attempt to satisfy $|f_k(x) - 0| < \varepsilon$ uniformly in $[0, 1]$. Since $\frac{1}{k} < 1$ for any $k > 2$ and $f_k(\frac{1}{k}) = \frac{1}{2}$, there is at least one x in $[0, 1]$ for which $|f_k(x) - 0| = \frac{1}{2} > \varepsilon$ for any f_k. The convergence is not uniform, and the functions $f_k(x)$ do not fit inside an ε-band about $f(x) = 0$. Figure 8.17 shows the first 10 members of the sequence and an ε-band corresponding to $\varepsilon = \frac{1}{3}$.   ❖

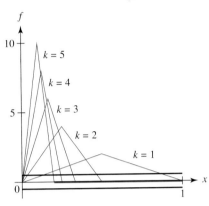

FIGURE 8.16 ε-band for sequence SEQ-7, Table 8.1

A classic theorem on uniform convergence is the proposition

the uniform limit of continuous functions is a continuous function.

Stated more formally, we have

THEOREM 8.1

1. $\{f_k(x)\}$ defined on the interval $[a, b]$.
2. $f_k(x)$ converge uniformly to $f(x)$ on $[a, b]$.
3.   Each $f_k(x)$ is continuous on $[a, b]$.

$\Longrightarrow$ $f(x)$ is continuous on $[a, b]$.

Our second classic theorem on uniform convergence and its consequences for sequences of functions is the proposition

the uniform limit of continuous functions can be integrated termwise.

A formal statement of this theorem is

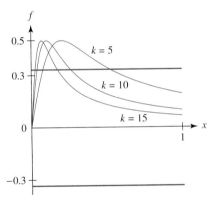

FIGURE 8.17 ε-band for sequence SEQ-3, Table 8.1

THEOREM 8.2

1. $\{f_k(x)\}$ converges uniformly to $f(x)$ on the interval $[a, b]$.
2.   Each $f_k(x)$ is continuous on $[a, b]$.

$$\Longrightarrow \lim_{k \to \infty} \int_a^b f_k(x)\,dx = \int_a^b \lim_{k \to \infty} f_k(x)\,dx = \int_a^b f(x)\,dx$$

Thus, the limit of the integrals is the integral of the limit, a result justified by uniform convergence. A proof of Theorem 8.2 is given after the following two examples.

EXAMPLE 8.8   The "moving tent" functions of SEQ-7, Table 8.1, converge pointwise, but not uniformly, to $f(x) = 0$ on the interval $[0, 1]$. The limit function is $f(x) = 0$, so the integral of the limit is likewise zero. But not so for the limit of the individual integrals. From Exercise A1, Section 8.1, the integral of each $f_k(x)$ is actually 1, so the limit of the integrals is 1, not zero.   ❖

EXAMPLE 8.9   Uniform convergence is sufficient, but not necessary, for the integral of the limit to be the limit of the integrals. SEQ-10, Table 8.1, is a modification of the previous sequence. Taking the heights of the tents to be $2\sqrt{k}$ instead of $2k$, we still have a sequence that converges to $f(x) = 0$ pointwise, but not uniformly. This time, however, the limit of the integrals will be the integral of the limit since $\int_0^1 f_k(x)\,dx = \frac{1}{\sqrt{k}} \to 0$.   ❖

PROOF OF THEOREM 8.2   To establish $\lim_{k\to\infty} \int_a^b f_k(x)\,dx = \int_a^b f(x)\,dx$, the conclusion of Theorem 8.2, it suffices to show that given an $\varepsilon > 0$, there is an N for which $k > N$ guarantees $|\int_a^b f_k(x)\,dx - \int_a^b f(x)\,dx| < \varepsilon$. The uniform convergence of $f_k(x)$ to $f(x)$ means there is an N for which $|f_k(x) - f(x)| < \frac{\varepsilon}{b-a}$ for all values of x in $[a,b]$, whenever $k > N$. Using this estimate we obtain

$$\left| \int_a^b f_k(x)\,dx - \int_a^b f(x)\,dx \right| = \left| \int_a^b f_k(x) - f(x)\,dx \right|$$

$$\leq \int_a^b |f_k(x) - f(x)|\,dx < \int_a^b \frac{\varepsilon}{b-a}\,dx = \varepsilon$$

Our third theorem on the consequences of uniform convergence concerns termwise differentiation of a sequence of functions. In a word, our theorem will state, roughly, that *if the derived (differentiated) sequence converges uniformly, then the limit of the derivatives is the derivative of the limit.* The formal statement of the theorem is

THEOREM 8.3

1. $\{f_k(x)\}$ is a sequence of functions differentiable on $[a, b]$.
2. $\{f_k(x_0)\}$ converges for some x_0 in $[a, b]$.
3. $\{f_k'(x)\}$ converges to $g(x)$ uniformly on $[a, b]$.

$\Longrightarrow$

1. $\{f_k(x)\}$ converges uniformly to some $f(x)$.
2. $f'(x)$ exists on $[a, b]$.
3. $f'(x) = \lim_{k\to\infty} f_k' = g(x)$

After considering the following two examples, we provide a proof of a slightly weaker form of Theorem 8.3.

EXAMPLE 8.10   As we saw in Example 8.5, the functions $f_k(x) = x^k/k$ from SEQ-6, Table 8.1, converge uniformly to $f(x) = 0$ on $[0, 1]$. The derivative of the limit function is $f'(x) = 0$. The derived sequence is $\{x^{k-1}\}$, which we know is pointwise convergent to $g(x) = \begin{cases} 0 & 0 \leq x < 1 \\ 1 & x = 1 \end{cases} \neq$ $f'(x) = 0$. The sequence of derivatives does not converge uniformly, so there is no guarantee that the limit of the derivatives is the derivative of the limit, even though the original sequence converges uniformly! The *derived* sequence must converge uniformly for the theorem to hold. ❖

EXAMPLE 8.11

The sequence defined by $F_k(x) = \frac{\ln(1+kx^2)}{2k}$ converges uniformly to $F(x) = 0$ on $[-1, 1]$. Figure 8.18 showing the first few members of the sequence suggests that each $F_k(x)$ exhibits even symmetry across the y-axis, and has a maximum at the endpoints of the interval $[-1, 1]$. The height of each maximum value is therefore $F_{\max} = \frac{\ln(1+k)}{2k}$, and this maximum height goes to zero as k increases. In essence, we have demonstrated the uniform convergence of $\{F_k(x)\}$ to $F(x) = 0$ because we have exhibited a uniform bound that itself goes to zero.

The derived sequence, given by $f_k(x) = \frac{x}{1+kx^2}$ is SEQ-11, Table 8.1, and it also converges uniformly to $f(x) = 0$. Indeed, Figure 8.11 suggests there might also be a uniform bound for the maximums of $|f_k(x)|$. In fact, we have $\left|f_k(\pm\frac{1}{\sqrt{k}})\right| = \frac{1}{2k}$. Since the maximum of each f_k clearly goes to zero as k increases, we have established uniform convergence to $f(x) = 0$. This is in agreement with the theorem on differentiation of sequences. Since the derived sequence converges uniformly, it converges to the derivative of the original sequence.

If we take the sequence $\{f_k(x)\}$ as the primary sequence and look at the sequence of *its* derivatives, we obtain the sequence $g_k(x) = \frac{1-kx^2}{(1+kx^2)^2}$, SEQ-12 in Table 8.1. Figure 8.12 suggests the derived sequence $g_k(x)$ converges pointwise to the discontinuous limit function $g(x) = \begin{cases} 1 & x = 0 \\ 0 & x \neq 0 \end{cases}$. If the convergence were uniform, the limit function would have to be continuous. Since that is not the case, the convergence could not have been uniform.

Even though the primary sequence $\{f_k(x)\}$ converges uniformly to $f(x) = 0$ so that $f'(x) = 0$, the derived sequence does not converge uniformly. Hence, there was no guarantee that the limit of the derivatives was the derivative of the limit. And, indeed, it was not. ❖

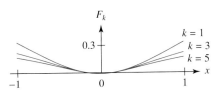

FIGURE 8.18 Members of the sequence $F_k(x)$ in Example 8.11

PROOF OF THEOREM 8.3 We prove Theorem 8.4, the following slightly weaker form of Theorem 8.3. It is "weaker" because we invoke the more restrictive hypotheses that f_k have continuous derivatives and that $\{f_k\}$ converges on the whole interval, not just at one point.

THEOREM 8.4

1. For each k, $f_k(x)$ has a continuous derivative on the interval $[a, b]$.
2. $\{f_k(x)\}$ converges to $f(x)$ on $[a, b]$.
3. $\{f_k'(x)\}$ converges uniformly to $g(x)$ on $[a, b]$.

$\Longrightarrow$

1. $f'(x)$ exists on $[a, b]$.
2. $f'(x) = \lim_{k \to \infty} f_k' = g(x)$

By hypothesis, each derivative $f_k'(x)$ is continuous. Since the $f_k'(x)$ converge to $g(x)$ uniformly, $g(x)$ is also continuous and hence integrable. Therefore,

$$\int_a^x g(t)\, dt = \lim_{k \to \infty} \int_0^x \frac{d}{dt} f_k(t)\, dt = \lim_{k \to \infty} [f_k(x) - f_k(a)] = f(x) - f(a)$$

and $f(x) = \int_a^x g(t)\, dt + f(a)$. The continuity of $g(x)$ makes $f(x)$ differentiable, and we have $f'(x) = g(x) = \lim_{k \to \infty} f_k'(x)$.

EXERCISES 8.3

1. For SEQ-8, Table 8.1:

  (a) Find $f(x)$, the pointwise limit.

  (b) Show that the convergence to $f(x)$ is not uniform by finding the maximum value of each $f_k(x)$.

  (c) Find the pointwise limit of $\{f_k'(x)\}$, the derived sequence, and observe that "the limit of the derivatives is not the derivative of the limit".

  (d) Show that $\{\int_0^x f_k(t)\,dt\}$ converges uniformly on $[0, 1]$. Is this a consequence of Theorem 8.2 or is it just a fortuitous accident?

2. Show that the sequence in Exercise B2, Section 8.1, converges uniformly. *Hint:* Use graphs to determine where the maximums of errors $\{|f(x) - f_k(x)|\}$ are occurring, and from this, exhibit a uniform bound on this sequence of errors.

3. Show that the sequence in Exercise B3, Section 8.1, converges uniformly. Then show that $f_k'(x) \to f'(x)$, but the convergence is not uniform.

4. Show that the convergence of the sequence in Exercise B4, Section 8.1, is not uniform by

  (a) showing there is no uniform (and decreasing) bound on the errors $\{|f(x) - f_k(x)|\}$.

  (b) using Theorem 8.1.

5. For the sequence of Exercise B5, Section 8.1:

  (a) Show that the convergence is uniform by finding a uniform (and decreasing) bound on the errors $\{|f(x) - f_k(x)|\}$.

  (b) Use Theorem 8.1 to show that the convergence of the sequence $\{f_k'(x)\}$ is not uniform.

  (c) Show that termwise integration is valid.

6. For the sequence of Exercise B6, Section 8.1:

  (a) Show that the convergence is not uniform by showing there is no uniform (and decreasing) bound on the errors $\{|f(x) - f_k(x)|\}$.

  (b) Show that $\{f_k'(x)\}$, the derived sequence, converges pointwise for $x \neq 0$, but the convergence on $[0, 2]$ cannot be uniform in light of Theorem 8.1.

  (c) Show that $\{\int_0^x f_k(t)\,dt\}$ converges pointwise on $[0, 2]$ but not uniformly, again because of Theorem 8.1.

7. For the sequence defined by $f_k(x) = \frac{k^2 x}{1+k^5 x}, -1 \le x \le 1$:

  (a) Show that the convergence is uniform by finding a uniform (and decreasing) bound on the errors $\{|f(x) - f_k(x)|\}$.

  (b) Show that $\{f_k'(x)\}$, the derived sequence, converges pointwise for $x \neq 0$, but the convergence on $[-1, 1]$ cannot be uniform in light of Theorem 8.1.

  (c) By finding a uniform (and decreasing) bound on the errors $\{|f(x) - f_k(x)|\}$, show that $\{\int_{-1}^x f_k(t)\,dt\}$ converges uniformly.

  (d) Why do the integrals of the antiderivatives converge to the antiderivative of $f(x) = \lim_{k\to\infty} f_k(x)$?

8. For the sequence of Exercise B8, Section 8.1:

  (a) Show that the convergence is pointwise, but, in light of Theorem 8.1, not uniform.

  (b) Show that $\{f_k'(x)\}$, the derived sequence, converges pointwise for $x \neq 0$, but the convergence on $[0, 1]$ cannot be uniform in light of Theorem 8.1.

  (c) By finding a uniform (and decreasing) bound on the errors $\{|f(x) - f_k(x)|\}$, show that $\{\int_0^x f_k(t)\,dt\}$ converges uniformly.

  (d) Show that $\lim_{k\to\infty} \int f_k(x)\,dx = \int f(x)\,dx$, where $f(x) = \lim_{k\to\infty} f_k(x)$. Is this a consequence of Theorem 8.2?

9. Show that SEQ-9, Table 8.1, converges uniformly to $f(x) = 0$. *Hint:* Find a uniform (and decreasing) bound on $f_k(x) - f(x)$.

8.4 Convergence in the Mean

DEFINITION 8.4

The sequence of functions $\{f_k(x)\}$ *converges in the mean* to $f(x)$ on the interval $[a, b]$ provided

$$\lim_{k\to\infty} \int_a^b |f_k(x) - f(x)|^2\,dx = 0 \qquad (8.1)$$

EXAMPLE 8.12   On the interval $[0, 2\pi]$, the sequence of functions $f_k(x) = \frac{\sin kx}{k}, k = 1, 2, \ldots,$ converges to $f(x) = 0$ both pointwise and in the mean because $\int_0^{2\pi} \frac{\sin^2 kx}{k^2}\,dx = \frac{\pi}{k^2}$. ❖

Discussion At each fixed x, the integrand in (8.1) is the square of the difference between $f_k(x)$ and $f(x)$. The integral sums this measure for all x in the interval $[a, b]$. If the difference were not squared, the integral would give the area between the functions, and this could be zero even when the functions were far apart. (See Exercise A1.)

For real-valued functions, the absolute value in (8.1) is superfluous when squaring the difference between f_k and f. For complex-valued functions, the absolute value is essential.

We also point out that some texts use the notation "l.i.m." for "limit in the mean."

Pointwise, Uniform, and Mean Convergence

Sequences of functions can now converge in the three ways: pointwise, uniformly, and in the mean. These are not independent concepts because some sequences can converge in more than one sense. Certainly, a sequence that converges uniformly also converges pointwise. The following theorem gives a connection between uniform convergence and convergence in the mean. The proof is direct and informative.

THEOREM 8.5

On a finite interval, uniform convergence to $f(x)$ implies convergence in the mean to $f(x)$.

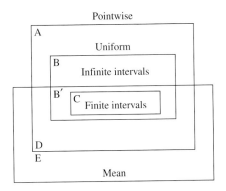

Pointwise

Uniform

Infinite intervals

Finite intervals

Mean

FIGURE 8.19 Relations between pointwise, uniform, and mean convergence

PROOF OF THEOREM 8.5 Assume the sequence $f_k(x)$ converges uniformly to $f(x)$ on the finite interval $[a, b]$. To show convergence in the mean, we have to show (8.1). Thus, we have to show that given an $\varepsilon > 0$, we can provide an N for which $\int_a^b |f_k(x) - f(x)|^2 \, dx < \varepsilon$ for every $k > N$. The uniform convergence of $f_k(x)$ to $f(x)$ means we can make $|f_k(x) - f(x)| < \sqrt{\frac{\varepsilon}{b-a}}$ for all $k > N(\varepsilon)$, independent of x. Consequently, $\int_a^b |f_k(x) - f(x)|^2 \, dx < \int_a^b \frac{\varepsilon}{b-a} \, dx = \varepsilon$, provided $k > N(\varepsilon)$, and we have established the desired limit.

The connections between pointwise, uniform, and mean convergence are summarized in Table 8.3 and illustrated by the Venn diagram in Figure 8.19.

Region	Convergence
A	Pointwise, but neither uniform nor mean
B	Uniform (hence pointwise), but not mean; infinite interval
B′	Uniform (hence pointwise), and mean; infinite interval
C	Uniform on finite interval; hence mean
D	Pointwise (but not uniform) and mean
E	Mean, but not pointwise

TABLE 8.3 Connections between pointwise, uniform, and mean convergence

Region E in Figure 8.19 requires the most explanation since it is possible for a sequence to converge in the mean to different limit functions. Thus, the set of functions that behaves as the zero function under mean convergence has many members. This difficulty is resolved in functional analysis courses by defining classes of functions equivalent under mean convergence, but this will not be discussed further here.

The following examples show that the regions in Figure 8.19 are not empty.

EXAMPLE 8.13   The "moving tent" functions from SEQ-7, Table 8.1, converge pointwise to $f(x) = 0$ on the interval $[0, 1]$, but the convergence is not uniform (see Section 8.3). Moreover, this sequence does not converge to $f(x) = 0$ in the mean because $\int_0^1 |f_k(x) - 0|^2 \, dx = \frac{4}{3}k$ for any of the $f_k(x)$. Clearly, the limit of the integrals is unbounded, so the sequence does not converge in the mean. This sequence is an example of a member of Region A in Figure 8.19.

❖

EXAMPLE 8.14   The functions from SEQ-9, Table 8.1, converge uniformly (see Exercise 9, Section 8.3) to $f(x) = 0$ on the semi-infinite interval $[0, \infty)$. The sequence does not converge in the mean because $\int_0^k (\frac{1}{\sqrt{k}} - 0)^2 \, dx = 1$ for each k, $k = 1, 2, \ldots$, and is therefore a member of Region B in Figure 8.19.

❖

EXAMPLE 8.15   The sequence $\{\frac{1}{kx^2}\}$, $k = 1, 2, \ldots$, defined on $[1, \infty)$, converges uniformly to $f(x) = 0$ because, on that interval, $\left|\frac{1}{kx^2}\right| \leq \frac{1}{k}$. Thus, the bound on $f_k(x)$ is uniform. The sequence also converges in the mean to $f(x) = 0$ because $\int_1^\infty (\frac{1}{kx^2} - 0)^2 \, dx = \frac{1}{3k^2}$ and, therefore, is a member of Region B' in Figure 8.19.

❖

EXAMPLE 8.16   The functions $f_k(x) = xe^{-kx}$ from SEQ-5, Table 8.1, converge uniformly to $f(x) = 0$ on the interval $[0, 1]$, as we saw in Section 8.3. The sequence also converges in the mean to $f(x) = 0$ because

$$\int_0^1 x^2 e^{-2kx} \, dx = \frac{1 - e^{-2k}(1 + 2k + 2k^2)}{4k^3}$$

which goes to zero as k goes to infinity. Hence, SEQ-5 is a member of Region C in Figure 8.19.

❖

EXAMPLE 8.17   The functions $f_k(x) = kxe^{-kx}$ from SEQ-8, Table 8.1, converge pointwise to $f(x) = 0$ on the interval $[0, 1]$. In Section 8.3 we saw that this sequence does not converge uniformly, but since

$$\int_0^1 k^2 x^2 e^{-2kx} \, dx = \frac{1 - e^{-2k}(1 + 2k + 2k^2)}{4k}$$

it converges in the mean to $f(x) = 0$. Thus, this sequence is a member of Region D in Figure 8.19.

❖

EXAMPLE 8.18   Consider the sequence $\{f_k(x)\}$ whose members are alternately $g_1(x)$ and $g_2(x)$, where

$$g_1(x) = \begin{cases} 3 & x = 1 \\ 0 & x \neq 1 \end{cases} \quad \text{and} \quad g_2(x) = \begin{cases} 3 & x = 2 \\ 0 & x \neq 2 \end{cases}$$

This sequence does not converge pointwise. However, since each function is zero except at a single point, their integrals, and integrals of their squares, are zero. The sequence therefore converges to zero in the mean but does not converge pointwise and so is a member of Region E in Figure 8.19.

❖

EXERCISES 8.4–Part A

A1. Sketch $f(x) = x$ and $g(x) = 8x^3 - 12x^2 + 5x$, and show that $\int_0^1 (f - g)\,dx = 0$. Find the maximum of the vertical separation between these two functions.

A2. Verify $\lim_{k\to\infty} \int_0^1 x^2 e^{-2kx}\,dx = 0$.

A3. If $f_k(x) = x^k/(1 + x^{2k})$, $0 \le x \le 1$, show that $\{f_k\}$ converges in the mean to $f(x) = 0$. Find the pointwise limit of this sequence. Is the convergence uniform?

A4. If $f_k(x) = \frac{x}{k}e^{-x/k}$, $0 \le x \le 10$, show that $\{f_k\}$ converges in the mean to $f(x) = 0$. Find the pointwise limit of this sequence. Is the convergence uniform?

A5. If $f_k(x) = \frac{1}{1+kx}$, $0 \le x \le 1$, show that $\{f_k\}$ converges in the mean to $f(x) = 0$. Find the pointwise limit of this sequence. Is the convergence uniform?

EXERCISES 8.4–Part B

B1. The recursively defined sequence ([78, p. 169]) $f_0 = 0$, $f_{k+1}(x) = f_k(x) + (x^2 - f_k^2(x))/2$, $k = 0, 1, \ldots$, is supposed to converge uniformly to $|x|$ on $[-1, 1]$. Test this claim graphically.

For the sequences in Exercises B2–10:

(a) Find the pointwise limit.

(b) Determine if the convergence is uniform or not.

(c) Find the limit in the mean.

(d) In which region of Figure 8.19 does the sequence fall?

B2. SEQ-3, Table 8.1   **B3.** SEQ-6, Table 8.1

B4. SEQ-10, Table 8.1   **B5.** SEQ-11, Table 8.1

B6. SEQ-12, Table 8.1   **B7.** Exercise B5, Section 8.1

B8. Exercise B6, Section 8.1   **B9.** Exercise 7, Section 8.3

B10. Exercise B8, Section 8.1

B11. For the sequence of Exercise B2, Section 8.1:

(a) Find the pointwise limit.

(b) Use numeric integration to find evidence that the sequence converges in the mean to the pointwise limit.

B12. For the sequence of Exercise B3, Section 8.1:

(a) Find the pointwise limit.

(b) Show the sequence converges in the mean to the pointwise limit.

(c) In which region of Figure 8.19 does this sequence fall?

The Stone–Weierstrass theorem [78] says that any function $f(x)$, continuous on the interval $[a, b]$, can be *uniformly* approximated by a sequence of polynomials that depend on $f(x)$. For example, given $f(x)$, $0 \le x \le 1$, the *Bernstein polynomials* $B_k(x) = \sum_{n=0}^{k} \frac{k!}{n!(k-n)!} x^n (1 - x)^{k-n} f\left(\frac{n}{k}\right)$ converge uniformly to $f(x)$ on the interval $[0, 1]$. For the functions $f(x)$, $0 \le x \le 1$, given in Exercises B13–16, obtain and plot the first five Bernstein polynomials. If your favorite computer algebra system has a built-in function for generating the Bernstein polynomials, compare your results to the polynomials it provides.

B13. $f(x) = \sqrt{x}$ **B14.** $f(x) = \sin 2\pi x$

B15. $f(x) = \cos 4\pi x$ **B16.** $f(x) = e^{-x^2}$

If $f(z)$ is defined (and continuous) on the interval $a \le z \le b$, then $g(x) = f((b - a)x + a)$, $0 \le x \le 1$, takes on the same values as $f(z)$, $a \le z \le b$. Use this change of variables to obtain and plot the first five Bernstein polynomials for the functions in Exercises B17–20.

B17. $f(z) = \sqrt{3 + z^2}$, $2 \le z \le 5$ **B18.** $f(z) = z \sin z$, $0 \le z \le 4\pi$

B19. $f(z) = \dfrac{1}{1 + z^2}$, $-2 \le z \le 3$ **B20.** $f(z) = \arctan z$, $-5 \le z \le 5$

B21. If $f_k(x) = k^{3/2} e^{-k^2 x^2}$, $-1 \le x \le 1$, show $\{f_k\}$ converges pointwise to 0, but not in the mean.

8.5   Series of Functions

DEFINITION 8.5

We say the infinite series $\sum_{k=0}^{\infty} f_k(x)$ converges to $f(x)$ *pointwise* if $\{\sum_{k=0}^{n} f_k(x)\}$, the sequence of partial sums, converges as a sequence of numbers at each point x.

Two things have happened here. First, we have defined the convergence of the series of functions in terms of the behavior of the sequence of partial sums. Second, we have reduced pointwise convergence to the behavior, at each x, of a series of just numbers.

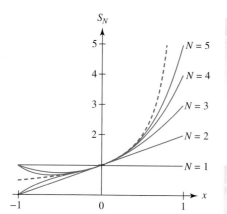

FIGURE 8.20   Partial sums of $\sum_{k=0}^{\infty} x^k$ converging to $\frac{1}{1-x}$ (dashed)

DEFINITION 8.6

If the convergence of the sequence of partial sums is uniform, then we say $\sum_{k=0}^{\infty} f_k(x)$ converges to $f(x)$ *uniformly*. Again, we have reduced the issue of uniform convergence for the series to one of uniform convergence of a sequence, the sequence of partial sums of the series.

DEFINITION 8.7

If the sequence of partial sums converges in the mean to $f(x)$, we say $\sum_{k=0}^{\infty} f_k(x)$ converges to $f(x)$ *in the mean*. The issue of convergence for the series is again determined by the convergence of the sequence of partial sums.

EXAMPLE 8.19   For $|x| < 1$, the function $f(x) = \frac{1}{1-x}$ can be represented by the geometric series $\sum_{k=0}^{\infty} x^k$. The first five partial sums, $S_N = \sum_{k=0}^{N} x^k$, are graphed in Figure 8.20. The graph of $f(x)$ is the dashed curve in the figure. The sequence of partial sums seems to converge pointwise to $f(x)$ in the open interval $(-1, 1)$. In fact, in the closed interval $|x| \leq \alpha < 1$, the convergence is uniform. However, at $x = 1$ the function $f(x)$ is undefined, having a vertical asymptote there. Determining convergence at the endpoints of the interval $[-1, 1]$ requires special care, and such matters will be discussed in Chapter 9, which specifically treats power series. ❖

EXAMPLE 8.20   The function $f(x) = x^2$ is well represented on the interval $[-\pi, \pi]$ by the series $\frac{\pi^2}{3} + 4\sum_{k=1}^{\infty}((-1)^k/k^2)\cos kx$. (This is an example of a *Fourier series*, to be studied in detail in Chapter 10.) Figure 8.21 shows the partial sum $S_2 = \frac{\pi^2}{3} - 4\cos x + \cos 2x$ (dashed), and the function $f(x)$ (solid) gives an indication of how quickly this series converges. ❖

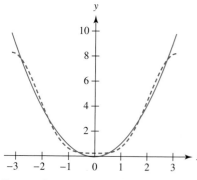

FIGURE 8.21   Partial sum S_2 (dashed) and $f(x) = x^2$ (solid) in Example 8.20

An nth Term Rule

In a convergent series of numbers, the nth term necessarily goes to zero (Theorem 7.1). However, this need not be true for a convergent series of functions, as the following series shows.

Define the functions

$$f_k(x) = \begin{cases} 0 & x \leq k \\ 1 & k < x \leq k+1 \\ 0 & x > k+1 \end{cases}$$

so that $f_k(x)$ is nothing more than a unit pulse of duration 1 starting at $x = k$. The sum of the series $\sum_{k=0}^{\infty} f_k(x)$ is the function $f(x) = 1$. However, $\lim_{k\to\infty} f_k(x) \neq 0$.

A condition ruling out this aberrant behavior is given by the following theorem.

THEOREM 8.6

If the series $\sum_{k=0}^{\infty} f_k(x)$ converges uniformly on the pointset A, then $f_k(x)$ converges uniformly to zero on the set A.

PROOF OF THEOREM 8.6   To prove $f_k(x)$ goes to zero uniformly in x, we need to show that given an $\varepsilon > 0$, we can produce an $N(\varepsilon)$ that is independent of x and for which $|f_k(x) - 0| < \varepsilon$ whenever $k > N$. Let the series converge uniformly so that $\sum_{k=0}^{\infty} f_k(x) = f(x)$, and represent the partial sums by $S_n = \sum_{k=0}^{n} f_k(x)$. Write

$$f_n(x) = S_n(x) - S_{n-1}(x) = S_n(x) - f(x) + f(x) - S_{n-1}(x)$$

so as to obtain

$$|f_n(x)| \leq |S_n(x) - f(x)| + |f(x) - S_{n-1}(x)|$$

Let $\varepsilon > 0$ be given, and use the uniform convergence of the sequence of partial sums $\{S_n(x)\}$ to $f(x)$ to determine an N for which both $|S_n(x) - f(x)| < \frac{\varepsilon}{2}$ and $|f(x) - S_{n-1}(x)| < \frac{\varepsilon}{2}$ hold whenever $n - 1 > N$. Clearly, this is the N for which $|f_n(x)| < \frac{\varepsilon}{2} + \frac{\varepsilon}{2} = \varepsilon$ likewise holds uniformly in x when $n - 1 > N$.

Determining Uniform Convergence

It is generally tedious to use the definition when verifying the uniform convergence of a series. The *comparison* test

domination by a uniformly convergent series implies uniform convergence

and its corollary, the *Weierstrass M-test* (domination by a constant series), provide two simpler criteria for determining whether or not a series converges uniformly. The precise meaning of *domination* is given by the following.

DEFINITION 8.8

The series $\sum_{k=0}^{\infty} u_k(x)$ is said to *dominate* the series $\sum_{k=0}^{\infty} v_k(x)$ if $|v_k(x)| \leq u_k(x)$ for every k and for every x in a specified domain.

A formal statement of the comparison test is given by the following theorem.

THEOREM 8.7   COMPARISON TEST

1. $\displaystyle\sum_{k=0}^{\infty} u_k(x) = u(x)$ uniformly on a pointset A.
2. $|f_k(x)| \leq u_k(x)$ (on A) for every k.

$\implies \displaystyle\sum_{k=0}^{\infty} f_k(x)$ converges uniformly on A to a function $f(x)$.

Theorem 8.7 simply says that if the series $\sum_{k=0}^{\infty} f_k(x)$ is dominated by a uniformly convergent series, then it is also uniformly convergent. Before proving this theorem, we state and illustrate the corollary called the Weierstrass M-test. The M-test arises when the dominating series is a series of constants M_k so that $\sum_{k=0}^{\infty} M_k$ dominates $\sum_{k=0}^{\infty} f_k(x)$.

WEIERSTRASS M-TEST   If $|f_k(x)| \leq M_k$ for each k and every x in a set A and if $\sum_{k=0}^{\infty} M_k$ converges, then $\sum_{k=0}^{\infty} f_k(x)$ converges uniformly.

EXAMPLE 8.21   Use the Weierstrass M-test to establish the uniform convergence of the series $\sum_{k=0}^{\infty} e^{-kx} \cos kx$ on the interval $[a, \infty)$, where $a > 0$. To this end, estimate $f_k(x)$ as

$$|f_k(x)| = |e^{-kx} \cos kx| \leq e^{-kx} \leq e^{-ka} = \left(\frac{1}{e^a}\right)^k = r^k = M_k$$

The first inequality holds because $|\cos t| \leq 1$, while the second holds because e^{-kx} attains its maximum at the left endpoint of the interval $[a, \infty)$. We then have $\sum_{k=0}^{\infty} M_k = 1/(1-e^{-a})$ as the sum of the geometric series. Since the series of dominating constants converges, the original series converges uniformly.   ❖

PROOF OF THEOREM 8.7   By the comparison test of Exercise B22, Section 7.3, the series $\sum_{k=0}^{\infty} |f_k(x)|$ converges pointwise, so $\sum_{k=0}^{\infty} f_k(x)$ converges pointwise to some function $f(x)$ because, from Section 7.4, absolute convergence implies convergence. Noting that necessarily $u_k(x) \geq 0$, we then follow [65] in writing

$$\left|\sum_{k=0}^{n} f_k(x) - f(x)\right| = \left|\sum_{k=n+1}^{\infty} f_k(x)\right| \leq \sum_{k=n+1}^{\infty} u_k(x) = \left|\sum_{k=0}^{n} u_k(x) - u(x)\right| \tag{8.2}$$

Since the term on the far right in (8.2) can be made uniformly small when $n > N(\varepsilon)$, so too will the term on the left be uniformly small, thus establishing the *uniform* convergence of $\sum_{k=0}^{\infty} f_k(x)$.

Further Consequences of Uniform Convergence

We next present three theorems that guarantee the continuity, integrability, and differentiability of the limit of a uniformly convergent series. These theorems allow the operations of calculus to be performed termwise on a uniformly convergent series.

As a straightforward consequence of the equivalent theorem for sequences, we have that

a uniformly convergent series of continuous functions converges to a continuous function.

Stated formally, we have the following theorem.

THEOREM 8.8   CONTINUITY

1. For each k, $f_k(x)$ is a continuous function of x on the set A.

2. $\sum_{k=0}^{\infty} f_k(x)$ converges uniformly to $f(x)$ on A.

$\implies f(x)$ is continuous on A.

We prove Theorem 8.8 by noting each partial sum $S_n(x) = \sum_{k=0}^{n} f_k(x)$ is a finite sum of continuous functions and, hence, is continuous. Since $f(x)$ is the limit of the sequence of partial sums and the partial sums themselves are a uniformly converging sequence of continuous functions, the limit function must be continuous.

Again paralleling the equivalent theorem for sequences, we have that

the uniform limit of continuous functions can be integrated termwise.

EXAMPLE 8.22 Termwise integration of the geometric series $\sum_{k=0}^{\infty} x^k = 1 + x + x^2 + x^3 + \cdots = \frac{1}{1-x}$ on $|x| \le r < 1$ is valid because of uniform convergence. Hence,

$$\sum_{k=0}^{\infty} \int_{-r}^{r} x^k \, dx = 2 \sum_{k=0}^{\infty} \frac{r^{2k+1}}{2k+1} = \int_{-r}^{r} = \int_{-r}^{r} \frac{1}{1-x} \, dx = \ln \frac{1+r}{1-r}$$

and we have obtained a series expansion for $\ln \frac{1+r}{1-r}$ by integration!   ❖

A formal statement of this theorem is as follows.

THEOREM 8.9   TERMWISE INTEGRATION

1. For each k, $f_k(x)$ is continuous on the interval $[a, b]$.

2. $\sum_{k=0}^{\infty} f_k(x)$ converges uniformly to $f(x)$ on $[a, b]$.

$$\implies \int_{a}^{b} f(x) \, dx = \int_{a}^{b} \sum_{k=0}^{\infty} f_k(x) \, dx = \sum_{k=0}^{\infty} \int_{a}^{b} f_k(x) \, dx$$

PROOF OF THEOREM 8.9   We prove this theorem by again noting each partial sum $S_n(x) = \sum_{k=0}^{n} f_k(x)$ is a finite sum of continuous functions and, hence, is continuous. Since $f(x)$ is the uniform limit of the sequence of (continuous) partial sums, $f(x)$ is itself continuous and hence integrable, with value $\int_{a}^{b} f(x) \, dx = \int_{a}^{b} \lim_{n \to \infty} S_n(x) \, dx = \lim_{n \to \infty} \int_{a}^{b} S_n(x) \, dx$. The second equality follows from the integrability of a uniformly converging sequence of integrable functions. Now each $S_n(x)$ is a finite sum, so

$$\lim_{n \to \infty} \int_{a}^{b} S_n(x) \, dx = \lim_{n \to \infty} \int_{a}^{b} \sum_{k=0}^{n} f_k(x) \, dx$$

$$= \lim_{n \to \infty} \sum_{k=0}^{n} \int_{a}^{b} f_k(x) \, dx = \sum_{k=0}^{\infty} \int_{a}^{b} f_k(x) \, dx$$

(8.3)

The term on the right in (8.3) is exactly the meaning of the limit on the left. Hence, the integral of the limit function $f(x)$ is the sum of the integrated terms of the original series, that is, $\int_{a}^{b} f(x) \, dx = \sum_{k=0}^{\infty} \int_{a}^{b} f_k(x) \, dx$.

As with sequences, if a series converges at least for a single point and the series formed by termwise differentiation converges uniformly, then the derived series converges to the derivative of the original series. In a word, uniform convergence of the derived series justifies termwise differentiation. This is formalized in the following theorem.

THEOREM 8.10   TERMWISE DIFFERENTIATION

1. For each k, $f_k(x)$ is differentiable on $[a, b]$.

2. $\displaystyle\sum_{k=0}^{\infty} f_k(x_0)$ converges for some x_0 in $[a, b]$.

3. $\displaystyle\sum_{k=0}^{\infty} f_k'$ converges uniformly on $[a, b]$.

$\Longrightarrow$

1. $\displaystyle\sum_{k=0}^{\infty} f_k(x)$ converges uniformly to $f(x)$.

2. $f'(x)$ exists.

3. $\displaystyle f'(x) = \sum_{k=0}^{\infty} f_k'$

A slightly weaker form of the theorem assumes that each function $f_k(x)$ has a *continuous* derivative. The conclusion is still that $f'(x) = \sum_{k=0}^{\infty} f_k'$. The formal statement of this weaker theorem is

1. For each k, $f_k(x)$ has a continuous derivative.

2. $\displaystyle\sum_{k=0}^{\infty} f_k(x)$ converges to $f(x)$.

3. $\displaystyle\sum_{k=0}^{\infty} f_k'$ converges uniformly.

$\displaystyle \Longrightarrow f'(x) = \sum_{k=0}^{\infty} f_k'$

The proof of this weaker form of the theorem on termwise differentiation of a series follows from the equivalent theorem on the termwise differentiation of a sequence. In particular, set $S_n(x) = \sum_{k=0}^{n} f_k(x)$, the nth partial sum of $\sum_{k=0}^{\infty} f_k(x)$.

EXERCISES 8.5–Part A

A1. For what values does the series (*) converge uniformly?

$$(*) \quad \sum_{k=1}^{\infty} \frac{1}{k^3} \frac{1 - x^{2k}}{1 + x^{2k}}$$

A2. If $\{c_k\}$ is a sequence of constants for which $\sum_{k=1}^{\infty} c_k$ converges absolutely, show that both $\sum_{k=1}^{\infty} c_k \sin kx$ and $\sum_{k=1}^{\infty} c_k \cos kx$ converge uniformly for $|x| < \infty$.

A3. If $f(x) = \sum_{k=1}^{\infty} \frac{\sin kx}{k^3}$, verify that $f'(x) = \sum_{k=1}^{\infty} \frac{\cos kx}{k^2}$ for all x.

A4. If $f(x)$ is the function in Exercise A3, obtain a series for $\int_0^x f(t)\, dt$.

A5. If $f(x) = \sum_{k=1}^{\infty} \frac{1}{k} \left(\frac{x}{1+2x} \right)^k$, for what values of x is $f'(x) = \frac{1}{(1+2x)^2} \sum_{k=0}^{\infty} \left(\frac{x}{1+2x} \right)^k$?

EXERCISES 8.5–Part B

B1. Let $f(x) = \begin{cases} x \ln x & x > 0 \\ 0 & x = 0 \end{cases}$. Show that $f(x)$ is continuous at $x = 0$. Determine R for which $\sum_{k=0}^{\infty} f(x)^k$ converges uniformly on $0 \le x \le r < R$.

In Exercises B2–8:

    (a) Show graphically that the sequence of partial sums seems to converge pointwise to the given function on the indicated domain.

    (b) Apply the Weierstrass M-test to determine uniform convergence.

    (c) Examine, at least empirically, the validity of $\int f \, dx = \sum \int f_k \, dx$.

    (d) Examine, at least empirically, the validity of $\sum f_k' = f'$.

B2. $e^x = \sum_{k=0}^{\infty} \frac{x^k}{k!}, \ 0 \le x \le 2$

B3. $\cos x = \sum_{k=0}^{\infty} \frac{(-1)^k x^{2k}}{(2k)!}, \ -\pi \le x \le \pi$

B4. $\sin x = \sum_{k=0}^{\infty} \frac{(-1)^k x^{2k+1}}{(2k+1)!}, \ -\pi \le x \le \pi$

B5. $\cosh x = \sum_{k=0}^{\infty} \frac{x^{2k}}{(2k)!}, \ -1 \le x \le 1$

B6. $\sinh x = \sum_{k=0}^{\infty} \frac{x^{2k+1}}{(2k+1)!}, \ -1 \le x \le 1$

B7. $\ln(1-x) = -\sum_{k=0}^{\infty} \frac{x^{k+1}}{k+1}, \ 0 \le x < 1$

B8. $\sqrt{1-x} = -\sum_{k=0}^{\infty} \frac{(2k)! x^k}{4^k (k!)^2 (2k-1)}, \ -1 < x < 1$

Exercises B9–16 each contain a Fourier series, of great important in science and engineering, and studied at length in Chapter 10. For each such series given, valid on the $0 \le x \le \pi$:

    (a) Show graphically that the sequence of partial sums seems to converge pointwise to the given function.

    (b) Examine, at least empirically, the validity of $\int f \, dx = \sum \int f_k \, dx$.

    (c) Examine, at least empirically, the validity of $\sum f_k' = f'$.

B9. $x = 2 \sum_{k=1}^{\infty} \frac{(-1)^{k+1}}{k} \sin kx$ **B10.** $x = \frac{\pi}{2} - \frac{4}{\pi} \sum_{k=1}^{\infty} \frac{\cos(2k-1)x}{(2k-1)^2}$

B11. $x^2 = \sum_{k=1}^{\infty} \left[\frac{2\pi}{k}(-1)^{k+1} + \frac{4}{\pi k^3}((-1)^k - 1) \right] \sin kx$

B12. $x^2 = \frac{\pi^2}{3} + 4 \sum_{k=1}^{\infty} \frac{(-1)^k}{k^2} \cos kx$

B13. $x^3 = 2 \sum_{k=1}^{\infty} \frac{(-1)^{k+1}(k^2\pi^2 - 6)}{k^3} \sin kx$

B14. $x^3 = \frac{\pi^3}{4} + \sum_{k=1}^{\infty} \left[\frac{6\pi}{k^2}(-1)^k - \frac{12}{\pi k^4}((-1)^k - 1) \right] \cos kx$

B15. $\cos(x) = \frac{8}{\pi} \sum_{k=1}^{\infty} \frac{k}{4k^2 - 1} \sin 2kx$

B16. $\sin(x) = \frac{2}{\pi} + \frac{4}{\pi} \sum_{k=1}^{\infty} \frac{\cos 2kx}{1 - 4k^2}$

B17. Use the series in Exercise B9 to establish $\frac{\pi}{4} = \sum_{k=1}^{\infty} (-1)^{k+1}/(2k-1)$. (*Hint:* Evaluate at $x = \frac{\pi}{2}$.)

B18. Use the series in Exercise B10 to establish $\frac{\pi^2\sqrt{2}}{16} = 1 - \frac{1}{3^2} - \frac{1}{5^2} + \frac{1}{7^2} + \frac{1}{9^2} - \frac{1}{11^2} - \frac{1}{13^2} + \frac{1}{15^2} + \frac{1}{17^2} - \cdots$. (*Hint:* Evaluate at $x = \frac{\pi}{4}$.)

B19. Use the series in Exercise B15 to establish $\frac{\pi\sqrt{2}}{16} = \sum_{k=0}^{\infty} (-1)^k \frac{2k+1}{4(2k+1)^2 - 1}$. (*Hint:* Evaluate at $x = \frac{\pi}{4}$.)

B20. The Legendre polynomials $P_k(x)$, $-1 \le x \le 1$, can be defined by the recursion $P_0 = 1$, $P_1 = x$, $P_k = \frac{2k-1}{k}x P_{k-1} - \frac{k-1}{k} P_{k-2}$. (Modern computer algebra systems have these polynomials built-in. The first five are listed in Section 10.7.)

    (a) Use the given recursion to generate $P_2, \ldots, P_{10}$, and compare the results to the built-in Legendre polynomials in your favorite computer algebra system.

    (b) Two functions $f(x)$ and $g(x)$ are said to be orthogonal on the interval $[a, b]$ if $\int_a^b f(x)g(x) \, dx = 0$. For $0 \le i, j \le 5$, show that the Legendre polynomials are orthogonal on the interval $[-1, 1]$ by showing
$$\int_{-1}^{1} P_k(x) P_j(x) \, dx = \begin{cases} 0 & j \ne k \\ \frac{2}{2k+1} & j = k \end{cases}$$

    (c) Let $f(x) = x$, and compute $a_k = \frac{2k+1}{2} \int_{-1}^{1} f(x) P_k(x) \, dx$, $k = 0, \ldots, 10$. Demonstrate that the series $\sum_{k=0}^{\infty} a_k P_k(x)$ converges to $f(x)$ by forming and graphing the sequence of partial sums $S_n(x) = \sum_{k=0}^{n} a_k P_k(x)$.

    (d) Repeat part (c) for $f(x) = x^2$.

    (e) Repeat part (c) for $f(x) = \cos \pi x$.

    (f) Repeat part (c) for $f(x) = \begin{cases} 1 & -1 \le x \le 0 \\ -1 & 0 < x \le 1 \end{cases}$.

B21. The Chebyshev polynomials $T_k(x)$, $-1 \le x \le 1$, can be defined by the recursion $T_0 = 1$, $T_1 = x$, $T_k = 2x T_{k-1} - T_{k-2}$. (Modern computer algebra systems have these polynomials built-in. The first eight are listed in Table 40.5 of Section 40.3.)

    (a) Use the given recursion to generate $T_2, \ldots, T_{10}$, and compare the results to the built-in Chebyshev polynomials in your favorite computer algebra system. (The spelling of this Russian name varies greatly in the literature. Some texts render it as Tchebyshev.)

(b) On the interval $[a, b]$, two functions $f(x)$ and $g(x)$ are orthogonal with respect to the weight function $w(x)$ if $\int_a^b w(x) f(x) g(x) \, dx = 0$. Show that on the interval $[-1, 1]$ the Chebyshev polynomials are *orthogonal with respect to the weight function* $w = \frac{1}{\sqrt{1-x^2}}$ by showing

$$\int_{-1}^1 w(x) T_k(x) T_j(x) \, dx = \begin{cases} \pi & k = j = 0 \\ \dfrac{\pi}{2} & k = j > 0 \\ 0 & k \neq j \end{cases}$$

(c) Let $f(x) = e^x$, and compute (via numerical integration) $a_k = \frac{2}{\pi} \int_{-1}^1 f(x) w(x) T_k(x) \, dx, k = 0, \ldots, 10$. Demonstrate that the series $\frac{a_0}{2} + \sum_{k=1}^\infty a_k T_k(x)$ converges to $f(x)$ by forming and graphing the sequence of partial sums $S_n(x) = \frac{a_0}{2} + \sum_{k=1}^n a_k T_k(x)$.

(d) Repeat part (c) for $f(x) = \sin \pi x$.

(e) Repeat part (c) for $f(x) = \cos \pi x$.

(f) Repeat part (c) for $f(x) = \begin{cases} 1 & -1 \leq x \leq 0 \\ -1 & 0 < x \leq 1 \end{cases}$.

Chapter Review

1. Define the pointwise convergence of a sequence of functions.

2. Give an example of a sequence of functions that converges pointwise. What is the limiting function?

3. Give an example of a sequence of continuous functions that converges pointwise to a discontinuous limit function.

4. Give an example of a sequence of functions that does not converge pointwise, that is, for which there is no limiting function. Explain why the sequence does not converge.

5. Give a verbal description of the concept of uniform convergence of a sequence of functions. Give an example.

6. State the mathematical definition of uniform convergence of a sequence of functions.

7. Does pointwise convergence imply uniform convergence? Explain, and illustrate with an example.

8. Does uniform convergence imply pointwise convergence?

9. Explain what it means for a sequence to fail to converge uniformly. Give an example.

10. Give an example of a sequence of continuous functions that converges uniformly. What is the limiting function? Is it continuous?

11. Give an example of a sequence of discontinuous functions that converges uniformly. What is the limiting function? Is it continuous? Can you find a sequence of discontinuous functions that converges uniformly to a discontinuous limit? Does this violate Theorem 8.1 on the uniform limit of continuous functions?

12. Use an example to illustrate Theorem 8.2, which states that a sequence of continuous functions converging uniformly can be integrated termwise.

13. Can a sequence of continuous functions that converges only pointwise and not uniformly be integrated termwise? Explain.

14. State a theorem justifying termwise integration of a sequence of functions. Illustrate it with an example. Give an example that violates one of the hypotheses of your theorem and where termwise differentiation fails. Can you find an example of a sequence that fails one of the hypotheses of your theorem, but for which termwise differentiation succeeds?

15. Give a verbal description of convergence in the mean for a sequence of functions. Give an example.

16. Give the precise mathematical definition of convergence in the mean for a sequence of functions.

17. Give an example of a sequence of functions that converges in the mean but does not converge pointwise.

18. Give an example of a sequence of functions that converges in the mean, converges pointwise, but does not converge uniformly.

19. Give an example of a sequence of functions that converges in the mean, converges pointwise, and converges uniformly.

20. Give an example of a sequence of functions that converges pointwise but does not converge in the mean.

21. Give an example of a sequence of functions that converges uniformly but does not converge in the mean.

22. Give a condition under which uniform convergence of a sequence of functions implies convergence in the mean. Give an example.

23. Give a verbal description of what it means for an infinite series of functions to converge pointwise. Be sure to include the concept of the partial sum. Give an example of a series of continuous functions that converges pointwise to a discontinuous limit function.

24. Give a verbal description of what it means for an infinite series of functions to converge uniformly. Give an example of a series of continuous functions that converges uniformly to a continuous function.

25. Give a verbal description of what it means for an infinite series of functions to converge in the mean. Give an example of a series of functions that converges both pointwise and in the mean. Give an example for which the convergence is in the mean but not pointwise.

26. Give an example of a convergent series of functions for which the nth term does not go to zero. Why is this surprising?

27. Give a sufficient condition under which a convergent series of functions will have its nth term go to zero.

28. State the Weierstrass M-test, and give an example of its applicability.

29. Give an example that illustrates the validity of termwise integration of a uniformly convergent series of continuous functions.

30. Give an example illustrating conditions under which a series of functions can be differentiated termwise.

Chapter 9

Power Series

INTRODUCTION   Chapter 9 discusses the *Taylor* (or *power*) *series,* and manipulations thereon, in anticipation of representing solutions of differential equations as power series. The Taylor series should be seen as but a first example of representing a function by an infinite series of "simpler" functions.

The calculus student should be familiar with *Taylor polynomials* and may even be familiar with the infinite Taylor series. For students so prepared, this chapter can be considered optional.

9.1 Taylor Polynomials

Polynomial Approximation

If a function $f(x)$ is difficult to evaluate or manipulate, it is often replaced by an approximating polynomial. Polynomials whose derivatives match those of $f(x)$ at a single fixed point $x = a$ are called *Taylor polynomials,* or *Maclaurin polynomials* if $a = 0$. Such approximations are typically studied in most calculus courses.

EXAMPLE 9.1   To construct a cubic polynomial that approximates the function $f(x) = \frac{1}{1+x}$ in a neighborhood of $x = 2$, write the approximating cubic as $Y = ax^3 + bx^2 + cx + d$ and determine the four constants a, b, c, d by the four conditions $f^{(k)}(2) = Y^{(k)}(2)$, $k = 0, 1, 2, 3$. Taking derivatives and evaluating at $x = 2$ gives the four equations

$$\tfrac{1}{3} = 8a + 4b + 2c + d \qquad -\tfrac{1}{9} = 12a + 4b + c \qquad \tfrac{2}{27} = 12a + 2b \qquad -\tfrac{2}{27} = 6a$$

from whose solution we find

$$Y(x) = \tfrac{65}{81} - \tfrac{11}{27}x + \tfrac{1}{9}x^2 - \tfrac{1}{81}x^3$$

Figure 9.1 illustrates the accuracy of the approximation, showing $f(x)$ as the solid curve and $Y(x)$ as the dashed.

In some neighborhood of $x = 2$ the approximation is good, but far enough away from $x = 2$ the approximation, an example of a Taylor polynomial, weakens.   ❖

Taylor's Theorem

The conditions under which such approximating polynomials exist are captured in theorems such as the following, taken essentially from [91].

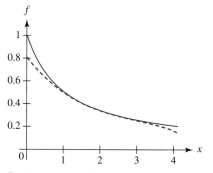

FIGURE 9.1 Cubic approximation (dashed) for $f(x) = \frac{1}{1+x}$ (solid) constructed at $x = 2$

THEOREM 9.1

1. $f(x), f'(x), \ldots, f^{(n+1)}(x)$ are continuous in a closed interval containing $x = a$ (either inside or at one end).

2. x is any point in the interval.

3. $R_{n+1} = \dfrac{1}{n!} \displaystyle\int_a^x (x-t)^n f^{(n+1)}(t)\, dt$

$\Longrightarrow f(x) = f(a) + \dfrac{f'(a)}{1!}(x-a) + \cdots + \dfrac{f^{(n)}(a)}{n!}(x-a)^n + R_{n+1}$

THEOREM 9.2

1. $f(x), f'(x), \ldots, f^{(n)}(x)$ are continuous in the closed interval $[a, x]$, with x fixed.

2. $f^{(n+1)}(x)$ exists on the open interval (a, x).

3. $R_{n+1}(c) = \dfrac{f^{(n+1)}(c)}{(n+1)!}(x-a)^{(n+1)}$

$\Longrightarrow$ in the open interval (a, x), there exists a value c for which the equality

$$f(x) = f(a) + \frac{f'(a)}{1!}(x-a) + \cdots + \frac{f^{(n)}(a)}{n!}(x-a)^n + R_{n+1}(c)$$

holds.

The nth-degree Taylor *polynomial* $p_n(x)$ is given by

$$p_n(x) = f(a) + \frac{f'(a)}{1!}(x-a) + \cdots + \frac{f^{(n)}(a)}{n!}(x-a)^n \tag{9.1}$$

It is a polynomial approximation to $f(x)$ constructed from values of the function f and its derivatives evaluated at $x = a$. Clearly, $p_n(a) = f(a)$, so at $x = a$ the polynomial is exact; but as x moves away from a, the approximation becomes poorer. The general coefficient of $(x-a)^k$ is $f^{(k)}(a)/k!$, the kth derivative evaluated at $x = a$ and divided by k factorial.

The remainder in Theorem 9.2 is ascribed to Lagrange and is more apt to be found in an elementary calculus text. Many other forms of the remainder exist, including forms attributed to Cauchy. The integral form of the remainder is an identity in x. It is exactly what remains if the Taylor polynomial is subtracted from the function. On the other hand, in the Lagrange form of the remainder, the constant c depends on x. For each value of x at which the function is to be approximated by a Taylor polynomial, the constant c is different. Hence, in the integral form of the remainder we have a difficult integral to evaluate, whereas in the Lagrange form we have the constant c, which cannot generally be found.

Theorem 9.1 is easier to prove than Theorem 9.2, requiring just repeated integration by parts for its derivation. In particular, writing $f(x) = f(a) + \int_a^x f'(t)\, dt$ and integrating by parts with $u = f'(t)$, we obtain

$$f(x) = f(a) + f'(a)(x-a) + \int_a^x f''(t)(x-t)\, dt$$

Again integrating by parts, this time with $u = f''(t)$, we find

$$f(x) = f(a) + f'(a)(x - a) + \frac{f''(a)}{2!}(x - a)^2 + \frac{1}{2!}\int_a^x f'''(t)(x - t)^2\, dt \qquad (9.2)$$

Repeated integration by parts, each time taking the derivative in the integrand as u, leads to the essence of an induction proof. Alternatively (see [65]), write the function as $f(x) = f(a) + \int_a^x f'(t)\, dt$ and make the change of variables $x - t = z$ to obtain $f(x) = f(a) + \int_0^{x-a} f'(x - z)\, dz$. Integration by parts with u taken as the derivative again yields the desired result.

EXERCISES 9.1–Part A

A1. Using the interpolation technique of Example 9.1, obtain a quadratic polynomial that approximates $f(x) = \frac{x}{1+2x}$ in a neighborhood of $x = 0$.

A2. Use (9.1) to obtain the quadratic polynomial of Exercise A1.

A3. For $f(x) = e^{2x}$, obtain $p_3(x)$, the third-degree Maclaurin polynomial, and then differentiate it termwise.

A4. For $f'(x)$, where $f(x)$ is the function in Exercise A3, obtain the second-degree Maclaurin polynomial. Compare the results of Exercises A3 and 4.

A5. For the polynomial $p_3(x)$ determined in Exercise A3, compute $\int_0^x p_3(t)\, dt$ and compare the result to the Maclaurin expansion of $F(x) = \int_0^x e^{2t}\, dt$.

EXERCISES 9.1–Part B

B1. For $f(x) = \frac{1}{1+x}$ and $a = 0$, apply the integration by parts that led to (9.2) and evaluate the remaining integral. Show that the result is an identity in x.

B2. For the function $f(x) = \sin\left(\frac{2x-3}{3x+2}\right)$:

  (a) Find $R_4(x) = \frac{f^{(4)}(c)}{4!}(x - 2)^4$, the Lagrange form of the remainder.

  (b) If $P_3(x)$ is the cubic Taylor polynomial constructed at $x = 2$, find c in R_4 by solving the equation $f(3) - P_3(3) = R_4(3)$ for c.

  (c) Repeat part (b) if $x = 4$.

  (d) Using a graph, find M, the maximum value attained by $|f^{(4)}(x)|$ on the interval $[1, 4]$.

  (e) Graphically compare $|R_4(x)|_{\max} = \frac{M}{4!}(x - 2)^4$, an upper bound for the error in the Taylor approximation, with $|f - P_3|$, the actual error in the Taylor approximation.

  (f) Find the largest interval centered at $x = 2$, over which $P_3(x)$ approximates $f(x)$ with an error no worse than 10^{-3}.

  (g) Find $P_k(x)$, the first Taylor polynomial constructed at $x = 2$ for which $|f(x) - P_k(x)| < 10^{-5}$ on the interval $|x - 2| \le 1$.

  (h) Find and plot the sequence of Taylor polynomials $P_k(x)$, $k = 1, \dots, 4$, if each polynomial is constructed at $x = 2$.

For the functions $f(x)$ in Exercises B3–7:

  (a) Using a computer algebra system, find and plot the sequence of Taylor polynomials $P_k(x)$, $k = 1, \dots, 4$, if each polynomial is constructed at $x = a$.

  (b) Find the largest interval centered at $x = a$, over which $P_3(x)$ approximates $f(x)$ with an error no worse than 10^{-3}.

  (c) Find $P_k(x)$, the first Taylor polynomial constructed at $x = a$ for which $|f(x) - P_k(x)| < 10^{-5}$ on the interval $|x - a| \le 1$.

  (d) Construct

$$P_3(x) = \sum_{k=0}^{3} \frac{f^{(k)}(a)}{k!}(x - a)^k$$

  by computing and evaluating the appropriate derivatives.

  (e) Construct $P_3(x)$ by determining the coefficients in the cubic polynomial $Y = ax^3 + bx^2 + cx + d$.

  (f) Find $R_4(x) = \frac{f^{(4)}(c)}{4!}(x - a)^4$, the Lagrange form of the remainder.

  (g) Find c in R_4 by solving the equation $f(a + 1) - P_3(a + 1) = R_4(a + 1)$ for c.

  (h) Repeat part (g) at $x = a + 2$.

  (i) Using a graph, find M, the maximum value attained by $|f^{(4)}(x)|$ on the interval $[a - 1, a + 2]$.

  (j) Graphically compare $|R_4(x)|_{\max} = \frac{M}{4!}(x - a)^4$, an upper bound for the error in the Taylor approximation, with $|f - P_3|$, the actual error in the Taylor approximation.

B3. $f(x) = \cos\left(\dfrac{5x - 4}{3x + 7}\right); a = 1$

B4. $f(x) = \dfrac{x + 1}{x^2 + 2x + 2}; a = 0$

B5. $f(x) = x\sqrt{x^2 + 2x + 3} - \sqrt{x + 4}; a = 3$

B6. $f(x) = e^{\sqrt{x}}; a = 4$ **B7.** $f(x) = \arctan x; a = 1$

For the function $f(x)$ and the value of a in each of Exercises B8–11:

(a) Find $R_4(x)$, the integral form of the remainder, and using a computer algebra system, evaluate the integral.

(b) Show that $P_3(x) + R_4(x) = f(x)$, exactly.

B8. $f(x)$ from Exercise B4   **B9.** $f(x)$ from Exercise B5

B10. $f(x)$ from Exercise B6   **B11.** $f(x)$ from Exercise B7

For the definite integrals given in Exercises B12–16:

(a) Obtain an accurate numeric evaluation via your favorite technology.

(b) Replace the integrand with a Taylor polynomial constructed at $x = 0$ and integrate termwise. The Taylor polynomial should be one for which termwise integration yields an answer accurate to 10^{-3}.

(c) Graph the Taylor polynomial used in part (b) and determine the maximum error it incurs as an approximation to the integrand on the interval $[0, 1]$.

B12. $\displaystyle\int_0^1 \frac{1 - x + x^4}{x^3 + 5}\, dx$ **B13.** $\displaystyle\int_0^1 \arctan^2 x\, dx$

B14. $\displaystyle\int_0^1 \sqrt{1 + x^5}\, dx$ **B15.** $\displaystyle\int_0^1 \sin\sqrt{1 + x^2}\, dx$

B16. $\displaystyle\int_0^1 e^x \sin x^2\, dx$

B17. For each of the following, show that termwise integration of the second-degree Taylor polynomial constructed at $x = 0$ gives the Taylor polynomial for an appropriate antiderivative of the function.

(a) $f(x) = \dfrac{1}{\sqrt{1 - x^2}}$ (b) $f(x) = \dfrac{1}{1 + x^2}$

(c) $f(x) = \dfrac{1}{\sqrt{1 + x^2}}$ (d) $f(x) = \dfrac{1}{1 - x^2}$

9.2 Taylor Series

Power Series

We saw in Section 9.1 that sufficiently well-behaved functions can be approximated by polynomials of arbitrarily high degree. Is, then, the representation

$$f(x) = \sum_{k=0}^{\infty} \frac{f^{(k)}(a)}{k!}(x - a)^k$$

meaningful? If it is, it will be called a *Taylor* series.

A series of the form $\sum_{k=0}^{\infty} c_k(x - a)^k$ is called a *power series,* that is, a series in powers of $x - a$. Since it can be shown that the coefficients in a Taylor series for $f(x)$ are unique, $c_k = f^{(k)}(a)/k!$ and a power series representation for a function $f(x)$ must be its Taylor series. From now on, we will treat the terms *Taylor series* and *power series* as equivalent.

From Chapter 8, an infinite series make sense if the associated sequence of partial sums converge. But the sequence of partial sums for this power series is just the sequence of Taylor polynomials. This sequence converges if the remainder term for the Taylor polynomials goes to zero.

EXAMPLE 9.2   Let $f(x) = e^x$ so that, at $a = 0$, the nth partial sum would be the polynomial $p_n(x) = \sum_{k=0}^{n} x^k/k!$ and the Lagrange form of the remainder would be $R = e^c x^{n+1}/(n + 1)!$, where c lies in the open interval $(0, x)$. In particular, c depends on x and n, so to show the remainder goes to zero for increasing n requires fixing x at M and noting that for the increasing functions e^x and x^{n+1} the maximum on $(0, x)$ will occur at $x = M$. Thus, the remainder is bounded by $R_1 = e^M M^{n+1}/(n + 1)!$ for which the limit is 0.   ❖

The appropriate arena for proving general results like this is the complex plane. Hence, we restrict ourselves here to a discussion of convergence for power series along the real line.

Obtaining Power Series

Power series representations for the elementary functions are well known and tabulated. In fact, most are met in an elementary calculus course. For example, the Taylor series representation of $\sin x$ is

$$\sum_{k=0}^{\infty} \frac{(-1)^k x^{2k+1}}{(2k+1)!}$$

Alternatively, some series can be obtained by manipulating other known series.

EXAMPLE 9.3 Obtain a power series representation of $f(x) = \frac{x}{x-5}$, valid in a neighborhood of $x = 2$.
Write $f(x)$ so that each x appears as $x - 2$, and exploit the similarity of $f(x)$ to $\frac{1}{1-x}$, the sum of the geometric series. Thus, obtain

$$f(x) = \frac{x}{x-5} = \frac{(x-2)+2}{(x-2)-3} = \frac{z+2}{-3\left(1-\frac{z}{3}\right)} = -\frac{z+2}{3}\sum_{k=0}^{\infty}\left(\frac{z}{3}\right)^k \tag{9.3}$$

where $z = x - 2$ and the geometric series has been used to expand $\frac{1}{1-z/3}$. Then, writing $u = \frac{z}{3}$, (9.3) leads to

$$-\left(u + u^2 + \cdots + \frac{2}{3}(1 + u + u^2 + \cdots)\right) = -\left(\frac{2}{3} + \frac{5}{3}(u + u^2 + \cdots)\right)$$

$$= -\left(-1 + \frac{5}{3} + \frac{5}{3}(u + u^2 + \cdots)\right)$$

$$= 1 - \frac{5}{3}\sum_{k=0}^{\infty} u^k$$

so $f(x) = 1 - \frac{5}{3}\sum_{k=0}^{\infty}\left(\frac{x-2}{3}\right)^k$, when u is replaced by $\frac{x-2}{3}$. ❖

Interval of Convergence

On the real line, power series of the form $\sum_{k=0}^{\infty} c_k(x-a)^k$ converge in $(a - R, a + R)$, an *interval of convergence* about $x = a$, with R being a real number to be defined. The series may or may not converge at the endpoints of this interval.

Radius of Convergence

The real number R is the *radius of convergence,* the distance from a to the closest point of non-convergence in the *complex plane*. Hence, it is possible for the power series on the *real line* to converge at both ends of the interval of convergence or at neither.

If $R = 0$, then the power series converges at just the single point $x = a$ and we say there is no interval of convergence. On the other hand, if $R = \infty$, then the power series converges everywhere and the interval of convergence is the whole real line.

The radius of convergence for a power series can usually be found by an application of the ratio test.

EXAMPLE 9.4 For the power series for $e^x = \sum_{k=0}^{\infty} x^k/k!$, the condition $\rho = \lim_{k\to\infty} |a_{k+1}/a_k| < 1$ in the ratio-test becomes $\lim_{k\to\infty}\left|\frac{x}{k+1}\right| < 1$, from which we obtain $|x| < \lim_{k\to\infty}(k+1) = \infty$. Thus R, the radius of convergence, is infinite and we have the interval of convergence defined by $|x| < \infty$. ❖

EXAMPLE 9.5   For the power series $\frac{1}{1-x} = \sum_{k=0}^{\infty} x^k$, the ratio-test condition $\rho = \lim_{k\to\infty} |a_{k+1}/a_k| < 1$ becomes $\lim_{k\to\infty} |x^{k+1}/x^k| < 1$, from which we obtain $|x| < 1$. Thus, the radius of convergence is $R = 1$, which is not surprising since this series is the familiar geometric series $\sum_{k=0}^{\infty} r^k$, known to converge for $|r| < 1$.   ❖

EXAMPLE 9.6   Find R, the radius of convergence for

$$\sin x = \sum_{k=0}^{\infty} \frac{(-1)^n x^{2k+1}}{(2k+1)!}$$

an example deliberately picked because the powers in successive terms increase by two, not one. Now, the ratio test leads to $R = \lim_{k\to\infty} \sqrt{2(k+1)(2k+3)} = \infty$, where the radical appears because the ratio a_{k+1}/a_k simplifies to a multiple of x^2, not just x.   ❖

EXERCISES 9.2–Part A

A1. Using just the geometric series $\frac{1}{1-x} = \sum_{k=0}^{\infty} x^k$, obtain the Taylor series for $f(x) = \frac{1}{2+x}$ about $x = -1$.

A2. Use the ratio test to determine the radius and interval of convergence for the series obtained in Exercise A1.

A3. A Maclaurin series is a Taylor series about $x = 0$. For example, the Maclaurin series for $f(x) = \frac{1}{x^2+3x+2} = \frac{1}{x+1} - \frac{1}{x+2}$ is

$$\sum_{k=0}^{\infty} (-1)^k \left(1 - \frac{1}{2^{k+1}}\right) x^k$$

Verify that the first few terms are correctly given by this sum.

A4. Obtain the Maclaurin series for the fractions $\frac{1}{x+1}$ and $\frac{1}{x+2}$, and use them to verify your work in Exercise A3.

A5. Since $f(x)$ in Exercise A3 factors to $\frac{1}{(x+1)(x+2)}$, the product of the Maclaurin series obtained in Exercise A4 should again reproduce the Maclaurin series of Exercise A3. Show that this is so for the first few terms.

EXERCISES 9.2–Part B

B1. Obtain the Taylor series for $f(x) = \frac{1}{x^2+2x+5}$, valid in a neighborhood of $x = -1$. *Hint:* Complete the square in the denominator, and use the technique of Example 9.3.

In Exercises B2–6:

  (a) Establish the given Taylor series.

  (b) Use the ratio test to find R, the radius of convergence.

B2. $\cos x = \sum_{k=0}^{\infty} \frac{(-1)^k x^{2k}}{(2k)!}$ **B3.** $\sqrt{1-x} = -\sum_{k=0}^{\infty} \frac{(2k)! x^k}{4^k (k!)^2 (2k-1)}$

B4. $\ln(1-x) = -\sum_{k=0}^{\infty} \frac{x^{k+1}}{k+1}$ **B5.** $\arctan x = \sum_{k=0}^{\infty} \frac{(-1)^k x^{2k+1}}{2k+1}$

B6. $\frac{1}{1+x^2} = \sum_{k=0}^{\infty} (-1)^k x^{2k}$

For the functions in Exercises B7–9:

  (a) Obtain the Maclaurin series, that is, the Taylor series at $x = 0$.

  (b) Use the ratio test to find R, the radius of convergence.

  (c) Decompose $f(x)$ by partial fractions, and obtain the Maclaurin series for each fraction, noting that the denominators of each fraction are linear, and hence, the associated series are just geometric series.

  (d) Add the series stemming from the partial fraction decomposition, and show that the original series is obtained.

  (e) Obtain the series for $f(x)$ by equating $f(x)$ to the formal sum $\sum_{k=0}^{\infty} a_k x^k$, multiplying through by the denominator of $f(x)$ and then matching coefficients of like powers of x on each side.

B7. $f(x) = \frac{5x-17}{4x^2+41x+45}$ **B8.** $f(x) = \frac{37x-12}{45x^2-61x+20}$

B9. $f(x) = \frac{7x-31}{6x^2+6x-12}$

For the functions $f(x)$ and $g(x)$ in Exercises B10–12:

  (a) Verify $f(g(x)) = g(f(x)) = x$, thereby showing that $f(x)$ and $g(x)$ are functional inverses.

  (b) Graph $f(x)$, $g(x)$, and $y = x$ all on the same set of axes.

  (c) Obtain, at $x = 0$, the Taylor series for $f(x)$ and $g(x)$.

(d) Take $g(x)$ as the formal sum $\sum_{k=0}^{\infty} a_k x^k$, and substitute it into $f(g(x)) = x$, matching coefficients of like powers of x on each side of the resulting equation. This should determine the coefficients a_k, and these coefficients should be the same as the coefficients in the Taylor series for $g(x)$.

B10. $f(x) = \sin x, -\dfrac{\pi}{2} \le x \le \dfrac{\pi}{2}; g(x) = \arcsin x$

B11. $f(x) = \tan x, -\dfrac{\pi}{2} \le x \le \dfrac{\pi}{2}; g(x) = \arctan x$

B12. $f(x) = \dfrac{x}{1+x}, x \ge -1; g(x) = \dfrac{x}{1-x}$

B13. Using just the geometric series $\frac{1}{1-x} = \sum_{k=0}^{\infty} x^k$, obtain the Taylor series of

(a) $f(x) = \dfrac{1}{2+x}$ about $x = 0$ **(b)** $f(x) = \dfrac{1}{2+x}$ about $x = 2$

(c) $f(x) = \dfrac{x}{2+x}$ about $x = 0$ **(d)** $f(x) = \dfrac{x}{2+x}$ about $x = -1$

(e) $f(x) = \dfrac{x}{2+x}$ about $x = 2$

B14. For each function in Exercise B13, determine the interval of convergence of the Taylor series.

9.3 Termwise Operations on Taylor Series

Introduction

Inside the interval of convergence, a Taylor series can be manipulated as if it were a polynomial. Termwise integration and differentiation, as well as termwise arithmetic, are valid. In this section we state and illustrate a number of specific theorems that justify termwise manipulations with Taylor series. We will state results for series in powers of x, since the generalizations to series in powers of $x - a$ are immediate.

Theorems on Convergence

Stated without proof, the following two theorems are the basis for most manipulations with power series. Inside the interval of convergence, a power series converges absolutely. In light of Theorem 8.7, the comparison theorem of Section 8.5, the convergence is also uniform. A formal statement of the theorems follows.

THEOREM 9.3   ABSOLUTE CONVERGENCE

The power series $\sum_{n=0}^{\infty} a_n x^n$ converges absolutely in the open interval $(-R, R)$.

THEOREM 9.4   UNIFORM CONVERGENCE

The power series $\sum_{n=0}^{\infty} a_n x^n$ converges uniformly in any closed interval completely contained inside the open interval $(-R, R)$. Thus, uniform convergence takes place in $[-r, r]$, where $0 < r < R$. In terms of x, the set on which uniform convergence takes place is $|x| \le r < R$, so x is *bounded away from* R.

Related to convergence of a power series is the concept of *analyticity* for the limit function.

ANALYTICITY   A function $f(x)$ defined on the real line is said to be *analytic* at $x = a$ if $f(x)$ has a Taylor series that converges in some *interval* about a. Typical real analytic functions are $\sin x$, $\cos x$, and e^x.

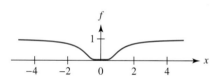

FIGURE 9.2 A function which, at $x = 0$, is not analytic, but is infinitely differentiable

Clearly, if a real function is analytic at $x = a$, then all derivatives of $f(x)$ exist at $x = a$. Oddly, the converse of this statement is not true. Real functions having all their derivatives at a point fail to be analytic at that point if there is no Taylor series converging to the function in an *interval* about the point. A classic example is the function

$$f(x) = \begin{cases} 0 & x = 0 \\ e^{-1/x^2} & x \neq 0 \end{cases}$$

whose graph, appearing in Figure 9.2, suggests examining the behavior at the origin. Since $\lim_{t \to \infty} e^{-t} = 0$, we accept $\lim_{x \to 0} f(x) = 0$ as true. The derivative at $x = 0$ is the limit of the difference quotient $e^{-1/h^2}/h$, likewise zero. The second derivative at $x = 0$ is the limit of the difference quotient $2e^{-1/h^2}/h^4$, again zero. Continuing in this manner, we can verify that all derivatives $f^{(k)}(0) = 0$. Hence, the Taylor series so generated for $f(x)$ will be identically zero and, therefore, converges to $f(x)$ just at the single point $x = 0$, not in an interval about $x = 0$. As predicted, this function has all its derivatives, but its Taylor series does not converge to the function in an interval.

Real functions for which all derivatives exist are said to be of class C^∞ (read "C infinity"). Real *analytic* functions are a subset of the class C^∞, the infinitely differentiable functions. For a real function, having one derivative does not guarantee existence of any others, and having derivatives of all orders does not guarantee the existence of a Taylor series converging in an interval.

Theorems on Continuity

Since the uniform limit of continuous functions is a continuous function, inside the interval of convergence a power series represents a continuous function. This is formalized as follows.

THEOREM 9.5   CONTINUITY

The function defined by the series $\sum_{n=0}^{\infty} a_n x^n$, provided $R \neq 0$, is continuous on the open interval $(-R, R)$.

At the endpoints of its interval of convergence, a power series may or may not converge. The question of continuity of the limit function at the endpoint of the interval of convergence is addressed by the following theorem of Abel.

THEOREM 9.6   ABEL'S THEOREM

If $\sum_{n=0}^{\infty} a_n x^n$ converges at $x = R$ and R is nonzero and finite, then the convergence is uniform at R. Hence, the limit function is continuous on the closed interval $[0, R]$.

EXAMPLE 9.7   The function $f(x) = \sqrt{1 + x}$ has the Taylor expansion

$$-\sum_{k=0}^{\infty} \frac{(-1)^k (2k)! x^k}{4^k (2k - 1)(k!)^2}$$

converging for $|x| < 1$, so the series has a radius of convergence $R = 1$. We can see this from the ratio test since the absolute value of the ratio of two successive terms is $\frac{1}{2}\left|\frac{(2k-1)x}{k+1}\right|$, from which we realize $R = 1$.

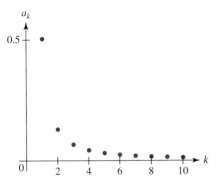

FIGURE 9.3 The sequence $a_k = \dfrac{(2k)!}{4^k(2k-1)(k!)^2}$ decreases monotonically to zero

Now suppose we are interested in using the series to compute the value of $f(1) = \sqrt{1+1} = \sqrt{2}$. Can we expect the series to converge to $\sqrt{2}$ at $x = 1$? According to Abel's theorem, if the series converges to anything at $x = 1$, then it represents a continuous function on $[0, 1]$ and must therefore give $f(1) = \sqrt{2}$ at $x = 1$. That the series converges at $x = 1$ can be shown as follows.

At $x = 1$ the series is alternating, and the absolute value of the general term is $(2k)!/4^k$ $(2k-1)(k!)^2$, which is monotone decreasing as a function of k, as seen in Figure 9.3. Theorem 7.3, Leibniz theorem for alternating series then assures convergence. Abel's theorem in turn assures continuity of the limit function; and since $f(1) = \sqrt{2}$, the series must likewise converge to $\sqrt{2}$ at $x = 1$. ❖

Uniqueness Theorems

We have already mentioned that the power series $f(x) = \sum_{k=0}^{\infty} a_k x^k$ is actually the Taylor series for $f(x)$, that is, $a_k = f^{(k)}(a)/k!$. Therefore, it is not surprising to have the following uniqueness theorem.

THEOREM 9.7 UNIQUENESS

If two power series converge to the same values on a common interval $|x| < r$ so that $\sum_{k=0}^{\infty} a_k x^k = \sum_{k=0}^{\infty} b_k x^k$ in this interval, then $a_k = b_k$ for all k. Thus, the coefficients of the power series for $f(x)$ are unique in a given interval.

The Arithmetic of Power Series

Power series behave like polynomials with respect to scalar multiplication and the four arithmetic operations of addition, subtraction, multiplication and division. The following theorem formalizes this claim. (See [65, p. 425] and [91, p. 676].)

THEOREM 9.8

1. $f_1(x) = \sum_{k=0}^{\infty} a_k x^k$ in R_1, its interval of convergence.
2. $f_2(x) = \sum_{k=0}^{\infty} b_k x^k$ in R_2, its interval of convergence.
3. $c_k = \sum_{n=0}^{k} a_n b_{k-n}, k = 0, 1, 2, \ldots$

$\Longrightarrow$

1. $\alpha f_1(x) = \sum_{k=0}^{\infty} \alpha a_k x^k$ in R_1, where α is any constant.
2. $\alpha f_1(x) + \beta f_2(x) = \sum_{k=0}^{\infty} (\alpha a_k + \beta b_k) x^k$ for all points common to R_1 and R_2, and any constants α and β.
3. $f_1(x) f_2(x) = \sum_{k=0}^{\infty} c_k x^k$ for all points interior to the intersection of R_1 and R_2.
4. $\frac{f_1(x)}{f_2(x)} = \sum_{k=0}^{\infty} d_k x^k$ for sufficiently small x, provided $b_0 \neq 0$. The coefficients d_k can be found by long division or by solving the equations $\sum_{n=0}^{k} b_n d_{k-n} = a_k$ successively for $d_0, d_1 d_2, \ldots$. The first few equations are $b_0 d_0 = a_0$, $b_0 d_1 + b_1 d_0 = a_1$, $b_0, d_2 + b_1 d_1 + b_2 d_0 = a_2$ and clearly arise by equating coefficients in the product $\left[\sum_{k=0}^{\infty} d_k x^k\right]\left[\sum_{k=0}^{\infty} b_k x^k\right] = \sum_{k=0}^{\infty} a_k x^k$.

The Algebra of Power Series

Our discussion of the algebra of power has two components. First, we discuss the relation between a Taylor series of the composition of two functions and the composition of the Taylor series of the individual functions. Then, we discuss the issue of *reversion of series*, that is, the process of generating from the power series of a function, the power series of the *inverse* of the function.

Composition of Power Series

Table 9.1 lists functions $f(x), g(x), h(x)$, the compositions $f(g(x))$ and $f(h(x))$, and Taylor polynomials for each. We are interested in the composition of the Taylor polynomials corresponding to the two composite functions. For $f(g(x))$ we find

$$1 + \frac{1}{2}\left(1 - \frac{x^2}{2!} + \frac{x^4}{4!}\right) - \frac{1}{8}\left(1 - \frac{x^2}{2!} + \frac{x^4}{4!}\right)^2 + \frac{1}{16}\left(1 - \frac{x^2}{2!} + \frac{x^4}{4!}\right)^3 = \frac{365}{256} - \frac{107}{512}x^2 + \cdots$$

but for $f(h(x))$ we find

$$1 + \frac{1}{2}\left(x - \frac{x^3}{3!} + \frac{x^5}{5!}\right) - \frac{1}{8}\left(x - \frac{x^3}{3!} + \frac{x^5}{5!}\right)^2 + \frac{1}{16}\left(x - \frac{x^3}{3!} + \frac{x^5}{5!}\right)^3 = 1 + \frac{x}{2} - \frac{x^2}{8} - \frac{x^3}{48}$$

The coefficients in the expansion of $f(g(x))$ are not obtained by composing partial sums of the series for $f(x)$ and $g(x)$, but for $f(h(x))$, they are. Hence, we have the following theorem [35].

Function	Taylor Polynomial	Function	Taylor Polynomial
$f(x) = \sqrt{1+x}$	$1 + \dfrac{x}{2} - \dfrac{x^2}{8} + \dfrac{x^3}{16}$	$f(g(x)) = \sqrt{1 + \cos x}$	$\sqrt{2}\left(1 - \dfrac{x^2}{8}\right)$
$g(x) = \cos x$	$1 - \dfrac{x^2}{2!} + \dfrac{x^4}{4!}$	$f(h(x)) = \sqrt{1 + \sin x}$	$1 + \dfrac{x}{2} - \dfrac{x^2}{8} - \dfrac{x^3}{48}$
$h(x) = \sin x$	$x - \dfrac{x^3}{3!} + \dfrac{x^5}{5!}$		

TABLE 9.1   Taylor series and the composition of functions

THEOREM 9.9   COMPOSITION

1. $f(y) = \displaystyle\sum_{k=0}^{\infty} a_k(y - y_0)^k$, convergent in $|y - y_0| < R_1$.

2. $g(x) = \displaystyle\sum_{k=0}^{\infty} b_k(x - x_0)^k$, convergent in $|x - x_0| < R_2$.

3. $g(x_0) = b_0 = y_0$

$\Longrightarrow$

There is an interval about x_0, namely $|x - x_0| < R_3$, in which $|g(x) - y_0| < R_1$ and in which the composition has a Taylor expansion given by

$$h(x) = f(g(x)) = \sum_{k=0}^{\infty} c_k(x - x_0)^k = \sum_{k=0}^{\infty} a_k \left[\sum_{n=1}^{\infty} b_n (x - x_0)^n\right]^k$$

In both examples above, the center for the expansions was $y_0 = 0$, so the crucial condition became whether or not the lead coefficient b_0 in the series for $g(x)$ was zero. In the first case, $g(0) = b_0 = 1 \neq y_0$ and composition of the individual Taylor series failed. In the second case, $g(0) = b_0 = 0$ and the composition of the individual series succeeded.

REVERSION OF POWER SERIES   If the function $y = f(x)$ has an inverse, when can we obtain the Taylor series of the inverse function from the Taylor series of the given function? For example, given the function $f(x) = \sin x$ with Taylor series $\sum_{k=0}^{\infty}(-1)^k x^{2k+1}/(2k+1)!$, what connection, if any, exists between its coefficients and the coefficients in the Taylor series for the inverse function, that is, with the coefficients in

$$\arcsin(x) = \sum_{k=0}^{\infty} \frac{(2k)! x^{2k+1}}{4^k (k!)^2 (2k+1)}$$

From [35], we have the following theorem, where the term *series reversion* refers to the process of computing the coefficients of the power series for an inverse function from the coefficients of the power series of the function.

THEOREM 9.10   REVERSION OF SERIES

1. $f(x) = \sum_{k=0}^{\infty} a_k (x - x_0)^k$ converges in $|x - x_0| < R$.

2. $f'(x_0) = a_1 \neq 0$

$\implies$

1. In a neighborhood of $f(x_0) = y_0 = a_0$, defined by $|y - a_0| < r$ for some real r, an inverse function ϕ exists.

2. In this neighborhood, the function ϕ satisfies $x = \phi(y) = x_0 + \sum_{k=1}^{\infty} b_k (y - a_0)^k$.

3. The coefficients b_k satisfy $x = x_0 + \sum_{k=1}^{\infty} b_k \left[\sum_{n=1}^{\infty} a_n (x - x_0)^n \right]^k$.

$1 = b_1$

$0 = b_2$

$0 = -\frac{1}{6} b_1 + b_3$

$0 = -\frac{1}{3} b_2 + b_4$

$0 = \frac{1}{120} b_1 - \frac{1}{2} b_3 + b_5$

TABLE 9.2   Coefficient equations which revert series for $\sin x$

The first hypothesis gives $y = f(x) = \sum_{k=0}^{\infty} a_k (x - x_0)^k = a_0 + \sum_{k=1}^{\infty} a_k (x - x_0)^k$, so $y - a_0 = \sum_{k=1}^{\infty} a_k (x - x_0)^k$. Since $y_0 = f(x_0) = a_0$, expanding $x = \phi(y)$ about y_0 gives $x = \phi(y) = x_0 + \sum_{k=1}^{\infty} b_k (y - a_0)^k$. Replacing $y - a_0$ with $\sum_{k=1}^{\infty} a_k (x - x_0)^k$ then gives the result in the third conclusion.

EXAMPLE 9.8   The first few terms of the Maclaurin expansion of $f(x) = \sin x$ are given by

$$F = x - \frac{1}{3!} x^3 + \frac{1}{5!} x^5 - \frac{1}{7!} x^7$$

Theorem 9.10 suggests we form the identity

$$x = x_0 + \sum_{k=1}^{5} b_k F^k$$

and match coefficients of like powers of x on each side. This generates the equations in Table 9.2, which are then solved for the unknown coefficients b_k, yielding $x + \frac{1}{6} x^3 + \frac{3}{40} x^5$ for an expansion of the inverse function. ❖

The Calculus of Power Series

Inside the interval of convergence of a power series, the operations of termwise differentiation and integration are valid. The following theorem from [65] justifies termwise differentiation of power series.

THEOREM 9.11 TERMWISE DIFFERENTIATION

The power series $f(x) = \sum_{k=0}^{\infty} a_k x^k$ and the *derived series* $\sum_{k=1}^{\infty} k a_k x^{k-1}$ have the same radius of convergence. Moreover, the derived series is the Taylor expansion of the derivative, $f'(x)$.

EXAMPLE 9.9 The function $\sin x$ with Maclaurin series

$$\sum_{k=0}^{\infty} \frac{(-1)^k x^{2k+1}}{(2k+1)!} = x - \frac{x^3}{3!} + \cdots$$

has derivative $\cos x$ with Maclaurin expansion

$$\sum_{k=0}^{\infty} \frac{(-1)^k x^{2k}}{(2k)!} = 1 - \frac{x^2}{2!} + \cdots$$

The second series is what is obtained if the first is differentiated termwise. ❖

Termwise integration of power series is justified by the following theorem from [65].

THEOREM 9.12 TERMWISE INTEGRATION

1. $f(x) = \sum_{k=0}^{\infty} a_k x^k$ with radius of convergence R.
2. α and β are within the interval of convergence.

$$\implies \int_{\alpha}^{\beta} f(x)\,dx = \sum_{k=0}^{\infty} a_k \int_{\alpha}^{\beta} x^k\,dx = \sum_{k=0}^{\infty} \frac{a_k}{k+1}(\beta^{(k+1)} - \alpha^{(k+1)})$$

EXAMPLE 9.10 Consider the function $f(x) = e^x$ whose Maclaurin series, convergent on the whole real line, is $\sum_{k=0}^{\infty} x^k/k!$. Termwise integration between $x = \alpha$ and $x = \beta$ yields the series expansion for $e^{\beta} - e^{\alpha}$, and integration of $f(x)$ yields precisely $e^{\beta} - e^{\alpha}$. ❖

Examples

The following examples illustrate valid and practical operations with series.

EXAMPLE 9.11 The series for e^{x^2} can be obtained from the series for $f(x) = e^x$ by substituting x^2 for x in the expansion for $f(x)$. ❖

EXAMPLE 9.12 The Maclaurin series for $\sinh(x) = (e^x - e^{-x})/2$, namely, $\sum_{k=0}^{\infty} x^{2k+1}/(2k+1)!$, can be obtained from the series $e^x = \sum_{k=0}^{\infty} x^k/k!$ and $e^{-x} = \sum_{k=0}^{\infty} (-x)^k/k!$ since the coefficient of x^k in the termwise sum, namely, $(1 - (-1)^k)/2k!$, is zero when k is even and 2 when k is odd. ❖

EXAMPLE 9.13 The Taylor series for $\cos(x - a)$ about $x = a$ is $\sum_{k=0}^{\infty}(-1)^k(x - a)^{2k}/(2k)!$, a result equivalent to $\sum_{k=0}^{\infty}(-1)^k x^{2k}/(2k)!$, the Maclaurin series for $\cos x$, with each x shifted to $x - a$. ❖

EXAMPLE 9.14 The Maclaurin expansion of the product $\sin x \cos x$ requires us to multiply the two series

$$\sum_{k=0}^{\infty}\frac{(-1)^k x^{2k+1}}{(2k + 1)!} \sum_{k=0}^{\infty}\frac{(-1)^k x^{2k}}{(2k)!}$$

The result should be $\sum_{k=0}^{\infty}(-4)^k x^{2k+1}/(2k + 1)!$, the Maclaurin series of $\frac{1}{2}\sin 2x = \sin x \cos x$. Working with partial sums and truncating, we obtain

$$\left(x - \frac{x^3}{3!} + \frac{x^5}{5!}\right)\left(1 - \frac{x^2}{2!} + \frac{x^4}{4!}\right) = x - \frac{2}{3}x^3 + \frac{2}{15}x^5 = \frac{1}{2}\left(2x - \frac{(2x)^3}{3!} + \frac{(2x)^5}{5!}\right)$$

which is the partial sum of the series for $\frac{1}{2}\sin 2x$. ❖

EXAMPLE 9.15 The nth-order Bessel functions of the first kind, $J_n(x)$, are "special functions," *defined* by power series and obtained as solutions of second-order linear differential equations with nonconstant coefficients. (Details appear in Section 16.2.) The graphs of

$$J_0(x) = \sum_{k=0}^{\infty}\frac{(-1)^k x^{2k}}{4^k (k!)^2} \text{ (solid)} \quad \text{and} \quad J_1(x) = \frac{1}{2}\sum_{k=0}^{\infty}\frac{(-1)^k x^{2k+1}}{4^k (k!)^2(k + 1)} \text{ (dashed)}$$

are shown in Figure 9.4. By termwise differentiation of the series representation for $J_0(x)$, valid within the interval of convergence, it is possible to verify that $J_0'(x) = -J_1(x)$. ❖

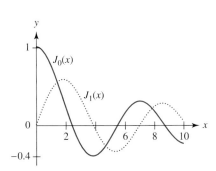

FIGURE 9.4   The Bessel functions $J_0(x)$ and $J_1(x)$, respectively the solid and dotted curves

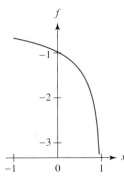

FIGURE 9.5   Graph of $f(x) = \frac{\ln(1-x)}{x}$

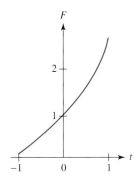

FIGURE 9.6   Graph of $F(t) = 1 - \int_0^t \frac{\ln(1-x)}{x}\,dx$

EXAMPLE 9.16 The function $f(x) = \frac{\ln(1-x)}{x}$ is undefined at $x = 0$, but $\lim_{x\to 0} f(x) = -1$, so at $(0, -1)$ the graph of $f(x)$ simply has a "hole" that can be "patched" be defining $f(0) = -1$. Thus, $f(x)$ is said to have a removable singularity at $x = 0$. The function $f(x)$ also has a vertical asymptote at $x = 1$ because $\lim_{x\to 1^-} f(x) = -\infty$. Both of these features are seen in Figure 9.5.

The function $F(t) = 1 - \int_0^t f(x)\,dx$, $0 \le t \le 1$, graphed in Figure 9.6, has the series expansion $1 + \sum_{k=0}^{\infty} t^{k+1}/(k + 1)^2$. Dividing the series $\ln(1 - x) = -\sum_{k=0}^{\infty} x^{k+1}/(k + 1)$ by x, and then integrating termwise give the series expansion for $F(t)$. ❖

EXERCISES 9.3–Part A

A1. Insight into why a power series converges uniformly on $|x| \leq r < R$, where R is the radius of convergence, can be obtained by examining the convergence of the geometric series $\sum_{k=0}^{\infty} x^k$. Use the Weierstrass M-test and assume $\sum_{k=0}^{\infty} r^k$ converges.

A2. Obtain the first few terms of the Maclaurin series of $\tan x$, $\sin x$, and $\cos x$. Long divide to test whether the division of the series for $\sin x$ and $\cos x$ gives the series for $\tan x$.

A3. Use the truncated Maclaurin series for $e^x \sin x$, e^x, and $\sin x$ to test whether the series of a product is the product of the series.

A4. Obtain $g(x)$, the function inverse to $f(x) = \frac{x}{1+x}$. Obtain the truncated Maclaurin series for both $f(x)$ and $g(x)$ and then apply the technique of series reversion to obtain the series for $g(x)$. Of course, the two series for $g(x)$ should agree.

A5. For the Bessel functions given in Example 9.15, verify that $J_0'(x) = -J_1(x)$.

EXERCISES 9.3–Part B

B1. Evaluate $F(a) = \int_1^2 \sqrt{a + x^2}\, dx$ and then obtain the Maclaurin series for $F(a)$. Expand the integrand in a Taylor series about $a = 0$, integrate termwise, and compare (analytically and graphically) the two expansions for $F(a)$.

For the functions given in Exercises B2–6:

   (a) Obtain the Maclaurin series.

   (b) Determine R, the radius of convergence.

   (c) Determine if the series converges at $x = R$ and $x = -R$.

   (d) If the series converges at an endpoint of the interval of convergence, obtain the sum of the series at that point and then discuss the relationship of this outcome to Abel's theorem.

   (e) Show graphically, by plotting the first few members of the sequence of partial sums, that convergence in the vicinity of the endpoint in part (d) is uniform.

B2. $\ln(1 + x)$ **B3.** $\ln(1 + x^2)$ **B4.** $\sqrt{1 + x^2}$
B5. $\ln(x^2 + \frac{13}{6}x + 1)$ **B6.** $\sqrt{1 - x}$

For the functions given in Exercises B7–16:

   (a) Obtain $\sum_{k=0}^{\infty} a_k x^k$ and $\sum_{k=0}^{\infty} b_k x^k$, the Maclaurin series for $f(x)$ and $g(x)$, respectively.

   (b) Obtain $\sum_{k=0}^{\infty} c_k x^k$, the Maclaurin series for the product $f(x)g(x)$.

   (c) Apply $c_k = \sum_{n=0}^{k} a_n b_{k-n}$, the product rule for coefficients, and show that these are indeed the coefficients found in part (b).

B7. $f(x) = \sin x$, $g(x) = e^x$ **B8.** $f(x) = \cos x$, $g(x) = \sin x$
B9. $f(x) = \ln(1 - x)$, $g(x) = \sqrt{1 + x}$
B10. $f(x) = \sqrt{1 - x}$, $g(x) = \sinh x$
B11. $f(x) = \dfrac{1}{1 + x^2}$, $g(x) = e^{-x^2}$
B12. $f(x) = \sqrt{1 + x^2}$, $g(x) = \ln(1 + x)$
B13. $f(x) = \sinh x$, $g(x) = \dfrac{x}{1 - x}$
B14. $f(x) = \cosh x$, $g(x) = \cos x$

B15. $f(x) = e^{-x^2}$, $g(x) = \cosh x$
B16. $f(x) = \dfrac{x}{1 + x}$, $g(x) = \dfrac{1}{2 + x}$

For the functions in Exercises B17–26:

   (a) Obtain $\sum_{k=0}^{\infty} a_k x^k$ and $\sum_{k=0}^{\infty} b_k x^k$, the Maclaurin series for $f(x)$ and $g(x)$, respectively.

   (b) Obtain $\sum_{k=0}^{\infty} d_k x^k$, the Maclaurin series for the quotient $\frac{f(x)}{g(x)}$.

   (c) Compute the coefficients d_k by solving the equations $\sum_{n=0}^{k} b_n d_{k-n} = a_k$, making sure that the results agree with those in part (b).

   (d) Obtain the Maclaurin series for the composition $f(g(x))$.

   (e) Compose partial sums of the Maclaurin series for $f(x)$ and $g(x)$, and determine if the results in parts (d) and (e) agree.

   (f) In either event, show that the expansion of the composition does or does not agree with the composition of partial sums of the expansions depending on the validity of $g(x_0) = y_0$, which for Maclaurin series, becomes $g(0) = 0$.

B17. $f(x) = \sin x$, $g(x) = e^x - 1$
B18. $f(x) = \sin x$, $g(x) = 1 - \cos x$
B19. $f(x) = \ln(2 - x)$, $g(x) = \sqrt{1 + x}$
B20. $f(x) = \sinh x$, $g(x) = \sqrt{1 - x}$
B21. $f(x) = \dfrac{1}{1 + x^2}$, $g(x) = e^{-x^2} - 1$
B22. $f(x) = \ln(1 + x)$, $g(x) = \sqrt{1 + x^2}$
B23. $f(x) = \cosh x$, $g(x) = 1 - \cos x$
B24. $f(x) = e^{-x^2}$, $g(x) = \cosh x$
B25. $f(x) = \dfrac{x}{1 + x}$, $g(x) = \dfrac{1}{2 + x}$
B26. $f(x) = \arctan x$, $g(x) = \dfrac{\pi}{2} - \arccos x$

In Exercises B27–33:

   (a) Obtain the Maclaurin series for $f(x)$ and for $g(x)$, the functional inverse of $f(x)$, defined by $f(g(x)) = x$.

(b) In accordance with Theorem 9.10, revert the series for $f(x)$ by solving for b_k, $k = 1, \ldots$, via the equations generated by equating coefficients of like-powers of x in $x = x_0 + \sum_{k=1}^{\infty} b_k [\sum_{n=1}^{\infty} (x - x_0)^n]^k$. The b_k should agree with the coefficients in the expansion for $g(x)$ obtained in part (a), and the calculations are simplified since $x_0 = 0$.

B27. $f(x) = \arccos x$ **B28.** $f(x) = \sinh x$ **B29.** $f(x) = e^x$

B30. $f(x) = \sqrt{1-x}$ **B31.** $f(x) = \tan x$ **B32.** $f(x) = \dfrac{x}{1-x}$

B33. $f(x) = \ln(1-x)$

In each of Exercises B34–43:

  (a) Obtain the Maclaurin series for both $f(x)$ and $f'(x)$.

  (b) Differentiate termwise the series for $f(x)$ and show the resulting series agrees with the series for $f'(x)$ found in part (a).

B34. $f(x) = \dfrac{\sin x}{1+x}$ **B35.** $f(x) = \dfrac{x}{1-x^2}$

B36. $f(x) = \sqrt{2 + \cos^2 x}$ **B37.** $f(x) = \tan x$

B38. $f(x) = e^{-2x} \cos x$ **B39.** $f(x) = \dfrac{x-1}{x+1}$

B40. $f(x) = \ln(1-x)$ **B41.** $f(x) = \sqrt{1+x^2}$

B42. $f(x) = \arctan x$ **B43.** $f(x) = \arccos x$

B44. Evaluate each of the following definite integrals either analytically or numerically. Then, expand the integrand in an appropriate Taylor series and integrate termwise. Compare the resulting answers.

  (a) $\displaystyle\int_0^2 \sqrt{1 + \sin x^2}\, dx$ **(b)** $\displaystyle\int_1^2 \ln x\, dx$ **(c)** $\displaystyle\int_0^5 e^x \sin e^x\, dx$

  (d) $\displaystyle\int_1^3 \dfrac{x}{1+x}\, dx$ **(e)** $\displaystyle\int_0^1 e^{\sin x} \cos x\, dx$

In Exercises B45 and 46, neither of the given integrals is easily expressed in terms of simple functions. For each:

  (a) Evaluate $F(a)$ numerically, exhibiting these values as a graph.

  (b) Expand the integrand in a Maclaurin series about $a = 0$ and integrate termwise. Compare graphically the two resulting representations of $F(a)$.

B45. $F(a) = \displaystyle\int_1^2 \sqrt{a + x^3}\, dx, 0 \le a \le 2$

B46. $F(a) = \displaystyle\int_0^\pi \sqrt{1 + a \cos x}\, dx, 0 \le a \le 1$

Chapter Review

1. Explain the difference between a Taylor polynomial and a Taylor series.

2. Give a verbal description of just what it is that Taylor's theorem provides.

3. For Taylor's theorem, illustrate the use of the Lagrange form of the remainder when estimating the error made by a quadratic Taylor polynomial. Use a function of your choice, expanded at a point of your choice.

4. What is the radius and interval of convergence for a power series? Illustrate the use of the ratio test to find both the radius and interval of convergence for one representative example.

5. Detail the complication that arises when applying the ratio test to a power series leads to an inequality of the form $\rho^2 = \lim_{n \to \infty} |a_{n+1}/a_n| < 1$.

6. Show that the power series $\sum_{n=0}^{\infty} n! x^n$ converges just at $x = 0$. Hence, $R = 0$ defines the radius of convergence.

7. What does *analyticity* mean for a real function? Give an example of a real analytic function.

8. Give an example of a real function that has all its derivatives at a point but does not have a Taylor series converging in a *neighborhood* of that point.

9. With an example, verify the absolute convergence of a power series inside its interval of convergence.

10. With an example, verify the uniform convergence of a power series inside its interval of convergence.

11. Use the Taylor series for $f(x) = \cos x$ and $g(x) = e^{-x}$ to verify that termwise addition and multiplication of power series are valid. Show that termwise division is valid for the fraction $\frac{f(x)}{g(x)}$.

12. Use the Taylor series for xe^{-x} to verify that inside the interval of convergence, a power series can be integrated or differentiated termwise.

13. Give an example when the statement "the Taylor series of a composition is the composition of the Taylor series" is true. Give an example when it is false. What is a sufficient condition that guarantees the statement to be true?

14. The term *series reversion* refers to the process of computing the coefficients of the power series for an inverse function from the coefficients of the power series of the function. If $f(x) = \cos x$, obtain the Maclaurin series for its inverse, $\arccos x$. Then, show how the process of series reversion can be used to construct the same series.

Chapter 10

Fourier Series

INTRODUCTION   A Taylor series for $f(x)$ is an infinite sum of powers of x, with coefficients determined by derivatives of $f(x)$. A Fourier series for $f(x)$ is an infinite series of sine and/or cosine terms, with coefficients determined by integrals of $f(x)$. For general categories of functions, the convergence of a Fourier series is *in the mean*. For smooth enough functions, the convergence is pointwise and, sometimes, even uniform. In the absence of uniform convergence such series, sums of continuous functions, can represent discontinuous functions. Until adequate definitions of limit, continuity, and convergence were developed in the nineteenth century, Fourier's use of these series in solving heat transfer problems was not well received in the early 1800s.

As with other series, we will examine the Fourier series with respect to integration and differentiation.

Section 10.5 gives an example where the Fourier series is used to solve an IVP for a driven damped oscillator. The driving term is a periodic function that is represented by a Fourier series, and partial sums of this series are used as the forcing function for the motion. Surprisingly, the convergence of the approximate solutions so generated is quite rapid.

The last two sections of the chapter are optional and theoretical but practical. They study the *optimizing property* of the Fourier series whereby its coefficients minimize the mean-square error between the function and the series. Central to this optimizing property is the *orthogonality* of sine and cosine functions. Other functions have similar orthogonality properties, and we show that Fourier–Legendre series can be constructed with *Legendre polynomials.*

10.1   General Formalism

Introduction

In the Taylor series

$$f(x) = \sum_{k=0}^{\infty} \frac{f^{(k)}(a)}{k!}(x-a)^k$$

the function $f(x)$ is represented as a sum of powers, with coefficients determined by derivatives evaluated at a single point $x = a$. In a *Fourier series,* $f(x)$ is represented as a sum of trigonometric terms, with coefficients determined by integrals of $f(x)$ over a fixed interval.

Before stating what a Fourier series *is*, we need language to describe what the function to which it converges looks like. Hence, we define what we mean by a *periodic extension* of the function $f(x)$ whose domain is the finite interval $[\alpha, \beta]$.

PERIODIC EXTENSION   The *periodic extension* of $f(x)$ whose domain is the finite interval $[\alpha, \beta]$ of length $2L$ is a periodic function $\hat{f}(x)$ having period $2L$ and agreeing with $f(x)$ on $[\alpha, \beta]$.

If the x-axis on either side of $[\alpha, \beta]$ were divided into contiguous subintervals of length $2L$, then over each such subinterval, $\hat{f}(x)$ would look like $f(x)$. For example, the periodic extension of $f(x) = x^2, 0 \leq x \leq 1$, is shown in Figure 10.1.

Discontinuities

In order to define classes of functions for which a Fourier series will converge to the function it is supposed to represent, we review the terminology for the kinds of discontinuities a function $f(x)$ can have.

REMOVABLE DISCONTINUITY   If $f(x)$ is continuous except at $x = a$ where $f(x)$ is not defined but $b = \lim_{x \to a} f(x)$ is a finite number, then $x = a$ is said to be a *removable discontinuity* of $f(x)$.

The function $\hat{f}(x) = \begin{cases} f(x) & x \neq a \\ b & x = a \end{cases}$ is then continuous at $x = a$ and agrees with $f(x)$ at all points where $f(x)$ is defined. In this sense, $\hat{f}(x)$ is the same as $f(x)$, except that the discontinuity of $f(x)$ has been "removed." (See Example 9.16, Section 9.3.)

JUMP DISCONTINUITY   If $f(x)$ is continuous except at $x = a$ where the one-sided limits $\lim_{x \to a^-} f(x)$ and $\lim_{x \to a^+} f(x)$ both exist but are unequal, then $x = a$ is said to be a *jump discontinuity* of $f(x)$.

At a jump discontinuity, the graph of the function literally jumps from one value to another, as, for example, in Figure 10.1.

Piecewise Continuity

A function is said to be *piecewise continuous* on the closed interval $[a, b]$ provided it is continuous there, with at most a finite number of exceptions where, at worst, we would find a removable or jump discontinuity. At both a removable and a jump discontinuity, the one-sided limits $f(t^+) = \lim_{x \to t^+} f(x)$ (the limit from the left) and $f(t^-) = \lim_{x \to t^-} f(x)$ (the limit from the right) exist and are finite.

This definition rules out, therefore, discontinuities where the function becomes unbounded or has infinite oscillation. We illustrate in the following examples.

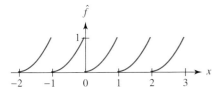

FIGURE 10.1   Periodic extension of $f(x) = x^2, 0 \leq x \leq 1$

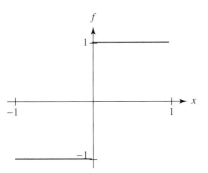

FIGURE 10.2   The piecewise continuous function in Example 10.1

EXAMPLE 10.1   The function $f(x) = \begin{cases} -1 & x < 0 \\ 1 & x > 0 \end{cases}$, graphed in Figure 10.2, is piecewise continuous on the interval $[-1, 1]$. It has a single jump discontinuity at $x = 0$. The limits from the left and right, at $x = 0$, are respectively, the finite values -1 and 1. ❖

EXAMPLE 10.2   The function $f(x) = \begin{cases} 1 & x < 0 \\ \frac{1}{x} & x > 0 \end{cases}$ is not piecewise continuous on any closed interval that includes $x = 0$. For example, Figure 10.3 shows its graph on the interval $[-1, 3]$. The y-axis is a vertical asymptote, so the one-sided limit $\lim_{x \to 0^+} f = \infty$ is not finite. ❖

EXAMPLE 10.3   The function $f(x) = \sin \frac{1}{x}$ is not piecewise continuous on any interval containing $x = 0$. The singularity is not removable since there is no limit at $x = 0$. Figure 10.4 suggests why.

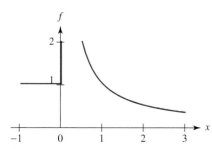

FIGURE 10.3 This function (Example 10.2) is not piecewise continuous

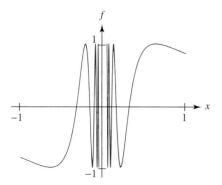

FIGURE 10.4 This function (Example 10.3) is not piecewise continuous

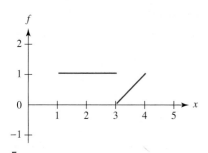

FIGURE 10.5 The piecewise continuous function of Example 10.4

The function $f(x)$ actually takes on every value in the interval $[-1, 1]$ an infinite number of times. Therefore, near $x = 0$, the values of $f(x)$ never "approach" a unique limit, so the limit fails to exist. ❖

First Fourier Theorem

The following theorem both *defines* a Fourier series, and gives conditions under which it converges to a reasonable facsimile of the function.

THEOREM 10.1

1. $f(x)$ and $f'(x)$ are both piecewise continuous on the closed interval $[\alpha, \beta]$ whose length is $2L$.

2. $a_n = \dfrac{1}{L} \displaystyle\int_\alpha^\beta f(x) \cos \dfrac{n\pi x}{L}\, dx \quad n = 0, 1, 2, \ldots$

3. $b_n = \dfrac{1}{L} \displaystyle\int_\alpha^\beta f(x) \sin \dfrac{n\pi x}{L}\, dx \quad n = 1, 2, \ldots$

$\Longrightarrow$

1. For x in the open interval (α, β) we have

$$\tilde f(x) = \tfrac12[f(x+) + f(x-)] = \frac{a_0}{2} + \sum_{n=1}^\infty \left(a_n \cos \frac{n\pi x}{L} + b_n \sin \frac{n\pi x}{L}\right)$$

2. $\displaystyle\lim_{n\to\infty} a_n = \lim_{n\to\infty} b_n = 0$

The *Fourier series* for $f(x)$ is the trigonometric sum in the conclusion of Theorem 10.1, provided the coefficients a_n and b_n are computed by the integrals stated in the second and third hypotheses. If the function obeys the first hypothesis, then Theorem 10.1 guarantees the Fourier series for $f(x)$ converges at each x to $\tilde f(x)$, the average value of the limit from the left and the limit from the right. At any point where $f(x)$ is continuous, $\tilde f(x) = f(x)$. At points where a jump discontinuity in $f(x)$ occurs, the Fourier series converges to the average of the left and right limits. No statement is made about convergence at the endpoints of the interval. However, outside the interval $[\alpha, \beta]$, the Fourier series converges to the periodic extension of $\tilde f(x)$, not $f(x)$.

The second conclusion of the theorem is generally called the *Riemann–Lebesgue lemma*, and it states that the Fourier coefficients a_n and b_n tend to zero as n increases.

◆ **EXAMPLE 10.4** We consider the function $f(x) = \begin{cases} 1 & 1 \le x \le 3 \\ x-3 & 3 < x \le 4 \end{cases}$ defined on the interval $[1, 4]$ and graphed in Figure 10.5.

THE FOURIER SERIES For the interval $[1, 4]$ we have $2L = 4 - 1$, so $L = \frac{3}{2}$. The first coefficient, $a_0 = \frac{2}{3}\int_1^4 f(x)\, dx = \frac{5}{3}$, is computed separately. Then,

$$a_n = \frac{2}{3}\int_1^4 f(x)\cos\frac{2}{3}n\pi x\, dx = \frac{2n\pi \sin\frac{8}{3}n\pi + 3\cos\frac{8}{3}n\pi - 3}{2n^2\pi^2} - \frac{\sin\frac{2}{3}n\pi}{n\pi}$$

$$b_n = \frac{2}{3}\int_1^4 f(x)\sin\frac{2}{3}n\pi x\, dx = \frac{3\sin\frac{8}{3}n\pi - 2\cos\frac{8}{3}n\pi}{2n^2\pi^2} + \frac{\cos\frac{2}{3}n\pi - 2}{n\pi}$$

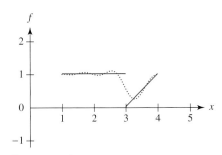

FIGURE 10.6 The partial sum S_3 and the function of Example 10.4

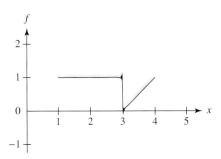

FIGURE 10.7 The partial sum S_{100} and the function of Example 10.4

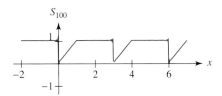

FIGURE 10.8 The partial sum S_{100} for Example 10.4

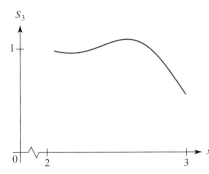

FIGURE 10.9 Example 10.4's Gibbs phenomenon illustrated by S_3

The n in the denominators of both a_n and b_n suggest they decrease in magnitude. The Riemann–Lebesgue lemma, in fact, guarantees that valid Fourier coefficients must tend to zero as n gets large. However, the more slowly the coefficients tend to zero, the more terms it will take for a partial sum of the series to represent the function adequately.

The kth partial sum is given by $S_k = \frac{a_0}{2} + \sum_{n=1}^{k}(a_n \cos\frac{n\pi x}{L} + b_n \sin\frac{n\pi x}{L})$; the partial sum corresponding to $k = 3$ and the function $f(x)$ are graphed in Figure 10.6, the approximation as the dotted curve and the function as the solid. As noted, the slow convergence of the Fourier coefficients to zero suggests we will need many terms to approximate $f(x)$ more closely. Thus, we draw Figure 10.7, showing $f(x)$ along with S_{100}.

Each partial sum of the Fourier series is a sum of continuous functions and is hence continuous. Thus, Figure 10.7 shows the partial sum, the dotted line, as a continuous function. A nearly vertical dotted line is drawn at $x = 3$. However, in conformity with Theorem 10.1, at $x = 3$ the Fourier series converges to the average value of the one-sided limits across the jump, namely, to $\frac{1+0}{2} = \frac{1}{2}$, whereas the partial sum has value 0.5015122679.

THE PERIODIC EXTENSION It is instructive to plot the partial sum of the Fourier series outside the interval over which it was calculated. Figure 10.8 shows S_{100} plotted on the interval $[-2, 7]$. The Fourier coefficients, and hence the Fourier series, reflect values of the function $f(x)$ on the interval for which the integrals were defined. If the Fourier series is evaluated outside this interval, the periodicity of the sine and cosine functions in the Fourier series yields a periodic function. The Fourier series converges to $\hat{f}(x)$, the periodic extension of $\tilde{f}(x)$, the limit of the Fourier series on $[1, 4]$.

THE GIBBS PHENOMENON In the neighborhood of a discontinuity of $f(x)$, the Fourier series cannot converge uniformly. Since each partial sum is a sum of continuous functions, if the convergence were uniform, then, by Theorem 8.8, Section 8.5, the limiting function would be continuous. But the limit function is not continuous at a jump discontinuity, so the convergence cannot be uniform in a neighborhood of such a discontinuity.

The nonuniformity of the convergence near a discontinuity is manifested by the "spikes" seen in the graphs of the partial sums. These "overshoots" persist for all values of n, with the location of the spike simply shifting closer to the point of discontinuity. In fact, the height of the overshoot tends to a value proportional to the jump at the discontinuity. The proportionality factor is dependent on the function.

Figure 10.9 illustrates this behavior, called the *Gibbs phenomenon,* for the partial sum S_3 from Example 10.4. The local maximum in Figure 10.9 occurs at (2.557942578, 1.083586807). Since the height of the jump is 1, the overshoot in S_3 is approximately 0.0835, or a bit less than 9%.

Repeating the experiment with the partial sum S_{100}, we find the local maximum closest to $x = 3$ in Figure 10.10 to be located at (2.985054260, 1.089318304). Here, the overshoot is approximately 0.089, or roughly 9% of the value of the jump in $f(x)$ at $x = 3$.

Figure 10.11 contains both S_3 (dashed) and S_{100} (solid), thereby showing how the overshoot retains a fairly constant height but moves closer to the point of discontinuity. ❖

The Usual Formalism

Typically, the defining interval for the Fourier series is taken as the symmetric interval, $[-L, L]$. The formulas for the Fourier coefficients are then

$$a_n = \frac{1}{L}\int_{-L}^{L} f(x)\cos\frac{n\pi x}{L}\,dx \qquad b_n = \frac{1}{L}\int_{-L}^{L} f(x)\sin\frac{n\pi x}{L}\,dx$$

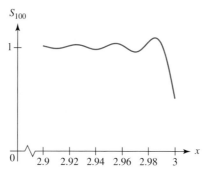

FIGURE 10.10   Example 10.4's Gibbs phenomenon illustrated by S_{100}

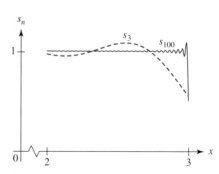

FIGURE 10.11 S_3 (dashed) and S_{100} (solid) for Example 10.4

with the Fourier series itself remaining

$$\tilde{f}(x) = \tfrac{1}{2}[f(x+) + f(x-)] = \frac{a_0}{2} + \sum_{n=1}^{\infty}\left(a_n \cos \frac{n\pi x}{L} + b_n \sin \frac{n\pi x}{L}\right)$$

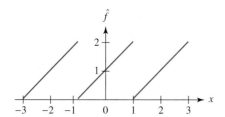

FIGURE 10.12   Periodic extension of the function in Example 10.5

EXAMPLE 10.5   On the interval $[-1, 1]$, let us obtain the Fourier series of the function $f(x) = x + 1$. For this interval, $L = 1$, so

$$a_0 = \int_{-1}^{1} (x + 1)\, dx = 2$$

$$a_n = \int_{-1}^{1} (x + 1) \cos n\pi x \, dx = 0$$

$$b_n = \int_{-1}^{1} (x + 1) \sin n\pi x \, dx = -2(-1)^n / n\pi$$

By inspection, we can see that $b_n \to 0$ slowly, as $\frac{1}{n}$, so it will take a large number of terms to obtain a good approximation to $f(x)$.

Anticipating a second Fourier theorem that concludes this section, we observe that at the endpoints, the series converges to the midpoint of the jump discontinuity in $\hat{f}(x)$, the *periodic extension* of $f(x)$, which is shown in Figure 10.12. The Fourier series converges to 1 at both endpoints where we find $y = 1$ to be the midpoint of the jump in the periodic extension.

The Gibbs phenomenon, the overshoot in the partial sum at both endpoints where there are discontinuities in the periodic extension, can be seen in Figure 10.13, which displays the partial sum S_{10}. The local maximum just to the left of $x = 1$ is at (0.9090909091, 2.086692509). Since $f(1) = 2$, the overshoot is roughly 4.33%.   ❖

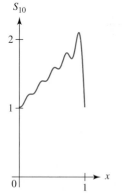

FIGURE 10.13   Example 10.5's Gibbs phenomenon illustrated by S_{10}

Second Fourier Theorem

ONE-SIDED DERIVATIVES   The function $f(x) = x^{3/2}$ is defined and differentiable for x in $[0, \infty)$. At $x = 0$ the derivative is taken via a limiting process that approaches from the right. Indeed, at $x = 0$ the difference quotient $\frac{f(0+h)-f(0)}{h}$ is $\sqrt{h}$ and the limit as h approaches zero must be taken from the right, through the positive values of h, giving zero. The derivative obtained by evaluating a one-sided limit of the difference quotient is called a one-sided derivative and is denoted by either f'_R or f'_L, depending on whether the derivative is from the right or left, respectively.

THEOREM 10.2

1. $f(x)$ is piecewise continuous on the closed interval $[-L, L]$.

$\Longrightarrow$

1. At any x in $(-L, L)$ where $f(x)$ has both a right and a left derivative, the Fourier series for $f(x)$ converges to $\frac{1}{2}[f(x+) + f(x-)]$.

2. If $f'_R(-L)$ and $f'_L(L)$ both exist, then at $-L$ and L the Fourier series for $f(x)$ converges to $\frac{1}{2}[f(-L+) + f(L-)]$.

Thus, if the function $f(x)$ is sufficiently well behaved, then its Fourier series converges at any jump in the *periodic extension* to the average value across that jump.

Fourier Series: A Perspective

The Fourier series is a remarkable object. A sum of continuous and periodic functions converges pointwise to a possibly discontinuous and nonperiodic function. This was a startling realization for mathematicians of the early nineteenth century. Even more startling is the realization that a smooth function can be expressed as a sum of the waveforms $\sin nx$ and $\cos nx$. Just as a musical note is composed of a sum of a fundamental note and its harmonics, a function is the sum of the same waveforms as the musical note!

This means a function, like a musical note, can be described by the strength, or amplitude, of each "harmonic" its Fourier series contains. For example, consider the function $f(x) = x$, $-\pi \le x \le \pi$, whose Fourier series is $-2 \sum_{n=1}^{\infty} ((-1)^n / n) \sin nx$. Once the frequencies $\omega_n = n$ are specified, the function is determined by the spectrum of amplitudes $b_n = -2(-1)^n / n$.

Hence, the *amplitude spectrum* shown in Figure 10.14, a bar graph of amplitudes for $f(x)$, is as much a representation of $f(x)$ as its ordinary graph. Each bar is centered on a frequency $\omega_n = n$ and its height is proportional to the coefficient of the corresponding "waveform" $\sin nx$ in the Fourier series.

In Section 30.3, the notion of the amplitude spectrum is generalized by the Fourier (integral) transform to the case of a continuous spectrum of frequencies.

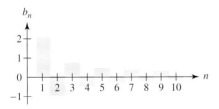

FIGURE 10.14   Amplitude spectrum for $f(x) = x$

EXERCISES 10.1–Part A

A1. Obtain the Fourier series for $f(x) = 1 - x$, $-1 \le x \le 1$, and show that a_n and b_n go to zero as n increases.

A2. Sketch $\tilde{f}(x)$, the function to which the Fourier series in Exercise A1 converges on $[-1, 1]$. Sketch $\hat{f}(x)$, the periodic extension of $\tilde{f}(x)$.

A3. Obtain the Fourier series for $f(x) = (x - 1)^2$, $1 \le x \le 2$, and show that a_n and b_n go to zero as n increases.

A4. Sketch $\tilde{f}(x)$, the function to which the Fourier series in Exercise A3 converges on $[1, 2]$. Sketch $\hat{f}(x)$, the periodic extension of $\tilde{f}(x)$.

A5. Obtain the Fourier series for $f(x) = \begin{cases} 0 & -\pi \le x < 0 \\ 1 - x & 0 \le x \le \pi \end{cases}$, and investigate the Gibbs phenomenon at $x = 0$.

EXERCISES 10.1–Part B

B1. According to [97], the series $\sum_{k=2}^{\infty} \frac{\sin kx}{\ln k}$ converges for all x. According to [93], it is not a Fourier series since there is no function whose Fourier coefficients are given by $b_k = \frac{1}{\ln k}$. Plot a few partial sums of this series to verify the claim of convergence.

In Exercises B2–17, a function and its domain is given. For each:

(a) Sketch $f(x)$ on the indicated domain.

(b) Obtain the Fourier series valid on the given interval $[\alpha, \beta]$ of length $2L$.

(c) Verify the Riemann–Lebesgue lemma, which states that $\lim_{k\to\infty} a_k = \lim_{k\to\infty} b_k = 0$.

(d) Obtain and plot enough partial sums of the Fourier series that convergence to the given function is apparent.

(e) Show that at points where the function has a jump discontinuity, its Fourier series converges to the average of the left and right limits at the jump.

(f) By plotting a partial sum on the interval $[-2L, 4L]$, show that the Fourier series converges to the periodic extension of the given function.

(g) At points where a jump discontinuity generates a Gibbs phenomenon, determine the maximum height of the Gibbs overshoot and express it as a percentage of the amount of the jump.

(h) Verify, analytically or numerically, the identity $\frac{a_0^2}{2} + \sum_{k=1}^{\infty}(a_k^2 + b_k^2) = \frac{1}{L}\int_0^{2L} f^2(x)\,dx$, known as Parseval's identity.

B2. $f(x) = \begin{cases} x & 0 \le x < 1 \\ 1 & 1 \le x \le 2 \end{cases}$, with domain $[0, 2]$

B3. $f(x) = \begin{cases} 1-x & 0 \le x < 1 \\ 0 & 1 \le x \le 2 \end{cases}$, with domain $[0, 2]$

B4. $f(x) = \begin{cases} (x-1)^2 & 0 \le x < 1 \\ 0 & 1 \le x \le 2 \end{cases}$, with domain $[0, 2]$

B5. $f(x) = \begin{cases} (x-1)^2 & 0 \le x < 1 \\ 1 & 1 \le x \le 2 \end{cases}$, with domain $[0, 2]$

B6. $f(x) = |x - 1|$, with domain $[0, 2]$

B7. $f(x) = \begin{cases} 1-x & 0 \le x < 1 \\ (x-1)(3-x) & 1 \le x \le 3 \end{cases}$, with domain $[0, 3]$

B8. $f(x) = \begin{cases} x & 0 \le x < 1 \\ 1 & 1 \le x < 2 \\ 3-x & 2 \le x \le 3 \end{cases}$, with domain $[0, 3]$

B9. $f(x) = \begin{cases} x & 0 \le x < 1 \\ 1 & 1 \le x < 2 \\ 0 & 2 \le x \le 3 \end{cases}$, with domain $[0, 3]$

B10. $f(x) = \begin{cases} 0 & 0 \le x < 1 \\ 1 & 1 \le x < 2 \\ 3-x & 2 \le x \le 3 \end{cases}$, with domain $[0, 3]$

B11. $f(x) = \begin{cases} 1-x & 0 \le x < 1 \\ 0 & 1 \le x < 2 \\ x-2 & 2 \le x \le 3 \end{cases}$, with domain $[0, 3]$

B12. $f(x) = \begin{cases} 0 & 0 \le x < 1 \\ (x-1)(3-x) & 1 \le x < 3 \\ 0 & 3 \le x \le 4 \end{cases}$, with domain $[0, 4]$

B13. $f(x) = \begin{cases} x^2 & 0 \le x < 1 \\ 1 & 1 \le x < 2 \\ (x-3)^2 & 2 \le x \le 3 \end{cases}$, with domain $[0, 3]$

B14. $f(x) = \begin{cases} 0 & 0 \le x < 1 \\ 1 & 1 \le x < 2 \\ 0 & 2 \le x \le 3 \end{cases}$, with domain $[0, 3]$

B15. $f(x) = \begin{cases} 0 & 0 \le x < 2 \\ 1 & 2 \le x < 3 \\ 0 & 3 \le x \le 4 \end{cases}$, with domain $[0, 4]$

B16. $f(x) = \begin{cases} x(2-x) & 0 \le x < 1 \\ 1 & 1 \le x < 2 \\ 4x-3-x^2 & 2 \le x \le 3 \end{cases}$, with domain $[0, 3]$

B17. $f(x) = \begin{cases} x(2-x) & 0 \le x < 1 \\ 1 & 1 \le x < 2 \\ (x-3)^2 & 2 \le x \le 3 \end{cases}$, with domain $[0, 3]$

In Exercises B18–37, a function and a domain is given. For each:

(a) Sketch $f(x)$ on the indicated domain.

(b) Obtain the Fourier series valid on the interval $-L \le x \le L$.

(c) Verify the Riemann–Lebesgue lemma, which states that $\lim_{k\to\infty} a_k = \lim_{k\to\infty} b_k = 0$.

(d) Obtain and plot enough partial sums of the Fourier series that convergence to the given function is apparent.

(e) Show that at points where the function has a jump discontinuity, its Fourier series converges to the average of the left and right limits at the jump.

(f) By plotting a partial sum on the interval $[-3L, 3L]$, show that the Fourier series converges to the periodic extension of the given function.

(g) At points where a jump discontinuity generates a Gibbs phenomenon, determine the maximum height of the Gibbs overshoot and express it as a percentage of the amount of the jump.

(h) Verify, analytically or numerically, the identity $\frac{a_0^2}{2} + \sum_{k=1}^{\infty}(a_k^2 + b_k^2) = \frac{1}{L}\int_{-L}^{L} f^2(x)\,dx$, known as Parseval's identity.

B18. $f(x) = |x|$, with domain $-1 \le x \le 1$

B19. $f(x) = x^2$, with domain $-1 \le x \le 1$

B20. $f(x) = x^3$, with domain $-1 \le x \le 1$

B21. $f(x) = 1 - |x|$, with domain $-1 \le x \le 1$

B22. $f(x) = \begin{cases} 1 & -1 \le x < 0 \\ -1 & 0 \le x \le 1 \end{cases}$, with domain $-1 \le x \le 1$

B23. $f(x) = \begin{cases} x+1 & -1 \le x < 0 \\ x-1 & 0 \le x \le 1 \end{cases}$, with domain $-1 \le x \le 1$

B24. $f(x) = \begin{cases} 0 & -1 \le x < 0 \\ x & 0 \le x \le 1 \end{cases}$, with domain $-1 \le x \le 1$

B25. $f(x) = \begin{cases} -x(x+\pi) & -\pi \le x < 0 \\ -x(x-\pi) & 0 \le x \le \pi \end{cases}$, with domain $-\pi \le x \le \pi$

B26. $f(x) = \begin{cases} 0 & -\pi \le x < 0 \\ -x(x-\pi) & 0 \le x \le \pi \end{cases}$, with domain $-\pi \le x \le \pi$

B27. $f(x) = \begin{cases} x(x+\pi) & -\pi \le x < 0 \\ -x(x-\pi) & 0 \le x \le \pi \end{cases}$, with domain $-\pi \le x \le \pi$

B28. $f(x) = \begin{cases} 0 & -2 \le x < 0 \\ 1-|x-1| & 0 \le x \le 2 \end{cases}$, with domain $-2 \le x \le 2$

B29. $f(x) = \begin{cases} 0 & -1 \le x < 0 \\ e^{-x} & 0 \le x \le 1 \end{cases}$, with domain $-1 \le x \le 1$

B30. $f(x) = \begin{cases} 1 & -2 \le x < -1 \\ x^2 & -1 \le x < 1 \\ 1 & 1 \le x \le 2 \end{cases}$, with domain $-2 \le x \le 2$

B31. $f(x) = \begin{cases} 2+x & -2 \le x < -1 \\ x^2 & -1 \le x < 1 \\ 1 & 1 \le x \le 2 \end{cases}$, with domain $-2 \le x \le 2$

B32. $f(x) = \begin{cases} 2+x & -2 \le x < -1 \\ x^2 & -1 \le x < 1 \\ 2-x & 1 \le x \le 2 \end{cases}$, with domain $-2 \le x \le 2$

B33. $f(x) = \begin{cases} 2+x & -2 \le x < -1 \\ x^2 & -1 \le x < 1 \\ x & 1 \le x \le 2 \end{cases}$, with domain $-2 \le x \le 2$

B34. $f(x) = \begin{cases} 0 & -2 \le x < -1 \\ 1-|x| & -1 \le x < 1 \\ 0 & 1 \le x \le 2 \end{cases}$, with domain $-2 \le x \le 2$

B35. $f(x) = \begin{cases} 1 & -2 \le x < -1 \\ 1-x & -1 \le x < 1 \\ 0 & 1 \le x \le 2 \end{cases}$, with domain $-2 \le x \le 2$

B36. $f(x) = \begin{cases} -2-x & -2 \le x < -1 \\ x & -1 \le x < 1 \\ 2-x & 1 \le x \le 2 \end{cases}$, with domain $-2 \le x \le 2$

B37. $f(x) = \begin{cases} -1 & -2 \le x < -1 \\ x & -1 \le x < 1 \\ 2-x & 1 \le x \le 2 \end{cases}$, with domain $-2 \le x \le 2$

B38. Any trigonometric series that converges uniformly represents a function whose Fourier series is the given trigonometric series [35, p. 466]. The trigonometric series $\sum_{k=1}^{\infty} \sin kx$ is not a Fourier series because the Riemann–Lebesgue lemma fails. Surprisingly, the trigonometric series $\sum_{k=1}^{\infty} \frac{\sin kx}{\sqrt{k}}$ is not a Fourier series of a Riemann-integrable function because $\sum_{k=1}^{\infty} b_k^2 = \sum_{k=1}^{\infty} \frac{1}{k}$ diverges. It can be shown (see [65, p. 533]) that for $x \ne n\pi, n = 1, \dots$, the series converges. By graphing some partial sums of this series, discover why convergence is to a function that is not Riemann integrable, that is, to a function whose Riemann integral does not exist. (The Riemann integral is the integral learned in elementary calculus.)

B39. One of the key steps in proving the convergence of the Fourier series is showing that $\frac{1}{2} + \sum_{k=1}^{n} \cos kx = \frac{\sin(n+1/2)x}{2\sin(x/2)} = D(x)$, the *Dirichlet kernel*. This equality can be established as follows.

(a) Since $2\cos A \sin B = \sin(A+b) - \sin(A-B)$, it follows that $\cos 2k\alpha = \frac{\sin(2k+1)\alpha - \sin(2k-1)\alpha}{2\sin\alpha}$.

(b) $\sum_{k=1}^{n} \cos 2k\alpha = \frac{1}{2\sin\alpha} \sum_{k=1}^{n} (\sin(2k+1)\alpha - \sin(2k-1)\alpha) = \frac{\sin(2k+1)\alpha - \sin(-\alpha)}{2\sin\alpha}$ because the sum of sine terms telescopes. The desired result is now immediate if $\alpha = \frac{x}{2}$.

(c) Graph the Dirichlet kernel for several values of n to see how it behaves.

(d) It can be shown that the partial sums of the Fourier series for $f(x)$ on $[-\pi, \pi]$ are given by $S_n(x) = \frac{1}{2\pi} \int_{-\pi}^{\pi} D(u) f(x+u)\, du = \frac{1}{2\pi} \int_{-\pi}^{\pi} \frac{\sin(n+1/2)u}{2\sin(u/2)} f(x+u)\, du$. Unfortunately, even for the function $f(x) = 1$, this integral cannot be evaluated analytically. (Its value is precisely the sum of terms found in the corresponding partial sum!) However, for any function in Exercises B2–37, it would be possible to fix n and compute $S_n(x)$ numerically for enough values of x that a graph could be drawn. Compare the graphs of the partial sums obtained in Exercise B7 with the numeric evaluation of the corresponding integral defining $S_n(x)$.

10.2   Termwise Integration and Differentiation

Integration

Integration is a smoothing process, and under very mild hypotheses it is valid to integrate a Fourier series termwise. In fact, we have the following theorem. (See, e.g., [35].)

> ### THEOREM 10.3   INTEGRATION
>
> 1. $f(x)$ is piecewise continuous on $[-L, L]$.
> 2. $\dfrac{a_0}{2} + \displaystyle\sum_{n=1}^{\infty} \left(a_n \cos \dfrac{n\pi x}{L} + b_n \sin \dfrac{n\pi x}{L} \right)$ is the Fourier series associated with $f(x)$ on $[-L, L]$.
> 3. x is in $[-L, L]$
>
> $\Longrightarrow$
>
> $$\int_{-L}^{x} f(t)\, dt = \frac{a_0}{2}(x + L) + \frac{L}{\pi} \sum_{n=1}^{\infty} \frac{1}{n} \left[a_n \sin \frac{n\pi x}{L} - b_n \left(\cos \frac{n\pi x}{L} - \cos n\pi \right) \right]$$

The expression on the right in the conclusion is obtained by termwise integrating the Fourier series. Under the assumption that $f(x)$ is merely piecewise continuous, this series need not converge to $f(x)$, yet the termwise integral of the series converges to the integral of the function.

EXAMPLE 10.6   Consider the function $f(x) = x + 1$ on the interval $[-1, 1]$. Its Fourier series, obtained in Section 10.1, is

$$f(x) = 1 - \frac{2}{\pi} \sum_{n=1}^{\infty} \frac{(-1)^n}{n} \sin n\pi x \qquad (10.1)$$

The integral of $f(x)$ is $F(x) = \int_{-1}^{x} f(t)\, dt = \frac{1}{2} + x + \frac{1}{2} x^2$, for which the Fourier series is

$$F(x) = \frac{2}{3} + \frac{2}{\pi^2} \sum_{n=1}^{\infty} \frac{(-1)^n}{n^2} \cos n\pi x - \frac{2}{\pi} \sum_{n=1}^{\infty} \frac{(-1)^n}{n} \sin n\pi x \qquad (10.2)$$

Termwise integration of the Fourier series for $f(x)$ yields

$$\int_{-1}^{x} \left(1 - \frac{2}{\pi} \sum_{n=1}^{\infty} \frac{(-1)^n}{n} \sin n\pi t \right) dt = x + 1 + \frac{2}{\pi^2} \sum_{n=1}^{\infty} \frac{(-1)^n \cos n\pi x - 1}{n^2} \qquad (10.3)$$

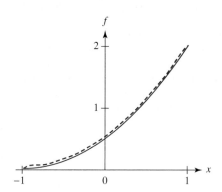

FIGURE 10.15 $F(x)$ (solid) and partial sum (dashed) in Example 10.6

Figure 10.15 shows a partial sum of the series on the right in (10.3) (dashed) and $F(x)$ (solid), graphically validating the termwise integration. To make the right side in (10.3) look like the Fourier series for $F(x)$ in (10.2), replace $x + 1$ with its Fourier series (10.1) and split the sum into two terms, obtaining

$$1 - \frac{2}{\pi} \sum_{n=1}^{\infty} \frac{(-1)^n}{n} \sin n\pi x + \frac{2}{\pi^2} \sum_{n=1}^{\infty} \frac{(-1)^n}{n^2} \cos n\pi x - \frac{2}{\pi^2} \sum_{n=1}^{\infty} \frac{1}{n^2} \qquad (10.4)$$

The rightmost term (10.4) evaluates to $-\frac{1}{3}$, so (10.2) and (10.4) are actually the same.   ❖

Differentiation

Termwise integration of a Fourier series is valid under mild restrictions, as we have just seen. Termwise differentiation requires as much more in the way of hypotheses as termwise integration required less. Indeed, a theorem validating termwise differentiation is the following. (See, e.g., [35].)

THEOREM 10.4   DIFFERENTIATION

1. $f(x)$ is continuous, and $f'(x)$ is piecewise continuous on $[-L, L]$.
2. $f(L) = f(-L)$
3. $f''(x)$ exists at x in $(-L, L)$.

$$\implies f'(x) = \frac{\pi}{L} \sum_{n=1}^{\infty} n \left(-a_n \sin \frac{n\pi x}{L} + b_n \cos \frac{n\pi x}{L} \right)$$

The series on the right is precisely the series obtained by termwise differentiation of the Fourier series for $f(x)$.

EXAMPLE 10.7   The function $f(x) = x^2$, $-\pi \le x \le \pi$, satisfying the hypotheses of Theorem 10.4, has

$$f(x) = x^2 = \frac{\pi^2}{3} + 4 \sum_{n=1}^{\infty} \frac{(-1)^n}{n^2} \cos nx \tag{10.5}$$

as its Fourier series. The Fourier series for $2x$, the derivative of $f(x)$, is

$$2x = -4 \sum_{n=1}^{\infty} \frac{(-1)^n}{n} \sin nx$$

which is exactly the termwise derivative of (10.5).   ❖

EXERCISES 10.2

1. Show formally that if $f(x)$ has a convergent Fourier series on $[-\pi, \pi]$, then $\frac{1}{\pi} \int_{-\pi}^{\pi} f^2 \, dx = \frac{a_0^2}{2} + \sum_{n=1}^{\infty} (a_n^2 + b_n^2)$. (This result is known as *Parseval's identity*.)

Exercises 2–17 use the functions of Exercises B2–17 in Section 10.1. For each:

(a) Obtain the function $F(x) = \int_0^x f(t) \, dt$, $0 \le x \le 2L$.

(b) Obtain the Fourier series for $F(x)$.

(c) Termwise integrate the Fourier series for $f(x)$.

(d) Use graphs of the partial sums of the series obtained in part (c) to determine if termwise integration of the Fourier series was valid.

(e) What can be said about convergence at points of discontinuity of $f(x)$?

(f) If the evidence in part (d) is positive, show analytically that the Fourier series for the antiderivative is the termwise antiderivative of the Fourier series for $f(x)$.

(g) Comment on the applicability of the theorem governing termwise integration of a Fourier series.

(h) Obtain the function $f'(x)$.

(i) Obtain the Fourier series for $f'(x)$.

(j) Termwise differentiate the Fourier series for $f(x)$.

(k) Use graphs of the partial sums of the series obtained in part (j) to determine if termwise differentiation of the Fourier series was valid.

(l) If the evidence in part (k) is positive, show analytically that the Fourier series for the derivative is the termwise derivative of the Fourier series for $f(x)$.

(m) Comment on the applicability of the theorem governing termwise differentiation of a Fourier series.

Exercises 18–37 use the functions of Exercises B18–37 in Section 10.1. For each:

(a) Obtain the function $F(x) = \int_{-L}^x f(t) \, dt$, $-L \le x \le L$.

(b) Obtain the Fourier series for $F(x)$.

(c) Termwise integrate the Fourier series for $f(x)$.

(d) Use graphs of the partial sums of the series obtained in part (c) to determine if termwise integration of the Fourier series was valid.

(e) What can be said about convergence at points of discontinuity of $f(x)$?

(f) If the evidence in part (d) is positive, show analytically that the Fourier series for the antiderivative is the termwise antiderivative of the Fourier series for $f(x)$.

(g) Comment on the applicability of the theorem governing termwise integration of a Fourier series.

(h) Obtain the function $f'(x)$.

(i) Obtain the Fourier series for $f'(x)$.

(j) Termwise differentiate the Fourier series for $f(x)$.

(k) Use graphs of the partial sums of the series obtained in part (j) to determine if termwise differentiation of the Fourier series was valid.

(l) If the evidence in part (k) is positive, show analytically that the Fourier series for the derivative is the termwise derivative of the Fourier series for $f(x)$.

(m) Comment on the applicability of the theorem governing termwise differentiation of a Fourier series.

10.3 Odd and Even Functions and Their Fourier Series

Even and Odd Symmetry

Functions with certain symmetry properties are called either *even* or *odd* functions. The Fourier series for such functions assume particularly simple forms.

EVEN FUNCTIONS   A function $f(x)$ is said to be *even* if it exhibits the symmetry expressed in the relationship $f(-x) = f(x)$. The prototypical even functions $\cos x$ and x^2 are shown in Figure 10.16.

ODD FUNCTIONS   A function $f(x)$ is said to be *odd* if it exhibits the symmetry expressed in the relationship $f(-x) = -f(x)$. The prototypical odd functions $\sin(x)$ and x^3 are shown in Figure 10.17.

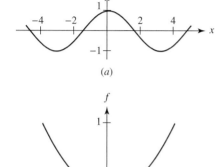

FIGURE 10.16   Prototypical even functions: $\cos x$ and x^2

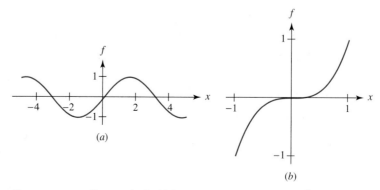

FIGURE 10.17   Prototypical odd functions: (*a*) $\sin x$ and (*b*) x^3

Fourier Series of an Even Function

Special things happen when you compute the Fourier series of an even function over $[-L, L]$. The integrals with $\cos \frac{n\pi x}{L}$ can be taken over half the interval and doubled. The integrals with $\sin \frac{n\pi x}{L}$ are all zero.

 EXAMPLE 10.8   On the interval $[-2, 2]$, let $f(x)$ be the even function $f(x) = x^2$. From Figure 10.18, which shows graphs of $f(x)$ times a cosine and times a sine, we suspect that *an even function times an even function yields an even function* but *an even function times an odd function yields an odd function.* A general proof is left to the Exercises.

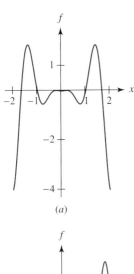

(a)

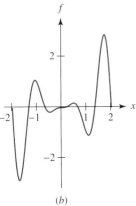

(b)

FIGURE 10.18 Graphs of (a) $x^2 \cos \frac{3}{2}x$, and (b) $x^2 \sin \frac{3}{2}x$

Moreover, we recall from calculus that when integrating functions with appropriate symmetry, we can use the symmetry to "balance areas" when setting up the integrals. When integrating the even function, we can use the rule $\int_{-r}^{r} even\, dx = 2\int_{0}^{r} even\, dx$ and when integrating the odd function we find $\int_{-r}^{r} odd\, dx = 0$.

Applying both these observations to the computation of the Fourier coefficients of the even function $f(x) = x^2$ we find

$$a_0 = \frac{1}{2}\int_{-2}^{2} x^2\, dx = \int_{0}^{2} x^2\, dx = \frac{8}{3}$$

$$a_n = \frac{1}{2}\int_{-2}^{2} x^2 \cos \frac{1}{2}n\pi x\, dx = \int_{0}^{2} x^2 \cos \frac{1}{2}n\pi x\, dx = 16\frac{(-1)^n}{n^2\pi^2}$$

$$b_n = \frac{1}{2}\int_{-2}^{2} x^2 \sin \frac{1}{2}n\pi x\, dx = 0$$ ❖

SUMMARY For an even function, only the even terms survive in the Fourier series. The coefficients of the odd terms are all zero. Even functions have a cosine series. Symmetry allows the a_n's to be computed by integrating over $[0, L]$. The Fourier series for the *even* function is

$$f(x) = \frac{a_0}{2} + \sum_{n=1}^{\infty} a_n \cos \frac{n\pi x}{L} \quad \text{where} \quad a_n = \frac{2}{L}\int_{0}^{L} f(x) \cos \frac{n\pi x}{L}\, dx, n = 0, 1, 2, \dots$$

Fourier Series of an Odd Function

Special things happen when you compute the Fourier series of an odd function over $[-L, L]$. The integrals with $\sin \frac{n\pi x}{L}$ can be done over half the interval and doubled. The integrals with $\cos \frac{n\pi x}{L}$ are all zero.

EXAMPLE 10.9 On the interval $[-2, 2]$, let $f(x)$ be the odd function $f(x) = x$. From Figure 10.19, which shows graphs of $f(x)$ times a cosine and times a sine, we suspect that *an odd function times an even function yields an odd function* but *an odd function times an odd function yields an even function*. Again, verification is left to the exercises.

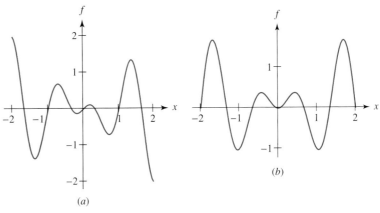

(a)

(b)

FIGURE 10.19 Graphs of (a) $x \cos \frac{3}{2}x$, and (b) $x \sin \frac{3}{2}x$

Further, recalling the "rules" $\int_{-r}^{r} even\,dx = 2\int_{0}^{r} even\,dx$ and $\int_{-r}^{r} odd\,dx = 0$, we find the Fourier coefficients of the odd function $f(x) = x$ to be

$$a_0 = \frac{1}{2}\int_{-2}^{2} x\,dx = 0$$

$$a_n = \frac{1}{2}\int_{-2}^{2} x\cos\frac{1}{2}n\pi x\,dx = 0$$

$$b_n = \frac{1}{2}\int_{-2}^{2} x\sin\frac{1}{2}n\pi x\,dx = \int_{0}^{2} x\sin\frac{1}{2}n\pi x\,dx = -4\frac{(-1)^n}{n\pi} \qquad ❖$$

SUMMARY   For an odd function, only the odd terms survive in the Fourier series. The coefficients of the even terms are all zero. Odd functions have a sine series. Symmetry allows each b_n to be computed by integrating over $[0, L]$. The Fourier series for the odd function is

$$f(x) = \sum_{n=1}^{\infty} b_n \sin\frac{n\pi x}{L} \quad \text{where} \quad b_n = \frac{2}{L}\int_{0}^{L} f(x)\sin\frac{n\pi x}{L}\,dx \qquad n = 1, 2, \dots$$

EXERCISES 10.3–Part A

A1. Justify analytically each of the following statements.

   (a) The product of two odd functions is an even function.

   (b) The product of two even functions is an even function.

   (c) The product of an odd and an even function is an odd function.

A2. Verify analytically that

   (a) $\sin\frac{k\pi x}{L}$, for any $k = 1, 2, \dots$, is an odd function.

   (b) $\cos\frac{k\pi x}{L}$, for any $k = 0, 1, 2, \dots$, is an even function.

A3. Determine whether each of the following functions is odd, even, or neither. Give evidence for your declaration.

   (a) $f(x) = e^x \sin x$ **(b)** $f(x) = e^{-|x|}\cos x$

   (c) $f(x) = \dfrac{1}{1-x^2}$ **(d)** $f(x) = \dfrac{x}{1-x}$ **(e)** $f(x) = \dfrac{x}{1+x^2}$

   (f) $f(x) = \sin(\cos x)$ **(g)** $f(x) = \cos(\sin x)$

   (h) $f(x) = e^{-\sin x}$ **(i)** $f(x) = e^{-|\sin x|}$ **(j)** $f(x) = \sin^2 x$

   (k) $f(x) = x + \sin x$ **(l)** $f(x) = x - \cos x$

   (m) $f(x) = \sin x^2$ **(n)** $f(x) = x^2 + \cos x$

   (o) $f(x) = \sin(x + e^x)$

A4. Determine if each of the following statements is true or false, and give justification for your conclusion.

   (a) The sum of two odd functions is an odd function.

   (b) The difference of two odd functions is an even function.

   (c) The sum of two even functions is an even function.

   (d) The difference of two even functions is an odd function.

   (e) The sum of an odd function and an even function is an odd function.

   (f) The sum of an odd function and an even function is an even function.

   (g) The difference of an odd function and an even function is an odd function.

   (h) The sum of two odd functions is an even function.

   (i) The difference of two odd functions is an odd function.

   (j) The sum of two even functions is an odd function.

   (k) The difference of two even functions is an even function.

EXERCISES 10.3–Part B

B1. Using numeric integration implemented with appropriate technology, obtain the Fourier series for $f(x) = e^{-x^2}$, $-1 \le x \le 1$. Graph $f(x)$ and a partial sum to verify the correctness of your calculations. Graph this same partial sum on $[-3, 3]$ to see $\hat{f}(x)$, the periodic extension of $\tilde{f}(x)$, the limit of the Fourier series on $[-1, 1]$.

The functions in Exercises B2–9 are either odd or even on the indicated domains. For each:

   (a) Obtain the Fourier series without using the relevant symmetry property.

(b) Identify the symmetry as odd or even, and obtain the Fourier series via the formulas for functions with such symmetry. Be sure the results in parts (a) and (b) agree.

(c) Sketch, on $[-3L, 3L]$, the function to which the Fourier series converges.

B2. $f(x) = 1 - x^2$, with domain $-1 \leq x \leq 1$

B3. $f(x) = \begin{cases} x^2 - 1 & -1 \leq x < 0 \\ 1 - x^2 & 0 \leq x \leq 1 \end{cases}$, with domain $-1 \leq x \leq 1$

B4. $f(x) = |\sin(x)|, -\pi \leq x \leq \pi$

B5. $f(x) = \begin{cases} 2 + x & -2 \leq x < -1 \\ 1 & -1 \leq x < 1 \\ 2 - x & 1 \leq x \leq 2 \end{cases}$, with domain $-2 \leq x \leq 2$

B6. $f(x) = \begin{cases} (x+1)(3+x) & -3 \leq x < -1 \\ 0 & -1 \leq x < 1 \\ (x-1)(3-x) & 1 \leq x \leq 3 \end{cases}$, with domain $-3 \leq x \leq 3$

B7. $f(x) = e^{-|x|}, -1 \leq x \leq 1$

B8. $f(x) = \begin{cases} -2 - x & -2 \leq x < -1 \\ -1 & -1 \leq x < 0 \\ 1 & 0 \leq x < 1 \\ 2 - x & 1 \leq x \leq 2 \end{cases}$, with domain $-2 \leq x \leq 2$

B9. $f(x) = \begin{cases} -3 - x & -3 \leq x < -1 \\ x - 1 & -1 \leq x < 0 \\ x + 1 & 0 \leq x < 1 \\ 3 - x & 1 \leq x \leq 3 \end{cases}$, with domain $-3 \leq x \leq 3$

Determine if each function given in Exercises B10–29 is an odd function, an even function, or neither an odd nor an even function.

For each odd function, show that $a_k = 0, k = 0, 1, \ldots$, and $b_k = \frac{2}{L} \int_0^L f(x) \sin \frac{n\pi x}{L} \, dx$. For each even function, show that $b_k = 0$, $k = 1, 2, \ldots$, and $a_k = \frac{2}{L} \int_0^L f(x) \cos \frac{n\pi x}{L} \, dx$.

B10. $f(x)$ from Exercise B18, Section 10.1
B11. $f(x)$ from Exercise B19, Section 10.1
B12. $f(x)$ from Exercise B20, Section 10.1
B13. $f(x)$ from Exercise B21, Section 10.1
B14. $f(x)$ from Exercise B22, Section 10.1
B15. $f(x)$ from Exercise B23, Section 10.1
B16. $f(x)$ from Exercise B24, Section 10.1
B17. $f(x)$ from Exercise B25, Section 10.1
B18. $f(x)$ from Exercise B26, Section 10.1
B19. $f(x)$ from Exercise B27, Section 10.1
B20. $f(x)$ from Exercise B28, Section 10.1
B21. $f(x)$ from Exercise B29, Section 10.1
B22. $f(x)$ from Exercise B30, Section 10.1
B23. $f(x)$ from Exercise B31, Section 10.1
B24. $f(x)$ from Exercise B32, Section 10.1
B25. $f(x)$ from Exercise B33, Section 10.1
B26. $f(x)$ from Exercise B34, Section 10.1
B27. $f(x)$ from Exercise B35, Section 10.1
B28. $f(x)$ from Exercise B36, Section 10.1
B29. $f(x)$ from Exercise B37, Section 10.1

10.4   Sine Series and Cosine Series

Motivation

If $f(x)$ is an even function on $[-L, L]$, its Fourier series will contain just cosine terms; whereas if $f(x)$ is an odd function on $[-L, L]$, its Fourier series will contain just sine terms.

Alternatively, suppose you have the function $f(x)$ defined on $[0, L]$ and you want a Fourier series for it containing just cosine terms. Extend $f(x)$ to a function $g(x)$ that is even on $[-L, L]$, and get the Fourier series of $g(x)$. It will contain only cosine terms and represent $f(x)$ on $[0, L]$.

Likewise, suppose you have the function $f(x)$ defined on $[0, L]$ and you want a Fourier series for it containing just sine terms. Extend $f(x)$ to a function $g(x)$ that is odd on $[-L, L]$, and get the Fourier series of $g(x)$. It will contain only sine terms and represent $f(x)$ on $[0, L]$.

In each case the function $g(x)$ is called an *extension* of $f(x)$.

General Extension

The function $g(x)$ is an *extension* of $f(x)$ if $g(x) = f(x)$ on the domain of $f(x)$, but $g(x)$ has a bigger domain.

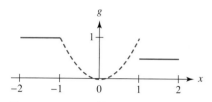

FIGURE 10.20 $f(x)$, $-1 \le x \le 1$, (dashed) extended to $g(x)$ (solid plus dashed)

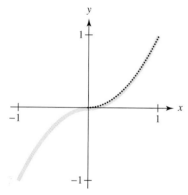

FIGURE 10.21 $g(x)$ (solid), the odd extension of $f(x)$ (dotted) in Example 10.10

For example, on the interval $[-1, 1]$ let $f(x)$ be the function $f(x) = x^2$ and define

$$g(x) = \begin{cases} 1 & x < -1 \\ x^2 & -1 \le x < 1 \\ \frac{1}{2} & 1 \le x \end{cases}$$

embedding $f(x) = x^2$ as part of its definition. Figure 10.20 shows a graph of $g(x)$, where the portion common with $f(x)$ is dashed. The function $g(x)$ agrees with $f(x)$ wherever $f(x)$ is defined, but the domain of $g(x)$ is larger than the domain of $f(x)$. Thus, $g(x)$ is *an extension* of $f(x)$. The functions agree where they have a common domain but $g(x)$ has a bigger domain.

EXAMPLE 10.10   **An odd extension**   Let $f(x) = x^2$ with domain $[0, 1]$. The goal is to extend $f(x)$ to a new function $G(x)$ that then has domain $[-1, 1]$ and is an odd function on $[-1, 1]$. The general recipe for extending $f(x)$, defined on $[0, L]$, to $G(x)$, an odd function on $[-L, L]$, is given on the left in (10.6). Applied to $f(x)$, it yields the rightmost member of (10.6), the extension $g(x)$ that is graphed in Figure 10.21

$$G(x) = \begin{cases} -f(-x) & -L \le x \le 0 \\ f(x) & 0 \le x \le L \end{cases} \quad \Rightarrow \quad g(x) = \begin{cases} -x^2 & -1 \le x \le 0 \\ x^2 & 0 \le x \le 1 \end{cases} \tag{10.6}$$

FOURIER SINE SERIES   The Fourier series of the odd function $G(x)$ is given by

$$G(x) = \sum_{n=1}^{\infty} b_n \sin \frac{n\pi x}{L} \quad \text{where} \quad b_n = \frac{2}{L} \int_0^L f(x) \sin \frac{n\pi x}{L} \, dx \quad n = 1, 2, \dots$$

It is not a mistake to put $f(x)$ into the integral for the coefficients b_n. Since $G(x)$ *is* $f(x)$ in the interval $[0, L]$, we can write $f(x)$ in place of $G(x)$ for that interval. Moreover, on the interval $[0, L]$, if the Fourier series represents $G(x)$, then it also represents $f(x)$ on that interval.

A Fourier sine series for the function $f(x) = x^2$, valid on the interval $[0, 1]$, is obtained by a computation of the coefficients that does not require us to use a representation of the extension $g(x)$. In fact, we find

$$b_n = \int_0^1 x^2 \sin n\pi x \, dx = 4\frac{(-1)^n - 1}{n^3 \pi^3} - 2\frac{(-1)^n}{n\pi}$$

CONCLUSION   Given $f(x)$ defined on $[0, L]$, you can obtain its Fourier sine series by extending $f(x)$ to the interval $[-L, L]$ as the odd function $G(x)$ and obtaining the Fourier series of $g(x)$. This Fourier series represents both $f(x)$ and $G(x)$ on the interval $[0, L]$. In practice, the extension to $G(x)$ need not be made because the sine series can be computed by integrating $f(x)$ over the interval $[0, L]$.   ❖

EXAMPLE 10.11   **An even extension**   Obtain the Fourier cosine series for the function

$$f(x) = \begin{cases} x & 0 \le x \le 1 \\ 2 - x & 1 < x \le 2 \end{cases}$$

defined on the interval $[0, 2]$ and graphed in Figure 10.22. To end up with a *cosine* series, form the Fourier series of an *even* function. Hence, extend $f(x)$ to $[-2, 2]$ as the *even* function $g(x)$. The general recipe for extending $f(x)$, defined on $[0, L]$, to $G(x)$, an even function on $[-L, L]$, is given on the left in (10.7). Applied to $f(x)$, it gives the rightmost member of (10.7), the extension $g(x)$, graphed in Figure 10.23

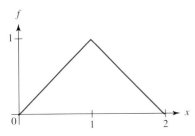

FIGURE 10.22 The function $f(x)$ of Example 10.11

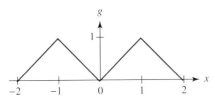

FIGURE 10.23 $g(x)$, the even extension of $f(x)$ in Example 10.11

$$G(x) = \begin{cases} f(-x) & -L \le x \le 0 \\ f(x) & 0 \le x \le L \end{cases} \Rightarrow g(x) = \begin{cases} 2+x & -2 \le x \le -1 \\ -x & -1 < x \le 0 \\ x & x < 0 \le 1 \\ 2-x & 1 < x \le 2 \end{cases} \quad (10.7)$$

Again, we repeat that it is not necessary to have an explicit formula for $G(x)$ in order to obtain the required Fourier cosine series for $f(x)$. The Fourier cosine series for $f(x)$ coincides with the series for $G(x)$, which can be obtained via the recipe

$$G(x) = \frac{a_0}{2} + \sum_{n=1}^{\infty} a_n \cos \frac{n\pi x}{L} \quad \text{where} \quad a_n = \frac{2}{L} \int_0^L f(x) \cos \frac{n\pi x}{L} \, dx \quad n = 1, 2, \ldots$$

On the interval $[0, L]$, the interval over which the integration is performed, $g(x)$ agrees with $f(x)$, so to get the Fourier series for $G(x)$ you can use the values provided to the integral by $f(x)$. We never needed an expression for $G(x)$ at all. Thus, the Fourier coefficients are

$$a_0 = \int_0^2 f(x)\,dx = 1 \quad a_n = \int_0^2 f(x) \cos \frac{n\pi x}{2}\,dx = \frac{4}{n^2\pi^2}\left(2\cos\frac{n\pi}{2} - (-1)^n - 1\right)$$

❖

EXERCISES 10.4–Part A

A1. Given $f(x) = x, 0 \le x \le 1$:

(a) Sketch $g(x)$, the even extension of $f(x)$.

(b) Sketch the even *periodic* extension of $f(x)$.

(c) Sketch the odd extension of $f(x)$.

(d) Sketch the odd *periodic* extension of $f(x)$.

(e) Obtain a Fourier cosine series for $f(x)$.

(f) On $[-3, 3]$, sketch the function to which the series in (e) converges.

(g) Obtain a Fourier sine series for $f(x)$.

(h) On $[-3, 3]$, sketch the function to which the series in (g) converges.

A2. Repeat Exercise A1 for the function $f(x) = \begin{cases} x & 0 \le x \le 1 \\ 1 & 1 < x \le 2 \end{cases}$.

A3. Obtain a Fourier cosine series for $f(x) = x^2$, valid on $[0, \pi]$. By evaluating the series at $x = 0$, show that $\sum_{k=1}^{\infty}(-1)^{k+1}/k^2 = \frac{\pi^2}{12}$.

A4. By evaluating the series in Exercise A3 at $x = \pi$, show that $\sum_{k=1}^{\infty} \frac{1}{k^2} = \frac{\pi^2}{6}$.

EXERCISES 10.4–Part B

B1. If $f(x)$ is the function given in Exercise A1 and $h(x)$ extends $f(x)$ to be zero on $-1 \le x \le 0$, show that the Fourier series for $h(x)$ is the average of the two series from parts (e) and (g) of Exercise A1.

In Exercises B2–17, the given functions have a domain of the form $[0, \lambda]$. For each:

(a) Find g_{odd}, the odd extension of $f(x)$.

(b) Using the formalism appropriate for the interval $[-\lambda, \lambda]$, obtain the Fourier series of g_{odd}.

(c) Obtain the Fourier sine series of $f(x)$, valid on $[0, \lambda]$.

(d) By graphing a partial sum of the sine series on $[-\lambda, \lambda]$, show that the series represents $f(x)$ on $[0, \lambda]$.

(e) Find g_{even}, the even extension of $f(x)$.

(f) By using the formalism appropriate for the interval $[-\lambda, \lambda]$, obtain the Fourier series of g_{even}.

(g) Obtain the Fourier cosine series of $f(x)$, valid on $[0, \lambda]$.

(h) By graphing a partial sum of the cosine series on $[-\lambda, \lambda]$, show that series represents $f(x)$ on $[0, \lambda]$.

(i) Show that the average of the Fourier sine and cosine series represents the function $\begin{cases} 0 & -\lambda \le x < 0 \\ f(x) & 0 \le x \le \lambda \end{cases}$.

(j) Extend $f(x)$ to a function $g_1(x) = \begin{cases} 1 & -\lambda \le x < 0 \\ f(x) & 0 \le x \le \lambda \end{cases}$ and obtain its Fourier series on $[-\lambda, \lambda]$.

(k) Graph a partial sum of the Fourier series of $g_1(x)$ on the interval $[-3\lambda, 3\lambda]$.

10.5 Periodically Driven Damped Oscillator

Damped Oscillator Driven Periodically

Consider the driven damped spring-mass system governed by the differential equation $y'' + 2y' + 10y = f(t)$, where $f(t)$ is the periodic extension of $g(t) = \begin{cases} t & 0 \le t \le \pi \\ 2\pi - t & \pi < t \le 2\pi \end{cases}$. Three cycles of the periodic forcing function $f(t)$ are shown in Figure 10.24.

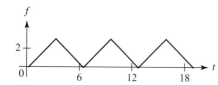

FIGURE 10.24   Forcing function for damped oscillator

Exact Solution by Laplace Transform

A graph of the exact motion of this oscillator, started with the inert initial conditions $y(0) = 0$, $y'(0) = 0$, can be found via the Laplace transform. To this end, we obtain $F(s)$, the transform of $f(t)$, with the formula for periodic functions from Section 6.7. Thus, with $p = 2\pi$, the period of $f(t)$, we obtain

$$F(s) = \frac{1}{1 - e^{-ps}} \int_0^p f(t)e^{-st}\, dt = \frac{1 - e^{-\pi s}}{s^2(1 + e^{-\pi s})}$$

Transforming the differential equation and solving for Y, the Laplace transform of $y(t)$, we get

$$Y = \frac{1}{s^2(s^2 + 2s + 10)} \frac{1 - e^{-\pi s}}{1 + e^{-\pi s}}$$

It is impossible to invert this transform with just a finite number of elementary functions, so we expand $(1 - e^{-\pi s})/(1 + e^{-\pi s})$ in powers of $e^{-\pi s}$, obtaining $-1 + 2\sum_{k=0}^{\infty}(-1)^k e^{-k\pi s}$. The inverse Laplace transform of the factor $\frac{1}{s^2(s^2+2s+10)}$ is the function

$$h(t) = \frac{e^{-t}}{25}\left(\frac{1}{2}\cos 3t - \frac{2}{3}\sin 3t\right) + \frac{5t - 1}{50}$$

so the solution to the differential equation is

$$y(t) = -h(t) + 2\sum_{k=0}^{\infty}(-1)^k h(t - k\pi)\mathrm{H}(t - k\pi)$$

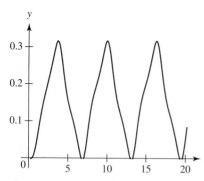

FIGURE 10.25   Exact solution for the driven damped oscillator

The graph of $\hat{y} = -h(t) + 2\sum_{k=0}^{7}(-1)^k h(t - k\pi)\mathrm{H}(t - k\pi)$ is given in Figure 10.25, where, on the interval $[0, 20]$, the solution is exact because $\mathrm{H}(t - k\pi)$, $k > 7$, is zero, since $8\pi > 20$.

Approximate Solution by Fourier Series

We next obtain a sequence of approximate solutions that rapidly converge to the exact solution. In the differential equation, the driving function $f(t)$ is replaced by $f_k(t)$, a partial sum of the Fourier series of $f(t)$, and the corresponding solution $y_k(t)$ computed. The sequence of approximate solutions $y_k(t)$ converges rapidly to the exact solution.

Using the general formalism of Section 10.1, we first obtain a Fourier series representation of $f(t)$. Hence, with the interval $[0, 2L]$ taken as $[0, 2\pi]$ so $L = \pi$, we obtain

$$a_0 = \frac{1}{\pi} \int_0^{2\pi} f(t)\,dt = \pi$$

$$a_n = \frac{1}{\pi} \int_0^{2\pi} f(t)\cos nt\,dt = 2\frac{(-1)^n - 1}{n^2\pi}$$

$$b_n = \frac{1}{\pi} \int_0^{2\pi} f(t)\sin nt\,dt = 0$$

The first three distinct partial sums of the resulting Fourier series are then

$$f_1 = \frac{\pi}{2} - \frac{4}{\pi}\cos t$$

$$f_2 = \frac{\pi}{2} - \frac{4}{\pi}\left(\cos t + \frac{1}{9}\cos 3t\right)$$

$$f_3 = \frac{\pi}{2} - \frac{4}{\pi}\left(\cos t + \frac{1}{9}\cos 3t + \frac{1}{25}\cos 5t\right)$$

Letting $y_k(t)$, $k = 1, 2, 3$, be the solutions of the original IVP when $f(t)$ is replaced by $f_k(t)$, we obtain Figure 10.26 wherein the approximations y_k are plotted against $\hat{y}$. The third partial sum approximates $f(t)$ well enough for the graph of the resulting approximation to $y(t)$ to be indistinguishable from the graph of the exact solution. In Chapter 14 we will study additional techniques for obtaining approximation analytic solutions of differential equations.

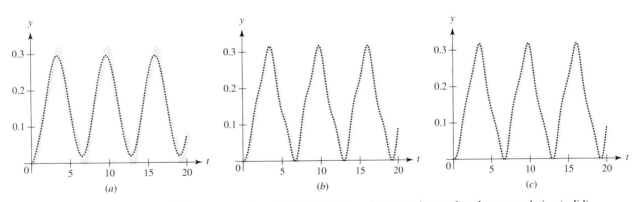

FIGURE 10.26 Approximate solutions y_1, y_2, and y_3, dotted in (a), (b), and (c), superimposed on the exact solution (solid).

EXERCISES 10.5–Part A

A1. The supposition in Section 10.5 is that solutions of initial value problems with "close" driving functions should themselves be "close." Use the initial value problem $y'' + 2y' + 10y = e^{-(1+a)t}$, $y(0) = y'(0) = 0$ to quantize this notion.

(a) Obtain $y_0(t)$, the solution of the initial value problem when $a = 0$.

(b) Obtain $y_a(t)$, the solution of the initial value problem when $a \neq 0$.

(c) Show that $\lim_{a\to 0} y_a = y_0$, making $y_a(t)$ continuous in a at $a = 0$.

(d) One measure of distance between $f_0(t) = e^{-t}$ and $f_a(t) = e^{-(1+a)t}$ is the mean-square norm, $\sqrt{\int_0^\infty (f_0 - f_a)^2\, dt}$. Show that by this measure, f_0 and f_a are close if a is small and that the distance goes to zero as a goes to zero.

(e) Show that by the same measure, y_0 and y_a are close and the distance between these two solutions go to zero as a goes to zero.

In Exercises A2–6, a function $g(t)$ is defined on an interval of the form $[0, \lambda]$. For each:

(a) Over the interval $[0, 3\lambda]$, graph the function $f(t)$, whose period is λ and which is the periodic extension of $g(t)$.

(b) Obtain $F(s)$, the Laplace transform of $f(t)$.

A2. $g(t) = t, 0 \le t \le 1$ **A3.** $g(t) = \begin{cases} t & 0 \le t < 1 \\ 1 & 1 \le t \le 2 \end{cases}$

A4. $g(t) = \begin{cases} 1 & 0 \le t < 1 \\ -1 & 1 \le t \le 2 \end{cases}$ **A5.** $g(t) = \begin{cases} t & 0 \le t < 1 \\ t-2 & 1 \le t \le 2 \end{cases}$

A6. $g(t) = \begin{cases} t & 0 \le t < 1 \\ 0 & 1 \le t \le 2 \end{cases}$

EXERCISES 10.5–Part B

B1. The sequence of functions $f_k(t) = kte^{-kt}, k = 1, 2, \ldots$, converges to zero both pointwise and in the mean (but not uniformly) for $t \ge 0$. Obtain $y_k(t)$, the solutions of the initial value problems $y'' + 2y' + 10y = f_k(t), y(0) = y'(0) = 0$, and show that they converge pointwise to zero, the solution of the completely homogeneous initial value problem.

In Exercises B2–6, $g(t)$ is prescribed and extended periodically to $f(t)$, which then drives the damped oscillator described by the IVP $y'' + 2y' + 5y = f(t), y(0) = y'(0) = 0$. In each case:

(a) Obtain $Y(s)$, the Laplace transform of $y(t)$, the solution of the given IVP.

(b) Approximate $Y(s)$ with a Taylor polynomial in powers of an appropriate exponential $e^{-\sigma s}$.

(c) Invert the approximation in part (b) and graph it over a domain for which it is the exact solution.

(d) Obtain $f_k(t), k = 1, \ldots, 5$, partial sums of a Fourier series for $f(t)$.

(e) Obtain $y_k(t), k = 1, \ldots, 5$, solutions of the initial value problem where $f(t)$ has been replaced by $f_k(t)$.

(f) Plot each solution $y_k(t)$ against the exact solution found in part (c).

B2. $g(t)$ from Exercise A2   **B3.** $g(t)$ from Exercise A3
B4. $g(t)$ from Exercise A4   **B5.** $g(t)$ from Exercise A5
B6. $g(t)$ from Exercise A6

In Exercises B7–11, $g(t)$ is prescribed and extended periodically to $f(t)$, which then drives the damped oscillator described by the IVP $y'' + 4y' + 13y = f(t), y(0) = y'(0) = 0$. In each case:

(a) Obtain $Y(s)$, the Laplace transform of $y(t)$, the solution of the given IVP.

(b) Approximate $Y(s)$ with a Taylor polynomial in powers of an appropriate exponential $e^{-\sigma s}$.

(c) Invert the approximation in part (b) and graph it over a domain for which it is the exact solution.

(d) Obtain $f_k(t), k = 1, \ldots, 5$, partial sums of a Fourier series for $f(t)$.

(e) Obtain $y_k(t), k = 1, \ldots, 5$, solutions of the initial value problem where $f(t)$ has been replaced by $f_k(t)$.

(f) Plot each solution $y_k(t)$ against the exact solution found in part (c).

B7. $g(t)$ from Exercise A2   **B8.** $g(t)$ from Exercise A3
B9. $g(t)$ from Exercise A4   **B10.** $g(t)$ from Exercise A5
B11. $g(t)$ from Exercise A6

In Exercises B12–16, $g(t)$ is prescribed and extended periodically to $f(t)$, which then drives the damped oscillator described by the IVP $y'' + 6y' + 13y = f(t), y(0) = y'(0) = 0$. In each case:

(a) Obtain $Y(s)$, the Laplace transform of $y(t)$, the solution of the given IVP.

(b) Approximate $Y(s)$ with a Taylor polynomial in powers of an appropriate exponential $e^{-\sigma s}$.

(c) Invert the approximation in part (b) and graph it over a domain for which it is the exact solution.

(d) Obtain $f_k(t), k = 1, \ldots, 5$, partial sums of a Fourier series for $f(t)$.

(e) Obtain $y_k(t), k = 1, \ldots, 5$, solutions of the initial value problem where $f(t)$ has been replaced by $f_k(t)$.

(f) Plot each solution $y_k(t)$ against the exact solution found in part (c).

B12. $g(t)$ from Exercise A2   **B13.** $g(t)$ from Exercise A3
B14. $g(t)$ from Exercise A4   **B15.** $g(t)$ from Exercise A5
B16. $g(t)$ from Exercise A6

B17. The solution of the *boundary value problem* $y'' + 2y' + 2y = 1$, $y(0) = y(1) = 0$, can be obtained from $y = \frac{1}{2} + e^{-t}(A\cos t + B\sin t)$, the general solution, by imposing the endpoint conditions to get the algebraic equations $0 = \frac{1}{2} + A$ and $0 = \frac{1}{2} + e^{-1}(A\cos 1 + B\sin 1)$ and solving for the constants A and B. (Use of a computer algebra system is highly recommended. For example, Maple's **dsolve** command solves this boundary value problem immediately, yielding $y = \frac{1}{2} - \frac{1}{2}e^{-t}\cos t + \frac{1}{2}(\cos 1 - \frac{e}{\sin 1})e^{-t}\sin t$.)

(a) Obtain $f_k(t), k = 1, 2, 3$, the first three distinct partial sums of the Fourier sine series for $f(t) = 1, 0 \le t \le 1$.

(b) Plot the three partial sums on the same set of axes.

(c) Solve the three boundary value problems $y'' + 2y' + 2y = f_k$, $y(0) = y(1) = 0$, for $y_k(t), k = 1, 2, 3$.

(d) Plot $y(t)$, the exact solution, and the three approximate solutions all on the same axes. Although the Fourier series converges slowly, the approximate solutions to the boundary value problem converge much more rapidly.

In Exercises B18–22, a driving function $f(x)$ and boundary conditions are given. These, along with the differential equation $y'' + 4y' + 5y = f(t)$, form a BVP. In each exercise:

(a) Solve the boundary value problem exactly by finding the general solution and applying the boundary conditions to determine the arbitrary constants.

(b) Solve the boundary problem directly with your favorite computer algebra system.

(c) Plot the exact solution.

(d) Approximate the driving function with a sequence of partial sums of a Fourier sine series.

(e) Plot these partial sums on the same set of axes.

(f) Obtain the approximate solutions which correspond to replacing $f(t)$ with one of its partial sums.

(g) Plot the approximate solutions and the exact solution on the same set of axes.

B18. $f(t) = t, y(0) = 1, y(1) = 2$

B19. $f(t) = e^{-t}, y(0) = 1, y(1) = -1$

B20. $f(t) = te^t, y(0) = -1, y(2) = 0$

B21. $f(t) = \cos t, y(0) = 2, y(1) = 1$

B22. $f(t) = t\sin t, y(0) = 0, y(1) = 1$

In Exercises B23–27, a driving function $f(x)$, and boundary conditions are given. These, along with the differential equation $y'' + 6y' + 18y = f(t)$, form a BVP. In each exercise:

(a) Solve the boundary value problem exactly by finding the general solution and applying the boundary conditions to determine the arbitrary constants.

(b) Solve the boundary problem directly with your favorite computer algebra system.

(c) Plot the exact solution.

(d) Approximate the driving function with a sequence of partial sums of a Fourier sine series.

(e) Plot these partial sums on the same set of axes.

(f) Obtain the approximate solutions that correspond to replacing $f(t)$ with one of its partial sums.

(g) Plot the approximate solutions, and the exact solution, on the same set of axes.

B23. $f(t) = t, y(0) = 1, y(1) = 2$

B24. $f(t) = e^{-t}, y(0) = 1, y(1) = -1$

B25. $f(t) = te^t, y(0) = -1, y(2) = 0$

B26. $f(t) = \cos t, y(0) = 2, y(1) = 1$

B27. $f(t) = t\sin t, y(0) = 0, y(1) = 1$

10.6   Optimizing Property of Fourier Series

Approximating $f(x)$ via Trigonometric Polynomials

A partial sum of a Fourier series is a trigonometric polynomial approximating the function to which the series converges. For the function $f(x) = x$ defined on the interval $[0, \pi]$, a two-term trigonometric polynomial that might approximate it is $g(x) = s_1 \sin x + s_2 \sin 2x$.

Judging the quality of the approximation of $f(x)$ by $g(x)$ requires a measure of the goodness of the fit of $g(x)$ to $f(x)$. In Section 8.4 we used the notion of convergence in the mean to measure convergence of a sequence of functions. The measure of performance for our trigonometric approximation will be

$$\|f(x) - g(x)\|_2^2 = \int_a^b (f(x) - g(x))^2\, dx \qquad (10.8)$$

the square of the measure used for convergence in the mean. The *subscript* 2 indicates that the difference between $f(x)$ and $g(x)$ in the integral has been *squared*. We illustrate this measure in the following example.

EXAMPLE 10.12   Assigning s_1 and s_2 the values 1 in $g(x)$ yields the approximating function $g_1(x) = \sin x + \sin 2x$ appearing in Figure 10.27 as the solid curve. The dashed curve is, of course, $f(x)$. A computation shows

$$\|f(x) - g_1(x)\|_2^2 = \int_0^\pi (f(x) - g_1(x))^2 \, dx \doteq 10.3 \qquad \diamondsuit$$

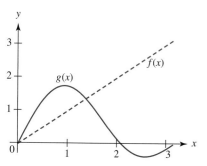

FIGURE 10.27   Approximating $f(x) = x$ (dashed) with $g(x) = \sin x + \sin 2x$ (solid)

OPTIMIZING THE FIT HEURISTICALLY   For our example, applying (10.8) to $f(x)$ and $g(x)$ would give

$$Q(s_1, s_2) = \int_0^\pi (x - s_1 \sin x - s_2 \sin 2x)^2 \, dx = \tfrac{1}{3}\pi^3 - 2\pi s_1 + \tfrac{1}{2}s_2^2\pi + \pi s_2 + \tfrac{1}{2}s_1^2\pi$$

a function of the two parameters s_1 and s_2, graphed in Figure 10.28. The surface determined by $Q(s_1, s_2)$ has a minimum point near $(s_1, s_2) = (2, -1)$. We next find the exact critical point by the techniques of calculus.

OPTIMIZING THE FIT ANALYTICALLY   Equating to zero the partial derivatives of $Q(s_1, s_2)$ taken with respect to s_1 and s_2 yields the equations

$$\pi(s_1 - 2) = 0 \quad \text{and} \quad \pi(s_2 + 1) = 0$$

whose solution is $(s_1, s_2) = (2, -1)$. Hence, the best fit to $f(x) = x$ that we can obtain under this measure, using the fitting function $g(x) = s_1 \sin x + s_2 \sin 2x$, is the function $G(x) = 2 \sin x - \sin 2x$, graphed in Figure 10.29 and for which $Q(2, -1) \doteq 2.418$.

Best Fit, Generalized

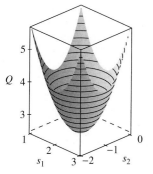

FIGURE 10.28 $\int_0^\pi (f - g)^2 \, dx$ as a function of s_1 and s_2

We next seek, for a general function $f(x)$ on the interval $[0, \pi]$ the best approximating trigonometric polynomial of the form $g(x) = \sum_{k=1}^\infty s_k \sin kx$. To minimize the measure of performance $Q = \int_0^\pi (f(x) - \sum_{k=1}^\infty s_k \sin kx)^2 \, dx$, solve the *normal equations*

$$\frac{1}{2}\frac{\partial Q}{\partial s_j} = \int_0^\pi \left(f(x) - \sum_{k=1}^\infty s_k \sin kx \right) \sin jx \, dx = 0 \qquad j = 1, \dots$$

for $s_k, k = 1, 2, \dots$. Because the integrals of the "cross terms" vanish, that is, because $\int_0^\pi \sin nx \sin mx \, dx = 0$ for $n \neq m$, these equations take the form

$$\int_0^\pi f(x) \sin kx \, dx = s_k \int_0^\pi \sin^2 kx \, dx = \frac{\pi}{2}s_k \tag{10.9}$$

from which we determine $s_k = \frac{2}{\pi}\int_0^\pi f(x) \sin kx \, dx$. These values for the s_k are precisely the coefficients of the Fourier sine series for $f(x)$ on the interval $[0, \pi]$. The Fourier series is the best trigonometric approximation to $f(x)$, provided the measure used to determine "best" is convergence in the mean.

Orthogonality of Functions

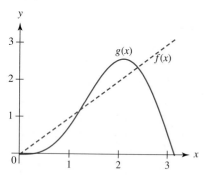

FIGURE 10.29   The best mean-square fit of $f(x) = x$ with $g(x) = s_1 \sin x + s_2 \sin 2x$

A key step in obtaining the Fourier coefficients by minimizing the mean-square norm Q is the vanishing of the "cross-terms," the integrals $\int_0^\pi \sin nx \sin mx \, dx, n \neq m$. This leads to a general definition of the orthogonality of functions.

DEFINITION 10.1

Two functions $f(x)$ and $g(x)$ are said to be orthogonal on the interval $[a, b]$ if $\int_a^b f(x)g(x)\,dx = 0$.

There are functions other than just sines and cosines with the same orthogonal, "minimizing" property, the basis for generalizations of the Fourier series. In Section 10.7 we will use orthogonality to obtain the Fourier–Legendre series, and in Section 16.2 we will again use orthogonality to obtain the Fourier–Bessel series. Similar techniques could be used to obtain series of Chebyshev polynomials, Laguerre polynomials, Hermite polynomials, and many other of the special functions of applied mathematics.

EXERCISES 10.6–Part A

A1. Verify that $\cos x$ and $\sin x$ are *orthogonal* on $[0, \pi]$ by showing the integral $\int_0^\pi \sin x \cos x\,dx$ evaluates to zero.

A2. Determine s_1 and s_2 so that $g(x) = s_1 \cos x + s_2 \sin x$ gives the best mean-square fit to $f(x) = x$ on $[0, \pi]$.

A3. Show that x and $\cos x$ are *not orthogonal* on $[0, \pi]$.

A4. Determine s_1 and s_2 so that $g(x) = s_1 \cos x + s_2 x$ gives the best mean-square fit to $f(x) = \sin x$ on $[0, \pi]$.

A5. Verify that the functions $\{\cos kx\}, k = 0, 1, \ldots$, are *mutually orthogonal* on the interval $[0, \pi]$ by showing that the integrals $\int_0^\pi \cos nx \cos mx\,dx = 0$ whenever $n \neq m$.

A6. Execute a best mean-square fit of $s_0 + \sum_{n=1}^\infty s_n \cos nx$ to $f(x)$ on $[0, \pi]$. Show that s_n are the coefficients of the Fourier cosine series for $f(x)$. What is the role of *orthogonality* in this calculation?

A7. Show that $\sin x$ and $\sin 3x$ are orthogonal on $[0, \pi]$. Graph these functions on the same set of axes and observe that, in spite of the terminology "orthogonal," there is nothing "geometricaly perpendicular" about these two curves. In fact, compute slopes at the points of intersection and show they are not negative reciprocals of each other.

EXERCISES 10.6–Part B

B1. Show that the functions $\{1, x, x^2, x^3\}$ are not mutually orthogonal on $[0, 1]$.

B2. Determine $s_k, k = 0, 1, 2$, so that $g(x) = \sum_{k=0}^3 s_k x^k$ gives the best mean-square fit to $f(x) = \sin \pi x$ on $[0, 1]$.

For the functions $f(x)$ given in Exercises B3–12:

(a) Obtain $b_k, k = 1, \ldots, 5$, the first five coefficients of the Fourier sine series on the interval $[0, \pi]$.

(b) Let $\Phi = \sum_{k=1}^5 b_k \sin kx$, obtain the normal equations in the process of minimizing $\int_0^\pi (f - \Phi)^2\,dx$, and show that the solution of these equations is the same set of coefficients as determined in part (a).

(c) Write the equation $f(x) = b_1 \sin x + b_2 \sin 2x + \cdots$, and multiply, in turn, by $\sin x, \sin 2x, \ldots, \sin 5x$, each time integrating with respect to x on the interval $[0, \pi]$. Show that the same five normal equations result.

(d) Let $p_5(x)$ be the value of the minimizing Φ. Plot $f(x)$ and $p_5(x)$ on the same set of axes.

(e) Compute the value of $\int_0^\pi (f - p_5)^2\,dx$.

(f) Show that $\int_0^\pi p_5^2(x)\,dx = \frac{\pi}{2}\sum_{k=1}^5 b_k^2 \le \int_0^\pi f^2(x)\,dx$. (The left half suggests Parseval's identity, first met in the exercises of Section 10.1; while the right half suggests *Bessel's inequality*, the relevant form of which is $\frac{\pi}{2}\sum_{k=1}^\infty b_k^2 \le \int_0^\pi f^2(x)\,dx$.)

(g) Determine the five coefficients b_k by *collocation*. Pick five points in the interval $(0, \pi)$ and at each point x_k require $f(x_k) = \Phi(x_k)$ to hold. Solve the resulting five equations for $b_k = B_k$, and call the ensuing approximation $P_5(x)$.

(h) Plot $f(x)$, $p_5(x)$, and $P_5(x)$ all on the same set of axes.

(i) Compute the value of $\int_0^\pi (f - P_5)^2\,dx$, and show that it is larger than the value computed in part (e).

B3. $f(x) = x$ **B4.** $f(x) = x^2$ **B5.** $f(x) = \pi - x$

B6. $f(x) = \pi^2 - x^2$ **B7.** $f(x) = 1$ **B8.** $f(x) = \cos x$

B9. $f(x) = \cos\frac{x}{2}$ **B10.** $f(x) = x^3$

B11. $f(x) = e^{-x}$ **B12.** $f(x) = x - \frac{\pi}{2}$

For the functions given in Exercises B13–22:

(a) Obtain a_k, $k = 0, 1, \ldots, 5$, the first six coefficients of the Fourier cosine series on the interval $[0, \pi]$.

(b) Let $\Phi = \frac{c_0}{2} + \sum_{k=1}^{5} c_k \cos kx$, obtain the normal equations in the process of minimizing $\int_0^\pi (f - \Phi)^2 \, dx$, and show that these equations determine $c_k = a_k$, where the a_k are determined in part (a).

(c) Write the equation $f(x) = \Phi$ and multiply, in turn, by $1, \cos x, \cos 2x, \ldots, \cos 5x$, each time integrating with respect to x on the interval $[0, \pi]$. Show that the same six normal equations result.

(d) Let $p_5(x)$ be the value of the minimizing Φ. Plot $f(x)$ and $p_5(x)$ on the same set of axes.

(e) Compute the value of $\int_0^\pi (f - p_5)^2 \, dx$.

(f) Show that $\int_0^\pi p_5^2(x) \, dx = \frac{\pi}{4} a_0^2 + \frac{\pi}{2} \sum_{k=1}^{5} a_k^2 \le \int_0^\pi f^2(x) \, dx$.

(g) Determine the six coefficients a_k by *collocation*. Pick six points in the interval $(0, \pi)$ and at each point x_k require $f(x_k) = \Phi(x_k)$ to hold. Solve the resulting six equations for c_k, and call the ensuing approximation $P_5(x)$.

(h) Plot $f(x)$, $p_5(x)$, and $P_5(x)$ all on the same set of axes.

(i) Compute the value of $\int_0^\pi (f - P_5)^2 \, dx$ and show that it is larger than the value computed in part (e).

B13. $f(x)$ from Exercise B3   **B14.** $f(x)$ from Exercise B4
B15. $f(x)$ from Exercise B5   **B16.** $f(x)$ from Exercise B6

B17. $f(x) = \left| x - \dfrac{\pi}{2} \right|$ **B18.** $f(x) = \sin x$

B19. $f(x) = \sin \dfrac{x}{2}$ **B20.** $f(x)$ from Exercise B10

B21. $f(x)$ from Exercise B11   **B22.** $f(x)$ from Exercise B12

For the functions in Exercises B23–27:

(a) Obtain the Fourier series coefficients a_k, $k = 0, 1, 2, 3$, and b_k, $k = 1, 2, 3$ for the interval $[-\pi, \pi]$.

(b) Let $\Phi = A + \sum_{k=1}^{3} u_k \cos(kx) + v_k \sin(kx)$, minimize $\int_{-\pi}^{\pi} (f - \Phi)^2 \, dx$ to obtain the normal equations, and show that the solution of these equations is the same set of coefficients as determined in part (a).

(c) Write the equation $f(x) = \Phi$ and multiply, in turn, by $1, \cos x, \cos 2x, \cos 3x, \sin x, \sin 2x, \sin 3x$, each time integrating with respect to x on the interval $[-\pi, \pi]$. Show that the normal equations of part (b) result.

(d) Let $p_3(x)$ by the value of the minimizing Φ. Plot $f(x)$ and $p_3(x)$ on the same set of axes.

(e) Compute the value of $\int_{-\pi}^{\pi} (f - p_3)^2 \, dx$.

(f) Show that $\int_{-\pi}^{\pi} p_3^2(x) \, dx = \frac{\pi}{2} a_0^2 + \pi \sum_{k=1}^{3} (a_k^2 + b_k^2) \le \int_{-\pi}^{\pi} f^2(x) \, dx$.

(g) Determine the seven coefficients $\{A, u_k, v_k\}$ by collocation. Pick seven points in the interval $(-\pi, \pi)$ and at each point x_k

require $f(x_k) = \Phi(x_k)$ to hold. Solve the resulting seven equations for $\{A, u_k, v_k\}$ and call the ensuing approximation $P_3(x)$.

(h) Plot $f(x)$, $p_3(x)$, and $P_3(x)$ all on the same set of axes.

(i) Compute the value of $\int_{-\pi}^{\pi} (f - P_3)^2 \, dx$ and show that it is larger than the value computed in part (e).

B23. $f(x) = \pi + x$ **B24.** $f(x) = e^x$ **B25.** $f(x) = x(1 - x)$
B26. $f(x) = x^3 - 7x - 6$

B27. Show that the basis functions $1, x, x^2, \ldots, x^5$ are not *orthogonal* on the interval $[0, 1]$. (See Exercises A1–6 for the notion of orthogonality.)

For the functions in Exercises B28–37:

(a) On the interval $[0, 1]$, obtain $p_5 = \sum_{k=1}^{5} b_k \sin k\pi x$, the fifth partial sum of the Fourier sine series, which the preceding exercises show to be the minimizer of the mean-square norm of the difference between $f(x)$ and $p_5(x)$.

(b) Graph $f(x)$ and $p_5(x)$ on the same set of axes.

(c) Obtain the value of $\int_0^1 (f - p_5)^2 \, dx$.

(d) Let $t_2(x) = \sum_{k=0}^{2} \beta_k x^k$, and then determine the coefficients in $t_2(x)$ by minimizing $\int_0^1 (f - t_2)^2 \, dx$ (obtain and solve the normal equations).

(e) Let $t_3(x) = \sum_{k=0}^{3} \beta_k x^k$, and then determine the coefficients in $t_3(x)$ by minimizing $\int_0^1 (f - t_3)^2 \, dx$.

(f) Let $t_4(x) = \sum_{k=0}^{4} \beta_k x^k$, and then determine the coefficients in $t_4(x)$ by minimizing $\int_0^1 (f - t_4)^2 \, dx$.

(g) Let $t_5(x) = \sum_{k=0}^{5} \beta_k x^k$, and then determine the coefficients in $t_5(x)$ by minimizing $\int_0^1 (f - t_5)^2 \, dx$.

(h) Observe that the normal equations in parts (d)–(g) are *coupled* and that adding a term to the approximating polynomial causes the preceding coefficients to change.

(i) Compare the graphs of the polynomials $t_k(x)$ with the graph of $f(x)$.

(j) Obtain the values of $\int_0^1 (f - t_k)^2 \, dx$, $k = 2, \ldots, 5$.

B28. $f(x) = x \sin \pi x$ **B29.** $f(x) = x \cos \pi x$
B30. $f(x) = e^x \sin \pi x$ **B31.** $f(x) = e^{-x} \cos \pi x$
B32. $f(x) = x e^{-x}$

B33. $f(x) = \begin{cases} 1 & 0 \le x < \frac{1}{2} \\ -1 & \frac{1}{2} \le x \le 1 \end{cases}$ **B34.** $f(x) = \begin{cases} x & 0 \le x < \frac{1}{2} \\ \frac{1}{2} & \frac{1}{2} \le x \le 1 \end{cases}$

B35. $f(x) = \begin{cases} x & 0 \le x < \frac{1}{2} \\ 1 - x & \frac{1}{2} \le x \le 1 \end{cases}$ **B36.** $f(x) = x(1 - x)$

B37. $f(x) = \begin{cases} 3x & 0 \le x < \frac{1}{3} \\ 1 & \frac{1}{3} \le x < \frac{2}{3} \\ 3(1 - x) & \frac{2}{3} \le x \le 1 \end{cases}$

B38. Use the following implementation of the *Gram–Schmidt orthogonalization process* to orthogonalize $1, x, x^2, \ldots, x^5$, thereby forming the set of polynomials $1, v_k(x), k = 1, \ldots, 5$, which, on the interval $[0, 1]$, are *orthogonal* in the sense that $\int_0^1 v_k(x)v_j(x)\,dx = 0$ when $k \neq j$. Take $v_0 = 1$ and $v_1 = 1 - \lambda_{11}x$. Determine λ_{11} by forcing v_1 to be orthogonal to its predecessor, $v_0 = 1$. Having determined v_1, write $v_2 = 1 - \lambda_{21}x - \lambda_{22}x^2$ and determine λ_{21} and λ_{22} by forcing v_2 to be orthogonal to its predecessors, $v_0 = 1$ and v_1. Having determined v_2, write $v_3 = 1 - \lambda_{31}x - \lambda_{32}x^2 - \lambda_{33}x^3$ and determine $\lambda_{3k}, k = 1, 2, 3$, by forcing v_3 to be orthogonal to its predecessors v_0, v_1, and v_2. Continue in this way until v_4 and v_5 have been determined.

For the functions in Exercises B39–48, find the best approximation in terms of the minimum mean-square norm, using the orthogonal basis polynomials $1, v_k(x), k = 1, \ldots, 5$, found in Exercise B38. Do this by

forming and solving the normal equations. Show that the normal equations are now uncoupled and that the resulting polynomial is still $t_5(x)$ as found in the earlier exercise with that function. Hence, orthogonality makes computing the best approximation about as easy as computing partial sums of Fourier series.

B39. $f(x)$ from Exercise B28 **B40.** $f(x)$ from Exercise B29
B41. $f(x)$ from Exercise B30 **B42.** $f(x)$ from Exercise B31
B43. $f(x)$ from Exercise B32 **B44.** $f(x)$ from Exercise B33
B45. $f(x)$ from Exercise B34 **B46.** $f(x)$ from Exercise B35
B47. $f(x)$ from Exercise B36 **B48.** $f(x)$ from Exercise B37

10.7 Fourier–Legendre Series

Orthogonality and Minimization

We wish to show that, in general, orthogonal functions lead to series with the minimization property of Fourier series. To do this, we carry out the following experiment in an arena less familiar than that of the trig functions [58].

On the interval $[-1, 1]$, consider the set of functions $p_k(x), k = 0, 1, \ldots$, with the two properties:

$$\int_{-1}^1 p_n(x)p_m(x)\,dx = 0, n \neq m \quad \text{and} \quad \int_{-1}^1 p_k^2(x)\,dx = \frac{2}{k+1} \quad (10.10)$$

The property on the left in (10.10) is orthogonality of the functions on the interval $[-1, 1]$, and the one on the right is the analog of $\int_0^\pi \sin^2 kx\,dx = \frac{\pi}{2}$, already used in (10.9) when developing the Fourier series. Are these two properties enough to reproduce the minimization property of the Fourier series?

The measure of performance, $Q = \int_{-1}^1 (f(x) - \sum_{k=0}^N s_k p_k(x))^2\,dx$, is the same as we used for the Fourier series. Again forming the normal equations

$$0 = \frac{1}{2}\frac{\partial Q}{\partial s_j} = \int_{-1}^1 \left(f(x) - \sum_{k=0}^N s_k p_k(x)\right) p_j(x)\,dx$$

we find from (10.10) that

$$\int_{-1}^1 f(x)p_j(x)\,dx = s_j \int_{-1}^1 p_j(x)^2\,dx = s_j\frac{2}{2j+1}$$

Hence, $s_j = \frac{2j+1}{2}\int_{-1}^1 f(x)p_j(x)\,dx$.

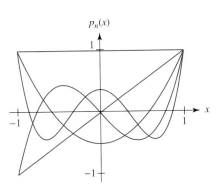

FIGURE 10.30 The first five Legendre polynomials

LEGENDRE POLYNOMIALS Do such functions exist? Are there actually functions that satisfy (10.10)? Yes, the Legendre polynomials $p_n(x)$, the first five of which, graphed in Figure 10.30, are $1, x, \frac{3}{2}x^2 - \frac{1}{2}, \frac{5}{2}x^3 - \frac{3}{2}x$, and $\frac{35}{8}x^4 - \frac{15}{4}x^2 + \frac{3}{8}$. As seen from the graph, $p_n(1) = 1$ for each Legendre polynomial $p_n(x)$. This is not the only normalization found in the literature, and the reader is advised to read any new text carefully.

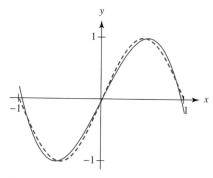

FIGURE 10.31   Fourier–Legendre approximation (solid) for $\sin \pi x$ (dashed)

Next, consider the function $f(x) = \sin \pi x$ on the interval $[-1, 1]$, and compute the coefficients

$$s_k = \frac{2k+1}{2} \int_{-1}^{1} \sin(\pi x) p_k(x)\, dx \qquad k = 0, \ldots, 4$$

obtaining $s_0 = s_2 = s_4 = 0$, $s_1 = \frac{3}{\pi}$, and $s_3 = 7\frac{\pi^2 - 15}{\pi^3}$, so that

$$g(x) = \frac{3}{\pi} x + \frac{7}{\pi^3}(\pi^2 - 15)\left(\frac{5}{2}x^3 - \frac{3}{2}x\right)$$

plotted as the solid curve in Figure 10.31, is a Fourier–Legendre approximation to $f(x)$, plotted as the dashed curve. The approximation $g(x)$ is a partial sum of the full Fourier–Legendre series $f(x) = \sum_{k=0}^{\infty} s_k p_k(x)$, where $s_k = \frac{2k+1}{2} \int_{-1}^{1} f(x) p_k(x)\, dx$. Likewise, $f(x)$ can be represented as infinite series in other orthogonal families such as the Hermite, Laguerre, or Chebyshev polynomials, developed in the exercises.

EXERCISES 10.7–Part A

A1. Obtain s_k, $k = 0, 1, 2$, the first three coefficients in the Fourier–Legendre series for $f(x) = e^{-x}$, $-1 \le x \le 1$.

A2. Obtain s_k, $k = 0, 1, 2$, the first three coefficients in the Fourier–Legendre series for $f(x) = \frac{1}{x+2}$, $-1 \le x \le 1$.

A3. Show that the functions $\{1, x, x^2\}$ are not mutually orthogonal on $[0, 1]$. Determine a, b, c so that, in the mean-square sense, $g(x) = a + bx + cx^2$ best approximates $f(x) = \frac{1}{1+x}$. How do the calculations differ from approximating with orthogonal functions?

EXERCISES 10.7–Part B

B1. Obtain s_k, $k = 0, 1, \ldots, 10$, the first eleven coefficients in the Fourier–Legendre series for $f(x) = x \cos \pi x$, $-1 \le x \le 1$. Show $\sum_{k=0}^{10} 2s_k^2/(2k+1) \le \int_{-1}^{1} f^2\, dx$.

B2. Account for the inequality in Exercise B1 by considering $\int_{-1}^{1} \left(\sum_{k=0}^{10} s_k p_k(x)\right)^2 dx$.

For the functions $f(x)$, $-1 \le x \le 1$, given in Exercises B3–14:

  (a) Obtain s_k, $k = 0, \ldots, 5$, the first six coefficients of the Fourier–Legendre series.

  (b) Let $\Phi(x) = \sum_{k=0}^{5} \sigma_k p_k(x)$, obtain the normal equations by minimizing $\int_{-1}^{1}(f - \Phi)^2\, dx$, and show that the solution of these equations is the same set of coefficients as determined in part (a).

  (c) Write the equation $f(x) = \Phi(x)$ and multiply, in turn, by $p_0, p_1, \ldots$, each time integrating with respect to x on the interval $[-1, 1]$. Show that the same six normal equations result.

  (d) Let $f_5(x)$ be the minimizing $\Phi(x)$, and plot $f(x)$ and $f_5(x)$ on the same set of axes.

  (e) Compute the value of $\int_{-1}^{1}(f - f_5)^2\, dx$.

  (f) Determine the six coefficients s_k by *collocation*. Pick six points in the interval $[-1, 1]$ and at each point x_k require $f(x_k) = \Phi(x_k)$ to hold. Solve the resulting six equations for $s_k = S_k$, and call the ensuing approximation $F_5(x)$.

  (g) Plot $f(x)$, $f_5(x)$, and $F_5(x)$ all on the same set of axes.

  (h) Compute the value of $\int_{-1}^{1}(f - F_5)^2\, dx$ and show that it is larger than the value computed in part (e).

B3. $f(x) = \sin \pi x$ **B4.** $f(x) = \cos \pi x$ **B5.** $f(x) = e^x$

B6. $f(x) = xe^{-x}$ **B7.** $f(x) = x \sin \pi x$ **B8.** $f(x) = x \cos \pi x$

B9. $f(x) = \begin{cases} 1 & -1 \le x < 0 \\ -1 & 0 \le x \le 1 \end{cases}$ **B10.** $f(x) = \begin{cases} 1 & -1 \le x < 0 \\ x & 0 \le x \le 1 \end{cases}$

B11. $f(x) = |x|$ **B12.** $f(x) = 1 - |x|$

B13. $f(x) = |\sin \pi x|$ **B14.** $f(x) = |\cos \pi x|$

B15. The Gram–Schmidt orthogonalization process was described in Exercise B38, Section 10.6. On the interval $[-1, 1]$, apply this process to x^k, $k = 0, 1, \ldots, 5$. Normalize the resulting polynomials so that they pass through $(1, 1)$, thereby making them the Legendre polynomials defined in Section 10.7.

B16. In Exercise B21, Section 8.5, the Chebyshev polynomials were defined by $T_0 = 1$, $T_1 = x$, $T_k = 2xT_{k-1} - T_{k-2}$, a definition under which $T_k(1) = 1$. In addition, on the interval $[-1, 1]$, the Chebyshev polynomials were shown to be orthogonal with respect to the weight function $w = \frac{1}{\sqrt{1-x^2}}$. Take 1 and x as the first two orthogonal polynomials, and show that using the *inner product* $\langle f, g \rangle_T = \int \frac{f(x)g(x)}{\sqrt{1-x^2}}\, dx$ and the Gram–Schmidt process applied to x^k, $k = 2, 3, \ldots$, the Chebyshev polynomials result if the polynomials are normalized to pass through the point $(1, 1)$.

For the functions $f(x)$, $-1 \le x \le 1$, given in Exercises B17–26:

(a) Obtain $f_3 = \frac{a_0}{2} + \sum_{k=1}^{3} a_k T_k(x)$, $a_k = \frac{2}{\pi} \langle f, T_k \rangle_T$, the third-degree partial sum of the Fourier–Chebyshev series. Don't be surprised if the integrations have to be carried out numerically. (Your favorite computer algebra system should have the Chebyshev polynomials built-in. Maple has them as $T(k, x)$ in its *orthopoly* package.)

(b) Obtain $\int_{-1}^{1} w(x)(f(x) - f_3(x))^2 \, dx$, the square of the minimum weighted distance between $f(x)$ and $f_3(x)$, where the least-squares norm is predicated on the inner product $\langle f, f_3 \rangle_T$.

(c) Write $\Phi = \sum_{k=0}^{3} s_k T_k(x)$ and determine the four coefficients s_k by *collocation*. Pick four points in the interval $[-1, 1]$ and at each point x_k require $f(x_k) = \Phi(x_k)$ to hold. Solve the resulting four equations for $s_k = S_k$, and call the ensuing approximation $F_3(x)$.

(d) Compute $\int_{-1}^{1} w(x)(f(x) - F_3(x))^2 \, dx$, which should be larger than the value computed in part (b).

(e) Plot $f(x)$, f_3, and F_3 all on the same set of axes.

B17. $f(x) = x \sin \pi x$ **B18.** $f(x) = x \cos \pi x$

B19. $f(x) = \arctan x$ **B20.** $f(x) = \dfrac{e^x}{1 + x^2}$

B21. $f(x) = |\sin \pi x|$ **B22.** $f(x) = 1 - |x|$

B23. $f(x) = e^{-|x|}$ **B24.** $f(x) = |x(1 - x)|$

B25. $f(x) = \dfrac{2x + 1}{x^2 + 2x + 2}$ **B26.** $f(x) = \sqrt{2 + x}$

B27. The Laguerre polynomials $L_0 = 1$, $L_1 = 1 - x$, $L_k = \frac{2k-1-x}{k} L_{k-1} - \frac{k-1}{k} L_{k-2}$, $k = 2, 3, \ldots$, are orthogonal on $[0, \infty)$ under the *inner product* $\langle f, g \rangle_L = \int_0^\infty e^{-x} f(x) g(x) \, dx$. Using this inner product, apply the Gram–Schmidt process to x^k, $k = 0, 1, \ldots$, to obtain the Laguerre polynomials. (Your favorite computer algebra system should have these polynomials built-in. Maple has them in the *orthopoly* package under the name $L(k, x)$.)

B28. Show that the Laguerre polynomials satisfy
$$\int_0^\infty e^{-x} L_k(x) L_j(x) \, dx = \begin{cases} 1 & j = k \\ 0 & j \neq k \end{cases}$$

For the functions $f(x)$, $0 \le x < \infty$, in Exercises B29–38:

(a) Obtain $f_5 = \sum_{k=0}^{5} a_k L_k(x)$, $a_k = \langle f, L_k \rangle_L$, the fifth-degree partial sum of the Fourier–Laguerre series.

(b) Obtain $\int_0^\infty e^{-x}(f(x) - f_5(x))^2 \, dx$, the square of the minimum weighted distance between $f(x)$ and $f_5(x)$, where the least-squares norm is predicated on the inner product $\langle f, f_5 \rangle_L$.

(c) If it exists, obtain F_5, the fifth-degree Taylor polynomial at $x = 0$, and compute $\int_0^\infty e^{-x}(f(x) - F_5(x))^2 \, dx$. It should be larger than the value found in part (b).

(d) Plot $f(x)$, $f_5(x)$, and $F_5(x)$ on the same set of axes.

(e) Explain why approximating with polynomials over an infinite interval *must* yield those results such as those seen in the graphs of part (d).

B29. $f(x) = \sqrt{x + 1}$ **B30.** $f(x) = \sin x$ **B31.** $f(x) = e^{-x}$

B32. $f(x) = xe^{-x}$ **B33.** $f(x) = x \sin x$

B34. $f(x) = 1 - H(x - 1)$ **B35.** $f(x) = (1 - H(x - 1)) \sin x$

B36. $f(x) = (1 - H(x - 1))e^x$ **B37.** $f(x) = (1 - H(x - 1))x^2$

B38. $f(x) = (1 - H(x - 1))(1 - x^2)$

B39. The Hermite polynomials $H_k(x)$ are orthogonal on $(-\infty, \infty)$ under the inner product $\langle f, g \rangle_H = \int_{-\infty}^{\infty} e^{-x^2} f(x) g(x) \, dx$, and satisfy the normalization condition $\int_{-\infty}^{\infty} e^{-x^2} H_k^2(x) \, dx = \sqrt{\pi} 2^k k!$. If $H_0(x) = 1$, obtain $H_1(x)$ and $H_2(x)$. *Hint:* To get H_2 from H_0 and H_1, set $p(x) = a + bx + cx^2$ and solve the equations $\langle H_0, p \rangle_H = \langle H_1, p \rangle_H = 0$, $\langle p, p \rangle_H = \sqrt{\pi} 2^2 2!$.

Chapter Review

1. What is a piecewise continuous function? Give an example of a piecewise continuous function and an example of a function that is not piecewise continuous.

2. Sketch the periodic extension of $f(x) = (x - 1)(3 - x)$, $1 \le x \le 2$. To what does the Fourier series of $f(x)$ converge? Does it converge to the periodic extension at every point?

3. State the formulas by which a Fourier series representing $f(x)$ on $[\alpha, \beta]$ can be constructed. To what does the Fourier series converge on the interval $[\alpha, \beta]$? To what does the Fourier series converge outside $[\alpha, \beta]$?

4. Obtain the Fourier series for $f(x) = x - 1$, $1 \le x \le 2$. Sketch the function to which the Fourier series converges for $0 \le x \le 3$. To what values does the Fourier series converge at $x = 1$ and at $x = 2$?

5. State the formulas by which a Fourier series representing $f(x)$ on $[-L, L]$ can be constructed. To what does the Fourier series converge on the interval $[-3L, 3L]$?

6. Obtain the Fourier series for $f(x) = (x + 1)(2 - x)$, $-1 \le x \le 1$. Sketch the function to which the Fourier series converges in $[-3, 3]$. Sketch the periodic extension of $f(x)$. Is this the same as the limit of the Fourier series?

7. Sketch the odd periodic extension of $f(x) = x, 0 \leq x \leq 2$. Obtain a Fourier sine series for $f(x)$, and on $[-6, 6]$ sketch the function to which it converges. Is this the odd periodic extension?

8. Sketch the even periodic extension of $f(x) = x, 0 \leq x \leq 2$. Obtain a Fourier cosine series for $f(x)$, and on $[-6, 6]$ sketch the function to which it converges. Is this the even periodic extension?

9. Use an example to verify that for a continuous function $f(x)$, $-2 \leq x \leq 2$, the termwise integral of its Fourier series is the Fourier series of $F(x) = \int_{-2}^{x} f(t) \, dt$.

10. Use an example to verify that for a smooth function $f(x)$, $-2 \leq x \leq 2$, the termwise derivative of its Fourier series is the Fourier series of $f'(x)$.

11. Describe the optimizing property of the Fourier series. What do the Fourier coefficients optimize?

12. What is the role of orthogonality in the optimization property of the Fourier series?

13. Obtain the Fourier–Legendre series for $f(x) = \sqrt{x+2}$, $-1 \leq x \leq 1$.

Chapter *11*

Asymptotic Series

INTRODUCTION   Series that do not converge are said to *diverge*. Some divergent series, called *asymptotic series,* can actually be used to approximate solutions of differential equations. Instead of adding terms until the error of the approximation decreases, the number of terms is fixed and a region for which the approximation is useful is determined. Clearly, this is an unusual representation for a function, and the examples in Section 11.1 are designed to take the mystery out of this concept. Section 11.2 makes precise the notion of the asymptotic series, and the optional Section 11.3 provides conditions under which termwise manipulations of asymptotic series are valid.

11.1 Computing with Divergent Series

Creating a Divergent Series

The point $x = 0$ is a *singular point* of the differential equation

$$x^2 y'' + (3x + 1)y' + y = 0 \tag{11.1}$$

because $y''(0)$ cannot be determined from (11.1) itself. However, for $x > 0$, (11.1) has $F(x) = \int_0^\infty \frac{e^{-t}}{1+xt}\, dt$ as a solution. (See Exercise A1.) The integral defining $F(x)$ cannot be evaluated in closed form in terms of elementary functions but can be evaluated numerically so that, for example, $F(0.2) = 0.852$. A graph of $F(x)$ is given in Figure 11.1.

Expanding the factor $\frac{1}{1+xt}$ in a power series about $t = 0$ and integrating termwise, we have the *formal* power series $F(x) = \int_0^\infty e^{-t} \sum_{k=0}^\infty (-t)^k x^k \, dt = \sum_{k=0}^\infty k!(-x)^k$, which, unfortunately, *diverges everywhere,* except at $x = 0$. Indeed, using the ratio test, form the ratio of two successive terms, obtaining $|(n + 1)x| < 1$ and $|x| < \frac{1}{n+1}$, from which the radius of convergence is seen to be $R = 0$.

Shortly, we will show that information about $F(x)$ can be obtained from this divergent series!

INTEGRATION BY PARTS   Sometimes, integration by parts can be used to create the divergent series. Thus, interpret $F(x)$ as $\int u\, dv$, choose u as the factor multiplying e^{-t}, and integrate by parts to obtain F_1 in Table 11.1. Again choosing u as the factor multiplying e^{-t} and integrating by parts, obtain F_2. One last repetition yields F_3. Not only does this process provide the terms of the divergent series, it also yields a useful remainder term. Thus,

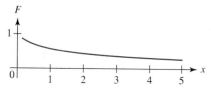

FIGURE 11.1   Graph of
$F(x) = \int_0^\infty \frac{e^{-t}}{1+xt} dt$

$$F_1 = 1 - \int_0^\infty \frac{xe^{-t}}{(1+xt)^2} dt$$

$$F_2 = 1 - x + \int_0^\infty \frac{2x^2 e^{-t}}{(1+xt)^3} dt$$

$$F_3 = 1 - x + 2x^2 - \int_0^\infty \frac{6x^3 e^{-t}}{(1+xt)^4} dt$$

TABLE 11.1   Repeated integration by parts
applied to $F(x) = \int_0^\infty \frac{e^{-t}}{1+xt} dt$

generalizing from these few cases, we write

$$F(x) = \sum_{k=0}^{n} k!(-x)^k + (n+1)! \int_0^\infty \frac{(-x)^{n+1} e^{-t}}{(1+xt)^{n+2}} dt$$

Computing with a Divergent Series

Surprisingly, it is possible to use such divergent series computationally. In essence, for each value of x, there is an optimal number of terms of this series for which the corresponding partial sum gives a best approximation to the function value $F(x)$. To see this, express $F(x)$ as $\frac{1}{x} \mathrm{Ei}(1, \frac{1}{x}) e^{1/x}$, where the exponential integral $\mathrm{Ei}(1, z) = \int_z^\infty e^{-t}/t \, dt$, while not elementary, is tabulated and available in computer programs such as Maple. Then, define the nth partial sum as $S_n(x) = \sum_{k=0}^{n} k!(-x)^k$ so that the error at any x and any number of terms n is defined by the difference $Error = |F(x) - \sum_{k=0}^{n} k!(-x)^k|$. A graph of the error as a function of both x and n is obtained as the surface plot shown in Figure 11.2.

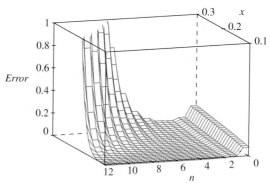

FIGURE 11.2
Graph of $Error = \left| F(x) - \sum_{k=0}^{n} k!(-x)^k \right|$

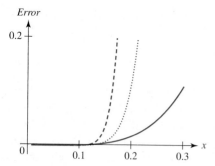

FIGURE 11.3   Partial sums $S_{12}(x)$ (solid), $S_9(x)$ (dotted), and $S_4(x)$ (dashed)

From this error surface it is possible to infer that for fixed n the error goes to zero as x goes to zero. Specifically, consider the curves in Figure 11.3. The solid line is the plot of the partial sum $S_{12}(x)$, the dotted line is the plot of the partial sum $S_9(x)$, and the dashed line is the plot of the partial sum $S_4(x)$. In each case, the error goes to zero as x goes to zero.

In addition, the surface plot of the error shows that for fixed x there is an optimal number of terms at which the error is a minimum. For example, at $x = 0.1$, we have the error as a function of the number of terms as shown in Figure 11.4. Taking three other plane sections corresponding to $x = 0.25, 0.2$, and 0.15, we have the three curves shown in Figure 11.5. In each case, we see a minimum error for some number of terms, n. In regions where a small number of terms yields accurate approximations, such divergent series are extremely useful representations of functions in applied mathematics.

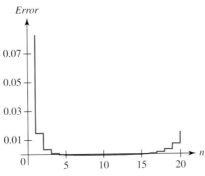

FIGURE 11.4   At $x = 0.1$, *Error* as a function of n

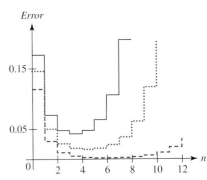

FIGURE 11.5   *Error* as a function of n for $x = 0.25, 0.2, 0.15$, solid, dotted, dashed, respectively

Standard references such as [1] give divergent series representations for many of the special functions of applied mathematics. These expressions are useful for determining the behavior of these functions in certain regions and are even used in computer library programs that evaluate these functions numerically.

EXERCISES 11.1–Part A

A1. By differentiating under the integral sign, show formally that $F(x) = \int_0^\infty \frac{e^{-t}}{1+xt} \, dt$ satisfies (11.1).

A2. Show formally that the series $\sum_{k=0}^\infty k!(-x)^k$ satisfies (11.1).

A3. Integrate

$$k! \int_0^\infty \frac{(-x)^k e^{-t}}{(1+xt)^{k+1}} \, dt$$

by parts once to generate the next term for the series given in Exercise A2.

EXERCISES 11.1–Part B

B1. The *error function* of applied mathematics is given by $\operatorname{erf}(x) = \frac{2}{\sqrt{\pi}} \int_0^x e^{-t^2} \, dt$.

   (a) Show that $\lim_{x \to \infty} \operatorname{erf}(x) = 1$.

   (b) Expand e^{-t^2} in a Maclaurin series and integrate termwise, obtaining a Maclaurin series for $\operatorname{erf}(x)$.

   (c) Graph $\operatorname{erf}(x)$.

B2. The Laplace transform of $f(t) = \frac{1}{\sqrt{1+t}}$ is $F(s) = \int_0^\infty e^{-st}/\sqrt{1+t} \, dt$.

   (a) Obtain $F(s) = e^s \sqrt{\frac{\pi}{s}}(1 - \operatorname{erf}(s))$, even if only by using a computer algebra system. (An alternative is to consult an extensive table of Laplace transforms such as found in [1].)

   (b) Expand $f(t)$ in a Maclaurin series and integrate termwise, obtaining

$$\sum_{k=0}^\infty \frac{(-1)^k (2k)!}{4^k k! s^{k+1}}$$

   (c) Use the ratio test to show that the series in part (b) diverges for $s \neq 0$.

   (d) Use repeated integration by parts to obtain this same divergent series but in the form

$$\sum_{k=0}^{N-1} \frac{(-1)^k (2k)!}{4^k k! s^{k+1}} + \frac{(-1)^N (2N)!}{4^N N! s^N} \int_0^\infty \frac{e^{-st}}{(1+t)^{N+1/2}} \, dt$$

(e) By simply creating a table of values, find the optimum number terms in the divergent series in part (d) that yields the smallest error at $s = 1$. What is that smallest error?

(f) Repeat part (e) for $s = 2, \ldots, 5$.

B3. The Laplace transform of $f(t) = \frac{1}{\sqrt{1+t^2}}$ is $F(s) = \int_0^\infty e^{-st}/\sqrt{1+t^2}\, dt$.

(a) Expand $f(t)$ in a Maclaurin series and integrate termwise, obtaining

$$\sum_{k=0}^\infty \frac{(-1)^k[(2k)!]^2}{4^k[k!]^2 s^{2k+1}}$$

(b) Use the ratio test to show that the series in part (b) diverges for $s \neq 0$.

(c) Use repeated integration by parts to obtain the first few terms of this divergent series. The remainder term becomes complicated after just a few terms.

(d) By simply creating a table of values, find the optimum number terms in the divergent series in part (c) that yields the smallest error at $s = 1$. What is that smallest error?

(e) Repeat part (e) for $s = 5, 10$, and 15.

B4. The integral $F(x) = \int_0^\infty e^{-xt}/\sqrt{x+t}\, dt$ is not a Laplace transform.

(a) Verify that $F(x) = e^{x^2}\sqrt{\frac{\pi}{x}}(1 - \mathrm{erf}(x))$. (You might find the result in a table of integrals, you might use a computer algebra system, or you might evaluate $F(x)$ numerically at a few points and compare to the result of numeric integration.)

(b) Expand $\frac{1}{\sqrt{x+t}}$ in a Maclaurin series and integrate termwise, obtaining

$$\sum_{k=0}^\infty \frac{(-1)^k(2k)!}{4^k k! x^{2k+3/2}}$$

(c) Use the ratio test to show the series in part (b) diverges for $x \neq 0$.

(d) Use repeated integration by parts to obtain the series in part (b) in the form

$$\sum_{k=0}^N \frac{(-1)^k(2k)!}{4^k k! x^{2k+3/2}} + \frac{(-1)^{N+1}(2N+2)!}{4^{N+1}(N+1)! x^{N+1}} \int_0^\infty \frac{e^{-xt}}{(x+t)^{N+1/2}}\, dt$$

(e) At $x = 1, 3$ and 5, find the optimum number of terms in the divergent series so as to best approximate $F(x)$.

(f) What is the error made in each case in part (e)?

B5. If $F(x) = \int_0^{\pi/4} e^{-x\sin t}\, dt$:

(a) Evaluate $F(1), F(5), F(10)$ by numerical integration.

(b) Make the change of variables $s = \sin t$ to obtain the integral $G(x) = \int_0^{1/\sqrt{2}} e^{-xs}/\sqrt{1-s^2}\, ds$.

(c) In the integrand for $G(x)$, expand $\frac{1}{\sqrt{1-s^2}}$ in a Maclaurin series and integrate termwise over $0 \leq s < \infty$, obtaining

$$\sum_{k=0}^\infty \frac{(2k)!]^2}{4^2[k!]^2 x^{2k+1}}$$

(While the integrand for $G(x)$ is not real for $s > 1$, Watson's lemma (Section 11.3) justifies the expansion and termwise integration!)

(d) Use the ratio test to show that the series in part (c) diverges for $x \neq 0$.

(e) Find the optimum number of terms that minimize the error when $F(5)$ is approximated by the divergent series. What is the error in the approximation?

B6. If $F(x) = \int_0^4 e^{-xt^2}t^2\sqrt{1+t}\, dt$, [17] suggests expanding $t^2\sqrt{1+t}$ in a Maclaurin series and integrating termwise on the interval $0 \leq t < \infty$.

(a) By integrating numerically, obtain accurate values for $F(x)$ at $x = 1.5, 2.5, 3.5$.

(b) The integral $\int_0^\infty e^{-xt}t^2\sqrt{1+t}\, dt$ is a Laplace transform with value

$$\frac{\sqrt{\pi}}{2}x^{-3/2}e^x(1 - \mathrm{erf}(\sqrt{x}))\left(1 - \frac{3}{x} + \frac{15}{4x^2}\right) - \frac{1}{2x^2} + \frac{15}{4x^3}$$

Obtain this transform, either using a computer algebra system or by looking in a table of transforms.

(c) Follow the suggestion of [17] to obtain

$$-\sum_{k=0}^\infty \frac{(-1)^k(2k)!(k+1)(k+2)}{k!4^k(2k-1)x^{k+3}}$$

(d) Use the ratio test to show the series in part (c) diverges.

(e) At each of $x = 1.5, 2.5$, and 3.5 find the optimum number of terms for the series in part (c) to best approximate $F(x)$.

(f) What is the actual error made for each of these approximations?

11.2 Definitions

EXAMPLE 11.1 Recall, from Section 11.1, the function $F(x) = \int_0^\infty e^{-t}/(1+xt)\, dt$ for which we computed the divergent series

$$\sum_{k=0}^\infty k!(-x)^k = \sum_{k=0}^n k!(-x)^k + (n+1)! \int_0^\infty \frac{(-x)^{n+1}e^{-t}}{(1+xt)^{n+2}}\, dt$$

We now make three observations to motivate key generalizations.

OBSERVATION 1   The magnitude of the remainder term can be estimated as

$$\left|(n+1)!\int_0^\infty \frac{(-x)^{n+1}e^{-t}}{(1+xt)^{n+2}}\,dt\right| \le (n+1)!x^{n+1}\int_0^\infty e^{-t}\,dt = (n+1)!x^{n+1}$$

since decreasing the denominator from $1+xt$ to 1 increases the value of the fraction in the integrand and then $\int_0^\infty e^{-t}\,dt = 1$. Consequently, $F(x)$ can be expressed as $\sum_{k=0}^n k!(-x)^k + R_n$, where $|R_n| \le (n+1)!x^{n+1}$. In fact, the remainder satisfies

$$\left|\frac{R_n}{x^{n+1}}\right| \le (n+1)! \quad \text{and} \quad \lim_{x\to 0}\frac{R_n}{x^n} = 0 \qquad (11.2)$$

The inequality on the left in (11.2) says that in a sufficiently small neighborhood of $x = 0$ the ratio of R_n to x^{n+1} is bounded by a constant, and the limit on the right in (11.2) says the remainder R_n goes to zero more quickly than x^n.

OBSERVATION 2   The series is a sum of increasing powers of x. For small values of x, each successive power of x is an order of magnitude smaller than its predecessor. Specifically, for each k we have $\lim_{x\to 0} x^{k+1}/x^k = 0$.

OBSERVATION 3   The coefficients a_n in the series $\sum_{k=0}^\infty a_k x^k$ can be computed by the recipe

$$a_n = \lim_{x\to 0}\frac{F - \sum_{k=0}^{n-1} a_k x^k}{x^n}$$

Computing the first few coefficients this way, we get $a_0 = \lim_{x\to 0} F(x) = 1, a_1 = \lim_{x\to 0}\frac{F-1}{x} = -1$, and $a_2 = \lim_{x\to 0}\frac{F-(1-x)}{x^2} = 2$. ❖

Bachman–Landau Order Symbols

THE "BIG OH" SYMBOL   A function $f(x)$ is said to be "big oh" of $g(x)$ in a region R, and we write $f(x) = O(g(x))$, provided $|f(x)|$ is bounded by a constant multiple of $|g(x)|$ in R, that is, the inequality $|f(x)| \le A|g(x)|$ holds in R, with A being a constant independent of x.

If $g(x) \ne 0$ in R, then we can more simply say that the ratio of $|f(x)|$ to $|g(x)|$ is bounded by some constant.

This definition makes precise our first observation about the remainder term for the function $F(x)$. For each n, $R_n = O(x^{n+1})$ because $|R_n| \le (n+1)!x^{n+1}$.

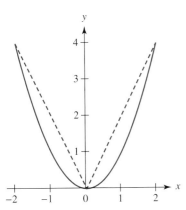

FIGURE 11.6   In Example 11.1, $2|x|$ (dashed) dominates $|x^2|$ (solid), on $|x| < 2$

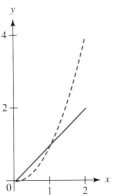

FIGURE 11.7   In Example 11.2, x^2 (dashed) dominates x (solid), for x large

EXAMPLE 11.2   The function $f(x) = x^2$ is $O(x)$ on $|x| < 2$ since $|x^2| \le 2|x|$ on that interval. Simply consider Figure 11.6, in which $2|x|$ (the dashed line) dominates $|x^2|$ (the solid line). ❖

EXAMPLE 11.3   The function $f(x) = x$ is $O(x^2)$ for x large enough. Indeed, Figure 11.7 shows, for $x > 1$, $|x| \le |x^2|$, that is, $|x|$ (solid) lies below $|x^2|$ (dashed). ❖

THE "LITTLE OH" SYMBOL   A function $f(x)$ is said to be $o(g(x))$, that is, "little oh" of $g(x)$ at $x = x_0$, provided $\lim_{x\to x_0}\left|\frac{f(x)}{g(x)}\right| = 0$.

This definition supposes that even if $g(x_0) = 0$, $g(x) \ne 0$ elsewhere in R, thereby simplifying the definition. Intuitively, if $f(x) = o(g(x))$, then, at $x = x_0$, $f(x)$ goes to zero "more rapidly" than $g(x)$.

In terms of this definition, we see that near $x = 0$, $x^{k+1} = o(x^k)$. Each power of x is smaller than its predecessor by a prescribed amount, namely, by an "order of magnitude." By Observation 2, namely, $\lim_{x\to 0} x^{k+1}/x^k = 0$, we can say the remainder in the series for $F(x)$ satisfies $R_n = o(x^n)$. In fact, this remainder satisfies both $R_n = O(x^{n+1})$ and $R_n = o(x^n)$.

Often, the context of the statement $f = o(g(x))$ provides ample indication of the point x_0 at which $f(x)$ is being compared to $g(x)$, or barring that, the statement is interpreted as being true at *some* specific x_0.

EXAMPLE 11.4   The function $f(t) = t^3$ is $o(e^t)$ as $t \to \infty$, since $\lim_{t\to\infty} |t|^3/e^t = 0$.   ❖

EXAMPLE 11.5   The function $f(x) = n!/x^{n+1}$ is $O(1/x^{n+1})$ and $o(1/x^n)$ as $x \to \infty$. The first claim is verified by writing $g(x) = 1/x^{n+1}$ and examining the ratio $\frac{f(x)}{g(x)} = n!$. Choosing A to be any constant greater than $n!$ means $|f(x)| \le A|g(x)|$. The second claim is valid since

$$\lim_{x\to\infty} \frac{f(x)}{1/x^n} = \lim_{x\to\infty} \frac{n!x^n}{x^{n+1}} = 0$$ ❖

Asymptotic Sequences

We have already noted that for small values of x and for each k we have $x^{k+1} = o(x^k)$, an observation generalized by the following definition.

> **DEFINITION 11.1**
>
> A sequence of functions $\{\phi_k(x)\}$ whose magnitudes decrease systematically according to the prescription $\phi_{k+1}(x) = o(\phi_k(x))$ is said to be an *asymptotic sequence*.

EXAMPLE 11.6   We have just seen that the sequence $\{x^k\}$ is an asymptotic sequence near $x = 0$. Additionally, $\{1/x^k\}$ is an asymptotic sequence for x large. Indeed, we have

$$\lim_{x\to\infty} \frac{1/x^{k+1}}{1/x^k} = \lim_{x\to\infty} \frac{x^k}{x^{k+1}} = 0$$ ❖

EXAMPLE 11.7   The functions $\phi_k(x) = (2 + \cos x^k)/x^k$ form an asymptotic sequence for large x because

$$\left|\frac{\phi_{k+1}}{\phi_k}\right| = \left|\frac{2 + \cos x^{k+1}}{2 + \cos x^k \|x\|}\right| \le \frac{3}{|x|} \to 0$$

The inequality follows upon maximizing the numerator and minimizing the denominator.   ❖

EXAMPLE 11.8   The functions $\phi_k(x) = (\sin kx)/x^k$ do *not* form an asymptotic sequence for x large because each fraction

$$\left|\frac{\phi_{k+1}}{\phi_k}\right| = \left|\frac{\sin(k+1)x}{x \sin kx}\right|$$

becomes unbounded infinitely often, at every zero of $\sin kx$.   ❖

Asymptotic Expansions

Once again consider the function $F(x)$ and its divergent series expansion $\sum_{k=0}^{n} k!(-1)^k x^k + R_n$, where we describe the remainder with the notation $R_n = o(x^n)$. Thus, we have

$$F(x) = \sum_{k=0}^{n} k!(-1)^k x^k + o(x^n) \qquad n = 0, 1, \ldots \tag{11.3}$$

as the meaning of the divergent series. In fact, we now define such an expansion as an *asymptotic (power) series* and write $F(x) \sim \sum_{k=0}^{\infty} k!(-1)^k x^k$ to mean the collection of equalities in (11.3). Hence, we have the following definitions.

ASYMPTOTIC SERIES   If

$$f(x) = \sum_{k=0}^{n} a_k x^k + o(x^n) \quad \text{or} \quad f(x) = \sum_{k=0}^{n} a_k x^k + O(x^{n+1}) \tag{11.4}$$

holds for each $n = 0, 1, \ldots$, then $\sum_{k=0}^{\infty} a_k x^k$ is called an *asymptotic series*, or *asymptotic power series*, for $f(x)$ and we write $f(x) \sim \sum_{k=0}^{\infty} a_k x^k$ to denote it.

The sets of equalities in (11.4) are equivalent.

GENERAL ASYMPTOTIC EXPANSION   If $\{\phi_k(x)\}$ is an asymptotic sequence for $x \to x_0$ and $f(x) = \sum_{k=0}^{n} a_k \phi_k(x) + o(\phi_n)$ holds for every $n = 0, 1, \ldots$, we call $\sum_{k=0}^{\infty} a_k \phi_k(x)$ an *asymptotic expansion* and write $f(x) \sim \sum_{k=0}^{\infty} a_k \phi_k(x)$ to denote it.

By virtue of Observation 3, we also have

$$a_n = \lim_{x \to x_0} \frac{f(x) - \sum_{k=0}^{n-1} a_k \phi_k(x)}{\phi_n(x)}$$

with $a_0 = \lim_{x \to x_0} \frac{f(x)}{\phi_0(x)}$.

Asymptotic Expansions to N Terms

Not every asymptotic expansion results in an infinite series. Some expansions terminate after a finite number of terms, in which case we make the following definition.

If $f(x) = \sum_{k=0}^{N-1} a_k \phi_k(x) + o(\phi_{N-1})$ but $f(x) \neq \sum_{k=0}^{N} a_k \phi_k(x) + o(\phi_N)$, then $f(x) \sim \sum_{k=0}^{N-1} a_k \phi_k(x)$ is an *asymptotic expansion to N terms*.

When the expansion at x_0 is to one term, namely, to $g(x)$, it is common to say "$f(x)$ is asymptotic to $g(x)$," denoted by $f(x) \sim g(x)$, and meaning $f(x) = g(x) + o(g(x))$ as $x \to x_0$, so that $\lim_{x \to x_0} \frac{f-g}{g} = 0$.

EXAMPLE 11.9   The asymptotic power series for the function

$$f(x) = \begin{cases} 0 & x \leq 0 \\ e^{-1/x} & x > 0 \end{cases}$$

is zero as $x \to 0$. Thus, we would write $f(x) \sim 0$ as $x \to 0$. Indeed, we have $\lim_{x \to 0} f = \lim_{x \to 0} \frac{f}{x} = \lim_{x \to 0} \frac{f}{x^2} = \cdots = 0$. Hence, with respect to the asymptotic sequence $\{x^k\}$, the asymptotic expansion of $f(x)$ is zero. Thus, two functions, $F(x)$ and $G(x) = F(x) + f(x)$, may have the same asymptotic expansion.   ❖

EXAMPLE 11.10   To expand the function $f(x) = \sqrt{1 - \sin x}$ in an asymptotic series with respect to the asymptotic sequence $\{\sin^k x\}$, $x \to 0$, first establish $\sin^{k+1} x = o(\sin^k x)$ via

$$\lim_{x \to 0} \frac{\sin^{k+1} x}{\sin^k x} = 0$$

Then, compute the coefficients of the asymptotic expansion by the recipe

$$a_n = \lim_{x \to x_0} \frac{f(x) - \sum_{k=0}^{n-1} a_k \phi_k(x)}{\phi_n(x)}$$

thereby obtaining the expansion $f(x) \sim 1 - \frac{1}{2} \sin x - \frac{1}{2^2 2!} \sin^2 x - \frac{3}{2^3 3!} \sin^3 x + \cdots$ as $x \to 0$.   ❖

EXAMPLE 11.11   Obtain the asymptotic expansion of $f(x) = \sqrt{1 - \sin x}$ with respect to the asymptotic sequence $\{\phi_k(x)\} = \{x^{2k}/\sqrt{1 + \sin x}\}$, $x \to 0$. First, show $\phi_{k+1} = o(\phi_k)$ by verifying $\lim_{x \to 0} \phi_{k+1}/\phi_k = \lim_{x \to 0} x^{2k+2}/x^{2k} = 0$. Then, compute the coefficients via the recipe

$$a_n = \lim_{x \to x_0} \frac{f(x) - \sum_{k=0}^{n-1} a_k \phi_k(x)}{\phi_n(x)}$$

thereby obtaining the expansion $f(x) \sim \sum_{k=0}^{\infty} (-1)^k x^{2k}/(2k)! \sqrt{1 + \sin x}$ as $x \to 0$. Of course, this expansion converges, and follows from the trigonometric identity $f(x) = \sqrt{1 - \sin x} = \frac{\cos x}{\sqrt{1 + \sin x}}$. The point being made here is that the asymptotic expansion of a given function is determined by the asymptotic sequence used.   ❖

EXAMPLE 11.12   This final example demonstrates that a finite asymptotic expansion actually can terminate. If we expand $F(x) = \int_0^\infty e^{-t}/(1 + xt)\, dt$ with respect to the asymptotic sequence $\{x^{2k}\}$, $x \to 0$, we find $a_0 = \lim_{x \to 0} F(x) = 1$ but $a_1 = \lim_{x \to 0} \frac{F(x)-1}{x^2}$ is undefined. Consequently, with respect to this asymptotic sequence, the most we can say about $F(x)$ is that $F(x) \sim 1$ to one term.   ❖

EXERCISES 11.2–Part A

A1. For each of the following, verify the indicated order relationship.

(a) $x^2 = O(x)$, $|x| < 2$ (b) $\sin x = O(1)$, $|x| < \infty$

(c) $\sin x = O(x)$, $|x| < \infty$ (d) $x^2 = O(\sin x)$, $x \to 0$

(e) $e^{-x} = O(1)$, $x \to \infty$ (f) $x = O(x^2)$, $x \to \infty$

(g) $\cos x = 1 + o(x)$, $x \to 0$

A2. Indicate whether each of the following is true or false, and give evidence for the determination.

(a) $e^{-x} = o(1)$, $x \to \infty$ (b) $e^{-x} = O(x)$, $x \to \infty$

(c) $\sinh x = O(x)$, $x \to 0$ (d) $\sinh x = o(x)$, $x \to 0$

(e) $\sinh x = O(e^x)$, $x \to \infty$ (f) $\sinh x = o(x)$, $x \to \infty$

(g) $\ln x = o(x)$, $x \to \infty$ (h) $\ln x = O(1)$, $x \to \infty$

A3. If $h = O(f)$ and $g = O(h)$, show that $g = O(f)$. Thus, $O(O(f)) = O(f)$.

A4. If $h = o(f)$ and $g = O(h)$, show that $g = o(f)$. Thus, $O(o(f)) = o(f)$.

A5. If $h = O(f)$ and $g = o(h)$, show that $g = o(f)$. Thus, $o(O(f)) = o(f)$.

A6. If $h = o(f)$ and $g = o(h)$, show that $g = o(f)$. Thus, $o(o(f)) = o(f)$.

A7. If $F = O(f)$ and $G = O(g)$, show that $FG = O(fg)$. Thus, $O(f)O(g) = O(fg)$.

A8. If $F = O(f)$ and $G = o(g)$, show that $FG = o(fg)$. Thus, $O(f)o(g) = o(fg)$.

A9. If $F = o(f)$ and $G = o(g)$, show that $FG = o(fg)$. Thus, $o(f)o(g) = o(fg)$.

A10. If $F_1 = O(f)$ and $F_2 = O(f)$, show that $F_1 + F_2 = O(f)$. Thus, $O(f) + O(f) = O(f)$.

A11. If $F_1 = O(f)$ and $F_2 = o(f)$, show that $F_1 + F_2 = O(f)$. Thus, $O(f) + o(f) = O(f)$.

A12. If $F_1 = o(f)$ and $F_2 = o(f)$, show that $F_1 + F_2 = o(f)$. Thus, $o(f) + o(f) = o(f)$.

A13. Show that $f \sim g \Rightarrow g \sim f$.

EXERCISES 11.2–Part B

B1. A function can have two different asymptotic expansions if two different asymptotic sequences are used.

(a) Show that $\frac{1+x}{1-x} \sim 1 + 2x + 2x^2 + \cdots$, $|x| < 1$.

(b) Show that $\sin^k x$, $k = 0, 1, \ldots$, is an asymptotic sequence for $x \to 0$.

(c) Show that $\frac{1+x}{1-x} \sim 1 + 2 \sin x + 2 \sin^2 x + \frac{7}{3} \sin^3 x + \cdots$, $|x| < 1$.

B2. Verify each of the following both analytically and graphically.

(a) $\sin x - x = O(x^3)$ as $x \to 0$ (b) $e^{\tan x} = O(1)$ as $x \to 0$

(c) $e^{\tan x} - 1 = O(x)$ as $x \to 0$

B3. Let $f(x) = \ln(1 + \sqrt{x}) \sim \sum_{k=1}^{\infty} a_k x^{k/2}$:

    (a) Show that $\{x^{k/2}\}$, $k = 1, 2, \ldots$, is an asymptotic sequence as $x \to 0$.

    (b) Obtain the coefficients

$$a_n = \lim_{x \to 0} \frac{f(x) - \sum_1^{n-1} a_k x^{k/2}}{x^{n/2}}$$

    (c) Show that the series actually converges and that the radius of convergence is $R = 1$.

B4. Show that $\{1, \ln(1 + x), \ln(1 + x^2), \ln(1 + x^3), \ldots, \}$ is an asymptotic sequence as $x \to 0$.

B5. Show that $(\tan x)^{k/2}$, $k = 1, 2, \ldots$, is an asymptotic sequence as $x \to 0$ from the right.

B6. Obtain an asymptotic expansion of e^x in the basis of Exercise B4.

B7. Obtain the asymptotic expansion of $f(x) = e^x - 1$ with respect to the basis of Exercise B5.

B8. If $f(x) = \sin(x + \sqrt{x})$:

    (a) Show that, with respect to the basis of Exercise B4, $f(x)$ has a terminating asymptotic expansion.

    (b) Obtain, if possible, an asymptotic expansion of $f(x)$ in the basis of Exercise B5.

B9. For each of the following, obtain the asymptotic power series as $x \to 0$. Test the validity and utility of the expansion by graphical and numerical techniques.

    (a) $\int_0^x e^{-t^2}\, dt$ **(b)** $e^{\sin x}$ **(c)** $e^{1/(1-x)}$

B10. For $f(x) = \dfrac{x^2}{\sqrt{x - x^2} + \sqrt{x}}$, obtain the asymptotic expansion with respect to the asymptotic sequence $\{x^{k/2}\}$, $k = 0, 1, \ldots$. On $[0, 1]$, graphically compare the expansion and $f(x)$.

11.3   Operations with Asymptotic Series

Overview

Theorems validating arithmetic with asymptotic power series parallel theorems validating arithmetic with convergent power series. Moreover, composition of asymptotic power series can be justified. Integration of asymptotic power series is likewise typically valid, but differentiation may not be. If the derivative is known to have an asymptotic expansion, then termwise differentiation yields the asymptotic expansion of the derivative.

    In the following sections, we give conditions under which various operations can be performed on asymptotic power series of the form $\sum a_k x^k$ and $\sum a_k x^{-k}$.

    Loosely speaking, the asymptotic expansion of a linear combination of two functions is the termwise linear combination of the terms of the asymptotic expansions of the two functions. Thus, we have the following theorem.

THEOREM 11.1

1. $\{\phi_k(x)\}$ is an asymptotic sequence as $x \to x_0$.

2. $f(x) \sim \sum_{k=0}^{\infty} a_k \phi_k(x)$ and $g(x) \sim \sum_{k=0}^{\infty} b_k \phi_k(x)$

$$\implies \alpha f(x) + \beta g(x) \sim \sum_{k=0}^{\infty} (\alpha a_k + \beta b_k) \phi_k(x)$$

EXAMPLE 11.13   Consider the functions

$$f(x) = \frac{1}{1 + x^2} \sim \frac{1}{x^2} - \frac{1}{x^4} + \frac{1}{x^6} - \frac{1}{x^8} + \cdots$$

$$g(x) = \frac{x}{2 - x^3} \sim -\frac{1}{x^2} - \frac{2}{x^5} - \frac{4}{x^8} + \cdots$$

(11.5)

The asymptotic expansion of $h(x) = 2f(x) + 3g(x)$ is $-\frac{1}{x^2} - \frac{2}{x^4} - \frac{6}{x^5} + \frac{2}{x^6} - \frac{14}{x^8} + \cdots$, which is just $2\left(\frac{1}{x^2} - \frac{1}{x^4} + \frac{1}{x^6} - \frac{1}{x^8} + \cdots\right) + 3\left(-\frac{1}{x^2} - \frac{2}{x^5} - \frac{4}{x^8} + \cdots\right)$.   ❖

As for multiplication of power series, multiplication of asymptotic power series is also valid. Thus, we have the following theorem.

THEOREM 11.2

1. Let the asymptotic sequence $\{w_k(x)\}$ be either $\{x^k\}$ or $\{x^{-k}\}$.

2. $f(x) \sim \displaystyle\sum_{k=0}^{\infty} a_k w_k(x)$ and $g(x) \sim \displaystyle\sum_{k=0}^{\infty} b_k w_k(x)$

3. $c_k = \displaystyle\sum_{n=0}^{k} a_n b_{k-n}$

$$\implies f(x)g(x) \sim \sum_{k=0}^{\infty} c_k w_k(x)$$

EXAMPLE 11.14   If $f(x)$ and $g(x)$ are given in (11.5), the asymptotic expansion of $h(x) = f(x)g(x)$ is

$$-\frac{1}{x^4} + \frac{1}{x^6} - \frac{2}{x^7} - \frac{1}{x^8} + \frac{2}{x^9} - \frac{3}{x^{10}} - \frac{2}{x^{11}} + \cdots$$

which matches, through terms of order 11, the product

$$\left(\frac{1}{x^2} - \frac{1}{x^4} + \frac{1}{x^6} - \frac{1}{x^8} + \cdots\right)\left(-\frac{1}{x^2} - \frac{2}{x^5} - \frac{4}{x^8} + \cdots\right)$$ ❖

THEOREM 11.3

1. $\{w_k\}$ is either $\{x^k\}$ or $\{x^{-k}\}$.

2. $f(x) \sim \displaystyle\sum_{k=0}^{\infty} c_k w_k$ with $c_0 \neq 0$

$$\implies \frac{1}{f(x)} \sim \sum_{k=0}^{\infty} d_k w_k \text{ for appropriate constants } d_k.$$

In each case, the restriction on c_0 prevents the quotient series from containing terms not in the asymptotic sequence for $f(x)$.

EXAMPLE 11.15   The reciprocal of neither series in (11.5) is an asymptotic power series for $x \to \infty$ since $\frac{1}{f(x)} = 1 + x^2$ and $\frac{1}{g(x)} = -x^2 + \frac{2}{x}$ introduce positive powers of x, terms not in the asymptotic basis for the original expansions.

On the other hand, at $x = 0$, we have

$$f(x) = 1 - x^2 + x^4 + O(x^6) \quad \text{and} \quad g(x) = \tfrac{1}{2}x + \tfrac{1}{4}x^4 + O(x^7)$$

Therefore, the reciprocal of $f(x)$ remains an asymptotic expansion with respect to the same asymptotic sequence, whereas the reciprocal of $g(x)$ does not.   ❖

Composition of asymptotic power series is not as well-behaved as it is for convergent power series. As for convergent power series, composition of asymptotic power series is valid at $x = 0$ under certain restrictions but is generally not valid for $x \to \infty$. A theorem governing the case at $x = 0$ reflects the condition imposed on convergent power series in Section 9.3.

THEOREM 11.4

1. $f(x) \sim \displaystyle\sum_{k=0}^{\infty} a_k x^k$ and $g(x) \sim \displaystyle\sum_{k=1}^{\infty} b_k x^k$

$$\implies f(g(x)) \sim \sum_{k=0}^{\infty} a_k \left(\sum_{n=1}^{k} b_n x^n \right)^k$$

This result is in complete accord with its analog for convergent power series wherein $g(x_0) = b_0 = y_0 = 0$.

Composition of series developed for $x \to \infty$ is not as well-behaved. Consider the following example.

EXAMPLE 11.16   For the composition $f(g(x))$ of the functions in (11.5), we have

$$f(g(x)) = \frac{(x^3 - 2)^2}{4 + x^2 - 4x^3 + x^6} \sim 1 - \frac{1}{x^4} - \frac{4}{x^7} + \frac{1}{x^8} + \cdots$$

On the other hand, composing the individual asymptotic series yields

$$\frac{1}{\left(-\frac{1}{x^2} - \frac{2}{x^5} - \frac{4}{x^8} \right)^2} - \frac{1}{\left(-\frac{1}{x^2} - \frac{2}{x^5} - \frac{4}{x^8} \right)^4} + \cdots$$

which does not even have a power series development! A moment's reflection reveals the difficulty. The series for $g(x)$ is a development for large values of x. In fact, $\lim_{x \to \infty} g(x) = 0$. The asymptotic power series for $f(x)$ is also a development for large x. But in the composite $f(g(x))$, when x is large, $g(x)$ is small, so $f(x)$ is not getting the large values of its argument that its series calls for.   ❖

Termwise integration of asymptotic series is generally valid. In fact, for sufficiently well-behaved functions $f(x)$ and $g(x)$, we have the following theorem.

THEOREM 11.5(a)

1. $f(x) \sim \displaystyle\sum_{k=0}^{\infty} a_k x^k$

$$\implies \int_{0}^{x} f(t) \, dt \sim \sum_{k=0}^{\infty} \frac{a_k}{k+1} x^{k+1}$$

THEOREM 11.5(b)

1. $g(x) \sim \displaystyle\sum_{k=2}^{\infty} b_k x^{-k}$

$$\implies \int_{x}^{\infty} g(t) \, dt \sim \sum_{k=2}^{\infty} \frac{b_k}{1-k} x^{1-k}$$

Each conclusion simply says the asymptotic expansion of the integral is obtained by the termwise integration of the expansion of the integrand.

EXAMPLE 11.17   Consider the function $f(x) = x \sin x$ and $\sum_{k=0}^{\infty} (-1)^k x^{2k+2}/(2k+1)!$, its asymptotic power series at $x = 0$. The asymptotic expansion of $\int_0^x f(t)\,dt = \sin x - x\cos x$ is

$$\sum_{k=0}^{\infty} \frac{(-1)^k x^{2k+3}}{(2k+3)(2k+1)!}$$

which is just the termwise integral of the expansion for $f(x)$.   ❖

EXAMPLE 11.18   Consider the function $g(x) = \frac{1}{x^2+1} \sim \frac{1}{x^2} - \frac{1}{x^4} + \frac{1}{x^6} + \cdots$ and its integral

$$G(x) = \int_x^{\infty} g(t)\,dt = \frac{1}{2}\pi - \arctan x \sim \frac{1}{x} - \frac{1}{3}\frac{1}{x^3} + \frac{1}{5}\frac{1}{x^5} + \cdots$$

Termwise integration of the expansion for $g(x)$ yields precisely the expansion for $G(x)$.   ❖

Differentiation of Asymptotic Power Series

Differentiation of asymptotic expansions, like composition, requires some care, since it is not always true that an asymptotic expansion can be differentiated. For example, the function $f(x) = e^{-1/x} \sin e^{1/x}$, graphed in Figure 11.8, is asymptotic to 0 since

$$\lim_{x\to 0^+} \frac{e^{-1/x} \sin e^{1/x}}{x^k} = 0$$

for each $k = 0, 1, 2, \ldots$. But the derivative $f'(x) = \frac{1}{x^2}(e^{-1/x} \sin e^{1/x} - \cos e^{1/x})$ has no asymptotic expansion at all since not even the limit of this derivative exists at the origin. However, if both $f(x)$ and $f'(x)$ have asymptotic power series expansions, then the expansion for $f'(x)$ is the termwise derivative of the expansion for $f(x)$.

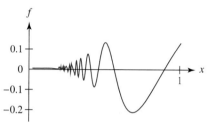

FIGURE 11.8   Graph of $f(x) = e^{-1/x} \sin e^{1/x}$ from Example 11.19

THEOREM 11.6(a)	THEOREM 11.6(b)
1. $f(x) \sim \displaystyle\sum_{k=0}^{\infty} a_k x^k$	**1.** $g(x) \sim \displaystyle\sum_{k=0}^{\infty} b_k x^{-k}$
2. $f'(x) \sim \displaystyle\sum_{k=0}^{\infty} c_k x^k$	**2.** $g'(x) \sim \displaystyle\sum_{k=2}^{\infty} d_k x^{-k}$
$\implies c_k = (k+1)a_{k+1}$	$\implies d_k = -(k-1)b_{k-1}$

EXAMPLE 11.19   For large x, consider

$$f(x) = \frac{x}{1+x} \sim 1 - \frac{1}{x} + \frac{1}{x^2} - \frac{1}{x^3} + \frac{1}{x^4} - \frac{1}{x^5} + \cdots$$

$$f'(x) = \frac{1}{(1+x)^2} \sim \frac{1}{x^2} - \frac{2}{x^3} + \frac{3}{x^4} - \frac{4}{x^5} + \cdots$$

Termwise differentiation of the series for $f(x)$ yields the asymptotic power series of the derivative.   ❖

WATSON'S LEMMA

1. $f(t) = \sum\limits_{k=0}^{\infty} a_k t^k$, with convergence in $|t| < R$.

2. $f(t) = O(e^{\alpha t}), \alpha > 0$, as $t \to \infty$, so that $f(t)$ is of exponential order.

$$\implies \int_0^{\infty} f(t) e^{-st}\, dt \sim \sum_{k=0}^{\infty} \frac{a_k}{s^{k+1}}$$

In essence, Watson's lemma says that the termwise Laplace transform of a series expansion of $f(t)$ is the asymptotic expansion of its Laplace transform.

EXAMPLE 11.20   Consider the function $f(t) = \frac{1}{1+t}$ whose nontrivial Laplace transform $F(s) = e^s \, \mathrm{Ei}(1,s)$ is given in terms of the tabulated exponential integral $\mathrm{Ei}(1,s) = \int_1^{\infty} e^{-xt}/t\, dt$. We then have

$$f(t) = \frac{1}{1+t} = \sum_{k=0}^{\infty} (-1)^k t^k \quad \text{and} \quad F(s) = \int_1^{\infty} \frac{e^{-xt}}{t}\, dt \sim \sum_{k=0}^{\infty} \frac{(-1)^k k!}{s^{k+1}}$$

The asymptotic series for $F(s)$ is the termwise transform of the Maclaurin series for $f(t)$. ❖

EXERCISES 11.3

1. In Exercise B2, Section 11.1, $F(s) = \int_0^{\infty} e^{-st}/\sqrt{1+t}\, dt$ is evaluated as a Laplace transform, from which an asymptotic series is developed for $s \to \infty$.

   (a) Termwise differentiate the series obtained for $F(s)$.

   (b) Evaluate $F'(s)$ by differentiating under the integral sign and recognizing another Laplace transform.

   (c) Obtain an asymptotic power series in the basis $\{s^{-k}\}$ for the function found in part (b).

   (d) Show that the series in part (a) is the same as the series in part (c). Hence, termwise differentiation of the asymptotic expansion of $F(s)$ is valid.

2. In Exercise B4, Section 11.1, $F(x) = \int_0^{\infty} e^{-xt}/\sqrt{x+t}\, dt$ is evaluated in closed form, and an asymptotic series is developed for $x \to \infty$.

   (a) Evaluate $F(10)$, expand $F(x)$ in an asymptotic series, evaluate the series at $x = 10$, and compare.

   (b) Obtain $F'(10)$ by differentiating under the integral sign, and integrating numerically. Then termwise differentiate the series for $F(x)$, evaluate the differentiated series at $x = 10$, and compare.

3. In Exercise B5, Section 11.1, an asymptotic series for $F(x) = \int_0^{\pi/4} e^{-x \sin t}\, dt$ is developed for large x.

   (a) Termwise differentiate the asymptotic series for $F(x)$.

   (b) Obtain an asymptotic expansion for $F'(x)$ by differentiating under the integral sign and expanding the integrand in the same way as was done for $F(x)$.

   (c) Show that the series in part (a) is the same as the series in part (b). Hence, termwise differentiation of the asymptotic expansion of $F(x)$ is valid.

4. In Exercise B6, Section 11.1, an asymptotic expansion for $F(x) = \int_0^4 e^{-xt} t^2 \sqrt{1+t}\, dt$ is developed for large x.

   (a) Obtain $F(10)$ by numeric integration. Approximate $F(x)$ with $G(x) = \int_0^{\infty} e^{-xt} t^2 \sqrt{1+t}\, dt$ as a Laplace transform, and obtain $G(10)$. Expand $G(x)$ in an asymptotic series and evaluate it at $x = 10$. Compare the three values.

   (b) Obtain $F'(10)$ by differentiating under the integral sign and integrating numerically. Obtain $G'(10)$. Termwise differentiate the expansion for $F(x)$ and evaluate the differentiated series at $x = 10$. Compare the three values.

For the pairs of functions $f(x)$ and $g(x)$ given in Exercises 5–9:

   (a) Obtain $F(x)$ and $G(x)$, the asymptotic expansions, for large x, of $f(x)$ and $g(x)$, respectively.

   (b) Obtain the asymptotic expansion of the product $f(x)g(x)$, and compare to $F(x)G(x)$, the product of the asymptotic expansions of $f(x)$ and $g(x)$.

(c) If the expansion of the product $f(x)g(x)$ matches the termwise product $FG \sim \sum c_k x^{-k}$, show that the coefficients c_k are given by $\sum_{n=0}^{k} a_n b_{k-n}$, where $f(x) \sim \sum a_k x^{-k}$ and $g(x) \sim \sum b_k x^{-k}$.

(d) Obtain the asymptotic expansion of the composite function $f(g(x))$, and compare to $F(G(x))$, the composition of the asymptotic expansions of expansions of $f(x)$ and $g(x)$.

(e) Obtain the asymptotic expansion of the composite function $g(f(x))$, and compare to $G(F(x))$, the composition of the asymptotic expansions of expansions of $f(x)$ and $g(x)$.

(f) Obtain the asymptotic expansion of the function $\frac{1}{f(x)}$ and compare it to the expansion of $\frac{1}{F(x)}$; in particular, note whether or not the expansions are asymptotic series even if they match.

(g) Obtain the asymptotic expansion of the function $\frac{1}{g(x)}$ and compare it to the expansion of $\frac{1}{G(x)}$; in particular, note whether or not the expansions are asymptotic series even if they match.

(h) Obtain the asymptotic expansion of the function $\frac{f(x)}{g(x)}$ and compare it to the expansion of $\frac{F(x)}{G(x)}$; in particular, note whether or not the expansions are asymptotic series even if they match.

(i) If it exists, obtain $f_1(x) = \int_x^{\infty} f(t)\, dt$ and expand it in an asymptotic series. Compare this result with the termwise integral of $F(x)$.

(j) If it exists, obtain $g_1(x) = \int_x^{\infty} g(t)\, dt$ and expand it in an asymptotic series. Compare this result with the termwise integral of $G(x)$.

5. $f(x) = \dfrac{x+1}{x^2 + 2x + 2}$, $g(x) = \dfrac{1}{x+1}$

6. $f(x) = \dfrac{1}{x^2 + 3x}$, $g(x) = \dfrac{x}{x+1}$

7. $f(x) = \dfrac{x+1}{x+2}$, $g(x) = \dfrac{x^2 + 1}{x^3 + x + 3}$

8. $f(x) = \dfrac{x}{x^3 + 2}$, $g(x) = \dfrac{x^2}{x^5 + x + 1}$

9. $f(x) = \dfrac{x+1}{x^2 + 2}$, $g(x) = \dfrac{x+1}{x^3 + x + 1}$

For the pairs of functions $f(x)$ and $g(x)$ given in Exercises 10–14:

(a) Obtain $F(x)$ and $G(x)$, the asymptotic expansions, for small x, of $f(x)$ and $g(x)$, respectively.

(b) Obtain the asymptotic expansion of the product $f(x)g(x)$, and compare to $F(x)G(x)$, the product of the asymptotic expansions of $f(x)$ and $g(x)$.

(c) If the expansion of the product $f(x)g(x)$ matches the termwise product $F(x)G(x) \sim \sum c_k x^k$, show that the coefficients c_k are given by $\sum_{n=0}^{k} a_n b_{k-n}$, where $f(x) \sim \sum a_k x^k$ and $g(x) \sim \sum b_k x^k$.

(d) Obtain the asymptotic expansion of the composite function $f(g(x))$, and compare to $F(G(x))$, the composition of the asymptotic expansions of expansions of $f(x)$ and $g(x)$.

(e) Obtain the asymptotic expansion of the composite function $g(f(x))$, and compare to $G(F(x))$, the composition of the asymptotic expansions of expansions of $f(x)$ and $g(x)$.

(f) If $f(0) \neq 0$, obtain the asymptotic expansion of the function $\frac{1}{f(x)}$ and compare it to the expansion of $\frac{1}{F(x)}$; in particular, note whether or not the expansions are asymptotic series even if they match.

(g) If $g(0) \neq 0$, obtain the asymptotic expansion of the function $\frac{1}{g(x)}$ and compare it to the expansion of $\frac{1}{G(x)}$; in particular, note whether or not the expansions are asymptotic series even if they match.

(h) If $g(0) \neq 0$, obtain the asymptotic expansion of the function $\frac{f(x)}{g(x)}$ and compare it to the expansion of $\frac{F(x)}{G(x)}$; in particular, note whether or not the expansions are asymptotic series even if they match.

(i) Obtain $f_1(x) = \int_0^x f(t)\, dt$ and expand it in an asymptotic series.

(j) Obtain the termwise integral of the expansion of $F(x)$ and compare it to the series developed in part (i).

(k) Obtain $g_1(x) = \int_0^x g(t)\, dt$ and expand it in an asymptotic series.

10. $f(x)$ and $g(x)$ as given in Exercise 5

11. $f(x)$ and $g(x)$ as given in Exercise 6

12. $f(x)$ and $g(x)$ as given in Exercise 7

13. $f(x)$ and $g(x)$ as given in Exercise 8

14. $f(x)$ and $g(x)$ as given in Exercise 9

Chapter Review

1. Explain the statement "near $x = 0$, $f(x) = O(g(x))$." Give an example.

2. Explain the statement "for large x, $f(x) = O(g(x))$." Give an example.

3. Explain the statement "at $x = 0$, $f(x) = o(g(x))$." Give an example.

4. Explain the statement "as $x \to \infty$, $f(x) = o(g(x))$." Give an example.

5. Explain the statement "x^k, $k = 0, 1, \ldots$, is an asymptotic sequence near $x = 0$."

6. Explain the statement "x^{-k}, $k = 0, 1, \ldots$, is an asymptotic sequence for large x."

7. Let $\phi_k = x^k$, $k = 0, 1, \ldots$, and let $f(x) = \sqrt{x+1}$. Use the formulas

$$a_0 = \lim_{x \to 0} \frac{f(x)}{\phi(x)} \quad \text{and} \quad a_n = \lim_{x \to 0} \frac{f(x) - \sum_{k=0}^{n-1} a_k \phi_k}{\phi_n}$$

to generate the asymptotic expansion $\sum_{k=0}^{N} a_k \phi_k$. Show that the expansion is just the Taylor series for $f(x)$.

8. Use the results of Question 7 to explain the statement "$f \sim 1$ as $x \to 0$."

9. Use the results of Question 7 to explain the statement "$f \sim 1 - \frac{x}{2}$ as $x \to 0$."

10. Explain the statement "$f(x) \sim g(x)$ for x large." Give an example.

11. Explain the statement "$O(g(x)) + O(g(x)) = O(g(x))$." Give an example.

12. Verify Watson's lemma for the function $f(t) = \sin t$. Thus, show that the asymptotic expansion of $F(s) = \frac{1}{s^2+1}$ is the termwise Laplace transform of the Maclaurin series for $f(t)$.

Ordinary Differential Equations—Part Two

Physical systems described by differential equations are often complex enough to require more than one differential equation in their models. Typically, these equations will be coupled, forming a system of differential equations. The simplest such system is the first-order constant-coefficient system, which we introduce and study in terms of interconnected mixing tanks.

These systems have a natural vector structure that prompts a study of enough matrix algebra to support a discussion of the eigenvalue problem for matrices. In addition to the applicability to such physical models as mixing tanks, the first-order linear system also captures the second-order linear ODE. The second-order linear ODE of Unit One can be expressed as a first-order linear system, and the characteristic roots of Unit One become the eigenvalues of Unit Three.

The behavior of a system for which an exact solution is available is readily determined. In particular, a system's stability, the propensity to return to an equilibrium, is important to the designers of control systems. The stability of nonlinear systems can often be determined by examining the stability of the associated linear system obtained from linearizing the original system.

The second-order, constant-coefficient, linear ODE can be solved exactly. Some second-order, linear ODEs with nonconstant coefficients also can be solved exactly. But the majority of such equations defy analytic solution, and numeric methods must be used. The Runge–Kutta–Nystrom method can be applied directly to the second-order equation. Alternatively, the second-order equation can be converted to a first-order system, and a vectorized version of one of the numeric methods of Chapter 4 can be applied. One such method, the vector version of rk4, is studied.

At regular points, second-order ODEs have solutions expressible as power series, which we learn to find directly from the DE itself. At regular singular points, the linear ODE has generalized series solutions that we also learn to obtain. Equations whose coefficients are not analytic may have solutions given in terms of asymptotic series that we learn to obtain as well.

Having accumulated a suite of tools for solving the second-order initial value problem, we turn our attention to the second-order boundary value problem. In the BVP, conditions are placed on the solution at the endpoints of an interval. Given a general solution containing two constants, the BVP is solved by writing and solving an appropriate pair of algebraic equations. Otherwise, we resort to numerical methods such as the shooting method and finite differences, and analytic approximation techniques such as the finite element method, or the methods of least squares, Rayleigh–Ritz, Galerkin, and collocation.

The ability to solve a BVP means we can now solve the eigenvalue problem for differential equations. This problem generally arises when solving partial differential equations. The Sturm–Liouville self-adjoint eigenvalue problem generates an infinite sequence of eigenvalues and a complete orthonormal set of eigenfunctions that behave in the infinite dimensional case, like the finite basis vectors $\{\mathbf{i}, \mathbf{j}, \mathbf{k}\}$ in three dimensions.

Two important cases that arise in the applications are Bessel's equation for problems in cylindrical coordinates and Legendre's equation for problems in spherical coordinates. In each case, an infinite set of mutually orthogonal eigenfunctions are generated, and these functions can be used to form Fourier-type series, just as we did with trig functions in Chapter 10.

The unit concludes with a finite-difference scheme for obtaining numeric solutions of eigenvalue problems.

Chapter *12*

Systems of First-Order ODEs

INTRODUCTION   Just as coupled systems of algebraic equations arise in the applications, so too do coupled systems of differential equations. Complex systems with many components will naturally be described by more than one interacting differential equation. The simplest systems are the first-order linear systems with constant coefficients. We use the example of interconnected mixing tanks to illustrate such systems.

Matrix and vector notation is the natural language for describing and solving first-order linear systems. After posing representative problems and solving them by Laplace transform techniques, we elucidate the vector nature of the solutions of these linear systems and proceed with a study of the requisite linear algebra In particular, the eigenvalue problem for matrices is examined. Since the second-order constant-coefficient linear equation of Unit One can be represented as a first-order system, the characteristic roots of Unit One become the eigenvalues of Chapter 12.

The matrix formalism introduced allows us to discuss the question of stability. Points at which the first derivatives in the first-order system simultaneously vanish are equilibrium points, and the system point in the phase plane will remain at rest at an equilibrium point if it once arrives there. However, physical systems are subject to disturbances, and the system point can be displaced from equilibrium. Stability of a system refers to its propensity to return to equilibrium after such a perturbation.

Not all physical systems are described by linear systems of differential equations. Most real-world systems are nonlinear. Nonlinear systems are rarely solvable in closed form and generally require numerical techniques for solution. However, establishing the stability of a nonlinear system is often as important as obtaining its solution. By the process of linearization, the stability of a nonlinear system can generally be determined by examining a related linear system.

The chapter concludes with a study of the nonlinear pendulum.

12.1   Mixing Tanks—Closed Systems

A Two-Tank Problem Without Flow-Through

We introduce the notion of a coupled set of first-order linear differential equations by setting up and solving the following mixing-tank problem associated with Figure 12.1. The tanks form a closed system because no matter enters or leaves the two tanks but only recirculates between them.

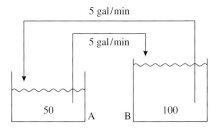

FIGURE 12.1 Schematic for mixing-tank problem

Tank A initially contains 50 gal of pure water, whereas tank B initially contains 100 gal of brine ($\frac{3}{4}$ lb of salt per gallon). Instantaneous mixing occurs when liquid is pumped back and forth between the tanks at 5 gal/min. To find the instantaneous amounts of salt in each tank proceed as follows.

Let the variables $x(t)$ and $y(t)$ represent the amount of salt in tanks A and B, respectively. Then, the initial amounts of salt in the tanks are given by $x(0) = 0$ and $y(0) = \frac{3}{4}(100) = 75$.

Before doing any further mathematics, we reason to the limiting outcome as follows. Eventually, the fluid in both tanks will be uniformly mixed. Thus, there will be 75 lb of salt mixed in 150 gal of water. There will be one-third of that in tank A and two-thirds of that in tank B. Hence, tank A will eventually contain 25 lb of salt, and tank B will eventually contain 50 lb of salt.

Differential equations governing $x(t)$ and $y(t)$ are written using the rate-balance principle

rate at which salt-content changes = rate at which salt enters − rate at which salt leaves

learned in Section 2.3. Thus, each minute, tank A loses 5/50 of its volume, so it loses 5/50 of its salt. Each minute, tank A gains salt from tank B, namely, 5/100 of the salt in tank B, so we have the equation $\frac{dx}{dt} = -\frac{5}{50}x(t) + \frac{5}{100}y(t)$.

Each minute, tank B gains salt from tank A, namely, 5/50 of the salt in tank A, and tank B loses salt to tank A, namely, 5/100 of its own salt. Thus, we have the equation $\frac{dy}{dt} = \frac{5}{50}x(t) - \frac{5}{100}y(t)$, and the IVP

$$\frac{dx}{dt} = -\frac{5}{50}x(t) + \frac{5}{100}y(t)$$
$$\frac{dy}{dt} = \frac{5}{50}x(t) - \frac{5}{100}y(t)$$
with $x(0) = 0, y(0) = 75$

These two differential equations are linear, first order, and coupled. Each variable appears in both equations, so the equations must be solved simultaneously. Fortunately, the equations have constant coefficients and are, therefore, amenable to solution by the Laplace transform.

Solution by Laplace Transform

Our solution starts by computing the Laplace transform of each differential equation. There are then two unknowns, $X = X(s)$ and $Y = Y(s)$, the transforms of $x(t)$ and $y(t)$, respectively. We thereby obtain

$$sX - x(0) = -\frac{1}{10}X + \frac{1}{20}Y \quad \text{and} \quad sY - y(0) = \frac{1}{10}X - \frac{1}{20}Y$$

supply the initial data to obtain

$$sX = -\frac{1}{10}X + \frac{1}{20}Y \quad \text{and} \quad sY - 75 = \frac{1}{10}X - \frac{1}{20}Y$$

and solve for $X = X(s)$ and $Y = Y(s)$ to obtain

$$X(s) = \frac{75}{s(20s + 3)} = \frac{25}{s} - \frac{25}{s + \frac{3}{20}} \quad \text{and} \quad Y(s) = 150\frac{10s + 1}{s(20s + 3)} = \frac{50}{s} + \frac{25}{s + \frac{3}{20}}$$

Inverting each transform, we find $x(t) = 25 - 25e^{-3/20t}$ and $y(t) = 50 + 25e^{-3/20t}$.

The behavior of these solutions can be seen in Figure 12.2, where $x(t)$ is the solid curve, $y(t)$, the dashed, and the dotted horizontal lines are asymptotes. As predicted, each

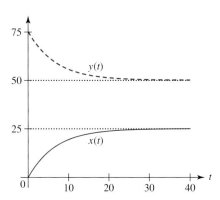

FIGURE 12.2 Asymptotes (dotted) approached by $x(t)$ (solid) and $y(x)$ (dashed)

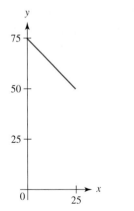

FIGURE 12.3   Phase-plane representation of solution

curve is asymptotic to the limiting values 25 and 50, respectively. Figure 12.3 shows an alternate representation of the relation between $y(t)$ and $x(t)$, generated by interpreting $x(t)$ and $y(t)$ as the parametric representation of a curve $y(x)$. As t increases from zero, the "system point" (x, y) travels along the curve $y(x)$, from $(0, 75)$ on the y-axis, toward the terminal point $(25, 50)$.

That the system point $(x(t), y(t))$ actually traverses a *line* can be verified by eliminating the parameter t from the equations for $x(t)$ and $y(t)$. Adding the two equations $x = 25 - 25e^{-3t/20}$ and $y = 50 + 25e^{-3t/20}$ gives $x + y = 75$, the explicit equation of the line, which is algebraically consistent with the realization that salt is conserved, neither entering nor leaving the *system*. This line is found in the *phase plane,* an xy-plane whose points represent system configurations (x, y).

As the state of the system varies in time, $x(t)$ and $y(t)$ change. A parametric plot of the points $(x(t), y(t))$ is a curve in this the phase plane, each point of which represents a potential state of the system's variables. The actual history of system states is the collection of phase-plane points constituting the parametric curve.

Alternatively, the equation of this line can be obtained by eliminating the parameter t from the differential equations themselves. This is done by recalling that the slope of the parametrically defined curve $y(x)$ is given by $\frac{dy}{dx} = \frac{dy/dt}{dx/dt} = -1$, from which $y(x) = 75 - x$ follows when the initial condition $(x(0), y(0)) = (0, 75)$ is imposed.

EXERCISES 12.1–Part A

A1. For the example described by Figure 12.1, show that the limiting values x_∞ and y_∞ can be found from the governing IVP by setting $x' = y' = 0$, then using the initial data and conservation of salt to get $x(0) + y(0) = x_\infty + y_\infty$.

A2. For the initial value problem $x'(t) = 5x + 6y$, $y'(t) = -7x - 8y$, $x(0) = -2$, $y(0) = 2$:

  (a) Obtain the solution by use of the Laplace transform.

  (b) Graph the curves $x(t)$ and $y(t)$.

  (c) Graph the parametric curve $y(x)$ determined by $x = x(t)$, $y = y(t)$.

A3. Assume that in Figure 12.1, tank A initially contains 200 gal of pure water, and tank B, 150 gal of brine in which 50 lb of salt has been dissolved. Instantaneous mixing occurs when liquid is pumped back and forth between the tanks at a rate of 6 gal/min.

  (a) Formulate an IVP whose solution gives, as a function of time, the amount of salt in each tank.

  (b) Reason to a limiting value of the amount of salt in each tank.

  (c) Use the method of Exercise A1 to calculate the limiting values for the salt in each tank.

A4. Use the Laplace transform to solve each of the following initial value problems.

  (a) $x'(t) = 3x + 6y$, $y'(t) = -5x + 14y$, $x(0) = 3$, $y(0) = -1$

  (b) $x'(t) = -38x + 21y$, $y'(t) = -70x + 39y$, $x(0) = -1$, $y(0) = 1$

  (c) $x'(t) = -27x + 12y$, $y'(t) = -72x + 33y$, $x(0) = 4$, $y(0) = -2$

  (d) $x'(t) = 31x + 45y$, $y'(t) = -18x - 26y$, $x(0) = -5$, $y(0) = 3$

  (e) $x'(t) = -30x - 32y$, $y'(t) = 37x + 39y$, $x(0) = 1$, $y(0) = -2$

Assume that in Figure 12.1, tank A initially contains α gallons of pure water whereas tank B initially contains β gallons of brine (b pounds of salt per gallon). Instantaneous mixing occurs when liquid is pumped back and forth between the tanks at r gallons per minute. For the parameters given in each of Exercises A5–9:

  (a) Let the variables $x(t)$ and $y(t)$ represent the amount of salt in tanks A and B, respectively, and formulate an appropriate initial value problem for the instantaneous amounts of salt in each tank.

  (b) Reason to the limiting values of $x(t)$ and $y(t)$.

A5. $\alpha = 75$, $\beta = 200$, $b = \frac{1}{10}$, $r = 3$

A6. $\alpha = 250$, $\beta = 150$, $b = 2$, $r = 4$

A7. $\alpha = 500$, $\beta = 300$, $b = 3$, $r = 7$

A8. $\alpha = 400$, $\beta = 125$, $b = 1$, $r = 10$

A9. $\alpha = 30$, $\beta = 80$, $b = \frac{1}{5}$, $r = 2$

EXERCISES 12.1–Part B

B1. Use the Laplace transform to solve the IVP $x'(t) = 27x - 72y$, $y'(t) = 16x - 41y$, $x(0) = a$, $y(0) = b$. Show that $\lim_{t\to\infty} x(t) = \lim_{t\to\infty} y(t) = 0$ for any values of a and b. What does this say about the parametric curve $y(x)$ in the phase plane?

Assume that in Figure 12.1, tank A initially contains α gallons of pure water whereas tank B initially contains β gallons of brine (b pounds of salt per gallon). Instantaneous mixing occurs when liquid is pumped back and forth between the tanks at r gallons per minute. For the parameters given in each of Exercises B2–6:

(a) Let the variables $x(t)$ and $y(t)$ represent the amount of salt in tanks A and B, respectively, and formulate an appropriate initial value problem for the instantaneous amounts of salt in each tank.

(b) As in the text, reason to the limiting values of $x(t)$ and $y(t)$.

(c) Use the method of Exercise A1 to calculate the limiting values of $x(t)$ and $y(t)$.

(d) Use the Laplace transform to solve the IVP formulated in part (a).

(e) Graph $x(t)$ and $y(t)$ on the same set of axes.

(f) Verify the limiting values obtained in parts (b) and (c).

(g) Plot $y(x)$, the trajectory in the phase plane.

(h) Eliminate the parameter t from the solutions obtained in part (d).

(i) Eliminate the parameter t by forming a differential equation for the function $y(x)$.

(j) Use a computer algebra system's differential equation solver to obtain the solution of the IVP, and compare to the Laplace transform solution found in part (d).

B2. $\alpha = 1000, \beta = 750, b = 3, r = 15$
B3. $\alpha = 225, \beta = 650, b = \frac{3}{10}, r = 8$
B4. $\alpha = 450, \beta = 350, b = \frac{5}{7}, r = 12$
B5. $\alpha = 100, \beta = 75, b = \frac{1}{2}, r = 5$
B6. $\alpha = 600, \beta = 800, b = 5, r = 20$

Assume that in Figure 12.1, tank A initially contains α gallons of brine (a pounds of salt per gallon) whereas tank B initially contains β gallons of brine (b pounds of salt per gallon). Instantaneous mixing occurs when liquid is pumped back and forth between the tanks at r gallons per minute. For the parameters given in Exercises B7–16:

(a) Let the variables $x(t)$ and $y(t)$ represent the amount of salt in tanks A and B, respectively, and formulate an appropriate initial value problem for the instantaneous amounts of salt in each tank.

(b) Reason to the limiting values of $x(t)$ and $y(t)$.

(c) Use the method of Exercise A1 to calculate the limiting values of $x(t)$ and $y(t)$.

(d) Use the Laplace transform to solve the IVP formulated in part (a).

(e) Graph $x(t)$ and $y(t)$ on the same set of axes.

(f) Verify the limiting values obtained in parts (b) and (c).

(g) Plot $y(x)$, the trajectory in the phase plane.

(h) Eliminate the parameter t from the solutions obtained in part (d).

(i) Eliminate the parameter t by forming a differential equation for the function $y(x)$.

(j) Use a computer algebra system's differential equation solver to obtain the solution of the IVP, and compare with the solution obtained in part (d).

B7. $\alpha = 85, \beta = 220, a = \frac{3}{4}, b = \frac{1}{5}, r = 5$
B8. $\alpha = 350, \beta = 250, a = 3, b = 4, r = 3$
B9. $\alpha = 50, \beta = 30, a = 2, b = 1, r = 6$
B10. $\alpha = 240, \beta = 325, a = 2, b = 3, r = 4$
B11. $\alpha = 70, \beta = 60, a = \frac{1}{7}, b = \frac{1}{2}, r = 1$
B12. $\alpha = 100, \beta = 75, a = \frac{2}{5}, b = \frac{1}{5}, r = \frac{3}{2}$
B13. $\alpha = 850, \beta = 550, a = \frac{2}{5}, b = \frac{3}{10}, r = 16$
B14. $\alpha = 275, \beta = 575, a = \frac{5}{9}, b = \frac{5}{7}, r = 14$
B15. $\alpha = 300, \beta = 175, a = \frac{3}{4}, b = \frac{1}{2}, r = 6$
B16. $\alpha = 450, \beta = 675, a = 3, b = 2, r = 8$

B17. Assume that in Figure 12.1, tank A contains 100 gal of 10% brine, and tank B, 200 gal of 15% brine. (Amounts of salt are therefore measured in *gallons,* not pounds.) Find r, the recirculation rate for which, at time $t = 10$, the amount of salt in tank B is within 5% of its limiting value.

B18. Assume that in Figure 12.1, tank A contains 300 gal of $16\frac{2}{3}\%$ brine, and tank B, 100 gal of 10% brine. (Amounts of salt are therefore measured in *gallons,* not pounds.) Find r, the recirculation rate for which, at time $t = 20$, the amount of salt in tank A is within 5% of its limiting value.

B19. Assume that in Figure 12.1, tank A contains 250 gal of 2% brine, and tank B, 400 gal of 20% brine. (Amounts of salt are therefore measured in *gallons,* not pounds.) Find r, the recirculation rate for which, at time $t = 15$, the amount of salt in tank A is at 50% of its limiting value.

In Exercises B20–29:

(a) Use the Laplace transform to obtain $x(t)$, $y(t)$, the solution of the given IVP.

(b) Plot $x(t)$ and $y(t)$ on the same set of axes.

(c) Plot $y(x)$, the trajectory in the phase plane.

(d) Use a computer algebra system's differential equation solver to obtain the solution of the IVP, and compare with the solution found in part (a).

B20. $x'(t) = 8x(t) + 4y(t)$, $y'(t) = -5x(t) - 5y(t)$, $x(0) = -2$,
$y(0) = 3$

B21. $x'(t) = -x(t) - y(t)$, $y'(t) = -4x(t) + 2y(t)$, $x(0) = 5$,
$y(0) = -3$

B22. $x'(t) = -4x(t) + y(t)$, $y'(t) = -9x(t) + 2y(t)$, $x(0) = 4$,
$y(0) = -5$

B23. $x'(t) = -4x(t) + 7y(t)$, $y'(t) = 3x(t) - 8y(t)$, $x(0) = -3$,
$y(0) = 5$

B24. $x'(t) = -\frac{16}{19}x(t) - \frac{260}{19}y(t)$, $y'(t) = -\frac{117}{38}x(t) - \frac{3}{19}y(t)$,
$x(0) = -5$, $y(0) = -1$

B25. $x'(t) = -\frac{89}{23}x(t) + \frac{72}{23}y(t)$, $y'(t) = \frac{66}{23}x(t) - \frac{95}{23}y(t)$, $x(0) = -1$,
$y(0) = 1$

B26. $x'(t) = -\frac{29}{5}x(t) + 2y(t)$, $y'(t) = \frac{12}{25}x(t) - \frac{31}{5}y(t)$, $x(0) = 2$,
$y(0) = 9$

B27. $x'(t) = \frac{72}{11}x(t) + \frac{6}{11}y(t)$, $y'(t) = \frac{5}{11}x(t) + \frac{71}{11}y(t)$, $x(0) = -7$,
$y(0) = 8$

B28. $x'(t) = -\frac{1}{5}x(t) + \frac{108}{5}y(t)$, $y'(t) = \frac{8}{5}x(t) + \frac{11}{5}y(t)$, $x(0) = 6$,
$y(0) = -4$

B29. $x'(t) = \frac{243}{49}x(t) + \frac{160}{49}y(t)$, $y'(t) = \frac{60}{49}x(t) + \frac{247}{49}y(t)$, $x(0) = 9$,
$y(0) = -7$

For the initial value problems in Exercises B30–39:

    (a) Use the Laplace transform to obtain the solution $x(t)$,
    $y(t)$, $z(t)$.

    (b) Plot $x(t)$, $y(t)$, and $z(t)$ on the same set of axes.

    (c) Plot the space curve determined parametrically by
    $(x(t), y(t), z(t))$.

    (d) Use a computer algebra system's differential equation solver
    to obtain the solution of the IVP, and compare to the solution
    obtained in part (a).

B30. $x' = -12x - 12y - 30z$, $y' = -44x - 19y - 60z$,
$z' = 30x + 15y + 42z$, $x(0) = -1$, $y(0) = 2$, $z(0) = 1$
Hint: Work in floating-point form.

B31. $x' = 82x + 108y + 72z$, $y' = -\frac{355}{3}x - 153y - 100z$,
$z' = 76x + 96y + 61z$, $x(0) = 3$, $y(0) = 0$, $z(0) = -2$

B32. $x' = 91x - 22y + 24z$, $y' = 656x - 161y + 168z$,
$z' = 280x - 70y + 69z$, $x(0) = -5$, $y(0) = 1$, $z(0) = 4$

B33. $x' = 16x + 2y - 4z$, $y' = 120x + 14y - 32z$, $z' = 120x + \frac{29}{2}y - 31z$, $x(0) = 7$, $y(0) = -3$, $z(0) = -1$

B34. $x' = 19x + 29y - 15z$, $y' = -93x - 139y + 75z$,
$z' = -147x - 219y + 119z$, $x(0) = y(0) = z(0) = 1$

B35. $x' = 63x + 232y + 80z$, $y' = -22x - 82y - 28z$,
$z' = 10x + 40y + 13z$, $x(0) = 1$, $y(0) = -2$, $z(0) = 3$

B36. $x' = 45x + 18y + 6z$, $y' = 24x + 3y + 6z$, $z' = -424x - 151y - 64z$, $x(0) = 4$, $y(0) = -7$, $z(0) = 2$

B37. $x' = 29x - y - 31z$, $y' = -72x + 7y + 92z$, $z' = 24x - y - 26z$,
$x(0) = -9$, $y(0) = 5$, $z(0) = 1$

B38. $x' = 155x + 30y - 39z$, $y' = 120x + 26y - 30z$, $z' = 732x + 144y - 184z$, $x(0) = 8$, $y(0) = -6$, $z(0) = -1$

B39. $x' = 189x + 376y - 80z$, $y' = -166x - 331y + 70z$,
$z' = -350x - 700y + 147z$, $x(0) = 2$, $y(0) = 3$, $z(0) = 5$

12.2 Mixing Tanks—Open Systems

Two-Tank Example with Flow-Through

We consider a second example of a two-tank mixing system and add "flow-through" to its structure. This makes the system "open" since matter can now enter and leave it. The goal is again to model the system as a coupled set of first-order linear differential equations, which we solve with the Laplace transform. The solution, plotted in the phase plane, reveals the typical complexity of the computations and casts further light on the notion of a *path* in this plane.

Consider, then, the two-tank system shown in Figure 12.4. Tank A contains 100 gal of brine (1 lb of salt per gallon), while tank B contains 100 gal of water. Fresh water flows into tank A at 2 gal/min and there is instantaneous mixing. Liquid flows from A to B at 3 gal/min, liquid flows from B to A at 1 gal/min, and liquid overflows from B at 2 gal/min. Find the amounts of salt present in tanks A and B at any time $t > 0$.

Fluid enters and leaves each tank at the net rate of 3 gal/min. The long-term outcome will be that both tanks contain pure water, since pure water flows into tank A and is then transferred to tank B by the recirculation pipes.

If we define $x(t)$ as the amount of salt in tank A and $y(t)$ as the amount of salt in tank B, the initial amounts of salt in the tanks will then be $x(0) = 100$, $y(0) = 0$. The rate of change

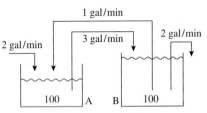

FIGURE 12.4   Schematic for mixing-tank problem

of the amount of salt in tank A is given by the differential equation $\frac{dx}{dt} = -\frac{3}{100}x(t) + \frac{1}{100}y(t)$ because tank A loses $\frac{3}{100}$ of its salt each minute (it goes to tank B) but gains $\frac{1}{100}$ of the salt in tank B each minute. Likewise, the rate of change of the amount of salt in tank B is given by the differential equation $\frac{dy}{dt} = \frac{3}{100}x(t) - \frac{1}{100}y(t) - \frac{2}{100}y(t)$ or $\frac{dy}{dt} = \frac{3}{100}x(t) - \frac{3}{100}y(t)$ because tank B gains, each minute, $\frac{3}{100}$ of the salt in tank A but loses, each minute, $\frac{1}{100}$ of its salt to tank A, and $\frac{2}{100}$ of its salt to the drain.

Solution by Laplace Transform

To solve the system of differential equations just derived, Laplace-transform each differential equation, obtaining

$$sX - x(0) = -\tfrac{3}{100}X + \tfrac{1}{100}Y \quad \text{and} \quad sY - y(0) = \tfrac{3}{100}X - \tfrac{3}{100}Y$$

where $X = X(s)$ and $Y = Y(s)$. Apply the initial data, finding

$$sX - 100 = -\tfrac{3}{100}X + \tfrac{1}{100}Y \quad \text{and} \quad sY = \tfrac{3}{100}X - \tfrac{3}{100}Y$$

the solution of which is

$$X = 5000\frac{100s + 3}{5000s^2 + 300s + 3} = 100\frac{s + \frac{3}{100}}{\left(s + \frac{3}{100}\right)^2 - \left(\frac{\sqrt{3}}{100}\right)^2}$$

and

$$Y = \frac{15000}{5000s^2 + 300s + 3} = \frac{3}{\left(s + \frac{3}{100}\right)^2 - \left(\frac{\sqrt{3}}{100}\right)^2}$$

Inversion of these transforms yields

$$x(t) = 100e^{-3/100t}\cosh\tfrac{\sqrt{3}}{100}t \quad \text{and} \quad y(t) = 100\sqrt{3}e^{-3/100t}\sinh\tfrac{\sqrt{3}}{100}t$$

the graphs of which appear in Figure 12.5, where the solid curve is $x(t)$ and the dotted, $y(t)$. Both curves seem to have a limiting value of zero, but $y(t)$ has a maximum value. Thus, tank B attains a maximum amount of salt before its contents asymptotically become pure water.

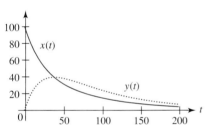

FIGURE 12.5 Solutions $x(t)$ (solid) and $y(t)$ (dotted)

Phase Plane Path

In Figure 12.6, we plot the path $y(x)$ by treating $x(t)$ and $y(t)$ as the parametric representation of a curve in the phase plane. The system point $(x(t), y(t))$ starts at $(100, 0)$ and traverses the path from right to left, heading toward the point $(0, 0)$. Obtaining the equation for the path is challenging.

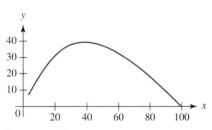

FIGURE 12.6 Phase-plane path, $y(x)$

ELIMINATING t FROM THE DIFFERENTIAL EQUATIONS First, we eliminate the parameter t from the differential equations, obtaining $\frac{dy}{dx} = -3\frac{x-y}{3x-y}$ and $y(100) = 0$ as an initial value problem for $y(x)$, the solution of which can be brought to the form $3x^2 - y^2 = 30{,}000e^{-2\sqrt{3}\,\text{arctanh}(y/x\sqrt{3})}$.

ELIMINATING t FROM THE PARAMETRIC REPRESENTATION Careful inspection of the solutions for $x(t)$ and $y(t)$ suggests that forming the ratio $\frac{y}{x}$ will allow us to solve for t, that is,

$$\frac{y}{x} = \frac{\sqrt{3}\sinh\frac{\sqrt{3}}{100}t}{\cosh\frac{\sqrt{3}}{100}t} = \sqrt{3}\tanh\frac{\sqrt{3}}{100}t \quad \text{gives} \quad t = t(x, y) = \frac{100}{\sqrt{3}}\text{arctanh}\frac{y}{x\sqrt{3}}$$

Since $x(t)$ is a hyperbolic cosine and $y(t)$ is a hyperbolic sine, we can use $\cosh^2 z - \sinh^2 z = 1$ so that we have $3x^2 - y^2 = 30{,}000e^{-3/50t} = 30{,}000e^{-2\sqrt{3}\arctanh(y/x\sqrt{3})}$.

EXERCISES 12.2–Part A

A1. For the example described by Figure 12.4, show that the limiting values $x_\infty = y_\infty = 0$ can be determined from the differential equations that govern the system.

A2. Use the Laplace transform to solve the IVP $x'(t) = 3 - 13x - 16y$, $y'(t) = 14x + 17y$, $x(0) = y(0) = 0$.

A3. Use the Laplace transform to solve the IVP $x'(t) = 1 + 3x + 3y$, $y'(t) = 2 - 24x - 14y$, $x(0) = y(0) = 0$.

A4. Modify the two-tank system shown in Figure 12.4 as follows. Tank A contains 400 gal of 35% brine, while tank B contains 125 gal of 40% brine. At a rate of 2 gal/min, 30% brine flows into tank A and there is instantaneous mixing. Liquid flows from tank A to tank B at 3 gal/min, liquid flows from tank B to tank A at 1 gal/min, and liquid overflows from tank B at 2 gal/min.

(a) Formulate an IVP for $x(t)$ and $y(t)$, the instantaneous amounts of salt in tanks A and B, respectively.

(b) Without solving the IVP in part (a), reason to x_∞ and y_∞, the limiting (steady-state) values of $x(t)$ and $y(t)$.

(c) Obtain x_∞ and y_∞ from the differential equations written in part (a) by noting that at steady state, $x' = y' = 0$.

A5. In Exercise A4, let pure water instead of brine enter tank A. Then, the parameter t can be eliminated from the resulting equations for $x(t)$, $y(t)$ by writing $\frac{dy}{dx} = \frac{y'(t)}{x'(t)}$ and making the change of variables $v(x) = \frac{y(x)}{x}$ as detailed in Section 3.2. Obtain an implicit representation of the phase plane trajectory $y = y(x)$ by this technique.

EXERCISES 12.2–Part B

B1. Modify the two-tank system shown in Figure 12.4 as follows. Tank A contains 450 gal of 17% brine, while tank B contains 200 gal of 14% brine. At a rate of 1 gal/min, 50% brine flows into tank A and there is instantaneous mixing. Liquid flows from tank A to tank B at 4 gal/min, liquid flows from tank B to tank A at 3 gal/min, and liquid overflows from tank B at 1 gal/min.

(a) Formulate an IVP for $x(t)$ and $y(t)$, the instantaneous amounts of salt in tanks A and B, respectively.

(b) Without solving the IVP in part (a), reason to x_∞ and y_∞, the limiting (steady-state) values of $x(t)$ and $y(t)$.

(c) Obtain x_∞ and y_∞ from the differential equations written in part (a) by noting that at steady state, $x' = y' = 0$.

(d) Use the Laplace transform to solve the IVP.

(e) Plot $x(t)$ and $y(t)$ on the same set of axes.

(f) From the solutions, compute $\lim_{t\to\infty} x(t)$ and $\lim_{t\to\infty} y(t)$ and compare to the answers in parts (b) and (c).

(g) Plot $y(x)$, the trajectory in the phase plane.

(h) Eliminate the parameter t in the differential equation by obtaining $\frac{dy}{dx} = \frac{y'(t)}{x'(t)}$, $y(x(0)) = y(0)$ and solving numerically, using a graph as the outcome of the computation.

(i) Solve the IVP with the differential equation solver of some computer algebra system.

For the nonhomogeneous IVPs in Exercises B2–11:

(a) Obtain the solution using the Laplace transform.

(b) Plot $x(t)$, $y(t)$, and $z(t)$ on the same set of axes.

(c) Plot the space curve defined parametrically by $(x(t), y(t), z(t))$.

(d) Use the differential equation solver of a computer algebra system to obtain the solution, and compare with the results in part (a).

B2. $x' = -343x + 300y + 432z$, $y' = 456x - 397y - 572z$, $z' = -588x + 513y + 739z$, $x(0) = -7$, $y(0) = 8$, $z(0) = 0$

B3. $x' = -384x - 840y + 240z$, $y' = 260x + 574y - 166z$, $z' = 286x + 644y - 191z$, $x(0) = -3$, $y(0) = 2$, $z(0) = 5$

B4. $x' = 703x - 324y - 606z$, $y' = 50x - 27y - 45z$, $z' = 790x - 362y - 680z$, $x(0) = 5$, $y(0) = 9$, $z(0) = -7$

B5. $x' = 58x + 166y + \frac{280}{3}z$, $y' = -240x - 666y - 400z$, $z' = 360x + 996y + 602z$, $x(0) = 10$, $y(0) = -7$, $z(0) = 4$

B6. $x' = -138x - 234y - 279z$, $y' = 405x + 699y + 837z$, $z' = -270x - 468y - 561z$, $x(0) = 0$, $y(0) = -1$, $z(0) = 2$

B7. $x' = 489x + 490y - 2456z$, $y' = 72x + 71y - 364z$, $z' = 112x + 112y - 563z$, $x(0) = 8$, $y(0) = -6$, $z(0) = 1$

B8. $x' = -109x - 45y + 5$, $y' = 252x + 104y - 5$, $z' = -63x - 27y - 4z$, $x(0) = 12$, $y(0) = 7$, $z(0) = -8$

B9. $x' = -45x + 10y + 4z$, $y' = -204x + 45y + 20z$, $z' = 48x - 10y - 9z$, $x(0) = 1$, $y(0) = -1$, $z(0) = 3$

B10. $x' = -135x + 32y + 492z - 2$, $y' = -43x + 12y + 156z$, $z' = -34x + 8y + 124z$, $x(0) = 0$, $y(0) = -3$, $z(0) = 5$

B11. $x' = 20x - 108y - 8z + 9$, $y' = -24x + 139y + 8z - 1$, $z' = 465x - 2670y - 159z$, $x(0) = 1$, $y(0) = -1$, $z(0) = 0$

In Figure 12.4, let r, in gallons per minute, represent the rate of influx of a $c\%$ brine solution, flowing into tank A holding α gallons of an $a\%$ brine solution. Let r_1 and r_2, in gallons per minute, be the recirculation rates from tank B to tank A and from tank A to tank B, respectively. Tank B initially contains β gallons of a $b\%$ brine solution, and the outflow from tank B to the drain is again at the rate r. For the parameters in Exercises B12–18, let $x(t)$ and $y(t)$ represent the amount of salt in tanks A and B, respectively, and

    (a) formulate the appropriate initial value problem.

    (b) reason to the limiting values of $x(t)$ and $y(t)$.

    (c) calculate the limiting values by setting $x' = y' = 0$ in the differential equations for the system, and then solving for $x(\infty)$, $y(\infty)$.

    (d) use the Laplace transform to solve the IVP.

    (e) plot $x(t)$ and $y(t)$ on the same set of axes.

    (f) from the solutions, compute $\lim_{t \to \infty} x(t)$ and $\lim_{t \to \infty} y(t)$ and compare to the answers in parts (b) and (c).

    (g) plot $y(x)$, the trajectory in the phase plane.

    (h) eliminate the parameter t in the differential equation by obtaining $\frac{dy}{dx} = \frac{y'(t)}{x'(t)}$, $y(x(0)) = y(0)$ and solving numerically, using a graph as the outcome of the computation.

    (i) for $x(t)$ or $y(t)$, determine any maximum that is not at an endpoint.

    (j) solve the IVP with the differential equation solver of some computer algebra system, and compare with the results in part (d).

B12. $r = 5, c = 80, \alpha = 625, a = 32, r_1 = 1, r_2 = 6, \beta = 725, b = 14$

B13. $r = 3, c = 0, \alpha = 775, a = 36, r_1 = 5, r_2 = 8, \beta = 450, b = 13$

B14. $r = 4, c = 25, \alpha = 800, a = 6, r_1 = 1, r_2 = 5, \beta = 800, b = 10$

B15. $r = 4, c = 30, \alpha = 550, a = 18, r_1 = 2, r_2 = 6, \beta = 675, b = 34$

B16. $r = 2, c = 0, \alpha = 725, a = 37, r_1 = 4, r_2 = 6, \beta = 975, b = 2$

B17. $r = 2, c = 55, \alpha = 325, a = 36, r_1 = 7, r_2 = 9, \beta = 450, b = 11$

B18. $r = 4, c = 0, \alpha = 500, a = 2, r_1 = 2, r_2 = 6, \beta = 225, b = 10$

In Exercises B13, 16, and 18, $c = 0$, so the resulting differential equations were homogeneous. For the systems in Exercises B19–21, write

$\frac{dy}{dx} = \frac{y'(t)}{x'(t)}$ and use the change of variables $v(x) = \frac{y(x)}{x}$ (see Section 3.2) to obtain an implicit representation of the phase plane trajectory $y = y(x)$.

B19. IVP of Exercise B13   **B20.** IVP of Exercise B16

B21. IVP of Exercise B18

B22. Modify the two-tank system shown in Figure 12.4 as follows. Tank A contains 200 gal of 15% brine while tank B contains 175 gal of 7% brine. At a rate of 4 gal/min, 10% brine flows into tank A and there is instantaneous mixing. Liquid flows from tank A to tank B at 9 gal/min, and liquid flows from tank B to tank A at 5 gal/min. In addition, at a rate of 3 gal/min, 20% brine flows directly into tank B and liquid overflows from tank B at 7 gal/min.

    (a) Formulate an IVP for $x(t)$ and $y(t)$, the instantaneous amounts of salt in tanks A and B, respectively.

    (b) Without solving the IVP in part (a), reason to x_∞ and y_∞, the limiting (steady-state) values of $x(t)$ and $y(t)$.

    (c) Obtain x_∞ and y_∞ from the differential equations written in part (a) by noting that at steady state, $x' = y' = 0$.

    (d) Use the Laplace transform to solve the IVP.

    (e) Plot $x(t)$ and $y(t)$ on the same set of axes.

    (f) From the solutions, compute $\lim_{t \to \infty} x(t)$ and $\lim_{t \to \infty} y(t)$ and compare to the answers in parts (b) and (c).

    (g) Plot $y(x)$, the trajectory in the phase plane.

    (h) Eliminate the parameter t in the differential equation by obtaining $\frac{dy}{dx} = \frac{y'(t)}{x'(t)}$, $y(x(0)) = y(0)$ and solving numerically, using a graph as the outcome of the computation.

    (i) Solve the IVP with the differential equation solver of some computer algebra system, and compare to the results in part (d).

In Exercises B23–27, solve the given IVP

    (a) with the Laplace transform.

    (b) with the differential equation solver of some computer algebra system, and compare to the solution in part (a).

B23. $x'(t) = 1 - 2x + 3y, \ y'(t) = 2 - 14x + 11y, \ x(0) = 9, \ y(0) = -5$

B24. $x'(t) = 2 - 3x + 4y, \ y'(t) = 5 - 12x + 11y, \ x(0) = -5, \ y(0) = 3$

B25. $x'(t) = 5 - 9x + 30y, \ y'(t) = 7 + 6x + 18y, \ x(0) = 3, \ y(0) = -2$

B26. $x'(t) = 4 + 5x - 24y, \ y'(t) = 2 + 3x - 12y, \ x(0) = -7, \ y(0) = 1$

B27. $x'(t) = 6 + 8x + 28y, \ y'(t) = 1 - 6x + 18y, \ x(0) = 8, \ y(0) = -3$

12.3   Vector Structure of Solutions

Mixing Tanks Revisited

Recall that for the closed-system mixing tanks in Section 12.1 we had the equations

$$x'(t) = -\tfrac{1}{10}x(t) + \tfrac{1}{20}y(t) \quad \text{and} \quad y(t) = \tfrac{1}{10}x(t) - \tfrac{1}{20}y(t) \tag{12.1}$$

and the solution $x(t) = 25 - 25e^{-3/20t}$, $y(t) = 50 + 25e^{-3/20t}$, which we now write as a sum of two vectors

$$\mathbf{X}(t) = \begin{bmatrix} x(t) \\ y(t) \end{bmatrix} = 25 \begin{bmatrix} 1 \\ 2 \end{bmatrix} + 25e^{(-3/20t)} \begin{bmatrix} -1 \\ 1 \end{bmatrix}$$

where $\mathbf{X}(t)$ is the vector whose components are $x(t)$ and $y(t)$. Similarly, we write the solution of the open-system mixing tanks in Section 12.2 as the sum

$$\mathbf{X}(t) = \begin{bmatrix} x(t) \\ y(t) \end{bmatrix} = 50e^{(-3+\sqrt{3})t/100} \begin{bmatrix} 1 \\ \sqrt{3} \end{bmatrix} + 50e^{(-3-\sqrt{3})t/100} \begin{bmatrix} 1 \\ -\sqrt{3} \end{bmatrix}$$

Vector Solutions, Directly

These examples suggest that the right way to look at a system of linear first-order ODEs is in vector-matrix form. For example, defining A to be the array

$$A = \begin{bmatrix} -\frac{1}{10} & \frac{1}{20} \\ \frac{1}{10} & -\frac{1}{20} \end{bmatrix}$$

allows us to write the equations in (12.1) in the form

$$\mathbf{X}'(t) = \begin{bmatrix} x'(t) \\ y'(t) \end{bmatrix} = \begin{bmatrix} -\frac{1}{10} & \frac{1}{20} \\ \frac{1}{10} & -\frac{1}{20} \end{bmatrix} \begin{bmatrix} x(t) \\ y(t) \end{bmatrix} = A\mathbf{X}(t) \tag{12.2}$$

where $\mathbf{X}'(t)$ is the vector whose components are the derivatives $x'(t)$ and $y'(t)$. Now implied in (12.2), which can be written compactly as $\mathbf{X}'(t) = A\mathbf{X}(t)$, is a multiplication between the matrix A and the vector $\mathbf{X}(t)$.

Indeed, momentarily in Example 12.1, we will study matrix-vector arithmetic in detail. However, before getting lost in the details of such formalism, we point out the reason for the digression. Having written (12.1) in the form (12.2), we will look for *vector* solutions of the form $\mathbf{X}(t) = \mathbf{B}e^{\lambda t}$, where $\mathbf{B}$ is a constant vector and λ is a constant to be determined. This is not a wild guess since $y' = ay$, the scalar case of the first-order linear homogeneous equation, has the exponential solution $y = be^{at}$.

EXAMPLE 12.1   The algebraic equations

$$3x + 4y = 7 \quad \text{and} \quad 2x - 5y = 1$$

are reduced to the matrix-vector form $A\mathbf{X} = \mathbf{B}$ by writing

$$A\mathbf{X} = \begin{bmatrix} 3 & 4 \\ 2 & -5 \end{bmatrix} \begin{bmatrix} x \\ y \end{bmatrix} = \begin{bmatrix} 7 \\ 1 \end{bmatrix} = \mathbf{B} \tag{12.3}$$

where in (12.3) A is the matrix, $\mathbf{X}$ is the vector of unknowns, and $\mathbf{B}$ is the vector of constants. Multiplication of a vector by a matrix is accomplished by computing the *dot product*

$$[a \quad b] \cdot \begin{bmatrix} x \\ y \end{bmatrix} = ax + by$$

of each row of the matrix with the vector. The first entry in the vector forming the product is the dot product of the first row of the matrix with the vector $\mathbf{X}$. The second entry in the product is the dot product of the second row of the matrix with $\mathbf{X}$, etc. ❖

Caution

Nonlinear equations cannot be put into matrix form. Matrix form can only be obtained for *linear* first-order equations. For example, $x'(t) = xy$, $y'(t) = xy - y$ cannot be put into matrix form since each equation is nonlinear.

Matrix Notation

Elements in the matrix are doubly subscripted, a_{rc}, with r, the first subscript, being the *row* index, and c, the second subscript, being the *column* index. In fact, for a 2×2 matrix the elements would be designated as $\begin{bmatrix} a_{11} & a_{12} \\ a_{21} & a_{22} \end{bmatrix}$.

Matrix Arithmetic

Matrices are added, subtracted, and multiplied by scalars *elementwise*. Thus, we illustrate the three computations of matrix addition, matrix subtraction, and multiplication of a matrix by a scalar. Note how addition and subtraction are accomplished by adding or subtracting corresponding entries of the matrices. Scalar multiplication requires that each entry of the matrix be multiplied by the scalar. For example, if

$$A = \begin{bmatrix} 1 & 2 \\ 7 & 9 \end{bmatrix} \quad \text{and} \quad B = \begin{bmatrix} 6 & 1 \\ 3 & 4 \end{bmatrix}$$

then we have

$$A + B = \begin{bmatrix} 1+6 & 2+1 \\ 7+3 & 9+4 \end{bmatrix} = \begin{bmatrix} 7 & 3 \\ 10 & 13 \end{bmatrix}$$

$$A - B = \begin{bmatrix} 1-6 & 2-1 \\ 7-3 & 9-4 \end{bmatrix} = \begin{bmatrix} -5 & 1 \\ 4 & 5 \end{bmatrix}$$

$$3A = \begin{bmatrix} 3 \times 1 & 3 \times 2 \\ 3 \times 7 & 3 \times 9 \end{bmatrix} = \begin{bmatrix} 3 & 6 \\ 21 & 27 \end{bmatrix}$$

Matrix Multiplication

To multiply matrix A against matrix B, dot row i of A with column j of B. This gives the ij-entry in the product matrix. Thus, using the matrices in (12.3), we obtain the product AB as

$$\begin{bmatrix} 1 & 2 \\ 7 & 9 \end{bmatrix} \begin{bmatrix} 6 & 1 \\ 3 & 4 \end{bmatrix} = \begin{bmatrix} 1 \times 6 + 2 \times 3 & 1 \times 1 + 2 \times 4 \\ 7 \times 6 + 9 \times 3 & 7 \times 1 + 9 \times 4 \end{bmatrix} = \begin{bmatrix} 12 & 9 \\ 69 & 43 \end{bmatrix}$$

Dimension Rule for Multiplication

A row in the first matrix has to have as many elements as a column in the second matrix. Hence, the rule for multiplying matrices requires $[n \times m][m \times k] = [n \times k]$. The number of columns in the first must equal the number of rows in the second, in which case the product matrix has as many rows as the first and as many columns as the second.

EXAMPLE 12.2   We distinguish between the product of two matrices and the product of a matrix and a vector, for which the dimension rules are

$$[2 \times 2][2 \times 2] = [2 \times 2] \quad \text{and} \quad [2 \times 2][2 \times 1] = [2 \times 1]$$

In the first case, we are multiplying two 2×2 matrices and thereby obtain another 2×2 matrix. The second case is illustrated by

$$\begin{bmatrix} 1 & 2 \\ 7 & 9 \end{bmatrix} \begin{bmatrix} 3 \\ 11 \end{bmatrix} = \begin{bmatrix} 1 \times 3 + 2 \times 11 \\ 7 \times 3 + 9 \times 11 \end{bmatrix} = \begin{bmatrix} 25 \\ 120 \end{bmatrix}$$

where the 2×2 matrix A multiplies a 2×1 column vector, giving a 2×1 column vector. ❖

Multiplicative Identity

The matrix that behaves like the number "1" for multiplication is called the multiplicative identity. In the 2×2 case, the multiplicative identity is the matrix $I = \begin{bmatrix} 1 & 0 \\ 0 & 1 \end{bmatrix}$. For the matrix A from (12.3), we verify that $IA = AI = A$ by the calculations

$$\begin{bmatrix} 1 & 0 \\ 0 & 1 \end{bmatrix} \begin{bmatrix} 1 & 2 \\ 7 & 9 \end{bmatrix} = \begin{bmatrix} 1 \times 1 + 0 \times 7 & 1 \times 2 + 0 \times 9 \\ 0 \times 1 + 1 \times 7 & 0 \times 2 + 1 \times 9 \end{bmatrix} = \begin{bmatrix} 1 & 2 \\ 7 & 9 \end{bmatrix}$$

$$\begin{bmatrix} 1 & 2 \\ 7 & 9 \end{bmatrix} \begin{bmatrix} 1 & 0 \\ 0 & 1 \end{bmatrix} = \begin{bmatrix} 1 \times 1 + 2 \times 0 & 1 \times 0 + 2 \times 1 \\ 7 \times 1 + 9 \times 0 & 7 \times 0 + 9 \times 1 \end{bmatrix} = \begin{bmatrix} 1 & 2 \\ 7 & 9 \end{bmatrix}$$

Matrix Multiplication is NOT Commutative

Matrix multiplication *is not commutative*. In general, $AB \neq BA$; that is, the order of the factors in a matrix product cannot usually be reversed. Indeed, for the matrices A and B in (12.3), we have

$$AB = \begin{bmatrix} 12 & 9 \\ 69 & 43 \end{bmatrix} \quad \text{but} \quad BA = \begin{bmatrix} 13 & 21 \\ 31 & 42 \end{bmatrix}$$

No Zero-Principle

If the product of two real numbers is zero, then at least one of the factors must itself be zero. This is the *zero-principle*. However, there is *no zero-principle* for matrices. Thus, you can have two matrices, neither of which is zero, whose product is itself zero. In the following example, two matrices, neither of which has only zero entries, are multiplied. The product is the zero matrix.

$$\begin{bmatrix} 1 & 0 \\ 0 & 0 \end{bmatrix} \begin{bmatrix} 0 & 0 \\ 0 & 1 \end{bmatrix} = \begin{bmatrix} 0 & 0 \\ 0 & 0 \end{bmatrix}$$

Consequently, if $\alpha \neq 0$ in the scalar equation $\alpha x - \alpha y = 0$ we can conclude that $x = y$ because one of the factors on the left in $\alpha(x - y) = 0$ must be zero. It isn't α, so it has to be $x - y = 0$ and that means $x = y$. The equivalent algebra does not hold for matrices. In fact, if $A \neq 0$ for the matrix A (that is, there is at least one nonzero entry in A), then the equation $AX - AY = 0$ does not guarantee $X = Y$ because neither factor in $A(X - Y)$ need itself be the zero matrix.

EXERCISES 12.3–Part A

A1. Write each solution in Exercise A4, Section 12.1, in the form $X = e^{\lambda_1 t} B_1 + e^{\lambda_2 t} B_2$, where B_1 and B_2 are constant vectors and X is as defined in (12.2).

A2. Write the linear algebraic equations $8x - 9y = 12$, $3x - 5y = 11$ in matrix-vector form.

A3. Write the linear algebraic equations $3x - 7y - 13z = 21$, $7x - 2y - z = 14$, $x - 6y - 7z = 3$ in matrix-vector form.

A4. Write the IVP in Exercise A3, Section 12.1, in matrix-vector form.

A5. Find two 2×2 matrices, neither of which contains any zeros but whose product is the zero matrix.

A6. Find two 3×3 matrices, neither of which is the zero matrix but whose product is the zero matrix.

A7. If $A = \begin{bmatrix} 2 & 3 \\ 1 & -1 \end{bmatrix}$ and $\mathbf{X} = e^{-2t} \begin{bmatrix} 5 \\ -4 \end{bmatrix}$, compute $\mathbf{X}'$ and $A\mathbf{X}$.

A8. Given the matrices $A = \begin{bmatrix} 3 & -5 \\ 2 & 7 \end{bmatrix}$, $B = \begin{bmatrix} 8 & 1 \\ -1 & 4 \end{bmatrix}$, $C = \begin{bmatrix} 6 & 0 \\ 5 & -2 \end{bmatrix}$, $D = \begin{bmatrix} \frac{1}{2} & \frac{6}{7} \\ \frac{2}{3} & \frac{5}{4} \end{bmatrix}$, $I = \begin{bmatrix} 1 & 0 \\ 0 & 1 \end{bmatrix}$ and the vectors $\mathbf{X} = \begin{bmatrix} 2 \\ -3 \end{bmatrix}$, $\mathbf{Y} = \begin{bmatrix} -1 \\ 4 \end{bmatrix}$,

$\mathbf{Z} = \begin{bmatrix} 5 \\ -2 \end{bmatrix}$, compute

(a) $A\mathbf{X}$

(b) det A, the determinant of A, defined as $a_{11}a_{22} - a_{12}a_{21}$.

(c) $A - 2B$ **(d)** AB, BA

(e) $(A + B)^2$, $A^2 + 2AB + B^2$, $A^2 + AB + BA + B^2$

(f) $A(tB), t(AB)$ **(g)** $(A - 2I)\mathbf{X}, A\mathbf{X} - 2\mathbf{X}$

(h) $\det(A - \lambda I)$, the determinant of the 2×2 matrix $A - \lambda I$.

EXERCISES 12.3–Part B

B1. Using the definitions in Exercise A8, obtain the indicated quantities.

   (a) $A\mathbf{Y}$ **(b)** $B + 3C$ **(c)** AC, CA

   (d) $(B - C)^2$, $B^2 - 2BC + C^2$, $B^2 - BC - CB + C^2$

   (e) $2A - 7B + 4C$ **(f)** $A(2B), 2(AB)$

   (g) $(B + 5I)\mathbf{Y}, B\mathbf{Y} + 5\mathbf{Y}$

B2. Using the definitions in Exercise A8, obtain the indicated quantities.

   (a) $A\mathbf{Z}$ **(b)** $A\mathbf{X} - B\mathbf{Y}$ **(c)** $\mathbf{X} - 2\mathbf{Y}$

   (d) AD, DA **(e)** BC, CB

   (f) $(A + 3C)^2$, $A^2 + 6AC + 9C^2$, $A^2 + 3(AC + CA) + 9C^2$

   (g) $A(tC), t(AC)$ **(h)** $AC\mathbf{X} - B\mathbf{Y}$

   (i) $(-A^2 - 4B + 7C^3)\mathbf{Z} - 2D\mathbf{Y} + 5B^2\mathbf{X}$ **(j)** $\det(C - \lambda I)$

B3. Using the definitions in Exercise A8, obtain the indicated quantities.

   (a) $\mathbf{Y} + 3\mathbf{Z}, 5\mathbf{X} - 7\mathbf{Z}$ **(b)** BD, DB **(c)** CD, DB

   (d) $(3A^2 - 6B^3)\mathbf{X}$ **(e)** $5\mathbf{X} - 7\mathbf{Y} + 8\mathbf{Z}$

   (f) $AB - C^2 - 6D$ **(g)** $I + A + \frac{1}{2}A^2 + \frac{1}{3!}A^3$

   (h) $\det(B - \lambda I)$ **(i)** $(C + 7I)\mathbf{Z}, C\mathbf{Z} + 7\mathbf{Z}$

In Exercises B4–9, rewrite the solution of the indicated IVP in the form $\mathbf{X} = e^{\lambda_1 t}\mathbf{B}_1 + e^{\lambda_2 t}\mathbf{B}_2$, where $\mathbf{B}_1$ and $\mathbf{B}_2$ are constant vectors.

B4. Exercise B20, Section 12.1   **B5.** Exercise B21, Section 12.1

B6. Exercise B23, Section 12.1   **B7.** Exercise B24, Section 12.1

B8. Exercise B25, Section 12.1   **B9.** Exercise B26, Section 12.1

In Exercises B10–15, rewrite the solution of the indicated IVP in the form $\mathbf{X} = e^{\lambda_1 t}\mathbf{B}_1 + e^{\lambda_2 t}\mathbf{B}_2 + e^{\lambda_3 t}\mathbf{B}_3$, where $\mathbf{B}_k$, $k = 1, 2, 3$, are constant vectors and $\mathbf{X} = \begin{bmatrix} x(t) \\ y(t) \\ z(t) \end{bmatrix}$.

B10. Exercise B2, Section 12.2   **B11.** Exercise B3, Section 12.2

B12. Exercise B4, Section 12.2   **B13.** Exercise B5, Section 12.2

B14. Exercise B7, Section 12.2   **B15.** Exercise B9, Section 12.2

In Exercises B16–25, write the given system of linear algebraic equations in the matrix-vector form $A\mathbf{x} = \mathbf{b}$, where $\mathbf{x} = \begin{bmatrix} x \\ y \end{bmatrix}$.

B16. $2x - 3y = 7$, $4x - 5y = 9$ **B17.** $6x - 5y = 2$, $x + 2y = 7$

B18. $-3x + y - 5 = 0$, $7x + 8y - 1 = 0$

B19. $7x = 5y + 4$, $2y = 9x - 5$

B20. $-13x = 12 - 9y$, $12y = x + 7$

B21. $3x - 5 = -8y$, $2y - 7 = -9x$

B22. $14y - 8 = 3x$, $21x - 23 = 13y$

B23. $56 - 31y = 11x$, $41y + 6 = 13x$

B24. $161x - 79 = 123y$, $37y + 5x - 4 = 0$

B25. $11.7x - 19.8y = 6.7$, $4.5x = 12.3y - 7.7$

In Exercises B26–35, write the given system of linear algebraic equations in the matrix form $A\mathbf{x} = \mathbf{b}$, where $\mathbf{x} = \begin{bmatrix} x \\ y \\ z \end{bmatrix}$.

B26. $2x - 3y + 5z = 7$, $-x + 4y - 8z = 9$, $12x + 7y - 13z = 17$

B27. $21x - 51y + 77z = 6$, $-x - y + 4z = 7$, $23x + 37y - 12z = 2$

B28. $11.3x - 72.1y = 55.4z$, $15.9x = 27.7z + 33.7z$, $71.8y = 81.1x + 64.9z$

B29. $67x - 113y + 107z = 91$, $123x + 231z = 13$, $321x + 55y - 76z = -98$

B30. $16y = 21x + 5z - 7$, $23z - 7y = 14 - 32x$, $17 - 3x = 4y + 2z$

B31. $171y - 75z - 6x - 4 = 0$, $5 - 4z - 3y + 13x = 0$, $8x - 11 = 7y + 12z$

B32. $34x - 23z = 9 - 29y$, $15 - 2x = 9z - 8y$, $43y + 7 - 6x = 65z$

B33. $1.2x - 4.3y - 5.7z = 9.8$, $6.2x - 7.3y - 5.4z = 2.2$, $7.7x - 8.3y + 9.5z = 6.8$

B34. $2x = 7y - 5z$, $8x - 5y = 12z$, $13x - 23y = 17z$

B35. $57y - 11 = 13x - 12z$, $8z - 19y = 24x$, $119x - 84z = 101y - 62$

In Exercises B36–40, write the indicated IVP in the matrix form $\mathbf{X}'(t) = A\mathbf{X}(t)$, $\mathbf{X}(0) = \mathbf{X}_0$.

B36. Exercise A5, Section 12.1   **B37.** Exercise A6, Section 12.1

B38. Exercise A7, Section 12.1   **B39.** Exercise A8, Section 12.1

B40. Exercise A9, Section 12.1

B41. If $A = \begin{bmatrix} 5 & 7 \\ -3 & 3 \end{bmatrix}$ and $\mathbf{X} = e^{-3t}\begin{bmatrix} 2 \\ 3 \end{bmatrix}$, compute $\mathbf{X}'$ and $A\mathbf{X}$.

B42. If $A = \begin{bmatrix} -8 & 1 \\ 7 & 9 \end{bmatrix}$ and $\mathbf{X} = e^{-5t}\begin{bmatrix} -7 \\ 8 \end{bmatrix}$, compute $\mathbf{X}'$ and $A\mathbf{X}$.

B43. If $A = \begin{bmatrix} 6 & -5 \\ 3 & 2 \end{bmatrix}$ and $\mathbf{X} = e^{-4t}\begin{bmatrix} 9 \\ -2 \end{bmatrix}$, compute $\mathbf{X}'$ and $A\mathbf{X}$.

B44. If $A = \begin{bmatrix} -1 & 1 \\ 4 & 5 \end{bmatrix}$ and $\mathbf{X} = e^{-7t}\begin{bmatrix} 6 \\ -1 \end{bmatrix}$, compute $\mathbf{X}'$ and $A\mathbf{X}$.

12.4   Determinants and Cramer's Rule

Computing Determinants

The determinant of an $n \times n$ matrix A is the scalar $\det(A)$ whose computation we prescribe as follows. The determinant of the 2×2 array A on the left in (12.4) is the scalar $ad - bc$. The determinant of the 3×3 array B in the center in (12.4) can be computed in several ways.

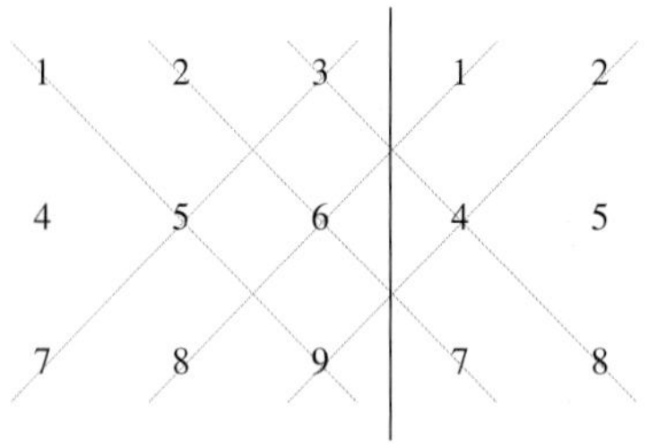

FIGURE 12.7   Computing the value of a 3×3 determinant

$$A = \begin{bmatrix} a & b \\ c & d \end{bmatrix} \qquad B = \begin{bmatrix} 1 & 2 & 3 \\ 4 & 5 & 6 \\ 7 & 8 & 9 \end{bmatrix} \qquad C = \begin{bmatrix} -1 & -2 & 5 & 1 \\ -6 & 1 & -2 & 1 \\ -6 & -3 & 3 & -4 \\ 5 & -2 & 7 & 3 \end{bmatrix} \tag{12.4}$$

For example, Figure 12.7 shows the array B with its first two columns written to its right. Each of the six slanted lines in the figure passes through three numbers. Each such slanted line represents the product of the three numbers through which it passes. Products associated with lines slanted upward (/) are taken as negative, while products associated with lines slanted downward (\) are taken as positive. The determinant is then the sum of the six products.

Here, the three positive products are $(1)(5)(9) = 45$, $(2)(6)(7) = 84$, $(3)(4)(8) = 96$. The three products that will be taken as negative are $(3)(5)(7) = 105$, $(1)(6)(8) = 48$, $(2)(4)(9) = 72$. The determinant is then $45 + 84 + 96 - 105 - 48 - 72 = 0$. Note well, however, that this method does not generalize to higher order determinants.

Laplace (Row or Column) Expansion of the Determinant

Higher order determinants can be evaluated via the Laplace expansion along any row or column. Given the matrix A with elements a_{ij}, define M_{ij}, the *minor* of element a_{ij}, as the determinant of the submatrix formed by deleting row i and column j from A. Then, define c_{ij}, the *cofactor* of element a_{ij}, as the signed minor of element a_{ij} so that $c_{ij} = (-1)^{i+j} M_{ij}$. The determinant of A is then the sum, along any row or column, of the products of each entry and its cofactor. For example, computing the determinant along row i is accomplished via the sum $\det A = \sum_{j=1}^{n} a_{ij} c_{ij}$.

EXAMPLE 12.3   The determinant of the matrix C on the right in (12.4) computed by a row-one expansion is $-1(-12) - (-2)(-275) + 5(127) - 93 = 4$.   ❖

EXAMPLE 12.4   From multivariable calculus, $\mathbf{a} \times \mathbf{b}$, the *cross-product* of the vectors $\mathbf{a} = a_1\mathbf{i} + a_2\mathbf{j} + a_3\mathbf{k}$ and $\mathbf{b} = b_1\mathbf{i} + b_2\mathbf{j} + b_3\mathbf{k}$, is the vector produced when the determinant on the left is formally evaluated by expanding the first row

$$
\det \begin{bmatrix} \mathbf{i} & \mathbf{j} & \mathbf{k} \\ a_1 & a_2 & a_3 \\ b_1 & b_2 & b_3 \end{bmatrix} = (a_2b_3 - a_3b_2)\mathbf{i} - (a_1b_3 - a_3b_1)\mathbf{j} + (a_1b_2 - a_2b_1)\mathbf{k} = \begin{bmatrix} a_2b_3 - a_3b_2 \\ a_3b_1 - a_1b_3 \\ a_1b_2 - a_2b_1 \end{bmatrix}
$$

$$(12.5)$$

We take this opportunity to remind the reader of the following useful properties of the cross-product $\mathbf{a} \times \mathbf{b}$, where θ in $[0, \pi]$ is the angle between $\mathbf{a}$ and $\mathbf{b}$. In Section 31.3 we show that the following first and third properties uniquely determine the cross-product.

1. $\mathbf{a}$, $\mathbf{b}$, and $\mathbf{a} \times \mathbf{b}$ form a *right-handed system* of mutually perpendicular vectors.
2. $\mathbf{a} \times \mathbf{b} = -\mathbf{b} \times \mathbf{a}$
3. $\|\mathbf{a} \times \mathbf{b}\| = \|\mathbf{a}\|\|\mathbf{b}\| \sin \theta$
4. $\mathbf{a} \times (\mathbf{b} \times \mathbf{c}) \neq (\mathbf{a} \times \mathbf{b}) \times \mathbf{c}$

The first property says that $\mathbf{a} \times \mathbf{b}$ is perpendicular to both $\mathbf{a}$ and $\mathbf{b}$ and is thus perpendicular to the plane containing $\mathbf{a}$ and $\mathbf{b}$. There are two possible directions for such a vector, and the *right-hand rule* selects a unique direction as follows. If the fingers of the right hand are wrapped around $\mathbf{a} \times \mathbf{b}$ in such a way as to point in the direction of rotation of $\mathbf{a}$ into $\mathbf{b}$ (through the angle θ), then the extended thumb of the right hand points in the direction of $\mathbf{a} \times \mathbf{b}$. This is equivalent to the statement that if $\mathbf{a}$ is rotated through the angle θ, into coincidence with $\mathbf{b}$, then a screw with a right-handed thread, turned in the same direction, would advance in the direction of $\mathbf{a} \times \mathbf{b}$.

The second property, an immediate consequence of the first, says that the cross-product is not *commutative*. In fact, it says the cross-product is *anticommutative*, reversing direction if the order of the vectors in the cross-product is reversed.

In the statement of the third property, the notation $\|\mathbf{a}\|$ indicates the length of the vector $\mathbf{a}$, so that

$$\|\mathbf{a}\| = \sqrt{a_1^2 + a_2^2 + a_3^2}$$

Therefore, the third property says that the length of the cross-product is equal to the product of the lengths of the factors, times the sine of the angle between the factors.

The fourth property says that the cross-product is not *associative*. Table 31.3 in Section 31.3 lists several other algebraic properties of the cross-product. In particular, it gives more detail on the nonassociativity described by the fourth property. ❖

PLANE AREA VIA DETERMINANT Again from multivariable calculus, $\|\mathbf{a} \times \mathbf{b}\|$ is the area of the parallelogram whose edges are the vectors $\mathbf{a}$ and $\mathbf{b}$. If $\mathbf{a} = a_1\mathbf{i} + a_2\mathbf{j} + 0\mathbf{k}$ and $\mathbf{b} = b_1\mathbf{i} + b_2\mathbf{j} + 0\mathbf{k}$ are plane vectors, then the area of the parallelogram they span is

$$\left| \det \begin{bmatrix} \mathbf{i} & \mathbf{j} & \mathbf{k} \\ a_1 & a_2 & 0 \\ b_1 & b_2 & 0 \end{bmatrix} \right| = \left| \det \begin{bmatrix} a_1 & a_2 \\ b_1 & b_2 \end{bmatrix} \right| = |a_1b_2 - a_2b_1|$$

The absolute value of the 2×2 determinant $\det(A)$ can be interpreted as the area of the parallelogram whose edges are the row vectors of the matrix A.

TRIPLE SCALAR (BOX) PRODUCT In multivariable calculus the absolute value of $\mathbf{A} \cdot (\mathbf{B} \times \mathbf{C})$, the *Triple Scalar* (or *Box*) *Product*, is found to be the volume of the parallelepiped formed by the three vectors **A**, **B**, and **C**. Indeed, given the vectors $\mathbf{A} = a_1\mathbf{i} + a_2\mathbf{j} + a_3\mathbf{k}$, $\mathbf{B} = b_1\mathbf{i} + b_2\mathbf{j} + b_3\mathbf{k}$, and $\mathbf{C} = c_1\mathbf{i} + c_2\mathbf{j} + c_3\mathbf{k}$, this volume is the absolute value of

$$\mathbf{A} \cdot (\mathbf{B} \times \mathbf{C}) = \det \begin{bmatrix} a_1 & a_2 & a_3 \\ b_1 & b_2 & b_3 \\ c_1 & c_2 & c_3 \end{bmatrix}$$

Determinant of a Product

The determinant of a product is the product of the determinants, as illustrated with the following example. Thus, for the matrices A, B, and the product AB in (12.6), the determinants are -134, -56, and $7504 = (-134)(-56)$, respectively.

$$A = \begin{bmatrix} -1 & -2 & 5 \\ 1 & -6 & 1 \\ -2 & 1 & -6 \end{bmatrix} \qquad B = \begin{bmatrix} -3 & 3 & -4 \\ 5 & -2 & 7 \\ 3 & -1 & -1 \end{bmatrix} \qquad AB = \begin{bmatrix} 8 & -4 & -15 \\ -30 & 14 & -47 \\ -7 & -2 & 21 \end{bmatrix}$$

(12.6)

Cramer's Rule

Cramer's rule represents the solution of a system of linear equations as ratios of determinants. For example, the Cramer's rule solution of the linear system $A\mathbf{x} = \mathbf{y}$

$$\begin{bmatrix} a_{11} & a_{12} & a_{13} \\ a_{21} & a_{22} & a_{23} \\ a_{31} & a_{32} & a_{33} \end{bmatrix} \begin{bmatrix} x_1 \\ x_2 \\ x_3 \end{bmatrix} = \begin{bmatrix} y_1 \\ y_2 \\ y_3 \end{bmatrix}$$

(12.7)

is given by

$$x_1 = \frac{\begin{vmatrix} y_1 & a_{12} & a_{13} \\ y_2 & a_{22} & a_{23} \\ y_3 & a_{32} & a_{33} \end{vmatrix}}{\begin{vmatrix} a_{11} & a_{12} & a_{13} \\ a_{21} & a_{22} & a_{23} \\ a_{31} & a_{32} & a_{33} \end{vmatrix}} \qquad x_2 = \frac{\begin{vmatrix} a_{11} & y_1 & a_{13} \\ a_{21} & y_2 & a_{23} \\ a_{31} & y_3 & a_{33} \end{vmatrix}}{\begin{vmatrix} a_{11} & a_{12} & a_{13} \\ a_{21} & a_{22} & a_{23} \\ a_{31} & a_{32} & a_{33} \end{vmatrix}} \qquad x_3 = \frac{\begin{vmatrix} a_{11} & a_{12} & y_1 \\ a_{21} & a_{22} & y_2 \\ a_{31} & a_{32} & y_3 \end{vmatrix}}{\begin{vmatrix} a_{11} & a_{12} & a_{13} \\ a_{21} & a_{22} & a_{23} \\ a_{31} & a_{32} & a_{33} \end{vmatrix}}$$

where $|A| = \det A$. Note that the array in the numerator of the fraction giving x_k, $k = 1, 2, 3$, is the system matrix A with the kth column replaced by the vector **y**.

Finally, we point out that Cramer's rule is not practical if $n > 2$. For example, if $n = 3$, it takes 12 multiplications and 5 additions to compute a single determinant. There are four different determinants to compute, and then three divisions. Hence, Cramer's rule requires 48 multiplications, 20 additions, and 3 divisions, for a total of 71 arithmetic operations. As we will learn in Unit Eight, a 3×3 linear system can be solved with 27 arithmetic operations using *Gaussian elimination,* a technique to be studied in Section 12.5.

EXERCISES 12.4–Part A

A1. Find the area of the parallelogram determined by the vectors $\mathbf{A} = 2\mathbf{i} + 3\mathbf{j}$, $\mathbf{B} = 5\mathbf{i} - 7\mathbf{j}$.

A2. Find the volume of the parallelepiped determined by the vectors $\mathbf{A} = 3\mathbf{i} - 7\mathbf{j} + 4\mathbf{k}$, $\mathbf{B} = 6\mathbf{i} - 9\mathbf{j} - 11\mathbf{k}$, $\mathbf{C} = 7\mathbf{i} - 12\mathbf{j} + 2\mathbf{k}$.

A3. Use Cramer's rule to find the solution of the system in Exercise A2, Section 12.3.

A4. Use Cramer's rule to find the solution of the system in Exercise A3, Section 12.3.

A5. Verify that $\det(A)\det(B) = \det(AB)$ for the matrices in (12.6).

EXERCISES 12.4–Part B

B1. Find the cofactors for the matrix A in (12.6).

In Exercises B2–5, find the area of the parallelogram determined by the given pairs of vectors.

B2. $A = -3\mathbf{i} + 5\mathbf{j}, B = 8\mathbf{i} - 13\mathbf{j}$ **B3.** $A = 9\mathbf{i} - 5\mathbf{j}, B = 6\mathbf{i} + 11\mathbf{j}$

B4. $A = 7\mathbf{i} + 13\mathbf{j}, B = -4\mathbf{i} - 17\mathbf{j}$ **B5.** $A = \mathbf{i} + \mathbf{j}, B = -\mathbf{i} + 3\mathbf{j}$

In Exercises B6–9, find the volume of the parallelepiped determined by each of the given triples of vectors.

B6. $A = -8\mathbf{i} + 5\mathbf{j} + 7\mathbf{k}, B = 2\mathbf{i} - 3\mathbf{j} - 5\mathbf{k}, C = 21\mathbf{i} + 2\mathbf{j} + 17\mathbf{k}$

B7. $A = 13\mathbf{i} + 9\mathbf{j} - 17\mathbf{k}, B = -3\mathbf{i} + 5\mathbf{j} + 23\mathbf{k}, C = -18\mathbf{i} + 13\mathbf{j} + 14\mathbf{k}$

B8. $A = -5\mathbf{i} - 12\mathbf{j} - 13\mathbf{k}, B = -\mathbf{i} + 7\mathbf{j} - 15\mathbf{k}, C = 10\mathbf{i} - 11\mathbf{j} + 13\mathbf{k}$

B9. $A = 2\mathbf{i} + 4\mathbf{j} + \mathbf{k}, B = 19\mathbf{i} - 11\mathbf{j} + 17\mathbf{k}, C = 25\mathbf{i} + 16\mathbf{j} - 17\mathbf{k}$

For the linear algebraic equations given in Exercises B10–19:

(a) Obtain a solution by Cramer's rule.

(b) Obtain a solution using the solver in some modern technological device and then compare answers.

B10. Exercise B16, Section 12.3   **B11.** Exercise B17, Section 12.3

B12. Exercise B18, Section 12.3   **B13.** Exercise B19, Section 12.3

B14. Exercise B20, Section 12.3   **B15.** Exercise B21, Section 12.3

B16. Exercise B22, Section 12.3   **B17.** Exercise B23, Section 12.3

B18. Exercise B24, Section 12.3   **B19.** Exercise B25, Section 12.3

For the linear algebraic equations given in Exercises B20–29:

(a) Obtain a solution by Cramer's rule.

(b) Obtain a solution using the solver in some modern technological device and then compare answers.

B20. Exercise B26, Section 12.3   **B21.** Exercise B27, Section 12.3

B22. Exercise B28, Section 12.3   **B23.** Exercise B29, Section 12.3

B24. Exercise B30, Section 12.3   **B25.** Exercise B31, Section 12.3

B26. Exercise B32, Section 12.3   **B27.** Exercise B33, Section 12.3

B28. Exercise B34, Section 12.3   **B29.** Exercise B35, Section 12.3

B30. If the matrices A, B, C, I are as defined in Exercise A8, Section 12.3, compute and comment on each of

(a) $\det(AB), \det(A)\det(B), \det(BA)$

(b) $\det(ABC), \det(A)\det(B)\det(C), \det(BCA), \det(ACB)$

(c) $\det(A - \lambda I)$ **(d)** $\det(A + B), \det(A) + \det(B)$

(e) $\det(B - 2I), \det(B) - 2$

B31. If the matrices A, B, C, D, I are as defined in Exercise A8, Section 12.3 compute and comment on each of

(a) $\det(AC), \det(A)\det(C), \det(CA)$

(b) $\det(ABD), \det(A)\det(B)\det(D), \det(DBA), \det(ADB)$

(c) $\det(B - \lambda I)$ **B32.** $\det(A - C), \det(A) - \det(C)$

(d) $\det(C + 3I), \det(C) + 3$

B33. If $A = \begin{bmatrix} -1 & 1 & 2 \\ -3 & 2 & 3 \\ 1 & 0 & 4 \end{bmatrix}, B = \begin{bmatrix} 2 & -2 & 1 \\ 3 & 0 & -1 \\ 1 & 1 & -3 \end{bmatrix}, C = \begin{bmatrix} 5 & 0 & -2 \\ 3 & 1 & -1 \\ 4 & 4 & 1 \end{bmatrix},$

$D = \begin{bmatrix} 7 & -1 & 1 \\ 2 & -3 & 0 \\ 0 & 3 & 2 \end{bmatrix}, I = \begin{bmatrix} 1 & 0 & 0 \\ 0 & 1 & 0 \\ 0 & 0 & 1 \end{bmatrix}$, compute and comment on each of

(a) $\det(AB), \det(A)\det(B), \det(BA)$

(b) $\det(ABC), \det(A)\det(B)\det(C), \det(BCA), \det(ACB)$

(c) $\det(ADC), \det(A)\det(D)\det(C), \det(DCA), \det(ACD)$

(d) $\det(A - \lambda I)$ **(e)** $\det(A + B), \det(A) + \det(B)$

(f) $\det(A - C), \det(A) - \det(C)$ **(g)** $\det(B - 2I), \det(B) - 2$

(h) $\det(C + 3I), \det(C) + 3$ **(i)** $\det(A - I), \det(A) - 1$

For the matrices in Exercises B33–36:

(a) Compute the determinant by some form of modern technology.

(b) Obtain the Laplace expansion by any one row and any one column.

(c) Obtain all sixteen cofactors.

B34. $\begin{bmatrix} 2 & -2 & 1 & 3 \\ -1 & -3 & 2 & 0 \\ 1 & 5 & -2 & 2 \\ 4 & 3 & -3 & 1 \end{bmatrix}$ **B35.** $\begin{bmatrix} -3 & 1 & -1 & 2 \\ -1 & 1 & -2 & 4 \\ 3 & 0 & 1 & 1 \\ 5 & 2 & -2 & -2 \end{bmatrix}$

B36. $\begin{bmatrix} 2 & 3 & 1 & -1 \\ -1 & 1 & -2 & 1 \\ 5 & 4 & -4 & -2 \\ -3 & 0 & 1 & 3 \end{bmatrix}$ **B37.** $\begin{bmatrix} 6 & 5 & -5 & 2 \\ 2 & -2 & 3 & 1 \\ 4 & -3 & 1 & -1 \\ 7 & 6 & 0 & -7 \end{bmatrix}$

12.5 Solving Linear Algebraic Equations

Introduction

A set of simultaneous linear algebraic equations such as

$$a + b + c = 2 \qquad 9a - 3b + c = 4 \qquad 25a + 5b + c = -5$$

can be solved in a variety of ways. We examine some of those ways.

Successive Elimination

Although somewhat tedious, successively eliminating variables in a set of linear equations will lead to a solution. The first equation is solved for the "first" unknown, and this value is substituted into all remaining equations, thereby eliminating the first unknown from the set of equations. The process is repeated with the two remaining equations, yielding a numeric value for the third unknown. By working back through the equations, values for the second and first unknowns can be found.

Solve the first equation for a, obtaining $a = -b - c + 2$, which is substituted into the second and third equations to yield

$$-12b - 8c = -14 \tag{12.8a}$$
$$-20b - 24c = -55 \tag{12.8b}$$

The set of three equations in three unknowns has been reduced to a set of two equations in two unknowns. The variable a has been eliminated from the system. Solve (12.8a) for $b = -\frac{2}{3}c + \frac{7}{6}$ and substitute this into (12.8b), producing

$$\frac{32}{3}c = -\frac{95}{3}$$

an equation in which just the variable c appears. There are now the three equations

$$a + b + c = 2 \tag{12.9}$$
$$-12b - 8c = -14 \tag{12.10}$$
$$\frac{32}{3}c = -\frac{95}{3} \tag{12.11}$$

We then determine c from (12.11) via simple arithmetic, obtaining $c = \frac{95}{32}$. Substitute this value for c into (12.10), thereby obtaining $b = -\frac{13}{16}$. Make substitutions for both b and c in (12.9), and find $a = -\frac{5}{32}$. Clearly, this successive elimination process that results in the "triangular" system (12.9)–(12.11) will be exceedingly simplified if the variables are not dragged along at each step.

Gauss Elimination

By working with just the array of numbers and suppressing the names of the unknowns, we can represent the set of equations via the augmented matrix

$$B = \begin{bmatrix} 1 & 1 & 1 & 2 \\ 9 & -3 & 1 & 4 \\ 25 & 5 & 1 & -5 \end{bmatrix}$$

The process of eliminating the unknown a from the second and third equations amounts to adding and subtracting multiples of row 1 in such a way as to create zeros in the second and third positions in column 1. This row elimination arithmetic is accomplished by adding to row 2, row 1 multiplied by -9, yielding the matrix

$$B_1 = \begin{bmatrix} 1 & 1 & 1 & 2 \\ 0 & -12 & -8 & -14 \\ 25 & 5 & 1 & -5 \end{bmatrix}$$

and then adding to row 3, row 1 multiplied by -25, yielding the matrix

$$B_2 = \begin{bmatrix} 1 & 1 & 1 & 2 \\ 0 & -12 & -8 & -14 \\ 0 & -20 & -24 & -55 \end{bmatrix}$$

Of course, the rows of matrix B_2 represent equations (12.8a) and (12.8b).

The equivalent of solving (12.8a) for b, and using it to eliminate b from (12.8b) is using the second row in matrix B_2 to create a zero at the 3-2 position. This is done by adding to row 3, row 2 multiplied by $-\frac{-20}{-12}$, yielding the matrix

$$B_3 = \begin{bmatrix} 1 & 1 & 1 & 2 \\ 0 & -12 & -8 & -14 \\ 0 & 0 & -\frac{32}{3} & -\frac{95}{3} \end{bmatrix}$$

The process of solving (12.11) for c, then using that value to eliminate c from (12.10), then using the values of b and c to eliminate b and c from (12.9) is called *back substitution*. Since we have already done that arithmetic, we'll not repeat it.

The process of row-reducing the augmented matrix B to the upper triangular form seen in B_3 is called *Gaussian elimination*. Thus, the matrix system $A\mathbf{x} = \mathbf{b}$ can be solved by forming the augmented matrix $[A, \mathbf{b}]$, row-reducing to an upper triangular form by Gaussian elimination, and finishing with back substitution.

Gauss–Jordan Reduction

The process of row reducing the augmented matrix until only 1s remain on the main diagonal is called Gauss–Jordan reduction. Requiring more computations, it is more "expensive" than Gaussian elimination followed by back substitution. Section 39.1 considers the number of arithmetic operations needed to complete each process and seeks ways to reduce this number to a minimum.

Gauss–Jordan reduction of the augmented matrix B yields

$$\begin{bmatrix} 1 & 0 & 0 & -\frac{5}{32} \\ 0 & 1 & 0 & -\frac{13}{16} \\ 0 & 0 & 1 & \frac{95}{32} \end{bmatrix} \tag{12.12}$$

from which the solution for the vector $\mathbf{x}$ is immediate. It's the rightmost column of (12.12).

Row-Echelon and Reduced Row-Echelon Forms

If Gaussian elimination steps lead to a matrix wherein the leftmost nonzero entry of each row is a 1, rows with all zeros are below all rows with nonzero entries, and the lead 1 in row $k + 1$ occurs in a column to the right of the column in which the lead 1 occurs in row k, then that form of the matrix is called a *row-echelon form*.

Reducing B to B_3 does not achieve an echelon form because the leftmost nonzero entries in rows 2 and 3 are not 1's. If we multiply row 2 by $-\frac{1}{12}$ and row 3 by $-\frac{3}{32}$, we achieve the echelon form

$$\begin{bmatrix} 1 & 1 & 1 & 2 \\ 0 & 1 & \frac{2}{3} & \frac{7}{6} \\ 0 & 0 & 1 & \frac{95}{32} \end{bmatrix}$$

The leftmost nonzero entry in each row is a 1, there are no zero rows, and each leading 1 occurs to the right of the leading 1 in the previous row. The row-echelon form of a matrix is not unique. A matrix has many row-echelon equivalents, depending on exactly how the Gaussian elimination steps were performed.

When each leading 1 in the nonzero rows is used to create as many zeros in the column above it as possible, the matrix is said to be in *reduced row-echelon form*. The reduced row-echelon form (12.12) of the matrix augmented B, denoted by rref(B), is unique.

The number of nonzero rows in the reduced row-echelon form of A is a unique number called the *rank* of A.

EXERCISES 12.5–Part A

A1. To solve the linear system $A\mathbf{x} = \mathbf{b}$, use Gaussian elimination to row-reduce the augmented matrix $[A, \mathbf{b}]$ to an upper triangular matrix. Is this necessarily a row-echelon matrix? Explain.

A2. Answer true or false: To use back substitution, the upper triangular matrix must be in row-echelon form. Explain your answer.

A3. Solve the linear system in Exercise B26, Section 12.3, by the method of successive elimination.

A4. Solve the system considered in Exercise A3 by writing the augmented matrix, row reducing to an upper triangular form, and using back substitution. Compare the solution process to that used in Exercise A3.

A5. Obtain the reduced row-echelon form for the augmented matrix in Exercise A4.

EXERCISES 12.5–Part B

B1. If A is given in Exercise B32, Section 12.4 and $\mathbf{Y} = \begin{bmatrix} -12 \\ 13 \\ 5 \end{bmatrix}$, solve $A\mathbf{X} = \mathbf{Y}$ by

  (a) an appropriate technological tool.

  (b) the method of successive elimination.

  (c) Gaussian elimination and back substitution.

  (d) Gauss–Jordan reduction to reduced row-echelon form.

B2. Repeat Exercise B1 for $B\mathbf{X} = \mathbf{Y}$, where B is given in Exercise B32, Section 12.4 and $\mathbf{Y}$ is given in Exercise B1.

B3. Repeat Exercise B1 for $C\mathbf{X} = \mathbf{Y}$, where C is given in Exercise B32, Section 12.4 and $\mathbf{Y}$ is given in Exercise B1.

B4. Repeat Exercise B1 for $D\mathbf{X} = \mathbf{Y}$, where D is given in Exercise B32, Section 12.4 and $\mathbf{Y}$ is given in Exercise B1.

For the linear systems designated in Exercises B5–13, solve by

  (a) the method of successive elimination.

  (b) Gaussian elimination and back substitution.

  (c) Gauss–Jordan reduction (which yields the reduced row-echelon form of the augmented system matrix).

B5. Exercise B27, Section 12.3   **B6.** Exercise B28, Section 12.3

B7. Exercise B29, Section 12.3   **B8.** Exercise B30, Section 12.3

B9. Exercise B31, Section 12.3   **B10.** Exercise B32, Section 12.3

B11. Exercise B33, Section 12.3   **B12.** Exercise B34, Section 12.3

B13. Exercise B35, Section 12.3

If $\mathbf{Y} = \begin{bmatrix} 3 \\ -5 \\ 7 \\ -8 \end{bmatrix}$ and M is the matrix specified in Exercises B14–17, solve each system $M\mathbf{X} = \mathbf{Y}$ by

  (a) an appropriate technological tool.

  (b) the method of successive elimination.

  (c) Gaussian elimination and back substitution.

  (d) Gauss–Jordan reduction to reduced row-echelon form.

B14. Exercise B33, Section 12.4   **B15.** Exercise B34, Section 12.4

B16. Exercise B35, Section 12.4   **B17.** Exercise B36, Section 12.4

For the augmented matrices in Exercises B18–27:

  (a) Write the set of equations represented by the matrix.

  (b) Obtain a row-echelon form.

  (c) Write the set of equations corresponding to the row-echelon form found in part (b).

  (d) If a solution to the system of equations exists, obtain it from part (b) and back substitution.

  (e) Obtain the reduced row-echelon form.

  (f) If a solution to the set of equations exists, use an appropriate technological tool to find it.

B18. $\begin{bmatrix} 0 & -2 & 11 & -12 \\ -10 & 9 & -4 & -8 \\ -4 & 1 & 10 & -6 \end{bmatrix}$ **B19.** $\begin{bmatrix} -4 & 10 & 12 & -7 \\ 9 & 6 & 8 & -4 \\ -6 & -3 & 6 & 10 \end{bmatrix}$

B20. $\begin{bmatrix} 6 & 6 & 6 & -8 \\ 7 & -7 & 0 & -8 \\ -6 & 3 & 9 & 7 \end{bmatrix}$ **B21.** $\begin{bmatrix} 5 & -12 & 11 & -2 \\ 5 & -10 & 11 & 9 \\ -11 & 4 & -12 & -9 \end{bmatrix}$

B22. $\begin{bmatrix} -8 & 2 & 6 & 8 \\ -3 & -9 & -2 & 6 \\ -3 & -2 & 1 & 12 \end{bmatrix}$ **B23.** $\begin{bmatrix} 4 & -6 & -7 & 1 & 1 \\ -2 & -5 & -1 & -2 & -5 \\ 9 & -3 & -9 & -3 & -10 \\ -12 & 0 & 6 & -2 & 1 \end{bmatrix}$

B24. $\begin{bmatrix} 8 & -4 & 1 & 5 & -10 \\ 11 & -5 & -2 & -2 & -2 \\ -10 & -5 & 6 & -5 & 10 \\ -9 & 5 & -10 & -2 & -12 \end{bmatrix}$ **B25.** $\begin{bmatrix} -7 & -6 & -3 & -3 & -7 \\ -12 & -10 & -6 & -4 & -9 \\ 0 & -11 & 9 & 10 & -9 \\ 8 & 4 & 11 & -2 & 7 \end{bmatrix}$

B26. $\begin{bmatrix} -3 & -3 & 3 & -8 & -3 \\ 7 & -9 & -1 & 6 & 7 \\ 5 & -4 & 8 & 0 & -12 \\ -10 & -8 & -11 & -7 & 9 \end{bmatrix}$ **B27.** $\begin{bmatrix} -1 & -10 & 0 & 4 & -2 \\ 8 & -9 & 7 & 3 & -6 \\ 5 & 3 & -4 & -11 & 1 \\ 10 & 2 & 8 & 10 & -1 \end{bmatrix}$

12.6 Homogeneous Equations and the Null Space

Homogeneous Equations

Homogeneous linear algebraic equations are of the form $\sum_{k=1}^{n} a_k x_k = 0$, where the a_k are coefficients and the x_k are variables. A pair of such homogeneous equations in two unknowns represents a pair of lines each passing through the origin. Three such equations in three unknowns represent three planes passing through the origin.

An $n \times n$ system of homogeneous linear equations of the form $A\mathbf{x} = \mathbf{0}$ either has $\mathbf{x} = \mathbf{0}$ as its only solution or infinitely many solutions. In the first case, det $A \neq 0$, and in the second, det $A = 0$. For clarity, we formalize these remarks as a theorem.

THEOREM 12.1

If det $A \neq 0$, then $A\mathbf{x} = \mathbf{0}$ has $\mathbf{x} = \mathbf{0}$ as its only solution.
If det $A = 0$, then $A\mathbf{x} = \mathbf{0}$ has infinitely many solutions.

Table 12.1 lists five examples that illustrate Theorem 12.1.

Example	Equations	det A	Solution
12.5	$2x + 3y = 0$ $5x - 7y = 0$	$\begin{vmatrix} 2 & 3 \\ 5 & -7 \end{vmatrix} = -29$	$\begin{bmatrix} 0 \\ 0 \end{bmatrix}$
12.6	$2x + 3y = 0$ $7x + \frac{21}{2}y = 0$	$\begin{vmatrix} 2 & 3 \\ 7 & \frac{21}{2} \end{vmatrix} = 0$	$t \begin{bmatrix} -\frac{3}{2} \\ 1 \end{bmatrix}$
12.7	$2x + 3y + 4z = 0$ $5x - y - 7z = 0$ $x - 2y + 6z = 0$	$\begin{vmatrix} 2 & 3 & 4 \\ 5 & -1 & -7 \\ 1 & -2 & 6 \end{vmatrix} = -187$	$\begin{bmatrix} 0 \\ 0 \\ 0 \end{bmatrix}$
12.8	$2x + 3y + 4z = 0$ $2x + 3y + 4z = 0$ $x - 2y + 6z = 0$	$\begin{vmatrix} 2 & 3 & 4 \\ 3 & \frac{9}{2} & 6 \\ 1 & -2 & 6 \end{vmatrix} = 0$	$t \begin{bmatrix} -\frac{26}{7} \\ \frac{8}{7} \\ 1 \end{bmatrix}$
12.9	$2x + 3y + 4z = 0$ $3x + \frac{9}{2}y + 6z = 0$ $5x + \frac{15}{2}y + 10z = 0$	$\begin{vmatrix} 2 & 3 & 4 \\ 3 & \frac{9}{2} & 6 \\ 5 & \frac{15}{2} & 10 \end{vmatrix} = 0$	$t_1 \begin{bmatrix} -\frac{3}{2} \\ 1 \\ 0 \end{bmatrix} + t_2 \begin{bmatrix} -2 \\ 0 \\ 1 \end{bmatrix}$

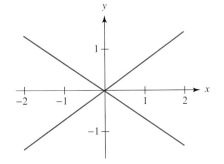

FIGURE 12.8   Homogeneous equations in Example 12.5

TABLE 12.1   Examples of homogeneous equations

EXAMPLE 12.5   The equations of Example 1, Table 12.1, implicitly define the lines graphed in Figure 12.8. The two lines are distinct and pass through the origin. The only solution to this set of equations is $(x, y) = (0, 0)$. In fact, the array of system coefficients has the nonzero determinant

−29, so we can use Cramer's rule to solve this (homogeneous) system, obtaining

$$x = \frac{\begin{vmatrix} 0 & 3 \\ 0 & -7 \end{vmatrix}}{\begin{vmatrix} 2 & 3 \\ 5 & -7 \end{vmatrix}} = \frac{0}{-29} = 0 \qquad y = \frac{\begin{vmatrix} 2 & 0 \\ 5 & 0 \end{vmatrix}}{\begin{vmatrix} 2 & 3 \\ 5 & -7 \end{vmatrix}} = \frac{0}{-29} = 0$$

Thus, for homogeneous equations, if the system determinant is nonzero, then the only solution for the system is zero. For each variable, the determinant in the numerator is of a matrix with a column of zeros. Such determinants are always zero.   ❖

EXAMPLE 12.6   The equations of Example 12.6, Table 12.1, are not distinct, as Figure 12.9 shows. In fact, multiplying the second equation by $\frac{2}{7}$ converts it to the first. Moreover, the determinant of the system array is now zero, so Cramer's rule does not apply. However, there are now an infinite number of solutions, obtained by solving the first equation for $x = -\frac{3}{2}y$. For each value of y, there is a corresponding value of x, namely, $x = -\frac{3}{2}y$. Alternatively, this solution can be written in terms of a parameter t as $x = -\frac{3}{2}t$, $y = t$. For every possible real number t, there is a corresponding pair $(x, y) = (-\frac{3}{2}t, t)$. We can also express these solutions in vector form, as listed in Table 12.1. Thus, for homogeneous equations, a zero determinant means many solutions; whereas a nonzero determinant means just one solution, the zero solution.   ❖

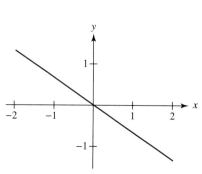

FIGURE 12.9   Homogeneous equations in Example 12.6

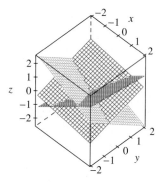

FIGURE 12.10   Homogeneous equations in Example 12.7

EXAMPLE 12.7   Each equation in Example 12.7, Table 12.1, represents a plane passing through the origin. Figure 12.10 suggests the three planes are distinct, so the only solution of these three equations is the zero solution, namely, $(x, y, z) = (0, 0, 0)$. Indeed, the system matrix has the nonzero determinant −187, so Cramer's rule applies and the only solution is the zero solution.   ❖

EXAMPLE 12.8   The equations of Example 12.8, Table 12.1, are graphed in Figure 12.11, which suggests two of the planes are coincident, so there might be only two distinct equations. The determinant of the system matrix is zero, so Cramer's rule does not apply. From the reduced row-echelon form of the system matrix in (12.13), we obtain the multiple solutions $(x, y, z) = (-\frac{26}{7}t, \frac{8}{7}t, t)$, which Table 12.1 and (12.13) give in vector form. From the reduced row-echelon form of the system matrix in (12.13), we see that it has two nonzero rows and, hence, has rank 2. The rank is consistent with the number of distinct equations we found the system to contain

$$\begin{bmatrix} 2 & 3 & 4 \\ 3 & \frac{9}{2} & 6 \\ 1 & -2 & 6 \end{bmatrix} \longrightarrow \begin{bmatrix} 1 & 0 & \frac{26}{7} \\ 0 & 1 & -\frac{8}{7} \\ 0 & 0 & 0 \end{bmatrix} \Rightarrow \begin{cases} x = -\frac{26}{7}z \\ y = \frac{8}{7}z \\ z = z \end{cases} \Rightarrow \mathbf{x} = t \begin{bmatrix} -\frac{26}{7} \\ \frac{8}{7} \\ 1 \end{bmatrix} \quad (12.13)$$

❖

EXAMPLE 12.9

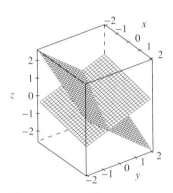

FIGURE 12.11 Homogeneous equations in Example 12.8

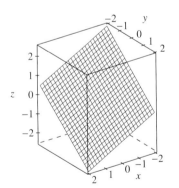

FIGURE 12.12 Homogeneous equations in Example 12.9

Figure 12.12, a graph of the equations in Example 12.9, Table 12.1, suggests there may be only one distinct plane and, hence, one distinct equation. The determinant of the system matrix is zero, so Cramer's rule cannot be used. From the reduced row-echelon form of the system matrix in (12.14), we obtain the solution $(x, y, z) = (-\frac{3}{2}t_1 - 2t_2, t_1, t_2)$, which now depends on the two arbitrary constants, t_1 and t_2. From the reduced row-echelon form of the system matrix in (12.14), we see it has only one nonzero row, so the rank is 1. The rank is consistent with the number of distinct nonzero equations we found the system to contain

$$\begin{bmatrix} 2 & 3 & 4 \\ 3 & \frac{9}{2} & 6 \\ 5 & \frac{15}{2} & 10 \end{bmatrix} \longrightarrow \begin{bmatrix} 1 & \frac{3}{2} & 2 \\ 0 & 0 & 0 \\ 0 & 0 & 0 \end{bmatrix} \Rightarrow \begin{cases} x = -\frac{3}{2}y - 2z \\ y = y \\ z = z \end{cases} \Rightarrow \mathbf{x} = t_1 \begin{bmatrix} -\frac{3}{2} \\ 1 \\ 0 \end{bmatrix} + t_2 \begin{bmatrix} -2 \\ 0 \\ 1 \end{bmatrix}$$

$$(12.14)$$

Table 12.1 and (12.14) express the *general solution* in terms of the two "basis" vectors

$$\mathbf{v}_5 = \begin{bmatrix} -\frac{3}{2} \\ 1 \\ 0 \end{bmatrix} \quad \text{and} \quad \mathbf{v}_6 = \begin{bmatrix} -2 \\ 0 \\ 1 \end{bmatrix}$$

Any member of the general solution can be realized as a sum of the form $a\mathbf{v}_5 + b\mathbf{v}_6$, where a and b are constants. Such a sum is known as a ***linear combination.*** Any such member of the general solution lies in the plane described by the one distinct equation to which the system reduces and that plane can be described by the single equation $x + \frac{3}{2}y + z = 0$. ❖

The Null Space

The collection of all solutions to a homogeneous linear system is called the *null space.* Vectors whose linear combinations give all members of the solution set are said to *span,* or generate, the null space. If a set which spans the null space is "minimal" because no vector in it is a linear combination of the others, it is called a *basis.* Table 12.2 lists bases for the null spaces of the examples in Table 12.1. In Examples 12.5 and 12.7, the only solution is $\mathbf{x} = \mathbf{0}$ so the bases for these null spaces are empty.

In Example 12.6, the null space consists of all multiples of a single vector and any multiple of that single vector can be considered a basis vector for the null space. This is again the case in Example 12.8. All solutions are a multiple of a single vector, so a basis for the null space consists of a single vector. That vector can be any multiple of the vector used to express the solution set, and often, it is convenient to use a multiple whose components have no fractions.

In Example 12.9, the general solution contains all vectors lying in the single distinct plane represented by the given equations. It takes two noncolinear vectors in a plane to express all other vectors in that plane. The reduced row-echelon form of the system matrix led to the vectors $\mathbf{v}_5$ and $\mathbf{v}_6$ in (12.14), and Table 12.2 lists these two vectors as a basis

	Solution	Basis for Null Space
12.5	$\mathbf{x} = \mathbf{0}$	$\{\mathbf{0}\}$
12.6	$\mathbf{x} = t \begin{bmatrix} -\frac{3}{2} \\ 1 \end{bmatrix}$	$\left\{ \begin{bmatrix} -\frac{3}{2} \\ 1 \end{bmatrix} \right\}$ or $\left\{ \begin{bmatrix} -3 \\ 2 \end{bmatrix} \right\}$
12.7	$\mathbf{x} = \mathbf{0}$	$\{\mathbf{0}\}$
12.8	$\mathbf{x} = t \begin{bmatrix} -\frac{26}{7} \\ \frac{8}{7} \\ 1 \end{bmatrix}$	$\left\{ \begin{bmatrix} -\frac{26}{7} \\ \frac{8}{7} \\ 1 \end{bmatrix} \right\}$ or $\left\{ \begin{bmatrix} -26 \\ 8 \\ 7 \end{bmatrix} \right\}$
12.9	$\mathbf{x} = t_1 \begin{bmatrix} -\frac{3}{2} \\ 1 \\ 0 \end{bmatrix} + t_2 \begin{bmatrix} -2 \\ 0 \\ 1 \end{bmatrix}$	$\left\{ \begin{bmatrix} -\frac{3}{2} \\ 1 \\ 0 \end{bmatrix}, \begin{bmatrix} -2 \\ 0 \\ 1 \end{bmatrix} \right\}$ or $\left\{ \begin{bmatrix} -3 \\ 2 \\ 0 \end{bmatrix}, \begin{bmatrix} -2 \\ 0 \\ 1 \end{bmatrix} \right\}$

TABLE 12.2   Null spaces for examples in Table 12.1

for the null space. Additionally, Table 12.2 lists fraction-free multiples of $\mathbf{v}_5$ and $\mathbf{v}_6$ in an alternative basis for the null space. In fact, any two linear combinations of $\mathbf{v}_5$ and $\mathbf{v}_6$ which result in noncolinear vectors can be taken as a basis for the null space. For example, the vectors $\mathbf{U} = 2\mathbf{v}_5 + \mathbf{v}_6$ and $\mathbf{V} = 4\mathbf{v}_5 - \mathbf{v}_6$, given in (12.15), can be taken as a basis for the null space in Example 12.9

$$\mathbf{U} = 2 \begin{bmatrix} -\frac{3}{2} \\ 1 \\ 0 \end{bmatrix} + \begin{bmatrix} -2 \\ 0 \\ 1 \end{bmatrix} = \begin{bmatrix} -5 \\ 2 \\ 1 \end{bmatrix} \quad \text{and} \quad \mathbf{V} = 4 \begin{bmatrix} -\frac{3}{2} \\ 1 \\ 0 \end{bmatrix} - \begin{bmatrix} -2 \\ 0 \\ 1 \end{bmatrix} = \begin{bmatrix} -4 \\ 4 \\ -1 \end{bmatrix} \quad \textbf{(12.15)}$$

Linear Dependence and Independence

Section 5.4 describes linear dependence and independence of functions. The functions $f(x)$ and $g(x)$ are linearly dependent if the equation $af(x) + bg(x) = 0$ has nonzero solutions for the constants a and b. If the only solution to this equation is $a = b = 0$, then the functions are linearly independent.

The concepts of linear dependence and independence also apply to vectors. A collection of vectors $\{\mathbf{v}_k\}, k = 1, \ldots, n$, is said to be *linearly independent* if the only solution of the vector equation $\sum_{k=1}^{n} s_k \mathbf{v}_k = \mathbf{0}$ is the zero solution $s_k = 0, k = 1, \ldots, n$. If there are nonzero solutions for the constants s_k, then the vectors $\{\mathbf{v}_k\}$ are said to be *linearly dependent*. $s_1 \neq 0$, then we can express $\mathbf{v}_1$ in terms of the other vectors in the collection by writing $\mathbf{v}_1 = -\sum_{k=2}^{n} (s_k/s_1) \mathbf{v}_k$.

In practice, the condition $\sum_{k=1}^{n} s_k \mathbf{v}_k = \mathbf{0}$ can be applied by writing the augmented matrix $A = [\mathbf{v}_1, \ldots, \mathbf{v}_n]$ and determining if the vector $\mathbf{s}$, whose components are $s_k, k = 1, \ldots, n$, is $\mathbf{0}$ or not. If the null space of A contains more than just $\mathbf{0}$, then a solution $\mathbf{s} \neq \mathbf{0}$ exists and the columns of A are linearly dependent. If the basis for the null space of A contains only $\mathbf{0}$, then $\mathbf{s} = \mathbf{0}$ and the columns of A are linearly independent.

EXAMPLE 12.10   Let $\mathbf{v}_k, k = 1, 2, 3$, in $\sum_{k=1}^{3} s_k \mathbf{v}_k = \mathbf{0}$ be the column vectors given in (12.16), and let $A = [\mathbf{v}_1, \mathbf{v}_2, \mathbf{v}_3]$ be the matrix whose columns are the vectors $\mathbf{v}_k$. Since $\det A = 0$, the system of homogeneous equations in (12.16) has multiple solutions, so there are nonzero solutions for the constants s_k. Hence, the given vectors are linearly dependent. (See Exercise A1.)

$$s_1 \begin{bmatrix} 1 \\ 2 \\ 3 \end{bmatrix} + s_2 \begin{bmatrix} -1 \\ 0 \\ 1 \end{bmatrix} + s_3 \begin{bmatrix} -2 \\ 2 \\ 4 \end{bmatrix} = \begin{bmatrix} 1 & -1 & -2 \\ 2 & 0 & 2 \\ 3 & 1 & 6 \end{bmatrix} \begin{bmatrix} s_1 \\ s_2 \\ s_3 \end{bmatrix} = \begin{bmatrix} 0 \\ 0 \\ 0 \end{bmatrix} \qquad (12.16)$$

EXERCISES 12.6–Part A

A1. Obtain at least one nonzero solution for the s_k in (12.16), thereby confirming the linear dependence of the vectors $\mathbf{v}_k$ in Example 12.10.

A2. Determine the rank of each of the following matrices. (See the end of Section 12.5.)

(a) $\begin{bmatrix} -24 & -36 & -20 \\ -30 & -45 & -25 \\ 48 & 72 & 40 \end{bmatrix}$ (b) $\begin{bmatrix} 6 & -10 & -1 \\ -3 & -8 & -12 \\ 8 & -9 & 1 \end{bmatrix}$ (c) $\begin{bmatrix} -26 & -25 & -31 \\ 70 & 71 & 89 \\ 26 & 17 & 19 \end{bmatrix}$

A3. Determine bases for the null spaces of the matrices in Exercise A2.

A4. The *dimension* of the null space of the matrix A is the number of (necessarily linearly independent) vectors in a basis for the null space. For each matrix A in Exercise A2, show that the dimension of the null space is $3 - \text{rank } A$.

A5. For each matrix A in Exercise A2, write the general solution of the homogeneous system $A\mathbf{x} = \mathbf{0}$ in vector form.

EXERCISES 12.6–Part B

B1. For each of the following matrices, determine the rank and a basis for the null space.

(a) $\begin{bmatrix} 70 & -64 & 62 \\ 15 & -6 & 3 \\ 0 & 18 & -24 \end{bmatrix}$ (b) $\begin{bmatrix} -68 & -50 & 6 & -56 \\ -46 & -37 & -15 & -22 \\ -2 & 9 & 63 & -54 \\ -12 & -22 & -78 & 56 \end{bmatrix}$

(c) $\begin{bmatrix} 26 & 62 & -4 & 77 \\ -96 & -36 & 40 & -57 \\ 57 & 81 & -136 & 57 \\ 70 & 26 & 16 & 59 \end{bmatrix}$

In Exercises B2–8, determine if the columns of the given matrices are linearly independent or dependent. Support your conclusions with appropriate evidence.

B2. $\begin{bmatrix} -10 & 26 & -24 \\ -12 & 13 & 57 \\ -45 & 96 & -9 \end{bmatrix}$ **B3.** $\begin{bmatrix} -9 & 12 & -12 \\ 6 & -4 & -7 \\ 3 & 5 & -9 \end{bmatrix}$

B4. $\begin{bmatrix} 27 & -27 & -6 \\ -63 & 63 & 14 \\ -81 & 81 & 18 \end{bmatrix}$ **B5.** $\begin{bmatrix} 5 & -10 & -11 & 6 \\ -2 & 12 & -9 & 10 \\ 11 & 1 & 6 & -10 \\ -12 & -9 & -3 & 3 \end{bmatrix}$

B6. $\begin{bmatrix} 36 & 81 & -103 \\ 10 & -2 & -26 \\ 29 & -28 & 0 \\ -20 & -13 & 59 \end{bmatrix}$ **B7.** $\begin{bmatrix} 3 & 4 & 2 \\ 12 & 16 & 8 \\ 27 & 36 & 18 \\ 6 & 8 & 4 \end{bmatrix}$

B8. $\begin{bmatrix} -35 & 23 & 40 \\ 8 & -19 & -62 \\ 59 & -15 & 24 \\ 51 & -22 & -14 \end{bmatrix}$

In Exercises B9–14, $n \times n$ matrices A of different rank are given. If each matrix represents a homogeneous system $A\mathbf{X} = \mathbf{0}$,

(a) compute the determinant of A.

(b) if $\det A \ne 0$, show that Cramer's rule gives the unique solution $\mathbf{X} = \mathbf{0}$.

(c) if $\det A = 0$, obtain the general solution, making free use of symbolic technology where appropriate.

(d) from the general solution, determine a basis for the null space of A.

(e) determine the dimension of the null space of A; that is, determine the number of independent vectors in a basis for the null space of A.

(f) obtain the reduced row-echelon form of A and, from it, construct the general solution to the homogeneous system of equations.

(g) obtain a basis for the null space of A by using a computer algebra system's built-in functionality.

(h) determine the *rank* of A; show that in each case it is equal to the number of independent rows in A.

(i) verify that the rank of A plus the dimension of the null space of A adds up to n.

B9. $\begin{bmatrix} 7 & 6 \\ -5 & 1 \end{bmatrix}$ **B10.** $\begin{bmatrix} 54 & 24 \\ 72 & 32 \end{bmatrix}$ **B11.** $\begin{bmatrix} -54 & -36 & -54 \\ 12 & 8 & 12 \\ 54 & 36 & 54 \end{bmatrix}$

B12. $\begin{bmatrix} 36 & 44 & 40 \\ -72 & 22 & 74 \\ 81 & -36 & -99 \end{bmatrix}$ **B13.** $\begin{bmatrix} -7 & 6 & 3 & -52 \\ 41 & 74 & 115 & -31 \\ -8 & 9 & 27 & -44 \\ 4 & 10 & 38 & 13 \end{bmatrix}$

B14. $\begin{bmatrix} 68 & 51 & -6 & 49 \\ -17 & 51 & 12 & 38 \\ 20 & -35 & -10 & -25 \\ -31 & 33 & 12 & 22 \end{bmatrix}$

If I is the identity matrix of the appropriate dimension, for each matrix A given in Exercises B15–19:

   (a) Obtain all values of λ that satisfy the equation $\det(A - \lambda I) = 0$.

   (b) For each value of $\lambda = \lambda_k$ found in part (a), form the matrix $C_k = A - \lambda_k I$.

   (c) Find a basis for the null space of each C_k found in part (b).

   (d) Show that any vector $\mathbf{v}_k$ in the null space of C_k satisfies the equation $A\mathbf{v}_k = \lambda_k \mathbf{v}_k$.

B15. $\begin{bmatrix} 6 & 5 \\ -7 & -11 \end{bmatrix}$ **B16.** $\begin{bmatrix} -3 & -5 & -65 \\ -5 & -7 & -41 \\ 2 & 3 & 22 \end{bmatrix}$

B17. $\begin{bmatrix} -7 & -3 & -16 \\ -32 & 1 & -100 \\ 0 & 1 & -2 \end{bmatrix}$ **B18.** $\begin{bmatrix} -177 & 116 & -524 & 112 \\ -14 & 7 & -35 & 5 \\ 77 & -51 & 230 & -50 \\ 96 & -63 & 286 & -62 \end{bmatrix}$

B19. $\begin{bmatrix} -17 & 18 & 19 & -91 \\ -15 & 15 & 18 & -80 \\ 32 & -30 & -26 & 142 \\ 6 & -6 & -5 & 28 \end{bmatrix}$

In each of Exercises B20–25, a matrix A and a vector $\mathbf{Y}$ is given. For each:

   (a) Show that $\det A = 0$.

   (b) Obtain a basis for the null space of A.

   (c) Using a computer algebra system, obtain $\mathbf{X}$, the general solution of the system $A\mathbf{X} = \mathbf{Y}$.

   (d) Show that $\mathbf{X} = \mathbf{X}_p + \mathbf{X}_h$, the sum of $\mathbf{X}_p$, a particular solution of the nonhomogeneous equation, and $\mathbf{X}_h$, an arbitrary member of the null space of A, which is therefore the general solution of the homogeneous equation $A\mathbf{X} = \mathbf{0}$.

B20. $\begin{bmatrix} -51 & -29 & -58 \\ -45 & -45 & -57 \\ 9 & -9 & 6 \end{bmatrix}, \begin{bmatrix} 305 \\ 339 \\ -3 \end{bmatrix}$

B21. $\begin{bmatrix} 14 & 35 & 3 \\ -32 & -47 & 23 \\ -63 & -63 & 72 \end{bmatrix}, \begin{bmatrix} -164 \\ 246 \\ 369 \end{bmatrix}$ **B22.** $\begin{bmatrix} 66 & 54 & 18 \\ 38 & 35 & 5 \\ 78 & 56 & 32 \end{bmatrix}, \begin{bmatrix} -60 \\ 4 \\ -148 \end{bmatrix}$

B23. $\begin{bmatrix} -48 & -10 & 56 & 44 \\ -12 & -10 & 14 & 6 \\ 36 & 12 & -42 & -30 \\ -6 & -23 & 7 & -9 \end{bmatrix}, \begin{bmatrix} 110 \\ 10 \\ -72 \\ -37 \end{bmatrix}$

B24. $\begin{bmatrix} -32 & 10 & -75 & -13 \\ 8 & -5 & 16 & 26 \\ -40 & -33 & 41 & -51 \\ 64 & 52 & -34 & 10 \end{bmatrix}, \begin{bmatrix} 503 \\ -145 \\ -12 \\ -38 \end{bmatrix}$

B25. $\begin{bmatrix} -39 & -35 & 18 & -65 \\ -39 & -35 & 18 & -65 \\ -32 & -32 & 8 & -64 \\ 23 & 19 & -14 & 33 \end{bmatrix}, \begin{bmatrix} 129 \\ 129 \\ 152 \\ -53 \end{bmatrix}$

12.7   Inverses

The Matrix Inverse

If A and B are square matrices ($n \times n$) and $AB = I$, the $n \times n$ identity, then B is called the *multiplicative inverse* of A, much like $\frac{1}{3}$ is the multiplicative inverse of 3. Not every matrix A has an inverse, but if A does have an inverse it is often denoted by the symbol A^{-1}. Matrices that have an inverse are called *nonsingular*, or *invertible*.

The Inverse by Brute Force

Given the matrix $A = \begin{bmatrix} 1 & 2 \\ 3 & 4 \end{bmatrix}$ we seek its multiplicative inverse using brute force. Hence, seek the matrix $B = \begin{bmatrix} a & b \\ c & d \end{bmatrix}$ for which the product $AB = I$, where I is the 2×2 identity matrix $I = \begin{bmatrix} 1 & 0 \\ 0 & 1 \end{bmatrix}$. If we equate the product AB with I, we form the four equations

$$\begin{array}{ll} a + 2c = 1 & b + 2d = 0 \\ 3a + 4c = 0 & \text{and} \quad 3b + 4d = 1 \end{array} \tag{12.17}$$

in the four unknowns a, b, c, and d. Solving these equations leads to the matrix $B = \begin{bmatrix} -2 & 1 \\ \frac{3}{2} & -\frac{1}{2} \end{bmatrix}$ for which $AB = BA = I$. Thus, we have computed B, the multiplicative inverse of the matrix A.

Inverse by Gaussian Elimination

As an alternative to the brute force method, we can find the inverse of A by augmenting A with I and then row reducing this augmented matrix so that the portion where A was becomes I. The portion where I was will be A^{-1}. Complete (Gauss–Jordan) row reduction of the augmented matrix $[A, I] = \begin{bmatrix} 1 & 2 & 1 & 0 \\ 3 & 4 & 0 & 1 \end{bmatrix}$ leads to $[I, A^{-1}] = \begin{bmatrix} 1 & 0 & -2 & 1 \\ 0 & 1 & \frac{3}{2} & -\frac{1}{2} \end{bmatrix}$. A glance at (12.17) shows that the product AB takes place by columns. Writing B as $[\mathbf{b}_1, \mathbf{b}_2]$, where $\mathbf{b}_1$ and $\mathbf{b}_2$ are the columns of B, shows that $AB = [A\mathbf{b}_1, A\mathbf{b}_2]$. Hence, solving for B amounts to solving the two systems $A\mathbf{b}_1 = \mathbf{e}_1$ and $A\mathbf{b}_2 = \mathbf{e}_2$, where $\mathbf{e}_1$ and $\mathbf{e}_2$ are the columns of I. Instead of forming two different augmented matrices, we have combined $[A, \mathbf{e}_1]$ and $[A, \mathbf{e}_2]$ to form the one augmented matrix $[A, \mathbf{e}_1, \mathbf{e}_2] = [A, I]$.

A Matrix with No Inverse

A square $(n \times n)$ matrix A having no inverse is said to be *singular*, or *noninvertible*.

Why would a matrix not have an inverse? What symptoms would indicate that a matrix would not be invertible? We will show that A is invertible if $\det A \neq 0$ and singular if $\det A = 0$.

For example, augmenting to $[A, I]$ the matrix A on the left in (12.18) and row-reducing lead to the three inconsistent equations $0 = 1, 0 = -2$, and $0 = 1$, shown in the bottom row of the reduced matrix on the right in (12.18)

$$A = \begin{bmatrix} 1 & 2 & 3 \\ 4 & 5 & 6 \\ 7 & 8 & 9 \end{bmatrix} \Rightarrow [A, I] = \begin{bmatrix} 1 & 2 & 3 & 1 & 0 & 0 \\ 4 & 5 & 6 & 0 & 1 & 0 \\ 7 & 8 & 9 & 0 & 0 & 1 \end{bmatrix} \Rightarrow \begin{bmatrix} 1 & 0 & -1 & 0 & -\frac{8}{3} & \frac{5}{3} \\ 0 & 1 & 2 & 0 & \frac{7}{3} & -\frac{4}{3} \\ 0 & 0 & 0 & 1 & -2 & 1 \end{bmatrix}$$

(12.18)

If $C = [\mathbf{c}_1, \mathbf{c}_2, \mathbf{c}_3]$ is a candidate for the inverse of A and the columns of I are $\mathbf{e}_k, k = 1, 2, 3$, then the equation $AC = I$ decomposes into the three sets of equations $A\mathbf{c}_k = \mathbf{e}_k, k = 1, 2, 3$. Solving any one of these sets of equations by Cramer's rule would require the determinant of A to be nonzero. However, we find that the determinant of A is 0, so Cramer's rule does not apply. The row reductions that lead to three zeros as the third row of the reduced A do not likewise lead to a zero as the third component in any of the vectors $\mathbf{e}_k$. The rank of A is not $n = 3$, the rank of I. When A row-reduces to an echelon matrix of rank less than n, the determinant will be zero and the equations in the reduced form of $[A, I]$ will be inconsistent. Thus, if $\det A = 0$, the inverse of A will not exist. Correspondingly, if $\det A \neq 0$, then Cramer's rule will lead to solutions for the vectors $\mathbf{c}_k$ and, hence, to $C = A^{-1}$.

EXERCISES 12.7–Part A

A1. If $A = \begin{bmatrix} 1 & 2 \\ 3 & 4 \end{bmatrix}$:

    (a) Show A is nonsingular.

    (b) Obtain A^{-1} by solving $AB = I$ for $B = A^{-1}$, element by element.

    (c) Obtain A^{-1} by row-reducing $[A, I]$ to $[I, A^{-1}]$.

A2. If $A = \begin{bmatrix} 9 & -63 \\ 8 & -56 \end{bmatrix}$:

    (a) Show $\det A = 0$.

    (b) Show that the reduced row-echelon form of A is not I.

    (c) Show the reduced row-echelon form of $[A, I]$ contains an inconsistent equation, so that A^{-1} does not exist.

A3. If $A = [a_1, \ldots, a_n]$ is an $n \times n$ (square) matrix whose kth column is $\mathbf{a}_k$, and $AB = I_n$, show that $BA = I_n$. *Hint:* Argue that $\det(B) \neq 0$ so that, by Cramer's rule, each $B\mathbf{a}_k = \mathbf{e}_k$, $k = 1, \ldots, n$, has a solution, where $\mathbf{e}_k$ is the kth column of I_n.

A4. Suppose $AB = I$, making $B = A^{-1}$ the inverse of A. Suppose further that C is proposed as a different candidate for the inverse of A. Show that $C = A^{-1}$, making the inverse unique. *Hint:* If C is an inverse of A, then $AC = I$.

A5. If $A = \begin{bmatrix} a & b \\ c & d \end{bmatrix}$ and $ad - bc \neq 0$, the inverse can be found by solving $A\mathbf{b}_1 = \mathbf{e}_1$ and $A\mathbf{b}_2 = \mathbf{e}_2$ for $B = [\mathbf{b}_1, \mathbf{b}_2]$, where $\mathbf{e}_1$ and $\mathbf{e}_2$

are the columns of the 2×2 identity matrix. Show this leads to $B = A^{-1} = \frac{1}{ad-bc}\begin{bmatrix} d & -b \\ -c & a \end{bmatrix}$ as a simple rule for finding the inverse of a 2×2 matrix.

A6. The *transpose* of the $n \times m$ matrix A, denoted A^{T}, is the $m \times n$ matrix formed by interchanging the rows and columns of A. Thus, the rows of A^{T} are the columns of A, and the columns of A^{T} are the rows of A. Let A_1 and A_2 be the matrices A in Exercises A1 and 2, respectively. Show that

    (a) $(A_1A_2)^{\mathrm{T}} = A_2^{\mathrm{T}}A_1^{\mathrm{T}}$ **(b)** $\det(A_1) = \det(A_1^{\mathrm{T}})$

EXERCISES 12.7–Part B

B1. If $A = \begin{bmatrix} -3 & -1 & 7 \\ -9 & 5 & 0 \\ 2 & -11 & -10 \end{bmatrix}$ and $\mathbf{Y} = \begin{bmatrix} 2 \\ -3 \\ 5 \end{bmatrix}$, obtain $\mathbf{X}$, the solution of $A\mathbf{X} = \mathbf{Y}$ as $\mathbf{X} = A^{-1}\mathbf{Y}$. Verify this solution by row-reducing $[A, \mathbf{Y}]$ to $[I, \mathbf{X}]$.

For the matrices A given in Exercises B2–7:

    (a) Obtain $\det A$, since A^{-1} exists only if $\det(A) \neq 0$.

    (b) Obtain A^{-1} using a modern technological device.

    (c) Verify that $AA^{-1} = A^{-1}A = I$.

    (d) Obtain A^{-1} by "brute force," writing and solving the n^2 equations arising from $AA^{-1} = I$.

    (e) Obtain $A^{-1} = [\mathbf{X}_1, \mathbf{X}_2, \ldots, \mathbf{X}_n]$ column by column by solving the n linear systems $A\mathbf{X}_k = \mathbf{e}_k$, $k = 1, \ldots, n$, where $\mathbf{e}_k$ is the kth column of I.

    (f) Obtain A^{-1} using Gaussian elimination to bring the augmented matrix $[A, I]$ to the form $[I, A^{-1}]$.

    (g) Obtain A^{-1} by the formula $A^{-1} = \frac{\mathrm{adj}\,A}{\det A}$, where adj A is the transpose of the matrix of cofactors. Thus, each element a_{ij} in A is replaced by $(-1)^{i+j}M_{ij}$, its signed minor, and the resulting matrix is transposed (see Exercise A6) to form adj A, the matrix called the *adjoint* (or conjoint, adjunct, adjugate, and a host of other names). The naming problem arises because *adjoint,* the most common name for this matrix, is already used in a different context in linear algebra and mathematics in general.

B2. $\begin{bmatrix} 11 & -12 & 7 \\ -11 & 12 & 10 \\ 1 & 3 & 7 \end{bmatrix}$ **B3.** $\begin{bmatrix} -1 & -1 & -3 \\ 4 & -6 & 8 \\ 11 & 12 & 6 \end{bmatrix}$

B4. $\begin{bmatrix} 7 & -11 & -12 \\ -6 & 8 & -2 \\ -7 & 4 & -1 \end{bmatrix}$ **B5.** $\begin{bmatrix} -12 & -2 & 9 & -8 \\ -6 & 12 & 5 & 10 \\ -3 & 6 & 7 & -4 \\ 10 & 2 & 2 & 9 \end{bmatrix}$

B6. $\begin{bmatrix} -9 & -2 & 7 & 4 \\ -10 & -7 & -7 & 5 \\ -1 & -5 & 7 & 11 \\ -2 & 0 & -2 & -2 \end{bmatrix}$ **B7.** $\begin{bmatrix} -1 & 4 & -5 & 7 \\ 6 & 4 & 10 & 4 \\ -12 & 1 & 7 & -3 \\ -6 & 5 & 7 & -9 \end{bmatrix}$

By the formula $A^{-1} = \frac{\mathrm{adj}\,A}{\det A}$, the inverse of the 2×2 matrix $A = \begin{bmatrix} a & b \\ c & d \end{bmatrix}$ is $\frac{1}{ad-bc}\begin{bmatrix} d & -b \\ -c & a \end{bmatrix}$ because adj A is formed by interchanging the two entries on the main diagonal and negating the other two entries. For each of the 2×2 matrices A in Exercises B8–10:

    (a) Obtain the inverse by this method.

    (b) Verify the inverse is correct by showing $AA^{-1} = I$.

B8. $\begin{bmatrix} -10 & 12 \\ -2 & -1 \end{bmatrix}$ **B9.** $\begin{bmatrix} 11 & -9 \\ 11 & 8 \end{bmatrix}$ **B10.** $\begin{bmatrix} 7 & -9 \\ 3 & -7 \end{bmatrix}$

For the matrix A and vector $\mathbf{Y}$ given in each of Exercises B11–20:

    (a) Use an efficient solver to obtain $\mathbf{X}$, the solution of $A\mathbf{X} = \mathbf{Y}$.

    (b) Calculate $\mathbf{X} = A^{-1}\mathbf{Y}$. (The reader is cautioned to accept this exercise as theoretical only and not to believe that computing an inverse is an efficient way of solving a linear system.)

B11. $\begin{bmatrix} -9 & -6 \\ 5 & -8 \end{bmatrix}, \begin{bmatrix} -4 \\ -10 \end{bmatrix}$ **B12.** $\begin{bmatrix} 11 & -12 \\ -10 & 9 \end{bmatrix}, \begin{bmatrix} -8 \\ -3 \end{bmatrix}$

B13. $\begin{bmatrix} 10 & -6 \\ -4 & 10 \end{bmatrix}, \begin{bmatrix} 6 \\ 4 \end{bmatrix}$ **B14.** $\begin{bmatrix} 8 & -4 & -6 \\ -3 & 6 & 10 \\ 6 & 6 & 6 \end{bmatrix}, \begin{bmatrix} 9 \\ -5 \\ -1 \end{bmatrix}$

B15. $\begin{bmatrix} -8 & 7 & -7 \\ 0 & -8 & -6 \\ 3 & 9 & 7 \end{bmatrix}, \begin{bmatrix} -3 \\ 4 \\ 11 \end{bmatrix}$ **B16.** $\begin{bmatrix} 4 & -12 & -9 \\ -8 & 2 & 6 \\ 8 & -3 & -9 \end{bmatrix}, \begin{bmatrix} -10 \\ 3 \\ 10 \end{bmatrix}$

B17. $\begin{bmatrix} -8 & 2 & -12 & 1 \\ 7 & 9 & -1 & -7 \\ 2 & -11 & 6 & -12 \\ -4 & -6 & 10 & -4 \end{bmatrix}, \begin{bmatrix} 1 \\ 6 \\ -10 \\ -12 \end{bmatrix}$

B18. $\begin{bmatrix} -4 & 4 & 11 & -4 \\ 1 & 4 & 9 & -8 \\ -9 & 10 & 5 & -5 \\ 8 & -7 & 4 & -9 \end{bmatrix}, \begin{bmatrix} -9 \\ -3 \\ 3 \\ -9 \end{bmatrix}$

B19. $\begin{bmatrix} 5 & -6 & -4 & -11 \\ -8 & -8 & -4 & -2 \\ -5 & -12 & -8 & 5 \\ 1 & -8 & 6 & -10 \end{bmatrix}, \begin{bmatrix} 12 \\ 2 \\ -3 \\ -5 \end{bmatrix}$

B20. $\begin{bmatrix} -1 & -3 & -8 & -12 \\ 8 & -9 & 1 & 0 \\ 3 & -11 & -10 & 9 \\ -4 & -7 & 11 & 10 \end{bmatrix}, \begin{bmatrix} 5 \\ 6 \\ 7 \\ 3 \end{bmatrix}$

B21. The geometric series $\sum_{k=0}^{\infty} x^k$ converges to $\frac{1}{1-x}$, provided $|x| < 1$. If $A = \begin{bmatrix} \frac{1}{2} & \frac{1}{3} \\ \frac{1}{4} & \frac{1}{5} \end{bmatrix}$ and $A^0 = I$, show computationally that $\sum_{k=0}^{n} A^k \to (I - A)^{-1}$ as n increases. Is this result true for an arbitrary matrix A?

12.8   Vectors and the Laplace Transform

EXAMPLE 12.11   Recall from Section 12.1 how an initial value problem such as

$$x'(t) = 9x - 14y$$
$$y'(t) = 4x - 6y \qquad \text{and} \quad x(0) = 2, \, y(0) = -1$$

can be solved with the Laplace transform. Denoting the transforms of $x(t)$, $y(t)$ by $X = X(s)$, $Y = Y(s)$, respectively, we obtain

$$\begin{array}{ccc} sX - 2 = 9X - 14Y & & (s-9)X + 14Y = 2 \\ sY + 1 = 4X - 6Y & \text{or} & -4X + (s+6)Y = -1 \end{array} \quad \text{or} \quad \begin{bmatrix} s-9 & 14 \\ -4 & s+6 \end{bmatrix} \begin{bmatrix} X \\ Y \end{bmatrix} = \begin{bmatrix} 2 \\ -1 \end{bmatrix}$$

$$(12.19)$$

and note that the matrix on the right in (12.19) is just $sI - A$. Solving (12.19), we get

$$X(s) = \frac{30}{s-2} - \frac{28}{s-1}$$
$$\qquad \qquad \text{and hence} \qquad \begin{aligned} x(t) &= 30e^{-2t} - 28e^{-t} \\ y(t) &= 15e^{-2t} - 16e^{-t} \end{aligned}$$
$$Y(s) = \frac{15}{s-2} - \frac{16}{s-1}$$ ❖

Vector Version of Laplace Transform Solution

We detail here a vector version of the Laplace transform solution to the initial value problem in Example 12.11. First, we fix notation. As usual, lower-case bold letters represent vectors. In particular, define

$$\mathbf{v} = \mathbf{v}(t) = \begin{bmatrix} x(t) \\ y(t) \end{bmatrix} \qquad \mathbf{v}_0 = \begin{bmatrix} 2 \\ -1 \end{bmatrix} \quad \text{and} \quad \mathbf{V} = \mathbf{V}(s) = \begin{bmatrix} X(s) \\ Y(s) \end{bmatrix} \qquad (12.20)$$

where an upper-case bold letter is the Laplace transform of a vector and $X(s)$, $Y(s)$ are the Laplace transforms of $x(t)$, $y(t)$, respectively. Thus, the Laplace transform of a vector is computed componentwise. If the IVP in Example 12.11 is written in matrix-vector form as

$$\mathbf{v}' = A\mathbf{v} \qquad \mathbf{v}(0) = \mathbf{v}_0 \qquad (12.21)$$

the Laplace transform of (12.21) gives

$$s\mathbf{V} - \mathbf{v}_0 = A\mathbf{V}$$

To solve for $\mathbf{V}$, the vector of unknown Laplace transforms, we first get

$$(sI - A)\mathbf{V} = \mathbf{v}_0$$

which is just the formal statement of the equations on the right in (12.19). Solving for $\mathbf{V}$ gives as an explicit representation for $\mathbf{V} = \mathbf{V}(s)$

$$\mathbf{V} = (sI - A)^{-1}\mathbf{v}_0 = \Phi\mathbf{v}_0$$

where $\Phi = (sI - A)^{-1}$ is the matrix inverse of the matrix $sI - A$. The solution $\mathbf{v} = \mathbf{v}(t)$ then follows by inverting the transforms in the vector $\mathbf{V}$. This solution is denoted by

$$\mathbf{v} = \mathbf{v}(t) = L^{-1}[(sI - A)^{-1}]\mathbf{v}_0 = \phi\mathbf{v}_0$$

where $\phi = L^{-1}[(sI - A)^{-1}]$.

EXAMPLE 12.12   For the IVP in Example 12.11, the vector $\mathbf{v}_0$ is defined in (12.20) and the system matrix is $A = \begin{bmatrix} 9 & -14 \\ 4 & -6 \end{bmatrix}$. The matrices $\Phi = (sI - A)^{-1}$ and $\phi = L^{-1}[(sI - A)^{-1}] = L^{-1}[\Phi]$ are then

$$\Phi = \frac{1}{s^2 - 3s + 2}\begin{bmatrix} s + 6 & -14 \\ 4 & s - 9 \end{bmatrix} \quad \text{and} \quad \phi = \begin{bmatrix} 8e^{2t} - 7e^t & -14e^{2t} + 14e^t \\ 4e^{2t} - 4e^t & -7e^{2t} + 8e^t \end{bmatrix} \tag{12.22}$$

and the solution to the IVP is

$$\mathbf{v} = \mathbf{v}(t) = \phi\mathbf{v}_0 = \begin{bmatrix} 8e^{2t} - 7e^t & -14e^{2t} + 14e^t \\ 4e^{2t} - 4e^t & -7e^{2t} + 8e^t \end{bmatrix}\begin{bmatrix} 2 \\ -1 \end{bmatrix} = \begin{bmatrix} 30e^{2t} - 28e^t \\ 15e^{2t} - 16e^t \end{bmatrix} \quad \text{❖}$$

The Fundamental Matrix

The matrix ϕ is called the *fundamental matrix* for the system $\mathbf{x}' = A\mathbf{x}$ and is determined by the matrix A. The recipe we used, namely, $\phi = L^{-1}[(sI - A)^{-1}]$, causes $\phi(0) = I$. The fundamental matrix has many other interesting properties, which will be explored in Section 12.9.

EXERCISES 12.8–Part A

A1. If $A = \begin{bmatrix} 3 & 6 \\ -5 & 14 \end{bmatrix}$, compute $\Phi = (sI - A)^{-1}$ out by the algorithm given in Exercise A5, Section 12.7.

A2. For Φ determined in Exercise A1, obtain $\phi = L^{-1}[\Phi]$. Show that $\phi(0) = I$.

A3. Use the Laplace transform to solve the IVP $x'(t) = 5x + 6y$, $y'(t) = -7x - 8y$, $x(0) = 3$, $y(0) = -2$.

A4. Cast the IVP in Exercise A3 into matrix-vector form, and obtain the fundamental matrix ϕ.

A5. Use the fundamental matrix ϕ computed in Exercise A4 to obtain the solution of the IVP in Exercise A3.

A6. For ϕ computed in Exercise A4, show that $\phi' = A\phi$, $\phi(0) = I$, where A is the system matrix for the IVP in Exercise A3.

EXERCISES 12.8–Part B

B1. Show that ϕ determined in Exercise A2 satisfies $\phi' = A\phi$, where A is the matrix in Exercise A1.

In Exercises B2–11, the given matrices A and initial vectors $\mathbf{v}_0$ define an IVP $\mathbf{v}' = A\mathbf{v}$, $\mathbf{v}(0) = \mathbf{v}_0$. For each:

(a) Obtain the solution $\mathbf{v}(t)$ using a computer algebra system.

(b) Plot each component of $\mathbf{v}(t)$ on the same set of axes.

(c) Write the solution in the vector form $\mathbf{v}(t) = \sum_{k=1}^{2} e^{\lambda_k t}\mathbf{B}_k$, where each $\mathbf{B}_k$, $k = 1, 2$, is a constant vector.

(d) Show that $A\mathbf{B}_k = \lambda_k\mathbf{B}_k$, $k = 1, 2$.

(e) Obtain the fundamental matrix $\phi = L^{-1}[(sI - A)^{-1}]$.

(f) Show that $\mathbf{v}(t) = \phi\mathbf{v}_0$.

(g) Verify that $\mathbf{v}' = A\mathbf{v}$ and that $\mathbf{v}(0) = \mathbf{v}_0$.

(h) Show that $\phi(0) = I$.

(i) If $\mathbf{X}_k$ is the kth column of ϕ, show that $\mathbf{X}'_k = A\mathbf{X}_k$, $k = 1, 2$, making each column of ϕ a solution of the given differential equation.

(j) Show that $\phi' = A\phi$.

(k) Show that $\mathbf{X}_k(0) = \mathbf{e}_k$, where $\mathbf{e}_k$ is the kth column of the 2×2 identity matrix.

B2. $\begin{bmatrix} 19 & -24 \\ 18 & -23 \end{bmatrix}$, $\begin{bmatrix} -2 \\ 6 \end{bmatrix}$ **B3.** $\begin{bmatrix} -35 & -16 \\ 72 & 33 \end{bmatrix}$, $\begin{bmatrix} 1 \\ 10 \end{bmatrix}$

B4. $\begin{bmatrix} -10 & 4 \\ -14 & 5 \end{bmatrix}$, $\begin{bmatrix} 12 \\ -11 \end{bmatrix}$ **B5.** $\begin{bmatrix} -11 & 6 \\ -12 & 6 \end{bmatrix}$, $\begin{bmatrix} -1 \\ -2 \end{bmatrix}$

B6. $\begin{bmatrix} 76 & -120 \\ 50 & -79 \end{bmatrix}$, $\begin{bmatrix} -4 \\ -5 \end{bmatrix}$ **B7.** $\begin{bmatrix} -1 & 3 \\ -6 & 8 \end{bmatrix}$, $\begin{bmatrix} 8 \\ -1 \end{bmatrix}$

B8. $\begin{bmatrix} -9 & -12 \\ 14 & 17 \end{bmatrix}, \begin{bmatrix} 6 \\ 5 \end{bmatrix}$ **B9.** $\begin{bmatrix} 35 & 36 \\ -32 & -33 \end{bmatrix}, \begin{bmatrix} -7 \\ -11 \end{bmatrix}$

B10. $\begin{bmatrix} 25 & 14 \\ -42 & -24 \end{bmatrix}, \begin{bmatrix} -11 \\ 2 \end{bmatrix}$ **B11.** $\begin{bmatrix} -26 & 12 \\ -60 & 28 \end{bmatrix}, \begin{bmatrix} 2 \\ 11 \end{bmatrix}$

B12. Discuss why the formalism $\mathbf{v}(t) = \phi \mathbf{v}_0$ should work for an $n \times n$ IVP.

In Exercises B13–22, the given matrices A and initial vectors $\mathbf{v}_0$ define an IVP $\mathbf{v}' = A\mathbf{v}, \mathbf{v}(0) = \mathbf{v}_0$. For each:

(a) Obtain the solution $\mathbf{v}(t)$ using a computer algebra system.

(b) Plot each component of $\mathbf{v}(t)$ on the same set of axes.

(c) Write the solution in the vector form $\mathbf{v}(t) = \sum_{k=1}^{3} e^{\lambda_k t} \mathbf{B}_k$, where each $\mathbf{B}_k, k = 1, 2, 3$, is a constant vector.

(d) Show that $A\mathbf{B}_k = \lambda_k \mathbf{B}_k, k = 1, 2, 3$.

(e) Obtain the fundamental matrix $\phi = L^{-1}[(sI - A)^{-1}]$.

(f) Show that $\mathbf{v}(t) = \phi \mathbf{v}_0$.

(g) Verify that $\mathbf{v}' = A\mathbf{v}$ and that $\mathbf{v}(0) = \mathbf{v}_0$.

(h) Show that $\phi(0) = I$.

(i) If $\mathbf{X}_k$ is the kth column of ϕ, show that $\mathbf{X}_k' = A\mathbf{X}_k$, $k = 1, 2, 3$, making each column of ϕ a solution of the given differential equation.

(j) Show that $\phi' = A\phi$.

(k) Show that $\mathbf{X}_k(0) = \mathbf{e}_k$, where $\mathbf{e}_k$ is the kth column of the 3×3 identity matrix.

B13. $\begin{bmatrix} -73 & 178 & 24 \\ -41 & 98 & 12 \\ 74 & -166 & -14 \end{bmatrix}, \begin{bmatrix} -11 \\ 12 \\ -6 \end{bmatrix}$ **B14.** $\begin{bmatrix} -114 & 108 & 36 \\ -135 & 129 & 45 \\ 54 & -54 & -24 \end{bmatrix}, \begin{bmatrix} -9 \\ 8 \\ 4 \end{bmatrix}$

B15. $\begin{bmatrix} -61 & -30 & -60 \\ 20 & 9 & 20 \\ 60 & 30 & 59 \end{bmatrix}, \begin{bmatrix} -11 \\ 9 \\ -12 \end{bmatrix}$ **B16.** $\begin{bmatrix} 151 & -34 & -122 \\ 482 & -113 & -374 \\ 50 & -10 & -45 \end{bmatrix}, \begin{bmatrix} 5 \\ 1 \\ 4 \end{bmatrix}$

B17. $\begin{bmatrix} -267 & 195 & -195 \\ 16 & -4 & 12 \\ 380 & -270 & 278 \end{bmatrix}, \begin{bmatrix} 5 \\ -10 \\ -6 \end{bmatrix}$ **B18.** $\begin{bmatrix} 155 & -204 & -41 \\ 120 & -158 & -30 \\ 12 & -16 & -12 \end{bmatrix}, \begin{bmatrix} -4 \\ -7 \\ 7 \end{bmatrix}$

B19. $\begin{bmatrix} 158 & -99 & -33 \\ 320 & -202 & -70 \\ -190 & 123 & 49 \end{bmatrix}, \begin{bmatrix} -6 \\ -4 \\ 11 \end{bmatrix}$ **B20.** $\begin{bmatrix} 335 & 184 & -252 \\ -530 & -289 & 390 \\ 70 & 40 & -59 \end{bmatrix}, \begin{bmatrix} 3 \\ 9 \\ -7 \end{bmatrix}$

B21. $\begin{bmatrix} -290 & 12 & 168 \\ -226 & 2 & 134 \\ -470 & 20 & 272 \end{bmatrix}, \begin{bmatrix} -11 \\ 11 \\ -10 \end{bmatrix}$ **B22.** $\begin{bmatrix} 7 & 8 & -2 \\ 134 & 97 & -22 \\ 591 & 444 & -102 \end{bmatrix}, \begin{bmatrix} 9 \\ 6 \\ 7 \end{bmatrix}$

12.9 The Matrix Exponential

Motivation

The initial value problem $x'(t) = ax, x(0) = x_0$, governs exponential growth or decay and has the solution $x(t) = x_0 e^{at}$. If A is a constant matrix, the IVP $\mathbf{x}' = A\mathbf{x}, \mathbf{x}(0) = \mathbf{x}_0$, has the solution $\mathbf{x}(t) = \phi(t)\mathbf{x}_0$, where $\phi(t) = L^{-1}[(sI - A)^{-1}]$ is the fundamental matrix we saw in Section 12.8.

Presently, we are going to see that the solution can also be represented as $\mathbf{x}(t) = e^{At}\mathbf{x}_0$, where e^{At} is called the *matrix exponential* and, by the uniqueness of the solution of the IVP, must clearly be the same as the fundamental matrix ϕ. We begin by defining the matrix exponential, e^A.

The Matrix Exponential

The exponential of the matrix A is defined as a power series in A. Thus, motivated by the power series for e^x in (12.23a), we have, with $A^0 = I$, the *definitions* in (12.23b) and (12.23c)

$$e^x = \sum_{k=0}^{\infty} \frac{x^k}{k!} = 1 + x + \frac{x^2}{2!} + \frac{x^3}{3!} + \cdots \tag{12.23a}$$

$$e^A = \sum_{k=0}^{\infty} \frac{A^k}{k!} = I + A + \frac{A^2}{2!} + \frac{A^3}{3!} + \cdots \tag{12.23b}$$

$$e^{At} = \sum_{k=0}^{\infty} \frac{(At)^k}{k!} = I + At + \frac{(At)^2}{2!} + \frac{(At)^3}{3!} + \cdots \tag{12.23c}$$

EXAMPLE 12.13   The following calculations yield e^D for the 2×2 *diagonal* matrix $D = \begin{bmatrix} a & 0 \\ 0 & b \end{bmatrix}$.

$$e^D = \begin{bmatrix} 1 & 0 \\ 0 & 1 \end{bmatrix} + \begin{bmatrix} a & 0 \\ 0 & b \end{bmatrix} + \frac{1}{2!}\begin{bmatrix} a^2 & 0 \\ 0 & b^2 \end{bmatrix} + \frac{1}{3!}\begin{bmatrix} a^3 & 0 \\ 0 & b^3 \end{bmatrix} + \cdots$$

$$= \begin{bmatrix} \sum_{k=0}^{\infty} \dfrac{a^k}{k!} & 0 \\ 0 & \sum_{k=0}^{\infty} \dfrac{b^k}{k!} \end{bmatrix} = \begin{bmatrix} e^a & 0 \\ 0 & e^b \end{bmatrix}$$

EXAMPLE 12.14   If D is the diagonal matrix of Example 12.13 and P is invertible, then for $A = PDP^{-1}$ we have $e^A = Pe^D P^{-1}$. We can show this as soon as we realize that for any power k we have

$$A^k = (PDP^{-1})^k = (PDP^{-1})(PDP^{-1})\cdots(PDP^{-1}) = PD^k P^{-1}$$

and for e^A we have

$$e^A = \sum_{k=0}^{\infty} \frac{(PDP^{-1})^k}{k!} = \sum_{k=0}^{\infty} P\frac{D^k}{k!}P^{-1} = P\left(\sum_{k=0}^{\infty} \frac{D^k}{k!}\right)P^{-1} = Pe^D P^{-1}$$

EXAMPLE 12.15   The matrices $A = PDP^{-1}$ in (12.24) determine $e^A = Pe^D P^{-1}$ by the calculation shown in Example 12.14.

$$P = \begin{bmatrix} 7 & -2 \\ 4 & -1 \end{bmatrix} \quad D = \begin{bmatrix} 1 & 0 \\ 0 & 2 \end{bmatrix} \quad A = \begin{bmatrix} 9 & -14 \\ 4 & -6 \end{bmatrix} \Rightarrow e^A = \begin{bmatrix} 8e^2 - 7e & -14e^2 + 14e \\ 4e^2 - 4e & -7e^2 + 8e \end{bmatrix}$$

$$(12.24)$$

EXAMPLE 12.16   If A is the matrix in (12.24), then e^{At} is the matrix ϕ in (12.22). This is obtained by using a built-in command from a computer algebra system. We will consider access to such a "black-box" algorithm as a *first recipe* for computing the matrix exponential.

e^{At} is the Fundamental Matrix

Our *second recipe* for computing the matrix exponential is $e^{At} = \phi = L^{-1}[(sI - A)^{-1}]$.

     Therefore, the matrix exponential is the fundamental matrix for the equation $\mathbf{x}' = A\mathbf{x}$, where A is a constant matrix.

The Columns of the Fundamental Matrix

The definition of e^{At} in (12.23c) assures us that $e^{At}|_{t=0} = I = \phi(0)$. A computation also based on the definition in (12.23c) leads to $\frac{d}{dt}e^{At} = Ae^{At}$ or $\phi'(t) = A\phi(t)$. (See Exercise A1.) Thus, each column of $\phi = e^{At}$ must itself be a solution of the differential equation $\mathbf{x}' = A\mathbf{x}$. In light of $\phi(0) = I$, $\boldsymbol{\phi}_k$, the kth column of ϕ, satisfies the initial condition $\boldsymbol{\phi}_k(0) = \mathbf{e}_k$, where $\mathbf{e}_k$ is the kth column of I. Consequently, we have the following theorem.

> **THEOREM 12.2**
>
> The columns of the fundamental matrix $\phi = e^{At}$ are each a solution of $\mathbf{x}' = A\mathbf{x}$; and $\boldsymbol{\phi}_k$, the kth column, satisfies the initial condition $\boldsymbol{\phi}_k(0) = \mathbf{e}_k$, where $\mathbf{e}_k$ is the kth column of I.

Fundamental Matrix: Recipe 3

We now have a *third recipe* for finding the fundamental matrix $\phi = e^{At}$ for a constant matrix A. We can solve the initial value problems $\mathbf{x}'_k = A\mathbf{x}_k$, $\mathbf{x}_k(0) = \mathbf{e}_k$, for the individual columns of the fundamental matrix and write $\phi = [\mathbf{x}_k, \ldots, \mathbf{x}_n]$.

EXAMPLE 12.17 We next demonstrate Recipe 3 for the matrix A given in (12.24). The differential equations represented by $\mathbf{x}' = A\mathbf{x}$ are $x'(t) = 9x - 14y$ and $y'(t) = 4x - 6y$. Solutions satisfying the initial conditions $\mathbf{x}_1(0) = \mathbf{e}_1$ and $\mathbf{x}_2(0) = \mathbf{e}_2$ are found to be

$$\mathbf{x}_1 = \begin{bmatrix} 8e^{2t} - 7e^t \\ 4e^{2t} - 4e^t \end{bmatrix} \quad \text{and} \quad \mathbf{x}_2 = \begin{bmatrix} -14e^{2t} + 14e^t \\ -7e^{2t} + 8e^t \end{bmatrix}$$

which are precisely the columns of $\phi = e^{At}$ computed in (12.22). A matrix with these solutions as its columns therefore must be the fundamental matrix. ❖

Inverse of the Fundamental Matrix

When A is a constant matrix, the inverse of the fundamental matrix $\phi(t) = e^{At}$ is $[\phi(t)]^{-1} = \phi(-t) = e^{-At}$. This is consistent with the scalar exponential function where $1/e^x = e^{-x}$. Intuitively, $e^{At}e^{-At} = e^O = I$, where O is the zero matrix, but a formal proof can be found in [40].

EXERCISES 12.9–Part A

A1. Differentiate (12.23c) termwise to show that $\frac{d}{dt}e^{At} = Ae^{At}$.

A2. Generally, two matrices A and B do not commute, that is, $AB \neq BA$. Show that for the constant matrix A, the fundamental matrix e^{At} commutes with A. *Hint:* Look at the series definition of e^{At}.

A3. If $A = \begin{bmatrix} 1 & t \\ t^2 & t^3 \end{bmatrix}$, show that A does not commute with A', suggesting that the result in Exercise A2 would generally not be valid for nonconstant matrices A.

A4. If $A = \begin{bmatrix} -3 & 4 \\ -12 & 11 \end{bmatrix}$, obtain $\phi = L^{-1}[(sI - A)^{-1}]$, the fundamental matrix.

A5. If A is the matrix in Exercise A4, obtain e^{At} by using the Laplace transform to solve $\mathbf{x}'_k = A\mathbf{x}_k$, $\mathbf{x}_k(0) = \mathbf{e}_k$, $k = 1, 2$. Compare to the results in Exercise A4.

EXERCISES 12.9–Part B

B1. Let $A = \begin{bmatrix} 1 & 1 \\ -1 & 3 \end{bmatrix}$. Verify that

$$e^{At} = \begin{bmatrix} e^{2t}(1-t) & te^{2t} \\ -te^{2t} & e^{2t}(1+t) \end{bmatrix}$$

For e^{At}, obtain the partial sum $\sum_{k=0}^{4}(At)^k/k!$ and, to the same order, the partial sums of the Maclaurin expansion of each component. Show that through terms of order four, the expansions match.

For the 2×2 matrix A and initial vector $\mathbf{v}_0$ given in each of Exercises B2–11:

(a) Use a computer algebra system to obtain ϕ, the fundamental matrix for the initial value problem $\mathbf{v}' = A\mathbf{v}$, $\mathbf{v}(0) = \mathbf{v}_0$.

(b) Verify that $\phi = L^{-1}[(sI - A)^{-1}]$ as found in Section 12.8.

(c) Construct ϕ columnwise as the matrix $[\mathbf{x}_1, \mathbf{x}_2]$, where $\mathbf{x}'_k = A\mathbf{x}_k$, $k = 1, 2$, and $\mathbf{x}_k(0) = \mathbf{e}_k$, the kth column of the identity matrix.

(d) Verify that $\sum_{k=0}^{n}(At)^k/k! \to e^{At} = \phi$ as n gets large by comparing a partial sum on the left with a comparable Maclaurin expansion of each component of the matrix on the right.

(e) Verify that $Ae^{At} = e^{At}A$, indicating that A commutes with the fundamental matrix it generates.

(f) Verify that $\phi' = Ae^{At}$.

(g) Verify that $[\phi(t)]^{-1} = \phi(-t)$.

B2. Exercise B2, Section 12.8   **B3.** Exercise B3, Section 12.8

B4. Exercise B4, Section 12.8   **B5.** Exercise B5, Section 12.8

B6. Exercise B6, Section 12.8   **B7.** Exercise B7, Section 12.8

B8. Exercise B8, Section 12.8   **B9.** Exercise B9, Section 12.8

B10. Exercise B10, Section 12.8   **B11.** Exercise B11, Section 12.8

For each 3×3 matrix A and initial vector $\mathbf{v}_0$ given in Exercises B12–21:

(a) Use a computer algebra system to obtain ϕ, the fundamental matrix for the initial value problem $\mathbf{v}' = A\mathbf{v}$, $\mathbf{v}(0) = \mathbf{v}_0$.

(b) Verify that $\phi = L^{-1}[(sI - A)^{-1}]$ as found in Section 12.8.

(c) Construct ϕ columnwise as the matrix $[\mathbf{x}_1, \mathbf{x}_2, \mathbf{x}_3]$, where $\mathbf{x}'_k = A\mathbf{x}_k$, $k = 1, 2, 3$, and $\mathbf{x}_k(0) = \mathbf{e}_k$, the kth column of the identity matrix.

(d) Verify that $Ae^{At} = e^{At}A$, indicating that A commutes with the fundamental matrix it generates.

(e) Verify that $\phi' = Ae^{At}$.

(f) Verify that $[\phi(t)]^{-1} = \phi(-t)$.

B12. Exercise B13, Section 12.8   **B13.** Exercise B14, Section 12.8

B14. Exercise B15, Section 12.8   **B15.** Exercise B16, Section 12.8

B16. Exercise B17, Section 12.8   **B17.** Exercise B18, Section 12.8

B18. Exercise B19, Section 12.8   **B19.** Exercise B20, Section 12.8

B20. Exercise B21, Section 12.8   **B21.** Exercise B22, Section 12.8

12.10 Eigenvalues and Eigenvectors

Vectors Hiding in a Solution

In (12.25) the matrix A and the vector $\mathbf{x}_0$ for the IVP $\mathbf{x}' = A\mathbf{x}$, $\mathbf{x}(0) = \mathbf{x}_0$, appears on the left, and the solution of the IVP appears on the right

$$A = \begin{bmatrix} 18 & -2 \\ 12 & 7 \end{bmatrix} \quad \mathbf{x}_0 = \begin{bmatrix} \alpha \\ \beta \end{bmatrix} \Rightarrow \mathbf{x}(t) = \begin{bmatrix} \left(-\frac{3}{5}\alpha + \frac{2}{5}\beta\right)e^{10t} + \left(\frac{8}{5}\alpha - \frac{2}{5}\beta\right)e^{15t} \\ \left(-\frac{12}{5}\alpha + \frac{8}{5}\beta\right)e^{10t} + \left(\frac{12}{5}\alpha - \frac{3}{5}\beta\right)e^{15t} \end{bmatrix}$$

$$(12.25)$$

With $a = -\frac{3}{5}\alpha + \frac{2}{5}\beta$ and $b = \frac{8}{5}\alpha - \frac{2}{5}\beta$, the solution becomes

$$\mathbf{x}(t) = \begin{bmatrix} ae^{10t} + be^{15t} \\ 4ae^{10t} + \frac{3}{2}be^{15t} \end{bmatrix} = ae^{10t}\begin{bmatrix} 1 \\ 4 \end{bmatrix} + be^{15t}\begin{bmatrix} 1 \\ \frac{3}{2} \end{bmatrix} \qquad (12.26)$$

where on the right, a and b have been factored out from $\mathbf{x}(t)$. In (12.27), the definitions on the left put $\mathbf{x}(t)$ into the form on the right

$$\mathbf{v}_1 = \begin{bmatrix} 1 \\ 4 \end{bmatrix} \quad \mathbf{v}_2 = \begin{bmatrix} 1 \\ \frac{3}{2} \end{bmatrix} \Rightarrow \mathbf{x}(t) = ae^{10t}\mathbf{v}_1 + be^{15t}\mathbf{v}_2 \qquad (12.27)$$

The vectors $\mathbf{v}_1, \mathbf{v}_2$ appearing in (12.26) and (12.27) are special, being closely related both to the system matrix A and to the 10 and 15 in the exponentials in the solution. Presently, we explore these vectors, the scalars 10 and 15, and their relationship to the matrix A.

Eigenvalues and Eigenvectors

The calculations

$$Av_1 = \begin{bmatrix} 18 & -2 \\ 12 & 7 \end{bmatrix} \begin{bmatrix} 1 \\ 4 \end{bmatrix} = \begin{bmatrix} 10 \\ 40 \end{bmatrix} = 10 \begin{bmatrix} 1 \\ 4 \end{bmatrix} = 10v_1$$

$$Av_2 = \begin{bmatrix} 18 & -2 \\ 12 & 7 \end{bmatrix} \begin{bmatrix} 1 \\ \frac{3}{2} \end{bmatrix} = \begin{bmatrix} 15 \\ \frac{45}{2} \end{bmatrix} = 15 \begin{bmatrix} 1 \\ \frac{3}{2} \end{bmatrix} = 15v_2$$

show that the relation between A, the vectors v_1 and v_2, and the numbers 10 and 15 is expressed by the equations

$$Av_1 = 10v_1 \quad \text{and} \quad Av_2 = 15v_2$$

Vectors such as v_1 and v_2 are called *eigenvectors,* while the numbers 10 and 15 are called *eigenvalues.*

Eigenvectors from the Start

Eigenvectors and eigenvalues capture the behavior of the linear system of differential equations. But we would like a more direct way of obtaining these eigenvalues and eigenvectors without having to extract them from a solution of $x' = Ax$. Therefore, look for this solution directly in the form of the vector $x = ve^{\lambda t}$, where

$$v = \begin{bmatrix} b_1 \\ b_2 \end{bmatrix}$$

and b_1, b_2, and λ are constants to be determined.

Substitution into $x' = Ax$ gives $\lambda ve^{\lambda t} = Ave^{\lambda t}$, and upon canceling the exponential term on both sides $\lambda v = Av$. Thus, the vector Av is a multiple of the special vector $v \neq 0$. Such a vector v is called an *eigenvector,* and the multiplier λ is called an *eigenvalue.* The equation

$$Av = \lambda v$$

determines both the eigenvalue and eigenvector.

The general solution of the system of differential equations $x' = Ax$ is known as soon as the eigenvalues and eigenvectors are known. In fact, if the eigenvalues are λ_1, λ_2 and the corresponding eigenvectors are v_1, v_2, then the general solution would be $x = ae^{\lambda_1 t} v_1 + be^{\lambda_2 t} v_2$.

EXERCISES 12.10–Part A

A1. The matrix $A = \begin{bmatrix} 26 & -18 \\ 21 & -13 \end{bmatrix}$ and initial vector $x_0 = \begin{bmatrix} \alpha \\ \beta \end{bmatrix}$ define the IVP $x' = Ax, x(0) = x_0$.

    (a) Using the Laplace transform, obtain the general solution.

    (b) Write the general solution in the form $x(t) = ae^{\lambda_1 t} v_1 + be^{\lambda_2 t} v_2$.

    (c) Show that $Av_k = \lambda_k v_k, k = 1, 2$.

    (d) Compute $\det(A - \lambda_k I), k = 1, 2$.

    (e) Obtain the null space for each of the matrices $A - \lambda_k I$, $k = 1, 2$.

EXERCISES 12.10–Part B

B1. The vector $\mathbf{v} = \mathbf{0}$ is never considered an eigenvector of a matrix A. However, $\lambda = 0$ can be an eigenvalue of A. For the matrix $A = \begin{bmatrix} 1 & 2 & 3 \\ 4 & 5 & 6 \\ 7 & 8 & 9 \end{bmatrix}$, show that $\lambda = 0$ is an eigenvalue and find the associated eigenvector. *Hint:* $A\mathbf{v} = \lambda\mathbf{v}$ becomes $A\mathbf{v} = 0\mathbf{v} = \mathbf{0}$.

In Exercises B2–11, the given matrix A determines a linear system $\mathbf{x}' = A\mathbf{x}$. For each:

(a) Obtain the general solution.

(b) Write the general solution in the form $ae^{\lambda_1 t}\mathbf{v}_1 + be^{\lambda_2 t}\mathbf{v}_2$.

(c) Show that $A\mathbf{v}_k = \lambda_k\mathbf{v}_k, k = 1, 2$.

(d) Compute $\det(A - \lambda_k I), k = 1, 2$.

(e) Obtain the null space for each of the matrices $A - \lambda_k I$, $k = 1, 2$.

(f) Using appropriate technology, obtain the eigenvalues and eigenvectors of A and compare to the λ_k and $\mathbf{v}_k$ determined in part (b).

(g) Plot $\mathbf{v} = \begin{bmatrix} \cos\theta \\ \sin\theta \end{bmatrix}$ and $A\mathbf{v}$ for several different values of θ.

B2. $\begin{bmatrix} 15 & 48 \\ -4 & -13 \end{bmatrix}$ **B3.** $\begin{bmatrix} -6 & -2 \\ 28 & 9 \end{bmatrix}$ **B4.** $\begin{bmatrix} 27 & -42 \\ 12 & -18 \end{bmatrix}$

B5. $\begin{bmatrix} 27 & 24 \\ -40 & -35 \end{bmatrix}$ **B6.** $\begin{bmatrix} 26 & 24 \\ -21 & -19 \end{bmatrix}$ **B7.** $\begin{bmatrix} 26 & -30 \\ 12 & -12 \end{bmatrix}$

B8. $\begin{bmatrix} 1 & -4 \\ 6 & -9 \end{bmatrix}$ **B9.** $\begin{bmatrix} 22 & 15 \\ -18 & -11 \end{bmatrix}$ **B10.** $\begin{bmatrix} -16 & -84 \\ 2 & 10 \end{bmatrix}$

B11. $\begin{bmatrix} -9 & -20 \\ 4 & 9 \end{bmatrix}$

12.11   Solutions by Eigenvalues and Eigenvectors

From Solution to Eigenvectors

In (12.28), the general solution of $\mathbf{x}' = A\mathbf{x}$ is given on the right for the matrix A given on the left

$$\begin{bmatrix} 9 & -5 & 2 \\ 14 & -8 & 4 \\ 10 & -7 & 5 \end{bmatrix} \Rightarrow \mathbf{x} = \begin{bmatrix} (-4c_1 + 3c_2 - c_3)e^t + (2c_1 - c_2)e^{2t} + (3c_1 - 2c_2 + c_3)e^{3t} \\ (-8c_1 + 6c_2 - 2c_3)e^t + (2c_1 - c_2)e^{2t} + (6c_1 - 4c_2 + 2c_3)e^{3t} \\ (-4c_1 + 3c_2 - c_3)e^t - (2c_1 - c_2)e^{2t} + (6c_1 - 4c_2 + 2c_3)e^{3t} \end{bmatrix}$$

(12.28)

If the constants a, b, c are introduced by the equations on the left in (12.29), the solution assumes the form shown on the right

$$\left. \begin{array}{r} -4c_1 + 3c_2 - c_3 = a \\ 2c_1 - c_2 = b \\ 3c_1 - 2c_2 + c_3 = c \end{array} \right\} \Rightarrow \mathbf{x} = a \begin{bmatrix} 1 \\ 2 \\ 1 \end{bmatrix} e^t + b \begin{bmatrix} 1 \\ 1 \\ -1 \end{bmatrix} e^{2t} + c \begin{bmatrix} 1 \\ 2 \\ 2 \end{bmatrix} e^{3t} \quad (12.29)$$

The equations on the left in (12.29) come from (arbitrarily) setting the coefficients of the exponentials in the first component of $\mathbf{x}$ to the single constants a, b, c. If the resulting equations are solved for $c_k, k = 1, 2, 3$, and the results substituted back into the general solution on the right in (12.28), then the form of $\mathbf{x}$ on the right in (12.29) results. Defining the vectors

$$\mathbf{v}_1 = \begin{bmatrix} 1 \\ 2 \\ 1 \end{bmatrix} \qquad \mathbf{v}_2 = \begin{bmatrix} 1 \\ 1 \\ -1 \end{bmatrix} \qquad \mathbf{v}_3 = \begin{bmatrix} 1 \\ 2 \\ 2 \end{bmatrix} \qquad (12.30)$$

means the general solution has the form

$$\mathbf{x} = ae^t\mathbf{v}_1 + be^{2t}\mathbf{v}_2 + ce^{3t}\mathbf{v}_3 \qquad (12.31)$$

and computation shows

$$A\mathbf{v}_1 = \begin{bmatrix} 1 \\ 2 \\ 1 \end{bmatrix} = \mathbf{v}_1 \qquad A\mathbf{v}_2 = \begin{bmatrix} 2 \\ 2 \\ -2 \end{bmatrix} = 2\mathbf{v}_2 \qquad A\mathbf{v}_3 = \begin{bmatrix} 3 \\ 6 \\ 6 \end{bmatrix} = 3\mathbf{v}_3$$

Hence, $\mathbf{v}_1, \mathbf{v}_2, \mathbf{v}_3$ are eigenvectors with eigenvalues 1, 2, 3, respectively.

From a general solution of the system we have extracted the eigenvalues and eigenvectors of the system matrix A.

From Eigenvectors to Solution

The general solution of $\mathbf{x}' = A\mathbf{x}$ can be constructed from the eigenvalues and eigenvectors. If the eigenvalues are determined to be 1, 2, 3, with the corresponding eigenvectors $\mathbf{v}_1, \mathbf{v}_2, \mathbf{v}_3$ given in (12.30), then the general solution can be written immediately as the linear combinationreproduces the solution on the right in (12.29).

Consequently, an efficient recipe for finding eigenvalues and eigenvectors is really a recipe for solving the linear system of differential equations $\mathbf{x}' = A\mathbf{x}$. Such a recipe is the subject of Sections 12.12, 12.14, and 12.15.

EXERCISES 12.11–Part A

A1. Solve the equations on the left in (12.29), thereby obtaining $c_k, k = 1, 2, 3$, in terms of a, b, c. Substitute these values into the general solution on the right in (12.28) and show that the form of the solution on the right in (12.29) results.

A2. If A is the matrix on the left in (12.28) and $\mathbf{x}_0 = \begin{bmatrix} -3 \\ 1 \\ 2 \end{bmatrix}$, use (12.30) and (12.31) to solve the IVP $\mathbf{x}' = A\mathbf{x}, \mathbf{x}(0) = \mathbf{x}_0$.

A3. Let $\Psi(t) = [\mathbf{v}_1 e^t, \mathbf{v}_2 e^{2t}, \mathbf{v}_3 e^{3t}]$, where the $\mathbf{v}_k$ are as given in (12.30). Show $\Psi(0) = [\mathbf{v}_1, \mathbf{v}_2, \mathbf{v}_3] \neq I$, so $\Psi(t)$ is not the fundamental matrix $\phi(t)$. Form the matrix $\Psi(t)[\Psi(0)]^{-1}$ and show that at $t = 0$ it is I, the identity matrix. Hence, it must be $\phi(t)$. To

support this, calculate $\Psi(t)\Psi^{-1}(0)\mathbf{x}_0$ and show this is the solution found in Exercise A2.

A4. If $A = \begin{bmatrix} -11 & 7 \\ -14 & 10 \end{bmatrix}$, use the Laplace transform to obtain the general solution of $\mathbf{x}' = A\mathbf{x}$. From the general solution, extract λ_k and $\mathbf{v}_k$, the eigenvalues and eigenvectors of A. Verify that $A\mathbf{v}_k = \lambda_k \mathbf{v}_k$, $k = 1, 2$.

A5. Show that $\mathbf{v}_1 = \begin{bmatrix} 7 \\ 4 \end{bmatrix}$ and $\mathbf{v}_2 = \begin{bmatrix} 5 \\ 3 \end{bmatrix}$ are eigenvectors, with respective eigenvalues $-1, 3$, for the matrix $A = \begin{bmatrix} -81 & 140 \\ -48 & 83 \end{bmatrix}$. From the eigenvalues and eigenvectors, construct the general solution of $\mathbf{x}' = A\mathbf{x}$.

EXERCISES 12.11–Part B

B1. Use appropriate technology to obtain the eigenvalues and eigenvectors of the matrix $\begin{bmatrix} -22 & -12 \\ 54 & 29 \end{bmatrix}$. Verify that the technology used has given correct answers.

In each of Exercises B2–13 a 3×3 matrix A and an initial vector $\mathbf{x}_0$ is given. For the IVP $\mathbf{x}' = A\mathbf{x}, \mathbf{x}(0) = \mathbf{x}_0$, thereby determined:

  (a) Use an appropriate technological device to determine the eigenvalues λ_k and the corresponding eigenvectors $\mathbf{v}_k$, $k = 1, \ldots, 3$.

  (b) Verify that $A\mathbf{v}_k = \lambda_k \mathbf{v}_k, k = 1, \ldots, 3$.

  (c) Write the general solution in the form $\mathbf{x}(t) = \sum_{k=1}^{3} a_k e^{\lambda_k t} \mathbf{v}_k$, where the $a_k, k = 1, 2, 3$, are arbitrary constants.

  (d) Apply the initial data to the general solution in part (c), forming and solving a set of algebraic equations for the constants $a_k, k = 1, 2, 3$, thereby obtaining the solution of the given IVP.

  (e) Obtain the fundamental matrix ϕ either using an appropriate technological device or by computing $L^{-1}[(sI - A)^{-1}]$.

  (f) Obtain the solution of the initial value problem as $\phi\mathbf{x}_0$ and compare to the result in part (d).

  (g) Obtain a solution of the initial value problem by a solver from a suitable technological device.

B2. $\begin{bmatrix} -2 & 14 & -1 \\ -1 & -11 & 0 \\ 6 & 78 & 2 \end{bmatrix}, \begin{bmatrix} -6 \\ 8 \\ 10 \end{bmatrix}$ **B3.** $\begin{bmatrix} -15 & -18 & -16 \\ 41 & 42 & 40 \\ -12 & -11 & -10 \end{bmatrix}, \begin{bmatrix} 1 \\ -11 \\ 1 \end{bmatrix}$ **B8.** $\begin{bmatrix} -35 & -23 & -48 \\ 48 & 31 & 63 \\ 4 & 3 & 7 \end{bmatrix}, \begin{bmatrix} -5 \\ 12 \\ -1 \end{bmatrix}$ **B9.** $\begin{bmatrix} 8 & 14 & 8 \\ 0 & 1 & 0 \\ -15 & -31 & -14 \end{bmatrix}, \begin{bmatrix} 11 \\ 4 \\ -3 \end{bmatrix}$

B4. $\begin{bmatrix} -72 & 64 & -75 \\ -11 & 6 & -12 \\ 63 & -61 & 65 \end{bmatrix}, \begin{bmatrix} -2 \\ 6 \\ 9 \end{bmatrix}$ **B5.** $\begin{bmatrix} -8 & 2 & -1 \\ -18 & 12 & -7 \\ -11 & 11 & -7 \end{bmatrix}, \begin{bmatrix} -3 \\ -7 \\ 1 \end{bmatrix}$ **B10.** $\begin{bmatrix} -10 & -8 & 7 \\ 21 & 22 & -22 \\ 14 & 16 & -17 \end{bmatrix}, \begin{bmatrix} 11 \\ -6 \\ 12 \end{bmatrix}$ **B11.** $\begin{bmatrix} 41 & -57 & 11 \\ 28 & -39 & 7 \\ -28 & 38 & -10 \end{bmatrix}, \begin{bmatrix} 4 \\ 11 \\ 5 \end{bmatrix}$

B6. $\begin{bmatrix} -40 & 39 & -38 \\ -94 & 94 & -95 \\ -50 & 52 & -55 \end{bmatrix}, \begin{bmatrix} -5 \\ -2 \\ 12 \end{bmatrix}$ **B7.** $\begin{bmatrix} 0 & 87 & 45 \\ 2 & -55 & -30 \\ -5 & 108 & 60 \end{bmatrix}, \begin{bmatrix} -2 \\ 4 \\ -1 \end{bmatrix}$ **B12.** $\begin{bmatrix} 32 & 97 & 32 \\ -7 & -24 & -8 \\ -7 & -14 & -4 \end{bmatrix}, \begin{bmatrix} 4 \\ 6 \\ -12 \end{bmatrix}$ **B13.** $\begin{bmatrix} -7 & -15 & -8 \\ 10 & 18 & 9 \\ -18 & -30 & -14 \end{bmatrix}, \begin{bmatrix} -1 \\ 1 \\ 1 \end{bmatrix}$

12.12 Finding Eigenvalues and Eigenvectors

Computing Eigenpairs

In (12.32), a matrix A and its eigenvectors $\mathbf{v}_1$ and $\mathbf{v}_2$ are listed

$$A = \begin{bmatrix} 1 & 2 \\ 4 & -1 \end{bmatrix} \Rightarrow \mathbf{v}_1 = \begin{bmatrix} 1 \\ 1 \end{bmatrix} \quad \mathbf{v}_2 = \begin{bmatrix} 1 \\ -2 \end{bmatrix} \tag{12.32}$$

Calculations show that $A\mathbf{v}_1 = 3\mathbf{v}_1$ and $A\mathbf{v}_2 = -3\mathbf{v}_2$, so the eigenvalues are $\lambda_1 = 3$ and $\lambda_2 = -3$. How are the *eigenpairs,* the pairs of eigenvalues and eigenvectors $(\lambda_k, \mathbf{v}_k)$, $k = 1, 2$, determined?

EIGENVALUES   The defining relationship for eigenvalues and eigenvectors is the equation $A\mathbf{v} = \lambda\mathbf{v}$, so if $\mathbf{v}$ is a constant vector, the substitution

$$\mathbf{v} = \begin{bmatrix} b_1 \\ b_2 \end{bmatrix} \quad \text{into} \quad A\mathbf{v} = \lambda\mathbf{v}$$

gives the equations

$$\begin{bmatrix} b_1 + 2b_2 \\ 4b_1 - b_2 \end{bmatrix} = \begin{bmatrix} \lambda b_1 \\ \lambda b_2 \end{bmatrix}$$

or

$$\begin{bmatrix} (1-\lambda)b_1 + 2b_2 \\ 4b_1 + (-1-\lambda)b_2 \end{bmatrix} = \begin{bmatrix} 1-\lambda & 2 \\ 4 & -1-\lambda \end{bmatrix} \begin{bmatrix} b_1 \\ b_2 \end{bmatrix} = \begin{bmatrix} 0 \\ 0 \end{bmatrix} \tag{12.33}$$

The matrix form of the homogeneous equations on the right of (12.33) is $(A - \lambda I)\mathbf{v} = \mathbf{0}$. If the determinant of the matrix $A - \lambda I$ is not zero, then Cramer's rule applies and the only solution for (12.33) is $b_1 = b_2 = 0$. (This is equivalent to applying Theorem 12.1 and concluding $\mathbf{v} = \mathbf{0}$.

To prevent $b_1 = b_2 = 0$ from being the only solution, the rows of $A - \lambda I$ must be proportional so that instead of two independent equations in (12.33) there is only one. The rows of $A - \lambda I$ will be proportional if λ satisfies

$$\frac{1-\lambda}{4} = \frac{2}{-1-\lambda} \Rightarrow (1-\lambda)(-1-\lambda) = 8 \Rightarrow \lambda^2 - 9 = 0 \tag{12.34}$$

The last equation on the right in (12.34) is just

$$\det(A - \lambda I) = \lambda^2 - 9 = 0$$

which is called the *characteristic equation,* and the two values of λ that satisfy it are called the *characteristic roots,* or *eigenvalues* of the matrix A. It is obvious from the characteristic equation $\lambda^2 - 9 = 0$ that the eigenvalues are 3 and -3. How do we then use that information to get the corresponding eigenvectors?

EIGENVECTORS An eigenvector is a nonzero solution of the homogeneous equations defined by $(A - \lambda_k I)\mathbf{v} = \mathbf{0}$. Hence, the eigenvector is a basis vector for the null space of the matrix $A - \lambda_k I$. The first column of Table 12.3 lists the eigenvalues of A, and opposite them in the second column the matrices $A - \lambda_k I$ are given. The third column lists the one independent equation in the homogeneous system $(A - \lambda_k I)\mathbf{v} = \mathbf{0}$. Solving this equation leads to the eigenvector in the fourth column.

λ	$A - \lambda I$	$(A - \lambda I)\mathbf{v} = \mathbf{0}$	**Basis for Null Space**
3	$\begin{bmatrix} -2 & 2 \\ 4 & -4 \end{bmatrix}$	$b_1 - b_2 = 0$	$\mathbf{v} = \begin{bmatrix} b_1 \\ b_1 \end{bmatrix} \Rightarrow \mathbf{v}_1 = \begin{bmatrix} 1 \\ 1 \end{bmatrix}$
-3	$\begin{bmatrix} 4 & 2 \\ 4 & 2 \end{bmatrix}$	$2b_1 + b_2 = 0$	$\mathbf{v} = \begin{bmatrix} b_1 \\ -2b_1 \end{bmatrix} \Rightarrow \mathbf{v}_2 = \begin{bmatrix} 1 \\ -2 \end{bmatrix}$

TABLE 12.3 Eigenvectors as bases of null spaces $A - \lambda I$

If $\lambda = 3$, the one independent equation in $(A - 3I)\mathbf{v} = \mathbf{0}$ can be taken as $b_1 - b_2 = 0$. Solving for $b_2 = b_1$ gives the vector $\mathbf{v}$, from which the basis vector $\mathbf{v}_1$ is determined when $b_1 = 1$

If $\lambda = -3$, the one independent equation in $(A + 3I)\mathbf{v} = \mathbf{0}$ can be taken as $2b_1 + b_2 = 0$. Solving for $b_2 = -2b_1$ gives the vector $\mathbf{v}$, from which the basis vector $\mathbf{v}_2$ is determined when $b_1 = 1$.

EXAMPLE 12.18 The characteristic equation for the matrix A in (12.28) is

$$\det(A - \lambda I) = \begin{vmatrix} 9 - \lambda & -5 & 2 \\ 14 & -8 - \lambda & 4 \\ 10 & -7 & 5 - \lambda \end{vmatrix} = (1 - \lambda)(2 - \lambda)(3 - \lambda) = 0 \qquad \textbf{(12.35)}$$

so the eigenvalues are 1, 2, 3. The corresponding eigenvectors are obtained by the manipulations summarized in Table 12.4. The first column lists the eigenvalues, and the second, the reduced row-echelon form of the matrix $A - \lambda_k I$. For each λ_k, the matrix $A - \lambda_k I$ contains two distinct rows and so has rank 2. Hence, the basis for each null space contains only

λ	$\text{rref}(A - \lambda I)$	$(A - \lambda I)\mathbf{v} = \mathbf{0}$	**Basis for Null Space**
1	$\begin{bmatrix} 1 & 0 & -1 \\ 0 & 1 & -2 \\ 0 & 0 & 0 \end{bmatrix}$	$\begin{aligned} b_1 - b_3 &= 0 \\ b_2 - 2b_3 &= 0 \end{aligned}$	$\mathbf{v} = \begin{bmatrix} b_3 \\ 2b_3 \\ b_3 \end{bmatrix} \Rightarrow \mathbf{v}_1 = \begin{bmatrix} 1 \\ 2 \\ 1 \end{bmatrix}$
2	$\begin{bmatrix} 1 & 0 & 1 \\ 0 & 1 & 1 \\ 0 & 0 & 0 \end{bmatrix}$	$\begin{aligned} b_1 + b_3 &= 0 \\ b_2 + b_3 &= 0 \end{aligned}$	$\mathbf{v} = \begin{bmatrix} -b_3 \\ -b_3 \\ b_3 \end{bmatrix} \Rightarrow \mathbf{v}_2 = \begin{bmatrix} 1 \\ 1 \\ -1 \end{bmatrix}$
3	$\begin{bmatrix} 1 & 0 & -\frac{1}{2} \\ 0 & 1 & -1 \\ 0 & 0 & 0 \end{bmatrix}$	$\begin{aligned} b_1 - \tfrac{1}{2}b_3 &= 0 \\ b_2 - b_3 &= 0 \end{aligned}$	$\mathbf{v} = \begin{bmatrix} \frac{1}{2}b_3 \\ b_3 \\ b_3 \end{bmatrix} \Rightarrow \mathbf{v}_3 = \begin{bmatrix} 1 \\ 2 \\ 2 \end{bmatrix}$

TABLE 12.4 Eigenvectors for Example 12.18

one vector. The third column lists the two distinct equations in the homogeneous system $(A + \lambda_k I)\mathbf{v} = \mathbf{0}$. When these equations are solved, the eigenvector in the fourth column is determined.

If $\lambda = 1$, the two distinct equations in the third column are solved for $b_1 = b_3$ and $b_2 = 2b_3$, giving $\mathbf{v}$. Setting $b_3 = 1$ then gives the basis vector $\mathbf{v}_1$.

If $\lambda = 2$, the two distinct equations in the third column are solved for $b_1 = -b_3$ and $b_2 = -b_3$, giving $\mathbf{v}$. Setting $b_3 = -1$ then gives the basis vector $\mathbf{v}_2$.

If $\lambda = 3$, the two distinct equations in the third column are solved for $b_1 = \frac{1}{2}b_3$ and $b_2 = b_3$, giving $\mathbf{v}$. Setting $b_3 = 2$ then gives the basis vector $\mathbf{v}_3$. ❖

EXAMPLE 12.19   For the matrix A given in (12.36), we will calculate the eigenvalues $\lambda = 2, 3, 3$ and the corresponding eigenvectors $\mathbf{v}_k, k = 1, 2, 3$, shown on the right in (12.36)

$$A = \begin{bmatrix} 4 & -1 & -1 \\ 1 & 2 & -1 \\ 1 & -1 & 2 \end{bmatrix} \Rightarrow \mathbf{v}_1 = \begin{bmatrix} 1 \\ 1 \\ 1 \end{bmatrix} \quad \mathbf{v}_2 = \begin{bmatrix} 1 \\ 1 \\ 0 \end{bmatrix} \quad \mathbf{v}_3 = \begin{bmatrix} 1 \\ 0 \\ 1 \end{bmatrix} \tag{12.36}$$

❖

EIGENVALUES   The characteristic equation for the matrix A is

$$\det(A - \lambda I) = \begin{vmatrix} 4-\lambda & -1 & -1 \\ 1 & 2-\lambda & -1 \\ 1 & -1 & 2-\lambda \end{vmatrix} = (2-\lambda)(3-\lambda)^2 = 0 \tag{12.37}$$

so the characteristic roots, or eigenvalues, are $\lambda = 2, 3, 3$.

EIGENVECTORS   Table 12.5 summarizes the calculations leading to the eigenvectors in (12.36). To find the eigenvector corresponding to $\lambda = 2$, obtain the reduced row-echelon form of $A - 2I$, listed in the second column. The rank of this matrix is 2 because there are two distinct rows. This means there are two distinct equations in the homogeneous system $(A - 2I)\mathbf{v} = \mathbf{0}$. Displayed in the third column, these two equations are solved for $b_1 = b_3$ and $b_2 = b_3$, giving the vector $\mathbf{v}$ in the fourth column. Setting $b_3 = 1$ then gives $\mathbf{v}_1$ as the basis vector for the null space of $A - 2I$, and this is the vector listed in (12.36) as the first eigenvector.

To find the eigenvector corresponding to $\lambda = 3$, obtain the reduced row-echelon form of the matrix $A - 3I$, listed in the second column. The rank of this matrix is 1 because there is only one nonzero row. The homogeneous system $(A - 3I)\mathbf{B} = \mathbf{0}$ will have just the one distinct equation $b_1 - b_2 - b_3 = 0$, which we write as $b_1 = b_2 + b_3$ in the third column. From this equation we obtain the vector $\mathbf{v}$ in the fourth column. The two free parameters b_2

λ	$\text{rref}(A - \lambda I)$	$(A - \lambda I)\mathbf{v} = \mathbf{0}$	**Basis for Null Space**
2	$\begin{bmatrix} 1 & 0 & -1 \\ 0 & 1 & -1 \\ 0 & 0 & 0 \end{bmatrix}$	$b_1 - b_3 = 0$ $b_2 - b_3 = 0$	$\mathbf{v} = \begin{bmatrix} b_3 \\ b_3 \\ b_3 \end{bmatrix} \Rightarrow \mathbf{v}_1 = \begin{bmatrix} 1 \\ 1 \\ 1 \end{bmatrix}$
3	$\begin{bmatrix} 1 & -1 & -1 \\ 0 & 0 & 0 \\ 0 & 0 & 0 \end{bmatrix}$	$b_1 = b_2 + b_3$	$\mathbf{v} = \begin{bmatrix} b_2 + b_3 \\ b_2 \\ b_3 \end{bmatrix} = b_2 \begin{bmatrix} 1 \\ 1 \\ 0 \end{bmatrix} + b_3 \begin{bmatrix} 1 \\ 0 \\ 1 \end{bmatrix}$

TABLE 12.5   Eigenvectors for Example 12.19

and b_3 are factored as shown in the fourth column, and the vectors they multiply are taken as v_2 and v_3, the second and third eigenvectors listed in (12.36).

The eigenvectors from the repeated eigenvalue $\lambda = 3$ all lie in a plane, and any two distinct (independent) vectors from this plane can be taken as "the" eigenvectors. Every vector in this plane will satisfy the equation $A\mathbf{v} = \lambda\mathbf{v}$.

EXERCISES 12.12–Part A

A1. Verify $A\mathbf{v}_1 = 3\mathbf{v}_1$ and $A\mathbf{v}_2 = -3\mathbf{v}_2$ for A and $\mathbf{v}_k$, $k = 1, 2$, given in (12.32).

A2. Verify $\det(A - \lambda I) = \lambda^2 - 9$ for A given in (12.32).

A3. Verify the entries in the second and third columns of Table 12.3.

A4. Verify the calculation of $\det(A - \lambda I)$ appearing in (12.35).

A5. Verify the entries in the second and third columns of Table 12.4.

A6. Verify the calculation of $\det(A - \lambda I)$ appearing in (12.37).

A7. Verify the entries in the second and third columns of Table 12.5.

EXERCISES 12.12–Part B

B1. For A and $\mathbf{v}_k$, $k = 2, 3$, appearing in (12.36), verify that $A\mathbf{x} = 3\mathbf{x}$ for $\mathbf{x} = \alpha\mathbf{v}_2 + \beta\mathbf{v}_3$, where α and β are arbitrary real numbers.

For the matrix A given in each of Exercises B2–11:

(a) Use appropriate technology to obtain the eigenvalues and eigenvectors.

(b) Obtain and solve the characteristic equation $\det(A - \lambda I) = 0$.

(c) Show that the characteristic equation is a direct consequence of making the rows of $A - \lambda I$ proportional.

(d) For each eigenvalue λ_k, obtain the corresponding eigenvector $\mathbf{v}_k$ as the solution of the homogeneous system of equations $(A - \lambda_k I)\mathbf{v}_k = \mathbf{0}$.

(e) For each eigenvalue λ_k, determine the null space of $A - \lambda_k I$.

(f) Verify that the rank of $A - \lambda_k I$ plus the dimension of the null space of $A - \lambda_k I$ equals n, which in this instance, is 2.

(g) Verify that $A\mathbf{v}_k = \lambda_k\mathbf{v}_k$, $k = 1, 2$.

(h) Form $P = [\mathbf{v}_1, \mathbf{v}_2]$, a matrix whose columns are the eigenvectors, and compute the product $P^{-1}AP$. This should be a diagonal matrix whose diagonal entries are the eigenvalues λ_k.

(i) If $\mathbf{u}$ is the unit vector $\begin{bmatrix} \cos\theta \\ \sin\theta \end{bmatrix}$, solve the pair of equations represented by $A\mathbf{u} = \lambda\mathbf{u}$ for θ and λ, from which the unit eigenvectors $\mathbf{u}_k$ can be obtained.

B2. $\begin{bmatrix} 9 & 26 \\ -4 & -12 \end{bmatrix}$ **B3.** $\begin{bmatrix} 18 & 55 \\ -8 & -24 \end{bmatrix}$ **B4.** $\begin{bmatrix} 33 & 35 \\ -48 & -49 \end{bmatrix}$ **B5.** $\begin{bmatrix} -18 & 18 \\ -8 & 7 \end{bmatrix}$

B6. $\begin{bmatrix} -5 & -6 \\ 4 & 6 \end{bmatrix}$ **B7.** $\begin{bmatrix} -8 & -3 \\ 7 & 2 \end{bmatrix}$ **B8.** $\begin{bmatrix} -12 & -1 \\ 56 & 3 \end{bmatrix}$ **B9.** $\begin{bmatrix} -43 & -64 \\ 33 & 49 \end{bmatrix}$

B10. $\begin{bmatrix} 1 & -10 \\ 1 & 8 \end{bmatrix}$ **B11.** $\begin{bmatrix} -19 & 20 \\ -12 & 12 \end{bmatrix}$

For the matrix A given in each of Exercises B12–21:

(a) Use appropriate technology to obtain the eigenvalues and any eigenvectors.

(b) Obtain and solve the characteristic equation $\det(A - \lambda I) = 0$.

(c) Observing that $\lambda_1 = \lambda_2$, show that the dimension of the null space of $A - \lambda_1 I$ is one.

(d) Find the single eigenvector that arises as the solution of the homogeneous system of equations $(A - \lambda_1 I)\mathbf{v}_1 = \mathbf{0}$.

(e) Verify that the rank of $A - \lambda_1 I$ plus the dimension of the null space of $A - \lambda_1 I$ equals n, which in this instance is 2.

B12. $\begin{bmatrix} 63 & 64 \\ -49 & -49 \end{bmatrix}$ **B13.** $\begin{bmatrix} 14 & 9 \\ -25 & -16 \end{bmatrix}$ **B14.** $\begin{bmatrix} 38 & -25 \\ 49 & -32 \end{bmatrix}$

B15. $\begin{bmatrix} -11 & 1 \\ -25 & -1 \end{bmatrix}$ **B16.** $\begin{bmatrix} -41 & 81 \\ -25 & 49 \end{bmatrix}$ **B17.** $\begin{bmatrix} 5 & -9 \\ 1 & -1 \end{bmatrix}$

B18. $\begin{bmatrix} -3 & -1 \\ 16 & 5 \end{bmatrix}$ **B19.** $\begin{bmatrix} -25 & -25 \\ 16 & 15 \end{bmatrix}$ **B20.** $\begin{bmatrix} -3 & 25 \\ -1 & 7 \end{bmatrix}$ **B21.** $\begin{bmatrix} 51 & -25 \\ 81 & -39 \end{bmatrix}$

For the matrix A given in each of Exercises B22–36:

(a) Use appropriate technology to obtain the eigenvalues and any eigenvectors.

(b) Obtain and solve the characteristic equation $\det(A - \lambda I) = 0$.

(c) For each distinct eigenvalue λ_k, determine the rank of $A - \lambda_k I$.

(d) By solving the homogeneous system $(A - \lambda_k I)\mathbf{v} = \mathbf{0}$, obtain the eigenvectors associated with each distinct eigenvalue λ_k.

(e) For each distinct eigenvalue λ_k, verify that the rank of $A - \lambda_k I$ plus the dimension of the null space of $A - \lambda_k I$ equals n, which in this instance is 3.

(f) If three linearly independent eigenvectors have been obtained, form $P = [\mathbf{v}_1, \mathbf{v}_2, \mathbf{v}_3]$, a matrix whose columns are the eigenvectors, and compute the product $P^{-1}AP$. This should be a diagonal matrix whose diagonal entries are the eigenvalues λ_k.

B22. $\begin{bmatrix} 14 & 4 & 8 \\ -20 & -4 & -15 \\ -12 & -3 & -8 \end{bmatrix}$ **B23.** $\begin{bmatrix} 1 & -4 & -1 \\ 4 & -7 & -1 \\ -1 & 1 & -3 \end{bmatrix}$ **B24.** $\begin{bmatrix} 3 & -4 & 18 \\ -2 & 1 & -9 \\ -1 & 0 & -4 \end{bmatrix}$

B25. $\begin{bmatrix} 31 & 3 & -35 \\ 25 & 3 & -30 \\ 38 & 4 & -43 \end{bmatrix}$
B26. $\begin{bmatrix} -49 & -18 & -6 \\ 96 & 35 & 12 \\ 48 & 18 & 5 \end{bmatrix}$
B27. $\begin{bmatrix} 3 & 10 & 130 \\ -3 & -9 & -40 \\ 1 & 3 & 4 \end{bmatrix}$
B31. $\begin{bmatrix} -19 & -22 & -20 \\ 27 & 28 & 26 \\ -17 & -16 & -15 \end{bmatrix}$
B32. $\begin{bmatrix} -3 & -12 & -6 \\ -2 & 7 & 6 \\ 4 & -24 & -17 \end{bmatrix}$
B33. $\begin{bmatrix} 0 & 5 & -3 \\ 1 & -41 & 21 \\ 2 & -65 & 33 \end{bmatrix}$

B28. $\begin{bmatrix} 39 & 33 & 75 \\ 42 & 43 & 83 \\ -46 & -44 & -90 \end{bmatrix}$
B29. $\begin{bmatrix} -5 & -1 & 0 \\ 3 & -8 & -1 \\ -9 & 6 & -2 \end{bmatrix}$
B30. $\begin{bmatrix} -16 & -9 & -15 \\ 18 & 11 & 15 \\ 12 & 6 & 12 \end{bmatrix}$
B34. $\begin{bmatrix} 8 & -5 & 4 \\ -4 & 10 & -7 \\ -5 & 7 & -3 \end{bmatrix}$
B35. $\begin{bmatrix} -51 & 90 & -60 \\ -15 & 24 & -20 \\ 15 & -30 & 14 \end{bmatrix}$
B36. $\begin{bmatrix} -2 & 0 & 1 \\ 61 & 29 & -31 \\ 39 & 21 & -20 \end{bmatrix}$

12.13   System versus Second-Order ODE

System to Second-Order

In (12.32), a matrix A and its eigenvectors $\mathbf{v}_1$ and $\mathbf{v}_2$ are given. In addition, this matrix has characteristic equation $\lambda^2 - 9 = 0$ and eigenvalues $\lambda = \pm 3$. Therefore, the system of differential equations $\mathbf{x}' = A\mathbf{x}$ on the left in (12.38) has the general solution given on the right

$$\begin{aligned} u'(t) &= u + 2v \\ v'(t) &= 4u - v \end{aligned} \quad \Rightarrow \mathbf{x} = ae^{3t}\begin{bmatrix} 1 \\ 1 \end{bmatrix} + be^{-3t}\begin{bmatrix} 1 \\ -2 \end{bmatrix} \tag{12.38}$$

The system of equations on the left in (12.38) can be converted to a single second-order ODE by the following procedure.

Pick one equation, say the first, and differentiate it with respect to t, obtaining $u'' = u' + 2v'$. Having introduced v' into this new equation, eliminate it using the second equation, obtaining $u'' = u' + 2(4u - v)$, which now contains v. Finally, eliminate v by again using the first equation, obtaining $u'' = u' + 8u - 2(\frac{u'-u}{2})$ or $u'' = 9u$.

This second-order ODE has the characteristic equation $m^2 - 9 = 0$ and characteristic roots $m = \pm 3$, so its general solution is $u(t) = c_1 e^{3t} + c_2 e^{-3t}$. To avoid introducing two more arbitrary constants, the solution for $v(t)$ is found from the first differential equation as $v(t) = \frac{1}{2}(u' - u) = c_1 e^{3t} - 2c_2 e^{-3t}$. Except for a renaming of the constants, this solution is the same as the one from eigenvalues and eigenvectors.

Second-Order to System

Given the second-order linear ordinary differential equation

$$y''(t) + p(t)y'(t) + q(t)y(t) = r(t)$$

it is useful to learn how to represent it as a first-order linear *system*. Do this by the following change of variables. Define u and v as on the left in (12.39) and then differentiate (and use the DE itself) to obtain the results on the right

$$\begin{aligned} u(t) &= y(t) \\ v(t) &= y'(t) \end{aligned} \quad \Rightarrow \quad \begin{aligned} u'(t) &= y'(t) = v(t) \\ v'(t) &= y''(t) = -p(t)y'(t) - q(t)y(t) + r(t) \\ &= -p(t)v(t) - q(t)u(t) + r(t) \end{aligned} \tag{12.39}$$

The matrix form of the resulting system is then

$$\begin{bmatrix} u \\ v \end{bmatrix}' = \begin{bmatrix} 0 & 1 \\ -q & -p \end{bmatrix}\begin{bmatrix} u \\ v \end{bmatrix} + \begin{bmatrix} 0 \\ r \end{bmatrix}$$

Third-Order to System

The third-order equation

$$y'''(t) + f(t)y''(t) + g(t)y'(t) + h(t)y(t) = r(t)$$

can be expressed as a linear system by the following steps. Define u, v, w as on the left in (12.40) then differentiate (and use the DE itself) to obtain the results on the right

$$
\begin{aligned}
&& u'(t) = y'(t) = v(t) \\
u(t) = y(t) && \\
v(t) = y'(t) \Rightarrow && v'(t) = y''(t) = w(t) \\
w(t) = y''(t) && w'(t) = y'''(t) = -f(t)y''(t) - g(t)y'(t) - h(t)y(t) + r(t) \\
&& w'(t) = -f(t)w(t) - g(t)v(t) - h(t)u(t) + r(t)
\end{aligned}
\tag{12.40}
$$

The matrix form of the resulting system is then

$$
\begin{bmatrix} u \\ v \\ w \end{bmatrix}' = \begin{bmatrix} 0 & 1 & 0 \\ 0 & 0 & 1 \\ -h & -g & -f \end{bmatrix} \begin{bmatrix} u \\ v \\ w \end{bmatrix} + \begin{bmatrix} 0 \\ 0 \\ r \end{bmatrix}
$$

EXAMPLE 12.20 The second-order differential equation $y''(t) + 4y'(t) + 13y(t) = 0$ has characteristic equation $m^2 + 4m + 13 = 0$ and characteristic roots $m = -2 \pm 3i$. To convert the equation to a first-order linear system, define u and v as on the left in (12.41) and then differentiate (and use the DE itself) to obtain the results on the right

$$
\begin{aligned}
&& u'(t) = y'(t) = v(t) \\
u(t) = y(t) && \\
v(t) = y'(t) \Rightarrow && v'(t) = y''(t) = -4y'(t) - 13y(t) \\
&& v'(t) = -4v(t) - 13u(t)
\end{aligned}
\tag{12.41}
$$

In matrix notation, the desired system is

$$
\begin{bmatrix} u \\ v \end{bmatrix}' = \begin{bmatrix} 0 & 1 \\ -13 & -4 \end{bmatrix} \begin{bmatrix} u \\ v \end{bmatrix}
\tag{12.42}
$$

The eigenvalues of the system matrix in (12.42) are $\lambda = -2 \pm 3i$, the same as the characteristic roots for the original second-order differential equation. However, the eigenvalues are a complex conjugate pair, dealt with at length in Section 12.14. ❖

EXERCISES 12.13–Part A

A1. If the system $\mathbf{x}' = A\mathbf{x}$ is defined by the matrix A in Exercise A4, Section 12.11, obtain an equivalent second-order equation. Solve the second-order equation by the methods of Chapter 5, and from this solution, construct the general solution of the given system.

A2. Repeat Exercise A1 for the matrix A of Exercise A5, Section 12.11.

A3. Repeat Exercise A1 for the matrix A of Exercise B1, Section 12.11.

A4. Cast into the form $\mathbf{v}' = M\mathbf{v}$, the second-order differential equation obtain in Exercise A1. Show that the eigenvalues of M are the same as those of A.

A5. Repeat Exercise A4 for the second-order differential equation obtained in Exercise A2.

A6. Repeat Exercise A4 for the second-order differential equation obtained in Exercise A3.

EXERCISES 12.13–Part B

B1. Let A be the matrix in Exercise A1 and M be the matrix in Exercise A4. Let $P = \begin{bmatrix} p_1 & p_2 \\ p_3 & p_4 \end{bmatrix}$ and set $\mathbf{v} = P\mathbf{x}$. Show that $\mathbf{v}' = M\mathbf{v}$ becomes $\mathbf{x} = P^{-1}MP\mathbf{x}$. By direct calculation, find all matrices P for which $P^{-1}MP = A$.

B2. Repeat Exercise B1 for the matrix A in Exercise A2 and the matrix M in Exercise A5.

B3. Repeat Exercise B1 for the matrix A in Exercise A3 and the matrix M in Exercise A6.

For the matrix A specified in each of Exercises B4–13:

(a) Write the general solution to $\mathbf{x}' = A\mathbf{x}$ as $\mathbf{x} = ae^{\lambda_1 t}\mathbf{v}_1 + be^{\lambda_2 t}\mathbf{v}_2$, where λ_k and $\mathbf{v}_k$, $k = 1, 2$, are the eigenvalues and eigenvectors, respectively, for A.

(b) By an appropriate combination of differentiations and eliminations, convert the system $\mathbf{x}' = A\mathbf{x}$ to a single second-order linear differential equation.

(c) Obtain the general solution of the second-order equation found in part (b), and show it is the same as the general solution found in part (a).

B4. Exercise B2, Section 12.12   **B5.** Exercise B3, Section 12.12

B6. Exercise B4, Section 12.12   **B7.** Exercise B5, Section 12.12

B8. Exercise B6, Section 12.12   **B9.** Exercise B7, Section 12.12

B10. Exercise B8, Section 12.12   **B11.** Exercise B9, Section 12.12

B12. Exercise B10, Section 12.12   **B13.** Exercise B11, Section 12.12

For Exercises B14–33, convert the given nth-order differential equations to a first-order system. For each *linear* differential equation, put the system into the matrix form $\mathbf{x}' = A\mathbf{x} + \mathbf{F}$. Be sure to specify $\mathbf{x}$ and $\mathbf{F}$.

B14. $3y'' + 2y' + 5y = \sin t$ **B15.** $t^2 y'' + 4ty' + 6y = 0$

B16. $y'' + e^{-2t}y' - \dfrac{1}{t}y = \cos 2t$ **B17.** $5y'' + 3yy' + 4y = te^{-t}$

B18. $2y''' + 7y'' - 9y' + 8y = 6$ **B19.** $y''' - (y')^2 y - \sin y = 13 \ln t$

B20. $6y''' + \sin y'' + y' \cos t = 7ye^{-y}$

B21. $2y'' + 3y' - 7y = e^{-t} \sin 2t$

B22. $\sqrt{t}\,y'' - 4y' + ty = \sqrt{1 - t^2}$ **B23.** $(1 + t^2)y''' + 5y'' - 13y = 1$

B24. $8y''' - 2y'' + 3y' + 9y = 25t$ **B25.** $\sin(y'y'') + 2y = \sqrt{t}$

B26. $y'' + y' + y = y^2$ **B27.** $y'' + y' + y = t^2$

B28. $ty''' + 2y'' - e^{-t}y' - 7y = 5$

B29. $12y'' - 7y' + 5y = 2 + 3t - t^2$

B30. $6y''' + 7ty'' - 5e^{2t}y' + 15y = e^{-3t}\cos 2t$

B31. $4y'' + 13y' + 6y = \dfrac{t}{1 + t}$ **B32.** $yy'' = y'$

B33. $3t^2 y'' + 2ty' - 5y = t \sin t$

12.14 Complex Eigenvalues

Introduction

When reduced to a system, the differential equation in Example 12.20, Section 12.13, led to the matrix A in (12.42) with complex conjugate eigenvalues. As a guide to solving the system with complex eigenvalues, we will use techniques from Chapter 5 to solve the second-order equation.

EXAMPLE 12.21 The second-order differential equation $y''(t) + 4y'(t) + 13y(t) = 0$ with characteristic equation $m^2 + 4m + 13 = 0$ and complex conjugate characteristic roots $-2 \pm 3i$ appeared in Example 12.20, Section 12.13. From our work in Chapter 5 we know that the fundamental set $\{e^{(-2+3i)t}, e^{(-2-3i)t}\}$ is equivalent to $\{e^{-2t}\cos 3t, e^{-2t}\sin 3t\}$, so the general solution is $y(t) = e^{-2t}(a \cos 3t + b \sin 3t)$.

Reduction to a system is accomplished in (12.41) and results in the equation $\mathbf{x}' = A\mathbf{x}$ detailed in (12.42). Using the known value of $y(t)$, the general solution of the system can then be written as

$$\mathbf{x} = \begin{bmatrix} u \\ v \end{bmatrix} = \begin{bmatrix} y \\ y' \end{bmatrix} = e^{-2t}\begin{bmatrix} a \cos 3t + b \sin 3t \\ a(-2\cos 3t - 3\sin 3t) + b(3\cos 3t - 2\sin 3t) \end{bmatrix} \quad \text{(12.43a)}$$

$$= ae^{-2t}\mathbf{x}_1 + be^{-2t}\mathbf{x}_2 = ae^{-2t}\begin{bmatrix} \cos 3t \\ -2\cos 3t - 3\sin 3t \end{bmatrix} + be^{-2t}\begin{bmatrix} \sin 3t \\ 3\cos 3t - 2\sin 3t \end{bmatrix}$$

$$\text{(12.43b)}$$

The system with complex eigenvalues appears to have a more complicated solution than the system with real eigenvalues. Our goal now is to see if this solution can be determined without first needing to solve the system as a second-order equation. ❖

Complex Case for Systems

The eigenvalues of the matrix A on the left in (12.44) are the complex conjugate pair $\lambda = -2 \pm 3i$, and the eigenvectors $\mathbf{v}_1$ and $\mathbf{v}_2$, complex conjugates, are on the right

$$A = \begin{bmatrix} 0 & 1 \\ -13 & -4 \end{bmatrix} \Rightarrow \mathbf{v}_1 = \begin{bmatrix} 1 \\ -2 + 3i \end{bmatrix} \quad \mathbf{v}_2 = \begin{bmatrix} 1 \\ -2 - 3i \end{bmatrix} \tag{12.44}$$

The real and imaginary parts of the one complex solution $e^{(-2+3i)t}\mathbf{v}_1$ are then two distinct real solutions of the first-order system $\mathbf{x}' = A\mathbf{x}$. In fact, applying Euler's formulas to both $e^{(-2+3i)t}\mathbf{v}_1$ and $e^{(-2-3i)t}\mathbf{v}_2$, we get

$$e^{(-2+3i)t} \begin{bmatrix} 1 \\ -2 + 3i \end{bmatrix} = \begin{bmatrix} e^{-2t}\cos 3t \\ -e^{-2t}(2\cos 3t + 3\sin 3t) \end{bmatrix} + i \begin{bmatrix} e^{-2t}\sin 3t \\ -e^{-2t}(3\cos 3t - 2\sin 3t) \end{bmatrix}$$

$$e^{(-2-3i)t} \begin{bmatrix} 1 \\ -2 - 3i \end{bmatrix} = \begin{bmatrix} e^{-2t}\cos 3t \\ -e^{-2t}(2\cos 3t + 3\sin 3t) \end{bmatrix} - i \begin{bmatrix} e^{-2t}\sin(3t) \\ -e^{-2t}(3\cos 3t - 2\sin 3t) \end{bmatrix}$$

so the linear combinations

$$\frac{1}{2}(e^{(-2+3i)t}\mathbf{v}_1 + e^{(-2-3i)t}\mathbf{v}_2) = \begin{bmatrix} e^{-2t}\cos 3t \\ -e^{-2t}(2\cos 3t + 3\sin 3t) \end{bmatrix} = \mathbf{x}_1$$

$$\frac{1}{2i}(e^{(-2+3i)t}\mathbf{v}_1 - e^{(-2-3i)t}\mathbf{v}_2) = \begin{bmatrix} e^{-2t}\sin 3t \\ -e^{-2t}(3\cos 3t - 2\sin 3t) \end{bmatrix} = \mathbf{x}_2$$

are simply the real and imaginary parts of the one complex solution $e^{(-2+3i)t}\mathbf{v}_1$.

The general solution of the linear system is then the linear combination $\mathbf{x} = ae^{-2t}\mathbf{x}_1 + be^{-2t}\mathbf{x}_2$ given in (12.43b).

Finding Complex Eigenvectors

Matrix arithmetic already learned will also yield complex eigenvectors, provided the rules of complex arithmetic are heeded. To begin, recall how looking for a solution of the form $\mathbf{v}e^{\lambda t}$ leads to the eigenvalue/eigenvector equation $A\mathbf{v} = \lambda\mathbf{v}$ or $(A - \lambda I)\mathbf{v} = \mathbf{0}$. The eigenvalues are the roots of the characteristic equation, $\det(A - \lambda I) = 0$. For the matrix in (12.44), the characteristic equation is

$$\det(A - \lambda I) = \begin{vmatrix} -\lambda & 1 \\ -13 & -4 - \lambda \end{vmatrix} = \lambda^2 + 4\lambda + 13 = 0 \tag{12.45}$$

so the eigenvalues are $-2 \pm 3i$.

Computation of the eigenvectors is summarized in Table 12.6. For each eigenvalue λ listed in the first column, the matrix $A - \lambda I$ is listed in the second column. In the third column, the one distinct equation in the homogeneous system $(A - \lambda I)\mathbf{v} = \mathbf{0}$ is given, and in the fourth column, the eigenvector is listed.

For example, when $\lambda = \lambda_1 = -2 + 3i$, the complex arithmetic masks the proportionality of the rows of $A - \lambda_1 I$, but multiplication of the first row by $-2 - 3i$ yields the second. Hence, there is but one equation, say $(2 - 3i)b_1 + b_2 = 0$, as shown. Setting the arbitrary parameter $b_1 = 1$ in $\mathbf{v}$ gives $\mathbf{v}_1$ in the fourth column.

EXAMPLE 12.22   A 3×3 system with real coefficients will have either one or three real eigenvalues. The matrix A on the left in (12.46) has the real eigenvalue $\lambda = 2$ and the pair of complex conjugate eigenvalues $\lambda = 2 \pm 3i$. The corresponding eigenvectors $\mathbf{v}_k, k = 1, 2, 3$, are

λ	$A - \lambda I$	$(A - \lambda I)\mathbf{v} = 0$	**Basis for Null Space**
$-2 + 3i$	$\begin{bmatrix} 2 - 3i & 1 \\ -13 & -2 - 3i \end{bmatrix}$	$(2 - 3i)b_1 + b_2 = 0$	$\mathbf{v} = \begin{bmatrix} b_1 \\ (-2 + 3i)b_1 \end{bmatrix} \Rightarrow \mathbf{v}_1 = \begin{bmatrix} 1 \\ -2 + 3i \end{bmatrix}$
$-2 - 3i$	$\begin{bmatrix} 2 + 3i & 1 \\ -13 & -2 + 3i \end{bmatrix}$	$(2 + 3i)b_1 + b_2 = 0$	$\mathbf{v} = \begin{bmatrix} b_1 \\ (-2 - 3i)b_1 \end{bmatrix} \Rightarrow \mathbf{v}_2 = \begin{bmatrix} 1 \\ -2 - 3i \end{bmatrix}$

TABLE 12.6   Eigenvectors for the matrix $A = \begin{bmatrix} 0 & 1 \\ -13 & -4 \end{bmatrix}$

shown on the right in (12.46)

$$A = \begin{bmatrix} -88 & 42 & -72 \\ 21 & -7 & 18 \\ 123 & -57 & 101 \end{bmatrix} \Rightarrow \mathbf{v}_1 = \begin{bmatrix} -\frac{3}{2} \\ -\frac{3}{2} \\ 1 \end{bmatrix} \quad \mathbf{v}_2 = \begin{bmatrix} -2i \\ 1 \\ \frac{1}{2} + \frac{5}{2}i \end{bmatrix} \quad \mathbf{v}_3 = \begin{bmatrix} 2i \\ 1 \\ \frac{1}{2} - \frac{5}{2}i \end{bmatrix}$$

$$(12.46)$$

The general solution of the first-order system $\mathbf{x}' = A\mathbf{x}$ is a linear combination of the three solutions $\mathbf{x}_1 = e^{2t}\mathbf{v}_1$, and $\mathbf{x}_2$ and $\mathbf{x}_3$, the real and imaginary parts of $e^{(2+3i)t}\mathbf{v}_2$, respectively. Thus, we have

$$\mathbf{x} = ae^{2t}\begin{bmatrix} -\frac{3}{2} \\ -\frac{3}{2} \\ 1 \end{bmatrix} + be^{2t}\begin{bmatrix} 2\sin 3t \\ \cos 3t \\ \frac{1}{2}\cos 3t - \frac{5}{2}\sin 3t \end{bmatrix} + ce^{2t}\begin{bmatrix} -2\cos 3t \\ \sin 3t \\ \frac{5}{2}\cos 3t + \frac{1}{2}\sin 3t \end{bmatrix}$$

If the initial data on the left in (12.47) were to be applied, the equations in the center would result. Solving these equations for $a = -26$, $b = -37$, and $c = 19$ then gives $\mathbf{x}$ on the right in (12.47) as the solution to the IVP $\mathbf{x}' = A\mathbf{x}$, $\mathbf{x}(0) = \mathbf{x}_0$.

$$\mathbf{x}_0 = \begin{bmatrix} 1 \\ 2 \\ 3 \end{bmatrix} \Rightarrow \begin{array}{l} -\frac{3}{2}a - 2c = 1 \\ -\frac{3}{2}a + b = 2 \\ a + \frac{1}{2}b + \frac{5}{2}c = 3 \end{array} \Rightarrow \mathbf{x} = \begin{bmatrix} e^{2t}(39 - 38\cos 3t - 74\sin 3t) \\ e^{2t}(39 - 37\cos 3t + 19\sin 3t) \\ e^{2t}(-26 + 29\cos 3t + 102\sin 3t) \end{bmatrix} \quad (12.47)$$

❖

EXERCISES 12.14–Part A

A1. Verify that the matrix A in (12.44) has eigenvalues $\lambda = -2 \pm 3i$.

A2. Multiply out $e^{3it}\mathbf{v}_1 = (\cos 3t + i\sin 3t)\mathbf{v}_1$ for the vector $\mathbf{v}_1$ defined in (12.44).

A3. Multiply out $e^{-3it}\mathbf{v}_2 = (\cos 3t - i\sin 3t)\mathbf{v}_2$ for the vector $\mathbf{v}_2$ defined in (12.44).

A4. Verify the determinant in (12.45).

A5. Show that the rows of $A - (-2 - 3i)I$ in Table 12.6 are indeed proportional and that the corresponding entry in the third column of the table is correct.

A6. Explain why a 3×3 matrix with real entries will have either one or three real eigenvalues but not two.

A7. Verify that the matrix A in (12.46) has the eigenvalues $2, 2 \pm 3i$.

A8. Verify that for A in (12.46), $\mathbf{v}_1$ is the eigenvector corresponding to $\lambda = 2$.

A9. Verify that for A in (12.46), $\mathbf{v}_2$ is the eigenvector corresponding to $\lambda = 2 + 3i$.

A10. Verify that for A in (12.46), $\mathbf{v}_3$ is the eigenvector corresponding to $\lambda = 2 - 3i$.

EXERCISES 12.14–Part B

B1. Solve the equations in the center of (12.47), and verify that (12.47) provides the correct solution to the IVP $\mathbf{x}' = A\mathbf{x}$, $\mathbf{x}(0) = \mathbf{x}_0$, where A is given in (12.46) and $\mathbf{x}_0$ is given in (12.47).

In Exercises B2–11, where a matrix A and initial vector $\mathbf{x}_0$ are given:

    (a) Use an appropriate technology to obtain the eigenvalues and eigenvectors of A.

    (b) Obtain the fundamental matrix $\phi = e^{At}$ and use it to solve the initial value problem $\mathbf{x}' = A\mathbf{x}$, $\mathbf{x}(0) = \mathbf{x}_0$, where $\mathbf{x} = \begin{bmatrix} x(t) \\ y(t) \end{bmatrix}$.

    (c) Convert the first-order IVP to a second-order IVP and solve, making sure the solution agrees with the one found in part (b).

    (d) Find $\mathbf{v}_1$, a solution to $(A - \lambda_1 I)\mathbf{v} = \mathbf{0}$.

    (e) Obtain $\mathbf{x}_1$ and $\mathbf{x}_2$, the real and imaginary parts, respectively, of $e^{\lambda_1 t}\mathbf{v}_1$.

    (f) Show that $\mathbf{x}'_k = A\mathbf{x}_k$, $k = 1, 2$, so the real and imaginary parts of a single complex solution are themselves solutions.

    (g) Write the general solution to the first-order system as $\mathbf{x} = a\mathbf{x}_1 + b\mathbf{x}_2$ and apply the initial data. Show that the resulting solution agrees with the solution found in part (b).

    (h) Plot $x(t)$ and $y(t)$ on the same set of axes.

    (i) Plot $y(x)$ as a trajectory in the phase plane.

B2. $\begin{bmatrix} -3 & -3 \\ 3 & -3 \end{bmatrix}, \begin{bmatrix} -6 \\ 8 \end{bmatrix}$ **B3.** $\begin{bmatrix} -13 & -20 \\ 10 & 15 \end{bmatrix}, \begin{bmatrix} 10 \\ 1 \end{bmatrix}$

B4. $\begin{bmatrix} 3 & -10 \\ 2 & -1 \end{bmatrix}, \begin{bmatrix} -11 \\ 1 \end{bmatrix}$ **B5.** $\begin{bmatrix} 5 & -5 \\ 10 & -9 \end{bmatrix}, \begin{bmatrix} -2 \\ 6 \end{bmatrix}$

B6. $\begin{bmatrix} 10 & 5 \\ -8 & -2 \end{bmatrix}, \begin{bmatrix} 9 \\ -3 \end{bmatrix}$ **B7.** $\begin{bmatrix} -9 & 3 \\ -30 & 9 \end{bmatrix}, \begin{bmatrix} -7 \\ 1 \end{bmatrix}$

B8. $\begin{bmatrix} 4 & -2 \\ 10 & 0 \end{bmatrix}, \begin{bmatrix} -5 \\ -2 \end{bmatrix}$ **B9.** $\begin{bmatrix} 6 & 2 \\ -13 & 4 \end{bmatrix}, \begin{bmatrix} 12 \\ -2 \end{bmatrix}$

B10. $\begin{bmatrix} 4 & -2 \\ 1 & 6 \end{bmatrix}, \begin{bmatrix} 4 \\ -1 \end{bmatrix}$ **B11.** $\begin{bmatrix} -2 & 10 \\ -4 & 10 \end{bmatrix}, \begin{bmatrix} -5 \\ 12 \end{bmatrix}$

In Exercises B12–21, where a matrix A and an initial vector $\mathbf{x}_0$ are given:

    (a) From the eigenvalues and eigenvectors of A, construct the general solution for the first-order equation $\mathbf{x}' = A\mathbf{x}$. If A has complex eigenvalues and eigenvectors, express the solution in terms of real functions.

    (b) Apply the initial data to obtain the unique solution of the IVP $\mathbf{x}' = A\mathbf{x}$, $\mathbf{x}(0) = \mathbf{x}_0$.

    (c) Obtain the fundamental matrix $\phi = e^{At}$ and use it to solve the IVP stated in part (b). Be sure the solution agrees with the one found in part (b).

B12. $\begin{bmatrix} 21 & 20 & 5 \\ -22 & -21 & -5 \\ 40 & 41 & 5 \end{bmatrix}, \begin{bmatrix} 12 \\ -3 \\ -2 \end{bmatrix}$ **B13.** $\begin{bmatrix} 16 & 12 & 39 \\ -6 & -6 & -13 \\ -6 & -4 & -15 \end{bmatrix}, \begin{bmatrix} -5 \\ 5 \\ 1 \end{bmatrix}$

B14. $\begin{bmatrix} -34 & -14 & 1 \\ 64 & 26 & -2 \\ -76 & -33 & 0 \end{bmatrix}, \begin{bmatrix} -10 \\ 12 \\ -2 \end{bmatrix}$ **B15.** $\begin{bmatrix} 49 & 66 & -22 \\ -44 & -61 & 22 \\ -22 & -33 & 16 \end{bmatrix}, \begin{bmatrix} -1 \\ 11 \\ -9 \end{bmatrix}$

B16. $\begin{bmatrix} 112 & 565 & 337 \\ -20 & -98 & -59 \\ -6 & -35 & -20 \end{bmatrix}, \begin{bmatrix} 11 \\ 8 \\ 11 \end{bmatrix}$ **B17.** $\begin{bmatrix} -23 & -46 & 4 \\ 10 & 20 & -2 \\ 10 & 24 & -4 \end{bmatrix}, \begin{bmatrix} 2 \\ 0 \\ -5 \end{bmatrix}$

B18. $\begin{bmatrix} 53 & 40 & -60 \\ -30 & -27 & 30 \\ 30 & 20 & -37 \end{bmatrix}, \begin{bmatrix} 12 \\ 7 \\ -9 \end{bmatrix}$ **B19.** $\begin{bmatrix} -51 & 39 & -55 \\ -28 & 26 & -30 \\ 20 & -10 & 22 \end{bmatrix}, \begin{bmatrix} 3 \\ -7 \\ -5 \end{bmatrix}$

B20. $\begin{bmatrix} -20 & -18 & 36 \\ -72 & -50 & 108 \\ -48 & -36 & 76 \end{bmatrix}, \begin{bmatrix} -11 \\ 0 \\ 10 \end{bmatrix}$ **B21.** $\begin{bmatrix} 24 & -18 & 12 \\ 33 & -25 & 18 \\ 11 & -9 & 8 \end{bmatrix}, \begin{bmatrix} 1 \\ 5 \\ 10 \end{bmatrix}$

12.15 **The Deficient Case**

Difficulty Articulated

When all the eigenvalues for a system of first-order ODEs are different, there will be one eigenvector for each eigenvalue and enough eigenpairs with which to construct the general solution. When there is a repeated eigenvalue, there may or may not be enough eigenvectors for constructing the general solution. Example 12.19, Section 12.12, contains in (12.36) the matrix A with a repeated eigenvalue but a full slate of eigenvectors. The case we study here is that of the repeated eigenvalue where there are not enough eigenvectors. This is called the *deficient case* for which we need to learn the form of the "missing solution."

EXAMPLE 12.23   Table 12.7 contains a *deficient* matrix A for which the eigenvalues are the repeated $\lambda = 2, 2$. There is just the one eigenvector $\mathbf{v}$, the vector $\mathbf{b}$ with $b_2 = 1$. That A will be deficient can be seen from the third column of the table where the reduced row-echelon form of $A - 2I$ has one nonzero row, so its rank is 1 and the dimension of its null space is $2 - 2 = 1$.

To solve the differential equation $\mathbf{x}' = A\mathbf{x}$, we have $\mathbf{x}_1(t) = e^{2t}\mathbf{v}$ as one solution but are in need of a second independent solution. If $y_1(t)$ is a solution of a second-order constant-

A	$A - 2I$	$\text{rref}(A - 2I)$	$(A - 2I)b = 0$	b	v
$\begin{bmatrix} -1 & -9 \\ 1 & 5 \end{bmatrix}$	$\begin{bmatrix} -3 & -9 \\ 1 & 3 \end{bmatrix}$	$\begin{bmatrix} 1 & 3 \\ 0 & 0 \end{bmatrix}$	$\begin{aligned} b_1 + 3b_2 &= 0 \\ \Rightarrow b_1 &= -3b_2 \end{aligned}$	$b_2 \begin{bmatrix} -3 \\ 1 \end{bmatrix}$	$\begin{bmatrix} -3 \\ 1 \end{bmatrix}$

TABLE 12.7 Calculations for Example 12.23

coefficient differential equation with repeated characteristic roots, Section 5.7 shows that a second independent solution is $y_2(t) = ty_1$. Unfortunately, the situation for the first-order linear system is a bit more complicated. Not only must the one available solution be multiplied by t, but a constant multiple of the associated exponential must be added as well. Thus, the form of the second solution $\mathbf{x}_2(t)$ is

$$\begin{aligned} \mathbf{x}_2(t) &= t\mathbf{x}_1 + \mathbf{c}e^{2t} \\ &= (t\mathbf{v} + \mathbf{c})e^{2t} \end{aligned} \quad \text{where } \mathbf{c} = \begin{bmatrix} c_1 \\ c_2 \end{bmatrix}$$

where the constant vector $\mathbf{c}$ is determined from the differential equation by the following calculation

$$\mathbf{x}_2' = A\mathbf{x}_2 \tag{12.48a}$$
$$\{(t\mathbf{v} + \mathbf{c})e^{2t}\}' = A\{(t\mathbf{v} + \mathbf{c})e^{2t}\} \tag{12.48b}$$
$$2(t\mathbf{v} + \mathbf{c})e^{2t} + \mathbf{v}e^{2t} = A\mathbf{c}e^{2t} + A(t\mathbf{v})e^{2t} \tag{12.48c}$$
$$2(t\mathbf{v} + \mathbf{c}) + \mathbf{v} = A\mathbf{c} + tA\mathbf{v} \tag{12.48d}$$
$$2t\mathbf{v} + 2\mathbf{c} + \mathbf{v} = A\mathbf{c} + t(2\mathbf{v}) \tag{12.48e}$$
$$2t\mathbf{v} + 2\mathbf{c} + \mathbf{v} = A\mathbf{c} + 2t\mathbf{v} \tag{12.48f}$$
$$2\mathbf{c} + \mathbf{v} = A\mathbf{c} \tag{12.48g}$$
$$\mathbf{v} = A\mathbf{c} - 2\mathbf{c} \tag{12.48h}$$
$$(A - 2I)\mathbf{c} = \mathbf{v} \tag{12.48i}$$

The derivative in (12.48b) leads to (12.48c), where every term contains e^{2t}, the removal of which gives (12.48d). On the right in (12.48d) we find $A\mathbf{v} = 2\mathbf{v}$ since $\mathbf{v}$ is an eigenvector with eigenvalue 2. Hence, we get (12.48e), on the right of which we find $t(2\mathbf{v}) = 2t\mathbf{v}$. Hence, we get (12.48f), which simplifies to (12.48g) because $2t\mathbf{v}$ appears on both sides of (12.48f). Simple transposition gives (12.48h), and factoring gives (12.48i).

The matrix $A - 2I$ in (12.48i) has rank 1 and only one independent row. Therefore, the vector $\mathbf{c}$ satisfies the single equation $c_1 + 3c_2 = 1$, which has the solution

$$\begin{bmatrix} 1 - 3c_2 \\ c_2 \end{bmatrix} = \begin{bmatrix} 1 \\ 0 \end{bmatrix} + c_2 \begin{bmatrix} -3 \\ 1 \end{bmatrix} = \begin{bmatrix} 1 \\ 0 \end{bmatrix} + c_2\mathbf{v} \tag{12.49}$$

Since we only need a single particular solution for building a fundamental set, we set c_2 equal to any convenient value. Here, the choice $c_2 = 0$ gives

$$\mathbf{c} = \begin{bmatrix} 1 \\ 0 \end{bmatrix} \quad \mathbf{x}_2 = e^{2t}\begin{bmatrix} 1 - 3t \\ 1 \end{bmatrix} \quad \text{and} \quad \mathbf{x} = a\mathbf{x}_1 + b\mathbf{x}_2 = e^{2t}\begin{bmatrix} -3a + b(1 - 3t) \\ a + bt \end{bmatrix} \tag{12.50}$$

EXERCISES 12.15–Part A

A1. Verify that the only eigenvalue of the matrix A in Table 12.7 is $\lambda = 2$.

A2. Verify the reduced row-echelon form for $A - 2I$ given in Table 12.7.

A3. Set $c_2 = 3$ in (12.49) and show that the resulting general solution $\mathbf{x} = a\mathbf{x}_1 + b\mathbf{x}_2$, where $\mathbf{x}_2 = (t\mathbf{v} + \mathbf{c})e^{2t}$, is really the same as the one in (12.50).

A4. Suppose c_2 remains arbitrary in (12.49). Show that the general solution $\mathbf{x} = a\mathbf{x}_1 + b\mathbf{x}_2$, where $\mathbf{x}_2 = (t\mathbf{v} + \mathbf{c})e^{2t}$, is really the same as in (12.50).

A5. If A is the matrix in Table 12.7 and $\mathbf{x}(t) = \begin{bmatrix} u(t) \\ v(t) \end{bmatrix}$:

  (a) Reduce the system $\mathbf{x}' = A\mathbf{x}$ to a single second-order equation in $u(t)$.

  (b) Obtain $u(t)$, the general solution of the equation determined in part (a).

  (c) Use $u(t)$ to determine $\mathbf{x}$, the general solution of the system in part (a).

  (d) Reconcile $\mathbf{x}$ found in part (c) with $\mathbf{x}$ given in (12.50).

EXERCISES 12.15–Part B

B1. Let $y(t)$ satisfy the third-order equation $y''' + 6y'' + 12y' + 8y = 0$, which we will call $(*)$.

  (a) Show that the characteristic equation for $(*)$ has a repeated eigenvalue λ of multiplicity three.

  (b) Obtain the general solution for $y(t)$.

  (c) Convert the third-order equation into a system $\mathbf{x}'(t) = A\mathbf{x}$, with $\mathbf{x}(t)$ having components $u(t)$, $v(t)$, $w(t)$.

  (d) Show that the eigenvalues of A are the characteristic roots for the third-order equation.

  (e) Show that the rank of $A - \lambda I$ is two so that there is only one linearly independent eigenvector associated with λ.

  (f) Find the one eigenvector of A.

  (g) Use the solution in part (b) to obtain a general solution for the system in part (c).

In Exercises B2–11 a deficient matrix A is given.

  (a) In each case, take $\mathbf{x}(0) = \mathbf{x}_0 = \begin{bmatrix} -3 \\ 2 \end{bmatrix}$ as the initial data in the initial value problem $\mathbf{x}' = A\mathbf{x}$, $\mathbf{x}(0) = \mathbf{x}_0$, and obtain the solution by finding and using the fundamental matrix ϕ.

  (b) The solution $\mathbf{v}_1 = e^{\lambda t}\mathbf{v}$, where $\mathbf{v}$ is an eigenvector, was obtained earlier. Now, obtain the second linearly independent solution $\mathbf{v}_2 = (\mathbf{c} + t\mathbf{v})e^{\lambda t}$, and write the general solution $\mathbf{x} = a\mathbf{v}_1 + b\mathbf{v}_2$.

  (c) Apply the initial condition, making sure the resulting solution agrees with the solution found in part (a).

B2. Exercise B12, Section 12.12   **B3.** Exercise B13, Section 12.12

B4. Exercise B14, Section 12.12   **B5.** Exercise B15, Section 12.12

B6. Exercise B16, Section 12.12   **B7.** Exercise B17, Section 12.12

B8. Exercise B18, Section 12.12   **B9.** Exercise B19, Section 12.12

B10. Exercise B20, Section 12.12   **B11.** Exercise B21, Section 12.12

In Exercises B12–26, the given matrix A may be deficient. Some of the deficient matrices are missing one eigenvector, and some are missing two.

  (a) For each matrix, take $\mathbf{x}(0) = \mathbf{x}_0 = \begin{bmatrix} -1 \\ 5 \\ 4 \end{bmatrix}$ as the initial data in the initial value problem $\mathbf{x}' = A\mathbf{x}$, $\mathbf{x}(0) = \mathbf{x}_0$, and obtain the solution by finding and using the fundamental matrix ϕ.

  (b) If the matrix A is missing one eigenvector, there are two solutions of the form $\mathbf{V}_1 = \mathbf{v}_1 e^{\lambda_1 t}$ and $\mathbf{V}_2 = \mathbf{v}_2 e^{\lambda_2 t}$, where $\mathbf{v}_k$, $k = 1, 2$, are eigenvectors. The third solution is of the form $\mathbf{V}_3 = (\mathbf{c} + t\mathbf{v}_2)e^{\lambda_2 t}$, where $\mathbf{v}_2$ is the one eigenvector from the repeated eigenvalue λ_2. Find this third vector, and write the general solution in the form $a\mathbf{V}_1 + b\mathbf{V}_2 + c\mathbf{V}_3$.

  (c) Apply the initial condition, making sure the resulting solution agrees with the solution found in part (a).

  (d) If the matrix A is missing two eigenvectors, there is just one solution of the form $\mathbf{V}_1 = \mathbf{v}e^{\lambda_1 t}$, where $\mathbf{v}$ is an eigenvector. The second solution is of the form $\mathbf{V}_2 = (\mathbf{c}_1 + t\mathbf{v})e^{\lambda_1 t}$, and the third solution is of the form $\mathbf{V}_3 = (\mathbf{c}_2 + t\mathbf{c}_1 + \frac{t^2}{2}\mathbf{v})e^{\lambda_1 t}$. The constant vector $\mathbf{c}_1$ is a single solution of the equation $(A - \lambda_1 I)\mathbf{c}_1 = \mathbf{v}$, whereas the constant vector $\mathbf{c}_2$ is a single solution of the equation $(A - \lambda_1 I)\mathbf{c}_2 = \mathbf{c}_1$. Obtain the solutions $\mathbf{V}_2$ and $\mathbf{V}_3$, and then write the general solution in the form $a\mathbf{V}_1 + b\mathbf{V}_2 + c\mathbf{V}_3$.

  (e) Apply the initial condition, making sure the resulting solution agrees with the solution found in part (a).

B12. $\begin{bmatrix} -28 & 18 & -41 \\ -35 & 23 & -47 \\ 5 & -3 & 9 \end{bmatrix}$ **B13.** Exercise B23, Section 12.12

B14. Exercise B24, Section 12.12   **B15.** Exercise B28, Section 12.12

B16. Exercise B29, Section 12.12   **B17.** Exercise B33, Section 12.12

B18. Exercise B34, Section 12.12

B19. $\begin{bmatrix} 6 & 11 & -12 \\ 14 & 20 & -23 \\ 16 & 22 & -26 \end{bmatrix}$ **B20.** $\begin{bmatrix} 8 & 10 & -13 \\ -1 & 1 & 2 \\ 1 & 2 & 0 \end{bmatrix}$ **B21.** $\begin{bmatrix} 3 & 5 & -50 \\ 5 & 7 & 50 \\ -2 & -3 & -3 \end{bmatrix}$

B22. $\begin{bmatrix} 2 & 1 & -1 \\ 1 & 2 & 0 \\ 4 & 4 & -1 \end{bmatrix}$ **B23.** $\begin{bmatrix} -3 & 0 & 1 \\ -21 & 3 & 11 \\ -4 & 0 & 1 \end{bmatrix}$ **B24.** $\begin{bmatrix} -2 & -1 & -1 \\ -1 & -3 & 0 \\ 6 & 3 & 2 \end{bmatrix}$

B25. $\begin{bmatrix} 20 & -20 & 20 \\ 85 & -84 & 84 \\ 59 & -56 & 55 \end{bmatrix}$ **B26.** $\begin{bmatrix} -51 & 18 & 17 \\ -35 & 12 & 11 \\ -98 & 35 & 33 \end{bmatrix}$

The cases considered in Exercises B22–36 do not exhaust the complexities that can be encountered. The astute reader will notice that there is one configuration that has not yet been presented. Consider the case of a deficient 3×3 that has a single eigenvalue of algebraic multiplicity 3 and *two* eigenvectors $\mathbf{v}_1$ and $\mathbf{v}_2$. The third solution of the differential equation $\mathbf{x}' = A\mathbf{x}$ has the form $\mathbf{z} = [(a\mathbf{v}_1 + b\mathbf{v}_2)t + \mathbf{c}]e^{\lambda t}$, where a and b are chosen to make the equations $(A - \lambda I)\mathbf{c} = a\mathbf{v}_1 + b\mathbf{v}_2$ consistent. To explore this case, take A as any of the matrices given in Exercises B27–34.

(a) Show that there is a single eigenvalue with algebraic multiplicity 3.

(b) Show that there are two eigenvectors $\mathbf{v}_1$ and $\mathbf{v}_2$.

(c) By row-reducing the augmented matrix $[A - \lambda I, a\mathbf{v}_1 + b\mathbf{v}_2]$ to an upper triangular form, find the constants a and b which make the resulting equations consistent. Then, using the resulting eigenvector $a\mathbf{v}_1 + b\mathbf{v}_2$, obtain the constant vector $\mathbf{c}$, and $\mathbf{z}$, the third solution of the differential equation.

B27. $\begin{bmatrix} 4 & -8 & -4 \\ 1 & -2 & -2 \\ -1 & 4 & 4 \end{bmatrix}$ **B28.** $\begin{bmatrix} 14 & 8 & -4 \\ -5 & 0 & 2 \\ 15 & 12 & -2 \end{bmatrix}$ **B29.** $\begin{bmatrix} 17 & -9 & 27 \\ 10 & -4 & 18 \\ -5 & 3 & -7 \end{bmatrix}$

B30. $\begin{bmatrix} 1 & -8 & -1 \\ 5 & 14 & 1 \\ -15 & -24 & 3 \end{bmatrix}$ **B31.** $\begin{bmatrix} -15 & -10 & 14 \\ 12 & 8 & -21 \\ 4 & 5 & -14 \end{bmatrix}$ **B32.** $\begin{bmatrix} 11 & 12 & -15 \\ -12 & -15 & 10 \\ 12 & 8 & -17 \end{bmatrix}$

B33. $\begin{bmatrix} 11 & 3 & -24 \\ 4 & 7 & -16 \\ 2 & 1 & -3 \end{bmatrix}$ **B34.** $\begin{bmatrix} -5 & 2 & -9 \\ 9 & -8 & 27 \\ 3 & -2 & 7 \end{bmatrix}$

12.16 Diagonalization and Uncoupling

Coupled Systems

We refer to the pair of equations

$$x'(t) = 18x(t) - 2y(t) \quad \text{and} \quad y'(t) = 12x(t) + 7y(t)$$

as *coupled* since both $x(t)$ and $y(t)$ appear in each equation. We apply the same term to the equivalent system $\mathbf{x}' = A\mathbf{x}$, where

$$A = \begin{bmatrix} 18 & -2 \\ 12 & 7 \end{bmatrix} \quad \text{and} \quad \mathbf{x} = \begin{bmatrix} x(t) \\ y(t) \end{bmatrix} \tag{12.51}$$

It is this coupling that makes it difficult to solve the system.

Uncoupled Systems

The pair of differential equations on the left in (12.52) forms a system of the form $\mathbf{u}' = B\mathbf{u}$, where the *diagonal* matrix B and the vector $\mathbf{u}$ are given on the right in (12.52). We refer to such a system as *uncoupled* since $u(t)$ appears just in the first equation and $v(t)$ appears just in the second

$$\begin{aligned} u'(t) &= 10u(t) \\ v'(t) &= 15v(t) \end{aligned} \qquad \mathbf{u}(t) = \begin{bmatrix} u(t) \\ v(t) \end{bmatrix} \qquad B = \begin{bmatrix} 10 & 0 \\ 0 & 15 \end{bmatrix} \tag{12.52}$$

Uncoupled equations of this type are easily solved by separating variables. Each of $u(t)$ and $v(t)$ are just constants times exponentials; specifically, they are $u(t) = c_1 e^{10t}$ and $v(t) = c_2 e^{15t}$.

Uncoupling Coupled Equations

Since uncoupled systems are so simple to solve, we ask if there is a way of uncoupling coupled systems. Consider the change of variables defined on the left in (12.53) and expressed in matrix form by $\mathbf{x} = P\mathbf{u}$, where the matrix P is defined on the right in (12.53)

$$
\begin{aligned}
x(t) &= au(t) + bv(t) \\
y(t) &= cu(t) + dv(t)
\end{aligned}
\qquad
P = \begin{bmatrix} a & b \\ c & d \end{bmatrix}
\tag{12.53}
$$

Making this change of variables in the original coupled system $\mathbf{x}' = A\mathbf{x}$, we obtain

$$(P\mathbf{u})' = A(P\mathbf{u}) \quad \text{or} \quad \mathbf{u}' = (P^{-1}AP)u$$

where we isolated $\mathbf{u}'$ by multiplying through by P^{-1}, the inverse of P. We want our change of variables to result in a system of the form $\mathbf{u}' = B\mathbf{u}$, where B is a diagonal matrix. If A is the matrix in (12.51), then we want

$$
P^{-1}AP = \frac{1}{ad-bc}
\begin{bmatrix}
18ad - 12ba - 2dc - 7bc & 11bd - 12b^2 - 2d^2 \\
-11ac + 12a^2 + 2c^2 & -18bc + 12ba + 2dc + 7ad
\end{bmatrix}
$$

to reduce to a diagonal of the form $C = \begin{bmatrix} \alpha & 0 \\ 0 & \beta \end{bmatrix}$ for some as-yet unknown values of α and β. The solutions of the four equations in the six unknowns a, b, c, d, α, and β contained in $P^{-1}AP = C$ are

$$
C_1 = \begin{bmatrix} 10 & 0 \\ 0 & 15 \end{bmatrix}
\quad
P_1 = \begin{bmatrix} \frac{1}{4}c & \frac{2}{3}d \\ c & d \end{bmatrix}
\quad \text{and} \quad
C_2 = \begin{bmatrix} 15 & 0 \\ 0 & 10 \end{bmatrix}
\quad
P_2 = \begin{bmatrix} \frac{2}{3}c & \frac{1}{4}d \\ c & d \end{bmatrix}
$$

The diagonal matrices C_1 and C_2 differ only in the order of the diagonal elements. Correspondingly, the matrices P_1 and P_2 are really the same, except for the order of the columns. If, for example, in P_1 we set $c = 4$ and $d = \frac{3}{2}$ and in P_2 set $c = \frac{3}{2}$ and $d = 4$, we obtain

$$
\begin{bmatrix} 1 & 1 \\ 4 & \frac{3}{2} \end{bmatrix}
\quad \text{and} \quad
\begin{bmatrix} 1 & 1 \\ \frac{3}{2} & 4 \end{bmatrix}
$$

(Since we solved four equations in six variables, there are two free variables left in the solution for P.) The matrix P_1 diagonalizes the coupled system, resulting in the diagonal matrix C_1; while the matrix P_2 also diagonalizes the coupled system, resulting in the diagonal matrix C_2.

The eigenvalues of A are 10 and 15, and the corresponding eigenvectors are

$$
\begin{bmatrix} 1 \\ 4 \end{bmatrix}
\quad \text{and} \quad
\begin{bmatrix} 1 \\ \frac{3}{2} \end{bmatrix}
$$

Thus, the columns of the matrix P_1 (or P_2) are the eigenvectors of A, and the diagonal entries of C_1 (or C_2) are the eigenvalues. Hence, knowledge of the eigenvalues and eigenvectors leads to an uncoupling, or diagonalization, of the system of differential equations. In fact, let us state this as a theorem.

THEOREM 12.3

1. P is a square matrix; its n columns are the n eigenvectors of the $n \times n$ matrix A.
2. D is a diagonal matrix; its entries are the corresponding eigenvalues of A.
3. $\mathbf{x} = P\mathbf{u}$

$\Longrightarrow$

1. $P^{-1}AP = D$
2. $\mathbf{u}' = D\mathbf{u}$ is the uncoupled version of $\mathbf{x}' = A\mathbf{x}$.

Of course, if we know the eigenvalues and eigenvectors of A, we can immediately write the general solution of $\mathbf{x}' = A\mathbf{x}$. We wouldn't find the eigenpairs just to uncouple the system for the purpose of *solving* it. However, we will see in Section 12.17 that uncoupling the equations describing oscillations of a mechanical system yields the *normal modes,* certain fundamental vibrations that characterize the system.

Similar Matrices

The matrix A is *similar* to the matrix B if there is an invertible matrix P for which $P^{-1}AP = B$. The matrix P is said to generate a *similarity transform* of A to B.

Theorem 12.3 says that if the $n \times n$ matrix A has n linearly independent eigenvectors, then it is similar to a diagonal matrix D whose entries are its eigenvalues. The similarity transform from A to D is accomplished by the matrix P whose columns are the eigenvectors of A ordered in correspondence with the eigenvalues in D.

The exercises contain additional information about similar matrices.

EXAMPLE 12.24 In (12.54), the coupled system of differential equations on the left has the system matrix A given on the right

$$\begin{aligned} x'(t) &= x(t) + 2y(t) \\ y'(t) &= 4x(t) - y(t) \end{aligned} \qquad A = \begin{bmatrix} 1 & 2 \\ 4 & -1 \end{bmatrix} \tag{12.54}$$

The change of variables $\mathbf{x} = P\mathbf{u}$, where P is the matrix of eigenvectors of A, results in the diagonal (uncoupled) system $\mathbf{u}' = D\mathbf{u}$, where D is a diagonal matrix whose entries are the eigenvalues of A. The matrices P and $P^{-1}AP = D$ are then

$$P = \begin{bmatrix} 1 & 1 \\ 1 & -2 \end{bmatrix} \quad \text{and} \quad P^{-1}AP = \begin{bmatrix} 3 & 0 \\ 0 & -3 \end{bmatrix} = D$$

Defining the vector $\mathbf{u}$ on the left in (12.55) gives the uncoupled system on the right

$$u = \begin{bmatrix} u(t) \\ v(t) \end{bmatrix} \Rightarrow \begin{cases} u'(t) = 3u(t) \\ v'(t) = -3v(t) \end{cases} \tag{12.55}$$

for which the solution is immediately $u(t) = c_1 e^{3t}$ and $v(t) = c_2 e^{-3t}$. The solution to the coupled system is then

$$\mathbf{x} = P\mathbf{u} = \begin{bmatrix} c_1 e^{3t} + c_2 e^{-3t} \\ c_1 e^{3t} - 2c_2 e^{-3t} \end{bmatrix} \qquad ❖$$

EXAMPLE 12.25 The matrix A on the left in (12.56) has eigenvalues $\lambda = -3, -2, -5$. The corresponding eigenvectors are the columns of the matrix P on the right in (12.56). The system $\mathbf{x}' = A\mathbf{x}$ is therefore coupled.

$$A = \begin{bmatrix} 358 & -447 & 384 \\ -228 & 279 & -242 \\ -606 & 750 & -647 \end{bmatrix} \qquad P = \begin{bmatrix} 1 & -\frac{5}{8} & -3 \\ \frac{5}{3} & 1 & 1 \\ 1 & \frac{7}{4} & 4 \end{bmatrix} \tag{12.56}$$

The change of variables $\mathbf{x} = P\mathbf{u}$ results in the diagonal (uncoupled) system $\mathbf{u}' = D\mathbf{u}$, where $D = P^{-1}AP$ is a diagonal matrix whose entries are the eigenvalues of A and $\mathbf{u}$ is the vector whose entries are $u(t)$, $v(t)$, and $w(t)$. In (12.57), the uncoupled system is on the left, its solution is in the center, and the solution for the coupled system is on the right

$$u'(t) = -3u(t) \qquad u(t) = c_1 e^{-3t}$$
$$v'(t) = -2v(t) \quad\Rightarrow\quad v(t) = c_2 e^{-2t} \quad\Rightarrow \mathbf{x} = P\mathbf{u} = \begin{bmatrix} c_1 e^{-3t} - \frac{5}{8} c_2 e^{-2t} - 3c_3 e^{-5t} \\[2mm] \frac{5}{3} c_1 e^{-3t} + c_2 e^{-2t} + c_3 e^{-5t} \\[2mm] c_1 e^{-3t} + \frac{7}{4} c_2 e^{-2t} + 4c_3 e^{-5t} \end{bmatrix}$$
$$w'(t) = -5w(t) \qquad w(t) = c_3 e^{-5t}$$

$$(12.57)$$

Normal Coordinates

It is sometimes convenient to think of the equation $\mathbf{x} = P\mathbf{u}$ as a change of coordinates in the phase plane. The coordinates x and y in the phase plane are changed to the new coordinates u and v. If the equations $\mathbf{x}' = A\mathbf{x}$ uncouple in the new coordinate system defined by $\mathbf{u} = P^{-1}\mathbf{x}$, we call the new coordinates the *normal coordinates*.

EXERCISES 12.16–Part A

A1. Compute the eigenpairs for the matrix A in (12.51).

A2. Show that if a matrix A is similar to a matrix B, then B is similar to A.

A3. Show that if A is similar to B and B is similar to C, then A is similar to C.

A4. Find a similarity transform that diagonalizes $A = \begin{bmatrix} -3 & -1 \\ 8 & 3 \end{bmatrix}$. Uncouple the system $\mathbf{x}' = A\mathbf{x}$, and use the solution of the uncoupled system to obtain the general solution of the coupled system.

A5. Show that $\begin{bmatrix} 33 & 16 \\ -49 & -23 \end{bmatrix}$ is not similar to a diagonal matrix.

EXERCISES 12.16–Part B

B1. Calculate the eigenpairs of the matrix A given in (12.56).

For the 2×2 matrix A given in each of Exercises B2–7:

  (a) Diagonalize A by applying the appropriate similarity transformation.

  (b) Uncouple the first-order linear system $\mathbf{x}' = A\mathbf{x}$ by setting $\mathbf{x} = P\mathbf{u}$, with $P = \begin{bmatrix} a & b \\ c & d \end{bmatrix}$ and $\mathbf{u} = \begin{bmatrix} u(t) \\ v(t) \end{bmatrix}$. Make the substitution $\mathbf{x} = P\mathbf{u}$ in $\mathbf{x}' = A\mathbf{x}$ so that $\mathbf{u}' = P^{-1}AP\mathbf{u}$ is obtained. Set up and solve four equations in six unknowns corresponding to $P^{-1}AP = D$, where $D = \begin{bmatrix} \alpha & 0 \\ 0 & \beta \end{bmatrix}$. Show that except for the two arbitrary constants that scale the eigenvectors in P, the solution of these four equations are the P and D matrices used in part (a).

  (c) Solve $\mathbf{u}' = D\mathbf{u}$ for $\mathbf{u}$, where $\mathbf{u}$ will contain two arbitrary constants of integration.

  (d) Setting the two free parameters in P to 1 (or any other convenient values), obtain $\mathbf{x} = P\mathbf{u}$ as the general solution to $\mathbf{x}' = A\mathbf{x}$.

  (e) Verify the general solution found in part (d) by computing $\phi\mathbf{c}$, where $\phi = e^{At}$, the fundamental matrix, and $\mathbf{c} = \begin{bmatrix} c_1 \\ c_2 \end{bmatrix}$ is a vector of arbitrary constants.

B2. Exercise B2, Section 12.10 **B3.** Exercise B3, Section 12.10

B4. Exercise B4, Section 12.10 **B5.** Exercise B5, Section 12.10

B6. Exercise B6, Section 12.10 **B7.** Exercise B7, Section 12.10

For the matrix A given in each of Exercises B8–13:

  (a) Diagonalize A by the similarity transform $P^{-1}AP = D$, where the columns of P are the eigenvectors of A.

  (b) Use the results in part (a) to uncouple the system $\mathbf{x}' = A\mathbf{x}$, thereby obtaining its general solution.

  (c) Verify the solution in part (b) by computing $\mathbf{x}(t) = e^{At}\mathbf{c}$ for an appropriate constant vector $\mathbf{c}$.

B8. Exercise B2, Section 12.11 **B9.** Exercise B3, Section 12.11

B10. Exercise B4, Section 12.11 **B11.** Exercise B5, Section 12.11

B12. Exercise B6, Section 12.11 **B13.** Exercise B7, Section 12.11

12.17   A Coupled Linear Oscillator

The Model

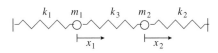

FIGURE 12.13   Coupled spring-mass system

Forces acting on mass m_1

$-k_1 x_1$

$-k_3(x_1 - x_2)$

Forces acting on mass m_2

$-k_2 x_2$

$-k_3(x_2 - x_1)$

TABLE 12.8   Forces acting on masses in Figure 12.13

Consider m_1 and m_2, two masses connected by three springs with spring constants k_1, k_2, k_3 as shown in Figure 12.13. The system thereby consists of two linear, undamped oscillators coupled together by the middle spring and held in place by the outside springs that are attached to vertical supports. Displacements of the masses from their respective equilibrium positions are, respectively, measured by the variables x_1 and x_2. At equilibrium, the two masses are L units apart.

By applying $\mathbf{F} = m\mathbf{a}$, Newton's Second Law of Motion, to the forces $\mathbf{F} = -kx\mathbf{i}$ exerted *by* the springs *on* their attached masses (see Table 12.8), we obtain

$$m_1 x_1'' + (k_1 + k_3)x_1 - k_3 x_2 = 0 \quad \text{and} \quad m_2 x_2'' - k_3 x_1 + (k_2 + k_3)x_2 = 0 \tag{12.58}$$

a coupled system of *second*-order differential equations as the equations of motion for the two masses. Since the equations are linear, we profit from a matrix formulation.

MATRIX FORMULATION   To put the two equations of motion into the form $M\mathbf{x}'' + K\mathbf{x} = \mathbf{0}$, define

$$M = \begin{bmatrix} m_1 & 0 \\ 0 & m_2 \end{bmatrix} \quad K = \begin{bmatrix} k_1 + k_3 & -k_3 \\ -k_3 & k_2 + k_3 \end{bmatrix} \quad \mathbf{x} = \begin{bmatrix} x_1(t) \\ x_2(t) \end{bmatrix} \tag{12.59}$$

as the mass-matrix, the stiffness matrix, and the vector of displacements, respectively. Initial conditions would consist of the initial displacements and initial velocities of each of the two masses. Thus, define

$$x_1(0) = a_1 \quad x_2(0) = a_2 \quad x_1'(0) = b_1 \quad x_2'(0) = b_2 \tag{12.60}$$

EXAMPLE 12.26   By picking values for the masses and spring constants, we realize a specific version of the coupled-oscillator system. Suppose, then, we specify the parameter values

$$m_1 = 1 \quad m_2 = 2 \quad k_1 = 1 \quad k_2 = 2 \quad k_3 = 3 \quad L = 4 \tag{12.61}$$

With these values, the equations of motion become

$$x_1''(t) + 4x_1(t) - 3x_2(t) = 0 \quad \text{and} \quad 2x_2''(t) - 3x_1(t) + 5x_2(t) = 0 \tag{12.62}$$

For this specific case of the model, the exact solution, obtained by Laplace transform (see Exercise B1), is

$$x_1(t) = \rho_1 \sin t + \rho_2 \cos t + 2\rho_3 \sin \tfrac{\sqrt{22}}{2}t + 2\rho_4 \cos \tfrac{\sqrt{22}}{2}t$$
$$x_2(t) = \rho_1 \sin t + \rho_2 \cos t - \rho_3 \sin \tfrac{\sqrt{22}}{2}t - \rho_4 \cos \tfrac{\sqrt{22}}{2}t \tag{12.63}$$

where

$$\rho_1 = \tfrac{1}{3}(b_1 + 2b_2) \qquad \rho_2 = \tfrac{1}{3}(a_1 + 2a_2)$$
$$\rho_3 = \tfrac{\sqrt{22}}{33}(b_1 - b_2) \qquad \rho_4 = \tfrac{1}{3}(a_1 - a_2) \tag{12.64}$$

Note the appearance of trig functions with two different angular frequencies. The motions will contain components with angular frequency 1 and components with angular frequency $\frac{\sqrt{22}}{2}$. The motion of the system is governed by the interaction of oscillations with these two different angular frequencies.

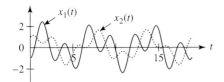

FIGURE 12.14 Solution curves $x_1(t)$ (solid) and $x_2(t)$ (dotted) for Example 12.25

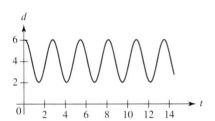

FIGURE 12.15 Physical separation of masses in Example 12.26

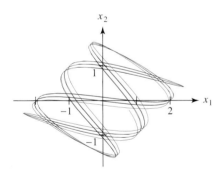

FIGURE 12.16 Lissajous figure for Example 12.26

Suppose we displace mass m_1 one unit to the left and mass m_2 one unit to the right and then impart an initial velocity of one unit to the right for both masses. For these specific initial conditions $a_1 = -1$, $a_2 = 1$, $b_1 = 1$, and $b_2 = 1$, so the solution becomes

$$x_1(t) = \sin t + \tfrac{1}{3}\cos t - \tfrac{4}{3}\cos\tfrac{\sqrt{22}}{2}t \quad \text{and} \quad x_2(t) = \sin t + \tfrac{1}{3}\cos t + \tfrac{2}{3}\cos\tfrac{\sqrt{22}}{2}t \quad \text{(12.65)}$$

❖

GRAPHS We try several different types of graphs in an effort to analyze and understand the motion of the system. In Figure 12.14 we plot the displacements, using a solid curve for the motion of mass m_1 and a dotted line for the motion of mass m_2. While there is apparent periodicity, we also note the motions are not synchronous. The first time mass m_1 attains a relative maximum displacement to the right, mass m_2 seems to experience a local minimum in its displacement to the left. Then, the second time mass m_1 attains a relative maximum displacement to the right, mass m_2 is moving toward a relative maximum displacement to the right also. Consequently, the motion is not simple.

In Figure 12.15 we graph $d = 4 + x_2 - x_1$, the physical distance between the two masses, assuming the equilibrium positions of the two masses are $L = 4$ units apart. In this graph we see what appears to be periodic behavior in the distance between the masses. So, the absolute motion seems to consist of oscillatory motion between the two masses, with translational motion superimposed.

Another way to study the motion of the coupled oscillator is to plot x_2 against x_1. Figure 12.16 shows the resulting parametric plot that forms a kind of phase plane diagram and the paths (orbits), called *Lissajous' figures,* after the person who first studied them.

Ultimately, to understand the interactions between the two masses, we need to uncouple the system and describe the motions in terms of the *normal coordinates.*

Diagonalization to Normal Coordinates

To uncouple the equations in (12.58), isolate $\mathbf{x}''$, the second-derivative term in $M\mathbf{x}'' + K\mathbf{x} = \mathbf{0}$, by multiplying through the equations of motion by M^{-1}, the inverse of the matrix M, to obtain $\mathbf{x}'' + (M^{-1}K)\mathbf{x} = \mathbf{0}$. Then, defining the matrix U as the matrix product $M^{-1}K$, we get the equations $\mathbf{x}'' + U\mathbf{x} = \mathbf{0}$. For the general system under consideration, the matrix U would be

$$U = \begin{bmatrix} \dfrac{k_1 + k_3}{m_1} & -\dfrac{k_3}{m_1} \\[2ex] -\dfrac{k_3}{m_2} & \dfrac{k_2 + k_3}{m_2} \end{bmatrix} \quad \text{(12.66)}$$

and the differential equations would now be

$$x_1''(t) + \frac{k_1 + k_3}{m_1}x_1(t) - \frac{k_3}{m_1}x_2(t) \quad \text{and} \quad x_2''(t) - \frac{k_3}{m_2}x_1(t) + \frac{k_2 + k_3}{m_2}x_2(t) \quad \text{(12.67)}$$

If P is a matrix whose columns are the eigenvectors of U, then $P^{-1}UP = D$, a diagonal matrix with the eigenvalues of U on its diagonal. Moreover, the change of variables defined by $\mathbf{x} = P\mathbf{r}$ gives

$$\mathbf{x}'' + U\mathbf{x} = \mathbf{0} \Rightarrow (P\mathbf{r})'' + UP\mathbf{r} = \mathbf{0} \Rightarrow \mathbf{r}'' + P^{-1}UP\mathbf{r} = \mathbf{0} \Rightarrow \mathbf{r}'' + D\mathbf{r} = \mathbf{0}$$

and the system $M\mathbf{x}'' + K\mathbf{x} = \mathbf{0}$ has been uncoupled. The uncoupled form $\mathbf{r}'' + D\mathbf{r} = \mathbf{0}$ consists of the two equations

$$r_1''(t) + \omega_1^2 r_1(t) = 0 \quad \text{and} \quad r_2''(t) + \omega_2^2 r_2(t) = 0 \quad \text{(12.68)}$$

each of which is the equation of simple harmonic motion in one variable. The coordinates defined by $r_1(t)$ and $r_2(t)$ are the normal coordinates, and the oscillations exhibited by the functions $r_1(t)$ and $r_2(t)$ are the *normal modes of vibration*.

EXAMPLE 12.27   For the parameter values specified in (12.61), the matrix U, given in (12.66), becomes U_1, given on the left in (12.69). In the center of (12.69) we have the matrix P whose columns are the eigenvectors of U_1, and on the right we have the diagonal matrix $D = P^{-1}U_1 P$

$$U_1 = \begin{bmatrix} 4 & -3 \\ -\frac{3}{2} & \frac{5}{2} \end{bmatrix} \Rightarrow P = \begin{bmatrix} 1 & -2 \\ 1 & 1 \end{bmatrix} \Rightarrow P^{-1}U_1 P = \begin{bmatrix} 1 & 0 \\ 0 & \frac{11}{2} \end{bmatrix} = D \qquad \textbf{(12.69)}$$

The uncoupled equations (12.68) are then

$$r_1''(t) + r_1(t) = 0 \quad \text{and} \quad r_2''(t) + \tfrac{11}{2} r_2''(t) = 0 \qquad \textbf{(12.70)}$$

Each equation in (12.70) describes an undamped harmonic oscillator, with fundamental sets $\{\cos t, \sin t\}$ and $\{\cos \sqrt{\tfrac{11}{2}} t, \sin \sqrt{\tfrac{11}{2}} t\}$, respectively. The general solution of $\mathbf{r}'' + D\mathbf{r} = \mathbf{0}$ is then

$$\mathbf{r} = \begin{bmatrix} c_1 \cos t + c_2 \sin t \\ c_3 \cos \sqrt{\tfrac{11}{2}} t + c_4 \sin \sqrt{\tfrac{11}{2}} t \end{bmatrix}$$

and harmonic motions with angular frequency 1 or $\sqrt{\tfrac{11}{2}}$ are called the *normal modes of vibration* for the physical system described by (12.62).

We want to know $c_k, k = 1, \ldots, 4$ (which belong to $\mathbf{r}$), in terms of the given initial conditions (12.60) on $\mathbf{x}$, since these latter quantities are the *physical* initial conditions over which control can be exerted. Using $\mathbf{x} = P\mathbf{r}$, we have the equations

$$\mathbf{r}(0) = \begin{bmatrix} c_1 \\ c_3 \end{bmatrix} = P^{-1}\mathbf{x}(0) = P^{-1} \begin{bmatrix} a_1 \\ a_2 \end{bmatrix} \qquad \textbf{(12.71a)}$$

$$\mathbf{r}'(0) = \begin{bmatrix} c_2 \\ \sqrt{\tfrac{11}{2}} c_4 \end{bmatrix} = P^{-1}\mathbf{x}'(0) = P^{-1} \begin{bmatrix} b_1 \\ b_2 \end{bmatrix} \qquad \textbf{(12.71b)}$$

from which we find

$$\mathbf{r} = \begin{bmatrix} \dfrac{2a_2 + a_1}{3} \cos t + \dfrac{2b_2 + b_1}{3} \sin t \\ \dfrac{a_2 - a_1}{3} \cos \sqrt{\tfrac{11}{2}} t + \dfrac{\sqrt{22}}{33}(b_2 - b_1) \sin \sqrt{\tfrac{11}{2}} t \end{bmatrix} \qquad \textbf{(12.72)}$$

The initial conditions $b_1 = b_2 = 0$ and $a_1 = a_2$ isolate the angular frequency $\omega_1 = 1$, whereas the initial conditions $b_1 = -2b_2$ and $a_1 = a_2 = 0$ isolate the angular frequency $\omega_2 = \sqrt{\tfrac{11}{2}} = \frac{\sqrt{22}}{2}$. Substitution into (12.72) confirms these claims, giving, respectively,

$$\mathbf{r}_1 = \begin{bmatrix} a_2 \cos t \\ 0 \end{bmatrix} \quad \text{and} \quad \mathbf{r}_2 = \begin{bmatrix} 0 \\ \frac{\sqrt{22}}{11} b_2 \sin \sqrt{\tfrac{11}{2}} t \end{bmatrix} \qquad \textbf{(12.73)}$$

NORMAL MODES REALIZED IN THE PHYSICAL COORDINATES   Again using $\mathbf{x} = P\mathbf{r}$ and applying it to $\mathbf{r}_k, k = 1, 2$, in (12.73), we find

$$\mathbf{x}_1 = P\mathbf{r}_1 = \begin{bmatrix} a_2 \cos t \\ a_2 \cos t \end{bmatrix} \quad \text{and} \quad \mathbf{x}_2 = P\mathbf{r}_2 = \begin{bmatrix} -\frac{2\sqrt{22}}{11} b_2 \sin \sqrt{\tfrac{11}{2}} t \\ \frac{\sqrt{22}}{11} b_2 \sin \sqrt{\tfrac{11}{2}} t \end{bmatrix} \qquad \textbf{(12.74)}$$

When the initial conditions are $b_1 = b_2 = 0$ and $a_1 = a_2$, the solution is $\mathbf{x}_1$ and just the angular frequency $\omega_1 = 1$ is observed, with the two masses moving as if the center spring were a rigid rod. If the initial conditions are $b_1 = -2b_2$ and $a_1 = a_2 = 0$, then the solution is $\mathbf{x}_2$ and the two masses oscillate with angular frequency $\omega_2 = \sqrt{\frac{11}{2}}$, but perfectly out of phase, the effect of the minus sign in the first component of $\mathbf{x}_2$.

EXERCISES 12.17–Part A

A1. Use Table 12.8 and Newton's Second Law ($\mathbf{F} = m\mathbf{a}$) to obtain the equations in (12.58).

A2. Show that (12.62) and (12.60) give (12.65) when $a_1 = -1$, $a_2 = b_1 = b_2 = 1$.

A3. If M and K are as given in (12.59), obtain U as given in (12.66).

A4. Using the parameters in (12.61), show that U in (12.66) becomes the matrix U_1 in (12.69).

A5. Solve the equations in (12.71a) and (12.71b) to obtain the vector $\mathbf{r}$ given in (12.72).

A6. For $\mathbf{r}_k, k = 1, 2$, given in (12.73), verify that $P\mathbf{r}_k$ gives $\mathbf{x}_k$ in (12.74).

EXERCISES 12.17–Part B

B1. Use the Laplace transform to obtain (12.63) and (12.64) from (12.62) and (12.60).

Exercises B2–9 use the data from one row in Table 12.9. For each:

   (a) Obtain $\mathbf{r}'' + D\mathbf{r} = \mathbf{0}$, the uncoupled differential equations that define the normal modes for the system.

   (b) Find initial conditions $x_1(0) = a_1$, $x_2(0) = a_2$, $x_1'(0) = b_1$, $x_2'(0) = b_2$, that separately excite each normal mode, that is, that cause the physical system to oscillate with one fundamental frequency or the other.

   (c) Solve the initial value problem determined by the appropriate entries in Table 12.9.

   (d) Plot $x_1(t)$ and $x_2(t)$ on the same set of axes.

   (e) Assuming the distance between the masses at rest is 4, plot the quantity $4 + x_2(t) - x_1(t)$, the dynamic distance between the moving masses.

   (f) Plot x_2 against x_1, thereby forming a Lissajous figure.

B2. Data from row 1 of Table 12.9

B3. Data from row 2 of Table 12.9

B4. Data from row 3 of Table 12.9

B5. Data from row 4 of Table 12.9

B6. Data from row 5 of Table 12.9

B7. Data from row 6 of Table 12.9

B8. Data from row 7 of Table 12.9

B9. Data from row 8 of Table 12.9

B10. If the left-hand mass in Figure 12.13 is driven by the external force $f(t) = \cos t$, and $m_1 = m_2 = k_1 = k_2 = k_3 = 1$,

   (a) write the governing differential equations for the motion of $x_1(t)$ and $x_2(t)$.

   (b) using the initial data $x_k(0) = x_k'(0) = 0, k = 1, 2$, solve for the motion.

   (c) Plot the resulting solutions as functions of time.

B11. For the system in Figure 12.13, set $k_1 = k_2 = 0, k_3 = k$, thereby making the masses free to translate horizontally.

   (a) Write the governing differential equations for the system.

   (b) Obtain and interpret the general solution. What are the natural frequencies for the system?

   (c) Uncouple the differential equations and determine the normal modes for the system.

B12. The general form for the equations governing harmonically driven coupled *damped* oscillators is $M\mathbf{x}'' + B\mathbf{x}' + K\mathbf{x} = \mathbf{F}$ where $M = \begin{bmatrix} m_{11} & m_{12} \\ m_{21} & m_{22} \end{bmatrix}$, $B = \begin{bmatrix} b_{11} & b_{12} \\ b_{21} & b_{22} \end{bmatrix}$, $K = \begin{bmatrix} k_{11} & k_{12} \\ k_{21} & k_{22} \end{bmatrix}$, $\mathbf{F} = \begin{bmatrix} F_1 \cos \omega t \\ F_2 \cos \omega t \end{bmatrix}$. Multiplying through by M^{-1} gives equations of the form $\mathbf{x}'' + A\mathbf{x}' + E\mathbf{x} = \mathbf{G}$. Show that if there exists a matrix P for which $P^{-1}AP = D_1$ and $P^{-1}EP = D_2$, then $\mathbf{x}'' + A\mathbf{x}' + E\mathbf{x} = \mathbf{G}$ becomes the uncoupled $\mathbf{r}'' + D_1\mathbf{r}' + D_2\mathbf{r} = P^{-1}\mathbf{G}$ upon

	k_1	k_2	k_3	m_1	m_2	a_1	a_2	b_1	b_2
1	2	2	5	1	3	-1	2	2	-1
2	2	4	5	1	3	-1	-2	-1	1
3	2	3	1	1	3	1	-2	0	-1
4	5	3	1	1	3	2	-2	-1	0
5	2	2	5	3	1	0	1	0	2
6	2	4	5	3	1	0	1	-1	1
7	2	3	1	3	1	$-\frac{3}{2}$	$\frac{1}{2}$	$-\frac{1}{2}$	$\frac{3}{2}$
8	5	3	1	3	1	-2	2	0	0

TABLE 12.9   Data for Exercises B2–9

making the change of variables $\mathbf{x} = P\mathbf{r}$. The matrices $D_1 = \begin{bmatrix} d_1 & 0 \\ 0 & d_2 \end{bmatrix}$ and $D_2 = \begin{bmatrix} d_3 & 0 \\ 0 & d_4 \end{bmatrix}$ are diagonal matrices, and the matrices A and E are said to be diagonalized *simultaneously*.

B13. If, in Exercise B12, $M = \begin{bmatrix} 1 & 2 \\ 3 & 4 \end{bmatrix}$, $B = \begin{bmatrix} 13 & 20 \\ 35 & 42 \end{bmatrix}$, $K = \begin{bmatrix} 14 & 19 \\ 36 & 41 \end{bmatrix}$, $F_1 = F_2 = 1, \omega = 1$:

   (a) Find A and E.

   (b) Find $P = \begin{bmatrix} p_1 & p_2 \\ p_3 & p_4 \end{bmatrix}$ by brute force, solving the eight equations $P^{-1}AP = D_1$, $P^{-1}EP = D_2$, for the eight unknowns p_k, d_k,

$k = 1, \ldots, 4$. The equations are nonlinear and do not have unique solutions. There are, in fact, two free parameters in P.

   (c) Fix the values of the two free parameters in P by requiring that the columns of P be vectors of length 1.

   (d) Use the results from parts (b) and (c) to uncouple the differential equations.

   (e) Solve the uncoupled equations for $\mathbf{r}$.

   (f) Obtain $\mathbf{x} = P\mathbf{r}$.

12.18   Nonhomogeneous Systems and Variation of Parameters

The First-Order Linear ODE

Consider, from Sections 3.4 and 3.5, the nonhomogeneous scalar differential equation

$$y'(t) = ay(t) + f(t) \tag{12.75}$$

where the coefficient a is a constant. Writing the equation as $y'(t) - ay(t) = f(t)$, the form of the first-order linear equation used in Section 3.4, and multiplying through by the "integrating factor" $e^{-\int a\,dt} = e^{-at}$, we obtain the general solution

$$y(t) = e^{at} \int e^{-at} f(t)\,dt + ce^{at}$$

Setting $c = 0$ leaves the particular solution

$$y_p = e^{at} \int e^{-at} f(t)\,dt \tag{12.76}$$

This same particular solution can also be obtained by the technique of variation of parameters. The homogeneous solution is $y_h(t) = ce^{at}$, and we look for a particular solution of the form $u(t)e^{at}$, with brute force leading to $u'(t) = e^{-at} f(t)$, and hence, $u(t) = \int e^{-at} f(t)\,dt$. Clearly, the resulting particular solution is again (12.76).

The Nonhomogeneous Linear System

The matrix analog to (12.75) is $\mathbf{x}'(t) = A\mathbf{x}(t) + \mathbf{F}(t)$, and the matrix analog to the particular solution (12.76) is

$$\mathbf{x}_p = e^{At} \int e^{-At} \mathbf{F}(t)\,dt \tag{12.77}$$

where e^{at} is replaced by e^{At}, the fundamental matrix ϕ of Section 12.9. The integral is applied to each component of the integrand, a vector.

EXAMPLE 12.28   The fundamental matrix for the nonhomogeneous first-order system

$$\begin{bmatrix} x(t) \\ y(t) \end{bmatrix}' = \begin{bmatrix} 1 & 2 \\ 4 & -1 \end{bmatrix} \begin{bmatrix} x(t) \\ y(t) \end{bmatrix} + \begin{bmatrix} t \\ e^t \end{bmatrix} \tag{12.78}$$

is on the left in (12.79), and its inverse, $\phi^{-1} = \phi(-t)$, is on the right.

$$\phi = \frac{1}{3}\begin{bmatrix} 2e^{3t} + e^{-3t} & e^{3t} - e^{-3t} \\ 2e^{3t} - 2e^{-3t} & e^{3t} + 2e^{-3t} \end{bmatrix} \Rightarrow \phi^{-1} = \frac{1}{3}\begin{bmatrix} 2e^{-3t} + e^{3t} & e^{-3t} - e^{3t} \\ 2e^{-3t} - 2e^{3t} & e^{-3t} + 2e^{3t} \end{bmatrix} \tag{12.79}$$

Consequently, if $\mathbf{F}(t)$ is the vector on the right in (12.78), the integrand in (12.77) becomes

$$e^{-At}\mathbf{F}(t) = \frac{1}{3}\begin{bmatrix} (2e^{-3t} + e^{3t})t + (e^{-3t} - e^{3t})e^t \\ (2e^{-3t} - 2e^{3t})t + (e^{-3t} + 2e^{3t})e^t \end{bmatrix} \quad (12.80)$$

Integrating, we obtain

$$\int e^{-At}\mathbf{F}(t)\,dt = \frac{1}{108}\begin{bmatrix} (12t-4)e^{3t} - (24t+8)e^{-3t} - 9e^{4t} - 18e^{-2t} \\ (-24t+8)e^{3t} - (24t+8)e^{-3t} + 18e^{4t} - 18e^{-2t} \end{bmatrix} \quad (12.81)$$

and after simplifying, we find

$$\mathbf{x}_p = e^{At}\int e^{-At}\mathbf{F}(t)\,dt = -\frac{1}{36}\begin{bmatrix} 9e^t + 4(t+1) \\ 16t \end{bmatrix} \quad (12.82)$$

❖

EXERCISES 12.18–Part A

A1. Verify that ϕ given in (12.79) is the fundamental matrix for the matrix A given in (12.78).

A2. For the fundamental matrix ϕ given in (12.79), verify that $[\phi(t)]^{-1} = \phi(-t)$.

A3. Verify the product $e^{-At}\mathbf{F}(t)$ given in (12.80) is correct.

A4. Verify the integration in (12.81).

A5. Show that $\mathbf{x}_p$ given in (12.82) satisfies (12.78).

EXERCISES 12.18–Part B

B1. Let $\mathbf{v} = \begin{bmatrix} x(t) \\ y(t) \end{bmatrix}$, and consider the initial value problem $\mathbf{v}'(t) = A\mathbf{v}(t) + \mathbf{f}(t)$, $\mathbf{v}(0) = \mathbf{v}_0 = \begin{bmatrix} a \\ b \end{bmatrix}$. Let $\mathbf{V}(s) = \begin{bmatrix} X(s) \\ Y(s) \end{bmatrix}$ be the Laplace transform of $\mathbf{v}$, and let $\mathbf{F}(s)$ be the Laplace transform of $\mathbf{f}$.

(a) Take the Laplace transform of the differential equation and show that $\mathbf{V}(s) = (sI - A)^{-1}\mathbf{v}_0 + (sI - A)^{-1}\mathbf{F}(s)$.

(b) Invert $\mathbf{V}(s)$, and show that $\mathbf{v}(t) = \phi\mathbf{v}_0 + L^{-1}[(sI - A)^{-1}\mathbf{F}(s)]$, where ϕ is the fundamental matrix e^{At}.

(c) Since $(sI - A)^{-1}\mathbf{F}(s)$ is the product of two Laplace transforms, we have $L^{-1}[(sI - A)^{-1}\mathbf{F}(s)] = \phi(t) * \mathbf{f}(t)$, the convolution of the fundamental matrix $\phi(t)$, and $\mathbf{f}(t)$. Show that $\phi * \mathbf{f} = e^{At}\int_0^t e^{-As}\mathbf{f}(s)\,ds = \mathbf{v}_p$, the particular solution.

(d) Show that $\mathbf{v} = \mathbf{v}_h + \mathbf{v}_p$, a sum of the homogeneous solution and the particular solution.

In Exercises B2–11, the matrix $A = \begin{bmatrix} -4 & 1 \\ 3 & -2 \end{bmatrix}$ and the given vectors $\mathbf{F}(t)$ and $\mathbf{x}_0$ determine an IVP of the form $\mathbf{x}'(t) = A\mathbf{x}(t) + \mathbf{F}(t)$, $\mathbf{x}(0) = \mathbf{x}_0$. For each:

(a) Solve by Laplace transform: compute the transform of each equation, solve for the unknown transforms, and invert each.

(b) Obtain the fundamental matrix $\phi = e^{At}$.

(c) Obtain the homogeneous solution $\mathbf{x}_h = \phi\mathbf{c}$, where $\mathbf{c}$ is a vector of arbitrary constants.

(d) Obtain the particular solution $\mathbf{x}_p = e^{At}\int e^{-At}\mathbf{F}(t)\,dt$.

(e) Obtain the general solution $\mathbf{x}_g = \mathbf{x}_h + \mathbf{x}_p$ and apply the data, making sure the resulting solution agrees with the result in part (a).

(f) On the same set of axes, plot $x(t)$ and $y(t)$, the components of $\mathbf{x}(t)$.

(g) Plot $y(x)$, the parametric curve determined by $(x(t), y(t))$.

B2. $\mathbf{F}(t) = \begin{bmatrix} 1 \\ t \end{bmatrix}$, $\mathbf{x}_0 = \begin{bmatrix} -6 \\ 8 \end{bmatrix}$ **B3.** $\mathbf{F}(t) = \begin{bmatrix} \cos t \\ \sin t \end{bmatrix}$, $\mathbf{x}_0 = \begin{bmatrix} 10 \\ 1 \end{bmatrix}$

B4. $\mathbf{F}(t) = \begin{bmatrix} e^{-t} \\ 0 \end{bmatrix}$, $\mathbf{x}_0 = \begin{bmatrix} -11 \\ 1 \end{bmatrix}$ **B5.** $\mathbf{F}(t) = \begin{bmatrix} 0 \\ e^{-2t} \end{bmatrix}$, $\mathbf{x}_0 = \begin{bmatrix} -2 \\ 6 \end{bmatrix}$

B6. $\mathbf{F}(t) = \begin{bmatrix} -3 \\ \cos 2t \end{bmatrix}$, $\mathbf{x}_0 = \begin{bmatrix} 9 \\ -3 \end{bmatrix}$ **B7.** $\mathbf{F}(t) = \begin{bmatrix} \sin 3t \\ 2 \end{bmatrix}$, $\mathbf{x}_0 = \begin{bmatrix} -5 \\ -2 \end{bmatrix}$

B8. $\mathbf{F}(t) = \begin{bmatrix} e^{-t}\cos t \\ 1 \end{bmatrix}$, $\mathbf{x}_0 = \begin{bmatrix} 4 \\ -1 \end{bmatrix}$ **B9.** $\mathbf{F}(t) = \begin{bmatrix} 2 \\ e^{-2t}\sin 3t \end{bmatrix}$, $\mathbf{x}_0 = \begin{bmatrix} -5 \\ 12 \end{bmatrix}$

B10. $\mathbf{F}(t) = \begin{bmatrix} t \\ t^2 \end{bmatrix}$, $\mathbf{x}_0 = \begin{bmatrix} -1 \\ 11 \end{bmatrix}$ **B11.** $\mathbf{F}(t) = \begin{bmatrix} t^2 \\ t \end{bmatrix}$, $\mathbf{x}_0 = \begin{bmatrix} 4 \\ -3 \end{bmatrix}$

In Exercises B12–21, the vector $\mathbf{F}(t) = \begin{bmatrix} \sin t \\ \cos t \end{bmatrix}$ and the given matrix A and initial vector $\mathbf{x}_0$ determine an IVP of the form $\mathbf{x}'(t) = A\mathbf{x}(t) + \mathbf{F}(t)$, $\mathbf{x}(0) = \mathbf{x}_0$. For each:

(a) Solve by Laplace transform: compute the transform of each equation, solve for the unknown transforms, and invert each.

(b) Obtain the fundamental matrix $\phi = e^{At}$.

(c) Obtain the homogeneous solution $\mathbf{x}_h = \phi\mathbf{c}$, where $\mathbf{c}$ is a vector of arbitrary constants.

(d) Obtain the particular solution $\mathbf{x}_p = e^{At}\int e^{-At}\mathbf{F}(t)\,dt$.

(e) Obtain the general solution $\mathbf{x}_g = \mathbf{x}_h + \mathbf{x}_p$ and apply the data, making sure the resulting solution agrees with the result in part (a).

(f) On the same set of axes, plot $x(t)$ and $y(t)$, the components of $\mathbf{x}(t)$.

(g) Plot $y(x)$, the parametric curve determined by $(x(t), y(t))$.

B12. Exercise B2, Section 12.14   **B13.** Exercise B3, Section 12.14

B14. Exercise B4, Section 12.14   **B15.** Exercise B5, Section 12.14

B16. Exercise B6, Section 12.14   **B17.** Exercise B7, Section 12.14

B18. Exercise B8, Section 12.14   **B19.** Exercise B9, Section 12.14

B20. Exercise B10, Section 12.14   **B21.** Exercise B11, Section 12.14

In Exercises B22–27, the vector $\mathbf{F}(t) = \begin{bmatrix} e^{-t} \\ \cos t \end{bmatrix}$, the initial vector $\mathbf{x}_0 = \begin{bmatrix} -3 \\ 1 \end{bmatrix}$, and the given matrix A determine an IVP of the form $\mathbf{x}'(t) = A\mathbf{x}(t) + \mathbf{F}(t)$, $\mathbf{x}(0) = \mathbf{x}_0$. For each:

(a) Solve by Laplace transform: compute the transform of each equation, solve for the unknown transforms, and invert each.

(b) Obtain the fundamental matrix $\phi = e^{At}$.

(c) Obtain the homogeneous solution $\mathbf{x}_h = \phi\mathbf{c}$, where $\mathbf{c}$ is a vector of arbitrary constants.

(d) Obtain the particular solution $\mathbf{x}_p = e^{At} \int e^{-At} \mathbf{F}(t)\, dt$.

(e) Obtain the general solution $\mathbf{x}_g = \mathbf{x}_h + \mathbf{x}_p$ and apply the data, making sure the resulting solution agrees with the result in part (a).

(f) On the same set of axes, plot $x(t)$ and $y(t)$, the components of $\mathbf{x}(t)$.

(g) Plot $y(x)$, the parametric curve determined by $(x(t), y(t))$.

B22. Exercise B6, Section 12.10   **B23.** Exercise B7, Section 12.10

B24. Exercise B8, Section 12.10   **B25.** Exercise B9, Section 12.10

B26. Exercise B10, Section 12.10   **B27.** Exercise B11, Section 12.10

In Exercises B28–33, the vector $\mathbf{F}(t) = \begin{bmatrix} -t \\ 2 \end{bmatrix}$, the initial vector $\mathbf{x}_0 = \begin{bmatrix} 5 \\ -4 \end{bmatrix}$, and the given matrix A determine an IVP of the form $\mathbf{x}'(t) = A\mathbf{x}(t) + \mathbf{F}(t)$, $\mathbf{x}(0) = \mathbf{x}_0$. For each:

(a) Solve by Laplace transform: compute the transform of each equation, solve for the unknown transforms, and invert each.

(b) Obtain the fundamental matrix $\phi = e^{At}$.

(c) Obtain the homogeneous solution $\mathbf{x}_h = \phi\mathbf{c}$, where $\mathbf{c}$ is a vector of arbitrary constants.

(d) Obtain the particular solution $\mathbf{x}_p = e^{At} \int e^{-At} \mathbf{F}(t)\, dt$.

(e) Obtain the general solution $\mathbf{x}_g = \mathbf{x}_h + \mathbf{x}_p$ and apply the data, making sure the resulting solution agrees with the result in part (a).

(f) On the same set of axes, plot $x(t)$ and $y(t)$, the components of $\mathbf{x}(t)$.

(g) Plot $y(x)$, the parametric curve determined by $(x(t), y(t))$.

B28. Exercise B3, Section 12.12   **B29.** Exercise B5, Section 12.12

B30. Exercise B7, Section 12.12   **B31.** Exercise B8, Section 12.12

B32. Exercise B10, Section 12.12   **B33.** Exercise B11, Section 12.12

In Exercises B34–39, the vector $\mathbf{F}(t) = \begin{bmatrix} 1+t \\ t \end{bmatrix}$, the initial vector $\mathbf{x}_0 = \begin{bmatrix} 2 \\ -1 \end{bmatrix}$, and the given matrix A determine an IVP of the form $\mathbf{x}'(t) = A\mathbf{x}(t) + \mathbf{F}(t)$, $\mathbf{x}(0) = \mathbf{x}_0$. For each:

(a) Solve by Laplace transform: compute the transform of each equation, solve for the unknown transforms, and invert each.

(b) Obtain the fundamental matrix $\phi = e^{At}$.

(c) Obtain the homogeneous solution $\mathbf{x}_h = \phi\mathbf{c}$, where $\mathbf{c}$ is a vector of arbitrary constants.

(d) Obtain the particular solution $\mathbf{x}_p = e^{At} \int e^{-At} \mathbf{F}(t)\, dt$.

(e) Obtain the general solution $\mathbf{x}_g = \mathbf{x}_h + \mathbf{x}_p$ and apply the data, making sure the resulting solution agrees with the result in part (a).

(f) On the same set of axes, plot $x(t)$ and $y(t)$, the components of $\mathbf{x}(t)$.

(g) Plot $y(x)$, the parametric curve determined by $(x(t), y(t))$.

B34. Exercise B12, Section 12.12   **B35.** Exercise B14, Section 12.12

B36. Exercise B15, Section 12.12   **B37.** Exercise B17, Section 12.12

B38. Exercise B19, Section 12.12   **B39.** Exercise B21, Section 12.12

12.19   Phase Portraits

The Phase Plane

A damped spring-mass system is governed by the second-order differential equation $my''(t) + by'(t) + ky(t) = 0$, which translates into a first-order system via the change of variables $u(t) = y(t)$, $v(t) = y'(t)$ since $u'(t) = y'(t) = v(t)$ and $v'(t) = y''(t) = -\frac{k}{m}y(t) -$

$\frac{b}{m} y'(t) = -\frac{k}{m} u(t) - \frac{b}{m} v(t)$, so

$$\begin{bmatrix} u(t) \\ v(t) \end{bmatrix}' = \begin{bmatrix} 0 & 1 \\ -\dfrac{k}{m} & -\dfrac{b}{m} \end{bmatrix} \begin{bmatrix} u(t) \\ v(t) \end{bmatrix}$$

A parametric plot of $v(t)$ against $u(t)$ is actually a plot of $y'(t)$ against $y(t)$. The system point $(y(t), y'(t))$ traces a history of position and velocity for the mechanical system. Such a curve is called an *orbit* (also, *path* or *trajectory*) and the uv-plane is called the *phase plane*. A graph of several orbits in the phase plane is called a *phase portrait* and is a characteristic geometric profile categorizing the behavior of the system.

A Phase Portrait

Consider the system of differential equations $x'(t) = x(t) + 2y(t)$ and $y'(t) = 4x(t) - y(t)$ for which Figure 12.17 shows solutions through the eight initial points $(1, 1)$, $(-1, -1)$, $(5, -10)$, $(-5, 10)$, $(6, -10)$, $(4, -10)$, $(-6, 10)$, and $(-4, 10)$. The directions on the trajectories correspond to the motion of the system point $(x(t), y(t))$ as the parameter t increases along the curve. These directions are obtained from the differential equations. Table 12.10 indicates how directions are obtained for paths through the four points labeled A, B, C, and D in Figure 12.17.

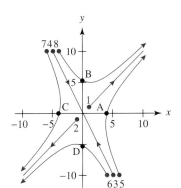

FIGURE 12.17 Phase portrait for $x' = x + 2y$, $y' = 4x - y$

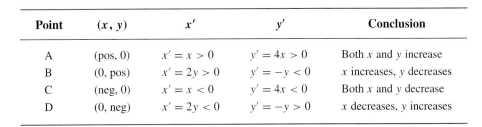

Point	(x, y)	x'	y'	Conclusion
A	(pos, 0)	$x' = x > 0$	$y' = 4x > 0$	Both x and y increase
B	(0, pos)	$x' = 2y > 0$	$y' = -y < 0$	x increases, y decreases
C	(neg, 0)	$x' = x < 0$	$y' = 4x < 0$	Both x and y decrease
D	(0, neg)	$x' = 2y < 0$	$y' = -y > 0$	x decreases, y increases

TABLE 12.10 Points on the trajectories in Figure 12.17

Equilibrium Points

An *equilibrium point* in the phase plane is a point at which $x'(t) = y'(t) = 0$. Such a point represents an *equilibrium solution,* a constant solution where the system point sits and does not travel along an orbit. For the typical linear system we find the only equilibrium point to be the origin, $(0, 0)$, provided the system matrix is not singular.

There are four kinds of equilibrium points; and they have the names *node, saddle, center,* and *spiral.* Table 12.11 summarizes these names and briefly characterizes each type of equilibrium point.

Name	Description
Node	All paths either enter or leave equilibrium point
Saddle	Two paths enter, two leave, all others "swoop by"
Center	Orbits are closed curves around equilibrium point
Spiral	Each path spirals around equilibrium point

TABLE 12.11 Equilibrium points

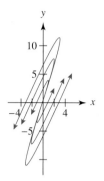

FIGURE 12.18   Phase portrait for the (outward) node of Example 12.29

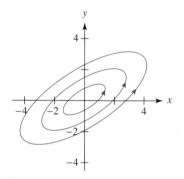

FIGURE 12.19   Phase portrait for the *center* in Example 12.31

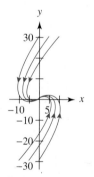

FIGURE 12.20   Phase portrait for the (inward) spiral of Example 12.32

Table 12.12 gives an example of each type of equilibrium point. A phase portrait for each example is given in the indicated figure. For system L_1 Table 12.12, the equilibrium point at the origin is a node "out," sometimes called a *repeller*. (See Figure 12.18.) Were the trajectories in this node to point inward toward the origin, the node would be a node "in," sometimes called an *attractor*.

Designation	Equations	Equilibrium Point	Figure
L_1	$x'(t) = 7x(t) - 2y(t)$ $y'(t) = 15x(t) - 4y(t)$	Node	Figure 12.18
L_2	$x'(t) = x(t) + 2y(t)$ $y'(t) = 4x(t) - y(t)$	Saddle	Figure 12.17
L_3	$x'(t) = x(t) - 2y(t)$ $y'(t) = x(t) - y(t)$	Center	Figure 12.19
L_4	$x'(t) = -x(t) - 2y(t)$ $y'(t) = x(t) - 3y(t)$	Spiral	Figure 12.20

TABLE 12.12   Examples of equilibrium points

For system L_2, Table 12.12, the equilibrium point at the origin is a saddle. This is the same system whose trajectories appear in Figure 12.17. A trajectory that enters or leaves the origin is called a *separatrix* (plural, *separatrices*). Trajectories starting on one side of a separatrix remain on that same side. The equations for the separatrices are considered in Exercise 1.

For system L_3, Table 12.12, the equilibrium point at the origin is a center. A center is surrounded by closed trajectories. (See Figure 12.19.)

For system L_4, Table 12.12, the equilibrium point at the origin is an inward spiral. Like the inward node, the inward spiral is sometimes called an attractor. (See Figure 12.20.) Clearly, if the trajectories spiral outward, the phase portrait would be that of an outward spiral and it would likewise be called a repeller.

Eigenvalues and the Phase Portrait

For the linear system $\mathbf{x}' = A\mathbf{x}$, the eigenvalues of the matrix A characterize the nature of the phase portrait at the origin. These relationships are summarized in Table 12.13.

Equilibrium Point	Eigenvalues
Node	Same sign
Saddle	Opposite signs
Center	Pure imaginary
Spiral	Complex conjugates

TABLE 12.13   Characterization of equilibrium points by eigenvalues

Representative Examples

Table 12.14 gives representative examples of the four types of equilibrium points and their eigenvalues. We distinguish between a *proper* node and an *improper* node. For the moment, the improper node is characterized by repeated eigenvalues. In Section 12.20, we will see that the improper node poses a minor difficulty when studying the *stability* of nonlinear systems. Nodes and saddles have exponential solutions that tend either to zero or infinity. Hence, trajectories for these systems will either tend toward or away from the origin. The center has trigonometric solutions that are the parametric representations of closed curves. The spirals have solutions with the trigonometric terms of the center but have exponential factors that force the solutions either toward or away from the origin.

Equilibrium Point	Eigenvalues	Solutions
Proper node in	$-1, -2$	e^{-t}, e^{-2t}
Improper node in	$-1, -1$	e^{-t}, te^{-t}
Proper node out	$1, 2$	e^{t}, e^{2t}
Improper node out	$1, 1$	e^{t}, te^{t}
Saddle	$-1, 2$	e^{-t}, e^{2t}
Center	$\pm 3i$	$\cos 3t, \sin 3t$
Spiral in	$-1 \pm 2i$	$e^{-t}\cos 2t, e^{-t}\sin 2t$
Spiral out	$1 \pm 2i$	$e^{t}\cos 2t, e^{t}\sin 2t$

TABLE 12.14   Eigenvalues and representative examples of equilibrium points

EXERCISES 12.19

1. Find the eigenvectors for the system matrix in Example 12.30, Table 12.12. Show that any solution $\mathbf{x}(t)$ starting at $\mathbf{x}_0$ on the line through the origin having direction of an eigenvector remains along that line. Hence, the separatrices are determined by the eigenvectors.

The 2×2 matrix A given in each of Exercises 2–21 determines a homogeneous system $\mathbf{x}' = A\mathbf{x}$, where the components of $\mathbf{x}(t)$ are $x(t)$ and $y(t)$. For each system:

  (a) Determine the nature of the equilibrium point $(0, 0)$ by computing the eigenvalues of A.

  (b) Obtain $\mathbf{x}_h$, the general homogeneous solution to the linear system.

  (c) Obtain $\mathbf{x}_k(t)$, $k = 1, \ldots, 4$, solutions through each of the four initial points $(x, y) = (1, 1), (-1, 1), (-1, -1), (1, -1)$.

  (d) In the phase plane, plot the trajectories determined by the four solutions found in part (c).

  (e) Use the differential equations to determine the direction of increasing t on each trajectory drawn in part (d).

  (f) Use an efficient technology to obtain a phase portrait for the system of differential equations.

  (g) If the origin is a saddle, show that the eigenvectors determine the separatrices.

2. $\begin{bmatrix} 9 & -3 \\ -7 & 1 \end{bmatrix}$ **3.** $\begin{bmatrix} 5 & 2 \\ -12 & 2 \end{bmatrix}$ **4.** $\begin{bmatrix} 3 & 3 \\ -6 & -3 \end{bmatrix}$ **5.** $\begin{bmatrix} 11 & 5 \\ 4 & 6 \end{bmatrix}$

6. $\begin{bmatrix} -12 & -1 \\ 1 & 1 \end{bmatrix}$ **7.** $\begin{bmatrix} 8 & 7 \\ 3 & 4 \end{bmatrix}$ **8.** $\begin{bmatrix} 8 & 12 \\ -3 & -2 \end{bmatrix}$ **9.** $\begin{bmatrix} -5 & 5 \\ 1 & -10 \end{bmatrix}$

10. $\begin{bmatrix} 12 & -2 \\ -1 & 11 \end{bmatrix}$ **11.** $\begin{bmatrix} -3 & 15 \\ -6 & 3 \end{bmatrix}$ **12.** $\begin{bmatrix} 2 & 0 \\ -5 & 12 \end{bmatrix}$ **13.** $\begin{bmatrix} 1 & 5 \\ -10 & -1 \end{bmatrix}$

14. $\begin{bmatrix} -5 & -11 \\ 0 & 10 \end{bmatrix}$ **15.** $\begin{bmatrix} 1 & 5 \\ 10 & 11 \end{bmatrix}$ **16.** $\begin{bmatrix} 10 & 10 \\ -12 & 12 \end{bmatrix}$ **17.** $\begin{bmatrix} 5 & 5 \\ 8 & 10 \end{bmatrix}$

18. $\begin{bmatrix} -2 & 5 \\ -1 & 2 \end{bmatrix}$ **19.** $\begin{bmatrix} 1 & -11 \\ -12 & 12 \end{bmatrix}$ **20.** $\begin{bmatrix} 18 & -30 \\ 12 & -18 \end{bmatrix}$ **21.** $\begin{bmatrix} 12 & 10 \\ 1 & 3 \end{bmatrix}$

12.20 Stability

Nonautonomous Systems

The general nonlinear system in two unknowns would typically be of the form

$$\frac{dx}{dt} = F(x(t), y(t), t) \qquad \frac{dy}{dt} = G(x(t), y(t), t)$$

When the independent variable t is explicitly present on the right-hand sides of the differential equations, the system is called *nonautonomous*. A time-varying breeze blowing the smoke coming from a chimney would give rise to a nonautonomous system. Table 12.15 lists several examples of both linear and nonlinear nonautonomous systems. In the table, system (b) is a linear nonautonomous system, while system (f) is a nonlinear nonautonomous system.

	Linear	Nonlinear
Autonomous	(a) $\begin{array}{l} x' = 2x + 3y \\ y' = 5x - 4y \end{array}$	(c) $\begin{array}{l} x' = x - \frac{1}{2}xy \\ y' = -2y + \frac{1}{4}xy \end{array}$
		(d) $\begin{array}{l} x' = x(60 - 3x - 4y) \\ y' = y(42 - 3y - 2x) \end{array}$
		(e) $\begin{array}{l} x' = y \\ y' = -\sin x \end{array}$
Nonautonomous	(b) $\begin{array}{l} x' = 2x + 3y + t \\ y' = 5x - 4y - e^t \end{array}$	(f)sf12.20 $\begin{array}{l} x' = xy + t \\ y' = x^2 + y \sin t \end{array}$

TABLE 12.15   Examples of autonomous and nonautonomous systems

Autonomous Systems

When the independent variable t is *not* explicitly found on the right-hand sides of the differential equations, the system is called *autonomous* and would be of the form

$$\frac{dx}{dt} = F(x(t), y(t)) \qquad \frac{dy}{dt} = G(x(t), y(t))$$

For the autonomous system, the notion of a fixed geometry in the phase plane, giving rise to time-invariant phase portraits, makes sense. Thus, we also have the notion of an equilibrium point, a point (x, y) in the phase plane where both $\frac{dx}{dt} = 0$ and $\frac{dy}{dt} = 0$. System (a) in Table 12.15 is a linear autonomous system, while systems (c)–(e) are nonlinear autonomous systems. System (a) is typical of the systems we have been studying in Sections 12.1–12.19. System (c) is a predator-prey model that we will examine in some detail in Section 12.21. System (d) is a competing-species model, also studied in Section 12.21. System (e) models the nonlinear pendulum, studied in Section 12.23.

Isolated Equilibrium Point

An *isolated equilibrium point* is one that can be surrounded by a circle containing no other equilibrium point. A linear autonomous system with constant coefficients and determinant zero would not have any of its equilibrium points isolated. For example, the system whose matrix is $A = \begin{bmatrix} 1 & 2 \\ 2 & 4 \end{bmatrix}$ has determinant zero, and every point on the line $x = -2y$ is an equilibrium point, none of which can be surrounded by a circle containing just the one equilibrium point.

Henceforth, we will assume each equilibrium point in our discussions is an isolated equilibrium point. In particular, the linear autonomous system for which det $A \neq 0$ has just one equilibrium point, the origin.

Stability of an Equilibrium Point

An isolated equilibrium point for an autonomous system of differential equations is said to be *orbitally (neutrally) stable* if all solutions sufficiently close to the equilibrium point remain close to that equilibrium point as $t \to \infty$, *asymptotically stable* if it is orbitally stable and, in addition, all solutions sufficiently close to the equilibrium point actually tend toward that equilibrium point as $t \to \infty$, and *unstable* if at least one solution near the equilibrium point at $t = t_1 > 0$ does not remain close to that equilibrium point as $t \to \infty$.

Table 12.16 lists Examples 12.29–12.31, which we will use to illustrate these terms and motivate precise mathematical definitions.

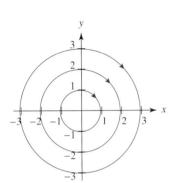

FIGURE 12.21   Example 12.29: Stable center

Example	System	A	Eigenvalues	Phase Portrait
(12.29)	$x'(t) = y(t)$ $y'(t) = -x(t)$	$\begin{bmatrix} 0 & 1 \\ -1 & 0 \end{bmatrix}$	$\pm i$	Figure 12.21
(12.30)	$x'(t) = -8x(t) + 3y(t)$ $y'(t) = -10x(t) + 3y(t)$	$\begin{bmatrix} -8 & 3 \\ -10 & 3 \end{bmatrix}$	$-3, -2$	Figure 12.22
(12.31)	$x'(t) = x(t)$ $y'(t) = 2x(t) - y(t)$	$\begin{bmatrix} 1 & 0 \\ 2 & -1 \end{bmatrix}$	± 1	Figure 12.23

TABLE 12.16   Examples of orbitally stable, asymptotically stable, and unstable systems

EXAMPLE 12.29   The system in Example 12.29, Table 12.16, has an *orbitally stable* equilibrium point at the origin. To see what this means, note the eigenvalues of the coefficient matrix A are the pure imaginary conjugate pair $\pm i$. Hence, the phase portrait is that of the center. The origin is the center of closed orbits that neither spread apart nor approach each other, the essential meaning of *orbital stability*. Any solution starting near the origin remains near the origin. It does not necessarily get arbitrarily close to the origin, but it does not get far from the origin either. The phase portrait in Figure 12.21 is useful in this regard because it shows that solutions starting near the origin remain near the origin, tending neither toward nor away from the origin. Hence, the following definition of orbital stability.

The equilibrium point P is *orbitally stable* if the solution $\mathbf{x}$ can be kept within ε of P, that is, $\|\mathbf{x}(t) - P\| < \varepsilon$, provided $\mathbf{x}$ initially starts out sufficiently close to P by satisfying $\|\mathbf{x}(0) - P\| < \delta$ for an appropriate $\delta > 0$.   ❖

EXAMPLE 12.30   The system in Example 12.30, Table 12.16, has an *asymptotically stable* equilibrium point at the origin. The eigenvalues of the system matrix A are -3, and -2. The eigenvalues are of the same sign and negative. Hence, the phase portrait is that of the inward node. That makes the origin an attractor; and hence, all solutions will be drawn into the origin, making the system asymptotically stable. Any solution that starts near the origin will tend toward the origin as $t \to \infty$. For example, the solution starting at (a, b), namely,

$$x(t) = (6a - 3b)e^{-3t} - (5a - 3b)e^{-2t} \quad \text{and} \quad y(t) = (10a - 5b)e^{-3t} - (10a - 6b)e^{-2t}$$

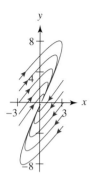

FIGURE 12.22   Example 12.30: Asymptotically stable (inward) node

satisfies $\lim_{t\to\infty} x(t) = 0$ and $\lim_{t\to\infty} y(t) = 0$. As promised, an arbitrary solution starting near the origin not only remains near the origin but tends to the origin as $t \to \infty$. Hence, *all* solutions, not only those near the origin, tend toward the origin with increasing t, as seen in the phase portrait in Figure 12.22.

In the nonlinear case where there may be other equilibrium points that attract some solutions starting some distance away from the origin, the more precise wording "starting near enough to the equilibrium point" is critical. Hence, we formulate the definition of asymptotic stability as follows.

If the equilibrium point P is a (neutrally or orbitally) stable equilibrium point for which $\lim_{t\to\infty} \mathbf{x}(t) = \mathbf{0}$ whenever any solution $\mathbf{x}$ satisfies $\|\mathbf{x}(0) - P\| < \delta$ for a sufficiently small and positive δ, then P is also *asymptotically stable*.

It is actually necessary to require that the equilibrium point be orbitally stable before checking that nearby solutions satisfy the limit condition. Examples can be constructed in which a sequence of solutions all starting near the equilibrium point successively tend arbitrarily far from the point before finally approaching it. Such nonuniformity in the limit is ruled out by first demanding that all solutions near enough to the critical point remain near the critical point, that is, by first demanding orbital stability.   ❖

EXAMPLE 12.31   The system in Example 12.31, Table 12.16, has an *unstable* equilibrium point at the origin. The eigenvalues of the system matrix A are ± 1, so the corresponding phase portrait is that of the saddle. There are two trajectories that will enter the origin, but all other solutions, no matter how close to the origin they begin, will tend toward infinity as t increases. Since at least one solution starting near the origin does not remain near the origin, the origin is an *unstable* equilibrium point. Solutions starting on the y-axis will tend toward the origin. In fact, such solutions are precisely $x(t) = 0$, $y(t) = ae^{-t}$, for which $\lim_{t\to\infty} x(t) = \lim_{t\to\infty} y(t) = 0$. Any other solution $x(t) = ae^t$, $y(t) = a(e^t - e^{-t})be^{-t}$ will not remain near the origin since, for $a > 0$, $\lim_{t\to\infty} x(t) = \lim_{t\to\infty} y(t) = \infty$, and for $a < 0$, $\lim_{t\to\infty} x(t) = \lim_{t\to\infty} y(t) = -\infty$. As promised, all other solutions, no matter how near the origin they may have started, tend toward infinity. As seen in Figure 12.23, the phase portrait, that of a saddle, makes the instability of this equilibrium clear.

An equilibrium point that is not stable is called *unstable*. Thus, if at least one solution starting near the equilibrium point does not remain near it, the equilibrium point is unstable.   ❖

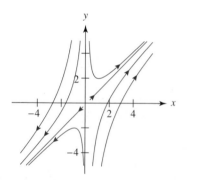

FIGURE 12.23   Example 12.31: Unstable saddle

Summary

Table 12.17 summarizes the relationships between phase portraits, eigenvalues, and stability.

Equilibrium Point	Eigenvalues	Conditions	Stability
Node (in)	Real and negative		Asymptotically stable
Spiral (in)	$a \pm bi$	$a < 0$	Asymptotically stable
Center	$\pm bi$	Pure imaginary	Orbitally stable
Node (out)	Real and positive		Unstable
Spiral (out)	$a \pm bi$	$a > 0$	Unstable
Saddle	Real, with opposite signs		Unstable

TABLE 12.17   Eigenvalues and the stability of equilibrium points

EXERCISES 12.20–Part A

For Exercises A1–5, the given matrix A determines a linear system $x' = Ax$. For each, classify the equilibrium point at the origin according to Table 12.17. Sketch a few representative trajectories that support your conclusion about stability.

A1. $\begin{bmatrix} 3 & 18 \\ -1 & -3 \end{bmatrix}$ **A2.** $\begin{bmatrix} 5 & -1 \\ -5 & 7 \end{bmatrix}$ **A3.** $\begin{bmatrix} 11 & -2 \\ 0 & -2 \end{bmatrix}$

A4. $\begin{bmatrix} 1 & -1 \\ 8 & 7 \end{bmatrix}$ **A5.** $\begin{bmatrix} 4 & -12 \\ 1 & 7 \end{bmatrix}$

EXERCISES 12.20–Part B

B1. Example 12.31, Table 12.12, is a system for which the origin is a center. Its phase portrait in Figure 12.19 shows ovals as the closed trajectories. Example 12.29 , Table 12.16, is a system for which the origin is a center. Its phase portrait in Figure 12.21 shows what appears to be circles as the closed trajectories. Investigate these two examples. Can centers have noncircular closed trajectories? How?

In Exercises B2–21, the given 2×2 matrix A determines the homogeneous system $\mathbf{x}' = A\mathbf{x}$, where $\mathbf{x}$ has components $x(t)$ and $y(t)$. For each:

   (a) Determine the nature of the equilibrium point $(0, 0)$ by computing the eigenvalues of A.

   (b) From the eigenvalues, determine if the equilibrium point is asymptotically stable, orbitally stable, or unstable.

   (c) Obtain $\mathbf{x}_h$, the general homogeneous solution to the linear system.

   (d) Obtain solutions through each of the four initial points $(x, y) = (1, 1), (-1, 1), (-1, -1), (1, -1)$.

   (e) For each solution in part (d), plot $x(t)$ and $y(t)$ and obtain $\lim_{t \to \infty} x(t)$ and $\lim_{t \to \infty} y(t)$, if they exist.

   (f) In the phase plane, plot the trajectories determined by the four solutions found in part (d).

   (g) Use the differential equations to determine the direction of increasing t on each trajectory drawn in part (f).

   (h) Use an efficient technology to obtain a phase portrait for the system of differential equations.

B2. $\begin{bmatrix} -6 & 10 \\ -2 & 9 \end{bmatrix}$ **B3.** $\begin{bmatrix} 8 & 10 \\ -7 & 1 \end{bmatrix}$ **B4.** $\begin{bmatrix} 8 & -3 \\ 12 & -2 \end{bmatrix}$ **B5.** $\begin{bmatrix} -8 & 7 \\ -11 & 0 \end{bmatrix}$

B6. $\begin{bmatrix} 9 & -2 \\ 11 & -7 \end{bmatrix}$ **B7.** $\begin{bmatrix} -9 & 4 \\ -6 & 12 \end{bmatrix}$ **B8.** $\begin{bmatrix} 3 & 10 \\ -3 & 2 \end{bmatrix}$ **B9.** $\begin{bmatrix} 11 & -9 \\ 2 & 12 \end{bmatrix}$

B10. $\begin{bmatrix} 10 & 5 \\ -25 & -10 \end{bmatrix}$ **B11.** $\begin{bmatrix} 3 & 11 \\ 12 & -1 \end{bmatrix}$ **B12.** $\begin{bmatrix} -4 & -5 \\ 3 & 5 \end{bmatrix}$

B13. $\begin{bmatrix} -11 & -17 \\ 10 & 11 \end{bmatrix}$ **B14.** $\begin{bmatrix} -5 & 8 \\ -5 & 3 \end{bmatrix}$ **B15.** $\begin{bmatrix} 6 & -9 \\ 8 & -6 \end{bmatrix}$ **B16.** $\begin{bmatrix} 4 & -10 \\ 2 & -4 \end{bmatrix}$

B17. $\begin{bmatrix} -8 & -6 \\ 12 & 5 \end{bmatrix}$ **B18.** $\begin{bmatrix} 10 & -3 \\ 6 & 7 \end{bmatrix}$ **B19.** $\begin{bmatrix} -4 & 10 \\ 2 & 2 \end{bmatrix}$

B20. $\begin{bmatrix} 9 & -9 \\ -2 & 7 \end{bmatrix}$ **B21.** $\begin{bmatrix} 7 & 6 \\ 4 & 10 \end{bmatrix}$

 ## Nonlinear Systems

Autonomous and Nonautonomous Systems

The general nonautonomous system is nonlinear and, by definition, contains the independent variable explicitly. The general autonomous system is also nonlinear but does not contain the independent variable explicitly. This "reversal" of negations can be confusing. The autonomous system does *not* contain, and the nonautonomous system *does* contain, the independent variable explicitly.

EXAMPLE 12.32   **Predator-prey model**   Lotka and Volterra proposed the following model to explain the relationship between two species, one a predator feeding off the other, the prey.

Let $x(t)$ and $y(t)$ be the numbers, respectively, of rabbits (prey) and foxes (predators) in an ecosystem. Assume that rabbits eat only clover and that the clover is an unlimited resource. Assume further that foxes eat only rabbits. Moreover, assume that if there were no foxes to limit the rabbits, the number of rabbits would grow exponentially, whereas if there were no rabbits to sustain the foxes, the foxes would die out exponentially.

Assume that the number of rabbits eaten by the foxes is proportional to the frequency of fox-rabbit encounters and that the number of fox-rabbit encounters is proportional to $x(t)y(t)$, the *product* of the population sizes. A particular example embodying these assumptions would be the initial value problem (see system (c), Table 12.15)

$$x'(t) = x - \tfrac{1}{2}xy$$
$$y'(t) = -2y + \tfrac{1}{4}xy \qquad \text{with} \quad x(0) = 10,\, y(0) = 5 \qquad (12.83)$$

Thus, if there were no foxes, $y(t)$ would be zero, in which case $x'(t) = x(t)$ and the rabbits grow exponentially. On the other hand, if there were no rabbits, $x(t)$ would be zero, in which case $y'(t) = -2y(t)$ and the foxes die off exponentially.

Unfortunately, there is no closed-form solution for this system of differential equations. The study of such systems must be done indirectly through numerical, graphical, and logical (analytical) investigations. As a first step in the investigation, we find all equilibrium points. For nonlinear systems, there may be equilibrium points at locations other than just the origin. For this system, there are two, $(0, 0)$ and $(8, 2)$, obtained as the solutions of the equations

$$x' = x - \tfrac{1}{2}xy = 0 \quad \text{and} \quad y' = -2y + \tfrac{1}{4}xy = 0$$

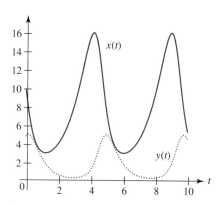

FIGURE 12.24 Example 12.32: $x(t)$ (solid), prey, and $y(t)$ (dotted), predator

It is instructive to obtain solutions both in the phase plane and in the time domain. Figure 12.24, computed numerically, shows time-domain graphs of $x(t)$ (solid curve) and $y(t)$ (dotted curve). The populations of the rabbits and the foxes vary periodically. As the number of foxes starts to increase, the number of rabbits will begin to decline. As the number of rabbits begins to increase, the population of foxes, though lagging the increase in the rabbits, begins increasing also.

Since $x(t)$ and $y(t)$ represent numbers of rabbits and foxes, respectively, only the first quadrant of the phase plane makes sense physically. Figure 12.25 shows the equilibrium point $(8, 2)$ inside closed orbits $y(x)$ in the phase plane. These orbits, traced counterclockwise, are consistent with periodic motion in the time domain since the system point keeps repeating values over and over again. Moreover, the phase portrait suggests $(8, 2)$ is a center. However, at this stage of our study, we cannot assert this with surety. It does appear, however, that the origin is a saddle point since there is one path (the y-axis) entering and one path (the x-axis) leaving. Section 12.22 discusses analytic techniques for determining the behavior of nonlinear systems in the vicinity of an equilibrium point. ❖

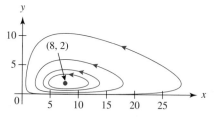

FIGURE 12.25 Phase portrait for Example 12.32

◆ **EXAMPLE 12.33** **Competing species** The equilibrium points of the system (see system (d), Table 12.15)

$$x'(t) = (60 - 3x)x - 4xy \quad \text{and} \quad y'(t) = (42 - 3y)y - 2xy \qquad (12.84)$$

are $(0, 0)$, $(20, 0)$, $(0, 14)$, and $(12, 6)$, obtained as solutions of the equations

$$x' = x(60 - 3x - 4y) = 0 \quad \text{and} \quad y' = y(42 - 3y - 2x) = 0 \qquad (12.85)$$

From (12.84), we can see that each species, in the absence of the other, obeys a logistic model. The interaction terms proportional to the product of the population sizes have negative coefficients, making interactions harmful to both species. Hence, the model represents competition between members of two species.

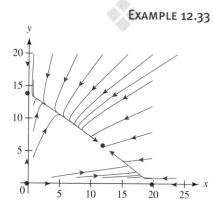

FIGURE 12.26 Phase portrait for Example 12.33

Figure 12.26 contains a phase portrait showing trajectories in the first quadrant of the phase plane. From the phase portrait, it would appear the equilibrium points are an outward node, a saddle, a saddle, and an inward node, respectively. However, these inferences require the analytical tools of Section 12.22 for verification. ❖

EXERCISES 12.21 –Part A

A1. The system $x' = x - xy$, $y' = -2y - xy$ can be thought of as a competing-species model where growth/decay for one species is exponential in the absence of the other. Find the equilibrium points, and describe the fate of the populations.

A2. The system $x' = (1 - x)x - xy$, $y' = -2y - xy$ can be thought of as a competing-species model where the first species obeys a logistic model in the absence of the second, but the second decays exponentially in the absence of the first. Find the equilibrium points, and describe the fate of the populations.

A3. The system $x' = x - xy$, $y' = (1 - y)y - xy$ can be thought of as a competing species model where the first species grows exponentially in the absence of the second, but the second obeys a

logistic model in the absence of the first. Find the equilibrium points, and describe the fate of the populations.

A4. Remove the interaction terms from the model in (12.84), leaving two uncoupled logistic models. Determine the carrying capacities for these two models.

A5. Write the first equation on the left in (12.83) as $\frac{x'}{x} = 1 - \frac{1}{2}y$, and integrate over $0 \le t \le T$, where T is the period for a cycle around one closed orbit in Figure 12.24. Let $\bar{y} = \frac{1}{T}\int_0^T y(t)\,dt$ be the average value of $y(t)$. Using $x(0) = x(T)$ (from periodicity), show that $\bar{y} = 2$.

A6. Repeat Exercise A5 for $x(t)$, obtaining $\bar{x} = 8$.

EXERCISES 12.21–Part B

B1. The system $x'(t) = (2 - x)x + \frac{1}{2}xy$, $y'(t) = (2 - y)y + \frac{1}{2}xy$ represents two species for which interactions are mutually beneficial. Each obeys a logistic model in the absence of the other. Obtain the equilibrium points. From a phase portrait or direction field drawn in the first quadrant, hypothesize about the nature of the equilibrium points.

For the autonomous systems in Exercises B2–21:

   (a) Find all equilibrium points.

   (b) Plot $x(t)$ and $y(t)$, numeric solutions passing through $(1, 1)$ at $t = 0$.

   (c) In the phase plane, plot $y(x)$, the trajectory corresponding to the solution found in part (b).

   (d) Using appropriate software, obtain a phase portrait or direction field, and from it, attempt to determine the nature of the equilibrium points found in part (a).

B2. $x' = x(-3 + x - y)$, $y' = y(-9 - 2x + 3y)$

B3. $x' = x(-9 + 3x + 2y)$, $y' = y(-1 + x + y)$

B4. $x' = x(5 + x)$, $y' = y(-5 + x - y)$

B5. $x' = y(-2 - x + 2y)$, $y' = x(14 + 3x - 7y)$

B6. $x' = x(3x + y + 6)$, $y' = y(3 + 4x + y)$

B7. $x' = x(-12 + 4x + 3y)$, $y' = y(-2 + x + y)$

B8. $x' = y(-1 + x + y)$, $y' = x(4 - 2x - 3y)$

B9. $x' = x(2 + x + y)$, $y' = y(7 - 8x - 7y)$

B10. $x' = x(4 - x - 2y)$, $y' = y(6 + x + 3y)$

B11. $x' = x(30 - 6x + 7y)$, $y' = y(24 - 5x + 6y)$

B12. $x' = y(3 + x - y)$, $y' = x(-20 - 4x + 3y)$

B13. $x' = x(15 + 3x + 2y)$, $y' = y(3 + 2x + y)$

B14. $x' = y(-3 + 8x + y)$, $y' = x(1 - x)$

B15. $x' = x(-2 + 2x - 3y)$, $y' = y(-1 - x + y)$

B16. $x' = x(3x + 4y + 9)$, $y' = y(-3 + 2x + 3y)$

B17. $x' = x(18 + 6x + 7y)$, $y' = y(-5 - x - y)$

B18. $x' = y(1 - x + y)$, $y' = x(-14 + 7x - 6y)$

B19. $x' = y(-1 + 4x - y)$, $y' = x(2 - 5x + y)$

B20. $x' = x(-5 + x)$, $y' = y(4 + 3x - y)$

B21. $x' = y(28 - 6x + 7y)$, $y' = x(-35 + 7x - 8y)$

12.22   Linearization

Tangent-Line Approximation

Near the point of contact, the curve $y = f(x)$ is approximated by its tangent line. For example, the function $f(x) = (x - 1)^2 + 2$ and its tangent at the point $(2, 3)$ appear in Figure 12.27. Near the point $(2, 3)$ the function $f(x)$ very much resembles its tangent line $y = 2x - 1$. Analytically, the tangent line is the Taylor polynomial of degree 1, $2x - 1$.

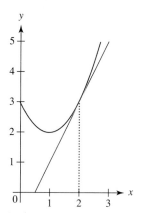

FIGURE 12.27   Tangent-line approximation

Tangent-Plane Approximation

From multivariable calculus, we know the tangent plane approximates a function of two variables. This is the two-dimensional version of linearization. For example, at the point where $(x, y) = (2, 2)$, the tangent plane approximating the function $f(x, y) = 3x^2 + 2y^2 - 5x - 7y - 9$ is $z = 7x + y - 29$. This first-degree polynomial approximation of $f(x, y)$ is just the multivariable Taylor polynomial of degree one.

Linearizing Autonomous Systems

Near an equilibrium point of the autonomous nonlinear system $\frac{dx}{dt} = F(x(t), y(t))$, $\frac{dy}{dt} = G(x(t), y(t))$, the behavior of solutions can be studied by linearizing the functions $F(x, y)$ and $G(x, y)$ at each equilibrium point. Indeed, let (a, b) be an equilibrium point determined by the equations $F(a, b) = G(a, b) = 0$. Then, lumping the higher order remainder terms into the functions $f(x, y)$ and $g(x, y)$, the first-degree Taylor expansions

$$F(x, y) = F(a, b) + F_x(a, b)(x - a) + F_y(a, b)(y - b) + f(x, y)$$
$$G(x, y) = G(a, b) + G_x(a, b)(x - a) + G_y(a, b)(y - b) + g(x, y)$$

become

$$F(x, y) = F_x(a, b)(x - a) + F_y(a, b)(y - b) + f(x, y)$$
$$G(x, y) = G_x(a, b)(x - a) + G_y(a, b)(y - b) + g(x, y)$$

because (a, b) is an equilibrium point at which $F(a, b) = G(a, b) = 0$. The nonlinear autonomous system now reads

$$\frac{dx}{dt} = F_x(a, b)(x - a) + F_y(a, b)(y - b) + f(x, y)$$
$$\frac{dy}{dt} = G_x(a, b)(x - a) + G_y(a, b)(y - b) + g(x, y)$$

Define the change of variables $u(t) = x(t) - a$, $v(t) = y(t) - b$ so that $\frac{du}{dt} = \frac{dx}{dt}$ and, $\frac{dv}{dt} = \frac{dy}{dt}$ and the system now reads

$$\mathbf{u}' = \begin{bmatrix} u \\ v \end{bmatrix}' = \begin{bmatrix} F_x(a, b) & F_y(a, b) \\ G_x(a, b) & G_y(a, b) \end{bmatrix} \begin{bmatrix} u \\ v \end{bmatrix} + \begin{bmatrix} f^*(u, v) \\ g^*(u, v) \end{bmatrix}$$

or $\mathbf{u}' = A\mathbf{u} + \mathbf{P}$, where the matrix

$$A = \begin{bmatrix} F_x(a, b) & F_y(a, b) \\ G_x(a, b) & G_y(a, b) \end{bmatrix}$$

is the *Jacobian matrix* and the functions $f^*(u, v)$ and $g^*(u, v)$ are just the functions $f(x, y)$ and $g(x, y)$ under the change of variables $x = u + a$, $y = v + b$.

The equilibrium point (a, b) in the xy-plane has been translated to the equilibrium point $(0, 0)$ in the uv-plane. If the perturbation term $\mathbf{P}$ is dropped, the resulting system $\mathbf{u}' = A\mathbf{u}$ is a linear system with an equilibrium point at $(0, 0)$. The behavior of the nonlinear xy-system at (a, b) can often be deduced by analyzing the behavior of the linear uv-system at $(0, 0)$, as we will shortly show.

EXAMPLE 12.34   Recall the competing-species system (d), Table 12.15, examined at some length in Example 12.33, Section 12.21, where the system equations written in (12.84) give $F(x, y) = x(60 - 3x - 4y)$ and $G(x, y) = y(42 - 3y - 2x)$. From Figure 12.26, the phase portrait for this system, we guessed the nature of the four equilibrium points $(0, 0)$, $(0, 14)$, $(20, 0)$, and $(12, 6)$ to be an outward node, a saddle, a saddle, and an inward node, respectively.

The Jacobian matrix is

$$A = \begin{bmatrix} 60 - 6x - 4y & -4x \\ -2y & 42 - 6y - 2x \end{bmatrix} \tag{12.86}$$

and at each of the equilibrium points this matrix evaluates to

$$\begin{bmatrix} 60 & 0 \\ 0 & 42 \end{bmatrix}, \begin{bmatrix} -60 & -80 \\ 0 & 2 \end{bmatrix}, \begin{bmatrix} 4 & 0 \\ -28 & -42 \end{bmatrix}, \begin{bmatrix} -36 & -48 \\ -12 & -18 \end{bmatrix} \tag{12.87}$$

respectively. The eigenvalues of these four matrices are 60, 42; -60, 2; 4, -42; and $-27 + 3\sqrt{73} \doteq -1.37$, $-27 - 3\sqrt{73} \doteq -52.6$. From the eigenvalues of the *linearized* systems, we have a node out, a saddle, a saddle, and a node in, respectively. But rules are needed for deducing the nature of the critical points of the nonlinear system from the information generated for the linear system. These rules appear in the next paragraph. ❖

Interpretation Rules

The following five rules summarize how to convert information about the linearized system into information about the nonlinear system. They form a hierarchy and are applied from top to bottom.

1. Equilibrium points that are unstable or asymptotically stable in the linearized system will be the same in the nonlinear system.

2. Saddles and spirals in the linearized system remain the same in the nonlinear system.

3. Proper nodes (λ's real and distinct) in the linearized system remain the same in the nonlinear system.

4. Improper nodes (λ's real and equal) in the linearized system can be nodes or spirals in the nonlinear system.

   Such nodes are called *improper* because of the slight ambiguity in passing from the linearized to the nonlinear system. However, instability and asymptotic stability will persist by Rule 1. Thus, an improper outward node will be either an outward node or an outward spiral. An improper inward node will be either an inward node or an inward spiral. So even though there is a slight ambiguity, it is still possible to determine whether the nonlinear system is unstable or asymptotically stable.

5. Centers in the linearized system can be centers or spirals in the nonlinear system.

   This is the difficult case. Since centers are not asymptotically stable to begin with, the center that becomes a spiral can become an inward or an outward spiral. Centers are neither unstable nor asymptotically stable, so Rule 1 does not apply. Thus, the very nature of the behavior of the nonlinear system is in complete doubt if the linearized system has a center. No decision can be made on the basis of the linearization, and far more powerful tools are then necessary to determine the behavior of the nonlinear system at such an equilibrium point.

It is often exceedingly difficult to distinguish between a spiral and a stable center in a nonlinear system. You can never be sure if the spiral behavior is an artifact of numerical (round-off) error in your numerical scheme. Analyzing the stability of such a point requires a great deal more mathematical skill than we have time to develop here.

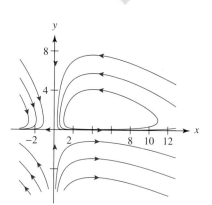

FIGURE 12.28 Phase portrait for Example 12.35

EXAMPLE 12.35 To analyze the stability of the nonlinear autonomous system given on the left in (12.88), obtain the equilibrium points $(0, 0)$ and $(4, 1)$ and the Jacobian matrix A given on the right

$$\begin{aligned} x' &= F(x, y) = 5x(1 - y) \\ y' &= G(x, y) = y(3x - 12) \end{aligned} \Rightarrow A = \begin{bmatrix} 5 - 5y & -5x \\ 3y & 3x - 12 \end{bmatrix} \quad (12.88)$$

Evaluate A at each equilibrium point, and compute the corresponding eigenvalues, obtaining the matrices

$$\begin{bmatrix} 5 & 0 \\ 0 & -12 \end{bmatrix} \text{ and } \begin{bmatrix} 0 & -20 \\ 3 & 0 \end{bmatrix} \quad (12.89)$$

and the respective eigenvalues $\{5, -12\}$, and $\pm 2i\sqrt{15}$. The point $(0, 0)$ is a saddle, and unstable, for the linearized system; therefore, it is a saddle, and unstable, for the nonlinear system. The point $(4, 1)$ is a center for the linearized system; therefore we cannot determine the behavior for the nonlinear system. The best we can do is try to infer the behavior from Figure 12.28, a phase portrait for the system. The phase portrait suggests that $(4, 1)$ may well be a center in the nonlinear case. ❖

EXERCISES 12.22–Part A

A1. Show that the origin is a center for the linearized version of the system $x' = y + 2x\sqrt{x^2 + y^2}$, $y' = -x + 2y\sqrt{x^2 + y^2}$ and is an unstable spiral for the nonlinear equations. *Hint:* The nonlinear equations can actually be solved by changing to polar coordinates. Make the change of variables $x(t) = r(t)\cos\theta(t)$, $y(t) = r(t)\sin\theta(t)$. Eliminate θ' to obtain an equation that can be solved for $r(t)$, then eliminate r' to obtain an equation that can be solved for $\theta(t)$.

A2. Show that the origin is a center for the linearized version of the system $x' = y - 2x\sqrt{x^2 + y^2}$, $y' = -x - 2y\sqrt{x^2 + y^2}$ and is an asymptotically stable spiral for the nonlinear equations. (See the Hint in Exercise A1.)

A3. Show that the origin is a center for the linearized, and nonlinearized, versions of the system $x' = y + y\sqrt{x^2 + y^2}$, $y' = -x - x\sqrt{x^2 + y^2}$. (See the Hint in Exercise A1.)

A4. For the competing-species model in Example 12.34, verify the Jacobian matrix given in (12.86).

A5. Obtain the four matrices in (12.87), and verify their eigenvalues have been correctly given in the text.

A6. Obtain the Jacobian matrix given in (12.88), verify that at the equilibrium points it becomes the matrices in (12.89), and then verify the eigenvalues have been correctly stated in the text.

EXERCISES 12.22–Part B

B1. The nonlinear system $x' = y + x(x^2 + y^2)$, $y' = -x + y(x^2 + y^2)$ is an example where the linearized system has a center that is not a center for the nonlinear system.

(a) Show that the origin is the only equilibrium point.

(b) At the origin, linearize and state the matrix A for the linearization.

(c) Obtain the eigenvalues for the matrix in part (b).

(d) On the basis of the eigenvalues of the linearization, categorize the equilibrium point for the linear system.

(e) Obtain a phase portrait for the nonlinear system.

(f) Use the phase portrait to determine the nature of the equilibrium point.

(g) Show that the change of variables $x(t) = r(t)\cos\theta(t)$, $y(t) = r(t)\sin\theta(t)$ converts the differential equations to $r'(t) = r^3$, $\theta'(t) = -1$. (*Hint:* Either compute $x' = \cos\theta - r\sin\theta\theta'$, etc., and combine the equations, or start with $r^2 = x^2 + y^2$ and obtain $rr' = xx' + yy'$, etc.)

(h) Solve the equations in part (g), and show that for any initial conditions, $r(t)$ becomes unbounded in finite time. Hence, the origin is an unstable equilibrium point.

B2–21. For the nonlinear system given in each of Exercises B2–21 in Section 12.21, linearize at each equilibrium point and determine the behavior of the nonlinear system in the neighborhood of the equilibrium point. If the equilibrium point is a center for the linearized system, use a phase portrait to predict the behavior of the nonlinear case.

12.23 The Nonlinear Pendulum

Pendulum Equations

SECOND-ORDER ODE   The differential equation for an undamped pendulum is

$$mL\theta''(t) + mg \sin \theta = 0 \tag{12.90}$$

where m is the mass of the bob, L is the length of the pendulum, g is the gravitational constant, and θ is the angle the pendulum makes with the vertical. (See Figure 12.29.) The equation is a nonlinear second-order differential equation for which the solution cannot be given in terms of elementary functions, so interpretation of the motion of the pendulum must come from numerical work and phase-plane analyses.

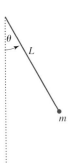

FIGURE 12.29   The plane pendulum

FIRST-ORDER SYSTEM   In anticipation of a phase-plane analysis, we rewrite the pendulum equation as a first-order system, a revision that does not remove the nonlinearity. The first-order system will also be nonlinear. As usual, define $x(t) = \theta(t)$ and $y(t) = \theta'(t)$, and then differentiate these definitions to obtain

$$x' = \theta' = y \quad \text{and} \quad y' = \theta'' = -\frac{mg \sin \theta}{mL} = -\frac{g}{L} \sin x$$

For simplicity, set $L = g$ so that the equations become

$$x'(t) = y \quad \text{and} \quad y'(t) = -\sin x$$

Equilibrium Points

The equilibrium points are clearly $(n\pi, 0)$, with $n = \ldots, -2, -1, 0, 1, 2, \ldots$. The nature of most of these equilibrium points can be determined by linearization. The Jacobian matrix, evaluated at the equilibrium points, is

$$\begin{bmatrix} 0 & 1 \\ -\cos x & 0 \end{bmatrix}_{x=n\pi} = \begin{bmatrix} 0 & 1 \\ -(-1)^n & 0 \end{bmatrix}$$

and has eigenvalues $\pm i$ when n is even and ± 1 when n is odd. At the even-integer multiples of π, the eigenvalues are pure imaginary, so the corresponding equilibrium points for the linearized system are centers. Unfortunately, this tells us nothing about the behavior of these points in the nonlinear case. At odd-integer multiples of π, the eigenvalues are real, but of opposite signs, so the corresponding equilibrium points for the linearized system are saddles. Hence, the corresponding points are also saddles in the nonlinear case. A phase portrait will be used to interpret the nonlinear system at the indeterminate equilibria.

Phase Portrait

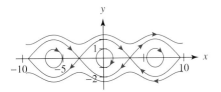

FIGURE 12.30   Phase portrait for the nonlinear plane pendulum

In the phase portrait shown in Figure 12.30, the points $(-2\pi, 0)$, $(0, 0)$, and $(2\pi, 0)$ appear to be centers in the nonlinear case also. The solution starting at $(0, 1)$ has initial displacement $\theta = 0$, $\theta' = 1$, so a pendulum hanging down at rest is given a positive initial (angular) velocity, causing the bob to swing counterclockwise. Moving counterclockwise, the bob rises, slowing down until it stops at its maximum displacement. The phase point on the closed orbit crosses the x-axis. As the bob begins swinging back down, its velocity becomes negative and its displacement is decreasing back to zero. When the bob again is vertical, it has negative velocity that will carry it in the clockwise direction. The return to vertical corresponds to the orbit crossing the y-axis at $y = -1$. The bob swings in the clockwise direction until it attains its maximal height. The system point on the orbit crosses the x-axis to the left of the origin. As the bob descends to the vertical, it moves counterclockwise. The system point on the orbit moves clockwise toward its initial point at $(0, 1)$.

The solution starting at $(0, 2.5)$ has initial displacement $\theta = 0, \theta' = 2.5$, so a pendulum hanging down at rest is given a positive initial (angular) velocity, causing the bob to swing counterclockwise. The initial velocity is large enough to cause the bob to "go over the top," rotating it in a complete circle.

The orbit starting at $(0, 2)$ in the phase plane ends at the saddle point $(\pi, 0)$. It takes an infinite amount of time to arrive at the saddle point. Correspondingly, the physical motion sees the bob move counterclockwise from the vertical, with just enough energy to coast to a stop $(y = 0)$ in the inverted position above the pivot $(x = \pi)$.

EXERCISES 12.23

1. Linearize the equations for the nonlinear pendulum by replacing $\sin\theta$ with θ. Obtain and interpret the phase portrait for the linearized equations.

2. Add linear damping to the equation for the nonlinear pendulum, obtaining the differential equation $mL\theta'' + b\theta' + mg\sin\theta = 0$. Obtain and interpret the phase portrait for this differential equation.

3. Derive (12.90), the equation for the nonlinear plane pendulum. *Hint:* Put the origin of an xy-coordinate system at the fulcrum of the pendulum so that the bob is located at $x = L\sin\theta, y = -L\cos\theta$. Apply Newton's Second Law in the form $\mathbf{F} = m\mathbf{a}$, where $\mathbf{F} = -mg\mathbf{j}$ and $\mathbf{a} = x''\mathbf{i} + y''\mathbf{j}$. Eliminate $(\theta')^2$ from the resulting equations.

Chapter Review

1. Formulate a two-tank mixing problem in which brine enters the first tank, is recirculated through the second tank, and leaves the second tank through a drain. Decide on flow rates, concentrations, recirculation rates, and drainage rates. Be sure that the fluid levels in each tank remain constant, and select meaningful initial conditions.

2. Write an IVP that describes the salt content of each tank in the system devised in Question 1.

3. Use the Laplace transform to obtain a solution to the IVP in Question 2. As a function of time, plot the salt content in each tank. Obtain a graph of the system's trajectory in phase space. Label the initial point on the trajectory, and indicate the direction of increasing time along the path.

4. Write the differential equations in Question 2 in matrix form. From the Laplace transform solution in Question 3, deduce the eigenvalues of the system matrix. Use a technological tool to obtain the eigenvalues of the system matrix, and compare.

5. Use a technological tool to obtain the eigenvectors of the system matrix. From the eigenvalues and eigenvectors, obtain the general solution of the first-order system of differential equations in Question 2, then apply the initial data, and reconstruct the solution found by the Laplace transform.

6. Solve $2x + 3y = 5, 4x - 7y = 1$ by Cramer's rule.

7. Obtain the general solution to the homogeneous equations
$x - 5y - z = 0, x + 3y - 3z = 0, -4x + 4y + 8z = 0$. Write the solution as the sum of homogeneous and particular solutions.

8. If $A = \begin{bmatrix} 2 & 6 \\ 6 & 3 \end{bmatrix}$, $B = \begin{bmatrix} 5 & 8 \\ 5 & 1 \end{bmatrix}$, and $\mathbf{x} = \begin{bmatrix} 7 \\ 9 \end{bmatrix}$, compute $(2A - 3B)^2\mathbf{x}$.

9. If $A = \begin{bmatrix} 4 & 9 \\ 1 & 6 \end{bmatrix}$, apply Gaussian elimination to the augmented matrix $[A, I]$ to obtain A^{-1}, its inverse. Verify that the inverse has indeed been found.

10. Given the IVP $x' = 7x + 4y, y' = -18x - 10y, x(0) = 2, y(0) = -3$:

   (a) Obtain the solution by Laplace transform.

   (b) Write the system as $\mathbf{x}' = A\mathbf{x}$, and find the eigenvalues and eigenvectors of the matrix A.

   (c) From the eigenvalues and eigenvectors, construct the general solution.

   (d) Apply the initial data to the general solution and thereby reconstruct the solution found in part (a).

   (e) Using the Laplace transform, find two solutions of the differential equations, one solution satisfying the initial data $x(0) = 1, y(0) = 0$, and the other solution satisfying $x(0) = 0, y(0) = 1$.

(f) From the solutions in part (e), obtain the fundamental matrix, and use it to solve the original IVP.

(g) Find a matrix P and a diagonal matrix D for which $A = PDP^{-1}$, define the new variable $\mathbf{r} = P^{-1}\mathbf{x}$, and thereby uncouple the original differential equations.

(h) Solve the uncoupled system found in part (g), and from this solution obtain the general solution of the original system.

(i) By differentiation and elimination, write the system as a single second-order differential equation.

(j) Obtain the solution of the equation in part (i), and from it, recreate the solution found in part (a).

(k) Obtain a phase portrait for the system $\mathbf{x}' = A\mathbf{x}$.

(l) Show the origin is the only equilibrium point, and categorize it as a node, saddle, center, or spiral.

(m) Make a definitive statement about the stability of this equilibrium point.

11. Given the IVP $x' = -x - 5y$, $y' = x + 3y$, $x(0) = -1$, $y(0) = 1$:

(a) Obtain the solution by Laplace transform.

(b) Write the system as $\mathbf{x}' = A\mathbf{x}$, and find the eigenvalues and eigenvectors of the matrix A.

(c) From one complex eigenpair, obtain two linearly independent solutions, and, from these two solutions, construct the general solution.

(d) Apply the initial data to the general solution and thereby reconstruct the solution found in part (a).

(e) Using the Laplace transform, find two solutions of the differential equations, one solution satisfying the initial data $x(0) = 1$, $y(0) = 0$, and the other solution satisfying $x(0) = 0$, $y(0) = 1$.

(f) From the solutions in part (e), obtain the fundamental matrix, and use it to solve the original IVP.

(g) Find a matrix P and a diagonal matrix D for which $A = PDP^{-1}$, define the new variable $\mathbf{r} = P^{-1}\mathbf{x}$, and thereby uncouple the original differential equations.

(h) Solve the uncoupled system found in part (g), and from this solution obtain the general solution of the original system.

(i) By differentiation and elimination, write the system as a single second-order differential equation.

(j) Obtain the solution of the equation in part (i), and from it, recreate the solution found in part (a).

(k) Obtain a phase portrait for the system $\mathbf{x}' = A\mathbf{x}$.

(l) Show the origin is the only equilibrium point, and categorize it as a node, saddle, center, or spiral.

(m) Make a definitive statement about the stability of this equilibrium point.

12. Given the IVP $x' = -6x + y$, $y' = -16x + 2y$, $x(0) = 3$, $y(0) = -2$:

(a) Obtain the solution by Laplace transform.

(b) Write the system as $\mathbf{x}' = A\mathbf{x}$, and find the eigenvalues and the single eigenvector of the matrix A.

(c) From the results in part (b), obtain one solution of the system of differential equations.

(d) Obtain a second linearly independent solution.

(e) From the solutions in parts (c) and (d), construct the general solution.

(f) Apply the initial data to the general solution and thereby reconstruct the solution found in part (a).

(g) Using the Laplace transform, find two solutions of the differential equations, one solution satisfying the initial data $x(0) = 1$, $y(0) = 0$, and the other solution satisfying $x(0) = 0$, $y(0) = 1$.

(h) From the solutions in part (g), obtain the fundamental matrix, and use it to solve the original IVP.

(i) By differentiation and elimination, write the system as a single second-order differential equation.

(j) Obtain the solution of the equation in part (i), and from it, recreate the solution found in part (a).

(k) Obtain a phase portrait for the system $\mathbf{x}' = A\mathbf{x}$.

(l) Show the origin is the only equilibrium point, and categorize it as a node, saddle, center, or spiral.

(m) Make a definitive statement about the stability of this equilibrium point.

13. Write $3y'' + 5y' + 7y = e^{-t}$, $y(0) = 1$, $y'(0) = 2$, as a first-order system of differential equations. Be sure to include the initial conditions.

14. Find and classify all equilibrium points of the system $x' = y(1 + x + y)$, $y' = (4 - 2x - 3y)$.

Chapter 13

Numerical Techniques: First-Order Systems and Second-Order ODEs

INTRODUCTION   The second-order differential equation can be solved numerically by specialized techniques such as the *Runge–Kutta–Nystrom* method. Alternatively, they can be solved by converting them to first-order systems and applying a vectorized version of one of the methods of Chapter 4. The vector version of the fourth-order Runge–Kutta technique is demonstrated.

13.1   Runge–Kutta–Nystrom

◆ EXAMPLE 13.1   **A second-order initial value problem**   In (13.1), the second-order initial value problem on the left has the exact solution shown on the right, the graph of which is seen as the solid curve in Figure 13.1. The dotted line in the figure is the horizontal asymptote $y_\infty = \lim_{t\to\infty} y(t) = \sqrt{2}$.

$$y''(t) + y'(t)y(t) = 0$$
$$y(0) = 0,\; y'(0) = 1 \quad \Rightarrow \quad y(t) = \sqrt{2}\tanh\frac{t}{\sqrt{2}} \qquad (13.1)$$

❖

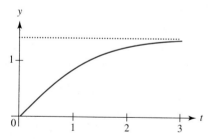

FIGURE 13.1   Example 13.1: Exact solution (solid) and asymptote $y = \sqrt{2}$ (dotted)

The rkn4 Algorithm

In 1925 the Finnish mathematician E. J. Nystrom proposed, for second-order equations, a variant of the fourth-order Runge–Kutta method. For an initial value problem whose differential equation is $y''(t) = f(t, y(t), y'(t))$, the following algorithm, the fourth-order Runge–Kutta–Nystrom (rkn4) method, yields a numeric solution

$$F_1 = hf(t_k, y_k, y'_k) \qquad\qquad F_a = h\left(y'_k + \tfrac{1}{2}F_1\right)$$

$$F_2 = hf\left(t_k + \tfrac{1}{2}h,\, y_k + \tfrac{1}{2}F_a,\, y'_k + \tfrac{1}{2}F_1\right) \qquad F_3 = hf\left(t_k + \tfrac{1}{2}h,\, y_k + \tfrac{1}{2}F_a,\, y'_k + \tfrac{1}{2}F_2\right)$$

$$F_b = h\left(y'_k + \tfrac{1}{2}F_3\right) \qquad\qquad F_4 = hf(t_{k+1},\, y_k + F_b,\, y'_k + F_3)$$

$$y_{k+1} = y_k + h\left(y'_k + \tfrac{1}{6}(F_1 + F_2 + F_3)\right) \qquad y'_{k+1} = y'_k + \tfrac{1}{6}(F_1 + 2F_2 + 2F_3 + F_4)$$

If the IVP in (13.1) is solved numerically with this algorithm, using a step size of $h = 0.1$, the solution at $t = 3$ is 1.374145628, as opposed to the exact solution of 1.374145962. The error of 3.34×10^{-7} is too small for a graph to illustrate.

EXERCISES 13.1

1. Using rkn4 implemented in a programming language of your choice, compute and plot the solution to each of the following initial value problems. Then, compare your solution with that generated by your favorite numeric differential equation solver. (Use $h = 0.1$ and integrate forward over an interval of length 3.)

(a) $t^2 y'' + y' \sin t - y \cos t = e^{\sqrt{t}}$, $y(1) = \pi$, $y'(1) = e$

(b) $yy'' - t(y' - \sqrt{y})^2 - \text{arcsinh}(ty) = 0$, $y(1) = 1$, $y'(1) = -2$

(c) $y'' + yy' + \sin y = e^{-y'}$, $y(0) = -1$, $y'(0) = 3$

(d) $y'y'' + \arctan y = e^t$, $y(1) = \sqrt{2}$, $y'(1) = \sqrt{3}$

(e) $y'' + y^2 y' - \dfrac{3}{1 + y^2} = \cosh y$, $y(0) = 2$, $y'(0) = -5$

In each of Exercises 2–6, use the given second-order IVP to test empirically whether rkn4 is a fourth-order method. As in Section 4.1 where this was done for rk4, integrate numerically from $t = 0$ to $t = t_f$, where the error in the computed solution is determined by comparing to the exact solution. Thus, in each case, carry out the following steps.

(a) Obtain y_{exact}, the exact solution of the initial value problem.

(b) Using rkn4, and step sizes $h_k = 1/2^k$, $k = 0, 1, 2, 3, 4$, obtain $y_{\text{approx}}(h_k)$, numeric approximations of $y_{\text{exact}}(t_f)$.

(c) Examine the constancy of
$$\frac{|y_{\text{exact}}(t_f) - y_{\text{approx}}(h_k)|}{h_k^4} \quad k = 0, \ldots, 4$$

2. $y'' + 3y' + 7y = t \cos 3t$, $y(0) = 2$, $y'(0) = -3$; $t_f = 5$

3. $4y'' + 5y' + 13y = e^t \cos 2t$, $y(0) = 1$, $y'(0) = -1$; $t_f = 5$

4. $y'' + \sqrt{2}y' + 10y = 5 \cos 3t$, $y(0) = 0$, $y'(0) = 0$; $t_f = 10$

5. $3y'' + 2y' + 8y = te^{-t} \cos 5t$, $y(0) = -2$, $y'(0) = 1$; $t_f = 5$

6. $y'' + 2y' + 18y = 3 \cos 4t$, $y(0) = 5$, $y'(0) = -2$; $t_f = 5$

13.2 rk4 for First-Order Systems

Conversion to First-Order System

A standard technique for obtaining a numeric solution of an initial value problem with a second-order differential equation is the conversion of the equation to a first-order system. The system is then solved by applying a method like the fourth-order Runge–Kutta algorithm adapted for such a system.

EXAMPLE 13.2 Using the technique of Section 12.13, the change of variables $u(t) = y(t)$, $v(t) = y'(t)$ converts the second-order IVP on the left in (13.2) to the first-order system on the right

$$
\begin{array}{lll}
y'' + y'y = 0 & & u'(t) = y' = v \\
y(0) = 0 & \text{becomes} & v'(t) = y'' = -y'y = -vu \qquad (13.2) \\
y'(0) = 1 & & u(0) = 0, v(0) = 1
\end{array}
$$

❖

Fourth-Order Runge–Kutta for Systems

The fourth-order Runge–Kutta algorithm of Section 4.4 generalizes to systems if each of the functions is interpreted as a vector. Specifically, for the initial value problem

$$ u'(t) = f(t, u(t), v(t)) \qquad v'(t) = g(t, u(t), v(t)) \qquad u(0) = u_0 \qquad v(0) = v_0 $$

the algorithm would be given by the formulae

$$ F_1 = hf(t_k, u_k, v_k) \qquad\qquad G_1 = hg(t_k, u_k, v_k) $$
$$ F_2 = hf\left(t_k + \tfrac{1}{2}h, u_k + \tfrac{1}{2}F_1, v_k + \tfrac{1}{2}G_1\right) \qquad G_2 = hg\left(t_k + \tfrac{1}{2}h, u_k + \tfrac{1}{2}F_1, v_k + \tfrac{1}{2}G_1\right) $$
$$ F_3 = hf\left(t_k + \tfrac{1}{2}h, u_k + \tfrac{1}{2}F_2, v_k + \tfrac{1}{2}G_2\right) \qquad G_3 = hg\left(t_k + \tfrac{1}{2}h, u_k + \tfrac{1}{2}F_2, v_k + \tfrac{1}{2}G_2\right) $$
$$ F_4 = hf(t_{k+1}, u_k + F_3, v_k + G_3) \qquad\qquad G_4 = hg(t_{k+1}, u_k + F_3, v_k + G_3) $$
$$ u_{k+1} = u_k + \tfrac{1}{6}(F_1 + 2F_2 + 2F_3 + F_4) \qquad v_{k+1} = v_k + \tfrac{1}{6}(G_1 + 2G_2 + 2G_3 + G_4) $$

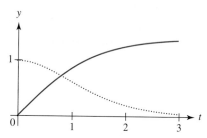

FIGURE 13.2   Example 13.2: $y(t) = u(t)$
(solid) and $y'(t) = v(t)$ (dotted)

Applied to the IVP in (13.2), the algorithm yields the graphs of $u(t)$ and $v(t)$ shown in Figure 13.2. The graph of $u(t)$ (solid) is recognized as the graph of $y(t)$ from Figure 13.1, while the graph of $v(t)$ (dotted) is the plot of $y'(t)$.

EXERCISES 13.2

1. Convert each initial value problem in Exercise 1, Section 13.1, into the form of a first-order system and obtain a numeric solution with a system-implementation of the fourth-order Runge–Kutta algorithm. Compare this numeric solution with a numeric solution generated by rkn4. (Use $h = 0.1$ and integrate forward over an interval of length 3.)

   In Exercises 2–6, convert the given IVP into a first-order system and obtain a numeric solution with a system-implementation of rk4. Compare

   this numeric solution with the numeric solution generated by rkn4 and with the exact solution. (Use $h = 0.1$ and integrate forward over an interval of length 3.)

2. Exercise 2, Section 13.1   3. Exercise 3, Section 13.1
4. Exercise 4, Section 13.1   5. Exercise 5, Section 13.1
6. Exercise 6, Section 13.1

Chapter Review

1. Write Euler's method of Section 4.2 for the IVP $x' = f(t, x, y)$, $y' = g(t, x, y)$, $x(0) = x_0$, $y(0) = y_0$.

2. Apply the results of Question 1 to the solution of $y'' + 3y' + 2y = e^{-t}$, $y(0) = 2$, $y'(0) = -3$. Obtain the exact solution of the IVP, and compare graphically with the numeric solution.

3. Solve the IVP of Question 2 by the Runge–Kutta–Nystrom algorithm, and graphically compare the solution with that generated by Euler's method.

Chapter 14

Series Solutions

INTRODUCTION   Second-order linear differential equations can be solved exactly if they have constant coefficients. Some equations with variable coefficients, like the Euler equation, can also be solved exactly. But in general, the solution of second-order ODEs with variable coefficients can only be expressed as power series. Hence, we consider techniques for obtaining power series solutions of ODEs directly from the differential equations themselves.

Points about which solutions have power series expansions are called *regular* points. All other points are then *singular* points. At *regular singular* points the linear ODE will have either two generalized power series solutions or a generalized power series solution and a series solution containing a logarithm. Techniques for obtaining these kinds of solutions are examined.

If the coefficients of the ODE are not analytic, its solution may be expressible only as an *asymptotic* series. This case is also examined in the chapter.

Nonlinear second-order ODEs containing a parameter and describing mechanical and electrical systems can sometimes be solved with *perturbation* series, which are a special kind of asymptotic expansion. The *Poincaré* perturbation scheme is described. However, the Poincaré series often fails because it can generate *secular terms* that become unbounded in finite time. Hence, we also discuss *Lindstedt's method* for generating perturbation solutions that have no such secular terms. When applied to the equation of a damped spring-mass system in which the spring is nonlinear, this method shows that the period of oscillation now depends on the initial amplitude, quite the contrary to the behavior of the equivalent system with a linear spring. Although Lindstedt's method can determine the dependence of frequency on initial amplitude and find periodic solutions of some nonlinear equations, it does not provide approximations valid away from the periodic solution itself. These solutions are better found by the method of *Krylov and Bogoliubov* appearing at the end of the chapter.

14.1 Power Series

Series Solutions at Regular Points

REGULAR POINT   The point t_0 is a *regular point* of the differential equation

$$a(t)y''(t) + b(t)y'(t) + c(t)y(t) = 0 \tag{14.1}$$

if a solution $y(t)$ can be obtained as a power series $\sum_{k=0}^{\infty} a_k (t - t_0)^k$, where each $a_k = y^k(t_0)/k!$ is computed directly from the differential equation itself.

343

If $a(t)$, $b(t)$, and $c(t)$ are analytic at $t = t_0$ and if $a(t_0) \neq 0$, then t_0 is a regular point for the differential equation, such a power series solution exists for $y(t)$, and its coefficients a_k are determined from (14.1) as follows. Write (14.1) as

$$ y''(t) = -\frac{b(t)}{a(t)} y'(t) - \frac{c(t)}{a(t)} y(t) = -p(t) y'(t) - q(t) y(t) $$

from which $y''(t_0)$ follows, provided initial values for $y'(t_0)$ and $y(t_0)$ are given and $a(t_0) \neq 0$. Additionally, higher derivatives such as $y'''(t_0)$ can be computed from $y''(t)$, since by our assumptions, the functions $p(t)$ and $q(t)$ possess the requisite derivatives.

EXAMPLE 14.1   In (14.2), the IVP on the left has the exact solution shown on the right. Also shown on the right are the first few terms of a Taylor series expansion of the exact solution

$$
\begin{aligned}
y'' + 4y' + 13y &= e^{-t} \\
y(0) = y'(0) &= 0
\end{aligned}
\quad \Rightarrow \quad
\begin{aligned}
y(t) &= \tfrac{1}{10}\left(e^{-t} - e^{-2t}\left(\cos 3t - \tfrac{1}{3}\sin 3t\right)\right) \\
&= \tfrac{1}{2}t^2 - \tfrac{5}{6}t^3 + \tfrac{1}{3}t^4 + \tfrac{4}{15}t^5 + \cdots
\end{aligned}
\tag{14.2}
$$

Table 14.1 shows how the differential equation itself is used to compute $y^{(k)}(0)$, and hence, $a_k = y^{(k)}(0)/k!$.

k	$y^{(k)}(t)$	$y^{(k)}(0)$	$\dfrac{y^{(k)}(0)}{k!} = a_k$
2	$y'' = e^{-t} - 4y' - 13y$	$1 - 4(0) - 13(0) = 1$	$\frac{1}{2!} = \frac{1}{2}$
3	$y''' = -e^{-t} - 4y'' - 13y'$	$-1 - 4(1) - 13(0) = -5$	$-\frac{5}{3!} = -\frac{5}{6}$
4	$y^{(4)} = e^{-t} - 4y''' - 13y''$	$1 - 4(-5) - 13(1) = 8$	$\frac{8}{4!} = \frac{1}{3}$
5	$y^{(5)} = -e^{-t} - 4y^{(4)} - 13y'''$	$-1 - 4(8) - 13(-5) = 34$	$\frac{34}{5!} = \frac{4}{15}$

TABLE 14.1   Using the IVP $y'' + 4y' + 13y = e^{-t}$, $y(0) = y'(0) = 0$ to generate the Taylor series for $y(t)$

A more practical technique for obtaining the series solution in (14.2) begins with the polynomial $\sum_{k=2}^{4} a_k t^k$, which already satisfies the initial conditions $y(0) = y'(0) = 0$. Substitution of this polynomial into the differential equation then gives

$$ 13a_4 t^4 + (16a_4 + 13a_3)t^3 + (12a_4 + 12a_3 + 13a_2)t^2 + (6a_3 + 8a_2)t + 2a_2 = \sum_{k=0}^{4} \frac{(-t)^k}{k!} \tag{14.3} $$

where, on the right, e^{-t} has been replaced by a Maclaurin polynomial. Matching coefficients for t^k, $k = 0, 1, 2$, on either side of (14.3) then gives, in (14.4), the equations on the left and the coefficients a_k, $k = 2, 3, 4$, on the right

$$
\begin{aligned}
2a_2 &= 1 & a_2 &= \tfrac{1}{2} \\
6a_3 + 8a_2 &= -1 \quad \Rightarrow \quad & a_3 &= -\tfrac{5}{6} \\
12a_4 + 12a_3 + 13a_2 &= \tfrac{1}{2} & a_4 &= \tfrac{1}{3}
\end{aligned}
\tag{14.4}
$$

The behavior of the approximate solution can be seen in Figure 14.1 where, as the dotted curve, it is compared to the exact solution, the solid curve. The approximate solution is accurate to about $t = 0.4$, whence it radically departs the oscillatory exact solution. Any order Taylor approximation will ultimately diverge from the exact solution since each Taylor approximation is but a polynomial that ultimately must become unbounded. ❖

Taylor Series Solution

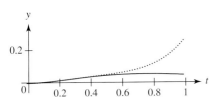

FIGURE 14.1 Example 14.1: Exact solution (solid) and Taylor approximation (dotted)

For the Taylor series solution to remain faithful to $y(t)$ for all time $t > 0$, the series would have to be an infinite sum, not just a polynomial (partial sum). The resulting manipulations with the formal sum $y(t) = \sum_{k=2}^{\infty} a_k t^k$ involve substitution, leading to

$$\sum_{k=2}^{\infty} k(k-1)a_k t^{k-2} + 4\sum_{k=2}^{\infty} k a_k t^{k-1} + 13\sum_{k=2}^{\infty} a_k t^k = \sum_{k=0}^{\infty} \frac{(-1)^k t^k}{k!}$$

where, on the right side, we wrote the Taylor series for e^{-t}. Next, bring all terms to the left side, writing all sums with like exponents. Changing the summation index on each of the first two sums on the left gives

$$\sum_{n=0}^{\infty}(n+2)(n+1)a_{n+2}t^n + 4\sum_{n=1}^{\infty}(n+1)a_{n+1}t^n + 13\sum_{n=2}^{\infty} a_n t^n - \sum_{n=0}^{\infty}\frac{(-1)^n t^n}{n!} = 0$$

Since the index in the third sum starts at $n = 2$, the other three sums must be made to start at that same index by extracting all terms prior to the $n = 2$ term. Thus, we have

$$(2a_2 - 1) + (6a_3 + 8a_2 + 1)t$$

$$+ \sum_{n=2}^{\infty}\left((n+2)(n+1)a_{n+2} + 4(n+1)a_{n+1} + 13a_n - \frac{(-1)^n}{n!}\right)t^n = 0$$

The coefficients of each power of t must vanish, giving us the equations $2a_2 - 1 = 0$, $6a_3 + 8a_2 + 1 = 0$, and

$$(2+n)(1+n)a_{2+n} + 4(1+n)a_{1+n} + 13a_n - \frac{(-1)^n}{n!} = 0$$

The first two equations determine a_2 and a_3, yielding $a_2 = \frac{1}{2}$ and $a_3 = -\frac{5}{6}$, whereas the third equation actually represents an infinite set of equations, one for each value of the index $n = 2, 3, \ldots$. Solving this two-term *recursion* (or *recurrence*) *relation* for a_{n+2} in terms of a_{n+1} and a_n, we find

$$a_{n+2} = \frac{-4(n+1)n!a_{n+1} - 13a_n n! + (-1)^n}{n!(n+1)(n+2)} \tag{14.5}$$

We don't anticipate obtaining a closed-form solution for the general coefficient a_{n+2}, but we can compute as many coefficients in succession as we need. Recalling that the initial conditions determined $a_0 = a_1 = 0$ and that the differential equation determined $a_2 = \frac{1}{2}$ and $a_3 = -\frac{5}{6}$, we can compute the next few coefficients as $a_4 = \frac{1}{3}$, $a_5 = \frac{4}{15}$, $a_6 = -\frac{77}{240}$, $a_7 = \frac{169}{1680}$, etc., thus yielding the corresponding terms of the series solution as

$$y(t) = \frac{1}{2}t^2 - \frac{5}{6}t^3 + \frac{1}{3}t^4 + \frac{4}{15}t^5 - \frac{77}{240}t^6 + \frac{169}{1680}t^7 + \frac{61}{2520}t^8 + O(t^9) \tag{14.6}$$

ALTERNATE ALGORITHM The following technique from [29] is an interesting alternative for obtaining the recurrence relation governing the coefficients in the Taylor series solution of certain differential equations. We illustrate the technique for

$$y''(t) + t^2 y'(t) + y(t) = 0$$

a differential equation with polynomial coefficients. Instead of using a full infinite sum for $y(t)$, write a finite sum, containing general terms that span the nth term, as in $a_{n-1}t^{n-1} + a_n t^n + a_{n+1}t^{n+1} + a_{n+2}t^{n+2}$. Substitute into the differential equation, and extract $2a_{n+2} + 3a_{n+2}n + a_{n+2}n^2 + a_{n-1}n - a_{n-1} + a_n$, the coefficient of the general term t^n. Setting this expression equal to zero, and solving for a_{n+2}, we find the recurrence relation

$$a_{n+2} = \frac{(1-n)a_{n-1} - a_n}{(n+1)(n+2)} \tag{14.7}$$

Define $a_{-1} = 0$ and compute the first few coefficients determined by (14.7), obtaining $a_2 = -\frac{1}{2}a_0$, $a_3 = -\frac{1}{6}a_1$, $a_4 = -\frac{1}{12}a_1 + \frac{1}{24}a_0$, and $a_5 = \frac{1}{20}a_0 + \frac{1}{120}a_1$. Each coefficient is given in terms of a_0 and a_1, the two arbitrary constants in the general solution of the differential equation.

Series Solutions at Regular Singular Points

REGULAR AND IRREGULAR SINGULAR POINTS   For the differential equation

$$y''(t) + p(t)y'(t) + q(t)y(t) = 0 \tag{14.8}$$

points that are not regular are called *singular points*.

A point t_0 can fail to be regular if either $p(t)$ or $q(t)$ fails to have a power series at $t = t_0$. For example, $t = 0$ would be a singular point if $p(t) = e^t/t$ since $p(t)$ does not have a *Taylor* series expansion about $t = 0$. In fact, the series representation for this function is

$$p(t) = \frac{e^t}{t} = \sum_{k=0}^{\infty} \frac{t^{k-1}}{k!} = t^{-1} + 1 + \tfrac{1}{2}t + \tfrac{1}{6}t^2 + \tfrac{1}{24}t^3 + \mathrm{O}(t^4)$$

Along with a "Taylor series" part, there is an additional term, namely $\frac{1}{t}$, that expresses the singularity of $p(t)$ at $t = 0$. This series expansion is called a *Laurent series* and is valid for $0 < |t| < R$ for some positive R. The function $p(t)$ is therefore said to have a *pole of order one* at $t = 0$.

Similarly, the function $q(t) = e^t/t^2$ has a *pole of order two* at $t = 0$ because its Laurent series

$$q(t) = \frac{e^t}{t^2} = \sum_{k=0}^{\infty} \frac{t^{k-2}}{k!} = t^{-2} + t^{-1} + \tfrac{1}{2} + \tfrac{1}{6}t + \tfrac{1}{24}t^2 + \tfrac{1}{120}t^3 + \mathrm{O}(t^4)$$

contains the term $\frac{1}{t^2}$. Consequently, we segregate singular points into two types, *regular* and *irregular*. The regular singular point is a singular point for which the singularities in $p(t)$ and $q(t)$ are within certain bounds, and the irregular singular point is one for which these singularities exceed the prescribed bounds. In particular, t_0 is a *regular singular point* provided that it is not a regular point of the differential equation and that $p(t)$ has no worse than a pole of order one at t_0 while $q(t)$ has no worse than a pole of order two at t_0.

SOLUTIONS AT REGULAR SINGULAR POINTS   Notation for the following discussion is greatly simplified if we assume the regular singular point $t = t_0$ has been translated to the origin. Thus, let $t = 0$ be a regular singular point of the differential equation (14.8). At $t = 0$, the function $p(t)$ has, at worst, a pole of order one and the representation shown on the left in (14.9), whereas $q(t)$ has, at worst, a pole of order two and the representation shown on the right in (14.9)

$$p(t) = \frac{p_{-1}}{t} + \sum_{k=0}^{\infty} p_k t^k \qquad q(t) = \frac{q_{-2}}{t^2} + \frac{q_{-1}}{t} + \sum_{k=0}^{\infty} q_k t^k \qquad (14.9)$$

Let α and β be the roots of the *indicial equation*

$$r^2 + (p_{-1} - 1)r + q_{-2} = 0$$

ordered so that $\mathrm{Re}(\beta) \geq \mathrm{Re}(\alpha)$. Then the two linearly independent solutions of (14.8) will be

$$Y_1 = t^{\beta} \sum_{k=0}^{\infty} a_k t^k \quad \text{and} \quad Y_2 = t^{\alpha} \sum_{k=0}^{\infty} b_k t^k \quad \text{or} \quad Y_3 = Y_1 \ln t + t^{\alpha} \sum_{k=0}^{\infty} d_k t^k \qquad (14.10)$$

There is always a solution of the form Y_1. If $\alpha = \beta$, the second solution will be of the form Y_3. If $\beta - \alpha = s$, a positive integer, then the second solution may be of the form Y_2 or Y_3. At great length, texts such as [74] develop criteria for distinguishing the nonlogarithmic case from the logarithmic case. We eschew the entanglement and instead rely on several examples to illustrate the two cases.

EXAMPLE 14.2   The differential equation

$$2t^2 y'' + t(5t + 3)y' + (7t - 1)y = 0 \qquad (14.11)$$

surely has $t = 0$ as a singular point, since the coefficient of $y''(t)$ vanishes at $t = 0$. The nature of this singular point follows from the identification and analysis of the coefficients

$$p(t) = \frac{5}{2} + \frac{3}{2}\frac{1}{t} \quad \text{and} \quad q(t) = \frac{7}{2}\frac{1}{t} - \frac{1}{2}\frac{1}{t^2}$$

Thus, $p(t)$ has a pole of order one at $t = 0$ and $p_{-1} = \frac{3}{2}$; while $q(t)$ has a pole of order two and $q_{-2} = -\frac{1}{2}$. Hence, the indicial equation is $r^2 + \frac{1}{2}r - \frac{1}{2} = 0$, with roots $\alpha = -1$ and $\beta = \frac{1}{2}$. Since the difference $\beta - \alpha = \frac{3}{2}$ is not an integer, the general solution is $y(t) = c_1 y_1(t) + c_2 y_2(t)$, where

$$y_1 = \sqrt{t}\left(1 - \frac{19}{10}t + \frac{551}{280}t^2 - \frac{7163}{5040}t^3 + \mathrm{O}(t^4)\right)$$

$$y_2 = t^{-1}\left(1 + 2t - 7t^2 + \frac{28}{3}t^3 + \mathrm{O}(t^4)\right) \qquad \text{❖}$$

EXAMPLE 14.3   The differential equation

$$2t^2 y'' + 5t^2 y' + \left(7t + \tfrac{1}{2}\right)y = 0 \qquad (14.12)$$

surely has $t = 0$ as a singular point, since the coefficient of $y''(t)$ vanishes at $t = 0$. The nature of this singular point follows from the identification and analysis of the coefficients

$$p(t) = \frac{5}{2} \quad \text{and} \quad q(t) = \frac{7}{2}\frac{1}{t} + \frac{1}{4}\frac{1}{t^2}$$

Thus, $p(t)$ does not have a pole at $t = 0$, so $p_{-1} = 0$; while $q(t)$ has a pole of order two and $q_{-2} = \frac{1}{4}$. Hence, the indicial equation is $r^2 - r + \frac{1}{4} = 0$, with the repeated roots $\alpha = \beta = \frac{1}{2}$. The general solution is therefore $y(t) = c_1 y_1(t) + c_3 y_3(t)$, where

$$y_1 = \sqrt{t}\left(1 - \frac{19}{4}t + \frac{551}{64}t^2 - \frac{7163}{768}t^3 + \mathrm{O}(t^4)\right)$$

$$y_3 = y_1 \ln t + \sqrt{t}\left(7t - \frac{1173}{64}t^2 + \frac{54{,}563}{2304}t^3 + \mathrm{O}(t^4)\right) \qquad \text{❖}$$

EXAMPLE 14.4   The differential equation

$$4t^2 y'' + 2t^2 y' - (t+3)y = 0 \tag{14.13}$$

has $t = 0$ as a singular point, since the coefficient of $y''(t)$ vanishes at $t = 0$. The nature of this singular point follows from the coefficients

$$p(t) = \frac{1}{2} \quad \text{and} \quad q(t) = -\frac{1}{4}\frac{1}{t} - \frac{3}{4}\frac{1}{t^2}$$

Thus, $p(t)$ does not have a pole $t = 0$, so $p_{-1} = 0$; while $q(t)$ has a pole of order two, and $q_{-2} = -\frac{3}{4}$. Hence, the indicial equation is $r^2 - r - \frac{3}{4} = 0$, with roots $\alpha = -\frac{1}{2}$ and $\beta = \frac{3}{2}$, differing by an integer. The general solution of the DE is $y(t) = c_1 y_1(t) + c_2 y_2(t)$, where

$$y_1 = t^{3/2}\left(1 - \tfrac{1}{6}t + \tfrac{1}{48}t^2 - \tfrac{1}{480}t^3 + O(t^4)\right) \quad \text{and} \quad y_2 = t^{-1/2}\left(-2 + t - \tfrac{1}{4}t^2 + \tfrac{1}{24}t^3 + O(t^4)\right) \tag{14.14}$$

so, although $\beta - \alpha = 2$, the second solution does not contain a logarithm. Surprisingly,

$$-\frac{1}{2}y_2 = f_2 = \frac{e^{-t/2}}{\sqrt{t}} \quad \text{and} \quad y_2 + \frac{1}{4}y_1 = f_1 = \frac{t-2}{\sqrt{t}} \tag{14.15}$$

so $\{f_1(t), f_2(t)\}$ is an alternate fundamental set for (14.13). (See Exercise A1.)   ❖

EXAMPLE 14.5   The differential equation

$$4t^2 y''(t) + 5t^2 y'(t) + (7t - 3)y(t) = 0 \tag{14.16}$$

surely has $t = 0$ as a singular point, since the coefficient of $y''(t)$ vanishes at $t = 0$. The nature of this singular point follows from the identification and analysis of the coefficients

$$p(t) = \frac{5}{4} \quad \text{and} \quad q(t) = \frac{7}{4}\frac{1}{t} - \frac{3}{4}\frac{1}{t^2}$$

Thus, $p(t)$ does not have a pole at $t = 0$, so $p_{-1} = 0$; while $q(t)$ has a pole of order two, and $q_{-2} = -\frac{3}{4}$. Hence, the indicial equation is the same as it was in Example 14.4, namely, $r^2 - r - \frac{3}{4} = 0$, with roots $\alpha = -\frac{1}{2}$ and $\beta = \frac{3}{2}$, roots that again differ by an integer. The general solution is $y(t) = c_1 y_1(t) + c_3 y_3(t)$, where

$$y_1 = t^{3/2}\left(1 - \tfrac{29}{24}t + \tfrac{377}{512}t^2 - \tfrac{18,473}{61,440}t^3 + O(t^4)\right)$$

$$y_3 = y_1 \ln t + \tfrac{64}{171}t^{-1/2}\left(-2 - \tfrac{9}{4}t + \tfrac{35}{8}t^2 - \tfrac{1609}{768}t^3 + O(t^4)\right)$$

so y_1 is of the form $Y_1 = t^\beta \sum_{k=0}^\infty a_k t^k$ while y_3 is of the form $Y_3 = Y_1 \ln t + t^\alpha \sum_{k=0}^\infty d_k t^k$.   ❖

EXERCISES 14.1–Part A

A1. Using (14.14), verify (14.15).

A2. Verify that the coefficients in (14.6) are given by (14.5).

A3. Substitute $y(t) = t^r$ into (14.11) and show that the indicial equation of Example 14.2 results when the coefficient of t^r is forced to vanish.

A4. With β determined as in Example 14.2, substitute $y(t) = t^\beta(1 + a_1 t + a_2 t^2)$ into (14.11) and verify a_1, a_2 in y_1.

A5. Substitute $y(t) = t^r$ into (14.12) and show that the indicial equation of Example 14.3 results when the coefficient of t^r is forced to vanish.

A6. With β determined as in Example 14.3, substitute $y(t) = t^\beta(1 + a_1 t + a_2 t^2)$ into (14.12) and verify a_1, a_2 in y_1.

A7. Substitute $y(t) = t^r$ into (14.13) and show that the indicial equation of Example 14.4 results when the coefficient of t^r is forced to vanish.

A8. With β determined as in Example 14.4, substitute $y(t) = t^{\beta}(1 + a_1 t + a_2 t^2)$ into (14.13) and verify a_1, a_2 in y_1.

A9. Substitute $y(t) = t^r$ into (14.16) and show that the indicial equation of Example 14.5 results when the coefficient of t^r is forced to vanish.

A10. With β determined as in Example 14.5, substitute $y(t) = t^{\beta}(1 + a_1 t + a_2 t^2)$ into (14.16) and verify a_1, a_2 in y_1.

EXERCISES 14.1–Part B

B1. The differential equation $t^2 y'' + t y' - y = 0$ is a Euler's equation with solutions of the form t^r. Show that $r = \pm 1$. Then, show $t = 0$ is a regular singular point of the equation, and find the indicial equation and its roots. Is the technique for solving the Euler equation in Section 5.12 at odds with the results in (14.10)? Explain.

In Exercises B2–7, $t = 0$ is an ordinary point of the given differential equation.

(a) Use a computer algebra system to generate $\sum_{k=0}^{5} b_k t^k$, a Taylor polynomial approximating the solution of the IVP consisting of the given ODE and the data $y(0) = 1$, $y'(0) = -2$.

(b) Obtain the exact solution to the initial value problem, and, from this solution, obtain an approximating Taylor polynomial that should agree with the polynomial found in part (a).

(c) Write the differential equation as $y'' = f(t, y, y')$ and use this to determine $b_k, k = 2, \dots, 5$, as was done in Table 14.1.

(d) Write $y = 1 - 2t + \sum_{k=2}^{5} a_k t^k$ and use the initial value problem to determine the coefficients $a_k, k = 2, \dots, 5$, which should agree with $b_k, k = 2, \dots, 5$.

(e) Write $y = \sum_{k=0}^{\infty} a_k t^k$ or $y = \sum_{k=n-p_1}^{n+p_2} a_k t^k$ and use the differential equation to obtain a recurrence relation for the coefficients a_k.

(f) Use the recurrence relation found in part (e) to determine $a_k, k = 0, \dots, 5$.

(g) On the same set of axes, graph the exact solution and the polynomial approximation found in part (a).

B2. $y'' + 2y' + 10y = \cos 2t$ *Hint:* $\cos(2t) = \sum_{k=0}^{\infty} \dfrac{\cos\left(\frac{k\pi}{2}\right) 2^k t^k}{k!}$.

B3. $y'' + 4y' + 13y = e^{-t} \sin t$ **B4.** $y'' + 5y' + 6y = \cos 5t$

B5. $y'' + 2y' + 5y = e^{-3t}$

B6. $y'' + 7y' + 10y = \sin t$ *Hint:* $\sin t = -\sum_{k=0}^{\infty} \dfrac{\cos\left(\frac{2k+1}{2}\pi\right) t^k}{k!}$.

B7. $y'' + 6y' + 18y = e^t \cos 2t$

In Exercises B8–11, $t = 0$ is a regular point of the given differential equation.

(a) Use a computer algebra system to generate $\sum_{k=0}^{5} b_k t^k$, a Taylor polynomial approximating the solution of the IVP

consisting of the given ODE and the data $y(0) = 2$, $y'(0) = -1$.

(b) Write $y = \sum_{k=n-p_1}^{n+p_2} a_k t^k$ and use the DE to obtain a recurrence relation for the coefficients a_k.

(c) Use the recurrence relation found in part (b) to determine $a_k, k = 0, \dots, 5$.

B8. $y'' + 10y' + 41y = \dfrac{1}{1+t}$ **B9.** $y'' + 7y' + 12y = \dfrac{1}{1+t^2}$

B10. $y'' + 8y' + 17y = \sqrt{1+t}$ **B11.** $y'' + 6y' + 9y = \cosh t$

In Exercises B12–26, $t = 0$ is a regular point of the given differential equation.

(a) Use a computer algebra system to generate $\sum_{k=0}^{5} b_k t^k$, a Taylor polynomial approximating the solution of the initial value problem for which the data is $y(0) = 2$, $y'(0) = -1$.

(b) Write the differential equation as $y'' = f(t, y, y')$ and use this to determine $b_k, k = 2, \dots, 5$, as was done in Table 14.1.

(c) Write $y = 2 - t + \sum_{k=2}^{5} a_k t^k$ and use the initial value problem to determine the coefficients $a_k, k = 2, \dots, 5$, which should agree with $b_k, k = 2, \dots, 5$.

(d) Write $y = \sum_{k=0}^{\infty} a_k t^k$ and use the differential equation to obtain a recurrence relation for the coefficients a_k.

(e) Write $y = \sum_{k=n-p_1}^{n+p_2} a_k t^k$ and use the differential equation to obtain a recurrence relation for the coefficients a_k.

(f) Use the recurrence relation found in either part (d) or (e) to determine $a_k, k = 0, \dots, 5$.

B12. $(5t^2 - 9t + 7)y'' + (11t - 17)y' + (18t^2 + 6)y = 0$

B13. $(18t^2 - 3)y'' + (4t - 7t^2)y' + (2 + 9t - 4t^2)y = 0$

B14. $(7t^2 - 13)y'' + (9t^2 + 10)y' + (18t - 4t^2)y = 0$

B15. $(31t - 3)y'' + (7t + 10t^2 + 4)y' + (11t^2 - 6t)y = 0$

B16. $(9 - 9t^2)y'' + (3t - 7t^2 - 7)y' + (14 - t)y = 0$

B17. $(t + 2t^2 + 2)y'' + (9 + 2t^2)y' + (10t + 4)y = 0$

B18. $(13 - 7t + t^2)y'' + 5y' + (5t + 3t^2 + 2)y = 0$

B19. $(5t^2 - 6 + 10t)y'' + -9ty' + (8t^2 - 10 - 2t)y = 0$

B20. $(16t^2 + 1)y'' + (t^2 - 5t + 2)y' + (3t - 5)y = 0$

B21. $(14t - 2)y'' + (4t^2 + 9)y' + (1 - 5t + 3t^2)y = 0$

B22. $(5t^2 + 8 - 2t)y'' + (14 + 6t)y' - 27t^2 y = 0$

B23. $(t^2 - 4 - 4t)y'' + (10t + t^2 + 4)y' + (1 + 8t^2)y = 0$
B24. $(3 + 17t^2)y'' + (8 - 13t)y' + (10 + 4t^2 - 4t)y = 0$
B25. $(t^2 + 11t + 10)y'' + (t^2 - 6t)y' + (11 - 8t)y = 0$
B26. $(5t - 11)y'' + (8t - 2 - 2t^2)y' + (11t^2 + 5 + 5t)y = 0$

In Exercises B27–41:

(a) Show $t = 0$ is a regular singular point of the given differential equation.

(b) Obtain and solve the indicial equation.

(c) Obtain the recurrence relation for each independent solution.

(d) Write the first few terms of each of the independent solutions.

(e) Use a computer algebra system to obtain a series solution, and compare this solution to the results in part (d).

B27. $-6t^2 y'' + t(-8t + 5)y' + (-4t + 7)y = 0$
B28. $5t^2 y'' + t(8t - 3)y' + (5t + 3)y = 0$
B29. $5t^2 y'' + t(-6t - 9)y' + 8y = 0$
B30. $(-2t^2 - 7t)y'' + (3t - 1)y' + 2y = 0$
B31. $6t^2 y'' + t(8t - 1)y' + 2y = 0$
B32. $-8t^2 y'' + 4ty' + (2t + 8)y = 0$
B33. $-4t^2 y'' + t(-5t - 7)y' + y = 0$
B34. $8t^2 y'' + t(2t^2 - 6)y' + 3y = 0$
B35. $(6t^2 + 7t)y'' + (-5t - 6)y' + 5y = 0$
B36. $9t^2 y'' + 9ty' + (-2t - 1)y = 0$
B37. $-7t^2 y'' + t(-8t - 2)y' + 2y = 0$
B38. $3t^2 y'' + t(3t + 2)y' - 2y = 0$
B39. $-8t^2 y'' + t(5t^2 - 2)y' + 5y = 0$
B40. $-8t^2 y'' + t(6t^2 + 4)y' + 8y = 0$
B41. $(4t^2 - 5t)y'' - (6t + 4)y' + 8y = 0$

In Exercises B42–46:

(a) Show $t = 0$ is a regular singular point of the given differential equation.

(b) Obtain and solve the indicial equation, noting that the roots, α and β, where $\text{Re}(\beta) \geq \text{Re}(\alpha)$, are real and differ by the positive integer $s = \beta - \alpha$.

(c) Using a computer algebra system, obtain a (truncated) series solution, thereby determining if the second solution contains $\ln t$.

(d) Obtain the recursion relation for the solution $y_1 = t^\beta \sum_{k=0}^{\infty} a_k t^k$.

(e) Use the recursion relation in part (d) to determine the first few terms of the solution y_1.

(f) Obtain the recursion relation for the solution $y_2 = t^\alpha \sum_{k=0}^{\infty} b_k t^k$.

(g) If part (c) suggests the second solution does not have a logarithmic term, use the recursion relation found in part (f) to determine the first few terms of the solution y_2.

(h) If part (c) suggests the second solution has a logarithmic term, use the recursion relation found in part (f) to show that it leads to the first solution, y_1.

(i) If part (c) suggests the second solution has a logarithmic term, use the differential equation and $y_2 = \ln t \sum_{k=0}^{n} a_k t^{k+\beta} + \sum_{k=0}^{n} d_k t^{k+\alpha}$, for some suitable value of n, to determine the first few coefficients d_k. Compare the resulting solution to the solution found in part (c).

B42. $2t^2 y'' + t(-3t - 4)y' + (9t + 4)y = 0$
B43. $t^2 y'' + t(5t - 7)y' + 7y = 0$
B44. $(t^2 - 3t)y'' + 3ty' - 8y = 0$
B45. $2t^2 y'' + t(3t + 6)y' - 6y = 0$
B46. $2t^2 y'' + t(8 - 5t^2)y' - 8y = 0$

14.2 Asymptotic Solutions

EXAMPLE 14.6   About the regular singular point $t = 0$, the differential equation

$$y''(t) - \left(1 + \frac{2}{t}\right)y(t) = 0 \tag{14.17}$$

has independent solutions $y_1 = te^t$ and $y_2 = y_1 \ln t + 1 - \frac{5}{2}t^2 - \frac{5}{3}t^3 + O(t^4)$, so there is indeed a logarithmic term as seen in Section 14.1. On the other hand, for large t, the differential equation appears to be $y''(t) - y(t) = 0$, which has the general solution $y(t) = c_1 e^t + c_2 e^{-t}$. This inspires the guess that for large t the solution of the original equation might be of the form $y(t) = e^{\lambda t} f(t)$, where perhaps we can take $f(t) = t^\alpha (1 + \sum_{k=1}^{\infty} a_k t^{-k})$. We therefore substitute the more pragmatic

$$y = e^{\lambda t} t^\alpha \left(1 + \frac{a_1}{t} + \frac{a_2}{t^2} + \frac{a_3}{t^3}\right) \tag{14.18}$$

for $y(t)$ in (14.17), obtaining $e^{\lambda t} \sum_{k=0}^{5} A_k t^{\alpha-5} = 0$, where

$$A_0 = \lambda^2 - 1$$
$$A_1 = -2 + a_1\lambda^2 + 2\lambda\alpha - a_1$$
$$A_2 = \lambda^2 a_2 + 2\lambda\alpha a_1 - 2\lambda a_1 - a_2 + \alpha^2 - \alpha - 2a_1$$
$$A_3 = \lambda^2 a_3 - 4\lambda a_2 + 2\lambda\alpha a_2 - a_3 - 2a_2 + 2a_1 + \alpha^2 a_1 - 3\alpha a_1 \qquad (14.19)$$
$$A_4 = -6\lambda a_3 + 2\lambda\alpha a_3 + 6a_2 - 2a_3 + \alpha^2 a_2 - 5\alpha a_2$$
$$A_5 = 12a_3 + \alpha^2 a_3 - 7\alpha a_3$$

Setting $A_0 = 0$ yields $\lambda = \pm 1$. If we pick $\lambda = 1$, then $A_1 = 2(\alpha - 1) = 0$, so $\alpha = 1$, and $A_2 = -2a_1 = 0$, $A_3 = -4a_2 = 0$, $A_4 = 2(a_2 - 3a_3) = 0$, and $A_5 = 6a_3$. Consequently, a_1 and a_2 must clearly vanish, and after a casual inspection of A_4 we see that $a_3 = 0$ also. In general, all the remaining coefficients are zero and one solution of the differential equation is te^t.

To find a second solution, pick $\lambda = -1$, in which case

$$A_1 = -2 - 2\alpha$$
$$A_2 = \alpha^2 - \alpha - 2\alpha a_1$$
$$A_3 = 2a_2 + 2a_1 + \alpha^2 a_1 - 3\alpha a_1 - 2\alpha a_2 \qquad (14.20)$$
$$A_4 = 6a_2 + 4a_3 + \alpha^2 a_2 - 5\alpha a_2 - 2\alpha a_3$$

Setting $A_1 = -2(1 + \alpha) = 0$ determines $\alpha = -1$, so $A_2 = 2(1 + a_1) = 0$, $A_3 = 4a_2 + 6a_1 = 0$, and $A_4 = 6(a_3 + 2a_2) = 0$. Hence, we have $a_1 = -1$, $a_2 = \frac{3}{2}$, $a_3 = -3, \ldots$. In general, the coefficients are determined by the recursion $a_{n-2} = \frac{1}{2}(n-1)a_{n-3}$, whose solution is $a_k = (-2)^{-k}(k+1)!$ ❖

This example illustrates the following more general theorem from [46].

THEOREM 14.1

1. $y''(t) + f(t)y'(t) + g(t)y(t) = 0$

2. $f(t) = \sum_{k=0}^{\infty} a_k t^{-k}$ and $g(t) = \sum_{k=0}^{\infty} b_k t^{-k}$, where a_0 and b_0 are not both zero.

3. λ_k and α_k, $k = 1, 2$ are defined by the equations
$$\lambda^2 + a_0\lambda + b_0 = 0 \quad \text{and} \quad (2\lambda + a_0)\alpha + \lambda a_1 + b_1 = 0$$

$\Longrightarrow$ There are solutions that, for large t, have asymptotic expansions of the form $e^{\lambda_k t} t^{\alpha_k}(1 + \sum_{k=1}^{\infty} c_k t^{-k})$, $k = 1, 2$.

Thus, this theorem prescribes, for large t, the asymptotic form of solutions to a linear second-order differential equation with a regular singular point at $t = 0$. Of course, for small values of t, the general solution may well be of the form $y(t) = c_1 y_1(t) + c_2(\ln(t)y_1(t) + y_2(t))$, that is, may well have a logarithmic term, as we learned in Section 14.1.

EXERCISES 14.2

1. Carry out the substitution of (14.18) into (14.17), obtaining (14.19).

2. Verify that when $\lambda = 1$, the equations in (14.19) determine $y(t) = te^t$.

3. Show that when $\lambda = -1$, the equations in (14.19) become the equations in (14.20).

4. For (14.17), write the solution for which $\lambda = \alpha = -1$.

In Exercises 5–14:

   (a) Write the given differential equation in the form $y'' + f(t)y' + g(t)y = 0$, and show that the hypotheses of Theorem 14.1 are met.

   (b) Obtain asymptotic expansions for $f(t)$ and $g(t)$, thereby determining a_0, b_0, a_1, b_1.

   (c) Obtain $\lambda_1, \lambda_2, \alpha_1, \alpha_2$.

   (d) Obtain the asymptotic solutions $y_j = e^{\lambda_j t} t^{\alpha_j} (1 + \sum_{k=1}^{3} c_{jk} t^{-k})$, $j = 1, 2$.

(e) Make the change of variables $t = \frac{1}{x}$ so that $y(t) = y(\frac{1}{x}) = w(x)$ leads to the differential equation $x^4 w'' + (2x^3 - x^2 f(\frac{1}{x}))w' + g(\frac{1}{x})w = 0$. If $x = 0$ is a regular (singular) point for the equation in $w(x)$, then $t = \infty$ is also a regular (singular) point for the equation in $y(t)$. Show that $t = \infty$ is an irregular singular point of the given equation.

5. $(4t^2 + 3t + 8)y'' + (6t^2 - 8t - 9)y' + (2t^2 + 9t - 1)y = 0$

6. $(2t^2 - 3t - 4)y'' + (6t^2 - t)y' + (4t^2 - 6t + 5)y = 0$

7. $(3t^2 + 5t - 4)y'' + (-2t^2 - 7t + 1)y' + (-t^2 + 4)y = 0$

8. $(2t^2 + 5t + 8)y'' + (3t^2 + 3t + 9)y' + (t^2 - t)y = 0$

9. $(t^2 - t - 8)y'' + (7t^2 - 9t + 3)y' - (8t^2 + 6t + 1)y = 0$

10. $(3t^2 - 7t + 6)y'' - (5t^2 - 2t)y' - (2t^2 - 3t - 5)y = 0$

11. $(2t^2 + 4t - 1)y'' - (3t^2 + t - 4)y' - (2t^2 + t + 4)y = 0$

12. $(3t^2 - 9t - 7)y'' - (2t^2 + t + 6)y' - (t^2 + 2t - 7)y = 0$

13. $(8t^2 + 4t + 2)y'' + (-2t^2 + 3t - 6)y' - 6(t^2 + 6t + 1)y = 0$

14. $(t^2 - 9t - 7)y'' + (-4t^2 + 9t + 2)y' + (3t^2 + 4t + 5)y = 0$

14.3   Perturbation Solution of an Algebraic Equation

Perturbation Solutions

A differential equation containing a parameter a will have a solution that depends on that parameter. For some equations, that solution can be computed exactly. For most equations of interest, an exact solution is not possible. In these cases it is sometimes useful to obtain the solution as a *perturbation series* $\sum_{k=0}^{\infty} x_k(t)a^k$, where the functions $x_k(t)$ are determined by the differential equation. For each fixed value of the independent variable t, such a series is generally an asymptotic power series in powers of a. Such solutions are the subject of Section 14.3.

   To clarify how a perturbation series provides a solution to an equation containing a parameter, we first consider the simpler case of algebraic equations.

EXAMPLE 14.7   **An algebraic equation**   The algebraic equation

$$ax^3 - x + 1 = 0 \qquad (14.21)$$

can be solved exactly for x, since it is a cubic. We will use (14.21) to illustrate a perturbation solution, using the exact solution to determine the accuracy of the approximate solution. Table 14.2 lists numeric values of x for two values of a. For a "small," there are three real solutions, one of which is close to $x = 1$. For a "not small," there is one real and a pair of complex conjugate roots. The real root is not near $x = 1$. Hence, the solutions appear to have a significant dependence on a.

   If $A = (12\sqrt{3}\sqrt{\frac{27a-4}{a}} - 108)a^2$, then

$$\frac{A^{1/3}}{6a} + \frac{2}{A^{1/3}} \quad \text{and} \quad -\frac{A^{1/3}}{12a} - \frac{1}{A^{1/3}} \pm \frac{i\sqrt{3}}{2}\left(\frac{A^{1/3}}{6a} - \frac{2}{A^{1/3}}\right) \qquad (14.22)$$

a	x
0.1	$-3.577089445, 1.153467305, 2.423622140$
0.2	$-2.627365085, 1.313682542 \pm 0.4210528070i$

TABLE 14.2   Roots of $ax^3 - x + 1 = 0$ for two values of a

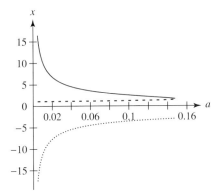

FIGURE 14.2   Example 14.7: f_1 (solid), f_2 (dotted), f_3 (dashed)

are the exact solutions of (14.21). In their present form, the expressions in (14.22) are difficult to interpret and/or use. We do see, however, that each solution depends on the parameter a. Since the quantity A containing $\sqrt{27a - 4}$ appears repeatedly in the three solutions, we are surprised to realize that for a "small," this radical is a complex number! However, there are other complex numbers in the expressions for the roots and, together, these roots can be real if $a < \frac{4}{27} = 0.\overline{148}$. Consequently, for $0 < a < \frac{4}{27}$ the three solutions are real and simplify to

$$f_1 = \frac{2}{\sqrt{3a}} \sin \lambda$$

$$f_2 = -\frac{1}{\sqrt{a}} \left(\frac{\sin \lambda}{\sqrt{3}} + \cos \lambda \right) \quad \text{where} \quad \lambda = \frac{1}{3} \arctan \left(\frac{\sqrt{4 - 27a}}{3\sqrt{3a}} + \frac{\pi}{6} \right) \quad \textbf{(14.23)}$$

$$f_3 = -\frac{1}{\sqrt{a}} \left(\frac{\sin \lambda}{\sqrt{3}} - \cos \lambda \right)$$

The dependence of the roots on a can also be seen in Figure 14.2, where the roots f_1, f_2, and f_3 are represented by the solid, dotted, and dashed curves, respectively. Equation (14.21) is more easily solved for $a = a(x) = \frac{x-1}{x^3}$, from which the curves $x = x(a)$ can be obtained by reflecting the graph of $a(x)$ across the line $a = x$. Figure 14.3, obtained in this way, clearly shows the behavior of the root represented by f_3, the root "near 1," which is continuously connected to $x = 1$, the solution of (14.21) when $a = 0$. Consequently, we seek a series expansion in powers of a for this root. ❖

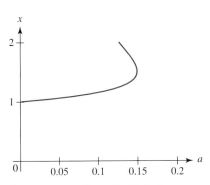

FIGURE 14.3   Example 14.7: f_3 as a function of a

Perturbation Series

A series expansion for the root near $a = 1$ gives $f_3 = F_3 + \mathrm{O}(a^6)$, where

$$F_3 = 1 + a + 3a^2 + 12a^3 + 55a^4 + 273a^5 \qquad \textbf{(14.24)}$$

Figure 14.4 shows the exact solution as a solid curve and the series approximation as a dotted curve, illustrating the validity of the series for small a.

The growth in the coefficients suggests the series is an asymptotic series that does not converge. The following construction obtains this same expansion as the perturbation series $\sum_{k=0}^{\infty} x_k a^k$. Substitute the truncated sum $\sum_{k=0}^{5} x_k a^k$ into the cubic equation, collect like powers of a, and obtain $\sum_{k=1}^{5} A_k a^k = 0$, where

$$A_1 = 1 - x_1$$

$$A_2 = 3x_1 - x_2$$

$$A_3 = 3x_1^2 + 3x_2 - x_3 \qquad \textbf{(14.25)}$$

$$A_4 = -x_4 + 4x_1 x_2 + 3x_3 + x_1(x_1^2 + 2x_2)$$

$$A_5 = -x_5 + x_2^2 + 3x_4 + 4x_1 x_3 + x_1(2x_3 + 2x_1 x_2) + x_2(x_1^2 + 2x_2)$$

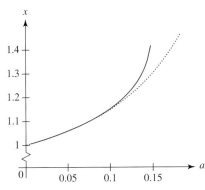

FIGURE 14.4   Example 14.7: f_3 (solid) and $\tilde{f}_3$ (dotted)

Then, set the coefficient of each power of a to zero, obtaining $x_1 = 1$, $x_2 = 3$, $x_3 = 12$, $x_4 = 55$, and $x_5 = 273$, giving, to degree five, the perturbation solution $x = F_3$. Alternatively, invoke the following theorem.

THEOREM 14.2   LAGRANGE EXPANSION THEOREM

1. $x = y + ag(x)$ $\qquad\qquad\qquad\qquad\qquad\qquad\qquad\qquad$ **(14.26)**

$$\Longrightarrow x = x(a, y) + \sum_{k=1}^{\infty} a^k \frac{1}{k!} \left[\left(\frac{d}{dx} \right)^{k-1} g(x)^k \right]_{x=y}$$ **(14.27)**

Equation (14.21) is a special case of (14.26) where $y = 1$ and $g(x) = x^3$. Theorem 14.2 then gives, by (14.27), $x = 1 + \sum_{k=1}^{\infty} a^k \left[\left(\frac{d}{dx} \right)^k x^{3k} \right]_{x=1}$. The differentiations are performed first, and then evaluated at $x = 1$, giving the first five coefficients as 1, 3, 12, 55, and 273, in agreement with our previous calculations.

The interested reader is referred to [8] for a complete statement of the Lagrange Expansion Theorem, and its generalization to the case where a perturbation series for $f(x)$, not just x, is to be computed.

EXERCISES 14.3–Part A

A1. Evaluate (14.22) for $a = 0.13$ and $a = 0.16$.

A2. Evaluate (14.23) for $a = 0.13$ and $a = 0.16$.

A3. Perform the calculations leading to (14.25).

A4. Verify that the solution of (14.25) reproduces (14.24).

A5. Use (14.27) to obtain (14.24).

EXERCISES 14.3–Part B

B1. Equation (14.21) prompts us to define the function $f(a, x) = ax^3 - x + 1$.

(a) Using appropriate technology, obtain a plot of the surface determined by $z = f(x, a)$. In this plot, include a representation of the plane $z = 0$. The surface and the plane intersect in the curve $x = x(a)$ defined implicitly by $f(a, x) = 0$.

(b) In the neighborhood of the point $(a, x) = (0, 1)$, the equation $f(a, x) = 0$ implicitly defines $x = x(a)$, a function that has a formal Maclaurin expansion in a. Obtain the coefficients of this series by subjecting $f(a, x) = 0$ to implicit differentiation, thereby computing $x = 1 + \sum_{k=1}^{\infty} x^{(k)}(0) a^k / k!$.

In Exercises B2–7, a function $f(a, x)$ is given.

(a) Determine σ_k, $k = 1, \ldots, 3$, then solutions of $f(x, 0) = 0$.

(b) Obtain a plot of the surface determined by $z = f(a, x)$. In this plot, include a representation of the plane $z = 0$. The

surface and the plane intersect in the curve $x = x(a)$ defined implicitly by $f(a, x) = 0$.

(c) Using appropriate technology, obtain a plot of the function $x(a)$ determined implicitly by the equation $f(a, x) = 0$.

(d) In the neighborhood of each point $(a, x) = (0, \sigma_k)$, the equation $f(a, x) = 0$ implicitly defines $x_k = x_k(a)$, functions that have formal Maclaurin expansions in a. Obtain the coefficients of these series by subjecting $f(a, x) = 0$ to implicit differentiation, thereby computing $x_k = \sigma_k + \sum_{n=1}^{\infty} x_k^{(n)}(0) a^n / n!$.

(e) Obtain the expansions found in part (d) by assuming series of the form $\sum_{n=0}^{\infty} c_n a^n$, substituting into the equation $f(a, x) = 0$, and matching coefficients of like powers of a in order to determine the coefficients c_n.

B2. $f(a, x) = x(x^2 - 1) + a$ **B3.** $f(a, x) = x^2 + 3x + 2 + a$

B4. $f(a, x) = x^2 - 3x + a$

B5. $f(a, x) = x^3 + 5x^2 - 2x - 24 + a$

B6. $f(a, x) = x^3 + x^2 - 2x + a$

B7. $f(a, x) = x^4 + 5x^3 - 22x^2 - 56x + a$

In Exercises B8–12, an equation of the form (14.26) is given. For each:

(a) Use the Lagrange expansion formula to obtain series representations of the solutions, $x = x(a)$.

(b) Graph the approximate solution found in part (a) along with $x = x(a)$ defined implicitly by the given equation.

B8. $x = 1 + ae^x$ **B9.** $x = 2 + a \cosh \left(\frac{x}{2} \right)$

B10. $x = 1 + ax \sin x$ **B11.** $x = 3 + ae^x \cos x$

B12. $x = 5 + ax^2 e^{-x}$

B13. Use implicit differentiation to obtain the formula for the Lagrange expansion of $x(a)$, the solution of the equation $x = y + ag(x)$. *Hint:* $x(a) = \sum_{k=0}^{\infty} x^{(k)}(0) a^k / k!$, and the derivatives $x^{(k)}(0)$ are computed by implicit differentiation.

14.4 Poincaré Perturbation Solution for Differential Equations

EXAMPLE 14.8   **Parameter-dependent damped oscillator**   For $0 \leq a < 1$, the damped linear oscillator whose governing differential equation is

$$x''(t) + 2ax'(t) + x(t) = 0 \tag{14.28}$$

is underdamped. Consequently, for a mass displaced one unit and released (therefore satisfying the initial conditions $x(0) = 1$, $x'(0) = 0$), the solution is given by

$$x(t) = \left(\frac{a \sin \left(\sqrt{1 - a^2} t \right)}{\sqrt{1 - a^2}} + \cos \left(\sqrt{1 - a^2} t \right) \right) e^{-at} \tag{14.29}$$

and the exponentially decaying oscillation exhibited by the solution is apparent. These motions were the concern of Chapters 5 and 6.

Presently, we seek to approximate the solution with a perturbation series of the form $x(t) = \sum_{k=0}^{\infty} x_k(t) a^k$, which for each fixed t is an asymptotic power series in powers of a. Obtained from a Maclaurin series expansion of (14.29) in powers of a, the first few terms of such an expansion are

$$\cos t + (\sin t - t \cos t)a - \tfrac{1}{2}(t \sin t - t^2 \cos t)a^2 + \tfrac{1}{6}(3 \sin t - t \cos t (3 + t^2))a^3 \tag{14.30}$$

In the coefficient of a, we find the term $t \cos t$, a bounded trig term multiplied by a power of t. Such terms are called *secular* terms, and they eventually become large, no matter how small a is taken. This behavior is shown in Figure 14.5, where the solid curve represents the exact solution and the dotted curve the approximate solution. Initially, and up to about $t = 9$, the two solutions are close, but thereafter, the approximation quickly becomes severely impaired. The secular terms dominate for large t and make the series an unbounded approximation to a solution that actually goes to zero. ❖

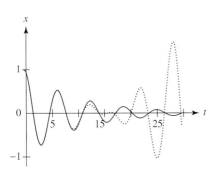

FIGURE 14.5   Example 14.8: Exact solution (solid) and series approximation (dotted)

Our goal, clearly, is to obtain useful approximations to solutions that could not otherwise be expressed in closed form. Thus, we first study a technique for generating perturbation solutions and then examine a technique for eliminating the secular terms.

The Regular Perturbation

The *regular perturbation* is developed by assuming a solution of the form $\sum_{k=0}^{\infty} x_k(t) a^k$, substituting it into (14.28), and letting the differential equation itself determine the functions $x_k(t)$ by equating to zero each coefficient of the powers of a. Substituting the truncated sum

$$x_0(t) + x_1(t)a + x_2(t)a^2 + x_3(t)a^3 \tag{14.31}$$

into (14.28), we obtain $\sum_{k=0}^{3} \phi_k(t)a^k = 0$, where

$$\phi_0 = x_0''(t) + x_0(t) \qquad\qquad \phi_1 = x_1''(t) + x_1(t) + 2x_0'(t)$$
$$\phi_2 = x_2''(t) + x_2(t) + 2x_1'(t) \qquad \phi_3 = x_3''(t) + x_3(t) + 2x'(t)$$

(14.32)

The functions $x_k(t)$ are determined by the requirement that each $\phi_k = 0$, the initial conditions $x_0(0) = 1$, $x_0'(0) = 0$, and the initial conditions $x_k(0) = x_k'(0) = 0$, $k \geq 1$.

The zeroth-order equation contains just $x_0(t)$. The solution satisfying the initial conditions corresponding to "displace one unit and release" is therefore given by $x_0(t) = \cos t$. Each of the other differential equations contains solutions of previous equations. Hence, the equations must be solved in sequence, with information propagated forward into successive equations. The solutions to the remaining equations, using the inert initial conditions of no displacement or velocity, are then

$$x_1(t) = \sin t - t \cos t$$
$$x_2(t) = -\tfrac{1}{2}(t \sin t - t^2 \cos t)$$
$$x_3(t) = \tfrac{1}{6}(3 \sin t - t(3 + t^2) \cos t)$$

(14.33)

Hence, to third order in a, the perturbation solution is precisely the same as (14.30), the Maclaurin expansion of the exact solution (14.29).

EXAMPLE 14.9   We give one further example of creating a regular perturbation solution for a differential equation, this time, one for which an exact solution is not known. We will illustrate how one might use a numeric solution to check the validity of the series solution.

Numeric solutions of the first-order IVP

$$x'(t) + x(t) = 1 + ax(t)^3 \qquad x(0) = 0$$

(14.34)

for three different values of a ($\frac{1}{27}$, $\frac{3}{27}$, 0.15) are shown in Figure 14.6. The differential equation has equilibrium solutions near $x(t) = 1$ for $a < \frac{4}{27}$ because the equilibrium solutions are solutions of the equation $x = 1 + ax^3$, the algebraic equation studied in Section 14.3. Since $a = 0.15 > \frac{4}{27} = 0.\overline{148}$, the corresponding curve does not at all resemble the curve $x(t) = 1$.

Lest the reader dismiss the utility of a regular perturbation solution, we instead obtain the Taylor series solution,

$$t - \frac{1}{2}t^2 + \frac{1}{6}t^3 + \left(\frac{a}{4} - \frac{1}{24}\right)t^4 + \left(\frac{1}{120} - \frac{7}{20}, a\right)t^5 + \cdots$$

(14.35)

that is, a solution of the form $x(t) = \sum_{k=0}^{\infty} c_k(a)t^k$. Figure 14.7, which shows a graph of the solution for $a = \frac{1}{27}$, indicates just how poor such an approximation actually is.

It is our intent to show that the regular perturbation solution is very much better at approximating solutions when $a < \frac{4}{27}$. To obtain this regular perturbation solution, substitute (14.31) into (14.34) and group terms of like powers of a, obtaining $\sum_{k=0}^{3} \sigma_k(t)a^k = 0$, where

$$\sigma_0 = x_0'(t) + x_0(t) - 1 \qquad\qquad \sigma_1 = x_1'(t) + x_1(t) - x_0(t)^3$$
$$\sigma_2 = x_2'(t) + x_2(t) - 3x_0(t)^2 x_1(t) \qquad \sigma_3 = x_3'(t) + x_3(t) - 3x_0(t)^2 x_2(t) - 3x_0(t)x_1(t)^2$$

(14.36)

Demanding that the coefficients of each power of a must vanish determines the $x_k(t)$ as solutions of the differential equations $\sigma_k = 0$ subject to the initial conditions $x_k(0) = 0$.

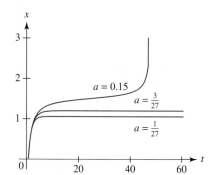

FIGURE 14.6   Example 14.9: Solutions for $a = \frac{1}{27}, \frac{3}{27}, 0.15$

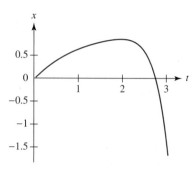

FIGURE 14.7   Example 14.9: Taylor series approximation for $a = \frac{1}{27}$

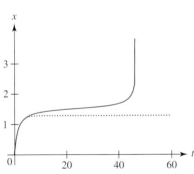

FIGURE 14.8   Example 14.9: for $a = 0.15$, numeric solution (solid) and perturbation solution (dotted)

Thus, we compute

$$x_0(t) = 1 - e^{-t}$$

$$x_1(t) = \tfrac{1}{2}(2 + (3 - 6t)e^{-t} - 6e^{-2t} + e^{-3t})$$

$$x_2(t) = 3 + \left(\tfrac{49}{8} - \tfrac{3}{2}t - \tfrac{9}{2}t^2\right)e^{-t} - (3 + 18t)e^{-2t} - \left(\tfrac{39}{4} - \tfrac{9}{2}t\right)e^{-3t} + 4e^{-4t} - \tfrac{3}{8}e^{-5t}$$

$$x_3(t) = 12 + \tfrac{51}{8}t + \left(\tfrac{2139}{80} - \tfrac{45}{4}t^2 - \tfrac{9}{2}t^3\right)e^{-t} + (3 - 54t - 54t^2)e^{-2t}$$
$$\qquad - \left(\tfrac{585}{16} + \tfrac{315}{4}t - \tfrac{81}{4}t^2\right)e^{-3t} - (23 - 48t)e^{-4t}$$
$$\qquad + \left(\tfrac{357}{16} - \tfrac{45}{8}t\right)e^{-5t} - \tfrac{24}{5}e^{-6t} + \tfrac{5}{16}e^{-7t}$$

Figure 14.8 shows a graph of the approximate solution (dotted), along with the numeric solution (solid) for $a = 0.15$. Even for $a = 0.15 > \tfrac{4}{27} = 0.\overline{148}$, the perturbation solution remains near the equilibrium solution $x = 1$, whereas the numeric solution departs from $x = 1$. In comparison, recall that the Taylor series solution, even for $a = \tfrac{1}{27}$, fails to represent the equilibrium solution $x = 1$ except very close to $t = 0$.   ❖

EXERCISES 14.4–Part A

A1. Use the Laplace transform (or other pertinent method) to obtain (14.29).

A2. Verify the expansion in (14.30).

A3. Verify that substituting (14.31) into (14.28) results in (14.32).

A4. Obtain the solutions listed in (14.33).

A5. Substitute $x(t) = \sum_{k=1}^{5} c_k(a)t^k$ into the DE in (14.34) and obtain (14.35).

A6. Substitute (14.31) into (14.34) to obtain (14.36).

EXERCISES 14.4–Part B

B1. Given the initial value problem $y'' + y' + ay = 0$, $y(0) = 1$, $y'(0) = 0$:

(a) Obtain the exact solution for $0 \le a < \tfrac{1}{4}$.

(b) For the exact solution found in part (a), obtain a Maclaurin expansion about $a = 0$.

(c) Obtain a regular perturbation solution of the form $y(t) = \sum_{k=0}^{\infty} y_k(t)a^k$, showing that the perturbation series agrees with the expansion found in part (b).

(d) On the same set of axes, and for $a = 0.1$, graph the exact solution, $y_0 + ay_1$, and $y_0 + ay_1 + a^2 y_2$.

(e) Repeat part (d) for $a = 0.2$.

B2. Given the initial value problem $y'' + 2y' + 2y = a\cos t$, $y(0) = 1$, $y'(0) = 1$:

(a) Obtain the exact solution.

(b) For the exact solution found in part (a), obtain a Maclaurin expansion about $a = 0$.

(c) Obtain a regular perturbation solution of the form $y(t) = \sum_{k=0}^{\infty} y_k(t)a^k$, showing that the perturbation series agrees with the expansion found in part (b).

(d) On the same set of axes, and for $a = 0.25$, graph the exact solution and y_0.

(e) Repeat part (d) for $a = 1$.

B3. Given the initial value problem $y'' + y' + ay^2 = 0$, $y(0) = 1$, $y'(0) = 0$:

(a) Obtain, and graph, a numeric solution for $a = 0.3$.

(b) Obtain a regular perturbation solution of the form $y(t) = \sum_{k=0}^{\infty} y_k(t)a^k$.

(c) On the same set of axes, graph the numeric solution, $y_0 + ay_1$, and $y_0 + ay_1 + a^2 y_2$.

B4. Given the initial value problem $y'' + y + ay^2 = 0$, $y(0) = 1$, $y'(0) = 0$:

(a) Obtain, and graph, a numeric solution for $a = 0.5$.

(b) Obtain a regular perturbation solution of the form $y(t) = \sum_{k=0}^{\infty} y_k(t)a^k$.

(c) On the same set of axes, graph the numeric solution, $y_0 + ay_1$, and $y_0 + ay_1 + a^2 y_2$.

B5. Given the initial value problem $y'' + y' + y + ay^2 = 0$, $y(0) = 1$, $y'(0) = 0$:

   (a) Obtain, and graph, a numeric solution for $a = 0.1$.

   (b) Obtain a regular perturbation solution of the form $y(t) = \sum_{k=0}^{\infty} y_k(t)a^k$.

   (c) On the same set of axes, graph the numeric solution and $y_0 + ay_1$.

B6. Given the boundary value problem $y'' + ay = 0$, $y(0) = 0$, $y(\pi) = 1$:

   (a) Obtain the exact solution for a small. *Hint:* Solution is a multiple of $\sin(\sqrt{a}t)$.

   (b) For the exact solution found in part (a), obtain a Maclaurin expansion about $a = 0$.

   (c) Obtain a regular perturbation solution of the form $y(t) = \sum_{k=0}^{\infty} y_k(t)a^k$, showing that the perturbation series agrees with the expansion found in part (b).

   (d) On the same set of axes, and for $a = 0.25$, graph the exact solution, $y_0 + ay_1$, and $y_0 + ay_1 + a^2y_2$.

   (e) Repeat part (d) for $a = 0.5$.

B7. Given the boundary value problem $y'' + ay' + 2y = 0$, $y(0) = 0$, $y(1) = 1$:

   (a) Obtain the exact solution for a small. *Hint:* Solution is a multiple of $e^{-at}\sin(\sqrt{1-a^2}t)$.

   (b) For the exact solution found in part (a), obtain a Maclaurin expansion about $a = 0$.

   (c) Obtain a regular perturbation solution of the form $y(t) = \sum_{k=0}^{\infty} y_k(t)a^k$, showing that the perturbation series agrees with the expansion found in part (b).

(d) On the same set of axes, and for $a = 0.2$, graph the exact solution, $y_0 + ay_1$, and $y_0 + ay_1 + a^2y_2$.

(e) Repeat part (d) for $a = 0.9$.

B8. Given the initial value problem $y' + y = ay^2$, $y(0) = 1$:

   (a) Obtain the exact solution for small a. (*Hint:* It's a Bernoulli equation, Section 3.6.)

   (b) For the exact solution found in part (a), obtain a Maclaurin expansion about $a = 0$.

   (c) Obtain a regular perturbation solution of the form $y(t) = \sum_{k=0}^{\infty} y_k(t)a^k$, showing that the perturbation series agrees with the expansion found in part (b).

   (d) On the same set of axes, and for $a = 0.5$, graph the exact solution, $y_0 + ay_1$, and $y_0 + ay_1 + a^2y_2$.

   (e) Repeat part (d) for $a = 0.75$.

B9. Given the initial value problem $y'' + a(y')^2 + y + ay = 0$, $y(0) = 1$, $y'(0) = 0$:

   (a) Obtain, and graph, a numeric solution for $a = 0.5$.

   (b) Obtain a regular perturbation solution of the form $y(t) = \sum_{k=0}^{\infty} y_k(t)a^k$.

   (c) On the same set of axes, graph the numeric solution, $y_0 + ay_1$, and $y_0 + ay_1 + a^2y_2$.

B10. Given the initial value problem $y'' + a(2(y')^3 - y') + y = 0$, $y(0) = 1$, $y'(0) = 0$:

   (a) Obtain, and graph, a numeric solution for $a = 0.5$.

   (b) Obtain a regular perturbation solution of the form $y(t) = \sum_{k=0}^{\infty} y_k(t)a^k$.

   (c) On the same set of axes, graph the numeric solution, $y_0 + ay_1$, and $y_0 + ay_1 + a^2y_2$.

14.5   The Nonlinear Spring and Lindstedt's Method

A Nonlinear Oscillator

The differential equation

$$x''(t) + x(t) + ax(t)^3 = 0 \tag{14.37}$$

governs an undamped oscillator whose spring obeys a nonlinear restoration law, $F = -(x + ax^3)$. For $a > 0$, the spring is called "hard," while for $a < 0$, it is called "soft." The magnitude of the restorative force for the hard spring is greater than that of the linear spring (whose restorative force is just $-kx$). Likewise, the magnitude of the restorative force for the soft spring is less than that of the linear spring for small enough displacements.

   When $a = 0$ the system is just a simple harmonic oscillator that clearly supports periodic solutions. We wish to show that for $a \neq 0$ the system also exhibits periodic solutions. We can show this by obtaining a first integral (a function constant on the trajectories in the phase plane) and showing that its level curves, the trajectories, are closed curves. To this

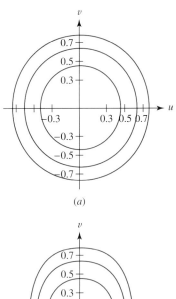

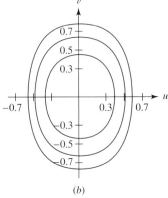

FIGURE 14.9 Level curves of the first integral $E(x, x')$

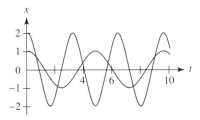

FIGURE 14.10 Frequency of the nonlinear oscillator depends on amplitude

end, multiply (14.37) by $x'(t)$ and integrate with respect to t, obtaining

$$\int x'(x'' + x + ax^3)\, dt = \int 0\, dt = E$$

The term on the right is the constant E, and each term on the left is an exact differential. Hence, we obtain

$$\frac{1}{2}(x')^2 + \frac{1}{2}x^2 + \frac{a}{4}x^4 = E(x, x')$$

For each value of a, the level curves of $E = E(x, x')$ define solutions of (14.37). Indeed, differentiating E with respect to t yields $x'(x'' + x + ax^3) = 0$, which is (14.37) multiplied by x'. Hence, a solution of (14.37) is a level curve of $E(x, x')$, and a level curve of $E = E(x, x')$, defines a solution of (14.37).

It is easier to plot the level curves of $E(x, x')$ if we change variables according to $u = x(t)$, $v = x'(t)$, so that $u' = x' = v$ and $v' = x'' = -u - au^3$, which then gives $E(u, v) = \frac{1}{2}v^2 + \frac{1}{2}u^2 + \frac{a}{4}u^4$. The contour plot in Figure 14.9(a) used $a = \frac{1}{5}$, while the contour plot in Figure 14.9(b) used $a = 5$. We rely on this figure to support the claim that the trajectories for this differential equation are closed curves. Hence, for $a \neq 0$ the undamped oscillator with the nonlinear spring supports periodic solutions.

Dependence of Frequency on Initial Displacement

In the nonlinear case, the (angular) frequency depends on the initial displacement. If we let that initial displacement be first 1, then 2 (along with an initial velocity $x'(0) = 0$), we obtain Figure 14.10, which shows two solutions corresponding to $a = 1$. The graph clearly indicates different periods, and hence frequencies, for the two solutions, which start with different initial displacements.

This is in distinction to the behavior of the linear oscillator, where for $a = 0$ the angular frequency is $\omega = \sqrt{\frac{k}{m}} = 1$, a constant depending on just the spring constant k and the mass m. Indeed, for $a = 0$ we have $x(t) = c_1 \cos t + c_2 \sin t$ and the angular frequency is $\omega = 1$.

In the nonlinear case, the dependence of the angular frequency on the initial displacement can be computed from the first integral $E(u, v)$. First, evaluate the constant E from the initial conditions $(u, v) = (x, x') = (A, 0)$, obtaining $E_0 = \frac{1}{2}A^2 + \frac{1}{4}A^4 a$. Next, solve the equation $E(u, v) = E_0$ for $v = \frac{dx}{dt}$, obtaining the two solutions

$$v(u) = \pm\frac{1}{2}\sqrt{2A^4 a - 4u^2 - 2u^4 a + 4A^2}$$

If we set $v(u) = \frac{dx}{dt} = \frac{du}{dt}$ and separate variables, we get

$$t = \int_A^{u_{\text{final}}} \frac{du}{v(u)}\, du$$

On a closed trajectory in the phase plane, motion is clockwise. At $(u, v) = (A, 0)$, the differential equations $u' = x' = v$ and $v' = x'' = -u - au^3$ give $v' = -A - aA^3 < 0$, so motion at this point is downward and, hence, clockwise around the trajectory. Thus, we pick the negative square root for $v(u)$ and integrate to $u_{\text{final}} = 0$, one-quarter of the way around the trajectory. Thus, the total time for one orbit is

$$T = -\int_A^0 \frac{8}{\sqrt{2A^4 a - 4u^2 - 2u^4 a + 4A^2}}\, du \tag{14.38}$$

To third order, a series expansion of $\omega = \frac{2\pi}{T}$ in powers of a is then

$$\omega = 1 + \frac{3}{8}A^2 a - \frac{21}{256}A^4 a^2 + \frac{81}{2048}A^6 a^3 \tag{14.39}$$

Thus, for the nonlinear oscillator, the angular frequency depends on the initial displacement, and we have obtained a series expansion of this dependence. Shortly, we will compute this same expression by Lindstedt's perturbation technique.

Regular Perturbation

As we saw in Section 14.4, a regular perturbation will lead to an expansion containing secular terms that are not periodic. Hence, the regular perturbation series of Poincaré cannot approximate the periodic solutions of the nonlinear oscillator. We therefore fail to learn about the dependence of the frequency on the initial displacement. However, for the sake of completeness, we will develop the regular perturbation. The purpose is to justify the claim that it contains secular terms that prevent it from approximating the periodic solutions of the nonlinear oscillator.

Assume a regular perturbation series of the form $x_0(t) + x_1(t)a + x_2(t)a^2$ and substitute this into (14.37), obtaining $\sum_{k=0}^{2} A_k a^k = 0$, where

$$A_0 = x_0''(t) + x_0(t) \qquad A_1 = x_1''(t) + x_1(t) + x_0(t)^3 \qquad A_2 = x_2''(t) + x_2(t) + 3x_0(t)^2 x_1(t)$$

The identical vanishing of the A_k gives differential equations for the x_k. The first equation depends only on $x_0(t)$ and so can be solved immediately. For simplicity, we choose the initial point as $(1, 0)$, thereby obtaining $x_0(t) = \cos t$. Each successive differential equation contains terms depending on the solutions of previous equations. We solve this sequence of equations, taking the initial point as $(0, 0)$ for each successive approximation, obtaining

$$x_1(t) = -\tfrac{1}{32} \cos t - \tfrac{3}{8} t \sin t + \tfrac{1}{32} \cos 3t$$

$$x_2(t) = \tfrac{1}{1024}(12t(8 \sin t - 3 \sin 3t - 6t \cos t) + 23 \cos t - 24 \cos 3t + \cos 5t)$$

The solutions for both $x_1(t)$ and $x_2(t)$ contain secular terms, as predicted. The regular perturbation series is not useful for studying the dependence of the frequency on the initial displacement. We therefore consider Lindstedt's method for generating a more useful asymptotic expansion.

Lindstedt's Method

In the Lindstedt method, the angular frequency is made to depend on a by assuming $\omega = \omega(a) = 1 + \sum_{k=1}^{\infty} \omega_k a^k$ and then carrying out a perturbation solution of (14.37) by writing $x(t) = \sum_{k=0}^{\infty} x_k(t)a^k$. Since the angular frequency ω is related to the period T by $\omega T = 2\pi$, the two expansions in the Lindstedt method are merged by the change of variables $\tau = \omega t$. In terms of τ, the solution assumes the form

$$x(t) = x\left(\frac{\tau}{\omega}\right) = U(\tau) = \sum_{k=0}^{\infty} u_k(\tau)a^k \tag{14.40}$$

and the differential equation (14.37) becomes

$$\omega^2 U''(\tau) + U(\tau) + aU(\tau)^3 = 0 \tag{14.41}$$

where now, primes denote differentiation with respect to τ. (The change of variables rests on the chain rule, which gives $\frac{dx}{dt} = \frac{dU}{d\tau}\frac{d\tau}{dt} = \frac{dU}{d\tau}\omega$. A second application of the chain rule then gives the ω^2 multiplying $U''(\tau)$.)

Write $U(\tau) = u_0(\tau) + u_1(\tau)a + u_2(\tau)a^2 + u_3(\tau)a^3$ and substitute the expansions for both $U(\tau)$ and $\omega(\tau)$ into the differential equation, collect coefficients of like powers of a, and thus obtain $\sum_{k=0}^{3} \sigma_k a^k = 0$, where σ_k, $k = 0, \ldots, 3$, are given in Table 14.3.

$$\sigma_0 = u_0''(\tau) + u_0(\tau)$$

$$\sigma_1 = u_1''(\tau) + u_1(\tau) + 2\omega_1 u_0''(\tau) + u_0(\tau)^3$$

$$\sigma_2 = u_2''(\tau) + u_2(\tau) + 2\omega_1 u_1''(\tau) + 3u_0(\tau)^2 u_1(\tau) + (\omega_1^2 + 2\omega_2)u_0''(\tau)$$

$$\sigma_3 = u_3'' + u_3 + (2\omega_2 + \omega_1^2)u_1'' + 3u_0^2 u_2 + 3u_0 u_1^2 + 2\omega_1 u_2'' + 2(\omega_3 + \omega_1\omega_2)u_0''$$

TABLE 14.3   Lindstedt's method applied to (14.37)

The identical vanishing of the σ_k give differential equations for the u_k. The first equation contains only $u_0(\tau)$, so this equation, and the initial conditions $u_0(0) = A, u_0'(0) = 0$, lead to the solution $u_0(\tau) = A\cos\tau$. The equation for $u_1(\tau)$ contains terms in $u_0(\tau)$. Incorporating this solution and the initial conditions $u_1(0) = u_0'(0) = 0$, we obtain

$$u_1(\tau) = \left(\omega_1 A - \tfrac{3}{8}A^3\right)\tau\sin\tau + \tfrac{1}{32}A^3\cos 3\tau - \tfrac{1}{32}A^3\cos\tau \qquad \textbf{(14.42)}$$

As with the regular perturbation, we have secular terms appearing. However, the coefficient ω_1 is as yet undetermined, so we choose a value for it that will guarantee the vanishing of the secular term $\omega_1 A - \tfrac{3}{8}A^3$, thus determining $\omega_1 = \tfrac{3}{8}A^2$ and consequently

$$u_1(\tau) = \tfrac{1}{32}A^3\cos 3\tau - \tfrac{1}{32}A^3\cos\tau \qquad \textbf{(14.43)}$$

The equation for $u_2(\tau)$ contains terms involving $u_0(\tau), u_1(\tau)$, and ω_1. Thus, making the appropriate substitutions, we solve for the function $u_2(\tau)$ satisfying the initial conditions $u_2(0) = u_2'(0) = 0$, obtaining

$$u_2(\tau) = \left(\tfrac{21}{256}A^5 + \omega_2 A\right)\tau\sin\tau + \tfrac{1}{1024}A^5\cos 5\tau - \tfrac{3}{128}A^5\cos 3\tau + \tfrac{23}{1024}A^5\cos\tau \quad \textbf{(14.44)}$$

We then determine ω_2 by the requirement that there be no secular terms in $u_2(\tau)$. Hence, the equation $\tfrac{21}{256}A^5 + \omega_2 A = 0$ yields $\omega_2 = -\tfrac{21}{256}A^4$ and thus

$$u_2(\tau) = \tfrac{1}{1024}A^5\cos 5\tau - \tfrac{3}{128}A^5\cos 3\tau + \tfrac{23}{1024}A^5\cos\tau \qquad \textbf{(14.45)}$$

The equation for $u_3(\tau)$ contains terms in $u_0(\tau), u_1(\tau), u_2(\tau), \omega_1$, and ω_2. Making the appropriate substitutions and solving for the $u_3(\tau)$ that satisfies the initial conditions $u_3(0) = u_3(0) = 0$, we obtain

$$u_3 = \left(-\tfrac{81}{2048}A^7 + \omega_3 A\right)\tau\sin\tau + \tfrac{1}{32,768}A^7\cos 7\tau - \tfrac{3}{2048}A^7\cos 5\tau \qquad \textbf{(14.46)}$$

$$+ \tfrac{297}{16,384}A^7\cos 3\tau - \tfrac{547}{32,768}A^7\cos\tau$$

To choose the ω_3 that eliminates the secular term from $u_3(\tau)$, solve $-\tfrac{81}{2048}A^7 + \omega_3 A = 0$ for $\omega_3 = \tfrac{81}{2048}A^6$, thereby obtaining

$$u_3(\tau) = \tfrac{1}{32,768}A^7\cos 7\tau - \tfrac{3}{2048}A^7\cos 5\tau + \tfrac{297}{16,384}A^7\cos 3\tau - \tfrac{547}{32,768}A^7\cos\tau$$

and, the first few terms of the expansion ω are those in (14.39).

The Lindstedt solution is then given by $U(\tau) = \sum_{k=0} u_k(\tau)a^k$, where $U(\tau) = U(\omega t) = x(t)$ and ω is as above. To judge the viability of the Lindstedt expansion, write the approximation as a function of t, obtaining

$$x(t) = \sum_{k=0} u_k(\omega t)a^k$$

where ω is given by (14.39). Since our test of the utility of the expansion will be a graph, assign the values $A = 2$ and $a = \tfrac{1}{10}$ to the initial displacement and spring parameter, respectively. Thus, the approximate solution is

$$x(t) = \frac{506{,}893}{256{,}000} \cos \frac{36{,}461}{32{,}000} t + \frac{2537}{128{,}000} \cos \frac{109{,}383}{32{,}000} t + \frac{1}{8000} \cos \frac{36{,}461}{6400} t + \frac{1}{256{,}000} \cos \frac{255{,}227}{32{,}000} t$$

Figure 14.11 shows the Lindstedt solution (dotted) and a numeric solution (solid) for $100 \le t \le 110$. In the neighborhood of $t = 105$, the two solutions are still in agreement. Figure 14.12 shows these two solutions for $150 \le t \le 160$. In the neighborhood of $t = 155$, the two solutions start separating. However, with $a = \frac{1}{10}$ and $t = 100$, the Lindstedt expansion gives remarkable accuracy.

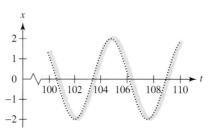

FIGURE 14.11   Lindstedt solution (dotted) and numeric solution (solid)

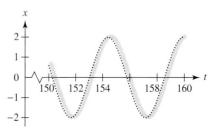

FIGURE 14.12   Lindstedt solution (dotted) and numeric solution (solid)

EXERCISES 14.5

1. On the same set of axes, plot $|x + x^3|$ and $|x|$, the magnitudes of the restorative forces for the hard spring and linear spring, respectively. Is the magnitude of the restorative force for the hard spring greater than that of the linear spring?

2. On the same set of axes, plot $|x - x^3|$ and $|x|$, the magnitudes of the restorative forces for the soft spring and linear spring, respectively. Is the magnitude of the restorative force for the soft spring less than that of the linear spring?

3. Carry out the change of variables $\tau = \omega t$ by which (14.37) becomes (14.41).

4. Set $A = 1$ and $a = \frac{1}{2}$ in (14.38) and evaluate numerically to determine $\omega = \frac{2\pi}{T}$. Compare with the value given by (14.39).

5. Obtain (14.39) from (14.38) as follows.

    (a) Expand the integrand of (14.38) in a Maclaurin series in powers of a and integrate termwise.

    (b) Write $\omega = \frac{2\pi}{T}$ and obtain a Maclaurin expansion, again in powers of a.

6. Solve the IVP defined by $\sigma_0 = 0$ (from Table 14.3) and $u_0(0) = A$, $u_0'(0) = 0$, obtaining $u_0(\tau) = A \cos \tau$.

7. Using $u_0(\tau) = A \cos \tau$, solve the IVP defined by $\sigma_1 = 0$ (from Table 14.3) and $u_1(0) = u_1'(0) = 0$ and obtain (14.42).

8. Using $u_0(\tau) = A \cos \tau$, $\omega_1 = \frac{3}{8} A^2$, and (14.43), solve the IVP defined by $\sigma_2 = 0$ (from Table 14.3) and $u_2(0) = u_2'(0) = 0$ and obtain (14.44).

9. Using $u_0(\tau) = A \cos \tau$, $\omega_1 = \frac{3}{8} A^2$, $\omega_2 = -\frac{21}{256} A^4$, and (14.45), solve the IVP defined by $\sigma_3 = 0$ (from Table 14.3) and $u_3(0) = u_3'(0) = 0$ and obtain (14.46).

In Exercises 10–24, the given differential equation and the initial conditions $y(0) = A$, $y'(0) = 0$, form an IVP. For each:

    (a) Show that the regular Poincaré perturbation scheme leads to secular terms.

    (b) Obtain a perturbation solution using the Lindstedt method, being sure to provide the frequency-amplitude relationship.

    (c) Write the solution found in part (b) as a function of t.

    (d) Obtain and plot numeric solutions for $a = 0.1$ and $A = 1, 2, 3$. Do the graphs show a dependence of the frequency on the initial conditions?

    (e) Along with the corresponding perturbation solution from part (c), plot, for $A = 2$, the numeric solution obtained in part (d).

10. $y'' + y = ay^2$ **11.** $y'' + y = a(y^2 + \frac{1}{5} y^3)$

12. $y'' + y + a(y')^2 = 0$ **13.** $(1 + ay')y'' + y = 0$

14. $y'' + y + a(y^3 + \frac{1}{2} y^5) = 0$ **15.** $y'' + y + ay^5 = 0$

16. $y'' + y + a((y')^2 + \frac{3}{4} y^3) = 0$ **17.** $(1 + y^2)y'' + y = 0$

18. $y'' + y - a((y')^2 + y^2) = 0$ **19.** $y'' + y = ayy'$

14.6 The Method of Krylov and Bogoliubov

Limit Cycles

Some nonlinear second-order differential equations have special periodic solutions whose trajectories in the phase plane are closed curves called limit cycles. A *limit cycle* is a closed trajectory for which no nearby trajectory is also closed. Hence, the nearby trajectories must spiral in toward, or away from, the limit cycle.

A limit cycle C is called *stable* if every trajectory starting sufficiently close to C approaches C as $t \to \infty$. If every trajectory starting sufficiently close to C approaches C as $t \to -\infty$, it is called *unstable*. If trajectories approach C on one side but depart from it on the other, then C is called *semistable*.

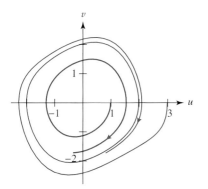

FIGURE 14.13   Phase portrait for the van der Pol equation

van der Pol's Equation

van der Pol's equation

$$x''(t) + x(t) = a(1 - x(t)^2)x'(t) \tag{14.47}$$

arising in the study of triode vacuum-tube oscillatory circuits, has a limit cycle throughout a range of values for the small parameter a. Setting $a = \frac{1}{4}$ and writing the system $u'(t) = v$, $v'(t) = \frac{1}{4}(1 - u^2)v - u$, we obtain the phase portrait in Figure 14.13. The limit cycle, representing a periodic solution of the differential equation, is the thick curve whose initial point is $(2, 0)$. The trajectory starting at $(1, 0)$ spirals out toward the limit cycle, whereas the trajectory starting at $(3, 0)$ spirals in toward the limit cycle. These nearby trajectories that spiral toward the limit cycle represent oscillations approaching the periodic solution. The time-domain view of these three solutions is provided by Figure 14.14.

The limit cycle supported by the van der Pol equation illustrates a second significant difference between the behaviors of the linear and nonlinear equations. In Section 14.5 we saw that the nonlinear system could have solutions whose frequency depended on the initial displacement. In other words, for a nonlinear spring, the larger the initial displacement, the faster the oscillation. For the linear spring, the frequency is determined by the mass, spring constant, and damping coefficients and not by the initial displacement.

In addition, we have just now seen that regardless of the initial displacement, the nonlinear system can have a system-determined amplitude of oscillation. Whether the initial displacement is 1 or 3, the solution of the van der Pol equation tends toward the periodic solution whose amplitude is determined by the solution that starts with initial displacement 2. This is in contradiction to the behavior of the damped linear oscillator for which all oscillations will gradually decay to zero.

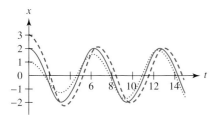

FIGURE 14.14   Solutions of the van der Pol equation corresponding to trajectories in Figure 14.13

Lindstedt's Solution to van der Pol's Equation

Lindstedt's method, studied in Section 14.5, finds periodic solutions of nonlinear equations. Applied to van der Pol's equation, it approximates the solution corresponding to the limit cycle of the equation. Using Lindstedt's method (with details to follow), we obtain the approximation

$$x = 2\cos \omega t + u_1(\omega t)a + u_2(\omega t)a^2 + u_4(\omega t)a^4 + \cdots$$

where $\omega = 1 - \frac{a^2}{16} - \frac{17}{1024}a^4 + \cdots$ and

$$u_1(\omega t) = \tfrac{1}{4}(3\sin\omega t - \sin 3\omega t)$$

$$u_2(\omega t) = -\tfrac{1}{96}(13\cos\omega t - 18\cos 3\omega t + 5\cos\omega t)$$

$$u_4(\omega t) = \tfrac{83}{11,520}\cos\omega t - \tfrac{15}{512}\cos 3\omega t + \tfrac{605}{18,432}\cos 5\omega t - \tfrac{455}{36,864}\cos 7\omega t + \tfrac{33}{20,480}\cos 9\omega t$$

The leading term in the Lindstedt expansion is $2\cos\omega t$, indicating that the method isolates the periodic solution of the van der Pol equation. At the moment, the point is not the computation of the expansion, but the need for a more robust method of approximating solutions of nonlinear equations. We find that method in the work of Krylov and Bogoliubov [62].

With $a = \tfrac{1}{4}$ the Lindstedt approximation (computed to terms in a^4) is

$$x = \tfrac{5,873,363}{2,949,120}\cos\tfrac{261,103}{262,144}t + \tfrac{3}{16}\sin\tfrac{261,103}{262,144}t - \tfrac{1}{16}\sin\tfrac{783,309}{262,144}t + \tfrac{1521}{131,072}\cos\tfrac{783,309}{262,144}t$$

$$- \tfrac{14,755}{4,718,592}\cos\tfrac{1,305,515}{262,144}t - \tfrac{455}{9,437,184}\cos\tfrac{1,827,721}{262,144}t + \tfrac{33}{5,242,880}\cos\tfrac{2,349,927}{262,144}t \qquad \textbf{(14.48)}$$

The phase-plane trajectory determined by this approximation is shown in Figure 14.15. The Lindstedt expansion approximates the limit cycle, the closed trajectory representing the periodic solution of the van der Pol equation.

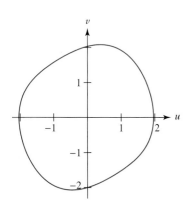

FIGURE 14.15   Lindstedt's approximation to the limit cycle for van der Pol's equation

DETAILS OF THE LINDSTEDT APPROXIMATION TO THE VAN DER POL EQUATION   Application of the Lindstedt method to the van der Pol equation presents a few challenges not met in our work in Section 14.5. However, we do begin with the same form for the angular frequency, namely,

$$\omega = 1 + \sum_{k=1}^{\infty}\omega_k a^k \qquad \textbf{(14.49)}$$

and make the change of variables $\tau = \omega t$ to obtain

$$\omega^2 U''(\tau) + U(\tau) = a(1 - U(\tau)^2)\omega U'(\tau) \qquad \textbf{(14.50)}$$

Fourth-order truncations of (14.49) and (14.40), substituted into (14.50), lead to the equation $\sum_{k=0}^{4}A_k a^k = 0$, where

$$A_0 = u_0''(\tau) + u_0(\tau)$$

$$A_1 = u_1''(\tau) + u_1(\tau) + 2\omega_1 u_0''(\tau) + (u_0(\tau)^2 - 1)u_0'(\tau)$$

$$A_3 = u_3'' + u_3(\tau) + 2\omega_1\omega_2 u_0'' + 2\omega_2 u_1'' + 2\omega_3 u_0'' + 2\omega_1 u_2'' + \omega_1^2 u_1'' + (u_0^2 - 1)u_2'$$
$$\quad + ((u_0^2 - 1)\omega_1 + 2u_0 u_1)u_1' + ((u_0^2 - 1)\omega_2 + 2u_0 u_1\omega_1 + 2u_0 u_2 + u_1^2)u_0'$$

$$A_4 = u_4'' + u_4 + 2\omega_4 u_0'' + 2\omega_1 u_3'' + 2\omega_2 u_2'' + \omega_1^2 u_2'' + 2\omega_1\omega_2 u_1'' + 2\omega_1\omega_3 u_0''$$
$$\quad + \omega_2^2 u_0'' + 2\omega_3 u_1'' + (u_0^2 - 1)u_3' + ((u_0^2 - 1)\omega_1 + 2u_0 u_1)u_2'$$
$$\quad + ((u_0^2 - 1)\omega_2 + 2u_0 u_1\omega_1 + 2u_0 u_2 + u_1^2)u_1'$$
$$\quad + ((u_0^2 - 1)\omega_3 + 2u_0 u_1\omega_2 + (2u_0 u_2 + u_1^2)\omega_1 + 2u_0 u_3 + 2u_1 u_2(\tau))u_0'$$

The identical vanishing of each A_k gives differential equations determining the $u_k(\tau)$. As in Section 14.5, we solve the first equation subject to the initial conditions $u_0(0) = A$, $u_0'(0) = 0$, obtaining $u_0(\tau) = A\cos\tau$. The solution $u_0(\tau)$ is needed in the differential equation for $u_1(\tau)$. Making the appropriate substitution and imposing the initial conditions $u_1(0) = u_1'(0) = 0$, we obtain

$$u_1(\tau) = \left(\left(\tfrac{1}{2}A - \tfrac{1}{8}A^3\right)\cos\tau + \omega_1 A\sin\tau\right)\tau - \tfrac{1}{32}A^3\sin 3\tau + \left(\tfrac{7}{32}A^3 - \tfrac{1}{2}A\right)\sin\tau$$

As expected, the solution for $u_1(\tau)$ contains secular terms that are eliminated by setting $\omega_1 = 0$ and $A = 2$, giving

$$u_1(\tau) = \tfrac{3}{4}\sin\tau - \tfrac{1}{4}\sin 3\tau$$

Even with $u(\tau) = u_0(\tau) = A$, Lindstedt's method forces us to set $A = 2$.

All this information is needed in the differential equation determining $u_2(\tau)$, whose solution is then

$$u_2(\tau) = \left(\tfrac{1}{8} + 2\omega_2\right)\tau\sin\tau - \tfrac{5}{96}\cos 5\tau + \tfrac{3}{16}\cos 3\tau - \tfrac{13}{96}\cos\tau$$

Again, there are secular terms that are eliminated by solving $\tfrac{1}{8} + 2\omega_2 = 0$ for $\omega_2 - \tfrac{1}{16}$, yielding

$$u_2(\tau) = -\tfrac{5}{96}\cos 5\tau + \tfrac{3}{16}\cos 3\tau - \tfrac{13}{96}\cos\tau$$

Passing all this information to the equation for $u_3(\tau)$, we compute that function to be

$$u_3(\tau) = \left(\tfrac{1}{96}\cos\tau + 2\omega_3\sin\tau\right)\tau + \tfrac{7}{576}\sin 7\tau - \tfrac{35}{576}\sin 5\tau + \tfrac{11}{128}\sin 3\tau - \tfrac{19}{384}\sin\tau$$

but another complication has arisen. The coefficient of τ in the secular term contains both $\sin\tau$ and $\cos\tau$. Hence, no choice of ω_3 will eliminate this secular term. Our only recourse is to set $u_3(\tau) = 0$ and take $\omega_3 = 0$.

Finally, the equation for $u_4(\tau)$ contains terms in each previous $u_k(\tau)$, so making the appropriate substitutions we obtain

$$u_4(\tau) = \left(2\omega_4 + \tfrac{17}{512}\right)\tau\sin\tau + \tfrac{33}{20{,}480}\cos 9\tau - \tfrac{455}{36{,}864}\cos 7\tau$$
$$+ \tfrac{605}{18{,}432}\cos 5\tau - \tfrac{15}{512}\cos 3\tau + \tfrac{83}{11{,}520}\cos\tau$$

and eliminate the secular terms by solving the equation $2\omega_4 + \tfrac{17}{512} = 0$ for $\omega_4 = -\tfrac{17}{1024}$ so that $u_4(\tau)$ is

$$u_4(\tau) = \tfrac{33}{20{,}480}\cos 9\tau - \tfrac{455}{36{,}864}\cos 7\tau + \tfrac{605}{18{,}432}\cos 5\tau - \tfrac{15}{512}\cos 3\tau + \tfrac{83}{11{,}520}\cos\tau$$

Thus, $\omega(a)$, to terms in a^4, is $\omega = 1 - \tfrac{1}{16}a^2 - \tfrac{17}{1024}a^4$ and $x(\tau)$ is then, to the same order, given by

$$U(\tau) = 2\cos\tau + \left(\tfrac{3}{4}\sin\tau - \tfrac{1}{4}\sin 3\tau\right)a - \left(\tfrac{5}{96}\cos 5\tau - \tfrac{3}{16}\cos 3\tau + \tfrac{13}{96}\cos\tau\right)a^2$$
$$+ \left(\tfrac{33}{20{,}480}\cos 9\tau - \tfrac{455}{36{,}864}\cos 7\tau + \tfrac{605}{18{,}432}\cos 5\tau - \tfrac{15}{512}\cos 3\tau + \tfrac{83}{11{,}520}\cos\tau\right)a^4$$

Finally, the Lindstedt approximation for $x(t)$ is

$$x(t) = 2\cos T + \left(\tfrac{3}{4}\sin T - \tfrac{1}{4}\sin 3T\right)a - \left(\tfrac{5}{96}\cos 5T - \tfrac{3}{16}\cos 3T + \tfrac{13}{96}\cos T\right)a^2$$
$$+ \left(\tfrac{33}{20{,}480}\cos 9T - \tfrac{455}{36{,}864}\cos 7T + \tfrac{605}{18{,}432}\cos 5T - \tfrac{15}{512}\cos 3T + \tfrac{83}{11{,}520}\cos T\right)a^4$$

where $T = t\left(1 - \tfrac{1}{16}a^2 - \tfrac{17}{1024}a^4\right)$.

The Method of Krylov and Bogoliubov

The first term of the Krylov–Bogoliubov (KB) approximation to the solution of the van der Pol equation is

$$x_{\text{KB}} = \frac{2A_0\sin(t + \phi_0)}{\sqrt{(4 - A_0^2)e^{-at} + A_0^2}} \tag{14.51}$$

where A_0 and ϕ_0 are determined by the initial conditions $x(0) = \alpha$ and $x'(0) = \beta$. Setting

$\phi_0 = \frac{\pi}{2}$ and $A_0 = A$ means the KB approximation will satisfy the initial conditions $x(0) = A$ and $x'(0) = \frac{1}{8}aA(4 - A^2)$. Shortly, we will show how to determine A_0 and ϕ_0 to ensure the KB approximation satisfies the arbitrary initial conditions $x(0) = \alpha$ and $x'(0) = \beta$.

From left to right in the top row of Figure 14.16 we have, for van der Pol's equation with $a = \frac{1}{4}$, a numerically computed trajectory in the phase plane, corresponding results for the Lindstedt approximation (14.48), and the results for the KB approximation. The bottom

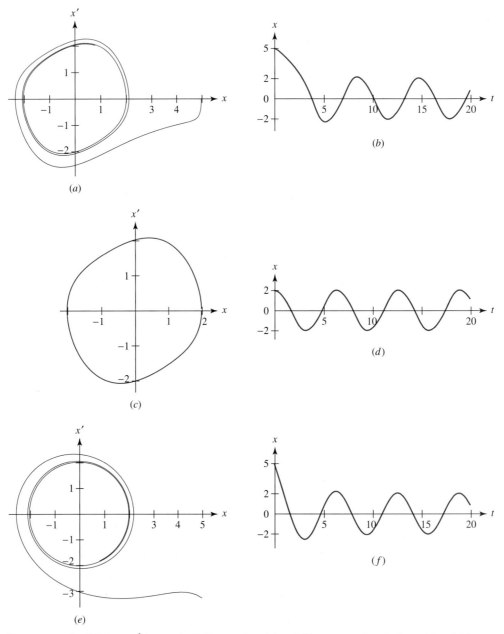

(a)

(b)

(c)

(d)

(e)

(f)

FIGURE 14.16   With $a = \frac{1}{4}$ in van der Pol's equation, (a) and (b) are numeric solutions, (c) and (d) are Lindstedt's solution, and (e) and (f) are the KB solution. For each pair, the first member is in the phase plane, and the second, in the time domain.

row shows the corresponding time-domain representations. The initial conditions for the numeric solution are $x(0) = A = 5$ and $x'(0) = 0$. The Lindstedt approximation depicts the limit cycle. The KB solution uses $x(0) = A = 5$ and $x'(0) = \frac{1}{8}aA(A^2 - 4) = \frac{105}{32}$. The KB approximation is "better" than the Lindstedt approximation because the former shows solutions *approaching* the limit cycle, whereas the Lindstedt approximation to the van der Pol equation only finds the periodic solution as represented by the limit cycle itself.

THE ALGORITHM   The Krylov–Bogoliubov approximation [45] to the differential equation

$$\frac{d^2x}{dt^2} + \omega^2 x + af\left(x, \frac{dx}{dt}\right) = 0 \tag{14.52}$$

for a small is of the form $x(t) = A(t)\sin(\omega t + \phi(t))$, where the functions $A(t)$ and $\phi(t)$ are determined by the differential equations

$$\frac{dA}{dt} = -\frac{a}{2\pi\omega}\int_0^{2\pi} \cos(\theta) f(A\sin\theta, A\omega\cos\theta)\, d\theta$$
$$\frac{d\phi}{dt} = \frac{a}{2\pi A\omega}\int_0^{2\pi} \sin(\theta) f(A\sin\theta, A\omega\cos\theta)\, d\theta \tag{14.53}$$

Each A in the *integrals* on the right-hand sides of these two differential equations is treated as a constant. The reader is cautioned that some texts [62] write the equation as $\frac{d^2x}{dt^2} + \omega^2 x = aF(x, \frac{dx}{dt})$ and the solution as $x(t) = A(t)\cos(\omega t + \phi(t))$, in which case $A(t)$ and $\phi(t)$ are determined by

$$\frac{dA}{dt} = -\frac{a}{2\pi\omega}\int_0^{2\pi} \sin(\theta) F(A\cos\theta, -A\omega\sin\theta)\, d\theta$$
$$\frac{d\phi}{dt} = -\frac{a}{2\pi A\omega}\int_0^{2\pi} \cos(\theta) F(A\cos\theta, -A\omega\sin\theta)\, d\theta$$

The reader is further advised that an extension of this method to terms of higher order in a is known as the method of Krylov, Bogoliubov, and Mitropolsky.

EXAMPLE 14.10   The van der Pol equation in (14.47), with $\omega = 1$ and $f(u,v) = -(1-u^2)v$, is a candidate for approximation by the method of Krylov and Bogoliubov. The integrals in (14.53) give

$$A'(t) = -\int_0^{2\pi} A\cos^2\theta(1 - A^2\sin^2\theta)\, d\theta = -\tfrac{1}{8}aA(-4 + A^2)$$

$$\phi'(t) = -\int_0^{2\pi} A\sin\theta\cos\theta(1 - A^2\sin^2\theta)\, d\theta = 0$$

so that

$$A(t) = \pm\frac{2A_0}{\sqrt{(4 - A_0^2)e^{-at} + A_0^2}} \quad\text{and}\quad \phi(t) = \phi_0$$

The KB approximation $A(t)\sin(t + \phi(t))$ therefore agrees with (14.51).

Finally, we consider the question of initial data. The KB approximation satisfies $x(0) = \alpha$, $x'(0) = \beta$ if the equations

$$A_0\sin\phi_0 = \alpha \quad\text{and}\quad A_0\cos\phi_0 + \frac{a}{2}A_0\sin\phi_0 - \frac{a}{8}A_0^3\sin\phi_0 = \beta$$

are satisfied. Since the equations are at least cubic in A_0, a closed-form solution for A_0 and ϕ_0 will be exceptionally complex. We demonstrate, instead, a numeric solution in the event

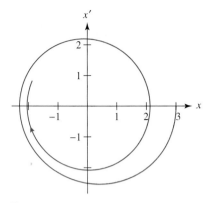

FIGURE 14.17 x_{KB} for $a = \frac{1}{4}, \alpha = 3, \beta = 0$

$a = \frac{1}{4}, \alpha = 3, \beta = 0$. Thus, we compute $A_0 = 3.039977166$, $\phi_0 = -4.874742954$; obtain the formula for the KB approximation

$$x_{KB} = 6.079954332 \frac{\sin(t - 4.874742954)}{\sqrt{9.241461170 - 5.241461170 e^{-t/4}}}$$

and graph its phase-plane trajectory in Figure 14.17.   ❖

DERIVATION   For the case $a = 0$, the differential equation (14.52) will have the solution $x(t) = A \sin(\omega t + \phi)$, which necessarily satisfies $\frac{dx}{dt} = \omega A \cos(\omega t + \phi)$, where A and ϕ are arbitrary constants of integration. For small values of a, Krylov and Bogoliubov assumed the solution satisfied $x(t) = A(t) \sin(\omega t + \phi(t))$ and $\frac{dx}{dt} = \omega A(t)\cos(\omega t + \phi(t))$. The alert reader will immediately notice that the formal derivative of $x(t)$ does not have the form assumed by Krylov and Bogoliubov. Hence, the original nonlinear differential equation and the two assumptions of Krylov and Bogoliubov are used to determine the functions $A(t)$ and $\phi(t)$.

Formally differentiate $x(t)$ and equate the result to the assumed form of $\frac{dx}{dt}$ resulting in the equation

$$A'(t) \sin(\omega t + \phi(t)) + A(t) \cos(\omega t + \phi(t))(\omega + \phi'(t)) = \omega A(t) \cos(\omega t + \phi(t))$$

Substitute into (14.52) the two assumptions of Krylov and Bogoliubov. Do this by differentiating the assumed form of $x'(t) = \frac{dx}{dt}$ to get $x''(t)$, then using the assumed forms for $x(t)$ and $x'(t) = \frac{dx}{dt}$ in each other instance where these terms are needed. The result is

$$\omega A'(t), \cos(\omega t + \phi(t)) - \omega A(t) \sin(\omega t + \phi(t))(\omega + \phi'(t)) + \omega^2 A(t) \sin(\omega t + \phi(t))$$

$$+ af(A(t) \sin(\omega t + \phi(t)), \omega A(t) \cos(\omega t + \phi(t))) = 0$$

We now have two equations in the derivatives A' and ϕ'. Solving for these derivatives, we obtain

$$A'(t) = -\frac{\cos(\omega t + \phi(t)) af(A(t) \sin(\omega t + \phi(t)), \omega A(t) \cos(\omega t + \phi(t)))}{\omega}$$

$$\phi'(t) = \frac{af(A(t) \sin(\omega t + \phi(t)), \omega A(t) \cos(\omega t + \phi(t))) \sin(\omega t + \phi(t))}{\omega A(t)}$$

A slight simplification follows if we define $\theta = \omega t + \phi(t)$, leading to

$$A'(t) = -\frac{\cos(\theta) af(A(t) \sin(\theta), \omega A(t) \cos(\theta))}{\omega}$$

$$\phi'(t) = \frac{af(A(t) \sin(\theta), \omega A(t) \cos(\theta)) \sin(\theta)}{\omega A(t)}$$

In the right-hand sides of these equations, we treat $A(t)$ as the constant A, yielding

$$A'(t) = -\frac{a \cos \theta f(A \sin \theta, \omega A \cos \theta)}{\omega}$$

$$\phi'(t) = \frac{a \sin \theta f(A \sin \theta, \omega A \cos \theta)}{\omega A}$$

Finally, the right-hand sides are replaced by their averages over one period, and the result is (14.53).

EXERCISES 14.6

1. Carry out the change of variables $\tau = \omega t$ by which (14.47) becomes (14.50).

2. Obtain the Lindstedt approximation to the van der Pol equation. Show where the Lindstedt approximation to this equation is forced to select the periodic solution having $x(0) = A = 2$.

3. For $y'' + ay' + y = 0$, $y(0) = 1$, $y'(0) = 1$, an initial value problem for a linear differential equation:

  (a) Obtain the exact solution.

  (b) Obtain the Krylov–Bogoliubov approximation.

  (c) Set $a = 0.1$ and graph, on the same set of axes, the solutions found in parts (a) and (b).

In Exercises 4–9 take the initial conditions as $y(0) = 1$, $y'(0) = 0$.

  (a) Write the differential equation as a first-order system and determine all equilibrium points.

  (b) By linearizing, determine (when possible) the nature of each equilibrium point found in part (a).

  (c) For a small, obtain a regular Poincaré perturbation expansion.

  (d) Obtain a Lindstedt expansion.

  (e) Obtain the Krylov–Bogoliubov approximation.

  (f) Set $a = 0.1$ and obtain a numeric solution.

  (g) Use appropriate graphs to compare the four solutions found in parts (c)–(f).

  (h) In the phase plane, plot the trajectory corresponding to the numeric solution found in part (d).

4. $y'' + y - a(1 + y)^2 = 0$ **5.** $y'' + y + ay^2 = 0$

6. $y'' + y + ay^3 = 0$ **7.** $y'' + y + a(y')^2 = 0$

8. $y'' + y + a((y')^2 + y^2) = 0$ **9.** $y'' + y + a((y')^2 + y^3) = 0$

In Exercises 10–17, take the initial conditions as $y(0) = 1$, $y'(0) = 0$.

  (a) Obtain the Krylov-Bogoliubov approximation.

  (b) Graphically compare the approximation in part (a) with a numeric solution of the given IVP.

10. $y'' + y - a(\frac{1}{5}(y')^3 - y') = 0$

11. $y'' + y + a(y' + y^2) = 0$, $y(0) = 1$, $y'(0) = -1$

12. $y'' + y + a(y' + y^3) = 0$

13. $y'' + y + a((y')^2 + y)y' = 0$, $y(0) = A$, $y'(0) = 0$

14. $y'' + y + a((y')^3 + y)y' = 0$ **15.** $y'' + y + a((y')^3 + y^2) = 0$

16. $y'' + y + a((y')^3 + y^3) = 0$ **17.** $y'' + y + a(y' + y)y' = 0$

Chapter Review

1. Given the IVP $y'' + 2y' + y = \cos t$, $y(0) = y'(0) = 0$:

  (a) Obtain the exact solution.

  (b) Expand the exact solution in a Maclaurin series.

  (c) Substitute $y = \sum_{k=0}^{5} a_k t^k$ into the differential equation, and by matching coefficients, recreate the first few terms of the series solution of the IVP.

  (d) Construct the first few terms of the series solution by repeatedly differentiating the differential equation itself.

2. For a differential equation of the form $p(t)y'' + q(t)y' + r(t)y = f(t)$:

  (a) What is a regular point, a regular singular point, and an irregular singular point?

  (b) For each, illustrate with a differential equation.

  (c) What can be said about the form of a solution at each such point?

3. Let $f(t) = \frac{t}{1+t}$ and $g(t) = \frac{1}{1+t^2}$.

  (a) Obtain the asymptotic expansions $f \sim \sum_{k=0}^{\infty} a_k t^{-k}$ and $g \sim \sum_{k=0}^{\infty} b_k t^{-k}$.

  (b) Determine λ and α by the equations $\lambda^2 + a_0\lambda + b_0 = 0$ and $(2\lambda + a_0)\alpha + \lambda a_1 + b_1 = 0$.

  (c) For the differential equation $y'' + f(t)y' + g(t)y = 0$, determine asymptotic solutions of the form $y \sim e^{\lambda t}t^\alpha(1 + \sum_{k=1}^{\infty} c_k t^{-k})$.

4. Obtain $x(a) \sim \sum_{k=0}^{\infty} x_k a^k$, a perturbation series solution of the equation $x = 1 + ae^x$.

5. For the IVP $y'' + 2ay' + y = \cos t$, $y(0) = A$, $y'(0) = 0$:

  (a) Obtain the exact solution and expand it in a Maclaurin series about $a = 0$.

  (b) Obtain a regular (Poincaré) perturbation solution of the form $\sum_{k=0}^{\infty} \phi_k(t)a^k$ and compare it with the series obtained in part (a).

  (c) Obtain a perturbation solution using Lindstedt's method, being sure to provide the frequency-amplitude relationship.

  (d) Obtain the Krylov–Bogoliubov approximation.

6. Explain the difference between the regular (Poincaré) perturbation, the Lindstedt perturbation, and the Krylov–Bogoliubov approximation. What are the distinguishing characteristics and ideas for each of these three approximations to the solution of an IVP?

Chapter 15

Boundary Value Problems

INTRODUCTION   Up to this point, the initial value problem has been the focus of our study. We now consider the *boundary value problem* in which restrictions are placed on the solution at the endpoints of an interval. Such problems arise, for example, if position at initial and terminal times is prescribed, or if temperatures are prescribed at both ends of an object like a rod. Techniques and theory for such problems will guide us when we study the related eigenvalue problem.

As with initial value problems, exact solutions are available for a small class of BVPs. In the general case, numeric methods and the analytic approximation technique are needed. Hence, we study the *shooting method* and the technique of *finite differences* to obtain numeric solutions of BVPs. We also study the analytic approximations generated by *least-squares, Rayleigh–Ritz, Galerkin,* and *collocation* techniques. These methods all begin with a solution containing undetermined parameters that the methods determine by minimizing the *residual* in some way. These four methods differ in how that residual is minimized.

A final analytic approximation is provided by the *finite element* method. If the Rayleigh–Ritz, Galerkin, or collocation techniques are used with functions that are nonzero only on subsets of the solution interval, the method becomes a finite element technique. Depending on the approximating functions used, the finite element method can generate a classical (*strong*) solution or a *weak* solution that, like the Green's function, does not possess enough derivatives to satisfy the DE pointwise.

15.1 Analytic Solutions

EXAMPLE 15.1   Suppose a shock absorber is being designed so that an attached unit mass, displaced one unit from rest, returns to the equilibrium position at $t = 1$. Suppose further that the mathematical model for the system uses

$$y''(t) + 4y'(t) + 13y(t) = 0 \tag{15.1}$$

as the equation of a damped oscillator. Thus, the two *boundary conditions* $y(0) = 0$ and $y(1) = 0$ must be imposed on the solution of (15.1). The differential equation and the two conditions, imposed at the endpoints of the interval $0 \le t \le 1$, constitute a *boundary value problem* that has the solution

$$y(t) = e^{-2t}(\cos 3t - \cot(3) \sin 3t) \tag{15.2}$$

whose graph is found in Figure 15.1.

It is interesting that the mass must be displaced more than three units in order for the displacement to be zero at $t = 1$. Thus, the behavior of solutions of boundary value problems (BVPs) will be at least as interesting as the behavior of initial value problems (IVPs) has proven to be. ❖

EXAMPLE 15.2 Suppose we wanted to know how the displacements in Example 15.1 would depend on a if the condition at $t = 1$ were changed to $y(1) = a$. The solution can be found by applying the two boundary conditions to the general solution

$$y(t) = e^{-2t}(c_1 \cos 3t + c_2 \sin 3t)$$

thereby producing the two equations

$$c_1 = 1 \quad \text{and} \quad e^{-2}(c_1 \cos 3 + c_2 \sin 3) = a$$

whose solution gives

$$y_a(t) = e^{-2t} \cos 3t + \frac{ae^2 - \cos 3}{\sin 3} e^{-2t} \sin 3t$$

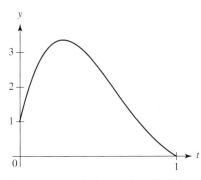

FIGURE 15.1 Solution of the BVP in Example 15.1

Figure 15.2(a) shows the displacement's dependence on the terminal value at $t = 1$ in a three-dimensional plot of the solution considered as a function $y(t, a)$. Each plane section $a = c$, constant, is a solution curve for the differential equation and the boundary conditions $y(0) = 1$, $y(1) = c$. Several such plane sections are shown in Figure 15.2(b). ❖

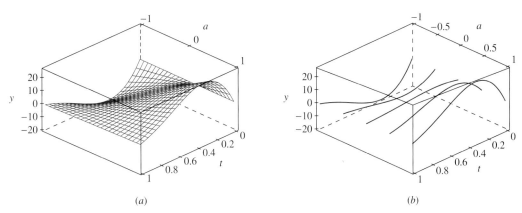

(a) (b)

FIGURE 15.2 Example 15.2: (a) shows $y_a(t)$ as the surface $y(t, a)$ while (b) shows as plane sections, solutions of BVPs determined by several values of a

EXAMPLE 15.3 The differential equation

$$y(x)y''(x) - y'(x)^2 - 1 = 0 \tag{15.3}$$

determines the curve that, when rotated about the x-axis, generates a surface of revolution of minimal surface area. (See Unit Nine.) If the curve must connect the points $(0, 1)$ and $(1, 2)$, we have the boundary conditions $y(0) = 1$ and $y(1) = 2$. The general solution of the nonlinear differential equation is $y(t) = \frac{1}{\alpha} \cosh(\alpha(x + \beta))$, and the solution satisfying the boundary conditions is

$$y(t) = 0.9499888271 \cosh(1.052643959x + 0.3230741021) \tag{15.4}$$

since the algebraic equations for the constants α and β virtually demand a numeric solution. The curve appears in Figure 15.3(a), and the resulting surface of minimal surface area appears in Figure 15.3(b). ❖

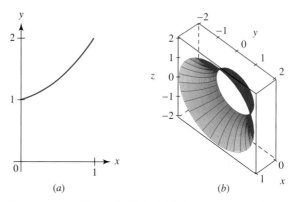

(a) $\qquad\qquad\qquad\qquad$ (b)

FIGURE 15.3   Example 15.3: (a) is the curve generating minimal surface of revolution while (b) is the generated surface

EXAMPLE 15.4   Some BVPs have only the trivial solution $y(x) = 0$, some have multiple solutions, and some have no solution. The only solution of the first BVP in Table 15.1 is $y(x) = 0$ because the two equations in the rightmost column, obtained by applying the boundary conditions to the DEs general solution in the middle column, determine the coefficients to be $c_1 = c_2 = 0$.

	BVP	General Solution $y(x)$	Equations
(1)	$y''(x) + y(x) = 0$ $y(0) = y(1) = 0$	$c_1 \cos x + c_2 \sin x$	$c_1 = 0$ $c_1 \cos 1 + c_2 \sin 1 = 0$
(2)	$y''(x) + y(x) = 0$ $y(0) = y(\pi) = 0$	$c_1 \cos x + c_2 \sin x$	$c_1 = 0$ $-c_1 = 0$
(3)	$y''(x) + y(x) = x$ $y(0) = y(\pi) = 0$	$x + c_1 \cos x + c_2 \sin x$	$c_1 = 0$ $\pi - c_1 = 0$

TABLE 15.1   Example 15.4: BVPs with one, many, and no solutions

The second BVP in Table 15.1 has infinitely many solutions of the form $y(x) = c_2 \sin x$ because the equations in the rightmost column both determine $c_1 = 0$.

The third BVP in Table 15.1 has no solution at all since the equations in the rightmost column place contradictory requirements on c_1. ❖

Fredholm Alternative

The results in Example 15.4 illustrate the *Fredholm alternative,* an existence-uniqueness theorem named after the Swedish mathematician Ivar Fredholm, who proved similar results for more general equations.

THEOREM 15.1

If $p(x)$ and $q(x)$ are continuous, the boundary value problem

$$y'' + p(x)y' + q(x)y = f(x) \quad \text{and} \quad y(a) = 0, \, y(b) = 0$$

has a solution for all continuous $f(x)$ precisely when the completely homogeneous BVP

$$y'' + p(x)y' + q(x)y = 0 \quad \text{and} \quad y(a) = 0, \, y(b) = 0$$

has only the trivial solution $y(x) = 0$. When the completely homogeneous BVP has multiple solutions, it clearly has at least one nontrivial solution. When the completely homogeneous BVP has multiple solutions, the nonhomogeneous BVP has no solution.

EXAMPLE 15.5   On the right in (15.5) is the general solution of the nonhomogeneous, but linear, DE on the left

$$x^2 y''(x) - 2xy'(x) + 2y(x) = x^2 \Rightarrow y(x) = c_1 x + c_2 x^2 + \tfrac{1}{2}x^3 \qquad \text{(15.5)}$$

The BVP consisting of the ODE in (15.5) and the mixed, but linear, boundary conditions

$$2y(1) - 3y'(1) + 5y(2) - 4y'(2) = 0 \qquad 3y(1) + 2y'(1) - 4y(2) + 5y'(2) = 4$$

is considered nonhomogeneous, both because of the ODE in (15.5) and because of the second boundary condition. Solving this BVP requires solving

$$5c_1 - \tfrac{15}{2} = 0 \quad \text{and} \quad 2c_1 + 11c_2 + \tfrac{37}{2} = 4$$

the algebraic equations obtained when applying the boundary conditions to the general solution in (15.5). A bit of algebra yields $c_1 = \tfrac{3}{2}$, $c_2 = -\tfrac{35}{22}$, and the solution

$$y(x) = \tfrac{3}{2}x + \tfrac{35}{22}x^2 + \tfrac{1}{2}x^3 \qquad \text{(15.6)}$$

whose graph is seen in Figure 15.4. The graph suggests, and calculation confirms, that the unspecified endpoint values are $y(1) = \tfrac{9}{22}$ and $y(2) = \tfrac{7}{11}$.   ❖

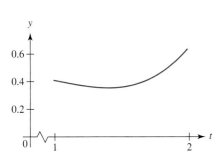

FIGURE 15.4   Solution of the BVP in Example 15.5

Thus, the general second-order linear BVP is of the form

$$a_0(x)y'' + a_1(x)y' + a_2(x)y = f(x)$$
$$B_1(y) = c_{11}y(a) + c_{12}y'(a) + d_{11}y(b) + d_{12}y'(b) = \alpha$$
$$B_2(y) = c_{21}y(a) + c_{22}y'(a) + d_{21}y(b) + d_{22}y'(b) = \beta$$

The BVP is called *regular* if $a_0(x) \neq 0$ in the interval $[a, b]$ and *singular* otherwise. Special cases of the general mixed boundary conditions are the *separated* boundary conditions

$$c_{11}y(a) + c_{12}y'(a) = \alpha \quad \text{and} \quad d_{21}y(b) + d_{22}y'(b) = \beta$$

and the *periodic* boundary conditions

$$y(a) = y(b) \quad \text{and} \quad y'(a) = y'(b)$$

Green's Functions

Given the inhomogeneous BVP whose differential equation is $y''(x) = x^2$ and whose boundary conditions are $y(0) = y(1) = 0$, we have the immediate solution $y(x) = \tfrac{1}{12}x^4 - \tfrac{1}{12}x$. Recalling Sections 5.13 and 6.11, we seek a Green's function for the inhomogeneous BVP. The Green's function will be a fundamental solution that also satisfies the given boundary conditions. A fundamental solution is a particular solution of the equation $y''(x) = \delta(x - t)$. Taking the Laplace transform with respect to x and writing $y'(0) = \alpha$, we obtain $Y(s) = \frac{\alpha}{s^2} + e^{-st}/s^2$ and therefore $y(x) = \alpha x + (x - t)H(x - t)$, where $H(x)$ is the Heaviside function. The endpoint condition $y(1) = 0$ then gives $\alpha = (t - 1)H(1 - t)$, so the Green's function is

$$G(x, t) = (x - t)H(x - t) + (t - 1)xH(1 - t) \qquad (15.7)$$

The solution of the inhomogeneous BVP then follows from the Green's function $G(x, t)$ via the integral $y(x) = \int_0^1 G(x, t) f(t)\, dt$, where $f(x) = x^2$ is the right-hand side of the differential equation. Hence, we again find $y(x) = \frac{1}{12}x^4 - \frac{1}{12}x$.

EXERCISES 15.1–Part A

A1. Verify that (15.2) solves the BVP of Example 15.1.

A2. Verify that $y(x) = \frac{1}{\alpha} \cosh(\alpha(x + \beta))$ is the general solution of (15.3). Apply the boundary conditions, and obtain the solution given in (15.4).

A3. Verify the general solution given in (15.5).

A4. Show that the general solution given in (15.6) satisfies the BVP of Example 15.5.

A5. Verify that as a function of x, the Green's function given in (15.7) satisfies the boundary conditions $y(0) = y(1) = 0$. Verify that $y(x) = \int_0^1 G(x, t) f(t)\, dt = -\frac{1}{12}x + \frac{1}{12}x^4$.

A6. An insulated rod is coincident with the interval $[0, 1]$ on the x-axis. Its left end is kept at a temperature of zero, and its right end at one. The steady-state temperatures in the rod are determined by the BVP $-u''(x) = 0$, $u(0) = 0$, $u(1) = 1$. Obtain $u(x)$, and explain its physical significance.

A7. The rod in Exercise A6 has both ends kept at a temperature of zero. However, there is a heat source internal to the rod so that the steady-state temperatures are now determined by the BVP $-u''(x) = x(1 - x)$, $u(0) = u(1) = 0$. Obtain and graph $u(x)$.

EXERCISES 15.1–Part B

B1. Carry out the steps of the following algorithm that finds the Green's function $G(x, t)$ for the BVP $y'' + 3y' + 2y = x$, $y(0) + y'(0) = 0$, $y(1) - y'(1) = 0$, and then gives the solution of the BVP.

  (a) Let $y_1(x)$ be a solution of the homogeneous DE $a_0 y'' + a_1 y' + a_2 y = 0$ satisfying the boundary condition $B_1(y) = 0$ at $x = a$.

  (b) Let $y_2(x)$ be a solution of the homogeneous DE satisfying the boundary condition $B_2(y) = 0$ at $x = b$. Be sure that y_1 and y_2 are linearly independent.

  (c) Let $W(x)$ be the Wronskian of $y_1(x)$ and $y_2(x)$.

  (d) The Green's function is

$$G(x, t) = \begin{cases} \dfrac{y_1(x)y_2(t)}{a_0(t)W(t)} & a \le x < t \\[2ex] \dfrac{y_1(t)y_2(x)}{a_0(t)W(t)} & t < x \le b \end{cases}$$

  (e) The solution of the BVP is $y(x) = \int_a^b G(x, t)t\, dt = \int_a^x \frac{y_1(t)y_2(x)}{a_0(t)W(t)} t\, dt + \int_x^b \frac{y_1(x)y_2(t)}{a_0(t)W(t)} t\, dt$

For the BVPs in Exercises B2–11:

  (a) Obtain an analytic solution with a suitable computer algebra system.

  (b) Graph the solution.

  (c) Obtain the general solution of the differential equation.

  (d) Form, and solve, the two algebraic equations that result from applying the boundary conditions to the general solution.

  (e) Write the solution of the BVP as per the results of part (d), comparing it to the solution found in part (a).

B2. $2y'' - 5y' + 3y = 0$, $y(0) = -1$, $y(1) = 1$

B3. $40y'' - 47y' + 12y = 0$, $y(0) = 2$, $y'(1) = 0$

B4. $28y'' - 41y' + 15y = 0$, $y(0) = -2$, $y(2) = 1$

B5. $6y'' - 13y' + 7y = 0$, $y(0) = 0$, $y'(1) = 1$

B6. $5y'' - 8y' + 3y = 0$, $y(0) = 5$, $y(3) = -1$

B7. $x^2 y'' + xy' - 25y = 0$, $y(1) = 1$, $y(3) = -1$

B8. $x^2 y'' - 2xy' - 4y = 0$, $y(1) = 0$, $y'(2) = 1$

B9. $x^2 y'' - 4xy' + 6y = 0$, $y'(1) = -1$, $y(3) = 0$

B10. $x^2 y'' + 7xy' + 8y = 0$, $y'(1) = 0$, $y(2) = 1$

B11. $x^2 y'' - 2xy' - 4y = 0$, $y(1) = -2$, $y(3) = 2$

For the BVPs in Exercises B12–21:

  (a) Obtain the general solution of the differential equation.

  (b) Form, and solve, the two algebraic equations which result from applying the boundary conditions to the general solution.

  (c) Write the solution of the BVP.

  (d) Graph the solution.

  (e) From the solution, determine the values of $y(x)$ at the endpoints.

  (f) Apply the differential equation solver in your favorite computer algebra system, and if it succeeds, compare the solution to that found in part (c).

B12. $16y'' - 86y' + 63y = 0$, $5y(0) + 9y'(0) + 7y(1) + 9y'(1) = 8$, $y(0) - 5y'(0) - 9y(1) - 2y'(1) = 4$

B13. $9y'' - 26y' + 16y = 0$, $8y(0) - 3y'(0) + 5y(1) + 3y'(1) = -9$, $3y(0) - 7y'(0) + 7y(1) + 4y'(1) = 7$

B14. $24y'' - 37y' + 14y = 0$, $3y(0) - 6y'(0) - 8y(1) + 3y'(1) = -5$, $3y(0) - 6y'(0) + 8y(1) - y'(1) = 2$

B15. $40y'' - 33y' - 18y = 0$, $4y(0) + 9y'(0) - 4y(1) = 3$, $4y(0) + 4y'(0) + 8y(1) + 9y'(1) = -2$

B16. $24y'' + 14y' - 5y = 0$, $7y(0) + 7y'(0) + y(1) - 6y'(1) = -9$, $6y(0) + 8y'(0) + 8y(1) - 6y'(1) = 4$

B17. $21y'' - 22y' - 63y = 0$, $y(0) + 7y'(0) - 5y(1) - 3y'(1) = 7$, $2y(0) + y'(0) + y(1) - 8y'(1) = -1$

B18. $y'' + 4y' + 13y = 0$, $6y(0) - 5y'(0) - 4y(1) + 9y'(1) = 6$, $5y(0) + 9y'(0) - 9y(1) - 9y'(1) = -4$

B19. $y'' - 10y' + 21y = 0$, $3y(0) - 8y'(0) + 6y(1) - 9y'(1) = -7$, $-3y(0) + 3y'(0) + 5y(1) - y'(1) = -8$

B20. $y'' - 4y' + 3y = 0$, $9y(0) + 8y'(0) - 7y(1) - 7y'(1) = 3$, $4y(0) - y'(0) + y(1) + 4y'(1) = -6$

B21. $40y'' + 57y' + 20y = 0$, $7y(0) + 6y'(0) - 8y(1) - 7y'(1) = -6$, $y(0) + 8y'(0) + 8y(1) - 2y'(1) = 2$

In Exercises B22–31, the given BVPs have a differential equation of the form $L[y(x)] = f(x)$, where L represents the operations on the left-hand side and have boundary conditions of the form $B_1(y) = \alpha$ and $B_2(y) = \beta$. For each:

  (a) Obtain an analytic solution with a suitable computer algebra system.

  (b) Graph the solution.

  (c) Obtain the general solution of the differential equation.

  (d) Form, and solve, the two algebraic equations that result from applying the boundary conditions to the general solution.

  (e) Write the solution of the BVP as per the results of part (d), comparing it to the solution found in part (a).

  (f) Obtain the Green's function $G(x, t)$ by the prescription of Exercise B1.

  (g) Obtain the solution of the boundary value problem as $y(x) = \int_0^1 G(x, t) f(t) \, dt$.

B22. $40y'' - 53y' + 9y = x$, $y(0) = 0$, $y(1) = 0$

B23. $7y'' - 50y' + 7y = 1 - x^2$, $y(0) = 0$, $y(1) = 0$

B24. $18y'' - 21y' + 5y = \cos x$, $y(0) = 0$, $y(1) = 0$

B25. $2y'' + 15y' - 27y = e^{-x}$, $y(0) = 0$, $y(1) = 0$

B26. $27y'' + 3y' - 10y = 1 - \sin x$, $y(0) = 0$, $y(1) = 0$

B27. $2y'' - y' - 3y = x + e^{-x}$, $y(0) = 0$, $y(1) = 0$

B28. $36y'' - 73y' - 18y = 2 + 3x$, $y'(0) = 0$, $y(1) = 0$

B29. $16y'' + 18y' + 5y = 3 - \cos 2x$, $y'(0) = 0$, $y'(1) = 0$

B30. $4y'' + 15y' + 9y = x + 2\sin x$, $y'(0) = 0$, $y(1) = 0$

B31. $5y'' + 24y' + 27y = 3x^2 + \sin \pi x$, $y'(0) = 0$, $y(1) = 0$

In Exercises B32–41, the given BVPs have a differential equation of the form $L[y(x)] = f(x)$, where L represents the operations on the left-hand side and have boundary conditions of the form $B_1(y) = \alpha$ and $B_2(y) = \beta$. For each:

  (a) Solve the completely homogeneous problem $L[y(x)] = 0$, $B_1(y) = 0$, $B_2(y) = 0$.

  (b) If the homogeneous problem in part (a) has only the trivial solution $y = 0$, solve the nonhomogeneous problem.

  (c) Obtain the general solution of the nonhomogeneous differential equation.

  (d) To the solution obtained in part (c), apply the nonhomogeneous boundary conditions, thereby obtaining two algebraic equations in two unknowns.

  (e) Show that the solution of the equations in part (d) leads to the solution of the nonhomogeneous BVP, provided the only solution found in part (b) is the trivial solution, $y = 0$. Otherwise, the algebraic equations formed in part (d) will have no solution and the nonhomogeneous BVP will also have no solution.

  (f) If the nonhomogeneous BVP has a solution, obtain its graph.

  (g) If the nonhomogeneous BVP has a solution, obtain it by the following extension of the method of Exercise B1. For the BVP with *homogeneous* boundary conditions where $B_1(y_1) = B_2(y_2) = 0$, obtain $G(x, t)$ as in Exercise B1. Then write $y(x) = \int_a^b G(x, t) f(t) \, dt + \frac{\beta}{B_2(y_1)} y_1(x) + \frac{\alpha}{B_1(y_2)} y_2(x)$, where $B_1(y) = \alpha$ and $B_2(y) = \beta$ are the given nonhomogeneous boundary conditions.

B32. $y'' + 2y' + 26y = 1$, $y(0) = 1$, $y\left(\frac{\pi}{2}\right) = -1$

B33. $y'' + 10y' + 41y = x$, $y(0) = -1$, $y\left(\frac{\pi}{2}\right) = 1$

B34. $y'' + 4y' + 20y = 1 + x$, $y(0) = 2$, $y(\pi) = -1$

B35. $y'' + 8y' + 20y = e^{-x}$, $y(0) = -2$, $y(\pi) = 1$

B36. $y'' + 8y' + 41y = x^2$, $y(0) = 0$, $y'\left(\frac{\pi}{2}\right) = 1$

B37. $y'' + 4y' + 13y = \cos x$, $y'(0) = 1$, $y(\pi) = -1$

B38. $y'' + 2y' + 5y = \sin x$, $y'(0) = -1$, $y\left(\frac{\pi}{2}\right) = 0$

B39. $y'' + 6y' + 18y = xe^{-x}$, $y'(0) = 2$, $y'(\pi) = 1$

B40. $y'' + 2y' + 17y = x\cos x$, $y'(0) = -2$, $y(\pi) = 1$

B41. $y'' + 10y' + 50y = x\sin x$, $y'(0) = 0$, $y\left(\frac{\pi}{2}\right) = 1$

15.2 Numeric Solutions

The Shooting Method

The *shooting method* is a numeric technique for solving a two-point boundary value problem. Based on the image of firing an artillery shell from a cannon, the method converts the boundary value problem into an initial value problem. The launch angle, $y'(0)$, is parametrized, and the solution is "launched" repeatedly until the "target is hit" at the endpoint of the interval.

EXAMPLE 15.6   Consider the BVP defined by

$$y''(x) + 2y'(x) + 10y(x) = \sin x \quad \text{and} \quad y(0) = 0, y(1) = 1 \tag{15.8}$$

Numeric solvers for ODEs actually solve initial value problems, so we convert (15.8) to the initial value problem

$$y''(x) + 2y'(x) + 10y(x) = \sin x \quad \text{and} \quad y(0) = 0, y'(0) = a \tag{15.9}$$

by replacing the boundary condition at $x = 1$ with the initial condition $y'(0) = a$. This IVP is solved for various values of a until a solution is found that attains $y(1) = 1$. The exact solution of (15.9) is

$$y_a(x) = \tfrac{1}{255}(27\sin x - 6\cos x + e^{-x}(6\cos 3x + (85a - 7)\sin 3x)) \tag{15.10}$$

and Figure 15.5 shows how $y_a(x)$ varies with a.

Of course, it would be inefficient to hunt at random for a value of a that puts the solution through the point $(1, 1)$. Analytically solving the equation $y_a(1) = 1$ for a gives $a = A$, where

$$A = \frac{e(255 + 6\cos 1 - 27\sin 1) - 6\cos 3 + 7\sin 3}{85\sin 3} \overset{\circ}{=} 53.95 \tag{15.11}$$

Figure 15.6 shows $y_A(x)$, the exact solution of (15.8).

However, we need to consider the case where $y_a(x)$ cannot be found analytically. Certainly, we don't want to engage in an inefficient search for the right value of a. We want to be able to determine a as the numeric solution of an "equation" $F(a) = y_a(1) - 1 = 0$. Each evaluation of $F(a)$ invokes a numeric differential equation solver that returns $y_a(1)$.

Newton's method from elementary calculus solves the equation $f(a) = 0$ by computing the iterates $a_{n+1} = a_n - f(a_n)/f'(a_n)$, starting from some initial a_0. To avoid computing $f'(a)$ required when solving $f(a) = 0$ by Newton's method, we use the *secant method*, which replaces the derivative with the difference quotient $(f(a_n) - f(a_{n-1}))/(a_n - a_{n-1})$, resulting in the iteration

$$a_{n+1} = a_n - \frac{f(a_n)(a_n - a_{n-1})}{f(a_n) - f(a_{n-1})} \tag{15.12}$$

Applying (15.12) to $F(a) = 0$ with $a_1 = 0$ and $a_2 = 1$ gives 53.95021153, 53.95018460, 53.95018462, 53.95018463, and 53.95018463 as the next five iterates. ❖

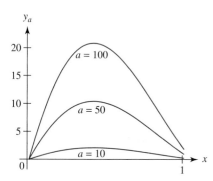

FIGURE 15.5   Several members of the family $x_a(t)$ in Example 15.6

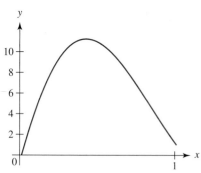

FIGURE 15.6   Exact solution of the BVP in Example 15.6

The Finite Difference Method

Replacing derivatives with approximating difference quotients leads to the *finite difference* method, an alternative for solving a boundary value problem. We will illustrate the method for a linear differential equation, but the method works for any equation that is easily put into the form $y''(x) = f(x, y(x), y'(x))$. Initially, we consider boundary conditions of the form $y(a) = \alpha$, $y(b) = \beta$ but later show how more general boundary conditions can be treated.

The interval $[a, b]$ is uniformly partitioned into N subintervals of length $h = \frac{b-a}{N}$, resulting in the partition points $x_k = a + kh, k = 0, 1, \ldots, N$. The derivatives at x_k are replaced with the $O(h^2)$ centered approximations

$$y''(x_k) = \frac{y_{k+1} - 2y_k + y_{k-1}}{h^2} \quad \text{and} \quad y'(x_k) = \frac{y_{k+1} - y_{k-1}}{2h}$$

where y_k approximates $y(x_k)$. (The formula for y'' is derived at the end of the section.) Substitution of these approximations into a linear differential equation yields a set of linear algebraic equations for $y_k, k = 0, 1, \ldots, N$. Nonlinear equations result in a set of nonlinear algebraic equations best solved with iterative methods such as Gauss–Seidel iteration (discussed in Section 39.5).

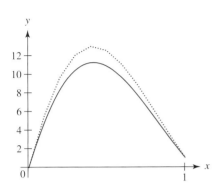

FIGURE 15.7 Exact (solid) and finite-difference (dotted) solutions of the BVP in Example 15.7

EXAMPLE 15.7 To implement the finite difference scheme for the BVP in (15.8), set $N = 10$ so that $h = \frac{1}{10}$ and $x_k = kh$. Next, define the function $f(x, y, y') = \sin x - 10x - 2x'$. At each "interior" node $x_k, k = 1, 2, \ldots, N - 1 = 9$, the discretization yields a single equation

$$100(y_{k+1} - 2y_k + y_{k-1}) = \sin\frac{k}{10} - 10(y_{k+1} + y_k - y_{k-1})$$

There are $N + 1 = 11$ unknowns, namely, $y_0, y_1, \ldots, y_{10}$, so we need two more equations. The endpoint conditions yield the remaining two equations $y_0 = 0$ and $y_{10} = 1$. Solving this set of 11 equations in 11 unknowns yields the values in Table 15.2 and the piecewise linear curve (dotted) in Figure 15.7. The solid curve is $y_A(x)$, the exact solution of (15.8). ❖

EXAMPLE 15.8 The finite-difference method can be adapted to more general boundary conditions. For example, in (15.8), change the condition at $x = 1$ to $y'(1) = 1$. The equations at the nine interior nodes remain the same as in Example 15.7. But we must write one more equation of the same type at the endpoint $x = x_{10}$. The resulting 10 equations are then

$$100(y_{k+1} - 2y_k + y_{k-1}) = \sin\frac{k}{10} - 10(y_{k+1} + y_k - y_{k-1}) \qquad k = 1, \ldots, 10$$

The 10th equation introduced the new unknown y_{11}. An 11th equation arises from the boundary condition at $x = 0$, and a 12th equation is generated by applying the finite-difference approximation to the derivative in the condition at $x = 1$. Thus, we have the two additional equations $y_0 = 0$ and $5y_{11} - 5y_9 = 1$, where the second of the two is actually $y'(x_{10}) = \frac{y_{11} - y_9}{2h} = 1$. The numeric solution of this set of 12 equations in 12 unknowns is given in Table 15.3.

The exact solution of the BVP is $y(x) = \lambda(\sigma_1 + e^{-x}\sigma_2 + e^{1-x}\sigma_3)$, where

$$\lambda = 510\cos 3 - 170\sin 3$$

$$\sigma_1 = 3\cos(x + 3) + 25\sin(x - 3) - 15\cos(x - 3) + 29\sin(x + 3)$$

$$\sigma_2 = 4\sin(3x - 3) + 12\cos(3x - 3)$$

$$\sigma_3 = 170\sin 3x - 9\sin(3x + 1) - 9\sin(3x - 1) - 2\cos(3x - 1) + 2\cos(3x + 1)$$

$y_0 = 0$
$y_1 = 5.578937329$
$y_2 = 9.637253870$
$y_3 = 12.08338677$
$y_4 = 12.98896509$
$y_5 = 12.55261795$
$y_6 = 11.05881799$
$y_7 = 8.836404040$
$y_8 = 6.220612425$
$y_9 = 3.521430481$
$y_{10} = 1$

TABLE 15.2 Finite-difference solution in Example 15.7

$y_0 = 0$	$y_3 = -0.4533344866$	$y_6 = -0.3947154904$	$y_9 = -0.0760352663$
$y_1 = -0.2109224910$	$y_4 = -0.4830077653$	$y_7 = -0.3004187839$	$y_{10} = 0.03132355666$
$y_2 = -0.3634130898$	$y_5 = -0.4598359387$	$y_8 = -0.1900996102$	$y_{11} = 0.1239647337$

TABLE 15.3   Finite-difference solution in Example 15.8

FIGURE 15.8   Exact (solid) and finite-difference (dotted) solutions of the BVP in Example 15.8

Figure 15.8 graphs the finite-difference solution (dotted) against the exact solution (solid), with the remarkable result that the finite-difference solution in Example 15.8 is much more accurate than the finite-difference solution in Example 15.7.   ❖

Derivation of the Central Difference Formula for f″

The central difference formula which replaced y'' is obtained by adding the following two Taylor series expansions and solving for $f''(x)$.

$$f(x+h) = f(x) + f'(x)h + \tfrac{1}{2}f''(x)h^2 + \tfrac{1}{6}f'''(x)h^3 + O(h^4)$$

$$f(x-h) = f(x) - f'(x)h + \tfrac{1}{2}f''(x)h^2 - \tfrac{1}{6}f'''(x)h^3 + O(h^4)$$

EXERCISES 15.2–Part A

A1. Using the methods of Chapter 5, obtain the exact solution of (15.8).

A2. Use the Laplace transform to obtain the exact solution of (15.8). Treat $y'(0) = \alpha$ as unknown, and use the condition $y(1) = 1$ to determine α.

A3. Show that the solution in Exercise A2 reflects the process by which $y_a(x)$ is obtained as the solution of (15.9) and by which $y_A(x)$ is obtained as the exact solution of (15.8).

A4. Starting with $x_0 = 3$ and $x_1 = 2.5$, obtain the next five secant-method iterates when $f(x) = x^2 - 3x + 2$.

A5. Obtain the exact solution of the BVP in Example 15.8.

EXERCISES 15.2–Part B

B1. To extend the shooting method to the BVP $y'' + 3y' + 2y = 0$, $y(0) + y(1) = 0$, $y'(0) - y'(1) = 1$, solve the IVP consisting of the same DE and the initial conditions $y(0) = a$, $y'(0) = b$. Solve this IVP analytically, and apply the boundary conditions to determine appropriate values for a and b. A numeric implementation of this process would lead to the desired extension of the shooting method.

For the BVP in each of Exercises B2–11:

(a) Obtain an analytic solution using a computer algebra system.

(b) Graph the solution.

(c) From the analytic solution, obtain the value of $y'(0)$.

(d) Obtain $G(x, t)$, the Green's function for the homogeneous problem.

(e) Obtain the solution as in Exercises B32–41, Section 15.1.

(f) Obtain the solution by the shooting method.

(g) Taking $h = \frac{1}{10}$, obtain the solution by the finite-difference method.

(h) Compare the two numeric solutions with the exact solution.

B2. $3y'' - 7y' + 4y = xe^{-2x}$, $y(0) = -2$, $y(1) = 1$

B3. $2y'' - 7y' + 3y = 1 - x + x\sin x$, $y(0) = -1$, $y(1) = 3$

B4. $21y'' - 8y' - 4y = 1 - 3x^2$, $y(0) = 1$, $y(1) = -2$

B5. $10y'' + y' - 3y = x$, $y(0) = 2$, $y(1) = 0$ on the right in B5, just x survives

B6. $3y'' + 5y' - 2y = x - 3\sin 2x$, $y(0) = -4$, $y(1) = 2$

B7. $15y'' + 52y' + 45y = x$, $y(0) = -7$, $y(1) = 1$

B8. $y'' + 2y' + 10y = 7e^{-x} + 4$, $y(0) = 2$, $y(1) = -5$

B9. $56y'' - 99y' + 40y = 1 + 3x^4$, $y(0) = \pi$, $y(1) = e$

B10. $y'' - 2y' - 3y = 6x^2 \sin 2x$, $y(0) = 2e$, $y(1) = 5$

B11. $y'' + 10y' + 26y = 2\cos(3x - 1)$, $y(0) = -6$, $y(1) = 5$

For the BVP in each of Exercises B12–26:

    (a) Obtain an analytic solution by using a computer algebra system

    (b) Graph the solution.

    (c) From the analytic solution, obtain the value of $y'(0)$.

    (d) Obtain $G(x, t)$, the Green's function for the homogeneous problem.

    (e) Obtain the solution as in Exercises B32–41, Section 15.1.

    (f) Obtain the solution by an extension of the shooting method. (Define a function $h(a)$ which, for each value $a = y'(0)$, returns a numerically computed $y'(1)$.)

    (g) Obtain the solution by the finite-difference method.

    (h) Compare the two numeric solutions with the exact solution.

B12. $12y'' - 8y' + y = e^{-x}$, $y(0) = 2$, $y'(1) = 0$

B13. $7y'' - 16y' + 4y = x$, $y(0) = -3$, $y'(1) = 2$

B14. $2y'' - 9y' + 7y = 1 - x^2$, $y(0) = 5$, $y'(1) = 3$

B15. $21y'' - 46y' - 7y = \cos 2x$, $y(0) = -4$, $y'(1) = 1$

B16. $21y'' - 2y' - 3y = 2 - 3\sin x$, $y(0) = 7$, $y'(1) = 0$

B17. $7y'' + 40y' - 63y = xe^x$, $y(0) = -1$, $y'(1) = 2$

B18. $3y'' + 11y' + 8y = 1 - 2e^x$, $y(0) = 1$, $y'(1) = -1$

B19. $18y'' + 39y' + 20y = xe^{-x}$, $y(0) = \pi$, $y'(1) = e$

B20. $3y'' + 5y' + 2y = 1 - x + x^2$, $y(0) = e$, $y'(1) = \pi$

B21. $9y'' + 56y' + 12y = 5$, $y(0) = 0$, $y'(1) = 3$

B22. $y'' + 8y' + 25y = x^3$, $y(0) = -6$, $y'(1) = 1$

B23. $y'' + 10y' + 41y = e^{-x}\cos x$, $y(0) = 3$, $y'(1) = -2$

B24. $y'' + 8y' + 20y = 1 - x^3$, $y(0) = 9$, $y'(1) = -1$

B25. $y'' + 2y' + 10y = x + e^x$, $y(0) = 11$, $y'(1) = 0$

B26. $y'' + 10y' + 41y = 2x + 1$, $y(0) = -8$, $y'(1) = 2$

For the BVP in each of Exercises B27–36:

    (a) Obtain an analytic solution using a computer algebra system.

    (b) Graph the solution.

    (c) From the analytic solution, obtain the value of $y'(1)$.

    (d) Obtain the general solution of the differential equation.

    (e) Form, and solve, the two algebraic equations resulting from applying the boundary conditions to the general solution found in part (d).

    (f) Write, and compare to part (a), the solution corresponding to the results of part (e).

    (g) Obtain $G(x, t)$, the Green's function for the homogeneous problem.

    (h) Obtain the solution as in Exercises B32–41, Section 15.1.

    (i) Obtain the solution by the finite-difference method.

    (j) Obtain the solution by the shooting method.

    (k) Compare the two numeric solutions with the analytic solution.

B27. $x^2 y'' + xy' - y = x$, $y(1) = -1$, $y(3) = 1$

B28. $x^2 y'' + 4xy' + 2y = x^2$, $y(1) = 2$, $y(2) = 3$

B29. $x^2 y'' - xy' - 15y = 1 + 2x$, $y(1) = 0$, $y(3) = 2$

B30. $x^2 y'' - 3xy' + 3y = x^{-1}$, $y(1) = -3$, $y'(2) = 1$

B31. $x^2 y'' + xy' - 4y = 5$, $y(1) = 7$, $y'(3) = 0$

B32. $x^2 y'' + 3xy' - 15y = 1 - 3x^2$, $y(1) = -5$, $y(2) = 7$

B33. $x^2 y'' + 7xy' + 5y = x^3$, $y(1) = 1$, $y(5) = -1$

B34. $x^2 y'' - 5xy' + 8y = x(2 + 3x)$, $y(1) = -4$, $y'(4) = 0$

B35. $x^2 y'' + 2xy' - 12y = x$, $y(1) = -2$, $y(2) = 1$

B36. $x^2 y'' + 10xy' + 20y = 7x$, $y(1) = 12$, $y'(3) = 0$

For the BVP in each of Exercises B37–40:

    (a) Obtain the general solution of the differential equation.

    (b) Form, and solve, the two algebraic equations that result from applying the boundary conditions to the general solution found in part (a).

    (c) Write, and graph, the solution of the BVP.

    (d) From the exact solution, obtain the values of $y(0)$ and $y'(0)$.

    (e) Try to obtain the solution directly with a computer algebra system.

    (f) Obtain the solution by the finite-difference method. (Partition $[0, 1]$ into $N = 10$ equal subintervals with the 11 nodes $x_k = \frac{k}{10}$, $k = 0, \ldots, 10$. Discretize the differential equation at each of the 11 nodes, thereby introducing the two additional unknowns y_{-1} and y_{11}. Obtain two additional equations from the boundary conditions, and solve for the 13 unknowns, y_k, $k = -1, \ldots, 11$. Graph the solution.)

B37. $3y'' - 25y' + 28y = \cos 2x$, $8y(0) + y'(0) + 8y(1) - 3y'(1) = 2$, $4y(0) - 7y'(0) - 2y(1) + 7y'(1) = 4$

B38. $9y'' - 86y' + 45y = x^2$, $3y(0) + 9y'(0) + 6y(1) + 5y'(1) = 6$, $8y(0) + 9y'(0) + 8y(1) - 8y'(1) = 1$

B39. $42y'' - 73y' + 28y = 1 - x$, $7y(0) + 6y'(0) + 6y(1) + 5y'(1) = 0$, $8y(0) - 3y'(0) + 8y(1) + y'(1) = 5$

B40. $2y'' + 7y' - 4y = 3$, $y(0) + 8y'(0) + y(1) + 2y'(1) = 2$, $y(0) + 2y'(0) + 5y(1) - 6y'(1) = -9$

15.3 Least-Squares, Rayleigh–Ritz, Galerkin, and Collocation Techniques

Introduction

The least-squares, Rayleigh–Ritz, Galerkin, and collocation techniques for approximating solutions of the differential equation

$$f(x, y, y', y'') = 0 \tag{15.13}$$

all stem from a central idea. An approximation in the form of a linear combination of some basis functions $\phi_k(x)$ is assumed to be a solution. Substitution into (15.13) results in either a zero or nonzero result. If the result is zero, then an extremely astute guess at the solution has been made and an exact solution has been found. In the more likely event that the result is not zero, we don't have an exact solution and the nonzero result is called the *residual, R*. However, the "closer" to zero we can make R, the "better" our guess approximates the true solution. What distinguishes one method from another is the manner in which "R close to zero" is defined and implemented.

EXAMPLE 15.9 We will illustrate the least-squares, Rayleigh–Ritz, Galerkin, and collocation techniques for (15.13), the BVP of Example 15.6 in Section 15.2, that is, for

$$y''(x) + 2y'(x) + 10y(x) = \sin x \quad y(0) = 0, \, y(1) = 1$$

The differential equation and the boundary conditions appear in (15.8), and the exact solution is given by (15.10) when $a = A$, given by (15.11). A graph of this exact solution is given in Figure 15.6. ❖

Common Preliminaries

For each method, we use the same approximation, namely,

$$\phi_0(x) + b\phi_1(x) + c\phi_2(x)$$

where $\phi_0(x) = x$ is introduced to satisfy the inhomogeneous boundary condition $y(1) = 1$ and $\phi_1(x) = x(x - 1)$ and $\phi_2(x) = x(x^2 - 1)$ both satisfy the homogeneous boundary conditions $y(0) = y(1) = 0$. Thus, $y(x)$ is approximated with the polynomial

$$y_p = x + bx(x - 1) + cx(x^2 - 1) \tag{15.14}$$

The residual is

$$R = 2 - 2c + (10 - 4c - 6b)x + (6c + 10b)x^2 + 10cx^3 - \sin x \tag{15.15}$$

The least-squares, Rayleigh–Ritz, Galerkin, and collocation techniques differ in how the residual is manipulated in an effort to determine values of the two constants b and c that optimize R in some sense.

Least-Squares Method

In the least-squares technique, b and c are chosen so as to minimize $Q = \int_0^1 R^2 \, dx$, a measure motivated by convergence in the mean, studied in Section 10.4. Thus, we seek the minimum of

$$Q = \int_0^1 R^2 \, dx = \tfrac{323}{6} + \tfrac{154}{3}b + \tfrac{160}{3}c + 2b^2 + 10cb + \tfrac{1346}{105}c^2$$
$$+ (44c - 28b - 20)\sin 1 + (24 - 32b - 124c)\cos 1 - \tfrac{1}{2}\cos(1)\sin(1)$$

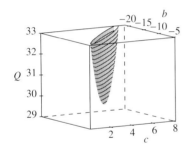

FIGURE 15.9 Least-squares method: $Q = \int_a^b R^2\,dx$

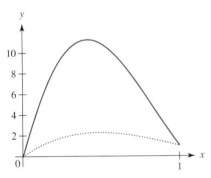

FIGURE 15.10 Least-squares solution (dotted) and exact solution (solid) for Example 15.9

a function of the two parameters b and c. Figure 15.9, showing Q as a surface over the bc-plane, suggests a minimum at some point with coordinates (b, c). The minimum sits at the bottom of a very narrow well.

In practice, finding such minimums using purely numerical techniques is a challenge. However, in this instance, experience in multivariable calculus suggests differentiation with respect to b and c, leading to the equations

$$\frac{154}{3} + 10c - 28\sin 1 - 32\cos 1 + 4b = 0$$

$$\frac{160}{3} + \frac{2692}{105}c + 10b - 124\cos 1 + 44\sin 1 = 0$$

whose solution is

$$b = -\frac{61{,}642}{201} + \frac{30{,}394}{67}\sin 1 - \frac{11{,}014}{67}\cos 1 \stackrel{\circ}{=} -13.77$$

$$c = \frac{7875}{67} - \frac{11{,}970}{67}\sin 1 + \frac{4620}{67}\cos 1 \stackrel{\circ}{=} 4.46$$

This gives the least-squares approximation

$$y_1 = 10.30992325x - 13.76946002x^2 + 4.45953677x^3$$

which is graphed against the exact solution in Figure 15.10. The least-squares solution (dotted) is not a very good approximation to the exact solution (solid) because the approximating polynomial with which we started was too limited. In fact, the minimum value of the residual is 29.6635146, which is not very close to zero.

LEAST-SQUARES AND ORTHOGONALITY For a linear differential equation of the form

$$L[y] = a_0 y'' + a_1 y' + a_2 y = f(x)$$

the least-squares algorithm assumes a solution of the form $y = \sum_{k=0}^{n} c_k \phi_k(x)$, where $c_0 = 1$, and generates the residual

$$R = L\left[\sum_{k=0}^{n} c_k \phi_k(x)\right] - f(x) = \sum_{k=0}^{n} c_k L[\phi_k(x)] - f(x) = \sum_{k=0}^{n} c_k \psi_k(x) - f(x)$$

where $\psi_k(x) = L[\phi_k(x)]$. Linearity allows the operations associated with L to pass through to the functions ϕ_k; and except for the nonhomogeneous term $f(x)$, the residual is a linear combination of L acting on these functions.

Now, when $Q(c_1, c_2, \ldots, c_n) = \int_a^b R^2\,dx$ is minimized, the equations

$$\frac{\partial}{\partial c_k} Q = 2\int_a^b R\left(\frac{\partial}{\partial c_k} R\right) dx = 0$$

determine the c_k. From the definition of R we find

$$\frac{\partial}{\partial c_k} R = \psi_k = L[\phi_k]$$

so the equations determining the c_k are really

$$\int_a^b R\psi_k\,dx = \int_a^b RL[\phi_k]\,dx = 0$$

from which we conclude that the residual R is orthogonal to each function that results when ϕ_k is acted on by L.

Thus, the least-squares algorithm sets out to minimize $Q = \int_0^1 R^2\,dx$ and ends up making the residual orthogonal to a set of functions. It is the *orthogonality* that generalizes to the Rayleigh–Ritz and Galerkin techniques.

Rayleigh–Ritz Technique

The Galerkin strategy generalizes the least-squares technique by making the residual orthogonal to each member of a set of functions called the *weighting* functions. Thus, Galerkin techniques are often called *weighted residual* techniques.

If the weighting functions are the same functions ϕ_k used in the approximation $y = \sum_{k=0}^{n} c_k \phi_k(x)$, the method is generally called the *Rayleigh–Ritz technique*. If the weighting functions are other than these functions, then the method is generally called a *Galerkin* technique. If the weighting functions are Dirac delta functions, then the method becomes a *collocation* method. The weighting functions are often taken as orthogonal.

A rich enough set of orthogonal functions behaves like the unit basis vectors $\mathbf{i}, \mathbf{j}$, and $\mathbf{k}$ in three-dimensional Cartesian space R^3. Just as a vector $\mathbf{v}$ in R^3 can be expressed as a sum of the form $\mathbf{v} = a\mathbf{i} + b\mathbf{j} + c\mathbf{k}$, so too can functions be expressed as linear combinations of orthogonal functions.

The Fourier series from Chapter 10 is an example. On the interval $[0, \pi]$, the Fourier sine series, namely $f(x) = \sum_{k=1}^{\infty} b_k \sin kx$, represents $f(x)$ in terms of the set of orthogonal functions $\phi_k(x) = \sin kx$. The functions $\phi_k(x)$ play the role that the vectors $\mathbf{i}, \mathbf{j}$, and $\mathbf{k}$ play in R^3.

The fundamental insight of the Rayleigh–Ritz and Galerkin techniques resides in the following observation. The only vector in R^3 that is orthogonal to all three basis vectors $\mathbf{i}, \mathbf{j}$, and $\mathbf{k}$ is the zero vector. There are only three mutually perpendicular directions in R^3, so there cannot be any new vectors perpendicular to all three basis vectors. A vector perpendicular to all three basis vectors must be the zero vector.

For functions, there are a countably infinite number of "basis vectors" and, correspondingly, a countably infinite number of mutually perpendicular "directions." The only function perpendicular to all of these directions is the zero function. The more basis vectors to which the residual is orthogonal, the closer the residual is to this zero function. Hence, the Galerkin methods seek to make the residual as close as possible to the zero function by making the residual orthogonal to as many of the members as possible in the set of weighting functions whether orthogonal or not.

Making the residual (15.15) orthogonal to $\phi_1 = x(x - 1)$ and $\phi_2 = x(x^2 - 1)$ gives the two equations

$$\int_0^1 R(x^2 - x)\,dx = \tfrac{1}{30}c + \tfrac{5}{6} - \sin 1 - 2\cos 1 = 0$$

$$\int_0^1 R(x^3 - x)\,dx = -\tfrac{4}{105}c - \tfrac{11}{6} - \tfrac{1}{30}b + 4\sin 1 - 6\cos 1 = 0$$

which then determine the parameters

$$b = -\tfrac{185}{7} + \tfrac{600}{7}\sin 1 - \tfrac{1740}{7}\cos 1 \quad \text{and} \quad c = -25 + 30\sin 1 + 60\cos 1$$

so that the Rayleigh–Ritz approximation is

$$y_2 = x - (x^2 - x)\left(\tfrac{185}{7} - \tfrac{600}{7}\sin 1 + \tfrac{1740}{7}\cos 1\right) - (x^3 - x)(25 - 30\sin 1 - 60\cos 1)$$

Figure 15.11 compares the exact solution (solid curve) and the Rayleigh–Ritz approximation (dotted curve); and we conclude that the Rayleigh–Ritz solution, for this problem, at least, appears to be "better" than the least-squares solution.

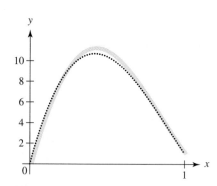

FIGURE 15.11 Rayleigh–Ritz solution (dotted) and exact solution (solid) for Example 15.9

Galerkin Method

In practice, the Galerkin technique differs from the Rayleigh–Ritz technique in the choice of weighting functions. There are also some additional differences in the nature of the differential equation being solved, with the Rayleigh–Ritz being restricted to a smaller class of problems. (See, e.g., [73] for more of the theory of the Rayleigh–Ritz method.)

Suppose we make the residual orthogonal to the weights $w_k = \sin k\pi x$, $k = 1, 2$. Then, the Galerkin orthogonality conditions are the two equations $\int_0^1 R w_k \, dx = 0, k = 1, 2$, which lead to

$$(\sin 1 - 4b - 8c - 14)\pi^4 + (44b + 92c + 14)\pi^2 - 40b - 84c = 0$$
$$(4\sin 1 - 16b - 48c - 40)\pi^4 + (4b + 72c + 10)\pi^2 - 15c = 0$$

Although an exact solution is possible, we settle for a numeric solution that then yields the Galerkin solution as the cubic polynomial

$$y_3 = 60.62190067x - 95.01456923x^2 + 35.39266856x^3$$

A graph of the exact solution (solid curve) and the Galerkin solution (dotted curve) is provided in Figure 15.12. Surprisingly, this Galerkin solution seems to be a better approximation than the Rayleigh–Ritz approximation.

Collocation

If the residual is made orthogonal to Dirac delta functions $\delta(x - x)_k$, where the x_k are in the interval $[a, b]$, we then have the equations $\int_a^b R(x)\delta(x - x_k) \, dx = R(x_k) = 0$. This is the method of collocation. The residual is made to vanish at enough points in the interval $[a, b]$ to generate the right number of equations for determining the coefficients in the approximation for $y(x)$. Typically, the points are taken as uniformly distributed, but that is not essential.

Evaluating the residual (15.15) at $x = \frac{1}{3}, \frac{2}{3}$ yields the collocation equations

$$-\frac{62}{27}c - \frac{8}{9}b + \frac{16}{3} - \sin\frac{1}{3} = 0 \quad \text{and} \quad \frac{26}{27}c + \frac{4}{9}b + \frac{26}{3} - \sin\frac{2}{3} = 0$$

whose solution

$$b = -\frac{1521}{10} + \frac{117}{20}\sin\frac{1}{3} + \frac{279}{20}\sin\frac{2}{3} \quad \text{and} \quad c = \frac{306}{5} - \frac{27}{10}\sin\frac{1}{3} - \frac{27}{5}\sin\frac{2}{3}$$

then gives the collocation approximation

$$y_4 = x - (x^2 - x)\left(\frac{1521}{10} - \frac{117}{20}\sin\frac{1}{3} - \frac{279}{20}\sin\frac{2}{3}\right) + (x^3 - x)\left(\frac{306}{5} - \frac{27}{10}\sin\frac{1}{3} - \frac{27}{5}\sin\frac{2}{3}\right)$$

Figure 15.13 compares the collocation solution (dotted) with the exact solution (solid).

Comparing Solutions

Figures 15.10–15.13 suggest that the Galerkin approximation of the solution to the BVP in Example 15.9 is best. Comparing the four approximations requires a way of assigning a number to the "difference" between two functions $f(x)$ and $g(x)$. We again employ (10.8) from Section 10.6 as the square of the "distance" between functions.

Evaluating the integrals

$$\int_0^1 (y_A - y_k)^2 \, dx \quad k = 1, \ldots, 4$$

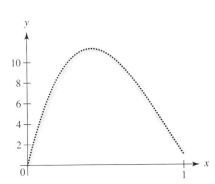

FIGURE 15.12 Galerkin solution (dotted) and exact solution (solid) for Example 15.9

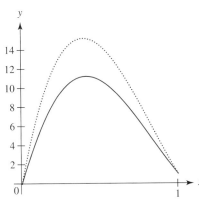

FIGURE 15.13 Collocation solution (dotted) and exact solution (solid) for Example 15.9

we obtain 39.48, 0.092, 0.049, 8.18, respectively, as the values of the square of the measure of difference between the exact solution and the least-squares, Rayleigh–Ritz, Galerkin, and collocation approximations. On the basis of this measure, the Galerkin technique produced the best of the four approximations.

A COLLOCATION EXPERIMENT   Let y_p in (15.14) be a polynomial approximation to $y_A(x)$, the exact solution of the BVP in Example 15.9. Determine the optimum values of b and c by minimizing

$$\|y_A(x) - y_p\|_2^2 = \int_0^1 (y_A(x) - y_p)^2 \, dx \tag{15.16}$$

This gives the very best mean-squared approximation possible with the polynomial y_p. Then, determine $x = u$ and $x = v$, two collocation points that generate these values of b and c by collocation.

The integral in (15.16) evaluates to

$$2.731555731b + 4.012228765c + 0.07619047607c^2 + 57.50018075 + \frac{b^2}{30} + \frac{bc}{10}$$

and has the minimum value 0.0187102577 when

$$b = -94.58938829 \quad \text{and} \quad c = 35.74403492 \tag{15.17}$$

Thus, the best approximation possible with the polynomial y_p in (15.14) is

$$y_5 = 59.84535337x - 94.58938829x^2 + 35.74403492x^3$$

Figure 15.14 compares the exact solution $y_A(x)$ (solid curve) with the best approximating cubic polynomial y_5 (dotted curve). The ratio $\frac{\|y_A(x) - y_3\|}{\|y_A(x) - y_5\|} = 1.62$ compares the performance of the Galerkin solution and the optimal solution, showing that the optimal solution is indeed closer to the exact solution than even the Galerkin solution.

Finally, we ask where are u and v, the two optimal collocating nodes that give us (15.17)? To obtain these nodes, we evaluate the residual (15.15) at $x = u$ and $x = v$ and then substitute for b and c using (15.17). The points u and v are determined by setting the residuals to zero, giving the two very symmetric equations

$$434.5601900u - 69.48806984 - 731.4296734u^2 + 357.4403492u^3 - \sin u = 0$$
$$434.5601900v - 69.48806984 - 731.4296734v^2 + 357.4403492v^3 - \sin v = 0$$

which we solve numerically for

$$u = 0.2593792693 \quad \text{and} \quad v = 0.6732659011$$

The two optimal collocation points, while symmetrically located in $[0, 1]$, are not uniformly spaced.

Our route to the optimum collocation points u and v was through the exact solution $y_A(x)$, which was used in (15.16). Although being able to locate collocation points optimally without using the exact solution would be valuable, we simply refer the reader to texts such as [73] for further study of this topic.

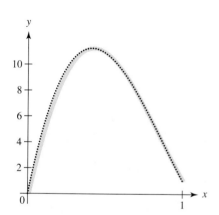

FIGURE 15.14   Optimal cubic solution (dotted) and exact solution (solid) for Example 15.9

EXERCISES 15.3–Part A

A1. Obtain y_e, the exact solution to the BVP $y'' = e^x$, $y(0) = 2$, $y(1) = e$.

A2. Find $\phi_0(x)$, a linear function that satisfies the two boundary conditions in Exercise A1.

A3. If $y_p = \phi_0 + b\phi_1 + c\phi_2$, where ϕ_0 is the function determined in Exercise A2 and ϕ_1, ϕ_2 are as in (15.14), obtain R, the residual y_p generates for the BVP in Exercise A1.

A4. Using R and y_p from Exercise A3, obtain y_1, the least-squares approximation to the solution of the BVP in Exercise A1.

A5. Using R and y_p from Exercise A3, obtain y_2, the Rayleigh–Ritz approximation to the solution of the BVP in Exercise A1.

A6. Using R and y_p from Exercise A3, obtain y_3, the Galerkin approximation to the solution of the BVP in Exercise A1. Use $w_k = \sin k\pi x$, $k = 1, 2$, for the weighting functions.

A7. Using R and y_p from Exercise A3, obtain y_4, the collocation approximation to the solution of the BVP in Exercise A1. Use $x = \frac{1}{3}, \frac{2}{3}$ for the collocation points.

A8. Evaluate the integrals $\int_0^1 (y_e - y_k)^2 \, dx$, $k = 1, \ldots, 4$.

A9. Graph y_e against each of y_k, $k = 1, \ldots, 4$.

EXERCISES 15.3–Part B

B1. In Exercise A6, obtain the Galerkin approximation using the weighting functions

  (a) $\sin x, \sin 2x$ **(b)** $1, x$ **(c)** $1, \cos x$

B2. Compare each approximation found in Exercise B1 with y_3, the Galerkin approximation found in Exercise A6. Explain your observations.

In Exercises B3–7, the given differential equation and the boundary conditions $y(0) = 0$, $y(1) = 1$, comprise a BVP. For each:

  (a) Obtain $y_e(x)$, the exact solution of the BVP. This can be done with a computer algebra system's differential-equation solver, with a Green's function, or with a general solution to which the boundary conditions have been applied.

  (b) Set $y_p = \sum_{k=0}^{2} c_k \phi_k(x)$, where $c_0 = 1$, $\phi_0 = x$, $\phi_1 = x(x - 1)$, $\phi_2 = x(x^2 - 1)$, and obtain $R(c_1, c_2) = L[y] - f(x)$, the residual for the differential equation whose form is $L[y(x)] = f(x)$.

  (c) Plot $Q(c_1, c_2) = \int_0^1 R^2 \, dx$ as a surface over the $c_1 c_2$-plane, being sure to capture the minimum in the graph.

  (d) Obtain a contour plot of $Q(c_1, c_2)$.

  (e) Using R, determine y_1, the least-squares approximation to $y_e(x)$.

  (f) Obtain $\psi_k = L[\phi_k]$, $k = 1, 2$, and show that the equations $\partial S/\partial c_k = \partial/\partial c_k \int_0^1 R^2 \, dx = 0$ are equivalent to $\int_0^1 R\psi_k \, dx = 0$.

  (g) Using R, determine y_2, the Rayleigh–Ritz approximation to $y_e(x)$.

  (h) Using R, and the weighting functions $w_k = \sin k\pi x$, $k = 1, 2$, determine y_3, the Galerkin approximation to $y_e(x)$.

  (i) Using R and two uniformly-spaced collocation points, determine y_4, the collocation approximation to $y_e(x)$.

  (j) Obtain y_5, the solution that minimizes $\|y_e(x) - y_p\|_2^2 = \int_0^1 (y_e(x) - y_p)^2 \, dx$.

  (k) Plot $y_e(x)$ and y_k, $k = 1, \ldots, 5$, all on the same axes.

  (l) Compute $\|y_e(x) - y_k\|_2^2$, $k = 1, \ldots, 5$.

  (m) Obtain R_u and R_v, the residual at $x = u$ and $x = v$, respectively. Using the optimal values of c_1 and c_2 determined in part (j), solve for u and v, the unevenly spaced collocation points that generate the same solution as y_5.

B3. $25y'' - 55y' + 24y = x^2$ **B4.** $y'' - 9y' + 14y = e^{-x}$

B5. $y'' + 2y' + 10y = \cos 2x$ **B6.** $y'' - 7y' + 6y = 1 - \sin x$

B7. $9y'' - 16y' - 4y = 1 + 2x^3$

B8. Show that the weighting functions $w_1 = x$ and $w_2 = x(1 - \frac{4}{3}x)$ are orthogonal on the interval $[0, 1]$.

In Exercises B9–13, the given differential equation and the boundary conditions $y(0) = 0$, $y(1) = 1$, comprise a BVP. For each, repeat the steps used for Exercises B3–7, but take $y_p = x + c_1 \sin \pi x + c_2 \sin 2\pi x$. For the Galerkin method, use the weighting functions w_1 and w_2 defined in Exercise B8.

B9. $27y'' + 60y' - 7y = xe^{-x}$ **B10.** $y'' + 6y' + 25y = 5$

B11. $5y'' + 12y' + 7y = x \cos x$

B12. $21y'' + 10y' + y = 2 - 3x$

B13. $y'' + 6y' + 10y = 2x$

In Exercises B14–18, the given differential equation and the boundary conditions $y(0) = 0$, $y(1) = 1$, comprise a BVP. For each, repeat the steps used for Exercises B3–7, but take $y_p = \sin \frac{\pi}{2}x + c_1 \sin \pi x + c_2 \sin 2\pi x$. For the Galerkin method, use the weighting functions w_1 and w_2 defined in Exercise B8.

B14. $y'' + 8y' + 17y = 1 - 2\sin \pi x$

B15. $y'' + 8y' + 25y = 1 + x - x^2$

B16. $28y'' - 67y' + 40y = x + 4x^3$

B17. $63y'' + 17y' - 10y = x^2 + 2e^x$

B18. $9y'' + 16y' + 7y = 3\cosh x$

B19. Starting with the polynomial $ax + bx^2 + cx^3$, obtain the form $c_0\phi_0(x) + c_1\phi_1(x) + c_2\phi_2(x)$ appropriate for the boundary conditions $y(0) = 0$, $y'(1) = \alpha$. Show that $c_0 = \alpha$, $c_1 = b$, $c_2 = c$, and $\phi_0 = x$, $\phi_1 = x(x - 2)$, $\phi_2 = x(x^2 - 3)$.

For the BVP given in each of Exercises B20–27, repeat the steps used for Exercises B3–7, but take $y_p = \sum_{k=0}^{2} c_k \phi_k(x)$, the polynomial determined in Exercise B19, with α chosen in accordance with the given boundary conditions. For the Galerkin method, use the weighting functions w_1 and w_2 defined in Exercise B1(c).

B20. $4y'' - 12y' + 9y = \cos x$, $y(0) = 0$, $y'(1) = 1$

B21. $9y'' - 61y' + 42y = \sin 2x$, $y(0) = 0$, $y'(1) = -1$

B22. $y'' + 2y' + 10y = e^{-x}$, $y(0) = 0$, $y'(1) = 2$

B23. $7y'' - 12y' - 4y = e^{-2x}$, $y(0) = 0$, $y'(1) = -2$

B24. $18y'' + 17y' - 15y = 5$, $y(0) = 0$, $y'(1) = 3$

B25. $y'' + 2y' + 2y = x^2$, $y(0) = 0$, $y'(1) = -3$

B26. $4y'' + 7y' + 3y = x$, $y(0) = 0$, $y'(1) = \pi$

B27. $y'' + 4y' + 29y = 3 - 2x$, $y(0) = 0$, $y'(1) = -1$

For the BVPs in each of Exercises B28–34, repeat the steps used in Exercises B20–27. However, this time, use $y_p = \alpha x + \sum_{k=1}^{2} c_k(1 - \cos k\pi x)$ where α is chosen in accordance with the given boundary conditions; for the Galerkin method, use the weighting functions $w_k = \sin k\pi x$, $k = 1, 2$.

B28. $8y'' + 26y' + 15y = 3x^2 - 5x$, $y(0) = 0$, $y'(1) = 1$

B29. $y'' + 8y' + 17y = xe^x$, $y(0) = 0$, $y'(1) = -2$

B30. $y'' + 8y' + 32y = x \sin \pi x$, $y(0) = 0$, $y'(1) = 2$

B31. $y'' + 4y' + 20y = x - e^x$, $y(0) = 0$, $y'(1) = -3$

B32. $2y'' - 9y' + 7y = 2e^{-x}$, $y(0) = 0$, $y'(1) = 3$

B33. $25y'' + 10y' - 24y = 2 + 3x - 5x^2$, $y(0) = 0$, $y'(1) = 4$

B34. $12y'' + 7y' + y = e^{-x} \cos \pi x$, $y(0) = 0$, $y'(1) = -4$

B35. Consider the nonlinear boundary value problem $y'' + (y')^2 + y^2 = \sin x$, $y(0) = 0$, $y(1) = 1$.

(a) Using y_p from Exercises B3–7, obtain the Rayleigh–Ritz approximation to the solution of the given BVP. Note that the equations for determining c_1 and c_2 are now quadratic. Show graphically that there are two solutions for the c_k and, consequently, there are two Rayleigh–Ritz solutions.

(b) Plot the two solutions found in part (a).

(c) Use the shooting method to obtain the two solutions of this BVP. Plot the two numeric solutions, and compare them to the Rayleigh–Ritz solutions.

(d) Using the weighting functions $w_k = \sin k\pi x$, $k = 1, 2$, obtain the two Galerkin solutions and compare them to the numeric (shooting method) solutions.

(e) Using collocation at two uniformly spaced nodes, obtain the two collocation solutions and compare them to the numeric (shooting method) solutions.

15.4   Finite Elements

Introduction

Historically, the approximation methods in Section 15.3 evolved into *finite element* methods, particularly important in solving boundary value problems involving partial differential equations. These methods are characterized by two features. First, the approximation is a linear combination of functions $\phi_k(x)$, typically splines (piecewise polynomials of sufficient smoothness), each having *compact support*. The *support* of a function is the set over which it is nonzero. Functions with compact support are nonzero on a closed finite interval.

The way the residual is made "small" is the second characteristic of finite element methods. One finds the names *Rayleigh–Ritz, Galerkin,* and *weighted residual* in the finite element literature. For some classes of problems, the Rayleigh–Ritz and Galerkin methods lead to exactly the same solution and to confusion when the names of the methods are used interchangeably. We will use the names as detailed in Section 15.3.

Thus, a finite element method is a Rayleigh–Ritz, Galerkin, or weighted-residual method where the approximating function is a linear combination of polynomial splines ϕ_k with compact support. When the functions ϕ_k are sufficiently differentiable, the method will be equivalent to the comparable method of Section 15.3. However, when the functions ϕ_k are not sufficiently differentiable to be substituted into the differential equation, the finite element method generates a *weak* solution, to be described.

The term finite element itself is elusive. For example, [7] identifies the finite element with a subregion of the domain over which the differential equation is solved, but [47] gives a more elaborate definition. The finite element is a triple consisting of a subregion, a collection of functions defined on the subregion, and a set of nodal values for the unknown and its relevant derivatives. In our examples where the finite element method will be applied just to ODEs, the subregions are subintervals along the real axis and the functions defined on these subintervals are the polynomials from which we construct the approximating splines.

Formulating the BVP

Let $L = a_0(x)\frac{d^2}{dx^2} + a_1(x)\frac{d}{dx} + a_2(x)$ represent the operations performed on $y(x)$ in the differential equation

$$L[y(x)] = a_0(x)y'' + a_1(x)y' + a_2(x)y = f(x) \qquad \textbf{(15.18)}$$

To change the BVP defined by (15.18) and the nonhomogeneous boundary conditions

$$y(a) = \alpha \qquad y(b) = \beta \qquad \textbf{(15.19)}$$

into a BVP with homogeneous boundary conditions, choose any function $u(x)$ satisfying $u(a) = \alpha, u(b) = \beta$, define $g(x) = f(x) - L[u(x)]$, and let $Y(x)$ be the solution of the BVP

$$L[Y(x)] = g(x) \quad \text{and} \quad Y(a) = Y(b) = 0$$

Then, $y(x) = Y(x) + u(x)$ satisfies the nonhomogeneous conditions (15.19) because

$$y(a) = 0 + \alpha = \alpha \quad \text{and} \quad y(b) = 0 + \beta = \beta$$

Moreover, $y(x)$ satisfies (15.18) because

$$L[y(x)] = L[Y(x) + u(x)] = L[Y(x)] + L[u(x)]$$
$$= g(x) + L[u(x)] = f(x) - L[u(x)] + L[u(x)] = f(x)$$

To convert the BVP of Example 15.9, Section 15.3, to one with homogeneous boundary conditions, write its differential equation as

$$y''(x) + 2y'(x) + 10y(x) = \sin x$$

define $f(x) = \sin x$, and set $u(x) = x$. Then, $g(x) = \sin x - 2 - 10x$, and the differential equation with homogeneous boundary conditions that we will solve by finite elements is $L[Y(x)] = g(x)$ or

$$Y''(x) + 2Y'(x) + 10Y(x) = \sin x - 2 - 10x$$

Strong Solution

A classical (or *strong*) solution of a differential equation is one that satisfies the differential equation *pointwise*. This means the solution is sufficiently differentiable so that when substituted into the equation, an identity results at each point in the relevant domain. Such solutions are precisely what we have been discussing to this point in the text. By contrast, we will shortly find *weak* solutions that do not satisfy the differential equation pointwise. The difference is necessitated by a choice of basis functions $\phi_k(x)$ from which to build an approximate solution.

First, we pick basis functions $\phi_k(x)$ having compact support and second derivatives. From such basis functions we can construct a classical (or strong) solution. After that, we will pick basis functions that do not possess two derivatives and will have to accept weak solutions as a result. In each case, our basis functions will be *B-splines,* special piecewise polynomial functions with compact support.

CUBIC B-SPLINES   A carpenter thinks of a *spline* as a strip of wood inserted into grooves in each of two pieces of wood whose joining is strengthened by the insertion. A draftsman thinks of a *spline* as a thin, flexible rod that can be bent into the shape of curves to be drawn. A numerical analyst thinks of a *spline* as a piecewise smooth polynomial. For example, a cubic spline is a collection of cubics defined on contiguous subintervals of the real axis and joined so that the resulting curve has a continuous second derivative. Such splines are discussed at length in Section 40.4.

A cubic B-spline is a cubic spline that is nonzero on a prescribed finite interval. The letter "B" designates "basis," and the cubic B-splines are used as a collection of basis or building-block functions when approximating solutions of differential equations. The following piecewise cubic polynomial is an example of a cubic B-spline of subinterval width $h = \frac{1}{5}$

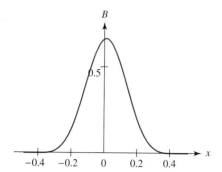

B

0.5

−0.4   −0.2   0   0.2   0.4   x

FIGURE 15.15   Cubic B-spline on $\left[-\frac{2}{5}, \frac{2}{5}\right]$

$$B = \begin{cases} 0 & x < -\frac{2}{5} \\ \frac{4}{3} + 10x + 25x^2 + \frac{125}{6}x^3 & -\frac{2}{5} \leq x < -\frac{1}{5} \\ \frac{2}{3} - 25x^2 - \frac{125}{2}x^3 & -\frac{1}{5} \leq x < 0 \\ \frac{2}{3} - 25x^2 + \frac{125}{2}x^3 & 0 \leq x < \frac{1}{5} \\ \frac{4}{3} - 10x + 25x^2 - \frac{125}{6}x^3 & \frac{1}{5} \leq x < \frac{2}{5} \\ 0 & \frac{2}{5} \leq x \end{cases}$$

Four distinct cubic polynomials, defined on four separate but contiguous subintervals, are joined together to form a function that is twice differentiable, as shown in Figure 15.15. The domain of this B-spline is the interval $[-\frac{2}{5}, \frac{2}{5}]$, and the endpoints of the subintervals over which the cubic polynomials are defined are $-\frac{2}{5}, -\frac{1}{5}, 0, \frac{1}{5}, \frac{2}{5}$. The values of the spline at these endpoints are $0, \frac{1}{6}, \frac{2}{3}, \frac{1}{6}, 0$. (The reader is cautioned that not all texts define the cubic B-spline the same. Some texts adopt a normalization that renders this sequence of values as $0, 1, 4, 1, 0$.)

The smoothness of the cubic B-spline can be seen from the graphs of the first and second derivatives shown in Figure 15.16. The cubic B-spline is therefore a piecewise-continuous function which is zero outside a closed and finite interval and which has a continuous second derivative. By translating and scaling such functions, we obtain a basis $\{\phi_k\}$ of twice-differentiable functions which are zero outside the closed finite intervals that are their domains.

The B-spline is nonzero over an interval of length $4h$, where h is the length of each subinterval over which one of the four cubic polynomials in the B-spline is defined. To illustrate the finite element technique, we will pick $n = \frac{1}{h}$ as the number of B-splines to use in the approximation. If we take $h = \frac{1}{5}$, we will need all appropriate B-splines that are nonzero over the interval $[0, 1]$. Thus, we define, for $n = 5$, the eight B-splines b_j, $j = -1, 0, 1, \ldots, 6$, whose graphs are shown in Figure 15.17. For those who find it difficult to keep track of each separate B-spline, we graph each one individually and also all at once.

There are eight cubic B-splines that are nonzero somewhere in the interval $[0, 1]$ if $h = \frac{1}{5}$. We next demonstrate that these functions can, indeed, be the basis of a representation of the form $h(x) = \sum_{k=-1}^{6} c_k b_k(x)$ for the exact solution of the BVP. Eight equations for the eight coefficients c_k, $k = -1, 1, 0, \ldots, 6$, are obtained by collocating at eight equispaced points in the interval $[0, 1]$. If $y_A(x)$ is the exact solution of the BVP of Example 15.9, Section 15.3, the eight equations $h(x_k) - y_A(x_k) = 0$, $x_k = \frac{k}{8}$, $k = 1, \ldots, 8$, written as

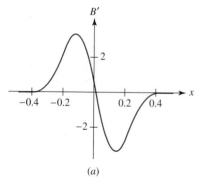

B′

2

−0.4   −0.2   0.2   0.4   x

−2

(a)

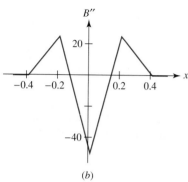

B″

20

−0.4   −0.2   0.2   0.4   x

−40

(b)

FIGURE 15.16   First derivative (*a*) and second derivative (*b*) of the cubic B-spline in Figure 15.15

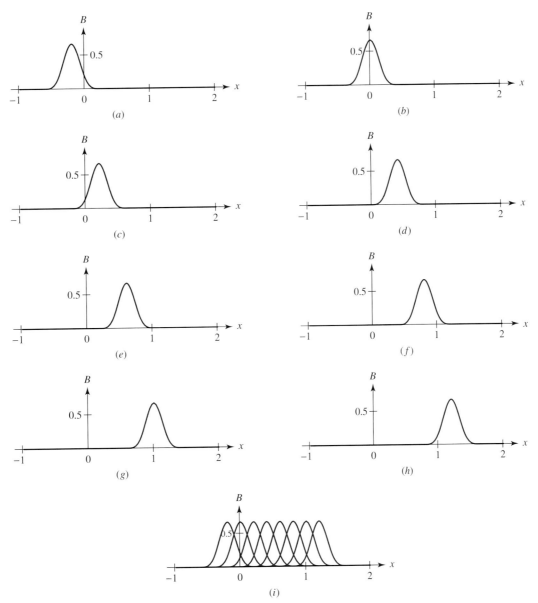

FIGURE 15.17   The eight cubic B-splines with support of length $\frac{4}{5}$ and which are nonzero somewhere in $[0, 1]$

$$0.1666666667c_{-1} + 0.6666666667c_0 + 0.1666666667c_1 = 0$$
$$0.008789062500c_{-1} + 0.3981119792c_0 + 0.5524088542c_1 + 0.04069010417c_2 = 5.813154399$$
$$0.07031250000c_0 + 0.6119791667c_1 + 0.3151041667c_2 + 0.002604166667c_3 = 9.548911635$$
$$0.0003255208333c_0 + 0.2360026042c_1 + 0.6520182292c_2 + 0.1116536458c_3 = 11.15867842$$
$$0.02083333333c_1 + 0.4791666667c_2 + 0.4791666667c_3 + 0.02083333333c_4 = 10.89467103$$
$$0.1116536458c_2 + 0.6520182292c_3 + 0.2360026042c_4 + 0.0003255208333c_5 = 9.208933635$$
$$0.002604166667c_2 + 0.3151041667c_3 + 0.6119791667c_4 + 0.07031250000c_5 = 6.647426235$$
$$0.04069010417c_3 + 0.5524088542c_4 + 0.3981119792c_5 + 0.008789062500c_6 = 3.754726150$$

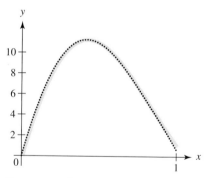

FIGURE 15.18   Exact solution (solid) of Example 15.9 and its B-spline approximation (dotted)

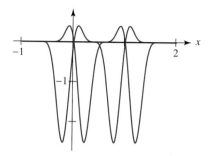

FIGURE 15.19   Adjusted B-splines $B_k, k = 0, 1, 4, 5$

have solution

$$c_{-1} = -12.37483789 \quad c_1 = 9.275318644 \quad c_3 = 10.05884153 \quad c_5 = 0.8341753103$$
$$c_0 = 0.7748798126 \quad c_2 = 12.03389744 \quad c_4 = 5.535894545 \quad c_6 = -5.090589006$$

so a graph of the resulting approximation (dotted curve) and the exact solution (solid curve) is shown in Figure 15.18, establishing that an extremely accurate approximation of the exact solution can be constructed from these cubic B-splines.

THE ADJUSTED SPLINE BASIS   We are particularly concerned with the leftmost and right-most B-splines, since these are not zero at the endpoints of the interval. The homogeneous boundary conditions we assumed dictate the need for picking a set of B-splines that satisfy the homogeneous endpoint conditons of our BVP. Hence, we write the following linear combinations $B_0 = b_0 - 4b_{-1}$, $B_1 = b_0 - 4b_1$, $B_4 = b_5 - 4b_4$, and $B_5 = b_5 - 4b_6$, which will reduce the number of basis splines to six from eight. The graphs of these four new basis functions are shown in Figure 15.19.

The remaining two B-splines are renamed as $B_2 = b_2$ and $B_3 = b_3$ so that we have a uniformly designated set of basis elements $B_0, B_1, \ldots, B_5$. Anticipating the inclusion of some matrix notation in our solution, we realize that a better choice of indexing would be $1, 2, \ldots, 6$ instead of $0, 1, \ldots, 5$, since most computer languages do not support matix elements indexed with zero indices. Thus, we again rename the basis elements, this time calling them $S_1, S_2, \ldots, S_6$ instead of $B_0, B_1, \ldots, B_5$.

FINITE ELEMENT SOLUTION   According to the nomenclature of Section 15.3, the following computations constitute a Rayleigh–Ritz method for approximating the solution of a boundary value problem. In the spirit of that section we seek to solve the differential equation $L[Y(x)] = g(x)$ by writing the solution in the form $Y(x) = \sum_{k=1}^{6} c_k S_k(x)$, which leads to $L[\sum_{k=1}^{6} c_k S_k(x)] = \sum_{k=1}^{6} c_k L[S_k(x)] = g(x)$ as a form of the residual. Then, minimizing the integral of the square of the residual leads to the orthogonality conditions of the Rayleigh–Ritz algorithm, namely,

$$\int_0^1 L\left[\sum_{k=1}^{6} c_k S_k(x)\right] S_j(x)\, dx = \sum_{k=1}^{6} c_k \int_0^1 L[S_k(x)] S_j(x)\, dx = \int_0^1 S_j(x) g(x)\, dx$$

The numbers $\int_0^1 L[S_k(x)] S_j(x)\, dx$ constitute the 6×6 matrix

$$A_{kj} = \begin{bmatrix} -3.1365079 & 4.2809524 & 0.73492063 & 0.039285714 & 0 & 0 \\ 1.32539682 & -46.012698 & -0.77619048 & -3.5289683 & 0.62857143 & 0 \\ 1.0238095 & -5.9095238 & -2.3746032 & 0.41706349 & -3.5289683 & 0.039285715 \\ 0.044841270 & -4.7678571 & 1.7781746 & -2.3746032 & -0.77619048 & 0.73492063 \\ 0 & 0.71746032 & -4.7678571 & -5.9095238 & -46.012698 & 4.2809524 \\ 0 & 0 & 0.044841270 & 1.0238095 & 1.3253968 & -3.1365079 \end{bmatrix}$$

and the six integrals $\int_0^1 S_j(x) g(x)\, dx$ constitute the vector **G** on the left in

$$\mathbf{G} = \begin{bmatrix} -0.2053713398 \\ 2.703416259 \\ -1.122634001 \\ -1.487822107 \\ 6.017227144 \\ -0.6687257465 \end{bmatrix} \quad \mathbf{C} = \begin{bmatrix} 3.040147671 \\ -2.268660776 \\ 11.63588707 \\ 9.456015632 \\ -1.186172705 \\ 0.9556239255 \end{bmatrix} \qquad \textbf{(15.20)}$$

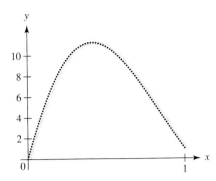

FIGURE 15.20 Exact solution (solid) and finite-element solution (dotted) of Example 15.9

Careful inspection of the left-hand side reveals the sum $\sum_{k=1}^{6} c_k A_{kj}$ is a sum on k, the *row*-index of the matrix A. To interpret the numbers c_k as members of a column vector $\mathbf{C}$, we would need to write the equations as $A^{\mathsf{T}}\mathbf{C} = \mathbf{G}$, where A^{T} is the *transpose* of A, the matrix obtained from A by making the ith row of A become the ith column of A^{T}.

Hence, the solution for the vector $\mathbf{C}$ is given on the right in (15.20) and $y(x) = \sum_{k=1}^{6} c_k S_k(x) = \mathbf{C} \cdot \mathbf{S}$, where $\mathbf{S}$ is the vector of basis functions $S_1, \ldots, S_6$. A graph of the solution $y(x) = Y(x) + u(x)$ (dotted) compared to the graph of the exact solution (solid) is given in Figure 15.20. It is gratifying to see how remarkably accurate the finite element solution actually is! The two curves are indistinguishable.

Weak Solution

Splines formed from piecewise-linear polynomials are also used in finite element methods. However, as we shall shortly see, such B-splines do not possess two derivatives and, hence, cannot be used for the pointwise approximation of solutions to second-order differential equations. The simplicity of the linear B-spline will require introducing the notion of the weak solution.

PIECEWISE-LINEAR SPLINES   The degree-one B-spline is a piecewise-linear function that consists of two linear functions defined on intervals of length h and is nonzero on an interval of length $2h$. For example, such a spline would be

$$B = \begin{cases} 0 & t < -\frac{1}{5} \\ 1 + 5t & -\frac{1}{5} \leq t < 0 \\ 1 - 5t & 0 \leq t < \frac{1}{5} \\ 0 & \frac{1}{5} \leq t \end{cases}$$

which is graphed in Figure 15.21. With $n = 5$ so that $h = \frac{1}{n} = \frac{1}{5}$ as with the cubic B-splines above, there are six piecewise-linear splines b_k, $k = 0, \ldots, 5$, whose values are nonzero somewhere on the interval $[0, 1]$. They are graphed in Figure 15.22.

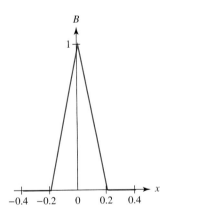

FIGURE 15.21 Degree-one B-spline of length $2h = \frac{2}{5}$

We again demonstrate that a linear combination of such B-splines can represent an arbitrary function and, in particular, $y_A(x)$ from Section 15.3. Thus, we form the sum $h(x) = \sum_{k=0}^{5} c_k b_k(x)$ and then use collocation at equispaced points to generate the six equations

$$c_0 = 0 \qquad c_2 - 11.24351066 = 0 \qquad c_4 - 5.501487521 = 0$$
$$c_1 - 11.24351066 = 0 \qquad c_3 - 9.634145453 = 0 \qquad c_5 - 1 = 0$$

There is an obvious and immediate solution leading to Figure 15.23, where a graph of the resulting approximation (dotted) along with the graph of $y_A(x)$ (solid) show that the piecewise-linear B-splines produce a piecewise-linear approximation!

More important, we realize from the values $c_0 = 0$, $c_5 = 1$ that these two basis elements are not needed for the approximation of a function satisfying *homogeneous* endpoint data. Consequently, for the solution of a BVP with homogeneous endpoint conditions, since the first and last linear B-splines do not vanish at the endpoints, we take as the basis elements of our piecewise-linear approximation just the four splines with support completely within the interval $[0, 1]$. In fact, these splines are shown in Figure 15.24.

GENERATING THE WEAK SOLUTION   If we follow the Rayleigh–Ritz procedure with a piecewise-linear approximation, we will have to take second derivatives of each of the

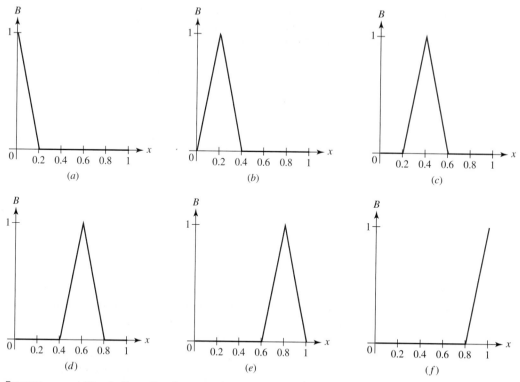

FIGURE 15.22   The six linear B-splines with support of length $\frac{2}{5}$ and which are nonzero somewhere in [0, 1]

FIGURE 15.23   Exact solution $y_A(x)$ (solid) and its approximation by linear B-splines (dotted)

FIGURE 15.24   Linear B-splines which are nonzero somewhere in [0, 1]

piecewise-linear B-splines. Recall that the Rayleigh–Ritz algorithm would have us compute the numbers $\int_0^1 L[b_k(x)]b_j(x)\,dx$. In particular, what happens when we try to compute the term $\int_0^1 b_1'' b_2\,dx$? Since

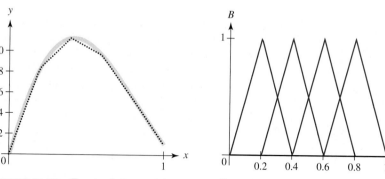

$$b_1'' = \begin{cases} \text{undefined} & t = 0 \\ \text{undefined} & t = \frac{1}{5} \\ \text{undefined} & t = \frac{2}{5} \\ 0 & \text{else} \end{cases}$$

the integral will vanish. To avoid this outcome, we integrate each such term by parts, shifting one derivative off the factor b_k'' and onto the factor b_j. Thus, $\int_0^1 b_k'' b_j\, dx = b_j b_k' |_0^1 - \int_0^1 b_k' b_j'\, dx = 5$.

Finally, we remind the reader that because the piecewise-linear approximation we are developing cannot be differentiated twice, it is not a pointwise (classical) solution. It will be the weak solution we are about to obtain.

FINITE ELEMENT SOLUTION   We are ready to implement the final stage of the finite element version of the weak solution for the problem $L[y(x)] = f(x)$, $y(0) = 0$, $y(1) = 1$. Set $u(x) = x$ and $g(x) = f(x) - 2 - 10x$. Then solve $L[Y(x)] = g(x)$, $Y(0) = Y(1) = 0$, after which $y(x) = Y(x) + u(x)$ is obtained. Writing $Y = \sum_{k=1}^4 c_k b_k$, $L[Y(x)] = g(x)$ becomes $\int_0^1 L[Y(x)] b_j\, dx = \int_0^1 g b_j\, dx$ or

$$\sum_{k=1}^4 c_k \int_0^1 (b_k'' + 2b_k' + 10 b_k) b_j\, dx = \int_0^1 g b_j\, dx$$

Integrate by parts in the second derivative term, obtaining

$$\sum_{k=1}^4 c_k \int_0^1 (-b_k' b_j' + 2 b_k' b_j + 10 b_k b_j)\, dx = \int_0^1 g b_j\, dx$$

Define the matrix $A_{kj} = \int_0^1 (-b_k' b_j' + 2 b_k' b_j + 10 b_k b_j)\, dx$, which evaluates to the array on the left in

$$A_{kj} = \begin{bmatrix} -\frac{26}{3} & \frac{13}{3} & 0 & 0 \\ \frac{19}{3} & -\frac{26}{3} & \frac{13}{3} & 0 \\ 0 & \frac{19}{3} & -\frac{26}{3} & \frac{13}{3} \\ 0 & 0 & \frac{19}{3} & -\frac{26}{3} \end{bmatrix} \quad \mathbf{F} = \begin{bmatrix} -\frac{4}{5} + 10\sin\frac{1}{5} - 5\sin\frac{2}{5} \\ -\frac{6}{5} - 5\sin\frac{1}{5} + 10\sin\frac{2}{5} - 5\sin\frac{3}{5} \\ -\frac{8}{5} - 5\sin\frac{2}{5} + 10\sin\frac{3}{5} - 5\sin\frac{4}{5} \\ -2 - 5\sin\frac{3}{5} + 10\sin\frac{4}{5} - 5\sin 1 \end{bmatrix} \quad \textbf{(15.21)}$$

The simplicity of the banded nature of this coefficient matrix is a strong reason why the theoretical entanglements of the weak solution are tolerated. That observation not withstanding, obtain $\mathbf{F}$, on the right in (15.21), as the vector of values for $\int_0^1 g b_j\, dx$, the right-hand side of the equation.

As noted in the case of the cubic B-spline solution, we now have the matrix equation $A^T \mathbf{C} = \mathbf{F}$, the solution of which is, in floating-point form,

$$\mathbf{C} = \begin{bmatrix} 6.932757137 \\ 9.366867891 \\ 7.897136620 \\ 4.162838280 \end{bmatrix}$$

and $y(x) = Y(x) + u(x)$ would be $\mathbf{C} \cdot \mathbf{B} + x$, where $\mathbf{B}$ is the vector of B-splines $b_1, \dots, b_4$. A graph of the exact solution (solid) and the weak solution (dotted) appears in Figure 15.25.

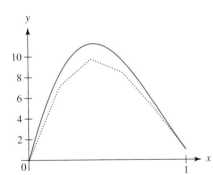

FIGURE 15.25   Exact solution $y_A(x)$ (solid) and weak linear B-spline solution (dotted) for Example 15.9

EXERCISES 15.4–Part A

A1. Verify the entries in the matrix A on the left in (15.21).

A2. Obtain b_k, $k = 1, 2$, the two linear B-splines with $h = \frac{1}{3}$ and support completely in the interval $[0, 1]$.

A3. If b_k, $k = 1, 2$, are as defined in Exercise A2, show that $\int_0^1 -b_1' b_2'\, dx = 3$.

A4. If $L = \frac{d^2}{dx^2} + 2\frac{d}{dx} + 10$ and b_k, $k = 1, 2$, are as defined in Exercise A2, obtain the 2×2 matrix A whose entries are given by $A_{kj} = \int_0^1 L[b_k]b_j \, dx = \int_0^1 (-b_k'b_j' + 2b_k'b_j + 10b_kb_j) \, dx$.

A5. Obtain the vector $\mathbf{F}$ whose entries are given by $\int_0^1 g(x)b_j \, dx$, where $g(x) = \sin x - 2 - 10x$.

A6. From Exercises A4 and A5, obtain the vector $\mathbf{C}$, the solution of $A^{\mathsf{T}}\mathbf{C} = \mathbf{F}$.

A7. If c_k, $k = 1, 2$, are the components of $\mathbf{C}$ obtain $Y = c_1b_1 + c_2b_2$ and $y(x) = Y + x$. Graph $y(x)$ and $y_A(x)$, the exact solution of (15.8), on the same set of axes. (See also, Example 15.9.)

EXERCISES 15.4–Part B

B1. Let $n = 4$ and $h = \frac{1}{4}$ and obtain b_k, $k = -1, \ldots, 5$, the seven cubic B-splines that are nonzero somewhere in the interval [0, 1].

B2. Form the adjusted B-splines $B_0 = b_0 - 4b_{-1}$, $B_1 = b_0 - 4b_1$, $B_3 = b_4 - 4b_3$, $B_4 = b_4 - 4b_5$. Let $B_2 = b_2$. Then renumber B_k, $k = 0, \ldots, 4$, as S_k, $k = 1, \ldots, 5$.

B3. Compute $A_{ij} = \int_0^1 L[S_i]S_j \, dx$, the entries of the 5×5 matrix A, if L is as given in Exercise A4.

B4. Compute $G_j = \int_0^1 g(x)S_j \, dx$, $k = 1, \ldots, 5$, the entries of the vector $\mathbf{G}$, if $g(x)$ is as given in Exercise A5.

B5. Obtain $\mathbf{C}$, the solution of the system $A^{\mathsf{T}}\mathbf{C} = \mathbf{G}$.

B6. If c_k, $k = 1, \ldots, 5$, are the components of $\mathbf{C}$, obtain $Y = \sum_{k=1}^{5} c_k S_k$ and $y(x) = Y + x$. Graph $y(x)$ and $y_A(x)$ on the same set of axes.

For the BVP in each of Exercises B7–11:

(a) Using a computer algebra system's built-in solver, a Green's function, or a general solution to which the boundary conditions have been applied, obtain the exact solution.

(b) Taking as $u(x)$ the lowest degree polynomial that works, convert to an equivalent BVP $L[Y] = g$, $Y(a) = Y(b) = 0$, where $Y(x) = y(x) - u(x)$.

(c) Obtain an $n = 5$ cubic B-spline approximation to $y(x)$ by finding a strong solution to the BVP formed in part (b).

(d) Graph the exact solution and the approximate solution found in part (c).

(e) Obtain an $n = 5$ linear B-spline approximation to $y(x)$ by finding a weak solution to the BVP formed in part (b).

(f) Obtain an $n = 10$ linear B-spline approximation to $y(x)$ by finding a weak solution to the BVP formed in part (b).

(g) Graph the exact solution and the two approximate solutions found in parts (f) and (g).

B7. $8y'' - 21y' + 10y = x$, $y(0) = -1$, $y(1) = 2$

B8. $9y'' - 52y' + 35y = 1 - x$, $y(0) = 1$, $y(2) = -3$

B9. $y'' + 2y' + 2y = x^2$, $y(0) = -2$, $y(1) = 5$

B10. $x^2y'' - 9xy' + 24y = 2 - x^2$, $y(1) = 3$, $y(3) = -4$

B11. $14y'' - 65y' + 9y = 3x + x^3$, $y(0) = 5$, $y(2) = -1$

B12. To solve a boundary value problem for which an endpoint condition is a derivative, the cubic B-spline basis and the function $u(x)$ must be adjusted accordingly. For example, if the condition

at the right endpoint is $y'(b) = \beta$, then define $Y(x) = y(x) - u(x)$, where $u(a) = \alpha$ but $u'(b) = \beta$. If $n = 5$ as in the example in the text, then $B_0 = b_0 - 4b_{-1}$, $B_1 = b_0 - 4b_1$, $B_2 = b_2$, $B_3 = b_3$, $B_4 = b_4 + b_6$, $B_5 = b_5$.

(a) Show that $B_4'(1) = b_4'(1) + b_6'(1) = 0$ and that $B_5'(1) = b_5'(1) = 0$.

(b) Show that the boundary conditions $y(a) = \alpha$, $y'(b) = \beta$ admit $u(x) = \beta x + (\alpha - \beta a)$.

For the BVP in each of Exercises B13–17:

(a) Using a computer algebra system's built-in solver, a Green's function, or a general solution to which the boundary conditions have been applied, obtain the exact solution.

(b) Taking $u(x)$ as detailed in Exercise B12(b), convert the given BVP to the equivalent problem of the form $L[Y] = g$, $Y(a) = Y'(b) = 0$, where $Y(t) = y(x) - u(x)$.

(c) Obtain an $n = 5$ cubic B-spline approximation to $y(x)$ by finding a strong solution to the BVP formed in part (b). Use a basis of cubic B-splines modified as per Exercise B12(a).

(d) Graph the exact solution and the approximate solution found in part (c).

B13. $y'' + 6y' + 18y = e^{-x}$, $y(0) = -5$, $y'(1) = 2$

B14. $4y'' - 8y' - 45y = 1 + x - 3x^2$, $y(0) = 7$, $y'(1) = 2$

B15. $x^2y'' + xy' - 64y = x^2$, $y(1) = 4$, $y'(3) = -1$

B16. $y'' + 6y' + 25y = 1 - e^x$, $y(0) = -1$, $y'(3) = 1$

B17. $7y'' - 20y' - 3y = x + e^{-x}$, $y(0) = 1$, $y'(1) = 3$

B18. Rework Exercise B12 for a derivative condition at the left endpoint where instead of the boundary condition $y(a) = \alpha$ the condition $y'(a) = \alpha$ is imposed.

For the BVP given in each of Exercises B19–23:

(a) Using a computer algebra system's built-in solver, a Green's function, or a general solution to which the boundary conditions have been applied, obtain the exact solution.

(b) Taking $u(x)$ as detailed in Exercise B18(b), convert the given BVP to the equivalent problem of the form $L[Y] = g$, $Y'(a) = Y(b) = 0$, where $Y(t) = y(x) - u(x)$.

(c) Obtain an $n = 5$ cubic B-spline approximation to $y(x)$ by finding a strong solution to the BVP formed in part (b). Use a basis of cubic B-splines modified as per Exercise B18(a).

(d) Graph the exact solution and the approximate solution found in part (c).

B19. $10y'' - 27y' - 81y = e^{-2x}$, $y'(0) = -2$, $y(3) = 1$

B20. $x^2 y'' + 10xy' + 14y = x^{-2}$, $y'(1) = 8$, $y(3) = 3$

B21. $56y'' + 67y' + 20y = x \sin x$, $y'(0) = -8$, $y(3) = -5$

B22. $y'' + 10y' + 24y = x \cos 2x$, $y'(0) = -3$, $y(2) = 1$

B23. $y'' + 6y' + 10y = 1 - \cos 3x$, $y'(0) = 3$, $y(3) = 1$

B24. Rework Exercise B2 so as to combine the results of Exercises B12 and 18, obtaining an adjusted B-spline basis suitable for a BVP with derivative conditions at both ends. In particular, instead of the boundary conditions $y(a) = \alpha$, $y(b) = \beta$, impose the conditions $y'(a) = \alpha$, $y'(b) = \beta$.

For the BVP given in each of Exercises B25–34:

(a) Using a computer algebra system's built-in solver, a Green's function, or a general solution to which the boundary conditions have been applied, obtain the exact solution.

(b) Taking $u(x)$ as detailed in Exercise B24(b), convert the given BVP to the equivalent problem of the form $L[Y] = g$, $Y'(a) = Y'(b) = 0$, where $Y(x) = y(x) - u(x)$.

(c) Obtain an $n = 5$ cubic B-spline approximation to $y(x)$ by finding a strong solution to the BVP formed in part (ii). Use a basis of cubic B-splines modified as per Exercise B24(a).

(d) Graph the exact solution and the approximate solution found in part (c).

B25. $y'' + 2y' + y = 2x + 3 \sin x$, $y'(0) = -2$, $y'(1) = 3$

B26. $24y'' + 85y' + 56y = xe^{-x} + 1$, $y'(0) = 2$, $y'(3) = -1$

B27. $y'' + 6y' + 10y = \cos \pi x$, $y'(0) = -1$, $y'(2) = 3$

B28. $y'' + 6y' + 25y = 5$, $y(0) = 1$, $y'(3) = 5$

B29. $24y'' + 50y' + 25y = xe^x \sin \pi x$, $y'(0) = -5$, $y'(2) = 1$

B30. $y'' + 2y' + 17y = 3x$, $y'(0) = 5$, $y'(1) = -2$

B31. $x^2 y'' - 2xy' - 28y = 1 + 3x$, $y'(1) = 2$, $y'(3) = -1$

B32. $x^2 y'' - 4xy' + 6y = x^{-3}$, $y'(1) = 3$, $y'(2) = 1$

B33. $x^2 y'' + xy' - 36y = 5x^2$, $y'(1) = -4$, $y'(3) = 2$

B34. $24y'' - 10y' + y = \sin x - 2 \cos x$, $y'(0) = 2$, $y'(1) = -1$

Chapter Review

1. Give an example of a boundary value problem for a second-order ODE. Explain why, in general, it is reasonable to impose two conditions on the general solution of the second-order differential equation.

2. Describe the process by which one might obtain the analytic solution to a BVP for a second-order ODE.

3. Give an example of a BVP that does not have a solution.

4. Give an example of a BVP for which the solution is not unique.

5. Give an example that illustrates the claim that when the completely homogeneous BVP has multiple solutions, the inhomogeneous BVP has no solution. Show, in your example, where the solution process described in Question 2 breaks down.

6. What is the difference between a regular and a singular BVP?

7. Give examples of separated boundary conditions and periodic boundary conditions.

8. Describe the essential idea behind the shooting method for numerically solving a BVP. Illustrate your discussion with an analytic example in which the numerical computations are replaced by algebraic calculations.

9. Describe the finite-difference technique for solving a BVP. Illustrate your discussion with an example.

10. Describe the least-squares technique for solving a BVP. Illustrate your discussion with an example.

11. Describe the Rayleigh–Ritz technique for solving a BVP. Illustrate your discussion with an example.

12. Describe the Galerkin technique for solving a BVP. Illustrate your discussion with an example.

13. Describe the collocation technique for solving a BVP. Illustrate your discussion with an example.

14. What is the unifying principle for the methods of Questions 10–13?

15. What are B-splines? Give examples of both linear and cubic B-splines.

16. Describe how the linear B-splines are adjusted to form a basis conformable to representing a solution to a BVP posed on the interval [0, 1].

17. Describe how the cubic B-splines are adjusted to form a basis conformable to representing a solution to a BVP posed on the interval [0, 1].

18. Describe the process by which a finite element solution of a BVP is obtained with linear B-splines. Illustrate with an example.

19. Describe the process by which a finite element solution of a BVP is obtained with cubic B-splines. Illustrate with an example.

20. Describe the difference between a strong or classical solution of a BVP and a weak solution. Why will the use of linear B-splines necessarily lead to a weak solution?

Chapter *16*

The Eigenvalue Problem

I N T R O D U C T I O N   The eigenvalues of a differential equation are very often of great physical significance. In mechanics, the eigenvalues and eigenfunctions determine mode shapes and buckling modes for structures such as plates, beams, and columns. In physics, the eigenvalues determine energy levels for electrons in atoms and molecules. Thus, the determination of eigenvalues for a differential equation has an inherent importance.

Unit Five details the method of *separation of variables,* the classical analytic technique for solving BVPs for partial differential equations. In this method, the PDE is reduced to several ODEs, one or more of which form an eigenvalue problem. Determining the eigenvalues and eigenvectors is then an essential step in solving the BVP for the partial differential equation.

The regular *Sturm–Liouville eigenvalue problem* for ODEs generates an infinite sequence of eigenvalues and a complete orthonormal set of associated eigenfunctions. The complete orthonormal set of eigenfunctions is an infinite "basis" for functions, much like $\{\mathbf{i}, \mathbf{j}, \mathbf{k}\}$ is a basis for vectors in three dimensions.

Unfortunately, many of the interesting problems of engineering science and mathematical physics lead to singular Sturm–Liouville problems. Two of the more important cases are Bessel's equation and Legendre's equation, each of which is studied in the chapter. Bessel's equation frequently arises when solving partial differential equations in cylindrical regions, and Legendre's equation generally arises when solving them in spherical domains.

Of course, analytic techniques must be supplemented by numerical methods, and we demonstrate how to find a finite-difference solution of an eigenvalue problem.

16.1 Regular Sturm–Liouville Problems

The Eigenvalue Problem for Differential Equations

An eigenpair for a matrix A consists of a scalar λ, called an eigenvalue, and a nonzero vector $\mathbf{x}$, called an eigenvector, satisfying the equation $A\mathbf{x} = \lambda\mathbf{x}$. Similarly, an eigenvalue problem can also be formulated for boundary value problems in differential equations. For example,

$$y''(x) = \lambda y(x) \quad \text{and} \quad y(0) = y(L) = 0 \tag{16.1}$$

constitutes an eigenvalue problem on the interval $0 \leq x \leq L$. A solution consists of a scalar λ, called an eigenvalue, and a nonzero function $y(x)$, called an *eigenfunction,* that satisfy (16.1). The parallel with the matrix eigenvalue problem is strengthened if the left-hand side of the DE in (16.1) is written as the operator A acting on $y(x)$, where $A = \frac{d^2}{dx^2}$, so that the differential equation becomes $Ay = \lambda y$.

The differential equation is linear with constant coefficients. Hence, we will make an exponential guess, trying $y(x) = e^{mx}$ and obtaining the characteristic equation $m^2 - \lambda = 0$. The characteristic roots are then $\pm\sqrt{\lambda}$. We must distinguish between the case of repeated roots and distinct roots. If $\lambda = 0$, the characteristic roots are $0, 0$ and the solution has a different form than in the distinct root case.

The Case $\lambda = 0$

When $\lambda = 0$, the solution is $y(x) = 0$ because the general solution is $y = c_1 + c_2 x$, and application of the boundary conditions leads to the algebraic equations $c_1 = 0$ and $c_1 + c_2 L = 0$, whose solution is $c_1 = c_2 = 0$. Hence, $\lambda = 0$ is *not* an eigenvalue because the corresponding solution is just $y(x) = 0$.

The Case $\lambda \neq 0$

When $\lambda \neq 0$, the general solution of the differential equation is

$$y(x) = c_1 e^{\sqrt{\lambda}x} + c_2 e^{-\sqrt{\lambda}x} \tag{16.2}$$

Applying the boundary conditions, we obtain the equations

$$c_1 + c_2 = 0 \quad \text{and} \quad c_1 e^{\sqrt{\lambda}L} + c_2 e^{-\sqrt{\lambda}L} = 0$$

These are homogeneous equations, so the only solution is $c_1 = c_2 = 0$ unless the determinant of the coefficient array vanishes, that is, unless $e^{-\sqrt{\lambda}L} - e^{\sqrt{\lambda}L} = 0$. Equivalently, the first equation gives $c_2 = -c_1$, so the second equation becomes $c_1(e^{\sqrt{\lambda}L} - e^{-\sqrt{\lambda}L}) = 0$, from which we have $e^{\sqrt{\lambda}L} - e^{-\sqrt{\lambda}L} = 0$. This difference of exponentials inspires us to write

$$\frac{e^{i\sqrt{\lambda}L/i} - e^{-i\sqrt{\lambda}L/i}}{2i} = \sin\frac{\sqrt{\lambda}L}{i} = 0$$

with $\frac{\sqrt{\lambda}L}{i}$ real, from which we conclude $\frac{\sqrt{\lambda}L}{i} = n\pi, n = 1, 2, \ldots$, so that $\lambda_n = -\frac{n^2\pi^2}{L^2}, n = 1, 2, \ldots$, as an infinite sequence of eigenvalues. The general solution, now in the form

$$y_n(x) = \alpha_n e^{in\pi x/L} + \beta_n e^{-in\pi x/L} \tag{16.3}$$

upon application of the left-point boundary condition, becomes

$$y_n(x) = \alpha_n(e^{in\pi x/L} - e^{-in\pi x/L}) = 2i\alpha_n \frac{e^{in\pi x/L} - e^{-in\pi x/L}}{2i} = 2i\alpha_n \sin\frac{n\pi x}{L} = a_n \sin\frac{n\pi x}{L}$$

so that $y_n(x) = a_n \sin\frac{n\pi x}{L}, n = 1, 2, \ldots$, is the corresponding infinite sequence of *eigenfunctions*.

The reader should see a connection between the eigenfunctions and the Fourier sine series of Section 10.4.

Additional Examples

We next obtain solutions to several important eigenvalue problems often met in the applications. We leave the details to the exercises.

EXAMPLE 16.1 Consider the eigenvalue problem

$$y''(x) = \lambda y(x) \quad \text{and} \quad y'(0) = y'(L) = 0 \tag{16.4}$$

When $\lambda = 0$, the differential equation becomes $y''(x) = 0$, the general solution is $y(x) =$

$a + bx$, and application of the boundary conditions gives $y_0(x) = a_0$ as an eigenfunction. Thus, $\lambda = 0$ is an eigenvalue!

When $\lambda \neq 0$, the general solution of the differential equation is again (16.2). The left-point boundary condition gives $c_1 = c_2$ and the right-point, $c_1(e^{\sqrt{\lambda}L} - e^{-\sqrt{\lambda}L}) = 0$, from which we conclude $\lambda_n = -\frac{n^2\pi^2}{L^2}$, $n = 1, 2, \ldots$, as an infinite sequence of eigenvalues. The general solution, again in the form (16.3), upon application of the left-point boundary condition, becomes

$$y_n(x) = \alpha_n(e^{in\pi x/L} + e^{-in\pi x/L}) = 2\alpha_n \frac{e^{in\pi x/L} + e^{-in\pi x/L}}{2} = 2\alpha_n \cos\frac{n\pi x}{L} = a_n \cos\frac{n\pi x}{L}$$

so that $y_n(x) = a_n \cos(\frac{n\pi x}{L})$, $n = 1, 2, \ldots$, is the corresponding infinite sequence of *eigenfunctions*.   ❖

Again, the reader should see a connection between these eigenfunctions and the Fourier cosine series from Section 10.4.

EXAMPLE 16.2   Consider the eigenvalue problem

$$y''(x) = \lambda y(x) \quad \text{and} \quad \begin{array}{l} y(-L) = y(L) \\ y'(-L) = y'(L) \end{array} \tag{16.5}$$

The constraints on the right in (16.5), called the *periodic* boundary conditions, arise in problems where $y(x)$ and $y'(x)$ must be continuous when $x = -L$ is identified with $x = L$. In particular, if x is the polar angle θ, then these boundary conditions impose smoothness across the negative x-axis.

When $\lambda = 0$ we again have the equation $y''(x) = 0$ and the general solution $y(x) = c_1 + c_2 x$. The periodic boundary conditions give the equations

$$c_1 - c_2 L = c_1 + c_2 L \quad \text{and} \quad c_2 = c_2$$

from which we conclude $c_2 = 0$, $\lambda = 0$ is an eigenvalue, and $y_0(x) = a_0$.

When $\lambda \neq 0$ we again have the general solution (16.2). The boundary conditions give

$$(e^{\sqrt{\lambda}L} - e^{-\sqrt{\lambda}L})(c_1 \pm c_2) = 0$$

from which we must conclude $e^{\sqrt{\lambda}L} - e^{-\sqrt{\lambda}L} = 0$. Hence, as before, $\lambda_n = -\frac{n^2\pi^2}{L^2}$, $n = 1, 2, \ldots$, but now, starting with (16.2), we get

$$y_n(x) = \alpha_n e^{in\pi x/L} + \beta_n e^{-in\pi x/L}$$

$$= \alpha_n\left(\cos\frac{n\pi x}{L} + i\sin\frac{n\pi x}{L}\right) + \beta_n\left(\cos\frac{n\pi x}{L} - i\sin\frac{n\pi x}{L}\right)$$

$$= (\alpha_n + \beta_n)\cos\frac{n\pi x}{L} + i(\alpha_n - \beta_n)\sin\frac{n\pi x}{L}$$

$$= a_n \cos\frac{n\pi x}{L} + b_n \sin\frac{n\pi x}{L}$$

For each nonzero eigenvalue, there are *two* eigenfunctions, $\cos\frac{n\pi x}{L}$ and $\sin\frac{n\pi x}{L}$.   ❖

Once again we point out the obvious connection with the Fourier series from Section 10.1.

EXAMPLE 16.3   Consider the eigenvalue problem

$$y''(x) = \lambda y(x) \quad \text{and} \quad y(0) = y'(L) = 0 \tag{16.6}$$

which arises in Chapter 25 for heat transfer calculations where one end of a rod or uniform slab is held at temperature $y = 0$ and the other end is simply insulated.

The case $\lambda = 0$ reduces the differential equation to $y''(x) = 0$, which has the solution $y_0(x) = a + bx$. The boundary conditions then give the equations $a = 0$ and $bL = 0$, so $a = b = 0$, $y_0(x) = 0$, and $\lambda = 0$ is not an eigenvalue.

For the case $\lambda \neq 0$, the general solution of the differential equation is again given by (16.2). The left-point boundary condition gives $c_2 = -c_1$, so the solution becomes $y(x) = c_1(e^{\sqrt{\lambda}x} - e^{-\sqrt{\lambda}x})$. The right-point boundary condition is applied to the derivative, so we have the equation $e^{\sqrt{\lambda}L} + e^{-\sqrt{\lambda}L} = 0$. The strategy in Example 16.1 now leads to $\cos \frac{\sqrt{\lambda}L}{i} = 0$, so $\frac{\sqrt{\lambda}L}{i} = \frac{2n+1}{2}\pi$ and $\lambda_n = -\left(\frac{2n+1}{2}\frac{\pi}{L}\right)^2$, $n = 1, 2, \ldots$ (The phrase "odd half-multiples of π" is useful when seeking the zeros of the cosine function.) The corresponding eigenfunctions are then $y_n(x) = a_n \sin(\frac{2n+1}{2}\frac{\pi x}{L})$. ❖

Sturm–Liouville Theory

The regular Sturm–Liouville (or self-adjoint) boundary value problem consists of

1. the finite interval $[a, b]$.

2. the functions p, p', q, w, all continuous on $[a, b]$, with p and w positive on $[a, b]$.

3. the differential equation

$$(py')' + (q + \lambda w)y = 0 \tag{16.7}$$

4. the unmixed boundary conditions $\alpha_1 y(a) + \alpha_2 y'(a) = 0$ and $\beta_1 y(b) + \beta_2 y'(b) = 0$ or the periodic boundary conditions $y(a) - y'(a) = 0$ and $y(b) - y'(b) = 0$

For this important class of eigenvalue problems it can be proved [87] that there is an infinite sequence of real eigenvalues, with the corresponding eigenfunctions $y_n(x)$ satisfying

$$\int_a^b w(x)y_n(x)y_m(x)\,dx = 0$$

whenever $n \neq m$. Thus, the eigenfunctions are said to be *orthogonal with respect to the weight function* $w(x)$. If the eigenfunctions are *normalized* so that

$$\int_a^b w(x)y_k^2(x)\,dx = 1$$

then the eigenfunctions are said to be *orthonormal,* that is, both orthogonal and normalized.

For the separated boundary conditions as in (16.1), (16.4), and (16.6), the eigenspaces are one-dimensional. For each eigenvalue, there is but one eigenfunction. (An eigenspace consists of all constant multiples of this one eigenfunction.) For periodic boundary conditions as in (16.5), the eigenspace can have dimension greater than one. In Example 16.3, each eigenvalue $\lambda_n = -\left(\frac{n\pi}{L}\right)^2$ has two independent eigenfunctions, $\cos \frac{n\pi x}{L}$ and $\sin \frac{n\pi x}{L}$.

In either event, the eigenfunctions form an *orthonormal basis* for function space, much like the unit vectors $\mathbf{i}$, $\mathbf{j}$, and $\mathbf{k}$ form a basis for vectors in three-dimensional Cartesian space. For each such basis of eigenfunctions, we can expand functions in Fourier-type series. In fact, in each of Examples 16.1–16.3, sets of orthonormal eigenfunctions resulted, and the theorems of Sturm and Liouville guarantee that these sets are *complete,* that is, contain enough members that Fourier-type series built from them converge to the functions they represent.

The differential equation (16.7) is the eigenvalue equation for the *formally self-adjoint* operator defined by $L[y(x)] = (py')' + qy$. We briefly consider the formal adjoint for a differential equation and the idea of a self-adjoint operator. In a series of papers (1836–7), Charles Sturm (1803–1855) and Joseph Liouville (1809–1882) published many more than the results summarized here. The concepts of the adjoint operator, and self-adjointness, are deep, interesting, and more significant than the following cursory introduction might suggest.

THE ADJOINT   Given the differential operator

$$L = a(x)\frac{d^2}{dx^2} + b(x)\frac{d}{dx} + c(x)$$

we define L^*, its *adjoint*, as

$$L^* = a(x)\frac{d^2}{dx^2} + (2a'(x) - b(x))\frac{d}{dx} + (a''(x) - b'(x) + c(x))$$

and $J(u, v)$, the *bilinear concomitant* (or *conjunct*) of u and v, as

$$J(u, v) = a(vu' - uv') + (b - a')uv$$

These definitions arise from *Green's formula*

$$\int \{vL[u] - uL^*[v]\}\,dx = J(u, v) \tag{16.8}$$

which is established by integrating $\int vL[u]\,dx$ by parts to obtain

$$\int vL[u]\,dx = \int v\{au'' + bu' + cu\}\,dx$$

$$= \int v\{au'' + bu' + cu\}\,dx$$

$$= \int u\{(av)'' - (bv)' + cv\}\,dx + a(vu' - uv') + uv(a - b')$$

$$= \int uL^*[v]\,dx + J(u, v)$$

We are particularly interested in the case where $b = a'$, since then,

$$L = \frac{d}{dx}\left(a\frac{d}{dx}\right) + b\frac{d}{dx} + c = L^* \tag{16.9}$$

and $J(u, v)$ simplifies to just $a(vu' - uv')$. The operator L is then called *symmetric*, or formally self-adjoint.

With these results at hand, we can show two things. First, we can define the "self-adjoint boundary value problem." Second, we can prove that the eigenfunctions from a self-adjoint eigenvalue problem are always orthogonal.

SELF-ADJOINT BOUNDARY VALUE PROBLEMS   For the formally self-adjoint operator L, (16.9), Green's formula (16.8) for the interval $\alpha \le x \le \beta$ becomes

$$\int_{\alpha}^{\beta} \{vL[u] - uL[v]\}\,dx = J(u, v)(\beta) - J(u, v)(\alpha) \tag{16.10}$$

If both u and v satisfy the same boundary conditions and these boundary conditions cause the right side of (16.10) to vanish, then we will say the *boundary value problem* is *self-adjoint*. For example, the boundary conditions $y(\alpha) = y(\beta) = 0$, if satisfied by both u and v, lead to a self-adjoint BVP, since $J(u, v) = a(vu' - uv')$ and

$$J(u, v)(\beta) - J(u, v)(\alpha) = a(\beta)(v(\beta)u' - u(\beta)v') - a(\alpha)(v(\alpha)u' - u(\alpha)v') = 0$$

In fact, for both separated and periodic boundary conditions, $J(u, v)(\beta) - J(u, v)(\alpha) = 0$, so that with these conditions, the formally self-adjoint operator L forms a self-adjoint BVP for which

$$\int_{\alpha}^{\beta} \{vL[u] - uL[v]\}\, dx = 0 \qquad (16.11)$$

ORTHOGONALITY OF EIGENFUNCTIONS   The symmetric operator (16.9), along with separated or periodic boundary conditions, form the eigenvalue problem $L[y] + \lambda wy = 0$ for which there is a countably infinite set of eigenfunctions orthonormal with respect to the weight function $w(x)$. The existence of these eigenfunctions is established in [87]. The orthogonality follows from (16.11).

Let $y_n(x)$ and $y_m(x)$ be eigenfunctions corresponding to the distinct eigenvalues λ_n and λ_m, respectively, where $\lambda_n \neq \lambda_m$. Then the two differential equations

$$L[y_n] + \lambda_n wy_n = 0 \quad \text{and} \quad L[y_m] + \lambda_m wy_m = 0$$

are identically satisfied. Multiply the first by y_m and the second by y_n. Subtract the second equation from the first and integrate over $\alpha \leq x \leq \beta$, obtaining

$$0 = \int_{\alpha}^{\beta} \{y_m L[y_n] - y_n L[y_m]\}\, dx + (\lambda_n - \lambda_m) \int_{\alpha}^{\beta} wy_n y_m\, dx$$

The first integral on the right vanishes in light of (16.11), leaving

$$0 = (\lambda_n - \lambda_m) \int_{\alpha}^{\beta} wy_n y_m\, dx$$

Since $\lambda_n \neq \lambda_m$, we must conclude that $0 = \int_{\alpha}^{\beta} wy_n y_m\, dx$, which establishes the orthogonality, with respect to the weight function $w(x)$, of eigenfunctions corresponding to different eigenvalues.

EXERCISES 16.1–Part A

A1. Supply the details for the eigenvalue problem solved in Example 16.1.

A2. Supply the details for the eigenvalue problem solved in Example 16.2.

A3. Supply the details for the eigenvalue problem solved in Example 16.3.

A4. Solve the eigenvalue problem defined by $y'' = \lambda y$, $y'(0) = y(L) = 0$.

A5. If $L[y] = 3y'' + 4y' + 5y$, obtain the formal adjoint L^* and the bilinear concomitant $J(u, v)$ and then verify Green's formula (16.8).

A6. Show that the only spring-mass systems $my'' + by' + ky = 0$ that are formally self-adjoint are those without damping. Hence, self-adjointness is lost in nonconservative systems.

EXERCISES 16.1–Part B

B1. The eigenvalues of $y'' - \lambda y = 0$, $y(0) + 3y'(0) = 0$, $y(\pi) + 2y'(\pi) = 0$ are solutions of a transcendental equation that must be solved numerically.

(a) Obtain, for the case $\lambda \neq 0$, a general solution of the differential equation. Following the examples in the text, this solution will be of the form (16.2).

(b) Apply each boundary condition, resulting in two homogeneous algebraic equations in the two arbitrary constants α and β.

(c) Write the two algebraic equations from part (b) in matrix form. Since the equations are homogeneous, the two arbitrary constants will be zero unless the determinant of the system matrix is zero. Thus, the condition that will determine the eigenvalues is obtained by setting the determinant of the system matrix to zero.

(d) Use a graph to determine if there are real solutions for λ.

(e) Replace $\sqrt{\lambda}$ with $i\mu$ and again use a graph to determine if there are solutions for real μ.

(f) Compute several solutions for either λ or μ numerically. From these solutions, obtain the corresponding eigenfunctions. If the eigenvalues arise from the case $\sqrt{\lambda} = i\mu$, it might be simpler to rewrite the general solution in terms of trig functions rather than continue to wrestle with imaginary exponentials. This will lead to simpler real solutions for the eigenfunctions.

(g) Determine if these several eigenfunctions are mutually orthogonal on the interval $[0, \pi]$.

(h) Graph the eigenfunctions found in part (f).

(i) Next, consider the case $\lambda = 0$ (the repeated-root case), again obtaining a general solution of the differential equation and applying the boundary conditions. Determine if there are nonzero solutions to these equations.

In Exercises B2–10, the differential equation $y'' - \lambda y = 0$ and the given boundary conditions determine an eigenvalue problem. In each:

(a) Obtain all eigenvalues and eigenfunctions. In some cases, the eigenvalues are solutions of transcendental equations and must be calculated numerically.

(b) By brute-force integration, determine whether or not the eigenfunctions from different eigenvalues are orthogonal on the interval $[0, \pi]$. Compare your conclusions with the predictions of the Sturm–Liouville theory.

(c) Graph two eigenfunctions, one from each of two different eigenvalues.

(d) Determine if the eigenspaces are one-dimensional or not. Sturm–Liouville theory guarantees that separated boundary conditions generate one-dimensional eigenspaces. However, the boundary conditions in B9 and 10 are not separated.

B2. $y'(0) = y(\pi) = 0$ **B3.** $y(0) = 0$, $y'(\pi) + y(\pi) = 0$

B4. $y'(0) + y(0) = 0$, $y(\pi) = 0$ **B5.** $y'(0) = 0$, $y'(\pi) + y(\pi) = 0$

B6. $y'(0) + y(0) = 0$, $y'(\pi) = 0$

B7. $y(0) + y'(0) = 0$, $y(\pi) + y'(\pi) = 0$

B8. $y(0) + 2y'(0) = 0$, $y(\pi) + 3y'(\pi) = 0$

B9. $y(0) + y'(\pi) = 0$, $y'(0) + y(\pi) = 0$

B10. $y(0) = y(\pi)$, $y'(0) = y'(\pi)$

B11. Show that the differential equation $y'' + y' + \lambda y = 0$ is not formally self-adjoint.

In Exercises B12–23, the differential equation $y'' + y' + \lambda y = 0$ and the given boundary conditions determine an eigenvalue problem. For each:

(a) Obtain all eigenvalues and eigenfunctions. In some cases, the eigenvalues are solutions of transcendental equations and must be calculated numerically.

(b) By brute-force integration, determine whether or not the eigenfunctions from different eigenvalues are orthogonal on the interval $[0, \pi]$. Compare your conclusions with the predictions of the Sturm–Liouville theory.

(c) Graph two eigenfunctions, one from each of two different eigenvalues.

(d) Determine if the eigenspaces are one-dimensional or not.

B12. $y(0) = y(\pi) = 0$ **B13.** $y'(0) = y'(\pi) = 0$

B14. $y(0) = y'(\pi) = 0$ **B15.** $y(0) = y(\pi)$, $y'(0) = y'(\pi)$

B16. $y'(0) = y(\pi) = 0$ **B17.** $y(0) = 0$, $y'(\pi) + y(\pi) = 0$

B18. $y'(0) + y(0) = 0$, $y(\pi) = 0$

B19. $y'(0) = 0$, $y'(\pi) + y(\pi) = 0$

B20. $y'(0) + y(0) = 0$, $y'(\pi) = 0$

B21. $y(0) + y'(0) = 0$, $y(\pi) + y'(\pi) = 0$

B22. $y(0) + 2y'(0) = 0$, $y(\pi) + 3y'(\pi) = 0$

B23. $y(0) + y'(\pi) = 0$, $y'(0) + y(\pi) = 0$

The differential equations given in each of Exercises B24–38 determine an operator L. For each:

(a) Obtain L^*, the adjoint differential operator, and $J(u, v)$, the bilinear concomitant.

(b) Determine if the operator is formally self-adjoint.

(c) Evaluate $J(u, v)\big|_{x=0}^{x=\pi} = J(u, v)(\pi) - J(u, v)(0)$.

(d) Apply the separated boundary conditions of Exercise B12 to both u and v in $J(u, v)\big|_{x=0}^{x=\pi}$.

(e) Apply the separated boundary conditions of Exercise B13 to both u and v in $J(u, v)\big|_{x=0}^{x=\pi}$.

(f) Apply the separated boundary conditions of Exercise B14 to both u and v in $J(u, v)\big|_{x=0}^{x=\pi}$.

(g) Apply the separated boundary conditions of Exercise B17 to both u and v in $J(u, v)\big|_{x=0}^{x=\pi}$.

(h) Apply the separated boundary conditions of Exercise B21 to both u and v in $J(u, v) \big|_{x=0}^{x=\pi}$.

(i) Apply the periodic boundary conditions of Exercise B15 to both u and v in $J(u, v) \big|_{x=0}^{x=\pi}$.

B24. $x^2 y'' + 3xy' - 5y = 0$ **B25.** $4y'' + 2y' + 7y = 0$

B26. $xy'' + y' + x^2 y = 0$ **B27.** $xy'' + 2y' + xy = 0$

B28. $y'' + e^x y' + 2y = 0$ **B29.** $(1 - x^2)y'' - 2xy' + 3y = 0$

B30. $\cos(x)y'' + xy' + y = 0$ **B31.** $e^x y'' + x^2 y' - y = 0$

B32. $3y'' + 2xy' + 5y = 0$ **B33.** $y'' - 2xy' + 3y = 0$

B34. $(1 - x^2)y'' - xy' + 9y = 0$ **B35.** $x^2(1 - x)y'' + 3xy' - 4y = 0$

B36. $2xy'' - 3(1 - x^2)y' + e^{-x}y = 0$

B37. $xy'' + (1 - x)y' + 6y = 0$

B38. $\sin(x)y'' + \cos(x)y' + e^x y = 0$

16.2 Bessel's Equation

Bessel's Equation of Order ν

The differential equation

$$x^2 y''(x) + xy'(x) + (x^2 - \nu^2)y(x) = 0 \tag{16.12}$$

known as Bessel's equation of order ν, occurs often in the applications. It typically arises in problems formulated in cylindrical coordinates.

The equation has variable coefficients, and its solutions cannot be given in terms of elementary functions. In fact, the solutions must be given as series. However, these series solutions occur so often that they have become "named functions." This equation, then, is the first instance of a new function being defined by the differential equation it satisfies. This is far from the exception and often a surprise for students whose experience is with just the elementary functions of the calculus.

In general, the "special functions" of engineering, science, and applied mathematics are defined as solutions of a differential equation. In addition to the hypergeometric function, we mention the orthogonal polynomials of Legendre, Hermite, Laguerre, Jacobi, Gegenbauer, Bernoulli, and Chebyshev and the functions named after Bessel, Legendre, Airy, Mathieu, Hankel, Struve, Kelvin, and Weber. Details of these and other special functions can be found in [1].

Preparing for a series solution of Bessel's equation, we notice that $x = 0$ is a regular singular point. For ν not an integer, there are two nonlogarithmic solutions; otherwise, the two solutions are generalized power series. (See Section 14.1.) Following [1], when ν is not an integer, we take the two linearly independent solutions of Bessel's equation as $J_\nu(x)$ and $J_{-\nu}(x)$, Bessel functions of the first kind, where

$$J_\nu(x) = \left(\frac{x}{2}\right)^\nu \sum_{k=0}^\infty \frac{\left(-\frac{x^2}{4}\right)^k}{k!\,\Gamma(\nu + k + 1)}$$

We remind the reader (see (6.5), Section 6.2) that the gamma function, generalizing the factorial function, satisfies the recursion $\Gamma(z + 1) = z\Gamma(z)$. Hence, if z is the integer n, then $\Gamma(z + 1) = n!$.

The Bessel function of the second kind is defined by

$$Y_\nu(x) = \frac{J_\nu(x)\cos(\nu\pi) - J_{-\nu}(x)}{\sin(\nu\pi)}$$

If ν is the integer n, the two linearly independent solutions of Bessel's equation are taken as $J_n(x)$ and $Y_n(x) = \lim_{\nu \to n} Y_\nu(x)$. The limiting process generates the expected logarithmic term, but the reader is referred to [1] for the results. Examples of series expansions of Bessel

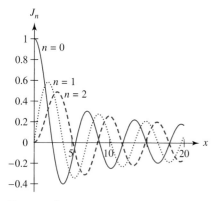

FIGURE 16.1   The Bessel functions $J_k(x), k = 0, 1, 2$

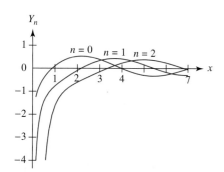

FIGURE 16.2   Bessel functions of the second kind, $Y_k(x), k = 0, 1, 2$

functions are

$$J_2(x) = \frac{1}{8}x^2 - \frac{1}{96}x^4 + \frac{1}{3072}x^6 - \frac{1}{184320}x^8 + O(x^{10})$$

$$Y_2(x) = -\frac{4}{\pi}x^{-2} - \frac{1}{\pi} + \frac{1}{4\pi}\left(\ln\frac{x}{2} + \gamma - \frac{3}{4}\right)x^2 + O(x^4)$$

In the expansion for $Y_2(x)$ we see the appearance of the logarithm. We also see Euler's constant,

$$\gamma = \lim_{n \to \infty}\left[\left(\sum_{i=1}^{n}\frac{1}{i}\right) - \ln n\right] \stackrel{\circ}{=} 0.5772156649$$

More interesting are the graphs of the Bessel functions $J_n(x), n = 0, 1, 2$, seen in Figure 16.1, illustrating that $J_0(0) = 1$ but, for $n \neq 0$, $J_n(0) = 0$. The oscillations are not periodic, but for large x, the asymptotic relation

$$J_n(x) = \sqrt{\frac{2}{\pi x}}\cos\left(x - \frac{n\pi}{2} - \frac{\pi}{4}\right) + O\left(|x|^{-1}\right)$$

reveals that Bessel functions of the first kind eventually resemble phase-shifted cosines enveloped by a multiple of $\frac{1}{\sqrt{x}}$.

Figure 16.2 shows graphs of $Y_n(x), n = 0, 1, 2$, indicating that these functions also oscillate; but in addition, they are all unbounded at the origin. For large x, they have the asymptotic representation

$$Y_n(x) = \sqrt{\frac{2}{\pi x}}\sin\left(x - \frac{n\pi}{2} - \frac{\pi}{4}\right) + O\left(|x|^{-1}\right)$$

By manipulating the full power series representations of

$$J_0(x) = \sum_{k=0}^{\infty}\frac{(-1)^k x^{2k}}{4^k(k!)^2} \quad \text{and} \quad J_1(x) = \frac{1}{2}\sum_{k=0}^{\infty}\frac{(-1)^k x^{2k+1}}{4^k(k!)^2(k+1)}$$

it is possible to establish the relation $\frac{dJ_0}{dx} = -J_1$. An example of a Bessel function of fractional order is

$$J_{1/3}(x) = \frac{3}{4\pi}2^{2/3}\sqrt{3}\Gamma\left(\frac{2}{3}\right)\sum_{k=0}^{\infty}\frac{(-1)^k x^{2k+1/3}}{4^k\left(\frac{4}{3}\right)_k k!}$$

where $(z)_a = \frac{\Gamma(z+a)}{\Gamma(z)}$ is called the *Pochhammer symbol*.

An Eigenvalue Problem

In Section 16.1 we defined the regular Sturm–Liouville boundary value problem as one with a formally self-adjoint differential equation

$$(py')' + (q + \lambda w)y = 0$$

where p and w are positive on the interval over which the equation is being solved. The self-adjoint form of (16.12), the Bessel equation, appears if it is divided by x, resulting in

$$xy''(x) + y'(x) + xy(x) - \frac{y(x)v^2}{x} = 0$$

Clearly, the first two terms can be written as $\frac{d}{dx}[x\frac{dy}{dx}]$, so $p(x) = x$. Restricting the equation to the case $v = 0$ and the interval to $[0, c]$ for c finite, we see that $p(x)$ is not positive on

[0, c]. Hence a boundary value problem for the equation

$$xy''(x) + y'(x) + xy(x) = 0 \qquad (16.13)$$

on this interval is not regular and is called *singular*. In place of a boundary condition at $x = 0$, we impose the physically based requirement of boundedness. The general solution of (16.13) is $y(x) = c_1 J_0(x) + c_2 Y_0(x)$, a linear combination of $J_0(x)$ and $Y_0(x)$. Remembering that $J_0(0) = 1$ but $Y_0(x)$ is unbounded at $x = 0$, we conclude that $c_2 = 0$, making the solution a multiple of just $J_0(x)$.

The typical appearance of Bessel's equation in applications arising from cylindrical coordinates has the form

$$R''(r) + \frac{1}{r}R'(r) + \mu^2 R(r) = 0 \qquad (16.14)$$

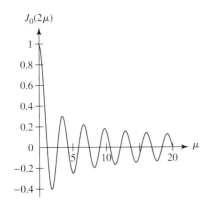

FIGURE 16.3 Zeros of $J_0(2\mu)$

where μ^2 plays the role of λ in the eigenvalue problem. The singular Sturm–Liouville boundary value problem for (16.14) on the finite interval $[0, c]$ simply has the conditions that $R(r)$ be bounded and $R(c) = 0$. Traditionally, the equation in $R(r)$ is explicitly converted to Bessel's equation by the change of variables $x = \mu r$. Then, $R(r) = R(\frac{x}{\mu}) = y(x)$ and $y(x)$ satisfies the equation

$$\mu^2 y''(x) + \frac{\mu^2}{x}y'(x) + \mu^2 y(x) = 0$$

which we recognize as Bessel's equation (16.13) as soon as we "divide out" the factor μ^2 and multiply through by x.

Under the boundedness requirement, the solution of (16.13) is $y(x) = C J_0(x)$ for some constant C, so can write $R(r) = y(x) = y(\mu r) = C J_0(\mu r)$ as the solution of (16.14). To apply the boundary condition $R(c) = 0$, we take the specific example $c = 2$ and solve the equation $R(2) = J_0(2\mu) = 0$ for μ. In fact, any value of μ satisfying the equation $J_0(2\mu) = 0$ is an eigenvalue for this singular BVP. In essence, the eigenvalues are the zeros of the Bessel function $J_0(2x)$, and these zeros can be seen from the graph in Figure 16.3. The first three eigenvalues are $m_1 = 1.202412779, m_2 = 2.760039055$, and $m_3 = 4.326863956$.

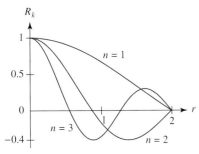

FIGURE 16.4 The eigenfunctions $R_k(r) = J_0(\mu_k r)$, $k = 1, 2, 3$

The corresponding eigenfunctions are then $R_k(r) = J_0(\mu_k r)$, plotted in Figure 16.4. For students with experience only with sines and cosines as eigenfunctions, the graph of the eigenfunctions is useful, if not essential. Each eigenfunction has one more oscillation than its predecessor, and between a pair of zeros of an eigenfunction there is a zero of the next eigenfunction. These are some of the observations proved by Sturm and Liouville as general properties of self-adjoint BVPs.

ORTHOGONALITY OF THE EIGENFUNCTIONS The eigenfunctions $R_k(r) = J_0(\mu_k r)$, where μ_k is the kth zero of $J_0(cx)$, are orthogonal with respect to the weight function $w(r) = r$. We can establish this by appeal to the general Sturm–Liouville theory from Section 16.1 or directly from the identity

$$\int_0^c r J_0(ar) J_0(br) \, dr = \frac{c}{a^2 - b^2}(a J_0(cb) J_1(ca) - b J_0(ca) J_1(cb)) \qquad (16.15)$$

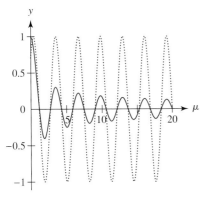

FIGURE 16.5 Comparison of $J_0(2\mu)$ (solid) and $\cos(2\mu - \frac{\pi}{4})$ (dotted)

If a and b are distinct zeros of $J_0(cr)$, then the right side of (16.15) vanishes, thus establishing the orthogonality of the eigenfunctions of (16.14).

ADDITIONAL EIGENVALUES Finally, we seek an efficient algorithm for computing at least the first 10 eigenvalues. Realizing that the zeros of $J_0(2\mu)$ rapidly approach the zeros of $\cos(2\mu - \frac{\pi}{4})$, we compare these two functions in Figure 16.5. Almost immediately, $\frac{n\pi}{2} - \frac{\pi}{8} = \frac{(4n-1)\pi}{8}$, $n = 1, 2, \ldots$, the zeros of $\cos(2\mu - \frac{\pi}{4})$, the dotted curve, are good

$\mu_1 = 1.202412779$
$\mu_2 = 2.760039055$
$\mu_3 = 4.326863956$
$\mu_4 = 5.895767220$
$\mu_5 = 7.465458854$
$\mu_6 = 9.035531984$
$\mu_7 = 10.60581831$
$\mu_8 = 12.17623577$
$\mu_9 = 13.74673957$
$\mu_{10} = 15.31730323$

TABLE 16.1   First ten zeros of $J_0(2\mu)$

approximations of the zeros of $J_0(2\mu)$, the solid curve. With this insight and an appropriate computer program, the first 10 eigenvalues, listed in Table 16.1, are easily computed.

A Fourier–Bessel Series   Sturm–Liouville theory tells us that the eigenfunctions of this eigenvalue problem form a complete set, orthogonal with respect to the weight function $w(r) = r$. Thus, the eigenfunctions are a basis and can be used to represent functions in a Fourier–Bessel series, much like the Fourier–Legendre series of Section 10.7. So, a function $f(r)$ has the series representation $f(r) = \sum_{k=1}^{\infty} c_k J_0(\mu_k r)$, where the coefficients c_k are given by

$$c_k = \frac{\int_0^c r f(r) J_0(\mu_k r)\, dr}{\int_0^c r J_0^2(\mu_k r)\, dr} = \frac{\int_0^c r f(r) J_0(\mu_k r)\, dr}{\frac{c^2}{2} J_1^2(c\mu_k)} \tag{16.16}$$

The integral in the denominator on the left in (16.16) is evaluated to the closed form shown on the right using the identity

$$\int_0^c r J_0^2(ar)\, dr = \frac{c^2}{2}(J_0^2(ac) + J_1^2(ac)) \tag{16.17}$$

where a and c are any constants. In particular, if $a = \mu_k$ is a zero of $J_0(c\mu)$, then the first term on the right in (16.17) is zero, by definition.

Taking $c = 2$ and $f(r) = r(2 - r)$ and then numerically evaluating the integrals in the numerator of (16.16), the coefficients $c_k, k = 1, \ldots, 10$, listed in Table 16.2, are readily obtained. Hence, a 10-term partial sum of the series expansion of $f(r)$ is given by $F_{10} = \sum_{k=1}^{10} c_k J_0(\mu_k r)$. Figure 16.6 shows both $f(r)$ and the partial sum F_{10}. Except near $r = 0$ (because $J_0(0) = 1$), the agreement between the two curves is substantial.

$c_1 = 1.294011356$
$c_2 = -0.8338654470$
$c_3 = -0.03055067335$
$c_4 = -0.1534142976$
$c_5 = -0.02137816209$
$c_6 = -0.06015106815$
$c_7 = -0.01339750016$
$c_8 = -0.03146792599$
$c_9 = -0.00902993004$
$c_{10} = -0.01918084051$

TABLE 16.2   Fourier–Bessel coefficients for $F(r) = r(2 - r)$

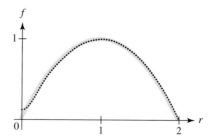

FIGURE 16.6   The function $f(r) = r(2 - r)$ (solid) and F_{10}, the partial sum of the Fourier–Bessel approximation (dotted)

EXERCISES 16.2–Part A

A1. Carry out the change of variables $x = \mu r$ by which (16.14) becomes (16.13).

A2. Show that $x = 0$ is a regular singular point of (16.12), Bessel's equation of order ν.

A3. Obtain the indicial equation and its roots for (16.12), Bessel's equation of order ν. (See Section 14.1.)

A4. Obtain $J_0(x)$ as the nonlogarithmic solution of (16.13), Bessel's equation of order zero.

A5. Evaluate the Pochhammer symbols $(5)_2$, $(3)_4$, and $\left(\frac{1}{2}\right)_3$.

EXERCISES 16.2–Part B

B1. If p_k is the kth zero of $J_0(x)$, give empirical evidence that each of the following may indeed be correct.

(a) $\sum_{k=1}^{\infty} \dfrac{1}{p_k^2} = \dfrac{1}{4}$ **(b)** $\sum_{k=1}^{\infty} \dfrac{1}{p_k^4} = \dfrac{1}{32}$

Exercises B2–21 specify the BVP $rR''(r) + R'(r) + r\mu^2 R(r) = 0$, $R(c) = 0$, $R(0) < \infty$, by giving $f(r)$ and c. For each:

(a) Transform the differential equation into the standard form of Bessel's equation $xy''(x) + y'(x) + xy(x) = 0$ by the change of variables $r = \frac{x}{\mu}$, $R(r) = R(\frac{x}{\mu}) = y(x)$, thereby obtaining $y(x) = J_0(x)$ so that $R(r) = J_0(\mu r)$.

(b) Obtain the first ten eigenvalues μ_k, $k = 1, \ldots, 10$.

(c) Obtain and graph the first three eigenfunctions $R_k(r) = J_0(\mu_k r)$, $k = 1, 2, 3$.

(d) Verify that on $0 \le r \le c$ these three eigenfunctions are mutually orthogonal with respect to the weighting function $w(r) = r$.

(e) Obtain $\sum_{k=1}^{10} A_k J_0(\mu_k r)$, the tenth partial sum of the Fourier–Bessel series of $f(r)$, and graph it along with $f(r)$. What is the effect on the approximation if $f(0) \ne 1$ or $f(c) \ne 0$?

B2. $f(r) = 1 - \dfrac{r^2}{9}, c = 3$ **B3.** $f(r) = \sqrt{r}, c = 4$

B4. $f(r) = \dfrac{1}{1 + r^2}, c = 5$ **B5.** $f(r) = e^{-r}, c = 2$

B6. $f(r) = re^{-r}, c = 3$ **B7.** $f(r) = \cos r, c = 2\pi$

B8. $f(r) = \dfrac{e^r}{1 + r^2}, c = \dfrac{3}{2}$ **B9.** $f(r) = \cos^2 r, c = 2\pi$

B10. $f(r) = \arctan r, c = 1$ **B11.** $f(r) = 1 - \sqrt{2r}, c = \frac{1}{2}$

B12. $f(r) = r \sin r, c = \pi$ **B13.** $f(r) = \begin{cases} 1 & 0 \le r < 1 \\ 0 & 1 \le r \le 2 \end{cases}, c = 2$

B14. $f(r) = \begin{cases} r & 0 \le r < 1 \\ 1 & 1 \le r \le 2 \end{cases}, c = 2$

B15. $f(r) = \begin{cases} r & 0 \le r < 1 \\ 1 - r & 1 \le r \le 2 \end{cases}, c = 2$

B16. $f(r) = \begin{cases} e^{-r} & 0 \le r < 1 \\ 1 & 1 \le r \le 2 \end{cases}, c = 2$

B17. $f(r) = \begin{cases} \sin r & 0 \le r < \dfrac{\pi}{2} \\ 1 & \dfrac{\pi}{2} \le r \le \pi \end{cases}, c = \pi$

B18. $f(r) = \begin{cases} r & 0 \le r < 1 \\ 1 & 1 \le r < 2 \\ 3 - r & 2 \le r \le 3 \end{cases}, c = 3$

B19. $f(r) = |(r - 1)(3 - r)|, c = 4$ **B20.** $f(r) = |r - 1|, c = 2$

B21. $f(r) = |\sin r|, c = 2\pi$

In Exercises B22–27, make the indicated change of variables to bring the given equation into the form (16.12), from which a solution for $w(z)$ can be given in terms of Bessel functions.

B22. $w''(z) + \left(\lambda^2 - \dfrac{v^2 - \frac{1}{4}}{z^2}\right) w(z) = 0$; $x = \lambda z$, $w(z) = \sqrt{\dfrac{x}{\lambda}} y(x)$

B23. $w''(z) + \left(\dfrac{\lambda^2}{4z} - \dfrac{v^2 - 1}{4z^2}\right) w(z) = 0$; $x = \lambda \sqrt{z}$, $w(z) = \dfrac{x}{\lambda} y(x)$

B24. $w''(z) + \lambda^2 z^{p-2} w(z) = 0$; $x = \left(\dfrac{2\lambda}{p}\right) z^{p/2}$, $w(z) = \left(\dfrac{px}{2\lambda}\right)^{1/p} y(x)$

B25. $w''(z) - \dfrac{2v - 1}{z} w'(z) + \lambda^2 w(z) = 0$; $x = \lambda z$, $w(z) = \left(\dfrac{x}{\lambda}\right)^v y(x)$

B26. $w''(z) + (\lambda^2 e^{2z} - v^2) w(z) = 0$; $x = \lambda e^z$, $w(z) = y(x)$

B27. $z^2 w''(z) + (1 - 2p)zw'(z) + (\lambda^2 q^2 z^{2q} + p^2 - v^2 q^2) w(z) = 0$; $x = \lambda z^q$, $w(z) = \left(\dfrac{x}{\lambda}\right)^{p/q} y(x)$ In particular, consider the case $p = 2, q = 4$.

B28. By evaluating at $x = 1, 2, 3$ and performing the appropriate numerical integration, give empirical evidence that the following formulas may indeed be correct.

(a) $J_0(x) = \dfrac{1}{\pi} \int_0^\pi \cos(x \sin \theta)\, d\theta$

(b) $J_0(x) = \dfrac{1}{\pi} \int_0^\pi \cos(x \cos \theta)\, d\theta$

(c) $J_0(x) = \dfrac{2}{\pi} \int_0^1 \dfrac{\cos sx}{\sqrt{1 - s^2}}\, ds$

(d) $J_n(x) = \dfrac{1}{\pi} \int_0^\pi \cos(x \sin \theta - n\theta)\, d\theta$, for $n = 1, 2$, and 3

B29. By differentiating under the integral sign, show formally that the integral in

(a) Exercise B28(a) is a solution of Bessel's equation of order zero.

(b) Exercise B28(b) is a solution of Bessel's equation of order zero.

(c) Exercise B28(c) is a solution of Bessel's equation of order zero.

B30. Show that if the integrands of the integrals in Exercise B28(a)–(c) are expanded in Maclaurin series at $x = 0$ and the results integrated termwise, the series for $J_0(x)$ is obtained in each case.

B31. For $n = 2$, show that if the integrand of the integral in Exercise B28(d) is expanded in a Maclaurin series at $x = 0$ and the result integrated termwise, the series for $J_n(x)$ is obtained.

B32. Evaluating numerically at $x = 1, 3, 5$, give empirical evidence that each of the following formulas may indeed be correct.

(a) $J_0(x) + 2 \sum_{k=1}^{\infty} J_{2k}(x) = 1$

(b) $J_0(x) + 2 \sum_{k=1}^{\infty} (-1)^k J_{2k}(x) = \cos x$

(c) $2\sum_{k=0}^{\infty}(-1)^k J_{2k+1}(x) = \sin x$

B33. For $\nu = 1, 3, 5$ and $x = 1.5, 3, 5.5$, give empirical evidence that each of the following recurrence relations may indeed by correct.

(a) $J_{\nu-1}(x) + J_{\nu+1}(x) = \dfrac{2\nu}{x}J_\nu(x)$

(b) $J_{\nu-1}(x) - J_{\nu+1}(x) = 2J'_\nu(x)$

(c) $J'_\nu(x) = J_{\nu-1}(x) - \dfrac{\nu}{x}J_\nu(x)$

(d) $J'_\nu(x) = -J_{\nu+1}(x) + \dfrac{\nu}{x}J_\nu(x)$

B34. It is asserted in [1, p. 369] that on the interval $-3 \le x \le 3$, the polynomial $f(x) = 1 - 2.2499997\left(\frac{x}{3}\right)^2 + 1.2656208\left(\frac{x}{3}\right)^4 - 0.3163866\left(\frac{x}{3}\right)^6 + 0.0444479\left(\frac{x}{3}\right)^8 - 0.0039444\left(\frac{x}{3}\right)^{10} + 0.0002100\left(\frac{x}{3}\right)^{12}$, approximates $J_0(x)$ with an error no worse than 5×10^{-8}. Give empirical evidence of a numeric and graphical nature that this claim may indeed be correct. For example, plot the difference between $f(x)$ and $J_0(x)$ and determine the maximum value of this difference.

B35. It is asserted in [1, p. 369] that if $f_0 = 0.79788456 - 0.00000077(\frac{3}{x}) - 0.00552740\left(\frac{3}{x}\right)^2 - 0.00009512\left(\frac{3}{x}\right)^3 + 0.00137237\left(\frac{3}{x}\right)^4 - 0.00072805\left(\frac{3}{x}\right)^5 + 0.00014476\left(\frac{3}{x}\right)^6$, and $\theta_0 = x - 0.78539816 - 0.04166397(\frac{3}{x}) - 0.00003954\left(\frac{3}{x}\right)^2 + 0.00262573\left(\frac{3}{x}\right)^3 - 0.00054125\left(\frac{3}{x}\right)^4 - 0.00029333\left(\frac{3}{x}\right)^5 + 0.00013558\left(\frac{3}{x}\right)^6$, then on the interval $3 \le x < \infty$, $\phi(x) = \dfrac{f_0 \cos\theta_0}{\sqrt{x}}$ is an approximation of $J_0(x)$ with errors no worse than 1.6×10^{-8} in f_0 and 7×10^{-8} in θ_0. Examine empirically just how well $\phi(x)$ approximates $J_0(x)$ in $3 \le x \le 10$.

B36. For $n = 1, 2, 3$, and $c = \frac{3}{2}, 3, 5$, test empirically

(a) $\displaystyle\int_0^c x^n J_{n-1}(x)\,dx = c^n J_n(c)$

(b) $\displaystyle\int_0^c x J_n^2(\mu_2 x)\,dx = \dfrac{c^2}{2}[J'_n(\mu_2 c)]^2$, where μ_2 is the second zero of $J_n(\mu c)$ (excluding $\mu = 0$).

B37. The differential equation $\frac{dy}{dx} + cy^2 + \frac{1}{x}y + \frac{1}{c} = 0$ is a Riccati equation.

(a) Show that the change of variables $y(x) = \frac{1}{cw(x)}\frac{dw}{dx}$ transforms the equation into Bessel's equation of order zero.

(b) Set $c = 3$ in the Ricatti equation, and obtain a numeric solution to the initial value problem consisting of this equation and the initial condition $y(1) = 1$.

(c) In terms of Bessel functions, obtain an exact solution to the initial value problem stated in part (b), and show graphically that it agrees with the numeric solution found in part (b).

B38. The differential equation $\frac{d^2y}{dz^2} + \frac{1}{z}\frac{dy}{dz} + \rho^2 y = 0$ is a disguised form of Bessel's equation of order zero.

(a) Make the change of variables $x = \rho z$ and transform the equation into a Bessel equation.

(b) Set $\rho = 2$ and obtain a numeric solution of the initial value problem consisting of the given differential equation and the initial conditions $y(1) = 1$, $y'(1) = -2$.

(c) In terms of Bessel functions, obtain an exact solution to the initial value problem stated in part (b), and show graphically that it agrees with the numeric solution found in part (b).

B39. Let $y(x)$ satisfy Bessel's equation of order $\nu = \frac{1}{2}$.

(a) Make the change of variables $y(x) = \frac{u(x)}{\sqrt{x}}$, obtaining $u'' + u = 0$, and thus, $y(x) = c_1\frac{\sin x}{\sqrt{x}} + c_2\frac{\cos x}{\sqrt{x}}$.

(b) Using the series for $J_\nu(x)$, show that $J_{1/2}(x) = \sqrt{\frac{2}{\pi x}}\sin x$ and $J_{-1/2}(x) = \sqrt{\frac{2}{\pi x}}\cos x$.

16.3 Legendre's Equation

A Singular Sturm–Liouville Eigenvalue Problem

Legendre's equation

$$(1 - x^2)y''(x) - 2xy'(x) + \lambda y(x) = 0 \qquad (16.18)$$

on the interval $-1 \le x \le 1$ typically arises in physical problems being solved in spherical coordinates. With $p(x) = 1 - x^2$, it's clear the equation is in the self-adjoint form $(py')' + (q + \lambda w)y = 0$, where $q = 0$ and $w = 1$. Although the values for q and the weighting function $w = 1$ are as simple as they could be, solving this equation presents unique challenges. In fact, we now realize that both $x = -1$ and $x = 1$ are regular singular points of the differential equation and the problem is singular because $p(x)$ is not positive on the interval $[-1, 1]$. Hence, there are no boundary conditions at the endpoints! Instead, the solution $y(x)$ must be bounded throughout the interval, especially at each endpoint.

Our boundary value problem then is the differential equation (16.18) along with the "boundary" conditions $y(-1)$ finite and $y(1)$ finite.

Solving the Differential Equation

The eigenvalue problem based on the equation $y'' = \lambda y$ presented some new challenges, but since the solutions were the familiar trig and exponential functions we ultimately determined both the eigenvalues and eigenfunctions. The Bessel equation presented some additional challenges since the functions that satisfy the equation, the Bessel functions, are not part of the suite of elementary functions studied in calculus. Legendre's equation poses further challenges because, unlike the Bessel equation, there are no "Legendre functions" with which we can ultimately construct the solution. Each of these three classic BVPs differ in key ways, making it seem to the beginning student that there is no coherence in the subject at all.

We avoided grinding out the power series solution of Bessel's equation because we ultimately relied on a computer algebra system such as Maple to generate the functions for us. We cannot do that with the solutions of Legendre's equation and now must revert to a series solution at a fundamental level.

POWER SERIES SOLUTION About $x = 0$, seek a Taylor series solution of the form $y(x) = \sum_{k=0}^{\infty} a_k x^k$. A seventh-order finite approximating sum satisfying the initial conditions $y(0) = A$ and $y'(0) = B$ is

$$\left(-\frac{\lambda(\lambda-6)(\lambda-20)}{720}x^6 + \frac{\lambda(\lambda-6)}{24}x^4 - \frac{\lambda}{2}x^2 + 1 \right) A$$
$$+ \left(-\frac{(\lambda-2)(\lambda-12)(\lambda-30)}{5040}x^7 + \frac{(\lambda-2)(\lambda-12)}{120}x^5 + \frac{2-\lambda}{6}x^3 + x \right) B$$

The solution appears to split into a series of even powers of x (multiplying $A = y(0)$) and a series of odd powers of x (multiplying $B = y'(0)$). Also, if λ is an integer such as 2, 6, 12, or 30, coefficients will be zero and the possibility exists that instead of an infinite series, the solution is just a polynomial. However, verification of these suspicions requires knowing more about the general form of the general coefficient a_k.

RECURSION RELATION To obtain a general recursion formula for the coefficients, seek a solution of the form

$$y = y(0) \sum_{n=0}^{\infty} a_{2n} x^{2n} + y'(0) \sum_{n=0}^{\infty} a_{2n+1} x^{2n+1}$$

As seen in Section 14.1, the sum $\sum_{n=k}^{k+2} a_n x^n$ suffices and substitution into the differential equation gives

$$(1-x^2)\left(\sum_{n=k}^{k+2} a_n x^n \right)'' - 2x \left(\sum_{n=k}^{k+2} a_n x^n \right)' + \lambda \sum_{n=k}^{k+2} a_n x^n = 0$$

Differentiating termwise leads to

$$a_k(k^2-k)x^{k-2} + a_{k+1}(k+k^2)x^{k-1} + (a_{k+2}(k^2+3k+2) - a_k(k^2+k-\lambda))x^k$$
$$a_{k+1}(\lambda - 3k - k^2 - 2)x^{k+1} + a_{k+2}(\lambda - 6 - k^2 - 5k)x^{k+2} = 0$$

where the coefficient of each power of x must separately vanish. Setting the coefficient of

x^k to zero yields

$$a_{k+2}(k^2 + 3k + 2) - a_k(k^2 + k - \lambda) = 0$$

and solving for a_{k+2} gives the desired recursion relationship

$$a_{k+2} = \frac{k(k+1) - \lambda}{(k+2)(k+1)} a_k = g(k)a_k$$

Applying the Boundary Conditions

The recurrence relation suggests that a_0 determines all other even-indexed coefficients and a_1 determines all other odd-indexed coefficients. Moreover, if λ is the integer $k(k+1)$, then $a_{k+2} = 0 = a_{k+4} = a_{k+6}$, etc., so the series starting with a_0 (if k is even) or with a_1 (if k is odd) becomes a simple polynomial, called a *Legendre polynomial*.

To determine the large-k behavior of the coefficients in either infinite series, express a_2 in terms of a_0, then a_4 in terms of a_2 and hence a_0, etc., obtaining the results in Table 16.3. Thus, with $a_0 = 1$, a_{2m} is the product $\prod_{n=0}^{m-1} \frac{2n(2n+1) - \lambda}{(2n+2)(2n+1)}$. A similar result holds when a_{k+2} is based on a_1. Evaluating the expression for a_{2m} gives the closed-form expression

$$a_{2m} = \frac{4^m \Gamma\left(m + \frac{1}{4} + \frac{1}{4}\sqrt{1+4\lambda}\right) \Gamma\left(m + \frac{1}{4} - \frac{1}{4}\sqrt{1+4\lambda}\right)}{\Gamma(2m+1)\Gamma\left(\frac{1}{4} + \frac{1}{4}\sqrt{1+4\lambda}\right)\Gamma\left(\frac{1}{4} - \frac{1}{4}\sqrt{1+4\lambda}\right)}$$

from which we can obtain

$$\lim_{m \to \infty} \frac{a_{2m}}{\frac{1}{m}} = c = \frac{\sqrt{\pi}}{\Gamma\left(\frac{1}{4} - \frac{1}{4}\sqrt{1+4\lambda}\right)\Gamma\left(\frac{1}{4} + \frac{1}{4}\sqrt{1+4\lambda}\right)} > 0$$

Thus, the coefficients grow at the same rate as the coefficients in the harmonic series $\sum_{k=1}^{\infty} \frac{1}{k}$, a series known to diverge. (See Section 7.2.) Hence, if this limit is true, solutions of Legendre's equation that are infinite series will diverge at $x = -1$ and 1. The only solutions that could then be bounded at the endpoints would be the polynomial solutions, those arising by choice of λ as one of the integers $k(k+1)$.

Numeric Solutions

Figure 16.7 shows a numeric solution of Legendre's equation with $\lambda = 3$ (which is not an eigenvalue) and the initial conditions $y(0) = 0$, $y'(0) = 1$ as a solid curve, and the derivative $y'(x)$ as the dotted curve. The solid curve suggests the values of $y(-1)$ and $y(1)$ are finite and not unbounded. However, the graph of the derivative shows that the slope at the endpoints is becoming large, and hence, there well may be vertical asymptotes at the endpoints.

Evaluating $y(x)$ near an endpoint reveals the computational difficulty.

$$y(0.99) = 0.4209825881239523$$
$$y(0.999) = 0.1148798427442815$$
$$y(0.9999) = -0.1934045835681208$$
$$y(0.99999) = -0.5015322459675798$$
$$y(0.999999) = -0.8096067971765240$$
$$y(0.9999999) = -1.117672295649649$$
$$y(0.99999999) = -1.425736522635797$$
$$y(0.999999999) = -1.425736522635797$$

The graph does not capture the actual endpoint behavior. Moreover, computing the numeric

---

$a_2 = g(0)a_0$

$a_4 = g(0)g(2)a_0$

$a_6 = g(0)g(2)g(4)a_0$

$a_8 = g(0)g(2)g(4)g(6)a_0$

$a_{10} = g(0)g(2)g(4)g(6)g(8)a_0$

TABLE 16.3   The recurrence relation for Legendre's equation

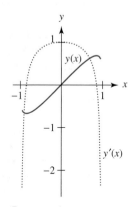

FIGURE 16.7   Solid curve is $y(x)$ and dotted curve is $y'(x)$ for Legendre's equation

solution of the differential equation near an endpoint becomes increasingly difficult. This observation is consistent with the slow convergence of the harmonic series, the series that models the behavior of $y(x)$ at the endpoint. The computational tools are just inadequate for convincingly demonstrating the unbounded behavior at the endpoint. Here is an example, therefore, where symbolic analysis and theoretical argument cannot be displaced by brute-force computation.

Legendre Polynomials

In Section 10.7 we showed that a function $f(x)$ can be represented on the interval $[-1, 1]$ by a Fourier–Legendre series built of the Legendre polynomials $p_k(x)$, where $p_k(x)$, the solution of Legendre's equation corresponding to the eigenvalue $\lambda = k(k + 1)$, is normalized so that $p_k(1) = 1$. The first six such polynomials are

$$p_0 = 1 \qquad p_2 = \tfrac{3}{2}x^2 - \tfrac{1}{2} \qquad p_4 = \tfrac{35}{8}x^4 - \tfrac{15}{4}x^2 + \tfrac{3}{8}$$
$$p_1 = x \qquad p_3 = \tfrac{5}{2}x^3 - \tfrac{3}{2}x \qquad p_5 = \tfrac{63}{8}x^5 - \tfrac{35}{4}x^3 + \tfrac{15}{8}x \tag{16.19}$$

The orthogonality of the Legendre polynomials has already been discussed in an earlier lesson, so we conclude this study of Legendre's equation and the resulting eigenvalue problem. Legendre polynomials and the Fourier–Legendre series will appear again in Sections 29.4 and 29.5 in the context of boundary value problems involving partial differential equations.

EXERCISES 16.3

1. Starting with $P_0(x) = 1$ and $P_1(x) = x$, use the recursion $P_{k+1}(x) = (\tfrac{2k+1}{k+1})x P_k(x) - (\tfrac{k}{k+1})P_{k-1}(x)$ to obtain $P_2(x), \ldots, P_8(x)$.

The eigenvalues of $(1 - x^2)y'' - 2xy' + \lambda y = 0$, Legendre's equation, are $\lambda = k(k + 1)$. For λ not an eigenvalue, a solution $y(x)$ will be unbounded at $x = \pm 1$, an outcome difficult to establish numerically. For the initial conditions and value of λ given in each of Exercises 2–6, obtain numeric solutions and plot both y and y'. The graphs should be consistent with the claim about endpoint behavior.

2. $y(0) = 1, y'(0) = 2; \lambda = 7$ **3.** $y(0) = 0, y'(0) = 1; \lambda = 11$

4. $y(\tfrac{1}{2}) = 1, y'(\tfrac{1}{2}) = -1; \lambda = 17$ **5.** $y(\tfrac{3}{4}) = 0, y'(\tfrac{3}{4}) = 2; \lambda = 29$

6. $y(\tfrac{9}{10}) = 1, y'(\tfrac{9}{10}) = -1; \lambda = 41$

The eigenfunctions of Legendre's equation are $P_k(x)$, the Legendre polynomials, for which $\int_{-1}^{1} P_k^2(x)\,dx = \tfrac{2}{2k+1}$. Let $\phi_k(x) = \sqrt{\tfrac{2k+1}{2}}P_k(x)$ be an orthonormal version of these polynomials for which the Fourier–Legendre series of $f(x)$ would be given by $\sum_{k=0}^{\infty} c_k \phi_k$, where $c_k = \int_{-1}^{1} f(x)\phi_k(x)\,dx = \sqrt{\tfrac{2k+1}{2}}\int_{-1}^{1} f(x)P_k(x)\,dx$. For each of the functions $f(x)$ given in Exercises 7–16, obtain the coefficients c_k and demonstrate empirically that $\sum_{k=0}^{\infty} c_k \phi_k$ converges to $f(x)$ and *Parseval's equality*, $\sum_{k=0}^{\infty} c_k^2 = \int_{-1}^{1} f^2(x)\,dx$, holds.

7. $f(x) = \sin x$ **8.** $f(x) = e^x$ **9.** $f(x) = e^{-2x}$

10. $f(x) = \cos \pi x$ **11.** $f(x) = \tan x$

12. $f(x) = \begin{cases} 1 & -1 \le x < 0 \\ 1 - x & 0 \le x \le 1 \end{cases}$

13. $f(x) = \begin{cases} 1 + x & -1 \le x < 0 \\ (x - 1)^2 & 0 \le x \le 1 \end{cases}$

14. $f(x) = \begin{cases} x^2 & -1 \le x < 0 \\ \sin \pi x & 0 \le x \le 1 \end{cases}$

15. $f(x) = |\sin \pi x|$

16. $f(x) = \begin{cases} x(x + 1) & -1 \le x < 0 \\ x & 0 \le x \le 1 \end{cases}$

The transformation $z(x) = \tfrac{a+b}{2} + (\tfrac{b-a}{2})x$ maps the interval $-1 \le x \le 1$ to $a \le z \le b$. If $F(z)$ is defined on $a \le z \le b$, then $f(x) = F(z(x))$ is defined and has a Fourier–Legendre series, $\Phi(x)$, on $-1 \le x \le 1$. Then, $\Phi(x(z))$ is a Fourier–Legendre series for $F(z)$ on $a \le z \le b$. In Exercises 17–21:

(a) Obtain a Fourier–Legendre series on the indicated interval.

(b) Plot at least one partial sum of $\Phi(x(z))$ and $F(z)$ on the same set of axes.

17. $F(z) = 4z - 3 - z^2, 1 \le z \le 3$

18. $F(z) = z^3 - 7z^2 + 14z - 8, 1 \le z \le 4$

19. $F(z) = \dfrac{2(z - 3)}{z^2 - 6z + 13}, 1 \le z \le 5$

20. $F(z) = \begin{cases} \frac{1}{2} & 2 \le z < 3 \\ 2 - \frac{z}{2} & 3 \le z < 4 \\ z - 4 & 4 \le z < 5 \\ 1 & 5 \le z \le 6 \end{cases}$, $2 \le z \le 6$

21. $F(z) = \begin{cases} \frac{z^2}{4} - \frac{5}{2}z + \frac{25}{4} & 3 \le z < 5 \\ -\frac{z^2}{4} + \frac{7}{2}z - \frac{45}{4} & 5 \le z \le 7 \end{cases}$, $3 \le z \le 7$

22. Use Rodrigues' formula,

$$P_k(x) = \frac{1}{2^k k!} \frac{d^k}{dx^k}(x^2 - 1)^k$$

to obtain $P_0(x), \ldots, P_8(x)$.

23. For $k = 0, \ldots, 10$, verify that

$$P_k(x) = \frac{1}{2^k} \sum_{n=0}^{[k/2]} \frac{(-1)^n (2k - 2n)!}{n!(k - 2n)!(k - n)!} x^{k-2n}$$

where $[\frac{k}{2}]$, the greatest integer in $\frac{k}{2}$, is $\frac{k}{2}$ when k is even and $\frac{k-1}{2}$ when k is odd.

24. Show $P_k^{(k)}(x) = (2k)!/2^k(k!)$ by using the formula in Exercise 23.

25. Using the binomial expansion, write

$$(x^2 - 1)^k = \sum_{n=0}^{k} (-1)^n \frac{k!}{n!(k-n)!} x^{2k-2n}$$

and differentiate it k times to obtain

$$\frac{d^k}{dx^k}(x^2 - 1)^k = \sum_{n=0}^{[k/2]} \frac{(-1)^n k!(2k - 2n)!}{n!(k - n)!(k - 2n)!} x^{k-2n}$$

from which, upon comparison to the formula in Exercise 23, Rodrigues' formula follows.

26. Obtain $\int_{-1}^{1} P_k^2(x)\, dx = \frac{2}{2k+1}$ from Rodrigues' formula, as follows.

(a) In

$$\int_{-1}^{1} P_k^2(x)\, dx = \int_{-1}^{1} P_k(x) \frac{1}{2^k k!} \frac{d^k}{dx^k}(x^2 - 1)^k\, dx$$

integrate by parts with $u = P_k(x)$, obtaining

$$\int_{-1}^{1} P_k^2(x)\, dx = -\frac{1}{2^k k!} \int_{-1}^{1} P_k'(x) \frac{d^{k-1}}{dx^{k-1}}(x^2 - 1)^k\, dx$$

(b) Continue integrating by parts for a total of $k - 1$ times, obtaining

$$\int_{-1}^{1} P_k^2(x)\, dx = \frac{(-1)^{k-1}}{2^k k!} \int_{-1}^{1} P_k^{(k-1)}(x) \frac{d}{dx}(x^2 - 1)^k\, dx$$

(c) Obtain $\int_{-1}^{1} P_k^2(x)\, dx = (-1)^k/2^k k! \int_{-1}^{1} P_k^{(k)}(x)(x^2 - 1)^k\, dx$ via an additional integration by parts.

(d) Establish $\int_{-1}^{1} (x^2 - 1)^k\, dx = (-1)^k 2^{2k+1}(k!)^2/(2k + 1)!$ by k applications of the reduction formula

$$\int (x^2 - 1)^k\, dx = \frac{1}{2k + 1}\left[x(x^2 - 1)^k - 2k \int (x^2 - 1)^k\, dx\right]$$

(e) Use the results in parts (c) and (d) and the result in Exercise 24 to deduce $\int_{-1}^{1} P_k^2(x)\, dx = \frac{2}{2k+1}$.

27. For $k \le 5$, verify empirically

(a) $P_{k+1}' - x P_k' = (k + 1)P_k$ **(b)** $x P_k' - P_{k-1}' = k P_k$

(c) $P_{k+1}' - P_{k-1}' = (2k + 1)P_k$ **(d)** $\int_{-1}^{1} x P_k^2(x)\, dx = 0$

(e) $\int_{x}^{1} P_k(x)\, dx = \frac{1}{2k + 1}(P_{k-1}(x) - P_{k+1}(x))$

(f) $P_k(1) = 1$ and $P_k(-1) = (-1)^k$

(g) $\int_{-1}^{1} P_k'(x) P_j(x)\, dx = \begin{cases} 1 - (-1)^{k+j} & k > j \\ 0 & k \le j \end{cases}$

28. When $\lambda = k(k + 1)$, Legendre's equation $(1 - x^2)y'' - 2xy' + k(k + 1)y = 0$ has one solution $y(x) = P_k(x)$, the Legendre polynomial. A second linearly independent solution is $Q_k(x) = P_k(x) \int dx/(1 - x^2)P_k^2(x)$, the *Legendre function of the second kind.*

(a) Calculate and graph $Q_0, \ldots, Q_3$.

(b) Show that $Q_k(x)$ satisfies Legendre's equation.

(c) Use reduction of order to obtain this second solution. Thus, set $y(x) = u(x)P_k(x)$ and obtain, from Legendre's equation, $2w(x)P_k'(x) + w'(x)P_k(x) = 0$, where $w(x) = (1 - x^2)u'(x)$. The equation for $w(x)$ is variable-separable, and $u(x)$ follows by an integration after finding $w(x)$.

29. Verify each of the following alternate representations of $Q_k(x)$, both taken from [1, pp. 334–5].

(a) $Q_k(x) = \frac{1}{2} \int_{-1}^{1} \frac{P_k(t)}{x - t}\, dt$

(b) $Q_k(x) = \frac{1}{2}P_k(x) \ln\left(\frac{1 + x}{1 - x}\right) - \sum_{n=1}^{k} \frac{1}{n} P_{n-1}(x)P_{k-n}(x)$

16.4 Solution by Finite Differences

The Finite-Difference Scheme

Section 15.2 implements a finite-difference solution of the boundary value problem

$$y''(x) = f(x, y, y') \quad \text{and} \quad y(a) = \alpha,\ y(b) = \beta$$

Recall that the interval $[a, b]$ is uniformly partitioned into N subintervals of length $h = \frac{b-a}{N}$, resulting in the partition points $x_k = a + kh$, $k = 0, 1, \ldots, N$. The derivatives at x_k are replaced with the $O(h^2)$ centered approximations

$$y''(x_k) = \frac{y_{k+1} - 2y_k + y_{k-1}}{h^2} \quad \text{and} \quad y'(x_k) = \frac{y_{k+1} - y_{k-1}}{2h} \qquad \textbf{(16.20)}$$

Substitution of these approximations into a differential equation yields a set of algebraic equations for $y(x_k)$, $k = 0, 1, \ldots, N$.

In extending these ideas to the eigenvalue problem

$$a(x)\frac{d^2 y}{dx^2} + b(x)\frac{dy}{dx} + (c(x) + \lambda w(x))y(x) = 0 \qquad y(\alpha) = y(\beta) = 0$$

we again use (16.20). Instead of a set of linear algebraic equations to solve for the values of the y_k, an eigenvalue problem of the form $A\mathbf{y} = -\lambda\mathbf{y}$ must be solved for both λ and $\mathbf{y}$, a vector whose components are the y_k. Since there are many eigenvalues and corresponding eigenfunctions, the approximation scheme must likewise generate these multiple solutions.

The following example illustrates these ideas.

EXAMPLE 16.4   Consider the eigenvalue problem defined, on the interval $[0, 1]$,

$$y''(x) + 2y'(x) + \lambda y(x) = 0 \quad \text{and} \quad y(0) = y(1) = 0$$

The methods of Section 16.1 lead to the exact solution $y_n(x) = a_n e^{-x} \sin n\pi x$ and the eigenvalues $\lambda_n = 1 + n^2\pi^2$, $n = 1, 2, \ldots$.

The finite-difference solution discretizes the interval $[0, 1]$ with a partition of N subintervals of length $h = \frac{1-0}{N}$, giving the nodes $x_k = 0 + kh$. If, for example, N is taken as 10, we then have $h = \frac{1}{10}$ and $x_k = kh$. The boundary conditions fix the values at the endpoints. Hence, make the assignments $u_0 = 0$ and $u_{10} = 0$, where $y(x_k) = u_k$. At each "interior" node x_k, $k = 1, 2, \ldots, N - 1 = 9$, the discretization yields a single equation, so we have the nine interior equations

$$110u_2 - 200u_1 + \lambda u_1 = 0 \qquad 110u_7 - 200u_6 + 90u_5 + \lambda u_6 = 0$$
$$110u_3 - 200u_2 + 90u_1 + \lambda u_2 = 0 \qquad 110u_8 - 200u_7 + 90u_6 + \lambda u_7 = 0$$
$$110u_4 - 200u_3 + 90u_2 + \lambda u_3 = 0 \qquad 110u_9 - 200u_8 + 90u_7 + \lambda u_8 = 0$$
$$110u_5 - 200u_4 + 90u_3 + \lambda u_4 = 0 \qquad -200u_9 + 90u_8 + \lambda u_9 = 0$$
$$110u_6 - 200u_5 + 90u_4 + \lambda u_5 = 0$$

The matrix form for these equations is $A\mathbf{u} = -\lambda\mathbf{u}$, where

$$A = \begin{bmatrix} -200 & 110 & 0 & 0 & 0 & 0 & 0 & 0 & 0 \\ 90 & -200 & 110 & 0 & 0 & 0 & 0 & 0 & 0 \\ 0 & 90 & -200 & 110 & 0 & 0 & 0 & 0 & 0 \\ 0 & 0 & 90 & -200 & 110 & 0 & 0 & 0 & 0 \\ 0 & 0 & 0 & 90 & -200 & 110 & 0 & 0 & 0 \\ 0 & 0 & 0 & 0 & 90 & -200 & 110 & 0 & 0 \\ 0 & 0 & 0 & 0 & 0 & 90 & -200 & 110 & 0 \\ 0 & 0 & 0 & 0 & 0 & 0 & 90 & -200 & 110 \\ 0 & 0 & 0 & 0 & 0 & 0 & 0 & 90 & -200 \end{bmatrix}$$

λ	$1 + k^2\pi^2$
389.2578571	800.4379567
360.9923492	632.6546819
316.9677883	484.6106158
261.4936055	356.3057585
200.0000000	247.7401101
138.5063945	158.9136705
83.03221165	89.82643964
39.00765084	40.47841762
10.74214286	10.86960440

TABLE 16.4   Eigenvalues of A and exact eigenvalues of BVP In Example 16.4

Table 16.4 lists λ_k, the negatives of the eigenvalues of A, along with the exact values $1 + k^2\pi^2$. The smallest eigenvalue is the most accurate. In fact, the relative error in the smallest eigenvalue is -0.01172641941.

APPROXIMATING THE EIGENFUNCTION   The eigenvector $\mathbf{y}$ corresponding to the eigenvalue λ, obtained from the matrix eigenvalue problem $A\mathbf{y} = -\lambda\mathbf{y}$, is a discrete approximation of the eigenfunction $y(x)$. For the case $N = 20$, the eigenvector corresponding to the smallest eigenvalue $\lambda = 10.8376345$ is the vector $\mathbf{y}$, whose 19 components are

$y_1 = 0.0751$	$y_5 = 0.278$	$y_9 = 0.318$	$y_{13} = 0.235$	$y_{17} = 0.0979$
$y_2 = 0.141$	$y_6 = 0.303$	$y_{10} = 0.306$	$y_{14} = 0.203$	$y_{18} = 0.0634$
$y_3 = 0.197$	$y_7 = 0.317$	$y_{11} = 0.288$	$y_{15} = 0.169$	$y_{19} = 0.0305$
$y_4 = 0.243$	$y_8 = 0.322$	$y_{12} = 0.263$	$y_{16} = 0.133$	

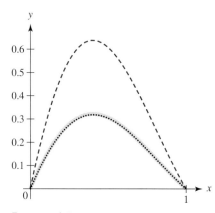

(Remember, $y_0 = y_{21} = 0$.) The dashed curve in Figure 16.8 is the exact solution $y_1(x) = e^{-x}\sin(\pi x)$, the solid curve is the piecewise-linear curve connection of the points (x_k, y_k) determined by $\mathbf{y}$, and the dotted curve is $\frac{1}{2}y_1(x)$. Since eigenfunctions are determined only up to an arbitrary multiple, $\mathbf{y}$ agrees with $\frac{1}{2}y_1(x)$.

FIGURE 16.8   Example 16.4: Exact solution $y_1(x)$ (dashed), finite-difference solution (solid), and $\frac{1}{2}y_1(x)$ (dotted)

Error Analysis

The relative error in the computed value of the smallest eigenvalue, for $N = 10$, was -0.01172641941. With $N = 20$, the error, now that h has been halved, is -0.002941220197. The ratio of errors is 3.986923326, which is close to 4, the square of the ratio of stepsizes. Since derivatives were replaced with $O(h^2)$ approximations, it is not unreasonable to find the same order persisting in subsequent computations.

Additional results, such as recipes for the matrix A in terms of the coefficients of the differential equation, can be found in [60]. ❖

EXERCISES 16.4

1. For the BVP $y'' + 3y' + \lambda y = 0$, $y(0) = y(1) = 0$:

(a) Obtain analytic solutions for the eigenvalues and eigenfunctions via the techniques of Section 16.1.

(b) Obtain numeric estimates of the eigenvalues and eigenfunctions via the finite-difference technique of this present section using $N = 10$.

(c) Compare graphically the first eigenfunctions computed in parts (a) and (b).

2. For the BVP $3y'' + 2y' + \lambda y = 0$, $y(0) = y(1) = 0$:

(a) Obtain analytic solutions for the eigenvalues and eigenfunctions via the techniques of Section 16.1.

(b) Obtain numeric estimates of the eigenvalues and eigenfunctions via the finite-difference technique of this present section using $N = 10$.

(c) Compare graphically the first eigenfunctions computed in parts (a) and (b).

In Exercises 3–7, the given differential equation and the boundary conditions $y(0) = y(1) = 0$ determine a BVP. For each:

(a) Obtain numeric estimates of the eigenvalues and eigenfunctions via the finite-difference technique of this present section using $N = 10$. The nodal equations will now contain terms that arise from evaluating the coefficient of y' at the nodes.

(b) Graph the first two eigenfunctions.

3. $y'' + 2xy' + \lambda y = 0$ **4.** $y'' + e^{-x}y' + \lambda y = 0$

5. $y'' + \sin(\pi x)y' + \lambda y = 0$ **6.** $y'' + x^2 y' + \lambda y = 0$

7. $y'' + x(x-1)y' + \lambda y = 0$

8. Consider the eigenvalue problem $y'' + \lambda y = 0$, $y'(0) = y'(1) = 0$.

(a) Obtain the exact solution.

(b) Using $N = 10$, discretize by central differences, obtaining nodal equations for the 11 nodes x_k, $k = 0, \ldots, 10$. The first and last equations will contain y_{-1} and y_{11}, respectively. Use the boundary conditions to eliminate these two unknowns, leaving an eigenvalue problem for an 11×11 matrix.

(c) Compare the first two eigenvalues and eigenfunctions computed numerically with their exact counterparts, graphing the eigenfunctions as part of the comparison.

9. Consider the eigenvalue problem $y'' + 2y' + \lambda y = 0$, $y'(0) = y'(1) = 0$.

(a) Obtain the exact solution.

(b) Using $N = 10$, discretize by central differences and obtain numeric solutions for the first two eigenvalues and eigenfunctions.

(c) Compare numerically and graphically the approximate and exact solutions.

For the given BVP in Exercises 10–17:

(a) Use $N = 10$ and obtain a solution by finite differences.

(b) Compare the finite-difference solution with the analytic solution obtained earlier.

10. Exercise B2, Section 16.1   **11.** Exercise B3, Section 16.1

12. Exercise B4, Section 16.1   **13.** Exercise B5, Section 16.1

14. Exercise B6, Section 16.1   **15.** Exercise B7, Section 16.1

16. Exercise B8, Section 16.1   **17.** Exercise B9, Section 16.1

For the given BVP in Exercises 18–27:

(a) Use $N = 10$ and obtain a solution by finite differences.

(b) Compare the finite-difference solution with the solution obtained earlier.

18. Exercise B12, Section 16.1   **19.** Exercise B13, Section 16.1
20. Exercise B14, Section 16.1   **21.** Exercise B16, Section 16.1
22. Exercise B17, Section 16.1   **23.** Exercise B18, Section 16.1
24. Exercise B19, Section 16.1   **25.** Exercise B20, Section 16.1
26. Exercise B21, Section 16.1   **27.** Exercise B22, Section 16.1

28. Consider the eigenvalue problem $y'' - \lambda y = 0$, $y(-1) = y(1) = 0$.

(a) Obtain the exact solution for the eigenvalues λ_k and the eigenfunctions $y_k(x)$.

(b) Approximate the solution with $\phi(x) = \sum_{k=0}^{2} c_k f_k(x)$, where $f_k(x) = (x^2 - 1)x^k$, obtaining R, the residual that arises from substituting this approximation into the differential equation.

The residual will be a function of the three coefficients c_0, c_1, c_2 and λ as well as x.

(c) Using a collocation technique, force the residual to be zero at the four equally spaced points $x_0 = -1, \ldots, 1 = x_3$, resulting in four equations in the four unknowns c_0, c_1, c_2 and λ. These equations are linear and homogeneous in c_0, c_1, c_2.

(d) Write the equations from part (c) in the matrix form $A(\lambda)\mathbf{C} = \mathbf{0}$, where A is a 4×3 matrix. For most values of λ, the solution for $\mathbf{C}$ is the trivial solution, $\mathbf{0}$.

(e) To obtain nontrivial solutions for $\mathbf{C}$, choose $\lambda = \hat{\lambda}$ so the rank of A is just 2.

(f) Set $\lambda = \hat{\lambda}$ in the matrix A, forming the homogeneous system $\hat{A}\mathbf{C} = \mathbf{0}$, and obtain $\hat{\mathbf{C}}$, the general solution.

(g) Write the function $\hat{\phi} = \sum_{k=0}^{2} \hat{c}_k f_k(x)$, where $\hat{c}_k$ is the kth entry in $\hat{\mathbf{C}}$. There will be a free parameter in this approximate solution.

(h) Normalize the solution found in part (g): determine the free parameter so that $\hat{\phi}(0) = 1$.

(i) Normalize the exact eigenfunction correspondingly, and plot both on one set of axes.

Chapter Review

1. What is a regular Sturm–Liouville eigenvalue problem? Give an example.

2. Reproduce the solution process for the eigenvalue problem $y'' - \lambda y = 0$, $y(0) = y(L) = 0$. Clearly state the resulting eigenvalues and the companion eigenfunctions.

3. Explain why, in answering Question 2, the cases $\lambda = 0$ and $\lambda \neq 0$ are treated separately.

4. Reproduce the solution process for the eigenvalue problem $y'' - \lambda y = 0$, $y'(0) = y'(L) = 0$. Clearly state the resulting eigenvalues and the companion eigenfunctions.

5. Reproduce the solution process for the eigenvalue problem $y'' - \lambda y = 0$, $y(0) = y(2\pi)$, $y'(0) = y'(2\pi)$. Clearly state the resulting eigenvalues and the companion eigenfunctions.

6. Reproduce the solution process for the eigenvalue problem $y'' - \lambda y = 0$, $y(0) = y'(L) = 0$. Clearly state the resulting eigenvalues and the companion eigenfunctions.

7. Why is the eigenvalue problem posed with $x^2 y'' + xy' + x^2 y = 0$, $0 \leq x \leq c$, Bessel's equation of order zero, and the conditions that $y(x)$ be bounded and satisfy $y(c) = 0$, called a *singular* Sturm–Liouville eigenvalue problem?

8. Describe the process by which the eigenvalues and eigenfunctions for the problem in Question 7 are determined.

9. Show that Bessel's equation is formally self-adjoint.

10. In what sense are the eigenfunctions determined by the problem in Question 7 orthogonal? Illustrate with the appropriate mathematical statements.

11. Reproduce the process by which the differential equation $R''(r) + \frac{1}{r}R'(r) + \mu^2 R(r) = 0$ is cast into the form of Bessel's equation of order zero. Show how $R(r)$ is determined from the solution of Bessel's equation.

12. Detail how to obtain $\sum_{k=0}^{\infty} c_k J_0(\mu_k r)$, the Fourier–Bessel series for the function $f(r)$, $0 \leq r \leq c$, given the eigenvalues μ_k and the eigenfunctions $J_0(\mu_k r)$.

13. Why is the eigenvalue problem posed with $(1 - x^2)y''(x) - 2xy'(x) + \lambda y(x) = 0$, $-1 \leq x \leq 1$, Legendre's equation, and the condition that $y(x)$ be bounded called a *singular* Sturm–Liouville eigenvalue problem?

14. Describe the process by which the eigenvalues and eigenfunctions for the problem in Question 13 are determined. Include a discussion of the series solution and the recursion relation that determines its coefficients.

15. Detail how to obtain $\sum_{k=0}^{\infty} c_k P_k(x)$, the Fourier–Legendre series for the function $f(x)$, $-1 \leq x \leq 1$, given that the problem in Question 13 determines the eigenvalues $\lambda = k(k + 1)$ and the corresponding eigenfunctions $P_k(x)$, the kth Legendre polynomial.

16. Solving Laplace's equation in which coordinate system typically leads to Bessel's equation and solution by Fourier–Bessel series?

17. Solving Laplace's equation in which coordinate system typically leads to Legendre's equation and solution by Fourier–Legendre series?

UNIT FOUR

Vector Calculus

The seven chapters of Unit Four can be summarized in the phrase "div, grad, curl, Green's, Stokes', and Divergence." The three differential operators of divergence, gradient, and curl provide information about vector fields and are related through three integration theorems attributed to Green, Stokes, and Gauss. In addition, the differential operator called the laplacian, expressible in terms of the divergence and gradient, is treated.

Vector fields describing fluid flows, electromagnetism, and heat transfer exist in space and can be tangent to curves or orthogonal to surfaces. The geometry of curves and surfaces is therefore background against which our study of vector calculus takes place.

The gradient vector arises in our attempt to define a rate of change for a scalar-valued function of several variables. This vector is orthogonal to the level sets of the scalar function and is tangent to the lines of flow of the gradient field. Gravitational and electrostatic potentials are examples of scalar fields for which the gradient field is physically significant. The electric field vector is obtained from the gradient of the electrostatic potential function and is therefore an example of a conservative force.

Properties of scalar and vector fields are distributed in space, so attempts at averaging or accumulating their values leads to line integrals and surface integrals. The line integral is a process of *accumulation* taking place along a space curve. Mechanical work, mathematically equivalent to *circulation,* is the line integral of the tangential component of a vector field, and flux is the line or surface integral of the normal component of a vector field. The divergence (or spread) of a vector field at a point is related to its flux per unit area, while its curl is related to its circulation per unit area.

The divergence theorem relates the volume integral of the divergence to the flux of the field through the enclosing surface. Stokes' theorem relates a curl's flux through a surface and the circulation around the curve bounding the surface. Green's theorem gives the planar version of these two theorems. Using these theorems, the differential operators div, grad, and curl can be expressed as surface integrals.

Although initially defined in Cartesian coordinates, the operators div, grad, curl, and laplacian are transformed to polar, cylindrical, and spherical coordinates. This makes it possible to solve problems of wave motions, heat transfer, and electrostatics in domains that have either circular or spherical symmetry.

The unit concludes with a brief look at the equation of continuity obeyed by continuous fields and Green's identities, the generalization to higher dimension of Green's formula from Section 16.1.

Chapter *17*

Space Curves

I N T R O D U C T I O N   A curve in space can be described either parametrically or in radius-vector form. For the radius-vector form of the curve, differentiation with respect to the curve's parameter yields a vector tangent to the curve. If the curve describes the motion of an object in time, then the tangent vector is a velocity vector. If the curve is a coordinate curve, then the normalized tangent vector is a unit basis vector.

The curvature of a plane curve is the rate at which the slope of the tangent line changes as arc length along the curve varies. If at a point the curve is approximated by a circle making second-order contact (the first and second derivatives for the curve and circle agree), the circle is the circle of curvature and its center is the center of curvature. In fact, the curvature of the curve is the reciprocal of the radius of curvature of the circle.

Curvature is generalized to space curves by defining a unit tangent vector $\mathbf{T}$ along the curve. Then, the rate at which $\mathbf{T}$ changes as arc length changes along the curve must be ascribed to the change in orientation of $\mathbf{T}$, since its length remains constant. The magnitude of the vector describing this change is called the curvature.

Along a general curve, we also define the principal normal vector $\mathbf{N}$, a unique normal along the curve. The vectors $\mathbf{T}$ and $\mathbf{N}$ then determine a unique plane, called the osculating plane. In this plane, a circle making second-order contact with the curve is called the circle of curvature, and the center of this circle is the center of curvature. The principal normal points toward the center of curvature.

The binormal vector $\mathbf{B}$, defined as the cross-product $\mathbf{T} \times \mathbf{N}$, is then the third vector of a moving orthonormal basis defined along a space curve. Thus, for a fighter pilot putting an aircraft through strenuous maneuvers, the cockpit contains the moving frame consisting of tangent, normal, and binormal vectors. To the pilot, the frame is fixed, but to the observer on the ground, the frame twists and turns along the path of the aircraft.

The osculating plane also contains the vector $\mathbf{R}''$ that satisfies the decomposition rule

$$\mathbf{R}'' = \rho'\mathbf{T} + \kappa\rho^2\mathbf{N}$$

where $\rho = \|\mathbf{R}'\|$. If the parameter along the curve is time, then $\mathbf{R}''$ is the acceleration vector and $\mathbf{R}'' = \frac{dv}{dt}\mathbf{T} + \kappa v^2\mathbf{N}$. For this case where the parameter along the curve is time, $\rho = v$, the speed. The decomposition then says the acceleration vector always lies in the plane formed by the tangent and normal vectors. Indeed, it says more than that since it links the dynamics on the left side of the equation and the geometry on the right. The acceleration vector is a dynamic quantity, but the tangent vector $\mathbf{T}$, the principal normal $\mathbf{N}$, the curvature κ, and v, the length of the tangent vector, are all geometric quantities. There is a link between the geometry of a space curve and the dynamics that can take place along it.

17.1 Curves and Their Tangent Vectors

Space Curves—Parametric Representation

A curve in space is determined by three equations of the form

$$x = x(p) \qquad y = y(p) \qquad z = z(p)$$

where p is a parameter specifying points $(x(p), y(p), z(p))$ along the curve. The equations in Table 17.1 determine two different space curves. The equations on the left determine the line shown in Figure 17.1, while the equations on the right determine the helix, or spiral, shown in Figure 17.2. In the equations for the spiral, $x = x(p)$ and $y = y(p)$ determine a plane circle of radius 6 while $z(p) = p$ causes the rotating system-point to rise at a constant rate, thus creating the spiral. Both examples should be familiar from a multivariable calculus course.

Space Curves—Radius-Vector Representation

A curve can also be represented by a vector drawn from the origin to a moving point on the curve. For example, we could write the helix as the column vector on the left in (17.1), or, in terms of the unit basis vectors $\{\mathbf{i}, \mathbf{j}, \mathbf{k}\}$, as on the right in (17.1). A single representative radius vector is shown in Figure 17.2.

$$\mathbf{R} = \begin{bmatrix} x(p) \\ y(p) \\ z(p) \end{bmatrix} = \begin{bmatrix} 6\cos p \\ 6\sin p \\ p \end{bmatrix} = 6\cos p\mathbf{i} + 6\sin p\mathbf{j} + p\mathbf{k} \qquad (17.1)$$

Tangent Vectors

The derivative of the radius-vector form of a curve is defined by

$$\mathbf{R}'(p) = \lim_{dp \to 0} \frac{\mathbf{R}(p + dp) - \mathbf{R}(p)}{dp}$$

and yields a vector tangent to the curve. Thus, if the radius-vector form of a curve, either in space or in the plane, is differentiated with respect to its parameter, the result is a vector tangent to the curve. We call $\mathbf{R}'$ the *natural* tangent vector for the curve $\mathbf{R}$. The following examples illustrate this result. The first is in space and the second in the plane.

EXAMPLE 17.1   If $\mathbf{R}(p)$ for the helix given on the right in Table 17.1 is differentiated with respect to its parameter p, we get the vector

$$\mathbf{R}' = \begin{bmatrix} -6\sin p \\ 6\cos p \\ 1 \end{bmatrix} = -6\sin p\mathbf{i} + 6\cos p\mathbf{j} + \mathbf{k}$$

At each point on the curve represented by $\mathbf{R}$, the vector $\mathbf{R}'$ is a vector tangent to $\mathbf{R}$, as illustrated in Figure 17.3. ❖

EXAMPLE 17.2   In (17.2) we have on the left, the parametric representation of a plane curve, in the center, $\mathbf{R}(p)$, its radius-vector form, and on the right, $\mathbf{R}'(p)$, the natural tangent vector along the

Line	Helix
$x(p) = 1 + 3p$	$x(p) = 6\cos p$
$y(p) = 2 - 5p$	$y(p) = 6\sin p$
$z(p) = 3 + 2p$	$z(p) = p$

TABLE 17.1   Examples of space curves

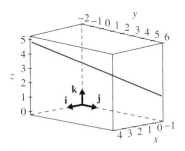

FIGURE 17.1   Line given on the left in Table 17.1

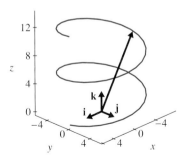

FIGURE 17.2   Helix given on the right in Table 17.1. Radius vector drawn from origin to (x, y, z) on the curve.

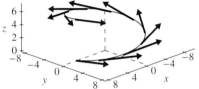

FIGURE 17.3   Vectors tangent to the helix given on the right in Table 17.1

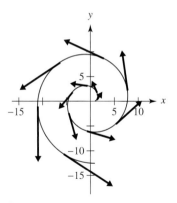

FIGURE 17.4   Curve and tangent vectors for Example 17.2

curve. Figure 17.4 shows the curve, an outward spiral, and some of its tangent vectors.

$$\left.\begin{array}{l} x(p) = \cos p + p \sin p \\ y(p) = \sin p - p \cos p \end{array}\right\} \Rightarrow \mathbf{R} = \begin{bmatrix} \cos p + p \sin p \\ \sin p - p \cos p \end{bmatrix} \qquad \mathbf{R}' = \begin{bmatrix} p \cos p \\ p \sin p \end{bmatrix} \qquad (17.2)$$

❖

Unit Tangent Vectors

A *unit tangent vector* **T** is obtained from the natural tangent vector $\mathbf{R}'(p)$ by normalization, that is, by dividing $\mathbf{R}'$ by its length. Thus, $\mathbf{R} = \mathbf{R}(p) \Rightarrow \frac{d\mathbf{R}}{dp} = \mathbf{R}'$ is tangent, but not, in general, of unit length (i.e., $\|\mathbf{R}'\| \neq 1$). The varying length of $\mathbf{R}'$ is $\rho = \|\mathbf{R}'\|$, so $\mathbf{T} = \frac{\mathbf{R}'}{\rho}$ is now a unit tangent vector. For the curve $\mathbf{R}(p)$ from Example 17.2, the tangent vector $\mathbf{R}'$ on the right in (17.2) has magnitude $\rho = \sqrt{p^2(\cos^2 p + \sin^2 p)} = p$, assuming $p \geq 0$. Dividing $\mathbf{R}'$ by ρ yields the *unit* tangent vector

$$\mathbf{T} = \frac{\mathbf{R}'}{\rho} = \frac{1}{p}\begin{bmatrix} p \cos p \\ p \sin p \end{bmatrix} = \begin{bmatrix} \cos p \\ \sin p \end{bmatrix} = \cos p\,\mathbf{i} + \sin p\,\mathbf{j}$$

EXERCISES 17.1–Part A

A1. If $P_k, k = 1, 2$, are the points $(3, -2)$ and $(7, 4)$, respectively, and $\mathbf{P}_k, k = 1, 2$, are vectors from the origin to these respective points, show that $\mathbf{R}(p) = \mathbf{P}_1 + p(\mathbf{P}_2 - \mathbf{P}_1), 0 \leq p \leq 1$, determines the line segment from P_1 to P_2.

A2. Let $P_k, k = 1, 2$, be the points $(3, -1, 2)$ and $(5, 7, -4)$, and generalize the results in Exercise A1 to three dimensions.

A3. Eliminate the parameter p from the representation of the line in Exercise A1, and show the result agrees with the point-slope form of the line in the plane.

A4. Derive $x(\theta) = a(\theta - \sin \theta)$, $y(\theta) = a(1 - \cos \theta)$, the parametric equations of a cycloid, the curve traced by a point on the circumference of a circle of radius a as it rolls along the x-axis from an initial position in which its center was at $(0, a)$. *Hint:* See Figure 17.5, where θ is the angle between the vertical and the radius to the point P, which was initially at the origin.

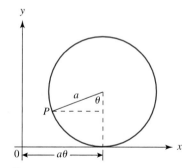

FIGURE 17.5   Hint for Exercise A4

A5. The plane curve $y = f(x)$ can be parametrized as $x(p) = p$, $y(p) = f(p)$. Verify this is so for $f(x) = x(1 - x)$.

EXERCISES 17.1–Part B

B1. The parametric equations $x(\theta) = 7 \cos \theta - \cos 7\theta$, $y(\theta) = 7 \sin \theta - \sin 7\theta$, $0 \leq \theta \leq 2\pi$, describe an epicycloid, the curve traced by a point on the circumference of a circle rolling around the outside of a second fixed circle. The angle θ is the angle from the horizontal to the line connecting the centers of the circles. Sketch this epicycloid.

B2. Let $A = a + b$ and $B = 1 + \frac{a}{b}$, and show that $x(\theta) = A \cos \theta - b \cos B\theta$, $y(\theta) = A \sin \theta - b \sin B\theta$, are the general equations for an epicycloid generated by a circle of radius b rolling on the outside of a fixed circle of radius a. What are the values of a and b for the epicycloid in Exercise B1?

In Exercises B3–12, a plane curve $y = y(x)$ and three values $x = x_k$, $k = 1, 2, 3$, are given. In each case:

(a) Obtain $\mathbf{R}(p)$, the radius vector form of the curve, by the parametrization $x = p, y = y(p)$.

(b) Find $\mathbf{R}'(p)$, the natural tangent vector at an arbitrary point along the curve.

(c) Obtain $\mathbf{R}'(p_k)$, the natural tangent vector at each $p_k = x_k$.

(d) Obtain $\rho(p_k) = \|\mathbf{R}'(p_k)\|$, the length of each tangent vector found in part (b).

(e) Obtain $\mathbf{T}(p_k)$, a unit tangent vector, for each $p_k = x_k$.

(f) Plot the curve, and on it draw the unit tangent vectors found in part (d).

(g) Find $\rho(p)$, the magnitude of $\mathbf{R}'(p)$.

(h) Plot $\rho(p)$.

(i) Determine p_m and p_M, the parameter values for which $\rho(p)$ attains, respectively, its absolute minimum and absolute maximum on the interval $[x_0, x_2]$. What are the coordinates of the corresponding points along the curve?

B3. $y = 2x^2 - 3x + 1$, $x_k = 0, 1, 2$

B4. $y = 5x^3 - 7x^2 + 2x + 3$, $x_k = -1, 0, 1$

B5. $y = \dfrac{1}{1 + x^2}$, $x_k = 0, 1, 2$ **B6.** $y = \dfrac{x}{1 + x^2}$, $x_k = -2, 0, 1$

B7. $y = \dfrac{1 - x^2}{1 + x^2}$, $x_k = 0, 1, 2$ **B8.** $y = xe^x$, $x_k = -1, 0, 1$

B9. $y = x^2 e^x$, $x_k = -1, 0, 1$ **B10.** $y = e^{-x^2}$, $x_k = -1, 0, 2$

B11. $y = \arctan x$, $x_k = 0, 1, 2$

B12. $y = \sqrt{1 + \sin^2 x}$, $x_k = 0, \dfrac{\pi}{4}, \dfrac{\pi}{2}$

In Exercises B13–17, a curve is given parametrically, along with several values p_k for the parameter p. In each case:

(a) Find $\mathbf{R}'(p)$, the natural tangent vector at an arbitrary point along the curve.

(b) Obtain $\mathbf{R}'(p_k)$, the natural tangent vector at each p_k.

(c) Obtain $\rho(p_k) = \|\mathbf{R}'(p_k)\|$, the length of each tangent vector found in part (b).

(d) Obtain $\mathbf{T}(p_k)$, a unit tangent vector, for each p_k.

(e) Plot the curve, and on it draw the unit tangent vectors found in part (d).

(f) Find $\rho(p)$, the magnitude of $\mathbf{R}'(p)$.

(g) Plot $\rho(p)$.

(h) Determine p_m and p_M, the parameter values for which $\rho(p)$ attains, respectively, its absolute minimum and absolute maximum on the smallest closed interval containing all the given p_k. What are the coordinates of the corresponding points along the curve?

B13. $x(p) = \sqrt{1 + p}$, $y(p) = \sqrt{1 + \frac{p^2}{10}}$, $p_1 = 1$, $p_2 = 3$

B14. $x(p) = p \cos p$, $y(p) = p^2 \sin p$, $p_k = 2, 5, 9$, $k = 1, 2, 3$, respectively

B15. $x(p) = 1 - p^2$, $1 + p^2$, $p_k = 1, 2$, $k = 1, 2$, respectively

B16. $x(p) = (1 + p^3)e^{-p}$, $y(p) = (1 - p^2)e^{-p}$, $p_k = 0, 1, 2$, $k = 1, 2, 3$, respectively

B17. $x(p) = 2 + 3p - 5p^2$, $y(p) = \cos p$, $p_k = 0, 1$, $k = 1, 2$, respectively

In Exercises B18–24, a curve is given parametrically, along with several values p_k for the parameter p. In each case:

(a) Find $\mathbf{R}'(p)$, the natural tangent vector at an arbitrary point along the curve.

(b) Obtain $\mathbf{R}'(p_k)$, the natural tangent vector at each p_k.

(c) Obtain $\rho(p_k) = \|\mathbf{R}'(p_k)\|$, the length of each tangent vector found in part (b).

(d) Obtain $\mathbf{T}(p_k)$, a unit tangent vector, for each p_k.

(e) Find $\rho(p)$, the magnitude of $\mathbf{R}'(p)$.

(f) Plot $\rho(p)$.

(g) Determine p_m and p_M, the parameter values for which $\rho(p)$ attains, respectively, its absolute minimum and absolute maximum on the smallest closed interval containing all the given p_k. What are the coordinates of the corresponding points along the curve?

B18. $x(p) = 3p^2$, $y(p) = 2p + 1$, $z(p) = p^3$, $p_k = 0, 1$, $k = 1, 2$, respectively

B19. $x(p) = p^2 \cos p$, $y(p) = p^3 \sin p$, $z(p) = p(1 - p)$, $p_k = 0, 1, 2$, $k = 1, 2, 3$, respectively

B20. $x(p) = \dfrac{1 - p}{1 + p}$, $y(p) = \dfrac{2 - p}{2 + p}$, $z(p) = \dfrac{3 - p}{3 + p}$, $p_k = 0, 1, 2$, $k = 1, 2, 3$, respectively

B21. $x(p) = 1 - p$, $y(p) = 1 + p^2$, $z(p) = \dfrac{p}{1 + p^2}$, $p_k = 0, 1$, $k = 1, 2$, respectively

B22. $x(p) = \cos 2p$, $y(p) = \sin 3p$, $z(p) = \cos p$, $p_k = 0, 1$, $k = 1, 2$, respectively

B23. $x(p) = (1 + p)^2$, $y(p) = (1 - p)^2$, $z(p) = p$, $p_k = 0, 1$, $k = 1, 2$, respectively

B24. $x(p) = p$, $y(p) = p^2$, $z(p) = p^3$, $p_k = 0, 1, 2$, $k = 1, 2, 3$, respectively

In Exercises B25–29, two plane curves $y = f(x)$ and $y = g(x)$ are given, along with a value for $x = x_0$. In each case:

(a) At $x = x_0$, obtain the equation of the line tangent to $y = f(x)$.

(b) Find the coordinates of the intersection of $y = g(x)$ and the tangent line found in part (a).

(c) Construct a vector from $(x_0, f(x_0))$ to the point found in part (b).

(d) Along $f(x)$, obtain $\mathbf{R}'(x_0)$, the natural tangent vector at $(x_0, f(x_0))$.

(e) Show that the vectors in parts (c) and (d) are parallel. (*Hint:* Show their components are proportional.)

(f) Draw both curves and the tangent vector found in part (d).

B25. $f(x) = \sin x$, $g(x) = 3$, $x_0 = \dfrac{\pi}{4}$

B26. $f(x) = x^2$, $g(x) = 8 - \left(\dfrac{x}{4}\right)^2$, $x \geq 0$, $x_0 = 1$

B27. $f(x) = x^3 + 3x^2 + 2x - 5$, $g(x) = 8 - x^2$, $x \geq 0$, $x_0 = 1$

B28. $f(x) = \dfrac{1}{1 + x^2}$, $g(x) = x + 2$, $x_0 = 2$

B29. $f(x) = \dfrac{1 - x^2}{1 + x^2}$, $g(x) = 2 - x$, $x_0 = -2$

17.2 Arc Length

Arc Length Formulas

Table 17.2 lists formulas for the arc length of a curve given in different formats. A plane curve can be given explicitly as $y = f(x)$ or parametrically as $x = x(p)$, $y = y(p)$. The radius-vector form $\mathbf{R}$ for each of these is shown in the table. In three dimensions, the curve is given parametrically and has the obvious radius-vector form $\mathbf{R}$.

Curve	R	R′	Arc Length
$y = f(x)$	$\begin{bmatrix} x \\ f(x) \end{bmatrix}$	$\begin{bmatrix} 1 \\ \frac{df}{dx} \end{bmatrix}$	$s = \int_a^b \sqrt{1 + \left(\frac{df}{dx}\right)^2}\, dx$
$\begin{cases} x = x(p) \\ y = y(p) \end{cases}$	$\begin{bmatrix} x(p) \\ y(p) \end{bmatrix}$	$\begin{bmatrix} \frac{dx}{dp} \\ \frac{dy}{dp} \end{bmatrix}$	$s = \int_a^b \sqrt{\left(\frac{dx}{dp}\right)^2 + \left(\frac{dy}{dp}\right)^2}\, dp$
$\begin{cases} x = x(p) \\ y = y(p) \\ z = z(p) \end{cases}$	$\begin{bmatrix} x(p) \\ y(p) \\ z(p) \end{bmatrix}$	$\begin{bmatrix} \frac{dx}{dp} \\ \frac{dy}{dp} \\ \frac{dz}{dp} \end{bmatrix}$	$s = \int_a^b \sqrt{\left(\frac{dx}{dp}\right)^2 + \left(\frac{dy}{dp}\right)^2 + \left(\frac{dz}{dp}\right)^2}\, dp$

TABLE 17.2   Arc length formulas

In each case, $\mathbf{R}'$, the derivative of $\mathbf{R}$, is listed; and Table 17.2 thereby shows that the integrand of the arc length integral is $\|\mathbf{R}'\|$, the length of the natural tangent vector. We state this observation as

$$ s = \int_a^b \|\mathbf{R}'\|\, dp = \int_a^b \rho(p)\, dp $$

EXAMPLE 17.3   For $\mathbf{R}$, the helix of Example 17.1, $\rho = \|\mathbf{R}'\| = \sqrt{37}$, so the arc length along one turn of the helix is

$$ s = \int_0^{2\pi} \sqrt{37}\, dp = 2\pi \sqrt{37} \qquad ❖ $$

EXAMPLE 17.4   **The arc length function**   From the starting point where $p = 0$, to the arbitrary terminal point p, compute the length of $\mathbf{R}$, the curve given on the left in

$$ \mathbf{R} = \begin{bmatrix} p^2 \\ p^3 \end{bmatrix} \Rightarrow \mathbf{R}' = \begin{bmatrix} 2p \\ 3p^2 \end{bmatrix} \tag{17.3} $$

For the natural tangent vector $\mathbf{R}'$, given on the right in (17.3), we have $\|\mathbf{R}'\| = \tau \sqrt{4 + 9\tau^2}$,

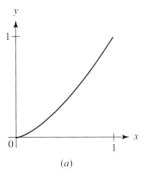

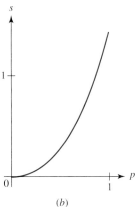

FIGURE 17.6   Example 17.4: The curve (*a*) and its arc length function (*b*)

where τ is a dummy variable of integration. The desired arc length is then

$$s(p) = \int_0^p \tau\sqrt{4 + 9\tau^2}\,d\tau = \tfrac{1}{27}\left(4 + 9p^2\right)^{3/2} - \tfrac{8}{27}$$

The curve $\mathbf{R}$ is shown in Figure 17.6(*a*), and $s(p)$, the arc length function, is shown in Figure 17.6(*b*). The vertical axis in Figure 17.6(*b*) exhibits the readings on an odometer in a car driven along the path $\mathbf{R}(p)$.   ❖

Arc Length as Parameter

THEOREM 17.1

If the parameter for a curve is s, the curve's arc length, the natural tangent vector $\mathbf{R}'(s)$ is automatically the *unit* vector $\mathbf{T}$.

The proof follows from an application of the chain rule by which we obtain

$$\frac{d\mathbf{R}}{ds} = \frac{d\mathbf{R}}{dp}\frac{dp}{ds} = \frac{d\mathbf{R}}{dp}\frac{1}{\frac{ds}{dp}} = \left(\frac{dx}{dp}\mathbf{i} + \frac{dy}{dp}\mathbf{j} + \frac{dz}{dp}\mathbf{k}\right)\frac{1}{\sqrt{\left(\frac{dx}{dp}\right)^2 + \left(\frac{dy}{dp}\right)^2 + \left(\frac{dz}{dp}\right)^2}}$$

Then, computing the length of $\frac{d\mathbf{R}}{ds}$, we have

$$\left\|\frac{d\mathbf{R}}{ds}\right\| = \left\|\frac{d\mathbf{R}}{dp}\right\| \frac{1}{\sqrt{\left(\frac{dx}{dp}\right)^2 + \left(\frac{dy}{dp}\right)^2 + \left(\frac{dz}{dp}\right)^2}} = \frac{\sqrt{\left(\frac{dx}{dp}\right)^2 + \left(\frac{dy}{dp}\right)^2 + \left(\frac{dz}{dp}\right)^2}}{\sqrt{\left(\frac{dx}{dp}\right)^2 + \left(\frac{dy}{dp}\right)^2 + \left(\frac{dz}{dp}\right)^2}} = 1$$

◆ **EXAMPLE 17.5**   Table 17.3 contains the parametric and radius-vector forms of a plane curve and its natural tangent vector $\mathbf{R}'(p)$. Figure 17.7, a graph of $\mathbf{R}'(p)$ at several points along $\mathbf{R}(p)$, shows how the lengths of the natural tangent vector vary with p.

Parametric Form	$\mathbf{R}(p)$	$\mathbf{R}'(p)$
$x(p) = \frac{3}{2}p^2 - \frac{3}{4}p + 1$ $y(p) = 2p^{3/2} + 1$	$\begin{bmatrix} \frac{3}{2}p^2 - \frac{3}{4}p + 1 \\ 2p^{3/2} + 1 \end{bmatrix}$	$\begin{bmatrix} 3p - \frac{3}{4} \\ 3\sqrt{p} \end{bmatrix}$

TABLE 17.3   Plane curves for Example 17.5

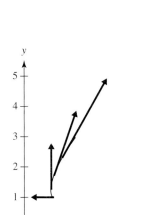

FIGURE 17.7   The curve and natural tangent vectors in Example 17.5

For the curve in Table 17.3, the following steps will show (a) how to parametrize the curve in terms of s, the arc length, and (b) that $\mathbf{R}'(s)$ is $\mathbf{T}$, the unit tangent vector.

1.   Get $s(p)$, arc length as a function of p, the stopping point on the curve, by computing $s(p) = \int_0^p \|\mathbf{R}'(\tau)\|\,d\tau$.

2.   Invert the function $s = s(p)$ to obtain $p = p(s)$.

3.   By writing the expressions $x = x(s) = x(p(s))$, $y = y(s) = y(p(s))$, obtain $\mathbf{R}(s)$, the curve expressed with parameter s.

4. Compute $\frac{d\mathbf{R}(s)}{ds}$ to get a tangent vector along $\mathbf{R}(s)$.

5. Show that $\|\mathbf{R}'(s)\| = 1$. Hence, $\mathbf{R}'(s) = \mathbf{T}$.

Step 1   Obtain the function $s(p)$ by the following calculation.

$$s(p) = \int_0^p \|\mathbf{R}'(\tau)\|\,d\tau = \int_0^p \sqrt{(x'(\tau))^2 + (y'(\tau))^2}\,d\tau = \int_0^p \left(3\tau + \tfrac{3}{4}\right) d\tau = \tfrac{3}{2}p^2 + \tfrac{3}{4}p$$

Step 2   Obtain the function $p = p(s)$ by solving the equation

$$s = s(p) = \tfrac{3}{2}p^2 + \tfrac{3}{4}p \quad \text{for} \quad p = -\tfrac{1}{4} \pm \tfrac{1}{12}\sqrt{9 + 96s}$$

Select the positive root because at $p = 0$, s must also be 0. Consequently, we have $p(s) = -\tfrac{1}{4} + \tfrac{1}{12}\sqrt{9 + 96s}$.

Step 3   Substitution yields

$$x = x(s) = x(p(s)) = \tfrac{11}{8} - \tfrac{1}{8}\sqrt{9 + 96s} + s$$
$$y = y(s) = y(p(s)) = 2\left(-\tfrac{1}{4} + \tfrac{1}{12}\sqrt{9 + 96s}\right)^{3/2} + 1$$

Step 4   Form $\mathbf{R}(s)$ and compute $\mathbf{R}'(s) = \frac{d\mathbf{R}(s)}{ds}$, obtaining

$$\mathbf{R}(s) = \begin{bmatrix} \tfrac{11}{8} - \tfrac{1}{8}\sqrt{9+96s} + s \\ 2\left(-\tfrac{1}{4} + \tfrac{1}{12}\sqrt{9+96s}\right)^{3/2} + 1 \end{bmatrix} \quad \text{and} \quad \mathbf{R}'(s) = \begin{bmatrix} \dfrac{-6 + \sqrt{9+96s}}{\sqrt{9+96s}} \\ 2\dfrac{\sqrt{-9 + 3\sqrt{9+96s}}}{\sqrt{9+96s}} \end{bmatrix}$$

Step 5   A straightforward but tedious computation now shows $\|\mathbf{R}'(s)\| = 1$, verifying the claim that $\mathbf{R}'(s) = \mathbf{T}$, the unit tangent vector along $\mathbf{R}(s)$.

A graph of the curve $\mathbf{R}(s)$ with (unit) tangent vectors $\mathbf{T} = \frac{d\mathbf{R}(s)}{ds}$ appears in Figure 17.8.   ❖

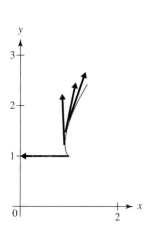

FIGURE 17.8   The curve and unit tangent vectors in Example 17.5

Generalization

1. Given $\mathbf{R}(p)$, the radius-vector form of a curve, $\mathbf{R}'(p) = \frac{d\mathbf{R}}{dp}$ is the natural tangent vector.

2. The length of the tangent vector is $\rho(p) = \|\mathbf{R}'(p)\| = \sqrt{\left(\frac{dx}{dp}\right)^2 + \left(\frac{dy}{dp}\right)^2 + \left(\frac{dz}{dp}\right)^2}$.

3. The arc length function is $s(p) = \int_0^p \rho(\tau)\,d\tau = \int_0^p \|\mathbf{R}'(\tau)\|\,d\tau$.

4. By the fundamental theorem of calculus, the derivative of the arc length function is

$$\frac{ds}{dp} = s'(p) = \rho(p) = \left\|\frac{d\mathbf{R}}{dp}\right\| = \sqrt{\left(\frac{dx}{dp}\right)^2 + \left(\frac{dy}{dp}\right)^2 + \left(\frac{dz}{dp}\right)^2}$$

5. In principle, the arc length function has an inverse $p = p(s)$.

6. By elementary calculus, $\frac{dp}{ds} = \frac{1}{ds/dp} = \frac{1}{\rho}$, that is, the derivative of an inverse function is the reciprocal of the derivative of the function. (See Exercises 29–34 for a refresher on this result from calculus.)

7. The unit tangent vector $\mathbf{T}$ can be computed by normalizing $\mathbf{R}' = \frac{d\mathbf{R}}{dp}$ via $\mathbf{T} = \frac{d\mathbf{R}/dp}{\|d\mathbf{R}/dp\|} = \frac{d\mathbf{R}/dp}{\rho(p)} = \frac{d\mathbf{R}}{dp}\frac{1}{\rho}$.

8. By the chain rule for differentiation, we can compute $\mathbf{T} = \frac{d\mathbf{R}}{ds}$ from $\mathbf{R}(p)$ as

$$\mathbf{T} = \frac{d\mathbf{R}}{ds} = \frac{d\mathbf{R}}{dp}\frac{dp}{ds} = \frac{d\mathbf{R}}{dp}\frac{1}{\frac{ds}{dp}} = \frac{d\mathbf{R}}{dp}\frac{1}{\rho}$$

There really aren't two different ways for computing $\mathbf{T}$. Given $\mathbf{R}(p)$, there is only one way to compute $\mathbf{T}$ and that is to compute $\mathbf{R}'(p)$ and divide by $\rho = \|\mathbf{R}'(p)\|$ to normalize $\mathbf{R}'$. Although the computation can be conceptualized in two different ways, the results are exactly the same.

EXERCISES 17.2

In Exercises 1–10, obtain the arc length for the given plane curve $y = y(x)$ between the given endpoints x_1 and x_3.

1. Exercise B3, Section 17.1   **2.** Exercise B4, Section 17.1

3. Exercise B5, Section 17.1   **4.** Exercise B6, Section 17.1

5. Exercise B7, Section 17.1   **6.** Exercise B8, Section 17.1

7. Exercise B9, Section 17.1   **8.** Exercise B10, Section 17.1

9. Exercise B11, Section 17.1   **10.** Exercise B12, Section 17.1

In Exercises 11–22:

(a) If $0 \le p \le 3$, find the length of the arc for the parametrically given curve.

(b) Obtain a graph of $s(p)$, the arc length function, if $0 \le p < 5$. Use numeric integration as needed. The arc length can be evaluated numerically for 10 values of p and the resulting points plotted.

(c) Obtain a graph of $p = p(s)$, the inverse function. If $s(p)$ was obtained as a discrete collection of points (p_k, s_k), simply plot the points (s_k, p_k) or reflect the graph of $p(s)$ about the line $p = s$.

11. Exercise B13, Section 17.1   **12.** Exercise B14, Section 17.1

13. Exercise B15, Section 17.1   **14.** Exercise B16, Section 17.1

15. Exercise B17, Section 17.1   **16.** Exercise B18, Section 17.1

17. Exercise B19, Section 17.1   **18.** Exercise B20, Section 17.1

19. Exercise B21, Section 17.1   **20.** Exercise B22, Section 17.1

21. Exercise B23, Section 17.1   **22.** Exercise B24, Section 17.1

In Exercises 23–28, a curve is defined parametrically for $0 \le p < \infty$. For each:

(a) Obtain a graph for $0 \le p \le 5$.

(b) Obtain, in closed form, $s = s(p)$, the arc length function.

(c) Compute $\frac{ds}{dp}$ and $\frac{dp}{ds}$ and show that these derivatives are actually reciprocals.

(d) Express the curve $\mathbf{R}(p) = x(p)\mathbf{i} + y(p)\mathbf{j}$ as $\mathbf{R}(s)$ by finding $x(p(s))$ and $y(p(s))$.

(e) Compute $\mathbf{R}'(s)$ and show it is a unit vector.

23. $x(p) = 1 + 9p - \frac{1}{2}p^2$, $y(p) = 1 + 4p^{3/2}$

24. $x(p) = 8p^2 + 16p + 5$, $y(p) = 4p^2 + 8p - 3$

25. $x(p) = 4p^2 + \frac{28}{5}p - 9$, $y(p) = 5p^2 + 7p$

26. $x(p) = 9p^2 + \frac{18}{5}p + 4$, $y(p) = 5p^2 + 2p - 5$

27. $x(p) = 7p^2 - \frac{49}{9}p + 4$, $y(p) = 5 + 7p - 9p^2$

28. $x(p) = 4p^2 + \frac{12}{5}p - 8$, $y(p) = 5p^2 + 3p - 9$

Two functions $f(x)$ and $g(x)$ are inverses if their compositions are related by $f(g(x)) = g(f(x)) = x$. Hence, the chain rule yields $f'(g(x))g'(x) = 1$, so $g'(x) = \frac{1}{f'(g(x))}$. For example, if $f(x) = \sin x$ and $g(x) = \arcsin x$, we know $f'(x) = \cos x$ and $g'(x) = \frac{1}{\sqrt{1-x^2}}$, but the second does not appear to be the reciprocal of the first. That is because f' has not been evaluated at the proper argument, namely, at $x = g(x)$. Doing that yields $g'(x) = \frac{1}{\cos(\arcsin x)} = \frac{1}{\sqrt{1-x^2}}$, which is the expected derivative of $g(x) = \arcsin x$.

In Exercises 29–34:

(a) For x in an appropriate domain, show that $f(x)$ and $g(x)$ are functional inverses by showing $f(g(x)) = x$ for x in the domain of g and $g(f(x)) = x$ for x in the domain of f.

(b) Obtain the derivatives $f'(x)$ and $g'(x)$.

(c) Show that f' and g' are actually reciprocals of each other. (*Hint:* Show, for example, $g'(x) = \frac{1}{f'(g(x))}$.)

29. $f(x) = \sinh x$, $g(x) = \text{arcsinh } x$

30. $f(x) = \tan x$, $g(x) = \arctan x$

31. $f(x) = e^x$, $g(x) = \ln x$ **32.** $f(x) = x^2$, $g(x) = \sqrt{x}$

33. $f(x) = \frac{2-x}{3+x}$, $g(x) = \frac{2-3x}{1+x}$

34. $f(x) = \frac{1}{1+x^2}$, $g(x) = \frac{\sqrt{x(1-x)}}{x}$

Curvature

Curvature of a Plane Curve

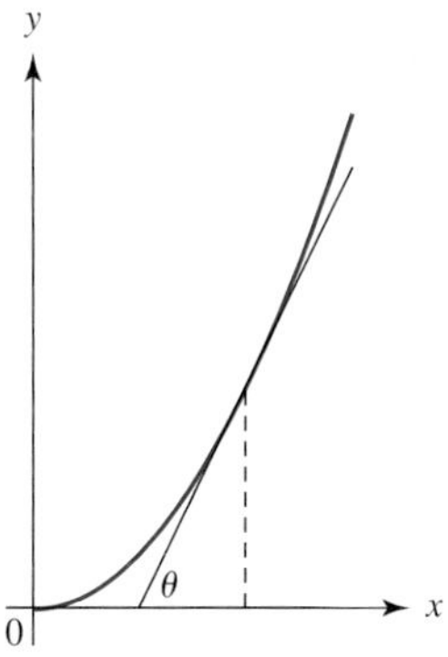

Given the curve whose equation is $y = f(x)$, we define κ, the *curvature* of this curve, measuring how "bent" the curve is at each point, to be the derivative, with respect to arc length, of the angle between the tangent line and the horizontal. This results in the formula

$$\kappa = \frac{|y''|}{(1 + (y')^2)^{3/2}} \tag{17.4}$$

whose derivation is based on Figure 17.9, which shows a curve, a tangent line, and the angle θ formed by the tangent line and the horizontal.

By the definition of the derivative, $y'(x) = \tan\theta$, so we have $\theta = \arctan y'$. The curvature κ is defined as the rate of change, with respect to arc length s, of the angle θ. Thus,

$$\kappa = \frac{d\theta}{ds} = \frac{d}{ds}\arctan(f'(x(s))) = \frac{f''(x(s))x'(s)}{1 + (f'(x(s)))^2}$$

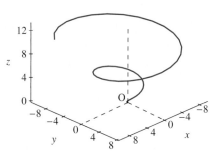

FIGURE 17.9 Curvature as the derivative of the angle θ

From Section 17.2, recall (a) the reciprocal rule for derivatives of inverse functions, namely, $\frac{dx}{ds} = \frac{1}{ds/dx}$ and (b), the definition of $s(x)$, namely, $s(x) = \int_0^x \sqrt{1 + (f'(\tau))^2}\, d\tau$. Hence, we have $\frac{ds}{dx} = \sqrt{1 + (f'(x))^2}$ and $\frac{dx}{ds} = \frac{1}{\sqrt{1+(f')^2}}$, so substitution for $\frac{dx}{ds}$ gives $\kappa = \frac{f''}{(1+(f')^2)^{3/2}}$. The important insight, here, is that the curvature formula follows from measuring the rate at which the tangent line rotates as the arc length increases.

Curvature of a Space Curve

The curvature of a space curve is defined as the magnitude of the rate of change (with respect to arc length) in the unit tangent vector $\mathbf{T}$. Since $\mathbf{T}$ always has length 1, any change in it must be caused by a change in its orientation. This again captures the idea of curvature as a measure of the turning of the tangent.

The formula $\kappa = \|\frac{d\mathbf{T}}{ds}\|$ captures the *definition* but is best avoided in practice. It requires parametrization by arc length, which, as Example 17.5 showed, is a complication best avoided when possible. To be derived shortly, a more practical formula is

$$\kappa = \frac{\|\mathbf{R}' \times \mathbf{R}''\|}{\rho^3} = \frac{\|\mathbf{R}' \times \mathbf{R}''\|}{\|\mathbf{R}'\|^3} \tag{17.5}$$

FIGURE 17.10 Expanding helix of Example 17.6

which works with whatever parameter is given for the curve. The derivatives $\mathbf{R}'$ and $\mathbf{R}''$ are simply the first and second derivatives of the curve's components and are taken with respect to the given parameter on the curve. The numerator is the length of the cross-product of these two vectors and the denominator is the length of $\mathbf{R}'$, a scalar, cubed.

◆ **EXAMPLE 17.6** In (17.6), we have $\mathbf{R}(p)$, the radius-vector form of the expanding helix shown in Figure 17.10. Also given in (17.6), we have $\mathbf{R}'(p)$ and $\mathbf{R}''(p)$, the first and second derivatives, respectively, along this curve.

$$\mathbf{R} = \begin{bmatrix} p\cos p \\ p\sin p \\ p \end{bmatrix} \quad \mathbf{R}' = \begin{bmatrix} \cos p - p\sin p \\ \sin p + p\cos p \\ 1 \end{bmatrix} \quad \mathbf{R}'' = \begin{bmatrix} -2\sin p - p\cos p \\ 2\cos p - p\sin p \\ 0 \end{bmatrix} \tag{17.6}$$

The cross-product is given by

$$\mathbf{R}' \times \mathbf{R}'' = \begin{vmatrix} \mathbf{i} & \mathbf{j} & \mathbf{k} \\ \cos p - p\sin p & \sin p + p\cos p & 1 \\ -2\sin p - p\cos p & 2\cos p - p\sin p & 0 \end{vmatrix} = \begin{bmatrix} -2\cos p + p\sin p \\ -2\sin p - p\cos p \\ p^2 + 2 \end{bmatrix}$$

and its length by $\|\mathbf{R}' \times \mathbf{R}''\| = \sqrt{8 + p^4 + 5p^2}$. The denominator contains $\rho = \|\mathbf{R}'\| = \sqrt{p^2 + 2}$, so the curvature along this space curve is given by $\kappa = \frac{\sqrt{8+p^4+5p^2}}{(p^2+2)^{3/2}}$. Figure 17.11 shows how $\kappa(p)$ varies along the curve. As the radius of the helix increases, the curvature decreases. The curvature is greatest when the helix is just beginning to unwind. ❖

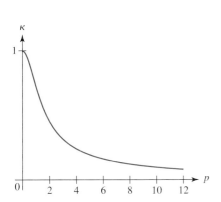

FIGURE 17.11   Curvature as a function of p along the expanding helix of Example 17.6

DERIVATION   Setting $\mathbf{R} = x(p)\mathbf{i} + y(p)\mathbf{j} + z(p)\mathbf{k}$, we can verify that $\kappa = \|\frac{d\mathbf{T}}{ds}\| = \frac{\|\mathbf{R}' \times \mathbf{R}''\|}{\rho^3}$ as follows. Write $\mathbf{T} = \frac{\mathbf{R}'}{\|\mathbf{R}'\|}$, and use the chain rule to obtain $\frac{d\mathbf{T}}{ds} = \frac{d\mathbf{T}}{dp}\frac{dp}{ds} = \frac{d\mathbf{T}}{dp}\frac{1}{ds/dp} = \frac{d\mathbf{T}}{dp}\frac{1}{\|\mathbf{R}'\|}$. Then,

$$\frac{d\mathbf{T}}{dp} = \frac{d}{dp}\left(\frac{\mathbf{R}'}{\|\mathbf{R}'\|}\right) = \frac{\mathbf{R}''}{\|\mathbf{R}'\|} + \mathbf{R}'\frac{d}{dp}\left(\frac{1}{\|\mathbf{R}'\|}\right) \qquad (17.7)$$

Now, $\|\mathbf{R}'\| = \sqrt{(x')^2 + (y')^2 + (z')^2}$, so

$$\frac{d}{dp}\left(\frac{1}{\|\mathbf{R}'\|}\right) = -\frac{1}{2}\left((x')^2 + (y')^2 + (z')^2\right)^{-3/2}(2x'x'' + 2y'y'' + 2z'z'') = -\frac{\mathbf{R}' \cdot \mathbf{R}''}{\|\mathbf{R}'\|^3}$$

$$(17.8)$$

Thus, using (17.7) and (17.8), $\frac{d\mathbf{T}}{ds}$ becomes

$$\frac{d\mathbf{T}}{ds} = \left(\frac{\mathbf{R}''}{\|\mathbf{R}'\|} - \mathbf{R}'\frac{\mathbf{R}' \cdot \mathbf{R}''}{\|\mathbf{R}'\|^3}\right)\frac{1}{\|\mathbf{R}'\|} = \left(\mathbf{R}'' - \mathbf{R}'\frac{\mathbf{R}' \cdot \mathbf{R}''}{\|\mathbf{R}'\|^2}\right)\frac{1}{\|\mathbf{R}'\|^2}$$

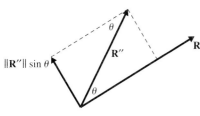

FIGURE 17.12   Components of $\mathbf{R}''$ along and perpendicular to $\mathbf{R}'$

Now, $\mathbf{R}'\frac{\mathbf{R}' \cdot \mathbf{R}''}{\|\mathbf{R}'\|^2}$ is the component of $\mathbf{R}''$ along $\mathbf{R}'$ (see Exercise A5 and Figure 17.12), so $\mathbf{R}'' - \mathbf{R}'\frac{\mathbf{R}' \cdot \mathbf{R}''}{\|\mathbf{R}'\|^2}$ is therefore the component of $\mathbf{R}''$ perpendicular to $\mathbf{R}'$ (see Exercise A6 and Figure 17.12) and is given by $\|\mathbf{R}''\|\sin\theta$, where θ is the angle from $\mathbf{R}'$ to $\mathbf{R}''$. Thus,

$$\frac{d\mathbf{T}}{ds} = \frac{\|\mathbf{R}'\|\|\mathbf{R}''\|\sin\theta}{\|\mathbf{R}'\|^3}$$

and since $\|\mathbf{R}' \times \mathbf{R}''\| = \|\mathbf{R}'\|\|\mathbf{R}''\|\sin\theta$, the desired result follows.

EXERCISES 17.3–Part A

A1. At $x = 1$, compute the curvature of $y = x^2$.

A2. Let $\mathbf{R} = \mathbf{i} + p\mathbf{j} + p^2\mathbf{k}$ define a space curve. Obtain its curvature at the point corresponding to $p = 2$.

A3. A plane curve is given parametrically by $x = x(p)$, $y = y(p)$. Write the curve as $\mathbf{R} = x(p)\mathbf{i} + y(p)\mathbf{j} + 0\mathbf{k}$, and use (17.5) to show that the curvature is given by $\kappa = \frac{|x'y'' - y'x''|}{((x')^2+(y')^2)^{3/2}}$.

A4. The vector component of the vector $\mathbf{B}$ along the vector $\mathbf{A}$, denoted $\mathbf{B}_\mathbf{A}$, is given by $\frac{\mathbf{B}\cdot\mathbf{A}}{\mathbf{A}\cdot\mathbf{A}}\mathbf{A}$, and the vector component of $\mathbf{B}$ perpendicular to $\mathbf{A}$, denoted $\mathbf{B}_{\perp\mathbf{A}}$, is given by $\mathbf{B} - \mathbf{B}_\mathbf{A}$. If $\mathbf{A} = \mathbf{i} - 2\mathbf{j}$, $\mathbf{B} = 2\mathbf{i} + 5\mathbf{j}$, obtain the vector components of $\mathbf{B}$ along $\mathbf{A}$, and perpendicular to $\mathbf{A}$.

A5. Use Figure 17.12 to show that the component of $\mathbf{R}''$ along $\mathbf{R}'$ is $\mathbf{R}'\frac{\mathbf{R}'\cdot\mathbf{R}''}{\|\mathbf{R}'\|^2} = \left(\frac{\mathbf{R}'}{\|\mathbf{R}'\|}\cdot\mathbf{R}''\right)\frac{\mathbf{R}'}{\|\mathbf{R}'\|}$.

A6. Use Figure 17.12 to show that the component of $\mathbf{R}''$ perpendicular to $\mathbf{R}'$ is $\mathbf{R}'' - \mathbf{R}'\frac{\mathbf{R}'\cdot\mathbf{R}''}{\|\mathbf{R}'\|^2} = \|\mathbf{R}''\|\sin\theta$.

EXERCISES 17.3–Part B

B1. For each of the following pairs of vectors **A** and **B**, obtain the vector component of **B** along, and perpendicular to, **A**.

    (a) $\mathbf{A} = 6\mathbf{i} - 7\mathbf{j} + 8\mathbf{k}$, $\mathbf{B} = 2\mathbf{i} + 3\mathbf{j} - 5\mathbf{k}$

    (b) $\mathbf{A} = 2\mathbf{i} - 3\mathbf{j} + 5\mathbf{k}$, $\mathbf{B} = -7\mathbf{i} + \mathbf{j} + 4\mathbf{k}$

In Exercises B2–6:

    (a) Use (17.4) to obtain, as a function of x, the curvature of the given curve.

    (b) Plot the curvature as a function of x.

    (c) Obtain $\kappa(x_0)$.

    (d) Write $\theta = \arctan y'$ and compute the curvature as $\kappa = \frac{d\theta}{ds}$.

    (e) Parametrize the curve with $x = p$, $y = y(p)$, and write the curve in radius-vector form $\mathbf{R}(p)$.

    (f) Obtain $\mathbf{T}(p)$, a unit tangent vector, and $\rho(p) = \|\mathbf{R}'(p)\|$.

    (g) Obtain $\kappa(x_0)$ by evaluating $\left\|\frac{d\mathbf{T}}{ds}\right\| = \left\|\frac{d\mathbf{T}}{dp}\frac{1}{\rho}\right\|$ at $p = x_0$.

B2. $y(x) = e^{2x}$, $x_0 = 0$ **B3.** $y(x) = x^3 - 4x^2 + 5x - 3$, $x_0 = 2$

B4. $y(x) = \sqrt{x}$, $x_0 = 1$ **B5.** $y(x) = \dfrac{1}{1 + x^2}$, $x_0 = 1$

B6. $y(x) = \dfrac{1 - x}{1 + x}$, $x_0 = 0$

In Exercises B7–16:

    (a) Use (17.4) to obtain, as a function of x, the curvature of the given curve.

    (b) Plot the curvature as a function of x.

    (c) Obtain $\kappa(x_1)$.

    (d) Determine where κ has its absolute maximum and minimum on the interval $[x_1, x_3]$.

    (e) Write $\theta = \arctan y'$ and compute the curvature as $\kappa = \frac{d\theta}{ds}$.

    (f) Parametrize the curve with $x = p$, $y = y(p)$, and write the curve in radius-vector form $\mathbf{R}(p)$.

    (g) Obtain $\mathbf{T}(p)$, a unit tangent vector, and $\rho(p) = \|\mathbf{R}'(p)\|$.

    (h) Obtain $\kappa(x_1)$ by evaluating $\left\|\frac{d\mathbf{T}}{ds}\right\| = \left\|\frac{d\mathbf{T}}{dp}\frac{1}{\rho}\right\|$ at $p = x_1$.

B7. Exercise B3, Section 17.1   **B8.** Exercise B4, Section 17.1

B9. Exercise B5, Section 17.1   **B10.** Exercise B6, Section 17.1

B11. Exercise B7, Section 17.1   **B12.** Exercise B8, Section 17.1

B13. Exercise B9, Section 17.1   **B14.** Exercise B10, Section 17.1

B15. Exercise B11, Section 17.1   **B16.** Exercise B12, Section 17.1

In Exercises B17–28:

    (a) Use (17.5) or the formula in Exercise A3, as appropriate, to obtain $\kappa(p_1)$.

    (b) Obtain $\mathbf{T}(p)$, a unit tangent vector, and $\rho(p) = \|\mathbf{R}'(p)\|$.

    (c) Obtain $\kappa(p_1)$ by evaluating $\left\|\frac{d\mathbf{T}}{ds}\right\| = \left\|\frac{d\mathbf{T}}{dp}\frac{1}{\rho}\right\|$ at $p = p_1$.

B17. Exercise B13, Section 17.1   **B18.** Exercise B14, Section 17.1

B19. Exercise B15, Section 17.1   **B20.** Exercise B16, Section 17.1

B21. Exercise B17, Section 17.1   **B22.** Exercise B18, Section 17.1

B23. Exercise B19, Section 17.1   **B24.** Exercise B20, Section 17.1

B25. Exercise B21, Section 17.1   **B26.** Exercise B22, Section 17.1

B27. Exercise B23, Section 17.1   **B28.** Exercise B24, Section 17.1

In Exercises B29–34, it is possible to obtain $R = R(s)$, where s is the arc length. For each:

    (a) Obtain $\mathbf{T}(s) = \mathbf{R}'(s)$.   **(b)** Obtain $\kappa = \|\mathbf{T}'(s)\|$.

    (c) Use the formula in Exercise A3 to obtain $\kappa(p)$.

    (d) Show that the two expressions for curvature computed in parts (b) and (c) agree, provided the change of variables $s = s(p)$ is made in the first expression for κ.

B29. Exercise 23, Section 17.2   **B30.** Exercise 24, Section 17.2

B31. Exercise 25, Section 17.2   **B32.** Exercise 26, Section 17.2

B33. Exercise 27, Section 17.2   **B34.** Exercise 28, Section 17.2

The *circle of curvature*, given by $(x - h)^2 + (y - k)^2 = R^2$, where $h = a - \frac{f'(a)(1 + f'(a)^2)}{f''(a)}$, $k = f(a) + \frac{1 + f'(a)^2}{f''(a)}$, $R = \frac{1}{\kappa} = \frac{(1 + f'(a)^2)^{3/2}}{f''(a)}$, is the unique circle that, at $x = a$, makes "second-order" contact with the curve $y = f(x)$. The point (h, k) is called the *center of curvature* at $x = a$. For the curve specified in each of Exercises B35–39:

    (a) Obtain the center of curvature at $a = x_0$.

    (b) Graph the circle of curvature along with the given curve.

    (c) Plot the locus of the centers of curvature. The resulting curve is called the *evolute* of the given curve.

    (d) Taking $a = x_0$, show that the circle of curvature touches the given curve at $(x_0, f(x_0))$; that is, show $(x_0, f(x_0))$ is a point whose coordinates satisfy the equation of the circle of curvature.

    (e) Show that at $(x_0, f(x_0))$ the slopes of the circle of curvature and of the curve $y = f(x)$ agree.

    (f) Show that at $(x_0, f(x_0))$ the second derivatives of the circle of curvature and the curve $y = f(x)$ agree.

    (g) Show that at $(x_0, f(x_0))$ the third derivatives of the circle of curvature and the curve $y = f(x)$ do not agree.

B35. Exercise B2   **B36.** Exercise B3   **B37.** Exercise B4

B38. Exercise B5   **B39.** Exercise B6

B40. Exercise B3, Section 17.1, with $a = x_1$

B41. Exercise B10, Section 17.1, with $a = x_1$

B42. Derive the formulas for the circle of curvature. Begin with $(x - h)^2 + (y - k)^2 = R^2$ and

    (a) write the equation that corresponds to having the point $(a, f(a))$ satisfy the equation of the circle.

(b) use implicit differentiation on the equation of the circle to obtain $y'(a)$ and then form the equation $y'(a) = f'(a)$.

(c) use implicit differentiation again to obtain $y''(a)$ and then form the equation $y''(a) = f''(a)$.

(d) solve, for h, k, and R^2, the three equations from parts (a)–(c).

The lines $y = \pm 1$ are *envelopes* of the family of circles described by $f(x, y, a) \equiv (x - a)^2 + y^2 - 1 = 0$. One member of the family is tangent to the envelope at each point on the envelope. In general, an envelope, if it exists, can be found by eliminating the parameter a from the two equations $f(x, y, a) = 0$ and $\frac{\partial f}{\partial a} = 0$. Sometimes, these two equations lead to $y = y(x)$, but more often, the envelope must be given parametrically by $x = x(a)$, $y = y(a)$. For the families in Exercises B43–47:

(a) Obtain any envelope that might exist.

(b) Graph the envelope, along with some representative members of the family.

B43. $x^3 - a(y - a) = 0$ **B44.** $2a(x - a) + a^2 - y = 0$

B45. $3a^2(x - a) + a^3 - y = 0$ **B46.** $y + 2a(x - a) + a^2 - 1 = 0$

B47. $x \cos a + y \sin a - 1 = 0$

The *evolute* of the curve C is the envelope of the family of normals along C. For the curves given in Exercises B48–52:

(a) Obtain the equation of the evolute, the locus of the centers of curvature.

(b) Obtain the equation for the family of normals along C.

(c) Obtain the envelope of the family of normals, and show that this envelope is the evolute found in part (a).

(d) Plot the curve C, the evolute, and several representative normals along C.

B48. $y = x^2$ **B49.** $y = x^3$ **B50.** $y = \sqrt{x}$

B51. $y = e^x$ **B52.** $y = \frac{1}{x}, x > 0$

B53. Show, in general, that for any smooth curve $y = f(x)$, the envelope of the family of normals is the evolute, defined as the locus of the centers of curvature.

17.4 Principal Normal and Binormal Vectors

Normal Vectors and the Principal Normal

A vector $\mathbf{n}$ is said to be *normal to a curve* if $\mathbf{n}$ is perpendicular to the natural tangent vector $\mathbf{R}'$ or to any of its multiples such as $\mathbf{T}$, the unit tangent vector. In the plane, there are two possible directions for the vector $\mathbf{n}$, as shown in Figure 17.13. In any event, the recipe

$$\mathbf{N} = \frac{1}{\kappa}\frac{d\mathbf{T}}{ds} = \frac{1}{\kappa}\frac{d}{dp}\mathbf{T}(p)\frac{1}{\rho} \qquad (17.9)$$

singles out $\mathbf{N}$, a unique unit normal vector called the *principal normal vector*.

NORMAL VECTORS IN THE PLANE   Evaluating (17.9) at $p = 2$ for the plane curve given on the left in (17.10) results in the principal normal given on the right. (See Exercise A1.)

$$\mathbf{R} = \begin{bmatrix} p^2 \\ p^3 \end{bmatrix} \qquad \mathbf{N} = \frac{1}{\sqrt{10}}\begin{bmatrix} -3 \\ 1 \end{bmatrix} \qquad (17.10)$$

In the plane a specific principal normal can be constructed as follows. From $\mathbf{T}$, the unit tangent vector, form two candidates for $\mathbf{N}$ by switching the two components of $\mathbf{T}$ and negating, in turn, each of the components in the new vectors. These three vectors are given in (17.11) and are the vectors graphed in Figure 17.13.

$$\mathbf{T} = \frac{1}{\sqrt{10}}\begin{bmatrix} 1 \\ 3 \end{bmatrix} \qquad \mathbf{n}_1 = \frac{1}{\sqrt{10}}\begin{bmatrix} 3 \\ -1 \end{bmatrix} \qquad \mathbf{n}_2 = \frac{1}{\sqrt{10}}\begin{bmatrix} -3 \\ 1 \end{bmatrix} \qquad (17.11)$$

The two vectors $\mathbf{n}_1$ and $\mathbf{n}_2$ point in opposite directions since they differ by a minus sign. By construction, they are both perpendicular to $\mathbf{T}$ since

$$\begin{bmatrix} a \\ b \end{bmatrix} \cdot \begin{bmatrix} b \\ -a \end{bmatrix} = ab - ba = 0$$

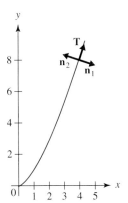

FIGURE 17.13   Two possible normals for a plane curve

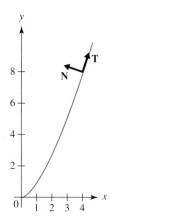

FIGURE 17.14   **T**(2) and **N**(2) for
$\mathbf{R} = p^2\mathbf{i} + p^3\mathbf{j}$

In this example, the vector $\mathbf{n}_2$ is the principal normal **N**. In general, the principal normal points inward toward the center of curvature (defined in Exercise B34, Section 17.3). It is helpful to observe that after switching the components in **T**, negating the first component (as in $\mathbf{n}_2$) yields a vector "to the left" of **T** whereas negating the second component (as in $\mathbf{n}_1$) yields a vector "to the right" of **T**. Thus, facing in the direction of **T**, the right arm extended out to the side would point in the direction of a vector that is to the right of **T**. Figure 17.14 shows the curve **R** and the vectors **T**(2) and **N**(2).

NORMAL VECTORS IN SPACE   In space, there are an infinite number of vectors perpendicular to the tangent vector $\mathbf{R}'$. These normal vectors form a plane to which the tangent vector is perpendicular. For example, evaluating (17.9) at $p = \frac{\pi}{2}$ for the helix given on the left in (17.12) results in the principal normal given on the right. The unit tangent vector **T** is given in the center. (See Exercise A2.)

$$\mathbf{R} = \begin{bmatrix} \cos p \\ \sin p \\ p \end{bmatrix} \qquad \mathbf{T}\left(\frac{\pi}{2}\right) = \frac{1}{\sqrt{2}}\begin{bmatrix} -1 \\ 0 \\ 1 \end{bmatrix} \qquad \mathbf{N}\left(\frac{\pi}{2}\right) = \begin{bmatrix} 0 \\ -1 \\ 0 \end{bmatrix} \qquad (17.12)$$

The point on **R** corresponding to $p = \frac{\pi}{2}$ is $(0, 1, \frac{\pi}{2})$. In Figure 17.15 the thickest arrow represents $\mathbf{T}(\frac{\pi}{2})$, while the thinner arrows are six of the infinite number of possible vectors normal to **R**. The figure also shows a portion of the plane which is perpendicular to $\mathbf{T}(\frac{\pi}{2})$ and contains the six normal vectors. The principal normal, also lying in this plane, is the unique vector **N** shown in the figure.

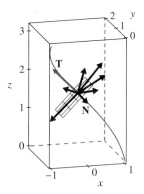

FIGURE 17.15 $\mathbf{T}(\frac{\pi}{2})$, $\mathbf{N}(\frac{\pi}{2})$, and several of the infinite normals at a point along a helix

Principal Normal Vector—Theory

Amongst all the possible normal vectors at a point on a space curve, the *principal normal* is the normalized version of the vector $\frac{d\mathbf{T}}{ds}$. Since the curvature κ is defined as the length of this vector, that is, since $\kappa = \left\| \frac{d\mathbf{T}}{ds} \right\|$, the principal normal, a unit vector, must then be $\mathbf{N} = \frac{1}{\kappa}\frac{d\mathbf{T}}{ds}$. Computing the derivative $\frac{d\mathbf{T}}{ds}$ by the chain rule leads to

$$\mathbf{N} = \frac{1}{\kappa}\left(\frac{d}{dp}\mathbf{T}\right)\frac{dp}{ds} = \frac{1}{\kappa}\frac{d\mathbf{T}}{dp}\frac{1}{\frac{ds}{dp}} = \frac{1}{\kappa}\frac{d\mathbf{T}}{dp}\frac{1}{\rho} = \frac{1}{\kappa\rho}\frac{d\mathbf{T}}{dp}$$

as a practical formula for computing **N**.

Since **T** is a unit vector, **T** and **N** must be perpendicular. The proof is essentially computational, starting with the identity $\mathbf{T} \cdot \mathbf{T} = 1$. Differentiating this dot product gives

$$\frac{d\mathbf{T}}{ds} \cdot \mathbf{T} + \mathbf{T} \cdot \frac{d\mathbf{T}}{ds} = 0$$

or $2\mathbf{T} \cdot \frac{d\mathbf{T}}{ds} = 0$, from which $\mathbf{T} \cdot \frac{d\mathbf{T}}{ds} = 0$ and $\mathbf{T} \cdot \frac{1}{\kappa}\frac{d\mathbf{T}}{ds} = 0$ clearly follow. But this last equation is just $\mathbf{T} \cdot \mathbf{N} = 0$, which shows **T** and **N** are perpendicular.

EXAMPLE 17.7   Table 17.4 lists a space curve **R** and vectors arising in the computation of **T**(1) and **N**(1). First, $\mathbf{R}'(p)$ is computed, and from this,

$$\rho = \frac{ds}{dp} = \|\mathbf{R}'(p)\| = \sqrt{1 + 4p^2 + 9p^4}$$

$$\mathbf{R} = \begin{bmatrix} p \\ p^2 \\ p^3 \end{bmatrix} \qquad \mathbf{R}' = \begin{bmatrix} 1 \\ 2p \\ 3p^2 \end{bmatrix} \qquad \mathbf{T} = \sqrt{\lambda}\begin{bmatrix} 1 \\ 2p \\ 3p^2 \end{bmatrix} \qquad \mathbf{T}(1) = \frac{1}{\sqrt{14}}\begin{bmatrix} 1 \\ 2 \\ 3 \end{bmatrix}$$

$$\mathbf{R}'' = \begin{bmatrix} 0 \\ 2 \\ 6p \end{bmatrix} \qquad \mathbf{R}' \times \mathbf{R}'' = \begin{bmatrix} 6p^2 \\ -6p \\ 2 \end{bmatrix} \qquad \frac{d\mathbf{T}}{dp} = 2\lambda^{3/2}\begin{bmatrix} -p(2+9p^2) \\ 1-9p^4 \\ 3p(1+2p^2) \end{bmatrix} \qquad \mathbf{N}(1) = \frac{1}{\sqrt{266}}\begin{bmatrix} -11 \\ 8 \\ 9 \end{bmatrix}$$

$$\lambda = \frac{1}{1+4p^2+9p^4}$$

TABLE 17.4   Vectors for Example 17.7

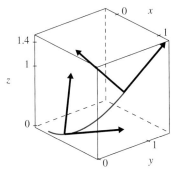

FIGURE 17.16 $\mathbf{T}$ and $\mathbf{N}$ at $p = \frac{3}{10}$ and 1 for the space curve in Example 17.7

and $\mathbf{T}(p) = \frac{1}{\rho}\mathbf{R}'(p)$. Evaluating at $p = 1$, we get $\mathbf{T}(1)$. Next, $\mathbf{R}''(p)$ is computed, and from this, $\|\mathbf{R}' \times \mathbf{R}''\| = 2\sqrt{9p^4 + 9p^2 + 1}$ and

$$\kappa = \frac{\|\mathbf{R}' \times \mathbf{R}''\|}{\rho^3} = 2\frac{\sqrt{9p^4 + 9p^2 + 1}}{(1+4p^2+9p^4)^{3/2}} \qquad (17.13)$$

Evaluating at $p = 1$, we have $\kappa(1) = \frac{\sqrt{266}}{98}$. Then, $\frac{d\mathbf{T}}{dp}$ is computed, and from this, the principal normal

$$\mathbf{N} = \frac{1}{\kappa\rho}\frac{d\mathbf{T}}{dp} = \frac{1}{2\sqrt{1+4p^2+9p^4}\sqrt{1+9p^2+9p^4}}\begin{bmatrix} -2p(2+9p^2) \\ 2(1-9p^4) \\ 6p(1+2p^2) \end{bmatrix} \qquad (17.14)$$

is obtained. Evaluating at $p = 1$ gives $\mathbf{N}(1)$.

Figure 17.16 shows the curve $\mathbf{R}$, along with $\mathbf{T}$ and $\mathbf{N}$ at $p = \frac{3}{10}$ and 1.   ❖

EXAMPLE 17.8   Table 17.5 contains the parametric equations of the plane curve shown in Figure 17.17. The table includes the radius vector $\mathbf{R}$ and the derivative $\mathbf{R}'$. Since

$$\rho = \|\mathbf{R}'\| = \frac{ds}{dp} = \sqrt{p^2 + 1}$$

the unit tangent vector $\mathbf{T}$ is readily obtained. The curvature, computed by the formula in

$$\begin{cases} x(p) = p\cos p \\ y(p) = p\sin p \end{cases} \qquad \mathbf{R}' = \begin{bmatrix} \cos p - p\sin p \\ \sin p + p\cos p \end{bmatrix} \qquad \frac{d\mathbf{T}}{dp} = -\frac{2+p^2}{(1+p^2)^{3/2}}\begin{bmatrix} \sin p + p\cos p \\ p\sin p - \cos p \end{bmatrix}$$

$$\mathbf{R} = \begin{bmatrix} p\cos p \\ p\sin p \end{bmatrix} \qquad \mathbf{T} = \frac{1}{\sqrt{p^2+1}}\begin{bmatrix} \cos p - p\sin p \\ \sin p + p\cos p \end{bmatrix} \qquad \mathbf{N} = \frac{1}{\sqrt{p^2+1}}\begin{bmatrix} -\sin p - p\cos p \\ \cos p - p\sin p \end{bmatrix}$$

TABLE 17.5   Vectors for Example 17.8

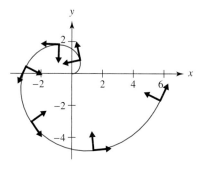

FIGURE 17.17 Example 17.8: The curve, and several sets of vectors **T** and **N**

Exercise A3, Section 17.3, is

$$\kappa = \frac{|x'y'' - y'x''|}{\rho^3} = \frac{2 + p^2}{(1 + p^2)^{3/2}} \tag{17.15}$$

A straightforward but tedious computation gives $\frac{d\mathbf{T}}{dp}$, from which the principal normal is obtained as $\mathbf{N} = \frac{1}{\kappa\rho}\frac{d\mathbf{T}}{dp}$. At each point on the curve **R**, there is a unit tangent vector **T** and a principal normal vector **N**, also a unit vector, as shown in Figure 17.17. ❖

The Binormal Vector

Given the unit vectors **T** and **N** along a space curve **R**, we define the binormal vector **B** = **T** × **N**, which is a unit vector perpendicular to both **T** and **N**. (Students who remember the word **BuTtoN** will be "right on the button" when computing the binormal vector.)

EXAMPLE 17.9 **Tangent, normal, binormal vectors** In Table 17.6, **R** defines a helix for which **R**′ and **R**″ are given. Since

$$\rho = \|\mathbf{R}'\| = \frac{ds}{dp} = \frac{1}{4}\sqrt{145} \tag{17.16}$$

we immediately have both **T** and $\frac{d\mathbf{T}}{dp}$. The curvature is given by

$$\kappa = \frac{\|\mathbf{R}' \times \mathbf{R}''\|}{\rho^3} = \frac{48}{145} \tag{17.17}$$

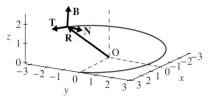

FIGURE 17.18 Example 17.9: The curve **R** and the vectors **B**, **T**, and **N**

so $\mathbf{N} = \frac{1}{\kappa\rho}\frac{d\mathbf{T}}{dp}$ readily follows. Finally, the binormal is computed as **B** = **T** × **N**. Figure 17.18 shows the helix, with a radius vector to a point on it at which the vectors **B**, **T**, and **N** are drawn. ❖

$$\mathbf{R} = \begin{bmatrix} 3\cos p \\ 3\sin p \\ \frac{1}{4}p \end{bmatrix} \quad \mathbf{R}' = \begin{bmatrix} -3\sin p \\ 3\cos p \\ \frac{1}{4} \end{bmatrix} \quad \mathbf{T} = \frac{12}{\sqrt{145}} \begin{bmatrix} -\sin p \\ \cos p \\ \frac{1}{12} \end{bmatrix} \quad \frac{d\mathbf{T}}{dp} = -\frac{12}{\sqrt{145}} \begin{bmatrix} \cos p \\ \sin p \\ 0 \end{bmatrix}$$

$$\mathbf{R}'' = \begin{bmatrix} -3\cos p \\ -3\sin p \\ 0 \end{bmatrix} \quad \mathbf{R}' \times \mathbf{R}'' = \frac{3}{4}\begin{bmatrix} \sin p \\ -\cos p \\ 12 \end{bmatrix} \quad \mathbf{N} = \begin{bmatrix} -\cos p \\ -\sin p \\ 0 \end{bmatrix} \quad \mathbf{B} = \frac{1}{\sqrt{145}}\begin{bmatrix} \sin p \\ -\cos p \\ 12 \end{bmatrix}$$

TABLE 17.6 Vectors for Example 17.9

EXERCISES 17.4–Part A

A1. Compute **N** for the plane curve **R** in (17.10).

A2. Compute **T** and **N** for the curve **R** in (17.12).

A3. Verify the computation of **T**, $\frac{d\mathbf{T}}{dp}$, and **N**(1) in Table 17.4.

A4. Verify κ in (17.13).

A5. Verify **N** in (17.14).

A6. Verify the computation of **R**′, **T**, and $\frac{d\mathbf{T}}{dp}$ in Table 17.5.

A7. Verify the computation of κ in (17.15).

A8. Verify the computation of **N** in Table 17.5.

A9. Verify (17.16) and (17.17) in Example 17.9.

A10. Verify **B** = **T** × **N** in Example 17.9.

EXERCISES 17.4–Part B

B1. Given $\mathbf{R}(p)$, an alternative recipe for computing $\mathbf{T}$, $\mathbf{N}$, and $\mathbf{B}$ begins with $\rho = \|\mathbf{R}'(p)\|$ and $\mathbf{T} = \frac{1}{\rho}\mathbf{R}'$. Then, $\mathbf{B} = \frac{\mathbf{R}' \times \mathbf{R}''}{\|\mathbf{R}' \times \mathbf{R}''\|}$ and $\mathbf{N} = \mathbf{B} \times \mathbf{T}$ follow. This recipe avoids differentiation by the chain rule.

    (a) Show that $\frac{\mathbf{R}' \times \mathbf{R}''}{\|\mathbf{R}' \times \mathbf{R}''\|} = \mathbf{T} \times \mathbf{N}$ and, hence, $\mathbf{B}$.

    (b) Show that $\mathbf{B} = \mathbf{T} \times \mathbf{N}$ implies $\mathbf{B} \times \mathbf{T} = \mathbf{N}$ since $(\mathbf{T} \times \mathbf{N}) \times \mathbf{T}$ can be simplified with the vector identity $(\mathbf{a} \times \mathbf{b}) \times \mathbf{c} = (\mathbf{a} \cdot \mathbf{c})\mathbf{b} - (\mathbf{b} \cdot \mathbf{c})\mathbf{a}$.

For the functions $y = y(x)$ specified in Exercises B2–11:

    (a) Graph the curve corresponding to the function.

    (b) Using the parametrization $x = p$, $y = y(p)$, represent the curve in radius-vector form.

    (c) Obtain $\mathbf{T}(1)$ from $\mathbf{T}(p) = \frac{1}{\rho}\mathbf{R}'(p)$.

    (d) Obtain $\mathbf{N}(1)$ from $\mathbf{T}(1)$ by interchanging the components of $\mathbf{T}$ and inserting a minus sign to make $\mathbf{N}(1)$ point inward toward the center of curvature.

    (e) Sketch $\mathbf{T}(1)$ and $\mathbf{N}(1)$ on the graph drawn in part (a).

    (f) Obtain $\kappa(1)$ from $\frac{|y''|}{(1+(y')^2)^{3/2}}$.

    (g) Obtain $\mathbf{T}'(p)$.

    (h) Show that $\mathbf{N}(1)$ agrees with $\frac{1}{\kappa\rho}\mathbf{T}'(p)$ when $p = 1$.

B2. Exercise B3, Section 17.1   **B3.** Exercise B4, Section 17.1

B4. Exercise B5, Section 17.1   **B5.** Exercise B6, Section 17.1

B6. Exercise B7, Section 17.1   **B7.** Exercise B8, Section 17.1

B8. Exercise B9, Section 17.1   **B9.** Exercise B10, Section 17.1

B10. Exercise B11, Section 17.1   **B11.** Exercise B12, Section 17.1

For the plane curves specified in Exercises B12–22:

    (a) Graph the curve determined by the given parametric equations.

    (b) Obtain $\mathbf{T}(1)$ from $\mathbf{T}(p) = \frac{1}{\rho}\mathbf{R}'(p)$.

    (c) Obtain $\mathbf{N}(1)$ from $\mathbf{T}(1)$ by interchanging the components of $\mathbf{T}$ and inserting a minus sign to make $\mathbf{N}(1)$ point inward toward the center of curvature.

    (d) Sketch $\mathbf{T}(1)$ and $\mathbf{N}(1)$ on the graph drawn in part (a).

    (e) Obtain $\kappa(1)$ from $\mathbf{T}'(p)$ and the chain rule.

    (f) Show that $\mathbf{N}(1)$ agrees with $\frac{1}{\kappa\rho}\mathbf{T}'(p)$ when $p = 1$.

B12. Exercise B13, Section 17.1   **B13.** Exercise B14, Section 17.1

B14. $x(p) = 1 - p^2$; $y(p) = 1 + p^3$

B15. Exercise B16, Section 17.1

B16. Exercise B17, Section 17.1   **B17.** Exercise B23, Section 17.2

B18. $x(p) = \arctan p$; $y(p) = p^2$ **B19.** $x(p) = p^3$; $y(p) = \sin p$

B20. $x(p) = \cos p$; $y(p) = e^{-p}$

B21. $x(p) = \ln(1 + p)$; $y(p) = 1 + 2p$

B22. $x(p) = \ln(1 + p)$; $y(p) = p^2$

For the space curves specified in each of Exercises B23–29:

    (a) Obtain $\mathbf{R}'$, $\mathbf{R}''$, and $\rho = \|\mathbf{R}'\|$.   **(b)** Obtain $\mathbf{T}(p)$ and $\mathbf{T}(1)$.

    (c) Obtain $\kappa(1)$ from $\frac{\|\mathbf{R}' \times \mathbf{R}''\|}{\rho^3}$.   **(d)** Obtain $\mathbf{T}'(p)$ and $\mathbf{T}'(1)$.

    (e) From part (d) and the chain rule, obtain $\mathbf{N}(1)$ from $\frac{1}{\kappa\rho}\mathbf{T}'(1)$.

    (f) Obtain $\mathbf{B}(1) = \mathbf{T}(1) \times \mathbf{N}(1)$.

    (g) Obtain $\mathbf{T}$, $\mathbf{N}$, and $\mathbf{B}$ by the recipe in Exercise B1 and test your results by evaluating at $p = 1$.

B23. Exercise B18, Section 17.1   **B24.** Exercise B19, Section 17.1

B25. Exercise B20, Section 17.1   **B26.** Exercise B21, Section 17.1

B27. Exercise B22, Section 17.1   **B28.** Exercise B23, Section 17.1

B29. Exercise B24, Section 17.1

As a result of the work of Frenet and Serret (1852 and 1851, respectively) it is known that the moving basis vectors $\mathbf{T}$, $\mathbf{N}$, and $\mathbf{B}$, and the two scalars κ, the curvature, and the torsion τ, completely determine a curve $\mathbf{R} = x\mathbf{i} + y\mathbf{j} + z\mathbf{k}$. The torsion is given by $\tau = -\mathbf{B}'(s) \cdot \mathbf{N} = \mathbf{N}'(s) \cdot \mathbf{B} = \frac{(\mathbf{R}' \times \mathbf{R}'') \cdot \mathbf{R}'''}{\|\mathbf{R}' \times \mathbf{R}''\|^2}$ and measures the twist of the curve $\mathbf{R}$ out of the *osculating plane*, the plane determined by $\mathbf{T}$ and $\mathbf{N}$. For each space curve specified in Exercises B30–36:

    (a) Calculate the torsion via $\tau = -\mathbf{B}'(s) \cdot \mathbf{N}$, where $\mathbf{B}'(s) = \frac{1}{\rho}\frac{d\mathbf{B}}{dp}$ by the chain rule.

    (b) Calculate the torsion via $\tau = \mathbf{N}'(s) \cdot \mathbf{B}$, where $\mathbf{N}'(s) = \frac{1}{\rho}\frac{d\mathbf{N}}{dp}$ by the chain rule.

    (c) Calculate the torsion via $\tau = \frac{(\mathbf{R}' \times \mathbf{R}'') \cdot \mathbf{R}'''}{\|\mathbf{R}' \times \mathbf{R}''\|^2}$.

    (d) Graph $\tau(p)$.

B30. Exercise B18, Section 17.1   **B31.** Exercise B20, Section 17.1

B32. Exercise B21, Section 17.1   **B33.** Exercise B22, Section 17.1

B34. Exercise B23, Section 17.1   **B35.** Exercise B24, Section 17.1

B36. $x(p) = \ln(1 + p^2)$; $y(p) = p$; $z(p) = p^2$

The differential equations $\mathbf{T}'(s) = \kappa\mathbf{N}$, $\mathbf{N}'(s) = -\kappa\mathbf{T} + \tau\mathbf{B}$, $\mathbf{B}'(s) = -\tau\mathbf{N}$ are known as the Frenet–Serret formulas for a space curve $\mathbf{R}$. For each space curve specified in Exercises B37–43, show that these differential equations are valid. Compute the derivatives on the left by the chain rule so that the variable of differentiation is p. Thus, $\mathbf{T}'(s) = \frac{1}{v}\frac{d\mathbf{T}}{dp}$, for example.

B37. Exercise B30   **B38.** Exercise B31   **B39.** Exercise B32

B40. $x(p) = \cos 2p$; $y(p) = \sin 2p$; $z(p) = \cos 2p$

B41. Exercise B34   **B42.** Exercise B35   **B43.** Exercise B36

B44. The Frenet–Serret formulas are nine differential equations for nine unknowns, three components each in the three vectors $\mathbf{T}$, $\mathbf{N}$, and $\mathbf{B}$. Take $\kappa = s$ and $\tau = s^2$, $\mathbf{T}(0) = \mathbf{i}$, $\mathbf{N}(0) = \mathbf{j}$, and $\mathbf{B}(0) = \mathbf{k}$. Solving the nine Frenet–Serret equations will give the moving basis vectors but not yet give the curve $\mathbf{R} = x(s)\mathbf{i} +$

$y(s)\mathbf{j} + z(s)\mathbf{k}$. The connection between the curve and $\mathbf{T}$ is $\mathbf{R'} = \mathbf{T}$. Hence, $\mathbf{T'} = \mathbf{R''} = x''(s)\mathbf{i} + y''(s)\mathbf{j} + z''(s)\mathbf{k}$, and the first three equations are second order, requiring such additional initial conditions as $x(0) = y(0) = z(0) = 0$. Now, the nine differential equations can be integrated numerically to give the moving basis vectors and the curve $\mathbf{R}$. Obtain this solution and plot the graph of $\mathbf{R}$ as a space curve.

17.5   Resolution of R″ into Tangential and Normal Components

Resolution of R″ *along* T *and* N

The vector $\mathbf{R''}(p)$ has components only along the moving basis vectors $\mathbf{T}$ and $\mathbf{N}$, a result expressed by the equation

$$\mathbf{R''} = \rho'\mathbf{T} + \kappa\rho^2\mathbf{N} \qquad (17.18)$$

where $\rho' = \frac{d\rho}{dp}$. Thus, $\mathbf{R''}(p)$ always remains in the *osculating plane,* the plane determined by the vectors $\mathbf{T}$ and $\mathbf{N}$, as we shall prove at the end of this section.

In the special case where the parameter p is the time t, the derivative $\mathbf{R'}(t)$ is $\mathbf{V}$, the velocity vector; $\mathbf{R''}(t)$ is $\mathbf{a}$, the acceleration vector; and $v = \|\mathbf{R'}(t)\|$ is the *speed,* the magnitude of the velocity vector. In this event, the decomposition of the acceleration vector into components along the tangent and normal vectors is written as

$$\mathbf{a} = \dot{v}\mathbf{T} + \kappa v^2\mathbf{N} \qquad (17.19)$$

where the raised dot on v denotes the derivative $\frac{dv}{dt}$. Thus, the formula for the acceleration vector reads aloud with a memorable "vee-dot tee," by which the author recalls the component that goes with the vector $\mathbf{T}$.

The term $\frac{dv}{dt}$, called *the rate of change of the speed,* is *not* the "scalar acceleration a," nor is it the magnitude of the acceleration vector. In fact, *there is no scalar associated with acceleration.* The length of the velocity vector $\mathbf{V}$ is the scalar speed v, but the length of the acceleration vector $\mathbf{a}$ is not a scalar of any dynamic significance. It is an egregious error to believe $\frac{dv}{dt}$ is a, the length of the acceleration vector. In fact, using raised dots to denote differentiation with respect to t, the acceleration vector is $\mathbf{a} = \ddot{x}\mathbf{i} + \ddot{y}\mathbf{j} + \ddot{z}\mathbf{k}$ and the speed v is given by $v = \sqrt{\dot{x}^2 + \dot{y}^2 + \dot{z}^2}$. The length of the acceleration vector, $\|\mathbf{a}\|$, and the rate of change of the speed, $\frac{dv}{dt}$, are given respectively by

$$\|\mathbf{a}\| = \sqrt{\ddot{x}^2 + \ddot{y}^2 + \ddot{z}^2} \quad \text{and} \quad \frac{dv}{dt} = \frac{\dot{x}\ddot{x} + \dot{y}\ddot{y} + \dot{z}\ddot{z}}{\sqrt{\dot{x}^2 + \dot{y}^2 + \dot{z}^2}}$$

It should be clear that $\frac{dv}{dt}$ and $\|\mathbf{a}\|$ are completely unrelated.

EXAMPLE 17.10   To demonstrate the validity of the decomposition (17.18) for the parabola curve $\mathbf{R}$, given in Table 17.7 and graphed in Figure 17.19, compute $\mathbf{R'}$ and $\mathbf{R''}$ as shown in Table 17.7.

$$\mathbf{R} = \begin{bmatrix} 2p+3 \\ p^2-1 \end{bmatrix} \qquad \mathbf{R'} = \begin{bmatrix} 2 \\ 2p \end{bmatrix} \qquad \mathbf{R''} = \begin{bmatrix} 0 \\ 2 \end{bmatrix} \qquad \mathbf{T} = \begin{bmatrix} \dfrac{1}{\sqrt{1+p^2}} \\ \dfrac{p}{\sqrt{1+p^2}} \end{bmatrix} \qquad \mathbf{N} = \begin{bmatrix} -\dfrac{p}{\sqrt{1+p^2}} \\ \dfrac{1}{\sqrt{1+p^2}} \end{bmatrix}$$

TABLE 17.7   Vectors for Example 17.10

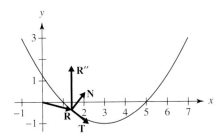

FIGURE 17.19   The parabola in Example 17.10

Then, $\|\mathbf{R}'\| = \rho = 2\sqrt{1 + p^2}$ and the curvature, given by the formula in Exercise A3, Section 17.3, is $\kappa = \frac{1}{2}\frac{1}{(1+p^2)^{3/2}}$. On the right in (17.18) there are four terms, namely, $\mathbf{T}$, $\mathbf{N}$, ρ', and $\kappa\rho^2$. The unit tangent vector is $\mathbf{T} = \frac{\mathbf{R}'}{\rho}$ and $\mathbf{N}$, the principal normal vector, is obtained from $\mathbf{T}$ by interchanging the components and negating the first (new) component, causing $\mathbf{N}$ to lie to the *left* of $\mathbf{T}$. (Of course, this alternative only exists for a planar curve.) The tangential component of $\mathbf{R}''$ is the derivative $\frac{dv}{dp} = \frac{2p}{\sqrt{1+p^2}}$, and its normal component is the product $\kappa\rho^2 = \frac{2}{\sqrt{1+p^2}}$. The sum $\rho'\mathbf{T} + \kappa\rho^2\mathbf{N}$ is then

$$\frac{2p}{\sqrt{1+p^2}}\frac{1}{\sqrt{1+p^2}}\begin{bmatrix}1\\p\end{bmatrix} + \frac{2}{\sqrt{1+p^2}}\frac{1}{\sqrt{1+p^2}}\begin{bmatrix}-p\\1\end{bmatrix} = \begin{bmatrix}0\\2\end{bmatrix} = \mathbf{R}'' \qquad \diamondsuit$$

EXAMPLE 17.11   The plane curve $\mathbf{R}$, given in Table 17.8, is a circle with radius 3 and center at the origin. Interpreting the parameter t as the time, $\mathbf{R}'$ is the velocity vector $\mathbf{V}$, the speed is $v = \|\mathbf{V}\| = 3$, and $\mathbf{R}''$ is the acceleration vector $\mathbf{a}$. The unit tangent vector is given by $\mathbf{T} = \frac{\mathbf{V}}{v}$, and $\mathbf{N}$, the principal normal vector, is most easily obtained from $\mathbf{T}$ by interchanging components and negating the second component. The curve $\mathbf{R}$ represents uniform motion around a circle, so $\dot{v} = 0$ and $\mathbf{a} = \mathbf{R}''$ has no tangential component. The curvature of a circle is the reciprocal of the radius, so $\kappa = \frac{1}{3}$; and the normal component of $\mathbf{a}$ is $\kappa v^2 = \frac{1}{3}3^2 = 3$, so $\mathbf{a} = 3\mathbf{N}$, verifying the decomposition $\mathbf{a} = \dot{v}\mathbf{T} + \kappa v^2\mathbf{N}$. Hence, the acceleration vector is directed completely along $\mathbf{N}$, the principal normal. $\diamondsuit$

$$\mathbf{R} = \begin{bmatrix}3\cos t\\3\sin t\end{bmatrix} \qquad \mathbf{V} = \begin{bmatrix}-3\sin t\\3\cos t\end{bmatrix} \qquad \mathbf{a} = \begin{bmatrix}-3\cos t\\-3\sin t\end{bmatrix} \qquad \mathbf{T} = \begin{bmatrix}-\sin t\\\cos t\end{bmatrix} \qquad \mathbf{N} = \begin{bmatrix}-\cos t\\-\sin t\end{bmatrix}$$

TABLE 17.8   Vectors for Example 17.11

EXAMPLE 17.12   Table 17.9 contains the radius vector $\mathbf{R}(p)$ defining a space curve, along with the vectors $\mathbf{R}'(p)$, $\mathbf{R}''(p)$, $\mathbf{T} = \frac{\mathbf{R}'}{\rho}$, and $\mathbf{T}' = \frac{d\mathbf{T}}{dp}$. These vectors are computed as functions of the parameter p, even though we want to verify (17.18) at $p = 0$. In addition, we must compute

$$\mathbf{R} = \begin{bmatrix}\dfrac{1}{1+p^2}\\[2mm]p\\[2mm]\dfrac{p}{1+p^2}\end{bmatrix} \qquad \mathbf{R}' = \begin{bmatrix}-2\dfrac{p}{(1+p^2)^2}\\[2mm]1\\[2mm]\dfrac{1-p^2}{(1+p^2)^2}\end{bmatrix} \qquad \mathbf{R}'' = \frac{2}{(1+p^2)^3}\begin{bmatrix}3p^2-1\\0\\p(p^2-3)\end{bmatrix}$$

$$\mathbf{T} = \frac{1}{\sqrt{p^4+2p^2+2}}\begin{bmatrix}\dfrac{-2p}{1+p^2}\\[2mm]1+p^2\\[2mm]\dfrac{1-p^2}{1+p^2}\end{bmatrix} \qquad \frac{d\mathbf{T}}{dp} = \frac{2}{(p^4+2p^2+2)^{3/2}(1+p^2)^2}\begin{bmatrix}3p^6+5p^4+2p^2-2\\p(1+p^2)^2\\p(p^6-p^4-5p^2-5)\end{bmatrix}$$

TABLE 17.9   Parameter-dependent vectors for Example 17.12

the scalar quantities

$$\rho = \|\mathbf{R}'\| = \frac{\sqrt{p^4 + 2p^2 + 2}}{1 + p^2} \quad \text{and} \quad \rho' = -\frac{2p}{(1 + p^2)^2 \sqrt{p^4 + 2p^2 + 2}} \qquad (17.20)$$

At $p = 0$, the curve passes through $(1, 0, 0)$, at which point the six vectors

$$\mathbf{R}'(0) = \mathbf{j} + \mathbf{k} \quad \mathbf{R}''(0) = -2\mathbf{i} \quad \mathbf{T}(0) = \frac{1}{\sqrt{2}}(\mathbf{j} + \mathbf{k}) \quad \mathbf{T}'(0) = -\sqrt{2}\mathbf{i} \qquad (17.21a)$$

$$\mathbf{N}(0) = \frac{1}{\kappa(0)\rho(0)}\mathbf{T}'(0) = -\mathbf{i} \quad \mathbf{B}(0) = \mathbf{T}(0) \times \mathbf{N}(0) = \frac{1}{\sqrt{2}}(\mathbf{k} - \mathbf{j}) \qquad (17.21b)$$

are computed, provided the scalars $\rho(0) = \sqrt{2}$, $\rho'(0) = 0$, and $\kappa(0) = \frac{\|\mathbf{R}'(0)\times\mathbf{R}''(0)\|}{\rho^3(0)} = 1$ are also obtained. Thus, (17.18) is verified by noting that $\mathbf{R}''(0) = 0\mathbf{T} + (\sqrt{2})^2\mathbf{N}(0)$. Moreover, $\mathbf{B}$ is orthogonal to the plane determined by $\mathbf{T}$ and $\mathbf{N}$. But that plane contains $\mathbf{R}''$, so $\mathbf{B}$ must be orthogonal to $\mathbf{R}''$. At $p = 0$ we see that, indeed, $\mathbf{B} \cdot \mathbf{R}'' = 0$. ❖

This example demonstrates just which quantities need to be computed as functions of the parameter p and which quantities can be evaluated at the specific argument $p = 0$.

Derivation

A general proof that $\mathbf{R}''$ lies in the plane of $\mathbf{T}$ and $\mathbf{N}$ follows immediately upon verifying the decomposition (17.18) because this formula plainly says that $\mathbf{R}''$ consists of components along just $\mathbf{T}$ and $\mathbf{N}$. Now, this formula can be derived from the two familiar results $\mathbf{T} = \frac{\mathbf{R}'}{\rho}$ and $\mathbf{N} = \frac{1}{\kappa}\frac{d\mathbf{T}}{ds}$ if they are written as $\mathbf{R}' = \rho\mathbf{T}$ and $\frac{d\mathbf{T}}{ds} = \kappa\mathbf{N}$. Differentiating, with respect to p, the first of these, gives $\mathbf{R}'' = \frac{d}{dp}[\rho\mathbf{T}]$, which, by the product rule for differentiation, becomes $\mathbf{R}'' = \frac{d\rho}{dp}\mathbf{T} + \rho\frac{d\mathbf{T}}{dp}$. By the chain rule, the second term on the right becomes $\frac{d\mathbf{T}}{dp} = \frac{d\mathbf{T}}{ds}\frac{ds}{dp} = \frac{d\mathbf{T}}{ds}\rho$. Hence, $\mathbf{R}''$ becomes $\mathbf{R}'' = \frac{d\rho}{dp}\mathbf{T} + \rho\frac{d\mathbf{T}}{ds}\rho$, which, upon replacing $\frac{d\mathbf{T}}{ds}$ with $\kappa\mathbf{N}$ and $\frac{d\rho}{dp}$ with ρ', becomes (17.18).

EXERCISES 17.5–Part A

A1. Show that $\frac{dv}{dt} = \mathbf{T} \cdot \mathbf{a}$.

A2. Obtain κ for Example 17.10.

A3. Verify that the curvature of a circle is the reciprocal of its radius.

A4. Verify the computations in (17.20).

A5. Verify the calculations in (17.21a) and (17.21b).

EXERCISES 17.5–Part B

B1. Verify the computations in Table 17.9.

For the curves specified in Exercises B2–11:

   (a) Obtain a graph of the curve.

   (b) Using the parametrization $x = p$, $y = y(p)$, obtain $\mathbf{R}''(p)$.

   (c) Obtain $\mathbf{T}(p)$, $\mathbf{N}(p)$, $\rho(p)$, $\rho'(p)$, and $\kappa(p)$.

   (d) Show that (17.18) is valid.

   (e) Using the projection operator $\mathbf{B}_\mathbf{A} = \frac{\mathbf{B}\cdot\mathbf{A}}{\mathbf{A}\cdot\mathbf{A}}\mathbf{A}$ for the component of $\mathbf{B}$ along $\mathbf{A}$ and working at $p = x_1$, obtain the components

of $\mathbf{R}''$ along $\mathbf{T}$ and along $\mathbf{N}$. Show that these are $\rho'\mathbf{T}$ and $\kappa\rho^2\mathbf{N}$, respectively.

   (f) On the graph of the curve, and at the specified point, draw the vectors $\rho'\mathbf{T}$, $\rho^2\kappa\mathbf{N}$, and $\mathbf{R}''$.

   (g) Show that $\frac{d\rho}{dp} = \frac{f'(p)f''(p)}{\sqrt{1+(f'(p))^2}}$ and $\kappa\rho^2 = \frac{|f''(p)|}{\sqrt{1+(f'(p))^2}}$.

B2. Exercise B3, Section 17.1   **B3.** $y = 7x^2 + 2x + 3$, $x_1 = -1$

B4. Exercise B5, Section 17.1   **B5.** Exercise B6, Section 17.1

B6. Exercise B7, Section 17.1   **B7.** Exercise B8, Section 17.1

B8. Exercise B9, Section 17.1   **B9.** Exercise B10, Section 17.1

B10. Exercise B11, Section 17.1   **B11.** $y = \ln(1 + x^2)$, $x_1 = 0$

B12. Show that, in general, for the plane curve $y = f(x)$, $\rho'(x) = \frac{f'(x)f''(x)}{\sqrt{1+(f'(x))^2}}$ and $\kappa\rho^2 = \frac{|f''(x)|}{\sqrt{1+(f'(x))^2}}$.

For the curves specified in Exercises B13–17:

    (a) Obtain a graph of the curve.

    (b) Obtain $\mathbf{R}''(p)$.

    (c) Obtain $\mathbf{T}(p)$, $\mathbf{N}(p)$, $\rho(p)$, $\rho'(p)$, and $\kappa(p)$.

    (d) Show that (17.18) is valid.

    (e) Using the projection operator $\mathbf{B_A} = \frac{\mathbf{B \cdot A}}{\mathbf{A \cdot A}}\mathbf{A}$ for the component of $\mathbf{B}$ along $\mathbf{A}$, and working at $p = 1$, obtain the components of $\mathbf{R}''$ along $\mathbf{T}$ and along $\mathbf{N}$. Show that these are $\rho'\mathbf{T}$ and $\kappa\rho^2\mathbf{N}$, respectively.

    (f) On the graph of the curve, at the point corresponding to $p = 1$, draw the vectors $\rho'\mathbf{T}$, $\rho^2\kappa\mathbf{N}$, and $\mathbf{R}''$.

B13. Exercise B13, Section 17.1   **B14.** Exercise B18, Section 17.4

B15. Exercise B14, Section 17.4   **B16.** Exercise B16, Section 17.1

B17. Exercise B22, Section 17.4

For the curves specified in Exercises B18–24:

    (a) Obtain $\mathbf{R}''(p)$.

    (b) Obtain $\mathbf{T}(p)$, $\mathbf{N}(p)$, $\rho(p)$, $\rho'(p)$, and $\kappa(p)$.

(c) Show that (17.18) is valid.

(d) Using the projection operator $\mathbf{B_A} = \frac{\mathbf{B \cdot A}}{\mathbf{A \cdot A}}\mathbf{A}$ for the component of $\mathbf{B}$ along $\mathbf{A}$, and working at $p = 1$, obtain the components of $\mathbf{R}''$ along $\mathbf{T}$ and along $\mathbf{N}$. Show that these are $\rho'\mathbf{T}$ and $\kappa\rho^2\mathbf{N}$, respectively.

B18. Exercise B18, Section 17.1   **B19.** Exercise B36, Section 17.4

B20. Exercise B20, Section 17.1   **B21.** Exercise B21, Section 17.1

B22. Exercise B22, Section 17.1   **B23.** Exercise B23, Section 17.1

B24. Exercise B24, Section 17.1

In Exercises B25–29, $\mathbf{V}(t_0)$ and $\mathbf{a}(t_0)$ are given, for the same curve, at time $t = t_0$.

    (a) Obtain $v(t_0)$.   **(b)** Obtain $\mathbf{T}(t_0)$.

    (c) Obtain $v'(t_0)$.   **(d)** Obtain $\kappa(t_0)$.

    (e) Obtain $\mathbf{B}(t_0)$. (*Hint:* See Exercise B1, Section 17.4.)

    (f) Obtain $\mathbf{N}(t_0)$.

B25. $\mathbf{V}(t_0) = 3\mathbf{i} - 4\mathbf{j} + 2\mathbf{k}$, $\mathbf{a}(t_0) = 13\mathbf{i} + 2\mathbf{j} + 5\mathbf{k}$

B26. $\mathbf{V}(t_0) = -5\mathbf{i} + 7\mathbf{j} - 8\mathbf{k}$, $\mathbf{a}(t_0) = 4\mathbf{i} + 17\mathbf{j} - 8\mathbf{k}$

B27. $\mathbf{V}(t_0) = 2\mathbf{i} - 5\mathbf{j} - 12\mathbf{k}$, $\mathbf{a}(t_0) = 5\mathbf{i} - 6\mathbf{k}$

B28. $\mathbf{V}(t_0) = 6\mathbf{i} + \mathbf{j} - 2\mathbf{k}$, $\mathbf{a}(t_0) = -9\mathbf{i} + 11\mathbf{j} + 15\mathbf{k}$

B29. $\mathbf{V}(t_0) = -\mathbf{i} - 7\mathbf{j} + 9\mathbf{k}$, $\mathbf{a}(t_0) = \frac{3}{2}\mathbf{i} + \frac{4}{5}\mathbf{j} - \mathbf{k}$

17.6 Applications to Dynamics

Geometry and Dynamics

If the parameter for a curve $\mathbf{R}(p)$ is actually the time, t, then, from Section 17.5, the decomposition formula (17.18) becomes

$$\mathbf{a} = \dot{v}\mathbf{T} + \kappa v^2\mathbf{N} \tag{17.22}$$

where $\mathbf{a}$ is the acceleration vector for a particle moving along the curve described by $\mathbf{R}(t)$. This is a remarkable result. It says that motion (as described on the left side by the acceleration vector $\mathbf{a}$) is determined by the geometry on the right side (as described by the geometric quantities $\mathbf{T}$, the unit tangent vector, $\mathbf{N}$, the principal normal, κ, the curvature, and v, the length of $\mathbf{R}'$).

Hence, this decomposition equation for the acceleration vector can be used to solve problems in dynamics where, primarily, the geometry of the path of motion is known. When (17.22) is combined with Newton's second law, $\mathbf{F} = m\mathbf{a}$, several interesting applications become accessible.

EXAMPLE 17.13 Together, all four tires of a car in uniform motion around a circular track can exert toward the center of the track, a total frictional force of no more than 450 lb. The radius of the track is 200 ft and the car weighs 3200 lb. What is the fastest constant speed the car can sustain without skidding?

Taking the gravitational constant to be 32 ft/s², the mass of the car is $m = \frac{w}{g} = \frac{3200}{32} = 100$ slugs. The curvature of the track is the reciprocal of its radius (Exercise A3, Section 17.5), so $\kappa = \frac{1}{r} = \frac{1}{200}$. The speed v is constant, so $\dot{v} = 0$. Hence, the acceleration becomes $\mathbf{a} = 0\mathbf{T} + \frac{1}{200}v^2\mathbf{N}$. The acceleration is directed along $\mathbf{N}$, which itself points toward the center of the circular track. The force of friction also points toward the center of the track, and so must be along $\mathbf{N}$. Hence, $\mathbf{F} = 450\mathbf{N}$. But $\mathbf{F} = m\mathbf{a}$ and $\mathbf{a} = \frac{v^2}{200}\mathbf{N}$, so $450\mathbf{N} = 100(\frac{v^2}{200})\mathbf{N}$ or $450 = \frac{1}{2}v^2$. Solving this for the speed yields $v = \pm 30$, with the positive answer, 30 ft/s, as the appropriate answer. Note that the units in the problem were fixed as soon as the gravitational constant g was taken as 32 ft/s². To obtain an answer in miles per hour, convert to 20.45 mph *afterward*. ❖

EXAMPLE 17.14 An object weighing 8 lb is moving from left to right along the curve $y = -x^2$. When it is at the origin, the total force on it is $\mathbf{F} = \mathbf{i} - 2\mathbf{j}$ lb. What is its speed, and at what rate is it speeding up or slowing down at that point?

The unknowns in the problem are v and $\dot{v}$. From the geometry, the vectors $\mathbf{T}(0)$ and $\mathbf{N}(0)$ can be found. From the force $\mathbf{F}$, Newton's second law ($\mathbf{F} = m\mathbf{a}$), and from (17.22), the equation

$$\mathbf{F} = \begin{bmatrix} 1 \\ -2 \end{bmatrix} = m[\dot{v}\mathbf{T}(0) + \kappa v^2 \mathbf{N}(0)] \tag{17.23}$$

is obtained. From the weight $w = 8$ lb, the mass $m = \frac{8}{32} = \frac{1}{4}$ slug can be determined, but the curvature κ at $x = 0$ must be computed from the equation of the curve. (The problem is predicated on the curvature being defined as a positive quantity.)

To write the curve $y = y(x)$ parametrically with parameter p take $x = x(p) = p$ and $y = y(p) = -p^2$. Then, we have

$$\mathbf{R}(p) = \begin{bmatrix} p \\ -p^2 \end{bmatrix} \qquad \mathbf{T}(0) = \begin{bmatrix} 1 \\ 0 \end{bmatrix} \qquad \mathbf{N}(0) = \begin{bmatrix} 0 \\ -1 \end{bmatrix}$$

The curvature at the origin is $\kappa(0) = 2$, so (17.23) becomes

$$\begin{bmatrix} 1 \\ -2 \end{bmatrix} = \frac{1}{4}\left\{ \dot{v}\begin{bmatrix} 1 \\ 0 \end{bmatrix} + 2v^2\begin{bmatrix} 0 \\ -1 \end{bmatrix} \right\} \tag{17.24}$$

or the pair of equations $1 = \frac{\dot{v}}{4}$ and $-2 = -\frac{v^2}{2}$, obtained by equating corresponding components in (17.24). Since $v = \|\mathbf{R}'(t)\|$ is a positive quantity, $v = 2$, and $\dot{v} = 4$, the object is speeding up at the rate of 4 ft/s, the units being determined by the choice of $g = 32$ ft/s². ❖

Note that $\mathbf{R}$ was parametrized with p to avoid the absolutely crushing error of writing the letter t and then believing it must therefore be the *time*. Nothing could be further from the truth. There is no indication in this problem as to how the object moves in time. The curve is given geometrically and cannot be parametrized in terms of the time. Consequently, computing from $\mathbf{R}(p)$ the magnitude $\|\frac{d}{dp}\mathbf{R}(p)\|$ and thinking it is $v(t) = \|\frac{d}{dt}\mathbf{R}(t)\|$ is precisely the error this problem is designed to detect.

EXERCISES 17.6

1. Find the force of friction required to keep a car on a circular track of radius 300 ft if the car weighs 3200 lb and travels at the constant speed of 200 ft/s.

2. A satellite is traveling at a constant speed in a circular orbit 400 mi above the surface of Earth where the acceleration of gravity is 30 ft/s^2. If the radius of Earth is taken as 4000 mi, what is the speed (in miles per hour) of the satellite?

3. A coin weighing λ lb remains on a turning phonograph record if the force on it is not more than $\frac{\lambda}{7}$ lb. How far (in inches) from the center of a revolving 45-rpm disk can this coin sit without sliding off?

4. What is the maximum magnitude of the force needed to cause an object weighing 2 lb to move at the constant speed of 3 ft/s along the parabola $y = x^2$?

5. An object of mass 3 kg is traveling counterclockwise around the ellipse $4x^2 + 9y^2 = 36$. When it reaches the point $(0, -2)$ the force on it is $3\mathbf{i} + 5\mathbf{j}$. At this point, what is its speed?

6. To create artificial gravity, a cylindrical satellite with radius 100 ft is given a constant rotation about its axis. How fast (in revolutions per minute) should it rotate if an astronaut whose feet are against the outer wall (head pointing toward the axis of rotation) is to experience, at foot-level, an acceleration of 32 ft/s^2? What would be the value of the acceleration experienced at head-level for an astronaut 6-ft tall?

7. What rotational speed (in revolutions per minute) imparted to a space station will cause the center of the crew's quarters, 8 m from the axis of rotation, to experience the same gravitational field as the Earth's surface?

8. A car moves with constant speed 20 ft/s along the curve $y = x^2$, in such a way that $x > 0$, $y > 0$, and y increases as x increases. At $(1, 1)$, find the acceleration vector.

9. Jogger X weighs 160 lb and runs clockwise around the circle $x^2 + y^2 = 25$. At the point $(3, 4)$ Jogger X is running 5 ft/s and is speeding up at the rate of 10 ft/s^2. Find X's acceleration vector at the moment X is at the point $(3, 4)$.

10. In a horizontal plane, a 1500-kg sports car enters a section of circular track and slows down to a constant speed of 100 m/s. The radius of curvature of the track in this curve is 80 m. Determine the horizontal force exerted on the tires by the road.

11. A 4-kg mass is in uniform motion around the inside of a plane vertical track whose radius of curvature is everywhere 1.8 m. Determine the minimum speed that the mass can have and still remain in contact with the track when it is at the top of the loop.

12. A 12-lb object traverses the curve $y = e^x$ from left to right. At $x = 0$, the force that is keeping the object moving is $-2\mathbf{i} + 3\mathbf{j}$. Find the speed, and determine if the object is speeding up or slowing down at this point by first finding $\mathbf{T}(0)$, $\mathbf{N}(0)$, $\kappa(0)$ from the curve and, from these, $\mathbf{a}(0)$. Now use $\mathbf{F} = m\mathbf{a}$, Newton's Second Law, to set up and solve two equations in the two unknowns $\dot{v}$ and v^2.

13. A 4-lb object traverses the curve $y = \frac{1}{1+x^2}$ from left to right. At $x = 1$, the force that is keeping the object moving is $-5\mathbf{i} + 7\mathbf{j}$. Find the speed, and determine if the object is speeding up or slowing down at this point.

14. Find the force $\mathbf{F}$ that will cause a particle of mass 3 to travel at constant speed $v = 7$ around the ellipse $x(p) = 5\cos p$, $y(p) = 12\sin p$.

15. An object of mass 2 kg is traveling from left to right along the curve $y = 4x^3 - 7x^2 + 2x - 1$. When it reaches the point $(1, -2)$ the force on it is $5\mathbf{i} + 2\mathbf{j}$. At this point, what is its speed?

Chapter Review

1. Let $\mathbf{R} = p^2\mathbf{i} + p^3\mathbf{j} + p\mathbf{k}$, $0 \le p \le 2$, define a space curve.

   (a) Obtain $\mathbf{R}'(p)$ and $\mathbf{R}''(p)$.

   (b) Obtain $v(p) = \frac{ds}{dp} = \|\mathbf{R}'(p)\|$.

   (c) Find the length of the curve.

   (d) Obtain $\mathbf{T}(p) = \frac{1}{v}\mathbf{R}'$, the unit tangent vector.

   (e) Obtain $\kappa = \frac{\|\mathbf{R}' \times \mathbf{R}''\|}{v^3}$, the curvature.

   (f) Obtain $\mathbf{N}(p) = \frac{1}{\kappa v}\frac{d\mathbf{T}}{dp}$, the principal normal.

   (g) Obtain $\mathbf{B}(1) = \mathbf{T}(1) \times \mathbf{N}(1)$, the binormal vector at $p = 1$.

   (h) Evaluate $v'\mathbf{T} + \kappa v^2\mathbf{N}$ at $p = 1$, and compare to $\mathbf{R}''(1)$.

   (i) Show that $\mathbf{R}''(1)$ lies in the plane determined by the unit tangent and principal normal vectors by verifying $\mathbf{R}'' \cdot \mathbf{B} = 0$ at $p = 1$.

2. Show that $\kappa = \frac{\|\mathbf{R}' \times \mathbf{R}''\|}{v^3}$ becomes $\kappa = \frac{|x'y'' - y'x''|}{v^3}$ for the plane curve $\mathbf{R} = x(p)\mathbf{i} + y(p)\mathbf{j} + 0\mathbf{k}$.

3. Let $\mathbf{R} = p\mathbf{i} - p^2\mathbf{j}$, $0 \le p \le 2$, define a plane curve.

   (a) Sketch the curve.

   (b) Obtain $\mathbf{R}'(1)$, $v(1)$, and $\mathbf{T}(1)$.

   (c) Obtain $\mathbf{N}(1)$ from $\mathbf{T}(1)$, being sure to have the principal normal point towards the center of curvature.

   (d) Using Question 2, obtain $\kappa(1)$, the curvature at $p = 1$.

   (e) Verify $\mathbf{R}'' = v'\mathbf{T} + \kappa v^2\mathbf{N}$, at least at $p = 1$.

   (f) Obtain the length of the curve.

4. State and illustrate the reciprocal rule for the derivative of an inverse function.

5. For the curvature of a plane curve $y(x)$, derive the formula $\kappa = \frac{|y''|}{(1+(y')^2)^{3/2}}$. Explain what the curvature measures.

6. Explain the formula $\kappa = \left\| \frac{d\mathbf{T}}{ds} \right\|$ for the curvature of a space curve.

7. Show that if $\mathbf{u}$ is a unit vector, it is necessarily orthogonal to $\mathbf{u}'$, its derivative.

8. Determine the force exerted on a particle of mass m if the particle undergoes uniform motion on a circle of radius r.

9. A particle of mass $m = 1$ traverses the curve $\mathbf{R} = 3t\mathbf{i} + 9t^2\mathbf{j}$, where the parameter t is time.

 (a) Show that the particle describes the parabola $y = x^2$.

 (b) By differentiation, obtain the acceleration vector $\mathbf{a} = \ddot{\mathbf{R}}$.

 (c) Using $\mathbf{R} = x\mathbf{i} + x^2\mathbf{j}$ as the trajectory, obtain $\mathbf{R}'' = v'\mathbf{T} + \kappa v^2\mathbf{N}$. Explain why this is not the acceleration vector $\mathbf{a}$. *Hint:* The chain rule for differentiation should enter the discussion.

10. A particle of mass $m = 1$ traverses the curve $\mathbf{R} = t^2\mathbf{i} + t^4\mathbf{j}$, where the parameter t is time.

 (a) Show that the particle describes the parabola $y = x^2$.

 (b) By differentiation, obtain the acceleration vector $\mathbf{a} = \ddot{\mathbf{R}}$.

 (c) Using $\mathbf{R} = x\mathbf{i} + x^2\mathbf{j}$ as the trajectory, obtain $\mathbf{R}'' = v'\mathbf{T} + \kappa v^2\mathbf{N}$. Again as in Question 9, $\mathbf{R}'' \neq \mathbf{a}$, and unlike the outcome of Question 9, $\mathbf{R}''$ is not even proportional to $\mathbf{a}$.

 (d) Deduce a condition under which $\mathbf{R}''$ would be proportional to $\mathbf{a}$.

Chapter 18

The Gradient Vector

INTRODUCTION   If $f(x, y, z)$ is a scalar-valued function describing, for example, temperatures in a room, then $D_{\mathbf{U}}f$, the rate of change of f along a fixed line whose direction is given by the unit vector $\mathbf{U}$, is called the *directional derivative* of f in the direction $\mathbf{U}$. This derivative is a spatial rate of change, not a temporal one. It gives the temperature gradient, the change of temperature per unit length in a given direction.

It turns out that the directional derivative can be expressed as the dot product of $\text{grad}(f) = f_x\mathbf{i} + f_y\mathbf{j} + f_z\mathbf{k}$, the *gradient vector*, and the vector $\mathbf{U}$. This gradient vector turns out to be orthogonal to the isotherms of f, the surfaces along which the temperatures described by f are constant. The gradient vector points in the direction of increasing temperatures, and the greatest rate of change in f is the length of $\text{grad}(f)$.

By defining the *del* or *nabla* operator $\nabla = \mathbf{i}\frac{\partial}{\partial x} + \mathbf{j}\frac{\partial}{\partial y} + \mathbf{k}\frac{\partial}{\partial z}$, the gradient vector can be expressed as ∇f. The operator ∇ will continue to appear throughout the rest of the unit.

The *flow lines* for f are curves whose tangent vectors are the gradient vector field $\text{grad}(f)$. They can be found as the solution of the system of differential equations $\frac{d\mathbf{R}}{dp} = \nabla f$, where p is a parameter along the flow line.

The Lagrange multiplier method is a technique for solving constrained optimization problems. The extrema are at points of tangency of the constraint and the level curves of the objective function. These points of tangency are obtained by locating points where, along the constraint and the level curves, the gradient vectors become parallel.

A force $\mathbf{F}$ is called *conservative* if $\mathbf{F} = -\nabla u$, the negative of the gradient of some scalar function $u(x, y, z)$ is called the *potential*. Several other characterizations of the conservative force are detailed in Section 21.5.

18.1   Visualizing Vector Fields and Their Flows

Vector Fields

Mathematically, a *vector field* is a function that assigns a vector to each point in its domain, which can be a plane or space curve or a region in two or three dimensions.

As an example of a vector field defined along a curve, recall the field of tangent vectors $\mathbf{R}'(p)$ defined along the curve $\mathbf{R}(p)$ in Example 17.2. This curve and its tangent vectors are shown in Figure 17.4. Such a field along a curve is, in general, represented formally by

$$\mathbf{F}(p) = \begin{bmatrix} f(p) \\ g(p) \end{bmatrix}$$

a vector-valued function of the scalar p.

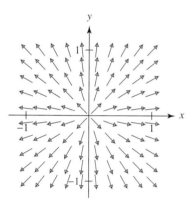

FIGURE 18.1   The planar vector field of Example 18.1

A vector-valued function of a vector-valued argument would define a vector at each point of a domain in two or three dimensions. Thus, the notation $\mathbf{F}(\mathbf{x})$, where $\mathbf{F}$ is a vector of functions and $\mathbf{x}$ is a vector of coordinates, represents the vector-valued function of a vector-valued argument. For example, a function of the form

$$\mathbf{F}(\mathbf{x}) = \begin{bmatrix} f(x, y) \\ g(x, y) \end{bmatrix} \tag{18.1}$$

is a vector field defining an arrow at each point in the xy-plane. An example of such a field is seen in Figure 18.1 appearing in Example 18.1.

Flow Lines

Given a vector field $\mathbf{F}(\mathbf{x})$, the *flow lines* are curves along which the arrows of the field are tangent. Thus, the flow lines are curves defined by the vector $\mathbf{R}(p)$. The given field $\mathbf{F}(\mathbf{x})$ is precisely the field of tangents to the curves $\mathbf{R}(p)$, so the flow lines are defined by the equations $\mathbf{R}'(p) = \mathbf{F}(\mathbf{x})$ along $\mathbf{x} = \mathbf{R}(p)$. Thus, given the field (18.1), write the equations of the flow lines in radius-vector form so that $\mathbf{R}(p)$ and $\mathbf{R}'(p)$ are given by

$$\mathbf{R}(p) = \begin{bmatrix} x(p) \\ y(p) \end{bmatrix} \Rightarrow \mathbf{R}'(p) = \begin{bmatrix} x'(p) \\ y'(p) \end{bmatrix}$$

The differential equations $\mathbf{R}'(p) = \mathbf{F}(\mathbf{x})$ then become $x'(p) = f(x, y)$, $y'(p) = g(x, y)$. Integrating these equations, that is, solving the system of differential equations $\mathbf{R}'(p) = \mathbf{F}(\mathbf{x})$, produces the *flow* of the field. On occasion, the flow lines can be found explicitly and exactly. Typically, however, the differential equations defining the flow of the field must be solved numerically.

EXAMPLE 18.1   The vector field

$$\mathbf{F}(\mathbf{x}) = \mathbf{F}(x, y) = \frac{1}{\left(x^2 + y^2\right)^{3/2}} \begin{bmatrix} x \\ y \end{bmatrix} \tag{18.2}$$

represents the force of a positive charge placed at the origin of the xy-plane. A plot of some of the arrows in this field appears in Figure 18.1. The flow lines, which we anticipate to be radial lines emanating from the origin, can be found analytically by solving the differential equations $\mathbf{R}'(p) = \mathbf{F}(x(p), y(p))$. Thus, the equations

$$x'(p) = \frac{x}{\left(x^2+y^2\right)^{3/2}} \quad \text{and} \quad y'(p) = \frac{y}{\left(x^2+y^2\right)^{3/2}} \tag{18.3}$$

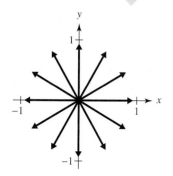

FIGURE 18.2   Flow lines of the planar field in Example 18.1

can be solved by writing $\frac{dy}{dx} = \frac{y'(p)}{x'(p)} = \frac{y}{x}$, from which we obtain $y = \alpha x$, radial lines through the origin. Figure 18.2 shows some of these radial lines and is, in fact, a phase portrait for the differential equations $\mathbf{R}'(p) = \mathbf{F}(x(p), y(p))$. ❖

EXERCISES 18.1–Part A

A1. Verify that the solution of (18.3) is $y = ax$.

A2. For the field $\mathbf{F} = x\mathbf{i} + y^2\mathbf{j}$, find and sketch the flow line through the point $(1, 1)$. Include the direction of the flow along this curve.

A3. At the grid points $(x_i, y_j) = (i, j), i = 0, \ldots, 3, j = 0, \ldots, 4$, sketch the arrows determined by the field in Exercise A2.

A4. For the field $\mathbf{F} = y\mathbf{i} + 2x\mathbf{j}$, find and sketch the flow line through the point $(1, 1)$. Include the direction of the flow along this curve.

A5. At the grid points $(x_i, y_j) = (i, j), i = 0, \ldots, 2, j = -2, \ldots, 2$, sketch the arrows determined by the field in Exercise A4.

EXERCISES 18.1–Part B

B1. The force due to a positive charge is supposedly inversely proportional to the square of the distance from the charge. Show that the field in (18.2) is indeed such a field. *Hint:* Define a *unit* vector from the charge to the point (x, y).

In Exercises B2–9:

(a) Find all critical points where $\mathbf{F} = \mathbf{0}$. *Hint:* If $\mathbf{F} = f(x, y)\mathbf{i} + g(x, y)\mathbf{j}$, then $f(x, y) = 0$ implicitly defines $y = y_1(x)$, whereas $g(x, y) = 0$ implicitly defines $y = y_2(x)$. Any intersections of the curves $y_1(x)$ and $y_2(x)$ are critical points.

(b) Obtain a field plot.

(c) Obtain a graph of the flow lines of $\mathbf{F}$, indicating the direction of increasing t along the flow lines. Such a graph is actually a phase portrait for the autonomous system $\mathbf{R}'(p) = \mathbf{F}(\mathbf{x}(p)) = \mathbf{F}(x(p), y(p))$, where $\mathbf{R} = x(p)\mathbf{i} + y(p)\mathbf{j}$. The critical points for the field are the equilibrium points for the autonomous system of differential equations defining the flow.

(d) Linearize at each equilibrium point (see Section 12.22) to determine, where possible, whether the point is ustable, orbitally stable, or asymptotically stable.

(e) Where possible, categorize each equilibrium point as a node, saddle, center, or spiral.

(f) Let C be the unit circle in the xy-plane; and let $p_k = (\cos\frac{k\pi}{5}, \sin\frac{k\pi}{5})$, $k = 0, 1, \ldots, 9$, be 10 equally spaced points around C. For this circle, let $\mathbf{T}_k$ be the unit tangent vector at p_k, and let $\mathbf{n}_k$ be the outward unit normal at p_k. For each p_k, obtain and graph the vector $\mathbf{F}(p_k)$, and its tangential and normal components, $\mathbf{F_T} = (\mathbf{F} \cdot \mathbf{T}_k)\mathbf{T}_k$ and $\mathbf{F_n} = (\mathbf{F} \cdot \mathbf{n}_k)\mathbf{n}_k$, respectively.

B2. $\mathbf{F} = x^2\mathbf{i} - y\mathbf{j}$ **B3.** $\mathbf{F} = y\mathbf{i} + x^2\mathbf{j}$ **B4.** $\mathbf{F} = y^2\mathbf{i} - x\mathbf{j}$

B5. $\mathbf{F} = xy\mathbf{i} + \cos y\mathbf{j}$ **B6.** $\mathbf{F} = \dfrac{1}{1+y^2}\mathbf{i} + \dfrac{1}{1+x^2}\mathbf{j}$

B7. $\mathbf{F} = \dfrac{x}{1+y^2}\mathbf{i} + \dfrac{y}{1+x^2}\mathbf{j}$ **B8.** $\mathbf{F} = \dfrac{y}{x^2+y^2}\mathbf{i} + \dfrac{x}{x^2+y^2}\mathbf{j}$

B9. $\mathbf{F} = x^2\mathbf{i} + y\mathbf{j}$

In Exercises B10–20, repeat the steps used for Exercises B2–9. However, in part (a) find only those critical points in the square $-2 \le x, y \le 2$. In part (f), draw the vectors just at the point $(\frac{1}{\sqrt{2}}, \frac{1}{\sqrt{2}})$ on the unit circle.

B10. $\mathbf{F} = (-5 - 3y - 4x^3 + 8y^3 + 8x^2y + 6x^2)\mathbf{i} + (7 - 6x^3 + 7x^2y + xy + 7xy^2)\mathbf{j}$

B11. $\mathbf{F} = (-3x - 5x^3 - 3y^3 - 2x^2y + xy^2 - 4x^2)\mathbf{i} + (-3x - 8y + 4y^2 + 6y^3 - 8x^2y + 3xy)\mathbf{j}$

B12. $\mathbf{F} = (4 - 6y + 2y^2 - 8x^3 - 6xy^2 - 9x^2)\mathbf{i} + (-6 - 9x + 9y + 6y^2 - 4y^3 + 2xy^2)\mathbf{j}$

B13. $\mathbf{F} = (9x + 8y - 7y^2 + 4x^3 + 5x^2y - 9x^2)\mathbf{i} + (4y + 9y^2 - 5x^3 + 7y^3 - 7x^2y - 9xy)\mathbf{j}$

B14. $\mathbf{F} = (9 + 6x^3 + 2y^3 + 8x^2y - 7xy + 6xy^2)\mathbf{i} + (4x - 4y - 6x^3 - 9x^2y + 2xy + 5x^2)\mathbf{j}$

B15. $\mathbf{F} = (-8x - 2y + 2y^2 + 2x^2y - 9xy^2 - 7x^2)\mathbf{i} + (8x + 9y - 3x^3 - 6y^3 + 3xy^2 - 3x^2)\mathbf{j}$

B16. $\mathbf{F} = (-2 + 4x + 3x^3 + 9y^3 - 7x^2y - 3xy^2)\mathbf{i} + (-8 - 9x + 6y^2 - 4x^3 - 7y^3)\mathbf{j}$

B17. $\mathbf{F} = (-8 + 5x - 5y^2 - 9y^3 - 3xy^2)\mathbf{i} + (-6 - 9y^3 + 5x^2y - 7xy^2)\mathbf{j}$

B18. $\mathbf{F} = (-8 - 3x - 6y^2 - 5x^3 - 3y^3 - 6x^2y)\mathbf{i} + (2 - 6x + 6y + 8x^2y + 5xy^2 - 9x^2)\mathbf{j}$

B19. $\mathbf{F} = (-3x - 5x^3 - 3y^3 - 2x^2y + xy^2 - 4x^2)\mathbf{i} + (-3x - 8y + 4y^2 + 6y^3 - 8x^2y + 3xy)\mathbf{j}$

B20. $\mathbf{F} = (-2x + y + 4y^2 - 7x^3 + xy^2 + 5x^2)\mathbf{i} + (-8 - y^2 - 6x^3 - 6x^2y + 7xy + 7xy^2)\mathbf{j}$

In Exercises B21–30, obtain and plot as a space curve, the flow line emanating from $(1, 1, 1)$ for the autonomous system of differential equations $\mathbf{R}' = \mathbf{F}(x(p), y(p), z(p))$ determined by the given field $\mathbf{F}(\mathbf{x})$. Along the solution curve, indicate the direction of increasing p.

B21. $\mathbf{F} = (-5 - 9x - 2xz)\mathbf{i} + (9yz - 5x - 5xz)\mathbf{j} + (9x + 2yz - 4z^2)\mathbf{k}$

B22. $\mathbf{F} = (2 + 4yz + 2z)\mathbf{i} + (x^2 - 5y^2 + 3xy)\mathbf{j} + (8xz - 3y + z)\mathbf{k}$

B23. $\mathbf{F} = (8y + 7xz + 8z^2)\mathbf{i} + (9 - 3x - 5y^2)\mathbf{j} + (6yz - 8x^2 - 9z)\mathbf{k}$

B24. $\mathbf{F} = (1 - 3y^2 - 6z)\mathbf{i} + (7yz - 9z^2 + 9z)\mathbf{j} + (8yz - 3x^2 - z^2)\mathbf{k}$

B25. $\mathbf{F} = (2y^2 - 8xz - 4z)\mathbf{i} + (1 - 3x - y)\mathbf{j} + (5x^2 - 4yz + 4z)\mathbf{k}$

B26. $\mathbf{F} = (2z^2 - 5y^2 - 7x^2)\mathbf{i} + (9 + 8xz + 6xy)\mathbf{j} + (4 - 9yz + 3x^2)\mathbf{k}$

B27. $\mathbf{F} = (4xz - 4xy - 9x^2)\mathbf{i} + (2xy - 7yz + 2z^2)\mathbf{j} + (5x + 2y - 9xy)\mathbf{k}$

B28. $\mathbf{F} = (6y - 4 - 2y^2)\mathbf{i} + (7y + 9xz - 8yz)\mathbf{j} + (-y - 2y^2 - x^2)\mathbf{k}$

B29. $\mathbf{F} = (yz - 2xz + 5y^2)\mathbf{i} + (2y + 6xz - 2x^2)\mathbf{j} + (-4x - 9y^2)\mathbf{k}$

B30. $\mathbf{F} = (8xz + 3z^2 - 8z)\mathbf{i} + (7xz - 3 - y)\mathbf{j} + (9xz - 2xy - 4z)\mathbf{k}$

18.2 The Directional Derivative and Gradient Vector

The Directional Derivative

Let $\mathbf{U} = a\mathbf{i} + b\mathbf{j} + c\mathbf{k}$ be a unit vector, and let $w = f(x, y, z)$ be a scalar-valued function defined at each point in a three-dimensional region. At the point $P_0 = (x_0, y_0, z_0)$ in the region, the rate of change of w in the direction of the vector $\mathbf{U}$ is called the *directional derivative* of $f(x, y, z)$ in the direction $\mathbf{U}$.

For example, w could give the temperature at each point in a room. The directional derivative of w would be the rate of change of temperature, computed at P_0, and in the direction of the vector $\mathbf{U}$. Although the word "rate" is used, the directional derivative is not a measurement of change in time. It is not a velocity. For a temperature, it would be a temperature gradient, measured in degrees per unit length, and would indicate the tendency for the temperature to change in a given direction.

In general, the equation of the line passing through point P_0 and having the direction of the vector $\mathbf{U}$ is

$$\mathbf{R}(p) = \mathbf{P}_0 + p\mathbf{U} = \begin{bmatrix} x_0 + pa \\ y_0 + pb \\ z_0 + pc \end{bmatrix} \tag{18.4}$$

The values of w along this line are a function of the parameter p used to mark progress along the line, so w becomes $w(p) = f(x(p), y(p), z(p))$ along this line. To determine the rate at which $w(p)$ is changing along this line, differentiate with respect to the parameter p, obtaining, by the chain rule,

$$\frac{dw}{dp} = f_x(x_0+pa, y_0+pb, z_0+pc)a + f_y(x_0+pa, y_0+pb, z_0+pc)b$$
$$+ f_z(x_0+pa, y_0+pb, z_0+pc)c$$

Since the rate of change at P_0 is the objective, evaluate this derivative at $p = 0$, which corresponds to the point P_0 on the line through P_0. This gives

$$\left.\frac{dw}{dp}\right|_{p=0} = f_x(x_0, y_0, z_0)a + f_y(x_0, y_0, z_0)b + f_z(x_0, y_0, z_0)c$$

The point (x_0, y_0, z_0) is arbitrary, so it can be replaced with just (x, y, z). If we also introduce the notation $D_\mathbf{U} f$ for the derivative $\left.\frac{dw}{dp}\right|_{p=0}$, we obtain

$$D_\mathbf{U} f = f_x(x, y, z)a + f_y(x, y, z)b + f_z(x, y, z)c \tag{18.5}$$

Careful inspection of (18.5) suggests it is a dot product between the vector $\mathbf{U}$ and a vector

$$\mathrm{grad}(f) = \nabla f = \left(\mathbf{i}\frac{\partial}{\partial x} + \mathbf{j}\frac{\partial}{\partial y} + \mathbf{k}\frac{\partial}{\partial z}\right) f = f_x\mathbf{i} + f_y\mathbf{j} + f_z\mathbf{k} = \begin{bmatrix} f_x(x, y, z) \\ f_y(x, y, z) \\ f_z(x, y, z) \end{bmatrix}$$

called the *gradient* vector. Notations for this vector include $\mathrm{grad}(f)$ and ∇f. In Cartesian coordinates, the *nabla* symbol ∇ stands for the *del* operator

$$\nabla = \mathbf{i}\frac{\partial}{\partial x} + \mathbf{j}\frac{\partial}{\partial y} + \mathbf{k}\frac{\partial}{\partial z} \tag{18.6}$$

As an operator, (18.6) is applied to a scalar-valued function *from the left*. It is common to read ∇f as "del f."

When the derivative of $f(x, y, z)$ in the direction $\mathbf{U}$ exists, it can be computed as the dot product $D_\mathbf{U} f = \nabla f \cdot \mathbf{U}$. The gradient ∇f is computed first, yielding a vector, which is then dotted with the unit vector $\mathbf{U}$. We illustrate this with two examples in three dimensions, leaving to the exercises the obvious modification to two dimensions.

EXAMPLE 18.2   Given the function $f(x, y, z) = x^3 - xy^2 - z$, we find, at the point $P_0 = (1, 1, 0)$, the directional derivative in the direction $\mathbf{u} = 2\mathbf{i} - 3\mathbf{j} + 6\mathbf{k}$. The direction vector $\mathbf{U}$ must be a unit vector. Given the direction $\mathbf{u}$, not a unit vector, we begin by normalizing $\mathbf{u}$ to form

$\mathbf{U} = \frac{1}{7}(2\mathbf{i} - 3\mathbf{j} + 6\mathbf{k})$. Next, we get

$$\nabla f = \begin{bmatrix} 3x^2 - y^2 \\ -2xy \\ -1 \end{bmatrix} = (3x^2 - y^2)\mathbf{i} - 2xy\mathbf{j} - \mathbf{k}$$

the gradient of the function $f(x, y, z)$, and evaluate the gradient at the point P_0, obtaining $\nabla f(P_0) = 2\mathbf{i} - 2\mathbf{j} - \mathbf{k}$. Finally, the desired directional derivative is the dot product $\nabla f(P_0) \cdot \mathbf{U} = \frac{4}{7}$.   ❖

Again, although the directional derivative is described as the *rate of change* in the direction $\mathbf{U}$, there is no time in this derivative. The directional derivative is a *spatial* rate of change that measures the change in function value per unit length in the direction of the vector $\mathbf{U}$. The gradient (or pitch) of a lawn, measured as a change in height per unit length, is a relevant image.

EXAMPLE 18.3   At the point $P_0 = (2, 0, 0)$ and in the direction of the point $P_1 = (4, 1, -2)$, find the directional derivative of the function $f(x, y, z) = xe^y + yz$. First, the vector *from* the point P_0 *to* the point P_1 is found by the difference $\mathbf{P}_1 - \mathbf{P}_0 = 2\mathbf{i} + \mathbf{j} - 2\mathbf{k}$. Then, the direction vector $\mathbf{U}$ must be a unit vector, so normalize to obtain $\mathbf{U} = \frac{1}{3}(2\mathbf{i} + \mathbf{j} - 2\mathbf{k})$. Compute

$$\nabla f = \begin{bmatrix} e^y \\ xe^y + z \\ y \end{bmatrix} = e^y\mathbf{i} + (xe^y + z)\mathbf{j} + y\mathbf{k}$$

the gradient of $f(x, y, z)$, and then evaluate at the point P_0 to obtain $\nabla f(P_0) = \mathbf{i} + 2\mathbf{j}$. Finally, obtain the dot product $\nabla f(P_0) \cdot \mathbf{U} = \frac{4}{3}$.   ❖

EXERCISES 18.2–Part A

A1. Show that the directional derivative of $f(x, y)$ in the direction of $\mathbf{i}$ is just f_x and in the direction of $\mathbf{j}$ is just f_y.

A2. At $(1, 1)$, and in the direction $\mathbf{u} = 2\mathbf{i} + 3\mathbf{j}$, obtain the directional derivative of $f(x, y) = xy^2$.

A3. At $(2, -1)$, and in the direction of the point $(-3, 5)$, obtain the directional derivative of $f(x, y) = x^2 y^3$.

A4. At $(3, -2, 1)$, and in the direction $\mathbf{u} = 4\mathbf{i} - 7\mathbf{j} + 2\mathbf{k}$, obtain the directional derivative of $f(x, y, z) = x^2 y - 3yz + 2xz$.

A5. At $(5, -1, 1)$, and in the direction of the point $(2, -3, 1)$, obtain the directional derivative of $f(x, y, z) = xy^2 z^3$.

EXERCISES 18.2–Part B

B1. If $\mathbf{F}(x, y, z) = f(x, y, z)\mathbf{i} + g(x, y, z)\mathbf{j} + h(x, y, z)\mathbf{k}$, define $\mathbf{F}(p)$ as the value of $\mathbf{F}(x, y, z)$ along (18.4). By computing $\frac{d\mathbf{F}}{dp}\big|_{p=0}$, show that the directional derivative of a vector field can formally be represented by the product $A\mathbf{U}$, where

$$A = \begin{bmatrix} f_x & f_y & f_z \\ g_x & g_y & g_z \\ h_x & h_y & h_z \end{bmatrix}$$

is the Jacobian matrix for $\mathbf{F}$. In Cartesian coordinates, A is the covariant derivative of $\mathbf{F}$, the natural generalization of the notion of a directional derivative.

In Exercises B2–11, a function $f(x, y)$ and a pair of points P and Q are given. For each:

(a) Obtain a graph of the surface defined by $z = f(x, y)$.

(b) Obtain a contour plot of $f(x, y)$.

(c) Obtain ∇f, the gradient vector field.

(d) Obtain a field plot of the arrows of the gradient vector field.

(e) Obtain a phase portrait for the autonomous system $\mathbf{R}' = \nabla f$ and obtain for the gradient field a graph of the flow lines superimposed on the contour plot of the level curves of $f(x, y)$. The flow lines should be orthogonal to the level curves.

(f) Obtain the coordinates of the tail and head of $\nabla f(P)$, the gradient vector constructed at point P.

(g) At the point P, obtain for $f(x, y)$, the directional derivative in the direction of the point Q.

B2. $f(x, y) = -7x^2y - 6xy^2 - 9x^2$, $P = (-2, -2)$, $Q = (-10, -5)$

B3. $f(x, y) = 7xy^2 + 4y^2 + y^3 + 5xy$, $P = (8, 4)$, $Q = (11, -2)$

B4. $f(x, y) = 3x^2 - 5x - 8y + 3x^2y$, $P = (-3, -3)$, $Q = (-7, -12)$

B5. $f(x, y) = 8x - 9y + 6y^2 - 2y^3$, $P = (-4, 8)$, $Q = (0, -12)$

B6. $f(x, y) = 7 - x + 9x^2y - 2xy$, $P = (4, -2)$, $Q = (8, -9)$

B7. $f(x, y) = y^2 - x^2 - 9x^3$, $P = (10, 2)$, $Q = (8, 10)$

B8. $f(x, y) = 7y^3 - 8 - 9x^2y - 9xy^2$, $P = (-6, 5)$, $Q = (-8, 1)$

B9. $f(x, y) = 9y + 4xy^2 - 4x^3 + 8y^3$, $P = (-8, -4)$, $Q = (1, 10)$

B10. $f(x, y) = xy - 5y - 5xy^2 - 4x^2$, $P = (-4, -6)$, $Q = (-3, 6)$

B11. $f(x, y) = 6 + 3y + 2x^2y + xy^2$, $P = (-6, 3)$, $Q = (9, 7)$

In Exercises B12–21, at the given point P, find the directional derivative of $f(x, y)$ in the direction of ∇g evaluated at point Q.

B12. $f(x, y) = 7y - 6x^2 - 9x^2y$, $g(x, y) = -3x + 9x^2 - 4xy$, $P = (8, 7)$, $Q = (-2, 1)$

B13. $f(x, y) = 8x - 2y^2 + 6x^2$, $g(x, y) = 3y + 6x^2y$, $P = (-7, 8)$, $Q = (5, -7)$

B14. $f(x, y) = 6x - 5x^2 - 6y^3$, $g(x, y) = 8 + 3y^2 - 7xy$, $P = (5, 8)$, $Q = (5, 11)$

B15. $f(x, y) = 2x - 6y - 9xy$, $g(x, y) = x - 8 - 2y^3$, $P = (-8, -9)$, $Q = (-3, -5)$

B16. $f(x, y) = 2y^2 - 8x - 2x^3$, $g(x, y) = 4xy^2 - 8xy + y^3$, $P = (3, 6)$, $Q = (1, 2)$

B17. $f(x, y) = 5x^2 - 2$, $g(x, y) = 4xy - 9x^2y$, $P = (-4, 0)$, $Q = (-10, 10)$

B18. $f(x, y) = 9x^3 - 8x^2 + 4y^3$, $g(x, y) = x^3 - 7x^2y$, $P = (1, -3)$, $Q = (4, -6)$

B19. $f(x, y) = 2x^2 + 3xy$, $g(x, y) = 8x^2 - 6y^3$, $P = (3, 7)$, $Q = (-12, -1)$

B20. $f(x, y) = 9x^3 - 6x - 7xy$, $g(x, y) = 7 - 5xy^2 - 8x^3$, $P = (-5, -3)$, $Q = (10, 11)$

B21. $f(x, y) = 3 + 8y^2$, $g(x, y) = 5 - 4xy^2 - 3x^2y$, $P = (3, 0)$, $Q = (7, 2)$

In Exercises B22–31, at the given point P, find the directional derivative of $f(x, y, z)$ in the direction of the point Q.

B22. $f(x, y, z) = 9xyz + 8x^3 + 4z^2$, $P = (-6, 12, -5)$, $Q = (-5, -10, 9)$

B23. $f(x, y, z) = 6z - 2yz$, $P = (-9, -10, 12)$, $Q = (-7, 9, -6)$

B24. $f(x, y, z) = 3yz + 5xz + 9yz^2$, $P = (-7, -5, -12)$, $Q = (2, 5, 2)$

B25. $f(x, y, z) = 6y - 5x^2 + 4z^2$, $P = (-5, 0, -6)$, $Q = (10, 7, 1)$

B26. $f(x, y, z) = 3y^2 - 8xz^2 - 3y^3$, $P = (-1, -2, 10)$, $Q = (6, -12, -2)$

B27. $f(x, y, z) = -7x^2 + 2x^2y - 7z^3$, $P = (2, -5, 3)$, $Q = (0, -2, 0)$

B28. $f(x, y, z) = 4x - yz - 7z^3$, $P = (4, 1, -4)$, $Q = (-1, 9, 7)$

B29. $f(x, y, z) = 4x^2y - 5x - 5z$, $P = (2, 8, -12)$, $Q = (11, -4, 1)$

B30. $f(x, y, z) = 2y + 2xy + 7z^3$, $P = (0, 8, -6)$, $Q = (4, -12, -2)$

B31. $f(x, y, z) = 4x^2 - 2z - 5z^3$, $P = (0, -8, 4)$, $Q = (1, 11, -4)$

In Exercises B32–41, at the given point P, find the directional derivative of $f(x, y, z)$ in the direction of ∇g evaluated at point Q.

B32. $f = -6y - 7xyz - 6x^2z$, $g = 3xy - 9yz^2$, $P = (-12, 11, -10)$, $Q = (9, 12, -3)$

B33. $f = 4y + 7xz^2 - 6z$, $g = 4xy - 8z^3 - 6xz^2$, $P = (12, -11, -8)$, $Q = (3, 10, -7)$

B34. $f = 2y + 5z^2 - 7xz^2$, $g = 3y^2 + 5y^2z + 2z$, $P = (3, -11, 7)$, $Q = (5, 3, -8)$

B35. $f = 6x^2y - 8xz - 8z^2$, $g = yz - 7z^3 - 7x^2$, $P = (8, 0, -9)$, $Q = (12, 7, 0)$

B36. $f = 2yz^2 - 9x^2z - 5y^3$, $g = 2x + 9x^2z - xy$, $P = (6, -9, -5)$, $Q = (8, -3, 12)$

B37. $f = xz - 7x^2z - 3y^2$, $g = 3xyz - 5yz - 8y^2$, $P = (2, 3, -7)$, $Q = (1, -12, -11)$

B38. $f = 4xz - 8xyz - 6y^3$, $g = 4y - 5xyz - 8xz^2$, $P = (10, -12, -4)$, $Q = (8, 2, 9)$

B39. $f = 5y^2 + 7z^2 - 7xyz$, $g = 6z^3 - 2x^2y - 7y^3$, $P = (2, 4, -1)$, $Q = (-11, 9, -9)$

B40. $f = -2yz + 8y + 8z^3$, $g = 3y - 4y^2z - y^3$, $P = (-2, -3, 2)$, $Q = (11, -3, -11)$

B41. $f = 4xy^2 - 8z^2 - 7xz^2$, $g = 3y - 4xz - xy^2$, $P = (12, 1, 3)$, $Q = (-6, -4, 12)$

B42. At $(1, 2)$ the directional derivative of $f(x, y)$ in the direction of the point $(5, 6)$ is 8 and in the direction of the point $(-4, 3)$ is 13. At $(1, 2)$, find the directional derivative in the direction of the point $(7, -5)$. (*Hint:* Use the two given directional derivatives to set up two equations in the two unknowns $f_x(1, 2)$ and $f_y(1, 2)$. Once these have been determined, the directional derivative in any direction can be calculated.)

B43. At $(1, 2, -3)$ the directional derivative of $f(x, y, z)$ in the direction of the point $(5, 6, -7)$ is 8, in the direction of the point $(-4, 3, -2)$ is 13, and in direction of point $(8, 0, -12)$ it is -3. At $(1, 2, -3)$, find the directional derivative in the direction of the point $(7, -5, 9)$.

B44. For $0 \le \theta \le 2\pi$, the unit vector $\mathbf{U} = \cos\theta\mathbf{i} + \sin\theta\mathbf{j}$, with tail at the origin, rotates around the unit circle. If $f(x, y) = \frac{xy^2}{x^3 + y^2}$:

 (a) Obtain, at the point $P = (1, 1)$, the directional derivative of f in the direction $\mathbf{U}$. This directional derivative will be a function of θ.

 (b) Plot the directional derivative as a function of θ.

 (c) Determine $\hat{\theta}$, the value of θ for which the directional derivative is a maximum.

 (d) Determine the vector $\mathbf{U}(\hat{\theta})$.

 (e) Normalize $\nabla f(P)$, the gradient evaluated at point P, and compare to $\mathbf{U}(\hat{\theta})$.

B45. Repeat Exercise B44 for $f(x, y) = \frac{3x^2 - 5y^3}{2x + 7y}$ and $P = (2, 3)$.

18.3 Properties of the Gradient Vector

Level Sets

The *level sets* for $f(x, y)$ are the curves $y = y(x)$ defined implicitly by an equation of the form $f(x, y) = c$, where c is a real constant. We will refer to these curves as the *level curves* of $f(x, y)$. The level curves for $f(x, y) = x^2 + y$ are the downward-opening parabolas $y = c - x^2$, some of which are shown in Figure 18.3. Note that these curves lie in the xy-plane.

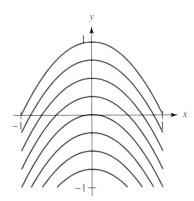

FIGURE 18.3 Level curves of $f(x, y) = x^2 + y$

The *level sets* for $g(x, y, z)$ are the surfaces $z = z(x, y)$ defined implicitly by an equation of the form $g(x, y, z) = c$, where c is a real constant. We will refer to these surfaces as the *level surfaces* of $g(x, y, z)$.

If the function $g(x, y, z)$ represents the temperature w at every point in a room, then it defines a number at each point in space inside the room. If each "air molecule" would sit still long enough for someone to paint on it the number representing the temperature at that point, a representation of this function of four variables would be obtained. Of course, air molecules, even if they existed, can't be recruited to this task, so imagine the room filled with ping-pong balls on which the appropriate temperatures have been painted. Again, a representation of this function of four variables has been obtained. Unfortunately, it would be difficult to extract information about the temperatures in the middle of the room!

Suppose further, however, that all the ping-pong balls whose temperatures were 68°F were painted blue and all those whose temperatures were 78° were painted red. Then, the collection of blue ping-pong balls would form a level surface, the surface $g(x, y, z) = 68$, and the collection of red ping-pong balls would form another level surface, the surface $g(x, y, z) = 78$. Thus, it is impossible to graph $w = g(x, y, z)$, a function of four variables, but $g(x, y, z) = c$ implicitly defines level surfaces of the form $z = z(x, y)$ and such surfaces can be graphed with three axes.

For example, the level surfaces of the function

$$w = g(x, y, z) = x^2 + y^2 - z \qquad (18.7)$$

are defined by the equation $x^2 + y^2 - z = c$, from which the surfaces $z = z(x, y)$ can be determined by $z = x^2 + y^2 - c$. For each value of the constant c, a level surface is determined, as shown in Figure 18.4, where c takes on the values $0, -1, -2$.

Four Properties of the Gradient

The following four properties of the gradient vector make it important, and useful, in the applications.

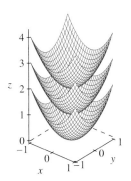

FIGURE 18.4 Level surfaces for $F(x, y, z) = x^2 + y^2 - z$

 1. ∇f is orthogonal to the level sets of f.

 2. ∇f points in the direction of increasing values of f.

3. The maximum rate of change in f is along ∇f.

4. The maximum rate of change in f is precisely the number $\|\nabla f\|$.

Property 1

The level sets for the function $z = f(x, y)$ are curves in the xy-plane, whereas the level sets for the function $w = g(x, y, z)$ are surfaces embedded in space. The gradient vectors in each case have different dimensions, and the geometric representations for each case are different. The wise student will pay close attention to the differences between the two cases.

TWO INDEPENDENT VARIABLES $z = f(x, y)$ The gradient vectors for $f(x, y)$ exist in the same space as the level curves of $f(x, y)$, namely, the xy-plane, the space of the independent variables. Even though the function $z = f(x, y)$ can be represented as a surface, the gradient vectors are found in the xy-plane.

For $f(x, y) = x^2 + y$, the gradient vector $\nabla f = 2x\mathbf{i} + \mathbf{j}$ has only two components and is defined only in the xy-plane. In fact, Figure 18.5 shows the two level curves $y = 1 - x^2$ and $y = 4 - x^2$ and some of the gradient vectors along them.

If the level curve $y = c - x^2$ is written as $\mathbf{R}(p) = p\mathbf{i} + (c - p^2)\mathbf{j}$, the tangent vector $\mathbf{R}'(p) = \mathbf{i} - 2p\mathbf{j}$ follows by differentiation with respect to the parameter on the curve. The induced parametrization of the gradient field along this level curve is $\nabla f = 2p\mathbf{i} + \mathbf{j}$ and $\mathbf{R}' \cdot \nabla f = 0$, showing that the gradient vector along a level curve is orthogonal to the tangent vector. Thus, the gradient is perpendicular to the level curve since it is perpendicular to a tangent to the level curve.

A general proof that the gradient of the function $f(x, y)$ is necessarily orthogonal to the level curves defined by $f(x, y) = c$ is easily given once it is recognized that this equation implicitly defines the level curve $y = y(x)$. Then, $f(x, y(x)) \equiv c$ is an identity, and differentiation with respect to x (by the chain rule) leads to $f_x + f_y y' = 0$, from which we get the derivative $y'(x) = -f_x/f_y$. The level curve $y = y(x)$ is written in vector form as $\mathbf{R} = x\mathbf{i} + y(x)\mathbf{j}$ by taking the parameter along the curve to be x. The tangent vector is then $\mathbf{R}' = \mathbf{i} + y'(x)\mathbf{j} = \mathbf{i} - (f_x/f_y)\mathbf{j}$ and the gradient vector is $\nabla f = f_x\mathbf{i} + f_y\mathbf{j}$. Consequently, $\mathbf{R}' \cdot \nabla f = 0$, so the gradient vector is orthogonal to the tangent vector, proving that, in general, gradients are orthogonal to the level curves.

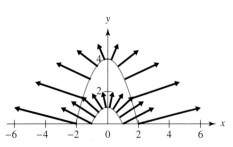

FIGURE 18.5   Gradient vectors on two level curves of $f(x, y) = x^2 + y$

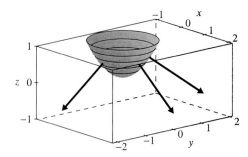

FIGURE 18.6   Three gradient vectors on the level surface $x^2 + y^2 - z = 0$

THREE INDEPENDENT VARIABLES $w = g(x, y, z)$ For $g(x, y, z)$ given in (18.7), the gradient is $\nabla g = 2x\mathbf{i} + 2y\mathbf{j} - \mathbf{k}$. It has three components and is attached to a level surface. Figure 18.6 shows three such gradient vectors emanating from the level surface $w = 0$.

The following construction demonstrates that the gradient vector is always perpendicular to the level surface. In fact, the construction shows more than that. It shows how to construct a normal to a surface, whether the surface is given implicitly in the form $g(x, y, z) = 0$ or explicitly in the form $z = f(x, y)$. In the implicit case, the normal vector will turn out to be the gradient vector, and in the explicit case, the normal vector will be a new formula.

Let the function $g(x, y, z) = 0$ implicitly define $z = f(x, y)$, and compute, by implicit differentiation based on the identity $g(x, y, f(x, y)) \equiv 0$, the derivatives

$$f_x = -\frac{g_x(x, y, z)}{g_z(x, y, z)} \quad \text{and} \quad f_y = -\frac{g_y(x, y, z)}{g_z(x, y, z)}$$

Next, on the surface $z = f(x, y)$ consider the two plane sections $y =$ constant and $x =$ constant, respectively, projecting to grid lines $y =$ constant and $x =$ constant. Along the grid line where y is constant, x varies and is the parameter on that space curve. Along the grid line where x is constant, y varies and is the parameter on the corresponding space curve.

Both space curves lying in the surface can be described in radius-vector notation by

$$\mathbf{R} = x\mathbf{i} + y\mathbf{j} + f(x, y)\mathbf{k}$$

where x is fixed on one curve and y on the other. Differentiating with respect to the parameter on each curve produces

$$\mathbf{R}_x = \mathbf{i} + f_x\mathbf{k} \quad \text{and} \quad \mathbf{R}_y = \mathbf{j} + f_y\mathbf{k}$$

vectors tangent to the respective curve. The cross-product

$$\mathbf{R}_x \times \mathbf{R}_y = -f_x\mathbf{i} - f_y\mathbf{j} + \mathbf{k}$$

is a vector normal to the surface. Thus, the normal to the surface given explicitly by $z = f(x, y)$ is

$$\mathbf{N} = -f_x\mathbf{i} - f_y\mathbf{j} + \mathbf{k}$$

Making the substitutions $f_x = -g_x/g_z$ and $f_y = -g_y/g_z$ in $\mathbf{N}$ gives $\mathbf{N} = (g_x/g_z)\mathbf{i} + (g_y/g_z)\mathbf{j} + \mathbf{k}$, and scaling this vector by the factor g_z yields $g_x\mathbf{i} + g_y\mathbf{j} + g_z\mathbf{k}$, which is precisely the gradient vector, ∇g. Thus, the gradient of $w = g(x, y, z)$ is normal to the level surface $g(x, y, z) = c$. If this result is taken as fundamental, then the expression for the normal to the surface given explicitly by $z = f(x, y)$ follows from the rearrangement $g(x, y, z) = z - f(x, y) = 0$. Since ∇g is normal to the surface defined by $g(x, y, z) = 0$, a normal to the original surface is then $-f_x\mathbf{i} - f_y\mathbf{j} + \mathbf{k}$, which recovers the result obtained earlier.

Property 2

The previous discussion detailed why the gradient is perpendicular to the level curves of $z = f(x, y)$ or to the level surfaces of $w = g(x, y, z)$. To show that the gradient points in the direction of increasing function values, compute the directional derivative along the direction defined by the gradient vector. The appropriate unit vector $\mathbf{U}$, obtained by normalizing ∇f, is $\mathbf{U} = \frac{\nabla f}{\|\nabla f\|}$. Then the directional derivative in this direction is $D_{\mathbf{U}} f = \nabla f \cdot \frac{\nabla f}{\|\nabla f\|} = \frac{\|\nabla f\|^2}{\|\nabla f\|} > 0$. Since the directional derivative of f in the direction of the gradient is positive, f must be increasing in the direction of the gradient vector. Of course, this demonstration is valid for $f(x, y)$ or $f(x, y, z)$.

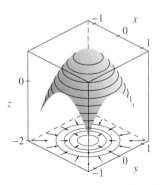

FIGURE 18.7   Example 18.4: Level curves and gradient vectors for $z = 1 - x^2 - y^2$

EXAMPLE 18.4   The function $z = f(x, y) = 1 - x^2 - y^2$ defines the surface in Figure 18.7 where the level curves (circles) are projected to the xy-plane on which gradient vectors are also drawn. Function values increase as the radii of the level curves decrease. The gradient vectors point "inward" since that is the direction of increasing function values.   ❖

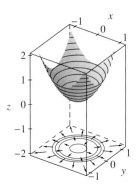

FIGURE 18.8   Example 18.5: Level curves and gradient vectors for $z = x^2 + y^2$

EXAMPLE 18.5   The function $g(x, y) = x^2 + y^2$ has the same contour plot as $f(x, y)$ from Example 18.4 but gradient vectors pointing in the opposite direction. Again, the gradient vectors point in the direction of increasing function values, but here, that direction is "outward." A plot showing the surface, the contours, and the outward-pointing gradient arrows is shown in Figure 18.8. ❖

Properties 3 and 4

The rate of change of the function $f(x, y, z)$, computed in the direction $\mathbf{U}$, is the directional derivative $D_{\mathbf{U}} f = \nabla f \cdot \mathbf{U}$. Expressing the dot product as

$$\nabla f \cdot \mathbf{U} = \|\nabla f\| \|\mathbf{U}\| \cos \theta = \|\nabla f\| 1 \cos \theta = \|\nabla f\| \cos \theta$$

shows the directional derivative is greatest when $\theta = 0$ so that $\cos \theta = 1$. But this means the directional derivative of f is greatest when it is taken exactly along the gradient vector. Hence, the maximal rate of change of f is along the gradient.

When $\theta = 0$ so that the directional derivative is being taken precisely along the gradient vector, the value of the maximal directional derivative is then $\nabla f \cdot \mathbf{U} = \|\nabla f\|$.

EXAMPLE 18.6   Consider the function $f(x, y) = 2x^2 + 5y^2$ whose level curves are ellipses and whose gradient vectors point "outward" in the direction of increasing values of f, as shown in Figure 18.9.

Now, pick a point on the level curve $f(x, y) = 2$. If $x = \frac{2}{3}$, the corresponding y coordinates are found by solving $f(\frac{2}{3}, y) = 2$ for y. This yields the two possibilities $y = \pm \frac{1}{3}\sqrt{2}$. Pick $(\frac{2}{3}, \frac{\sqrt{2}}{3})$, and at this point compute the directional derivative of f, using an arbitrary unit vector $\mathbf{U}$ for the direction. First, compute $\nabla f = 4x\mathbf{i} + 10y\mathbf{j}$, and evaluate it at the fixed point on the level curve $f = 2$, obtaining $\nabla f(\frac{2}{3}, \frac{\sqrt{2}}{3}) = \frac{1}{3}(8\mathbf{i} + 10\sqrt{2}\mathbf{j})$. Then, define the unit vector $\mathbf{U} = \mathbf{i} \cos t + \mathbf{j} \sin t$. As t varies through the interval $[0, 2\pi]$, $\mathbf{U}$ points in every direction around the compass. The directional derivative in any of these arbitrary directions is therefore $D_{\mathbf{U}} f = \frac{8}{3} \cos t + \frac{10}{3}\sqrt{2} \sin t$.

Figure 18.10, a graph of the directional derivative as a function of t, shows the dependence of the directional derivative on the direction of $\mathbf{U}$. Moreover, the directional derivative clearly has a maximum value that can be found by differentiation. Thus, the value of t at which this maximal value occurs is $t_{\max} = \arctan \frac{5}{4}\sqrt{2} \doteq 1.056$. The direction vector

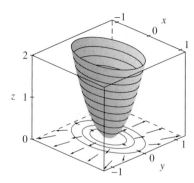

FIGURE 18.9   Example 18.6: Level curves and gradient vectors for $z = 2x^2 + 5y^2$

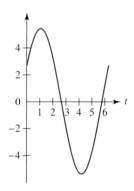

FIGURE 18.10   Example 18.6: At $(\frac{2}{3}, \frac{\sqrt{2}}{3})$, the directional derivative as a function of the angle t

$\mathbf{U} = \frac{1}{\sqrt{33}}(2\sqrt{2}\mathbf{i} + 5\mathbf{j})$ corresponding to this maximizing direction is supposed to be the direction of the gradient vector. However, $\mathbf{U}_{max}$ is a unit vector while the gradient vector is not. Therefore, normalize the gradient, obtaining $\frac{\nabla f}{\|\nabla f\|} = \frac{1}{\sqrt{33}}(2\sqrt{2}\mathbf{i} + 5\mathbf{j}) = \mathbf{U}$, as expected.

❖

EXERCISES 18.3–Part A

A1. Let P be the point $(x, y) = (1, 2)$ and define $f(x, y) = 3x^2 + 2xy + 5y^2$.

   (a) Obtain a formula for, and a sketch of, the level curve of $f(x, y)$ that passes through the point P.

   (b) At P, obtain a vector tangent to the level curve found in part (a).

   (c) At P, obtain the gradient vector, and show it is perpendicular to the tangent vector found in part (b).

   (d) Find the coordinates of the tip of the gradient vector.

   (e) By comparing the values of f at both the tail and tip of the gradient vector obtained in part (c), show that this vector points in the direction of increasing values of f.

   (f) At P, obtain the value of the directional derivative in the direction of the gradient.

   (g) At the point corresponding to P, find a vector normal to the surface $z = f(x, y)$.

A2. Let P be the point $(2, 1, -1)$ and define $g(x, y, z) = 2x^2 + 3y^2 + 4z^2$.

   (a) Obtain a formula for, and a sketch of, the level surface of $g(x, y, z)$ that passes through the point P.

   (b) At P, obtain the gradient vector. Find the coordinates of the tip of this vector.

   (c) Show that the gradient in part (b) points in the direction of increasing values of g.

   (d) At P, obtain the value of the directional derivative in the direction of the gradient.

EXERCISES 18.3–Part B

B1. If $\mathbf{R} = x\mathbf{i} + y(x)\mathbf{j}$ is an implicitly defined level curve of $f(x, y)$, show that a tangent vector along $\mathbf{R}$ is given by $\mathbf{R}' = \mathbf{i} - (f_x/f_y)\mathbf{j}$. *Hint:* Differentiate $f(x, y(x)) = c$ implicitly to get y'.

In Exercises B2–11, a function $f(x, y)$ and a point P are given. For each:

   (a) Obtain a contour plot of $f(x, y)$, $-2 \le x, y \le 2$.

   (b) On the contour plot found in part (a), superimpose a field plot of ∇f, the gradient field.

   (c) Obtain the coordinates of the tail and head of $\nabla f(P)$.

   (d) Obtain the vector $\mathbf{R}'(P)$, tangent to the level curve through P. (See Exercise B1.)

   (e) Show that $\nabla f(P)$ is perpendicular to $\mathbf{R}'(P)$.

   (f) Obtain, at point P, and in the direction of $\mathbf{U}(\theta) = \cos\theta\mathbf{i} + \sin\theta\mathbf{j}$, $0 \le \theta \le 2\pi$, $\delta(\theta) = D_{\mathbf{U}}(f)(P)(\theta)$, the directional derivative of f.

   (g) Plot as a function of θ, the directional derivative, $\delta(\theta)$.

   (h) Calculate $\hat{\theta}$, the value of θ for which $\delta(\theta)$, the directional derivative, is a maximum.

   (i) Obtain $\mathbf{U}(\hat{\theta})$, the unit vector in the direction of maximum value for the directional derivative.

   (j) Normalize $\nabla f(P)$ and compare to $\mathbf{U}(\hat{\theta})$.

   (k) Compute $\delta(\hat{\theta})$, the maximum value of the directional derivative $\delta(\theta)$, and compare to $\|\nabla f(P)\|$.

B2. $f(x, y) = 4x - 4x^2y + 2x^3$, $P = (8, -10)$

B3. $f(x, y) = 6y - y^3 - 6x^3$, $P = (0, 1)$

B4. $f(x, y) = 3x + y^2$, $P = (-9, -9)$

B5. $f(x, y) = 9 + 2x^2 - 7y$, $P = (8, -11)$

B6. $f(x, y) = 2x^2y - 2 - 8y$, $P = (11, -5)$

B7. $f(x, y) = \dfrac{7x^2 - 9y^3 + xy}{2 - 2x^2y + xy}$, $P = (4, -1)$

B8. $f(x, y) = \dfrac{8x - 2x^2y + 2x^3}{9x^3 - 5y}$, $P = (3, -2)$

B9. $f(x, y) = \dfrac{8x^2 - 4y^2}{6xy^2 - 9y^2 + 3x^3}$, $P = (-4, -2)$

B10. $f(x, y) = \dfrac{2y - 9x^2y - 4xy^2}{9x^2 - x^2y - y^3}$, $P = (-1, 7)$

B11. $f(x, y) = \dfrac{2 - 5x^2 - 9xy}{5(y - x^2 - y^3)}$, $P = (5, 5)$

B12. Let $g(x, y, z) = 0$ implicitly define a surface $z = z(x, y)$. On this surface, the plane section $x = \alpha$ is the curve $\mathbf{R}_y = \alpha\mathbf{i} + y\mathbf{j} + z(\alpha, y)\mathbf{k}$, while the plane section $y = \beta$ is the plane section $\mathbf{R}_x = x\mathbf{i} + \beta\mathbf{j} + z(x, \beta)\mathbf{k}$. Show that $\mathbf{T}_x = \mathbf{i} - (g_x/g_z)\mathbf{k}$ and $\mathbf{T}_y = \mathbf{j} - (g_y/g_z)\mathbf{k}$ are vectors tangent to $\mathbf{R}_x$ and $\mathbf{R}_y$, respectively.

B13. Show that $\mathbf{U}(\theta, \phi) = \cos\theta \sin\phi\mathbf{i} + \sin\theta \sin\phi\mathbf{j} + \cos\phi\mathbf{k}$,
$0 \le \theta \le 2\pi, 0 \le \phi \le \pi$, is a unit vector with tail at the origin and head on the unit sphere.

In Exercises B14–25:

   (a) Obtain $\nabla g(P)$.

   (b) Obtain the coordinates of the tail and head of $\nabla g(P)$.

   (c) Using the result of Exercise B12, obtain $\mathbf{T}_x(P)$ and $\mathbf{T}_y(P)$, independent vectors tangent to the surface at point P.

   (d) Show that $\nabla g(P)$ is orthogonal to both $\mathbf{T}_x(P)$ and $\mathbf{T}_y(P)$.

   (e) With $\mathbf{U}(\theta, \phi)$ as defined in Exercise B13, obtain, at point P, the directional derivative $\delta(\theta, \phi) = D_{\mathbf{U}}(g)(P)(\theta, \phi)$.

   (f) Plot, as a surface over $0 \le \theta \le 2\pi, 0 \le \phi \le \pi$, the directional derivative $\delta(\theta, \phi)$.

   (g) Calculate, using the techniques of multivariable calculus, $\hat{\theta}$ and $\hat{\phi}$, the values of θ and ϕ that maximize $|\delta(\theta, \phi)|$.

   (h) Obtain $\mathbf{U}(\hat{\theta}, \hat{\phi})$, the unit vector in the direction of maximum absolute value for the directional derivative.

   (i) Normalize $\nabla g(P)$ and compare to $\mathbf{U}(\hat{\theta}, \hat{\phi})$.

   (j) Compute $|\delta(\hat{\theta}, \hat{\phi})|$, the maximum absolute value of the directional derivative $\delta(\theta, \phi)$, and compare to $\|\nabla g(P)\|$.

B14. $g(x, y, z) = 6y^2z + 8xy^2 + 4xz, P = (-11, -2, 3)$

B15. $g(x, y, z) = 9x^2 + 8yz + 9yz^2, P = (6, -9, -9)$

B16. $g(x, y, z) = 5z - 4x^2 + 3xy^2, P = (3, 1, 2)$

B17. $g(x, y, z) = 4y - 2z + 4z^3, P = (11, -6, 3)$

B18. $g(x, y, z) = y + 6x^2 + 5x^2z, P = (7, 8, -5)$

B19. $g(x, y, z) = 3y^2 + 7yz^2 + 9xz^2, P = (0, 3, -8)$

B20. $g(x, y, z) = \dfrac{x^2y + 3x^3}{4z + 6z^2 - 7yz}, P = (-5, -1, -1)$

B21. $g(x, y, z) = \dfrac{2xy^2 + 8yz^2}{6y^2z - 7y^3 - 9x^2z}, P = (-3, 7, 4)$

B22. $g(x, y, z) = \dfrac{4y - 2xy^2 + 3xy}{3y^2 + 4z^2 - 8yz^2}, P = (4, 8, -3)$

B23. $g(x, y, z) = \dfrac{8x^2 + xy + 2yz^2}{2y^3 - 4z}, P = (9, 2, 10)$

B24. $g(x, y, z) = \dfrac{4 + 4xy + xz}{2z^2 - 6z^3 + 6xz^2}, P = (-5, 2, 9)$

B25. $g(x, y, z) = \dfrac{7x - 9x^2 + 2xz^2}{6x^2z - y^2 + yz}, P = (3, 11, -6)$

18.4 | Lagrange Multipliers

Constrained Optimization

Elementary calculus considers constrained optimization problems of the form

find the rectangular box of maximum area if the perimeter is fixed at 100.

Initially, these problems are solved by eliminating one of the variables. For example, with the area of the box written as $A = xy$ and the perimeter constraint as $2x + 2y = 100$, the constraint can be solved for $y = 50 - x$ so that $A = A(x) = x(50 - x)$. The ordinary techniques of differentiation (or plotting) now lead to the maximum of 625 at $x = y = 25$.

The Lagrange Multiplier Method

The Lagrange multiplier technique is an alternate method for solving constrained optimization problems. By making an elegant application of the gradient vector, it avoids the algebra of using the constraint to eliminate one of the variables. The method of Lagrange multipliers is illustrated through the following five examples.

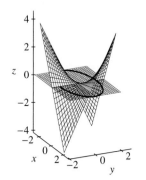

FIGURE 18.11   Objective function and constraint in Example 18.7

EXAMPLE 18.7   In this first example, we show how the Lagrange multiplier method selects points at which the level curves of the objective function $f(x, y)$ are tangent to the constraint curve. To do this, we will find the extreme value of $f(x, y) = xy$ along the constraining ellipse $g(x, y) = x^2 + 4y^2 - 8 = 0$.

Figure 18.11 shows the constraint ellipse in the xy-plane, along with the surface $z = f(x, y)$. By way of interpretation, imagine being restricted to walking on this ellipse in the xy-plane. Overhead, the function $f(x, y)$ determines the shape of the ceiling. You want to

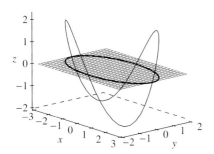

FIGURE 18.12 Example 18.7: Constraint ellipse and curve of intersection of the surface $z = f(x, y)$ and the cylinder whose cross-section is the constraint

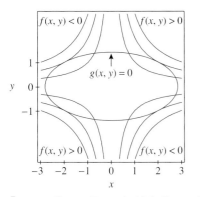

FIGURE 18.13 Example 18.7: Constraint ellipse and level curves of $f(x, y)$

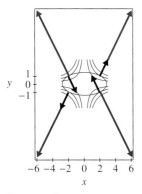

FIGURE 18.14 Example 18.7: Gradients (in color) along constraint ellipse, and gradients (in black) along level curves of $f(x, y)$ that are tangent to the constraint

know where, on the elliptic path being walked, you will find the highest and lowest points of the ceiling.

The intersection of the cylinder $g(x, y) = 0$ and the surface $z = f(x, y)$ is the space curve (thin black) shown in Figure 18.12. Included is a rendition of the ellipse (thick black) in the xy-plane. Clearly, there are two maxima and two minima.

The contour map in Figure 18.13 is yet another way of looking at this problem. The level curves of $f(x, y)$ and the constraint ellipse are all drawn in the xy-plane. Where a level curve of $f(x, y)$ cuts through the graph of $g = 0$, there can't be a stationary value. Where the level curve of $f(x, y)$ is tangent to the graph of $g = 0$, the value of $f(x, y)$ becomes stationary because it "pauses" momentarily.

The method of *Lagrange multipliers* seeks points where the level curves of f are tangent to the constraint curve $g = 0$. It does this by looking for points where the gradient vectors of f and g are colinear. This happens where the gradient $\nabla f = y\mathbf{i} + x\mathbf{j}$ is a multiple of the gradient $\nabla g = 2x\mathbf{i} + 8y\mathbf{j}$. Figure 18.14 shows the gradient vectors at the four points where they are colinear. The companion Maple worksheet contains a complete animation of the gradient vectors as the constraint ellipse is traversed. The animation shows two things. First, ∇f (in black) never equals ∇g (in color), which is considerably longer than ∇f. Second, at two points of tangency, ∇f points in the same direction as ∇g, but at the other two points of tangency, the two gradients point in exactly opposite directions.

The analytic condition that expresses the colinearity of the gradients is $\nabla f = \lambda \nabla g$, which stands for the pair of component equations

$$f_x = \lambda g_x \quad \text{and} \quad f_y = \lambda g_y \tag{18.8}$$

The factor of proportionality, λ, is called the *Lagrange multiplier*. There are three unknowns, namely, the coordinates x and y and the Lagrange multiplier λ. These are determined by the two equations in (18.8) and the constraint equation $g(x, y) = 0$. Table 18.1 lists the equations, the solutions, and the value of $f(x, y)$ at each extreme point. Hence, there are two places on the constraint curve $g(x, y) = 0$ where $f(x, y)$ attains a maximum and two places on the constraint curve where $f(x, y)$ attains a minimum. ❖

Equations	λ	(x, y)	$f(x, y)$
$x = 8\lambda y$	$\frac{1}{4}$	$(2, 1)$	2
$y = 2\lambda x$	$\frac{1}{4}$	$(-2, -1)$	2
$x^2 + 4y^2 = 8$	$-\frac{1}{4}$	$(2, -1)$	-2
	$-\frac{1}{4}$	$(-2, 1)$	-2

TABLE 18.1 Example 18.7: Equations $\nabla f = \lambda \nabla g$ and $g = 0$, critical points, and value of f at the critical points

The solutions in Table 18.1 can be obtained by dividing the left and right sides of the first two equations on the left in Table 18.1. This eliminates λ and gives $x^2 = 4y^2$. If this equation is now used to eliminate x in the constraint equation, we obtain $y^2 = 1$, so $y = \pm 1$ and, hence, $x = \pm 2$.

EXAMPLE 18.8 In this second example, we will find the Lagrange multiplier is zero, so an extreme point occurs where the constraint is not operative. The extreme point is on the constraint, but it is also a critical point for the unconstrained objective function. To illustrate this, we find the extreme values of $f(x, y) = x^2 y$ subject to the constraint $g(x, y) = x + y - 3 = 0$.

Figure 18.15 shows the surface $z = f(x, y)$, the xy-plane, constraint (line) $g(x, y) = 0$, and the intersection of the plane $x + y = 3$ with the surface $z = f(x, y)$. The space curve in which the plane $x + y = 3$ intersects the surface $z = f(x, y)$ is shown in Figure 18.16. It suggests $f(x, y)$ has extreme values at $x = 0$ and $x = 2$ along the constraint line. Figure 18.17 shows a contour plot of the surface $z = f(x, y)$ and a graph of the constraint line $g(x, y) = 0$. This view is significant since it shows but a single point of tangency at about $(2, 1)$.

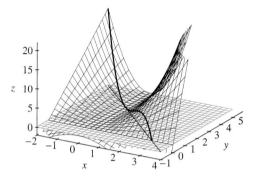

FIGURE 18.15   Example 18.8: Constraint line, surface $z = f(x, y)$, and intersection of plane $x + y = 3$ with this surface

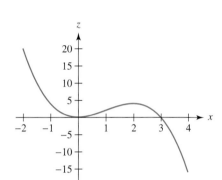

FIGURE 18.16   Example 18.8: Intersection of plane $x + y = 3$ and $z = f(x, y)$ drawn in the plane $x + y = 3$

Table 18.2 lists the equations $\nabla f = \lambda \nabla g$ and $g(x, y) = 0$, the solutions for λ, x, and y, and the values of $f(x, y)$ at the extreme points. When $\lambda = 0$, the constraint does not apply. The corresponding extreme point is a critical point for the unconstrained optimization problem and would have been found without the constraint. That is why Figure 18.17 shows but one point of tangency between the constraint line and the level curves of the objective function $f(x, y)$.   ❖

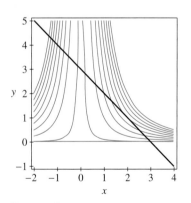

FIGURE 18.17   Example 18.8: Constraint line and level curves of $f(x, y)$

Equations	λ	(x, y)	$f(x, y)$
$2xy = \lambda$	4	$(2, 1)$	4
$x^2 = \lambda$	0	$(0, 3)$	0
$x + y = 3$			

TABLE 18.2   Example 18.8: Equations $\nabla f = \lambda \nabla g$ and $g = 0$, critical points, and value of f at the critical points

◆◆ **EXAMPLE 18.9**   In this third example, the objective function must be constructed from a verbal description provided by the problem statement, namely, the requirement to find the (shortest) distance from the origin to the plane $2x + y - z = 5$.

Thus, the quantity to be minimized is the distance from the origin to a point (x, y, z) on the plane. It's generally easier, however, to minimize the *square* of the distance. We therefore have $f(x, y, z) = x^2 + y^2 + z^2$, with $g(x, y, z) = 2x + y - z - 5 = 0$, the equation of the plane, as the constraint. There are three equations in $\nabla f = \lambda \nabla g$ and four unknowns, namely, x, y, z, and λ. The fourth equation is the constraint equation itself. Table 18.3 lists the four equations and their single solution.

Equations	λ	(x, y, z)	$\sqrt{f(x, y, z)}$
$2x = 2\lambda$	$\frac{5}{3}$	$\left(\frac{5}{3}, \frac{5}{6}, -\frac{5}{6}\right)$	$\frac{5}{\sqrt{6}}$
$2y = \lambda$			
$2z = -\lambda$			
$2x + y - z = 5$			

TABLE 18.3 Example 18.9: Equations $\nabla f = \lambda \nabla g$ and $g = 0$, critical point, and value of $\sqrt{f}$ at the critical point

EXAMPLE 18.10 A pentagon is formed from a rectangle surmounted by an isosceles triangle. What dimensions give the pentagon least perimeter if the area is fixed at the value a? (See Figure 18.18.) In this fourth example, the constraint contains a symbolic parameter and the algebra becomes significantly more complicated. The point of the example is to show how to navigate through such complexity.

From Figure 18.18 the area of the rectangle is $2xy$, the area of the triangle is $2(\frac{1}{2}xy)$, the total area is $g(x, y, z) = 2xy + xz = a$, and the perimeter is $f(x, y, z) = 2\sqrt{x^2 + z^2} + 2x + 2y$. The objective function is the perimeter. The constraint is the fixed area. Table 18.4 lists the four equations arising from $\nabla f = \lambda \nabla g$ and the constraint $g = a$. There are four possible solutions for x, y, z, and λ; computed exactly in Maple, they are listed in floating-point form in Table 18.4. Only one solution gives all three dimensions as positive.

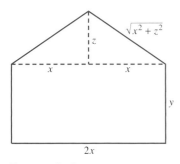

FIGURE 18.18 Example 18.10: Pentagon whose perimeter is to be minimized while its area remains fixed at a

Equations	λ	(x, y, z)	$f(x, y, z)$
$\dfrac{2x}{\sqrt{x^2 + z^2}} + 2 = \lambda(2y + z)$	$\dfrac{0.52}{\sqrt{a}}$	$\left(1.9\sqrt{a}, 0.82\sqrt{a}, -1.1\sqrt{a}\right)$	
$2 = 2\lambda x$	$-\dfrac{0.52}{\sqrt{a}}$	$\left(-1.9\sqrt{a}, -0.82\sqrt{a}, 1.1\sqrt{a}\right)$	
$\dfrac{2z}{\sqrt{x^2 + z^2}} = \lambda x$	$\dfrac{1.9}{\sqrt{a}}$	$\left(0.52\sqrt{a}, 0.82\sqrt{a}, 0.30\sqrt{a}\right)$	$3.86\sqrt{a}$
$2xy + xz = a$	$-\dfrac{1.9}{\sqrt{a}}$	$\left(-0.52\sqrt{a}, -0.82\sqrt{a}, -0.30\sqrt{a}\right)$	

TABLE 18.4 Example 18.10: Equations $\nabla f = \lambda \nabla g$ and $g = 0$, critical points, and value of f at the one feasible critical point

The third solution, given exactly as $(\beta, (1 + \frac{1}{\sqrt{3}})\beta, \frac{\beta}{\sqrt{3}})$, with $\beta = \sqrt{a(2 - \sqrt{3})}$, is the physically meaningful one. The minimum value of the perimeter can also be given exactly as $(1 + \sqrt{3})\sqrt{2a}$. It is surprising how much the presence of the symbolic parameter a complicates the algebra. But the reader should not let the additional complexity in the algebra obscure the underlying simplicity of the basic technique of Lagrange multipliers.

EXAMPLE 18.11   In this fifth example we extend the method to the case of two constraints by finding the shortest distance from the origin to the intersection of the two planes

$$g_1 = 2x - 3y + 5z = 9 \quad \text{and} \quad g_2 = 6x + y - 7z = 12$$

We give two solutions, the first with the use of Lagrange multipliers, and the second, without. For both solutions, the objective function is $f(x, y, z) = x^2 + y^2 + z^2$, the square of the distance from the origin to the point (x, y, z). As we will explain, the Lagrange multiplier method generalizes to $\nabla f = \lambda_1 \nabla g_1 + \lambda_2 \nabla g_2$, so now, there are five equations in the five unknowns x, y, z, λ_1, and λ_2. Alternatively, the function $F = f - \sum_{k=1}^{2} \lambda_k g_k$ could be defined and the same set of equations obtained from $\nabla F = \mathbf{0}$. The equations and their single solution are listed in Table 18.5. The point closest to the origin is denoted by P.

Equations	(λ_1, λ_2)	(x, y, z)	$\sqrt{f(x, y, z)}$
$2x = 2\lambda_1 + 6\lambda_2$	$\left(\frac{181}{216}, \frac{115}{216}\right)$	$\left(\frac{263}{108}, -\frac{107}{108}, \frac{25}{108}\right)$	$\frac{\sqrt{1003}}{12}$
$2y = -3\lambda_1 + \lambda_2$			
$2z = 5\lambda_1 - 7\lambda_2$			
$2x - 3y + 5z = 9$			
$6x + y - 7z = 12$			

TABLE 18.5   Example 18.11: Equations $\nabla f = \lambda_1 \nabla g_1 + \lambda_2 \nabla g_2$, $g_1 = 0$ and $g_2 = 0$, the critical point, and value of $\sqrt{f}$ at the critical point

A second solution to this problem can be framed within the confines of simple multivariable calculus operations. First, a vector along the line of intersection of the two constraint planes is given by the cross-product of normals to the two planes. Thus, we compute the vector $\mathbf{V} = \nabla g_1 \times \nabla g_2 = 16\mathbf{i} + 44\mathbf{j} + 20\mathbf{k}$, which lies along the line of intersection of the constraint planes. This vector is perpendicular to ∇f at the point P.

Hence, the line of intersection of the constraint planes is tangent to the level surface determined by the objective function, $f(x, y, z)$. This is the analog of the geometry that applies in the case of the objective function $f(x, y)$ constrained by a single constraint curve $g(x, y) = 0$.

To find a point on the line of intersection of the two constraint planes, set $z = 0$ in both planes to find where the line passes through the xy-plane. The solution of the equations

$$2x - 3y = 9 \qquad 6x + y = 12$$

is found to be $x = \frac{9}{4}$, $y = -\frac{3}{2}$, so the line of intersection can be given parametrically by the vector

$$\mathbf{R}(t) = \left(\tfrac{9}{4} + 16t\right)\mathbf{i} + \left(44t - \tfrac{3}{2}\right)\mathbf{j} + 20t\mathbf{k}$$

If we evaluate $f(x, y, z)$ along this line, we obtain $f(t) = \frac{117}{16} - 60t + 2592t^2$, and the techniques of elementary differential calculus, setting the derivative to zero and solving, gives the critical value of t as $\frac{5}{432}$ and $\mathbf{R}\left(\frac{5}{432}\right) = P$.

Figure 18.19 shows a sketch of the sphere $x^2 + y^2 + z^2 = \frac{1003}{144}$, a segment of the line $\mathbf{R}(t)$, and the vectors $\nabla f, \nabla g_1$, and ∇g_2, all drawn at point P. (The segment of the line $\mathbf{R}(t)$ is drawn in color.) The line of intersection of the two constraining planes is tangent to the level surface determined by the objective function $f(x, y, z)$. The plane spanned by the

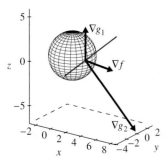

FIGURE 18.19   Example 18.11: Sphere $x^2 + y^2 + z^2 = \frac{1003}{144}$, $\mathbf{R}(t)$ (the line of intersection of the constraints $g_1 = 0$ and $g_2 = 0$), and the vectors $\nabla f, \nabla g_1$, and ∇g_2

gradients ∇g_1 and ∇g_2 is orthogonal to this line. The vector ∇f lies in the plane spanned by the gradients ∇g_1 and ∇g_2. The Lagrange multiplier condition $\nabla f = \lambda_1 \nabla g_1 + \lambda_2 \nabla g_2$ finds points at which the gradient of f lies in the plane spanned by the gradients of g_1 and g_2. At such points, the curve of intersection of the constraints is tangent to a level surface of the objective function f. ❖

EXERCISES 18.4–Part A

A1. Use the Lagrange multiplier technique to find the shortest distance from the point $(2, -1, 3)$ to the plane $3x - 5y + 7z = 15$.

A2. Use the Lagrange multiplier technique to find the shortest distance from the point $(5, 2, -4)$ to the intersection of the planes $x - 2y + 5z = 12$ and $5x + 8y - 2z = 9$.

EXERCISES 18.4–Part B

B1. Using the Lagrange multiplier technique, find the minimum distance from the point $(1, -2, 3)$ to the plane $3x - 5y + 7z = 4$ and find the point on the plane where this minimum distance occurs. Show that the vector from $(1, -2, 3)$ to the distance-minimizing point on the plane is perpendicular to the plane.

B2. Find the minimum distance from the point $(-3, 1, 5)$ to the line $x(t) = 3 - 4t$, $y(t) = -2 + 5t$, $z(t) = 6 + 11t$, and find the coordinates of the point on the given line at which this minimum distance occurs. Show that the vector from $(-3, 1, 5)$ to the distance-minimizing point on the line is perpendicular to the line.

B3. Find the minimum distance between the two skew lines $x(t) = 7 + 4t$, $y(t) = -3 + 2t$, $z(t) = -1 - 5t$, and $x(t) = -\frac{1}{2}(5 + 3t)$, $y(t) = \frac{1}{3}(7 + 8t)$, $z(t) = -3 + 5t$; find the points on the lines where this minimum distance occurs. Show that the vector connecting the distance-minimizing points on the two lines is simultaneously perpendicular to each line.

B4. Using the Lagrange multiplier technique, find the point on the curve $x^2 y = 2$ that is closest to the origin. Find the minimum distance. Obtain a graph of the level curves of the objective function $f(x, y)$ that is to be minimized, and on this graph superimpose the graph of the constraint curve.

B5. Using the Lagrange multiplier technique, find the points on the curve $x^2 + xy + y^2 = 1$ closest to and farthest from the origin.

Obtain a graph of the level curves of the objective function $f(x, y)$ that is to be minimized, and on this graph superimpose the graph of the constraint curve.

B6. Using the Lagrange multiplier technique, find the point on the surface $xy + 1 - z = 0$ that is closest to the origin.

B7. Find the extrema of $f(x, y) = 4x + y$ on the constraint $g(x, y) = 2x^2 + 3y^2 - 1 = 0$. Sketch the level curves of $f(x, y)$ and superimpose the graph of $g(x, y) = 0$.

B8. Find the distance from the point $(1, 2, -3)$ to the plane $3x + 2y - 3z = 5$.

B9. Find the extrema of $f(x, y, z) = 2x + y + 2z$ subject to the constraints $g_1(x, y, z) = x^2 + y^2 - 4 = 0$ and $g_2(x, y, z) = x + z - 2 = 0$.

B10. Find the extrema of $f(x, y, z) = 2x + y^2 + 5z$ subject to the constraint $g(x, y, z) = 9x^2 + 4y^2 - 25z^2 - 16 = 0$.

B11. Find the point(s) on $3x^2 + 2y^2 + 2xy + x + y = 16$ closest to the origin.

B12. Find the points on the intersection of $2x - y = 1$ and $2x^2 + y^2 + 2z^2 = 1$ that are closest to and farthest from the origin.

B13. Find the distance from $(-1, 2, 3)$ to the line of intersection of the planes $x + 2y - 3z = 4$ and $2x - y + 2z = 5$.

18.5 Conservative Forces and the Scalar Potential

The Conservative Force

A force $\mathbf{F}$ is called *conservative* if $\mathbf{F} = -\nabla u$ for a scalar function u called the *scalar potential*.

  Inclusion of the minus sign makes the scalar potential u agree with the potential energy function in physics.

  Section 21.5 presents several other definitions of the conservative force and shows the properties that characterize them to be equivalent.

EXAMPLE 18.12   Let the force field $\mathbf{F} = \frac{A}{x^2}\mathbf{i}$, $A > 0$, be generated by a unit positive charge at the origin on an x-axis. The potential energy function is defined as the work it takes to move another unit positive charge from infinity to the point $x > 0$ and is given by

$$u(x) = \int_{\infty}^{x} -\|\mathbf{F}\|\, d\tau = \int_{\infty}^{x} \left(-\frac{A}{\tau^2}\right) d\tau = \frac{A}{x} \qquad (18.9)$$

The force generated by the charge at the origin is directed radially outward. Along the axis, the force that the fixed charge exerts on the moving charge is directed to the right, so the force needed to push the moving charge toward the origin points to the left. That is the reason for the minus sign in the integral in (18.9). Thus, the field $\mathbf{F}$ arises not as $\nabla u = -\frac{A}{x^2}\mathbf{i}$, but rather, as $-\nabla u = \frac{A}{x^2}\mathbf{i}$. ❖

Often, it is mathematically convenient to characterize the conservative field $\mathbf{F}$ as one that is "the gradient of a scalar potential," without explicit mention of the physicist's minus sign. The resolution is that the physicist's potential u could easily be the mathematician's potential $U = -u$. Then, the physicist's $\mathbf{F} = -\nabla u$ is the same as the mathematician's $\mathbf{F} = \nabla U$.

EXAMPLE 18.13   In addition to the positive charge at the origin in Example 18.12, assume a similar charge, placed at the point $(5, 5)$ in the xy-plane. In two dimensions, the charges at the origin and at $(5, 5)$ have the respective potentials

$$u(x, y) = \frac{1}{\sqrt{x^2 + y^2}} \quad \text{and} \quad v(x, y) = \frac{1}{\sqrt{(x-5)^2 + (y-5)^5}}$$

The potential for the system is then the sum

$$U(x, y) = u(x, y) + v(x, y) = \frac{1}{\sqrt{x^2 + y^2}} + \frac{1}{\sqrt{(x-5)^2 + (y-5)^2}}$$

The potential surface for this system is seen in Figure 18.20; and the equipotentials, the level curves of this surface, appear in Figure 18.21. The contours show that at a sufficient distance away from the two charges, they begin to resemble one charge.

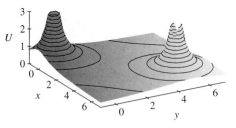

FIGURE 18.20   Potential function for Example 18.13

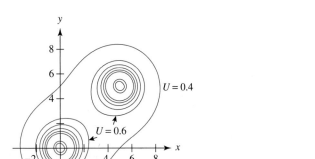

FIGURE 18.21   Equipotentials for Example 18.13. The values on the contours are $U = 0.4, 0.6, 0.7, 0.8, 0.9, 1, 2, 3$, with the smallest value occurring on the outermost contour.

The force field generated by the two charges is then

$$F = -\nabla U = \begin{bmatrix} \dfrac{x}{(x^2+y^2)^{3/2}} + \dfrac{x-5}{\left((x-5)^2+(y-5)^2\right)^{3/2}} \\[2ex] \dfrac{y}{(x^2+y^2)^{3/2}} + \dfrac{y-5}{\left((x-5)^2+(y-5)^2\right)^{3/2}} \end{bmatrix}$$

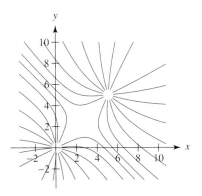

FIGURE 18.22   Flow lines for the field in Example 18.13

and if $\mathbf{R} = x(p)\mathbf{i} + y(p)\mathbf{j}$, solving the system of equations $\mathbf{R}'(p) = \mathbf{F}(x(p), y(p))$ gives the phase portrait seen in Figure 18.22. The points $(0, 0)$ and $(5, 5)$, where the charges are located, are singular points of the differential equations. The derivatives x' and y' are infinite at those two points, making the system of differential equations very difficult to solve numerically.

There is an equilibrium point at $(\frac{5}{2}, \frac{5}{2})$ since $\mathbf{F}(\frac{5}{2}, \frac{5}{2}) = \mathbf{0}$. The linearization technique of Section 12.22 can be applied here to verify that $(\frac{5}{2}, \frac{5}{2})$ is a saddle point. At $(\frac{5}{2}, \frac{5}{2})$, the Jacobian matrix for the system is $-\frac{\sqrt{2}}{25}\begin{bmatrix} 2 & 6 \\ 6 & 2 \end{bmatrix}$, with eigenvalues $-\frac{8}{125}\sqrt{2}$, $\frac{4}{125}\sqrt{2}$. Since the eigenvalues are of different signs, the equilibrium point is indeed a saddle.   ❖

Recovering the Potential

It is sometimes possible to construct the scalar potential from a given vector field. In two dimensions, the test for the existence of a scalar potential is essentially the straightforward test for the exactness of a first-order differential equation. (See Section 3.3.) In three dimensions, the full characterization of conservative fields is given in Section 21.5.

If $\mathbf{F} = f(x, y)\mathbf{i} + g(x, y)\mathbf{j} = \nabla U$, then, by the equality of the mixed second partial derivatives of U, that is, since $U_{xy} = U_{yx}$ must hold, then so too must $f_y = g_x$ also hold. Applying that test to the scalar potential of Example 18.13 gives

$$-3\frac{xy}{\left(x^2+y^2\right)^{5/2}} - 3\frac{(x-5)(y-5)}{\left((x-5)^2 + (y-5)^2\right)^{5/2}}$$

for both f_y and g_x, so a potential function exists.

To find the scalar potential $U(x, y)$, start with the equations

$$U_x = f(x, y) \quad \text{and} \quad U_y = g(x, y)$$

Then, determine $U(x, y)$ by integrating with respect to x, the left-hand equation, holding y fixed, thereby making the constant of integration a function of y. This yields

$$\hat{U} = -\frac{1}{\sqrt{x^2+y^2}} - \frac{1}{\sqrt{(x-5)^2 + (y-5)^2)}} + C(y)$$

If this candidate for $U(x, y)$ is differentiated with respect to y, $g(x, y)$ should result. Hence,

$$\hat{U}_y = \frac{y}{\left(x^2 + y^2\right)^{3/2}} + \frac{y-5}{\left((x-5)^2 + (y-5)^2\right)^{3/2}} + C'(y)$$

Matching that with the known result for $g(x, y)$ gives $C'(y) = 0$, so $C(y)$ is at most a constant. The scalar potential is, at best, determined up to an additive constant, and with $C(y) = 0$, $\hat{U} = U$.

Not every vector field comes from a scalar potential. For example, form the vector field

$$\mathbf{G} = \begin{bmatrix} \dfrac{y}{\left(x^2 + y^2\right)^{3/2}} + \dfrac{y-5}{\left((x-5)^2 + (y-5)^2\right)^{3/2}} \\[4mm] -\dfrac{x}{\left(x^2 + y^2\right)^{3/2}} - \dfrac{x-5}{\left((x-5)^2 + (y-5)^2\right)^{3/2}} \end{bmatrix}$$

everywhere orthogonal to the vector field $\mathbf{F}$, by interchanging the components of $\mathbf{F}$ and introducing one minus sign. This new field $\mathbf{G}$ will have its vectors perpendicular to the flow lines of $\mathbf{F}$, so if a scalar potential exists for this new field, the equipotentials of $U(x, y)$ will be its flow lines. That sometimes does happen, but not here, since the "cross-partials" test shows

$$f_y - g_x = -\frac{\left(x^2+y^2\right)^{3/2} + \left((x-5)^2 + (y-5)^2\right)^{3/2}}{\left(x^2+y^2\right)^{3/2}\left((x-5)^2 + (y-5)^2\right)^{3/2}} \neq 0$$

so no scalar potential exists for the field **G**.

A Convention

Usually, the electric force field in physics is normalized so that it gives the force on a unit positive test charge. In physics, the convention for gravitational fields seems to be that each field is defined in terms of the force exerted on the specific mass in question. However, for the sake of uniformity, we will normalize all fields in this text so that a field always gives the force on a unit particle or unit mass.

EXERCISES 18.5–Part A

A1. Use the technique in (18.9) to obtain the potential energy function for the force $\mathbf{F} = x^{-3}\mathbf{i}$.

A2. Determine which, if any, of the following forces are conservative. Find a scalar potential for each conservative force.

   (a) $\mathbf{F} = x^2 y\mathbf{i} - xy^2\mathbf{j}$ **(b)** $\mathbf{F} = (1 + xy)e^{xy}(y\mathbf{i} + x\mathbf{j})$

   (c) $\mathbf{F} = y^2 \cos xy\mathbf{i} + (\sin xy + xy \cos xy)\mathbf{j}$

A3. If $r = \sqrt{x^2 + y^2}$ and $\mathbf{r} = \frac{1}{r}(x\mathbf{i} + y\mathbf{j})$, show that $\mathbf{F} = f(r)\mathbf{r}$ is conservative for any differentiable function $f(r)$.

A4. The force exerted by a spring on an attached object is $\mathbf{F} = -kx\mathbf{i}$, where x is the displacement of the object from the equilibrium position of the spring. Find $w(x)$, the work done in stretching the spring x units from its equilibrium length. Show that $w(x)$ is the potential function for **F**.

A5. A unit positive charge is placed at the origin in an xy-plane. If at (x, y) the charge exerts a force **F** inversely proportional to r^2, find **F** and its scalar potential.

EXERCISES 18.5–Part B

B1. An object weighing 4 lb is raised x ft. Find the work done, and show this gives the scalar potential for the gravitational force acting on the object.

For the scalar potentials given in each of Exercises B2–11, let R be the region $-1 \leq x, y \leq 1$. In R:

   (a) Obtain a plot of the potential surface $z = u(x, y)$.

   (b) Obtain a contour plot of the equipotentials (level curves) of $u(x, y)$.

   (c) Obtain the conservative field $\mathbf{F} = -\nabla u$.

   (d) Find all critical points where $\mathbf{F} = \mathbf{0}$. *Hint:* Since $\mathbf{F} = -u_x(x, y)\mathbf{i} - u_y(x, y)\mathbf{j}, u_x(x, y) = 0$ implicitly defines $y = y_1(x)$, whereas $u_y(x, y) = 0$ implicitly defines $y = y_2(x)$. Any intersections of the curves $y_1(x)$ and $y_2(x)$ are critical points.

   (e) Obtain a field plot for **F**.

   (f) Obtain a graph of the flow lines of **F**, indicating the direction of increasing t along the flow lines. Such a graph is actually a phase portrait for the autonomous system $\mathbf{R}'(p) = \mathbf{F}(\mathbf{x}(p)) = $

$\mathbf{F}(x(p), y(p))$, where $\mathbf{R} = x(p)\mathbf{i} + y(p)\mathbf{j}$. The critical points for the field are the equilibrium points for the autonomous system of differential equations defining the flow.

   (g) By linearizing at each equilibrium point, determine whether the point is asymptotically stable or unstable.

   (h) Where possible, categorize each equilibrium point as a node, saddle, center, or spiral.

   (i) Superimpose the flow lines on the contour plot found in part (b).

   (j) By appropriately integrating **F**, reconstruct the potential function $u(x, y)$. Compare your result with what your favorite computer algebra system gives if it has a built-in facility for finding the scalar potential.

B2. $u(x, y) = 2y^3 + 11x^2 y - 12xy^2$

B3. $u(x, y) = 10y + 12xy^2 - 7x^3$

B4. $u(x, y) = 6y + 6x^2 - 8xy^2$

B5. $u(x, y) = 12y^3 + 11x^2 y - 2x^3$

B6. $u(x, y) = 6x^2 + 8y^3 - 3xy$

B7. $u(x, y) = \dfrac{10x^2 + 8xy + 4xy^2}{7 + 6xy^2}$

B8. $u(x, y) = \dfrac{12x - 5x^2 + 4y^3}{15 - 12y^3 + x^3}$ **B9.** $u(x, y) = \dfrac{2x - 11y^3 + 6y^2}{10 - 4x^2 - 4x^3}$

B10. $u(x, y) = \dfrac{4xy + 6y^2 - 5xy^2}{4 + 11xy + 10x^2}$ **B11.** $u(x, y) = \dfrac{6 + 12y}{1 + x^2 y^2}$

B12. Let $u(x, y)$ be the potential of a unit positive charge placed at $(0, 0)$ in the xy-plane. Let $v(x, y)$ be the potential of a unit positive charge placed at $(2, 0)$, and let $w(x, y)$ be the potential of a unit negative charge placed at $(1, 1)$. Then, the potential for the system of three charges is $U(x, y) = u + v + w$.

(a) Write the potential function $U(x, y)$.

(b) Obtain a graph of the surface $U(x, y)$.

(c) Obtain a contour plot of the equipotentials (level curves) of $U(x, y)$.

(d) Obtain a field plot of $\mathbf{F} = -\nabla U$, the associated force field.

(e) Obtain a graph of the two flow lines that start at $(-0.1, 0.1)$ and $(0.1, 0)$.

(f) Find the critical point where $\mathbf{F} = \mathbf{0}$.

(g) The critical point found in part (f) is an equilibrium point for the differential equation $\mathbf{R}' = \mathbf{F}(x(t), y(t))$. By linearization, categorize this equilibrium point as a node, saddle, center, or spiral.

In Exercises B13–27, the vector field $\mathbf{F} = M(x, y)\mathbf{i} + N(x, y)\mathbf{j}$ is conservative and has a potential function $u(x, y)$ if the condition $M_y = N_x$ is satisfied. For each, determine if a potential function exists. If it does, find a scalar function $u(x, y)$ for which $\mathbf{F} = -\nabla u$.

B13. $M = 11y^2 + 24xy,\ N = 3 + 22xy + 12x^2$

B14. $M = -18x - 18xy,\ N = 9x^2$

B15. $M = -22x - 6xy,\ N = -9 - 3x^2$

B16. $M = 4y^2 - 4xy,\ N = 8xy - 24y^2 - 2x^2$

B17. $M = 10x - 6y,\ N = 6x + 2y$

B18. $M = 6xy + 3y^2,\ N = 3x^2 + 6xy$

B19. $M = y \arctan \dfrac{y}{x} + \dfrac{xy^2}{x^2 + y^2},\ N = x \arctan \dfrac{y}{x} + \dfrac{x^2 y}{x^2 + y^2}$

B20. $M = \dfrac{x^2 - 2xy - 2y^2}{(x - y)^2},\ N = \dfrac{2x^2 + 2xy - y^2}{(x - y)^2}$

B21. $M = \dfrac{y(x^2 + y^4)}{(x^2 - y^4)^2},\ N = -\dfrac{x(x^2 + 3y^4)}{(x^2 - y^4)^2}$

B22. $M = 6\dfrac{x(2x - 1)}{4 + 3y^3},\ N = 3\dfrac{y^2(10 - 9x^2 + 12x^3)}{(4 + 3y^3)^2}$

B23. $M = \dfrac{1}{2}\dfrac{7xy^2 - 6 + 12y}{x^3},\ N = -\dfrac{3 + 7xy}{x^2}$

B24. $M = \dfrac{1}{5}\dfrac{4x^3 + 7y}{x^3},\ N = -\dfrac{7}{10}\dfrac{1}{x^2}$

B25. $M = \dfrac{y^2 - 11}{(x - y)^2},\ N = \dfrac{11 - 10xy + 5y^2 + 4x^2}{(x - y)^2}$

B26. $M = \dfrac{x^3 + 1 + y^2}{(y^2 - 1)x^2},\ N = -\dfrac{y(x^3 - 4)}{x(y^2 - 1)^2}$

B27. $M = 4\dfrac{2 + 6x + 9x^2}{y(1 + 3x)^2},\ N = 2\dfrac{6x^2 - 3 - 5x}{y^2(1 + 3x)}$

For the potential functions $u(x, y, z)$ given in each of Exercises B28–39:

(a) Obtain the corresponding conservative force field $\mathbf{F} = -\nabla u$.

(b) Graph solutions of the differential equation $\mathbf{R}' = \mathbf{F}(x(p), y(p), z(p))$ to obtain, as space curves, the flow lines for the field $\mathbf{F}$.

B28. $u(x, y, z) = 5x - 11y^2 z - 8xz^2 + 8yz$

B29. $u(x, y, z) = 9xy - 11y^2 - 9x^2 - 11yz^2$

B30. $u(x, y, z) = 10xy + 2z^3 + 3y^2 z + 11xy^2$

B31. $u(x, y, z) = 3y^2 z - 7x^2 z + 8xz^2 - 12xz$

B32. $u(x, y, z) = 2y - x^2 z + 3xz^2$

B33. $u(x, y, z) = 10 + 12xz^2 - 12y^3 + 8yz$

B34. $u(x, y, z) = 2yz^2 - 8x^2 z + 2y^3$

B35. $u(x, y, z) = \dfrac{9xyz - 3xy - 6y^2}{1 + 2y + z^3}$

B36. $u(x, y, z) = \dfrac{9y^2 z + 12yz^2 - 7xz}{11x - 10yz}$

B37. $u(x, y, z) = \dfrac{5xy + 5z^3}{5y^2 z - 5xz^2 + x}$

B38. $u(x, y, z) = \dfrac{8x^2 z + xz^2 + 11y^3}{11y^2 - 6z^3}$

B39. $u(x, y, z) = \dfrac{5y^2 - 5xz - z^2}{11yz - 7x}$

The vector field $\mathbf{F} = f(x, y, z)\mathbf{i} + g(x, y, z)\mathbf{j} + h(x, y, z)\mathbf{k}$ is conservative and has a potential function $u(x, y, z)$ if the conditions $h_y = g_z, f_z = h_x$, and $g_x = f_y$ are satisfied. In Section 21.5, $u(x, y, z) = -\int_a^x f(t, b, c)\,dt - \int_b^y g(x, t, c)\,dt - \int_c^{zd} h(x, y, t)\,dt$ is then shown to be a potential for $\mathbf{F}$. The potential so constructed is determined up to an arbitrary constant via the arbitrary point (a, b, c). For each of Exercises B40–51, determine if a potential function exists, using the given recipe to find it, should it exist.

B40. $\mathbf{F} = (8xz + 3z^2)\mathbf{i} - 7z^2\mathbf{j} + (4x^2 - 14yz + 6xz)\mathbf{k}$

B41. $\mathbf{F} = (7 - 21x^2 - z)\mathbf{i} - x\mathbf{k}$

B42. $\mathbf{F} = (2xz + 2y)\mathbf{i} + (2x - 2z)\mathbf{j} + (x^2 - 2y)\mathbf{k}$

B43. $\mathbf{F} = 10y^2\mathbf{i} + (4z^2 + 20xy)\mathbf{j} + 8yz\mathbf{k}$

B44. $\mathbf{F} = (12y + z)\mathbf{i} + (12x - 14y)\mathbf{j} + x\mathbf{k}$

B45. $\mathbf{F} = (3x^2 - 7y^2 + 4z)\mathbf{i} - 14xy\mathbf{j} + 4x\mathbf{k}$

B46. $\mathbf{F} = 3yz\mathbf{i} - 3xz\mathbf{j} + (3xy - 12z^2)\mathbf{k}$

B47. $\mathbf{F} = 5\dfrac{x}{y(6y+5)}\mathbf{i} - \dfrac{5}{2}\dfrac{4y^2 + 12x^2y + 5x^2}{y^2(6y+5)^2}\mathbf{j}$

B48. $\mathbf{F} = -14\dfrac{y}{z(8z^2-9)}\mathbf{j} + 21\dfrac{y^2(8z^2-3)}{z^2(8z^2-9)^2}\mathbf{k}$

B49. $\mathbf{F} = 4\dfrac{y}{y^2-8z}\mathbf{i} - 2\dfrac{2xy^2 + 16xz + 11yz}{(y^2-8z)^2}\mathbf{j} + \dfrac{y(11y+32x)}{(y^2-8z)^2}\mathbf{k}$

B50. $\mathbf{F} = -\dfrac{y}{z}\mathbf{i} - \dfrac{1}{11}\dfrac{11x-2z^2}{z}\mathbf{j} + \dfrac{1}{11}\dfrac{y(2z^2+11x)}{z^2}\mathbf{k}$

B51. $\mathbf{F} = \dfrac{1+y}{z(y-z)x^2}\mathbf{i} + \dfrac{z+2}{xz(y-z)^2}\mathbf{j} + \dfrac{(2+y)(y-2z)}{xz^2(y-z)^2}\mathbf{k}$

Chapter Review

1. Describe the calculation by which the flow lines for a vector field are obtained. Give an example.

2. Describe the gradient vector and state its most interesting properties.

3. Give an example of a gradient vector field in two dimensions. To what are the gradient vectors orthogonal?

4. Give an example of a gradient vector field in three dimensions. To what are the gradient vectors orthogonal?

5. Define the concept of the directional derivative. Show how to obtain the expression $\nabla f \cdot \mathbf{u}$ for the directional derivative of the scalar $f(x, y, z)$.

6. Describe what is meant by the level curves of the function $z = f(x, y)$. Give an example.

7. Describe what is meant by the level surfaces of the function $w = f(x, y, z)$. Give an example.

8. Give an example that demonstrates that the gradient vector ∇f points in the direction of increasing values of f.

9. Give an example that demonstrates that the directional derivative, taken in the direction of the gradient vector, has the value $\|\nabla f\|$.

10. Use the Lagrange multiplier technique to find the minimum distance from the point $(3, 7, -2)$ to the plane $6x - 4y + z = 12$.

11. Use the Lagrange multiplier technique to find the minimum distance from the point $(4, -5, 2)$ to the intersection of the planes $2x - 3y + 7z = 9$ and $3x + 5y - 9z = 16$.

12. Define a conservative force. Give an example.

13. A unit positive charge at the origin of the real line exerts its influence along that line as a *field*. A second unit positive charge is at $x = +\infty$ on the line. The potential energy $u(x) = \int_{-\infty}^{x} \mathbf{F}(r)\,dr$ is defined as the mechanical work required to move the particle from $+\infty$ to a new location x along the line. Since the two charges repel each other, the rightmost charge moves toward the origin only if it experiences a force pointed to the left. For electrostatic charges, this force is inversely proportional to the square of the distance between the charges. Hence, the force on the rightmost charge is $-\dfrac{k}{x^2}$, where k is a constant of proportionality.

    (a) Show that the work done in moving the rightmost charge from infinity to x is $u(x) = \dfrac{k}{x}$.

    (b) Show that $\mathbf{F} = -\nabla u = -\dfrac{du}{dx}\mathbf{i}$ is the force exerted *by* the field (from the charge at the origin) *on* the particle to its right.

Chapter **19**

Line Integrals in the Plane

INTRODUCTION   Students of general physics know that work is defined as the product of a force and the distance through which it acts. If the force is not directed along the line of motion, its component along the path is used. If the force varies along the path, or if the path is curved, the component of the force on the tangent to the path is summed along the path. This results in a *line integral*, which happens to be the same, mathematically, as the *circulation integral* for a moving fluid. For work, the *tangential component* of a force is integrated along a path. For circulation, the tangential component of the velocity vector is integrated along a closed path such as a circle.

If a fluid flows across a curve, there is a net transport of mass across that curve. The *flux,* or transport across the curve, is determined by a line integral of the *normal component* of the velocity vector along the curve. Thus, flow through a boundary is related to the flux, a line integral of the normal component of the field crossing the boundary.

These two concepts of work and flux are fundamental in understanding the vector calculus in the following two chapters. Although initially defined for plane curves, both work and flux exist in three dimensions. The work integral generalizes to a line integral in three dimensions, but the flux integral generalizes to a *surface integral* in Section 21.2, where flow through a *surface* is defined.

19.1   Work and Circulation

Integration as Accumulation

Integrating to find area under a curve is a summation process. Rectangular bits of area are accumulated, and as the rectangles become thinner and more numerous, the approximate area becomes the total area sought. For the function $f(x) = x^2 + 1$, Figure 19.1 suggests how 10 rectangles begin to approximate the area under $f(x)$ on the interval $[0, 1]$.

Of course, this intuition was well established in the integral calculus course, as was the following analysis of the summation process that resulted in the definition of the integral. The sum of n equally spaced rectangles, each having its height determined by the function value at the left endpoint of the subinterval over which the rectangle sits is

$$\frac{1}{n}\sum_{i=0}^{n-1}\left(\frac{i^2}{n^2} + 1\right) = \frac{1}{n}\left(\frac{4}{3}n - \frac{1}{2} + \frac{1}{6n}\right)$$

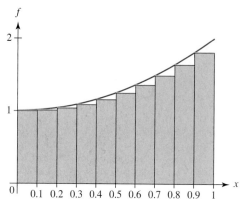

FIGURE 19.1   Integration as accumulation of area

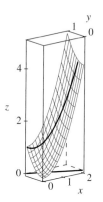

FIGURE 19.2   The curve
$y = 1 - \frac{x}{2}$ "lifted" to the

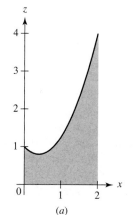

(*a*)

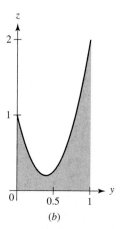

(*b*)

FIGURE 19.3   Areas under the
projections of the curve in Figure 19.2 to
(*a*) the *xz*-plane and (*b*) the *yz*-plane

which, in the limit as *n* becomes infinite, is $\frac{4}{3}$, recognized as the value of the definite integral $\int_0^1 (x^2 + 1)\,dx$.

Line Integrals

INTRODUCTION   A similar accumulation process for a quantity $f(x, y)$ defined along a curve *C* lying in the *xy*-plane results in a *line integral*, written with one of the notation $\int_C f(x, y)\,dx$ or $\int_C f(x, y)\,dy$. The subscript *C* indicates that the integration is along some curve *C*.

Suppose $f(x, y) = x^2 + y^2$ and *C* is the curve $x + 2y = 1$. A graph of the surface $z = f(x, y)$, the curve *C*, and the "lift" of *C* onto the surface is shown in Figure 19.2. Figure 19.3(*a*) shows $z = f(x, y(x))$, the lift projected onto the *xz*-plane; while Figure 19.3(*b*) shows $z = f(x(y), y)$, the lift projected onto the *yz*-plane. The area under each projection is also shown. Hence the line integrals

$$\int_C f(x, y)\,dx = \int_{x=a}^{x=b} f(x, y(x))\,dx = \int_{y=y^{-1}(a)}^{y=y^{-1}(b)} f(x(y), y)x'(y)\,dy \quad \textbf{(19.1a)}$$

$$\int_C f(x, y)\,dy = \int_{y=y^{-1}(a)}^{y=y^{-1}(b)} f(x(y), y)\,dy = \int_{x=a}^{x=b} f(x, y(x))y'(x)\,dx \quad \textbf{(19.1b)}$$

compute areas analogous to the shaded regions in Figures 19.3(*a*) and 19.3(*b*), respectively. The middle integral in (19.1a) is used if *C* is parametrized by $y = y(x)$, whereas the rightmost integral is used if *C* is parametrized by $x = x(y)$. The middle integral in (19.1b) is used if *C* is parametrized by $x = x(y)$, while the rightmost integral is used if *C* is parametrized by $y = y(x)$. If the curve *C* is defined parametrically by expressions of the form $x = x(p)$ and $y = y(p)$, then the line integrals are evaluated as

$$\int_C f(x, y)\,dx = \int_{p=p_0}^{p=p_1} f(x(p), y(p))x'(p)\,dp$$

$$\int_C f(x, y)\,dy = \int_{p=p_0}^{p=p_1} f(x(p), y(p))y'(p)\,dp$$

EXAMPLE 19.1   Let the curve C be the upper semicircle of radius 3, centered at the origin, and parametrized by p via the equations $x(p) = 3\cos p$ and $y(p) = 3\sin p$; let the function be $f(x, y) = x^2 + y^3$. The function must inherit the parametrization given the curve C so that $f = f(x(p), y(p)) = 9\cos^2 p + 27\sin^3 p$. Hence, we have the two line integrals

$$\int_C f(x, y)\, dx = \int_0^\pi (9\cos^2 p + 27\sin^3 p)(-3\sin p)\, dp = -18 - \tfrac{243}{4}\pi \qquad \textbf{(19.2a)}$$

$$\int_C f(x, y)\, dy = \int_0^\pi (9\cos^2 p + 27\sin^3 p)(3\cos p)\, dp = 0 \qquad \textbf{(19.2b)}$$

❖

CLOSED LINE INTEGRALS   If the line integral is taken around a curve C forming a closed loop, it is called a *closed* line integral and is often denoted by an integral sign decorated with a closed loop: $\oint$. Since the direction in which the curve C is traversed matters, some texts also decorate the little loop with an arrow to indicate whether a simple (non-self-intersecting) loop is traversed in the clockwise or in the counterclockwise sense. Moreover, the applications most often contain closed line integrals of the form

$$\oint_C f(x, y)\, dx + g(x, y)\, dy = \oint_C f(x, y)\, dx + \oint_C g(x, y)\, dy$$

which arise as *work, circulation,* and *flux* calculations.

WORK AS A LINE INTEGRAL   Mechanical work done by a force acting on a particle is defined in physics as the product of the component of the force in the direction of motion times the distance through which that component acts. If the force varies along the path, or the path is curved, then this definition is generalized as the line integral of the component of the force tangent to the path. This definition gives the work done *on* the particle *by* the force.

In particular, let $\mathbf{F}$ be the force acting on the particle, and C be the curve along which the particle moves under the action of the force $\mathbf{F}$, and ds be the element of arc length along C. Further, let $\mathbf{T}$ be a unit tangent vector along the curve C. Then define

$$\text{Work} = \int_C \mathbf{F} \cdot \mathbf{T}\, ds = \int_C \mathbf{F} \cdot d\mathbf{r}$$

Figure 19.4 shows why, along the curve C, $\mathbf{F} \cdot \mathbf{T}$ represents the tangential component of the force $\mathbf{F}$. Mathematically, the projection of $\mathbf{F}$ onto the unit vector $\mathbf{T}$ is determined by simple right triangle trigonometry and the standard formula for interpreting the dot product, namely, $\mathbf{F} \cdot \mathbf{T} = \|\mathbf{F}\|\|\mathbf{T}\|\cos\theta = \|\mathbf{F}\|1\cos\theta = \|\mathbf{F}\|\cos\theta$.

Next, examine why $\mathbf{F} \cdot \mathbf{T}\, ds = \mathbf{F} \cdot d\mathbf{r}$, or equivalently, why $\mathbf{T}\, ds = d\mathbf{r}$. The radius vector $\mathbf{r} = x(p)\mathbf{i} + y(p)\mathbf{j}$ describes the curve C. Along C, the unit tangent vector is $\mathbf{T} = \frac{d\mathbf{r}}{ds} = \frac{dx}{ds}\mathbf{i} + \frac{dy}{ds}\mathbf{j}$. Thus, $\mathbf{T}\, ds = \frac{d\mathbf{r}}{ds}\, ds = d\mathbf{r} = dx\,\mathbf{i} + dy\,\mathbf{j}$. Finally, if $\mathbf{F} = f(x, y)\mathbf{i} + g(x, y)\mathbf{j}$, then $\mathbf{F} \cdot \mathbf{T}\, ds = \mathbf{F} \cdot d\mathbf{r} = f\, dx + g\, dy$ is the tangential component of $\mathbf{F}$ along C.

Hence, the line integral for *work* is precisely the line integral of the tangential component of the field. *Circulation* is also defined as the closed line integral (around the curve C) of the tangential component of the velocity field of a fluid flow. Therefore, the same structure expressing mechanical work also expresses the notion of circulation, which plays an important role in Sections 36.9 and 36.10 where problems in planar fluid flow are solved.

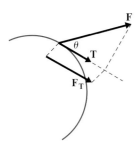

FIGURE 19.4   Tangential component of $\mathbf{F}$ along the curve C

EXAMPLE 19.2   Find the work done by the force field $\mathbf{F} = (3x - 4y)\mathbf{i} + (4x + 2y)\mathbf{j}$ on a particle of mass $m = 3$ as the particle is moved around the ellipse C defined by $x = 4\cos p$, $y = 3\sin p$. According to the convention stated at the end of Section 18.5, the field $\mathbf{F}$ is defined as the force exerted on a particle of *unit* mass.

The work done *by* the field $\mathbf{F}$ *on* the particle as it moves along the closed curve C is the closed line integral of the tangential component of $\mathbf{F}$, namely, $\mathbf{F} \cdot d\mathbf{r}$. If the curve C is expressed in radius-vector form, we have

$$\mathbf{r} = 4\cos p\,\mathbf{i} + 3\sin p\,\mathbf{j} \quad \text{and} \quad d\mathbf{r} = -4\sin p\,dp\,\mathbf{i} + 3\cos p\,dp\,\mathbf{j}$$

Expressing x and y as functions of p then gives

$$3\int_C f\,dx + g\,dy = 3\int_{t=0}^{2\pi}[-4(12\cos p - 12\sin p)\sin p$$
$$+ 3(16\cos p + 6\sin p)\cos p]\,dp = 288\pi \qquad ❖$$

EXAMPLE 19.3   When the curve C is parametrized by one explicitly given set of equations, computing a line integral whose integrand is of the form $f(x, y)\,dx + g(x, y)\,dy$ is straightforward. If, however, the curve C is defined in segments, with different sets of formulas for different segments, the line integral becomes more of a challenge. In fact, if no parametrization for C is given at all, then the user must supply that all-important parametrization before any line integration can take place. Consider, then, the following example.

Let C be the triangle with vertices $(1, 0)$, $(1, 1)$, and $(0, 0)$. Let $f(x, y) = y^2$ and $g(x, y) = x^2$. Find the value of the line integral of $f\,dx + g\,dy$ taken in the counterclockwise sense around the triangle C. Note that traversing the triangle in the counterclockwise sense means that a person walking the path would have the interior of the triangle to the left as the boundary was traversed. (See Figure 19.5.)

Parametrize segment (a) with $x = p$, $y = 0$, segment (b) with $x = 1$, $y = p$, and segment (c) with $x = p$, $y = p$, giving

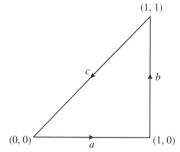

FIGURE 19.5   The triangular path for Example 19.3

$$\int_C f\,dx + g\,dy = \int_{t=0}^{1} 0\,dp + \int_0^1 1\,dp + \int_1^0 2p^2\,dp = 0 + 1 - \tfrac{2}{3} = \tfrac{1}{3}$$

Note that segment (c) must be traversed from $(1, 1)$ to $(0, 0)$. ❖

EXERCISES 19.1–Part A

A1. Evaluate the line integrals $\int_C f(x, y)\,dx$ and $\int_C f(x, y)\,dy$ if $f(x, y) = 2x^2 - 3xy + 5y^2$ and C is the curve $y = 2x - 3$, $0 \le x \le 1$.

A2. Evaluate the line integrals $\int_C f(x, y)\,dx$ and $\int_C f(x, y)\,dy$ if $f(x, y) = 3x^2 + 2y^2$ and C is a circle with radius 1 and center at the origin.

A3. Compute the work done when an object of mass 2 is moved from $(0, 0)$ to $(2, 4)$ along the curve $y = x^2$ if it is subjected to the field $\mathbf{F} = x^2y\mathbf{i} + xy^2\mathbf{j}$.

A4. Compute the work done in Exercise A3 if the path is the straight line connecting the two given points. Is $\mathbf{F}$ a conservative field?

A5. Let $\mathbf{F} = f(x, y)\mathbf{i} + g(x, y)\mathbf{j}$ be a force field. Show that $\int_C \mathbf{F} \cdot d\mathbf{r} = \int_C f(x, y)\,dx + \int_C g(x, y)\,dy$.

EXERCISES 19.1–Part B

B1. Graph the integrand in (19.2b), Example 19.1, and argue from symmetry that the value given is correct. Graph the integrand in (19.2a) and explain why the graph does not contradict the given value of the integral.

In Exercises B2–11, evaluate the line integrals $\int_C f(x, y)\, dx$ and $\int_C f(x, y)\, dy$, where C is the curve given by $y = y(x)$ for x in the indicated interval.

B2. $f(x, y) = -6 + 7y - 7y^2$, $y = -7x^4 - 4x^3 - 14x^2 - 4x + 9$, $[-3, 6]$

B3. $f(x, y) = 8 - 6x^2y - 9xy^2$, $y = 14x^4 + 8x^3 + 1$, $[-9, 7]$

B4. $f(x, y) = -3 + 4x^2y + 9y^2$, $y = 8x^4 - x^3 + 8x^2 + 6x$, $[-6, 6]$

B5. $f(x, y) = -8x - 4y + xy$, $y = 9x^4 - 6x^3 + 5x^2 - 6x + 1$, $[-7, 9]$

B6. $f(x, y) = 8 + 8x^2y + x^2$, $y = -7x^3 + 4x^2 + 14x + 3$, $[-6, 1]$

B7. $f(x, y) = xy^2 - 9xy + 8x^2$, $y = 6x^3 - 18x$, $[-2, 6]$

B8. $f(x, y) = 3x + 9xy^2 - 3xy$, $y = x^4 - 3x^3 - 2x - 8$, $[-2, 6]$

B9. $f(x, y) = -4x + 5xy^2 + 2xy$, $y = 12x^4 - x^3 - 5x^2 + 6$, $[-5, 7]$

B10. $f(x, y) = 9y + 7x^2y - 3y^3$, $y = -4x^4 + 7x^3 - 3x^2 - 3x + 1$, $[-3, 5]$

B11. $f(x, y) = 8 - 7y + 3y^2$, $y = -2x^3 + 12x - 8$, $[-4, 8]$

In Exercises B12–21, evaluate the line integrals $\int_C f(x, y)\, dx$ and $\int_C f(x, y)\, dy$, where C is the curve given by $x = x(t)$, $y = y(t)$ for t in the indicated interval.

B12. $f(x, y) = 6x - 4xy^2 - 3x^2$, $x = -9t^2 - 5$, $y = -7t^3 + 3t^2 + 3t$, $[-8, 8]$

B13. $f(x, y) = 7x + 3y^2 + 2y$, $x = 8t^3 + 8t$, $y = 2t^4 + 9t^3 - 3$, $[-2, 0]$

B14. $f(x, y) = 3 + 6y^2 - x^2$, $x = 13t^3 - 4$, $y = -6t^4 + 12t - 2$, $[-1, 1]$

B15. $f(x, y) = 6 - 8x^2y + xy^2$, $x = -t^2 - 1$, $y = -2t^4 + 8t^3 + 2t^2 + 3t$, $[1, 2]$

B16. $f(x, y) = 7xy^2 + 8xy + 6x^3$, $x = -8t^3 + 6t - 2$, $y = -5t^4 - 5t^3 - 3t^2$, $[0, 1]$

B17. $f(x, y) = x + 8y - 7y^3$, $x = 9t^3 - 2t^2 + 8t + 2$, $y = -6t^4 + t^3 - 18t$, $[-\frac{1}{2}, 1]$

B18. $f(x, y) = y - 7x^2y + 9y^3$, $x = 3t^3 + 6t^2 - 8t$, $y = 7t^4 - 3t^2 - 4$, $[0, \frac{1}{5}]$

B19. $f(x, y) = 3y + 4y^3 + 5xy$, $x = 3t^3 - 4t + 5$, $y = -t^2 - 7$, $[-1, 1]$

B20. $f(x, y) = 6x + 9y^3$, $x = -2t^3 - 7t^2 - 5t$, $y = 6t^4 + 3t^2 + 9t + 2$, $[0, \frac{1}{10}]$

B21. $f(x, y) = 6 - 5x^2y + 5xy$, $x = -5t^2 - 2t + 12$, $y = -8t^4 + t^2 + 3$, $[-1, 2]$

In Exercises B22–31:

  (a) Graph the path C, the curve given by $y = y(x)$, $-2 \le x \le 3$.

  (b) Add to the graph of C a direction field for $\mathbf{G} = \mathbf{F}/\|\mathbf{F}\|$. Calculate both $\mathbf{F}$ and $\mathbf{F_T}$, the tangential component of $\mathbf{F}$, at $x = 1$.

  (c) Find the work done by the field $\mathbf{F}$, acting on a particle of mass m, as the particle moves from left to right along the curve C.

B22. $\mathbf{F} = (1 - 7y - 4y^2)\mathbf{i} + (9y^3 + 7x^3 - 6xy^2)\mathbf{j}$, $m = 7$, $y = 12x^2 + 9$

B23. $\mathbf{F} = (3 + 5x^3 + 9xy)\mathbf{i} + (5y - 9x^3 - 2x^2)\mathbf{j}$, $m = 1$, $y = 6x^4 - 3x^3 - 9x + 5$

B24. $\mathbf{F} = (2y^2 + 7x^3 + 2xy^2)\mathbf{i} + (8 + xy^2 - 5x^2y)\mathbf{j}$, $m = 8$, $y = 5x^3 + 3x^2 + 3x + 1$

B25. $\mathbf{F} = (5y - 6y^3 + 3x^2)\mathbf{i} + (5 - y + 4x^2y)\mathbf{j}$, $m = 3$, $y = -5x^4 - 3x^3 + 7$

B26. $\mathbf{F} = (7 - 4y - 5xy^2)\mathbf{i} + (6 - 4x^2 + 3xy^2)\mathbf{j}$, $m = 1$, $y = 7x^2 + 15x + 9$

B27. $\mathbf{F} = (4x^3 - 3xy^2)\mathbf{i} + (6 + y^3 - xy)\mathbf{j}$, $m = 9$, $y = -3x^4 + 2x^3 - 7$

B28. $\mathbf{F} = (8y + 7xy - 7xy^2)\mathbf{i} + (3 + 2x^2y)\mathbf{j}$, $m = 4$, $y = 2x^4 + x^3 - 3x^2 - 5x$

B29. $\mathbf{F} = 3xy^2\mathbf{i} + (2y^3 - 6x^2)\mathbf{j}$, $m = 4$, $y = 3x^3 - 11x - 11$

B30. $\mathbf{F} = (4xy - 2y^2 - 5x^2)\mathbf{i} + (4 + 3x - 7y^2)\mathbf{j}$, $m = 8$, $y = -7x^4 + 3x^3 - 4x^2$

B31. $\mathbf{F} = (7 + 4x^2y)\mathbf{i} + (2 + 2xy^2)\mathbf{j}$, $m = 4$, $y = 11x^4 + 8x^2$

In Exercises B32–43:

  (a) Graph the given closed path C formed by y_1 and y_2.

  (b) Add to the graph of C a direction field for $\mathbf{G} = \mathbf{F}/\|\mathbf{F}\|$. Calculate both $\mathbf{F}$ and $\mathbf{F_T}$, the tangential component of $\mathbf{F}$, at $P_0 = (x_0, y_0)$, the lowest point on y_1.

  (c) Find the work done by the field $\mathbf{F}$, acting on a particle of mass m, as the particle moves once counterclockwise around the curve C.

B32. $\mathbf{F} = (8y^2 - 6x^2y - 7x^3)\mathbf{i} + (2 - 2xy^2)\mathbf{j}$, $m = 6$, $y_1 = 2x^2 - 8x$, $y_2 = -2x + 56$

B33. $\mathbf{F} = (1 + 2x - 6x^2y)\mathbf{i} + (7x + 8y - 7xy^2)\mathbf{j}$, $m = 1$, $y_1 = 7x^2 - 7x - 3$, $y_2 = 7x + \frac{23}{4}$

B34. $\mathbf{F} = (6xy - 7x^2y - 4)\mathbf{i} + (8x + 3y + 5xy)\mathbf{j}$, $m = 1$, $y_1 = x^2 + 2x + 2$, $y_2 = 2x + 27$

B35. $\mathbf{F} = (2x + 8y + 4x^2y)\mathbf{i} + (x + 9y^2 - 5x^3)\mathbf{j}$, $m = 8$, $y_1 = 5x^2 - 7$, $y_2 = 5x + 93$

B36. $\mathbf{F} = (5y + 4xy - 2x^2)\mathbf{i} + (3x - 2y + 4x^2)\mathbf{j}$, $m = 3$, $y_1 = 2x^2 - 4x - 4$, $y_2 = 10x - 4$

B37. $\mathbf{F} = (8xy - 9x^2y - 6)\mathbf{i} + (1 - 5x^2y)\mathbf{j}$, $m = 8$, $y_1 = 2x^2 - 8x - 2$, $y_2 = 2x + 26$

B38. $\mathbf{F} = (9y + 8y^3 - 3x^3)\mathbf{i} + (x - 3y)\mathbf{j}$, $m = 6$, $y_1 = 4x^2 - 4x - 9$, $y_2 = 16x + 2$

B39. $\mathbf{F} = (2y + 3xy + 8x^3)\mathbf{i} + (5x - 6xy - y^2)\mathbf{j}$, $m = 6$, $y_1 = 9x^2 + 8$, $y_2 = 9x + 26$

B40. $\mathbf{F} = (6x + 4y^3)\mathbf{i} + (5 + 8y^2 + 3x^3)\mathbf{j}$, $m = 8$, $y_1 = x^2 + 8x + 5$, $y_2 = x + 13$

B41. $\mathbf{F} = (3 - 8x + 9y^2)\mathbf{i} + (4xy - 4x^3 + 6y)\mathbf{j}$, $m = 7$, $y_1 = 4x^2 + 8x - 2$, $y_2 = \frac{23}{8}x^2 + 8x + \frac{65}{8}$

B42. $\mathbf{F} = (4y + 5y^3 + 9x^2)\mathbf{i} + (5x^2y + 4y^3 + 8x^2)\mathbf{j}$, $m = 7$, $y_1 = 3x^2 + 6x + 1$, $y_2 = \frac{3}{2}x^2 - \frac{9}{2}x + 1$

B43. $\mathbf{F} = (8 - 7xy - 8x^2)\mathbf{i} + (6x + y - 8y^2)\mathbf{j}$, $m = 2$, $y_1 = 3x^2 - 9x - 9$, $y_2 = \frac{5}{2}x^2 - 6x - \frac{83}{8}$

In Exercises B44–55:

(a) Graph the closed polygonal path C determined by the given vertices.

(b) Add to the graph of C a direction field for $\mathbf{G} = \mathbf{F}/\|\mathbf{F}\|$. Calculate both $\mathbf{F}$ and $\mathbf{F_T}$, the tangential component of $\mathbf{F}$, each computed at $x = 0$. (The path C is closed, with two points on it corresponding to $x = 0$.)

(c) Find the work done by the field $\mathbf{F}$, acting on a particle of mass $m = 1$, as the particle moves once counterclockwise around the curve C.

B44. $\mathbf{F} = (4 - 5x + 2x^2y)\mathbf{i} + (1 + 2x^2y + 7xy)\mathbf{j}$, $\{(10, 1), (-11, 4), (-10, -3)\}$

B45. $\mathbf{F} = (3x + 6y^2 + 9x^2)\mathbf{i} + (9x^3 - 9y^3)\mathbf{j}$, $\{(9, 6), (-11, 9), (-1, -10)\}$

B46. $\mathbf{F} = (4y + x^2 + 3xy)\mathbf{i} + (3x^2y + 4xy^2 - 3)\mathbf{j}$, $\{(4, 2), (-7, 11), (-4, -3)\}$

B47. $\mathbf{F} = (3x^2y - 6x)\mathbf{i} + (7 - 3xy^2)\mathbf{j}$, $\{(12, 9), (-2, 3), (-4, -1)\}$

B48. $\mathbf{F} = (1 - 3y - x^3)\mathbf{i} + (2x + 4y + x^3)\mathbf{j}$, $\{(7, 10), (-7, 10), (-10, -7)\}$

B49. $\mathbf{F} = (8y - 6x^3 + 7xy)\mathbf{i} + (5x - 9y + 7x^3)\mathbf{j}$, $\{(2, 10), (-3, 6), (-6, -2)\}$

B50. $\mathbf{F} = (7xy^2 + 4)\mathbf{i} + (9 - 6x + 3xy)\mathbf{j}$, $\{(2, 9), (-2, 9), (-10, -8)\}$

B51. $\mathbf{F} = (2xy - 6y^2 + x^2)\mathbf{i} + (4y + 8xy^2 - 3x)\mathbf{j}$, $\{(5, 10), (-11, 4), (-7, -12)\}$

B52. $\mathbf{F} = 9xy^2\mathbf{i} + (7x^2y + 5xy^2 + 3)\mathbf{j}$, $\{(5, 5), (-9, 11), (-1, -8), (12, -10)\}$

B53. $\mathbf{F} = (2 + 8x + 6xy)\mathbf{i} + (7x - 2x^3 - 8y^3)\mathbf{j}$, $\{(15, 9), (-5, 7), (-1, -10), (12, -11)\}$

B54. $\mathbf{F} = (y + 9x^3 + 6y^3)\mathbf{i} + (4x - x^2y + 5)\mathbf{j}$, $\{(2, 4), (-12, 10), (-3, -8), (5, -9)\}$

B55. $\mathbf{F} = (xy^2 + x - y)\mathbf{i} + (2y - 8x^2y + 6x)\mathbf{j}$, $\{(7, 6), (-4, 9), (-1, -9), (11, -2)\}$

19.2 Flux Through a Plane Curve

DEFINITION 19.1

The *flux* of the vector field $\mathbf{F} = f\mathbf{i} + g\mathbf{j}$ through a plane curve C is defined as the line integral, along the curve C, of the normal component of $\mathbf{F}$ on the curve, that is, by $\int_C \mathbf{F} \cdot \mathbf{N}\, ds$.

If at a point on C the two normals are denoted by $\mathbf{N}$ and $-\mathbf{N}$, then the flux in the direction of $-\mathbf{N}$ is the negative of the flux in the direction of $\mathbf{N}$. Shortly, we will describe a standard convention for specifying which unit normal vector is usually chosen when computing flux through a plane curve.

The flux of $\mathbf{F}$ through C is a measure of how much of $\mathbf{F}$ flows along the normal to C. This is most easily seen by examining the units of the field $\mathbf{F} = \rho\mathbf{v}$, where ρ is the planar density (mass per unit area) and $\mathbf{v}$ is the velocity field of a planar fluid flow. The units for $\mathbf{F}$ are then $\frac{\text{mass}}{\text{area}}\frac{\text{length}}{\text{time}} = \frac{\text{mass}}{\text{length time}}$, indicating that $\mathbf{F}$ is the mass flow-rate per unit length along the curve C. The flux of this field through C is then a measure of the mass of fluid flowing per unit time through (across, not along) the curve.

If C is given in vector form by the radius vector $\mathbf{r}$ and $\mathbf{T}$ and $\mathbf{N}$ are unit tangent and normal vectors, respectively, along C, with $\mathbf{N}$ to the right of $\mathbf{T}$, then the following comparison of the work and flux integrals is useful.

$$\text{Work} = \int_C \mathbf{F} \cdot \mathbf{T}\, ds = \int_C \mathbf{F_T}\, ds = \int_C f\, dx + g\, dx = \int_C \mathbf{F} \cdot d\mathbf{r} \qquad \textbf{(19.3a)}$$

$$\text{Flux} = \int_C \mathbf{F} \cdot \mathbf{N}\, ds = \int_C \mathbf{F_N}\, ds = \int_C f\, dy - g\, dx \qquad \textbf{(19.3b)}$$

It is not an oversight that flux has no counterpart for the rightmost expression in (19.3a). The closest we come to filling that gap is (23.17) in Section 23.2.

The Unit Normal Vector N

A standard convention for specifying the normal vector **N** is to pick it "to the right of the unit tangent vector **T**." This means that when walking along the curve C facing the direction in which the tangent vector **T** points, the normal vector points to the right. Hence, this choice of **N** implies $\mathbf{T} \times \mathbf{N} = -\mathbf{k}$.

Put **N** to the right of **T** as follows. Let $\mathbf{T} = \cos p\mathbf{i} + \sin p\mathbf{j}$, making **T** a unit vector and $\mathbf{N} = \sin p\mathbf{i} - \cos p\mathbf{j}$. Compute $\mathbf{T} \times \mathbf{N}$ and notice it will necessarily be $-\mathbf{k}$. Thus, by interchanging the components of **T** and negating the second component, the normal vector **N** is to the right of the tangent vector **T**. If **N** is to the right of **T**, then turning **T** toward **N** advances a right-handed thread into the page, so $\mathbf{T} \times \mathbf{N} = -\mathbf{k}$.

If the curve C is closed, it is usual to select the outward normal. Hence, the closed curve must be traversed in a counterclockwise sense for the normal to be both outward and to the right of **T**. If C is not closed, the choice of parametrization determines the direction of **T**, so that the direction of **N** will depend on the parametrization of C.

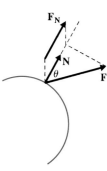

FIGURE 19.6 The normal component of **F** along the curve C

Normal Component of F

Figure 19.6 helps show why $\mathbf{F} \cdot \mathbf{N}$ is the component of **F** along the unit normal **N**. From the definition of the dot product we have, for the normal component of **F** along the curve C,

$$\mathbf{F} \cdot \mathbf{N} = \|\mathbf{F}\|\|\mathbf{N}\| \cos\theta = \|\mathbf{F}\| 1 \cos\theta = \|\mathbf{F}\| \cos\theta$$

Consequently, $\mathbf{F} \cdot \mathbf{N}$ is the component of **F** along the unit normal **N**.

To obtain

$$\mathbf{F} \cdot \mathbf{N}\, ds = f\, dy - g\, dx$$

recall that the radius vector $\mathbf{r} = x\mathbf{i} + y\mathbf{j}$ gives $\mathbf{T} = \frac{d\mathbf{r}}{ds} = \frac{dx}{ds}\mathbf{i} + \frac{dy}{ds}\mathbf{j}$, the unit tangent vector. Then $\mathbf{N} = \frac{dy}{ds}\mathbf{i} - \frac{dx}{ds}\mathbf{j}$ is formed from **T** by interchanging components and negating the second, so that $\mathbf{N} ds = dy\, \mathbf{i} - dx\, \mathbf{j}$, and finally, $\mathbf{F} \cdot \mathbf{N}\, ds = f\, dy - g\, dx$.

It might prove useful for the reader to notice that the word *flux* begins with the letter f and ends with the letter x, just like the flux expression $f\, dy - g\, dx$.

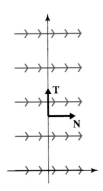

FIGURE 19.7 The field and path in Example 19.4

EXAMPLE 19.4 Using Figure 19.7, compute the flux of the field **F** through C if $\mathbf{F} = 3\mathbf{i}$ is a constant field of strength 3 pointing horizontally left to right and C is that portion of the y-axis where y is in $[0, 5]$. This makes $\mathbf{T} = \mathbf{j}$ and $\mathbf{N} = \mathbf{i}$. Parametrize C as $x(p) = 0$, $y(p) = p$, p in $[0, 5]$. The flux of **F** through the curve C is computed by evaluating the line integral

$$\text{Flux} = \int_C \mathbf{F} \cdot \mathbf{N}\, ds = \int_C f\, dy - g\, dx = \int_{p=0}^{p=5} 3\, dp - \int_{p=0}^{p=5} 0 = 15$$

The flux is positive. Note that the flow of the field **F** is in the same direction as **N**. Positive flux indicates flow along the direction of the normal **N**. ❖

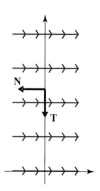

EXAMPLE 19.5   Repeat the calculation in Example 19.4, but parametrize the curve C as $x(p) = 0$, $y(p) = 5 - p$ so that the direction of travel is downward from $(0, 5)$ to the origin. This reverses the direction of **T**, making $\mathbf{T} = -\mathbf{j}$, and **N**, to the right of **T**, must be taken as $\mathbf{N} = -\mathbf{i}$. The field **F** now opposes the normal **N**, as seen in Figure 19.8.

The flux through the curve C is then calculated by evaluating the line integral

$$\text{Flux} = \int_C \mathbf{F} \cdot \mathbf{N} \, ds = \int_C f \, dy - g \, dx = \int_{p=0}^{p=5} 3(-dp) - \int_{p=0}^{p=5} 0 = -15$$

FIGURE 19.8   The field and path in Example 19.5

The negative value of the flux indicates that the flow is opposed to the direction of the normal **N**. Depending on the parametrization of the curve C, the flux can be computed as ± 15, even though the normal vector **N** was chosen to the right of **T**. What this means is that there is no way for selecting a unique normal along a segment of the curve C. Hence, when computing the value of the flux through such a curve, it is essential that the normal **N** be reported along with the value computed for the flux of **F** through C. Positive flux indicates a net flow along the chosen normal, and negative flux indicates a net flow against the chosen normal. ❖

EXAMPLE 19.6   Find the flux of the field $\mathbf{F} = (\frac{3}{8}x - \frac{1}{2}y)\mathbf{i} + (\frac{1}{2}x + \frac{1}{4}y)\mathbf{j}$ through the ellipse C given parametrically by $x(p) = 5\cos p$, $y(p) = 3\sin p$. A graph of the field **F** superimposed on a plot of the ellipse appears in Figure 19.9. The field **F** seems to be spreading, passing out through the curve C, so there should be some net flux of **F** passing through this curve C. Figure 19.9 also shows, at the point $(\frac{5}{2}, \frac{3}{2}\sqrt{3})$, a representative tangent and normal vector. Along C,

$$\mathbf{F} = \mathbf{F}(x(p), y(p)) = \left(\tfrac{15}{8}\cos p - \tfrac{3}{2}\sin p\right)\mathbf{i} + \left(\tfrac{5}{2}\cos p + \tfrac{3}{4}\sin p\right)\mathbf{j}$$

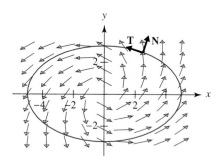

FIGURE 19.9   The field and path in Example 19.6

so the flux is given by the integral

$$\int_C f \, dy - g \, dx = \int_0^{2\pi} \left[3\left(\tfrac{15}{8}\cos p - \tfrac{3}{2}\sin p\right)\cos p \right.$$
$$\left. + 5\left(\tfrac{5}{2}\cos p + \tfrac{3}{4}\sin p\right)\sin p\right] dp = \tfrac{75}{8}\pi$$

Since the ellipse C is traced in the counterclockwise sense by the parametrization used, the normal vector **N** is both "to the right of **T**" and "outward" with respect to the closed curve C. Hence, the positive value of the flux confirms that there is a net outward flow, through the boundary of C, for the vector field **F**. ❖

EXERCISES 19.2–Part A

A1. Compute the flux of $\mathbf{F} = x^2 y\mathbf{i} + xy^2\mathbf{j}$ through the curve $y = x^2$, $0 \le x \le 1$.

A2. Compute the flux of $\mathbf{F} = xy^3\mathbf{i} + x^3 y\mathbf{j}$ through the triangle whose vertices are $(0, 0)$, $(1, 0)$, $(0, 2)$.

A3. Predict, then compute, the flux of $\mathbf{F} = 3\mathbf{i}$ through the square whose vertices are $(0, 0)$, $(1, 0)$, $(1, 1)$, $(0, 1)$.

A4. Predict, then compute, the flux of $\mathbf{F} = 3\mathbf{i}$ through the triangle whose vertices are $(0, 0)$, $(1, 0)$, $(1, 1)$.

A5. Find the flux of $\mathbf{F} = (3x + 4y)\mathbf{i} + (5x - 7y)\mathbf{j}$ through a circle with radius 1 and center at the origin.

EXERCISES 19.2–Part B

B1. Let $u(x, y) = -xy$ so that $\mathbf{F} = -\nabla u$ is conservative. If C is a circle with radius $c > 0$ and center at (a, b),

   (a) obtain a field plot for $\mathbf{F}$.

   (b) obtain a contour plot for $u(x, y)$.

   (c) find, and categorize as node, saddle, center, or spiral, all equilibrium points for the differential equation $\mathbf{R}' = \mathbf{F}(x, y)$. For nodes and spirals, determine if stable or unstable.

   (d) obtain the flow lines for the field $\mathbf{F}$ and superimpose them on the contour plot found in part (b).

   (e) show that the flux through C is zero for all such circles. *Hint:* $x = a + c\cos\theta$, $y = b + c\sin\theta$, and integrate counterclockwise around C so that the normal, to the right of the tangent vector $\mathbf{T}$, is outward along C.

B2. Repeat Exercise B1 for $u(x, y) = -\frac{1}{2}(x^2 + y^2)$. In part (e), show that the flux is $2\pi c^2$ for all circles C. Thus, vanishing of the flux must be attributed to a property of the field other than being conservative. See Section 20.1.

In Exercises B3–12:

   (a) Obtain a field plot for $\mathbf{F}$.

   (b) Find, and categorize as node, saddle, center, or spiral, all equilibrium points for the differential equation $\mathbf{R}' = \mathbf{F}(x, y)$.

   (c) Obtain the flow lines for the field $\mathbf{F}$.

   (d) Let C be the circle with center at (a, b) and with radius c, as given. At $(a, b + c)$, the point at the "top" of the circle, plot $\mathbf{F}$, and also $\mathbf{F_N}$, the component of $\mathbf{F}$ along $\mathbf{N}$, the outward unit normal.

   (e) Compute the flux of $\mathbf{F}$ through C.

B3. $\mathbf{F} = (4x - 4y^3)\mathbf{i} + (8y^2 - 5x - 6y^3)\mathbf{j}$, $(-8, -3)$, $c = 7$

B4. $\mathbf{F} = (9y^2 + 3xy^2 - 4x)\mathbf{i} + (10 - 10y^2 + 4y)\mathbf{j}$, $(-8, 8)$, $c = 4$

B5. $\mathbf{F} = (12x^2 - 5y + 11x^2y)\mathbf{i} + (2xy + 10x - 7)\mathbf{j}$, $(-1, 2)$, $c = 5$

B6. $\mathbf{F} = (9x - 2xy^2 + 9y)\mathbf{i} + (4x^2y - 10xy - 8x^2)\mathbf{j}$, $(-1, 3)$, $c = 9$

B7. $\mathbf{F} = (xy^2 - 8y + 3x^2)\mathbf{i} + (11xy^2 + 4x - 8y)\mathbf{j}$, $(9, 4)$, $c = 1$

B8. $\mathbf{F} = (7x^2 - 4y - 5)\mathbf{i} + (9x - 7y^2 + 3x^2)\mathbf{j}$, $(6, 3)$, $c = 6$

B9. $\mathbf{F} = (9y^2 + 8x + 5x^2)\mathbf{i} + (11xy + 8y^2 - 4x^3)\mathbf{j}$, $(-1, -5)$, $c = 6$

B10. $\mathbf{F} = (4y^3 - 10x + y)\mathbf{i} + (6xy^2 - 12y^3 + 10x^2)\mathbf{j}$, $(2, 8)$, $c = 4$

B11. $\mathbf{F} = (12x^2 - 3xy^2 - 5y^2)\mathbf{i} + (9x^2y - 7x^2 - 9)\mathbf{j}$, $(8, -9)$, $c = 2$

B12. $\mathbf{F} = (xy + 5y - x^2)\mathbf{i} + (x^3 - 3x + 4y)\mathbf{j}$, $(-5, -5)$, $c = 5$

In Exercises B13–22:

   (a) Obtain a field plot for $\mathbf{F}$.

   (b) Find, and categorize as node, saddle, center, or spiral, all equilibrium points for the differential equation $\mathbf{R}' = \mathbf{F}(x, y)$. For nodes and spirals, determine if stable or unstable.

   (c) Obtain the flow lines for the field $\mathbf{F}$.

   (d) Let C be the given ellipse that has been parametrized with p, $0 \leq p \leq 2\pi$. At the highest point on the ellipse, plot $\mathbf{F}$, and also $\mathbf{F_N}$, the component of $\mathbf{F}$ along $\mathbf{N}$, the outward unit normal.

   (e) Compute the flux of $\mathbf{F}$ through C.

B13. $\mathbf{F} = (2x - x^2 + y)\mathbf{i} + (1 - 2y - 2x^2)\mathbf{j}$, $x = 3\cos p$, $y = \sin p$

B14. $\mathbf{F} = (8x - xy + 4x^2y)\mathbf{i} + (3xy + 6y + x^3)\mathbf{j}$, $x = 3 + 6\cos p$, $y = 6 + 2\sin p$

B15. $\mathbf{F} = (11x^2y + 11x + 3)\mathbf{i} + (8xy + 5x^2 + 9)\mathbf{j}$, $x = -5 + 8\cos p$, $y = -6 + 6\sin p$

B16. $\mathbf{F} = (3x - y^2 - 12)\mathbf{i} + (x^2 - 3y^2)\mathbf{j}$, $x = 1 + 4\cos p$, $y = 2 + 7\sin p$

B17. $\mathbf{F} = (12y^3 - 1 - 5x^3)\mathbf{i} + 10y\mathbf{j}$, $x = 4 + 3\cos p$, $y = 2\sin p$

B18. $\mathbf{F} = (9x^2y + 5)\mathbf{i} + (12x^2 - 10xy)\mathbf{j}$, $x = 8 + 6\cos p$, $y = 4 + 2\sin p$

B19. $\mathbf{F} = (10xy^2 - 2 - 2y)\mathbf{i} + (6x^2y - 7y^2 - 2)\mathbf{j}$, $x = 4 + \cos p$, $y = 6 + 3\sin p$

B20. $\mathbf{F} = (7y^3 - 5y - 8x^2)\mathbf{i} + (5y^3 + 10 - 5x^3)\mathbf{j}$, $x = 5 + 9\cos p$, $y = 6 + 4\sin p$

B21. $\mathbf{F} = (3x^3 - 5)\mathbf{i} + (4x^2y + 4y^2 - 10y)\mathbf{j}$, $x = 3 + 9\cos p$, $y = -9 + 3\sin p$

B22. $\mathbf{F} = (10xy - 9)\mathbf{i} + (2x + 4y^2)\mathbf{j}$, $x = 9 + 8\cos p$, $y = 4 + 3\sin p$

In Exercises B23–32:

   (a) Obtain a field plot for $\mathbf{F}$.

   (b) Graph C, the closed curve formed by y_1 and y_2.

   (c) At the lowest point on y_1, plot $\mathbf{F}$, and also $\mathbf{F_N}$, the component of $\mathbf{F}$ along $\mathbf{N}$, the outward unit normal.

   (d) Compute the flux of $\mathbf{F}$ through C.

B23. $\mathbf{F} = (5x^2 + 2 - 6y^2)\mathbf{i} + (2x^2 + 5x + 6y^2)\mathbf{j}$, $y_1 = 2x^2 + 8x + 7$, $y_2 = 19 - 2x$

B24. $\mathbf{F} = (5y^3 + 5x + 5y^2)\mathbf{i} + (5x^3 - 2y^3 - 4y)\mathbf{j}$, $y_1 = x^2 + 3x + 4$, $y_2 = 3x + \frac{97}{4}$

B25. $\mathbf{F} = (x + y)\mathbf{i} + (5x - 8y - 9)\mathbf{j}$, $y_1 = 6x^2 + 4x - 7$, $y_2 = \frac{79}{3} - 6x$

B26. $\mathbf{F} = (4x^3 - 4 - 2y)\mathbf{i} + (1 + y - 2x)\mathbf{j}$, $y_1 = 3x^2 - 3x - 9$, $y_2 = \frac{7}{3}x^2 + \frac{1}{3}x - \frac{5}{2}$

B27. $\mathbf{F} = (3xy - 7xy^2 + 8x)\mathbf{i} + (8x^2y + 6 - y)\mathbf{j}$, $y_1 = 8x^2 - 4x + 3$, $y_2 = 7x^2 - \frac{5}{2}x + \frac{75}{16}$

B28. $\mathbf{F} = (7x^3 - 5x^2 + 9y^3)\mathbf{i} + (6x^2y + 8y^3 - x)\mathbf{j}$, $y_1 = 2x^2 - 4x - 4$, $y_2 = \frac{4}{5}x^2 + \frac{16}{5}x - 4$

B29. $\mathbf{F} = (4x^3 + 5 + x)\mathbf{i} + (5x - y)\mathbf{j}$, $y_1 = 2x^2 - 6x - 9$, $y_2 = \frac{3}{2}x^2 - \frac{5}{2}x - \frac{57}{8}$

B30. $\mathbf{F} = (4x^2 - y)\mathbf{i} + (y^2 - 7x)\mathbf{j}$, $y_1 = 4x^2 - 6x - 6$, $y_2 = \frac{7}{2}x^2 - \frac{21}{4}x + \frac{55}{32}$

B31. $\mathbf{F} = (2 - 8y^2 + 9x)\mathbf{i} + (x^2y - y - 8xy)\mathbf{j}$, $y_1 = x^2 + 9x + 5$, $y_2 = \frac{5}{6}x^2 + \frac{22}{3}x + \frac{47}{8}$

B32. $\mathbf{F} = (7x^3 - 8 + 7y^2)\mathbf{i} + (4x^3 - 8xy^2 - 2xy)\mathbf{j}$, $y_1 = 6x^2 - 4x + 8$, $y_2 = 4x^2 + \frac{4}{3}x + \frac{112}{9}$

In Exercises B33–42:

   (a) Obtain a field plot for $\mathbf{F}$.

   (b) Graph C, the triangle whose vertices {A, B, C} are given.

   (c) At the midpoint of segment AB, plot $\mathbf{F}$, and also $\mathbf{F_N}$, the component of $\mathbf{F}$ along $\mathbf{N}$, the outward unit normal.

   (d) Compute the flux of $\mathbf{F}$ through C.

B33. $\mathbf{F} = (3x^3 + 7y^3 + 4y)\mathbf{i} + (3x^3 + 7)\mathbf{j}$, {(4, 4), (−9, 4), (−10, −4)}

B34. $\mathbf{F} = (6x - 5x^3 - y)\mathbf{i} + 6\mathbf{j}$, {(6, 3), (−9, 5), (−5, −1)}

B35. $\mathbf{F} = (9x^2 - 3xy^2)\mathbf{i} + (2 + 5y^2 - 4x)\mathbf{j}$, {(3, 12), (−10, 8), (−5, −11)}

B36. $\mathbf{F} = (5y - 2xy - x^2y)\mathbf{i} + (2x^3 - 7x^2y + 4y^2)\mathbf{j}$, {(8, 1), (−4, 4), (−9, −7)}

B37. $\mathbf{F} = (x^3 - 7xy + 9xy^2)\mathbf{i} + (3xy + 2 + 4x)\mathbf{j}$, {(5, 4), (−11, 5), (−5, −1)}

B38. $\mathbf{F} = (6x^3 - 4y^3 - 6y^2)\mathbf{i} + (3xy - 4x^3 + 9x^2)\mathbf{j}$, {(9, 1), (−11, 8), (−5, −2)}

B39. $\mathbf{F} = (5x^2 - 4y)\mathbf{i} + (7xy^2 - y + 3)\mathbf{j}$, {(1, 8), (−10, 6), (−5, −11)}

B40. $\mathbf{F} = (4x^3 + 3x^2y - 8x)\mathbf{i} + (5xy + 2x^3 + 2xy^2)\mathbf{j}$, {(1, 9), (−12, 7), (−1, −8)}

B41. $\mathbf{F} = (3y^3 - 2xy - y)\mathbf{i} + (2y^3 - 6xy - 2y^2)\mathbf{j}$, {(2, 11), (−5, 5), (−8, −6)}

B42. $\mathbf{F} = (5xy + 7y^2 + 9x)\mathbf{i} + (5 + 6x + y)\mathbf{j}$, {(7, 11), (−9, 7), (−3, −8)}

In Exercises B43–52:

   (a) Obtain a field plot for $\mathbf{F}$.

   (b) Graph C, the quadrilateral whose vertices {A, B, C, D} are given.

   (c) At the midpoint of segment AB, plot $\mathbf{F}$, and also $\mathbf{F_N}$, the component of $\mathbf{F}$ along $\mathbf{N}$, the outward unit normal.

   (d) Compute the flux of $\mathbf{F}$ through C.

B43. $\mathbf{F} = (-5x^2y + 9y^3 - 8)\mathbf{i} + (-6x^3 - y^2 - 3x)\mathbf{j}$, {(7, 6), (−1, 6), (−12, −11), (6, −11)}

B44. $\mathbf{F} = (7x^3 - 2y^2 - 3xy^2)\mathbf{i} + (x^2 + 7y - 6xy^2)\mathbf{j}$, {(11, 1), (−3, 4), (−12, −6), (6, −4)}

B45. $\mathbf{F} = (7 + 7y^2)\mathbf{i} + (-3x^2 - 7x^2y + 3y^2)\mathbf{j}$, {(12, 12), (−11, 4), (−1, −6), (3, −8)}

B46. $\mathbf{F} = (9x^2y + 6 - 6y^2)\mathbf{i} + (6y^2 - 8x + 3xy^2)\mathbf{j}$, {(7, 6), (−5, 5), (−7, −1), (4, −9)}

B47. $\mathbf{F} = (x^3 - 6x^2 + 8x^2y)\mathbf{i} + (x + y)\mathbf{j}$, {(1, 7), (−12, 2), (−4, −3), (5, −2)}

B48. $\mathbf{F} = (5xy + x^2 - 4y)\mathbf{i} + (8x^3 + 7y^2 + 3y)\mathbf{j}$, {(11, 2), (−7, 6), (−10, −10), (2, −8)}

B49. $\mathbf{F} = (2x^2 + 8 - 5y^2)\mathbf{i} + (x^2y - 5x - 8y)\mathbf{j}$, {(5, 7), (−12, 2), (−2, −2), (1, −8)}

B50. $\mathbf{F} = (7x^3 - 5y^3 + 6y)\mathbf{i} + (5x^2y - 9x + 2y)\mathbf{j}$, {(9, 6), (−6, 10), (−1, −5), (5, −11)}

B51. $\mathbf{F} = (3y^3 - 8x^3 + 6x)\mathbf{i} + (8x^2 - 5y)\mathbf{j}$, {(3, 5), (−10, 10), (−11, −3), (3, −5)}

B52. $\mathbf{F} = (8x^3 + 3y^2 - 2y)\mathbf{i} + (2 - 7x^3 - 8xy)\mathbf{j}$, {(7, 7), (−10, 1), (−1, −12), (1, −9)}

Chapter Review

1. The line integral of $f(x, y)$ is to be evaluated along a curve C parametrized by $x = x(t)$, $y = y(t)$. How are the line integrals $\int_C f(x, y)\,dx$ and $\int_C f(x, y)\,dy$ evaluated? Give an example of each.

2. The line integral of $f(x, y)$ is to be evaluated along a curve C given by $y = y(x)$. How are the line integrals $\int_C f(x, y)\,dx$ and $\int_C f(x, y)\,dy$ evaluated? Give an example of each.

3. The line integral of $f(x, y)$ is to be evaluated along a curve C given by $x = x(y)$. How are the line integrals $\int_C f(x, y)\,dx$ and $\int_C f(x, y)\,dy$ evaluated? Give an example of each.

4. Evaluate the line integrals $\int_C f(x, y)\,dx$ and $\int_C f(x, y)\,dy$, where $f(x, y) = x^2y^3$ and C is the counterclockwise path formed by the edges of the triangle whose vertices are (1, 1), (4, 5), and (9, 8).

5. A particle is moved from point A to point B along the curve C in a field $\mathbf{F}$. The work done is computed by the line integral

$W = \int_C \mathbf{F} \cdot d\mathbf{r}$. If $W > 0$, did the field move the particle or did an outside agent have to push the particle against the field? What if $W < 0$?

6. If $\mathbf{F} = x \sin y\mathbf{i} + y \cos x\mathbf{j}$ is a field and C is the line segment from (2, 3) to (−3, 2):

   (a) Find the work done when a unit particle is moved through the field $\mathbf{F}$ and along C.

   (b) Find the flux of $\mathbf{F}$ through C. Clearly state the normal direction, and interpret the value obtained for the flux.

7. Obtain the circulation of the field $\mathbf{F}$ from Question 6 if C is the unit circle centered at the origin.

8. Sketch representative arrows for the field $\mathbf{F}$ in Question 6.

9. Sketch representative flow lines for the field $\mathbf{F}$ in Question 6.

Chapter 20

Additional Vector
Differential Operators

INTRODUCTION   The *divergence* and *curl* of a vector field are introduced as
two additional vector differential operators. The divergence of a vector field is defined for
a vector with any number of components, but the curl is defined only for a vector in three
dimensions.

The divergence is interpreted physically as the limiting value of the ratio of the flux
through the boundary of a curve that shrinks to a point and the area enclosed by that curve.
As such, the divergence is a measure of the pointwise dispersion of a vector field.

The curl is interpreted physically as the limiting ratio of the circulation around a closed
plane curve that shrinks to a point and the area enclosed by that curve. As such, the curl is
a measure of the pointwise rotation in a vector field. This interpretation is generalized in
Section 21.6.

There are now three vector operators, the divergence, the gradient, and the curl. A
fourth operator, the laplacian, can be expressed in terms of the divergence and the gradient.
We compute the divergence and curl of a vector field and the gradient of a scalar field. The
divergence yields a scalar, but the curl yields another vector. The laplacian of a scalar field
is again a scalar field.

We present 11 formulas or identities relating these vector operators to one another. Six
contain a single del operator that acts on a product of two operands. The other 5 contain
two del operators, del acting on the result of another del applied to a single operand.

20.1   Divergence and Its Meaning

DEFINITION 20.1   DIVERGENCE

Given the vector field $\mathbf{F} = u(x, y, z)\mathbf{i} + v(x, y, z)\mathbf{j} + w(x, y, z)\mathbf{k}$, the *divergence* of
$\mathbf{F}$ at the point (x, y, z) is the scalar

$$\nabla \cdot \mathbf{F} = \mathrm{div}(\mathbf{F}) = u_x(x, y, z) + v_y(x, y, z) + w_z(x, y, z)$$

The divergence of the field **F** measures the "spread" of the field at each point. Notational clarifications, as well as examples and calculations designed to illuminate this claim about divergence measuring "spread" of the field at a point, follow.

EXAMPLE 20.1   The field $\mathbf{F} = x\mathbf{i} + y\mathbf{j}$, graphed in Figure 20.1, appears to be "flowing" radially outward, with flow lines $y = \alpha x$, the solution of the set differential equations $x'(p) = x$, $y'(p) = y$, contained in $\mathbf{R}'(p) = \mathbf{F}$. The parameter p is eliminated from these equations by writing $\frac{dy}{dx} = \frac{y'}{x'} = \frac{y}{x}$, and the solution is immediate. The field arises from the scalar potential $u(x, y) = \frac{1}{2}(x^2 + y^2)$, where the minus sign has been ignored. The level curves of $u(x, y)$ are concentric circles orthogonal to the flow lines of the field **F**, as seen in Figure 20.1. The divergence is $\nabla \cdot \mathbf{F} = 2$. The constant and positive value of the divergence signals the uniform spread of the field, evidenced by the flow pointing radially outward from the origin.   ❖

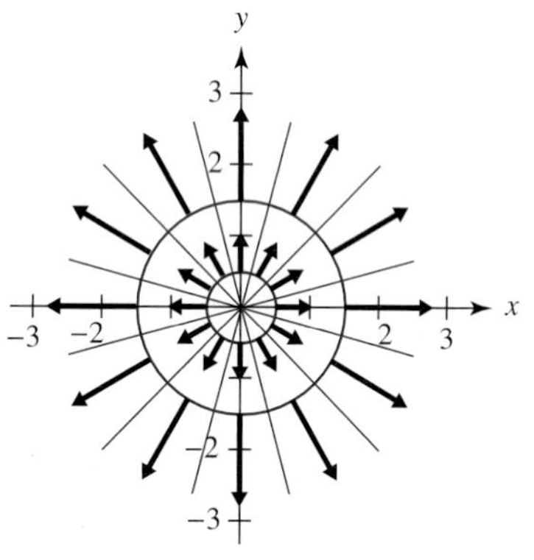

FIGURE 20.1   The vector field, flow lines, and equipotentials for Example 20.1

Divergence as Limiting Value of Flux per Unit Area

We next give a physical interpretation of the divergence based on the notion of flux through a closed curve C. It demonstrates that in the limit as the curve C shrinks to a point P, the ratio of the flux of the field **F** through C divided by the area enclosed in C becomes the divergence of the field **F** evaluated at P.

To compute the flux of the arbitrary plane vector $\mathbf{F} = f(x, y)\mathbf{i} + g(x, y)\mathbf{j}$ through C, a circle of radius a, centered at the point P whose coordinates are (x_0, y_0), write the parametric representation of the circle as $x(p) = x_0 + a \cos p$, $y(p) = y_0 + a \sin p$, so that on C we have $\mathbf{F} = \mathbf{F}(x_0 + a \cos p, y_0 + a \sin p)$. As the circle C shrinks to the point P, the radius a goes to zero. Hence, expand **F** in a Taylor series about $a = 0$, obtaining

$$F = \begin{bmatrix} f(x_0, y_0) + (f_x(x_0, y_0) \cos p + f_y(x_0, y_0) \sin p)a + \mathrm{O}(a^2) \\ g(x_0, y_0) + (g_x(x_0, y_0) \cos p + g_y(x_0, y_0) \sin p)a + \mathrm{O}(a^2) \end{bmatrix}$$

The flux integral $\oint_C f \, dy - g \, dx$ is then

$$\int_0^{2\pi} a[(f_0 + (f_x \cos p + f_y \sin p)a) \cos p$$
$$+ (g_0 + (g_x \cos p + g_y \sin p)a) \sin p] \, dp + \mathrm{O}(a^3)$$

so dividing by πa^2, the area enclosed by the circle C, gives $f_x + g_y + \mathrm{O}(a)$. Thus, in the limit as a goes to zero, the limiting ratio of flux to area becomes $f_x + g_y$, the divergence of **F**.

Solenoidal Fields

Because the magnetic field set up by a solenoid (coil of wire) has zero divergence in the space surrounded by the coil, physicists use the term *solenoidal* for fields whose divergence is everywhere zero. For example, the field $\mathbf{F} = x\mathbf{i} - y\mathbf{j}$ is solenoidal since its divergence is $1 - 1 = 0$. On the other hand, the field $\mathbf{F} = x\mathbf{i} + y\mathbf{j}$ is not solenoidal, since the divergence of **F** is $1 + 1 = 2$, as seen previously. Connections between conservative fields and solenoidal fields are detailed in Section 21.5.

If **F** is the field of velocity vectors for a steady fluid flow with constant density, then its divergence must be zero, as we discuss in Section 23.3. But a steady flow with constant density is *incompressible*, so the analog of the solenoidal field is the incompressible flow.

The Laplacian

The *laplacian* of the scalar $u(x, y, z)$ is the scalar field $\nabla \cdot (\nabla u) = \nabla^2 u$, the divergence of the gradient field of u.

In Cartesian coordinates, the del operator (Section 18.2) is $\nabla = \mathbf{i}\frac{\partial}{\partial x} + \mathbf{j}\frac{\partial}{\partial y} + \mathbf{k}\frac{\partial}{\partial z}$, and the gradient of the scalar $u(x, y, z)$ is the vector $\nabla u = u_x \mathbf{i} + u_y \mathbf{j} + u_z \mathbf{k}$. The divergence of $\mathbf{F}$ is

$$\nabla \cdot \mathbf{F} = \left(\mathbf{i}\frac{\partial}{\partial x} + \mathbf{j}\frac{\partial}{\partial y} + \mathbf{k}\frac{\partial}{\partial z} \right) \cdot (f\mathbf{i} + g\mathbf{j} + h\mathbf{k}) = f_x + g_y + h_z$$

so the laplacian of $u(x, y, z)$ is

$$\nabla \cdot (\nabla u) = \left(\mathbf{i}\frac{\partial}{\partial x} + \mathbf{j}\frac{\partial}{\partial y} + \mathbf{k}\frac{\partial}{\partial z} \right) \cdot \left(\mathbf{i}\frac{\partial}{\partial x} + \mathbf{j}\frac{\partial}{\partial y} + \mathbf{k}\frac{\partial}{\partial z} \right) u$$

$$= \left(\frac{\partial^2}{\partial x^2} + \frac{\partial^2}{\partial y^2} + \frac{\partial^2}{\partial z^2} \right) u = \frac{\partial^2 u}{\partial x^2} + \frac{\partial^2 u}{\partial y^2} + \frac{\partial^2 u}{\partial z^2}$$

$$= u_{xx} + u_{yy} + u_{zz} = \nabla^2 u$$

Thus, the laplacian operator is generally called "del-squared" and in Cartesian coordinates is written $\nabla^2 = \frac{\partial^2}{\partial x^2} + \frac{\partial^2}{\partial y^2} + \frac{\partial^2}{\partial z^2}$.

Generalizations of the Directional Derivative

FIRST GENERALIZATION Texts on fluid dynamics or electromagnetic theory generalize $D_{\mathbf{U}} f = \nabla f \cdot \mathbf{U}$, the directional derivative of the scalar $f(x, y, z)$, by introducing the operator

$$\mathbf{V} \cdot \nabla = (u\mathbf{i} + v\mathbf{j} + w\mathbf{k}) \cdot \left(\mathbf{i}\frac{\partial}{\partial x} + \mathbf{j}\frac{\partial}{\partial y} + \mathbf{k}\frac{\partial}{\partial z} \right) \tag{20.1a}$$

$$= u\frac{\partial}{\partial x} + v\frac{\partial}{\partial y} + w\frac{\partial}{\partial z} \tag{20.1b}$$

The directional derivative of Section 18.2 requires that the vector $\mathbf{U}$ be a unit vector. In this first generalization of the directional derivative, the vector $\mathbf{V}$ need not be a unit vector. When applied to a scalar $f(x, y, z)$, this operator is just the gradient dotted with $\mathbf{V}$, giving for f a derivative in the direction $\mathbf{V}$.

SECOND GENERALIZATION Between two neighboring points P_1 and P_2, a change in the vector field

$$\mathbf{F} = f(x, y, z)\mathbf{i} + g(x, y, z)\mathbf{j} + h(x, y, z)\mathbf{k} \tag{20.2}$$

is again a vector. If $\mathbf{V}$ is proportional to $\mathbf{P}_2 - \mathbf{P}_1$, then at P_1 the directional derivative of $\mathbf{F}$ in the direction $\mathbf{V}$ is given by

$$(\mathbf{V} \cdot \nabla)\mathbf{F} = \begin{bmatrix} \mathbf{V} \cdot \nabla f \\ \mathbf{V} \cdot \nabla g \\ \mathbf{V} \cdot \nabla h \end{bmatrix} = \begin{bmatrix} uf_x + vf_y + wf_z \\ ug_x + vg_y + wg_z \\ uh_x + vh_y + wh_z \end{bmatrix} = \begin{bmatrix} f_x & f_y & f_z \\ g_x & g_y & g_z \\ h_x & h_y & h_z \end{bmatrix} \begin{bmatrix} u \\ v \\ w \end{bmatrix} \tag{20.3}$$

where the matrix on the right in (20.3) is the Jacobian matrix for $\mathbf{F}$ and the operator $\mathbf{V} \cdot \nabla$ is given by (20.1a) and (20.1b). The formalism in (20.3) is established in Exercise B1 of Section 18.2. The idea behind the formalism is a generalization of the directional derivative.

Think of $\mathbf{F}$ as a force field in which $\mathbf{R}(p) = \mathbf{P}_1 + p\mathbf{V}$ is a line through P_1 with direction $\mathbf{V}$. Evaluating $\mathbf{F}$ along the line gives $\mathbf{F}(p)$, and $\frac{d\mathbf{F}}{dp}\big|_{p=0}$ is the directional derivative of the field measured at P_1 in the direction of $\mathbf{V}$.

When this process was carried out in Section 18.2 for $D_{\mathbf{U}} f = \nabla f \cdot \mathbf{U}$, the directional derivative of the scalar field f, the gradient vector ∇f appeared as a new object dotted with the direction vector $\mathbf{U}$. The matrix that appears in (20.3) is a new object, one "rank" higher than the vector field $\mathbf{F}$, called the *covariant derivative*. Multiplication of the direction vector $\mathbf{V}$ by this object gives the desired rate of change.

EXERCISES 20.1–Part A

A1. Compute the divergence of

   (a) $\mathbf{F} = xy^2\mathbf{i} + x^2y\mathbf{j}$ **(b)** $\mathbf{F} = xy\mathbf{i} + yz^2\mathbf{j} + x^3z\mathbf{k}$

A2. Let $\mathbf{F} = x\mathbf{i} + y\mathbf{j}$ and let C be a circle with radius a and center at the origin. Obtain the flux of F through C, and compute the limit, as a goes to zero, of the ratio of the flux to the area enclosed by C. Show that this limit is the divergence of F evaluated at the origin.

A3. Show that $\mathbf{F} = yz\mathbf{i} + xz\mathbf{j} + xy\mathbf{k}$ is solenoidal.

A4. Let $\mathbf{F}$ be as in (20.2), where f, g, h, all have continuous second partial derivatives. Show that the field $(h_y - g_z)\mathbf{i} + (f_z - h_x)\mathbf{j} + (g_x - f_y)\mathbf{k}$ is solenoidal.

A5. Obtain the laplacian of $u(x, y, z) = x^2y^3z^4$.

EXERCISES 20.1–Part B

B1. For each of the vector fields $\mathbf{F}_1 = x\mathbf{i} + y\mathbf{j}$ and $\mathbf{F}_2 = \frac{x}{x^2+y^2}\mathbf{i} + \frac{y}{x^2+y^2}\mathbf{j}$:

   (a) Obtain a field plot.

   (b) Obtain the flow lines for the field $\mathbf{F}_k, k = 1, 2$.

   (c) Obtain the divergence of $\mathbf{F}_k, k = 1, 2$. Both fields "appear" to be diverging, but one actually has positive divergence and the other has zero divergence. Divergence is a "local" or "point" property of a vector field, whereas the field plot and the flow lines exhibit more "global" behavior of the field. Caution, therefore, should be exercised when interpreting graphs of vector fields and their flows.

B2. If $f(x, y) = \arctan \frac{y}{x}$ and $g(x, y) = x^2 - y^2$, show that the vector field $\mathbf{F} = \nabla f \times \nabla g$ is solenoidal.

B3. Show that, in general, for scalar-valued functions $f(x, y, z)$ and $g(x, y, z)$ with continuous second partial derivatives, the vector field $\mathbf{F} = \nabla f \times \nabla g$ is solenoidal.

In Exercises B4–13, a pair of scalar-valued functions $f(x, y, z)$ and $g(x, y, z)$ and a point P are given. For each, compute (at P) the divergence of the field $\mathbf{F} = \|\nabla f\| \nabla g$.

B4. $f(x, y, z) = -8xy^2 - 6xyz - 9y^3$, $g(x, y, z) = 4x^2y + z^3 + 2x$, $P = (0, -5, 4)$

B5. $f(x, y, z) = 3x^2 - 7xz^2 + yz^2$, $g(x, y, z) = -z^2 + 7x^2z + 2y^3$, $P = (4, -7, -2)$

B6. $f(x, y, z) = z^2 + 8y^2z + 9z$, $g(x, y, z) = -3xz^2 - 2xyz + 6y^3$, $P = (9, -3, 8)$

B7. $f(x, y, z) = -yz^2 - 7$, $g(x, y, z) = y - 8y^2z + 8xyz$, $P = (-7, -4, -6)$

B8. $f(x, y, z) = 3z^3 - 3x^3 - yz$, $g(x, y, z) = -4xy^2 + 7x^3 + 3yz$, $P = (5, 0, 9)$

B9. $f(x, y, z) = 2y^2z - 7 + 2y$, $g(x, y, z) = 8x^2 + 9xz^2 + 4y$, $P = (-5, 9, 1)$

B10. $f(x, y, z) = 3x^2 + 4x^2y$, $g(x, y, z) = 5x^2 - 8y^2$, $P = (-9, -7, -1)$

B11. $f(x, y, z) = 2 - 8xz^2 - 5x^3$, $g(x, y, z) = 8x^2z + 3x + 3xz$, $P = (3, -9, 4)$

B12. $f(x, y, z) = 5 - 7x^3 - 3y$, $g(x, y, z) = 4x^2y - 8xy^2 - 4y^2z$, $P = (-6, -6, -2)$

B13. $f(x, y, z) = 5z - 7xy^2 + 5xyz$, $g(x, y, z) = 7 - 9x + 3z$, $P = (-8, -4, -7)$

For the vector field $\mathbf{F}$ given in each of Exercises B14–25:

   (a) Obtain a field plot.

   (b) Obtain a graph of the flow lines.

   (c) Compute $\nabla \cdot \mathbf{F}$, the divergence of $\mathbf{F}$. Is $\mathbf{F}$ solenoidal?

B14. $\mathbf{F} = (4xy^2 + 5)\mathbf{i} + (8x^3 - 3x^2)\mathbf{j}$ **B15.** $\mathbf{F} = -18y^2\mathbf{i} - 6x^2\mathbf{j}$

B16. $\mathbf{F} = (y - 2x)\mathbf{i} - (9 + 7y)\mathbf{j}$

B17. $\mathbf{F} = -9x^2\mathbf{i} + (5xy - 9y^2)\mathbf{j}$

B18. $\mathbf{F} = (3x^2 - 9x)\mathbf{i} - 9y(1 - y)\mathbf{j}$

B19. $\mathbf{F} = (-3y^2 - 5xy)\mathbf{i} + (1 - 2x^2)\mathbf{j}$

B20. $\mathbf{F} = (7x - 8x^2)\mathbf{i} + (2y - 2)\mathbf{j}$

B21. $\mathbf{F} = (4y - 7x^2)\mathbf{i} + (4x^2 - 7y)\mathbf{j}$

B22. $\mathbf{F} = (xy - 6y^2)\mathbf{i} - 3x^2\mathbf{j}$ **B23.** $\mathbf{F} = -12xy^2\mathbf{i} + (4y^3 - 16x)\mathbf{j}$

B24. $\mathbf{F} = (2y + 4xy)\mathbf{i} + (8 + 5y^2)\mathbf{j}$

B25. $\mathbf{F} = (6y^2 - 8)\mathbf{i} + (7x + 5y)\mathbf{j}$

For each vector field $\mathbf{F}$ and point $P = (a, b)$ given in Exercises B26–35:

(a) Compute $\nabla \cdot \mathbf{F}$ at P.

(b) Obtain a field plot of $\mathbf{F}$.

(c) Obtain a graph of the flow lines of $\mathbf{F}$.

(d) Obtain $\mathcal{F}$, the flux of $\mathbf{F}$ through a circle with radius ρ and center at P.

(e) Obtain $\lim_{\rho \to 0} \frac{\mathcal{F}}{\pi\rho^2}$, that is, compute, as the circle shrinks to a point, the limit of the ratio of the flux to the enclosed area. It should agree with the divergence computed in part (a).

(f) Compute $\mathcal{F}$, the flux of $\mathbf{F}$ through a square with side 2ρ and center at P.

(g) Obtain $\lim_{\rho \to 0} \frac{\mathcal{F}}{4\rho^2}$, that is, compute, as the square shrinks to a point, the limit of the ratio of the flux to the enclosed area. It should agree with the divergence computed in part (a).

B26. $\mathbf{F} = (7x^3 + 4x - y^2)\mathbf{i} + (-9x - 4y^2 - 7xy)\mathbf{j}$, $P = (2, 2)$

B27. $\mathbf{F} = (9x^2 - 3x^3 - 9y)\mathbf{i} + (7y - 8y^3 - 8)\mathbf{j}$, $P = (7, 4)$

B28. $\mathbf{F} = (3x - 3x^3 - 2xy)\mathbf{i} + (2x^2 - 2 - 7x^2y)\mathbf{j}$, $P = (-3, 9)$

B29. $\mathbf{F} = (7y + 2x^2y - 2)\mathbf{i} + (4xy^2 + 9x)\mathbf{j}$, $P = (8, -4)$

B30. $\mathbf{F} = (2x^2 + 7x^3 + 3)\mathbf{i} + (x^2y - 2x + 6y^2)\mathbf{j}$, $P = (-3, 2)$

B31. $\mathbf{F} = (2x^2y + 7xy - 4)\mathbf{i} + (5 - 2x + 6x^2y)\mathbf{j}$, $P = (3, -7)$

B32. $\mathbf{F} = (xy^2 + 9x + 3y)\mathbf{i} + (2x^2 + 9y^3 - 6xy)\mathbf{j}$, $P = (-3, 5)$

B33. $\mathbf{F} = (9x + 2y^3 - 9x^2y)\mathbf{i} + (9x^2 + 2xy^2 - x)\mathbf{j}$, $P = (-5, 3)$

B34. $\mathbf{F} = (y - 6x^2y - 6y^2)\mathbf{i} + (y^2 + 6xy - 3)\mathbf{j}$, $P = (-5, 2)$

B35. $\mathbf{F} = (3xy^2 - 5x^2y)\mathbf{i} + (3 + 7y^2 - 8x^2y)\mathbf{j}$, $P = (-5, -8)$

B36. Show that the area enclosed by the ellipse $\frac{x^2}{a^2} + \frac{y^2}{b^2} = 1$ is πab.

B37. Show that the area enclosed by the ellipse $x = a + \rho\cos t$, $y = b + \kappa\rho\sin t$, $0 \le t \le 2\pi$, is $\kappa\pi\rho^2$. *Hint:* Arrange the parametric equations for the ellipse so that the identity $\cos^2\theta + \sin^2\theta = 1$ can be used.

For the vector field $\mathbf{F}$ and point $P = (a, b)$ given in each of Exercises B38–47:

(a) Compute $\nabla \cdot \mathbf{F}$ at point P.

(b) Obtain a field plot of $\mathbf{F}$.

(c) Obtain a graph of the flow lines of $\mathbf{F}$.

(d) Compute $\mathcal{F}$, the flux of $\mathbf{F}$ through the ellipse $x = a + \rho\cos t$, $y = b + \kappa\rho\sin t$, $0 \le t \le 2\pi$, where κ is as given in each case.

(e) Obtain $\lim_{\rho \to 0} \frac{\mathcal{F}}{\kappa\pi\rho^2}$, that is, compute, as the ellipse shrinks to a point, the limit of the ratio of the flux to the enclosed area. It should agree with the divergence computed in part (a).

B38. $\mathbf{F} = (8x^2 + 8 - y)\mathbf{i} + (4x^2 + 4y)\mathbf{j}$, $P = (6, -6)$, $\kappa = 4$

B39. $\mathbf{F} = (6 - 4y - 8xy)\mathbf{i} + (5y - 8y^2 - 5x)\mathbf{j}$, $P = (1, -2)$, $\kappa = 6$

B40. $\mathbf{F} = (7x^2 + 9y^2 + 2y)\mathbf{i} + (6x^3 - 5y^2)\mathbf{j}$, $P = (7, -5)$, $\kappa = 3$

B41. $\mathbf{F} = (8x - 5y^3 - 4y^2)\mathbf{i} + (5x^2 - 6x^2y - 8xy)\mathbf{j}$, $P = (1, 1)$, $\kappa = 7$

B42. $\mathbf{F} = (8xy - 2x^3 - 8y^3)\mathbf{i} + (5xy^2 - 7y^3 - x^2y)\mathbf{j}$, $P = (-2, -6)$, $\kappa = 7$

B43. $\mathbf{F} = (8 - 6y^3 - 7xy^2)\mathbf{i} + (8y - 8x^3 - 5y^3)\mathbf{j}$, $P = (5, -6)$, $\kappa = 6$

B44. $\mathbf{F} = (5x^2y - 4xy^2 - 3y^2)\mathbf{i} + (2x^2 + 8y)\mathbf{j}$, $P = (-3, -1)$, $\kappa = 3$

B45. $\mathbf{F} = (9x^3 - 8y + 7y^2)\mathbf{i} + (7x^2 + 6x^3 - 5y^2)\mathbf{j}$, $P = (-2, -2)$, $\kappa = 7$

B46. $\mathbf{F} = (2 - 2x^2 + xy)\mathbf{i} + (5xy - 8y)\mathbf{j}$, $P = (-5, -5)$, $\kappa = 5$

B47. $\mathbf{F} = (9x^2y - 9x - 4xy)\mathbf{i} + (5x^3 - 8x^2y + 4xy)\mathbf{j}$, $P = (-1, -8)$, $\kappa = 7$

For the vector field $\mathbf{V}$, scalar-valued function $f(x, y)$, and point P given in each of Exercises B48–57, compute $(\mathbf{V} \cdot \nabla)f$ at P.

B48. $\mathbf{V} = (6y^3 - 3x)\mathbf{i} + (3y - 2x^2y + 8xy)\mathbf{j}$, $f(x, y) = \frac{x^3 - 5xy}{8 + 5xy}$, $P = (4, -7)$

B49. $\mathbf{V} = (x - 3xy^2 - 7x^2y)\mathbf{i} + (5x - 4)\mathbf{j}$, $f(x, y) = \frac{x^2 + y^3}{2x^3 - 2x^2y}$, $P = (5, -3)$

B50. $\mathbf{V} = (6x - 3x^3 + 2x^2y)\mathbf{i} + (9y^2 - 7x - 9x^2y)\mathbf{j}$, $f(x, y) = \frac{5x^2}{4x^3 - x}$, $P = (9, -7)$

B51. $\mathbf{V} = (7x^2 + 9y)\mathbf{i} + (3y^3 + 2x + y^2)\mathbf{j}$, $f(x, y) = \frac{x^2 - 8x^3}{8y^3 - 9}$, $P = (-1, 7)$

B52. $\mathbf{V} = (5 - 4x - 9y^2)\mathbf{i} + (2x^3 - 6x - 9x^2y)\mathbf{j}$, $f(x, y) = \frac{3xy^2 + 7xy}{3 - 7xy}$, $P = (-4, 1)$

B53. $\mathbf{V} = (7x^2 - 9x^2y + 7xy)\mathbf{i} + (8 - 4x^3 - 3y^2)\mathbf{j}$, $f(x, y) = \frac{9y + 9xy}{3x^3 + 2y}$, $P = (1, 6)$

B54. $\mathbf{V} = (7x^2 - 6xy^2 - 5y^2)\mathbf{i} + (6xy - y^3 - 3xy^2)\mathbf{j}$, $f(x, y) = \frac{7 - 2x^2y}{8xy^2 + 7y}$, $P = (2, -2)$

B55. $\mathbf{V} = (8x^2y - 6y^3 - 6xy)\mathbf{i} + (6x^2y - 3x^2 - xy)\mathbf{j}$, $f(x, y) = \frac{x^2}{y^2}$, $P = (-5, 2)$

B56. $\mathbf{V} = (1 - 8y^3 + 5x^2y)\mathbf{i} + (3y^3 - 6x - 2y^2)\mathbf{j}$, $f(x, y) = \frac{9x^3 + 6}{8x^2 + 3xy}$, $P = (-9, 5)$

B57. $\mathbf{V} = (9x^3 - 4y + xy)\mathbf{i} + (3x^2 - 5x + 2y)\mathbf{j}$,

$f(x, y) = \dfrac{6x^2 - 3}{4x^2 - 6y^3}$, $P = (-5, 5)$

In Exercises B58–67, given the vector fields $\mathbf{F}$ and $\mathbf{V}$, compute, at point P, the generalized directional derivative, $(\mathbf{V} \cdot \nabla)\mathbf{F}$.

B58. $\mathbf{F} = (7xy^2 - 8)\mathbf{i} + (8y^3 + 3xy)\mathbf{j}$, $\mathbf{V} = (x^2 + 2)\mathbf{i} + (2y^3 - 8x^2y)\mathbf{j}$,
$P = (5, 3)$

B59. $\mathbf{F} = (6xy^2 + 5)\mathbf{i} + (7x^2 + 3xy)\mathbf{j}$, $\mathbf{V} = (2 + 9xy)\mathbf{i} + (5y^3 - 9x)\mathbf{j}$,
$P = (8, -3)$

B60. $\mathbf{F} = 2y\mathbf{i} + (6x^3 - 7y)\mathbf{j}$, $\mathbf{V} = (7x^2 + 9y^2)\mathbf{i} + (xy^2 + 8y)\mathbf{j}$,
$P = (9, 5)$

B61. $\mathbf{F} = (6x^2 + y^2)\mathbf{i} + (6 - 3xy)\mathbf{j}$, $\mathbf{V} = (9y^3 + 4x)\mathbf{i} + (4y^3 + 9x)\mathbf{j}$,
$P = (1, -7)$

B62. $\mathbf{F} = 4xy^2\mathbf{i} + (x + 4x^2y)\mathbf{j}$, $\mathbf{V} = (xy^2 - 2y)\mathbf{i} + (3x^2 - 4xy^2)\mathbf{j}$,
$P = (-2, -5)$

B63. $\mathbf{F} = (5x^2 - 2y^3)\mathbf{i} + (7x^3 + 4y^2)\mathbf{j}$, $\mathbf{V} = -3y^2\mathbf{i} + (6y^2 + xy)\mathbf{j}$,
$P = (-4, 2)$

B64. $\mathbf{F} = (y^3 + 8xy^2)\mathbf{i} + (6x^2 + 3y)\mathbf{j}$, $\mathbf{V} = -9xy\mathbf{i} + (2x^3 - x^2y)\mathbf{j}$,
$P = (6, 8)$

B65. $\mathbf{F} = (9x + 9x^2y)\mathbf{i} + (3x^3 - 8y^3)\mathbf{j}$, $\mathbf{V} = (2x + 3xy)\mathbf{i} + (9x^3 - 2x^2y)\mathbf{j}$, $P = (7, 2)$

20.2   Curl and Its Meaning

> ### DEFINITION 20.2   CURL
>
> A third operator of vector calculus is the *curl*, taking a vector field $\mathbf{F}$ into another vector field $\nabla \times \mathbf{F}$, which measures "local rotation" in the vector field $\mathbf{F}$. As suggested by the notation, the formalism for computing the curl looks like the cross-product of the del operator $\nabla = \mathbf{i}\frac{\partial}{\partial x} + \mathbf{j}\frac{\partial}{\partial y} + \mathbf{k}\frac{\partial}{\partial z}$ and the vector field $\mathbf{F} = u(x, y, z)\mathbf{i} + v(x, y, z)\mathbf{j} + w(x, y, z)\mathbf{k}$.
>
> $$\text{curl}(\mathbf{F}) = \nabla \times \mathbf{F} = \begin{vmatrix} \mathbf{i} & \mathbf{j} & \mathbf{k} \\ \dfrac{\partial}{\partial x} & \dfrac{\partial}{\partial y} & \dfrac{\partial}{\partial z} \\ u & v & w \end{vmatrix} = \begin{bmatrix} w_y - v_z \\ -(w_x - u_z) \\ v_x - u_y \end{bmatrix} = \begin{bmatrix} w_y - v_z \\ u_z - w_x \\ v_x - u_y \end{bmatrix}$$

The determinant on the left is a mnemonic, a memory device, whereas the vector on the right is the definition of the curl in Cartesian coordinates.

EXAMPLE 20.2   The velocity field $\mathbf{V} = -y\mathbf{i} + x\mathbf{j} + 0\mathbf{k}$ arises from a flow in uniform rotation about the z-axis, and Figure 20.2 shows flow lines and field vectors for one representative z-plane. A computation shows this field has curl $\nabla \times \mathbf{V} = 2\mathbf{k}$. The following argument shows the angular velocity vector for the field is $\Omega = \mathbf{k}$, and we conclude that for the uniformly rotating field, the curl is twice the angular velocity vector.

The angular velocity vector lies along the z-axis, the axis of rotation, and points upward by the right-hand rule. Its magnitude is $\frac{d\theta}{dt}$, where θ is the angle that measures the rotation in radians. For $\mathbf{V}$, the magnitude of $\frac{d\theta}{dt}$ is 1 because

$$\|\mathbf{V}\| = \sqrt{x^2 + y^2} = r = \frac{ds}{dt} = r\frac{d\theta}{dt}$$

where $s = r\theta$ measures arc length around a circle of radius r. Hence, $\Omega = \mathbf{k}$.   ❖

Curl as Limiting Value of Circulation per Unit Area

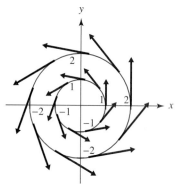
FIGURE 20.2   Example 20.2: The velocity field for uniform rotation about the z-axis

Another classic demonstration of the curl as a measure of rotation is the computation for the closed curve C and the planar field $\mathbf{F} = f(x, y)\mathbf{i} + g(x, y)\mathbf{j}$ of the limiting value of the ratio of circulation to area enclosed. Circulation of $\mathbf{F}$ about C is the line integral of the tangential component of the field **F**. The expression for circulation is therefore the same as that for mechanical work. This circulation is divided by the area enclosed by the curve C, and in the limit as C shrinks to a point P, the ratio becomes the z-component of $\nabla \times \mathbf{F}$ at P. Thus, taking C as a circle of radius a, we will show

$$(\nabla \times \mathbf{F}) \cdot \mathbf{k} = \lim_{a \to 0} \frac{1}{\pi a^2} \oint_C \mathbf{F} \cdot d\mathbf{r} = g_x - f_y$$

Let the circle C, with radius a and center (x_0, y_0), be given by $x(p) = x_0 + a \cos p$, $y(p) = y_0 + a \sin p$. On C, $\mathbf{F} = \mathbf{F}(x_0 + a \cos p, y_0 + a \sin p)$. As the circle C shrinks to the point P, the radius a goes to zero. Hence, expand **F** in a Taylor series about $a = 0$, obtaining

$$F = \begin{bmatrix} f(x_0, y_0) + (f_x(x_0, y_0) \cos p + f_y(x_0, y_0) \sin p)a + O(a^2) \\ g(x_0, y_0) + (g_x(x_0, y_0) \cos p + g_y(x_0, y_0) \sin p)a + O(a^2) \end{bmatrix}$$

The circulation integral $\oint_C f\, dx + g\, dy$ becomes

$$\int_0^{2\pi} a[-(f_0 + (f_x \cos p + f_y \sin p)a) \sin p$$
$$+ (g_0 + (g_x \cos p + g_y \sin t)a) \cos p]\, dp + O(a^3)$$

so dividing by πa^2, the area enclosed by the circle C, gives $g_x - f_y + O(a)$. Thus, in the limit as a goes to zero, the ratio of circulation to area becomes $g_x - f_y$, the z-component of the curl of **F**.

Irrotational Fields

Fields for which the curl vanishes at all points are called *irrotational*. Section 21.5 explores the important connections between *irrotational* fields and *conservative* fields.

EXAMPLE 20.3 The field $\mathbf{V} = (4xy - z^3)\mathbf{i} + 2x^2\mathbf{j} - 3xz^2\mathbf{k}$ is irrotational since $\nabla \times \mathbf{V} = \mathbf{0}$. (Once again, notice how we distinguish between **0**, the zero vector, and 0, the zero scalar.) ❖

EXERCISES 20.2–Part A

A1. Obtain the curl of
  (a) $\mathbf{F} = xe^y\mathbf{i} + y \sin x\mathbf{j}$ (b) $\mathbf{F} = xz\mathbf{i} + yz\mathbf{j} + xy\mathbf{k}$

A2. Verify $\nabla \times \mathbf{V} = 2\mathbf{k}$ for **V** given in Example 20.2.

A3. If $\mathbf{F} = y^2\mathbf{i} + x^2\mathbf{j}$ and C is the circle with radius a and center at $P = (2, 3)$, obtain the curl of **F** at P. Then, compute, as $a \to 0$, the limiting ratio of the circulation of **F** around C and the area enclosed by C.

A4. Determine whether the following fields are irrotational.
  (a) $\mathbf{F} = 3x^2y^2\mathbf{i} + 2x^3y\mathbf{j}$ (b) $\mathbf{F} = y^2z^3\mathbf{i} + 2xyz^3\mathbf{j} + 3xy^2z^2\mathbf{k}$

A5. Let $\mathbf{F} = \nabla u$, where $u(x, y, z)$ has continuous second partial derivatives. Show **F** is irrotational, that is, show $\nabla \times \mathbf{F} = \mathbf{0}$.

EXERCISES 20.2–Part B

B1. For each of the vector fields $\mathbf{F}_1 = y\mathbf{i} - x\mathbf{j}$ and $\mathbf{F}_2 = \frac{y}{x^2+y^2}\mathbf{i} - \frac{x}{x^2+y^2}\mathbf{j}$:
  (a) Obtain a field plot.
  (b) Obtain the flow lines for the field $\mathbf{F}_k$, $k = 1, 2$.
  (c) Obtain the curl of $\mathbf{F}_k$, $k = 1, 2$. Both fields seem to have rotation about the origin, but one actually has nonzero curl and the other has zero curl. Curl is a "local" or "point"

property of a vector field, whereas the field plot and the flow lines exhibit more "global" behavior of the field. Caution, therefore, should be exercised when interpreting graphs of vector fields and their flows.

B2. Give an example that shows $\mathbf{F}$ and $\nabla \times \mathbf{F}$ are not necessarily orthogonal.

B3. If $\mathbf{F} = f\nabla g$ for scalar-valued functions f and g, show that $\mathbf{F}$ and $\nabla \times \mathbf{F}$ are orthogonal.

B4. If $\mathbf{A}$ and $\mathbf{B}$ are irrotational, show that the cross-product $\mathbf{A} \times \mathbf{B}$ is solenoidal.

B5. Find solenoidal fields $\mathbf{A}$ and $\mathbf{B}$ for which the cross-product $\mathbf{A} \times \mathbf{B}$ is not irrotational.

B6. This exercise is designed to determine if the concepts of solenoidal and irrotational are mutually exclusive.

(a) Find a vector field that is solenoidal but not irrotational. Verify your claims.

(b) Find a vector field that is irrotational but not solenoidal. Verify your claims.

(c) Find a vector field that is *both* solenoidal *and* irrotational. Verify your claims.

(d) Show that for every solution $u(x, y, z)$ of Laplace's equation $\nabla^2 u = 0$, the gradient vector ∇u is both solenoidal and irrotational.

In Exercises B7–16:

(a) Categorize the given field $\mathbf{F}$ as neither solenoidal nor irrotational, just solenoidal, just irrotational, or both solenoidal and irrotational.

(b) Obtain a field plot of $\mathbf{F}$.

(c) Obtain a graph of the flow lines of $\mathbf{F}$.

(d) Evaluate $\nabla \times \mathbf{F}$, the curl of $\mathbf{F}$, at the given point P.

(e) Compute Λ_1, the circulation around C_1, a circle with radius ρ and center at point P.

(f) Compute $\lim_{\rho \to 0} \frac{\Lambda_1}{\pi\rho^2}$, the limiting value of the ratio of circulation to enclosed area. It should agree with the value of curl($\mathbf{F}$) at point P, as obtained in part (d).

(g) Compute Λ_2, the circulation around C_2, a square with side 2ρ and center at point P.

(h) Compute $\lim_{\rho \to 0} \frac{\Lambda_2}{4\rho^2}$, the limiting value of the ratio of circulation to enclosed area. It should again agree with the value of curl($\mathbf{F}$) at point P, as obtained in part (d).

B7. $\mathbf{F} = (5xy - y^3)\mathbf{i} - (x + 4x^2 y)\mathbf{j}$, $P = (-8, -2)$

B8. $\mathbf{F} = (24y^3 + 6x^2)\mathbf{i} - 12xy\mathbf{j}$, $P = (6, -8)$

B9. $\mathbf{F} = (32x^3 + 14xy)\mathbf{i} + (7x^2 - 1)\mathbf{j}$, $P = (-8, -4)$

B10. $\mathbf{F} = (5x^2 - 5xy)\mathbf{i} + (3 - 4y^3)\mathbf{j}$, $P = (-1, 2)$

B11. $\mathbf{F} = (3y - 6xy^2)\mathbf{i} + (5y + 9xy)\mathbf{j}$, $P = (-8, 6)$

B12. $\mathbf{F} = (6y^3 - 12xy)\mathbf{i} + (18xy^2 - 18y^2 - 6x^2)\mathbf{j}$, $P = (-4, 2)$

B13. $\mathbf{F} = (15xy^2 - 3x)\mathbf{i} + (3y - 5y^3)\mathbf{j}$, $P = (-3, 3)$

B14. $\mathbf{F} = (14xy - 3x^2)\mathbf{i} + (6xy - 7y^2)\mathbf{j}$, $P = (-1, -1)$

B15. $\mathbf{F} = (10x - 27x^2 y)\mathbf{i} - (16y^3 + 9x^3)\mathbf{j}$, $P = (-1, 2)$

B16. $\mathbf{F} = (8xy^2 - 4x^2)\mathbf{i} + (8y^3 - 7x^2 y)\mathbf{j}$, $P = (-9, -3)$

In Exercises B17–26:

(a) Categorize the given field $\mathbf{F}$ as neither solenoidal nor irrotational, just solenoidal, just irrotational, or both solenoidal and irrotational.

(b) Obtain a field plot of $\mathbf{F}$.

(c) Obtain a graph of the flow lines of F.

(d) Evaluate $\nabla \times \mathbf{F}$, the curl of $\mathbf{F}$, at the given point $P = (a, b)$.

(e) Compute Λ, the circulation around C, the ellipse given parametrically by $x = a + \rho \cos t$, $y = b + \kappa\rho \sin t$, $0 \le t \le 2\pi$.

(f) Compute $\lim_{\rho \to 0} \frac{\Lambda}{\pi\kappa\rho^2}$, the limiting value of the ratio of circulation to enclosed area. It should agree with the value of curl($\mathbf{F}$) at point P, as obtained in part (d).

B17. $\mathbf{F} = (5 + 12x + 5y + 27x^2 - 27y^2)\mathbf{i} + (8 + 5x - 12y - 54xy)\mathbf{j}$, $P = (1, 2)$, $\kappa = 8$

B18. $\mathbf{F} = (16x^2 y - 8x^3)\mathbf{i} + (24x^2 y - 16xy^2)\mathbf{j}$, $P = (-2, -1)$, $\kappa = 4$

B19. $\mathbf{F} = (2y^2 + 10xy^2)\mathbf{i} + (12y^2 + 4xy + 10x^2 y)\mathbf{j}$, $P = (0, -1)$, $\kappa = 4$

B20. $\mathbf{F} = (2 - 3xy^2)\mathbf{i} + (2y^2 - 8y)\mathbf{j}$, $P = (-1, 3)$, $\kappa = 7$

B21. $\mathbf{F} = (2 - 4x + 2y - 6x^2 + 6y^2)\mathbf{i} + (9 + 2x + 4y + 12xy)\mathbf{j}$, $P = (8, 2)$, $\kappa = 3$

B22. $\mathbf{F} = (20x^3 + 3y)\mathbf{i} + (12y + 3x)\mathbf{j}$, $P = (-1, 1)$, $\kappa = 6$

B23. $\mathbf{F} = (3 - 6x + 6y + 12xy)\mathbf{i} + (6x + 6y - 6y^2 + 6x^2 - 4)\mathbf{j}$, $P = (3, 6)$, $\kappa = 6$

B24. $\mathbf{F} = (9x^2 y - 2y)\mathbf{i} + (8 - 5x^2 y)\mathbf{j}$, $P = (4, -4)$, $\kappa = 5$

B25. $\mathbf{F} = (8xy + 12x^2 y)\mathbf{i} + (4x^3 - 6y^2 + 4x^2)\mathbf{j}$, $P = (1, 2)$, $\kappa = 3$

B26. $\mathbf{F} = (xy^2 - 6x^2 y)\mathbf{i} + (8 - 5x)\mathbf{j}$, $P = (2, 0)$, $\kappa = 9$

In each of Exercises B27–36, compute, at point P, and in the direction of $\mathbf{V}$, the directional derivative of the divergence of $\mathbf{F}$.

B27. $\mathbf{F} = (3x^3 + 8x^2)\mathbf{i} + (9xy^2 - xy)\mathbf{j}$, $\mathbf{V} = (8 + 7y)\mathbf{i} + (4 - 9x)\mathbf{j}$, $P = (9, -4)$

B28. $\mathbf{F} = (8y + 5x^2 y)\mathbf{i} + (6x^3 - 3x^2)\mathbf{j}$, $\mathbf{V} = (y - 3xy)\mathbf{i} + (9y^2 + 8x^2 y)\mathbf{j}$, $P = (7, -6)$

B29. $\mathbf{F} = (8y^2 - 5x^2 y)\mathbf{i} + (7x + 7x^3)\mathbf{j}$, $\mathbf{V} = (6y - 8xy^2)\mathbf{i} + (9y + 5y^3)\mathbf{j}$, $P = (8, -2)$

B30. $\mathbf{F} = 8x\mathbf{i} + (xy^2 - 7x^2 y)\mathbf{j}$, $\mathbf{V} = (2 - 5x^2)\mathbf{i} + (7x + 3y^3)\mathbf{j}$, $P = (-8, -3)$

B31. $\mathbf{F} = (6 + 2xy)\mathbf{i} + (6y^3 - 3x^2 y)\mathbf{j}$, $\mathbf{V} = (2 - 8xy)\mathbf{i} + (5x^3 + 6y^2)\mathbf{j}$, $P = (-9, -4)$

B32. $\mathbf{F} = (y + 4x^2)\mathbf{i} + (9x + 6xy^2)\mathbf{j}$, $\mathbf{V} = 8y^2\mathbf{i} + (6y^2 + 3xy)\mathbf{j}$, $P = (6, 1)$

B33. $\mathbf{F} = (6y - 4y^2)\mathbf{i} + (4x^2 + 7y^3)\mathbf{j}$, $\mathbf{V} = (9y^3 + 3x^2y)\mathbf{i} + (x - 3x^3)\mathbf{j}$, $P = (7, -3)$

B34. $\mathbf{F} = (7xy^2 - 5y^3)\mathbf{i} + (6x^3 - 9xy^2)\mathbf{j}$, $\mathbf{V} = (9x^3 - 3y^3)\mathbf{i} + (2x - 4y^2)\mathbf{j}$, $P = (3, 4)$

B35. $\mathbf{F} = (2x^2 + 8x^2y)\mathbf{i} + (5xy^2 + 7y^2)\mathbf{j}$, $\mathbf{V} = (5x - 8y)\mathbf{i} + (3x + 5xy^2)\mathbf{j}$, $P = (-7, 3)$

B36. $\mathbf{F} = (9x + 8xy^2)\mathbf{i} - 8x^2y\mathbf{j}$, $\mathbf{V} = (4x^2 + y^3)\mathbf{i} + (5 - 6y^3)\mathbf{j}$, $P = (4, 8)$

20.3 Products—One ∇ and Two Operands

Basic Tableau

Let $f(x, y, z)$ and $g(x, y, z)$ be scalar functions, and let $\mathbf{u}$ and $\mathbf{v}$ be vector fields. Table 20.1 lists all possible combinations of one del operator acting on a product of two factors as shown in

$$\nabla \text{ acting via } \begin{Bmatrix} * \\ \cdot \\ \times \end{Bmatrix} \text{ on } \begin{Bmatrix} fg \\ f\mathbf{v} \\ \mathbf{u} \times \mathbf{v} \\ \mathbf{u} \cdot \mathbf{v} \end{Bmatrix} \qquad (20.4)$$

where here, $*$ represents ordinary operator application. Of the $1 \times 3 \times 4 = 12$ possible combinations of these symbols, only the six numbered expressions are defined.

∇	Ordinary Operator Application	$\nabla(fg)$	(1)
		$\nabla(f\mathbf{v})$	
		$\nabla(\mathbf{u} \times \mathbf{v})$	
		$\nabla(\mathbf{u} \cdot \mathbf{v})$	(2)
∇	$\cdot$	$\nabla \cdot (fg)$	
		$\nabla \cdot (f\mathbf{v})$	(3)
		$\nabla \cdot (\mathbf{u} \times \mathbf{v})$	(4)
		$\nabla \cdot (\mathbf{u} \cdot \mathbf{v})$	
∇	$\times$	$\nabla \times (fg)$	
		$\nabla \times (f\mathbf{v})$	(5)
		$\nabla \times (\mathbf{u} \times \mathbf{v})$	(6)
		$\nabla \times (\mathbf{u} \cdot \mathbf{v})$	

TABLE 20.1   All possible combinations of ∇ acting on products

The six numbered entries in Table 20.1 are listed as identities in Table 20.2.

(1) $\nabla(fg) = f\nabla g + g\nabla f$

(2) $\nabla(\mathbf{u} \cdot \mathbf{v}) = (\mathbf{u} \cdot \nabla)\mathbf{v} + (\mathbf{v} \cdot \nabla)\mathbf{u} + \mathbf{u} \times (\nabla \times \mathbf{v}) + \mathbf{v} \times (\nabla \times \mathbf{u})$

(3) $\nabla \cdot (f\mathbf{v}) = f\nabla \cdot \mathbf{v} + \mathbf{v} \cdot \nabla f$

(4) $\nabla \cdot (\mathbf{u} \times \mathbf{v}) = \mathbf{v} \cdot \nabla \times \mathbf{u} - \mathbf{u} \cdot \nabla \times \mathbf{v}$

(5) $\nabla \times (f\mathbf{v}) = f\nabla \times \mathbf{v} + (\nabla f) \times \mathbf{v}$

(6) $\nabla \times (\mathbf{u} \times \mathbf{v}) = (\mathbf{v} \cdot \nabla)\mathbf{u} - (\mathbf{u} \cdot \nabla)\mathbf{v} + \mathbf{u}\nabla \cdot \mathbf{v} - \mathbf{v}\nabla \cdot \mathbf{u}$

TABLE 20.2   Six vector identities where ∇ acts on products

Notation

In addition to the scalars $f(x, y, z)$ and $g(x, y, z)$, we will use the vectors $\mathbf{u} = a\mathbf{i} + b\mathbf{j} + c\mathbf{k}$ and $\mathbf{v} = \alpha\mathbf{i} + \beta\mathbf{j} + \gamma\mathbf{k}$ while exploring the six identities in Table 20.2.

Identity (1)

Identity (1) in Table 20.2 says the gradient operator is well-behaved with respect to the product rule. The gradient of the product fg is "the first times the gradient of the second plus the second times the gradient of the first."

The vector on each side of the first identity is

$$\nabla(fg) = \begin{bmatrix} f_x g + f g_x \\ f_y g + f g_y \\ f_z g + f g_z \end{bmatrix}$$

Identity (2)

Identity (2) in Table 20.2 is complex, indeed. It says that the gradient of a dot product does not behave like the derivative of a product. In addition to the first two terms that seem to follow a product-like rule (differentiate each factor and multiply by the partner), there are two other terms involving curls and cross-products. The terms involving the dot products are generalized directional derivatives. Where multiplication by the undifferentiated factor enters, it is a dot product. Where differentiation occurs, it is by the gradient operator. In the other two terms, where multiplication occurs, it is a cross-product. Where differentiation occurs, it is a curl.

The vector on each side is

$$\nabla(\mathbf{u} \cdot \mathbf{v}) = \begin{bmatrix} a\alpha_x + \alpha a_x + b\beta_x + \beta b_x + c\gamma_x + \gamma c_x \\ a\alpha_y + \alpha a_y + b\beta_y + \beta b_y + c\gamma_y + \gamma c_y \\ a\alpha_z + \alpha a_z + b\beta_z + \beta b_z + c\gamma_z + \gamma c_z \end{bmatrix}$$

Identity (3)

Identity (3) in Table 20.2 again nearly behaves according to the product rule. The divergence of this product sees the del operator applied, once to the scalar and once to the vector. When applied to the vector, a divergence is computed, and that is multiplied by the scalar factor. When applied to the scalar, a gradient is computed, and that is multiplied by the vector via a dot product.

The scalar on each side of this identity is

$$\nabla \cdot (f\mathbf{v}) = f_x \alpha + f\alpha_x + f_y \beta + f\beta_y + f_z \gamma + f\gamma_z$$

Identity (4)

Identity (4) in Table 20.2 is again close to behaving like a product rule. The divergence of the cross-product expands to two terms, each containing one of the factors multiplying a derivative of the other. The differentiations are curls, and the multiplications are dot products. The only quirk is the minus sign appearing in the second term.

The scalar on each side of this identity is

$$\nabla \cdot (\mathbf{u} \times \mathbf{v}) = \gamma b_x + b\gamma_x - \beta c_x - c\beta_x + \alpha c_y + c\alpha_y - \gamma a_y - a\gamma_y$$
$$+ \beta a_z + a\beta_z - \alpha b_z - b\alpha_z$$
$$= (b\gamma)_x - (c\beta)_x + (c\alpha)_y - (a\gamma)_y + (a\beta)_z - (b\alpha)_z$$

Identity (5)

Identity (5) in Table 20.2 expands the curl of a product and behaves much like a product rule. There are two terms, each containing one factor multiplying a derivative of the other. The derivative of the vector is a curl, while the derivative of the scalar is a gradient. Multiplication of the curl is ordinary scalar multiplication, whereas multiplication of the gradient is a cross-product. Note the order in which that cross-product is formed in the second term.

The vector on each side of this identity is

$$\nabla \times (f\mathbf{v}) = \begin{bmatrix} f_y\gamma + f\gamma_y - f_z\beta - f\beta_z \\ f_z\alpha + f\alpha_z - f_x\gamma - f\gamma_x \\ f_x\beta + f\beta_x - f_y\alpha - f\alpha_y \end{bmatrix} = \begin{bmatrix} (f\gamma)_y - (f\beta)_z \\ (f\alpha)_z - (f\gamma)_x \\ (f\beta)_x - (f\alpha)_y \end{bmatrix}$$

Identity (6)

Identity (6) in Table 20.2 gives a complicated expression for the curl of a cross-product. Since there are four terms, it is not akin to a product rule. The first two terms on the right are generalized directional derivatives, whereas the second two are simply the product of a vector by a divergence. The first two terms have the reciprocity of a product rule, and so too, the last two.

The vector on each side of this identity is

$$\nabla \times (\mathbf{u} \times \mathbf{v}) = \begin{bmatrix} \beta a_y + a\beta_y - \alpha b_y - b\alpha_y - \alpha c_z - c\alpha_z + \gamma a_z + a\gamma_z \\ \gamma b_z + b\gamma_z - \beta c_z - c\beta_z - \beta a_x - a\beta_x + \alpha b_x + b\alpha_x \\ \alpha c_x + c\alpha_x - \gamma a_x - a\gamma_x - \gamma b_y - b\gamma_y + \beta c_y + c\beta_y \end{bmatrix}$$

$$= \begin{bmatrix} (a\beta)_y - (b\alpha)_y - (c\alpha)_z + (a\gamma)_z \\ (b\gamma)_z - (c\beta)_z - (a\beta)_x + (b\alpha)_x \\ (c\alpha)_x - (a\gamma)_x - (b\gamma)_y + (c\beta)_y \end{bmatrix}$$

EXERCISES 20.3

1. Explain why each unnumbered entry in Table 20.1 is undefined.

In Exercises 2–11, using the given pairs of vectors and scalars, verify each of the six vector identities in Table 20.2.

2. $f = -6xz^2 - 9x^2z$, $\mathbf{u} = \begin{bmatrix} xy^2 + xz^2 \\ 9x^2 - 3xz \\ 4xz^2 + 4xyz \end{bmatrix}$, $\mathbf{v} = \begin{bmatrix} 9xy - 9y^2 \\ 4xyz - 8x \\ 8z^3 - 3xz^2 \end{bmatrix}$

$g = \dfrac{8xz - 2x^3}{4z - 3z^2}$

3. $f = 4yz - 4xz$, $\mathbf{u} = \begin{bmatrix} 2y^3 - xyz \\ 6z^3 - 3 \\ 7z^2 - 7x^2y \end{bmatrix}$, $\mathbf{v} = \begin{bmatrix} 4z^3 + 4xy^2 \\ 9 - 3z^3 \\ 2yz \end{bmatrix}$

$g = \dfrac{7xz + 9y^2z}{9x^2z + 2y^2}$

4. $f = z^3 - 7yx^2$, $\mathbf{u} = \begin{bmatrix} 6yz^2 - x^2 \\ 6xz - 5y^2 \\ 7y - 2 \end{bmatrix}$, $\mathbf{v} = \begin{bmatrix} 3x - 2yz \\ 3x^2y - 2y^2z \\ 3xz + 9y^2 \end{bmatrix}$

$g = \dfrac{4xy^2 - y^2z}{5x^3 + 8x^2z}$

5. $f = 2yz - x^2y$, $\mathbf{u} = \begin{bmatrix} 2yz^2 + 8x^2 \\ 5xy + 6z^2 \\ x^2 + 6y^2z \end{bmatrix}$, $\mathbf{v} = \begin{bmatrix} 2 - 7y^3 \\ 7yz + 4xyz \\ 5xy + 2z \end{bmatrix}$

$g = \dfrac{7z + 3}{5y + 7x^2z}$

6. $f = 3yz - 6x^3$, $\mathbf{u} = \begin{bmatrix} 5xz - 8y^3 \\ 4xy^2 - 7z \\ 6xz^2 - 5y \end{bmatrix}$, $\mathbf{v} = \begin{bmatrix} 7xy \\ 9z - x \\ 6z^2 + 4xy \end{bmatrix}$

$g = \dfrac{5z^3 - 4x}{5xy^2 - 3xyz}$

7. $f = xz - 7y^2$, $\mathbf{u} = \begin{bmatrix} yz - 5x^3 \\ 8x^2y + 5z^2 \\ 9yz - 7x \end{bmatrix}$, $\mathbf{v} = \begin{bmatrix} 2z^3 - 9xy^2 \\ 8xz - 3y \\ 5xy + 2yz \end{bmatrix}$

$g = \dfrac{9z^3 + 4x^2}{4z^3 - 3xy^2}$

8. $f = y^3 - 9xz^2$, $\mathbf{u} = \begin{bmatrix} xy^2 - 3x^2 \\ 9x^2y + 7z \\ 8y^3 - 6z^2 \end{bmatrix}$, $\mathbf{v} = \begin{bmatrix} 3x^3 - 3z^2 \\ 2z^3 - 7y^2 \\ 4x^2y - y \end{bmatrix}$

$g = \dfrac{9xy^2 - y^3}{3z^3 + 5}$

9. $f = 2z^2 + 6y^2$, $\mathbf{u} = \begin{bmatrix} xyz - 7 \\ 7x^2y - 6z^2 \\ x^2 - 7y^3 \end{bmatrix}$, $\mathbf{v} = \begin{bmatrix} 8xz^2 - 8y^3 \\ z \\ 8x^2 + 7y^2z \end{bmatrix}$

$g = \dfrac{2z^3 - 6y}{9x^3 + 2y}$

10. $f = yz^2 - 7x^2$, $\mathbf{u} = \begin{bmatrix} x^2 \\ 8xz^2 - 9y^2z \\ 8y^2 - 6x^2z \end{bmatrix}$, $\mathbf{v} = \begin{bmatrix} 3z^3 - 8xy \\ 3y^2 - 7x^2 \\ x^2y - 6z \end{bmatrix}$

$g = \dfrac{8y^3 + 5x}{9z^2 - 6xy}$

11. $f = z^3 + 9xy$, $\mathbf{u} = \begin{bmatrix} z^3 - 3y^2 \\ 7z - 2xy \\ 4xz + 9y^3 \end{bmatrix}$, $\mathbf{v} = \begin{bmatrix} 3z - 4y^2x \\ 8z^2 - 9x^3 \\ 7yz^2 + 4x \end{bmatrix}$

$g = \dfrac{6xy^2 - 7z^2}{8 - yz^2}$

12. A particle of mass m moving under the governance of Newton's Second Law, $\mathbf{F} = m\mathbf{a}$, follows a path $\mathbf{x}(t)$ that is constrained to lie in an equipotential surface of the conservative force $\mathbf{F}$. Show that the particle moves with constant speed. *Hint:* See Section 17.5.

13. If, in $\mathbf{R}^3$, the vectors $\mathbf{u}(t)$, $\mathbf{v}(t)$, and $\mathbf{w}(t)$ are differentiable, with $\mathbf{u}' = \frac{d\mathbf{u}}{dt}$, $\mathbf{v}' = \frac{d\mathbf{v}}{dt}$, and $\mathbf{w}' = \frac{d\mathbf{w}}{dt}$, show

(a) $(\mathbf{u} \cdot \mathbf{v})' = \mathbf{u} \cdot \mathbf{v}' + \mathbf{v} \cdot \mathbf{u}'$

(b) $(\mathbf{u} \times \mathbf{v})' = \mathbf{u}' \times \mathbf{v} + \mathbf{u} \times \mathbf{v}'$

(c) $[\mathbf{u} \cdot (\mathbf{v} \times \mathbf{w})]' = \mathbf{u}' \cdot (\mathbf{v} \times \mathbf{w}) + \mathbf{u} \cdot (\mathbf{v}' \times \mathbf{w}) + \mathbf{u} \cdot (\mathbf{v} \times \mathbf{w}')$

20.4   Products—Two ∇'s and One Operand

Basic Tableau

Of the 18 possible ways two del operators, the three operations of multiply, dot, and cross, and a single operand can be arranged, only 5 are valid. These 5 valid expressions are numbered in Table 20.3, and studied later in this section. We assume throughout that the operands f and $\mathbf{v}$ have continuous second partial derivatives.

	$\nabla(\nabla f)$	(2) $\nabla \cdot (\nabla f)$	(4) $\nabla \times (\nabla f)$
	$\nabla(\nabla \mathbf{v})$	$\nabla \cdot (\nabla \mathbf{v})$	$\nabla \times (\nabla \mathbf{v})$
	$\nabla(\nabla \cdot f)$	$\nabla \cdot (\nabla \cdot f)$	$\nabla \times (\nabla \cdot f)$
(1)	$\nabla(\nabla \cdot \mathbf{v})$	$\nabla \cdot (\nabla \cdot \mathbf{v})$	$\nabla \times (\nabla \cdot \mathbf{v})$
	$\nabla(\nabla \times f)$	$\nabla \cdot (\nabla \times f)$	$\nabla \times (\nabla \times f)$
	$\nabla(\nabla \times \mathbf{v})$	(3) $\nabla \cdot (\nabla \times \mathbf{v})$	(5) $\nabla \times (\nabla \times \mathbf{v})$

TABLE 20.3   All possible combinations of ∇, acting on another ∇ that acts on a single operand

The 13 expressions that are not numbered are operationally undefined. For example, at the top of the leftmost column, one cannot compute a gradient of a gradient since a gradient can be computed only for a scalar-valued function. Immediately following is the attempt to compute the gradient of the vector $\mathbf{v}$, again undefined. In like manner, the remaining 11 unnumbered entries should be examined and the reason why they are undefined determined.

For the 5 numbered entries, capsule comments appear in Table 20.4. Detailed discussions constitute the essence of this section.

(1)	$\nabla(\nabla \cdot \mathbf{v})$ grad(div $\mathbf{v}$)		Appears in formula (5)
(2)	$\nabla \cdot \nabla f$ div(grad f)	laplacian	$\nabla^2 f = 0 \Rightarrow f$ called harmonic function
(3)	$\nabla \cdot (\nabla \times \mathbf{v})$ div(curl $\mathbf{v}$)	always 0	Curls don't spread (curl fields are solenoidal)
(4)	$\nabla \times (\nabla f)$ curl(grad f)	always $\mathbf{0}$	Gradients don't twist (conservative $\leftrightarrow$ irrotational)
(5)	$\nabla \times \nabla \times \mathbf{v}$ curl(curl $\mathbf{v}$)	$\nabla(\nabla \cdot \mathbf{v}) - \nabla^2 \mathbf{v}$	laplacian applied to each component of $\mathbf{v}$

TABLE 20.4   Five meaningful cases of two ∇'s and a single operand

Expressions (3) and (4) are the most important of all the results. Each states that some combination of vector differential operators is zero, in one case the scalar zero and in the other the zero vector. The implications of each are so significant that a mnemonic phrase has been provided.

Expression (1)

The first expression, namely, grad(div $\mathbf{v}$), is defined but will not play a significant role in our work. If $\mathbf{v} = u\mathbf{i} + v\mathbf{j} + w\mathbf{k}$, then

$$\nabla(\nabla \cdot \mathbf{v}) = \begin{bmatrix} u_{xx} + v_{xy} + w_{xz} \\ u_{yx} + v_{yy} + w_{yz} \\ u_{zx} + v_{zy} + w_{zz} \end{bmatrix}$$

Expression (2)

VERIFICATION   The second expression, namely, div(grad f) = laplacian(f), was discussed formally in Section 20.1 where the notation $\nabla \cdot (\nabla f) = \nabla^2 f$ was first met. The laplacian operator is of particular importance in the applications since each of the three partial differential equations of mathematical physics, equations governing wave motion, heat transfer, and electric potential all contain this operator.

FICK'S LAW   As just noted, many of the laws of engineering and the physical sciences are expressed in partial differential equations. In almost all cases, the partial differential equation contains the laplacian. Unit Five is devoted to a study of special cases of the following three partial differential equations of mathematical physics, each of which contains the laplacian

$$\begin{aligned} \text{Wave equation} \quad & u_{tt} = c^2(u_{xx} + u_{yy} + u_{zz}) \\ \text{Heat equation} \quad & u_t = \kappa(u_{xx} + u_{yy} + u_{zz}) \\ \text{Potential equation} \quad & 0 = u_{xx} + u_{yy} + u_{zz} \end{aligned}$$

Here's a glimpse why. Fick's Law states that *flux is proportional to the gradient,* that is, the amount of flow down a hill is proportional to the steepness of the hill. For example, when discussing heat flow, it is assumed that heat flows the instant a temperature gradient is established and, as in Newton's law of cooling, with a rate proportional to the difference in temperatures. This difference in temperature between an object and its surroundings is precisely the temperature gradient. For temperature distributions at equilibrium, we would have $\mathbf{F} = -k\nabla f$, where $\mathbf{F}$ is the flux (or flow) of heat and f is the temperature. Computing the divergence of the flux vector $\mathbf{F}$ (to determine whether heat flows into or out of the region) immediately yields the laplacian $\nabla \cdot \mathbf{F} = -k\nabla^2 f$.

AVERAGING PROPERTY OF THE LAPLACIAN   Another glimpse of the importance of the laplacian operator is reflected in its averaging property. For example, suppose the values of a function $f(x, y)$ and its derivatives are known at (a, b), the center of a square of side $2h$. Using a Taylor expansion computed at (a, b), the values of $f(x, y)$ can be approximated at the four corners of the square, yielding

$$f(a+h, b-h) = f(a, b) + f_x(a, b)h - f_y(a, b)h$$
$$+ \tfrac{1}{2}f_{xx}(a, b)h^2 - f_{xy}(a, b)h^2 + \tfrac{1}{2}f_{yy}(a, b)h^2$$

$$f(a+h, b+h) = f(a, b) + f_x(a, b)h + f_y(a, b)h$$
$$+ \tfrac{1}{2}f_{xx}(a, b)h^2 + f_{xy}(a, b)h^2 + \tfrac{1}{2}f_{yy}(a, b)h^2$$

$$f(a-h, b+h) = f(a, b) - f_x(a, b)h + f_y(a, b)h$$
$$+ \tfrac{1}{2}f_{xx}(a, b)h^2 - f_{xy}(a, b)h^2 + \tfrac{1}{2}f_{yy}(a, b)h^2$$

$$f(a-h, b-h) = f(a, b) - f_x(a, b)h - f_y(a, b)h$$
$$+ \tfrac{1}{2}f_{xx}(a, b)h^2 + f_{xy}(a, b)h^2 + \tfrac{1}{2}f_{yy}(a, b)h^2$$

The average of the four corner-values is then

$$f_{av} = f(a, b) + \tfrac{1}{2}f_{xx}(a, b)h^2 + \tfrac{1}{2}f_{yy}(a, b)h^2$$

and one interpretation of this result is that the laplacian is proportional to the difference between $f(a, b)$ and the average value at the corners.

Expression (3)

The third expression is the vector identity div(curl $\mathbf{v}$) = 0, which says that the divergence of the curl is always zero. Since the divergence measures spread, the phrase "curls don't spread" is provided in Table 20.4. (Visualize an elegant hair-do of long, softly wound curls, bouncing, but not coming undone. The *curls don't spread*.)

That $\nabla \cdot (\nabla \times \mathbf{v})$ is zero follows from the equality of the mixed second partial derivatives. Let $\mathbf{v} = u\mathbf{i} + v\mathbf{j} + w\mathbf{k}$, and compute

$$\nabla \cdot (\nabla \times \mathbf{v}) = (w_y - v_z)_x + (u_z - w_x)_y + (v_x - u_y)_z$$
$$= w_{yx} - v_{zx} + u_{zy} - w_{xy} + v_{xz} - u_{yz} = 0$$

This actually characterizes fields with zero divergence. Any field that arises as the curl of some other field is necessarily solenoidal. In Section 21.5 a proof of the converse is given whereby it is shown that a field whose divergence vanishes must necessarily be the curl of another field called the *vector potential*. Thus, any curl field is necessarily a solenoidal field, a characterization of solenoidal fields. A field is *solenoidal* if and only if it is a curl field (that is, the solenoidal field arises as the curl of some other field). This is discussed further in Section 21.5.

Expression (4)

The fourth expression is the vector identity curl(grad f) = **0**, which says the curl of the gradient is always the zero vector. Since the curl measures rotation (or twist), the phrase "gradients don't twist" is provided in Table 20.4. (It looks like this section is a frightening lesson, so your hair is standing straight out from your head. Each hair is a gradient vector on the surface of your scalp. Now, grab a handful and try to twist it. You'll tear your scalp off, so don't twist the gradients, and *gradients don't twist*.)

That $\nabla \times (\nabla f)$ is the zero vector follows from the equality of the mixed second partial derivatives of $f(x, y, z)$, since

$$\nabla \times (\nabla f) = \begin{vmatrix} \mathbf{i} & \mathbf{j} & \mathbf{k} \\ \dfrac{\partial}{\partial x} & \dfrac{\partial}{\partial y} & \dfrac{\partial}{\partial z} \\ f_x & f_y & f_z \end{vmatrix} = \begin{bmatrix} f_{zy} - f_{yz} \\ f_{xz} - f_{zx} \\ f_{yx} - f_{xy} \end{bmatrix} = \mathbf{0}$$

This actually characterizes irrotational fields. Any gradient field is irrotational. In Section 21.5 a proof of the converse is given whereby it is shown that an irrotational field is the gradient of some scalar, called the scalar potential. Thus, gradient fields are always irrotational, and a field is *irrotational* if and only if it is the gradient of some scalar potential. Consequently, a vector field has a scalar potential if and only if its curl vanishes. That means the irrotational fields are those arising as gradients of scalar potentials, so the *conservative* fields are precisely the *irrotational* fields.

The existence of a scalar potential for a vector field is established by the vanishing of the curl of the vector field. If the curl of the field vanishes, then the field is irrotational, is therefore conservative, and hence arises as the gradient of some scalar potential. This is examined in detail in Section 21.5.

Expression (5)

The fifth expression, namely, curl(curl **v**) = grad(div **v**) − laplacian(**v**), is an identity of use in working with Maxwell's equations in electromagnetic theory. The first term on the right, namely grad(div **v**), is the quantity in expression (1).

EXERCISES 20.4–Part A

A1. Explain why each of the 13 unnumbered entries in Table 20.3 is undefined.

A2. For what value of the constant α is the vector field $\mathbf{V} = (x + 4y)\mathbf{i} + (y - 3z)\mathbf{j} + \alpha z\mathbf{k}$ the curl of some other vector field?

A3. Compute grad(div **v**) $= \nabla(\nabla \cdot \mathbf{v})$ for $\mathbf{v} = xy^2\mathbf{i} + x^2 y\mathbf{j}$.

A4. Verify div(curl **v**) $= \nabla \cdot (\nabla \times \mathbf{v}) = 0$ for $\mathbf{v} = y^2 z^2\mathbf{i} + xyz\mathbf{j} + xy^2\mathbf{k}$.

A5. Verify curl(grad f) $= \nabla \times (\nabla f) = \mathbf{0}$ for $f(x, y, z) = xyz$.

A6. Verify $\nabla \times \nabla \times \mathbf{v} = \nabla(\nabla \cdot \mathbf{v}) - \nabla^2 \mathbf{v}$ for **v** given in Exercise A4.

EXERCISES 20.4–Part B

B1. Given the vectors $\mathbf{u} = a(x, y, z)\mathbf{i} + b(x, y, z)\mathbf{j} + c(x, y, z)\mathbf{k}$ and $\mathbf{v} = f(x, y, z)\mathbf{i} + g(x, y, z)\mathbf{j} + h(x, y, z)\mathbf{k}$, define the operator

$$(\mathbf{u} \times \nabla) = \begin{vmatrix} \mathbf{i} & \mathbf{j} & \mathbf{k} \\ a & b & c \\ \dfrac{\partial}{\partial x} & \dfrac{\partial}{\partial y} & \dfrac{\partial}{\partial z} \end{vmatrix} = \begin{bmatrix} b\dfrac{\partial}{\partial z} - c\dfrac{\partial}{\partial y} \\ c\dfrac{\partial}{\partial x} - a\dfrac{\partial}{\partial z} \\ a\dfrac{\partial}{\partial y} - b\dfrac{\partial}{\partial x} \end{bmatrix}$$ and show that

$(\mathbf{u} \times \nabla) \cdot \mathbf{v} = \mathbf{u} \cdot (\nabla \times \mathbf{v})$.

In Exercises B2–11, a scalar $f(x, y, z)$ and a point P are given. For each:

(a) Evaluate both the divergence of the gradient of f and the laplacian of f. Show that the two numbers are the same.

(b) Verify that the curl of the gradient is the zero vector.

(c) For each scalar function $f(x, y, z)$ and $g(x, y, z) = \frac{x+yz}{z-xy^2}$, verify that, at P, $(\nabla f) \times (\nabla g) = \nabla \times (f\nabla g) = -\nabla \times (g\nabla f)$.

B2. $f(x, y, z) = \dfrac{1}{x^2 + y^2 + z^2}, P = (1, 2, 3)$

B3. $f(x, y, z) = \dfrac{yz^2 + 9y^3}{3z + 4}, P = (-3, 1, 5)$

B4. $f(x, y, z) = \dfrac{5y^2 - 4z^2}{x^2 y}, P = (4, -1, 3)$

B5. $f(x, y, z) = \dfrac{x + 3y^3}{5x^2 - 6xy^2}, P = (-2, 2, 3)$

B6. $f(x, y, z) = \dfrac{xz}{6xy + 8y^3}, P = (-3, -5, 2)$

B7. $f(x, y, z) = \dfrac{2y^2 + y}{7xz^2 + y^3}, P = (3, 2, 1)$

B8. $f(x, y, z) = \dfrac{9 - 2xz}{9x^2 + 4xz^2 + 9yz^2}, P = (3, 1, 7)$

B9. $f(x, y, z) = \dfrac{x^2}{3yz - x^3 + 3}, P = (1, 2, 5)$

B10. $f(x, y, z) = \dfrac{6y^2 z - 3x}{9x^2 - 4x^2 y - 6yz^2}, P = (1, 1, -1)$

B11. $f(x, y, z) = \dfrac{9z^3 - 4}{3xy - 9y^2 z - 8xz^2}, P = (2, 5, 1)$

For the vectors **F** given in Exercises B12–16, verify that the divergence of the curl is zero.

B12. $\mathbf{F} = x^2 yz\mathbf{i} + xy^2 z\mathbf{j} + xyz^2\mathbf{k}$

B13. $\mathbf{F} = xyz\mathbf{i} + z\mathbf{j} + y\mathbf{k}$

B14. $\mathbf{F} = (1 - z)^2\mathbf{i} + \dfrac{xz}{x + y}\mathbf{j} + z^2\mathbf{k}$

B15. $\mathbf{F} = 2x^3\sqrt{yz}\,\mathbf{i} + ye^z\mathbf{j} + \sin(xy)\mathbf{k}$

B16. $\mathbf{F} = z\cos y\mathbf{i} + x\sin z\mathbf{j} + yz\tan(xz)\mathbf{k}$

For the vector fields **F** given in each of Exercises B17–23, determine a function $u(x, y, z)$ that makes **F** the curl of some other vector field.

B17. $\mathbf{F} = u(x, y, z)\mathbf{i} + (6y + 3)\mathbf{j} + (-3x + 5xz)\mathbf{k}$

B18. $\mathbf{F} = u(x, y, z)\mathbf{i} + (9xy - z^2)\mathbf{j} + (3z - 3)\mathbf{k}$

B19. $\mathbf{F} = u(x, y, z)\mathbf{i} + (2y - 5x^2)\mathbf{j} + (8y^2 - 4x)\mathbf{k}$

B20. $\mathbf{F} = u(x, y, z)\mathbf{i} + (8xy - 6y)\mathbf{j} + (8z - 7x)\mathbf{k}$

B21. $\mathbf{F} = u(x, y, z)\mathbf{i} + (9xz - 3x^2)\mathbf{j} - 7xz\mathbf{k}$

B22. $\mathbf{F} = u(x, y, z)\mathbf{i} + (7y - 5)\mathbf{j} + (9x - 4y^2)\mathbf{k}$

B23. $\mathbf{F} = u(x, y, z)\mathbf{i} + (4xz - z)\mathbf{j} + (7xz - 7)\mathbf{k}$

In Exercises B24–33, determine a value of the constant α for which the given vector field is a gradient field.

B24. $\mathbf{F} = (9x^2 - 5)\mathbf{i} + \alpha yz\mathbf{j} - 5y^2\mathbf{k}$

B25. $\mathbf{F} = 14xz\mathbf{i} + (\alpha z^2 - 16yz)\mathbf{j} + (7x^2 - 16yz - 8y^2)\mathbf{k}$

B26. $\mathbf{F} = (8xz + 2y^2)\mathbf{i} + (7z + 4xy)\mathbf{j} + (\alpha y + 4x^2)\mathbf{k}$

B27. $\mathbf{F} = \alpha xy\mathbf{i} - 9x^2\mathbf{j} + (14z - 21z^2)\mathbf{k}$

B28. $\mathbf{F} = (\alpha y^2 - 9z)\mathbf{i} - 10xy\mathbf{j} + (8 - 9x)\mathbf{k}$

B29. $\mathbf{F} = (\alpha xy - 4yz + 8y^2)\mathbf{i} + (16xy - 2x^2 - 4xz)\mathbf{j} - 4xy\mathbf{k}$

B30. $\mathbf{F} = (4yz + 8y^2 + 2z^2)\mathbf{i} + (4xz + \alpha xy)\mathbf{j} + (4xy + 4xz)\mathbf{k}$

B31. $\mathbf{F} = (\alpha yz + 8z^2)\mathbf{i} - 8xz\mathbf{j} + (16xz - 8xy)\mathbf{k}$

B32. $\mathbf{F} = (\alpha yz + 3x^2)\mathbf{i} + 3xz\mathbf{j} + (3xy - 9z^2)\mathbf{k}$

B33. $\mathbf{F} = 5yz\mathbf{i} + (6z + 5xz)\mathbf{j} + (6y + \alpha xy + 7)\mathbf{k}$

Chapter Review

1. If $\mathbf{F} = u\mathbf{i} + v\mathbf{j} + w\mathbf{k}$:

   (a) Write the formula for the divergence, $\nabla \cdot \mathbf{F}$.

   (b) Write the formula for the curl, $\nabla \times \mathbf{F}$.

2. Give a verbal description of what the divergence measures.

3. Give a verbal description of what the curl measures.

4. Give an example that illustrates the meaning of the statement "the divergence is the limiting value of flux per unit area enclosed."

5. Give an example that illustrates the meaning of the statement "the curl is the limiting value of the circulation per unit area enclosed."

6. If $F = xy^2 - yz^2 + zx^3$:

   (a) Compute the divergence.   **(b)** Compute the curl.

7. What is a solenoidal field? Give an example.

8. What is the divergence of the gradient field? Give an example.

9. Is the laplacian of a scalar field necessarily zero?

10. Write the formula for computing the laplacian of the scalar $u(x, y, z)$ in Cartesian coordinates.

11. Compute the laplacian of $u(x, y) = \dfrac{x - y^2}{\sqrt{x^2 + y^2}}$.

12. What is an irrotational field? Give an example.

13. Of the 12 possible expressions of the form $\nabla \begin{Bmatrix} * \\ \cdot \\ \times \end{Bmatrix} \begin{Bmatrix} fg \\ f\mathbf{v} \\ \mathbf{u} \times \mathbf{v} \\ \mathbf{u} \cdot \mathbf{v} \end{Bmatrix}$, state which 6 are valid. For those that are not valid, explain why.

14. Of the 18 possible expressions of the form $\nabla \begin{Bmatrix} * \\ \cdot \\ \times \end{Bmatrix} \nabla \begin{Bmatrix} * \\ \cdot \\ \times \end{Bmatrix} \begin{Bmatrix} f \\ \mathbf{v} \end{Bmatrix}$,

   state which 5 are valid. For those that are not valid, explain why. For those that are valid, state the significance of any associated identity.

15. What is a harmonic function. Give an example, and demonstrate that it is indeed harmonic.

16. Characterize conservative fields.

17. Characterize solenoidal fields.

18. Characterize irrotational fields.

19. What is a field called if it is the curl of some other field?

Chapter 21

Integration

INTRODUCTION   A procedure for computing surface area of a surface of revolution is available in elementary integral calculus. We generalize this to the computation of surface area of a general surface. We then define the surface integral that accumulates the values of a scalar on a surface. For example, the scalar can be a surface density giving either charge or mass per unit area. The surface integral would then yield the total charge or mass for the surface.

If the component of a vector field normal to a surface is integrated over that surface, the resulting number is called the surface flux of the field and is a measure of how much of the field "passes through" the surface.

If the divergence of a vector field is integrated over the interior of a region R, the result must match the surface flux of the field through the boundary of R. This law of balance is the *divergence theorem of Gauss* and simply says that a net change in the content inside the region must be balanced by an appropriate flow through the boundary.

If a field passes through an open surface such as a hemisphere, the surface integral of the normal component of the curl must balance the circulation of the field around the curve that bounds the open surface. For the hemisphere, the bounding curve would be the "equator." This law of balance is *Stokes' Theorem.*

There are two forms of *Green's Theorem*, one that looks like the planar version of the divergence theorem, and one that looks like the planar version of Stokes' Theorem.

Section 21.5 collects and connects alternate definitions of a conservative force. In fact, the equivalence of the various definitions is rigorously established. The *solenoidal* (divergence-free) field is characterized in terms of conservative fields for which the potential function is *harmonic*. It is also characterized in terms of a field that is itself the curl of another field called the *vector potential*. The *irrotational* (curl-free) field is characterized as a conservative field, and vice-versa.

Finally, we see the integral equivalents for the vector differential operators of divergence, gradient, and curl. On the basis of the formulas relating the differential operators to integrals, richer interpretations of divergence and curl are possible.

21.1 Surface Area

Surface Area—Surfaces of Revolution

Surface area for surfaces of revolution is a staple of the integral calculus course. After a brief review, the question of finding the surface area of a surface that does not necessarily have rotational symmetry becomes the chief focus.

If a curve given by $y = f(x)$ is rotated about the x-axis to form a solid of revolution, then an integral that expresses the surface area of the solid is $\int_a^b 2\pi\rho\,ds$, where ρ is the radius of rotation and ds is the element of arc length. Corresponding to Figure 21.1, this integral becomes

$$\int_a^b 2\pi f(x)\sqrt{1 + (f'(x))^2}\,dx$$

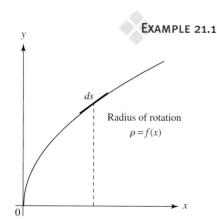

y

ds

Radius of rotation
$\rho = f(x)$

0

x

FIGURE 21.1   Surface area for a surface of revolution

EXAMPLE 21.1   **Sphere of radius a**   Rotate, about the x-axis, a semi-circle of radius a, thereby forming a sphere of radius a. Thus, the curve being rotated is $f(x) = \sqrt{a^2 - x^2}$, so the element of arc length is $ds = \frac{a}{\sqrt{a^2 - x^2}}$. The surface area sought is therefore $A = 2\pi \int_{-a}^{a} a\,dx = 4\pi a^2$, the well-known value for the surface area of a sphere of radius a.   ❖

Surface Area—General Case

The element of surface area in the general case is approximated by a parallelogram, one corner of which is attached to the surface in question. The area of a parallelogram whose edges are the vectors $\mathbf{U}$ and $\mathbf{V}$ is $\|\mathbf{U} \times \mathbf{V}\|$. A brief reminder of this result from multivariable calculus is based on Figure 21.2. The dotted line forms a right triangle where the hypotenuse has length $\|\mathbf{V}\|$ and the leg opposite angle t has length $\|\mathbf{V}\| \sin t$. The height of the parallelogram is therefore $\|\mathbf{V}\| \sin t$ and its area is $\|\mathbf{U}\|\|\mathbf{V}\| \sin t$, which is precisely the length of the vertical segment, the cross-product of $\mathbf{U}$ and $\mathbf{V}$. Thus, the area of a parallelogram of sides $\mathbf{U}$ and $\mathbf{V}$ is $\|\mathbf{U} \times \mathbf{V}\|$.

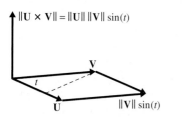

$\|\mathbf{U} \times \mathbf{V}\| = \|\mathbf{U}\|\,\|\mathbf{V}\|\sin(t)$

$\mathbf{V}$

t

$\mathbf{U}$ $\|\mathbf{V}\|\sin(t)$

FIGURE 21.2   Area of a parallelogram

On the surface $z = f(x, y)$ the surface area element is the parallelogram formed by two vectors tangent to plane sections $x = \alpha$ and $y = \beta$, as shown in Figure 21.3. The plane sections are

$$\mathbf{R}_y = \alpha\mathbf{i} + y\mathbf{j} + f(\alpha, y)\mathbf{k} \quad \text{and} \quad \mathbf{R}_x = x\mathbf{i} + \beta\mathbf{j} + f(x, \beta)\mathbf{k}$$

respectively. Infinitesimal tangents to these curves are then

$$\mathbf{T}_x = (\mathbf{i} + f_x\mathbf{k})\,dx \quad \text{and} \quad \mathbf{T}_y = (\mathbf{j} + f_y\mathbf{k})\,dy$$

so the area of the parallelogram they form is

$$\|\mathbf{T}_x \times \mathbf{T}_y\| = \sqrt{1 + f_x^2 + f_y^2}\,dx\,dy$$

Hence, the element of surface area for the surface $z = f(x, y)$ that projects to the xy-plane is

$$d\sigma = \sqrt{1 + f_x^2 + f_y^2}\,dx\,dy$$

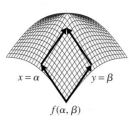

$x = \alpha$ $y = \beta$

$f(\alpha, \beta)$

FIGURE 21.3   Element of surface area approximated by parallelogram

Of course, if the surface were projected to a different coordinate plane, the expression for $d\sigma$ would change accordingly.

EXAMPLE 21.2   Find the surface area of the "North Polar Cap" cut off the sphere $x^2 + y^2 + z^2 = a^2$ by the plane $z = \frac{2a}{3}$.

Solution   As shown in Figure 21.4, the required surface area is given by an integral of the form $\iint_Q d\sigma$, where Q is the disk in the xy-plane above which sits the North Polar Cap. To find the radius of the disk Q, intersect the sphere with the cutting plane by solving

$$x^2 + y^2 + z^2 = a^2 \quad \text{and} \quad z = \frac{2a}{3}$$

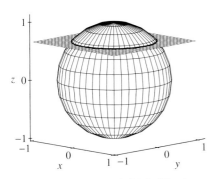

FIGURE 21.4   Example 21.2: "North Polar Cap"

simultaneously. Thus, solve

$$r^2 + \tfrac{4}{9}a^2 = a^2$$

for r, calling the positive value $r_1 = \frac{\sqrt{5}}{3}a$. Then write the upper hemisphere as

$$z = f(x, y) = \sqrt{a^2 - x^2 - y^2}$$

the positive square root being used when finding the area of the *North* Polar Cap. Hence,

$$d\sigma = \sqrt{1 + f_x^2 + f_y^2}\, dx\, dy = \frac{a}{\sqrt{a^2 - x^2 - y^2}}\, dx\, dy$$

The surface-area integral will be a double integral over a disk of radius r_1. Set this up in Cartesian coordinates, but evaluate it in polar coordinates, obtaining

$$\int_{-\sqrt{5}a/3}^{\sqrt{5}a/3} \int_{-\sqrt{5a^2-9x^2}/3}^{\sqrt{5a^2-9x^2}/3} \frac{a}{\sqrt{a^2 - x^2 - y^2}}\, dy\, dx = \int_0^{2\pi} \int_0^{\sqrt{5}a/3} \frac{ra}{\sqrt{a^2 - r^2}}\, dr\, d\theta = \tfrac{2}{3}\pi a^2 \quad ❖$$

EXERCISES 21.1–Part A

A1. Find the surface area of the solid of revolution formed when $y = x^2, 0 \le x \le 1$, is rotated about the x-axis.

A2. Find the area of the parallelogram that has $\mathbf{u} = 3\mathbf{i} - 5\mathbf{j} + 7\mathbf{k}$ and $\mathbf{v} = 2\mathbf{i} + \mathbf{j} - 4\mathbf{k}$ as adjacent sides.

A3. Write $d\sigma$ for the surface:

   (a) $x = g(y, z)$ **(b)** $y = h(x, z)$

A4. The first-octant triangle T is cut from the plane $5x + 7y + 9z = 315$ by the coordinate planes. Obtain the area of T by

   (a) using the formula $\frac{1}{2}\|\mathbf{u} \times \mathbf{v}\|$, where $\mathbf{u}$ and $\mathbf{v}$ are vectors that form the triangle.

   (b) projecting T to the xy-plane and using integration.

   (c) projecting T to the xz-plane and using integration.

   (d) projecting T to the yz-plane and using integration.

A5. Calculate the surface area of the sphere $x^2 + y^2 + z^2 = a^2$ by

   (a) projecting to the xy-plane and integrating.

   (b) projecting to the yz-plane and integrating.

   (c) projecting to the xz-plane and integrating.

EXERCISES 21.1–Part B

B1. Find the surface area of the surface of revolution formed when $y = \sin x, 0 \le x \le \frac{\pi}{2}$, is rotated about the x-axis.

B2. Find the surface area of that portion of the plane in Exercise A4 that lies inside the cylinder

   (a) $(x - 3)^2 + (y - 5)^2 = 4$

   (b) whose footprint, in the xy-plane, is the triangle with vertices $(1, 3), (8, 5), (13, 22)$.

   (c) whose walls are the curves $y = x^2 - 8x + 21$ and $y = x + 13$.

   (d) whose walls are the curves $y = x^2 - 8x + 21$ and $y = \frac{1}{12}(100 + 75x - 7x^2)$.

   (e) $6x^2 + 15y^2 = 90$ **(f)** $(x - 6)^2 + (z - 7)^2 = 9$

   (g) $(y - 8)^2 + (z - 10)^2 = 25$

   (h) whose footprint, in the xz-plane, is the triangle with vertices $(3, 4), (9, 12), (6, 17)$.

   (i) whose footprint, in the yz-plane, is the triangle with vertices $(6, 1), (1, 7), (12, 13)$.

B3. Find the surface area of that portion of the surface $z = 36 - 4x^2 - 9y^2$ that is inside the cylinder $2x^2 + 3y^2 = 12$.

B4. Find the surface area of that portion of the surface $z = 1 + 2x^2 + 3y^2$ that lies below the plane $z = 5$.

B5. Find the surface area of that portion of the surface $z = \sqrt{x^2 + y^2}$ that lies in the first octant and below the plane $z = 1$.

B6. Find the surface area of that portion of the surface $z = \sqrt{x^2 + y^2 - 1}$ that lies in the first octant and below the plane $z = 3$.

B7. Find the surface area of that portion of the cylinder $x^2 + y^2 = 1$ cut out by the cylinder $x^2 + z^2 = 1$.

B8. Find the surface area of that portion of the surface $z = \sqrt{x^2 + y^2}$ that lies within the cylinder $x^2 + y^2 - y = 0$.

B9. Find the surface area of that portion of the cylinder $x = y^2$ cut off by the planes $z = 0$, $y = z$, and $x = 2$ and for which $y \geq 0$.

B10. Find the surface area of that portion of the cylinder $x^2 + 2z^2 = 2$ that lies within the cylinder $x^2 + y^2 = 1$.

B11. Find the area of that portion of the plane $3x + 4y = 12$ that lies within the ellipsoid $\frac{x^2}{16} + \frac{y^2}{49} + \frac{z^2}{25} = 1$.

B12. Find the area of that portion of the plane $6x + 4y + 3z = 12$ that lies within the ellipsoid $\frac{x^2}{9} + \frac{y^2}{16} + \frac{z^2}{25} = 1$.

B13. Obtain the surface area of the surface of revolution formed when the curve $y = f(x) = x^3$, $0 \leq x \leq 1$, is rotated about the x-axis.

B14. Obtain the surface area of the surface of revolution formed when the curve $y = f(x) = \tan x$, $0 \leq x \leq 1$, is rotated about the x-axis.

B15. When the curve $y = f(x)$, $a \leq x \leq b$, is rotated about the x-axis, the surface of revolution so formed is given implicitly by $z^2 + y^2 = f^2(x)$. To substantiate this claim, obtain graphs of the implicit surfaces corresponding to the surfaces in Exercises B13 and 14.

B16. Writing the implicit surface in Exercise B15 as $z(x, y) = \sqrt{f^2(x) - y^2}$:

(a) Obtain $d\sigma$, the element of surface area.

(b) Set up a double integral for the surface area for this surface of revolution.

(c) Show that evaluation of the inner integral in part (b) leads to $2\pi \int_a^b f(x)\sqrt{1 + (f')^2}\, dx$.

21.2   Surface Integrals and Surface Flux

Surface Integrals

The surface integral $\iint_S f(x, y, z)\, d\sigma$ accumulates the values of the function $f(x, y, z)$ on a surface S just the way the line integrals $\int_C f(x, y)\, dx$ and $\int_C f(x, y)\, dy$ accumulate values of the function $f(x, y)$ along a curve C. For the line integral, the parametrization of the path C is passed to the integrand to guarantee that $f(x, y)$ assumes values just on C. Similarly, for a surface integral, the integrand must be restricted to assume values just on the surface S. But there are two ways to express the surface S.

The surface S can be given either explicitly as $z = z(x, y)$ or parametrically via formulas of the form $x = x(u, v)$, $y = y(u, v)$, $z = z(u, v)$. When the surface S is given explicitly as $z = z(x, y)$, the surface area element is $d\sigma = \sqrt{1 + z_x^2 + z_y^2}\, dx\, dy$. Section 23.2 considers the more complicated expression for the surface integral when the surface is given parametrically.

EXAMPLE 21.3   **Surface integral**   Suppose the functions $f(x, y, z) = x$ and $g(x, y, z) = xz$ give the charge per unit area on the surface S, defined by $z = x^2 + y$, with x in $[0, 1]$ and y in $[-1, 1]$. The element of surface area is $d\sigma = \sqrt{2 + 4x^2}\, dx\, dy$. The double integral that accumulates the total charge due to $f(x, y, z) = x$ on the surface S is

$$\int_{-1}^{1}\int_{0}^{1} x\sqrt{2 + 4x^2}\, dx\, dy = \sqrt{6} - \tfrac{1}{3}\sqrt{2}$$

whereas the double integral that accumulates the total charge due to $g(x, y, z) = xz$ on the surface S is

$$\int_{-1}^{1}\int_{0}^{1} x(x^2 + y)\sqrt{2 + 4x^2}\, dx\, dy = \tfrac{2}{5}\sqrt{6} + \tfrac{1}{15}\sqrt{2}$$

The explicit z in $g(x, y, z) = xz$ is replaced by $z = x^2 + y$, the value of z on the surface S. ❖

Surface Flux

CONCEPT   In analogy to the notion of flux through a curve C, the *surface flux* of the vector field

$$\mathbf{F} = f(x, y, z)\mathbf{i} + g(x, y, z)\mathbf{j} + h(x, y, z)\mathbf{k}$$

through a surface S is given by the surface integral of the normal component of $\mathbf{F}$ over the surface. The normal component of $\mathbf{F}$ is $\mathbf{F} \cdot \mathbf{N}$, where $\mathbf{N}$ is a unit normal vector on the surface S. Once this normal component has been obtained, the surface flux integral becomes a *surface integral,* that is, it becomes the integral of some scalar-valued function over a surface S. Thus, the surface flux integral is

$$\iint_S \mathbf{F} \cdot \mathbf{N}\, d\sigma$$

Choosing the "right" normal takes more than a few words. For the flux through a curve C there were two possible normals to consider and the convention adopted was to take the normal to the "right" of the tangent vector $\mathbf{T}$. On a closed curve, this normal is outward for traverse counterclockwise. There is also a choice of two normals on a surface, but unfortunately, for a surface there is no recipe by which every user will end up with the same normal. Hence, the choice of normal on a surface is arbitrary, and that choice will have to be stated clearly in every case.

There is one case where the choice of surface normal permits a convention that removes the ambiguity. The **closed surface** (like a sphere or an ellipsoid) permits the convention "**outward normal.**" That's the easy case. Look at the surface, determine the inside and outside of the closed surface, and select the outward-pointing normal.

EXAMPLE 21.4   **Closed surface**   We will find the flux of the field $\mathbf{F} = x\mathbf{i} + 2y\mathbf{j} + 3z\mathbf{k}$ through the unit sphere given implicitly by

$$f(x, y, z) \equiv x^2 + y^2 + z^2 - 1 = 0 \tag{21.1}$$

Since the divergence $\nabla \cdot \mathbf{F} = 6$, the net flow should be outward and positive.

Determining the outward normal belongs in the arena of the sphere. It has nothing to do with the field $\mathbf{F}$. The sphere is given implicitly by (21.1). For a surface defined implicitly by a formula of the form $f(x, y, z) = 0$, the gradient operator produces a normal vector. Whether or not this normal vector is inward or outward for the closed surface cannot be determined a priori. We must examine the gradient so calculated to see which way it points.

From (21.1) we have $\nabla f = 2(x\mathbf{i} + y\mathbf{j} + z\mathbf{k})$, which points in the direction of increasing f. For $w = f(x, y, z)$, that makes ∇f outward but not unit. A *unit* normal is $\frac{x\mathbf{i}+y\mathbf{j}+z\mathbf{k}}{\sqrt{x^2+y^2+z^2}}$, and a unit normal field on the *surface* of the sphere is $\mathbf{N} = x\mathbf{i} + y\mathbf{j} + z\mathbf{k}$, since there, $x^2 + y^2 + z^2 = 1$.

Next, obtain $\mathbf{F} \cdot \mathbf{N} = x^2 + 2y^2 + 3z^2$, which for either branch of $z = \pm\sqrt{1 - x^2 - y^2}$ becomes $3 - 2x^2 - y^2$. The surface area element is $d\sigma = \frac{dx\, dy}{\sqrt{1-x^2-y^2}}$ for both the upper and lower hemispheres, each of which project to R_{xy}, the unit disk in the xy-plane. The desired flux integral is then

$$2\iint_{R_{xy}} \frac{3 - 2x^2 - y^2}{\sqrt{1 - x^2 - y^2}}\, dx\, dy = 2\int_0^{2\pi}\int_0^1 \frac{r(3 - r^2(1 + \cos^2 t))}{\sqrt{1 - r^2}}\, dr\, dt = 8\pi \tag{21.2}$$

where the integral on the left is evaluated by changing to polar coordinates. The flux is positive, indicating the net flow of the field **F** is along the (outward) normal **N**, and hence, is itself outward, consistent with the divergence of **F** being 6.   ❖

NONCLOSED SURFACES   A nonclosed surface such as a plane or a hemisphere does not have an "inside" or "outside," so a new convention for selecting a unique normal is needed. Note that the word "upward" is also ambiguous because it depends on perspective. For example, consider the plane $x - y + z = 1$. Write the plane as $z = 1 - x + y$, and the normal $-z_x \mathbf{i} - z_y \mathbf{j} + \mathbf{k} = \mathbf{i} - \mathbf{j} + \mathbf{k}$ will point upward for an observer who thinks z is inherently upward. Write the plane as $y = y(x, z) = x + z - 1$ and $\mathbf{N} = -y_x \mathbf{i} + \mathbf{j} - y_z \mathbf{k} = -\mathbf{i} + \mathbf{j} - \mathbf{k}$, a vector that is "upward" to an observer whose north is the positive y-direction but is "downward" to observers who think the z-direction is upward. So, the recipe for finding a normal for an explicitly given surface is not fool-proof. It works only for surfaces described explicitly by a formula of the form $z = z(x, y)$. Then, the recipe yields an "upward" normal, provided everyone is in agreement that the z-direction is "upward."

ORIENTING SURFACES   Some texts describe for a surface, the process of *orientation* (sketched below), but leave the impression that giving a surface a right-handed orientation solves the problem of picking a unique normal. This is not true. Orienting a surface insures that a continuous normal field has been selected over all the surface, not that a unique normal has been selected.

The orientation process starts with the arbitrary choice of a normal, then covers the whole surface with a field of normals continuous with respect to the original arbitrary choice. It thus guarantees that when constructing surfaces, even closed surfaces, from nonclosed parts, the complete surface ends up with a consistent (continuous) normal field.

Right-handed orientation is accomplished by walking along the bounding edge of a nonclosed surface in such a way that the surface is to the left, the head pointing in the direction of the normal field. Advance along the bounding edge, walking on the surface itself, with the surface to the left. The body is the normal vector.

Unfortunately, a right-handed orientation can be given to the same surface with the opposite normal being selected. For example, consider the upper hemisphere. Walk around the equator while standing on the outside of the hemisphere, left hand toward the surface, and head pointing "outward." Having walked west-to-east, the body is the "outward" normal. Alternatively, stand *inside* the surface, walk along the equator on the inside of the hemisphere, left hand pointing into the surface. The journey around the equator is now east-to-west, and the body is now the "inward" normal. Thus, orienting a surface merely assures continuity in the normal field, it does not select one of the two possible normal fields.

There are surfaces, however, that cannot be oriented. The classic example is the Moebius band formed by gluing together the ends of a strip of paper after the strip has been given a one-half turn, or twist. The Moebius band, shown in Figure 21.5, is not orientable because a continuous normal field cannot be prescribed on it. The half-twist in the strip rotates the normal field as the field is followed around the band. The initial and terminal points of the circuit around the band should be identified since the surface has no thickness. Therefore, the normals at the initial and terminal points should be the same, but instead, they are 180° out of phase. Hence, the normal field is inherently discontinuous on the band, and the band is not orientable.

FIGURE 21.5   Moebius band with its discontinuous normal field

EXAMPLE 21.5   **Nonclosed surface**   We will find the flux of the field $\mathbf{F} = 18z\mathbf{i} - 12\mathbf{j} + 3y\mathbf{k}$ through the surface S, which is the first-octant portion of the plane $2x + 3y + 6z = 12$.

A unit normal field on this plane is $\mathbf{N} = \frac{2\mathbf{i}+3\mathbf{j}+6\mathbf{k}}{7}$, as per the sketch in Figure 21.6. The normal component of $\mathbf{F}$ is $\mathbf{F} \cdot \mathbf{N} = \frac{18}{7}(2z + y - 2)$, and writing the plane in the explicit form $z = 2 - \frac{y}{2} - \frac{x}{3}$ we get the element of surface area $d\sigma = \frac{7}{6}$. Evaluating $\mathbf{F} \cdot \mathbf{N}\, d\sigma$ on the surface by substituting $z = z(x, y)$, we get the integrand $6 - 2x$. The domain of integration is a triangle in the xy-plane, the first-quadrant shadow of the plane onto the xy-plane. The bounding curves are $x = 0$, $y = 0$, and the line $y = -\frac{2}{3}x + 4$. Hence, the required flux is $\int_0^6 \int_0^{-2x/3+4}(6 - 2x)\,dy\,dx = 24$. ❖

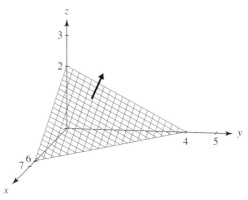

FIGURE 21.6   Example 21.5: The surface and a representative normal vector

EXAMPLE 21.6   **Nonclosed surface**   We will find the flux of the field $\mathbf{F} = z\mathbf{i} + x\mathbf{j} - 3y^2 z\mathbf{k}$ through the surface S, which is the first-octant part of the curved wall of the cylinder $x^2 + y^2 = 16$, between $z = 0$ and $z = 5$.

The surface S is given implicitly by $f(x, y, z) \equiv x^2 + y^2 - 16 = 0$, so a normal vector is $\nabla f = 2(x\mathbf{i} + y\mathbf{j})$, a unit normal field is $\frac{x\mathbf{i}+y\mathbf{j}}{\sqrt{x^2+y^2}}$, and a unit normal field *on the surface* where $x^2 + y^2 = 16$ is $\mathbf{N} = \frac{x\mathbf{i}+y\mathbf{j}}{4}$. The surface S is a cylinder with axis parallel to the z-axis and, hence, has no projection to the xy-plane. The surface cannot be expressed in the form $z = z(x, y)$. Instead, project the cylinder onto the xz-plane, and solve $f(x, y, z) = 0$ for $y = y(x, z) = \sqrt{16 - x^2}$, where the positive root has been chosen because S lies in the first octant. Next, compute

$$d\sigma = \sqrt{1 + y_x^2 + y_z^2}\,dx\,dz = \frac{4}{\sqrt{16 - x^2}}\,dx\,dz \qquad (21.3)$$

then form the integrand

$$\mathbf{F} \cdot \mathbf{N}\, d\sigma = \frac{zx + xy}{\sqrt{16 - x^2}}\,dx\,dz = \left(\frac{zx}{\sqrt{16 - x^2}} + x\right)dx\,dz \qquad (21.4)$$

where y has been replaced with $y(x, z)$. The projection of the surface $y = y(x, z)$ onto the xz-plane is the rectangle $0 \le x \le 4, 0 \le z \le 5$. Hence, $\int_0^5 \int_0^4 (\frac{zx}{\sqrt{16-x^2}} + x)\,dx\,dz = 90$ is the flux through S. ❖

EXAMPLE 21.7   **Closed surface**   Find the flux of the field $\mathbf{F} = \mathbf{k}$ passing through the surface S, which is just the unit sphere centered at the origin. Note that the field $\mathbf{F}$ is constant and upward, so that whatever passes into S on the bottom leaves again through the top. We would therefore expect that the flux through S is 0.

As in Example 21.4, represent the unit sphere implicitly by (21.1) so that on the whole surface, the unit normal field is $\mathbf{N} = x\mathbf{i} + y\mathbf{j} + z\mathbf{k}$ and $d\sigma = \frac{dx\,dy}{\sqrt{1-x^2-y^2}} = \frac{dx\,dy}{\sqrt{z^2}} = \frac{dx\,dy}{|z|}$. If $\sqrt{z^2}$ is erroneously simplified to z instead of $|z|$, the net flux will be the counterintuitive 2π. Since $\mathbf{F} \cdot \mathbf{N}\,d\sigma = \frac{z}{|z|}$ and the upper and lower hemispheres both project to R_{xy}, the unit disk in the xy-plane, the total flux is given by

$$\iint_{R_{xy}} dx\,dy + \iint_{R_{xy}} -dx\,dy = 0$$

On the upper hemisphere, $\frac{z}{|z|} = 1$, but on the lower hemisphere, $\frac{z}{|z|} = -1$; so the two double integrals differ in sign. Incorrectly simplifying $|z|$ to z gives two identical integrals, each giving π, the area of the unit disk.   ❖

EXERCISES 21.2–Part A

A1. Show how the unit disk with center at the origin can have a right-handed orientation with the normal field pointed upward or downward.

A2. Evaluate $\iint_S f(x, y, z)\,d\sigma$ if $f(x, y, z) = x^2 y^2$ and S is the surface $z = 1 - x^2 - y^2, z \geq 0$.

A3. Evaluate $\iint_S f(x, y, z)\,d\sigma$ if $f(x, y, z) = x^2 z$ and S is the surface $z = 1 - x^2 - y^2, z \geq 0$.

A4. Compute the flux of $\mathbf{F} = x^2\mathbf{i} + 2y^2\mathbf{j} + 3z\mathbf{k}$ through S, the surface $z = 1 - x^2 - y^2, z \geq 0$.

A5. Compute the flux of $\mathbf{F} = 3x^2\mathbf{i} + 5y^2\mathbf{j} + z^2\mathbf{k}$ through S, the surface $z = 1 - x^2 - y^2, z \geq 0$.

EXERCISES 21.2–Part B

B1. For Example 21.4, verify the integration in (21.2).

B2. For Example 21.6, verify $d\sigma$ given in (21.3).

B3. For Example 21.6, verify the integrand in (21.4).

B4. For Example 21.6, verify the integral
$\int_0^5 \int_0^4 (\frac{zx}{\sqrt{16-x^2}} + x)\,dx\,dz = 90$.

In Exercises B5–14, compute $\iint_S f(x, y, z)\,d\sigma$ for the given $f(x, y, z)$ and surface S.

B5. $f(x, y, z) = (2x^2 - 3y^2)z^3$; S is that part of the cylinder $x^2 + y^2 = 9$ between $z = 0$ and $z = 4$.

B6. $f(x, y, z) = x^2 y z^2$; S is that part of the cylinder $2x^2 + 3z^2 = 5$ between $y = 0$ and $y = 2$.

B7. $f(x, y, z) = 5x^2 - 4y^2$; S is that part of the sphere $x^2 + y^2 + z^2 = 49$ lying between $z = 2$ and $z = 6$.

B8. $f(x, y, z) = 5x^2 y - 3z + 4$; S is the first-octant portion of the plane $2x + 5y + 9z = 45$.

B9. $f(x, y, z) = 3x + 4y - 5z^2$; S is the first-octant portion of the cylinder $y = x^2$ lying below $z = 5$ and bounded by the plane $x = 1$.

B10. $f(x, y, z) = 4x^3 + 8y^3 + 8x^2 z$; S is that part of the plane $7x - 3y + 5z = 21$ interior to the cylinder $4x^2 + 3y^2 - 2y = 1$.

B11. $f(x, y, z) = 4xz - 4xy - 9x^2$; $S = \{z = 10 - x^2 - 4y^2, z \geq 0\}$

B12. $f(x, y, z) = 7y + 9xz - 8yz$; S is that part of $z = \sqrt{x^2 + y^2}$ interior to the cylinder $x^2 + 2x + y^2 = 2$.

B13. $f(x, y, z) = 9xz - 2xy^2 - 4z$; S is that part of $z = 4 - y^2$ inside $R = \{0 \leq y \leq 4 - x^2\}$.

B14. $f(x, y, z) = 6yz - 8x^2 - 9z^3$; S is that part of $z = xy$ defined on $R = \{0 \leq y \leq 3, 0 \leq x \leq y\}$.

In Exercises B15–24, compute $\iint_S \mathbf{F} \cdot \mathbf{N}\,d\sigma$, the flux of $\mathbf{F}$ through the given surface S, where $\mathbf{N}$ is the upward unit normal on S.

B15. $\mathbf{F} = 5xy\mathbf{i} - 3yz\mathbf{j} + 13xz\mathbf{k}$; S is the first-octant portion of the plane $3x + 5y + 7z = 105$.

B16. $\mathbf{F} = y\mathbf{i} + z^2\mathbf{j} + x^3\mathbf{k}$; S is that part of $z = 125 - 2x^2 - 3y^2$ lying above the triangle whose vertices are $(1, 1, 0)$, $(2, 5, 0)$, $(-3, 4, 0)$.

B17. $\mathbf{F} = yz\mathbf{i} - 2x^2\mathbf{j} + 3xz\mathbf{k}$; S is that part of $z = xy$ defined on $0 \leq y \leq 1 - x^2$.

B18. $\mathbf{F} = 3z\mathbf{i} + 2xj - 5y\mathbf{k}$; S is that part of $z = 1 + xy^2$ defined on $R = \{0 \leq x \leq \pi, 0 \leq y \leq \sin x\}$.

B19. $\mathbf{F} = y^2\mathbf{i} - z^2\mathbf{j} + x^2\mathbf{k}$; S is that part of $z = x^2 + y^2$ satisfying $0 \leq z \leq 1$.

B20. $\mathbf{F} = (x - y)\mathbf{i} + (y + z)\mathbf{j} + xz\mathbf{k}$; $S = \{z = \sqrt{1 - x^2 - y^2}\}$

B21. $\mathbf{F} = (yz - x)\mathbf{i} + (xz - y)\mathbf{j} + (xy - z^2)\mathbf{k}$; S is that part of $x^2 + y^2 - z^2 = 9$ satisfying $-4 \leq z \leq 4$. (Use an outward normal.)

B22. $\mathbf{F} = x\mathbf{i} + y^2\mathbf{j} - z^3\mathbf{k}$; S is the first-octant portion of $x^2 + z^2 = 1$ satisfying $0 \leq x \leq y$.

B23. $\mathbf{F} = \nabla f$, $f = 9xz - 9xy^2 + 5z$; S is that part of the cylinder $z = \sin y$ satisfying $0 \le y \le \frac{\pi}{2}$, $0 \le x \le 1$.

B24. $\mathbf{F} = \nabla f$, $f = 8yz - 8x^2 + 7z^3$; S is that part of the cylinder $z = x^3$ satisfying $0 \le x \le 1$, $0 \le y \le x$.

In Exercises B25–34, compute $\iint_S \mathbf{F} \cdot \mathbf{N}\, d\sigma$, the flux of $\mathbf{F}$ through the given closed surface S, where $\mathbf{N}$ is the outward unit normal on S.

B25. $\mathbf{F} = x^2\mathbf{i} + y\mathbf{j} - z\mathbf{k}$; $S = \{z = \sqrt{1 - x^2 - y^2}, z = 0\}$

B26. $\mathbf{F} = yz\mathbf{i} - xz\mathbf{j} + xy\mathbf{k}$; $S = \{z = x^2 + y^2, z = 1\}$

B27. $\mathbf{F} = z\mathbf{i} - 2x\mathbf{j} + y\mathbf{k}$; $S = \{z = 1 - x^2 - y^2, z = 0\}$

B28. $\mathbf{F} = y^2\mathbf{i} + z\mathbf{j} - x^2\mathbf{k}$; $S = \{2x + 3y + 5z = 30, 0 \le x, y, z\}$

B29. $\mathbf{F} = x\mathbf{i} - yz\mathbf{j} + y^2\mathbf{k}$; $S = \{x^2 + y^2 = 1, z = 0, z = 1\}$

B30. $\mathbf{F} = \nabla f$, $f = x^2yz$; S is the rectangular box whose faces are $x = \pm 1$, $y = \pm 2$, $z = \pm 3$.

B31. $\mathbf{F} = \nabla f$, $f = yz^2 - 3xy^2$; S is the boundary of the region enclosed by $\{x^2 + y^2 = 1, z = 0, z = 2 - x\}$.

B32. $\mathbf{F} = \nabla f$, $f = 2xz^3 + 5y^2$; S is the boundary of the region enclosed by $z = \sqrt{2x^2 + 5y^2}$ and $z = 1$.

B33. $\mathbf{F} = xz\mathbf{i} - y^2\mathbf{j} + yz\mathbf{k}$; $S = \{2x^2 + 3y^2 + 4z^2 = 24\}$

B34. $\mathbf{F} = (y + z)\mathbf{i} + (x - z)\mathbf{j} + 2xy\mathbf{k}$; S is the triangular cylinder with vertices $(1, 3)$, $(-4, 5)$, $(3, -9)$ in the xy-plane, bounded above by the plane $5x + 7y + 9z = 315$ and below by $z = 0$.

21.3 The Divergence Theorem and the Theorems of Green and Stokes

Statement of the Theorems

There are connections between the vector operators div, grad, and curl and the integrals defining flux and circulation. Three classic theorems attributed to Green, Stokes, and Gauss connect these notions and are an essential part of vector calculus.

The three theorems are stated first so that they may be compared and contrasted. The divergence theorem of Gauss is the most intuitive and is stated before Stokes' Theorem. Green's Theorem has two forms, one resembling the divergence theorem and one resembling Stokes' Theorem, each form being the planar counterpart of its analog. In fact, some texts confuse the issue by calling Green's Theorem "the divergence theorem in the plane."

For simple regions, both the divergence theorem and Green's Theorem are proven by straightforward integration ideas. The proof of Stokes' Theorem generally invokes Green's Theorem. All three theorems are extended to more complicated regions by arguments based on subdivision to the simpler cases. Details of such proofs are found in texts such as [84], [92], [91], [35], [97], and [65].

THE DIVERGENCE THEOREM OF GAUSS   Alternatively called the "divergence theorem," the divergence theorem of Gauss should not be confused with "Gauss' Theorem," which appears in Section 23.1. Gauss' Theorem is a consequence of the divergence theorem, but not conversely.

The essence of the divergence theorem is the integration formula

$$\iiint_V \nabla \cdot \mathbf{F}\, dv = \iint_S \mathbf{F} \cdot \mathbf{N}\, d\sigma \tag{21.5}$$

and appropriate hypotheses for making (21.5) valid. Conditions on the vector field $\mathbf{F}$ are just that its components have continuous derivatives inside and on the surface S. The surface S, bounding the open connected set V, must be piecewise smooth, closed, and oriented. The vector $\mathbf{N}$ is the unit outward normal on S. The volume represented by V is connected, but not necessarily simply connected. Thus, V can be the interior of a sphere (simply connected) or of a torus (donut; not simply connected). It can even be the (simply connected) interior of a sphere that has had a smaller sphere extracted from its interior. In this case, S would be the union of the inner and outer boundaries.

STOKES' THEOREM   The essence of Stokes' Theorem is the integration formula

$$\iint_S [\nabla \times \mathbf{F}] \cdot \mathbf{N}\, d\sigma = \oint_C \mathbf{F} \cdot d\mathbf{r} = \oint_C \mathbf{F} \cdot \mathbf{T}\, ds \qquad (21.6)$$

with sufficient hypotheses for making (21.6) valid. For example, the vector field **F** should have continuously differentiable components. The surface S should be oriented, its unit normal field **N** consistent with the orientation. In addition, S should be described by piecewise smooth functions and bounded by a piecewise smooth, simple closed curve C, itself oriented consistently with the orientation of S.

GREEN'S THEOREM IN THE PLANE   Green's Theorem consists of an integration formula that is valid in the plane and resembles either the divergence theorem or Stokes' Theorem. One form of the theorem can be transformed into the other by suitably defining the vector field **F** appearing in the theorem. In Table 21.1, each form is given in both vector and component forms. Both forms are valid under the general hypotheses [35] that R is a bounded closed region whose boundary consists of a finite number of simple, closed, rectifiable curves and that the components of **F** have continuous partial derivatives. A *rectifiable curve* is one that has finite arc length.

Form	Green's Theorem Integral Formula
Divergence	$\iint_R \nabla \cdot \mathbf{F}\, dx\, dy = \oint_C \mathbf{F} \cdot \mathbf{N}\, ds$ $\iint_R (f_x + g_y)\, dx\, dy = \oint_C f\, dy - g\, dx$
Stokes'	$\iint_R [\nabla \times \mathbf{F}] \cdot \mathbf{k}\, dx\, dy = \oint_C \mathbf{F} \cdot d\mathbf{r} = \oint_C \mathbf{F} \cdot \mathbf{T}\, ds$ $\iint_R (g_x - f_y)\, dx\, dy = \oint_C f\, dx + g\, dy$

TABLE 21.1   Alternative forms of Green's Theorem

If the divergence form of the theorem is applied to the vector field $\mathbf{F} = g\mathbf{i} - f\mathbf{j}$, the resulting component form will read precisely as the Stokes' form. If the Stokes' form of the theorem is applied to the field $\mathbf{F} = -g\mathbf{i} + f\mathbf{j}$, the resulting component form will read precisely as the divergence form. A more complete discussion of Green's Theorem is the subject of Section 21.4.

Discussion of the Divergence Theorem

In the integration formula $\iiint_V \nabla \cdot \mathbf{F}\, dv = \iint_S \mathbf{F} \cdot \mathbf{N}\, d\sigma$ that is the essence of the divergence theorem, V is a three-dimensional region in space, representing a volume, and S is its bounding surface, necessarily closed. A vector field **F** is defined throughout the region V, and $\nabla \cdot \mathbf{F}$ is its divergence. The volume element in V is dv, which is $dx\, dy\, dz$ in Cartesian coordinates or $r\, dr\, d\theta\, dz$ in cylindrical coordinates.

The left side of the divergence theorem is the volume integral of the divergence. We saw earlier that this yields the net spread of the field **F** inside the region V. The divergence theorem is a *law of balance,* so whatever enters or leaves the region V must have passed through the boundaries of the region. This boundary is the surface S, and on the right side of the formula we find the surface integral of $\mathbf{F} \cdot \mathbf{N}$, where $\mathbf{N}$ is the outward unit normal on S. Thus, the integrand on the right is the surface integral of the normal component of the field $\mathbf{F}$, namely, the flux of $\mathbf{F}$ through the surface S.

The divergence theorem is intuitively appealing, simply claiming that the net divergence (spread) of the field $\mathbf{F}$ inside a volume V must be accounted for by the flux of $\mathbf{F}$ through the bounding surface S enclosing V.

EXAMPLE 21.8

For the field $\mathbf{F} = x\mathbf{i} + y\mathbf{j} + z\mathbf{k}$ and the region V taken as the interior of the unit sphere, verify that the divergence theorem is valid.

Define S implicitly by (21.1), so a unit normal field on S is $\mathbf{N} = x\mathbf{i} + y\mathbf{j} + z\mathbf{k}$ and $\mathbf{F} \cdot \mathbf{N} = x^2 + y^2 + z^2 = 1$ on S. The divergence of $\mathbf{F}$ is $\nabla \cdot \mathbf{F} = 3$, so the volume integral on the left in the divergence theorem, $\iiint_V 3 \, dv$, is just three times the volume of the unit sphere, or $3(\frac{4}{3}\pi 1^3) = 4\pi$. The flux integral on the right is $\iint_S 1 \, d\sigma = 4\pi$ because it is just the surface area of the unit sphere. ❖

EXAMPLE 21.9

Verify the validity of the divergence theorem for the field of Example 21.8, and V', the region between concentric spheres with radii $\frac{1}{2}$ and 1. Thus, S has two pieces, the outer surface, which is the sphere of radius 1, and the inner surface, which is the sphere of radius $\frac{1}{2}$. A cut-away sketch of the region V', along with representative outward normal vectors on the inner and outer surfaces, are shown in Figure 21.7.

From Example 21.8, the field is $\mathbf{F} = x\mathbf{i} + y\mathbf{j} + z\mathbf{k}$, with divergence $\nabla \cdot \mathbf{F} = 3$. The integral of the divergence over the region V' can be realized by subtracting from 4π (the value over V computed in Example 21.8), the integral of the divergence over V'', the interior of the inner sphere. Thus, the left side of the divergence theorem will be

$$4\pi - \iiint_{V''} 3 \, dx \, dy \, dz = \frac{7}{2}\pi$$

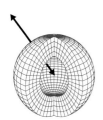

FIGURE 21.7 Example 21.9: Region between concentric spheres

using $\frac{4}{3}\pi \left(\frac{1}{2}\right)^3 = \frac{\pi}{6}$ for the volume of the inner sphere of radius $\frac{1}{2}$.

The unit outward normal field on the outer surface is still $\mathbf{N} = x\mathbf{i} + y\mathbf{j} + z\mathbf{k}$, but on the inner surface it is $\mathbf{n} = -2(x\mathbf{i} + y\mathbf{j} + z\mathbf{k})$, the minus sign because the normal field must be *outward* with respect to the region V', which is therefore *inward* with respect to s, the smaller sphere. The value of $\mathbf{F} \cdot \mathbf{N}$ is 1 on the outer surface and $-\frac{1}{2}$ on the inner sphere. The value of the flux through the surface of the larger sphere is 4π, as computed in Example 21.8. The flux through the surface of the smaller sphere is

$$\iint_s -\frac{1}{2} \, d\sigma = -\frac{1}{2}4\pi \left(\frac{1}{2}\right)^2 = -\frac{\pi}{2}$$

so the total flux through V' is $\frac{7}{2}\pi$.

Discussion of Stokes' Theorem

The essential formula in Stokes' Theorem is (21.6). The surface S, usually not closed, is called a *capping surface* for the bounding curve C. For example, a hemisphere is the capping surface for the circle C, which is the equator. Alternatively, a thimble is the capping surface for the circle that defines its opening.

The integrand in the double integral on the left is the flux, through the surface S, of the curl field of $\mathbf{F}$. On the right, the tangential component of $\mathbf{F}$ is integrated around the bounding curve C, yielding the circulation of $\mathbf{F}$ around C.

Stokes' Theorem balances the field's net *vorticity* (circulation) flowing through the surface and the average circulation of **F** around the bounding curve C. Vorticity (or twist, rotation) of **F** on the surface S is measured by the curl of **F**. Integrating the normal component of the curl of **F** on the surface S allows local swirls that oppose each other to cancel out, leaving just the uncanceled parts on the boundary curve C to reckon with. This residual vorticity is along the bounding curve C and accumulates along C in the line integral on the right side of (21.6).

Figure 21.8 illustrates how local rotation of the normal components of $\nabla \times \mathbf{F}$ "cancel" where two such vectors are contiguous, leaving just the points along the boundary to accumulate circulation. Thus, the net circulation as measured by the flux of the curl field can be measured by summing the tangential component of **F** along the bounding curve C.

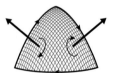

FIGURE 21.8   Flux of the curl field passing through a surface

EXAMPLE 21.10   If the surface S is the upper unit hemisphere, verify Stokes' Theorem for the field $\mathbf{F} = -y\mathbf{i} + x\mathbf{j} + z\mathbf{k}$. Let the capping surface S be defined by (21.1), so that the unit normal field on S is $\mathbf{N} = x\mathbf{i} + y\mathbf{j} + z\mathbf{k}$. The curl of **F** is $\nabla \times \mathbf{F} = 2\mathbf{k}$, from which follows $[\nabla \times \mathbf{F}] \cdot \mathbf{N} = 2z$. The element of surface area, computed from $z = z(x, y) = \sqrt{1 - x^2 - y^2}$, is

$$d\sigma = \frac{dx\,dy}{\sqrt{1 - x^2 - y^2}} = \frac{dx\,dy}{\sqrt{z^2}} = \frac{dx\,dy}{z}$$

valid on the upper hemisphere. The surface integral on the left of (21.6) is $\iint_{R_{xy}} 2\,dx\,dy$, where R_{xy} is the unit disk in the xy-plane. The flux integral, twice the area of the unit disk, evaluates to 2π.

For the line integral on the right side of (21.6), the parametrization of the bounding curve C must be consistent with the orientation induced by the choice of normal field used in the flux integral on the left. The outward normal was used. Walk the equator with the left hand pointing inward, toward the surface S, and the head in the direction of the normal. If the hemisphere is set on the xy-plane, the equator is the unit circle centered at the origin. The orientation induced by the outward normal means the curve C is traversed in the counterclockwise direction. A parametrization that accomplishes this is $x(p) = \cos p$, $y(p) = \sin p$. Since $dz = 0$ in the xy-plane,

$$\mathbf{F} \cdot \mathbf{T}\,ds = \mathbf{F} \cdot d\mathbf{r} = F_1\,dx + F_2\,dy + F_3\,dz = 1$$

and the closed line integral on the right reduces to $\oint_C 1\,d\sigma = 2\pi$, the circumference of C.

STOKES' THEOREM ON A CLOSED SURFACE   If Stokes' Theorem is applied to a *closed* surface S, the result is always 0. The simplest way to see this is to look at both Stokes' Theorem and the divergence theorem

$$\iint_S [\nabla \times \mathbf{F}] \cdot \mathbf{N}\,d\sigma = \oint_C \mathbf{F} \cdot d\mathbf{r} \qquad \iiint_V \nabla \cdot \mathbf{A}\,dv = \iint_S \mathbf{A} \cdot \mathbf{N}\,d\sigma$$

In the divergence theorem take $\mathbf{A} = \nabla \times \mathbf{F}$, obtaining

$$\iiint_V \nabla \cdot [\nabla \times \mathbf{F}]\,dv = \iint_S [\nabla \times \mathbf{F}] \cdot \mathbf{N}\,d\sigma$$

The divergence of the curl is necessarily zero since *curls don't spread*. That makes the left side of the divergence theorem zero. But with $\mathbf{A} = \nabla \times \mathbf{F}$, the right side of the divergence theorem reads the same as the left side of Stokes' Theorem. Hence, applying the surface integral in Stokes' Theorem to a *closed* surface S yields zero.

EXERCISES 21.3–Part A

A1. If $\mathbf{F} = y\mathbf{i} + z^2\mathbf{j} + x^2\mathbf{k}$, verify the divergence theorem for the region bounded by $z = \sqrt{1 - x^2 - y^2}$ and $z = 0$.

A2. If $\mathbf{F} = xy\mathbf{i} + yz\mathbf{j} + xz\mathbf{k}$, verify Stokes' Theorem for the surface $z = \sqrt{1 - x^2 - y^2}$.

A3. If C is a simple closed curve and $f(x, y, z)$ and $g(x, y, z)$ are scalar-valued functions, apply Stokes' Theorem to $\mathbf{F} = \nabla(fg)$ to prove that $\oint_C f\nabla g \cdot d\mathbf{r} = -\oint_C g\nabla f \cdot d\mathbf{r}$.

A4. If V is the region internal to the closed surface S, $\mathbf{N}$ is the outward unit normal field on S, and $f(x, y, z)$ is a scalar-valued function, apply the divergence theorem to the vector $f\mathbf{F}$ to prove that $\iint_S f\mathbf{F} \cdot \mathbf{N} \, d\sigma = \iiint_V (f\nabla \cdot \mathbf{F} + \mathbf{F} \cdot \nabla f) \, dv$.

A5. For any surface S, the surface area is given by $\iint_S d\sigma$. If S is a closed surface with unit outward normal $\mathbf{N}$, prove that $\iint_S \mathbf{N} \, d\sigma = \mathbf{0}$. *Hint:* Let $\mathbf{B}$ be a constant vector, consider $\mathbf{B} \cdot \iint_S \mathbf{N} \, d\sigma = \iint_S \mathbf{B} \cdot \mathbf{N} \, d\sigma$, and apply the divergence theorem.

EXERCISES 21.3–Part B

B1. If v is the volume internal to the closed surface S, show that $v = \frac{1}{6} \iint_S \nabla(r^2) \cdot \mathbf{N} \, d\sigma$, where $\mathbf{N}$ is the outward unit normal field on S, and $r = \|x\mathbf{i} + y\mathbf{j} + z\mathbf{k}\|$.

In Exercises B2–8, if $\mathbf{F} = 2y^2\mathbf{i} + 3z^2\mathbf{j} + 4x\mathbf{k}$, verify the divergence theorem for the region bounded by the given surfaces.

B2. $z = \sqrt{1 - x^2 - y^2}$ and $z = 0$.

B3. $x = \pm 5$, $y = \pm 3$, and $z = \pm 1$

B4. $z = x^2 + y^2$ and $z = 4$ **B5.** $z = \sqrt{x^2 + y^2}$ and $z = 1$

B6. $z = x^2 + y^2$ and $z = 3 - 2\sqrt{x^2 + y^2}$

B7. $z = 10 - 2x^2 - 3y^2$ and $z = 0$

B8. $x^2 + 4x + y^2 + 4y = 8$, $z = 0$, and $2x + 3y + 5z = 30$

In Exercises B9–15, if $\mathbf{F} = 2x^2\mathbf{i} + 3y^2\mathbf{j} + 4z\mathbf{k}$, verify the divergence theorem for the region bounded by the given surfaces.

B9. $z = \sqrt{4 - x^2 - y^2}$ and $z = 0$

B10. $x = \pm 2$, $y = \pm 3$ and $z = \pm 4$

B11. $z = x^2 + y^2$ and $z = 1$ **B12.** $z = \sqrt{x^2 + y^2}$ and $z = 3$

B13. $z = x^2 + y^2$ and $z = 3 - 2\sqrt{x^2 + y^2}$

B14. $z = 9 - 3x^2 - 4y^2$ and $z = 0$

B15. $x^2 + 6x + y^2 + 4y = 3$, $z = 0$, and $5x + 3y + 6z = 90$

In Exercises B16–22, if $\mathbf{F} = 2xy\mathbf{i} + 3yz\mathbf{j} + 5xz\mathbf{k}$, verify the divergence theorem for the region bounded by the given surfaces.

B16. $z = x^2 + y^2$ and $z = 3 - 2\sqrt{x^2 + y^2}$

B17. $x = \pm 3$, $y = \pm 1$, and $z = \pm 5$

B18. $z = x^2 + y^2$ and $z = 4$ **B19.** $z = \sqrt{x^2 + y^2}$ and $z = 1$

B20. $z = \sqrt{1 - x^2 - y^2}$ and $z = 0$

B21. $z = 16 - 4x^2 - 2y^2$ and $z = 0$

B22. $x^2 + y^2 = 1$, $y + z = 4$, and $z = 0$. *Hint:* Project the cylinder onto the yz-plane.

In Exercises B23–29, if $\mathbf{F} = yz\mathbf{i} - x^2z\mathbf{j} + xy^2\mathbf{k}$, verify the divergence theorem for the region bounded by the given surfaces.

B23. $x = \pm 1$, $y = \pm 2$, and $z = \pm 3$

B24. $z = x^2 + y^2$ and $z = 3 - 2\sqrt{x^2 + y^2}$

B25. $z = x^2 + y^2$ and $z = 4$ **B26.** $z = 10 - 2x^2 - 3y^2$ and $z = 0$

B27. $z = \sqrt{x^2 + y^2}$ and $z = 1$

B28. $z = \sqrt{1 - x^2 - y^2}$ and $z = 0$

B29. $3x + 5y + 7z = 95$ and lying in the first octant.

In Exercises B30–36, if $\mathbf{F} = (3x^2 - 2yz)\mathbf{i} + (5yz - 2x)\mathbf{j} + (7xz + 6y^2)\mathbf{k}$, verify Stokes' Theorem for the given surface.

B30. $y = \sqrt{9 - x^2 - z^2}$ **B31.** $x^2 + z^2 = 9$, $0 \le y \le 1$, and $y = 1$

B32. $x^2 + z^2 = 9$, $0 \le y \le 5$, and $y = 5$ **B33.** $z = \sqrt{1 - x^2 - y^2}$

B34. $z = 9 - x^2 - y^2$, $z \ge 0$

B35. $x^2 + y^2 = 9$, $0 \le z \le 1$, and $z = 1$

B36. $x^2 + y^2 = 9$, $0 \le z \le 5$, and $z = 5$

In Exercises B37–41, if $\mathbf{F} = 3yz\mathbf{i} + 5xz^2\mathbf{j} - 7xy^2\mathbf{k}$, verify Stokes' Theorem for the given surface.

B37. The surface bounded by the planes $x = 0$, $y = 0$, $4x + 5y + 6z = 120$, and lying in the first octant.

B38. $z = x^2 + y^2$, $0 \le z \le 4$ **B39.** $z = \sqrt{x^2 + y^2}$, $0 \le z \le 5$

B40. $x + y + z = 4$, inside the cylinder $x^2 + y^2 = 1$.

21.4 # Green's Theorem

Integral Formulas of Green's Theorem

Table 21.1 in Section 21.3 lists the integral formulas for the two forms of Green's Theorem. The following two examples illustrate both forms of the theorem.

EXAMPLE 21.11   Taking the region R as the unit disk in the xy-plane, we verify both forms of Green's Theorem for the vector field $\mathbf{F} = f(x, y)\mathbf{i} + g(x, y)\mathbf{j} = x\mathbf{i} + xy\mathbf{j}$. The boundary of R is the unit circle, parametrized with $x(p) = \cos p$, $y(p) = \sin p$. The curl of $\mathbf{F}$ is $\nabla \times \mathbf{F} = y\mathbf{k}$, and the divergence of $\mathbf{F}$ is $\nabla \cdot \mathbf{F} = 1 + x$. Except on the line $x = -1$, $\nabla \cdot \mathbf{F} \neq 0$. Therefore, the question "What is the net divergence within the unit disk?" becomes an interesting question.

DIVERGENCE FORM OF GREEN'S THEOREM   In the divergence form of Green's Theorem, $\nabla \cdot \mathbf{F}$ is integrated over the region R. Since here R is the unit disk, a change to polar coordinates is appropriate, so the double integral on the left becomes

$$\int_0^{2\pi} \int_0^1 (1 + r\cos t)r\, dr\, dt = \pi \tag{21.7}$$

The closed line integral on the right, the net flux of $\mathbf{F}$ through the boundary of the region R, is around a circle. The parametrization of C given previously traces the unit circle in the counterclockwise sense, putting the normal $\mathbf{N}$ to the right of the tangent vector $\mathbf{T}$. This is the positive orientation of the curve C. The line integral around C is then

$$\oint_C f\, dy - g\, dx = \int_0^{2\pi} \cos^2 t\, dt - \int_0^{2\pi} -\cos t \sin^2 t\, dt = \pi \tag{21.8}$$

STOKES' FORM OF GREEN'S THEOREM   In Stokes' form of Green's Theorem, the integrand of the double integral on the left is $[\nabla \times \mathbf{F}] \cdot \mathbf{k} = y$, and in polar coordinates, that integral is

$$\int_0^{2\pi} \int_0^1 r^2 \sin t\, dr\, dt = 0$$

Note that the value of this double integral is not the same as the value of the double integral in the divergence form of the theorem. There are two different versions of Green's Theorem. One form computes the net divergence of $\mathbf{F}$ throughout the region R, and the other computes the net curl. There is no reason to believe that these quantities should be the same. What the theorem does guarantee, however, is that in each case the corresponding line integral taken around the boundary of R will agree with the respective double integral.

For the Stokes' form, the line integral on the right is the circulation integral, the integral of the tangential component of $\mathbf{F}$. Thus,

$$\oint_C f\, dx + g\, dy = \int_0^{2\pi} -\cos t \sin t\, dt + \int_0^{2\pi} \cos^2 t \sin t\, dt = 0 \tag{21.9}$$

and both sides in Stokes' form of Green's Theorem agree.   ❖

EXAMPLE 21.12   We verify both forms of Green's Theorem for the field $\mathbf{F} = f(x, y)\mathbf{i} + g(x, y)\mathbf{j} = xy^2\mathbf{i} + (x + y)\mathbf{j}$, and R, the region inside the curves $y_1 = x^{1/4}$ and $y_2 = x^4$ but outside the square defined by the inequalities $\frac{3}{10} \leq x \leq \frac{6}{10}$ and $\frac{3}{10} \leq y \leq \frac{6}{10}$. A sketch of the region R appears in Figure 21.9.

A double integral over the region R requires that R be subdivided into four parts, $R_k, k = 1, \ldots, 4$. Each edge of the interior square will require a separate parametrization, so each edge is separately labeled. The boundary of R is in two parts. One part consists of the curves y_1 and y_2, forming the outer boundary. The other part consists of the four edges of the square, forming the inner boundary. A positive orientation for the boundary is counterclockwise around the outer boundary and clockwise around the inner boundary. All these features are shown in Figure 21.9.

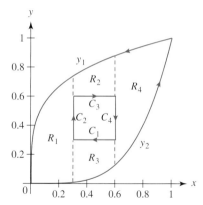

FIGURE 21.9 The region for Example 21.12

The curl of $\mathbf{F}$ is $\nabla \times \mathbf{F} = (1 - 2xy)\mathbf{k}$, and the divergence of $\mathbf{F}$ is $\nabla \cdot \mathbf{F} = 1 + y^2 = \zeta \neq 0$. ❖

DIVERGENCE FORM OF GREEN'S THEOREM In the divergence form of Green's Theorem, the double integral on the left is

$$
\int_0^{3/10} \int_{x^4}^{x^{1/4}} \zeta \, dy \, dx + \int_{3/10}^{3/5} \int_{3/5}^{x^{1/4}} \zeta \, dy \, dx + \int_{3/10}^{3/5} \int_{x^4}^{3/10} \zeta \, dy \, dx
$$

$$
+ \int_{3/5}^{1} \int_{x^4}^{x^{1/4}} \zeta \, dy \, dx = \frac{596{,}901}{910{,}000} \tag{21.10}
$$

the sum of four integrals over the subregions $R_k, k = 1, \ldots, 4$. The line integral around R will take six integrals. First, on the outer boundary, the integral

$$
\oint f \, dy - g \, dx = \int_{y_1} f \, dy - g \, dx + \int_{y_2} f \, dy - g \, dx
$$

will be

$$
\int_0^1 4x^{12} \, dx - \int_0^1 (x + x^4) \, dx + \int_1^0 \tfrac{1}{4} x^{3/4} \, dx - \int_1^0 (x + x^{1/4}) \, dx = \frac{348}{455} \tag{21.11}
$$

Then, on the inner boundary, there are four different line segments $C_k, k = 1, \ldots, 4$, parametrized as shown in Table 21.2. The integral $\oint f \, dy - g \, dx$ is now $-\frac{1089}{10{,}000}$, the sum of the four line integrals listed in Table 21.2. The flux through the complete boundary of R is then $\frac{348}{455} - \frac{1089}{10{,}000} = \frac{596{,}901}{910{,}000}$, which agrees with the double integral of the divergence of $\mathbf{F}$ throughout R.

C_k	Parametrization	$\oint f \, dy - g \, dx$	Value
C_1	$x_1 = t, \, y_1 = \tfrac{3}{10}, \, \tfrac{3}{5} \leq t \leq \tfrac{3}{10}$	$\int_{3/5}^{3/10} 0 \, dt - \int_{3/5}^{3/10} \left(t + \tfrac{3}{10}\right) dt$	$\tfrac{9}{40}$
C_2	$x_2 = \tfrac{3}{10}, \, y_2 = t, \, \tfrac{3}{10} \leq t \leq \tfrac{3}{5}$	$\int_{3/10}^{3/5} \tfrac{3}{10} t^2 \, dt - \int_{3/10}^{3/5} 0 \, dt$	$\tfrac{189}{10000}$
C_3	$x_3 = t, \, y_3 = \tfrac{3}{5}, \, \tfrac{3}{10} \leq t \leq \tfrac{3}{5}$	$\int_{3/10}^{3/5} 0 \, dt - \int_{3/10}^{3/5} \left(t + \tfrac{3}{5}\right) dt$	$-\tfrac{63}{200}$
C_4	$x_4 = \tfrac{3}{5}, \, y_4 = t, \, \tfrac{3}{5} \leq t \leq \tfrac{3}{10}$	$\int_{3/5}^{3/10} \tfrac{3}{5} t^2 \, dt - \int_{3/5}^{3/10} 0 \, dt$	$-\tfrac{189}{5000}$

TABLE 21.2 Example 21.12: Line integrals in divergence form of Green's Theorem

STOKES' FORM OF GREEN'S THEOREM For Stokes' form of Green's Theorem, the double integral on the left contains $[\nabla \times \mathbf{F}] \cdot \mathbf{k} = 1 - 2xy = \rho$ and is

$$
\int_0^{3/10} \int_{x^4}^{x^{1/4}} \rho \, dy \, dx + \int_{3/10}^{3/5} \int_{3/5}^{x^{1/4}} \rho \, dy \, dx + \int_{3/10}^{3/5} \int_{x^4}^{3/10} \rho \, dy \, dx
$$

$$
+ \int_{3/5}^{1} \int_{x^4}^{x^{1/4}} \rho \, dy \, dx = \frac{4929}{20{,}000} \tag{21.12}
$$

the sum of four separate integrals over the subregions $R_k, k = 1, \ldots, 4$.

The circulation integral around the boundary will contain six separate line integrals. Using the parameter x, the line integrals on the separate curves forming the outer boundary sum to

$$\left(\int_0^1 x^9 \, dx + \int_0^1 4(x + x^4)x^3 \, dx \right) + \left(\int_1^0 x^{3/2} \, dx + \int_1^0 \frac{x + x^{1/4}}{4x^{3/4}} \, dx \right) = \frac{3}{10} \quad (21.13)$$

The four integrals on the inner boundary, listed in Table 21.3, sum to $-\frac{1071}{20,000}$, so the net circulation around the boundary of R is $\frac{3}{10} - \frac{1071}{20,000} = \frac{4929}{20,000}$, in agreement with the value of the double integral on the left in Stokes' form of Green's Theorem.

C_k	$\oint f \, dx + g \, dy$	Value
C_1	$\int_{3/5}^{3/10} \frac{9}{100} t \, dt + \int_{3/5}^{3/10} 0 \, dt$	$-\frac{243}{20000}$
C_2	$\int_{3/10}^{3/5} 0 \, dt + \int_{3/10}^{3/5} \left(t + \frac{3}{10} \right) dt$	$\frac{9}{40}$
C_3	$\int_{3/10}^{3/5} \frac{9}{25} t \, dt + \int_{3/10}^{3/5} 0 \, dt$	$\frac{243}{5000}$
C_4	$\int_{3/5}^{3/10} 0 \, dt + \int_{3/5}^{3/10} \left(t + \frac{3}{5} \right) dt$	$-\frac{63}{200}$

TABLE 21.3 Example 21.12: Line integrals in Stokes' form of Green's Theorem

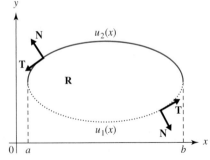

FIGURE 21.10 Proof of Green's Theorem: simple region, bounding curves $y = y(x)$

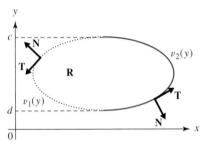

FIGURE 21.11 Proof of Green's Theorem: simple region, bounding curves $x = x(y)$

Proof of Green's Theorem

An outline of a proof of the divergence form of Green's Theorem for a *simple* region R is given. The region is shown in Figure 21.10 where the boundary curves are written as functions of x and Figure 21.11 where the boundary curves are written as functions of y. (A nonsimple region would be broken into simple regions where the following analysis would be carried out, and the results combined for the larger region.)

The double integral $\iint_R (f_x + g_y) \, dx \, dy$ is equivalent to the sum of the iterated integrals $\int_a^b \int_{u_1(x)}^{u_2(x)} g_y \, dy \, dx$ and $\int_c^d \int_{v_1(y)}^{v_2(y)} f_x \, dx \, dy$. The first iterated integral is guided by Figure 21.10, where the boundary consists of the two curves $y = u_1(x)$ and $y = u_2(x)$ and the second by Figure 21.11, where the boundary consists of the two curves $x = v_1(y)$ and $x = v_2(y)$. These iterated integrals evaluate, respectively, to

$$\int_a^b g(x, u_2(x)) \, dx - \int_a^b g(x, u_1(x)) \, dx \quad \text{and} \quad \int_c^d f(v_2(y), y) \, dy - \int_c^d f(v_1(y), y) \, dy \quad (21.14)$$

Reversing the limits of integration in the integrals containing $u_2(x)$ and $v_1(y)$ gives

$$-\int_a^b g(x, u_1(x)) \, dx - \int_b^a g(x, u_2(x)) \, dx \quad \text{and} \quad \int_c^d f(v_2(y), y) \, dy + \int_d^c f(v_1(y), y) \, dy \quad (21.15)$$

which sum to $\oint_C f \, dy - g \, dx$.

EXERCISES 21.4–Part A

A1. Use Green's Theorem and an appropriate field **F** to establish that the area inside the simple closed curve C is given by each of the integrals $\oint_C x\,dy = -\oint_C y\,dx = \frac{1}{2}\oint_C x\,dy - y\,dx$.

In Exercises A2–6, let R be the given region. If $\mathbf{F} = (y^3 - 3x^2)\mathbf{i} + (x^2 + 5y^2)\mathbf{j}$:

    (a) Verify the divergence form of Green's Theorem for the region R.

    (b) Verify the Stokes' form of Green's Theorem for the region R.

    (c) Use the third integral in Exercise A1 to find the area of R.

A2. the triangle whose vertices are $(0, 0)$, $(5, 3)$, $(1, 4)$

A3. the region bounded by the parabola $y = x^2$ and the line $y = x + 1$

A4. the region inside the ellipse $\frac{x^2}{9} + \frac{y^2}{16} = 1$, best parametrized as $x = 3\cos t, y = 4\sin t, 0 \le t \le 2\pi$

A5. the region bounded by the parabolas $y = x^2$ and $y = 8 - x^2$

A6. the region inside $x = 2 + \sin t$, $y = (2 + \sin t)\cos t, 0 \le t \le 2\pi$

A7. Verify the integrals in (21.14).

A8. Verify the integrals in (21.15).

EXERCISES 21.4–Part B

B1. Verify the integral in (21.7).

B2. Verify the line integral in (21.8).

B3. Verify the line integral in (21.9).

B4. Verify the results in (21.10).

B5. Verify the results in (21.11).

B6. Verify the entries in Table 21.2.

B7. Verify the results in (21.12).

B8. Verify the results in (21.13).

B9. Verify the results in Table 21.3.

In Exercises B10–21, a region R is given. For each:

    (a) Use the third integral in Exercise A1 to find the area of R.

    (b) Verify both the divergence and Stokes' form of Green's Theorem for R and the field $\mathbf{F} = (x^2 - 3y^2)\mathbf{i} + (y^3 + 5x^2)\mathbf{j}$.

    (c) Verify both the divergence and Stokes' form of Green's Theorem for R and the field $\mathbf{F} = xy^2\mathbf{i} - (3x^2 + 2y)\mathbf{j}$.

    (d) Verify both the divergence and Stokes' form of Green's Theorem for R and the field $\mathbf{F} = (x^2y + 1)\mathbf{i} + (xy^3 - x + y)\mathbf{j}$.

    (e) Verify both the divergence and Stokes' form of Green's Theorem for R and the field $\mathbf{F} = x^2y^3\mathbf{i} - (1 + xy)\mathbf{j}$.

    (f) Verify both the divergence and Stokes' form of Green's Theorem for R and the field $\mathbf{F} = (x^2 - 3y)\mathbf{i} + (y^3 + 2x)\mathbf{j}$.

B10. the region inside $x = \cos t$, $y = \sin 2t$, $-\frac{\pi}{2} \le t \le \frac{\pi}{2}$

B11. the region inside the loop in the curve given by $x = t^2 - 1$, $y = t(t^2 - 1), -1 \le t \le 1$

B12. the region inside the curve defined implicitly by $|x| + |y| = 1$

B13. the region between the two circles $x^2 + y^2 = 1$ and $x^2 + y^2 = 4$

B14. the region between the square $x = \pm 1$, $y = \pm 1$ and the circle $x^2 + y^2 = 4$

B15. the region between the triangle with vertices $(1, 1)$, $(-2, 3)$, $(4, -5)$ and the square $x = \pm 7$, $y = \pm 7$

B16. the region between the circle $x^2 + y^2 = 1$ and the square $x = \pm 3$, $y = \pm 3$

B17. the region between the circle $x^2 + y^2 = 1$ and the ellipse $x = 3\cos t$, $y = 4\sin t, 0 \le t \le 2\pi$

B18. the region between $|x| + |y| = 1$ and the ellipse $x = 3\cos t$, $y = 4\sin t, 0 \le t \le 2\pi$

B19. the region inside the square $x = \pm 5$, $y = \pm 5$ but outside the circles $(x - 2)^2 + (y - 2)^2 = 1$ and $(x + 2)^2 + (y + 2)^2 = 1$

B20. the region inside the hypocycloid $x = \cos^3 t$, $y = \sin^3 t$, $0 \le t \le 2\pi$

B21. the region between $x = \sin t$, $y = t(1 - t), 0 \le t \le 1$, and the x-axis

In Exercises B22–31, use Green's Theorem to evaluate the given line integrals.

B22. $\oint_C (3x^2 - 5y^2)\,dx + x^2y^2\,dy$; C is the square $x = \pm 1$, $y = \pm 1$

B23. $\oint_C xy^2\,dx + (2x - 3y)\,dy$; C is the triangle with vertices $(2, 3)$, $(7, 5)$, $(4, -2)$

B24. $\oint_C (3x - 4y^2)\,dx + x^2y\,dy$; C is the circle $x^2 + y^2 = 1$

B25. $\oint_C x\cos(y)\,dy - y\sin(x)\,dx$; C is the triangle with vertices $(-2, 5)$, $(5, 2)$, $(9, 8)$

B26. $\oint_C ye^x\,dy - xe^y\,dx$; C is the rectangle $x = \pm 1$, $y = \pm 2$

B27. $\oint_C xy^4\,dy - x^4y\,dx$; C bounds the region enclosed by $y = x^2$ and $y = x^3$

B28. $\oint_C x^3y\,dx + (2y^2 - x)\,dx$; C is the cardioid $r = 1 + \cos\theta$

B29. $\oint_C (3xy^2 - 2x^3)\,dx + (x + 5y)\,dy$; C is the first-quadrant loop of the rose $r = \sin 3\theta$

B30. $\oint_C (x^2y^4 + x)\,dy - (x^4y^2 - y)\,dx$; C is the first-quadrant loop of the rose $r = \sin 2\theta$

B31. $\oint_C (5x^2 - 4y^2)\,dy - (3xy + x^2)\,dx$; C is the outer loop of the limaçon $r = 1 + 2\sin\theta$

B32. Consider the field $\mathbf{F} = \frac{x\mathbf{i}}{x^2+y^2} + \frac{y\mathbf{j}}{x^2+y^2}$, and recall that one hypothesis of Green's Theorem is that $f(x, y)$ and $g(x, y)$, the components of $\mathbf{F} = f\mathbf{i} + g\mathbf{j}$, have continuous first partial derivatives on the region R over which the double integral is computed.

(a) Show graphically that f_x is not continuous at $(x, y) = (0, 0)$.

(b) Show analytically that f_x is not continuous at $(x, y) = (0, 0)$ by approaching the origin, for example, along the x-axis.

(c) Show that in any region that does not include the origin, both the curl and the divergence of $\mathbf{F}$ are zero.

(d) For this field, verify both Stokes' and the divergence forms of Green's Theorem if C is the circle with radius 1 and with center at $(2, 2)$. Thus, the singularity in F is not inside C, Green's Theorem applies, and $0 = 0$ results for each form of the theorem.

(e) Let C be the unit circle centered at the origin. The singularity in the field is now inside the region R and Green's Theorem does not apply. Show that for this field, the Stokes' form of the theorem seems to "work," but the divergence form yields the contradictory $0 = 2\pi$.

(f) Let C' be a small circle with radius a and with center at the origin. Let R' be the region between C and C'. Green's Theorem is valid on R'. The boundary of R' now consists of C and C' (both oriented in the counterclockwise sense). Hence, the following extended forms of Green's Theorem are suggested, namely, $\oint_C \mathbf{F} \cdot \mathbf{T}\, ds = \oint_{C'} \mathbf{F} \cdot \mathbf{T}\, ds + \iint_{R'} \nabla \cdot \mathbf{F}\, dA$ and $\oint_C \mathbf{F} \cdot \mathbf{N}\, ds = \oint_{C'} \mathbf{F} \cdot \mathbf{N}\, ds + \iint_{R'} \nabla \times \mathbf{F}\, dA$. Verify both forms of the extended theorem for this field $\mathbf{F}$. Show that the line integral around C' is independent of the value of a.

In Exercises B33–37, analyze the given field in a manner analogous to the discussion in Exercise B32. However, in part (c) the divergence and/or the curl may not vanish, and in part (e), the outcomes for the two forms of Green's theorem may not be the same as in Exercise 68.

B33. $\mathbf{F} = -\dfrac{y}{x^2+y^2}\mathbf{i} + \dfrac{x}{x^2+y^2}\mathbf{j}$

B34. $\mathbf{F} = \dfrac{x}{(x^2+y^2)^{3/2}}\mathbf{i} + \dfrac{y}{(x^2+y^2)^{3/2}}\mathbf{j}$

B35. $\mathbf{F} = -\dfrac{y}{(x^2+y^2)^{3/2}}\mathbf{i} + \dfrac{x}{(x^2+y^2)^{3/2}}\mathbf{j}$

B36. $\mathbf{F} = -\dfrac{y}{\sqrt{x^2+y^2}}\mathbf{i} + \dfrac{x}{\sqrt{x^2+y^2}}\mathbf{j}$

B37. $\mathbf{F} = \dfrac{x}{\sqrt{x^2+y^2}}\mathbf{i} + \dfrac{y}{\sqrt{x^2+y^2}}\mathbf{j}$

B38. If F is singular at points P_k, $k = 1, \ldots, N$, inside C, then each singularity is surrounded by small circles C_k, $k = 1, \ldots, N$, respectively, so that Green's Theorem extends to either $\oint_C \mathbf{F} \cdot \mathbf{T}\, ds = \sum_{k=1}^{N} \oint_{C_k} \mathbf{F} \cdot \mathbf{T}\, ds + \iint_{R'} \nabla \cdot \mathbf{F}\, dA$ or $\oint_C \mathbf{F} \cdot \mathbf{N}\, ds = \sum_{k=1}^{N} \oint_{C_k} \mathbf{F} \cdot \mathbf{N}\, ds + \iint_{R'} \nabla \times \mathbf{F}\, dA$. Of course, R' is that portion of the interior of C exclusive of the C_k and their interiors.

(a) Apply this formalism to the field $F = x\mathbf{i} + y\mathbf{j}$, where C is a circle with center at the origin and radius 5, whereas C_1 and C_2 are circles of radius 1, centered at $(2, 0)$ and $(0, 2)$, respectively.

(b) Apply this formalism to the same field, but take C to be the square $x = \pm 4$, $y = \pm 4$, and take C_1 and C_2 as squares with side 1 and with centers at $(2, 0)$ and $(0, 2)$, respectively.

21.5   Conservative, Solenoidal, and Irrotational Fields

Introduction

A conservative force was initially defined as $\mathbf{F} = -\nabla u$, that is, as a gradient of a scalar potential $u(x, y, z)$. Moreover, it was intimated that there are other, equivalent definitions of a conservative force. These alternative definitions, and their consequences and connections, listed in Table 21.4, are the substance of this section. We will assume the domains for the functions in Table 21.4 are *simply connected,* that is, have no "holes," and that the functions themselves have continuous partial derivatives. Exercise A10 shows the difficulty one encounters if the domain is not simply connected.

Conservative and Irrotational Fields—A Discussion of Theorem 21.1

Theorem 21.1 is stated in Section 20.4 where we showed that for $\mathbf{F}$ conservative, $\mathrm{curl}(\mathbf{F}) = -\mathrm{curl}(\mathrm{grad}\, u) = \mathbf{0}$ and created the mnemonic "gradients don't twist" to capture this identity. To prove the converse, we have to exhibit the scalar potential for the irrotational field $\mathbf{F}$. We provide two constructive recipes for this.

Theorem 21.1	$\mathbf{F}$ is conservative $\Leftrightarrow$ $\mathbf{F}$ is *irrotational*
Theorem 21.2	$\mathbf{F}$ is conservative $\Leftrightarrow$ $\mathbf{F} \cdot d\mathbf{r}$ *exact*
Theorem 21.3	$\mathbf{F}$ is conservative $\Leftrightarrow$ $\oint_C \mathbf{F} \cdot d\mathbf{r} = 0$ for all closed paths C
Theorem 21.4	$\mathbf{F}$ is conservative $\Leftrightarrow$ $\int_A^B \mathbf{F} \cdot d\mathbf{r}$ independent of path from A to B
Theorem 21.5	The *conservative* $\mathbf{F}$ is also *solenoidal* $\Leftrightarrow$ $\mathbf{F}$ has *harmonic* scalar potential
Theorem 21.6	$\mathbf{F}$ is *solenoidal* $\Leftrightarrow$ $\mathbf{F} = \nabla \times \mathbf{A}$ for some vector $\mathbf{A}$, the *vector potential* for $\mathbf{F}$

TABLE 21.4   Characterizations and properties of conservative forces

RECIPE 1   If $\mathbf{F} = f(x, y, z)\mathbf{i} + g(x, y, z)\mathbf{j} + h(x, y, z)\mathbf{k}$, its curl is $\nabla \times \mathbf{F} = (h_y - g_z)\mathbf{i} + (f_z - h_x)\mathbf{j} + (g_x - f_y)\mathbf{k}$. The working hypothesis is that the curl of $\mathbf{F}$ is the zero vector, so $h_y = g_z$, $h_x = f_z$, and $g_x = f_y$. Next, define the three integrals $\int_a^x f(t, b, c)\, dt$, $\int_b^y g(x, t, c)\, dt$, and $\int_c^z h(x, y, t)\, dt$ along straight lines parallel to the axes. The first integral is from (a, b, c) to (x, b, c) along a line parallel to the x-axis. The second integral is from (x, b, c) to (x, y, c) along a line parallel to the y-axis. The third integral is from (x, y, c) to (x, y, z) along a line parallel to the z-axis. Define $U(x, y, z)$ to be the sum of these three integrals and $u(x, y, z) = -U(x, y, z)$. Thus,

$$-u(x, y, z) = U(x, y, z) = \int_a^x f(t, b, c)\, dt + \int_b^y g(x, t, c)\, dt + \int_c^z h(x, y, t)\, dt$$

$$(21.16)$$

works because $-u_x = U_x = f$, $-u_y = U_y = g$, $-u_z = U_z = h$ follow from (21.16) by direct computation, provided $h_y = g_z$, $h_x = f_z$, and $g_x = f_y$, from $\nabla \times \mathbf{F} = \mathbf{0}$, are used. Indeed, $U_z = h(x, y, z)$ is immediate, and

$$U_y = g(x, y, c) + \int_c^z h_y(x, y, t)\, dt = g(x, y, c) + \int_c^z g_z(x, y, t)\, dt = g(x, y, z)$$

follows. Finally, we obtain

$$U_x = f(x, b, c) + \int_b^y g_x(x, t, c)\, dt + \int_c^z h_x(x, y, t)\, dt$$

$$= f(x, b, c) + \int_b^y f_y(x, t, c)\, dt + \int_c^z f_z(x, y, t)\, dt = f(x, y, z)$$

EXAMPLE 21.13   The curl of the vector

$$\mathbf{F} = (12xy + yz)\mathbf{i} + (6x^2 + xz)\mathbf{j} + xy\mathbf{k} \qquad (21.17)$$

is zero. Using (21.16), construct

$$-u = U = \int_a^x (12tb + bc)\, dt + \int_b^y (6x^2 + xc)\, dt + \int_c^z xy\, dt$$

$$= -6a^2 b - abc + 6x^2 y + xyz$$

$$(21.18)$$

Setting $(a, b, c) = (0, 0, 0)$ gives $u = -(6x^2 y + xyz)$, since the terms in a, b, and c constitute an additive constant, and the scalar potential is determined up to an additive constant by any recipe. Verification that $-\nabla u$ recovers $\mathbf{F}$ is left to Exercise A2.   ❖

RECIPE 2   The radius-vector form of the line connecting (a, b, c) with (x, y, z) is given by $\mathbf{r}(t)$, where

$$\mathbf{R0} = a\mathbf{i} + b\mathbf{j} + c\mathbf{k} \tag{21.19a}$$

$$\mathbf{R} = x\mathbf{i} + y\mathbf{j} + z\mathbf{k} \tag{21.19b}$$

$$\mathbf{r}(t) = \mathbf{R0} + t(\mathbf{R} - \mathbf{R0}) = [a + t(x-a)]\mathbf{i} + [b + t(y-b)]\mathbf{j} + [c + t(z-c)]\mathbf{k} \tag{21.19c}$$

The parameter on the line is t, and the segment is traced as t varies in the interval $[0, 1]$. In fact, $\mathbf{r}(0) = \mathbf{R0}$ and $\mathbf{r}(1) = \mathbf{R}$. Then, the potential function is

$$-u(x, y, x) = U(x, y, z) = \int_0^1 \mathbf{F}(\mathbf{r}(t)) \cdot \mathbf{r}'\, dt$$

EXAMPLE 21.14   For the field (21.17), with $(a, b, c) = (0, 0, 0)$, we have $\mathbf{r}(t) = t\mathbf{R}$, $\mathbf{r}' = \mathbf{R}$, and $\mathbf{F}(\mathbf{r}(t)) = t^2\mathbf{F}(x, y, z)$. Hence, we again find

$$-u = U = \int_0^1 (18x^2 y + 3xyz)\, dt = 6x^2 y + xyz \tag{21.20}$$

❖

CONSERVATIVE FIELDS, AND EXACT DIFFERENTIALS—A DISCUSSION OF THEOREM 21.2
Given the scalar function $f(x, y, z)$, the *exact* or *total differential* df is defined by

$$df = f_x\, dx + f_y\, dy + f_z\, dz$$

where the subscripts denote partial derivatives. The symbols dx, dy, and dz denote "increments" and can just as well be expressed as $dx = x - x_0$, $dy = y - y_0$, and $dz = z - z_0$. If the partial derivatives are evaluated at a point, then $df = a(x - x_0) + b(y - y_0) + c(z - z_0)$, which is just a linearization of the function f.

But given three expressions $A(x, y, z)$, $B(x, y, z)$, and $C(x, y, z)$, interpreting them as the partial derivatives of some function $f(x, y, z)$ isn't necessarily possible. The equations $f_x = A(x, y, z)$, $f_y = B(x, y, z)$, and $f_z = C(x, y, z)$ just may not have a solution for some function $f(x, y, z)$. Thus, the form $df = A\, dx + B\, dy + C\, dz$ may or may not be the total differential of some function $f(x, y, z)$. If it is, df is called an exact (or total) differential. This definition generalizes to three dimensions, the notion of exactness first seen in Section 3.3.

EXAMPLE 21.15   Given $f(x, y, z) = x \sin \frac{y}{z}$, the total differential is

$$df = \sin\left(\frac{y}{z}\right) dx + \frac{x \cos(\frac{y}{z})}{z}\, dy - \frac{x \cos(\frac{y}{z}) y}{z^2}\, dz$$

Note that the same expression can be constructed from

$$\nabla f = \sin\left(\frac{y}{z}\right)\mathbf{i} + \frac{x \cos(\frac{y}{z})}{z}\mathbf{j} - \frac{x \cos(\frac{y}{z}) y}{z^2}\mathbf{k}$$

dotted with $d\mathbf{r} = dx\mathbf{i} + dy\mathbf{j} + dz\mathbf{k}$.   ❖

The example shows why an arbitrary declaration of df may not be a total differential. There may not be a function whose gradient consists of the three components A, B, C. Thus, the problem of determining if a differential is exact is the same as determining if a force field comes from a potential by a gradient operation. It's the same problem. So, there is clearly a connection between a conservative force field $\mathbf{F} = A\mathbf{i} + B\mathbf{j} + C\mathbf{k}$ and finding a scalar potential $f(x, y, z)$ for which $-\text{grad}(f) = \mathbf{F}$.

Theorem 21.2 in Table 21.4 characterizes as conservative, fields $\mathbf{F} = A\mathbf{i} + B\mathbf{j} + C\mathbf{k}$ for which $\mathbf{F} \cdot d\mathbf{r}$ is an exact or total) differential df. To see why this is so, assume $\mathbf{F}$ is conservative so that $\mathbf{F} = -\nabla u = \nabla U$. Then,

$$\mathbf{F} \cdot d\mathbf{r} = A\,dx + B\,dy + C\,dz = U_x\,dx + U_y\,dy + U_z\,dz$$

is the (total) differential dU. On the other hand, if $A\,dx + B\,dy + C\,dz$ is known to be the exact differential df, then $\mathbf{F} = A\mathbf{i} + B\mathbf{j} + C\mathbf{k}$ is a gradient vector and hence, $A = f_x$, $B = f_y$, $C = f_z$, so $\mathbf{F} = \text{grad}(f)$ and is conservative.

Conservative Fields, and Work on Closed Paths—A Discussion of Theorem 21.3

A field $\mathbf{F}$ for which $\oint_C \mathbf{F} \cdot d\mathbf{r}$, the work done by the field as a closed path is traversed, vanishes for every closed path C is said to have the *closed-loop property*. Theorem 21.3 in Table 21.4 characterizes conservative fields as exactly those that have the closed-loop property. To see this, suppose $\mathbf{F}$ is conservative. Then, $\mathbf{F} \cdot d\mathbf{r}$ is an exact differential, df, and

$$\mathbf{F} \cdot d\mathbf{r} = A\,dx + B\,dy + C\,dz = f_x\,dx + f_y\,dy + f_z\,dz = df$$

The work integral then evaluates to $\oint_C \mathbf{F} \cdot d\mathbf{r} = \int_{P_0}^{P_0} df = f\big|_{P_0}^{P_0} = f(P_0) - f(P_0) = 0$.

On the other hand, if $\mathbf{F}$ has the closed-loop property, let C be any closed curve and let S be any open capping surface with C as its boundary. Invoke Stokes' Theorem to obtain $\iint_S [\nabla \times \mathbf{F}] \cdot \mathbf{N}\,d\sigma = \oint_C \mathbf{F} \cdot d\mathbf{r} = 0$ for every possible closed path C. Since the integral $\iint_S [\nabla \times \mathbf{F}] \cdot \mathbf{N}\,d\sigma$ vanishes for *arbitrary* closed curves C and their equally arbitrary capping surfaces S, the conclusion $\nabla \times \mathbf{F} = \mathbf{0}$ follows, thereby making $\mathbf{F}$ irrotational and, hence, conservative.

Conservative Fields, and Path Independence—A Discussion of Theorem 21.4

If the work done by a field $\mathbf{F}$ is independent of which path connects the starting and ending points, then the field is said to be *path independent*.

Theorem 21.4 in Table 21.4 characterizes conservative fields as those that are *path independent*. To see this, suppose $\mathbf{F}$ is conservative so that $\mathbf{F} \cdot d\mathbf{r}$ is an exact differential df. Thus, the work integral becomes

$$\text{Work} = \int_{P_0}^{P_1} \mathbf{F} \cdot d\mathbf{r} = \int_{P_0}^{P_1} df = f(P_1) - f(P_0)$$

a number that depends only on the points P_0 and P_1 and not on the path connecting them.

On the other hand, consider Figure 21.12, which depicts a closed loop composed from two paths connecting points A and B. The left path is traversed is from A to B in the counterclockwise sense, and the right path is traversed from A to B in the clockwise sense. Let the value of the work integral on the left path be u, and the value on the right path, v. If the right path is traversed in the counterclockwise sense, the work integral is then negated to $-v$. The assumption of path independence is captured in the equality $u = v$ or $u - v = 0$. The complete counterclockwise loop from A back to A is a closed path along which the work integral is the sum of the two values $u + (-v) = u - v = 0$. Hence, path independence is equivalent to the work vanishing on any closed loop, and by Theorem 21.3, $\mathbf{F}$ is conservative.

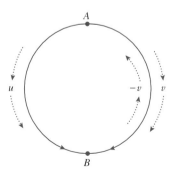

FIGURE 21.12   A closed loop constructed from two paths connecting A and B

EXAMPLE 21.16

A path-dependent field   The field $\mathbf{F} = f\mathbf{i} + g\mathbf{j} = (x^2 - y)\mathbf{i} + (y^3 + 2x)\mathbf{j}$ is one for which the work integral is not path independent. Since $\nabla \times \mathbf{F} = 3\mathbf{k} \neq \mathbf{0}$, $\mathbf{F}$ is not conservative. If $\mathbf{F}$ *were* conservative, the work integral *would* be path independent, but we will show the work integral for $\mathbf{F}$ to depend on the curve connecting starting and ending points. This will be demonstrated by computing the work integral for at least two different paths connecting $(0, 0)$, with $(1, 1)$. First, let the path be the line $y = x$ and compute

$$\int_{C_1} f \, dx + g \, dy = \int_0^1 (x^2 - x) \, dx + \int_0^1 (x^3 + 2x) \, dx = \tfrac{13}{12} \tag{21.21}$$

along this path. Next, take the path to be the line segments $x = t$, $y = 0$, then $x = 1$, $y = t$. The work along this second path is

$$\int_0^1 t^2 \, dt + 2\int_0^1 0 \, dt + \int_0^1 (t^3 + 2) \, dt = \tfrac{31}{12} \tag{21.22}$$

These two paths produced different amounts of work. The field $\mathbf{F}$ is not irrotational, is not conservative, and is not path independent.   ❖

Conservative, Solenoidal Fields, and Harmonic Potentials—A Discussion of Theorem 21.5

A function $u(x, y, z)$ with continuous second partial derivatives is said to be *harmonic* if it satisfies Laplace's equation

$$\nabla^2 u = u_{xx} + u_{yy} + u_{zz} = 0 \tag{21.23}$$

Some fields $\mathbf{F}$ are conservative, some are solenoidal, some are both conservative *and* solenoidal, and the rest are neither conservative nor solenoidal. The relationships are captured graphically in the Venn diagram in Figure 21.13. The circle $\mathbf{C}$ represents all conservative forces, whereas the ellipse $\mathbf{S}$ represents all solenoidal forces. In the overlap, $\mathbf{H}$, are forces that are both conservative *and* solenoidal. These forces are conservative forces with a harmonic scalar potential. The region $\mathbf{N}$ represents the remaining forces that are neither conservative nor solenoidal.

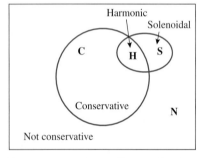

FIGURE 21.13   Venn diagram showing relationships between fields for which the potential function is *harmonic* (**H**) and fields that are *conservative* (**C**), *solenoidal* (**S**), and *nonconservative* (**N**)

These relationships are captured analytically in Theorem 21.5 in Table 21.4. In particular, the forces in region $\mathbf{H}$ in Figure 21.13 are those conservative forces for which the scalar potential u satisfies Laplace's equation (21.23).

To see why this is so, let $\mathbf{F}$ be conservative so that $\mathbf{F} = -\text{grad}(u)$ for some scalar function $u(x, y, z)$. If $\mathbf{F}$ is also solenoidal (divergence free), then $\text{div}(\mathbf{F}) = -\text{div}(\text{grad } u) = 0$. But the divergence of the gradient is the laplacian, so the scalar potential u satisfies $\nabla^2 u = 0$, making it harmonic. On the other hand, if u is harmonic, then $\text{div } \mathbf{F} = -\nabla^2 u = 0$, so $\mathbf{F}$ is solenoidal.

EXAMPLE 21.17

The function $u = \sin x \cosh y$ is harmonic because

$$\nabla^2 u = u_{xx} + u_{yy} = -\sin x \cosh y + \sin x \cosh y = 0$$

Hence, the field $\mathbf{F} = -\nabla u = -\cos x \cosh y\mathbf{i} - \sin x \sinh y\mathbf{j}$ must be conservative, since it arises from a scalar potential as a gradient. This field is also solenoidal since the divergence $\nabla \cdot \mathbf{F} = 0$. Thus, fields with a harmonic scalar potential are both conservative and solenoidal.   ❖

EXAMPLE 21.18

There are conservative forces for which the scalar potential function is not harmonic. In fact, it is more likely that the scalar potential is *not* harmonic. Just start with an arbitrary scalar function $u(x, y, z)$ and set $\mathbf{F} = -\nabla u$. It is unlikely that $\mathbf{F}$ is divergence-free since it is unlikely that the function $u(x, y, z)$ will satisfy Laplace's equation.

Consider the following scalar function $u(x, y, z) = x \sin \frac{y}{z}$, and compute

$$\mathbf{F} = -\nabla u = -\sin \frac{y}{z} \mathbf{i} - \frac{x}{z} \cos \frac{y}{z} \mathbf{j} + \frac{xy}{z^2} \cos \frac{y}{z} \mathbf{k} \qquad (21.24)$$

The field $\mathbf{F}$ just created is conservative because it was obtained as a gradient. However,

$$\nabla \cdot \mathbf{F} = -\nabla^2 u = z^{-4} \left(x(y^2 + z^2) \sin \frac{y}{z} - 2xyz \cos \frac{y}{z} \right) \neq 0 \qquad (21.25)$$

so $\mathbf{F}$ is not solenoidal and $u(x, y, z)$ is not harmonic. The field $\mathbf{F}$ is conservative but not solenoidal. The scalar potential $u(x, y, z)$ is not harmonic.   ❖

Solenoidal Fields and the Vector Potential—A Discussion of Theorem 21.6

If the field $\mathbf{F}$ arises from the field $\mathbf{A}$ as its curl, that is, if $\mathbf{F} = \nabla \times \mathbf{A}$, then $\mathbf{A}$ is called the *vector potential* for $\mathbf{F}$. Any vector $\mathbf{F} = \nabla \times \mathbf{A}$ is necessarily solenoidal as we saw from the vector identity $\operatorname{div}(\operatorname{curl} \mathbf{A}) = 0$ (*curls don't spread*) in Section 20.4.

EXAMPLE 21.19   The field $\mathbf{A} = (\frac{z^2}{2} - xy)\mathbf{i} - yz\mathbf{j}$ is the vector potential for

$$\mathbf{F} = \nabla \times \mathbf{A} = y\mathbf{i} + z\mathbf{j} + x\mathbf{k} \qquad (21.26)$$

and a computation verifies that $\nabla \cdot s\mathbf{F} = 0$.   ❖

Theorem 21.6 in Table 21.4 characterizes solenoidal fields with continuous partial derivatives as those that arise as the curl of a vector potential. To see why this is so, we have only to exhibit a recipe for finding the vector potential for a solenoidal field $\mathbf{F}$. (We have already verified the vector identity $\operatorname{div}(\operatorname{curl} \mathbf{A}) = 0$.)

RECIPE 3   If $\mathbf{F} = u(x, y, z)\mathbf{i} + v(x, y, z)\mathbf{j} + w(x, y, z)\mathbf{k}$ is a solenoidal vector, then a vector potential for it is

$$\mathbf{A} = \begin{bmatrix} 0 \\ \displaystyle\int_a^x w(t, y, z)\, dt - \int_c^z u(a, y, t)\, dt \\ \displaystyle -\int_a^x v(t, y, z)\, dt \end{bmatrix} \qquad (21.27)$$

Indeed, computing the curl of (21.27) we get

$$\mathbf{F} = \operatorname{curl}(\mathbf{A}) = \left[u(a, y, z) - \int_a^x (v_y(t, y, z) + w_z(t, y, z))\, dt \right] \mathbf{i}$$
$$+ v(x, y, z)\mathbf{j} + w(x, y, z)\mathbf{k}$$

From $\nabla \cdot \mathbf{F} = u_x + v_y + w_z = 0$, obtain $-v_y(t, y, z) - w_z(t, y, z) = u_x(t, y, z)$, so the integral in the first component of $\operatorname{curl}(\mathbf{A})$ reduces to $u(x, y, z) - u(a, y, z)$ and $\operatorname{curl}(\mathbf{A}) = \mathbf{F}$.

EXAMPLE 21.20   With $(a, b, c) = (0, 0, 0)$, (21.27) gives the vector potential for (21.26), a solenoidal field, as

$$\mathbf{A}_1 = \left(\int_0^x t\, dt - \int_0^z y\, dt \right)\mathbf{j} - \int_0^x z\, dt\, \mathbf{k} = \left(\frac{x^2}{2} - yz \right)\mathbf{j} - xz\mathbf{k} \qquad (21.28)$$

❖

RECIPE 4   An alternate recipe for constructing a vector potential of $\mathbf{F}$ is $\mathbf{A} = \int_0^1 t\mathbf{F} \times d\mathbf{r}$, where $\mathbf{R0}, \mathbf{R}$, and $\mathbf{r}(t)$ are given by (21.19a), (21.19b), and (21.19c), respectively, and $d\mathbf{r}$ is given by $(\mathbf{R} - \mathbf{R0})\, dt$. The vector field $\mathbf{F} = u(x, y, z)\mathbf{i} + v(x, y, z)\mathbf{j} + w(x, y, z)\mathbf{k}$, evaluated along this line segment, becomes $\mathbf{F}(\mathbf{r}(t))$.

Verification of this second recipe in the general case is computationally intensive. The reader is referred to the accompanying Maple worksheet or to [22] for the details.

EXAMPLE 21.21   For the solenoidal field (21.26) and $(a, b, c) = (0, 0, 0)$,

$$\mathbf{A} = \int_0^1 t^2[(z^2 - xy)\mathbf{i} + (x^2 - yz)\mathbf{j} + (y^2 - xz)\mathbf{k}]\, dt = \frac{1}{3}\begin{bmatrix} z^2 - xy \\ x^2 - yz \\ y^2 - xz \end{bmatrix}$$

However, (21.27) gave (21.28), a different vector potential. The vector potential is not unique, but any two differ by the gradient of a scalar. Indeed,

$$\mathbf{A}_1 - \mathbf{A} = -\tfrac{1}{3}(z^2 - xy)\mathbf{i} + \tfrac{1}{6}(x^2 - 4yz)\mathbf{j} - \tfrac{1}{3}(y^2 + 2xz)\mathbf{k} \qquad \textbf{(21.29)}$$

is the gradient of $\frac{1}{6}(x^2 y - 2xz^2 - 2y^2 z)$.   ❖

EXERCISES 21.5–Part A

A1. Show that $\nabla \times \mathbf{F} = \mathbf{0}$ for $\mathbf{F}$ given in (21.17).

A2. Show that $-\nabla u = \mathbf{F}$ for u given by (21.18) and $\mathbf{F}$ in (21.17).

A3. Verify the integration in (21.20).

A4. Verify the line integral in (21.21).

A5. Verify the results in (21.22).

A6. In Example 21.17, show that $\nabla \cdot \mathbf{F} = 0$.

A7. For u given in Example 21.18, verify $\mathbf{F} = -\nabla u$ as in (21.24).

A8. Verify (21.25) in Example 21.18.

A9. Show $\nabla \cdot \mathbf{F} = 0$ for $\mathbf{F}$ in Example 21.19.

A10. Let $\mathbf{F} = -\frac{y}{x^2+y^2}\mathbf{i} + \frac{x}{x^2+y^2}\mathbf{j}$.

   (a) Show that for $(x, y) \neq (0, 0)$, $\mathbf{F}$ is conservative.

   (b) Show that $\oint_C \mathbf{F} \cdot d\mathbf{r} = 2\pi \neq 0$ for C, the unit circle centered at the origin.

   (c) Explain why this conservative field does not have the closed-loop property and therefore, is, not path independent.

EXERCISES 21.5–Part B

B1. Verify that $u(x, y, z) = \frac{1}{6}(x^2 y - 2xz^2 - 2y^2 z)$ is a scalar potential for (21.29).

Exercises B2–11 give a potential function $u(x, y, z)$ and a point $P = (a, b, c)$. In each:

   (a) Obtain the conservative field $\mathbf{F} = -\nabla u$.

   (b) Obtain du, the total differential of u.

   (c) Show that $\mathbf{F} \cdot d\mathbf{r}$ is exact by showing it is precisely $-du$.

   (d) Show that $\nabla \times \mathbf{F} = \mathbf{0}$.

   (e) Reconstruct the potential $u(x, y, z)$ from $\mathbf{F}$ using Recipe 1.

   (f) Reconstruct the potential $v(x, y, z)$ from $\mathbf{F}$ using Recipe 2.

   (g) Show that $u(x, y, z)$ and $v(x, y, z)$ differ by a constant.

   (h) Compare u and v to the output of an appropriate command in your favorite computer algebra system.

   (i) Show $\int_{C_1} \mathbf{F} \cdot d\mathbf{r} = \int_{C_2} \mathbf{F} \cdot d\mathbf{r}$, where C_1 is the straight line from $(0, 0, 0)$ to $P = (a, b, c)$ and C_2 is the segmented linear path from $(0, 0, 0)$ to $(a, 0, 0)$ to $(a, b, 0)$ to (a, b, c).

   (j) Compute $\oint_C \mathbf{F} \cdot d\mathbf{r}$, where C is the closed space curve $x(t) = \cos t, y(t) = \sin t, z(t) = \sin t, 0 \leq t \leq 2\pi$. Of course, the value is zero since $\mathbf{F}$ is conservative.

   (k) Compute $\oint_C \mathbf{F} \cdot d\mathbf{r}$, where C is the closed space curve $x(t) = \cos t, y(t) = \sin t, z(t) = \sin 2t, 0 \leq t \leq 2\pi$. Again, the value is zero since $\mathbf{F}$ is conservative.

   (l) Compute $\oint_C \mathbf{F} \cdot d\mathbf{r}$, where C is the closed space curve consisting of $x(t) = t^2 + t - 1, y(t) = t^3 + t^2 - t - 3$, $z(t) = t^3 + 2, -1 \leq t \leq 1$, and the straight line connecting the initial and terminal points. Again, the value is zero since $\mathbf{F}$ is conservative.

B2. $u(x, y, z) = 8z^3 - xy^2$, $P = (-4, 8, -7)$

B3. $u(x, y, z) = 4z^3 - 6xz - 5y^3$, $P = (8, -7, -7)$

B4. $u(x, y, z) = 2z^3 + 6y + 5xz$, $P = (-8, 5, 6)$

B5. $u(x, y, z) = 6x^2z - 5xy - 2y^3$, $P = (-7, -9, -6)$

B6. $u(x, y, z) = 5yz - 6z^2 - 5x$, $P = (8, 7, -6)$

B7. $u(x, y, z) = 2x^2y - y^2z + 3z^2$, $P = (-4, 8, 7)$

B8. $u(x, y, z) = 5yz - 6z^3 - 8x$, $P = (8, 4, -6)$

B9. $u(x, y, z) = x^2z - 9y^2z + 3x$, $P = (-4, 8, -2)$

B10. $u(x, y, z) = 8x^2z - 8xyz$, $P = (-2, -6, 3)$

B11. $u(x, y, z) = 3y^2z + 5xz - 9x^2y$, $P = (-8, 4, 6)$

For each vector field **F** given in Exercises B12–21:

(a) Show **F** is irrotational, so that **F** is thereby conservative.

(b) Use Recipe 1 to find $u(x, y, z)$, a scalar potential for **F**.

(c) Use Recipe 2 to find $v(x, y, z)$, a scalar potential for **F**.

(d) Show that $u(x, y, z)$ and $v(x, y, z)$ differ by a constant.

(e) Compare u and v to the output of an appropriate command in your favorite computer algebra system.

(f) Show that **F** is solenoidal.

(g) Use Recipe 3 to find **A**, a vector potential for **F**.

(h) Use a computer algebra system to find **B**, a vector potential for **F**.

(i) Show that **A** and **B** differ by a gradient vector.

B12. $\mathbf{F} = (-5y - 9z)\mathbf{i} - (8z + 5x)\mathbf{j} + (9 - 8y - 9x)\mathbf{k}$

B13. $\mathbf{F} = 2z\mathbf{i} + (5z - 9)\mathbf{j} + (5y + 2x + 2)\mathbf{k}$

B14. $\mathbf{F} = (yz + x)\mathbf{i} + xz\mathbf{j} + (xy - z)\mathbf{k}$

B15. $\mathbf{F} = (4yz + 3)\mathbf{i} + (4xz - 8z - 8)\mathbf{j} + (4xy - 8y)\mathbf{k}$

B16. $\mathbf{F} = (8z - y)\mathbf{i} + (7z - x)\mathbf{j} + (7y + 8x)\mathbf{k}$

B17. $\mathbf{F} = (3y - 7yz)\mathbf{i} + (3x - 7xz + 9)\mathbf{j} - 7xy\mathbf{k}$

B18. $\mathbf{F} = (9y + 5yz - 5)\mathbf{i} + (9x + 5xz + 7)\mathbf{j} + 5xy\mathbf{k}$

B19. $\mathbf{F} = (5y + 16xy)\mathbf{i} + (8x^2 - 8z^2 + 5x)\mathbf{j} - 16yz\mathbf{k}$

B20. $\mathbf{F} = -8xy\mathbf{i} + (4z^2 - 4x^2 + 6z)\mathbf{j} + (6y + 4 + 8yz)\mathbf{k}$

B21. $\mathbf{F} = (9yz + x)\mathbf{i} + 9xz\mathbf{j} + (9xy - z + 8)\mathbf{k}$

In Exercises B22–36, determine if the given expression is an exact differential. (Write du as $\mathbf{F} \cdot d\mathbf{r}$ and examine $\nabla \times \mathbf{F}$.) If du is not exact, find C_1 and C_2, two paths connecting $(0, 0, 0)$ with $(1, 1, 1)$, for which $\int_{C_1} du \neq \int_{C_2} du$. If du is exact, find $u(x, y, z)$ using a computer algebra system by Recipe 1 and by Recipe 2. Show that the resulting functions differ by a constant, and verify that $(\nabla u) \cdot d\mathbf{r} = du$.

B22. $du = -4xz\,dx + (14yz + 7z^2)\,dy + (7y^2 + 14yz - 2x^2 - 6z^2)\,dz$

B23. $du = (3yz + 12xy - 2xz)\,dx + (3xz + 6x^2)\,dy + (18z - 3xy - x^2)\,dz$

B24. $du = (3 - 16xy)\,dx - (16yz + 8x^2)\,dy - (8y^2 + 8z)\,dz$

B25. $du = (5y^2 - 4yz + 12xy)\,dx + (10xy - 14yz - 4xz + 6x^2)\,dy - (7y^2 + 4xy)\,dz$

B26. $du = (1 - 12xy + 10xz)\,dx - 6x^2dy + 5x^2\,dz$

B27. $du = (6x + y^2 + 6yz - 12xz)\,dx + (2xy + 6xz)\,dy + (6xy - 6x^2)\,dz$

B28. $du = (4z + 2y^2 - 8xz)\,dx + (2yz + 4xy)\,dy + (y^2 + 4x - 4x^2)\,dz$

B29. $du = (7z^2 - 7y)\,dx - (7x + 24y^2)\,dy + (14xz - 12z)\,dz$

B30. $du = (4z^2 - 2yz)\,dx - (18yz + 2xz)\,dy + (8xz - 9y^2 - 2xy)\,dz$

B31. $du = (8z - 2y^2)\,dx + (18y^2 - 4xy)\,dy + (16z - 8x)\,dz$

B32. $du = (4z + y)\,dx + (7 - 2z + x)\,dy + (4x - 2y)\,dz$

B33. $du = (2z - 2xy)\,dx + (8 - x^2)\,dy + (18z - 2x)\,dz$

B34. $du = (18x^2 + 10xz)\,dx - (6 + 12y^2)\,dy + 5x^2\,dz$

B35. $du = (21x^2 + 10xy)\,dx + (5x^2 - 8 - 5z)\,dy - 5y\,dz$

B36. $du = 12xy\,dx + (5z^2 - 6x^2 - 3z)\,dy + (10yz - 3y)\,dz$

Let $u(x, y, z)$ be harmonic inside and on S, a sphere with radius a and center at $P = (x_0, y_0, z_0)$. Then it can be shown (see, e.g., [77, p. 593]) that $u(x_0, y_0, z_0) = \frac{1}{4\pi a^2} \iint_S u(x, y, z)\,d\sigma$. In Exercises B37–39:

(a) Verify that the given function is harmonic.

(b) At the given point P, verify the *mean-value property*.

B37. $u(x, y, z) = 3yz - 6y^3 + 9yz^2 + 9x^2y$, $P = (2, -3, 5)$

B38. $u(x, y, z) = -9y^2 + 9z^2 + 3xz^2 - 3xy^2$, $P = (1, 3, 7)$

B39. $u(x, y, z) = 6xz + xyz + y^2z - x^2z$, $P = (-5, 8, 2)$

B40. Let $u(x, y, z)$ be harmonic on R, a closed, bounded region with piecewise smooth boundary S, oriented by the outward unit normal field **N**. The directional derivative of u, in the direction of **N**, evaluated on S is called the *normal derivative* and is generally denoted by $\frac{\partial u}{\partial n}$; thus, $(\nabla u) \cdot \mathbf{N} = \frac{\partial u}{\partial n}$. Use the divergence theorem to show that $\iint_S \frac{\partial u}{\partial n}\,d\sigma = 0$.

In Exercises B41–44, verify the given function is harmonic, and show $\iint_S \frac{\partial u}{\partial n}\,d\sigma = 0$ for S:

(a) The unit sphere centered at the origin.

(b) A cube with side $2a$ and center at the origin.

(c) The closed cylinder $x^2 + y^2 = 1$, $z = \pm 1$.

B41. $u(x, y, z) = 7y^3 - xyz - 21x^2y$

B42. $u(x, y, z) = -3xz^2 + 3xy^2$

B43. $u(x, y, z) = -4xy + 7z^3 - 21x^2z$

B44. $u(x, y, z) = 4x^2 - 4y^2 - 2yz$

Let $u(x, y, z)$ be harmonic on R, a closed, bounded region with piecewise smooth boundary S, oriented by the outward unit normal field **N**. Then $\iiint_R \|\nabla u\|^2\,dv = \iint_S u\frac{\partial u}{\partial n}\,d\sigma$, where $\|\nabla u\|^2 = (\nabla u) \cdot (\nabla u)$, S can be established using Green's First Identity, to be seen in Section 23.4. In Exercises B45–47, verify the given function is harmonic, and then verify $\iiint_R \|\nabla u\|^2\,dv = \iint_S u\frac{\partial u}{\partial n}\,d\sigma$ for R:

(a) The unit sphere centered at the origin.

(b) A cube with side $2a$ and center at the origin.

(c) The closed cylinder $x^2 + y^2 = 1$, $z = \pm 1$.

B45. $u(x, y, z) = -9xyz + 3xz^2 - 3xy^2$

B46. $u(x, y, z) = 2xy - 6yz - 9$ **B47.** $u(x, y, z) = 6xy - 2xz + 6y$

Integral Equivalents of div, grad, and curl

The Tableau

Let $\otimes$ represent any of the three operations: "ordinary operator application," $\cdot$, and $\times$. Let $d\sigma$ be the element of surface area on the closed surface S bounding the three-dimensional region ΔV. Let $\mathbf{N}$ be the unit outward normal on this closed surface S. Then the operators div, grad, and curl can be expressed as integrals according to the template

$$\nabla \otimes \Psi = \lim_{\Delta V \to 0} \frac{\iint \mathbf{N} \otimes \Psi \, d\sigma}{\Delta V}$$

As the operator $\otimes$ ranges through its three values, the quantity Ψ it acts on changes from scalar to vector, but that still gives only three values for the formula. These three values correspond to expressions for the divergence, gradient, and curl shown in Table 21.5.

div	$\nabla \cdot \mathbf{F}$	$\displaystyle\lim_{\Delta V \to 0} \frac{\iint \mathbf{N} \otimes \mathbf{F} \, d\sigma}{\Delta V}$	$= \displaystyle\lim_{\Delta V \to 0} \frac{\iint \mathbf{N} \cdot \mathbf{F} \, d\sigma}{\Delta V}$	$= \displaystyle\lim_{\Delta V \to 0} \frac{\iint \mathbf{F} \cdot \mathbf{N} \, d\sigma}{\Delta V}$
grad	∇f	$\displaystyle\lim_{\Delta V \to 0} \frac{\iint \mathbf{N} \otimes f \, d\sigma}{\Delta V}$	$= \displaystyle\lim_{\Delta V \to 0} \frac{\iint \mathbf{N} f \, d\sigma}{\Delta V}$	$= \displaystyle\lim_{\Delta V \to 0} \frac{\iiint f \mathbf{N} \, d\sigma}{\Delta V}$
curl	$\nabla \times \mathbf{F}$	$\displaystyle\lim_{\Delta V \to 0} \frac{\iint \mathbf{N} \otimes \mathbf{F} \, d\sigma}{\Delta V}$	$= \displaystyle\lim_{\Delta V \to 0} \frac{\iint \mathbf{N} \times \mathbf{F} \, d\sigma}{\Delta V}$	$= \displaystyle\lim_{\Delta V \to 0} -\frac{\iint \mathbf{F} \times \mathbf{N} \, d\sigma}{\Delta V}$

TABLE 21.5   Integral equivalents of divergence, gradient, and curl

DIVERGENCE   The integral formula for the divergence is the easiest of the three to explain. In Section 20.1 the divergence for a planar vector field was shown to be the limiting value of the "flux per unit area" for flux through the boundary of a plane curve. The generalization of this result to fields in space is precisely the first formula in Table 21.5.

Since the integrand of the surface integral is $\mathbf{F} \cdot \mathbf{N}$, the integral computes the flux of $\mathbf{F}$ through the surface S. The divergence of $\mathbf{F}$ is then the limiting ratio of the flux through S divided by the volume enclosed. The intuition developed in the planar case is simply extended by this new formula.

GRADIENT   Surprisingly, the most challenging formula for which to provide insight is the gradient formula. However, if the surface S is taken as a cube of side $2h$, centered at the point (a, b, c), the following computations show why the right side of the middle formula in Table 21.5 produces the partial derivatives found in the gradient vector.

In Figure 21.14, a representative cube is seen, along with the unit normals $\mathbf{N}$ on each face. These normals are just the unit basis vectors $\mathbf{i}, \mathbf{j}, \mathbf{k}$, and their negatives. On the face to which $\mathbf{N} = \mathbf{i}$ is attached, $(x, y, z) = (a + h, y, z)$; while on the opposite face where $\mathbf{N} = -\mathbf{i}$ is attached, $(x, y, z) = (a - h, y, z)$. Hence, $f\mathbf{N}$ on these two faces contributes $f(a + h, y, z) - f(a - h, y, z)$ to the surface integral. This contribution, made in the first component of the vector of surface integrals, is

$$\int_{c-h}^{c+h} \int_{b-h}^{b+h} [f(a + h, y, z) - f(a - h, y, z)] \, dy \, dz$$

where the integration in the variables y and z are over the two faces in question. By the Mean Value theorem for integrals [91, p. 383], there is a point (y, z) for which the integral

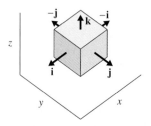

FIGURE 21.14   Integral formula for gradient: cube with outward normals

is equal to

$$[f(a+h, y, z) - f(a-h, y, z)] \int_{b-h}^{b+h} \int_{c-h}^{c+h} dy\, dz$$

The double integral is $(2h)^2$, the surface area of one of the faces, so this pair of faces contributes

$$4[f(a+h, y, z) - f(a-h, y, z)]h^2$$

Dividing this component by the volume in the cube yields

$$\frac{f(a+h, y, z) - f(a-h, y, z)}{2h}$$

recognized as a form of the difference quotient for the derivative f_x. Similar calculations applied to the other two pairs of faces yield f_y and f_z for the remaining two components, respectively. (See Exercises A1 and A2.)

CURL   In Section 20.2, the curl of a planar vector field was extracted from the limiting ratio of the field's circulation around a closed curve to the area enclosed by the curve. The circulation around a curve is the net tangential component of the field.

The surface integral for the curl in the third formula of Table 21.5 is actually the generalization of the circulation around a closed curve. The cross-product in the surface integral yields a vector in the plane tangent to the surface S, a vector equal in length to the component of $\mathbf{F}$ orthogonal to $\mathbf{N}$. However, $\mathbf{N} \times \mathbf{F}$ is not simply the projection of the component of $\mathbf{F}$ normal to $\mathbf{N}$ but is rotated a quarter-turn to the left of that projection.

To illustrate these remarks, consider a surface S aligned so that $\mathbf{N} = \mathbf{k}$, and take $\mathbf{F} = u\mathbf{i} + v\mathbf{j} + w\mathbf{k}$. From elementary multivariable calculus, the projection of $\mathbf{F}$ onto $\mathbf{N}$, since $\mathbf{N}$ is a unit vector, is given by $\mathbf{F_N} = (\mathbf{F} \cdot \mathbf{N})\mathbf{N} = w\mathbf{k}$, and the component of $\mathbf{F}$ orthogonal to $\mathbf{N}$ is given by the difference $\mathbf{F_{\perp N}} = \mathbf{F} - \mathbf{F_N} = u\mathbf{i} + v\mathbf{j}$. The cross-product, $\mathbf{N} \times \mathbf{F} = -v\mathbf{i} + u\mathbf{j}$, is in the plane orthogonal to $\mathbf{N}$, here, the xy-plane. However, $\mathbf{N} \times \mathbf{F}$ is not the vector $\mathbf{F_{\perp N}}$, the component of $\mathbf{F}$ orthogonal to $\mathbf{N}$, projected to the tangent plane, but is that projection rotated a quarter-turn to the left.

Figure 21.15 shows the relationships between the five vectors $\mathbf{N}, \mathbf{F}, \mathbf{F_N}, \mathbf{F_{\perp N}}$, and $\mathbf{N} \times \mathbf{F}$ when $\mathbf{F} = \mathbf{i} + 2\mathbf{j} + 3\mathbf{k}$.

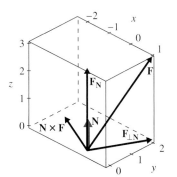

FIGURE 21.15   The vector $\mathbf{N} \times \mathbf{F}$ is orthogonal to $\mathbf{N}$, rotated a quarter-turn to the left of $\mathbf{F_{\perp N}}$

EXAMPLE 21.22

Divergence   Using $\mathbf{F} = x\mathbf{i} + 2y\mathbf{j} + 3z\mathbf{k}$ and S, the surface of a sphere of radius ρ centered at the origin, let's verify the divergence formula in Table 21.5, computing $\nabla \cdot \mathbf{F} = 6$ at the origin.

Describe S by the equation $x^2 + y^2 + z^2 = \rho^2$, so the enclosed volume is $V = \frac{4}{3}\pi\rho^3$. A unit (outward) normal field on S is $\mathbf{N} = \frac{1}{\rho}(x\mathbf{i} + y\mathbf{j} + z\mathbf{k})$, the sphere is explicitly given by $z = \pm\sqrt{\rho^2 - x^2 - y^2}$, the surface area element is $d\sigma = \frac{\rho}{|z|}dx\,dy$, and

$$\mathbf{F} \cdot \mathbf{N}\, d\sigma = \frac{x^2 + 2y^2 + 3z^2}{|z|}dx\,dy = \frac{3\rho^2 - 2x^2 - y^2}{\sqrt{\rho^2 - x^2 - y^2}}dx\,dy \qquad (21.30)$$

Consequently, the surface integral $\iint_{R_{xy}} \mathbf{F} \cdot \mathbf{N}\, d\sigma$, where R_{xy} is the unit disk in the xy-plane, is the sum of two integrals, both the same, so changing to polar coordinates we have

$$2\int_0^{2\pi}\int_0^{\rho}\frac{3\rho^2 - r^2(1 + \cos^2 t)}{\sqrt{\rho^2 - r^2}}r\,dr\,dt = 8\pi\rho^3 \qquad (21.31)$$

which, divided by $V = \frac{4}{3}\pi\rho^3$, yields 6.

EXAMPLE 21.23   **Gradient**   Using the scalar function $f(x, y, z) = x + 2y + 3z$ and the same surface S as in Example 21.22, let's use the second formula in Table 21.5 to obtain $\nabla f = \mathbf{i} + 2\mathbf{j} + 3\mathbf{k}$ at the origin.

The integrand of the surface integral in the formula for the gradient is $f\mathbf{N}\,d\sigma$, with $\mathbf{N}$ being the unit (outward) normal vector. This is interpreted to mean three surface integrals, one for each component of the vector $\mathbf{N}$. The quantities $\mathbf{N} = \frac{1}{\rho}(x\mathbf{i} + y\mathbf{j} + z\mathbf{k})$ and $d\sigma = \frac{\rho}{|z|}\,dx\,dy$ were obtained in Example 21.22. Each component of

$$f\mathbf{N}\,d\sigma = \frac{x + 2y + 3z}{|z|}(x\mathbf{i} + y\mathbf{j} + z\mathbf{k})\,dx\,dy \tag{21.32}$$

will contain two surface integrals, one for the upper hemisphere where $z = \sqrt{\rho^2 - x^2 - y^2}$ and one for the lower hemisphere where $z = -\sqrt{\rho^2 - x^2 - y^2}$. For $|z|$ in the denominator of $d\sigma$ the sign will not matter. But where z appears to an odd power in the numerator, the sign will indeed matter. On the upper hemisphere, $z > 0$, so $\frac{z}{|z|} = 1$, and $f\mathbf{N}\,d\sigma$ is the vector on the left in (21.33); on the lower hemisphere, $z < 0$, so $\frac{z}{|z|} = -1$, and $f\mathbf{N}\,d\sigma$ is the vector on the right in (21.33)

$$\begin{bmatrix} \dfrac{(x + 2y + 3\sqrt{\rho^2 - x^2 - y^2})x}{\sqrt{\rho^2 - x^2 - y^2}} \\[2mm] \dfrac{(x + 2y + 3\sqrt{\rho^2 - x^2 - y^2})y}{\sqrt{\rho^2 - x^2 - y^2}} \\[2mm] x + 2y + 3\sqrt{\rho^2 - x^2 - y^2} \end{bmatrix} dx\,dy \qquad \begin{bmatrix} \dfrac{(x + 2y - 3\sqrt{\rho^2 - x^2 - y^2})x}{\sqrt{\rho^2 - x^2 - y^2}} \\[2mm] \dfrac{(x + 2y - 3\sqrt{\rho^2 - x^2 - y^2})y}{\sqrt{\rho^2 - x^2 - y^2}} \\[2mm] -x - 2y + 3\sqrt{\rho^2 - x^2 - y^2} \end{bmatrix} dx\,dy$$

$$\tag{21.33}$$

The domain of integration in each case is R_{xy}, a disk of radius ρ in the xy-plane. Hence, adding each pair of integrals gives the integrand on the left in (21.34). Conversion to polar coordinates gives the vector in the middle in (21.34), and evaluating the integrals gives the vector on the right in (21.34)

$$\begin{bmatrix} 2\dfrac{x(x + 2y)}{\sqrt{\rho^2 - x^2 - y^2}} \\[2mm] 2\dfrac{y(x + 2y)}{\sqrt{\rho^2 - x^2 - y^2}} \\[2mm] 6\sqrt{\rho^2 - x^2 - y^2} \end{bmatrix} dx\,dy \quad \begin{bmatrix} \displaystyle\int_0^{2\pi}\!\!\int_0^{\rho} 2\dfrac{\cos t(\cos t + 2\sin t)}{\sqrt{\rho^2 - r^2}}r^3\,dr\,dt \\[2mm] \displaystyle\int_0^{2\pi}\!\!\int_0^{\rho} 2\dfrac{\cos t\sin t + 2\sin^2 t}{\sqrt{\rho^2 - r^2}}r^3\,dr\,dt \\[2mm] \displaystyle\int_0^{2\pi}\!\!\int_0^{\rho} 6r\sqrt{\rho^2 - r^2}\,dr\,dt \end{bmatrix} \quad \begin{bmatrix} \frac{4}{3}\pi\rho^3 \\[2mm] \frac{8}{3}\pi\rho^3 \\[2mm] 4\pi\rho^3 \end{bmatrix}$$

$$\tag{21.34}$$

When divided by $V = \frac{4}{3}\pi\rho^3$, the vector on the right in (21.34) becomes $\nabla f = \mathbf{i} + 2\mathbf{j} + 3\mathbf{k}$.   ❖

EXAMPLE 21.24   **Curl**   Using the vector field $\mathbf{F} = z\mathbf{i} + x\mathbf{j} + y\mathbf{k}$ and the same surface S as in Examples 21.22 and 21.23, let's use the third formula in Table 21.5 to compute $\nabla \times \mathbf{F} = \mathbf{i} + \mathbf{j} + \mathbf{k}$ at the origin.

The integrand needed is $\mathbf{N} \times \mathbf{F}\,d\sigma$. The quantities $\mathbf{N} = \frac{1}{\rho}(x\mathbf{i} + y\mathbf{j} + z\mathbf{k})$ and $d\sigma = \frac{\rho}{|z|}$ have already been obtained. As in Example 21.23, the integrand is a vector, so a surface integral for each component of this vector must be formulated and evaluated. Thus,

$$\mathbf{N} \times \mathbf{F}\, d\sigma = \frac{1}{|z|}[(y^2 - xz)\mathbf{i} + (z^2 - xy)\mathbf{j} + (x^2 - yz)\mathbf{k}]\, dx\, dy \qquad (21.35)$$

on the upper hemisphere where z is positive and $\frac{z}{|z|} = 1$, becomes the vector on the left in (21.36); while on the lower hemisphere, where z is negative and $\frac{z}{|z|} = -1$, it becomes the vector on the right in

$$\begin{bmatrix} \dfrac{y^2}{\sqrt{\rho^2 - x^2 - y^2}} - x \\[4mm] \sqrt{\rho^2 - x^2 - y^2} - \dfrac{xy}{\sqrt{\rho^2 - x^2 - y^2}} \\[4mm] \dfrac{x^2}{\sqrt{\rho^2 - x^2 - y^2}} - y \end{bmatrix} dx\, dy \qquad \begin{bmatrix} \dfrac{y^2}{\sqrt{\rho^2 - x^2 - y^2}} + x \\[4mm] \sqrt{\rho^2 - x^2 - y^2} - \dfrac{xy}{\sqrt{\rho^2 - x^2 - y^2}} \\[4mm] \dfrac{x^2}{\sqrt{\rho^2 - x^2 - y^2}} + y \end{bmatrix} dx\, dy$$

$$(21.36)$$

The upper and lower hemispheres both project to R_{xy}, a disk of radius ρ in the xy-plane. Each component of the integrand in the surface integral consists of two terms, one for the upper hemisphere and one for the lower. Adding the integrands yields the vector on the left in (21.37). Conversion to polar coordinates leads to the vector in the middle in (21.37), the value of which is given by the vector on the right in

$$\begin{bmatrix} 2\dfrac{y^2}{\sqrt{\rho^2 - x^2 - y^2}} \\[4mm] 2\sqrt{\rho^2 - x^2 - y^2} - 2\dfrac{xy}{\sqrt{\rho^2 - x^2 - y^2}} \\[4mm] 2\dfrac{x^2}{\sqrt{\rho^2 - x^2 - y^2}} \end{bmatrix} dx\, dy \quad \begin{bmatrix} \displaystyle\int_0^{2\pi}\int_0^{\rho} 2\dfrac{\sin^2 t}{\sqrt{\rho^2 - r^2}}r^3\, dr\, dt \\[4mm] \displaystyle\int_0^{2\pi}\int_0^{\rho} 2\dfrac{\rho^2 - r^2(1 + \cos t \sin t)}{\sqrt{\rho^2 - r^2}}r\, dr\, dt \\[4mm] \displaystyle\int_0^{2\pi}\int_0^{\rho} 2\dfrac{\cos^2 t}{\sqrt{\rho^2 - r^2}}r^3\, dr\, dt \end{bmatrix} \quad \begin{bmatrix} \frac{4}{3}\pi\rho^3 \\[2mm] \frac{4}{3}\pi\rho^3 \\[2mm] \frac{4}{3}\pi\rho^3 \end{bmatrix}$$

$$(21.37)$$

When divided by volume, $V = \frac{4}{3}\pi\rho^3$, the vector on the right in (21.37) becomes $\nabla \times \mathbf{F} = \mathbf{i} + \mathbf{j} + \mathbf{k}$. ❖

Derivations

DIVERGENCE   To obtain the first formula in Table 21.5, start with the divergence theorem

$$\iiint_{\Delta V} \nabla \cdot \mathbf{F}\, dv = \iint_S \mathbf{F} \cdot \mathbf{N}\, d\sigma$$

Invoke the Mean Value theorem for integrals by which, for $\mathbf{F}$ sufficiently continuous, there exists some point Q in the region of integration for which the *equality*

$$[\nabla \cdot \mathbf{F}]_Q \iiint_{\Delta V} dv = \iint_S \mathbf{F} \cdot \mathbf{N}\, d\sigma$$

holds. The volume integral on the left now evaluates to the volume inside S, leading to

$$[\nabla \cdot \mathbf{F}]_Q = \frac{1}{\Delta V} \iint_S \mathbf{F} \cdot \mathbf{N}\, d\sigma$$

In the limit as the surface S shrinks to point P, the point Q must approach P and the formula $[\nabla \cdot \mathbf{F}]_P = \lim_{\Delta \to 0} \iint \mathbf{F} \cdot \mathbf{N}\, d\sigma / \Delta V$ is obtained.

GRADIENT   To obtain the second formula in Table 21.5, start with the divergence theorem and set $\mathbf{F} = f\mathbf{C}$, where $\mathbf{C}$ is an arbitrary constant vector, thus establishing $\iiint_{\Delta V} \nabla f\, dv = \iint_S f\mathbf{N}\, d\sigma$. The Mean Value Theorem for integrals then guarantees the existence of a point Q for which the equality

$$[\nabla f]_Q \iiint_{\Delta V} dv = \iint_S f\mathbf{N}\, d\sigma$$

holds. Evaluating the volume integral on the left leads to

$$[\nabla f]_Q = \frac{1}{\Delta V} \iint_S f\mathbf{N}\, d\sigma$$

and in the limit to $[\nabla f]_P = \lim_{\Delta \to 0} \frac{\iint f\mathbf{N}\, d\sigma}{\Delta V}$.

To establish the all-important starting point $\iiint_{\Delta V} \nabla f\, dv = \iint_S f\mathbf{N}\, d\sigma$, manipulate the divergence theorem as follows.

$$\iiint_{\Delta V} \nabla \cdot \mathbf{F}\, dv = \iint_S \mathbf{F} \cdot \mathbf{N}\, d\sigma$$

$$\iiint_{\Delta V} \nabla \cdot [f\mathbf{C}]\, dv = \iint_S [f\mathbf{C}] \cdot \mathbf{N}\, d\sigma$$

$$\iiint_{\Delta V} [\nabla f \cdot \mathbf{C} + f \nabla \cdot \mathbf{C}]\, dv = \iint_S \mathbf{C} \cdot [f\mathbf{N}]\, d\sigma$$

$$\iiint_{\Delta V} [\nabla f \cdot \mathbf{C} + f 0]\, dv = \iint_S \mathbf{C} \cdot [f\mathbf{N}]\, d\sigma$$

$$\iiint_{\Delta V} [\mathbf{C} \cdot \nabla f]\, dv = \iint_S \mathbf{C} \cdot [f\mathbf{N}]\, d\sigma$$

$$\iiint_{\Delta V} \nabla f\, dv = \iint_S f\mathbf{N}\, d\sigma$$

The final equality follows because $\mathbf{C}$, being arbitrary, could be one of the unit vectors $\mathbf{i}$, $\mathbf{j}$, or $\mathbf{k}$, in which case one of the *components* of the vectors in the integrals is isolated. If the result is then true for each component of the vectors, it is true for the vectors themselves.

CURL   To obtain the third formula in Table 21.5, start with the divergence theorem and set $\mathbf{F} = \mathbf{B} \times \mathbf{C}$, with $\mathbf{C}$ an arbitrary constant vector, to establish

$$\iiint_{\Delta V} \nabla \times \mathbf{B}\, dv = \iint_S \mathbf{N} \times \mathbf{B}\, d\sigma$$

By the Mean Value Theorem for integrals, there is then a point Q for which the equality

$$[\nabla \times \mathbf{B}]_Q \iiint_{\Delta V} dv = \iint_S \mathbf{N} \times \mathbf{B}\, d\sigma$$

holds. The triple integral on the left evaluates to the volume ΔV, leading to

$$[\nabla \times \mathbf{B}]_Q = \frac{1}{\Delta V} \iint_S \mathbf{N} \times \mathbf{B}\, d\sigma$$

and in the limit to $[\nabla \times \mathbf{B}]_P = \lim_{\Delta \to 0} \iint \mathbf{N} \times \mathbf{B}\, d\sigma / \Delta V$.

To establish the all-important starting point $\iiint_{\Delta V} \nabla \times \mathbf{B} \, dv = \iint_S \mathbf{N} \times \mathbf{B} \, d\sigma$, manipulate the divergence theorem as follows.

$$\iiint_{\Delta V} \nabla \cdot \mathbf{F} \, dv = \iint_S \mathbf{F} \cdot \mathbf{N} \, d\sigma$$

$$\iiint_{\Delta V} \nabla \cdot [\mathbf{B} \times \mathbf{C}] \, dv = \iint_S [\mathbf{B} \times \mathbf{C}] \cdot \mathbf{N} \, d\sigma$$

$$\iiint_{\Delta V} [\mathbf{C} \cdot \nabla \times \mathbf{B} - \mathbf{B} \cdot \nabla \times \mathbf{C}] \, dv = \iint_S \mathbf{N} \cdot [\mathbf{B} \times \mathbf{C}] \, d\sigma$$

$$\iiint_{\Delta V} [\mathbf{C} \cdot \nabla \times \mathbf{B} - \mathbf{B} \cdot \mathbf{0}] \, dv = \iint_S \mathbf{N} \cdot [\mathbf{B} \times \mathbf{C}] \, d\sigma$$

$$\iiint_{\Delta V} [\mathbf{C} \cdot \nabla \times \mathbf{B}] \, dv = \iint_S \mathbf{C} \cdot [\mathbf{N} \times \mathbf{B}] \, d\sigma$$

$$\iiint_{\Delta V} \nabla \times \mathbf{B} \, dv = \iint_S \mathbf{N} \times \mathbf{B} \, d\sigma$$

As before, the final equality follows by choosing the arbitrary vector $\mathbf{C}$ as each of $\mathbf{i}$, $\mathbf{j}$, and $\mathbf{k}$, which makes the equation true for each *component* of the vectors and hence for the vectors themselves.

In the surface integral on the right, the triple scalar product was manipulated as follows

$$\mathbf{N} \cdot [\mathbf{B} \times \mathbf{C}] = \begin{vmatrix} n_1 & n_2 & n_3 \\ b_1 & b_2 & b_3 \\ c_1 & c_2 & c_3 \end{vmatrix} = - \begin{vmatrix} n_1 & n_2 & n_3 \\ c_1 & c_2 & c_3 \\ b_1 & b_2 & b_3 \end{vmatrix} = \begin{vmatrix} c_1 & c_2 & c_3 \\ n_1 & n_2 & n_3 \\ b_1 & b_2 & b_3 \end{vmatrix} = \mathbf{C} \cdot [\mathbf{N} \times \mathbf{B}]$$

EXERCISES 21.6–Part A

A1. Using the cube in Figure 21.14, obtain $f\mathbf{N}$ for the two faces orthogonal to $\mathbf{j}$, and show how this leads to f_y.

A2. Using the cube in Figure 21.14, obtain $f\mathbf{N}$ for the two faces orthogonal to $\mathbf{k}$, and show how this leads to f_z.

A3. Verify the calculations leading to (21.30).

A4. Verify the integral in (21.31).

A5. Verify the calculations leading to (21.32).

A6. Verify the vectors in (21.33).

A7. Verify the calculations represented by (21.34).

A8. Verify the calculations leading to (21.35).

A9. Verify the vectors in (21.36).

A10. Verify the calculations represented by (21.37).

EXERCISES 21.6–Part B

B1. Verify that interchanging two rows in a 3×3 determinant negates the determinant.

In Exercises B2–4, at the given point $P = (a, b, c)$ and for the given field $\mathbf{F}$, verify the integral formula for $\nabla \cdot \mathbf{F}$ using

    (a) S, a sphere of radius ρ centered at P.

    (b) S, the cube $x = a \pm \rho$, $y = b \pm \rho$, $z = c \pm \rho$.

    (c) S, a cylindrical "can" with radius ρ and height 2ρ, centered at P. Thus, the cylinder will have equation $(x - a)^2 + (y - b)^2 = \rho^2$ and "lids" in the planes $z = c \pm \rho$.

B2. $\mathbf{F} = x^2 y \mathbf{i} + y^2 z \mathbf{j} + xz^2 \mathbf{k}$, $P = (2, 3, 5)$

B3. $\mathbf{F} = (5y + xyz)\mathbf{i} - 4x^2 z\mathbf{j} + (y - 9y^2 z)\mathbf{k}$, $P = (1, -7, -3)$

B4. $\mathbf{F} = (4z + 6xy^2)\mathbf{i} + (8xy^2 + 3z^3)\mathbf{j} + (3y^2 - 7yz^2)\mathbf{k}$, $P = (-8, 9, 4)$

In Exercises B5–7, at the given point $P = (a, b, c)$ and for the given field $\mathbf{F}$, verify the integral formula for $\nabla \times \mathbf{F}$ using

    (a) S, a sphere of radius ρ centered at P.

    (b) S, the cube $x = a \pm \rho$, $y = b \pm \rho$, $z = c \pm \rho$.

    (c) S, a cylindrical "can" with radius ρ and height 2ρ, centered at P. Thus, the cylinder will have equation $(x - a)^2 + (y - b)^2 = \rho^2$ and "lids" in the planes $z = c \pm \rho$.

B5. $\mathbf{F} = (5y + 2xz^2)\mathbf{i} + (7yz + 4y^3)\mathbf{j} + (7z + 6yz)\mathbf{k}$, $P = (6, 8, 1)$

B6. $\mathbf{F} = (3z + 8x^2y)\mathbf{i} + (9xz^2 - 3y)\mathbf{j} + (9 - 3z^3)\mathbf{k}$, $P = (9, -9, -8)$

B7. $\mathbf{F} = xz\mathbf{i} + (4x^2 + 8y^3)\mathbf{j} - yz^2\mathbf{k}$, $P = (8, 9, 3)$

In Exercises B8–11, at the given point $P = (a, b, c)$ and for the given scalar field $\phi(x, y, z)$, verify the integral formula for $\nabla\phi$ using

(a) S, a sphere of radius ρ centered at P.

(b) S, the cube $x = a \pm \rho$, $y = b \pm \rho$, $z = c \pm \rho$.

(c) S, a cylindrical "can" with radius ρ and height 2ρ, centered at P. Thus, the cylinder will have equation $(x - a)^2 + (y - b)^2 = \rho^2$ and "lids" in the planes $z = c \pm \rho$.

B8. $\phi(x, y, z) = x^2y^2z$, $P = (2, -3, -5)$

B9. $\phi(x, y, z) = 3z^3 + 5x^2$, $P = (7, -7, -1)$

B10. $\phi(x, y, z) = 8xy - 7z^2$, $P = (9, -9, -8)$

B11. $\phi(x, y, z) = 6x^2 + 2xyz$, $P = (2, 9, -3)$

In Exercises B12–13, for the given fields $\mathbf{F}$, verify the integral formula $\iint_S \mathbf{N} \times \mathbf{F} \, d\sigma = \iiint_R \nabla \times \mathbf{F} \, dv$, where R is the interior of the closed surface S, $\mathbf{N}$ is the outward unit normal field on S, and S is

(a) a sphere with radius 1 and center at the origin.

(b) a cube $x = \pm 1$, $y = \pm 1$, $z = \pm 1$.

(c) the closed cylinder $x^2 + y^2 = 1$, $z = \pm 1$.

B12. $\mathbf{F} = 7xyz\mathbf{i} + (5x^2 - 5y^3)\mathbf{j} + (yz - x^3)\mathbf{k}$

B13. $\mathbf{F} = (5y + 4z^3)\mathbf{i} + (4xz - x^3)\mathbf{j} - 9xk$

In Exercises B14–16, for the given scalar field ϕ, verify the integral formula $\iint_S \phi\mathbf{N} \, d\sigma = \iiint_R \nabla\phi \, dv$, where R is the interior of the closed surface S, $\mathbf{N}$ is the outward unit normal field on S, and S is

(a) a sphere with radius 1 and center at the origin.

(b) a cube $x = \pm 1$, $y = \pm 1$, $z = \pm 1$.

(c) the closed cylinder $x^2 + y^2 = 1$, $z = \pm 1$.

B14. $\phi(x, y, z) = 6z^3 - xyz$ **B15.** $\phi(x, y, z) = xz - x^2y$

B16. $\phi(x, y, z) = 4xy + 3z^3$

In Exercises B17–19, for the given scalar field $\phi(x, y, z)$, verify the integral formula $\oint_C \phi \, d\mathbf{r} = \iint_S \mathbf{N} \times \nabla\phi \, d\sigma$, where S is an open "capping" surface for the simple closed curve C lying in the plane $z = -1$, $\mathbf{N}$ is

the unit normal field on S (positively oriented with respect to C), and S is taken to be

(a) the unit upper hemisphere with center at the origin.

(b) the cube $x = \pm 1$, $y = \pm 1$, $z = \pm 1$ and open in the plane $z = -1$.

(c) the cylinder $x^2 + y^2 = 1$, $z = \pm 1$, capped in the plane $z = 1$ and open in the plane $z = -1$.

B17. $\phi(x, y, z) = x^2 + 5y^2z$ **B18.** $\phi(x, y, z) = 2xz - 2y^2z$

B19. $\phi(x, y, z) = 4xy^2z^3$

In Exercises B20–21, for the given vector field $\mathbf{F}$, verify the integral formula $\oint_C d\mathbf{r} \times \mathbf{F} = \iint_S (\mathbf{N} \times \nabla) \times \mathbf{F} \, d\sigma$, where S is an open "capping" surface for the simple closed curve C lying in the plane $z = -1$, $\mathbf{N}$ is the unit normal field on S (positively oriented with respect to C), and S is taken to be

(a) the unit upper hemisphere with center at the origin.

(b) the open cube $x = \pm 1$, $y = \pm 1$, $z = 1$.

(c) the cylinder $x^2 + y^2 = 1$, capped in the plane $z = 1$.

B20. $\mathbf{F} = 6y^2z\mathbf{i} + (6z^3 - 9x^3)\mathbf{j} + (8xy + 7yz)\mathbf{k}$

B21. $\mathbf{F} = (4y + 6xyz)\mathbf{i} + (4xz^2 + x^2z)\mathbf{j} + (9xz - 3y^3)\mathbf{k}$

B22. Obtain the integral formula used in Exercises B20 and 21 by applying Stokes' Theorem to the vector $\mathbf{F} \times \mathbf{B}$, where $\mathbf{B}$ is an arbitrary constant vector.

B23. Obtain the integral formula used in Exercises B17–19 by applying Stokes' Theorem to the vector $\phi\mathbf{B}$, where $\mathbf{B}$ is an arbitrary constant vector.

B24. Obtain the integral formula used in Exercises B14–16 by applying the divergence theorem to the vector $\phi\mathbf{B}$, where $\mathbf{B}$ is an arbitrary constant vector.

B25. Obtain the integral formula used in Exercises B12 and 13 by applying the divergence theorem to the vector $\mathbf{F} \times \mathbf{B}$, where $\mathbf{B}$ is an arbitrary constant vector.

B26. If $\mathbf{B}$ is a constant vector, show that $\iint_S \mathbf{N} \times (\mathbf{B} \times \mathbf{r}) \, d\sigma = 2\mathbf{B}v$, where S is a closed surface, $\mathbf{N}$ is the outward unit normal field on S, $\mathbf{r}$ is the position vector to a point on S, and v is the volume enclosed by S. *Hint:* Use the integral formula from Exercises B12 and 13 and a vector identity from Table 20.2.

Chapter Review

1. What is $d\sigma$, the element of surface area, for a surface described by

(a) $z = f(x, y)$ (b) $y = g(x, z)$ (c) $x = h(y, z)$

2. Show that $d\sigma = \|\nabla f\|/|f_z| \, dx \, dy$ for a surface that is defined implicitly by $f(x, y, z) = 0$ and has a projection on the xy-plane. *Hint:* Use Question 1(a) and implicit differentiation.

3. What is meant by the phrase "surface integral?" Give an example.

4. If S is a surface described by $z = f(x, y)$, write a mathematical recipe by which the flux of the field $\mathbf{F}$ through S is computed. State how one obtains the unit normal $\mathbf{N}$. Give a verbal explanation of the prescription.

5. What is the usual convention for selecting a normal on a closed surface when computing the flux through the surface? With that

convention, what physical significance attaches to a positive value of the flux? To negative value?

6. Let S be a cylinder with a closed top and an open bottom. Let $\mathbf{N}_0$ be a unit normal attached to the top of the surface, and let it point into the cylinder. Use the device of orienting the surface to embed $\mathbf{N}_0$ in a continuous normal field on S.

7. Repeat Question 6 for $\mathbf{N}_0$ pointing in the opposite direction.

8. Let S be the first octant portion of the plane $2x + 3y + 4z = 5$. Obtain the flux of $\mathbf{F} = 7x\mathbf{i} + 8y\mathbf{j} + 9z\mathbf{k}$ through S.

9. State the divergence theorem. Give a verbal description of its conclusion.

10. State Stokes' Theorem. Give a verbal description of its conclusion.

11. State the "divergence form" of Green's Theorem. Write out its essential formula for the vector $\mathbf{F} = f\mathbf{i} + g\mathbf{j}$.

12. State the "Stokes' form" of Green's Theorem. Write out its essential formula for the vector $\mathbf{F} = f\mathbf{i} + g\mathbf{j}$.

13. Show how one form of Green's Theorem can be transformed into the other by an appropriate choice of vector field $\mathbf{F}$.

14. Demonstrate the application of the divergence theorem to the vector field $\mathbf{F} = 2x\mathbf{i} + 3y\mathbf{j} + 4z\mathbf{k}$ for the region V' that lies between concentric spheres of radii $\frac{1}{3}$ and 1.

15. Demonstrate the application of Stokes' Theorem to the field of Question 14 for the surface S, the upper unit hemisphere with center at the origin.

16. Demonstrate both forms of Green's Theorem for the field $\mathbf{F} = 5x\mathbf{i} - 3y\mathbf{j}$ and the region R between concentric circles with radii 1 and 2 and centers at the origin.

17. Explain why the application of Stokes' Theorem to a *closed* surface always results in the identity $0 = 0$.

18. State five definitions for a conservative force.

19. What is an exact differential? Give an example. Give an example of a differential that is not exact.

20. Explain the phrase "path independent." What is its relationship with conservative forces?

21. What is meant by the "closed path" property? What is its relationship with conservative forces?

22. Present the argument that shows "path independence" is equivalent to the "closed path" property.

23. What can be said about a force whose scalar potential is harmonic?

24. What can be said about the scalar potential of a force that is both conservative and solenoidal?

25. Give an example of a solenoidal field that is not conservative.

26. Give an example of a field that is conservative but not solenoidal.

27. Give an example of a field that is both conservative and solenoidal.

28. State a formula for constructing the scalar potential from the components of an irrotational force field.

29. If $u(x, y, z) = xyz$, obtain $\mathbf{F} = \nabla u$ and show it is irrotational. Then, determine the extent to which the formula in Question 28 reconstructs u.

30. State a formula for constructing the vector potential from the components of a solenoidal field.

31. If $\mathbf{A} = 2xy\mathbf{i} - 3yz^2\mathbf{j} + 4xz\mathbf{k}$, obtain $\mathbf{F} = \nabla \times \mathbf{A}$ and show it is solenoidal. Then, determine the extent to which the formula in Question 30 reconstructs $\mathbf{A}$. If $\mathbf{A}$ itself is not obtained, show that the difference is a gradient field.

32. The integral equivalent of $\nabla \times \mathbf{F}$ contains the surface integral of $\mathbf{N} \times \mathbf{F}$. The integral is a generalization of the circulation if $\mathbf{N} \times \mathbf{F}$ is the component of $\mathbf{F}$ tangential to the surface. Show that $\mathbf{N} \times \mathbf{F}$ lies in the plane tangent to $\mathbf{N}$ and, hence, in the surface to which $\mathbf{N}$ is normal. Further show that the length of $\mathbf{N} \times \mathbf{F}$ is the length of the component of $\mathbf{F}$ orthogonal to $\mathbf{N}$.

33. Describe how the integral equivalent of $\nabla \cdot \mathbf{F}$ is the direct generalization of "the limiting ratio of the flux per unit area enclosed."

34. Use a cube centered at (a, b, c) and having its faces parallel to the coordinate planes to argue the validity of the integral equivalent to ∇f. The argument for the faces orthogonal to the yz-plane is given in the text. Provide the arguments for the faces parallel to the xz- and xy-planes.

Chapter 22

NonCartesian Coordinates

INTRODUCTION   Up to this point in the unit, vector calculus has taken place in Cartesian coordinates. We now move into polar, cylindrical, and spherical coordinates, systems that are orthogonal but nonCartesian. We first learn the relationship between a change of coordinates and a mapping. For example, in elementary calculus, polar coordinates are typically seen as a change of coordinates. The points in the xy-plane remain fixed, but their coordinates, or "addresses," change. As a mapping, the coordinates in the xy-plane remain fixed, but the point is moved to the $r\theta$-plane, a separate plane with its own rectangular grid.

Interpreting a change of coordinates as a mapping allows us to determine how to transform a multiple integral from one coordinate system to another.

We then derive expressions for the gradient, divergence, and laplacian operators in polar coordinates. Each result is derived by mapping the calculation into the Cartesian plane where the operators are known, to the polar plane where the equivalent calculations are to be determined.

Similar formulas for gradient, divergence, curl, and laplacian in cylindrical and spherical coordinates appear in two separate tables. The table for spherical coordinates lists two common forms for this system found in the literature. Although these two versions of spherical coordinates differ only by an interchange in the names of two of its coordinates, it can be very confusing to read results in one system thinking they are given in the other. The results for cylindrical coordinates are not derived because the calculations would be almost identical to those for polar coordinates. Although the gradient in spherical coordinates is derived, the remaining formulas are left as exercises.

22.1   Mappings and Changes of Coordinates

Motivation—Double Integrals in Polar Coordinates

More than once, double integrals in Cartesian coordinates have been converted to double integrals in polar coordinates. On each such occasion the conversion was done by declaring $dx\, dy \to r\, dr\, d\theta$. Typically, in the multivariable calculus course an explanation, based on Figure 22.1, is given. The shaded region is roughly a rectangle with edges dr and $r\, d\theta$, from which it is argued that the element of area in the polar plane is $r\, dr\, d\theta$. This is intuitively appealing, but not a proof.

Under the more general coordinate change $x = \alpha(u, v)$, $y = \beta(u, v)$, a double integral changes according to the rule

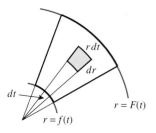

FIGURE 22.1   Element of area in polar coordinates

$$\iint_R f(x, y)\, dx\, dy = \iint_{R'} f(\alpha(u, v), \beta(u, v)) \left| \frac{\partial(x, y)}{\partial(u, v)} \right| du\, dv$$

In the integrand's $f(x, y)$, each x is replaced by $\alpha(u, v)$ and each y by $\beta(u, v)$. But $dx\, dy$ is replaced by the absolute value of the Jacobian $\frac{\partial(x,y)}{\partial(u,v)}$, the determinant of the Jacobian matrix

$$\begin{bmatrix} \alpha_u(u, v) & \alpha_v(u, v) \\ \beta_u(u, v) & \beta_v(u, v) \end{bmatrix}$$

Heuristically, imagine the "denominator" of the Jacobian "cancels" with $du\, dv$, so the integral on the right is left with $\partial(x, y)$, which, if interpreted as $dx\, dy$, "makes" the integral on the right "equivalent" to the integral on the left. To understand why the Jacobian appears when changing variables in a multiple integral, a sound understanding of coordinate changes, per se, is necessary.

Coordinate Change vs. Mapping

POLAR COORDINATES AS COORDINATE CHANGE   The typical treatment of polar coordinates in elementary calculus is slanted toward "change of coordinates." Thus, the point $(x, y) = (1, 1)$ in Cartesian coordinates is given by the polar coordinates $(r, \theta) = (\sqrt{2}, \frac{\pi}{4})$. The "physical point" has remained fixed, but its "house number" has been changed. The house didn't move; the number attached to the front door changed.

How is this interpretation as "change of coordinates" played out in the multivariable calculus course? When curves given in polar coordinates are plotted in the Cartesian plane, the interpretation is that the points remain fixed in the Cartesian plane but their identifying names are changed.

For example, consider the Cartesian curve defined implicitly by the equation $x^2 + y^2 + x = \sqrt{x^2 + y^2}$, graphed in Figure 22.2. The curve looks suspiciously like a cardioid, and its "formula" would be more readily recognized if expressed in polar coordinates as $r = 1 - \cos\theta$. Figure 22.2 reinforces the interpretation of polar coordinates as a change of coordinates. The points remain in the xy-plane. Just their names change.

But what happens if, in $r = 1 - \cos\theta$, the variables r and θ are treated as belonging to a rectangular grid? What does it mean to plot $r = f(\theta)$ in the $r\theta$-plane, a plane in which the grid lines $\theta = $ constant and $r = $ constant are themselves rectangular? Figure 22.3 results.

POLAR COORDINATES AS A MAPPING   Plotting $r = f(\theta)$ as an equation in the $r\theta$-plane means the polar plane is not identified with the xy-plane. The planes are treated as distinct worlds, with communication via the transformation equations, $x = r\cos\theta$, $y = r\sin\theta$, and $r = \sqrt{x^2 + y^2}$, $\theta = \arctan(\frac{y}{x})$.

The graph in Figure 22.3 now hardly resembles the cardioid. Yet, it is the same curve, only mapped onto the rectangular $r\theta$-plane. In this interpretation of polar coordinates, imagine that the points in the xy-plane are moved to a new neighborhood. The rectangular xy-neighborhood is transported, being distorted as it goes through space, to a new planet, the $r\theta$-planet, where lines that were straight in the xy-world are curved to fit into the $r\theta$-world.

MAPPING GRID LINES   In elementary calculus, the rectangular grid lines in the $r\theta$-plane are mapped back to curves in the xy-plane by the action of the transformations $x = r\cos\theta$, $y = r\sin\theta$. For example, the images of the $r\theta$-plane's grid lines $r = 1$ and $\theta = 1$ are a circle and a radial line through the origin in the xy-plane, as shown in Figure 22.4.

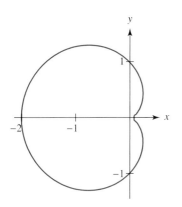

FIGURE 22.2   Graph of $x^2 + y^2 + x = \sqrt{x^2 + y^2}$ in the xy-plane

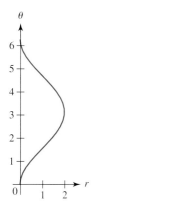

FIGURE 22.3   Graph of $r = 1 - \cos\theta$ in a Cartesian $r\theta$-plane

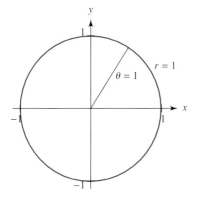

FIGURE 22.4   Rectangular grid lines from $r\theta$-plane mapped back to xy-plane

Elementary calculus treats polar coordinates as if they existed in the Cartesian xy-plane, giving the impression the $r\theta$-plane is "distorted." The truth of the matter is that neither the xy-plane nor the $r\theta$-plane is distorted. The "distortion" arises only when one plane is mapped onto the other.

When the rectangular $r\theta$-plane is mapped onto the xy-plane, the familiar picture of circles and radial lines is obtained. That represents only one view of the mapping induced by the transformations defining polar coordinates. The other view would be based on an image of the xy-plane's grid lines over in the $r\theta$-plane. For example, Figure 22.5(b) is the image in the $r\theta$-plane of the square $1 \leq x, y \leq 2$, in xy-plane. Figure 22.5(c) is the image in the xy-plane of the square $1 \leq r, \theta \leq 2$, in the $r\theta$-plane.

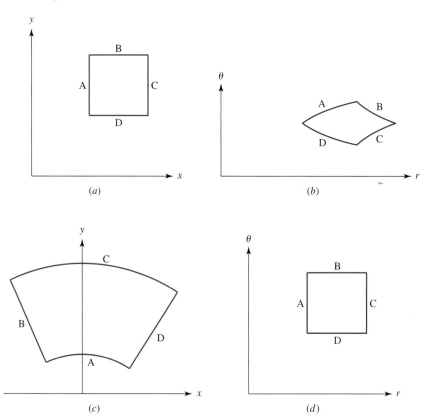

FIGURE 22.5   Square in xy-plane (a) mapped to $r\theta$-plane (b); square in $r\theta$-plane (d) mapped to xy-plane (c)

The starting point has been regained. This investigation was begun by asking why the Jacobian appears when changing coordinates in a multiple integral. The answer is in the picture just drawn. A unit square in the $r\theta$-plane maps to the region shown in Figure 22.5(c). A unit area in the $r\theta$-plane changes to some other amount of area when mapped to the xy-plane. If the factor by which this unit area changed can be deduced, the reason why the Jacobian appears when changing variables in a multiple integral will have been discovered.

AREA   The following calculation has appeared earlier. Represent the radius vector to a general point, use it to define coordinate curves, then differentiate the coordinate curves to obtain vectors tangent to the coordinate curves. This is, in fact, how the surface area element was found and how the formula for a normal to a surface was determined.

The xy-plane is mapped to the uv-plane by transformation equations of the form $x = x(u, v)$, $y = y(u, v)$, with inversion formulae of the form $u = u(x, y)$, $v = v(x, y)$.

It is not known how to measure area in the uv-plane, so bring back to the xy-plane images of rectangular grid lines from the uv-plane. This puts into the xy-plane, images of coordinate curves from the uv-plane. Differentiation along these curves gives tangent vectors expressed in the xy-system. The area of the parallelogram spanned by these tangent vectors is the approximate area, expressed in xy-terms, of a unit rectangle in the uv-plane.

The radius vector in the xy-plane is $\mathbf{R} = x(u, v)\mathbf{i} + y(u, v)\mathbf{j}$. Images, in the xy-plane, of grid lines $v = \beta$, and $u = \alpha$ in the uv-plane are given, respectively, by

$$\mathbf{R}_u = x(u, \beta)\mathbf{i} + y(u, \beta)\mathbf{j} \quad \text{and} \quad \mathbf{R}_v = x(\alpha, v)\mathbf{i} + y(\alpha, v)\mathbf{j}$$

Infinitesimal tangent vectors on these curves in the xy-plane are

$$\mathbf{T}_u = (x_u\mathbf{i} + y_u\mathbf{j})\, du \quad \text{and} \quad \mathbf{T}_v = (x_v\mathbf{i} + y_v\mathbf{j})\, dv$$

which lead to

$$\|\mathbf{T}_u \times \mathbf{T}_v\| = \|(x_u y_v - x_v y_u)\, du\, dv\mathbf{k}\| = |x_u y_v - x_v y_u|\, du\, dv = \left|\frac{\partial(x, y)}{\partial(u, v)}\right| du\, dv$$

The symbol $\frac{\partial(x,y)}{\partial(u,v)}$ represents the determinant of the Jacobian matrix

$$\mathbf{J}_{\Phi^{-1}} = \begin{bmatrix} x_u & x_v \\ y_u & y_v \end{bmatrix}$$

for Φ^{-1}, the inverse transformation from the uv-plane to the xy-plane. The Jacobian matrix for Φ, the forward transformation from the xy-plane to the uv-plane, is

$$\mathbf{J}_{\Phi} = \begin{bmatrix} u_x & u_y \\ v_x & v_y \end{bmatrix}$$

and its determinant is $\frac{\partial(u,v)}{\partial(x,y)}$. The two Jacobian matrices are related as multiplicative inverses, so that

$$\mathbf{J}_{\Phi^{-1}}\mathbf{J}_{\Phi} = \begin{bmatrix} 1 & 0 \\ 0 & 1 \end{bmatrix}$$

and the two Jacobians are related as reciprocals, so that $\frac{\partial(x,y)}{\partial(u,v)}\frac{\partial(u,v)}{\partial(x,y)} = 1$.

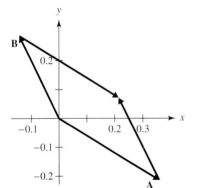

FIGURE 22.6 Example 22.1: In xy-plane, inverse image of unit square in uv-plane

◆ **EXAMPLE 22.1** For the forward mapping Φ (given by $u = 4x + 2y$, $v = 3x + 5y$), taking the xy-plane to the uv-plane,

$$\mathbf{J}_{\Phi} = \begin{bmatrix} u_x & u_y \\ v_x & v_y \end{bmatrix} = \begin{bmatrix} 4 & 2 \\ 3 & 5 \end{bmatrix}$$

and $|\mathbf{J}_{\Phi}| = \frac{\partial(u,v)}{\partial(x,y)} = 14$. For the inverse map Φ^{-1} (given by $x = \frac{1}{14}(5u - 2v)$, $y = \frac{1}{14}(4v - 3u)$), taking the uv-plane to the xy-plane,

$$\mathbf{J}_{\Phi^{-1}} = \begin{bmatrix} x_u & x_v \\ y_u & y_v \end{bmatrix} = \frac{1}{14}\begin{bmatrix} 5 & -2 \\ -3 & 4 \end{bmatrix}$$

and $|\mathbf{J}_{\Phi^{-1}}| = \frac{\partial(x,y)}{\partial(u,v)} = \frac{1}{14}$. The points $(1, 0)$ and $(0, 1)$ in the uv-plane are taken by Φ^{-1} to $(\frac{5}{14}, -\frac{3}{14})$ and $(-\frac{2}{14}, \frac{4}{14})$, respectively, in the xy-plane. The vectors $\mathbf{A} = \frac{1}{14}(5\mathbf{i} - 3\mathbf{j})$ and $\mathbf{B} = \frac{1}{14}(-2\mathbf{i} + 4\mathbf{j})$ span a parallelogram in the xy-plane, as shown in Figure 22.6. The area of this parallelogram is $\|\mathbf{A} \times \mathbf{B}\| = \frac{1}{14} = \frac{\partial(x,y)}{\partial(u,v)}$. The unit square in the uv-plane is pulled back to a parallelogram with area $\frac{\partial(x,y)}{\partial(u,v)} = \frac{1}{14}$ in the xy-plane. Thus, to measure this parallelogram's image in the uv-plane, the unit square's area of 1 must be multiplied by $\frac{1}{14} = \frac{\partial(x,y)}{\partial(u,v)}$. Consequently, $\left|\frac{\partial(x,y)}{\partial(u,v)}\right| du \, dv$ replaces $dx \, dy$ as the element of area when changing a double integral from one in the xy-plane to one in the uv-plane.   ❖

EXERCISES 22.1–Part A

A1. Give the polar coordinates of $(x, y) = (2, -3)$. Plot its image in a rectangular $r\theta$-plane.

A2. Plot $r(\theta) = 1 - 2\cos\theta$, $0 \le \theta \le 2\pi$, as a polar-coordinate curve in the xy-plane.

A3. Plot $r(\theta) = 1 - 2\cos\theta$, $0 \le \theta \le 2\pi$, in a rectangular $r\theta$-plane.

A4. Let $x = 2u + 3v$, $y = 5u - 7v$, determine a change of coordinates, and let R be the triangular region bounded by $y = x$, $x = 1$,

$y = 0$. Evaluate $\iint_R x^2 y^3 \, dy \, dx$, transform the integral to uv-coordinates, and evaluate again.

A5. Cylindrical coordinates are defined by $x = r\cos\theta$, $y = r\sin\theta$, $z = z$. Obtain $\left|\frac{\partial(x,y,z)}{\partial(r,\theta,z)}\right|$ and transform $\iiint_R f(x, y, z) \, dz \, dy \, dx$ to cylindrical coordinates.

EXERCISES 22.1–Part B

B1. Spherical coordinates are sometimes defined by $x = \rho \sin\phi \cos\theta$, $y = \rho \sin\phi \sin\theta$, $z = \rho \cos\phi$. (The angle ϕ is measured from the vertical z-axis to the radius vector, the angle θ measures counterclockwise rotation around the z-axis, starting from the x-axis and ρ is the distance to the origin.) Obtain $\left|\frac{\partial(x,y,z)}{\partial(\rho,\phi,\theta)}\right|$ and transform $\iiint_R f(x, y, z) \, dz \, dy \, dx$ to this version of spherical coordinates.

B2. Spherical coordinates are sometimes defined by $x = \rho \sin\theta \cos\phi$, $y = \rho \sin\theta \sin\phi$, $z = \rho \cos\theta$. (The angle θ is measured from the vertical z-axis to the radius vector, the angle ϕ measures counterclockwise rotation around the z-axis, starting from the x-axis and ρ is the distance to the origin.) Obtain $\left|\frac{\partial(x,y,z)}{\partial(\rho,\theta,\phi)}\right|$ and transform $\iiint_R f(x, y, z) \, dz \, dy \, dx$ to this version of spherical coordinates.

In Exercises B3–12:

(a) Plot $y(x)$ implicitly, using a computer algebra system.

(b) Convert the given equation to polar coordinates, writing the result in the form $r = f(\theta)$.

(c) Plot, in polar coordinates, the result found in part (b). The resulting graph should agree with the one found in part (a).

(d) Plot $r = f(\theta)$ as a curve in a rectangular $r\theta$-plane.

B3. $2x^2 + 2y^2 = \sqrt{x^2 + y^2} + 2x$

B4. $2x^2 + 2y^2 = 3\sqrt{x^2 + y^2} + 2y$

B5. $x^2 + y^2 = 2\sqrt{x^2 + y^2} + y$ **B6.** $(x^2 + y^2)^3 = (x^2 - y^2)^2$

B7. $(x^2 + y^2)^3 = 4x^2 y^2$ **B8.** $x^2 + y^2 = 2x$

B9. $(x^2 + y^2)^2 = x(x^2 - 3y^2)$ **B10.** $8x^2 + 9y^2 - 1 + 2x = 0$

B11. $y^2 = 1 - 2x$ **B12.** $y^2 = 8x^2 - 6x + 1$

B13. Let $u = 3x + 2y$, $v = 5x - 7y$ define a mapping of the xy-plane to the uv-plane.

(a) In the uv-plane, find and graph the images of the grid lines $x = 1, 2, 3$.

(b) In the uv-plane, find and graph the images of the grid lines $y = 1, 2, 3$.

(c) Solve the transformation equations for $x = x(u, v)$, $y = y(u, v)$.

(d) In the xy-plane, find and graph the inverse images of the grid lines $u = 1, 2, 3$.

(e) In the xy-plane, find and graph the inverse images of the grid lines $v = 1, 2, 3$.

(f) In the xy-plane, find and graph the inverse image of the square $u = 0$, $u = 1$, $v = 0$, $v = 1$. Find the area within this inverse image (it's a parallelogram) and show it is equal to $\left|\frac{\partial(x,y)}{\partial(u,v)}\right|$.

(g) In the uv-plane, find and graph the image of the square $x = \pm 1$, $y = \pm 1$.

(h) In the uv-plane, find and graph the image of the circle $x^2 + y^2 = 1$.

(i) In the uv-plane, find and graph the image of the circle $(x - 3)^2 + (y - 2)^2 = 1$.

(j) In the uv-plane, find and graph the image of the parabola $y = x^2$.

(k) In your favorite computer algebra system, install the coordinates found in part (c) as a new coordinate system called *slant*. Convert $x^2 + y^2 = 1$ to the *slant* coordinates, and use an implicit plotting device that knows the *slant* coordinate system to obtain its graph in the *xy*-plane. This process is the analog of what was done with polar coordinates in Exercises B3–12. (The accompanying Maple worksheet demonstrates the installation of new coordinates in Maple.)

(l) Convert $y = x^2$ to the *slant* coordinates, and use an implicit plotting device that knows the *slant* coordinate system to obtain its graph in the *xy*-plane.

B14. The mapping $x = \frac{1}{2}(u^2 - v^2)$, $y = uv$, with inversion equations $u = \pm \frac{y}{\sqrt{\sqrt{x^2+y^2}-x}}$, $v = \pm \sqrt{\sqrt{x^2 + y^2} - x}$, defines the *parabolic* coordinate system, known to some computer algebra systems.

(a) In the *xy*-plane, find and graph the inverse images of the grid lines $u = 1, 2, 3$.

(b) In the *xy*-plane, find and graph the inverse images of the grid lines $v = 1, 2, 3$.

(c) In the *uv*-plane, find and graph the images of the grid lines $x = 1, 2, 3$.

(d) In the *uv*-plane, find and graph the images of the grid lines $y = 1, 2, 3$.

(e) In the *uv*-plane, find and graph the image of the square $1 \le x, y \le 2$.

(f) In the *uv*-plane, find and graph the image of the circle $(x - 3)^2 + (y - 2)^2 = 1$.

(g) In the *uv*-plane, find and graph the image of the parabola $y = x^2$.

(h) Convert $x^2 + y^2 = 1$ to *uv*-coordinates, and use an implicit plotting device that knows the *parabolic* coordinate system to obtain its graph in the *xy*-plane.

(i) Convert $y = x^2$ to *uv*-coordinates, and use an implicit plotting device that knows the *parabolic* coordinate system to obtain its graph in the *xy*-plane.

B15. The mapping $x = \frac{u}{u^2+v^2}$, $y = \frac{v}{u^2+v^2}$, with inversion equations $u = \frac{x}{x^2+y^2}$, $v = \frac{y}{x^2+y^2}$, defines the *tangent* coordinate system that is known to some computer algebra systems.

(a) In the *xy*-plane, find and graph the inverse images of the grid lines $u = 1, 2, 3$.

(b) In the *xy*-plane, find and graph the inverse images of the grid lines $v = 1, 2, 3$.

(c) In the *uv*-plane, find and graph the images of the grid lines $x = 1, 2, 3$.

(d) In the *uv*-plane, find and graph the images of the grid lines $y = 1, 2, 3$.

(e) In the *uv*-plane, find and graph the image of the square $1 \le x, y \le 2$.

(f) In the *uv*-plane, find and graph the image of the circle $(x - 3)^2 + (y - 2)^2 = 1$.

(g) In the *uv*-plane, find and graph the image of the parabola $y = x^2$.

(h) Convert $x^2 + y^2 = 1$ to *uv*-coordinates, and use an implicit plotting device that knows the *tangent* coordinate system to obtain its graph in the *xy*-plane. The unit circle should result. Once again, this process mirrors exactly what was done with polar coordinates in Exercise B3–12.

(i) Convert $y = x^2$ to *uv*-coordinates, and use an implicit plotting device that knows the *tangent* coordinate system to obtain its graph in the *xy*-plane.

B16. Consider the parabolic coordinate system from Exercise B14.

(a) In the *xy*-plane, obtain the inverse image of the grid line $u = \alpha$. Differentiate $\mathbf{R}(x(u, v), y(u, v))$ to obtain a vector tangent to this curve.

(b) Repeat part (a) for the grid line $v = \beta$.

(c) Find the area of the parallelogram determined by the tangent vectors in parts (a) and (b).

(d) Show that the area found in part (c) is $\left| \frac{\partial(x,y)}{\partial(u,v)} \right|$.

B17. Repeat Exercise B16 for the tangent coordinate system.

B18. Let R_{xy} denote, in the *xy*-plane, the region bounded by $y = 1 - x^2$ and the *x*-axis.

(a) Obtain R_{uv}, the image of R_{xy} under the mapping $u = 4x - 7y$, $v = 3x + 5y$, being sure to indicate the new bounding curves and their intersections.

(b) Obtain as $\iint_{R_{xy}} dy\, dx$, the area of R_{xy}.

(c) Change variables in the double integral of part (b), writing it as $\iint_{R_{uv}} \left| \frac{\partial(x,y)}{\partial(u,v)} \right| du\, dv$, an integral in *u* and *v*. Evaluate this integral, obtaining the same value as found in part (b).

B19. Let $x = u^2 + v^2$, $y = \frac{u}{v}$, define a mapping from the *uv*-plane to the *xy*-plane.

(a) Obtain $u = u(x, y)$, $v = v(x, y)$, the formulas for mapping the *xy*-plane back to the *uv*-plane.

(b) In the *uv*-plane, obtain T', the image of triangle T whose vertices are $(1, 1)$, $(3, 5)$, $(2, 7)$, being sure to write the equations of the curves bounding the mapped region. *Hint:* It is easier to describe the boundaries as $u = u(v)$.

(c) Find the area of triangle T.

(d) Obtain the area of T' using double integration in the *uv*-plane.

B20. Let R_{xy} be the region in the *xy*-plane bounded by the curves $y_1 = x^2$, $y_2 = x^2 + 1$, $y_3 = 2x + 2$, $y_4 = 2x$. (One point of the boundary is the origin.)

(a) Use double integrals to obtain the area of R_{xy}. (It takes three such integrals whether the integration is done $dy\, dx$ or $dx\, dy$.)

(b) Change coordinates to $u = y - x^2$, $v = y - 2x$, so that the region of integration becomes a rectangle in the *uv*-plane. Obtain $x = x(u, v)$, $y = y(u, v)$, and also $\left| \frac{\partial(x,y)}{\partial(u,v)} \right|$.

(c) Compute the area as a double integral in the uv-plane. Clearly, the answer should match that found in part (a).

B21. Let R_{xy} by the region in the xy-plane bounded by the curves $y_1 = x$, $y_2 = x + 2$, $y_3 = \frac{1}{x}$, $y_4 = \frac{5}{x}$.

  (a) Use double integrals to obtain the area of R_{xy}. (It takes three such integrals whether the integration is done $dy\,dx$ or $dx\,dy$.)

(b) Change coordinates to $u = y - x$, $v = xy$, so that the region of integration becomes a rectangle in the uv-plane. Obtain $x = x(u, v)$, $y = y(u, v)$, and also $\left|\frac{\partial(x,y)}{\partial(u,v)}\right|$.

(c) Compute the area as a double integral in the uv-plane. Clearly, the answer should match that found in part (a).

22.2 Vector Operators in Polar Coordinates

Gradient in Polar Coordinates

In Cartesian coordinates, the gradient of the scalar $u(x, y)$ is the vector $\nabla u = u_x \mathbf{i} + u_y \mathbf{j}$, where $\mathbf{i}$ and $\mathbf{j}$ are unit basis vectors for the Cartesian system. What does it mean to compute the gradient of the scalar $u(r, \theta)$ in polar coordinates?

GRADIENT IN POLAR COORDINATES—DERIVATION   Recall the equations defining polar coordinates, $x = r\cos\theta$, $y = r\sin\theta$, and the equations defining the inverse transformation, $r = \sqrt{x^2 + y^2}$, $\theta = \arctan\frac{y}{x}$. Write the radius vector as $\mathbf{R} = x\mathbf{i} + y\mathbf{j}$, with $x = x(r, \theta)$ and $y = y(r, \theta)$. If θ is held constant and r allowed to vary, $\mathbf{R} = \mathbf{R}(r)$, so an r-coordinate curve is traced in the xy-plane. For example, $\theta = 1$ defines a radial line emanating from the origin, while $r = 1$ defines the θ-coordinate curve that is a circle about the origin.

A fundamental idea of vector calculus, now long familiar, is that differentiation of a curve with respect to its parameter yields a vector tangent to that curve. Apply that principle to the radius vector $\mathbf{R}(r, \theta)$, first differentiating with respect to r and then with respect to θ. In the first case, obtain a vector tangent to an r-coordinate curve (a radial line) and in the second a vector tangent to a θ-coordinate curve, a circle about the origin. See Figure 22.7 for a sketch of the vectors $\frac{\partial \mathbf{R}}{\partial r}$ and $\frac{\partial \mathbf{R}}{\partial \theta}$ drawn at $(x, y) = (\sqrt{2}, \sqrt{2})$.

The vector tangent to the r-coordinate curve turns out to be a unit vector. The vector tangent to the θ-coordinate curve needs to be normalized since it does not automatically have unit length. Typical notation for these unit vectors in polar coordinates is

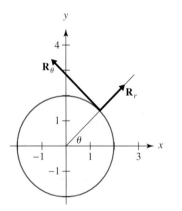

$$\mathbf{e}_r = \cos\theta\,\mathbf{i} + \sin\theta\,\mathbf{j} \quad \text{and} \quad \mathbf{e}_\theta = \frac{1}{r}\mathbf{R}_\theta = -\sin\theta\,\mathbf{i} + \cos\theta\,\mathbf{j} \tag{22.1}$$

FIGURE 22.7   The vector $\frac{\partial \mathbf{R}}{\partial r}$ and $\frac{\partial \mathbf{R}}{\partial \theta}$ in polar coordinates

Solving for $\mathbf{i}$ and $\mathbf{j}$ in terms of $\mathbf{e}_r$ and $\mathbf{e}_\theta$, we have

$$\mathbf{i} = \cos\theta\,\mathbf{e}_r - \sin\theta\,\mathbf{e}_\theta \quad \text{and} \quad \mathbf{j} = \sin\theta\,\mathbf{e}_r + \cos\theta\,\mathbf{e}_\theta \tag{22.2}$$

The point here is to discover the proper expressions for the gradient and laplacian of a scalar-valued function and the divergence of a vector-valued function, when each is given in polar coordinates. Since the expression for the gradient in Cartesian coordinates is already known, use that knowledge to deduce the result for polar coordinates. Start with the expression for the gradient in Cartesian coordinates, replacing $\mathbf{i}$ and $\mathbf{j}$ with their equivalents in terms of $\mathbf{e}_r$ and $\mathbf{e}_\theta$ and transforming the partial derivatives using the chain rule.

The second step expresses $u(x, y)$ as $U(r(x, y), \theta(x, y))$ prior to using the chain rule for determining the equivalents of the partial derivatives u_x and u_y in polar coordinates. For insight, consider the function $h(x, y) = xy^2$ and its conversion to polar coordinates, $h(x(r, \theta), y(r, \theta)) = r^3 \cos\theta \sin^2\theta = H(r, \theta)$. The introduction of $H(r, \theta)$ is essential, since it is a different function of r and θ than $h(x, y)$ was a function of x and y. In $h(x, y)$

the first variable, x, simply multiplies the square of the second. In $H(r, \theta)$, the first variable, r, is cubed. Thus, two different letters, h and H, are more than appropriate, they are necessary.

Now, go back the other way by writing $H(r(x, y), \theta(x, y)) = h(x, y)$, the starting point for the partial differentiations about to take place. Thus, start with

$$u(x, y) = U(r(x, y), \theta(x, y))$$

and by the chain rule obtain

$$u_x = U_r r_x + U_\theta \theta_x \quad \text{and} \quad u_y = U_r r_y + U_\theta \theta_y \qquad (22.3)$$

Since $r_x = \cos\theta$, $\theta_x = -\frac{\sin\theta}{r}$, $r_y = \sin\theta$, and $\theta_y = \frac{\cos\theta}{r}$, the gradient in Cartesian coordinates becomes

$$\left[U_r \cos\theta - U_\theta \frac{\sin\theta}{r} \right] [\cos\theta \mathbf{e}_r - \sin\theta \mathbf{e}_\theta] + \left[U_r \sin\theta + U_\theta \frac{\cos\theta}{r} \right] [\sin\theta \mathbf{e}_r + \cos\theta \mathbf{e}_\theta]$$

$$= U_r \mathbf{e}_r + \frac{1}{r} U_\theta \mathbf{e}_\theta \qquad (22.4)$$

Divergence in Polar Coordinates

In Cartesian coordinates, the vector field $\mathbf{F} = f(x, y)\mathbf{i} + g(x, y)\mathbf{j}$ has as its divergence, the scalar field $\nabla \cdot \mathbf{F} = f_x + g_y$. The divergence of the vector $\mathbf{B} = U(r, \theta)\mathbf{e}_r + V(r, \theta)\mathbf{e}_\theta$ is the scalar $U_r + \frac{1}{r} U + \frac{1}{r} V_\theta$. Notice the term without any derivative! There is something to discover here!

DIVERGENCE IN POLAR COORDINATES—DERIVATION To derive the expression for the divergence of the polar vector $\mathbf{B} = U(r, \theta\mathbf{e}_r + V(r, \theta)\mathbf{e}_\theta$, transform $\mathbf{B}$ completely to Cartesian coordinates, compute the divergence in Cartesian coordinates, then transform that result back to polar coordinates. Therefore, in terms of $\mathbf{i}$ and $\mathbf{j}$, we have $\mathbf{B} = F(r, \theta)\mathbf{i} + G(r, \theta)\mathbf{j}$, where

$$F = U \cos\theta - V \sin\theta \quad \text{and} \quad G = U \sin\theta + V \cos\theta \qquad (22.5)$$

Since F and G are functions of r and θ, define

$$f(x, y) = F(r(x, y), \theta(x, y)) \quad \text{and} \quad g(x, y) = G(r(x, y), \theta(x, y))$$

so that the Cartesian version of $\mathbf{B}$ is really $f(x, y)\mathbf{i} + g(x, y)\mathbf{j}$. Then, the partial derivatives f_x and g_y, computed by the chain rule, are precisely what is needed to determine the divergence of $\mathbf{B}$.

By the chain rule, $f_x = F_r r_x + F_\theta \theta_x$ and $g_y = G_r r_y + G_\theta \theta_y$, so

$$f_x = [U_r \cos\theta - V_r \sin\theta] \cos\theta$$
$$+ [U_\theta \cos\theta - U \sin\theta - V_\theta \sin\theta - V \cos\theta] \left(-\frac{\sin\theta}{r} \right) \qquad (22.6\text{a})$$

$$g_y = [U_r \sin\theta + V_r \cos\theta] \sin\theta$$
$$+ [U_\theta \sin\theta + U \cos\theta + V_\theta \cos\theta - V \sin\theta] \left(\frac{\cos\theta}{r} \right) \qquad (22.6\text{b})$$

The sum $f_x + g_y = U_r + \frac{1}{r} U + \frac{1}{r} V_\theta = \nabla \cdot \mathbf{B}$ is the divergence in polar coordinates.

laplacian in Polar Coordinates

In rectangular coordinates, the laplacian of $f(x, y)$ is $\nabla^2 f = f_{xx} + f_{yy}$. In polar coordinates, the laplacian is $F_{rr} + \frac{1}{r} F_r + \frac{1}{r^2} F_{\theta\theta}$, where $F = F(r, \theta)$.

EXAMPLE 22.2   For example, the laplacian of the function $f(x, y) = x + y^3$ is $\nabla^2 f = 6y$. Converting the function $f(x, y) = x + y^3$ to polar coordinates gives

$$F(r, \theta) = f(x(r, \theta), y(r, \theta)) = r \cos\theta + r^3 \sin^3\theta$$

In polar coordinates, the laplacian of $F(r, \theta)$ is $6r \sin\theta$, the same as if $\nabla^2 f = 6y$ were converted to polar coordinates.

TRADITIONAL CHAIN RULE DERIVATION   The starting point for the differentiations is the statement $f(x, y) = F(r(x, y), \theta(x, y))$. An application of the chain rule gives $f_x = F_r r_x + F_\theta \theta_x$ and $f_y = F_r r_y + F_\theta \theta_y$. Noting that F_r and F_θ depend on $r(x, y)$ and $\theta(x, y)$, second derivatives computed by the chain rule are

$$f_{xx} = \left(\frac{\partial}{\partial x} F_r\right) r_x + F_r r_{xx} + \left(\frac{\partial}{\partial x} F_\theta\right)\theta_x + F_\theta \theta_{xx} \tag{22.7a}$$

$$= (F_{rr} r_x + F_{r\theta}\theta_x)r_x + F_r r_{xx} + (F_{\theta r}r_x + F_{\theta\theta}\theta_x)\theta_x + F_\theta\theta_{xx} \tag{22.7b}$$

and

$$f_{yy} = \left(\frac{\partial}{\partial y} F_r\right) r_y + F_r r_{yy} + \left(\frac{\partial}{\partial y} F_\theta\right)\theta_y + F_\theta \theta_{yy} \tag{22.8a}$$

$$= (F_{rr} r_y + F_{r\theta}\theta_y)r_y + F_r r_{yy} + (F_{\theta r}r_y + F_{\theta\theta}\theta_y)\theta_y + F_\theta\theta_{yy} \tag{22.8b}$$

Then, substituting the derivatives listed in Table 22.1, into the sum $f_{xx} + f_{yy}$ and using $F_{r\theta} = F_{\theta r}$, we get

$$\nabla^2 F = F_{rr} + \frac{1}{r} F_r + \frac{1}{r^2} F_{\theta\theta} \tag{22.9}$$

❖

$r_x = \cos\theta$	$r_{xx} = \dfrac{\sin^2\theta}{r}$
$r_y = \sin\theta$	$r_{yy} = \dfrac{\cos^2\theta}{r}$
$\theta_x = -\dfrac{\sin\theta}{r}$	$\theta_{xx} = 2\dfrac{\sin\theta\cos\theta}{r^2}$
$\theta_y = \dfrac{\cos\theta}{r}$	$\theta_{yy} = -2\dfrac{\sin\theta\cos\theta}{r^2}$

TABLE 22.1   First and second partial derivatives of $r(x, y)$ and $\theta(x, y)$

EXERCISES 22.2–Part A

A1. Obtain the vectors in (22.1).   **A2.** Obtain the vectors in (22.2).

A3. Obtain the derivatives in (22.3).

A4. Verify the calculations in (22.4).

A5. Verify the representations in (22.5).

A6. Verify the differentiations in (22.6a) and (22.6b).

A7. Verify the calculations in (22.7a) and (22.7b).

A8. Verify the calculations in (22.8a) and (22.8b).

A9. Using (22.7a), (22.7b), (22.8a), (22.8b), and Table 22.1, obtain (22.9).

EXERCISES 22.2–Part B

B1. Obtain the derivatives in Table 22.1.

For the scalar-valued functions $f(x, y)$ in Exercises B2–6:

(a) Obtain ∇f, the gradient, in Cartesian coordinates.

(b) Transform ∇f from part (a) to polar coordinates by substituting both for the variables x and y and for the basis vectors $\mathbf{i}$ and $\mathbf{j}$.

(c) Convert $f(x, y)$ to polar coordinates by substituting for the variables x and y. (Since u is a scalar, there is no problem with changing basis vectors.) Call the transformed $f(x, y)$ by a new name such as $F(r, \theta)$.

(d) Obtain ∇F in polar coordinates, and compare with the vector found in part (b).

B2. $f(x, y) = 7x^2y + 2y$ **B3.** $f(x, y) = 3xy + 7x$

B4. $f(x, y) = x^3 - 5y^2$ **B5.** $f(x, y) = 7y^2 - 9x^2y$

B6. $f(x, y) = 9x^3 + 6xy^2$

For the vector-valued functions $\mathbf{F}(x, y)$ in Exercises B7–11:

(a) Obtain $\nabla \cdot \mathbf{F}$, the divergence, in Cartesian coordinates.

(b) Transform $\nabla \cdot \mathbf{F}$ to polar coordinates. (Since the divergence is a scalar, simply substitute for the variables x and y.)

(c) Change **F** to polar coordinates by substituting both for the variables x and y and for the basis vectors **i** and **j**. Call the transformed vector by the new name $\mathbf{G}(r, \theta)$.

(d) Compute $\nabla \cdot \mathbf{G}$ in polar coordinates, and compare with the result in part (b).

B7. $\mathbf{F} = (9x^2 - 4y)\mathbf{i} + (9y^2 - 3x)\mathbf{j}$ **B8.** $\mathbf{F} = 5y^2\mathbf{i} + (5y^3 + 4x^2)\mathbf{j}$

B9. $\mathbf{F} = (5y^2 + 6x)\mathbf{i} + (5xy^2 + 2y)\mathbf{j}$

B10. $\mathbf{F} = (4x^2y + 7x)\mathbf{i} + 7x^3\mathbf{j}$ **B11.** $\mathbf{F} = (6y^3 + 9x^2)\mathbf{i} + (8y^2 + 9)\mathbf{j}$

For the scalar-valued functions $f(x, y)$ in Exercises B12–16:

(a) Obtain $\nabla^2 f$, the laplacian, in Cartesian coordinates.

(b) Transform $\nabla^2 f$ to polar coordinates. (Since the laplacian is a scalar, simply substitute for the variables x and y.)

(c) Convert $f(x, y)$ to polar coordinates by substituting for the variables x and y. (Since f is a scalar, there is no problem with changing basis vectors.) Call the transformed $f(x, y)$ by a new name such as $F(r, \theta)$.

(d) Compute $\nabla^2 F$ in polar coordinates, and compare with the result in part (b).

B12. $f(x, y) = 9x^2y - 8x^3y$ **B13.** $f(x, y) = 8x^2y + 4x^4$

B14. $f(x, y) = 6x^3 + 7y^3$ **B15.** $f(x, y) = 2xy^2 - 8x^4$

B16. $f(x, y) = 8xy^2 - 7y^3$

B17. Let $u = 3x - 2y$, $v = 5x + 7y$, define a change of coordinates from (x, y) to (u, v). If $F(u, v)$ is a scalar-valued function, implement the ensuing steps to derive an expression for ∇F in the skewed (uv) coordinate system.

(a) Obtain the inversion formulas $x = x(u, v)$, $y = y(u, v)$.

(b) From $\mathbf{R} = x(u, v)\mathbf{i} + y(u, v)\mathbf{j}$, obtain the tangent vectors $\frac{\partial \mathbf{R}}{\partial u}$ and $\frac{\partial \mathbf{R}}{\partial v}$.

(c) Obtain the unit tangent vectors $\hat{\mathbf{e}}_u$ and $\hat{\mathbf{e}}_v$, then express the unit basis vectors **i** and **j** in terms of $\hat{\mathbf{e}}_u$ and $\hat{\mathbf{e}}_v$.

(d) Write $f(x, y) = F(u(x, y), v(x, y))$, and use the chain rule to obtain f_x and f_y.

(e) Convert $\nabla f = f_x\mathbf{i} + f_y\mathbf{j}$ to its equivalent in the uv-coordinate system.

(f) Apply the result in part (e) to $f = x^2y^3$. (Transform $f(x, y)$ to $F(u, v)$, then compute the gradient in uv-coordinates.)

(g) Compute ∇f in Cartesian coordinates, and, using the results of part (c), transform this gradient to skewed coordinates. Compare to the result obtained in part (f).

B18. Add the skewed coordinates of Exercise B17 to the list of known coordinate systems in your favorite computer algebra system. Obtain a graph showing, in the xy-plane, the inverse images of the grid lines of the uv-plane.

B19. Using the skewed coordinates defined in Exercise B17:

(a) Obtain, in the uv-coordinate system, an expression for $\nabla^2 F$, the laplacian of the arbitrary scalar $F(u, v)$. (Write

$f(x, y) = F(u(x, y), v(x, y))$, and use the chain rule to obtain f_x, f_y, f_{xx}, and f_{yy}. Convert $\nabla^2 f = f_{xx} + f_{yy}$ to the desired expression in the uv-coordinate system.)

(b) Apply the result in part (a) to $f = x^3y^4$.

(c) Compute $\nabla^2 f$ in Cartesian coordinates.

(d) Transform the result in part (c) to uv-coordinates. Compare to the result in part (b).

B20. Using the skewed coordinates defined in Exercise B17:

(a) Obtain, in the uv-coordinate system, an expression for $\nabla \cdot \mathbf{F}$, the divergence of the arbitrary vector $\mathbf{F}(u, v) = f(u, v)\hat{\mathbf{e}}_u + g(u, v)\hat{\mathbf{e}}_v$. (Convert **F** to the form $A(u, v)\mathbf{i} + B(u, v)\mathbf{j}$, and apply the chain rule to $a(x, y) = A(u(x, y), v(x, y))$ and $b(x, y) = B(u(x, y), v(x, y))$.)

(b) Apply the result in part (a) to the vector $\mathbf{V} = (2x^2y - 3y^2)\mathbf{i} + (5xy^2 + 4x^2)\mathbf{j}$. (Convert $\mathbf{V}(x, y)$ to $\mathbf{F}(u, v) = f(u, v)\hat{\mathbf{e}}_u + g(u, v)\hat{\mathbf{e}}_v$, and compute the divergence in the uv-coordinate system.)

(c) Compute $\nabla \cdot \mathbf{V}$ in Cartesian coordinates.

(d) Transform the result in part (c) to uv-coordinates. Compare to the result in part (b).

B21. The equations $x = \frac{u}{u^2 + v^2}$, $y = \frac{v}{u^2 + v^2}$, define the *tangent* coordinate system. If $F(u, v)$ is a scalar, use the following steps to derive an expression for ∇F in the tangent (uv) coordinate system.

(a) Obtain the inversion formulas $u = u(x, y)$, $v = v(x, y)$.

(b) From $\mathbf{R} = x(u, v)\mathbf{i} + y(u, v)\mathbf{j}$, obtain the tangent vectors $\frac{\partial \mathbf{R}}{\partial u}$ and $\frac{\partial \mathbf{R}}{\partial v}$.

(c) Obtain the unit tangent vectors $\hat{\mathbf{e}}_u$ and $\hat{\mathbf{e}}_v$, then express the unit basis vectors **i** and **j** in terms of $\hat{\mathbf{e}}_u$ and $\hat{\mathbf{e}}_v$.

(d) Write $f(x, y) = F(u(x, y), v(x, y))$, and use the chain rule to obtain f_x and f_y.

(e) Convert $\nabla f = f_x\mathbf{i} + f_y\mathbf{j}$ to its equivalent in the uv-coordinate system.

(f) Apply the result in part (e) to $f = x^2y^3$. (Transform $f(x, y)$ to $F(u, v)$, and then compute the gradient in uv-coordinates.)

(g) Compute ∇f in Cartesian coordinates and, using the results of part (c), transform this gradient to skewed coordinates. Compare to the result obtained in part (f).

B22. Using the tangent coordinates defined in Exercise B21:

(a) Obtain, in the uv-coordinate system, an expression for $\nabla^2 F$, the laplacian of the arbitrary scalar-valued function $F(u, v)$. (Write $f(x, y) = F(u(x, y), v(x, y))$, and use the chain rule to obtain f_x, f_y, f_{xx}, and f_{yy}. Convert $\nabla^2 f = f_{xx} + f_{yy}$ to the desired expression in the uv-coordinate system.)

(b) Apply the result in part (a) to $f = x^3y^4$. (Transform $f(x, y)$ to $F(u, v)$, and then compute the gradient in uv-coordinates.)

(c) Compute $\nabla^2 f$ in Cartesian coordinates.

(d) Transform the result in part (c) to uv-coordinates. Compare to the result in part (b).

(e) Check parts (a) and (b) with a computer algebra system in which the tangent coordinate system is known.

B23. Using the tangent coordinates defined in Exercise B21:

(a) Obtain, in the uv-coordinate system, an expression for $\nabla \cdot \mathbf{F}$, the divergence of the arbitrary vector $\mathbf{F}(u, v) = f(u, v)\hat{\mathbf{e}}_u + g(u, v)\hat{\mathbf{e}}_v$. (Convert $\mathbf{F}$ to the form $A(u, v)\mathbf{i} + B(u, v)\mathbf{j}$, and apply the chain rule to $a(x, y) = A(u(x, y), v(x, y))$ and $b(x, y) = B(u(x, y), v(x, y))$.)

(b) Apply the result in part (a) to the vector $\mathbf{V} = (2x^2 y - 3y^2)\mathbf{i} + (5xy^2 + 4x^2)\mathbf{j}$. (Convert $\mathbf{V}(x, y)$ to $\mathbf{F}(u, v) = f(u, v)\hat{\mathbf{e}}_u + g(u, v)\hat{\mathbf{e}}_v$, and compute the divergence in the uv-coordinate system.)

(c) Compute $\nabla \cdot \mathbf{V}$ in Cartesian coordinates.

(d) Transform the result in part (c) to uv-coordinates. Compare to the result in part (b).

(e) Check parts (a) and (b) with a computer algebra system in which the tangent coordinate system is known.

B24. The parabolic coordinate system is defined by $x = \frac{1}{2}(u^2 - v^2)$, $y = uv$, but the inversion formulas $u = u(x, y)$ and $v = v(x, y)$ are sufficiently complex as to warrant computational adjustments. For example, to obtain an expression for the gradient of the scalar $F(u, v)$, the chain rule is applied to $f(x, y) = F(u(x, y), v(x, y))$ to obtain $f_x = F_u u_x + F_v v_x$ and $f_y = F_u u_y + F_v v_y$.

(a) Using $x = \frac{1}{2}(u^2 - v^2)$, $y = uv$, and implicit differentiation, obtain, in terms of u and v, expressions for u_x, u_y, v_x, v_y. (Computer algebra systems typically have commands for implicit differentiation.)

(b) From $\mathbf{R} = x(u, v)\mathbf{i} + y(u, v)\mathbf{j}$ and part (a), obtain the tangent vectors $\frac{\partial \mathbf{R}}{\partial u}$ and $\frac{\partial \mathbf{R}}{\partial v}$.

(c) Obtain the unit tangent vectors $\hat{\mathbf{e}}_u$ and $\hat{\mathbf{e}}_v$, and then express the unit basis vectors $\mathbf{i}$ and $\mathbf{j}$ in terms of $\hat{\mathbf{e}}_u$ and $\hat{\mathbf{e}}_v$.

(d) Convert $\nabla f = f_x \mathbf{i} + f_y \mathbf{j}$ to its equivalent in the parabolic coordinate system.

(e) Check the result in part (d) with a computer algebra system in which the parabolic coordinate system is known.

(f) Apply the result in part (d) to $f = x^2 y^3$. (Transform $f(x, y)$ to $F(u, v)$, and then compute the gradient in the parabolic coordinate system.)

(g) Compute ∇f in Cartesian coordinates and, using the results of part (c), transform this gradient to parabolic coordinates. Compare to the result obtained in part (f).

B25. Using the parabolic coordinate system from Exercise B24:

(a) Obtain, in the uv-coordinate system, an expression for $\nabla \cdot \mathbf{F}$, the divergence of the arbitrary vector-valued function $\mathbf{F}(u, v) = f(u, v)\hat{\mathbf{e}}_u + g(u, v)\hat{\mathbf{e}}_v$. (Convert $\mathbf{F}$ to the form $A(u, v)\mathbf{i} + B(u, v)\mathbf{j}$, and apply the chain rule to $a(x, y) = A(u(x, y), v(x, y))$ and $b(x, y) = B(u(x, y), v(x, y))$. Again, use implicit differentiation to obtain expressions for u_x, u_y, v_x, v_y.)

(b) Apply the result in part (a) to the vector $\mathbf{V} = (2x^2 y - 3y^2)\mathbf{i} + (5xy^2 + 4x^2)\mathbf{j}$. (Convert $\mathbf{V}(x, y)$ to $\mathbf{F}(u, v) = f(u, v)\hat{\mathbf{e}}_u + g(u, v)\hat{\mathbf{e}}_v$, and compute the divergence in the parabolic coordinate system.)

(c) Compute $\nabla \cdot \mathbf{V}$ in Cartesian coordinates.

(d) Transform the result in part (c) to parabolic coordinates. Compare to the result in part (b).

(e) Check parts (a) and (b) with a computer algebra system in which the parabolic coordinate system is known.

22.3 Vector Operators in Cylindrical and Spherical Coordinates

Introduction

Tangent to a coordinate curve, a *natural* tangent basis vector is formed by differentiating $\mathbf{R}$, the radius vector, with respect to the parameter defining the coordinate curve. For example, $\frac{\partial \mathbf{R}}{\partial \theta}$ in polar coordinates is a natural tangent basis vector but is not the unit vector $\mathbf{e}_\theta$. Although natural tangent basis vectors are generally not unit vectors, they are the basis vectors used when generalizing vector calculus to the domain of tensor calculus.

In the *orthogonal* coordinate systems considered in this text, polar, cylindrical, and spherical, the basis vectors are unit vectors. The formulas developed in Section 22.2 for polar coordinates, and those to be developed in this section for the differential operators div, grad, curl, and laplacian in cylindrical and spherical coordinates, all use unit basis vectors. Based on these coordinate systems, it should be possible for the reader to decipher the vector operators in other *orthogonal* coordinate systems. The reader who confronts these operators in coordinate systems that are not orthogonal must therefore be careful about the basis vectors used.

Unit Basis Vectors

The notation $\mathbf{e}_r$ and $\mathbf{e}_\theta$ was introduced for *unit* basis vectors in the polar coordinate system. With respect to these *unit* basis vectors in polar coordinates, a vector is expressed as $\mathbf{F} = f\mathbf{e}_r + g\mathbf{e}_\theta$ and the gradient vector as $\mathrm{grad}(u) = u_r\mathbf{e}_r + (u_\theta/r)\mathbf{e}_\theta$. For the vector $\mathbf{F}$ given in polar coordinates, the divergence was found to be $\mathrm{div}(F) = f_r + \frac{1}{r}f + \frac{1}{r}g_\theta$.

All these results are predicated on the *unit* basis vectors $\mathbf{e}_r$ and $\mathbf{e}_\theta$. Using this convention for basis vectors, Table 22.2 lists expressions for div, grad, curl, and laplacian in Cartesian and cylindrical coordinates, where $\mathbf{e}_z = \mathbf{k}$. Expressions for these operators in polar coordinates can be deduced from those in cylindrical coordinates. In Table 22.3, these same operators are listed for the two different versions of spherical coordinates found in the literature.

System	Cartesian	Cylindrical
Basis	$\mathbf{i}, \mathbf{j}, \mathbf{k}$	$\mathbf{e}_r, \mathbf{e}_\theta, \mathbf{e}_z = \mathbf{k}$
Coordinates	x y z	$x = r\cos(\theta)$ $y = r\sin(\theta)$ $z = z$
grad ∇u	$\begin{bmatrix} u_x \\ u_y \\ u_z \end{bmatrix}$	$\begin{bmatrix} u_r \\ \frac{1}{r}u_\theta \\ u_z \end{bmatrix}$
div $\nabla \cdot \mathbf{F}$	$f_x + g_y + h_z$	$\frac{1}{r}(rf)_r + \frac{1}{r}g_\theta + h_z$
curl $\nabla \times \mathbf{F}$	$\begin{vmatrix} \mathbf{i} & \mathbf{j} & \mathbf{k} \\ \frac{\partial}{\partial x} & \frac{\partial}{\partial y} & \frac{\partial}{\partial z} \\ f & g & h \end{vmatrix} = \begin{bmatrix} h_y - g_z \\ f_z - h_x \\ g_x - f_y \end{bmatrix}$	$\begin{vmatrix} \frac{\mathbf{e}_r}{r} & \mathbf{e}_\theta & \frac{\mathbf{e}_z}{r} \\ \frac{\partial}{\partial r} & \frac{\partial}{\partial \theta} & \frac{\partial}{\partial z} \\ f & rg & h \end{vmatrix} = \begin{bmatrix} \frac{h_\theta - (rg)_z}{r} \\ f_z - h_r \\ \frac{(rg)_r - f_\theta}{r} \end{bmatrix}$
laplacian $\nabla^2 u$	$u_{xx} + u_{yy} + u_{zz}$	$\frac{(ru_r)_r}{r} + \frac{u_{\theta\theta}}{r^2} + u_{zz}$

TABLE 22.2   Vector operators in Cartesian and cylindrical coordinates

Symbols for the *unit* basis vectors in each system are provided *in the order in which the basis vectors are used*. Hence, in each coordinate system, vectors are of the form $\mathbf{F} = f\mathbf{e}_1 + g\mathbf{e}_2 + h\mathbf{e}_3$, where the ordering 1, 2, 3 is determined by the way the basis vectors are listed in the tables. For each system the gradient vector is given as a column vector. It is therefore essential that the order of the basis vectors be known. For each system the curl

System	Spherical ϕ measured down from z-axis	Spherical θ measured down from z-axis
Basis	$\mathbf{e}_\rho, \mathbf{e}_\phi, \mathbf{e}_\theta$	$\mathbf{e}_\rho, \mathbf{e}_\theta, \mathbf{e}_\phi$
Coordinates	$x = \rho\cos\theta\sin\phi$ $y = \rho\sin\theta\sin\phi$ $z = \rho\cos\phi$	$x = \rho\cos\phi\sin\theta$ $y = \rho\sin\phi\sin\theta$ $z = \rho\cos\theta$
grad ∇u	$\begin{bmatrix} u_\rho \\ \dfrac{1}{\rho}u_\phi \\ \dfrac{u_\theta}{\rho\sin\phi} \end{bmatrix}$	$\begin{bmatrix} u_\rho \\ \dfrac{1}{\rho}u_\theta \\ \dfrac{u_\phi}{\rho\sin\theta} \end{bmatrix}$
div $\nabla\cdot\mathbf{F}$	$\dfrac{(\rho^2 f)_\rho}{\rho^2} + \dfrac{(g\sin\phi)_\phi}{\rho\sin\phi} + \dfrac{h_\theta}{\rho\sin\phi}$	$\dfrac{(\rho^2 f)_\rho}{\rho^2} + \dfrac{(g\sin\theta)_\theta}{\rho\sin\theta} + \dfrac{h_\phi}{\rho\sin\theta}$
curl $\nabla\times\mathbf{F}$	$\begin{vmatrix} \dfrac{\mathbf{e}_\rho}{\rho^2\sin\phi} & \dfrac{\mathbf{e}_\phi}{\rho\sin\phi} & \dfrac{\mathbf{e}_\theta}{\rho} \\ \dfrac{\partial}{\partial\rho} & \dfrac{\partial}{\partial\phi} & \dfrac{\partial}{\partial\theta} \\ f & \rho g & \rho h\sin\phi \end{vmatrix}$ $=\begin{bmatrix} \dfrac{h\cos\phi}{\rho\sin\phi}+\dfrac{h_\phi}{\rho}-\dfrac{g_\theta}{\rho\sin\phi} \\ \dfrac{f_\theta}{\rho\sin\phi}-\dfrac{h}{\rho}-h_\rho \\ \dfrac{g}{\rho}+g_\rho-\dfrac{f_\phi}{\rho} \end{bmatrix}$	$\begin{vmatrix} \dfrac{\mathbf{e}_\rho}{\rho^2\sin\theta} & \dfrac{\mathbf{e}_\theta}{\rho\sin\theta} & \dfrac{\mathbf{e}_\phi}{\rho} \\ \dfrac{\partial}{\partial\rho} & \dfrac{\partial}{\partial\theta} & \dfrac{\partial}{\partial\phi} \\ f & \rho g & \rho h\sin\theta \end{vmatrix}$ $=\begin{bmatrix} \dfrac{h\cos\theta}{\rho\sin\theta}+\dfrac{h_\theta}{\rho}-\dfrac{g_\phi}{\rho\sin\theta} \\ \dfrac{f_\phi}{\rho\sin\theta}-\dfrac{h}{\rho}-h_\rho \\ \dfrac{g}{\rho}+g_\rho-\dfrac{f_\theta}{\rho} \end{bmatrix}$
laplacian $\nabla^2 u$	$\dfrac{(\rho^2 u_\rho)_\rho}{\rho^2} + \dfrac{(u_\phi\sin\phi)_\phi}{\rho^2\sin\phi} + \dfrac{u_{\theta\theta}}{\rho^2\sin^2\phi}$	$\dfrac{(\rho^2 u_\rho)_\rho}{\rho^2} + \dfrac{(u_\theta\sin\theta)_\theta}{\rho^2\sin\theta} + \dfrac{u_{\phi\phi}}{\rho^2\sin^2\theta}$

TABLE 22.3   Vector operators in spherical coordinates

is given both as a determinant and as a column vector. The top row in the matrix for each determinant shows the order of the basis vectors. The column vectors are consistent with the determinants expanded by the top rows of these matrices. In addition, the order of the basis vectors stated at the top of each column corresponds to the order of the components in the column vectors for the gradient and curl vectors.

The two definitions considered for spherical coordinates differ by an interchange of the angles θ and ϕ. The resulting formulas are similar enough that not having a complete reference to both systems can be very confusing. Hence, we have chosen to list all formulas in both systems. Note, however, that in both versions of spherical coordinates, the second component belongs to the angle measured downward from the z-axis.

Cartesian to Cylindrical Coordinates

Table 22.1 collects the relationships between the gradient, divergence, curl, and laplacian in Cartesian and cylindrical coordinates.

The results in this table can be reproduced by the techniques of Section 22.2. However, this table is of little value if its entries are not understood.

GRADIENT   The gradient in cylindrical coordinates is the vector

$$\nabla u = \begin{bmatrix} u_r(r,\theta,z) \\ \dfrac{u_r(r,\theta,z)}{r} \\ u_z(r,\theta,z) \end{bmatrix} = u_r(r,\theta,z)\mathbf{e}_r + \frac{1}{r}u_\theta(r,\theta,z)\mathbf{e}_\theta + u_z(r,\theta,z)\mathbf{e}_z$$

DIVERGENCE   A general vector in cylindrical coordinates is

$$\mathbf{F} = f(r,\theta,z)\mathbf{e}_r + g(r,\theta,z)\mathbf{e}_\theta + h(r,\theta,z)\mathbf{e}_z$$

for which the divergence is the scalar

$$\nabla \cdot \mathbf{F} = f_r(r,\theta,z) + \frac{f(r,\theta,z)}{r} + \frac{g_\theta(r,\theta,z)}{r} + h_z(r,\theta,z)$$

The first two terms are often combined into the single term $\frac{1}{r}(rf)_r = \frac{1}{r}(rf_r + f) = f_r + \frac{1}{r}f$.

CURL   When the curl is learned in Cartesian coordinates, its determinant form is proffered as a mnemonic, or memory, aid. The determinant form in cylindrical coordinates should be interpreted in the same light.

The determinant, expanded along the first row, yields the vector

$$\nabla \times \mathbf{F} = \begin{bmatrix} \dfrac{h_\theta - (rg)_z}{r} \\ f_z - h_r \\ \dfrac{(rg)_r - f_\theta}{r} \end{bmatrix} = \frac{\mathbf{e}_r}{r}(h_\theta - (rg)_z) - \mathbf{e}_\theta(h_r - f_z) + \frac{\mathbf{e}_z}{r}((rg)_r - f_\theta)$$

LAPLACIAN   In cylindrical coordinates, the laplacian is the scalar $\nabla^2 u = (ru_r)_r/r + u_{\theta\theta}/r^2 + u_{zz}$, in which the first term can be expanded to $\frac{1}{r}(ru_{rr} + u_r) = u_{rr} + \frac{1}{u}u_r$.

Spherical Coordinates

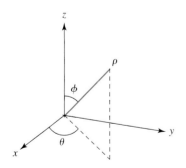

FIGURE 22.8   Spherical coordinates where ϕ is measured down from the z-axis

Consider, first, spherical coordinates listed in the center column in Table 22.3. The angle θ measures rotation around the z-axis and lies in the interval $[0, 2\pi]$. The angle ϕ ("phi") measures rotation down from the vertical and lies in the interval $[0, \pi]$. The basis vector $\mathbf{e}_\phi$ is considered the second basis vector. Figure 22.8 illustrates this definition of spherical coordinates for which (22.10) lists the formulas for mapping from and to spherical coordinates, respectively.

$$\begin{aligned} x &= \rho\cos\theta\sin\phi \\ y &= \rho\sin\theta\sin\phi \\ z &= \rho\cos\phi \end{aligned} \quad\Longleftrightarrow\quad \begin{aligned} \rho &= \sqrt{x^2 + y^2 + z^2} \\ \phi &= \arccos\frac{z}{\sqrt{x^2 + y^2 + z^2}} \\ \theta &= \arctan\frac{y}{x} \end{aligned} \qquad (22.10)$$

Cartesian to Spherical Coordinates

Table 22.3 collects the relationships between the gradient, divergence, curl, and laplacian in Cartesian and spherical coordinates. In addition, results are listed for both definitions of spherical coordinates prevalent in the literature. For both systems, always take the angle that is measured down from the z-axis as the **second coordinate.** Thus, the permanent is that the second coordinate in the list of coordinates always refers to the *same physical angle,* regardless of the name used for it.

GRADIENT   The gradient vector is the vector

$$\nabla u = \begin{bmatrix} u_\rho \\ \dfrac{1}{\rho}u_\phi \\ \dfrac{u_\theta}{\rho \sin \phi} \end{bmatrix} = u_\rho(\rho,\phi,\theta)\mathbf{e}_\rho + \frac{u_\phi(\rho,\phi,\theta)}{\rho}\mathbf{e}_\phi + \frac{u_\theta(\rho,\phi,\theta)}{\rho \sin \phi}\mathbf{e}_\theta$$

DIVERGENCE   The divergence of the vector

$$\mathbf{F} = f(\rho,\phi,\theta)\mathbf{e}_\rho + g(\rho,\phi,\theta)\mathbf{e}_\phi + h(\rho,\phi,\theta)\mathbf{e}_\theta$$

is the scalar

$$\nabla \cdot \mathbf{F} = \frac{(\rho^2 f)_\rho}{\rho^2} + \frac{(g \sin \phi)_\phi}{\rho \sin \phi} + \frac{h_\theta}{\rho \sin \phi}$$

in which the first and second terms, respectively, expand to

$$\frac{1}{\rho^2}(\rho^2 f_\rho + 2\rho f) = f_\rho + \frac{2}{\rho}f \quad \text{and} \quad \frac{1}{\rho \sin \phi}(g_\phi \sin \phi + g \cos \phi) = \frac{1}{\rho}g_\phi + \frac{g \cos \phi}{\rho \sin \phi} \tag{22.11}$$

CURL   The determinant that expresses the curl in spherical coordinates is just a mnemonic, or memory device. It does not have conceptual content. Expanding the determinant by the first row yields the vector

$$\frac{\mathbf{e}_\rho}{\rho^2 \sin \phi}(\rho(h \sin \phi)_\phi - \rho g_\theta) - \frac{\mathbf{e}_\phi}{\rho \sin \phi}(\sin \phi(\rho h)_\rho - f_\theta) + \frac{\mathbf{e}_\theta}{\rho}((\rho g)_\rho - f_\phi) \tag{22.12}$$

Carrying out the differentiations and adopting subscript notation for the resulting partial derivatives then yields

$$\mathbf{e}_\rho\left(\frac{h \cos \phi}{\rho \sin \phi} + \frac{h_\phi}{\rho} - \frac{g_\theta}{\rho \sin \phi}\right) + \mathbf{e}_\phi\left(\frac{f_\theta}{\rho \sin \phi} - \frac{h}{\rho} - h_\rho\right) + \mathbf{e}_\theta\left(\frac{g}{\rho} + g_\rho - \frac{f_\phi}{\rho}\right) \tag{22.13}$$

LAPLACIAN   The laplacian of the scalar $u(\rho,\phi,\theta)$ is the scalar

$$\nabla^2 u = \frac{(\rho^2 u_\rho)_\rho}{\rho^2} + \frac{(u_\phi \sin \phi)_\phi}{\rho^2 \sin \phi} + \frac{u_{\theta\theta}}{\rho^2 \sin^2 \phi}$$

where the first and second terms, respectively, expand to

$$\frac{1}{\rho^2}(\rho^2 u_{\rho\rho} + 2\rho u_\rho) = u_{\rho\rho} + \frac{2}{\rho}u_\rho \quad \text{and} \quad \frac{1}{\rho^2 \sin \phi}(u_{\phi\phi} \sin \phi + u_\phi \cos \phi) \tag{22.14}$$

Gradient—From Cartesian to Spherical

As an illustration, we derive one of the entries in Table 22.3. As was done for the gradient in polar coordinates, the gradient in spherical coordinates can be deduced by transforming the result from Cartesian coordinates to spherical coordinates.

In Cartesian coordinates, the radius vector is $\mathbf{R} = x\mathbf{i} + y\mathbf{j} + z\mathbf{k}$. In spherical coordinates it becomes $\mathbf{R} = x(\rho, \phi, \theta)\mathbf{i} + y(\rho, \phi, \theta)\mathbf{j} + z(\rho, \phi, \theta)\mathbf{k}$ or, better still,

$$\mathbf{R} = \rho \cos \theta \sin \phi \mathbf{i} + \rho \sin \theta \sin \phi \mathbf{j} + \rho \cos \phi \mathbf{k}$$

Unit vectors tangent to the coordinate curves are obtained by differentiating $\mathbf{R}$ with respect to each of ρ, θ, and ϕ and then normalizing. The results are

$$\mathbf{e}_\rho = \cos \theta \sin \phi \mathbf{i} + \sin \theta \sin \phi \mathbf{j} + \cos \phi \mathbf{k} \tag{22.15a}$$

$$\mathbf{e}_\phi = \cos \theta \cos \phi \mathbf{i} + \sin \theta \cos \phi \mathbf{j} - \sin \phi \mathbf{k} \tag{22.15b}$$

$$\mathbf{e}_\theta = -\sin \theta \mathbf{i} + \cos \theta \mathbf{j} \tag{22.15c}$$

which, when solved for $\mathbf{i}, \mathbf{j}, \mathbf{k}$, give

$$\mathbf{i} = -\sin \theta \mathbf{e}_\theta + \cos \theta \sin \phi \mathbf{e}_\rho + \cos \theta \cos \phi \mathbf{e}_\phi \tag{22.16a}$$

$$\mathbf{j} = \cos \theta \mathbf{e}_\theta + \sin \theta \sin \phi \mathbf{e}_\rho + \sin \theta \cos \phi \mathbf{e}_\phi \tag{22.16b}$$

$$\mathbf{k} = \cos \phi \mathbf{e}_\rho - \sin \phi \mathbf{e}_\phi \tag{22.16c}$$

In Cartesian coordinates $\mathrm{grad}(u) = u_x \mathbf{i} + u_y \mathbf{j} + u_z \mathbf{k}$, so we need to invoke the chain rule to express the derivatives in terms of spherical coordinates. For this, start from the identity

$$u(x, y, z) = U(\rho(x, y, z), \theta(x, y, z), \phi(x, y, z))$$

and differentiating by the chain rule, obtain

$$u_x = U_\rho \rho_x + U_\phi \phi_x + U_\theta \theta_x \tag{22.17a}$$

$$u_y = U_\rho \rho_y + U_\phi \phi_y + U_\theta \theta_y \tag{22.17b}$$

$$u_z = U_\rho \rho_z + U_\phi \phi_z + U_\theta \theta_z \tag{22.17c}$$

Replacing the derivatives with their equivalents from Table 22.4 and simplifying the assemblage of all the terms, we find

$$\nabla U = U_\rho(\rho, \phi, \theta)\mathbf{e}_\rho + \frac{U_\phi(\rho, \phi, \theta)}{\rho}\mathbf{e}_\phi + \frac{U_\theta(\rho, \phi, \theta)}{\rho \sin \phi}\mathbf{e}_\theta \tag{22.18}$$

$\rho_x = \sin \phi \cos \theta$	$\rho_y = \sin \phi \sin \theta$	$\rho_z = \cos \phi$
$\phi_x = \dfrac{\cos \phi \cos \theta}{\rho}$	$\phi_y = \dfrac{\cos \phi \sin \theta}{\rho}$	$\phi_z = -\dfrac{\sin \phi}{\rho}$
$\theta_x = -\dfrac{\sin \theta}{\rho \sin \phi}$	$\theta_y = \dfrac{\cos \theta}{\rho \sin \phi}$	$\theta_z = 0$

TABLE 22.4   First parital derivatives of $\rho(x, y, z)$, $\phi(x, y, z)$, and $\theta(x, y, z)$

EXERCISES 22.3–Part A

A1. Verify the simplifications in (22.11).

A2. Verify that a row-one expansion of the determinant in the middle column of Table 22.3 leads to (22.12).

A3. Show that (22.13) is the simplification of (22.12).

A4. Verify the simplifications in (22.14).

A5. Show that (22.15a), (22.15b), and (22.15c) indeed lead to (22.16a), (22.16b), and (22.16c).

A6. Verify the derivatives listed in Table 22.4.

EXERCISES 22.3–Part B

B1. Use Table 22.4, along with (22.17a–c), and (22.16a–c), to show (22.18) is grad(u) in spherical coordinates.

For the scalar $f(x, y, z)$ and point $p = (x_0, y_0, z_0)$ given in each of Exercises B2–11:

    (a) Compute $\nabla^2 f$, the laplacian, and evaluate it at p.

    (b) Transform f to spherical coordinates by writing $F(\rho, \phi, \theta) = f(x(\rho, \phi, \theta), y(\rho, \phi, \theta), z(\rho, \phi, \theta))$. Express p as a point P in spherical coordinates.

    (c) Compute $\nabla^2 F$, the laplacian in spherical coordinates, and evaluate it at P. Since this is a scalar, the value should be exactly the one found in part (a).

    (d) Compute ∇f, the gradient, and evaluate it at p.

    (e) Compute ∇F, the gradient in spherical coordinates, and evaluate it at P. This vector will *not* match the result in part (d) until a change of basis is effected.

    (f) Evaluate the unit basis vectors $\mathbf{e}_\rho, \mathbf{e}_\phi, \mathbf{e}_\theta$ at P and express ∇F in terms of $\mathbf{i}, \mathbf{j}$, and $\mathbf{k}$. Agreement with ∇f at p should now be achieved.

B2. $f(x, y, z) = x^2 z + 6y^3$, $p = (1, 7, -9)$

B3. $f(x, y, z) = 2yz - 7xz^2$, $p = (7, -6, 6)$

B4. $f(x, y, z) = xy^2 + z^2$, $p = (6, 9, -1)$

B5. $f(x, y, z) = 2xz^2 - 8y^3$, $p = (2, -4, -5)$

B6. $f(x, y, z) = 6y^2 z - 4x^3$, $p = (4, 2, -6)$

B7. $f(x, y, z) = 8x^2 y + 9xz^2$, $p = (2, 2, 7)$

B8. $f(x, y, z) = 7x^2 y - 2yz^2$, $p = (1, 7, -5)$

B9. $f(x, y, z) = 8x^2 y - (yz)^3$, $p = (1, -5, -3)$

B10. $f(x, y, z) = 3y^2 + x^2 z$, $p = (2, 2, 8)$

B11. $f(x, y, z) = 4yz + 8x^2 z^2$, $p = (6, -7, 8)$

For the vector $\mathbf{f}(x, y, z)$ and point $p = (x_0, y_0, z_0)$ given in each of Exercises B12–21:

    (a) Compute $\nabla \cdot \mathbf{f}$, the divergence, and evaluate it at p.

    (b) Transform $\mathbf{f}$ in Cartesian coordinates to $\mathbf{F}$ in spherical coordinates: transform the component functions by coordinate-substitution, and change the basis vectors from $\mathbf{i}, \mathbf{j}, \mathbf{k}$, to $\mathbf{e}_\rho, \mathbf{e}_\phi, \mathbf{e}_\theta$.

    (c) Express p as a point P in spherical coordinates.

    (d) Compute $\nabla \cdot \mathbf{F}$, the divergence in spherical coordinates, and evaluate it at P. Since this is a scalar, the value should be exactly the one found in part (a).

    (e) Compute $\nabla \times \mathbf{f}$, the curl, and evaluate it at p.

    (f) Compute $\nabla \times \mathbf{F}$, the curl in spherical coordinates, and evaluate it at P. This vector will not match the result in part (e) until a change of basis is effected.

    (g) Evaluate the unit basis vectors $\mathbf{e}_\rho, \mathbf{e}_\phi, \mathbf{e}_\theta$ at P, and use these results to express $\nabla \times \mathbf{F}$ in terms of $\mathbf{i}, \mathbf{j}$, and $\mathbf{k}$. Agreement with $\nabla \times \mathbf{f}$ at p should now be achieved.

B12. $\mathbf{f} = (6xz + 5x)\mathbf{i} + (5z^2 - 6y^2)\mathbf{j} + (9y^2 z - xy)\mathbf{k}$, $p = (6, 5, -6)$

B13. $\mathbf{f} = (8 + 6x^2)\mathbf{i} + (9z + 2xy^2)\mathbf{j} + (z^2 - 9xy)\mathbf{k}$, $p = (2, -9, 1)$

B14. $\mathbf{f} = (2y^3 - 7y)\mathbf{i} + (9x^2 y - 3x^2)\mathbf{j} + (8y^2 z - xyz)\mathbf{k}$, $p = (6, -3, 9)$

B15. $\mathbf{f} = (9x^2 y + 2z^3)\mathbf{i} + (y^2 - 7xyz)\mathbf{j} + (4 + 8x)\mathbf{k}$, $p = (7, -7, 5)$

B16. $\mathbf{f} = (5xy^2 + 8x^2 y)\mathbf{i} + (2x^2 y + 4x^2)\mathbf{j} + (9z^2 - y)\mathbf{k}$, $p = (9, -3, 6)$

B17. $\mathbf{f} = (9xz^2 + x^2 y)\mathbf{i} + (4y + 6x^3)\mathbf{j} + (3y^3 + 6xz)\mathbf{k}$, $p = (5, 3, 4)$

B18. $\mathbf{f} = (2xy^2 + 8xyz)\mathbf{i} + (6 + 5x^3)\mathbf{j} + (8z^2 + 4x^2 z)\mathbf{k}$, $p = (1, 3, -1)$

B19. $\mathbf{f} = (8xy^2 - xy)\mathbf{i} + (7xz^2 - 2yz^2)\mathbf{j} + (5z^3 - 7yz^2)\mathbf{k}$, $p = (3, 2, 5)$

B20. $\mathbf{f} = (6y^2 - z^3)\mathbf{i} + (x^2 y - 8xy^2)\mathbf{j} + (3y^2 - 2xy)\mathbf{k}$, $p = (9, -2, 4)$

B21. $\mathbf{f} = (6x^2 y - x^2 z)\mathbf{i} + (2xy - 9x^2 z)\mathbf{j} + (y^2 z - x^2)\mathbf{k}$, $p = (3, -3, -2)$

B22. Using a computer algebra system, derive the expression for the laplacian in spherical coordinates.

    (a) Obtain expressions for the nine derivatives $\rho_x, \ldots, \theta_z$, where, for example, $\rho_x = \dfrac{x}{\sqrt{x^2+y^2+z^2}}$.

    (b) Apply the chain rule to $f(x, y, z) = F(\rho(x, y, z), \phi(x, y, z), \theta(x, y, z))$ to obtain the first partial derivatives f_x, f_y, f_z.

    (c) Apply the chain rule to the derivatives found in part (b), obtaining the second derivatives f_{xx}, f_{yy}, f_{zz}.

    (d) Obtain the spherical-coordinate equivalent of $f_{xx} + f_{yy} + f_{zz}$.

B23. Using a computer algebra system, derive the expression for the divergence in spherical coordinates.

    (a) Start with the vector $\mathbf{F} = U(\rho, \phi, \theta)\mathbf{e}_\rho + V(\rho, \phi, \theta)\mathbf{e}_\phi + W(\rho, \phi, \theta)\mathbf{e}_\theta$ and convert it to Cartesian coordinates, obtaining $\mathbf{f} = a(\rho, \phi, \theta)\mathbf{i} + b(\rho, \phi, \theta)\mathbf{j} + c(\rho, \phi, \theta)\mathbf{k}$.

    (b) Write $A(x, y, z) = a(\rho(x, y, z), \phi(x, y, z), \theta(x, y, z))$, with similar expressions for b and c, and use the chain rule to obtain the derivatives A_x, B_y, and C_z.

    (c) Obtain the spherical-coordinate equivalent of $A_x + B_y + C_z$.

B24. Using a computer algebra system, derive the expression for the curl in spherical coordinates.

    (a) Start with the vector in Exercise B23, part (a). As in part (b) of that exercise, obtain the derivatives $A_y, A_z, B_x, B_z, C_x, C_y$.

    (b) In $\nabla \times \mathbf{f} = (C_y - B_z)\mathbf{i} + (A_z - C_x)\mathbf{j} + (B_x - A_y)\mathbf{k}$, replace the derivatives with their values from part (a) and replace the vectors $\mathbf{i}, \mathbf{j}, \mathbf{k}$ with their equivalents in terms of $\mathbf{e}_\rho, \mathbf{e}_\phi, \mathbf{e}_\theta$, thereby obtaining $\nabla \times \mathbf{F}$ in spherical coordinates.

B25. The laplacian in spherical coordinates can be derived with a computer algebra system as follows.

   (a) Instead of starting with the generic $f(x, y, z) = F(\rho(x, y, z), \phi(x, y, z), \theta(x, y, z))$, write $f(x, y, z) = F(\sqrt{x^2 + y^2 + z^2}, \arctan \frac{z}{\sqrt{x^2+y^2+z^2}}, \arccos \frac{y}{x})$ and compute $f_{xx} + f_{yy} + f_{zz}$.

   (b) Restore the variables ρ, ϕ, θ and simplify, obtaining the expression for the laplacian in spherical coordinates.

B26. Using the technique of Exercise 25, obtain the divergence in spherical coordinates.

   (a) Start with $\mathbf{F}$ as in Exercise 23, part (a), and obtain $\mathbf{f}$, as indicated. Instead of writing $A(x, y, z) = a(\rho(x, y, z), \phi(x, y, z), \theta(x, y, z))$, write $A(x, y, z) = a(\sqrt{x^2 + y^2 + z^2}, \arctan \frac{z}{\sqrt{x^2+y^2+z^2}}, \arccos \frac{y}{x})$. The component A contains the coefficients U, V, W from $\mathbf{F}$ and the variables ρ, ϕ, θ that are present explicitly, having been inherited from the change of basis vectors from $\mathbf{e}_\rho, \mathbf{e}_\phi, \mathbf{e}_\theta$, to $\mathbf{i}, \mathbf{j}, \mathbf{k}$. So also with B and C. Now compute $A_x + B_y + C_z$.

   (b) Restore the variables ρ, ϕ, θ and simplify, obtaining the expression for the divergence in spherical coordinates.

B27. Using the technique of Exercise 25, obtain the curl in spherical coordinates.

   (a) Proceed as in Exercise 26, part (a). Compute the derivatives $A_y, A_z, B_x, B_z, C_x, C_y$.

   (b) In $\nabla \times \mathbf{f} = (C_y - B_z)\mathbf{i} + (A_z - C_x)\mathbf{j} + (B_x - A_y)\mathbf{k}$, replace the derivatives with their values from part (a) and the vectors $\mathbf{i}, \mathbf{j}, \mathbf{k}$ with their equivalents in terms of $\mathbf{e}_\rho, \mathbf{e}_\phi, \mathbf{e}_\theta$, thereby obtaining $\nabla \times \mathbf{F}$ in spherical coordinates.

B28. The equations $u_k = u_k(x, y, z)$, $k = 1, 2, 3$, define an orthogonal coordinate system, if, for $\mathbf{R} = u_1\mathbf{i} + u_2\mathbf{j} + u_3\mathbf{k}$, the tangent vectors $\partial \mathbf{R}/\partial u_k$, $k = 1, 2, 3$, are mutually perpendicular. For such a system, define the scale factors $h_k = \|\partial \mathbf{R}/\partial u_k\|$ and obtain the unit basis vectors $\mathbf{e}_k = (1/h_k)(\partial \mathbf{R}/\partial u_k)$, $k = 1, 2, 3$. Then, as shown in [84],

   (i) $\nabla \phi(u_1, u_2, u_3) = \frac{\mathbf{e}_1}{h_1}\frac{\partial \phi}{\partial u_1} + \frac{\mathbf{e}_2}{h_2}\frac{\partial \phi}{\partial u_2} + \frac{\mathbf{e}_3}{h_3}\frac{\partial \phi}{\partial u_3}$

   (ii) $\nabla \cdot \mathbf{A}(u_1, u_2, u_3) =$
   $\frac{1}{h_1 h_2 h_3}\left[\frac{\partial}{\partial u_1}(A_1 h_2 h_3) + \frac{\partial}{\partial u_2}(A_2 h_1 h_3) + \frac{\partial}{\partial u_3}(A_3 h_1 h_2)\right]$

   (iii) $\nabla \times \mathbf{A}(u_1, u_2, u_3) = \frac{1}{h_1 h_2 h_3}\begin{vmatrix} h_1\mathbf{e}_1 & h_2\mathbf{e}_2 & h_3\mathbf{e}_3 \\ \frac{\partial}{\partial u_1} & \frac{\partial}{\partial u_2} & \frac{\partial}{\partial u_3} \\ A_1 h_1 & A_2 h_2 & A_3 h_3 \end{vmatrix}$

   (iv) $\nabla^2 \phi(u_1, u_2, u_3) =$
   $\frac{1}{h_1 h_2 h_3}\left[\frac{\partial}{\partial u_1}\left(\frac{h_2 h_3}{h_1}\frac{\partial \phi}{\partial u_1}\right) + \frac{\partial}{\partial u_2}\left(\frac{h_3 h_1}{h_2}\frac{\partial \phi}{\partial u_2}\right) + \frac{\partial}{\partial u_3}\left(\frac{h_1 h_2}{h_3}\frac{\partial \phi}{\partial u_3}\right)\right]$

   (a) Obtain the scale factors for the spherical coordinate system.

   (b) Show that (i) gives the gradient in spherical coordinates.

   (c) Show that (ii) gives the divergence in spherical coordinates.

   (d) Show that (iii) gives the curl in spherical coordinates.

   (e) Show that (iv) gives the laplacian in spherical coordinates.

B29. The *6-sphere* coordinate system is defined by the equations $x = \frac{u}{u^2+v^2+w^2}$, $y = \frac{v}{u^2+v^2+w^2}$, $z = \frac{w}{u^2+v^2+w^2}$.

   (a) From $\mathbf{R} = x(u, v, w)\mathbf{i} + y(u, v, w)\mathbf{j} + z(u, v, w)\mathbf{k}$, obtain the tangent vectors $\frac{\partial \mathbf{R}}{\partial u}, \frac{\partial \mathbf{R}}{\partial v}, \frac{\partial \mathbf{R}}{\partial w}$ and verify that the coordinate system is orthogonal.

   (b) Obtain as space curves, the coordinate curves defined by this system.

   (c) Obtain the scale factors for this coordinate system.

   (d) Using (i) in Exercise B28, obtain the gradient in these coordinates and check via a computer algebra system.

   (e) Using (ii) in Exercise B28, obtain the divergence in these coordinates and check via a computer algebra system.

   (f) Using (iii) in Exercise B28, obtain the curl in these coordinates and check via a computer algebra system.

   (g) Using (iv) in Exercise B28, obtain the laplacian in these coordinates and check via a computer algebra system.

B30. One definition of the *paraboloidal* coordinate system is given by the equations $x = uv \cos w$, $y = uv \sin w$, $z = \frac{1}{2}(u^2 - v^2)$. Repeat the steps of Exercise B29 for these coordinates.

B31. The *bispherical* coordinate system is given by the equations $x = \frac{1}{d}\sin u \cos w$, $y = \frac{1}{d}\sin u \sin w$, $z = \frac{1}{d}\sinh v$, where $d = \cosh v - \cos u$. Repeat the steps of Exercise B29 for these coordinates.

B32. For the 6-sphere coordinate system of Exercise B29:

   (a) Obtain the equations $u = u(x, y, z)$, $v = v(x, y, z)$, $w = w(x, y, z)$.

   (b) Obtain graphs of the coordinate surfaces, $u = $ constant, $v = $ constant, $w = $ constant.

   (c) Use the built-in functionality of your favorite computer algebra system to draw these coordinate surfaces without the need for the explicit formulas found in part (a).

B33. Obtain graphically, the coordinate surfaces for the paraboloidal coordinate system in Exercise B30.

B34. Obtain graphically, the coordinate surfaces for the bispherical coordinate system in Exercise B31.

B35. Repeat the steps of Exercises B2–11 for the 6-sphere coordinate system and the data in Exercise B2.

B36. Repeat the steps of Exercises B2–11 for the paraboloidal coordinate system and the data in Exercise B3.

B37. Repeat the steps of Exercises B2–11 for the bispherical coordinate system and the data in Exercise B4 but with $P = (\frac{1+\sqrt{2}}{2}, \frac{\sqrt{3}}{2}(1 + \sqrt{2}), 0)$.

Chapter Review

1. Let Φ, given by $u = 7x - 3y$, $v = 5x + 2y$, define a coordinate change on the xy-plane. Demonstrate why, in a double integral, $dx\,dy$ must be replaced with $\left|\frac{\partial(x,y)}{\partial(u,v)}\right| du\,dv$ when invoking this change of coordinates. This demonstration might include the following steps.

   (a) Obtain $x = x(u, v)$ and $y = y(u, v)$.

   (b) Find the coordinates of the points in the xy-plane to which Φ^{-1} maps $(0, 0)$, $(1, 0$, $(1, 1)$, $(0, 1)$.

   (c) Find the area of the parallelogram in the xy-plane formed by the points determined in part (b).

   (d) Find the value of the Jacobian $\frac{\partial(x,y)}{\partial(u,v)}$ and relate the area found in part (c) to the value of the Jacobian found in part (d).

2. Explain the difference between mapping the xy-plane to the uv-plane and changing coordinates from x and y to u and v.

3. Use an example to illustrate that the Jacobians $\frac{\partial(x,y)}{\partial(u,v)}$ and $\frac{\partial(u,v)}{\partial(x,y)}$ are reciprocals.

4. Use an example to illustrate that the Jacobian matrices $\mathbf{J}_\Phi$ and $\mathbf{J}_{\Phi^{-1}}$ are multiplicative inverses of each other.

5. Compute the Jacobian $\frac{\partial(x,y,z)}{\partial(u,v,w)}$ for spherical coordinates. Take $(u, v, w) = (\rho, \phi, \theta)$ in a mathematical context and (ρ, θ, ϕ) in an applied context, but in either case, the second variable should refer to the angle between the z-axis and the radius vector.

6. Let $u(x, y) = x^2 y^3$.

   (a) Obtain ∇u in Cartesian coordinates.

   (b) Transform ∇u to a vector in polar coordinates, being sure to change basis vectors from $\{\mathbf{i}, \mathbf{j}\}$ to the unit vectors $\{\hat{\mathbf{e}}_r, \hat{\mathbf{e}}_\theta\}$.

   (c) Obtain $U(r, \theta) = u(x(r, \theta), y(r, \theta))$ and, from it, the gradient in polar coordinates. Compare to the result in part (b).

   (d) Obtain $\nabla^2 u$, the laplacian of u in Cartesian coordinates.

   (e) Transform $\nabla^2 u$ to polar coordinates.

   (f) Obtain the laplacian of $U(r, \theta)$ directly in polar coordinates, and compare to the result in part (e).

7. Let $\mathbf{v} = \frac{x}{y}\mathbf{i} + \frac{y}{x}\mathbf{j}$.

   (a) Obtain $\nabla \cdot \mathbf{v}$, the divergence of $\mathbf{v}$, in Cartesian coordinates.

   (b) Transform the divergence to polar coordinates.

   (c) Transform $\mathbf{v}$ to $\mathbf{V}$, its equivalent in polar coordinates.

   (d) Obtain the divergence of $\mathbf{V}$ directly in polar coordinates, and compare to the result in part (b).

8. Let $u(x, y, z) = xyz$.

   (a) Obtain ∇u in cylindrical coordinates.

   (b) Transform ∇u to a vector in cylindrical coordinates, being sure to change basis vectors from $\{\mathbf{i}, \mathbf{j}, \mathbf{k}\}$ to the unit vectors $\{\hat{\mathbf{e}}_r, \hat{\mathbf{e}}_\theta, \hat{\mathbf{e}}_z\}$.

   (c) Obtain $U(r, \theta, z) = u(x(r, \theta), y(r, \theta), z)$, and from it, the gradient in cylindrical coordinates. Compare to the result in part (b).

   (d) Obtain $\nabla^2 u$, the laplacian of u in Cartesian coordinates.

   (e) Transform $\nabla^2 u$ to cylindrical coordinates.

   (f) Obtain the laplacian of $U(r, \theta)$ directly in cylindrical coordinates, and compare to the result in part (e).

9. Let $\mathbf{v} = xy\mathbf{i} + yz\mathbf{j} + zx\mathbf{k}$.

   (a) Obtain $\nabla \cdot \mathbf{v}$, the divergence of $\mathbf{v}$, in Cartesian coordinates.

   (b) Transform the divergence to cylindrical coordinates.

   (c) Transform $\mathbf{v}$ to $\mathbf{V}$, its equivalent in cylindrical coordinates.

   (d) Obtain the divergence of $\mathbf{V}$ directly in cylindrical coordinates, and compare to the result in part (b).

   (e) Obtain $\nabla \times \mathbf{v}$, the curl in Cartesian coordinates.

   (f) Transform the curl to a vector in cylindrical coordinates.

   (g) Obtain the curl of $\mathbf{V}$ in cylindrical coordinates and compare to the result in part (f).

10. Let $u(x, y, z) = xyz$.

    (a) Obtain ∇u in spherical coordinates.

    (b) Transform ∇u to a vector in spherical coordinates, being sure to change basis vectors from $\{\mathbf{i}, \mathbf{j}, \mathbf{k}\}$ to the unit vectors $\{\hat{\mathbf{e}}_\rho, \hat{\mathbf{e}}_\phi, \hat{\mathbf{e}}_\theta\}$, where ϕ is the angle from the z-axis to the radius vector.

    (c) Obtain $U(\rho, \phi, \theta) = u(x(\rho, \phi, \theta), y(\rho, \phi, \theta), z(\rho, \phi, \theta))$ and, from it, the gradient in spherical coordinates. Compare to the result in part (b).

    (d) Obtain $\nabla^2 u$, the laplacian of u in spherical coordinates.

    (e) Transform $\nabla^2 u$ to spherical coordinates.

    (f) Obtain the laplacian of $U(r, \theta)$ directly in spherical coordinates, and compare to the result in part (e).

11. Let $\mathbf{v} = xy\mathbf{i} + yz\mathbf{j} + zx\mathbf{k}$.

    (a) Obtain $\nabla \cdot \mathbf{v}$, the divergence of $\mathbf{v}$, in Cartesian coordinates.

    (b) Transform the divergence to spherical coordinates.

    (c) Transform $\mathbf{v}$ to $\mathbf{V}$, its equivalent in spherical coordinates.

    (d) Obtain the divergence of $\mathbf{V}$ directly in spherical coordinates, and compare to the result in part (b).

    (e) Obtain $\nabla \times \mathbf{v}$, the curl in Cartesian coordinates.

    (f) Transform the curl to a vector in spherical coordinates.

    (g) Obtain the curl of $\mathbf{V}$ in spherical coordinates and compare to the result in part (f).

Chapter 23

Miscellaneous Results

INTRODUCTION   This chapter contains a miscellaneous mix of four ideas. First, we see *Gauss' Theorem,* which appears repeatedly in courses on electricity and magnetism. Then, we show how to compute the surface integral for a surface given parametrically. Next, we derive the *equation of continuity* and gives some of its consequences; finally, we state, illustrate, and derive *Green's first and second identities.*

23.1 Gauss' Theorem

EXAMPLE 23.1   Let S_1 be a closed cylinder of radius 1 and height 2, with axis along the z-axis and situated between $z = 1$ and $z = 3$. Let S_2 be the same cylinder but situated between $z = -1$ and $z = 1$. The origin is not inside S_1 but is inside S_2. We will show that the surface integral

$$\iint_S \mathbf{N} \cdot \left(\frac{\mathbf{r}}{r^3}\right) d\sigma \tag{23.1}$$

is zero for S_1 and 4π for S_2.

A sketch of S_1, along with several outward normal vectors, is shown in Figure 23.1. To construct the integrand, first define the vector

$$\mathbf{F} = \frac{\mathbf{r}}{r^3} = \left(x^2 + y^2 + z^2\right)^{-3/2} (x\mathbf{i} + y\mathbf{j} + z\mathbf{k}) \tag{23.2}$$

The unit normal on top of the cylinder is $\mathbf{k}$ and on the bottom is $-\mathbf{k}$. On the top and bottom of the cylinder, $d\sigma = dx\, dy$. The sum of $\mathbf{F} \cdot \mathbf{k}\, d\sigma$ for $z = 3$ and $z = 1$ is

$$\left(\frac{3}{\left(x^2 + y^2 + 9\right)^{3/2}} - \frac{1}{\left(x^2 + y^2 + 1\right)^{3/2}}\right) dx\, dy \tag{23.3}$$

The combined surface integral for the top and bottom of the cylinder is taken over the unit disk. Hence, change to polar coordinates to obtain

$$\int_0^{2\pi} \int_0^1 \frac{\left(3\left(1 + r^2\right)^{3/2} - \left(9 + r^2\right)^{3/2}\right)}{\left(9 + r^2\right)^{3/2} \left(1 + r^2\right)^{3/2}} r\, dr\, dt = \pi\sqrt{2}\left(1 - \frac{3}{\sqrt{5}}\right) \tag{23.4}$$

The curved wall of the cylinder is described implicitly by $g(x, y, z) \equiv x^2 + y^2 - 1 = 0$, which has no projection onto the xy-plane. Therefore, project to the xz-plane and solve for

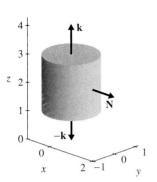

FIGURE 23.1   The surface S_1 in Example 23.1

541

$y = y(x, z) = \pm\sqrt{1 - x^2}$, giving

$$d\sigma = \sqrt{1 + y_x^2 + y_z^2}\, dx\, dz = \frac{dx\, dz}{\sqrt{1 - x^2}} \tag{23.5}$$

On the curved wall of the cylinder the unit normal vector $\mathbf{N} = x\mathbf{i} + y\mathbf{j}$ can be obtained from the gradient vector $\nabla g = 2(x\mathbf{i} + y\mathbf{j})$, so

$$\mathbf{F} \cdot \mathbf{N}\, d\sigma = \frac{x^2 + y^2}{\sqrt{1 - x^2}\,\left(x^2 + y^2 + z^2\right)^{3/2}}\, dx\, dz \tag{23.6}$$

arises for both projections onto the rectangle $R_{xz} = \{-1 \le x \le 1, 1 \le z \le 3\}$. Since $y(x, z)$ appears only as y^2, we can double the contribution of one projection, obtaining

$$2\int_1^3 \int_{-1}^1 \frac{1}{\sqrt{1 - x^2}\,\left(1 + z^2\right)^{3/2}}\, dx\, dz = \pi\sqrt{2}\left(\frac{3}{\sqrt{5}} - 1\right) \tag{23.7}$$

and the total contribution of all the surface integrals is then 0.

Next, slide the cylinder down two units so that it lies between $z = -1$ and $z = 1$. Then, the unit normals on top, bottom, and side of S_2 are still $\mathbf{k}$, $-\mathbf{k}$, and $\mathbf{N}$, respectively. The functions $g(x, y, z)$ and $y = y(x, z)$ are also the same, as is $d\sigma$ on the curved wall of the cylinder. Thus, the combined integrands for the top and bottom surfaces will be $2(1 + x^2 + y^2)^{-3/2}$; and the surface integral, in polar coordinates, will be

$$\int_0^{2\pi} \int_0^1 \frac{2}{\left(1 + r^2\right)^{3/2}}\, r\, dr\, dt = 2\pi(2 - \sqrt{2}) \tag{23.8}$$

The integrand for the surface integral on the curved wall of the cylinder, namely, $\mathbf{F} \cdot \mathbf{N}\, d\sigma$, is again the same, so the surface integral itself is

$$2\int_{-1}^1 \int_{-1}^1 \frac{1}{\sqrt{1 - x^2}\,\left(1 + z^2\right)^{3/2}}\, dx\, dz = 2\pi\sqrt{2} \tag{23.9}$$

Hence, the total of the surface integrals is the predicted 4π and we have shown that (23.1) evaluates to 0 for a surface S_1 not enclosing the origin, and to 4π for a surface S_2 which does enclose the origin. ❖

This result is true in general and is known as *Gauss' Theorem*. This should not be confused with the *divergence theorem*, sometimes called the *Gauss' divergence theorem*, which is used in the proof of Gauss' Theorem, but is not equivalent to it.

THEOREM 23.1   GAUSS' THEOREM

Let S be a closed surface with unit outward normal $\mathbf{N}$ and surface area element $d\sigma$. Let $\mathbf{r} = x\mathbf{i} + y\mathbf{j} + z\mathbf{k}$ be the radius vector from an origin O to the point (x, y, z), and let $r = \sqrt{x^2 + y^2 + z^2}$ be the scalar magnitude of the radius vector. Then

$$\iint_S \mathbf{N} \cdot \left(\frac{\mathbf{r}}{r^3}\right) d\sigma = \begin{cases} 0 & \text{if } O \text{ lies outside } S \\ 4\pi & \text{if } O \text{ lies inside } S \end{cases}$$

PROOF OF GAUSS' THEOREM   **Case (1)** Let O lie outside the surface S, which encloses the volume V. As long as the tip of the radius vector $\mathbf{r}$ lies outside S, r cannot be zero in

V, so the integrand in the volume integral remains finite and we can apply the divergence theorem $\iiint_V \nabla \cdot \mathbf{F}\, dv = \iint_S \mathbf{F} \cdot \mathbf{N}\, d\sigma$, with $\mathbf{F}$ given by (23.2), for which $\nabla \cdot \mathbf{F} = 0$, as a routine calculation shows. Hence, we get $\iint_S \mathbf{N} \cdot \left(\frac{\mathbf{r}}{r^3}\right) d\sigma = \iiint_V 0\, dv = 0$.

Case (2) Let O lie inside S so that there is one point inside S, namely O itself, at which $r = 0$. Because of the singularity so introduced, the divergence theorem cannot be applied to V, the region inside S. However, if O is surrounded by a small sphere s of radius a, the region inside S but outside s is one for which r isn't zero. The boundary of this closed region V' is $S + s$. Apply the divergence theorem to the closed surface $S + s$ that encloses the volume V' and obtain

$$\iint_{S+s} \mathbf{N} \cdot \left(\frac{\mathbf{r}}{r^3}\right) d\sigma = \iint_S \mathbf{N} \cdot \left(\frac{\mathbf{r}}{r^3}\right) d\sigma + \iint_s \mathbf{N} \cdot \left(\frac{\mathbf{r}}{r^3}\right) d\sigma = \iiint_V \nabla \cdot \left(\frac{\mathbf{r}}{r^3}\right) dv = 0$$

The surface integral over $S + s$ is zero because V' does not contain O. The significant conclusion is that

$$\iint_S \mathbf{N} \cdot \left(\frac{\mathbf{r}}{r^3}\right) d\sigma = -\iint_s \mathbf{N} \cdot \left(\frac{\mathbf{r}}{r^3}\right) d\sigma$$

The surface integral over S can't be computed because S is a general surface, but that surface integral is equivalent to one over a small sphere of radius a. Now that surface integral can be computed.

The origin O from which the vector $\mathbf{r}$ emanates is now at the center of the small sphere s. Describe this small sphere implicitly with $f(x, y, z) \equiv x^2 + y^2 + z^2 - a^2 = 0$. To perform the surface integration on this small sphere, a unit normal vector $\mathbf{N}$, a radius vector $\mathbf{r}$ with tip on the surface of the sphere s, and a surface area element $d\sigma$ are all needed. The unit normal vector is obtained by normalizing the gradient vector computed for f. However, this normal vector must point outward on the surface $S + s$, which is "outward" on S but "inward" on the small sphere s! (This calculation is much like Example 21.9 in Section 21.3.) The gradient vector for the small sphere is $\nabla f = 2(x\mathbf{i} + y\mathbf{j} + z\mathbf{k})$, so an outward unit normal on s, pointing toward the void inside S is $N = -\frac{1}{a}(x\mathbf{i} + y\mathbf{j} + z\mathbf{k})$. As for Case (1), both the surface and volume integrals require use of the vector $\mathbf{F}$ given by (23.2) and $\mathbf{F} \cdot \mathbf{N} = -\frac{1}{a^2}$ on the surface of the small sphere s. Hence, the surface integral on S becomes

$$\iint_S \mathbf{N} \cdot \left(\frac{\mathbf{r}}{r^3}\right) d\sigma = -\iint_s \mathbf{N} \cdot \left(\frac{\mathbf{r}}{r^3}\right) d\sigma = -\iint_s -\frac{1}{a^2}\, d\sigma$$

$$= \frac{1}{a^2} \iint_s d\sigma = \frac{1}{a^2} 4\pi a^2 = 4\pi$$

THEOREM 23.2   GAUSS' THEOREM IN THE PLANE

Let C be a simple closed plane curve with unit outward normal $\mathbf{N}$ and arc-length element ds. Let $\mathbf{r} = x\mathbf{i} + y\mathbf{j}$ be the radius vector from an origin O to the point (x, y). Then $r = \sqrt{x^2 + y^2}$ and

$$\oint_C \mathbf{N} \cdot \left(\frac{\mathbf{r}}{r^2}\right) ds = \begin{cases} 0 & \text{if } O \text{ lies outside } C \\ 2\pi & \text{if } O \text{ lies inside } C \end{cases} \tag{23.10}$$

The two-dimensional version of Gauss' Theorem differs slightly from the three-dimensional case. In the plane, both the integrand and the computed value differ and, instead of a surface integral, there is a line integral.

Writing the vector $\frac{\mathbf{r}}{r^2}$ as

$$\mathbf{F} = \frac{\mathbf{r}}{r^2} = f\mathbf{i} + g\mathbf{j} = \frac{x}{x^2 + y^2}\mathbf{i} + \frac{y}{x^2 + y^2}\mathbf{j}$$

and noting that the line integral in (23.10) is a flux integral, it can be written in the more familiar form

$$\oint_C \mathbf{N} \cdot \left(\frac{\mathbf{r}}{r^2}\right) ds = \oint_C f\, dy - g\, dx \qquad (23.11)$$

(See, e.g., Section 19.2, where the flux integral in the plane was first developed.)

The proof parallels exactly the proof seen for the three-dimensional case, provided the divergence form of Green's Theorem is used instead of the divergence theorem. The proof is left as an exercise, but the following example is provided.

EXAMPLE 23.2   Let C be a circle of radius 1, with center at $(0, 2)$ on the y-axis. If the origin O is taken as $(0, 0)$, then O lies outside of C. Parametrize C via the obvious $x = \cos p$, $y = 2 + \sin p$ so that (23.11) becomes

$$\oint_C f\, dy - g\, dx = \int_0^{2\pi} \frac{2\sin p + 1}{5 + 4\sin p}\, dp = 0 \qquad (23.12)$$

Now, move the circle so it encloses O. For example, let the circle have its center at the origin so that it is parametrized by $x = \cos p$, $y = \sin p$, in which case (23.11) becomes $\oint_C f\, dy - g\, dx = \int_0^{2\pi} 1\, dp = 2\pi$. ❖

EXERCISES 23.1–Part A

A1. Verify that the sum of $\mathbf{F} \cdot \mathbf{N}\, d\sigma$ on the top and bottom surfaces of S_1 in Example 23.1 is given by (23.3).

A2. Verify that when evaluated on the top and bottom surfaces of S_1 in Example 23.1, (23.1) becomes (23.4).

A3. Verify that (23.5) gives $d\sigma$ for the curved wall of S_1 in Example 23.1.

A4. Verify that (23.6) gives $\mathbf{F} \cdot \mathbf{N}\, d\sigma$ for the curved wall of S_1 in Example 23.1.

A5. Verify the integral and its value in (23.7).

A6. Verify the integral and its value in (23.8).

A7. Verify the integral and its value in (23.9).

A8. Verify the integral and its value in (23.12).

A9. Prove Gauss' Theorem in the plane.

EXERCISES 23.1–Part B

B1. If the origin is $O = (0, 0, 0)$, verify Gauss' Theorem for the cube

(a) $1 \le x \le 3, 1 \le y \le 3, 1 \le z \le 3$

(b) $|x| \le 1, |y| \le 1, |z| \le 1$

B2. If the origin is $O = (0, 0, 0)$, verify Gauss' Theorem for the cylindrical "can" bounded by

(a) $x^2 + y^2 = 1, z = 1, z = 3$ (b) $x^2 + y^2 = 1, z = \pm 3$

B3. If the origin is $O = (0, 0, 0)$, verify Gauss' Theorem for the closed surface bounded by

(a) $z = 1 - x^2 - y^2$ and $z = \frac{1}{2}$

(b) $z = 1 - x^2 - y^2$ and $z = x^2 + y^2 - 1$

B4. If the origin is $O = (0, 0)$, verify Gauss' Theorem in the plane for the closed curve C determined by

(a) the parabola $y = 1 + x^2$ and the line $y = 2x + 9$.

(b) the parabola $y = x^2 - 1$ and the line $y = 2x + 7$.

B5. If the origin is $O = (0, 0)$, verify Gauss' Theorem in the plane for the closed curve C determined by

(a) the parabolas $y = x^2 - 4$ and $y = \frac{1}{4}x^2 - 1$.

(b) the parabolas $y = x^2 - 4$ and $y = \frac{1}{3}x^2 + 2$.

B6. If the origin is $O = (0, 0)$, verify Gauss' Theorem in the plane for C, the ellipse

(a) $x = 3 \cos \theta$, $y = 2 \sin \theta$ (b) $x = 3 \cos \theta$, $y = 5 + 2 \sin \theta$

B7. A key step in the derivation of Gauss' Theorem is discovering that $\mathbf{F} = \frac{\mathbf{r}}{r^3}$ and $\mathbf{F} = \frac{\mathbf{r}}{r^2}$ are solenoidal fields in three dimensions and two dimensions, respectively.

(a) Show that in three dimensions, the only value of n for which $\mathbf{r}/r^n$ is solenoidal is $n = 3$.

(b) Show that in two dimensions, the only value of n for which $\mathbf{r}/r^n$ is solenoidal is $n = 2$.

B8. Show that in three dimensions, the only solenoidal field of the form $\mathbf{F} = \mathbf{r} f(r)$ is $\mathbf{F} = c \frac{\mathbf{r}}{r^3}$.

B9. Show that in two dimensions, the only solenoidal field of the form $\mathbf{F} = \mathbf{r} f(r)$ is $\mathbf{F} = c \frac{\mathbf{r}}{r^2}$.

23.2 Surface Area for Parametrically Given Surfaces

Explicitly Given Surface

Recall that for the surface given explicitly by $z = z(x, y)$ the element of surface area is given by

$$d\sigma = \sqrt{1 + z_x^2 + z_y^2}\, dx\, dy \tag{23.13}$$

obtained as the area of the parallelogram spanned by infinitesimal vectors tangent to the two coordinate curves

$$\mathbf{R}_x = x\mathbf{i} + \beta\mathbf{j} + z(x, \beta)\mathbf{k} \quad \text{and} \quad \mathbf{R}_y = \alpha\mathbf{i} + y\mathbf{j} + z(\alpha, y)\mathbf{k}$$

Thus, $d\sigma = \|\mathbf{T}_x \times \mathbf{T}_y\|$, where $\mathbf{T}_x = (\mathbf{i} + z_x\mathbf{k})\, dx$ and $\mathbf{T}_y = (\mathbf{j} + z_y\mathbf{k})\, dy$.

Implicitly Given Surface

If the surface $z = z(x, y)$ is defined implicitly by $f(x, y, z) = 0$, the derivatives

$$z_x = -\frac{f_x}{f_z} \quad \text{and} \quad z_y = -\frac{f_y}{f_z} \tag{23.14}$$

can be computed by implicit differentiation. Substituting (23.14) into (23.13) yields

$$d\sigma = \sqrt{\frac{f_x^2 + f_y^2 + f_z^2}{f_z^2}}\, dx\, dy = \frac{\|\nabla f\|}{|f_z|}\, dx\, dy \tag{23.15}$$

Parametrically Given Surface

Surface Area Element   If the surface is given parametrically in terms of the two parameters u and v by $x = x(u, v)$, $y = y(u, v)$, $z = z(u, v)$, then the element of surface area is given by $d\sigma = \sqrt{J_1^2 + J_2^2 + J_3^2}\, du\, dv$, where

$$J_1 = \frac{\partial(y, z)}{\partial(u, v)} = \begin{vmatrix} y_u & y_v \\ z_u & z_v \end{vmatrix} = y_u z_v - y_v z_u$$

$$J_2 = \frac{\partial(z, x)}{\partial(u, v)} = \begin{vmatrix} z_u & z_v \\ x_u & x_v \end{vmatrix} = z_u x_v - z_v x_u$$

$$J_3 = \frac{\partial(x, y)}{\partial(u, v)} = \begin{vmatrix} x_u & x_v \\ y_u & y_v \end{vmatrix} = x_u y_v - x_v y_u$$

are Jacobians.

EXAMPLE 23.3   Parametric equations for the surface of a sphere of radius a and the resulting Jacobians J_k, $k = 1, 2, 3$, are

$$x = a\cos\theta\sin\phi \qquad J_1 = -a^2\cos\theta\sin^2\phi$$
$$y = a\sin\theta\sin\phi \quad\Rightarrow\quad J_2 = -a^2\sin^2\phi\sin\theta \qquad (23.16)$$
$$z = a\cos\phi \qquad J_3 = -a^2\sin\phi\cos\phi$$

Then, $d\sigma = a^2\sin\phi\,d\theta\,d\phi$ and the surface area of the sphere is given by $\int_0^\pi\int_0^{2\pi} a^2\sin\phi\,d\theta\,d\phi = 4\pi a^2$.

It is also instructive to obtain from first principles, the expression for $d\sigma$ in this example. Thus, form coordinate curves on the sphere, differentiate to form tangent vectors, and find $d\sigma$ as the length of the cross-product of these tangent vectors.

On the surface of the sphere, let a curve for which $\phi = \beta$ be $\mathbf{R}_\theta$ and let a curve for which $\theta = \alpha$ be $\mathbf{R}_\phi$, as listed in Table 23.1. Differentiating with respect to the curves' parameters yields the tangent vectors $\mathbf{T}_\theta$, and $\mathbf{T}_\phi$, listed in Table 23.1 for the generic point (θ, ϕ). The area spanned by these tangent vectors is $\|\mathbf{T}_\theta \times \mathbf{T}_\phi\| = a^2\sin\phi$, from which the expression for $d\sigma$ is obtained. ❖

$\mathbf{R}_\theta = a(\cos\theta\sin\beta\mathbf{i} + \sin\theta\sin\beta\mathbf{j} + \cos\beta\mathbf{k})$ $\qquad \mathbf{T}_\theta = a(-\sin\theta\sin\phi\mathbf{i} + \cos\theta\sin\phi\mathbf{j})$

$\mathbf{R}_\phi = a(\cos\alpha\sin\phi\mathbf{i} + \sin\alpha\sin\phi\mathbf{j} + \cos\phi\mathbf{k})$ $\qquad \mathbf{T}_\phi = a(\cos\theta\cos\phi\mathbf{i} + \sin\theta\cos\phi\mathbf{j} - \sin\phi\mathbf{k})$

TABLE 23.1   Coordinate curves and their tangent vectors on a sphere

GENERAL DERIVATION   Using the same technique of forming coordinate curves, differentiating to form tangent vectors, and computing the area of the parallelogram so spanned, we can derive the general result for $d\sigma$ for the surface given parametrically by $x = x(u, v)$, $y = y(u, v)$, $z = z(u, v)$. The radius vector to a point on the surface is then

$$\mathbf{R} = x(u, v)\mathbf{i} + y(u, v)\mathbf{j} + z(u, v)\mathbf{k}$$

On the surface, let

$$\mathbf{R}_u = x(u, \beta)\mathbf{i} + y(u, \beta)\mathbf{j} + z(u, \beta)\mathbf{k} \quad\text{and}\quad \mathbf{R}_v = x(\alpha, v)\mathbf{i} + y(\alpha, v)\mathbf{j} + z(\alpha, v)\mathbf{k}$$

be curves for which $v = \beta$ and $u = \alpha$, respectively. Differentiate with respect to u and v to form vectors tangent to these curves, obtaining

$$\mathbf{T}_u = x_u\mathbf{i} + y_u\mathbf{j} + z_u\mathbf{k} \quad\text{and}\quad \mathbf{T}_v = x_v\mathbf{i} + y_v\mathbf{j} + z_v\mathbf{k}$$

The cross-product of these two tangent vectors must be a normal vector $\mathbf{N} = \mathbf{T}_u \times \mathbf{T}_v$ whose value we obtain from a row-one expansion of the determinant

$$\mathbf{N} = \begin{vmatrix} \mathbf{i} & \mathbf{j} & \mathbf{k} \\ x_u & y_u & z_u \\ x_v & y_v & z_v \end{vmatrix}$$

This gives

$$\mathbf{N} = \begin{vmatrix} y_u & z_u \\ y_v & z_v \end{vmatrix} \mathbf{i} - \begin{vmatrix} x_u & z_u \\ x_v & z_v \end{vmatrix} \mathbf{j} + \begin{vmatrix} x_u & y_u \\ x_v & y_v \end{vmatrix} \mathbf{k}$$

$$= \begin{vmatrix} y_u & y_v \\ z_u & z_v \end{vmatrix} \mathbf{i} - \begin{vmatrix} x_u & x_v \\ z_u & z_v \end{vmatrix} \mathbf{j} + \begin{vmatrix} x_u & x_v \\ y_u & y_v \end{vmatrix} \mathbf{k}$$

$$= \begin{vmatrix} y_u & y_v \\ z_u & z_v \end{vmatrix} \mathbf{i} + \begin{vmatrix} z_u & z_v \\ x_u & x_v \end{vmatrix} \mathbf{j} + \begin{vmatrix} x_u & x_v \\ y_u & y_v \end{vmatrix} \mathbf{k}$$

$$= \frac{\partial(y, z)}{\partial(u, v)} \mathbf{i} + \frac{\partial(z, x)}{\partial(u, v)} \mathbf{j} + \frac{\partial(x, y)}{\partial(u, v)} \mathbf{k} = J_1 \mathbf{i} + J_2 \mathbf{j} + J_3 \mathbf{k}$$

with length $\|\mathbf{N}\| = \sqrt{J_1^2 + J_2^2 + J_3^2}$.

FLUX THROUGH PARAMETRICALLY GIVEN SURFACE   The flux of the vector field

$$\mathbf{F} = P(x, y, z)\mathbf{i} + Q(x, y, z)\mathbf{j} + R(x, y, z)\mathbf{k}$$

through the surface S is

$$\iint_{R_{yz}} P \, dy \, dz + \iint_{R_{zx}} Q \, dz \, dx + \iint_{R_{xy}} R \, dx \, dy \qquad \text{(23.17)}$$

In the leftmost integral, the region R_{yz} is the projection of the surface S onto the yz-plane and the integration, merely a double integral (not a surface integral), is carried out over R_{yz}. In the middle integral, the region R_{zx} is the projection of the surface S onto the zx-plane and the integration, merely a double integral (not a surface integral), is carried out over R_{zx}. In the rightmost integral, the region R_{xy} is the projection of the surface S onto the xy-plane and the integration, merely a double integral (not a surface integral), is carried out over R_{xy}.

As strange as this seems, this form of the flux integral is our best analog for the work integral. Recall that work is defined as the line integral of the component of the field tangential to the path of integration, that is, as

$$\int_C \mathbf{F} \cdot \mathbf{T} \, ds = \int_C \mathbf{F} \cdot d\mathbf{r} = \int_C f \, dx + g \, dy + h \, dz$$

The flux, however, is the surface integral of the normal component of the field, that is, $\iint_S \mathbf{F} \cdot \mathbf{N} \, d\sigma$. There is no perfect analog to $f \, dx + g \, dy + h \, dz$ for the flux integral, but (23.17) is close to being $\mathbf{F}$ dotted with something.

We conclude this discussion by showing how this formula can be obtained from the expression for flux through a parametrically given surface. But first, an example.

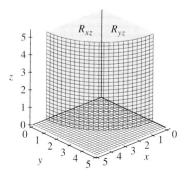

FIGURE 23.2   Example 23.4: The cylindrical surface S and its projections onto the xy-plane and yz-plane

EXAMPLE 23.4   Find the flux of the field

$$\mathbf{F} = P\mathbf{i} + Q\mathbf{j} + R\mathbf{k} = z\mathbf{i} + x\mathbf{j} - 3y^2 z\mathbf{k}$$

through the surface S, the first octant portion of the cylinder $g(x, y, z) \equiv x^2 + y^2 - 16 = 0$ lying between $z = 0$ and $z = 5$, shown in Figure 23.2. The projections of S onto the xz-plane and the yz-plane are rectangles. There is no projection of S onto the xy-plane.

The Old Way   Compute the flux through S by integrating $\mathbf{F} \cdot \mathbf{N}\,d\sigma$ on S. From $\nabla g = 2(x\mathbf{i} + y\mathbf{j})$, obtain $N = \frac{1}{4}(x\mathbf{i} + y\mathbf{j})$, a unit normal field on S. Since the surface S is parallel to the z-axis, $g(x, y, z) = 0$ cannot be solved for $z = z(x, y)$. Instead, solve for $y = y(x, z) = \sqrt{16 - x^2}$. The surface area element is $d\sigma = \frac{4\,dx\,dz}{\sqrt{16-x^2}}$, so the flux is

$$\int_0^4 \int_0^5 \left(\frac{xz}{\sqrt{16 - x^2}} + x \right) dz\,dx = 90 \qquad (23.18)$$

The New Way   Now, compute the flux by (23.17). Since S has no projection onto the xy-plane, the rightmost integral vanishes. Hence, the flux is

$$\int_0^5 \int_0^4 z\,dy\,dz + \int_0^5 \int_0^4 x\,dx\,dz = 90 \qquad \text{❖}$$

The General Case   In the general case, the surface S is given parametrically by $x = x(u, v)$, $y = y(u, v)$, $z = z(u, v)$; the field is $\mathbf{F} = P(x, y, z)\mathbf{i} + Q(x, y, z)\mathbf{j} + R(x, y, z)\mathbf{k}$; and a normal to S is $J_1\mathbf{i} + J_2\mathbf{j} + J_3\mathbf{k}$, which comes from $\mathbf{n} = -z_x\mathbf{i} - z_y\mathbf{j} + \mathbf{k}$ and

$$z_x = -\frac{\frac{\partial(y,z)}{\partial(u,v)}}{\frac{\partial(x,y)}{\partial(u,v)}} = -\frac{J_1}{J_3} \quad \text{and} \quad z_y = -\frac{\frac{\partial(z,x)}{\partial(u,v)}}{\frac{\partial(x,y)}{\partial(u,v)}} = -\frac{J_2}{J_3} \qquad (23.19)$$

so that we get $\mathbf{n} = (\frac{J_1}{J_3})\mathbf{i} + (\frac{J_2}{J_3})\mathbf{j} + \mathbf{k}$ or $J_1\mathbf{i} + J_2\mathbf{j} + J_3\mathbf{k}$. A *unit* normal, and the integrand of the flux integral become, respectively,

$$\mathbf{N} = \frac{J_1\mathbf{i} + J_2\mathbf{j} + J_3\mathbf{k}}{\sqrt{J_1^2 + J_2^2 + J_3^2}} \quad \text{and} \quad \mathbf{F} \cdot \mathbf{N}\,d\sigma = \frac{PJ_1 + QJ_2 + RJ_3}{\sqrt{J_1^2 + J_2^2 + J_3^2}} \sqrt{J_1^2 + J_2^2 + J_3^2}\,du\,dv$$

That gives the flux integral as

$$\iint_S \mathbf{F} \cdot \mathbf{N}\,d\sigma = \iint_{R_{uv}} [PJ_1 + QJ_2 + RJ_3]\,du\,dv$$

$$= \iint_{R_{uv}} P\frac{\partial(y, z)}{\partial(u, v)}\,du\,dv + \iint_{R_{uv}} Q\frac{\partial(z, x)}{\partial(u, v)}\,du\,dv + \iint_{R_{uv}} R\frac{\partial(x, y)}{\partial(u, v)}\,du\,dv$$

$$= \iint_{R_{yz}} P\,dy\,dz + \iint_{R_{zx}} Q\,dz\,dx + \iint_{R_{xy}} R\,dx\,dy$$

$$= \iint_S P\,dy\,dz + Q\,dz\,dx + R\,dx\,dy$$

The third equality follows from the second by recognizing the formula for changing variables in a double integral.

EXERCISES 23.2–Part A

A1. If $f(x, y, z) = 0$ defines $y = g(x, z)$ implicitly, obtain the analogs of (23.14) and (23.15).

A2. If $f(x, y, z) = 0$ defines $x = h(y, z)$ implicitly, obtain the analogs of (23.14) and (23.15).

A3. Verify the Jacobians in (23.16).

A4. Show that $\sqrt{J_1^2 + J_2^2 + J_3^2} = a^2 \sin\phi$ for the Jacobians J_k, $k = 1, 2, 3$, given in (23.16).

A5. Show that $\|\mathbf{T}_\theta \times \mathbf{T}_\phi\| = a^2 \sin\phi$ for $\mathbf{T}_\theta$ and $\mathbf{T}_\phi$ given in Table 23.1.

A6. Verify the integral and its value in (23.18).

A7. Establish the results in (23.19) as follows. Start with $z(u, v) = z(x(u, v), y(u, v))$, and use the chain rule to obtain the derivatives $\frac{\partial z}{\partial u}$ and $\frac{\partial z}{\partial v}$. Use Cramer's rule to solve for z_x and z_y.

EXERCISES 23.2–Part B

B1. The surface described by $z = z(x, y) = 12x^2 + 4xy + 9y^2$ has a minimum at the origin. The portion of the surface satisfying $z(x, y) \leq 26$ projects to a rotated ellipse in the xy-plane.

  (a) Obtain a graph of the surface.

  (b) Verify that $z(x, y)$ has a minimum at the origin.

  (c) Graph the ellipse $z = 26$.

  (d) Obtain the element of surface area as
$$d\sigma = \sqrt{1 + z_x^2 + z_y^2}\, dy\, dx.$$

  (e) Obtain the surface area of that portion of the surface satisfying $z(x, y) \leq 26$. (Numeric integration may be necessary.)

  (f) Parametrize the surface via the equations $x = 7u + 5v$, $y = 6u - 3v$. (Of course, $z(x(u, v), y(u, v)) = Z(u, v)$.)

  (g) Graph the ellipse $Z(u, v) = 26$.

  (h) Obtain the element of surface area in parametric form.

  (i) Use the result in part (h) to obtain the surface area of that portion of the surface satisfying $Z(u, v) \leq 26$.

  (j) To the double integral used in part (e), apply the change of coordinates defined by the equations in part (f). Show that the double integral used in part (i) results.

B2. Let $f(x, y, z) = 5xy + 8xz - 4yz$, $\mathbf{F} = xz\mathbf{i} - y^2\mathbf{j} + yz\mathbf{k}$; and let S be the surface described parametrically by $x = 3u - 4v$, $y = u^2 + v$, $z = 10u + 7v$, $0 \leq u \leq 2$, $0 \leq v \leq 3$.

  (a) Obtain a graph of S.

  (b) Evaluate the surface integral of $f(x, y, z)$ on S.

  (c) Obtain, in the direction of the upward normal, the flux of $\mathbf{F}$ through S.

B3. Let $f(x, y, z) = 8xy + 5yz + 9x^2y$, $\mathbf{F} = yz\mathbf{i} - x^2y\mathbf{j} + xz\mathbf{k}$; and let S be the surface described parametrically by $x = 8v + 4u$, $y = 7v^2 + 9uv$, $z = 8v + 9u$, $3 \leq u \leq 4$, $-4 \leq v \leq -3$.

  (a) Obtain a graph of S.

  (b) Evaluate the surface integral of $f(x, y, z)$ on S.

  (c) Obtain, in the direction of the upward normal, the flux of $\mathbf{F}$ through S.

B4. Let $f(x, y, z) = 3x^2z - 5yz^2 + 3xyz$, $\mathbf{F} = z^2\mathbf{i} + y^2\mathbf{j} - xz\mathbf{k}$; and let S be the surface described parametrically by $x = 7u + 8uv$, $y = 4v - 5u$, $z = 2v - 4u$, $-5 \leq u \leq 5$, $-5 \leq v \leq 4$.

  (a) Obtain a graph of S.

  (b) Evaluate the surface integral of $f(x, y, z)$ on S.

  (c) Obtain, in the direction of the upward normal, the flux of $\mathbf{F}$ through S.

B5. Let $f(x, y, z) = 2z^2 + 5xy^2 + xyz$, $\mathbf{F} = x^2\mathbf{i} + z^2\mathbf{j} - xy\mathbf{k}$; and let S be the surface described parametrically by $x = 7v - 5u$, $y = 4v - 4u$, $z = 9v + 6v^2 + 6u^2$, $-2 \leq u \leq 5$, $-6 \leq v \leq 4$.

  (a) Obtain a graph of S.

  (b) Evaluate the surface integral of $f(x, y, z)$ on S.

  (c) Obtain, in the direction of the upward normal, the flux of $\mathbf{F}$ through S.

B6. Let $\mathbf{F} = (9z^3 + 4y^3)\mathbf{i} + (2x^3 + 7xy^2)\mathbf{j} - 4xyz\mathbf{k}$; and let S be the first-octant portion of the plane $-12 + 84x + 23y + 46z = 0$. Obtain, in the direction of the upward normal, the flux of $\mathbf{F}$ through S using

  (a) one double integral with $d\sigma = \sqrt{1 + z_x^2 + z_y^2}\, dy\, dx$.

  (b) formula (23.17).

  (c) the parametrization $x = 4u - 3v$, $y = 3u - v$.

B7. Let $\mathbf{F} = (3z^2 - 8y^2z)\mathbf{i} + (9x^3 + 8x^2y)\mathbf{j} + (6x^2z - 4x^2y)\mathbf{k}$; and let S be the first-octant portion of the plane $-52 + 86x + 41y + 81z = 0$. Obtain, in the direction of the upward normal, the flux of $\mathbf{F}$ through S using

  (a) one double integral with $d\sigma = \sqrt{1 + z_x^2 + z_y^2}\, dy\, dx$.

  (b) formula (23.17).

  (c) the parametrization $x = 12u + 9v$, $y = 3u - 5v$.

B8. Let $\mathbf{F} = (8x^2y - 3xz^2)\mathbf{i} + (6xyz + x^3)\mathbf{j} + (7xy^2 - 5xy)\mathbf{k}$; and let S be the first-octant portion of the plane $-12 + 28x + 27y + 79z = 0$. Obtain, in the direction of the upward normal, the flux of $\mathbf{F}$ through S using

  (a) one double integral with $d\sigma = \sqrt{1 + z_x^2 + z_y^2}\, dy\, dx$.

  (b) formula (23.17).

  (c) the parametrization $x = 9u - 5v$, $y = 12u + v$.

B9. Let $\mathbf{F} = (7x^2 + 4xz)\mathbf{i} + (8x + 8xz^2)\mathbf{j} + (5y^2z - 8xz^2)\mathbf{k}$; and let S be the first-octant portion of the plane $-46 + 31x + 74y + 6z = 0$. Obtain, in the direction of the upward normal, the flux of $\mathbf{F}$ through S using

  (a) one double integral with $d\sigma = \sqrt{1 + z_x^2 + z_y^2}\, dy\, dx$.

  (b) formula (23.17).

  (c) the parametrization $x = 7u - v$, $y = 3u - 6v$.

B10. Let $\mathbf{F} = (x^2y + 8y^2z)\mathbf{i} + (7 - 9xy)\mathbf{j} + (x^2 - 8yz)\mathbf{k}$; and let S be the first-octant portion of the plane $-100 + 25x + 85y + 65z = 0$. Obtain, in the direction of the upward normal, the flux of $\mathbf{F}$ through S using

  (a) one double integral with $d\sigma = \sqrt{1 + z_x^2 + z_y^2}\, dy\, dx$.

  (b) formula (23.17).

  (c) the parametrization $x = 12u + 7v$, $y = u - 8v$.

B11. Let $\mathbf{F} = (5y^3 - 4xy^2)\mathbf{i} + (x^3 - y^2z)\mathbf{j} + (z + 8x^2y)\mathbf{k}$; and let S be the upper hemisphere of the unit sphere. Obtain, in the direction of the upward normal, the flux of $\mathbf{F}$ through S using

  (a) one double integral with $d\sigma = \sqrt{1 + z_x^2 + z_y^2}\, dy\, dx$.

  (b) formula (23.17).   **(c)** cylindrical coordinates.

  (d) spherical coordinates.

B12. Let $\mathbf{F} = (5x + 2x^3)\mathbf{i} + (6x^2y + 6yz)\mathbf{j} + (3x^2 - 2z^2)\mathbf{k}$; and let S be the cone $z = \sqrt{x^2 + y^2}$, $0 \le z \le 1$. Obtain, in the direction of the upward normal, the flux of $\mathbf{F}$ through S using

    (a) one double integral with $d\sigma = \sqrt{1 + z_x^2 + z_y^2}\ dy\,dx$.

    (b) formula (23.17).   **(c)** cylindrical coordinates.

    (d) spherical coordinates. *Hint:* On S, $z = \frac{\rho}{\sqrt{2}}$.

B13. Let $\mathbf{F} = (7xyz - 6x^2y)\mathbf{i} + (6x^2z + 3xz)\mathbf{j} + (3x - 4y^2z)\mathbf{k}$; and let S be the paraboloid $z = x^2 + y^2$, $0 \le z \le 1$. Obtain, in the direction of the upward normal, the flux of $\mathbf{F}$ through S using

    (a) one double integral with $d\sigma = \sqrt{1 + z_x^2 + z_y^2}\ dy\,dx$.

    (b) formula (23.17).   **(c)** cylindrical coordinates.

    (d) spherical coordinates. *Hint:* On S, $z = \cot^2 \phi$.

B14. Let $\mathbf{F} = (x^2z + 7y^2)\mathbf{i} + (9xyz + 8x^3)\mathbf{j} + (6z - 8xy^2)\mathbf{k}$; and let S be the paraboloid $z = 1 - x^2 - y^2$, $0 \le z$. Obtain, in the direction

of the upward normal, the flux of $\mathbf{F}$ through S using

    (a) one double integral with $d\sigma = \sqrt{1 + z_x^2 + z_y^2}\ dy\,dx$.

    (b) formula (23.17).   **(c)** cylindrical coordinates.

B15. Obtain the element of surface area for the coordinate surface $\phi = a$ in the spherical coordinate system.

B16. Obtain the element of surface area for the coordinate surface $w = a$ in the 6-sphere coordinate system defined in Exercise B29, Section 22.3.

B17. Obtain the element of surface area for the coordinate surface $v = a$ in the paraboloidal coordinate system defined in Exercise B30, Section 22.3.

B18. Obtain the element of surface area for the coordinate surface $v = a$ in the bispherical coordinate system defined in Exercise B31, Section 22.3.

23.3   The Equation of Continuity

Equation of Continuity—Statement

The partial differential equation

$$\frac{\partial \rho}{\partial t} + \nabla \cdot (\rho \mathbf{v}) = 0 \tag{23.20}$$

called the *equation of continuity,* is a balance law for a substance of density ρ flowing with velocity $\mathbf{v}$. It captures the reality that the substance can be neither created nor destroyed. The temporal rate of change of the density is balanced by the divergence of the product $\rho\mathbf{v}$, where the divergence is computed with respect to the spatial variables x, y, z.

    In fact, consider a fluid of density $\rho(x, y, z, t)$, moving so that at every point (x, y, z) in a fixed coordinate system an external observer sees fluid flowing with a velocity

$$\mathbf{v} = f(x, y, z, t)\mathbf{i} + g(x, y, z, t)\mathbf{j} + h(x, y, z, t)\mathbf{k}$$

For example, think of water flowing through a glass trough, on the walls of which are painted the grid lines of an *xyz*-coordinate system. At any particular point inside the trough, fluid flowing with velocity $\mathbf{v}(x, y, z, t)$ will be observed.

    The product $\rho\mathbf{v}$ is a measure of mass-flow per unit time, per unit area, as the following examination of physical units suggests.

$$\rho\mathbf{v} \Rightarrow \frac{\text{mass}}{\text{volume}}\frac{\text{distance}}{\text{time}} = \frac{\text{mass}}{\text{area time}}$$

Derivation

An insightful derivation based on the divergence theorem is easily given. Let S be a fixed, closed, but arbitrary surface enclosing a geometric region V inside the fluid. Fluid is imagined to flow through the surface S, which, again, does not move in time. Let $M(t)$ be the total mass of fluid inside S at any time t. Then

$$M(t) = \iiint_V \rho\, dv$$

where the t-dependence of $M(t)$ is through the density $\rho(x, y, z, t)$ but not through the region V since the region V is fixed in time. Further note that dv is the volume element, not the velocity vector $\mathbf{v}$ that we typically write as boldfaced since it is a vector.

The (time) rate of increase of mass inside S is

$$\frac{dM}{dt} = \frac{d}{dt} \iiint_V \rho\, dv = \iiint_V \frac{\partial \rho}{\partial t}\, dv$$

which, by its very definition as a derivative, is positive if the mass inside S increases. The outward flux of the quantity $\rho\mathbf{v}$ passing out along $\mathbf{N}$, an outward unit normal on S, is given by $\iint_S \rho\mathbf{v} \cdot \mathbf{N}\, d\sigma$. But this flux is the amount of mass lost from V, through the walls of S, per unit time. Hence,

$$\frac{dM}{dt} = \iiint_V \frac{\partial \rho}{\partial t}\, dv = -\iint_S \rho\mathbf{v} \cdot \mathbf{N}\, d\sigma = -\iiint_V \nabla \cdot (\rho\mathbf{v})\, dv \qquad \textbf{(23.21)}$$

The middle equality in (23.21) was explained previously: the flux of mass outward is the surface integral without the minus sign. The derivative $\frac{dM}{dt}$ measures the rate of mass increase (mass passing inward) in V. The rightmost equality in (23.21) results from an application of the divergence theorem to the surface integral.

Combining the two triple integrals over the arbitrary volume V gives

$$\iiint_V \left[\frac{\partial \rho}{\partial t} + \nabla \cdot (\rho\mathbf{v}) \right] dv = 0$$

Since V is arbitrary, this equality must hold for all possible regions V, and that can only happen if the integrand itself is zero. Hence, the equation of continuity (23.20) results.

Some practitioners call the equation of continuity an *accounting principle* since it accounts for mass under transport. Others characterize the equation of continuity as a law of conservation, or mass-balance. In either event, it states that mass can't be created or destroyed without violating the continuity of the flow.

Three Applications

1. In the context of fluid flow, the equation of continuity leads to new conclusions about the flow. If ρ, the density of a fluid, is constant, then $\rho_t = 0$ and $\nabla \cdot (\rho\mathbf{v}) = \rho \nabla \cdot \mathbf{v}$. Hence, the equation of continuity yields $0 + \rho \nabla \cdot \mathbf{v} = 0$, so $\nabla \cdot \mathbf{v} = 0$. Therefore, a constant density means the flow field $\mathbf{v}$ is solenoidal, or divergence-free.

2. Surprisingly, the equation of continuity arises in contexts other than fluid flows. For example, in electromagnetic theory where ρ is charge density, the quantity $\mathbf{J} = \rho\mathbf{v}$ is current density obeying, as a consequence of Maxwell's equations, the continuity equation $\nabla \cdot \mathbf{J} + \rho_t = 0$.

3. In traffic flow theory for long lines of cars moving on a one-lane road with no exits or entrances, at location x and time t, let $u(x, t)$ be the observed car-velocity and $\rho(x, t)$ be the traffic density in number of cars per mile. Then the conservation of cars demands the one-dimensional equation of continuity $\rho_t + (\rho u)_x = 0$ be satisfied. In fact, such traffic flows have been successfully modeled as if the cars were gas particles in a tube. One of the results of such models is that the bunching-up braking causes travels back through the line of traffic like a shock wave in a gas-filled tube.

EXERCISES 23.3

1. If $\rho(x, t)$ is prescribed, show that the one-dimensional equation of continuity determines the velocity field as $u(x, t) = \frac{1}{\rho}(f(t) - \int \rho_t \, dx)$, where $f(t)$ is an arbitrary function. Thus, the velocity $u(x, t)$ is constrained, but not uniquely determined, by the equation of continuity.

2. In the model for single-lane traffic flow, let $\rho(x, t) = e^{-(x-t)^2}$, measured in number of cars per mile.

   (a) Show that the equation of continuity then yields $u(x, t) = 1 + f(t)e^{(x-t)^2}$, where $f(t)$ is an arbitrary function.

   (b) Take $f(t) = 1$ and graph, either as an animation or as a series of images, $\rho(x, t)$ along with $u(x, t)$, thereby showing the time-history of the traffic flow. (A localized region of higher density cars moves to the right along the x-axis. Where the density is high, the velocity is low, leading to the hypothesis that perhaps the equation of continuity always gives this result. The next two problems show this to be a false premise.)

3. In the model for single-lane traffic flow, let $\rho(x, t) = 1 + \frac{1}{1+(x-t)^2}$ be the density of cars per unit length of highway.

   (a) Show that ρ represents a localized region of high-density cars moving uniformly along the x-axis.

   (b) Use the equation of continuity to obtain $u(x, t) = \frac{1+f(t)(1+(x-t)^2)}{2+(x-t)^2}$, the velocity of the cars at location x.

   (c) Show that $u_x(t, t) = 0$, making $x = t$ a critical value for u. Show further that $u_{xx}(t, t) = \frac{f(t)-1}{2}$, so that at $x = t$, $u(x, t)$ will have, for example, a maximum if $f(t) = \frac{1}{2}$ and a minimum, if $f(t) = 2$.

   (d) Let $u_1 = \frac{1}{2}\left[\frac{3+(x-t)^2}{2+(x-t)^2}\right]$, the solution with $f(t) = \frac{1}{2}$. Along with ρu_1, the number of cars per unit time that pass location x, graph ρ and u_1, showing either by an animation or by multiple images the time-history of the traffic flow.

   (e) Plot as a function of time, $\int_{x=1}^{x=2} \rho \, dx$, the number of cars between $x = 1$ and $x = 2$.

   (f) Plot as a function of time, $(\rho u_1)_{x=1} - (\rho u_1)_{x=2}$, the net change in the number of cars between $x = 1$ and $x = 2$. Is this graph consistent with that found in part (e)?

   (g) Let $u_2 = \frac{3+2(x-t)^2}{2+(x-t)^2}$, the solution with $f(t) = 2$. Along with ρu_2, the number of cars per unit time that pass location x, graph ρ and u_2, showing either by an animation or by multiple images the time-history of the traffic flow.

   (h) Plot as a function of time, $(\rho u_2)_{x=1} - (\rho u_2)_{x=2}$, the net change in the number of cars between $x = 1$ and $x = 2$. Is this graph consistent with that found in part (e)?

4. Repeat Exercise 3 for the density function $\rho(x, t) = 1 + \frac{(x-t)^2}{1+(x-t)^2}$. Show that the continuity equation gives $u(x, t) = \frac{f(t)(1+(x-t)^2)-1}{1+2(x-t)^2}$ and that $u_{xx}(t, t) = 4 - 2f(t)$. In part (d) take $f(t) = 3$, and in part (g), take $f(t) = 1$.

5. Still working in one dimension, determine $u(x, t)$ when:

   (a) $\rho = \rho(x)$ (b) $\rho = \rho(t)$

6. Let $\rho(x, t) = e^{-(x-1)^2}$, and obtain $u(x, t)$. In succession, take $f(t)$, the arbitrary function in $u(x, t)$, as t, $\sin t$, and e^{-t}. Analyze the traffic flow in each case.

7. Let $\rho(x, t) = e^{-t}$, and obtain $u(x, t)$, taking $f(t)$, the arbitrary function, to be zero. Analyze the traffic flow.

8. Ohm's law, $V = IR$, linearly relates the *voltage* V to the *current* I via the proportionality constant R, the *resistance*. This direct proportionality is also expressed by the equivalent form of Ohm's law, $\mathbf{E} = r\mathbf{J}$, where $\mathbf{E}$ is the electric field vector and r is the resistivity. Thus, the electric field is directly proportional to the current density. (See [25].) Recalling $\mathbf{E} = -\nabla V$, use the equation of continuity to show that for a steady DC current where ρ is constant, we obtain Laplace's equation $\nabla^2 V = 0$.

23.4 Green's Identities

Statement of Green's Identities

Green's first and second identities, in which V is a suitable well-behaved volume in space and S is its bounding surface, are given.

GREEN'S FIRST IDENTITY

$$\iiint_V [f\nabla^2 g + (\nabla f) \cdot (\nabla g)]dv = \iint_S (f\nabla g) \cdot \mathbf{N} \, d\sigma$$

GREEN'S SECOND IDENTITY

$$\iiint_V [f\nabla^2 g - g\nabla^2 f]dv = \iint_S [f\nabla g - g\nabla f] \cdot \mathbf{N} \, d\sigma$$

In Green's second identity, pay particular attention to the symmetry of the operators in the integrand on the left. That will be the key to understanding why such a formula might be of interest.

Adjoint Operator in Linear Algebra

Some simple matrix algebra provides the motivation for Green's identities. The dot product between vectors is the basic mathematics being examined. For example, consider the vectors and matrix

$$
\mathbf{x} = \begin{bmatrix} -4 \\ 7 \\ 8 \end{bmatrix} \qquad \mathbf{y} = \begin{bmatrix} 10 \\ -6 \\ -8 \end{bmatrix} \qquad A = \begin{bmatrix} -5 & 7 & 6 \\ -6 & 0 & 5 \\ -10 & 1 & 1 \end{bmatrix}
$$

for which $\mathbf{x} \cdot \mathbf{y} = \mathbf{x}^T \mathbf{y} = -146$, where $\mathbf{x}^T = [-4, 7, 8]$ is the transpose of $\mathbf{x}$, and $\mathbf{x}^T \mathbf{y}$ is a matrix product between the row vector $\mathbf{x}^T$ and the column vector $\mathbf{y}$. The dot product of the vector $A\mathbf{x}$ with the vector $\mathbf{y}$ is $(A\mathbf{x}) \cdot \mathbf{y} = 346$. This new dot product is equiva-lent to the matrix product

$$
(A\mathbf{x})^T \mathbf{y} = \mathbf{x}^T A^T \mathbf{y} = \mathbf{x}^T (A^T \mathbf{y}) = \mathbf{x} \cdot (A^T \mathbf{y}) = \mathbf{x} \cdot (A^* \mathbf{y})
$$

where the matrix A^* is just A^T when A is real and $\bar{A}^T$, the complex conjugate, transposed, when A is complex. It is typically called the *adjoint* of A.

Write the calculation as $(A\mathbf{x}) \cdot \mathbf{y} - \mathbf{x} \cdot (A^* \mathbf{y}) = 0$, which becomes $(A\mathbf{x}) \cdot \mathbf{y} - \mathbf{x} \cdot (A\mathbf{y}) = 0$ for a *Hermitian* matrix, that is, one for which $A = A^* = \bar{A}^T$. Another name for the Hermitian matrix is "self-adjoint," a term first encountered for differential equations in Section 16.1.

The left side in Green's second identity has this same structure. The matrix operator A becomes the laplacian. The vector $\mathbf{x}$ becomes the function g. The vector $\mathbf{y}$ becomes the function f. The dot product is the integral.

In fact, the laplacian is just one example of a differential operator, the analog of the operator "multiply by the matrix A." In general, if the differential operator is denoted by the symbol L and its adjoint is denoted by L^*, then Green's second identity is the multivariable generalization of

$$
\int [vLy - yL^*v]\, dx = J(u, v)
$$

where $J(u, v)$ is the bilinear concomitant. This is just Green's formula, equation (16.8), from Section 16.1.

Deriving Green's Identities

GREEN'S FIRST IDENTITY   To obtain Green's first identity, set $\mathbf{F} = f\nabla g$ in the divergence theorem, that is, in $\iiint_V \nabla \cdot \mathbf{F}\, dv = \iint_S \mathbf{F} \cdot \mathbf{N}\, d\sigma$, to obtain

$$
\iiint_V \nabla \cdot (f\nabla g)\, dv = \iint_S (f\nabla g) \cdot \mathbf{N}\, d\sigma
$$

$$
\iiint_V [f\nabla \cdot (\nabla g) + (\nabla f) \cdot (\nabla g)]\, dv = \iint_S (f\nabla g) \cdot \mathbf{N}\, d\sigma
$$

$$
\iiint_V [f\nabla^2 g + (\nabla f) \cdot (\nabla g)]\, dv = \iint_S (f\nabla g) \cdot \mathbf{N}\, d\sigma
$$

GREEN'S SECOND IDENTITY   Green's second identity is obtained by reversing the roles of f and g in the first identity, giving

$$\iiint_V [f\nabla^2 g + (\nabla f) \cdot (\nabla g)] \, dv = \iint_S (f\nabla g) \cdot \mathbf{N} \, d\sigma \qquad \textbf{(23.22a)}$$

$$\iiint_V [g\nabla^2 f + (\nabla g) \cdot (\nabla f)] \, dv = \iint_S (g\nabla f) \cdot \mathbf{N} \, d\sigma \qquad \textbf{(23.22b)}$$

Then, (23.22b) is subtracted from (23.22a), causing the product of the gradients to cancel in the integrands on the left and resulting in Green's second identity.

EXERCISES 23.4

1. Green's identities are often described as nothing more than multidimensional versions of integration by parts. The validity of this claim is easily established for the first identity if the region of integration is taken as the box $V = \{x_0 \le x \le x_1, y_0 \le y \le y_1, z_0 \le z \le z_1\}$, whose sides are parallel to the coordinate planes. Start with $\phi = \iiint_V f\nabla^2 g \, dv$, and

   (a) write ϕ as an iterated triple integral, taking the order of integration as, for example, $dz \, dy \, dx$.

   (b) since $f\nabla^2 g = f(g_{xx} + g_{yy} + g_{zz})$, write ϕ as a sum of three triple integrals where the inner integral in each corresponds to the variable with respect to which the derivatives on g are taken.

   (c) apply integration by parts to the inner integral in each term.

   (d) obtain $\phi + \iiint_V (\nabla f) \cdot (\nabla g) \, dv = \int_{z_0}^{z_1} \int_{y_0}^{y_1} (fg_x)]_{x_0}^{x_1} dy \, dz + \int_{z_0}^{z_1} \int_{x_0}^{x_1} (fg_y)]_{y_0}^{y_1} dx \, dz + \int_{x_0}^{x_1} \int_{y_0}^{y_1} (fg_z)]_{z_0}^{z_1} dx \, dy$.

   (e) show that the three double integrals on the right constitute $\iint_S (f\nabla g) \cdot \mathbf{N} \, d\sigma$, where S is the surface of the box V and $\mathbf{N}$ is the unit outward normal on S.

For the region V and functions f, g given in each of Exercises 2–6, verify Green's first and second identities.

2. $f = 7x^2$, $g = 9z^2$; first octant, bounded by $x = 5$ and $25z^2 + 36y^2 = 144$

3. $f = 3xy - 4y^2 z$, $g = 5yz + 7x^2 z$; the cube $x = \pm 1$, $y = \pm 1$, $z = \pm 1$

4. $f = xyz^2$, $g = x^3 y - z^2$; the unit sphere centered at the origin

5. $f = xy^2 z + 5xz^3$, $g = 2yz - 3x^3$; the cylindrical "can" $x^2 + y^2 = 1$, $z = \pm 1$

6. $f = 4x^2 y^3 + 9z^3$, $g = z^2 - 5xy^3$; first octant, under the plane $12x + 23y + 37z = 100$

For the matrix A and vectors $\mathbf{x}$ and $\mathbf{y}$ given in each of Exercises 7–11, show that $(A\mathbf{x}) \cdot \mathbf{y} - \mathbf{x} \cdot (A^T\mathbf{y}) = 0$.

7. $A = \begin{bmatrix} -6 & 8 & 10 \\ 1 & -11 & 1 \\ -2 & 6 & 9 \end{bmatrix}$, $\mathbf{x} = \begin{bmatrix} 5 \\ 1 \\ -10 \end{bmatrix}$, $\mathbf{y} = \begin{bmatrix} 7 \\ -9 \\ 3 \end{bmatrix}$

8. $A = \begin{bmatrix} 3 & 7 & -1 \\ 5 & 2 & -12 \\ 2 & -4 & 1 \end{bmatrix}$, $\mathbf{x} = \begin{bmatrix} 12 \\ -2 \\ -1 \end{bmatrix}$, $\mathbf{y} = \begin{bmatrix} 7 \\ 5 \\ 11 \end{bmatrix}$

9. $A = \begin{bmatrix} -5 & 12 & -1 \\ 11 & 4 & -3 \\ 11 & -6 & 12 \end{bmatrix}$, $\mathbf{x} = \begin{bmatrix} 11 \\ -9 \\ 11 \end{bmatrix}$, $\mathbf{y} = \begin{bmatrix} 3 \\ 10 \\ 1 \end{bmatrix}$

10. $A = \begin{bmatrix} 4 & 11 & 5 \\ 4 & 6 & -12 \\ -1 & 1 & 1 \end{bmatrix}$, $\mathbf{x} = \begin{bmatrix} 8 \\ 11 \\ 2 \end{bmatrix}$, $\mathbf{y} = \begin{bmatrix} 5 \\ 10 \\ 11 \end{bmatrix}$

11. $A = \begin{bmatrix} 8 & 7 & 3 \\ 4 & 8 & 12 \\ -3 & -2 & -5 \end{bmatrix}$, $\mathbf{x} = \begin{bmatrix} 7 \\ -5 \\ 12 \end{bmatrix}$, $\mathbf{y} = \begin{bmatrix} 4 \\ 3 \\ 5 \end{bmatrix}$

12. Let $A = \begin{bmatrix} 12 & 5 & 5 \\ 8 & 10 & 8 \\ 2 & -8 & -11 \end{bmatrix}$, $M = \begin{bmatrix} m_{11} & m_{12} & m_{13} \\ m_{21} & m_{22} & m_{23} \\ m_{31} & m_{32} & m_{33} \end{bmatrix}$, $\mathbf{x} = \begin{bmatrix} x_1 \\ x_2 \\ x_3 \end{bmatrix}$, $\mathbf{y} = \begin{bmatrix} y_1 \\ y_2 \\ y_3 \end{bmatrix}$, and form the equation $(A\mathbf{x}) \cdot \mathbf{y} - \mathbf{x} \cdot (M\mathbf{y}) = 0$. By treating this as an identity in the x_k and y_k, $k = 1, 2, 3$, show that M must be A^T, the transpose of A.

Chapter Review

1. State Gauss' Theorem in three dimensions.

2. State Gauss' Theorem in two dimensions.

3. Give the expression for $d\sigma$, the surface area element, when the surface is given parametrically by equations of the form $x = x(u, v)$, $y = y(u, v)$, $z = z(u, v)$.

4. Let the surface S be defined parametrically by $x = 2u - 3v$, $y = 5u + 7v$, $z = u + 4v$.

   (a) Obtain $d\sigma$ in parametric form.

   (b) Eliminate the parameters u and v, and obtain $d\sigma$ in Cartesian coordinates.

5. State the formula for the flux of $\mathbf{F} = P(x, y, z)\mathbf{i} + Q(x, y, z)\mathbf{j} + R(x, y, z)\mathbf{k}$ through a parametrically given surface.

6. If S is as given in Question 4 and $0 \le u, v \le 1$, obtain the flux of $F = y\mathbf{i} - z\mathbf{j} + x\mathbf{k}$ through S using the formula in Question 5.

7. State the equation of continuity and provide a derivation based on the divergence theorem.

8. Give at least one application of the equation of continuity.

9. State Green's identities, and illustrate each with an example.

Boundary Value Problems for PDEs

The second-order partial differential equations referred to as the wave equation, the heat equation, and Laplace's equation are archetypes of the important equations of engineering, science, and applied mathematics. These PDEs are usually posed in the context of side conditions, both initial and boundary, so we will actually study *boundary value problems*.

The wave and heat equations are called *evolution equations* since they model how initial states propagate into the future. The wave equation governs the propagation of disturbances that, in the absence of losses through dissipation, retain their initial shapes—thus, the notion of wave motion.

The heat equation models diffusion processes in which initial configurations are immediately dissipated in an irreversible manner. The diffusion of dye in water, the spread of a rumor on a college campus, the spread of potato beetles through a farm, and the dispersal of heat in a conducting medium can all be modeled by the heat equation. In distinction to the wave equation, the heat equation immediately smooths out discontinuities in initial data. Heat at one end of a rod diffuses through the rod and cannot be resequestered into its original configuration. Yet, an electric pulse sent out along a wire travels along the wire as a pulse, retaining its shape in the absence of losses.

We study the wave equation for a finite string in which disturbances take place in a single vertical plane. Modeling the string in a musical instrument, this BVP allows us to describe the solution surface and its role in understanding a solution. Plane sections of this solution surface are images of the moving string, and a sequence of such plane sections constitute a movie of the motion of the string.

The solution of the one-dimensional wave equation can be given by an infinite series, or it can be given by D'Alembert's formula. We examine both and give a numerical solution computed by finite differences.

We also study longitudinal vibrations in an elastic rod. The same PDE governing displacements in the stretched string describe the displacements of cross-sections in the elastic rod. Visualizing the vibrations in the rod is a significantly more difficult task, but one that appropriate computer simulations can help resolve.

In two spatial dimensions, the simplest geometry for studying waves is that of the rectangle. We solve for the vibrations of a rectangular plate with clamped edges. We also consider vibrations in a circular disk and solve the wave equation in polar coordinates. This is the problem of the circular drumhead, a disk with clamped edges.

The one-dimensional heat equation describes heat transfer along the axis of a rod that is wrapped in insulation. If both ends of the rod are kept at the same constant temperature, a series solution for the time-dependent temperatures in the rod is readily given. If the endpoint temperatures differ, or vary in time, a series solution is a much more complicated computation. However, the finite-difference solution we study is readily adapted to varying endpoint temperatures, and we solve the thermal diffusion problem for the wall of a house air-conditioned on the inside and subjected to the daily cycle of heating and cooling during a Midwestern summer.

Laplace's equation, also called the *potential equation,* models either the electrostatic potential or the steady-state temperatures of a plane region. The same BVP describes the physics in both cases. The *equipotentials* or the *isotherms* are the level curves of the harmonic function that satisfies Laplace's equation. The negative of the gradient of this function gives the electric field vector or indicates the flux field for heat flow in the region. The field lines (lines of force) for the electric field are the flow lines for the thermal process.

We solve Laplace's equation on a rectangle, on a disk, in a cylinder, and in a sphere. On the rectangle or disk, the solutions are given in terms of an infinite series akin to a Fourier series. In the cylinder, the solution is a series in Bessel functions; while in the sphere, it is a series in Legendre polynomials. For both the cylinder and the sphere, obtaining the series solutions is interesting and reducible to an algorithm. Visualizing the solutions is a greater challenge because in both cases the quantity to be visualized, either temperature or electric potential, exists in the interior of a solid object. Part of our study of these problems will involve the graphics options available for visualization.

The classic solutions for Laplace's equation are series of eigenfunctions. The geometry of the region over which the equation is being solved determines the eigenfunctions. Hence, the series solution is feasible only for geometries in which the eigenfunctions are readily computed. In more complex geometries, numeric solutions are required, and we give an example of a finite-difference solution for the rectangle.

If the spatial region for the wave, heat, or potential equation is infinite, then series solutions must give way to transform techniques. We will see how to use the Laplace transform to solve the heat equation on a semi-infinite rod. We must then introduce the Fourier transform to solve the wave and heat equations on an infinite domain, the wave equation on the semi-infinite domain, and Laplace's equation on quarter-planes and half-planes. For odd and even functions, the Fourier transform collapses to the Fourier sine and cosine transforms, respectively, and with these integral transforms, we can solve Laplace's equation on semi-infinite strips.

Chapter 24

Wave Equation

I N T R O D U C T I O N   The one-dimensional wave equation is a partial differential equation describing how a disturbance propagates in time along a coordinate axis. It models the motion of a taut, finite string with fixed endpoints and constrained to move in a single plane. If put into motion by imparting an initial shape, we say the string has been *plucked*. Alternatively, the string can be put into motion by imparting an initial velocity. If this initial velocity is achieved by striking the string, we say the string has been *struck*. Therefore, our prototypical examples are the harpsichord, where the string is plucked, and the piano, where the string is struck. However, we do admit that in these two musical instruments, nothing constrains the strings to vibrate in a single plane.

We solve both problems by the technique of separation of variables, which reduces the PDE to two ODEs. One ODE is part of a Sturm–Liouville eigenvalue problem whose eigenfunctions are the fundamental shapes, or *modes,* for oscillation in the string. The eigenvalues are then the harmonic frequencies supported by the string. Just the way a musical sound is a blend, or sum, of harmonics of different intensities, so too is the shape of the vibrating string a sum of fundamental mode shapes with distinctive intensities. This sum, a consequence of the principle of superposition, is the infinite series that represents the solution of the one-dimensional wave equation.

By studying the solution of the plucked string, we discover that the initial disturbance in the string travels along the string with no distortion in its shape. Called a *wave,* this initial shape splits into two similar disturbances of half the initial amplitude, with one copy moving to the right and one to the left. Since the string is finite, the propagated waves are absorbed by the fixed endpoints and are returned *reflected,* to travel the string in the opposite direction. With no provision for energy loss built into our model of the wave equation, these waves continue to traverse the string indefinitely.

This analysis of the motion of waves in a finite string is even more readily seen from D'Alembert's solution, a "closed-form" solution that does not require summing an infinite series.

Using Newton's Second Law and some simple physics, we derive the wave equation. The marvel is that the resulting equation gives such an accurate account of wave motion.

The very same PDE that describes the plucked string also describes the longitudinal vibrations in an elastic rod. For the string, the deflection of the string is the unknown in the governing BVP, and a graph of the deflection at a fixed time simulates the shape of the string at that instant. However, for the elastic rod, the unknown deflections are those of cross-sections of the rod, and a graph of such a cross-section is not a direct visualization of the motion of the rod. Consequently, while the solution process for this BVP poses no new challenges, interpretation becomes the dominant issue, one for which computer simulation is a most welcomed partner.

Finally, we study a finite-difference numeric scheme for solving the wave equation in one dimension. Although our analytic techniques are restricted to problems with homogeneous Dirichlet ($u = 0$) or Neumann ($u_x = 0$) conditions, the numeric solution is sufficiently flexible that either type boundary condition can be nonhomogeneous. This means we can study how waves travel in a string that has one endpoint fixed, and the other made to execute a prescribed motion in time, much like a string that is tied to a post and "snapped" at its other end.

24.1   The Plucked String

Wave Equation on the Finite String

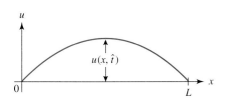

FIGURE 24.1   The finite string with displacement $u(x, \hat{t})$

An idealized string, uniform and without internal friction, is stretched along an x-axis from $x = 0$ to $x = L$. The ends are fixed, the string is taut, and it is constrained to move in a vertical plane. The equilibrium position for the string coincides with the interval $[0, L]$ on the x-axis, and displacements in the string are measured vertically by u. At time t, and at location x in $[0, L]$, the displacement of the string from equilibrium is $u(x, t)$. For fixed $t = \hat{t}$, the shape of the string is given by $u(x, \hat{t})$. Figure 24.1 shows the shape of the string at a particular instant $\hat{t} > 0$.

As we will derive in Section 24.4, displacements in the string are governed by $u_{tt} = c^2 u_{xx}$ the one-dimensional *wave equation* (1) in Table 24.1. This is a *partial differential equation* in the two independent variables x and t. The parameter c will turn out to be the *wave speed,* the speed at which a disturbance can propagate horizontally along the string. The boundary conditions (2) and (3) in Table 24.1 assure that the endpoints of the string remain fixed. Because they prescribe values of u at the endpoints, they are called *Dirichlet* conditions. Since the prescribed values are zero, these conditions are *homogeneous* Dirichlet conditions. The initial conditions (4) and (5) in Table 24.1 give the initial shape and velocity of the string. At first, we will take $g(x) = 0$, so that the string is given a (small) initial displacement $f(x)$ and released. This corresponds to a gentle plucking of the string, much like the action in a harpsichord.

Wave Equation (1) $u_{tt} = c^2 u_{xx}$ for $t > 0$			
Fixed left end	(2) $u(0, t) = 0$	Initial shape	(4) $u(x, 0) = f(x)$
Fixed right end	(3) $u(L, t) = 0$	Initial velocity	(5) $u_t(x, 0) = g(x)$

TABLE 24.1   BVP for wave equation on a finite string

Together, the five equations in Table 24.1 constitute an initial boundary value problem (or BVP) for the finite string. Our goal is to obtain a solution to this BVP and to understand how the function $u(x, t)$ so determined describes the physical motion of the string.

◆ **EXAMPLE 24.1**   We will show that $u(x, t) = \sin x \cos 3t$ is a solution to the BVP in Table 24.1 when $c = 3$, $L = \pi$, $g(x) = 0$, and $f(x) = \sin x$. A calculation shows

$$u_{tt} = -9 \sin x \cos 3t \quad \text{and} \quad c^2 u_{xx} = 3^2(-\sin x)\cos 3t$$

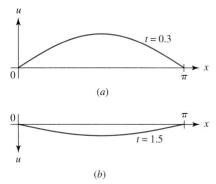

FIGURE 24.2   Two snapshots of the moving string in Example 24.1

Moreover, $u(0, t) = u(\pi, t) = 0$ since $\sin 0 = \sin \pi = 0$. Finally, $u(x, 0) = \sin(x) = f(x)$. Figure 24.2 shows the shape of the string at two different times. Figure 24.3 shows the solution surface whereby $u(x, t)$ is graphed as a surface in the xt-plane. Figure 24.4 shows several snapshots of the string superimposed on the solution surface. The solution surface is traced out in time by a succession of moving images of the dynamic string, or equivalently, the plane sections of the solution surface are snapshots in time of the physical motion of the string. Figures 24.3 and 24.4 show that the motion of the string is an example of a *standing wave* because the points of zero displacement do not translate along the string. (The accompanying Maple worksheet contains complete animations corresponding to Figures 24.3 and 24.4.)   ❖

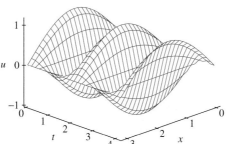

FIGURE 24.3   The solution surface for Example 24.1

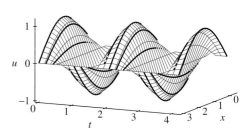

FIGURE 24.4   Snapshots of the string in Example 24.1 superimposed on the solution surface

EXAMPLE 24.2

A wave free to move *along* the string is called a *traveling wave*. Suppose the initial disturbance in the string has the shape shown in Figure 24.5. Under the action of the wave equation, that initial disturbance would travel along the string, reflecting at the endpoints and bouncing back and forth between them. Figure 24.6 shows the traveling wave at two successive times after $t = 0$. The initial energy splits into two equal parts, with each new wave traveling in opposite directions. When the wave reaches the boundary, it is absorbed and then retransmitted, only now, with a reflection. Figure 24.7 shows the solution surface for the traveling wave. The initial disturbance splits into two waves of half the initial height, one traveling left and one traveling right. Without changing shape, these waves move in the xt-plane along *characteristics,* the lines $x \pm ct =$ constant, which are therefore said to "carry the initial information." (Animations of the traveling wave are available in the accompanying Maple worksheet.)

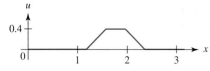

FIGURE 24.5   Initial disturbance in Example 24.2

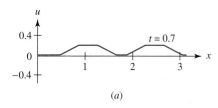

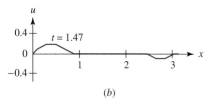

FIGURE 24.6   Two snapshots of the traveling wave in Example 24.2

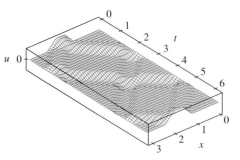

FIGURE 24.7   Solution surface for the traveling wave in Example 24.2

Solution by Separation of Variables

The classical method of solution for such BVPs is the method of *separation of variables.* Assume a solution of the *separated* form $u(x, t) = X(x)T(t)$, a product of one function just containing x and one function just containing t. Recall the initial example, $u(x, t) = \sin x \cos 3t$. Now, generalize this to the product $u(x, t) = X(x)T(t)$, to which we next apply the fixed endpoint conditions, obtaining $X(0)T(t) = 0$ and $X(L)T(t) = 0$. Ruling out the choice $T(t) = 0$ because that implies $u(x, t) \equiv 0$, we conclude $X(0) = X(L) = 0$.

Under the assumption of a separated solution, the initial conditions become $u(x, 0) = X(x)T(0) = f(x)$, and $u_t(x, 0) = X(x)T'(0) = g(x)$. Unless either $f(x)$ or $g(x)$ is zero, no conclusion can be drawn from these two equations. Assuming, for simplicity, that the initial velocity is zero, so that $g(x) = 0$, we conclude

$$u_t(x, 0) = X(x)T'(0) = 0 \Rightarrow T'(0) = 0$$

The initial condition $u(x, 0) = f(x)$ can be satisfied with the aid of a Fourier series. Before we can see that, we must apply the separation assumption to the PDE itself, giving

$$X(x)T''(t) = c^2 X''(x)T(t)$$

We have just deliberately committed a notational no-no. The derivatives on the left are with respect to t, while the derivatives on the right are with respect to x. Yet, we have used the *same* symbol for each differentiation, a mathematical imprecision and ambiguity that is too convenient to avoid.

Next, divide by $c^2 u(x, t)$, with $u(x, t) = X(x)T(t)$. This general strategy, which seems to work with linear second-order PDEs, gives

$$\frac{T''(t)}{c^2 T(t)} = \frac{X''(x)}{X(x)}$$

the *separated form* of the PDE wherein each side of the equation is a function of only one of the variables. Since on the left there is a function purely of t and, on the right, a function purely of x, and these two functions must be equal for **all** values of x and t, each must be equal to the same constant.

This argument is generally not transparent to the novice. Consider the following. Student A is asked to state a "favorite function of t." Student B is asked to state a "favorite function of x." Typically, the functions now written on the board are something like t^2 and $\sin x$. Student C is asked to select a "favorite value of t," and student D is asked for a "favorite value of x." The functions t^2 and $\sin x$ are now computed at values such as $t = 1$ and $x = \pi$. The students then realize that because $1^2 \neq \sin \pi = 0$, the only way for a function of t to equal, consistently, a function of x is for the two functions to be the same *constant* function.

This constant is typically taken as λ and is called the *Bernoulli separation constant.* Hence, we have the two ordinary differential equations $\frac{T''(t)}{c^2 T(t)} = \lambda$ and $\frac{X''(x)}{X(x)} = \lambda$, which, together with the relevant data, form the two problems

$$\begin{array}{ccc} X''(x) - \lambda X(x) = 0 & & T''(t) - c^2 \lambda T(t) = 0 \\ X(0) = X(L) = 0 & \text{and} & T'(0) = 0 \end{array} \qquad \textbf{(24.1)}$$

The problem on the left in (24.1) is a Sturm–Liouville eigenvalue problem, while the problem on the right is a modified initial value problem.

The solution of the Sturm–Liouville eigenvalue problem, obtained in Section 16.1, is $X_n(x) = B_n \sin \frac{n\pi x}{L}$, with eigenvalues $\lambda_n = -\frac{n^2 \pi^2}{L^2}$, $n = 1, 2, \ldots$. For each value of n, the function $T(t)$ is the solution of $T'' + c^2 \frac{n^2 \pi^2}{L^2} T = 0$, $T'(0) = 0$, so we write each solution as $T_n(t) = C_n \cos \frac{cn\pi t}{L}$, $n = 1, 2, \ldots$.

A single eigensolution is the product

$$u_n(x, t) = B_n \sin\left(\frac{n\pi x}{L}\right) C_n \cos\left(\frac{cn\pi t}{L}\right) = b_n \sin\frac{n\pi x}{L} \cos\frac{cn\pi t}{L}$$

where we replaced the product $B_n C_n$ with the single constant b_n. The set of eigensolutions $\{u_n\}$ can be thought of as the analog of a fundamental set for a linear ODE. The general solution of a linear ODE is a linear combination of all the members of the fundamental set. Similarly, the most general solution of the linear PDE is a linear combination of all the members of its "fundamental set," namely, the set of eigensolutions $\{u_n(x, t)\}$. Since there are an infinite number of eigensolutions, our general solution is the infinite sum

$$u(x, t) = \sum_{n=1}^{\infty} u_n(x, t) = \sum_{n=1}^{\infty} b_n \sin\frac{n\pi x}{L} \cos\frac{cn\pi t}{L} \tag{24.2}$$

The final condition to be applied in this BVP is the initial shape requirement

$$u(x, 0) = \sum_{n=1}^{\infty} b_n \sin\frac{n\pi x}{L} = f(x)$$

The coefficients b_n must therefore be the Fourier sine series coefficients for the function $f(x)$. There is no other choice for these constants if this final condition is to be satisfied. If $f(x)$ is to be equal to a sum of sine terms, then the coefficients in that sum must be the Fourier sine series coefficients. Therefore, the b_n's are determined by the Fourier sine series formulas

$$b_n = \frac{2}{L} \int_0^L f(x) \sin\frac{n\pi x}{L} \, dx \qquad n = 1, 2, \ldots \tag{24.3}$$

and the solution $u(x, t)$ is given by (24.2). This infinite sum, along with the defining integrals for the b_n, constitutes the formal solution to the given BVP for the finite plucked string. ❖

EXAMPLE 24.3   A string of length $L = \pi$ has its ends fixed at $x = 0$ and $x = \pi$ on an x-axis. The string is under tension, so when it is given the initial shape $f(x) = \frac{x(\pi - x)}{10}$, as shown in Figure 24.8, and released, it oscillates. We solve for the motion of the string at any time $t > 0$.

The associated BVP that models the motion of this string is given in Table 24.2. The solution, given by (24.2) and (24.3), is then $u(x, t) = \sum_{n=1}^{\infty} b_n \sin nx \cos cnt$, where

$$b_n = \frac{2}{\pi} \int_0^\pi \frac{x(\pi - x)}{10} \sin nx \, dx = \frac{2}{5} \frac{1 - (-1)^n}{\pi n^3}$$

Table 24.3 shows the first 10 coefficients both exactly and in floating-point form. Only the odd-indexed coefficients are nonzero, and these rapidly become zero since $b_n = O(\frac{1}{n^3})$.

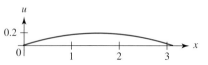

FIGURE 24.8   Initial shape of the string in Example 24.3

Wave Equation (1) $u_{tt} = c^2 u_{xx}$ for $t > 0$			
Fixed left end	(2) $u(0, t) = 0$	Initial shape	(4) $u(x, 0) = \dfrac{x(\pi - x)}{10}$
Fixed right end	(3) $u(\pi, 0) = 0$	Initial velocity	(5) $u_t(x, 0) = 0$

TABLE 24.2   BVP for Example 24.3

n	1	2	3	4	5	6	7	8	9	10
b_n	$\dfrac{4}{5\pi}$	0	$\dfrac{4}{5\pi}\dfrac{1}{27}$	0	$\dfrac{4}{5\pi}\dfrac{1}{125}$	0	$\dfrac{4}{5\pi}\dfrac{1}{343}$	0	$\dfrac{4}{5\pi}\dfrac{1}{729}$	0
b_n	0.255	0	0.00943	0	0.00204	0	0.000742	0	0.0003493	0

TABLE 24.3   Values of b_n, $n = 1, \ldots, 10$, in Example 24.3

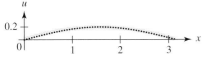

FIGURE 24.9   Example 24.3: $f(x)$ (solid) and p_1 (dotted)

Hence, a very few terms will be needed to get a good approximation to $u(x, t)$. We can test this by examining how many terms it takes to get the Fourier series of $f(x)$ to approximate $f(x)$ well. The first two distinct partial sums in the Fourier series for $f(x)$ are

$$p_1 = \frac{4}{5\pi}\sin x \quad \text{and} \quad p_3 = \frac{4}{5\pi}\sin x + \frac{4}{135\pi}\sin 3x$$

with $f(x)$ (solid) and p_1 (dotted) shown in Figure 24.9. The experiment suggests that two terms of the series for $u(x, t)$ might yield a reasonably good approximation, so write

$$U = \sum_{n=1}^{3} b_n \sin nx \cos cnt = \frac{4}{135\pi}(27\sin x \cos ct + \sin 3x \cos 3ct)$$

as an approximation to the solution. Set $c = 1$ to obtain the graph of the solution surface seen in Figure 24.10. The plane sections $t = t_k$ are snapshots in time, freezing the physical motion of the string for each value of t_k. Exhibiting these snapshots in succession forms a movie, or animation, of the physical motion of the string. (Such an animation can be found in the accompanying Maple worksheet.)   ❖

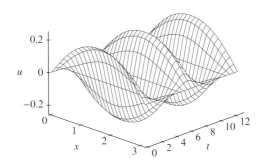

FIGURE 24.10   Solution surface for Example 24.3

EXERCISES 24.1–Part A

A1. Let $c = 2$, $L = 1$, $f(x) = x^2(1 - x)$, and $g(x) = 0$ in Table 24.1. Use (24.2) and (24.3) to write the solution $u(x, t)$ for the resulting BVP.

A2. Starting with $u(x, t) = X(x)T(t)$, provide all the details leading to the solution given in Exercise A1.

For the BVPs given in Exercises A3–8, write $u(x, t)$ as the separated product $X(x)T(t)$. Apply this assumption about $u(x, t)$ to the given initial and boundary data, being sure to identify the condition that leads to the Fourier series. Apply the separation assumption to the PDE itself and obtain the resulting separated ODEs. Do not attempt to find any eigenvalues or to solve any of the ODEs. For your answers, group the appropriate conditions with each separated ODE.

A3. $u_{tt} = 4u_{xx} + 3u$, $u(0, t) = 0$, $u(x, 0) = f(x)$, $u(\pi, t) = 0$, $u_t(x, 0) = 0$

A4. $u_{tt} = 3u_{xx} + 5u_x$, $u_x(0, t) = 0$, $u(x, 0) = 0$, $u_x(7, t) = 0$, $u_t(x, 0) = g(x)$

A5. $u_t = 2u_{xx} + u, u(0, t) = 0, u(x, 0) = f(x), u_x(1, t) = 0$

A6. $3u_{xx} + 2u_x + 5u_t = 0, u(0, t) = 0, u(x, 0) = f(x),$
$u_x(4, t) = 0$

A7. $2u_{tt} - u_t - u_x + 3u_{xx} = 0, u_x(1, t) = 0, u(x, 0) = 0, u(3, t) = 0,$
$u_t(x, 0) = g(x)$

A8. $6u_t + 5u_{xx} - 2u = 0, u_x(0, t) = 0, u_x(x, 0) = h(x), u(1, t) = 0$

EXERCISES 24.1–Part B

B1. If $f(x)$ is twice differentiable, show that $u(x, t) = f(x + ct)$ is a solution of the wave equation $u_{tt} = c^2 u_{xx}$.

B2. Repeat Exercise B1 for $u(x, t) = f(x - ct)$.

For the functions $f(x)$ given in Exercises B3–10:

    (a) Show that $f(x + ct)$ and $f(x - ct)$ are solutions of the wave equation $u_{tt} = c^2 u_{xx}$.

    (b) Obtain graphs of $f(x + t)$ for $t = 0, 1, 2, 3$.

    (c) Obtain graphs of $f(x - t)$ for $t = 0, 1, 2, 3$.

    (d) Graph the surface $z = f(x + t)$ for $t \geq 0$.

    (e) Graph the surface $z = f(x - t)$ for $t \geq 0$.

B3. $f(x) = e^{-x^2}$ **B4.** $f(x) = \dfrac{1}{1 + x^2}$ **B5.** $f(x) = \dfrac{1 + x}{1 + x^2}$

B6. $f(x) = \dfrac{1 - x^2}{1 + x^4}$ **B7.** $f(x) = x^2 e^{-x^2}$

B8. $f(x) = e^{-x^2} \arctan x$ **B9.** $f(x) = \sin 2x$

B10. $f(x) = \cos 3x$

In Exercises B11–13, for the given value of the wave speed c:

    (a) Show that the given function $u(x, t)$ is a solution of $u_{tt} = c^2 x_{xx}, u(0, t) = u(L, t) = 0, u(x, 0) = f(x), u_t(x, 0) = 0$, the BVP governing wave motion in a finite string.

    (b) From the given solution, determine $f(x)$, the initial shape of the string.

    (c) Obtain a plot of the solution surface.

    (d) Obtain graphs of the "snapshots" $u(x, t), t = 0, 1, 2, 3$.

    (e) Using appropriate software, animate the motion of the string governed by the given BVP.

    (f) Plot the surface $u_t(x, t)$ and analyze the velocity of a finite string governed by the given BVP.

    (g) Use an animation utility to explore how, for the finite string governed by the given BVP, the velocity propagates in time.

B11. $u(x, t) = \sin 2x \cos 3t, L = \pi, c = \frac{3}{2}$

B12. $u(x, t) = \sin \dfrac{\pi x}{2} \cos \pi t, L = 2, c = 2$

B13. $u(x, t) = \sin \dfrac{2\pi x}{3} \cos 2\pi t, L = 3, c = \frac{3}{2}$

For the data given in each of Exercises B14–23:

    (a) Solve $u_{tt} = c^2 x_{xx}, u(0, t) = u(L, t) = 0, u(x, 0) = f(x),$
$u_t(x, 0) = 0$, the BVP for one-dimensional wave motion in a finite string supporting the wave speed c.

    (b) Choose $U(x, t)$, a partial sum of the infinite series representing the solution, so that $U(x, 0)$ gives a reasonable approximation to $f(x)$, and graph $U(x, 0)$ along with $f(x)$.

    (c) Plot $U(x, t)$, the approximate solution surface.

    (d) Using $U(x, t)$, animate the motion, studying how the initial disturbance propagates in time.

    (e) Plot the surface determined by $U_t(x, t)$, the approximate velocity along the string.

    (f) In an attempt to study the propagation of the string's velocity in time, animate $U_t(x, t)$ and compare with the displacements given by $U(x, t)$.

B14. $c = 1, L = \pi, f(x) = \begin{cases} x & 0 \leq x < \dfrac{\pi}{2} \\ \pi - x & \dfrac{\pi}{2} \leq x \leq \pi \end{cases}$

B15. $c = 2, L = 4, f(x) = \begin{cases} 0 & 0 \leq x < 1 \\ x - 1 & 1 \leq x < 2 \\ 3 - x & 2 \leq x < 3 \\ 0 & 3 \leq x \leq 4 \end{cases}$

B16. $c = 1, L = 5, f(x) = \begin{cases} 0 & 0 \leq x < 1 \\ x - 1 & 1 \leq x < 2 \\ 1 & 2 \leq x < 3 \\ 4 - x & 3 \leq x < 4 \\ 0 & 4 \leq x \leq 5 \end{cases}$

B17. $c = \sqrt{2}, L = 3, f(x) = \begin{cases} x & 0 \leq x < 1 \\ 1 & 1 \leq x < 2 \\ 3 - x & 2 \leq x \leq 3 \end{cases}$

B18. $c = 1, L = \pi, f(x) = \begin{cases} 0 & 0 \leq x < \dfrac{\pi}{3} \\ \left(x - \dfrac{\pi}{3}\right)\left(\dfrac{2\pi}{3} - x\right) & \dfrac{\pi}{3} \leq x < \dfrac{2\pi}{3} \\ 0 & \dfrac{2\pi}{3} \leq x \leq \pi \end{cases}$

B19. $c = 2, L = 2\pi, f(x) = \dfrac{1 - \cos x}{4}$

B20. $c = \sqrt{3}, L = 3, f(x) = \begin{cases} x & 0 \leq x < 1 \\ 2 - x & 1 \leq x < 2 \\ 0 & 2 \leq x \leq 3 \end{cases}$

B21. $c = \sqrt{3}$, $L = 4$, $f(x) = \begin{cases} x & 0 \le x < 1 \\ 1 & 1 \le x < 2 \\ 3 - x & 2 \le x < 3 \\ 0 & 3 \le x \le 4 \end{cases}$

B22. $c = \pi$, $L = 4$, $f(x) = \begin{cases} x & 0 \le x < 1 \\ 1 & 1 \le x < 2 \\ 2 - \frac{x}{2} & 2 \le x \le 4 \end{cases}$

B23. $c = 1$, $L = 4$, $f(x) = \begin{cases} 0 & 0 \le x < 1 \\ (x-1)(2-x) & 1 \le x < 2 \\ 0 & 2 \le x \le 4 \end{cases}$

B24. Modify the wave equation to $u_{tt} = c^2 u_{xx} - 2\alpha u_t$, thereby including linear damping in the model of the motion of the finite string. Show that the assumption $u(x,t) = X(x)T(t)$, in the BVP $u_{tt} = c^2 x_{xx} - 2\alpha u_t$, $u(0,t) = u(L,t) = 0$, $u(x,0) = f(x)$, $u_t(x,0) = 0$, yields

(a) the same eigenvalues and eigenfunctions as for the undamped model.

(b) $T'' + 2\alpha T' + \left(\frac{cn\pi}{L}\right)^2 T = 0$, $T'(0) = 0$, for determining $T(t)$.

B25. For the model proposed in Exercise B24, let $c = 1$, $L = \pi$, $\alpha = \frac{1}{2}$, and $f(x) = \frac{x(\pi - x)}{10}$. Then:

(a) Determine $T_n(t)$ according to part (b) of Exercise B24.

(b) Obtain $u(x,t)$, the solution of the modified BVP.

(c) Choose $U(x,t)$, a partial sum of the infinite series representing the solution, so that $U(x,0)$ gives a reasonable approximation to $f(x)$, and graph $U(x,0)$ along with $f(x)$.

(d) Plot $U(x,t)$, the approximate solution surface.

(e) Using $U(x,t)$, animate the motion, studying how the initial disturbance propagates in time.

B26. Change the model in Exercise B25 so that $\alpha = \frac{3}{2}$. Notice that $T_1(t)$ now contains just real exponentials. Obtain a solution of this BVP and comment on the effect of $T_1(t)$ in the solution.

For the data given in each of Exercises B27–36, let $\alpha = \frac{c\pi}{2L}$ in the model of Exercise B24.

(a) Solve the resulting BVP.

(b) Choose $U(x,t)$, a partial sum of the infinite series representing the solution, so that $U(x,0)$ gives a reasonable approximation to $f(x)$, and graph $U(x,0)$ along with $f(x)$.

(c) Plot $U(x,t)$, the approximate solution surface.

B27. the data in Exercise B14 **B28.** the data in Exercise B15

B29. the data in Exercise B16 **B30.** the data in Exercise B17

B31. the data in Exercise B18 **B32.** the data in Exercise B19

B33. the data in Exercise B20 **B34.** the data in Exercise B21

B35. the data in Exercise B22 **B36.** the data in Exercise B23

B37. For the model proposed in Exercise B24, discuss what happens to both $T_1(t)$ and $u(x,t)$ if $\alpha = \frac{c\pi}{L}$. ($T_1(t)$ is the solution to the IVP in Exercise B24(b) when $n = 1$.)

B38. For the model proposed in Exercise B24, discuss what happens to both $T_1(t)$ and $u(x,t)$ if $\alpha = \frac{3c\pi}{2L}$. ($T_1(t)$ is the solution to the IVP in Exercise B24(b) when $n = 1$.)

24.2 The Struck String

The BVP

An idealized string, uniform and having no internal friction, is stretched taut between $x = 0$ and $x = L$ on an x-axis. The ends of the string are fixed; the equilibrium position of the string is horizontal, along the axis; and the string is constrained to move only in a vertical plane. As in Section 24.1, let $u(x,t)$ measure the deflection of the string from equilibrium.

At time $t = 0$, the string is struck, thereby acquiring, at all its points, the initial velocity distribution $u_t(x,0) = g(x)$. The boundary value problem modeling the motion of this struck string is given in Table 24.4. For $t > 0$, the string obeys the wave equation (1). We take the initial shape (4) to be the equilibrium configuration, $u(x,0) = 0$.

In Section 24.1 we considered the wave equation on a finite string that had been plucked, suggesting the action in a harpsichord as a realization of the mathematical model. The struck string is found in a piano where a felt hammer hitting the string puts it into motion.

The Solution by Separation of Variables

Assuming a solution of the form $u(x,t) = X(x)T(t)$ will again result in the Sturm–Liouville eigenvalue problem on the left in (24.4). The differential equation for $T(t)$ on the right in (24.4) is the same as it was for the plucked string, but now, the initial condition is $T(0) = 0$

Wave Equation (1) $u_{tt} = c^2 u_{xx}$ **for** $t > 0$			
Fixed left end	(2) $u(0, t) = 0$	Initial shape	(4) $u(x, 0) = 0$
Fixed right end	(3) $u(L, t) = 0$	Initial velocity	(5) $u_t(x, 0) = g(x)$

TABLE 24.4   BVP for the struck string

instead of $T'(0) = 0$.

$$X''(x) - \lambda X(x) = 0 \quad \text{and} \quad T''(t) - c^2 \lambda T(t) = 0$$
$$X(0) = X(L) = 0 \qquad\qquad T(0) = 0 \tag{24.4}$$

As before, the Sturm–Liouville problem has eigenfunctions $X_n = B_n \sin \frac{n\pi x}{L}$ and eigenvalues $\lambda_n = -\frac{n^2 \pi^2}{L^2}, n = 1, 2, \ldots$.

For each value of n, the function $T(t)$, satisfying the differential equation $T''(t) + c^2 \frac{n^2 \pi^2}{L^2} T(t) = 0$ and the condition $T(0) = 0$, is $T_n(t) = C_n \sin \frac{cn\pi t}{L}$. Consequently, the general eigensolution is

$$u_n(x, t) = b_n \sin \frac{n\pi x}{L} \sin \frac{cn\pi t}{L}$$

where the product of constants $B_n C_n$ has been replaced with the single constant b_n. As in Section 24.1, the general solution of the PDE will be a linear combination of all such eigensolutions. Hence, take the most general possible linear combination of all the eigensolutions. Since there are an infinite number of eigensolutions, our linear combination is the infinite sum

$$u(x, t) = \sum_{n=1}^{\infty} b_n \sin \frac{n\pi x}{L} \sin \frac{cn\pi t}{L} \tag{24.5}$$

The final condition to be applied in this BVP is the initial velocity requirement

$$u_t(x, 0) = \sum_{n=1}^{\infty} \frac{cn\pi}{L} b_n \sin \frac{n\pi x}{L} = \sum_{n=1}^{\infty} B_n \sin \frac{n\pi x}{L} = g(x)$$

where $B_n = \frac{cn\pi}{L} b_n$ must be the Fourier sine series coefficient of $g(x)$. Thus, we must have

$$B_n = \frac{n\pi c}{L} b_n = \frac{2}{L} \int_0^L g(x) \sin \frac{n\pi x}{L} \, dx$$

so $b_n = \frac{2}{cn\pi} \int_0^L g(x) \sin \frac{n\pi x}{L} \, dx$. The complete solution is then

$$u(x, t) = \sum_{n=1}^{\infty} b_n \sin \frac{n\pi x}{L} \sin \frac{cn\pi t}{L} \qquad b_n = \frac{2}{cn\pi} \int_0^L g(x) \sin \frac{n\pi x}{L} \, dx$$

EXAMPLE 24.4   A finite string of length $L = \pi$, with fixed endpoints at $x = 0$ and $x = \pi$, and $c = 1$, struck at its center, acquires the velocity profile $g(x) = -\frac{1}{10} x(\pi - x)$, graphed in Figure 24.11. The coefficients are

$$b_n = -\frac{2}{n\pi} \int_0^\pi \frac{1}{10} x(\pi - x) \sin nx \, dx = \frac{2}{5} \frac{(-1)^n - 1}{n^4 \pi} \tag{24.6}$$

Observing the n^4 in the denominator, we realize the b_n rapidly decrease in magnitude, so the series for $u(x, t)$ will converge rapidly. Thus, just several terms will be needed to

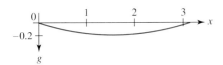

FIGURE 24.11   Initial velocity in Example 24.4

approximate this solution and we approximate $u(x, t)$ with

$$U = \sum_{n=1}^{3} b_n \sin nx \sin nt = -\frac{4}{5\pi}\left(\sin x \sin t + \frac{1}{81}\sin 3x \sin 3t\right) \quad (24.7)$$

Figure 24.12 holds a representation of the solution surface. The plane sections $t = t_k$ are snapshots in time, freezing the physical motion of the string for each value of t_k. Animating these plane sections simulates the physical motion of the string. Figures 24.13(a) and (b) show snapshots of the string at $t = \frac{\pi}{2}$ and $t = \frac{3\pi}{2}$, respectively. As the string moves in space-time, its history generates the solution surface. This and the physical motion of the string appear as animations in the companion Maple worksheet.   ❖

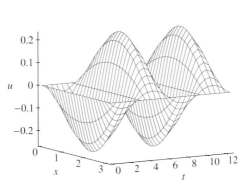

FIGURE 24.12   Solution surface for Example 24.4

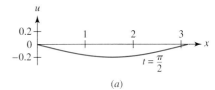

(a)

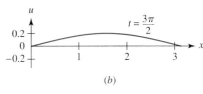

(b)

FIGURE 24.13   Example 24.4: Snapshots of the string taken at $t = \frac{\pi}{2}$ and $t = \frac{3\pi}{2}$

EXERCISES 24.2–Part A

A1. Verify the calculations in (24.6).

A2. Provide a complete solution to $u_{tt} = 4x_{xx}, u(0, t) = u(2\pi, t) = 0, u(x, 0) = 0, u_t(x, 0) = 1 - \cos x$. Separate variables, solve the Sturm–Liouville problem for the eigenfunctions and eigenvalues,

solve the equation for $T_n(t)$, find the eigensolutions $u_n(x, t)$, write $u(x, t)$ in the form (24.5), and obtain b_n.

A3. Give an example of an initial velocity $u_t(x, 0) = g(x)$ that causes standing waves, not traveling waves, to exist in the finite string.

EXERCISES 24.2–Part B

B1. Show that the error made using (24.7) instead of the infinite sum is no worse than 0.0006.

In Exercises B2–8:

(a) Solve $u_{tt} = c^2 u_{xx}, u(0, t) = u(L, t) = 0, u(x, 0) = 0$, $u_t(x, 0) = g(x)$, the BVP for one-dimensional wave motion in a finite string.

(b) Choose $U(x, t)$, a partial sum of the infinite series representing the solution, so that $U_t(x, 0)$ gives a reasonable approximation to $g(x)$, and graph $U(x, 0)$.

(c) Plot $U(x, t)$, the approximate solution surface.

(d) Using $U(x, t)$, animate the motion, studying how the displacements caused by the initial velocity propagate in time.

(e) Plot the surface determined by $U_t(x, t)$, the approximate velocity along the string.

(f) In an attempt to study the propagation of the string's velocity in time, animate $U_t(x, t)$ and compare with the displacements given by $U(x, t)$.

B2. $c = 1, L = 5, g(x) = \begin{cases} 0 & 0 \leq x < 1 \\ (x-1)^2 & 1 \leq x < 2 \\ 1 & 2 \leq x < 3 \\ (x-4)^2 & 3 \leq x < 4 \\ 0 & 4 \leq x \leq 5 \end{cases}$

B3. $c = 2, L = 3, g(x) = \begin{cases} x^2 & 0 \le x < 1 \\ 1 & 1 \le x < 2 \\ (x-3)^2 & 2 \le x \le 3 \end{cases}$

B4. $c = 1, L = 1, g(x) = \begin{cases} x & 0 \le x < \frac{1}{2} \\ 1-x & \frac{1}{2} \le x \le 1 \end{cases}$

B5. $c = 2, L = 4, g(x) = \begin{cases} 0 & 0 \le x < 1 \\ x-1 & 1 \le x < 2 \\ 3-x & 2 \le x < 3 \\ 0 & 3 \le x \le 4 \end{cases}$

B6. $c = 1, L = 5, g(x) = \begin{cases} 0 & 0 \le x < 1 \\ x-1 & 1 \le x < 2 \\ 1 & 2 \le x < 3 \\ 4-x & 3 \le x < 4 \\ 0 & 4 \le x \le 5 \end{cases}$

B7. $c = \sqrt{2}, L = 3, g(x) = \begin{cases} x & 0 \le x < 1 \\ 1 & 1 \le x < 2 \\ 3-x & 2 \le x \le 3 \end{cases}$

B8. $c = 1, L = \pi, g(x) = \begin{cases} 0 & 0 \le x < \frac{\pi}{3} \\ \left(x - \frac{\pi}{3}\right)\left(\frac{2\pi}{3} - x\right) & \frac{\pi}{3} \le x < \frac{2\pi}{3} \\ 0 & \frac{2\pi}{3} \le x \le \pi \end{cases}$

In Exercises B9–15, add linear damping to the wave equation so that it becomes $u_{tt} = c^2 u_{xx} - 2\alpha u_t$ with $\alpha = \frac{c\pi}{2L}$. Then:

(a) Solve $u_{tt} = c^2 u_{xx} - 2\alpha u_t$, $u(0, t) = u(L, t) = 0$, $u(x, 0) = 0$, $u_t(x, 0) = g(x)$, the BVP for damped one-dimensional wave motion in a finite string.

(b) With $\alpha = \frac{c\pi}{2L}$, choose $U(x, t)$, a partial sum of the infinite series representing the solution, so that $U_t(x, 0)$ gives a reasonable approximation to $g(x)$, and graph $U(x, 0)$.

(c) Plot $U(x, t)$, the approximate solution surface.

(d) Use an animation of $U(x, t)$ to study how the displacements caused by the initial velocity propagate in time.

(e) Plot the surface determined by $U_t(x, t)$, the approximate velocity along the string.

(f) In an attempt to study the propagation of the string's velocity in time, animate $U_t(x, t)$ and compare with the displacements given by $U(x, t)$.

B9. $c = 3, L = 3, g(x) = \begin{cases} 0 & 0 \le x < 1 \\ x-1 & 1 \le x < 2 \\ 3-x & 2 \le x \le 3 \end{cases}$

B10. $c = 1, L = 4, g(x) = \begin{cases} 0 & 0 \le x < 1 \\ x-1 & 1 \le x < 2 \\ 1 & 2 \le x < 3 \\ 4-x & 3 \le x \le 4 \end{cases}$

B11. $c = 1, L = 4, g(x) = \begin{cases} \frac{x}{2} & 0 \le x < 2 \\ 1 & 2 \le x < 3 \\ 4-x & 3 \le x \le 4 \end{cases}$

B12. $c = 2, L = 4, g(x) = \begin{cases} 0 & 0 \le x < 1 \\ (x-1)(2-x) & 1 \le x < 2 \\ 0 & 2 \le x \le 4 \end{cases}$

B13. $c = 2, L = \pi, g(x) = \begin{cases} 0 & 0 \le x < \frac{3\pi}{4} \\ \left(x - \frac{3\pi}{4}\right)(\pi - x) & \frac{3\pi}{4} \le x \le \pi \end{cases}$

B14. $c = 3, L = 1, g(x) = \begin{cases} 10x & 0 \le x < \frac{1}{10} \\ 1 & \frac{1}{10} \le x < \frac{9}{10} \\ 10(1-x) & \frac{9}{10} \le x \le 1 \end{cases}$

B15. $c = \sqrt{2}, L = 1$,

$g(x) = \begin{cases} -2000x^3 + 300x^2 & 0 \le x < \frac{1}{10} \\ 1 & \frac{1}{10} \le x < \frac{9}{10} \\ 2000x^3 - 5700x^2 + 5400x - 1700 & \frac{9}{10} \le x \le 1 \end{cases}$

24.3 D'Alembert's Solution

Separation of Variables Solution

If the initial velocity is zero, that is, if $u_t(x, 0) = g(x) = 0$, then the solution of the BVP for the finite string with fixed endpoints (see Table 24.1) is

$$u_f(x, t) = \sum_{n=1}^{\infty} f_n \sin \frac{n\pi x}{L} \cos \frac{cn\pi t}{L} \tag{24.8a}$$

$$f_n = \frac{2}{L} \int_0^L f(x) \sin \frac{n\pi x}{L} \, dx \tag{24.8b}$$

Alternatively, if $f(x) = 0$, the BVP in Table 24.1 will have the solution

$$u_g(x, t) = \sum_{n=1}^{\infty} s_n \sin \frac{n\pi x}{L} \sin \frac{cn\pi x}{L}$$

with

$$s_n = \frac{2}{cn\pi} \int_0^L g(x) \sin \frac{n\pi x}{L} \, dx = \frac{L}{cn\pi} \frac{2}{L} \int_0^L g(x) \sin \frac{n\pi x}{L} \, dx = \frac{L}{cn\pi} g_n$$

where

$$g_n = \frac{2}{L} \int_0^L g(x) \sin \frac{n\pi x}{L} \, dx$$

is the Fourier coefficient for the Fourier sine series of $g(x)$. If neither $f(x)$ nor $g(x)$ is zero, then the solution will be the sum of the two separate solutions u_f and u_g.

D'Alembert's Solution

The function

$$u(x, t) = \frac{1}{2}[F(x + ct) + F(x - ct)] + \frac{1}{2c}\left[\int_0^{x+ct} G(y)\, dy - \int_0^{x-ct} G(y)\, dy\right] \quad \textbf{(24.9)}$$

where $F(x)$ and $G(x)$ are the odd periodic-extensions of $f(x)$ and $g(x)$, respectively, turns out to be a solution called *D'Alembert's solution*. In the absence of an initial velocity, the solution reduces to $u(x, t) = \frac{1}{2}[F(x + ct) + F(x - ct)]$, which justifies earlier claims that the initial disturbance splits into two waves of half the initial energy, one traveling to the left and one to the right, along the string. In the xt-plane, these waves travel along $x \pm ct =$ constant, the *characteristics* of Section 24.1.

D'Alembert's solution can be derived from the separation-of-variables solution. Consider, first, the case where $g(x) = 0$, so the solution is u_f given in (24.8a). The coefficients f_n, given in (24.8b), are the coefficients of the series $\sum_{n=1}^{\infty} f_n \sin \frac{n\pi x}{L}$ that converges to $F(x)$, the odd periodic extension of $f(x)$. This is a crucial point and will have great significance shortly.

Transform the product of trig terms $\sin A \cos B$ with the identity

$$\sin A \cos B = \tfrac{1}{2} \sin(A + B) + \tfrac{1}{2} \sin(A - B)$$

Impose this identity on the terms in u_f, so

$$u_f(x, t) = \frac{1}{2}\sum_{n=1}^{\infty} f_n \sin \frac{n\pi(x + ct)}{L} + \frac{1}{2}\sum_{n=1}^{\infty} f_n \sin \frac{n\pi(x - ct)}{L} \quad \textbf{(24.10)}$$

The sum $\sum_{n=1}^{\infty} f_n \sin \frac{n\pi\sigma}{L}$ converges to $F(\sigma)$, the odd periodic-extension evaluated at σ. Consequently, the first sum in (24.10) converges to $\frac{1}{2}F(x + ct)$ and the second to $\frac{1}{2}F(x - ct)$.

In the case where $f(x) = 0$, transforming u_g, the separation-of-variables solution, to D'Alembert's solution

$$u_g(x, t) = \frac{1}{2c}\left[\int_0^{x+ct} G(y)\, dy - \int_0^{x-ct} G(y)\, dy\right] = \frac{1}{2c}\int_{x-ct}^{x+ct} G(y)\, dy \quad \textbf{(24.11)}$$

requires one additional but subtle step. First, apply the trig formula

$$\sin A \sin B = \tfrac{1}{2} \cos(A - B) - \tfrac{1}{2} \cos(A + B)$$

so that the separation-of-variables solution becomes

$$-\frac{1}{2}\sum_{n=1}^{\infty} s_n \left[\cos\frac{n\pi(x+ct)}{L} - \cos\frac{n\pi(x-ct)}{L}\right] \tag{24.12}$$

from which we conclude that

$$u(x,t) = \varphi(x+ct) - \varphi(x-ct)$$

Applying the initial condition $u_t(x,0) = g(x)$, we get $2c\varphi'(x) = g(x)$, and

$$\varphi(x) = \frac{1}{2c}\int_{\alpha}^{x} G(y)\,dy \tag{24.13}$$

where α arises, as a constant of integration. Hence, $u(x,t)$ is given by (24.11).

EXAMPLE 24.5   Even when $g(x) = 0$ so there is no initial velocity, it is not a trivial task implementing D'Alembert's solution to the wave equation. Let $c = 1$, define the initial displacement function on the interval $[0,3]$ by (24.14), and graph it in Figure 24.14.

$$f(x) = \begin{cases} 0 & 0 \le x \le 1 \\ (x-1)(2-x) & 1 \le x \le 2 \\ 0 & 2 \le x \le 3 \end{cases} \tag{24.14}$$

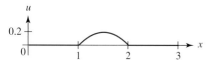

FIGURE 24.14   Initial shape function $f(x)$ in Example 24.5

The odd periodic extension of $f(x)$, namely, $F(x)$, is graphed in Figure 24.15.

Forming an *analytic* representation of $F(x)$ is the challenge. Once an expression for $F(x)$ exists, the solution $u(x,t) = \frac{1}{2}[F(x+t) + F(x-t)]$ is easily formed and can be graphed as in Figure 24.16 or used to create a simulation of the string's motion. (See the accompanying Maple worksheet for a representation of $F(x)$ and an animation that simulates the motion of the string.) ❖

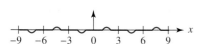

FIGURE 24.15   Odd periodic extension of $f(x)$ in Figure 24.14

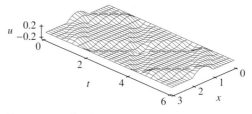

FIGURE 24.16   Solution surface to Example 24.5

EXERCISES 24.3–Part A

A1. Obtain the series in (24.12) and determine the coefficients s_n.

By Leibniz' rule, the derivative of the function $F(x) = \int_a^{u(x)} f(x,t)\,dt$ is $F'(x) = \int_a^{u(x)} f_x(x,t)\,dt + f(x,u(x))u'(x)$. In Exercises A2 and 3:

  (a) Obtain $F(x)$ explicitly by evaluating the given integral.

  (b) Obtain $F'(x)$ from the representation obtained in part (a).

  (c) Obtain $F'(x)$ by applying Leibniz' rule, and then compare to part (b).

A2. $F(x) = \int_1^{\sin x}(2x+t)^2\,dt$ **A3.** $F(x) = \int_0^{x^2}\sin xt\,dt$

A4. Use direct substitution to verify that (24.9) satisfies the wave equation. *Hint:* Use Leibniz' rule to differentiate the integrals.

EXERCISES 24.3–Part B

B1. If $f(x) = 0$ and $g(x) = \sin x$, use (24.11) to obtain $u(x, t)$, the solution of the wave equation on $[0, \pi]$. Why is (24.11) so easy to apply in this case?

In each of Exercises B2–11, $g(x) = 0$ and c, L, and $f(x)$ are given. For each:

    (a) Obtain D'Alembert's solution.

    (b) Using D'Alembert's solution, plot the solution surface. This requires finding a representation of the odd periodic extension of $f(x)$, the initial-shape function. (See the accompanying Maple worksheet for software tools that yield a graphical representation of a periodic extension.)

    (c) Animate D'Alembert's solution and compare with the animation based on the approximation used in Section 24.1.

B2. Exercise B14, Section 24.1   **B3.** Exercise B15, Section 24.1

B4. Exercise B16, Section 24.1   **B5.** Exercise B17, Section 24.1

B6. Exercise B18, Section 24.1   **B7.** Exercise B19, Section 24.1

B8. Exercise B20, Section 24.1   **B9.** Exercise B21, Section 24.1

B10. Exercise B22, Section 24.1   **B11.** Exercise B23, Section 24.1

Given the initial velocity $u_t(x, t) = g(x)$, integration of $G(y)$, the odd periodic extension of $g(x)$, can become a difficult tangle of conditional statements. However, let $\hat{g}(x)$ be the odd extension of $g(x)$ to the interval $[-L, L]$; let $\hat{G}(x) = \int \hat{g}(x)\, dx$ be the indefinite integral of $\hat{g}(x)$, meaningful on the interval $[-L, L]$; and let $\tilde{G}(x)$ be the periodic extension of $\hat{G}(x)$. Then (24.11) becomes $u(x, t) = \frac{1}{2c}[\tilde{G}(x + ct) - \tilde{G}(x - ct)]$. In Exercises B12–18, use software to evaluate $\tilde{G}(x \pm ct)$ and graph $u(x, t)$, showing that this solution replicates the solution obtained earlier.

B12. Exercise B2, Section 24.2   **B13.** Exercise B3, Section 24.2

B14. Exercise B4, Section 24.2   **B15.** Exercise B5, Section 24.2

B16. Exercise B6, Section 24.2   **B17.** Exercise B7, Section 24.2

B18. Exercise B8, Section 24.2

24.4   Derivation of the Wave Equation

Derivation

The wave equation is a consequence of Newton's Second Law, $\mathbf{F} = m\mathbf{a}$. Just what forces are exerted on the string are established by some simple assumptions and the diagram in Figure 24.17.

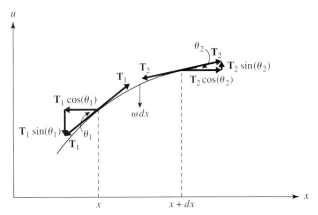

FIGURE 24.17   Forces on a segment of an ideal string under tension

    The string is ideal, so it is uniform and frictionless. It is at equilibrium, and under tension, so every point that is not an endpoint is subjected to equal and opposite forces whose common magnitude we denote by τ. The string is subjected to a "gentle plucking" $f(x)$ so that it experiences a *small* deflection from equilibrium, one small enough that the magnitude of the tension remains τ. We also assume that the tension in the deformed string acts tangential to the string and that individual particles in the string move only vertically. The string experiences no lateral translation and remains in horizontal equilibrium.

At some fixed time t, consider a small segment of the string lying between x and $x + dx$. Let ω be the weight per unit length of the string. Let $m = \frac{\omega}{g}$ be the mass per unit length of the string, where g is the gravitational constant. The segment of string in $[x, x + dx]$ in Figure 24.17 has total weight $\omega\,dx$ and total mass $m\,dx$. The horizontal components of the tension are equal in magnitude, so we have

$$T_1 \cos \theta_1 = T_2 \cos \theta_2 = \tau \tag{24.15}$$

Applying Newton's Second Law ($\mathbf{F} = m\mathbf{a}$) to the vertical direction, with the acceleration $a = u_{tt}(x, t) = u_{tt}$ and the net vertical force

$$\mathbf{F} = (T_2 \sin \theta_2 - T_1 \sin \theta_1 - \omega\,dx)\mathbf{i}$$

we obtain

$$T_2 \sin \theta_2 - T_1 \sin \theta_1 - \omega\,dx = \frac{\omega}{g}\,dx\,u_{tt}$$

From (24.15) we have, upon multiplication by $\tan \theta_k$,

$$T_1 \sin \theta_1 = \tau \tan \theta_1 \quad \text{and} \quad T_2 \sin \theta_2 = \tau \tan \theta_2$$

Hence, the equation $\mathbf{F} = m\mathbf{a}$ becomes

$$\tau[\tan \theta_2 - \tan \theta_1] - \omega\,dx = \frac{\omega}{g}\,dx\,u_{tt}$$

Divide by dx, obtaining

$$\tau\left[\frac{\tan \theta_2 - \tan \theta_1}{dx}\right] - \omega = \frac{\omega}{g}u_{tt} \tag{24.16}$$

Now make the obvious geometric interpretation of $\tan \theta_1$ and $\tan \theta_2$ as derivatives. From elementary calculus, the derivative is the slope of the tangent line and the slope of the tangent line is the tangent of the angle it makes with the horizontal. Thus, the slopes at the two endpoints are

$$\tan \theta_2 = u_x(x + dx, t) \quad \text{and} \quad \tan \theta_1 = u_x(x, t)$$

so (24.16) becomes

$$\tau\left[\frac{u_x(x + dx, t) - u_x(x, t)}{dx}\right] - \omega = \frac{\omega}{g}u_{tt}$$

The fraction in the bracket on the left is a difference quotient for the x-derivative of $u_x(x, t)$. Hence, in the limit as the segment of string becomes very small ($dx \to 0$) we have

$$\tau u_{xx}(x, t) - \omega = \frac{\omega}{g}u_{tt}(x, t)$$

Multiply through by $\frac{g}{\omega}$ and write u_{tt} as the left side of the equation, giving

$$u_{tt} = \frac{\tau g}{\omega}u_{xx} - g$$

Define $c^2 = \frac{\tau g}{\omega}$, leading to

$$u_{tt} = c^2 u_{xx} - g$$

Assume that the tension τ is great enough that the term $c^2 u_{xx}$ is large in comparison to g, resulting in $u_{tt} = c^2 u_{xx}$, the wave equation.

The constant $c^2 = \frac{\tau g}{\omega}$ has dimensions of velocity squared, so c is the *wave speed*. the speed with which the initial disturbance (or shape) travels along the string.

EXERCISES 24.4–Part A

A1. Show that $c = \sqrt{\frac{\tau g}{\omega}}$ has the units of velocity.

A2. In three dimensions, the Cartesian form of the wave equation would be $u_{tt} = c^2 \nabla^2 u = c^2(u_{xx} + u_{yy} + u_{zz})$, where $u = u(x, y, z, t)$.

(a) Show that $u = 2\sin(x + ct) - \cos(x - ct) + e^{y+ct} + \arctan(y - ct) - \frac{1}{1+(z+ct)^2}$ is a solution of the three-dimensional wave equation.

(b) Show that for any twice-differentiable functions $f_k(\sigma)$, $k = 1, \ldots, 6$, the sum $u = f_1(x + ct) + f_2(x - ct) + f_3(y + ct) + f_4(y - ct) + f_5(z + ct) + f_6(z - ct)$ is a solution of the wave equation.

EXERCISES 24.4–Part B

B1. The telegrapher's equation, $u_{tt} = c^2 u_{xx} - 2\alpha u_t - ku$, models the transmission of electrical signals along a wire. The term $-2\alpha u_t$ models a linear viscous damping, and the term $-ku$ models an elastic restoring force akin to the Hooke's law term in the damped oscillator of Unit One.

(a) Verify that inclusion of a linear elastic restoring force in the model of the finite string results in the modification seen in the telegrapher's equation.

(b) Obtain a formal solution to the undamped telegrapher's equation ($\alpha = 0$), $u(0, t) = u(L, t) = 0, u(x, 0) = f(x)$, and $u_t(x, 0) = 0$.

(c) For the solution in part (b), take $k = \sqrt{2}, L = \pi, c = 1$, and
$$f(x) = \begin{cases} 20x\left(\frac{\pi}{10} - x\right) & 0 \le x < \frac{\pi}{10} \\ 0 & \frac{\pi}{10} \le x \le \pi \end{cases} \quad \text{and determine } u(x,t).$$

(d) Obtain a plot of the solution surface for the BVP solved in part (c).

(e) Plot time-slices for the solution found in part (c), and animate them if possible.

(f) Compare the solution found in part (c) with the solution for which $k = 0$. Determine what physical effects inclusion of the elastic restorative force has on the model of the finite string.

24.5 Longitudinal Vibrations in an Elastic Rod

Introduction

A rod of length L and density ρ (mass per unit volume) is deformed by pulling or pushing in opposite directions at its ends. The rod is said to stretch or compress *linearly* along its axis if the restorative forces induced in the rod obey Hooke's law. In effect, the rod behaves like a spring that is both extensible and compressible. Such a rod is called *elastic* if after a deformation, it returns to its original state.

Under compression, the distance between contiguous cross-sections in the rod decreases, while under expansion (rarefaction) it increases. Compression or expansion experienced by neighboring plane cross-sections can propagate through the rod as a wave that obeys the wave equation $u_{tt} = c^2 u_{xx}$. Since the disturbance travels along the axis of the rod, the wave is said to be *longitudinal*. Thus, a steel rod, hammered on its end, will support longitudinal waves of compression and rarefaction traveling the length of the rod. Similarly, the rubber engine mounts supporting the engine in an automobile, when subjected to shock, experience the same passage of distortions progressing as plane waves through the medium.

Although such longitudinal vibrations in a rod are governed by the same wave equation as the vibrating string, the meaning of the displacement $u(x, t)$ is different and subtly so. In the rod, $u(x, t)$ measures the displacement, at time t, of the plane in the rod that was at x initially. Since these displacements are relative to where the plane section was located initially, there is an additivity that makes the complete motion much harder to fathom than the vibrations in the finite string where the displacement was directly observable.

In addition, the free endpoint condition whereby an end is free to move is formalized in terms of the *strain,* a new concept for the reader not familiar with the theory of elasticity. Hence, our approach will be to state an appropriate boundary value problem, nurture intuition by illustrating the solution, and finally develop the theory and derive the governing partial differential equation.

EXAMPLE 24.6   **PROBLEM STATEMENT**   Formulate and solve an appropriate boundary value problem for the longitudinal elastic vibrations in a rod of length π whose left and right ends are both free. Assume $c = 1$ as the wave speed, and uniformly stretch the rod by one unit, releasing it without imparting any initial velocity.

PROBLEM FORMULATION   If the rod is placed between $x = 0$ and $x = \pi$ on an x-axis and a copy of the x-axis is etched on the rod at equilibrium, then the desired boundary value problem consists of the wave equation $u_{tt} = u_{xx}$, the boundary conditions on the left in (24.17), and the initial conditions on the right.

$$\begin{aligned} u_x(0, t) = 0 &\qquad u(x, 0) = f(x) = \frac{x}{\pi} \\ u_x(\pi, t) = 0 &\qquad u_t(x, 0) = 0 \end{aligned} \qquad \textbf{(24.17)}$$

The function $u(x, t)$ measures the displacement, at time t, of the plane section in the rod that was at location x initially. The endpoint conditions are derivatives because, as we will see later when we derive the governing equations, "free ends" mean no forces act on the ends. In fact, we will show the elastic force on a face at location x is proportional to $u_x(x, t)$.

To determine the initial distribution of displacements in the rod, we must first obtain the constant value of the *strain,* defined as the change in length, per unit length. Thus, we compute the constant strain as

$$\frac{\Delta L}{L} = \frac{(\pi + 1) - \pi}{\pi} = \frac{1}{\pi}$$

The initial distribution of displacements is then found by accumulating the strain along the rod, obtaining

$$u(x, 0) = \int_0^x \frac{1}{\pi} \, ds = \frac{x}{\pi}$$

Finally, the initial velocity, namely, $u_t(x, 0)$, is zero. This boundary value problem is new because we have not yet solved the wave equation with homogeneous Neumann conditions (as derivative conditions at the endpoints are called).

SOLUTION BY SEPARATION OF VARIABLES   Assume $u(x, t) = X(x)T(t)$ as the separated form of the solution. As in Sections 24.1 and 24.2, the wave equation then yields the two ODEs

$$X''(x) - \lambda X(x) = 0 \quad \text{and} \quad T''(t) - \lambda T(t) = 0$$

However, applying the separation assumption to the endpoint conditions now gives

$$u_x(0, t) = 0 \Rightarrow X'(0)T(t) = 0 \Rightarrow X'(0) = 0$$
$$u_x(\pi, t) = 0 \Rightarrow X'(\pi)T(t) = 0 \Rightarrow X'(\pi) = 0$$

In addition, applying the separation assumption to the initial velocity condition yields

$$u_t(x, 0) = 0 \Rightarrow X(x)T'(0) = 0 \Rightarrow T'(0) = 0$$

a result we obtained each time we solved the wave equation with a similarly vanishing initial velocity.

We therefore face the Sturm–Liouville system on the left in (24.18) and the modified IVP on the right.

$$\begin{matrix} X''(x) - \lambda X(x) = 0 & T'' - \lambda T = 0 \\ X'(0) = X'(\pi) = 0 & T'(0) = 0 \end{matrix} \qquad (24.18)$$

From Section 16.1, the Sturm–Liouville BVP for $X(x)$ has the eigenvalues $\lambda = -n^2, n = 0, 1, 2, \ldots$, and the corresponding eigenfunctions $X_n = A_n \cos nx, n = 0, 1, 2, \ldots$. The corresponding solutions for $T(t)$ are then $T_n = B_n \cos nt, n = 0, 1, 2, \ldots$. A typical eigen-solution is therefore

$$u_n(x, t) = a_n \cos nx \cos nt$$

and $u(x, t)$ is a linear combination of all such eigensolutions. Evaluating at $t = 0$ tells us the coefficients in the sum are the Fourier coefficients for $f(x)$, so we have the solution

$$u(x, t) = \frac{a_0}{2} + \sum_{n=1}^{\infty} a_n \cos nx \cos nt \qquad a_n = \frac{2}{\pi} \int_0^\pi f(x) \cos nx \, dx$$

For $f(x) = \frac{x}{\pi}$, the Fourier coefficients are

$$a_0 = \frac{2}{\pi} \int_0^\pi \frac{x}{\pi} \, dx = 1 \quad \text{and} \quad a_n = \frac{2}{\pi} \int_0^\pi \frac{x}{\pi} \cos nx \, dx = 2 \frac{(-1)^n - 1}{\pi^2 n^2}$$

The partial sum $u_{20}(x, t) = \frac{1}{2} + 2 \sum_{n=1}^{20} [((-1)^n - 1)/\pi^2 n^2] \cos nx \cos nt$ is graphed as a surface in Figure 24.18.

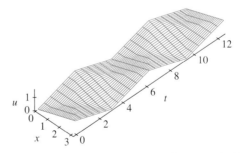

FIGURE 24.18   Solution surface for Example 24.6

The wave-like properties of the solution are apparent, but the detailed interpretation of the meaning of these waves requires more investigation. Remember, $u(x, t)$ gives the displacement, at time t, of the plane section that was at x initially. What we want to see is a representation of the physical motion of the elastic rod, and unlike the transverse waves on the finite string, $u(x, t)$ does not directly show these physical motions.

An animation of the solution $u(x, t)$ is a dynamic interpretation of the solution surface but is still not a representation of the physical motions in the rod. It shows the displacements $u(x, t)$ that each face that was at location x is now experiencing at time t. It tells us that whatever is happening is happening linearly, and it takes place from the endpoints in and

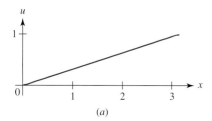

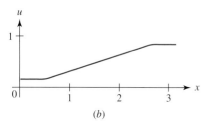

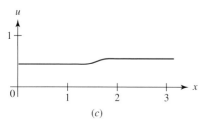

FIGURE 24.19   Example 24.6: $u(x, t)$ at $t = 0, 0.5, 1.4$

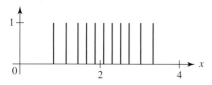

FIGURE 24.20   Example 24.6: Snapshot of the moving elastic rod at $t = 3.7$

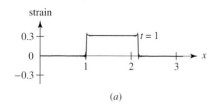

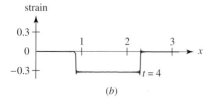

FIGURE 24.21   Example 24.6: The strains $u_x(x, t)$ at $t = 1$ and $t = 4$

back with a wave-front like a shock wave. Several frames of this animation appear in Figure 24.19.

A more revealing animation in the accompanying software simulates the actual rod by imagining a set of 11 equispaced scratch marks on the rod and following the motion of these scratch marks in real time. Figure 24.20 shows the bar at $t = 3.7$.

The uniformly stretched rod relaxes to its natural length, with the collapse occurring from the "outside, in." The center of the rod is the only part of the rod that does not move. Then, the collapse continues past the point of reaching natural length, and the rod contracts by an amount equal to the original stretch. This continuing contraction now takes place from "inside, out" as the center of the rod first experiences the compression, with the outside of the rod feeling the compression last. The rod then expands from this state of full compression to natural length, from "outside, in." Having achieved natural length, the rod continues expanding to its initial stretched state, this time from the "inside, out."

In addition to this simulation of the rod itself, it is also instructive to examine an animation of $u_x(x, t)$, which is the localized (per-face) value of the strain, $\frac{\Delta L}{L}$. This animation shows more clearly how the compression and rarefaction waves travel with a "front" through the rod. To compute the derivative $u_x(x, t)$ we have to differentiate the infinite sum representing the solution $u(x, t)$. The derivative of a Fourier series converges even more slowly than the original series. Where we obtained acceptable results with a 20-term approximation for $u(x, t)$ itself, it will take 100 terms for the derived series to converge. This translates into a modest increase in computation time.

Figure 24.21 shows, at times $t = 1$ and $t = 4$, the strain in the rod. The strain moves longitudinally through the rod, as a shock wave. It starts out uniform across the rod and drops to zero with a shock-front. When the strain is completely gone, the rod has reached its natural length. As the rod then enters a compressive mode, the strain becomes negative and the region of compression expands from the center out to the ends. At this juncture the strain graph is a constant negative amount. The compression then starts to vanish from the ends, inward, until the rod again reaches its natural length. This corresponds to the vanishing of the strain. As the rod enters an expansion phase, the strain appears above the x-axis, growing, with a shock-front, outward from the center. The expansion continues until the rod returns to its original stretched state.

The accompanying Maple worksheet contains animations of the displacement $u(x, t)$, the physical motion $x + u(x, t)$, and the strain $u_x(x, t)$. A simultaneous side-by-side display of these three animations is also available in the worksheet.   ❖

Theory and Derivation

A rod that obeys a linear law of elasticity is said to obey *Hooke's law,* wherein forces are proportional to displacement. This linear relationship between force and displacement, expressed by the formula $F = kx$, is a staple of elementary calculus and general physics, especially in any discussion of the behavior of a spring. In fact, the linear spring was an essential element in Chapters 5 and 6. Thus, F, the force required to stretch a spring, is proportional to x, the amount of the stretch. The constant of proportionality, k, is called the spring constant.

In the context of elasticity, Hooke's law is sometimes stated as "*stress* is proportional to the *strain*," where stress is the force per unit cross-sectional area and

$$\text{strain} = \frac{\Delta L}{L}$$

is defined as the elongation per unit length. In the context of elasticity, Hooke's law is then stress = E strain, where E is Young's modulus, the constant of proportionality between stress and strain. This gives Young's modulus the units of

$$E \sim \frac{\text{stress}}{\text{strain}} \sim \frac{\frac{\text{force}}{\text{area}}}{\frac{\text{length}}{\text{length}}} = \frac{\text{force}}{\text{area}}$$

In fact, starting with $F = kx$ and dividing through by A, the cross-sectional area, we have $\frac{F}{A} = \frac{k}{A}x$. The displacement x is the elongation ΔL from the definition of strain, namely, strain $= \frac{\Delta L}{L}$. Hence, the spring law becomes

$$\text{stress} = \frac{F}{A} = \frac{k}{A}\Delta L = \frac{k}{A}L \text{ strain } = E \text{ strain}$$

where E, Young's modulus, is $\frac{kL}{A}$. Hence, the two statements of Hooke's law are equivalent.

Since longitudinal waves travel in the direction of the rod's axis, we want a coordinate system along that axis. However, we must assume a linear elasticity that keeps plane sections both plane and parallel. Thus, forces within the rod obey the linear Hooke's law and preserve the orientation of all plane sections that are perpendicular to the axis of the rod.

Put the rod (whose equilibrium length is L) on a ruler, aligning the left end of the rod with the left end of the ruler. At equilibrium, paint a copy of the ruler's coordinates on the rod. For clarity, take this linear coordinate system to be an x-coordinate system. Identify a face (i.e., a plane section) within the rod by the coordinate it had at equilibrium. If the equilibrium coordinates are painted on the rod, then these coordinates travel with the rod should the rod be stretched or compressed. In addition, leave the ruler on the table so that the original meaning of an x-coordinate is preserved.

Consider R, a segment of the rod bounded at equilibrium by faces at x and $x + dx$. This segment has length dx and is shown in Figure 24.22. Let $u(x, t)$ denote the displacement experienced, at time t, by the face, which at equilibrium, was located at x. It helps to remember that this face still bears the coordinate x. The displacement undergone by this face is the quantity $u(x, t)$ and is measured along the axis of the rod. Unlike the vibrating string where the displacements are very obvious, the lateral displacements of faces in the rod are much more difficult to visualize.

The partial differential equation governing longitudinal vibrations in the rod will be a consequence of $\mathbf{F} = m\mathbf{a}$, Newton's Second Law of Motion. It's not hard to guess that $\mathbf{a}$, the acceleration of the face bearing the coordinate x, is given by $\mathbf{a} = u_{tt}(x, t)\mathbf{i}$. Determining $\mathbf{F}$, the net force acting on this face, is more subtle.

At time t, the left and right faces of the segment R are now located, respectively, at

$$X_{\text{left}} = x + u(x, t) \quad \text{and} \quad X_{\text{right}} = x + dx + u(x + dx, t)$$

The length of the deformed segment is then

$$\Delta = dx + u(x + dx, t) - u(x, t)$$

The *change* in length of this segment is therefore

$$\Delta - dx = u(x + dx, t) - u(x, t)$$

the relative displacement between the left and right faces of the segment. The strain associated with the face that, at equilibrium, was at location x, is given by the limiting ratio $\lim_{dx \to 0} \frac{\Delta - dx}{dx}$. Notice that strain, an extensive property, is ascribed to a point by this limit. This definition parallels the definition of pressure, which is also an extensive property, force per unit area, ascribed to a point.

Thus, the strain on the single face bearing the coordinate x is $\lim_{dx \to 0} \frac{u(x + dx, t) - u(x, t)}{dx}$, so the value of the strain is strain $= u_x(x, t)$. This is a fundamental result in the theory of elasticity. Getting this correct at the beginning is well worth the time invested. Thus, at

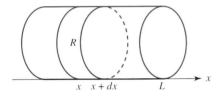

FIGURE 24.22 A segment of an elastic rod

time t and location x, the strain caused by the displacement $u(x, t)$ is positive if the applied force tends to increase the size of the region R and negative otherwise.

The force on a plane section can be obtained from the strain, using Hooke's law, stress $= E$ strain. With A as the cross-sectional area of the rod, we have

$$\frac{\text{force}}{\text{area}} = \text{stress} = E \text{ strain}$$

so that

$$\text{force} = EA \text{ strain} = EAu_x(x, t)$$

Thus, the elastic force on the face whose equilibrium coordinate was x is given by $F(x, t) = EAu_x(x, t)$.

The net force on segment R is given by the sum of the forces acting on the left and right faces of R, namely, by

$$EAu_x(x + dx, t) - EAu_x(x, t)$$

The total mass of segment R is $m = \rho A\, dx$ since the volume of the segment is A, the cross-sectional area, times the length, dx. Taking the acceleration of a face as $u_{tt}(x, t)$, we can write the **i**-component of Newton's Second Law, $\mathbf{F} = m\mathbf{a}$, as

$$EAu_x(x + dx, t) - EAu_x(x, t) = \rho A\, dx\, u_{tt}(x, t)$$

Divide by $A\rho\, dx$ to get

$$\frac{E}{\rho} \frac{u_x(x + dx, t) - u_x(x, t)}{dx} = u_{tt}(x, t)$$

which, in the limit as $dx \to 0$, becomes

$$\frac{E}{\rho} u_{xx}(x, t) = u_{tt}(x, t) \tag{24.19}$$

Upon defining $c^2 = \frac{E}{\rho}$, (24.19) becomes $u_{tt} = c^2 u_{xx}$, the very same wave equation we derived for the vibrating string. The difference here is that $u(x, t)$ represents the displacement of the face that was at x at time t, and this displacement is lateral, along the axis of the rod.

EXERCISES 24.5

1. An elastic rod of length 10 whose ends are free is uniformly stretched by two units and released without imparting an initial velocity.

(a) Formulate an appropriate BVP.

(b) Solve the BVP by separating variables and obtain the relevant Fourier series.

(c) Set $c = 1$ and graph, as a function of time, the displacements of the free right end of the rod.

(d) With $c = 1$, graph, as a function of time, the displacements of the free left end of the rod.

(e) With $c = 1$, obtain a plot of the surface $x + u(x, t)$ for $0 \le x \le 10$ and $0 \le t \le 20$. Each plane section $x = x_0$ depicts the motion undergone by the face initially located at $x = x_0$.

(f) With $c = 1$, graph the curves $x_0 + u(x_0, t)$, where $0 \le t \le 20$ and x_0 assumes the values 1, 4, and 7. Again, these plane

sections depict the actual motions undergone by the corresponding three faces in the rod.

2. An elastic rod of length π whose left end is held fixed and whose right end is free is uniformly stretched by one unit and released without imparting an initial velocity.

(a) Formulate an appropriate BVP.

(b) Solve the BVP by separation of variables and obtain the relevant Fourier series.

(c) Set $c = 1$ and graph, as a function of time, the displacements of the free right end of the rod.

(d) With $c = 1$, obtain a plot of the surface $x + u(x, t)$ for $0 \le x \le \pi$ and $0 \le t \le 20$.

(e) With $c = 1$, graph the curves $x_0 + u(x_0, t)$, where $0 \le t \le 20$ and x_0 assumes the values 1, 2, and 3.

3. An elastic rod of length π whose ends are free is stretched by one unit and released without imparting an initial velocity. The initial strain is not uniform but is proportional to location. Thus, it is given by $u_x(x, 0) = \frac{2}{\pi^2}x$, so the initial displacement is given by $u(x, 0) = \int_0^x u_x(s, 0)\, ds = \frac{1}{\pi^2}x^2$. Note that $u(\pi, 0) = 1$, which is consistent with the hypothesis of a total stretch of one unit.

   (a) Formulate an appropriate BVP.

   (b) Solve the BVP by separating variables and obtain the relevant Fourier series.

   (c) Set $c = 1$ and graph, as a function of time, the displacements of the free right end of the rod.

   (d) With $c = 1$, graph, as a function of time, the displacements of the free left end of the rod.

   (e) With $c = 1$, obtain a plot of the surface $x + u(x, t)$ for $0 \le x \le \pi$ and $0 \le t \le 20$.

   (f) With $c = 1$, graph the curves $x_0 + u(x_0, t)$, where $0 \le t \le 20$ and x_0 assumes the values 1, 2, and 3.

4. The elastic rod of Exercise 3 has its left end fixed.

   (a) Formulate and solve the relevant BVP.

   (b) Set $c = 1$ and graph, as a function of time, the displacements of the free right end of the rod.

   (c) With $c = 1$, obtain a plot of the surface $x + u(x, t)$ for $0 \le x \le \pi$ and $0 \le t \le 20$.

   (d) With $c = 1$, graph the curves $x_0 + u(x_0, t)$, where $0 \le t \le 20$ and x_0 assumes the values 1, 2, and 3.

5. An elastic rod of length 5 whose left end is at the origin of an x-axis, is nonuniformly stretched a total of one unit. The strain in the rod is proportional to the square root of the coordinate of the location. Both ends of the rod are free, and after stretching, the rod is released without imparting an initial velocity.

   (a) Obtain $u(x, 0)$, the initial displacement function for the rod.

   (b) Formulate and solve the relevant BVP.

   (c) Set $c = 1$ and graph, as a function of time, the displacements of the free right end of the rod.

   (d) With $c = 1$, graph, as a function of time, the displacements of the free left end of the rod.

   (e) With $c = 1$, obtain a plot of the surface $x + u(x, t)$ for $0 \le x \le 5$ and $0 \le t \le 20$.

   (f) With $c = 1$, graph the curves $x_0 + u(x_0, t)$, where $0 \le t \le 20$ and x_0 assumes the values 1, $\frac{5}{2}$, and 4.

6. The elastic rod of Exercise 5 has its left end fixed.

   (a) Formulate and solve the relevant BVP.

   (b) Set $c = 1$ and graph, as a function of time, the displacements of the free right end of the rod.

   (c) With $c = 1$, obtain a plot of the surface $x + u(x, t)$ for $0 \le x \le 5$ and $0 \le t \le 20$.

   (d) With $c = 1$, graph the curves $x_0 + u(x_0, t)$, where $0 \le t \le 20$ and x_0 assumes the values 1, $\frac{5}{2}$, and 4.

7. An elastic rod of length π whose ends are free is uniformly compressed one unit and released without imparting an initial velocity.

   (a) Formulate and solve the relevant BVP.

   (b) Set $c = 1$ and graph, as a function of time, the displacements of the free right end of the rod.

   (c) With $c = 1$, graph, as a function of time, the displacements of the free left end of the rod.

   (d) With $c = 1$, obtain a plot of the surface $x + u(x, t)$ for $0 \le x \le \pi$ and $0 \le t \le 20$.

   (e) With $c = 1$, graph the curves $x_0 + u(x_0, t)$, where $0 \le t \le 20$ and x_0 assumes the values 1, 2, and 3.

8. The elastic rod of Exercise 7 has its left end fixed.

   (a) Formulate and solve the relevant BVP.

   (b) Set $c = 1$ and graph, as a function of time, the displacements of the free right end of the rod.

   (c) With $c = 1$, obtain a plot of the surface $x + u(x, t)$ for $0 \le x \le \pi$ and $0 \le t \le 20$.

   (d) With $c = 1$, graph the curves $x_0 + u(x_0, t)$, where $0 \le t \le 20$ and x_0 assumes the values 1, 2, and 3.

9. An elastic rod of length π whose ends are free is compressed by one unit and released without imparting an initial velocity. The initial strain is not uniform but is proportional to location. Thus, it is given by $u_x(x, 0) = -\frac{2}{\pi^2}x$, so the initial displacement is given by $u(x, 0) = \int_0^x u_x(s, 0)\, ds = -\frac{1}{\pi^2}x^2$. Note that $u(\pi, 0) = -1$, which is consistent with the hypothesis of a total compression of one unit.

   (a) Formulate and solve the relevant BVP.

   (b) Set $c = 1$ and graph, as a function of time, the displacements of the free right end of the rod.

   (c) With $c = 1$, graph, as a function of time, the displacements of the free left end of the rod.

   (d) With $c = 1$, obtain a plot of the surface $x + u(x, t)$ for $0 \le x \le \pi$ and $0 \le t \le 20$.

   (e) With $c = 1$, graph the curves $x_0 + u(x_0, t)$, where $0 \le t \le 20$ and x_0 assumes the values 1, 2, and 3.

10. The elastic rod of Exercise 9 has its left end fixed.

    (a) Formulate and solve the relevant BVP.

    (b) Set $c = 1$ and graph, as a function of time, the displacements of the free right end of the rod.

    (c) With $c = 1$, obtain a plot of the surface $x + u(x, t)$ for $0 \le x \le \pi$ and $0 \le t \le 20$.

    (d) With $c = 1$, graph the curves $x_0 + u(x_0, t)$, where $0 \le t \le 20$ and x_0 assumes the values 1, 2, and 3.

11. An elastic rod of length π is nonuniformly compressed a total of one unit. The strain in the rod is proportional to the square of the location. Both ends of the rod are free, and after compression, the rod is released without imparting an initial velocity.

(a) Obtain $u(x, 0)$, the initial displacement function for the rod.

(b) Formulate and solve the relevant BVP.

(c) Set $c = 1$ and graph, as a function of time, the displacements of the free right end of the rod.

(d) With $c = 1$, graph, as a function of time, the displacements of the free left end of the rod.

(e) With $c = 1$, obtain a plot of the surface $x + u(x, t)$ for $0 \le x \le \pi$ and $0 \le t \le 20$.

(f) With $c = 1$, graph the curves $x_0 + u(x_0, t)$, where $0 \le t \le 20$ and x_0 assumes the values 1, 2, and 3.

12. The elastic rod of Exercise 11 has its left end fixed.

(a) Formulate and solve the relevant BVP.

(b) Set $c = 1$ and graph, as a function of time, the displacements of the free right end of the rod.

(c) With $c = 1$, obtain a plot of the surface $x + u(x, t)$ for $0 \le x \le \pi$ and $0 \le t \le 20$.

(d) With $c = 1$, graph the curves $x_0 + u(x_0, t)$, where $0 \le t \le 20$ and x_0 assumes the values 1, 2, and 3.

24.6   Finite-Difference Solution of the One-Dimensional Wave Equation

Problem Statement

A numeric solution is sought for the boundary value problem consisting of the one-dimensional wave equation $u_{tt} = c^2 u_{xx}$ in a string of length L, the boundary conditions on the left in (24.20), and the initial conditions on the right.

$$u(0, t) = \psi_1(t) \qquad u(x, 0) = f(x)$$
$$u(L, t) = \psi_2(t) \qquad u_t(x, 0) = g(x) \tag{24.20}$$

Notice that the boundary conditions are not homogeneous.

We will describe and implement a finite-difference technique in which derivatives in the wave equation are replaced by the $O(h^2)$ central difference approximations given in (24.22) and derived in Exercises A1 and A2.

An exact solution of this problem would require analytic techniques similar to those developed in Section 27.2. Of course, if $\psi_1(t) = \psi_2(t) = 0$, the problem reduces to one for which we can compute an exact solution.

An Explicit Scheme

First, a grid is defined on that portion of the xt-plane corresponding to the extension of the string in space-time. Let the x-interval $[0, L]$ be divided into N equal subintervals of length $\Delta x = \frac{L}{N}$ by the $N + 1$ points $x_n = x_0 + n\Delta x, n = 0, 1, \ldots, N$. Similarly, discretize the t-axis with the partition $t_m = t_0 + m\Delta t, m = 0, 1, \ldots, M$, where $\Delta t \le \frac{\Delta x}{c}$. For the explicit method being described, this is a very real restriction on the size of the time-step permitted. It is derived in a subsequent discussion.

Let $u(x_n, t_m)$ be the exact value of the solution at the point (x_n, t_m). Let $v_{n,m}$ be the numeric approximation to that exact value, where $v_{n,m}$ is computed from

$$v_{n,m+1} = \lambda^2(v_{n-1,m} + v_{n+1,m}) + 2(1 - \lambda^2)v_{n,m} - v_{n,m-1} \tag{24.21}$$

where $\lambda = \frac{c\Delta t}{\Delta x} \le 1$. (The restriction on λ is derived next.) Then, except for round-off error, the difference $u(x_n, t_m) - v_{n,m}$ goes to zero as x_n and t_m themselves go to zero. (See [18] for details.)

Theoretical Considerations

DERIVATION   The explicit method given by (24.21) is obtained by making the $O(h^2)$ central difference replacements

$$u_{xx}(x_n, t_m) = \frac{v_{n+1,m} - 2v_{n,m} + v_{n-1,m}}{(\Delta x)^2} \quad \text{and} \quad u_{tt}(x_n, t_m) = \frac{v_{n,m+1} - 2v_{n,m} + v_{n,m-1}}{(\Delta t)^2}$$

(24.22)

in the wave equation, multiplying through by $(\Delta t)^2$, introducing $\lambda = \frac{c\Delta t}{\Delta x}$, and solving for $v_{n,m+1}$. Values at "interior points" are computed with (24.21).

IMPLEMENTATION   A careful examination of the terms on the right in (24.21) suggests the schematic, or computational molecule, shown in Figure 24.23. The solid dots represent known values that are used to compute the unknown $v_{n,m+1}$, represented by the open dot in the schematic. Moreover, the diagram shows that to compute $v_{n,m+1}$, v must be known for the previous *two* time-steps.

The very first time-step forward from the initial data at time $t = 0$ is taken via the equation

$$v_{n,1} = \lambda^2(v_{n-1,0} + v_{n+1,0}) + 2(1 - \lambda^2)v_{n,0} - v_{n,-1}$$

(24.23)

When computing $v_{n,1}$, the index is $m = 0$, so the row with three black dots is the very first row consisting of the values of $u(x,0) = f(x)$. Thus, with the values of $v_{n,0} = f(x_n) = f_n$, the equation for the very first row of new values becomes

$$v_{n,1} = \lambda^2(f_{n-1} + f_{n+1}) + 2(1 - \lambda^2)f_n - v_{n,-1}$$

(24.24)

The value $v_{n,-1}$, represented by the lone black dot in the row beneath, has to be obtained from the other initial condition $u_t(x,0) = g(x)$. For this, solve the $O(h^2)$ central-difference representation

$$u_t(x_n, t_0) = \frac{v_{n,1} - v_{n,-1}}{2\Delta t} = g(x_n) \equiv g_n$$

for $v_{n,-1} = -2g_n\Delta t + v_{n,1}$, which, when inserted into (24.24), yields

$$v_{n,1} = (1 - \lambda^2)f_n + \frac{\lambda^2}{2}(f_{n+1} + f_{n-1}) + g_n\Delta t$$

(24.25)

STABILITY ANALYSIS   The following technique for determining the stability of a finite-difference method such as (24.21) is due to John von Neumann. At time $t_m = t_0 + m\Delta t$, assume $v_{n,m}$ can be expanded in Fourier series, the terms of which are of the form $\phi(t_m)\cos(\omega x_n)$. Equation (24.21) is then transformed by

1.   using dx and dt in place of Δx and Δt, respectively when defining λ.

2.   writing $\phi(t_k) = \phi_k$.

3.   noting $\cos(\omega x_{n\pm1}) = \cos(\omega(x_n \pm dx)) = \cos(\omega x_n)\cos(\omega\,dx) \mp \sin(\omega x_n)\sin(\omega\,dx)$.

4.   dividing through by $\phi_m\cos(\omega x_n)$.

5.   defining $R = \phi_{m+1}/\phi_m$ and noting it is the *amplification factor* by which a given component in the numeric solution grows in time.

6.   noting that $\phi_{m-1}/\phi_m = \frac{1}{R}$ because it is the reciprocal of $R = \phi_m/\phi_{m-1}$.

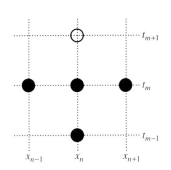

FIGURE 24.23   Computational molecule for the explicit finite-difference scheme for the wave equation

These steps result in $R = 2\lambda^2(\cos(\omega\,dx) - 1) + 2 - \frac{1}{R}$. Writing $A = \cos(\omega\,dx) - 1$ yields, for $R = R(A, \lambda)$, the two roots

$$R_\pm = 1 + \lambda^2 A \pm \lambda\sqrt{2A + \lambda^2 A^2}$$

Stability requires $|R_\pm| \le 1$ for both roots with $-2 \le A \le 0$, achieved for $0 < \lambda \le 1$. The bounds on A follow from the behavior of the cosine function, and the bounds on λ can be deduced from Figure 24.24, where $|R_+(A, \lambda)|$ is graphed, Figure 24.25, where $|R_-(A, \lambda)|$ is graphed, and Figure 24.26, where $R_+(0, \lambda) = R_-(0, \lambda) = 1$ and $R_\pm(-2, \lambda)$, constituting the curve, are graphed.

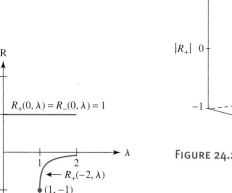

FIGURE 24.26   Graphs of $R_+(0, \lambda) = R_-(0, \lambda) = 1$ and $R_\pm(-2, \lambda)$

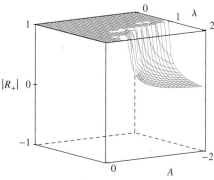

FIGURE 24.24   Graph of $|R_+(A, \lambda)|$

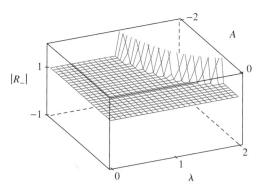

FIGURE 24.25   Graph of $|R_-(A, \lambda)|$

If $1 < \lambda$, at least one branch of the amplification factor R has a magnitude greater than 1. The gap in the graph over the interval $0 < \lambda \le 1$ suggests that at least one branch of R is complex in that range. Thus, the *amplification factor* $R = \phi(t_{m+1})/\phi(t_m)$ has magnitude 1 for any value of $\cos(\omega x_n)$ as long as $0 < \lambda \le 1$. For these values of λ, the explicit scheme (24.21) is stable for the one-dimensional wave equation. Readers interested in alternate approaches to this stability analysis should consult more specialized texts such as [18] or [90].

EXAMPLE 24.7   Let $L = 1$, $c = 1$, $\psi_1 = \sin t$, $\psi_2 = \cos t$, $f(x) = x$, and $g(x) = x - 1$, and take $\Delta x = \frac{1}{10}$ corresponding to the choice $N = 10$. Let $\Delta t = \frac{1}{10}$ so $\lambda = c\frac{\Delta t}{\Delta x} = 1$, and define the nodes $(x_n, t_m) = (\frac{n}{10}, \frac{m}{10})$. At the first time-step, compute the $N - 1 = 9$ values $v_{n,1}$ with (24.25), which becomes

$$v_{n,1} = \frac{11}{100}n - \frac{1}{10} \qquad (24.26)$$

The values $v_{n,m+1}$ for $m = 1, 2, \ldots$ are given by (24.21), which becomes

$$v_{n,m+1} = v_{n-1,m} + v_{n+1,m} - v_{n,m-1} \qquad (24.27)$$

Ordinarily, from three contiguous values at time t_m and one value at time t_{m-1}, one new value of the approximate solution can be computed at time t_{m+1} using (24.21). The choice of $\lambda = 1$ eliminates $v_{n,m}$, so (24.27) is particularly simple. The open dot in Figure 24.23 moves from left to right between x_1 and x_9, so the values of $v_{0,m+1}$ and $v_{10,m+1}$ are computed from the endpoint data, not from the same formula used to compute at the $N - 1 = 9$ interior points.

For the first "row," the values of $v_{n,0}$ are obtained from the initial function, $f(x)$. Hence, to begin the calculation, the initialization step defines values for $v_{n,0}$, $n = 0, \ldots, N$. Then, row 1 is computed from (24.26), after which each new row can be computed from the previous two rows provided (24.27) is used for the $N - 1 = 9$ interior points, and the endpoints are computed from the boundary functions.

If only the numeric values of the approximate solution are desired, there is no need to store the values for every row during the computation. Only three rows are needed during the computation, the last two known rows and the new row being computed. After a complete row has been computed, the values in that row can be printed or stored for further processing and graphical displays. Then, the values in the row m are shifted to become the oldest row, the most recently computed row becomes row m, and the iteration repeats.

The companion Maple worksheet contains code embodying the algorithm just enunciated. The results of the calculations appear as the solution surface in Figure 24.27. Three plane sections $t = t_m$, $m = 10, 15, 40$, are shown in Figure 24.28. A complete animation of the motion of the string is given in the companion Maple worksheet. ❖

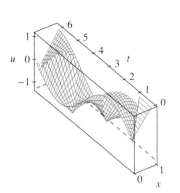

FIGURE 24.27   Solution surface for Example 24.7

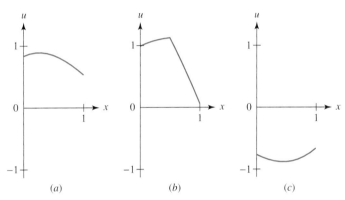

FIGURE 24.28   Example 24.7: Three snapshots of the moving string taken at times (a) $t_{10} = 1$; (b) $t_{15} = 1.5$; (c) $t_{40} = 4$

EXERCISES 24.6–Part A

A1. By adding the Taylor expansions
$$u(x + h, t) = u(x, t) + u_x(x, t)h + \tfrac{1}{2}u_{xx}(x, t)h^2$$
$$+ \tfrac{1}{6}u_{xxx}(x, t)h^3 + O(h^4)$$
$$u(x - h, t) = u(x, t) - u_x(x, t)h + \tfrac{1}{2}u_{xx}(x, t)h^2$$
$$- \tfrac{1}{6}u_{xxx}(x, t)h^3 + O(h^4)$$

and solving for u_{xx}, obtain the $O(h^2)$ central difference formula appearing on the left in (24.22).

A2. Obtain the $O(h^2)$ central difference formula appearing on the right in (24.22).

A3. Use (24.22) to obtain (24.21).   **A4.** Derive (24.25).

A5. Obtain (24.26).

EXERCISES 24.6–Part B

B1. In Example 24.7, take $c = \tfrac{1}{2}$ so $\lambda = \tfrac{1}{2}$. Keeping all other data the same, apply the explicit finite-difference method, obtaining a graph of the solution surface.

B2. In Example 24.7, take $c = 2$ so $\lambda = 2$. Keeping all other data the same, apply the explicit finite-difference method. Since $\lambda > 1$, your results should show the computation becomes unstable, generating meaningless large values for the displacement u.

A string of length L has its endpoints fixed. For that string and the data given in each of Exercises B3–8:

(a) Obtain the exact solution by the methods of Sections 24.1 and 24.2.

(b) Letting $N = 10$ so $\Delta x = \tfrac{L}{10}$ and choosing $\Delta t \leq \tfrac{\Delta x}{c}$, obtain a numeric solution.

(c) Compare the exact and numeric solutions by graphing the resulting solution surfaces.

(d) Compare the exact and numeric solutions by graphing the time-dependent displacement at the midpoint of the string.

B3. $u(x, 0) = x(\pi - x), u_t(x, 0) = 0, c = 1, L = \pi$

B4. $u(x, 0) = \begin{cases} x & 0 \le x < 1 \\ 2 - x & 1 \le x < 2, \ u_t(x, 0) = 0, c = 1, L = 3 \\ 0 & 2 \le x \le 3 \end{cases}$

B5. $u(x, 0) = \begin{cases} 0 & 0 \le x < 1 \\ (x - 1)(2 - x) & 1 \le x < 2, \ u_t(x, 0) = 0, \\ 0 & 2 \le x \le 3 \end{cases}$

$c = 2, L = 3$

B6. $u(x, 0) = 0, u_t(x, 0) = \dfrac{1 - \cos x}{4}, c = 1, L = 2\pi$

B7. $u(x, 0) = 0, u_t(x, 0) = \begin{cases} x(x - 1) & 0 \le x < 1 \\ 0 & 1 \le x \le 3 \end{cases}, c = 1, L = 3$

B8. $u(x, 0) = \begin{cases} x(1 - x) & 0 \le x < 1 \\ 0 & 1 \le x \le 3 \end{cases}$

$u_t(x, 0) = \begin{cases} 0 & 0 \le x < 2 \\ (x - 2)(x - 3) & 2 \le x \le 3 \end{cases}, c = 2, L = 3$

The displacements $u(x, t)$ in a string of length 10 obey the differential equation $u_{tt} = u_{xx}$, the endpoint data $u(0, t) = \psi_1(t), u(10, t) = \psi_2(t)$, and the initial conditions $u(x, 0) = f(x), u_t(x, 0) = g(x)$. Using the additional data provided in each of Exercises B9–17:

(a) Let $N = 20$ so $\Delta x = \frac{L}{20}$, choose $\Delta t \le \frac{\Delta x}{c} = \Delta x$, and obtain a numeric solution.

(b) Plot the solution surface.

(c) Animate the motion of the string, or minimally, plot several snapshots of the shape of the string for successive times.

B9. $\psi_1(t) = \sin t, \psi_2(t) = \sin 2t, f(x) = g(x) = 0$

B10. $\psi_1(t) = \sin t, \psi_2(t) = 0, f(x) = g(x) = 0$

B11. $\psi_1(t) = \sin^2 t, \psi_2(t) = 0, f(x) = g(x) = 0$

B12. $\psi_1(t) = |\sin t|, \psi_2(t) = 0, f(x) = g(x) = 0$

B13. $\psi_1(t) = \begin{cases} \sin t & 0 \le t \le \pi \\ 0 & t > \pi \end{cases}, \psi_2(t) = 0, f(x) = g(x) = 0$

B14. $\psi_1(t) = \begin{cases} \sin t & 0 \le t \le \pi \\ 0 & t > \pi \end{cases}$

$\psi_2(t) = \begin{cases} 0 & 0 \le t \le \pi \\ -\sin t & \pi \le t \le 2\pi, \ f(x) = g(x) = 0 \\ 0 & t > 2\pi \end{cases}$

B15. $\psi_1(t) = \begin{cases} \sin t & 0 \le t \le \pi \\ 0 & t > \pi \end{cases}, \psi_2(t) = 0,$

$f(x) = \begin{cases} 0 & 0 \le x < 8 \\ (x - 8)(10 - x) & 8 \le x \le 10 \end{cases}, g(x) = 0$

B16. $\psi_1(t) = \begin{cases} \sin t & 0 \le t < \pi \\ 0 & t \ge \pi \end{cases}, \psi_2(t) = 0,$

$f(x) = \begin{cases} 0 & 0 \le x < 8 \\ (x - 8)(10 - x) & 8 \le x \le 10 \end{cases}$

$g(x) = \begin{cases} 0 & 0 \le x < 4 \\ (x - 4)(x - 6) & 4 \le x < 6 \\ 0 & 6 \le x \le 10 \end{cases}$

B17. $\psi_1(t) = \begin{cases} \sin t & 0 \le t < \pi \\ 0 & t \ge \pi \end{cases}$

$\psi_2(t) = \begin{cases} 0 & 0 \le t \le \pi \\ -\sin t & \pi \le t \le 2\pi \\ 0 & t > 2\pi \end{cases}$

$f(x) = \begin{cases} 0 & 0 \le x < 8 \\ (x - 8)(10 - x) & 8 \le x \le 10 \end{cases}$

$g(x) = \begin{cases} 0 & 0 \le x < 4 \\ (x - 4)(x - 6) & 4 \le x < 6 \\ 0 & 6 \le x \le 10 \end{cases}$

B18. Add viscous damping to the model of the finite string so that the partial differential equation becomes $u_{tt} = c^2 u_{xx} - bu_t$. Reformulate the explicit finite-difference scheme to account for this additional term, and test the new algorithm on the BVP for which $u(0, t) = u(1, t) = 0, u(x, 0) = f(x) = x(1 - x)$, $u_t(x, 0) = g(x) = 0, c = 1, b = 2, L = 1$.

(a) Approximate u_t with the $O(h^2)$ central-difference formula $u_t(x_n, t_m) = (v_{n,m+1} - v_{n,m-1})/2\Delta t$.

(b) Obtain, in place of (24.21), the equation $v_{n,m+1} = [\lambda^2(v_{n-1,m} + v_{n+1,m}) + 2(1 - \lambda^2)v_{n,m} - v_{n,m-1} + \frac{b\Delta t}{2} v_{n,m-1}] \frac{1}{1+(b\Delta t)/2}$.

(c) Obtain, in place of (24.25), the equation $v_{n,1} = (1 - \lambda^2)f_n + \frac{\lambda^2}{2}(f_{n+1} + f_{n-1}) + g_n \Delta t - \frac{b(\Delta t)^2}{2} g_n$.

(d) Obtain a plot of the solution surface.

(e) Animate the motion of the string, being sure to observe damping of the oscillations.

B19. Take the damping coefficient as $b = 2$ and numerically solve the damped wave equation for the system in Exercise B17. Obtain both a solution surface and an animation of the motion of the string. If damping isn't observed, adjust b until damping is visible.

In Exercises B20–22, a time-dependent wave speed is given. For each:

(a) Modify the explicit algorithm for the undamped wave equation to account for the variable wave speed. In particular, let $c_m = c(t_m)$ and $\lambda_m = c_m \frac{\Delta t}{\Delta x}$. Apply the modified algorithm to the undamped wave equation on a string of length $L = 10$, for which $u(0, t) = u(10, t) = 0, u(x, 0) =$

$$f(x) = \begin{cases} 4x(1-x) & 0 \le x < 1 \\ 0 & 1 \le x \le 10 \end{cases}, u_t(x, 0) = g(x) = 0.$$ Obtain a plot of the solution surface, and animate the motion of the string.

(b) Compare the solution in part (a) with the solution to the same problem for which the wave speed remains constant at $c = 1$.

B20. $c(t) = \left| \sin \dfrac{t}{10} \right|$ **B21.** $c(t) = \dfrac{1 - (\cos t)}{2}$

B22. $c(t) = \dfrac{1}{1 + t^2}$

In each of Exercises B23–25, a position-dependent wave speed is given. For each:

(a) Modify the explicit algorithm for the undamped wave equation to account for the variable wave speed. In particular, let $c_n = c(x_n)$ and $\lambda_n = c_n \frac{\Delta t}{\Delta x}$. Apply the modified algorithm

to the undamped wave equation on a string of length $L = 2$, for which $u(0, t) = u(2, t) = 0, u(x, 0) = f(x) = x(2-x)$, $u_t(x, 0) = g(x) = 0$. Obtain a plot of the solution surface, and animate the motion of the string.

(b) Compare the solution in part (a) with the solution to the same problem for which the wave speed remains constant at $c = 1$.

(c) In part (a), take $f(x) = \begin{cases} x(1-x) & 0 \le x \le 1 \\ 0 & 1 < x \le 5 \end{cases}$, and $L = 5$; solve the wave equation with position-dependent wave speed, and graphically examine the behavior of the solution.

B23. $c(x) = 1 + x$ **B24.** $c(x) = \dfrac{1}{1 + x^2}$

B25. $c(x) = 1 + (x - 1)^2$

Chapter Review

1. Formulate a boundary value problem whose solution describes the planar motion of a taut string of length L, stretched between two fixed endpoints, and whose initial shape and velocity are $f(x)$ and $g(x)$, respectively.

2. In Question 1, let $L = \pi, g(x) = 0$, and $f(x) = \frac{1}{4}x(\pi - x)$. Show how to obtain the solution by the technique of separation of variables. Discuss the Sturm–Liouville eigenvalue problem that arises from the separation process, and show how the eigenvalues and eigenfunctions are obtained. Also, show how the coefficients in the infinite series solution are determined as coefficients of a Fourier sine series.

3. In Question 1, let $L = \pi, f(x) = 0$, and $g(x) = \frac{1}{4}x(\pi - x)$. Show how to solve the resulting BVP by separation of variables.

4. State D'Alembert's solution to the wave equation. In the case $g(x) = 0$, use D'Alembert's formula to explain what happens to an initial disturbance. Explain how to use D'Alembert's formula for wave motion on a finite string given that $g(x) = 0$.

5. Summarize the derivation of the wave equation for the finite string.

6. Formulate a boundary value problem for the longitudinal vibrations in a finite elastic rod whose ends are fixed and whose plane sections that start parallel remain parallel during the motion.

7. Formulate a boundary value problem for the longitudinal vibrations in a finite elastic rod whose ends are free and whose plane sections that start parallel remain parallel during the motion.

8. If $u(x, t)$ represents the longitudinal displacements of plane sections in an elastic rod, explain why $EAu_x(a, t)$ should represent the elastic force on the face that was at $x = a$ at equilibrium.

9. Justify calling $u_x(x, t)$ the *strain* in the elastic rod.

10. Demonstrate how to obtain a finite-difference solution to the wave equation on a finite string with fixed endpoints and with initial shape and velocity given by the functions $f(x)$ and $g(x)$, respectively.

Chapter 25

Heat Equation

INTRODUCTION   Heat transfer along a finite rod whose curved surface is insulated is modeled by the *one-dimensional heat equation* $u_t = \kappa u_{xx}$. We obtain analytic solutions of three BVPs posed for this rod. First, we maintain the temperatures at each end at the same constant temperature, which we take to be zero. Then, we insulate each end so no heat can enter or leave the rod. Then, we consider the problem of time-varying temperatures at the endpoints.

In the first case, that of zero endpoint temperatures, the steady-state solution for the rod is a temperature of zero throughout. In the second, the steady-state temperature is the average value of the initial temperature profile. In both cases, we use the technique of separation of variables, converting the PDE into two ODEs, one of which forms a Sturm–Liouville eigenvalue problem. The problem with fixed endpoint temperatures results in the same eigenfunctions and eigenvalues as we had for the plucked string. Thus, the eigenfunctions are sine functions. The problem with insulated ends has cosines for eigenfunctions, like the elastic rod with free ends.

The analytic solution of the problem with time-varying endpoint conditions is significantly more complex than either of the first two BVPs. The difficulties of this analytic approach are circumvented by a finite-difference numeric solution that easily implements a solution for the BVP with time-varying endpoint temperatures. An analytic solution is postponed for a later chapter.

25.1 One-Dimensional Heat Diffusion

Problem Statement

If an insulated rod of length L has its end faces maintained at a temperature of $0°$, then heat dispersion in the rod is governed by the boundary value problem in Table 25.1, where (1) is the heat equation that we will derive in Section 25.2, (2) and (3) are homogeneous Dirichlet boundary conditions, and (4) is an initial condition. The function $u(x, t)$ gives, at time t, the temperature of the rod at location x (measured from the left end where $x = 0$). Heat can enter and leave the rod only through the ends, not through the round lateral surface. The constant κ is called the *thermal diffusivity*, has units $\frac{\text{length}^2}{\text{time}}$, and is large in metal, which is a good conductor of heat, but small in wool, which is a good insulator. Since the units of κ are essentially $\frac{\text{area}}{\text{time}}$, κ can be interpreted as the amount of area to which heat can spread per unit time.

Heat equation	(1) $u_t = \kappa u_{xx}$	$t > 0$
Boundary condition	(2) $u(0, t) = 0$	Left end at $0°$
Boundary condition	(3) $u(L, t) = 0$	Right end at $0°$
Initial condition	(4) $u(x, 0) = f(x)$	Initial temperature profile

TABLE 25.1   BVP for heat transfer in an insulated rod

EXAMPLE 25.1   The function $u(x, t) = e^{-t} \sin x$ is the solution of the heat equation in a rod of length π for which the initial temperature was $u(x, 0) = \sin x$ and for which the thermal diffusivity is $\kappa = 1$. Direct substitution into the heat equation, the partial differential satisfied by many diffusion processes, of which heat transfer is one familiar example, gives the identity $-e^{-t} \sin x = -e^{-t} \sin x$.

Figure 25.1 is a plot of $u(x, t)$ as a surface over the xt-plane. The plane section $t = 0$ is the initial temperature profile, the graph of $u(x, 0) = \sin x$, $0 \leq x \leq \pi$. The edges of the solution surface corresponding to $x = 0$ and $x = \pi$ remain at $u = 0$. The plane sections $t = c$ for $c > 0$ show a decrease in temperature throughout the rod. This is because heat escapes through the ends, which are held at temperature $u = 0$. A collection of such plane sections are shown in Figure 25.2.

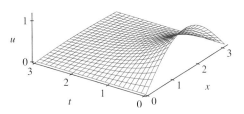

FIGURE 25.1   Solution surface for Example 25.1

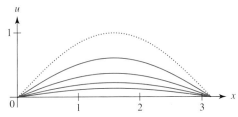

FIGURE 25.2   Example 25.1: Several temperature profiles for $t = \frac{k}{4}$, $k = 0, \ldots, 4$. The dotted curve is $u(x, 0)$

Figure 25.2 shows graphs of temperature profiles at various times $t > 0$. Each plane section $t = t^*$ is the plot of $u(x, t^*)$, the temperature profile in the rod at time $t = t^*$. The initial temperature profile is the dotted curve. The other curves, decreasing in height because the rod is cooling, correspond to profiles at increasing times. Were these profiles to be viewed in sequence, an animation of the temperature history in the rod would result. This more dynamic view would show the decrease in temperatures as a function of time and would illustrate how heat is extracted from the rod by the ice-bath at each end. The ultimate state of the rod, its steady state, is a uniform temperature of $0°$. The accompanying Maple worksheet contains such an animation.   ❖

Solution by Separation of Variables

We next apply the technique of separation of variables to the BVP in Table 25.1, the problem of heat conduction in an insulated rod of length L. According to the separation assumption, set $u(x, t) = X(x)T(t)$ and apply this separation assumption to the endpoint data, obtaining $X(0)T(t) = 0$ and $X(L)T(t) = 0$, from which we conclude, as in Section 24.1, that $X(0) = X(L) = 0$.

Next, apply the separation assumption to the heat equation, which, after division by $\kappa u(x, t) = \kappa X(x) T(t)$, gives

$$\frac{T'(t)}{\kappa T(t)} = \frac{X''(x)}{X(x)} = \lambda$$

The variables are separated, and the two ordinary differential equations

$$X''(x) - \lambda X(x) = 0 \quad \text{and} \quad T'(t) - \lambda \kappa T(t) = 0 \tag{25.1}$$

follow. The Sturm–Liouville eigenvalue problem that results is the same one we saw for the wave equation on the finite string. We know, therefore, that the solution to the BVP

$$X''(x) - \lambda X(x) = 0 \qquad X(0) = X(L) = 0$$

is $X_n = A_n \sin \frac{n\pi x}{L}$, $\lambda_n = -\frac{n^2\pi^2}{L^2}$, $n = 1, 2, \dots$. We can now solve the ODE

$$T'(t) + \kappa \frac{n^2\pi^2}{L^2} T(t) = 0 \tag{25.2}$$

for $T_n(t) = B_n e^{-\kappa n^2 \pi^2 t / L^2}$.

Next, we form the general eigensolution

$$u_n(x, t) = X_n(x) T_n(t) = b_n \sin \frac{n\pi x}{L} e^{-\kappa n^2 \pi^2 t / L^2}$$

A linear combination of all such eigensolutions,

$$u(x, t) = \sum_{n=1}^{\infty} b_n \sin \frac{n\pi x}{L} e^{-\kappa n^2 \pi^2 t / L^2}$$

is the most general solution possible for the original BVP. Application of the initial condition $u(x, 0) = f(x)$, the condition on the initial temperature profile in the rod, leads to the Fourier series. Because

$$u(x, 0) = \sum_{n=1}^{\infty} b_n \sin \frac{n\pi x}{L} = f(x)$$

means the coefficients b_n in the series for $u(x, t)$ must therefore be the Fourier sine series coefficients for representing $f(x)$ on the interval $[0, L]$, we have $b_n = \frac{2}{L} \int_0^L f(x) \sin \frac{n\pi x}{L} dx$.

EXAMPLE 25.2   Let the thermal diffusivity be $\kappa = \frac{1}{2}$, and let the rod have length π. If the initial temperature distribution $u(x, 0)$ is given by

$$f(x) = \begin{cases} 0 & 0 \le x < 1 \\ 1 & 1 \le x \le 2 \\ 0 & 2 < x \le \pi \end{cases}$$

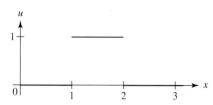

FIGURE 25.3   Initial termperature profile for Example 25.2

we may obtain the complete solution to the heat transfer BVP on this rod. Figure 25.3 shows the physically impossible discontinuities in $f(x)$. This discontinuous temperature profile might be achieved by cutting the rod into three segments, cooling the outer two segments, and heating the center section. Then, at the instant the clock starts ticking, the three segments are slammed together with enough pressure that the thermal interfaces between the segments disappear.

Applying ice cubes to the ends of a warm rod would generate the same discontinuities at $t = 0$. The purpose of this example is to demonstrate that the heat equation instantly smooths out these initial discontinuities because it embodies Newton's law of cooling whereby heat flows as soon as a temperature gradient is established.

For the Fourier sine series for $f(x)$, we compute the coefficients

$$b_n = \frac{2}{\pi} \int_0^\pi f(x) \sin nx \, dx = \frac{2}{n\pi} (\cos n - \cos 2n) \tag{25.3}$$

Then the solution to the boundary value problem is the infinite series

$$u(x, t) = \frac{2}{\pi} \sum_{n=1}^\infty \frac{1}{n} (\cos n - \cos 2n) \sin(nx) e^{-n^2 t/2}$$

At $t = 0$ the series is the Fourier sine series for $f(x)$, a series whose coefficients slowly decrease in magnitude since $b_n = O(\frac{1}{n})$. However, for $t > 0$, the exponential term rapidly becomes small with increasing n. Hence, for $t > 0$ only a few terms of this series are needed for an accurate approximation, whereas for $t = 0$ a very large number of terms are needed to represent the discontinuity in $f(x)$. Approximating the exact solution with the finite sum

$$u_{50}(x, t) = \frac{2}{\pi} \sum_{n=1}^{50} \frac{1}{n} (\cos n - \cos 2n) \sin(nx) e^{-n^2 t/2} \tag{25.4}$$

we obtain Figure 25.4, a graph of the solution surface. First, note the difficulty the discontinuity in the initial temperature profile causes. The approximate solution does not represent the discontinuity very well and tends to show a smooth profile instead of a discontinuous one. Second, note that for $t > 0$, the discontinuity is indeed smoothed out since heat flows instantaneously and everywhere.

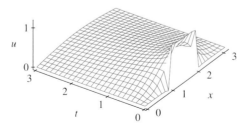

FIGURE 25.4   Solution surface in Example 25.2

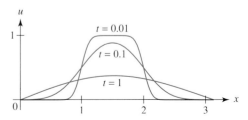

FIGURE 25.5   Example 25.2: Temperature profiles at $t = 0.01, 0.1$, and 1.0

n	$e^{-n^2(0.1)/2}$	$e^{-n^2(1.0)/2}$
1	0.951	0.607
2	0.819	0.135
3	0.638	0.0111
4	0.449	$0.335 10^{-3}$
5	0.287	$0.373 10^{-5}$
6	0.165	$0.152 10^{-7}$
7	0.0863	$0.229 10^{-10}$
8	0.0408	$0.127 10^{-13}$

TABLE 25.2   At $t = 0.1$ and 1.0, the exponential factors for the first eight terms in the solution for Example 25.2

Figure 25.5 shows temperature profiles observed at times $t = 0.01, 0.1$, and 1.0. Again, temperatures change everywhere instantaneously. Every portion of the cool segments instantly experiences a rise in temperature, and every portion of the heated segment instantaneously experiences a drop in temperature. A complete animation of the cooling of this cooling process is found in the accompanying Maple worksheet.

Table 25.2 lists values for the exponential factor $e^{-n^2 t/2}$ for $t = 0.1$ and 1.0. For the larger value of t, the exponential decays more rapidly. Consequently, for $t > 1$, $u(x, t)$ can be well approximated by a few terms of the series. However, as t gets smaller, it requires more and more terms of the series to approximate the solution accurately. The rapid decay of the exponential term $e^{-\kappa(n\pi/L)^2 t}$ in the solution of the heat equation is responsible for the smoothing seen in solutions of this equation. This smoothing property is built into the equation, as the derivation in the Section 25.2 will show. ❖

EXERCISES 25.1–Part A

A1. Verify the derivation of the equations in (25.1).

A2. Obtain the solution of (25.2).

A3. Verify the integration in (25.3).

A4. Evaluate $e^{-n^2(0.01)/2}$ for $n = 5, 10, 20$.

A5. Show that the method of separation of variables fails for the BVP in Table 25.1 if the Dirichlet conditions at the endpoints are not homogeneous.

EXERCISES 25.1–Part B

B1. Use (25.4) to obtain a graph of $u_x(0, t)$, the heat flux at the left end of the rod in Example 25.2. Verify that this partial sum suffices by comparing this graph with the one generated by a partial sum of 100 terms. Explain the physics this graph represents.

B2. Repeat Exercise B1 at $x = 1$, the point where the initial temperature was discontinuous.

In Exercises B3–7:

   (a) Formulate and solve $u_t = \kappa u_{xx}$, $u(0, t) = u(L, t) = 0$, $u(x, 0) = f(x)$, the BVP for heat transfer in an insulated rod with its ends kept at a temperature of zero.

   (b) For the series found in part (a), let $U(x, t)$ be a partial sum with enough terms so that $U(x, 0)$ is a reasonable approximation to $f(x)$. Plot $U(x, t)$ as an approximation to the solution surface.

   (c) Animate $U(x, t)$, or at least plot time-slices at a sequence of increasing times.

   (d) Plot $U_x(0, t)$ and $U_x(L, t)$. These are the temperature gradients at the end faces and are proportional to the heat flowing through the respective ends of the rod. Give a physical interpretation of the graphs.

B3. $L = \pi, \kappa = \frac{1}{2}, f(x) = x(\pi - x)$

B4. $L = \pi, \kappa = \frac{1}{4}$,

$$f(x) = \begin{cases} 0 & 0 \le x < \frac{\pi}{3} \\ 5\left(x - \frac{\pi}{3}\right)\left(\frac{2\pi}{3} - x\right) & \frac{\pi}{3} \le x < \frac{2\pi}{3} \\ 0 & \frac{2\pi}{3} \le x \le \pi \end{cases}$$

B5. $L = 4, \kappa = \frac{2}{3}, f(x) = \begin{cases} 0 & 0 \le x < 2 \\ 5 & 2 \le x < 3 \\ 0 & 3 \le x \le 4 \end{cases}$

B6. $L = 3, \kappa = \frac{3}{4}, f(x) = \begin{cases} 0 & 0 \le x < 1 \\ (x-1)(3-x) & 1 \le x \le 3 \end{cases}$

B7. $L = 5, \kappa = \frac{2}{5}, f(x) = \begin{cases} 0 & 0 \le x < 1 \\ 1 & 1 \le x < 2 \\ 0 & 2 \le x < 3 \\ 1 & 3 \le x < 4 \\ 0 & 4 \le x \le 5 \end{cases}$

B8. For $0 < \kappa \le 1$, what is the maximum temperature at $x = \frac{5}{2}$ in the rod of Exercise B7? *Hint:* Obtain an approximate solution $U(x, t)$ in which κ is a parameter, then examine graphs of $U(2.5, t)$ for different values of κ. An efficient way to do this is to animate on the parameter κ.

B9. The partial differential equation $u_t = \kappa u_{xx} + hu$, $h > 0$, models one-dimensional heat transfer in the presence of a source term. A finite rod made of heat-generating radioactive material might be so modeled.

   (a) Let $L = \pi$, and solve the BVP $u_t = u_{xx} + hu$, $u(0, t) = u(\pi, t) = 0$, $u(x, 0) = x(\pi - x)$. Note that $T_n(t)$ is a constant multiple of $e^{(h-n^2)t}$, so the solution will behave differently depending on the magnitude of h.

   (b) Let $h = \frac{1}{2}$, and plot the resulting solution surface. Also, plot the rod's temperature at a succession of several times. If possible, obtain an animation of the time-history of the temperatures in the rod. Plot $u_x(0, t)$, the temperature gradient at the left end of the rod.

   (c) Let $h = 1$ and repeat part (b).

   (d) Let $h = \frac{3}{2}$ and repeat part (b).

   (e) In each case, explain the observed behavior of the solution.

B10. In the belief that the term u_x adds viscous resistance to the flow of heat in a rod of length $L = \pi$, the partial differential equation $u_t = \kappa u_{xx} + 2\alpha u_x$ is used to model heat flow in this rod whose ends are kept at a temperature of zero.

   (a) Show that the differential equation is separable, with eigenvalues $\lambda_n = -n^2 - \frac{\alpha^2}{\kappa^2}$, $n = 1, \ldots$ and eigenfunctions $X_n(x) = b_n e^{-\alpha x/\kappa} \sin nx$, where b_n is a constant.

   (b) Show that $T_n(t)$ is a constant multiple of $e^{-(\alpha^2/\kappa + \kappa n^2)t}$.

   (c) Show that on the interval $[0, \pi]$, the eigenfunctions $X_n(x)$ are orthogonal with respect to the weight function $w(x) = e^{2\alpha x/\kappa}$.

   (d) The solution of the BVP can be written as $u(x, t) = \sum_{n=0}^{\infty} b_n e^{-\alpha x/\kappa} \sin(nx) e^{-(\alpha^2/\kappa + \kappa n^2)t}$, where $b_n = \frac{2}{\pi} \int_0^\pi f(x) w(x) X_n(x) \, dx$. Obtain $U(x, t)$, an approximating partial sum for the BVP with $f(x) = x(\pi - x)$ and $\alpha = \frac{1}{2}$, $\kappa = \frac{1}{3}$.

   (e) Animate $U(x, t)$, or at least plot time-slices at a sequence of increasing times. Is the viscous damping independent of x?

Derivation of the One-Dimensional Heat Equation

Derivation of the Heat Equation

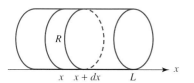

FIGURE 25.6   Segment of an insulated rod

Consider a rod of length L on an x-axis, the left end being at the origin. In this rod imagine R, a segment of length dx, lying between x and $x + dx$. (See Figure 25.6.) Let $Q(t)$ be the total quantity of heat, measured in calories, contained in the region R; let $u(x, t)$ be the temperature at location x at time t; and let M be a constant that converts temperatures (in degrees) to heat (in calories) via the integral

$$Q(t) = \int_x^{x+dx} Mu(x, t)\, dx$$

The constant M has units $\frac{\text{calories}}{\text{degrees length}}$. This is not unreasonable, since specific heat is defined as $\frac{\text{calories}}{\text{degrees mass}}$, so M is a "linear" specific heat that gives the number of calories required to raise the temperature of a one-unit length of rod one degree.

The rate of change of the amount of heat in the region R is then

$$\frac{dQ}{dt} = \int_x^{x+dx} Mu_t(x, t)\, dx$$

For a small segment R (dx is small) we can argue heuristically that $u_t(x, t)$ is roughly constant over the interval $[x, x + dx]$. Hence, the integral is approximately

$$\frac{dQ}{dt} = Mu_t(x, t) \int_x^{x+dx} dx = Mu_t(x, t)\, dx$$

For there to be a change in heat content in R, heat must pass through one of the faces of R, faces located at x and $x + dx$. Now heat will flow across one of these faces if there is a difference of temperature on either side of such a face. A spatial region in which there is a difference in temperature is said to have a *temperature gradient*, much like land has a gradient wherever it has a difference in height. Let $u_x(x, t)$ be the *temperature gradient* across the face at location x, and let $u_x(x + dx, t)$ be the *temperature gradient* across the face at location $x + dx$. Newton's law of cooling (from Section 2.4) says that the flow of heat is proportional to the difference in temperature. Make that same assumption here. It amounts to the claim that heat has no inertia. Hence, there is **instantaneous** heat flow the instant a temperature gradient exists. (This assumption is responsible for the smoothing property of the heat equation.)

Let N be a constant of proportionality between the temperature gradient, which itself has units $\frac{\text{degrees}}{\text{length}}$, and the resulting amount of heat that flows, which therefore has units $\frac{\text{calories}}{\text{time}}$. The constant N will consequently have the units $\frac{\text{calories length}}{\text{time degrees}}$ so that $Nu_x(x, t)$ is the amount of heat flowing per second across the face at x and $Nu_x(x + dx, t)$ is the amount of heat flowing per second across the face at $x + dx$. The net heat flowing in/out of R, per second, which is actually the same quantity as $\frac{dQ}{dt}$, is therefore $N[u_x(x + dx, t) - u_x(x, t)]$.

Thus, the equation for monitoring heat flow into and out of R is

$$Mu_t(x, t)\, dx = N[u_x(x + dx, t) - u_x(x, t)]$$

Dividing by both dx and M we have

$$u_t(x, t) = \frac{N}{M}\left[\frac{u_x(x + dx, t) - u_x(x, t)}{dx}\right]$$

In the limit as $dx \to 0$, the difference quotient on the right becomes $u_{xx}(x, t)$. The constant

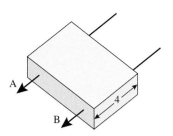

FIGURE 25.7   Uniform slab supporting one-dimensional heat flow in Example 25.3

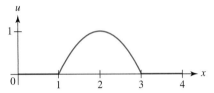

FIGURE 25.8   Example 25.3: $u(x, 0) = f(x)$, the initial temperature distribution

$\frac{N}{M}$ is the *thermal diffusivity* κ with units

$$\frac{\frac{\text{calories length}}{\text{time degrees}}}{\frac{\text{calories}}{\text{degrees length}}} = \frac{\text{length}^2}{\text{time}}$$

The reader is cautioned that there is another, similar constant used in heat transfer. The *thermal conductivity* K has units $\frac{\text{calories}}{\text{time degrees length}}$ and is related to the thermal diffusivity by $\kappa = \frac{K}{c\rho}$, where ρ is the density, with units $\frac{\text{mass}}{\text{volume}}$, and c, the specific heat with units $\frac{\text{calories}}{\text{degrees mass}}$, is the amount of heat required to raise the temperature of a unit mass by one degree. A moment's notice suggests that both κ and K will be "big" in metals and "small" in insulators.

EXAMPLE 25.3   The rectangular slab in Figure 25.7 has width 4, and, except for its left and right faces (which are kept at a temperature of zero degrees), is coated with insulation. Assuming $\kappa = \frac{1}{3}$, and that heat flow is one-dimensional (strictly horizontal, as suggested by arrows A and B in Figure 25.7), solve for the temperatures in the slab if the initial temperature distribution along any horizontal line is given by the function $f(x)$ on the right in (25.5). Figure 25.8 contains a graph of $f(x)$, and $x = 0$ corresponds to the left edge of the slab. The boundary value problem to be solved is on the left in

$$u_t(x, t) = \tfrac{1}{3}u_{xx}(x, t), t > 0 \quad u(0, t) = 0$$
$$u(x, 0) = f(x) \qquad\qquad u(4, t) = 0$$

with $f(x) = \begin{cases} 0 & 0 \le x < 1 \\ (1 - x)(x - 3) & 1 \le x \le 3 \\ 0 & 3 < x \le 4 \end{cases}$

(25.5)

From our work in Section 25.1, the solution of (25.5) is

$$u(x, t) = \sum_{n=1}^{\infty} b_n \sin\left(\frac{n\pi x}{4}\right) e^{-n^2\pi^2 t/48} \qquad \textbf{(25.6a)}$$

$$b_n = \frac{1}{2} \int_0^4 f(x) \sin\frac{n\pi x}{4} \, dx \qquad \textbf{(25.6b)}$$

$$= \frac{16}{n^3\pi^3}\left(4\cos\frac{n\pi}{4} - n\pi \sin\frac{n\pi}{4} - 4\cos\frac{3n\pi}{4} - n\pi \sin\frac{3n\pi}{4}\right) \qquad \textbf{(25.6c)}$$

Table 25.3 lists the first 16 coefficients, giving some sense of how fast these coefficients

n	b_n		n	b_n	
1	$-16\dfrac{\sqrt{2}(-4 + \pi)}{\pi^3}$	0.626	9	$-\dfrac{16}{729}\dfrac{\sqrt{2}(-4 + 9\pi)}{\pi^3}$	-0.0243
2	0	-0.194×10^{-9}	10	0	-0.155×10^{-10}
3	$-\dfrac{16}{27}\dfrac{\sqrt{2}(4 + 3\pi)}{\pi^3}$	-0.363	11	$-\dfrac{16}{1331}\dfrac{\sqrt{2}(4 + 11\pi)}{\pi^3}$	-0.0211
4	0	0.165×10^{-9}	12	0	0.100×10^{-9}
5	$\dfrac{16}{125}\dfrac{\sqrt{2}(-4 + 5\pi)}{\pi^3}$	0.0684	13	$\dfrac{16}{2197}\dfrac{\sqrt{2}(-4 + 13\pi)}{\pi^3}$	0.0122
6	0	0	14	0	0.752×10^{-11}
7	$\dfrac{16}{343}\dfrac{\sqrt{2}(4 + 7\pi)}{\pi^3}$	0.0553	15	$\dfrac{16}{3375}\dfrac{\sqrt{2}(4 + 15\pi)}{\pi^3}$	0.0111
8	0	0.178×10^{-10}	16	0	-0.796×10^{-10}

TABLE 25.3   Example 25.3: Coefficients b_n in exact and floating-point form

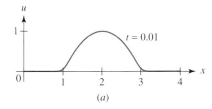

(a)

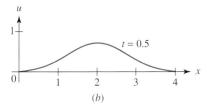

(b)

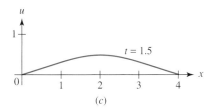

(c)

FIGURE 25.10 Example 25.3: Temperature profiles at times $t = 0.01$, 0.5, and 1.5

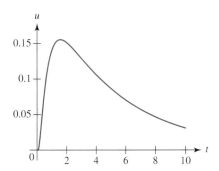

FIGURE 25.11 A graph of $u(\frac{1}{2}, t)$ in Example 25.3

get small. The even-indexed coefficients are zero, and the odd-indexed coefficients slowly decay. Hence, it will take a partial sum with more than just a few terms to approximate the solution well in any region which includes the x-axis.

The 50-term approximation $u_{50}(x, t) = \sum_{n=1}^{50} b_n \sin(\frac{n\pi x}{4})e^{-n^2\pi^2 t/48}$ is graphed in Figure 25.9. For $t > 0$, the temperatures change everywhere and instantaneously. Where $u = 0$ initially, the temperature instantly rises. Where the temperatures were positive initially, they decrease instantly.

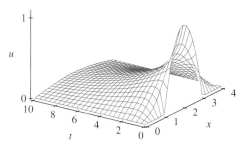

FIGURE 25.9 Solution surface for Example 25.3

The dynamic behavior of the temperature profiles is better seen by examining the plane sections $t = c$ representing instantaneous temperature profiles. Temperatures in the intervals [0, 1] and [3, 4] instantly rise from zero to some positive temperature, then later decrease back to zero. This happens because of the assumption that heat flows the instant there is a temperature gradient (Newton's law of cooling). Instantly, the temperatures in those two segments rise. Heat flow is everywhere, instantaneous, and has no inertia. The three plane sections $t = 0.01, 0.5$, and 1.5 are shown in Figure 25.10, and an animation of the cooling process is found in the accompanying Maple worksheet.

In fact, if the temperatures at $x = \frac{1}{2}$ are plotted as a function of time, Figure 25.11 results, showing that for at least one point outside the initially heated region, the temperature does immediately increase. ❖

EXERCISES 25.2

1. Verify that (25.6c) gives the correct evaluation of (25.6b).

In each of Exercises 2–6, a rectangular slab of width L has its front and back faces, as well as its top and bottom edges, insulated. The left and right edges are kept at a temperature of zero degrees, the flow of heat is one-dimensional, and the initial temperature distribution along any horizontal line is given by $u(x, 0) = f(x)$.

    (a) Formulate and solve a relevant BVP for the temperatures in the slab.

(b) For the series found in part (a), let $U(x, t)$ be a partial sum with enough terms so that $U(x, 0)$ is a reasonable approximation to $f(x)$. Plot $U(x, t)$ as an approximation to the solution surface.

(c) Animate $U(x, t)$, or at least plot time-slices at a sequence of increasing times.

(d) On the same set of axes, plot $U_x(0, t)$ and $U_x(L, t)$. These are the temperature gradients at the end faces and are proportional to the heat flowing through the respective ends of the rod.

2. $L = 3, \kappa = \frac{4}{3}, f(x) = 3x^2 - x^3$

3. $L = 5, \kappa = \frac{1}{2}, f(x) = \begin{cases} 0 & 0 \le x < 1 \\ 1 & 1 \le x < 2 \\ 0 & 2 \le x < 3 \\ 3 & 3 \le x < 4 \\ 0 & 4 \le x \le 5 \end{cases}$

4. $L = 5, \kappa = 2, f(x) = \begin{cases} 0 & 0 \le x < 1 \\ x - 1 & 1 \le x < 2 \\ 1 & 2 \le x < 3 \\ \dfrac{5 - x}{2} & 3 \le x \le 5 \end{cases}$

5. $L = 4, \kappa = 3, f(x) = \begin{cases} 0 & 0 \le x < 1 \\ 1 & 1 \le x < 2 \\ 2 & 2 \le x < 3 \\ 0 & 3 \le x \le 4 \end{cases}$

6. $L = 4, \kappa = \frac{3}{5}, f(x) = \begin{cases} x & 0 \le x < 1 \\ 3 & 1 \le x < 3 \\ 4 - x & 3 \le x \le 4 \end{cases}$

For each of the indicated heat-flow problems in Exercises 7–11:

(a) Calculate $Q_0 = M \int_0^L f(x)\, dx$, the initial heat content, where M, constant, is unknown.

(b) Calculate $F_0 = N U_x(0, t)$ and $F_L = N U_x(L, t)$, the heat flux at the left and right faces, respectively. The constant $N = \kappa M$ is unknown. Having used $U(x, t)$ instead of $u(x, t)$, these time-dependent fluxes are approximate, not exact.

(c) Calculate $\Delta Q_{\text{total}} = \int_0^\infty (F_0 - F_L)\, dt$, an approximation to the total heat that has flowed from the slab.

(d) Using $N = \kappa M$, show that $Q_0 = \Delta Q_{\text{total}}$, thus showing that the total initial heat content has dissipated through the two ends.

(e) Calculate $Q(t) = M \int_0^L U(x, t)\, dx$, the time-dependent heat content in the slab.

(f) Plot $\frac{Q(t)}{M}$ and $\frac{Q'(t)}{M} = \int_0^L U_t(x, t)\, dx$, where the latter is proportional to the change in heat content in the slab.

(g) Show that $Q'(t) = F_L - F_0$ by showing
$$[U_x(L, t) - U_x(0, t)]/\int_0^L U_t(x, t)\, dx = \tfrac{1}{\kappa}.$$

7. Exercise 2   **8.** Exercise 3   **9.** Exercise 4

10. Exercise 5   **11.** Exercise 6

When computing the Fourier sine series for $f(x) = \cos 3x, 0 \le x \le \pi$, the coefficients are given by

$$b_n = \frac{2}{\pi} \int_0^\pi \cos 3x \sin nx\, dx = \begin{cases} \dfrac{2n(1 + (-1)^n)}{\pi(n^2 - 9)} & n \ne 3 \\ 0 & n = 3 \end{cases}$$

The integral formula that yields $2n(1 + (-1)^n)/\pi(n^2 - 9)$ is not valid when $n = 3$. A different integration formula must be used to compute b_3. When instructing a computer program to construct a partial sum of this Fourier series, it is best to obtain $b_1 = 0$ and $b_2 = -\frac{8}{5\pi}$ first. Then, the partial sum $-\frac{8}{5\pi} + \sum_{n=4}^N b_n \sin nx$ can be written. In Exercises 12–14:

(a) Solve the BVP $u_t = \kappa u_{xx}, u(0, t) = u(\pi, t) = 0,$ $u(x, 0) = f(x).$

(b) For the series found in part (a), let $U(x, t)$ be a partial sum with enough terms so that $U(x, 0)$ is a reasonable approximation to $f(x)$. Plot $U(x, t)$ as an approximation to the solution surface.

(c) Animate $U(x, t)$, or at least plot time-slices at a sequence of increasing times.

12. $\kappa = 2, f(x) = 3 \cos 2x$ **13.** $\kappa = \frac{1}{2}, f(x) = 5 \cos 4x$

14. $\kappa = \sqrt{2}, f(x) = \pi \cos 5x$

25.3 Heat Flow in a Rod with Insulated Ends

General Solution

Consider the heat equation in a rod whose length is L and whose ends are insulated. The associated boundary value problem consists of the heat equation $u_t(x, t) = \kappa u_{xx}$, an initial temperature distribution $u(x, 0) = f(x)$, and homogeneous Neumann boundary conditions $u_x(0, t) = 0$ and $u_x(L, t) = 0$ at the ends of the rod. Requiring the temperature gradient at the ends of the rod to vanish means there is no temperature gradient at those faces. Hence, there can be no heat flow across the ends of the rod, so "insulation" has been achieved. Where there is no temperature gradient, there can be no heat flow.

To solve this problem by the technique of separation of variables, assume a solution of the form $u(x, t) = X(x)T(t)$, a product of one function just containing x and one function just containing t. Then, apply the separation assumption to the initial data, obtaining

$$u_x(0, t) = 0 \Rightarrow X'(0)T(t) = 0 \Rightarrow X'(0) = 0$$
$$u_x(L, t) = 0 \Rightarrow X'(L)T(t) = 0 \Rightarrow X'(L) = 0$$

The consequences of the separation assumption and a division by $\kappa u(x,t) = \kappa X(x)T(t)$ are the separated form

$$\frac{T'(t)}{\kappa T(t)} = \frac{X''(x)}{X(x)} = \lambda$$

and the two ODEs

$$X''(x) - \lambda X(x) = 0 \quad \text{and} \quad T'(t) - \lambda \kappa T(t) = 0$$

These ODEs are not solved in isolation. There is data belonging to each equation. Hence, in (25.7) we have two problems to solve, namely,

$$\begin{aligned} X''(x) - \lambda X(x) &= 0 \\ X'(0) = X'(L) &= 0 \end{aligned} \quad \text{and} \quad T'(t) - \lambda k T(t) = 0 \tag{25.7}$$

a Sturm–Liouville eigenvalue problem on the left and an ODE on the right.

As seen in Section 16.1, the solution of this Sturm–Liouville eigenvalue problem is not $\sin(\frac{n\pi x}{L})$. For this problem, $\lambda = 0$ is an eigenvalue, with corresponding eigenfunction $X_0(x) = $ constant. In addition, the other eigenvalues are $\lambda = -\left(\frac{n\pi}{L}\right)^2$, $n = 1, 2, \ldots$, with corresponding eigenfunctions $X_n(x) = c_n \cos(\frac{n\pi x}{L})$. The functions $T_n(t)$ now satisfy

$$T_n'(t) + \kappa \left(\frac{n\pi}{L}\right)^2 T_n(t) = 0$$

so $T_0(t) = $ constant and $T_n(t) = d_n e^{-\kappa(n\pi/L)^2 t}$. The general eigensolution is then

$$u_n(x,t) = X_n(x)T_n(t) = a_n \cos\left(\frac{n\pi x}{L}\right) e^{-\kappa(n\pi/L)^2 t}$$

The complete solution of the partial differential equation will be a linear combination of all such eigensolutions. Hence, we take the most general possible linear combination of all the eigensolutions. Since there are an infinite number of eigensolutions, our linear combination is the infinite sum

$$u(x,t) = \frac{a_0}{2} + \sum_{n=1}^{\infty} a_n \cos\left(\frac{n\pi x}{L}\right) e^{-\kappa(n\pi/L)^2 t} \tag{25.8}$$

The final condition to be applied is the initial shape requirement, $u(x,0) = f(x)$, giving

$$u(x,0) = \frac{a_0}{2} + \sum_{n=1}^{\infty} a_n \cos \frac{n\pi x}{L} = f(x)$$

The coefficients must be the Fourier cosine series coefficients $a_n = \frac{2}{L}\int_0^L f(x)\cos(\frac{n\pi x}{L})\,dx$, $n = 0, 1, 2, \ldots$, for the function $f(x)$.

EXAMPLE 25.4 Let a rod with insulated ends have length π, thermal diffusivity $\kappa = \frac{1}{4}$, and initial temperature distribution

$$f(x) = \begin{cases} 0 & 0 \le x < \frac{\pi}{3} \\ 4\left(\frac{\pi}{3} - x\right)\left(x - \frac{2\pi}{3}\right) & \frac{\pi}{3} \le x \le \frac{2\pi}{3} \\ 0 & \frac{2\pi}{3} < x \le \pi \end{cases}$$

FIGURE 25.12   Example 25.4: $u(x,0) = f(x)$, the initial temperature distribution

graphed in Figure 25.12. This puts heat in the center of the rod, heat that cannot escape out the insulated ends. Hence, the long-term outcome will be to even out the temperature throughout the rod.

The coefficients for the Fourier cosine series of $f(x)$ are

$$a_0 = \frac{2}{\pi} \int_0^\pi f(x)\, dx = \frac{4}{81}\pi^2 \tag{25.9a}$$

$$a_n = \frac{2}{\pi} \int_0^\pi f(x) \cos nx\, dx \tag{25.9b}$$

$$= -\frac{8}{3\pi n^3}\left(n\pi \cos \frac{2}{3} n\pi - 6 \sin \frac{2}{3} n\pi + n\pi \cos \frac{n\pi}{3} + 6 \sin \frac{n\pi}{3} \right) \tag{25.9c}$$

These coefficients are $O(n^{-2})$, so they decrease at a satisfactory rate. Figure 25.13 contains a graph of the solution surface as approximated by

$$u_{50}(x, t) = \frac{a_0}{2} + \sum_{n=1}^{50} a_n \cos(nx) e^{-n^2 t/4}$$

the 50-term partial sum. Again, there is instantaneous heat flow everywhere. In the portions of the rod where $u = 0$ initially, there is an immediate rise in temperature for $t > 0$ and an immediate reduction in temperature where initially the temperature was positive. In addition, along the insulated edges $x = 0$ and $x = \pi$, the temperature rises since heat can no longer escape from the ends of the rod. In fact, the temperature of the rod approaches a constant value for sufficiently large t. It's as if the heat content were water enclosed in a vessel, and the water flowed out to a uniform level.

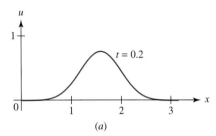

(a)

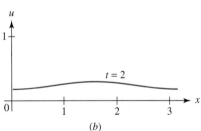

(b)

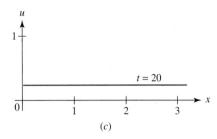

(c)

FIGURE 25.14 Example 25.4: Temperature profiles at times $t = 0.2$, 2.0, and 20

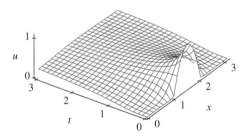

FIGURE 25.13   Solution surface in Example 25.4

Some of this time-dependent behavior can be seen in Figure 25.14, which contains the plane sections $t = 0.2$, 2.0, and 20. The steady-state solution is not $u(x, t) = 0$, but rather, some nonzero constant. The total initial caloric content of the rod becomes evened out, much like water seeking its own level. The uniform level representing the steady-state temperature in the rod can be guessed from $\frac{a_0}{2} = 0.2436939359$. In fact, if $t \to \infty$ in (25.8), it's clear that the only term surviving at steady state is the constant term outside the sum, the only term not multiplied by an exponential function. That surviving constant is $\frac{a_0}{2}$.

Evolution to steady state can be seen in an animation available with the accompanying software. ◆

EXERCISES 25.3

1. Verify the integration in (25.9a) and verify that (25.9c) gives the correct evaluation of the integral in (25.9b).

In Exercises 2–16:

(a) Formulate and solve the BVP governing heat transfer in the indicated rod whose ends are insulated and whose initial temperature profile is given by the function $f(x)$.

(b) For the series found in part (a), let $U(x, t)$ be a partial sum with enough terms so that $U(x, 0)$ is a reasonable approximation to $f(x)$. Plot $U(x, t)$ as an approximation to the solution surface.

(c) Animate $U(x, t)$, or at least plot time-slices at a sequence of increasing times.

(d) Obtain the steady-state temperature of the rod.

(e) Since $u_x(x, t)$ is proportional to the heat flow at x, obtain and plot $U_x(j\frac{L}{4}, t)$, $j = 1, 2, 3$.

2. $L = \pi, \kappa = \frac{1}{2}, f(x) = x(\pi - x)$

3. $L = \pi, \kappa = 2, f(x) = 3 \sin 2x$

4. $L = 3, \kappa = \frac{2}{3}, f(x) = \begin{cases} 0 & 0 \le x < 1 \\ 4(x-1)(2-x) & 1 \le x < 2 \\ 0 & 2 \le x \le 3 \end{cases}$

5. $L = \pi, \kappa = \sqrt{2}, f(x) = 5 \sin 3x$

6. $L = 4, \kappa = \frac{3}{2}, f(x) = \begin{cases} 0 & 0 \le x < 2 \\ 5 & 2 \le x < 3 \\ 0 & 3 \le x \le 4 \end{cases}$

7. $L = 3, \kappa = \frac{4}{3}, f(x) = \begin{cases} 0 & 0 \le x < 1 \\ 2(x-1)(3-x) & 1 \le x \le 3 \end{cases}$

8. $L = \pi, \kappa = 3, f(x) = \sin^2 x$

9. $L = 5, \kappa = 2, f(x) = \begin{cases} 0 & 0 \le x < 1 \\ 1 & 1 \le x < 2 \\ 0 & 2 \le x < 3 \\ 1 & 3 \le x < 4 \\ 0 & 4 \le x \le 5 \end{cases}$

10. $L = 3, \kappa = \sqrt{3}, f(x) = 3x^2 - x^3$

11. $L = 5, \kappa = 2, f(x) = \begin{cases} 0 & 0 \le x < 1 \\ 1 & 1 \le x < 2 \\ 0 & 2 \le x < 3 \\ 3 & 3 \le x < 4 \\ 0 & 4 \le x \le 5 \end{cases}$

12. $L = 5, \kappa = 2, f(x) = \begin{cases} 0 & 0 \le x < 1 \\ x - 1 & 1 \le x < 2 \\ 1 & 2 \le x < 3 \\ \frac{5-x}{2} & 3 \le x \le 5 \end{cases}$

13. $L = 4, \kappa = \frac{1}{5}, f(x) = \begin{cases} 0 & 0 \le x < 1 \\ 1 & 1 \le x < 2 \\ 2 & 2 \le x < 3 \\ 3 & 3 \le x \le 4 \end{cases}$

14. $L = 1, \kappa = \frac{1}{3}, f(x) = x$

15. $L = 1, \kappa = \frac{1}{3}, f(x) = x^2$

16. $L = 4, \kappa = \frac{3}{5}, f(x) = \begin{cases} x & 0 \le x < 1 \\ 3 & 1 \le x < 3 \\ 4 - x & 3 \le x \le 4 \end{cases}$

25.4   Finite-Difference Solution of the One-Dimensional Heat Equation

Problem Statement

As in Section 24.6 we seek a numeric solution by the finite-difference technique for the boundary value problem consisting of the one-dimensional heat equation $u_t = \kappa u_{xx}$ in a rod of length L, the boundary conditions $u(0, t) = \psi_1(t)$, $u(L, t) = \psi_2(t)$, and the initial condition $u(x, 0) = f(x)$. Notice that the boundary conditions are not homogeneous, so an exact solution of this problem must be postponed until Section 27.2. Of course, if $\psi_1(t) = \psi_2(t) = 0$, the problem reduces to one for which we can compute an exact solution with the tools presently at hand.

An Explicit Scheme

First, a grid is defined on the rectangle $0 \le x \le L, 0 \le t \le T$, in the xt-plane, with T being the length of time for which the solution is to be computed. Let the x-interval $[0, L]$ be divided into N equal subintervals of length $\Delta x = \frac{L}{N}$ by the $N+1$ points $x_n = x_0 + n\Delta x$,

$n = 0, 1, \ldots, N$. Similarly, discretize the t-axis with the partition $t_m = t_0 + m\Delta t$, $m = 0, 1, \ldots, M$, where $\Delta t \leq \frac{(\Delta x)^2}{2\kappa}$. For the explicit method being described, this is a very real restriction on the size of the time-step permitted and is derived.

Let $u(x_n, t_m)$ be the exact value of the solution at the point (x_n, t_m). Let $v_{n,m}$ be the numeric approximation to that exact value, where $v_{n,m}$ is computed from

$$v_{n,m+1} = \lambda v_{n-1,m} + (1 - 2\lambda)v_{n,m} + \lambda v_{n+1,m} \qquad (25.10)$$

with $\lambda = \frac{\kappa \Delta t}{(\Delta x)^2} \leq \frac{1}{2}$. (Both (25.10) and the restriction on λ are derived here.) Then, except for round-off error, the difference $u(x_n, t_m) - v_{n,m}$ goes to zero as Δx and Δt themselves go to zero.

EXAMPLE 25.5 Let $L = 1$; and define $\psi_1(t) = \sin t$, $\psi_2(t) = \cos t$. Take $\kappa = \frac{1}{2}$ and pick $\Delta x = \frac{1}{5}$ corresponding to the choice $N = 5$. Let $\Delta t = 0.03 < \frac{(1/5)^2}{2(1/2)} = 0.04$, and take $f(x) = x$ as the initial temperature in the rod. Define the nodes $(x_n, t_m) = (\frac{n}{5}, \frac{3}{100}m)$. Then, with $\lambda = \frac{3}{8}$, define $\sigma = 1 - 2\lambda = \frac{1}{4}$. There are $N - 1 = 4$ interior points at which new values $v_{n,m+1}$ are computed from old values by the equation

$$v_{n,m+1} = \lambda v_{n-1,m} + \sigma v_{n,m} + \lambda v_{n+1,m} \qquad (25.11)$$

as per the computational molecule (schematic) in Figure 25.15. The solid dots represent known values that are used to compute the unknown $v_{n,m+1}$ represented by the open dot in the schematic.

From three contiguous values at time t_m, one new value of $v_{n,m+1}$ can be computed at time t_{m+1}. The "vertical riser" connecting (x_n, t_m) with (x_n, t_{m+1}) moves from left to right between x_1 and x_{N-1}, so the values of $v_{0,m+1}$ and $v_{N,m+1}$ are computed from the endpoint data, not from (25.11).

For the first "row," the values of $v_{n,0}$ are obtained from the initial function, $f(x)$. Hence, to begin the calculation, the initialization step defines values for $v_{n,0}$, $n = 0, \ldots, N$. Then, each new "row" can be computed from the previous "row" provided (25.11) is used for the $N - 1$ interior points, and the endpoints are computed from the boundary functions.

The solution surface in Figure 25.16 and the three temperature profiles in Figure 25.17 are obtained by graphing the results of these calculations. ❖

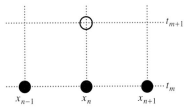

FIGURE 25.15 Computational molecule for the explicit finite-difference scheme for the heat equation

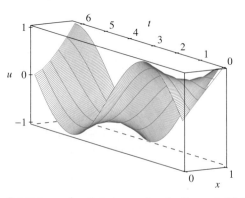

FIGURE 25.16 Solution surface for Example 25.5

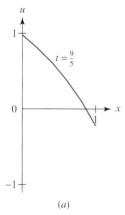

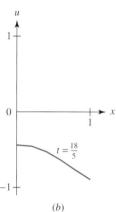

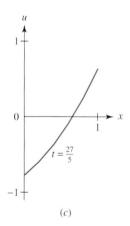

FIGURE 25.17   Example 25.5: Temperature profiles at times $t = \frac{9}{5}, \frac{18}{5},$ and $\frac{27}{5}$

Derivation of the Explicit Method

The explicit method (25.10) is obtained by making the $O(h^2)$ central-difference replacement

$$u_{xx}(x_n, t_m) = \frac{v_{n-1,m} - 2v_{n,m} + v_{n+1,m}}{(\Delta x)^2}$$

and the $O(h)$ forward-difference replacement

$$u_t(x_n, t_m) = \frac{v_{n,m+1} - v_{n,m}}{\Delta t}$$

in the heat equation, defining $\lambda = \frac{\kappa \Delta t}{(\Delta x)^2}$, and solving

$$\frac{v_{n,m+1} - v_{n,m}}{\Delta t} = \kappa \frac{v_{n-1,m} - 2v_{n,m} + v_{n+1,m}}{(\Delta x)^2} \qquad (25.12)$$

for $v_{n,m+1}$. (See Exercise 1.)

DERIVATION OF THE RESTRICTION $0 < \lambda \leq \frac{1}{2}$ The following technique for determining the stability of a finite-difference scheme is due to John von Neumann. At time $t_m = t_0 + m\Delta t$, assume $v_{n,m}$ can be expanded in Fourier series, the terms of which are of the form $\phi(t_m)\cos(\omega x_n)$. Equation (25.10) is then transformed by

1. using dx and dt in place of Δx and Δt, respectively when defining λ.

2. writing $\phi(t_k) = \phi_k$.

3. noting that $\cos(\omega x_{n\pm 1}) = \cos(\omega(x_n \pm dx)) = \cos(\omega x_n)\cos(\omega\, dx) \mp \sin(\omega x_n)\sin(\omega\, dx)$.

4. dividing through by $\phi_m \cos(\omega x_n)$.

5. defining $R = \phi_{m+1}/\phi_m$ and noting it is the *amplification factor* by which a given component in the numeric solution grows in time.

These steps result in $R = 2\lambda(\cos(\omega\, dx) - 1) + 1$, which must satisfy $|R| \leq 1$ if the scheme is to be stable. Hence, we are interested in the values of λ for which the nonlinear inequality

$$|2\lambda(\cos(\omega\, dx) - 1) + 1| \leq 1 \qquad (25.13)$$

holds for all values of $\cos(\omega\, dx)$ in $[-1, 1]$. Under the substitution $A = \cos(\omega\, dx)$ we have

$$R = R(A, \lambda) = 2\lambda(A - 1) + 1$$

Figure 25.18 contains a graph of the surface $|R(A, \lambda)|$ plotted over the λA-plane. It shows a surface lying between $R = 0$ and $R = 1$ for only a portion of the rectangular domain. Rotating Figure 25.18 so that it appears viewed from directly above yields Figure 25.19, in which the shaded region represents points (λ, A) satisfying $|R(A, \lambda)| \leq 1$. For this inequality to hold for $-1 \leq A \leq 1$, we can see from Figure 25.19 that $0 < \lambda \leq \frac{1}{2}$.

Alternatively, solve $R = 2\lambda(A - 1) + 1$ for $A = \frac{2\lambda - 1 + R}{2\lambda}$, and plot $A(\lambda)$ for several values of R in $[-1, 1]$. Figure 25.20 shows these level curves, along with cross-hatching where the inequality is satisfied for all A in $[-1, 1]$. The inequality $|R| \leq 1$ is valid at the solid dot but not for all A in $[-1, 1]$.

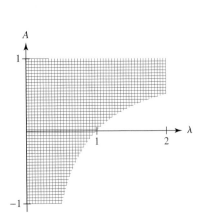

FIGURE 25.19   Region in λA-plane determined by $|R(A, \lambda)| \leq 1$

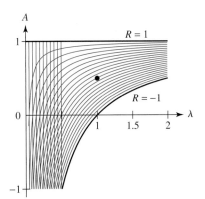

FIGURE 25.20   Region in λA-plane where $|R(A, \lambda)| \leq 1$ holds for all A in $[-1, 1]$

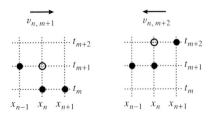

FIGURE 25.21   Computational molecule for Saul'yev's alternating direction scheme

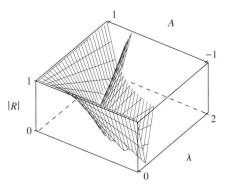

FIGURE 25.18   Graph of $|R(A, \lambda)|$

Proceeding analytically, we note the inequality $|2\lambda(A - 1) + 1| \leq 1$ is really the two inequalities

$$-1 \leq 1 + 2\lambda(A - 1) \leq 1$$

Subtracting 1 from each side of each of the two inequalities, we get

$$-2 \leq 2\lambda(A - 1) \leq 0$$

Dividing by 2 gives

$$-1 \leq \lambda(A - 1) \leq 0$$

Since both inequalities must hold for all A in $[-1, 1]$, in particular when $A = -1$, we have $-1 \leq -2\lambda$, from which follows the restriction $\lambda \leq \frac{1}{2}$. No matter how this inequality is obtained, it implies (25.10) is only *conditionally* stable, that is, it is stable only if λ is suitably restricted.

An Explicit Unconditionally Stable Method

The method expressed by (25.10) has the advantage of being explicit, that is, (25.10) directly and explicitly gives the new function value $v_{n,m+1}$. Unfortunately, (25.10) is only conditionally stable, as we have just seen. There are several explicit unconditionally stable finite-difference schemes for solving the one-dimensional heat equation. These methods remove the restriction on step-size but remain explicit. There is some additional computational complexity, however, but that is a small price to pay for being able to take larger steps in time. We will consider the method of Saul'yev and only mention the names Larkin, Barakat and Clark, and DuFort–Frankel. Texts such as [90] and [18] give the details of these other methods.

The method of Saul'yev is an example of an *alternating direction* schemewherein the computation proceeds from left to right and from right to left in alternating "rows." The values $v_{n,m+1}, n = 1, \ldots, N - 1$, are computed from left to right by the formula

$$v_{n,m+1} = \frac{1}{1 + \lambda}[v_{n,m} + \lambda(v_{n-1,m+1} - v_{n,m} + v_{n+1,m})] \qquad \textbf{(25.14)}$$

whereas the values $v_{n,m+2}, n = 1, \ldots, N - 1$, are computed from right to left by the formula

$$v_{n,m+2} = \frac{1}{1 + \lambda}[v_{n,m+1} + \lambda(v_{n-1,m+1} - v_{n,m+1} + v_{n+1,m+2})] \qquad \textbf{(25.15)}$$

The schematic in Figure 25.21 highlights how these alternating formulas are related to the alternating directions. The solid dots represent known values, whereas the open dots represent the unknown values. In each case, the dots correspond to the terms appearing in (25.14) and (25.15).

EXERCISES 25.4

1. Solve (25.12) for $n_{n,m+1}$, thereby obtaining (25.10).

2. Verify that (25.13) is obtained via the five steps given in the text.

In Exercises 3–8, the given data specified a BVP of the form $u_t = \kappa u_{xx}$, $u(0,t) = \psi_1(t), u(L,t) = \psi_2(t), u(x,0) = f(x)$. For each:

    (a) Use the explicit finite-difference scheme to obtain a plot of the solution surface.

    (b) Animate the temperature profiles as they vary in time.

3. $L = \pi, \kappa = \frac{1}{3}, \psi_1(t) = \sin t, \psi_2(t) = \sin 2t, f(x) = x(\pi - x)$

4. $L = 3, \kappa = 2, \psi_1(t) = 4(1 - e^{-2t}), \psi_2(t) = 0, f(x) = 0$

5. $L = \pi, \kappa = \sqrt{2}, \psi_1(t) = \sin t, \psi_2(t) = 0, f(x) = x^2(\pi - x)$

6. $L = \pi, \kappa = 3, \psi_1(t) = |\sin t|, \psi_2(t) = 0, f(x) = \sin x$

7. $L = 5, \kappa = \frac{3}{5}, \psi_1(t) = \sin 2t,$
$$\psi_2(t) = \begin{cases} 0 & 0 \le t < 1 \\ \cos 2(t-1) & 1 \le t \end{cases}, \quad f(x) = \sin^2 \frac{\pi x}{5}$$

8. $L = 4, \kappa = \frac{1}{2}, \psi_1(t) = \sin^2 t, \psi_2(t) = \dfrac{t^2}{1+t^2},$
$$f(x) = \begin{cases} 0 & 0 \le x < 1 \\ 5 & 1 \le x < 2 \\ 2 & 2 \le x < 3 \\ 0 & 3 \le x \le 4 \end{cases}$$

9. Modify the explicit finite-difference scheme so that it applies to the heat equation with insulated ends.

    (a) Discretize $u_x(0, t_m) = 0$ with the $O(h^2)$ approximation $(v_{1,m} - v_{-1,m})/2\Delta x = 0$ so that $v_{-1,m} = v_{1,m}$.

    (b) Apply (25.10) at $x_0 = 0$, replacing $v_{-1,m}$ with $v_{1,m}$, obtaining $v_{0,m+1} = (1 - 2\lambda)v_{0,m} + 2\lambda v_{1,m}$.

    (c) Apply (25.10) at $x_N = L$, making the appropriate replacements, and obtaining $v_{N,m+1} = (1 - 2\lambda)v_{N,m} + 2\lambda v_{N-1,m}$.

    (d) Apply (25.10) to the $N - 1$ interior nodes to obtain $v_{k,m+1}$, $k = 1, \ldots, N - 1$.

In Exercises 10–15, the given data specifies a BVP of the form $u_t = \kappa u_{xx}$, $u_x(0,t) = u_x(L,t) = 0, u(x,0) = f(x)$. For each, test the algorithm developed in Exercise 9. In particular:

    (a) Obtain the exact solution using the techniques of Section 25.3.

    (b) Obtain a numeric solution with $\lambda = \frac{\kappa \Delta t}{(\Delta x)^2} = \frac{3}{8}$, plotting the solution surface and animating the temperature profiles.

    (c) Compare the exact and numeric solutions at the point $(x, t) = (\frac{L}{2}, t_5)$.

10. $L = 1, \kappa = 2, f(x) = e^x$

11. $L = 3, \kappa = \sqrt{2}, f(x) = 5x(3 - x)$

12. $L = 5, \kappa = \frac{2}{3}, f(x) = x^2$ **13.** $L = \pi, \kappa = 3, f(x) = \sin^2 x$

14. $L = 4, \kappa = \frac{3}{10}, f(x) = \sin^2 \frac{\pi}{2}x$ **15.** $L = 2, \kappa = \frac{5}{3}, f(x) = x^3$

16. Modify the explicit finite-difference algorithm to solve the BVP $u_t = \kappa u_{xx} + \alpha u_x, u(0,t) = \psi_1(t), u(L,t) = \psi_2(t), u(x,0) = f(x)$. *Hint:* Discretize $u_x(x_n, t_m)$ with the $O(h^2)$ approximation $(v_{n+1,m} - v_{n-1,m})/2\Delta x$, obtaining, in place of (25.10), the equation $v_{n,m+1} = \lambda v_{n-1,m} + (1 - 2\lambda)v_{n,m} + \lambda v_{n+1,m} + \frac{\alpha \Delta t}{2\Delta x}(v_{n+1,m} - v_{n-1,m})$.

In Exercises 17–21, apply the algorithm developed in Exercise 16. Plot the solution surface and animate the temperature profiles.

17. $L = 3, \kappa = \frac{1}{3}, \alpha = 1, \psi_1(t) = \sin t, \psi_2(t) = 1 - e^{-t},$
$f(x) = x^3(3 - x)$

18. $L = \pi, \kappa = 3, \alpha = \frac{1}{2}, \psi_1(t) = \cos t, \psi_2(t) = 1, f(x) = \cos^2 x$

19. $L = 2, \kappa = \frac{5}{3}, \alpha = 2, \psi_1(t) = 1, \psi_2(t) = 2e^{-t}, f(x) = 1 + \frac{19}{2}x - \frac{9}{2}x^2$

20. $L = 4, \kappa = 2, \alpha = 3, \psi_1(t) = -2, \psi_2(t) = 3 + \cos t,$
$f(x) = -\frac{13}{6}x^2 + \frac{61}{6}x - 2$

21. $L = 5, \kappa = 1, \alpha = \sqrt{2}, \psi_1(t) = 1 - e^{-t}, \psi_2(t) = \dfrac{1}{1 + t^2},$
$f(x) = \dfrac{x}{10}(47 - 9x)$

22. In the programming language of your choice, implement the method of Saul'yev.

In Exercises 23–28, apply the algorithm developed in Exercise 22 to each of the following BVPs. Plot the solution surface and animate the temperature profiles. Caution: Even though Saul'yev's method is unconditionally stable, its time-step is still restricted by $0 < \lambda \le \frac{1}{2}$, although only for accuracy. See [42] for a similar outcome with the DuFort–Frankel method, another unconditionally stable explicit method.

23. the BVP determined by Exercise 3

24. the BVP determined by Exercise 4

25. the BVP determined by Exercise 5

26. the BVP determined by Exercise 6

27. the BVP determined by Exercise 7

28. the BVP determined by Exercise 8

29. Modify the explicit finite difference method to provide a solution to the BVP $u_t = \kappa u_{xx}, u(0,t) = 0, u_x(L,t) = 0, u(x,0) = f(x)$. Thus, the left end of a rod of length L is kept at a temperature of zero, whereas the right end is insulated. See Exercise 9 where an explicit finite-difference scheme was developed for a finite rod with *both* ends insulated.

In Exercises 30–32, apply the algorithm of Exercise 29 to the given BVP, graphing the solution surface and animating the temperature profiles.

30. $L = \pi, \kappa = 2, f(x) = x^3(\pi - x)$ **31.** $L = 3, \kappa = \frac{2}{3}, f(x) = x^3$

32. $L = 5, \kappa = \frac{1}{2}, f(x) = x$

33. Modify the explicit finite-difference method to provide a solution to the BVP $u_t = \kappa u_{xx}, u(0, t) = \psi_1(t), u_x(L, t) = 0,$ $u(x, 0) = f(x)$. Thus, the left end of a rod of length L is maintained at the varying temperature $\psi_1(t)$. Clearly, a solution to this exercise provides a solution to Exercise 29.

In Exercises 34–36, apply the algorithm of Exercise 33 to the given BVP, graphing the solution surface and animating the temperature profiles.

34. $L = \pi, \kappa = 2, \psi_1(t) = \sin t, f(x) = x^3(\pi - x)$

35. $L = 3, \kappa = \frac{1}{3}, \psi_1(t) = \cos^2 t, f(x) = 0$

36. $L = 5, \kappa = 3, \psi_1(t) = 2(1 - e^{-t}), f(x) = \left(\frac{x}{5}\right)^3$

37. Using separation of variables, obtain a series solution of the BVP given in Exercise 29.

In Exercises 38–40, use the results of Exercise 37 to obtain an analytic solution. Compare to the numeric solution.

38. the BVP of Exercise 30 **39.** the BVP of Exercise 31

40. the BVP of Exercise 32

Chapter Review

1. Formulate a boundary value problem describing one-dimensional heat flow in a homogeneous solid of length L that is kept at a temperature of zero at each end and has an initial temperature profile of $f(x)$.

2. Describe the separation of variables process by which a series solution to the BVP in Question 1 can be obtained. Include a discussion of the Sturm–Liouville eigenvalue problem that arises from separating variables, especially its solution for eigenvalues and eigenfunctions. Also, describe how the coefficients of the series solution are found to be the Fourier sine coefficients of the initial temperature distribution function.

3. Formulate a boundary value problem describing one-dimensional heat flow in a homogeneous solid of length L that has its ends insulated and an initial temperature profile of $f(x)$.

4. Describe the separation of variables process by which a series solution of the BVP in Question 3 can be obtained. Include a discussion of the Sturm–Liouville eigenvalue problem that arises

from separating variables, especially its solution for eigenvalues and eigenfunctions. Also, describe how the coefficients of the series solution are found to be the Fourier cosine coefficients of the initial temperature distribution function.

5. Using the solution detailed in Question 4, determine the steady-state temperature for the solid in Question 3. Explain why this result is intuitive and perfectly reasonable.

6. Using separation of variables, solve the BVP consisting of the one-dimensional heat equation for a finite rod of length L if the left end is insulated, the right end is kept at a temperature of zero, and the initial temperature distribution is $u(x, 0) = f(x)$.

7. Solve the BVP in Question 6 if $L = \pi, \kappa = \frac{1}{3}$, and $f(x) = x(\pi - x)$.

8. Sketch a derivation of the one-dimensional heat equation.

9. Sketch a finite-difference solution for the BVP in Question 1.

10. Sketch a finite-difference solution for the BVP in Question 3.

Chapter 26

Laplace's Equation in a Rectangle

I N T R O D U C T I O N   Laplace's equation on a region R can be solved if the values of the dependent variable or its normal derivative are prescribed on the boundary of the region. If the values of the dependent variable are prescribed, the conditions are called Dirichlet conditions and the BVP is called the Dirichlet problem for the region R.

The normal derivative along the boundary is the directional derivative in the direction of the normal on the boundary. If this derivative is prescribed, the condition is called a Neumann condition. If only Neumann conditions are prescribed for Laplace's equation on the region R, then the resulting BVP is called the Neumann problem for R.

The simplest region on which to solve either the Dirichlet or Neumann problems is the rectangle. Working in Cartesian coordinates, we use separation of variables to reduce Laplace's PDE to two ODEs, one of which forms a Sturm–Liouville eigenvalue problem. If homogeneous Dirichlet conditions have been prescribed, then the eigenfunctions are sines; if Neumann conditions, then cosines. In either event, the solution is an infinite series of eigensolutions.

The Dirichlet problem for the rectangle models the electric potential or the steady-state temperatures in a rectangular plate. For the potential problem, the edges are grounded, that is, maintained at a zero potential. For the thermal problem, the edges are maintained at a temperature of zero. The level curves of the solution are equipotentials in the potential problem or isotherms in the thermal problem. In either case, the gradient vectors are orthogonal to the level curves, so the flow lines that have these gradients as tangents are also orthogonal to the level curves. In the potential problem, the negative of the gradient vector is the electric field vector and the flow lines are the field lines or lines of force. In the thermal problem, the flow lines indicate the path of the heat energy flowing through the plate.

The Neumann problem models the rectangular plate with insulated edges, either electrostatic or thermal insulation. The homogeneous Neumann condition in the first case indicates the gradient of the potential is zero at the boundary. Thus, no charges can flow across the boundary, the meaning of electrostatic insulation. For the thermal problem, the vanishing of the temperature gradient at the boundary means no heat can flow across it, so thermal insulation is achieved.

Various combinations of nonhomogeneous Dirichlet conditions on different edges lead to a variety of problems that can be solved by superposition of the solutions of simpler problems. We can even combine Dirichlet and Neumann on different edges.

Finally, we obtain a finite-difference solution of the Dirichlet problem on the rectangle. The difference between this numeric scheme and the ones for the wave and heat equations is that now, the unknown values at all the interior nodes must be solved for simultaneously. For the evolution equations (wave and heat), the solution is computed at each time-step. For

Laplace's equation, the solution must be computed for the whole region R simultaneously. Thus, solving Laplace's equation requires hardware and software capable of solving large systems of simultaneous algebraic equations.

26.1   Nonzero Temperature on the Bottom Edge

Steady-State Temperatures in a Rectangle

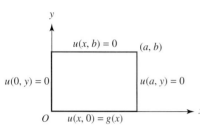

FIGURE 26.1   A Dirichlet problem for a rectangle

The time-dependent heat equation in one spatial dimension is $u_t(x, t) = \kappa u_{xx}(x, t)$. In two spatial dimensions (such as in a rectangular slab), this equation generalizes to $u_t(x, y, t) = \kappa(u_{xx}(x, y, t) + u_{yy}(x, y, t))$. At steady state when the temperatures have stopped changing, $u_t = 0$, so the steady-state heat equation in two dimensions becomes $u_{xx}(x, y) + u_{yy}(x, y) = 0$. This is Laplace's equation, $\nabla^2 u = 0$, the same equation that governs the electrostatic potential in a rectangular dielectric (insulator). It is this equation we now embed in a boundary value problem and then solve via separation of variables.

With respect to Figure 26.1, the boundary value problem to consider consists of Laplace's equation $u_{xx} + u_{yy} = 0$ inside the rectangle $0 \le x \le a, 0 \le y \le b$ and the boundary conditions $u(x, 0) = g(x), u(0, y) = 0, u(a, y) = 0, u(x, b) = 0$. A BVP such as this, consisting of Laplace's equation and Dirichlet conditions, is called a *Dirichlet problem*.

EXAMPLE 26.1

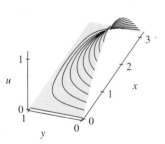

FIGURE 26.2   Solution surface for Example 26.1

The solution of the Dirichlet problem corresponding to Figure 26.1 when $u(x, 0) = g(x) = \sin \frac{\pi x}{a}$ along the bottom edge is

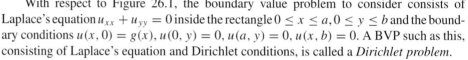

$$u(x, y) = -\frac{\sin \frac{\pi x}{a} \sinh \frac{\pi(y-b)}{a}}{\sinh \frac{\pi b}{a}} \tag{26.1}$$

A straightforward computation shows $\nabla^2 u = 0$. Setting $a = \pi$ and $b = 1$ we draw the solution surface shown in Figure 26.2. The rectangle over which the surface is plotted is the physical slab whose temperatures are being measured. Thus, the heights plotted above this rectangle are representative of the actual temperatures in the slab.

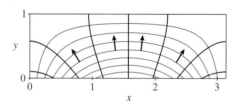

FIGURE 26.3   Isotherms (in color), flow lines (in black), and several arrows from the negative gradient field for the solution in Example 26.1

The contour lines (level curves) on the solution surface (26.1) are the *isotherms*, curves of constant temperature. A contour plot, the equivalent of looking straight down at Figure 26.2, appears in Figure 26.3. Along with the isotherms, Figure 26.3 also contains several arrows of the gradient field,

$$-\nabla u = \frac{1}{\sinh(1)}(\cos x \sinh(y - 1)\mathbf{i} + \sin x \cosh(y - 1)\mathbf{j})$$

which is orthogonal to the isotherms. The minus sign is included because here, $u(x, y)$ represents temperatures and heat flows from *hotter* temperatures to *cooler* temperatures, and we want arrows pointing in the direction of *decreasing*, not *increasing*, values of $u(x, y)$. Figure 26.3 also shows the flow lines, the paths along which heat would flow if it were a fluid. Since the flow is along the gradient vectors, these lines are solutions of the differential equations $\mathbf{R}' = -\nabla(u)$. In component form, these equations are

$$x'(t) = -u_x(x(t), y(t)) \quad \text{and} \quad y'(t) = -u_y(x(t), y(t))$$

The gradient vectors are tangent to the flow lines, so the flow lines are orthogonal to the isotherms.   ❖

Separation of Variables

We now apply the separation of variables technique to obtain a series solution of this boundary value problem. First, the separation assumption $u(x, y) = X(x)Y(y)$ is applied to the homogeneous (Dirichlet) boundary data, resulting in

$$X(0)Y(y) = 0 \qquad X(a)Y(y) = 0 \qquad X(x)Y(b) = 0$$

from which we conclude

$$X(0) = X(a) = Y(b) = 0 \tag{26.2}$$

Next, apply the separation assumption to the partial differential equation, and, after dividing by $u(x, y) = X(x)Y(y)$, obtain

$$\frac{X''(x)}{X(x)} = -\frac{Y''(y)}{Y(y)} = \lambda$$

and the two ODEs

$$X''(x) - \lambda X(x) = 0 \quad \text{and} \quad Y''(y) + \lambda Y(y) = 0 \tag{26.3}$$

(Again, the primes on X and Y denote differentiation with respect to different independent variables.)

The first equation in (26.3) and the first two conditions in (26.2) form the familiar Sturm–Liouville BVP on the left in (26.4), the solution of which appears on the right in

$$
\begin{aligned}
X'' - \lambda X &= 0 \\
X(0) = X(a) &= 0
\end{aligned}
\quad \Rightarrow \quad
\begin{aligned}
X_n(x) &= B_n \sin \frac{n\pi x}{a} \\
\lambda_n &= -\frac{n^2\pi^2}{a^2} \qquad n = 1, 2, \ldots
\end{aligned}
\tag{26.4}
$$

With λ determined, the equation for $Y(y)$ becomes $Y_n'' - \frac{n^2\pi^2}{a^2}Y_n = 0$, with solution

$$Y_n(y) = c_n e^{n\pi y/a} + d_n e^{n\pi y/a} \tag{26.5}$$

The condition $Y_n(b) = 0$ from (26.2) gives $d_n = -c_n e^{2n\pi b/a}$, and consequently

$$
\begin{aligned}
Y_n(y) &= c_n(e^{n\pi y/a} - e^{2(n\pi b/a} e^{-n\pi y/a}) \\
&= c_n e^{n\pi b/a}(e^{-n\pi b/a} e^{(n\pi y/a} - e^{n\pi b/a} e^{-n\pi y/a}) \\
&= c_n e^{n\pi b/a}(e^{n\pi(y-b)/a} - e^{-(n\pi(y-b)/a}) \\
&= C_n \sinh\left(\frac{n\pi}{a}(y - b)\right)
\end{aligned}
$$

Therefore we have the eigensolutions

$$u_n(x, y) = A_n \sin \frac{n\pi x}{a} \sinh\left(\frac{n\pi}{a}(y - b)\right)$$

where $A_n = B_n C_n$. A linear combination of all possible eigensolutions gives the general solution

$$u(x, y) = \sum_{n=1}^{\infty} A_n \sin \frac{n\pi x}{a} \sinh \left(\frac{n\pi}{a}(y - b) \right)$$

Application of the inhomogeneous Dirichlet condition on the edge $y = 0$ yields the equation

$$u(x, 0) = g(x) = \sum_{n=1}^{\infty} A_n \sin \frac{n\pi x}{a} \sinh \left(\frac{n\pi}{a}(-b) \right)$$

Unlike the solutions of the wave and heat equation in which $T_n(0) = 1(\cos(0))$, and e^0, respectively), for Laplace's equation $Y_n(0) = \sinh(\frac{n\pi}{a}(-b)) \neq 1$. Thus, match the coefficients in the Fourier sine series

$$g(x) = \sum_{n=1}^{\infty} b_n \sin \frac{n\pi x}{a}$$

$$b_n = \frac{2}{a} \int_0^a g(x) \sin \frac{n\pi x}{a} \, dx$$

and $g(x) = \sum_{n=1}^{\infty} \left[A_n \sinh \left(\frac{n\pi}{a}(-b) \right) \right] \sin \frac{n\pi x}{a}$

concluding $[A_n \sinh(\frac{n\pi}{a}(-b))] = b_n$, so

$$A_n = -\frac{b_n}{\sinh \frac{n\pi b}{a}} = -\frac{2}{a \sinh \frac{n\pi b}{a}} \int_0^a g(x) \sin \frac{n\pi x}{a} \, dx$$

Hence, the complete solution to the given Dirichlet problem is the series

$$u(x, y) = \sum_{n=1}^{\infty} A_n \sin \frac{n\pi x}{a} \sinh \left(\frac{n\pi}{a}(y - b) \right) \tag{26.6a}$$

$$A_n = -\frac{2}{a \sinh \frac{n\pi b}{a}} \int_0^a g(x) \sin \frac{n\pi x}{a} \, dx \tag{26.6b}$$

EXERCISES 26.1–Part A

A1. Verify that (26.1) satisfies Laplace's equation.

A2. Separate variables in Laplace's equation and obtain (26.3).

A3. Separate variables in the Dirichlet problem corresponding to Figure 26.1 and obtain (26.2).

A4. Obtain (26.5).

A5. Verify the identity $\cosh(u + v) = \cosh u \cosh v + \sinh u \sinh v$.

A6. Verify the identity $\sinh(u - v) = \sinh u \cosh v - \cosh u \sinh v$.

EXERCISES 26.1–Part B

B1. Write the solution of $Y_n'' - \frac{n^2\pi^2}{a^2} Y_n = 0$ as $Y(y) = \alpha_n \cosh \frac{n\pi}{a} + \beta_n \sinh \frac{n\pi}{a}$ and apply the condition $Y(b) = 0$. Use Exercise A6 to obtain $Y_n(y) = C_n \sinh(\frac{n\pi}{a}(y - b))$.

In Exercises B2–21:

  (a) Solve the Dirichlet problem corresponding to Figure 26.1.

  (b) For the series solution found in part (a), let $U(x, y)$ be a partial sum for which $U(x, 0)$ is a reasonable approximation to $g(x)$. Use $U(x, y)$ to approximate $u(\frac{a}{2}, \frac{b}{2})$, the value of u at the center of the rectangle.

  (c) Graph the solution surface.

  (d) Obtain a contour plot showing the isotherms (interpreting $u(x, y)$ as temperatures) or equipotentials (interpreting $u(x, y)$ as a potential).

  (e) Obtain a plot of the gradient field arising from $u(x, y)$.

  (f) Calculate and plot the flow lines of the gradient field.

  (g) Superimpose the flow lines on the contour plot of the level curves of u.

B2. $(a, b) = (\pi, 1)$, $g(x) = x(\pi - x)$

B3. $(a, b) = (\pi, 1)$, $g(x) = x^2(\pi - x)$

B4. $(a, b) = (\pi, 1)$, $g(x) = x^3(\pi - x)$

B5. $(a, b) = (\pi, 1)$, $g(x) = \sin 2x$

B6. $(a, b) = (\pi, 1)$, $g(x) = \sin^2 x$

B7. $(a, b) = (\pi, 1)$, $g(x) = \sin^2 2x$

B8. $(a, b) = (\pi, 1)$, $g(x) = \cos x$

B9. $(a, b) = (\pi, 1)$, $g(x) = \cos^2 x$

B10. $(a, b) = (1, 1)$, $g(x) = x$ **B11.** $(a, b) = (1, 1)$, $g(x) = x^2$

B12. $(a, b) = (1, 1)$, $g(x) = x^3$ **B13.** $(a, b) = (2, 3)$, $g(x) = e^{-x}$

B14. $(a, b) = (4, 4)$, $g(x) = \begin{cases} 0 & 0 \le x < 1 \\ 1 & 1 \le x < 2 \\ 2 & 2 \le x < 3 \\ 0 & 3 \le x \le 4 \end{cases}$

B15. $(a, b) = (4, 4)$, $g(x) = \begin{cases} 0 & 0 \le x < 1 \\ x - 1 & 1 \le x < 2 \\ 3 - x & 2 \le x < 3 \\ 0 & 3 \le x \le 4 \end{cases}$

B16. $(a, b) = (2, 3)$, $g(x) = \begin{cases} x & 0 \le x < 1 \\ 2 - x & 1 \le x \le 2 \end{cases}$

B17. $(a, b) = (4, 4)$, $g(x) = \begin{cases} x & 0 \le x < 1 \\ 1 & 1 \le x < 3 \\ 4 - x & 3 \le x \le 4 \end{cases}$

B18. $(a, b) = (4, 4)$, $g(x) = \begin{cases} x^2 & 0 \le x < 1 \\ 1 & 1 \le x < 3 \\ (4 - x)^2 & 3 \le x \le 4 \end{cases}$

B19. $(a, b) = (4, 4)$, $g(x) = \begin{cases} x(2 - x) & 0 \le x < 1 \\ 1 & 1 \le x < 3 \\ (x - 2)(4 - x) & 3 \le x \le 4 \end{cases}$

B20. $(a, b) = (3, 2)$, $g(x) = \begin{cases} x & 0 \le x < 1 \\ \frac{1}{2}(3 - x) & 1 \le x \le 3 \end{cases}$

B21. $(a, b) = (4, 4)$, $g(x) = \begin{cases} 2 & 0 \le x < 1 \\ 0 & 1 \le x < 3 \\ 3 & 3 \le x \le 4 \end{cases}$

26.2 **Nonzero Temperature on the Top Edge**

Nonzero Temperature on Bottom Edge

In Section 26.1 we learned how to solve the Dirichlet problem for a rectangle on which a nonhomogeneous Dirichlet condition is prescribed on the *bottom* edge. Our immediate goal is to solve the Dirichlet problem for a rectangle with a nonhomogeneous Dirichlet condition prescribed on the *top* edge. Our strategy for solving this new problem will be to parallel, as much as possible, our solution to the old problem. Hence, we start with an example of the Dirichlet problem for a rectangle with a nonhomogeneous Dirichlet condition on the bottom edge.

EXAMPLE 26.2   In the BVP summarized in Figure 26.1, let $a = 4$, $b = 4$, and

$$g(x) = \begin{cases} 0 & 0 \le x < 1 \\ (1 - x)(x - 3) & 1 \le x \le 3 \\ 0 & 3 < x \le 4 \end{cases}$$

Graph $g(x)$ in Figure 26.4, and interpret the boundary value problem as the steady-state temperatures in a square plate. Obtain the solution $u(x, y)$ by computing the coefficients

$$A_n = -\frac{1}{2 \sinh n\pi} \int_0^4 g(x) \sin \frac{n\pi x}{4} \, dx \tag{26.7a}$$

$$= -\frac{16}{n^3 \pi^3 \sinh n\pi} \left(4 \cos \frac{n\pi}{4} - n\pi \sin \frac{n\pi}{4} - 4 \cos \frac{3n\pi}{4} - n\pi \sin \frac{3n\pi}{4} \right) \tag{26.7b}$$

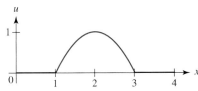

FIGURE 26.4   Graph of $u(x, 0) = g(x)$ in Example 26.2

the first 10 of which appear in Table 26.1, whose entries are consistent with $A_n = \mathrm{O}(n^{-2}e^{-n\pi})$.

In the interior of the rectangle, away from $y = 0$, the denominators contain $\sinh n$, which gets large quickly, so the coefficients themselves equally quickly get small. Hence, the partial sum $u_7(x, y) = \sum_{n=1}^7 A_n \sin \frac{n\pi x}{4} \sinh(\frac{n\pi}{4}(y - 4))$, graphed in Figure 26.5, is a satisfactory approximation to the solution.

n	A_n	$\tilde{A}_n$	n	A_n	$\tilde{A}_n$
1	$16\dfrac{\sqrt{2}(\pi-4)}{\pi^3\sinh\pi}$	-0.0542	6	0	0
2	0	0.723×10^{-12}	7	$-\dfrac{16}{343}\dfrac{\sqrt{2}(7\pi+4)}{\pi^3\sinh 7\pi}$	-0.311×10^{-10}
3	$\dfrac{16}{27}\dfrac{\sqrt{2}(3\pi+4)}{\pi^3\sinh 3\pi}$	0.586×10^{-4}	8	0	-0.433×10^{-21}
4	0	-0.115×10^{-14}	9	$\dfrac{16}{729}\dfrac{\sqrt{2}(9\pi-4)}{\pi^3\sinh 9\pi}$	0.255×10^{-13}
5	$-\dfrac{16}{125}\dfrac{\sqrt{2}(5\pi-4)}{\pi^3\sinh 5\pi\pi}$	-0.206×10^{-7}	10	0	0.703×10^{-24}

TABLE 26.1   Coefficients A_n in Example 26.2. The numbers $\tilde{A}_n$ are the floating-point values produced by (26.7b)

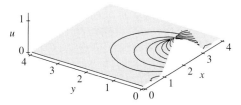

FIGURE 26.5   Solution surface in Example 26.2

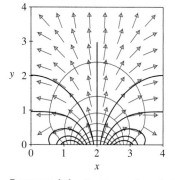

FIGURE 26.6   Isotherms (in color), flow lines (in black), and the negative of the gradient field for the solution in Example 26.2

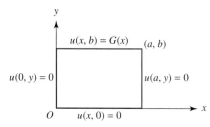

FIGURE 26.7   Dirichlet problem for rectangle where top edge supports nonhomogeneous Dirichlet condition

The contours on the surface in Figure 26.5 are the isotherms, or level curves, shown as a contour plot in Figure 26.6. In addition, Figure 26.6 shows the vectors $-\nabla u$, the minus sign because heat flows from higher to lower temperatures, but gradients point from lower to higher values. Thus, the vector field plotted shows the direction of heat flow. Finally, Figure 26.6 also shows the flow lines along which heat would flow if heat were water and the solution surface, a hill. These lines are orthogonal to the isotherms, the level curves of $u(x, y)$. Thus, the flow lines are *along* the vectors $-\nabla u$; that is, the flow lines have as their tangent vectors the negatives of the gradient vectors of $u(x, y)$ and are obtained as solutions to the differential equations $x'(t) = -u_x(x(t), y(t))$, $y'(t) = -u_y(x(t), y(t))$.   ❖

Nonzero Temperature on Top Edge

Consider the following slight variation of the boundary value problem in Example 26.2. Instead of prescribing a nonzero temperature on the edge $y = 0$, prescribe it on the edge $y = b$. Thus, the boundary value problem now consists of Laplace's equation in the rectangular region shown in Figure 26.7, with the top edge maintained at a temperature of $G(x)$ and the bottom edge maintained at a temperature of zero. The two edges $x = 0$ and $x = a$ are still held at a temperature of zero, so the complete statement of the problem consists of Laplace's equation $u_{xx} + u_{yy} = 0$ and the Dirichlet boundary conditions $u(x, 0) = 0$, $u(0, y) = 0$, $u(a, y) = 0$, $u(x, b) = G(x)$.

A little thought will help avoid repeating many of the computations in Section 26.1. First, separation of variables will still yield the Sturm–Liouville eigenvalue problem

$$X''(x) - \lambda X(x) = 0 \quad \text{and} \quad X(0) = X(a) = 0$$

because the homogeneous Dirichlet conditions on the opposing edges $x = 0$ and $x = a$

haven't changed. Hence, the eigenfunctions and eigenvalues

$$X_n(x) = B_n \sin \frac{n\pi x}{a} \qquad \lambda_n = -\frac{n^2\pi^2}{a^2} \qquad n = 1, 2, \ldots$$

will be the same as before. The only part that will be different is the solution for $Y(y)$, and that, not by much. The function $Y(y)$ must still be a hyperbolic sine because it must be zero exactly once. In fact, $Y_n(y)$ will satisfy the same ODE as before,

$$Y_n''(y) - \frac{n^2\pi^2}{a^2} Y_n(y) = 0 \tag{26.8}$$

because the separation of variables steps will be the same as before. However, there will now be the different initial condition $Y_n(0) = 0$, so that $Y_n(y) = C_n \sinh \frac{n\pi y}{a}$ and the solution to the boundary value problem will be

$$u(x, y) = \sum_{n=1}^{\infty} A_n \sin \frac{n\pi x}{a} \sinh \frac{n\pi y}{a} \quad \text{where} \quad A_n = \frac{2}{a \sinh \frac{bn\pi}{a}} \int_0^a G(x) \sin \frac{n\pi x}{a}\, dx$$

EXAMPLE 26.3 As an example of a problem with a nonzero temperature on the top edge, solve Laplace's equation for the steady-state temperatures in the plate of Example 26.2 (where $a = 4$, $b = 4$) if $G(x) = \frac{x(4-x)}{4}$, graphed in Figure 26.8.

The coefficients,

$$A_n = \frac{1}{2 \sinh n\pi} \int_0^4 \frac{1}{4} x(4-x) \sin \frac{n\pi x}{4}\, dx = 16 \frac{1 - (-1)^n}{n^3\pi^3 \sinh n\pi} \tag{26.9}$$

are $O(n^{-3}e^{-n\pi})$, so the partial sum

$$u_5(x, y) = \frac{32}{\pi^3} \left(\frac{\sin \frac{\pi x}{4} \sinh \frac{\pi y}{4}}{\sinh \pi} + \frac{\sin \frac{3\pi x}{4} \sinh \frac{3\pi y}{4}}{27 \sinh 3\pi} + \frac{\sin \frac{5\pi x}{4} \sinh \frac{5\pi y}{4}}{125 \sinh 5\pi} \right)$$

graphed in Figure 26.9, will be an adequate approximation of the solution.

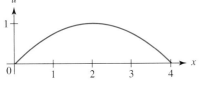

FIGURE 26.8 Graph of $G(x)$ in Example 26.3

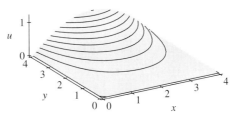

FIGURE 26.9 Solution surface in Example 26.3

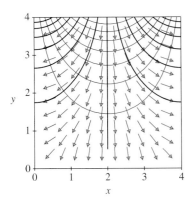

FIGURE 26.10 Isotherms (in color), flow lines (in black), and the negative of the gradient field for the solution in Example 26.3

The contours on the surface in Figure 26.9 are the isotherms, or level curves. They again appear in Figure 26.10 where, in addition, the flow lines and vectors $-\nabla u$ also appear. The flow lines are orthogonal to the isotherms, that is, they are along the negatives of the gradient vectors of $u(x, y)$. The vectors $-\nabla u$ are tangent vectors along the flow lines, which themselves are obtained by integrating the differential equations $x'(t) = -u_x(x(t), y(t))$, $y'(t) = -u_y(x(t), y(t))$. The minus sign is present because heat flows from higher to lower temperatures, but gradient vectors point from lower to higher function values. ❖

Nonzero Temperatures on Top and Bottom Edges

The solution to Laplace's equation on a rectangle that has two nonhomogeneous Dirichlet conditions can be obtained by adding the solutions to two simpler problems, each of which has only one nonhomogeneous Dirichlet condition.

◆ **EXAMPLE 26.4**   For example, suppose $u(x, 0) = g(x)$ and $u(x, b) = G(x)$, as shown in Figure 26.11. The solution of Laplace's equation on this rectangle, with the top and bottom edges held at the respective temperature profiles $G(x)$ and $g(x)$, is the sum of the solutions of the problems in Examples 26.2 and 26.3. Hence, we merely have to add the solutions $u_g(x, y)$ and $u_G(x, y)$ found, respectively, in Examples 26.2 and 26.3, and we will have the solution to the problem in Example 26.4. Linearity is the key here, since the sum $u_{gG}(x, y) = u_g(x, y) + u_G(x, y)$ will satisfy Laplace's equation and the homogeneous Dirichlet conditions on the edges $x = 0$ and $x = a$ because u_g and u_G both do. The verifications are

$$u_{gG}(0, y) = u_g(0, y) + u_G(0, y) = 0 + 0 = 0$$

$$u_{gG}(a, y) = u_g(a, y) + u_G(a, y) = 0 + 0 = 0$$

and

$$u_{gG}(x, 0) = u_g(x, 0) + u_G(x, 0) = g(x) + 0 = g(x)$$

$$u_{gG}(x, b) = u_g(x, b) + u_G(x, b) = 0 + G(x) = G(x)$$

Figure 26.12 contains the graph of the solution surface $u_{gG}(x, y) = u_g(x, y) + u_G(x, y)$. Figure 26.13 combines images of the flow lines (thick curves), isotherms (thin curves), and the arrows of the field $-\nabla u_{gG}$. The flow lines are interesting. Those contributed by $G(x)$ do the expected. They show heat flows from the edge $y = 4$ to the two sides $x = 0$ and $x = 4$. However, the flow lines contributed by $g(x)$ show heat flowing into the plate along this heated edge, then back out at a cooler spot on the same heated edge! Not all the heat escapes to the edges $x = 0$ and $x = 4$. Some also flows out in the two outer segments on the edge $y = 0$, where u is maintained at temperature zero.   ❖

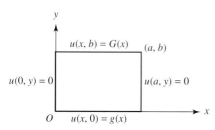

FIGURE 26.11   Dirichlet problem for rectangle where both top and bottom edges support nonhomogeneous Dirichlet conditions

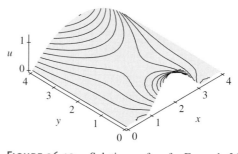

FIGURE 26.12   Solution surface for Example 26.4

Neumann Boundary Conditions

In Section 25.3 the heat equation was solved for a finite rod with insulated ends. Thus, the endpoint conditions $u_x(0, t) = 0$ and $u_x(L, t) = 0$ were imposed. Boundary conditions such as these which prescribe the value of the derivative are examples of *Neumann conditions*.

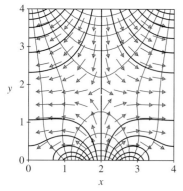

FIGURE 26.13   Isotherms (in color), flow lines (in black), and the negative of the gradient field in Example 26.4

The general Neumann boundary condition prescribes the *normal derivative* $u_n = \nabla(u) \cdot \mathbf{n}$, where $\mathbf{n}$ is a unit normal vector on the boundary of the region R over which the BVP is posed. The homogeneous Neumann condition $u_n = 0$ corresponds to thermal insulation in heat-transfer problems, and electrostatic grounding in potential problems. Boundary value problems with Neumann conditions are left to the exercises.

EXERCISES 26.2–Part A

A1. Verify that the integral in (26.7a) evaluates to the result shown in (26.7b).

A2. Show that A_n in (26.7b) is $O(n^{-2}e^{-n\pi})$. For example, show graphically that the ratio $|A_n/n^{-2}e^{-n\pi}|$ is bounded by a constant.

A3. Show that (26.8) and $Y_n(0) = 0$ are satisfied by $Y_n(y) = C_n \sinh\frac{n\pi y}{a}$.

A4. Verify the integration in (26.9).

A5. Show that A_n in (26.9) is $O(n^{-3}e^{-n\pi})$.

EXERCISES 26.2–Part B

B1. Using $u_7(x, y)$ as the solution to the Dirichlet problem in Example 26.2, show that the net heat flow through the rectangle is zero by evaluating the flux of ∇u through the boundary.

B2. Using $u_5(x, y)$ as the solution to the Dirichlet problem in Example 26.3, show that the net heat flow through the rectangle is zero by evaluating the flux of ∇u through the boundary.

In Exercises B3–22:

    (a) Solve the Dirichlet problem corresponding to Figure 26.7.

    (b) For the series solution found in part (a), let $U(x, y)$ be a partial sum for which $U(x, b)$ is a reasonable approximation to $G(x)$. Use $U(x, y)$ to approximate $u(\frac{a}{2}, \frac{b}{2})$, the value of u at the center of the rectangle.

    (c) Graph the solution surface.

    (d) Obtain a contour plot showing the isotherms (interpreting $u(x, y)$ as temperatures) or equipotentials (interpreting $u(x, y)$ as a potential).

    (e) Obtain a plot of the gradient field arising from $u(x, y)$.

    (f) Calculate and plot the flow lines of the gradient field.

    (g) Superimpose the flow lines on the contour plot of the level curves of u.

B3. $(a, b) = (\pi, 1), G(x) = x(\pi - x)^3$

B4. $(a, b) = (\pi, 1), G(x) = x(\pi - x)^2$

B5. $(a, b) = (\pi, 1), G(x) = x(\pi - x)$

B6. $(a, b) = (\pi, 1), G(x) = \cos 2x$

B7. $(a, b) = (\pi, 1), G(x) = \cos^2 x$

B8. $(a, b) = (\pi, 1), G(x) = \cos^2 2x$

B9. $(a, b) = (\pi, 1), G(x) = \cos\frac{x}{2}$

B10. $(a, b) = (\pi, 1), G(x) = \sin^2 x$

B11. $(a, b) = (1, 1), G(x) = x^3$ **B12.** $(a, b) = (1, 1), G(x) = x$

B13. $(a, b) = (1, 1), G(x) = x^2$

B14. $(a, b) = (2, 3), G(x) = 1 - e^{-x}$

B15. $(a, b) = (4, 4), G(x) = \begin{cases} 0 & 0 \le x < 1 \\ 3 & 1 \le x < 2 \\ 1 & 2 \le x < 3 \\ 0 & 3 \le x \le 4 \end{cases}$

B16. $(a, b) = (4, 4), G(x) = \begin{cases} 1 & 0 \le x < 1 \\ 2 - x & 1 \le x < 2 \\ x - 2 & 2 \le x < 3 \\ 1 & 3 \le x \le 4 \end{cases}$

B17. $(a, b) = (2, 3), G(x) = \begin{cases} 1 - x & 0 \le x < 1 \\ x - 1 & 1 \le x \le 2 \end{cases}$

B18. $(a, b) = (4, 4), G(x) = \begin{cases} x & 0 \le x < 1 \\ 1 & 1 \le x < 2 \\ 2 - \frac{x}{2} & 2 \le x \le 4 \end{cases}$

B19. $(a, b) = (4, 4), G(x) = \begin{cases} x^3 & 0 \le x < 1 \\ 1 & 1 \le x < 3 \\ (4 - x)^3 & 3 \le x \le 4 \end{cases}$

B20. $(a, b) = (4, 4), G(x) = \begin{cases} x(2 - x)^2 & 0 \le x < 1 \\ 1 & 1 \le x < 3 \\ (x - 2)^2(4 - x) & 3 \le x \le 4 \end{cases}$

B21. $(a, b) = (3, 2), G(x) = \begin{cases} \dfrac{x}{2} & 0 \le x < 2 \\ 3 - x & 2 \le x \le 3 \end{cases}$

B22. $(a, b) = (4, 4), G(x) = \begin{cases} 5 & 0 \le x < 1 \\ 0 & 1 \le x < 3 \\ 2 & 3 \le x \le 4 \end{cases}$

In Exercises B23–32, functions $g(x)$ and $G(x)$ are given for the Dirichlet problem corresponding to Figure 26.11. In each case, the rectangles for which $g(x)$ and $G(x)$ are given have the same dimensions. The solution is the sum of Dirichlet problems corresponding to Figures 26.1 and 26.7.

    (a) Graph the solution surface.

    (b) Obtain a contour plot showing the isotherms (interpreting $u(x, y)$ as temperatures) or equipotentials (interpreting $u(x, y)$ as a potential).

(c) Obtain a plot of the gradient field arising from $u(x, y)$.

(d) Calculate and plot the flow lines of the gradient field.

(e) Superimpose the flow lines on the contour plot of the level curves of u.

B23. $g(x)$ from Exercise B2, Section 26.1, and $G(x)$ from Exercise B3, Section 26.2

B24. $g(x)$ from Exercise B3, Section 26.1, and $G(x)$ from Exercise B4, Section 26.2

B25. $g(x)$ from Exercise B4, Section 26.1, and $G(x)$ from Exercise B5, Section 26.2

B26. $g(x)$ from Exercise B5, Section 26.1, and $G(x)$ from Exercise B6, Section 26.2

B27. $g(x)$ from Exercise B6, Section 26.1, and $G(x)$ from Exercise B7, Section 26.2

B28. $g(x)$ from Exercise B12, Section 26.1, and $G(x)$ from Exercise B13, Section 26.2

B29. $g(x)$ from Exercise B13, Section 26.1, and $G(x)$ from Exercise B14, Section 26.2

B30. $g(x)$ from Exercise B14, Section 26.1, and $G(x)$ from Exercise B15, Section 26.2

B31. $g(x)$ from Exercise B15, Section 26.1, and $G(x)$ from Exercise B16, Section 26.2

B32. $g(x)$ from Exercise B16, Section 26.1, and $G(x)$ from Exercise B17, Section 26.2

B33. In Figure 26.1, change the homogeneous Dirichlet conditions on the left and right edges to the homogeneous Neumann conditions $u_x(0, y) = u_x(a, y) = 0$, thereby applying insulation to the pair of opposing edges. Obtain the general solution for this BVP.

For the data in Exercises B34–43, apply the results of Exercise B33 to solve the corresponding BVP.

(a) Graph the solution surface.

(b) Obtain a contour plot showing the isotherms (interpreting $u(x, y)$ as temperatures) or equipotentials (interpreting $u(x, y)$ as a potential).

(c) Obtain a plot of the gradient field arising from $u(x, y)$.

(d) Calculate and plot the flow lines of the gradient field.

(e) Superimpose the flow lines on the contour plot of the level curves of u.

(f) Compute the flux of ∇u through the boundary of the rectangle.

B34. the data in Exercise B7, Section 26.1

B35. the data in Exercise B8, Section 26.1

B36. the data in Exercise B9, Section 26.1

B37. the data in Exercise B10, Section 26.1

B38. the data in Exercise B11, Section 26.1

B39. the data in Exercise B17, Section 26.1

B40. the data in Exercise B18, Section 26.1

B41. the data in Exercise B19, Section 26.1

B42. the data in Exercise B20, Section 26.1

B43. the data in Exercise B21, Section 26.1

B44. In Figure 26.7, change the homogeneous Dirichlet conditions on the left and right edges to the homogeneous Neumann conditions $u_x(0, y) = u_x(a, y) = 0$, and obtain the general solution for this BVP.

For the data in Exercises B45–46, use the results of Exercise B44 to solve the corresponding BVP.

(a) Plot the solution surface.

(b) Obtain level curves and flow lines.

(c) Compute the flux of ∇u through the boundary of the rectangle.

B45. $(a, b) = (\pi, 1)$, $G(x) = x^2(\pi - x)$

B46. the data in Exercise B4, Section 26.2

B47. In Figure 26.11, change the homogeneous Dirichlet conditions on the left and right edges to the homogeneous Neumann conditions $u_x(0, y) = u_x(a, y) = 0$, and obtain the general solution for this BVP.

For the data in Exercises B48 and 49, use the results of Exercise B47 to solve the corresponding BVP. The given functions are defined for rectangles with the same dimensions.

(a) Plot the solution surface.

(b) Obtain level curves and flow lines.

(c) Compute the flux of ∇u through the boundary of the rectangle.

B48. $g(x)$ from Exercise B7, Section 26.1, and $G(x)$ from Exercise B8, Section 26.2

B49. $g(x)$ from Exercise B8, Section 26.1, and $G(x)$ from Exercise B9, Section 26.2

B50. In Figure 26.1, change the homogeneous Dirichlet condition on the top edge to the homogeneous Neumann condition $u_y(x, b) = 0$, and obtain the general solution to the resulting BVP.

For the data in Exercises B51 and 52, use the results of Exercise B50 to solve the corresponding BVP.

(a) Plot the solution surface.

(b) Obtain level curves and flow lines.

(c) Compute the flux of ∇u through the boundary of the rectangle.

B51. the data from Exercise B16, Section 26.1

B52. the data from Exercise B17, Section 26.1

B53. In Figure 26.1, change the homogeneous Dirichlet conditions on the top and right edges to homogeneous Neumann conditions, and obtain the general solution to the resulting BVP.

For the data in Exercises B54 and 55, use the results of Exercise B53 to solve the corresponding BVP.

(a) Plot the solution surface.

(b) Obtain level curves and flow lines.

(c) Compute the flux of ∇u through the boundary of the rectangle.

B54. the data from Exercise B18, Section 26.1

B55. the data from Exercise B19, Section 26.1

B56. In Figure 26.1, change all three homogeneous Dirichlet conditions to homogeneous Neumann conditions, and solve the corresponding BVP.

For the data in Exercises B57 and 58, use the results of Exercise B56 to solve the corresponding BVP.

(a) Plot the solution surface.

(b) Obtain level curves and flow lines.

(c) Compute the flux of ∇u through the boundary of the rectangle.

B57. the data from Exercise B20, Section 26.1

B58. the data from Exercise B21, Section 26.1

B59. In Figure 26.7, change the homogeneous Dirichlet condition on the bottom edge to the homogeneous Neumann condition $u_y(x, 0) = 0$, and obtain the general solution of the resulting BVP.

For the data in Exercises B60 and 61, use the results of Exercise B59 to solve the corresponding BVP.

(a) Plot the solution surface.

(b) Obtain level curves and flow lines.

(c) Compute the flux of ∇u through the boundary of the rectangle.

B60. the data from Exercise B19, Section 26.2

B61. the data from Exercise B20, Section 26.2

B62. In Figure 26.7, change the homogeneous Dirichlet conditions on the bottom and left edges to homogeneous Neumann conditions, and obtain the general solution of the resulting BVP.

For the data in Exercises B63 and 64, use the results of Exercise B62 to solve the corresponding BVP.

(a) Plot the solution surface.

(b) Obtain level curves and flow lines.

(c) Compute the flux of ∇u through the boundary of the rectangle.

B63. the data from Exercise B21, Section 26.2

B64. the data from Exercise B22, Section 26.2

B65. In Figure 26.11, change the homogeneous Dirichlet condition on the left edge to the homogeneous Neumann condition $u_x(0, y) = 0$, and obtain the general solution to the resulting BVP.

For the data in Exercises B66 and 67, use the results of Exercise B65 to solve the corresponding BVP.

(a) Plot the solution surface.

(b) Obtain level curves and flow lines.

(c) Compute the flux of ∇u through the boundary of the rectangle.

B66. the data from Exercise B9, Section 26.1, and Exercise B10, Section 26.2

B67. the data from Exercise B10, Section 26.1, and Exercise B11, Section 26.2

B68. In Figure 26.11, change the homogeneous Dirichlet condition on the right edge to the homogeneous Neumann condition $u_x(a, y) = 0$, and obtain the general solution to the resulting BVP.

For the data in Exercises B69 and 70, use the results of Exercise B68 to solve the corresponding BVP.

(a) Plot the solution surface.

(b) Obtain level curves and flow lines.

(c) Compute the flux of ∇u through the boundary of the rectangle.

B69. the data from Exercise B11, Section 26.1, and Exercise B12, Section 26.2

B70. the data from Exercise B12, Section 26.1, and Exercise B13, Section 26.2

26.3   Nonzero Temperature on the Left Edge

Problem Formulation

Suppose the one nonhomogeneous Dirichlet condition on the rectangle in Sections 26.1 or 26.2 were imposed on the left edge, namely, on the edge $x = 0$. The temperature on that edge would be a function of y, as summarized in Figure 26.14. A formal statement of this boundary value problem would include Laplace's equation $u_{xx} + u_{yy} = 0$ in the rectangle

$0 \le x \le a, 0 \le y \le b$ and the Dirichlet boundary conditions $u(x, 0) = 0, u(0, y) = f(y)$, $u(a, y) = 0, u(x, b) = 0$.

Solution by Symmetry

If the variables x and y and parameters a and b are interchanged in (26.6a) and (26.6b), the solution of the Dirichlet problem where the bottom edge is maintained at a nonzero temperature, we obtain the solution of the Dirichlet problem associated with Figure 26.14. In fact, that interchange of variables leads to the solution

$$u(x, y) = \sum_{n=1}^{\infty} A_n \sin \frac{n\pi y}{b} \sinh \left(\frac{n\pi}{b}(x - a) \right) \tag{26.10a}$$

$$A_n = -\frac{2}{b \sinh \frac{n\pi a}{b}} \int_0^b f(y) \sin \frac{n\pi y}{b} \, dy \tag{26.10b}$$

where $f(y)$ replaces $g(x)$ in (26.6b).

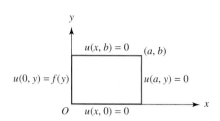

FIGURE 26.14   Dirichlet problem on rectangle whose left edge satisfies the nonhomogeneous Dirichlet condition

EXAMPLE 26.5   Let $a = 4, b = 4$, and define

$$f(y) = \begin{cases} y^2 & 0 \le y < 1 \\ 1 & 1 \le y \le 3 \\ (y - 4)^2 & 3 < y \le 4 \end{cases}$$

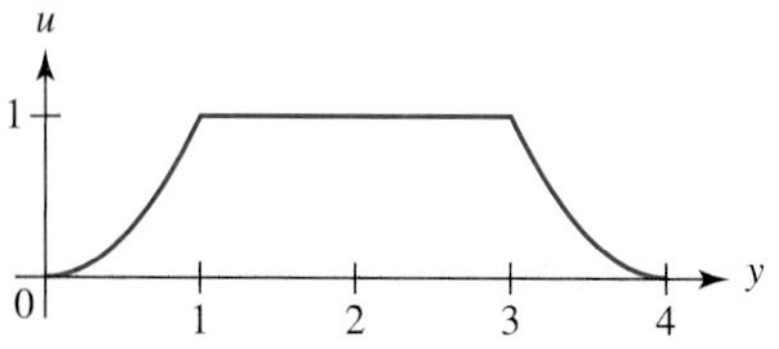

graphed in Figure 26.15. The series coefficients (26.10b) evaluate to

$$\frac{16}{n^3 \pi^3 \sinh n\pi} \left[4 \left(\cos \frac{3n\pi}{4} - \cos \frac{n\pi}{4} + 1 - (-1)^n \right) - n\pi \left(\sin \frac{3n\pi}{4} + \sin \frac{n\pi}{4} \right) \right]$$

$$\tag{26.11}$$

and are $O(n^{-2} e^{-n\pi})$, so the partial sum

$$u_9(x, y) = \sum_{n=1}^{9} A_n \sin \frac{n\pi y}{4} \sinh \left(\frac{n\pi}{4}(x - 4) \right)$$

graphed in Figure 26.16, will adequately approximate the solution on the interior of the square.

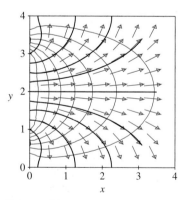

FIGURE 26.15   Dirichlet condition $u(0, y) = f(y)$

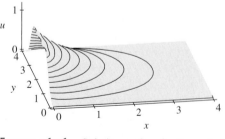

FIGURE 26.16   Solution surface for Example 26.5

FIGURE 26.17   Isotherms (in color), flow lines (in black), and vectors from the field $-\nabla u$ for Example 26.5

Figure 26.17 shows, along with the contours on the surface in Figure 26.16, the vectors of the field $-\nabla u$ and the flow lines for which these vectors are tangent. The contours are the thin curves, and the flow lines, the thick. There is just a hint that heat flows into the slab from warmer portions of the heated edge and then flows back out, not only on the three edges kept at temperature zero but also on a cooler portion of the heated edge!   ❖

Solution by Separation of Variables

First, the separation assumption $u(x, y) = X(x)Y(y)$ is applied to the boundary data to conclude $Y(0) = 0$, $X(a) = 0$, and $Y(b) = 0$. Then, Laplace's equation is separated to

$$\frac{Y''(y)}{Y(y)} = -\frac{X''(x)}{X(x)} = \lambda$$

from which we obtain the ODEs

$$Y''(y) - \lambda Y(y) = 0 \quad \text{and} \quad X''(x) + \lambda X(x) = 0$$

The first equation for $Y(y)$ is incorporated into the Sturm–Liouville BVP on the left in (26.12); the eigenfunctions and eigenvalues are shown on the right.

$$\begin{aligned} Y''(y) - \lambda Y(y) &= 0 \\ Y(0) = Y(b) &= 0 \end{aligned} \Rightarrow \begin{aligned} Y_n(y) &= C_n \sin \frac{n\pi y}{b} \\ \lambda_n &= -\frac{n^2\pi^2}{b^2} \quad n = 1, 2, \dots \end{aligned} \tag{26.12}$$

That leaves us with the ODE

$$X_n''(x) - \frac{n^2\pi^2}{b^2} X_n(x) = 0$$

for which the general solution is $X_n(x) = c_n e^{n\pi x/b} + d_n e^{-n\pi x/b}$. Application of the condition $X_n(a) = 0$ gives $d_n = -c_n e^{2n\pi a/b}$ so that

$$\begin{aligned} X_n(x) &= c_n(e^{n\pi x/b} - e^{2n\pi a/b} e^{-n\pi x/b}) \\ &= c_n e^{n\pi a/b}(e^{-n\pi a/b} e^{n\pi x/b} - e^{n\pi a/b} e^{-n\pi x/b}) \\ &= c_n e^{n\pi a/b}(e^{n\pi(x-a)/b} - e^{-n\pi(x-a)/b}) \\ &= B_n \sinh\left(\frac{n\pi}{b}(x - a)\right) \end{aligned}$$

Therefore we have the general eigensolution

$$u_n(x, y) = A_n \sin \frac{n\pi y}{b} \sinh\left(\frac{n\pi}{b}(x - a)\right)$$

where $A_n = B_n C_n$. The general solution is then (26.10a), a linear combination of all possible eigensolutions. To this solution we apply the final boundary condition, obtaining

$$u(0, y) = f(y) = \sum_{n=1}^{\infty}\left[-A_n \sinh \frac{n\pi a}{b}\right] \sin \frac{n\pi y}{b}$$

This is a Fourier sine series for $f(y) = \sum_{n=1}^{\infty} b_n \sin \frac{n\pi y}{b}$, so

$$\left[-A_n \sinh \frac{n\pi a}{b}\right] = b_n = \frac{2}{b}\int_0^b f(y) \sin \frac{n\pi y}{b}\, dy$$

and A_n is given by (26.10b). Hence, the complete solution to the given BVP is the series given by (26.10a) and (26.10b).

Extensions

A companion Dirichlet problem has the nonhomogeneous Dirichlet condition on the right edge of the rectangle in Figure 26.14. As we saw in the exercises in Section 26.2, we are now in a position to solve the Dirichlet problem for a rectangle where any two opposing edges support a nonhomogeneous Dirichlet condition. Using the same strategy of superposition, we can also solve Dirichlet problems where any three edges, or even all four edges, support nonhomogeneous Dirichlet conditions. As we also saw in the exercises in Section 26.2, we can introduce nonhomogeneous Neumann conditions to insulate one or more edges of the rectangle. These and other variations of the basic problem for Laplace's equation on the rectangle are left to the exercises.

EXERCISES 26.3–Part A

A1. Verify that A_n in Example 26.5 is given by (26.11).

A2. Verify graphically that in Example 26.5, A_n, given by (26.11), is $O(n^{-2}e^{-n\pi})$.

A3. Obtain the general solution to the Dirichlet problem for the rectangle in Figure 26.14 if it has three homogeneous Dirichlet conditions and one nonhomogeneous Dirichlet condition $u(a, y) = F(y)$ on the right edge.

A4. Write the general solution to the Dirichlet problem for the rectangle in Figure 26.14 if it has two homogeneous Dirichlet conditions and two nonhomogeneous Dirichlet conditions $u(0, y) = f(y)$ and $u(a, y) = F(y)$ on the left and right edges, respectively.

A5. Write the general solution of the Dirichlet problem for the rectangle in Figure 26.14 if it has the nonhomogeneous Dirichlet conditions $u(0, y) = f(y)$ and $u(a, y) = F(y)$ on the left and right edges, respectively, and $u(x, 0) = g(x)$ and $u(x, b) = G(x)$ on the bottom and top edges, respectively.

EXERCISES 26.3–Part B

B1. Obtain the general solution to the BVP on the rectangle in Figure 26.14 if the solution satisfies the homogeneous Neumann condition $u_x(0, y) = 0$ on the left edge, homogeneous Dirichlet conditions on the top and bottom edges, and the nonhomogeneous Dirichlet condition $u(a, y) = F(y)$ on the right edge.

B2. Obtain the general solution to the BVP on the rectangle in Figure 26.14 if the solution satisfies the homogeneous Neumann condition $u_x(a, y) = 0$ on the right edge, homogeneous Dirichlet conditions on the top and bottom edges, and the nonhomogeneous Dirichlet condition $u(0, y) = f(y)$ on the left edge.

B3. Obtain the general solution to the BVP on the rectangle in Figure 26.14 if the solution satisfies the homogeneous Neumann condition $u_y(x, b) = 0$ on the top edge, a homogeneous Dirichlet condition on the bottom, and the nonhomogeneous Dirichlet conditions $u(0, y) = f(y)$ and $u(a, y) = F(y)$ on the left and right edges, respectively.

B4. Obtain the general solution to the BVP on the rectangle in Figure 26.14 if the solution satisfies the homogeneous Neumann condition $u_y(x, 0) = 0$ on the bottom edge, a homogeneous Dirichlet condition on the top, and the nonhomogeneous Dirichlet conditions $u(0, y) = f(y)$ and $u(a, y) = F(y)$ on the left and right edges, respectively.

B5. Obtain the general solution to the BVP on the rectangle in Figure 26.14 if the solution satisfies the homogeneous Neumann conditions $u_y(x, b) = 0$ and $u_y(x, 0) = 0$, respectively, on the top and bottom edges, and the nonhomogeneous Dirichlet conditions $u(0, y) = f(y)$ and $u(a, y) = F(y)$ on the left and right edges, respectively.

B6. Obtain the general solution to the BVP on the rectangle in Figure 26.14 if the solution satisfies the homogeneous Neumann

conditions $u_x(a, y) = 0$ and $u_y(x, 0) = 0$, respectively, on the right and bottom edges, the homogeneous Dirichlet condition $u(x, b) = 0$ on the top, and the nonhomogeneous Dirichlet condition $u(0, y) = f(y)$ on the left.

B7. Obtain the general solution to the BVP on the rectangle in Figure 26.14 if the solution satisfies the homogeneous Neumann conditions $u_x(a, y) = 0$ and $u_y(x, b) = 0$, respectively, on the right and top edges, the homogeneous Dirichlet condition $u(x, 0) = 0$ on the bottom, and the nonhomogeneous Dirichlet condition $u(0, y) = f(y)$ on the left.

B8. Obtain the general solution to the BVP on the rectangle in Figure 26.14 if the solution satisfies the homogeneous Neumann conditions $u_x(0, y) = 0$ and $u_y(x, 0) = 0$, respectively, on the left and bottom edges, the homogeneous Dirichlet condition $u(x, b) = 0$ on the top, and the nonhomogeneous Dirichlet condition $u(a, y) = F(y)$ on the right.

B9. Obtain the general solution to the BVP on the rectangle in Figure 26.14 if the solution satisfies the homogeneous Neumann conditions $u_x(0, y) = 0$ and $u_y(x, b) = 0$, respectively, on the left and top edges, the homogeneous Dirichlet condition $u(x, 0) = 0$ on the bottom, and the nonhomogeneous Dirichlet condition $u(a, y) = F(y)$ on the right.

B10. Solve the BVP in Exercise A5 if the Dirichlet condition on the left edge is replaced with a homogeneous Neumann condition.

B11. Solve the BVP in Exercise A5 if the Dirichlet condition on the right edge is replaced with a homogeneous Neumann condition.

B12. Solve the BVP in Exercise A5 if the Dirichlet condition on the bottom edge is replaced with a homogeneous Neumann condition.

B13. Solve the BVP in Exercise A5 if the Dirichlet condition on the top edge is replaced with a homogeneous Neumann condition.

For the data given in each of Exercises B14–18:

(a) Solve the Dirichlet problem corresponding to Figure 26.14.

(b) For the series solution found in part (a), let $U(x, y)$ be a partial sum for which $U(0, y)$ is a reasonable approximation to $f(y)$. Use $U(x, y)$ to approximate $u(\frac{a}{2}, \frac{b}{2})$, the value of u at the center of the rectangle.

(c) Graph the solution surface.

(d) Obtain a contour plot showing the isotherms (interpreting $u(x, y)$ as temperatures) or equipotentials (interpreting $u(x, y)$ as a potential).

(e) Obtain a plot of the gradient field arising from $u(x, y)$.

(f) Calculate and plot the flow lines of the gradient field.

(g) Superimpose the flow lines on the contour plot of the level curves of u.

B14. $(a, b) = (\pi, 1)$, $f(y) = y(1 - y)$

B15. $(a, b) = (\pi, 1)$, $f(y) = y^2(1 - y)$

B16. $(a, b) = (\pi, 1)$, $f(y) = y^3(1 - y)$

B17. $(a, b) = (\pi, 1)$, $f(y) = \sin 2\pi y$

B18. $(a, b) = (\pi, 1)$, $f(y) = \sin^2 \pi y$

For the data given in each of Exercises B19–23:

(a) Use Exercise A3 to solve the Dirichlet problem $u_{xx} + u_{yy} = 0$, $u(x, 0) = u(x, b) = u(0, y) = 0$, $u(a, y) = F(y)$.

(b) For the series solution found in part (a), let $U(x, y)$ be a partial sum for which $U(0, y)$ is a reasonable approximation to $f(y)$. Use $U(x, y)$ to approximate $u(\frac{a}{2}, \frac{b}{2})$, the value of u at the center of the rectangle.

(c) Graph the solution surface.

(d) Obtain a contour plot showing the isotherms (interpreting $u(x, y)$ as temperatures) or equipotentials (interpreting $u(x, y)$ as a potential).

(e) Obtain a plot of the gradient field arising from $u(x, y)$.

(f) Calculate and plot the flow lines of the gradient field.

(g) Superimpose the flow lines on the contour plot of the level curves of u.

B19. $(a, b) = (\pi, 1)$, $F(y) = y(1 - y)^2$

B20. $(a, b) = (\pi, 1)$, $F(y) = y(1 - y)^3$

B21. $(a, b) = (\pi, 1)$, $F(y) = y(1 - y)$

B22. $(a, b) = (\pi, 1)$, $F(y) = \cos 2\pi y$

B23. $(a, b) = (\pi, 1)$, $F(y) = \cos^2 \pi y$

In Exercises B24 and 25, apply the results in Exercise A4 to the given data, obtaining

(a) the solution surface.

(b) the flow lines superimposed on the level curves.

B24. $(a, b) = (\pi, 1)$, $f(y) = \sin^2 2\pi y$, $F(y) = \cos^2 2\pi y$

B25. $(a, b) = (\pi, 1)$, $f(y) = \cos \pi y$, $F(y) = \sin \pi y$

In Exercises B26 and 27, apply the results of Exercise B1 to the given data, obtaining

(a) the solution surface.

(b) the flow lines superimposed on the level curves.

B26. $(a, b) = (\pi, 1)$, $F(y) = \sin^2 \pi y$

B27. $(a, b) = (1, 1)$, $F(y) = 1 - y^3$

In Exercises B28 and 29, apply the results of Exercise B2 to the given data, obtaining

(a) the solution surface.

(b) the flow lines superimposed on the level curves.

B28. $(a, b) = (\pi, 1)$, $f(y) = \cos^2 \pi y$

B29. $(a, b) = (1, 1)$, $f(y) = 1 - y$

B30. Apply the results of Exercise B3 to the data $(a, b) = (1, 1)$, $f(y) = 1 - y^2$, $F(y) = 1 - y$, obtaining the solution surface, the level curves, and the flow lines. Compute the flux of ∇u through the boundary of the rectangle.

B31. Apply the results of Exercise B4 to the data $(a, b) = (1, 1)$, $f(y) = 1 - y^3$, $F(y) = 1 - y^2$, obtaining the solution surface, the level curves, and the flow lines. Compute the flux of ∇u through the boundary of the rectangle.

B32. Apply the results of Exercise B5 to the data $(a, b) = (2, 3)$, $f(y) = ye^{-y}$, $F(y) = y^2e^{-y}$, obtaining the solution surface, the level curves, and the flow lines. Compute the flux of ∇u through the boundary of the rectangle.

B33. Apply the results of Exercise B6 to the data
$(a, b) = (4, 4)$, $f(y) = \begin{cases} 0 & 0 \le y < 1 \\ 5 & 1 \le y < 2 \\ 4 & 2 \le y < 3 \\ 0 & 3 \le y \le 4 \end{cases}$, obtaining the solution surface, the level curves, and the flow lines. Compute the flux of ∇u through the boundary of the rectangle.

B34. Apply the results of Exercise B7 to the data
$(a, b) = (4, 4)$, $f(y) = \begin{cases} 0 & 0 \le y < 1 \\ (y-1)^2 & 1 \le y < 2 \\ 3 - y & 2 \le y < 3 \\ 0 & 3 \le y \le 4 \end{cases}$, obtaining the solution surface, the level curves, and the flow lines. Compute the flux of ∇u through the boundary of the rectangle.

B35. Apply the results of Exercise B8 to the data
$(a, b) = (4, 4)$, $F(y) = \begin{cases} -1 & 0 \le y < 1 \\ 1 & 1 \le y < 2 \\ 2 & 2 \le y < 3 \\ 0 & 3 \le y \le 4 \end{cases}$, obtaining the solution surface, the level curves, and the flow lines. Compute the flux of ∇u through the boundary of the rectangle.

B36. Apply the results of Exercise B9 to the data
$(a, b) = (4, 4)$, $F(y) = \begin{cases} 0 & 0 \le y < 1 \\ y - 1 & 1 \le y < 2 \\ (3-y)^2 & 2 \le y < 3 \\ 0 & 3 \le y \le 4 \end{cases}$, obtaining the solution surface, the level curves, and the flow lines. Compute the flux of ∇u through the boundary of the rectangle.

B37. Apply the results of Exercise B10 to the data
$$(a, b) = (4, 4), \quad F(y) = \begin{cases} y^3 & 0 \le y < 1 \\ 1 & 1 \le y < 3 \\ (4 - y)^2 & 3 \le y \le 4 \end{cases}, \quad g(x) \text{ as given in}$$
Exercise B17, Section 26.1, and $G(x)$ as given in Exercise B18, Section 26.2. Obtain the solution surface, the level curves, and the flow lines. Compute the flux of ∇u through the boundary of the rectangle.

B38. Apply the results of Exercise B11 to the data
$$(a, b) = (4, 4), \quad f(y) = \begin{cases} y^2 & 0 \le y < 1 \\ 1 & 1 \le y < 3 \\ 4 - y & 3 \le y \le 4 \end{cases}, \quad g(x) \text{ as given in}$$
Exercise B19, Section 26.1, and $G(x)$ as given in Exercise B20, Section 26.2. Obtain the solution surface, the level curves, and the flow lines. Compute the flux of ∇u through the boundary of the rectangle.

B39. Apply the results of Exercise B12 to the data
$$(a, b) = (4, 4), \quad f(y) = \begin{cases} y & 0 \le y < 1 \\ 1 & 1 \le y < 3 \\ (4 - y)^2 & 3 \le y \le 4 \end{cases},$$
$$F(y) = \begin{cases} y^4 & 0 \le y < 1 \\ 1 & 1 \le y < 3 \\ (4 - y)^4 & 3 \le y \le 4 \end{cases}, \text{ and } G(x) \text{ as given in Exercise B22,}$$
Section 26.2. Obtain the solution surface, the level curves, and the flow lines.

B40. Apply the results of Exercise B13 to the data
$$(a, b) = (4, 4), \quad f(y) = \begin{cases} y(2 - y)^3 & 0 \le y < 1 \\ 1 & 1 \le y < 3 \\ (y - 2)(4 - y) & 3 \le y \le 4 \end{cases},$$
$$F(y) = \begin{cases} 3 & 0 \le y < 1 \\ 0 & 1 \le y < 3 \\ -2 & 3 \le y \le 4 \end{cases}, \text{ and } g(x) \text{ as given in Exercise B21,}$$
Section 26.1. Obtain the solution surface, the level curves, and the flow lines. Compute the flux of ∇u through the boundary of the rectangle.

26.4   Finite-Difference Solution of Laplace's Equation in a Rectangle

Dirichlet Problem on a Rectangle

In accord with Figure 26.18, we derive a finite-difference solution of the Dirichlet problem consisting of Laplace's equation in the rectangle $0 \le x \le a, 0 \le y \le b$ and the (nonhomogeneous) Dirichlet boundary conditions $u(0, y) = f(y), u(a, y) = F(y), u(x, 0) = g(x), u(x, b) = G(x)$.

Discretization

First, define a grid on the rectangle. Let the x-interval $[0, a]$ be divided by the $N + 1$ points $x_n = n\Delta x, n = 0, 1, \ldots, N$, into N equal subintervals of length $\Delta x = \frac{a}{N}$. Similarly, discretize the y-interval $[0, b]$ with the partition $y_m = m\Delta y, m = 0, 1, \ldots, M$, where $\Delta y = \frac{b}{M}$. Let $u(x_n, y_m)$ be the exact value of the solution at the point (x_n, y_m). Let $v_{n,m}$ be the numeric approximation to that exact value, obtained by replacing $u_{xx}(x_n, y_m)$ and $u_{yy}(x_n, y_m)$ with the second-order central-difference formulas

$$u_{xx}(x_n, y_m) = \frac{v_{n+1,m} - 2v_{n,m} + v_{n-1,m}}{(\Delta x)^2} \quad \text{and} \quad u_{yy}(x_n, y_m) = \frac{v_{n,m+1} - 2v_{n,m} + v_{n,m-1}}{(\Delta y)^2}$$

$$(26.13)$$

in Laplace's equation, giving

$$\frac{v_{n+1,m} - 2v_{n,m} + v_{n-1,m}}{(\Delta x)^2} + \frac{v_{n,m+1} - 2v_{n,m} + v_{n,m-1}}{(\Delta y)^2} = 0 \qquad (26.14)$$

Figure 26.19 shows the computational "molecule" expressing the discretized Laplacian at each interior node (x_n, y_m), $1 \le n \le N - 1, 1 \le m \le M - 1$. The open dots represent unknown values of v. At each node, the equation to be formed is the sum of the "north, east, south, and west" nodal values minus four times the value at that node. There are no "known values" in the interior of the rectangle! All the $N_1 \times M_1 = (N - 1) \times (M - 1)$ nodal equations are coupled. Boundary data is propagated to the interior through the interlocking

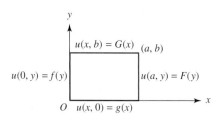

FIGURE 26.18   Dirichlet problem on a rectangle where all four edges support nonhomogeneous Dirichlet conditions

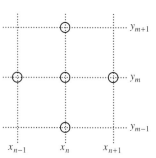

FIGURE 26.19   Computational molecule for Laplace's equation

of these nodal equations, and a fine mesh generates large systems of algebraic equations whose solutions often require sophisticated techniques from numerical linear algebra rather than from the theory of partial differential equations.

EXAMPLE 26.6   For simplicity, take $a = b = 1, N = M = 5$ so that $\Delta x = \Delta y = \frac{1}{5}$. Then define the boundary functions $f(y) = y(1 - y)$, $F(y) = \sin \pi y$, $g(x) = \sin 2\pi x$, and $G(x) = x(x - 1)$. Figure 26.20 shows these functions and their relative orientation on the square.

The grid imposed on the rectangle contains six rows of six nodes, for a total of 36 nodes. There are $N_1 \times M_1 = (N - 1) \times (M - 1) = 16$ interior unknown values to compute, so we will need to formulate and solve a system of 16 equations in 16 unknowns. Rather than work with doubly indexed variables, label the unknowns according the pattern shown in the schematic in Figure 26.21, where the letter b indicates a value determined from a boundary function.

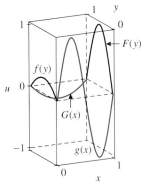

FIGURE 26.20   Boundary functions for Example 26.6

b	b	b	b	b	b
b	v_1	v_2	v_3	v_4	b
b	v_5	v_6	v_7	v_8	b
b	v_9	v_{10}	v_{11}	v_{12}	b
b	v_{13}	v_{14}	v_{15}	v_{16}	b
b	b	b	b	b	b

FIGURE 26.21   Schematic for the nodal values in Example 26.6

For the grid chosen, the discretized Laplacian becomes

$$v_{n+1,m} - 4v_{n,m} + v_{n-1,m} + v_{n,m+1} + v_{n,m-1} = 0 \qquad (26.15)$$

The coordinates at the nodes are $(x_n, y_m) = (\frac{n}{N}, \frac{m}{M}) = (\frac{n}{5}, \frac{m}{5})$ and a mapping from the nodal point (x_n, y_m) to the corresponding variable v_s, $s = 1, 2, \ldots, 16$, is achieved with the help of the function

$$s(n, m) = n + (M_1 - m)N_1 \qquad (26.16)$$

where $M_1 = M - 1$ and $N_1 = N - 1$. The values of (n, m) corresponding to the grid points (x_n, y_m) do not readily correlate with (r, c), the row-column indices on a matrix. The index n on x_n is the "column" index on the rectangle's grid. The index m on y_m is the "row" index, but in the reverse order, for the rows in the rectangle's grid. Thus, to map the points (x_n, y_m) to the array of nodal values v_s, there must be two "reversals" in the indices. First, the row and column indices must be interchanged; then, the column index must run in the opposite order. The matrix with entries

$$A_{ij} = s(j, M - i) = j + (i - 1)N_1 = j + (i - 1)(N - 1) \qquad (26.17)$$

is therefore

$$\begin{bmatrix} 1 & 2 & 3 & 4 \\ 5 & 6 & 7 & 8 \\ 9 & 10 & 11 & 12 \\ 13 & 14 & 15 & 16 \end{bmatrix} \qquad (26.18)$$

Equation (26.15) becomes, at each grid point,

$$\text{West} + \text{East} + \text{North} + \text{South} - 4v_s = 0$$

When West, East, North, or South represent a node on the boundary, the value must be obtained from the appropriate function $f(y)$, $F(y)$, $G(x)$, $g(x)$, respectively. Thus, formation of the equations for this system requires a distinction between nodes that are next to no boundary points, one boundary point, or two boundary points (at the four corners). For example, at the point (x_4, y_4), the variable v_4 represents $v_{4,4}$. The neighbor to the west is v_3 and to the south is v_8. The neighbor to the east is $F(y_4)$ and the neighbor to the north is $G(x_4)$. Code for obtaining the nodal equations is given in the accompanying Maple worksheet.

If the 16 equations are written in the form $A\mathbf{v} = \mathbf{c}$, the matrix A containing the coefficients of all 16 equations is given in Table 26.2.

−4	1	0	0	1	0	0	0	0	0	0	0	0	0	0	0
1	−4	1	0	0	1	0	0	0	0	0	0	0	0	0	0
0	1	−4	1	0	0	1	0	0	0	0	0	0	0	0	0
0	0	1	−4	0	0	0	1	0	0	0	0	0	0	0	0
1	0	0	0	−4	1	0	0	1	0	0	0	0	0	0	0
0	1	0	0	1	−4	1	0	0	1	0	0	0	0	0	0
0	0	1	0	0	1	−4	1	0	0	1	0	0	0	0	0
0	0	0	1	0	0	1	−4	0	0	0	1	0	0	0	0
0	0	0	0	1	0	0	0	−4	1	0	0	1	0	0	0
0	0	0	0	0	1	0	0	1	−4	1	0	0	1	0	0
0	0	0	0	0	0	1	0	0	1	−4	1	0	0	1	0
0	0	0	0	0	0	0	1	0	0	1	−4	0	0	0	1
0	0	0	0	0	0	0	0	1	0	0	0	−4	1	0	0
0	0	0	0	0	0	0	0	0	1	0	0	1	−4	1	0
0	0	0	0	0	0	0	0	0	0	1	0	0	1	−4	1
0	0	0	0	0	0	0	0	0	0	0	1	0	0	1	−4

TABLE 26.2   Coefficient matrix A for Example 26.6

The vector $\mathbf{c}$ and the solution vector $\mathbf{v}$ are

$$
\mathbf{c} = \begin{bmatrix}
0 \\
0.2400000000 \\
0.2400000000 \\
-0.4277852520 \\
-0.2400000000 \\
0 \\
0 \\
-0.9510565160 \\
-0.2400000000 \\
0 \\
0 \\
-0.9510565160 \\
-1.111056516 \\
-0.5877852520 \\
0.5877852520 \\
0.3632712640
\end{bmatrix}
\qquad
\mathbf{v} = \begin{bmatrix}
0.04610335322 \\
0.00533911906 \\
0.05477087277 \\
0.2337890290 \\
0.1790742938 \\
0.1604822503 \\
0.2199553429 \\
0.4525999913 \\
0.2697115717 \\
0.2375602453 \\
0.2119682571 \\
0.4055990772 \\
0.4222117474 \\
0.3080789020 \\
-0.01524163722 \\
0.00677154400
\end{bmatrix}
\qquad (26.19)
$$

A graph of the solution surface appears in Figure 26.22.

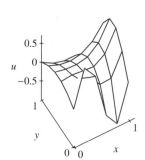

FIGURE 26.22   Example 26.6: Solution surface computed with $M = N = 5$

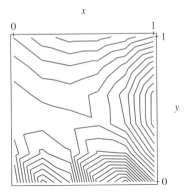

FIGURE 26.23 Example 26.6: Contours generated by a 5 × 5 grid

OBSERVATIONS AND COMMENTS The matrix A is very regular and very sparse. It is regular because it is essentially *banded* and *sparse* because it consists mostly of zeros. Thus, it could have been constructed without ever writing a single nodal equation. Deducing the pattern is left to the exercises. By the same token, deducing the pattern for the vector $\mathbf{c}$ in the system $A\mathbf{v} = \mathbf{c}$ is also left to the exercises.

Banded matrices can be stored as lists of diagonal elements. Diagonals of zeros need not be stored. Arithmetic consisting of adding, subtracting, and multiplying zeros can also be avoided by special algorithms for solving systems with banded matrices. Thus, the effort it took to organize the variables and establish patterns makes it possible to solve much larger problems. These matters are typically dealt with in texts on numerical linear algebra.

CONTOURS Figure 26.23 is a contour plot obtained from the data upon which Figure 26.22 is based. The contours are not smooth because the grid was a course 5 × 5. If N and M are both increased to 20, there will be 361 equations in 361 unknowns to solve, a task taking several minutes to complete. The resulting solution surface, pleasant to view, is given in Figure 26.24. Figure 26.25 contains a smoother contour plot. ❖

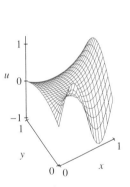

FIGURE 26.24
Example 26.6: Solution surface generated on a 20 × 20 grid

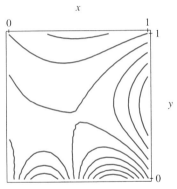

FIGURE 26.25 Example 26.6: Contours generated by a 20 × 20 grid

EXERCISES 26.4–Part A

A1. Show that for equal grid spacing ($\Delta x = \Delta y$), (26.14) becomes (26.15).

A2. Obtain the analog of the array in Figure 26.21 when $N = M = 6$.

A3. Show that (26.16) correctly maps $v_{n,m}$ to v_s when $N = M = 6$.

A4. If $N = M = 6$, show that (26.17) correctly maps v_s to a 5 × 5 matrix analogous to the array in (26.18).

A5. Modify the finite-difference scheme for the Dirichlet problem to account for homogeneous Neumann conditions. For example, suppose $u_x(0, y) = 0$ is imposed on the left edge of the rectangle in Figure 26.18. The $O(h^2)$ central-difference approximation $\frac{1}{2h}(u(0 + h, y) - u(0 - h, y)) = 0$ leads to $u(-h, y) = u(h, y)$. For each interior node along the edge $x = 0$, the Laplacian is

evaluated. Each such interior node becomes an unknown in the problem, and function values at nodes outside the rectangle are replaced by function values along the line $x = h$.

A6. In Figure 26.18, replace the Dirichlet condition on the right edge with a homogeneous Neumann condition. Obtain a finite-difference algorithm for the resulting BVP.

A7. In Figure 26.18, replace the Dirichlet condition on the top edge with a homogeneous Neumann condition. Obtain a finite-difference algorithm for the resulting BVP.

A8. In Figure 26.18, replace the Dirichlet condition on the bottom edge with a homogeneous Neumann condition. Obtain a finite-difference algorithm for the resulting BVP.

EXERCISES 26.4–Part B

B1. In Figure 26.18, replace the Dirichlet conditions on the top and bottom edges with homogeneous Neumann conditions. Obtain a finite-difference algorithm for the resulting BVP.

B2. In Figure 26.18, replace the Dirichlet conditions on the bottom and right edges with homogeneous Neumann conditions. Obtain a finite-difference algorithm for the resulting BVP.

B3. In Figure 26.18, replace the Dirichlet conditions on the top and right edges with homogeneous Neumann conditions. Obtain a finite-difference algorithm for the resulting BVP.

B4. In Figure 26.18, replace the Dirichlet conditions on the left and bottom edges with homogeneous Neumann conditions. Obtain a finite-difference algorithm for the resulting BVP.

B5. In Figure 26.18, replace the Dirichlet conditions on the left and top edges with homogeneous Neumann conditions. Obtain a finite-difference algorithm for the resulting BVP.

In Exercises B6–10, solve the indicated BVP using finite differences. Take $N = M = 5$, and plot the solution surface and the level curves. Compare to the analytic solution found in Section 26.3.

B6. Exercise B14, Section 26.3

B7. Exercise B15, Section 26.3

B8. Exercise B16, Section 26.3

B9. Exercise B17, Section 26.3

B10. Exercise B18, Section 26.3

B11. Deduce the pattern of entries in the matrix shown in Table 26.2, verifying the deductions by comparing their consequences to the coefficient matrices obtained in any of Exercises B6–10.

B12. Deduce the pattern of entries in the vector **c** found in (26.19), verifying the deductions by comparing their consequences to the corresponding vectors obtained in any of Exercises B6–10.

In Exercises B13–17, solve the indicated BVP using finite differences. Take $N = M = 10$, and plot the solution surface and the level curves. Compare to the analytic solution found in Section 26.3.

B13. Exercise B19, Section 26.3   **B14.** Exercise B20, Section 26.3

B15. Exercise B21, Section 26.3   **B16.** Exercise B22, Section 26.3

B17. Exercise B23, Section 26.3

In Exercises B18 and 19, solve the indicated BVP using finite differences. Take $N = M = 20$, and plot the solution surface and level curves. Compare to the analytic solution found in Section 26.3.

B18. Exercise B24, Section 26.3   **B19.** Exercise B25, Section 26.3

In Exercises B20 and 21, use the results in Exercise A5 to solve the indicated BVP by finite differences. Take $N = M = 15$, and plot the solution surface and level curves. Compare to the analytic solution found in Section 26.3.

B20. Exercise B26, Section 26.3   **B21.** Exercise B27, Section 26.3

In Exercises B22 and 23, use the results in Exercise A6 to solve the indicated BVP by finite differences. Take $N = M = 15$, and plot the solution surface and level curves. Compare to the analytic solution found in Section 26.3.

B22. Exercise B28, Section 26.3   **B23.** Exercise B29, Section 26.3

B24. Use the results in Exercise A7 to solve Exercise B30, Section 26.3, by finite differences. Take $N = M = 10$, and plot the solution surface and level curves. Compare to the analytic solution found in Section 26.3.

B25. Use the results in Exercise A8 to solve Exercise B31, Section 26.3, by finite differences. Take $N = M = 12$, and plot the solution surface and level curves. Compare to the analytic solution found in Section 26.3.

B26. Use the results in Exercise B1 to solve Exercise B32, Section 26.3, by finite differences. Take $N = M = 20$, and plot the solution surface and level curves. Compare to the analytic solution found in Section 26.3.

B27. Use the results in Exercise B2 to solve Exercise B33, Section 26.3, by finite differences. Take $N = M = 10$, and plot the solution surface and level curves. Compare to the analytic solution found in Section 26.3.

B28. Use the results in Exercise B3 to solve Exercise B34, Section 26.3, by finite differences. Take $N = M = 15$, and plot the solution surface and level curves. Compare to the analytic solution found in Section 26.3.

B29. Use the results in Exercise B4 to solve Exercise B35, Section 26.3, by finite differences. Take $N = M = 12$, and plot the solution surface and level curves. Compare to the analytic solution found in Section 26.3.

B30. Use the results in Exercise B5 to solve Exercise B36, Section 26.3, by finite differences. Take $N = M = 10$, and plot the solution surface and level curves. Compare to the analytic solution found in Section 26.3.

B31. Obtain a finite-difference solution for the BVP in Exercise B37, Section 26.3. Take $N = M = 15$, and plot the solution surface and level curves. Compare to the analytic solution found in Section 26.3.

B32. Obtain a finite-difference solution for the BVP in Exercise B38, Section 26.3. Take $N = M = 15$, and plot the solution surface and level curves. Compare to the analytic solution found in Section 26.3.

B33. Obtain a finite-difference solution for the BVP in Exercise B39, Section 26.3. Take $N = M = 20$, and plot the solution surface and level curves. Compare to the analytic solution found in Section 26.3.

B34. Obtain a finite-difference solution for the BVP in Exercise B40, Section 26.3. Take $N = M = 20$, and plot the solution surface and level curves. Compare to the analytic solution found in Section 26.3.

Chapter Review

1. The Dirichlet problem for a rectangle consists of Laplace's equation $\nabla^2 u = 0$ in the rectangle and the specification of u on the boundary of the rectangle. Write the solution for the Dirichlet problem on a rectangle that has one corner at the origin of the xy-plane and the opposite corner at the point (a, b). On the bottom edge, u is $g(x)$; and on the other three edges, it is zero.

2. Write the solution for the Dirichlet problem on the rectangle of Question 1 if u is $G(x)$ on the top edge and is zero on the other three edges.

3. Write the solution for the Dirichlet problem on the rectangle of Question 1 if u is $f(y)$ on the left edge and is zero on the other three edges.

4. Write the solution for the Dirichlet problem on the rectangle of Question 1 if u is $F(y)$ on the right edge and is zero on the other three edges.

5. Show how to use the results of Questions 1–4 to solve the Dirichlet problem on the same rectangle if the boundary conditions $u(x, 0) = g(x)$, $u(x, b) = G(x)$, $u(0, y) = f(y)$, and $u(a, y) = F(y)$ are imposed.

6. Detail the separation of variables process by which the solution stated in Question 1 is obtained.

7. Detail the separation of variables process by which the solution stated in Question 2 is obtained.

8. Detail the separation of variables process by which the solution stated in Question 3 is obtained.

9. Detail the separation of variables process by which the solution stated in Question 4 is obtained.

10. Obtain the separation of variables solution for Laplace's equation on the rectangle of Question 1, if the conditions $u_x(a, y) = u_x(b, y) = 0$ are imposed on the left and right edges, $u(x, b) = 0$ is imposed on the top edge, and $u(x, 0) = f(x)$ is imposed on the bottom edge.

11. Obtain the separation of variables solution for Laplace's equation on the rectangle of Question 1, if, on the left edge u satisfies the homogeneous Neumann condition $u_x(0, y) = 0$, on the bottom edge it satisfies the nonhomogeneous Dirichlet condition $u(x, 0) = f(x)$, and on the other two edges it is zero.

12. Obtain the solution for the BVP described in Question 11 if $(a, b) = (\pi, 1)$ and $f(x) = x(\pi - x)$. Obtain a plot of the level curves of $u(x, y)$.

Chapter 27

Nonhomogeneous Boundary Value Problems

INTRODUCTION   The nonhomogeneous Dirichlet condition in either the heat or wave equation poses a severe challenge for the method of separation of variables. One of the ODEs generated by separating variables must have two boundary conditions to become a Sturm–Liouville eigenvalue problem. Unless this happens, there can be no solution in terms of eigenfunctions.

A thermal transfer problem in a rod with one end subject to a nonhomogeneous but constant-valued Dirichlet condition is readily solved analytically by superposition. The desired solution is a sum of two parts, one of which satisfies a BVP with homogeneous Dirichlet conditions.

However, the BVP governing heat transfer in the wall of a house whose interior is air-conditioned and whose exterior is subjected to the time-varying temperatures of the daily heating-cooling cycle of a Midwestern summer can only be solved analytically by a significant extension of the process of separating variables.

By superposition, the BVP must be split into two problems, one of which has homogeneous Dirichlet conditions. This problem will lead to a Sturm–Liouville eigenvalue problem that has a full set of eigenfunctions for building a series solution. However, the PDE for this problem is now the nonhomogeneous heat equation whose solution can be given only as an eigenfunction expansion with time-dependent coefficients. Thus, the nonhomogeneity in the boundary data is exchanged for a nonhomogeneity in the PDE, an exchange resulting in a much more difficult solution process for the PDE.

We do discover, however, that while the temperatures outside the house vary sinusoidally during the course of the daily cycle, the wall retains a residual heat content and its temperature profile varies in a decidedly nonlinear way. We even discover how the solution depends on the thermal diffusivity. Although we could have obtained these insights with a numerical solution of this BVP, there is merit in exposure to the strengths and weaknesses of analytic and numeric techniques.

27.1 One-Dimensional Heat Equation with Different Endpoint Temperatures

Problem Statement

If the boundary conditions in a boundary value problem are not homogeneous, the method of separation of variables cannot be used without modification. Consider one-dimensional heat flow in an insulated rod with left end at temperature zero but right end at some nonzero, but constant, temperature ϕ. In particular, consider the boundary value problem

$$u_t = \kappa u_{xx}, t > 0 \qquad \text{and} \qquad \begin{matrix} u(0, t) = 0 \\ u(L, t) = \phi \end{matrix} \qquad (27.1)$$
$$u(x, 0) = f(x)$$

Separation of variables fails, not because of the differential equation but because of the boundary conditions. Separation of the partial differential equation, as we have seen earlier, leads, as expected, to the ODEs $X''(x) - \lambda X(x) = 0$ and $T'(t) - \lambda \kappa T(t) = 0$. However, trouble arises when separating the data. The homogeneous boundary condition on the left leads to a conclusion since $u(0, t) = 0 \Rightarrow X(0)T(t) = 0 \Rightarrow X(0) = 0$. No conclusion can be drawn from the nonhomogeneous boundary condition on the right since $u(L, t) = \phi \Rightarrow X(L)T(t) = \phi \neq 0$. We no longer can conclude that $X(L) = 0$, so there is no Sturm–Liouville boundary value problem for determining the eigenvalues λ. Separation of variables fails.

Homogenizing Substitution

We consider the substitution $u(x, t) = v(x, t) + \psi(x)$ in an attempt to reformulate (27.1) as a BVP with homogeneous boundary conditions. Applied to the heat equation, we obtain $v_t = \kappa(v_{xx} + \psi'')$. Demanding $\psi'' = 0$ restores this differential equation to the simple form of the heat equation. Applying the endpoint conditions to the hypothesized solution $u(x, t) = v(x, t) + \psi(x)$ leads to

$$v(0, t) + \psi(0) = 0 \quad \text{and} \quad v(L, t) + \psi(L) = \phi$$

Now, if we demand that $\psi(0) = 0$ and $\psi(L) = \phi$, then $v(0, t) = 0$ and $v(L, t) = 0$.

Thus, if we demand $\psi'' = 0$, along with $\psi(0) = 0$ and $\psi(L) = \phi$, then $v(x, t)$ satisfies the homogeneous heat equation with homogeneous boundary conditions. The initial condition $v(x, t)$ must now satisfy is $v(x, 0) = f(x) - \psi(x)$, since

$$u(x, 0) = f(x) \Rightarrow v(x, 0) + \psi(x) = f(x) \Rightarrow v(x, 0) = f(x) - \psi(x)$$

If we can find such a function $\psi(x)$, then $v(x, t)$ satisfies the homogeneous BVP

$$\begin{matrix} v_t(x, t) = \kappa v_{xx}(x, t) \\ v(x, 0) = f(x) - \psi(x) \end{matrix} \qquad \text{and} \qquad \begin{matrix} v(0, t) = 0 \\ v(L, t) = 0 \end{matrix}$$

and the solution to (27.1) will be $u(x, t) = v(x, t) + \psi(x)$.

A simple integration gives $\psi(x) = a + bx$, but the conditions $\psi(0) = 0$ and $\psi(L) = \phi$ determine a and b so that $\psi(x) = \frac{\phi}{L}x$. We can now solve for $v(x, t)$, obtaining

$$v(x, t) = \sum_{n=1}^{\infty} b_n \sin\left(\frac{n\pi x}{L}\right) e^{-\kappa(n\pi/L)^2 t}$$

$$b_n = \frac{2}{L} \int_0^L [f(x) - \psi(x)] \sin\frac{n\pi x}{L} \, dx$$

and consequently, $u(x, t) = v(x, t) + \psi(x)$. Since $v(x, t)$ tends to zero for large values of t, the steady-state solution is $u_\infty = \psi(x) = \frac{\phi}{L}x$.

EXAMPLE 27.1   Consider a rod of length π, with left end held at temperature zero and right end maintained at a temperature of $10°$. The thermal diffusivity is $\kappa = \frac{1}{2}$, and the initial temperature profile is $f(x) = x(\pi - x)$.

From the preceding calculations, we find $\psi(x) = \frac{\phi}{L}x = \frac{10}{\pi}x$, then write $u(x, t) = v(x, t) + \frac{10}{\pi}x$, and solve for $v(x, t)$ by computing the Fourier coefficients for the function $F(x) = f(x) - \psi(x) = x(\pi - x) - \frac{10}{\pi}x$. We get

$$b_n = \frac{2}{\pi}\int_0^\pi \left(x(\pi - x) - \frac{10}{\pi}x\right)\sin nx\, dx = \frac{4}{\pi n^3}(1 + (-1)^n(5n^2 - 1)) \qquad (27.2)$$

coefficients that are $O(\frac{1}{n})$ and therefore slowly decaying.

Taking the partial sum $v_{30}(x, t) = \sum_{n=1}^{30} b_n \sin(nx)e^{-n^2t/2}$ as an approximation to $v(x, t)$, we then have $u_{30} = v_{30} + \frac{10}{\pi}x$ as an approximation to the solution $u(x, t)$. Figure 27.1 shows this approximate solution surface. The temperature is initially discontinuous at $x = \pi$, but that discontinuity is immediately smoothed out by heat flowing into the rod from the heated right end.

Progress to the ultimate steady state of a linear temperature distribution across the rod can be better seen via plane sections $t = c$. Figure 27.2 contains the plane sections $t = 0$, 0.01, and 0.1. The approximate solution exhibits the Gibbs phenomenon at $x = \pi$ because of the discontinuity in the initial temperatures there. Immediately after, the solution begins smoothing out, the steady state being the linear temperature profile $u_\infty = \frac{10}{\pi}x$. The progression to equilibrium can be seen as an animation in the accompanying Maple worksheet. ❖

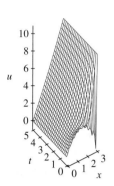

FIGURE 27.1   Solution surface for Example 27.1

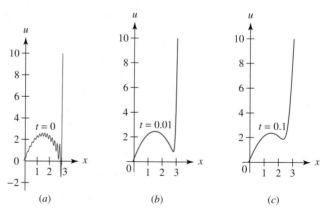

FIGURE 27.2   Example 27.1: At times $t = 0, 0.01$, and 0.1, plane sections determined by a partial sum of the series solution

EXERCISES 27.1

1. Verify the calculations in (27.2).

2. Demonstrate graphically that b_n in (27.2) is indeed $O(\frac{1}{n})$.

In Exercises 3–7:

  (a) Obtain an analytic solution of the BVP (27.1).

  (b) Graph the solution surface.

  (c) Graph $u(\frac{L}{2}, t)$.

  (d) Obtain and graph the steady-state solution.

  (e) Animate the temperature profiles as they change in time.

  (f) Obtain a finite-difference solution via the scheme of Section 25.4. Compare the numerical solution with the exact solution.

3. $L = \pi, \kappa = 2, \phi = 1, f(x) = \sin x$

4. $L = 3, \kappa = \frac{3}{5}, \phi = 25, f(x) = \frac{25}{27}x^3$

5. $L = 10, \kappa = \sqrt{2}, \phi = 75, f(x) = \frac{3}{400}x^4$

6. $L = 5, \kappa = 3, \phi = 50, f(x) = \frac{2}{125}x^5$

7. $L = 1, \kappa = \frac{3}{2}, \phi = 30, f(x) = 30x^7$

8. Obtain an analytic solution of the BVP $u_t = \kappa u_{xx}, u(0, t) = \sigma,$ $u(L, t) = 0, u(x, 0) = f(x)$.

In Exercises 9–13, the given data specifies a BVP of the type solved in Exercise 8. For each:

    (a) Use the results of Exercise 8 to solve the resulting BVP.

    (b) Graph the solution surface.

    (c) Graph $u(\frac{L}{2}, t)$.

    (d) Obtain and graph the steady-state solution.

    (e) Animate the temperature profiles as they change in time.

    (f) Obtain a finite-difference solution via the scheme of Section 25.4. Compare the numerical solution with the exact solution.

9. $L = 10, \kappa = 2, \sigma = 20, f(x) = \dfrac{1000 - x^3}{50}$

10. $L = 15, \kappa = \frac{2}{3}, \sigma = 100, f(x) = 100\left(1 - \dfrac{x}{15}\right)^5$

11. $L = 4, \kappa = 3, \sigma = 12, f(x) = \frac{3}{4}(16 - x^2)$

12. $L = 5, \kappa = \sqrt{3}, \sigma = -1, f(x) = \left(\dfrac{x}{5}\right)^5 - 1$

13. $L = 3, \kappa = \frac{1}{2}, \sigma = 15, f(x) = \frac{5}{27}(81 - x^4)$

27.2   One-Dimensional Heat Equation with Time-Varying Endpoint Temperatures

Problem Formulation

We will model the time-varying temperature profile within the wall of a mid-western house during the summer air-conditioning season. The wall has thickness $L = \pi$ and is insulated. Since the wall of a frame house is approximately 6-in thick, our units of length are each about 2 in. Our coordinate system (an x-axis) has its origin on the surface of the wall inside the house and extends into the wall. The outside surface of the wall (the part receiving the beating rays of the sun) is located at $x = \pi$.

The surface of the wall facing the living quarters is maintained at 70°F, and we take this temperature as the reference temperature $u(0, t) = 0$. The time-varying temperature of the outside wall is modeled as $\phi(t) = 15(1 + \sin\frac{\pi}{12}t)$, graphed in Figure 27.3. This represents a sinusoidal variation over a 24-hr period, with a maximum of 30 and a minimum of 0. That range corresponds to a mid-day temperature of 100°F and a night-time low of 70°F.

A boundary value problem governing this physical system is given in Table 27.1. The initial condition suggests that at time $t = 0$ the air-conditioner is turned on after the house and its walls heated up to the maximum outside temperature during, say, a power outage. We are also assuming that as soon as the air-conditioner is turned on, the surface of the wall inside the house is instantly at 70°F.

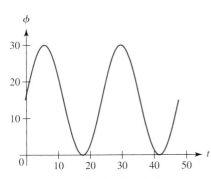

FIGURE 27.3   Graph of $\phi(t)$, the time-varying temperature on the outer wall for the BVP in Table 27.1

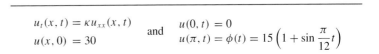

TABLE 27.1   BVP governing time-varying temperatures in a wall

Solution

The time-varying endpoint temperature at the right end leads to a much more complex solution than the nonhomogeneous, but constant, Dirichlet condition generated in Section 27.1. The solution to the BVP in Table 27.1 is obtained by executing, in the order given, the calculations in Table 27.2. After implementing this solution in Example 27.2 and exploring its implications, we will derive it in the last part of this section.

---

(1) $F(x) = 30 - \dfrac{x}{\pi}\phi(0) = \dfrac{15}{\pi}(2\pi - x)$

(2) $b_n = \dfrac{2}{\pi}\displaystyle\int_0^\pi F(x)\sin nx\,dx = \dfrac{30}{\pi n}(2 - (-1)^n)$

(3) $G(x, t) = -\dfrac{x}{\pi}\phi'(t) = -\dfrac{5}{4}x\cos\dfrac{\pi}{12}t$

(4) $G_n(t) = \dfrac{2}{\pi}\displaystyle\int_0^\pi G(x, t)\sin nx\,dx = \dfrac{5}{2n}(-1)^n\cos\dfrac{\pi}{12}t$

(5) $A_n(t) = \displaystyle\int_0^t G_n(s)e^{-\kappa n^2(t-s)}\,ds = \dfrac{30(-1)^n\left(12\kappa n^2\cos\frac{\pi}{12}t + \pi\sin\frac{\pi}{12}t - 12\kappa n^2 e^{-\kappa n^2 t}\right)}{(144\kappa^2 n^4 + \pi^2)n}$

(6) $v(x, t) = \displaystyle\sum_{n=1}^\infty A_n(t)\sin nx + \sum_{n=1}^\infty b_n\sin(nx)e^{-\kappa n^2 t}$

(7) $u(x, t) = v(x, t) + \dfrac{x}{\pi}\phi(t)$

TABLE 27.2   Calculations which solve BVP in Table 27.1

EXAMPLE 27.2   If we take $\kappa = \frac{1}{10}$, execute the calculations in Table 27.2, and then approximate the solution $u = v + \frac{x}{\pi}\phi(t)$ with

$$u_{\text{approx}} = \sum_{n=1}^{10} A_n(t)\sin nx + \sum_{n=1}^{5} b_n\sin(nx)e^{-\kappa n^2 t} + \frac{x}{\pi}\phi(t) \qquad (27.3)$$

we get, in Figure 27.4, the surface plot for $0.1 \le t \le 5$. Even though the graph starts six minutes into the first hour, there is still rippling in the surface initially because of the slow convergence of the series solution.

Looking beyond the transient phase of this thermal process, a graph starting after the second day and extending for two days is then given in Figure 27.5. The temperatures

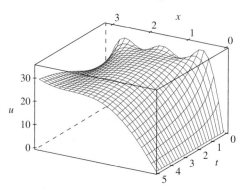

FIGURE 27.4   Example 27.2: Approximate solution surface for $0.1 \le t \le 5$

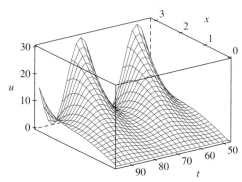

FIGURE 27.5   Example 27.2: Solution surface for $48 \le t \le 96$

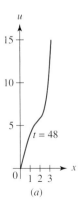

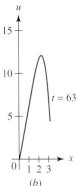

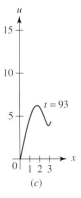

FIGURE 27.6   Plane sections at times $t = 48, 63$, and 96 hours

inside the wall rise and fall in step with the variation of the outdoor temperatures, but in diminishing proportions.

The temperature profiles $t = c$ are not linear, as Figure 27.6 attests. It depicts plane sections at $t = 48, 63$, and 93 hours, frames from a complete animation included in the accompanying Maple worksheet. The temperature profiles do not simply rise and fall linearly but undergo a decidedly nonlinear variation. It turns out that at the coolest part of the outdoor cycle, $t = 18 + 24k, k = 0, 1, \ldots$, the temperature in the wall never goes to zero. There is always some residual heat trapped in the wall.   ❖

Derivation

Given the nonhomogeneous boundary value problem

$$u_t(x, t) = \kappa u_{xx}(x, t) \quad \text{and} \quad \begin{aligned} u(0, t) &= \phi_0(t) \\ u(x, 0) &= f(x) \end{aligned} \quad u(L, t) = \phi_1(t)$$

define the functions

$$\psi(x, t) = \phi_0(t) + \frac{x}{L}[\phi_1(t) - \phi_0(t)] \quad \text{and} \quad G(x, t) = -\psi_t(x, t)$$

Let $v(x, t)$ be the solution of the boundary value problem

$$v_t(x, t) = \kappa v_{xx}(x, t) + G(x, t) \quad \text{and} \quad \begin{aligned} v(0, t) &= 0 \\ v(x, 0) &= f(x) - \psi(x, 0) \equiv F(x) \end{aligned} \quad v(L, t) = 0$$

If $u(x, t)$ is defined by $u(x, t) = v(x, t) + \psi(x, t)$, then $u(x, t)$ satisfies the original nonhomogeneous boundary value problem.

Before investigating how to find $v(x, t)$, the solution to a nonhomogeneous partial differential equation, we verify the claims made about the utility of $\psi(x, t)$ and $G(x, t)$. Since $v(0, t) = 0$ and $\psi(0, t) = \phi_0(t)$, we clearly have $u(0, t) = \phi_0(t)$. Similarly, since $v(L, t) = 0$ and $\psi(L, t) = \phi_1(t)$, we clearly have $u(L, t) = \phi_1(t)$. Finally, if

$$v(x, 0) = f(x) - \psi(x, 0) \quad \text{and} \quad u(x, t) = v(x, t) + \psi(x, t)$$

then $u(x, 0) = v(x, 0) + \psi(x, 0) = f(x) - \psi(x, 0) + \psi(x, 0) = f(x)$.

To verify that $v(x, t)$ should satisfy the partial differential equation

$$v_t(x, t) = \kappa v_{xx}(x, t) + G(x, t) = \kappa v_{xx}(x, t) - \psi_t(x, t) \tag{27.4}$$

define $u(x, t) = v(x, t) + \psi(x, t)$ and compute $u_t = \kappa u_{xx}$, obtaining

$$v_t + \psi_t = \kappa(v_{xx} + \psi_{xx})$$

Moving $\psi_t(x, t)$ from the left side to the right yields

$$v_t = \kappa(v_{xx} + \psi_{xx}) - \psi_t$$

Noting that $\psi(x, t) = \phi_0(t) + \frac{x}{L}[\phi_1(t) - \phi_0(t)]$ depends linearly on x, we realize that $\psi_{xx}(x, t) = 0$. Hence, the partial differential equation that $v(x, t)$ must satisfy becomes $v_t = \kappa v_{xx} - \psi_t$.

Now we face the serious challenge of solving this nonhomogeneous partial differential equation for which separation of variables fails. Were $v(x, t)$ to satisfy the homogeneous heat equation, we could immediately write

$$v(x, t) = \sum_{n=0}^{\infty} b_n \sin\left(\frac{n\pi x}{L}\right) e^{-\kappa(n\pi/L)^2 t}$$

Hence, for the nonhomogeneous equation, we are moved to seek a solution of the form

$$v(x, t) = \sum_{n=1}^{\infty} B_n(t) \sin \frac{n\pi x}{L}$$

This will require that $G(x, t) = -\psi_t(x, t)$ also be expanded in an eigenfunction expansion of the form

$$G(x, t) = \sum_{n=1}^{\infty} G_n(t) \sin \frac{n\pi x}{L} \quad \text{where} \quad G_n(t) = \frac{2}{L} \int_0^L G(x, t) \sin \frac{n\pi x}{L} \, dx \quad \text{(27.5)}$$

To compute the coefficients $B_n(t)$, set $\lambda_n = \frac{n\pi}{L}$ and substitute $v = \sum_{n=1}^{\infty} B_n(t) \sin \lambda_n x$ into the differential equation on the left in (27.4), obtaining

$$\sum_{n=1}^{\infty} \left(B_n'(t) + \kappa \lambda_n^2 B_n(t) - G_n(t) \right) \sin \lambda_n x = 0$$

from which the differential equations

$$B_n'(t) + \kappa \lambda_n^2 B_n(t) = G_n(t) \quad n = 1, 2, \ldots \quad \text{(27.6)}$$

follow. These first-order linear equations are readily solved for

$$B_n(t) = e^{-\kappa \lambda_n^2 t} \int_0^t e^{\kappa \lambda_n^2 u} G_n(u) \, du + e^{-\kappa \lambda_n^2 t} b_n \quad \text{(27.7)}$$

the additive constant of integration having been chosen so that at $t = 0$, $B_n(0) = b_n$.

In Table 27.2, we wrote this coefficient in the form

$$B_n(t) = \int_0^t G_n(s) e^{-\kappa \lambda_n^2 (t-s)} \, ds + b_n e^{-\kappa \lambda_n^2 t}$$

giving

$$v(x, t) = \sum_{n=1}^{\infty} \left[\int_0^t G_n(s) e^{-\kappa \lambda_n^2 (t-s)} \, ds + b_n e^{-\kappa \lambda_n^2 t} \right] \sin \lambda_n x$$

Since $v(x, 0) = F(x)$, we then find

$$F(x) = \sum_{n=1}^{\infty} \left[\int_0^0 G_n(s) e^{-\kappa \lambda_n^2 (-s)} \, ds + b_n e^0 \right] \sin \lambda_n x$$

$$= \sum_{n=1}^{\infty} [0 + b_n] \sin \lambda_n x = \sum_{n=1}^{\infty} b_n \sin \frac{n\pi x}{L}$$

Thus, the b_n must be the Fourier sine series coefficients for $F(x)$ and are thus given by

$$b_n = \frac{2}{L} \int_0^L F(x) \sin \frac{n\pi x}{L} \, dx$$

EXERCISES 27.2–Part A

A1. Verify the results in (27.6).

A2. Verify the results in (27.7).

A3. Obtain an analytic solution of the BVP $u_t = \kappa u_{xx}$, $u(0, t) = \sigma(t)$, $u(L, t) = 0$, $u(x, 0) = f(x)$.

EXERCISES 27.2–Part B

B1. Execute the calculations in Table 27.2 to obtain (27.3).

B2. Use (27.5) to obtain the eigenfunction expansion of $\rho(x, t) = x(\pi - x)e^{-xt}$ on $0 \le x \le \pi$. Use a partial sum and a three-dimensional graph to demonstrate the validity of the expansion.

In Exercises B3–7:

    (a) Obtain an analytic solution of the BVP $u_t = \kappa u_{xx}$, $u(0, t) = 0, u(L, t) = \phi_1(t), u(x, 0) = f(x)$.

    (b) Graph the solution surface.

    (c) Graph $u(\frac{L}{2}, t)$.

    (d) Animate the temperature profiles as they change in time.

    (e) Obtain a finite-difference solution via the scheme of Section 25.4. Compare the numerical solution with the exact solution.

B3. $L = \pi, \kappa = 2, \phi_1(t) = \cos t, f(x) = \dfrac{x}{\pi}$

B4. $L = 5, \kappa = \frac{2}{3}, \phi_1(t) = 3\cos^2 t, f(x) = \frac{3}{125}x^3$

B5. $L = 3, \kappa = \sqrt{2}, \phi_1(t) = 5(1 - e^{-t}), f(x) = x(3 - x)$

B6. $L = 1, \kappa = \frac{3}{5}, \phi_1(t) = \sin 24t, f(x) = x(1 - x)^2$

B7. $L = 4, \kappa = \frac{7}{10}, \phi_1(t) = 2te^{-t/5}, f(x) = x(4 - x)^3$

In Exercises B8–12:

    (a) Using the results of Exercise A3, solve the BVP $u_t = \kappa u_{xx}$, $u(0, t) = \phi_0(t), u(L, t) = 0, u(x, 0) = f(x)$.

    (b) Graph the solution surface.

    (c) Graph $u(\frac{L}{2}, t)$.

    (d) Animate the temperature profiles as they change in time.

    (e) Obtain a finite-difference solution via the scheme of Section 25.4. Compare the numerical solution with the exact solution.

B8. $L = \pi, \kappa = \frac{1}{2}, \phi_0(t) = \sin t, f(x) = x^2(\pi - x)$

B9. $L = 3, \kappa = \frac{5}{2}, \phi_0(t) = 2\sin^2 2t, f(x) = x(3 - x)^3$

B10. $L = 10, \kappa = 3, \phi_0(t) = 4(1 - e^{-t/2}), f(x) = 5\sin\dfrac{\pi}{10}x$

B11. $L = 1, \kappa = \frac{1}{10}, \phi_0(t) = 1 + \cos 2t, f(x) = 2(1 - x^5)$

B12. $L = 4, \kappa = \sqrt{2}, \phi_0(t) = 3e^{-0.1t}, f(x) = \frac{3}{4}(4 - x)$

Chapter Review

1. Given the BVP consisting of the one-dimensional heat equation, the endpoint conditions $u(0, t) = A, u(L, t) = 0$, and the initial condition $u(x, 0) = f(x)$, demonstrate how to use the substitution $u(x, t) = v(x, t) + \psi(x)$ to convert it to one with homogeneous boundary conditions.

2. Obtain the separation of variables solution for the BVP in Question 1 if $A = 10, L = 1, \kappa = \frac{1}{2}$, and $f(x) = 10(x - 1)$.

3. Solve the one-dimensional heat equation on the interval $0 \le x \le L$ if $u(0, t) = A, u(L, t) = B$, and $u(x, 0) = f(x)$.

4. Solve the one-dimensional heat equation on the interval $0 \le x \le L$ if $u_x(0, t) = A, u(L, t) = 0$, and $u(x, 0) = f(x)$.

5. Given the BVP consisting of the one-dimensional heat equation, the endpoint conditions $u(0, t) = \phi_0(t), u(L, t) = \phi_1(t)$, and the initial condition $u(x, 0) = f(x)$, show how the substitution $u(x, t) = v(x, t) + \psi(x, t)$, where $\psi(x, t) = \phi_0(t) + \frac{x}{L}[\phi_1(t) - \phi_0(t)]$, converts it to a new BVP with homogeneous boundary conditions, the nonhomogeneous PDE $v_t(x, t) = \kappa v_{xx}(x, t) - \psi_t(x, t)$, and the new initial condition $v(x, 0) = f(x) - \psi(x, 0)$.

Chapter 28

Time-Dependent Problems in Two Spatial Dimensions

I N T R O D U C T I O N   The wave and heat equations are evolution equations for which one independent variable is time. So far, we have solved BVPs for these equations in which there has been but a single spatial variable. In this chapter we consider problems with three independent variables, time, and two spatial variables.

The wave equation in two dimensions is most easily solved on a rectangle. The BVP consisting of the wave equation on a rectangle with homogeneous Dirichlet conditions on the edges models the vibrations in a rectangular plate with clamped edges. Thus, the oscillations in a plate in a ship's hull, in a window in a sky scraper, or in a panel in an aircraft's wing are all examples of physical systems governed by the two-dimensional wave equation. Surprisingly, where the vibrating string can support nodes, that is, points that remain at rest throughout the motion, the plate can exhibit *nodal curves* along which no motion occurs.

If these same rectangular plates are the object of thermal processes, then we are interested in solving the time-dependent heat equation with two spatial variables. For simplicity, we consider only the case of homogeneous Dirichlet conditions.

For either equation, separation of variables now generates two Sturm–Liouville eigenvalue problems, and the solution to the original BVPs will be double sums of eigensolutions. Hence, to solve these problems, we must develop a theory of two-dimensional Fourier series. For the wave equation we discover that there is a doubly indexed set of fundamental frequencies, but they are not integer multiples of a smallest frequency. Hence, the oscillations of the rectangular plate are not musical, in distinction to the oscillations of the finite string.

28.1   Oscillations of a Rectangular Membrane

Problem Formulation

A rectangular membrane such as a steel plate in a ship's hull or a plate of glass in a sky-scraper's window supports oscillations governed by the two-dimensional wave equation. A boundary value problem for the wave equation on a rectangular membrane with all four edges clamped consists of the partial differential equation on the left in Table 28.1, the

$$u_{tt} = c^2(u_{xx} + u_{yy})$$

$$u(0, y, t) = 0$$
$$u(x, 0, t) = 0 \qquad u(x, y, 0) = f(x, y)$$
$$u(a, y, t) = 0 \qquad u_t(x, y, 0) = g(x, y)$$
$$u(x, b, t) = 0$$

TABLE 28.1   BVP for the two-dimensional wave equation on a rectangular plate with clamped edges

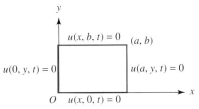

FIGURE 28.1   Rectangular domain with homogeneous Dirichlet conditions on each edge

boundary conditions in the middle, and the initial conditions on the right. Figure 28.1 is an icon for this BVP.

Statement of the Solution

Taking the initial velocity $g(x, y)$ to be zero and seeking a solution by the method of separation of variables, assume $u(x, y, t) = X(x)Y(y)T(t)$. Then, the solution is given by

$$u(x, y, t) = \sum_{m=1}^{\infty} \sum_{n=1}^{\infty} d_{nm} \sin \frac{n\pi x}{a} \sin \frac{m\pi y}{b} \cos \left(c\pi \sqrt{\frac{n^2}{a^2} + \frac{m^2}{b^2}} \, t \right)$$

$$d_{nm} = \frac{4}{ab} \int_0^b \int_0^a f(x, y) \sin \frac{n\pi x}{a} \sin \frac{m\pi y}{b} \, dx \, dy$$

where the two-dimensional Fourier series, discussed below, has been used. In the simplest case, the rectangle is a square with sides $a = b = \pi$, so the solution is

$$u(x, y, t) = \sum_{m=1}^{\infty} \sum_{n=1}^{\infty} d_{nm} \sin nx \sin my \cos \left(c\sqrt{n^2 + m^2} t \right)$$

$$d_{nm} = \frac{4}{\pi^2} \int_0^\pi \int_0^\pi f(x, y) \sin nx \sin my \, dx \, dy$$

In either event, several interesting features are exhibited by such solutions of the wave equation. First, the natural frequencies $\omega_{nm} = \pi \sqrt{\frac{n^2}{a^2} + \frac{m^2}{b^2}}$ are not integer multiples of a smallest fundamental frequency. Hence, the sound made by a vibrating rectangular membrane is not "musical." Second, while the vibrating string can exhibit nodal points where $u = 0$, the vibrating membrane can exhibit *nodal curves,* curves along which $u = 0$.

EXAMPLE 28.1   COMPUTING $u(x, y, t)$ Take the rectangle as a square of side $a = b = \pi$, the wave speed as $c = 1$, and $f(x, y) = \frac{1}{8}xy(x - \pi)(y - \pi)$ as the initial displacement function. As an initial displacement, $f(x, y)$ represents a symmetric "dimpling up" of the surface, as seen in Figure 28.2. The Fourier coefficients are

$$d_{nm} = \frac{4}{\pi^2} \int_0^\pi \int_0^\pi f(x, y) \sin nx \sin my \, dx \, dy = 2 \frac{(-1)^{m+n} - (-1)^m - (-1)^n + 1}{\pi^2 m^3 n^3}$$

$$(28.1)$$

To see how many of these coefficients are zero and how big the nonzero coefficients might be, the first 25 are arranged in Table 28.2. The coefficients are zero unless both n and m are odd integers, in which case it appears that $d_{nm} = \frac{8}{n^3 m^3 \pi^2}$.

Some values of $\omega_{nm} = \sqrt{n^2 + m^2}$ appearing in the function $\cos(\sqrt{n^2 + m^2} t)$ are listed in Table 28.3. We see in this table the reason why a drum is not "musical." Unlike the

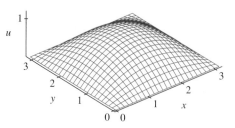

FIGURE 28.2   Initial displacement function for Example 28.1

$n \backslash m$	$m = 1$	$m = 2$	$m = 3$	$m = 4$	$m = 5$
$n = 1$	$\dfrac{8}{\pi^2}$	0	$\dfrac{8}{27}\dfrac{8}{\pi^2}$	0	$\dfrac{8}{\pi^2}$
$n = 2$	0	0	0	0	0
$n = 3$	$\dfrac{8}{27}\dfrac{8}{\pi^2}$	0	$\dfrac{8}{729}\dfrac{8}{\pi^2}$	0	$\dfrac{8}{3375}\dfrac{8}{\pi^2}$
$n = 4$	0	0	0	0	0
$n = 5$	$\dfrac{8}{125}\dfrac{8}{\pi^2}$	0	$\dfrac{8}{3375}\dfrac{8}{\pi^2}$	0	$\dfrac{8}{15625}\dfrac{8}{\pi^2}$

TABLE 28.2   Coefficients d_{nm} in Example 28.1

$n \backslash m$	$m = 1$	$m = 2$	$m = 3$	$m = 4$	$m = 5$
$n = 1$	$\sqrt{2}$	$\sqrt{10}$	$\sqrt{26}$	$\sqrt{50}$	$\sqrt{82}$
$n = 2$	$\sqrt{10}$	$\sqrt{18}$	$\sqrt{34}$	$\sqrt{58}$	$\sqrt{90}$
$n = 3$	$\sqrt{26}$	$\sqrt{34}$	$\sqrt{50}$	$\sqrt{74}$	$\sqrt{106}$
$n = 4$	$\sqrt{50}$	$\sqrt{58}$	$\sqrt{74}$	$\sqrt{98}$	$\sqrt{130}$
$n = 5$	$\sqrt{82}$	$\sqrt{90}$	$\sqrt{106}$	$\sqrt{130}$	$\sqrt{162}$

TABLE 28.3   Values of ω_{nm} in Example 28.1

vibrating string that supports harmonics (frequencies that are multiples of a fundamental note), the membrane does not have higher frequencies that are multiples of some lowest fundamental frequency. Remarkably, the Caribbean steel drum, hammered out of a 50-gal steel drum, actually sounds more musical than would be expected for a vibrating membrane. Special shaping and tempering of the steel account for the drum's unique properties.

Taking the wave speed as $c = 1$, and noting how quickly the magnitude of the Fourier coefficients decreases, we approximate $u(x, y, t)$ with

$$u_{10}(x, y, t) = \sum_{m=1}^{10} \sum_{n=1}^{10} \frac{8}{n^3 m^3 \pi^2} \sin nx \sin my \cos\left(\sqrt{n^2 + m^2}\, t\right)$$

Figure 28.3 shows $u_{10}(x, y, 2)$, the solution surface at time $t = 2$.

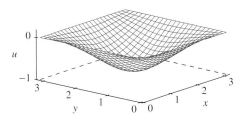

FIGURE 28.3   Example 28.1: Solution surface
at time $t = 2$

NODAL ANALYSIS   The vibrating string supports both traveling waves and standing waves. The standing wave exhibits nodes, points at which the string is at rest. Such points are found as solutions of the equation

$$u_n(x, t) = \sin \frac{n\pi x}{L} \cos \frac{cn\pi t}{L} = 0 \qquad (28.2)$$

The points x in $[0, L]$ satisfying this equation for all values of t are solutions of $\sin \frac{n\pi x}{L} = 0$, and the nodes are therefore solutions of $\frac{n\pi x}{L} = k\pi$, $k = 0, 1, 2, \ldots$. Thus, the nodes are located at $x = \frac{kL}{n}$, $k = 0, 1, 2, \ldots, n$. For example, if $L = 5$ and $n = 7$, the nodes are at $x = 0$, $\frac{5}{7}, 2(\frac{5}{7}), 3(\frac{5}{7}), 4(\frac{5}{7}), 5(\frac{5}{7}), 6(\frac{5}{7}), 7(\frac{5}{7}) = 5$, as confirmed by Figure 28.4, in which $D = \frac{5}{7}$.

The situation with the vibrating membrane is more complex. Instead of nodal points, there are nodal curves, curves along which it is possible to have no displacement. In fact, sand spread on a membrane driven by the vibrations of an acoustic speaker will show the patterns of nodal curves. For the present example where $a = b = \pi$, the mathematical equivalent are nodal curves defined implicitly by the vanishing of an eigenfunction

$$\psi(n, m) = \sin nx \sin my$$

Each such eigenfunction generates its own nodal curves. For $n \neq m$ the equation $\psi(n, m) = 0$ has solutions $x = \frac{k\pi}{n}$, $y = \frac{k\pi}{m}$, $k = 0, 1, \ldots$. These are just straight lines parallel to the axes.

However, on the square, two eigenfunctions can have the same frequency. In fact, the eigenfunctions

$$\psi(n, m) = \sin nx \sin my \quad \text{and} \quad \psi(m, n) = \sin mx \sin ny$$

both have the equal frequencies $\omega_{nm} = \omega_{mn} = \sqrt{n^2 + m^2} = \omega$, so the terms

$$d_{nm} \sin nx \sin my \cos c\omega_{nm} t \quad \text{and} \quad d_{mn} \sin mx \sin ny \cos c\omega_{mn} t$$

can be grouped into the equation

$$(\alpha \sin nx \sin my + \beta \sin mx \sin ny) \cos c\omega t = 0$$

Since the equation must hold for any time t, we are led to

$$\alpha \sin nx \sin my + \beta \sin mx \sin ny = 0$$

Letting $\frac{\beta}{\alpha} = h$, we finally get

$$\sin nx \sin my + h \sin mx \sin ny = 0$$

as an equation defining nodal curves for the plate under consideration.

The nodal curves defined implicitly by the equations in Table 28.4 appear in Figure 28.5. Figure 28.6 represents the nodal curves $y(x)$ determined by

$$\sin x \sin 3y + h \sin 3x \sin y = 0 \qquad (28.3)$$

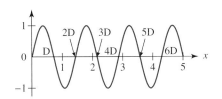

FIGURE 28.4   Nodal points determined by $\sin \frac{7\pi}{5} x = 0$

(a) $\sin x \sin 2y + \frac{1}{3} \sin 2x \sin y = 0$

(b) $\sin x \sin 2y + \frac{1}{\sqrt{2}} \sin 2x \sin y = 0$

(c) $\sin x \sin 2y + \frac{3}{2} \sin 2x \sin y = 0$

(d) $\sin x \sin 3y + \sin 3x \sin y = 0$

TABLE 28.4   Equations that implicitly define the nodal curves in Figure 28.5

when values of h are fixed. Alternatively, (28.3) can be interpreted as a definition of the implicit function $h(x, y)$ whose level curves are the nodal curves. The accompanying Maple worksheet animates a sequence of these curves for $0 < h < 2$. ❖

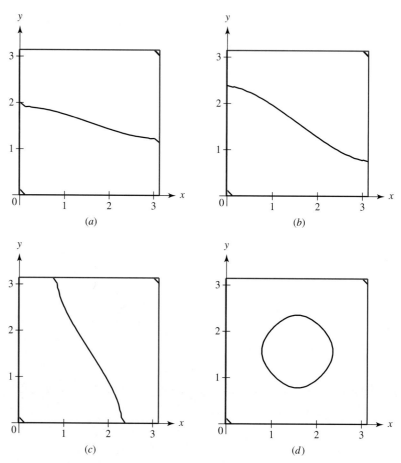

(a)

(b)

(c)

(d)

FIGURE 28.5   Nodal curves defined implicitly by the equations in Table 28.4

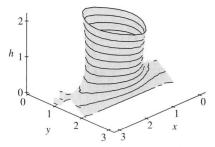

FIGURE 28.6   Level curves of $h(x, y)$ defined implicitly by $\sin x \sin 3y + h \sin 3x \sin y = 0$

Solution by Separation of Variables

We will next develop the solution of the two-dimensional wave equation by separation of variables. To this end, assume $u(x, y, t) = X(x)Y(y)T(t)$ and apply the separation assumption to the boundary conditions, obtaining

$$X(0)Y(y)T(t) = X(a)Y(y)T(t) = X(x)Y(0)T(t) = X(x)Y(b)T(t) = 0$$

from which we conclude $X(0) = X(a) = Y(0) = Y(b) = 0$. In addition, apply the separation assumption to the initial velocity condition, finding

$$u_t(x, y, 0) = X(x)Y(y)T'(0) = 0 \Rightarrow T'(0) = 0$$

Apply the separation assumption to the equation itself, divide by $c^2 u(x, y, t)$, and obtain

$$\frac{T''(t)}{c^2 T(t)} = \frac{X''(x)}{X(x)} + \frac{Y''(y)}{Y(y)}$$

Rearrange this to

$$\frac{T''(t)}{c^2 T(t)} - \frac{Y''(y)}{Y(y)} = \frac{X''(x)}{X(x)} \qquad (28.4)$$

a function of t and y on the left and a function of just x on the right. Hence, each side is a constant, say λ, so

$$\frac{X''(x)}{X(x)} = \lambda \quad \text{and} \quad \frac{T''(t)}{c^2 T(t)} - \frac{Y''(y)}{Y(y)} = \lambda$$

From the first equation, the familiar ODE $X''(x) - \lambda X(x) = 0$ results. From the second, obtain

$$\frac{Y''(y)}{Y(y)} = -\lambda + \frac{T''(t)}{c^2 T(t)}$$

where, on the left we find a function of just y and on the right a function of just t. Hence, each side is a constant, say μ, and we have the equations

$$\frac{Y''(y)}{Y(y)} = \mu \quad \text{and} \quad -\lambda + \frac{T''(t)}{c^2 T(t)} = \mu$$

The y-equation becomes the familiar $Y''(y) - \mu Y(y) = 0$, and the t-equation can be put into the form $T''(t) - c^2(\lambda + \mu)T(t) = 0$.

The three separated ODEs, with their accompanying conditions, are therefore

$$X''(x) - \lambda X(x) = 0, \, X(0) = X(a) = 0$$

$$Y''(y) - \mu Y(y) = 0, \, Y(0) = Y(b) = 0$$

$$T''(t) - c^2(\lambda + \mu)T(t) = 0, \, T'(0) = 0$$

The functions $X(x)$ and $Y(y)$ each satisfy identical Sturm-Liouville boundary value problems, problems for which we have already found the solution. The solutions for the eigenfunctions and eigenvalues are therefore

$$X_n(x) = A_n \sin \frac{n\pi x}{a} \qquad\qquad Y_m(y) = B_m \sin \frac{m\pi y}{b}$$

$$\text{and}$$

$$\lambda_n = -\left(\frac{n\pi}{a}\right)^2 \quad n = 1, 2, \ldots \qquad \mu_m = -\left(\frac{m\pi}{b}\right)^2 \quad m = 1, 2, \ldots$$

There are two different sets of eigenvalues, each indexed by an integer. As we went from one spatial dimension to two, we correspondingly went from one set of eigenvalues to two.

Finally, we solve the remaining ODE for $T(t)$, which now reads

$$T''(t) + c^2 \pi^2 \left(\frac{n^2}{a^2} + \frac{m^2}{b^2}\right) T(t) = 0$$

Were the equation $T'' + \omega^2 T = 0$, the solution would be

$$T = c_1 \cos \omega t + c_2 \sin \omega t$$

and the condition $T'(0) = 0$ would imply $c_2 = 0$ so that $T = c_1 \cos \omega t$. Thus, define

$$\omega_{nm} = \pi \sqrt{\frac{n^2}{a^2} + \frac{m^2}{b^2}}$$

and the desired solution becomes

$$T_{nm}(t) = C_{nm} \cos c\omega_{nm}t = C_{nm} \cos \left(c\pi \sqrt{\frac{n^2}{a^2} + \frac{m^2}{b^2}} \, t \right)$$

A single eigensolution is the product

$$u_{nm}(x, y, t) = d_{nm} \sin \frac{n\pi x}{a} \sin \frac{m\pi y}{b} \cos \left(c\pi \sqrt{\frac{n^2}{a^2} + \frac{m^2}{b^2}} \, t \right)$$

where $d_{nm} = A_n B_m C_{nm}$. The general solution is then a linear combination of all such eigenfunctions, a double sum over the two indices n and m, so

$$u(x, y, t) = \sum_{m=1}^{\infty} \sum_{n=1}^{\infty} d_{nm} \sin \frac{n\pi x}{a} \sin \frac{m\pi y}{b} \cos \left(c\pi \sqrt{\frac{n^2}{a^2} + \frac{m^2}{b^2}} \, t \right)$$

Applying the initial-shape condition yields the equation

$$f(x, y) = \sum_{m=1}^{\infty} \sum_{n=1}^{\infty} d_{nm} \sin \frac{n\pi x}{a} \sin \frac{m\pi y}{b} \tag{28.5}$$

from which we obtain the constants

$$d_{nm} = \frac{4}{ab} \int_0^b \int_0^a f(x, y) \sin \frac{n\pi x}{a} \sin \frac{m\pi y}{b} \, dx \, dy \tag{28.6}$$

by the usual Fourier techniques. The details are as follows.

THE TWO-DIMENSIONAL FOURIER SERIES   The equation (28.5) and the orthogonality of the eigenfunctions

$$\psi_{nm} = \sin \frac{n\pi x}{a} \sin \frac{m\pi y}{b}$$

on the rectangle $R = \{0 \le x \le a, 0 \le y \le b\}$ lead to the definition of the Fourier coefficients d_{nm}. The orthogonality of the eigenfunctions is realized in the vanishing of the integrals

$$\int_0^b \int_0^a \psi_{nm} \psi_{NM} \, dx \, dy = \int_0^b \int_0^a \sin \frac{n\pi x}{a} \sin \frac{m\pi y}{b} \sin \frac{N\pi x}{a} \sin \frac{M\pi y}{b} \, dx \, dy$$

when $n \ne N, m \ne M$, and N and M are integers. When $N = n$ and $M = m$, the integral becomes

$$\int_0^b \int_0^a \sin^2 \frac{n\pi x}{a} \sin^2 \frac{m\pi y}{b} \, dx \, dy = \frac{ab}{4} \tag{28.7}$$

If (28.5) is multiplied by the eigenfunction ψ_{NM} and integrated over the rectangle R, the left side is just

$$\int_0^b \int_0^a f(x, y) \sin \frac{N\pi x}{a} \sin \frac{M\pi y}{b} \, dx \, dy$$

but the right side contains an infinite sum of integrals of the products of eigenfunctions. All but one of these integrals on the right vanish by orthogonality of the eigenfunctions. The one integral that does not vanish is the one in which $n = N$ and $m = M$, whence the value is $\frac{ab}{4}d_{NM}$. Thus, from the equation

$$\int_0^b \int_0^a f(x, y) \sin\frac{N\pi x}{a} \sin\frac{M\pi y}{b} \, dx \, dy = \frac{ab}{4}d_{NM}$$

results the Fourier coefficient $d_{NM} = \frac{4}{ab}\int_0^b\int_0^a f(x, y) \sin\frac{N\pi x}{a} \sin\frac{M\pi y}{b}\, dx\, dy$.

EXERCISES 28.1–Part A

A1. Verify the integration in (28.1).

A2. If $n = 5$ and $L = 3$, find the first six nonnegative nodal points determined by (28.2).

A3. Explain how (28.3) determines a function $h(x, y)$ whose level curves are nodal curves $y(x)$.

A4. Starting with the wave equation on the left in Table 28.1, obtain the results in (28.4).

A5. Verify the integration in (28.7).

EXERCISES 28.1–Part B

B1. Investigate the nodal curves determined by $\sin 2x \sin 3y + h \sin 3x \sin 2y = 0$, where $0 < h < 2$, and $0 \le x, y \le \pi$.

Exercises B2–5 give an eigensolution of the BVP for the vibrating string of length L. For each:

(a) Find the nodes.

(b) Give the value of c, the wave speed.

(c) Give the period of the motion.

(d) Animate the motion of string oscillating in the given mode shape.

B2. $L = \pi, u(x, t) = \sin 2x \cos 3t$

B3. $L = 2\pi, u(x, t) = \sin 3x \cos 4t$

B4. $L = 3, u(x, t) = \sin 2\pi x \cos 5t$

B5. $L = 5, u(x, t) = \sin \pi x \cos \sqrt{2}t$

In Exercises B6–10, the data pertains to a $\pi \times \pi$ plate with clamped edges. For each, graph the nodal curves generated by $\sin nx \sin my + h \sin mx \sin ny = 0$.

B6. $n = 2, m = 3, h = \frac{3}{2}$ **B7.** $n = 3, m = 1, h = 2$

B8. $n = 4, m = 2, h = \frac{1}{2}$ **B9.** $n = 4, m = 3, h = 2$

B10. $n = 5, m = 2, h = \frac{3}{2}$

For $g(x, y) = 0$, and the data given in each of Exercises B11–15:

(a) Solve the BVP in Table 28.1.

(b) Plot $f(x, y)$.

(c) For the series found in part (a), let $U(x, y, t)$ be a partial sum for which $U(x, y, 0)$ is a reasonable approximation to $f(x, y)$. Obtain graphs of the solution surface at times $t = 1, 2, 3$.

(d) Graph the motion of the center of the plate.

(e) Animate the solution surface using the appropriate software.

B11. $(a, b) = (\pi, \pi)$, $c = \frac{1}{2}$, $f(x, y) = \frac{1}{48}x^3 y^2(x - \pi)(y - \pi)$

B12. $(a, b) = (\pi, 1)$, $c = \frac{1}{3}$, $f(x, y) = \frac{1}{2}x^3 y(x - \pi)(y - 1)^3$

B13. $(a, b) = (3, 3)$, $c = \frac{3}{2}$, $f(x, y) = \frac{1}{16}xy(x - 3)^2(y - 3)^2$

B14. $(a, b) = (3, 5)$, $c = \frac{2}{3}$, $f(x, y) = \frac{1}{160}xy^2(x - 3)^3(y - 5)$

B15. $(a, b) = (5, 4)$, $c = \frac{3}{5}$, $f(x, y) = \frac{1}{64}xy(x - 5)^2(4 - y)$

In Exercises B16–20:

(a) On the rectangle $0 \le x \le a, 0 \le y \le b$, obtain the double Fourier series (28.5), where d_{nm} are given by (28.6).

(b) On the rectangle $0 \le x \le a, 0 \le y \le b$, plot $F(x, y)$, a partial sum that gives a reasonable approximation to $f(x, y)$.

(c) Plot $F(x, y)$ on the rectangle $-a \le x \le a, -b \le y \le b$. Observe that this Fourier series converges to a function that is the extension of $f(x, y)$ odd in both x and y.

B16. $(a, b) = (\pi, \pi)$, $f(x, y) = xy(x - \pi)(y - \pi)e^{-y}$

B17. $(a, b) = (\pi, 1)$, $f(x, y) = x^2 y(x - \pi)(y - 1)^2$

B18. $(a, b) = (3, 3)$, $f(x, y) = (x - 3)(y - 3)^3 \sin x \sin y$

B19. $(a, b) = (3, 5)$, $f(x, y) = (1 - e^{-x/2})(1 - e^{-y/2})(x - 3)(y - 5)$

B20. $(a, b) = (5, 4)$, $f(x, y) = xy(x - 5)^2(y - 4)^3$

For $f(x, y)$ defined on the rectangle $R = \{-a \le x \le a, -b \le y \le b\}$, the general Fourier series representation is given by $f(x, y) = d_{00} + A + B + C$, where

$$A = \sum_{m=1}^{\infty}\sum_{n=1}^{\infty}\left(d_{nm}^1 \sin\frac{n\pi x}{a}\sin\frac{m\pi y}{b} + d_{nm}^2 \cos\frac{n\pi x}{a}\cos\frac{m\pi y}{b}\right.$$

$$\left. + d_{nm}^3 \sin\frac{n\pi x}{a}\cos\frac{m\pi y}{b} + d_{nm}^4 \cos\frac{n\pi x}{a}\sin\frac{m\pi y}{b}\right)$$

$$B = \sum_{n=1}^{\infty}\left(c_n^1 \cos\frac{n\pi x}{a} + c_n^2 \sin\frac{n\pi x}{a}\right)$$

$$C = \sum_{m=1}^{\infty}\left(c_m^3 \cos\frac{m\pi y}{b} + c_m^4 \sin\frac{m\pi y}{b}\right)$$

$$d_{00} = \frac{1}{4ab}\iint_R f\, dx\, dy$$

$$d_{nm}^1 = \frac{1}{ab}\iint_R f \sin\frac{n\pi x}{a}\sin\frac{m\pi y}{b}\, dx\, dy$$

$$d_{nm}^2 = \frac{1}{ab}\iint_R f \cos\frac{n\pi x}{a}\cos\frac{m\pi y}{b}\, dx\, dy$$

$$d_{nm}^3 = \frac{1}{ab}\iint_R f \sin\frac{n\pi x}{a}\cos\frac{m\pi y}{b}\, dx\, dy$$

$$d_{nm}^4 = \frac{1}{ab}\iint_R f \cos\frac{n\pi x}{a}\sin\frac{m\pi y}{b}\, dx\, dy$$

$$c_n^1 = \frac{1}{2ab}\iint_R f \cos\frac{n\pi x}{a}\, dx\, dy$$

$$c_n^2 = \frac{1}{2ab}\iint_R f \sin\frac{n\pi x}{a}\, dx\, dy$$

$$c_m^3 = \frac{1}{2ab}\iint_R f \cos\frac{m\pi y}{b}\, dx\, dy$$

$$c_m^4 = \frac{1}{2ab}\iint_R f \sin\frac{m\pi y}{b}\, dx\, dy$$

In Exercises B21–25:

   (a) Obtain the general Fourier series on R.

   (b) Plot a reasonably accurate partial sum.

   (c) If $f(x, y)$ is odd in both x and y, show that the general Fourier series collapses to the case determined by (28.5) and (28.6).

B21. $a = 1, b = 1, f(x, y) = \dfrac{(x^2 + 1)(y^3 + 2y^2 - y + 1)}{30}$

B22. $a = 3, b = 4, f(x, y) = x^2 + y^2 - 25$

B23. $a = 2, b = 3, f(x, y) = x^2 + y^2$

B24. $a = \pi, b = 1, f(x, y) = (y^2 - 1)\cos(\sqrt{2}x)$

B25. $a = \pi, b = \pi, f(x, y) = \dfrac{xy(x^2 - \pi^2)(y^2 - \pi^2)}{100}$

B26. Specialize the general Fourier series of Exercises B21–25 for a function that is even in both x and y.

B27. Specialize the general Fourier series of Exercises B21–25 for a function that is even in x but odd in y.

B28. Specialize the general Fourier series of Exercises B21–25 for a function that is odd in x but even in y.

28.2 Time-Varying Temperatures in a Rectangular Plate

Problem Formulation

Suppose the rectangular plate in Figure 28.1 has its flat faces insulated and its edges kept at a temperature of zero. If initially the temperatures in the plate are $u(x, y, 0) = f(x, y)$, the steady-state temperatures will eventually be zero. However a boundary value problem describing the time-dependent temperatures throughout the plate consists of the two-dimensional heat equation on the left in Table 28.5, the boundary conditions in the center, and the initial condition on the right.

$$u_t = \kappa(u_{xx} + u_{yy})$$

$$u(0, y, t) = 0$$
$$u(x, 0, t) = 0$$
$$u(a, y, t) = 0$$
$$u(x, b, t) = 0$$

$$u(x, y, 0) = f(x, y)$$

TABLE 28.5   BVP for two-dimensional heat equation on a rectangular plate whose edges are maintained at temperature zero

Statement of the Solution

Separation of variables leads to the solution

$$u(x, y, t) = \sum_{m=1}^{\infty} \sum_{n=1}^{\infty} d_{nm} \sin \frac{n\pi x}{a} \sin \left(\frac{m\pi y}{b} \right) e^{-\kappa \pi^2 (n^2/a^2 + m^2/b^2)t} \qquad (28.8a)$$

$$d_{nm} = \frac{4}{ab} \int_0^b \int_0^a f(x, y) \sin \frac{n\pi x}{a} \sin \frac{m\pi y}{a} \, dx \, dy \qquad (28.8b)$$

EXAMPLE 28.2   As in Example 28.1, take the rectangle to be a square of side $a = b = \pi$ but take the initial temperature profile to be the function $f(x, y) = \frac{1}{8}x, y^2(x - \pi)(y - \pi)$, plotted in Figure 28.7, which suggests the plate has initially been heated in an asymmetric way. The Fourier coefficients are

$$d_{nm} = \frac{4}{\pi^2} \int_0^\pi \int_0^\pi \frac{1}{8} xy^2(x - \pi)(y - \pi) \sin nx \sin my \, dx \, dy$$

$$= 2\frac{2(-1)^{m+n} - 2(-1)^m + (-1)^n - 1}{\pi m^3 n^3} \qquad (28.9)$$

Table 28.6, listing the first 25 coefficients, shows the coefficients are zero when n is even and rapidly decrease in magnitude with increasing n or m.

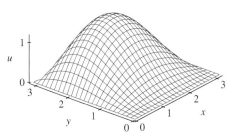

FIGURE 28.7   Initial temperature distribution in Example 28.2

$n \backslash m$	$m = 1$	$m = 2$	$m = 3$	$m = 4$	$m = 5$
$n = 1$	$\dfrac{4}{\pi}$	$-\dfrac{3}{2}\dfrac{1}{\pi}$	$\dfrac{4}{27}\dfrac{1}{\pi}$	$-\dfrac{3}{16}\dfrac{1}{\pi}$	$\dfrac{4}{125}\dfrac{1}{\pi}$
$n = 2$	0	0	0	0	0
$n = 3$	$\dfrac{4}{27}\dfrac{1}{\pi}$	$-\dfrac{1}{18}\dfrac{1}{\pi}$	$\dfrac{4}{729}\dfrac{1}{\pi}$	$-\dfrac{1}{144}\dfrac{1}{\pi}$	$\dfrac{4}{3375}\dfrac{1}{\pi}$
$n = 4$	0	0	0	0	0
$n = 5$	$\dfrac{4}{125}\dfrac{1}{\pi}$	$-\dfrac{3}{250}\dfrac{1}{\pi}$	$\dfrac{4}{3375}\dfrac{1}{\pi}$	$-\dfrac{3}{2000}\dfrac{1}{\pi}$	$\dfrac{4}{15625}\dfrac{1}{\pi}$

TABLE 28.6   Coefficients d_{nm} in Example 28.2

Taking the thermal diffusivity as $\kappa = 1$ and noting how quickly the Fourier coefficients decay, we approximate $u(x, y, t)$ with the partial sum

$$u_{10}(x, y, t) = \sum_{m=1}^{10} \sum_{n=1}^{10} d_{nm} \sin nx \sin(my)e^{-(n^2+m^2)t}$$

Figure 28.8 shows $u_{10}(x, y, 0.5)$, but the accompanying Maple worksheet contains an animation of the cooling process as all heat escapes from the plate and its temperature approaches the steady-state $u = 0$. ❖

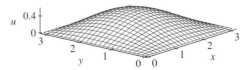

FIGURE 28.8 Example 28.2: Solution surface at time $t = \frac{1}{2}$

Solution by Separation of Variables

Separation of variables proceeds as in the solution of the wave equation on the rectangular membrane. First, apply the separation assumption, namely, $u(x, y, t) = X(x)Y(y)T(t)$, to the boundary conditions, obtaining

$$X(0)Y(y)T(t) = X(a)Y(y)T(t) = X(x)Y(0)T(t) = X(x)Y(b)T(t) = 0$$

and hence $X(0) = X(a) = Y(0) = Y(b) = 0$. Applying the separation assumption to the heat equation, dividing by $\kappa u(x, y, t)$, and rearranging lead to

$$\frac{T'(t)}{\kappa T(t)} - \frac{Y''(y)}{Y(y)} = \frac{X''(x)}{X(x)} = \lambda$$

where λ is the Bernoulli separation constant introduced because the function of t and y on the left must equal the function of x on the right. The self-adjoint equation $X''(x) - \lambda X(x) = 0$ follows, as does

$$\frac{T'(t)}{\kappa T(t)} - \lambda = \frac{Y''(y)}{Y(y)} = \mu$$

where μ is a second separation constant introduced because the function of t on the left must equal the function of y on the right. Consequently, we have the additional two differential equations

$$Y''(y) - \mu Y(y) = 0 \quad \text{and} \quad T'(t) - \kappa(\lambda + \mu)T(t) = 0$$

The equation for $T(t)$ is solved after obtaining the eigenfunctions of the two Sturm–Liouville BVPs

$$\begin{array}{cc} X''(x) - \lambda X(x) = 0 & Y''(y) - \mu Y(y) = 0 \\ X(0) = X(a) = 0 & \text{and} \quad Y(0) = Y(b) = 0 \end{array}$$

Just as for the wave equation in Section 28.1, we have

$$X_n(x) = A_n \sin \frac{n\pi x}{a} \qquad\qquad Y_m(y) = B_m \sin \frac{m\pi y}{b}$$

$$\lambda_n = -\frac{n^2\pi^2}{a^2} \quad n = 1, 2, \dots \qquad \text{and} \qquad \mu_m = -\frac{m^2\pi^2}{b^2} \quad m = 1, 2, \dots$$

As with the wave equation on the rectangular membrane, there are two different sets of integer-indexed eigenvalues. They make the equation and solution for $T(t)$ become

$$T'_{nm}(t) + \kappa \pi^2 \left(\frac{n^2}{a^2} + \frac{m^2}{b^2} \right) T_{nm}(t) = 0 \Rightarrow T_{nm}(t) = C_{nm} e^{-\kappa \pi^2 (n^2/a^2 + m^2/b^2)t} \quad \textbf{(28.10)}$$

A single eigensolution is

$$u_{nm} = d_{nm} \sin \frac{n\pi x}{a} \sin \left(\frac{m\pi y}{b} \right) e^{-\kappa \pi^2 (n^2/a^2 + m^2/b^2)t}$$

where d_{nm} replaces the product $A_n B_m C_{nm}$. The general solution is then given by (28.8a), the sum over all such eigensolutions. Application of the initial condition leads to the equation

$$f(x, y) = \sum_{m=1}^{\infty} \sum_{n=1}^{\infty} d_{nm} \sin \frac{n\pi x}{a} \sin \frac{m\pi y}{b} \quad \textbf{(28.11)}$$

from which Fourier techniques identical to those of Section 28.1 determine the constants d_{nm} to be given by (28.8b).

EXERCISES 28.2–Part A

A1. Verify the integration in (28.9).

A2. Verify the DE and its solution in (28.10).

A3. Show how (28.11) leads to (28.8b) for the coefficients d_{nm} in (28.8a).

A4. Insulate the left and right edges of the plate in Figure 28.1 and solve the BVP $u_t = \kappa(u_{xx} + u_{yy}), u_x(0, y, t) = u_x(a, y, t) = u(x, 0, t) = u(x, b, t) = 0, u(x, y, 0) = f(x, y)$. See Exercise

B27, Section 28.1, for details on the appropriate Fourier series in two variables.

A5. Insulate the top and bottom edges of the plate in Figure 28.1 and solve the BVP $u_t = \kappa(u_{xx} + u_{yy}), u(0, y, t) = u(a, y, t) = u_y(x, 0, t) = u_y(x, b, t) = 0, u(x, y, 0) = f(x, y)$. See Exercise B28, Section 28.1, for details on the appropriate Fourier series in two variables.

EXERCISES 28.2–Part B

B1. Insulate the top and right edges of the plate in Figure 28.1 and solve the BVP $u_t = \kappa(u_{xx} + u_{yy}), u(0, y, t) = u_x(a, y, t) = u(x, 0, t) = u_y(x, b, t) = 0, u(x, y, 0) = f(x, y)$. Develop the appropriate Fourier series in two variables.

Using the data given in Exercises B2–6:

   (a) Solve the BVP in Table 28.5.

   (b) Using a suitable partial sum of the series developed in part (a), obtain a plot of the solution surface at times $t = 1, 2, 3$.

   (c) Obtain contour plots (graphs of the isotherms) at times $t = 1, 2, 3$.

   (d) Using appropriate software, animate the solution surface as it varies in time.

B2. $a = \pi, b = \pi, \kappa = 2, f(x, y) = \frac{1}{10} xy(x - \pi)^2(\pi - y)$

B3. $a = 1, b = \pi, \kappa = \frac{1}{3}, f(x, y) = 3xy^2(x - 1)^3(y - \pi)$

B4. $a = 3, b = 3, \kappa = \frac{5}{3}, f(x, y) = \frac{1}{16} xy(x - 3)^2(y - 3)^2$

B5. $a = 5, b = 3, \kappa = \sqrt{3}, f(x, y) = \frac{1}{250} x^3 y(x - 5)(y - 3)^3$

B6. $a = 4, b = 5, \kappa = \frac{1}{2}, f(x, y) = \frac{1}{250} x^3 y^2(x - 4)(y - 5)$

In Exercises B7–11, use the given data and the results of Exercise A4. For each:

   (a) Solve the BVP detailed in Exercise A4.

   (b) Using a suitable partial sum of the series developed in part (a), obtain a plot of the solution surface at times $t = 1, 2, 3$.

   (c) Obtain contour plots (graphs of the isotherms) at times $t = 1, 2, 3$.

   (d) Using appropriate software, animate the solution surface as it varies in time.

B7. the data from Exercise B2   **B8.** the data from Exercise B3

B9. the data from Exercise B4   **B10.** the data from Exercise B5

B11. the data from Exercise B6

In Exercises B12–16, use the given data and the results of Exercise A5. For each:

(a) Solve the BVP detailed in Exercise A5.

(b) Using a suitable partial sum of the series developed in part (a), obtain a plot of the solution surface at times $t = 1, 2, 3$.

(c) Obtain contour plots (graphs of the isotherms) at times $t = 1, 2, 3$.

(d) Using appropriate software, animate the solution surface as it varies in time.

B12. the data from Exercise B2   **B13.** the data from Exercise B3

B14. the data from Exercise B4   **B15.** the data from Exercise B5

B16. the data from Exercise B6

In Exercises B17–18, use the given data and the results of Exercise B1. For each:

(a) Solve the BVP detailed in Exercise B1.

(b) Using a suitable partial sum of the series developed in part (a), obtain a plot of the solution surface at times $t = 1, 2, 3$.

(c) Obtain contour plots (graphs of the isotherms) at times $t = 1, 2, 3$.

(d) Obtain graphs of $u(\frac{a}{2}, b, t)$ and $u(a, \frac{b}{2}, t)$.

B17. $a = \pi, b = \pi, \kappa = 2, f(x, y) = xy^2$

B18. $a = 1, b = 2, \kappa = 1, f(x, y) = x^2 y$

Chapter Review

1. Expand $f(x, y) = (x - 1)^2(y - 2)^2, 0 \le x \le a = 2, 0 \le y \le b = 4$, in a double Fourier series whose terms are of the form $\sin \frac{n\pi x}{a} \sin \frac{m\pi y}{b}$, with n and m positive integers.

2. Sketch the separation of variables solution of the two-dimensional wave equation for a rectangular plate with clamped edges and an initial displacement $u(x, y, 0) = f(x, y)$.

3. What are nodal lines, and how do they occur?

4. Sketch the separation of variables solution of the two-dimensional heat equation for a rectangular plate with each of its edges kept at a temperature of zero and an initial temperature distribution given by $u(x, y, 0) = f(x, y)$. Let one corner of the plate be at the origin in the xy-plane and the diagonally opposite corner be at (a, b).

5. Sketch the separation of variables solution of the two-dimensional heat equation for a rectangular plate situated with one corner at the origin in the xy-plane and the diagonally opposite corner at (a, b). The left and right edges are insulated, and the top and bottom edges are kept at a temperature of zero. The initial temperature distribution is given by $u(x, y, 0) = f(x, y)$. The eigenfunctions in x will be cosines, while the eigenfunctions in y will be sines. This means a new double Fourier series must be devised and justified.

Chapter 29

Separation of Variables
in NonCartesian Coordinates

I N T R O D U C T I O N   So far, the technique of separation of variables has been demonstrated for a variety of BVPs but always in Cartesian coordinates. The technique does extend to other coordinate systems, but with an increasing complexity for the resulting Sturm–Liouville eigenvalue problems.

To solve Laplace's equation on a disk, we use polar coordinates because the disk's boundary, a circle, is a grid line in that system. Separation of variables leads to a Sturm–Liouville eigenvalue problem in which each nonzero eigenvalue has two associated eigenfunctions. In this case, solutions of BVPs rest on the theory of the general Fourier series.

To solve Laplace's equation in a cylinder, we use cylindrical coordinates. Separation of variables leads to a Sturm–Liouville eigenvalue problem in which the ODE is Bessel's equation and for which the eigenfunctions are Bessel functions. In this case, solutions of BVPs rest on the theory of Fourier–Bessel series.

To solve Laplace's equation in a sphere, we use spherical coordinates. Separation of variables leads to a Sturm–Liouville eigenvalue problem in which the ODE is Legendre's equation and for which the eigenfunctions are the Legendre polynomials. In this case, solutions of BVPs rest on the theory of Fourier–Legendre series.

With these tools, we are able to solve such problems as the Dirichlet problem on the disk and in a sphere and a heat transfer problem in a cylinder. For the cylinder, although we impose both Dirichlet and Neumann conditions, the bulk of the analysis is graphical because the challenge is to visualize and represent temperatures inside a solid cylinder. For the sphere, we solve the Dirichlet problem but face the same challenge for visualization and interpretation as we faced for the cylinder.

Finally, we solve Laplace's equation for the potential inside and outside a spherical dielectric shell. This is a thin hollow nonconducting sphere placed in a uniform electric field. We find that inside the sphere, the field is not distorted, but outside the sphere, there is local distortion which diminishes with distance away from the shell.

29.1 Laplace's Equation in a Disk

laplacian in Polar Coordinates

In rectangular coordinates, the laplacian of $f(x, y)$ is $\nabla^2 f(x, y) = f_{xx}(x, y) + f_{yy}(x, y)$. In polar coordinates, the laplacian of $u(r, \theta)$ is $\nabla^2 u(r, \theta) = u_{rr}(r, \theta) + \frac{1}{r} u_r(r, \theta) + \frac{1}{r^2} u_{\theta\theta}(r, \theta)$, as was derived in Section 22.2.

Problem Formulation

The Dirichlet problem on the disk of radius σ consists of Laplace's equation $\nabla^2 u(r, \theta) = 0$ inside the disk and a Dirichlet boundary condition of the form $u(\sigma, \theta) = f(\theta)$ on the circumference of the disk. The domain for the angle θ is the interval $(-\pi, \pi]$.

Implicit in this boundary value problem is the requirement that inside the disk, $u(r, \theta)$ be continuous with a continuous derivative. These continuity conditions require, for each value of $r < \sigma$, $u(r, -\pi) = u(r, \pi)$ and $u_\theta(r, -\pi) = u_\theta(r, \pi)$, conditions that guarantee continuity across the negative portion of the x-axis where θ jumps from $-\pi$ to π.

The problem models the steady-state temperatures in a disk around the boundary of which the temperatures have been prescribed by the function $f(\theta)$.

Statement of the Solution

The solution of the Dirichlet problem on the disk is

$$u(r, \theta) = \frac{A_0}{2} + \sum_{n=1}^{\infty} (A_n r^n \cos n\theta + B_n r^n \sin n\theta) \tag{29.1a}$$

$$A_n = \frac{1}{\pi \sigma^n} \int_{-\pi}^{\pi} f(\theta) \cos n\theta \, d\theta \qquad n = 0, 1, 2, \ldots \tag{29.1b}$$

$$B_n = \frac{1}{\pi \sigma^n} \int_{-\pi}^{\pi} f(\theta) \sin n\theta \, d\theta \qquad n = 1, 2, \ldots \tag{29.1c}$$

Some texts give this solution in the form

$$u(r, \theta) = \frac{a_0}{2} + \sum_{n=1}^{\infty} \left(a_n \left(\frac{r}{\sigma} \right)^n \cos n\theta + b_n \left(\frac{r}{\sigma} \right)^n \sin n\theta \right)$$

$$a_n = \frac{1}{\pi} \int_{-\pi}^{\pi} f(\theta) \cos n\theta \, d\theta \qquad n = 0, 1, 2, \ldots$$

$$b_n = \frac{1}{\pi} \int_{-\pi}^{\pi} f(\theta) \sin n\theta \, d\theta \qquad n = 1, 2, \ldots$$

In this alternate form, the coefficients a_n and b_n are the Fourier coefficients of the boundary function $f(\theta)$.

EXAMPLE 29.1   Take the radius of the circle to be $\sigma = 2$, and let the boundary function be

$$f(\theta) = \begin{cases} 0 & -\pi < \theta \le 0 \\ 3 & 0 < \theta \le \pi \end{cases}$$

as depicted in Figure 29.1. Using (29.1b) and (29.1c), we obtain

$$A_0 = \frac{1}{\pi} \int_{-\pi}^{\pi} f(\theta) \, d\theta = 3 \tag{29.2a}$$

$$A_n = \frac{1}{2^n \pi} \int_{-\pi}^{\pi} f(\theta) \cos n\theta \, d\theta = 0 \tag{29.2b}$$

$$B_n = \frac{1}{2^n \pi} \int_{-\pi}^{\pi} f(\theta) \sin n\theta \, d\theta = \frac{3}{2^n \pi n} (1 - (-1)^n) \tag{29.2c}$$

Using (29.1a), a partial sum of the series is then $u_{30}(r, \theta) = \frac{3}{2} + \sum_{n=1}^{30} B_n r^n \sin(n\theta)$, graphed in Figure 29.2. At $\theta = 0$ and $\theta = \pi$, the solution is discontinuous on the boundary. It is difficult to represent these discontinuities with the finite sum used to approximate the solution.

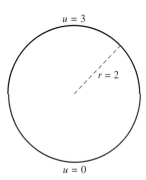

$u = 3$

$r = 2$

$u = 0$

FIGURE 29.1   Boundary function $f(\theta)$ in Example 29.1

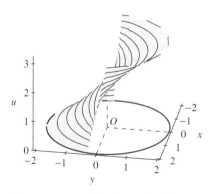

FIGURE 29.2   Solution surface in Example 29.1

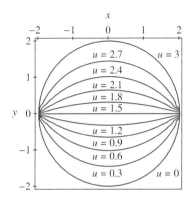

FIGURE 29.3   Level curves in Example 29.1

An alternative view of the difficulties the discontinuities cause is the solution's contour plot contained in Figure 29.3, which shows all contour lines converging at the two points of discontinuity on the boundary where the temperature jumps from $u = 0$ to $u = 3$. Corresponding to these jumps, the solution surface takes on all intermediate temperatures along two vertical lines that are nearly discernible in Figure 29.2.   ❖

Solution by Separation of Variables

The solution by separation of variables for Laplace's equation in a disk of radius σ begins with the separated form of the solution, $u(r, \theta) = R(r)\Theta(\theta)$. The boundary conditions are the implicit continuity conditions $u(r, -\pi) = u(r, \pi)$ and $u_\theta(r, -\pi) = u_\theta(r, \pi)$, which, by the obvious substitutions, lead to the *periodic* boundary conditions $\Theta(-\pi) = \Theta(\pi)$ and $\Theta'(-\pi) = \Theta'(\pi)$. Applying the separation assumption to Laplace's equation in polar coordinates, multiplying by r^2, dividing by $u = R\Theta$, rearranging, and introducing the Bernoulli separation constant, we get

$$\frac{\Theta''(\theta)}{\Theta(\theta)} = -\frac{r R'(r)}{R(r)} - \frac{r^2 R''(r)}{R(r)} = \lambda$$

which then gives the two ODEs

$$\Theta''(\theta) - \lambda\Theta(\theta) = 0 \quad \text{and} \quad r^2 R''(r) + r R'(r) + \lambda R(r) = 0 \tag{29.3}$$

The Θ-equation belongs to the Sturm–Liouville eigenvalue problem on the left in (29.4). The eigenfunctions and eigenvalues on the right in (29.4) were derived in Section 16.1.

$$\begin{array}{l} \Theta''(\theta) - \lambda\Theta(\theta) = 0 \\ \quad \Theta(-\pi) = \Theta(\pi) \quad \Rightarrow \\ \quad \Theta'(-\pi) = \Theta'(\pi) \end{array} \qquad \begin{array}{l} \Theta_n(\theta) = \alpha_n \cos n\theta + \beta_n \sin n\theta \\ \quad \lambda_n = -n^2 \quad n = 0, 1, 2, \ldots \end{array} \tag{29.4}$$

This is a case where the single eigenvalue $-n^2$ has associated with it the two eigenfunctions $\cos n\theta$ and $\sin n\theta$, except for $n = 0$ when the only eigenfunction is $\Theta_0(\theta) = \text{constant}$.

To solve for $R(r)$, consider first the case $\lambda_0 = 0$ where we find

$$r^2 R''(r) + r R'(r) = 0 \Rightarrow R_0(r) = c_1 + c_2 \ln r \tag{29.5}$$

The solution must remain bounded at $r = 0$, so we set $c_2 = 0$, effectively eliminating the log term and leaving just $R_0(r) = \alpha$ for some constant α.

For the case $\lambda_n = -n^2 \neq 0$, the differential equation on the right in (29.3) is the Euler equation (Section 5.12)

$$r^2 R''(r) + r R'(r) - n^2 R(r) = 0 \tag{29.6}$$

with solution $R_n(r) = C_1 r^n + C_2 r^{-n}$. For the solution to remain bounded at $r = 0$, we must take $C_2 = 0$, so $R_n(r) = c_n r^n$.

The eigensolutions are then

$$u_0(r, \theta) = R_0(r)\Theta_0(\theta) = C \quad \text{and} \quad u_n(r, \theta) = (A_n \cos n\theta + B_n \sin n\theta)r^n$$

where we have redefined the constants $A_n = \alpha_n c_n$ and $B_n = \beta_n c_n$. The general solution,

$$u(r, \theta) = C + \sum_{n=1}^{\infty} (A_n \cos n\theta + B_n \sin n\theta)r^n$$

is a sum over all possible eigensolutions.

Applying the Dirichlet boundary condition yields the equation

$$f(\theta) = C + \sum_{n=1}^{\infty} (\sigma^n A_n \cos n\theta + \sigma^n B_n \sin n\theta)$$

but a Fourier series expansion of $f(\theta)$ is

$$f(\theta) = \frac{a_0}{2} + \sum_{n=1}^{\infty} (a_n \cos n\theta + b_n \sin n\theta)$$

$$a_n = \frac{1}{\pi} \int_{-\pi}^{\pi} f(\theta) \cos n\theta \, d\theta$$

$$b_n = \frac{1}{\pi} \int_{-\pi}^{\pi} f(\theta) \sin n\theta \, d\theta$$

By comparison, then, we conclude $C = \frac{a_0}{2}$ and, from

$$\sigma^n A_n = a_n = \frac{1}{\pi} \int_{-\pi}^{\pi} f(\theta) \cos n\theta \, d\theta \quad \text{and} \quad \sigma^n B_n = b_n = \frac{1}{\pi} \int_{-\pi}^{\pi} f(\theta) \sin n\theta \, d\theta$$

that (29.1b), (29.1c), and (29.1a) must hold.

EXERCISES 29.1–Part A

A1. Verify the integrations in (29.2a), (29.2b), and (29.2c).

A2. Starting with the laplacian in polar coordinates, separate variables and show the equations in (29.3) result.

A3. Obtain the eigenfunctions and eigenvalues on the right in (29.4).

A4. Verify the solution in (29.5).

A5. Solve the Euler equation in (29.6).

EXERCISES 29.1–Part B

B1. The flow lines for $u(r, \theta)$, a solution of Laplace's equation $\nabla^2 u = 0$, can be obtained in polar coordinates by solving (numerically) the pair of ODEs $\mathbf{R}'(t) = -\nabla u$. Working in polar coordinates, $\mathbf{R}'(t) = r'(t)\mathbf{e}_r + r(t)\theta'(t)\mathbf{e}_\theta$, whereas $\nabla u = u_r \mathbf{e}_r + (u_\theta/r)\mathbf{e}_\theta$, so the equations to solve are $r'(t) = -u_r$ and $\theta'(t) = -u_\theta/r^2$. Take initial conditions of the form $r(0) = \sigma, \theta(0) = \theta_0$; and assume that as t, the parameter along the flow line, increases from 0 to t_f, the flow line moves through the interior of the disk. Almost any numeric solver will provide $r(t)$ and $\theta(t)$. The challenge is to merge the output of the solver with a graphics utility that will plot the flow line *in polar coordinates*. Obtain graphs of several flow lines for Example 29.1.

In Exercises B2–11:

(a) Using polar coordinates where $-\pi < \theta \leq \pi$, solve Laplace's equation in the disk of radius σ. Included in this BVP is the requirement that $u(r, \theta)$ is smooth across the negative x-axis.

(b) Plot the solution as a surface over the appropriate disk.

(c) Plot the level curves (isotherms or equipotentials) in the given disk.

(d) Obtain, and superimpose on the level curves, graphs of the flow lines.

B2. $\sigma = 2$, $f(\theta) = \sin\theta$ **B3.** $\sigma = 2$, $f(\theta) = \cos^2\theta$

B4. $\sigma = \frac{1}{2}$, $f(\theta) = \cos 2\theta$ **B5.** $\sigma = \frac{1}{2}$, $f(\theta) = \sin^2 2\theta$

B6. $\sigma = 3$, $f(\theta) = \sin 3\theta$ **B7.** $\sigma = 3$, $f(\theta) = \cos^2 3\theta$

B8. $\sigma = \frac{5}{2}$, $f(\theta) = 1 - \left(\frac{\theta}{\pi}\right)^2$ **B9.** $\sigma = \frac{3}{2}$, $f(\theta) = 1 - \left|\frac{\theta}{\pi}\right|$

B10. $\sigma = \frac{2}{3}$, $f(\theta) = \left|\frac{\theta}{\pi}\right|$

B11. $\sigma = \frac{5}{4}$, $f(\theta) = \begin{cases} \frac{2}{\pi}(\pi + \theta) & -\pi < \theta \leq -\frac{\pi}{2} \\ 1 & -\frac{\pi}{2} < \theta \leq \frac{\pi}{2} \\ \frac{2}{\pi}(\pi - \theta) & \frac{\pi}{2} < \theta \leq \pi \end{cases}$

B12. Solve the nonhomogeneous Neumann problem on the disk of radius σ. This problem consists of Laplace's equation in the disk, a smoothness requirement across the negative x-axis, and the prescription $u_r(\sigma, \theta) = F(\theta)$. In the context of steady-state temperatures on the disk, $u_r(\sigma, \theta)$, the directional derivative in the normal direction, is the flux of heat across the boundary of the disk. In the context of electrostatics, it would be the normal component of the electric field.

(a) Make a formal statement of the BVP, including Laplace's equation in polar coordinates, the induced periodic boundary conditions arising from the smoothness requirement, and the imposed Neumann condition.

(b) Separate variables and obtain the solution $u(r, \theta) = \frac{a_0}{2} + \sum_{n=1}^{\infty} r^n (A_n \cos n\theta + B_n \sin n\theta)$.

(c) Compute $u_r(r, \theta)$, the normal derivative, and evaluate it on the boundary where $r = \sigma$.

(d) Impose the Neumann condition, obtaining $u_r(\sigma, \theta) = \sum_{n=1}^{\infty} n\sigma^{n-1}(A_n \cos n\theta + B_n \sin n\theta) = F(\theta)$, and deduce

(i) $A_n = \dfrac{1}{n\pi\sigma^{n-1}} \int_{-\pi}^{\pi} F(\theta) \cos n\theta \, d\theta, \ n \geq 1$

(ii) $B_n = \dfrac{1}{n\pi\sigma^{n-1}} \int_{-\pi}^{\pi} F(\theta) \sin n\theta \, d\theta, \ n \geq 1$

(iii) $0 = \int_{-\pi}^{\pi} F(\theta) \, d\theta$. Since the derived series has "no constant term," the integral normally giving this constant term must be zero. Hence, the average value of the imposed boundary flux must be zero or there is no solution to this BVP. Laplace's equation arises, for example, from the steady state heat transfer problem, so if the flux is not in balance, there can be no steady state and a solution to the BVP cannot exist.

In Exercises B13–17, with $-\pi < \theta \leq \pi$:

(a) Verify that the average value of $F(\theta)$ vanishes.

(b) Apply the results of Exercise 12 and obtain a solution to the indicated nonhomogeneous Neumann problem.

(c) Plot the solution as a surface over the appropriate disk.

(d) Plot the level curves (isotherms or equipotentials) in the given disk.

(e) Obtain, and superimpose on the level curves, graphs of the flow lines.

B13. $\sigma = 2, F(\theta) = \sin\theta$ **B14.** $\sigma = 3, F(\theta) = \cos\theta$

B15. $\sigma = \frac{3}{2}, F(\theta) = \theta$

B16. $\sigma = \frac{3}{5}, F(\theta) = \begin{cases} 1 & -\pi < \theta \leq 0 \\ -1 & 0 < \theta \leq \pi \end{cases}$

B17. $\sigma = 4, F(\theta) = \begin{cases} -1 & -\pi < \theta \leq -\frac{\pi}{3} \\ -2 & -\frac{\pi}{3} < \theta \leq \frac{\pi}{3} \\ 3 & \frac{\pi}{3} < \theta \leq \pi \end{cases}$

B18. For a disk with radius σ, find the steady-state temperatures in the sector $0 \leq \theta \leq \omega \leq \pi$ if the radial boundaries are kept at a temperature of zero and the temperature along the circumference

is given by $f(\theta)$. Thus, in $0 < r < \sigma, 0 < \theta < \omega$, solve $\nabla^2 u(r, \theta) = 0$ subject to the conditions $u(r, 0) = u(r, \omega) = 0$, $u(\sigma, \theta) = f(\theta)$.

(a) The boundary conditions along the radii are $u(r, 0) = u(r, \omega) = 0$. Assume the separated solution $u(r, \theta) = R(r)\Theta(\theta)$ and find the boundary conditions $\Theta(0) = \Theta(\omega) = 0$.

(b) Obtain the separated ODEs $r^2 R''(r) + r R'(r) + \lambda R(r) = 0$ and $\Theta''(\theta) - \lambda\Theta(\theta) = 0$.

(c) Obtain the eigenvalues $\lambda_n = -(\frac{n\pi}{\omega})^2$ and the eigenfunctions $\Theta_n = B_n \sin\frac{n\pi}{\omega}\theta, n = 1, \ldots$.

(d) Show that Θ_n and Θ_m are orthogonal for $n \neq m$ by showing $\int_0^{\omega} \sin\frac{n\pi}{\omega}\theta \sin\frac{m\pi}{\omega}\theta \, d\theta = 0$.

(e) Obtain $R_n = r^{n\pi/\omega}$.

(f) Obtain $u(r, \theta) = \sum_{n=1}^{\infty} B_n r^{n\pi/\omega} \sin\frac{n\pi}{\omega}\theta$, and use $u(\sigma, \theta) = f(\theta)$ to determine B_n.

In Exercises B19–28:

(a) Apply the results of Exercise B18 to obtain the solution of the Dirichlet problem given for the sector.

(b) Plot the solution surface over the sector.

(c) Plot the isotherms (level curves or contours) in the sector.

(d) Obtain, and superimpose on the level curves, graphs of the flow lines.

B19. $\sigma = 2, \omega = \dfrac{\pi}{4}, f(\theta) = 6\theta\left(\dfrac{\pi}{4} - \theta\right)$

B20. $\sigma = 2, \omega = \dfrac{\pi}{4}, f(\theta) = \sin 4\theta$

B21. $\sigma = 3, \omega = \dfrac{\pi}{3}, f(\theta) = 4\theta\left(\dfrac{\pi}{3} - \theta\right)^2$

B22. $\sigma = 3, \omega = \dfrac{\pi}{3}, f(\theta) = \sin 3\theta$

B23. $\sigma = 4, \omega = \dfrac{\pi}{2}, f(\theta) = 2\theta\left(\dfrac{\pi}{2} - \theta\right)^3$

B24. $\sigma = 4, \omega = \dfrac{\pi}{2}, f(\theta) = \sin 2\theta$

B25. $\sigma = \dfrac{3}{2}, \omega = \dfrac{\pi}{6}, f(\theta) = 16\theta\left(\dfrac{\pi}{6} - \theta\right)$

B26. $\sigma = \dfrac{7}{2}, \omega = \dfrac{2\pi}{3}, f(\theta) = \theta\left(\dfrac{2\pi}{3} - \theta\right)^3$

B27. $\sigma = \dfrac{7}{2}, \omega = \dfrac{2\pi}{3}, f(\theta) = \sin\dfrac{3}{2}\theta$

B28. $\sigma = \dfrac{7}{2}, \omega = \dfrac{2\pi}{3}, f(\theta) = \sin^2 3\theta$

B29. For an annulus with radii σ_1 and σ_2, find the steady-state temperatures in the sector $0 \leq \theta \leq \omega \leq \pi$ if the radial boundaries and the inner circumference are kept at a temperature of zero and the temperature along the outer circumference is given by $f(\theta)$. Thus, in $\sigma_1 < r < \sigma_2, 0 < \theta < \omega$, solve $\nabla^2 u(r, \theta) = 0$ subject to the conditions $u(r, 0) = u(r, \omega) = u(\sigma_1, \theta) = 0$, $u(\sigma_2, \theta) = f(\theta)$.

(a) The boundary conditions along the radii are $u(r, 0) = u(r, \omega) = 0$. As in Exercise B18, assume the separated solution $u(r, \theta) = R(r)\Theta(\theta)$ and find the boundary conditions $\Theta(0) = \Theta(\omega) = 0$ and $R(\sigma_1) = 0$.

(b) Obtain the separated ODEs $r^2 R''(r) + r R'(r) + \lambda R(r) = 0$ and $\Theta''(\theta) - \lambda\Theta(\theta) = 0$.

(c) Obtain the eigenvalues $\lambda_n = -\left(\frac{n\pi}{\omega}\right)^2$ and the eigenfunctions $\Theta_n = \sin\frac{n\pi}{\omega}\theta$, $n = 1, \ldots$.

(d) Show that Θ_n and Θ_m are orthogonal for $n \neq m$ by showing $\int_0^\omega \sin\frac{n\pi}{\omega}\theta \sin\frac{m\pi}{\omega}\theta \, d\theta = 0$.

(e) Obtain $R_n = A_n r^{n\pi/\omega} + B_n r^{-n\pi/\omega}$, apply $R_n(\sigma_1) = 0$, and obtain

$$R_n = C_n\left[\left(\frac{r}{\sigma_1}\right)^{n\pi/\omega} - \left(\frac{\sigma_1}{r}\right)^{n\pi/\omega}\right]$$

Note that when $\sigma_1 = 1$, the simpler $R_n = C_n[r^{n\pi/\omega} - r^{-n\pi/\omega}]$ results.

(f) Obtain

$$u(r, \theta) = \sum_{n=1}^\infty C_n\left[\left(\frac{r}{\sigma_1}\right)^{n\pi/\omega} - \left(\frac{\sigma_1}{r}\right)^{n\pi/\omega}\right]\sin\frac{n\pi}{\omega}\theta$$

and use $u(\sigma_2, \theta) = f(\theta)$ to determine C_n.

In Exercises B30–39:

(a) Apply the results of Exercise B29 to obtain the solution of the Dirichlet problem given for the sector of the annulus.

(b) Plot the solution surface over the sector.

(c) Plot the isotherms (level curves or contours) in the sector.

(d) Obtain, and superimpose on the level curves, graphs of the flow lines.

B30. $\sigma_1 = 1, \sigma_2 = 2, \omega = \frac{\pi}{4}, f(\theta) = 60\left(\frac{\pi}{4} - \theta\right)$

B31. $\sigma_1 = 1, \sigma_2 = 2, \omega = \frac{\pi}{4}, f(\theta) = \sin 4\theta$

B32. $\sigma_1 = 2, \sigma_2 = 3, \omega = \frac{\pi}{3}, f(\theta) = 40\left(\frac{\pi}{3} - \theta\right)^2$

B33. $\sigma_1 = 2, \sigma_2 = 3, \omega = \frac{\pi}{3}, f(\theta) = \sin 3\theta$

B34. $\sigma_1 = 2, \sigma_2 = 4, \omega = \frac{\pi}{2}, f(\theta) = 20\left(\frac{\pi}{2} - \theta\right)^3$

B35. $\sigma_1 = \frac{1}{2}, \sigma_2 = 3, \omega = \frac{\pi}{6}, f(\theta) = 160\left(\frac{\pi}{6} - \theta\right)^3$

B36. $\sigma_1 = \frac{3}{2}, \sigma_2 = 4, \omega = \frac{3\pi}{4}, f(\theta) = \sin\frac{4}{3}\theta$

B37. $\sigma_1 = 1, \sigma_2 = \frac{5}{2}, \omega = \frac{2\pi}{3}, f(\theta) = \theta\left(\frac{2\pi}{3} - \theta\right)^2$

B38. $\sigma_1 = 1, \sigma_2 = \frac{5}{2}, \omega = \frac{2\pi}{3}, f(\theta) = \sin\frac{3}{2}\theta$

B39. $\sigma_1 = 1, \sigma_2 = \frac{5}{2}, \omega = \frac{2\pi}{3}, f(\theta) = \sin^2 3\theta$

B40. For an annulus with radii σ_1 and σ_2, find the steady-state temperatures in the sector $0 \leq \theta \leq \omega \leq \pi$ if the radial boundaries are kept at a temperature of zero, the inner circumference is insulated, and the outer circumference is kept at temperature

$f(\theta)$. Thus, in $\sigma_1 < r < \sigma_2, 0 < \theta < \omega$, solve $\nabla^2 u(r, \theta) = 0$ subject to the conditions $u(r, 0) = u(r, \omega) = 0$, $u_r(\sigma_1, \theta) = 0$, $u(\sigma_2, \theta) = f(\theta)$.

(a) The boundary conditions along the radii are $u(r, 0) = u(r, \omega) = 0$. As in Exercise B29, assume the separated solution $u(r, \theta) = R(r)\Theta(\theta)$ and find the boundary conditions $\Theta(0) = \Theta(\omega) = 0$ and $R'(\sigma_1) = 0$.

(b) Obtain the separated ODEs $r^2 R''(r) + r R'(r) + \lambda R(r) = 0$ and $\Theta''(\theta) - \lambda\Theta(\theta) = 0$.

(c) Obtain the eigenvalues $\lambda_n = -\left(\frac{n\pi}{\omega}\right)^2$ and the eigenfunctions $\Theta_n = \sin\frac{n\pi}{\omega}\theta$, $n = 1, \ldots$.

(d) Show that Θ_n and Θ_m are orthogonal for $n \neq m$ by showing $\int_0^\omega \sin\frac{n\pi}{\omega}\theta \sin\frac{m\pi}{\omega}\theta \, d\theta = 0$.

(e) Obtain $R_n = A_n r^{n\pi/\omega} + B_n r^{-n\pi/\omega}$, apply $R'_n(\sigma_1) = 0$, and obtain

$$R_n = C_n\left[\left(\frac{r}{\sigma_1}\right)^{n\pi/\omega} + \left(\frac{\sigma_1}{r}\right)^{n\pi/\omega}\right]$$

Note that when $\sigma_1 = 1$, the simpler $R_n = C_n[r^{n\pi/\omega} + r^{-n\pi/\omega}]$ results.

(f) Obtain

$$u(r, \theta) = \sum_{n=1}^\infty C_n\left[\left(\frac{r}{\sigma_1}\right)^{n\pi/\omega} + \left(\frac{\sigma_1}{r}\right)^{n\pi/\omega}\right]\sin\frac{n\pi}{\omega}\theta$$

and use $u(\sigma_2, \theta) = f(\theta)$ to determine C_n.

In Exercises B41–50:

(a) Apply the results of Exercise B40 to obtain the solution of the BVP given for the sector of the annulus.

(b) Plot the solution surface over the sector.

(c) Plot the isotherms (level curves or contours) in the sector.

(d) Obtain, and superimpose on the level curves, graphs of the flow lines.

B41. $\sigma_1 = 1, \sigma_2 = 2, \omega = \frac{\pi}{4}, f(\theta) = 60\left(\frac{\pi}{4} - \theta\right)$

B42. $\sigma_1 = 1, \sigma_2 = 2, \omega = \frac{\pi}{4}, f(\theta) = \sin 4\theta$

B43. $\sigma_1 = 2, \sigma_2 = 3, \omega = \frac{\pi}{3}, f(\theta) = 40\left(\frac{\pi}{3} - \theta\right)^2$

B44. $\sigma_1 = 2, \sigma_2 = 3, \omega = \frac{\pi}{3}, f(\theta) = \sin 3\theta$

B45. $\sigma_1 = 2, \sigma_2 = 4, \omega = \frac{\pi}{2}, f(\theta) = 20\left(\frac{\pi}{2} - \theta\right)^3$

B46. $\sigma_1 = 2, \sigma_2 = 4, \omega = \frac{\pi}{2}, f(\theta) = \sin 2\theta$

B47. $\sigma_1 = \frac{1}{2}, \sigma_2 = 3, \omega = \frac{\pi}{6}, f(\theta) = \sin 6\theta$

B48. $\sigma_1 = \frac{3}{2}, \sigma_2 = 4, \omega = \frac{3\pi}{4}, f(\theta) = \theta\left(\frac{3\pi}{4} - \theta\right)^3$

B49. $\sigma_1 = 1, \sigma_2 = \frac{5}{2}, \omega = \frac{2\pi}{3}, f(\theta) = \sin\frac{3}{2}\theta$

B50. $\sigma_1 = 1, \sigma_2 = \frac{5}{2}, \omega = \frac{2\pi}{3}, f(\theta) = \sin^2 3\theta$

29.2 Laplace's Equation in a Cylinder

Problem Statement and Formulation

Find the steady-state, bounded temperatures inside a solid cylinder of radius σ and height h if the base is insulated and the curved lateral surface and the top are maintained at temperatures of zero and $f(r)$, respectively.

Because none of the data depends on the angle θ, the temperature will be a function of only the height z and radius r. Thus, the temperature function, which can be written as $u(r, z)$ in cylindrical coordinates, satisfies Laplace's equation inside the cylinder. The insulation condition on the bottom of the cylinder, a homogeneous Neumann condition, is expressed by the vanishing of the temperature gradient, a partial derivative in the z-direction, on that face. The complete boundary value problem determining $u(r, z)$ therefore consists of the partial differential equation

$$\nabla^2 u(r, z) = u_{rr}(r, z) + \frac{1}{r}u_r(r, z) + u_{zz}(r, z) = 0 \tag{29.7}$$

and the boundary conditions $u(\sigma, z) = 0, u_z(r, 0) = 0, u(r, h) = f(r)$, and $u(r, z)$ bounded.

Statement and Analysis of the Solution

The solution is given by

$$u(r, z) = \sum_{k=1}^{\infty} A_k J_0(\lambda_k r) \cosh(\lambda_k z) \tag{29.8a}$$

$$A_k = \frac{\int_0^\sigma r f(r) J_0(\lambda_k r)\, dr}{\cosh(\lambda_k h) \int_0^\sigma r J_0^2(\lambda_k r)\, dr} = \frac{2 \int_0^\sigma r f(r) J_0(\lambda_k r)\, dr}{\sigma^2 \cosh(\lambda_k h) J_1^2(\lambda_k \sigma)} \tag{29.8b}$$

where $J_0(x)$ is the Bessel function of order zero and λ_k is the kth zero of $J_0(\sigma x)$. The functions $J_\nu(x)$ satisfy Bessel's equation

$$x^2 y''(x) + x y'(x) + (x^2 - \nu^2) y(x) = 0 \tag{29.9}$$

and are called Bessel functions of order ν. As seen in Section 16.2, two linearly independent solutions of Bessel's equation are $J_\nu(x)$ and $Y_\nu(x)$. Only one of these solutions, $J_\nu(x)$, is bounded at the origin. The Bessel function of the second kind, $Y_\nu(x)$, has a logarithmic singularity at $x = 0$.

Series expansions for the Bessel functions of order zero and one are, respectively,

$$J_0(x) = \sum_{k=0}^{\infty} \frac{(-1)^k x^{2k}}{4^k (k!)^2} \quad \text{and} \quad J_1(x) = \frac{1}{2} \sum_{k=0}^{\infty} \frac{(-1)^k x^{2k+1}}{4^k (k!)^2 (k+1)}$$

FIGURE 29.4   The Bessel functions $J_0(x)$ (solid) and $J_1(x)$ (dotted)

In Figure 29.4, the solid curve is $J_0(x)$, the dotted, $J_1(x)$. It should be recalled from Section 16.2 that if λ_k, $k = 1, 2, \ldots$, are the roots of $J_0(\sigma x)$, then the functions $J_0(\lambda_k x)$ are, on the interval $0 \le x \le \sigma$, orthogonal with respect to the weight function x. Thus, for $\lambda_k \ne \lambda_j$, we have $\int_0^\sigma x J_0(\lambda_k x) J_0(\lambda_j x)\, dx = 0$.

EXAMPLE 29.2   For the cylinder described earlier, let $\sigma = 2$, $h = 3$, and $f(r) = 1$. Then, the first 10 eigenvalues λ_k are the first 10 zeros of $J_0(2x)$, listed as μ_k in Table 16.1 in Section 16.2. The coefficients

$$A_k = \frac{\int_0^2 r J_0(\lambda_k r)\, dr}{2 \cosh(3\lambda_k) J_1^2(2\lambda_k)}$$

can be evaluated with numeric integration, yielding, for example, $A_1 = 0.08684853220$. Alternatively, because $f(r) = 1$, the integral in the numerator can be evaluated *exactly* via the formula

$$\int_0^\alpha x J_0(x)\, dx = \alpha J_1(\alpha)$$

if the change of variables $\lambda_k r = x$ is made in the integral in the numerator of A_k, as summarized by

$$\left\{ \int_0^2 r J_0(\lambda_k r)\, dr \right\}\Bigg|_{x = \lambda_k r} = \frac{1}{\lambda_k^2} \int_0^{2\lambda_k} x J_0(x)\, dx = \frac{2}{\lambda_k} J_1(2\lambda_k)$$

Consequently, we find

$$A_k = \frac{\operatorname{sech}(3\lambda_k)}{\lambda_k J_1(2\lambda_k)}$$

which we use to compute A_k, $k = 1, \ldots, 10$, listed in Table 29.1. The coefficients rapidly decrease in magnitude, so the convergence is rapid, and the partial sum

$$u_{10}(r, z) = \sum_{k=1}^{10} A_k J_0(\lambda_k r) \cosh(\lambda_k z)$$

is adequate, except on the boundary $z = 3$ where a much slower convergence is evident, as seen in Figure 29.5.

$A_1 = 8.68 \times 10^{-2}$
$A_2 = -5.40 \times 10^{-4}$
$A_3 = 3.92 \times 10^{-6}$
$A_4 = -3.04 \times 10^{-8}$
$A_5 = 2.43 \times 10^{-10}$
$A_6 = -2.00 \times 10^{-12}$
$A_7 = 1.65 \times 10^{-14}$
$A_8 = -1.39 \times 10^{-16}$
$A_9 = 1.18 \times 10^{-18}$
$A_{10} = -1.99 \times 10^{-20}$

TABLE 29.1   Coefficients A_k, $k = 1, \ldots, 10$, in Example 29.2

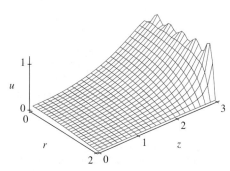

FIGURE 29.5   Example 29.2: Approximation of the solution surface $u(r, 2)$

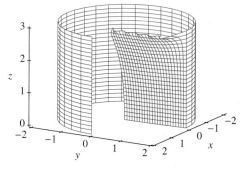

FIGURE 29.6   Example 29.2: Representation of temperatures on the axial cross-section inside the cylinder

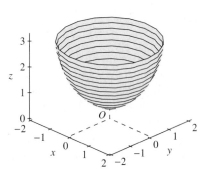

FIGURE 29.7   Example 29.2: Isothermal surface $u(r, z) = 0.1$

Temperatures on the rectangle $R = \{0 \le r \le 2, 0 \le z \le 3\}$, half of a plane section through the axis of the cylinder, are represented in Figure 29.5. It is impossible to draw a graph of the temperatures *inside* the cylinder. Every representation of these temperatures will require some visual compromise. In Figure 29.6, the temperatures are graphed on the rectangle R, shown supporting the temperature surface *inside* the cylinder. The rectangle R rotates about the axis of the cylinder, supporting the same temperatures for all values of the angle θ.

Figure 29.7 gives another view of the temperatures inside the cylinder, the isothermal surface on which $u(r, z) = 0.1$. In fact, the level surfaces $u(r, z) = \frac{k}{10}$, $k = 1, 2, \ldots, 9$, are each bowl-shaped, with the same circle bounding the open top. This corresponds to the discontinuity in temperatures at the junction of the top of the cylinder with the curved lateral wall of the cylinder. At this juncture, all level surfaces intersect. The accompanying Maple worksheet presents this sequence of isothermal surfaces as an animation.   ❖

Solution by Separation of Variables

Assuming a separated solution of the form $u(r, z) = R(r)Z(z)$ and substituting it into the boundary data $u(\sigma, z) = 0$ and $u_z(r, 0) = 0$ lead to

$$R(\sigma)Z(z) = 0 \Rightarrow R(\sigma) = 0 \quad \text{and} \quad R(r)Z'(0) = 0 \Rightarrow Z'(0) = 0$$

Application of the separation assumption to Laplace's equation, division by $u = RZ$, rearrangement, and introduction of the Bernoulli separation constant μ lead to

$$\frac{R''(r)}{R(r)} + \frac{R'(r)}{rR(r)} = -\frac{Z''(z)}{Z(z)} = \mu \tag{29.10}$$

Overlooking the abuse of notation whereby primes denote differentiation with respect to both r and z, we have the two ODEs

$$rR''(r) + R'(r) - \mu r R(r) = 0 \quad \text{and} \quad Z''(z) + \mu Z(z) = 0 \tag{29.11}$$

The r-equation is a modified form of Bessel's equation. In fact, the solution is

$$R(r) = c_1 J_0(\sqrt{-\mu}\,r) + c_2 Y_0(\sqrt{-\mu}\,r)$$

The appearance of $\sqrt{-\mu}$ suggests renaming the separation constant to $-\lambda^2$ so that the two ordinary differential equations become

$$rR''(r) + R'(r) + \lambda^2 rR(r) = 0 \quad \text{and} \quad Z''(z) - \lambda^2 Z(z) = 0 \tag{29.12}$$

Then, the solution of the r-equation is

$$R(r) = c_1 J_0(\lambda r) + c_2 Y_0(\lambda r)$$

Since we know $Y_0(x)$, the zero-order Bessel function of the second kind, is not bounded at the origin, the constant c_2 is taken as zero and the solution for $R(r)$ is a multiple of $J_0(\lambda r)$.

To transform the r-equation to Bessel's equation (29.9), make the change of variables $r = \frac{x}{\lambda}$ so that $R(r)$ becomes $R(\frac{x}{\lambda}) = y(x)$ and the r-equation becomes

$$\lambda[xy''(x) + y'(x) + xy(x)] = 0 \tag{29.13}$$

Except for the factor λ, the equation is precisely Bessel's equation of order zero.

Incidentally, the change of variables was accomplished by a careful use of the chain rule starting with the definitions $x = \lambda r$ and

$$R(r) = R\left(\frac{x}{\lambda}\right) = y(x) = y(\lambda r)$$

Then, we obtain

$$\frac{dR}{dr} = \frac{dy}{dx}\frac{dx}{dr} = \frac{dy}{dx}\lambda \quad \text{and} \quad \frac{d^2 R}{dr^2} = \frac{d}{dx}\left(\frac{dy}{dx}\right)\frac{dx}{dr}\lambda = \frac{d^2 y}{dx^2}\lambda^2 \tag{29.14}$$

so that the equation on the left in (29.12) becomes (29.13) when r is replaced by $\frac{x}{\lambda}$.

Since $R(r) = J_0(\lambda r)$, the boundary condition $R(\sigma) = 0$ becomes $J_0(\lambda \sigma) = 0$, the equation whose roots give the eigenvalues λ_k. The z-equation on the right in (29.12) now inherits the known eigenvalues and becomes

$$Z_k''(z) - \lambda_k^2 Z_k(z) = 0$$

which has exponential solutions. The solution satisfying the Neumann boundary condition $Z'(0) = 0$ is

$$Z_k(z) = C_k \cosh(\lambda_k z)$$

Consequently, a single eigensolution is

$$u_k(r, z) = A_k J_0(\lambda_k r) \cosh(\lambda_k z)$$

and the general solution, a sum over all such eigensolutions, is (29.8a).

Application of $u(r, h) = f(r)$, the final nonhomogeneous Dirichlet boundary condition at the top of the cylinder, leads to the equation

$$f(r) = \sum_{k=1}^{\infty} A_k J_0(\lambda_k r) \cosh(\lambda_k h) \tag{29.15}$$

from which the coefficients A_k are determined via multiplication by $r J_0(\lambda_j r)$ and integration with respect to r over the interval $0 \le r \le \sigma$. The orthogonality of the eigenfunctions $J_0(\lambda_k r)$ with respect to the weighting function r causes all terms but one on the right side to vanish. The one surviving term is the one for which $k = j$. Hence, what results is the equation

$$\int_0^{\sigma} r f(r) J_0(\lambda_j r) \, dr = A_j \cosh(\lambda_j h) \int_0^{\sigma} r J_0^2(\lambda_j r) \, dr \tag{29.16}$$

from which we obtain (29.8b).

EXERCISES 29.2

1. Start with $\nabla^2 u(r, z) = 0$ as in (29.7), and show how to obtain (29.10) and (29.11).

2. Make the change of variables $r = \frac{x}{\lambda}$ in (29.12), use (29.14), and show that (29.13) results.

3. Show how (29.15) leads to (29.16).

In Exercises 4–13:

(a) Solve Laplace's equation $\nabla^2 u = 0$ for the steady-state temperatures $u(r, z)$ in the cylinder $0 \le r \le \sigma, 0 \le z \le h$ if the curved wall $(r = \sigma)$ is kept at a temperature of zero, the bottom $(z = 0)$ is insulated, and the top $(z = h)$ is maintained at the temperature $f(r)$.

(b) Graph $u(r, z)$ as a surface over the rectangle $0 \le r \le \sigma$, $0 \le z \le h$.

(c) Rotate the surface in part (b) so that it stands vertical inside a partly open cylinder. Thus, the rectangle $0 \le r \le \sigma, 0 \le z \le h$ should be the right-half of the rectangle exposed if the cylinder is cut with a plane passing through the z-axis.

(d) If $\rho = u_{\max} - u_{\min}$, plot the three isothermal surfaces $u = u_{\min} + \frac{k}{4}\rho, k = 1, 2, 3$.

(e) Treating r and z as Cartesian variables, obtain a contour plot of $u(r, z)$ on the rectangle $0 \le r \le \sigma, 0 \le z \le h$. Each level curve is the intersection of an isothermal surface and a vertical plane passing through the z-axis.

(f) For each fixed value of $\theta = \theta_0$, there is a flow line that starts at the point (r_0, θ_0, h) and remains in the vertical plane $\theta = \theta_0$. Obtain these flow lines by solving the ODEs $r'(t) = -u_r$, $z'(t) = -u_z$ and graphing the solutions as Cartesian curves in the rectangle $0 \le r \le \sigma, 0 \le z \le h, \theta = 0$.

4. $\sigma = 2, h = 3, f(r) = 2 - r$

5. $\sigma = 2, h = 3, f(r) = \frac{1}{2}(4 - r^2)$

6. $\sigma = 2, h = 3, f(r) = 2\cos\frac{\pi}{4}r$

7. $\sigma = 3, h = 5, f(r) = r^2(3 - r)$

8. $\sigma = 3, h = 5, f(r) = r(3 - r)^2$

9. $\sigma = 3, h = 5, f(r) = r^3(3 - r)$

10. $\sigma = 3, h = 5, f(r) = r(3 - r)^3$

11. $\sigma = 5, h = 4, f(r) = 10\sin^2\frac{2\pi}{5}r$

12. $\sigma = 5, h = 4, f(r) = 10\sin\frac{2\pi}{5}r$

13. $\sigma = 5, h = 4, f(r) = 10\sin\frac{\pi}{5}r$

14. Solve Laplace's equation $\nabla^2 u = 0$ for the steady-state temperatures $u(r, z)$ in the cylinder $0 \le r \le \sigma, 0 \le z \le h$ if the curved wall $(r = \sigma)$ and the bottom $(z = 0)$ are kept at a temperature of zero and the top $(z = h)$ is maintained at the temperature $f(r)$. Thus, solve the BVP $\nabla^2 u = 0, u(\sigma, z) = u(r, 0) = 0, u(r, h) = f(r)$ for the *bounded* temperatures $u(r, z)$ inside the cylinder $0 \le r \le \sigma$, $0 \le z \le h$.

(a) Using cylindrical coordinates, write $u(r, z) = R(r)Z(z)$ and obtain $rR''(r) + R'(r) + \lambda^2 rR(r) = 0, R(\sigma) = 0$ and $Z''(z) - \lambda^2 Z(z) = 0, Z(0) = 0$.

(b) Obtain the eigenvalues $\lambda_k, k = 1, \ldots$, where λ_k is the kth zero of the Bessel function $J_0(\sigma x)$.

(c) Obtain the eigenfunctions $R_k(r) = J_0(\lambda_k r)$.

(d) Obtain $Z_k(z) = \sinh \lambda_k z$.

(e) Obtain $u(r, z) = \sum_{k=1}^{\infty} A_k J_0(\lambda_k r) \sinh \lambda_k z$, and determine A_k by the condition $u(r, h) = f(r)$.

For the data in each of Exercises 15–24:

(a) Use the results of Exercise 14 to solve Laplace's equation $\nabla^2 u = 0$ for the steady-state temperatures $u(r, z)$ in the cylinder $0 \le r \le \sigma, 0 \le z \le h$ if the curved wall ($r = \sigma$) and the bottom ($z = 0$) are kept at a temperature of zero and the top ($z = h$) is maintained at the temperature $f(r)$.

(b) Graph $u(r, z)$ as a surface over the rectangle $0 \le r \le \sigma$, $0 \le z \le h$.

(c) Rotate the surface in part (b) so that it stands vertical inside a partly open cylinder. Thus, the rectangle $0 \le r \le \sigma, 0 \le z \le h$ should be the right-half of the rectangle exposed if the cylinder is cut with a plane passing through the z-axis.

(d) If $\rho = u_{max} - u_{min}$, plot the three isothermal surfaces $u = u_{min} + \frac{k}{4}\rho, k = 1, 2, 3$.

(e) Treating r and z as Cartesian variables, obtain a contour plot of $u(r, z)$ on the rectangle $0 \le r \le \sigma, 0 \le z \le h$. Each level curve is the intersection of an isothermal surface and a vertical plane passing through the z-axis.

(f) For each fixed value of $\theta = \theta_0$, there is a flow line that starts at the point (r_0, θ_0, h) and remains in the vertical plane $\theta = \theta_0$. Obtain these flow lines by solving the ODEs $r'(t) = -u_r$, $z'(t) = -u_z$ and graphing the solutions as Cartesian curves in the rectangle $0 \le r \le \sigma, 0 \le z \le h, \theta = 0$.

15. the data from Exercise 4   16. the data from Exercise 5

17. the data from Exercise 6   18. the data from Exercise 7

19. the data from Exercise 8   20. the data from Exercise 9

21. the data from Exercise 10   22. the data from Exercise 11

23. the data from Exercise 12   24. the data from Exercise 13

25. Solve Laplace's equation $\nabla^2 u = 0$ for the steady-state temperatures $u(r, z)$ in the cylinder $0 \le r \le \sigma, 0 \le z \le h$ if the curved wall ($r = \sigma$) is insulated, the bottom ($z = 0$) is kept at a temperature of zero, and the top ($z = h$) is maintained at the temperature $f(r)$. Thus, solve the BVP $\nabla^2 u = 0, u_r(\sigma, z) = u(r, 0) = 0$, $u(r, h) = f(r)$ for the *bounded* temperatures $u(r, z)$ inside the cylinder $0 \le r \le \sigma, 0 \le z \le h$.

(a) Using cylindrical coordinates, write $u(r, z) = R(r)Z(z)$ and obtain (29.12), $R'(\sigma) = 0$ and $Z(0) = 0$.

(b) Obtain the eigenvalues $\lambda_k, k = 0, 1, \ldots$, where λ_k is the kth zero of the Bessel function $J_0'(\sigma x) = -J_1(\sigma x)$. Note that $\lambda_0 = 0$ is an eigenvalue.

(c) Obtain the eigenfunctions $R_k(r) = J_0(\lambda_k r), k = 0, \ldots$. Note that $R_0(r) = A_0$, a constant.

(d) Obtain $Z_k(z) = \sinh \lambda_k z, k = 0, \ldots$, but obtain $Z_0(z)$ by a separate calculation.

(e) Obtain $u(r, z) = A_0 Z_0 + \sum_{k=1}^{\infty} A_k J_0(\lambda_k r) \sinh \lambda_k z$, and determine $A_k, k = 0, 1, \ldots$, by the condition $u(r, h) = f(r)$.

For the data in each of Exercises 26–35:

(a) Use the results of Exercise 25 to solve Laplace's equation $\nabla^2 u = 0$ for the steady-state temperatures $u(r, z)$ in the cylinder $0 \le r \le \sigma, 0 \le z \le h$ if the curved wall ($r = \sigma$) is insulated, the bottom ($z = 0$) is kept at a temperature of zero, and the top ($z = h$) is maintained at the temperature $f(r)$.

(b) Graph $u(r, z)$ as a surface over the rectangle $0 \le r \le \sigma$, $0 \le z \le h$.

(c) Rotate the surface in part (b) so that it stands vertical inside a partly open cylinder. Thus, the rectangle $0 \le r \le \sigma, 0 \le z \le h$ should be the right-half of the rectangle exposed if the cylinder is cut with a plane passing through the z-axis.

(d) If $\rho = u_{max} - u_{min}$, plot the three isothermal surfaces $u = \frac{k}{4}\rho$, $k = 1, 2, 3$.

(e) Treating r and z as Cartesian variables, obtain a contour plot of $u(r, z)$ on the rectangle $0 \le r \le \sigma, 0 \le z \le h$. Each level curve is the intersection of an isothermal surface and a vertical plane passing through the z-axis.

(f) For each fixed value of $\theta = \theta_0$, there is a flow line that starts at the point (r_0, θ_0, h) and remains in the vertical plane $\theta = \theta_0$. Obtain these flow lines by solving the ODEs $r'(t) = -u_r$, $z'(t) = -u_z$ and graphing the solutions as Cartesian curves in the rectangle $0 \le r \le \sigma, 0 \le z \le h, \theta = 0$.

26. the data from Exercise 4   27. the data from Exercise 5

28. the data from Exercise 6   29. the data from Exercise 7

30. the data from Exercise 8   31. the data from Exercise 9

32. the data from Exercise 10   33. the data from Exercise 11

34. the data from Exercise 12   35. the data from Exercise 13

29.3 The Circular Drumhead

Problem Statement and Formulation

A circular drumhead consisting of a membrane with radius $r = \sigma$ that is clamped on its circumference at time $t = 0$ is given a radially symmetric displacement and no initial velocity. If the motions of the membrane are governed by the two-dimensional wave equation $u_{tt} = c^2(u_{xx} + u_{yy})$, formulate an appropriate boundary value problem for this drumhead.

The circular domain suggests the two-dimensional wave equation is best given in polar coordinates as

$$u_{tt} = \frac{c^2}{r^2} \left(r(ru_r)_r + u_{\theta\theta} \right)$$

The radial symmetry of the initial disturbance suggests $u(r, \theta, t)$ is independent of the angle θ and becomes just $u(r, t)$. Thus, the boundary value problem we seek consists of the partial differential equation

$$u_{tt}(r, t) = c^2 \left(u_{rr}(r, t) + \frac{1}{r} u_r(r, t) \right)$$

the homogeneous Dirichlet condition $u(\sigma, t) = 0$ that expresses the clamping of the rim and the homogeneous initial condition $u_t(r, 0) = 0$ that expresses the zero initial velocity. The initial displacement is given by $u(r, 0) = f(r)$, and the physical nature of the system requires that the displacement function $u(r, t)$ be continuous and, hence, bounded.

Solution and Analysis

The solution is given by the infinite series

$$u(r, t) = \sum_{k=1}^{\infty} A_k J_0(\lambda_k r) \cos(c\lambda_k t) \tag{29.17a}$$

$$A_k = \frac{\int_0^{\sigma} r f(r) J_0(\lambda_k r)\, dr}{\int_0^{\sigma} r J_0^2(\lambda_k r)\, dr} = \frac{2}{\sigma^2 J_1^2(\lambda_k \sigma)} \int_0^{\sigma} r f(r) J_0(\lambda_k r)\, dr \tag{29.17b}$$

where A_k are the Fourier–Bessel coefficients for the function $f(r)$. Of course, the functions $J_0(x)$ and $J_1(x)$ are Bessel functions of order zero and one, respectively, and λ_k, the eigenvalues, are the zeros of $J_0(\sigma x)$.

EXAMPLE 29.3   Let the radius of the membrane be $\sigma = 5$, the wave speed be $c = 1$, and the initial displacement $u(r, 0)$ be the smooth function

$$f(r) = \begin{cases} \frac{3}{4} - r^2 & 0 \leq r < \frac{1}{2} \\ \frac{1}{8}(4r^2 - 12r + 9) & \frac{1}{2} \leq r < \frac{3}{2} \\ 0 & \frac{3}{2} \leq r \leq 5 \end{cases} \tag{29.18}$$

graphed in Figure 29.8, which shows the initial shape's profile along any radius of the membrane. The actual shape of the membrane during this initial displacement is shown in Figure 29.9.

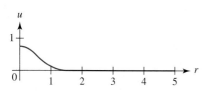

FIGURE 29.8   Initial function $f(r)$ in Example 29.3

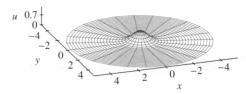

FIGURE 29.9   Initial shape of drumhead in Example 29.3

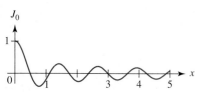

FIGURE 29.10   Graph of $J_0(5\lambda)$ whose zeros are the eigenvalues λ_k in Example 29.3

The eigenvalues λ_k are the zeros of the Bessel function $J_0(5\lambda)$ whose graph appears in Figure 29.10. Since, as seen in Section 16.2, $J_0(5x)$ is asymptotic to $\sqrt{\frac{2}{5\pi x}} \cos(5x - \frac{\pi}{4})$, we are guided to the location of the first 24 eigenvalues, which we compute numerically and display in Table 29.2.

$\lambda_1 = 0.481$	$\lambda_5 = 2.986$	$\lambda_9 = 5.499$	$\lambda_{13} = 8.0117$	$\lambda_{17} = 10.525$	$\lambda_{21} = 13.038$
$\lambda_2 = 1.104$	$\lambda_6 = 3.614$	$\lambda_{10} = 6.127$	$\lambda_{14} = 8.640$	$\lambda_{18} = 11.153$	$\lambda_{22} = 13.666$
$\lambda_3 = 1.731$	$\lambda_7 = 4.242$	$\lambda_{11} = 6.755$	$\lambda_{15} = 9.268$	$\lambda_{19} = 11.781$	$\lambda_{23} = 14.295$
$\lambda_4 = 2.358$	$\lambda_8 = 4.870$	$\lambda_{12} = 7.383$	$\lambda_{16} = 9.897$	$\lambda_{20} = 12.410$	$\lambda_{24} = 14.923$

TABLE 29.2 First 24 eigenvalues in Example 29.3

Then, with $\sigma = 5$ and $f(r)$ given by (29.18), numeric integration of (29.17b) provides the coefficients listed in Table 29.3.

$A_1 = 0.0587$	$A_5 = 0.124$	$A_9 = 0.00414$	$A_{13} = 0.000889$	$A_{17} = -0.00416$	$A_{21} = -0.000620$
$A_2 = 0.122$	$A_6 = 0.0831$	$A_{10} = 0.00132$	$A_{14} = -0.00247$	$A_{18} = -0.00213$	$A_{22} = -0.000794$
$A_3 = 0.154$	$A_7 = 0.0442$	$A_{11} = 0.00252$	$A_{15} = -0.00505$	$A_{19} = -0.000701$	$A_{23} = -0.000397$
$A_4 = 0.153$	$A_8 = 0.0171$	$A_{12} = 0.00293$	$A_{16} = -0.00555$	$A_{20} = -0.000337$	$A_{24} = 0.000454$

TABLE 29.3 Example 29.3: Coefficients A_k, $k = 1, \ldots, 24$

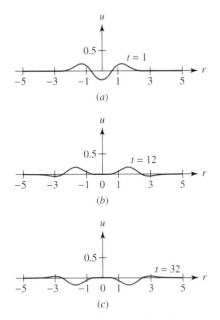

FIGURE 29.11 Example 29.3: Graph of $u(r, t)$ for $t = 1$, 12, and 32

In Figure 29.11, the partial sum $u_{24}(r, t) = \sum_{k=1}^{24} A_k J_0(\lambda_k r) \cos(c\lambda_k t)$ approximating $u(r, t)$ is graphed for $t = 1$, 12, and 32. The companion Maple worksheet provides animations for the motion of both the profiles $u(r, t)$ and the drumhead itself. ❖

Solution by Separation of Variables

Assuming a separated solution of the form $u(r, t) = R(r)T(t)$ and substituting it into the boundary data $u(\sigma, z) = 0$ and initial condition $u_t(r, 0) = 0$ lead to

$$R(\sigma)T(t) = 0 \Rightarrow R(\sigma) = 0 \quad \text{and} \quad R(r)T'(0) = 0 \Rightarrow T'(0) = 0$$

Application of the separation assumption to the wave equation, division by $c^2 RT$, mild rearrangement, and introduction of the Bernoulli separation constant μ yield

$$\frac{R''(r)}{R(r)} + \frac{R'(r)}{rR(r)} = \frac{T''(t)}{c^2 T(t)} = \mu$$

from which we get the two ODEs

$$rR''(r) + R'(r) - \mu rR(r) = 0 \quad \text{and} \quad T''(t) - c^2\mu T(t) = 0 \qquad \textbf{(29.19)}$$

As in Section 29.2, we replace μ with $-\lambda^2$ so that (29.19) becomes

$$rR''(r) + R'(r) + \lambda^2 rR(r) = 0 \quad \text{and} \quad T''(t) + c^2\lambda^2 T(t) = 0 \qquad \textbf{(29.20)}$$

Then, the solution of the r-equation in (29.20) is $R(r) = c_1 J_0(\lambda r) + c_2 Y_0(\lambda r)$. Since we know $Y_0(x)$, the zero-order Bessel function of the second kind is not bounded at the origin, the constant c_2 is taken as zero, and the solution for $R(r)$ is a multiple of $J_0(\lambda r)$. The details are precisely the same as in Section 29.2.

Since $R(r) = J_0(\lambda r)$, the boundary condition $R(\sigma) = 0$ becomes $J_0(\lambda\sigma) = 0$, the equation whose roots give the eigenvalues λ_k. The t-equation now inherits the known eigenvalues and becomes

$$T_k''(t) + c^2\lambda_k^2 T_k(t) = 0$$

and has trigonometric solutions. The solution satisfying the initial condition $T_k'(0) = 0$ is

$$T_k(t) = C_k \cos(c\lambda_k t)$$

Consequently, a single eigensolution is

$$u_k(r, t) = A_k J_0(\lambda_k r) \cos(c\lambda_k t)$$

and the general solution is given by (29.17a), a sum over all such eigensolutions. Application of the final initial condition $u(r, 0) = f(r)$ leads to the equation

$$f(r) = \sum_{k=1}^{\infty} A_k J_0(\lambda_k r)$$

The coefficients A_k are the Fourier–Bessel coefficients for the function $f(r)$ and are determined by multiplying this equation by $r J_0(\lambda_j r)$ and integrating with respect to r over the interval $0 \le r \le \sigma$, just as in Section 29.2. The orthogonality of the eigenfunctions $J_0(\lambda_k r)$ with respect to the weighting function r causes all terms but one on the right side to vanish. The one surviving term is the one for which $k = j$. Hence, what results is the equation

$$\int_0^{\sigma} r f(r) J_0(\lambda_j r)\, dr = A_j \int_0^{\sigma} r J_0^2(\lambda_j r)\, dr$$

from which we obtain (29.17b).

EXERCISES 29.3

1. Verify the entries in Table 29.2.

2. Verify the entries in Table 29.3.

In Exercises 3–7:

(a) Solve the BVP $u_{tt} = c^2 \nabla^2 u(r, t)$ on the circular domain $0 \le r < \sigma, -\pi < \theta \le \pi$ for a membrane with clamped edges, no initial velocity, and an initial displacement $f(r)$.

(b) For $t = 1, 2, 3$, graph the profile of the membrane along the line $\theta = 0, -\sigma \le r \le \sigma$.

(c) Animate the motion of the cross-section $\theta = 0$.

(d) Graph the solution surface at times $t = 1, 2, 3$.

3. $\sigma = 5, c = 1, f(r) = \begin{cases} 1 - r & 0 \le r < 1 \\ 0 & 1 \le r \le 5 \end{cases}$

4. $\sigma = 4, c = 2, f(r) = \begin{cases} 1 & 0 \le r < 1 \\ 2 - r & 1 \le r < 2 \\ 0 & 2 \le r \le 4 \end{cases}$

5. $\sigma = 8, c = \frac{2}{3}, f(r) = \begin{cases} 1 + \cos r & 0 \le r < \pi \\ 0 & \pi \le r \le 8 \end{cases}$

6. $\sigma = 9, c = \frac{5}{4}, f(r) = \begin{cases} \sin^2 r & 0 \le r < \pi \\ 0 & \pi \le r \le 9 \end{cases}$

7. $\sigma = 1, c = \frac{1}{2}, f(r) = 1 - r$

8. Solve the wave equation on a circular membrane of radius σ if the edges are clamped, there is no initial velocity, and the initial displacement is a function of both r and θ. Thus, the BVP consists of the wave equation $u_{tt} = c^2 \nabla^2 u(r, \theta, t)$, the boundary condition $u(\sigma, \theta, t) = 0$, the initial conditions $u_t(r, \theta, 0) = 0, u(r, \theta, 0) = F(r, \theta)$, and, since $-\pi < \theta \le \pi$, the requirement that u and u_θ be continuous across the negative x-axis.

(a) In cylindrical coordinates, the wave equation is $\frac{1}{c^2} u_{tt} = u_{rr} + \frac{1}{r} u_r + \frac{1}{r^2} u_{\theta\theta}$. Write $u = R(r)\Theta(\theta)T(t)$ and obtain, after introducing the two separation constants λ^2 and μ,

  (i) $r^2 R'' + r R' + (\lambda^2 r^2 + \mu)R = 0, R(\sigma) = 0$;

  (ii) $\Theta'' - \mu\Theta = 0$, with the periodic boundary conditions (from continuity) $\Theta(-\pi) = \Theta(\pi), \Theta'(-\pi) = \Theta'(\pi)$;

  (iii) $T'' + \lambda^2 c^2 T = 0, T'(0) = 0$.

(b) Solve the Θ-equation (it's a Sturm–Liouville problem solved previously), to obtain $\mu = -n^2, n = 0, 1, \ldots$, with $\Theta_n = a_n \cos n\theta + b_n \sin n\theta, n = 1, \ldots$, and $\Theta_0 = a_0$.

(c) Solve the R-equation, which now reads $r^2 R'' + r R' + (\lambda^2 r^2 - n^2)R = 0, R(\sigma) = 0$, a (singular) Sturm–Liouville problem whose solution is $R_{nk}(r) = J_n(\lambda_{nk} r)$, Bessel functions of order n. The eigenvalue λ_{nk} is the kth zero of $J_n(\sigma x)$. Thus, λ_{nk} is a doubly indexed collection of eigenvalues that must be computed numerically.

(d) Solve the T-equation to obtain $T_{nk} = \cos \lambda_{nk} ct$. Hence, there are eigensolutions of the form $A_{nk} J_n(\lambda_{nk} r) \cos n\theta \cos \lambda_{nk} ct$, $B_{nk} J_n(\lambda_{nk} r) \sin n\theta \cos \lambda_{nk} ct$, and $A_{0k} J_0(\lambda_{0k} r) \cos \lambda_{0k} ct$. The solution $u(r, \theta, t)$ is a sum over all such eigensolutions. Hence, $u = \sum_{k=1}^{\infty} A_{0k} J_0(\lambda_{0k} r) \cos \lambda_{0k} ct + \sum_{k=1}^{\infty} \sum_{n=1}^{\infty} J_n(\lambda_{nk} r) \cos \lambda_{nk} ct (A_{nk} \cos n\theta + B_{nk} \sin n\theta)$.

(e) Show that the constants A_{nk} and B_{nk} are determined by the initial condition $F(r, \theta) = \sum_{k=1}^{\infty} A_{0k} J_0(\lambda_{0k} r) + \sum_{k=1}^{\infty} \sum_{n=1}^{\infty} J_n(\lambda_{nk} r) (A_{nk} \cos n\theta + B_{nk} \sin n\theta)$ and are therefore

$$A_{0k} = \frac{\int_{-\pi}^{\pi} \int_0^{\sigma} F(r, \theta) r J_0(\lambda_{0k} r) \, dr \, d\theta}{2\pi \int_0^{\sigma} r J_0^2(\lambda_{0k} r) \, dr}$$

$$A_{nk} = \frac{\int_{-\pi}^{\pi} \int_0^{\sigma} F(r, \theta) r J_n(\lambda_{nk} r) \cos n\theta \, dr \, d\theta}{\pi \int_0^{\sigma} r J_n^2(\lambda_{nk} r) \, dr}$$

$$B_{nk} = \frac{\int_{-\pi}^{\pi} \int_0^{\sigma} F(r, \theta) r J_n(\lambda_{nk} r) \sin n\theta \, dr \, d\theta}{\pi \int_0^{\sigma} r J_n^2(\lambda_{nk} r) \, dr}$$

9. Apply the results of Exercise 8 to a circular membrane with radius $\sigma = 2$, taking $c = 1$, and $F(r, \theta) = f(r)g(\theta)$, where

$$f(r) = \begin{cases} 0 & 0 \le r < \frac{1}{2} \\ 40(r - \frac{1}{2})^2(1 - r)^2 & \frac{1}{2} \le r < 1 \\ 0 & 1 \le r \le 2 \end{cases}$$

and

$$g(\theta) = \begin{cases} 0 & -\pi < \theta \le 0 \\ 40\theta^2 \left(\frac{\pi}{3} - \theta\right)^2 & 0 < \theta \le \frac{\pi}{3} \\ 0 & \frac{\pi}{3} < \theta \le \pi \end{cases}$$

(a) Obtain the first four zeros of each of $J_n(2x)$, $n = 0, \ldots, 5$. This gives λ_{nk} for $k = 1, \ldots, 5$ and $n = 0, \ldots, 5$.

(b) Obtain A_{0k}, $k = 1, \ldots, 5$.

(c) Obtain A_{nk}, $k = 1, \ldots, 5$, $n = 1, \ldots, 5$.

(d) Obtain B_{nk}, $k = 1, \ldots, 5$, $n = 1, \ldots, 5$.

(e) Form U, a partial sum containing the terms for which the coefficients have just been computed. Compare the plot of $U(r, \theta, 0)$ with that of $F(r, \theta)$.

(f) Plot $U(r, \theta, t)$ at $t = 1, 2, 3$. If possible (it may be computationally taxing on older computers), animate the solution surface as a function of time.

29.4 Laplace's Equation in a Sphere

Problem Statement and Formulation

Find u, the steady-state temperature inside a sphere of radius $\rho = \sigma$ if the temperature on the boundary, the surface of the sphere, is prescribed.

Steady-state temperatures satisfy Laplace's equation, and in a sphere, it makes most sense to work in spherical coordinates. Using the definitions in the middle column of Table 22.3, Section 22.3, and assuming the prescribed temperatures on the surface of the sphere are given by $f(\phi)$, then they are constant on latitudes and $u(\sigma, \phi, \theta) = f(\phi)$ on the surface. Hence, u is independent of the angle θ, the boundary condition is the homogeneous Dirichlet condition $u(\sigma, \phi) = f(\phi)$, and Laplace's equation in the sphere gives

$$\rho^2 \nabla^2 u = \rho^2 u_{\rho\rho}(\rho, \phi) + 2\rho u_\rho(\rho, \phi) + u_{\phi\phi}(\rho, \phi) + \cot(\phi) u_\phi(\rho, \phi) = 0 \quad (29.21)$$

The only other condition on $u(\rho, \phi)$ is boundedness, as dictated by the physics of the problem.

Solution

The solution is given by

$$u(\rho, \phi) = \sum_{k=0}^{\infty} A_k \rho^k P_k(\cos \phi) \quad (29.22a)$$

$$A_k = \frac{2k+1}{2\sigma^k} \int_{-1}^{1} f(\arccos z) P_k(z) \, dz = \frac{2k+1}{2\sigma^k} \int_0^{\pi} f(\phi) P_k(\cos \phi) \sin \phi \, d\phi$$

$$(29.22b)$$

where $P_k(x)$ is the kth Legendre polynomial. The Legendre polynomials are detailed in Section 16.3. Incidentally, the solution to the *exterior* problem in which the temperature is again prescribed on the surface of a sphere but the domain of interest is the region *outside* the sphere is similar, being given by

$$u(\rho, \phi) = \sum_{k=0}^{\infty} A_k \rho^{(-k-1)} P_k(\cos\phi) \tag{29.23a}$$

$$A_k = \frac{2k+1}{2\sigma^{(-k-1)}} \int_{-1}^{1} f(\arccos z) P_k(z)\, dz = \frac{2k+1}{2\sigma^{(-k-1)}} \int_{0}^{\pi} f(\phi) P_k(\cos\phi) \sin\phi\, d\phi \tag{29.23b}$$

◆ **EXAMPLE 29.4**   Consider a sphere of radius $\sigma = 2$, on the surface of which the temperatures are prescribed by the function $f(\phi) = \frac{\phi}{\pi}$. Thus, if the sphere were the earth, the temperatures would vary linearly from zero at the North Pole, to 1 at the South Pole. To find the temperatures inside the "earth," use (29.22b) to compute the first 11 coefficients, listing the nonzero coefficients in Table 29.4. The even-indexed coefficients A_{2k}, $k \geq 1$, vanish ($P_{2k}(\cos\phi)$ is even and $f(\phi)$ is odd), and the surviving odd-indexed coefficients clearly go to zero rapidly. Hence, the series for $u(\rho, \phi)$ will converge rapidly. A finite approximation to the solution is then the partial sum

$$u_{10}(r, \phi) = \sum_{k=0}^{10} A_k r^k P_k(\cos\phi)$$

for which $u_{10}(2, \phi)$ should reproduce $f(\phi) = \frac{\phi}{\pi}$. Figure 29.12 shows that except near $\phi = 0$ and $\phi = \pi$, this approximation is good.

The temperatures inside the solid sphere cannot be "seen" and cannot be readily plotted. The rest of this example demonstrates a variety of techniques for visualizing and interpreting the temperatures inside the sphere.

To begin, we first ask for the level surfaces inside the sphere. These are the isothermal surfaces, the surfaces on which the temperature remains constant inside the sphere. Because of the symmetry with respect to the angle θ, any vertical plane section through the z-axis supports the same temperature profile. In particular, right along the z-axis, the temperatures are a function of ρ, and these temperatures are plotted in Figure 29.13. The dotted curve represents temperatures along the positive z-axis, while the solid curve represents temperatures along the negative z-axis.

Is it possible that for each plane section $z =$ constant, the temperatures in the sphere are just the boundary temperatures at that height? If so, the isothermal surfaces would be disks, slices of the sphere parallel to the xy-plane. To test this hypothesis, compute the temperature on the z-axis at height $z = 1.9$, obtaining 0.068, and compare this value to the temperature on the boundary at the same height. Since the temperatures on the boundary are given by $f(\phi) = \frac{\phi}{\pi}$, the appropriate angle ϕ is given by $\phi = \arccos\frac{z}{\rho} = \arccos\frac{1.9}{2}$. The corresponding temperature is then $f(\arccos 0.95) = 0.101$, and the hypothesis of "slices" is ruled out. The isothermal surfaces are not simply disks.

Exploiting the solution's symmetry with respect to angle θ, a vertical plane section through the z-axis slices an isothermal surface in a curve. Imagine the sphere sliced by the yz-plane, exposing the circle of radius 2 shown in Figure 29.14. As angle ϕ varies from 0 to $\frac{\pi}{2}$, angle ω varies from $\frac{\pi}{2}$ to 0. Hence, set $\phi = \frac{\pi}{2} - \omega$ and interpret ω as the polar angle in the yz-plane. The curves in which this plane cuts the isothermal surfaces are shown in Figure 29.15.

If the graph in Figure 29.15 were rotated about the z-axis, the isothermal surfaces inside the sphere would be generated as surfaces of revolution. The isothermal surface

$$A_0 = \tfrac{1}{2}$$

$$A_1 = -\tfrac{3}{16}$$

$$A_3 = -\tfrac{7}{1024}$$

$$A_5 = -\tfrac{11}{16384}$$

$$A_7 = -\tfrac{375}{4194304}$$

$$A_9 = -\tfrac{931}{67108864}$$

TABLE 29.4   Coefficients A_k in Example 29.4

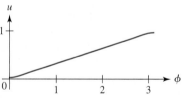

FIGURE 29.12   Example 29.4: Approximation of $f(\phi)$ by $u_{10}(2, \phi)$

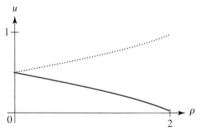

FIGURE 29.13   In Example 29.4, temperatures along the z-axis; for $z > 0$, they are given by the dotted curve, and for $z < 0$, by the solid.

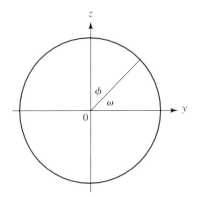

FIGURE 29.14   Example 29.4: Geometry of a cross-section that contains the z-axis

$u(\rho, \phi) = 0.3$ is graphed, but not to scale, in Figure 29.16. Inside the sphere the temperatures can be represented as a stack of nested isothermal surfaces of a similar type, as an animation in the accompanying Maple worksheet shows.

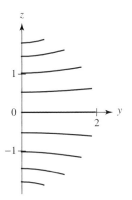

FIGURE 29.15   Example 29.4: Intersection of isothermal surfaces and cross-section in Figure 29.14

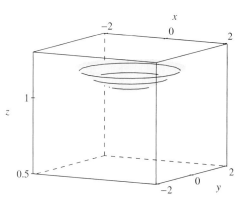

FIGURE 29.16   Example 29.4: A magnified view of the isothermal surface $u(\rho, \phi) = 0.3$

The gradient field

$$\nabla u(\rho, \phi) = u_\rho \mathbf{e}_\rho + \frac{1}{\rho} u_\phi \mathbf{e}_\phi$$

indicates how heat is transported across the level surfaces of $u(\rho, \phi)$. In particular, $-u_\rho$, the normal component of $-\nabla u$, indicates the heat flux across the surface $\rho = \sigma = 2$. Figure 29.17 shows a graph of $-u_\rho(2, \phi)$, the heat flux across the surface of the sphere. It was computed from the partial sum $u_{25}(r, \phi)$ since computing a gradient involves differentiation, and the termwise derivative of a series generally converges more slowly than the series itself. The graph agrees with our intuition. At the North Pole, in the vicinity of $\phi = 0$, the heat flow is out of the sphere since the temperature is lowest there. The graph shows a positive flux, indicating the heat flow is along the outward normal. At the South Pole, in the vicinity of $\phi = \pi$, the heat flow is into the sphere since that is the warmest part of the sphere. The graph shows a negative flux, indicating the heat flow is into the sphere and against the outward normal.

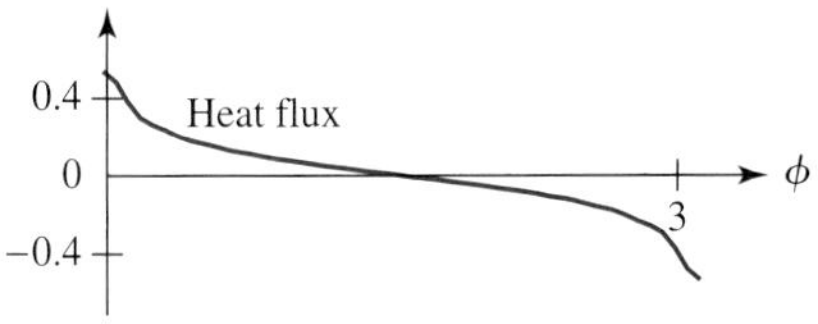

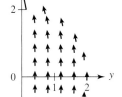

FIGURE 29.17   Example 29.4: Heat flux across the surface of the sphere

To visualize the flow of heat inside the sphere, look at the plane section $x = 0$, the yz-plane inside the sphere. Figure 29.18 shows arrows of the field $-\nabla u$ for $y \geq 0$. The flow of heat in the fourth quadrant is upward and to the right, whereas in the first quadrant it is upward and to the left. Clearly, heat flows from the warmer southern hemisphere to the cooler northern hemisphere. Above the equator there is also a component of heat flow through the boundary and out of the sphere, while below the equator there is a component of the heat flow into the sphere through the boundary. ❖

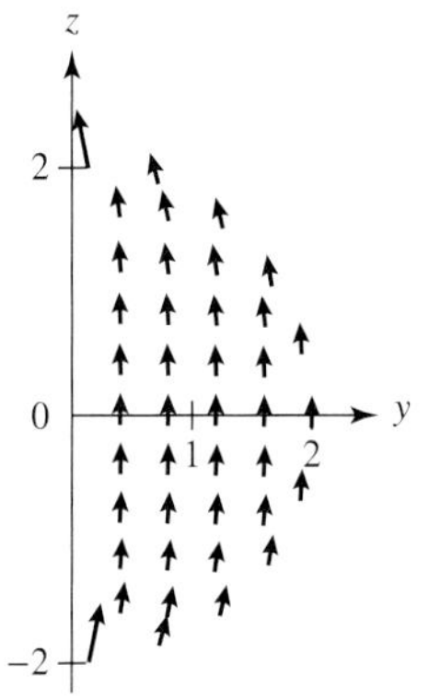

Solution by Separation of Variables

Apply the separation assumption $u(\rho, \phi) = R(\rho)\Phi(\phi)$ to Equation (29.21), divide by $u = R\Phi$, rearrange, and introduce the Bernoulli separation constant λ to obtain

$$\rho^2 \frac{R''(\rho)}{R(\rho)} + 2\rho \frac{R'(\rho)}{R(\rho)} = -\frac{\Phi''(\phi)}{\Phi(\phi)} - \cot(\phi)\frac{\Phi'(\phi)}{\Phi(\phi)} = \lambda \qquad \textbf{(29.24)}$$

FIGURE 29.18   Example 29.4: The field $-\nabla u$ in the cross-section from Figure 29.14

yielding the two ODEs

$$\rho^2 R''(\rho) + 2\rho R'(\rho) - \lambda R(\rho) = 0 \tag{29.25a}$$

$$\sin(\phi)\Phi''(\phi) + \cos(\phi)\Phi'(\phi) + \lambda\sin(\phi)\Phi(\phi) = 0 \tag{29.25b}$$

The ρ-equation is an Euler equation, solvable when λ is known. The ϕ-equation becomes

$$(1 - x^2)y''(x) - 2xy'(x) + \lambda y(x) = 0 \tag{29.26}$$

Legendre's equation, upon making the change of variables $x = \cos(\phi)$.

The change of variables is performed analytically by setting $x = x(\phi) = \cos\phi$ so that

$$\Phi(\phi) = \Phi(\arccos x) = y(x) = y(\cos\phi)$$

Then, starting with $\Phi(\phi) = y(x(\phi))$, the chain rule gives

$$\Phi' = \frac{d\Phi}{d\phi} = \frac{dy}{dx}\frac{dx}{d\phi} = \frac{dy}{dx}[-\sin\phi]$$

and

$$\Phi'' = \frac{d}{d\phi}\Phi' = \frac{d^2y}{dx^2}\frac{dx}{d\phi}[-\sin\phi] + \frac{dy}{dx}\frac{d}{d\phi}[-\sin\phi]$$

$$= \frac{d^2y}{dx^2}[-\sin\phi][-\sin\phi] + \frac{dy}{dx}[-\cos\phi]$$

$$= y''\sin^2\phi - y'\cos\phi$$

$$= [1 - \cos^2\phi]y'' - y'\cos\phi$$

$$= (1 - x^2)y'' - xy'$$

Then (29.25b) becomes

$$\sin(\phi)[(1 - x^2)y'' - xy'] + \cos(\phi)[-\sin(\phi)y'] + \lambda\sin(\phi)y = 0$$

Factoring $\sin\phi$ and grouping the first-derivative terms, we finally have

$$\sin(\phi)[(1 - x^2)y'' - 2xy' + \lambda y] = 0$$

and therefore Legendre's equation (29.26).

The interval $0 \le \phi \le \pi$ becomes $-1 \le x \le 1$ under the change of variables $x = \cos\phi$. We saw in Section 16.3 that the condition of boundedness on the interval $[-1, 1]$ leads to the eigenvalues $\lambda_k = k(k+1), k = 0, 1, \ldots$, and to the corresponding eigenfunctions $P_k(x)$, the Legendre polynomials $1, x, \frac{3}{2}x^2 - \frac{1}{2}, \frac{5}{2}x^3 - \frac{3}{2}x, \ldots$. If $[x]$ is the greatest integer function, then the Legendre polynomials can be represented as

$$P_k(x) = \sum_{n=0}^{[k/2]} \frac{(-1)^n(2k-2n)!x^{k-2n}}{2^k n!(k-2n)!(k-n)!} \tag{29.27}$$

We also saw in Section 10.7 that the Legendre polynomials are orthogonal on the interval $[-1, 1]$. In fact,

$$\int_{-1}^{1} P_k(x)P_j(x)\,dx = \begin{cases} 0 & k \ne j \\ \dfrac{2}{2k+1} & k = j \end{cases}$$

The solution of the ρ-equation (29.25a) is much more straightforward, since it is just an Euler equation (see Section 5.12). Restricting λ to the eigenvalues, the equation becomes

$$\rho^2 R_k''(\rho) + 2\rho R_k'(\rho) - k(k+1)R_k(\rho) = 0 \qquad \textbf{(29.28)}$$

with solution $R_k(\rho) = c_1\rho^k + c_2\rho^{-k-1}$. Boundedness on the interior of the sphere demands $c_2 = 0$, giving $R_k(\rho) = \rho^k$, $k = 0, 1, \ldots$, in the interior. Obviously, if solving Laplace's equation on the exterior of the sphere, we would set $c_1 = 0$, giving $R_k(\rho) = \rho^{-k-1}$, $k = 0, 1, \ldots$, in the region exterior to the sphere.

We can now write a single eigensolution as

$$u_k(\rho, \phi) = \rho^k P_k(\cos\phi)$$

and the general solution as the infinite sum (29.22a). The final boundary condition, the Dirichlet condition $u(\sigma, \phi) = f(\phi)$, leads to

$$f(\phi) = \sum_{k=0}^{\infty} A_k \sigma^k P_k(\cos\phi)$$

Since $f(\phi)$ has the Fourier–Legendre series

$$f(\phi) = \sum_{k=0}^{\infty} \alpha_k P_k(\cos\phi) \quad \text{where} \quad \alpha_k = \frac{2k+1}{2}\int_0^\pi f(\phi)P_k(\cos\phi)\sin\phi\,d\phi$$

we conclude that $A_k\sigma^k = \alpha_k$, so A_k is given by (29.22b).

EXERCISES 29.4–Part A

A1. Show the two integrals in (29.22b) are equivalent.

A2. Show that the composition of an even function with an even function is again an even function. Hence, verify that $A_{2k}, k \geq 1$, vanish, in Example 29.4.

A3. Set $u(\rho, \phi) = R(\rho)\Phi(\phi)$ in (29.21), and verify that (29.24) results.

A4. Obtain (29.25a) and (29.25b) from (29.24).

A5. Solve (29.28).

EXERCISES 29.4–Part B

B1. Verify the entries in Table 29.4.

B2. Verify that (29.27) reproduces $P_k(x)$, $k = 0, \ldots, 10$.

In Exercises 3–12:

(a) Solve the Dirichlet problem on the sphere of radius $\rho = \sigma$ if $u(\sigma, \phi) = f(\phi)$.

(b) Obtain graphs of the level surfaces $u = f_{\min} + \frac{k}{4}(f_{\max} - f_{\min})$, $k = 1, 2, 3$.

(c) Obtain graphs of the curves in which the yz-plane intersects the level surfaces obtained in part (b).

(d) Obtain a graph of the normal component of the surface flux of u.

(e) Obtain, in the yz-plane, a plot of the arrows of ∇u, the gradient field inside the sphere.

B3. $\sigma = 3$, $f(\phi) = \left(\dfrac{\phi}{\pi}\right)^2$ **B4.** $\sigma = 1$, $f(\phi) = \sin\phi$

B5. $\sigma = 1$, $f(\phi) = \sin 2\phi$ **B6.** $\sigma = 2$, $f(\phi) = \cos^2\phi$

B7. $\sigma = 5$, $f(\phi) = \phi(\pi - \phi)^2$ **B8.** $\sigma = 5$, $f(\phi) = \phi^2(\pi - \phi)$

B9. $\sigma = \frac{1}{2}$, $f(\phi) = \dfrac{2}{\pi^3}(\pi^3 - \phi^3)$ **B10.** $\sigma = 4$, $f(\phi) = 8\phi^2 e^{-2\phi}$

B11. $\sigma = 6$, $f(\phi) = \begin{cases} \left(\phi - \dfrac{\pi}{2}\right)^2 & 0 \leq \phi < \dfrac{\pi}{2} \\[2mm] \phi - \dfrac{\pi}{2} & \dfrac{\pi}{2} \leq \phi \leq \pi \end{cases}$

B12. $\sigma = 6, f(\phi) = \begin{cases} \left(\phi - \dfrac{\pi}{2}\right)^3 & 0 \le \phi < \dfrac{\pi}{2} \\ \left(\phi - \dfrac{\pi}{2}\right)^2 & \dfrac{\pi}{2} \le \phi \le \pi \end{cases}$

In Exercises 13–22:

(a) Use (29.23a) and (29.23b) to solve the Dirichlet problem on the *exterior* of the sphere of radius $\rho = \sigma$ if $u(\sigma, \phi) = f(\phi)$.

(b) Obtain graphs of the level surfaces $u = \frac{k}{4}(f_{max} - f_{min})$, $k = 1, 2, 3$.

(c) Obtain graphs of the curves in which the yz-plane intersects the level surfaces obtained in part (b).

(d) Obtain a graph of the normal component of the surface flux of u, interpreting u as temperature, and flux, as heat flow.

(e) Obtain, in the yz-plane, a plot of the arrows of ∇u, the gradient field outside the sphere.

B13. $\sigma = 3, f(\phi) = \left(\dfrac{\phi}{\pi}\right)^3$ **B14.** $\sigma = 1, f(\phi) = \cos\phi$

B15. $\sigma = 1, f(\phi) = \cos 2\phi$ **B16.** $\sigma = 2, f(\phi) = \sin^2\phi$

B17. $\sigma = 5, f(\phi) = \phi(\pi - \phi)$ **B18.** $\sigma = 5, f(\phi) = \phi(\pi - \phi)^3$

B19. $\sigma = 5, f(\phi) = \phi^3(\pi - \phi)$ **B20.** $\sigma = \frac{1}{2}, f(\phi) = \dfrac{2}{\pi^2}(\pi^2 - \phi^2)$

B21. $\sigma = 4, f(\phi) = 8\phi e^{-2\phi}$

B22. $\sigma = 6, f(\phi) = \begin{cases} \left(\phi - \dfrac{\pi}{2}\right)^3 & 0 \le \phi < \dfrac{\pi}{2} \\ \phi - \dfrac{\pi}{2} & \dfrac{\pi}{2} \le \phi \le \pi \end{cases}$

29.5 The Spherical Dielectric

Problem Statement and Formulation

A dielectric (nonconductor of electricity) in the shape of a sphere of radius σ is placed in a uniform electric field $\mathbf{E} = \eta\mathbf{k}$, where $\mathbf{k}$ is the unit vector in the z-direction. Determine $v(\rho, \phi)$, the potential both inside and outside the sphere.

The potentials inside and outside of the sphere, $u(\rho, \phi)$ and $U(\rho, \phi)$, respectively, must each satisfy Laplace's equation and, in addition, the continuity conditions

$$u(\sigma, \phi) = U(\sigma, \phi) \quad \text{and} \quad \kappa u_\rho(\sigma, \phi) = U_\rho(\sigma, \phi)$$

for the positive constant κ. The second condition expresses the equality of the normal components of the electric displacement vector $\mathbf{D} = \varepsilon\mathbf{E}$ across the dielectric surface. Far from the dielectric, the exterior potential must approach the potential for the uniform field $\mathbf{E}$, so that $\lim_{\rho \to \infty} U(\rho, \phi) = -\eta z = -\eta\rho\cos(\phi)$.

Solution

The solution, derived later in this section, is given by

$$v(\rho, \phi) = \begin{cases} u(\rho, \phi) = -\dfrac{3\eta}{\kappa + 2}\rho\cos\phi & 0 \le \rho \le \sigma \\[2mm] U(\rho, \phi) = -\eta\rho\cos\phi + \eta\sigma^3\dfrac{\kappa - 1}{\kappa + 2}\dfrac{\cos\phi}{\rho^2} & \sigma \le \rho < \infty \end{cases} \tag{29.29}$$

EXAMPLE 29.5   Setting $\eta = 1, \sigma = 1$, and $\kappa = 4$, the solution becomes

$$v_1(\rho, \phi) = \begin{cases} u_1(\rho, \phi) = -\dfrac{1}{2}\rho\cos\phi & 0 \le \rho \le 1 \\[2mm] U_1(\rho, \phi) = -\rho\cos\phi + \dfrac{1}{2}\dfrac{\cos\phi}{\rho^2} & 1 \le \rho < \infty \end{cases}$$

a solution we will now proceed to interpret graphically. As in Section 29.4, slicing the sphere through the yz-plane allows us to change to polar coordinates with the polar angle taken as $\omega = \frac{\pi}{2} - \phi$, as shown in Figure 29.14.

The solution, with respect to the polar variables ρ and ω, is then

$$
V(\rho, \omega) = \begin{cases} -\dfrac{1}{2}\rho \sin \omega & 0 \le \rho \le 1 \\[2ex] -\dfrac{2\rho^3 - 1}{2\rho^2}\sin \omega & 1 \le \rho < \infty \end{cases}
$$

A graph of the equipotentials for the solution would consist of a set of surfaces. The intersection of these surfaces with the yz-plane is the set of curves shown in Figure 29.19.

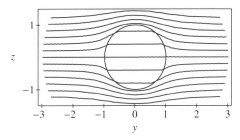

FIGURE 29.19 Example 29.5: Intersection of equipotential surfaces with the yz-plane

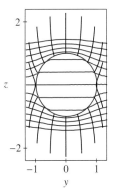

The flow lines of the electric field are generally called the *lines of force,* or *field lines.* We can obtain a graph of these lines by integrating the gradient field for the exterior potential. Since this will be easier in Cartesian coordinates, convert the polar coordinates on the yz-plane back to the Cartesian variables y and z using the transformations $\rho = \sqrt{y^2 + z^2}$ and $\omega = \arctan \frac{z}{y}$. Thus, the exterior potential, restricted to the yz-plane, becomes

$$
\frac{z}{2}\left(y^2 + z^2\right)^{-3/2} - z
$$

and the differential equations for the flow lines are then

$$
y'(t) = \tfrac{3}{2}yz\left(y^2 + z^2\right)^{-5/2} \quad \text{and} \quad z'(t) = 1 + \frac{2z^2 - y^2}{2\left(y^2 + z^2\right)^{5/2}}
$$

The resulting flow lines are shown in Figure 29.20, superimposed on the equipotentials of Figure 29.19. Of course, inside the circle, the flow lines (not shown) are just vertical line segments. Under the symmetry in the angle θ, the diagram can be rotated about the z-axis. Inside the circle, the vertical field lines become cylinders with axis parallel the z-axis, and the equipotentials become segments of planes parallel to the xy-plane. ❖

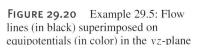

FIGURE 29.20 Example 29.5: Flow lines (in black) superimposed on equipotentials (in color) in the yz-plane

Derivation of the Solution

From Section 29.4, the interior solution to Laplace's equation in a sphere is of the form

$$
u(\rho, \phi) = \sum_{k=0}^{\infty} a_k \rho^k P_k(\cos \phi)
$$

while the exterior solution is of the form

$$
U(\rho, \phi) = \sum_{k=0}^{\infty} (\alpha_k \rho^k + \beta_k \rho^{-k-1}) P_k(\cos \phi)
$$

Because the electric field does not vanish at infinity, we did not take $\alpha_k = 0, k = 0, 1, \ldots$. In fact, for ρ large, $U(\rho, \phi)$ must tend to $-\eta \rho \cos \phi$. Since $P_1(x) = x$, we recognize $\cos \phi$ as $P_1(\cos \phi)$ and realize we should take $\alpha_0 = 0$, $\alpha_1 = -\eta$, $\alpha_k = 0, k = 2, 3, \ldots$. Thus, the exterior solution reduces to

$$U(\rho, \phi) = -\eta \rho \, P_1(\cos \phi) + \sum_{k=0}^{\infty} \beta_k \rho^{-k-1} P_k(\cos \phi)$$

The coefficients a_k and β_k are determined by the continuity conditions on the surface of the sphere. The condition $u(\sigma, \phi) = U(\sigma, \phi)$ requires the coefficients satisfy

$$\sum_{k=0}^{\infty} a_k \sigma^k P_k = -\eta \sigma P_1 + \sum_{k=0}^{\infty} \beta_k \sigma^{-k-1} P_k \qquad (29.30)$$

If the sums in (29.30) are truncated at $k = 5$, we obtain

$$u(\sigma, \phi) = a_0 P_0 + a_1 \sigma P_1 + a_2 \sigma^2 P_2 + a_3 \sigma^3 P_3 + a_4 \sigma^4 P_4 + a_5 \sigma^5 P_5$$

$$U(\sigma, \phi) = -\eta \sigma P_1 + \frac{\beta_0 P_0}{\sigma} + \frac{\beta_1 P_1}{\sigma^2} + \frac{\beta_2 P_2}{\sigma^3} + \frac{\beta_3 P_3}{\sigma^4} + \frac{\beta_4 P_4}{\sigma^5} + \frac{\beta_5 P_5}{\sigma^6}$$

Coefficients of like Legendre polynomials must match on the left and right sides of the equation $u(\sigma, \phi) = U(\sigma, \phi)$, so subtract and group terms according to

$$\left(a_0 - \frac{\beta_0}{\sigma} \right) P_0 + \left(a_1 \sigma + \eta \sigma - \frac{\beta_1}{\sigma^2} \right) P_1 + \left(a_2 \sigma^2 - \frac{\beta_2}{\sigma^3} \right) P_2 + \left(a_3 \sigma^3 - \frac{\beta_3}{\sigma^4} \right) P_3$$

$$+ \left(a_4 \sigma^4 - \frac{\beta_4}{\sigma^5} \right) P_4 + \left(a_5 \sigma^5 - \frac{\beta_5}{\sigma^6} \right) P_5 = 0$$

Thus, each coefficient of a Legendre polynomial must vanish identically, giving rise to the equations

$$a_0 - \frac{\beta_0}{\sigma} = 0 \qquad a_1 \sigma + \eta \sigma - \frac{\beta_1}{\sigma^2} = 0 \qquad a_2 \sigma^2 - \frac{\beta_2}{\sigma^3} = 0$$

$$a_3 \sigma^3 - \frac{\beta_3}{\sigma^4} = 0 \qquad a_4 \sigma^4 - \frac{\beta_4}{\sigma^5} = 0 \qquad a_5 \sigma^5 - \frac{\beta_5}{\sigma^6} = 0$$

Except for the case $k = 1$, each equation appears to be of the form $a_n \sigma^n - \beta_n / \sigma^{n+1} = 0$.

The second continuity condition, $\kappa u_\rho(\sigma, \phi) = U_\rho(\sigma, \phi)$, requires that we differentiate with respect to ρ and, then evaluate on the surface. Truncating as earlier, we obtain

$$u_n = a_1 P_1 + 2a_2 \sigma P_2 + 3a_3 \sigma^2 P_3 + 4a_4 \sigma^3 P_4 + 5a_5 \sigma^4 P_5$$

$$U_n = -\eta P_1 - \frac{\beta_0 P_0}{\sigma^2} - 2\frac{\beta_1 P_1}{\sigma^3} - 3\frac{\beta_2 P_2}{\sigma^4} - 4\frac{\beta_3 P_3}{\sigma^5} - 5\frac{\beta_4 P_4}{\sigma^6} - 6\frac{\beta_5 P_5}{\sigma^7}$$

Again, to match like-coefficients of Legendre polynomials we subtract to form the equation

$$\frac{\beta_0 P_0}{\sigma^2} + \left(\eta + \kappa a_1 + 2\frac{\beta_1}{\sigma^3} \right) P_1 + \left(2\kappa a_2 \sigma + 3\frac{\beta_2}{\sigma^4} \right) P_2 + \left(3\kappa a_3 \sigma^2 + 4\frac{\beta_3}{\sigma^5} \right) P_3$$

$$+ \left(4\kappa a_4 \sigma^3 + 5\frac{\beta_4}{\sigma^6} \right) P_4 + \left(5\kappa a_5 \sigma^4 + 6\frac{\beta_5}{\sigma^7} \right) P_5 = 0$$

The identical vanishing of each coefficient of a Legendre polynomial leads to the equations

$$\frac{\beta_0}{\sigma^2} = 0 \qquad \eta + \kappa a_1 + 2\frac{\beta_1}{\sigma^3} = 0 \qquad 2\kappa a_2 \sigma + 3\frac{\beta_2}{\sigma^4} = 0$$

$$3\kappa a_3 \sigma^2 + 4\frac{\beta_3}{\sigma^5} = 0 \qquad 4\kappa a_4 \sigma^3 + 5\frac{\beta_4}{\sigma^6} = 0 \qquad 5\kappa a_5 \sigma^4 + 6\frac{\beta_5}{\sigma^7} = 0$$

Except for the case $k = 1$, it appears the general equation is of the form $n\kappa a_n \sigma^{n-1} + (n+1)\beta_n/\sigma^{n+2} = 0$.

Now, consider the equations in pairs, the first pair being

$$a_0 - \frac{\beta_0}{\sigma} = 0 \quad \text{and} \quad \frac{\beta_0}{\sigma^2} = 0$$

Being homogeneous, the equations indicate $a_0 = 0$, $\beta_0 = 0$. However, the next pair,

$$a_1 \sigma + \eta \sigma - \frac{\beta_1}{\sigma^2} = 0 \quad \text{and} \quad \eta + \kappa a_1 + 2\frac{\beta_1}{\sigma^3} = 0$$

is not homogeneous and need not have the trivial solution. The remaining pairs are again homogeneous, as we can see from, for example,

$$a_2 \sigma^2 - \frac{\beta_2}{\sigma^3} = 0 \quad \text{and} \quad 2\kappa a_2 \sigma + 3\frac{\beta_2}{\sigma^4} = 0$$

The general pair is

$$a_n \sigma^n - \frac{\beta_n}{\sigma^{n+1}} = 0 \quad \text{and} \quad n\kappa a_n \sigma^{n-1} + \frac{(n+1)\beta_n}{\sigma^{n+2}} = 0$$

with solution $a_n = 0$, $\beta_n = 0$. These equations are homogeneous with nonzero determinant, as can be seen from

$$\begin{vmatrix} \sigma^n & -\dfrac{1}{\sigma^{n+1}} \\ \kappa n \sigma^{n-1} & \dfrac{n+1}{\sigma^{n+2}} \end{vmatrix} = \frac{n+1+\kappa n}{\sigma^2}$$

Hence, it seems as though only a_1 and β_1 are nonzero, with all the remaining coefficients equal to zero.

Thus, the interior and exterior solutions, inheriting this information about the coefficients, indeed become (29.29).

EXERCISES 29.5

1. The BVP of this lesson prescribes at infinity, the potential exterior to a spherical dielectric. Alternatively, suppose the exterior potential were prescribed on a surrounding sphere of radius $\rho = S$. Then, the Dirichlet problem would be $\nabla^2 u = 0$ inside $\rho = \sigma$, $\nabla^2 U = 0$ for $\sigma < \rho < S$, and $U(S, \phi) = F(\phi)$, $u(\sigma, \phi) = U(\sigma, \phi)$, $\kappa u_\rho(\sigma, \phi) = U_\rho(\sigma, \phi)$. Solve this BVP by completing the following steps.

(a) Write $u(\rho, \phi) = \sum_{k=0}^{\infty} a_k \rho^k P_k(\cos \phi)$ and $U(\rho, \phi) = \sum_{k=0}^{\infty}(\alpha_k \rho^k + \beta_k \rho^{-k-1}) P_k(\cos \phi)$ for the interior and exterior potentials, respectively. Compute $F(\phi) = \sum_{k=0}^{\infty} f_k P_k(\cos \phi)$, the Fourier–Legendre expansion of $F(\phi)$. Hence, obtain $f_k = \frac{2k+1}{2} \int_0^\pi F(\phi) P_k(\cos \phi) \sin \phi \, d\phi, k = 0, 1, \ldots$.

(b) Form Equation 1 by equating coefficients of Legendre polynomials in $U(S, \phi) = F(\phi)$.

(c) Form Equation 2 by equating coefficients of Legendre polynomials in $u(\sigma, \phi) = U(\sigma, \phi)$.

(d) Form Equation 3 by equating coefficients of Legendre polynomials in $\kappa u_\rho(\sigma, \phi) = U_\rho(\sigma, \phi)$.

(e) Solve Equations 1–3 for a_k, α_k, and β_k.

(f) Write $v(\rho, \phi) = \begin{cases} u(\rho, \phi) & 0 \le \rho < \sigma \\ U(\rho, \phi) & \sigma \le \rho \le S \end{cases}$ as the solution of the BVP.

In Exercises 2–13:

(a) Obtain, by the results of Exercise 1, $v(\rho, \phi)$.

(b) For the series in part (a), let $V(\rho, \phi)$ be a partial sum for which $V(S, \phi)$ is a reasonable approximation to $F(\phi)$. This approximation can be checked graphically by plotting both $V(S, \phi)$ and $F(\phi)$ as functions of ϕ.

(c) In the polar coordinates $\rho, \omega, -\pi < \omega \le \pi$, obtain (in the yz-plane) level curves (equipotentials) of $\tilde{V}(\rho, \omega) = V(\rho, \frac{\pi}{2} - \omega)$.

(d) Obtain $\hat{V}(y, z) = \tilde{V}(\sqrt{y^2 + z^2}, \arctan \frac{z}{y})$, and solve the differential equations $y'(t) = -\hat{V}_y(y, z)$, $z'(t) = -\hat{V}_z(y, z)$ for the flow lines between the two spheres. Take initial points around the dielectric sphere and produce a graph of flow (field) lines superimposed on the equipotentials. (If you have access to software that will numerically generate a phase portrait in polar coordinates, it would certainly be easier to work in the coordinates ρ and ω.)

2. $\sigma = 1$, $S = 5$, $\kappa = 3$, $F(\phi) = 10\left(\dfrac{\phi}{\pi}\right)$

3. $\sigma = 1$, $S = 5$, $\kappa = 3$, $F(\phi) = 10\left(\dfrac{\phi}{\pi}\right)^2$

4. $\sigma = 1$, $S = 5$, $\kappa = 3$, $F(\phi) = 10\left(\dfrac{\phi}{\pi}\right)^3$

5. $\sigma = 2$, $S = 4$, $\kappa = 5$, $F(\phi) = 10\sin\phi$

6. $\sigma = 2$, $S = 4$, $\kappa = 5$, $F(\phi) = 10\cos^2\phi$

7. $\sigma = 2$, $S = 4$, $\kappa = 5$, $F(\phi) = 10\sin 2\phi$

8. $\sigma = \frac{1}{2}$, $S = 2$, $\kappa = \frac{1}{5}$, $F(\phi) = 8\phi e^{-2\phi}$

9. $\sigma = \frac{1}{2}$, $S = 2$, $\kappa = \frac{1}{5}$, $F(\phi) = 8\phi^2 e^{-2\phi}$

10. $\sigma = \frac{1}{2}$, $S = 2$, $\kappa = \frac{1}{5}$, $F(\phi) = 5\phi(\pi - \phi)$

11. $\sigma = \frac{1}{2}$, $S = 2$, $\kappa = \frac{1}{5}$, $F(\phi) = 5\phi^2(\pi - \phi)$

12. $\sigma = \frac{1}{2}$, $S = 2$, $\kappa = \frac{1}{5}$, $F(\phi) = \phi(\pi - \phi)^3$

13. $\sigma = \frac{1}{2}$, $S = 2$, $\kappa = \frac{1}{5}$, $F(\phi) = \phi^2(\pi - \phi)^3$

14. Repeat Exercise 1 for the case in which $F(\phi)$ is the normal component of ∇U on the surface $\rho = S$. The condition $U_\rho(S, \phi) = F(\phi)$ means a_0 and α_0 are indeterminate. Thus, the potential $v(\rho, \phi)$ is determined up to an additive constant for the Neumann problem. Show, in fact, that if $f_0 = 0$, then $\beta_0 = 0$ and $a_0 = \alpha_0$ and that otherwise, there is no solution to the problem.

In Exercises 15–22:

   (a) Obtain, by the results of Exercise 14, $v(\rho, \phi)$.

(b) For the series in part (a), let $V(\rho, \phi)$ be a partial sum for which $V_\rho(S, \phi)$ is a reasonable approximation to $F(\phi)$. This approximation can be checked graphically by plotting both $V_\rho(S, \phi)$ and $F(\phi)$ as functions of ϕ.

(c) In the polar coordinates $\rho, \omega, -\pi < \omega \le \pi$, obtain (in the yz-plane) level curves (equipotentials) of $\tilde V(\rho, \omega) = V(\rho, \frac{\pi}{2} - \omega)$.

(d) Obtain $\hat V(y, z) = \tilde V(\sqrt{y^2 + z^2}, \arctan\frac{z}{y})$, and solve the differential equations $y'(t) = -\hat V_y(y, z)$, $z'(t) = -\hat V_z(y, z)$ for the flow lines between the two spheres. Take initial points around the dielectric sphere and produce a graph of flow (field) lines superimposed on the equipotentials. (See the comment in part (d) for Exercises B2–13.)

15. $\sigma = 1$, $S = 4$, $\kappa = 2$, $F(\phi) = 3\sin 2\phi$

16. $\sigma = 1$, $S = 4$, $\kappa = 2$, $F(\phi) = 3\cos\phi$

17. $\sigma = \dfrac{2}{3}$, $S = 3$, $\kappa = \dfrac{1}{4}$, $F(\phi) = 3\left(\phi - \dfrac{\pi}{2}\right)$

18. $\sigma = \dfrac{2}{3}$, $S = 3$, $\kappa = \dfrac{1}{4}$, $F(\phi) = 3\left(\phi - \dfrac{\pi}{2}\right)^3$

19. $\sigma = 2$, $S = 5$, $\kappa = 3$, $F(\phi) = 3\phi^2 + 2\phi + 6 - \pi(\frac{3}{2}\pi + 1)$

20. $\sigma = 2$, $S = 5$, $\kappa = 3$, $F(\phi) = \dfrac{12}{\pi^2}\phi^2 - \dfrac{10}{\pi}\phi + \dfrac{24}{\pi^2} - 1$

21. $\sigma = 2$, $S = 5$, $\kappa = 3$,
$$F(\phi) = 2\phi^3 + \left(\dfrac{6}{\pi^2} - 3\pi\right)\phi^2 + \left(\pi^2 - \dfrac{6}{\pi}\right)\phi + \dfrac{12}{\pi^2}$$

22. $\sigma = 2$, $S = 5$, $\kappa = 3$,
$$F(\phi) = 2\phi^3 + \left(\dfrac{12}{\pi^2} - 3\pi\right)\phi^2 + \left(\pi^2 - \dfrac{10}{\pi}\right)\phi + \dfrac{24}{\pi^2} - 1$$

Chapter Review

1. Working in polar coordinates, obtain the separation of variables solution for the Dirichlet problem on a disk of radius σ. The Dirichlet problem consists of Laplace's equation on the disk, and the prescription of $u(\sigma, \theta) = f(\theta)$ on the circumference of the disk.

   (a) State the appropriate BVP. Show how the requirement of smoothness leads to the periodic boundary conditions.

   (b) Write Laplace's equation in polar coordinates, and carry out the separation of variables. Show how the separated boundary conditions lead to the eigenvalues and eigenfunctions.

   (c) Show how the coefficients in the infinite series solution are determined through the medium of a Fourier series for the boundary function.

2. Interpreting the Dirichlet problem in Question 1 as determining the steady-state temperatures on the disk, what is the solution if $f(\theta) = 1$? Before doing any computing, think of the physics!

3. Solve the problem in Question 1 if $\sigma = 2$ and $f(\theta) = \pi^2 - \theta^2$.

4. Use separation of variables to solve the Dirichlet problem for the exterior of the disk of radius σ wherein the value of u is prescribed on the circumference of the disk and Laplace's equation is to hold outside the disk.

5. Answer Question 4 if $\sigma = 2$ and $f(\theta) = \pi^2 - \theta^2$ on the circumference of the disk.

6. Working in polar coordinates, obtain the separation of variables solution for the Neumann problem on a disk of radius σ. The Neumann problem consists of Laplace's equation on the disk, and the prescription of the directional derivative normal to the boundary. The geometry of a disk renders $u_r(\sigma, \theta) = F(\theta)$ as the appropriate Neumann condition. Explain the physical meaning of the additional condition $\int_{-\pi}^{\pi} F(\theta)\, d\theta = 0$, which must be imposed in order for the Neumann problem to have a solution.

7. Working in cylindrical coordinates, obtain the separation of variables solution to the Dirichlet problem in a cylinder of height h and radius σ, with u prescribed as a function of r on top of the cylinder and zero on all other surfaces of the cylinder. Show how separation of variables leads to Bessel's equation as part of a singular Sturm–Liouville eigenvalue problem. Show how the coefficients in the infinite series solution are determined via the medium of a Fourier–Bessel expansion of $f(r)$, the function giving the values on the top surface of the cylinder.

8. Sketch the separation of variables solution for the wave equation on the disk of radius σ having a clamped circumference and an initial displacement that is just a function of the radius r.

9. Sketch the separation of variables solution for the Dirichlet problem in a sphere of radius σ if the value of u prescribed on the surface of the sphere is a function only of the angle between the z-axis and the radius vector. Show how separation of variables leads to Legendre's equation as part of a singular Sturm–Liouville eigenvalue problem. Describe how the eigenvalues and eigenfunctions are determined by Legendre's equation and the requirement of boundedness. Show how the coefficients in the infinite series representing the solution are determined from the Fourier–Legendre expansion of the function giving the surface values.

Chapter 30

Transform Techniques

INTRODUCTION   Boundary value problems posed on infinite domains cannot be solved with series techniques. For such problems, we use integral transforms such as the Laplace or Fourier transform. For example, we use the Laplace transform to solve the heat equation in a semi-infinite rod. It is here that the notion of a partial transform of a function of several variables is developed. And it is here that the roles played by the data and the PDE are examined.

We then introduce the Fourier transform for problems whose spatial domain is the whole real line. The theoretical basis for the Fourier transform is the Fourier integral theorem by which a function is represented as a double integral of itself. The inner integral defines the transform, and the outer, the inverse transform. The operational properties of the Fourier transform and its inverse, as well as convolution, are important here.

The Fourier transform is used to solve the one-dimensional wave equation on an infinite domain, the outcome being *D'Alembert's solution*. It is also used to solve the one-dimensional heat equation on the infinite rod and Laplace's equation on the infinite strip.

Restricted to odd functions, the Fourier transform becomes the Fourier sine transform, which can be used to solve such problems as Laplace's equation on the quarter-plane. Restricted to even functions, the Fourier transform becomes the Fourier cosine transform, which can be used to solve such problems as Laplace's equation on the semi-infinite strip.

30.1 Solution by Laplace Transform

Partial Laplace Transforms

We call the Laplace transform with respect to one of the variables in $u(x, y)$ or $u(x, t)$ the *partial* Laplace transform [43]. To indicate which variable has been "transformed," we adopt the notation

$$L_x[u(x, y)] = U(s, y) = \int_0^\infty u(x, y)e^{-sx}\,dx$$

for the partial transform with respect to x and

$$L_y[u(x, y)] = U(x, s) = \int_0^\infty u(x, y)e^{-sy}\,dy$$

for the partial transform with respect to y. In the first case, the variable y is held constant; while in the second, x is held fixed. Of course, the notation would change accordingly, if instead of y, the second variable were t.

EXAMPLE 30.1 The partial Laplace transforms of the function $u(x, y) = x^2 + 2y$ can both be obtained using the linearity and ordinary operational rules of the Laplace transform. The results are

$$L_x[u(x, y)] = U(s, y) = \frac{2}{s^3} + 2\frac{y}{s} \quad \text{and} \quad L_y[u(x, y)] = U(x, s) = \frac{x^2}{s} + \frac{2}{s^2} \quad \text{❖}$$

PARTIAL LAPLACE TRANSFORM OF PARTIAL DERIVATIVES   Since our intent is the use of the Laplace transform to solve partial differential equations, it behooves us to determine how partial derivatives transform under the partial Laplace transform. In a differential equation containing both $u_{xx}(x, y)$ and $u_{yy}(x, y)$, the partial transform, with respect to x or y, and of either derivative, may occur. Hence,

$$L_x[u_x(x, y)] = sU(s, y) - u(0, y)$$
$$L_x[u_{xx}(x, y)] = s^2 U(s, y) - su(0, y) - u_x(0, y)$$

and

$$L_y[u_y(x, y)] = sU(x, s) - u(x, 0)$$
$$L_y[u_{yy}(x, y)] = s^2 U(x, s) - su(x, 0) - u_y(x, 0)$$

but

$$L_x[u_y(x, y)] = \int_0^\infty u_y(x, y)e^{-sx}\, dx = \frac{\partial}{\partial y} \int_0^\infty u(x, y)e^{-sx}\, dx = \frac{\partial}{\partial y} U(s, y)$$

$$L_x[u_{yy}(x, y)] = \int_0^\infty u_{yy}(x, y)e^{-sx}\, dx = \frac{\partial^2}{\partial y^2} \int_0^\infty u(x, y)e^{-sx}\, dx = \frac{\partial^2}{\partial y^2} U(s, y)$$

and

$$L_y[u_x(x, y)] = \int_0^\infty u_x(x, y)e^{-sy}\, dy = \frac{\partial}{\partial x} \int_0^\infty u(x, y)e^{-sy}\, dy = \frac{\partial}{\partial x} U(x, s)$$

$$L_y[u_{xx}(x, y)] = \int_0^\infty u_{xx}(x, y)e^{-sy}\, dy = \frac{\partial^2}{\partial x^2} \int_0^\infty u(x, y)e^{-sy}\, dy = \frac{\partial^2}{\partial x^2} U(x, s)$$

Applying the partial Laplace transform to a partial differential equation may convert the partial differential equation into an ordinary differential equation in the unknown transform. It may also lead to a more difficult equation!

EXAMPLE 30.2 The Laplace transform can be used to solve the boundary value problem consisting of the partial differential equation $u_y(x, y) + xu_x(x, y) = x$ on the first quadrant ($0 \le x < \infty$, $0 \le y < \infty$) subject to the boundary conditions $u(x, 0) = 0$ and $u(0, y) = 0$.

If x is held fixed, the term $xu_x(x, y)$ will transform most simply as a constant multiple of a derivative. The less desirable alternative is the transform of a derivative multiplied by a variable. Taking the partial Laplace transform with respect to y gives

$$sU(x, s) - u(x, 0) + xU_x(x, s) = \frac{x}{s} \tag{30.1}$$

The first boundary condition gives $u(x, 0) = 0$, so (30.1) becomes

$$sU(x, s) + xU_x(x, s) = \frac{x}{s} \tag{30.2}$$

This is an ordinary differential equation in the function $U(x, s)$, where s is simply a parameter. The independent variable is x and the dependent variable U. Hence, let $U(x, s) = F(x)$, and rewrite (30.2) as

$$xF'(x) + sF(x) = \frac{x}{s} \tag{30.3}$$

The other boundary condition contains $u(0, y)$, so take the partial Laplace transform to obtain $U(0, s) = 0$, equivalent to the initial condition $F(0) = 0$. The differential equation (30.3) is a first-order linear equation, but the initial condition is prescribed at $x = 0$, a singular point for the differential equation. Hence, the general solution, after invocation of the integrating factor $e^{\int (s/x)\,dx} = x^s$, is

$$x^s F(x) = \frac{x^{s+1} + c_1 s(s + 1)}{s(s + 1)} \tag{30.4}$$

Applying the initial condition leads to $c_1 = 0$ and $F(x) = \frac{x}{s(s+1)}$ so that $U(x, s) = F(x)$ and $u(x, y) = x(1 - e^{-y})$, a portion of which is shown in Figure 30.1.

Finally, note that taking the partial Laplace transform with respect to x yields

$$\frac{\partial}{\partial y} U(s, y) - U(s, y) - s \frac{\partial}{\partial s} U(s, y) = \frac{1}{s^2}$$

still a partial differential equation!

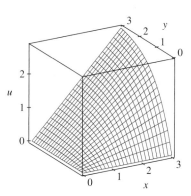

FIGURE 30.1   Solution surface for Example 30.2

THEOREM 30.1

If the transformable function $f(t)$ is bounded ($|f(t)| < M$), then $\lim_{s\to\infty} |F(s)| = 0$.

This follows easily upon the obvious estimate

$$|F(s)| = \left| \int_0^\infty f(t)e^{-st}\,dt \right| \leq \int_0^\infty |f(t)e^{-st}|\,dt < M \int_0^\infty e^{-st}\,dt = \frac{M}{s}$$

Thus, the Laplace transforms of bounded functions will go to zero as s goes to infinity. For example, $f(t) = \sin t$ is bounded by $M = 1$, and $L[f] = \frac{1}{s^2+1}$ clearly goes to zero as s goes to infinity.

Equivalently, if a Laplace transform does *not* go to zero as s goes to infinity, then the function from which it came is *not* bounded. However, avoid the temptation to argue that since $L[\delta(t)] = 1$, which certainly does not go to zero as s goes to infinity, the Dirac delta function $\delta(t)$ is not bounded. The Dirac delta is not a "function," and Theorem 30.1 really does not apply to it.

EXAMPLE 30.3   Solve the heat equation for a semi-infinite rod if the rod is initially cold, the left end remains warm, and the solution remains bounded. Thus, we pose the boundary value problem consisting of the one-dimensional heat equation $u_t(x, t) = \kappa u_{xx}(x, t)$ on the first quadrant $(0 < x < \infty, 0 < t < \infty)$ of the xt-plane, the boundary condition $u(0, t) = A$, and the initial condition $u(x, 0) = 0$. Boundedness of the solution plays a vital role because of Theorem 30.1.

The partial Laplace transform with respect to t gives

$$sU(x, s) - u(x, 0) = \kappa U_{xx}(x, s)$$

which the initial condition $u(x, 0) = 0$ simplifies to

$$sU(x, s) = \kappa U_{xx}(x, s)$$

Again writing $F(x) = U(x, s)$, we have the ODE

$$sF(x) = \kappa F''(x)$$

with general solution

$$F(x) = c_1 e^{x\sqrt{s/\kappa}} + c_2 e^{-x\sqrt{s/\kappa}}$$

For $F(x) = U(x, s)$ to invert back to a bounded function $u(x, t)$, the transform must vanish as s goes to infinity. This suggests setting $c_1 = 0$, giving $F(x) = c_2 e^{-x\sqrt{s/\kappa}}$. The boundary condition $u(0, t) = A$ transforms to $U(0, s) = \frac{A}{s} = F(0)$. Hence, $c_2 = \frac{A}{s}$, so we have

$$U(x, s) = \frac{A}{s} e^{-x\sqrt{s/\kappa}}$$

and using a sufficiently robust table of transforms (or using a suitable computer algebra system), we find

$$u(x, t) = A\left(1 - \operatorname{erf}\left(\frac{x}{2\sqrt{\kappa t}}\right)\right)$$

where $\operatorname{erf}(z)$, the *error function*, is defined as

$$\operatorname{erf}(z) = \frac{2}{\sqrt{\pi}} \int_0^z e^{-\sigma^2} d\sigma$$

Since $\operatorname{erf}(-z) = -\operatorname{erf}(z)$, the error function is odd and $\lim_{z \to \infty} \operatorname{erf}(z) = 1$, as seen in Figure 30.2. For $z > 0$, the value of $\operatorname{erf}(z)$ is the shaded area in Figure 30.3, a graph of $f(\sigma) = \frac{2}{\pi} e^{-\sigma^2}$.

Finally, if in $u(x, t)$ we set $A = 2$ and $\kappa = 1$, we get the function

$$u(x, t) = 2\left(1 - \operatorname{erf}\left(\frac{x}{2\sqrt{t}}\right)\right) \tag{30.5}$$

graphed as the surface in Figure 30.4. The instantaneous rise in temperature everywhere is suggested both in the graph of the solution surface and in the animation of the plane sections $t = $ constant available in the accompanying Maple worksheet.

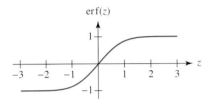

FIGURE 30.2 The error function, $\operatorname{erf}(z)$

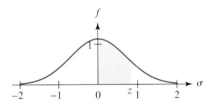

FIGURE 30.3 Shaded area represents value of $\operatorname{erf}(z)$, $z > 0$

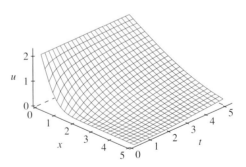

FIGURE 30.4 Solution surface for Example 30.3

For every x, the temperature instantly rises, no matter how small t might be. For fixed x and small enough t the argument of the error function in $u(x, t)$ is large. In (30.5), replacing $\frac{x}{2\sqrt{t}}$ with z and expanding for large z gives

$$2(1 - \operatorname{erf}(z)) = \frac{e^{-z^2}}{\sqrt{\pi}}\left(\frac{2}{z} - \frac{1}{z^3}\right) + O\left(z^{-5}e^{-z^2}\right) \tag{30.6}$$

If $x = 10$, then $z = \frac{5}{\sqrt{t}}$ and (30.6) becomes

$$u(10, t) = \frac{\sqrt{t}e^{-25/t}}{5\sqrt{\pi}}\left(2 - \frac{t}{25}\right) + O\left(t^{5/2}e^{-25/t}\right) \tag{30.7}$$

At $t = 0.01$, the first term on the right in (30.7) evaluates to 0.414×10^{-1087}, a small but positive temperature.

The limiting temperature at any given x is $u = 2$ since $\lim_{t \to \infty} 2(1 - \operatorname{erf}(\frac{x}{2\sqrt{t}})) = 2$. Every point in the rod experiences an instantaneous rise in temperature, and the whole rod will eventually be at temperature $u = 2$.   ❖

EXERCISES 30.1–Part A

A1. Verify that (30.4) is the general solution of (30.3).

If, in Exercises A2–6, a and b are constants:

  (a) Show that $L_x[f_x(x, y)] = sF(s, y) - f(0, y)$.

  (b) Show that $L_y[f_x(x, y)] = F_x(x, s)$.

  (c) Show that $L_x[f_y(x, y)] = F_y(s, y)$.

  (d) Show that $L_y[f_y(x, y)] = sF(x, s) - f(x, 0)$.

  (e) Show that $L_x[f_{xx}(x, y)] = s^2 F(s, y) - sf(0, y) - f_x(0, y)$.

  (f) Show that $L_x[f_{yy}(x, y)] = F_{yy}(s, y)$.

  (g) Show that $L_y[f_{xx}(x, y)] = F_{xx}(x, s)$.

  (h) Show that $L_y[f_{yy}(x, y)] = s^2 F(x, s) - sf(x, 0) - f_y(x, 0)$.

A2. $f(x, y) = a \sin x + be^{-y}$ **A3.** $f(x, y) = ae^{2x} \sin by$

A4. $f(x, y) = ax \sin y + by^2 \cos x$ **A5.** $f(x, y) = ax^3 y^2 - be^{xy}$

A6. $f(x, y) = g(ax - by)$, where $g(x)$ is twice differentiable

EXERCISES 30.1–Part B

B1. How accurate is (30.7) when it is used to make the approximation $u(10, 0.01) \doteq 0.414 \times 10^{-1087}$? *Hint:* In the error term in (30.6), substitute $z = \frac{5}{\sqrt{0.01}}$.

In Exercises B2–11:

  (a) Solve by Laplace transform.

  (b) Graph the solution surface.

B2. $xu_y + 3u_x = 0, u(x, 0) = 0, u(0, y) = y^2$

B3. $xu_y + u_x = 0, u(x, 0) = 1, u(0, y) = y^3$

B4. $xu_y + 5u_x = 0, u(x, 0) = 7, u(0, y) = \sin y$

B5. $xu_y + 2u_x = 0, u(x, 0) = 4, u(0, y) = \cos y$

B6. $2xu_y + 7u_x = 3x, u(x, 0) = 2, u(0, y) = y$

B7. $2xu_y + 3u_x = 5x, u(x, 0) = 4, u(0, y) = \sin y$

B8. $5xu_y + 2u_x = 4x, u(x, 0) = 1, u(0, y) = \cos y$

B9. $3u_y + yu_x = 0, u(0, y) = 1, u(x, 0) = e^{-x}$

B10. $5u_y + yu_x = 0, u(0, y) = 2, u(x, 0) = x \sin x$

B11. $4u_y + yu_x = 0, u(0, y) = 3, u(x, 0) = e^x \cos 2x$

B12. For the BVP $u_t = u_{xx}, 0 \le x < \infty, t > 0, u(x, 0) = 0$, $u(0, t) = 1$, the heat equation on a semi-infinite rod:

  (a) Use the Laplace transform to obtain a bounded solution.

  (b) Plot the solution surface.

  (c) Plot temperature profiles at several increasing times, or create an animation if suitable software is available.

B13. For the BVP $u_t = 2u_{xx}, 0 \le x < \infty, t > 0, u(x, 0) = 0$, $u(0, t) = 10 \sin t$, the heat equation on a semi-infinite rod,

use the Laplace transform and the following steps to obtain the solution $u(x, t) = \frac{5x}{\sqrt{2\pi}} \int_0^t e^{-x^2/8\tau} \sin(t - \tau)\tau^{-3/2} d\tau$.

  (a) Apply L_t to the heat equation, taking the initial data as $u(x, 0) = 0$, obtaining an ODE in $U(x, s)$.

  (b) Solve the ODE along with the boundary conditions $U(0, s) = 0, U(p, s) = 0, p > 0$.

  (c) Obtain $u(x, t)$ by letting $p \to \infty$ in $U(x, s)$ and inverting.

  (d) Numerically evaluate $u(5, t)$ for a sequence of times t in an attempt to discover if the temperatures at $x = 5$ oscillate, increase, or remain zero.

B14. For the BVP $u_t = u_{xx}, 0 \le x < \infty, t > 0, u(x, 0) = 0$, $u_x(0, t) = -1$, the heat equation on a semi-infinite rod:

  (a) Apply L_t to the heat equation, taking the initial data as $u(x, 0) = 0$, obtaining an ODE in $U(x, s)$.

  (b) Solve the ODE along with the boundary conditions $U'(0, s) = -\frac{1}{s}, U(p, s) = 0, p > 0$.

  (c) Obtain $u(x, t)$ by letting $p \to \infty$ in $U(x, s)$ and inverting.

  (d) Plot the solution surface.

  (e) Plot temperature profiles at several increasing times, or create an animation if suitable software is available.

  (f) Since heat is being continuously pumped into the left end of the rod, what is the ultimate fate of the rod's temperature?

B15. Repeat Exercise B14 for the boundary condition $u(x, 0) = 1$, $u_x(0, t) = 1$.

B16. For the BVP $u_t = u_{xx}, 0 \le x < \infty, t > 0, u(x, 0) = 1$, $u_x(0, t) = \frac{1}{2}u(0, t)$, the heat equation on a semi-infinite rod whose left end is subject to a homogeneous Robin condition (after the French mathematician Victor G. Robin, 1855–1897):

(a) Apply L_t to the heat equation, taking the initial data as $u(x, 0) = 1$, obtaining an ODE in $U(x, s)$.

(b) Solve the ODE along with the boundary condition $U(p, s) = 0, p > 0$.

(c) Let $p \to \infty$ in $U(x, s)$, leaving one constant of integration that is determined by applying $L_x[u_x(0, t) = \frac{1}{2}u(0, t)]$.

(d) Invert $U(x, s)$ using $L^{-1}\left[ae^{-k\sqrt{s}}/s(a + \sqrt{s})\right] =$ $-e^{ak}e^{a^2 t}\text{erfc}\left(a\sqrt{t} + \frac{k}{2\sqrt{t}}\right) + \text{erfc}\frac{k}{2\sqrt{t}}$, where $\text{erfc}(z) = 1 - \text{erf}(z)$. (This transform is entry 29.3.89, p. 1027, in [1].)

(e) Plot the solution surface.

(f) Plot temperature profiles at several increasing times, or create an animation if suitable software is available.

B17. For the BVP $u_t = u_{xx}, 0 \le x < \infty, t > 0, u(x, 0) = 0$, $u(0, t) = e^{-t}$, the heat equation on a semi-infinite rod, use the Laplace transform and numeric evaluation of the convolution integral to obtain a graph of $u(1, t)$.

B18. For the BVP $u_t = \frac{1}{16}u_{xx}, 0 \le x < \infty, t > 0, u(x, 0) = 0$, $u_x(0, t) = -10\delta(t)$, the heat equation on a semi-infinite rod with an impulsive flow of heat at the left end:

(a) Use the Laplace transform to obtain a bounded solution.

(b) Plot the solution surface.

(c) Plot temperature profiles at several increasing times, or create an animation if suitable software is available. Note that the initial bulge of raised temperatures disperses as the heat progresses along the rod.

B19. For the BVP $u_t = \frac{1}{16}u_{xx}, 0 \le x \le 1, t > 0, u(x, 0) = 0$, $u(0, t) = u(1, t) = 1$, the heat equation on a finite rod with both ends kept at the same temperature:

(a) Apply L_t to the heat equation and solve the resulting ODE to obtain $U(x, s)$.

(b) Expand $U(x, s)$ in powers of $e^{-4\sqrt{s}}$, and invert termwise to obtain a partial sum of a series solution to the BVP.

(c) Plot the solution surface. Note that the initial region of lowered temperatures disperses as the heat progresses along the rod.

(d) Determine the time at which the lowest temperature in the rod (clearly at $x = \frac{1}{2}$) is within 10% of the steady-state temperature of 1.

B20. For the BVP $u_{tt} = c^2 u_{xx}, 0 \le x < \infty, t > 0, u(x, 0) =$ $u_t(x, 0) = 0, u(0, t) = \begin{cases} \sin t & 0 \le t < \pi \\ 0 & t \ge \pi \end{cases}$, the wave equation on a semi-infinite string:

(a) Use the Laplace transform to obtain a bounded solution.

(b) Take $c = 1$ and plot the solution surface.

(c) Animate the motion in the string, or graph snapshots of the solution for a succession of times t.

In Exercises B21–B23, use the Laplace transform to obtain a bounded solution of $u_{tt} = u_{xx}, 0 \le x < \infty, t > 0, u(x, 0) = f(x), u_t(x, 0) = 0$, $u(0, t) = 0, \lim_{x \to \infty} u(x, t) = 0$, the BVP for the wave equation on a semi-infinite string. In particular:

(a) Apply L_t to the wave equation, using $u(x, 0) = f(x)$ and $u_t(x, 0) = 0$ as the initial data.

(b) Solve the resulting ODE, using $U(0, s) = 0$ and $U(p, s) = 0, p > 0$, as boundary data.

(c) Obtain $U(x, s)$ by letting $p \to \infty$.

(d) Invert to obtain $u(x, t)$.

(e) Plot the solution surface.

(f) Graph the solution for several increasing values of t, or animate the displacements in the string if suitable software is available.

B21. $f(x) = e^{-x}$ **B22.** $f(x) = xe^{-x}$ **B23.** $f(x) = x^2 e^{-x}$

B24. For the BVP $u_{tt} = u_{xx}, 0 \le x \le 1, t > 0, u(x, 0) = u_t(x, 0) = 0$, $u_x(1, t) = 1$, modeling an elastic bar of length 1 with left end fixed and right end pulled by a constant force:

(a) Obtain $U(x, s)$, the Laplace transform $L_t[u(x, t)]$.

(b) Expand $U(x, s)$ in powers of e^{-2s}, invert termwise, and obtain a partial sum of a series solution.

(c) Plot the solution surface.

(d) If appropriate software is available, animate the displacements in the bar.

(e) If appropriate software is available, animate the actual motion of the bar.

(f) Explain why the bar supports oscillations if there is a constant force pulling on the right end.

B25. For the BVP $u_{tt} = u_{xx}, 0 \le x \le 1, t > 0, u(x, 0) = u_t(x, 0) = 0$, $u_x(1, t) = \frac{1}{5}\sin 5t$, modeling an elastic bar of length 1 with left end fixed and right end subjected to a periodic force:

(a) Obtain $U(x, s)$, the Laplace transform $L_t[u(x, t)]$.

(b) Expand $U(x, s)$ in powers of e^{-2s}, invert termwise, and obtain a partial sum of a series solution.

(c) Plot the solution surface.

(d) If appropriate software is available, animate the displacements in the bar.

(e) Plot $\frac{1}{2} + u(\frac{1}{2}, t)$ to visualize the actual motion of the center of the bar.

B26. For the BVP $u_{tt} = u_{xx}, 0 \le x \le 1, t > 0, u(x, 0) = u_t(x, 0) = 0$, $u_x(1, t) = -\frac{1}{10}\delta(t)$, modeling an elastic bar of length 1 with left end fixed and right end subjected to an impulsive force:

(a) Obtain $U(x, s)$, the Laplace transform $L_t[u(x, t)]$.

(b) Expand $U(x, s)$ in powers of e^{-2s}, invert termwise, and obtain a partial sum of a series solution.

(c) Plot the solution surface.

(d) If appropriate software is available, animate the displacements in the bar.

(e) Plot $\frac{1}{2} + u(\frac{1}{2}, t)$ to visualize the actual motion of the center of the bar.

B27. The BVP $u_{tt} = u_{xx}, 0 \le x < \infty, t > 0, u(x, 0) = 0,$ $u_t(x, 0) = -\frac{1}{2}, \lim_{x \to \infty} u_x(x, t) = 0$ models the motion of a semi-infinite elastic bar moving right to left with velocity $-\frac{1}{2}$ and hitting a solid wall at $x = 0$. To solve this problem:

(a) Apply L_t to the wave equation, using $u(x, 0) = 0$ and $u_t(x, 0) = -\frac{1}{2}$ as the initial data.

(b) Solve the resulting ODE, using $U(0, s) = 0$ and $U'(p, s) = 0, p > 0$, as boundary data.

(c) Obtain $U(x, s)$ by letting $p \to \infty$.

(d) Invert to obtain $u(x, t)$.

(e) Plot the solution surface.

(f) Plot $x + u(x, t)$ for several values of x.

(g) Describe (in words) what happens to the bar.

(h) Change the initial velocity to -1 and again solve, describing in words what now happens to the bar. Is this physically reasonable?

30.2 The Fourier Integral Theorem

Exponential Form of the Fourier Integral Theorem

The remarkable Fourier integral theorem expresses, under suitable conditions, the function $f(x)$ as a double integral with itself. There are two equivalent ways of writing the exponential form of this double integral and several ways of writing the trigonometric forms. With $i = \sqrt{-1}$, we have, for the exponential form of the theorem,

THEOREM 30.2

1. $f(x)$ and $f'(x)$ are piecewise continuous in every finite interval

2. $\int_{-\infty}^{\infty} |f(x)| \, dx < \infty$

$$\Longrightarrow f(x) = \frac{1}{2\pi} \int_{-\infty}^{\infty} \int_{-\infty}^{\infty} e^{i\alpha x} f(\beta) e^{-i\alpha\beta} \, d\beta \, d\alpha$$

The outer integral is really a *Cauchy principal value* (for further background, see [65]), and should be interpreted as

$$\lim_{\lambda \to \infty} \int_{-\lambda}^{\lambda} e^{i\alpha x} \left[\int_{-\infty}^{\infty} f(\beta) e^{-i\alpha\beta} \, d\beta \right] d\alpha \qquad (30.8)$$

The inner integral, that performed first in (30.8), contains the exponential with the minus sign, a convention followed by texts such as [51] and [66]. Texts such as [87], [84], and [100] reverse the signs on the exponential terms, so an alternate but equivalent statement of the Fourier integral theorem reads

$$f(x) = \frac{1}{2\pi} \int_{-\infty}^{\infty} \int_{-\infty}^{\infty} e^{-i\alpha x} f(\beta) e^{i\alpha\beta} \, d\beta \, d\alpha \qquad (30.9)$$

where now, the *outer* integral has the exponential term with the minus sign. Some texts such as [51] combine the exponentials, so the theorem reads

$$f(x) = \frac{1}{2\pi} \int_{-\infty}^{\infty} \int_{-\infty}^{\infty} f(\beta) e^{i\alpha(x-\beta)} \, d\beta \, d\alpha$$

At points of discontinuity, the double integral converges to $\frac{1}{2}[f(x+0) + f(x-0)]$, the average of the left-hand and right-hand limits at x. Finally, a function $f(x)$ satisfying the second hypothesis is generally called an *absolutely integrable* function.

EXAMPLE 30.4 Using the function

$$f(x) = e^{-|x|} = \begin{cases} e^x & x < 0 \\ e^{-x} & x \geq 0 \end{cases}$$

whose graph is given in Figure 30.5, verify the Fourier integral theorem.

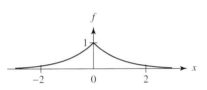

The function is continuous, and its derivative is piecewise continuous. It is also absolutely integrable, since

$$\int_{-\infty}^{\infty} |e^{-|x|}|\, dx = \int_{-\infty}^{\infty} e^{-|x|}\, dx = 2\int_{0}^{\infty} e^{-x}\, dx = 2$$

FIGURE 30.5 The function $f(x)$ in Example 30.4

The inner integral in the Fourier integral theorem is

$$\int_{-\infty}^{\infty} e^{-|\beta|} e^{-i\alpha\beta}\, d\beta = \frac{2}{1+\alpha^2} \tag{30.10}$$

The outer integral is

$$\frac{1}{2\pi}\int_{-\infty}^{\infty} 2\frac{e^{i\alpha x}}{1+\alpha^2}\, d\alpha = \begin{cases} e^x & x < 0 \\ e^{-x} & x \geq 0 \end{cases} \tag{30.11}$$

after an appropriate tussle with a table of integrals or a consultation with a computer algebra system. ❖

EQUIVALENCE OF EXPONENTIAL FORMS The two forms of the Fourier integral theorem found in the literature are equivalent under the change of variables $a = -\alpha$. Indeed, writing the inner integral as

$$\int_{-\infty}^{\infty} f(\beta)e^{-i\alpha\beta}\, d\beta$$

and applying this change of variables, we get

$$F(\beta) = \int_{-\infty}^{\infty} f(\beta)e^{ia\beta}\, d\beta$$

The outer integral is then

$$\int_{-\infty}^{\infty} F(\beta)e^{i\alpha x}\, d\alpha$$

Making the same change of variables in the outer integral gives

$$\int_{-\infty}^{\infty} F(\beta)e^{-ixa}\, da$$

Since a is just a "dummy" variable of integration, it can be replaced with the letter α, so the outer integral reads

$$\int_{-\infty}^{\infty} F(\beta)e^{-i\alpha x}\, da$$

Inserting the inner integral for $F(\beta)$ (with a replaced by α), we get

$$\int_{-\infty}^{\infty}\int_{-\infty}^{\infty} f(\beta)e^{i\alpha\beta}\, d\beta\, e^{-i\alpha x}\, d\alpha$$

which, except for some simple rearrangement and a factor of $\frac{1}{2\pi}$, is (30.9), the alternate exponential form of the Fourier integral theorem.

Trigonometric Form of the Fourier Integral Theorem

The trigonometric equivalent of $f(x) = \frac{1}{2\pi} \int_{-\infty}^{\infty} \int_{-\infty}^{\infty} e^{i\alpha x} f(\beta) e^{-i\alpha \beta} \, d\beta \, d\alpha$ is

$$f(x) = \frac{1}{\pi} \int_0^{\infty} \int_{-\infty}^{\infty} f(\beta) \cos(\alpha(\beta - x)) \, d\beta \, d\alpha \tag{30.12}$$

obtained by writing the Cauchy principal value in the exponential form

$$\frac{1}{2\pi} \lim_{\lambda \to \infty} \int_{-\lambda}^{\lambda} \int_{-\infty}^{\infty} f(\beta) e^{i\alpha(x-\beta)} \, d\beta \, d\alpha$$

as

$$\frac{1}{2\pi} \lim_{\lambda \to \infty} \left(\int_{-\lambda}^{0} \int_{-\infty}^{\infty} f(\beta) e^{i\alpha(x-\beta)} \, d\beta \, d\alpha + \int_{0}^{\lambda} \int_{-\infty}^{\infty} f(\beta) e^{i\alpha(x-\beta)} \, d\beta \, d\alpha \right)$$

In the first integral, changing variables from α to $-\alpha$ gives

$$\frac{1}{2\pi} \left(\lim_{\lambda \to \infty} \int_{0}^{\lambda} \int_{-\infty}^{\infty} f(\beta) e^{-i\alpha(x-\beta)} \, d\beta \, d\alpha + \int_{0}^{\lambda} \int_{-\infty}^{\infty} f(\beta) e^{i\alpha(x-\beta)} \, d\beta \, d\alpha \right)$$

which can be written as

$$\frac{1}{\pi} \int_{0}^{\lambda} \int_{-\infty}^{\infty} f(\beta) \left[\frac{e^{i\alpha(x-\beta)} + e^{-i\alpha(x-\beta)}}{2} \right] d\beta \, d\alpha$$

and, hence, as

$$\frac{1}{\pi} \int_{0}^{\infty} \int_{-\infty}^{\infty} f(\beta) \cos(\alpha(x - \beta)) \, d\beta \, d\alpha = \frac{1}{\pi} \int_{0}^{\infty} \int_{-\infty}^{\infty} f(\beta) \cos(\alpha(\beta - x)) \, d\beta \, d\alpha$$

If, in addition, the cosine is expanded to $\cos \alpha\beta \cos \alpha x + \sin \alpha\beta \sin \alpha x$, then an alternate trigonometric form is

$$f(x) = \frac{1}{\pi} \int_{0}^{\infty} \left(\int_{-\infty}^{\infty} f(\beta) \cos \alpha\beta \, d\beta \right) \cos \alpha x \, d\alpha$$

$$+ \frac{1}{\pi} \int_{0}^{\infty} \left(\int_{-\infty}^{\infty} f(\beta) \sin \alpha\beta \, d\beta \right) \sin \alpha x \, d\alpha$$

which is typically written as

$$f(x) = \frac{1}{\pi} \int_{0}^{\infty} A(\alpha) \cos \alpha x \, d\alpha + \frac{1}{\pi} \int_{0}^{\infty} B(\alpha) \sin \alpha x \, d\alpha \tag{30.13}$$

where

$$A(\alpha) = \int_{-\infty}^{\infty} f(x) \cos \alpha x \, dx \quad \text{and} \quad B(\alpha) = \int_{-\infty}^{\infty} f(x) \sin \alpha x \, dx \tag{30.14}$$

Notice how β, the variable of integration in $A(\alpha)$ and $B(\alpha)$, has been switched to x. Note, too, that there is no uniformity on the placement of the factor $\frac{1}{\pi}$. Texts such as [100] and [66] follow our usage, whereas texts such as [84] and [51] put the factor with $A(\alpha)$ and $B(\alpha)$.

◆◆ **EXAMPLE 30.5**   Use the function

$$f(x) = \begin{cases} 0 & x < -1 \\ 1 & -1 \le x \le 1 \\ 0 & x > 1 \end{cases}$$

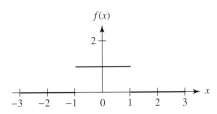

$f(x)$

FIGURE 30.6   The function $f(x)$ in Example 30.5

graphed in Figure 30.6, to verify (30.12). Since $f(x)$ is nonzero only in the interval $[-1, 1]$, we obtain

$$\frac{1}{\pi} \int_0^\infty \int_{-1}^1 \cos(\alpha(\beta - x)) \, d\beta \, d\alpha = \frac{1}{\pi} \int_0^\infty \frac{1}{\alpha} [\sin(\alpha(1 + x)) + \sin(\alpha(1 - x))] \, d\alpha \tag{30.15}$$

The tabulated integral

$$\int_0^\infty \frac{\sin \lambda u}{u} \, du = \begin{cases} -\dfrac{\pi}{2} & \lambda < 0 \\ 0 & \lambda = 0 \\ \dfrac{\pi}{2} & \lambda > 0 \end{cases} \tag{30.16}$$

can be used to evaluate the remaining integral in (30.15). If $x \neq \pm 1$, then (30.15) is $\frac{\pi}{\pi} = 1$. At $x = \pm 1$, the integrand on the right in (30.15) is $\frac{\sin 2\alpha}{\alpha}$, so the Fourier integral converges to $\frac{1}{2}$, the average value across each of the jumps at those points.   ❖

◆◆ **EXAMPLE 30.6**   Compute $A(\alpha)$ and $B(\alpha)$ for the function $f(x)$ in Example 30.5.

Since $f(x)$ is nonzero only in the interval $[-1, 1]$, we obtain

$$A = \int_{-1}^1 \cos \alpha \beta \, d\beta = \frac{2}{\alpha} \sin \alpha \quad \text{and} \quad B = \int_{-1}^1 \sin \alpha \beta \, d\beta = 0 \tag{30.17}$$

The Fourier integral representation of $f(x)$ is then

$$f(x) = \frac{1}{\pi} \left[\int_0^\infty A(\alpha) \cos \alpha x \, d\alpha + \int_0^\infty B(\alpha) \sin \alpha x \, d\alpha \right]$$

$$= \frac{1}{\pi} \int_0^\infty \frac{2}{\alpha} \sin \alpha \cos \alpha x \, d\alpha$$

$$= \frac{1}{\pi} \int_0^\infty \frac{1}{\alpha} [\sin(\alpha(1 + x)) + \sin(\alpha(1 - x))] \, d\alpha$$

the integral obtained in Example 30.5.   ❖

Solving BVPs with the Fourier Integral Theorem

Heuristic reasoning and the Fourier integral theorem lead to a solution of Laplace's equation on the upper half-plane if the potential, prescribed on the x-axis, remains bounded. Thus, Laplace's equation holds for $y > 0$, $u(x, y)$ is bounded, and $u(x, 0) = f(x)$.

Were the partial Laplace transform taken with respect to y, the result would be

$$\frac{d^2}{dx^2} U(x, s) + s^2 U(x, s) - s u(x, 0) - u_y(x, 0) = 0 \tag{30.18}$$

The boundary condition $u(x, 0) = 0$ eliminates the third term on the left, but the fourth brings this idea up short. Knowing the potential function on the x-axis is not enough information for us to determine the normal derivative along the boundary. That would require knowledge of the potential inside the region, but we don't have that. Hence, we cannot solve this problem with the Laplace transform.

Instead, we seek solutions of the form $u(x, y) = e^{px}e^{qy}$. The laplacian of such a solution, set equal to zero, is $(p^2 + q^2)e^{px}e^{qy} = 0$ or $p^2 + q^2 = 0$, from which we conclude $p = \pm iq$. Thus, u will be either $e^{iqx}e^{qy}$ or $e^{-iqx}e^{qy}$. Boundedness for $y > 0$ suggests we take $q < 0$, which we accomplish by writing $q = -\alpha$, with $\alpha > 0$. Then, u will be either of the forms $e^{i\alpha x}e^{-\alpha y}$ or $e^{-i\alpha x}e^{-\alpha y}$, which we choose to write as $\cos(\alpha x)e^{-\alpha y}$ and $\sin(\alpha x)e^{-\alpha y}$. The most general potential $u(x, y)$ we can construct from these functions would be the linear combination

$$u_\alpha(x, y) = A(\alpha)\cos(\alpha x)e^{-\alpha y} + B(\alpha)\sin(\alpha x)e^{-\alpha y}$$

The sum of all such solutions $u_\alpha(x, y)$ is achieved by integrating over α, resulting in

$$u(x, y) = \frac{1}{\pi}\int_0^\infty A(\alpha)\cos(\alpha x)e^{-\alpha y}\, d\alpha + \frac{1}{\pi}\int_0^\infty B(\alpha)\sin(\alpha x)e^{-\alpha y}\, d\alpha$$

Applying the boundary condition $u(x, 0) = f(x)$ means we must have

$$f(x) = \frac{1}{\pi}\int_0^\infty A(\alpha)\cos\alpha x\, d\alpha + \frac{1}{\pi}\int_0^\infty B(\alpha)\sin\alpha x\, d\alpha$$

which, by the Fourier integral theorem, we can have if we set

$$A(\alpha) = \int_{-\infty}^\infty f(x)\cos\alpha x\, dx \quad \text{and} \quad B(\alpha) = \int_{-\infty}^\infty f(x)\sin\alpha x\, dx$$

We can then write

$$
\begin{aligned}
u(x, y) &= \frac{1}{\pi}\int_0^\infty \left(\int_{-\infty}^\infty f(\beta)\cos\alpha\beta\, d\beta\right)\cos(\alpha x)e^{-\alpha y}\, d\alpha \\
&\quad + \frac{1}{\pi}\int_0^\infty \left(\int_{-\infty}^\infty f(\beta)\sin\alpha\beta\, d\beta\right)\sin(\alpha x)e^{-\alpha y}\, d\alpha \\
&= \frac{1}{\pi}\int_0^\infty \int_{-\infty}^\infty f(\beta)[\cos\alpha\beta\cos\alpha x - \sin\alpha\beta\sin\alpha x]e^{-\alpha y}\, d\beta\, d\alpha \\
&= \frac{1}{\pi}\int_0^\infty \int_{-\infty}^\infty f(\beta)\cos(\alpha(\beta - x))e^{-\alpha y}\, d\beta\, d\alpha
\end{aligned}
$$

(30.19)

thereby achieving a closed-form solution for this boundary value problem!

EXAMPLE 30.7   It is possible to evaluate (30.19) and determine the behavior of the potential $u(x, y)$ when $f(x)$ is the function

$$f(x) = \begin{cases} 0 & x < 0 \\ 1 & x > 0 \end{cases}$$

Since $f(x)$ is nonzero only when $x > 0$, we obtain

$$u(x, y) = \frac{1}{\pi}\int_0^\infty \int_0^\infty \cos(\alpha(\beta - x))e^{-\alpha y}\, d\beta\, d\alpha$$

A change of variables from β to $\sigma = \beta - x$ in the inner integral leads to

$$
\int_0^\infty \int_{-x}^\infty \cos(\alpha\sigma)e^{-\alpha y}\, d\sigma\, d\alpha = \int_{-x}^\infty \int_0^\infty \cos(\alpha\sigma)e^{-\alpha y}\, d\alpha\, d\sigma
$$

$$
= \int_{-x}^\infty \frac{y}{y^2 + \sigma^2}\, d\sigma = \frac{1}{2} + \frac{1}{\pi}\arctan\frac{x}{y}
$$

(30.20)

where reversing the order of integration gives the middle integral.

To verify that $u(x, y)$ satisfies the boundary condition $u(x, 0) = f(x)$, approach the x-axis by a limiting processes from above, first, on the positive x-axis, obtaining $\lim_{y \to 0^+} u = 1$, and then on the negative x-axis, obtaining $\lim_{y \to 0^+} u = 0$.

The equipotentials, radial lines emanating from the origin, are shown in Figure 30.7. The electric field is

$$-\nabla u = \mathbf{E} = \frac{1}{\pi(x^2 + y^2)}(-y\mathbf{i} + x\mathbf{j})$$

The field lines, the flow lines for the gradient field, are the solution of the differential equations

$$x'(t) = -\frac{y}{\pi(x^2 + y^2)} \quad \text{and} \quad y'(t) = \frac{x}{\pi(x^2 + y^2)}$$

An exact solution for the field lines can be obtained using the derivative formula

$$\frac{dy}{dx} = \frac{\frac{dy}{dt}}{\frac{dx}{dt}} = \frac{E_2}{E_1} = -\frac{x}{y}$$

leading to $x^2 + y^2 = r^2$, the equation of a circle having radius r and center at the origin. In Figure 30.7, the field arrows and the flow lines are superimposed on the equipotentials. ❖

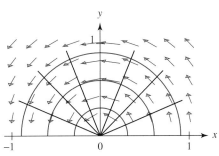

FIGURE 30.7   Example 30.7: Equipotentials (in black), field lines (in color), and electric field vectors

EXERCISES 30.2–Part A

A1. Verify the integration in (30.10).

A2. Verify the integration in (30.15).

A3. Use (30.16) to verify that (30.15) converges to $\begin{cases} 0 & |x| > 1 \\ \frac{1}{2} & x = \pm 1 \\ 1 & |x| < 1 \end{cases}$

A4. Verify the integrations in (30.17).

A5. Verify the Laplace transform in (30.18).

A6. Verify the integration in (30.20).

EXERCISES 30.2–Part B

B1. Verify the integration in (30.11).

In Exercises B2–13, where H(x) is the function Heaviside(x):

 (a) Sketch $f(x)$ and select x_0, an x for which $f(x_0) \neq 0$.

 (b) At x_0 verify the exponential form of the Fourier integral theorem.

 (c) At x_0 verify (30.12), the trigonometric form of the Fourier integral theorem.

 (d) Compute $A(\alpha)$ and $B(\alpha)$ as defined in (30.14), then at x_0 verify $f(x) = \frac{1}{\pi}\int_0^\infty A(\alpha) \cos \alpha x \, d\alpha + \frac{1}{\pi}\int_0^\infty B(\alpha) \sin \alpha x \, d\alpha$.

B2. $f(x) = xe^{-x}H(x)$ **B3.** $f(x) = x^2e^{-x}H(x)$

B4. $f(x) = e^{-x}\cos x H(x)$ **B5.** $f(x) = \sin x(H(x) - H(x - \pi))$

B6. $f(x) = \cos x(H(x) - H(x - 2\pi))$

B7. $f(x) = x(H(x) - H(x - 1))$

B8. $f(x) = x^3(H(x) - H(x - 1))$

B9. $f(x) = (1 - x)(H(x) - H(x - 1))$

B10. $f(x) = H(x + 1) - H(x - 2)$

B11. $f(x) = x^2(H(x + 1) - H(x - 1))$

B12. $f(x) = (1 - |x|)(H(x + 1) - H(x - 1))$

B13. $f(x) = (1 - x^2)(H(x + 1) - H(x - 1))$

In Exercises B14–23, $f(x)$ is given, thereby determining the BVP consisting of $\nabla^2 u = 0$ on the upper half-plane, where $u(x, y)$ is bounded and satisfies $u(x, 0) = f(x)$. For each such BVP:

(a) Sketch $f(x)$.

(b) Obtain the solution using the Fourier integral theorem and (30.19).

(c) Interchange the order of integration and evaluate the resulting integrals.

(d) Plot the solution surface.

(e) Obtain a contour plot.

(f) Obtain, and superimpose on the contour plot, a graph of the flow lines.

B14. $f(x) = x(H(x) - H(x - 1))$

B15. $f(x) = x^2(H(x) - H(x - 1))$

B16. $f(x) = x(1 - x)(H(x) - H(x - 1))$

B17. $f(x) = x(1 - x)^3(H(x) - H(x - 1))$

B18. $f(x) = x^2(1 - x)(H(x) - H(x - 1))$

B19. $f(x) = (1 - x)(H(x) - H(x - 1))$

B20. $f(x) = (1 - x^2)(H(x + 1) - H(x - 1))$

B21. $f(x) = (1 - |x|)(H(x + 1) - H(x - 1))$

B22. $f(x) = H(x + 1) - H(x - 3)$

B23. $f(x) = \begin{cases} 0 & x < -1 \\ x + 1 & -1 \le x < 0 \\ 1 & 0 \le x < 2 \\ 3 - x & 2 \le x < 3 \\ 0 & x \ge 3 \end{cases}$

B24. Use the Fourier integral theorem to solve Laplace's equation on the infinite horizontal strip $-\infty < x < \infty, 0 \le y \le b$, where $u(x, b) = f(x)$ on the top of the strip, $u(x, 0) = 0$ on the bottom, and u remains bounded for $|x|$ large.

(a) Assume $u(x, y) = e^{px+qy}$ and use physical reasoning to arrive at the possible solutions $\{\sinh \alpha y \cos \alpha x, \sinh \alpha y \sin \alpha x\}$, where $\alpha > 0$.

(b) Write $u(x, y)$ as the most general linear combination of these two solutions, obtaining $u(x, y) = \frac{1}{\pi}\int_0^\infty A(\alpha)\sinh \alpha y \cos \alpha x \, d\alpha + \frac{1}{\pi}\int_0^\infty B(\alpha)\sinh \alpha y \sin \alpha x \, d\alpha$.

(c) Apply the boundary condition $u(x, b) = f(x)$ to determine $A(\alpha)$ and $B(\alpha)$ via the Fourier integral theorem, and obtain $u(x, y) = \frac{1}{\pi}\int_0^\infty \int_{-\infty}^\infty f(\beta)\frac{\sinh \alpha y}{\sinh \alpha b}\cos \alpha(\beta - x) \, d\beta \, d\alpha$.

In Exercises B25–34, take $b = 1$. For each $f(x)$ given:

(a) Sketch $f(x)$.

(b) Evaluate the inner integral in the solution for $u(x, y)$ obtained in Exercise B24.

(c) Show that $u(x, 1)$ evaluates to $f(x)$.

(d) Evaluate $u(0, \frac{1}{2})$ by integrating numerically, after having graphed the integrand.

(e) Evaluate $u(5, \frac{1}{2})$ by integrating numerically, after having graphed the integrand.

B25. $f(x) = H(x + 1) - H(x - 1)$

B26. $f(x) = x(H(x) - H(x - 1))$

B27. $f(x) = (1 - x^2)(H(x + 1) - H(x - 1))$

B28. $f(x) = x(1 - x)(H(x) - H(x - 1))$

B29. $f(x) = x(1 - x)^2(H(x) - H(x - 1))$

B30. $f(x) = x(1 - x)^3(H(x) - H(x - 1))$

B31. $f(x) = x^2(1 - x)(H(x) - H(x - 1))$

B32. $f(x) = x^3(1 - x)(H(x) - H(x - 1))$

B33. $f(x) = (H(x) - H(x - 1))\sin \pi x$

B34. $f(x) = (\cos \pi x + 1)(H(x + 1) - H(x - 1))$

B35. Use the Fourier integral theorem to solve Laplace's equation on the infinite horizontal strip $-\infty < x < \infty, 0 \le y \le b$, where $u(x, b) = 0$ on the top of the strip, $u(x, 0) = g(x)$ on the bottom, and u remains bounded for $|x|$ large. A construction similar to that in Exercise B24 leads to $u(x, y) = \frac{1}{\pi}\int_0^\infty \int_{-\infty}^\infty g(\beta)\frac{\sinh \alpha(y-b)}{\sinh(-\alpha b)}\cos \alpha(\beta - x) \, d\beta \, d\alpha$.

B36. Use the Fourier integral theorem to solve Laplace's equation on the infinite horizontal strip $-\infty < x < \infty, 0 \le y \le b$, where $u(x, b) = f(x)$ on the top of the strip, $u_y(x, 0) = 0$ on the bottom, and u remains bounded for $|x|$ large. A construction similar to that in Exercise B24 leads to $u(x, y) = \frac{1}{\pi}\int_0^\infty \int_{-\infty}^\infty f(\beta)\frac{\cosh \alpha y}{\cosh \alpha b}\cos \alpha(\beta - x) \, d\beta \, d\alpha$ for the steady-state temperatures on the infinite strip with prescribed temperatures on top and insulation on the bottom.

In Exercises B37–46, set $b = 1$ in Exercise B36. For each $f(x)$ given:

(a) Sketch $f(x)$.

(b) Evaluate the inner integral in the solution for $u(x, y)$.

(c) Show that $u(x, 1)$ evaluates to $f(x)$.

(d) Evaluate $u(0, \frac{1}{2})$ by integrating numerically, after having graphed the integrand.

(e) Evaluate $u(5, \frac{1}{2})$ by integrating numerically, after having graphed the integrand.

(f) By integrating numerically, obtain a graph of $u(x, 0)$, the temperatures on the insulated (bottom) edge.

B37. $f(x) = H(x) - H(x-1)$

B38. $f(x) = x^2(H(x) - H(x-1))$

B39. $f(x) = x(1-x)^2(H(x) - H(x-1))$

B40. $f(x) = x(1-x)^3(H(x) - H(x-1))$

B41. $f(x) = x^2(1-x)(H(x) - H(x-1))$

B42. $f(x) = x^3(1-x)(H(x) - H(x-1))$

B43. $f(x) = (\cos \pi x + 1)(H(x+1) - H(x-1))$

30.3   The Fourier Transform

DEFINITION 30.1

The Fourier integral theorem

$$f(x) = \frac{1}{2\pi} \int_{-\infty}^{\infty} \int_{-\infty}^{\infty} e^{i\alpha x} f(\beta) e^{-i\alpha \beta} \, d\beta \, d\alpha \qquad (30.21)$$

suggests defining the inner integral as the *Fourier transform*

$$F(\alpha) = \int_{-\infty}^{\infty} f(\beta) e^{-i\alpha \beta} \, d\beta \qquad (30.22)$$

so that the outer integral supplies the inversion formula

$$f(x) = \frac{1}{2\pi} \int_{-\infty}^{\infty} F(\alpha) e^{i\alpha x} \, d\alpha \qquad (30.23)$$

The integral in (30.22) exists for absolutely integrable functions $f(x)$ that are piecewise continuous on every interval of the form $[-L, L]$. Corresponding to the alternate form of the Fourier integral theorem, there is an alternate definition of the Fourier transform.

EXAMPLE 30.8   Show that the Fourier transform of the function $f(x) = e^{-|x|}$ is $F(\alpha) = \frac{2}{1+\alpha^2}$.

In Example 30.4, Section 30.2, the Fourier integral theorem was illustrated for this function. The inner integral, the one equivalent to the Fourier transform, was found to be

$$\int_{-\infty}^{\infty} e^{-|\beta|} e^{-i\alpha \beta} \, d\beta = \frac{2}{1 + \alpha^2}$$

The inversion integral can be evaluated as

$$\frac{1}{2\pi} \int_{-\infty}^{\infty} \frac{2}{1 + \alpha^2} e^{i\alpha x} \, d\alpha = e^{-x} H(x) + e^x H(-x)$$

where $H(x)$ is the Heaviside function. When x is negative, $H(x) = 0$, $H(-x) = 1$, and the inverse is e^x; but when x is positive, $H(x) = 1$, $H(-x) = 0$, and the inverse is $e^{(-x)}$. On rare occasions it is possible to evaluate the inversion integral directly. Most computations of the inversion integral are done as contour integrals in the complex plane, as we will see in Section 36.3. ❖

Operational Properties of the Fourier Transform

Just as with the Laplace transform, the operational properties of the Fourier transform are useful.

DIFFERENTIATION   Letting $\mathcal{F}$ represent the Fourier transform operator, the operational rule for derivatives is

$$\mathcal{F}[f^{(n)}(x)] = (i\alpha)^n F(\alpha)$$

where $F(\alpha)$ is the Fourier transform of $f(x)$. In particular, for the cases $n = 1$ and 2, we have

$$\mathcal{F}[f'(x)] = i\alpha F(\alpha) \quad \text{and} \quad F[f''(x)] = (i\alpha)^2 F(\alpha) = -\alpha^2 F(\alpha)$$

Conditions under which this result is valid are:

1. $f^{(n)}(x)$ is piecewise continuous on every interval of the form $[-L, L]$.
2. $f^{(n-1)}(x)$ is absolutely integrable.
3. $\lim_{x \to \pm\infty} f^{(k)}(x) = 0$ for $k = 0, 1, 2, \ldots, n-1$.

EXAMPLE 30.9   The Fourier transform of the function $f(x) = \frac{1}{1+x^2}$ is $F(\alpha) = e^{\alpha}\pi H(-\alpha) + e^{-\alpha}\pi H(\alpha)$. The Fourier transform of the derivative $f'(x) = -\frac{2x}{(1+x^2)^2}$, related to $F(\alpha)$ through the multiplication by the factor $i\alpha$, is $\mathcal{F}[f'(x)] = i\alpha\pi(e^{\alpha}H(-\alpha) + H(\alpha)e^{-\alpha})$. ❖

MULTIPLICATION   Differentiation of a function maps to multiplication of the transform by $i\alpha$. Multiplication of the function by x maps to differentiation of the transform, much like the duality found for the Laplace transform. In particular, if $f(x)$ is piecewise continuous on every interval $[-L, L]$, and $x^n f(x)$ is absolutely integrable, we have the operational rule

$$\mathcal{F}[x^n f(x)] = i^n F^{(n)}(\alpha)$$

which, in the case $n = 1$, becomes $\mathcal{F}[xf(x)] = iF'(\alpha)$, or $\mathcal{F}[-ixf(x)] = F'(\alpha)$.

EXAMPLE 30.10   The Fourier transform of the function $f(x) = \frac{1}{1+x^2}$ is $F(\alpha) = \pi(e^{\alpha}H(-\alpha) + e^{-\alpha}H(\alpha))$. Computation then shows $\mathcal{F}[-ixf(x)] = F'(\alpha) = \pi(e^{\alpha}H(-\alpha) - e^{-\alpha}H(\alpha))$. ❖

INTEGRATION   If $f(x)$ is piecewise continuous on every interval $[-L, L]$, is absolutely integrable, and has a transform $F(\alpha)$ for which $F(0) = 0$, then, as with the Laplace transform, integration on $f(x)$ maps to division in the frequency domain. In particular, we have

$$\mathcal{F}\left[\int_{-\infty}^{x} f(\sigma)\, d\sigma\right] = \frac{F(\alpha)}{i\alpha}$$

EXAMPLE 30.11   The Fourier transform of the function $f(x) = \frac{1}{x^2}$ is $F(\alpha) = \pi\alpha(H(-\alpha) - H(\alpha))$. Integrating $f(x)$ we get $\int_{-\infty}^{x} \frac{1}{\sigma^2}\, d\sigma = -\frac{1}{x}$ whose Fourier transform is $-i\pi(H(-\alpha) - H(\alpha)) = \frac{F(\alpha)}{i\alpha}$. ❖

FIRST SHIFTING LAW   Just like for the Laplace transform, multiplication by an exponential maps to shifting in the frequency domain. Formally, we have

$$\mathcal{F}[e^{iax} f(x)] = F(\alpha - a)$$

EXAMPLE 30.12   For the function $f(x) = \frac{1}{1+x^2}$ with Fourier transform $F(\alpha) = \pi(e^{\alpha}H(-\alpha) + e^{-\alpha}H(\alpha))$, multiplication by an exponential creates a function whose Fourier transform is now $\pi(e^{\alpha-a}H(-\alpha+a) + e^{-\alpha+a}H(\alpha-a))$. Clearly, this is just shifting in the frequency domain, as seen by direct substitution. ❖

Second Shifting Law   Again, in analogy with the Laplace transform, there is a "dual" shifting law for multiplication by an exponential in the frequency domain. The formal result is

$$\mathcal{F}[f(x-a)] = e^{-i\alpha a}F(\alpha)$$

Thus, the transform of a shifted function is the transform of the unshifted function but multiplied by an exponential, provided the shift is by a real constant a.

EXAMPLE 30.13   If the function $f(x) = \frac{1}{1+x^2}$ with Fourier transform $F(\alpha) = \pi(e^{\alpha}H(-\alpha) + e^{-\alpha}H(\alpha))$ is shifted to $\frac{1}{1+(x-a)^2}$, then the Fourier transform of the shifted function is $e^{-i\alpha a}\pi(e^{\alpha}H(-\alpha) + e^{-\alpha}H(\alpha))$, which is clearly just the product of $e^{-i\alpha a}$ and $F(\alpha)$.   ❖

Modulation   For a real, the Fourier transforms of $f(x)$ multiplied by either $\cos a\alpha$ or $\sin a\alpha$ are given by

$$\mathcal{F}[f(x)\cos ax] = \frac{1}{2}[F(\alpha+a) + F(\alpha-a)]$$

$$\mathcal{F}[f(x)\sin ax] = \frac{i}{2}[F(\alpha+a) - F(\alpha-a)]$$

Both results are a consequence of the first shifting law if we write $\cos ax$ as $(e^{iax} + e^{-iax})/2$ and $\sin ax$ as $(e^{iax} - e^{-iax})/2i$ and recall that $\frac{1}{i} = -i$.

EXAMPLE 30.14   The function $f(x) = \frac{1}{x}$ has the Fourier transform $F(\alpha) = i\pi(H(-\alpha) - H(\alpha))$, whereas the function $g(x) = \frac{\cos ax}{x}$ has Fourier transform $G(\alpha) = \frac{\pi i}{2}(H(-\alpha+a) - H(\alpha-a) + H(-\alpha-a) - H(\alpha+a))$, which is just $\frac{1}{2}[F(\alpha+a) + F(\alpha-a)]$.   ❖

Symmetry   The transform of a transform is "almost" the original function, a result not seen for the Laplace transform. In particular, if $\mathcal{F}[f(x)] = F(\alpha)$, then

$$\mathcal{F}[F(x)] = 2\pi f(-\alpha) \tag{30.24}$$

The function $f(x)$ is not quite recovered; there is a factor of 2π, and the argument has been negated.

EXAMPLE 30.15   The function $f(x) = \frac{1}{x}$ has the transform $F(\alpha) = i\pi(H(-\alpha) - H(\alpha))$. The Fourier transform of $F(x) = i\pi(H(-x) - H(x))$ is $-2\frac{\pi}{\alpha}$, which is just $2\pi f(-\alpha)$.   ❖

Convolution   In the context of the Laplace transform, the convolution of two functions $f(t)$ and $g(t)$ was defined (Section 6.8) as

$$f(t) * g(t) = \int_0^t f(x)g(t-x)\,dx = \int_0^t f(t-x)g(x)\,dx$$

In the context of the Fourier transform, the convolution of two functions $f(x)$ and $g(x)$ is defined as

$$f(x) * g(x) = \int_{-\infty}^{\infty} f(\sigma)g(x-\sigma)\,d\sigma = \int_{-\infty}^{\infty} f(x-\sigma)g(\sigma)\,d\sigma$$

The difference in the definition of convolution for functions with Fourier transforms means that in addition to an analog for the convolution theorem for Laplace transforms, there is a second convolution theorem unique to the Fourier transform. Thus, if two functions

$f(x)$ and $g(x)$ have Fourier transforms $F(\alpha)$ and $G(\alpha)$, respectively, then

$$\mathcal{F}[f(x) * g(x)] = F(\alpha)G(\alpha) \quad \text{and} \quad \mathcal{F}[f(x)g(x)] = \frac{1}{2\pi}F(\alpha) * G(\alpha)$$

The first theorem says the Fourier transform of the convolution is the (expected) product of the transforms. The second theorem says the Fourier transform of an ordinary product is the (unexpected) convolution of the transforms, divided by 2π.

EXAMPLE 30.16   To demonstrate $\mathcal{F}[f(x) * g(x)] = F(\alpha)G(\alpha)$, the first convolution theorem, let $f(x)$ be the function $f(x) = \frac{1}{1+x^2}$ and take $g(x) = f(x)$. Then, we have the convolution

$$f(x) * g(x) = \int_{-\infty}^{\infty} \frac{1}{(1 + \sigma^2)(1 + (x - \sigma)^2)}\, d\sigma = \frac{2\pi}{4 + x^2}$$

obtained from a table or a computer algebra system. The Fourier transform of this convolution is then

$$\mathcal{F}[f(x) * g(x)] = \mathcal{F}\left[\frac{2\pi}{4 + x^2}\right] = \pi^2(e^{2\alpha}H(-\alpha) + e^{-2\alpha}H(\alpha)) = \pi^2 e^{-2|\alpha|}$$

Next, obtain $F(\alpha)G(\alpha) = F^2(\alpha)$, the square of the Fourier transform of $f(x) = \frac{1}{1+x^2}$. Then, compute

$$F(\alpha) = \pi(e^{\alpha}H(-\alpha) + e^{-\alpha}H(\alpha)) = \pi e^{-|\alpha|}$$

$$\Rightarrow F^2(\alpha) = \left(\pi e^{-|\alpha|}\right)^2 = \pi^2 e^{-2|\alpha|} = \mathcal{F}[f(x) * g(x)]$$

demonstrating the convolution theorem $\mathcal{F}[f(x) * g(x)] = F(\alpha)G(\alpha)$.   ❖

EXAMPLE 30.17   To demonstrate $\mathcal{F}[f(x)g(x)] = \frac{1}{2\pi}F(\alpha) * G(\alpha)$, the second convolution theorem, again use $f(x) = g(x) = \frac{1}{1+x^2}$ and obtain the Fourier transform

$$\mathcal{F}[f(x)g(x)] = \mathcal{F}[f^2(x)] = \frac{\pi}{2}[(1 - \alpha)H(-\alpha)e^{\alpha} + (1 + \alpha)H(\alpha)e^{-\alpha}] = \frac{\pi}{2}(1 + |\alpha|)e^{-|\alpha|}$$

To obtain $\frac{1}{2\pi}F(\alpha) * G(\alpha) = \frac{1}{2\pi}F(\alpha) * F(\alpha)$, evaluate the integral

$$\pi \int_{-\infty}^{\infty} (e^{\sigma}H(-\sigma) + e^{-\sigma}H(\sigma))(e^{\alpha-\sigma}H(-\alpha+\sigma) + e^{-\alpha+\sigma}H(\alpha - \sigma))\, d\sigma$$

and divide by 2π. For α negative, get $-\frac{\pi}{2}e^{\alpha}(\alpha - 1)$ and for α positive, get $\frac{\pi}{2}(1 + \alpha)e^{-\alpha}$, equivalent to $\frac{\pi}{2}(1+|\alpha|)e^{-|\alpha|}$. Hence, we have demonstrated the second convolution theorem, $\mathcal{F}[f(x)g(x)] = \frac{1}{2\pi}F(\alpha) * G(\alpha)$.   ❖

EXAMPLE 30.18   The convolution of $f(x) = (H(x) - H(x - 3))x$ and $g(x) = (H(x - 1) - H(x - 5))x^2$ is given by the integral

$$\int_{-\infty}^{\infty} (H(\sigma) - H(\sigma - 3))(H(x - \sigma - 1) - H(x - \sigma - 5))\sigma(x - \sigma)^2\, d\sigma$$

the integrand of which is nonzero over the shaded region in Figure 30.8. Corresponding to the three intervals $1 \le x \le 4, 4 \le x \le 5, 5 \le x \le 8$, we therefore have the three integrals

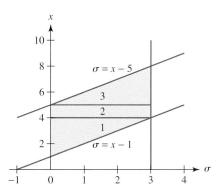

FIGURE 30.8 Domain of integration in Example 30.18

$\int_0^{x-1} \sigma (x - \sigma)^2 \, d\sigma$, $\int_0^3 \sigma (x - \sigma)^2 \, d\sigma$, $\int_{x-5}^3 \sigma (x - \sigma)^2 \, d\sigma$ and, hence, the piecewise-defined

$$f * g = \begin{cases} 0 & x < 1 \\ \frac{1}{12}(x^4 - 4x + 3) & 1 \le x \le 4 \\ \frac{9}{4}(2x^2 - 8x + 9) & 4 \le x \le 5 \\ -136 + \frac{71}{3}x + \frac{9}{2}x^2 - \frac{1}{12}x^4 & 5 \le x \le 8 \\ 0 & x > 8 \end{cases} \tag{30.25}$$

Amplitude Spectrum

The Fourier (sine) *series* for $f(t) = t$, $0 \le t \le \pi$, has coefficients $b_n = \frac{2}{\pi} \int_0^\pi t \sin nt \, dt = -\frac{2}{n}(-1)^n$, whose magnitudes are graphed in Figure 30.9. The graph represents the contribution of each harmonic $\sin nt$ in the complete series expansion for the function $f(t) = t$. Such a graph represents the *discrete amplitude spectrum* for the function $f(t)$ and gives an idea of the contribution made by each frequency n.

The Fourier transform of $f(x) = \sin(x)(\mathrm{H}(x) - \mathrm{H}(x - 2\pi))$, the function whose graph is shown in Figure 30.10, is

$$F(\alpha) = \int_0^{2\pi} \sin(\beta) e^{-i\alpha\beta} \, d\beta = \frac{e^{-2i\pi\alpha} - 1}{\alpha^2 - 1} \tag{30.26}$$

Incidentally, the expression for $F(\alpha)$ is not defined at $\alpha = \pm 1$. However, the singularities are removable since the limits at these points are $\mp \pi i$, respectively, matching the values of the integrals

$$\int_0^{2\pi} \sin(\beta) e^{i\beta} \, d\beta = \pi i \quad \text{and} \quad \int_0^{2\pi} \sin(\beta) e^{-i\beta} \, d\beta = -\pi i \tag{30.27}$$

The Fourier transform is typically a complex-valued function of the real variable α, as this example shows. A complex number $z = u + iv$ has magnitude $|z| = \sqrt{u^2 + v^2}$. The magnitude of the Fourier transform is called the *amplitude spectrum* for $f(x)$ and shows the contributions of the continuous spectrum of frequencies α. For the present example, the frequency spectrum, graphed in Figure 30.11, is

$$|F(\alpha)| = \sqrt{\frac{2(1 - \cos 2\pi\alpha)}{(\alpha^2 - 1)^2}} \tag{30.28}$$

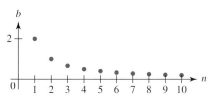

FIGURE 30.9 Amplitude spectrum for $f(t) = t$

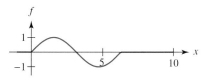

FIGURE 30.10 The function $f(x) = \sin(x)(H(x) - H(x - 2\pi))$

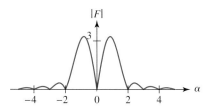

FIGURE 30.11 Amplitude spectrum for the function in Figure 30.10

EXERCISES 30.3–Part A

A1. In Example 30.11, verify the transforms of $f(x) = \frac{1}{x^2}$ and its integral, $-\frac{1}{x}$.

A2. In Example 30.12, verify the transform of $f(x) = \frac{1}{1+x^2}$.

A3. In Example 30.13, verify the transform of $f(x - a) = \frac{1}{1+(x-a)^2}$.

A4. In Example 30.14, verify the transform of $g(x) = \frac{\cos ax}{x}$.

A5. Verify the integrations leading to (30.25).

A6. Verify the integration in (30.26).

A7. Verify the integrations in (30.27).

EXERCISES 30.3–Part B

Throughout these exercises, H(x) is the Heaviside function.

B1. For $F(\alpha)$ in (30.26), show that $|F(\alpha)|$ is given by (30.28). Why is it incorrect to simplify (30.28) to $\frac{\sqrt{2(1-\cos 2\pi\alpha)}}{(\alpha^2-1)}$? What is the correct simplification?

In Exercises B2–11:

   (a) Sketch $f(x)$.

   (b) Compute $\mathcal{F}[f(x)]$ by the definition of the Fourier transform.

   (c) Obtain $\mathcal{F}[f(x)]$ by an appropriate software tool such as Maple's **fourier** command.

   (d) Recover $f(x)$ from its transform by evaluating the inversion integral.

   (e) Recover $f(x)$ from its transform by an appropriate software tool such as Maple's **invfourier** command.

   (f) Plot the amplitude spectrum of $f(x)$.

B2. $f(x) = H(x) - H(x - 1)$ **B3.** $f(x) = (H(x) - H(x - 1))x$

B4. $f(x) = (H(x) - H(x - 1))x^3$

B5. $f(x) = (H(x) - H(x - 1))(1 - x)$

B6. $f(x) = (H(x) - H(x - 1))x(1 - x)$

B7. $f(x) = (H(x) - H(x - 1))x(1 - x)^2$

B8. $f(x) = (H(x) - H(x - 1))x^2(1 - x)$

B9. $f(x) = (H(x) - H(x - 1)) \cos \pi x$

B10. $f(x) = H(x)e^{-x}$ **B11.** $f(x) = H(x)xe^{-x}$

In Exercises B12–21:

   (a) Obtain $F(\alpha) = \mathcal{F}[f(x)]$ and $\mathcal{F}[f'(x)]$, either by a software tool or by looking up the transforms in a table.

   (b) Verify $\mathcal{F}[f'(x)] = i\alpha F(\alpha)$, the rule for the transforms of derivatives.

B12. $f(x) = (H(x) - H(x - 2))x$

B13. $f(x) = (H(x) - H(x - 2))x^2$

B14. $f(x) = (H(x) - H(x - 2))x^3$

B15. $f(x) = (H(x) - H(x - 2))e^{-2x}$

B16. $f(x) = (H(x) - H(x - 2)) \sin \pi x$

B17. $f(x) = (H(x) - H(x - 2)) \cos^2 \pi x$

B18. $f(x) = (H(x) - H(x - 2))x(2 - x)$

B19. $f(x) = (H(x) - H(x - 2))x^2(2 - x)$

B20. $f(x) = H(x)xe^{-x}$ **B21.** $f(x) = H(x)x^2e^{-x}$

In Exercises B22–29:

   (a) Obtain $F(\alpha) = \mathcal{F}[f(x)]$ and $\mathcal{F}[xf(x)]$, either by a software tool or by use of a table of transforms.

   (b) Verify $\mathcal{F}[xf(x)] = iF'(\alpha)$, the rule for differentiating transforms.

B22. $f(x) = (H(x) - H(x - \pi))x \sin x$

B23. $f(x) = (H(x) - H(x - \pi))x \cos x$

B24. $f(x) = (H(x) - H(x - \pi))x \sin^2 x$

B25. $f(x) = (H(x) - H(x - \pi))x \cos^2 x$

B26. $f(x) = (H(x) - H(x - 3))x$

B27. $f(x) = (H(x) - H(x - 3))x^2$

B28. $f(x) = (H(x) - H(x - 3))x^3$ **B29.** $f(x) = H(x)e^{-x}$

In Exercises B30–39:

   (a) Obtain $F(\alpha) = \mathcal{F}[f(x)]$.

   (b) Obtain $g(x) = \int_{-\infty}^{x} f(\sigma)\, d\sigma$.

   (c) Obtain $\mathcal{F}[g(x)]$.

   (d) Verify $\mathcal{F}[\int_{-\infty}^{x} f(\sigma)\, d\sigma] = \frac{1}{i\alpha}F(\alpha)$, the rule for Fourier transforms of integrals.

B30. $f(x) = (H(x) - H(x - \pi)) \sin 2x$

B31. $f(x) = (H(x) - H(x - \pi)) \cos 2x$

B32. $f(x) = (H(x) - H(x - \pi)) \sin^2 2x$

B33. $f(x) = (H(x) - H(x - \pi)) \cos^2 2x$

B34. $f(x) = H(x) - H(x - 5)$ **B35.** $f(x) = (H(x) - H(x - 5))x$

B36. $f(x) = (H(x) - H(x - 5))x^2$

B37. $f(x) = (H(x) - H(x - 5))x^3$

B38. $f(x) = H(x)e^{-x}$ **B39.** $f(x) = H(x)xe^{-x}$

In Exercises B40–49:

   (a) If $a > 0$, obtain $F(\alpha) = \mathcal{F}[f(x)]$, $\mathcal{F}[e^{iax}f(x)]$, and $\mathcal{F}[f(x - a)]$.

   (b) Verify $\mathcal{F}[e^{iax}f(x)] = F(\alpha - a)$, the first shifting rule for Fourier transforms.

   (c) Verify $\mathcal{F}[f(x - a)] = e^{-i\alpha a}F(\alpha)$, the second shifting rule for Fourier transforms.

   (d) Obtain $\mathcal{F}[f(x) \cos ax]$ and verify $\mathcal{F}[f(x) \cos ax] = \frac{1}{2}[F(\alpha + a) + F(\alpha - a)]$.

   (e) Obtain $\mathcal{F}[f(x) \sin ax]$ and verify $\mathcal{F}[f(x) \sin ax] = \frac{i}{2}[F(\alpha + a) - F(\alpha - a)]$.

   (f) Obtain $\mathcal{F}[F(\alpha)]$ and verify (30.24), the symmetry law for Fourier transforms.

B40. $f(x) = H(x) - H(x - 4)$ **B41.** $f(x) = (H(x) - H(x - 4))x$

B42. $f(x) = (H(x) - H(x - 4))x^2$

B43. $f(x) = (H(x) - H(x - 4)) \sin \frac{\pi}{2}x$

B44. $f(x) = (H(x) - H(x - 4)) \cos^2 \frac{\pi}{2}x$

B45. $f(x) = (H(x) - H(x - 4))x \sin \frac{\pi}{2}x$

B46. $f(x) = (\mathrm{H}(x) - \mathrm{H}(x - 4))x^2 \cos \frac{\pi}{2}x$

B47. $f(x) = (\mathrm{H}(x) - \mathrm{H}(x - 4))x^2 \sin^2 \frac{\pi}{2}x$

B48. $f(x) = \mathrm{H}(x)e^{-x}$ **B49.** $f(x) = \mathrm{H}(x)xe^{-x}$

In Exercises B50–54, obtain $\mathcal{F}[F(\alpha)]$ and verify (30.24), the symmetry law for Fourier transforms.

B50. $f(x) = e^{-x^2}$ **B51.** $f(x) = e^{-x^2}\sin x$

B52. $f(x) = e^{-x^2}\cos x$ **B53.** $f(x) = xe^{-x^2}\sin x$

B54. $f(x) = xe^{-x^2}\cos x$

B55. If $f(x) = g(x) = e^{-x^2}$:

  (a) Obtain $F(\alpha) = G(\alpha) = \mathcal{F}[f(x)]$.

  (b) Obtain $f * g$, the convolution of $f(x)$ with itself.

  (c) Obtain $\mathcal{F}[f * g]$, the transform of the convolution, and show that it equals $F(\alpha)G(\alpha)$. This verifies the convolution theorem for the time domain.

  (d) Obtain $\mathcal{F}[f(x)g(x)]$, the convolution of the product $fg = f^2$.

  (e) Obtain $\frac{1}{2\pi}F * G = \frac{1}{2\pi}F * F$, the convolution of the transform with itself (divided by 2π), and show it equals $\mathcal{F}[f(x)g(x)]$. This verifies the convolution for the frequency domain.

In Exercises B56–61:

  (a) Obtain $F(\alpha) = \mathcal{F}[f(x)]$ and $G(\alpha) = \mathcal{F}[f(x)]$, the transforms of $f(x)$ and $g(x)$.

  (b) Obtain the convolution $f * g$ and draw its graph.

  (c) Obtain $\mathcal{F}[f * g]$, the transform of the convolution, and show that it equals $F(\alpha)G(\alpha)$, thus verifying the convolution theorem for the time domain.

  (d) Graph the product $f(x)g(x)$, and obtain $\mathcal{F}[f(x)g(x)]$, its Fourier transform.

  (e) Obtain $F * G$, the convolution of the transforms, using, for example, $F * G = \mathcal{F}^{-1}[\mathcal{F}[F]\mathcal{F}[G]]$.

  (f) Show $\mathcal{F}[f(x)g(x)] = \frac{1}{2\pi}F * G$, thereby verifying the convolution for the frequency domain.

B56. $f(x) = (\mathrm{H}(x) - \mathrm{H}(x - 1))x$, $g(x) = \mathrm{H}(x)e^{-x}$

B57. $f(x) = (\mathrm{H}(x) - \mathrm{H}(x - 1))x^2$, $g(x) = \mathrm{H}(x)e^{-x}$

B58. $f(x) = (\mathrm{H}(x) - \mathrm{H}(x - 1))\sin \pi x$, $g(x) = \mathrm{H}(x)e^{-x}$

B59. $f(x) = (\mathrm{H}(x) - \mathrm{H}(x - 1))\cos \pi x$, $g(x) = \mathrm{H}(x)e^{-x}$

B60. $f(x) = g(x) = \mathrm{H}(x)e^{-x}$ **B61.** $f(x) = g(x) = e^{-|x|}$

In Exercises B62–81:

  (a) Sketch $f(x)$ and $g(x)$.

  (b) Obtain $F(\alpha) = \mathcal{F}[f(x)]$ and $G(\alpha) = \mathcal{F}[f(x)]$, the transforms of $f(x)$ and $g(x)$.

  (c) Obtain the convolution $f * g$ and draw its graph.

  (d) Obtain $\mathcal{F}[f * g]$, the transform of the convolution, and show that it equals $F(\alpha)G(\alpha)$, thus verifying the convolution theorem for the time domain.

  (e) Use the frequency domain convolution theorem to obtain the convolution of the transforms by computing $F * G = 2\pi\mathcal{F}[f(x)g(x)]$.

  (f) Use the time domain convolution theorem to obtain the convolution of the transforms by treating $F(\alpha)$ and $G(\alpha)$ as "functions," and inverting the product of their transforms. Thus, compute $F * G = \mathcal{F}^{-1}[\mathcal{F}[F(\alpha)]\mathcal{F}[G(\alpha)]]$.

  (g) In part (f), use the symmetry relation to transform the transforms, thus computing the convolution as $F * G = \mathcal{F}^{-1}[2\pi f(-\alpha)2\pi g(-\alpha)]$.

B62. $f(x) = \mathrm{H}(x) - \mathrm{H}(x - 2)$, $g(x) = \mathrm{H}(x - 1) - \mathrm{H}(x - 3)$

B63. $f(x) = \mathrm{H}(x) - \mathrm{H}(x - 2)$, $g(x) = \mathrm{H}(x - 3) - \mathrm{H}(x - 5)$

B64. $f(x) = (\mathrm{H}(x) - \mathrm{H}(x - 3))x$, $g(x) = (\mathrm{H}(x - 2) - \mathrm{H}(x - 5))x^2$

B65. $f(x) = (\mathrm{H}(x-1) - \mathrm{H}(x-3))x$, $g(x) = (\mathrm{H}(x-2) - \mathrm{H}(x-4))e^{-x}$

B66. $f(x) = (\mathrm{H}(x-1) - \mathrm{H}(x-3))x^2$, $g(x) = (\mathrm{H}(x-2) - \mathrm{H}(x-4))e^{-x}$

B67. $f(x) = (\mathrm{H}(x-1) - \mathrm{H}(x-3))x^3$, $g(x) = (\mathrm{H}(x-2) - \mathrm{H}(x-4))e^{-x}$

B68. $f(x) = (\mathrm{H}(x) - \mathrm{H}(x - 3))x$, $g(x) = (\mathrm{H}(x - 2) - \mathrm{H}(x - 4))\sin x$

B69. $f(x) = (\mathrm{H}(x) - \mathrm{H}(x - 3))x^2$, $g(x) = (\mathrm{H}(x - 2) - \mathrm{H}(x - 4))\sin x$

B70. $f(x) = (\mathrm{H}(x) - \mathrm{H}(x - 3))x^3$, $g(x) = (\mathrm{H}(x - 2) - \mathrm{H}(x - 4))\sin x$

B71. $f(x) = (\mathrm{H}(x - 2) - \mathrm{H}(x - 4))e^{-x}$, $g(x) = (\mathrm{H}(x - 3) - \mathrm{H}(x - 5))\sin x$

B72. $f(x) = \mathrm{H}(x - 1) - \mathrm{H}(x - 5)$, $g(x) = \mathrm{H}(x - 2) - \mathrm{H}(x - 3)$

B73. $f(x) = (\mathrm{H}(x-1) - \mathrm{H}(x-5))x$, $g(x) = (\mathrm{H}(x-2) - \mathrm{H}(x-3))x^2$

B74. $f(x) = (\mathrm{H}(x-1) - \mathrm{H}(x-5))x$, $g(x) = (\mathrm{H}(x-2) - \mathrm{H}(x-3))x^3$

B75. $f(x) = (\mathrm{H}(x) - \mathrm{H}(x - 4))x$, $g(x) = (\mathrm{H}(x - 1) - \mathrm{H}(x - 3))e^{-x}$

B76. $f(x) = (\mathrm{H}(x) - \mathrm{H}(x - 4))x^2$, $g(x) = (\mathrm{H}(x - 1) - \mathrm{H}(x - 3))e^{-x}$

B77. $f(x) = (\mathrm{H}(x) - \mathrm{H}(x - 4))x^3$, $g(x) = (\mathrm{H}(x - 1) - \mathrm{H}(x - 3))e^{-x}$

B78. $f(x) = (\mathrm{H}(x-2) - \mathrm{H}(x-6))x$, $g(x) = (\mathrm{H}(x-3) - \mathrm{H}(x-4))\sin x$

B79. $f(x) = (\mathrm{H}(x - 2) - \mathrm{H}(x - 6))x^2$, $g(x) = (\mathrm{H}(x - 3) - \mathrm{H}(x - 4))\sin x$

B80. $f(x) = (\mathrm{H}(x - 2) - \mathrm{H}(x - 6))x^3$, $g(x) = (\mathrm{H}(x - 3) - \mathrm{H}(x - 4))\sin x$

B81. $f(x) = (\mathrm{H}(x - 2) - \mathrm{H}(x - 6))\cos x$, $g(x) = (\mathrm{H}(x - 3) - \mathrm{H}(x - 4))\sin x$

30.4 Wave Equation on the Infinite String—Solution by Fourier Transform

Problem Statement

The BVP consisting of the one-dimensional wave equation $u_{tt}(x, t) = c^2 u_{xx}(x, t)$ on the infinite domain $-\infty < x < \infty$, the initial conditions $u(x, 0) = f(x)$, $u_t(x, 0) = g(x)$, and the boundary conditions $\lim_{x \to \pm\infty} u(x, t) = 0$ models the loss-less propagation of a disturbance along the real axis. We call this the propagation of a wave along an infinitely long string.

D'Alembert's Solution

Momentarily, we will use the Fourier transform to derive (30.29), D'Alembert's solution to this BVP.

$$u(x, t) = \frac{1}{2}[f(x - ct) + f(x + ct)] + \frac{1}{2c}\int_{x-ct}^{x+ct} g(\sigma)\, d\sigma \qquad (30.29)$$

EXAMPLE 30.19   Consider the particular initial shape

$$f_1(x) = \begin{cases} 0 & x < -1 \\ 1 - x^2 & -1 \le x \le 1 \\ 0 & x > 1 \end{cases}$$

and the initial velocity $u_t(x, 0) = 0$. The Fourier transform of the initial conditions gives $U(\alpha, 0) = F_1(\alpha)$ and $U_t(\alpha, 0) = 0$, where

$$F_1(\alpha) = \int_{-1}^{1} (1 - \beta^2)e^{-i\alpha\beta}\, d\beta = \frac{4}{\alpha^3}(\sin\alpha - \alpha\cos\alpha) \qquad (30.30)$$

is the Fourier transform of $f_1(x)$. The function $f_1(x)$ is the localized disturbance shown in Figure 30.12. The partial Fourier transform of the wave equation, taken with respect to x, yields

$$U_{tt}(\alpha, t) = -c^2\alpha^2 U(\alpha, t)$$

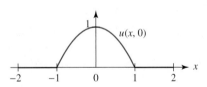

FIGURE 30.12   Initial disturbance in Example 30.19

If $Y(t) = U(\alpha, t)$, the IVP to solve becomes $Y''(t) = -c^2\alpha^2 Y(t)$, $Y(0) = F_1(\alpha)$, $Y'(0) = 0$, for which the solution is $Y(t) = U(\alpha, t) = F_1(\alpha)\cos\alpha ct$. Inverting, we find $u(x, t) = \frac{1}{4}\sum_{k=1}^{4} h_k$, where

$$h_1 = (x + 1 + ct)(x - 1 + ct)[\mathrm{H}(x - 1 + ct) - \mathrm{H}(-x + 1 - ct)]$$

$$h_2 = (-x - 1 + ct)(-x + 1 + ct)[\mathrm{H}(x - 1 - ct) - \mathrm{H}(-x + 1 + ct)]$$

$$h_3 = (-x - 1 + ct)(-x + 1 + ct)[\mathrm{H}(-x - 1 + ct) - \mathrm{H}(x + 1 - ct)]$$

$$h_4 = (x + 1 + ct)(x - 1 + ct)[\mathrm{H}(-x - 1 - ct) - \mathrm{H}(x + 1 + ct)]$$

and $\mathrm{H}(x)$ is the Heaviside function.

Figure 30.13, showing the solution surface for $c = 1$, suggests that the initial disturbance splits into two equal parts, each part traveling in opposite directions along the string. Figure 30.14 shows $u(x, 2)$, the shape of the string (in color), and $u_t(x, 2)$, the velocity profile (in black). The velocity decreases linearly from the front of the traveling wave to the rear. Figure 30.15 shows the shape and velocity at times $t = \frac{1}{5}$, 1, and $\frac{8}{5}$. When the wave traveling to the right hits an $x > 0$, the vertical velocity of that point jumps to 1. The veloc-

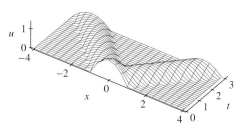

FIGURE 30.13 Solution surface for Example 30.19

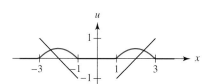

FIGURE 30.14 Example 30.19: Shape (in color) and velocity (in black) of the string at time $t = 2$

ities of points to the left of that x decrease linearly since the wave has already passed over them. Where the displacement is a maximum, the velocity is zero. At the rear of the wave, the velocity is negative, since back there, the string is on its way back toward equilibrium. See the accompanying Maple worksheet for a complete animation of this motion. ❖

EXAMPLE 30.20

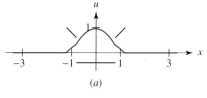

(a)

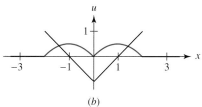

(b)

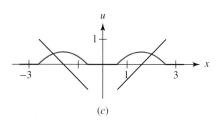

(c)

FIGURE 30.15 Example 30.19: In color and black, respectively, the shape and velocity of the string at times $t = 0.2$, 1.0, 1.6

We find the velocity profile induced in Example 30.19 to be sufficiently interesting that we next solve another special case of the wave equation. This time, we take the initial shape to be $u(x, 0) = 0$ and set the string into motion by imposing the initial velocity $u_t(x, 0) = f_1(x)$. Thus, $U(\alpha, t)$, the Fourier transform of $u(x, t)$, is $Y(t) = F_1(\alpha)\frac{\sin \alpha ct}{\alpha c}$, the solution of the initial value problem $Y''(t) = -c^2\alpha^2 Y(t)$, $Y(0) = 0$, $Y'(0) = F_1(\alpha)$. Inversion of this transform yields $u(x, t) = \frac{1}{12c}\sum_{k=1}^{4}\eta_k$, where

$$\eta_1 = (ct + x + 2)(x - 1 + ct)^2[\mathrm{H}(x - 1 + ct) - \mathrm{H}(-x + 1 - ct)]$$

$$\eta_2 = (ct - x - 2)(-x + 1 + ct)^2[\mathrm{H}(x - 1 - ct) - \mathrm{H}(-x + 1 + ct)]$$

$$\eta_3 = (ct - x + 2)(-x - 1 + ct)^2[\mathrm{H}(-x - 1 + ct) - \mathrm{H}(x + 1 - ct)]$$

$$\eta_4 = (ct + x - 2)(x + 1 + ct)^2[\mathrm{H}(-x - 1 - ct) - \mathrm{H}(x + 1 + ct)]$$

Figure 30.16 shows the solution surface for the shape of the string when $c = 1$. The shape of the string is arresting, being displaced a uniform amount and remaining displaced. This certainly does not fit with our intuition of rubber-bands, guitar strings, and piano strings subjected to a tap that imparts an initial velocity.

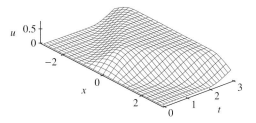

FIGURE 30.16 Solution surface for Example 30.20

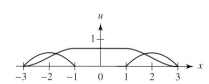

FIGURE 30.17 Example 30.20: Shape (in color) and velocity (in black) of the string at time $t = 2$

Figure 30.17 shows, for $t = 2$, the velocity wave as a thin curve and the displacement as a thick curve. Indeed, the initial velocity splits and travels in both directions as a wave. The displacement of the string becomes $u = \frac{2}{3}$ after the velocity wave has passed. The companion Maple worksheet provides a complete animation for this motion. ❖

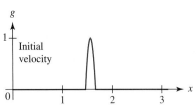

FIGURE 30.18 Initial velocity profile for the finite string in Example 30.21

FIGURE 30.19 Example 30.21: In color and black, respectively, the shape and velocity of the string at time $t = 0.4$

EXAMPLE 30.21 Do we see the same behavior in the finite string? To find out, consider an undisturbed string of length π, set into motion with an initial velocity described by

$$
g(x) = \begin{cases}
0 & 0 \le x < \dfrac{\pi}{2} - \dfrac{1}{10} \\[2mm]
100\left(x - \dfrac{\pi}{2} + \dfrac{1}{10}\right)\left(\dfrac{\pi}{2} + \dfrac{1}{10} - x\right) & \dfrac{\pi}{2} - \dfrac{1}{10} \le x \le \dfrac{\pi}{2} + \dfrac{1}{10} \\[2mm]
0 & \dfrac{\pi}{2} + \dfrac{1}{10} < x \le \pi
\end{cases}
$$

and graphed in Figure 30.18. If we take the wave speed to be $c = 1$, the solution to the wave equation on the finite string is the infinite series

$$
u(x, y) = \sum_{n=1}^{\infty} B_n \sin nx \sin nt \tag{30.31a}
$$

$$
B_n = \frac{2}{n\pi} \int_0^{\pi} g(x) \sin nx \, dx = \frac{80}{\pi n^4}\left(10 \sin \frac{n}{10} - n \cos \frac{n}{10}\right) \sin \frac{n\pi}{2} \tag{30.31b}
$$

Approximating $u(x, t)$ with the partial sum $u_{100}(x, t) = \sum_{n=1}^{100} B_n \sin nx \sin nt$ we obtain Figure 30.19, which shows, for $t = \frac{2}{5}$, the velocity wave (in black) and the string itself (in color). Indeed, until the traveling waves (see the accompanying Maple worksheet for an animation) reach the fixed endpoints, the same behavior in the shape of the string is observed. If that's true, then why does our intuition, derived from finite strings, not reflect this insight? Perhaps it is because we are never able to set a string into motion with a localized velocity profile. ❖

EXAMPLE 30.22 Typically, striking a string with a sharp instrument gives an initial velocity to the *whole* string, as modeled by the function $g(x) = \frac{1}{8}x(\pi - x)$, graphed in Figure 30.20. Again, the solution has the form (30.31a) with

$$
B_n = \frac{2}{n\pi} \int_0^{\pi} g(x) \sin nx \, dx = \frac{(1 - (-1)^n)}{2\pi n^4} \tag{30.32}
$$

The partial sum $\sum_{n=1}^{100} B_n \sin nx \sin nt$ yields Figure 30.21, the shape (in color) and velocity (in black) at a typical time $t > 0$. Imparting a nonzero initial velocity to the whole string results in a "familiar" motion for the finite string. ❖

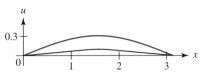

FIGURE 30.20 Initial velocity profile for the finite string in Example 30.22

Derivation of D'Alembert's Solution

The Fourier transform of $u_{tt}(x, t) = c^2 u_{xx}(x, t)$ is $U_{tt}(\alpha, t) = -c^2 \alpha^2 U(\alpha, t)$, written as the ODE $Y''(t) = -c^2 \alpha^2 Y(t)$. The initial conditions $u(x, 0) = f(x)$ and $u_t(x, 0) = g(x)$ transform to $Y(0) = U(\alpha, 0) = F(\alpha)$ and $Y'(0) = U_t(\alpha, 0) = G(\alpha)$. The solution is

$$
Y(t) = U(\alpha, t) = F(\alpha) \cos \alpha ct + \frac{1}{\alpha c} G(\alpha) \sin \alpha ct \tag{30.33}
$$

and if written in exponential form

$$
U(\alpha, t) = \frac{1}{2}\left(F(\alpha) - \frac{i}{\alpha c} G(\alpha)\right) e^{i\alpha ct} + \frac{1}{2}\left(F(\alpha) + \frac{i}{\alpha c} G(\alpha)\right) e^{-i\alpha ct}
$$

inverts to (30.29), D'Alembert's solution of the wave equation.

FIGURE 30.21 Example 30.22: At time $t > 0$, shape and velocity of the finite string in Example 30.22

Inversion by Convolution

The reader may have seen in (30.33) the opportunity to invert a Fourier transform by applying a convolution result, namely, the product of two transforms inverts back to the convolution of the inverses. There are two products in (30.33), one involving $F(\alpha)$ and one involving $G(\alpha)$. The companion factors are $\cos \alpha ct$ and $\frac{\sin \alpha ct}{\alpha c}$, respectively. We know that the inverses of $F(\alpha)$ and $G(\alpha)$ are $f(x)$ and $g(x)$, respectively. The inverses of the corresponding companion factors are, respectively,

$$\tfrac{1}{2}[\delta(-x - ct) + \delta(-x + ct)]$$

$$\frac{1}{4c}[H(x + ct) - H(-x - ct) - H(x - ct) + H(-x + ct)]$$

The inverse of the first product is the convolution

$$\frac{1}{2}\int_{-\infty}^{\infty}[\delta(-x - ct) + \delta(-x + ct)]f(x - \sigma)\,d\sigma = \tfrac{1}{2}[f(x + ct) + f(x - ct)]$$

The inverse of the second product is the convolution

$$\frac{1}{4c}\int_{-\infty}^{\infty}[H(x + ct) - H(-x - ct) - H(x - ct) + H(-x + ct)]g(x - \rho)\,d\rho$$

$$= \frac{1}{2c}\int_{x-ct}^{x+ct} g(\sigma)\,d\sigma \tag{30.34}$$

EXERCISES 30.4–Part A

A1. Verify the integration in (30.30).

A2. Verify the integration in (30.31b).

A3. Verify the integration in (30.32).

EXERCISES 30.4–Part B

Throughout these exercises, $H(x)$ is the Heaviside function.

B1. Verify the calculation in (30.34).

In Exercises B2–9:

  (a) Use the Fourier transform to solve $u_{tt} = c^2 u_{xx}$, $-\infty < x < \infty$, $u(x, 0) = f(x)$, $u_t(x, 0) = 0$, the wave equation on an infinite string for which the initial shape is prescribed.

  (b) Obtain D'Alembert's solution for the same problem.

  (c) Graph the solution surface.

  (d) Where possible, animate the deformations in the string, or at least graph the shape of the string at a succession of three moments in time.

B2. $c = 1$, $f(x) = H(x + 1) - H(x - 1)$

B3. $c = 2$, $f(x) = (H(x + 1) - H(x - 1))x$

B4. $c = \frac{1}{2}$, $f(x) = (H(x + 1) - H(x - 1))x^2$

B5. $c = 3$, $f(x) = (H(x + 1) - H(x - 1))(x + 1)(1 - x)$

B6. $c = \frac{2}{3}$, $f(x) = (H(x + 1) - H(x - 1))(x + 1)(1 - x)^2$

B7. $c = \frac{5}{4}$, $f(x) = (H(x + 1) - H(x - 1))\sin \pi x$

B8. $c = \frac{4}{3}$, $f(x) = (H(x + 1) - H(x - 1))\sin^2 \frac{\pi}{2}x$

B9. $c = \frac{3}{4}$, $f(x) = (H(x + 1) - H(x - 1))(1 - x^2)$

In Exercises B10–21:

  (a) Use the Fourier transform to solve $u_{tt} = c^2 u_{xx}$, $-\infty < x < \infty$, $u(x, 0) = 0$, $u_t(x, 0) = g(x)$, the wave equation on an infinite string for which the initial velocity is prescribed.

  (b) Obtain D'Alembert's solution for the same problem.

(c) Graph the solution surface.

(d) Where possible, animate the deformations in the string, or at least graph the shape of the string at a succession of three moments in time.

B10. $c = 1, g(x) = e^{-x^2}$ **B11.** $c = 2, g(x) = e^{-|x|}$

B12. $c = 3, g(x) = H(x) - H(x - 1)$

B13. $c = \frac{1}{2}, g(x) = (H(x) - H(x - 1))x(1 - x)$

B14. $c = \frac{2}{3}, g(x) = (H(x) - H(x - 1))x(1 - x)^2$

B15. $c = \frac{5}{3}, g(x) = (H(x) - H(x - 1))x^2(1 - x)$

B16. $c = \sqrt{2}, g(x) = (H(x + 1) - H(x - 1))(x + 1)(1 - x)$

B17. $c = \frac{6}{5}, g(x) = (H(x + 1) - H(x - 1))(x + 1)(1 - x)^2$

B18. $c = 2, g(x) = (H(x) - H(x - 1)) \sin \pi x$

B19. $c = 1, g(x) = (H(x + 1) - H(x - 1)) \sin \pi x$

B20. $c = \sqrt{3}, g(x) = \begin{cases} 0 & x < -1 \\ -x^2 & -1 \leq x < 0 \\ x^2 & 0 \leq x < 1 \\ 0 & x \geq 1 \end{cases}$

B21. $c = \frac{1}{3}, g(x) = \begin{cases} 0 & x < -2 \\ -2 - x & -2 \leq x < -1 \\ x & -1 \leq x < 1 \\ 2 - x & 1 \leq x < 2 \\ 0 & x \geq 2 \end{cases}$

In Exercises B22–28, use the Fourier transform to solve the BVP $u_{tt} = u_{xx} + u_x, -\infty < x < \infty, u(x, 0) = f(x), u_t(x, 0) = 0$, expressing the solution as an integral, the inversion integral for the Fourier transform.

B22. $f(x) = H(x + 1) - H(x - 1)$

B23. $f(x) = (H(x + 1) - H(x - 1))x$

B24. $f(x) = (H(x + 1) - H(x - 1)) \sin \pi x$

B25. $f(x) = (H(x + 1) - H(x - 1))(1 - x^2)$

B26. $f(x) = e^{-x^2}$ **B27.** $f(x) = e^{-|x|}$ **B28.** $f(x) = \dfrac{1}{1 + x^2}$

In Exercises B29–35, use the Fourier transform to solve the BVP $u_{tt} + u_t = u_{xx}, -\infty < x < \infty, u(x, 0) = f(x), u_t(x, 0) = 0$, expressing the solution as an integral, the inversion integral for the Fourier transform.

B29. $f(x)$ from Exercise B22   **B30.** $f(x)$ from Exercise B23

B31. $f(x)$ from Exercise B24   **B32.** $f(x)$ from Exercise B25

B33. $f(x)$ from Exercise B26   **B34.** $f(x)$ from Exercise B27

B35. $f(x)$ from Exercise B28

B36. Use the Fourier transform to solve Laplace's equation on the upper half-plane. In particular, the BVP is $u_{xx} + u_{yy} = 0$, $-\infty < x < \infty, y > 0, u(x, 0) = f(x), u(x, y) \to 0$ as $y \to \infty$. It can be solved as follows.

(a) Compute the Fourier transform (with respect to x) of Laplace's equation, obtaining an ODE in $U(\alpha, y) = \mathcal{F}[u(x, y)]$.

(b) Solve the ODE along with the boundary conditions $U(\alpha, 0) = F(\alpha) = \mathcal{F}[f(x)]$ and $U(\alpha, R) = 0, R > 0$. Obtain $U_R(\alpha, y)$.

(c) Obtain $\lim_{R \to \infty} U_R(\alpha, y)$ for the two cases $\alpha > 0$ and $\alpha < 0$. This gives $U(\alpha, y) = F(\alpha)e^{-|\alpha|y}$.

(d) Invert $e^{-|\alpha|y}$ and obtain $u(x, y) = \frac{1}{\pi} \int_{-\infty}^{\infty} \frac{yf(\sigma)}{(x-\sigma)^2+y^2} d\sigma$ by convolution. This is known as the *Poisson integral formula for the half-plane* $y > 0$ or the *Schwarz integral formula*.

In Exercises B37–46:

(a) Evaluate the Schwarz integral formula in Exercise B36.

(b) Obtain a graph of the solution surface.

(c) Obtain a contour plot showing the equipotentials (level curves) of $u(x, y)$.

(d) Obtain, and superimpose on the contour plot, a graph of the flow lines of $u(x, y)$.

B37. $f(x) = H(x)$ **B38.** $f(x) = \dfrac{1}{x + 1} H(x)$

B39. $f(x) = H(x + 1) - H(x - 1)$

B40. $f(x) = (H(x + 1) - H(x - 1))x$

B41. $f(x) = (H(x + 1) - H(x - 1))x^2$

B42. $f(x) = (H(x + 1) - H(x - 1))(1 - x^2)$

B43. $f(x) = (H(x + 1) - H(x - 1))(x + 1)(1 - x)$

B44. $f(x) = (H(x + 1) - H(x - 1))(x + 1)(1 - x)^2$

B45. $f(x) = (H(x + 1) - H(x - 1))(x + 1)^2(1 - x)$

B46. $f(x) = \dfrac{1}{1 + x^2}$

30.5 Heat Equation on the Infinite Rod—Solution by Fourier Transform

Statement of the Problem and Its Solution

The heat equation $u_t(x, t) = \kappa u_{xx}(x, t) = 0$ on the infinite rod $(-\infty < x < \infty)$, along with the initial condition $u(x, 0) = f(x)$ and the boundary conditions $\lim_{x \to \pm\infty} u(x, t) = 0$, is satisfied by the function

$$u(x, t) = \frac{1}{2\sqrt{\kappa \pi t}} \int_{-\infty}^{\infty} f(\sigma) e^{-(\sigma - x)^2 / 4\kappa t} \, d\sigma \tag{30.35}$$

as we derive in this section.

Solution by Fourier Transform

The partial Fourier transform of the heat equation, taken with respect to x, yields $U_t(\alpha, t) = -\kappa \alpha^2 U(\alpha, t)$, which we write as the ODE $Y'(t) = -\kappa \alpha^2 Y(t)$ through the definition $Y(t) = U(\alpha, t)$. The initial condition $u(x, 0) = f(x)$ transforms to $Y(0) = U(\alpha, 0) = F(\alpha)$, so $Y(t) = e^{-\kappa \alpha^2 t} F(\alpha)$. The inverse of $Y(t) = U(\alpha, t)$ is (30.35), as a table of transforms or computer algebra system will reveal.

Alternatively, seeing a product of two Fourier transforms on the right side of $Y(t) = e^{-\kappa \alpha^2 t} F(\alpha)$, we reach for the convolution theorem whereby the inverse of a product of two transforms is the convolution of the inverses of the individual factors. The inverse of the factor $F(\alpha)$ is clearly $f(x)$. Its companion factor $e^{-\kappa \alpha^2 t}$ has inverse $h(x) = \frac{1}{2\sqrt{\pi \kappa t}} e^{-x^2 / 4\kappa t}$, so $u(x, t)$ in (30.35) is the convolution $f(x) * h(x)$.

EXAMPLE 30.23   If $f(x) = 3(H(x + 1) - H(x - 1))$ and $\kappa = 1$, then (30.35) gives

$$u(x, t) = \frac{3}{2} \left(\mathrm{erf} \left(\frac{x + 1}{2\sqrt{t}} \right) - \mathrm{erf} \left(\frac{x - 1}{2\sqrt{t}} \right) \right) \tag{30.36}$$

where $\mathrm{erf}\, z = \frac{2}{\sqrt{\pi}} \int_0^z e^{-\sigma^2} \, d\sigma$. The solution surface is seen in Figure 30.22.   ❖

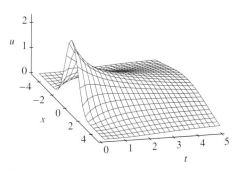

FIGURE 30.22   Solution surface in Example 30.23

EXERCISES 30.5

Throughout these exercises, $H(x)$ is the Heaviside function.

1. Obtain the solution in (30.36). Show $\lim_{t \to 0^+} u(x, t) = f(x)$.

In Exercises 2–11:

(a) Solve the BVP $u_t = \kappa u_{xx}$, $-\infty < x < \infty, t > 0$, $u(x, 0) = f(x)$, $\lim_{|x| \to \infty} u(x, t) = 0$, for the heat equation on an infinite rod.

(b) Plot the solution surface.

(c) Obtain a contour plot showing the isotherms.

(d) Either animate the temperature profiles as they vary in time or graph the temperature at a succession of times.

2. $\kappa = 2$, $f(x) = (H(x + 1) - H(x - 1))x$

3. $\kappa = 3$, $f(x) = (H(x + 1) - H(x - 1))(1 - x^2)$

4. $\kappa = \frac{1}{2}$, $f(x) = (H(x+1) - H(x-1))(x+1)(1-x)$

5. $\kappa = \frac{2}{3}$, $f(x) = (H(x+1) - H(x-1))(x+1)(1-x)^2$

6. $\kappa = \sqrt{2}$, $f(x) = (H(x+1) - H(x-1))(x+1)^3(1-x)$

7. $\kappa = \sqrt{3}$, $f(x) = (H(x) - H(x-1))e^{-x/10}$

8. $\kappa = 1$, $f(x) = (H(x) - H(x-1))10x(1-x)^3$

9. $\kappa = \frac{2}{5}$, $f(x) = e^{-x^2}$ **B10.** $\kappa = \frac{5}{2}$, $f(x) = e^{-|x|}$

11. $\kappa = 2$, $f(x) = \delta(x)$

12. For $u_t = ru_x$, $-\infty < x < \infty$, $t > 0$, $u(x, 0) = f(x)$, $r > 0$:

    (a) Verify that $f(x + rt)$ is the solution.

    (b) Use the Fourier transform to obtain $u(x, t) = f(x + rt)$ as the solution.

In Exercises 13–20:

    (a) Solve
      $u_t = ru_x$, $-\infty < x < \infty$, $t > 0$, $\lim_{t \to 0^+} u(x, t) = f(x)$.

    (b) With $r = 1$, plot the solution surface.

    (c) With $r = 1$, use either an animation or a succession of plots to show that $f(x)$, the initial disturbance, is propagated to the left, without distortion and with "speed" r, by $u(x, t)$.

13. $f(x) = (H(x+1) - H(x-1)) \sin \pi x$

14. $f(x) = (H(x+1) - H(x-1)) \sin^2 \pi x$

15. $f(x) = (H(x+1) - H(x-1))(1 - x^2)$

16. $f(x) = (H(x) - H(x-1))x(1-x)$

17. $f(x) = (H(x) - H(x-1))x(1-x)^2$

18. $f(x) = (H(x) - H(x-1))x(1-x)^3$

19. $f(x) = e^{-x^2}$ **20.** $f(x) = \dfrac{1}{1+x^2}$

In Exercise 21, do all parts. In Exercises 22–30, do parts (a)–(d).

    (a) Solve the BVP $u_t = \kappa u_{xx} + ru_x$, $-\infty < x < \infty$, $t > 0$, r real, $u(x, 0) = f(x)$, $\lim_{x \to \pm\infty} u(x, t) = 0$, which for $r = 0$ is the heat equation on the infinite rod.

    (b) Plot the solution surfaces corresponding to $r = 0$ and $r = 1$.

    (c) For $r = 1$, obtain a contour plot showing the isotherms.

    (d) For $r = 1$, animate the temperature profiles as they vary in time or plot a succession of temperature profiles at increasing times.

    (e) For $r = 1$, determine the location of the maximum temperature in the rod.

    (f) Plot the maximum temperature as a function of time.

    (g) Describe how the term u_x changes the behavior of the solution of the heat equation. *Hint:* Consider Exercise 12.

    (h) What happens when $r < 0$?

21. $\kappa = \frac{1}{2}$, $f(x) = H(x+1) - H(x-2)$

22. $\kappa = \frac{2}{3}$, $f(x) = (H(x+1) - H(x-2))x$

23. $\kappa = \frac{2}{3}$, $f(x) = (H(x+1) - H(x-2))x^2$

24. $\kappa = 2$, $f(x) = (H(x+1) - H(x-2))(x+1)(2-x)$

25. $\kappa = 3$, $f(x) = (H(x+1) - H(x-2))(x+1)(2-x)^2$

26. $\kappa = \sqrt{2}$, $f(x) = (H(x) - H(x-3))e^{-x}$

27. $\kappa = 1$, $f(x) = (H(x) - H(x-2))10x^2(2-x)$

28. $\kappa = \frac{3}{2}$, $f(x) = e^{-x^2}$ **29.** $\kappa = \frac{3}{5}$, $f(x) = e^{-|x|}$

30. $\kappa = 2$, $f(x) = \delta(x)$

31. The time-dependent heat equation in two spatial variables can be solved by repeated use of the Fourier transform. Solve the BVP $u_t = \kappa(u_{xx} + u_{yy})$, $-\infty < x, y < \infty$, $u(x, y, 0) = f(x, y)$, by implementing the following steps.

    (a) Let $\mathcal{F}_x$ and $\mathcal{F}_y$ represent Fourier transforms with respect to x and y, respectively, so that $F(\alpha, \beta) = \mathcal{F}_y[\mathcal{F}_x[f(x, y)]]$ is the y-transform of the x-transform of $f(x, y)$. Apply $\mathcal{F}_x$ to the PDE, obtaining $U_t(\alpha, y, t) = \kappa(-\alpha^2 U(\alpha, y, t) + U_{yy}(\alpha, y, t))$.

    (b) Apply $\mathcal{F}_y$ to the PDE obtain in part (a), obtaining the ODE $U_t(\alpha, \beta, t) = \kappa(-\alpha^2 U(\alpha, \beta, t) - \beta^2 U(\alpha, \beta, t))$.

    (c) The ODE from part (b), along with the initial condition from part (a), is solved for $U(\alpha, \beta, t) = F(\alpha, \beta)e^{-\kappa(\alpha^2 + \beta^2)t} = e^{-\kappa\alpha^2 t}e^{-\kappa\beta^2 t}$.

    (d) Apply $\mathcal{F}_y^{-1}$ using the convolution theorem.

    (e) Apply $\mathcal{F}_x^{-1}$ again using the convolution theorem, so that $u(x, y, t) =$

$$\frac{1}{4\pi\kappa t} \int_{-\infty}^{\infty} \int_{-\infty}^{\infty} f(\sigma_1, \sigma_2)e^{-(x-\sigma_1)^2/4\kappa t}e^{-(y-\sigma_2)^2/4\kappa t}\, d\sigma_1\, d\sigma_2$$

In Exercises 32–41:

    (a) Sketch $f(x, y)$.

    (b) Solve $u_t = \kappa(u_{xx} + u_{yy})$, $-\infty < x, y < \infty$, $u(x, y, 0) = f(x, y)$ by evaluating the integral obtained in part (e) of Exercise 31.

    (c) With $\kappa = 1$, draw the solution surface at several instants in time, or animate the time-dependent temperature surfaces using an appropriate software tool.

32. $f(x, y) = e^{-x^2-y^2}$ **33.** $f(x, y) = \delta(x)\delta(y)$

34. $f(x, y) = [H(x+1) - H(x-1)][H(y+1) - H(y-1)]$

35. $f(x, y) = [H(x+1) - H(x-1)][H(y+1) - H(y-1)]x$

36. $f(x, y) = [H(x+1) - H(x-1)][H(y+1) - H(y-1)]xy$

37. $f(x, y) = [H(x+1) - H(x-1)][H(y+1) - H(y-1)](1 - x^2)$

38. $f(x, y) = [H(x+1) - H(x-1)][H(y+1) - H(y-1)](1 - x^2)$
    $(1 - y^2)$

Laplace's Equation on the Infinite Strip—Solution by Fourier Transform

Problem Statement

In this section we will consider the Dirichlet problem consisting of Laplace's equation $u_{xx}(x, y) + u_{yy}(x, y) = 0$ in the infinite (horizontal) strip $-\infty < x < \infty, 0 \le y \le b$, $0 < b$, and the boundary conditions $u(x, 0) = 0$, $u(x, b) = f(x)$, $\lim_{x \to \pm\infty} u(x, y) < \infty$. The conditions at $x = \pm\infty$ simply require $u(x, y)$ to remain bounded.

Solution

The solution of this Dirichlet problem, derived later in this section, is

$$u(x, y) = \frac{1}{\pi} \int_0^\infty \int_{-\infty}^\infty f(\beta) \frac{\sinh \alpha y}{\sinh \alpha b} \cos(\alpha(\beta - x)) \, d\beta \, d\alpha$$

EXAMPLE 30.24   For even the simplest of functions $f(x)$, the integrals in the solution given previously are difficult to evaluate. For example, with $b = 1$ and

$$f(x) = \begin{cases} 1 & |x| \le 1 \\ 0 & |x| > 1 \end{cases}$$

the inner integral becomes

$$\phi(\alpha, x, y) = \int_{-1}^1 \frac{\sinh \alpha y \cos(\alpha(\beta - x))}{\sinh \alpha} \, d\beta = \frac{\sinh \alpha y}{\alpha \sinh \alpha}[\sin(\alpha(x+1)) - \sin(\alpha(x - 1))]$$

and the outer integral is then the intractable $u(x, y) = \frac{1}{\pi} \int_0^\infty \phi(\alpha, x, y) \, d\alpha$.

Evaluating $u(x, y)$ at specific points is still a challenge, since for every such point a finite approximation for $\alpha = \infty$ must be determined. The integrand oscillates, adding to the difficulty of numeric evaluation. Figure 30.23 shows the integrands $\phi(\alpha, \frac{1}{2}, \frac{1}{2})$ and $\phi(\alpha, 3, \frac{1}{2})$ as the solid and dotted curves, respectively. As x increases, the integrand oscillates more rapidly. Numerically evaluating the integral gives, for example, $u(0.5, 0.5) = 0.4318996631$, a computation still in need of an error analysis. Although this was a simply stated boundary value problem, its solution, either exact or approximate, involves more than naive mathematics. ❖

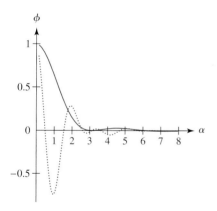

FIGURE 30.23   Example 30.24: Integrands $\phi(\alpha, \frac{1}{2}, \frac{1}{2})$ (solid) and $\phi(\alpha, 3, \frac{1}{2})$ (dotted)

Solution by Fourier Transform

The partial Fourier transform of Laplace's equation, taken with respect to x, yields

$$-\alpha^2 U(\alpha, y) + U_{yy}(\alpha, y) = 0$$

which we write, with the definition $V(y) = U(\alpha, y)$, as the ODE

$$-\alpha^2 V(y) + V''(y) = 0$$

The boundary conditions $u(x, 0) = 0$ and $u(x, b) = f(x)$ transform to

$$V(0) = U(\alpha, 0) = 0 \quad \text{and} \quad V(b) = U(\alpha, b) = F(\alpha)$$

The solution is

$$Y(y) = U(\alpha, y) = \frac{\sinh \alpha y}{\sinh \alpha b} F(\alpha)$$

and the Fourier inversion integral gives

$$u(x, y) = \frac{1}{2\pi} \int_{-\infty}^{\infty} F(\alpha) \frac{\sinh \alpha y}{\sinh \alpha b} e^{i\alpha x} \, d\alpha$$

Recognizing that $F(\alpha) = \int_{-\infty}^{\infty} f(\beta) e^{-i\alpha\beta} \, d\beta$ is the Fourier transform of $f(x)$, we have

$$u(x, y) = \frac{1}{2\pi} \int_{-\infty}^{\infty} \int_{-\infty}^{\infty} f(\beta) \frac{\sinh \alpha y}{\sinh \alpha b} e^{-i\alpha(\beta - x)} \, d\beta \, d\alpha$$

where the outer integral is a Cauchy principal value.

The exponential form of what is essentially the Fourier integral theorem can be brought into the trigonometric form as follows. Write the inner integral as

$$g(\alpha, \beta) = \int_{-\infty}^{\infty} f(\beta) \frac{\sinh \alpha y}{\sinh \alpha b} e^{-i\alpha(\beta - x)} \, d\beta$$

so that

$$u(x, y) = \frac{1}{2\pi} \lim_{\rho \to \infty} \left(\int_{-\rho}^{0} g(\alpha, \beta) \, d\alpha + \int_{0}^{\rho} g(\alpha, \beta) \, d\alpha \right)$$

Make the change of variables $\alpha = -\sigma$ in the first integral and $\alpha = \sigma$ in the second, obtaining

$$u(x, y) = \frac{1}{2\pi} \lim_{\rho \to \infty} \left(\int_{\rho}^{0} -g(-\sigma, \beta) \, d\sigma + \int_{0}^{\rho} g(\sigma, \beta) \, d\sigma \right)$$

Renaming σ as α, we actually have

$$
\begin{aligned}
u(x, y) &= \frac{1}{2\pi} \int_{0}^{\infty} \int_{-\infty}^{\infty} f(\beta) \frac{\sinh \alpha y}{\sinh \alpha b} e^{-i\alpha(\beta - x)} \, d\beta \, d\alpha \\
&\quad + \frac{1}{2\pi} \int_{0}^{\infty} \int_{-\infty}^{\infty} f(\beta) \frac{\sinh(-\alpha y)}{\sinh(-\alpha b)} e^{i\alpha(\beta - x)} \, d\beta \, d\alpha \\
&= \frac{1}{2\pi} \int_{0}^{\infty} \int_{-\infty}^{\infty} f(\beta) \frac{\sinh \alpha y}{\sinh \alpha b} \left(e^{-i\alpha(\beta - x)} + e^{i\alpha(\beta - x)} \right) \, d\beta \, d\alpha \\
&= \frac{1}{\pi} \int_{0}^{\infty} \int_{-\infty}^{\infty} f(\beta) \frac{\sinh \alpha y}{\sinh \alpha b} \cos(\alpha(\beta - x)) \, d\beta \, d\alpha
\end{aligned}
\tag{30.37}
$$

EXERCISES 30.6

1. If $\mathcal{F}_x^{-1}\left[\frac{\sinh \alpha y}{\sinh \alpha b}\right] = \frac{1}{2b} \frac{\sin(\pi y/b)}{\cosh(\pi x/b) + \cos(\pi y/b)}$, invert $U(\alpha, y) = \frac{\sinh \alpha y}{\sinh \alpha b} F(\alpha)$ using the convolution theorem.

2. If the boundary conditions on the infinite strip $-\infty < x < \infty$, $0 \le y \le b, b > 0$, are $u(x, b) = 0$, $u(x, 0) = g(x)$, $\lim_{x \to \pm\infty} u(x, y) < \infty$, solve the Dirichlet problem by the Fourier transform and

(a) the technique leading to (30.37).

(b) the convolution theorem. (*Hint:* See Exercise 1.)

(c) explain where $\lim_{|x| \to \infty} u(x, y) < \infty$, the boundary conditions at infinity, are used in the solution process.

3. If the boundary conditions on the infinite strip $-\infty < x < \infty$, $0 \le y \le b, b > 0$, are $u_y(x, 0) = 0$, $u(x, b) = f(x)$,

$\lim_{x \to \pm\infty} u(x, y) < \infty$, solve Laplace's equation by the Fourier transform and

(a) the technique leading to (30.37).

(b) the convolution theorem, given that $\mathcal{F}_x^{-1}\left[\frac{\cosh \alpha y}{\cosh \alpha b}\right] = \frac{1}{b} \frac{\cos(\pi y/2b) \cosh(\pi x/2b)}{\cosh(\pi x/b) + \cos(\pi y/b)}$.

4. If the boundary conditions on the infinite strip $-\infty < x < \infty$, $0 \le y \le b, b > 0$, are $u(x, 0) = g(x)$, $u_y(x, b) = 0$, $\lim_{|x| \to \infty} u(x, y) < \infty$, solve Laplace's equation by the Fourier transform and

(a) the technique leading to (30.37).

(b) the convolution theorem. (*Hint:* See Exercise 3.)

(c) give a physical interpretation of this BVP.

5. If the boundary conditions on the infinite strip $-\infty < x < \infty$, $0 \le y \le b, b > 0$, are $u(x, 0) = 0, u_y(x, b) = f(x)$, $\lim_{x \to \pm\infty} u(x, y) < \infty$, solve Laplace's equation by the Fourier transform and

(a) the technique leading to (30.37).

(b) the convolution theorem, given that $\mathcal{F}_x^{-1}\left[\frac{\sinh \alpha y}{\alpha \cosh \alpha b}\right] = \frac{1}{2\pi} \ln\left[\frac{\cosh(\pi x/2b)+\sin(\pi y/2b)}{\cosh(\pi x/2b)-\sin(\pi y/2b)}\right]$.

(c) give a physical interpretation of this BVP.

6. If the boundary conditions on the infinite strip $-\infty < x < \infty$, $0 \le y \le b, b > 0$, are $u(x, b) = 0, u_y(x, 0) = g(x)$, $\lim_{x \to \pm\infty} u(x, y) < \infty$, solve Laplace's equation by the Fourier transform and

(a) the technique leading to (30.37).

(b) the convolution theorem. (*Hint:* See Exercise 5.)

(c) give a physical interpretation of this BVP.

7. If the boundary conditions on the infinite strip $-\infty < x < \infty$, $0 \le y \le b, b > 0$, are $u_y(x, b) = f(x), u_y(x, 0) = 0$, $\lim_{x \to \pm\infty} u(x, y) < \infty$, solve Laplace's equation by the Fourier transform and

(a) the technique leading to (30.37).

(b) the convolution theorem, given that $\mathcal{F}_x^{-1}\left[\frac{\cosh \alpha y}{\alpha \sinh \alpha b}\right] = -\frac{1}{2\pi} \ln(\cosh \frac{\pi x}{b} + \cos \frac{\pi y}{b})$.

(c) give a physical interpretation of this BVP.

8. If the boundary conditions on the infinite strip $-\infty < x < \infty$, $0 \le y \le b, b > 0$, are $u_y(x, b) = 0, u_y(x, 0) = g(x)$, $\lim_{x \to \pm\infty} u(x, y) < \infty$, solve Laplace's equation by the Fourier transform and

(a) the technique leading to (30.37).

(b) the convolution theorem. (*Hint:* See Exercise 7.)

(c) give a physical interpretation of this BVP.

30.7 The Fourier Sine Transform

The Fourier Integral Theorem for Odd Functions

If $f(x)$ is an odd function whereby $f(-x) = -f(x)$, then (30.13) and (30.14), the trigonometric form of the Fourier integral theorem, reduces to

$$f(x) = \frac{2}{\pi} \int_0^\infty \int_0^\infty f(\beta) \sin \alpha\beta \sin \alpha x \, d\beta \, d\alpha \tag{30.38}$$

because $A(\alpha) = 0$ and $B(\alpha) = \int_{-\infty}^\infty f(\beta) \sin \alpha\beta \, d\beta = 2\int_0^\infty f(\beta) \sin \alpha\beta \, d\beta$.

The Fourier Sine Transform

The form (30.38) taken by the Fourier integral theorem for odd functions inspires the definition of the self-inverting *Fourier sine transform*

$$\mathcal{F}_s[f(x)] = F_s(\alpha) = \sqrt{\frac{2}{\pi}} \int_0^\infty f(\beta) \sin \alpha\beta \, d\beta \tag{30.39}$$

for which the inversion integral is $f(x) = \sqrt{\frac{2}{\pi}} \int_0^\infty F_s(\alpha) \sin \alpha x \, d\alpha$, the same integral as for the transform.

EXAMPLE 30.25 The Fourier sine transform of $f(x) = e^{-x}$ with domain $[0, \infty)$ is

$$F_s(\alpha) = \sqrt{\frac{2}{\pi}} \int_0^\infty e^{-\beta} \sin \alpha\beta \, d\beta = \sqrt{\frac{2}{\pi}} \frac{\alpha}{(1 + \alpha^2)} \tag{30.40}$$

The inversion integral for this Fourier sine transform is $f(x) = \frac{2}{\pi} \int_0^\infty \frac{\alpha \sin \alpha x}{(1+\alpha^2)} \, d\alpha = e^{-x}$, $x \ge 0$. ❖

> **THEOREM 30.3   OPERATIONAL PROPERTY OF THE FOURIER SINE TRANSFORM**
>
> The one operational property of consequence in applying the Fourier sine transform to differential equations is the derivative rule, stated in the following theorem.
>
> 1. f, f', f'' are all continuous on the interval $[0, \infty)$.
> 2. f and f' both go to zero as x approaches ∞.
> 3. f is absolutely integrable, that is, $\int_0^\infty |f(x)|\, dx < \infty$.
>
> $$\Longrightarrow \mathcal{F}_s[f''(x)] = -\alpha^2 \mathcal{F}_s[f(x)] + \alpha \sqrt{\frac{2}{\pi}} f(0)$$

Notice that the operational law for derivatives is stated in terms of the *second* derivative. The Fourier sine transform of the *first* derivative cannot be expressed in terms of the sine transform itself, so it is not useful to consider an operational law for first derivatives.

EXAMPLE 30.26   Since $f(x) = e^{-x} = f''(x)$, the operational rule gives

$$\mathcal{F}_s[f''(x)] = -\alpha^2 \left(\sqrt{\frac{2}{\pi}} \frac{\alpha}{1+\alpha^2} \right) + \alpha \sqrt{\frac{2}{\pi}} = \sqrt{\frac{2}{\pi}} \frac{\alpha}{1+\alpha^2} = \mathcal{F}_s[e^{-x}] \qquad \textbf{(30.41)}$$

In fact, $\mathcal{F}_s[e^{-x}]$ can be obtained from the operational rule by solving $\mathcal{F}_s[e^{-x}] = \mathcal{F}_s[(e^{-x})''] = -\alpha^2 \mathcal{F}_s[e^{-x}] + \alpha \sqrt{\frac{2}{\pi}}$ for $\mathcal{F}_s[e^{-x}] = \sqrt{\frac{2}{\pi}} \frac{\alpha}{1+\alpha^2}$. ❖

PROOF OF THEOREM 30.3   The operational rule for derivatives is established via integration by parts. Integrating to a finite upper limit and taking $u = \sin \alpha\beta$ in the integration-by-parts formula $\int u\, dv = uv - \int v\, du$, the transform $\mathcal{F}_s[f''(x)]$ becomes

$$\sqrt{\frac{2}{\pi}} \int_0^\sigma f''(\beta) \sin \alpha\beta\, d\beta = \sqrt{\frac{2}{\pi}} \sin \alpha\sigma f'(\sigma) - \alpha \sqrt{\frac{2}{\pi}} \int_0^\sigma f'(\beta) \cos \alpha\beta\, d\beta \qquad \textbf{(30.42)}$$

Integrating by parts a second time with $u = \cos \alpha\beta$, we then obtain for the terms on the right in (30.42)

$$\sqrt{\frac{2}{\pi}} \sin \alpha\sigma f'(\sigma) - \alpha \sqrt{\frac{2}{\pi}} f(\sigma) \cos \alpha\sigma + \alpha \sqrt{\frac{2}{\pi}} f(0) - \alpha^2 \sqrt{\frac{2}{\pi}} \int_0^\sigma f(\beta) \sin \alpha\beta\, d\beta$$

$$\textbf{(30.43)}$$

Now let σ become infinite. The first term in (30.43) goes to zero because f' does. The second term in (30.43) goes to zero because f does. The remaining two terms in (30.43) then become the operational rule we set out to derive.

Solving the Dirichlet Problem with the Fourier Sine Transform

The Fourier sine transform, applied to the Dirichlet problem on the first quadrant ($0 \le x < \infty, 0 \le y < \infty$) with boundary conditions $u(0, y) = 0, u(x, 0) = f(x), \lim_{y \to \infty} u(x, y) = 0$, gives the solution

$$u(x, y) = \frac{1}{\pi} \int_0^\infty f(\beta) \left(\frac{y}{y^2 + (\beta - x)^2} - \frac{y}{y^2 + (\beta + x)^2} \right) d\beta \qquad \textbf{(30.44)}$$

◆ **EXAMPLE 30.27**   For the pulse

$$f(x) = 1 - H(x - 1) = H(1 - x) \qquad x \geq 0 \tag{30.45}$$

where $H(x)$ is the Heaviside function, the solution given by (30.44) and graphed in Figure 30.24 is

$$u(x, y) = \frac{1}{\pi} \left(2 \arctan \frac{x}{y} - \arctan \frac{x-1}{y} - \arctan \frac{x+1}{y} \right) \tag{30.46}$$

If $u(x, y)$ is interpreted as a potential, the equipotentials are then the level curves shown in color in Figure 30.25. The field lines, shown in black in Figure 30.25, are the integral curves for the differential equations $\mathbf{r}' = -\nabla u$, which, in component form, are $x'(t) = -u_x(x(t), y(t))$ and $y'(t) = -u_y(x(t), y(t))$. The arrows in Figure 30.25 represent the electric field $\mathbf{E} = -\nabla u$.   ❖

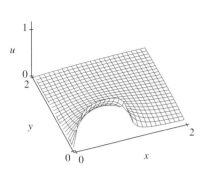

FIGURE 30.24   Solution surface in Example 30.27

DERIVATION OF THE SOLUTION (30.44)   For Laplace's equation, the Fourier sine transform, taken with respect to x, yields

$$\alpha \sqrt{\frac{2}{\pi}} u(0, y) - \alpha^2 U(\alpha, y) + U_{yy}(\alpha, y) = 0$$

and for the boundary condition on the x-axis, yields $U(\alpha, 0) = F(\alpha)$. Then, since $u(0, y) = 0$, the transformed differential equation, with $V(y) = U(\alpha, y)$, becomes

$$V''(y) - \alpha^2 V(y) = 0$$

The solution is $V(y) = c_1 e^{\alpha y} + c_2 e^{-\alpha y}$.

For $u(x, y)$ to go to zero far from the origin, the unbounded positive exponential must not appear in $V(y) = U(\alpha, y)$. This determines $c_1 = 0$. Then, apply the boundary condition $V(0) = U(\alpha, 0) = F(\alpha)$ to determine that $c_2 = F(\alpha)$. Hence, we have

$$V(y) = U(\alpha, y) = F(\alpha)e^{-\alpha y} \quad \text{and} \quad u(x, y) = \sqrt{\frac{2}{\pi}} \int_0^\infty F(\alpha)e^{-\alpha y} \sin \alpha x \, d\alpha$$

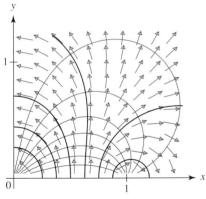

FIGURE 30.25   Example 30.27: Equipotentials (in color), field lines (in black), and electric field vectors

Replacing $F(\alpha)$, the Fourier sine transform of $f(x)$, with its defining integral from (30.39), the solution becomes

$$u(x, y) = \frac{2}{\pi} \int_0^\infty \int_0^\infty f(\beta) \sin \alpha \beta \, e^{-\alpha y} \sin \alpha x \, d\beta \, d\alpha$$

Since $\sin \alpha \beta \sin \alpha x = \frac{1}{2}[\cos(\alpha(\beta - x)) - \cos(\alpha(\beta + x))]$, we can reverse the order of integration, obtaining for the new inner integral,

$$\frac{1}{2} \int_0^\infty e^{-\alpha y} [\cos(\alpha(\beta - x)) - \cos(\alpha(\beta + x))] \, d\alpha = \frac{1}{2} \left(\frac{y}{y^2 + (\beta - x)^2} - \frac{y}{y^2 + (\beta + x)^2} \right) \tag{30.47}$$

so the solution can be represented as

$$u(x, y) = \frac{2}{\pi} \int_0^\infty f(\beta) \frac{1}{2} \left(\frac{y}{y^2 + (\beta - x)^2} - \frac{y}{y^2 + (\beta + x)^2} \right) d\beta$$

which is just (30.44).

EXERCISES 30.7–Part A

Throughout these exercises, H(x) is the Heaviside function.

A1. Verify the integration in (30.40).

A2. Verify the transform in (30.41).

A3. Verify the calculation in (30.42).

A4. Verify the calculation in (30.43).

A5. Verify that $1 - H(x - 1) = H(1 - x)$.

A6. Verify that (30.44) applied to $f(x) = H(1 - x)$ gives (30.46).

A7. Verify the integration in (30.47).

EXERCISES 30.7–Part B

B1. Show that the Fourier sine transform of $f(x)$ is given by $F_s(\alpha) = \frac{i}{\sqrt{2\pi}}\mathcal{F}[f(x)]$ when $f(x)$ is odd.

If $a > 0$, in Exercises B2–16:

   (a) Use a software tool with a built-in Fourier sine transform to obtain $F_s(\alpha)$.

   (b) Where possible, obtain the Fourier sine transform by integration.

   (c) Where possible, obtain the Fourier sine transform from $\mathcal{F}[f(x)]$, the exponential Fourier transform. (See Exercise B1.)

B2. $f(x) = \dfrac{1}{x}$ **B3.** $f(x) = \dfrac{1}{\sqrt{x}}$ **B4.** $f(x) = x^{-1/3}$

B5. $f(x) = \dfrac{x}{a^2 + x^2}$ **B6.** $f(x) = \dfrac{x}{(a^2 + x^2)^2}$

B7. $f(x) = \dfrac{1}{x(a^2 + x^2)}$ **B8.** $f(x) = e^{-ax}$ **B9.** $f(x) = xe^{-ax}$

B10. $f(x) = \dfrac{1}{x}e^{-ax}$ **B11.** $f(x) = xe^{-a^2x^2}$

B12. $f(x) = \arctan\dfrac{a}{x}$

B13. $f(x) = H(a - x)$

B14. $f(x) = \dfrac{x}{4 + x^4}$ **B15.** $f(x) = e^{-ax}\sin ax$

B16. $f(x) = \text{erfc}(ax) = 1 - \text{erf}(ax)$

In Exercises B17–28:

   (a) Sketch $f(x)$.

   (b) Obtain the solution of the BVP $u_{xx} + u_{yy} = 0$, in the first quadrant, $0 < x, y < \infty$, $u(0, y) = 0$, $u(x, 0) = f(x)$.

   (c) Plot the solution surface.

   (d) Obtain a contour plot showing the level curves (isotherms or equipotentials).

   (e) Obtain, and superimpose on the contour plot, a graph of the flow lines.

   (f) Show that $u \to 0$ as either x or y become infinite.

B17. $f(x) = \begin{cases} x & 0 \le x < 1 \\ 2 - x & 1 \le x < 2 \\ 0 & x \ge 2 \end{cases}$

B18. $f(x) = \begin{cases} x & 0 \le x < 1 \\ 1 & 1 \le x < 2 \\ 3 - x & 2 \le x < 3 \\ 0 & x \ge 3 \end{cases}$

B19. $f(x) = \begin{cases} x^2 & 0 \le x < 1 \\ 1 & 1 \le x < 2 \\ 3 - x & 2 \le x < 3 \\ 0 & x \ge 3 \end{cases}$

B20. $f(x) = \begin{cases} \dfrac{x}{5} & 0 \le x < 5 \\ 6 - x & 5 \le x < 6 \\ 0 & x \ge 6 \end{cases}$

B21. $f(x) = (1 - H(x - 1))x$ **B22.** $f(x) = (1 - H(x - 1))x^2$

B23. $f(x) = (1 - H(x - 1))(1 - x)$

B24. $f(x) = (1 - H(x - 1))(1 - x^2)$

B25. $f(x) = (1 - H(x - 1))x(1 - x)$

B26. $f(x) = (1 - H(x - 1))x(1 - x)^2$

B27. $f(x) = (1 - H(x - 1))x^2(1 - x)$

B28. $f(x) = (1 - H(x - 1))x^3(1 - x)$

B29. Use the Fourier sine transform (taken with respect to y) to solve the BVP $u_{xx} + u_{yy} = 0$, $0 < x, y < \infty$, $u(0, y) = g(y)$, $u(x, 0) = 0$.

In Exercises B30–35:

   (a) Using superposition, solve the BVP $u_{xx} + u_{yy} = 0$, in the first quadrant, $0 < x, y < \infty$, $u(0, y) = g(y)$, $u(x, 0) = f(x)$.

   (b) Plot the solution surface.

   (c) Obtain a contour plot showing the level curves (isotherms or equipotentials).

   (d) Obtain, and overlay on the contour plot, a graph of the flow lines.

B30. $f(x) = \begin{cases} x & 0 \le x < 1 \\ 2 - x & 1 \le x < 2, \\ 0 & x \ge 2 \end{cases}$ $g(y) = \begin{cases} \dfrac{y}{5} & 0 \le y < 5 \\ 6 - y & 5 \le y < 6 \\ 0 & y \ge 6 \end{cases}$

B31. $f(x) = \begin{cases} x & 0 \le x < 1 \\ 1 & 1 \le x < 2 \\ 3 - x & 2 \le x < 3 \\ 0 & x \ge 3 \end{cases}$, $g(y) = \begin{cases} \dfrac{y}{5} & 0 \le y < 5 \\ 6 - y & 5 \le y < 6 \\ 0 & y \ge 6 \end{cases}$

B32. $f(x) = (1 - H(x - 1))x$, $g(y) = (1 - H(y - 1))y^3(1 - y)$

B33. $f(x) = (1 - H(x - 1))x^2$, $g(y) = (1 - H(y - 1))y^2(1 - y)$

B34. $f(x) = (1 - H(x - 1))(1 - x)$, $g(y) = (1 - H(y - 1))y(1 - y)^2$

B35. $f(x) = (1 - H(x - 1))(1 - x^2)$, $g(y) = (1 - H(y - 1))y(1 - y)$

B36. Use the Fourier sine transform (taken with respect to x) to solve the BVP $u_{xx} + u_{yy} = 0, 0 < x < \infty, 0 < y < b, u(0, y) = 0$, $u(x, 0) = f(x), u(x, b) = 0$, obtaining

$$u(x, y) = \sqrt{\frac{2}{\pi}} \int_0^\infty F(\alpha) \frac{\sinh \alpha(b - y)}{\sinh \alpha b} \sin \alpha x \, d\alpha$$

$$= \frac{2}{\pi} \int_0^\infty \int_0^\infty f(\beta) \frac{\sinh \alpha(b - y)}{\sinh \alpha b} \sin \alpha \beta \sin \alpha x \, d\beta \, d\alpha$$

Show that, in the limit as $b \to \infty$, the second form of the solution approaches that in (30.44).

B37. Use the Fourier sine transform (taken with respect to x) to solve the BVP $u_{xx} + u_{yy} = 0, 0 < x < \infty, 0 < y < b, u(0, y) = 0$, $u(x, 0) = 0, u(x, b) = g(x)$. What additional conditions must $u(x, y)$ and $g(x)$ satisfy in order for the solution to exist?

B38. Use the Fourier sine transform (taken with respect to y) to solve the BVP $u_{xx} + u_{yy} = 0, 0 < y < \infty, 0 < x < b, u(0, y) = f(y)$, $u(x, 0) = 0, u(b, y) = 0$. What additional conditions must $u(x, y)$ and $f(y)$ satisfy in order for the solution to exist?

B39. Use the Fourier sine transform (taken with respect to y) to solve the BVP $u_{xx} + u_{yy} = 0, 0 < y < \infty, 0 < x < b, u(0, y) = 0$, $u(x, 0) = 0, u(b, y) = g(y)$. What additional conditions must $u(x, y)$ and $g(y)$ satisfy in order for the solution to exist?

In Exercises B40–42:

   (a) Use the Fourier sine transform (taken with respect to y) to obtain a formal solution of the BVP $u_{xx} + u_{yy} = 0$, $0 < y < \infty, 0 < x < \pi, u(0, y) = 0, u(x, 0) = f(x)$, $u(\pi, y) = 0$.

   (b) If possible, obtain an approximate value for $u(1, 1)$ via numeric integration.

B40. $f(x) = 1$ **B41.** $f(x) = \sin x$

B42. $f(x) = x(\pi - x)$

In Exercises B43–45, use the Fourier sine transform (taken with respect to y) to obtain a formal solution to the BVP $u_{xx} + u_{yy} = 0, 0 < y < \infty$, $0 < x < 1, u(0, y) = 0, u(x, 0) = f(x), u(1, y) = 0$.

B43. $f(x) = x(1 - x)$ **B44.** $f(x) = e^{-x}$ **B45.** $f(x) = 1$

In Exercises B46–48, obtain a finite-difference solution. The numerical solution requires imposing a boundary condition at $y = \infty$, which can be approximated using $u(x, b) = 0$ for a suitable value of b such as $b = 10$ or $b = 20$.

B46. the BVP of Exercise B40   **B47.** the BVP of Exercise B41

B48. the BVP of Exercise B42

B49. Use the Fourier sine transform (taken with respect to x) to obtain a formal solution to the BVP $u_{xx} + u_{yy} = 0, 0 < x < \infty$, $0 < y < 1, u(0, y) = y^2, u(x, 0) = 0, u(1, y) = 0$.

B50. For the BVP $u_{xx} + u_{yy} = 0, 0 < x < \infty, 0 < y < b$, $u(0, y) = A, u_y(x, 0) = f(x), u_y(x, b) = g(x)$:

   (a) Give a physical interpretation.

   (b) Use the Fourier sine transform (taken with respect to x) to obtain a formal solution.

In Exercises B51–55, use the Fourier sine transform (taken with respect to x) to obtain a formal solution of the BVP $u_{xx} + u_{yy} = 0, 0 < x < \infty$, $0 < y < 1, u(0, y) = h(y), u_y(x, 0) = f(x), u_y(x, 1) = g(x)$. Interpret each BVP in terms of the steady-state temperatures in a semi-infinite strip.

B51. $h(y) = 1, f(x) = 1 - H(x - 1), g(x) = H(x - 1) - H(x - 2)$

B52. $h(y) = y(1 - y), f(x) = (H(x - 1) - H(x - 2))(x - 1)$ $(2 - x), g(x) = 0$

B53. $h(y) = \sin \pi y, f(x) = e^{-x}, g(x) = \arctan \dfrac{1}{x}$

B54. $h(y) = 0, f(x) = (H(x - 1) - H(x - 2))(x - 1)(2 - x)$, $g(x) = (H(x - 3) - H(x - 4))(x - 3)(4 - x)$

B55. $h(y) = 0, f(x) = xe^{-x}, g(x) = \dfrac{x}{1 + x^2}$

B56. For the BVP $u_{xx} + u_{yy} = 0, 0 < x < \infty, 0 < y < b$, $u(0, y) = A, u(x, 0) = f(x), u_y(x, b) = g(x)$:

   (a) Give a physical interpretation.

   (b) Use the Fourier sine transform (taken with respect to x) to obtain a formal solution.

In Exercises B57–61, use the Fourier sine transform (taken with respect to x) to obtain a formal solution of the BVP $u_{xx} + u_{yy} = 0, 0 < x < \infty, 0 < y < 1, u(0, y) = h(y), u(x, 0) = f(x), u_y(x, 1) = g(x)$. Interpret each BVP in terms of the steady-state temperatures in a semi-infinite strip.

B57. $h(y) = 1, f(x) = 1 - H(x - 1), g(x) = 1 - H(x - 2)$

B58. $h(y) = y(1 - y), f(x) = (1 - H(x - 1))x(1 - x), g(x) = e^{-x}$

B59. $h(y) = \sin \pi y, f(x) = 0, g(x) = \dfrac{x}{1 + x^2}$

B60. $h(y) = 0, f(x) = xe^{-x}, g(x) = 0$

B61. $h(y) = \sin 2\pi y, f(x) = \dfrac{x}{4 + x^4}, g(x) = xe^{-x}$

B62. Use the Fourier sine transform (taken with respect to x) to obtain

$$u(x, t) = \sqrt{\frac{2}{\pi}} \int_0^\infty F_s(\alpha) e^{-\alpha^2 \kappa t} \sin \alpha x \, d\alpha$$

$$= \frac{2}{\pi} \int_0^\infty \int_0^\infty f(\beta) \sin \alpha \beta e^{-\alpha^2 \kappa t} \sin \alpha x \, d\beta \, d\alpha$$

as the formal solution of the BVP $u_t = \kappa u_{xx}, 0 < x < \infty$, $u(0, t) = 0, u(x, 0) = f(x)$. The solution represents the time-dependent temperatures in a semi-infinite rod.

In Exercises B63 and 64, use numeric integration and the results of Exercise B62 to approximate $u(0.5, 0.1)$ when $\kappa = 1$.

B63. $f(x) = 1 - H(x - 1)$ **B64.** $f(x) = e^{-x}$

B65. Use the Fourier sine transform (taken with respect to x) to obtain

$$u(x, t) = \sqrt{\frac{2}{\pi}} \int_0^\infty F_s(\alpha) e^{-\alpha^2 \kappa t - t^2/2} \sin \alpha x \, d\alpha$$

$$= \frac{2}{\pi} \int_0^\infty \int_0^\infty f(\beta) \sin \alpha \beta e^{-\alpha^2 \kappa t - t^2/2} \sin \alpha x \, d\beta \, d\alpha$$

as the formal solution of the BVP $u_t + tu = \kappa u_{xx}, 0 < x < \infty$, $u(0, t) = 0, u(x, 0) = f(x)$.

In Exercises B66 and 67, use numeric integration and the results of Exercise B65 to approximate $u(0.1, 0.5)$ when $\kappa = 1$.

B66. $f(x) = 1 - H(x - 1)$ **B67.** $f(x) = H(x - 1) - H(x - 2)$

In Exercises B68 and 69:

(a) Use the Fourier sine transform (taken with respect to x) to obtain a formal solution of the BVP $u_t + hu = u_{xx}$, $0 < x < \infty, u(0, t) = 0, u(x, 0) = f(x)$.

(b) With $h = 1$, and using numeric integration, compute and plot temperature profiles for $t = \frac{1}{2}, 1, \frac{3}{2}$.

(c) Repeat part (b) with $h = -1$.

B68. $f(x) = 1 - H(x - 1)$ **B69.** $f(x) = H(x - 1) - H(x - 2)$

In Exercises B70–72:

(a) Use the Fourier sine transform (taken with respect to x) to obtain a formal solution of the BVP $u_{tt} = u_{xx}, 0 < x < \infty$, $u(0, t) = f(t), u(x, 0) = 0, u_t(x, 0) = 0$.

(b) Give a physical interpretation of the BVP.

(c) Obtain a graph of $u(\frac{1}{2}, t)$ for $0 < t \leq 3$.

B70. $f(t) = 1 - H(t - 1)$ **B71.** $f(t) = (1 - H(t - 1))t(1 - t)$

B72. $f(t) = \begin{cases} t & 0 \leq t < 1 \\ 2 - t & 1 \leq t < 2 \\ 0 & t \geq 2 \end{cases}$

B73. Give a physical interpretation of the BVP $u_{tt} + u_t = u_{xx}$, $0 < x < \infty, u(0, t) = 0, u(x, 0) = f(x), u_t(x, 0) = 0$, and then use the Fourier sine transform (taken with respect to x) to obtain a formal solution.

B74. Give a physical interpretation of the BVP $u_{tt} + u_t = u_{xx}$, $0 < x < \infty, u(0, t) = f(t), u(x, 0) = 0, u_t(x, 0) = 0$, then take $f(t) = te^{-t}$ and use the Fourier sine transform (taken with respect to x) to obtain a formal solution. Don't be surprised at the algebraic complexity of the solution.

30.8   The Fourier Cosine Transform

The Fourier Integral Theorem for Even Functions

If $f(x)$ has the even symmetry $f(-x) = f(x)$, then (30.13) and (30.14), the trigonometric form of the Fourier integral theorem, reduces to

$$f(x) = \frac{2}{\pi} \int_0^\infty \int_0^\infty f(\beta) \cos \alpha \beta \cos \alpha x \, d\beta \, d\alpha \qquad (30.48)$$

because $B(\alpha) = 0$ and $A(\alpha) = \int_{-\infty}^\infty f(\beta) \cos \alpha \beta \, d\beta = 2 \int_0^\infty f(\beta) \cos \alpha \beta \, d\beta$.

The Fourier Cosine Transform

The form (30.48) taken by the Fourier integral theorem for even functions inspires the definition of the self-inverting *Fourier cosine transform*

$$\mathcal{F}_c[f(x)] = F_c(\alpha) = \sqrt{\frac{2}{\pi}} \int_0^\infty f(\beta) \cos \alpha \beta \, d\beta \qquad (30.49)$$

Inversion is by the same integral, namely, $f(x) = \sqrt{\frac{2}{\pi}} \int_0^\infty F_c(\alpha) \cos \alpha x \, d\alpha$.

EXAMPLE 30.28   The Fourier cosine transform of $f(x) = e^{-x}$ with domain $[0, \infty)$ is

$$F_c(\alpha) = \sqrt{\frac{2}{\pi}} \int_0^\infty e^{-\alpha} \cos \alpha \beta \, d\beta = \sqrt{\frac{2}{\pi}} \frac{1}{1 + \alpha^2}$$

The inversion integral for this Fourier cosine transform is then $f(x) = \frac{2}{\pi} \int_0^\infty \frac{\cos \alpha x}{1+\alpha^2} \, d\alpha = e^{-x}, x \geq 0$. ❖

THEOREM 30.4   OPERATIONAL PROPERTY OF THE FOURIER COSINE TRANSFORM

The one operational property of consequence in applying the Fourier cosine transform to differential equations is the derivative rule, stated in the following theorem.

1. f, f', f'' are all continuous on the interval $[0, \infty)$.
2. f and f' both go to zero as x approaches ∞.
3. f is absolutely integrable, that is, $\int_0^\infty |f(x)| \, dx < \infty$.

$$\implies \mathcal{F}_c[f''(x)] = -\alpha^2 \mathcal{F}_c[f(x)] - \sqrt{\frac{2}{\pi}} f'(0)$$

Notice that the operational law for derivatives is stated in terms of the *second* derivative. The Fourier cosine transform of the *first* derivative cannot be expressed in terms of the cosine transform itself, so it is not useful to consider an operational law for first derivatives.

EXAMPLE 30.29 Since $f(x) = e^{-x} = f''(x)$, the operational rule gives $\mathcal{F}_c[f''(x)] = -\alpha^2 \left(\sqrt{\frac{2}{\pi}} \frac{1}{1+\alpha^2} \right) - \sqrt{\frac{2}{\pi}} = \sqrt{\frac{2}{\pi}} \frac{1}{1+\alpha^2} = \mathcal{F}_c[e^{-x}]$. As in Example 30.26, $\mathcal{F}_c[e^{-x}]$ can be obtained from the operational rule by solving $\mathcal{F}_c[e^{-x}] = \mathcal{F}_c[(e^{-x})''] = -\alpha^2 \mathcal{F}_c[e^{-x}] + \sqrt{\frac{2}{\pi}}$ for $\mathcal{F}_c[e^{-x}] = \sqrt{\frac{2}{\pi}} \frac{1}{1+\alpha^2}$. ❖

PROOF OF THEOREM 30.4   The operational rule for derivatives is established via integration by parts. Integrating to a finite upper limit and taking $u = \cos \alpha \beta$ in the integration-by-parts formula $\int u \, dv = uv - \int v \, du$, the transform $\mathcal{F}_c[f''(x)]$ becomes

$$\sqrt{\frac{2}{\pi}} f'(\sigma) \cos \alpha \sigma - \sqrt{\frac{2}{\pi}} f'(0) + \alpha \sqrt{\frac{2}{\pi}} \int_0^\sigma f'(\beta) \sin \alpha \beta \, d\beta$$

Integrating by parts a second time with $u = \sin \alpha \beta$, we then obtain

$$\sqrt{\frac{2}{\pi}} f'(\sigma) \cos \alpha \sigma - \sqrt{\frac{2}{\pi}} f'(0) + \alpha \sqrt{\frac{2}{\pi}} f(\sigma) \sin \alpha \sigma - \alpha^2 \sqrt{\frac{2}{\pi}} \int_0^\sigma f(\beta) \cos \alpha \beta \, d\beta$$

$$(30.50)$$

Now let σ become infinite. The first term in (30.50) goes to zero because f' does. The third term in (30.50) goes to zero because f does. The remaining two terms in (30.50) then become the operational rule we set out to prove.

Solving Boundary Value Problems with the Fourier Cosine Transform

The Fourier cosine transform, applied to the boundary value problem consisting of Laplace's equation $u_{xx}(x, y) + u_{yy}(x, y) = 0$ in the semi-infinite (horizontal) strip $0 < x < \infty$, $0 < y < b$ and the boundary conditions $u_x(0, y) = 0$, $u(x, 0) = f(x)$, $u(x, b) = 0$, gives the solution in the form

$$u(x, y) = -\sqrt{\frac{2}{\pi}} \int_0^\infty F(\alpha) \frac{\sinh(\alpha(y-b))}{\sinh \alpha b} \cos \alpha x \, d\alpha \qquad (30.51a)$$

$$= -\frac{2}{\pi} \int_0^\infty \int_0^\infty f(\beta) \frac{\sinh(\alpha(y-b))}{\sinh \alpha b} \cos \alpha \beta \cos \alpha x \, d\beta \, d\alpha \qquad (30.51b)$$

The homogeneous Neumann boundary condition $u_x(0, y) = 0$ represents insulation, either thermal or electrical, on the left edge.

EXAMPLE 30.30   The pulse (30.45) has the Fourier cosine transform $F(\alpha) = \sqrt{\frac{2}{\pi}} \frac{\sin \alpha}{\alpha}$. With $b = 1$, the integrand in (30.51a) becomes

$$\phi(\alpha, x, y) = \frac{\sin \alpha \, \sinh(\alpha(y-1))}{\alpha \sinh \alpha} \cos \alpha x$$

and the solution itself becomes the intractable $u(x, y) = -\frac{2}{\pi} \int_0^\infty \phi(\alpha, x, y) \, d\alpha$. Explicit evaluation must yield to numeric computation, but because of the oscillatory nature of the integrand and the infinite upper limit of integration, even this task is not simple. Figure 30.26 shows $\phi(\alpha, \frac{1}{2}, \frac{1}{2})$ as the solid curve and $\phi(\alpha, 3, \frac{1}{2})$ as the dotted curve. As x increases, the integrand oscillates more rapidly, making numeric integration even more difficult.

Graphs of $u(x, \frac{1}{4})$, $u(x, \frac{1}{2})$, and $u(x, \frac{3}{4})$, computed numerically, are shown as solid, dotted, and dashed curves, respectively, in Figure 30.27. The solution surface corresponding to $u(x, y)$ is given in Figure 30.28. ❖

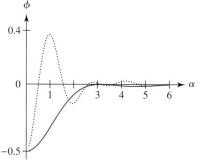

FIGURE 30.26   Example 30.30: Integrands $\phi(\alpha, \frac{1}{2}, \frac{1}{2})$ (solid) and $\phi(\alpha, 3, \frac{1}{2})$ (dotted)

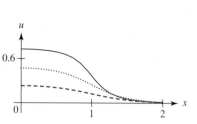

FIGURE 30.27   Example 30.30: Graphs of $u(x, \frac{1}{4})$, $u(x, \frac{1}{2})$, and $u(x, \frac{3}{4})$ are solid, dotted, and dashed curves, respectively

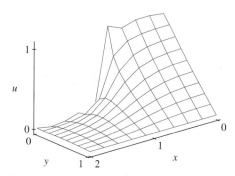

FIGURE 30.28   Solution surface in Example 30.30

DERIVATION OF FORMULAS (30.51a) AND (30.51b)   The Fourier cosine transform, taken with respect to x, yields

$$-\sqrt{\frac{2}{\pi}} u_x(0, y) - \alpha^2 U(\alpha, y) + U_{yy}(\alpha, y) = 0$$

for Laplace's equation and $U(\alpha, 0) = F(\alpha)$ for the boundary condition on the x-axis. Then, since $u_x(0, y) = 0$, the transformed differential equation becomes, with $V(y) = U(\alpha, y)$,

$$V''(y) - \alpha^2 V(y) = 0$$

This equation, along with the boundary conditions $V(0) = F(\alpha)$, $V(b) = 0$, has solution

$$V(y) = U(\alpha, y) = -F(\alpha) \frac{\sinh(\alpha(y-b))}{\sinh \alpha b}$$

Applying the inversion integral for the Fourier cosine transform then gives the solution as (30.51a). Replacing $F(\alpha)$, the Fourier cosine transform of $f(x)$, with (30.49), the solution then becomes (30.51b). Writing

$$\cos\alpha\beta\cos\alpha x = \tfrac{1}{2}[\cos(\alpha(\beta - x)) + \cos(\alpha(\beta + x))]$$

does not lead to a simplification because reversing the order of integration just gives the intractable integrals

$$\int_0^\infty \frac{\sinh(\alpha(y - b))}{\sinh\alpha b}\cos(\alpha(x - \beta))\,d\alpha \text{ and } \int_0^\infty \frac{\sinh(\alpha(y - b))}{\sinh\alpha b}\cos(\alpha(x + \beta))\,d\alpha$$

for the new inner integral.

EXERCISES 30.8

Throughout these exercises, $H(x)$ is the Heaviside function.

1. Show that the Fourier cosine transform of $f(x)$ is given by $F_c(\alpha) = \frac{1}{\sqrt{2\pi}}\mathcal{F}[f(x)]$ when $f(x)$ is even.

2. Verify the transform and its inversion in Example 30.28.

If $a, b > 0$ in Exercises 3–22:

(a) Use a software tool with a built-in Fourier cosine transform to obtain $F_c(\alpha)$.

(b) Where possible, obtain the Fourier cosine transform by integration.

(c) Where possible, obtain the Fourier cosine transform from $\mathcal{F}[f(x)]$, the exponential Fourier transform.

(d) Where possible, verify the differentiation rule $\mathcal{F}_c[f''] = -\alpha^2 F_c(\alpha) - \sqrt{\frac{2}{\pi}}f'(0)$; where $\mathcal{F}_c[f'']$ cannot be computed directly, use the differentiation rule to obtain its value.

(e) Where possible, verify the differentiation rule $\mathcal{F}_s[f''] = -\alpha^2 F_s(\alpha) + \alpha\sqrt{\frac{2}{\pi}}f(0)$; where $\mathcal{F}_s[f'']$ cannot be computed directly, use the differentiation rule to obtain its value.

(f) Where possible, verify the scaling rules $\mathcal{F}_c[f(\frac{x}{\lambda})] = \lambda F_c(\lambda\alpha)$ and $\mathcal{F}_s[f(\frac{x}{\lambda})] = \lambda F_s(\lambda\alpha)$, both subject to the restriction $\lambda > 0$.

(g) Where possible, verify the multiplication rules $\mathcal{F}_c[xf(x)] = F_s'(\alpha)$ and $\mathcal{F}_s[xf(x)] = -F_c'(\alpha)$; where the transform of the product cannot be computed directly, use the appropriate multiplication rule to obtain its value.

(h) Where possible, verify the shifting rules $\mathcal{F}_c[f(x)\cos\omega x] = \frac{1}{2}[F_c(\alpha + \omega) + F_c(\alpha - \omega)]$ and $\mathcal{F}_s[f(x)\cos\omega x] = \frac{1}{2}[F_s(\alpha + \omega) + F_s(\alpha - \omega)]$.

(i) Where possible, verify the shifting rules $\mathcal{F}_c[f(x)\sin\omega x] = \frac{1}{2}[F_s(\alpha + \omega) - F_s(\alpha - \omega)]$ and $\mathcal{F}_s[f(x)\sin\omega x] = \frac{1}{2}[F_c(\alpha - \omega) - F_c(\alpha + \omega)]$.

(j) Where possible, verify the integration rule $\mathcal{F}_c\left[\int_x^\infty f(\sigma)\,d\sigma\right] = \frac{1}{\alpha}F_s(\alpha)$.

(k) Where possible, verify the differentiation rule $\mathcal{F}_c[f'(x)] = -\sqrt{\frac{2}{\pi}}f(0) + \alpha F_s(\alpha)$; where $\mathcal{F}_c[f'(x)]$ cannot be obtained directly, use the differentiation rule to obtain its value.

3. $f(x) = e^{-ax}$ **4.** $f(x) = xe^{-ax}$

5. $f(x) = (1 - x)e^{-ax}$ **6.** $f(x) = x^2 e^{-ax}$

7. $f(x) = (H(x) - H(x - 1))(2x^3 - 3x^2 + 1)$

8. $f(x) = \dfrac{1}{\sqrt{x}}$ **9.** $f(x) = x^{-1/3}$ **10.** $f(x) = x^{-2/3}$

11. $f(x) = e^{-ax}\cos bx$ **12.** $f(x) = e^{-ax}\sin bx$

13. $f(x) = H(a - x)$ **14.** $f(x) = \dfrac{1}{a^2 + x^2}$

15. $f(x) = (H(x) - H(x - 1))x^2$

16. $f(x) = \cos a^2 x^2$

17. $f(x) = (H(x) - H(x - 1))x^3$

18. $f(x) = H(a - x)(a^2 - x^2)$ **19.** $f(x) = \ln\left(1 + \dfrac{a^2}{x^2}\right)$

20. $f(x) = (H(x) - H(x - 1))x^4$

21. $f(x) = \arctan\dfrac{a}{x}$ **22.** $f(x) = \arctan\dfrac{4}{x^2}$

In Exercises 23–31:

(a) Obtain the formal solution of the BVP $u_{xx} + u_{yy} = 0$, $0 < x < \infty, 0 < y < 1, u_x(0, y) = 0, u(x, 0) = f(x)$, $u(x, 1) = 0$.

(b) By integrating numerically over the finite domain $0 \le \alpha \le 20$, obtain a graph of the solution surface.

(c) From the graph of the solution surface, obtain a contour plot of the level curves (isotherms or equipotentials) determined by the BVP.

(d) Estimate the error made when computing $u(2, \frac{1}{2})$ through integration over $0 \le \alpha \le 20$ instead of over $0 \le \alpha < \infty$.

23. $f(x) = H(1 - x)$ **24.** $f(x) = H(1 - x)x(1 - x)$

25. $f(x) = H(1 - x)x(1 - x)^2$ **26.** $f(x) = H(1 - x)\sin\pi x$

27. $f(x) = H(1 - x) \sin 2\pi x$ **28.** $f(x) = e^{-x^2}$

29. $f(x) = (2x^3 - 3x^2 + 1)H(1 - x)$ **30.** $f(x) = \arctan \dfrac{1}{x^2}$

31. For the BVP $u_{xx} + u_{yy} = 0$, $0 < x < \infty$, $0 < y < b$, $u_x(0, y) = 0$, $u(x, 0) = 0$, $u(x, b) = f(x)$:

(a) Give a physical interpretation.

(b) Use the Fourier cosine transform (taken with respect to x) to obtain the formal solution $u(x, y) = \sqrt{\frac{2}{\pi}} \int_0^\infty F_c(\alpha) \frac{\sinh \alpha y}{\sinh \alpha b} \cos \alpha x \, d\alpha = \frac{2}{\pi} \int_0^\infty \int_0^\infty f(\beta) \cos \alpha \beta \frac{\sinh \alpha y}{\sinh \alpha b} \cos \alpha x \, d\beta \, d\alpha$, where $F_c(\alpha) = \mathcal{F}_c[f(x)]$.

In Exercises 32–39:

(a) Obtain the formal solution of the BVP $u_{xx} + u_{yy} = 0$, $0 < x < \infty$, $0 < y < 1$, $u_x(0, y) = 0$, $u(x, 0) = 0$, $u(x, 1) = f(x)$.

(b) By integrating numerically over the finite domain $0 \le \alpha \le 20$, obtain a graph of the solution surface.

(c) From the graph of the solution surface, obtain a contour plot of the level curves (isotherms or equipotentials) determined by the BVP.

(d) Estimate the error made when computing $u(2, \frac{1}{2})$ through integration over $0 \le \alpha \le 20$ instead of over $0 \le \alpha < \infty$.

32. $f(x) = H(1 - x)$ **33.** $f(x) = H(1 - x)x(1 - x)$

34. $f(x) = H(1 - x)x^2(1 - x)$ **35.** $f(x) = H(1 - x) \sin \pi x$

36. $f(x) = H(1 - x) \sin 2\pi x$ **37.** $f(x) = e^{-x}$

38. $f(x) = xe^{-x}$ **39.** $f(x) = \dfrac{1}{1 + x^2}$

40. For the BVP $u_{xx} + u_{yy} = 0$, $0 < x < \infty$, $0 < y < b$, $u_x(0, y) = A$, $u(x, 0) = 0$, $u(x, b) = f(x)$, where A is a positive constant:

(a) Give a physical interpretation.

(b) Use the Fourier cosine transform (taken with respect to x) to obtain the formal solution $u(x, y) = \sqrt{\frac{2}{\pi}} \int_0^\infty (F_c(\alpha) \frac{\sinh \alpha y}{\sinh \alpha b} + \sqrt{\frac{2}{\pi}} A \frac{\sinh \alpha y - \sinh \alpha b - \sinh \alpha (y - b)}{\alpha^2 \sinh \alpha b}) \cos \alpha x \, d\alpha$, where $F_c(\alpha) = \mathcal{F}_c[f(x)]$.

In Exercises 41–50:

(a) Obtain the formal solution of the BVP $u_{xx} + u_{yy} = 0$, $0 < x < \infty$, $0 < y < 1$, $u_x(0, y) = \sqrt{\frac{\pi}{2}}$, $u(x, 0) = 0$, $u(x, 1) = f(x)$.

(b) By integrating numerically over the finite domain $0 \le \alpha \le 20$, obtain a graph of the solution surface.

(c) From the graph of the solution surface, obtain a contour plot of the level curves (isotherms or equipotentials) determined by the BVP.

(d) Estimate the error made when computing $u(2, \frac{1}{2})$ through integration over $0 \le \alpha \le 20$ instead of over $0 \le \alpha < \infty$.

41. $f(x) = H(1 - x)$ **42.** $f(x) = (1 - x)H(1 - x)$

43. $f(x) = x(1 - x)H(1 - x)$ **44.** $f(x) = x^2(1 - x)H(1 - x)$

45. $f(x) = x(x - 1)(2 - x)H(2 - x)$

46. $f(x) = H(x - 1) - H(x - 2)$

47. $f(x) = (H(x - 1) - H(x - 2))(x - 1)(2 - x)$

48. $f(x) = (H(x - 1) - H(x - 2))(x - 1)^2(2 - x)$

49. $f(x) = -\sin \pi x H(2 - x)$ **50.** $f(x) = xe^{-x}$

In Exercises 51–55:

(a) Use the Fourier cosine transform (taken with respect to x) to solve the BVP $u_{xx} + u_{yy} = 0$, $0 < x < \infty$, $0 < y < 1$, $u_x(0, y) = h(y)$, $u_y(x, 0) = 0$, $u(x, 1) = f(x)$.

(b) Give a physical interpretation of the BVP.

(c) By integrating numerically over the finite domain $0 \le \alpha \le 20$, obtain a graph of the solution surface.

(d) From the graph of the solution surface, obtain a contour plot of the level curves (isotherms or equipotentials) determined by the BVP.

51. $f(x) = H(x - 1) - H(x - 2)$, $h(y) = y(1 - y)$

52. $f(x) = (H(x - 1) - H(x - 2))(x - 1)(2 - x)$, $h(y) = \sin \pi y$

53. $f(x) = \sin \pi x \, H(1 - x)$, $h(y) = y^2(1 - y)$

54. $f(x) = (H(x - 1) - H(x - 3))(x - 1)(x - 2)(3 - x)$, $h(y) = y$

55. $f(x) = (H(x - 1) - H(x - 3))(x - 1)(x - 2)(3 - x)$, $h(y) = 1 - y$

56. For the BVP $u_{xx} + u_{yy} = 0$, $0 < x < \infty$, $0 < y < b$, $u_x(0, y) = 0$, $u_y(x, 0) = 0$, $u(x, b) = f(x)$:

(a) Give a physical interpretation.

(b) Use the Fourier cosine transform (taken with respect to x) to obtain the formal solution $u(x, y) = \sqrt{\frac{2}{\pi}} \int_0^\infty F_c(\alpha) \frac{\cosh \alpha y}{\cosh \alpha b} \cos \alpha x \, d\alpha$, where $F_c(\alpha) = \mathcal{F}_c[f(x)]$.

In Exercises 57–61:

(a) Obtain the formal solution of the BVP $u_{xx} + u_{yy} = 0$, $0 < x < \infty$, $0 < y < 1$, $u_x(0, y) = 0$, $u_y(x, 0) = 0$, $u(x, 1) = f(x)$.

(b) By integrating numerically over the finite domain $0 \le \alpha \le 20$, obtain a graph of the solution surface.

(c) From the graph of the solution surface, obtain a contour plot of the level curves (isotherms or equipotentials) determined by the BVP.

(d) Estimate the error made when computing $u(2, \frac{1}{2})$ through integration over $0 \le \alpha \le 20$ instead of over $0 \le \alpha < \infty$.

57. $f(x) = H(1 - x)$ **58.** $f(x) = H(x - 1) - H(x - 2)$

59. $f(x) = (H(x - 1) - H(x - 2))(x - 1)(2 - x)$

60. $f(x) = (H(x - 1) - H(x - 2))(x - 1)(2 - x)^2$

61. $f(x) = \begin{cases} 0 & 0 \le x < 1 \\ x - 1 & 1 \le x < 2 \\ 3 - x & 2 \le x < 3 \\ 0 & x \ge 3 \end{cases}$

62. For the BVP $u_{xx} + u_{yy} = 0, 0 < x < \infty, 0 < y < b$,
$u_x(0, y) = 0, u_y(x, 0) = g(x), u(x, b) = f(x)$:

(a) Give a physical interpretation.

(b) Use the Fourier cosine transform (taken with respect to x) to obtain the formal solution $u(x, y) = \sqrt{\frac{2}{\pi}} \int_0^\infty F_c(\alpha) \frac{\cosh \alpha y}{\cosh \alpha b}$
$\cos \alpha x \, d\alpha + \sqrt{\frac{2}{\pi}} \int_0^\infty G_c(\alpha) \frac{\sinh \alpha(y-b)}{\alpha \cosh \alpha b} \cos \alpha x \, dx$ where
$F_c(\alpha) = \mathcal{F}_c[f(x)]$ and $G_c(\alpha) = \mathcal{F}_c[g(x)]$.

In Exercises 63–70:

(a) Obtain the formal solution of the BVP $u_{xx} + u_{yy} = 0$,
$0 < x < \infty, 0 < y < 1, u_x(0, y) = 0, u_y(x, 0) = g(x)$,
$u(x, 1) = f(x)$.

(b) By integrating numerically over the finite domain $0 \le \alpha \le 20$, obtain a graph of the solution surface.

(c) From the graph of the solution surface, obtain a contour plot of the level curves (isotherms or equipotentials) determined by the BVP.

(d) Estimate the error made when computing $u(2, \frac{1}{2})$ through integration over $0 \le \alpha \le 20$ instead of over $0 \le \alpha < \infty$.

63. $f(x) = xH(1 - x), g(x) = H(1 - x)$

64. $f(x) = H(1 - x), g(x) = xH(1 - x)$

65. $f(x) = (1 - x)H(1 - x), g(x) = H(1 - x) \sin \pi x$

66. $f(x) = x(1 - x)H(1 - x), g(x) = H(x - 1) - H(x - 2)$

67. $f(x) = H(x - 1) - H(x - 2)$,
$g(x) = (H(x - 1) - H(x - 2))(x - 1)(2 - x)$

68. $f(x) = (H(x - 1) - H(x - 2))(x - 1)(2 - x)$,
$g(x) = H(x - 1) - H(x - 2)$

69. $f(x) = (1 - x^2)H(1 - x), g(x) = -xH(1 - x)$

70. $f(x) = 2x(x - 1)H(1 - x), g(x) = -xH(1 - x)$

Chapter Review

1. Use the Laplace transform to solve the one-dimensional heat on the semi-infinite rod if the rod is initially at a temperature of zero, the left end, at the origin, is maintained at a temperature of A and the temperature in the rod must remain bounded.

2. Use the Laplace transform to solve the one-dimensional wave equation on the semi-infinite string for which the initial shape and velocity are zero and the left end satisfies the condition $u(0, t) = (1 - \text{Heaviside}(t - 1)) \sin t$.

3. State the exponential form of the Fourier integral theorem. Give an example of its validity.

4. State the trigonometric form of the Fourier integral theorem. Give an example of its validity.

5. Define the Fourier transform and give its inversion integral. Give an example of a function for which the transform and its inversion can be computed.

6. State the operational rule for the Fourier transform of a derivative and the rule for the derivative of a Fourier transform. Give examples illustrating each.

7. State the First Shifting law for the Fourier transform and illustrate it with an example.

8. State the Second Shifting law for the Fourier transform and illustrate it with an example.

9. State each of the two convolution theorems for Fourier transforms, and given an example of each.

10. Explain the term *amplitude spectrum*.

11. Demonstrate how the Fourier transform is used to obtain D'Alembert's solution of the one-dimensional wave equation on an infinite domain.

12. Demonstrate how the Fourier transform is used to solve the one-dimensional heat equation on an infinite rod.

13. Demonstrate how the Fourier transform is used to solve Laplace's equation on the infinite strip $-\infty < x < \infty, 0 \le y \le b$, on which the conditions $u(x, 0) = 0, u(x, b) = f(x)$, and $\lim_{x \to \pm\infty} u(x, y) < \infty$ are imposed.

14. State the rule for the Fourier sine transform and give an example of a sine transform and its inversion.

15. State the operational rule for the Fourier sine transform of a second derivative and give an example illustrating its validity.

16. State the rule for the Fourier cosine transform and give an example of a cosine transform and its inversion.

17. State the operational rule for the Fourier cosine transform of a second derivative, and give an example illustrating its validity.

18. Demonstrate the use of the Fourier sine transform in solving the BVP consisting of Laplace's equation in the first quadrant of the xy-plane, and the boundary conditions $u(0, y) = 0$, $u(x, 0) = f(x), \lim_{y \to \infty} u(x, y) = 0$.

19. Demonstrate the use of the Fourier sine transform (taken with respect to x) in solving the BVP consisting of the one-dimensional heat equation on the semi-infinite rod whose temperatures also satisfy the conditions $u(0, t) = 0, u(x, 0) = f(x)$.

20. Demonstrate the use of the Fourier cosine transform in solving the BVP consisting of Laplace's equation in the semi-infinite strip $0 < x < \infty, 0 < y < b$ and the boundary conditions $u_x(0, y) = 0$, $u(x, 0) = f(x)$, and $u(x, b) = 0$. (If u is the steady-state temperature in the strip, the left edge is insulated, the temperature on the bottom edge is $f(x)$, and the temperature on the top edge is zero.)

Matrix Algebra

Facility with the concepts and operations of matrix algebra is of increasing importance to the modern engineer and scientist. This unit discusses the vector and matrix manipulations most often met in the applications. With an emphasis on vector and matrix computations, it treats vector space, subspace, basis, and dimension intuitively, exemplifying them in the three-dimensional space familiar from multivariable calculus.

Vectors are introduced as arrows, based on a familiarity with the orthogonal unit basis vectors {**i**, **j**, **k**}, and then generalized to a column of real numbers giving the components with respect to the basis vectors. The distinction between row and column vectors is maintained because we treat the contravariant and covariant transformation laws. In Unit Four we saw how tangent vectors along a curve arise by differentiation, the model being the velocity vector for a moving particle. We associate tangent vectors with column vectors and show why a gradient vector, also seen in Unit Four, should be a row vector.

We discover how the components of a vector change when the basis used to express the vector changes. The change of basis relationship allows us to characterize rigid rotations and leads us to the orthogonal matrix. Then, after defining the reciprocal basis, we obtain the contravariant and covariant transformation laws for column (tangent) vectors and row (gradient) vectors.

A great deal of matrix arithmetic was introduced in Unit Three, where it was used to study systems of differential equations. After a brief review of these skills and ideas, we consider the concept of a projection. We give a prescription for forming the matrix of a projection of a vector onto another vector or a space spanned by a collection of vectors. Finding the orthogonal component of the vector is the key idea in the Gram–Schmidt orthogonalization process that transforms a set of vectors into an "equivalent" set of mutually orthogonal vectors.

Eigenvalues and eigenvectors for a matrix were also introduced and used in Unit Three when solving systems of differential equations. In Unit Six, the eigenpairs of a matrix are used in an analysis of quadratic forms, allowing us to generalize the "second derivative test" for functions of several variables.

Our discussion of matrix norms also relies on a familiarity with eigenpairs of a matrix. Vector norms measure the length or "magnitude" of a vector. A matrix norm is a function that assigns a real number to a matrix in a similar attempt to assign a "size" to the matrix. One common norm, the 2-norm, is computed from eigenvalues, while other norms can be determined from the matrix entries themselves.

Using the least-squares technique to fit a curve to data points is a very common operation in the applications. Data points come from observations and experiments; curves of fit come from theory. Least squares is a technique for obtaining the best fit of the curve to the data. Our matrix formulation of the calculation is an entry point for understanding the geometry underlying the linear system $A\mathbf{x} = \mathbf{y}$. Thus, we tie the row space, null space, and column space to a discussion of what really happens when a least-squares solution of an overdetermined system of equations is calculated.

A matrix written as a product of special factors is said to be factorized or decomposed. There are a number of useful such factorizations that arise in the applications, and we discuss four of them. We treat the *LU* decomposition, which is very important for solving linear systems $A\mathbf{x} = \mathbf{y}_k$, $k = 1, \ldots, n$. The similarity transform from Unit Three is seen as the factorization $A = PDP^{-1}$. It is not possible to diagonalize every matrix in this fashion. For some matrices, the diagonal matrix must be replaced by a Jordan matrix J, which is almost diagonal and gives the factorization $A = PJP^{-1}$.

Far more important is the *QR* decomposition in which A is expressed as a product of an orthogonal matrix and an upper triangular factor. This factorization is the basis of the very powerful *QR* algorithm for finding eigenvalues. We treat the underlying concept in this unit and revisit the algorithm in Unit Eight, where we study numeric methods.

The fourth factorization is the SVD, the singular value decomposition, which relates to computing the 2-norm of a matrix and provides a basis for approximating a matrix with "nearby" matrices. Finally, the singular value decomposition is related to the general least-squares problem and the pseudoinverse, a matrix that behaves like an inverse when a true inverse does not exist.

Chapter 31

Vectors as Arrows

I N T R O D U C T I O N   Vectors as arrows, and vectors as columns of numbers—both views are useful, and often at the same time. Vectors as columns of numbers were used in Chapter 12 when solving systems of linear differential equations. Vectors as arrows were used in the vector calculus of Unit Four.

This chapter reviews the algebra and geometry of vector addition and subtraction. The dot product, used as early as Chapter 12, is generalized to the inner product and is defined for vectors with complex numbers as components. The cross-product, introduced in Section 12.4 and used in Unit Four is revisited. We derive the determinantal form of $\mathbf{A} \times \mathbf{B}$ from the assumptions that the cross-product forms a right-handed orthogonal system with $\mathbf{A}$ and $\mathbf{B}$ and has length $\|\mathbf{A}\|\|\mathbf{B}\| \sin \theta$, where θ is the angle between $\mathbf{A}$ and $\mathbf{B}$.

31.1 The Algebra and Geometry of Vectors

Directed Line Segments

In a Cartesian (i.e., Euclidean) space, such as the familiar three-dimensional rectangular space with mutually perpendicular x-, y-, and z-axes, a vector $\mathbf{P}$ is the directed line segment connecting the origin O to a point p. It is said to run *from* the origin *to* the point p. The origin is then the *tail* of the vector, and the point p is its *head*. When an arrow is used to represent the vector, the point of the arrow becomes the head of the vector.

The length and direction of the directed line segment are the two essential pieces of information carried by the vector. The length of a vector is the length of the line segment connecting the origin to the point p, while the direction of the vector is the direction of this line segment taken from O to p.

If $-p$ is the point whose coordinates are the negatives of the coordinates of p, then the vector from O to $-p$ is $-\mathbf{P}$, the *negative* of the vector $\mathbf{P}$. The vectors $-\mathbf{P}$ and $\mathbf{P}$ are colinear but point in opposite directions.

Unit Basis Vectors

Vectors of length 1, directed along the coordinate axes, are called *unit basis vectors*. In three dimensions, these vectors are typically taken as $\mathbf{i}$, $\mathbf{j}$, $\mathbf{k}$ and are directed along the x-, y-, and z-axes, respectively. Consequently, the coordinates of the heads of these vectors are then $(1, 0, 0)$, $(0, 1, 0)$, and $(0, 0, 1)$, respectively. Figure 31.1 shows these three vectors.

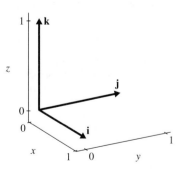
FIGURE 31.1   Unit basis vectors $\mathbf{i}, \mathbf{j}, \mathbf{k}$

$$\|\mathbf{P}\|_1 = |p_1| + |p_2| + |p_3|$$

$$\|\mathbf{P}\|_2 = \sqrt{|p_1|^2 + |p_2|^2 + |p_3|^2}$$

$$\|\mathbf{P}\|_3 = \sqrt[3]{|p_1|^3 + |p_2|^3 + |p_3|^3}$$

$$\|\mathbf{P}\|_4 = \sqrt[4]{|p_1|^4 + |p_2|^4 + |p_3|^4}$$

$$\|\mathbf{P}\|_5 = \sqrt[5]{|p_1|^5 + |p_2|^5 + |p_3|^5}$$

$$\|\mathbf{P}\|_\infty = \max(|p_1|, |p_2|, |p_3|)$$

TABLE 31.1   Different norms for measuring the length of a vector

Coordinates and Components

If the point p has coordinates (p_1, p_2, p_3), then p_1, p_2, p_3 are the *components*, with respect to $\mathbf{i}$, $\mathbf{j}$, and $\mathbf{k}$, of the vector

$$\mathbf{P} = p_1\mathbf{i} + p_2\mathbf{j} + p_3\mathbf{k} = \begin{bmatrix} p_1 \\ p_2 \\ p_3 \end{bmatrix}$$

where, on the right, we have used the column-vector notation for $\mathbf{P}$ and in the middle, we have written $\mathbf{P}$ in terms of $p_1\mathbf{i}$, $p_2\mathbf{j}$, and $p_3\mathbf{k}$, the *vector components* of $\mathbf{P}$. Typographically, the column vector looks like a matrix with one column, and indeed, mathematics blurs the distinction between the two.

Length of a Vector

The length of the real vector $\mathbf{P}$, also called its *magnitude* or *norm*, is ordinarily taken as

$$\|\mathbf{P}\| = \sqrt{p_1^2 + p_2^2 + p_3^2}$$

the square root of the sum of the squares of the components of $\mathbf{P}$. Hence, this is also called the "two-norm," denoted $\|\mathbf{P}\|_2$.

As in the definitions in Table 31.1, the "2" can be replaced with any positive integer k or, in the limit as k becomes arbitrarily large, by the symbol "∞." (See Exercise A1.) The 1-norm is just the sum of the absolute values of the components, while the infinity-norm is the largest magnitude amongst the components. For vectors with real components, the absolute values can be dropped in $\|\mathbf{P}\|_{2k}$, $k \geq 1$.

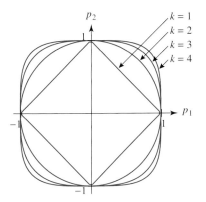

FIGURE 31.2   Loci of (p_1, p_2) satisfying $\|\mathbf{P}\|_k = 1$, $k = 1, \ldots, 4$, where $\mathbf{P} = p_1\mathbf{i} + p_2\mathbf{j}$

$\|\mathbf{P}\|_1$	$\|\mathbf{P}\|_2$	$\|\mathbf{P}\|_3$	$\|\mathbf{P}\|_4$	$\|\mathbf{P}\|_5$	$\|\mathbf{P}\|_\infty$
15	$\sqrt{83} = 9.11$	$\sqrt[3]{495} = 7.91$	$\sqrt[4]{3107} = 7.47$	$\sqrt[5]{20175} = 7.26$	7

TABLE 31.2   Vector norms for $\mathbf{P} = 5\mathbf{i} + 3\mathbf{j} - 7\mathbf{k}$

For example, if $\mathbf{P} = 5\mathbf{i} + 3\mathbf{j} - 7\mathbf{k}$, then we have the norms listed in Table 31.2. The norms form a decreasing sequence converging to the infinity-norm. The *Euclidean norm* (2-norm) is but the second member of the sequence.

Figure 31.2 shows the loci of (p_1, p_2), the heads of vectors $\mathbf{P} = p_1\mathbf{i} + p_2\mathbf{j}$ for which $\|\mathbf{P}\|_k = 1$, $k = 1, \ldots, 4$. The innermost curve, the square, is the locus of the tips of all plane vectors whose 1-norm equals 1. The curve just outside the square is the unit circle, the locus of the tips of all plane vectors whose 2-norm equals 1. Next beyond the circle is the locus for the 3-norm; then finally, the locus for the 4-norm is the outermost curve.

Direction of a Vector

In the plane, the direction of a vector can be given by the angle formed by the positive x-axis and the vector. In three dimensions, the direction of a vector can be given by the *direction angles* α, β, γ, which the vector respectively makes with the three coordinate axes, as shown in Figure 31.3. The cosines of these angles are called the *direction cosines*.

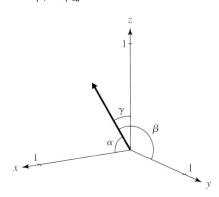

FIGURE 31.3   The direction angles α, β, γ

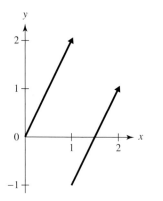

FIGURE 31.4   Parallel translation of a plane vector

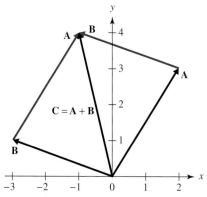

FIGURE 31.5   Parallelogram of vector addition

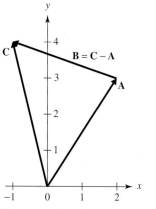

FIGURE 31.6   The vector from the head of **A** to the head of **C** is the difference **C** − **A**

Addition, Subtraction, and Scalar Multiplication

Vectors are added and subtracted componentwise. Multiplication by a scalar is accomplished by multiplying all components of the vector by the scalar. Thus, addition, subtraction, and scalar multiplication of vectors are given by

$$\begin{bmatrix} a_1 \\ a_2 \\ a_3 \end{bmatrix} \pm \begin{bmatrix} b_1 \\ b_2 \\ b_3 \end{bmatrix} = \begin{bmatrix} a_1 \pm b_1 \\ a_2 \pm b_2 \\ a_3 \pm b_3 \end{bmatrix} \qquad c \begin{bmatrix} a_1 \\ a_2 \\ a_3 \end{bmatrix} = \begin{bmatrix} ca_1 \\ ca_2 \\ ca_3 \end{bmatrix}$$

Multiplication of a vector by a positive scalar $c \neq 1$ changes the length of the vector without changing its direction. Multiplying by a negative scalar $c \neq -1$ changes the length of the vector and reverses its direction.

Parallel Translation of Vectors

Consider the collection of all possible smooth curves passing through point p in Euclidean space. At this point, each curve has a tangent vector. The collection of all such tangent vectors is the *tangent space* T_p at p. A tangent space for a point in Euclidean space is again a "copy" of Euclidean space, so we can imagine each point to be the origin O for these tangent vectors. The distinctive property of Euclidean space is that the unit basis vectors $\{\mathbf{i}, \mathbf{j}, \mathbf{k}\}$ in the tangent spaces remain parallel to their counterparts in the original space.

The parallelism of the unit basis vectors allows us to translate an arbitrary vector parallel to itself so that its tail is no longer the origin, but some other point. The vector still has the same components, but the tail rests at a point other than the origin. When a vector is attached to a point, it is sometimes called a *bound* vector. While the vector is being translated, or if its tail is attached to no specific point, it is sometimes called a *free* vector.

Figure 31.4 shows the vector $\mathbf{P} = \mathbf{i} + 2\mathbf{j}$ drawn both at the origin and at $(1, -1)$. In either case, the vector representation is the same, namely, $\mathbf{P} = \mathbf{i} + 2\mathbf{j}$. To see this, consider the parallel lines described by

$$\mathbf{R}_1 = t\mathbf{i} + 2t\mathbf{j} \quad \text{and} \quad \mathbf{R}_2 = (t+1)\mathbf{i} + 2(t-1)\mathbf{j}$$

the first passing through $(0, 0)$, the second through $(1, -1)$ at $t = 0$. Differentiation with respect to the parameter t leads to $\mathbf{R}_1' = \mathbf{R}_2' = \mathbf{P}$. Thus, under parallel translation in Cartesian space, a vector and its parallel translate have the same magnitude and direction and, therefore, are considered to be equivalent.

The Geometry of Vector Addition

The vector $\mathbf{A} + \mathbf{B}$ is the diagonal of the parallelogram whose sides are $\mathbf{A}$ and $\mathbf{B}$. This diagonal, sometimes called the *resultant* of $\mathbf{A}$ and $\mathbf{B}$, is illustrated in Figure 31.5 for the vectors $\mathbf{A} = 2\mathbf{i} + 3\mathbf{j}$, $\mathbf{B} = -3\mathbf{i} + \mathbf{j}$, $\mathbf{C} = \mathbf{A} + \mathbf{B} = -\mathbf{i} + 4\mathbf{j}$. The resultant is the diagonal of the parallelogram or the third side of the triangle formed by $\mathbf{A}$ and the translation of $\mathbf{B}$, which puts the tail of $\mathbf{B}$ at the tip of $\mathbf{A}$. Since $\mathbf{C} = \mathbf{A} + \mathbf{B}$, it follows that $\mathbf{B} = \mathbf{C} - \mathbf{A}$, so the difference of two vectors with the same initial points ($\mathbf{A}$ and $\mathbf{C}$) is a vector ($\mathbf{B}$) pointing from the tip of the first to the tip of the second. Isolating the triangle formed by $\mathbf{A}$, $\mathbf{B}$, and $\mathbf{C}$ gives Figure 31.6. Consequently, the vector *from* point p_1 *to* the point p_2 is the vector $\mathbf{P}_2 - \mathbf{P}_1$.

EXAMPLE 31.1   The vector *from* the point $(4, -3, 2)$ *to* the point $(5, 1, -2)$ is the vector $\mathbf{i} + 4\mathbf{j} - 4\mathbf{k}$. ❖

EXERCISES 31.1–Part A

A1. Obtain $\|\mathbf{P}\|_k$ for the vector $\mathbf{P} = 2\mathbf{i} + 3\mathbf{j} - 4\mathbf{k}$, and compute $\lim_{k \to \infty} \|\mathbf{P}\|_k$. Show this is 4, and then construct a more general argument justifying the claim about $\|\mathbf{P}\|_\infty$ in the lower right corner of Table 31.1.

A2. If $\mathbf{P} = a\mathbf{i} + b\mathbf{j} + c\mathbf{k}$ and α, β, γ are its direction angles, show $\frac{a}{\|\mathbf{P}\|_2} = \cos\alpha$, $\frac{b}{\|\mathbf{P}\|_2} = \cos\beta$, $\frac{c}{\|\mathbf{P}\|_2} = \cos\gamma$. Consequently, if $\mathbf{P}$ is a unit vector, then its components are its direction cosines.

A3. Show that the direction cosines always satisfy $\cos^2\alpha + \cos^2\beta + \cos^2\gamma = 1$.

A4. Find a unit vector from the tip of $\mathbf{A} = 5\mathbf{i} - 7\mathbf{j} + 8\mathbf{k}$ toward the tip of $\mathbf{B} = 6\mathbf{i} + \mathbf{j} - \mathbf{k}$.

A5. For $\mathbf{A}$ and $\mathbf{B}$ given in Exercise A4, obtain $\mathbf{A} + 2\mathbf{B}$.

EXERCISES 31.1–Part B

B1. Obtain the direction cosines for the vectors $\mathbf{A}$ and $\mathbf{B}$ given in Exercise A4.

In Exercises B2–11:

    (a) Find $\|\mathbf{A}\|_1, \|\mathbf{A}\|_2, \|\mathbf{A}\|_3, \|\mathbf{A}\|_\infty$.

    (b) Find $\|\mathbf{B}\|_1, \|\mathbf{B}\|_2, \|\mathbf{B}\|_3, \|\mathbf{B}\|_\infty$.

    (c) Plot $\mathbf{A}, \mathbf{B}, \mathbf{A} + \mathbf{B}, \mathbf{A} - \mathbf{B}, \frac{1}{2}(\mathbf{A} + \mathbf{B})$.

    (d) Show that $\|\mathbf{A} + \mathbf{B}\|^2 + \|\mathbf{A} - \mathbf{B}\|^2 = 2(\|\mathbf{A}\|^2 + \|\mathbf{B}\|^2)$, the parallelogram law, holds.

    (e) Obtain a unit vector from $\mathbf{A}$ to $\mathbf{B}$.

    (f) Determine the locus of the tip of $\mathbf{A} + t\mathbf{B}, 0 \le t < \infty$.

    (g) Determine the values of α, β, γ for which $\mathbf{A} = \alpha\mathbf{C}_1 + \beta\mathbf{C}_2 + \gamma\mathbf{C}_3$ if $\mathbf{C}_1 = \mathbf{i} + 2\mathbf{j} - 3\mathbf{k}$, $\mathbf{C}_2 = 2\mathbf{i} - \mathbf{j} - \mathbf{k}$, $\mathbf{C}_3 = -\mathbf{i} + 3\mathbf{j} + 2\mathbf{k}$.

    (h) In the parallelogram formed by $\mathbf{A}$ and $\mathbf{B}$, show that the diagonals $\mathbf{A} + \mathbf{B}$ and $\mathbf{A} - \mathbf{B}$ bisect each other.

    (i) The third side of the triangle formed by $\mathbf{A}$ and $\mathbf{B}$ is $\mathbf{B} - \mathbf{A}$. Show that $\frac{1}{2}(\mathbf{A} + \mathbf{B})$ is a vector from the origin to the midpoint of the third side of the triangle. Show also that the medians of the triangle intersect at the tip of the vector $\frac{1}{3}(\mathbf{A} + \mathbf{B})$.

    (j) In the triangle formed by $\mathbf{A}$ and $\mathbf{B}$, show that the line segment connecting the midpoints of sides $\mathbf{A}$ and $\mathbf{B}$ is parallel to the third side and half its length.

B2. $\mathbf{A} = \mathbf{i} + 8\mathbf{j} - 2\mathbf{k}, \mathbf{B} = 7\mathbf{i} - 7\mathbf{j} + 3\mathbf{k}$

B3. $\mathbf{A} = 7\mathbf{i} - 3\mathbf{j}, \mathbf{B} = -9\mathbf{i} + 2\mathbf{j} + 9\mathbf{k}$

B4. $\mathbf{A} = 5\mathbf{i} + \mathbf{j} - 2\mathbf{k}, \mathbf{B} = -\mathbf{i} + 3\mathbf{j} - 8\mathbf{k}$

B5. $\mathbf{A} = 7\mathbf{i} - 2\mathbf{j} + \mathbf{k}, \mathbf{B} = -8\mathbf{i} + 7\mathbf{j} + 3\mathbf{k}$

B6. $\mathbf{A} = 2\mathbf{i} - 5\mathbf{j} - 4\mathbf{k}, \mathbf{B} = \mathbf{i} - 4\mathbf{j} - \mathbf{k}$

B7. $\mathbf{A} = -4\mathbf{i} - 7\mathbf{j} - 4\mathbf{k}, \mathbf{B} = 2\mathbf{i} + 2\mathbf{j} + 4\mathbf{k}$

B8. $\mathbf{A} = \mathbf{i} + 2\mathbf{j} - 2\mathbf{k}, \mathbf{B} = \mathbf{i} + 9\mathbf{j} - 7\mathbf{k}$

B9. $\mathbf{A} = 9\mathbf{i} - 6\mathbf{j} - 9\mathbf{k}, \mathbf{B} = 5\mathbf{i} - 2\mathbf{j} + 8\mathbf{k}$

B10. $\mathbf{A} = -2\mathbf{i} + 6\mathbf{j} + 8\mathbf{k}, \mathbf{B} = -6\mathbf{i} + 9\mathbf{j} - 2\mathbf{k}$

B11. $\mathbf{A} = -3\mathbf{i} + 9\mathbf{j} - 9\mathbf{k}, \mathbf{B} = 7\mathbf{i} - 3\mathbf{j} + 8\mathbf{k}$

In Exercises B12–16, two points p_1 and p_2 in the (vertical) xz-plane are given. Also given is the value of ω, the weight in pounds of an object suspended at the origin by two ropes connected to the points p_1 and p_2. Find the magnitude of the tension in each rope. *Hint:* If $\mathbf{u}_1$ and $\mathbf{u}_2$ are unit vectors from the origin toward p_1 and p_2, respectively, find α and β for which $\mathbf{F} = \alpha\mathbf{u}_1 + \beta\mathbf{u}_2 = \omega\mathbf{k}$.

B12. $p_1 = (10, 0, 11), p_2 = (-19, 0, 18), \omega = 195$

B13. $p_1 = (21, 0, 23), p_2 = (-11, 0, 10), \omega = 115$

B14. $p_1 = (17, 0, 13), p_2 = (-30, 0, 22), \omega = 345$

B15. $p_1 = (27, 0, 30), p_2 = (-23, 0, 16), \omega = 130$

B16. $p_1 = (26, 0, 11), p_2 = (-29, 0, 17), \omega = 380$

In Exercises B17–21, the given points $p_k, k = 1, \ldots, 4$, form a quadrilateral in which A, B, C, D, the midpoints of consecutive sides, are joined to form the quadrilateral $ABCD$. Show that $ABCD$ is a parallelogram. *Hint:* Show the vector equalities $\mathbf{AB} = \mathbf{DC}$ and $\mathbf{BC} = \mathbf{AD}$.

B17. $p_1 = (18, 28), p_2 = (20, -29), p_3 = (-17, -14),$ $p_4 = (-29, 20)$

B18. $p_1 = (25, 26), p_2 = (24, -30), p_3 = (-25, -28),$ $p_4 = (-18, 29)$

B19. $p_1 = (22, 24), p_2 = (16, -17), p_3 = (-19, -20),$ $p_4 = (-25, 21)$

B20. $p_1 = (29, 10), p_2 = (25, -12), p_3 = (-24, -21),$ $p_4 = (-20, 14)$

B21. $p_1 = (17, 10), p_2 = (19, -26), p_3 = (-10, -25),$ $p_4 = (-23, 15)$

In Exercises B22–26, if the points $p_k, k = 1, \ldots, 3$, form a triangle and $\mathbf{P}_k, k = 1, \ldots, 3$, are vectors from the origin to the respective points p_k,

    (a) Show that the medians of the triangle intersect at the tip of the vector $\frac{1}{3}(\mathbf{P}_1 + \mathbf{P}_2 + \mathbf{P}_3)$. (A median connects a vertex with the midpoint of the opposite side.)

    (b) Show that the line segment connecting the midpoints of the two sides meeting at $\mathbf{P}_1$, and the median to the remaining side bisect each other.

B22. $p_1 = (21, 20, -9)$, $p_2 = (24, 12, -18)$, $p_3 = (11, -19, -15)$ **B25.** $p_1 = (14, 1, 20)$, $p_2 = (-10, 13, 16)$, $p_3 = (20, 14, 19)$

B23. $p_1 = (20, -13, -19)$, $p_2 = (-18, 9, 9)$, $p_3 = (-23, -1, 25)$ **B26.** $p_1 = (19, 12, 18)$, $p_2 = (-21, -24, -7)$, $p_3 = (2, -12, 8)$

B24. $p_1 = (-5, 16, -9)$, $p_2 = (1, 19, -11)$, $p_3 = (-9, 25, 3)$

31.2   Inner and Dot Products

The Dot Product

The dot product between the vectors

$$\mathbf{A} = a_1\mathbf{i} + a_2\mathbf{j} + a_3\mathbf{k} \quad \text{and} \quad \mathbf{B} = b_1\mathbf{i} + b_2\mathbf{j} + b_3\mathbf{k}$$

is the real number

$$\mathbf{A} \cdot \mathbf{B} = a_1 b_1 + a_2 b_2 + a_3 b_3 \tag{31.1}$$

Alternatively, the dot product can be expressed by

$$\mathbf{A} \cdot \mathbf{B} = \mathbf{A}^{\mathsf{T}}\mathbf{B}$$

in conformity with the process of matrix multiplication. If $\mathbf{A}$ and $\mathbf{B}$ are column vectors, then $\mathbf{A}^{\mathsf{T}}$, the transpose of $\mathbf{A}$, is a row vector that multiplies $\mathbf{B}$ from the left.

EXAMPLE 31.2   The dot product of the vectors

$$3\mathbf{i} - 2\mathbf{j} + 5\mathbf{k} \quad \text{and} \quad 7\mathbf{i} + 3\mathbf{j} - 4\mathbf{k}$$

is the real number $\mathbf{A} \cdot \mathbf{B} = \mathbf{A}^{\mathsf{T}}\mathbf{B} = (3)(7) + (-2)(3) + (5)(-4) = -5$.   ❖

From the expression for $\mathbf{A} \cdot \mathbf{B}$, we see immediately that

$$\mathbf{A} \cdot \mathbf{A} = a_1 a_1 + a_2 a_2 + a_3 a_3 = a_1^2 + a_2^2 + a_3^2 = \|\mathbf{A}\|^2$$

and $\|\mathbf{A}\| = \sqrt{\mathbf{A} \cdot \mathbf{A}}$. (Henceforth, unless stated otherwise, $\|\mathbf{A}\|$ will denote the 2-norm. The expression

$$\mathbf{A} \cdot \mathbf{B} = \|\mathbf{A}\|\|\mathbf{B}\| \cos\theta \tag{31.2}$$

where θ is the angle between $\mathbf{A}$ and $\mathbf{B}$ satisfying $0 \le \theta \le \pi$, is derived after Example 31.3.

EXAMPLE 31.3   The angle between the vectors $\mathbf{P}_1 = 5\mathbf{i}$ and $\mathbf{P}_2 = 3\mathbf{i} + 3\mathbf{j}$ is clearly $\theta = \frac{\pi}{4}$ since the first vector is along the positive x-axis and the second along the line $y = x$. Thus, we calculate $\mathbf{P}_1 \cdot \mathbf{P}_2 = (5)(3) + (0)(3) = 15$, and since $\|\mathbf{P}_1\| = 5$, $\|\mathbf{P}_2\| = 3\sqrt{2}$, and $\cos\frac{\pi}{4} = \frac{1}{\sqrt{2}}$, we get $\mathbf{P}_1 \cdot \mathbf{P}_2 = 5(3\sqrt{2})\frac{1}{\sqrt{2}} = 15$, thereby verifying (31.2).   ❖

DERIVATION OF EQUATION (31.2)   The equality of the right-hand sides of (31.1) and (31.2) hinges on the law of cosines, $w^2 = u^2 + v^2 - 2uv\cos\theta$, where u, v, w, θ are as in Figure 31.7. Identify the vector $\mathbf{B} = b_1\mathbf{i} + b_2\mathbf{j} + b_3\mathbf{k}$ with side u and $\mathbf{A} = a_1\mathbf{i} + a_2\mathbf{j} + a_3\mathbf{k}$ with side v. Then $\mathbf{C}$, the vector from $\mathbf{B}$ to $\mathbf{A}$, is $\mathbf{C} = \mathbf{A} - \mathbf{B} = (a_1 - b_1)\mathbf{i} + (a_1 - b_1)\mathbf{j} + (a_1 - b_1)\mathbf{k}$, and we identify $\mathbf{C}$ with side w. Thus, we have the triangle in Figure 31.8.

The law of cosines gives

$$\cos\theta = \frac{w^2 - u^2 - v^2}{-2uv} = \frac{\|\mathbf{C}\|^2 - \|\mathbf{A}\|^2 - \|\mathbf{B}\|^2}{\|\mathbf{A}\|\|\mathbf{B}\|}$$

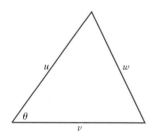

FIGURE 31.7   Triangle for illustrating the law of cosines

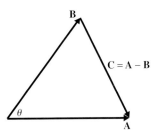

FIGURE 31.8   Vector version of the law of cosines

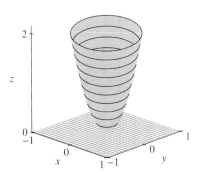

FIGURE 31.9   The function $f(x, y) = 3x^2 + 2xy + 3y^2$ has a minimum of zero at $(0, 0)$

and multiplication by $\|\mathbf{A}\|\|\mathbf{B}\|$ gives

$$\mathbf{A} \cdot \mathbf{B} = -\tfrac{1}{2}(\|\mathbf{C}\|^2 - \|\mathbf{A}\|^2 - \|\mathbf{B}\|^2)$$
$$= -\tfrac{1}{2}[(a_1 - b_1)^2 + (a_2 - b_2)^2 + (a_3 - b_3)^2] + \tfrac{1}{2}(a_1^2 + a_2^2 + a_3^2) + \tfrac{1}{2}(b_1^2 + b_2^2 + b_3^2)$$
$$= a_1 b_1 + a_2 b_2 + a_3 b_3$$

ANGLE BETWEEN TWO VECTORS   Solving (31.2) for $\cos\theta = \frac{\mathbf{A}\cdot\mathbf{B}}{\|\mathbf{A}\|\|\mathbf{B}\|}$ gives a simple way for determining the angle between the vectors $\mathbf{A}$ and $\mathbf{B}$. If $\mathbf{A}$ and $\mathbf{B}$ are perpendicular, $\theta = \frac{\pi}{2}$ and $\cos\theta = 0$, so $\mathbf{A} \cdot \mathbf{B} = 0$. Orthogonality is therefore characterized by the vanishing of the dot product.

Positive Definite Matrices

A symmetric matrix A with positive eigenvalues is said to be *positive definite*. The strict positivity of the eigenvalues guarantees $\mathbf{x}^\mathrm{T} A \mathbf{x} > 0$ for $\mathbf{x} \neq \mathbf{0}$. For example, the matrix

$$A = \begin{bmatrix} 3 & 1 \\ 1 & 3 \end{bmatrix} \tag{31.3}$$

is symmetric and has eigenvalues $\lambda = 2, 4$. If $\mathbf{x} = x\mathbf{i} + y\mathbf{j}$, the quadratic function

$$\mathbf{x}^\mathrm{T} A \mathbf{x} = 3x^2 + 2xy + 3y^2 = f(x, y) \tag{31.4}$$

has a minimum of zero at $(x, y) = (0, 0)$, as suggested by Figure 31.9 and confirmed by the second derivative test. (See Exercise A1.) Section 33.4 deals at greater length with the connection between positive definite matrices A and the quadratic functions $\mathbf{x}^\mathrm{T} A \mathbf{x}$ they define.

Real Inner Products

The dot product is actually a function that, for each pair of real vectors, returns a real number. This number, $\mathbf{A} \cdot \mathbf{B} = \langle \mathbf{A}, \mathbf{B} \rangle$, satisfies the conditions

1. $\langle \mathbf{A}, \mathbf{A} \rangle \geq 0$
2. $\langle \mathbf{A}, \mathbf{A} \rangle = 0$ just when $\mathbf{A} = \mathbf{0}$.
3. $\langle \mathbf{A}, \mathbf{B} \rangle = \langle \mathbf{B}, \mathbf{A} \rangle$
4. $\langle a\mathbf{A} + b\mathbf{B}, \mathbf{C} \rangle = a\langle \mathbf{A}, \mathbf{C} \rangle + b\langle \mathbf{B}, \mathbf{C} \rangle$, for real scalars a and b.

Any function satisfying these conditions is called a (real) *inner product*. Hence, the dot product is an example of a real inner product.

The first condition, requiring the inner product of a vector with itself to be nonnegative, makes sense since the dot product of a vector with itself is the length squared. A vector should have positive length unless, of course, it is the zero vector. That is the content of the second condition, whereby the only vector with zero length is the zero vector.

The third condition is symmetry and holds for real vectors, as does the fourth condition, which requires the real inner product to be linear in the first entry. In light of the third condition, the real inner product is also linear in the second entry.

EXAMPLE 31.4   If A is the positive definite matrix in (31.3), then $\langle \mathbf{x}, \mathbf{y} \rangle = \mathbf{x}^\mathrm{T} A \mathbf{y}$ defines an inner product. The first and second properties for an inner product hold because the function $f(x, y)$ defined in (31.4) has a minimum of zero at the origin. The third and fourth properties are a consequence of the linearity of matrix multiplication and are left to Exercises A2 and A3.   ❖

Of course, the dot product is an inner product for which $A = I$, the identity matrix. (See Exercise A4.)

Complex Inner Products

The definition of the dot product must be changed for complex vectors. If a real inner product is applied to vectors containing complex components, the following anomalous result can occur. Using the dot product and the vector $\mathbf{A} = \mathbf{i} + i\mathbf{j}$, we would get $\mathbf{A} \cdot \mathbf{A} = 1^2 + i^2 = 1 - 1 = 0$, thereby violating the requirement that only the zero vector have zero length. This problem can be avoided if one of the vectors in the dot product is conjugated. (If $z = x + iy$ is a complex number, then $\bar{z} = x - iy$ is the *complex conjugate* of z. The superimposed "bar" in $\bar{z}$ typically denotes the complex conjugate in the literature, and $\bar{\mathbf{z}}$, the conjugate of a vector, is obtained by conjugating each component of $\mathbf{z}$.)

Unfortunately, the literature is about evenly split on the issue of which vector to conjugate. Texts such as [52] conjugate the second member of the pair, while texts such as [64] conjugate the first. We will follow the convention of conjugating the second vector so, in three dimensions, $\mathbf{v} \cdot \mathbf{w} = \sum_{k=1}^{3} v_k \bar{w}_k$. As a consequence, the properties of the (complex) inner product must be stated as

1. $\langle \mathbf{AA} \rangle \geq 0$
2. $\langle \mathbf{A}, \mathbf{A} \rangle = 0$ just when $\mathbf{A} = \mathbf{0}$.
3. $\langle \mathbf{A}, \mathbf{B} \rangle = \overline{\langle \mathbf{B}, \mathbf{A} \rangle}$
4. $\langle a\mathbf{A} + b\mathbf{B}, \mathbf{C} \rangle = a\langle \mathbf{A}, \mathbf{C} \rangle + b\langle \mathbf{B}, \mathbf{C} \rangle$

The first two conditions still require that only the zero vector have zero length and all other vectors have positive length. The third condition modifies the symmetry exhibited by the real inner product. Here, interchanging the order of the vectors conjugates the value of the inner product.

The fourth condition still requires the inner product to be linear in the first entry but, in light of the third condition, no longer yields linearity in the second entry. A linear combination of vectors in the second entry would now simplify to

$$\langle \mathbf{A}, b\mathbf{B} + c\mathbf{C} \rangle = \bar{b}\langle \mathbf{A}, \mathbf{B} \rangle + \bar{c}\langle \mathbf{A}, \mathbf{C} \rangle$$

Scalar multiples of the second vector "come out of" the inner product as conjugates. Texts that conjugate the first vector will still impose the third condition but preserve linearity in the second entry and conjugate-linearity in the first entry.

EXAMPLE 31.5   Given the vectors

$$\mathbf{A} = (1 + 2i)\mathbf{i} + 3\mathbf{j} + (4 - 5i)\mathbf{k} \quad \text{and} \quad \mathbf{B} = (3 - i)\mathbf{i} + (2 + 3i)\mathbf{j} + (5 + 6i)\mathbf{k}$$

we find

$$\|\mathbf{A}\| = \sqrt{\langle \mathbf{A}, \mathbf{A} \rangle} = \sqrt{55} \qquad \langle \mathbf{A}, \mathbf{B} \rangle = -3 - 51i \qquad \langle \mathbf{B}, \mathbf{A} \rangle = -3 + 51i$$

Furthermore, if

$$\mathbf{C} = (5 - 7i)\mathbf{i} - 2i\mathbf{j} + (9 + 8)\mathbf{k} \qquad z_1 = 7 - 3i \qquad z_2 = 6 - 4i$$

then

$$\langle z_1\mathbf{A} + z_2\mathbf{B}, \mathbf{C} \rangle = 537 - 571i = z_1\langle \mathbf{A}, \mathbf{C} \rangle + z_2\langle \mathbf{B}, \mathbf{C} \rangle$$

and

$$\langle \mathbf{A}, z_1 \mathbf{B} + z_2 \mathbf{C} \rangle = 270 - 742i = \bar{z}_1 \langle \mathbf{A}, \mathbf{B} \rangle + \bar{z}_2 \langle \mathbf{A}, \mathbf{C} \rangle$$

Without conjugating the scalar multipliers in the second expression, the incorrect $z_1 \langle \mathbf{A}, \mathbf{B} \rangle + z_2 \langle \mathbf{A}, \mathbf{C} \rangle = -468 - 620i$ would result. ❖

EXERCISES 31.2–Part A

A1. Use the "second derivative test" of multivariable calculus to classify the critical point of $f(x, y)$ given by (31.4). (At the critical point, $f_{xx} f_{yy} - f_{xy}^2 > 0$ and $f_{xx} > 0$ indicate a minimum.)

A2. Define the inner product $\langle \mathbf{x}, \mathbf{y} \rangle$ by $\mathbf{x}^T A \mathbf{y}$, where A is given by (31.3). Show that $\langle \mathbf{x}, \mathbf{y} \rangle = \langle \mathbf{y}, \mathbf{x} \rangle$.

A3. For the inner product defined in Exercise A2, show $\langle a\mathbf{x} + b\mathbf{y}, \mathbf{z} \rangle = a \langle \mathbf{x}, \mathbf{z} \rangle + b \langle \mathbf{y}, \mathbf{z} \rangle$, where a and b are real scalars.

A4. Show that $\mathbf{x} \cdot \mathbf{y}$, the ordinary dot product, is an inner product with $A = I$, the identity matrix.

A5. Show that $A = \begin{bmatrix} 5 & 3 \\ 3 & 2 \end{bmatrix}$ is positive definite. Verify that $\mathbf{x}^T A \mathbf{x} > 0$ for $\mathbf{x} \neq \mathbf{0}$.

EXERCISES 31.2–Part B

B1. Verify that $A = \begin{bmatrix} 8 & -5 & 4 \\ -5 & 4 & -2 \\ 4 & -2 & 9 \end{bmatrix}$ is positive definite so that $\mathbf{x}^T A \mathbf{y}$ defines an inner product. Compute this inner product for $\mathbf{x} = 2\mathbf{i} - 3\mathbf{j} + 5\mathbf{k}$ and $\mathbf{y} = -4\mathbf{i} + \mathbf{j} + 7\mathbf{k}$.

B2. If $\langle \mathbf{x}, \mathbf{y} \rangle = 0$, the vectors $\mathbf{x}$ and $\mathbf{y}$ are said to be *conjugate*. (Think of this as a generalization of orthogonality.) For the inner product defined in Exercise B1, find all vectors conjugate to $\mathbf{x}$.

In Exercises B3–7:

(a) Write $\mathbf{V} = \alpha \mathbf{C}_1 + \beta \mathbf{C}_2 + \gamma \mathbf{C}_3$ and solve for α, β, γ by equating components.

(b) Write $\mathbf{V} = \alpha \mathbf{C}_1 + \beta \mathbf{C}_2 + \gamma \mathbf{C}_3$ and solve for α, β, γ by computing $\mathbf{V} \cdot \mathbf{C}_k, k = 1, 2, 3$.

(c) Write $\mathbf{V} = \alpha \mathbf{C}_1 + \beta \mathbf{C}_2 + \gamma \mathbf{C}_3$ and solve for α, β, γ by computing $\langle \mathbf{V}, \mathbf{C}_k \rangle, k = 1, 2, 3$, where the inner product is defined in Exercise B1.

B3. $\mathbf{V} = \begin{bmatrix} -6 \\ 8 \\ 10 \end{bmatrix}, \mathbf{C}_1 = \begin{bmatrix} -2 \\ 4 \\ -1 \end{bmatrix}, \mathbf{C}_2 = \begin{bmatrix} 4 \\ 6 \\ -12 \end{bmatrix}, \mathbf{C}_3 = \begin{bmatrix} 5 \\ 1 \\ -10 \end{bmatrix}$

B4. $\mathbf{V} = \begin{bmatrix} 1 \\ -11 \\ 1 \end{bmatrix}, \mathbf{C}_1 = \begin{bmatrix} -5 \\ 12 \\ -1 \end{bmatrix}, \mathbf{C}_2 = \begin{bmatrix} -1 \\ 1 \\ 1 \end{bmatrix}, \mathbf{C}_3 = \begin{bmatrix} 12 \\ -2 \\ -1 \end{bmatrix}$

B5. $\mathbf{V} = \begin{bmatrix} -2 \\ 6 \\ 9 \end{bmatrix}, \mathbf{C}_1 = \begin{bmatrix} 11 \\ 4 \\ -3 \end{bmatrix}, \mathbf{C}_2 = \begin{bmatrix} 8 \\ 7 \\ 3 \end{bmatrix}, \mathbf{C}_3 = \begin{bmatrix} 11 \\ -9 \\ 11 \end{bmatrix}$

B6. $\mathbf{V} = \begin{bmatrix} -3 \\ -7 \\ 1 \end{bmatrix}, \mathbf{C}_1 = \begin{bmatrix} 11 \\ -6 \\ 12 \end{bmatrix}, \mathbf{C}_2 = \begin{bmatrix} 4 \\ 8 \\ 12 \end{bmatrix}, \mathbf{C}_3 = \begin{bmatrix} 8 \\ 11 \\ 2 \end{bmatrix}$

B7. $\mathbf{V} = \begin{bmatrix} -5 \\ -2 \\ 12 \end{bmatrix}, \mathbf{C}_1 = \begin{bmatrix} 4 \\ 11 \\ 5 \end{bmatrix}, \mathbf{C}_2 = \begin{bmatrix} -3 \\ -2 \\ -5 \end{bmatrix}, \mathbf{C}_3 = \begin{bmatrix} 0 \\ -5 \\ 12 \end{bmatrix}$

In Exercises B8–17, graph $\mathbf{A}$ and $\mathbf{B}$, use a ruler to measure their lengths, use a protractor to measure the angle θ between A and B, and compute $\|\mathbf{A}\| \|\mathbf{B}\| \cos \theta$. Then, compare to the computed value of $\mathbf{A} \cdot \mathbf{B}$.

B8. $\mathbf{A} = 7\mathbf{i} - 8\mathbf{j}, \mathbf{B} = 3\mathbf{i} + 5\mathbf{j}$　　**B9.** $\mathbf{A} = -5\mathbf{i} + 5\mathbf{j}, \mathbf{B} = -7\mathbf{i} + \mathbf{j}$

B10. $\mathbf{A} = -\mathbf{i} + 3\mathbf{j}, \mathbf{B} = 8\mathbf{i} + 3\mathbf{j}$　　**B11.** $\mathbf{A} = -\mathbf{i} - 4\mathbf{j}, \mathbf{B} = 2\mathbf{i} + 9\mathbf{j}$

B12. $\mathbf{A} = -2\mathbf{i} - 7\mathbf{j}, \mathbf{B} = 8\mathbf{i} + 5\mathbf{j}$　　**B13.** $\mathbf{A} = -3\mathbf{i} - 6\mathbf{j}, \mathbf{B} = -8\mathbf{i} + 2\mathbf{j}$

B14. $\mathbf{A} = 5\mathbf{i} + \mathbf{j}, \mathbf{B} = 3\mathbf{i} + 4\mathbf{j}$　　**B15.** $\mathbf{A} = 9\mathbf{i} - 8\mathbf{j}, \mathbf{B} = \mathbf{i} + 9\mathbf{j}$

B16. $\mathbf{A} = 9\mathbf{i} - 4\mathbf{j}, \mathbf{B} = -5\mathbf{i} + 6\mathbf{j}$　　**B17.** $\mathbf{A} = 5\mathbf{i} - 7\mathbf{j}, \mathbf{B} = -3\mathbf{i} - 8\mathbf{j}$

In Exercises B18–22:

(a) Show that $\mathbf{A} \cdot \mathbf{B} = \mathbf{B} \cdot \mathbf{A}$.

(b) Show that $(r\mathbf{A} + s\mathbf{B}) \cdot \mathbf{C} = r(\mathbf{A} \cdot \mathbf{C}) + s(\mathbf{B} \cdot \mathbf{C})$.

(c) Show that $\mathbf{A} \cdot (r\mathbf{B} + s\mathbf{C}) = r(\mathbf{A} \cdot \mathbf{B}) + s(\mathbf{A} \cdot \mathbf{C})$.

(d) Show that $\mathbf{A} \cdot \mathbf{B} = \frac{1}{4}(\|\mathbf{A} + \mathbf{B}\|^2 - \|\mathbf{A} - \mathbf{B}\|^2)$.

B18. $\mathbf{A} = \begin{bmatrix} 7 \\ -9 \\ 3 \end{bmatrix}, \mathbf{B} = \begin{bmatrix} -12 \\ -5 \\ -5 \end{bmatrix}, \mathbf{C} = \begin{bmatrix} 7 \\ -11 \\ 12 \end{bmatrix}, r = 2, s = -3$

B19. $\mathbf{A} = \begin{bmatrix} -7 \\ -5 \\ -11 \end{bmatrix}, \mathbf{B} = \begin{bmatrix} -8 \\ -10 \\ -8 \end{bmatrix}, \mathbf{C} = \begin{bmatrix} 10 \\ 1 \\ 3 \end{bmatrix}, r = -7, s = 5$

B20. $\mathbf{A} = \begin{bmatrix} 0 \\ 10 \\ 1 \end{bmatrix}, \mathbf{B} = \begin{bmatrix} -2 \\ 8 \\ 11 \end{bmatrix}, \mathbf{C} = \begin{bmatrix} 7 \\ -1 \\ -1 \end{bmatrix}, r = 11, s = -4$

B21. $\mathbf{A} = \begin{bmatrix} 5 \\ 10 \\ 11 \end{bmatrix}, \mathbf{B} = \begin{bmatrix} 1 \\ -11 \\ -12 \end{bmatrix}, \mathbf{C} = \begin{bmatrix} -3 \\ 4 \\ -6 \end{bmatrix}, r = -9, s = 2$

B22. $\mathbf{A} = \begin{bmatrix} -10 \\ -10 \\ 12 \end{bmatrix}, \mathbf{B} = \begin{bmatrix} 12 \\ 11 \\ -12 \end{bmatrix}, \mathbf{C} = \begin{bmatrix} 8 \\ 11 \\ 12 \end{bmatrix}, r = 6, s = -13$

In Exercises B23–27:

(a) Show that $\mathbf{A} \cdot \mathbf{B} = \overline{\mathbf{B} \cdot \mathbf{A}}$.

(b) Show that $(z_1\mathbf{A} + z_2\mathbf{B}) \cdot \mathbf{C} = z_1(\mathbf{A} \cdot \mathbf{C}) + z_2(\mathbf{B} \cdot \mathbf{C})$.

(c) Show that $\mathbf{A} \cdot (z_1\mathbf{B} + z_2\mathbf{C}) = \bar{z}_1(\mathbf{A} \cdot \mathbf{B}) + \bar{z}_2(\mathbf{A} \cdot \mathbf{C})$.

(d) Show that $\mathbf{A} \cdot \mathbf{B} = \frac{1}{4}(\|\mathbf{A} + \mathbf{B}\|^2 - \|\mathbf{A} - \mathbf{B}\|^2) + \frac{i}{4}(\|\mathbf{A} + i\mathbf{B}\|^2 - \|\mathbf{A} - i\mathbf{B}\|^2)$.

B23. $\mathbf{A} = \begin{bmatrix} -1 - 6i \\ 6i \\ -8 + 9i \end{bmatrix}, \mathbf{B} = \begin{bmatrix} -3 - 7i \\ -4 - 7i \\ -1 - 6i \end{bmatrix}, \mathbf{C} = \begin{bmatrix} 9 + 2i \\ -7 - 2i \\ 6 + 8i \end{bmatrix},$
$z_1 = 7 - 5i, z_2 = -8 - 5i$

B24. $\mathbf{A} = \begin{bmatrix} 2 - 3i \\ 5 + 2i \\ 8 + 2i \end{bmatrix}, \mathbf{B} = \begin{bmatrix} 3 + 9i \\ 8 - i \\ -7i \end{bmatrix}, \mathbf{C} = \begin{bmatrix} -5 - 8i \\ -4 - 6i \\ -3 + 8i \end{bmatrix},$
$z_1 = 9 - 2i, z_2 = -9 + i$

B25. $\mathbf{A} = \begin{bmatrix} -3 - 5i \\ -3 + 6i \\ -2 - 7i \end{bmatrix}, \mathbf{B} = \begin{bmatrix} -4 - 4i \\ -5 - 4i \\ 8i \end{bmatrix}, \mathbf{C} = \begin{bmatrix} -5 - i \\ 5 - 3i \\ 1 + 9i \end{bmatrix},$
$z_1 = -4 + 5i, z_2 = 1 + 6i$

B26. $\mathbf{A} = \begin{bmatrix} -i \\ 9 - i \\ -3 + 6i \end{bmatrix}, \mathbf{B} = \begin{bmatrix} 2 + 6i \\ -3 - i \\ -4 \end{bmatrix}, \mathbf{C} = \begin{bmatrix} 4 - 2i \\ -6 + 2i \\ 5 + i \end{bmatrix},$
$z_1 = 3 - 7i, z_2 = -2 - i$

B27. $\mathbf{A} = \begin{bmatrix} -9 + 6i \\ -2 + 3i \\ 8 + 5i \end{bmatrix}, \mathbf{B} = \begin{bmatrix} -4 + i \\ -2 - i \\ -7 \end{bmatrix}, \mathbf{C} = \begin{bmatrix} 4 + 7i \\ 1 + 9i \\ 8 + 6i \end{bmatrix},$
$z_1 = -5 - i, z_2 = 3 - 6i$

In Exercises B28–37:

(a) Use the dot product to find α so that $\mathbf{B} - \alpha\mathbf{A}$ is perpendicular to $\mathbf{A}$.

(b) Use calculus to minimize $\|\mathbf{B} - \alpha\mathbf{A}\|^2$, thereby recomputing α from an alternative perspective.

(c) Sketch $\mathbf{A}, \mathbf{B}, \mathbf{X} = \alpha\mathbf{A}$, and $\mathbf{Y} = \mathbf{B} - \mathbf{X}$.

(d) Use the diagram in part (c) to explain why α is the same in parts (a) and (b).

B28. $\mathbf{A} = 8\mathbf{i} + 12\mathbf{j}, \mathbf{B} = -3\mathbf{i} - 2\mathbf{j}$ **B29.** $\mathbf{A} = -5\mathbf{i} + 5\mathbf{j}, \mathbf{B} = \mathbf{i} - 10\mathbf{j}$

B30. $\mathbf{A} = 12\mathbf{i} - 2\mathbf{j}, \mathbf{B} = -\mathbf{i} + 11\mathbf{j}$ **B31.** $\mathbf{A} = \mathbf{i} + 5\mathbf{j}, \mathbf{B} = 10\mathbf{i} + 11\mathbf{j}$

B32. $\mathbf{A} = -9\mathbf{i} + 11\mathbf{j}, \mathbf{B} = 8\mathbf{i} + 3\mathbf{j}$ **B33.** $\mathbf{A} = 7\mathbf{i} - 9\mathbf{j}, \mathbf{B} = 3\mathbf{i} - 7\mathbf{j}$

B34. $\mathbf{A} = 7\mathbf{i} - \mathbf{j}, \mathbf{B} = -\mathbf{i} - 3\mathbf{j}$ **B35.** $\mathbf{A} = 11\mathbf{i} - 12\mathbf{j}, \mathbf{B} = 7\mathbf{i} - 11\mathbf{j}$

B36. $\mathbf{A} = -8\mathbf{i} - 2\mathbf{j}, \mathbf{B} = 7\mathbf{i} + 11\mathbf{j}$ **B37.** $\mathbf{A} = 2\mathbf{i} + \mathbf{j}, \mathbf{B} = \mathbf{i} + 3\mathbf{j}$

In Exercises B38–47:

(a) Use the dot product to find α and β for which $\mathbf{C} - (\alpha\mathbf{A} + \beta\mathbf{B})$ is perpendicular to the plane spanned by $\mathbf{A}$ and $\mathbf{B}$.

(b) Use calculus to minimize $\|\mathbf{C} - (\alpha\mathbf{A} + \beta\mathbf{B})\|^2$, thereby recomputing α and β from an alternative perspective.

(c) Sketch $\mathbf{A}, \mathbf{B}$, and the plane in which they lie, and then include $\mathbf{C}, \mathbf{X} = \alpha\mathbf{A} + \beta\mathbf{B}$, and $\mathbf{Y} = \mathbf{C} - \mathbf{X}$.

B38. $\mathbf{A} = 4\mathbf{i} - 6\mathbf{j} + 8\mathbf{k}, \mathbf{B} = 11\mathbf{i} + 12\mathbf{j} + 6\mathbf{k}, \mathbf{C} = 7\mathbf{i} - 11\mathbf{j} - 12\mathbf{k}$

B39. $\mathbf{A} = -\mathbf{i} - 8\mathbf{j} - 12\mathbf{k}, \mathbf{B} = -6\mathbf{i} + 10\mathbf{j} - 2\mathbf{k}, \mathbf{C} = 9\mathbf{i} + 8\mathbf{j} + 10\mathbf{k}$

B40. $\mathbf{A} = -7\mathbf{i} + \mathbf{j} + 8\mathbf{k}, \mathbf{B} = -3\mathbf{i} + 12\mathbf{j} - 2\mathbf{k}, \mathbf{C} = -8\mathbf{i} + 7\mathbf{j} - 11\mathbf{k}$

B41. $\mathbf{A} = 9\mathbf{j} - 2\mathbf{k}, \mathbf{B} = 11\mathbf{i} - 7\mathbf{j} - 9\mathbf{k}, \mathbf{C} = 4\mathbf{i} - 6\mathbf{j} + 12\mathbf{k}$

B42. $\mathbf{A} = -3\mathbf{i} - 10\mathbf{j} + 3\mathbf{k}, \mathbf{B} = -2\mathbf{i} - 11\mathbf{j} + 9\mathbf{k}, \mathbf{C} = -2\mathbf{i} - 12\mathbf{j} + 6\mathbf{k}$

B43. $\mathbf{A} = -5\mathbf{i} - \mathbf{j} + 5\mathbf{k}, \mathbf{B} = 3\mathbf{i} + 11\mathbf{j} + 12\mathbf{k}, \mathbf{C} = -\mathbf{i} - 4\mathbf{j} - 5\mathbf{k}$

B44. $\mathbf{A} = 3\mathbf{i} + 5\mathbf{j} + 3\mathbf{k}, \mathbf{B} = 7\mathbf{i} - 9\mathbf{j} - 6\mathbf{k}, \mathbf{C} = -5\mathbf{i} + 8\mathbf{j} - 5\mathbf{k}$

B45. $\mathbf{A} = 3\mathbf{i} - 5\mathbf{j} + \mathbf{k}, \mathbf{B} = -3\mathbf{i} - 4\mathbf{j} + \mathbf{k}, \mathbf{C} = -12\mathbf{i} - 2\mathbf{j} + 9\mathbf{k}$

B46. $\mathbf{A} = -8\mathbf{i} - 6\mathbf{j} + 12\mathbf{k}, \mathbf{B} = 5\mathbf{i} + 10\mathbf{j} - 3\mathbf{k}, \mathbf{C} = 6\mathbf{i} + 7\mathbf{j} - 4\mathbf{k}$

B47. $\mathbf{A} = 10\mathbf{i} + 2\mathbf{j} + 2\mathbf{k}, \mathbf{B} = 9\mathbf{i} - 9\mathbf{j} - 2\mathbf{k}, \mathbf{C} = 7\mathbf{i} + 4\mathbf{j} - 10\mathbf{k}$

B48. Using vectors, prove that the median to the base of an isosceles triangle is also the altitude to that side.

31.3   The Cross-Product

Derivation of Formula (12.5) for the Cross-Product

Formula (12.5) in Section 12.4 can be derived from the three properties that $\mathbf{A} \times \mathbf{B}$, the cross-product of the vectors $\mathbf{A}$ and $\mathbf{B}$, is orthogonal to both $\mathbf{A}$ and $\mathbf{B}$, forms a right-handed system with $\mathbf{A}$ and $\mathbf{B}$, and has length given by

$$\|\mathbf{A} \times \mathbf{B}\| = \|\mathbf{A}\|\|\mathbf{B}\| \sin\theta \qquad (31.5)$$

where θ in $[0, \pi]$ is the angle between $\mathbf{A}$ and $\mathbf{B}$. To do this, let the vectors $\mathbf{A}, \mathbf{B}$, and $\mathbf{C}$ be given, respectively, by

$$\mathbf{A} = a_1\mathbf{i} + a_2\mathbf{j} + a_3\mathbf{k} \qquad \mathbf{B} = b_1\mathbf{i} + b_2\mathbf{j} + b_3\mathbf{k} \qquad \mathbf{C} = \mathbf{A} \times \mathbf{B} = c_1\mathbf{i} + c_2\mathbf{j} + c_3\mathbf{k}$$

The condition that $\mathbf{C}$ be perpendicular to both $\mathbf{A}$ and $\mathbf{B}$ yields the two equations

$$\mathbf{A} \cdot \mathbf{C} = a_1c_1 + a_2c_2 + a_3c_3 = 0 \quad \text{and} \quad \mathbf{B} \cdot \mathbf{C} = b_1c_1 + b_2c_2 + b_3c_3 = 0$$

Then, (31.5) yields the third equation

$$\sqrt{c_1^2 + c_2^2 + c_3^2} = \sqrt{a_1^2 + a_2^2 + a_3^2}\sqrt{b_1^2 + b_2^2 + b_3^2}\sqrt{1 - \frac{(a_1b_1 + a_2b_2 + a_3b_3)^2}{(a_1^2 + a_2^2 + a_3^2)(b_1^2 + b_2^2 + b_3^2)}}$$

Solving these three equations for c_1, c_2, c_3 gives the two solutions

$$\begin{bmatrix} a_2b_3 - a_3b_2 \\ a_3b_1 - a_1b_3 \\ a_1b_2 - a_2b_1 \end{bmatrix} \quad \text{and} \quad -\begin{bmatrix} a_2b_3 - a_3b_2 \\ a_3b_1 - a_1b_3 \\ a_1b_2 - a_2b_1 \end{bmatrix}$$

A simple test with the unit vectors $\{\mathbf{i}, \mathbf{j}, \mathbf{k}\}$ shows the first solution obeys the right-hand rule and points in the appropriate direction. The other solution is its negative. Consequently, orthogonality, the length condition (31.5), and the right-hand rule determine the cross-product, which we write more compactly as

$$\mathbf{A} \times \mathbf{B} = \det \begin{bmatrix} \mathbf{i} & \mathbf{j} & \mathbf{k} \\ a_1 & a_2 & a_3 \\ b_1 & b_2 & b_3 \end{bmatrix} \tag{31.6}$$

in terms of the determinant of a 3×3 matrix. (Determinants are discussed in Section 12.4.)

EXAMPLE 31.6   If $\mathbf{A} = 2\mathbf{i} - 2\mathbf{j} - \mathbf{k}$ and $\mathbf{B} = \mathbf{i} + \mathbf{j} + \mathbf{k}$, then

$$\mathbf{A} \times \mathbf{B} = \begin{vmatrix} \begin{bmatrix} \mathbf{i} & \mathbf{j} & \mathbf{k} \\ 2 & -2 & -1 \\ 1 & 1 & 1 \end{bmatrix} \end{vmatrix} = \begin{bmatrix} -1 \\ -3 \\ 4 \end{bmatrix}$$

with $\mathbf{A}$, $\mathbf{B}$, and $\mathbf{A} \times \mathbf{B}$ shown in Figure 31.10.   ❖

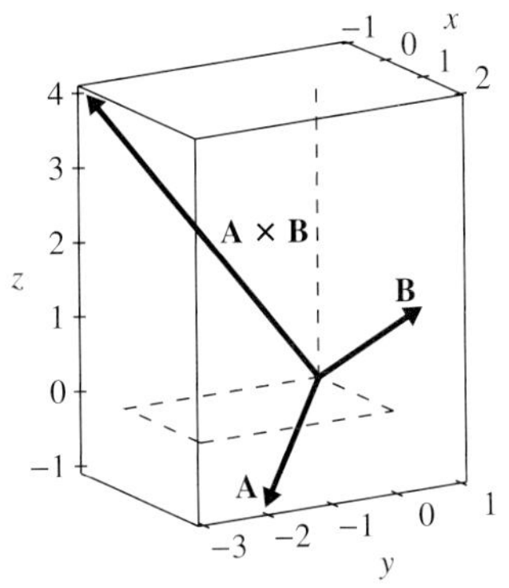

FIGURE 31.10   The vectors $\mathbf{A}$, $\mathbf{B}$, and $\mathbf{A} \times \mathbf{B}$ in Example 31.6

Additional Properties of the Cross-Product

A compendium of formulas for the cross-product includes those listed in Table 31.3. The first two laws suggest the cross-product obeys some of the rules of ordinary algebra. However, the failure of the commutative law (3) and of the associative laws (4) points out the singular nature of the cross-product.

(1) Scalar multiplication	$(a\mathbf{A}) \times (b\mathbf{B}) = ab(\mathbf{A} \times \mathbf{B})$
(2) Distributive laws	$\mathbf{A} \times (\mathbf{B} + \mathbf{C}) = \mathbf{A} \times \mathbf{B} + \mathbf{A} \times \mathbf{C}$ $(\mathbf{B} + \mathbf{C}) \times \mathbf{A} = \mathbf{B} \times \mathbf{A} + \mathbf{C} \times \mathbf{A}$
(3) Anticommutation	$\mathbf{B} \times \mathbf{A} = -\mathbf{A} \times \mathbf{B}$
(4) Nonassociativity	$\mathbf{A} \times (\mathbf{B} \times \mathbf{C}) = (\mathbf{A} \cdot \mathbf{C})\mathbf{B} - (\mathbf{A} \cdot \mathbf{B})\mathbf{C}$ $(\mathbf{A} \times \mathbf{B}) \times \mathbf{C} = (\mathbf{A} \cdot \mathbf{C})\mathbf{B} - (\mathbf{B} \cdot \mathbf{C})\mathbf{A}$

TABLE 31.3   Additional formulas for the cross-product

EXERCISES 31.3–Part A

In Exercises A1–5, let $\mathbf{A} = \mathbf{i} + \mathbf{j} + \mathbf{k}$, $\mathbf{B} = \mathbf{i} - \mathbf{j} - \mathbf{k}$, $\mathbf{C} = \mathbf{i} - \mathbf{j} + \mathbf{k}$.

A1. Verify $\mathbf{A} \times (\mathbf{B} + \mathbf{C}) = \mathbf{A} \times \mathbf{B} + \mathbf{A} \times \mathbf{C}$.

A2. Verify $(\mathbf{B} + \mathbf{C}) \times \mathbf{A} = \mathbf{B} \times \mathbf{A} + \mathbf{C} \times \mathbf{A}$.

A3. Verify $\mathbf{B} \times \mathbf{A} = -\mathbf{A} \times \mathbf{B}$.

A4. Verify $\mathbf{A} \times (\mathbf{B} \times \mathbf{C}) = (\mathbf{A} \cdot \mathbf{C})\mathbf{B} - (\mathbf{A} \cdot \mathbf{B})\mathbf{C}$.

A5. Verify $(\mathbf{A} \times \mathbf{B}) \times \mathbf{C} = (\mathbf{A} \cdot \mathbf{C})\mathbf{B} - (\mathbf{B} \cdot \mathbf{C})\mathbf{A}$.

EXERCISES 31.3–Part B

B1. The triple scalar product, $\mathbf{A} \cdot (\mathbf{B} \times \mathbf{C}) = \det \begin{bmatrix} a_1 & a_2 & a_3 \\ b_1 & b_2 & b_3 \\ c_1 & c_2 & c_3 \end{bmatrix}$, appeared in a proof in Section 21.6. Show that $|\mathbf{A} \cdot (\mathbf{B} \times \mathbf{C})|$ represents the volume of the parallelepiped whose edges are the vectors $\mathbf{A}, \mathbf{B}, \mathbf{C}$, and their translates.

B2. Verify the identity $(\mathbf{A} \times \mathbf{B}) \times (\mathbf{C} \times \mathbf{D}) = (\mathbf{A} \cdot \mathbf{B} \times \mathbf{D})\mathbf{C} - (\mathbf{A} \cdot \mathbf{B} \times \mathbf{C})\mathbf{D}$.

B3. Verify the identity $(\mathbf{A} \times \mathbf{B}) \times (\mathbf{C} \times \mathbf{D}) = (\mathbf{A} \cdot \mathbf{C} \times \mathbf{D})\mathbf{B} - (\mathbf{B} \cdot \mathbf{C} \times \mathbf{D})\mathbf{A}$.

B4. Verify the identity $(\mathbf{A} \times \mathbf{B}) \cdot (\mathbf{C} \times \mathbf{D}) = (\mathbf{A} \cdot \mathbf{C})(\mathbf{B} \cdot \mathbf{D}) - (\mathbf{A} \cdot \mathbf{D})(\mathbf{B} \cdot \mathbf{C})$.

B5. Verify the identity $\mathbf{A} \times (\mathbf{B} \times \mathbf{C}) + \mathbf{B} \times (\mathbf{C} \times \mathbf{A}) + \mathbf{C} \times (\mathbf{A} \times \mathbf{B}) = \mathbf{0}$.

B6. Verify the identity $\|\mathbf{A} \times \mathbf{B}\|^2 + (\mathbf{A} \cdot \mathbf{B})^2 - \|\mathbf{A}\|^2 \|\mathbf{B}\|^2 = 0$.

B7. Verify the identity $[\mathbf{A} \times (\mathbf{A} \times \mathbf{B})] \times \mathbf{A} \cdot \mathbf{C} = \|\mathbf{A}\|^2 (\mathbf{A} \cdot \mathbf{B} \times \mathbf{C})$.

B8. Verify the identity $(\mathbf{A} \times \mathbf{B}) \cdot (\mathbf{B} \times \mathbf{C}) \times (\mathbf{C} \times \mathbf{A}) = (\mathbf{A} \cdot \mathbf{B} \times \mathbf{C})^2$.

In Exercises B9–18:

(a) Obtain $\mathbf{A} \times \mathbf{B}$ via the determinant formula, (31.6), and then compute its length.

(b) Compare part (a) to $\|\mathbf{A} \times \mathbf{B}\| = \|\mathbf{A}\| \|\mathbf{B}\| \sin \theta$, where θ is obtained from $\mathbf{A} \cdot \mathbf{B} = \|\mathbf{A}\| \|\mathbf{B}\| \cos \theta$.

B9. $\mathbf{A} = 7\mathbf{i} - 9\mathbf{j} + 3\mathbf{k}$, $\mathbf{B} = -7\mathbf{i} - 5\mathbf{j} - 11\mathbf{k}$

B10. $\mathbf{A} = -10\mathbf{i} - 10\mathbf{j} + 12\mathbf{k}$, $\mathbf{B} = -12\mathbf{i} - 5\mathbf{j} - 5\mathbf{k}$

B11. $\mathbf{A} = -8\mathbf{i} - 10\mathbf{j} - 8\mathbf{k}$, $\mathbf{B} = -2\mathbf{i} + 8\mathbf{j} + 11\mathbf{k}$

B12. $\mathbf{A} = 7\mathbf{i} - 11\mathbf{j} + 12\mathbf{k}$, $\mathbf{B} = 10\mathbf{i} + \mathbf{j} + 3\mathbf{k}$

B13. $\mathbf{A} = 7\mathbf{i} - \mathbf{j} - \mathbf{k}$, $\mathbf{B} = -3\mathbf{i} + 4\mathbf{j} - 6\mathbf{k}$

B14. $\mathbf{A} = -12\mathbf{i} - 6\mathbf{j} + 8\mathbf{k}$, $\mathbf{B} = -2\mathbf{i} - 7\mathbf{j} + 4\mathbf{k}$

B15. $\mathbf{A} = 10\mathbf{i} - 7\mathbf{j} + \mathbf{k}$, $\mathbf{B} = 8\mathbf{i} - 3\mathbf{j} + 12\mathbf{k}$

B16. $\mathbf{A} = -2\mathbf{i} + 11\mathbf{j} - 7\mathbf{k}$, $\mathbf{B} = -9\mathbf{i} + 4\mathbf{j} - 6\mathbf{k}$

B17. $\mathbf{A} = 9\mathbf{i} - 2\mathbf{j} - 12\mathbf{k}$, $\mathbf{B} = 6\mathbf{i} - 5\mathbf{j} - \mathbf{k}$

B18. $\mathbf{A} = 5\mathbf{i} + 3\mathbf{j} + 11\mathbf{k}$, $\mathbf{B} = 12\mathbf{i} - \mathbf{j} - 4\mathbf{k}$

For each pair of vectors $\mathbf{A}$ and $\mathbf{B}$ in Exercises B19–26, let $\mathbf{C} = c_1\mathbf{i} + c_2\mathbf{j} + c_3\mathbf{k}$ and determine $\mathbf{C} = \mathbf{A} \times \mathbf{B}$ from the three conditions $\mathbf{A} \cdot \mathbf{C} = 0$, $\mathbf{B} \cdot \mathbf{C} = 0$, and $\|\mathbf{C}\| = \|\mathbf{A}\| \|\mathbf{B}\| \sin \theta$. Verify the result by computing $\mathbf{A} \times \mathbf{B}$.

B19. $\mathbf{A} = \mathbf{i} - 11\mathbf{j} - 12\mathbf{k}$, $\mathbf{B} = 12\mathbf{i} + 11\mathbf{j} - 12\mathbf{k}$

B20. $\mathbf{A} = 8\mathbf{i} + 11\mathbf{j} + 12\mathbf{k}$, $\mathbf{B} = 6\mathbf{i} + 7\mathbf{j} - 11\mathbf{k}$

B21. $\mathbf{A} = -12\mathbf{i} - 6\mathbf{j} + 10\mathbf{k}$, $\mathbf{B} = -2\mathbf{i} + 9\mathbf{j} + 8\mathbf{k}$

B22. $\mathbf{A} = 12\mathbf{i} - 3\mathbf{j} - 10\mathbf{k}$, $\mathbf{B} = 3\mathbf{i} - 2\mathbf{j} - 11\mathbf{k}$

B23. $\mathbf{A} = -5\mathbf{i} + 3\mathbf{j} + 5\mathbf{k}$, $\mathbf{B} = 3\mathbf{i} + 7\mathbf{j} - 9\mathbf{k}$

B24. $\mathbf{A} = -6\mathbf{i} - 5\mathbf{j} + 8\mathbf{k}$, $\mathbf{B} = -5\mathbf{i} + 3\mathbf{j} - 5\mathbf{k}$

B25. $\mathbf{A} = \mathbf{i} - 3\mathbf{j} - 4\mathbf{k}$, $\mathbf{B} = \mathbf{i} - 12\mathbf{j} - 2\mathbf{k}$

B26. $\mathbf{A} = -2\mathbf{i} - 8\mathbf{j} + 7\mathbf{k}$, $\mathbf{B} = 5\mathbf{i} + 10\mathbf{j} + 11\mathbf{k}$

In Exercises B27–36:

(a) Verify $\mathbf{A} \times (\mathbf{B} \times \mathbf{C}) = (\mathbf{A} \cdot \mathbf{C})\mathbf{B} - (\mathbf{A} \cdot \mathbf{B})\mathbf{C}$.

(b) Sketch $\mathbf{B}, \mathbf{C}$, and $\mathbf{A} \times (\mathbf{B} \times \mathbf{C})$.

(c) Verify $(\mathbf{A} \times \mathbf{B}) \times \mathbf{C} = (\mathbf{A} \cdot \mathbf{C})\mathbf{B} - (\mathbf{B} \cdot \mathbf{C})\mathbf{A}$.

(d) Find the triple scalar product $\mathbf{A} \cdot (\mathbf{B} \times \mathbf{C})$ and the volume of the related parallelepiped.

(e) Sketch $\mathbf{A}, \mathbf{B}$, and $(\mathbf{A} \times \mathbf{B}) \times \mathbf{C}$.

B27. $\mathbf{A} = 9\mathbf{i} - 8\mathbf{j} - 6\mathbf{k}$, $\mathbf{B} = 12\mathbf{i} + 5\mathbf{j} + 10\mathbf{k}$, $\mathbf{C} = -3\mathbf{i} + 6\mathbf{j} + 7\mathbf{k}$

B28. $\mathbf{A} = -4\mathbf{i} + 10\mathbf{j} + 2\mathbf{k}$, $\mathbf{B} = 2\mathbf{i} + 9\mathbf{j} - 9\mathbf{k}$, $\mathbf{C} = -2\mathbf{i} + 7\mathbf{j} + 4\mathbf{k}$

B29. $\mathbf{A} = -10\mathbf{i} - 7\mathbf{j} - 7\mathbf{k}$, $\mathbf{B} = 5\mathbf{i} - \mathbf{j} - 5\mathbf{k}$, $\mathbf{C} = 7\mathbf{i} + 11\mathbf{j} - 2\mathbf{k}$

B30. $\mathbf{A} = 10\mathbf{i} + 4\mathbf{j} - 12\mathbf{k}$, $\mathbf{B} = \mathbf{i} + 7\mathbf{j} - 3\mathbf{k}$, $\mathbf{C} = -6\mathbf{i} + 5\mathbf{j} + 7\mathbf{k}$

B31. $\mathbf{A} = -9\mathbf{i} + 12\mathbf{j} + 8\mathbf{k}$, $\mathbf{B} = -10\mathbf{i} - 7\mathbf{k}$, $\mathbf{C} = 3\mathbf{i} + 4\mathbf{j} - 6\mathbf{k}$

B32. $\mathbf{A} = -7\mathbf{i} + \mathbf{j} + \mathbf{k}$, $\mathbf{B} = -2\mathbf{i} - 5\mathbf{j} - \mathbf{k}$, $\mathbf{C} = -2\mathbf{i} - 5\mathbf{j} + 9\mathbf{k}$

B33. $\mathbf{A} = -3\mathbf{i} - 9\mathbf{j} - 3\mathbf{k}$, $\mathbf{B} = -10\mathbf{i} - 12\mathbf{j}$, $\mathbf{C} = 6\mathbf{i} - 2\mathbf{j} + \mathbf{k}$

B34. $\mathbf{A} = 8\mathbf{i} - 4\mathbf{j} + \mathbf{k}$, $\mathbf{B} = 5\mathbf{i} - 10\mathbf{j} + 11\mathbf{k}$, $\mathbf{C} = -5\mathbf{i} - 2\mathbf{j} - 2\mathbf{k}$

B35. $\mathbf{A} = -2\mathbf{i} - 10\mathbf{j} - 5\mathbf{k}$, $\mathbf{B} = 6\mathbf{i} - 5\mathbf{j} + 10\mathbf{k}$, $\mathbf{C} = -9\mathbf{i} + 5\mathbf{j} - 10\mathbf{k}$

B36. $\mathbf{A} = -2\mathbf{i} - 12\mathbf{j} - 7\mathbf{k}$, $\mathbf{B} = -6\mathbf{i} - 3\mathbf{j} - 3\mathbf{k}$, $\mathbf{C} = -7\mathbf{i} - 12\mathbf{j} - 10\mathbf{k}$

In Exercises B37–41, find the area of the triangle determined by the given points. *Hint:* See Figure 21.2, Section 21.1, and the related discussion.

B37. $p_1 = (9, 6, 8)$, $p_2 = (-4, -6, -3)$, $p_3 = (6, 10, 6)$

B38. $p_1 = (9, 7, 5)$, $p_2 = (-12, 11, -2)$, $p_3 = (5, -10, 11)$

B39. $p_1 = (9, -11, 4)$, $p_2 = (-12, -9, -8)$, $p_3 = (2, 6, 8)$

B40. $p_1 = (6, -5, 1)$, $p_2 = (-8, 4, 12)$, $p_3 = (-5, 4, -2)$

B41. $p_1 = (10, 5, -5)$, $p_2 = (8, -7, 4)$, $p_3 = (-9, 5, -6)$

In Exercises B42–49:

(a) Show that the four given points are coplanar.

(b) Show that the four given points determine a parallelogram, and sketch the parallelogram.

(c) Find the area of the parallelogram.

(d) Find the lengths of the diagonals of the parallelogram.

(e) Find the coordinates of the intersection of the diagonals.

B42. $p_1 = (-3, -7, 1)$, $p_2 = (-5, -2, 12)$, $p_3 = (-2, 4, -1)$, $p_4 = (0, -1, -12)$

B43. $p_1 = (-5, 12, -1)$, $p_2 = (11, 4, -3)$, $p_3 = (27, -14, 10)$, $p_4 = (11, -6, 12)$

B44. $p_1 = (4, 11, 5)$, $p_2 = (4, 6, -12)$, $p_3 = (-1, -4, -16)$, $p_4 = (-1, 1, 1)$

B45. $p_1 = (8, 7, 3)$, $p_2 = (4, 8, 12)$, $p_3 = (-7, -1, 4)$, $p_4 = (-3, -2, -5)$

B46. $p_1 = (5, 1, -10)$, $p_2 = (12, -2, -1)$, $p_3 = (18, -12, 20)$, $p_4 = (11, -9, 11)$

B47. $p_1 = (7, -11, 12)$, $p_2 = (10, 1, 3)$, $p_3 = (10, 11, -10)$, $p_4 = (7, -1, -1)$

B48. $p_1 = (8, 11, 2)$, $p_2 = (0, -5, 12)$, $p_3 = (-1, -25, 13)$, $p_4 = (7, -9, 3)$

B49. $p_1 = (-2, 8, 11)$, $p_2 = (1, -11, -12)$, $p_3 = (15, -8, -35)$, $p_4 = (12, 11, -12)$

When a force **F** is applied at point Q in an object, the tendency of the force to generate a rotation about the point P is called the *torque* or moment of the force about P. If **r** is the vector from P to Q, this torque is given by $\mathbf{T} = \mathbf{r} \times \mathbf{F}$. The magnitude of **T** measures the tendency of the vector **r** to be rotated counterclockwise about an axis lying along **T**. In Exercises B50–59:

(a) Obtain **T** and $\|\mathbf{T}\|$.

(b) Sketch **r**, **F**, and **T**.

B50. $P = (7, -11, 12)$, $Q = (10, 1, 3)$, $\mathbf{F} = 7\mathbf{i} - 11\mathbf{j} + 12\mathbf{k}$

B51. $P = (-3, 4, -6)$, $Q = (8, 11, 12)$, $\mathbf{F} = -3\mathbf{i} + 4\mathbf{j} - 6\mathbf{k}$

B52. $P = (-12, -6, 8)$, $Q = (-2, -7, 4)$, $\mathbf{F} = -12\mathbf{i} - 6\mathbf{j} + 8\mathbf{k}$

B53. $P = (10, -7, 1)$, $Q = (8, -3, 12)$, $\mathbf{F} = 10\mathbf{i} - 7\mathbf{j} + \mathbf{k}$

B54. $P = (6, -5, -1)$, $Q = (5, 3, 11)$, $\mathbf{F} = 6\mathbf{i} - 5\mathbf{j} - \mathbf{k}$

B55. $P = (-5, 3, 5)$, $Q = (3, 7, -9)$, $\mathbf{F} = -5\mathbf{i} + 3\mathbf{j} + 5\mathbf{k}$

B56. $P = (9, -8, -6)$, $Q = (12, 5, 10)$, $\mathbf{F} = 9\mathbf{i} - 8\mathbf{j} - 6\mathbf{k}$

B57. $P = (-4, 10, 2)$, $Q = (2, 9, -9)$, $\mathbf{F} = -4\mathbf{i} + 10\mathbf{j} + 2\mathbf{k}$

B58. $P = (10, 4, -12)$, $Q = (1, 7, -3)$, $\mathbf{F} = 10\mathbf{i} + 4\mathbf{j} - 12\mathbf{k}$

B59. $P = (-7, 1, 1)$, $Q = (-2, -5, -1)$, $\mathbf{F} = -7\mathbf{i} + \mathbf{j} + \mathbf{k}$

Chapter Review

Let the vectors **A**, **B**, **C** be given by
$\mathbf{A} = 2\mathbf{i} + 3\mathbf{j} - 5\mathbf{k}$, $\mathbf{B} = 5\mathbf{i} - 2\mathbf{j} + 7\mathbf{k}$, $\mathbf{C} = 4\mathbf{i} - 5\mathbf{j} - 6\mathbf{k}$.

1. What are the components of **A**?

2. What are the vector components of **A**?

3. What is $\|\mathbf{A}\|$, the length of **A**?

4. Obtain $\mathbf{A} + \mathbf{B}$, and draw the parallelogram that represents this sum graphically.

5. Obtain $\mathbf{A} - \mathbf{B}$, and draw the triangle that represents this sum graphically.

6. Compute $\mathbf{A} \cdot \mathbf{B}$ and verify that $\mathbf{A} \cdot \mathbf{B} = \|\mathbf{A}\|\|\mathbf{B}\| \cos\theta$, where θ is the angle between **A** and **B**.

7. Compute $\mathbf{A} \times \mathbf{B}$ and verify that $\|\mathbf{A} \times \mathbf{B}\| = \|\mathbf{A}\|\|\mathbf{B}\| \sin\theta$.

8. Verify that $\mathbf{A} \times \mathbf{B}$ is orthogonal to both **A** and **B**.

9. Compute $\mathbf{A} \cdot \mathbf{B} \times \mathbf{C}$ and interpret its value.

10. Compute $\mathbf{A} \times (\mathbf{B} \times \mathbf{C})$ and $(\mathbf{A} \times \mathbf{B}) \times \mathbf{C}$.

11. Verify that $\mathbf{A} \times (\mathbf{B} + \mathbf{C}) = \mathbf{A} \times \mathbf{B} + \mathbf{A} \times \mathbf{C}$.

12. Verify that $(\mathbf{B} + \mathbf{C}) \times \mathbf{A} = \mathbf{B} \times \mathbf{A} + \mathbf{C} \times \mathbf{A}$.

Let $M = \begin{bmatrix} -1 & 9 \\ -2 & 8 \end{bmatrix}$ and define $\langle \mathbf{x}, \mathbf{y} \rangle = \mathbf{x}^{\mathrm{T}} M \mathbf{y}$, where $\mathbf{x} = \begin{bmatrix} x_1 \\ x_2 \end{bmatrix}$ and $\mathbf{y} = \begin{bmatrix} y_1 \\ y_2 \end{bmatrix}$. Show that $\langle \mathbf{x}, \mathbf{y} \rangle$ defines an inner product by answering Questions 13–16.

13. Show that $\langle \mathbf{x}, \mathbf{x} \rangle > 0$ for $\mathbf{x} \neq \mathbf{0}$.

14. Show that $\langle \mathbf{x}, \mathbf{x} \rangle = 0 \Rightarrow \mathbf{x} = \mathbf{0}$, and conversely.

15. Show that $\langle \mathbf{x}, \mathbf{y} \rangle = \langle \mathbf{y}, \mathbf{x} \rangle$.

16. Show that $\langle a\mathbf{x} + b\mathbf{y}, \mathbf{z} \rangle = a\langle \mathbf{x}, \mathbf{z} \rangle + b\langle \mathbf{y}, \mathbf{z} \rangle$, where $\mathbf{z} = \begin{bmatrix} z_1 \\ z_2 \end{bmatrix}$.

17. If $\mathbf{x} = 2\mathbf{i} - 3\mathbf{j}$ and $\mathbf{y} = 5\mathbf{i} + 7\mathbf{j}$, compute $\mathbf{x} \cdot \mathbf{y}$ and $\langle \mathbf{x}, \mathbf{y} \rangle$.

18. When is a square matrix A said to be positive definite? Give an example of a positive definite 2×2 matrix which contains no zeros.

Chapter 32

Change of Coordinates

INTRODUCTION   In the vector calculus of Unit Four, vectors tangent to curves arose by differentiation of the radius-vector form of the curve and vectors orthogonal to coordinates surfaces arose as gradient vectors on those surfaces. In the orthogonal coordinates systems of Unit Four, the essential difference between tangent and gradient vectors is masked by the orthonormal bases used. This difference is in the way components change when the coordinates in the space are changed.

Therefore, to understand the contravariant and covariant transformation laws for obtaining the components of tangent and gradient vectors under a change of coordinates, we begin the chapter with an examination of a change of basis for vectors thought of as arrows. From this perspective, we then discuss rotations and reflections in space and the orthogonal matrices that express them.

A change to polar or spherical coordinates induces a change in basis for vectors in those systems. The new basis vectors can be taken as either tangents to coordinate curves or normals to coordinate surfaces. These sets of basis vectors are related as reciprocal sets of vectors. In coordinate systems that have orthonormal bases, the two sets of reciprocal vectors are coincident. However, in nonorthogonal coordinates systems or in orthogonal systems where the basis vectors are not normalized, the reciprocal bases are different. Vectors expressed in terms of these two different types of basis vectors transform differently. Vectors expressed in terms of a basis of vectors tangent to coordinate curves are represented as column vectors, and transform contravariantly. Vectors expressed in terms of a basis of vectors normal to coordinate surfaces are represented as row vectors, and transform covariantly.

32.1   Change of Basis

Basis

We have informally used the notion of a *basis* for a null space in Sections 12.6 and 12.12 and, in Section 31.1, have used $\{\mathbf{i}, \mathbf{j}, \mathbf{k}\}$ as the set of *unit basis* vectors for three-dimensional Euclidean space (R^3). A more formal definition is the following.

A *basis* for a space of vectors is a "minimal" set of linearly independent vectors whose linear combinations generate all members of the space. By minimal, we mean that the number of vectors in the set is as small as possible. This unique number is the *dimension* of the space.

For example, the *xy*-plane is a two-dimensional space in its own right. When it sits inside three-dimensional Euclidean space, it is considered a *subspace* of the larger space. In fact, any plane through the origin of R^3 is a two-dimensional subspace of R^3. Although every vector in the *xy*-plane can be represented in terms of $\{\mathbf{i}, \mathbf{j}, \mathbf{k}\}$ by taking the component of $\mathbf{k}$ to be zero, a basis for the *xy*-plane is $\{\mathbf{i}, \mathbf{j}\}$, a set of just *two* linearly independent vectors.

Components of a Vector

A vector, defined as a directed line segment, has been represented geometrically by an arrow and analytically by its components with respect to the orthonormal set $\{\mathbf{i}, \mathbf{j}, \mathbf{k}\}$, a basis of mutually perpendicular vectors of unit length. Typographically, the vector $\mathbf{V}$ whose components are the numbers v^1, v^2, v^3 can be written explicitly in terms of $\{\mathbf{i}, \mathbf{j}, \mathbf{k}\}$ or as a column, as in

$$\mathbf{V} = v^1\mathbf{i} + v^2\mathbf{j} + v^3\mathbf{k} = \begin{bmatrix} v^1 \\ v^2 \\ v^3 \end{bmatrix}$$

In the course of Chapter 32, the formalism that pairs superscripted indices with column vectors will eventually be matched with a formalism that pairs subscripted indices with row vectors. Both are related to the type of basis used to represent the vector, and it is to this question we now turn. In particular, we wish to learn how a vector's components, given with respect to one basis, are related to its components given with respect to another basis.

New Basis Vectors

Deliberately written with subscripts, the three vectors

$$\mathbf{e}_1 = -8\mathbf{i} + 8\mathbf{j} - 5\mathbf{k} \qquad \mathbf{e}_2 = -3\mathbf{i} - 5\mathbf{j} + 3\mathbf{k} \qquad \mathbf{e}_3 = -8\mathbf{i} + 3\mathbf{j} - \mathbf{k}$$

are linearly independent, in accord with the definition in Section 12.6. We wish to express the vector $\mathbf{V} = v^1\mathbf{i} + v^2\mathbf{j} + v^3\mathbf{k}$ in terms of the new basis $\{\mathbf{e}_1, \mathbf{e}_2, \mathbf{e}_3\}$, that is, we wish to know $'v^1, 'v^2, 'v^3$, the components of $\mathbf{V}$ with respect to the new basis vectors. The vector $\mathbf{V}$ will still be the same arrow, even though its representation will be

$$\mathbf{V} = {'v^1}\mathbf{e}_1 + {'v^2}\mathbf{e}_2 + {'v^3}\mathbf{e}_3 = \sum_{k=1}^{3} {'v^k}\mathbf{e}_k = {'v^k}\mathbf{e}_k$$

where the *Einstein summation convention* (sum on the repeated index k appearing once raised and once lowered) allows us to drop the explicit sigma notation.

To find the new components, solve the equations

$$\mathbf{e}_1 = -8\mathbf{i} + 8\mathbf{j} - 5\mathbf{k} \qquad \mathbf{e}_2 = -3\mathbf{i} - 5\mathbf{j} + 3\mathbf{k} \qquad \mathbf{e}_3 = -8\mathbf{i} + 3\mathbf{j} - \mathbf{k}$$

for $\{\mathbf{i}, \mathbf{j}, \mathbf{k}\}$ in terms of $\{\mathbf{e}_1, \mathbf{e}_2, \mathbf{e}_3\}$, obtaining

$$\mathbf{i} = -\tfrac{4}{61}\mathbf{e}_1 - \tfrac{7}{61}\mathbf{e}_2 - \tfrac{1}{61}\mathbf{e}_3 \qquad \mathbf{j} = -\tfrac{27}{61}\mathbf{e}_1 - \tfrac{32}{61}\mathbf{e}_2 + \tfrac{39}{61}\mathbf{e}_3 \qquad \mathbf{k} = -\tfrac{49}{61}\mathbf{e}_1 - \tfrac{40}{61}\mathbf{e}_2 + \tfrac{64}{61}\mathbf{e}_3$$

from which we get

$$'\mathbf{V} = \left(-\tfrac{4}{61}v^1 - \tfrac{27}{61}v^2 - \tfrac{49}{61}v^3\right)\mathbf{e}_1 + \left(-\tfrac{7}{61}v^1 - \tfrac{32}{61}v^2 - \tfrac{40}{61}v^3\right)\mathbf{e}_2 + \left(-\tfrac{1}{61}v^1 + \tfrac{39}{61}v^2 + \tfrac{64}{61}v^3\right)\mathbf{e}_3$$

For insight into the structure of this result, let A be the matrix of coefficients for the equations relating the old basis to the new. In particular, let the columns of A be the coefficients of the new basis vectors in terms of the old. Then, A and A^{-1} are given by

$$A = \begin{bmatrix} -8 & -3 & -8 \\ 8 & -5 & 3 \\ -5 & 3 & -1 \end{bmatrix} \qquad A^{-1} = \begin{bmatrix} -\frac{4}{61} & -\frac{27}{61} & -\frac{49}{61} \\ -\frac{7}{61} & -\frac{32}{61} & -\frac{40}{61} \\ -\frac{1}{61} & \frac{39}{61} & \frac{64}{61} \end{bmatrix}$$

and inspection shows that if $\mathbf{V}$ and $'\mathbf{V}$ are written as column vectors, then $'\mathbf{V} = A^{-1}\mathbf{V}$ is the transformation law for the change in coordinates induced by a change of basis.

EXAMPLE 32.1   In the xy-plane, change bases from $\{\mathbf{i}, \mathbf{j}\}$ to $\mathbf{e}_1 = -\mathbf{i} - 4\mathbf{j}$, $\mathbf{e}_2 = -5\mathbf{i} + 3\mathbf{j}$. To find the components of $\mathbf{V} = 2\mathbf{i} + 3\mathbf{j}$ with respect to the new basis vectors, solve the equations $-\mathbf{i} - 4\mathbf{j} = \mathbf{e}_1$, $-5\mathbf{i} + 3\mathbf{j} = \mathbf{e}_2$ for $\{\mathbf{i}, \mathbf{j}\}$ in terms of $\{\mathbf{e}_1, \mathbf{e}_2\}$, obtaining $\mathbf{i} = -\frac{3}{23}\mathbf{e}_1 - \frac{4}{23}\mathbf{e}_2$, $\mathbf{j} = -\frac{5}{23}\mathbf{e}_1 + \frac{1}{23}\mathbf{e}_2$. Then, by direct substitution, obtain $'\mathbf{V} = -\frac{21}{23}\mathbf{e}_1 - \frac{5}{23}\mathbf{e}_2$.

Alternatively, proceed algorithmically by forming A, the matrix whose columns are the coefficients of the new basis with respect to the old, and then computing A^{-1} and $'\mathbf{V} = A^{-1}\mathbf{V}$, thereby obtaining

$$A = \begin{bmatrix} -1 & -5 \\ -4 & 3 \end{bmatrix} \qquad A^{-1} = \begin{bmatrix} -\frac{3}{23} & -\frac{5}{23} \\ -\frac{4}{23} & \frac{1}{23} \end{bmatrix} \Rightarrow \ '\mathbf{V} = A^{-1}\mathbf{V} = \begin{bmatrix} -\frac{21}{23} \\ -\frac{5}{23} \end{bmatrix} = -\frac{21}{23}\mathbf{e}_1 - \frac{5}{23}\mathbf{e}_2$$

Figure 32.1 shows $\mathbf{V}$, $\mathbf{e}_1$, $\mathbf{e}_2$, and the components of $\mathbf{V}$ along $\mathbf{e}_1$ and $\mathbf{e}_2$. In addition, the parallelogram of addition formed by the components of $\mathbf{V}$ along $\mathbf{e}_1$ and $\mathbf{e}_2$ is also shown. ❖

EXAMPLE 32.2   For the vectors

$$\mathbf{V} = \begin{bmatrix} -5 \\ -6 \\ -8 \end{bmatrix} \qquad \mathbf{e}_1 = \begin{bmatrix} 4 \\ 8 \\ 2 \end{bmatrix} \qquad \mathbf{e}_2 = \begin{bmatrix} -3 \\ 7 \\ -6 \end{bmatrix} \qquad \mathbf{e}_3 = \begin{bmatrix} 2 \\ 8 \\ 0 \end{bmatrix}$$

find the components of $\mathbf{V}$ with respect to $\{\mathbf{e}_1, \mathbf{e}_2, \mathbf{e}_3\}$. Thus, form A, the transition matrix whose columns are components of the new basis vectors in terms of the old. Then, the components of $\mathbf{V}$ with respect to the new basis are given by $'\mathbf{V} = A^{-1}\mathbf{V}$ or

$$'\mathbf{V} = \begin{bmatrix} 4 & -3 & 2 \\ 8 & 7 & 8 \\ 2 & -6 & 0 \end{bmatrix}^{-1} \begin{bmatrix} -5 \\ -6 \\ -8 \end{bmatrix} = \frac{1}{10} \begin{bmatrix} 68 \\ 36 \\ -107 \end{bmatrix} \qquad ❖$$

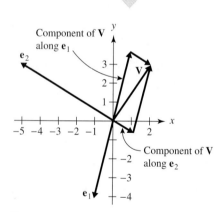

FIGURE 32.1   Example 32.1: The vectors v, $\mathbf{e}_1$, and $\mathbf{e}_2$ and the vector components of v along $\mathbf{e}_1$ and $\mathbf{e}_2$

EXERCISES 32.1

In Exercises 1–10:

(a) Show $\mathbf{e}_1$ and $\mathbf{e}_2$ are linearly independent.

(b) Obtain $'\mathbf{V}$, the representation of $\mathbf{V}$ with respect to $\mathbf{e}_1$ and $\mathbf{e}_2$, by solving for $\mathbf{i}$ and $\mathbf{j}$ in terms of $\mathbf{e}_1$ and $\mathbf{e}_2$ and then substituting into $\mathbf{V}$.

(c) Obtain $'\mathbf{V}$ using the transition matrix $A = [\mathbf{e}_1, \mathbf{e}_2]$.

1. $\mathbf{V} = -2\mathbf{i} - 12\mathbf{j}$, $\mathbf{e}_1 = -7\mathbf{i} - 6\mathbf{j}$, $\mathbf{e}_2 = -3\mathbf{i} - 3\mathbf{j}$

2. $\mathbf{V} = -7\mathbf{i} - 12\mathbf{j}$, $\mathbf{e}_1 = -10\mathbf{i} - 6\mathbf{j}$, $\mathbf{e}_2 = -4\mathbf{i} - 9\mathbf{j}$

3. $\mathbf{V} = -4\mathbf{i} - 11\mathbf{j}$, $\mathbf{e}_1 = \mathbf{i} + 10\mathbf{j}$, $\mathbf{e}_2 = 2\mathbf{i} + 8\mathbf{j}$

4. $\mathbf{V} = 4\mathbf{i} + 11\mathbf{j}$, $\mathbf{e}_1 = -2\mathbf{i} + 7\mathbf{j}$, $\mathbf{e}_2 = -3\mathbf{i} - 3\mathbf{j}$

5. $\mathbf{V} = 3\mathbf{i} - 8\mathbf{j}$, $\mathbf{e}_1 = -3\mathbf{i} + 7\mathbf{j}$, $\mathbf{e}_2 = -9\mathbf{i} - \mathbf{j}$

6. $\mathbf{V} = -12\mathbf{i} - 10\mathbf{j}$, $\mathbf{e}_1 = -8\mathbf{i} - 11\mathbf{j}$, $\mathbf{e}_2 = -7\mathbf{i} + 9\mathbf{j}$

7. $\mathbf{V} = -9\mathbf{i} + 7\mathbf{j}$, $\mathbf{e}_1 = 3\mathbf{i} - 6\mathbf{j}$, $\mathbf{e}_2 = 5\mathbf{i} + 3\mathbf{j}$

8. $\mathbf{V} = 10\mathbf{i} - \mathbf{j}$, $\mathbf{e}_1 = 8\mathbf{i} - 2\mathbf{j}$, $\mathbf{e}_2 = -9\mathbf{i} + 11\mathbf{j}$

9. $\mathbf{V} = 3\mathbf{i} - 7\mathbf{j}$, $\mathbf{e}_1 = -9\mathbf{i} - 6\mathbf{j}$, $\mathbf{e}_2 = 5\mathbf{i} - 8\mathbf{j}$

10. $\mathbf{V} = 10\mathbf{i} - 6\mathbf{j}$, $\mathbf{e}_1 = -4\mathbf{i} + 10\mathbf{j}$, $\mathbf{e}_2 = 12\mathbf{i} - 7\mathbf{j}$

In Exercises 11–20:

    (a) Show $\mathbf{e}_k$, $k = 1, 2, 3$, are linearly independent.

    (b) Obtain $'\mathbf{V}$, the representation of $\mathbf{V}$ with respect to $\mathbf{e}_k$, $k = 1, 2, 3$, by solving for $\mathbf{i}$, $\mathbf{j}$, and $\mathbf{k}$ in terms of $\mathbf{e}_1$, $\mathbf{e}_2$, and $\mathbf{e}_3$ and then substituting in $\mathbf{V}$.

    (c) Obtain $'\mathbf{V}$ using the transition matrix $A = [\mathbf{e}_1, \mathbf{e}_2, \mathbf{e}_3]$.

11. $\mathbf{V} = -12\mathbf{i} - 9\mathbf{j} - 8\mathbf{k}$, $\mathbf{e}_1 = 2\mathbf{i} + 6\mathbf{j} + 8\mathbf{k}$, $\mathbf{e}_2 = -3\mathbf{i} - 9\mathbf{j} - 2\mathbf{k}$, $\mathbf{e}_3 = 6\mathbf{i} - 3\mathbf{j} - 2\mathbf{k}$

12. $\mathbf{V} = \mathbf{i} + 12\mathbf{j} - 8\mathbf{k}$, $\mathbf{e}_1 = -9\mathbf{i} + 11\mathbf{j} + 10\mathbf{k}$, $\mathbf{e}_2 = 8\mathbf{i} + 4\mathbf{j} + 10\mathbf{k}$, $\mathbf{e}_3 = -8\mathbf{i} + 6\mathbf{j} + 7\mathbf{k}$

13. $\mathbf{V} = 6\mathbf{i} - 5\mathbf{j} + \mathbf{k}$, $\mathbf{e}_1 = -8\mathbf{i} + 4\mathbf{j} + 12\mathbf{k}$, $\mathbf{e}_2 = -5\mathbf{i} + 4\mathbf{j} - 2\mathbf{k}$, $\mathbf{e}_3 = -6\mathbf{i} - 2\mathbf{j} - 8\mathbf{k}$

14. $\mathbf{V} = 2\mathbf{i} - 12\mathbf{j} + \mathbf{k}$, $\mathbf{e}_1 = 7\mathbf{i} + 9\mathbf{j} - \mathbf{k}$, $\mathbf{e}_2 = -7\mathbf{i} + 2\mathbf{j} - 11\mathbf{k}$, $\mathbf{e}_3 = 6\mathbf{i} - 12\mathbf{j} - 4\mathbf{k}$

15. $\mathbf{V} = -6\mathbf{i} + 10\mathbf{j} - 4\mathbf{k}$, $\mathbf{e}_1 = -4\mathbf{i} + 4\mathbf{j} + 11\mathbf{k}$, $\mathbf{e}_2 = -4\mathbf{i} + \mathbf{j} + 4\mathbf{k}$, $\mathbf{e}_3 = 9\mathbf{i} - 8\mathbf{j} - 9\mathbf{k}$

16. $\mathbf{V} = 10\mathbf{i} + 5\mathbf{j} - 5\mathbf{k}$, $\mathbf{e}_1 = 8\mathbf{i} - 7\mathbf{j} + 4\mathbf{k}$, $\mathbf{e}_2 = -9\mathbf{i} + 5\mathbf{j} - 6\mathbf{k}$, $\mathbf{e}_3 = -4\mathbf{i} - 11\mathbf{j} - 8\mathbf{k}$

17. $\mathbf{V} = -8\mathbf{i} - 4\mathbf{j} - 2\mathbf{k}$, $\mathbf{e}_1 = -5\mathbf{i} - 12\mathbf{j} - 8\mathbf{k}$, $\mathbf{e}_2 = 5\mathbf{i} + \mathbf{j} - 8\mathbf{k}$, $\mathbf{e}_3 = 6\mathbf{i} - 10\mathbf{j} - \mathbf{k}$

18. $\mathbf{V} = -7\mathbf{i} + 11\mathbf{j} + 10\mathbf{k}$, $\mathbf{e}_1 = 5\mathbf{i} + 12\mathbf{j} + 11\mathbf{k}$, $\mathbf{e}_2 = 12\mathbf{i} + 10\mathbf{j} + 3\mathbf{k}$, $\mathbf{e}_3 = 12\mathbf{i} - \mathbf{j} - 7\mathbf{k}$

19. $\mathbf{V} = -12\mathbf{i} + 12\mathbf{j} + 11\mathbf{k}$, $\mathbf{e}_1 = -4\mathbf{i} + 2\mathbf{j} - 3\mathbf{k}$, $\mathbf{e}_2 = -12\mathbf{i} + \mathbf{j} + 11\mathbf{k}$, $\mathbf{e}_3 = -8\mathbf{j} - 4\mathbf{k}$

20. $\mathbf{V} = 12\mathbf{i} + 6\mathbf{j} + 4\mathbf{k}$, $\mathbf{e}_1 = 6\mathbf{i} - 5\mathbf{j} + 9\mathbf{k}$, $\mathbf{e}_2 = -5\mathbf{i} - \mathbf{j} - 3\mathbf{k}$, $\mathbf{e}_3 = 4\mathbf{i} + 11\mathbf{j} + 10\mathbf{k}$

If L is an $n \times n$ square matrix, then $L\mathbf{V} = \mathbf{Y}$ defines a (linear) transformation on vectors $\mathbf{V}$. If the columns of $A = [\mathbf{e}_1, \ldots, \mathbf{e}_n]$ are a new set of basis vectors so that $\mathbf{V}$ and $\mathbf{Y}$ undergo a change of basis defined by $A\,'\mathbf{V} = \mathbf{V}$ and $A\,'\mathbf{Y} = \mathbf{Y}$, then $L(A\,'\mathbf{V}) = A\,'\mathbf{Y}$ and $A^{-1}LA\,'\mathbf{V} = '\mathbf{Y}$. Hence, the matrix $B = A^{-1}LA$, similar to A, represents the action of L in the new coordinate system. Thus, similar matrices represent the same transformation in different coordinate systems. In Exercises 21–30, using the given L, $\mathbf{V}$, and $A = [\mathbf{e}_1, \mathbf{e}_2]$:

    (a) Obtain $\mathbf{Y} = L\mathbf{V}$ and $'\mathbf{Y} = A^{-1}\mathbf{Y}$.

    (b) Obtain $B = A^{-1}LA$ and show $B\,'\mathbf{V} = '\mathbf{Y}$.

21. $L = \begin{bmatrix} -6 & 8 \\ 10 & 1 \end{bmatrix}$ and A, $\mathbf{V}$ from Exercise 1

22. $L = \begin{bmatrix} -11 & 1 \\ -2 & 6 \end{bmatrix}$ and A, $\mathbf{V}$ from Exercise 2

23. $L = \begin{bmatrix} 9 & -3 \\ -7 & 1 \end{bmatrix}$ and A, $\mathbf{V}$ from Exercise 3

24. $L = \begin{bmatrix} -5 & -2 \\ 12 & -2 \end{bmatrix}$ and A, $\mathbf{V}$ from Exercise 4

25. $L = \begin{bmatrix} 4 & -1 \\ -5 & 12 \end{bmatrix}$ and A, $\mathbf{V}$ from Exercise 5

26. $L = \begin{bmatrix} -1 & 11 \\ 4 & -3 \end{bmatrix}$ and A, $\mathbf{V}$ from Exercise 6

27. $L = \begin{bmatrix} 11 & -6 \\ 12 & 4 \end{bmatrix}$ and A, $\mathbf{V}$ from Exercise 7

28. $L = \begin{bmatrix} 11 & 5 \\ 4 & 6 \end{bmatrix}$ and A, $\mathbf{V}$ from Exercise 8

29. $L = \begin{bmatrix} -12 & -1 \\ 1 & 1 \end{bmatrix}$ and A, $\mathbf{V}$ from Exercise 9

30. $L = \begin{bmatrix} 8 & 7 \\ 3 & 4 \end{bmatrix}$ and A, $\mathbf{V}$ from Exercise 10

In Exercises 31–40, using the given $A = [\mathbf{e}_1, \mathbf{e}_2, \mathbf{e}_3]$, $\mathbf{V}$, and L:

    (a) Obtain $\mathbf{Y} = L\mathbf{V}$ and $'\mathbf{Y} = A^{-1}\mathbf{Y}$.

    (b) Obtain $B = A^{-1}LA$ and show $B\,'\mathbf{V} = '\mathbf{Y}$.

31. $L = \begin{bmatrix} 8 & 12 & -3 \\ -2 & -5 & 5 \\ 1 & -10 & 12 \end{bmatrix}$ and A, $\mathbf{V}$ from Exercise 11

32. $L = \begin{bmatrix} -2 & -1 & 11 \\ -9 & 11 & 8 \\ 11 & 2 & 0 \end{bmatrix}$ and A, $\mathbf{V}$ from Exercise 12

33. $L = \begin{bmatrix} -5 & 12 & 7 \\ -9 & 3 & -7 \\ -5 & -11 & 0 \end{bmatrix}$ and A, $\mathbf{V}$ from Exercise 13

34. $L = \begin{bmatrix} 10 & 1 & 5 \\ 10 & 11 & -10 \\ -10 & 12 & -12 \end{bmatrix}$ and A, $\mathbf{V}$ from Exercise 14

35. $L = \begin{bmatrix} -5 & -5 & -8 \\ -10 & -8 & -2 \\ 8 & 11 & 1 \end{bmatrix}$ and A, $\mathbf{V}$ from Exercise 15

36. $L = \begin{bmatrix} -11 & -12 & 12 \\ 11 & -12 & 7 \\ -11 & 12 & 10 \end{bmatrix}$ and A, $\mathbf{V}$ from Exercise 16

37. $L = \begin{bmatrix} 1 & 3 & 7 \\ -1 & -1 & -3 \\ 4 & -6 & 8 \end{bmatrix}$ and A, $\mathbf{V}$ from Exercise 17

38. $L = \begin{bmatrix} 11 & 12 & 6 \\ 7 & -11 & -12 \\ -6 & 8 & -2 \end{bmatrix}$ and A, $\mathbf{V}$ from Exercise 18

39. $L = \begin{bmatrix} -7 & 4 & -1 \\ 4 & 0 & 0 \\ -1 & -8 & -12 \end{bmatrix}$ and A, $\mathbf{V}$ from Exercise 19

40. $L = \begin{bmatrix} -6 & 10 & -2 \\ 9 & 8 & 10 \\ -7 & 1 & 8 \end{bmatrix}$ and A, $\mathbf{V}$ from Exercise 20

The dot product $\mathbf{V} \cdot \mathbf{W}$ can be written as the ordinary matrix multiplication $\mathbf{V}^{\mathrm{T}}\mathbf{W}$ wherein a row "vector" multiplies a "column" vector, yielding a scalar. If the columns of $A = [\mathbf{e}_1, \ldots, \mathbf{e}_n]$ are a new set of basis vectors so that $\mathbf{V}$ and $\mathbf{W}$ undergo a change of basis defined by $A\,'\mathbf{V} = \mathbf{V}$ and $A\,'\mathbf{W} = \mathbf{W}$, then $\mathbf{V} \cdot \mathbf{W} = \mathbf{V}^{\mathrm{T}}\mathbf{W} = (A\,'\mathbf{V})^{\mathrm{T}}(A\,'\mathbf{W}) = '\mathbf{V}^{\mathrm{T}}A^{\mathrm{T}}A\,'\mathbf{W}$. Thus, in the new basis, the dot product is obtained by multiplying the symmetric matrix $A^{\mathrm{T}}A$ from the left by a row vector and from the right by a column vector. In Exercises 41–50, using the given $\mathbf{V}$, $A = [\mathbf{e}_1, \mathbf{e}_2]$, and $\mathbf{W}$:

    (a) Obtain $\mathbf{V} \cdot \mathbf{W}$ and the matrix $C = A^{\mathrm{T}}A$.

    (b) Obtain $'\mathbf{W}$.   **(c)** compute $'\mathbf{V}^{\mathrm{T}}C\,'\mathbf{W}$, and compare to $\mathbf{V} \cdot \mathbf{W}$.

41. $\mathbf{W} = 5\mathbf{i} - 11\mathbf{j}$ and A, $\mathbf{V}$ from Exercise 1

42. $\mathbf{W} = -11\mathbf{i} + 2\mathbf{j}$ and A, $\mathbf{V}$ from Exercise 2

43. $\mathbf{W} = 2\mathbf{i} + 11\mathbf{j}$ and A, $\mathbf{V}$ from Exercise 3

44. $\mathbf{W} = -11\mathbf{i} + 12\mathbf{j}$ and A, $\mathbf{V}$ from Exercise 4

45. $\mathbf{W} = 3\mathbf{i} - 9\mathbf{j}$ and A, $\mathbf{V}$ from Exercise 5

46. $\mathbf{W} = 8\mathbf{i} + 4\mathbf{j}$ and A, $\mathbf{V}$ from Exercise 6

47. $\mathbf{W} = \mathbf{i} + 4\mathbf{j}$ and A, $\mathbf{V}$ from Exercise 7

48. $\mathbf{W} = -\mathbf{i} - 6\mathbf{j}$ and A, $\mathbf{V}$ from Exercise 8

49. $\mathbf{W} = 7\mathbf{i} - 11\mathbf{j}$ and A, $\mathbf{V}$ from Exercise 9

50. $\mathbf{W} = 9\mathbf{i} + 6\mathbf{j}$ and A, $\mathbf{V}$ from Exercise 10

In Exercises 51–60, using the given $\mathbf{V}$, $A = [\mathbf{e}_1, \mathbf{e}_2, \mathbf{e}_3]$, and $\mathbf{W}$:

(a) Obtain $\mathbf{V} \cdot \mathbf{W}$ and the matrix $C = A^{\mathsf{T}}A$.

(b) Obtain $'\mathbf{W} = A^{-1}\mathbf{W}$.

(c) Compute $'\mathbf{V}^{\mathsf{T}} C \, '\mathbf{W}$, where $'\mathbf{V} = A^{-1}\mathbf{V}$, and compare to $\mathbf{V} \cdot \mathbf{W}$.

51. $\mathbf{W} = -12\mathbf{i} + 5\mathbf{j} + 8\mathbf{k}$, $\mathbf{V}$ and A from Exercise 11

52. $\mathbf{W} = 10\mathbf{i} - 3\mathbf{j} - 11\mathbf{k}$, $\mathbf{V}$ and A from Exercise 12

53. $\mathbf{W} = 9\mathbf{i} + 12\mathbf{j} + 3\mathbf{k}$, $\mathbf{V}$ and A from Exercise 13

54. $\mathbf{W} = -8\mathbf{i} - 11\mathbf{j} - 12\mathbf{k}$, $\mathbf{V}$ and A from Exercise 14

55. $\mathbf{W} = 6\mathbf{i} + \mathbf{j} - 9\mathbf{k}$, $\mathbf{V}$ and A from Exercise 15

56. $\mathbf{W} = \mathbf{i} - 4\mathbf{j} + 3\mathbf{k}$, $\mathbf{V}$ and A from Exercise 16

57. $\mathbf{W} = 12\mathbf{i} + 7\mathbf{j} + 3\mathbf{k}$, $\mathbf{V}$ and A from Exercise 17

58. $\mathbf{W} = 9\mathbf{i} - 8\mathbf{j} - 3\mathbf{k}$, $\mathbf{V}$ and A from Exercise 18

59. $\mathbf{W} = 5\mathbf{i} + 11\mathbf{j} - 7\mathbf{k}$, $\mathbf{V}$ and A from Exercise 19

60. $\mathbf{W} = \mathbf{i} + 7\mathbf{j} - 9\mathbf{k}$, $\mathbf{V}$ and A from Exercise 20

32.2   Rotations and Orthogonal Matrices

Rotation as a Change of Basis

Consider a change of basis to $\{\mathbf{e}_1, \mathbf{e}_2\}$,

$$\mathbf{e}_1 = \begin{bmatrix} \cos\theta \\ \sin\theta \end{bmatrix} \qquad \mathbf{e}_2 = \begin{bmatrix} -\sin\theta \\ \cos\theta \end{bmatrix} \tag{32.1}$$

where $\mathbf{e}_1$ is a unit vector rotated an angle θ counterclockwise from the x-axis and $\mathbf{e}_2$ is a unit vector rotated $\frac{\pi}{2}$ counterclockwise from $\mathbf{e}_1$. Figure 32.2 shows $\mathbf{e}_1$ and $\mathbf{e}_2$ for $\theta = \frac{\pi}{6}$. From the figure, we see that $\mathbf{e}_1 = \cos\theta\mathbf{i} + \sin\theta\mathbf{j}$. Then, reversing the components and inserting a minus sign gives $\mathbf{e}_2$, a vector orthogonal and to the left of $\mathbf{e}_1$.

The matrix A whose columns are the vectors $\mathbf{e}_1$ and $\mathbf{e}_2$ is the transition matrix for a change of basis. Hence, $'\mathbf{V} = A^{-1}\mathbf{V}$, where $\mathbf{V}$ is a vector referred to the unrotated basis and $'\mathbf{V}$ is that same vector given in terms of the rotated basis. Thought of this way, the vector $\mathbf{V}$ is an arrow that does not move. Just the sets of basis vectors used to express it are rotated with respect to each other.

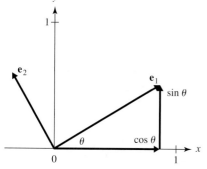

FIGURE 32.2   Vectors $\mathbf{e}_1$ and $\mathbf{e}_2$ are the vectors $\mathbf{i}$ and $\mathbf{j}$ rotated counterclockwise through the angle θ

Rotation as an Operation

We continue to think of the basis $\{\mathbf{i}, \mathbf{j}\}$ as fixed while we have $'\mathbf{V} = a\mathbf{e}_1 + b\mathbf{e}_2$ and $A = [\mathbf{e}_1, \mathbf{e}_2]$, where the rotated basis vectors $\mathbf{e}_k$, $k = 1, 2$, are given by (32.1). The arrow represented by $'\mathbf{V} = A^{-1}\mathbf{V}$ is fixed in the plane described by $\{\mathbf{i}, \mathbf{j}\}$. The vector $\mathbf{Y} = a\mathbf{i} + b\mathbf{j}$ has the same geometric relationship to the axes in the fixed plane as $'\mathbf{V}$ has to the rotated axes. Since the two sets of axes are related by a rotation through an angle θ, the vectors $'\mathbf{V}$ and $\mathbf{Y}$ must also be related by a rotation through an angle θ. Thus, $\mathbf{V} = A\,'\mathbf{V}$ must be rotated through an angle θ with respect to $\mathbf{Y}$, so $A\mathbf{Y}$ must be the rotation of $\mathbf{Y}$ through the angle θ. Indeed, if

$$A\mathbf{Y} = \begin{bmatrix} \cos\theta & -\sin\theta \\ \sin\theta & \cos\theta \end{bmatrix} \begin{bmatrix} a \\ b \end{bmatrix} = \begin{bmatrix} a\cos\theta - b\sin\theta \\ a\sin\theta + b\cos\theta \end{bmatrix} \tag{32.2}$$

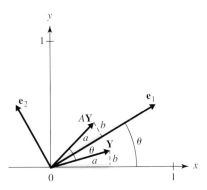

FIGURE 32.3 $A\mathbf{Y}$ as a rotation of $\mathbf{Y}$ through the angle θ

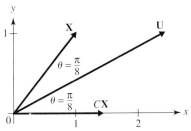

FIGURE 32.4 Reflection of $\mathbf{X} = \mathbf{i} + \mathbf{j}$ across $\mathbf{U} = (1 + \sqrt{2})\mathbf{i} + \mathbf{j}$

then the angle between $\mathbf{Y}$ and $A\mathbf{Y}$ is θ because the cosine of that angle is given by

$$\frac{\mathbf{Y} \cdot A\mathbf{Y}}{\|\mathbf{Y}\|\|A\mathbf{Y}\|} = \frac{(a^2 + b^2)\cos\theta}{(a^2 + b^2)} = \cos\theta$$

Hence, as shown in Figure 32.3, A rotates $\mathbf{Y}$ through the same angle as $\mathbf{e}_1$ and $\mathbf{e}_2$ have been "rotated."

Properties of a Rotation Matrix

The matrix A is called a *rotation matrix* since the vector $A\mathbf{Y}$ is rotated by an angle θ with respect to the vector $\mathbf{Y}$. Such a matrix has the property that the columns are *orthonormal* (length 1 and mutually perpendicular); consequently, the rows are orthonormal, as we will verify.

The orthonormality of the rows and columns of A imply that the transpose of A is its inverse, so that $A^{\mathrm{T}} = A^{-1}$. Showing $A^{\mathrm{T}}A = I$ for A given in (32.2) is left to Exercise A1, but not before observing that $\det A = 1$.

Reflections and Reflection Matrices

The matrices

$$B(\theta) = \begin{bmatrix} \cos\theta & \sin\theta \\ \sin\theta & -\cos\theta \end{bmatrix} \quad \text{and} \quad C = B\left(\frac{\pi}{4}\right) = \frac{1}{\sqrt{2}}\begin{bmatrix} 1 & 1 \\ 1 & -1 \end{bmatrix} \tag{32.3}$$

are examples of *reflection matrices,* matrices that have orthonormal rows and orthonormal columns, inverses that are the transpose, but determinant -1. Given a vector $\mathbf{Y}$, the vector $B\mathbf{Y}$ is the reflection of $\mathbf{Y}$ across a line that passes through the origin and has direction $\mathbf{U}$. Figure 32.4 illustrates the reflection of $\mathbf{X} = \mathbf{i} + \mathbf{j}$ for the matrix C given in (32.3).

EXAMPLE 32.3 Let $\theta = \angle(\mathbf{u}, \mathbf{v})$, $0 \le \theta \le \pi$, denote the angle between the vectors $\mathbf{u}$ and $\mathbf{v}$. If $\mathbf{Y} = x\mathbf{i} + y\mathbf{j}$ and $\mathbf{U} = a\mathbf{i} + b\mathbf{j}$, then the equation $\angle(\mathbf{Y}, \mathbf{U}) = \angle(\mathbf{U}, C\mathbf{Y})$ where C is given by (32.3) should be an identity in x and y for appropriate values of a and b. Indeed, this equation is

$$\arccos\left(\frac{ax + by}{\|\mathbf{Y}\|\|\mathbf{U}\|}\right) = \arccos\left(\frac{(a + b)x + (a - b)y}{\sqrt{2}\|\mathbf{Y}\|\|\mathbf{U}\|}\right)$$

and the obvious manipulations give

$$ax + by = \frac{(a + b)x + (a - b)y}{\sqrt{2}} \tag{32.4}$$

For (32.4) to be an identity in x and y, we must then have

$$(2 - \sqrt{2})a - \sqrt{2}b = 0 \quad \text{and} \quad \sqrt{2}a - (2 + \sqrt{2})b = 0 \tag{32.5}$$

for which a solution is $a = 1 + \sqrt{2}$, $b = 1$. With $\mathbf{U} = (1 + \sqrt{2})\mathbf{i} + \mathbf{j}$ and $\mathbf{YX} = \mathbf{i} + \mathbf{j}$, the angle θ in Figure 32.4 is $\frac{\pi}{8}$, a calculation we leave for Exercise A3. ❖

Rotations in Three Dimensions

The matrices

$$R_x = \begin{bmatrix} 1 & 0 & 0 \\ 0 & \cos\alpha & -\sin\alpha \\ 0 & \sin\alpha & \cos\alpha \end{bmatrix} \quad R_y = \begin{bmatrix} \cos\beta & 0 & -\sin\beta \\ 0 & 1 & 0 \\ \sin\beta & 0 & \cos\beta \end{bmatrix} \quad R_z = \begin{bmatrix} \cos\gamma & -\sin\gamma & 0 \\ \sin\gamma & \cos\gamma & 0 \\ 0 & 0 & 1 \end{bmatrix}$$

represent counterclockwise rotations through angles α, β, γ about the x-, y-, and z-axes, respectively.

A rotation of $\alpha = \frac{\pi}{3}$ about the x-axis, followed by a rotation of $\beta = \frac{\pi}{4}$ about the y-axis, followed by a rotation of $\gamma = \frac{\pi}{6}$ about the z-axis is represented by the product

$$R_z\left(\frac{\pi}{6}\right) R_y\left(\frac{\pi}{4}\right) R_x\left(\frac{\pi}{3}\right) = \begin{bmatrix} \frac{1}{4}\sqrt{6} & -\frac{1}{4} - \frac{3}{8}\sqrt{2} & \frac{1}{4}\sqrt{3} - \frac{1}{8}\sqrt{6} \\ \frac{1}{4}\sqrt{2} & \frac{1}{4}\sqrt{3} - \frac{1}{8}\sqrt{6} & -\frac{3}{4} - \frac{1}{8}\sqrt{2} \\ \frac{1}{2}\sqrt{2} & \frac{1}{4}\sqrt{6} & \frac{1}{4}\sqrt{2} \end{bmatrix} = R_{zyx} \quad (32.6)$$

Not only is the order in which the individual rotation matrices are multiplied important, more importantly, the transpose of R_{zyx} is the inverse, as a straightforward computation of $R_{zyx}^{\mathrm{T}} R_{zyx} = I$ shows.

Orthogonal Matrices

A square matrix P with real entries is called an *orthogonal* matrix if its transpose is its inverse, that is, if $P^{\mathrm{T}} = P^{-1}$, or equivalently, if $P^{\mathrm{T}} P = P P^{\mathrm{T}} = I$.

The properties of a real orthogonal matrix P include the following.

1. The columns (rows) of P are orthonormal and a matrix with orthonormal columns (rows) is orthogonal.

2. Lengths and angles are preserved. Thus, $\|P\mathbf{u}\|_2 = \|\mathbf{u}\|_2$ and $\angle(P\mathbf{u}, P\mathbf{v}) = \angle(\mathbf{u}, \mathbf{v})$.

3. The determinant of P is ± 1. (When $\det P = 1$, P is a rotation matrix, otherwise it is a reflection matrix.)

4. The eigenvalues of P have magnitude 1.

5. If $P \neq I$ is 3×3 and P is a rotation, then $\lambda = 1$ is its one real eigenvalue and P represents a rotation about an axis whose direction is the eigenvector associated with this eigenvalue.

Property 1

If Q is a square matrix with orthonormal columns, then the following schematic, drawn for the 3×3 case where $Q = [\mathbf{u}, \mathbf{v}, \mathbf{w}]$, with $\{\mathbf{u}, \mathbf{v}, \mathbf{w}\}$ being orthonormal, shows why $Q^{\mathrm{T}} Q = I$.

$$\begin{bmatrix} \leftarrow \mathbf{u} \rightarrow \\ \leftarrow \mathbf{v} \rightarrow \\ \leftarrow \mathbf{w} \rightarrow \end{bmatrix} \begin{bmatrix} \uparrow & \uparrow & \uparrow \\ \mathbf{u} & \mathbf{v} & \mathbf{w} \\ \downarrow & \downarrow & \downarrow \end{bmatrix} = \begin{bmatrix} \mathbf{u} \cdot \mathbf{u} & \mathbf{u} \cdot \mathbf{v} & \mathbf{u} \cdot \mathbf{w} \\ \mathbf{v} \cdot \mathbf{u} & \mathbf{v} \cdot \mathbf{v} & \mathbf{v} \cdot \mathbf{w} \\ \mathbf{w} \cdot \mathbf{u} & \mathbf{w} \cdot \mathbf{v} & \mathbf{w} \cdot \mathbf{w} \end{bmatrix} = \begin{bmatrix} 1 & 0 & 0 \\ 0 & 1 & 0 \\ 0 & 0 & 1 \end{bmatrix}$$

Showing $Q Q^{\mathrm{T}} = I$, which is equivalent to showing the rows of Q are orthonormal, is more difficult. To begin, you need to prove that Q^{-1}, the inverse of Q, exists. Once you know this inverse exists, you can show it is Q^{T}. If Q^{T} is Q^{-1}, then clearly $Q Q^{\mathrm{T}} = I$. The complete proof is given next.

Take the determinant of both sides of $Q^{\mathrm{T}} Q = I$ to obtain $\det Q^{\mathrm{T}} Q = \det I = 1$. Since the determinant of a product is the product of the determinants, we have $\det Q^{\mathrm{T}} \det Q = 1$. Since the determinant of the transpose is the determinant of the matrix, we have $\det Q \det Q = 1$, from which we get $(\det Q)^2 = 1$, so $\det Q \neq 0$ and Q^{-1} exists.

Once the existence of the inverse is established, the inverse can be used in the remainder of the proof. Again, take the self-evident statement $Q^{\mathrm{T}} Q = I$ and multiply by the inverse of Q, obtaining $(Q^{\mathrm{T}} Q = I) Q^{-1}$. Clearing parentheses, we have $Q^{\mathrm{T}} Q Q^{-1} = I Q^{-1}$, which reduces to $Q^T = Q^{-1}$. We now know the transpose is the inverse, so we can write $QQ^T = QQ^{-1} = I$, which is what we wished to show.

If W is a square matrix with orthonormal *rows*, then $WW^T = I$ is self-evident and $\det WW^T = 1$ establishes the existence of $W^{-1} = W^T$, from which we then get $W^T W = I$. Thus, a square matrix with orthonormal rows must have orthonormal columns and is an orthogonal matrix.

Property 2

Preservation of length by an orthogonal matrix P is established by the following calculation that uses the 2-norm.

$$\|P\mathbf{u}\|^2 = (P\mathbf{u}) \cdot (P\mathbf{u}) = (P\mathbf{u})^T P\mathbf{u} = \mathbf{u}^T P^T P\mathbf{u} = \mathbf{u}^T I\mathbf{u} = \mathbf{u}^T\mathbf{u} = \mathbf{u} \cdot \mathbf{u} = \|\mathbf{u}\|^2 \quad \textbf{(32.7)}$$

Preservation of angle by an orthogonal matrix P follows from (32.7), the calculation

$$(P\mathbf{u}) \cdot (P\mathbf{v}) = (P\mathbf{u})^T P\mathbf{v} = \mathbf{u}^T P^T P\mathbf{v} = \mathbf{u}^T I\mathbf{v} = \mathbf{u}^T\mathbf{v} = \mathbf{u} \cdot \mathbf{v}$$

and the observation that

$$\angle(P\mathbf{u}, P\mathbf{v}) = \arccos\left(\frac{(P\mathbf{u}) \cdot (P\mathbf{v})}{\|P\mathbf{u}\|\|P\mathbf{v}\|}\right) = \arccos\left(\frac{\mathbf{u} \cdot \mathbf{v}}{\|\mathbf{u}\|\|\mathbf{v}\|}\right) = \angle(\mathbf{u}, \mathbf{v})$$

Property 3

If P is orthogonal, then $P^T P = I$, so $\det P^T P = (\det P)^2 = 1$, as we obtained when proving Property 1. Hence, we must have $\det P = \pm 1$.

Property 4

Let $(\lambda, \mathbf{v})$ be an eigenpair for the orthogonal matrix P so that $P\mathbf{v} = \lambda\mathbf{v}$. Then, using a calculation similar to that in (32.7), we have

$$\|P\mathbf{v}\| = \|\mathbf{v}\| \quad \text{and} \quad \|P\mathbf{v}\| = \|\lambda\mathbf{v}\| = |\lambda|\|\mathbf{v}\|$$

from which $\|\mathbf{v}\| = |\lambda|\|\mathbf{v}\|$ and $|\lambda| = 1$ follow.

Property 5

If A is an $n \times n$ matrix with eigenvalues λ_k, $k = 1, \ldots, n$, then its characteristic polynomial is the nth-degree polynomial

$$p(\lambda) = \det(A - \lambda I) = (-1)^n \prod_{k=1}^{n}(\lambda - \lambda_k) \quad \textbf{(32.8)}$$

and $\det A = p(0) = \prod_{k=1}^{n} \lambda_k$. For a 3×3 rotation matrix P where $\det P = 1$, we have $\lambda_1 \lambda_2 \lambda_3 = 1$, with one or three eigenvalues being real, since they are roots of a cubic equation with real coefficients. In light of Property 4, if the three eigenvalues are real, they are 1, 1, 1 or 1, -1, -1. In the first case, $P = I$; in the second, we have established the existence of the single eigenvalue $\lambda = 1$.

If there is a single real eigenvalue, it must be ± 1 and the other two eigenvalues must be complex conjugates whose product, by Property 4, is 1. Finally, if two of the eigenvalues are complex conjugates, then the real eigenvalue must be 1 to satisfy $\lambda_1 \lambda_2 \lambda_3 = 1$. We have therefore established that a 3×3 rotation matrix always has the eigenvalue $\lambda = 1$ and, except for the two cases where $P = I$ or the eigenvalues are 1, -1, -1, the other two eigenvalues are complex conjugates with magnitude 1.

EXAMPLE 32.4   The matrix R_{zyx} in (32.6) has the eigenvalues 1 and $0.04638 \pm 0.9989i$. The eigenvector corresponding to the eigenvalue 1, spanning the null space of the matrix $R_{zyx} - 1I$, is

$$\mathbf{v} = \begin{bmatrix} 5\sqrt{3} - 4\sqrt{6} - 10 + 6\sqrt{2} \\ 1 \\ 2\sqrt{6} - 3\sqrt{3} + 4 - 4\sqrt{2} \end{bmatrix} = \begin{bmatrix} -2.652423558 \\ 1.0 \\ -1.954027186 \end{bmatrix}$$

The plane for which this eigenvector is the normal passes through the origin and has an equation of the form

$$(5\sqrt{3} - 4\sqrt{6} - 10 + 6\sqrt{2})x + y + (2\sqrt{6} - 3\sqrt{3} + 4 - 4\sqrt{2})z = 0$$

If multiplied by R_{zyx}, a vector in this plane will remain in this plane but will be rotated about the axis defined by the eigenvector that is normal to the plane.

For example, two vectors in the plane of rotation are

$$\mathbf{a} = \begin{bmatrix} 1 \\ 0 \\ -\frac{14}{23}\sqrt{2} + \frac{1}{23} - \frac{10}{23}\sqrt{3} + \frac{2}{23}\sqrt{6} \end{bmatrix} \qquad \mathbf{b} = \begin{bmatrix} 0 \\ 1 \\ -\frac{6}{23}\sqrt{6} + \frac{7}{23}\sqrt{3} + \frac{20}{23} - \frac{4}{23}\sqrt{2} \end{bmatrix}$$

and the angle between these two vectors is found to be $111.52°$. The angle between the vectors $\mathbf{a}' = R_{zyx}\mathbf{a}$ and $\mathbf{b}' = R_{zyx}\mathbf{b}$ is likewise found to be $111.52°$. The actual angle of rotation about the axis of rotation can be determined by calculating θ', the angle between $\mathbf{a}$, a vector in the plane, and $\mathbf{a}'$. This angle turns out to be 1.5244 radians, or about $87.342°$. Thus, the combined rotation about the coordinate axes is equivalent to a single rotation through the angle θ', about the axis defined by the eigenvector whose eigenvalue is 1.

A sketch of the pair of vectors before and after rotation and the plane and its normal (the axis of rotation) is given in the Figure 32.5. The plane has been displaced slightly from the origin so that the vectors may be seen more clearly. Additional views and animations of the rotation are found in the accompanying Maple worksheet. ❖

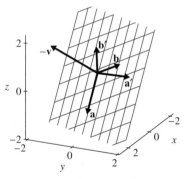

FIGURE 32.5   Example 32.4: The axis about which R_{zyx} rotates vectors has direction $\mathbf{v}$. A plane orthogonal to $\mathbf{v}$ contains vectors $\mathbf{a}$ and $\mathbf{b}$ that are both rotated the same amount θ' by R_{zyx}.

Real Symmetric Matrices

Continued experimentation with orthogonal matrices would be expedited if we had some way to generate them. One approach is to use a matrix whose columns are orthogonal vectors that can be normalized. The real symmetric matrix will turn out to have eigenvectors that either are, or can be made, orthogonal.

Typically, the term "symmetric matrix" refers to a matrix whose entries are real; but on occasion, the word "real" is included for absolute clarity. Such real symmetric matrices have the following four very special properties.

1. Symmetric matrices always have the maximal number of eigenvectors and, hence, are diagonalizable.

2. The eigenvalues of a symmetric matrix are always real.

3. In a symmetric matrix, the eigenvectors corresponding to distinct eigenvalues are orthogonal.

4. Consequently, symmetric matrices are diagonalizable via an orthogonal transition matrix.

The first property is a theorem that can be found in linear algebra texts such as [64] or [52]. After Example 32.5, we give proofs of the second and third properties. The fourth property is actually a consequence of the first three and an orthogonalization algorithm we will study in Section 33.3. (See Section 12.16 for details on diagonalizing a matrix.)

EXAMPLE 32.5 Consider the following (real) symmetric matrix A with eigenvalues $2, 5, -1$ and corresponding eigenvectors $\mathbf{v}_1, \mathbf{v}_2, \mathbf{v}_3$, where

$$A = \begin{bmatrix} 3 & 1 & -2 \\ 1 & 0 & 1 \\ -2 & 1 & 3 \end{bmatrix} \quad \mathbf{v}_1 = \begin{bmatrix} 1 \\ 1 \\ 1 \end{bmatrix} \quad \mathbf{v}_2 = \begin{bmatrix} -1 \\ 0 \\ 1 \end{bmatrix} \quad \mathbf{v}_3 = \begin{bmatrix} 1 \\ -2 \\ 1 \end{bmatrix}$$

Since $\mathbf{v}_1 \cdot \mathbf{v}_2 = \mathbf{v}_1 \cdot \mathbf{v}_3 = \mathbf{v}_2 \cdot \mathbf{v}_3 = 0$, the eigenvectors are mutually orthogonal. If, as in Section 12.16, the eigenvectors are made the columns of a matrix S, we see that S is the transition matrix for a similarity transformation of A to a diagonal form because

$$S^{-1}AS = \begin{bmatrix} 1 & -1 & 1 \\ 1 & 0 & -2 \\ 1 & 1 & 1 \end{bmatrix}^{-1} \begin{bmatrix} 3 & 1 & -2 \\ 1 & 0 & 1 \\ -2 & 1 & 3 \end{bmatrix} \begin{bmatrix} 1 & -1 & 1 \\ 1 & 0 & -2 \\ 1 & 1 & 1 \end{bmatrix} = \begin{bmatrix} 2 & 0 & 0 \\ 0 & 5 & 0 \\ 0 & 0 & -1 \end{bmatrix}$$

However, S will not necessarily be an orthogonal transition matrix because its columns will not have been normalized, and therefore its transpose is not its inverse. If P is the matrix consisting of the normalized columns from S, then $\det P = 1$, the rows of P each have length 1, and the rows of P are orthogonal. In fact, we can show that the transpose of P is its inverse by showing

$$P^{\mathrm{T}}P = \begin{bmatrix} \frac{1}{\sqrt{3}} & -\frac{1}{\sqrt{2}} & \frac{1}{\sqrt{6}} \\ \frac{1}{\sqrt{3}} & 0 & -\frac{2}{\sqrt{6}} \\ \frac{1}{\sqrt{3}} & \frac{1}{\sqrt{2}} & \frac{1}{\sqrt{6}} \end{bmatrix}^{\mathrm{T}} \begin{bmatrix} \frac{1}{\sqrt{3}} & -\frac{1}{\sqrt{2}} & \frac{1}{\sqrt{6}} \\ \frac{1}{\sqrt{3}} & 0 & -\frac{2}{\sqrt{6}} \\ \frac{1}{\sqrt{3}} & \frac{1}{\sqrt{2}} & \frac{1}{\sqrt{6}} \end{bmatrix} = I$$

In addition, P is the transition matrix for a similarity transformation of A to diagonal form, with the entries on the diagonal being the eigenvalues of A. Thus, we find

$$P^{\mathrm{T}}AP = \begin{bmatrix} \frac{1}{\sqrt{3}} & -\frac{1}{\sqrt{2}} & \frac{1}{\sqrt{6}} \\ \frac{1}{\sqrt{3}} & 0 & -\frac{2}{\sqrt{6}} \\ \frac{1}{\sqrt{3}} & \frac{1}{\sqrt{2}} & \frac{1}{\sqrt{6}} \end{bmatrix}^{\mathrm{T}} \begin{bmatrix} 3 & 1 & -2 \\ 1 & 0 & 1 \\ -2 & 1 & 3 \end{bmatrix} \begin{bmatrix} \frac{1}{\sqrt{3}} & -\frac{1}{\sqrt{2}} & \frac{1}{\sqrt{6}} \\ \frac{1}{\sqrt{3}} & 0 & -\frac{2}{\sqrt{6}} \\ \frac{1}{\sqrt{3}} & \frac{1}{\sqrt{2}} & \frac{1}{\sqrt{6}} \end{bmatrix} = \begin{bmatrix} 2 & 0 & 0 \\ 0 & 5 & 0 \\ 0 & 0 & -1 \end{bmatrix}$$

❖

Consequently, for the (real) symmetric matrix A with distinct eigenvalues, there is an orthogonal transition matrix P taking A to diagonal form by a similarity transformation. The matrix A is said to be orthogonally similar to a diagonal matrix and is, therefore, said to be *orthogonally diagonalizable*.

After we see the Gram–Schmidt orthogonalization algorithm in Section 33.3, we can consider the real symmetric matrix with repeated eigenvalues. In that case, the eigenspace associated with the repeated eigenvalue has a basis of eigenvectors that can be orthogonalized by the Gram–Schmidt process and then normalized.

In Section 34.2 we will consider the diagonalization of matrices that are not symmetric.

Proof: Symmetric Matrices Have Real Eigenvalues

Let A be a real symmetric matrix for which $A\mathbf{x} = \lambda\mathbf{x}$, making $\mathbf{x}$ an eigenvector with eigenvalue λ. Compute

$$\langle A\mathbf{x}, \mathbf{x} \rangle = (A\mathbf{x})^T\bar{\mathbf{x}} = \mathbf{x}^T A^T \bar{\mathbf{x}} = \mathbf{x}^T A\bar{\mathbf{x}} = \mathbf{x}^T \overline{A\mathbf{x}} = \mathbf{x}^T \overline{\lambda\mathbf{x}} = \bar{\lambda}\mathbf{x}^T\bar{\mathbf{x}} = \bar{\lambda}\langle \mathbf{x}, \mathbf{x} \rangle = \bar{\lambda}\|\mathbf{x}\|^2$$

$$\langle A\mathbf{x}, \mathbf{x} \rangle = (A\mathbf{x})^T\bar{\mathbf{x}} = (\lambda\mathbf{x})^T\bar{\mathbf{x}} = \lambda\mathbf{x}^T\bar{\mathbf{x}} = \lambda\langle \mathbf{x}, \mathbf{x} \rangle = \lambda\|\mathbf{x}\|^2$$

then equate each result to obtain $(\bar{\lambda} - \lambda)\|\mathbf{x}\|^2 = 0$. But by definition of an eigenvector, $\mathbf{x} \neq \mathbf{0}$, so $\bar{\lambda} - \lambda = 0$, or $\bar{\lambda} = \lambda$, making λ real.

Proof: Symmetric Matrices Have Orthogonal Eigenvectors

Let A be a real symmetric matrix for which $A\mathbf{x} = \lambda\mathbf{x}$, making $\mathbf{x}$ an eigenvector with eigenvalue λ, and for which $A\mathbf{y} = \mu\mathbf{y}$, making $\mathbf{y}$ an eigenvector with eigenvalue $\mu \neq \lambda$. Since A is real and we now know λ and μ are real, we also know $\mathbf{x}$ and $\mathbf{y}$ are real. Hence, we compute

$$(A\mathbf{x}) \cdot \mathbf{y} = (\lambda\mathbf{x}) \cdot \mathbf{y} = \lambda\mathbf{x} \cdot \mathbf{y} \quad \text{and} \quad (A\mathbf{x}) \cdot \mathbf{y} = (A\mathbf{x})^T\mathbf{y} = \mathbf{x}^T A\mathbf{y} = \mathbf{x}^T\mu\mathbf{y} = \mu\mathbf{x} \cdot \mathbf{y}$$

Equate to obtain $(\lambda - \mu)(\mathbf{x} \cdot \mathbf{y}) = 0$. Since $\lambda \neq \mu$, we must have $\mathbf{x} \cdot \mathbf{y} = 0$, making $\mathbf{x}$ and $\mathbf{y}$ orthogonal.

EXERCISES 32.2–Part A

A1. For the matrix rotation A given in (32.2), verify $A^T A = I$.

A2. From (32.4), obtain the equations in (32.5).

A3. Obtain the angle between the vectors $\mathbf{U}$ and $\mathbf{X}$ in Example 32.3.

A4. Show that if $\mathbf{u}$ is a unit vector, the matrix $H = I - 2\mathbf{u}\mathbf{u}^T$ is symmetric and orthogonal.

A5. Let $\mathbf{y} = -2\mathbf{i} + 3\mathbf{j}$ and $\mathbf{u} = \frac{\mathbf{y}}{\|\mathbf{y}\|}$. Form H by the prescription in Exercise A4, and show that H is a reflection, reflecting vectors across a line that is orthogonal to $\mathbf{u}$.

EXERCISES 32.2–Part B

B1. For the matrix R_{zyx} given in (32.6), show that $R_{zyx}^T R_{zyx} = I$.

In Exercises B2–11:

    (a) Show that P is orthogonal by verifying $P^T P = PP^T = I$.

    (b) Verify that $\det P = 1$.

    (c) Show that the columns of P are orthonormal vectors.

    (d) Show that the rows of P are orthonormal vectors.

    (e) Find the angle between $\mathbf{x}$ and $P\mathbf{x}$.

    (f) Show $\|\mathbf{x}\| = \|P\mathbf{x}\|$.

    (g) Graph the vectors $\mathbf{x}$ and $P\mathbf{x}$.

    (h) If $\mathbf{v} = a\mathbf{i} + b\mathbf{j}$, show that $\|\mathbf{v}\| = \|P\mathbf{v}\|$.

B2. $P = \frac{1}{\sqrt{37}}\begin{bmatrix} 6 & 1 \\ -1 & 6 \end{bmatrix}$, $\mathbf{x} = \begin{bmatrix} 10 \\ 1 \end{bmatrix}$

B3. $P = -\frac{1}{\sqrt{41}}\begin{bmatrix} 5 & 4 \\ -4 & 5 \end{bmatrix}$, $\mathbf{x} = \begin{bmatrix} -11 \\ 1 \end{bmatrix}$

B4. $P = \frac{1}{\sqrt{29}}\begin{bmatrix} 5 & 2 \\ -2 & 5 \end{bmatrix}$, $\mathbf{x} = \begin{bmatrix} -2 \\ 6 \end{bmatrix}$

B5. $P = \frac{1}{\sqrt{17}}\begin{bmatrix} 4 & 1 \\ -1 & 4 \end{bmatrix}$, $\mathbf{x} = \begin{bmatrix} 9 \\ -3 \end{bmatrix}$

B6. $P = \frac{1}{\sqrt{5}}\begin{bmatrix} 1 & -2 \\ 2 & 1 \end{bmatrix}$, $\mathbf{x} = \begin{bmatrix} -7 \\ 1 \end{bmatrix}$

B7. $P = \frac{1}{\sqrt{2}}\begin{bmatrix} 1 & 1 \\ -1 & 1 \end{bmatrix}$, $\mathbf{x} = \begin{bmatrix} -5 \\ -2 \end{bmatrix}$

B8. $P = \frac{1}{\sqrt{13}}\begin{bmatrix} 2 & -3 \\ 3 & 2 \end{bmatrix}$, $\mathbf{x} = \begin{bmatrix} 12 \\ -2 \end{bmatrix}$

B9. $P = -\frac{1}{\sqrt{82}}\begin{bmatrix} 1 & -9 \\ 9 & 1 \end{bmatrix}$, $\mathbf{x} = \begin{bmatrix} 4 \\ -1 \end{bmatrix}$

B10. $P = \frac{1}{\sqrt{85}}\begin{bmatrix} 9 & -2 \\ 2 & 9 \end{bmatrix}$, $\mathbf{x} = \begin{bmatrix} -5 \\ 12 \end{bmatrix}$

B11. $P = -\frac{1}{\sqrt{97}}\begin{bmatrix} 4 & 9 \\ -9 & 4 \end{bmatrix}$, $\mathbf{x} = \begin{bmatrix} -1 \\ 11 \end{bmatrix}$

In Exercises B12–19, given the matrix P:

    (a) Show that P is orthogonal by verifying $P^T P = PP^T = I$.

    (b) Verify that $\det P = 1$.

    (c) Show that the columns of P are orthonormal vectors.

(d) Show that the rows of P are orthonormal vectors.

(e) Find the axis about which P rotates vectors.

(f) Find the equation of the plane orthogonal to the axis of rotation.

(g) Find $\mathbf{v}_1$ and $\mathbf{v}_2$, two vectors in the plane found in part (f).

(h) Find the angle between $\mathbf{v}_1$ and $\mathbf{v}_2$.

(i) Find the angle between $P\mathbf{v}_1$ and $P\mathbf{v}_2$.

(j) Find the length of $\mathbf{v}_1$ and $P\mathbf{v}_1$.

(k) Find the length of $\mathbf{v} = a\mathbf{i} + b\mathbf{j} + c\mathbf{k}$ and $P\mathbf{v}$.

B12. $\begin{bmatrix} \frac{8}{\sqrt{114}} & -\frac{82}{\sqrt{17670}} & \frac{3}{\sqrt{155}} \\ \frac{7}{\sqrt{114}} & \frac{85}{\sqrt{17670}} & -\frac{5}{\sqrt{155}} \\ \frac{1}{\sqrt{114}} & \frac{61}{\sqrt{17670}} & \frac{11}{\sqrt{155}} \end{bmatrix}$
B13. $\begin{bmatrix} -\frac{3}{\sqrt{11}} & -\frac{13}{3\sqrt{110}} & -\frac{1}{3\sqrt{10}} \\ \frac{1}{\sqrt{11}} & -\frac{25}{3\sqrt{110}} & \frac{5}{3\sqrt{10}} \\ -\frac{1}{\sqrt{11}} & \frac{14}{3\sqrt{110}} & \frac{8}{3\sqrt{10}} \end{bmatrix}$

B14. $\begin{bmatrix} -\frac{3}{5\sqrt{2}} & \frac{6}{5\sqrt{17}} & \frac{5}{\sqrt{34}} \\ \frac{1}{\sqrt{2}} & \frac{2}{\sqrt{17}} & \frac{3}{\sqrt{34}} \\ -\frac{4}{5\sqrt{2}} & \frac{17}{5\sqrt{17}} & 0 \end{bmatrix}$
B15. $\begin{bmatrix} \frac{8}{\sqrt{89}} & -\frac{13}{\sqrt{1869}} & -\frac{2}{\sqrt{21}} \\ \frac{4}{\sqrt{89}} & \frac{38}{\sqrt{1869}} & \frac{1}{\sqrt{21}} \\ -\frac{3}{\sqrt{89}} & -\frac{16}{\sqrt{1869}} & \frac{4}{\sqrt{21}} \end{bmatrix}$

B16. $\begin{bmatrix} -\frac{2}{\sqrt{14}} & \frac{58}{\sqrt{4886}} & -\frac{3}{\sqrt{349}} \\ \frac{3}{\sqrt{14}} & \frac{39}{\sqrt{4886}} & \frac{4}{\sqrt{349}} \\ \frac{1}{\sqrt{14}} & -\frac{1}{\sqrt{4886}} & -\frac{18}{\sqrt{349}} \end{bmatrix}$
B17. $\begin{bmatrix} -\frac{2}{7} & -\frac{18}{119}\sqrt{17} & \frac{3}{\sqrt{17}} \\ \frac{3}{7} & -\frac{22}{119}\sqrt{17} & -\frac{2}{\sqrt{17}} \\ \frac{6}{7} & \frac{5}{119}\sqrt{17} & \frac{2}{\sqrt{17}} \end{bmatrix}$

B18. $\begin{bmatrix} -\frac{1}{\sqrt{11}} & -\frac{3}{\sqrt{22}} & -\frac{1}{\sqrt{2}} \\ -\frac{3}{\sqrt{11}} & \frac{2}{\sqrt{22}} & 0 \\ \frac{1}{\sqrt{11}} & \frac{3}{\sqrt{22}} & -\frac{1}{\sqrt{2}} \end{bmatrix}$
B19. $\begin{bmatrix} -\frac{1}{\sqrt{11}} & -\frac{13}{\sqrt{1166}} & \frac{9}{\sqrt{106}} \\ \frac{1}{\sqrt{11}} & -\frac{31}{\sqrt{1166}} & -\frac{3}{\sqrt{106}} \\ \frac{3}{\sqrt{11}} & \frac{6}{\sqrt{1166}} & \frac{4}{\sqrt{106}} \end{bmatrix}$

In Exercises B20–29, orthogonally diagonalize the given symmetric matrix A, that is, find an orthogonal matrix P for which $P^{\mathrm{T}}AP = D$, where D is a diagonal matrix whose diagonal elements are the eigenvalues of A.

B20. $\begin{bmatrix} 2 & -8 \\ -8 & -10 \end{bmatrix}$
B21. $\begin{bmatrix} 3 & 6 \\ 6 & -6 \end{bmatrix}$
B22. $\begin{bmatrix} 6 & -4 \\ -4 & 12 \end{bmatrix}$

B23. $\begin{bmatrix} 7 & 4 \\ 4 & -8 \end{bmatrix}$
B24. $\begin{bmatrix} 2 & -8 \\ -8 & 2 \end{bmatrix}$
B25. $\begin{bmatrix} 8 & 6 \\ 6 & 8 \end{bmatrix}$
B26. $\begin{bmatrix} 0 & 8 \\ 8 & -12 \end{bmatrix}$

B27. $\begin{bmatrix} 6 & 12 \\ 12 & -4 \end{bmatrix}$
B28. $\begin{bmatrix} -5 & -4 \\ -4 & 10 \end{bmatrix}$
B29. $\begin{bmatrix} -4 & 2 \\ 2 & -7 \end{bmatrix}$

In Exercises B30–39, orthogonally diagonalize the given symmetric matrix A, that is, find an orthogonal matrix P for which $P^{\mathrm{T}}AP = D$, where D is a diagonal matrix whose diagonal elements are the eigenvalues of A. If A has a repeated eigenvalue, the corresponding eigenvectors $\{\mathbf{x}, \mathbf{y}\}$, which initially need not be orthogonal, can be orthonormalized to $\{\mathbf{w}_1, \mathbf{w}_2\}$ by taking $\mathbf{w}_1 = \frac{\mathbf{x}}{\|\mathbf{x}\|}$ and $\mathbf{w}_2 = \frac{\mathbf{y}_{\perp\mathbf{x}}}{\|\mathbf{y}_{\perp\mathbf{x}}\|}$, where $\mathbf{y}_{\perp\mathbf{x}}$ is the component of $\mathbf{y}$ orthogonal to $\mathbf{x}$.

B30. $\begin{bmatrix} -1 & -8 & -3 \\ -8 & 7 & -1 \\ -3 & -1 & 2 \end{bmatrix}$
B31. $\begin{bmatrix} 3 & 2 & 3 \\ 2 & 6 & 6 \\ 3 & 6 & -5 \end{bmatrix}$

B32. $\begin{bmatrix} 4 & 1 & -1 \\ 1 & 4 & -1 \\ -1 & -1 & 4 \end{bmatrix}$
B33. $\begin{bmatrix} -5 & -9 & -4 \\ -9 & 8 & 9 \\ -4 & 9 & -5 \end{bmatrix}$

B34. $\begin{bmatrix} 8 & 6 & 2 \\ 6 & -8 & -6 \\ 2 & -6 & 8 \end{bmatrix}$
B35. $\begin{bmatrix} 0 & 2 & 4 \\ 2 & 0 & -4 \\ 4 & -4 & 6 \end{bmatrix}$

B36. $\begin{bmatrix} -6 & 5 & -2 \\ 5 & 2 & 2 \\ -2 & 2 & 3 \end{bmatrix}$
B37. $\begin{bmatrix} 1 & -6 & 0 \\ -6 & 1 & 0 \\ 0 & 0 & 7 \end{bmatrix}$

B38. $\begin{bmatrix} 8 & -3 & 1 \\ -3 & 0 & 3 \\ 1 & 3 & 2 \end{bmatrix}$
B39. $\begin{bmatrix} 5 & -3 & 3 \\ -3 & 4 & 2 \\ 3 & 2 & 4 \end{bmatrix}$

The following recipe for constructing an orthogonal matrix that rotates objects counterclockwise about a specified axis and through a specified angle is found in texts on computer graphics. Let the axis of rotation be specified by the vector $\mathbf{v} = a\mathbf{i} + b\mathbf{j} + c\mathbf{k}$, and let the angle of counterclockwise rotation about this axis be θ. If $L = \sqrt{a^2 + b^2 + c^2}$ and $p = \sqrt{b^2 + c^2}$, then the desired orthogonal matrix is $R = [R_x(\alpha)]^{\mathrm{T}} [R_y(\beta)]^{\mathrm{T}} R_z(\theta) R_y(\beta) R_x(\alpha)$, where $R_x(\alpha)$ rotates the object (and the vector $\mathbf{v}$) through the angle α counterclockwise about the x-axis until $\mathbf{v}$ lies in the xz-plane; $R_y(\beta)$ rotates the object (and the vector $\mathbf{v}$) through the angle β counterclockwise about the y-axis until $\mathbf{v}$ lies along the z-axis; $R_z(\theta)$ rotates the object through the angle θ counterclockwise about the z-axis, the direction in which $\mathbf{v}$, the axis of rotation is now pointing; and

$$R_x(\alpha) = \frac{1}{p}\begin{bmatrix} p & 0 & 0 \\ 0 & c & -b \\ 0 & b & c \end{bmatrix} \qquad R_y(\beta) = \frac{1}{L}\begin{bmatrix} p & 0 & -a \\ 0 & L & 0 \\ a & 0 & p \end{bmatrix}$$

$$R_z(\theta) = \begin{bmatrix} \cos\theta & -\sin\theta & 0 \\ \sin\theta & \cos\theta & 0 \\ 0 & 0 & 1 \end{bmatrix}$$

This step actually accomplishes the intended rotation. However, the rotation about the y-axis must be undone by $[R_y(\beta)]^{-1} = [R_y(\beta)]^{\mathrm{T}}$ and the rotation about the x-axis must be undone by $[R_x(\alpha)]^{-1} = [R_x(\alpha)]^{\mathrm{T}}$. Since rotation matrices are orthogonal, the inverses are the transposes. For v and θ given in each of Exercises B40–44:

(a) Construct R, an orthogonal matrix corresponding to a counterclockwise rotation of θ about the line along the vector $\mathbf{v}$.

(b) Verify that R has an eigenpair $(1, \mathbf{V})$, where $\mathbf{V}$ is proportional to $\mathbf{v}$.

(c) Obtain the equation of the plane perpendicular to $\mathbf{V}$ (or $\mathbf{v}$).

(d) Find $\mathbf{u}$, a vector in this plane; verify that it is orthogonal to $\mathbf{v}$; and compute the angle between $\mathbf{u}$ and $R\mathbf{u}$. *Hint:* It should be θ.

B40. $\mathbf{v} = \mathbf{i} + 5\mathbf{j} + 2\mathbf{k}, \theta = \dfrac{\pi}{3}$
B41. $\mathbf{v} = 3\mathbf{i} - 2\mathbf{j} + 4\mathbf{k}, \theta = \dfrac{\pi}{4}$

B42. $\mathbf{v} = -2\mathbf{i} + 7\mathbf{j} + 6\mathbf{k}, \theta = \dfrac{\pi}{6}$
B43. $\mathbf{v} = 4\mathbf{i} - \mathbf{j} + 3\mathbf{k}, \theta = \dfrac{\pi}{12}$

B44. $\mathbf{v} = 5\mathbf{i} + 9\mathbf{j} - 8\mathbf{k}, \theta = \dfrac{\pi}{5}$

32.3 Change of Coordinates

Coordinate Curves

When the Cartesian plane is "changed to polar coordinates," a grid of circles and radial lines is superimposed on the rectangular grid of the xy-plane. These circles and radial lines are called *coordinate curves*. Points labeled (x, y) now acquire new labels, namely, (r, θ), but remain in the same location in the plane. (All the houses and streets in town remain in place, but the city declares new street names and house numbers.)

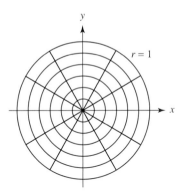

FIGURE 32.6   Coordinate curves for polar coordinates

POLAR COORDINATES—AN ORTHOGONAL COORDINATE SYSTEM   Polar coordinates are defined by the familiar equations $x = r\cos\theta$, $y = r\sin\theta$. Therefore, the radius-vector form for a curve in the xy-plane, namely, $\mathbf{R} = x\mathbf{i} + y\mathbf{j}$, becomes $\mathbf{R}(r, \theta) = x(r, \theta)\mathbf{i} + y(r, \theta)\mathbf{j}$. Curves for which θ is held constant, that is, along which r varies, are radial lines. Curves for which r is held constant, that is, along which θ varies, are concentric circles centered at the origin. (See Figure 32.6.) Since the radius is orthogonal to the circle, polar coordinates are an example of an orthogonal coordinate system.

We next examine a coordinate system that is not orthogonal.

◆ **EXAMPLE 32.6**   We will show that the change of coordinates defined by the following equations is not orthogonal.

$$x = \tfrac{1}{2}(u^2 - v^2) \qquad y = 2uv \tag{32.9}$$

Grid lines on which v is constant and u varies are parabolas opening to the right, while grid lines on which u is constant and v varies are parabolas opening to the left, as shown in Figure 32.7.   ❖

Notation

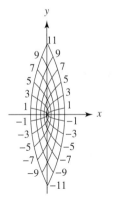

FIGURE 32.7   Coordinate curves for the nonorthogonal system $x = \tfrac{1}{2}(u^2 - v^2)$, $y = 2uv$

THE INVERSE MAP   If the xy-plane is taken as "fundamental," then equations of the form $x = x(u, v)$, $y = y(u, v)$, which define a mapping from the uv-plane to the xy-plane, comprise a vector-valued function of the form $\mathbf{x} = \mathbf{x}(\mathbf{u})$. Here, the vector $\mathbf{x}$ is the vector of coordinates (x, y) and the vector $\mathbf{u}$ is the vector of coordinates (u, v). We will call this the function Φ^{-1}, where the designation "inverse" suggests the mapping is "to" the xy-plane "from" the uv-plane.

THE FORWARD MAP   Again taking the xy-plane as fundamental, equations of the form $u = u(x, y)$, $v = v(x, y)$, which define a "forward" map from the xy-plane to the uv-plane, comprise a vector-valued function of the form $\mathbf{u} = \mathbf{u}(\mathbf{x})$. We will call this the function Φ, and for polar coordinates the function Φ would be defined by the equations $r = \sqrt{x^2 + y^2}$, $\theta = \arctan\frac{y}{x}$.

Natural Basis of Tangent Vectors

If the radius-vector form of a curve is differentiated with respect to the curve's parameter, a vector tangent to the curve results. (See Section 17.1 for details.) At each point in the xy-plane for which they are nonzero, the vectors tangent to the intersecting coordinate curves form the *natural* basis of tangent vectors. These vectors are not necessarily unit vectors and are not normalized.

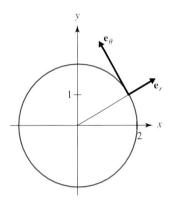

FIGURE 32.8   Polar coordinates: natural basis of tangent vectors $\mathbf{e}_r$ and $\mathbf{e}_\theta$ at $(x, y) = (\sqrt{3}, 1)$

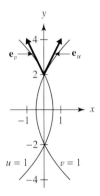

FIGURE 32.9   At $(x, y) = (0, 2)$, natural tangent vectors for the coordinates defined in Example 32.6

POLAR COORDINATES   The backward map Φ^{-1} for polar coordinates is defined by $x = r\cos\theta$, $y = r\sin\theta$. The radius-vector form for coordinate curves is then $\mathbf{R} = r\cos\theta\,\mathbf{i} + r\sin\theta\,\mathbf{j}$, and derivatives with respect to r and θ yield the tangent vectors

$$\mathbf{e}_r = \cos\theta\,\mathbf{i} + \sin\theta\,\mathbf{j} \quad\text{and}\quad \mathbf{e}_\theta = -r\sin\theta\,\mathbf{i} + r\cos\theta\,\mathbf{j} \tag{32.10}$$

These vectors are orthogonal, but $\|\mathbf{e}_\theta\| = r$, so $\mathbf{e}_\theta$ is not a unit vector. Figure 32.8 shows $\mathbf{e}_r$ and $\mathbf{e}_\theta$ at $(x, y) = (\sqrt{3}, 1)$, where $r = 2$ and $\theta = \frac{\pi}{6}$.

NONORTHOGONAL COORDINATES   The radius-vector form for coordinate curves of the coordinate system given in (32.9) is $\mathbf{R} = \frac{1}{2}(u^2 - v^2)\mathbf{i} + 2uv\,\mathbf{j}$, and derivatives with respect to u and v yield the tangent vectors

$$\mathbf{e}_u = u\mathbf{i} + 2v\mathbf{j} \quad\text{and}\quad \mathbf{e}_v = -v\mathbf{i} + 2u\mathbf{j}$$

Since $\|\mathbf{e}_u\| = \sqrt{u^2 + 4v^2}$ and $\|\mathbf{e}_v\| = \sqrt{v^2 + 4u^2}$, these vectors are not unit. Moreover, $\mathbf{e}_u \cdot \mathbf{e}_v = 3uv$ shows that, in general, the vectors are not orthogonal which verifies the nonorthogonality claimed in Example 32.6. Figure 32.9 shows these natural tangent basis vectors at the point $(x, y) = (0, 2)$ where $u = v = 1$.

GENERAL CASE   In general, if the equations defining the change of coordinates in three dimensions are

$$x^1 = x^1(u^1, u^2, u^3) \quad x^2 = x^2(u^1, u^2, u^3) \quad x^3 = x^3(u^1, u^2, u^3)$$

(where the convention is to use superscripts as indices on coordinates), then the Cartesian radius vector $\mathbf{R} = x\mathbf{i} + y\mathbf{j} + z\mathbf{k}$ becomes

$$\mathbf{R}(u^1, u^2, u^3) = x^1(u^1, u^2, u^3)\mathbf{i} + x^2(u^1, u^2, u^3)\mathbf{j} + x^3(u^1, u^2, u^3)\mathbf{k}$$

(Of course, the three-dimensional case includes the two-dimensional case by restriction to a single plane.)

The u^1-, u^2-, and u^3-coordinate curves through $(u^1, u^2, u^3) = (\alpha, \beta, \gamma)$ are given, respectively, by $\mathbf{R}(u^1, \beta, \gamma)$, $\mathbf{R}(\alpha, u^2, \gamma)$, and $\mathbf{R}(\alpha, \beta, u^3)$. Vectors tangent to these coordinate curves are obtained by differentiation, yielding

$$\mathbf{e}_1 = \frac{\partial \mathbf{R}}{\partial u^1} = \frac{\partial x^1}{\partial u^1}\mathbf{i} + \frac{\partial x^2}{\partial u^1}\mathbf{j} + \frac{\partial x^3}{\partial u^1}\mathbf{k}$$

$$\mathbf{e}_2 = \frac{\partial \mathbf{R}}{\partial u^2} = \frac{\partial x^1}{\partial u^2}\mathbf{i} + \frac{\partial x^2}{\partial u^2}\mathbf{j} + \frac{\partial x^3}{\partial u^2}\mathbf{k}$$

$$\mathbf{e}_3 = \frac{\partial \mathbf{R}}{\partial u^3} = \frac{\partial x^1}{\partial u^3}\mathbf{i} + \frac{\partial x^2}{\partial u^3}\mathbf{j} + \frac{\partial x^3}{\partial u^3}\mathbf{k}$$

or, as column vectors, $\mathbf{e}_1$, $\mathbf{e}_2$, $\mathbf{e}_3$ would be, respectively,

$$\begin{bmatrix} \dfrac{\partial x^1}{\partial u^1} \\[2ex] \dfrac{\partial x^2}{\partial u^1} \\[2ex] \dfrac{\partial x^3}{\partial u^1} \end{bmatrix} \quad \begin{bmatrix} \dfrac{\partial x^1}{\partial u^2} \\[2ex] \dfrac{\partial x^2}{\partial u^2} \\[2ex] \dfrac{\partial x^3}{\partial u^2} \end{bmatrix} \quad \begin{bmatrix} \dfrac{\partial x^1}{\partial u^3} \\[2ex] \dfrac{\partial x^2}{\partial u^3} \\[2ex] \dfrac{\partial x^3}{\partial u^3} \end{bmatrix}$$

These tangent (basis) vectors are just the columns of the Jacobian matrix

$$\mathbf{J}_{\Phi^{-1}} = \begin{bmatrix} \dfrac{\partial x^1}{\partial u^1} & \dfrac{\partial x^1}{\partial u^2} & \dfrac{\partial x^1}{\partial u^3} \\[2ex] \dfrac{\partial x^2}{\partial u^1} & \dfrac{\partial x^2}{\partial u^2} & \dfrac{\partial x^2}{\partial u^3} \\[2ex] \dfrac{\partial x^3}{\partial u^1} & \dfrac{\partial x^3}{\partial u^2} & \dfrac{\partial x^3}{\partial u^3} \end{bmatrix}$$

whose general element is $A_c^r = \partial x^r / \partial u^c$. The upper index is the row index, and the lower index is the column index. The columns of $\mathbf{J}_{\Phi^{-1}}$ are the tangent vectors along the coordinate curves for a curvilinear coordinate system pulled over to the Cartesian plane.

The Jacobian matrix for Φ, the forward mapping whose vector form is $\mathbf{u} = \mathbf{u}(\mathbf{x})$, is

$$\mathbf{J}_{\Phi} = \begin{bmatrix} \dfrac{\partial u^1}{\partial x^1} & \dfrac{\partial u^1}{\partial x^2} & \dfrac{\partial u^1}{\partial x^3} \\[2ex] \dfrac{\partial u^2}{\partial x^1} & \dfrac{\partial u^2}{\partial x^2} & \dfrac{\partial u^2}{\partial x^3} \\[2ex] \dfrac{\partial u^3}{\partial x^1} & \dfrac{\partial u^3}{\partial x^2} & \dfrac{\partial u^3}{\partial x^3} \end{bmatrix}$$

where the general element is $\partial u^r / \partial x^c$.

The Jacobian matrices $\mathbf{J}_{\Phi}$ and $\mathbf{J}_{\Phi^{-1}}$ are inverse matrices, as we now demonstrate for polar coordinates. The equations of the backward map Φ^{-1} are $x = r \cos \theta$, $y = r \sin \theta$; and the equations of the forward map Φ are $r = \sqrt{x^2 + y^2}$, $\theta = \arctan \frac{y}{x}$. The matrices $\mathbf{J}_{\Phi^{-1}}$, $\mathbf{J}_{\Phi}$, and $\mathbf{J}_{\Phi(\mathbf{u})}$, where $\mathbf{J}_{\Phi(\mathbf{u})}$ is $\mathbf{J}_{\Phi}$ expressed in terms of polar coordinates, are, respectively,

$$\begin{bmatrix} \cos \theta & -r \sin \theta \\ \sin \theta & r \cos \theta \end{bmatrix} \qquad \begin{bmatrix} \dfrac{x}{\sqrt{x^2 + y^2}} & \dfrac{y}{\sqrt{x^2 + y^2}} \\[2ex] -\dfrac{y}{x^2 + y^2} & \dfrac{x}{x^2 + y^2} \end{bmatrix} \qquad \begin{bmatrix} \cos \theta & \sin \theta \\[2ex] -\dfrac{\sin \theta}{r} & \dfrac{\cos \theta}{r} \end{bmatrix}$$

from which we can verify that $\mathbf{J}_{\Phi^{-1}} \mathbf{J}_{\Phi(\mathbf{u})} = I = \mathbf{J}_{\Phi(\mathbf{u})} \mathbf{J}_{\Phi^{-1}}$.

More generally, differentiating the identities

$$x^1 = x^1(u^1, u^2) = x^1(u^1(x^1, x^2), u^2(x^1, x^2))$$
$$x^2 = x^2(u^1, u^2) = x^2(u^1(x^1, x^2), u^2(x^1, x^2))$$

gives

$$\frac{\partial x^1}{\partial x^1} = 1 = \frac{\partial x^1}{\partial u^1} \frac{\partial u^1}{\partial x^1} + \frac{\partial x^1}{\partial u^2} \frac{\partial u^2}{\partial x^1} = \sum_{k=1}^{2} \frac{\partial x^1}{\partial u^k} \frac{\partial u^k}{\partial x^1}$$

$$\frac{\partial x^1}{\partial x^2} = 0 = \frac{\partial x^1}{\partial u^1} \frac{\partial u^1}{\partial x^2} + \frac{\partial x^1}{\partial u^2} \frac{\partial u^2}{\partial x^2} = \sum_{k=1}^{2} \frac{\partial x^1}{\partial u^k} \frac{\partial u^k}{\partial x^2}$$

$$\frac{\partial x^2}{\partial x^1} = 0 = \frac{\partial x^2}{\partial u^1} \frac{\partial u^1}{\partial x^1} + \frac{\partial x^2}{\partial u^2} \frac{\partial u^2}{\partial x^1} = \sum_{k=1}^{2} \frac{\partial x^2}{\partial u^k} \frac{\partial u^k}{\partial x^1}$$

$$\frac{\partial x^2}{\partial x^2} = 1 = \frac{\partial x^2}{\partial u^1} \frac{\partial u^1}{\partial x^2} + \frac{\partial x^2}{\partial u^2} \frac{\partial u^2}{\partial x^2} = \sum_{k=1}^{2} \frac{\partial x^2}{\partial u^k} \frac{\partial u^k}{\partial x^2}$$

Since $\mathbf{J}_{\Phi^{-1}} = [\partial x^i / \partial u^j]$ and $\mathbf{J}_{\Phi} = [\partial u^j / \partial x^k]$, the ik-element of the product $\mathbf{J}_{\Phi^{-1}}\mathbf{J}_{\Phi}$ would be

$$\sum_{j=1}^{2} \frac{\partial x^i}{\partial u^j} \frac{\partial u^j}{\partial x^k} = \delta_k^i = \begin{cases} 1 & i = k \\ 0 & i \neq k \end{cases}$$

because when multiplying two matrices P_{rc} and Q_{rc}, the product matrix M_{rc} is defined element-by-element via the sums $m_{rc} = \sum_{k=1}^{n} p_{rk}q_{kc}$. The summation varies over the column index of the first factor and over the row index of the second factor. This identity property of the Jacobian matrices for the mappings Φ and Φ^{-1} is normally expressed without the sums, with summation on a repeated index implied. Thus, texts on tensor analysis typically write this last result as $(\partial x^i / \partial u^j)(\partial u^j / \partial x^k) = \delta_k^i$, where the Kronecker delta is used to represent the identity matrix.

Alternatively, if we start with the identities

$$u^1 = u^1(x^1, x^2) = u^1(x^1(u^1, u^2), x^2(u^1, u^2))$$
$$u^2 = u^2(x^1, x^2) = u^2(x^1(u^1, u^2), x^2(u^1, u^2))$$

and again differentiate by the chain rule, we would obtain

$$\frac{\partial u^1}{\partial u^1} = 1 = \frac{\partial u^1}{\partial x^1} \frac{\partial x^1}{\partial u^1} + \frac{\partial u^1}{\partial x^2} \frac{\partial x^2}{\partial u^1} = \sum_{k=1}^{2} \frac{\partial u^1}{\partial x^k} \frac{\partial x^k}{\partial u^1}$$

$$\frac{\partial u^1}{\partial u^2} = 0 = \frac{\partial u^1}{\partial x^1} \frac{\partial x^1}{\partial u^2} + \frac{\partial u^1}{\partial x^2} \frac{\partial x^2}{\partial u^2} = \sum_{k=1}^{2} \frac{\partial u^1}{\partial x^k} \frac{\partial x^k}{\partial u^2}$$

$$\frac{\partial u^2}{\partial u^1} = 0 = \frac{\partial u^2}{\partial x^1} \frac{\partial x^1}{\partial u^1} + \frac{\partial u^2}{\partial x^2} \frac{\partial x^2}{\partial u^1} = \sum_{k=1}^{2} \frac{\partial u^2}{\partial x^k} \frac{\partial x^k}{\partial u^1}$$

$$\frac{\partial u^2}{\partial u^2} = 1 = \frac{\partial u^2}{\partial x^1} \frac{\partial x^1}{\partial u^2} + \frac{\partial u^2}{\partial x^2} \frac{\partial x^2}{\partial u^2} = \sum_{k=1}^{2} \frac{\partial u^2}{\partial x^k} \frac{\partial x^k}{\partial u^2}$$

If we now multiply the Jacobian matrices in the opposite order, we obtain $\mathbf{J}_{\Phi}\mathbf{J}_{\Phi^{-1}} = [\partial u^i / \partial x^k][\partial x^k / \partial u^j]$ and the ij-element of the product is $\sum_{k=1}^{2} (\partial u^i / \partial x^k)(\partial x^k / \partial u^j) = \delta_j^i$.

Except for the tedium of writing the sums in the case of n variables, the general case would proceed exactly as for the case $n = 2$.

General Tangent Vectors and the Contravariant Law

If $x = x(p)$, $y = y(p)$ are the parametric equations of a curve in the xy-plane and $\mathbf{R} = x(p)\mathbf{i} + y(p)\mathbf{j}$ is its radius-vector form, then

$$\frac{d}{dp}\mathbf{R} = \mathbf{R}'(p) = x'(p)\mathbf{i} + y'(p)\mathbf{j}$$

the derivative of $\mathbf{R}$ with respect to the curve's parameter, is a vector tangent to the curve. What is the equivalent expression for this tangent vector if coordinates in the xy-plane are changed via the mapping Φ^{-1} defined by $x = x(u, v)$, $y = y(u, v)$? Answering this question requires the functions $u = u(p)$, $v = v(p)$. For example, if a curve is given by $x = p$, $y = p^2$, then, for polar coordinates, we solve the equations $p = r\cos\theta$, $p^2 = r\sin\theta$ for $r = r(p) = p\sqrt{p^2 + 1}$ and $\theta = \theta(p) = \arctan p$.

The radius-vector form of the curve, namely, $\mathbf{R} = x(p)\mathbf{i} + y(p)\mathbf{j}$, becomes, under the change of coordinates $x = x(u, v)$, $y = y(u, v)$,

$$\mathbf{R} = x(u(p), v(p))\mathbf{i} + y(u(p), v(p))\mathbf{j}$$

Computing $\mathbf{R}'(p)$ by the chain rule yields

$$\mathbf{R}'(p) = (x_u u' + x_v v')\mathbf{i} + (y_u u' + y_v v')\mathbf{j} = u'(x_u\mathbf{i} + y_u\mathbf{j}) + v'(x_v\mathbf{i} + y_v\mathbf{j}) \qquad (32.11)$$

Recognizing that the terms in parentheses on the right in (32.11) are the natural basis vectors $\mathbf{e}_u$ and $\mathbf{e}_v$, respectively, we have obtained

$$\mathbf{R}'(p) = u'\mathbf{e}_u + v'\mathbf{e}_v$$

as the expression for a tangent vector in the new coordinate system. Notice that the components are just the derivatives of the coordinate functions, just as in the xy-plane. However, our use of the chain rule has brought about a change of basis vectors to the new (natural) basis.

Next, examine the derivatives $u'(p)$ and $v'(p)$ under the aegis of the chain rule. Begin with

$$u = u(x(p), y(p)) \qquad v = v(x(p), y(p))$$

and obtain

$$u' = \frac{du}{dp} = u_x x' + u_y y' \quad \text{and} \quad v' = \frac{dv}{dp} = v_x x' + v_y y'$$

Since u' and v' are components of a (column) vector, we are inspired to write

$$\begin{bmatrix} u' \\ v' \end{bmatrix} = \begin{bmatrix} u_x & u_y \\ v_x & v_y \end{bmatrix} \begin{bmatrix} x' \\ y' \end{bmatrix}$$

where the matrix on the right is the Jacobian matrix $\mathbf{J}_\Phi$. Thus, the transition matrix connected with the change of basis vectors from $\{\mathbf{i}, \mathbf{j}\}$ to $\{\mathbf{e}_u, \mathbf{e}_v\}$ can be applied to the *components* of the tangent vector. This is the *contravariant transformation law* for tangent vectors,

$$v^{k'} = \frac{\partial u^{k'}}{\partial x^k} v^k$$

In the original frame, the components of the contravariant vector are v^k and are with respect to a set of basis vectors such as the $\mathbf{i}, \mathbf{j}, \mathbf{k}$ system of the Cartesian plane. In the new frame, the transformed components of the contravariant vector are $v^{k'}$ and are with respect to the set of new basis vectors $\{\mathbf{e}_u, \mathbf{e}_v\}$. By the Einstein summation convention, the repeated index k is a summation index. The symbol $\partial u^{k'}/\partial x^k$ represents the elements of $\mathbf{J}_\Phi$, the Jacobian matrix for the forward transformation Φ, where Φ is given in the form $\mathbf{u} = \mathbf{u}(\mathbf{x})$. Quantities with primed indices are in the new coordinate system, while quantities with unprimed indices are in the old coordinate system.

The product of the Jacobian matrix $\mathbf{J}_\Phi$ and the column vector of components v^k is ordinary matrix multiplication with the matrix on the left and the column vector on the right. The sum on the index k is a sum on the column index in the matrix, and that corresponds to the matrix product against a column vector from the left.

EXAMPLE 32.7   In the xy-plane, the parabola $x(p) = p$, $y(p) = p^2$ has the tangent vector $\mathbf{i} + 2p\mathbf{j}$. The contravariant transformation law for transforming to polar coordinates requires multiplication by the Jacobian matrix $\mathbf{J}_\Phi$, evaluated along the parabola. Consequently, the transformed vector is then

$$'\mathbf{V} = \begin{bmatrix} \dfrac{1}{\sqrt{p^2+1}} & \dfrac{p}{\sqrt{p^2+1}} \\[2ex] -\dfrac{1}{p^2+1} & \dfrac{1}{p(p^2+1)} \end{bmatrix} \begin{bmatrix} 1 \\ 2p \end{bmatrix} = \begin{bmatrix} \dfrac{1+2p^2}{\sqrt{p^2+1}} \\[2ex] \dfrac{1}{p^2+1} \end{bmatrix} = \dfrac{1+2p^2}{\sqrt{p^2+1}}\mathbf{e}_r + \dfrac{1}{p^2+1}\mathbf{e}_\theta \qquad (32.12)$$

To obtain corroboration of this result, solve (32.10) for

$$\mathbf{e}_r = \frac{1}{\sqrt{p^2+1}}\mathbf{i} + \frac{p}{\sqrt{p^2+1}}\mathbf{j} \quad \text{and} \quad \mathbf{e}_\theta = -p^2\mathbf{i} + p\mathbf{j} \qquad (32.13)$$

along the parabola. Hence, $'\mathbf{V}$ transforms back to $\mathbf{i} + 2p\mathbf{j}$, thereby verifying the transformation law for this example. ❖

EXERCISES 32.3

1. Verify that (32.13) substituted into (32.12) restores the tangent vector $\mathbf{i} + 2p\mathbf{j}$.

In Exercises 2–7:

    (a) Sketch the given curve.

    (b) Obtain the tangent vector $\mathbf{V} = \mathbf{R}'(p)$.

    (c) Transform $\mathbf{V}$ to polar coordinates, obtaining $'\mathbf{V}$ by applying the contravariant law as per (32.12).

    (d) Obtain $'\mathbf{V}$ by expressing $\mathbf{i}$ and $\mathbf{j}$ in terms of $\mathbf{e}_r$ and $\mathbf{e}_\theta$, and substituting in $\mathbf{V}$.

    (e) Change $'\mathbf{V}$ back to $\mathbf{V}$ by replacing $\mathbf{e}_r$ and $\mathbf{e}_\theta$ with their equivalents in terms of $\mathbf{i}$ and $\mathbf{j}$.

2. $x(p) = p \cos p,\ y(p) = p \sin p,\ 0 \le p \le 4\pi$

3. $x(p) = \sqrt{1+p^2},\ y(p) = \sqrt{1-p^2},\ -1 < p < 1$

4. $x(p) = \dfrac{1-p}{1+p},\ y(p) = pe^{-p},\ -1 < p < 1$

5. $x(p) = p^2 + 3p - 1,\ y(p) = p^3 - p + 1,\ -1 \le p \le 1$

6. $x(p) = 3\cos 2p,\ y(p) = 4\sin p,\ 0 \le p \le 2\pi$

7. $x(p) = \dfrac{1-p^3}{1+p^2},\ y(p) = \dfrac{1-p^2}{1+p^3},\ -1 \le p \le 1$

In Exercises 8–15, for the coordinate system defined by the given inverse map Φ^{-1}:

    (a) Obtain a labeled plot of the coordinate curves.

    (b) Obtain $\{\mathbf{e}_u, \mathbf{e}_v\}$, the natural basis of tangent vectors.

    (c) Obtain $\mathbf{J}_{\Phi^{-1}}$, the Jacobian matrix of the inverse map.

    (d) Obtain $\mathbf{J}_{\Phi(u)} = [\mathbf{J}_{\Phi^{-1}}]^{-1}$, the Jacobian matrix for Φ, the forward map.

8. $x = \frac{1}{2}(u^2 - v^2),\ y = uv$, the parabolic coordinate system

9. $x = \dfrac{u}{u^2 + v^2},\ y = \dfrac{v}{u^2 + v^2}$, the tangent coordinate system

10. $x = \cosh u \cos v,\ y = \sinh u \sin v$, the elliptic coordinate system

11. $x = \sqrt{\sqrt{u^2 + v^2} + u},\ y = \sqrt{\sqrt{u^2 + v^2} - u}$, the hyperbolic coordinate system

12. $x = \dfrac{\sinh v}{\cosh v - \cos u},\ y = \dfrac{\sin u}{\cosh v - \cos u}$, the bipolar coordinate system

13. $x = 3u + 2v,\ y = 5u - 7v$, an undistinguished skew coordinate system

14. $x = 2u^2 - 5v,\ y = 3u + 4v^2$

15. $x = 5u^2 + 3v^2,\ y = 4u^2 - 9v^2$

For the curves given in Exercises 16–21:

    (a) Obtain the tangent vector $\mathbf{V} = \mathbf{R}'(p)$.

    (b) Transform $\mathbf{V}$ to the coordinate system $x = \frac{1}{2}(u^2 - v^2)$, $y = 2uv$, wherein Φ, the forward map, is given by $u = \frac{y}{d}$, $v = \frac{d}{2},\ d = \sqrt{2\sqrt{4x^2 + y^2} - 4x}$, obtaining $'\mathbf{V}$ by applying the contravariant law as per (32.12).

    (c) Obtain $'\mathbf{V}$ by expressing $\mathbf{i}$ and $\mathbf{j}$ in terms of $\mathbf{e}_u$ and $\mathbf{e}_v$ and substituting in $\mathbf{V}$.

    (d) Change $'\mathbf{V}$ back to $\mathbf{V}$ by replacing $\mathbf{e}_u$ and $\mathbf{e}_v$ with their equivalents in terms of $\mathbf{i}$ and $\mathbf{j}$.

16. the curve of Exercise 2   **B17.** the curve of Exercise 3

18. the curve of Exercise 4   **B19.** the curve of Exercise 5

20. the curve of Exercise 6   **B21.** the curve of Exercise 7

In Exercises 22–27:

    (a) Obtain the tangent vector $\mathbf{V} = \mathbf{R}'(p)$.

    (b) Transform $\mathbf{V}$ to the parabolic coordinate system wherein Φ, the forward map, is given by $u = \frac{y}{s}$, $v = s,\ s = \sqrt{\sqrt{x^2 + y^2} - x}$, obtaining $'\mathbf{V}$ by applying the contravariant law as per (32.12).

(c) Obtain $'\mathbf{V}$ by expressing $\mathbf{i}$ and $\mathbf{j}$ in terms of $\mathbf{e}_u$ and $\mathbf{e}_v$ and substituting in $\mathbf{V}$.

(d) Change $'\mathbf{V}$ back to $\mathbf{V}$ by replacing $\mathbf{e}_u$ and $\mathbf{e}_v$ with their equivalents in terms of $\mathbf{i}$ and $\mathbf{j}$.

22. the curve of Exercise 2   **B23.** the curve of Exercise 3

24. the curve of Exercise 4   **B25.** the curve of Exercise 5

26. the curve of Exercise 6   **B27.** the curve of Exercise 7

In Exercises 28–33:

(a) Obtain the tangent vector $\mathbf{V} = \mathbf{R}'(p)$.

(b) Transform $\mathbf{V}$ to the tangent coordinate system wherein Φ, the forward map, is given by $u = \frac{x}{x^2+y^2}$, $v = \frac{y}{x^2+y^2}$, obtaining $'\mathbf{V}$ by applying the contravariant law as per (32.12).

(c) Obtain $'\mathbf{V}$ by expressing $\mathbf{i}$ and $\mathbf{j}$ in terms of $\mathbf{e}_u$ and $\mathbf{e}_v$ and substituting in $\mathbf{V}$.

(d) Change $'\mathbf{V}$ back to $\mathbf{V}$ by replacing $\mathbf{e}_u$ and $\mathbf{e}_v$ with their equivalents in terms of $\mathbf{i}$ and $\mathbf{j}$.

28. the curve of Exercise 2   **B29.** the curve of Exercise 3

30. the curve of Exercise 4   **B31.** the curve of Exercise 5

32. the curve of Exercise 6   **B33.** the curve of Exercise 7

In Exercises 34–39:

(a) Obtain the tangent vector $\mathbf{V} = \mathbf{R}'(p)$.

(b) Transform $\mathbf{V}$ to the hyperbolic coordinate system wherein Φ, the forward map, is given by $u = \frac{1}{2}(x^2 - y^2)$, $v = xy$, obtaining $'\mathbf{V}$ by applying the contravariant law as per (32.12).

(c) Obtain $'\mathbf{V}$ by expressing $\mathbf{i}$ and $\mathbf{j}$ in terms of $\mathbf{e}_u$ and $\mathbf{e}_v$ and substituting in $\mathbf{V}$.

(d) Change $'\mathbf{V}$ back to $\mathbf{V}$ by replacing $\mathbf{e}_u$ and $\mathbf{e}_v$ with their equivalents in terms of $\mathbf{i}$ and $\mathbf{j}$.

34. the curve of Exercise 2   **B35.** the curve of Exercise 3

36. the curve of Exercise 4   **B37.** the curve of Exercise 5

38. the curve of Exercise 6   **B39.** the curve of Exercise 7

In Exercises 40–45:

(a) Obtain the tangent vector $\mathbf{V} = \mathbf{R}'(p)$.

(b) Transform $\mathbf{V}$ to the coordinate system $x = 3u + 2v$, $y = 5u - 7v$ where Φ, the forward map, is simple enough to

find, obtaining $'\mathbf{V}$ by applying the contravariant law as per (32.12).

(c) Obtain $'\mathbf{V}$ by expressing $\mathbf{i}$ and $\mathbf{j}$ in terms of $\mathbf{e}_u$ and $\mathbf{e}_v$ and substituting in $\mathbf{V}$.

(d) Change $'\mathbf{V}$ back to $\mathbf{V}$ by replacing $\mathbf{e}_u$ and $\mathbf{e}_v$ with their equivalents in terms of $\mathbf{i}$ and $\mathbf{j}$.

40. the curve of Exercise 2   **B41.** the curve of Exercise 3

42. the curve of Exercise 4   **B43.** the curve of Exercise 5

44. the curve of Exercise 6   **B45.** the curve of Exercise 7

46. Define spherical coordinates by $x = \rho \cos\theta \sin\phi$, $y = \rho \sin\theta \sin\phi$, $z = \rho \cos\phi$, where ϕ is the angle between the radius vector and the z-axis.

(a) Obtain the forward map Φ whose equations would be $\rho = \rho(x, y, z)$, $\phi = \phi(x, y, z)$, $\theta = \theta(x, y, z)$.

(b) Sketch the three coordinate surfaces $\rho = 1$, $\phi = \frac{\pi}{6}$, $\theta = \frac{\pi}{3}$.

(c) Obtain a plot of the three coordinate curves through point P whose coordinates are $(x, y, x) = (\frac{1}{2}, \frac{1}{2}, \frac{1}{\sqrt{2}})$.

(d) Obtain $\{\mathbf{e}_\rho, \mathbf{e}_\phi, \mathbf{e}_\theta\}$, the natural basis of vectors tangent to coordinate curves.

(e) Evaluate $\{\mathbf{e}_\rho, \mathbf{e}_\phi, \mathbf{e}_\theta\}$ at point P, and add their graphs to the sketch drawn in part (c).

In Exercises 47–50:

(a) Sketch the given curve.

(b) Obtain the tangent vector $\mathbf{V} = \mathbf{R}'(p)$.

(c) Transform $\mathbf{V}$ to spherical coordinates, obtaining $'\mathbf{V}$ by applying the contravariant law as per (32.12).

(d) Obtain $'\mathbf{V}$ by expressing $\mathbf{i}$, $\mathbf{j}$, and $\mathbf{k}$ in terms of $\mathbf{e}_\rho$, $\mathbf{e}_\phi$, and $\mathbf{e}_\theta$ and substituting in $\mathbf{V}$.

(e) Change $'\mathbf{V}$ back to $\mathbf{V}$ by replacing $\{\mathbf{e}_\rho, \mathbf{e}_\phi, \mathbf{e}_\theta\}$ with their equivalents in terms of $\mathbf{i}$, $\mathbf{j}$, and $\mathbf{k}$.

47. $x(p) = p$, $y(p) = p^2$, $z = p^3$, $-2 \le p \le 2$

48. $x(p) = \cos p$, $y(p) = \sin p$, $z(p) = p$, $0 \le p \le 4\pi$

49. $x(p) = p\cos p$, $y(p) = p\sin p$, $z(p) = p$, $0 \le p \le 4\pi$

50. $x(p) = \cos 2p$, $y(p) = \sin p$, $z(p) = p^2$, $0 \le p \le 2\pi$

32.4 Reciprocal Bases and Gradient Vectors

Reciprocal Vectors

A set of vectors $\{\mathbf{e}^1, \ldots, \mathbf{e}^n\}$ is *reciprocal* to the set of tangent basis vectors $\{\mathbf{e}_1, \ldots, \mathbf{e}_n\}$ if the dot products between the vectors obey $\mathbf{e}^i \cdot \mathbf{e}_j = \delta^i_j$.

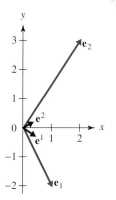

FIGURE 32.10   Tangent vectors $\{\mathbf{e}_1, \mathbf{e}_2\}$ and the reciprocal vectors $\{\mathbf{e}^1, \mathbf{e}^2\}$ in Example 32.8

EXAMPLE 32.8   Let

$$\mathbf{e}_1 = \mathbf{i} - 2\mathbf{j} \quad\text{and}\quad \mathbf{e}_2 = 2\mathbf{i} + 3\mathbf{j}$$

be a set of tangent basis vectors in the plane. The vectors

$$\mathbf{e}^1 = a_1\mathbf{i} + a_2\mathbf{j} \quad\text{and}\quad \mathbf{e}^2 = b_1\mathbf{i} + b_2\mathbf{j}$$

will be reciprocal if the equations

$$\mathbf{e}^1 \cdot \mathbf{e}_1 = a_1 - 2a_2 = 1 \quad \mathbf{e}^1 \cdot \mathbf{e}_2 = 2a_1 + 3a_2 = 0$$
$$\mathbf{e}^2 \cdot \mathbf{e}_1 = b_1 - 2b_2 = 0 \quad \mathbf{e}^2 \cdot \mathbf{e}_2 = 2b_1 + 3b_2 = 1$$

are satisfied. The unique solution is

$$\mathbf{e}^1 = \tfrac{1}{7}(3\mathbf{i} - 2\mathbf{j}) \quad\text{and}\quad \mathbf{e}^2 = \tfrac{1}{7}(2\mathbf{i} + \mathbf{j})$$

shown in Figure 32.10. The tangent vectors $\mathbf{e}_1$ and $\mathbf{e}_2$ are not perpendicular to each other, and neither are the reciprocal vectors $\mathbf{e}^1$ and $\mathbf{e}^2$. However, $\mathbf{e}_1$ is perpendicular to $\mathbf{e}^2$ and $\mathbf{e}_2$ is perpendicular to $\mathbf{e}^1$.   ❖

EXAMPLE 32.9   To test if two sets of vectors are mutually reciprocal, take all possible dot products between members of the two sets. For example, given the vectors $\{\mathbf{v}_1, \mathbf{v}_2, \mathbf{v}_3\}$ and the set of vectors $\{\mathbf{v}^1, \mathbf{v}^2, \mathbf{v}^3\}$, testing for reciprocity requires forming the nine dot products $\mathbf{v}^i \cdot \mathbf{v}_j, i = 1, 2, 3, j = 1, 2, 3$. This can be done by forming matrices $M_1 = [\mathbf{v}_1, \mathbf{v}_2, \mathbf{v}_3]$ and $M_2 = [\mathbf{v}^1, \mathbf{v}^2, \mathbf{v}^3]$ and computing

$$M_2^{\mathrm{T}} M_1 = \begin{bmatrix} \longleftarrow \mathbf{v}^1 \longrightarrow \\ \longleftarrow \mathbf{v}^2 \longrightarrow \\ \longleftarrow \mathbf{v}^3 \longrightarrow \end{bmatrix} \begin{bmatrix} \uparrow & \uparrow & \uparrow \\ \mathbf{v}_1 & \mathbf{v}_2 & \mathbf{v}_3 \\ \downarrow & \downarrow & \downarrow \end{bmatrix} = \begin{bmatrix} \mathbf{v}^1 \cdot \mathbf{v}_1 & \mathbf{v}^1 \cdot \mathbf{v}_2 & \mathbf{v}^1 \cdot \mathbf{v}_3 \\ \mathbf{v}^2 \cdot \mathbf{v}_1 & \mathbf{v}^2 \cdot \mathbf{v}_2 & \mathbf{v}^2 \cdot \mathbf{v}_3 \\ \mathbf{v}^3 \cdot \mathbf{v}_1 & \mathbf{v}^3 \cdot \mathbf{v}_2 & \mathbf{v}^3 \cdot \mathbf{v}_3 \end{bmatrix} \qquad (32.14)$$

If the product is the identity matrix, then the sets of vectors are mutually reciprocal.   ❖

EXAMPLE 32.10   The following sets of vectors, namely, $\{\mathbf{v}_1, \mathbf{v}_2, \mathbf{v}_3\}$ and $\{\mathbf{v}^1, \mathbf{v}^2, \mathbf{v}^3\}$, are mutually reciprocal. The $\mathbf{v}_k$ are unit vectors but are not orthogonal. The $\mathbf{v}^k$ are neither unit nor orthogonal. Indeed, the norms of the $\mathbf{v}^k$ are 1.27, 1.31, 1.12, respectively. The relative orientations of these vectors can be seen in Figure 32.11 on page 744.

$$\left\{ \begin{bmatrix} \frac{1}{\sqrt{26}} \\ -\frac{5}{\sqrt{26}} \\ 0 \end{bmatrix}, \begin{bmatrix} \frac{5}{\sqrt{66}} \\ -\frac{4}{\sqrt{66}} \\ \frac{5}{\sqrt{66}} \end{bmatrix}, \begin{bmatrix} -\frac{5}{\sqrt{30}} \\ \frac{1}{\sqrt{30}} \\ \frac{2}{\sqrt{30}} \end{bmatrix} \right\} \quad \left\{ \begin{bmatrix} -\frac{13\sqrt{26}}{162} \\ -\frac{35\sqrt{26}}{162} \\ -\frac{5\sqrt{26}}{54} \end{bmatrix}, \begin{bmatrix} \frac{5\sqrt{66}}{81} \\ \frac{\sqrt{66}}{81} \\ \frac{4\sqrt{66}}{27} \end{bmatrix}, \begin{bmatrix} -\frac{25\sqrt{30}}{162} \\ -\frac{5\sqrt{30}}{162} \\ \frac{7\sqrt{30}}{54} \end{bmatrix} \right\} \qquad ❖$$

EXAMPLE 32.11   Orthonormal vectors are *self-reciprocal*. For example, if the tangent vectors $S = \{\mathbf{v}_1, \mathbf{v}_2, \mathbf{v}_3\}$ are *orthonormal*, that is, are each of length 1 and mutually perpendicular, then they are their own reciprocal vectors. Indeed, taking $M_2 = M_1$ in (32.14) yields, by the orthonormality of S, $M_1^{\mathrm{T}} M_1 = I$.

Readers who have worked only in Cartesian, polar, spherical, and cylindrical coordinates will not have experienced reciprocal bases as distinct from a basis of tangent vectors. Each of these orthogonal coordinate systems has an *orthonormal* basis for which the reciprocal vectors are always the same as the tangent basis vectors. It is only when the tangent basis is not orthonormal that a difference between the basis and its reciprocal basis is observed.   ❖

EXAMPLE 32.12   Impose polar coordinates on the *xy*-plane, and use the natural tangent basis vectors

$$\mathbf{e}_r = \cos\theta\mathbf{i} + \sin\theta\mathbf{j} \quad \text{and} \quad \mathbf{e}_\theta = -r\sin\theta\mathbf{i} + r\cos\theta\mathbf{j}$$

instead of the more familiar unit basis vectors. By brute force, the reciprocal vectors are found to be

$$\mathbf{e}^r = \cos\theta\mathbf{i} + \sin\theta\mathbf{j} = \mathbf{e}_r \quad \text{and} \quad \mathbf{e}^\theta = \frac{1}{r}(-\sin\theta\mathbf{i} + \cos\theta\mathbf{j}) = \frac{1}{r^2}\mathbf{e}_\theta$$

Notice that the reciprocal vectors are not unit vectors, but they are mutually orthogonal, just like the vectors $\{\mathbf{e}_r, \mathbf{e}_\theta\}$. Notice further that the reciprocal vectors are the rows of the Jacobian matrix $\mathbf{J}_{\Phi(u)}$. Indeed, since $\mathbf{J}_{\Phi(u)}$ and $\mathbf{J}_{\Phi^{-1}}$ are inverse matrices, we have

$$[\mathbf{J}_{\Phi^{-1}}]^{-1} = \begin{bmatrix} \cos\theta & -r\sin\theta \\ \sin\theta & r\cos\theta \end{bmatrix}^{-1} = \begin{bmatrix} \cos\theta & \sin\theta \\ -\dfrac{\sin\theta}{r} & \dfrac{\cos\theta}{r} \end{bmatrix} = J_{\Phi(u)} \qquad ❖$$

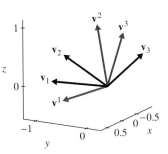

FIGURE 32.11   Reciprocal vectors $\{\mathbf{v}^k\}$ and $\{\mathbf{v}_k\}$, $k = 1, 2, 3$, in Example 32.10

Why Reciprocal Vectors?

Example 32.12 suggests the first of two reasons why reciprocal vectors are of importance. Because $\mathbf{J}_{\Phi(u)}\mathbf{J}_{\Phi^{-1}} = I$, the rows of $\mathbf{J}_{\Phi(u)}$ are *always* the vectors reciprocal to the columns of $\mathbf{J}_{\Phi^{-1}}$, that is, to the natural basis of tangent vectors. The basis of vectors tangent to the coordinate curves arise naturally by differentiating the radius vector with respect to the new coordinates. The formalism of the Jacobian matrix means these tangent vectors are the columns of $\mathbf{J}_{\Phi^{-1}}$.

The rows of $\mathbf{J}_{\Phi(u)}$ are reciprocal to the natural tangent basis, but what meaning do these row vectors have in their own right? We again illustrate with polar coordinates.

$$\mathbf{J}_\Phi\mathbf{J}_{\Phi^{-1}} = \left[\frac{\partial u^i}{\partial x^k}\right]\left[\frac{\partial x^k}{\partial u^j}\right] = \begin{bmatrix} \left(\dfrac{\partial u^1}{\partial x^1} \quad \dfrac{\partial u^1}{\partial x^2}\right) \\ \left(\dfrac{\partial u^2}{\partial x^1} \quad \dfrac{\partial u^2}{\partial x^2}\right) \end{bmatrix} \begin{bmatrix} \left(\dfrac{\partial x^1}{\partial u^1}\right) & \left(\dfrac{\partial x^1}{\partial u^2}\right) \\ \left(\dfrac{\partial x^2}{\partial u^1}\right) & \left(\dfrac{\partial x^2}{\partial u^2}\right) \end{bmatrix} = \begin{bmatrix} 1 & 0 \\ 0 & 1 \end{bmatrix}$$

$$= \begin{bmatrix} \left(\dfrac{\partial r}{\partial x} \quad \dfrac{\partial r}{\partial y}\right) \\ \left(\dfrac{\partial \theta}{\partial x} \quad \dfrac{\partial \theta}{\partial y}\right) \end{bmatrix} \begin{bmatrix} \left(\dfrac{\partial x}{\partial r}\right) & \left(\dfrac{\partial x}{\partial \theta}\right) \\ \left(\dfrac{\partial y}{\partial r}\right) & \left(\dfrac{\partial y}{\partial \theta}\right) \end{bmatrix} = \begin{bmatrix} 1 & 0 \\ 0 & 1 \end{bmatrix}$$

The rows of $\mathbf{J}_\Phi$, the Jacobian matrix for $\mathbf{u} = \mathbf{u}(\mathbf{x})$, the forward mapping Φ, are gradient vectors!

Gradient Vectors

For polar coordinates, the rows of the Jacobian matrix corresponding to the forward mapping Φ, the mapping whose equations are $\mathbf{u} = \mathbf{u}(\mathbf{x})$, are the gradient vectors

$$\mathbf{e}^r = \frac{\partial r}{\partial x}\mathbf{i} + \frac{\partial r}{\partial y}\mathbf{j} = \nabla r(x, y) = \frac{x}{\sqrt{x^2 + y^2}}\mathbf{i} + \frac{y}{\sqrt{x^2 + y^2}}\mathbf{j} = \cos\theta\mathbf{i} + \sin\theta\mathbf{j} \quad \textbf{(32.15a)}$$

$$\mathbf{e}^\theta = \frac{\partial \theta}{\partial x}\mathbf{i} + \frac{\partial \theta}{\partial y}\mathbf{j} = \nabla\theta(x, y) = -\frac{y}{x^2 + y^2}\mathbf{i} + \frac{x}{x^2 + y^2}\mathbf{j} = -\frac{\sin\theta}{r}\mathbf{i} + \frac{\cos\theta}{r}\mathbf{j} \quad \textbf{(32.15b)}$$

computed from the functions $r = \sqrt{x^2 + y^2}$, $\theta = \arctan\frac{y}{x}$. Since gradient vectors are orthogonal to their level sets, the gradient vectors (32.15a) and (32.15b) are orthogonal to the level curves $r = $ constant and $\theta = $ constant.

Conclusions

The backward map Φ^{-1}, a vector-valued function of the form $\mathbf{x} = \mathbf{x}(\mathbf{u})$, gives rise to the u^k-coordinate curves determined in the space of $\mathbf{x}$ by varying u^k while holding the other u's fixed. The vectors $\mathbf{e}_k$, tangent vectors along these coordinate curves, are the columns of the Jacobian matrix

The forward map Φ, a vector-valued function of the form $\mathbf{u} = \mathbf{u}(\mathbf{x})$, gives rise to level things determined in the space of $\mathbf{x}$ by implicit functions of the form $u^k(\mathbf{x}) = $ constant. The vectors $\mathbf{e}^k$, gradients of these functions, are vectors normal to the level sets; these gradients are the rows of the Jacobian matrix

Together, the tangent vectors and gradient vectors constitute sets of reciprocal vectors.

EXERCISES 32.4–Part A

A1. If $\mathbf{e}_1 = 3\mathbf{i} + 2\mathbf{j}$ and $\mathbf{e}_2 = 5\mathbf{i} - 7\mathbf{j}$, obtain $\{\mathbf{e}^1, \mathbf{e}^2\}$, the vectors reciprocal to $\{\mathbf{e}_1, \mathbf{e}_2\}$.

A2. Given the linearly independent vectors $\{\mathbf{e}_1, \mathbf{e}_2, \mathbf{e}_3\}$ in R^3, show that

$$\left\{\mathbf{e}^1 = \frac{\mathbf{e}_2 \times \mathbf{e}_3}{\mathbf{e}_1 \cdot \mathbf{e}_2 \times \mathbf{e}_3}, \mathbf{e}^2 = \frac{\mathbf{e}_3 \times \mathbf{e}_1}{\mathbf{e}_1 \cdot \mathbf{e}_2 \times \mathbf{e}_3}, \mathbf{e}^3 = \frac{\mathbf{e}_1 \times \mathbf{e}_2}{\mathbf{e}_1 \cdot \mathbf{e}_2 \times \mathbf{e}_3}\right\}$$

are reciprocal. *Hint:* Verify $\mathbf{e}^i \cdot \mathbf{e}_j = \delta^i_j$.

A3. For the vectors in Exercise A2, show that the numbers $\mathbf{e}^1 \cdot \mathbf{e}^2 \times \mathbf{e}^3$ and $\mathbf{e}_1 \cdot \mathbf{e}_2 \times \mathbf{e}_3$ are reciprocals. *Hint:* Use Exercise B8, Section 31.3 to obtain

$$\mathbf{e}^1 \cdot \mathbf{e}^2 \times \mathbf{e}^3 = \frac{(\mathbf{e}_3 \times \mathbf{e}_1) \times (\mathbf{e}_1 \times \mathbf{e}_2) \cdot (\mathbf{e}_2 \times \mathbf{e}_3)}{(\mathbf{e}_1 \cdot \mathbf{e}_2 \times \mathbf{e}_3)^3} = \frac{(\mathbf{e}_1 \cdot \mathbf{e}_2 \times \mathbf{e}_3)^2}{(\mathbf{e}_1 \cdot \mathbf{e}_2 \times \mathbf{e}_3)^3}$$

A4. Derive the recipe in Exercise A2. *Hint:* The equations $\mathbf{e}^i \cdot \mathbf{e}_j = \delta^i_j$ say that $\mathbf{e}^1$ is perpendicular to $\mathbf{e}_2$ and $\mathbf{e}_3$, so write $\mathbf{e}^1 = \alpha \mathbf{e}_2 \times \mathbf{e}_3$.

Determine α by the condition $\mathbf{e}_1 \cdot \mathbf{e}^1 = 1$. Similar calculations yield $\mathbf{e}^2$ and $\mathbf{e}^3$.

A5. Fill in the details of the following proof that a set of reciprocal vectors is unique. Let $S = \{\mathbf{e}_1, \mathbf{e}_2, \mathbf{e}_3\}$ be linearly independent in R^3, and assume both $\{\mathbf{e}^k\}$ and $\{\mathbf{v}^k\}$, $k = 1, 2, 3$, are sets of vectors reciprocal to S. The conditions $\mathbf{e}^i \cdot \mathbf{e}_j = \delta^i_j$ would hold, as would $\mathbf{e}_k \cdot \mathbf{v}^k = 1$, $k = 1, 2, 3$, and $\mathbf{e}_i \cdot \mathbf{v}^j = 0$, $i \neq j$. For all values of i and j, write out the equations $\mathbf{e}_i \cdot (\mathbf{e}^j - \mathbf{v}^j) = 0$ which are obtained by subtraction. Show that if one vector $\mathbf{e}^j - \mathbf{v}^j$ were nonzero, it would be orthogonal to three linearly independent vectors, which is impossible. Hence, $\mathbf{e}^j - \mathbf{v}^j = \mathbf{0}$ and a reciprocal set of vectors is unique.

EXERCISES 32.4–Part B

B1. Obtain $\mathbf{N}$, a vector normal to the plane through the points $p_1 = (1, 2, 3)$, $p_2 = (5, 3, -4)$, $p_3 = (2, 7, 6)$. Then, let $\mathbf{P}_k$ be the vector from the origin to the point p_k, $k = 1, 2, 3$. Find $\{\mathbf{e}^k\}$, the vectors reciprocal to $\{\mathbf{P}_k\}$. Show that $\mathbf{e}^1 + \mathbf{e}^2 + \mathbf{e}^3$ is along $\mathbf{N}$.

In Exercises B2–11:

    (a) Find the vectors $\{\mathbf{e}^1, \mathbf{e}^2\}$ reciprocal to the given $\{\mathbf{e}_1, \mathbf{e}_2\}$.

    (b) Sketch the given vectors and the reciprocal set.

B2. $\mathbf{e}_1 = -6\mathbf{i} + 8\mathbf{j}, \mathbf{e}_2 = 10\mathbf{i} + \mathbf{j}$

B3. $\mathbf{e}_1 = -11\mathbf{i} + \mathbf{j}, \mathbf{e}_2 = -2\mathbf{i} + 6\mathbf{j}$

B4. $\mathbf{e}_1 = 9\mathbf{i} - 3\mathbf{j}, \mathbf{e}_2 = -7\mathbf{i} + \mathbf{j}$ **B5.** $\mathbf{e}_1 = -5\mathbf{i} - 2\mathbf{j}, \mathbf{e}_2 = 12\mathbf{i} - 2\mathbf{j}$

B6. $\mathbf{e}_1 = 4\mathbf{i} - \mathbf{j}, \mathbf{e}_2 = -5\mathbf{i} + 12\mathbf{j}$ **B7.** $\mathbf{e}_1 = 11\mathbf{i} - 6\mathbf{j}, \mathbf{e}_2 = 12\mathbf{i} + 4\mathbf{j}$

B8. $\mathbf{e}_1 = 8\mathbf{i} + 7\mathbf{j}, \mathbf{e}_2 = 3\mathbf{i} + 4\mathbf{j}$ **B9.** $\mathbf{e}_1 = 8\mathbf{i} + 12\mathbf{j}, \mathbf{e}_2 = -3\mathbf{i} - 2\mathbf{j}$

B10. $\mathbf{e}_1 = -9\mathbf{i} + 11\mathbf{j}, \mathbf{e}_2 = 8\mathbf{i} + 11\mathbf{j}$ **B11.** $\mathbf{e}_1 = 7\mathbf{i} - 9\mathbf{j}, \mathbf{e}_2 = 3\mathbf{i} - 7\mathbf{j}$

In Exercises B12–21:

    (a) Use "brute force" to find the vectors $\{\mathbf{e}^1, \mathbf{e}^2, \mathbf{e}^3\}$ reciprocal to the given $\{\mathbf{e}_1, \mathbf{e}_2, \mathbf{e}_3\}$.

    (b) Use the recipe in Exercise A2 to find the vectors reciprocal to the given $\{\mathbf{e}_1, \mathbf{e}_2, \mathbf{e}_3\}$.

    (c) Show that the numbers $\mathbf{e}^1 \cdot \mathbf{e}^2 \times \mathbf{e}^3$ and $\mathbf{e}_1 \cdot \mathbf{e}_2 \times \mathbf{e}_3$ are reciprocals.

    (d) Sketch the given vectors and the reciprocal set.

B12. $\mathbf{e}_1 = -8\mathbf{i} - 2\mathbf{j} + 8\mathbf{k}, \mathbf{e}_2 = 11\mathbf{i} + \mathbf{j} - 11\mathbf{k}, \mathbf{e}_3 = -12\mathbf{i} + 12\mathbf{j} + 11\mathbf{k}$

B13. $\mathbf{e}_1 = -12\mathbf{i} + 7\mathbf{j} - 11\mathbf{k}, \mathbf{e}_2 = 12\mathbf{i} + 10\mathbf{j} + \mathbf{k}, \mathbf{e}_3 = 3\mathbf{i} + 7\mathbf{j} - \mathbf{k}$

B14. $\mathbf{e}_1 = -\mathbf{i} - 3\mathbf{j} + 4\mathbf{k}, \mathbf{e}_2 = -6\mathbf{i} + 8\mathbf{j} + 11\mathbf{k}, \mathbf{e}_3 = 12\mathbf{i} + 6\mathbf{j} + 7\mathbf{k}$

B15. $e_1 = -11i - 12j - 6k$, $e_2 = 8i - 2j - 7k$, $e_3 = 4i - j + 4k$

B16. $e_1 = 8i + 10j - 7k$, $e_2 = i + 8j - 3k$, $e_3 = 12i - 2j - 8k$

B17. $e_1 = -6i + 12j - 3k$, $e_2 = -10i + 3j - 2k$, $e_3 = -11i + 9j - 2k$

B18. $e_1 = -12i + 6j - 5k$, $e_2 = -i + 5j + 3k$, $e_3 = 11i + 12j - k$

B19. $e_1 = -4i - 5j + 3k$, $e_2 = 5i + 3j + 7k$, $e_3 = -9i - 6j - 5k$

B20. $e_1 = 8i - 5j + 3k$, $e_2 = -5i + j - 3k$, $e_3 = -4i + j - 12k$

B21. $e_1 = -2i + 9j - 8k$, $e_2 = -6i + 12j + 5k$, $e_3 = 10i - 3j + 6k$

B22. For the map Φ^{-1} defined by $x = \frac{1}{2}(u^2 - v^2)$, $y = 2uv$, with Φ given by $u = \frac{y}{d}$, $v = \frac{d}{2}$, $d = \sqrt{2\sqrt{4x^2 + y^2} - 4x}$, the natural basis of tangent vectors $\{e_u, e_v\}$ was obtained in Section 32.3.

    (a) Obtain the Jacobian matrices $J_{\Phi^{-1}}$ and $J_{\Phi(u)}$.

    (b) Show that the columns of $J_{\Phi^{-1}}$ are the vectors e_u and e_v.

    (c) Obtain a labeled plot of the level curves of $u(x, y)$ and $v(x, y)$.

    (d) Obtain the gradient vectors orthogonal to the level curves of $u(x, y)$ and $v(x, y)$, expressing them in terms of the variables u and v.

    (e) Show that the rows of $J_{\Phi(u)}$ are the gradient vectors found in part (c).

    (f) Show that the tangent vectors $\{e_u, e_v\}$ and the gradient vectors in parts (c) and (d) are reciprocal vectors.

    (g) Sketch the tangent and gradient vectors at the point where $u = v = 1$.

B23. In Exercise B8, Section 32.3, the natural basis of tangent vectors $\{e_u, e_v\}$ and the Jacobian matrices $J_{\Phi^{-1}}$ and $J_{\Phi(u)}$ were obtained for the parabolic coordinate system. The forward map Φ is given by $u = \frac{y}{s}$, $v = s$, $s = \sqrt{\sqrt{x^2 + y^2} - x}$.

    (a) Show that the columns of $J_{\Phi^{-1}}$ are the vectors e_u and e_v.

    (b) Obtain a labeled plot of the level curves of $u(x, y)$ and $v(x, y)$.

    (c) Obtain the gradient vectors orthogonal to the level curves of $u(x, y)$ and $v(x, y)$, expressing them in terms of the variables u and v.

    (d) Show that the rows of $J_{\Phi(u)}$ are the gradient vectors found in part (c).

    (e) Show that the tangent vectors $\{e_u, e_v\}$, and the gradient vectors in parts (c) or (d) are reciprocal vectors.

    (f) Sketch the tangent and gradient vectors at the point where $u = v = 1$.

B24. In Exercise B9, Section 32.3, the natural basis of tangent vectors $\{e_u, e_v\}$ and the Jacobian matrices $J_{\Phi^{-1}}$ and $J_{\Phi(u)}$ were obtained for the tangent coordinate system. The forward map Φ is given by $u = \frac{x}{x^2+y^2}$, $v = \frac{y}{x^2+y^2}$.

    (a) Show that the columns of $J_{\Phi^{-1}}$ are the vectors e_u and e_v.

    (b) Obtain a labeled plot of the level curves of $u(x, y)$ and $v(x, y)$.

    (c) Obtain the gradient vectors orthogonal to the level curves of $u(x, y)$ and $v(x, y)$, expressing them in terms of the variables u and v.

    (d) Show that the rows of $J_{\Phi(u)}$ are the gradient vectors found in part (c).

    (e) Show that the tangent vectors $\{e_u, e_v\}$ and the gradient vectors in parts (c) and (d) are reciprocal vectors.

    (f) Sketch the tangent and gradient vectors at the point where $u = v = 1$.

B25. In Exercise B11, Section 32.3, the natural basis of tangent vectors $\{e_u, e_v\}$ and the Jacobian matrices $J_{\Phi^{-1}}$ and $J_{\Phi(u)}$ were obtained for the hyperbolic coordinate system. The forward map Φ is given by $u = \frac{1}{2}(x^2 - y^2)$, $v = xy$.

    (a) Show that the columns of $J_{\Phi^{-1}}$ are the vectors e_u and e_v.

    (b) Obtain a labeled plot of the level curves of $u(x, y)$ and $v(x, y)$.

    (c) Obtain the gradient vectors orthogonal to the level curves of $u(x, y)$ and $v(x, y)$, expressing them in terms of the variables u and v.

    (d) Show that the rows of $J_{\Phi(u)}$ are the gradient vectors found in part (c).

    (e) Show that the tangent vectors $\{e_u, e_v\}$ and the gradient vectors in parts (c) and (d) are reciprocal vectors.

    (f) Sketch the tangent and gradient vectors at the point where $u = v = 1$.

B26. In Exercise B13, Section 32.3, the natural basis of tangent vectors $\{e_u, e_v\}$ and the Jacobian matrices $J_{\Phi^{-1}}$ and $J_{\Phi(u)}$ were obtained for a skew coordinate system.

    (a) Show that the columns of $J_{\Phi^{-1}}$ are the vectors e_u and e_v.

    (b) Obtain a labeled plot of the level curves of $u(x, y)$ and $v(x, y)$.

    (c) Obtain the gradient vectors orthogonal to the level curves of $u(x, y)$ and $v(x, y)$, expressing them in terms of the variables u and v.

    (d) Show that the rows of $J_{\Phi(u)}$ are the gradient vectors found in part (c).

    (e) Show that the tangent vectors $\{e_u, e_v\}$ and the gradient vectors in parts (c) and (d) are reciprocal vectors.

    (f) Sketch the tangent and gradient vectors at the point where $u = v = 1$.

B27. In Exercise B10, Section 32.3, the Jacobian matrices $J_{\Phi^{-1}}$ and $J_{\Phi(u)}$ for the elliptic coordinate system were obtained. The tangent vectors $\{e_u, e_v\}$ are the columns of $J_{\Phi^{-1}}$, and the gradient vectors $\{e^u, e^v\}$ are the rows of $J_{\Phi(u)}$.

    (a) Show that the tangent and gradient vectors are reciprocal sets of vectors.

    (b) Sketch the tangent and gradient vectors at the point where $u = v = 1$.

B28. In Exercise B12, Section 32.3, the Jacobian matrices $\mathbf{J}_{\Phi^{-1}}$ and $\mathbf{J}_{\Phi(u)}$ for the bipolar coordinate system were obtained. The tangent vectors $\{\mathbf{e}_u, \mathbf{e}_v\}$ are the columns of $\mathbf{J}_{\Phi^{-1}}$, and the gradient vectors $\{\mathbf{e}^u, \mathbf{e}^v\}$ are the rows of $\mathbf{J}_{\Phi(u)}$.

  (a) Show that the tangent and gradient vectors are reciprocal sets of vectors.

  (b) Sketch the tangent and gradient vectors at the point where $u = v = 1$.

B29. In Exercise B14, Section 32.3, the Jacobian matrices $\mathbf{J}_{\Phi^{-1}}$ and $\mathbf{J}_{\Phi(u)}$ for the coordinate system $x = 2u^2 - 5v$, $y = 3u + 4v^2$ were obtained. The tangent vectors $\{\mathbf{e}_u, \mathbf{e}_v\}$ are the columns of $\mathbf{J}_{\Phi^{-1}}$, and the gradient vectors $\{\mathbf{e}^u, \mathbf{e}^v\}$ are the rows of $\mathbf{J}_{\Phi(u)}$.

  (a) Show that the tangent and gradient vectors are reciprocal sets of vectors.

  (b) Sketch the tangent and gradient vectors at the point where $u = v = 1$.

B30. In Exercise B15, Section 32.3, the Jacobian matrices $\mathbf{J}_{\Phi^{-1}}$ and $\mathbf{J}_{\Phi(u)}$ for the coordinate system $x = 5u^2 + 3v^2$, $y = 4u^2 - 9v^2$ were obtained. The tangent vectors $\{\mathbf{e}_u, \mathbf{e}_v\}$ are the columns of $\mathbf{J}_{\Phi^{-1}}$, and the gradient vectors $\{\mathbf{e}^u, \mathbf{e}^v\}$ are the rows of $\mathbf{J}_{\Phi(u)}$.

  (a) Show that the tangent and gradient vectors are reciprocal sets of vectors.

  (b) Sketch the tangent and gradient vectors at the point where $u = v = 1$.

32.5 ## Gradient Vectors and the Covariant Transformation Law

The General Normal Vector

In Section 32.3, the contravariant transformation law for a vector tangent to a curve was obtained. In this section we will explore the covariant transformation law for a gradient vector. Remember, in Section 32.4 we discovered that the rows of the Jacobian matrix $\mathbf{J}_\Phi$ are vectors reciprocal to the natural basis of vectors tangent to the coordinate curves. Moreover, these row vectors are the gradients of the coordinate functions $u^k(\mathbf{x})$ and, therefore, are orthogonal to the coordinate surfaces $u^k = $ constant. (Chapter 18 details the orthogonality of the gradient vector with respect to level curves and surfaces.)

We denote the basis of gradient vectors, orthogonal to the coordinate surfaces, with superscripts, as in $\{\mathbf{e}^1, \mathbf{e}^2, \mathbf{e}^3\}$, for example. Again, these are just the basis vectors reciprocal to the natural basis of tangent vectors, $\{\mathbf{e}_1, \mathbf{e}_2, \mathbf{e}_3\}$.

If the function $f(x, y, z) = 0$ implicitly defines a surface $z = z(x, y)$, then the gradient vector $\nabla f = \frac{\partial f}{\partial x}\mathbf{i} + \frac{\partial f}{\partial y}\mathbf{j} + \frac{\partial f}{\partial z}\mathbf{k}$ is perpendicular to this surface. We wish to determine how this gradient vector transforms under a change of coordinates.

The Coordinate Change

Change coordinates via the backward map Φ^{-1} defined by the vector-valued function $\mathbf{x} = \mathbf{x}(\mathbf{u})$. Specifically, let the coordinate change be given by

$$x = x(u^1, u^2, u^3) \qquad y = y(u^1, u^2, u^3) \qquad z = z(u^1, u^2, u^3)$$

The corresponding forward map Φ is defined by the vector-valued function $\mathbf{u} = \mathbf{u}(\mathbf{x})$, whose equations would then be

$$u^1 = u^1(x, y, z) \qquad u^2 = u^2(x, y, z) \qquad u^3 = u^3(x, y, z)$$

Change of Coordinates in $f(x, y, z)$

Apply the mapping Φ^{-1} to the function $f(x, y, z)$, obtaining

$$f(x(u^1, u^2, u^3), y(u^1, u^2, u^3), z(u^1, u^2, u^3)) \equiv \hat{f}(u^1, u^2, u^3) \tag{32.16}$$

The notation signals that the transformed function is a different combination of the new variables. For example, if the function were $f(x, y, z) = x + y^2 + \frac{1}{z}$, and the new coordinates were spherical, then the transformed function would be

$$\hat{f}(\rho, \phi, \theta) = \rho \cos \theta \sin \phi + \rho^2 \sin^2 \theta \sin^2 \phi + \frac{1}{\rho \cos \phi}$$

In the function $f(x, y, z)$, the first variable is x, and that appears in f as itself. In the transformed function, the first variable is ρ, but it certainly does not appear just as itself. It appears multiplied by various trig functions, squared, and in the denominator of a fraction. Hence, the notation $f(x, y, z)$ instructs the user to perform certain operations on the variables x, y, z; whereas the symbol $\hat{f}(\rho, \phi, \theta)$ instructs the user to perform very different operations on the three variables ρ, ϕ, θ.

If we start with the transformed function, and restore the original variables via the map Φ, we get

$$\hat{f}(u^1(x, y, z), u^2(x, y, z), u^3(x, y, z)) = f(x, y, z) \tag{32.17}$$

The Transformed Gradient Vector

Under the change of coordinates, the gradient vector $\nabla f = \frac{\partial f}{\partial x}\mathbf{i} + \frac{\partial f}{\partial y}\mathbf{j} + \frac{\partial f}{\partial z}\mathbf{k}$ becomes

$$\frac{\partial \hat{f}}{\partial u^1}\mathbf{e}^1 + \frac{\partial \hat{f}}{\partial u^2}\mathbf{e}^2 + \frac{\partial \hat{f}}{\partial u^3}\mathbf{e}^3 \tag{32.18}$$

where $\mathbf{e}^1, \mathbf{e}^2, \mathbf{e}^3$ are the gradient vectors orthogonal to the coordinate surfaces defined by $\Phi = $ constant. In fact, these gradient vectors are reciprocal to the vectors tangent to the coordinate curves defined by Φ^{-1}. We prove these claims using (32.17) and the chain rule for differentiation, thereby obtaining

$$\nabla f = \frac{\partial f}{\partial x}\mathbf{i} + \frac{\partial f}{\partial y}\mathbf{j} + \frac{\partial f}{\partial z}\mathbf{k}$$

$$= \left(\frac{\partial \hat{f}}{\partial u^1}\frac{\partial u^1}{\partial x} + \frac{\partial \hat{f}}{\partial u^2}\frac{\partial u^2}{\partial x} + \frac{\partial \hat{f}}{\partial u^3}\frac{\partial u^3}{\partial x} \right)\mathbf{i} + \left(\frac{\partial \hat{f}}{\partial u^1}\frac{\partial u^1}{\partial y} + \frac{\partial \hat{f}}{\partial u^2}\frac{\partial u^2}{\partial y} + \frac{\partial \hat{f}}{\partial u^3}\frac{\partial u^3}{\partial y} \right)\mathbf{j}$$

$$+ \left(\frac{\partial \hat{f}}{\partial u^1}\frac{\partial u^1}{\partial z} + \frac{\partial \hat{f}}{\partial u^2}\frac{\partial u^2}{\partial z} + \frac{\partial \hat{f}}{\partial u^3}\frac{\partial u^3}{\partial z} \right)\mathbf{k}$$

$$= \frac{\partial \hat{f}}{\partial u^1}\left(\frac{\partial u^1}{\partial x}\mathbf{i} + \frac{\partial u^1}{\partial y}\mathbf{j} + \frac{\partial u^1}{\partial z}\mathbf{k} \right) + \frac{\partial \hat{f}}{\partial u^2}\left(\frac{\partial u^2}{\partial x}\mathbf{i} + \frac{\partial u^2}{\partial y}\mathbf{j} + \frac{\partial u^2}{\partial z}\mathbf{k} \right)$$

$$+ \frac{\partial \hat{f}}{\partial u^3}\left(\frac{\partial u^3}{\partial x}\mathbf{i} + \frac{\partial u^3}{\partial y}\mathbf{j} + \frac{\partial u^3}{\partial z}\mathbf{k} \right)$$

$$= \frac{\partial \hat{f}}{\partial u^1}\mathbf{e}^1 + \frac{\partial \hat{f}}{\partial u^2}\mathbf{e}^2 + \frac{\partial \hat{f}}{\partial u^3}\mathbf{e}^3$$

The Covariant Transformation Law

The result in (32.18) is the structure of the gradient vector under a change of coordinates. However, it is expressed in terms of derivatives of the transformed function. In general, we don't have the transformed function. What we have are the derivatives of $f(x, y, z)$ since what we really begin with are the coefficients of the gradient vector computed in Cartesian coordinates. Thus, we seek the law that details how to convert these derivatives in Cartesian coordinates to their appropriate counterparts in the new coordinate system.

This time, start with (32.16) and compute the components of the transformed gradient vector, as required in (32.18). It is tempting, but not useful, to write

$$
\begin{bmatrix} \dfrac{\partial \hat{f}}{\partial u^1} \\[2ex] \dfrac{\partial \hat{f}}{\partial u^2} \\[2ex] \dfrac{\partial \hat{f}}{\partial u^3} \end{bmatrix} = \begin{bmatrix} \dfrac{\partial f}{\partial x}\dfrac{\partial x}{\partial u^1} + \dfrac{\partial f}{\partial y}\dfrac{\partial y}{\partial u^1} + \dfrac{\partial f}{\partial z}\dfrac{\partial z}{\partial u^1} \\[2ex] \dfrac{\partial f}{\partial x}\dfrac{\partial x}{\partial u^2} + \dfrac{\partial f}{\partial y}\dfrac{\partial y}{\partial u^2} + \dfrac{\partial f}{\partial z}\dfrac{\partial z}{\partial u^2} \\[2ex] \dfrac{\partial f}{\partial x}\dfrac{\partial x}{\partial u^3} + \dfrac{\partial f}{\partial y}\dfrac{\partial y}{\partial u^3} + \dfrac{\partial f}{\partial z}\dfrac{\partial z}{\partial u^3} \end{bmatrix} = \begin{bmatrix} \dfrac{\partial x}{\partial u^1} & \dfrac{\partial y}{\partial u^1} & \dfrac{\partial z}{\partial u^1} \\[2ex] \dfrac{\partial x}{\partial u^2} & \dfrac{\partial y}{\partial u^2} & \dfrac{\partial z}{\partial u^2} \\[2ex] \dfrac{\partial x}{\partial u^3} & \dfrac{\partial y}{\partial u^3} & \dfrac{\partial z}{\partial u^3} \end{bmatrix} \begin{bmatrix} \dfrac{\partial f}{\partial x} \\[2ex] \dfrac{\partial f}{\partial y} \\[2ex] \dfrac{\partial f}{\partial z} \end{bmatrix}
$$

since the matrix on the right is not a Jacobian matrix. It is the *transpose* of the Jacobian matrix

$$
\mathbf{J}_{\Phi^{-1}} = \begin{bmatrix} \dfrac{\partial x^i}{\partial u^j} \end{bmatrix} = \begin{bmatrix} \dfrac{\partial}{\partial u^1}\begin{pmatrix} x \\ y \\ z \end{pmatrix}, & \dfrac{\partial}{\partial u^2}\begin{pmatrix} x \\ y \\ z \end{pmatrix}, & \dfrac{\partial}{\partial u^3}\begin{pmatrix} x \\ y \\ z \end{pmatrix} \end{bmatrix}
$$

so, instead we write

$$
\begin{bmatrix} \dfrac{\partial \hat{f}}{\partial u^1} & \dfrac{\partial \hat{f}}{\partial u^2} & \dfrac{\partial \hat{f}}{\partial u^3} \end{bmatrix} = \begin{bmatrix} \dfrac{\partial f}{\partial x} & \dfrac{\partial f}{\partial y} & \dfrac{\partial f}{\partial z} \end{bmatrix} \begin{bmatrix} \dfrac{\partial x}{\partial u^1} & \dfrac{\partial x}{\partial u^2} & \dfrac{\partial x}{\partial u^3} \\[2ex] \dfrac{\partial y}{\partial u^1} & \dfrac{\partial y}{\partial u^2} & \dfrac{\partial y}{\partial u^3} \\[2ex] \dfrac{\partial z}{\partial u^1} & \dfrac{\partial z}{\partial u^2} & \dfrac{\partial z}{\partial u^3} \end{bmatrix} \tag{32.19}
$$

and, using the Einstein summation convention, enunciate the *covariant law*

$$
v_{k'} = v_k \frac{\partial x^k}{\partial u^{k'}}
$$

for the transformation of gradient vectors v_k.

Contravariant vs. Covariant

The contravariant law for transforming tangent vectors is $v^{k'} = (\partial u^{k'}/\partial x^k)v^k$, whereas the covariant law for transforming gradient vectors is $v_{k'} = v_k(\partial x^k/\partial u^{k'})$. The tangent vector is written as a column vector, but the gradient vector is written as a row vector. The distinction between column and row is reinforced using upper indices on the components of a tangent vector but lower indices on the components of a gradient vector. Moreover, for the contravariant law, the components v^k are written on the right of the Jacobian matrix in agreement with the formalism for multiplying a column vector by a matrix. In contrast, for the covariant law, the components v_k are written on the left of the Jacobian matrix in conformity with the formalism for multiplying a row vector by a matrix. In each case, if the summation is viewed from the perspective of the Jacobian matrix, the contravariant law sums "uphill" against (contra) gravity, whereas the covariant law sums "downhill" with (co) gravity.

Tangent vectors are expressed via $\mathbf{e}_1, \mathbf{e}_2, \mathbf{e}_3$, a basis of vectors tangent to the coordinate curves; and components of tangent vectors are denoted with superscripts, as in v^k. Gradient vectors are expressed via $\mathbf{e}^1, \mathbf{e}^2, \mathbf{e}^3$, a basis of gradient vectors normal to the coordinate surfaces; and components of gradient vectors are denoted with subscripts, as in v_k.

The contravariant components v^k and the covariant components v_k are two distinct representations of the same geometric arrow. The two sets of basis vectors are mutually reciprocal and can be thought of as existing in the same space. Hence, it should be possible to convert a contravariant representation to a covariant one and vice versa. This is indeed the case and is known in the literature as "lowering and raising indices."

EXAMPLE 32.13   In Cartesian coordinates, the gradient of $f(x, y) = xy$ is $\nabla f = y\mathbf{i} + x\mathbf{j}$. To transform this vector to polar coordinates, we compute

$$[r \sin\theta \quad r \cos\theta]\mathbf{J}_{\Phi^{-1}} = [r \sin 2\theta \quad r^2 \cos 2\theta] = r \sin 2\theta \mathbf{e}^r + r^2 \cos 2\theta \mathbf{e}^\theta \qquad (32.20)$$

To corroborate this result, we solve

$$\mathbf{e}^r = \cos\theta\mathbf{i} + \sin\theta\mathbf{j} \quad \text{and} \quad \mathbf{e}^\theta = -\frac{\sin\theta}{r}\mathbf{i} + \frac{\cos\theta}{r}\mathbf{j} \qquad (32.21)$$

for

$$\mathbf{i} = \cos\theta\mathbf{e}^r - r\sin\theta\mathbf{e}^\theta \quad \text{and} \quad \mathbf{j} = \sin\theta\mathbf{e}^r + r\cos\theta\mathbf{e}^\theta \qquad (32.22)$$

and substitute into $\nabla f = y\mathbf{i} + x\mathbf{j} = r\sin\theta\mathbf{i} + r\cos\theta\mathbf{j}$. ❖

EXERCISES 32.5

1. Verify the calculations in (32.20).

2. Using (32.21) and (32.22), verify that $\nabla f = y\mathbf{i} + x\mathbf{j} = r\sin\theta\mathbf{i} + r\cos\theta\mathbf{j}$ becomes (32.20).

In Exercises 3–12:

(a) Obtain ∇f, the gradient of the given function $f(x, y)$.

(b) Convert ∇f to polar coordinates via the covariant transformation law as embodied in (32.19).

(c) Convert ∇f to polar coordinates by obtaining $\mathbf{i}$ and $\mathbf{j}$ in terms of $\mathbf{e}^r$ and $\mathbf{e}^\theta$, making the appropriate substitutions and changing coordinates in the expressions for $\frac{\partial f}{\partial x}$ and $\frac{\partial f}{\partial y}$.

(d) Obtain $\hat{f}(r, \theta) = f(x(r, \theta), y(r, \theta))$ and the derivatives $\frac{\partial \hat{f}}{\partial r}$ and $\frac{\partial \hat{f}}{\partial \theta}$.

(e) Since the gradient in polar coordinates is $\frac{\partial \hat{f}}{\partial r}\mathbf{e}^r + \frac{\partial \hat{f}}{\partial \theta}\mathbf{e}^\theta$, show that the derivatives obtained in part (d) are the components of the gradient in polar coordinates as found in parts (b) and (c).

(f) Obtain the *physical* components of the gradient in polar coordinates. (The physical components of a vector are the components with respect to *unit* basis vectors.)

3. $f(x, y) = 3 + 9xy - 7x^2$ **4.** $f(x, y) = 8 - 6x - 9y^2$

5. $f(x, y) = 3 + 7x + 7y - 8x^2 + 9xy^2$

6. $f(x, y) = 7xy - 3x^2 + 3y^3$

7. $f(x, y) = 2xy + 5x^2 + y^3$ **8.** $f(x, y) = 4x + 8xy - 4y^2$

9. $f(x, y) = 9x^2 - 3x^2y - 7x^3$

10. $f(x, y) = 1 + 6x + 9y - y^2 - 2x^2y + 2y^3$

11. $f(x, y) = 9x - 2xy + y^2 - 3x^2y - 4xy^2 + 8y^3$

12. $f(x, y) = 1 - 8xy^2 - 7x^3$

13. For the nonorthogonal coordinate system defined by $x = \frac{1}{2}(u^2 - v^2)$, $y = 2uv$, obtain $\{\mathbf{e}^u, \mathbf{e}^v\}$ from the rows of the Jacobian matrix $\mathbf{J}_{\Phi(u)} = [\mathbf{J}_{\Phi^{-1}}]^{-1}$.

In Exercises 14–23:

(a) Obtain ∇f, the gradient of the given function $f(x, y)$.

(b) Convert ∇f to uv-coordinates via the covariant transformation law as embodied in (32.19).

(c) Convert ∇f to uv-coordinates by obtaining $\mathbf{i}$ and $\mathbf{j}$ in terms of $\mathbf{e}^u$ and $\mathbf{e}^v$, making the appropriate substitutions and changing coordinates in the expressions for $\frac{\partial f}{\partial x}$ and $\frac{\partial f}{\partial y}$.

(d) Obtain $\hat{f}(u, v) = f(x(u, v), y(u, v))$ and the derivatives $\frac{\partial \hat{f}}{\partial u}$ and $\frac{\partial \hat{f}}{\partial v}$.

(e) Since the gradient in uv-coordinates is $\frac{\partial \hat{f}}{\partial u}\mathbf{e}^u + \frac{\partial \hat{f}}{\partial v}\mathbf{e}^v$, show that the derivatives obtained in part (d) are the components of the gradient in uv-coordinates as found in parts (b) and (c).

(f) Obtain the *physical* components of the gradient in uv-coordinates.

14. $f(x, y) = 6 + y + 10x^2y + 9x^3 + 12y^3$

15. $f(x, y) = 9x + 8y + y^2$ **16.** $f(x, y) = 5 - xy + 5x^2$

17. $f(x, y) = 4 - 2xy - 3y^3$ **18.** $f(x, y) = 4 + 8x^2y - xy^2$

19. $f(x, y) = 1 - 12x + 5y^2 + 8x^2 + 10x^2y - 3y^3$

20. $f(x, y) = 2x - 11y - 10xy$ **21.** $f(x, y) = y - 9x^2 - 10xy^2$

22. $f(x, y) = 2x - 8x^2 + y^3$ **23.** $f(x, y) = 6x - 4xy - 7y^2 + 7x^2$

24. For the parabolic coordinate system defined by $x = \frac{1}{2}(u^2 - v^2)$, $y = uv$, obtain $\{\mathbf{e}^u, \mathbf{e}^v\}$ from the rows of the Jacobian matrix $\mathbf{J}_{\Phi(u)} = [\mathbf{J}_{\Phi^{-1}}]^{-1}$.

In Exercises 25–27:

(a) Obtain ∇f, the gradient of the given function $f(x, y)$.

(b) Convert ∇f to parabolic coordinates via the covariant transformation law as embodied in (32.19).

(c) Convert ∇f to parabolic coordinates by obtaining $\mathbf{i}$ and $\mathbf{j}$ in terms of $\mathbf{e}^u$ and $\mathbf{e}^v$, making the appropriate substitutions and changing coordinates in the expressions for $\frac{\partial f}{\partial x}$ and $\frac{\partial f}{\partial y}$.

(d) Obtain $\hat{f}(u, v) = f(x(u, v), y(u, v))$ and the derivatives $\frac{\partial \hat{f}}{\partial u}$ and $\frac{\partial \hat{f}}{\partial v}$.

(e) Since the gradient in parabolic coordinates is $\frac{\partial \hat{f}}{\partial u}\mathbf{e}^u + \frac{\partial \hat{f}}{\partial v}\mathbf{e}^v$, show that the derivatives obtained in part (d) are the components of the gradient in parabolic coordinates as found in parts (b) and (c).

(f) Obtain the physical components of the gradient in parabolic coordinates.

25. $f(x, y) = 5y + xy - 2x^2 + 3x^3 + 3xy^2 + 6y^3$

26. $f(x, y) = 10 + 4x - 3y^2$ 27. $f(x, y) = 2 - 9xy + 7x^3$

28. For the tangent coordinate system defined by $x = \frac{u}{u^2+v^2}$, $y = \frac{v}{u^2+v^2}$, obtain $\{\mathbf{e}^u, \mathbf{e}^v\}$ from the rows of the Jacobian matrix $\mathbf{J}_{\Phi(u)} = [\mathbf{J}_{\Phi^{-1}}]^{-1}$.

In Exercises 29–31:

(a) Obtain ∇f, the gradient of the given function $f(x, y)$.

(b) Convert ∇f to tangent coordinates via the covariant transformation law as embodied in (32.19).

(c) Convert ∇f to tangent coordinates by obtaining $\mathbf{i}$ and $\mathbf{j}$ in terms of $\mathbf{e}^u$ and $\mathbf{e}^v$, making the appropriate substitutions and changing coordinates in the expressions for $\frac{\partial f}{\partial x}$ and $\frac{\partial f}{\partial y}$.

(d) Obtain $\hat{f}(u, v) = f(x(u, v), y(u, v))$ and the derivatives $\frac{\partial \hat{f}}{\partial u}$ and $\frac{\partial \hat{f}}{\partial v}$.

(e) Since the gradient in tangent coordinates is $\frac{\partial \hat{f}}{\partial u}\mathbf{e}^u + \frac{\partial \hat{f}}{\partial v}\mathbf{e}^v$, show that the derivatives obtained in part (d) are the components of the gradient in tangent coordinates as found in parts (b) and (c).

(f) Obtain the physical components of the gradient in tangent coordinates.

29. $f(x, y) = 7 + 7xy + 7y^2 + 9x^3 - 3xy^2$

30. $f(x, y) = x + 10y - 10xy$ 31. $f(x, y) = x^2 - 4xy - 10y^3$

32. For the elliptic coordinate system defined by $x = \cosh u \cos v$, $y = \sinh u \sin v$, obtain $\{\mathbf{e}^u, \mathbf{e}^v\}$ from the rows of the Jacobian matrix $\mathbf{J}_{\Phi(u)} = [\mathbf{J}_{\Phi^{-1}}]^{-1}$.

In Exercises 33–35:

(a) Obtain ∇f, the gradient of the given function $f(x, y)$.

(b) Convert ∇f to elliptic coordinates via the covariant transformation law as embodied in (32.19).

(c) Convert ∇f to elliptic coordinates by obtaining $\mathbf{i}$ and $\mathbf{j}$ in terms of $\mathbf{e}^u$ and $\mathbf{e}^v$, making the appropriate substitutions and changing coordinates in the expressions for $\frac{\partial f}{\partial x}$ and $\frac{\partial f}{\partial y}$.

(d) Obtain $\hat{f}(u, v) = f(x(u, v), y(u, v))$ and the derivatives $\frac{\partial \hat{f}}{\partial u}$ and $\frac{\partial \hat{f}}{\partial v}$.

(e) Since the gradient in elliptic coordinates is $\frac{\partial \hat{f}}{\partial u}\mathbf{e}^u + \frac{\partial \hat{f}}{\partial v}\mathbf{e}^v$, show that the derivatives obtained in part (d) are the components of the gradient in elliptic coordinates as found in parts (b) and (c).

(f) Obtain the physical components of the gradient in elliptic coordinates.

33. $f(x, y) = 5 - 11x + 4y + 11x^3 - 11x^2y + 5y^3$

34. $f(x, y) = 11x - 7y + 8xy$ 35. $f(x, y) = 9 - 2x^2 - 9x^2y$

36. For the coordinate system defined by $x = 5u^2 + 3v^2$, $y = 4u^2 - 9v^2$, obtain $\{\mathbf{e}^u, \mathbf{e}^v\}$ from the rows of the Jacobian matrix $\mathbf{J}_{\Phi(u)} = [\mathbf{J}_{\Phi^{-1}}]^{-1}$.

In Exercises 37–39:

(a) Obtain ∇f, the gradient of the given function $f(x, y)$.

(b) Convert ∇f to uv-coordinates via the covariant transformation law as embodied in (32.19).

(c) Convert ∇f to uv-coordinates by obtaining $\mathbf{i}$ and $\mathbf{j}$ in terms of $\mathbf{e}^u$ and $\mathbf{e}^v$, making the appropriate substitutions and changing coordinates in the expressions for $\frac{\partial f}{\partial x}$ and $\frac{\partial f}{\partial y}$.

(d) Obtain $\hat{f}(u, v) = f(x(u, v), y(u, v))$ and the derivatives $\frac{\partial \hat{f}}{\partial u}$ and $\frac{\partial \hat{f}}{\partial v}$.

(e) Since the gradient in uv-coordinates is $\frac{\partial \hat{f}}{\partial u}\mathbf{e}^u + \frac{\partial \hat{f}}{\partial v}\mathbf{e}^v$, show that the derivatives obtained in part (d) are the components of the gradient in uv-coordinates as found in parts (b) and (c).

(f) Obtain the physical components of the gradient in uv-coordinates.

37. $f(x, y) = 11y + xy - 7x^2 + 9xy^2$ 38. $f(x, y) = 9y + 3x + 8y^2$

39. $f(x, y) = 4xy + 11y^2 - 2x^2y$

40. For the bipolar coordinate system defined by $x = \frac{\sinh v}{\cosh v - \cos u}$, $y = \frac{\sin u}{\cosh v - \cos u}$, obtain $\{\mathbf{e}^u, \mathbf{e}^v\}$ from the rows of the Jacobian matrix $\mathbf{J}_{\Phi(u)} = [\mathbf{J}_{\Phi^{-1}}]^{-1}$.

In Exercises 41–43:

(a) Obtain ∇f, the gradient of the given function $f(x, y)$.

(b) Convert ∇f to bipolar coordinates via the covariant transformation law as embodied in (32.19).

(c) Convert ∇f to bipolar coordinates by obtaining $\mathbf{i}$ and $\mathbf{j}$ in terms of $\mathbf{e}^u$ and $\mathbf{e}^v$, making the appropriate substitutions and changing coordinates in the expressions for $\frac{\partial f}{\partial x}$ and $\frac{\partial f}{\partial y}$.

(d) Obtain $\hat{f}(u, v) = f(x(u, v), y(u, v))$ and the derivatives $\frac{\partial \hat{f}}{\partial u}$ and $\frac{\partial \hat{f}}{\partial v}$.

(e) Since the gradient in bipolar coordinates is $\frac{\partial \hat f}{\partial u}\mathbf{e}^u + \frac{\partial \hat f}{\partial v}\mathbf{e}^v$, show that the derivatives obtained in part (d) are the components of the gradient in bipolar coordinates as found in parts (b) and (c).

(f) Obtain the physical components of the gradient in bipolar coordinates.

41. $f(x, y) = 4 - 6x^2 - 2x^2 y - 5xy^2 + 3y^3$

42. $f(x, y) = 8y + 5y^2 + 11x^2$ **43.** $f(x, y) = 4x - 6xy^2 - y^3$

44. For the skew coordinate system defined by $x = 3u + 2v$, $y = 5u - 7v$, obtain $\{\mathbf{e}^u, \mathbf{e}^v\}$ from the rows of the Jacobian matrix $\mathbf{J}_{\Phi(u)} = [\mathbf{J}_{\Phi^{-1}}]^{-1}$.

In Exercises 45–47:

(a) Obtain ∇f, the gradient of the given function $f(x, y)$.

(b) Convert ∇f to skew coordinates via the covariant transformation law as embodied in (32.19).

(c) Convert ∇f to skew coordinates by obtaining $\mathbf{i}$ and $\mathbf{j}$ in terms of $\mathbf{e}^u$ and $\mathbf{e}^v$, making the appropriate substitutions and changing coordinates in the expressions for $\frac{\partial f}{\partial x}$ and $\frac{\partial f}{\partial y}$.

(d) Obtain $\hat f(u, v) = f(x(u, v), y(u, v))$ and the derivatives $\frac{\partial \hat f}{\partial u}$ and $\frac{\partial \hat f}{\partial v}$.

(e) Since the gradient in skew coordinates is $\frac{\partial \hat f}{\partial u}\mathbf{e}^u + \frac{\partial \hat f}{\partial v}\mathbf{e}^v$, show that the derivatives obtained in part (d) are the components of the gradient in skew coordinates as found in parts (b) and (c).

(f) Obtain the physical components of the gradient in skew coordinates.

45. $f(x, y) = 10x + 6y - 4xy + 12x^2 - 4x^2 y$

46. $f(x, y) = 9x - 3y - 5xy$ **47.** $f(x, y) = 4 + 3xy^2 - x^3$

48. For the coordinate system defined by $x = 2u^2 - 5v$, $y = 3u + 4v^2$, obtain $\{\mathbf{e}^u, \mathbf{e}^v\}$ from the rows of the Jacobian matrix $\mathbf{J}_{\Phi(u)} = [\mathbf{J}_{\Phi^{-1}}]^{-1}$.

In Exercises 49–51:

(a) Obtain ∇f, the gradient of the given function $f(x, y)$.

(b) Convert ∇f to uv-coordinates via the covariant transformation law as embodied in (32.19).

(c) Convert ∇f to uv-coordinates by obtaining $\mathbf{i}$ and $\mathbf{j}$ in terms of $\mathbf{e}^u$ and $\mathbf{e}^v$, making the appropriate substitutions and changing coordinates in the expressions for $\frac{\partial f}{\partial x}$ and $\frac{\partial f}{\partial y}$.

(d) Obtain $\hat f(u, v) = f(x(u, v), y(u, v))$ and the derivatives $\frac{\partial \hat f}{\partial u}$ and $\frac{\partial \hat f}{\partial v}$.

(e) Since the gradient in uv-coordinates is $\frac{\partial \hat f}{\partial u}\mathbf{e}^u + \frac{\partial \hat f}{\partial v}\mathbf{e}^v$, show that the derivatives obtained in part (d) are the components of the gradient in uv-coordinates as found in parts (b) and (c).

(f) Obtain the physical components of the gradient in uv-coordinates.

49. $f(x, y) = 8x + 3y - 4xy^2$ **B50.** $f(x, y) = 12y + 7y^2 + 12x^2$

51. $f(x, y) = 10x + 8xy + 11y^2$

Chapter Review

1. What is an orthogonal matrix? What are its properties?

2. Let $\mathbf{x} = 2\mathbf{i} + 3\mathbf{j}$, and define $\mathbf{e}_1 = 3\mathbf{i} - 2\mathbf{j}$, $\mathbf{e}_2 = 4\mathbf{i} + 5\mathbf{j}$.

(a) Find $'\mathbf{x}$, the representation of $\mathbf{x}$ with respect to the new basis $\{\mathbf{e}_1, \mathbf{e}_2\}$.

(b) If $\mathbf{X}$ and $'\mathbf{X}$ are the column-vector representations of $\mathbf{x}$ and $'\mathbf{x}$, respectively, obtain the transition matrix M for which $'\mathbf{X} = M\mathbf{X}$ holds.

3. Given $M = \begin{bmatrix} \cos a & -\sin a \\ \sin a & \cos a \end{bmatrix}$, where $a = \frac{\pi}{5}$ and $\mathbf{x} = \mathbf{i} + 2\mathbf{j}$:

(a) Show that M is an orthogonal matrix.

(b) If $\mathbf{y} = M\mathbf{x}$, show $\|\mathbf{y}\| = \|\mathbf{x}\|$.

(c) Show that $\mathbf{y}$ is the counterclockwise rotation of $\mathbf{x}$ through the angle a.

(d) Let the columns of M define new basis vectors $\mathbf{e}_1$ and $\mathbf{e}_2$, respectively. Express $\mathbf{x}$ with respect to this new basis, and show that $'\mathbf{x} = \mathbf{y}$.

(e) Show that $\mathbf{e}_1$ and $\mathbf{e}_2$ are, respectively, the counterclockwise rotation of $\mathbf{i}$ and $\mathbf{j}$, through the angle a.

4. Obtain an orthogonal matrix R for which the product $\mathbf{y} = R\mathbf{x}$ is the result of applying to $\mathbf{x}$ a sequence of three counterclockwise rotations about the coordinate axes through angles of $\frac{\pi}{5}, \frac{\pi}{8}$, and $\frac{\pi}{6}$, respectively.

5. If R is the matrix in Question 4:

(a) Show that $\lambda = 1$ is the only real eigenvalue and that the associated eigenvector is the axis of rotation for a single rotation moving $\mathbf{x}$ to $\mathbf{y}$.

(b) Obtain an equation for the plane whose normal is the eigenvector in part (a).

(c) Find a vector $\mathbf{z}$ that lies in the plane determined in part (b), and show that $R\mathbf{z}$ remains in that plane.

(d) Find the angle between $\mathbf{z}$ and $R\mathbf{z}$.

6. What is meant by the term *real symmetric matrix,* and what are the properties of such a matrix?

7. Give an example of a real symmetric 3×3 matrix and show that its eigenvalues are real and that its eigenvectors are orthogonal.

8. The matrix $A = \begin{bmatrix} i & 2 \\ 2 & 1 \end{bmatrix}$ is symmetric but not real. Show it does not have real eigenvalues.

9. Sketch the proof that the eigenvalues of a real symmetric matrix are real.

10. Sketch the proof that, for a real symmetric matrix, the eigenvectors of distinct eigenvalues are orthogonal.

11. What is *orthogonal diagonalization* of a matrix? Give an example.

12. Verify that the columns of $M = \begin{bmatrix} \frac{1}{\sqrt{3}} & \frac{1}{\sqrt{2}} \\ \frac{1}{\sqrt{3}} & -\frac{1}{\sqrt{2}} \\ \frac{1}{\sqrt{3}} & 0 \end{bmatrix}$ are orthogonal and each of unit length. Show that $M^\mathsf{T} M$ is the 2×2 identity matrix but that $M M^\mathsf{T}$ is *not* an identity matrix.

13. If $\mathbf{e}_1 = 2\mathbf{i} + 3\mathbf{j}$ and $\mathbf{e}_2 = -5\mathbf{i} + 4\mathbf{j}$, obtain $\{\mathbf{e}^1, \mathbf{e}^2\}$, the basis reciprocal to $\{\mathbf{e}_1, \mathbf{e}_2\}$.

14. For polar coordinates, the forward map Φ is given by $r = \sqrt{x^2 + y^2}$, $\theta = \arctan \frac{y}{x}$, whereas the inverse map Φ^{-1} is given by $x = r\cos\theta$, $y = r\sin\theta$.

  (a) Obtain the Jacobian matrix $\mathbf{J}_{\Phi(\mathbf{x})}$.

  (b) Obtain the Jacobian matrix $\mathbf{J}_{\Phi^{-1}(\mathbf{u})}$.

  (c) Show that $\mathbf{J}_{\Phi(\mathbf{u})} = [\mathbf{J}_{\Phi^{-1}(\mathbf{u})}]^{-1}$.

  (d) Show that the columns of $\mathbf{J}_{\Phi^{-1}(\mathbf{u})}$ are the natural tangent basis vectors.

  (e) Show that the rows of $\mathbf{J}_{\Phi(\mathbf{x})}$ are the natural gradient basis vectors.

  (f) Show that the sets of basis vectors in parts (d) and (e) are reciprocal vectors. *Hint:* Use part (c).

15. Let $\mathbf{R}(p) = p^2\mathbf{i} + p^3\mathbf{j}$ define a plane curve in the Cartesian plane.

  (a) Obtain $\mathbf{V} = \mathbf{R}'(p)$, the tangent vector along the given curve.

  (b) Evaluate $\mathbf{J}_{\Phi(\mathbf{x})}$, determined for polar coordinates in Question 14(a), along the given curve.

  (c) Use the contravariant transformation law for tangent vectors to obtain $'\mathbf{V}$, the representation of $\mathbf{V}$ in polar coordinates. Schematically, this calculation can be represented as $'\mathbf{V} = \mathbf{J}_{\Phi(\mathbf{x})}\mathbf{V}$, where $\mathbf{V}$ is written as a column vector.

  (d) Express the basis vectors $\{\mathbf{i}, \mathbf{j}\}$ in terms of $\{\mathbf{e}_r, \mathbf{e}_\theta\}$, the natural tangent basis vectors for polar coordinates.

  (e) Use the expressions in part (d) to transform $\mathbf{V}$ to $'\mathbf{V}$, and compare to the result in part (c).

16. Let $f(x, y) = x^2 y^2$.

  (a) Obtain the row vector $\mathbf{X} = \nabla f$, the gradient of f.

  (b) Use the covariant transformation law for gradient vectors to obtain $'\mathbf{X}$, the representation of ∇f in polar coordinates. Schematically, this calculation can be represented as $'\mathbf{X} = \mathbf{X}\mathbf{J}_{\Phi^{-1}(\mathbf{u})}$, provided the variables are all changed to polar coordinates.

  (c) Using the result in Question 15(d), change $\nabla f = f_x\mathbf{i} + f_y\mathbf{j}$ to polar coordinates and compare to $'\mathbf{X}$ computed in part (b).

Chapter *33*

Matrix Computations

I N T R O D U C T I O N   Basic matrix arithmetic and the eigenvalue problem were covered in Chapter 12. In the present chapter, we continue with a discussion of common, but more advanced, matrix and vector manipulations. Vector projection, and its cousin, the Gram–Schmidt orthogonalization process, the matrix norm, and its relation to the eigenvalue problem, the least-squares problem for fitting a curve to a set of data, and a study of quadratic forms are the framework on which we hang the details of more sophisticated matrix computations than were needed in Chapter 12.

After a brief review of Gaussian elimination, determinants, Cramer's rule, and the eigenvalue problem for vectors, carried out mainly by a set of review exercises, we show how to obtain the components of a vector along, and orthogonal to, a given direction. These components are called the projections on, and orthogonal to, a given vector. This operation of projecting a vector onto a given direction is cast into a matrix formalism that then generalizes to a projection of a vector onto, and orthogonal to, a subspace spanned by a collection of vectors.

This ability to project vectors onto a subspace and perpendicular to the subspace is at the heart of the Gram–Schmidt process for transforming a set of nonorthogonal vectors into an "equivalent" set of mutually orthogonal vectors. Interestingly enough, if the given set of vectors are not linearly independent, the Gram–Schmidt process will yield orthogonal vectors that are linearly independent.

Quadratic forms are the basis for the second derivative test of optimality for functions of several variables. The test for a function of two variables, learned in multivariable calculus, is explained and then generalized to the case of n variables using the theory of quadratic forms.

The length or magnitude of a vector is a single number measuring the "size" of the vector. We likewise assign a "sizing" number to a matrix and call it a matrix norm. There is more than one way to assign such a number, so we examine the Frobenius norm as well as the 1-norm, the 2-norm, and the infinity-norm for a matrix. Computing a 2-norm for a matrix requires solving an eigenvalue problem.

Finally, we conclude with an analytic and algebraic approach to the *least squares* fit of the general linear model to data. The overdetermined set of algebraic equations that is at the heart of this problem is solved both analytically and algebraically for the solution that minimizes the sum of squares of the error between the data points and the fitting curve. A geometric interpretation of this problem illustrates the role of the fundamental subspaces of a matrix in determining the solution of a system of linear equations.

33.1 Summary

Review

Chapter 33 details a variety of matrix manipulations and calculations. Many of these have been seen earlier in the context of a matrix formulation for systems of linear ordinary differential equations. Thus, we merely recall that we have already seen basic matrix arithmetic in Section 12.3; the determinant and Cramer's rule in Section 12.4; Gaussian elimination in Sections 12.5 and 12.6; matrix inverse in Section 12.7; the eigenvalue problem in Sections 12.10, 12.12, 12.14, and 12.15; and the matrix exponential in Sections 12.8 and 12.9.

The following exercises will serve as a review of most of the concepts and calculations seen in earlier parts of the text.

EXERCISES 33.1

In Exercises 1–5, compute the following quantities.

(a) $2A + 3B$, $A(5B - 7C)$, ABC

(b) $AB - BA$, $AC - CA$, $BC - CB$

(c) $I + A + A^2 + A^3$

(d) $\det AC$, $\det(AB)^{\mathrm{T}} - \det A \det B$, $\det A^{-1} - \frac{1}{\det A}$

(e) $A^{-1}BA$, $A^{\mathrm{T}}B$

(f) $(AB)^{-1} - B^{-1}A^{-1}$, $(AC)^{\mathrm{T}} - C^{\mathrm{T}}A^{\mathrm{T}}$

(g) $\mathrm{rref}(A)$, $\mathrm{rref}(B)$, $\mathrm{rref}(C)$

1. $A = \begin{bmatrix} -6 & 8 \\ 10 & 1 \end{bmatrix}$, $B = \begin{bmatrix} -11 & 1 \\ -2 & 6 \end{bmatrix}$, $C = \begin{bmatrix} 9 & -3 \\ -7 & 1 \end{bmatrix}$

2. $A = \begin{bmatrix} -5 & -2 & 12 \\ -2 & 4 & -1 \\ -5 & 12 & -1 \end{bmatrix}$, $B = \begin{bmatrix} 11 & 4 & -3 \\ 11 & -6 & 12 \\ 4 & 11 & 5 \end{bmatrix}$, $C = \begin{bmatrix} 4 & 6 & -12 \\ -1 & 1 & 1 \\ 8 & 7 & 3 \end{bmatrix}$

3. $A = \begin{bmatrix} 4 & 8 & 12 \\ -3 & -2 & -5 \\ 5 & 1 & -10 \end{bmatrix}$, $B = \begin{bmatrix} 12 & -2 & -1 \\ 11 & -9 & 11 \\ 8 & 11 & 2 \end{bmatrix}$, $C = \begin{bmatrix} 0 & -5 & 12 \\ 7 & -9 & 3 \\ -7 & -5 & -11 \end{bmatrix}$

4. $A = \begin{bmatrix} 0 & 10 & 1 \\ 5 & 10 & 11 \\ -10 & -10 & 12 \end{bmatrix}$, $B = \begin{bmatrix} -12 & -5 & -5 \\ -8 & -10 & -8 \\ -2 & 8 & 11 \end{bmatrix}$, $C = \begin{bmatrix} 1 & -11 & -12 \\ 12 & 11 & -12 \\ 7 & -11 & 12 \end{bmatrix}$

5. $A = \begin{bmatrix} 10 & 1 \\ 3 & 7 \end{bmatrix}$, $B = \begin{bmatrix} -1 & -1 \\ -3 & 4 \end{bmatrix}$, $C = \begin{bmatrix} -6 & 8 \\ 11 & 12 \end{bmatrix}$

For the system $A\mathbf{x} = \mathbf{b}$ given in each of Exercises 6–10:

(a) Obtain $\mathbf{x}$ using Cramer's rule.

(b) Obtain $\mathbf{x}$ using Gauss elimination and back substitution.

(c) Obtain $\mathbf{x}$ by computing and using A^{-1}.

(d) Obtain $\mathbf{x}$ by Gauss–Jordan reduction of $[A, \mathbf{b}]$ to the reduced row-echelon form.

6. $A = \begin{bmatrix} 6 & 7 \\ -11 & -12 \end{bmatrix}$, $\mathbf{b} = \begin{bmatrix} -11 \\ 9 \end{bmatrix}$ **7.** $A = \begin{bmatrix} -6 & 8 \\ -2 & -7 \end{bmatrix}$, $\mathbf{b} = \begin{bmatrix} -2 \\ -12 \end{bmatrix}$

8. $A = \begin{bmatrix} 12 & -2 & -8 \\ 7 & -11 & 0 \\ 9 & -2 & 11 \end{bmatrix}$, $\mathbf{b} = \begin{bmatrix} 6 \\ -5 \\ -1 \end{bmatrix}$

9. $A = \begin{bmatrix} 10 & -2 & 9 \\ 8 & 10 & -7 \\ 1 & 8 & -3 \end{bmatrix}$, $\mathbf{b} = \begin{bmatrix} 5 \\ 3 \\ 11 \end{bmatrix}$

10. $A = \begin{bmatrix} -7 & -9 & 4 \\ -6 & 12 & -3 \\ -10 & 3 & -2 \end{bmatrix}$, $\mathbf{b} = \begin{bmatrix} 12 \\ -1 \\ -4 \end{bmatrix}$

For the matrix A in each of Exercises 11–20:

(a) Obtain $\det A$, and obtain A^{-1} if $\det A \neq 0$.

(b) Obtain $\det A^{\mathrm{T}}$, and obtain $[A^{\mathrm{T}}]^{-1}$ if $\det A^{\mathrm{T}} \neq 0$.

(c) Obtain the eigenvalues and corresponding eigenvectors.

(d) Show that each eigenpair $(\lambda, \mathbf{v})$ satisfies $A\mathbf{v} = \lambda\mathbf{v}$.

(e) Obtain e^{At}.

11. $\begin{bmatrix} -7 & 3 \\ 4 & -6 \end{bmatrix}$ **12.** $\begin{bmatrix} 6 & -5 \\ 10 & -9 \end{bmatrix}$ **13.** $\begin{bmatrix} -7 & 9 \\ 6 & 8 \end{bmatrix}$

14. $\begin{bmatrix} 3 & 9 \\ 7 & 5 \end{bmatrix}$ **15.** $\begin{bmatrix} 6 & -3 \\ -2 & 1 \end{bmatrix}$ **16.** $\begin{bmatrix} 4 & 9 & 3 \\ 6 & 1 & 2 \\ -12 & -1 & -11 \end{bmatrix}$

17. $\begin{bmatrix} -12 & 10 & 3 \\ -6 & 4 & 3 \\ -2 & -4 & 7 \end{bmatrix}$ **18.** $\begin{bmatrix} -9 & 6 & 12 \\ 6 & -2 & 5 \\ 8 & 6 & -5 \end{bmatrix}$

19. $\begin{bmatrix} 11 & 2 & -2 \\ -11 & -5 & 9 \\ -11 & -3 & 7 \end{bmatrix}$ **20.** $\begin{bmatrix} 2 & -7 & 7 \\ -9 & 4 & 2 \\ 9 & 7 & -9 \end{bmatrix}$

33.2   Projections

Resolution of a Vector into Orthogonal Components

If neither of the vectors **B** and **A** are **0**, the zero vector, we can express **B** in terms of its vector components along **A** and perpendicular to **A**. Each of these vector components are called *projections*, either $\mathbf{B_A}$, the projection of **B** on or along **A**, and $\mathbf{B_{\perp A}}$, the projection (or component) of **B** orthogonal to **A**.

PROJECTION OF B ALONG A   The vector $\mathbf{B_A}$ ("**B** on **A**"), the projection of **B** on or along **A**, is given by

$$\mathbf{B_A} = \frac{\mathbf{B} \cdot \mathbf{A}}{\mathbf{A} \cdot \mathbf{A}} \mathbf{A} \qquad (33.1)$$

As an example, consider the two vectors $\mathbf{A} = 2\mathbf{i}$ and $\mathbf{B} = \mathbf{i} + \mathbf{j}$, graphed along with $\mathbf{B_A}$ in Figure 33.1. From the figure, the angle between **A** and **B** is $\frac{\pi}{4}$, so the length of $\mathbf{B_A}$ is found to be $\|\mathbf{B}\| \cos\frac{\pi}{4} = \sqrt{2}\frac{1}{\sqrt{2}} = 1$. A vector of this length in the direction of **A** is $1(\frac{\mathbf{A}}{\|\mathbf{A}\|}) = \mathbf{i}$. Alternatively, $\mathbf{B_A} = \frac{\mathbf{B} \cdot \mathbf{A}}{\mathbf{A} \cdot \mathbf{A}} \mathbf{A} = \frac{2}{4}(2\mathbf{i}) = \mathbf{i}$.

PROJECTION OF B ORTHOGONAL TO A   The projection, or component, of **B** perpendicular to **A** is $\mathbf{B_{\perp A}} = \mathbf{B} - \mathbf{B_A} = (\mathbf{i} + \mathbf{j}) - \mathbf{i} = \mathbf{j}$, as shown in Figure 33.1.

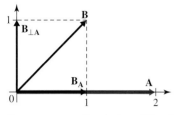

FIGURE 33.1   Resolution of $\mathbf{B} = \mathbf{i} + \mathbf{j}$ into components along and perpendicular to $\mathbf{A} = 2\mathbf{i}$

Projections—A Matrix Formulation

We next recast in matrix form our work on projections along and perpendicular to a vector.

PROJECTION OF B ALONG A   The matrix

$$P = \frac{\mathbf{A}\mathbf{A}^{\mathrm{T}}}{\mathbf{A}^{\mathrm{T}}\mathbf{A}} \qquad (33.2)$$

will project any vector onto the vector **A**. When applied to a vector **B**, we can write

$$P\mathbf{B} = \frac{\mathbf{A}\mathbf{A}^{\mathrm{T}}}{\mathbf{A}^{\mathrm{T}}\mathbf{A}}\mathbf{B} = \frac{(\mathbf{A}^{\mathrm{T}}\mathbf{B})}{\mathbf{A}^{\mathrm{T}}\mathbf{A}}\mathbf{A} = \frac{\mathbf{B} \cdot \mathbf{A}}{\mathbf{A} \cdot \mathbf{A}}\mathbf{A}$$

Remember, **A** is a vector, so $\mathbf{A}^{\mathrm{T}}\mathbf{B} = \mathbf{A} \cdot \mathbf{B}$ is a scalar that commutes with **A**, thereby validating the middle fraction.

EXAMPLE 33.1   The matrix that projects any vector **B** onto the vector $\mathbf{A} = \mathbf{i} + 2\mathbf{j} + 2\mathbf{k}$ is

$$P = \frac{\mathbf{A}\mathbf{A}^{\mathrm{T}}}{\mathbf{A}^{\mathrm{T}}\mathbf{A}} = \frac{\begin{bmatrix} 1 \\ 2 \\ 2 \end{bmatrix}\begin{bmatrix} 1 & 2 & 2 \end{bmatrix}}{\begin{bmatrix} 1 & 2 & 2 \end{bmatrix}\begin{bmatrix} 1 \\ 2 \\ 2 \end{bmatrix}} = \frac{1}{9}\begin{bmatrix} 1 & 2 & 2 \\ 2 & 4 & 4 \\ 2 & 4 & 4 \end{bmatrix} \qquad (33.3)$$

Applied to the vector $\mathbf{B} = \mathbf{i} + \mathbf{j} + \mathbf{k}$ we get $P\mathbf{B} = \mathbf{B_A} = \frac{5}{9}(\mathbf{i} + 2\mathbf{j} + 2\mathbf{k})$. Of course, $\mathbf{B_A}$ is also given by

$$\frac{\mathbf{B} \cdot \mathbf{A}}{\mathbf{A} \cdot \mathbf{A}}\mathbf{A} = \frac{\begin{bmatrix} 1 \\ 1 \\ 1 \end{bmatrix} \cdot \begin{bmatrix} 1 \\ 2 \\ 2 \end{bmatrix}}{\begin{bmatrix} 1 \\ 2 \\ 2 \end{bmatrix} \cdot \begin{bmatrix} 1 \\ 2 \\ 2 \end{bmatrix}}\begin{bmatrix} 1 \\ 2 \\ 2 \end{bmatrix} = \frac{5}{9}\begin{bmatrix} 1 \\ 2 \\ 2 \end{bmatrix} = \frac{5}{9}(\mathbf{i} + 2\mathbf{j} + 2\mathbf{k}) \qquad \text{❖}$$

PROJECTION OF B ORTHOGONAL TO A   The projection, or component, of **B** perpendicular to **A** is given by

$$\mathbf{B}_{\perp\mathbf{A}} = \mathbf{B} - \mathbf{B}_{\mathbf{A}} = \mathbf{B} - P\mathbf{B} = (I - P)\mathbf{B} = \tfrac{1}{9}(4\mathbf{i} - \mathbf{j} - \mathbf{k})$$

since

$$(I - P)\mathbf{B} = \begin{bmatrix} \frac{8}{9} & -\frac{2}{9} & -\frac{2}{9} \\ -\frac{2}{9} & \frac{5}{9} & -\frac{4}{9} \\ -\frac{2}{9} & -\frac{4}{9} & \frac{5}{9} \end{bmatrix} \begin{bmatrix} 1 \\ 1 \\ 1 \end{bmatrix} = \frac{1}{9} \begin{bmatrix} 4 \\ -1 \\ -1 \end{bmatrix}$$

In general, if P projects onto the vector **A**, then $I - P$ projects onto the direction orthogonal to **A**. Figure 33.2 shows the vectors **B**, **A**, $\mathbf{B}_{\mathbf{A}}$, and $\mathbf{B}_{\perp\mathbf{A}}$.

PROJECTION MATRICES   A matrix P is called a *projection matrix* if it is symmetric and $P^2 = P$. (If $P\mathbf{B}$ is the projection of **B** onto **A**, then $P(P\mathbf{B})$, the projection of the projection onto **A**, is still $P\mathbf{B}$, so $P^2 = P$.) The matrix P in (33.3) is symmetric, and a computation shows $P^2 = P$. Hence, it is a projection matrix.

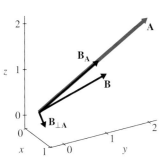

FIGURE 33.2   Resolution of **B** into components along and perpendicular to **A** in Example 33.1

Projection onto a Subspace

Next, we consider the concept of projecting a vector onto a subspace generated by the basis vectors $\mathbf{A}_1, \ldots, \mathbf{A}_k$. For example, the projection of **B** onto the subspace spanned by $\mathbf{A}_1$ and $\mathbf{A}_2$ would be $\mathbf{B}_{[\mathbf{A}_1,\mathbf{A}_2]} = \mathbf{B}_{[A]}$, the component of **B** lying in the plane formed by $\mathbf{A}_1$ and $\mathbf{A}_2$.

By calling the vectors $\mathbf{A}_1, \ldots, \mathbf{A}_k$ a *basis*, we are stating that they span the subspaceall possible linear combinations of the basis vectors) and that they are linearly independent. This last point is very important.

Let A be the matrix whose columns are the vectors $\mathbf{A}_1, \ldots, \mathbf{A}_k$ so that $A = [\mathbf{A}_1, \ldots, \mathbf{A}_k]$. A recipe for a matrix P that projects vectors onto this subspace is

$$P = A(A^{\mathrm{T}}A)^{-1}A^{\mathrm{T}} \tag{33.4}$$

The matrix in (33.2) can be written as $\mathbf{A}(\mathbf{A}^{\mathrm{T}}\mathbf{A})^{-1}\mathbf{A}^{\mathrm{T}}$ because $\mathbf{A}^{\mathrm{T}}\mathbf{A}$ is a scalar. This suggests (33.4) is an appropriate generalization of (33.2).

EXAMPLE 33.2   Form a matrix that will project vectors onto the subspace spanned by the vectors $\mathbf{A}_1 = \mathbf{i} + \mathbf{j}$ and $\mathbf{A}_2 = \mathbf{j}$, two vectors that constitute a basis for the *xy*-plane. To form P, a matrix that projects onto this plane, write

$$A = \begin{bmatrix} 1 & 0 \\ 1 & 1 \\ 0 & 0 \end{bmatrix} \qquad P = \begin{bmatrix} 1 & 0 \\ 1 & 1 \\ 0 & 0 \end{bmatrix} \begin{bmatrix} 2 & 1 \\ 1 & 1 \end{bmatrix}^{-1} \begin{bmatrix} 1 & 1 & 0 \\ 0 & 1 & 0 \end{bmatrix} = \begin{bmatrix} 1 & 0 & 0 \\ 0 & 1 & 0 \\ 0 & 0 & 0 \end{bmatrix}$$

The projection of $\mathbf{B} = a\mathbf{i} + b\mathbf{j} + c\mathbf{k}$ onto the plane spanned by $\mathbf{A}_1$ and $\mathbf{A}_2$ is $P\mathbf{B} = a\mathbf{i} + b\mathbf{j}$, a result that should not surprise us. ❖

EXAMPLE 33.3   To project the vector **B** onto the subspace spanned by $\mathbf{A}_1$ and $\mathbf{A}_2$, where

$$\mathbf{B} = \begin{bmatrix} 1 \\ 2 \\ 3 \\ 4 \end{bmatrix} \qquad \mathbf{A}_1 = \begin{bmatrix} 1 \\ 0 \\ -1 \\ 1 \end{bmatrix} \qquad \mathbf{A}_2 = \begin{bmatrix} 0 \\ 1 \\ 1 \\ 2 \end{bmatrix}$$

write

$$A = \begin{bmatrix} 1 & 0 \\ 0 & 1 \\ -1 & 1 \\ 1 & 2 \end{bmatrix} \qquad P = \begin{bmatrix} 1 & 0 \\ 0 & 1 \\ -1 & 1 \\ 1 & 2 \end{bmatrix} \begin{bmatrix} 3 & 1 \\ 1 & 6 \end{bmatrix}^{-1} \begin{bmatrix} 1 & 0 & -1 & 1 \\ 0 & 1 & 1 & 2 \end{bmatrix} = \frac{1}{17} \begin{bmatrix} 6 & -1 & -7 & 4 \\ -1 & 3 & 4 & 5 \\ -7 & 4 & 11 & 1 \\ 4 & 5 & 1 & 14 \end{bmatrix}$$

and compute $\mathbf{B}_{[\mathbf{A}_1,\mathbf{A}_2]} = \mathbf{B}_{[A]} = P\mathbf{B}$. The component of $\mathbf{B}$ perpendicular to this subspace is then $\mathbf{B}_{\perp[\mathbf{A}_1,\mathbf{A}_2]} = B_{\perp[A]} = \mathbf{B} - P\mathbf{B} = (I - P)\mathbf{B}$. Both components of $\mathbf{B}$ are given as

$$\mathbf{B}_{[A]} = \frac{1}{17}\begin{bmatrix} -1 \\ 37 \\ 38 \\ 73 \end{bmatrix} \qquad \mathbf{B}_{\perp[A]} = \frac{1}{17}\begin{bmatrix} 18 \\ -3 \\ 13 \\ -5 \end{bmatrix}$$ ❖

FINAL COMMENTS   If A is an $r \times c$ matrix whose columns are independent, then (33.4) projects onto the *column space* of A. (The column space of A is the span of the columns, that is, the set of all possible linear combinations of the columns of A.) The matrix $A^{\mathrm{T}}A$ will be an invertible $c \times c$ matrix and P can be constructed from (33.4).

If A has rank $k < c$, then $A^{\mathrm{T}}A$ will be singular and not invertible and (33.4) fails. The redundant columns in A must be deleted before the projection matrix P can be constructed with (33.4). For example,

1. The $r \times c$ matrix A is 4×5, so its rank is at most $k = 4$, and the 5×5 matrix $A^{\mathrm{T}}A$ is definitely singular.

2. The $r \times c$ matrix A is 5×4 and has maximal rank $k = 4$, so the 4×4 matrix $A^{\mathrm{T}}A$ is invertible.

3. The $r \times c$ matrix A is 5×4 and has rank $k < 4$, so the 4×4 matrix $A^{\mathrm{T}}A$ has rank $k < 4$ and is singular.

Minimizing Property of Projection

The projection of $\mathbf{B}$ onto the subspace spanned by $\mathbf{A}_1, \ldots, \mathbf{A}_k$ is the component of $\mathbf{B}$ in the subspace spanned by $\mathbf{A}_1, \ldots, \mathbf{A}_k$. The component orthogonal to this subspace is the shortest vector from the tip of $\mathbf{B}$ to the subspace formed by $\mathbf{A}_1, \ldots, \mathbf{A}_k$. Figure 33.3 illustrates this for the vectors

$$\mathbf{B} = \begin{bmatrix} -5 \\ 7 \\ 6 \end{bmatrix} \qquad \mathbf{A}_1 = \begin{bmatrix} 1 \\ 2 \\ 1 \end{bmatrix} \qquad \mathbf{A}_2 = \begin{bmatrix} -1 \\ 0 \\ 2 \end{bmatrix}$$

The matrices $A = [\mathbf{A}_1, \mathbf{A}_2]$ and $P = A(A^{\mathrm{T}}A)^{-1}A^{\mathrm{T}}$ are given by

$$A = \begin{bmatrix} 1 & -1 \\ 2 & 0 \\ 1 & 2 \end{bmatrix} \qquad P = \begin{bmatrix} 1 & -1 \\ 2 & 0 \\ 1 & 2 \end{bmatrix}\begin{bmatrix} 6 & 1 \\ 1 & 5 \end{bmatrix}^{-1}\begin{bmatrix} 1 & 2 & 1 \\ -1 & 0 & 2 \end{bmatrix} = \frac{1}{29}\begin{bmatrix} 13 & 12 & -8 \\ 12 & 20 & 6 \\ -8 & 6 & 25 \end{bmatrix}$$

The projection of $\mathbf{B}$ onto the span of $\mathbf{A}_1$ and $\mathbf{A}_2$ is the vector $P\mathbf{B} = \mathbf{B}_{[A]} = -\mathbf{i} + 4\mathbf{j} + 8\mathbf{k}$, whereas the component of $\mathbf{B}$ orthogonal to the subspace spanned by $\mathbf{A}_1$ and $\mathbf{A}_2$ is $\mathbf{B} - P\mathbf{B} = \mathbf{B}_{\perp[A]} = -4\mathbf{i} + 3\mathbf{j} - 2\mathbf{k}$. The vector $\mathbf{B}_{\perp[A]} = (I - P)\mathbf{B}$ is the shortest vector from the plane formed by $\mathbf{A}_1$ and $\mathbf{A}_2$ to the tip of $\mathbf{B}$.

To see this, let

$$\mathbf{Y} = c_1\mathbf{A}_1 + c_2\mathbf{A}_2 = (c_1 - c_2)\mathbf{i} + 2c_1\mathbf{j} + (c_1 + 2c_2)\mathbf{k}$$

be a general vector in the plane of $\mathbf{A}_1$ and $\mathbf{A}_2$. By varying c_1 and c_2 the vector $\mathbf{Y}$ moves in the plane. Correspondingly, the vector

$$\mathbf{R} = \mathbf{B} - \mathbf{Y} = (-5 - c_1 + c_2)\mathbf{i} + (7 - 2c_1)\mathbf{j} + (6 - c_1 - 2c_2)\mathbf{k}$$

from the tip of $\mathbf{Y}$ to the tip of $\mathbf{B}$ varies. We then seek values of c_1 and c_2 that minimize

$$f(c_1, c_2) = \|\mathbf{R}\|^2 = (-5 - c_1 + c_2)^2 + (7 - 2c_1)^2 + (6 - c_1 - 2c_2)^2$$

(Minimizing $\|\mathbf{R}\|^2$ is equivalent to, but simpler than, minimizing $\|\mathbf{R}\|$.)

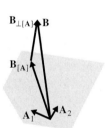

FIGURE 33.3   Projection of **B** onto plane spanned by $\mathbf{A}_1$ and $\mathbf{A}_2$. Component of **B** orthogonal to this plane is the shortest vector from the plane to the head of **B**.

The techniques of calculus are sufficient for finding the desired minimizing values of c_1 and c_2. They must satisfy the equations

$$\frac{\partial f}{\partial c_1} = -30 + 12c_1 + 2c_2 = 0 \quad \text{and} \quad \frac{\partial f}{\partial c_2} = -34 + 2c_1 + 10c_2 = 0$$

and are therefore $c_1 = 2$, $c_2 = 3$. The minimizing $\mathbf{Y}$ is then

$$\mathbf{Y} = 2\mathbf{A}_1 + 3\mathbf{A}_2 = -\mathbf{i} + 4\mathbf{j} + 8\mathbf{k} = P\mathbf{B}$$

The shortest vector $\mathbf{R}$ is then

$$\mathbf{R} = \mathbf{B} - \mathbf{Y} = \mathbf{B} - P\mathbf{B} = \mathbf{B}_{\perp[A]} = -4\mathbf{i} + 3\mathbf{j} - 2\mathbf{k}$$

in agreement with the original calculation.

EXERCISES 33.2–Part A

A1. If $\mathbf{A} = 2\mathbf{i} + 3\mathbf{j}$ and $\mathbf{B} = \mathbf{i} + 5\mathbf{j}$, obtain $\mathbf{B}_{\mathbf{A}}$ and $\mathbf{B}_{\perp\mathbf{A}}$ via the formalism in (33.1).

A2. With $\mathbf{A}$ as in Exercise A1, obtain P by the formalism in (33.2). Show $P^2 = P$.

A3. Use P from Exercise A2 and $\mathbf{B}$ from Exercise A1 to obtain $\mathbf{B}_{\mathbf{A}}$ and $\mathbf{B}_{\perp\mathbf{A}}$.

A4. For $\mathbf{A}$ and $\mathbf{B}$ given in Exercise A1, obtain $\mathbf{B}_{\perp\mathbf{A}}$ using the techniques of calculus to find the value of α that minimizes $\|\mathbf{B} - \alpha\mathbf{A}\|^2$.

A5. If $\mathbf{A}_1 = \mathbf{i} - \mathbf{k}$ and $\mathbf{A}_2 = \mathbf{i} + \mathbf{j}$, obtain P, the matrix that projects onto the plane determined by $\mathbf{A}_1$ and $\mathbf{A}_2$. Show $P^2 = P$.

A6. If $\mathbf{A}_1$ and $\mathbf{A}_2$ are as given in Exercise A5 and $\mathbf{B} = 2\mathbf{i} + 3\mathbf{j} - 5\mathbf{k}$, obtain $\mathbf{B}_{[A]}$ and $\mathbf{B}_{\perp[A]}$.

A7. For the vectors used in Exercise A6, obtain $\mathbf{B}_{\perp[A]}$ using the techniques of calculus to find the values of α_k, $k = 1, 2$, that minimize $\|\mathbf{B} - \sum_{k=1}^{2} \alpha_k\mathbf{A}_k\|^2$, where $\mathbf{A}_k$, $k = 1, 2$, are the two (independent) columns of A.

EXERCISES 33.2–Part B

B1. From $P^2 = P$ we easily see that $\det P$ must be either 0 or 1. Explain why $\det P = 0$ and not 1 when projecting onto a subspace whose dimension is smaller than that of the space in which it is contained.

In Exercises B2–11:

(a) Obtain $\mathbf{B}_{\mathbf{A}} = \frac{\mathbf{B}\cdot\mathbf{A}}{\mathbf{A}\cdot\mathbf{A}}\mathbf{A}$, the vector projection of $\mathbf{B}$ along $\mathbf{A}$.

(b) Obtain $\mathbf{B}_{\perp\mathbf{A}} = \mathbf{B} - \mathbf{B}_{\mathbf{A}}$, the vector component of $\mathbf{B}$ orthogonal to $\mathbf{A}$.

(c) Obtain the projection matrix $P = \frac{\mathbf{A}\mathbf{A}^{\mathsf{T}}}{\mathbf{A}^{\mathsf{T}}\mathbf{A}}$ and show $P^2 = P$.

(d) Show $\mathbf{B}_{\mathbf{A}} = P\mathbf{B}$ and $\mathbf{B}_{\perp\mathbf{A}} = (I - P)\mathbf{B}$.

(e) Obtain $\mathbf{B}_{\perp\mathbf{A}}$ using the techniques of calculus to find the value of α that minimizes $\|\mathbf{B} - \alpha\mathbf{A}\|^2$.

(f) Sketch $\mathbf{A}$, $\mathbf{B}$, $\mathbf{B}_{\mathbf{A}}$, and $\mathbf{B}_{\perp\mathbf{A}}$.

B2. $\mathbf{A} = -8\mathbf{i} - 2\mathbf{j}$, $\mathbf{B} = 8\mathbf{i} + 11\mathbf{j}$ **B3.** $\mathbf{A} = 12\mathbf{i} + 10\mathbf{j}$, $\mathbf{B} = \mathbf{i} + 3\mathbf{j}$

B4. $\mathbf{A} = -7\mathbf{i} - 9\mathbf{j}$, $\mathbf{B} = 4\mathbf{i} - 6\mathbf{j}$ **B5.** $\mathbf{A} = 5\mathbf{i} + 3\mathbf{j}$, $\mathbf{B} = 11\mathbf{i} + 12\mathbf{j}$

B6. $\mathbf{A} = -12\mathbf{i} - 6\mathbf{j}$, $\mathbf{B} = 10\mathbf{i} - 2\mathbf{j}$

B7. $\mathbf{A} = -\mathbf{i} - 4\mathbf{j} - 5\mathbf{k}$, $\mathbf{B} = 3\mathbf{i} + 5\mathbf{j} + 3\mathbf{k}$

B8. $\mathbf{A} = 7\mathbf{i} - 9\mathbf{j} - 6\mathbf{k}$, $\mathbf{B} = -5\mathbf{i} + 8\mathbf{j} - 5\mathbf{k}$

B9. $\mathbf{A} = 5\mathbf{i} + 10\mathbf{j} - 3\mathbf{k}$, $\mathbf{B} = 6\mathbf{i} + 7\mathbf{j} - 4\mathbf{k}$

B10. $\mathbf{A} = 10\mathbf{i} + 2\mathbf{j} + 2\mathbf{k}$, $\mathbf{B} = 9\mathbf{i} - 9\mathbf{j} - 2\mathbf{k}$

B11. $\mathbf{A} = 6\mathbf{i} + 4\mathbf{j} + 10\mathbf{k}$, $\mathbf{B} = 4\mathbf{i} - 12\mathbf{j} + \mathbf{k}$

In Exercises B12–21:

(a) Show that the columns of A are independent.

(b) Obtain the matrix $P = A(A^{\mathsf{T}}A)^{-1}A^{\mathsf{T}}$ that projects vectors onto the subspace spanned by the columns of A.

(c) Show $P^2 = P$.

(d) Obtain $\mathbf{B}_{[A]} = P\mathbf{B}$, the projection of $\mathbf{B}$ onto the column space of A.

(e) Obtain $\mathbf{B}_{\perp[A]} = (I - P)\mathbf{B}$, the component of $\mathbf{B}$ orthogonal to the column space of A.

(f) Obtain $\mathbf{B}_{\perp[A]}$ using the techniques of calculus to find the values of α_k, $k = 1, \ldots, n$, that minimize $\|\mathbf{B} - \sum_{k=1}^{n} \alpha_k\mathbf{A}_k\|^2$, where $\mathbf{A}_k$, $k = 1, \ldots, n$, are the n (independent) columns of A.

(g) Sketch a representative diagram showing $\mathbf{B}$, $\mathbf{B}_{[A]}$, $\mathbf{B}_{\perp[A]}$, and the column space of A represented as a plane.

B12. $A = \begin{bmatrix} -5 & 3 \\ 5 & 3 \\ 7 & -9 \end{bmatrix}$, $\mathbf{B} = \begin{bmatrix} -8 \\ -4 \\ 1 \end{bmatrix}$ **B13.** $A = \begin{bmatrix} -6 & -5 \\ 8 & -5 \\ 3 & -5 \end{bmatrix}$, $\mathbf{B} = \begin{bmatrix} 10 \\ -6 \\ -4 \end{bmatrix}$

B14. $A = \begin{bmatrix} -4 & 10 \\ 2 & 9 \\ -2 & 7 \end{bmatrix}$, $\mathbf{B} = \begin{bmatrix} -4 \\ -6 \\ -3 \end{bmatrix}$ **B15.** $A = \begin{bmatrix} -2 & -2 \\ 4 & -5 \\ 6 & 4 \\ 4 & -12 \end{bmatrix}$, $\mathbf{B} = \begin{bmatrix} 6 \\ 10 \\ 6 \\ 6 \end{bmatrix}$

B16. $A = \begin{bmatrix} 5 & -10 \\ -12 & -7 \\ -3 & -3 \\ -12 & -10 \end{bmatrix}$, $B = \begin{bmatrix} 0 \\ -8 \\ -6 \\ 3 \end{bmatrix}$ **B17.** $A = \begin{bmatrix} -3 & -9 \\ -10 & -12 \\ 6 & -2 \\ 8 & -4 \end{bmatrix}$, $B = \begin{bmatrix} 9 \\ 7 \\ 5 \\ -12 \end{bmatrix}$

B18. $A = \begin{bmatrix} 5 & -10 \\ -5 & -2 \\ -2 & -10 \\ 6 & -5 \end{bmatrix}$, $B = \begin{bmatrix} 11 \\ -2 \\ 5 \\ -10 \end{bmatrix}$

B19. $A = \begin{bmatrix} -6 & -4 & -9 \\ 0 & -11 & 9 \\ 10 & -9 & 8 \\ 4 & 11 & -2 \end{bmatrix}$, $B = \begin{bmatrix} 11 \\ 9 \\ -11 \\ 4 \end{bmatrix}$

B20. $A = \begin{bmatrix} 7 & -3 & -3 \\ 3 & -8 & -3 \\ 7 & -9 & -1 \\ 6 & 7 & 5 \end{bmatrix}$, $B = \begin{bmatrix} -12 \\ -9 \\ -8 \\ 2 \end{bmatrix}$

B21. $A = \begin{bmatrix} -6 & 5 & -8 \\ 1 & 12 & 0 \\ -2 & 11 & -12 \\ -10 & 9 & -4 \end{bmatrix}$, $B = \begin{bmatrix} 1 \\ 12 \\ -8 \\ -9 \end{bmatrix}$

B22. Find P, the matrix that projects vectors in the plane onto the line $2x + 3y = 0$, and then obtain the matrix $H = I - 2P$. Apply H to several vectors in the plane and deduce its general effect. Show that $H^2 = I$, and explain why this should be true.

B23. Given that the inverse of an invertible symmetric matrix is symmetric, show that $P = A(A^T A)^{-1} A^T$ is symmetric by computing P^T.

B24. If A is a square matrix of maximal rank (so it is invertible), show that $P = A(A^T A)^{-1} A^T = I$. Explain why this makes sense.

B25. Show that if $P = A(A^T A)^{-1} A^T$, then $P^2 = P$, necessarily.

In Exercises B25–30, find a matrix that projects onto the given plane by

    (a) forming $A(A^T A)^{-1} A^T$, where A is a matrix whose columns are any two independent vectors in the plane.

    (b) calculating $(I - P)$, where P is a matrix that projects onto the *normal* to the plane. (*Hint:* A vector perpendicular to the plane $ax + by + cz = 0$ is $a\mathbf{i} + b\mathbf{j} + c\mathbf{k}$.)

B26. $7x + 9y + 10z = 0$ **B27.** $11x - 2y + 7z = 0$

B28. $8x - 3y + 7z = 0$ **B29.** $10x - 8y - 11z = 0$

B30. $2x + 8y - 9z = 0$

33.3 **The Gram–Schmidt Orthogonalization Process**

The Gram–Schmidt Process

We seek an algorithm for deriving an orthonormal collection $\{\mathbf{w}_k\}$, $k = 1, \ldots, r \leq n$, whose span is the same as the span of the given vectors, from a given set of vectors $\{\mathbf{v}_k\}$, $k = 1, \ldots, n$. We can work in stages, first transforming the given vectors into an orthogonal collection and then normalizing each of the orthogonal vectors to produce a collection of orthonormal vectors.

The following prescription is a form of the Gram–Schmidt orthogonalization process, which yields a set of mutually orthogonal vectors that can then be normalized to form a set of orthonormal vectors

$$\mathbf{w}_1 = \mathbf{v}_1 \qquad \mathbf{w}_j = (I - P_{[\mathbf{w}_1, \ldots, \mathbf{w}_{j-1}]})\mathbf{v}_j \qquad j > 1$$

EXAMPLE 33.4   Given the four vectors $\mathbf{v}_k$, $k = 1, \ldots, 4$, shown respectively in (33.5),

$$\begin{bmatrix} -3 \\ -2 \\ 3 \\ -1 \end{bmatrix} \quad \begin{bmatrix} -1 \\ -3 \\ 2 \\ 3 \end{bmatrix} \quad \begin{bmatrix} -1 \\ -2 \\ 1 \\ -3 \end{bmatrix} \quad \begin{bmatrix} 2 \\ 3 \\ -2 \\ -2 \end{bmatrix} \tag{33.5}$$

we build $\mathbf{w}_k$, $k = 1, \ldots, 4$, a set of mutually orthogonal vectors as follows. Begin by accepting $\mathbf{v}_1$ as the first "new" vector, calling it $\mathbf{w}_1$. Next, take as $\mathbf{w}_2$ the component of $\mathbf{v}_2$ that is perpendicular to $\mathbf{v}_1$. Thus,

$$\mathbf{w}_2 = \mathbf{v}_2 - P_{[\mathbf{w}_1]}\mathbf{v}_2 = (I - P_{[\mathbf{w}_1]})\mathbf{v}_2$$

is $\mathbf{v}_2$ less the projection of $\mathbf{v}_2$ on $\mathbf{w}_1$, as shown in Figure 33.4. To obtain $\mathbf{w}_2$, form $P_{[\mathbf{w}_1]} = A_1(A_1^T A_1)^{-1} A_1^T$ from the matrix $A_1 = [\mathbf{w}_1]$ and compute $\mathbf{w}_2 = (I - P_{[\mathbf{w}_1]})\mathbf{v}_2$.

Take as $\mathbf{w}_3$ the component of $\mathbf{v}_3$ that is orthogonal to the plane spanned by $\mathbf{w}_1$ and $\mathbf{w}_2$. To obtain $\mathbf{w}_3$, form $P_{[\mathbf{w}_1, \mathbf{w}_2]} = A_2(A_2^T A_2)^{-1} A_2^T$, the matrix that projects onto the sub-

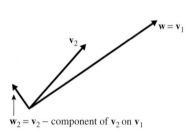

$\mathbf{w}_2 = \mathbf{v}_2$ – component of $\mathbf{v}_2$ on $\mathbf{v}_1$

FIGURE 33.4   Example 33.4: The vector $\mathbf{w}_2$ is the component of $\mathbf{v}_2$ orthogonal to $\mathbf{w}_1 = \mathbf{v}_1$

space spanned by $\mathbf{w}_1$ and $\mathbf{w}_2$, from the matrix $A_2 = [\mathbf{w}_1, \mathbf{w}_2]$. Then, compute $\mathbf{w}_3 = (I - P_{[\mathbf{w}_1,\mathbf{w}_2]})\mathbf{v}_3$.

Finally, take as $\mathbf{w}_4$ the component of $\mathbf{v}_4$ that is orthogonal to the subspace spanned by $\mathbf{w}_1$, $\mathbf{w}_2$, and $\mathbf{w}_3$. To obtain $\mathbf{w}_4$, form $P_{[\mathbf{w}_1,\mathbf{w}_2,\mathbf{w}_3]} = A_3(A_3^{\mathsf{T}} A_3)^{-1} A_3^{\mathsf{T}}$, the matrix that projects onto the subspace spanned by $\mathbf{w}_1$, $\mathbf{w}_2$, and $\mathbf{w}_3$, from the matrix $A_3 = [\mathbf{w}_1, \mathbf{w}_2, \mathbf{w}_3]$. Then, compute $\mathbf{w}_4 = (I - P_{[\mathbf{w}_1,\mathbf{w}_2,\mathbf{w}_3]})\mathbf{v}_4$. The projection matrices $P_{[\mathbf{w}_1]}$, $P_{[\mathbf{w}_1,\mathbf{w}_2]}$, $P_{[\mathbf{w}_1,\mathbf{w}_2,\mathbf{w}_3]}$ are, respectively,

$$\frac{1}{23}\begin{bmatrix} 9 & 6 & -9 & 3 \\ 6 & 4 & -6 & 2 \\ -9 & -6 & 9 & -3 \\ 3 & 2 & -3 & 1 \end{bmatrix}$$

$$\frac{1}{385}\begin{bmatrix} 158 & 75 & -145 & 96 \\ 75 & 155 & -120 & -125 \\ -145 & -120 & 155 & -15 \\ 96 & -125 & -15 & 302 \end{bmatrix} \quad \frac{1}{118}\begin{bmatrix} 69 & -14 & -56 & 7 \\ -14 & 114 & -16 & 2 \\ -56 & -16 & 54 & 8 \\ 7 & 2 & 8 & 117 \end{bmatrix}$$

and the orthogonal vectors $\mathbf{w}_k$, $k = 1, \ldots, 4$, are, respectively,

$$\begin{bmatrix} -3 \\ -2 \\ 3 \\ -1 \end{bmatrix} \quad \frac{1}{23}\begin{bmatrix} 13 \\ -45 \\ 10 \\ 81 \end{bmatrix} \quad \frac{1}{385}\begin{bmatrix} 356 \\ -640 \\ -200 \\ -388 \end{bmatrix} \quad \frac{1}{59}\begin{bmatrix} 21 \\ 6 \\ 24 \\ -3 \end{bmatrix}$$

Normalizing the vectors $\mathbf{w}_k$, $k = 1, \ldots, 4$, we get

$$\frac{1}{\sqrt{23}}\begin{bmatrix} -3 \\ -2 \\ 3 \\ -1 \end{bmatrix} \quad \frac{1}{\sqrt{8855}}\begin{bmatrix} 13 \\ -45 \\ 10 \\ 81 \end{bmatrix} \quad \frac{1}{\sqrt{45430}}\begin{bmatrix} 89 \\ -160 \\ -50 \\ -97 \end{bmatrix} \quad \frac{1}{\sqrt{118}}\begin{bmatrix} 7 \\ 2 \\ 8 \\ -1 \end{bmatrix} \quad ❖$$

Gram–Schmidt and Dependent Vectors

What happens if a set of *dependent* vectors is given to the Gram–Schmidt process? For example, the vectors $\mathbf{V}_k$, $k = 1, \ldots, 4$, where $\mathbf{V}_1 = \mathbf{v}_1$, $\mathbf{V}_2 = \mathbf{v}_2$, $\mathbf{V}_3 = 3\mathbf{v}_1 - 7\mathbf{v}_2$, $\mathbf{V}_4 = 5\mathbf{v}_1 + 2\mathbf{v}_2$, are linearly dependent because the second two vectors are linear combinations of the first two. Thus, applying the Gram–Schmidt process to the four linearly dependent vectors

$$\begin{bmatrix} -3 \\ -2 \\ 3 \\ -1 \end{bmatrix} \quad \begin{bmatrix} -1 \\ -3 \\ 2 \\ 3 \end{bmatrix} \quad \begin{bmatrix} -2 \\ 15 \\ -5 \\ -24 \end{bmatrix} \quad \begin{bmatrix} -17 \\ -16 \\ 19 \\ 1 \end{bmatrix}$$

yields the two orthonormal vectors

$$\frac{1}{\sqrt{23}}\begin{bmatrix} -3 \\ -2 \\ 3 \\ -1 \end{bmatrix} \quad \frac{1}{\sqrt{8855}}\begin{bmatrix} 13 \\ -45 \\ 10 \\ 81 \end{bmatrix}$$

The redundancy is eliminated, a useful feature of the orthogonalization process.

A moment's thought reveals how the redundancy is detected and eliminated. When $\mathbf{V}_3$ is projected onto the subspace spanned by $\mathbf{W}_1$ and $\mathbf{W}_2$, the projection is $\mathbf{V}_3$ itself. There is no component perpendicular to the subspace, so $\mathbf{W}_3 = \mathbf{0}$. The same happens with $\mathbf{V}_4$. There is no component orthogonal to the space spanned by $\mathbf{W}_1$, $\mathbf{W}_2$, and $\mathbf{W}_3$. Hence, $\mathbf{W}_4 = \mathbf{0}$.

Thus, the Gram–Schmidt process can be used as a filter for sifting out a maximal set of independent vectors from amongst a set of linearly dependent vectors.

EXERCISES 33.3

In Exercises 1–9, apply the Gram–Schmidt process to each of the following sets of vectors $\{\mathbf{v}_1, \ldots, \mathbf{v}_k\}$, obtaining the orthonormal sets $\{\mathbf{w}_1, \ldots, \mathbf{w}_r\}$, where $r < k$ if $\{\mathbf{v}_1, \ldots, \mathbf{v}_k\}$ is a linearly dependent set of vectors.

1. $\mathbf{v}_1 = 10\mathbf{i} - 9\mathbf{j} - 9\mathbf{k}$, $\mathbf{v}_2 = -3\mathbf{i} - 10\mathbf{j} - 8\mathbf{k}$, $\mathbf{v}_3 = -4\mathbf{i} - 12\mathbf{j} + 12\mathbf{k}$

2. $\mathbf{v}_1 = 8\mathbf{i} + 4\mathbf{j} - \mathbf{k}$, $\mathbf{v}_2 = 8\mathbf{i} + 3\mathbf{j} - \mathbf{k}$, $\mathbf{v}_3 = -2\mathbf{i} - 10\mathbf{j} + \mathbf{k}$

3. $\mathbf{v}_1 = 6\mathbf{i} + 10\mathbf{j} + 12\mathbf{k}$, $\mathbf{v}_2 = -10\mathbf{i} + 5\mathbf{j} - 7\mathbf{k}$, $\mathbf{v}_3 = -4\mathbf{i} + 15\mathbf{j} + 5\mathbf{k}$

4. $\mathbf{v}_1 = -12\mathbf{i} - 6\mathbf{j} - 2\mathbf{k}$, $\mathbf{v}_2 = -11\mathbf{i} - 9\mathbf{j} - 10\mathbf{k}$, $\mathbf{v}_3 = -3\mathbf{i} + 2\mathbf{j} + 9\mathbf{k}$, $\mathbf{v}_4 = 7\mathbf{i} + 7\mathbf{j} + 12\mathbf{k}$

5. $\mathbf{v}_1 = \mathbf{i} + 4\mathbf{j} + 11\mathbf{k}$, $\mathbf{v}_2 = -9\mathbf{i} - 4\mathbf{j} + 6\mathbf{k}$, $\mathbf{v}_3 = 7\mathbf{i} - 3\mathbf{j} - 10\mathbf{k}$, $\mathbf{v}_4 = 2\mathbf{i} - 2\mathbf{j} - \mathbf{k}$

6. $\mathbf{v}_1 = \begin{bmatrix} 7 \\ 7 \\ -1 \\ 4 \end{bmatrix}$, $\mathbf{v}_2 = \begin{bmatrix} -1 \\ -7 \\ -9 \\ -2 \end{bmatrix}$, $\mathbf{v}_3 = \begin{bmatrix} 12 \\ -8 \\ -8 \\ 11 \end{bmatrix}$

7. $\mathbf{v}_1 = \begin{bmatrix} 11 \\ -11 \\ 11 \\ -8 \end{bmatrix}$, $\mathbf{v}_2 = \begin{bmatrix} 2 \\ 4 \\ -6 \\ -1 \end{bmatrix}$, $\mathbf{v}_3 = \begin{bmatrix} -9 \\ 15 \\ -17 \\ 7 \end{bmatrix}$

8. $\mathbf{v}_1 = \begin{bmatrix} -6 \\ 6 \\ 8 \\ 9 \end{bmatrix}$, $\mathbf{v}_2 = \begin{bmatrix} 10 \\ 5 \\ -11 \\ 1 \end{bmatrix}$, $\mathbf{v}_3 = \begin{bmatrix} -1 \\ -4 \\ 0 \\ 6 \end{bmatrix}$, $\mathbf{v}_4 = \begin{bmatrix} 3 \\ -3 \\ 1 \\ 5 \end{bmatrix}$

9. $\mathbf{v}_1 = \begin{bmatrix} 3 \\ 7 \\ -3 \\ -3 \end{bmatrix}$, $\mathbf{v}_2 = \begin{bmatrix} -1 \\ 11 \\ -11 \\ 2 \end{bmatrix}$, $\mathbf{v}_3 = \begin{bmatrix} 5 \\ -3 \\ -11 \\ -8 \end{bmatrix}$, $\mathbf{v}_4 = \begin{bmatrix} 0 \\ 14 \\ -3 \\ 2 \end{bmatrix}$

The Gram–Schmidt process can be implemented without constructing projection matrices. Given the set of vectors $\{\mathbf{v}_1, \ldots, \mathbf{v}_k\}$, create the orthogonal set $\{\mathbf{w}_1, \ldots, \mathbf{w}_r\}$, $r \leq k$ by defining $\mathbf{w}_1 = \mathbf{v}_1$ and

$$\mathbf{w}_i = \mathbf{v}_i - \sum_{j=1}^{i-1} \frac{\langle \mathbf{v}_i, \mathbf{w}_j \rangle}{\langle \mathbf{w}_j, \mathbf{w}_j \rangle} \mathbf{w}_j \qquad i = 2, \ldots, k$$

If $\langle \mathbf{u}, \mathbf{v} \rangle = \mathbf{u} \cdot \mathbf{v} = \mathbf{u}^T \mathbf{v}$, apply this version of the Gram–Schmidt algorithm to each set of vectors $\{\mathbf{v}_1, \ldots, \mathbf{v}_k\}$ given in Exercises 10–17.

10. $\mathbf{v}_1 = -\mathbf{i} + \mathbf{j} - 3\mathbf{k}$, $\mathbf{v}_2 = 9\mathbf{i} - 12\mathbf{j} + 7\mathbf{k}$, $\mathbf{v}_3 = -2\mathbf{i} - 10\mathbf{j} + 11\mathbf{k}$

11. $\mathbf{v}_1 = 10\mathbf{i} + 4\mathbf{j} - 9\mathbf{k}$, $\mathbf{v}_2 = 8\mathbf{i} + 11\mathbf{j} + 9\mathbf{k}$, $\mathbf{v}_3 = 6\mathbf{i} - 8\mathbf{j} - 27\mathbf{k}$

12. $\mathbf{v}_1 = -6\mathbf{i} - 6\mathbf{j} + \mathbf{k}$, $\mathbf{v}_2 = 12\mathbf{i} - 4\mathbf{k}$, $\mathbf{v}_3 = 6\mathbf{i} + 18\mathbf{j} + \mathbf{k}$

13. $\mathbf{v}_1 = -2\mathbf{i} - 7\mathbf{j} - 6\mathbf{k}$, $\mathbf{v}_2 = 7\mathbf{i} + 17\mathbf{j} + 9\mathbf{k}$, $\mathbf{v}_3 = 5\mathbf{i} + 10\mathbf{j} + 3\mathbf{k}$, $\mathbf{v}_4 = -11\mathbf{i} - 16\mathbf{j} + 3\mathbf{k}$

14. $\mathbf{v}_1 = 4\mathbf{i} + 12\mathbf{j} - 8\mathbf{k}$, $\mathbf{v}_2 = 13\mathbf{i} + 12\mathbf{j} - \mathbf{k}$, $\mathbf{v}_3 = -11\mathbf{i} - 6\mathbf{j} - 3\mathbf{k}$, $\mathbf{v}_4 = 7\mathbf{i} - 6\mathbf{j} + 11\mathbf{k}$

15. $\mathbf{v}_1 = \begin{bmatrix} 7 \\ 11 \\ -9 \\ 4 \end{bmatrix}$, $\mathbf{v}_2 = \begin{bmatrix} 2 \\ -3 \\ -3 \\ -1 \end{bmatrix}$, $\mathbf{v}_3 = \begin{bmatrix} 7 \\ 5 \\ 10 \\ 0 \end{bmatrix}$

16. $\mathbf{v}_1 = \begin{bmatrix} -8 \\ -7 \\ 11 \\ 12 \end{bmatrix}$, $\mathbf{v}_2 = \begin{bmatrix} -16 \\ -10 \\ 12 \\ 15 \end{bmatrix}$, $\mathbf{v}_3 = \begin{bmatrix} -8 \\ -3 \\ 1 \\ 3 \end{bmatrix}$, $\mathbf{v}_4 = \begin{bmatrix} 0 \\ -4 \\ 10 \\ 9 \end{bmatrix}$

17. $\mathbf{v}_1 = \begin{bmatrix} 1 \\ 12 \\ -2 \\ -9 \end{bmatrix}$, $\mathbf{v}_2 = \begin{bmatrix} -8 \\ 4 \\ 10 \\ 5 \end{bmatrix}$, $\mathbf{v}_3 = \begin{bmatrix} 10 \\ -10 \\ 11 \\ 11 \end{bmatrix}$, $\mathbf{v}_4 = \begin{bmatrix} -1 \\ -9 \\ 6 \\ 11 \end{bmatrix}$

The finite-precision numerical implementation of the Gram–Schmidt process can suffer loss of orthogonality through round-off error. Greater accuracy is achieved by the following modification wherein, at each stage, all "old vectors" are made orthogonal to each "new vector." In particular, $\mathbf{w}_1 = \frac{\mathbf{v}_1}{\|\mathbf{v}_1\|}$, but each of the remaining vectors $\mathbf{v}_2, \ldots, \mathbf{v}_k$, are made orthogonal to $\mathbf{w}_1$ by subtracting from each $\mathbf{v}_i$, its projection along $\mathbf{w}_1$ by means of $\mathbf{v}_i^{(1)} = \mathbf{v}_i - (\mathbf{w}_1 \cdot \mathbf{v}_i)\mathbf{w}_1$, $i = 2, \ldots, k$. Then, $\mathbf{w}_2 = \frac{\mathbf{v}_2^{(1)}}{\|\mathbf{v}_2^{(1)}\|}$ is already orthogonal to $\mathbf{w}_1$, so the remaining $\mathbf{v}_3^{(1)}, \ldots, \mathbf{v}_k^{(1)}$ are made orthogonal to $\mathbf{w}_2$ by means of $\mathbf{v}_i^{(2)} = \mathbf{v}_i^{(1)} - (\mathbf{w}_2 \cdot \mathbf{v}_i^{(1)})\mathbf{w}_2$, $i = 3, \ldots, k$. In general, $\mathbf{w}_s = \mathbf{v}_s^{(s-1)}/\|\mathbf{v}_s^{(s-1)}\|$ and $\mathbf{v}_i^{(s)} = \mathbf{v}_i^{(s-1)} - (\mathbf{w}_s \cdot \mathbf{v}_i^{(s-1)})\mathbf{w}_s$, $i = s+1, \ldots, k$. Apply this modification of the Gram–Schmidt process to each set of vectors $\{\mathbf{v}_1, \ldots, \mathbf{v}_k\}$ given in Exercises 18–25.

18. $\mathbf{v}_1 = -11\mathbf{i} - \mathbf{j} + 6\mathbf{k}$, $\mathbf{v}_2 = 5\mathbf{i} - 4\mathbf{k}$, $\mathbf{v}_3 = 3\mathbf{i} - 2\mathbf{j} + 5\mathbf{k}$

19. $\mathbf{v}_1 = -7\mathbf{i} + 9\mathbf{j} - 7\mathbf{k}$, $\mathbf{v}_2 = -9\mathbf{i} + 11\mathbf{j} + 10\mathbf{k}$, $\mathbf{v}_3 = 2\mathbf{i} - 2\mathbf{j} - 17\mathbf{k}$

20. $\mathbf{v}_1 = 9\mathbf{i} - 12\mathbf{j} + 3\mathbf{k}$, $\mathbf{v}_2 = 13\mathbf{i} - \mathbf{j} - 7\mathbf{k}$, $\mathbf{v}_3 = -7\mathbf{i} - 7\mathbf{j} + 9\mathbf{k}$, $\mathbf{v}_4 = 16\mathbf{i} - 5\mathbf{j} - 6\mathbf{k}$

21. $\mathbf{v}_1 = -11\mathbf{i} + 3\mathbf{j} - 5\mathbf{k}$, $\mathbf{v}_2 = -10\mathbf{i} - 4\mathbf{j} - 2\mathbf{k}$, $\mathbf{v}_3 = 5\mathbf{i} - 9\mathbf{j} - 11\mathbf{k}$, $\mathbf{v}_4 = 3\mathbf{i} - 8\mathbf{j} - 4\mathbf{k}$

22. $\mathbf{v}_1 = -9\mathbf{i} + 4\mathbf{j} + \mathbf{k}$, $\mathbf{v}_2 = -5\mathbf{i} + 6\mathbf{j} + 4\mathbf{k}$, $\mathbf{v}_3 = 8\mathbf{i} + 4\mathbf{j} + 6\mathbf{k}$, $\mathbf{v}_4 = \mathbf{i} - 8\mathbf{j} - 7\mathbf{k}$

23. $\mathbf{v}_1 = \begin{bmatrix} 4 \\ 12 \\ 7 \\ -1 \end{bmatrix}$, $\mathbf{v}_2 = \begin{bmatrix} 5 \\ 2 \\ 5 \\ -7 \end{bmatrix}$, $\mathbf{v}_3 = \begin{bmatrix} 9 \\ 14 \\ 12 \\ -8 \end{bmatrix}$

24. $\mathbf{v}_1 = \begin{bmatrix} 7 \\ 10 \\ 2 \\ -12 \end{bmatrix}$, $\mathbf{v}_2 = \begin{bmatrix} 8 \\ 6 \\ 4 \\ 7 \end{bmatrix}$, $\mathbf{v}_3 = \begin{bmatrix} -10 \\ -3 \\ -6 \\ 6 \end{bmatrix}$, $\mathbf{v}_4 = \begin{bmatrix} -7 \\ -5 \\ -4 \\ -39 \end{bmatrix}$

25. $\mathbf{v}_1 = \begin{bmatrix} 9 \\ -9 \\ -8 \\ -8 \end{bmatrix}$, $\mathbf{v}_2 = \begin{bmatrix} 14 \\ -6 \\ -9 \\ -17 \end{bmatrix}$, $\mathbf{v}_3 = \begin{bmatrix} -4 \\ 12 \\ 7 \\ -1 \end{bmatrix}$, $\mathbf{v}_4 = \begin{bmatrix} 5 \\ 3 \\ -1 \\ -9 \end{bmatrix}$

In Exercises 26–35, orthogonalize the functions x^n, $n = 0, 1, \ldots, 5$, by applying the Gram–Schmidt process with the given inner product.

26. $\langle f(x), g(x) \rangle = \displaystyle\int_0^1 f(x)g(x)\, dx$

27. $\langle f(x), g(x) \rangle = \displaystyle\int_{-1}^1 f(x)g(x)\, dx$

28. $\langle f(x), g(x) \rangle = \displaystyle\int_{-1}^1 f(x)g(x)\sqrt{1 - x^2}\, dx$

29. $\langle f(x), g(x) \rangle = \displaystyle\int_0^1 f(x)g(x)\sin \pi x\, dx$

30. $\langle f(x), g(x) \rangle = \displaystyle\int_0^1 f(x)g(x)\cos \frac{\pi}{2}x\, dx$

31. $\langle f(x), g(x) \rangle = \displaystyle\int_0^1 f(x)g(x)\sqrt{x}\, dx$

32. $\langle f(x), g(x) \rangle = \displaystyle\int_0^1 f(x)g(x)\sqrt[3]{x}\, dx$

33. $\langle f(x), g(x) \rangle = \displaystyle\sum_{k=0}^4 f(x_k)g(x_k)$, where $x_i = \dfrac{i}{4}$, $i = 0, \ldots, 4$

34. $\langle f(x), g(x) \rangle = \displaystyle\sum_{k=0}^4 f(x_k)g(x_k)x_k$, where $x_i = \dfrac{i}{4}$, $i = 0, \ldots, 4$

35. $\langle f(x), g(x) \rangle = \displaystyle\sum_{k=0}^4 f(x_k)g(x_k)x_k^2$, where $x_i = \dfrac{i}{4}$, $i = 0, \ldots, 4$

Let $\langle \mathbf{u}, \mathbf{v} \rangle = \mathbf{u}^T M \mathbf{v}$, where $M = \begin{bmatrix} 5 & -2 & -3 & -3 \\ -2 & 3 & 2 & 2 \\ -3 & 2 & 8 & 3 \\ -3 & 2 & 3 & 9 \end{bmatrix}$. To the vectors given in Exercises 36–41, apply the Gram–Schmidt algorithm $\mathbf{w}_1 = \mathbf{v}_1$,

$$\mathbf{w}_i = \mathbf{v}_i - \sum_{j=1}^{i-1} \frac{\langle \mathbf{v}_i, \mathbf{w}_j \rangle}{\langle \mathbf{w}_j, \mathbf{w}_j \rangle} \mathbf{w}_j \qquad i > 1$$

36. the vectors in Exercise 8   **37.** the vectors in Exercise 9

38. the vectors in Exercise 16   **39.** the vectors in Exercise 17

40. the vectors in Exercise 24   **41.** the vectors in Exercise 25

33.4 Quadratic Forms

Second Derivative Test in One Variable

The following is a brief review of the second-derivative test for extrema of functions of a single variable.

Let $f'(c) = 0$ so that c is a critical value for $f(x)$. To test if the critical point $(c, f(c))$ is a maximum, minimum, or neither, examine $f''(c)$. If $f''(c) > 0$, the critical point is a minimum; if $f''(c) < 0$, a maximum; if $f''(c) = 0$, no decision. To tell if the critical point is a "saddle," check if $f''(x)$ changes sign across $x = c$.

This second derivative test flows from the local behavior of $f(x)$ near $x = c$. This local behavior, which can be studied via the Taylor expansion of $f(x)$ up through quadratic terms, namely,

$$f(x) = f(c) + f'(c)(x - c) + \tfrac{1}{2}f''(c)(x - c)^2 + \mathrm{O}(x^3)$$

Since $f'(c) = 0$, this expansion becomes (with $x - c = h$, so $x = c + h$)

$$f(c + h) = f(c) + f''(c)\frac{h^2}{2} + \mathrm{O}(h^3)$$

Near $x = c$ a function value $f(x)$ is the function value at $x = c$ altered by an amount that is approximately $f''(c)\frac{h^2}{2}$, a quantity whose sign is determined solely by the sign of $f''(c)$. If $f''(c) > 0$, then nearby values are greater than $f(c)$, so $f(c)$ is a minimum. If $f''(c) < 0$, then nearby values are less than $f(c)$, so $f(c)$ is a maximum.

Second Derivative Test in Two Variables

A point p at which $\nabla f = \mathbf{0}$ is called a *critical point* for the multivariable function $f(\mathbf{x})$. In two variables, the nature of the critical point p can be determined by evaluating $T = f_{xx}f_{yy} - f_{xy}^2$ at p. If $T < 0$, then p is a saddle point. If $T > 0$, then p is a minimum when

$f_{xx} > 0$ and a maximum when $f_{xx} < 0$. If $T = 0$, then T does not reveal the nature of p and the test fails.

EXAMPLE 33.5   Consider the function $f(x, y) = x^3 - 6xy + y^3$ whose critical points $(0, 0)$ and $(2, 2)$ are found by computing the first partial derivatives, setting them equal to zero, and solving for x and y, resulting in the pair of equations

$$3x^2 - 6y = 3y^2 - 6x = 0$$

Next, compute the second partial derivatives $f_{xx} = 6x$, $f_{yy} = 6y$, and $f_{xy} = -6$; then form the test expression

$$T = f_{xx}f_{yy} - f_{xy}^2 = 36xy - 36$$

and evaluate T at each critical point. Thus, $T(0, 0) = -36$ and $T(2, 2) = 108$. At $(0, 0)$ the test number $T(0, 0)$ is negative and we conclude that $(0, 0)$ is a saddle. At $(2, 2)$ the test number $T(2, 2)$ is positive. Since f_{xx} is also positive, we conclude that $(2, 2)$ is a minimum. ❖

Quadratic Forms and the Second Derivative Test

Let $F(x, y)$ have a critical point at $(0, 0)$ so that $F_x(0, 0) = 0$ and $F_y(0, 0) = 0$. Then, consider

$$F(x, y) = F(0, 0) + F_x(0, 0)x + F_y(0, 0)y + \tfrac{1}{2}F_{xx}(0, 0)x^2$$
$$+ F_{xy}(0, 0)xy + \tfrac{1}{2}F_{yy}(0, 0)y^2 + R(x, y)$$

a Taylor expansion of $F(x, y)$ with expansion point at $(0, 0)$. The remainder term $R(x, y)$ is $O(r^3)$, where $r = \sqrt{x^2 + y^2}$. The first derivative terms vanish at the critical point. The three second derivative terms in

$$F(x, y) = F(0, 0) + \tfrac{1}{2}[F_{xx}x^2 + 2F_{xy}xy + F_{yy}y^2] + R(x, y) \qquad \textbf{(33.6)}$$

suggest defining the symmetric matrix Q, called the *Hessian* of $F(x, y)$, and the vector V in (33.7)

$$Q = \begin{bmatrix} F_{xx} & F_{xy} \\ F_{yx} & F_{yy} \end{bmatrix} \qquad V = \begin{bmatrix} x \\ y \end{bmatrix} \qquad \textbf{(33.7)}$$

(Of course, for functions with continuous second partial derivatives, $F_{yx} = F_{xy}$.) Except for the multiplicative factor of $\tfrac{1}{2}$, the second derivative terms in (33.6) are given by the product $V^T Q V$, an expression called a *quadratic form*. The matrix Q is the *matrix of the quadratic form*.

The classification of the critical point $(0, 0)$ as a maximum, minimum, or saddle amounts to determining if the quadratic form $V^T Q V$ is always positive, always negative, or sometimes positive and sometimes negative. Since, to quadratic terms, $F(x, y) = F(0, 0) + \tfrac{1}{2}V^T Q V$, if the quadratic form is always positive, the critical point is a minimum since positive values are being added to $F(0, 0)$ in the neighborhood of the critical point. If the quadratic form is always negative, the critical point is then a maximum. If the quadratic form is sometimes positive and sometimes negative, the critical point is a saddle.

If the matrix Q were in diagonal form we could tell much more easily if the form were positive, negative, or both, because the associated quadratic form would contain just sums of squares and no cross-terms such as the product xy. Our strategy, then, is to transform the quadratic form to a strict sum of squares.

Since Q is a symmetric matrix, it is certainly diagonalizable with real eigenvalues. (See Section 32.2.) The columns of the transition matrix P are the eigenvectors that either are, or can be chosen as, orthogonal. If these eigenvectors are then normalized, P will be an orthogonal matrix for which $P^{-1} = P^{\mathrm{T}}$. (See Section 32.2.) Hence, the quadratic form becomes

$$\mathbf{V}^{\mathrm{T}}Q\mathbf{V} = \mathbf{V}^{\mathrm{T}}P^{\mathrm{T}}DP\mathbf{V} = (P\mathbf{V})^{\mathrm{T}}D(P\mathbf{V}) = \mathbf{W}^{\mathrm{T}}D\mathbf{W}$$

where D is a diagonal matrix of eigenvalues λ_1, λ_2 and $\mathbf{W} = u\mathbf{i} + v\mathbf{j}$ is an arbitrary vector. A sum of squares, the quadratic form can now be written as

$$\mathbf{V}^{\mathrm{T}}Q\mathbf{V} = \lambda_1 u^2 + \lambda_2 v^2$$

Clearly, if both eigenvalues are positive, the quadratic form is positive for every vector $\mathbf{W}$. If both eigenvalues are negative, then the quadratic form is negative for every vector $\mathbf{W}$. If the eigenvalues are of different signs, then the quadratic form can be either negative or positive, depending on the vector $\mathbf{W}$. Finally, if one of the eigenvalues is zero, then this analysis fails.

The second-derivative test for functions of two variables is based on an analysis of the eigenvalues in terms of the second partial derivatives. In light of the equality of the mixed second partials, these eigenvalues are

$$\lambda_\pm = \tfrac{1}{2}\left[F_{xx} + F_{yy} \pm \sqrt{(F_{xx} - F_{yy})^2 + 4F_{xy}^2}\right]$$

and $\lambda_+\lambda_-$, a product of the form $(a+b)(a-b) = a^2 - b^2$, is

$$\lambda_+\lambda_- = F_{xx}F_{yy} - F_{xy}^2 \tag{33.8}$$

This is the "test number" T of the second-derivative test. If the product of the eigenvalues is negative, the eigenvalues are of opposite signs; so there are vectors $\mathbf{V}_+$ for which the quadratic form is positive and vectors $\mathbf{V}_-$ for which it is negative. This means the second-derivative term in the Taylor expansion can either add to $F(0,0)$ or subtract from it, so the critical point must be a saddle.

If the product of the eigenvalues is positive, then the eigenvalues are of the same sign. The two eigenvalues are either both positive or both negative. If they are both positive, then for all vectors $\mathbf{V}$, the second-derivative term adds to $F(0,0)$ so the critical point is a minimum. If both eigenvalues are negative, then for all vectors $\mathbf{V}$, the second-derivative term subtracts from $F(0,0)$, so the critical point is a maximum.

Now, what determines if the two eigenvalues are both positive or both negative? First, the condition

$$\lambda_+\lambda_- = F_{xx}F_{yy} - F_{xy}^2 > 0$$

requires F_{xx} and F_{yy} to be nonzero and of the same sign. Certainly, the condition can't be true if either F_{xx} or F_{yy} is zero. If $F_{xy} = 0$ then $F_{xx}F_{yy} > 0$, so F_{xx} and F_{yy} are of the same sign. If $F_{xy} \neq 0$ then $F_{xx}F_{yy} > F_{xy}^2 > 0$, so again, $F_{xx}F_{yy} > 0$ and F_{xx} and F_{yy} are of the same sign.

Next, consider the sum of the eigenvalues, $\lambda_+ + \lambda_- = F_{xx} + F_{yy}$. Since F_{xx} and F_{yy} are of the same sign and the eigenvalues themselves are of the same sign, if F_{xx} is positive, both eigenvalues are positive, and if F_{xx} is negative, both eigenvalues are negative. Thus, when (33.8) is positive and F_{xx} is positive, the second-derivative term is positive and adds to the value $F(0,0)$, making the critical point a minimum, and when F_{xx} is negative, the second-derivative term is negative and subtracts from $F(0,0)$, making the critical point a maximum.

EXAMPLE 33.6   For the function $f(x,y)$ of Example 33.5, we would have, at the critical point $(0,0)$, the Taylor expansion $f(x,y) = -6xy$ and the Hessian Q in Table 33.1. The eigenvalues of Q

are $\lambda = \pm 6$, and the normalized eigenvectors of Q are the columns of P, an orthogonal transition matrix that diagonalizes Q via the similarity transform P^TQP, as given in Table 33.1. Thus, there is a coordinate system in which the quadratic form of second partials reduces to a sum of squares; but since the coefficients on these square terms are both positive and negative, it is possible to find points near the critical point $(0, 0)$ so that this quadratic form is positive and points near $(0, 0)$ where the form is negative. This is the essence of a saddle.

$$Q = \begin{bmatrix} 0 & -6 \\ -6 & 0 \end{bmatrix} \qquad P = \frac{1}{\sqrt{2}}\begin{bmatrix} 1 & 1 \\ -1 & 1 \end{bmatrix} \qquad P^TQP = \begin{bmatrix} 6 & 0 \\ 0 & -6 \end{bmatrix}$$

TABLE 33.1 Example 33.6: Testing the critical point $(0, 0)$

The same analysis at $(2, 2)$, the second critical point, gives the Taylor expansion $f(x, y) = -8 + 6X^2 - 6XY + 6Y^2$, where $X = x - 2$ and $Y = y - 2$, in which case the matrix of the quadratic form is Q, which has eigenvalues 6 and 18, and is given in Table 33.2. The eigenvalues are both positive. There is a basis in which Q is diagonal and the quadratic form is a sum of squares with positive coefficients. As before, the similarity transformation taking Q to its diagonal form can be constructed from its eigenvectors. Thus, P and P^TQP, given in Table 33.2, show $(2, 2)$ is a minimum.   ❖

$$Q = \begin{bmatrix} 12 & -6 \\ -6 & 12 \end{bmatrix} \qquad P = \frac{1}{\sqrt{2}}\begin{bmatrix} -1 & 1 \\ 1 & 1 \end{bmatrix} \qquad P^TQP = \begin{bmatrix} 18 & 0 \\ 0 & 6 \end{bmatrix}$$

TABLE 33.2 Example 33.6: Testing the critical point $(2, 2)$

Classification by Eigenvalues

On the basis of Example 33.6, we realize that we only need to know the eigenvalues of Q to determine the nature of the critical point p. We therefore have the terminology in Table 33.3. If the eigenvalues are all positive, then Q is positive definite and p is a minimum. If the eigenvalues are all negative, then Q is negative definite and p is a maximum. If there are positive and negative eigenvalues, then Q is indefinite and p is a saddle. Surprisingly, if zero is an eigenvalue so that Q is semidefinite, then p is minimum in the positive semidefinite case and a maximum in the negative semidefinite case.

Eigenvalues		Quadratic Form	Terminology
Positive	(all $\lambda_k > 0$)	$\mathbf{x}^TQ\mathbf{x} > 0$	Positive definite
Negative	(all $\lambda_k < 0$)	$\mathbf{x}^TQ\mathbf{x} < 0$	Negative definite
Nonnegative	(all $\lambda_k \geq 0$)	$\mathbf{x}^TQ\mathbf{x} \geq 0$	Positive semidefinite
Nonpositive	(all $\lambda_k \leq 0$)	$\mathbf{x}^TQ\mathbf{x} \leq 0$	Negative semidefinite
Positive and negative		Positive and negative	Indefinite

TABLE 33.3 Terminology for quadratic forms

Second-Derivative Test for Three Variables

Consider the function

$$f(x, y, z) = x^3 - 6xy - 6xz - 6yz + y^3 + z^3 \qquad (33.9)$$

with critical points $(0, 0, 0)$ and $(4, 4, 4)$. At $(4, 4, 4)$, a Taylor expansion gives

$$f = -96 + 12X^2 + 12Y^2 + 12Z^2 - 6YZ - 6XZ - 6XY$$

where $X = x - 4$, $Y = y - 4$, $Z = z - 4$. The (Hessian) matrix of second partials is then

$$Q = \begin{bmatrix} 24 & -6 & -6 \\ -6 & 24 & -6 \\ -6 & -6 & 24 \end{bmatrix} \qquad (33.10)$$

with eigenvalues $12, 30, 30$. The eigenvalues are all positive, so the quadratic form of second partials is positive definite (strictly positive) and the critical point $(4, 4, 4)$ is a minimum.

EXERCISES 33.4–Part A

A1. Obtain the critical point for the function $f(x, y) = 2x^2 + 3y^2 + 2x - 5y + 7xy$, and determine its nature using $T = f_{xx} f_{yy} - f_{xy}^2$.

A2. For the function in Exercise A1, determine the Hessian and its eigenvalues at the critical point. Is the information provided by the eigenvalues consistent with the conclusions in Exercise A1?

A3. Obtain the critical points for the function in (33.9).

A4. For the function in (33.9), obtain the Hessian matrix and its eigenvalues at $(0, 0, 0)$.

A5. Verify that (33.10) is the correct Hessian matrix at $(4, 4, 4)$ and that its eigenvalues are $\lambda = 12, 30, 30$.

EXERCISES 33.4–Part B

B1. With a graph, show that $f(x, y) = x^2$ has a minimum all along the y-axis. Evaluate $T = f_{xx} f_{yy} - f_{xy}^2$ at any point on the y-axis. Does the test succeed? Obtain the Hessian and its eigenvalues at any point on the y-axis. Is the information from the eigenvalues consistent with the geometric information obtained from the graph of $f(x, y)$?

In Exercises B2–11:

(a) Find all real critical points.

(b) Test each critical point with the second-derivative test based on $T = f_{xx} f_{yy} - f_{xy}^2$.

(c) Test each critical point by finding the eigenvalues of the corresponding Hessian.

B2. $f(x, y) = -y^2 + 12x^2 + 10 - 9x + 4y$

B3. $f(x, y) = -4y^2 - x^2 + 9 + 7xy + 12x + 9y$

B4. $f(x, y) = 8y^2 + 12x^2 - 10 - 7xy + 7x - 10y$

B5. $f(x, y) = -8y^2 + 4x^2 + 1 + 11xy - 4x + 12y$

B6. $f(x, y) = -y^2 - 11x^2 - 11 - 8xy - 4xy^2 - 11y^3$

B7. $f(x, y) = -2y^2 + 12x^2 - 3 + 10xy - 5x - 6y$

B8. $f(x, y) = 3y^2 + 9x^2 - 3 + 7xy + 10x$

B9. $f(x, y) = -10y^2 + 9x^2 - 1 - 8xy - 4x - 9y$

B10. $f(x, y) = -2y^2 + 11x^2 - 5 + 3xy + 11x - 8y$

B11. $f(x, y) = -10x^2 y + 3x^2 + 4 + 2xy + x^3 + 4x$

In Exercises B12–21:

(a) Find all real critical points.

(b) Test each critical point by finding the eigenvalues of the corresponding Hessian.

B12. $f(x, y, z) = 12y^2 - 10z^2 + 12 - 11x - 10x^2 y - 2z^3$

B13. $f(x, y, z) = 9y^2 - 6z^2 + 10z - 6x^2 + 6 - 7x$

B14. $f(x, y, z) = y^2 + 3xz - 1 + 3x^2 z + 10xz^2 - 2y^3$

B15. $f(x, y, z) = 6xz - 10z^2 + 10xy + 7x + 12y$

B16. $f(x, y, z) = 12z^2 - 10x^2 - 4yz + 9x^2 z - 2xyz + 7xz^2$

B17. $f(x, y, z) = -5y^2 - z^2 - 2z - 3x^2 - 7 + yz$

B18. $f(x, y, z) = -9xz - 3xz^2 + 11xy - 5x - 4y^3 + 3xy^2$

B19. $f(x, y, z) = 11xz + 4z^2 - 9z + 10x^2 - 12xy + 11x$

B20. $f(x, y, z) = -9y^2 + 9x^2 - 11 - 9yz - 11xy + 2x^2 z$

B21. $f(x, y, z) = xz + 3z^2 - 6z - 4 + 12yz - 8xy$

In Exercises B22–31, find the Hessian matrix.

B22. $f(x, y) = 7xy^3 - 9x^3 y + 5x^2 + 2x^2 y^2 - 11x$

B23. $f(x, y) = 6y^2 + 6xy^2 + 6x^2 y^2 - 3x^4 - 12x^3 - 7$

B24. $f(x, y) = 4x^4 + 8x^3 y^2 - x^3 + 7y^3 + 4y^4 - 3y$

B25. $f(x, y) = 4x^2 y + 5x^3 y + 4xy^2 - 6x^3 - 2y^3$

B26. $f(x, y) = 10xy + 4y^2 - 12x^3 - 2y + 8 + 3x$

B27. $f(x, y, z) = 8xy + 9x^2 yz - y^2 z + 9y + 9x$

B28. $f(x, y, z) = 5xy - 2x^2 yz + 4y^2 z^2 + 8yz^2 + 5xz^3 + 6yz$

B29. $f(x, y, z) = 3yz^3 - 9x^2 y^2 z - 8y^2 z^3 - 4y^2 z^2 - 2x$

B30. $f(x, y, z) = z^5 - 9y^2 z^3 + 6z^2 + 4x^3 yz - 2x^2 yz^2 + 9x$

B31. $f(x, y, z) = 9x^3 y - 4x^3 z - xy^2 - 5x^2 yz + 4x^3 + 3yz^2$

In Exercises B32–41:

   (a) Find Q, the (symmetric) matrix of the given quadratic form.

   (b) Verify that $\mathbf{x}^T Q\mathbf{x}$ generates the given quadratic form.

   (c) Find P, an orthogonal transition matrix that diagonalizes Q, that is, a matrix for which $Q = PDP^T$, where D is a diagonal matrix of eigenvalues.

   (d) Find a change of variables that reduces the quadratic form to a "sum of squares".

   (e) Determine if Q is positive definite ($\mathbf{x}^T Q\mathbf{x} > 0$), negative definite ($\mathbf{x}^T Q\mathbf{x} < 0$), positive semidefinite ($\mathbf{x}^T Q\mathbf{x} \geq 0$), negative semidefinite ($\mathbf{x}^T Q\mathbf{x} \leq 0$), or indefinite.

   (f) If Q is semidefinite, either positive or negative, find a nonzero vector $\mathbf{x}_1$ for which $\mathbf{x}_1^T Q\mathbf{x}_1 = 0$. (*Hint:* The null space of Q is the eigenspace belonging to the eigenvalue $\lambda = 0$.)

   (g) If Q is indefinite (the quadratic form has both positive and negative values), find a nonzero vector $\mathbf{x}_1$ for which $\mathbf{x}_1^T Q\mathbf{x}_1 < 0$ and another nonzero vector $\mathbf{x}_2$ for which $\mathbf{x}_2^T Q\mathbf{x}_2 > 0$.

B32. $x^2 - 6xy + 18xz + y^2 - 6yz + 9z^2$

B33. $-7x^2 - 10xy - 7y^2 - 2z^2$

B34. $8x^2 + 6xy + 10xz - 4y^2 - 6yz + 8z^2$

B35. $6x^2 - 4xy + 5y^2 - 4yz + 9z^2$

B36. $-8x^2 - 16xy - 8xz - 7y^2 - 12yz + 2z^2$

B37. $4x^2 - 8xy - 16xz + 4y^2 - 16yz + 8z^2$

B38. $9x^2 + 12xy - 12xz + 6y^2 - 8yz + 6z^2$

B39. $4x^2 + 14xy + 4xz + 4y^2 + 4yz + 4z^2$

B40. $2x^2 + 2xy - 18xz + 2y^2 + 18yz - 8z^2$

B41. $x^2 - 18xy - 6xz + 3y^2 + 6yz - 7z^2$

The *Rayleigh quotient* for the (real) symmetric $n \times n$ matrix A is the function $\rho_A(\mathbf{x}) = \frac{\mathbf{x}^T A\mathbf{x}}{\mathbf{x}^T \mathbf{x}}$ defined for $\mathbf{x} \neq \mathbf{0}$. If each matrix A in Exercises B42–51 has its eigenvalues ordered $\lambda_1 \leq \lambda_2 \leq \lambda_3$,

   (a) show that $\rho_A(\mathbf{x})$ assumes its maximum value at $\mathbf{x} = \mathbf{x}_3$, the eigenvector corresponding to λ_3, the largest eigenvalue, and

that the maximum value of the Rayleigh quotient is λ_3. *Hint:* If $\mathbf{v} = a\mathbf{i} + b\mathbf{j} + c\mathbf{k}$, then $\rho_A(\mathbf{v})$ is a function of a, b, c. Use the techniques of multivariable calculus to maximize this function.

   (b) show that $\rho_A(\mathbf{x})$ assumes its minimum value at $\mathbf{x} = \mathbf{x}_1$, the eigenvector corresponding to λ_1, the smallest eigenvalue, and that the minimum value of the Rayleigh quotient is λ_1. The maximization process used in part (a) should yield both the maximum and the minimum.

B42. $\begin{bmatrix} 12 & 8 & 8 \\ 8 & 6 & -2 \\ 8 & -2 & 6 \end{bmatrix}$ **B43.** $\begin{bmatrix} -1 & -8 & -3 \\ -8 & 7 & -1 \\ -3 & -1 & 2 \end{bmatrix}$ **B44.** $\begin{bmatrix} 3 & 2 & 3 \\ 2 & 6 & 6 \\ 3 & 6 & -5 \end{bmatrix}$

B45. $\begin{bmatrix} -5 & -9 & -4 \\ -9 & 8 & 9 \\ -4 & 9 & -5 \end{bmatrix}$ **B46.** $\begin{bmatrix} -6 & 5 & -2 \\ 5 & 2 & 2 \\ -2 & 2 & 3 \end{bmatrix}$ **B47.** $\begin{bmatrix} 0 & -1 & 1 \\ -1 & 5 & 4 \\ 1 & 4 & 5 \end{bmatrix}$

B48. $\begin{bmatrix} 2 & -1 & 6 \\ -1 & 9 & -1 \\ 6 & -1 & 2 \end{bmatrix}$ **B49.** $\begin{bmatrix} -7 & 4 & -2 \\ 4 & 9 & -6 \\ -2 & -6 & 0 \end{bmatrix}$

B50. $\begin{bmatrix} 1 & 4 & 6 \\ 4 & -11 & -9 \\ 6 & -9 & 1 \end{bmatrix}$ **B51.** $\begin{bmatrix} -6 & 6 & 8 \\ 6 & 7 & -6 \\ 8 & -6 & -6 \end{bmatrix}$

For the (real) symmetric $n \times n$ matrices of Exercises B42–51, the eigenvalues λ_s, $s = 2, \ldots, n - 1$, are characterized by the min-max principle. To pair the next-larger and next-smaller eigenvalues, let $r = 1, \ldots, n - 2$ so that λ_{n-r} and λ_{r+1} are, respectively, the rth eigenvalue smaller than the largest and the rth eigenvalue larger than the smallest. For each r, let $\{\mathbf{v}_1, \ldots, \mathbf{v}_r\}$ be a set of exactly r arbitrary vectors. Then, $\lambda_{n-r} = \min_{\mathbf{v}_1, \ldots, \mathbf{v}_r} \max_{\mathbf{x} \cdot \mathbf{v}_k = 0} \rho_A(\mathbf{x})$ and $\lambda_{r+1} = \max_{\mathbf{v}_1, \ldots, \mathbf{v}_r} \min_{\mathbf{x} \cdot \mathbf{v}_k = 0} \rho_A(\mathbf{x})$, where the optimizations are restricted by $\mathbf{x} \neq \mathbf{0}$. The extreme values occur for the eigenvectors corresponding to the eigenvalues. For λ_{n-r}, the "inner" maximum is taken over all vectors $\mathbf{x} \neq \mathbf{0}$ orthogonal to each fixed set of $\{\mathbf{v}_1, \ldots, \mathbf{v}_r\}$. Hence, the optimal vector $\mathbf{x}$ is a (complicated) function of the vectors $\mathbf{v}_k$. The "outer" minimization is then taken over all possible sets of vectors $\{\mathbf{v}_1, \ldots, \mathbf{v}_r\}$, with a similar explanation for λ_{r+1}. (See [64] or [44] for proofs.) For the matrices in Exercises B52–61, where $n = 3$, $r = 1$, and $n - r = r + 1 = 2$, so that the two conditions merge, apply the min-max principle to find the "middle" eigenvalue and its corresponding eigenvector. The "inner" extreme of $\rho_A(\mathbf{x})$ is taken over all vectors $\mathbf{x} = x\mathbf{i} + y\mathbf{j} + z\mathbf{k}$ that are orthogonal to $\mathbf{v}_1 = c_1\mathbf{i} + c_2\mathbf{j} + c_3\mathbf{k}$, so the optimization in x, y, z, constrained by $c_1 x + c_2 y + c_3 z = 0$, could be performed by the Lagrange multiplier technique. The optimal x, y, z are thereby computed as functions of c_1, c_2, c_3; and the "outer" optimization is computed with respect to the variables c_1, c_2, c_3.

B52. A is the matrix of Exercise B42

B53. A is the matrix of Exercise B43

B54. A is the matrix of Exercise B44

B55. A is the matrix of Exercise B45

33.5 Vector and Matrix Norms

Vector Norms

As seen in Section 31.1, the vector $\mathbf{P} = x_1\mathbf{i} + x_2\mathbf{j} + x_3\mathbf{k}$ can be "measured" with a variety of vector norms, some of which were given in Table 31.1. Corresponding to each vector norm, we can define a matrix norm, a real number that measures some "size" property of the matrix. How these numbers are assigned and what they measure are the subjects of this section.

Matrix Norms

A matrix norm is a function that assigns to the matrix A, the real number $\|A\|$ with the properties

1. $\|A\| \geq 0$.
2. $\|A\| = 0$ just when (if and only if) A is the zero matrix.
3. $\|\alpha A\| = |\alpha| \|A\|$ for any real or complex scalar α.
4. $\|A + B\| \leq \|A\| + \|B\|$ (the triangle inequality).
5. $\|AB\| \leq \|A\| \|B\|$.

Strictly speaking, property 5 is convenient but not necessary. We will include it in our requirements for a matrix norm because one of its consequences is that $\|A^k\| \leq \|A\|^k$. Thus, if $\|A\| < 1$, then $\lim_{k \to \infty} \|A^k\| = 0$.

EXAMPLE 33.7   The Frobenius norm of the $r \times c$ matrix A whose entries are a_{ij} is the number

$$\|A\|_{\mathrm{F}} = \sqrt{\sum_{j=1}^{c} \sum_{i=1}^{r} |a_{ij}|^2}$$

This is just the square root of the sum of the squares of the magnitude of each entry in the matrix. Of course, if A contains only real entries, the absolute values are not necessary. As an example, the Frobenius norm of the matrix A in (33.11) is $\|A\|_{\mathrm{F}} = \sqrt{313}$.

$$A = \begin{bmatrix} -6 & 8 & -9 \\ 9 & -3 & 5 \\ 3 & -2 & 2 \end{bmatrix} \tag{33.11}$$

❖

Compatible Matrix Norms

A matrix norm $\| \cdot \|_{\mathrm{M}}$ is said to be *compatible* with a vector norm $\| \cdot \|_{\mathrm{V}}$ if the inequality (33.12) holds for every vector $\mathbf{x}$ in R^n, the space of column vectors with n real components.

$$\|A\mathbf{x}\|_{\mathrm{V}} \leq \|A\|_{\mathrm{M}} \|\mathbf{x}\|_{\mathrm{V}} \tag{33.12}$$

Subordinate Matrix Norms

Each vector norm in Table 31.1 defines a *subordinate* (or *induced*) matrix norm by the prescription

$$\|A\|_k = \max_{\|\mathbf{x}\|_k = 1} \|A\mathbf{x}\|_k \tag{33.13}$$

That (33.13) defines a compatible norm (satisfying properties 1–5) is the content of theorems whose proofs can be found in [52], for example. That (33.13) is equivalent to

$$\|A\|_k = \max \frac{\|A\mathbf{x}\|_k}{\|\mathbf{x}\|_k} \quad \mathbf{x} \neq \mathbf{0}$$

is a consequence of property 3 for a matrix norm. Another consequence of (33.13) is the very useful

$$\|A\mathbf{x}\|_k \leq \|A\|_k \|\mathbf{x}\|_k$$

in which the norm on the left is the same vector norm as is used on the right (for $\|\mathbf{x}\|_k$) so that the matrix norm $\|A\|_k$ is a bound on the factor by which the matrix A "stretches" any vector $\mathbf{x}$.

When the context is clear, we will drop the subscript k.

The Frobenius norm is not a subordinate norm because there is no vector norm from which it is derived.

THEOREM 33.1

1. $\|A\|_1 = \max\limits_c \sum_{r=1}^{n} |a_{rc}|$, largest column-sum of absolute values in $A = \{a_{ij}\}$.

2. $\|A\|_\infty = \max\limits_r \sum_{c=1}^{n} |a_{rc}|$, largest row-sum of absolute values in $A = \{a_{ij}\}$.

3. $\|A\|_2 = \sqrt{\hat{\lambda}}$, where $\hat{\lambda}$ is the largest eigenvalue of the symmetric matrix $A^{\mathsf{T}}A$.

As a memory aid, we suggest the following. The "1" in $\|A\|_1$ is a vertical stroke, suggestive of "sum the elements down the columns," whereas the "∞" in $\|A\|_\infty$ appears to be the number 8 lying horizontally, suggestive of "sum across each row." For a proof of this theorem, we suggest a text such as [52] or [64].

EXAMPLE 33.8   In addition to $\|A\|_F = \sqrt{313} \doteq 17.7$ for the Frobenius norm, the matrix (33.11), has, among others, the norms $\|A\|_1 = 18$, $\|A\|_2 = 17.08$, and $\|A\|_\infty = 23$. The norms $\|A\|_1$ and $\|A\|_\infty$ are available by inspection. The 2-norm requires computing the eigenvalues of the (necessarily) symmetric matrix

$$A^{\mathsf{T}}A = \begin{bmatrix} 126 & -81 & 105 \\ -81 & 77 & -91 \\ 105 & -91 & 110 \end{bmatrix} \tag{33.14}$$

These eigenvalues of $A^{\mathsf{T}}A$ are 0.1783, 20.93, 291.89; and their square roots are 0.422, 4.575, 17.085. The square root of the largest eigenvalue of $A^{\mathsf{T}}A$ is indeed the 2-norm of A. ❖

EXAMPLE 33.9   Let A be the matrix on the left in (33.15), so that $A^{\mathsf{T}}A$ is the matrix on the right

$$A = \begin{bmatrix} 1 & 2 \\ 3 & 4 \end{bmatrix} \quad A^{\mathsf{T}}A = \begin{bmatrix} 10 & 14 \\ 14 & 20 \end{bmatrix} \tag{33.15}$$

We can compute

$$\|A\|_2 = \sqrt{\hat{\lambda}} = \sqrt{15 + \sqrt{221}} = \tfrac{1}{2}\left[\sqrt{34} + \sqrt{26}\right] \doteq \sqrt{29.866} \doteq 5.465 \tag{33.16}$$

where $\hat{\lambda}$ is the largest eigenvalue of $A^{\mathrm{T}}A$. Alternatively, we can compute $\|A\|_2$ by carrying out the optimization in (33.13) for the 2-norm.

For the unit vector $\mathbf{v} = \cos t\,\mathbf{i} + \sin t\,\mathbf{j}$, form the vector

$$A\mathbf{v} = (\cos t + 2\sin t)\mathbf{i} + (3\cos t + 4\sin t)\mathbf{j}$$

then compute and plot

$$\|A\mathbf{v}\|_2^2 = (\cos t + 2\sin t)^2 + (3\cos t + 4\sin t)^2 \tag{33.17}$$

obtaining Figure 33.5. Using the techniques of elementary calculus, the angle at which the first maximum occurs is found to be $t = \arctan\frac{5+\sqrt{221}}{14} \stackrel{\circ}{=} 0.9569101336$, and the corresponding maximum value of $\|A\mathbf{v}\|_2$ is then $\sqrt{15 + \sqrt{221}}$, in agreement with (33.16). The maximum in $\|A\mathbf{v}\|_2$ occurs for $\mathbf{v}_1 = 0.576\mathbf{i} + 0.817\mathbf{j}$, so $A\mathbf{v}_1 = 2.211\mathbf{i} + 4.998\mathbf{j}$. Figure 33.6 shows both $\mathbf{v}_1$ and $A\mathbf{v}_1$. (See the accompanying Maple worksheet for an animation showing $\mathbf{v}$ and $A\mathbf{v}$ as $\mathbf{v}$ rotates around the unit circle.) ❖

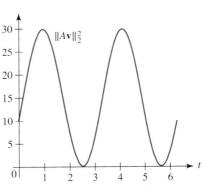

FIGURE 33.5 Graph of $\|A\mathbf{v}\|_2^2$ for a unit vector $\mathbf{v}$ and matrix A in Example 33.9

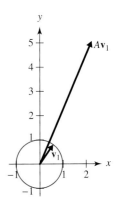

FIGURE 33.6 The vectors $\mathbf{v}_1$ and $A\mathbf{v}_1$ in Example 33.9

EXERCISES 33.5–Part A

A1. For A given in (33.11), verify that $\|A\|_{\mathrm{F}} = \sqrt{331}$.

A2. For A in (33.11), verify that $\|A\|_1 = 18$ and that $\|A\|_\infty = 23$.

A3. For A in (33.11), verify that $A^{\mathrm{T}}A$ is given by (33.14).

A4. Obtain the eigenvalues and their square roots for $A^{\mathrm{T}}A$ given in (33.14).

A5. For A in (33.15), verify that the largest eigenvalue of $A^{\mathrm{T}}A$ is $15 + \sqrt{221}$.

EXERCISES 33.5–Part B

B1. Show that $\sqrt{15 + \sqrt{221}} = \sqrt{\frac{17}{2}} + \sqrt{\frac{13}{2}} = \frac{1}{2}(\sqrt{34} + \sqrt{26})$. *Hint:* Write $\sqrt{15 + \sqrt{221}} = \sqrt{a} + \sqrt{b}$, square both sides, and match like terms.

B2. Show that the first maximum in (33.17) occurs at $\hat{t} = \arctan\frac{5+\sqrt{221}}{14}$.

B3. Show that at $\hat{t}$ given in Exercise B2, (33.17) becomes $15 + \sqrt{221}$.

B4. Obtain $\mathbf{v}_1(\hat{t})$ and $A\mathbf{v}_1(\hat{t})$, where $\hat{t}$ is given in Exercise B2.

For the vector $\mathbf{v}$ given in each of Exercises B5–14, obtain $\|\mathbf{v}\|_1$, $\|\mathbf{v}\|_2$, $\|\mathbf{v}\|_3$, and $\|\mathbf{v}\|_\infty$.

B5. $\mathbf{v} = 8\mathbf{i} + 12\mathbf{j}$ **B6.** $\mathbf{v} = 12\mathbf{i} - 2\mathbf{j}$ **B7.** $\mathbf{v} = 8\mathbf{i} + 11\mathbf{j}$

B8. $\mathbf{v} = -3\mathbf{i} - 2\mathbf{j}$ **B9.** $\mathbf{v} = -5\mathbf{i} + 12\mathbf{j}$ **B10.** $\mathbf{v} = 7\mathbf{i} - 9\mathbf{j} + 3\mathbf{k}$

B11. $\mathbf{v} = 5\mathbf{i} + 10\mathbf{j} + 11\mathbf{k}$ **B12.** $\mathbf{v} = \mathbf{i} - 11\mathbf{j} - 12\mathbf{k}$

B13. $\mathbf{v} = 6\mathbf{i} - 5\mathbf{j} - 2\mathbf{k}$ **B14.** $\mathbf{v} = -2\mathbf{i} + 8\mathbf{j} + 4\mathbf{k}$

For matrix A given in each of Exercises B15–24:

(a) Obtain $\|A\|_1$, $\|A\|_2$, $\|A\|_f$, and $\|A\|_\infty$. Use an appropriate software tool to determine $\|A\|_2$.

(b) Show that $\|A\|_2 = \sqrt{\lambda}$, where λ is the largest eigenvalue of $A^T A$.

(c) Show that $\|A\|_2 = \max_{\|\mathbf{x}\|_2=1} \|A\mathbf{x}\|_2$. *Hint:* Treat this as a constrained optimization problem, where the constraint is $\|\mathbf{x}\|_2^2 = 1$ and the objective function to be maximized is $\|A\mathbf{x}\|_2^2$. The Lagrange multiplier technique works, but a computer algebra system with a command for finding extrema will be useful here.

B15. $\begin{bmatrix} -6 & 8 \\ 10 & 1 \end{bmatrix}$ **B16.** $\begin{bmatrix} -11 & 1 \\ -2 & 6 \end{bmatrix}$ **B17.** $\begin{bmatrix} 9 & -3 \\ -7 & 1 \end{bmatrix}$

B18. $\begin{bmatrix} -5 & -2 \\ 12 & -2 \end{bmatrix}$ **B19.** $\begin{bmatrix} 4 & -1 \\ -5 & 12 \end{bmatrix}$ **B20.** $\begin{bmatrix} -1 & 11 & 4 \\ -3 & 11 & -6 \\ 12 & 4 & 11 \end{bmatrix}$

B21. $\begin{bmatrix} 5 & 4 & 6 \\ -12 & -1 & 1 \\ 1 & 8 & 7 \end{bmatrix}$ **B22.** $\begin{bmatrix} 3 & 4 & 8 \\ 12 & -3 & -2 \\ -5 & 5 & 1 \end{bmatrix}$

B23. $\begin{bmatrix} -10 & 12 & -2 \\ -1 & 11 & -9 \\ 11 & 8 & 11 \end{bmatrix}$ **B24.** $\begin{bmatrix} 2 & 0 & -5 \\ 12 & 7 & -9 \\ 3 & -7 & -5 \end{bmatrix}$

B25. Show that $\|A\|_2^2 = \max_{\mathbf{x}\neq 0} \frac{\|A\mathbf{x}\|_2^2}{\|\mathbf{x}\|_2^2} = \max_{\mathbf{x}\neq 0} \rho_{A^T A}(\mathbf{x})$, where $\rho_{A^T A}(\mathbf{x})$, the Rayleigh quotient for $A^T A$, defined for Exercises B42–51 in Section 33.4, has for its maximum value, the largest eigenvalue of $A^T A$. Hence, $\|A\|_2$ is the square root of the largest eigenvalue of $A^T A$.

For the matrix A in each of Exercises B26–28:

(a) Obtain $\|A\|_1$.

(b) Sketch the locus of points satisfying $\|\mathbf{x}\|_1 = |x| + |y| = 1$.

(c) Obtain $\|A\mathbf{x}\|_1$ for $\mathbf{x}$ along each segment of the locus sketched in part (b).

(d) Show that the maximum of $\|A\mathbf{x}\|$ is the largest sum of absolute values for the columns of A.

B26. $\begin{bmatrix} 5 & 10 \\ 11 & -10 \end{bmatrix}$ **B27.** $\begin{bmatrix} -10 & 12 \\ -12 & -5 \end{bmatrix}$ **B28.** $\begin{bmatrix} a & b \\ c & d \end{bmatrix}$

For the matrix A given in each of Exercises B29–32:

(a) Obtain $\|A\|_1$.

(b) Sketch the level curves of $\|\mathbf{x}\|_1 = |x| + |y| = c$, for $0 < c \leq 10$, being sure to include $c = 1$.

(c) Let $\mathbf{x}_k$, $k = 1, \ldots, 12$, be equally spaced along the level curve $c = 1$, and plot, on the sketch made in part (a), each of the vectors $A\mathbf{x}_k$, $k = 1, \ldots, 12$.

(d) Is the largest value for $\|A\mathbf{x}\|_1$ (as determined by the level curves from part (b)) close to the value of $\|A\|_1$?

B29. $\begin{bmatrix} -5 & 2 \\ 4 & -6 \end{bmatrix}$ **B30.** $\begin{bmatrix} 5 & 2 \\ 0 & 4 \end{bmatrix}$ **B31.** $\begin{bmatrix} -4 & 1 \\ -3 & -2 \end{bmatrix}$ **B32.** $\begin{bmatrix} -5 & -1 \\ 2 & -1 \end{bmatrix}$

For the matrix A given in each of Exercises B33–36:

(a) Obtain $\|A\|_\infty$.

(b) Show that the locus of points satisfying $\|\mathbf{x}\|_\infty = 1$ is a square with side 2 and center at the origin.

(c) Show that the vectors $\mathbf{x}_1, \ldots, \mathbf{x}_4 = \begin{bmatrix} 1 \\ t \end{bmatrix}, \begin{bmatrix} -1 \\ t \end{bmatrix}, \begin{bmatrix} t \\ 1 \end{bmatrix}, \begin{bmatrix} t \\ -1 \end{bmatrix}$, $-1 \leq t \leq 1$, trace the sides of the square in part (b).

(d) Obtain $A\mathbf{x}_k$ and $\|A\mathbf{x}_k\|_\infty$ for $k = 1, \ldots, 4$; show that $\|A\mathbf{x}_k\|_\infty$ is maximized only when $t = \pm 1$, with the sign dependent on the signs of the entries of the matrix A, and chosen so the vector norm is a maximal sum of absolute values across a row of A.

(e) Show that the largest of the four norms in part (d) is the largest row-sum of absolute values.

B33. $\begin{bmatrix} 10 & 1 \\ 3 & 7 \end{bmatrix}$ **B34.** $\begin{bmatrix} -1 & -1 \\ -3 & 4 \end{bmatrix}$ **B35.** $\begin{bmatrix} -6 & 8 \\ 11 & 12 \end{bmatrix}$ **B36.** $\begin{bmatrix} 6 & 7 \\ -11 & -12 \end{bmatrix}$

For the matrix A given in each of Exercises B37–40:

(a) Obtain $\|A\|_4$ by applying the definition $\|A\|_4 = \max_{\|\mathbf{x}\|_4=1} \|A\mathbf{x}\|_4$. This is a constrained optimization problem, a candidate for the Lagrange multiplier technique, where the constraint can be taken as $g(x, y) = x^4 + y^4 = 1$ and the objective function as $f(x, y) = \|A\mathbf{x}\|_4^4$.

(b) Obtain $\|A\|_4$ by applying the definition $\|A\|_4 = \max_{\mathbf{x}\neq 0} \frac{\|A\mathbf{x}\|_4}{\|\mathbf{x}\|_4}$. This calculation can be approached as an unconstrained optimization problem and yields to the techniques of multivariable calculus.

B37. $\begin{bmatrix} -6 & 8 \\ -2 & -7 \end{bmatrix}$ **B38.** $\begin{bmatrix} 8 & 10 \\ -7 & 1 \end{bmatrix}$ **B39.** $\begin{bmatrix} 8 & -3 \\ 12 & -2 \end{bmatrix}$ **B40.** $\begin{bmatrix} -6 & 10 \\ -2 & 9 \end{bmatrix}$

33.6   Least Squares

Interpolation

Passing a curve through a prescribed set of n points is called *interpolation*. The number of points matches the number of free parameters that define the curve, so the interpolation problem requires solving a set of n equations in n unknowns.

EXAMPLE 33.10 Consider the problem of passing the parabola $y = ax^2 + bx + c$ through the three points $(1, 2)$, $(-3, 4)$, $(5, -5)$. Since the coordinates of each point must satisfy the equation of the parabola, substitution of these coordinates into the equation of the parabola leads to the equations on the left in Table 33.4. This set of three equations in the three unknowns a, b, and c has the solution given on the right in Table 33.4, so the interpolating parabola is $y = -\frac{5}{32}x^2 - \frac{13}{16}x + \frac{95}{32}$. ❖

$$\left.\begin{array}{r} a + b + c = 2 \\ 9a - 3b + c = 4 \\ 25a + 5b + c = -5 \end{array}\right\} \Rightarrow \left\{\begin{array}{l} a = -\frac{5}{32} \\ b = -\frac{13}{16} \\ c = \frac{95}{32} \end{array}\right.$$

TABLE 33.4 Interpolation equations and their solution in Example 33.10

A Least-Squares Fit

Far more common than interpolation is the least-squares fitting problem arising from experimental data. Data points are acquired by experiment and some curve is passed through the data points in an optimal fashion. There are more points than parameters for the curve, so the set of equations that result are *overdetermined*, that is, there are more equations than unknowns. In fact, there is generally no solution to the set of equations, so a "solution" that minimizes the error of the fit is accepted as the "best fit" of the curve to the data.

EXAMPLE 33.11 As an example, we will fit the parabola

$$y(x) = ax^2 + bx + c \tag{33.18}$$

to the nine data points (x_k, y_k) in Table 33.5. Figure 33.7 contains a plot of these points that seem to fit a parabolic shape.

x	-4	-3	-2	-1	0	1	2	3	4
y	-60	-44	-21	-4	10	3	-5	-15	-39

TABLE 33.5 Data points for Example 33.11

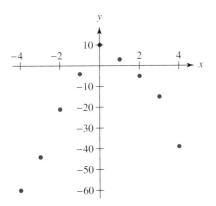

FIGURE 33.7 The data points from Table 33.4 in Example 33.11

For each point, we substitute the corresponding x and y values into (33.18), thereby generating the nine equations

$$\begin{array}{lll} 16a - 4b + c = -60 & a - b + c = -4 & 4a + 2b + c = -5 \\ 9a - 3b + c = -44 & c = 10 & 9a + 3b + c = -15 \\ 4a - 2b + c = -21 & a + b + c = 3 & 16a + 4b + c = -39 \end{array} \tag{33.19}$$

in the three unknowns a, b, and c. In general, we do not expect a set of nine equations in only three unknowns to have a solution. In fact, if the equations are written as $A\mathbf{x} = \mathbf{y}$, the augmented matrix $[A, \mathbf{y}]$ can be reduced to the row-echelon form $\text{rref}(A)$ in (33.20). The fourth row represents the equation $0a + 0b + 0c = 1$ for which there is no solution.

$$[A, \mathbf{y}] = \begin{bmatrix} 16 & -4 & 1 & -60 \\ 9 & -3 & 1 & -44 \\ 4 & -2 & 1 & -21 \\ 1 & -1 & 1 & -4 \\ 0 & 0 & 1 & 10 \\ 1 & 1 & 1 & 3 \\ 4 & 2 & 1 & -5 \\ 9 & 3 & 1 & -15 \\ 16 & 4 & 1 & -39 \end{bmatrix} \quad \text{rref}([A, \mathbf{y}]) = \begin{bmatrix} 1 & 0 & 0 & 0 \\ 0 & 1 & 0 & 0 \\ 0 & 0 & 1 & 0 \\ 0 & 0 & 0 & 1 \\ 0 & 0 & 0 & 0 \\ 0 & 0 & 0 & 0 \\ 0 & 0 & 0 & 0 \\ 0 & 0 & 0 & 0 \\ 0 & 0 & 0 & 0 \end{bmatrix} \tag{33.20}$$

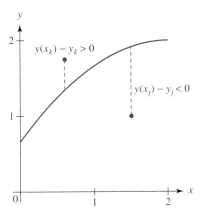

FIGURE 33.8 Deviation of data points from a least-squares curve

Hence, we seek a curve that is as close as possible to the points and that minimizes the vertical deviations from the points to the curve. (See Figure 33.8.) These deviations are

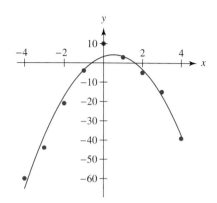

FIGURE 33.9 Data points and least-squares parabola in Example 33.11

squared (so that positive and negative deviations don't cancel out) and added. The curve that minimizes this sum of squares of the deviations is called the least-squares fit of the curve to the data points.

For the nine data points of Table 33.5, let $S = \sum_{k=1}^{9}[y(x_k) - y_k]^2$ be the sum of squares of the deviations from the hypothesized parabola $y(x)$. This sum of squares, which simplifies to

$$S = 4440a - 420b + 9c^2 + 60b^2 + 7873 + 120ac + 708a^2 + 350c \qquad (33.21)$$

is minimized by solving the three *normal* equations on the left in (33.22) for the solution on the right. This minimization process amounts to finding the critical point (a, b, c) for S and gives the least-squares curve $y = -\frac{790}{231}x^2 + \frac{7}{2}x + \frac{775}{231}$, graphed in Figure 33.9.

$$\left. \begin{array}{r} \frac{1}{2}\frac{\partial S}{\partial a} = 708a + 60c + 2220 = 0 \\[2mm] \frac{1}{2}\frac{\partial S}{\partial b} = 60b - 210 = 0 \\[2mm] \frac{1}{2}\frac{\partial S}{\partial c} = 60a + 9c + 175 = 0 \end{array} \right\} \Rightarrow \left\{ \begin{array}{l} a = -\dfrac{790}{231} \\[2mm] b = \dfrac{7}{2} \\[2mm] c = \dfrac{775}{231} \end{array} \right. \qquad (33.22)$$

The minimum value of S, the sum of squares of the deviations, is then $S_{\min} = \frac{30703}{231} = 132.91$. ❖

Matrix Formulation

The original nine equations in Example 33.11 are linear, as are the three normal equations, suggesting a matrix formulation of the least-squares problem. Let A be the matrix of coefficients in the original set of nine equations, and let **y** be the vector of right-hand sides in these equations. (See (33.20).) Let M be the matrix of coefficients in the three normal equations, and let **b** be the vector of right-hand sides, provided the equations are rearranged as

$$708a + 60c = -2220 \qquad 60b = 210 \qquad 60a + 9c = -175$$

Finally, define the vector **u** as the vector of unknowns a, b, c. Then the normal equations can be written as $M\mathbf{u} = \mathbf{b}$, where

$$M = \begin{bmatrix} 708 & 0 & 60 \\ 0 & 60 & 0 \\ 60 & 0 & 9 \end{bmatrix} \qquad \mathbf{u} = \begin{bmatrix} a \\ b \\ c \end{bmatrix} \qquad \mathbf{b} = \begin{bmatrix} -2220 \\ 210 \\ -175 \end{bmatrix} \qquad (33.23)$$

A calculation would show that $M = A^{\mathrm{T}}A$ and $\mathbf{b} = A^{\mathrm{T}}\mathbf{y}$. Thus, the original set of equations $A\mathbf{u} = \mathbf{y}$, multiplied through by A^{T}, become $M\mathbf{u} = \mathbf{b}$, the *normal equations*, since $A^{\mathrm{T}}[A\mathbf{u} = \mathbf{y}]$ leads to $A^{\mathrm{T}}A\mathbf{u} = A^{\mathrm{T}}\mathbf{y}$, or $M\mathbf{u} = \mathbf{b}$. The assumption that the columns of A are independent means M is invertible. In agreement with (33.22), the solution of $M\mathbf{u} = \mathbf{b}$ is found to be

$$\mathbf{u} = \begin{bmatrix} -\frac{790}{231} \\[2mm] \frac{7}{2} \\[2mm] \frac{775}{231} \end{bmatrix} \qquad (33.24)$$

Geometric Analysis of the Matrix A

The matrix A has dimensions 9×3, nine rows and three columns. It multiplies vectors in R^3 and sends them to R^9, the space of column vectors with nine real components. Matrix multiplication therefore defines a function from a domain in R^3 to a range in R^9. The domain is called the *row space*, a basis of which is given by $\{\mathbf{i}, \mathbf{j}, \mathbf{k}\}$, as can be seen from rref(A), the reduced row-echelon form of A given in (33.25).

$$\begin{bmatrix} 1 & 0 & 0 \\ 0 & 1 & 0 \\ 0 & 0 & 1 \\ 0 & 0 & 0 \\ 0 & 0 & 0 \\ 0 & 0 & 0 \\ 0 & 0 & 0 \\ 0 & 0 & 0 \\ 0 & 0 & 0 \end{bmatrix} \tag{33.25}$$

In (33.25), the three rows with the lead 1s are a basis for the row space. Here, that row space is all of R^3.

The range is that part of R^9 spanned ("reached") by the columns of A. This part of R^9 consists of all possible linear combinations of the columns of A and is called the *column space* of A. The reduced row-echelon for A shows all three columns of A are independent. Hence, the three columns of A are a basis for the column space of A.

A solution to the system $A\mathbf{u} = \mathbf{y}$ exists if the vector $\mathbf{y}$ lies in the column space of A. Thus, if $\mathbf{y}$ is a linear combination of the columns of A, then you can calculate a solution $\mathbf{u}$. If not, then there is no solution for $\mathbf{u}$. From (33.20), we know that $\mathbf{y}$ is not in the column space of A.

Rank of a Matrix

The number of independent rows or columns in a matrix is called its *rank*. Generally, texts define "row rank" as the number of independent rows and then prove that the "column rank" is the same. Hence, the "rank" of the matrix is a unique number.

The rank is therefore the number of basis elements in a basis for the row space. A basis for the column space would have the same number of elements. Thus, the rank of the matrix A is 3 because rref(A) has three nonzero rows. If the rank of A is 3 but the rank of the augmented matrix $[A, \mathbf{y}]$ is 4, then it is clear that one equation in the reduced row-echelon form of the augmented matrix must be the inconsistent equation $0 = 1$. This is a clear signal that the system $A\mathbf{u} = \mathbf{y}$ is inconsistent and, therefore, has no solution. In other words, $\mathbf{y}$ is not in the column space of A, so no $\mathbf{u}$ can be found that A sends to $\mathbf{y}$.

EXAMPLE 33.12   The matrix

$$A = \begin{bmatrix} -85 & -55 & -37 & -35 & 97 & 50 & 79 \\ 56 & 49 & 63 & 57 & -59 & 45 & -8 \\ -93 & 92 & 43 & -62 & 77 & 66 & 54 \end{bmatrix}$$

has rank 3. Since A has three rows and seven columns, the rank is as large as it can possibly be and A is said to be of *maximal rank*. The reduced row-echelon form of A is

$$\begin{bmatrix} 1 & 0 & 0 & \frac{172316}{409017} & -\frac{423736}{409017} & -\frac{340237}{409017} & -\frac{124109}{136339} \\ 0 & 1 & 0 & -\frac{318673}{409017} & -\frac{130444}{409017} & -\frac{515992}{409017} & \frac{139681}{136339} \\ 0 & 0 & 1 & \frac{464750}{409017} & \frac{95063}{409017} & \frac{995915}{409017} & \frac{201647}{136339} \end{bmatrix} \tag{33.26}$$

so a basis for the row space consists of the vectors forming the three rows of rref(A), or the three rows of A itself.

The column space is all of R^3. A basis for it is given by $\{\mathbf{i}, \mathbf{j}, \mathbf{k}\}$, as can be seen from the first three columns of rref(A) given in (33.26). ❖

EXAMPLE 33.13   The matrix A on the left in (33.27) has rank 2, as seen from its reduced row-echelon form given on the right in (33.27).

$$
\begin{bmatrix}
-28 & 28 & -15 & 5 & -10 & -9 & -6 \\
31 & -31 & 20 & -10 & 15 & 13 & 7 \\
24 & -24 & 21 & -15 & 18 & 15 & 6 \\
0 & 0 & -19 & 25 & -22 & -17 & -2 \\
6 & -6 & -9 & 15 & -12 & -9 & 0
\end{bmatrix}
\qquad
\begin{bmatrix}
1 & -1 & 0 & \frac{10}{19} & -\frac{5}{19} & -\frac{3}{19} & \frac{3}{19} \\
0 & 0 & 1 & -\frac{25}{19} & \frac{22}{19} & \frac{17}{19} & \frac{2}{19} \\
0 & 0 & 0 & 0 & 0 & 0 & 0 \\
0 & 0 & 0 & 0 & 0 & 0 & 0 \\
0 & 0 & 0 & 0 & 0 & 0 & 0
\end{bmatrix}
$$

$$\text{(33.27)}$$

The reduced row-echelon form of A tells us that there are two independent rows in A. Since we don't know which two of the five rows of A are independent, we select the two nonzero rows of rref(A) The reduced row-echelon form of A also says the first and third columns of A are a basis for the column space of A. The first and third columns of rref(A) are *not* a basis for this column space.

Null Space   The matrix A on the left in (33.27) is 5×7, that is, it has five rows and seven columns. Not all the rows are linearly independent. In fact, since A has rank 2, there are only two independent rows. Now A maps R^7 to R^5. The domain space for A is a two-dimensional subspace in R^7 spanned by, for example, the basis vectors $\mathbf{r}_1$ and $\mathbf{r}_2$ given on the left in (33.28). The matrix A maps the remaining portion of R^7 to $\mathbf{0}$, the zero vector in R^5. This part of R^7 is called the *null space* of A. Any vector $\mathbf{x}$ in the null space of A satisfies the equation $A\mathbf{x} = \mathbf{0}$. A basis for the null space of A, $\{\mathbf{n}_k, k = 1, \ldots, 5\}$, is given on the right

$$
\left\{
\begin{bmatrix} 0 \\ 0 \\ 1 \\ -\frac{25}{19} \\ \frac{22}{19} \\ \frac{17}{19} \\ \frac{2}{19} \end{bmatrix}
\begin{bmatrix} 1 \\ -1 \\ 0 \\ \frac{10}{19} \\ -\frac{5}{19} \\ -\frac{3}{19} \\ \frac{3}{19} \end{bmatrix}
\right\}
\left\{
\begin{bmatrix} -\frac{5}{2} \\ 0 \\ 0 \\ 1 \\ 0 \\ 0 \\ \frac{25}{2} \end{bmatrix}
\begin{bmatrix} 2 \\ 0 \\ 0 \\ 0 \\ 1 \\ 0 \\ -11 \end{bmatrix}
\begin{bmatrix} \frac{3}{2} \\ 0 \\ 0 \\ 0 \\ 0 \\ 1 \\ -\frac{17}{2} \end{bmatrix}
\begin{bmatrix} \frac{3}{2} \\ 0 \\ 1 \\ 0 \\ 0 \\ 0 \\ -\frac{19}{2} \end{bmatrix}
\begin{bmatrix} 1 \\ 1 \\ 0 \\ 0 \\ 0 \\ 0 \\ 0 \end{bmatrix}
\right\}
\quad \text{(33.28)}
$$

A calculation shows that $A\mathbf{n}_k = \mathbf{0}, k = 1, \ldots, 5$; and another calculation shows that for $j = 1, 2$, the dot products $\mathbf{r}_j \cdot \mathbf{n}_k, k = 1, \ldots, 5$, all vanish. Hence, the vectors in the row space are perpendicular to the vectors in the null space. (Since the basis vectors for the row space are orthogonal to the basis vectors for the null space, the row space itself is orthogonal to the null space.) The action of A on vectors in R^7 is shown in Figure 33.10.

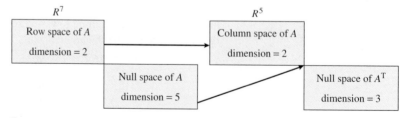

FIGURE 33.10   The four fundamental subspaces for A in Example 33.13

In the sense just discussed, the row space of A, of dimension 2, is orthogonal to the null space of A, of dimension 5. The matrix A maps R^7 to R^5 and takes vectors in the null space of A to the "origin" in R^5. The row space of A is mapped to the column space of A in R^5. The column space, of dimension 2, is orthogonal to the rest of R^5, which turns out to be the null space of the transpose of A. Clearly, if the column space in R^5 has dimension 2, then the orthogonal complement of the column space must be of dimension $5 - 2 = 3$.

These four subspaces are called "the four fundamental subspaces of A" in the literature.

❖

Geometric Interpretation of Least Squares

The least-squares problem $A\mathbf{u} = \mathbf{y}$, having the 9×3 matrix A defined in (33.20), has an interesting geometric interpretation. The matrix A maps R^3 to R^9. The vector $\mathbf{u}$ is in R^3, and the vector $\mathbf{y}$ is in R^9. Because $\mathbf{y}$ is not in the column space of A, there is no solution $\mathbf{u}$. What the least-squares solution provides, however, is that vector $\mathbf{u}$ that maps to the projection of $\mathbf{y}$ onto the column space of A. Because A is of maximal rank 3, the row space of A is all of R^3 and the null space of A is empty. All of R^3 maps to the three-dimensional column space of A over in R^9. The vector $\mathbf{y}$, in R^9, has a vector component in the column space of A and a vector component in the orthogonal complement. Figure 33.11 summarizes these observations.

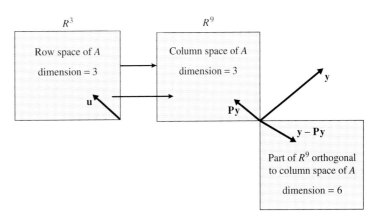

FIGURE 33.11 Fundamental subspaces for A in Example 33.11. Also shown are the projections of $\mathbf{y}$ onto, and orthogonal to, the column space of A

Analytic justification for the observations captured in Figure 33.11 begins with forming the projection matrix $P = A(A^TA)^{-1}A^T$, which projects vectors in R^9 onto the three-dimensional column space of A.

$$
P = \begin{bmatrix}
\frac{109}{165} & \frac{21}{55} & \frac{9}{55} & \frac{1}{165} & -\frac{1}{11} & -\frac{7}{55} & -\frac{17}{165} & -\frac{1}{55} & \frac{7}{55} \\
\frac{21}{55} & \frac{46}{165} & \frac{21}{110} & \frac{13}{110} & \frac{2}{33} & \frac{1}{55} & -\frac{1}{110} & -\frac{7}{330} & -\frac{1}{55} \\
\frac{9}{55} & \frac{21}{110} & \frac{232}{1155} & \frac{149}{770} & \frac{13}{77} & \frac{293}{2310} & \frac{26}{385} & -\frac{1}{110} & -\frac{17}{165} \\
\frac{1}{165} & \frac{13}{110} & \frac{149}{770} & \frac{268}{1155} & \frac{18}{77} & \frac{153}{770} & \frac{293}{2310} & \frac{1}{55} & -\frac{7}{55} \\
-\frac{1}{11} & \frac{2}{33} & \frac{13}{77} & \frac{18}{77} & \frac{59}{231} & \frac{18}{77} & \frac{13}{77} & \frac{2}{33} & -\frac{1}{11} \\
-\frac{7}{55} & \frac{1}{55} & \frac{293}{2310} & \frac{153}{770} & \frac{18}{77} & \frac{268}{1155} & \frac{149}{770} & \frac{13}{110} & \frac{1}{165} \\
-\frac{17}{165} & -\frac{1}{110} & \frac{26}{385} & \frac{293}{2310} & \frac{13}{77} & \frac{149}{770} & \frac{232}{1155} & \frac{21}{110} & \frac{9}{55} \\
-\frac{1}{55} & -\frac{7}{330} & -\frac{1}{110} & \frac{1}{55} & \frac{2}{33} & \frac{13}{110} & \frac{21}{110} & \frac{46}{165} & \frac{21}{55} \\
\frac{7}{55} & -\frac{1}{55} & -\frac{17}{165} & -\frac{7}{55} & -\frac{1}{11} & \frac{1}{165} & \frac{9}{55} & \frac{21}{55} & \frac{109}{165}
\end{bmatrix}
$$

The projection of $\mathbf{y}$ onto the column space of A is then $P\mathbf{y}$, and the vector component of $\mathbf{y}$ orthogonal to the column space of A is $\mathbf{y}_{\perp[A]} = \mathbf{y} - P\mathbf{y} = (I - P)\mathbf{y}$.

$$
P\mathbf{y} = \begin{bmatrix} -\frac{719}{11} \\ -\frac{2503}{66} \\ -\frac{1334}{77} \\ -\frac{549}{154} \\ \frac{775}{231} \\ \frac{529}{154} \\ -\frac{256}{77} \\ -\frac{1117}{66} \\ -\frac{411}{11} \end{bmatrix}
\qquad
\mathbf{y}_{\perp[A]} = \begin{bmatrix} \frac{59}{11} \\ -\frac{401}{66} \\ -\frac{283}{77} \\ \frac{67}{154} \\ \frac{1535}{231} \\ -\frac{67}{154} \\ -\frac{129}{77} \\ \frac{127}{66} \\ -\frac{18}{11} \end{bmatrix}
$$

The minimum value of S, the sum of squares found by the least-squares technique, was $\frac{30703}{231}$, which is also the value of $\|\mathbf{y}_{\perp[A]}\|^2$. The least-squares solution (33.24) is the vector in R^3 that A maps to $P\mathbf{y}$, the component of $\mathbf{y}$ lying in the column space of A over in R^9. This is verified by showing $A\mathbf{u} = P\mathbf{y}$. We summarize the solution process as follows.

1. The least-squares problem consists of the overdetermined system $A\mathbf{u} = \mathbf{y}$.
2. Multiplication by A^T gives the normal equations $A^T A\mathbf{u} = A^T\mathbf{y}$ or $M\mathbf{u} = \mathbf{b}$.
3. The 3×3 matrix $M = A^T A$ is invertible, so $\mathbf{u} = (A^T A)^{-1} A^T\mathbf{y}$ is the least-squares solution.

We summarize the analysis of this solution process as follows.

1. The matrix A maps $\mathbf{u} = (A^T A)^{-1} A^T\mathbf{y}$ to $A\mathbf{u} = A(A^T A)^{-1} A^T\mathbf{y} = P\mathbf{y}$.
2. The vector $P\mathbf{y}$ is the projection of $\mathbf{y}$ onto the column space of A over in R^9.
3. The least-squares solution $\mathbf{u}$ is the vector in R^3 mapped by A to $P\mathbf{y}$, not $\mathbf{y}$.
4. The vector $\mathbf{b} = A^T\mathbf{y}$ is the pre-image in R^3 of $P\mathbf{y}$.
5. The matrix $M = A^T A$ maps $\mathbf{u}$ in R^3 to $\mathbf{b}$ in R^3.
6. Solving the normal equations $M\mathbf{u} = \mathbf{b}$ finds the $\mathbf{u}$ in R^3 that A maps to $P\mathbf{y}$ in the column space of A over in R^9.

EXERCISES 33.6–Part A

A1. Verify the equations and solution in Table 33.4.

A2. Verify the equations listed in (33.19).

A3. Verify the equations and solution in (33.22).

A4. For Example 33.11, verify that the solution in (33.22) gives $S_{\min} = \frac{30703}{231}$.

A5. Verify that the solution of $M\mathbf{u} = \mathbf{b}$ determined by (33.23) is given by (33.24).

EXERCISES 33.6–Part B

B1. Obtain the reduced row-echelon matrix in (33.20).

B2. Obtain S in (33.21).

B3. For the 9×3 matrix A in Example 33.11, obtain the projection matrix $P = A(A^TA)^{-1}A^T$, and show that $A^T\mathbf{y} = A^T(P\mathbf{y}) = \mathbf{b}$.

In Exercises B4–8, interpolate the parabola $y = ax^2 + bx + c$ through the given set of points. Plot the points and the interpolating parabola.

B4. $\{(-8, 9), (-6, 7), (3, 5)\}$ **B5.** $\{(-12, 5), (-2, -10), (11, 11)\}$

B6. $\{(2, -3), (6, -9), (8, -2)\}$ **B7.** $\{(-3, 1), (-2, 12), (6, -8)\}$

B8. $\{(-8, 6), (6, -5), (7, 1)\}$

In Exercises B9–13, interpolate the cubic $y = ax^3 + bx^2 + cx + d$ through the given set of points. Plot the points and the interpolating cubic.

B9. $\{(-11, -8), (-10, 1), (0, -8), (9, 4)\}$

B10. $\{(-12, 12), (-9, 11), (6, -4), (11, 2)\}$

B11. $\{(-12, 0), (-3, -8), (1, -4), (11, -2)\}$

B12. $\{(-1, 2), (2, -12), (7, -10), (10, -6)\}$

B13. $\{(-11, -4), (-8, -2), (-6, -5), (-4, -12)\}$

For the set of data points given in each of Exercises B14–23:

(a) Write the equations that arise from fitting the parabola $y = ax^2 + bx + c$ to the data.

(b) Put the equations in part (a) into matrix form $A\mathbf{u} = \mathbf{y}$, where A is an $r \times c$ matrix.

(c) Use the reduced row-echelon form of the augmented matrix $[A, \mathbf{y}]$ to show that the equations in part (a) are inconsistent.

(d) Find the rank of A and the rank of $[A, \mathbf{y}]$.

(e) Obtain $S = \sum_{k=1}^{n}[y(x_k) - y_k]^2$, the sum of the squares of the deviations of the data from the target parabola.

(f) Using calculus, minimize S to obtain the least-squares solution.

(g) Obtain $S_{\min}$, the minimum value of S.

(h) Obtain the normal equations $A^TA\mathbf{u} = A^T\mathbf{y} = \mathbf{b}$ and, from them, the least-squares solution.

(i) Obtain $P = A(A^TA)^{-1}A^T$, the matrix that projects vectors in R^r onto the column space of A.

(j) Use the reduced row-echelon form of $[A, P\mathbf{y}]$ to show that the equations $A\mathbf{u} = P\mathbf{y}$ are consistent.

(k) Obtain the least-squares solution by solving the system $A\mathbf{u} = P\mathbf{y}$.

(l) Find $\mathbf{y}_{\perp[A]} = \mathbf{y} - P\mathbf{y}$, the component of $\mathbf{y}$ orthogonal to the column space of A.

(m) Show that $\|\mathbf{y} - P\mathbf{y}\|^2 = S_{\min}$.

(n) Sketch the data points and the least-squares parabola.

(o) Sketch a figure analogous to Figure 33.11.

B14. $\{(-4, 85), (-3, 55), (-2, 14), (-1, 25), (0, 23), (1, 45), (2, 30), (4, 176), (5, 228)\}$

B15. $\{(-4, 149), (-2, 26), (-1, -5), (0, -18), (2, 6), (3, 54), (4, 109), (5, 218)\}$

B16. $\{(-5, 217), (-4, 179), (-3, 80), (-1, 14), (0, -26), (2, 39), (3, 35)\}$

B17. $\{(-5, -7), (-4, 1), (-3, 7), (0, 8), (1, 1), (4, -23), (5, -27)\}$

B18. $\{(-5, 162), (-3, 54), (-2, 36), (-1, 20), (2, 12), (3, 62), (4, 78), (5, 159)\}$

B19. $\{(-5, 6), (-4, 11), (-2, 13), (0, 6), (1, 4), (4, -21), (5, -47)\}$

B20. $\{(-5, 147), (-2, 44), (-1, -26), (0, 8), (2, 48), (3, 95), (4, 183), (5, 277)\}$

B21. $\{(-5, 45), (-2, -4), (-1, -5), (1, 16), (2, 13), (3, 25), (5, 60)\}$

B22. $\{(-4, -64), (-3, -36), (-2, -14), (-1, -15), (2, 5), (3, 6), (4, 16), (5, -4)\}$

B23. $\{(-5, -105), (-2, -17), (-1, -12), (0, -22), (1, -16), (2, -26), (4, -102)\}$

For the set of data points given in each of Exercises B24–33:

(a) Write the equations that arise from fitting the cubic $y = ax^3 + bx^2 + cx + d$ to the data.

(b) Put the equations in part (a) into matrix form $A\mathbf{u} = \mathbf{y}$, where A is an $r \times c$ matrix.

(c) Use the reduced row-echelon form of the augmented matrix $[A, \mathbf{y}]$ to show that the equations in part (a) are inconsistent.

(d) Find the rank of A and the rank of $[A, \mathbf{y}]$.

(e) Obtain $S = \sum_{k=1}^{n}[y(x_k) - y_k]^2$, the sum of the squares of the deviations of the data from the target cubic.

(f) Using calculus, minimize S to obtain the least-squares solution.

(g) Obtain $S_{\min}$, the minimum value of S.

(h) Obtain the normal equations $A^TA\mathbf{u} = A^T\mathbf{y} = \mathbf{b}$ and, from them, the least-squares solution.

(i) Obtain $P = A(A^TA)^{-1}A^T$, the matrix that projects vectors in R^r onto the column space of A.

(j) Use the reduced row-echelon form of $[A, P\mathbf{y}]$ to show that the equations $A\mathbf{u} = P\mathbf{y}$ are consistent.

(k) Obtain the least-squares solution by solving the system $A\mathbf{u} = P\mathbf{y}$.

(l) Find $\mathbf{y}_{\perp[A]} = \mathbf{y} - P\mathbf{y}$, the component of $\mathbf{y}$ orthogonal to the column space of A.

(m) Show that $\|\mathbf{y} - P\mathbf{y}\|^2 = S_{\min}$.

(n) Sketch the data points and the least-squares cubic.

(o) Sketch a figure analogous to Figure 33.11.

B24. $\{(-2, 194), (-1, 133), (0, 81), (1, 169), (2, 181), (3, -156),$
$(4, -449), (5, -693)\}$

B25. $\{(-5, -828), (-4, -499), (-1, 178), (0, -196), (1, 34),$
$(2, 183), (4, 414), (5, 1098)\}$

B26. $\{(-5, 345), (-4, 269), (-2, 32), (-1, 7), (1, 55), (3, -26),$
$(5, -571)\}$

B27. $\{(-5, 450), (-4, 416), (-3, 275), (0, 93), (3, -144),$
$(4, -450), (5, -919)\}$

B28. $\{(-3, 56), (-2, 35), (-1, -28), (0, -40), (1, 42), (2, 14),$
$(5, 481)\}$

B29. $\{(-5, 1022), (-4, 594), (-1, 25), (0, 91), (2, -19),$
$(3, -243), (4, -478), (5, -766)\}$

B30. $\{(-5, -339), (-4, -65), (-1, -64), (1, -111), (4, 470),$
$(5, 806)\}$

B31. $\{(-3, -443), (-1, -161), (0, -236), (1, -185), (2, 29),$
$(3, 88), (4, 511), (5, 888)\}$

B32. $\{(-4, 575), (-3, 271), (-2, 245), (-1, 186), (0, -122),$
$(1, 135), (4, -396), (5, -919)\}$

B33. $\{(-5, 913), (-4, 752), (-3, 256), (-2, -137), (0, -125),$
$(1, -179), (2, 43), (3, 48)\}$

For the $r \times c$ matrix A in each of Exercises B34–43:

(a) Find the rank.

(b) Obtain a basis for the row space.

(c) Obtain a basis for the null space.

(d) Obtain a basis for the column space.

(e) Obtain a basis for the null space of A^T, the orthogonal complement of the column space of A.

(f) State the dimensions of the four fundamental subspaces.

(g) Show that the dimension of the row space, plus the dimension of the null space, equals c, the number of columns in A.

(h) Show that the dimension of the column space, plus the dimension of the null space of A^T, equals r, the number of rows in A.

(i) Show that the row space is orthogonal to the null space, that is, show the vectors in part (b) are orthogonal to the vectors in part (c).

(j) Show that the column space is orthogonal to the null space of A^T, that is, show the vectors in part (d) to be orthogonal to the vectors in part (e).

(k) Sketch a figure analogous to Figure 33.10.

B34. $\begin{bmatrix} 9 & 7 & 1 & -8 & -3 \\ -3 & -2 & -5 & 6 & -7 \\ -4 & -5 & 0 & -4 & -4 \\ 8 & -5 & 5 & 1 & -1 \\ -3 & 9 & -4 & 5 & 1 \\ 6 & 9 & 1 & 2 & -2 \end{bmatrix}$

B35. $\begin{bmatrix} -17 & 7 & -25 & -20 & 13 \\ -29 & 36 & 36 & 7 & -39 \\ -30 & 25 & 59 & 35 & 12 \\ -6 & 37 & 56 & -34 & -12 \\ 3 & -14 & -1 & 20 & 25 \\ 27 & -23 & -25 & -8 & -1 \end{bmatrix}$

B36. $\begin{bmatrix} 9 & 3 & 5 & 7 & 7 & 2 \\ -2 & -1 & -7 & -9 & 4 & 1 \\ 3 & -2 & 7 & 9 & 5 & -4 \\ 0 & 9 & -3 & -1 & -1 & 6 \\ 2 & -3 & -4 & 6 & -1 & 0 \end{bmatrix}$

B37. $\begin{bmatrix} 13 & -25 & 15 & 2 & 41 & -18 \\ 29 & -27 & -1 & 12 & 51 & -32 \\ -42 & -41 & 55 & -57 & -17 & 5 \\ -21 & 13 & -13 & -22 & -57 & 2 \\ -36 & 79 & -15 & 9 & -79 & 47 \end{bmatrix}$

B38. $\begin{bmatrix} 30 & 7 & -36 & -8 & 60 \\ -7 & 31 & -33 & -33 & 1 \\ 25 & 4 & -39 & -11 & 63 \\ 29 & 35 & -51 & -27 & 47 \\ 46 & 21 & -30 & -2 & 50 \\ -55 & -44 & 39 & 11 & -43 \end{bmatrix}$

B39. $\begin{bmatrix} -31 & 14 & 8 & -17 & 23 \\ -5 & 7 & 1 & -7 & 1 \\ 26 & -7 & -7 & 10 & -22 \\ -10 & 14 & 2 & -14 & 2 \\ 0 & 49 & -3 & -44 & -28 \\ -3 & -35 & -3 & 31 & 23 \end{bmatrix}$

B40. $\begin{bmatrix} 25 & 22 & 1 & -20 & -30 & -7 \\ -29 & 8 & -79 & -13 & -15 & -18 \\ -16 & -31 & 64 & 52 & 24 & -18 \\ 33 & 8 & 51 & -3 & -9 & 6 \\ -15 & 18 & -71 & -20 & -30 & -23 \end{bmatrix}$

B41. $\begin{bmatrix} 32 & -6 & 8 & -30 & 26 & -18 \\ 19 & 4 & 35 & -13 & 34 & -43 \\ 11 & -11 & -33 & -16 & -13 & 32 \\ 25 & -15 & -35 & -30 & -5 & 30 \\ 17 & -8 & -15 & -19 & 2 & 11 \end{bmatrix}$

B42. $\begin{bmatrix} 46 & -1 & 27 & 11 \\ -45 & -9 & -29 & -3 \\ 13 & 32 & 16 & -22 \\ 2 & -20 & -4 & 16 \\ -6 & -21 & -9 & 15 \\ 23 & 40 & 24 & -26 \end{bmatrix}$

B43. $\begin{bmatrix} -9 & -10 & -12 & -3 & 1 & 16 \\ -13 & -26 & -32 & 1 & 1 & 32 \\ -37 & 8 & 13 & -35 & 6 & 28 \\ 34 & 6 & 5 & 26 & -5 & -36 \end{bmatrix}$

In Exercises B44–48:

(a) Show that the equations $A\mathbf{u} = \mathbf{y}$ are inconsistent.

(b) Obtain the least-squares solution by solving the normal equations $A^T A\mathbf{u} = A^T\mathbf{y} = \mathbf{b}$.

(c) Show that the equations $A\mathbf{u} = A^T\mathbf{y}$ are consistent and have for their solution the same vector $\mathbf{u}$ as found in part (b).

B44. $A = \begin{bmatrix} 12 & -2 & -1 & 11 \\ -9 & 11 & 8 & 11 \\ 2 & 0 & -5 & 12 \\ 7 & -9 & 3 & -7 \\ -5 & -11 & 0 & 10 \end{bmatrix}$ $\mathbf{y} = \begin{bmatrix} 1 \\ 5 \\ 10 \\ 11 \\ -10 \end{bmatrix}$

B45. $A = \begin{bmatrix} -10 & 12 & -12 & -5 \\ -5 & -8 & -10 & -8 \\ -2 & 8 & 11 & 1 \\ -11 & -12 & 12 & 11 \\ -12 & 7 & -11 & 12 \\ 10 & 1 & 3 & 7 \end{bmatrix}$ $\mathbf{y} = \begin{bmatrix} -1 \\ -1 \\ -3 \\ 4 \\ -6 \\ 8 \end{bmatrix}$

B46. $A = \begin{bmatrix} 11 & 12 & 6 \\ 7 & -11 & -12 \\ -6 & 8 & -2 \\ -7 & 4 & -1 \end{bmatrix}$ $\mathbf{y} = \begin{bmatrix} -8 \\ -12 \\ -6 \\ 10 \end{bmatrix}$

B47. $A = \begin{bmatrix} -2 & 9 \\ 8 & 10 \\ -7 & 1 \\ 8 & -3 \end{bmatrix}$ $\mathbf{y} = \begin{bmatrix} 12 \\ -2 \\ -8 \\ 7 \end{bmatrix}$

B48. $A = \begin{bmatrix} -11 & 0 & 9 \\ -2 & 11 & -7 \\ -9 & 4 & -6 \\ 12 & -3 & -10 \\ 3 & -2 & -11 \end{bmatrix}$ $\mathbf{y} = \begin{bmatrix} 9 \\ -2 \\ -12 \\ 6 \\ -5 \end{bmatrix}$

Chapter Review

1. If $A = \begin{bmatrix} 1 & 2 \\ 3 & 4 \end{bmatrix}$, obtain det A, A^{-1}, and the eigenvalues of A. Write the characteristic polynomial for A. What is the rank of A?

2. For each, answer true or false. If false, explain why, and where possible, correct the statement.

   (a) The product of symmetric matrices is symmetric.

   (b) The product of orthogonal matrices is orthogonal.

   (c) The product of nonsingular matrices is nonsingular.

   (d) The product of invertible matrices is invertible.

   (e) The determinant of a product is the product of the determinants.

   (f) The transpose of a product is the product of the transposes.

   (g) The transpose of a sum is the sum of the transposes.

   (h) For matrices, the inverse of a sum is the sum of the inverses.

   (i) The transpose of the inverse is the inverse of the transpose, provided the inverses exist.

   (j) For matrices, the inverse of a product is the product of the inverses in the reverse order.

3. Find the eigenpairs of $A = \begin{bmatrix} 1 & 2 \\ 2 & 1 \end{bmatrix}$. Orthogonally diagonalize A. Write the characteristic polynomial for A. If $\mathbf{y} = [2 \ 3]^T$, use Cramer's rule to solve $A\mathbf{x} = \mathbf{y}$ for $\mathbf{x}$.

4. Given $\mathbf{A} = 3\mathbf{i} + 2\mathbf{j}$ and $\mathbf{B} = 7\mathbf{i} - 5\mathbf{j}$, find the components of $\mathbf{A}$ along and orthogonal to $\mathbf{B}$.

5. Given $\mathbf{A} = [3 \ -2 \ 4]^T$, $\mathbf{B} = [6 \ -1 \ 1]^T$, and $\mathbf{C} = [5 \ 3 \ -2]^T$, find the projection of $\mathbf{C}$ onto the subspace spanned by $\mathbf{A}$ and $\mathbf{B}$. What is the component of $\mathbf{C}$ orthogonal to this subspace?

6. Are the vectors $\mathbf{A}$, $\mathbf{B}$, $\mathbf{C}$ of Question 5 linearly dependent or independent? Why?

7. Apply the Gram–Schmidt process to orthogonalize $\mathbf{A}$, $\mathbf{B}$, and $\mathbf{C}$, the vectors in Question 5.

8. What is the quadratic form determined by the matrix A in Question 3? Is this form positive definite? Why?

9. Let (a, b, c) be a critical point of the function $f(x, y, z)$. Describe a "second-derivative test" for determining if the critical point is a maximum, minimum, or saddle point.

10. Apply criterion in Question 9 to the critical point of the function $f(x, y, z) = 2x^2 - 3y^2 + 5z^2 - 7x + 6y + 2z - 1$.

11. Define the two-norm for the matrix in Question 1. Describe how to compute this norm.

12. Obtain the one-norm and the infinity-norm of the matrix A in Question 1.

13. Obtain the frobenius norm of the matrix A in Question 1.

14. Given the matrix $A = \begin{bmatrix} 39 & -49 & -19 & -52 \\ 14 & 29 & -10 & -47 \\ 26 & -51 & 24 & -13 \\ 57 & -54 & -3 & -79 \end{bmatrix}$:

   (a) Obtain the reduced row-echelon form of A.

   (b) Find the rank of A.

   (c) Find a basis for its null space.

   (d) What is the most general solution of the equation $A\mathbf{x} = \mathbf{0}$?

   (e) Find a basis for its column space.

   (f) Find a basis for its row space.

   (g) Draw a diagram to show how the row space, null space, and column space are related.

   (h) Obtain the least-squares solution to $A\mathbf{x} = \mathbf{y}$, where $\mathbf{y} = [-1 \ 2 \ 4 \ 3]^T$.

   (i) Demonstrate the relationship between the least-squares solution found in part (h) and the null space.

15. Obtain the least-squares quadratic function that best fits the points $\{(-2, 3), (-1, 0), (0, -1), (2, 0)\}$.

16. For the data in Question 15, obtain the least-squares fit using the functions $\{1, \cos x, \cos 2x\}$.

Chapter *34*

Matrix Factorizations

INTRODUCTION   The organizational principle of *matrix factorization* structures this chapter. Integers can be uniquely factored as a product of prime numbers. Matrices can be factored in a variety of ways as a product of matrices with different properties. These different factorizations, or decompositions, reveal different aspects of matrix algebra and are useful in different computational arenas.

The factorization called the *LU decomposition* can be obtained as a by-product of Gaussian elimination. The row reductions that yield the upper triangular factor U also yield the lower triangular factor L. This decomposition is an efficient way to solve systems of the form $A\mathbf{x} = \mathbf{y}$, where the vector $\mathbf{y}$ could be one of a number of right-hand sides. In fact, the Doolittle, Crout, and Cholesky variations of the decomposition are important algorithms for the numerical solution of systems of linear equations.

In Sections 12.16 and 12.17 we uncoupled systems of differential equations by diagonalizing a matrix, obtaining the similarity transformation $A = PDP^{-1}$. This is another example of a factorization, one for which the most general form is $A = PJP^{-1}$, where J is a Jordan matrix rather than a diagonal matrix D. The Jordan matrix is a diagonal matrix with some additional 1's on the superdiagonal, the one above the main diagonal. For some matrices, the Jordan matrix is as close to diagonalization as can be achieved.

The *QR decomposition* factors a matrix into a product of an orthogonal matrix Q and an upper triangular matrix R. It is an important ingredient of powerful numeric methods for finding eigenvalues and for solving the least-squares problem. The *QR algorithm* for finding eigenvalues is discussed in Section 34.4 and also in Sections 45.3 and 45.4 where some of its more numerical aspects are treated.

The *singular value decomposition* factors a matrix as a product of three factors, two being orthogonal matrices and one being diagonal. The columns in one orthogonal factor are left singular vectors, and the columns in the other orthogonal factor are the right singular vectors. The matrix itself can be represented in outer product form in terms of the left and right singular vectors. One use of this representation is in digital image processing.

Finally, we use the least-squares problem to motivate the *pseudoinverse*. As seen in Section 12.7, not every square matrix has an inverse. And no inverse has been defined for matrices that are not square. Yet, the equation $A\mathbf{x} = \mathbf{y}$ represents a mapping of vectors from one space to another, and the question of reversing the map is certainly valid. The pseudoinverse inverts these maps in the sense of least squares and can be constructed from the factors of the singular value decomposition.

34.1 LU Decomposition

Decompositions of LU Type

The similarity transform that diagonalizes the matrix A is represented by the equality $P^{-1}AP = D$, where the transition matrix P has the eigenvectors of A for its columns and the entries on the diagonal of the diagonal matrix D are the corresponding eigenvalues. If this equation is rewritten as $A = PDP^{-1}$, then A is said to be *factored* into the product of the three matrices P, D, and P^{-1}. This is but one of a number of ways the matrix A can be factored, or *decomposed*.

In this section we study several LU decompositions in which the matrix A is factored into the product of L, a lower triangular matrix, and U, an upper triangular matrix. There are at least four different versions of this decomposition, and often the phrase "LU" can refer to any one of the four. The four possible decompositions to which the generic "LU" might refer, along with the related Cholesky decomposition, are collected in Table 34.1.

Doolittle	L_1U	1's on main diagonal of L
Crout	LU_1	1's on main diagonal of U
LDU	L_1DU_1	1's on main diagonals of L and U D is a diagonal matrix
Gauss	$L_1DL_1^T$	A is symmetric, 1's on main diagonal of L D is a diagonal matrix
Cholesky	RR^T	A is symmetric, positive definite $R = L_1D$, with D a diagonal matrix

TABLE 34.1   A spectrum of "LU" decompositions

EXAMPLE 34.1   **Doolittle decomposition**   The matrix A and its Doolittle factors L_1 and U are given in (34.1), where we see that $A = L_1U$.

$$A = \begin{bmatrix} -3 & -1 & 1 & 5 \\ -3 & -4 & 5 & -2 \\ -1 & 1 & 3 & 5 \\ 2 & -4 & 4 & -3 \end{bmatrix} = \begin{bmatrix} 1 & 0 & 0 & 0 \\ 1 & 1 & 0 & 0 \\ \frac{1}{3} & -\frac{4}{9} & 1 & 0 \\ -\frac{2}{3} & \frac{14}{9} & -\frac{7}{20} & 1 \end{bmatrix} \begin{bmatrix} -3 & -1 & 1 & 5 \\ 0 & -3 & 4 & -7 \\ 0 & 0 & \frac{40}{9} & \frac{2}{9} \\ 0 & 0 & 0 & \frac{113}{10} \end{bmatrix} \quad (34.1)$$

By definition, the lower triangular matrix L_1 has 1's on and 0's above the main diagonal. Again by definition, the upper triangular factor U has 0's below the main diagonal. (A memory aid: *Doolittle* contains the letter "*l*" and the factor L_1. *Crout* contains the letter "*u*" and the factor U_1.)   ❖

EXAMPLE 34.2   **Use of the *LU* decomposition**   Let A and its Doolittle decomposition be as given in (34.1), and consider the system $A\mathbf{x} = \mathbf{y}$ where $\mathbf{y}$ is the vector $\mathbf{y} = [2\ 0\ -2\ 1]^T$. Write $A\mathbf{x} = \mathbf{y}$ as $L_1U\mathbf{x} = \mathbf{y}$. Let $\mathbf{c} = [c_1\ c_2\ c_3\ c_4]^T$, and set $\mathbf{c} = U\mathbf{x}$, the original system reads $L_1\mathbf{c} = \mathbf{y}$, as seen in (34.2).

$$
\begin{bmatrix}
1 & 0 & 0 & 0 \\
1 & 1 & 0 & 0 \\
\frac{1}{3} & -\frac{4}{9} & 1 & 0 \\
-\frac{2}{3} & \frac{14}{9} & -\frac{7}{20} & 1
\end{bmatrix}
\begin{bmatrix}
c_1 \\ c_2 \\ c_3 \\ c_4
\end{bmatrix}
=
\begin{bmatrix}
2 \\ 0 \\ -2 \\ 1
\end{bmatrix}
\tag{34.2}
$$

The rows of (34.2) are solved from the top down, in a forward sense, as shown in (34.3). The kth equation yields c_k because at that point, $c_i, i = 1, \ldots, k - 1$, are known. This process is called *forward substitution*.

$$
\left.
\begin{aligned}
c_1 &= 2 \\
c_1 + c_2 &= 0 \\
\tfrac{1}{3}c_1 - \tfrac{4}{9}c_2 + c_3 &= -2 \\
-\tfrac{2}{3}c_1 + \tfrac{14}{9}c_2 - \tfrac{7}{20}c_3 + c_4 &= 1
\end{aligned}
\right\}
\Rightarrow
\left\{
\begin{aligned}
c_1 &= 2 \\
c_2 &= -c_1 = -2 \\
c_3 &= -2 + \tfrac{4}{9}c_2 - \tfrac{1}{3}c_1 = -\tfrac{32}{9} \\
c_4 &= 1 + \tfrac{7}{20}c_3 - \tfrac{14}{9}c_2 + \tfrac{2}{3}c_1 = \tfrac{21}{5}
\end{aligned}
\right.
\tag{34.3}
$$

Once the vector $\mathbf{c}$ is known, $\mathbf{x}$ is obtained as the solution of the system $U\mathbf{x} = \mathbf{c}$ in (34.4).

$$
\begin{bmatrix}
-3 & -1 & 1 & 5 \\
0 & -3 & 4 & -7 \\
0 & 0 & \frac{40}{9} & \frac{2}{9} \\
0 & 0 & 0 & \frac{113}{10}
\end{bmatrix}
\begin{bmatrix}
x_1 \\ x_2 \\ x_3 \\ x_4
\end{bmatrix}
=
\begin{bmatrix}
2 \\ -2 \\ -\frac{32}{9} \\ \frac{21}{5}
\end{bmatrix}
\tag{34.4}
$$

This time, solve the equations in the reverse order, from bottom to top. The equations, and the solution process that solves them, is shown in (34.5). The solution process is called *back substitution* and was seen in Section 12.5.

$$
\left.
\begin{aligned}
-3x_1 - x_2 + x_3 + 5x_4 &= 2 \\
-3x_2 + 4x_3 - 7x_4 &= -2 \\
\tfrac{40}{9}x_3 + \tfrac{2}{9}x_4 &= -\tfrac{32}{9} \\
\tfrac{113}{10}x_4 &= \tfrac{21}{5}
\end{aligned}
\right\}
\Rightarrow
\left\{
\begin{aligned}
x_1 &= \frac{2 + x_2 - x_3 - 5x_4}{-3} = \frac{25}{226} \\
x_2 &= \frac{-2 - 4x_3 + 7x_4}{-3} = -\frac{292}{226} \\
x_3 &= \frac{-\tfrac{32}{9} - \tfrac{2}{9}x_4}{\tfrac{40}{9}} = -\frac{185}{226} \\
x_4 &= \frac{\tfrac{21}{5}}{\tfrac{113}{10}} = \frac{84}{226}
\end{aligned}
\right.
\tag{34.5}
$$

❖

Crout Decomposition

The Crout decomposition can be obtained from the Doolittle decomposition by factoring out a diagonal matrix D from U to obtain the factorization $L_1 D U_1$. The factor D is combined with L_1 to form $L = L_1 D$, and the Crout factorization is then

$$
A = L_1 U = L_1(DU_1) = L_1 DU_1 = (L_1 D)U_1 = LU_1
$$

The diagonal matrix D has U_{ii} as its diagonal elements, and $U_1 = D^{-1}U$. The product $U = DU_1$ is

$$
\begin{bmatrix}
-3 & -1 & 1 & 5 \\
0 & -3 & 4 & -7 \\
0 & 0 & \frac{40}{9} & \frac{2}{9} \\
0 & 0 & 0 & \frac{113}{10}
\end{bmatrix}
=
\begin{bmatrix}
-3 & 0 & 0 & 0 \\
0 & -3 & 0 & 0 \\
0 & 0 & \frac{40}{9} & 0 \\
0 & 0 & 0 & \frac{113}{10}
\end{bmatrix}
\begin{bmatrix}
1 & \frac{1}{3} & -\frac{1}{3} & -\frac{5}{3} \\
0 & 1 & -\frac{4}{3} & \frac{7}{3} \\
0 & 0 & 1 & \frac{1}{20} \\
0 & 0 & 0 & 1
\end{bmatrix}
$$

The complete Crout decomposition is then

$$
\begin{bmatrix}
-3 & -1 & 1 & 5 \\
-3 & -4 & 5 & -2 \\
-1 & 1 & 3 & 5 \\
2 & -4 & 4 & -3
\end{bmatrix}
=
\begin{bmatrix}
-3 & 0 & 0 & 0 \\
-3 & -3 & 0 & 0 \\
-1 & \frac{4}{3} & \frac{40}{9} & 0 \\
2 & -\frac{14}{3} & -\frac{14}{9} & \frac{113}{10}
\end{bmatrix}
\begin{bmatrix}
1 & \frac{1}{3} & -\frac{1}{3} & -\frac{5}{3} \\
0 & 1 & -\frac{4}{3} & \frac{7}{3} \\
0 & 0 & 1 & \frac{1}{20} \\
0 & 0 & 0 & 1
\end{bmatrix}
$$

Cholesky Decomposition

When A is symmetric and positive definite, it has a *Cholesky decomposition* of the form $A = RR^{\mathrm{T}}$ where R is lower triangular, so R^{T} is upper triangular.

One way to build a symmetric positive definite matrix A is to start with the symmetric matrix M and compute $A = MM^{\mathrm{T}}$, as shown in (34.6).

$$
A =
\begin{bmatrix}
3 & -3 & 4 & 4 \\
-3 & -5 & 0 & -1 \\
4 & 0 & -1 & 0 \\
4 & -1 & 0 & -5
\end{bmatrix}
\begin{bmatrix}
3 & -3 & 4 & 4 \\
-3 & -5 & 0 & -1 \\
4 & 0 & -1 & 0 \\
4 & -1 & 0 & -5
\end{bmatrix}
=
\begin{bmatrix}
50 & 2 & 8 & -5 \\
2 & 35 & -12 & -2 \\
8 & -12 & 17 & 16 \\
-5 & -2 & 16 & 42
\end{bmatrix}
\tag{34.6}
$$

We see by inspection that A is symmetric, and since the eigenvalues are 3.77, 35.15, 51.61, 53.47, we are assured that A is also positive definite. The Cholesky factorization is then $A = RR^{\mathrm{T}}$, where R is given by

$$
R =
\begin{bmatrix}
5\sqrt{2} & 0 & 0 & 0 \\
\frac{\sqrt{2}}{5} & \frac{3\sqrt{97}}{5} & 0 & 0 \\
\frac{4\sqrt{2}}{5} & -\frac{308\sqrt{97}}{1455} & \frac{\sqrt{963,113}}{291} & 0 \\
-\frac{1}{\sqrt{2}} & -\frac{3}{\sqrt{97}} & \frac{4704}{\sqrt{963,113}} & \frac{605}{\sqrt{19,858}}
\end{bmatrix}
$$

The Cholesky decomposition can be obtained from the L_1DU_1 factorization if D is written as $\sqrt{D}\sqrt{D}$. (If the diagonal elements of D are U_{ii}, then the diagonal elements of $\sqrt{D}$ are $\sqrt{U_{ii}}$.) Hence, we have

$$
A = L_1DU_1 = L_1\sqrt{D}\sqrt{D}U_1 = (L_1\sqrt{D})(\sqrt{D}U_1) = (L_1\sqrt{D})(U_1^{\mathrm{T}}\sqrt{D})^{\mathrm{T}}
$$

and since $U_1^{\mathrm{T}} = L_1$ when A is symmetric, we finally have $A = RR^{\mathrm{T}}$, where $R = L_1\sqrt{D}$.

We leave it to the reader to verify that

$$
L_1\sqrt{D} =
\begin{bmatrix}
1 & 0 & 0 & 0 \\
\frac{1}{25} & 1 & 0 & 0 \\
\frac{4}{25} & -\frac{308}{873} & 1 & 0 \\
-\frac{1}{10} & -\frac{5}{97} & \frac{14,112}{9929} & 1
\end{bmatrix}
\begin{bmatrix}
5\sqrt{2} & 0 & 0 & 0 \\
0 & \frac{3\sqrt{97}}{5} & 0 & 0 \\
0 & 0 & \frac{\sqrt{963,113}}{291} & 0 \\
0 & 0 & 0 & \frac{605}{\sqrt{19,858}}
\end{bmatrix}
= R
$$

When A is symmetric, $U_1 = L_1^{\mathrm{T}}$, so $A = L_1 D U_1 = L_1 D L_1^{\mathrm{T}}$, and we have the Gauss decomposition of Table 34.1.

Decomposition by Brute Force

Some insight into the nature of matrix factorizations is obtained by factoring A into the product $L_1 U$ by the following "brute-force" technique. Write A as the product

$$
A =
\begin{bmatrix}
1 & 0 & 0 & 0 \\
L_{21} & 1 & 0 & 0 \\
L_{31} & L_{32} & 1 & 0 \\
L_{41} & L_{42} & L_{43} & 1
\end{bmatrix}
\begin{bmatrix}
U_{11} & U_{12} & U_{13} & U_{14} \\
0 & U_{22} & U_{23} & U_{24} \\
0 & 0 & U_{33} & U_{34} \\
0 & 0 & 0 & U_{44}
\end{bmatrix}
\tag{34.7}
$$

which yields 16 equations in the 16 unknowns $\{L_{ij}, i = 2, \ldots, 4, j = 1, \ldots, i-1\} \cup \{U_{ij}, i = 1, \ldots, 4, j = i, \ldots, 4\}$. For example, if A is the symmetric matrix in (34.6), the solution of these equations yields the factorization

$$
A =
\begin{bmatrix}
1 & 0 & 0 & 0 \\
\frac{1}{25} & 1 & 0 & 0 \\
\frac{4}{25} & -\frac{308}{873} & 1 & 0 \\
-\frac{1}{10} & -\frac{5}{97} & \frac{14,112}{9929} & 1
\end{bmatrix}
\begin{bmatrix}
50 & 2 & 8 & -5 \\
0 & \frac{873}{25} & -\frac{308}{25} & -\frac{9}{5} \\
0 & 0 & \frac{9929}{873} & \frac{1568}{97} \\
0 & 0 & 0 & \frac{366,025}{19,858}
\end{bmatrix}
$$

We also see why there is a multiplicity of LU decompositions. If $n = 4$, there are $\frac{n(n+1)}{2} = 10$ nonzero entries in an upper triangular or lower triangular matrix. The product in (34.7) yields $n^2 = 16$ equations, but there could have been as many as $10 + 10 = 20$ unknowns. We obtained a determinate system by choosing the diagonal elements in L to be 1, reducing the number of unknowns in L_1 to $\frac{n(n-1)}{2} = 6$. Without such a choice, the solution for L and U would have contained arbitrary parameters.

Elementary Matrices

We next consider *elementary matrices* as devices for recording the individual transformations applied to a matrix A during Gaussian elimination. Letting r_k represent row k, we list the following three types of elementary transformations used in row-reducing a matrix A by Gaussian elimination.

1. Interchange two rows of A.
2. Replace r_k with cr_k, that is, replace row k with c times row k.
3. Replace r_j with the sum $r_j + cr_i$, that is, add to row j, row i times the scalar c, called the *multiplier*.

Each of these three elementary operations can be respectively represented by a matrix $E_k, k = 1, 2, 3$, so that the matrix product $E_k A$ gives A subjected to the corresponding elementary Gaussian transformation. For each type of elementary transformation, the elementary matrix E_k is just an identity matrix with the elementary operation performed on *it*. For added clarity, we write E_3 as $E_3(i, j, c)$. We illustrate using the matrix

$$A = \begin{bmatrix} 1 & 4 & 6 & -3 & -2 \\ 3 & 5 & 4 & -6 & 2 \\ 5 & -1 & 5 & 4 & -5 \\ 4 & 0 & -4 & -3 & 0 \\ 0 & -1 & 2 & 1 & 5 \end{bmatrix}$$

The elementary matrix E_1 that records the interchange of rows 2 and 5 and its effect on A are

$$E_1 A = \begin{bmatrix} 1 & 0 & 0 & 0 & 0 \\ 0 & 0 & 0 & 0 & 1 \\ 0 & 0 & 1 & 0 & 0 \\ 0 & 0 & 0 & 1 & 0 \\ 0 & 1 & 0 & 0 & 0 \end{bmatrix} \begin{bmatrix} 1 & 4 & 6 & -3 & -2 \\ 3 & 5 & 4 & -6 & 2 \\ 5 & -1 & 5 & 4 & -5 \\ 4 & 0 & -4 & -3 & 0 \\ 0 & -1 & 2 & 1 & 5 \end{bmatrix} = \begin{bmatrix} 1 & 4 & 6 & -3 & -2 \\ 0 & -1 & 2 & 1 & 5 \\ 5 & -1 & 5 & 4 & -5 \\ 4 & 0 & -4 & -3 & 0 \\ 3 & 5 & 4 & -6 & 2 \end{bmatrix}$$

The elementary matrix E_2 that records multiplying the third row of A by s and its effect on A are

$$E_2 A = \begin{bmatrix} 1 & 0 & 0 & 0 & 0 \\ 0 & 1 & 0 & 0 & 0 \\ 0 & 0 & s & 0 & 0 \\ 0 & 0 & 0 & 1 & 0 \\ 0 & 0 & 0 & 0 & 1 \end{bmatrix} \begin{bmatrix} 1 & 4 & 6 & -3 & -2 \\ 3 & 5 & 4 & -6 & 2 \\ 5 & -1 & 5 & 4 & -5 \\ 4 & 0 & -4 & -3 & 0 \\ 0 & -1 & 2 & 1 & 5 \end{bmatrix} = \begin{bmatrix} 1 & 4 & 6 & -3 & -2 \\ 3 & 5 & 4 & -6 & 2 \\ 5s & -s & 5s & 4s & -5s \\ 4 & 0 & -4 & -3 & 0 \\ 0 & -1 & 2 & 1 & 5 \end{bmatrix}$$

The elementary matrix $E_3(1, 2, -3)$ that adds -3 times row 1 to row 2 and its effect on A are

$$E_3(1, 2, -3)A = \begin{bmatrix} 1 & 0 & 0 & 0 & 0 \\ -3 & 1 & 0 & 0 & 0 \\ 0 & 0 & 1 & 0 & 0 \\ 0 & 0 & 0 & 1 & 0 \\ 0 & 0 & 0 & 0 & 1 \end{bmatrix} \begin{bmatrix} 1 & 4 & 6 & -3 & -2 \\ 3 & 5 & 4 & -6 & 2 \\ 5 & -1 & 5 & 4 & -5 \\ 4 & 0 & -4 & -3 & 0 \\ 0 & -1 & 2 & 1 & 5 \end{bmatrix} = \begin{bmatrix} 1 & 4 & 6 & -3 & -2 \\ 0 & -7 & -14 & 3 & 8 \\ 5 & -1 & 5 & 4 & -5 \\ 4 & 0 & -4 & -3 & 0 \\ 0 & -1 & 2 & 1 & 5 \end{bmatrix}$$

The elementary matrix $E_3(1, 2, -3)$ pivots on the element a_{11}, thereby replacing a_{21} by zero. The multiplier is -3, and the notation $r_2 \to r_2 - 3r_1$ captures the effect of $E_3(1, 2, -3)$ on A.

EXAMPLE 34.3 The matrix $A = \begin{bmatrix} 1 & 3 & 4 \\ 2 & -1 & 1 \\ 3 & 0 & 2 \end{bmatrix}$ can be row-reduced to an upper triangular matrix U by applying the three elementary transformations contained in the matrices $E_3(1, 2, -2)$, $E_3(1, 3, -3)$, and $E_3(2, 3, -\frac{9}{7})$. Indeed, $E_3(2, 3, -\frac{9}{7})E_3(1, 3, -3)E_3(1, 2, -2)A = TA = U$, as shown.

$$\begin{bmatrix} 1 & 0 & 0 \\ 0 & 1 & 0 \\ 0 & -\frac{9}{7} & 1 \end{bmatrix} \begin{bmatrix} 1 & 0 & 0 \\ 0 & 1 & 0 \\ -3 & 0 & 1 \end{bmatrix} \begin{bmatrix} 1 & 0 & 0 \\ -2 & 1 & 0 \\ 0 & 0 & 1 \end{bmatrix} \begin{bmatrix} 1 & 3 & 4 \\ 2 & -1 & 1 \\ 3 & 0 & 2 \end{bmatrix} = \begin{bmatrix} 1 & 3 & 4 \\ 0 & -7 & -7 \\ 0 & 0 & -1 \end{bmatrix}$$

The matrix $T = E_3(2, 3, -\frac{9}{7})E_3(1, 3, -3)E_3(1, 2, -2)$, capturing the totality of the reduction of A to U, the matrix T^{-1}, and L, a 3×3 identity matrix in which zeros have been replaced with the negatives of the corresponding multipliers used in the reduction of A to U are

$$T = \begin{bmatrix} 1 & 0 & 0 \\ -2 & 1 & 0 \\ -\frac{3}{7} & -\frac{9}{7} & 1 \end{bmatrix} \quad T^{-1} = \begin{bmatrix} 1 & 0 & 0 \\ 2 & 1 & 0 \\ 3 & \frac{9}{7} & 1 \end{bmatrix} \quad L = \begin{bmatrix} 1 & 0 & 0 \\ 2 & 1 & 0 \\ 3 & \frac{9}{7} & 1 \end{bmatrix}$$

Since $TA = U$, we have $A = T^{-1}U$, but $T^{-1} = L$, so $A = LU$. However, $L = L_1$, so in reality, we have $A = L_1U$, the Doolittle factorization of A. The Gaussian arithmetic whereby A is row-reduced to the upper triangular matrix U provides L_1 by merely storing the negatives of the multipliers in the locations where pivoting creates zeros. Thus, the arithmetic of Gaussian elimination provides both factors in the Doolittle decomposition and does it in a storage-efficient manner.   ❖

Computer Implementation of the Doolittle Decomposition

The following algorithm would implement, on a digital computer, the Doolittle decomposition of the $n \times n$ matrix A.

for $m = 1, \ldots, n$, execute each of:

for $j = m, \ldots, n$, compute $U_{mj} = A_{mj} - \sum_{k=1}^{m-1} L_{mk}U_{kj}$

for $i = m + 1, \ldots, n$, compute $L_{im} = \frac{1}{U_{mm}}\left[A_{im} - \sum_{k=1}^{m-1} L_{ik}U_{km} \right]$

The decomposition is obtained by computing the kth row of U, followed by the kth column of L_1. The left schematic in (34.8) illustrates the "row over column" strategy of the algorithm. The "vector" **a** is computed first and is essentially the first row of A. The "column vector" **b** is computed next, followed by the pair **c** and **d**, then the pair **e** and **f**, then the pair **g** and **h**, and finally, the last calculation yields **i**. The algorithm fails if any of the $U_{mm} = 0$.

$$\begin{bmatrix} a_1 & a_2 & a_3 & a_4 & a_5 \\ b_1 & c_1 & c_2 & c_3 & c_4 \\ b_2 & d_1 & e_1 & e_2 & e_3 \\ b_3 & d_2 & f_1 & g_1 & g_2 \\ b_4 & d_3 & f_2 & h_1 & i_1 \end{bmatrix} \quad \begin{bmatrix} a_1 & b_1 & b_2 & b_3 & b_4 \\ a_2 & c_1 & d_1 & d_2 & d_3 \\ a_3 & c_2 & e_1 & f_1 & f_2 \\ a_4 & c_3 & e_2 & g_1 & h_1 \\ a_5 & c_4 & e_3 & g_2 & i_1 \end{bmatrix} \quad (34.8)$$

Computer Implementation of the Crout Decomposition

The following algorithm would implement, on a digital computer, the Crout decomposition of the $n \times n$ matrix A.

for $m = 1, \ldots, n$, execute each of:

for $i = m, \ldots, n$, compute $L_{im} = A_{im} - \sum_{k=1}^{m-1} L_{ik}U_{km}$

for $j = m + 1, \ldots, n$, compute $U_{mj} = \frac{1}{L_{mm}}\left[A_{mj} - \sum_{k=1}^{m-1} L_{mk}U_{kj} \right]$

The decomposition is obtained by computing the kth column of L, followed by the kth row of U_1. The right schematic in (34.8) illustrates the "column before row" strategy of the algorithm. The "vector" **a** is computed first and is essentially the first column of A. The "row vector" **b** is computed next, followed by the pair **c** and **d**, then the pair **e** and **f**, then the pair **g** and **h**, and finally, the last calculation yields **i**. The algorithm fails if any of the $L_{mm} = 0$.

Because the algorithm for the Crout decomposition calculates a column of L before it calculates a row of U_1, it is possible to implement a pivoting strategy whereby rows are interchanged to minimize the round-off error introduced when dividing by small values of L_{mm}. Thus, prior to computing column k, the row with the largest remaining element in column k can be moved to become row k.

EXERCISES 34.1–Part A

A1. Obtain the Doolittle decomposition of $A = \begin{bmatrix} 2 & -18 & 10 \\ 8 & -69 & 64 \\ -6 & 78 & 160 \end{bmatrix}$.

Obtain U by upper triangularizing A via Gaussian elimination, and obtain L_1 by storing the negatives of the multipliers in a copy of the 3×3 identity matrix. Verify that the product of L_1 and U is A.

A2. From the $L_1 U$ decomposition in Exercise A1, obtain the factorization $L_1 D U_1$.

A3. From Exercise A2, obtain the Crout decomposition of the matrix A given in Exercise A1.

A4. For the matrix A in Exercise A1, obtain L_1 and U by the "brute-force" technique illustrated in (34.7).

A5. Modify the calculation in Exercise A4 to obtain the Crout decomposition of the matrix A in Exercise A1.

A6. If $\mathbf{y} = -40\mathbf{i} - 178\mathbf{j} - 22\mathbf{k}$ and A is the matrix in Exercise A1, solve $A\mathbf{x} = \mathbf{y}$ by solving $L_1 \mathbf{c} = \mathbf{y}$ by forward substitution and $U\mathbf{x} = \mathbf{c}$ by back substitution. Verify that your solution is correct.

A7. Obtain L_1 from the three elementary matrices $E_3(1, 2, m_{12})$, $E_3(1, 3, m_{13})$, $E_3(2, 3, m_{23})$, which record the Gaussian arithmetic taking A to upper triangular form.

EXERCISES 34.1–Part B

B1. Verify $A = \begin{bmatrix} 4 & 16 & -12 \\ 16 & 73 & 24 \\ -12 & 24 & 628 \end{bmatrix}$ is positive definite, and write it as

$A = L_1 D L_1^{\mathrm{T}} = L_1 \sqrt{D} \sqrt{D} L_1^{\mathrm{T}} = R R^{\mathrm{T}}$, thereby obtaining its Cholesky decomposition.

For the matrix A given in each of Exercises B2–11:

   (a) Use a software tool to obtain the Doolittle ($L_1 U$) decomposition of A.

   (b) From the Doolittle decomposition, obtain the $L_1 D U_1$ factorization.

   (c) From the $L_1 D U_1$ factorization, obtain the Crout (LU_1) factorization of A.

   (d) If A is positive-definite, obtain the Cholesky ($R R^{\mathrm{T}}$) decomposition from the $L_1 D U_1$ factorization found in part (b). Check the result with an appropriate software tool.

$$\textbf{B2.}\ \begin{bmatrix} -6 & 8 & 10 \\ 1 & -11 & 1 \\ -2 & 6 & 9 \end{bmatrix} \quad \textbf{B3.}\ \begin{bmatrix} 9 & -6 & -1 \\ -6 & 10 & -5 \\ -1 & -5 & 8 \end{bmatrix}$$

$$\textbf{B4.}\ \begin{bmatrix} 7 & -7 & 8 \\ -7 & -2 & -9 \\ 8 & -9 & -8 \end{bmatrix} \quad \textbf{B5.}\ \begin{bmatrix} 4 & -12 & -12 & -4 \\ -12 & 40 & 44 & 12 \\ -12 & 44 & 61 & -6 \\ -4 & 12 & -6 & 47 \end{bmatrix}$$

$$\textbf{B6.}\ \begin{bmatrix} -5 & 12 & -1 & 11 \\ 4 & -3 & 11 & -6 \\ 12 & 4 & 11 & 5 \\ 4 & 6 & -12 & -1 \end{bmatrix} \quad \textbf{B7.}\ \begin{bmatrix} -20 & 7 & -24 & -18 \\ 7 & -30 & -7 & -9 \\ -24 & -7 & -19 & 11 \\ -18 & -9 & 11 & -27 \end{bmatrix}$$

$$\textbf{B8.}\ \begin{bmatrix} 5 & 15 & 10 & 35 & -10 \\ 15 & 49 & 30 & 109 & -30 \\ 10 & 30 & 23 & 76 & -26 \\ 35 & 109 & 76 & 263 & -88 \\ -10 & -30 & -26 & -88 & 56 \end{bmatrix} \quad \textbf{B9.}\ \begin{bmatrix} -1 & 11 & -9 & 11 & 8 \\ 11 & 2 & 0 & -5 & 12 \\ 7 & -9 & 3 & -7 & -5 \\ -11 & 0 & 10 & 1 & 5 \\ 10 & 11 & -10 & -10 & 12 \end{bmatrix}$$

$$\textbf{B10.}\ \begin{bmatrix} -18 & -20 & -27 & -30 & -13 \\ -20 & -16 & 26 & 26 & 6 \\ -27 & 26 & -27 & -26 & -18 \\ -30 & 26 & -26 & 21 & -6 \\ -13 & 6 & -18 & -6 & -26 \end{bmatrix} \quad \textbf{B11.}\ \begin{bmatrix} -12 & -5 & -5 & -8 & -10 \\ -8 & -2 & 8 & 11 & 1 \\ -11 & -12 & 12 & 11 & -12 \\ 7 & -11 & 12 & 10 & 1 \\ 3 & 7 & -1 & -1 & -3 \end{bmatrix}$$

In Exercises B12–21:

   (a) Obtain the Doolittle ($L_1 U$) decomposition of A.

   (b) Use $A = L_1 U$ and forward and back substitution to solve $A\mathbf{x}_k = \mathbf{e}_k$, $k = 1, \ldots, n$, where n is 3 or 4 as appropriate and $\mathbf{e}_k$ is the kth column of the $n \times n$ identity matrix.

   (c) Verify that $[\mathbf{x}_1, \ldots, \mathbf{x}_n] = A^{-1}$.

   (d) Obtain the Doolittle decomposition by setting up and solving n^2 equations in the n^2 unknown entries in L_1 and U.

   (e) Obtain the Crout decomposition by setting up and solving n^2 equations in the n^2 unknown entries in L and U_1.

   (f) Use Gaussian elimination to row-reduce A to an upper triangular matrix U. Record the transformations in a sequence of elementary matrices. Show that $L_1 = T^{-1}$, where T is the product of the elementary matrices. Show that $TA = U$ so that $A = L_1 U$.

$$\textbf{B12.}\ A = \begin{bmatrix} 7 & -11 & 0 \\ 9 & -2 & 11 \\ -7 & -9 & 4 \end{bmatrix} \quad \textbf{B13.}\ A = \begin{bmatrix} -6 & 12 & -3 \\ -10 & 3 & -2 \\ -11 & 9 & -2 \end{bmatrix}$$

$$\textbf{B14.}\ A = \begin{bmatrix} -12 & 6 & -5 \\ -1 & 5 & 3 \\ 11 & 12 & -1 \end{bmatrix} \quad \textbf{B15.}\ A = \begin{bmatrix} -4 & -5 & 3 \\ 5 & 3 & 7 \\ -9 & -6 & -5 \end{bmatrix}$$

$$\textbf{B16.}\ A = \begin{bmatrix} 8 & -5 & 3 \\ -5 & 1 & -3 \\ -4 & 1 & -12 \end{bmatrix} \quad \textbf{B17.}\ A = \begin{bmatrix} -2 & 9 & -8 & -6 \\ 12 & 5 & 10 & -3 \\ 6 & 7 & -4 & 10 \\ 2 & 2 & 9 & -9 \end{bmatrix}$$

B18. $A = \begin{bmatrix} -2 & 7 & 4 & -10 \\ -7 & -7 & 5 & -1 \\ -5 & 7 & 11 & -2 \\ 0 & -2 & -2 & -1 \end{bmatrix}$ **B19.** $A = \begin{bmatrix} 4 & -5 & 7 & 6 \\ 4 & 10 & 4 & -12 \\ 1 & 7 & -3 & -6 \\ 5 & 7 & -9 & 12 \end{bmatrix}$ **B24.** $A = \begin{bmatrix} 8 & -2 & -7 & 4 & -1 \\ 4 & 0 & 0 & -1 & -8 \\ -12 & -6 & 10 & -2 & 9 \\ 8 & 10 & -7 & 1 & 8 \\ -3 & 12 & -2 & -8 & 7 \end{bmatrix}$

B20. $A = \begin{bmatrix} 8 & -10 & 0 & -7 \\ 3 & 4 & -6 & -7 \\ 1 & 1 & -2 & -5 \\ -1 & -2 & -5 & 9 \end{bmatrix}$ **B21.** $A = \begin{bmatrix} -3 & -9 & -3 & -10 \\ -12 & 0 & 6 & -2 \\ 1 & 8 & -4 & 1 \\ 5 & -10 & 11 & -5 \end{bmatrix}$

In Exercises B25–27:

In Exercises B22–24:

(a) Obtain $A = L_1 D U_1$ as the product of the three given factors.

(b) Obtain the Doolittle decomposition of A from the given factors.

(a) In the computer language of your choice, implement the computer algorithm for the Doolittle decomposition. Apply it to the given matrix, and verify the correctness of your results.

(c) Apply the numeric procedure of developed for part (a) of Exercises B22–24. Does it succeed? Why?

(b) In the computer language of your choice, implement the computer algorithm for the Crout decomposition. Apply it to the given matrix, and verify the correctness of your results.

B25. $\begin{bmatrix} 1 & 0 & 0 \\ -4 & 1 & 0 \\ -9 & -2 & 1 \end{bmatrix}, \begin{bmatrix} 5 & 0 & 0 \\ 0 & 0 & 0 \\ 0 & 0 & -8 \end{bmatrix}, \begin{bmatrix} 1 & 9 & 3 \\ 0 & 1 & 8 \\ 0 & 0 & 1 \end{bmatrix}$

(c) Modify the code in part (b) so as to include the pivoting strategy described at the end of this section. Polished computer code would not literally move rows but would instead use a *pointer file* to keep track of the row-interchanges that are called for by the pivoting strategy. (See, e.g., [60] for details.) On a finite-precision machine, how much difference in accuracy is achieved by pivoting?

B26. $\begin{bmatrix} 1 & 0 & 0 & 0 \\ 3 & 1 & 0 & 0 \\ -9 & 5 & 1 & 0 \\ -3 & 6 & 4 & 1 \end{bmatrix}, \begin{bmatrix} -9 & 0 & 0 & 0 \\ 0 & 3 & 0 & 0 \\ 0 & 0 & 0 & 0 \\ 0 & 0 & 0 & 6 \end{bmatrix}, \begin{bmatrix} 1 & 5 & 7 & -9 \\ 0 & 1 & 4 & 0 \\ 0 & 0 & 1 & -1 \\ 0 & 0 & 0 & 1 \end{bmatrix}$

B27. $\begin{bmatrix} 1 & 0 & 0 & 0 & 0 \\ -9 & 1 & 0 & 0 & 0 \\ 1 & 8 & 1 & 0 & 0 \\ -8 & 3 & -9 & 1 & 0 \\ 0 & 3 & 4 & 6 & 1 \end{bmatrix}, \begin{bmatrix} -7 & 0 & 0 & 0 & 0 \\ 0 & -2 & 0 & 0 & 0 \\ 0 & 0 & 0 & 0 & 0 \\ 0 & 0 & 0 & -5 & 0 \\ 0 & 0 & 0 & 0 & -5 \end{bmatrix},$

B22. $A = \begin{bmatrix} -7 & 1 & -5 \\ -2 & 12 & -2 \\ 4 & -1 & -5 \end{bmatrix}$ **B23.** $A = \begin{bmatrix} 1 & -10 & 12 & -2 \\ -1 & 11 & -9 & 11 \\ 8 & 11 & 2 & 0 \\ -5 & 12 & 7 & -9 \end{bmatrix}$ $\begin{bmatrix} 1 & -8 & 1 & -9 & 5 \\ 0 & 1 & -8 & -5 & -8 \\ 0 & 0 & 1 & -6 & -2 \\ 0 & 0 & 0 & 1 & 7 \\ 0 & 0 & 0 & 0 & 1 \end{bmatrix}$

34.2 PJP^{-1} and Jordan Canonical Form

Similar Matrices

Two square matrices A and B are said to be *similar* if an invertible matrix P can be found for which $A = PBP^{-1}$. (See Sections 12.16 and 12.17.) Clearly, this equality is equivalent to $B = P^{-1}AP$. If A and B are thought of as transformations on a space of vectors, then, if they are similar they represent the same transformation but only in different bases.

Given the matrices A and P, the matrix $B = P^{-1}AP$ is then similar to A, as for example, with

$$A = \begin{bmatrix} 1 & -1 \\ 2 & 4 \end{bmatrix} \quad B = \begin{bmatrix} 4 & -1 \\ -4 & -1 \end{bmatrix}^{-1} \begin{bmatrix} 1 & -1 \\ 2 & 4 \end{bmatrix} \begin{bmatrix} 4 & -1 \\ -4 & -1 \end{bmatrix} = \begin{bmatrix} 2 & \frac{3}{4} \\ 0 & 3 \end{bmatrix}$$

The eigenvalues of A and B are 2, 3; and the determinant of both A and B is 6. Similar matrices have the same eigenvalues and determinants, but not necessarily the same eigenvectors. (The eigenvectors of A are $\mathbf{i} - \mathbf{j}$ and $\mathbf{i} - 2\mathbf{j}$, whereas the eigenvectors of B are $\mathbf{i}$ and $3\mathbf{i} + 4\mathbf{j}$.)

The equality of the eigenvalues is not that hard to prove. The eigenvalues of A and B satisfy characteristic equations of the form

$$\det(A - \lambda I) = 0 \quad \text{and} \quad \det(B - \lambda I) = 0$$

Multiply the first equation from the left by $\det(P^{-1})$ and from the right by $\det(P)$ to get

$$\det(P^{-1})\det(A - \lambda I)\det(P) = 0$$

But the product of determinants is the determinant of the product. Hence

$$\det(P^{-1}[A - \lambda I]P) = 0 \quad \text{or} \quad \det(P^{-1}AP - \lambda P^{-1}P) = 0$$

which is $\det(B - \lambda I) = 0$. Any value of λ that satisfies the characteristic equation for A also satisfies the characteristic equation for B.

The equality of the determinants of similar matrices also hinges on the property of determinants whereby the product of determinants is the determinant of the product. In fact,

$$\det(A) = [1]\det(A) = [\det(I)]\det(A) = [\det(P^{-1}P)]\det(A)$$

$$= [\det(P^{-1})\det(P)]\det(A) = \det(P^{-1})\det(A)\det(P) = \det(P^{-1}AP) = \det(B)$$

Notice that in going from the fourth step to the fifth, we were able to commute the *determinants* of P and A, since determinants are scalars.

Similarity to a Diagonal Matrix

In Section 12.16 we found, by brute force, a transition matrix P for which $P^{-1}AP = D$, where D was a diagonal matrix. In effect, we showed A was similar to the diagonal matrix D. The columns of P were the eigenvectors of A, and the diagonal elements of D were the corresponding eigenvalues.

In (34.9), the columns of P are the eigenvectors of A, the corresponding eigenvalues of A are the diagonal entries in D, and $P^{-1}AP = D$.

$$A = \begin{bmatrix} 1 & 1 \\ -2 & 4 \end{bmatrix} \quad P = \begin{bmatrix} 1 & 1 \\ 2 & 1 \end{bmatrix} \quad D = \begin{bmatrix} 3 & 0 \\ 0 & 2 \end{bmatrix} \tag{34.9}$$

Motivation was provided in Sections 12.16 and 12.17 where coupled sets of differential equations were uncoupled by finding a similarity transform taking the system matrix to diagonal form. The system of differential equations $\mathbf{x}' = A\mathbf{x}$, where A is given in (34.9), uncouples from

$$\begin{array}{cc} x' = x + y & \\ y' = -2x + 4y & \end{array} \quad \text{to} \quad \begin{array}{c} y_1' = 3y_1 \\ y_2' = 2y_2 \end{array}$$

under the diagonalization induced by P. In fact, write $A = PDP^{-1}$ to obtain $\mathbf{x}' = PDP^{-1}\mathbf{x}$; then multiply by P^{-1} (from the left) to obtain $P^{-1}\mathbf{x}' = DP^{-1}\mathbf{x}$. Defining $\mathbf{y} = P^{-1}\mathbf{x}$ gives the uncoupled system $\mathbf{y}' = D\mathbf{y}$, with $\mathbf{y}$ obtained by inspection, and $\mathbf{x} = P\mathbf{y}$, as shown in (34.10).

$$\mathbf{y} = \begin{bmatrix} ae^{3t} \\ be^{2t} \end{bmatrix} \Rightarrow \mathbf{x} = P\mathbf{y} = \begin{bmatrix} ae^{3t} + be^{2t} \\ 2ae^{3t} + be^{2t} \end{bmatrix} \tag{34.10}$$

Jordan Canonical Form

The $n \times n$ matrix A is diagonalizable if it has n linearly independent eigenvectors. Not every matrix has this maximal number of eigenvectors. In that case, A is similar to J, its *Jordan form*, a matrix that is as close to a diagonal as it can be, in a sense to be made precise by the following examples.

A matrix is always similar to its Jordan form, so that $P^{-1}AP = J$ or $A = PJP^{-1}$, where J is the Jordan form. The matrix J is either a diagonal matrix (with the eigenvalues on the main diagonal) or is nearly a diagonal matrix with the eigenvalues on the main diagonal and at least one 1 in the diagonal immediately above the main diagonal. Thus, the similarity transformation taking A to its Jordan form yields another factorization of the matrix.

We will consider three matrices A, B, and C, each of which fails to have a maximal number of eigenvectors. In each case J, the Jordan form of the matrix, is given. Although this canonical form J will have the eigenvalues of the matrix on its main diagonal, it will not be a diagonal matrix. Above the diagonal there will be at least one 1, the location being determined by inherent properties of the matrix.

The matrix A in (34.11) has eigenvalues 3, 2, 2, with only one eigenvector corresponding to the repeated eigenvalue 2. The first two columns of the transition matrix P_A are eigenvectors, and J_A is the Jordan form of A. The action of A on the columns of the transition matrix gives some insight into the meaning of the 1 in the super diagonal. The product AP_A shows the first column behaves as an eigenvector with eigenvalue 3, the second column behaves as an eigenvector with eigenvalue 2, but the third column behaves in a novel way. Denoting the columns of P_A by $\mathbf{p}_1$, $\mathbf{p}_2$, $\mathbf{p}_3$, we have the third column of the product AP_A behaving as a *generalized eigenvector*, that is, $A\mathbf{p}_3 = \mathbf{p}_2 + 2\mathbf{p}_3$.

$$A = \begin{bmatrix} 6 & 9 & -3 \\ -1 & 0 & 1 \\ 2 & 5 & 1 \end{bmatrix} \quad P_A = \begin{bmatrix} 1 & 3 & 0 \\ 0 & -1 & 0 \\ 1 & 1 & -1 \end{bmatrix} \quad J_A = \begin{bmatrix} 3 & 0 & 0 \\ 0 & 2 & 1 \\ 0 & 0 & 2 \end{bmatrix} \quad AP_A = \begin{bmatrix} 3 & 6 & 3 \\ 0 & -2 & -1 \\ 3 & 2 & -1 \end{bmatrix}$$
(34.11)

The matrix B in (34.12) has eigenvalues 2, 2, 2, with but two corresponding eigenvectors, the first and third columns of the transition matrix P_B. The Jordan form of B is J_B. Denoting the columns of P_B by $\mathbf{p}_1$, $\mathbf{p}_2$, $\mathbf{p}_3$, we see from the product BP_B that $\mathbf{p}_1$ and $\mathbf{p}_3$ are eigenvectors but that $\mathbf{p}_2$ is a generalized eigenvector because $A\mathbf{p}_2 = \mathbf{p}_1 + 2\mathbf{p}_2$.

$$B = \begin{bmatrix} 2 & -1 & 0 \\ 0 & 2 & 0 \\ 0 & -1 & 2 \end{bmatrix} \quad P_B = \begin{bmatrix} -1 & 0 & 0 \\ 0 & 1 & 0 \\ -1 & 1 & 1 \end{bmatrix} \quad J_B = \begin{bmatrix} 2 & 1 & 0 \\ 0 & 2 & 0 \\ 0 & 0 & 2 \end{bmatrix} \quad BP_B = \begin{bmatrix} -2 & -1 & 0 \\ 0 & 2 & 0 \\ -2 & 1 & 2 \end{bmatrix}$$
(34.12)

The matrix C in (34.13) has eigenvalues 2, 2, 2, with but one corresponding eigenvector, the first column of the transition matrix P_C. The Jordan form of C is J_C, and the two 1's on the superdiagonal indicate there are two generalized eigenvectors associated with the one eigenvector. Denoting the columns of P_C by $\mathbf{p}_1$, $\mathbf{p}_2$, $\mathbf{p}_3$, we see from the product CP_C that $\mathbf{p}_1$ is an eigenvector but that $\mathbf{p}_2$ and $\mathbf{p}_3$ are generalized eigenvectors because $C\mathbf{p}_2 = \mathbf{p}_1 + 2\mathbf{p}_2$, and $C\mathbf{p}_3 = \mathbf{p}_2 + 2\mathbf{p}_3$.

$$C = \begin{bmatrix} 5 & 5 & -3 \\ -1 & 0 & 1 \\ 1 & 1 & 1 \end{bmatrix} \quad P_C = \begin{bmatrix} 1 & 3 & 1 \\ 0 & -1 & 0 \\ 1 & 1 & 0 \end{bmatrix} \quad J_C = \begin{bmatrix} 2 & 1 & 0 \\ 0 & 2 & 1 \\ 0 & 0 & 2 \end{bmatrix} \quad CP_C = \begin{bmatrix} 2 & 7 & 5 \\ 0 & -2 & -1 \\ 2 & 3 & 1 \end{bmatrix}$$
(34.13)

EXERCISES 34.2–Part A

A1. Show that $A = \begin{bmatrix} -8 & 3 \\ -18 & 7 \end{bmatrix}$ and $B = \begin{bmatrix} -8 & 6 \\ -9 & 7 \end{bmatrix}$ have the same eigenvalues and determinant.

A2. For A from Exercise A1, find a transition matrix P for which $P^{-1}AP = D$, where D is a diagonal matrix.

A3. For B from Exercise A1, find a transition matrix Q for which $Q^{-1}BQ = D$, where D is from Exercise A2.

A4. For A and B from Exercise A1, find a transition matrix R for which $A = RBR^{-1}$.

A5. Answer *true* or *false,* and give a reason for your answer: Two matrices with the same determinant and eigenvalues are similar.

EXERCISES 34.2–Part B

B1. Show that $A = \begin{bmatrix} -1 & -3 & -6 \\ -4 & -2 & -8 \\ 4 & 4 & 10 \end{bmatrix}$ and $B = \begin{bmatrix} 11 & 5 & -2 \\ 4 & 4 & -1 \\ 44 & 24 & -8 \end{bmatrix}$ have the same eigenvalues and determinant.

B2. For A from Exercise B1, find a transition matrix for which $P^{-1}AP = D$, where D is a diagonal matrix.

B3. Show that B from Exercise B1 is not similar to a diagonal matrix. Are A and B from Exercise B1 similar?

B4. Answer *true* or *false*, and give a reason for your answer: If an eigenvalue of the matrix A is repeated (has algebraic multiplicity greater than one), then its Jordan form cannot be diagonal.

For the pair of matrices A and B given in each of Exercises B5–14:

(a) Obtain transition matrices P_A and P_B for which $P_A^{-1}AP_A = P_B^{-1}BP_B = D$, thereby showing A is similar to B.

(b) From P_A and P_B, construct a transition matrix P for which $B = P^{-1}AP$.

(c) Use "brute force," solving four equations in four unknowns, to find a transition matrix P for which $B = P^{-1}AP$.

B5. $A = \begin{bmatrix} 2 & 6 \\ -1 & 9 \end{bmatrix}$, $B = \begin{bmatrix} 19 & -44 \\ 4 & -8 \end{bmatrix}$

B6. $A = \begin{bmatrix} -2 & -9 \\ -7 & -4 \end{bmatrix}$, $B = \begin{bmatrix} -145 & -335 \\ 60 & 139 \end{bmatrix}$

B7. $A = \begin{bmatrix} -6 & -2 \\ -3 & -5 \end{bmatrix}$, $B = \begin{bmatrix} 8 & -22 \\ 8 & -19 \end{bmatrix}$

B8. $A = \begin{bmatrix} 0 & -6 \\ 2 & -7 \end{bmatrix}$, $B = \begin{bmatrix} -5 & -2 \\ 1 & -2 \end{bmatrix}$

B9. $A = \begin{bmatrix} -2 & 0 \\ 1 & -5 \end{bmatrix}$, $B = \begin{bmatrix} -57 & -65 \\ 44 & 50 \end{bmatrix}$

B10. $A = \begin{bmatrix} 7 & -2 \\ 0 & 3 \end{bmatrix}$, $B = \begin{bmatrix} 55 & 32 \\ -78 & -45 \end{bmatrix}$

B11. $A = \begin{bmatrix} -7 & -2 \\ 5 & 4 \end{bmatrix}$, $B = \begin{bmatrix} -13 & -2 \\ 56 & 10 \end{bmatrix}$

B12. $A = \begin{bmatrix} 8 & -2 \\ 4 & -1 \end{bmatrix}$, $B = \begin{bmatrix} 55 & 40 \\ -66 & -48 \end{bmatrix}$

B13. $A = \begin{bmatrix} 4 & 1 \\ 2 & 3 \end{bmatrix}$, $B = \begin{bmatrix} 5 & -16 \\ 0 & 2 \end{bmatrix}$

B14. $A = \begin{bmatrix} -1 & 0 \\ -2 & -3 \end{bmatrix}$, $B = \begin{bmatrix} -3 & -16 \\ 0 & -1 \end{bmatrix}$

For the Jordan matrix J in Exercises B15–24:

(a) Find the eigenvectors.

(b) Find the generalized eigenvectors, and give the algebraic equation, involving J, that each generalized eigenvector satisfies.

B15. $\begin{bmatrix} 2 & 1 & 0 & 0 \\ 0 & 2 & 1 & 0 \\ 0 & 0 & 2 & 1 \\ 0 & 0 & 0 & 2 \end{bmatrix}$ **B16.** $\begin{bmatrix} 2 & 1 & 0 & 0 \\ 0 & 2 & 1 & 0 \\ 0 & 0 & 2 & 0 \\ 0 & 0 & 0 & 2 \end{bmatrix}$

B17. $\begin{bmatrix} 2 & 1 & 0 & 0 \\ 0 & 2 & 0 & 0 \\ 0 & 0 & 2 & 1 \\ 0 & 0 & 0 & 2 \end{bmatrix}$ **B18.** $\begin{bmatrix} 2 & 1 & 0 & 0 \\ 0 & 2 & 0 & 0 \\ 0 & 0 & 2 & 0 \\ 0 & 0 & 0 & 2 \end{bmatrix}$

B19. $\begin{bmatrix} 2 & 1 & 0 & 0 & 0 \\ 0 & 2 & 1 & 0 & 0 \\ 0 & 0 & 2 & 1 & 0 \\ 0 & 0 & 0 & 2 & 1 \\ 0 & 0 & 0 & 0 & 2 \end{bmatrix}$ **B20.** $\begin{bmatrix} 2 & 1 & 0 & 0 & 0 \\ 0 & 2 & 1 & 0 & 0 \\ 0 & 0 & 2 & 0 & 0 \\ 0 & 0 & 0 & 2 & 1 \\ 0 & 0 & 0 & 0 & 2 \end{bmatrix}$

B21. $\begin{bmatrix} 2 & 1 & 0 & 0 & 0 \\ 0 & 2 & 1 & 0 & 0 \\ 0 & 0 & 2 & 1 & 0 \\ 0 & 0 & 0 & 2 & 0 \\ 0 & 0 & 0 & 0 & 2 \end{bmatrix}$ **B22.** $\begin{bmatrix} 2 & 1 & 0 & 0 & 0 \\ 0 & 2 & 1 & 0 & 0 \\ 0 & 0 & 2 & 0 & 0 \\ 0 & 0 & 0 & 2 & 0 \\ 0 & 0 & 0 & 0 & 2 \end{bmatrix}$

B23. $\begin{bmatrix} 2 & 1 & 0 & 0 & 0 \\ 0 & 2 & 0 & 0 & 0 \\ 0 & 0 & 2 & 0 & 0 \\ 0 & 0 & 0 & 2 & 1 \\ 0 & 0 & 0 & 0 & 2 \end{bmatrix}$ **B24.** $\begin{bmatrix} 2 & 1 & 0 & 0 & 0 \\ 0 & 2 & 0 & 0 & 0 \\ 0 & 0 & 2 & 0 & 0 \\ 0 & 0 & 0 & 2 & 0 \\ 0 & 0 & 0 & 0 & 2 \end{bmatrix}$

For the pair of matrices given in each of Exercises B25–29:

(a) Show the matrices have the same eigenvalues.

(b) Show the matrices have different Jordan forms and, therefore, are not similar.

B25. $\begin{bmatrix} 53 & 257 & 38 \\ -4 & -26 & -2 \\ -41 & -187 & -31 \end{bmatrix}$, $\begin{bmatrix} 107 & 275 & 110 \\ -10 & -28 & -10 \\ -80 & -200 & -83 \end{bmatrix}$

B26. $\begin{bmatrix} -17 & -30 & 20 \\ 0 & -2 & 0 \\ -15 & -30 & 18 \end{bmatrix}$, $\begin{bmatrix} -49 & -14 & 52 \\ -4 & 0 & 4 \\ -45 & -15 & 48 \end{bmatrix}$

B27. $\begin{bmatrix} -37 & -140 & 80 \\ 10 & 38 & -20 \\ -2 & -7 & 7 \end{bmatrix}$, $\begin{bmatrix} -53 & -204 & 80 \\ 14 & 54 & -20 \\ -3 & -11 & 7 \end{bmatrix}$

B28. $\begin{bmatrix} 18 & 170 & -59 \\ 1 & 7 & -3 \\ 9 & 81 & -29 \end{bmatrix}$, $\begin{bmatrix} -1 & 189 & -21 \\ 0 & 8 & -1 \\ 0 & 90 & -11 \end{bmatrix}$

B29. $\begin{bmatrix} -52 & 79 & 72 \\ -54 & 77 & 60 \\ 15 & -26 & -30 \end{bmatrix}$, $\begin{bmatrix} -130 & 105 & -84 \\ -126 & 101 & -84 \\ 42 & -35 & 24 \end{bmatrix}$

B30. Show that $(P^{-1}AP)^k = P^{-1}A^kP$ for conformable matrices A and P.

B31. If $A = PJP^{-1}$, where J is the Jordan form for A, show $e^{At} = e^{PJP^{-1}t} = Pe^{Jt}P^{-1}$. *Hint:* $e^{At} = \sum_{k=0}^{\infty} \frac{1}{k!}(At)^k$, and apply Exercise B30.

For the matrix A given in each of Exercises B32–41:

(a) Find P and J for which $A = PJP^{-1}$, where J is the Jordan form of A.

(b) Compute e^{At} either with a computer algebra system or with the Laplace transform as detailed in Section 12.8.

(c) Compute e^{Jt}, as per part (b).

(d) Verify that $e^{At} = Pe^{Jt}P^{-1}$.

B32. $\begin{bmatrix} -8 & -4 \\ 9 & 4 \end{bmatrix}$ **B33.** $\begin{bmatrix} -9 & -4 \\ 9 & 3 \end{bmatrix}$ **B34.** $\begin{bmatrix} -6 & -64 \\ 1 & 10 \end{bmatrix}$ **B38.** $\begin{bmatrix} 26 & -9 \\ 64 & -22 \end{bmatrix}$ **B39.** $\begin{bmatrix} -53 & -49 \\ 64 & 59 \end{bmatrix}$

B35. $\begin{bmatrix} -25 & -16 \\ 49 & 31 \end{bmatrix}$ **B36.** $\begin{bmatrix} -20 & 4 \\ -81 & 16 \end{bmatrix}$ **B37.** $\begin{bmatrix} 27 & 25 \\ -36 & -33 \end{bmatrix}$ **B40.** $\begin{bmatrix} 26 & -36 \\ 18 & -25 \end{bmatrix}$ **B41.** $\begin{bmatrix} -26 & 9 \\ -72 & 25 \end{bmatrix}$

34.3 QR Decomposition

The QR Decomposition of A

Every real matrix A can be factored into the product of Q, a matrix with orthonormal columns, and R, an upper triangular matrix R, so that $A = QR$. The factors Q and R are unique once the otherwise arbitrary signs on the diagonal of R are fixed. (See [4].)

For the normalized QR decomposition in [64], if A is a real $r \times c$ matrix, then Q is the matrix whose columns are the orthonormalized columns of A and the upper triangular factor R is given by $R = Q^T A$. If A has rank k, Q will be $r \times k$ and R will be $k \times c$. If $k = c$, then $|R_{ii}|$ equals the distance the ith column of A is from the space spanned by the first $i - 1$ columns of A.

EXAMPLE 34.4   The matrix A on the left in (34.14) has dimensions $r = 4, c = 4$ and rank $k = 3$. The columns of the matrix Q given in the center of (34.14) are the orthonormalized columns of A. (There are only three independent columns in A, so Q has just three columns.) The product $Q^T A$ gives the factor R, shown on the right in (34.14).

$$A = \begin{bmatrix} 1 & 2 & 0 & -1 \\ 1 & -1 & 3 & 2 \\ 1 & -1 & 3 & 2 \\ -1 & 1 & -3 & 1 \end{bmatrix} \quad Q = \begin{bmatrix} \frac{1}{2} & \frac{\sqrt{3}}{2} & 0 \\ \frac{1}{2} & -\frac{\sqrt{3}}{6} & \frac{1}{\sqrt{6}} \\ \frac{1}{2} & -\frac{\sqrt{3}}{6} & \frac{1}{\sqrt{6}} \\ -\frac{1}{2} & \frac{\sqrt{3}}{6} & \frac{2}{\sqrt{3}} \end{bmatrix} \quad R = \begin{bmatrix} 2 & -\frac{1}{2} & \frac{9}{2} & 1 \\ 0 & \frac{3\sqrt{3}}{2} & -\frac{3\sqrt{3}}{2} & -\sqrt{3} \\ 0 & 0 & 0 & \sqrt{6} \end{bmatrix}$$

$$(34.14)$$

Because of the orthonormality of the columns of Q, a computation would show $Q^T Q = I$. However,

$$QQ^T = \begin{bmatrix} 1 & 0 & 0 & 0 \\ 0 & \frac{1}{2} & \frac{1}{2} & 0 \\ 0 & \frac{1}{2} & \frac{1}{2} & 0 \\ 0 & 0 & 0 & 1 \end{bmatrix} \ne I \qquad (34.15)$$

a matrix for which an interpretation will be given shortly.   ❖

EXAMPLE 34.5   The matrix A on the left in (34.16) has dimensions $r = 4, c = 3$ and is of maximal rank $k = 3 = c$. The three columns of Q, shown in the center of (34.16), are the orthonormalized columns of A, which are independent. The product $Q^T A$ gives the factor R, shown on the right in (34.16).

$$A = \begin{bmatrix} 1 & -1 & 4 \\ 1 & 4 & -2 \\ 1 & 4 & 2 \\ 1 & -1 & 0 \end{bmatrix} \quad Q = \begin{bmatrix} \frac{1}{2} & -\frac{1}{2} & \frac{1}{2} \\ \frac{1}{2} & \frac{1}{2} & -\frac{1}{2} \\ \frac{1}{2} & \frac{1}{2} & \frac{1}{2} \\ \frac{1}{2} & -\frac{1}{2} & -\frac{1}{2} \end{bmatrix} \quad R = \begin{bmatrix} 2 & 3 & 2 \\ 0 & 5 & -2 \\ 0 & 0 & 4 \end{bmatrix} \qquad (34.16)$$

Because A is of maximal rank $k = 3$, the diagonal entries of R can be interpreted as lengths of orthogonal projections on subspaces of the column space of $A = [\mathbf{A}_1, \mathbf{A}_2, \mathbf{A}_3]$. The vector component of $\mathbf{A}_2$ orthogonal to $\mathbf{A}_1$, namely, $\mathbf{A}_2 - \frac{\mathbf{A}_2 \cdot \mathbf{A}_1}{\mathbf{A}_1 \cdot \mathbf{A}_1} \mathbf{A}_1$, has length $\left\| \left[-\frac{5}{2} \ \frac{5}{2} \ \frac{5}{2} \ -\frac{5}{2} \right]^T \right\| = 5 = |R_{22}|$. The vector component of $\mathbf{A}_3$ orthogonal to the subspace spanned by $\mathbf{A}_1$ and $\mathbf{A}_2$ is given by $(I - P)\mathbf{A}_3$, where $P = M(M^T M)^{-1} M^T$ and $M = [\mathbf{A}_1, \mathbf{A}_2]$. It's length is $\|[2 \ -2 \ 2 \ -2]^T\| = 4$, agreeing with $|R_{33}| = 4$. ❖

Interpreting QQ^T

The product $Q^T Q$ is I, but the product QQ^T need not be the identity I, as in Example 34.4 where the columns of A were not linearly independent. Since the columns of Q are the orthonormalized columns of A (with dependent columns filtered out), we can form $P = Q(Q^T Q)^{-1} Q^T$, the matrix that projects onto the column space of A. This projection matrix can be based on Q since its columns are independent, whereas it cannot always be based on A, whose columns might not be independent. We claim that $QQ^T = P = Q(Q^T Q)^{-1} Q^T$.

For the matrix A in Example 34.4, computing $P = Q(Q^T Q)^{-1} Q^T$ yields the matrix in (34.15), namely, QQ^T. When R^{-1} exists (as in Example 34.5 where A has maximal rank), QQ^T is the projection matrix $A(A^T A)^{-1} A^T$. This is demonstrated, in general, by the following calculation, valid when R^{-1} exists.

$$
\begin{aligned}
QQ^T &= QIIQ^T \\
&= Q[RR^{-1}][(R^T)^{-1} R^T] Q^T \\
&= QR[R^T R]^{-1} R^T Q^T \\
&= QR[R^T I R]^{-1} R^T Q^T \\
&= QR[R^T Q^T QR]^{-1} R^T Q^T \\
&= (QR)[(QR)^T (QR)]^{-1} (QR)^T \\
&= A(A^T A)^{-1} A^T
\end{aligned}
$$

Brief Summary

For the normalized decomposition in [64], when A, square ($r = c$) or "tall and thin" ($r > c$), is also of maximal rank, then R^{-1} exists, $A^T A$ is invertible so that $QQ^T = A(A^T A)^{-1} A^T$, and $|R_{ii}| = \|\mathbf{A}_{i \perp [\mathbf{A}_1, \dots, \mathbf{A}_{i-1}]}\|$. In all other cases, QQ^T projects onto the column space of A, and R (with dimensions $k \times c$) has full rank but more columns than rows. Table 34.2 illustrates these relationships.

$A = [\mathbf{A}_1, \dots, \mathbf{A}_c]$	$r = c$ (square) or $r < c$ (tall & thin)	$r < c$ (low & wide)
Full rank	R^{-1} exists $QQ^T = A(A^T A)^{-1} A^T$ $\|R_{ii}\| = \|\mathbf{A}_{i \perp [\mathbf{A}_1, \dots, \mathbf{A}_{i-1}]}\|$	R: low & wide $QQ^T = P$
Not full rank	R: low & wide $QQ^T = P$	R: low & wide $QQ^T = P$

TABLE 34.2   Characteristics of the normalized QR decomposition for different matrices A

796   Chapter 34   Matrix Factorizations

EXERCISES 34.3

1. For the matrices in (34.14), verify $A = QR$.

2. For the matrices in (34.14), verify that $Q^T Q = I$ and that $Q Q^T$ is given by (34.15).

3. For the matrices given in (34.14), let B be a matrix whose columns are a basis for the column space of A. This can be found from $\text{rref}(A)$, the reduced row-echelon form of A. Then show that $P = B(B^T B)^{-1} B^T = Q Q^T$.

4. For the matrices in (34.16), verify $A = QR$.

5. For the matrices in (34.16), verify $Q^T Q = I$ and compute $Q Q^T$.

6. For the matrices in (34.16), show that $P = A(A^T A)^{-1} A^T = Q Q^T$.

In Exercises 7–15, matrices of the same dimension, but different ranks, are given. In each case:

(a) Determine the rank of each given matrix A.

(b) Use an appropriate software tool to obtain $\tilde{Q}$ and $\tilde{R}$, the factors of the QR decomposition.

(c) Verify that $\tilde{Q}^T \tilde{Q} = I$.

(d) Orthonormalize the columns of A, and call the resulting matrix Q.

(e) Obtain $R = Q^T A$, and compare to part (a).

(f) If R^{-1} exists, show $Q Q^T = A(A^T A)^{-1} A^T$.

(g) If R^{-1} exists, verify $|R_{ii}| = \|\mathbf{A}_{i \perp [\mathbf{A}_1, \ldots, \mathbf{A}_{i-1}]}\|$, $i = 2, \ldots, c$.

(h) If R^{-1} does not exist, let B be a matrix whose columns are a basis for the column space of A. This can be found from $\text{rref}(A)$, the reduced row-echelon form of A. Then show that $P = B(B^T B)^{-1} B^T = Q Q^T$.

(i) Use Table 34.2 to explain the difference in outcomes for the given matrices.

7. $\begin{bmatrix} -52 & -63 & -7 \\ -7 & -3 & -7 \\ -7 & -60 & 56 \\ 26 & 3 & 35 \end{bmatrix}$, $\begin{bmatrix} -1 & 11 & -9 \\ 11 & 8 & 11 \\ 2 & 0 & -5 \\ 12 & 7 & -9 \end{bmatrix}$

8. $\begin{bmatrix} -85 & 9 & 32 \\ 5 & 2 & 1 \\ 21 & -26 & -35 \end{bmatrix}$, $\begin{bmatrix} 4 & -3 & 8 \\ -9 & 7 & 5 \\ 0 & 4 & 4 \end{bmatrix}$

9. $\begin{bmatrix} -44 & -2 & 73 & 18 \\ 14 & 18 & 3 & -31 \\ -33 & -22 & 13 & 49 \\ -37 & -20 & 20 & 49 \\ 21 & -8 & 2 & -22 \end{bmatrix}$, $\begin{bmatrix} 49 & -34 & 32 & 56 \\ -40 & 10 & -5 & -35 \\ -10 & 46 & -53 & -35 \\ 28 & 22 & -31 & 7 \\ -8 & 2 & -1 & -7 \end{bmatrix}$

10. $\begin{bmatrix} 40 & -6 & -21 & 41 \\ -4 & 124 & -88 & -1 \\ -35 & 97 & -57 & 35 \\ 64 & 10 & -49 & 75 \end{bmatrix}$, $\begin{bmatrix} 1 & -1 & 3 & 8 \\ 3 & 3 & -9 & -6 \\ 3 & 9 & -9 & -5 \\ 6 & 0 & 1 & -7 \end{bmatrix}$

11. $\begin{bmatrix} -9 & 4 & -6 & 12 \\ -3 & -10 & 3 & -2 \\ -11 & 9 & -2 & -12 \end{bmatrix}$, $\begin{bmatrix} 44 & -40 & -20 & 88 \\ -88 & -9 & 67 & -99 \\ 6 & 35 & -15 & -23 \end{bmatrix}$

12. $\begin{bmatrix} 6 & -5 & -1 & 5 & 3 \\ 11 & 12 & -1 & -4 & -5 \\ 3 & 5 & 3 & 7 & -9 \end{bmatrix}$, $\begin{bmatrix} -14 & 20 & -46 & 49 & -49 \\ -18 & 26 & -54 & 58 & -58 \\ 21 & -31 & 51 & -56 & 56 \end{bmatrix}$

13. $\begin{bmatrix} -1 & 29 & -48 & -9 & -92 \\ -9 & 49 & 34 & -42 & -72 \\ 22 & 27 & -35 & -25 & 11 \\ 57 & 82 & -58 & -20 & -78 \\ -10 & -17 & 29 & 6 & 25 \end{bmatrix}$, $\begin{bmatrix} 1 & 1 & 1 & -2 & 9 \\ -2 & 6 & -7 & -9 & 4 \\ 9 & 5 & -9 & -1 & 0 \\ 3 & -5 & 2 & -3 & 3 \\ -1 & -3 & -3 & 5 & 5 \end{bmatrix}$

14. $\begin{bmatrix} 18 & 8 & 2 \\ -69 & -25 & 14 \\ 75 & 22 & -35 \\ -40 & -31 & -55 \\ -48 & -27 & -27 \end{bmatrix}$, $\begin{bmatrix} 6 & 4 & 10 \\ 4 & -12 & 1 \\ 7 & -3 & -6 \\ 5 & 7 & -9 \\ 12 & 8 & -10 \end{bmatrix}$

15. $\begin{bmatrix} -4 & -2 & -22 & -59 & -28 \\ -8 & -160 & -172 & 104 & -60 \\ -76 & -6 & -14 & -69 & -88 \\ -21 & -87 & -105 & 12 & -63 \\ -13 & 133 & 126 & -149 & 10 \end{bmatrix}$, $\begin{bmatrix} 31 & -51 & -19 & 9 & -9 \\ -19 & 33 & 43 & -36 & 36 \\ -14 & 20 & -46 & 49 & -49 \\ -18 & 26 & -54 & 58 & -58 \\ 21 & -31 & 51 & -56 & 56 \end{bmatrix}$

16. Given the least-squares problem $A\mathbf{u} = \mathbf{y}$, the normal equations $A^T A\mathbf{u} = A^T\mathbf{y} = \mathbf{b}$ are consistent, but numeric computation with the matrix $M = A^T A$ can lead to large errors in the computed solution $\mathbf{x}$. Show that replacing A with QR leads to $R^T R\mathbf{u} = R^T Q^T\mathbf{y}$.

Since R always has maximal rank and at least as many columns as rows, R^T has maximal rank and no more columns than rows. By a theorem in [64], R^T then has a left inverse, that is, there exists a matrix L for which $L R^T = I_k$, the $k \times k$ identity. Thus, by Exercise 16, the equations $R^T R\mathbf{u} = R^T Q^T\mathbf{y}$ become $R\mathbf{u} = Q^T\mathbf{y}$, a triangular system easily solved by back substitution. For each of the least-squares problems $A\mathbf{u} = \mathbf{y}$ given in Exercises 17–25:

(a) Obtain $M = A^T A$ and find its condition number $C(M) = \|M\| \|M^{-1}\|$, where any matrix norm will do. (The condition number estimates the factor by which errors in the data, either M or $\mathbf{y}$, may be magnified in the solution $\mathbf{x}$.)

(b) Obtain $\mathbf{u}$ by solving the normal equations numerically.

(c) Obtain the QR decomposition of A.

(d) Obtain $\mathbf{u}$ using back substitution to solve $R\mathbf{u} = Q^T\mathbf{y}$.

(e) Obtain L, a left inverse of R^T, by solving the set of equations generated by $L R^T = I_k$, where L is a $k \times c$ matrix of unknowns and R^T is $c \times k$. Show that $L R^T = I_k$ for your L.

17. $A = \begin{bmatrix} 9.75 & 5.66 & 8.37 \\ 8.25 & -6.48 & 3.32 \\ -8.43 & -5.74 & -4.49 \\ 3.74 & 2.45 & -4.47 \\ -6.88 & 2.24 & -9.33 \end{bmatrix}$ $y = \begin{bmatrix} 0.28 \\ -7.94 \\ 7.68 \\ -1.41 \\ 5.48 \end{bmatrix}$

18. $A = \begin{bmatrix} 4.69 & -3.34 & 9.06 & 5.57 \\ 9.43 & -2.99 & 9.94 & 8.25 \\ -6.71 & 8.66 & -8.12 & -4.36 \\ -6.16 & 2.45 & 9.22 & 5.39 \\ 6.32 & 7.81 & -6.16 & 4.80 \end{bmatrix}$ $y = \begin{bmatrix} -4.95 \\ 8.31 \\ -5.48 \\ -5.57 \\ 3.46 \end{bmatrix}$

19. $A = \begin{bmatrix} -9.80 & 7.21 & 7.21 \\ 8.19 & -5.10 & 6.40 \\ 9.38 & 5.29 & -4.9 \\ 9.90 & -6.48 & 2.83 \end{bmatrix}$ $y = \begin{bmatrix} -8.31 \\ 8.54 \\ 8.31 \\ -9.22 \end{bmatrix}$

20. $A = \begin{bmatrix} -7.99 & 8.83 \\ 2.65 & -6.63 \\ -3.74 & 9.80 \\ 0.49 & -4.24 \\ 9.85 & 6.08 \end{bmatrix}$ $y = \begin{bmatrix} 8.77 \\ -5.57 \\ 8.43 \\ -5.50 \\ 7.62 \end{bmatrix}$

21. $A = \begin{bmatrix} -9.54 & 4.24 \\ 9.11 & -8.94 \\ -4.47 & 3.87 \\ 9.17 & -7.28 \end{bmatrix}$ $y = \begin{bmatrix} -5.10 \\ 6.40 \\ 7.35 \\ -4.80 \end{bmatrix}$

22. $A = \begin{bmatrix} 6.16 & 1.41 & 9.06 & 8.77 & 5.74 \\ 8.31 & 8.54 & 6.24 & 2.84 & 2.65 \\ 7.75 & 8.25 & 4.58 & 5.74 & 5.83 \\ 8.31 & 8.83 & 2.83 & 3.87 & 9.85 \\ 5.64 & 7.14 & 9.38 & 5.66 & 8.37 \\ 2.24 & 2.83 & 3.87 & 2.65 & 6.24 \end{bmatrix}$ $y = \begin{bmatrix} 4.24 \\ 5.57 \\ 3.74 \\ 7.94 \\ 5.10 \\ 7.87 \end{bmatrix}$

23. $A = \begin{bmatrix} 7.94 & 4.40 & 5.10 & 4.69 \\ -4.12 & 8.06 & 4.36 & -3.74 \\ 9.54 & 5.66 & -3.47 & 7.75 \\ 8.91 & -9.06 & 9.49 & 0.39 \\ 2.24 & 8.77 & 9.54 & 9.54 \\ 8.83 & 7.07 & -7.75 & 1.41 \end{bmatrix}$ $y = \begin{bmatrix} 7.75 \\ 4.40 \\ -0.23 \\ 6.78 \\ -8.77 \\ -6.16 \end{bmatrix}$

24. $A = \begin{bmatrix} -5.39 & 2.83 & 4.58 \\ 5.10 & -8.72 & 6.16 \\ 7.42 & -1.41 & -7.21 \\ -3.32 & 7.75 & 4.99 \\ 7.21 & -4.90 & 8.89 \\ -7.07 & 4.36 & -4.69 \end{bmatrix}$ $y = \begin{bmatrix} -8.72 \\ 0.90 \\ 9.22 \\ -7.62 \\ 7.55 \\ 8.60 \end{bmatrix}$

25. $A = \begin{bmatrix} 5.66 & -4.85 \\ 7.28 & 1.73 \\ -6.78 & 6.48 \\ 7.48 & 9.17 \\ 5.10 & -6.78 \\ 8.25 & 4.47 \end{bmatrix}$ $y = \begin{bmatrix} 5.57 \\ 9.54 \\ -5.92 \\ 6.78 \\ 9.49 \\ -7.48 \end{bmatrix}$

34.4 QR Algorithm for Finding Eigenvalues

The QR Algorithm for Finding Eigenvalues

Modern computational algorithms for finding eigenvalues numerically use some version of the QR algorithm. The following treatment will illustrate the idea behind the QR algorithm but will not show all the many refinements that have been added to speed up the computation and make it more stable. But with this introduction, the algorithm will be understandable when met, for example, in a numerical analysis course.

The QR algorithm finds the eigenvalues of $A = A_0$ by factoring A_0 to $Q_0 R_0$ and generating the sequences of matrices Q_k, R_k, and $A_k, k = 1, 2, \ldots,$ according to the prescription $A_{k+1} = R_k Q_k$, where $A_k = Q_k R_k$. Thus, at each step, the QR factorization of A_k is obtained and the factors are reversed to generate a new matrix A_{k+1}. At this point it is helpful to visualize the first few steps of the algorithm.

$$A_0 = Q_0 R_0 \qquad A_1 = R_0 Q_0 = Q_1 R_1 \qquad A_2 = R_1 Q_1 = Q_2 R_2 \qquad A_3 = R_2 Q_2 = Q_3 R_3$$

If A is real and no two eigenvalues have equal magnitude, that is, if the eigenvalues satisfy the condition

$$0 < |\lambda_n| < |\lambda_{n-1}| < \cdots < |\lambda_2| < |\lambda_1|$$

then the sequence of matrices A_k converges to an upper triangular matrix with the eigenvalues of A_0 on the main diagonal. If, in addition, A is symmetric, then A_k converges to a diagonal matrix.

If A is real, with at least two eigenvalues of equal magnitude, then A_k converges to a quasi-triangular matrix such as

$$
\begin{bmatrix}
x & a_{12} & a_{13} & a_{14} & a_{15} & a_{16} \\
0 & y_1 & y_2 & a_{24} & a_{25} & a_{26} \\
0 & y_3 & y_4 & a_{34} & a_{35} & a_{36} \\
0 & 0 & 0 & z & a_{45} & a_{46} \\
0 & 0 & 0 & 0 & w_1 & w_2 \\
0 & 0 & 0 & 0 & w_3 & w_4
\end{bmatrix}
\qquad (34.17)
$$

where all entries below the subdiagonal are zero and no two consecutive subdiagonal entries $a_{j+1,j}$ are nonzero. Going down the diagonal there are $n \times n$ submatrices, $n = 1, 2$, called, respectively, blocks of order 1 or 2. In (34.17), for example, x and z sit in blocks of order 1 and are eigenvalues. The submatrices

$$
\begin{bmatrix} y_1 & y_2 \\ y_3 & y_4 \end{bmatrix}
\quad
\begin{bmatrix} w_1 & w_2 \\ w_3 & w_4 \end{bmatrix}
$$

are blocks of order 2 and generate the usual pair of eigenvalues for a 2×2 matrix. Of course, if A is symmetric (with eigenvalues of equal magnitude), then A_k converges to a block-diagonal matrix with an interpretation similar to that of the quasi-triangular matrix.

The off-diagonal elements in A_k decrease by one of the factors $|\lambda_{k+1}/\lambda_k|, k = 1, \ldots, n-1$. The largest ratio of the magnitudes of successive eigenvalues determines how quickly the basic QR algorithm converges to a diagonal matrix.

At each step, the QR factorization can be obtained by applying Gram–Schmidt orthogonalization to the columns of A_k to form Q_k and then determining $R_k = Q_k^{\mathrm{T}} A_k$. In practice, much more sophisticated numerical techniques are used to obtain the QR decompositions required. (See, Chapter 45, [4] or [60].)

EXAMPLE 34.6   We illustrate the QR algorithm for finding the eigenvalues $-6, -\frac{1}{2}(1 \pm \sqrt{145})$ of the symmetric matrix

$$
A = \begin{bmatrix} -3 & 5 & 1 \\ 5 & 2 & 2 \\ 1 & 2 & -6 \end{bmatrix}
\qquad (34.18)
$$

The sorted absolute values of these eigenvalues are $5.520797290, 6, 6.520797290$, so the ratios of successive magnitudes are 0.92 and 0.92, ratios close enough to 1 to signal slow convergence. To see this, begin the QR algorithm by factoring $A_0 = A$ as $A_0 = Q_0 R_0$, where the factors are

$$
Q_0 = \begin{bmatrix}
-\dfrac{3}{\sqrt{35}} & \dfrac{166}{\sqrt{40{,}110}} & \dfrac{8}{\sqrt{1146}} \\[8pt]
\dfrac{5}{\sqrt{35}} & \dfrac{85}{\sqrt{40{,}110}} & \dfrac{11}{\sqrt{1146}} \\[8pt]
\dfrac{1}{\sqrt{35}} & \dfrac{73}{\sqrt{40{,}110}} & -\dfrac{31}{\sqrt{1146}}
\end{bmatrix}
\qquad
R_0 = \begin{bmatrix}
\sqrt{35} & -\dfrac{3}{\sqrt{35}} & \dfrac{1}{\sqrt{35}} \\[8pt]
0 & \dfrac{1146}{\sqrt{40{,}110}} & -\dfrac{102}{\sqrt{40{,}110}} \\[8pt]
0 & 0 & \dfrac{216}{\sqrt{1146}}
\end{bmatrix}
$$

then obtaining

$$
A_1 = R_0 Q_0 = \begin{bmatrix}
-\dfrac{17}{5} & \dfrac{5628}{\sqrt{1{,}403{,}850}} & \dfrac{216}{\sqrt{40{,}110}} \\[8pt]
\dfrac{5628}{\sqrt{1{,}403{,}850}} & \dfrac{2142}{955} & \dfrac{2628}{\sqrt{1{,}276{,}835}} \\[8pt]
\dfrac{216}{\sqrt{40{,}110}} & \dfrac{2628}{\sqrt{1{,}276{,}835}} & -\dfrac{1116}{191}
\end{bmatrix}
$$

To avoid excessive "expression swell," we can work in floating-point arithmetic when factoring A_1 to $Q_1 R_1$ and obtaining $A_2 = R_1 Q_1$. Rounding entries to three digits, we have

$$R_1 = \begin{bmatrix} 5.94 & -0.503 & 0.182 \\ 0 & 5.72 & -0.552 \\ 0 & 0 & 6.35 \end{bmatrix} \quad Q_1 = \begin{bmatrix} -0.572 & 0.780 & 0.254 \\ 0.800 & 0.462 & 0.383 \\ 0.182 & 0.422 & -0.888 \end{bmatrix}$$

$$A_2 = \begin{bmatrix} -3.77 & 4.48 & 1.15 \\ 4.48 & 2.41 & 2.68 \\ 1.15 & 2.68 & -5.64 \end{bmatrix}$$

Factoring A_2 to $Q_2 R_2$ and forming $A_3 = R_2 Q_2$, we find

$$R_2 = \begin{bmatrix} 5.96 & -0.499 & 0.193 \\ 0 & 5.73 & -0.595 \\ 0 & 0 & 6.32 \end{bmatrix} \quad Q_2 = \begin{bmatrix} -0.632 & 0.726 & 0.250 \\ 0.750 & 0.487 & 0.447 \\ 0.193 & 0.485 & -0.853 \end{bmatrix}$$

$$A_3 = \begin{bmatrix} -4.11 & 4.18 & 1.22 \\ 4.18 & 2.50 & 3.07 \\ 1.22 & 3.07 & -5.39 \end{bmatrix}$$

Factoring A_3 to $Q_3 R_3$ and forming $A_4 = R_3 Q_3$, we obtain

$$R_3 = \begin{bmatrix} 5.99 & -0.497 & 0.204 \\ 0 & 5.74 & -0.639 \\ 0 & 0 & 6.29 \end{bmatrix} \quad Q_3 = \begin{bmatrix} -0.686 & 0.670 & 0.285 \\ 0.699 & 0.496 & 0.516 \\ 0.204 & 0.553 & -0.808 \end{bmatrix}$$

$$A_4 = \begin{bmatrix} -4.41 & 3.88 & 1.28 \\ 3.88 & 2.49 & 3.48 \\ 1.28 & 3.48 & -5.08 \end{bmatrix}$$

Continuing in this way, we obtain

$$R_9 = \begin{bmatrix} 6.14 & -0.513 & 0.225 \\ 0 & 5.88 & -0.787 \\ 0 & 0 & 5.99 \end{bmatrix} \quad Q_9 = \begin{bmatrix} -0.897 & 0.319 & 0.306 \\ 0.381 & 0.206 & 0.901 \\ 0.225 & 0.925 & -0.306 \end{bmatrix}$$

$$A_{10} = \begin{bmatrix} -5.65 & 2.06 & 1.35 \\ 2.06 & 0.483 & 5.54 \\ 1.35 & 5.54 & -1.84 \end{bmatrix}$$

along with

$$A_{20} = \begin{bmatrix} -6.35 & 0.492 & 0.563 \\ 0.492 & -4.29 & 4.24 \\ 0.563 & 4.24 & 3.64 \end{bmatrix} \quad A_{40} = \begin{bmatrix} -6.51 & 0.0591 & 0.0253 \\ 0.0591 & -5.94 & 0.885 \\ 0.0253 & 0.885 & 5.45 \end{bmatrix}$$

Since the eigenvalues are approximately $-6, 5.521, -6.521$, we can see the slow convergence of the sequence $\{A_k\}$ to a diagonal matrix of eigenvalues. After 10 iterations, the off-diagonal elements in A_{10} have been reduced only by a factor of $0.92^{10} = 0.43$. In A_{20}, this factor is $0.92^{20} = 0.19$; and in A_{40}, it is $0.92^{40} = 0.036$. We leave to the exercises computation of A_{100}, significantly closer to a diagonal matrix for which $0.92^{100} = 0.00024$.

Theory Behind the QR Algorithm

If the $n \times n$ matrix A, of rank k, is QR-factored, the columns of the $n \times k$ matrix Q are orthonormal, so $Q^\mathrm{T}Q = I_k$ and $Q^\mathrm{T}A = I_k R = R$ since R is $k \times n$. Therefore, in the QR algorithm,

$$A_{k+1} = R_k Q_k = Q_k^\mathrm{T} A_k Q_k$$

so A_{k+1} is similar to A_k and has the same eigenvalues as A_k. By induction, A_{k+1} is similar to $A_0 = A$. The eigenvalues of A_{k+1} are therefore the same as the eigenvalues of A.

Indeed, starting the process from $A_0 = A$, we find

$$A_0 = Q_0 R_0 \qquad A_1 = R_0 Q_0 = [Q_0^\mathrm{T} A_0] Q_0 \Rightarrow A_1 \text{ similar to } A_0 = A$$

for the first step and

$$A_1 = Q_1 R_1 \qquad A_2 = R_1 Q_1 = [Q_1^\mathrm{T} A_1] Q_1 \Rightarrow A_2 \text{ similar to } A_1 \text{ similar to } A_0 = A$$

for the second. At each stage of the algorithm, the matrix A_{k+1} is orthogonally similar to the preceding A_k and, hence, has the same eigenvalues. In fact, generalizing from

$$
\begin{aligned}
A_3 &= R_2 Q_2 \\
&= Q_2^\mathrm{T} A_2 Q_2 \\
&= Q_2^\mathrm{T} R_1 Q_1 Q_2 \\
&= Q_2^\mathrm{T} Q_1^\mathrm{T} A_1 Q_1 Q_2 \\
&= Q_2^\mathrm{T} Q_1^\mathrm{T} R_0 Q_0 Q_1 Q_2 \\
&= Q_2^\mathrm{T} Q_1^\mathrm{T} Q_0^\mathrm{T} A_0 Q_0 Q_1 Q_2
\end{aligned}
$$

we have $A_{k+1} = (Q_k^\mathrm{T} \cdots Q_0^\mathrm{T}) A_0 (Q_0 \cdots Q_k) = (Q_0 \cdots Q_k)^\mathrm{T} A (Q_0 \cdots Q_k) = P_k^\mathrm{T} A P_k$.

The convergence of $\{A_k\}$ to a diagonal matrix with eigenvalues on the main diagonal is discussed, for example, in [4] or [60]. This convergence is related to the behavior of the *power* *method* for finding eigenvalues, seen in Section 45.1. In essence, $A^{k+1} = (Q_0 \cdots Q_k)(R_k \cdots R_0) = P_k U_k$, giving the QR factorization of A^{k+1}, established by an induction based on the following sample calculation.

$$
\begin{aligned}
A_0^3 &= A_0 A_0 A_0 \\
&= (Q_0 R_0)(Q_0 R_0)(Q_0 R_0) \\
&= Q_0 (R_0 Q_0)(R_0 Q_0) R_0 \\
&= Q_0 A_1 A_1 R_0 \\
&= Q_0 (Q_1 R_1)(Q_1 R_1) R_0 \\
&= Q_0 Q_1 (R_1 Q_1) R_1 R_0 \\
&= Q_0 Q_1 A_2 R_1 R_0 \\
&= Q_0 Q_1 (Q_2 R_2) R_1 R_0 \\
&= (Q_0 Q_1 Q_2)(R_2 R_1 R_0)
\end{aligned}
$$

Exercise 3 empirically explores this relation and also the identity $A_{k+1} = P_k^\mathrm{T} A^{k+2} U_k^{-1}$, which relates A_{k+1} to A^{k+2}.

EXERCISES 34.4

1. Using the software of your choice, implement a version of the QR algorithm.

2. Using the code from Exercise 1, continue the QR algorithm for the matrix in (34.18), obtaining A_{100}.

For the matrices given in Exercises 3–12:

  (a) Use an appropriate software tool to find the eigenvalues.

  (b) Sort the eigenvalues by magnitudes, and obtain the ratios of successive magnitudes.

  (c) Use the QR algorithm from Exercise 1 to obtain numerical approximations to the eigenvalues.

  (d) Test the decrease in the off-diagonal elements of $\{A_k\}$, comparing to the ratios in part (b).

  (e) With $P_k = Q_0 \cdots Q_k$ and $U_k = R_k \cdots R_0$, show, for $k = 5$, 10, 15, that $P_k U_k = A^{k+1}$ and that $A_{k+1} = P_k^{\mathrm{T}} A^{k+2} U_k^{-1}$. (See, e.g., [52] or [88] for induction proofs.)

3. $\begin{bmatrix} -3 & 4 & -6 \\ 8 & 11 & 12 \\ 6 & 7 & -11 \end{bmatrix}$ **4.** $\begin{bmatrix} 7 & -11 & 12 \\ 10 & 1 & 3 \\ 7 & -1 & -1 \end{bmatrix}$ **5.** $\begin{bmatrix} 0 & 0 & -12 \\ 1 & 0 & 8 \\ 0 & 1 & 1 \end{bmatrix}$

6. $\begin{bmatrix} -9 & -3 & -8 \\ -3 & -12 & 11 \\ -8 & 11 & -5 \end{bmatrix}$ **7.** $\begin{bmatrix} -12 & -6 & 8 \\ -2 & -7 & 4 \\ -1 & 4 & 0 \end{bmatrix}$

8. $\begin{bmatrix} -3 & 10 & 7 & -2 \\ -2 & 6 & 8 & -5 \\ -4 & 10 & 3 & 3 \\ -5 & -9 & -2 & 12 \end{bmatrix}$ **9.** $\begin{bmatrix} -1 & 12 & -1 & -7 \\ 0 & 12 & 4 & -8 \\ -6 & 3 & -11 & -2 \\ 5 & 12 & -4 & 11 \end{bmatrix}$

10. $\begin{bmatrix} -10 & -1 & 3 & 11 \\ -1 & -1 & 2 & -12 \\ 3 & 2 & 10 & 2 \\ 11 & -12 & 2 & 0 \end{bmatrix}$ **11.** $\begin{bmatrix} 0 & 0 & 0 & 48 \\ 1 & 0 & 0 & 44 \\ 0 & 1 & 0 & 4 \\ 0 & 0 & 1 & -5 \end{bmatrix}$

12. $\begin{bmatrix} -3 & 8 & -3 & -4 \\ 5 & 1 & 6 & -12 \\ 8 & -6 & 6 & 3 \\ 2 & 11 & 2 & 12 \end{bmatrix}$

For the vector $\mathbf{w}$ given in each of Exercises 13–22:

  (a) If I_4 is the 4×4 identity matrix, form $H_{\mathbf{w}} = I_4 - \frac{2}{\|\mathbf{w}\|_2^2}\mathbf{w}\mathbf{w}^{\mathrm{T}}$, the *Householder reflection matrix* that will be used for numeric computations in Section 45.2.

  (b) Let $\mathbf{x} = [1\ 2\ 3\ 4]^{\mathrm{T}}$, and obtain $H_{\mathbf{w}}\mathbf{x} = \mathbf{y}$.

  (c) Show that $\mathbf{x}_{\mathbf{w}}$ the projection of $\mathbf{x}$ along $\mathbf{w}$, is the negative of $\mathbf{y}_{\mathbf{w}}$, the projection of $\mathbf{y}$ along $\mathbf{w}$.

  (d) Show that $\mathbf{x}_{\perp\mathbf{w}}$, the component of $\mathbf{x}$ orthogonal to $\mathbf{w}$, and $\mathbf{y}_{\perp\mathbf{w}}$, the component of $\mathbf{y}$ orthogonal to $\mathbf{w}$, are equal. Hence, $\mathbf{y}$ is the reflection of $\mathbf{x}$ across the subspace orthogonal to $\mathbf{w}$.

  (e) Show that $H_{\mathbf{w}}$ is a symmetric orthogonal matrix.

  (f) Show that $\det(H_{\mathbf{w}}) = -1$.

13. $[-6\ 8\ 10\ 1]^{\mathrm{T}}$ **14.** $[-11\ 1\ -2\ 6]^{\mathrm{T}}$

15. $[9\ -3\ -7\ 1]^{\mathrm{T}}$ **16.** $[-5\ -2\ 12\ -2]^{\mathrm{T}}$

17. $[11\ -6\ 12\ 4]^{\mathrm{T}}$ **18.** $[11\ 5\ 4\ 6]^{\mathrm{T}}$

19. $[8\ 7\ 3\ 4]^{\mathrm{T}}$ **20.** $[8\ 12\ -3\ -2]^{\mathrm{T}}$

21. $[7\ -9\ 3\ -4]^{\mathrm{T}}$ **22.** $[5\ -7\ -8\ 10]^{\mathrm{T}}$

For each matrix A given in Exercises 23–27:

  (a) Let $\mathbf{x}$ be the first column of A, define $\mathbf{w} = \mathbf{x} - \|\mathbf{x}\|_2\mathbf{e}_1$, where $\mathbf{e}_1 = [1\ 0\ 0\ 0]^{\mathrm{T}}$; and obtain the Householder matrix $H_{\mathbf{w}}$.

  (b) Show that $H_{\mathbf{w}}\mathbf{x} = \|\mathbf{x}\|_2\mathbf{e}_1 = [\|\mathbf{x}\|_2\ 0\ 0\ 0]^{\mathrm{T}}$.

  (c) Show that $H_{\mathbf{w}}A$ is a matrix whose first column is $[\|\mathbf{x}\|_2\ 0\ 0\ 0]^{\mathrm{T}}$.

23. $\begin{bmatrix} -6 & -9 & -6 & -9 \\ -4 & 8 & 9 & -7 \\ -1 & -3 & -9 & -8 \\ 9 & -4 & 1 & -9 \end{bmatrix}$ **24.** $\begin{bmatrix} 2 & 9 & -5 & -2 \\ -6 & -2 & 6 & 1 \\ 4 & -6 & 8 & -9 \\ 9 & -3 & 5 & 3 \end{bmatrix}$

25. $\begin{bmatrix} -2 & 2 & -2 & 0 \\ 7 & 3 & -7 & -6 \\ 6 & -7 & 6 & -3 \\ -9 & -4 & -1 & -2 \end{bmatrix}$ **26.** $\begin{bmatrix} 1 & 7 & 5 & 2 \\ -1 & -6 & 4 & 4 \\ 6 & 6 & -8 & 1 \\ 7 & -9 & 5 & -4 \end{bmatrix}$

27. $\begin{bmatrix} 7 & 3 & -8 & -5 \\ 4 & 9 & 5 & 6 \\ 7 & 3 & 4 & -9 \\ 2 & -1 & -2 & 5 \end{bmatrix}$

The QR decomposition of a matrix can be obtained by multiplying it by a sequence of Householder matrices. For each matrix A given in Exercises 28–32:

  (a) Let $\mathbf{u}_1$ be the first column of A, and obtain $\mathbf{w}_1 = \mathbf{u}_1 - [\|\mathbf{u}_1\|_2\ 0\ 0\ 0]^{\mathrm{T}}$, $H_1 = I_4 - \frac{2}{\|\mathbf{w}_1\|_2^2}\mathbf{w}_1\mathbf{w}_1^{\mathrm{T}}$, and $A_1 = H_1 A$. The first column of A_1 will have three zeros for its last three elements. *Hint:* Compute with floating-point numbers or expression-swell will become a significant challenge.

  (b) Let $\mathbf{U}$ be the second column of A_1, set $\mathbf{u}_2 = [U_2\ U_3\ U_4]^{\mathrm{T}}$, and obtain $\mathbf{w}_2 = \mathbf{u}_2 - [\|\mathbf{u}_2\|_2\ 0\ 0]^{\mathrm{T}}$, $h_2 = I_3 - \frac{2}{\|\mathbf{w}_2\|_2^2}\mathbf{w}_2\mathbf{w}_2^{\mathrm{T}}$, $H_2 = \begin{bmatrix} 1 & 0 \\ 0 & h_2 \end{bmatrix}$, and $A_2 = H_2 A_1$. In the first and second columns of A_2 there should be only zeros below the main diagonal.

  (c) Let $\mathbf{V}$ be the third column of A_2, set $\mathbf{u}_3 = [V_3\ V_4]^{\mathrm{T}}$, and obtain $\mathbf{w}_3 = \mathbf{u}_3 - [\|\mathbf{u}_3\|_2\ 0]^{\mathrm{T}}$, $h_3 = I_2 - \frac{2}{\|\mathbf{w}_3\|_2^2}\mathbf{w}_3\mathbf{w}_3^{\mathrm{T}}$, $H_3 = \begin{bmatrix} I_2 & 0_2 \\ 0_2 & h_3 \end{bmatrix}$, and $A_3 = H_3 A_2$. The matrix A_3 is an upper triangular matrix and is, in fact, the factor R in the QR decomposition.

  (d) Obtain $H = H_3 H_2 H_1$, and show H is orthogonal. In fact, $H = Q^{\mathrm{T}}$, where Q is the factor companion to R in the QR decomposition.

  (e) Show $HA = A_3 = R$.

  (f) Show that $A = H^{\mathrm{T}}A_3$. This verifies that the QR decomposition can be achieved by Householder reflection matrices. This technique obtains R and along the way gets $Q = H^{\mathrm{T}}$, in contrast to the computation of Q by the Gram–Schmidt process, leading to $R = Q^{\mathrm{T}}A$. Computing QR by Householder matrices is "economical" and numerically stable. We will see this in Sections 45.2 and 45.3, where we consider the numerical calculation of eigenvalues.

28. $\begin{bmatrix} 4 & -6 & 8 & 11 \\ 12 & 6 & 7 & -11 \\ -12 & -6 & 8 & -2 \\ -7 & 4 & -1 & 4 \end{bmatrix}$
29. $\begin{bmatrix} 0 & 0 & -1 & -8 \\ -12 & -6 & 10 & -2 \\ 9 & 8 & 10 & -7 \\ 1 & 8 & -3 & 12 \end{bmatrix}$

30. $\begin{bmatrix} -2 & -8 & 7 & -11 \\ 0 & 9 & -2 & 11 \\ -7 & -9 & 4 & -6 \\ 12 & -3 & -10 & 3 \end{bmatrix}$
31. $\begin{bmatrix} 1 & 5 & 10 & 11 \\ -10 & -10 & 12 & -12 \\ -5 & -5 & -8 & -10 \\ -8 & -2 & 8 & 11 \end{bmatrix}$

32. $\begin{bmatrix} 1 & -11 & -12 & 12 \\ 11 & -12 & 7 & -11 \\ 12 & 10 & 1 & 3 \\ 7 & -1 & -1 & -3 \end{bmatrix}$

34.5   SVD, The Singular Value Decomposition

Singular Values, Singular Vectors and the SVD

A square matrix is similar to either a diagonal matrix or to a Jordan matrix. (See Section 34.2.) Thus, we have $A = PDP^{-1}$ in the case of the diagonal matrix and $A = PJP^{-1}$ in the case of the Jordan matrix. In either event, the eigenvalues of A appear on the diagonal of D or J, and the eigenvectors of A appear in the transition matrix P.

A square *symmetric* matrix is *orthogonally* similar to a diagonal matrix, in which case $A = PDP^T$ since now P can be chosen to be an orthogonal matrix. The eigenvalues of A are on the diagonal of D, and the columns of P are an orthonormal set of eigenvectors of A.

The $r \times c$ matrix A of rank k can be written as $A = U\Sigma V^T$, a factorization called the *singular value decomposition* (SVD) of A. In particular, for A real:

1. $A^T A$ is a $c \times c$ symmetric matrix of rank k, whose k nonzero eigenvalues are positive, as shown in (34.21).

2. AA^T is an $r \times r$ symmetric matrix of rank k, whose k nonzero eigenvalues are the same as those of $A^T A$. (See Exercise A1.)

3. Ordering the positive eigenvalues of $A^T A$ or AA^T in the descending order $\lambda_1 \geq \lambda_2 \geq \cdots \geq \lambda_k > 0$, the *singular values* of A are $\sigma_i = \sqrt{\lambda_i}, i = 1, \ldots, k$. (We follow [4] in taking singular values as nonzero, noting that texts such as [52] allow singular values to be zero.)

4. Σ is an $r \times c$ matrix whose only nonzero entries are $\Sigma_{ii} = \sigma_i, i = 1, \ldots, k$. (See (34.19).)

5. $U = [U_1, U_2]$ is an orthogonal matrix whose columns are the orthonormal eigenvectors of the symmetric $r \times r$ matrix AA^T. The columns of U are called the *left singular vectors* of A.

6. The columns of the matrix $U_1 = [\mathbf{u}_1, \ldots, \mathbf{u}_k]$ are an orthonormal basis for the column space of A.

7. The columns of the matrix $U_2 = [\mathbf{u}_{k+1}, \ldots, \mathbf{u}_r]$ are an orthonormal basis for the null space of A^T.

8. $V = [V_1, V_2]$ is an orthogonal matrix whose columns are the orthonormal eigenvectors of the symmetric $c \times c$ matrix $A^T A$. The columns of V are called the *right singular vectors* of A.

9. The columns of the matrix $V_1 = [\mathbf{v}_1, \ldots, \mathbf{v}_k]$ are an orthonormal basis for the row space of A.

10. The columns of the matrix $V_2 = [\mathbf{v}_{k+1}, \ldots, \mathbf{v}_c]$ are an orthonormal basis for the null space of A.

These relationships are summarized in (34.19).

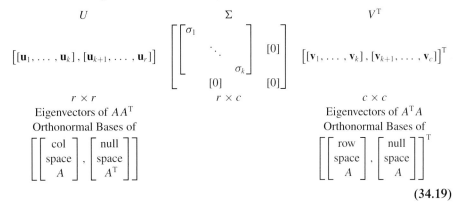

$$(34.19)$$

Figure 34.1 shows the relationship of the singular value decomposition to the four fundamental subspaces for A.

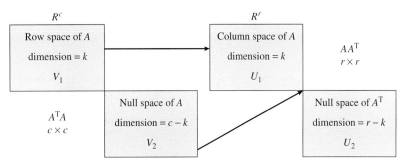

FIGURE 34.1 The fundamental subspaces of the matrix $A = U\Sigma V^T$ and their relationship to the SVD for A

EXAMPLE 34.7 Obtained as the product of 5×2 and 2×4 matrices of maximal rank, the 5×4 matrix A in (34.20) has rank $k = 2$.

$$A = \begin{bmatrix} 45 & -108 & 36 & -45 \\ 21 & -68 & 26 & -33 \\ 72 & -32 & -16 & 24 \\ -56 & 64 & -8 & 8 \\ 50 & -32 & -6 & 10 \end{bmatrix} \qquad (34.20)$$

THE MATRICES $A^T A$ AND $A A^T$ Figure 34.2 illustrates the action of A on a vector in $R^c = R^4$. Since the rank of A is $k = 2$, both the row space and the column space will have dimension 2. By subtraction, we find the dimensions of the spaces orthogonal to the row and column spaces. The space orthogonal to the row space of A is the null space of A. The space orthogonal to the column space of A is the null space of A^T.

The matrices $A^T A$ and $A A^T$ and their eigenvalues are given in Table 34.3. The matrix $A^T A$ is 4×4. In general, if A is $r \times c$, then $A^T A$ is $c \times c$. The matrix $A A^T$ is 5×5. In general, if A is $r \times c$, then $A A^T$ is $r \times r$. Observe, in addition, that these matrices are both

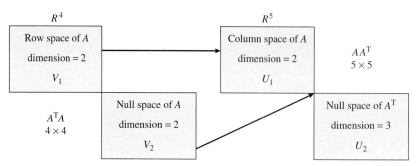

FIGURE 34.2   Fundamental subspaces for the matrix A in Example 34.7

$$A^{\mathrm{T}}A = \begin{bmatrix} 13{,}286 & -13{,}776 & 1162 & -938 \\ -13{,}776 & 22{,}432 & -5464 & 6528 \\ 1162 & -5464 & 2328 & -2986 \\ -938 & 6528 & -2986 & 3854 \end{bmatrix} \qquad AA^{\mathrm{T}} = \begin{bmatrix} 17{,}010 & 10{,}710 & 5040 & -10{,}080 & 5040 \\ 10{,}710 & 6830 & 2480 & -6000 & 2740 \\ 5040 & 2480 & 7040 & -5760 & 4960 \\ -10{,}080 & -6000 & -5760 & 7360 & -4720 \\ 5040 & 2740 & 4960 & -4720 & 3660 \end{bmatrix}$$

Eigenvalues: $0, 0, 20{,}950 \pm 30\sqrt{204{,}983}$ Eigenvalues: $0, 0, 0, 20{,}950 \pm 30\sqrt{204{,}983}$

TABLE 34.3   Example 34.7: The matrices $A^{\mathrm{T}}A$ and AA^{T} and their eigenvalues

symmetric. They have to be, since

$$(A^{\mathrm{T}}A)^{\mathrm{T}} = A^{\mathrm{T}}(A^{\mathrm{T}})^{\mathrm{T}} = A^{\mathrm{T}}A \quad \text{and} \quad (AA^{\mathrm{T}})^{\mathrm{T}} = (A^{\mathrm{T}})^{\mathrm{T}}A^{\mathrm{T}} = AA^{\mathrm{T}}$$

Because A has rank 2, $A^{\mathrm{T}}A$ and AA^{T} also have rank $k = 2$.

Each matrix has *nearly* the same set of eigenvalues. They each have the same nonzero eigenvalues, and the only difference is in the number of times $\lambda = 0$ is an eigenvalue. Clearly, $\lambda = 0$ has to be an eigenvalue because each matrix is not of full rank, so the determinant of each matrix is zero.

The eigenvalues of $A^{\mathrm{T}}A$ are necessarily real since $A^{\mathrm{T}}A$ is symmetric. (See Section 32.2.) Moreover, the eigenvalues are nonnegative. To see this, let ρ be an eigenvalue of $A^{\mathrm{T}}A$ and let $\mathbf{x}$ be the corresponding eigenvector. Then we have

$$\|A\mathbf{x}\|^2 = (A\mathbf{x}) \cdot (A\mathbf{x}) = (A\mathbf{x})^{\mathrm{T}}A\mathbf{x} = \mathbf{x}^{\mathrm{T}}A^{\mathrm{T}}A\mathbf{x} = \mathbf{x}^{\mathrm{T}}\rho\mathbf{x} = \rho\|\mathbf{x}\|^2 \qquad (34.21)$$

and consequently, $0 \le \rho = \frac{\|A\mathbf{x}\|^2}{\|\mathbf{x}\|^2}$. (A similar statement holds for AA^{T}. See Exercise A2.)

THE SINGULAR VALUES OF A   The $k = \mathrm{rank}(A) = 2$ singular values of A, the square roots of the nonzero eigenvalues of $A^{\mathrm{T}}A$ or AA^{T}, are

$$\sigma_1 = \sqrt{20{,}950 + 30\sqrt{204{,}983}} \quad \text{and} \quad \sigma_2 = \sqrt{20{,}950 - 30\sqrt{204{,}983}}$$

THE MATRIX V   To obtain the matrix $V = [V_1, V_2]$, we find the eigenvectors of $A^{\mathrm{T}}A$ and sort them in the order corresponding to decreasing eigenvalues, as shown in (34.22). Since $A^{\mathrm{T}}A$ is symmetric, the eigenvectors from distinct eigenvalues are necessarily orthogonal. Two eigenvectors are from the single eigenvalue zero; hence, they need not be orthogonal to each other. In fact, they are not since their dot product is -1300.

$$\begin{bmatrix} \frac{3277}{4434} + \frac{23\sqrt{204,983}}{4434} \\ -\frac{5666}{2217} - \frac{10\sqrt{204,983}}{2217} \\ 1 \\ -\frac{5641}{4434} + \frac{\sqrt{204,983}}{4434} \end{bmatrix} \quad \begin{bmatrix} \frac{3277}{4434} - \frac{23\sqrt{204,983}}{4434} \\ -\frac{5666}{2217} + \frac{10\sqrt{204,983}}{2217} \\ 1 \\ -\frac{5641}{4434} - \frac{\sqrt{204,983}}{4434} \end{bmatrix} \quad \begin{bmatrix} 0 \\ 1 \\ 28 \\ 20 \end{bmatrix} \quad \begin{bmatrix} 1 \\ 0 \\ -30 \\ -23 \end{bmatrix} \qquad \textbf{(34.22)}$$

The first two eigenvectors in (34.22) need only be normalized and then placed into the matrix

$$V_1 = \begin{bmatrix} \frac{739}{2}\frac{3277 + 23\sqrt{204,983}}{\sqrt{\alpha}} & -\frac{739}{2}\frac{-3277 + 23\sqrt{204,983}}{\sqrt{\beta}} \\ -1478\frac{2833 + 5\sqrt{204,983}}{\sqrt{\alpha}} & 1478\frac{-2833 + 5\sqrt{204,983}}{\sqrt{\beta}} \\ \frac{1,638,363}{\sqrt{\alpha}} & \frac{1,638,363}{\sqrt{\beta}} \\ \frac{739}{2}\frac{-5641 + \sqrt{204,983}}{\sqrt{\alpha}} & -\frac{739}{2}\frac{5641 + \sqrt{204,983}}{\sqrt{\beta}} \end{bmatrix}$$

$$\alpha = 52,054,667,238,495 + 80,926,940,385\sqrt{204,983}$$
$$\beta = 52,054,667,238,495 - 80,926,940,385\sqrt{204,983}$$

The remaining two eigenvectors need to be orthonormalized and then placed into the matrix

$$V_2 = \begin{bmatrix} 0 & \frac{237}{\sqrt{215,670}} \\ \frac{1}{\sqrt{1185}} & \frac{260}{\sqrt{215,670}} \\ \frac{28}{\sqrt{1185}} & \frac{170}{\sqrt{215,670}} \\ \frac{20}{\sqrt{1185}} & -\frac{251}{\sqrt{215,670}} \end{bmatrix}$$

Finally, we form $V = [V_1, V_2]$, obtaining

$$V = \begin{bmatrix} \frac{739}{2}\frac{3277 + 23\sqrt{204,983}}{\sqrt{\alpha}} & -\frac{739}{2}\frac{-3277 + 23\sqrt{204,983}}{\sqrt{\beta}} & 0 & \frac{237}{\sqrt{215,670}} \\ -1478\frac{2833 + 5\sqrt{204,983}}{\sqrt{\alpha}} & 1478\frac{-2833 + 5\sqrt{204,983}}{\sqrt{\beta}} & \frac{1}{\sqrt{1185}} & \frac{260}{\sqrt{215,670}} \\ \frac{1,638,363}{\sqrt{\alpha}} & \frac{1,638,363}{\sqrt{\beta}} & \frac{28}{\sqrt{1185}} & \frac{170}{\sqrt{215,670}} \\ \frac{739}{2}\frac{-5641 + \sqrt{204,983}}{\sqrt{\alpha}} & -\frac{739}{2}\frac{5641 + \sqrt{204,983}}{\sqrt{\beta}} & \frac{20}{\sqrt{1185}} & -\frac{251}{\sqrt{215,670}} \end{bmatrix}$$

$$\textbf{(34.23)}$$

THE MATRIX U Although the vectors $\mathbf{u}_j$, $j = 1, \ldots, k$, are eigenvectors of the matrix AA^T, we obtain them from the equation

$$A\mathbf{v}_j = \sigma_j\mathbf{u}_j \qquad j = 1, \ldots, k \qquad \textbf{(34.24)}$$

thereby linking them to the $\mathbf{v}_j$'s. Thus, compute $\mathbf{u}_j = A\mathbf{v}_j/\sigma_j$, making each such $\mathbf{u}_j$ an eigenvector of AA^T because multiplying $A\mathbf{v}_j = \sigma_j\mathbf{u}_j$ by A^T yields $A^\mathsf{T}A\mathbf{v}_j = \sigma_j A^\mathsf{T}\mathbf{u}_j$. But

$A^T A \mathbf{v}_j = \lambda_j \mathbf{v}_j$, so equating the two right sides yields $\sigma_j A^T \mathbf{u}_j = \lambda_j \mathbf{v}_j$, which we multiply by A (from the left) to produce $\sigma_j A A^T \mathbf{u}_j = \lambda_j A \mathbf{v}_j$. But $A \mathbf{v}_j = \sigma_j \mathbf{u}_j$, so

$$\sigma_j A A^T \mathbf{u}_j = \lambda_j A \mathbf{v}_j = \lambda_j \sigma_j \mathbf{u}_j \Rightarrow A A^T \mathbf{u}_j = \lambda_j \mathbf{u}_j$$

making $\mathbf{u}_j$ an eigenvector of $A A^T$. Incidentally, computing the $\mathbf{u}_j$'s with (34.24) yields the matrix relation $AV = U\Sigma$, from which it follows that $A = U\Sigma V^T$.

Using (34.24) to find the vectors $\mathbf{u}_1$ and $\mathbf{u}_2$, we get

$$U_1 = \begin{bmatrix} 232{,}785 \dfrac{2833 + 5\sqrt{204{,}983}}{\sqrt{\mu_1 \mu_2}} & -232{,}785 \dfrac{-2833 + 5\sqrt{204{,}983}}{\sqrt{\mu_3 \mu_4}} \\[2mm] 3695 \dfrac{114{,}083 + 181\sqrt{204{,}983}}{\sqrt{\mu_1 \mu_2}} & -3695 \dfrac{-114{,}083 + 181\sqrt{204{,}983}}{\sqrt{\mu_3 \mu_4}} \\[2mm] 29{,}560 \dfrac{4903 + 29\sqrt{204{,}983}}{\sqrt{\mu_1 \mu_2}} & -29{,}560 \dfrac{-4903 + 29\sqrt{204{,}983}}{\sqrt{\mu_3 \mu_4}} \\[2mm] -29{,}560 \dfrac{12{,}367 + 32\sqrt{204{,}983}}{\sqrt{\mu_1 \mu_2}} & 29{,}560 \dfrac{-12{,}367 + 32\sqrt{204{,}983}}{\sqrt{\mu_3 \mu_4}} \\[2mm] 22{,}170 \dfrac{7391 + 30\sqrt{204{,}983}}{\sqrt{\mu_1 \mu_2}} & -22{,}170 \dfrac{-7391 + 30\sqrt{204{,}983}}{\sqrt{\mu_3 \mu_4}} \end{bmatrix}$$

$$\mu_1 = 52{,}054{,}667{,}238{,}495 + 80{,}926{,}940{,}385\sqrt{204{,}983}$$
$$\mu_2 = 20{,}950 + 30\sqrt{204{,}983}$$
$$\mu_3 = 52{,}054{,}667{,}238{,}495 - 80{,}926{,}940{,}385\sqrt{204{,}983}$$
$$\mu_4 = 20{,}950 - 30\sqrt{204{,}983}$$

The vectors in U_2 are an orthonormal basis for the null space of A^T. Finding and normalizing such a basis gives

$$U_2 = \begin{bmatrix} 0 & 0 & \sqrt{\dfrac{5393}{13{,}979}} \\[2mm] 0 & \dfrac{168}{\sqrt{113{,}253}} & -\dfrac{5859}{\sqrt{75{,}388{,}747}} \\[2mm] -\dfrac{2}{\sqrt{21}} & \dfrac{143}{\sqrt{113{,}253}} & \dfrac{1080}{\sqrt{75{,}388{,}747}} \\[2mm] \dfrac{1}{\sqrt{21}} & \dfrac{254}{\sqrt{113{,}253}} & \dfrac{3276}{\sqrt{75{,}388{,}747}} \\[2mm] \dfrac{4}{\sqrt{21}} & \dfrac{8}{\sqrt{113{,}253}} & -\dfrac{279}{\sqrt{75{,}388{,}747}} \end{bmatrix}$$

The matrix $U = [U_1, U_2]$ is then

$$U = \begin{bmatrix} 232{,}785 \dfrac{2833 + 5\sqrt{204{,}983}}{\sqrt{\mu_1 \mu_2}} & -232{,}785 \dfrac{-2833 + 5\sqrt{204{,}983}}{\sqrt{\mu_3 \mu_4}} & 0 & 0 & \sqrt{\dfrac{5393}{13{,}979}} \\[2mm] 3695 \dfrac{114{,}083 + 181\sqrt{204{,}983}}{\sqrt{\mu_1 \mu_2}} & -3695 \dfrac{-114{,}083 + 181\sqrt{204{,}983}}{\sqrt{\mu_3 \mu_4}} & 0 & \dfrac{168}{\sqrt{113{,}253}} & -\dfrac{5859}{\sqrt{75{,}388{,}747}} \\[2mm] 29{,}560 \dfrac{4903 + 29\sqrt{204{,}983}}{\sqrt{\mu_1 \mu_2}} & -29{,}560 \dfrac{-4903 + 29\sqrt{204{,}983}}{\sqrt{\mu_3 \mu_4}} & -\dfrac{2}{\sqrt{21}} & \dfrac{143}{\sqrt{113{,}253}} & \dfrac{1080}{\sqrt{75{,}388{,}747}} \\[2mm] -29{,}560 \dfrac{12{,}367 + 32\sqrt{204{,}983}}{\sqrt{\mu_1 \mu_2}} & 29{,}560 \dfrac{-12{,}367 + 32\sqrt{204{,}983}}{\sqrt{\mu_3 \mu_4}} & \dfrac{1}{\sqrt{21}} & \dfrac{254}{\sqrt{113{,}253}} & \dfrac{3276}{\sqrt{75{,}388{,}747}} \\[2mm] 22{,}170 \dfrac{7391 + 30\sqrt{204{,}983}}{\sqrt{\mu_1 \mu_2}} & -22{,}170 \dfrac{-7391 + 30\sqrt{204{,}983}}{\sqrt{\mu_3 \mu_4}} & \dfrac{4}{\sqrt{21}} & \dfrac{8}{\sqrt{113{,}253}} & -\dfrac{279}{\sqrt{75{,}388{,}747}} \end{bmatrix}$$

$$\text{(34.25)}$$

THE MATRIX Σ Because Σ has dimension $r \times c$, it need not be a square matrix. In this case, Σ is 5×4, so the "main diagonal" has four entries, the two singular values, and two additional zeros.

$$\Sigma = \begin{bmatrix} \sqrt{20{,}950 + 30\sqrt{204{,}983}} & 0 & 0 & 0 \\ 0 & \sqrt{20{,}950 - 30\sqrt{204{,}983}} & 0 & 0 \\ 0 & 0 & 0 & 0 \\ 0 & 0 & 0 & 0 \\ 0 & 0 & 0 & 0 \end{bmatrix} \quad (34.26)$$

❖

Numeric Determination of the SVD

A numeric computation of the singular values of A in Example 34.7 would produce a list such as $185.8292617, 85.83405796, 0.1836522528 \times 10^{-13}, 0$. The third value should be zero but is computed as a small positive number. In practice, a threshold must be declared, below which values are declared to be zero.

A numerical computation of the SVD of A could produce

$$U_f = \begin{bmatrix} -0.6779323847 & -0.3932107662 & 0.5266768065 & -0.3291715495 & 0.007113908434 \\ -0.4138821428 & -0.3523417895 & -0.8289320969 & -0.05560144243 & -0.1197367678 \\ -0.3045822425 & 0.7216102297 & -0.05703451753 & -0.3371828652 & -0.5191946775 \\ 0.4535944034 & -0.1860447743 & -0.09626016074 & -0.8649313450 & 0.04761521857 \\ -0.2656897221 & 0.4073163084 & -0.1515393199 & -0.1635727801 & 0.8448563053 \end{bmatrix}$$

$$(34.27)$$

and

$$V_f = \begin{bmatrix} -0.5371284133 & 0.6716051721 & 0.4964801281 & -0.1180975988 \\ 0.7998709493 & 0.2142912742 & 0.5513842068 & -0.1012974370 \\ -0.1739650811 & -0.4172909716 & 0.5443539482 & 0.7065997273 \\ 0.2035575661 & 0.5734927965 & -0.3913588097 & 0.6902960319 \end{bmatrix}$$

$$(34.28)$$

as floating-point versions of U and V, respectively. Notice that U_f and V_f are not just the matrices U and V converted to floating-point form. The matrices U and V are not unique, and the numeric algorithm that produced U_f and V_f was not the same one we used "by hand" to obtain U and V. The singular values are unique, but not U and V.

If we convert Σ to floating-point form and call it Σ_f, thereby imposing a cut-off threshold of 10^{-10} on the singular values, the product $U_f \Sigma_f V_f^{\mathrm{T}}$ yields

$$A_f = \begin{bmatrix} 45.00000000 & -108.0000000 & 35.99999999 & -45.00000000 \\ 21.00000002 & -68.00000001 & 26.00000001 & -33.00000000 \\ 71.99999999 & -32.00000000 & -15.99999999 & 23.99999999 \\ -56.00000000 & 64.00000002 & -8.000000014 & 8.000000020 \\ 49.99999999 & -32.00000001 & -5.999999996 & 9.99999999 \end{bmatrix}$$

which, rounded to three significant figures, is just A.

The Outer Product Expansion of A

From the singular value decomposition of A we can obtain the *outer product expansion*

$$A = \sigma_1 \mathbf{u}_1 \mathbf{v}_1^{\mathrm{T}} + \sigma_2 \mathbf{u}_2 \mathbf{v}_2^{\mathrm{T}} + \cdots + \sigma_k \mathbf{u}_k \mathbf{v}_k^{\mathrm{T}} \quad (34.29)$$

For A in Example 34.7, $k = 2$, each vector $\mathbf{u}_j$ has dimension 5×1, and each vector $\mathbf{v}_j^{\mathrm{T}}$ has dimension 1×5. Hence, each $\mathbf{u}_j \mathbf{v}_j^T$ is a 5×4 matrix of rank 1. If we drop the term containing the smallest singular value, we obtain the matrix $B = \sigma_1 \mathbf{u}_1 \mathbf{v}_1^T$ as a rank-1 approximation of A. In floating-point form, we find

$$B = \begin{bmatrix} 67.6672627 & -100.7674819 & 21.9160643 & -25.6441159 \\ 41.3113053 & -61.5192049 & 13.3799002 & -15.6559000 \\ 30.4016257 & -45.2729303 & 9.8464746 & -11.5214179 \\ -45.2751518 & 67.4220126 & -14.6637103 & 17.1580938 \\ 26.5196008 & -39.4919683 & 8.5891649 & -10.0502324 \end{bmatrix}$$

The difference between A and B is measured by the 2-norm of their difference, $\|A - B\|_2 = 85.83405797$, which is just the value of σ_2. (See Exercises A3 and A4.)

APPLICATION: DIGITAL IMAGE PROCESSING   A video image or photograph can be digitized by breaking it up into a rectangular array of cells (or pixels) and measuring the gray level of each cell. This information can be stored and transmitted using a matrix A. The entries of A are nonnegative numbers corresponding to the measures of the gray levels. Because the gray levels of any one cell generally turn out to be close to the gray levels of its neighboring cells, it is possible to reduce the amount of storage by expressing A as an outer product expansion and dropping terms corresponding to small singular values. Since the measure of the approximation being made is the size of the singular value dropped, all terms corresponding to small singular values can be dropped without appreciably changing the quality of the image being transmitted. (See [52] for an example, pictures included, of such image compression.)

EXERCISES 34.5–Part A

A1. If λ is an eigenvalue for $A^T A$, show it is an eigenvalue for $A A^T$. *Hint:* Start with $A^T A \mathbf{x} = \lambda \mathbf{x}$.

A2. Construct an argument similar to that in (34.21) to show that the nonzero eigenvalues of AA^T are positive.

A3. Let A in (34.20) be given as its outer product expansion (34.29), and let $B = \sigma_1 \mathbf{u}_1 \mathbf{v}_1^T$. Then, $\|A - B\|_2 = \|\sigma_2 \mathbf{u}_2 \mathbf{v}_2^T\|_2 = \sigma_2 \|\mathbf{u}_2 \mathbf{v}_2^T\|_2$.

To compute the 2-norm of the matrix $C = \mathbf{u}_2 \mathbf{v}_2^T$, the largest eigenvalue of $C C^T$ must be computed. Show that $C C^T = \mathbf{u}_2 \mathbf{u}_2^T$.

A4. Show that $(1, \mathbf{u}_2)$ is an eigenpair for $\mathbf{u}_2 \mathbf{u}_2^T$. (If this is the only eigenpair, then $\|\mathbf{u}_2 \mathbf{v}_2^T\|_2 = 1$ and $\|A - B\|_2 = \sigma_2$.)

A5. Show that if $\mathbf{u}$ is a unit vector, then $(1, \mathbf{u})$ is the only eigenpair for $\mathbf{u}\mathbf{u}^T$. *Hint:* Show that $\mathbf{u}\mathbf{u}^T \mathbf{y} = \mu \mathbf{y}$ implies $\mathbf{y}$ is along $\mathbf{u}$.

EXERCISES 34.5–Part B

B1. Let $A = \begin{bmatrix} \frac{3}{5} & 1 \\ \frac{4}{5} & 0 \end{bmatrix}$ and let $\sigma_1 \geq \sigma_2$ be its singular values. Show $\|A\|_2 = \sigma_1$ and $\|A\|_F = \sqrt{\sigma_1^2 + \sigma_2^2}$, where F indicates the Frobenius norm.

B2. If A is the matrix in Exercise B1, obtain its singular value decomposition as follows.

(a) Obtain $A^T A$, its eigenvalues $\lambda_1 \geq \lambda_2$, and its corresponding normalized eigenvectors $\mathbf{v}_1$ and $\mathbf{v}_2$. The singular values are $\sigma_k = \sqrt{\lambda_k}, k = 1, 2$.

(b) Obtain $\mathbf{u}_k = A\mathbf{v}_k / \sigma_k, k = 1, 2$.

(c) Form the matrices $U = [\mathbf{u}_1, \mathbf{u}_2]$, $V = [\mathbf{v}_1, \mathbf{v}_2]$, and Σ, a diagonal matrix with σ_k on the diagonal.

(d) Show that $A = U \Sigma V^T$.

B3. If A is the matrix in Exercise B1, obtain the matrix $W = U V^T$ and compute $\|A - W\|_F$. Show that W is an orthogonal matrix for which $\det(W) = -1$.

B4. Let $M = \begin{bmatrix} a & b \\ c & d \end{bmatrix}$ be an orthogonal matrix, and let A be the matrix in Exercise B1. Minimize $\|A - M\|_F$ by the Lagrange multiplier method, taking as constraints the three distinct equations in $M^T M = I$, the condition that M be orthogonal. Compare the minimizing M with the matrix W in Exercise B3.

B5. Let A be the matrix in Exercise B1, and let $B = \sigma_1 \mathbf{u}_1 \mathbf{v}_1^{\mathrm{T}}$. Show $\|A - B\|_2 = \|A - B\|_{\mathrm{F}} = \sigma_2$. Compare $\|A - B\|_{\mathrm{F}}$ to $\|A - W\|_{\mathrm{F}}$. Remember, B is singular (rank 1) but W, an orthogonal matrix, is not.

For the $r \times c$ matrices A in Exercises B6–20:

(a) Find k, the rank of A.

(b) Use the software of your choice to obtain the singular value decomposition, $A = U \Sigma V^{\mathrm{T}}$.

(c) Obtain the outer product expansion, $A = \sum_{i=1}^{k} \sigma_i \mathbf{u}_i \mathbf{v}_i^{\mathrm{T}}$, where k is the rank of A, and the vectors $\mathbf{u}_i$, $\mathbf{v}_i$ are as defined by (34.19).

(d) Obtain $B = \sum_{i=1}^{k-1} \sigma_i \mathbf{u}_i \mathbf{v}_i^{\mathrm{T}}$, and show that $\|A - B\|_2 = \sigma_k$.

(e) Show that the nonzero eigenvalues of $A^{\mathrm{T}}A$ and AA^{T} are just σ_i^2, $i = 1, \ldots, k$, the squares of the singular values.

(f) Obtain the $r \times c$ matrix Σ from the singular values.

(g) Obtain the eigenvectors of $A^{\mathrm{T}}A$, normalize and/or orthonormalize as necessary, and form the matrix $V = [V_1, V_2]$.

(h) Obtain a basis of the row space of A by selecting the appropriate rows of the reduced row-echelon form of A.

(i) Verify that the columns of V_1 are a basis for the row space of A by showing that each column in V_1 is a linear combination of the basis vectors found in part (h).

(j) Verify that the columns of V_2 are a basis for the null space of A by showing $A\mathbf{v}_i = \mathbf{0}$ for each $\mathbf{v}_i$ in V_2 and then arguing on the grounds of independence and dimension.

(k) Using the results of part (g), calculate $\mathbf{u}_j = A\mathbf{v}_j / \sigma_j A$, $j = 1, \ldots, k$, and from these vectors, obtain U_1.

(l) Find a basis for the null space of A^{T} and orthonormalize it, thereby forming U_2 and $U = [U_1, U_2]$.

(m) Verify that $U \Sigma V^{\mathrm{T}} = A$.

(n) Show that the columns of U_1 are eigenvectors of AA^{T} and that the eigenvalues are the squares of the singular values.

(o) Show that U and V are orthogonal matrices.

(p) Sketch a relevant diagram similar to the one in Figure 34.2.

B6. $\begin{bmatrix} 2 & 5 & -3 \\ -1 & 1 & -2 \\ 7 & 3 & 4 \end{bmatrix}$ **B7.** $\begin{bmatrix} -3 & -2 & -7 & -8 \\ 2 & -6 & 1 & 9 \\ 5 & -1 & 5 & 11 \\ -2 & 3 & -1 & -6 \end{bmatrix}$

B8. $\begin{bmatrix} -5 & -1 & -5 & -1 \\ 3 & 1 & -7 & 11 \\ -4 & -5 & 2 & -11 \\ 2 & -2 & -4 & 4 \end{bmatrix}$ **B9.** $\begin{bmatrix} 3 & 21 & 6 & -12 \\ -2 & -16 & -5 & 11 \\ -5 & -5 & 5 & -25 \\ 5 & 7 & -4 & 22 \end{bmatrix}$

B10. $\begin{bmatrix} -3 & -15 & -3 & 3 \\ -3 & -15 & -3 & 3 \\ 3 & 21 & 6 & -12 \\ 2 & -6 & -6 & 22 \end{bmatrix}$ **B11.** $\begin{bmatrix} -2 & -6 & -4 & 3 \\ 3 & 5 & 2 & 6 \\ 1 & -3 & 5 & -7 \end{bmatrix}$

B12. $\begin{bmatrix} -5 & -3 & -5 & 0 \\ -6 & -4 & 4 & 2 \\ 6 & 6 & 1 & -2 \end{bmatrix}$ **B13.** $\begin{bmatrix} 7 & 4 & -2 & -5 & -6 \\ 4 & 2 & 6 & 5 & -6 \\ 3 & -5 & -7 & -4 & -6 \end{bmatrix}$

B14. $\begin{bmatrix} -5 & -6 & 7 & -3 & -7 \\ 4 & 3 & 3 & -1 & -3 \\ -7 & -3 & -4 & 4 & 4 \end{bmatrix}$

B15. $\begin{bmatrix} -4 & 1 & 5 & -3 & -22 \\ -7 & -4 & 1 & -15 & -21 \\ 6 & 6 & -4 & 4 & 30 \\ 4 & 4 & -1 & 6 & 15 \end{bmatrix}$

B16. $\begin{bmatrix} 4 & -6 & 5 & 28 & -13 \\ 2 & 6 & -5 & -10 & 25 \\ 6 & 0 & -2 & 14 & 18 \\ 0 & -1 & -7 & -13 & 20 \end{bmatrix}$ **B17.** $\begin{bmatrix} -2 & 4 & -3 & -4 \\ -3 & -7 & -4 & 6 \\ 2 & 1 & 2 & 1 \\ 5 & 6 & 2 & 5 \\ -6 & -5 & -4 & -5 \end{bmatrix}$

B18. $\begin{bmatrix} -5 & 2 & -2 & -13 \\ -4 & 2 & 1 & -16 \\ 1 & 7 & 2 & -8 \\ -4 & -5 & -2 & -3 \\ -5 & 5 & 1 & -22 \end{bmatrix}$ **B19.** $\begin{bmatrix} -6 & -28 & 5 & -27 \\ 3 & 9 & 0 & 6 \\ 5 & 17 & -1 & 13 \\ 1 & 1 & 1 & -1 \\ -6 & -4 & -7 & 9 \end{bmatrix}$

B20. $\begin{bmatrix} 5 & 29 & -7 & 31 \\ 2 & 2 & 2 & -2 \\ 4 & 4 & 4 & -4 \\ -1 & 11 & -7 & 19 \\ 7 & 9 & 6 & -4 \end{bmatrix}$

34.6   Minimum-Length Least-Squares Solution, and the Pseudoinverse

Least-Squares Problem: Formulation

In this section we again solve a least-squares problem as an example of an over-determined system. However, we will look at the general case where the rank of the coefficient matrix is less than maximal. Since the system is over-determined, there are more rows than columns in the system matrix, that is, $r > c$. If the rank of the system matrix is $k < c$, the columns of the matrix are not independent. (In Section 33.6 wesolved the least-squares problem for a system whose matrix was of maximal rank, that is, whose columns were linearly independent.)

The system to be solved is $A\mathbf{x} = \mathbf{y}$, where A is the rank-2 matrix given in (34.20), and listed (again), along with $\mathbf{y}$ and rref($[A, \mathbf{y}]$), the reduced row-echelon form of $[A, \mathbf{y}]$, in Table 34.4.

$$
A = \begin{bmatrix} 45 & -108 & 36 & -45 \\ 21 & -68 & 26 & -33 \\ 72 & -32 & -16 & 24 \\ -56 & 64 & -8 & 8 \\ 50 & -32 & -6 & 10 \end{bmatrix} \qquad \mathbf{y} = \begin{bmatrix} -2 \\ 4 \\ 4 \\ 6 \\ -4 \end{bmatrix} \qquad \text{rref}([A, \mathbf{y}]) = \begin{bmatrix} 1 & 0 & -\frac{5}{11} & \frac{7}{11} & 0 \\ 0 & 1 & -\frac{23}{44} & \frac{15}{22} & 0 \\ 0 & 0 & 0 & 0 & 1 \\ 0 & 0 & 0 & 0 & 0 \\ 0 & 0 & 0 & 0 & 0 \end{bmatrix}
$$

TABLE 34.4 A least-squares problem where the rank of A is not maximal

Least-Squares Problem: Solution

The third row of rref($[A, \mathbf{y}]$) given in Table 34.4 indicates an inconsistent equation $0 = 1$, so we know there is no solution to the over-determined system of linear equations $A\mathbf{x} = \mathbf{y}$. Hence, we seek a least-squares "best fit" solution.

The normal equations are given by $A^\mathrm{T} A \mathbf{x} = A^\mathrm{T} \mathbf{y}$, where $A^\mathrm{T} A$, $\mathbf{b} = A^\mathrm{T} \mathbf{y}$, and the solution $\mathbf{x} = \mathbf{X}$, are

$$
A^\mathrm{T} A = \begin{bmatrix} 13{,}286 & -13{,}776 & 1162 & -938 \\ -13{,}776 & 22{,}432 & -5464 & 6528 \\ 1162 & -5464 & 2328 & -2986 \\ -938 & 6528 & -2986 & 3854 \end{bmatrix}
$$

$$
\mathbf{b} = \begin{bmatrix} -254 \\ 328 \\ -56 \\ 62 \end{bmatrix} \qquad \mathbf{X} = \begin{bmatrix} c_1 \\ c_2 \\ -30c_1 + 28c_2 - \frac{7673}{13{,}979} \\ -23c_1 + 20c_2 - \frac{5720}{13{,}979} \end{bmatrix}
$$

The least-squares solution $\mathbf{X}$ contains two free parameters because $A^\mathrm{T} A$, like A, has rank 2. There are only two independent equations in the set of four normal equations. Note, however, that the four normal equations are now consistent, but under-determined. There is a solution, but not a unique solution.

Next, we write the solution as $\mathbf{X} = \mathbf{v}_1 + c_1 \mathbf{v}_2 + c_2 \mathbf{v}_3$, where $\mathbf{v}_k$, $k = 1, 2, 3$, are given by

$$
\mathbf{v}_1 = \begin{bmatrix} 0 \\ 0 \\ -\frac{7673}{13{,}979} \\ -\frac{5720}{13{,}979} \end{bmatrix} \qquad \mathbf{v}_2 = \begin{bmatrix} 1 \\ 0 \\ -30 \\ -23 \end{bmatrix} \qquad \mathbf{v}_3 = \begin{bmatrix} 0 \\ 1 \\ 28 \\ 20 \end{bmatrix}
$$

The vectors $\mathbf{v}_2$ and $\mathbf{v}_3$ are in the null space of A because $A\mathbf{v}_k = \mathbf{0}$, $k = 2, 3$. The vector $\mathbf{v}_1$ is *not* orthogonal to $\mathbf{v}_2$ and $\mathbf{v}_3$ because $\mathbf{v}_1 \cdot \mathbf{v}_2 = \frac{361{,}750}{13{,}979} \neq 0$ and $\mathbf{v}_1 \cdot \mathbf{v}_3 = -\frac{329{,}244}{13{,}979} \neq 0$. Thus, $\mathbf{v}_1$ is not orthogonal to the null space, so $\mathbf{v}_1$ is *not* in the row space of A. The vector $\mathbf{v}_1$ has a component in the row space of A, but it also has a component in the null space of A.

Minimum-Length Solution

When A has less than maximal rank, the solution to the least-squares problem is not unique. One way to select a unique solution from the resulting family of least-squares solutions is to pick the least-squares solution of minimum length. Therefore, minimize

$$\|\mathbf{X}\|^2 = 1430c_2^2 + 1185c_1^2 - 2600c_1c_2 + \frac{723,500}{13,979}c_2 - \frac{658,488}{13,979}c_1 + \frac{91,593,329}{195,412,441}$$

with respect to the two parameters c_1 and c_2. The usual techniques of multivariable calculus, namely, equating partial derivatives to zero and solving simultaneously, yield $c_1 = \frac{4184}{489,265}$, $c_2 = -\frac{13131}{1,272,089}$, and

$$\mathbf{X}^+ = \begin{bmatrix} -\frac{13,131}{1,272,089} \\ \frac{4184}{489,265} \\ \frac{1411}{6,360,445} \\ -\frac{939}{1,272,089} \end{bmatrix}$$

where $\mathbf{X}^+$ is the least-squares solution of minimum length.

Adjusting the component of $\mathbf{X}$ in the null space of A to minimize the length of $\mathbf{X}$ is equivalent to projecting $\mathbf{X}$ onto the row space of A. But the only part of $\mathbf{X}$ having a component in the row space of A is $\mathbf{v}_1$. Hence, the minimization process we just went through is nothing more than finding the projection of $\mathbf{v}_1$ onto the row space of A. We can demonstrate this claim by forming P', a matrix that projects onto the row space of A. (Distinguish between P' and P, the matrix that projects vectors in $R^r = R^5$ onto the column space of A.) If the columns of the matrix C are a basis for the row space of A, then

$$C = \begin{bmatrix} 0 & 1 \\ 1 & 0 \\ -\frac{23}{44} & -\frac{5}{11} \\ \frac{15}{22} & \frac{7}{11} \end{bmatrix} \qquad P' = C(C^TC)^{-1}C^T = \begin{bmatrix} \frac{673}{910} & -\frac{2}{7} & -\frac{17}{91} & \frac{251}{910} \\ -\frac{2}{7} & \frac{24}{35} & -\frac{8}{35} & \frac{2}{7} \\ -\frac{17}{91} & -\frac{8}{35} & \frac{93}{455} & -\frac{25}{91} \\ \frac{251}{910} & \frac{2}{7} & -\frac{25}{91} & \frac{337}{910} \end{bmatrix}$$

and $P'\mathbf{X} = P'\mathbf{v}_1 = \mathbf{X}^+$. Figure 34.3 shows the relationships between the vectors and the four fundamental subspaces.

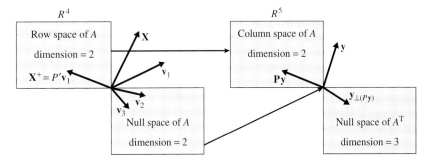

FIGURE 34.3 Fundamental subspaces and $\mathbf{X}^+$ for the least-squares problem in Table 34.4

Pseudoinverse

The over-determined system we are solving is $A\mathbf{x} = \mathbf{y}$. There is no solution—the equations are inconsistent because $\mathbf{y}$ is not in the column space of A. We settled for a least-squares solution $\mathbf{X}$ that was not unique. To extract a unique solution from the two-parameter family of solutions contained in **X**, we found **X**$^+$, the least-squares solution of minimum length, and confirmed that this unique solution resides in the row space of A and is the projection of **X** onto that row space.

If the system $A\mathbf{x} = \mathbf{y}$ were a square system of maximal rank, there would be an inverse matrix A^{-1}, multiplication by which would yield the solution $\mathbf{x} = A^{-1}\mathbf{y}$ via $A^{-1}A\mathbf{x} = A^{-1}\mathbf{y} = \mathbf{x}$. For the over-determined case, we seek a matrix A^+ with which **y** could be multiplied so as to yield the minimum-length least-squares solution $\mathbf{X}^+ = A^+\mathbf{y}$. Such a matrix will be called a *pseudoinverse,* and we now give two prescriptions for constructing such a matrix.

Pseudoinverse by Curious Formula

We seek a matrix, the application of which to **y** yields the minimum-norm least-squares solution $\mathbf{X}^+$. The matrix A^+ can be computed by the formula

$$A^+ = \lim_{s \to 0}[(A^{\mathrm{T}}A + s^2 I)^{-1} A^{\mathrm{T}}] \tag{34.30}$$

discussed briefly in [68]. When applied to the matrix A in Table 34.4, it yields

$$A^+ = \begin{bmatrix} -\dfrac{14{,}211}{12{,}720{,}890} & -\dfrac{1418}{908{,}635} & \dfrac{41{,}512}{6{,}360{,}445} & -\dfrac{2514}{908{,}635} & \dfrac{50{,}311}{12{,}720{,}890} \\[2mm] -\dfrac{1908}{489{,}265} & -\dfrac{186}{69{,}895} & \dfrac{48}{97{,}853} & \dfrac{104}{69{,}895} & \dfrac{62}{489{,}265} \\[2mm] \dfrac{32{,}391}{12{,}720{,}890} & \dfrac{3817}{1{,}817{,}270} & -\dfrac{4100}{1{,}272{,}089} & \dfrac{436}{908{,}635} & -\dfrac{11{,}013}{6{,}360{,}445} \\[2mm] -\dfrac{42{,}867}{12{,}720{,}890} & -\dfrac{2551}{908{,}635} & \dfrac{28{,}544}{6{,}360{,}445} & -\dfrac{678}{908{,}635} & \dfrac{30{,}917}{12{,}720{,}890} \end{bmatrix}$$

and a calculation shows $\mathbf{X}^+ = A^+\mathbf{y}$. Moreover, it can also be shown that the rank of A^+ is 2. Subsequently, we will show how A^+ is related to $U\Sigma V^{\mathrm{T}}$, the singular value decomposition of A.

The matrix A^+ behaves as an inverse on the column space of A, acting as follows. Vectors in $R^r = R^5$ are projected onto the column space of A, and those projections are "pulled back" to pre-images in $R^c = R^4$. These pre-images are then projected onto the row space of A. A vector **y** in $R^r = R^5$ is projected onto the two-dimensional column space of A, and that projection is inverted to a pre-image in $R^c = R^4$ that is not unique. By projecting this pre-image onto the row space of A, the unique, shortest least-squares solution, $\mathbf{X}^+ = A^+\mathbf{y}$ is obtained.

Pseudoinverse from SVD

If the singular value decomposition of A is given by $A = U\Sigma V^{\mathrm{T}}$, the pseudoinverse is given by $A^+ = V\Sigma^+U^{\mathrm{T}}$, where Σ^+ is the transpose of that version of Σ, in which each nonzero singular value has been replaced by its reciprocal. Since Σ, like A, has dimensions $r \times c = 5 \times 4$, then Σ^+ has dimensions $c \times r = 4 \times 5$. For our example, we have

$$\Sigma^+ = \begin{bmatrix} \dfrac{1}{\sqrt{20{,}950+30\sqrt{204{,}983}}} & 0 & 0 & 0 & 0 \\[2mm] 0 & \dfrac{1}{\sqrt{20{,}950-30\sqrt{204{,}983}}} & 0 & 0 & 0 \\[2mm] 0 & 0 & 0 & 0 & 0 \\[2mm] 0 & 0 & 0 & 0 & 0 \end{bmatrix}$$

and a calculation then shows $A^+ = V\Sigma^+U^{\mathrm{T}}$.

Numeric SVD and the Pseudoinverse

In practice, the singular value decomposition is computed numerically, not exactly. We therefore consider how A^+ might be assembled from the factors U_f and V_f that were obtained in floating-point form in (34.27) and (34.28). If Σ_f, a numerically computed version of Σ, is obtained from the singular values 185.8292617, 85.83405796, and $0.1836522528 \times 10^{-13}$, the reciprocals that appear in Σ^+ would include a term with magnitude 10^{13}. Clearly, the numerically determined singular values must be filtered through a threshold barrier to eliminate values that are probably zero. If this is done and Σ_f^+ is constructed from Σ, then $A_f^+ = V_f \Sigma_f^+ U_f^T$ is the matrix

$$\begin{bmatrix} -0.001117138816 & -0.001560582632 & 0.006526587369 & -0.002766787544 & 0.003954990570 \\ -0.003899727142 & -0.002661134558 & 0.000490531716 & 0.001487946205 & -0.000126720694 \\ 0.002546284104 & 0.002100403352 & -0.003223044929 & 0.000479840640 & -0.001731482624 \\ -0.003369811389 & -0.002807507965 & 0.004487736313 & -0.000746174206 & 0.002430411708 \end{bmatrix}$$

which agrees to eight significant figures with A^+ expressed in floating-point form. Moreover, $\mathbf{X}_f^+$, the floating-point form of the exact solution, and $A_f^+\mathbf{y}_f$, the minimum-norm least-squares solution computed entirely in floating-point arithmetic, can be seen to agree to eight significant figures.

$$\mathbf{X}_f^+ = \begin{bmatrix} -0.01032239096 \\ 0.00855160292 \\ 0.00022183982 \\ -0.00073815590 \end{bmatrix} \qquad A_f^+\mathbf{y}_f = \begin{bmatrix} -0.01032239096 \\ 0.00855160292 \\ 0.00022183982 \\ -0.00073815590 \end{bmatrix}$$

EXERCISES 34.6–Part A

A1. Let A be a square matrix of maximal rank. Show that $A^+ = A^{-1}$. *Hint:* Show $A^+A = AA^+ = I$.

A2. Use (34.30) to obtain A^+ for the matrix A of Exercise B1, Section 34.5. Show that $A^+ = A^{-1}$.

A3. For the matrix A in Exercise A2, use the factors U and V from Exercise B2, Section 34.5, to construct $A^+ = U\Sigma^+V^T$. It should agree with the results in Exercise A2.

EXERCISES 34.6–Part B

B1. If A is an $r \times c$ matrix of rank k so that there are k nonzero singular values $\sigma_1, \ldots, \sigma_k$, show each of the following.

(a) $(\Sigma^+)^+ = \Sigma$ (b) $(\Sigma\Sigma^+)^2 = \Sigma\Sigma^+$ (c) $(\Sigma^+\Sigma)^2 = \Sigma^+\Sigma$
(d) $(A^+)^+ = A$ (e) $(AA^+)^2 = AA^+$ (f) $(A^+A)^2 = A^+A$

In Exercises B2–11:

(a) Obtain k, the rank of A, and show that the columns of A are not linearly independent.

(b) Use (34.30) to obtain A^+, the pseudoinverse of A.

(c) Show that $(A^+)^T = (A^T)^+$.

(d) Show that $(A^+)^+ = A$.

(e) Obtain $\mathbf{X}^+ = A^+\mathbf{y}$, the minimum-norm least-squares solution of $A\mathbf{x} = \mathbf{y}$.

(f) Show $\mathbf{X}^+$ is in the row space of A by obtaining a basis for the null space of A and showing that $\mathbf{X}^+$ is orthogonal to each vector in this basis.

(g) Obtain the singular value decomposition of A, thereby factoring A to $U\Sigma V^T$.

(h) Compute $V\Sigma^+U^T$, and compare to A^+.

(i) Apply the Gram–Schmidt process to the columns of A, letting Q be the resulting matrix.

(j) From Q, build $P = Q(Q^TQ)^{-1}Q^T = QQ^T$, which projects onto the column space of A.

(k) Obtain $P\mathbf{y}$, the projection of $\mathbf{y}$ onto the column space of A.

(l) Solve $A\mathbf{x} = P\mathbf{y}$ for $\mathbf{x}$, and call this the general least-squares solution $\mathbf{v}$; if available, compare this to a least-squares solution provided by an appropriate software package.

(m) Obtain $\mathbf{X}^+$ from $\mathbf{v}$ using calculus to minimize $\|\mathbf{v}\|_2^2$.

(n) Obtain a basis for the row space of A, and let C be the matrix whose columns are these basis vectors.

(o) Construct $P' = C(C^{\mathrm{T}}C)^{-1}C^{\mathrm{T}}$, a matrix that projects onto the row space of A.

(p) Show that $P'\mathbf{v} = \mathbf{X}^+$, thereby verifying that projecting $\mathbf{v}$, the general least-squares solution onto the row space of A, gives $\mathbf{X}^+$, the minimum-norm least-squares solution.

(q) Obtain $\mathbf{V} = (I - P')\mathbf{v}$, the component of $\mathbf{v}$ orthogonal to the row space of A.

(r) Show $A\mathbf{V} = \mathbf{0}$, indicating that the component of $\mathbf{v}$ orthogonal to the row space of A is actually in the null space of A.

(s) Sketch a relevant diagram similar to the one in Figure 34.3.

B2. $A = \begin{bmatrix} 2 & -12 & 5 \\ 17 & 24 & -16 \\ -27 & -20 & 17 \\ -11 & -4 & 5 \end{bmatrix}$ $\mathbf{y} = \begin{bmatrix} -6 \\ 8 \\ 10 \\ 1 \end{bmatrix}$

B3. $A = \begin{bmatrix} 21 & -18 & -18 & 3 \\ 20 & -18 & -24 & 5 \\ -8 & 10 & 32 & -9 \\ 16 & -12 & 0 & -2 \\ 22 & -18 & -12 & 1 \end{bmatrix}$ $\mathbf{y} = \begin{bmatrix} -3 \\ -7 \\ 1 \\ -5 \\ -2 \end{bmatrix}$

B4. $A = \begin{bmatrix} -9 & 19 & 12 & -24 & 23 \\ 5 & -9 & 7 & 20 & -14 \\ -1 & 16 & 13 & 17 & 23 \\ 13 & -20 & -7 & 58 & -19 \\ 7 & -23 & -8 & 39 & -21 \\ -8 & -14 & 0 & -10 & -10 \end{bmatrix}$ $\mathbf{y} = \begin{bmatrix} 12 \\ -1 \\ 11 \\ 4 \\ -3 \\ 11 \end{bmatrix}$

B5. $A = \begin{bmatrix} -2 & -7 & 8 & 5 & -37 \\ -3 & -21 & -21 & 6 & 27 \\ 31 & -10 & 35 & -43 & -4 \\ -6 & -25 & -24 & 11 & 21 \\ 15 & -28 & -3 & -19 & 30 \\ 17 & -19 & 13 & -22 & 1 \end{bmatrix}$ $\mathbf{y} = \begin{bmatrix} -6 \\ 12 \\ 4 \\ 11 \\ 5 \\ 4 \end{bmatrix}$

B6. $A = \begin{bmatrix} -4 & 10 & 6 & -8 & -20 \\ 11 & -20 & -27 & 13 & 55 \\ -3 & 0 & 15 & 3 & -15 \\ -7 & 35 & -14 & -35 & -35 \\ -3 & 30 & -27 & -33 & -15 \\ -6 & 5 & 23 & 0 & -30 \end{bmatrix}$ $\mathbf{y} = \begin{bmatrix} 6 \\ -12 \\ -1 \\ 1 \\ 1 \\ 8 \end{bmatrix}$

B7. $A = \begin{bmatrix} -12 & -17 & 1 & -13 \\ -12 & 15 & -15 & 27 \\ -24 & -2 & -14 & 14 \\ 24 & 10 & 10 & -4 \\ 44 & 17 & 19 & -9 \\ -20 & -19 & -3 & -10 \end{bmatrix}$ $\mathbf{y} = \begin{bmatrix} -2 \\ -5 \\ 5 \\ 1 \\ -10 \\ 12 \end{bmatrix}$

B8. $A = \begin{bmatrix} -3 & 5 & 5 & -10 \\ -7 & 7 & -7 & 5 \\ 8 & 5 & 11 & 12 \end{bmatrix}$ $\mathbf{y} = \begin{bmatrix} -5 \\ 12 \\ 7 \end{bmatrix}$

B9. $A = \begin{bmatrix} -1 & -4 & -11 & 11 & -5 \\ -12 & 7 & 12 & -7 & 3 \\ -8 & 11 & -8 & -9 & -3 \end{bmatrix}$ $\mathbf{y} = \begin{bmatrix} -9 \\ 3 \\ -7 \end{bmatrix}$

B10. $A = \begin{bmatrix} -5 & -6 & 0 & 12 & -11 \\ 0 & -6 & -12 & -1 & -3 \\ 0 & 4 & 9 & 3 & 6 \\ 1 & 2 & -12 & -1 & -11 \end{bmatrix}$ $\mathbf{y} = \begin{bmatrix} 1 \\ 5 \\ 10 \\ 11 \end{bmatrix}$

B11. $A = \begin{bmatrix} -6 & 4 & -2 & -3 & 2 & -9 \\ 7 & 10 & 2 & -3 & -4 & 0 \\ -10 & 10 & -12 & 1 & -9 & -2 \\ -9 & -12 & 5 & 7 & 8 & -2 \end{bmatrix}$ $\mathbf{y} = \begin{bmatrix} -10 \\ -10 \\ 12 \\ -12 \end{bmatrix}$

B12. A matrix M that satisfies the four *Penrose conditions* $AMA = A$, $MAM = M$, $(AM)^{\mathrm{T}} = AM$, $(MA)^{\mathrm{T}} = MA$, can be shown to be A^+, the pseudoinverse of A.

(a) If A is an $r \times c$ matrix of rank k with the k nonzero singular values $\sigma_1, \ldots, \sigma_k$, show that Σ^+ satisfies the four Penrose conditions by showing $\Sigma\Sigma^+\Sigma = \Sigma$, $\Sigma^+\Sigma\Sigma^+ = \Sigma^+$, $(\Sigma\Sigma^+)^{\mathrm{T}} = \Sigma\Sigma^+$, and $(\Sigma^+\Sigma)^{\mathrm{T}} = \Sigma^+\Sigma$.

(b) Show that $A^+ = V\Sigma^+U^{\mathrm{T}}$ satisfies the four Penrose conditions.

For the pairs of matrices F and G given in Exercises B13–22:

(a) Show $F^{\mathrm{T}}F$ and GG^{T} are nonsingular.

(b) Obtain $A = FG$.

(c) Obtain A^+.

(d) Obtain $C = G^{\mathrm{T}}(GG^{\mathrm{T}})^{-1}(F^{\mathrm{T}}F)^{-1}F^{\mathrm{T}}$, and show $C = A^+$.

B13. $F = \begin{bmatrix} 3 & -5 & -4 \\ 4 & -7 & 3 \\ 10 & 11 & -7 \end{bmatrix}$ $G = \begin{bmatrix} 1 & -11 & -7 & 1 & 10 \\ 2 & 9 & -7 & 1 & 4 \\ -2 & -12 & -1 & -2 & 10 \end{bmatrix}$

B14. $F = \begin{bmatrix} -1 & -5 \\ -8 & 2 \end{bmatrix}$ $G = \begin{bmatrix} -1 & -1 & 0 & 4 & -6 & -11 \\ 12 & -7 & 12 & -8 & 3 & -2 \end{bmatrix}$

B15. $F = \begin{bmatrix} 1 & 1 & -11 \\ 9 & -11 & -4 \\ 5 & 10 & -9 \\ 8 & -12 & 8 \end{bmatrix}$ $G = \begin{bmatrix} -8 & -3 & 10 & -6 & -3 \\ 6 & 3 & -11 & -11 & -4 \\ 12 & 0 & -7 & 3 & -7 \end{bmatrix}$

B16. $F = \begin{bmatrix} -7 & -9 & 10 \\ -11 & 4 & 3 \\ -12 & 4 & -1 \\ 6 & -4 & -9 \end{bmatrix}$ $G = \begin{bmatrix} -7 & -4 & 11 & 12 & -2 \\ -3 & -6 & 2 & 11 & 2 \\ 6 & 6 & -5 & -2 & -4 \end{bmatrix}$

B17. $F = \begin{bmatrix} -7 & 0 & 1 & 2 \\ 8 & -9 & 8 & 0 \\ 2 & 8 & -1 & -2 \\ 8 & 10 & -5 & -12 \end{bmatrix}$ $G = \begin{bmatrix} 7 & -11 & 5 & -11 & -1 \\ 2 & 8 & 0 & 2 & 6 \\ 7 & 0 & 6 & 11 & -10 \\ -10 & -6 & 3 & -3 & -5 \end{bmatrix}$

B18. $F = \begin{bmatrix} 0 & 12 \\ 12 & -10 \end{bmatrix}$ $G = \begin{bmatrix} -7 & 2 & 10 \\ 1 & -11 & 0 \end{bmatrix}$

B19. $F = \begin{bmatrix} 6 & -6 \\ -8 & -9 \\ 1 & -5 \end{bmatrix}$ $G = \begin{bmatrix} 5 & 2 & -9 & -6 \\ 10 & -8 & 9 & -2 \end{bmatrix}$

B20. $F = \begin{bmatrix} -5 & 7 & -8 & 5 \\ -3 & 6 & 2 & 2 \\ -9 & 3 & -7 & -5 \\ 10 & -8 & -1 & 11 \\ 9 & -8 & 2 & 7 \\ 12 & -7 & -7 & -11 \end{bmatrix}$ $G = \begin{bmatrix} -1 & -11 & 4 & 11 & 12 \\ 2 & 1 & -5 & -12 & 10 \\ -9 & 0 & 5 & -10 & -2 \\ 7 & -5 & 5 & -1 & 5 \end{bmatrix}$

B21. $F = \begin{bmatrix} -12 & 1 \\ -3 & -5 \\ 1 & -3 \\ 9 & 5 \\ 5 & 4 \end{bmatrix}$ $G = \begin{bmatrix} -2 & -1 & -10 & 1 \\ -1 & -7 & -2 & -4 \end{bmatrix}$

B22. $F = \begin{bmatrix} 11 & -2 \\ -9 & -11 \\ 0 & -11 \end{bmatrix}$ $G = \begin{bmatrix} 1 & -5 & -1 & 3 & -4 \\ -9 & 6 & -5 & 2 & -10 \end{bmatrix}$

If the $r \times c$ real matrix A of rank k is factored into its normalized QR form, the factors are $r \times k$ and $k \times c$, respectively, and both are of rank k. The product $Q^T Q = I_k$ is necessarily nonsingular. The product RR^T is $k \times k$ and nonsingular because it is of rank k. Thus, Q and R are the factors F and G of Exercise B13–22, and $A^+ = R^T(RR^T)^{-1}(Q^TQ)^{-1}Q^T = R^T(RR^T)^{-1}Q^T$. In Exercises B23–32:

(a) Obtain the QR decomposition of the given matrix A.

(b) Obtain $A^+ = R^T(RR^T)^{-1}Q^T$.

(c) Obtain A^+ by (34.30).

B23. $\begin{bmatrix} -17 & -4 & 4 & -8 \\ 3 & 16 & -3 & -7 \\ 4 & 28 & -5 & -13 \end{bmatrix}$ **B24.** $\begin{bmatrix} -5 & -8 & -10 & -8 \\ -2 & 8 & 11 & 1 \\ -11 & -12 & 12 & 11 \end{bmatrix}$

B25. $\begin{bmatrix} 1 & -7 & -3 \\ 20 & -4 & -12 \\ 16 & 7 & -6 \\ 11 & 8 & -3 \end{bmatrix}$ **B26.** $\begin{bmatrix} 11 & 12 & 6 \\ 7 & -11 & -12 \\ -6 & 8 & -2 \\ -7 & 4 & -1 \end{bmatrix}$

B27. $\begin{bmatrix} -14 & 4 & -8 & 0 & -5 \\ -32 & 6 & -14 & -4 & -7 \\ 7 & -13 & 19 & -14 & 18 \end{bmatrix}$

B28. $\begin{bmatrix} 12 & -2 & -8 & 7 & -11 \\ 0 & 9 & -2 & 11 & -7 \\ -9 & 4 & -6 & 12 & -3 \end{bmatrix}$

B29. $\begin{bmatrix} 25 & 20 & 5 \\ 27 & 16 & 9 \\ 3 & 8 & -3 \\ -22 & -12 & -8 \\ -27 & -16 & -9 \end{bmatrix}$ **B30.** $\begin{bmatrix} -5 & 3 & 5 \\ 3 & 7 & -9 \\ -6 & -5 & 8 \\ -5 & 3 & -5 \\ 1 & -3 & -4 \end{bmatrix}$

B31. $\begin{bmatrix} -13 & -25 & -18 & -17 & 13 \\ 11 & -25 & 6 & -41 & -11 \\ -8 & 20 & -4 & 32 & 8 \\ 10 & 10 & 12 & 2 & -10 \end{bmatrix}$

B32. $\begin{bmatrix} -37 & -37 & -24 & 4 \\ -27 & -27 & -18 & 6 \\ 7 & 7 & 6 & -10 \\ -26 & -26 & -18 & 10 \\ -13 & -13 & -6 & -14 \end{bmatrix}$

B33. Prove that the matrix AA^+ projects onto the column space of A. *Hint:* See the calculations of Exercises B23–32.

B34. Let $\mathbf{y} = (I - A^+A)\mathbf{x}$, and prove that the matrix $I - A^+A$ projects onto the null space of A by showing

(a) $A\mathbf{y} = \mathbf{0}$

(b) $(I - A^+A)^2 = I - A^+A$

(c) $(A^+A)^T = A^+A$, so $(I - A^+A)^T = I - A^+A$ *Hint:* Express A and A^+ in terms of the factors of the singular value decomposition, and see Exercise B12.

Chapter Review

1. Describe the factorization of A into $L_1 U$, the Doolittle decomposition.

2. Show how the Doolittle factorization is used to solve the linear system $A\mathbf{x} = \mathbf{y}$ via forward and back substitutions.

3. Describe how the Doolittle decomposition can be obtained as a by-product of the Gaussian arithmetic, which row-reduces A to an upper triangular form.

4. Obtain the Doolittle decomposition of the matrix $\begin{bmatrix} 4 & 7 & -9 \\ 5 & 7 & -8 \\ 3 & 5 & -9 \end{bmatrix}$.
   Show how to use it to solve the system $A\mathbf{x} = \mathbf{y}$ if $\mathbf{y} = [4 \ -3 \ 5]^T$.

5. Convert the Doolittle decomposition in Question 4 to the Crout decomposition LU_1.

6. Obtain the Cholesky decomposition of $A = \begin{bmatrix} 8 & 2 \\ 2 & 2 \end{bmatrix}$ from its Doolittle decomposition by first factoring to the form L_1DU_1.

7. Describe what it means for a matrix A to be similar to a matrix B.

8. For each of the following, answer true or false. If true, verify. If false, give a reason for your answer.

(a) If A is similar to B and B is similar to C, then A is similar to C.

(b) If A is similar to B, then B is similar to A.

(c) Similar matrices have the same eigenvalues.

(d) Similar matrices have the same eigenvectors.

(e) If two matrices have the same eigenvalues, they are similar.

(f) If two matrices have the same eigenvectors, they are similar.

(g) If $\lambda = 0$ is an eigenvalue of the matrix A, then $\det A = 0$ and the matrix is not invertible.

9. Find the matrix P that defines a similarity transform of $A = \begin{bmatrix} 1 & 7 \\ 6 & 2 \end{bmatrix}$ to a diagonal matrix D.

10. Find an orthogonal matrix P that defines a similarity transform of $A = \begin{bmatrix} 1 & 2 \\ 2 & 1 \end{bmatrix}$ to a diagonal matrix D.

11. The matrix $A = \begin{bmatrix} 2 & 1 & 0 \\ 0 & 2 & 0 \\ 0 & 0 & 2 \end{bmatrix}$ is in Jordan form. Exhibit its two eigenvectors and its one generalized eigenvector. What equation does the generalized eigenvector satisfy?

12. The matrix $A = \begin{bmatrix} 2 & 1 & 0 \\ 0 & 2 & 1 \\ 0 & 0 & 2 \end{bmatrix}$ is in Jordan form. Exhibit its one eigenvector and its two generalized eigenvectors. What equations do the generalized eigenvectors satisfy?

13. Determine the rank of $A = \begin{bmatrix} -1 & -1 & 0 & 5 \\ -1 & -6 & 6 & 0 \\ -2 & -2 & 2 & -5 \\ -1 & 2 & -2 & -4 \end{bmatrix}$. Apply the Gram–Schmidt *orthonormalization* process to the columns of A, thereby forming the factor Q in the QR decomposition of A. Obtain $R = Q^T A$. What is the rank of R? What is the rank of Q? Show that $Q^T Q = I_3$, the 3×3 identity matrix. Find QQ^T and show it is the matrix that projects vectors onto the column space of A.

14. Describe the QR algorithm for finding eigenvalues. Consider just the case where A is real and no two eigenvalues have equal magnitude. Explain why the eigenvalues of A are preserved in the matrices A_k.

15. Describe the singular value decomposition of A, that is, the factorization $A = U \Sigma V^T$. What are the left singular vectors of A? What are the right singular vectors of A? What are the singular values of A?

16. Describe an algorithm by which the singular value decomposition of A might be found.

17. Given the singular value decomposition of A, state the outer product expansion of A. How can the outer product expansion of A be used to obtain the "best" lower rank approximation of A?

18. Show that the rank of $A = \begin{bmatrix} 1 & 2 \\ 2 & 4 \end{bmatrix}$ is 1. Obtain the general least-squares solution of $A\mathbf{x} = \mathbf{y}$, where $\mathbf{y} = [1 \ 2]^T$. Obtain $\mathbf{x}^+$, the least-squares solution of minimum length. Use the formula $A^+ = \lim_{s \to 0}[(A^T A + s^2 I)^{-1} A^T]$ to find the pseudoinverse A^+. Show that $A^+ \mathbf{y} = \mathbf{x}^+$.

19. Describe how to obtain the pseudoinverse A^+ from the singular value decomposition of A.

Unit Seven

Complex Variables

Chapter 35 **Fundamentals**
Chapter 36 **Applications**

Unit Seven has an unusual structure. It consists of two chapters, one on "fundamentals" and one on applications. Chapter 35 provides the tools needed for using functions of a complex variable in the applications. Chapter 36 presents a spectrum of applications.

Starting at the beginning, we revisit complex arithmetic and provide geometric insight for working with complex numbers in both rectangular and polar forms. Then, a systematic study of complex functions follows. First, we examine z^2 and z^3 and discover the notion of the Riemann surface, on the basis of which we can define inverse functions. Hence, we define the square root and cube root functions as examples of multivalued complex functions.

The exponential function is central to the study of complex variables. Since $w = e^z$ maps horizontal strips of height 2π to the whole w-plane, we have our first example of a complex function that is "infinitely many"-to-one. The Riemann surface it generates has an infinite number of "sheets"; and the inverse function, the complex logarithm, is our first example of a complex function that, for each w in its domain, assumes an infinite number of values.

Since the complex trigonometric functions and the complex hyperbolic functions are defined in terms of the complex exponential function, understanding the behavior of the exponential function is essential for understanding the behavior of this class of elementary functions. The inverses of these functions can all be expressed in terms of the complex logarithm. Hence, a thorough understanding of the differences between the real and complex logarithms is essential for effective use of complex functions in the applications.

Once the behavior of the complex version of the elementary functions is established, we extend the differential calculus to these functions. We de-fine the derivative of a complex function, obtain the Cauchy–Riemann partial differential equations, and specify the category of analytic functions. With complex differentiation, we can obtain the Taylor and Laurent series expansion of a complex function.

Integration of a complex function takes place in the complex plane, as a contour integral, the natural analog of the line integral of Unit Four. Surprisingly, there is an interesting connection between integration and differentiation in the complex plane. The coefficients of the Taylor and Laurent series can be obtained as *integrals* of the expanded function! This leads to the Cauchy residue theorem, which allows us to evaluate contour integrals by obtaining one particular coefficient in a Laurent series.

By way of applications, we study the evaluation of contour integrals, including the contour integrals that yield the inverses of Laplace and Fourier transforms. The complex form of the Fourier series is shown to be equivalent to the trigonometric form of Chapter 10. The root locus diagram and the Nyquist stability criterion, two elements of control theory, are examined.

Conformal mapping and its use in solving certain boundary value problems in the plane lead to a study of planar fluid flow. Complicated flows in the z-plane are mapped to simpler flows in the w-plane, solved there, and mapped back to the z-plane. However, we also study simple flows in the z-plane, which are then mapped to complicated flows in the w-plane. The flow past a cylinder in the z-plane, mapped to flow past the Joukowski airfoil in the w-plane, is an example.

Chapter 35

Fundamentals

INTRODUCTION   This chapter is designed to provide the student with a working suite of tools for understanding and manipulating complex quantities, that is, quantities arising from and associated with the complex plane. Some familiarity with the behavior of complex numbers and the complex exponential function was obtained in Chapters 5 and 12. The present chapter expands those ideas and presents new ones.

In *rectangular form,* the complex number $z = a + bi$ is a relatively tame construct. In *polar form,* $z = \rho(\cos\theta + i \sin\theta)$ is a construct that will confuse and bedevil the unwary. Each point in the complex plane has a unique representation in rectangular form but an infinite number of representations in polar form. Arithmetic in polar form is best done by converting to rectangular form. But the question of which polar form to give for the result is the heart of the problem. Hence, special rules for θ, the *argument* of the complex number, must be observed and, therefore, understood and learned. Thus, we begin by making clear the algebra and geometry of working with complex numbers.

The complex-valued functions $w = z^2$ and $w = z^3$ are studied as examples of a function of a complex variable. Such functions can be interpreted as maps from the z-plane to the w-plane. Hence, no simple graph easily represents such a mapping. The special techniques available for visualizing and representing the action of a complex function are explored, and this includes the *Riemann surface.*

The most important complex function, the exponential function, is discussed. Because of Euler's formulas, the polar form of a complex number is actually $z = \rho e^{i\theta}$. Hence, the polar representation is replaced by the exponential representation, but the difficulties of the multivalued representation are not avoided. Hence, the mapping $w = e^{i\theta}$ takes horizontal strips of width 2π in the z-plane to the whole w-plane, and the Riemann surface for the exponential function has an infinite number of sheets.

Consequently, the inverse function, the complex logarithm, is a multivalued complex function. Thus, $z = \ln w$ takes each point in the w-plane to an infinite number of points in the z-plane, requiring users to define appropriate branches of the logarithm in practical work. Because the complex logarithm is multivalued, all of the familiar laws of logarithms learned in elementary algebra and calculus must be revised. The familiar properties are generally not true for arbitrary complex numbers. Hence, this section pays particular attention to a precise statement of the *rules of logarithms.*

The familiar rules for working with real exponents are generally not true in the complex plane. Hence, the properties of exponents in the complex plane are carefully stated and illustrated. If the properties of exponents and logarithms in the complex plane are not understood and respected, there is little chance any arithmetic in the complex plane will ever be correct if implemented without the guiding hand of technology. And if it is implemented in technology, then it will not be understood if the rules of exponents and logarithms remain a mystery.

In the complex plane, the *trigonometric functions* are defined, by Euler's formula, in terms of the complex exponential function. They create periodic maps of the z-plane to the w-plane, and their Riemann surfaces are infinitely sheeted; likewise for the *hyperbolic functions*. The *inverse trigonometric functions*, like the complex logarithm, are all multivalued. Working with these functions again requires selection of appropriate branches. The *inverse hyperbolic functions* are also multivalued and can, in fact, be expressed in terms of the logarithm.

At this point, complex arithmetic has been seen and the elementary functions familiar from calculus extended to the complex plane. The building blocks and tools are all in place, so now, analysis can begin. Hence, we introduce the *derivative of a complex function* and deduce the *Cauchy–Riemann partial differential equations* that must be obeyed by the real and imaginary parts of complex functions that have a derivative.

The consequences of having one derivative in a neighborhood in the complex plane are remarkable. Such *analytic functions* have all their derivatives and a convergent Taylor series to boot. In addition, the real and imaginary parts of such functions are actually *harmonic functions,* solutions of Laplace's equation!

Some of the remarkable properties of analytic functions are proved as a consequence of theorems about integration in the complex plane. Line integrals along well-behaved paths called *contours* are fundamental here. The *Cauchy–Goursat theorem* shows that closed line integrals of analytic functions are always zero. We also have the *Cauchy integral formulas* for representing a complex function and its derivatives as line integrals. It is by these theorems that properties of analytic functions are proved.

We then treat power series in the complex plane and detail the ideas of the *radius and circle of convergence.* Functions that have isolated singularities of a benign type, called *poles,* have *Laurent series* rather than Taylor series. Because of the Cauchy integral formulas, the coefficients of both the Taylor and the Laurent series can be expressed as closed line integrals. One coefficient in particular, the *residue,* is the closed line integral of the function whose expansion is being obtained. Hence, we detail the *calculus of residues,* the devices for obtaining the values of closed line integrals in the complex plane via manipulations of Laurent series expansions.

Chapter 35 provides a robust set of tools for working in the complex plane. Nearly all of these tools are implemented in modern software packages like computer algebra systems. Hence, much of the tedious and detailed labor of working in the complex plane can be automated. However, without a sound knowledge of the tools, the results generated by computer programs will be neither understood nor appreciated. Because of modern software, it might not be essential to master all the manipulations represented in this chapter, but it is certainly wise to understand all of the concepts.

35.1 Complex Numbers

Imaginary Numbers

The equation $x^2 + 1 = 0$ does not have real solutions. Formally, we might write $x = \pm\sqrt{-1}$, "numbers" for which the squares are -1. Since no real number has a square that is negative, these two formal solutions cannot be real.

Thus, the real numbers are made a subset of a larger set of numbers called the *complex numbers* by defining the *imaginary* number $i = \sqrt{-1}$.

Rectangular Form of Complex Numbers

If a and b are real numbers, then $z = a + bi$ is the *rectangular* form of the complex number z. The real number $a = \mathcal{R}(z)$ is called the *real part* of z, and the real number $b = \mathcal{I}(z)$ is called the *imaginary part* of z. Surprisingly, the imaginary part of the complex number z is a real number! Examples of complex numbers would be

$$z_1 = 2 + 3i \quad \text{and} \quad z_2 = 4 - 7i \tag{35.1}$$

The real and imaginary parts of these complex numbers are, respectively, $\mathcal{R}(z_1) = 2$, $\mathcal{I}(z_1) = 3$, $\mathcal{R}(z_2) = 4$, and $\mathcal{I}(z_2) = -7$.

The Complex Plane

A complex number can be represented as a point in the complex plane. A plot of a complex number in this plane is often called an *Argand diagram* in which the horizontal axis is the real part of the complex number and the vertical axis is the imaginary part of the complex number. For example, Figure 35.1 shows both z_1 and z_2. If the point in the complex plane is connected to the origin with a line segment, the complex number $a + bi$ represents the planar vector $a\mathbf{i} + b\mathbf{j}$, where $\{\mathbf{i}, \mathbf{j}\}$ are unit vectors along the x- and y-axes, respectively.

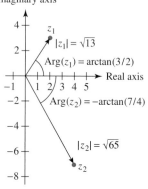

FIGURE 35.1   Argand diagram of $z_1 = 2 + 3i$ and $z_2 = 4 - 7i$

Magnitude and Argument

The analogy between the complex number and the vector in the plane suggests defining the length of the line segment from the origin to the complex number as the *magnitude* (or *modulus*) of the complex number. The magnitude of the complex number $z = a + bi$ is denoted by $|z| = \sqrt{a^2 + b^2}$. For example, $|z_1| = \sqrt{13}$ and $|z_2| = \sqrt{65}$.

The angle this line segment makes with the x-axis is called the *argument* of the complex number and is typically denoted by $\arg(z)$. Just like with real numbers in polar coordinates, the angle, measured from the positive real axis, can lie in $[0, 2\pi)$ or in $(-\pi, \pi]$, for example. The *principal argument*, $\text{Arg}(z)$, lies in $(-\pi, \pi]$, so that $\text{Arg}(z_1) = \arctan(\frac{3}{2})$ and $\text{Arg}(z_2) = -\arctan(\frac{7}{4})$.

Polar Form

The *polar* form of the complex number $z = a + bi$ is

$$z = r(\cos\theta + i\sin\theta)$$

where $r = |z|$ and $\theta = \arg(z)$. Occasionally, this is shortened to $z = r\,\text{cis}(\theta)$, where cis is an obvious shortening of *cosine-i-sine*. Thus, for z_1 and z_2 in (35.1) we would write

$$z_1 = \sqrt{13}\,\text{cis}\left(\arctan\tfrac{3}{2}\right) = \sqrt{13}\left(\tfrac{2}{\sqrt{13}} + i\tfrac{3}{\sqrt{13}}\right) = 2 + 3i \tag{35.2a}$$

$$z_2 = \sqrt{65}\,\text{cis}\left(-\arctan\tfrac{7}{4}\right) = \sqrt{65}\left(\tfrac{4}{\sqrt{65}} - i\tfrac{7}{\sqrt{65}}\right) = 4 - 7i \tag{35.2b}$$

Addition and Subtraction

Complex numbers are added and subtracted by adding and subtracting the real and imaginary parts, respectively. For the complex numbers in (35.1), we have $z_1 + z_2 = 6 - 4i$ and $z_1 - z_2 = -2 + 10i$. Addition and subtraction of complex numbers is modeled geometrically by the addition and subtraction of vectors in the plane.

Multiplication

Multiplication of two complex numbers in rectangular form follows the algebraic pattern used for multiplication of two binomials. For the complex numbers in (35.1) we have $z_1 z_2 = 29 - 2i$, and in general, we have

$$(a + bi)(c + di) = (ac - bd) + i(ad + bc)$$

Geometric Interpretation of Multiplication

A useful geometric interpretation of multiplication is revealed by expressing the factors and product in polar form, as

$$[r_1 \operatorname{cis} \theta_1][r_2 \operatorname{cis} \theta_2] = r_1 r_2 (\cos \theta_1 + i \sin \theta_1)(\cos \theta_2 + i \sin \theta_2)$$

$$= r_1 r_2 [(\cos \theta_1 \cos \theta_2 - \sin \theta_1 \sin \theta_2) + i(\sin \theta_1 \cos \theta_2 + \cos \theta_1 \sin \theta_2)]$$

$$= r_1 r_2 [\cos(\theta_1 + \theta_2) + i \sin(\theta_1 + \theta_2)]$$

$$= r_1 r_2 \operatorname{cis}(\theta_1 + \theta_2)$$

The magnitude of the product is the product of the magnitudes of the factors, and the argument of the product is the sum of the arguments, expressed as an angle in the range $(-\pi, \pi]$.

EXAMPLE 35.1 For the complex numbers given in (35.1), $z_1 z_2 = 29 - 2i = 13\sqrt{5} \operatorname{cis}(-\arctan \frac{2}{29})$, where clearly, $|z_1 z_2| = \sqrt{13}\sqrt{65} = 13\sqrt{5}$. It is more difficult to show that

$$\operatorname{Arg}(z_1 z_2) = -\arctan \frac{2}{29} = \operatorname{Arg}(z_1) + \operatorname{Arg}(z_2) = \arctan \frac{3}{2} - \arctan \frac{7}{4} \qquad (35.3)$$

In floating-point arithmetic, however, we have $-0.069 = 0.983 - 1.052$. (See Exercise A1.)

On the other hand, the arguments of $z_4 = -3 + i$ and $z_5 = -4 + \frac{i}{2}$ are approximately 2.820 and 3.017, respectively, but the principal argument of the product $z_4 z_5$ is -0.446, not the sum 5.837, because $5.837 - 2\pi = -0.446$. ❖

The Complex Conjugate

Given a complex number $a + bi$, the complex number $\overline{a + bi} = a - bi$ is the *complex conjugate*. Geometrically, a complex number and its conjugate are mirror images across the x-axis. The product of a complex number $z = a + bi$ and its complex conjugate $\bar{z} = a - bi$ is $z\bar{z} = |z|^2 = a^2 + b^2$. Thus, the product $z\bar{z}$ is always real.

Division

The mechanics of dividing complex numbers in rectangular form hinges on using the *complex conjugate* in a most judicious way: multiply numerator and denominator by the conjugate of the *denominator*. Thus, to evaluate the ratio $\frac{z_1}{z_2}$, where z_1 and z_2 are given in (35.1), we would compute

$$\frac{2 + 3i}{4 - 7i} = \frac{(2 + 3i)(4 + 7i)}{(4 - 7i)(4 + 7i)} = \frac{(8 - 21) + (12 + 14)i}{4^2 + 7^2} = \frac{-13 + 26i}{65} = -\frac{1}{5} + \frac{2}{5}i$$

The general case therefore follows the pattern

$$\frac{z_1}{z_2} = \frac{z_1 \bar{z}_2}{z_2 \bar{z}_2} = \frac{z_1 \bar{z}_2}{|z_2|^2}$$

In particular, note that the reciprocal of i is the conjugate of i since

$$\frac{1}{i} = \frac{1\bar{i}}{i\bar{i}} = -\frac{i}{1} = -i$$

The same device is used to evaluate the reciprocal of a complex number, as in

$$\frac{1}{a+ib} = \frac{a}{a^2+b^2} - \frac{ib}{a^2+b^2}$$

since

$$\frac{1}{z} = \frac{\bar{z}}{z\bar{z}} = \frac{\bar{z}}{|z|^2}$$

where $z\bar{z} = |z|^2 = (a+bi)(a-bi) = a^2 + b^2$.

Geometric Interpretation of Division

The quotient of two complex numbers in polar form is evaluated as

$$\frac{r_1 \operatorname{cis}\theta_1}{r_2 \operatorname{cis}\theta_2} = \frac{r_1}{r_2}\left[\frac{\cos\theta_1 + i\sin\theta_1}{\cos\theta_2 + i\sin\theta_2}\right]\left[\frac{\cos\theta_2 - i\sin\theta_2}{\cos\theta_2 - i\sin\theta_2}\right]$$

$$= \frac{r_1}{r_2}[(\cos\theta_1\cos\theta_2 + \sin\theta_1\sin\theta_2) + i(\sin\theta_1\cos\theta_2 - \cos\theta_1\sin\theta_2)]$$

$$= \frac{r_1}{r_2}[\cos(\theta_1 - \theta_2) + i\sin(\theta_1 - \theta_2)]$$

$$= \frac{r_1}{r_2}\operatorname{cis}(\theta_1 - \theta_2)$$

The magnitude of a quotient is the quotient of the magnitudes of the numbers being divided, whereas the argument of the quotient is the difference of the arguments expressed as an angle in the range $(-\pi, \pi]$.

EXAMPLE 35.2   Using the numbers in (35.1), we obtain the quotient

$$\frac{z_1}{z_2} = \frac{2+3i}{4-7i} = -\frac{1}{5} + \frac{2}{5}i = \frac{1}{\sqrt{5}}\operatorname{cis}(\pi - \arctan 2)$$

From (35.2a) and (35.2b), we see that $\frac{|z_1|}{|z_2|} = \sqrt{\frac{13}{65}} = \frac{1}{\sqrt{5}}$. Using floating-point arithmetic, we also see that

$$\operatorname{Arg}\left(\frac{z_1}{z_2}\right) = 2.034 = 0.9828 - (-1.0517) = \operatorname{Arg}(z_1) - \operatorname{Arg}(z_2)$$

If we divide two complex numbers that are near the negative real axis, one above and one below, the arguments will be close to π and $-\pi$, respectively. The difference of the arguments will be close to 2π, and the resulting angle will have to be expressed as an angle in the range $(-\pi, \pi]$. For example, the quotient of

$$z_6 = -1 + \frac{i}{10} = \frac{\sqrt{101}}{10}\operatorname{cis}\left(\pi - \arctan\frac{1}{10}\right) \quad \text{and} \quad \bar{z}_6 = \frac{\sqrt{101}}{10}\operatorname{cis}\left(\arctan\frac{1}{10} - \pi\right)$$

is $\operatorname{cis}(-\arctan\frac{20}{99})$, for which the argument $-\arctan\frac{20}{99} = -0.2$ is not just $\operatorname{Arg}(z_6) - \operatorname{Arg}(\bar{z}_6) = 3.04 - (-3.04) = 6.08$, but rather, $6.08 - 2\pi = -0.2$.   ❖

EXERCISES 35.1–Part A

A1. Assuming all angles remain in $(-\frac{\pi}{2}, \frac{\pi}{2}]$, establish $\arctan A - \arctan B = \arctan(\frac{A-B}{1+AB})$ as follows. Write $X = \arctan A - \arctan B = u - v$, and apply to $\tan X$ the formula for tangent of a difference.

A2. Use the result in Exercise A1 to show that (35.3) is exactly $-\arctan\frac{2}{29}$.

A3. If $\arctan A$ and $\arctan B$ are both in the first quadrant but $\arctan A + \arctan B$ is in the second, then $\arctan A + \arctan B = \pi + \arctan(\frac{A+B}{1-AB})$. Use this result to show that $\arctan\frac{3}{2} - \arctan(-\frac{7}{4}) = \pi - \arctan(2)$.

A4. Write $z = 5 - 4i$ in polar form.

A5. Write $\frac{2-7i}{5+3i}$ in rectangular form.

EXERCISES 35.1–Part B

B1. If $z_1 = a + bi$ and $z_2 = c + di$, show that $\mathcal{R}(\frac{z_1}{z_1+z_2}) + \mathcal{R}(\frac{z_2}{z_1+z_2}) = 1$.

For the complex numbers z given in Exercises B2–11:

    (a) Obtain $|z|$ and $\operatorname{Arg} z$.

    (b) Obtain $\bar{z}$ and then plot both z and $\bar{z}$.

    (c) Verify that $z + \bar{z} = 2\mathcal{R}(z)$ and that $z - \bar{z} = 2i\mathcal{I}(z)$.

    (d) Verify that $z\bar{z} = |z|^2$.

    (e) Express z in polar form.

    (f) Obtain $z/\bar{z}$ and $\bar{z}/z$.

B2. $8 + 4i$ **B3.** $3 - 5i$ **B4.** $-1 - 9i$ **B5.** $7 + 5i$

B6. $8 - 9i$ **B7.** $4 - 3i$ **B8.** $-5 - 2i$ **B9.** $8 + i$

B10. $6 + 2i$ **B11.** $9 - 7i$

For the pairs of complex numbers z_1, z_2 given in Exercises B12–21:

    (a) Obtain $z_1 + z_2$ and $z_1 - z_2$ in the form $a + bi$ and in the form $\rho \operatorname{cis}\theta$.

    (b) Obtain $z_1 z_2$ and $\frac{z_1}{z_2}$, both in rectangular form.

    (c) Verify that $|z_1 z_2| \operatorname{cis}(\operatorname{Arg} z_1 z_2) = |z_1||z_2| \operatorname{cis}(\operatorname{Arg} z_1 + \operatorname{Arg} z_2 + 2k\pi)$, for some $k = 0, \pm 1, \ldots$, which keeps the argument of cis in the range $(-\pi, \pi]$.

    (d) Verify that $|\frac{z_1}{z_2}| \operatorname{cis}(\operatorname{Arg}\frac{z_1}{z_2}) = \frac{|z_1|}{|z_2|} \operatorname{cis}(\operatorname{Arg} z_1 - \operatorname{Arg} z_2 + 2k\pi)$, for some $k = 0, \pm 1, \ldots$, which keeps the argument of cis in the range $(-\pi, \pi]$.

    (e) Verify that $|z_1 - z_2|^2 = |z_1|^2 + |z_2|^2 - 2\mathcal{R}(z_1\bar{z}_2)$.

    (f) Verify that $|z_1 + z_2| \leq |z_1| + |z_2|$.

    (g) Verify that $||z_1| - |z_2|| \leq |z_1 - z_2|$.

    (h) Verify that $\mathcal{R}(\frac{z_1}{z_1+z_2}) + \mathcal{R}(\frac{z_2}{z_1+z_2}) = 1$, the identity from Exercise B1.

B12. $4 + 7i, -9 + 5i$ **B13.** $7 - 8i, 3 + 5i$ **B14.** $-5 + 5i, -7 + i$

B15. $-1 + 3i, 8 + 3i$ **B16.** $-1 - 4i, 2 + 9i$ **B17.** $-2 - 7i, 8 + 5i$

B18. $3 - 6i, -8 + 2i$ **B19.** $5 + i, 3 + 4i$

B20. $2 - 9i, -5 + 6i$ **B21.** $9 - 8i, 1 + 9i$

B22. If $z_1 = x_1 + iy_1$ and $z_2 = x_2 + iy_2$, show that $|z_1 + z_2|^2 + |z_1 - z_2|^2 = 2(|z_1|^2 + |z_2|^2)$. Interpret geometrically in terms of the diagonals and sides of a parallelogram.

B23. If $z_1 = x_1 + iy_1$ and $z_2 = x_2 + iy_2$, show in general that

    (a) $\mathcal{R}(z_1 + z_2) = \mathcal{R}(z_1) + \mathcal{R}(z_2)$

    (b) $\mathcal{I}(z_1 + z_2) = \mathcal{I}(z_1) + \mathcal{I}(z_2)$

    (c) $\mathcal{R}(z_1 z_2) = \mathcal{R}(z_1)\mathcal{R}(z_2) - \mathcal{I}(z_1)\mathcal{I}(z_2)$

    (d) $\mathcal{I}(z_1 z_2) = \mathcal{R}(z_1)\mathcal{I}(z_2) + \mathcal{I}(z_1)\mathcal{R}(z_2)$

In Exercises B24–38, reduce the given complex number to rectangular form.

B24. $(-7 + i) + (1 + i) - (-2 + 9i)$

B25. $(-2 + 6i) + (-7 - 9i) + (4 + 9i)$

B26. $(2 - 3i) - (3 - i) + (-3 - 3i)$

B27. $(4 - 6i) - (8 - 9i) - (9 - 3i)$

B28. $(6 - 3i) + (-9 - 4i) - (-1 - 2i)$

B29. $(1 + 7i)(5 + 2i)(-1 - 6i)$ **B30.** $(-9i)(5 - 4i)(7 + 3i)$

B31. $(7 + 3i)(4 - 9i)(2 - i)$ **B32.** $(5 + 2i)(6 - 7i)(9 + 3i)$

B33. $(-8 + 3i)(3 + 7i)(6 + 5i)$ **B34.** $\dfrac{1 - 6i}{-9 + 8i} + \dfrac{3 - 7i}{-7 - 3i}$

B35. $\dfrac{2 + 9i}{-6 + 9i} + \dfrac{-2 - i}{4 + 5i}$ **B36.** $\dfrac{1 - 9i}{-2 + i} + \dfrac{-3 - 4i}{8 + 7i}$

B37. $\dfrac{-5 + 7i}{7 - 8i} + \dfrac{8 - 5i}{-9 + 2i}$ **B38.** $\dfrac{1 - 7i}{9 - 2i} + \dfrac{-5 + 4i}{3 + i}$

In Exercises B39–48, graph and/or describe the set of complex numbers that satisfy the given condition.

B39. $\mathcal{I}(z) = 3$ **B40.** $1 \leq \mathcal{I}(z) \leq 3$ **B41.** $|3z - 4| = 5$

B42. $|3z - 4| \leq 5$ **B43.** $|z| = |z + i|$ **B44.** $|z| = 2|z + 3|$

B45. $|z| = \mathcal{I}(z) - 3$ **B46.** $|z + 5 - 7i| = |z - i|$

B47. $|z - 2 + 3i| = |z + 1| + 8$ **B48.** $|z - 2i| = |z + 1| + 4\mathcal{R}(z)$

35.2 The Function $w = f(z) = z^2$

Complex Functions

An important component in any study of complex variables is a study of the behavior of complex functions of the form $w = f(z)$. Such functions map the complex z-plane to the complex w-plane.

If $z = x + iy$ is a complex variable, then the function $f(z)$ can be written as $f(z) = u(x, y) + iv(x, y)$, where $u(x, y)$ is called the *real part* of f and $v(x, y)$ is called the *imaginary part* of f. In this and the following six sections, we will explore properties of the elementary functions of the calculus extended to the complex plane and illustrate the decomposition of these functions into their real and imaginary parts.

Real and Imaginary Parts of $f(z) = z^2$

The real and imaginary parts of $f(z) = z^2$ are found by writing $f(x + iy) = (x + iy)^2$ as $x^2 - y^2 + 2xyi$, from which it follows that $u(x, y) = x^2 - y^2$ and $v(x, y) = 2xy$.

Graphing $f(z) = z^2$

Because $w = f(z)$ is a mapping of the z-plane to the w-plane, it is difficult to draw a simple graph representing the function. Hence, we will draw several different plots, each of which captures some aspect or other of the action of a complex function. Figure 35.2 shows a graph of the magnitude-surface

$$|w| = |f(z)| = \sqrt{u^2 + v^2} = x^2 + y^2$$

Figure 35.3 shows (in the plane of $w = u + iv$) images of the grid lines in the z-plane, namely, the lines $x = $ constant and $y = $ constant, under the mapping induced by the function $w = f(z)$. The curves opening to the left are the images of $x = $ constant, while the curves opening to the right are the images of $y = $ constant. For reasons to be revealed in Section 36.6, this graph is called a *conformal plot* for $f(z) = z^2$.

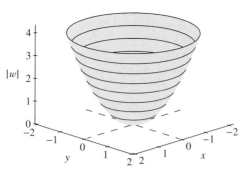

FIGURE 35.2   Magnitude surface for $f(z) = z^2$

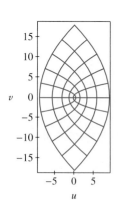

FIGURE 35.3   Conformal plot for $f(z) = z^2$

Figure 35.4 shows, in the xy-plane, the inverse images of the rectangular grid lines in the $w = uv$-plane. This graph therefore shows the contours (level curves) for $u(x, y) = $ constant and $v(x, y) = $ constant, contour lines that "live" in the xy-plane. The (solid) hyperbolas $x^2 - y^2 = $ constant are the contours of $u(x, y)$, while the (dotted) hyperbolas $xy = $ constant are the contours of $v(x, y)$.

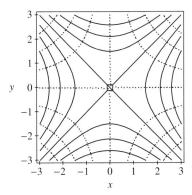

FIGURE 35.4   Inverse images of $u = $ constant (solid) and $v = $ constant (dotted) for $f(z) = z^2$

Action of the Mapping Induced by $f(z) = z^2$

If the function $f(z) = z^2$ is expressed in polar notation, its action as a mapping of the z-plane to the w-plane has a simple geometric interpretation. Since the complex number $z = x + iy$ can be written as $z = r(\cos\theta + i\sin\theta)$, we have for $f(z) = z^2$ the corresponding form

$$w = (r(\cos\theta + i\sin\theta))^2 = r^2[(\cos^2\theta - \sin^2\theta) + 2i\cos\theta\sin\theta] = r^2(\cos 2\theta + i\sin 2\theta)$$

Thus, the magnitude of z is squared and the argument of z is doubled, in conformity with our earlier discussion on multiplication.

For example, the complex number $z_1 = -10 + i$ has a magnitude of $\sqrt{101}$ and an argument, in degrees, of 174.29. The magnitude of $w_1 = f(z_1)$ is then 101, which is $|z_1|^2$. The argument of $w_1 = f(z_1)$, in degrees, is -11.42. The function z^2 doubles that to 348.58, an angle in the fourth quadrant, converted to an angle in the range $(-\pi, \pi]$ by the arithmetic $348.58 - 360 = -11.42$.

Riemann Surface for z^2

Because of the "doubling" property of $w = z^2$, points in the upper half of the z-plane where $0 \le \mathrm{Arg}\, z < \pi$ have their argument doubled and are mapped onto all the w-plane. Similarly, the lower half of the z-plane will also map onto all of the w-plane. Thus, the z-plane is mapped onto the w-plane *twice*. Two different points in the z-plane are thus mapped to the same point in the w-plane. The two points $r\,\mathrm{cis}\,\theta$ and $r\,\mathrm{cis}(\theta + \pi)$ are both mapped to

$$r^2\,\mathrm{cis}\,2\theta = r^2\,\mathrm{cis}(2\theta + 2\pi)$$

The polar forms of two complex numbers with the same magnitude and arguments differing by 2π evaluate to the same complex number in rectangular form. Hence, the rectangular form of a complex number has a primacy that allows us to think of the point as the complex number $a + bi$ and the numbers $\rho\,\mathrm{cis}(\phi + 2k\pi)$, $k = 1, 2, \ldots$, as *copies* of the primary point.

To account for the numbers $r^2\,\mathrm{cis}\,2\theta$ and $r^2\,\mathrm{cis}(2\theta + 2\pi)$ in the mapping $w = z^2$, create two copies of the w-plane and stack them one above the other. As z ranges over the upper half of the z-plane, let the images fall in the first copy of the w-plane. As z ranges over the lower half of the z-plane, let the images fall in the second copy of the w-plane. If crossing the negative real axis in the z-plane corresponds to passing from one copy of the w-plane to the other, the range of $w = z^2$ will be an *open connected set*. (An open set is connected if each pair of points in it can be connected by a polygonal path lying completely in the set. The interior of a disk is an example of an open connected set.) A graph of this set is called the *Riemann surface* generated by z^2 and is shown in Figure 35.5.

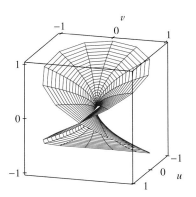

FIGURE 35.5   Riemann surface generated by $f(z) = z^2$

The Inverse Map $z = \sqrt{w}$

Since "squaring" amounts to squaring the magnitude and doubling the argument, finding the square root amounts to halving the argument and taking the square root of the magnitude. However, the function $w = f(z) = z^2$ is a two-to-one mapping of the z-plane onto the w-plane. Consequently, the mapping back from the w-plane to the z-plane is not a function since it is one-to-two. However, by defining two distinct branches of the square root function and associating each branch with a sheet of the Riemann surface generated by z^2, two single-valued functions for $z = \sqrt{w}$ are obtained.

A "geometric" point in the w-plane has two "numeric" representations, one above the other, on the two sheets of the Riemann surface generated by $w = z^2$. The inverse image of the point on the upper sheet lies in the upper half of the z-plane, and the inverse image of the point on the lower sheet lies in the lower half of the z-plane.

For example, if $w_1 = 4 + 3i = 5\,\text{cis}(\arctan\frac{3}{4})$, then $z_1 = \sqrt{w_1}$ should be the complex number $\sqrt{5}\,\text{cis}(\frac{1}{2}\arctan\frac{3}{4})$, for which the magnitude is the square root of $|w_1|$, and the argument, $\frac{1}{2}\text{Arg}(w_1)$. Thus, we have

$$z_1 = \sqrt{5}\left[\cos\left(\tfrac{1}{2}\arctan\tfrac{3}{4}\right) + i\,\sin\left(\tfrac{1}{2}\arctan\tfrac{3}{4}\right)\right] \tag{35.4a}$$

$$= \sqrt{5}\left[\sqrt{\frac{1 + \cos\left(\arctan\frac{3}{4}\right)}{2}} + i\sqrt{\frac{1 - \cos\left(\arctan\frac{3}{4}\right)}{2}}\right] \tag{35.4b}$$

$$= \sqrt{5}\left[\sqrt{\frac{1 + \frac{4}{5}}{2}} + i\sqrt{\frac{1 - \frac{4}{5}}{2}}\right] \tag{35.4c}$$

$$= \frac{(3 + i)}{\sqrt{2}} \tag{35.4d}$$

What about the square root that lies in the lower half of the z-plane? How is that to be found? Since the Riemann surface identifies the two numbers $r^2\,\text{cis}\,2\theta$ and $r^2\,\text{cis}(2\theta + 2\pi)$, the inverse image of the first is $r\,\text{cis}\,\theta$, just computed as $z_1 = \frac{(3+i)}{\sqrt{2}}$, whereas the inverse image of the second is $r\,\text{cis}(\theta + \pi)$. Thus, the arguments of the two square roots of w_1 differ by π, and the second square root of w_1 is $-z_1$.

EXERCISES 35.2–Part A

A1. Verify the calculations in (35.4a)–(35.4d).

A2. Let $z_1 = -1 + i$, and express $\tilde{w} = z_1^2$ in rectangular form and polar form $\tilde{w} = r\,\text{cis}\,\tilde{\theta}$.

A3. Find $z_2 \neq z_1$ for which $z_2^2 = \tilde{w}$, where $\tilde{w}$ is as computed in Exercise A2. How are z_1 and z_2 related?

A4. If $\tilde{w}$ is as computed in Exercise A2, obtain $\sqrt{\tilde{w}} = \sqrt{r}\,\text{cis}(\tilde{\theta}/2)$. Is this z_1 or z_2?

A5. Obtain $\sqrt{r}\,\text{cis}\,\frac{\tilde{\theta} + 2\pi}{2}$. How is this related to $\sqrt{\tilde{w}}$?

EXERCISES 35.2–Part B

B1. The function $w = z^2$ takes a point on the circle $|z| = 1$ to a point on the Riemann surface in Figure 35.5. The circle $x = \cos t$, $y = \sin t$, $0 \leq t \leq 2\pi$, can be mapped onto this surface by drawing the space curve $u = \cos 2t$, $v = \sin 2t$, $h = y = \sin t$, $0 \leq t \leq 2\pi$, where h is "height" on the Riemann surface. Using appropriate technology, obtain a graph of this path on the Riemann surface generated by z^2. Why did we choose h proportional to y?

For the complex numbers z in Exercises B2–11:

(a) Obtain $|z|$ and $\theta = \text{Arg}\,z$.

(b) Obtain z^2 in the form $\sigma = a + bi$, and then write σ in the form $r\,\text{cis}\,\omega$, where $\omega = \text{Arg}\,\sigma$.

(c) Show that $r = |z|^2$ and $\omega = 2\theta + 2k\pi$, where k is chosen so $-\pi < \omega \leq \pi$.

(d) Plot both z and z^2 on the same Argand diagram.

(e) Obtain $r_1 = \sqrt{|z|}\,\text{cis}\,\frac{\theta}{2}$ and $r_2 = \sqrt{|z|}\,\text{cis}(\frac{\theta}{2} + \pi)$, the two square roots of z.

(f) Plot z and its two square roots on the same Argand diagram.

(g) Compute $z^{3/2} = |z|^{3/2}\,\text{cis}(\frac{3}{2}\text{Arg}\,z)$, r_1^3, r_2^3, and $\sqrt{z^3} = |z|^{3/2}\,\text{cis}(\frac{1}{2}\text{Arg}\,z^3)$. If the first number is the principal value of $z^{3/2}$, which of the other three numbers is in agreement?

(h) Compute $z^{5/2} = |z|^{5/2}\,\text{cis}(\frac{5}{2}\text{Arg}\,z)$, r_1^5, r_2^5, and $\sqrt{z^5} = |z|^{5/2}\,\text{cis}(\frac{1}{2}\text{Arg}\,z^5)$. If the first number is the principal value of $z^{5/2}$, which of the other three numbers is in agreement?

B2. $-1 - 2i$ **B3.** $-4 - 5i$ **B4.** $8 - i$ **B5.** $6 + 5i$

B6. $-7 - 11i$ **B7.** $-11 + 2i$ **B8.** $-11 + 12i$ **B9.** $12 - 11i$

B10. $1 + 4i$ **B11.** $-1 - 6i$

For each pair of complex numbers z_1 and z_2 given in Exercises B12–21:

(a) Compute $\sqrt{z_1 z_2} = \sqrt{|z_1 z_2|}\operatorname{cis}\frac{\omega}{2}$, where $\omega = \operatorname{Arg} z_1 + \operatorname{Arg} z_2 + 2k\pi$, with $k = 0, \pm1, \dots$, chosen so that $-\pi < \omega \le \pi$.

(b) Compute $\sqrt{z_1}\sqrt{z_2} = \sqrt{|z_1|}\operatorname{cis}(\frac{1}{2}\operatorname{Arg} z_1)\sqrt{|z_2|}\operatorname{cis}(\frac{1}{2}\operatorname{Arg} z_2)$, and compare to part (a).

(c) Compute $\sqrt{\frac{z_1}{z_2}} = \sqrt{\left|\frac{z_1}{z_2}\right|}\operatorname{cis}\frac{\tau}{2}$, where $\tau = \operatorname{Arg} z_1 - \operatorname{Arg} z_2 + 2k\pi$, with $k = 0, \pm1, \dots$, chosen so that $-\pi < \tau \le \pi$.

(d) Compute $\frac{\sqrt{z_1}}{\sqrt{z_2}} = \frac{\sqrt{|z_1|}\operatorname{cis}(\frac{1}{2}\operatorname{Arg} z_1)}{\sqrt{|z_2|}\operatorname{cis}(\frac{1}{2}\operatorname{Arg} z_2)}$, and compare to part (c).

B12. $8 + 6i, -10 + i$ **B13.** $-10 + 2i, 3 - 5i$

B14. $-7 - 7i, 4 - 6i$ **B15.** $-7 - 7i, 3 + 3i$

B16. $1 + 11i, -8 + 12i$ **B17.** $3 - 8i, -12 + 5i$

B18. $-1 - i, 3 - 7i$ **B19.** $3 - 6i, -9 + 8i$

B20. $3 - 8i, -6 - 5i$ **B21.** $2 - 10i, -11 + 2i$

35.3 The Function $w = f(z) = z^3$

Real and Imaginary Parts of $f(z) = z^3$

Separation of $f(z) = (x + iy)^3$ into its real and imaginary parts gives, respectively,

$$u(x, y) = x^3 - 3xy^2 \quad \text{and} \quad v(x, y) = 3x^2 y - y^3 \tag{35.5}$$

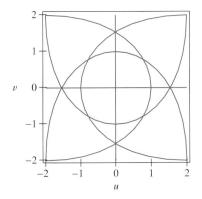

FIGURE 35.7   Conformal plot for $f(z) = z^3$

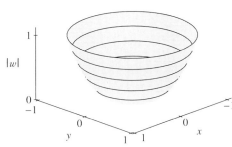

FIGURE 35.6   Magnitude surface for $f(z) = z^3$

Graphing $f(z) = z^3$

Figure 35.6 contains a plot of $|w| = \sqrt{u^2 + v^2} = (x^2 + y^2)^{3/2}$. Figure 35.7 contains a conformal plot showing how grid lines in the z-plane map into the w-plane. Curves opening left and right are the images of $x = $ constant, whereas curves opening toward the top or bottom of the graph are images of $y = $ constant. Figure 35.8 shows, in the xy-plane, the inverse images of the rectangular grid lines in the w-plane where $w = u + iv$. The (solid) curves are the contours (level curves) for $u(x, y) = $ constant and the (dotted) curves are the level curves for $v(x, y) = $ constant.

Action of the Mapping Induced by $f(z) = z^3$

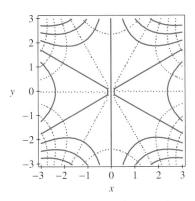

FIGURE 35.8   Inverse images of $u = $ constant (solid) and $v = $ constant (dotted) for $f(z) = z^3$

Writing $z = x + iy$ as $r \operatorname{cis}\theta$, we find that

$$w = z^3 = r^3(\cos 3\theta + i \sin 3\theta)$$

Thus, under the mapping induced by the function $f(z) = z^3$, the magnitude of z is cubed and the argument of z is tripled. For example, given the complex number $z_1 = -10 + i = $

$\sqrt{101}$ cis 3.042, the magnitude of $w_1 = f(z_1)$ is then $|z_1|^3 = 101\sqrt{101}$ and the argument of w_1 is $2.842 = 3(3.042) - 2\pi$.

Riemann Surface for z^3

On the unit circle $|z| = 1$, the arcs determined by

$$\left\{0 \le \theta < \frac{2\pi}{3}\right\} \qquad \left\{\frac{2\pi}{3} \le \theta < \frac{4\pi}{3}\right\} \qquad \left\{\frac{4\pi}{3} \le \theta < 2\pi\right\}$$

are each mapped by $w = f(z) = z^3$ to the complete circle $|w| = 1$ in the w-plane. The w-plane is covered *three* times by the mapping $w = z^3$. Three different points in the z-plane are thus mapped to the same point in the w-plane. For example, the three points

$$r\,\text{cis}\,\theta \qquad r\,\text{cis}\left(\theta + \frac{2\pi}{3}\right) \qquad r\,\text{cis}\left(\theta + \frac{4\pi}{3}\right)$$

are all mapped to $r^3\,\text{cis}(3\theta) = r^3\,\text{cis}(3\theta + 2\pi) = r^3\,\text{cis}(3\theta + 4\pi)$.

To account for these numbers in the mapping $w = z^3$, create three copies of the w-plane and stack them one above the other. As z ranges over the sector of the z-plane given by $0 \le \theta < \frac{2\pi}{3}$, let the images fall in the first copy of the w-plane. As z ranges over the sector of the z-plane given by $\frac{2\pi}{3} \le \theta < \frac{4\pi}{3}$, let the images fall in the second copy of the w-plane. As z ranges over the sector of the z-plane given by $\frac{4\pi}{3} \le \theta < 2\pi$, let the images fall in the third copy of the w-plane. If the three copies of the w-plane are connected along the positive real axis, the result is the Riemann surface seen in Figure 35.9.

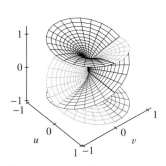

FIGURE 35.9   Riemann surface generated by $f(z) = z^3$

The Cube-Root Function

The function $w = f(z) = z^3$ is a three-to-one mapping of the z-plane onto the w-plane. Consequently, the mapping back from the w-plane to the z-plane is not a function since it is one-to-three. However, by defining three distinct branches of the cube-root function and associating each branch with a sheet of the Riemann surface for z^3, three single-valued functions for $z = w^{1/3}$ are obtained.

A "geometric" point in the w-plane has three "numeric" representations, one above the other, on the three sheets of the Riemann surface for $w = z^3$. The inverse images of three points lying one above the other in the w-plane will fall into the separate sectors $0 \le \theta < \frac{2\pi}{3}, \frac{2\pi}{3} \le \theta < \frac{4\pi}{3}, \frac{4\pi}{3} \le \theta < 2\pi$ in the z-plane.

The cube root of $w_1 = 4 + 3i = 5\,\text{cis}(\arctan\frac{3}{4})$ is $w_1^{1/3} = 5^{1/3}\,\text{cis}(\frac{1}{3}\arctan\frac{3}{4})$, for example. Since mapping the z-plane to the w-plane by $w = z^3$ cubed the modulus and tripled the argument, the inverse mapping consists of taking the cube root of the modulus and dividing the argument by three.

What about the cube root that lies in the sector $\frac{2\pi}{3} \le \theta < \frac{4\pi}{3}$ of the z-plane? How is that to be found? Since the Riemann surface identifies the three numbers $r^3\,\text{cis}(3\theta + 2k\pi)$, $k = 0, 1, 2$, the inverse image of the first is $r\,\text{cis}(\theta)$, as just seen, whereas the inverse image of the second is $r\,\text{cis}(\theta + \frac{2\pi}{3})$. Similarly, the cube root that lies in the sector $\frac{4\pi}{3} \le \theta < 2\pi$ of the z-plane is given by $r\,\text{cis}(\theta + \frac{4\pi}{3})$. These three cube roots are shown in Figure 35.10, along with the rays $\theta = 0, \frac{2\pi}{3}, \frac{4\pi}{3}$.

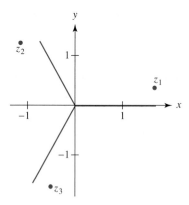

FIGURE 35.10   The three cube roots of $w_1 = 4 + 3i$

DeMoivre's Laws

A generalization of the process of finding square roots and cube roots of complex numbers is given by *DeMoivre's law* (Abraham DeMoivre, 1667–1754). For an integer $n > 0$ we have

$$w^{1/n} = [r \operatorname{cis} \theta]^{1/n} = [r \operatorname{cis}(\theta + 2k\pi)]^{1/n} = r^{1/n} \operatorname{cis}\left(\frac{\theta}{n} + \frac{2k\pi}{n}\right) \qquad k = 0, 1, \ldots, n-1$$

Thus, to obtain an nth root of a complex number, realize first that the function z^n maps the z-plane onto n copies of the w-plane. For example, the function z^3 takes the wedge $0 \le \theta < \frac{2\pi}{3}$ onto the whole w-plane. Hence, if we let z^3 map the whole z-plane to the w-plane, each point in the w-plane is a recipient of three different points in the z-plane. The sectors $0 \le \theta < \frac{2\pi}{3}$, $\frac{2\pi}{3} \le \theta < \frac{4\pi}{3}$, and $\frac{4\pi}{3} \le \theta < 2\pi$ each map onto the whole w-plane. To keep these three images in the w-plane distinct, three copies of the w-plane are stacked over each other and the w-points whose arguments differ by 2π are identified with each other. Then, when it's time to invert, each of the three points sitting one above the other is independently pulled back to the z-plane, one to each of the three sectors of size $\frac{2\pi}{3}$.

A generalization of DeMoivre's law, for any integer $m > 0$, is

$$w^m = [r \operatorname{cis} \theta]^m = r^m \operatorname{cis} m\theta$$

Combining both laws we would have $w^{m/n} = r^{m/n} \operatorname{cis}(\frac{m}{n}\theta + \frac{m}{n}2k\pi)$, $k = 0, 1, \ldots, n-1$, where both m and n are integers.

EXERCISES 35.3–Part A

A1. Verify the results in (35.5).

A2. If $z = x + yi$, show that $|z^3| = \sqrt{u^2 + v^2} = (x^2 + y^2)^{3/2}$.

A3. Express the cube roots of $w_1 = 4 + 3i$ in rectangular form.

A4. If $z_1 = 1 + i$, obtain $w_1 = z_1^3$ in both rectangular and polar forms.

A5. If z_1 and w_1 are as in Exercise A4, find z_2 and z_3, both different from z_1 and each satisfying $z_k^3 = w_1$, $k = 2, 3$.

EXERCISES 35.3–Part B

B1. Obtain the three cube roots of i.

For the complex number z in each of Exercises B2–11:

(a) Obtain $|z|$ and $\theta = \operatorname{Arg} z$.

(b) Obtain z^3 in the form $\sigma = a + bi$, and then write σ in the form $r \operatorname{cis} \omega$, where $\omega = \operatorname{Arg} \sigma$.

(c) Show that $r = |z|^3$ and that $\omega = 3\theta + 2k\pi$, where k is chosen so $-\pi < \omega \le \pi$.

(d) Using DeMoivre's law, compute the three cube roots of z and plot them on the same Argand diagram.

(e) Let r_k, $k = 1, 2, 3$, be the three cube roots of z. Compute $z^{2/3} = |z|^{2/3} \operatorname{cis}(\frac{2}{3} \operatorname{Arg} z)$, r_1^2, r_2^2, r_3^2, and $\sqrt[3]{z^2} = |z|^{2/3} \operatorname{cis}(\frac{1}{3} \operatorname{Arg} z^2)$. If the first number is the principal value of $z^{2/3}$, which of the other four numbers is in agreement?

(f) Compute $z^{5/3} = |z|^{5/3} \operatorname{cis}(\frac{5}{3} \operatorname{Arg} z)$, r_1^5, r_2^5, r_3^5, and $\sqrt[3]{z^5} = |z|^{5/3} \operatorname{cis}(\frac{1}{3} \operatorname{Arg} z^5)$. If the first number is the principal value of $z^{5/3}$, which of the other four numbers is in agreement?

B2. $3 - 5i$ **B3.** $-\frac{7}{2} - \frac{1}{5}i$ **B4.** $8 - 9i$ **B5.** $7 + 5i$

B6. $\frac{1}{4} - \frac{7}{10}i$ **B7.** $-9 + 5i$ **B8.** $7 - 8i$ **B9.** $\frac{8}{7} - \frac{9}{2}i$

B10. $-5 + 5i$ **B11.** $-\frac{7}{5} + \frac{1}{4}i$

For each pair of complex numbers z_1 and z_2 given in Exercises B12–21:

(a) Use DeMoivre's law to compute $\sqrt[3]{z_1 z_2}$ and $\sqrt[3]{z_1}\sqrt[3]{z_2}$. Do the results necessarily agree?

(b) Use DeMoivre's law to compute $\sqrt[3]{\frac{z_1}{z_2}}$ and $\frac{\sqrt[3]{z_1}}{\sqrt[3]{z_2}}$. Do the results necessarily agree?

B12. $4 + i, 5 - 3i$ **B13.** $-4 - 8i, \frac{9}{4} + \frac{4}{5}i$ **B14.** $\frac{6}{5} + \frac{1}{3}i, 8 + 5i$

B15. $9 - 2i, 5 + 4i$ **B16.** $-8 + 3i, \frac{4}{3} - \frac{5}{7}i$ **B17.** $-\frac{3}{7} - \frac{5}{4}i, 8 - 4i$

B18. $-2 - 7i, 8 + 3i$ **B19.** $\frac{1}{11} - \frac{3}{4}i, 3 - 7i$

B20. $9 + 9i, 2 + 9i$ **B21.** $2 - 5i, \frac{6}{7} - \frac{4}{5}i$

B22. Use DeMoivre's law to compute

(a) the four fourth roots of $1, -1$, and i.

(b) the five fifth roots of $1, -1$, and i.

B23. Show that a circle with center at z_0 has an equation of the form $z\bar{z} - z_0\bar{z} - \bar{z}_0 z + a = 0$, where a is real. What is the radius of the circle?

B24. Show that the line joining the points z_1 and z_2 is perpendicular to the line joining the points z_3 and z_4 if the number $\lambda = \frac{z_1 - z_2}{z_3 - z_4}$ is pure imaginary.

B25. Using $(\operatorname{cis}\theta)^m = \operatorname{cis} m\theta$, show that $\cos 5\theta = 16\cos^5\theta - 20\cos^3\theta + 5\cos\theta$.

In Exercises B27–30, derive the indicated trigonometric formulas.

B26. $\cos^3\theta + \sin^3\theta = \frac{1}{4}[\cos 3\theta + 3\cos\theta - \sin 3\theta + 3\sin\theta]$

B27. $\cos^4\theta + \sin^4\theta = \frac{1}{4}[\cos 4\theta + 3]$

B28. $\cos^5\theta + \sin^5\theta = \frac{1}{16}[\cos 5\theta + 5\cos 3\theta + 10\cos\theta + \sin 5\theta - 5\sin 3\theta + 10\sin\theta]$

B29. $\cos^6\theta + \sin^6\theta = \frac{1}{8}[3\cos 4\theta + 5]$

B30. If z_1, z_2, and z_3 are the vertices of a triangle in the complex plane, show that

(a) the centroid, the intersection of the medians, is located at $\frac{1}{3}(z_1 + z_2 + z_3)$.

(b) the centroid divides each median in the ratio $1:2$.

In Exercises B31–35, the given values of p, q, and r determine the cubic equation $z^3 - 3pz^2 + 3qz - r = 0$, whose three roots are the vertices of a triangle in the complex plane. Show that the centroid of the triangle is the point corresponding to p.

B31. $p = -\frac{1}{3} + 2i, q = 3 + \frac{31}{3}i, r = 12 - 44i$

B32. $p = \frac{1}{3} - \frac{5}{3}i, q = \frac{4}{3} + 6i, r = 26 - 22i$

B33. $p = \frac{1}{3} + i, q = 7 + \frac{4}{3}i, r = 39 - 91i$

B34. $p = \frac{2}{3} + 2i, q = \frac{4}{3} + \frac{31}{3}i, r = 17 - 33i$

B35. $p = -1 + i, q = -\frac{28}{3} + 9i, r = 60 - 40i$

35.4 The Exponential Function

Euler's Formulas

The exponential function, e^z, is a very important component of complex function theory. This importance derives, in part, from Euler's formulas, which state $e^{i\theta} = \operatorname{cis}\theta$. In greater detail, we have $e^{i\theta} = \cos\theta + i\sin\theta$ and $e^{-i\theta} = \cos\theta - i\sin\theta$. Consequently, the complex number $e^{i\theta}$ has magnitude 1, and argument θ. The points that satisfy $e^{i\theta} = 1$ comprise a circle with radius 1 and center at the origin.

Another consequence of Euler's formulas is

$$e^{x+iy} = e^x e^{iy} = e^x(\cos\theta + i\sin\theta)$$

Hence, the polar form of a complex number z is easily replaced by the exponential form $z = |z|e^{i\,\mathrm{Arg}(z)}$.

Euler's proof of the identity $e^{i\theta} = \cos\theta + i\sin\theta$ was based on the Taylor series expansion $e^x = \sum_{k=0}^{\infty} x^k/k!$. Just as this series was used to define e^{At} for a matrix A, so too is it used to define e^z. In particular, $e^{i\theta} = \sum_{k=0}^{\infty}(i\theta)^k/k!$, where

$$i^{4k+1} = i \qquad i^{4k+2} = i^2 = -1 \qquad i^{4k+3} = i^3 = -i \qquad i^{4k+4} = i^4 = 1$$

The four values $i, i^2 = -1, i^3 = -i, i^4 = 1$ are cyclic, and the terms in the series alternate between real and imaginary. Hence, the series can be rearranged to

$$e^{i\theta} = \sum_{k=0}^{\infty} \frac{(-1)^k \theta^{2k}}{(2k)!} + i\sum_{k=0}^{\infty} \frac{(-1)^k \theta^{2k+1}}{(2k+1)!}$$

On the right, the first series is the expansion of $\cos\theta$ and the second of $\sin\theta$. Of course, we have worked *formally* and have not considered questions of convergence. Typically, texts in complex variables *define* e^z by Euler's formula and then later in the course show the series representation is valid.

Real and Imaginary Parts of e^z

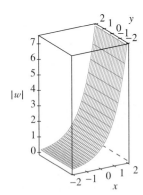

FIGURE 35.11 Magnitude surface for $f(z) = e^z$

The function $w = e^z$ can be written in the form $u(x,y) + iv(x,y)$, where $u(x,y) = e^x\cos y$ and $v(x,y) = e^x\sin y$.

Magnitude Surface for e^z

Figure 35.11 exhibits a graph of $|e^z| = |e^{x+iy}| = |e^x e^{-iy}| = e^x |\operatorname{cis} y| = e^x$, the magnitude of $w = e^z$.

Conformal Plot for e^z

The conformal plot for $w = e^z$ in Figure 35.12 shows, in the w-plane, the images of the grid lines of the z-plane. This graph suggests that the fundamental strip $-\pi < y \le \pi$ in the z-plane is mapped by $w = e^z$ to all of the w-plane less the origin. Indeed, the grid line $x = 1$, $-\pi < y \le \pi$, maps to a circle with radius e and center at the origin. As x varies, the circle increases or decreases in radius. Consequently, as x varies in the z-plane, w attains all values except $w = 0$. Indeed, every horizontal strip of width 2π in the z-plane will map onto the w-plane less its origin. With $\operatorname{Arg}(z)$ in the range $(-\pi, \pi]$, the horizontal strips of width 2π are bounded by the lines $y = (2k+1)\pi$, $k = 0, \pm 1, \pm 2, \ldots$.

Therefore, the map $w = e^z$ is "infinitely many"-to-one, and the Riemann surface for the exponential function has an infinite number of sheets. Figure 35.13 shows a small portion of this Riemann surface.

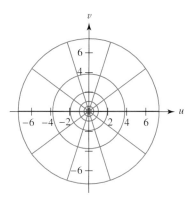

FIGURE 35.12 Conformal plot for $f(z) = e^z$

Law of Exponents

The "law of exponents," familiar from real-variable calculus, also holds for complex numbers. Thus, we have

$$e^{z_1 + z_2} = e^{z_1} e^{z_2} \tag{35.6}$$

EXAMPLE 35.3 If $z_1 = 2 + 3i$ and $z_2 = 4 - 7i$, then $e^{z_1 + z_2} = e^6 \cos 4 - i e^6 \sin 4 = e^{z_1} e^{z_2}$. ❖

Derivation of the Law of Exponents

We wish to show $e^{z_1} e^{z_2} = e^{x_1 + i y_1} e^{x_2 + i y_2}$ can be transformed into $e^{z_1 + z_2}$. Euler's formula gives

$$e^{x_1} e^{x_2} (\cos y_1 \cos y_2 - \sin y_1 \sin y_2) + i e^{x_1} e^{x_2} (\sin y_1 \cos y_2 + \cos y_1 \sin y_2)$$

and combining the trigonometric terms using the standard identities for the sum of two angles gives

$$e^{x_1} e^{x_2} \cos(y_1 + y_2) + i e^{x_1} e^{x_2} \sin(y_1 + y_2)$$

Combining the real exponentials and factoring then gives

$$e^{x_1 + x_2} [\cos(y_1 + y_2) + i \sin(y_1 + y_2)] = e^{z_1 + z_2}$$

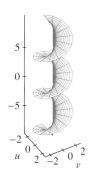

FIGURE 35.13 Portion of the Riemann surface for $f(z) = e^z$

The Reciprocal Rule

The reciprocal rule $e^{-z} = 1/e^z$ follows from (35.6) by setting $z_1 = z$ and $z_2 = -z$ since $e^{z-z} = e^0 = 1 = e^z e^{-z}$.

Conjugation

The exponential of the conjugate is the conjugate of the exponential, as made precise by the identity $e^{\bar{z}} = \overline{e^z}$. In fact, if $z = x + iy$, then $e^{\bar{z}} = \overline{e^z} = e^x \cos y - i e^x \sin y$.

The Equation $e^z = 1$

The solution set for the equation $e^z = 1$ is precisely the set $z = \{2k\pi i\}$, where k is an integer. Clearly, $z = 0$ satisfies the equation. By the following argument, the remaining solutions are found on the imaginary axis in the complex plane. Write

$$e^z = e^{x+iy} = e^x \cos y + i e^x \sin y = 1 + 0i$$

and equate real and imaginary parts. From $e^x \sin y = 0$, we determine $y = k\pi$, where k is an integer. From $e^x \cos y = 1$, we then obtain $\pm e^x = 1$, which forces us to choose $x = 0$ and k even.

If $e^{z_1} = e^{z_2}$, then $e^{z_1 - z_2} = 1$, so $z_1 - z_2 = 2k\pi i$, that is, the complex exponents differ by an integer multiple of $2\pi i$.

EXAMPLE 35.4   If $z = 2 + 3i = \sqrt{13}e^{i \arctan(3/2)}$, then $z^5 = |z|^5 e^{5i \arctan(3/2)} = |z|^5 \operatorname{cis}(5 \arctan \frac{3}{2}) = 169\sqrt{13}$ $[\cos(5 \arctan \frac{3}{2}) + i \sin(5 \arctan \frac{3}{2})] = 122 - 597i$.   ❖

DeMoivre's Laws

Example 35.4 uses DeMoivre's law, namely, the rule $[e^{i\theta}]^n = e^{in\theta}$, where n is an integer. Since $e^{ix} = \operatorname{cis} x$, the exponential function therefore subsumes DeMoivre's law.

EXERCISES 35.4–Part A

Exercises A1–6 use $z_1 = 1 + i$ and $z_2 = 1 + i\sqrt{3}$.

A1. Write z_1 in exponential form.

A2. Verify that $e^{-z_1} = 1/e^{z_1}$.

A3. Verify that $e^{z_1}/e^{z_2} = e^{z_1 - z_2}$.

A4. Verify that $e^{\bar{z}_1} = \overline{e^{z_1}}$.

A5. Using the exponential formalism, obtain both square roots of z_1.

A6. Using the exponential formalism, obtain all three cube roots of z_1.

EXERCISES 35.4–Part B

B1. By computation, show that $e^{(2k+1)\pi i} = -1$ for $k = 0, \pm1, \pm2, \pm3$.

For each $z = x + iy$ given in Exercises B2–11:

(a) Verify computationally that $e^{kz} = (e^z)^k$ for $k = 2, \ldots, 5$.

(b) Verify computationally that $|e^{iz^2}| = e^{-2xy}$.

(c) Compute $e^{z+2k\pi i}$ for $k = 0, \pm1, \pm2, \pm3$.

(d) Obtain $e^{e^{z^2}} = \exp(\exp(z^2))$ and $(e^{e^z})^2 = e^{2e^z} = \exp(\exp(z))^2$. Are these the same?

(e) Obtain $z = |z|e^{i \operatorname{Arg} z}$.

(f) Use the exponential form to obtain both square roots of z, and express them in rectangular form.

(g) Use the exponential form to obtain all three cube roots of z, and express them in rectangular form.

B2. $3 - 5i$ **B3.** $-\frac{7}{2} - \frac{1}{5}i$ **B4.** $8 - 9i$ **B5.** $7 + 5i$

B6. $\frac{1}{4} - \frac{7}{10}i$ **B7.** $-9 + 5i$ **B8.** $7 - 8i$ **B9.** $\frac{8}{7} - \frac{9}{2}i$

B10. $-5 + 5i$ **B11.** $-\frac{7}{5} + \frac{1}{4}i$

For each pair of complex numbers z_1 and z_2 given in Exercises B12–21, obtain, in rectangular form:

(a) $\dfrac{e^{z_1}}{e^{z_2}}$ (b) $\dfrac{z_1 z_2}{e^{z_1} + e^{z_2}}$ (c) $z_1 e^{\bar{z}_2} + z_2 e^{\bar{z}_1}$ (d) $\dfrac{e^{z_1 \bar{z}_2}}{e^{\bar{z}_1 z_2}}$

B12. $4 + i, 5 - 3i$ **B13.** $-4 - 8i, \frac{9}{4} + \frac{4}{5}i$ **B14.** $\frac{6}{5} + \frac{1}{3}i, 8 + 5i$

B15. $9 - 2i, 5 + 4i$ **B16.** $-8 + 3i, \frac{4}{3} - \frac{5}{7}i$ **B17.** $-\frac{3}{7} - \frac{5}{4}i, 8 - 4i$

B18. $-2 - 7i, 8 + 3i$ **B19.** $\frac{1}{11} - \frac{3}{4}i, 3 - 7i$

B20. $9 + 9i, 2 + 9i$ **B21.** $2 - 5i, \frac{6}{7} - \frac{4}{5}i$

In Exercises B22–31:

(a) Write each of the given functions in the form $u + iv$, thereby obtaining $u(x, y)$ and $v(x, y)$, the real and imaginary parts, respectively.

(b) Obtain in the w-plane, a graph of the images of grid lines in the z-plane.

(c) Obtain in the z-plane, a graph of the pre-images of grid lines in the w-plane. (*Hint:* Obtain contour plots of $u = $ constant and $v = $ constant.)

B22. $f(z) = ze^z$ **B23.** $f(z) = z^2 e^z$ **B24.** $f(z) = z^3 e^z$

B25. $f(z) = \dfrac{e^z}{z}$ **B26.** $f(z) = \dfrac{e^z}{z^2}$ **B27.** $f(z) = \dfrac{e^z}{z^3}$

B28. $f(z) = ze^{z^2}$ **B29.** $f(z) = z^2 e^{z^2}$

B30. $f(z) = e^{e^z}$ **B31.** $f(z) = \dfrac{e^{z^2}}{z^2}$

35.5 The Complex Logarithm

Complex Logarithm as Inverse to $w = e^z$

We will define the complex logarithm $\ln z$ as the function inverse to $f(z) = e^z$. However, e^z, the exponential function, is "infinitely many"-to-one. Hence, the complex logarithm will be one-to-"infinitely many" and will require careful definition of branches to yield a "function."

It is usual to consider two versions of the complex logarithm, $\log z$ and $\operatorname{Log} z$, as

$$\log z = \ln |z| + i\operatorname{Arg} z + 2k\pi i \qquad \operatorname{Log} z = \ln |z| + i\operatorname{Arg} z$$

where k is an integer, $\ln x$ is the real natural logarithm from elementary calculus, and $\operatorname{Arg} z$, lying in the interval $(-\pi, \pi]$, denotes the principal value of the argument of z. The second formula defines $\operatorname{Log} z$ as the *principal value of the logarithm* and is the single-valued branch. The first formula defines $\log z$ as a multivalued logarithm. Specification of a particular value of the integer k then yields a single branch of this multivalued function.

EXAMPLE 35.5 The principal logarithm maps the points $w_k, k = 1, 2, 3$, in Table 35.1 back to z_k in the z-plane. The arguments of $z_k, k = 1, 2, 3$, all lie in the *fundamental strip* $-\pi < \arg z \leq \pi$. ❖

w	$z = \operatorname{Log} w$	$= \ln |w| + i\operatorname{Arg} w$
$w_1 = i$	$z_1 = \operatorname{Log} i$	$= \frac{\pi}{2}i$
$w_2 = 5 + i$	$z_2 = \operatorname{Log}(5 + i)$	$= \ln \sqrt{26} + i\arctan \frac{1}{5}$
$w_3 = 2 - 13i$	$z_3 = \operatorname{Log}(2 - 13i)$	$= \ln \sqrt{173} - i\arctan \frac{13}{2}$

TABLE 35.1 Principal value of the logarithm for the three points in Example 35.5

EXAMPLE 35.6 The points $z_4 = 1, z_5 = 1 + 2\pi i, z_6 = 1 + 4\pi i, z_7 = 1 + 6\pi i$, lie on a vertical line segment in the z-plane, each in a successive strip of width 2π. The function $w = e^z$ maps each of these points to e. Consequently, the principal logarithm will return each image back to 1. To obtain the preimages from which each copy of e came, the multivalued logarithm must be used. Thus, using $\log e = \ln |e| + i\operatorname{Arg} e + 2k\pi i = 1 + 2k\pi i, k = 1, \ldots, 3$, we get

$$z_5 = \ln e + i\operatorname{Arg} e + 2\pi i \qquad z_6 = \ln e + i\operatorname{Arg} e + 4\pi i \qquad z_7 = \ln e + i\operatorname{Arg} e + 6\pi i$$

❖

Laws of Logarithms

With a real and x and y both positive, Table 35.2 lists the "usual" rules for working with the real logarithm. If a is not real, or if x or y fail to be positive, and the logarithm is Log,

$a \ln x = \ln x^a$	Multipliers become exponents
$\ln xy = \ln x + \ln y$	log of a product is the sum of the logs
$\ln \frac{x}{y} = \ln x - \ln y$	log of a quotient is the difference of the logs
$\ln e^x = e^{\ln x} = x$	e^x and $\ln x$ are inverse functions

TABLE 35.2 Properties of the real logarithm, where a is real, and x, y, are positive

the principal complex logarithm, then four of the five "rules" in the table fail to be true, in general. (We have counted the two expressions for inverse functions separately, since $e^{\text{Log} z} = z$ is always true but $\text{Log}\, e^z \neq z$ in general.)

The following examples verify the failure of the familiar "laws" of logs learned for real logarithms in calculus and precalculus.

EXAMPLE 35.7   For any z, we always have $e^{\text{Log} z} = z$. However, if $z_1 = 1 + 10i$, then $|e^{z_1}| = e$ but $\text{Arg}\, e^{z_1} = 10 - 4\pi$ because the 10 in e^{10i} means 10 radians, an argument that must be expressed as an angle in the interval $(-\pi, \pi]$. Hence,

$$\text{Log}\, e^{z_1} = \ln e + i(10 - 4\pi) = 1 + i(10 - 4\pi)$$

so $\text{Log}\, e^{1+10i} \neq 1 + 10i$. In general, therefore, $\text{Log}\, e^z \neq z$.   ❖

This result should not be so surprising since its analog exists even for the real trigonometric functions of calculus and precalculus. For example, the functions $\tan x$ and $\arctan x$ are inverse to each other. But because $\tan x$ is "infinitely many"-to-one, $\arctan x$ is one-to-"infinitely many," thereby requiring the same definition of a principal branch as does the complex logarithm. Consequently, while it is true that $\tan(\arctan x) = x$, it is also the case that $\arctan(\tan x) \neq x$. For example, $\arctan(\tan 10) = 10 - 3\pi$, not 10.

Thus, for any "infinitely many"-to-one function $f(x)$, the inverse $g(x)$ will always require defining a principal branch, and this same phenomenon, namely, $f(g(x)) = x$, but $g(f(x)) \neq x$, will occur.

EXAMPLE 35.8   If the "familiar" rule $\text{Log}\, e^z = z$ fails for the principal complex logarithm, then there is no reason to expect the "rule" $\text{Log}\, z^w = w\, \text{Log}\, z$ to hold either. For example, we have $\text{Log}\, i^4 = \text{Log}\, 1 = 0$, but for the alternative we have $4\,\text{Log}\, i = 4(\ln 1 + \frac{\pi}{2}i) = 2\pi i$. Thus, even in the simplest of cases, for the principal complex logarithm, "multipliers become exponents" is not a valid operational rule.   ❖

EXAMPLE 35.9   If $z_1 = -5 + 2i$ and $z_2 = -6 + i$ so that $z_1 z_2 = 28 - 17i$, then $\text{Log}\, z_1 = 1.68 + 2.76i$, $\text{Log}\, z_2 = 1.81 + 2.98i$, and $\text{Log}(z_1 z_2) = 3.49 - 0.55i$. "The Log of the product equals the sum of the Logs" fails for these numbers since

$$\text{Log}\, z_1 + \text{Log}\, z_2 = 3.49 + 5.74i \neq 3.5 - 0.55i = \text{Log}(z_1 z_2)$$

and, hence, is not true in general for the principal complex logarithm.   ❖

EXAMPLE 35.10   To demonstrate the failure of the rule "the Log of a quotient is the difference of the Logs," consider the complex number $z_1 = -1 + i$ and its complex conjugate $z_2 = -1 - i$. Then, we have the principal logarithms

$$\text{Log}\, z_1 = \frac{1}{2}\ln 2 + \frac{3}{4}\pi i \qquad \text{Log}\, z_2 = \frac{1}{2}\ln 2 - \frac{3}{4}\pi i \qquad \text{Log}\,\frac{z_1}{z_2} = \text{Log}(-i) = -\frac{\pi}{2}i$$

for which we see $\text{Log}\, z_1 - \text{Log}\, z_2 = \frac{3}{2}\pi i \neq -\frac{\pi}{2}i = \text{Log}\,\frac{z_1}{z_2}$. Thus, for the principal complex logarithm, it is not in general true that "the Log of a quotient is the difference of the Logs."   ❖

Salvaging the Laws of Logarithms

The following are valid for the complex logarithm, provided they are interpreted to hold for *some* branch.

$$\log e^z = z \qquad\qquad \log z^w = w \log z$$

$$\log(z_1 z_2) = \log z_1 + \log z_2 \qquad \log\frac{z_1}{z_2} = \log z_1 - \log z_2$$

When working with the principal complex logarithm, the following are valid.

$$\text{Log}\, e^z = z + (2\pi i)n_1 \tag{35.7a}$$

$$\text{Log}\, z^w = w\,\text{Log}\, z + (2\pi i)n_2 \tag{35.7b}$$

$$\text{Log}(z_1 z_2) = \text{Log}\, z_1 + \text{Log}\, z_2 + (2\pi i)n_3 \tag{35.7c}$$

$$\text{Log}\, \frac{z_1}{z_2} = \text{Log}\, z_1 - \text{Log}\, z_2 + (2\pi i)n_4 \tag{35.7d}$$

In each case there is an integer n_k for which the statement is true. More precisely, $n_1 = n_1(z)$, that is, n_1 depends on the complex number z; $n_2 = n_2(z, w)$, that is, n_2 depends on both w and z; and similarly, both n_3 and n_4 will depend on z_1 and z_2. Clearly, the n_k must be chosen so that the complex numbers on the right sides have arguments in the interval $(-\pi, \pi]$, just like the principal logarithms on the left sides. Explicit formulas for n_k, $k = 1, \ldots, 4$, are explored in the exercises.

EXAMPLE 35.11 In Example 35.7 we saw that for $z_1 = 1 + 10i$, the law $\text{Log}\, e^{z_1} = z_1$ failed because the left side evaluates to $1 + i(10 - 4\pi)$. Consequently, we must have $n_1 = -2$ in order for (35.7a) to be valid in this case. ❖

EXAMPLE 35.12 In Example 35.8 we saw that for $z = i$ and $w = 4$, the law $\text{Log}\, z^w = w\,\text{Log}\, z$ failed because the left side evaluates to 0 but the right side evaluates to $2\pi i$. Hence, $n_2 = -1$ makes (35.7b) valid in this case. ❖

EXAMPLE 35.13 In Example 35.9 we saw that for $z_1 = -5 + 2i$ and $z_2 = -6 + i$, the rule "the Log of the product is the sum of the Logs" failed because the Log of the product is $3.49 - 0.55i$ but the sum of the Logs is $3.49 + 5.74i$. Choosing $n_3 = -1$ makes (35.7c) valid in this case. ❖

EXAMPLE 35.14 In Example 35.10 we saw that for $z_1 = -1 + i$ and its conjugate $z_2 = -1 - i$, the rule "the Log of the quotient is the difference of the Logs" failed because the Log of the quotient is $-\frac{\pi}{2}i$ but the difference of the Logs is $\frac{3}{2}\pi i$. Hence, taking $n_4 = -1$ makes (35.7d) valid in this case. ❖

EXERCISES 35.5–Part A

Exercises A1–6 use $z_1 = 1 + i$, $z_2 = -2 + i$, and $z_3 = -10 + i$.

A1. Obtain $\text{Log}\, z_1$, $\text{Log}\, z_2$, $\text{Log}(z_1 z_2)$, $\text{Log}\, \frac{z_1}{z_3}$.

A2. Obtain $\log z_1$, $\log z_2$, $\log(z_1 z_2)$, $\log \frac{z_1}{z_3}$.

A3. Obtain $\text{Log}\, z_1 + \text{Log}\, z_2$ and compare to $\text{Log}(z_1 z_2)$.

A4. Determine the branch (that is, determine the integer n_3 in (35.7c) for which $\text{Log}\, z_1 + \text{Log}\, z_2 = \text{Log}(z_1 z_2)$.

A5. Obtain $\text{Log}\, z_1 - \text{Log}\, z_2$ and compare to $\text{Log}\, \frac{z_1}{z_3}$.

A6. Determine the branch (i.e., determine the integer k) for which $\log z_1 - \log z_2 = \log \frac{z_1}{z_3}$.

EXERCISES 35.5–Part B

B1. Compare $\text{Log}\, e^z$ to z for $z = -2 + 3i$ and $z = -2 + 4i$.

Define the function $N_1(z) = [\frac{1}{2} - \frac{\mathcal{I}(z)}{2\pi}]$, where $[\,\cdot\,]$ is the greatest integer function. Then, for z given in each of Exercises B2–11:

  (a) Compute $\text{Log}\, e^z$ and compare to z, determining empirically the value of the integer n_1 for which the law $\text{Log}\, e^z = z + (2\pi i)n_1$ is valid.

  (b) Compute $N_1(z)$ and compare to n_1.

  (c) Compute $z_k = z + 2k\pi i$, $k = \pm1, \pm2, \pm3$, and show $e^{z_k} = e^z = w$. Then, obtain each z_k from $\log w$, the multivalued logarithm.

B2. $-2 + 8i$ **B3.** $3 - 6i$ **B4.** $-4 - 11i$ **B5.** $1 + 10i$

B6. $10 - i$ **B7.** $-9 + 11i$ **B8.** $-10 + 9i$ **B9.** $5 - 8i$

B10. $8 - 2i$ **B11.** $11 - 12i$

Define the function $N_2(z, w) = [\frac{1}{2} - \frac{\mathcal{I}(w)\ln|z| + \mathcal{R}(w)\operatorname{Arg} z}{2\pi}]$, where again, $[\cdot]$ is the greatest integer function. Then, for z and w given in each of Exercises B12–21:

    (a) Compute $\operatorname{Log} z^w$ and compare to $w \operatorname{Log} z$, determining empirically the value of the integer n_2 for which the law $\operatorname{Log} z^w = w \operatorname{Log} z + (2\pi i)n_2$ is valid.

    (b) Compute $N_2(z, w)$ and compare to n_2.

B12. $z = -12 - 9i, w = -5 + 4i$ **B13.** $z = 6 + 8i, w = -2 - 7i$

B14. $z = 11 + 10i, w = 4 - 3i$ **B15.** $z = -2 - 6i, w = -9 + 8i$

B16. $z = 2 - 12i, w = -7 - 3i$ **B17.** $z = -6 + 10i, w = -9 - 8i$

B18. $z = 4 + 11i, w = 9 + 9i$ **B19.** $z = -6 - 4i, w = 7 - 4i$

B20. $z = -11 - 10i, w = 2 + 3i$ **B21.** $z = -8 - 4i, w = 2 - 4i$

Define the function $N_3(z_1, z_2) = \begin{cases} -1 & \pi < \operatorname{Arg} z_1 + \operatorname{Arg} z_2 \le 2\pi \\ 0 & -\pi < \operatorname{Arg} z_1 + \operatorname{Arg} z_2 \le \pi \\ 1 & -2\pi < \operatorname{Arg} z_1 + \operatorname{Arg} z_2 \le -\pi \end{cases}$

Then for the complex numbers z_1 and z_2 given in each of Exercises B22–31:

    (a) Compute $\operatorname{Log}(z_1 z_2)$ and compare to the sum $\operatorname{Log} z_1 + \operatorname{Log} z_2$, determining empirically the value of the integer n_3 for which the law $\operatorname{Log}(z_1 z_2) = \operatorname{Log} z_1 + \operatorname{Log} z_2 + (2\pi i)n_3$ is valid.

    (b) Compute $N_3(z_1, z_2)$ and compare to n_3.

B22. $z_1 = 2 - 5i, z_2 = -5 - 4i$ **B23.** $z_1 = -9 + 12i, z_2 = 2 + 5i$

B24. $z_1 = 3 - i, z_2 = 6 - 7i$ **B25.** $z_1 = -9 + 8i, z_2 = -3 + 7i$

B26. $z_1 = -12 - 9i, z_2 = -7 - 7i$ **B27.** $z_1 = 7 + 3i, z_2 = -1 - 8i$

B28. $z_1 = -3 + 2i, z_2 = 8 + 6i$ **B29.** $z_1 = -6 - 9i, z_2 = -2 - 4i$

B30. $z_1 = -6 - 4i, z_2 = -4 - 8i$ **B31.** $z_1 = -11 + 9i, z_2 = -3 - i$

Define the function $N_4(z_1, z_2) = \begin{cases} -1 & \pi < \operatorname{Arg} z_1 - \operatorname{Arg} z_2 \le 2\pi \\ 0 & -\pi < \operatorname{Arg} z_1 - \operatorname{Arg} z_2 \le \pi \\ 1 & -2\pi < \operatorname{Arg} z_1 - \operatorname{Arg} z_2 \le -\pi \end{cases}$.

Then for the complex numbers z_1 and z_2 given in each of Exercises B32–41:

    (a) Compute $\operatorname{Log} \frac{z_1}{z_2}$ and compare to the difference $\operatorname{Log} z_1 - \operatorname{Log} z_2$, determining empirically the value of the integer n_4 for which the law $\operatorname{Log} \frac{z_1}{z_2} = \operatorname{Log} z_1 - \operatorname{Log} z_2 + (2\pi i)n_4$ is valid.

    (b) Compute $N_4(z_1, z_2)$ and compare to n_4.

B32. $z_1 = 12 - 10i, z_2 = -9 + 5i$ **B33.** $z_1 = -4 + 3i, z_2 = 2 - 2i$

B34. $z_1 = 5 + 11i, z_2 = 4 + 6i$ **B35.** $z_1 = 1 + 11i, z_2 = -4 - 9i$

B36. $z_1 = -7 - 7i, z_2 = -1 + 6i$ **B37.** $z_1 = 5 + 9i, z_2 = -7 - 4i$

B38. $z_1 = 12 + 5i, z_2 = -4 - i$ **B39.** $z_1 = -11 + 5i, z_2 = -7 + i$

B40. $z_1 = -8 - 3i, z_2 = -2 + 9i$ **B41.** $z_1 = -1 - 3i, z_2 = -8 + i$

For each of Exercises B42–51, form the complex number $z = \rho e^{i\theta}$ from the given pair of real numbers (ρ, θ). Show that $\operatorname{Arg} z = \theta + 2\pi[\frac{1}{2} - \frac{\theta}{2\pi}]$, where $[\cdot]$ is the greatest integer function.

B42. $(11, -12)$ **B43.** $(8, -9)$ **B44.** $(10, 4)$ **B45.** $(8, -3)$

B46. $(11, -5)$ **B47.** $(6, -2)$ **B48.** $(5, 7)$ **B49.** $(10, -8)$

B50. $(9, -7)$ **B51.** $(11, 4)$

35.6 Complex Exponents

> **DEFINITION 35.1**
>
> Complex exponentiation is defined in terms of the logarithm by the equations
> $$z^c = e^{c \log z} \quad \text{or} \quad z^c = e^{c \operatorname{Log} z}$$

The first is the general multivalued exponentiation, while the second is the principal value for exponentiation. In a purely formal sense, if the logarithmic rule for moving multipliers to exponents were valid, the definition of complex exponentiation would reduce to the tautology $z^c = e^{c \log z} = e^{\log z^c} = z^c$.

The following examples explore the range of possibilities for complex exponentiation.

EXAMPLE 35.15   To compute $(1 + i)^5$ by the definition $z^c = e^{c \log z} = e^{c[\ln|z| + i\operatorname{Arg} z + 2k\pi i]}$, write

$$e^{5[\ln\sqrt{2} + \pi i/4 + 2k\pi i]} = e^{\ln 2^{5/2}} e^{5\pi i/4} e^{10k\pi i} = 2^{5/2} \operatorname{cis}\left(\frac{5}{4}\pi\right)$$

$$= 4\sqrt{2}\left(-\frac{1}{\sqrt{2}} - \frac{i}{\sqrt{2}}\right) = -4 - 4i$$

This is nothing more than evaluating the polynomial z^5 at $z = 1 + i$. We should not expect multiple values, and we should not expect polynomial arithmetic to yield a different result.

❖

EXAMPLE 35.16   To compute $(1 + i)^{1/5}$ by the definition $z^c = e^{c \log z} = e^{c[\ln|z| + i \, \text{Arg} \, z + 2k\pi i]}$, write

$$e^{1/5[\ln \sqrt{2} + (\pi i/4)i + 2k\pi i]} = e^{\ln 2^{1/10}} e^{\pi i/20 + 2k\pi i/5}$$

$$= 2^{1/10}\left[\cos\left(\frac{\pi}{20} + \frac{2}{5}k\pi\right) + i \sin\left(\frac{\pi}{20} + \frac{2}{5}k\pi\right)\right] \tag{35.8}$$

As k varies from 0 to 4, the five fifth-roots of $1 + i$ that emerge from (35.8) are listed in Table 35.3.

❖

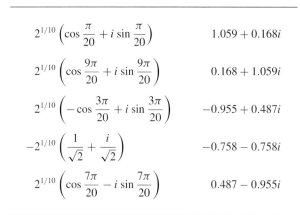

$2^{1/10}\left(\cos\dfrac{\pi}{20} + i\sin\dfrac{\pi}{20}\right)$	$1.059 + 0.168i$
$2^{1/10}\left(\cos\dfrac{9\pi}{20} + i\sin\dfrac{9\pi}{20}\right)$	$0.168 + 1.059i$
$2^{1/10}\left(-\cos\dfrac{3\pi}{20} + i\sin\dfrac{3\pi}{20}\right)$	$-0.955 + 0.487i$
$-2^{1/10}\left(\dfrac{1}{\sqrt{2}} + \dfrac{i}{\sqrt{2}}\right)$	$-0.758 - 0.758i$
$2^{1/10}\left(\cos\dfrac{7\pi}{20} - i\sin\dfrac{7\pi}{20}\right)$	$0.487 - 0.955i$

TABLE 35.3   The five fifth-roots of $1 + i$

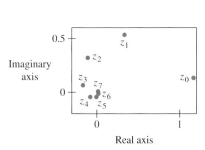

FIGURE 35.14   Argand diagram of the first 8 values of $3^{1/7 + i/10}$

EXAMPLE 35.17   If $z = 3$ and $c = \frac{1}{7} + \frac{i}{10}$, then computing $z^c = e^{c \log z} = e^{c[\ln|z| + i \, \text{Arg} \, z + 2k\pi i]}$ yields infinitely many values z_k. Indeed, since $\text{Arg} \, 3 = 0$, we have

$$3^c = e^{(1/7 + i/10)(\ln 3 + 2k\pi i)} = e^{(\ln 3)/7 - k\pi/5} \text{cis}\left(\frac{\ln 3}{10} + \frac{2k\pi}{7}\right)$$

In floating-point form, the first several values of 3^c are $1.163 + 0.128i$, $0.333 + 0.528i$, $-0.109 + 0.315i$, $-0.168 + 0.059i$, and $-0.080 - 0.050i$. Figure 35.14 is a plot of the values corresponding to $k = 0, 1, \ldots, 7$ and shows that as k increases, 3^c tends to spiral in toward the origin because 3^c contains the factor $e^{-k\pi/5}$, which tends to zero with increasing k.

❖

EXAMPLE 35.18   Surprisingly, i^i is real, as we see from

$$i^i = e^{i(\ln 1 + i\pi/2 + 2k\pi i)} = e^{-\pi/2} e^{-2k\pi} \qquad k = 0, \pm 1, \ldots$$

❖

Laws of Exponents

Because complex exponentiation is defined in terms of the complex logarithm, the rules for complex exponents and for powers of complex numbers seem to be the most difficult. Many of the familiar properties of real exponents of positive real quantities fail in the complex case. The leftmost column of Table 35.4 lists five valid rules for exponents. The first and fifth have already been seen but are listed here for the sake of completeness. The middle column of Table 35.4 lists five rules that, in general, are not valid. The rightmost column of Table 35.4 lists valid counterparts for three of the invalid operations in the center column.

Valid	Invalid	Valid
$e^{\operatorname{Log} z} = z$	$(zw)^a \neq z^a w^a$	$(zw)^a = z^a w^a e^{2\pi i a n_3}$
$z^c z^d = z^{c+d}$	$\left(\dfrac{z}{w}\right)^a \neq \dfrac{z^a}{w^a}$	$\left(\dfrac{z}{w}\right)^a = \dfrac{z^a}{w^a} e^{2\pi i a n_4}$
$\dfrac{z^c}{z^d} = z^{c-d}$	$(z^a)^b \neq z^{ab}$	$(z^a)^b = z^{ab} e^{2\pi i b n_2}$
$(z^a)^b \neq (z^b)^a$ Integer $n > 0$	$(z^{1/n})^n = z$	
$z^{-a} = \dfrac{1}{z^a},\ z \neq 0$	$(z^n)^{1/n} \neq z$ Integer $n > 0$	

TABLE 35.4   Valid and invalid manipulations with complex exponentiation

From [69], the integer $n_2(z, a)$ is given by

$$n_2(z, a) = \left[\frac{1}{2} - \frac{\mathcal{I}(a) \ln |z| + \mathcal{R}(a) \operatorname{Arg} z}{2\pi}\right] \tag{35.9}$$

where $[\,\cdot\,]$ is the greatest integer function, and (35.9) is $N_2(z, a)$, defined for Exercises B12–21, in Section 35.5.

The integer n_3 is chosen so that $\operatorname{Arg} z + \operatorname{Arg} w + 2\pi n_3$ lies in the interval $(-\pi, \pi]$. The integer n_4 is chosen so that $\operatorname{Arg} z - \operatorname{Arg} w + 2\pi n_4$ lies in the interval $(-\pi, \pi]$. Hence, $n_3(z, w) = N_3(z, w)$ and $n_4(z, w) = N_4(z, w)$, where N_3 and N_4 were defined for Exercises B22–31 and B32–41, respectively, in Section 35.5. Thus, we have

$$n_3(z, w) = \begin{cases} -1 & \pi < \operatorname{Arg} z + \operatorname{Arg} w \leq 2\pi \\ 0 & -\pi < \operatorname{Arg} z + \operatorname{Arg} w \leq \pi \\ 1 & -2\pi < \operatorname{Arg} z + \operatorname{Arg} w \leq -\pi \end{cases} \tag{35.10a}$$

$$n_4(z, w) = \begin{cases} -1 & \pi < \operatorname{Arg} z - \operatorname{Arg} w \leq 2\pi \\ 0 & -\pi < \operatorname{Arg} z - \operatorname{Arg} w \leq \pi \\ 1 & -2\pi < \operatorname{Arg} z - \operatorname{Arg} w \leq -\pi \end{cases} \tag{35.10b}$$

EXAMPLE 35.19   If $z = -10 + i$ and $w = -11 + i$, then

$$(zw)^i = -0.00341 - 1.210i \neq z^i w^i = -0.637 \times 10^{-5} - 0.00226i$$

The incorrect number on the right can be rendered correct by multiplying by $e^{2\pi i^2 n_3}$, where n_3 is determined by examining

$$\operatorname{Arg} z + \operatorname{Arg} w + 2\pi n_3 = 2\pi - \arctan \tfrac{1}{10} - \arctan \tfrac{1}{11} + 2\pi n_3 = 6.093 + 6.28 n_3$$

from which we determine $n_3 = -1$. Hence, the appropriate multiplicative factor is $e^{2\pi}$ and we find $z^i w^i e^{2\pi} = -0.00341 - 1.210i = (zw)^i$.   ❖

EXAMPLE 35.20   If $z = -10 + i$ and $w = \bar{z} = -10 - i$, then $\left(\tfrac{z}{w}\right)^i = e^{\arctan(20/99)} = 1.221$, and $z^i/w^i = e^{2\arctan(1/10) - 2\pi} = 0.00228$. The first number is correct. The incorrect second number is rendered correct if multiplied by the factor $e^{2\pi i^2 n_4}$, where n_4 is chosen so that

$$\operatorname{Arg} z - \operatorname{Arg} w + 2\pi n_4 = 2\pi - 2\arctan \tfrac{1}{10} + 2\pi n_4 = 6.08 + 6.28 n_4$$

lies in the interval $(-\pi, \pi]$. Hence, $n_4 = -1$ and the appropriate multiplicative factor is $e^{2\pi}$, so $z^i/w^i = e^{2\arctan(1/10) - 2\pi}e^{2\pi} = 1.221$. ❖

EXAMPLE 35.21   If $z = -10 + i$, $a = 3 + 2i$, and $b = 2 - i$, then

$$s_1 = \left(z^a\right)^b = 1.049 + 17.297i$$
$$s_2 = \left(z^b\right)^a = 10^{10}(8.63 + 142i)$$
$$s_3 = z^{ab} = 10^5(3.0 + 49.6i)$$

Each of these three numbers is different, whereas for real exponents and positive z they would have been the same.

To test the equality $(z^a)^b = z^{ab}e^{2\pi i b n_2}$, determine n_2 by (35.9), obtaining $[-1.69] = -2$, so that $s_3e^{-4\pi bi} = s_1$. To test the equality $\left(z^b\right)^a = z^{ab}e^{2\pi i a n_2}$, determine n_2 by (35.9), obtaining $[-0.101] = -1$, so that $s_3e^{-2\pi ai} = s_2$. ❖

EXAMPLE 35.22   As a more practical example of the interplay between roots and powers in the complex plane, let $z = -1$, $a = 2$, $b = \frac{1}{2}$ and observe that $z^{ab} = (-1)^{[2(1/2)]} = -1$. Note further that $(z^a)^b \neq z^{ab} = \left(z^b\right)^a$ because

$$\left[(-1)^2\right]^{1/2} = \sqrt{1} = 1 \neq \left[(-1)^{1/2}\right]^2 = i^2 = -1$$ ❖

Consequently, we infer that if n is a positive integer, then $\left(z^{1/n}\right)^n = z \neq (z^n)^{1/n}$ and articulate this with the phrase *roots before powers*. Thus, if the root is taken before the power, then the exponents may be multiplied. If the power precedes the root, then the exponents cannot be multiplied.

EXAMPLE 35.23   As a final example of the interplay between roots and powers, let $n = 2$ and take $z = -1 + i$. Then, direct computation gives

$$\left[z^{1/2}\right]^2 = \left[\frac{1}{2}\left(\sqrt{2\sqrt{2} - 2} + i\sqrt{2\sqrt{2} + 2}\right)\right]^2 = -1 + i = z$$
$$\left[z^2\right]^{1/2} = \sqrt{-2i} = 1 - i = -z \neq z$$

which is consistent with the adage *roots before powers*. ❖

EXERCISES 35.6–Part A

A1. Obtain all values of 2^i.   **A2.** Obtain all values of $(2i)^i$.   **A5.** Evaluate $\left((-1 - i)^3\right)^{1/3}$.

A3. Obtain all values of $\left(\frac{2}{i}\right)^i$.   **A4.** Obtain all values of $\left(2^i\right)^i$.

EXERCISES 35.6–Part B

B1. Let $a = -1 + i$, $z = -1 - i$, and $w = 2 - 5i$. Compare the principal values of $(zw)^a$ and $z^a w^a$.

For each complex number z given in Exercises B2–11:

   (a) Verify that $e^{\text{Log }z} = z$, where $\text{Log }z = \ln|z| + i\,\text{Arg }z$.

(b) Verify that $\left(z^{1/n}\right)^n = z$ for $n = 2, \ldots, 5$. Also, show that for the majority of these examples, $(z^n)^{1/n} \neq z$ for the principal value of the nth-root function.

B2. $3 - 5i$ **B3.** $-\frac{7}{2} - \frac{1}{5}i$ **B4.** $8 - 9i$ **B5.** $7 + 5i$

B6. $\frac{1}{4} - \frac{7}{10}i$ **B7.** $-9 + 5i$ **B8.** $7 - 8i$ **B9.** $\frac{8}{7} - \frac{9}{2}i$

B10. $-5 + 5i$ **B11.** $-\frac{7}{5} + \frac{1}{4}i$

For the complex numbers z, c and d given in Exercises B12–21:

    (a) Verify that $z^c z^d = z^{c+d}$.   **(b)** Verify that $z^c / z^d = z^{c-d}$.

B12. $z = -4 - 8i, c = 9 - 7i, d = 4 - 6i$

B13. $z = -2 + 5i, c = -8 + 9i, d = \frac{3}{7} - \frac{3}{8}i$

B14. $z = 6 + 9i, c = 4 - 4i, d = 8 + 5i$

B15. $z = -8 + 8i, c = 2 - 5i, d = \frac{3}{4} - \frac{2}{3}i$

B16. $z = 1 + 5i, c = -3 + 5i, d = -5 + 4i$

B17. $z = 8 - 4i, c = -2 - 7i, d = 8 + 3i$

B18. $z = -5 + 7i, c = 7 - 8i, d = \frac{8}{5} - \frac{3}{2}i$

B19. $z = -3 + 3i, c = 9 - 6i, d = 1 - 2i$

B20. $z = 9 - 5i, c = -7 + 3i, d = 7 - 8i$

B21. $z = -4 - 7i, c = -6 + 6i, d = 3 + 2i$

For the complex numbers z, w, and a given in each of Exercises B22–31:

    (a) Compare $(zw)^a$ to $z^a w^a$.

    (b) Determine empirically the integer n_3 for which the law $(zw)^a = z^a w^a e^{2\pi i a n_3}$ is valid.

    (c) Using (35.10a), compute $n_3(z, w)$ and compare to n_3 determined in part (b).

    (d) Compare $\left(\frac{z}{w}\right)^a$ to z^a / w^a.

    (e) Determine empirically the integer n_4 for which the law $\left(\frac{z}{w}\right)^a = z^a e^{2\pi i a n_4} / w^a$ is valid.

    (f) Using (35.10b), compute $n_4(z, w)$ and compare to n_4 determined in part (e).

B22. $z = 2 - 7i, w = -7 - 3i, a = 3 - 3i$

B23. $z = -5 - 4i, w = -4 + 9i, a = -5 + 3i$

B24. $z = -7 - 7i, w = 1 + 2i, a = 2 + 5i$

B25. $z = -4 - 5i, w = 8 + 6i, a = 9 + 5i$

B26. $z = 7 - 5i, w = -8 - 6i, a = -2 - 9i$

B27. $z = 7 + 4i, w = -6 - 4i, a = 3 - 6i$

B28. $z = 4 + 2i, w = 2 - 6i, a = -7 - 9i$

B29. $z = -6 + 4i, w = -7 - 4i, a = 2 - 4i$

B30. $z = -5 - 9i, w = 7 - 5i, a = 8 - 7i$

B31. $z = -2 + 3i, w = 4 - 9i, a = 5 - i$

For the complex numbers z, a, and b given in each of Exercises B32–41:

    (a) Compare z^{ab} to both $(z^a)^b$ and $(z^b)^a$.

    (b) Determine empirically the integer n_2 for which $(z^a)^b = z^{ab} e^{2\pi i b n_2}$ holds.

    (c) Using (35.9), determine $n_2(z, a)$ and compare to part (b).

    (d) Determine empirically the integer n_2 for which $(z^b)^a = z^{ab} e^{2\pi i a n_2}$ holds.

    (e) Using (35.9), determine $n_2(z, b)$ and compare to part (d).

B32. $z = -1 - 5i, a = 2 + 6i, b = 1 - 3i$

B33. $z = 8 - 4i, a = \frac{1}{2} - i, b = -5 - 9i$

B34. $z = -5 + 8i, a = -9 + 5i, b = -2 + i$

B35. $z = -4 + 9i, a = 9 - 3i, b = \frac{3}{2} + \frac{5}{6}i$

B36. $z = -3 + 5i, a = 6 + 4i, b = -9 + 3i$

B37. $z = 3 - 2i, a = \frac{4}{3} - \frac{7}{6}i, b = -5 - 5i$

B38. $z = -\frac{8}{7} + i, a = -\frac{3}{4} - \frac{1}{6}i, b = -\frac{7}{8} + \frac{9}{5}i$

B39. $z = 9 - 4i, a = -1 - 5i, b = 2 + \frac{1}{2}i$

B40. $z = -1 - i, a = \frac{1}{2} + \frac{3}{7}i, b = \frac{7}{6} + \frac{5}{8}i$

B41. $z = -3 + i, a = \frac{7}{4} + \frac{4}{5}i, b = -1 - 4i$

35.7 Trigonometric and Hyperbolic Functions

The Trigonometric Functions

In the complex plane, the elementary trigonometric functions are defined in terms of the exponential function. Because the exponential function defines an "infinitely many"-to-one map of the z-plane to the w-plane, the elementary trigonometric functions will do likewise. Any complications in the behavior of these functions can be traced to the behavior of the complex exponential function.

The Function $w = \sin z$

The real and imaginary parts of $\sin z = u + iv$ are found from the identity

$$\sin(x + iy) = \sin x \cos(iy) + \cos x \sin(iy)$$

after using the simplifications

$$\sin(iy) = \frac{1}{2i}\left(e^{i(iy)} - e^{-i(-iy)}\right) = i\sinh y \quad \text{and} \quad \cos(iy) = \frac{1}{2}\left(e^{i(iy)} + e^{-i(-iy)}\right) = \cosh y$$

Consequently, we have $u(x, y) = \sin x \cosh y$ and $v(x, y) = \cos x \sinh y$.

Graphing $w = \sin z$

Figure 35.15 shows a graph of $|\sin z| = \sqrt{\sin^2 x + \sinh^2 y}$, the magnitude of $\sin z$. Along $y = 0$ we have $|\sin z| = |\sin x|$, so the zeros of $\sin z$ are just the real zeros of $\sin x$ along the real axis. Figure 35.16 is a conformal plot for $\sin z$, showing in the w-plane, the images of grid lines in the z-plane. The hyperbolas are the images of the lines $x = $ constant, and the ellipses are the images of the lines $y = $ constant.

Figure 35.17 contains a portion of the Riemann surface for $\sin z$. The Riemann surface is "climbed" as a horizontal line in the z-plane is traversed from left to right. A vertical strip of width 2π in the z-plane maps to a sheet of the Riemann surface in the w-plane. The Riemann surface is drawn as a parametric surface whose height is determined by the variable in which the function $w = f(z)$ is periodic. For $w = e^z$ the height on the Riemann surface was determined by y, whereas for $w = \sin z$ the height is determined by x.

The Function $w = \cos z$

The real and imaginary parts of $\cos z = u + iv$ are found from the identity

$$\cos(x + iy) = \cos x \cos(iy) - \sin x \sin(iy)$$

after using the simplifications

$$\sin(iy) = i\sinh y \quad \text{and} \quad \cos(iy) = \cosh y \qquad \text{(35.11)}$$

Consequently, we have $u(x, y) = \cos x \cosh y$ and $v(x, y) = -\sin x \sinh y$.

Graphing $w = \cos z$

Figure 35.18 shows a graph of $|\cos z| = \sqrt{\cos^2 x + \sinh^2 y}$, the magnitude of $\cos z$. Along $y = 0$ we have $|\cos z| = |\cos x|$, so the zeros of $\cos z$ are just the real zeros of $\cos x$ along the real axis. Figure 35.19 is a conformal plot for $\cos z$, showing in the w-plane, the images of grid lines in the z-plane. The hyperbolas are the images of the lines $x = $ constant, and the ellipses are the images of the lines $y = $ constant. Figure 35.20 contains a portion of the

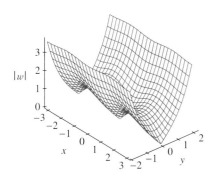

FIGURE 35.15 Magnitude surface for $f(z) = \sin z$

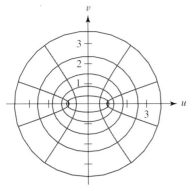

FIGURE 35.16 Conformal plot for $f(z) = \sin z$

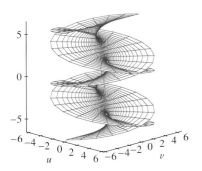

FIGURE 35.17 Portion of the Riemann surface generated by $f(z) = \sin z$

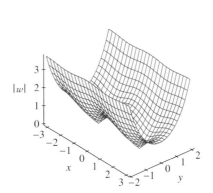

FIGURE 35.18 Magnitude surface for $f(z) = \cos z$

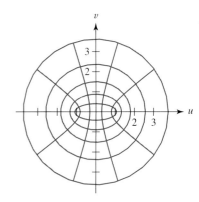

FIGURE 35.19 Conformal plot for $f(z) = \cos z$

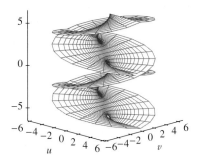

FIGURE 35.20   Portion of the Riemann surface generated by $f(z) = \cos z$

Riemann surface for $\cos z$. The Riemann surface is "climbed" as a horizontal line in the z-plane is traversed from left to right. A vertical strip of width 2π in the z-plane maps to a sheet of the Riemann surface in the w-plane.

The Function $w = \tan z$

If k is an integer and $z \neq \frac{2k+1}{2}\pi$, we can define $\tan z = \frac{\sin z}{\cos z}$. By the "usual" techniques, we can obtain the real and imaginary parts of the function $f(z) = \tan z$, finding

$$u(x, y) = \frac{\sin x \cos x}{\cos^2 x + \sinh^2 y} \quad \text{and} \quad v(x, y) = \frac{\sinh y \cosh y}{\cos^2 x + \sinh^2 y} \qquad (35.12)$$

Graphing $w = \tan z$

Figure 35.21 shows a graph of $|\tan z| = \sqrt{\frac{\cos^2 x - \cosh^2 y}{1 - \cos^2 x - \cosh^2 y}}$, the magnitude of $\tan z$. Along $y = 0$ we have $|\tan z| = |\tan x|$, so the zeros of $\tan z$ are just the real zeros of $\tan x$ along the real axis. Interspersed with these zeros are the singularities of $\tan z$ occurring at the zeros of $\cos z$ on the real axis. Thus, $\tan z$ is periodic across vertical strips of width π. Figure 35.22 is a conformal plot for $\tan z$, showing in the w-plane, the images of grid lines in the z-plane. The lines that look like longitudes on a globe are the images of the lines $x = $ constant, whereas the lines that look like latitudes are the images of the lines $y = $ constant.

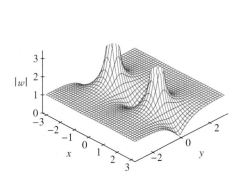

FIGURE 35.21   Magnitude surface for $f(z) = \tan z$

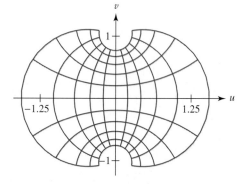

FIGURE 35.22   Conformal plot for $f(z) = \tan z$

Hyperbolic Functions

Like the trig functions, the hyperbolic functions $\sinh z$, $\cosh z$, and $\tanh z$ are defined in terms of the complex exponential function. Hence, they will be "infinitely many"-to-one functions, with inverses that are one-to-"infinitely many."

The Function $w = \sinh z$

To determine the real and imaginary parts of $\sinh z = u + iv$, use the expansion formula

$$\sinh(A + B) = \sinh A \cosh B + \cosh A \sinh B$$

to write

$$\sinh(x + iy) = \sinh x \cos y + i \cosh x \sin y = u(x, y) + iv(x, y)$$

From the definition of $\sinh z$, we also have $\sinh(iz) = \frac{1}{2}(e^{iz} - e^{-iz}) = i(e^{iz} - e^{-iz})/2i = i \sin z$.

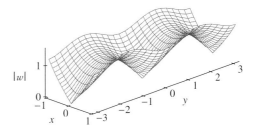

FIGURE 35.23   Magnitude surface for $f(z) = \sinh z$

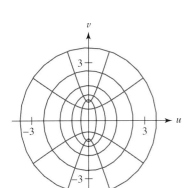

FIGURE 35.24   Conformal plot for $f(z) = \sinh z$

Graphing $w = \sinh z$

Figure 35.23 shows a graph of $|\sinh z| = \sqrt{\sin^2 y + \sinh^2 x}$. Along $x = 0$, we have $|\sinh z| = |\sin y|$, so the zeros of $\sinh z$ are the real zeros of $\sin y$ but located along the imaginary axis. Figure 35.24 is a conformal plot for $\sinh z$, showing in the w-plane, the images of grid lines in the z-plane. The ellipses are the images of the lines $x = $ constant, and the hyperbolas are the images of the lines $y = $ constant. Figure 35.25 contains a portion of the Riemann surface for $\sinh z$. The Riemann surface is "climbed" as a vertical line in the z-plane is traversed from bottom to top. A horizontal strip of width 2π in the z-plane maps to a sheet of the Riemann surface in the w-plane.

The Function $w = \cosh z$

To determine the real and imaginary parts of $\cosh z = u + iv$, use the expansion formula

$$\cosh(A + B) = \cosh A \cosh B + \sinh A \sinh B$$

to write

$$\cosh(x + iy) = \cosh x \cos y + i \sinh x \sin y$$

From the definition of $\cosh z$, we also have $\cosh(iz) = \frac{1}{2}(e^{iz} + e^{-iz}) = \cos z$.

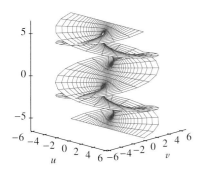

FIGURE 35.25   Portion of the Riemann surface generated by $f(z) = \sinh z$

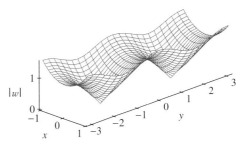

FIGURE 35.26   Magnitude surface for $f(z) = \cosh z$

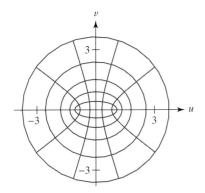

FIGURE 35.27   Conformal plot for $f(z) = \cosh z$

Graphing $w = \cosh z$

Figure 35.26 shows a graph of $|\cosh z| = \sqrt{\cos^2 y + \sinh^2 x}$. Along $x = 0$, we have $|\cosh z| = |\cos y|$, so the zeros of $\cosh z$ are the real zeros of $\cos y$ but located along the imaginary axis. Figure 35.27 is a conformal plot for $\cosh z$, showing in the w-plane, the images of grid lines in the z-plane. The ellipses are the images of the lines $x = $ constant, and the hyperbolas are the images of the lines $y = $ constant.

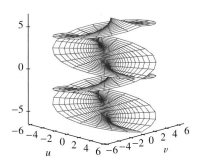

FIGURE 35.28   Portion of the Riemann surface generated by $f(z) = \cosh z$

Figure 35.28 contains a portion of the Riemann surface for $\cosh z$. The Riemann surface is "climbed" as a vertical line in the z-plane is traversed from bottom to top. A horizontal strip of width 2π in the z-plane maps to a sheet of the Riemann surface in the w-plane.

The Function $w = \tanh z$

If k is an integer and $z \neq \frac{2k+1}{2}\pi i$, we can define $\tanh z$ by $\tanh z = \frac{\sinh z}{\cosh z}$. By the "usual" techniques, we can obtain the real and imaginary parts of the function $f(z) = \tanh z$, finding

$$u(x, y) = \frac{\sinh x \cosh x}{\sinh^2 x + \cos^2 y} \quad \text{and} \quad v(x, y) = \frac{\sin y \cos y}{\sinh^2 x + \cos^2 y} \tag{35.13}$$

Graphing $w = \tanh z$

Figure 35.29 shows a graph of $|\tanh z| = \sqrt{\frac{\cosh 2x - \cos 2y}{\cosh 2x + \cos 2y}}$, the magnitude of $\tanh z$. The zeros of $\tanh z$ are along $x = 0$, coincident with the zeros of $\sinh y$. Interspersed with these zeros are the singularities of $\tanh z$, occurring at the zeros of $\cosh y$ on the imaginary axis. Thus, $\tanh z$ is periodic across horizontal strips of width π.

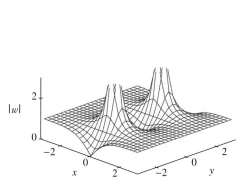

FIGURE 35.29   Magnitude surface for $f(z) = \tanh z$

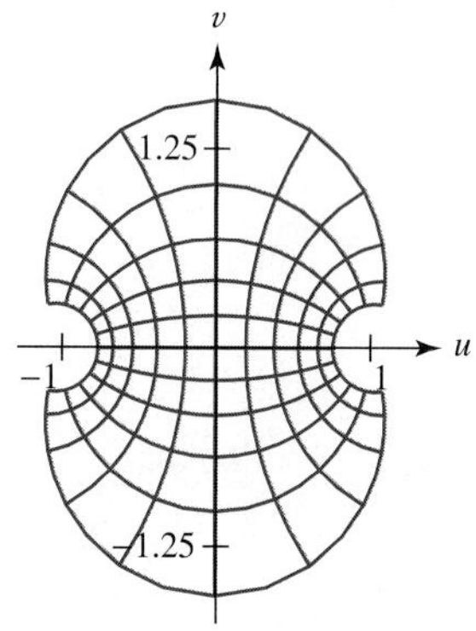

FIGURE 35.30   Conformal plot for $f(z) = \tanh z$

Figure 35.30 is a conformal plot for $\tanh z$, showing in the w-plane, the images of grid lines in the z-plane. The curves that open toward the left and right are the images of the lines $x = $ constant, whereas the curves that arc from left to right are the images of the lines $y = $ constant.

EXERCISES 35.7–Part A

A1. Verify the identities in (35.11).   **A2.** Verify the results in (35.12).   **A4.** $\sin \bar{z} = \overline{\sin z}$ **A5.** $\cos \bar{z} = \overline{\cos z}$

A3. Verify the results in (35.13).   **A6.** $\sinh \bar{z} = \overline{\sinh z}$ **A7.** $\cosh \bar{z} = \overline{\cosh z}$

Write $z = x + iy$ and establish the identities in Exercises A4–7.

EXERCISES 35.7–Part B

B1. Use (35.12) to establish $|\tan z| = \sqrt{\frac{\cos^2 x - \cosh^2 y}{\sin^2 x - \cosh^2 y}}$.

For the magnitudes given in each of Exercises B2–5, determine, on $|z| = 1$, the extreme values and their locations. *Hint:* Use the parametrization $z = \cos\theta + i\sin\theta$.

B2. $|\sin z|$ **B3.** $|\cos z|$ **B4.** $|\sinh z|$ **B5.** $|\cosh z|$

In Exercises B6–17, use the exponential definitions of the relevant functions to verify the given identity.

B6. $\cos^2 z + \sin^2 z = 1$

B7. $\sin(z_1 \pm z_2) = \sin z_1 \cos z_2 \pm \cos z_1 \sin z_2$

B8. $\cos(z_1 \pm z_2) = \cos z_1 \cos z_2 \mp \sin z_1 \sin z_2$

B9. $\sin 2z = 2 \sin z \cos z$

B10. $\cos 2z = \cos^2 z - \sin^2 z = 2\cos^2 z - 1 = 1 - 2\sin^2 z$

B11. $\cosh^2 z - \sinh^2 z = 1$

B12. $\sinh(z_1 \pm z_2) = \sinh z_1 \cosh z_2 \pm \cosh z_1 \sinh z_2$

B13. $\cosh(z_1 \pm z_2) = \cosh z_1 \cosh z_2 \pm \sinh z_1 \sinh z_2$

B14. $\sinh 2z = 2 \sinh z \cosh z$

B15. $\cosh 2z = \cosh^2 z + \sinh^2 z = 2\cosh^2 z = 1 + 2\sinh^2 z$

B16. $\sinh \dfrac{z}{2} = \sqrt{\dfrac{\cosh z - 1}{2}}$ **B17.** $\cosh \dfrac{z}{2} = \sqrt{\dfrac{\cosh z + 1}{2}}$

In Exercises B18–21, obtain a graph of the w-plane image, under the given mapping $w = f(z)$, of the z-plane curve

   (a) $y = x^2$ **(b)** $9x^2 + 4y^2 = 9$

B18. $f(z) = \sin z$ **B19.** $f(z) = \cos z$

B20. $f(z) = \sinh z$ **B21.** $f(z) = \cosh z$

If $\lambda = 2 + 3i$, evaluate in rectangular form, the expressions given in Exercises B22–31.

B22. $\sin(\text{Log }\lambda)$ **B23.** $\text{Log}(\sin\lambda)$ **B24.** $\cos(\sin\lambda^2)$

B25. $\sqrt{\lambda \sin\lambda}$ **B26.** $\cosh^\lambda \frac{1}{2}$ **B27.** $\sinh(\cos\lambda)$ **B28.** $\tanh(\lambda^\lambda)$

B29. $\cosh(\lambda \sin\lambda)$ **B30.** $\tan\sqrt{\lambda^2 - 1}$ **B31.** $\sinh e^{\cos\lambda}$

Inverses of Trigonometric and Hyperbolic Functions

Inverse Trigonometric Functions

Along the real line, the inverse trig functions on the complex plane must reduce to the familiar real-valued inverse trig functions. However, the complex trig functions are defined in terms of the "infinitely many"-to-one exponential function. The inverses are then defined in terms of the complex logarithm. The energy expended in understanding the Riemann surface for the exponential function and in mastering the complex logarithm was the appropriate capital investment for this examination of the inverse trig functions.

As with the complex logarithm, there will be a multivalued inverse trig function and a principal inverse function. A further complication arises from the dependence of these functions on an appropriate branch of the square root function. Although it's only along the real line, the following example is revealing.

EXAMPLE 35.24   Find all real solutions of the equation $\sin x = \frac{1}{2}$.

Past exposure to trigonometry and calculus suggests computing, even with a calculator, $s_1 = \arcsin \frac{1}{2} = \frac{\pi}{6}$, the first-quadrant solution from the principal branch of the arcsine function. Another solution, $s_2 = \pi - \frac{\pi}{6} = \frac{5\pi}{6}$, is in the second quadrant. Figure 35.31 shows where to look for additional solutions. It contains graphs of $\sin x$, the line $y = \frac{1}{2}$, and, as a black segment, that portion of $\sin x$ corresponding to the principal branch of the arcsine function. The first solution, s_1, is labeled a at the intersection of the graphs of $\sin x$ and the line $y = \frac{1}{2}$. The second solution, s_2, is labeled b. Additional solutions are labeled from c to f. Solutions f, a, and c are all generated by s_1 according to the recipe $s_1 + 2k\pi$, where k is an integer. Solutions e, b, and d are all generated by s_2 according to the recipe $s_2 + 2k\pi$.

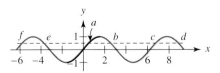

FIGURE 35.31   Solving the equation $\sin x = \frac{1}{2}$ for x

Thus, there are two infinite sets of solutions, one generated by s_1 and one by s_2. Within each set, the members are separated by 2π, corresponding to the periodicity of $\sin x$.

The Function $w = \arcsin z$

The complex function $w = \sin z$ takes a vertical strip of width 2π in the z-plane to the whole w-plane. (Recall the Riemann surface for $\sin z$, Section 35.7.) If $z_1 = \frac{\pi}{6} + i = 0.524 + i$ is mapped by $\sin z$ to the w-plane, we obtain $w_1 = \sin z_1 = \frac{1}{2}(\cosh 1 + i\sqrt{3}\sinh 1)$. In fact, because of the periodicity of $\sin z$, we also have $\sin(z_1 + 2k\pi) = w_1$ for any integer k. Consequently, the arcsine function must map w_1 back to the infinite collection $z_1 + 2k\pi$, $k = 0, \pm 1, \pm 2, \ldots$.

To visualize the inversion, write $\sin z = \sin(x + iy) = u + iv$ and simultaneously solve the equations

$$u(x, y) = \mathcal{R}(w_1) = \frac{\cosh 1}{2} \quad \text{and} \quad v(x, y) = \mathcal{I}(w_1) = \frac{\sqrt{3}\sinh 1}{2}$$

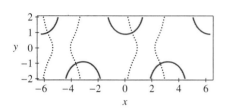

FIGURE 35.32   Solving the equation $\sin z = \frac{1}{2}(\cosh 1 + i\sqrt{3}\sinh 1)$ for z in the complex plane

Figure 35.32 contains graphs of the curves defined implicitly by each of these two equations. The dotted curves are defined by the first equation; the solid by the second. The intersection points are solutions of the equation $\sin z_1 = w_1$; and it appears that, just as in the real case, there are two sets of solutions, one of the form $z_1 + 2k\pi$ and one of the form $z_2 + 2k\pi$, where k is an integer.

The standard definition of the arcsine function is

$$\arcsin z = -i \log \left(iz + \sqrt{1 - z^2} \right) \tag{35.14}$$

as derived shortly. There are two possible values for the square root and an infinite number of values for the multivalued logarithm. The square root is given by

$$\left(1 - z^2\right)^{1/2} = e^{(1/2)\log(1-z^2)} = e^{1/2[\ln|1-z^2| + i\,\mathrm{Arg}(1-z^2) + 2k\pi i]} = \left|1 - z^2\right|^{1/2} e^{(i/2)\,\mathrm{Arg}(1-z^2)} e^{k\pi i}$$

where the first two factors on the right define the principal square root and the third factor is either 1 or -1 depending on whether k is even or odd, respectively. Thus, we can replace the definition of $\arcsin z$ with the two functions

$$\mathrm{asin}_1 z = -i \log \left(iz + \sqrt{1 - z^2} \right) \quad \text{and} \quad \mathrm{asin}_2 z = -i \log \left(iz - \sqrt{1 - z^2} \right) \tag{35.15}$$

Each function generates an infinite set of values when the multivalued logarithm is used. If the principal complex logarithm is used, we can also define the two functions

$$\mathrm{Asin}_1 z = -i \,\mathrm{Log} \left(iz + \sqrt{1 - z^2} \right) \quad \text{and} \quad \mathrm{Asin}_2 z = -i \,\mathrm{Log} \left(iz - \sqrt{1 - z^2} \right)$$

with which we obtain $z_1 = \mathrm{Asin}_1 w_1 = 0.524 + i$ and $z_2 = \mathrm{Asin}_2 w_1 = 2.618 - i$. The first is the principal value, and the second is the analog of $\frac{5\pi}{6}$, the second-quadrant solution corresponding to $\arcsin \frac{1}{2} = \frac{\pi}{6}$ when working on the real line. Corresponding to z_1 there would be an infinite set of solutions of the form $z_1 + 2k\pi$, and corresponding to z_2 there would be another infinite set of solutions of the form $z_2 + 2k\pi$. These infinite sets of solutions arise from the use of the multivalued logarithm in the definition of the arcsine function. Thus, we would have $\mathrm{asin}_1 w_1 = z_1 + 2k\pi$ and $\mathrm{asin}_2 w_1 = z_2 + 2k\pi$, where k is an integer.

Derivation of the arcsine Function

We conclude this discussion of $\arcsin z$ with a derivation of the expression for computing the arcsine function. Start with the equation $w = \sin z = \frac{1}{2i}(e^{iz} - e^{-iz})$ and solve for $z = z(w) = \arcsin w$. To do this, multiply through by $2i$, and set $r = e^{iz}$ so that $e^{-iz} = \frac{1}{r}$. The equation becomes $2iw = r - \frac{1}{r}$, which is actually the quadratic equation $r^2 - 2iwr - 1 = 0$. The solution for r is

$$r = e^{iz} = \frac{1}{2}\left(2iw \pm \sqrt{-4w^2 + 4}\right) = iw \pm \sqrt{1 - w^2}$$

The two radicals represented by $\pm\sqrt{1 - w^2}$ are expressed with the single term $\left(1 - w^2\right)^{1/2}$, where we have already seen how this evaluates to two separate complex numbers. Then, solving for z, we have

$$z = \frac{1}{i}\log\left(i\left(1 - w^2\right)^{1/2}\right) \quad \text{or} \quad z = -i\log\left(iw + \left(1 - w^2\right)^{1/2}\right)$$

Interchanging the variables z and w then gives (35.15), which is what (35.14) really means.

The Function $w = \arccos z$

The inverse cosine function, defined by

$$\arccos z = -i\log\left(z + i\sqrt{1 - z^2}\right) \tag{35.16}$$

can be extracted from the solution of the equation $w = \frac{1}{2}(e^{iz} + e^{-iz})$. (See Exercise A1.)

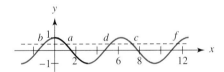

FIGURE 35.33 Solving the equation for $\cos x = \frac{1}{2}$ for x

EXAMPLE 35.25   Let $w_2 = \cos z_1 = \frac{1}{2}(\sqrt{3}\cosh 1 - i\sinh 1)$. Then, by the periodicity of the cosine, $\cos(z_1 + 2k\pi) = \cos z_1 = w_2$ for every integer k. To find all possible solutions of the equation $w_2 = \cos z$, compute

$$\arccos w_2 = -i\left[\mathrm{Log}\left(w_2 \pm i\sqrt{1 - w_2^2}\right) + 2k\pi i\right] = 2k\pi \pm z_1$$

consistent with the behavior of the cosine function on the real line. Indeed, Figure 35.33 shows graphs of $\cos x$ and the line $y = \frac{1}{2}$, the intersections of which are solutions of $\cos x = \frac{1}{2}$. The black portion of the cosine curve is the part used in the definition of the real principal arccosine function. For the intersection at a we write $x_1 = \frac{\pi}{3}$, the principal value of $\arccos\frac{1}{2}$, and for b write $x_2 = -\frac{\pi}{3}$, the fourth-quadrant companion. The intersections at b, d, f on the "back side" of the wave are of the form $x_2 + 2k\pi$, while the intersections at a, c on the "front edge" of the wave are of the form $x_1 + 2k\pi$. Finally, note that the set of solutions $\{x_2 + 2k\pi\}$ is equivalent to the set $\{2k\pi - x_1\}$ because $x_2 = -x_1$. ❖

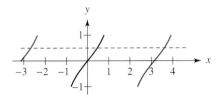

FIGURE 35.34 Solving the equation $\tan x = \frac{1}{2}$ in Example 35.26

The Function $w = \arctan z$

The inverse tangent function, defined by

$$\arctan z = \frac{i}{2}\log\left(\frac{i + z}{i - z}\right) \tag{35.17}$$

can be extracted from the solution of the equation $w = \tan z = \frac{1}{i}(e^{iz} - e^{-iz})/(e^{iz} + e^{-iz})$ by solving for e^{iz} and, hence, for z. (See Exercise A2.)

EXAMPLE 35.26   Let $w_3 = \tan z_1 = \frac{\sqrt{3}}{3 + \sinh^2 1} + \frac{4\sinh 1\cosh 1}{3 + \sinh^2 1}i$. Then, by the periodicity of $\tan z$, we have $\tan(z_1 + k\pi) = w_3$ for any integer k. To find all possible solutions of the equation $w_3 = \tan z$, namely, the infinite set $z_1 + k\pi$, use the multivalued logarithm to write $\arctan w_3 = \frac{i}{2}[\mathrm{Log}(\frac{i + w_3}{i - w_3}) + 2n\pi i] = z_1 + k\pi$, where $k = -n$ is any integer.

    Comparing the behavior on the real line with the behavior in the complex plane, Figure 35.34 shows graphs of $\tan x$ and the line $y = \frac{1}{2}$, the intersections of which are solutions of $\tan x = \frac{1}{2}$. The line $y = \frac{1}{2}$ intersects each branch of the tangent function just once, and the intersections are an integral multiple of π apart. The black branch is the one used in the definition of the principal branch of the arctangent function.   ❖

Inverse Hyperbolic Functions

The complex hyperbolic functions are defined in terms of the complex exponential function. Hence, they, too, are "infinitely many"-to-one functions. Their inverses are then one-to-"infinitely many" mappings, just like the complex inverse trig functions. However, for the inverse hyperbolic functions, the periodicity is across horizontal, not vertical, strips. And like the inverse trig functions, the inverses of the hyperbolic functions are given in terms of the complex logarithm.

The Function $w = \operatorname{arcsinh} z$

The complex function $w = \sinh z$ takes a horizontal strip of width 2π in the z-plane to the whole w-plane. (Recall the Riemann surface for $\sinh z$ given in Section 35.7.) If $z_4 = 1 + i$ is mapped by $\sinh z$ to the w-plane, we obtain $w_4 = \sinh 1 \cos 1 + i \cosh 1 \sin 1$. In fact, for $z_4 + 2k\pi i$, k an integer, we have, because of the periodicity of $\sinh z$, $\sinh(z_4 + 2k\pi i) = \sinh z_4 = w_4$. Consequently, the inverse hyperbolic sine function must map w_4 back to the infinite collection $z_4 + 2k\pi i$, $k = 0, \pm 1, \pm 2, \ldots$.

Figure 35.35 helps with visualizing the inversion process since it shows the intersection of the curves defined implicitly by the equations

$$u(x, y) = \sinh x \cos y = \mathcal{R}(w_4) \quad \text{and} \quad v(x, y) = \cosh x \sin y = \mathcal{I}(w_4)$$

The dotted curves are defined by the first equation, the solid by the second. The points of intersection are solutions of the equation $\sinh z = w_4$. It appears that, just as in the real case, there are two sets of solutions, one of the form $z_4 + 2k\pi i$ and one of the form $2k\pi i - z_4$, where k is an integer.

The standard definition of $\operatorname{arcsinh} z$ is given by

$$\operatorname{arcsinh} z = \log\left(z + \sqrt{z^2 + 1}\right) \tag{35.18}$$

There are two possible values for the square root and an infinite number of values for the multivalued logarithm. The square root is given by

$$\left(z^2 + 1\right)^{1/2} = e^{(1/2)\log(z^2+1)} = e^{1/2[\operatorname{Log}(z^2+1) + 2k\pi i]} = e^{(1/2)\operatorname{Log}(z^2+1)} e^{k\pi i}$$

The first factor on the right is the principal square root of $z^2 + 1$, and the second factor is 1 for even integers k and -1 for odd integers k. Thus, we can replace the definition of $\operatorname{arcsinh} z$ with the two functions

$$\operatorname{asinh}_1 z = \log\left(z + \sqrt{z^2 + 1}\right) \quad \text{and} \quad \operatorname{asinh}_2 z = \log\left(z - \sqrt{z^2 + 1}\right)$$

Each expression generates an infinite set of values when the multivalued log is used. If the principal complex logarithm is used, we can define the two functions

$$\operatorname{Asinh}_1 z = \operatorname{Log}\left(z + \sqrt{z^2 + 1}\right) \quad \text{and} \quad \operatorname{Asinh}_2 z = \operatorname{Log}\left(z - \sqrt{z^2 + 1}\right)$$

with which we obtain $z_4 = \operatorname{Asinh}_1 w_4 = 1 + i$ and $z_5 = \operatorname{Asinh}_2 w_4 = 2\pi i - z_4$. Corresponding to the principal value z_4 there would be the infinite set of solutions $\operatorname{asinh}_1 w_4 = z_4 + 2k\pi i$, and corresponding to z_5 there would be another infinite set of solutions $\operatorname{asinh}_2 w_4 = 2k\pi i - z_4$.

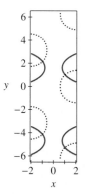

FIGURE 35.35   Solving the equation $\sinh z = \sinh 1 \cos 1 + i \cosh 1 \sin 1$ for z in the complex plane

Derivation of the Hyperbolic arcsine Function

To derive (35.18), start with the equation $w = \sinh z = \frac{1}{2}(e^z - e^{-z})$ and solve for $z = z(w) = \text{arcsinh } w$, obtaining $\log(w \pm \sqrt{w^2 + 1})$. An interchange of w and z and a proper understanding of the square root yield (35.18).

The Function $w = \text{arccosh } z$

The inverse hyperbolic cosine, defined by

$$\text{arccosh } z = \log\left(z + \sqrt{z^2 - 1}\right)$$

can be extracted from the solution of the equation $w = \frac{1}{2}(e^z + e^{-z})$. (See Exercise A3.)

EXAMPLE 35.27   If $w_5 = \cosh z_4 = \cosh 1 \cos 1 + i \sinh 1 \sin 1$, then, for any integer k, $\cosh(z_4 + 2k\pi i) = \cosh z_4 = w_5$ by the known periodicity of $\cosh z$. To find all possible solutions of the equation $w_5 = \cosh z$, write

$$\log\left(w_5 \pm \sqrt{w_5^2 - 1}\right) = \text{Log}\left(w_5 \pm \sqrt{w_5^2 - 1}\right) + 2k\pi i = 2k\pi i \pm z_4$$

where k is any integer.   ❖

The Function $w = \text{arctanh } z$

The inverse hyperbolic tangent is defined by

$$\text{arctanh } z = \frac{1}{2}\log\left(\frac{1+z}{1-z}\right) \qquad\qquad (35.19)$$

which can be extracted from the solution of the equation $w = (e^z - e^{-z})/(e^z + e^{-z})$. (See Exercise A4.)

EXAMPLE 35.28   If $w_6 = \tanh z_4 = \frac{\sinh 1 \cosh 1}{\sinh^2 1 + \cos^2 1} + \frac{\sin 1 \cos 1}{\sinh^2 1 + \cos^2 1} i$, then, for any integer k, $\tanh(z_4 + k\pi i) = w_6$ by the known periodicity of $\tanh z$. To find all possible solutions to the equation $w_6 = \tanh z$, write

$$\frac{1}{2}\log\left(\frac{1+w_6}{1-w_6}\right) = \frac{1}{2}\left[\text{Log}\left(\frac{1+w_6}{1-w_6}\right) + 2k\pi i\right] = z_4 + k\pi i$$

where k is any integer.   ❖

EXERCISES 35.8–Part A

A1. Solve $w = \cos z = \frac{1}{2}(e^{iz} + e^{-iz})$ for $r = e^{iz}$ and, hence, for z, thereby obtaining (35.16). *Hint:* $e^{-iz} = \frac{1}{r}$.

A2. Solve $w = \tan z = \frac{1}{i}(e^{iz} - e^{-iz})/(e^{iz} + e^{-iz})$ for $r = e^{iz}$ and, hence, for z, thereby obtaining (35.17).

A3. Solve $w = \cosh z = \frac{1}{2}(e^z + e^{-z})$ for $r = e^z$ and, hence for z, thereby obtaining $\text{arccosh } z = \log(z + \sqrt{z^2 - 1})$.

A4. Solve $w = \tanh z = (e^z - e^{-z})/(e^z + e^{-z})$ for $r = e^z$ and, hence for z, thereby obtaining (35.19).

A5. Find all solutions of $\sin x = \frac{\sqrt{3}}{2}$.

EXERCISES 35.8–Part B

B1. Find all solutions of $\cos x = \frac{1}{4}$.

Given the complex number ζ in Exercises B2–9, obtain all solutions of the equation $f(z) = \zeta$ if $f(z)$ is

(a) $\sin z$ (b) $\cos z$ (c) $\tan z$

(d) $\sinh z$ (e) $\cosh z$ (f) $\tanh z$

B2. $1 - 3i$ **B3.** $\frac{1}{2} + \frac{4}{3}i$ **B4.** $7 - 9i$ **B5.** $\frac{2}{3} + i$

B6. $-5 + 8i$ **B7.** $\frac{1}{2} + \frac{9}{7}i$ **B8.** $8 + 3i$ **B9.** $-\frac{6}{5} + \frac{9}{2}i$

B10. Define $\cot z = \frac{1}{\tan z}$ and obtain

(a) the real and imaginary parts of $\cot z$.   (b) $|\cot z|$

B11. Define $\sec z = \frac{1}{\cos z}$ and obtain

(a) the real and imaginary parts of $\sec z$.   (b) $|\sec z|$

B12. Define $\csc z = \frac{1}{\sin z}$ and obtain

(a) the real and imaginary parts of $\csc z$.   (b) $|\csc z|$

B13. Define $\coth z = \frac{1}{\tanh z}$ and obtain

(a) the real and imaginary parts of $\coth z$.   (b) $|\coth z|$

B14. Define $\operatorname{sech} z = \frac{1}{\cosh z}$ and obtain

(a) the real and imaginary parts of $\operatorname{sech} z$.   (b) $|\operatorname{sech} z|$

B15. Define $\operatorname{csch} z = \frac{1}{\sinh z}$ and obtain

(a) the real and imaginary parts of $\operatorname{csch} z$.   (b) $|\operatorname{csch} z|$

In Exercises B16–21, express, in terms of the complex logarithm, the inverse functions.

B16. $\operatorname{arccot} z$ **B17.** $\operatorname{arcsec} z$ **B18.** $\operatorname{arccsc} z$

B19. $\operatorname{arccoth} z$ **B20.** $\operatorname{arcsech} z$ **B21.** $\operatorname{arccsch} z$

B22. Obtain the real and imaginary parts of

(a) $\operatorname{arccot} z$ (b) $\operatorname{arccoth} z$

Given the complex number ζ in Exercises B23–30, obtain all solutions of the equation $f(z) = \zeta$ if $f(z)$ is

(a) $\cot z$ (b) $\sec z$ (c) $\csc z$

(d) $\coth z$ (e) $\operatorname{sech} z$ (f) $\operatorname{csch} z$

B23. $6 + 5i$ **B24.** $-\frac{1}{2} - \frac{7}{6}i$ **B25.** $\frac{3}{2} - \frac{1}{3}i$ **B26.** $-7 + 6i$

B27. $9 - 5i$ **B28.** $\frac{2}{3} + 2i$ **B29.** $-2 + 2i$ **B30.** $7 + 3i$

35.9   **Differentiation and the Cauchy–Riemann Equations**

Differentiation

A complex function $f(z)$ is said to be differentiable and has derivative $f'(z_0)$ at z_0 provided $f(z)$ is defined in a neighborhood of z_0 and $\lim_{h \to 0} \frac{f(z_0 + h) - f(z_0)}{h}$ exists as a limit in the plane (independent of direction of approach). The complex number h approaches 0 *in the plane,* so the direction of approach must not affect the value of the limit.

The differentiation rules of the calculus (linearity, the power, product, quotient, and chain rules) all apply. In addition, the formulas for the derivatives of the trig functions and their inverses, and the hyperbolic functions and their inverses, *almost* agree with their counterparts from elementary calculus. However, there are simple complex functions for which no derivative exists.

**EXAMPLE 35.29**   The function $f(x) = \bar{z}$, the conjugate of z, has no derivative To verify this claim, let $z_0 = x_0 + iy_0$ be a fixed point in the complex plane and attempt to compute, by formal definition, the derivative of $\bar{z}$ at this point. Thus, evaluate the limit

$$\lim_{h \to 0} \frac{\overline{x_0 + iy_0 + h} - \overline{x_0 + iy_0}}{h} \tag{35.20}$$

Now take the limit from two different directions. In the plane, for a limit to exist, it must be directionally independent. If the limit is different along two different paths, it does not exist. (This is akin to the behavior of a limit on the real line when the limits from the left and right do not agree and we conclude that the limit does not exist.)

Along a line parallel to the x-axis we have $h = t$ and (35.20) evaluates to 1. Along a line parallel to the y-axis we have $h = it$, and (35.20) evaluates to -1. Since the limit is not path independent, it does not exist and $f(x) = \bar{z}$ is not differentiable. ❖

Differentiating the Elementary Functions

POWERS   Formally, differentiation of polynomials proceeds exactly as in real-variable calculus. Thus, $\frac{d}{dz}z^k = kz^{k-1}$ is valid for integers k. It is also true that $\frac{d}{dz}z^c = cz^{c-1}$ for an arbitrary complex number c. Since z^c, defined in terms of the logarithm, is multivalued, the differentiation formula requires that the same branch of the function z^c be used on both sides of the equal sign.

THE EXPONENTIAL FUNCTION   The derivative of e^z, as in elementary calculus, is again e^z. In fact, the differentiation rules $\frac{d}{dz}e^{g(z)} = e^{g(z)}g'(z)$ and $\frac{d}{dz}c^z = c^z \log c$ both hold. In the second formula, c^z is defined in terms of a branch of the logarithm. Hence, the same branch of the logarithm must be used on both sides of the equal sign.

THE LOGARITHM   The derivative of any logarithm obeys the familiar rule, namely, $\frac{d}{dz} \log z = \frac{1}{z}$. Thus, for any branch of the logarithm, and in particular the principal branch, the "rate of change" of the logarithm is the same. It is the familiar $\frac{1}{z}$.

THE TRIG AND HYPERBOLIC TRIG FUNCTIONS   The first column of Table 35.5 lists the six complex trigonometric functions and the second, their derivatives. The third column lists the six complex hyperbolic functions and the fourth, their derivatives. The results in the complex plane agree exactly with the real-variable results for elementary calculus on the real line.

THE INVERSE TRIG AND HYPERBOLIC FUNCTIONS   Columns 2 and 3 in Table 35.6 list the derivatives of the inverse trigonometric functions. Column 2, headed by the symbol $f'(z)$, gives the form of the derivative valid over the whole complex plane; and column 3, headed by the symbol $f'(x)$, gives the form of the derivative found in the typical real-variable calculus text. On the real line, the domain of validity of the derivatives of arcsin x and arccos x would be $|x| < 1$, but for arcsec x and arccsc x it would be $|x| > 1$. For the derivatives of arcsec z and arccsc z, the square roots are not simplified because they require more care than on the real line. For example, evaluating the derivative of arcsec z at $z = 1 + i$ gives the complex number $-\frac{\sqrt{5}}{10}(\sqrt{\sqrt{5}-2} + i\sqrt{\sqrt{5}+2})$, whereas evaluating the expression for the derivative of arcsec x at the same point gives $\frac{\sqrt{10}}{20}(\sqrt{\sqrt{5}-2} - i\sqrt{\sqrt{5}+2})$. As floating-point numbers, the first is $-0.109 - 0.460i$, whereas the second is $0.249 - 0.402i$.

Columns 5 and 6 in Table 35.6 list the derivatives for the inverse hyperbolic functions. Column 5, headed by the symbol $g'(z)$, gives the form of the derivative valid over the whole complex plane, and column 6, headed by the symbol $g'(x)$, gives the form of the derivative found in the typical real-variable calculus text. In the standard calculus text, the domain for the derivative of arccosh x is $1 < x$; for the derivative of arctanh x, $|x| < 1$; and for the derivative of arcsech x, $0 < x < 1$. Column 5 is correct for the whole complex plane.

For example, consider the derivative of arcsech z at $z = -3$. This is clearly outside the domain for the formula from elementary calculus, but if working in the complex plane, the formula must be replaced with, for example, the one from column 5, which then yields $\frac{\sqrt{2}}{12}i$ as opposed to $-\frac{\sqrt{2}}{12}i$ from the formula in column 6.

Trigonometric	
f	f'
$\sin z$	$\cos z$
$\cos z$	$-\sin z$
$\tan z$	$\sec^2 z$
$\cot z$	$-\csc^2 z$
$\sec z$	$\sec z \tan z$
$\csc z$	$-\csc z \cot z$

Hyperbolic	
f	f'
$\sinh z$	$\cosh z$
$\cosh z$	$\sinh z$
$\tanh z$	$\text{sech}^2 z$
$\coth z$	$-\text{csch}^2 z$
$\text{sech } z$	$-\text{sech } z \tanh z$
$\text{csch } z$	$-\text{csch } z \coth z$

TABLE 35.5   Derivatives of the complex trigonometric and hyperbolic functions

Inverse Trigonometric Functions				Inverse Hyperbolic Functions		
f	$f'(z)$	$f'(x)$	g	$g'(z)$	$g'(x)$
arcsin z	$\dfrac{1}{\sqrt{1-z^2}}$	$\dfrac{1}{\sqrt{1-x^2}}$	arcsinh z	$\dfrac{1}{\sqrt{z^2+1}}$	$\dfrac{1}{\sqrt{x^2+1}}$
arccos z	$-\dfrac{1}{\sqrt{1-z^2}}$	$-\dfrac{1}{\sqrt{1-x^2}}$	arccosh z	$\dfrac{1}{\sqrt{z-1}\sqrt{z+1}}$	$\dfrac{1}{\sqrt{x^2-1}}$
arctan z	$\dfrac{1}{1+z^2}$	$\dfrac{1}{1+x^2}$	arctanh z	$\dfrac{1}{1-z^2}$	$\dfrac{1}{1-x^2}$
arccot z	$-\dfrac{1}{1+z^2}$	$-\dfrac{1}{1+x^2}$	arccoth z	$\dfrac{1}{1-z^2}$	$\dfrac{1}{1-x^2}$
arcsec z	$\dfrac{1}{z^2\sqrt{1-\frac{1}{z^2}}}$	$\dfrac{1}{|x|\sqrt{x^2-1}}$	arcsech z	$-\dfrac{1}{z^2\sqrt{\frac{1}{z}-1}\sqrt{\frac{1}{z}+1}}$	$-\dfrac{1}{x\sqrt{1-x^2}}$
arccsc z	$-\dfrac{1}{z^2\sqrt{1-\frac{1}{z^2}}}$	$-\dfrac{1}{|x|\sqrt{x^2-1}}$	arccsch z	$-\dfrac{1}{z^2\sqrt{1+\frac{1}{z^2}}}$	$-\dfrac{1}{|x|\sqrt{1-x^2}}$

TABLE 35.6   Derivatives of the real and complex inverse trigonometric and hyperbolic functions

The Cauchy–Riemann Equations

We next state two important theorems about complex differentiation. The first theorem gives some consequences when a function $f(z)$ has a derivative, whereas the second theorem is nearly a converse to the first.

THEOREM 35.1

At any point $z = x + iy$ at which $f(z) = u(x, y) + iv(x, y)$ is differentiable,

1. the partial derivatives u_x, v_x, u_y, v_y exist.
2. the derivative is given by $f'(z) = u_x(x, y) + iv_x(x, y) = v_y(x, y) - iu_y(x, y)$.
3. the Cauchy–Riemann equations $u_x = v_y$, $u_y = -v_x$ hold.

The Cauchy–Riemann partial differential equations (CRPDEs) are a consequence of the two expressions for the derivative in the second conclusion of the theorem. The equality of the real and imaginary parts of the expressions for the derivative yields the CRPDEs.

The CRPDEs alone are not enough to guarantee the existence of the derivative. For example, consider the function

$$f(z) = \begin{cases} 0 & z = 0 \\ \dfrac{z^5}{|z|^4} & z \neq 0 \end{cases}$$

for which the CRPDEs are satisfied at $z = 0$ but the derivative $f'(0)$ does not exist. The derivative at $z = 0$ is determined by $\lim_{h\to 0} \frac{f(0+h)-f(0)}{h} = \lim_{h\to 0} \frac{h^5}{|h|^4 h} = \lim_{h\to 0} \frac{h^4}{|h|^4}$. The limit must be path independent, but for $h = te^{i\theta}$, t real, the limit is the complex number

$e^{4i\theta}$, which is not independent of direction. Hence, the limit does not exist and $f(z)$ is not differentiable at the origin.

To show the CRPDEs are satisfied at $z = 0$, obtain the real and imaginary parts of $f(z)$ as

$$u(x, y) = \frac{x(x^4 - 10x^2y^2 + 5y^4)}{\left(x^2 + y^2\right)^2} \quad \text{and} \quad v(x, y) = \frac{y(5x^4 - 10x^2y^2 + y^4)}{\left(x^2 + y^2\right)^2}$$

respectively. At $z = 0$, the partial derivatives appearing in the CRPDEs are then

$$u_x(0, 0) = \lim_{h \to 0} \frac{u(0 + h, 0) - u(0, 0)}{h} = \lim_{h \to 0} \frac{h}{h} = 1$$

$$v_y(0, 0) = \lim_{h \to 0} \frac{v(0, 0 + h) - v(0, 0)}{h} = \lim_{h \to 0} \frac{h}{h} = 1$$

$$u_y(0, 0) = \lim_{h \to 0} \frac{u(0, 0 + h) - u(0, 0)}{h} = \lim_{h \to 0} \frac{0}{h} = 0$$

$$v_x(0, 0) = \lim_{h \to 0} \frac{v(0 + h, 0) - v(0, 0)}{h} = \lim_{h \to 0} \frac{0}{h} = 0$$

Thus, the CRPDEs are satisfied at $z = 0$ but $f(z)$ does not have a derivative at $z = 0$.

EXAMPLE 35.30   Consider the polynomial function $f(z) = z^2$ whose derivative is simply $2z = 2x + 2yi$ and whose real and imaginary parts are

$$u(x, y) = x^2 - y^2 \quad \text{and} \quad v(x, y) = 2xy$$

respectively. Then,

$$u_x = 2x \quad v_x = 2y \quad u_y = -2y \quad v_y = 2x$$

from which we see that the CRPDEs are satisfied. Using

$$f'(z) = u_x(x, y) + i v_x(x, y) = v_y(x, y) - i u_y(x, y) \tag{35.21}$$

to find the derivative again gives $2x + 2yi$.   ❖

EXAMPLE 35.31   The derivative of $f(z) = \sin z = u + iv$, where

$$u(x, y) = \cos x \cosh y \quad \text{and} \quad v(x, y) = \cos x \sinh y$$

is

$$\cos z = \cos x \cosh y - i \sin x \sinh y \tag{35.22}$$

The derivatives appearing in the CRPDEs are then

$$u_x = \cos x \cosh y \quad v_x = -\sin x \sinh y \quad u_y = \sin x \sinh y \quad v_y = \cos x \cosh y$$

from which we can see that $u_x = v_y$ and $u_y = -v_x$. Thus, the CRPDEs are satisfied. Using (35.21) to find the derivative again gives (35.22), so once more we see that the real and imaginary parts of a differentiable function satisfy the Cauchy–Riemann equations.   ❖

Derivation of the CRPDEs

The derivation of the Cauchy–Riemann equations is mechanical. If $f(z)$ is differentiable at z, then the limit of the difference quotient, $f'(z) = \lim_{h \to 0} \frac{f(z+h) - f(z)}{h}$, exists and is independent of direction of approach. Evaluating this limit along two different paths, one parallel to the x-axis ($h = t$) and the other parallel to the y-axis ($h = it$), gives two different but equal expressions for the derivative. If we write $f(z) = u(x, y) + iv(x, y)$, then the difference quotient along a horizontal line ($h = t$, t real) becomes

$$\frac{u(x+t, y) - u(x, y)}{t} + \frac{v(x+t, y) - v(x, y)}{t}i,$$

which, in the limit as t goes to zero, gives $f'(z) = u_x + iv_x$.

Repeating the process along the vertical path defined by $h = it$, we have the difference quotient

$$\frac{u(x, y+t) - u(x, y)}{it} + \frac{v(x, y+t) - v(x, y)}{it}i,$$

which, in the limit as t becomes zero, gives $f'(z) = v_y - iu_y$. Equating the real and imaginary parts of these two representations of f', we then get the Cauchy–Riemann equations $u_x = v_y, u_y = -v_x$.

THEOREM 35.2

In an open connected set, the Cauchy–Riemann equations, plus continuity of the partial derivatives, guarantees $f(z)$ is differentiable.

Theorem 35.1 showed that the real and imaginary parts of a differentiable function necessarily satisfied the Cauchy–Riemann equations. We also saw a counterexample that showed the CRPDEs alone are not sufficient to guarantee the existence of the derivative. Theorem 35.2 gives a sufficient condition, which, along with the CRPDEs, guarantees a complex function has a derivative. A proof of this theorem can be found in almost any complex variables text. See, for example, [69].

EXERCISES 35.9–Part A

A1. If $f(z)$ and $\overline{f(z)}$ are both differentiable on the open connected set R, apply the Cauchy–Riemann equations to each to show that $f(z)$ must be constant on R.

In Exercises A2–5, show that the real and imaginary parts of the given function $f(z)$ satisfy the Cauchy–Riemann equations.

A2. z^3 **A3.** $z + \dfrac{k^2}{z}$ **A4.** $\cos z$ **A5.** $\sinh z$

EXERCISES 35.9–Part B

B1. Show that in polar coordinates, the Cauchy–Riemann equations become $\frac{\partial u}{\partial r} = \frac{1}{r}\frac{\partial v}{\partial \theta}$ and $\frac{\partial v}{\partial r} = -\frac{1}{r}\frac{\partial u}{\partial \theta}$.

In Exercises B2–4, apply the result of Exercise B1 to verify that the given function is differentiable for $r > 0, 0 < \theta \le 2\pi$.

B2. $f(z) = \sqrt[n]{z} = \sqrt[n]{r}\left(\cos\dfrac{\theta}{n} + i\sin\dfrac{\theta}{n}\right)$

B3. $f(z) = \dfrac{\bar{z}}{|z|^2} = \dfrac{x - iy}{x^2 + y^2} = \dfrac{\cos\theta}{r} - i\dfrac{\sin\theta}{r}$

B4. $f(z) = z^i = e^{-\theta}(\cos \mathrm{Log}\, r + i\sin \mathrm{Log}\, r)$

In each of Exercises B5–20, show that the real and imaginary parts of the given function $f(z)$ satisfy the Cauchy–Riemann equations.

B5. $\sqrt{z}$ **B6.** $\sqrt[3]{z}$ **B7.** $\dfrac{2z - 3}{5z + 7}$ **B8.** $\tan z$

B9. $\cot z$ **B10.** $\sec z$ **B11.** $\csc z$ **B12.** $\cosh z$

B13. $\tanh z$ **B14.** $\coth z$ **B15.** $\mathrm{sech}\, z$ **B16.** $\mathrm{csch}\, z$

B17. $\arctan z$ **B18.** $\mathrm{arccot}\, z$ **B19.** $\mathrm{arctanh}\, z$ **B20.** $\mathrm{arccoth}\, z$

35.10 Analytic and Harmonic Functions

Analytic Functions—Definitions

In Section 35.9 we defined *differentiability* of $f(z)$, calling $f(z)$ differentiable at z if $f'(z)$, the derivative, exists at z. We showed that the function $f(z) = \bar{z}$ is not differentiable for any value of z, so there exist nondifferentiable functions. The function $f(z) = |z|^2 = x^2 + y^2$ is differentiable only at $z = 0$ because that is the only point at which the Cauchy–Riemann equations are satisfied. Examples like this suggest we need a condition stronger than pointwise differentiability, and that condition is *analyticity*. However, we first need the following.

DEFINITION 35.2   NEIGHBORHOOD

A *neighborhood* of a point z_0 in the complex plane is an open set containing z_0.

It suffices to think of a neighborhood of z_0 as an open disk with center at z_0 and radius $\varepsilon > 0$.

DEFINITION 35.3   DOMAIN

A *domain* is an open connected set.

As we saw in Section 35.2, the interior of a disk is an open connected set.

DEFINITION 35.4

A function $f(z)$ is *analytic at a point z_0* if $f(z)$ is differentiable in a neighborhood of the point z_0.

Thus, analyticity is a point property, but it requires differentiability in a *neighborhood* of the point. A function can have an isolated point of differentiability but not of analyticity. The very definition of analyticity at z_0 completely surrounds z_0 with enough points of differentiability that close neighbors of z_0 can, in turn, be surrounded by a neighborhood of differentiability.

DEFINITION 35.5

A function $f(z)$ is *analytic in a domain D* if it is analytic at every point in D.

This definition is equivalent to saying that if $f(z)$ is differentiable in a domain, then it is analytic in that domain. Thus, the intermediate concept of *analytic at a point* is handy but not strictly necessary.

Analytic functions are also called *regular* or *holomorphic*.

DEFINITION 35.6

A function $f(z)$ is *analytic at $z = \infty$* if $F(\zeta) = f(\frac{1}{\zeta})$ is analytic at $\zeta = 0$.

Thus, the function $z = \frac{1}{\zeta}$ maps the origin to infinity and the point at infinity to the origin. Behavior at $z = \infty$ is determined by behavior at $\zeta = 0$.

DEFINITION 35.7

Functions that are analytic (or equivalently, differentiable) on all of the finite complex plane are called *entire*.

EXAMPLE 35.32   The functions $f(z) = \bar{z}$ and $|z|$ are not analytic anywhere because they are not differentiable anywhere.   ❖

EXAMPLE 35.33   The function $f(z) = |z|^2$ is not analytic anywhere because it is differentiable only at the single point $z = 0$.   ❖

EXAMPLE 35.34   The function $f(z) = x^2 + iy^2$ is not analytic anywhere even though it is differentiable all along the line $y = x$. The points of differentiability do not form an open set in the plane, so there is no point of analyticity. To see why this function is differentiable only along the line $y = x$, write $u = x^2$ and $v = y^2$ and apply the CRPDEs, obtaining $2x = 2y$ and $0 = 0$. Thus, the CRPDEs are only satisfied on the line $y = x$.   ❖

EXAMPLE 35.35   The function $f(z) = z^2$ is analytic everywhere in the finite complex plane and is thus an entire function. For each finite z, the derivative is simply $f'(z) = 2z$. This function is not analytic at $z = \infty$. To see why, "invert" the complex plane by mapping z to $\frac{1}{\zeta}$, thereby obtaining $F(\zeta) = f(\frac{1}{\zeta}) = \frac{1}{\zeta^2}$. Clearly, $F(\zeta)$ is not differentiable at $\zeta = 0$, so $f(z)$ is deemed not to be differentiable at $z = \infty$.   ❖

EXAMPLE 35.36   The function $f(z) = \frac{1}{z}$ is analytic for $z \neq 0$, since, for each such z, the derivative is the familiar $f'(z) = -\frac{1}{z^2}$.   ❖

EXAMPLE 35.37   The function $f(z) = e^z$ is entire because it is analytic (differentiable) everywhere in the finite complex plane. This function is not analytic at $z = \infty$ because $F(\zeta) = f(\frac{1}{\zeta}) = e^{1/\zeta}$ is neither defined nor differentiable at $\zeta = 0$.   ❖

THEOREM 35.3   LIOUVILLE'S THEOREM

Functions that are both bounded on the finite complex plane, and entire, are necessarily constant.

The proof of this theorem is best given after obtaining the profound results of integration theory. (For example, see [69].) Presently, we simply contrast the situation in the complex plane with that on the real line. The real function $f(x) = \sin x$ is both differentiable and bounded everywhere, yet is not constant. The complex function $f(z) = \sin z$ is entire but not bounded. Off the real line, $\sin z$ becomes unbounded as $|z| \to \infty$.

Harmonic Functions

As we saw in Section 21.5, a scalar function $\Phi(x, y)$ with continuous second partial derivatives satisfying Laplace's equation, namely, $\nabla^2 \Phi = \Phi_{xx} + \Phi_{yy} = 0$, on a domain D, is called a harmonic function. The Cauchy–Riemann equations tell us that differentiable functions in the complex plane cannot have their real and imaginary parts chosen arbitrarily. The real and imaginary parts of an analytic function are linked by the Cauchy–Riemann equations. The resulting constraints force each part to satisfy Laplace's equation. Thus, the real and imaginary parts of an analytic function are necessarily harmonic. We formalize this observation in the following theorem.

THEOREM 35.4

Both the real and the imaginary parts of the analytic function $f(z) = u(x, y) + iv(x, y)$ are harmonic. The function $v(x, y)$ is called the *harmonic conjugate* of $u(x, y)$.

This result follows from the Cauchy–Riemann equations. Indeed, if $f = u + iv$ is analytic, then $u_x = v_y$ so $u_{xx} = v_{yx}$ and $u_y = -v_x$ so $u_{yy} = -v_{xy}$. Continuity of the second partial derivatives means $v_{yx} = v_{xy}$, so $u_{xx} + u_{yy} = v_{xy} - v_{xy} = 0$.

EXAMPLE 35.38   For the analytic function $f(z) = z^2$, we have $u(x, y) = x^2 - y^2$ and $v(x, y) = 2xy$. For both the real and imaginary parts we have $\nabla^2 u = \nabla^2 v = 0$. Each is harmonic, and the real and imaginary parts of an analytic function satisfy Laplace's equation.

The CRPDEs so constrain the relationship between $u(x, y)$ and $v(x, y)$ that the level curves of each are mutually orthogonal. We can see this in Figure 35.36, where the level curves of $u(x, y) = x^2 - y^2$ (solid lines) and the level curves of $v(x, y) = 2xy$ (dotted lines) are shown. That these curves are orthogonal is easily demonstrated by showing that the gradient vectors $\nabla u = 2x\mathbf{i} - 2y\mathbf{j}$ and $\nabla v = 2y\mathbf{i} + 2x\mathbf{j}$ are themselves orthogonal since the dot product $\nabla u \cdot \nabla v$ is zero. (See Exercise A2 for a general proof of this result.)

Finally, we show that $v(x, y) = 2xy$ can be recovered from $u(x, y) = x^2 - y^2$ by a variety of techniques. First, if $u(x, y)$ is harmonic at (x_0, y_0), then $v(x, y)$ is given by the formula

$$v(x, y) = c + \int_{y_0}^{y} u_x(x, t)\, dt - \int_{x_0}^{x} u_y(t, y_0)\, dt \qquad (35.23)$$

where c is a constant. (The harmonic conjugate is always determined up to an additive constant.) In this example, we would have

$$c + \int_{y_0}^{y} 2x\, dt - \int_{x_0}^{x} -2y_0\, dt = c + 2xy - 2x_0 y_0$$

If the constant c is taken as $2x_0 y_0$, then $v(x, y) = 2xy$ will have been recovered.   ❖

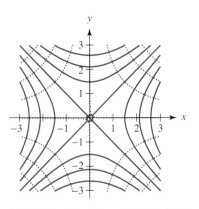

FIGURE 35.36   Example 35.38: The level curves of $u = x^2 - y^2$ are orthogonal to the level curves of $v = 2xy$

This formula is equivalent to the following algorithm, which starts with

$$v(x, y) = \int v_y(x, y)\, dy + g(x)$$

where, because x is held constant during the integration, the additive constant becomes $g(x)$, a function of x. Now, by the Cauchy–Riemann equation $u_x = v_y$, we have

$$v(x, y) = \int u_x(x, y)\, dy + g(x) = h(x, y) + g(x) \qquad (35.24)$$

for some function $h(x, y)$. By the Cauchy–Riemann equation $v_x = -u_y$, we have $v_x = h_x + g'(x) = -u_y$, from which $g(x)$ is then determined.

EXAMPLE 35.39   Consider the analytic function $f(z) = \sin z$ so that the real and imaginary parts are

$$u = \sin x \cosh y \quad \text{and} \quad v = \cos x \sinh y \qquad (35.25)$$

Each of the real and imaginary parts are harmonic since $\nabla^2 u = \nabla^2 v = 0$. The real and imaginary parts of the analytic function $f(z) = u(x, y) + iv(x, y)$ both satisfy Laplace's equation and are therefore called *harmonic* functions. The mutually orthogonal level curves of $u(x, y)$ and $v(x, y)$ are the dashed and solid curves, respectively, in Figure 35.37. The gradient vectors $\nabla u = \cos x \cosh y \mathbf{i} + \sin x \sinh y \mathbf{j}$ and $\nabla v = -\sin x \sinh y \mathbf{i} + \cos x \cosh y \mathbf{j}$ are therefore mutually orthogonal since $\nabla u \cdot \nabla v = 0$. ❖

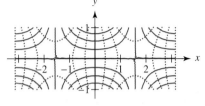

FIGURE 35.37   Example 35.39: The level curves of $u = \sin x \cosh y$ are orthogonal to the level curves of $v = \cos x \sinh y$

EXERCISES 35.10–Part A

A1. Show that $\lim_{z \to \infty} |\sin z| = \infty$ along the path $z = iy$, $y > 0$. Hence, $\sin z$ is analytic for all finite z but is not bounded.

A2. Let $f(z) = u + iv$ be analytic. Use the Cauchy–Riemann equations to show that $\nabla u \cdot \nabla v = 0$.

A3. Show that the real and imaginary parts of $f(z) = z^3$ are harmonic.

A4. Let $u(x, y)$ be the real part of $f(z) = z^3$. Use the algorithm embodied in (35.23) to recover $v(x, y)$, the harmonic conjugate.

A5. Repeat Exercise A4 with the algorithm embodied in (35.24).

EXERCISES 35.10–Part B

B1. The real and imaginary parts of $f(z) = \sin z$ are given by (35.25). The function $u(x, y)$ solves the Dirichlet problem $\nabla^2 u = 0$, $u(0, y) = u(\pi, y) = u(x, 1) = 0, u(x, 0) = \sin x$. The flow lines for u are the solutions of $\frac{dy}{dx} = \frac{u_y}{u_x}$. Show that the level curves of $v(x, y)$ satisfy this equation for the flow lines of u. *Hint:* Use implicit differentiation and the CRPDEs.

For the functions $f(z)$ given in Exercises B2–20:

    (a) Verify that the real and imaginary parts satisfy Laplace's equation and are, therefore, harmonic.

    (b) Superimpose graphs of the level curves of the real and imaginary parts of each function. Use a scaling that

demonstrates the orthogonality of the level curves of $u(x, y)$ and its harmonic conjugate, $v(x, y)$.

    (c) Construct the imaginary part from the real part using the algorithm contained in (35.23).

    (d) Construct the real part from the imaginary part using the alternative algorithm embodied in (35.24).

B2. $z + \dfrac{1}{z}$ **B3.** $\sqrt{z}$ **B4.** $\sqrt[3]{z}$ **B5.** $\dfrac{2z - 3}{5z + 7}$ **B6.** $\cos z$

B7. $\tan z$ **B8.** $\cot z$ **B9.** $\sec z$ **B10.** $\csc z$ **B11.** $\sinh z$

B12. $\cosh z$ **B13.** $\tanh z$ **B14.** $\coth z$ **B15.** $\operatorname{sech} z$

B16. $\operatorname{csch} z$ **B17.** $\arctan z$ **B18.** $\operatorname{arccot} z$

B19. $\operatorname{arctanh} z$ **B20.** $\operatorname{arccoth} z$

35.11 **Integration**

Indefinite Integrals

On the real line, the indefinite integral is often identified with the antiderivative. We review the distinction.

A function $F(x)$ is said to be an *antiderivative* of $f(x)$ if $F'(x) = f(x)$. The definite integral of $f(x)$, interpreted as the area between the graph of $f(x)$ and the x-axis, is given by the limit of a sum of areas of rectangular areas lying in $a \leq x \leq b$. Appropriately specifying x_k, the evaluation points, Δx_k, the widths of the rectangles, and the way in which the limit is to be taken, the notation and definition of the definite integral is typically written as

$$\int_a^b f(x)\,dx = \lim_{N \to \infty} \sum_{k=1}^N f(x_k)\Delta x_k$$

The Fundamental Theorem of Calculus declares $\int_a^b f(x)\,dx = F(b) - F(a)$, where $F(x)$ is an antiderivative of $f(x)$. Because of this theorem, the notation $\int f(x)\,dx$ is used to represent an antiderivative of $f(x)$. Moreover, if b, the right endpoint, is allowed to vary, then the definite integral defines the "area function" $F(x) = \int_a^x f(t)\,dx$ whose derivative, $F'(x) = f(x)$, is given by a modification of the Fundamental Theorem of Calculus. There are then two interpretations of these results.

First, to find an area, find an appropriate antiderivative. Second, to construct an antiderivative, evaluate the "area function." The first interpretation leads to drill and practice in applying the rules of "integration," which are really the "rules" of antidifferentiation. The second interpretation leads to "functions defined by integrals," functions such as $\Gamma(z)$, the gamma function; $\operatorname{erf}(x)$, the error function, and $\operatorname{sn}(u)$, $\operatorname{cn}(u)$, and $\operatorname{dn}(u)$, the Jacobian elliptic functions.

For complex functions, antiderivatives clearly exist. In fact, the typical theorem on the existence of antiderivatives will state something like

> *if f is defined and analytic on a simply connected region A, then there is an analytic function F, defined on A, and unique up to an additive constant, for which $F'(z) = f(z)$ for all z in A. We call F the antiderivative of f on A.*

However, without an appropriate "Fundamental Theorem," there is no justification for writing $\int f(z)\,dz$ for the antiderivative of $f(z)$. What is available, though, is the *contour integral* $\int_C f(z)\,dz$, which is actually a line integral along the curve C in the complex plane.

Contour Integration

Line Integrals in the Plane—A Review   Recall from Section 19.1 that the work done by the force $\mathbf{F}$, as it acts *on* a particle of unit mass moving along the curve C, is given by the *line integral* $\int_C \mathbf{F} \cdot d\mathbf{r}$, where $\mathbf{r}$ is the radius vector describing the curve. For example, if $\mathbf{F} = x\mathbf{i} + y^2\mathbf{j}$ and the curve C, a circle with radius 2 and center at the origin, is described parametrically by the equations

$$x(t) = 2\cos t \qquad y(t) = 2\sin t \qquad 0 \leq t \leq 2\pi \tag{35.26}$$

then

$$\int_C \mathbf{F} \cdot d\mathbf{r} = \int_0^{2\pi} (xx' + y^2 y')\,dt = \int_0^{2\pi} 4\sin t \cos t (2\sin t - 1)\,dt = 0 \tag{35.27}$$

CONTOURS   Paths in the complex plane along which line integrals are computed are called *contours,* and the resulting line integrals, *contour integrals.* A *contour* is a piecewise-smooth curve, defined as follows. A curve is a continuous map of a real interval $[a, b]$ into the plane, realized by $z(t) = x(t) + iy(t)$. The curve is *piecewise-smooth* if $[a, b]$ can be divided into a finite number of subintervals $a = a_0 < a_1 < \cdots < a_n = b$ for which $z'(t)$ exists on the open subintervals (a_k, a_{k+1}) and is continuous on the closed subintervals $[a_k, a_{k+1}]$. Continuity of $z'(t)$ on $[a_k, a_{k+1}]$ means the one-sided limits $z(a_k+)$ and $z(a_{k+1}-)$ exist. If $z'(t)$ is continous on $[a, b]$, then the curve is called *smooth.*

Curves that never cross themselves are called *simple.* Specifically, for a simple curve, when $t_1 \neq t_2$ are in (a, b) then $z(t_1) \neq z(t_2)$, so that different parameter values guarantee different points in the plane. If the initial and terminal points on the contour coincide, the contour is called *closed.* Thus, for a closed contour defined by $z(t), a \leq t \leq b$, we would have $z(a) = z(b)$.

Closed contours are said to have *positive orientation* if an increase in the parameter value causes the path to be traversed with "the left hand pointing to the interior." For a circle, the positive orientation is consistent with traverse in the counterclockwise direction. (The same convention was adopted in Section 19.2 for line integrals defining flux.)

We rely on Figure 35.38 to clarify the interactions of these terms. From top to bottom, the first contour is simple and smooth. The second contour is simple, closed, and also smooth. The third contour is neither simple nor closed. However, it is smooth. The fourth contour is closed but not simple. It, too, is smooth. The fifth contour is simple, not closed, and not smooth. The two arcs that comprise it form a cusp at their junction where the tangent, instead of turning continuously, jumps from one orientation to another. Hence, the contour is piecewise-smooth.

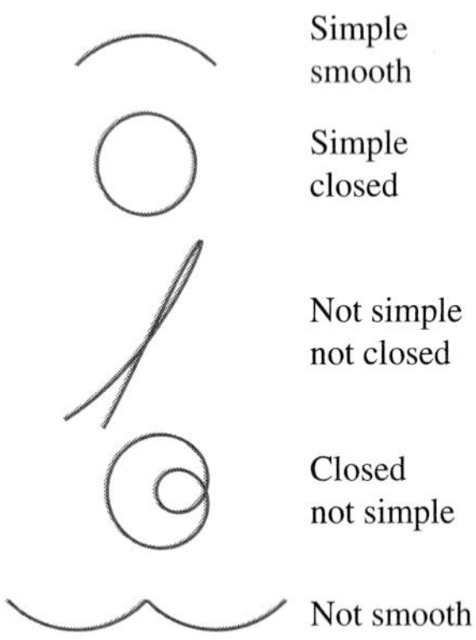

Simple
smooth

Simple
closed

Not simple
not closed

Closed
not simple

Not smooth

FIGURE 35.38   Examples of contours

CONTOUR INTEGRALS IN THE COMPLEX PLANE   If $f(z) = u(x, y) + iv(x, y)$ and $z(t) = x(t) + iy(t), a \leq t \leq b$, define a contour C in the complex plane, the *contour integral of $f(z)$ along C* is given by

$$\int_C f(z)\, dz = \int_a^b f(z(t)) z'(t)\, dt = \int_a^b [u(x(t), y(t)) + iv(x(t), y(t))](x' + iy')\, dt$$

$$= \int_a^b (ux' - vy') + i(vx' + uy')\, dt$$

We illustrate with the following examples.

EXAMPLE 35.40   Let $f(z) = e^z$, and take the contour C_1 as the straight line from the origin to the point $2 + \frac{\pi}{4}i$. Parametrize C_1 as $z(t) = (2 + \frac{\pi}{4}i)t, 0 \leq t \leq 1$, and obtain

$$\int_{C_1} f(z)\, dz = \int_a^b f(z(t)) z'(t)\, dt = \int_0^1 e^{(2+\pi i/4)t} \left(2 + \frac{\pi}{4}i\right) dt = \frac{e^2}{\sqrt{2}}(1 + i) - 1$$

(35.28)

❖

EXAMPLE 35.41   Using the function $f(z) = e^z$ and the endpoints from Example 35.40, we obtain the contour integral along the parabola $y = \frac{\pi}{16}x^2$. For this contour C_2, a parametrization of the form $z = x + iy(x)$ is realized by $z(t) = t + \frac{\pi i}{16}t^2, 0 \leq t \leq 2$, where x, the obvious parameter, has been renamed t. The desired contour integral is then

$$\int_{C_2} f(z)\, dz = \int_a^b f(z(t)) z'(t)\, dt = \int_0^2 e^{t + \pi i t^2/16} \left(1 + \frac{\pi i}{8}t\right) dt = \frac{e^2}{\sqrt{2}}(1 + i) - 1$$

(35.29)

which just happens to be the same value as the integral along the contour C_1. We will have more to say about this shortly.   ❖

EXAMPLE 35.42 Let $f(z) = z^3$ and C_3 be the contour $z(t) = t + \frac{i}{2}(t - 1)$, $-1 \le t \le 3$, a straight line connecting $z_1 = -1 - i$ to $z_2 = 3 + i$. The contour integral $\int_{C_3} f(z)\,dz$ is then given by

$$\int_a^b f(z(t))z'(t)\,dt = \int_{-1}^3 \left(t + \frac{i}{2}(t - 1)\right)^3 \left(1 + \frac{i}{2}\right)dt = 8 + 24i \qquad (35.30)$$

❖

EXAMPLE 35.43 With $f(z) = z^3$ as in Example 35.42, let the contour C_4 be the parabolic path defined by $x = y^2 + 2y$. The parametrization $x = t^2 + 2t$, $y = t$, $-1 \le t \le 1$, gives $z(t) = x(t) + iy(t)$, so that $z(-1) = z_1 = -1 - i$ and $z(1) = z_2 = 3 + i$. Thus, C_4 connects the same two endpoints as C_3 and the contour integral $\int_{C_4} f(z)\,dz = \int_a^b f(z(t))z'(t)\,dt$ becomes

$$\int_{-1}^1 \left(t^2 + 2t + it\right)^3 (2t + 2 + i)\,dt = 8 + 24i \qquad (35.31)$$

As with Examples 35.40 and 35.41, the contour integrals connecting the same endpoints with different contours are equal. We will have more to say about this momentarily.   ❖

EXAMPLE 35.44 If $f(z) = \bar{z}$ and the contour C_5 is the (lower) semicircle connecting the points $(-1, 0)$ and $(1, 0)$, the contour integral $\int_{C_5} f(z)\,dz$ can be computed by parametrizing C_5 with $z(t) = \cos t + i \sin t$, $\pi \le t \le 2\pi$. Then we have

$$\int_a^b f(z(t))z'(t)\,dt = \int_{\pi}^{2\pi} (\cos t - i \sin t)(-\sin t + i \cos t)\,dt = \pi i \qquad (35.32)$$

❖

EXAMPLE 35.45 With $f(z) = \bar{z}$ as in Example 35.44, let the contour C_6 be a parabola through $(-1, 0)$ and $(1, 0)$, the same endpoints as in Example 35.44. The parabolic contour $y = 1 - x^2$ can be parametrized by $z(t) = t + (1 - t^2)i$, $-1 \le t \le 1$, so that the contour integral $\int_{C_6} f(z)\,dz$ becomes

$$\int_a^b f(z(t))z'(t)\,dt = \int_1^{-1} \left(t - i(1 - t^2)\right)(1 - 2it)\,dt = \tfrac{8}{3}i \qquad (35.33)$$

The value along the parabola differs from the value along the semicircle!   ❖

EXAMPLE 35.46 Let $f(z) = \frac{1}{z - z_0}$ and the contour C_7 be a circle with radius R and center at z_0. To evaluate the contour integral of $f(z)$ about the closed contour C_7, parametrize the curve by $z(t) = z_0 + Re^{it}$, $0 \le t \le 2\pi$, so that we have

$$\int_{C_7} f(z)\,dz = \int_a^b f(z(t))z'(t)\,dt = \int_0^{2\pi} i\,dt = 2\pi i$$

Note that in this example, the contour integral around a closed contour did not evaluate to zero.   ❖

EXAMPLE 35.47 With the contour still as the circle C_7 from Example 35.46, take $f(z) = 1/(z - z_0)^n$, where n is an integer other than 1. Then the contour integral of $f(z)$ around the closed contour C_7 becomes

$$\int_{C_7} f(z)\,dz = \int_a^b f(z(t))z'(t)\,dt = \int_0^{2\pi} \frac{iRe^{it}}{\left(Re^{it}\right)^n}\,dt = iR^{1-n}\int_0^{2\pi} e^{i(1-n)t}\,dt = 0 \quad (35.34)$$

❖

CAUCHY–GOURSAT THEOREM   The various behaviors seen in Examples 35.40–35.47 are illuminated by the historic Cauchy–Goursat theorem. (Some texts call this the Cauchy Integral theorem; others, just the Cauchy theorem.) The following statement of the theorem is based on [69].

THEOREM 35.5   CAUCHY–GOURSAT

1. $f(z)$ is analytic in the interior of the open connected set D.

2. C is a closed contour lying in D.

$$\implies \int_C f(z)\, dz = 0$$

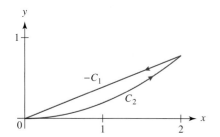

FIGURE 35.39   Example 35.48: The closed contour formed by the curves in Examples 35.40 and 35.41

Thus, analytic functions have a "closed contour integral" property whereby the contour integral of an analytic function is zero on a closed contour. Recall Section 21.5 where we saw that for conservative vector fields, the "closed line integral" property was equivalent to "path independence." By a similar argument based on the Cauchy–Goursat theorem, we can show that if $f(z)$ is analytic in D and if C_1 and C_2 are contours connecting z_1 and z_2 in D, then $\int_{C_1} f(z)\, dz = \int_{C_2} f(z)\, dz$. Thus, a contour can be deformed without changing the value of a contour integral, provided the deformation process does not carry the contour through a point where $f(z)$ is not analytic.

The Cauchy–Goursat theorem can be thought of as giving the "closed contour integral" property for analytic functions $f(z)$, a property that is equivalent to "path independence" for contours that remain in the domain of analyticity of $f(z)$.

EXAMPLE 35.48   The contours C_1 and C_2 from Examples 35.40 and 35.41 connect the same initial and terminal points. Hence, together they form a closed contour C, provided the direction on C_1 is reversed, as shown in Figure 35.39. The value for the contour integral along the parabolic arc in Example 35.41 was $\frac{e^2}{\sqrt{2}}(1+i) - 1$. Along the straight line in Example 35.40, the contour integral had the same value because of path independence for analytic functions. As for line integrals in Unit Four, reversing directions on the contour C_1 gives $-\frac{e^2}{\sqrt{2}}(1+i) + 1$ for this integral, and the sum is zero.   ❖

EXAMPLE 35.49   The contours C_3 and C_4 connect the same initial and terminal points in Examples 35.42 and 35.43, so together, provided the direction is reversed on C_4, they form the closed contour C shown in Figure 35.40. The value for the contour integral along both C_3 and C_4 was $8 + 24i$, as predicted by the path-independence property of analytic functions. If the direction of integration along C_4 is reversed, then the value of the contour integral is $-8 - 24i$ and the integral around the closed contour C is zero.   ❖

EXAMPLE 35.50   The contours C_5 and C_6 connect the same initial and terminal points in Examples 35.44 and 35.45, so together, provided the direction is reversed on C_6, they form the closed contour C shown in Figure 35.41. The value for the contour integral along C_5 was πi, whereas the value along C_6 was $\frac{8}{3}i$. Reversing the direction of integration on C_6 negates the value of the contour integral, so the integral around the closed contour C would be $(\pi - \frac{8}{3})i \neq 0$. The contour integral around C is not zero because $f(z) = \bar{z}$ is not analytic anywhere.   ❖

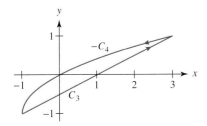

FIGURE 35.40 Example 35.49: The closed contour formed by the curves in Examples 35.42 and 35.43

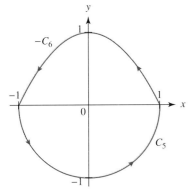

FIGURE 35.41 Example 35.50: The closed contour formed by the curves in Examples 35.44 and 35.45

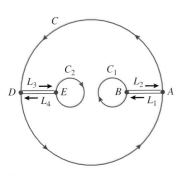

FIGURE 35.42 Contours C_1 and C_2 surrounding isolated singularities of $f(z)$ lying inside the contour C

DEFINITION 35.8

If N is a neighborhood of z_0 from which z_0 itself has been removed, then N is called a *deleted neighborhood* of z_0.

An example of a deleted neighborhood of the origin would be the points satisfying $0 < |z| < \varepsilon$ for ε positive.

DEFINITION 35.9

An *isolated singularity* of $f(z)$ is a point z_0 at which $f(z)$ is not analytic but for which $f(z)$ is analytic in a deleted neighborhood of z_0.

In Example 35.46, the function $f(z) = \frac{1}{z - z_0}$ is analytic everywhere except at $z = z_0$. The circular contour C_7 encloses the isolated singular point z_0, so the Cauchy–Goursat theorem does not apply. The contour integral around C_7 evaluates to $2\pi i \neq 0$.

In Example 35.47, the function $f(z) = 1/(z - z_0)^n$ is analytic everywhere except at $z = z_0$. The circular contour C_7 encloses the isolated singular point z_0, so again the Cauchy–Goursat theorem does not apply. That the contour integral around C_7 evaluates to zero is, in this case, a coincidence, not a consequence of the Cauchy–Goursat theorem.

THEOREM 35.6   GENERALIZED CAUCHY–GOURSAT THEOREM

1. $C, C_1, C_2, \ldots, C_k$ are $k + 1$ positively oriented, simple closed contours.

2. $C_1, \ldots, C_k$ are contained inside C in such a way that no C_i is inside or on any other C_j for $i \neq j$.

3. $f(z)$ is analytic on C, on all the $C_i, i = 1, \ldots, k$, and at each point that is inside C but exterior to all the C_i.

$$\implies \int_C f(z)\, dz = \sum_{i=1}^{k} \int_{C_i} f(z)\, dz$$

Loosely speaking, this theorem says the contour integral around a "large" contour inside of which there are small contours surrounding isolated singularities of $f(z)$ is the sum of the contour integrals around each of the small contours.

The following sketch of the proof for the case $k = 2$ shows how the Cauchy–Goursat theorem is used to establish the general result. In Figure 35.42, C contains two smaller contours C_1 and C_2. Points A and D on C are connected to points B and E on C_1 and C_2, respectively, by the paths L_1 and L_3. The points B and A are connected by another path L_2 and E and D are connected by L_4. The contour integrals inward along L_1 and outward along L_2 have the values I_1 and $-I_1$, respectively. Similarly, the contour integrals inward along L_3 and outward along L_4 have the values I_2 and $-I_2$, respectively.

The positive orientation around C_1 and C_2 is counterclockwise. Consider the contour Γ that consists of L_1 from A to B, C_1 in the clockwise sense, L_2 from B to A, along C to D, along L_3 to E, around C_2 in the clockwise sense, along L_4 from E to D, and around C to

A. Inside and along this contour, $f(z)$ is analytic, so the Cauchy–Goursat theorem applies. It gives

$$\int_\Gamma f\,dz = 0 = \int_C f\,dz + \int_{L_1} f\,dz + \int_{L_2} f\,dz + \int_{L_3} f\,dz + \int_{L_4} f\,dz - \int_{C_1} f\,dz - \int_{C_2} f\,dz$$

$$0 = \int_C f\,dz + I_1 - I_1 + I_2 - I_2 - \int_{C_1} f\,dz - \int_{C_2} f\,dz$$

$$0 = \int_C f\,dz - \int_{C_1} f\,dz - \int_{C_2} f\,dz$$

from which we get $\int_C f\,dz = \int_{C_1} f\,dz + \int_{C_2} f\,dz$.

EXAMPLE 35.51   To evaluate $\int_C \frac{2z}{z^2+2}\,dz$, where C is the circle $|z| = 2$, note that the integrand has two isolated singularities, namely, the points $(0, i\sqrt{2})$ and $(0, -i\sqrt{2})$, and both lie within the contour C. The parametrization $z(t) = 2\cos t + 2i\sin t, 0 \le t \le 2\pi$, leads to

$$4\int_0^{2\pi} \frac{i(2\cos^2 t + 1) - 2\cos t \sin t}{8\cos^2 t + 1}\,dt = 4\pi i$$

The generalized Cauchy–Goursat theorem suggests surrounding $(0, i\sqrt{2})$ with C_1, a small circle of radius, say $\frac{1}{2}$, and surrounding $(0, -i\sqrt{2})$ with C_2, also a circle of radius $\frac{1}{2}$. The parametrizations

$$z_a(t) = \tfrac{1}{2}\cos t + i\left(\sqrt{2} + \tfrac{1}{2}\sin t\right) \quad \text{and} \quad z_b(t) = \tfrac{1}{2}\cos t + i\left(-\sqrt{2} + \tfrac{1}{2}\sin t\right)$$

and Theorem 35.6 lead to $\int_C f(z)\,dz = \int_{C_1} f(z)\,dz + \int_{C_2} f(z)\,dz$, where

$$\int_{C_1} f(z)\,dz = 2\int_0^{2\pi} \frac{(15 + 2\sqrt{2}\sin t)i + 6\sqrt{2}\cos t}{31 - 8\sqrt{2}i\cos t}\,dt = 2\pi i$$

$$\int_{C_2} f(z)\,dz = 2\int_0^{2\pi} \frac{(15 - 2\sqrt{2}\sin t)i - 6\sqrt{2}\cos t}{31 + 8\sqrt{2}i\cos t}\,dt = 2\pi i$$

Thus, by the generalized Cauchy–Goursat theorem, we again obtain $4\pi i$ for the value of the original contour integral.

However, a more elegant solution can be constructed from the partial fraction decomposition

$$f(z) = \frac{1}{z - i\sqrt{2}} + \frac{1}{z + i\sqrt{2}}$$

Instead of integrating $f(z)$ around C_1 and C_2, integrate $f_1(z) + f_2(z)$ around each of C_1 and C_2. Inside C_1, $f_2(z)$ is analytic and does not contribute to the integral. Inside C_2, $f_1(z)$ is analytic and does not contribute to the integral. Hence, we need only evaluate the integrals

$$\int_{C_1} \frac{1}{z - i\sqrt{2}}\,dz = 2\pi i \quad \text{and} \quad \int_{C_2} \frac{1}{z + i\sqrt{2}}\,dz = 2\pi i$$

where the evaluations are transparent in light of Example 35.46.   ❖

EXAMPLE 35.52   Evaluate the contour integral $\int_C \frac{16z-1}{2z^2-z-1}\,dz$, where the contour C, shown in Figure 35.43, consists of the line segments connecting points P_1, P_3, P_2, P_4, in that order. Thus, the lines cross at P_5, thereby forming two simple contours, C_a and C_b. The contour C_a is the triangle $P_2 P_5 P_3$, whereas contour C_b is the triangle $P_1 P_5 P_4$. Contour C_a has a positive orientation,

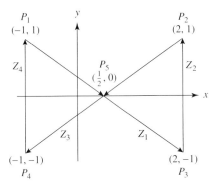

FIGURE 35.43   The contour C of Example 35.52

but contour C_b has a negative orientation. The equation of the line $P_1 P_3$ is $y = \frac{1-2x}{3}$, whereas the equation of the line $P_2 P_4$ is $y = \frac{2x-1}{3}$.

Using the parametric representations

$$z_1(t) = t + \frac{i}{3}(1 - 2t) \qquad -1 \le t \le 2 \qquad z_3(t) = t + \frac{i}{3}(2t - 1) \qquad -1 \le t \le 2$$

$$z_2(t) = 2 + it \qquad -1 \le t \le 1 \qquad z_4(t) = it - 1 \qquad -1 \le t \le 1$$

the contour integral around C is given by

$$\int_{-1}^{2} f(z_1(t)) z_1'(t)\, dt + \int_{-1}^{1} f(z_2(t)) z_2'(t)\, dt$$

$$+ \int_{2}^{-1} f(z_3(t)) z_3'(t)\, dt + \int_{-1}^{1} f(z_4(t)) z_4'(t)\, dt = 4\pi i$$

Confirmation of this result can be obtained if the contour C is decomposed into the two simple closed contours C_a and C_b. Then, these two triangles can be deformed continuously to the disjoint circles $z_5(t) = \frac{1}{2}(e^{-it} - 1)$ and $z_6(t) = 1 + \frac{1}{2}e^{it}$, where, for $0 \le t \le 2\pi$, each retains the orientation induced by C and one is centered at the singularity $z = -\frac{1}{2}$ while the other is centered at the singularity $z = 1$. The contour integral on C is then given by

$$\int_{0}^{2\pi} [f(z_5(t)) z_5'(t) + f(z_6(t)) z_6'(t)]\, dt = 4\pi i$$

Each of these two solutions rely on brute-force integrations that are *extremely* tedious to carry out "by hand." A simpler solution follows from the partial-fraction decomposition

$$f(z) = \frac{3}{z + \frac{1}{2}} + \frac{5}{z - 1} = f_1(z) + f_2(z)$$

Again, deform the triangular contours C_a and C_b into disjoint, positively oriented (counterclockwise) circles C_1 and C_2, centered respectively at the isolated singularities $z = -\frac{1}{2}$ and $z = 1$. In the contour integral of $f(z)$ around C_1, $f_2(z)$ makes no contribution; and in the contour integral around C_2, $f_1(z)$ makes no contribution. The required contour integral is then given by

$$\int_C \frac{16z - 1}{2z^2 - z - 1}\, dz = -\int_{C_1} f_1(z)\, dz + \int_{C_2} f_2(z)\, dz = -3(2\pi i) + 5(2\pi i) = 4\pi i$$

The minus sign in front of the integral over C_1 arises from the clockwise orientation imposed by C.

The intended conclusion for this example is the realization that careful analysis and clever technique can sometimes be far superior to brute-force computation.   ❖

Indefinite Integrals and Antiderivatives

Indefinite integrals can be defined for analytic functions since contour integrals of analytic functions are path independent. This means the value of a contour integral of an analytic function depends only on the endpoints and not on the path. Provided the path does not go through any singularities, we can use the definite integral to construct an antiderivative.

If $f(z)$ is analytic in a simply connected domain D and if z and a are points in the region where $f(z)$ is analytic, we can define $F(z) = \int_a^z f(\zeta)\, d\zeta$, where the integration is on any contour lying completely in D and connecting the fixed point a with the varying point z. Since the integral is path independent, the value of $F(z)$ is well defined. In addition, it can be shown that $F'(z) = f(z)$, so $F(z)$ is an antiderivative for the function $f(z)$.

EXAMPLE 35.53   The definite integral $\int_1^z \cos \zeta \, d\zeta$ defines the function $F(z) = \sin z - \sin 1$ whose derivative is $\cos z$. Thus, $F(z) = \sin z - \sin 1$, *an antiderivative of* $f(z) = \cos z$, is constructed, at least in principle, by a contour integral. ❖

EXAMPLE 35.54   The definite integral $\sigma = \int_4^z \frac{1}{\sqrt{\zeta}} \, d\zeta$ defines the function $F(z) = 2\sqrt{z} - 4$ whose derivative is $\frac{1}{\sqrt{z}}$. Thus, $F(z) = 2\sqrt{z} - 4$, *an antiderivative of* $f(z) = \frac{1}{\sqrt{z}}$, is constructed, at least in principle, by evaluating a contour integral. The sense in which $F(z)$ is an antiderivative is shown in the evaluation of the integral $\lambda = \int_{z_1}^{z_2} \frac{dz}{\sqrt{z}}$, where $z_1 = e^{-3\pi i/4} = -\frac{i}{\sqrt{2}}(1+i)$, $z_2 = e^{3\pi i/4} = \frac{1}{\sqrt{2}}(1 - i)$, and C is the arc $z(t) = e^{it}$, $-\frac{3}{4}\pi i \le t \le \frac{3}{4}\pi i$.

First, evaluate λ via the antiderivative $F(z)$, obtaining $F(z_2) - F(z_1) = 2i\sqrt{2 + \sqrt{2}} = s_1$. Then, evaluate λ as a contour integral on C, obtaining $\lambda = \int_{-3\pi/4}^{3\pi/4} i\sqrt{e^{it}} \, dt = s_1$. Thus, $F(z)$ can be used to evaluate a contour integral that does not cross a branch cut. On the other hand, if the formula for the antiderivative were not known, the path-independent contour integral σ *defines* the antiderivative.

Finally, we force the contour integration to cross the branch cut by integrating along $z(t) = z_1 + (z_2 - z_1)t$, $0 \le t \le 1$, the straight line from z_1 to z_2. Interrupting the integral at the branch cut, we have

$$\int_0^{1/2} \frac{z'(t)}{\sqrt{z(t)}} \, dt + \int_{1/2}^1 \frac{z'(t)}{\sqrt{z(t)}} \, dt = 2i\left(\sqrt{2 + \sqrt{2}} - 2^{3/4}\right) \ne s_1$$

Deliberately crossing the branch cut results in a value that is not consistent with the value obtained with the antiderivative. ❖

EXAMPLE 35.55   Let z_1 and z_2 be the same points, on the same circular arc C, as in Example 35.54. Let $z(t)$ be the same straight line connecting the points z_1 and z_2. An antiderivative for the integral $\sigma = \int_2^z \frac{d\zeta}{\zeta}$ is $F(z) = \text{Log } z - \text{Log } 2$, where $\text{Log } z$ is the principal logarithm. The sense in which $F(z)$ is an antiderivative is shown in the evaluation of the integral

$$\lambda = \int_{z_1}^{z_2} \frac{dz}{z} = F(z_2) - F(z_1) = \tfrac{3}{2}\pi i = s_1$$

Then, evaluate λ as a contour integral on C, obtaining $\lambda = \int_{-3\pi/4}^{3\pi/4} i \, dt = s_1$. Thus, $F(z)$ can be used to evaluate a contour integral that does not cross a branch cut. On the other hand, if the formula for the antiderivative were not known, the path-independent contour integral σ *defines* the antiderivative.

Finally, we force the contour integration to cross the branch cut by integrating along $z(t)$, the straight line from z_1 to z_2. Interrupting the integral at the branch cut, we have

$$\int_0^{1/2} \frac{z'(t)}{z(t)} \, dt + \int_{1/2}^1 \frac{z'(t)}{z(t)} \, dt = -\frac{\pi}{2}i \ne s_1$$

Deliberately crossing the branch cut results in a value that is not consistent with the value obtained with the antiderivative. ❖

Cauchy Integral Formulas

If $f(z)$ is analytic inside and on C—a positively oriented, simple, closed contour—and z_0 is any point inside C, then the *Cauchy integral formula*

$$f(z_0) = \frac{1}{2\pi i} \int_C \frac{f(z)}{z - z_0} \, dz \tag{35.35}$$

expresses the value of $f(z_0)$ in terms of the values of $f(z)$ along the enveloping contour C.

This single formula is one case of the *Cauchy integral formulae for derivatives*

$$f^{(n)}(z_0) = \frac{n!}{2\pi i} \int_C \frac{f(z)}{(z-z_0)^{n+1}} \, dz \tag{35.36}$$

The Cauchy Integral Formula

The Cauchy integral formula (35.35) is typically used to obtain the value of a contour integral without actually evaluating the integral. Thus, the value of the contour integral $\int_C \frac{f(z)}{z-z_0} \, dz$ is $2\pi i f(z_0)$. The contour integrals in Examples 35.56 and 35.57 will be evaluated numerically and then exactly by the Cauchy integral formula. The contour integral in Example 35.58 will be evaluated exactly by contour integration and by the Cauchy integral formula.

EXAMPLE 35.56   Let C be the contour given by $z(t) = 2e^{it}$, $-\pi \le t \le \pi$, a circle $|z| = 2$, having center at the origin and radius 2. Then, by numeric integration we get

$$\int_C \frac{e^z}{z-1} \, dz = 2i \int_{-\pi}^{\pi} \frac{e^{2e^{it}} e^{it}}{2e^{it} - 1} \, dt \overset{\circ}{=} 17.079i$$

Since the circle C encloses $z_0 = 1$, if we take $f(z) = e^z$, we can apply the Cauchy integral formula, obtaining

$$\int_C \frac{f(z)}{z-z_0} \, dz = 2\pi i f(1) = 2\pi i e \overset{\circ}{=} 17.079i \qquad \qquad ❖$$

EXAMPLE 35.57   Let C be the contour $z(t) = e^{it}$, $-\pi \le t \le \pi$, a circle centered at the origin and having radius 1. Then, by numeric integration we get

$$\int_C \frac{e^z}{2z^2 - 5z + 2} \, dz = i \int_{-\pi}^{\pi} \frac{e^{e^{it}} e^{it}}{2e^{2it} - 5e^{it} + 2} \, dt \overset{\circ}{=} -3.453i \tag{35.37}$$

If we write $f(z) = e^z/(2(z-2))$ and $F(z) = \frac{f(z)}{z-1/2}$, then $F(z)$ has two isolated singularities, one at $z = 2$ and another at $z = \frac{1}{2}$. The singularity at $z = 2$ is outside the circle $|z| = 1$, and hence, the factor $(z-2)$ can be "grouped with" the numerator. Thus, we let $z_0 = \frac{1}{2}$ and apply the Cauchy integral formula to the contour integral in (35.37), obtaining

$$\int_C \frac{e^z}{2z^2 - 5z + 2} \, dz = \int_C F(z) \, dz = \int_C \frac{f(z)}{z - \frac{1}{2}} \, dz = 2\pi i f\left(\tfrac{1}{2}\right) = -\tfrac{2}{3} i \pi e^{1/2} \overset{\circ}{=} -3.453i$$

$$❖$$

EXAMPLE 35.58   Let C be the contour $z(t) = 2e^{it}$, $-\pi \le t \le \pi$, a circle centered at the origin and having radius 2. Then, by direct contour integration we find

$$\int_C \frac{\sin z}{z-1} \, dz = 2i \int_{-\pi}^{\pi} \frac{\sin(2e^{it}) e^{it}}{2e^{it} - 1} \, dt = 2\pi i \sin 1 \tag{35.38}$$

To evaluate (35.38) with the Cauchy integral formula, let $f(z) = \sin z$. Since $z_0 = 1$ is within the circle $|z| = 2$, the Cauchy integral formula applies, giving $2\pi i f(1)$ as the value of the integral. With $f(z) = \sin z$, the value for the contour integral is clearly $2\pi i \sin 1$. ❖

The Cauchy Integral Formulae for Derivatives

The hypotheses required for the derivative formulae are the same as those needed for the Cauchy integral formula, namely, that $f(z)$ is analytic within and on the positively oriented, simple closed contour C surrounding z_0.

As an illustration of these formulae, verify that for $f(z) = e^{z^2}$, the contour integral $\frac{3!}{2\pi i} \int_C \frac{f(z)}{(z-i)^4}\, dz$ gives $f'''(i) = 4ie^{-1} \stackrel{\circ}{=} 1.4715$, if the contour C, a circle with radius 1 and center at $z = i$, is parametrized by $z(t) = i + e^{it}$, $-\pi \le t \le \pi$. By numeric integration we find

$$\frac{3!}{2\pi i} \int_C \frac{f(z)}{(z-i)^4}\, dz = \frac{3}{\pi}\int_{-\pi}^{\pi} e^{(i+e^{it})^2} e^{-3it}\, dt \stackrel{\circ}{=} 1.4715$$

EXERCISES 35.11–Part A

A1. Verify the result in (35.27).   **A2.** Verify the result in (35.28).   **A5.** Verify the result in (35.31).

A3. Verify the result in (35.29).   **A4.** Verify the result in (35.30).

EXERCISES 35.11–Part B

B1. Verify the result in (35.32).   **B2.** Verify the result in (35.33).

B3. Verify the result in (35.34).

In Exercises B4–18, apply the generalized Cauchy–Goursat theorem and the Cauchy integral formula to evaluate each of the given contour integrals, where

(a) C is the circle $z(t) = 15e^{it}, 0 \le t \le 2\pi$.

(b) C is a square with side 10 and center at the origin.

(c) possible, obtain corroboration by integrating along the contour in part (a), directly, either exactly or numerically.

B4. $\int_C \frac{dz}{z^2 + 9}$ **B5.** $\int_C \frac{\sin z}{z^2 + 25}\, dz$ **B6.** $\int_C \frac{e^{-z}}{z^2 + 16}\, dz$

B7. $\int_C \frac{dz}{z^2 - 18z + 85}$ **B8.** $\int_C \frac{dz}{z^2 + 8z + 41}$

B9. $\int_C \frac{dz}{z^2 - 6z + 34}$ **B10.** $\int_C \frac{dz}{z^2 + 10z + 61}$

B11. $\int_C \frac{dz}{z^2 + 4z + 29}$ **B12.** $\int_C \frac{6\cos z}{z^2 - 2z + 2}\, dz$

B13. $\int_C \frac{16\sinh z}{z^2 - 6z + 25}\, dz$ **B14.** $\int_C \frac{\cosh 2z}{z^2 + 14z + 98}\, dz$

B15. $\int_C \frac{e^z \sin z}{z^2 + 8z + 32}\, dz$ **B16.** $\int_C \frac{6z - 4 + 12i}{z^2 - 3z + 32 - 30i}\, dz$

B17. $\int_C \frac{z + 5 + 4i}{z^2 + (-2 + 5i)z - 9 - 7i}\, dz$

B18. $\int_C \frac{6z - 8 + 40i}{z^2 - 8z + 37 - 20i}\, dz$

In Exercises B19–23, let C be a circle with appropriate radius and center at the appropriate z_0, and use (35.36) to determine the indicated derivative. Corroborate your answer by direct computation.

B19. $f''(2 + 3i)$, $f(z) = \dfrac{z}{z^2 + 3}$

B20. $f'''(-1 - i)$, $f(z) = \dfrac{z + 1}{z^2 + 2z + 3}$

B21. $f^{(4)}(i)$, $f(z) = \dfrac{z - i}{z + i}$ **B22.** $f'''\left(\frac{1}{2} - \frac{1}{3}i\right)$, $f(z) = z^6$

B23. $f'\left(\frac{2}{3} + \frac{5}{4}i\right)$, $f(z) = z\sin z$

For the indefinite integrals $\int_a^z f(\zeta)\, d\zeta$ given in Exercises B24–28, find an antiderivative and use it to evaluate the definite integral $\int_{z_1}^{z_2} f(\zeta)\, d\zeta$, where z_1 and z_2 are as given. Then, being careful about singularities and branch cuts, connect z_1 and z_2 with a line segment or a polygonal path that does not cross the branch cut and evaluate the definite integral as a contour integral along this path. (*Hint:* The line segment from ζ_1 to ζ_2 can be parametrized as $z(t) = \zeta_1 + (\zeta_2 - \zeta_1)t, 0 \le t \le 1$.)

B24. $\int_2^z \zeta e^\zeta\, d\zeta$; $z_1 = 1 + 2i, z_2 = -1 - 3i$

B25. $\int_i^z \sqrt{\zeta}\, d\zeta$; $z_1 = -1 - i, z_2 = -1 + i$

B26. $\int_{-i}^z \frac{d\zeta}{\zeta}$; $z_1 = -1 - i, z_2 = -2 + 3i$

B27. $\int_1^z \frac{d\zeta}{\sqrt{\zeta}}$; $z_1 = 1 - i, z_2 = -7 + 2i$

B28. $\int_{3i}^z \text{Log}\,\zeta\, d\zeta$; $z_1 = 2 - 3i, z_2 = -12 + i$

35.12   **Series in Powers of z**

Power Series

A *power series* is a sum of the form

$$\sum_{k=0}^{\infty} a_k (z - z_0)^k \qquad (35.39)$$

where the complex numbers a_k are the *coefficients*, the complex number z_0 is the fixed point about which the series is constructed, and z is a complex variable. If this sum is evaluated at $z = z_0$, the first term is $a_0 0^0$, which we interpret as a_0.

The following statements summarize some of the important properties of a power series.

1. If $\lim_{n=\infty} \sum_{k=0}^{n} a_k (z - z_0)^k = f(z)$ exists pointwise, we say (35.39) converges to $f(z)$.

2. If (35.39) converges for $z_1 \neq z_0$, then it converges absolutely for all z in the disk $|z - z_0| < |z_1 - z_0| = r$.

3. The radius of the largest disk for which (35.39) converges is R, the *radius of convergence*.

4. The radius of convergence can be found by the ratio test.

5. The power series (35.39) converges absolutely in $|z - z_0| < R$ and uniformly in $|z - z_0| \leq r < R$.

6. The limit function $f(z)$ is analytic.

7. If (35.39) converges to $f(z)$, then $a_k = f^{(k)}(z_0)/k!$.

Thus, if a power series converges, it converges inside a disk called its *circle of convergence*, and converges to an analytic function. Conversely, the analytic function $f(z)$ has a power series expansion about any point of analyticity. This is the content of Taylor's theorem, which establishes the existence of Taylor series, the name of power series created under its aegis.

THEOREM 35.7   TAYLOR'S THEOREM

If $f(z)$ is analytic in an open connected set D and z_0 is in D, then

$$f(z) = \sum_{k=0}^{\infty} \frac{f^{(k)}(z_0)}{k!} (z - z_0)^k$$

converges absolutely inside the largest circle about z_0 for which $f(z)$ is analytic.

Thus, if you start with an analytic function and apply Taylor's theorem, a power series, called the Taylor series, results. If you start with a power series that converges, the limit function is analytic and the power series is the Taylor series for that function.

EXAMPLE 35.59   At $z_0 = 0$, the Taylor series for the complex function $f(z) = \frac{1}{1-z}$ is $\sum_{k=0}^{\infty} z^k$. It is just the geometric series of elementary algebra.   ❖

EXAMPLE 35.60   At $z_0 = 0$, the Taylor series for the function $g(z) = \sin z$ is $\sum_{k=0}^{\infty}(-1)^k z^{2k+1}/(2k+1)!$. ❖

EXAMPLE 35.61   At $z_0 = 0$, the Taylor series for the function $h(z) = \cos z$ is $\sum_{k=0}^{\infty}(-1)^k z^{2k}/(2k)!$. ❖

EXAMPLE 35.62   At $z_0 = 0$, the Taylor series for the function $w(z) = e^z$ is $\sum_{k=0}^{\infty} z^k/k!$. ❖

EXAMPLE 35.63   At $z = \frac{1}{2}(1+i)$, the Taylor series $f(z) = \frac{1}{1-z} = \sum_{k=0}^{\infty} z^k$ becomes $\sum_{k=0}^{\infty}(1+i)^k/2^k$, which converges to $1+i$, which is the value of $f(\frac{1+i}{2})$. On the other hand, if $z = 1+i$, to what will the partial sums $S_n = \sum_{k=0}^{n}(1+i)^k$ converge? The first few partial sums, $1, 2+i, 2+3i, 5i, -4+5i, -8+i, -8-7i, -15i, 16-15i, 32+i, 32+33i$, graphed in Figure 35.44, seem to be moving away from the origin, suggesting that perhaps this series does not converge for $z = 1+i$.

Thus, for some values of z such as $z = \frac{1+i}{2}$, the power series $\sum_{k=0}^{\infty} z^k$ converges. For other values such as $z = 1+i$, the series diverges. In general, convergence takes place within a circle called the *circle of convergence*. The circle is centered at $z = z_0$, where z_0 is the point about which the Taylor series was computed. The radius of this circle is R, the *radius of convergence*. The circle of convergence is typically denoted by $|z - z_0| < R$.

The *circle of convergence* expands outward from z_0 and grows until it hits the first point for which $f(z)$ is not analytic. For the function $f(z) = \frac{1}{1-z}$, this first singularity is at $(1, 0)$ on the real axis. That $z = \frac{1+i}{2}$ is inside $|z| = 1$ while $z = 1+i$ is outside this circle can be seen by computing the magnitudes $\frac{1}{\sqrt{2}}$ and $\sqrt{2}$, respectively. At the first point, within the circle of convergence, we observed convergence. At the second, outside the circle of convergence, we saw evidence of divergence. This information is summarized in Figure 35.45, which shows the circle of convergence, the point $z = 1$ where $f(z)$ is not analytic, and the two points $z = \frac{1+i}{2}$ and $z = 1+i$, one inside, and one outside, the circle. ❖

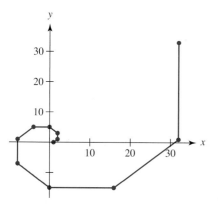

FIGURE 35.44   Example 35.63: The partial sums $\sum_{k=0}^{n}(1+i)^k$

Ratio Test

As in Section 9.2, we can use the ratio test to determine the radius of convergence of a power series.

EXAMPLE 35.64   For the geometric series $\sum_{k=0}^{\infty} z^k$, the coefficients are $a_k = 1, k = 0, 1, \ldots$. Hence, we examine the absolute value of the ratio of two successive terms but carry through the powers of z, obtaining $|z^{k+1}/z^k| = |z|$. The "test number" in the ratio test, namely, L, must be less than 1 for convergence. The ratio of two successive terms is $|z|$ in this example. Hence, we seek those values of z for which $L = \lim_{k \to \infty} |z| < 1$ and conclude that the geometric series converges for z satisfying $|z| < 1$. This is what we mean when we say R, the radius of convergence, is 1 and that the circle of convergence is given by $|z| < 1$. ❖

EXAMPLE 35.65   Although the coefficients in $\sum_{k=0}^{\infty} z^k/k!$, the power series expansion of $f(z) = e^z$, are $a_k = \frac{1}{k!}$, the ratio test is to be executed with $z^k/k!$ as the general term. The absolute value of the ratio of two such successive terms is then

$$\left| \frac{z^{k+1}k!}{z^k(k+1)!} \right| = \left| \frac{z}{k+1} \right|$$

The ratio test says the series converges provided $L = \lim_{k\to\infty}\left|\frac{z}{k+1}\right| < 1$. Isolating $|z|$, we find the condition for convergence to be $|z| < \lim_{k\to\infty} k + 1 = \infty$. Thus R, the radius of convergence, is infinite and we have the circle of convergence defined by $|z| < \infty$.   ❖

EXAMPLE 35.66   Find R, the radius of convergence for

$$\sin z = \sum_{k=0}^{\infty} \frac{(-1)^k}{(2k+1)!} z^{2k+1}$$

The general term of this power series is $(-1)^k z^{2k+1}/(2k+1)!$, so the absolute value of the ratio of two successive terms is

$$\left|\frac{(-1)^{k+1}z^{2k+3}(2k+1)!}{(-1)^k z^{2k+1}(2k+3)!}\right| = \left|\frac{z^2}{2(k+1)(2k+3)}\right|$$

We are looking for those values of z for which $L = \lim_{k\to\infty}\left|\frac{z^2}{2(k+1)(2k+3)}\right| < 1$. Isolating $|z^2| = |z|^2$ before computing the limit yields $|z|^2 < \lim_{k\to\infty} 2(k+1)(2k+3)$. The circle of convergence is therefore defined by

$$|z| < \lim_{k\to\infty}\sqrt{2(k+1)(2k+3)} = R = \infty$$

The series for $\sin z$ converges in the whole finite plane, and the radius of convergence is described by $R = \infty$.   ❖

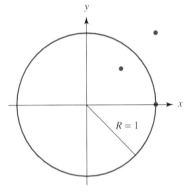

FIGURE 35.45   Example 35.63: The circle of convergence, along with the points $z = 1, \frac{1}{2}(1+i)$, and $1+i$

Why Are Power Series So Useful?

Inside the circle of convergence, a power series represents an analytic function and the operations of termwise addition, subtraction, multiplication, division, differentiation, and integration are all valid. Moreover, the coefficients in a power series are unique. Hence, inside the circle of convergence, an infinite power series can be manipulated like a polynomial.

Taylor Series and the Cauchy Integral Formulae

Taylor's theorem tells us that the coefficients a_k in the power series $f(z) = \sum_{k=0}^{\infty} a_k(z - z_0)^k$ are given by $a_k = f^{(k)}(z_0)/k!$. We also know from Section 35.11 that the Cauchy integral formulae for derivatives give $f^{(k)}(z_0) = \frac{k!}{2\pi i}\int_C (f(z)/(z - z_0)^{k+1})\, dz$, where C is an appropriate simple closed positively oriented contour surrounding z_0. Thus, the coefficients in the power series are now given, not by derivatives, but by integrals! Indeed, the coefficients are now

$$a_k = \frac{1}{k!}\left\{\frac{k!}{2\pi i}\int_C \frac{f(z)}{(z - z_0)^{k+1}}\, dz\right\} = \frac{1}{2\pi i}\int_C \frac{f(z)}{(z - z_0)^{k+1}}\, dz$$

where the contour C can be taken as a circle about z_0.

EXAMPLE 35.67   The function $f(z) = \frac{z+1}{z^2+2z-3}$ is analytic everywhere except where the denominator is zero, that is, except at the points $z = -3, 1$. If we expand $f(z)$ in a Taylor series about $z_0 = 0$, we obtain

$$\frac{1}{6}\sum_{k=0}^{\infty}\frac{1 - 3(-3)^k}{(-3)^k}z^k = -\frac{1}{3} - \frac{5}{9}z - \frac{13}{27}z^2 - \frac{41}{81}z^3 - \frac{121}{243}z^4 + O(z^5)$$

Since the singularity closest to $z_0 = 0$ occurs at $z = 1$, we choose for the contour C, the circle $z = e^{it}/2$, $0 \le t \le 2\pi$, which has radius $\frac{1}{2}$ and center at the origin. Thus, C does not enclose any singularity of $f(z)$ and we compute the value of the integrals

$$\frac{1}{2\pi i} \int_C \frac{f(z)}{(z - z_0)^{k+1}} \, dz = \int_0^{2\pi} \frac{i\left(\frac{1}{2}e^{it} + 1\right)e^{it}}{2\left(\frac{1}{4}e^{2it} + e^{it} - 3\right)\left(\frac{1}{2}e^{it}\right)^{k+1}} \, dt \quad k = 0, 1, \ldots, 4$$

to be $-\frac{1}{3}, -\frac{5}{9}, -\frac{13}{27}, -\frac{41}{81}$, and $-\frac{121}{243}$.   ❖

Analytic Continuation

We next investigate the relationship between two power series for $f(z)$ developed in two overlapping circles of convergence. The transition from one circle to the other is known as *analytic continuation*.

EXAMPLE 35.68   Inside the circle $|z| < 1$ we have $f(z) = \frac{1}{1-z} = \sum_{k=0}^{\infty} z^k$, a power series about the point $z_0 = 0$. Expanding about $z_0 = \frac{i}{2}$ gives the series

$$\sigma_1 = \sum_{k=0}^{\infty} \left(\frac{4}{5} + \frac{2}{5}i\right)^{k+1} \left(z - \frac{i}{2}\right)^k$$

The circle of convergence "expands out" from $z_0 = \frac{i}{2}$ until it hits the singularity at $z = 1$, as confirmed by the ratio test. The ratio of two successive terms simplifies to $\frac{2}{\sqrt{5}}\left|1 - \frac{i}{2}\right|$, from which we obtain $\left|1 - \frac{i}{2}\right| < \frac{\sqrt{5}}{2} = R$. In fact, $\left|1 - \frac{i}{2}\right| = \frac{\sqrt{5}}{2}$.

Expanding $f(z)$ about $z_0 = \frac{9}{10}i$ gives the series

$$\sigma_2 = \sum_{k=0}^{\infty} \left(\frac{100}{181} + \frac{90}{181}i\right)^{k+1} \left(z - \frac{9}{10}i\right)^k$$

Again, the circle of convergence expands out from $z_0 = \frac{9}{10}i$ until it hits the singularity at $z = 1$. Repeating the ratio test, we find $R = \frac{\sqrt{181}}{10} = \left|\frac{9}{10}i - 1\right|$.

Now, the point $z_1 = \frac{3}{2}i$ is outside the circle $|z| = 1$, but inside the circle $\left|z - \frac{9}{10}i\right| = \frac{\sqrt{181}}{10}$, as shown in Figure 35.46. Computation then shows $f(z_1) = \sigma_j(z_1) = \frac{1}{13}(4 + 6i)$, $j = 1, 2$, so $f(z)$ is represented by a host of Taylor series. Some of these series have circles of convergence that overlap. If two different series are valid at a point, both series have to produce the same value for $f(z)$. However, there may well be points not common to the different circles of convergence. The "new" series is then used to compute values outside the "old" circle of convergence but inside the "new" circle of convergence. That is how the values for a function are "continued analytically" beyond the original circle of convergence.   ❖

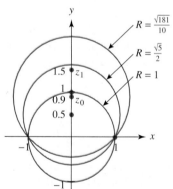

FIGURE 35.46   Analytic continuation in Example 35.68

Laurent Series

EXAMPLE 35.69   The function $f(z) = e^z/z^3$ does not have a Taylor series about $z_0 = 0$ where $f(z)$ is clearly not analytic. However, expanding e^z at $z_0 = 0$ and dividing termwise by z^3 gives the "series"

$$\sum_{k=0}^{\infty} \frac{z^{k-3}}{k!} = \frac{1}{z^3} + \frac{1}{z^2} + \frac{1}{2}\frac{1}{z} + \frac{1}{3!} + \frac{1}{4!}z + \frac{1}{5!}z^2 + \cdots$$

While the function $f(z)$ is not analytic in a disk $|z| < R$, it *is* analytic in the annulus (washer) $r < |z| < R$. A series of this form, with negative integer powers of z, is called a Laurent series, for which it is useful to distinguish between the two cases $r = 0$ and $r > 0$.   ❖

THEOREM 35.8   LAURENT'S THEOREM

1. $f(z)$ is analytic in the annulus $r < |z - z_0| < R$.

$\Longrightarrow$

1. Inside this annulus $f(z)$ has the unique representation

$$f(z) = \sum_{k=-\infty}^{\infty} c_k(z - z_0)^k = \sum_{k=1}^{\infty} b_k(z - z_0)^{-k} + \sum_{k=0}^{\infty} a_k(z - z_0)^k$$

2. $a_k = \dfrac{f^{(k)}(z_0)}{k!} = \dfrac{1}{k!}\left[\dfrac{k!}{2\pi i} \int_C \dfrac{f(z)}{(z - z_0)^{k+1}}\, dz \right] = \dfrac{1}{2\pi i} \int_C \dfrac{f(z)}{(z - z_0)^{k+1}}\, dz,$

$k = 0, 1, \dots$

3. $b_k = \dfrac{1}{2\pi i} \int_C \dfrac{f(z)}{(z - z_0)^{1-k}}\, dz \quad k = 1, 2, \dots$

The coefficients b_k are interesting, especially $b_1 = \frac{1}{2\pi i} \int_C f(z)\, dz$, which is a multiple of the contour integral of $f(z)$ around the positively oriented, simple closed contour C lying inside the appropriate annulus and containing z_0 in its interior. The coefficient b_1 is called the *residue* of $f(z)$ at z_0 and is typically found, not by integrating, but by other devices. The theory of using Laurent series to evaluate contour integrals is called the *residue calculus;* and Section 35.13 is devoted to its study.

When $r = 0$ so that the annulus is $0 < |z - z_0| < R$, the following hold.

1. The series $\sum_{k=1}^{\infty} b_k(z - z_0)^{-k}$ is called the *principal part* of $f(z)$ at z_0.
2. If the "last b" is b_N, the function $f(z)$ is said to have a *pole of order N* at z_0. If $N = 1$, $f(z)$ is said to have a *simple pole* at z_0.
3. If there are an infinite number of b's, then $f(z)$ has an *essential singularity* at z_0. An essential singularity has unique properties, captured in the following two theorems that we merely state. (See [61] for a proof of the first theorem and [93] or [23] for a proof of the second.)

THEOREM 35.9   CASORATI–WEIERSTRASS THEOREM

In any neighborhood of an isolated essential singularity z_0, the function $f(z)$ comes arbitrarily close to every value.

THEOREM 35.10   PICARD'S LITTLE THEOREM

In any neighborhood of an isolated essential singularity, $f(z)$ attains every possible complex value with at most one exception.

A function that has an essential singularity at $z = 0$ is $f(z) = e^{1/z} = \cos \frac{1}{z} + i \sin \frac{1}{z}$. The oscillatory behavior of $\cos \frac{1}{z}$ and $\sin \frac{1}{z}$ on the real line should make it plausible that in the complex plane, this function would oscillate wildly enough for both these characterizing theorems to be realized.

EXAMPLE 35.70   The Laurent series representation of $f(z) = \frac{1+\sin z}{z^3}$, valid in the annulus $0 < |z|$, is

$$z^{-3} + z^{-2} + \sum_{k=0}^{\infty} \frac{(-1)^{k+1}z^{2k}}{(2k+3)!} = z^{-3} + z^{-2} - \frac{1}{3!} + \frac{z^2}{5!} - \frac{z^4}{7!} + \cdots$$

found by termwise dividing the Taylor series for $1 + \sin z$ by z^3. Thus, $z = 0$ is a pole of order 3 and the principal part of $f(z)$ at $z = 0$ is $z^{-3} + z^{-2}$.   ❖

EXAMPLE 35.71   In the annulus $0 < |z|$, the Laurent series for the function $f(z) = e^{-z^{-2}}$ is $\sum_{k=0}^{\infty}(-1)^k/k!z^{2k}$, obtained by replacing x in $e^x = \sum_{k=0}^{\infty} x^k/k!$ with $-\frac{1}{z^2}$. Thus, $z = 0$ is an essential singularity of $f(z)$.   ❖

EXAMPLE 35.72   The function $f(z) = \frac{1}{(1-z)(2+z)}$ has singularities at $z = 1$ and $z = -2$. Hence, series expansions for $f(z)$ can be constructed in powers of z in each of the three regions $A = \{|z| < 1\}$, $B = \{1 < |z| < 2\}$, and $C = \{|z| > 2\}$ shown in Figure 35.47. In region A, $f(z)$ is given by the Taylor series $\sum_{k=0}^{\infty} z^k((-2)^{-k} + 2)/6$. In region C, $f(z)$ is given by the asymptotic series

$$\frac{1}{2} - \frac{1}{6}\sum_{k=0}^{\infty}((-2)^k + 2)z^{-k} = -z^{-2} + z^{-3} - 3z^{-4} + \cdots$$

a Taylor series constructed about the point $z_0 = \infty$.

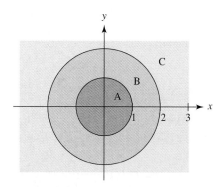

Obtaining a series valid in region B, the annulus $1 < |z| < 2$, requires the most work. Write $f_1(z) = \frac{1}{3}\frac{1}{1-z}$ and $f_2(z) = \frac{1}{3}\frac{1}{2+z}$, so $f(z) = f_1(z) + f_2(z)$, a partial fraction decomposition. A Taylor series would represent f_1 *inside* the circle $|z| < 1$, but we want a representation for $f(z)$ valid *outside* this circle. Hence, the expansion representing f_1 must be the asymptotic expansion $-\frac{1}{3}\sum_{k=0}^{\infty} z^{-k-1}$. The Taylor series $\frac{1}{6}\sum_{k=0}^{\infty}\left(\frac{-z}{2}\right)^k$ represents f_2 inside the circle $|z| < 2$. Consequently, the sum $\frac{1}{6}\sum_{k=0}^{\infty}\left(\frac{-z}{2}\right)^k - \frac{1}{3}\sum_{k=0}^{\infty} z^{-k-1}$ represents $f(z)$ in the annulus B.

Finally, note carefully that the annulus $1 < |z| < 2$ is of the form $r < |z - z_0| < R$ and $r \neq 0$. So, although the resulting Laurent series has an infinite number of b's, the origin is *not* an essential singularity for $f(z)$. The terms *pole of order N* and *essential singularity* are reserved for cases where the annulus is of the form $0 < |z - z_0| < R$.   ❖

FIGURE 35.47   The three regions of Example 35.72

EXERCISES 35.12

1. Obtain the Taylor expansion valid in region A in Example 35.72. *Hint:* Look for the geometric series in the partial fraction decomposition of $f(z)$.

2. Obtain the series expansion for $f(z)$ in region C in Example 35.72. *Hint:* Look for the geometric series in the partial fraction decomposition of $f(z)$.

In Exercises 3–12:

(a) Obtain a power series representation of $f(z)$ at the indicated z_0.

(b) Determine the radius of convergence geometrically by finding the distance to the singularity nearest to z_0.

(c) Where possible, use the ratio test to confirm the value of the radius of convergence.

3. $f(z) = \dfrac{z+1}{z^2 + 2z + 10}$; $z_0 = 0$ **4.** $f(z) = \dfrac{ze^z}{z^2 + 9}$; $z_0 = 0$

5. $f(z) = \dfrac{z}{z^2 + 16}$; $z_0 = 2i$ **6.** $f(z) = \dfrac{z+3}{z-5}$; $z_0 = 1$

7. $f(z) = \dfrac{z+3}{z-5}$; $z_0 = 10$ **8.** $f(z) = \dfrac{z^2 - 1}{z^2 + 1}$; $z_0 = 0$

9. $f(z) = \dfrac{z^2 - 1}{z^2 + 1}$; $z_0 = 3$ **10.** $f(z) = \sqrt{z}$; $z_0 = i$

11. $f(z) = \dfrac{z + 2}{z^2 + 4z + 13}$; $z_0 = 1$ **12.** $f(z) = \dfrac{z}{z^2 - z - 6}$; $z_0 = i$

In Exercises 13–22, determine all zeros and poles of the given functions. Indicate the multiplicity of the zeros and the order of the poles.

13. $f(z) = \dfrac{7z^2 - z - 1}{z^3 + (7 - 21i)z^2 - (149 + 114i)z - 507 + 279i}$

14. $f(z) = \dfrac{9z^2 + 5z + 1}{z^3 + (4 + 3i)z^2 + (69 - 42i)z + 378 + 171i}$

15. $f(z) = \dfrac{5z^2 - 6z - 5}{z^3 + (1 + i)z^2 + (7 - 6i)z + 15 + 5i}$

16. $f(z) = \dfrac{-6z^2 - 9z - 6}{z^3 - (10 + 13i)z^2 + (12 + 116i)z + 264 - 372i}$

17. $f(z) = \dfrac{2z^3 - z^2 - 7z + 6}{z^4 + (19 + i)z^3 + (162 - 6i)z^2 + (604 - 172i)z + 728 - 504i}$

18. $f(z) = \dfrac{3z^2 + 5z + 5}{z^4 + 4iz^3 + (8 + 48i)z^2 - (96 - 24i)z + 288i - 540}$

19. $f(z) = \dfrac{4z^3 + 9z^2 + 5z + 6}{z^4 + (20 - 18i)z^3 - (66 + 246i)z^2 - (1932 - 296i)z - 3648 + 7468i}$

20. $f(z) = \dfrac{5z^2 - 7z - 5}{z^4 + (4 - 10i)z^3 - (159 + 46i)z^2 - (406 - 638i)z + 4592 + 1794i}$

21. $f(z) = \dfrac{z^3 - 2z^2 - 5z + 6}{z^5 + (6 + 5i)z^4 + (17 + 12i)z^3 + (74 + 33i)z^2 + (96 + 40i)z + 184 - 36i}$

22. $f(z) = \dfrac{9z^2 - 3z + 5}{z^5 - (9 + 2i)z^4 + (74 - 18i)z^3 - (378 - 46i)z^2 + (965 - 630i)z - 2925 + 1100i}$

The functions in Exercises 23–30 have simple poles at z_1 and z_2. For each k, $k = 1, 2$, obtain a Laurent series expansion valid in $0 < |z - z_k| < R_k$ and determine the largest possible value of R_k.

23. $f(z) = \dfrac{z - 4}{z^2 - 6z + 8}$ **24.** $f(z) = \dfrac{z - 5}{z^2 - 4z - 45}$

25. $f(z) = \dfrac{z + 9}{z^2 - 4z}$ **26.** $f(z) = \dfrac{z + 8}{z^2 + z - 20}$

27. $f(z) = \dfrac{z}{z^2 + 6z + 5}$ **28.** $f(z) = \dfrac{z - 1}{z^2 + 4}$

29. $f(z) = \dfrac{1}{z^2 - 4z + 13}$ **30.** $f(z) = \dfrac{1}{z^2 + 6z + 25}$

For each function $f(z)$ and each region given in Exercises 31–36, obtain a Laurent series expansion in powers of z.

31. $f(z) = \dfrac{1}{z^2 + z - 12}$; $A = \{|z| < 3\}$; $B = \{3 < |z| < 4\}$; $C = \{|z| > 4\}$

32. $f(z) = \dfrac{1}{z^2 - 13z + 40}$; $A = \{|z| < 5\}$; $B = \{5 < |z| < 8\}$; $C = \{|z| > 8\}$

33. $f(z) = \dfrac{1}{z^2 - 7z + 10}$; $A = \{|z| < 2\}$; $B = \{2 < |z| < 5\}$; $C = \{|z| > 5\}$

34. $f(z) = \dfrac{z + 9}{z^2 + 2z - 8}$; $A = \{|z| < 2\}$; $B = \{2 < |z| < 4\}$; $C = \{|z| > 4\}$

35. $f(z) = \dfrac{z - 7}{z^2 + 9z + 14}$; $A = \{|z| < 2\}$; $B = \{2 < |z| < 7\}$; $C = \{|z| > 7\}$

36. $f(z) = \dfrac{z + 1}{z^2 - 14z + 45}$; $A = \{|z| < 5\}$; $B = \{5 < |z| < 9\}$; $C = \{|z| > 9\}$

35.13 The Calculus of Residues

What is a Residue?

DEFINITION 35.10   RESIDUE

Let $f(z)$ be analytic, except for a pole of order m at $z = z_0$. The *residue* of $f(z)$ at z_0 is the complex number b_1, the coefficient of the $\frac{1}{z}$-term in the Laurent series for $f(z)$ constructed at $z = z_0$.

As we saw in Section 35.12, if $f(z)$ has a pole of order m at $z = z_0$, then it has a Laurent series of the form

$$f(z) = \sum_{k=1}^{\infty} b_k (z - z_0)^{-k} + \sum_{k=0}^{\infty} a_k (z - z_0)^k$$

valid in the annulus $0 < |z - z_0| < R$ for some $R > 0$. The coefficients in the Laurent series are given by the formulas

$$a_k = \frac{1}{2\pi i} \int_C \frac{f(z)}{(z - z_0)^{k+1}} \, dz \quad k = 0, 1, \dots$$

$$b_k = \frac{1}{2\pi i} \int_C \frac{f(z)}{(z - z_0)^{1-k}} \, dz \quad k = 1, 2, \dots$$

where C is a positively oriented, simple closed contour lying in the annulus $0 < |z - z_0| < R$ and having z_0 in its interior. In particular, the coefficient

$$b_1 = \frac{1}{2\pi i} \int_C f(z) \, dz$$

called the residue of $f(z)$ at $z = z_0$, is denoted $\mathrm{Res}(f(z), z_0)$.

If the Laurent series is known by some means other than integration, then knowledge of b_1 yields the value of the integral $\int_C f(z) \, dz$ as $2\pi i b_1$. In fact, we have

$$\int_C f(z) \, dz = 2\pi i b_1 = 2\pi i \, \mathrm{Res}(f(z), z_0)$$

Thus, we have the remarkable result that an integral can be evaluated by obtaining a coefficient in a series expansion of the integrand.

EXAMPLE 35.73   Let C be the positively oriented contour $z(t) = \frac{1}{2}e^{it}$, $-\pi \le t \le \pi$, and $f(z) = \frac{\cot z}{z^2}$. To evaluate $\sigma = \int_C \frac{\cot z}{z^2} \, dz$ by the use of residues, note that $\cot z = \frac{\cos z}{\sin z}$, so the only singularity of $f(z)$ inside C is $z_0 = 0$. The Laurent series for $f(z) = z^{-2} \cot z$ is $z^{-3} - \frac{1}{3}z^{-1} - \frac{1}{45}z + \cdots$, so $z = 0$ is a pole of order 3, $b_1 = \mathrm{Res}(f(z), 0) = -\frac{1}{3}$, and $\sigma = 2\pi i(-\frac{1}{3}) = -\frac{2}{3}\pi i \doteq -2.0944i$. A corroborating $\int_{-\pi}^{\pi} 2i \cot(\frac{1}{2}e^{it})e^{-it} \, dt \doteq -2.0944i$ is obtained by numeric evaluation of the contour integral.   ❖

The Cauchy Residue Theorem

The generalized Cauchy–Goursat theorem of Section 35.11 concludes with the integration formula

$$\int_C f(z) \, dz = \sum_{j=1}^{k} \int_{C_i} f(z) \, dz$$

where, loosely speaking, $C_i, i = 1, \dots, k$, are "small loops" encircling singularities of $f(z)$ inside the larger positively oriented, simple closed contour C. If each integral on the right is evaluated using the appropriate residue, we then have the following theorem.

THEOREM 35.11   CAUCHY RESIDUE THEOREM

1.   Except for singularities at $z_1, z_2, \dots, z_k$, $f(z)$ is analytic inside and on the positively oriented, simple closed contour C

$$\implies \int_C f(z) \, dz = 2\pi i \sum_{i=1}^{k} \mathrm{Res}(f(z), z_i)$$

EXAMPLE 35.74   If C is the circle $z(t) = 5e^{it}$, $-\pi \leq t \leq \pi$, and $f(z) = \frac{z^2+2z-3}{z^3+2z^2-11z-12}$, evaluate the contour integral $\lambda = \int_C f(z)\,dz$. Factoring the denominator of $f(z)$, we find $(z+1)(z-3)(z+4)$, so $f(z)$ has simple poles at $-1, 3, -4$, all lying inside C. A direct (but tedious) computation gives

$$\lambda = 5i \int_{-\pi}^{\pi} \frac{(25e^{2it} + 10e^{it} - 3)e^{it}}{125e^{3it} + 50e^{2it} - 55e^{it} - 12}\,dt = 2\pi i$$

At $z = -1, 3, -4$, we have the Laurent series

$$f(z) = \tfrac{1}{3}(z+1)^{-1} - \tfrac{1}{36} + \cdots$$

$$f(z) = \tfrac{3}{7}(z-3)^{-1} + \tfrac{23}{196} - \tfrac{141}{5488}(z-3) + \cdots$$

$$f(z) = \tfrac{5}{21}(z+4)^{-1} - \tfrac{76}{441} - \tfrac{424}{9261}(z+4) + \cdots$$

respectively. Hence, the residues are $r_1 = \tfrac{1}{3}$, $r_2 = \tfrac{3}{7}$, and $r_3 = \tfrac{5}{21}$, respectively, and the Cauchy residue theorem gives the value of the contour integral as $2\pi i(\tfrac{1}{3} + \tfrac{3}{7} + \tfrac{5}{21}) = 2\pi i$. ❖

A Formula for Residues

If $f(z)$ is known to have a pole of order k at $z = z_0$, it is possible to compute a residue without resorting to the Laurent series. A formula for doing this is

$$\text{Res}(f(z), z_0) = \frac{1}{(k-1)!} \lim_{z \to z_0} \left[(z - z_0)^k f(z)\right]^{(k-1)} \qquad (35.40)$$

When $k = 1$, the formula indicates we are to take the derivative of order 0, which is interpreted to mean "take no derivative." We illustrate all this with an example.

EXAMPLE 35.75   Find the poles and residues of the function $f(z) = \frac{1}{(z-1)(z-3)^2(z+4)^3}$. Thus, $f(z)$ has a simple pole at $z = 1$, a pole of order 2 at $z = 3$, and a pole of order 3 at $z = -4$. The computation of the residues at these three poles is summarized in Table 35.7. ❖

Pole z_0	Principal Part of $f(z)$ at z_0	Residue b_1
1	$\dfrac{(z-1)^{-1}}{500}$	$\lim\limits_{z\to 1}(z-1)f(z) = \tfrac{1}{500}$
3	$\dfrac{(z-3)^{-2}}{686} - \dfrac{13(z-3)^{-1}}{9604}$	$\lim\limits_{z\to 3}[(z-3)^2 f(z)]' = -\tfrac{13}{9604}$
-4	$-\dfrac{(z+4)^{-3}}{245} - \dfrac{17(z+4)^{-2}}{8575} - \dfrac{194(z+4)^{-1}}{300125}$	$\tfrac{1}{2}\lim\limits_{z\to -4}[(z+4)^3 f(z)]'' = -\tfrac{194}{300125}$

TABLE 35.7   Residues in Example 35.75

EXERCISES 35.13–Part A

A1. Verify the residues given in the rightmost column of Table 35.7.

In Exercises A2–5, let C be the contour $z(t) = 3e^{it}$, $0 \le t \le 2\pi$, and, for the given function $f(z)$, evaluate $\int_C f(z)\,dz$ by

    (a) the Cauchy integral formula(s) applied to the first form of the function.

    (b) the Cauchy integral formula(s) applied to the second form of the function.

    (c) the Cauchy residue theorem.

A2. $f(z) = \dfrac{7 - z}{(z - 1)(z + 2)} = \dfrac{2}{z - 1} - \dfrac{3}{z + 2}$

A3. $f(z) = \dfrac{11 + 5z + 2z^2}{(z - 1)(z + 2)^2} = \dfrac{2}{z - 1} - \dfrac{3}{(z + 2)^2}$

A4. $f(z) = \dfrac{1 + 8z - 3z^2}{(z - 1)^2(z + 2)} = \dfrac{2}{(z - 1)^2} - \dfrac{3}{z + 2}$

A5. $f(z) = \dfrac{5 + 14z - z^2}{(z - 1)^2(z + 2)^2} = \dfrac{2}{(z - 1)^2} - \dfrac{3}{(z + 2)^2}$

EXERCISES 35.13–Part B

B1. Determine the poles of $f(z) = \frac{1}{(z+2)(z+5)^2(z-1)^3}$ and then use (35.40) to obtain the residues at each singularity.

In Exercises B2–16:

    (a) Evaluate the given contour integral using the Cauchy residue theorem, computing the residues with (35.40) or with an appropriate software tool.

    (b) Compute each residue by obtaining an appropriate Laurent series and selecting the coefficient b_1.

B2. Exercise B4, Section 35.11   **B3.** Exercise B5, Section 35.11

B4. Exercise B6, Section 35.11   **B5.** Exercise B7, Section 35.11

B6. Exercise B8, Section 35.11   **B7.** Exercise B9, Section 35.11

B8. Exercise B10, Section 35.11   **B9.** Exercise B11, Section 35.11

B10. Exercise B12, Section 35.11   **B11.** Exercise B13, Section 35.11

B12. Exercise B14, Section 35.11   **B13.** Exercise B15, Section 35.11

B14. Exercise B16, Section 35.11   **B15.** Exercise B17, Section 35.11

B16. Exercise B18, Section 35.11

For each function given in Exercises B17–26:

    (a) Obtain the complete Taylor expansion about $z_0 = 1$.

    (b) Obtain the Taylor coefficients at z_0 using the Cauchy integral formulae for derivatives, (35.36), of Section 35.11. Evaluate the integrals in (35.36) by the Cauchy residue theorem.

B17. $f(z) = 3z^3 + 5z^2 + 5z + 5$ **B18.** $f(z) = 6z^3 - 3z^2 + 3z + 2$

B19. $f(z) = 8z^3 - 9z^2 - 4z - 6$ **B20.** $f(z) = 4z^3 + z^2 - 9z + 2$

B21. $f(z) = 2z^3 + 6z^2 + z + 4$

B22. $f(z) = 7z^4 + 6z^3 - 3z^2 - 9z - 4$

B23. $f(z) = 9z^4 + 5z^3 - 4z^2 + 7z + 3$

B24. $f(z) = 4z^4 - 3z^3 - 7z^2 - z - 9$

B25. $f(z) = 8z^4 - 5z^3 + 4z^2 + 9z + 5$

B26. $f(z) = 9z^4 + z^3 + 3z^2 + 7z - 5$

In Exercises B27–36, use the Cauchy integral formulae for derivatives, (35.36), Section 35.11, and the Cauchy residue theorem, to obtain $f^{(n)}(z_0)$ for the given function $f(z)$ and the indicated n and z_0. Verify the result by evaluating the appropriate derivative.

B27. $f(z) = ze^{-z}$; $n = 2$; $z_0 = 2$

B28. $f(z) = \sin(z^2 - 1)$; $n = 3$; $z_0 = 2i$

B29. $f(z) = \tanh z^2$; $n = 2$; $z_0 = i$

B30. $f(z) = \arctan \sin z$; $n = 1$; $z_0 = \dfrac{\pi}{4}$

B31. $f(z) = \ln \cos z$; $n = 3$; $z_0 = \dfrac{\pi}{4}$

B32. $f(z) = \dfrac{\tan z}{1 + z}$; $n = 2$; $z_0 = 1 + i$

B33. $f(z) = \dfrac{z^2 - 1}{z^2 + 1}$; $n = 3$; $z_0 = 2$

B34. $f(z) = z \operatorname{arcsinh} z$; $n = 2$; $z_0 = 2$

B35. $f(z) = \sqrt{z^2 + 1}$; $n = 3$; $z_0 = 0$

B36. $f(z) = \cos^2 \sin z$; $n = 1$; $z_0 = \dfrac{\pi}{3}$

Chapter Review

Let $z_1 = 2 + 3i$, $z_2 = -3 + i$, $z_3 = -5 - i$.

1. Obtain $z_1 + z_2$, $z_2 - z_3$, $z_1 z_3$, $\frac{z_2}{z_3}$.

2. Express z_k, $k = 1, 2, 3$, in exponential form.

3. Obtain z_1^3, z_2^2, and z_3^5.

4. Obtain both square roots of z_1, z_2, and z_3.

5. Obtain the three cube roots of z_1, z_2, and z_3.

Write each of the following functions in the form $u(x, y) + iv(x, y)$.

6. $f(z) = \sin z$ **7.** $f(z) = \cos z$

8. $f(z) = \sinh z$ **9.** $f(z) = \cosh z$

Use the results of Questions 6–9 to obtain the rectangular form of each of Questions 10 and 11.

10. $\sin z_1$, $\cos z_2$ **11.** $\sinh z_1$, $\cosh z_3$

Obtain the rectangular form of each of Questions 12 and 13 using the appropriate formulas.

12. $\arcsin z_2$, $\arccos z_3$ **13.** $\text{arcsinh}\, z_1$, $\text{arccosh}\, z_2$

14. Define both the multivalued $\log z$ and $\text{Log}\, z$, the principal value of the complex logarithm.

15. In what sense is a formula like $\log z_1 z_2 = \log z_1 + \log z_2$ "true" in the complex plane?

16. Evaluate $\text{Log}\, z_1$.

17. Evaluate $\text{Log}\, z_1 z_3$ and compare it to $\text{Log}\, z_1 + \text{Log}\, z_3$.

18. Evaluate $\text{Log}\, z_1^{z_2}$ and compare it to $z_2 \,\text{Log}\, z_1$.

19. Evaluate $\text{Log}\, \frac{z_2}{z_3}$ and compare it to $\text{Log}\, z_2 - \text{Log}\, z_3$.

20. Evaluate $\text{Log}\, e^{z_2}$ and compare it to $z_2 \,\text{Log}\, e = z_2$.

21. Evaluate $(z_1 z_2)^{z_3}$ and compare it to $z_1^{z_3} z_2^{z_3}$.

22. Evaluate $\left(\frac{z_1}{z_2}\right)^{z_3}$ and compare it to $z_1^{z_3}/z_2^{z_3}$.

23. Evaluate $\left(z_1^{z_2}\right)^{z_3}$ and compare it to $z_1^{z_2 z_3}$.

24. Evaluate $\left(z_1^{z_2}\right)^{z_3}$ and compare it to $\left(z_1^{z_3}\right)^{z_2}$.

25. Evaluate $\left(z_1^3\right)^{1/3}$, $\left(z_2^3\right)^{1/3}$, and $\left(z_3^3\right)^{1/3}$.

26. Evaluate $\left(z_1^{1/3}\right)^3$, $\left(z_2^{1/3}\right)^3$, and $\left(z_3^{1/3}\right)^3$.

27. Let $f(z) = z^2$.

(a) Write $f(z)$ in the form $u(x, y) + iv(x, y)$.

(b) Verify that u and v satisfy the Cauchy–Riemann equations.

(c) Show that u and v are harmonic functions.

28. What is an analytic function of a real variable? What is an analytic function of a complex variable? How do these differ?

29. If $f(z) = z^2$, obtain the value of the contour integral $\int_C f(z)\, dz$ when

(a) C is the parabola $y = x^2$, $-1 \le x \le 1$.

(b) C is the straight line connecting $(-1, 1)$ with $(1, 1)$.

30. Explain why the two contour integrals in Question 29 have the same value.

31. What does the Cauchy–Goursat theorem say?

32. What does the generalized Cauchy–Goursat say?

33. What is the Cauchy integral formula?

34. What are the Cauchy integral formulas for derivatives?

35. What are the integral formulas for the coefficients in a Laurent series?

36. Use the generalized Cauchy–Goursat theorem to evaluate the contour integral $\int_C e^z/(2z^2 - 5z + 2)\, dz$, if C is the contour $z = 3e^{it}$, $-\pi \le t \le \pi$.

37. Evaluate the integral in Question 36 by using Cauchy's residue theorem.

38. State the content of Laurent's theorem for Laurent series expansions.

39. Find all zeros and poles of $f(z) = \frac{9z+7}{z^3 - (16+13i)z^2 + 12(3+11i)z + 72(1-4i)}$. Indicate the multiplicity of the zeros and the order of the poles.

40. If $f(z) = \frac{z^2 - 2z - 1}{z^2 - 4z + 3}$, obtain a Laurent series in powers of z, valid in the region

(a) $|z| < 1$ (b) $1 < |z| < 3$ (c) $|z| > 3$

Chapter 36

Applications

INTRODUCTION   This chapter contains a representative collection of practical applications of complex function theory. The suite of tools and techniques upon which this chapter rests was assembled in Chapter 35.

A variety of real integrals can be evaluated using the complex plane as an ally. Generally, the real integral is interpreted as a contour integral, a closed contour is formed in the complex plane, the contribution from the added contour is shown to be negligible, and the resulting integral is evaluated by Cauchy's residue theorem. Although it might be tempting to call this technique "contour integration," the reality is that all integrals in the complex plane are contour integrals. Certain contours can be deformed so that the Cauchy residue theorem applies, and it is residue theory that really evaluates the integrals.

The *Bromwich integral formula* for inverting Laplace transforms is a contour integral in the complex plane, with the contour being a vertical line. In some cases, the integral can be evaluated directly, and in other cases the contour must be closed so that Cauchy's residue theorem can be applied. This is the analog of the inversion formula for the Fourier transform in Section 30.3. In that section we inverted some Fourier transforms by brute-force integration. We can also evaluate these inversion integrals by contour integration and the Cauchy residue theorem.

The *root locus* is a graph used in the design of *feedback control systems*. In Chapter 12, the stability of a system of linear differential equations is seen to depend on the location of the characteristic roots. If all the characteristic roots have negative real parts, the roots lie in the left half of the complex plane. The roots of a control system under design typically depend on a parameter, called the *gain*. Thus, the location of the characteristic roots depends on this parameter. A graph of the location of the characteristic roots as a function of the gain is called the root locus.

The *Nyquist stability criterion,* based on two concepts, the *winding number* and the *principle of the argument,* is another concept arising from the design of feedback control systems. The winding number of a contour is the number of times it winds around a given point. The principle of the argument relates the number of poles and zeros a function has inside a contour to the winding number of the contour. Hence, by looking at the deformation of a contour under the action of a complex map, the winding number can be determined, and from that, the number of poles and zeros the mapping has inside the contour. Articulated as a way of determining the number of zeros in the left half-plane, the extended principle of the argument becomes the Nyquist stability criterion.

Roughly speaking, the conformal property of analytic functions amounts to the local preservation of angles. If two curves that intersect at a particular angle in the z-plane are mapped to the w-plane by an analytic function, their images in the w-plane will intersect at the same angle. Consequently, the mappings induced by analytic functions are called *conformal maps*. The mapping action of a number of simple analytic functions is cataloged. Because complex functions map the plane, it is not possible to describe their action by a simple graph. Various devices are used to show how the function maps regions in the z-plane to regions in the w-plane. These studies are collected and go under the heading of *conformal mapping*.

The important conformal map called the *Joukowski map* transforms certain circles in the z-plane to shapes in the w-plane, which resemble airfoils, the cross-sectional view of an airplane wing—hence, the phrase *Joukowski airfoil* and the interest this particular map generates.

The Dirichlet problem in the plane is solved by transforming it with a conformal map to an equivalent problem on a simpler region. The *Dirichlet problem* consists of Laplace's equation on a region on whose boundary the value of the unknown function has been prescribed. Using separation of variables, this problem has been solved for the rectangle in Chapter 26 and for the disk in Section 29.1. Chapter 30 contains examples of the Dirichlet problem being solved by integral transforms. In the conformal mapping method a conformal map is used to transform the given region to a more tractable one, one for which the solution of Laplace's equation is more readily obtained.

The first of two sections devoted to solving *fluid-flow problems* via complex variables begins with a brief formulation of the planar fluid-flow problem and then details several simple flows in the z-plane. Several flow problems are solved by conformal mapping techniques. In some cases, a complex flow in the z-plane is mapped to the w-plane, where it admits a simpler solution that can then be mapped back to the z-plane. In other cases, this technique fails because of the algebraic complexity of the inverse map. Hence, a known flow in the z-plane is mapped to a complex flow in the w-plane, and the flow in w-plane is declared to be solved by the simpler flow in the z-plane. This subtle reversal of the roles of the two planes must be recognized, lest the reader misunderstand some of the applications.

36.1 Evaluation of Integrals

Introduction

After being converted to contour integrals in the complex plane, a number of definite integrals along the real axis can be evaluated by the Cauchy residue theorem. In this section we will examine several classes of such integrals and the conditions under which they can be evaluated by contour integration.

Integrals of the Form $\int_0^{2\pi} F(\cos\theta, \sin\theta)\, d\theta$

Let $F(u, v)$ be a rational function, that is, a ratio of polynomials in u and v. The change of variables $z = e^{i\theta}$ will convert integrals of the form $\int_0^{2\pi} F(\cos\theta, \sin\theta)\, d\theta$ to contour integrals of the form $\int_C f(z)\, dz$, where C is the positively oriented circle $|z| = 1$. The contour integrals can then be evaluated using residues, and

$$\int_0^{2\pi} F(\cos\theta, \sin\theta)\, d\theta = 2\pi i \sum_{n=1}^{k} \operatorname{Res}(f(z), z_n)$$

where $z_n, n = 1, \ldots, k$, are the poles of $f(z)$ inside the contour C. The change of variables $z = e^{i\theta}$ is implemented using $\cos\theta = \frac{1}{2}(z + \frac{1}{z})$, $\sin\theta = \frac{1}{2i}(z - \frac{1}{z})$, and $d\theta = \frac{dz}{iz}$.

EXAMPLE 36.1   To obtain

$$\int_0^{2\pi} \frac{\sin^2\theta}{5 + 2\cos 2\theta}\, d\theta = \frac{\pi}{6}(\sqrt{21} - 3) \overset{\circ}{=} 0.828635$$

make the change of variables $z = e^{i\theta}$, and use $z^2 = e^{2i\theta} = \cos 2\theta + i\sin 2\theta$, so $\cos 2\theta = \frac{1}{2}(z^2 + \frac{1}{z^2})$, to obtain the integrand $f(z) = \frac{i(z^2-1)^2}{4z(z^4+5z^2+1)}$. The simple poles, their magnitudes, and the associated residues are listed in Table 36.1. The first three poles are inside $|z| = 1$, and the sum of the associated residues is $\frac{\pi}{6}(\sqrt{21} - 3) \overset{\circ}{=} 0.828635$.   ❖

k	Pole at $z = z_k$	$|z_k|$	$\mathrm{Res}(f(z), z_k)$
1	0	0	$\dfrac{i}{4}$
2	$\dfrac{i}{2}(\sqrt{7} - \sqrt{3})$	0.46	$-i\dfrac{\sqrt{21}}{24}$
3	$-\dfrac{i}{2}(\sqrt{7} - \sqrt{3})$	0.46	$-i\dfrac{\sqrt{21}}{24}$
4	$\dfrac{i}{2}(\sqrt{7} + \sqrt{3})$	2.2	$i\dfrac{\sqrt{21}}{24}$
5	$-\dfrac{i}{2}(\sqrt{7} + \sqrt{3})$	2.2	$i\dfrac{\sqrt{21}}{24}$

TABLE 36.1   Poles and residues of $f(z)$ in Example 36.1

EXAMPLE 36.2   The Poisson integral formula

$$u(r, \theta) = \frac{c^2 - r^2}{2\pi} \int_0^{2\pi} \frac{f(\sigma)}{c^2 - 2rc\cos(\theta - \sigma) + r^2}\, d\sigma$$

provides the solution to the Dirichlet problem $\nabla^2 u(r, \theta) = 0$, $u(c, \theta) = f(\theta)$ on the disk of radius c. (See Section 29.1.) Unfortunately, this is a difficult integral to evaluate. For example, if we take $c = 1$ and $f(\theta) = \sin^2\theta$, the integrand in the Poisson integral formula would be $F = \frac{\sin^2\sigma}{1+r^2-2r\cos(\theta-\sigma)}$. The change of variables $z = e^{i\sigma}$ leads to the new integrand

$$f(z) = \frac{i}{4}\frac{(z^2 - 1)^2}{z^4 r e^{-i\theta} - z^3(1 + r^2) + z^2 r e^{i\theta}}$$

which has a pole of order 2 at $z = 0$ and simple poles at $z = r e^{i\theta}$ and $\frac{1}{r}e^{i\theta}$. Since $0 \le r < 1$ inside the disk, only the first two poles are inside the contour. Adding residues at these two poles, multiplying by the factors $2\pi i$ and $\frac{1-r^2}{2\pi}$, and simplifying, we obtain

$$u(r, \theta) = -\frac{i}{4}\left[\frac{1+r^2}{r^2 e^{2i\theta}} + \frac{(r^2 e^{2i\theta} - 1)^2}{r^2 e^{2i\theta}(r^2 - 1)}\right](2\pi i)\left(\frac{1 - r^2}{2\pi}\right) = \frac{1}{2}(1 - r^2\cos 2\theta)$$ ❖

Integrals of the Form $\int_{-\infty}^{\infty} f(x)\, dx$

If $f(x)$ has no singularities on the real axis, then in $\int_{-\infty}^{\infty} f(x)\, dx = \lim_{\beta \to \infty} \lim_{\alpha \to -\infty} \int_{\alpha}^{\beta} f(x)\, dx$, the limit on the right *defines* the improper integral on the left. It can be shown that if the limit on the right exists, then so does the less restrictive *Cauchy principal value,* namely, $\lim_{R \to \infty} \int_{-R}^{R} f(x)\, dx$, and the two limits are equal. So, any condition that guarantees the existence of the first limit then allows us to evaluate it by computing the second "easier" limit. The sufficient condition we will impose on $f(x)$ will allow us to "complete (or close) the contour" in the complex plane and evaluate the improper integral by invoking Cauchy's residue theorem.

Let C_1 be the positively oriented contour consisting of the interval $[-R, R]$ on the real axis and C_R, the upper half of the circle $|z| = R$ in the complex plane. (See Figure 36.1(*a*).) In the upper half-plane, let $f(z)$ be analytic, except for isolated poles, and let $|f(z)|$ decay sufficiently rapidly on C_R so that $\lim_{R \to \infty} \int_{C_R} f(z)\, dz = 0$. Then $\int_{-\infty}^{\infty} f(x)\, dx$ is given by

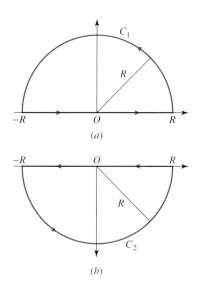

FIGURE 36.1   The positively oriented contours C_1 and C_2 are shown in (*a*) and (*b*), respectively

$$\lim_{R \to \infty} \int_{-R}^{R} f(x)\, dx + \lim_{R \to \infty} \int_{C_R} f(z)\, dz = \lim_{R \to \infty} \int_{C_1} f(z)\, dz = 2\pi i \sum_{n=1}^{k} \operatorname{Res}(f(z), z_n)$$

where $z_1, \ldots, z_k$ are the poles of $f(z)$ in the upper half-plane.

Precisely, if as $|z| \to \infty$ we have $f(z) = O(|z|^{-p})$, with $p > 1$, so that $|f(z)| < K/|z|^p$ for some positive constant K and large enough $|z|$, then the improper integral exists and $f(z)$ makes a negligible contribution to the contour integral along C_R. (A formal proof can be found in [69].)

Finally, note that if the contour is completed in the lower half-plane, then $\int_{-\infty}^{\infty} f(x)\, dx = -2\pi i \sum_{n=0}^{j} \operatorname{Res}(f(z), z_n)$, where $z_1, \ldots, z_j$ are the poles of $f(z)$ in the lower half-plane. To understand the minus sign on the right, let C_2 be the positively oriented contour consisting of the real interval $[-r, r]$ traversed right to left and C_r, the lower half of the circle $|z| = r$ in the complex plane. (See Figure 36.1(*b*).) The positive (counterclockwise) orientation on C_2 means that the integration along the real axis is from right to left. Hence, the contour integral gives the *negative* of the desired real improper integral.

EXAMPLE 36.3   Let $f(x) = \frac{1}{(x^2+1)(x+2i)}$, with simple poles at $\pm i$ and $-2i$. There are no poles on the real axis, one pole in the upper half-plane and two poles in the lower half-plane. The integral $\int_{-\infty}^{\infty} f(x)\, dx$ evaluates to $-\frac{\pi i}{3}$ using the antiderivative

$$F(x) = \frac{1}{6} \ln \frac{x^2+1}{x^2+4} - \frac{2i}{3} \arctan x + \frac{i}{3} \arctan \frac{x}{2}$$

obtained by writing $f(x)$ as

$$\frac{1}{3}\left(\frac{x-2i}{x^2+1} - \frac{x-2i}{x^2+4} \right)$$

Since $f(z) = O(|z|^{-3})$, the real axis may be closed to the contour C_R in the upper half-plane and the Cauchy residue theorem applied, giving $2\pi i \operatorname{Res}(f(z), i) = 2\pi i (-\frac{1}{6}) = -\frac{\pi i}{3}$, with $z = i$ being the single pole in the upper half-plane.

It would also be possible to close the contour in the lower half-plane by adjoining to the real axis the lower semicircle $|z| = r$. The positively oriented contour C_r consisting of the real axis between $x = r$ and $x = -r$, and the semicircle, is traversed counterclockwise. However, that causes the x-axis to be traversed from right to left and, therefore, induces a minus sign in front of the sum of residues when applying the Cauchy residue theorem. The residue at $z = -i$ is $\frac{1}{2}$, whereas the residue at $z = -2i$ is $-\frac{1}{3}$. The value of the integral is then $-2\pi i$ times the sum of the residues in the lower half-plane, or $-\frac{\pi i}{3}$, as computed earlier. ❖

EXAMPLE 36.4   If $f(x)$ is an even function, then

$$\int_0^\infty f(x)\, dx = \frac{1}{2}\int_{-\infty}^\infty f(x)\, dx = \frac{2\pi i}{2}\sum_{n=1}^k \operatorname{Res}(f(z), z_n) = \pi i \sum_{n=1}^k \operatorname{Res}(f(z), z_n)$$

where again, $z_1, \ldots, z_k$ are the poles of $f(z)$ in the upper half-plane. For example, if $f(x) = \frac{1}{1+x^6}$, with antiderivative

$$F(x) = \tfrac{1}{3}\arctan x + \tfrac{1}{6}\arctan(2x - \sqrt{3}) + \tfrac{1}{6}\arctan(2x + \sqrt{3}) + \frac{\sqrt{3}}{12}\ln \frac{\sqrt{3}x + x^2 + 1}{\sqrt{3}x - x^2 - 1}$$

then $\int_0^\infty f(x)\, dx = \lim_{x\to\infty} F(x) - F(0) = \frac{\pi}{3}$. On the other hand, $f(z)$ has poles at the six roots of $z = -1$, namely, at $\pm i$, $e^{\pm\pi i/6}$, $e^{\pm 5\pi i/6}$. At the three poles in the upper half-plane we then find $\operatorname{Res}(f(z), i) = -\frac{i}{6}$, $\operatorname{Res}(f(z), e^{\pi i/6}) = -\frac{1}{12}(\sqrt{3}+i)$, and $\operatorname{Res}(f(z), e^{5\pi i/6}) = \frac{1}{12}(\sqrt{3} - i)$, so πi times the sum of the residues is again $\frac{\pi}{3}$. ❖

Integrals of the Form $\int_{-\infty}^\infty e^{imx} f(x)\, dx$

Let $f(z)$ be a *meromorphic function,* that is, a function that is analytic, except for isolated poles. In addition, let $f(z) = O(|z|^{-p})$, $p > 0$, hold for $|z|$ large. If $z_1, \ldots, z_k$ and $\zeta_1, \ldots, \zeta_j$ are the poles of $f(z)$ in the upper and lower half-planes, respectively, then

$$\int_{-\infty}^\infty e^{imx} f(x)\, dx = \begin{cases} 2\pi i \displaystyle\sum_{n=1}^k \operatorname{Res}(e^{imz} f(z), z_n) & m > 0 \\[2ex] -2\pi i \displaystyle\sum_{n=1}^j \operatorname{Res}(e^{imz} f(z), \zeta_n) & m < 0 \end{cases}$$

When $m \neq 0$, the behavior of e^{imz} in the complex plane allows $f(z)$ to obey a less restrictive condition than in the case $m = 0$. If $m > 0$, let C_1 be the positively oriented contour consisting of the interval $[-R, R]$ on the real axis and C_R, the upper half of the circle $|z| = R$ in the complex plane. On C_R where $z = Re^{it}$, the magnitude of e^{imz} is

$$\left| e^{imRe^{it}} \right| = \left| e^{imR(\cos t + i\sin t)} \right| = \left| e^{-mR\sin t} \right| \left| e^{imR\cos t} \right| = e^{-mR\sin t}$$

Thus, if $m > 0$ and $0 \leq \sin t$ because $0 \leq t \leq \pi$ on C_R, the exponential will go to zero as $R \to \infty$.

On the other hand, if $m < 0$, the contour is completed to C_2, consisting of the real interval $[-r, r]$ traversed right to left, and C_r, the lower half of the circle $|z| = r$ in the complex plane. Then on C_r where $z = re^{it}$, the magnitude of e^{imz} is

$$\left| e^{imre^{it}} \right| \left| e^{imr(\cos t + i\sin t)} \right| = \left| e^{-mr\sin t} \right| \left| e^{imr\cos t} \right| = e^{-mr\sin t}$$

Because $m < 0$, to have decay of the exponential term, we must have $\sin t \leq 0$ on C_r, which is exactly what happens for $\pi \leq t \leq 2\pi$. If the contour is completed in the lower half-plane, the positive (counterclockwise) orientation of C_2 means the integral along the real axis is from right to left. Thus, the contour integral will yield the negative of the real improper integral, thereby explaining the minus sign in front of the sum of the lower half-plane's residues.

EXAMPLE 36.5   To evaluate the integral $\int_{-\infty}^{\infty} e^{3ix}/(x^3 - 8i)\,dx$, determine the poles of $f(z) = \frac{1}{z^3-8i}$ to be $z = -2i$ and $z = i \pm \sqrt{3}$. Since $m = 3 > 0$, close the contour in the upper half-plane, and find the residues of the two poles in that half-plane. Thus, the value of the integral is

$$2\pi i \left(\operatorname{Res}\left(\frac{e^{3iz}}{z^3 - 8i}, i + \sqrt{3} \right) + \operatorname{Res}\left(\frac{e^{3iz}}{z^3 - 8i}, i - \sqrt{3} \right) \right)$$

$$= 2\pi i \frac{e^{-3}}{24} \left((1 - i\sqrt{3})e^{3i\sqrt{3}} + (1 + i\sqrt{3})e^{-3i\sqrt{3}} \right)$$

$$= \frac{\pi i e^{-3}}{6}(\cos 3\sqrt{3} + \sqrt{3}\sin 3\sqrt{3}) \overset{\circ}{=} -0.0278i \qquad \blacklozenge$$

Integrals of the Form $\int_{-\infty}^{\infty} f(x) \cos mx\,dx$ and $\int_{-\infty}^{\infty} f(x) \sin mx\,dx$

Since $e^{imx} = \cos mx + i \sin mx$, integrals of the form

$$\int_{-\infty}^{\infty} f(x) \cos mx\,dx \quad \text{and} \quad \int_{-\infty}^{\infty} f(x) \sin mx\,dx$$

can be evaluated by taking the real and imaginary parts of $\int_{-\infty}^{\infty} f(x)e^{imx}\,dx$.

EXAMPLE 36.6   To obtain

$$\int_{-\infty}^{\infty} \frac{x \sin 5x}{x^2 + 9} = \pi e^{-15}$$

determine the poles of $f(z) = \frac{z}{z^2+9}$ to be $z = \pm 3i$. The required integral is the imaginary part of $\int_{-\infty}^{\infty} e^{5iz} f(z)\,dz$ in which $m = 5 > 0$ suggests completing the contour in the upper half-plane where $z = 3i$ is the only pole of $f(z)$. Since

$$2\pi i \operatorname{Res}\left(\frac{z e^{5iz}}{z^2 + 9}, 3i \right) = 2\pi i \left(\frac{1}{2} e^{-15} \right)$$

the value of the integral is πe^{-15}. $\qquad \blacklozenge$

EXAMPLE 36.7   To obtain

$$\int_0^{\infty} \frac{2\cos^2 x}{\left(4 + x^2\right)^2}\,dx = \frac{\pi}{32}(1 + 5e^{-4})$$

use $\cos^2 x = \frac{1}{2}(1 + \cos 2x)$ to write $\int_0^{\infty} \frac{dx}{(4+x^2)^2} + \int_0^{\infty} \frac{\cos 2x\,dx}{(4+x^2)^2}$. Each of these two integrals has even integrands, so we have $\frac{1}{2}\int_{-\infty}^{\infty} \frac{dx}{(4+x^2)^2} + \frac{1}{2}\int_{-\infty}^{\infty} \frac{\cos 2x\,dx}{(4+x^2)^2}$. In each of *these* two integrals, close the contour in the upper half-plane thereby forming C_R, and evaluate $\frac{1}{2}\int_{C_R} \frac{dz}{(4+z^2)^2} + \frac{1}{2}\int_{C_R} \frac{\cos 2z\,dz}{(4+z^2)^2}$. For the first integral we have, by the Cauchy residue theorem,

$$\frac{1}{2}(2\pi i) \operatorname{Res}\left(\frac{1}{\left(4 + z^2\right)^2}, 2i \right) = \pi i \left(-\frac{i}{32} \right) = \frac{\pi}{32}$$

where $z = 2i$ is the only pole the integrand has in the upper half-plane. For the second integral, we have

$$\frac{1}{2}\mathcal{R} \left(2\pi i \operatorname{Res}\left(\frac{e^{2iz}}{\left(4 + z^2\right)^2}, 2i \right) \right) = \frac{1}{2}\mathcal{R}\left(-\frac{5i}{32} e^{-4} 2\pi i \right) = \frac{5\pi}{32} e^{-4}$$

and for the original integral, we have $\frac{\pi}{32} + \frac{5\pi}{32} e^{-4} = \frac{\pi}{32}(1 + 5e^{-4})$. $\qquad \blacklozenge$

Indented Contours

Suppose $f(z)$ satisfies the following conditions:

1. On the real axis, $f(z)$ is analytic except for *simple* poles at $a_1, \ldots, a_j$.
2. In the upper half-plane, $f(z)$ is analytic except for isolated poles at $z_1, \ldots, z_k$.
3. $f(z) = O(|z|^{-p})$, $p > 1$, for $|z|$ large.

Then, by the Cauchy residue theorem, we have

$$\int_{-\infty}^{\infty} f(x)\, dx = 2\pi i \sum_{n=1}^{k} \operatorname{Res}(f(z), z_n) + \pi i \sum_{n=1}^{j} \operatorname{Res}(f(z), a_n) \tag{36.1}$$

Along the real axis a semicircular detour is made around each singularity. In the limit as the radius of the detour goes to zero, the contribution to the contour integral is half the residue ordinarily expected. This is illustrated in the following example.

EXAMPLE 36.8   If $f(x) = \frac{2}{(x-1)(x^2+1)}$, verify that $\lambda = \int_{-\infty}^{\infty} f(x)\, dx = -\pi$. The integrand has a simple pole at $x = 1$ on the real axis and at $z = i$ in the upper half-plane. Accordingly,

$$\lambda = 2\pi i \operatorname{Res}(f(z), i) + \pi i \operatorname{Res}(f(z), 1) = \tfrac{1}{2}(i - 1)2\pi i + \pi i = -\pi$$

The integral is evaluated by closing the contour in the upper half-plane but indenting it around the simple pole at $x = 1$. Thus, consider the contour C, shown in Figure 36.2 and consisting of the interval $[-R, 1-r]$ on the real axis, the semicircular arc C_r with center at $(1, 0)$ and radius r, the interval $[1 + r, R]$ on the real axis, and the semicircular arc C_R with center at $(0, 0)$ and radius R.

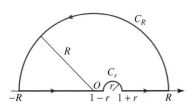

FIGURE 36.2   Indented contour for Example 36.8

Except for the poles at $z_1, \ldots, z_k$, $f(z)$ is analytic inside C. Thus, the Cauchy residue theorem applies and we have

$$\int_{-\infty}^{\infty} f(x)\, dx = \lim_{r \to 0} \lim_{R \to \infty} \left[\int_{-R}^{1-r} f(x)\, dx + \int_{C_r} f(z)\, dz + \int_{1+r}^{R} f(x)\, dx + \int_{C_R} f(z)\, dz \right]$$

$$= 2\pi i \sum_{n=1}^{k} \operatorname{Res}(f(z), z_n)$$

Since for $|z|$ large, $f(z) = O(|z|^{-p})$, $p > 1$, holds, we again have $\lim_{R \to \infty} \int_{C_R} f(z)\, dz = 0$. On C_r, the small "indented" contour, we will show that we have $\lim_{r \to 0} \int_{C_r} f(z)\, dz = -\pi i \operatorname{Res}(f(z), 1)$. Consequently, we then have

$$\int_{-\infty}^{\infty} f(x)\, dx - \pi i \operatorname{Res}(f(z), 1) = 2\pi i \sum_{n=1}^{k} \operatorname{Res}(f(z), z_n)$$

from which we conclude

$$\int_{-\infty}^{\infty} f(x)\, dx = 2\pi i \sum_{n=1}^{k} \operatorname{Res}(f(z), z_n) + \pi i \operatorname{Res}(f(z), 1)$$

By generalizing the process of contour-indentation, if there are simple poles $a_1, \ldots, a_j$ along the real axis, then we sum the residues at these poles and obtain the formula (36.1).

The integral on C_r is evaluated as follows. First, isolate the part of $f(x)$ that is singular at $x = 1$. This can be done by computing a Laurent series about $x = 1$ and getting

$$f(x) = (x - 1)^{-1} - 1 + \tfrac{1}{2}(x - 1) + \cdots$$

or by computing the partial fraction expansion of $f(x)$ and getting

$$f(x) = \frac{1}{x-1} - \frac{x+1}{x^2+1} = f_1(x) + f_2(x) \tag{36.2}$$

In either case, we can see that the residue of $f(z)$ at $z = 1$ is 1. For our purposes, it is more convenient to work with the partial fraction decomposition (36.2) because that keeps the part analytic at $x = 1$ "together" as a single fraction.

The contribution of $f_1 = \frac{1}{x-1}$ to the integral on C_r, parametrized as $z(t) = 1 + re^{it}, 0 \le t \le \pi$, is obtained directly as $\int_\pi^0 i\, dt = -i\pi$. Of course, if $\text{Res}(f(z), 1)$ were something other than 1, the value of this integral would be $-\pi i\, \text{Res}(f(z), 1)$.

We next argue that since $f_2(z) = -\frac{z+1}{z^2+1}$ is analytic (and therefore continuous) in a neighborhood of $z = 1$, it is bounded, say by M, in that neighborhood. Then, on C_r, the integral of f_2 can be estimated as $M\int_\pi^0 r\, dt = -M\pi r$, a bound that goes to zero as r goes to zero. ❖

Integrals of Multiple-Valued Functions

For integrals of the form $\int_0^\infty x^\beta f(x)\, dx$, where β is not an integer, we state the following theorem taken primarily from [69].

THEOREM 36.1

1. β is a real constant different from an integer.
2. $\lambda = \beta + 1$
3. $f(z)$ is analytic except for *simple* poles at $a_1, \ldots, a_j$ on the positive real axis, and poles at $z_1, \ldots, z_k$ anywhere in the complex plane exclusive of the nonnegative real axis.
4. $f(z) = O(|z|^{-p}), p > \lambda$, for $|z| \to \infty$, or equivalently, $\lim_{|z|\to\infty} |z|^\lambda |f(z)| = 0$.
5. $f(z) = O(|z|^p), p > -\lambda$, for $|z| \to 0$, or equivalently, $\lim_{|z|\to 0} |z|^\lambda |f(z)| = 0$.
6. $F_0(z) = e^{\beta \text{Log}\, z} f(z)$, where $\text{Log}\, z = \ln|z| + i\, \text{Arg}\, z$ is the principal logarithm, and $-\pi < \text{Arg}\, z \le \pi$.
7. $F(z) = e^{\beta[\text{Log}(-z)+\pi i]} f(z) = \begin{cases} F_0(z) & 0 < \mathcal{I}(z) \\ e^{2\pi i \beta} F_0(z) & \mathcal{I}(z) < 0 \end{cases}$

$\Longrightarrow$

$$CPV \int_0^\infty x^\beta f(x)\, dx = -\frac{\pi e^{-\lambda\pi i}}{\sin\lambda\pi} \sum_{n=1}^k \text{Res}(F(z), z_n) - \pi \cot\lambda\pi \sum_{n=1}^j \text{Res}(F_0(z), a_n)$$

where *CPV* denotes the Cauchy principal value of the integral on the left.

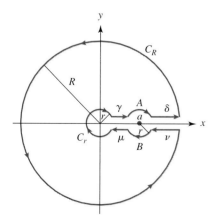

FIGURE 36.3   The contour C for Theorem 36.1

The *CPV* arises when each indentation circle such as those illustrated in Figure 36.3 are all taken to have the same radius r. Of course, whenever the integral exists as an improper integral, the principal value also exists and is the same number. Figure 36.3, which shows the positively oriented, closed contour C, will help in our discussion of Theorem 36.1.

When β is not an integer, the origin is a branch point for x^β. Since x^β is defined in terms of the logarithm, the behavior of the logarithm is crucial to the choice of contour along which to integrate. Closing the contour to C so as to indent around the representative singularity

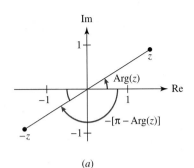

(a)

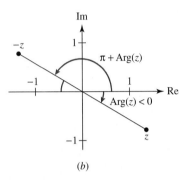

(b)

FIGURE 36.4   If $\mathcal{I}(z) > 0$ as in (a), then $\text{Arg}(-z) = \text{Arg}(z) - \pi$. If $\mathcal{I}(z) < 0$ as in (b), then $\text{Arg}(-z) = \text{Arg}(z) + \pi$.

at $x = a$ on the positive real axis, we include the poles of $f(x)$ inside C. Therefore, the logarithm used to define x^β must be analytic inside C, a region that includes the negative real axis. Thus, we must have the discontinuity in the argument of the logarithm occur across the positive real axis.

The principal logarithm, $\text{Log } z = \ln|z| + i\,\text{Arg } z$, is analytic for $-\pi < \theta \le \pi$. Thus, $F_0(z) = e^{\beta \, \text{Log } z} f(z)$ uses the principal logarithm and is analytic across the positive real axis. At simple poles along the positive real axis, residues are computed using this function.

Inside C, $F(z)$, as given by hypothesis 7, is analytic for $0 \le \theta < 2\pi$. There are two claims being made here. First, we need to verify that the second equality in hypothesis 7 is indeed valid; and second, we need to check the analyticity across the negative real axis.

To show the validity of the equality, let z be in the upper half-plane so that $\mathcal{I}(z) > 0$ and let $\text{Arg}(-z) = \text{Arg } z - \pi$. For example, if $z = \sqrt{3}+i$, then $\text{Arg } z = \frac{\pi}{6}$ and $\text{Arg}(-z) = -\frac{5}{6}\pi = \text{Arg}(z) - \pi$, as illustrated in Figure 36.4(a). On the other hand, let z be in the lower half-plane, so $\text{Arg}(z) < 0$ and $\text{Arg}(-z) = \text{Arg}(z) + \pi$. Again using an example, let $z = \sqrt{3} - i$ so that $\text{Arg } z = -\frac{\pi}{6}$ and $\text{Arg}(-z) = \text{Arg } z + \pi = \frac{5}{6}\pi$, as illustrated in Figure 36.4(b).

Since $\text{Log } z = \ln|z| + i\,\text{Arg } z$ defines the principal logarithm, for $\mathcal{I}(z) > 0$ we have

$$\begin{aligned} \text{Log}(-z) + \pi i &= \ln|-z| + i\,\text{Arg}(-z) + \pi i \\ &= \ln|z| + i(\text{Arg } z - \pi) + \pi i \\ &= \ln|z| + i\,\text{Arg } z \\ &= \text{Log } z \end{aligned}$$

but for $\mathcal{I}(z) < 0$ we have

$$\begin{aligned} \text{Log}(-z) + \pi i &= \ln|-z| + i\,\text{Arg}(-z) + \pi i \\ &= \ln|z| + i(\text{Arg } z + \pi) + \pi i \\ &= \ln|z| + i\,\text{Arg } z + 2\pi i \\ &= \text{Log } z + 2\pi i \end{aligned}$$

Thus, when $\mathcal{I}(z) < 0$, $e^{\beta[\text{Log}(-z)+\pi i]} = e^{\beta \, \text{Log } z} e^{2\pi i \beta}$ and $F(z) = e^{\beta[\text{Log}(-z)+\pi i]} f(z) = e^{\beta \, \text{Log } z} f(z) e^{2\pi i \beta} = F_0(z)e^{2\pi i \beta}$.

If $a < 0$, then

$$\lim_{b \to 0^+} F(a + ib) = \lim_{b \to 0^-} F(a + ib)$$

showing $F(z)$ is continuous across the negative real axis. If $a > 0$, then

$$\lim_{b \to 0^+} F(a + ib) - \lim_{b \to 0^-} F(a + ib) = a^\beta f(a)(1 - e^{2\pi i \beta})$$

so $F(z)$ is continuous across the positive real axis only if β is an integer, whereby $e^{2\pi i \beta} = 1$.

EXAMPLE 36.9   To obtain

$$\int_0^\infty \frac{dx}{\sqrt{x}(x + c)} = \frac{\pi}{\sqrt{c}} \quad c > 0$$

note that $\beta = -\frac{1}{2}$ so that $\beta + 1 = \frac{1}{2} = \lambda$. Since $f(z) = \frac{1}{z+c} = O(|z|^{-p})$, $|z| \to \infty$, with $p = 1 > \lambda = \frac{1}{2}$, and since $c > 0$, the only pole of $f(z)$ inside the contour C is at $z = -c$. There are no poles on the positive real axis. Define $F(z) = e^{(-1/2)[\text{Log}(-z)+\pi i]}/(z + c)$ and obtain the value of the integral as $-\frac{\pi e^{-\pi i/2}}{\sin(\pi/2)}\text{Res}(F(z), -c) = (\pi i)(-\frac{i}{\sqrt{c}}) = \frac{\pi}{\sqrt{c}}$.   ❖

EXAMPLE 36.10   The integral $\int_0^\infty \frac{dx}{\sqrt{x}(x+4)(x-5)}$ does not exist as an improper integral because of the simple pole at $x = 5$ on the real axis. However, the *CPV* is $-\frac{\pi}{18}$, as can be determined from the antiderivative

$$g(x) = -\frac{1}{9}\arctan\frac{\sqrt{x}}{2} - \frac{2\sqrt{5}}{45}\operatorname{arctanh}\sqrt{\frac{x}{5}}$$

by computing

$$\lim_{x\to\infty} g(x) - \lim_{x\to 0^+} g(x) + \lim_{c\to 0^+}[g(5-c) - g(5+c)] = -\frac{\pi}{18}$$

Alternatively, note that since $\beta = -\frac{1}{2}$, we have $\beta + 1 = \frac{1}{2} = \lambda$. Then define the functions

$$F(z) = \frac{e^{(-1/2)[\operatorname{Log}(-z)+\pi i]}}{(z+4)(z-5)} \quad \text{and} \quad F_0(z) = \frac{1}{\sqrt{z}(z+4)(z-5)}$$

and compute the residues

$$r_1 = \operatorname{Res}(F(z), -4) = \frac{i}{18} \quad \text{and} \quad r_2 = \operatorname{Res}(F_0(z), 5) = \frac{\sqrt{5}}{45}$$

The *CPV* of the integral is then $-\pi r_1 \frac{e^{-\pi i/2}}{\sin(\pi/2)} - \pi r_2 \cot\frac{\pi}{2} = -\pi(\frac{i}{18})(-i) = -\frac{\pi}{18}$.   ❖

EXERCISES 36.1–Part A

A1. Make the change of variables $z = e^{i\theta}$ in $\int_0^{2\pi} \frac{1+\sin^2\theta}{2+\cos\theta}\,d\theta$ to obtain $\int_C f(z)\,dz$, where C is the contour defined by $|z| = 1$.

A2. Use residues to evaluate $\int_{-\infty}^{\infty} \frac{1+x^2}{4+x^4}\,dx$.

A3. Evaluate $\int_0^{\infty} \frac{1+x^2}{4+x^4}\,dx$.

A4. Evaluate $\int_{-\infty}^{\infty} e^{2ix}\frac{1+x}{1+x^2}\,dx$.

A5. Evaluate $\int_{-\infty}^{\infty} \frac{1+x}{1+x^2}\cos 2x\,dx$ and $\int_{-\infty}^{\infty} \frac{1+x}{1+x^2}\sin 2x\,dx$.

EXERCISES 36.1–Part B

B1. Evaluate $\int_{-\infty}^{\infty} \frac{1}{1+x^2}\,dx$ by completing the contour as in Figure 36.1(*a*) and as in Figure 36.1(*b*). Of course, the value should be the same in each case.

In Exercises B2–16, evaluate by conversion to contour integrals. Where possible, verify by consulting a table of integrals, using a computer algebra system or integrating numerically.

B2. $\displaystyle\int_0^{2\pi} \frac{d\theta}{1 + a\cos\theta}, a^2 < 1$ **B3.** $\displaystyle\int_0^{2\pi} \frac{d\theta}{1 + a\sin\theta}, a^2 < 1$

B4. $\displaystyle\int_0^{2\pi} \frac{d\theta}{6 + 3\cos\theta}$ **B5.** $\displaystyle\int_0^{2\pi} \frac{d\theta}{7 + 2\sin\theta}$

B6. $\displaystyle\int_0^{2\pi} \frac{d\theta}{a + b\cos\theta}, a^2 > b^2$ **B7.** $\displaystyle\int_0^{2\pi} \frac{d\theta}{a + b\sin\theta}, a^2 > b^2$

B8. $\displaystyle\int_0^{2\pi} \frac{d\theta}{3 + 2\cos^2\theta}$ **B9.** $\displaystyle\int_0^{2\pi} \frac{d\theta}{2 + 5\sin^2\theta}$

B10. $\displaystyle\int_0^{2\pi} \frac{d\theta}{a + b\cos^2\theta}$ **B11.** $\displaystyle\int_0^{2\pi} \frac{d\theta}{a + b\sin^2\theta}$

B12. $\displaystyle\int_0^{2\pi} \frac{\sin^2\theta\,d\theta}{3 + 2\cos\theta}$ **B13.** $\displaystyle\int_0^{2\pi} \frac{\sin^2\theta\,d\theta}{a + b\cos\theta}, a > |b| > 0$

B14. $\displaystyle\int_0^{2\pi} \frac{\cos^2 5\theta\,d\theta}{7 - 3\cos 4\theta}$ (*Hint:* If $z = e^{i\theta}$, then $e^{5i\theta} = z^5$, etc.)

B15. $\displaystyle\int_0^{\pi} \frac{d\theta}{3 + 2\cos\theta}$ (*Hint:* Use symmetry or $\theta = 2\pi - \tau$ to establish $\int_0^{\pi} \frac{d\theta}{3+2\cos\theta} = \int_\pi^{2\pi} \frac{d\theta}{3+2\cos\theta}$.)

B16. $\displaystyle\int_0^{\pi} \frac{d\theta}{(a + \cos^2\theta)^2}$

In Exercises B17–26, evaluate by conversion to contour integrals. Where possible, verify by consulting a table of integrals, using a computer algebra system or integrating numerically.

B17. $\displaystyle\int_{-\infty}^{\infty} \frac{8x - 9}{(x^2 - 8x + 25)(x + 7 + 5i)}\,dx$

B18. $\displaystyle\int_{-\infty}^{\infty} \frac{4x - 2}{(x^2 + 25)(x - 7i)}\,dx$

B19. $\displaystyle\int_{-\infty}^{\infty} \frac{6x^2 - 4x + 5}{(x^2 + 2x + 50)(x^2 - 6x + 34)}\,dx$

B20. $\displaystyle\int_0^{\infty} \frac{x^2 + 7}{x^4 + 18x^2 + 162}\,dx$

B21. $\displaystyle\int_0^\infty \frac{dx}{x^2+9}$ **B22.** $\displaystyle\int_0^\infty \frac{x^2+8}{x^4-8x^2+32}\,dx$

B23. $\displaystyle\int_{-\infty}^\infty \frac{x}{(x^2-18x+162)(x-8-5i)}\,dx$

B24. $\displaystyle\int_{-\infty}^\infty \frac{x-2}{(x^2-2x+2)(x+9i)}\,dx$

B25. $\displaystyle\int_{-\infty}^\infty \frac{7x^2+6x+5}{(x^2+10x+89)(x^2-6x+18)}\,dx$

B26. $\displaystyle\int_0^\infty \frac{x^2-1}{(x^2+2)^2}\,dx$

In Exercises B27–36, evaluate by conversion to contour integrals. Where possible, verify by consulting a table of integrals, using a computer algebra system or integrating numerically.

B27. $\displaystyle\int_{-\infty}^\infty \frac{(2x^2-x+4)e^{5ix}}{(x^2+12x+117)(x+i)}\,dx$

B28. $\displaystyle\int_{-\infty}^\infty \frac{(3x^2+8x-3)e^{-ix}}{(x^2+16x+145)(x-5+7i)}\,dx$

B29. $\displaystyle\int_{-\infty}^\infty \frac{(3x^3-2x^2+4x+1)e^{-9ix}}{(x^2+12x+52)(x^2-6x+45)}\,dx$

B30. $\displaystyle\int_0^\infty \frac{(x^2+9)e^{-5ix}}{x^4+14x^2+85}\,dx$ **B31.** $\displaystyle\int_0^\infty \frac{x^3e^{6ix}}{x^4-16x^2+89}\,dx$

B32. $\displaystyle\int_{-\infty}^\infty \frac{(2x^2+2x-4)e^{-7ix}}{(x^2-18x+82)(x-2+5i)}\,dx$

B33. $\displaystyle\int_{-\infty}^\infty \frac{(8x^2-5x-9)e^{2ix}}{(x^2-14x+113)(x+9i)}\,dx$

B34. $\displaystyle\int_{-\infty}^\infty \frac{(9x^3+5x^2-x+4)e^{-3ix}}{(x^2+12x+40)(x^2-6x+25)}\,dx$

B35. $\displaystyle\int_0^\infty \frac{(x^2-5)e^{8ix}}{x^4-6x^2+18}\,dx$ **B36.** $\displaystyle\int_0^\infty \frac{(x^3-5x)e^{-4ix}}{x^4+18x^2+145}\,dx$

In Exercises B37–46, evaluate by conversion to contour integrals. Where possible, verify by consulting a table of integrals, using a computer algebra system or integrating numerically.

B37. $\displaystyle\int_{-\infty}^\infty \frac{(9x^2-4x-6)\cos(9x)}{(x^2-4x+68)(x-6-9i)}\,dx$

B38. $\displaystyle\int_{-\infty}^\infty \frac{(7x^2-6x+6)\sin(7x)}{(x^2-14x+58)(x+6-3i)}\,dx$

B39. $\displaystyle\int_{-\infty}^\infty \frac{(7x^3-5x^2+8x+3)\cos(x)}{(x^2-4x+5)(x^2+4x+29)}\,dx$

B40. $\displaystyle\int_{-\infty}^\infty \frac{(6x^3+7x^2+4x+1)\sin(5x)}{(x^2-8x+52)(x^2-2x+37)}\,dx$

B41. $\displaystyle\int_0^\infty \frac{\cos(2x)}{x^2+2}\,dx$ **B42.** $\displaystyle\int_0^\infty \frac{x\sin x}{x^2+5}\,dx$

B43. $\displaystyle\int_0^\infty \frac{(x^2+9)\cos(4x)}{x^4+18x^2+130}\,dx$ **B44.** $\displaystyle\int_0^\infty \frac{x\sin(3x)}{x^4-2x^2+26}\,dx$

B45. $\displaystyle\int_{-\infty}^\infty \frac{(x-4)\sin(3x)}{x^2-10x+89}\,dx$ **B46.** $\displaystyle\int_{-\infty}^\infty \frac{(x+9)\cos(2x)}{x^2-8x+65}\,dx$

Obtain the Cauchy principal value of the integrals in Exercises B47–56.

B47. $\displaystyle\int_{-\infty}^\infty \frac{7x-3}{(x^2-4x+85)(x-7)}\,dx$

B48. $\displaystyle\int_{-\infty}^\infty \frac{5x^2-9x-1}{(x^2+14x+130)(x^2-8x+97)x}\,dx$

B49. $\displaystyle\int_{-\infty}^\infty \frac{9x^3-8x^2+9x-4}{(x^2-18x+130)(x^2+2x+10)(x-1)}\,dx$

B50. $\displaystyle\int_{-\infty}^\infty \frac{6x^2-4x+4}{(x^2-4x+5)(x^2-4)}\,dx$

B51. $\displaystyle\int_{-\infty}^\infty \frac{5x-8}{(x^2+4x+53)(x-6)}\,dx$

B52. $\displaystyle\int_{-\infty}^\infty \frac{3x^2-x-3}{(x^2-6x+34)(x^2-4x+13)(x+3)}\,dx$

B53. $\displaystyle\int_{-\infty}^\infty \frac{9x^3-3x^2+5x+3}{(x^2-8x+52)(x^2-16x+145)(x+2)}\,dx$

B54. $\displaystyle\int_{-\infty}^\infty \frac{4x^2+9x+5}{(x^2+16x+89)(x^2-9)}\,dx$

B55. $\displaystyle\int_{-\infty}^\infty \frac{4x^2+7x+3}{(x^2+18x+106)^2(x^2-16)}\,dx$

B56. $\displaystyle\int_{-\infty}^\infty \frac{3x^2+4x-9}{(x^4-12x^2+85)(x^2-1)}\,dx$

Obtain the Cauchy principal value of the integrals in Exercises B57–66.

B57. $\displaystyle\int_0^\infty \frac{x^{-1/3}}{(x^2-16x+89)(x-3)}\,dx$

B58. $\displaystyle\int_0^\infty \frac{x^{-1/2}(3x^2-5x-2)}{(x^4-4x^2+85)(x-2)}\,dx$

B59. $\displaystyle\int_0^\infty \frac{x^{8/3}}{(x^2+8x+25)(x^2-8x+7)}\,dx$

B60. $\displaystyle\int_0^\infty \frac{x^{-1/4}}{(x^2+12x+100)(x-2)}\,dx$

B61. $\displaystyle\int_0^\infty \frac{x^{1/4}(5x^2-6x-2)}{(x^4-12x^2+37)(x-9)}\,dx$

B62. $\displaystyle\int_0^\infty \frac{x^{7/4}}{(x^2-12x+85)(x^2-12x+27)}\,dx$

B63. $\displaystyle\int_0^\infty \frac{x^{5/4}}{(x^2+2x+5)(x-3)}\,dx$

B64. $\displaystyle\int_0^\infty \frac{x^{6/5}(x^2-4x+3)}{(x^4-18x^2+90)(x-2)}\,dx$

B65. $\displaystyle\int_0^\infty \frac{x^{13/5}}{(x^2-6x+13)(x^2-2x+1)}\,dx$

B66. $\displaystyle\int_0^\infty \frac{x^{\sqrt2}(7x^2-5x-6)}{(x^4+2x^2+5)(x-2)}\,dx$

36.2 The Laplace Transform

The Laplace Transform Revisited

The Laplace transform, $F(s) = L[f(t)] = \int_0^\infty f(t)e^{-st}\,dt$, an integral transform that maps a well-behaved function $f(t)$ to $F(s)$, its Laplace transform, was studied in Chapter 6. For example, if $f(t) = t$, the transform, an integration along the real line, is given by $F(s) = \int_0^\infty te^{-st}\,dt = \frac{1}{s^2}$. In Chapter 6, the inversion of Laplace transforms was done with partial fractions, the four operational laws, and pattern-recognition.

Bromwich Integral Formula

The following theorem taken from [21] provides the *Bromwich integral formula* by means of which $F(s)$, the Laplace transform of $f(t)$, is "inverted" or mapped back to $f(t)$.

THEOREM 36.2

1. $f(t)$ is of exponential order $O(e^{\alpha t})$.
2. $f'(t)$ is piecewise continuous over each interval $0 < t < T$.
3. $c > \alpha$ and $s = c + iy$.

$\implies$

1. $f(t) = \frac{1}{2\pi i}\int_{c-i\infty}^{c+i\infty} F(s)e^{st}\,ds$, where this improper integral along the vertical line $s = c + iy$ is interpreted as the Cauchy principal value $\lim_{b\to\infty}\int_{c-ib}^{c+ib} F(s)e^{st}\,ds$ and is called the Bromwich integral formula.
2. If $f(t)$ is discontinuous at $t_0 > 0$, then the Bromwich integral converges to $\frac{1}{2}[f(t_0 + 0) + f(t_0 - 0)]$.
3. At $t = 0$, the Bromwich integral converges to $\frac{f(+0)}{2}$.
4. For $t < 0$, the Bromwich integral converges to 0.
   Also found in [21], the following alternate theorem places sufficient conditions on $F(s)$ rather than on $f(t)$.

THEOREM 36.3

1. $F(s) = O(s^{-k})$, $k > 1$, and is analytic for $s = x + iy$ in the half-plane $\beta \le x$.
2. $F(x)$ is real when $x \ge \beta$.
3. $\gamma \ge \beta$
4. t is real.

$\implies$

1. $\frac{1}{2\pi i}\int_{\gamma-i\infty}^{\gamma+i\infty} F(s)e^{st}\,ds$ converges to a continuous function $f(t)$ that is of exponential order $O(e^{\beta t})$ and vanishes for $t \le 0$.
2. $L[f(t)] = F(s)$

In practice, if $F(s)$ is analytic to the right of $x = \beta$, then the Bromwich integral can be calculated along any vertical line $x = c > \beta$.

EXAMPLE 36.11   We verify the inversion integral for the function $f(t) = 1$ for which the Laplace transform is $F(s) = \frac{1}{s}$. Since the only singularity of $F(s) = \frac{1}{s}$ is at $s = 0$, choose $c = 1$ and parametrize C, the vertical line, as $s = 1 + iy$. The Bromwich contour integral,

$$\frac{1}{2\pi i} \int_C F(s)e^{st}\, ds = \lim_{b \to \infty} \frac{1}{2\pi i} \int_{-b}^{b} i \frac{e^{(1+iy)t}}{1+iy}\, dy = \begin{cases} 1 & t > 0 \\ \frac{1}{2} & t = 0 \\ 0 & t < 0 \end{cases} \tag{36.3}$$

is evaluated with great difficulty using an antiderivative found, for example, by Maple. (See Exercises A1–5.) Shortly, we will learn how to evaluate such integrals by the Cauchy residue theorem.   ❖

EXAMPLE 36.12   We verify the Bromwich inversion integral for $f(t) = t$, for which the Laplace transform is $F(s) = \frac{1}{s^2}$. Again, the only singularity of $F(s)$ is $s = 0$, so we can use for the contour C, the same vertical line $s = 1 + iy$ as we used in Example 36.11. Thus, we have

$$\frac{1}{2\pi i} \int_C F(s)e^{st}\, ds = \lim_{b \to \infty} \frac{1}{2\pi i} \int_{-b}^{b} i \frac{e^{(1+iy)t}}{(1+iy)^2}\, dy = \begin{cases} t & t > 0 \\ 0 & t < 0 \end{cases} \tag{36.4}$$

where again the integral is evaluated with an antiderivative provided, for example, by Maple. (See Exercise B1.) Since $f(+0) = f(0) = 0$, there was no need to examine the integral for the case $t = 0$.   ❖

Evaluating the Bromwich Integral by the Cauchy residue theorem

Examples 36.11 and 36.12, in which direct evaluation of the Bromwich integral could only be done by recourse to a computer algebra system, motivate the more productive strategy of closing the contour so that Cauchy's residue theorem can be used.

Suppose the vertical line $s = c + iy$, $|y| \le b$, became part of a closed path, a path for which the additional contribution to the Bromwich integral along the new part of the contour went to zero as $b \to \infty$. Then, by evaluating this *closed* contour integral, in the limit as $b \to \infty$, we'd simply be computing the value of the integral along the original vertical line.

One such closure for the contour C is shown in Figure 36.5, where the curve is an arc from a circle with radius R and center at the origin. As the radius $R \to \infty$, with c held fixed, we have to have $b \to \infty$ also. So having $R \to \infty$ is equivalent to having $b \to \infty$, provided c remains fixed.

If the value of the Bromwich integral along the curved arc tends to zero as $R \to \infty$, then we can complete the contour as in Figure 36.5, and evaluate the integral as a closed contour integral via an appeal to residues. Any singularities of $F(s)$ are to the left of the vertical line and, thus,

$$f(t) = \frac{1}{2\pi i} \lim_{b \to \infty} \int_{c-ib}^{c+ib} F(s)e^{st}\, ds$$

$$= \frac{1}{2\pi i}(2\pi i) \sum_{n=1}^{k} \operatorname{Res}(e^{st}F(s), s_n) = \sum_{n=1}^{k} \operatorname{Res}(e^{st}F(s), s_n)$$

where $s_1, \ldots, s_k$ are the poles of $F(s)$ in the half-plane $\mathcal{R}(s) < c$.

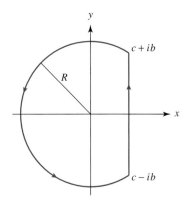

FIGURE 36.5   Closed contour for evaluating the Bromwich integral by the Cauchy residue theorem

EXAMPLE 36.13 To invert $F(s) = \frac{1}{s}$, note that $s = 0$ is a simple pole, and the only pole, of $e^{st} F(s)$. The residue at $s = 0$ is 1, the value of the Bromwich integral when $t > 0$. ❖

EXAMPLE 36.14 To invert $F(s) = \frac{1}{s^2}$, note that $s = 0$ is a pole of order 2, and the only pole, for $e^{st} F(s)$. The residue at $s = 0$ is t, the value of the Bromwich integral when $t > 0$. ❖

EXAMPLE 36.15 To invert $F(s) = \frac{s}{s^2 + \omega^2}$, note that there are simple poles at $s_1 = i\omega$ and $s_2 = -i\omega$. Hence,

$$f(t) = \sum_{n=1}^{2} \text{Res}(e^{st} F(s), s_n) = \tfrac{1}{2} e^{i\omega t} + \tfrac{1}{2} e^{-i\omega t} = \cos \omega t$$ ❖

EXAMPLE 36.16 To invert the Laplace transform $F(s) = \frac{1}{s} \tanh \frac{\pi}{2} s$, the poles, $s = 0$ and the zeros of $\cosh \frac{\pi}{2} s$, must be determined as $0, \pm i, \pm 3i, \pm 5i, \ldots$. Since $\text{Res}(e^{st} F(s), 0) = 0$ and $\text{Res}(e^{st} F(s), \pm ni) = \frac{2}{n\pi}(\mp i \cos nt + \sin nt), n = 1, 3, 5, \ldots$, we find

$$f(t) = \frac{2}{\pi} \sum_{n=1,3,5,\ldots}^{\infty} \frac{1}{n} [(-i \cos nt + \sin nt) + (i \cos nt + \sin nt)]$$

$$= \frac{4}{\pi} \sum_{n=1,3,5,\ldots}^{\infty} \frac{1}{n} \sin nt = \frac{4}{\pi} \sum_{k=0}^{\infty} \frac{\sin(2k+1)t}{2k+1}$$

This is just the Fourier series representation of the square wave $f(t) = 1 + 2 \sum_{n=1}^{\infty}(-1)^n$ $H(t - n\pi)$, the odd periodic extension of $g(t) = 1, 0 < t < \pi$, where $H(t)$ is the Heaviside function. Indeed, the Fourier coefficients

$$b_n = \frac{2}{\pi} \int_0^{\pi} \sin nt \, dt = \begin{cases} 0 & n \text{ even} \\ \dfrac{4}{n\pi} & n \text{ odd} \end{cases}$$

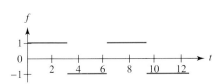

FIGURE 36.6 The graph of $f(t)$ in Example 36.16

are exactly the coefficients in the Fourier sine series for $f(t)$. Figure 36.6 contains a graph of $f(t)$. ❖

EXERCISES 36.2–Part A

A1. Show that $\int_{-b}^{b} e^{(1+iy)t}/(1+iy) \, dy = e^t \int_{-b}^{b} \frac{1}{1+y^2} [\cos ty + y \sin ty + i(\sin ty - y \cos ty)] \, dy$.

A2. Without computing, reason to the conclusion that the imaginary part of the right-hand integral in Exercise A1 is zero.

A3. If $t > 0$ implies that $\int_{-\infty}^{\infty} \frac{\cos ty}{1+y^2} \, dy = \int_{-\infty}^{\infty} \frac{y \sin ty}{1+y^2} \, dy = \pi(\cosh t - \sinh t)$, obtain the result for $t > 0$ in (36.3).

A4. If $t < 0$ implies that $\int_{-\infty}^{\infty} \frac{\cos ty}{1+y^2} \, dy = -\int_{-\infty}^{\infty} \frac{y \sin ty}{1+y^2} \, dy = \pi(\cosh t + \sinh t)$, obtain the result for $t < 0$ in (36.3).

A5. Obtain the result for $t = 0$ in (36.3).

EXERCISES 36.2–Part B

B1. Obtain the result in (36.4) as follows. First, integrate by parts once, taking $u = e^{ity}$ in $\int u \, dv = uv - \int v \, du$. Then use the results and techniques in Exercises A1–4.

For each Laplace transform $F(s)$ given in Exercises B2–11, obtain $f(t)$ using the Bromwich inversion formula. Evaluate the relevant integrals

using the Cauchy residue theorem. Where possible, verify the result by consulting a table of transforms or using a computer algebra system.

B2. $F(s) = \dfrac{1}{(s+2)^2}$

B3. $F(s) = \dfrac{s+3}{s^2 + 6s + 13}$

B4. $F(s) = \dfrac{1 - e^{-s}(s-1)}{s^2}$ **B5.** $F(s) = 18\dfrac{s^2-3}{(s^2+9)^3}$

B6. $F(s) = \dfrac{e^{-s+2} - e^{-3s+6}}{s-2}$

B7. $F(s) = 6\dfrac{s^4 - 8s^3 - 126s^2 + 568s + 41}{(s^2-4s+29)^4}$

B8. $F(s) = \dfrac{s^2+4}{(s^2-4)^2}$

B9. $F(s) = \dfrac{2(s^2-13)}{(s^2-4s+13)(s^2+4s+13)}$

B10. $F(s) = \dfrac{s - e^{-2s}(e^{-8}(s-4) + e^8(s+4))}{2(s^2-16)}$

B11. $F(s) = \dfrac{s^2+2a^2}{s(s^2+4a^2)}$

For the Laplace transforms given in Exercises B12–17:

(a) Obtain $f(t)$ by using the Bromwich inversion integral and the Cauchy residue theorem, summing an infinite series of residues.

(b) Expand $F(s)$ in powers of e^{-2s}, and invert termwise to obtain $f(t)$.

B12. $F(s) = \dfrac{1}{s(1+e^{2s})}$ **B13.** $F(s) = \dfrac{1}{s(e^{2s}-1)}$

B14. $F(s) = \dfrac{1}{s(1-e^{-2s})}$ **B15.** $F(s) = \dfrac{1}{s}\coth 2s$

B16. $F(s) = \dfrac{1}{s}\operatorname{sech}2s$ **B17.** $F(s) = \dfrac{1}{s}\operatorname{csch}2s$

36.3   Fourier Series and the Fourier Transform

The Fourier Series

Trigonometric Form   In Chapter 10 we saw the trigonometric form of the Fourier series. Recall that for a function $f(t)$ defined on the interval $[-L, L]$, the Fourier trig series is given by

$$f(t) = \frac{a_0}{2} + \sum_{n=1}^{\infty} a_n \cos \frac{n\pi t}{L} + b_n \sin \frac{n\pi t}{L}$$

where

$$a_n = \frac{1}{L} \int_{-L}^{L} f(t) \cos \frac{n\pi t}{L}\, dt \quad n = 0, 1, 2, \ldots$$

$$b_n = \frac{1}{L} \int_{-L}^{L} f(t) \sin \frac{n\pi t}{L}\, dt \quad n = 1, 2, \ldots$$

Exponential Form   In the applications, it is common to use the exponential form of the Fourier series. Thus, if $f(t)$ is defined on the interval $[-L, L]$, it has the representation

$$f(t) = \sum_{n=-\infty}^{\infty} c_n e^{in\pi t/L} \quad n = 0, \pm 1, \pm 2, \ldots$$

where

$$c_n = \frac{1}{2L} \int_{-L}^{L} f(t) e^{-in\pi t/L}\, dt \quad n = 0, \pm 1, \pm 2, \ldots$$

In light of Euler's formulas, namely, $e^{\pm ix} = \cos x \pm i \sin x$, the trig and exponential forms of the Fourier series are ultimately identical, as we will shortly show.

EXAMPLE 36.17   The exponential form of the Fourier series for

$$f(t) = \begin{cases} 1+t & -1 \le t \le 0 \\ 1 - t^2 & 0 \le t \le 1 \end{cases}$$

valid on the interval $[-1, 1]$, is $\sum_{n=-\infty}^{\infty} c_n e^{int}$, where

$$c_0 = \frac{1}{2} \int_{-1}^{1} f(t)\, dt = \frac{7}{12}$$

$$c_n = \frac{1}{2} \int_{-1}^{1} f(t) e^{-int}\, dt = \frac{1 - 3(-1)^n}{2n^2 \pi^2} + i\frac{(-1)^n - 1}{n^3 \pi^3} \qquad n \neq 0$$

Applying Euler's formulas to $\sum_{n=-\infty}^{\infty} c_n e^{int}$ gives

$$f(t) = \frac{7}{12} + \sum_{n=1}^{\infty} \left(\frac{1 - 3(-1)^n}{n^2 \pi^2} \cos n\pi t + \frac{2}{n^3 \pi^3}(1 - (-1)^n) \sin n\pi t \right)$$

the Fourier trig series for $f(t)$.

❖

EQUIVALENCE OF FORMS   To show that the exponential form of the Fourier series is equivalent to the trig form write

$$c_n = \frac{1}{2L} \int_{-L}^{L} f(t) e^{-in\pi t/L}\, dt$$

$$= \frac{1}{2L} \int_{-L}^{L} f(t) \cos \frac{n\pi t}{L}\, dt - \frac{i}{2L} \int_{-L}^{L} f(t) \sin \frac{n\pi t}{L}\, dt = \frac{1}{2}(a_n - ib_n)$$

Since n is both positive and negative, that is, $n = \pm k, k > 0$, write $c_n e^{in\pi t/L}$ as the two terms

$$c_k e^{ik\pi t/L} + c_{-k} e^{-ik\pi t/L} = \frac{1}{2}(a_k - ib_k) e^{ik\pi t/L} + \frac{1}{2}(a_k + ib_k) e^{-ik\pi t/L}$$

$$= \frac{a_k}{2}(e^{ik\pi t/L} + e^{-ik\pi t/L}) + \frac{b_k}{2i}(e^{ik\pi t/L} - e^{-ik\pi t/L})$$

$$= a_k \cos \frac{k\pi t}{L} + b_k \sin \frac{k\pi t}{L}$$

where $a_{-k} = a_k$ because the cosine is even and $b_{-k} = -b_k$ because the sine is odd. For the case $n = 0$ we have $c_0 = \frac{1}{2L} \int_{-L}^{L} f(t)\, dt = \frac{a_0}{2}$.

FOURIER TRANSFORM   In Section 30.2 the Fourier Integral Theorem $f(x) = \frac{1}{2\pi} \int_{-\infty}^{\infty} \int_{-\infty}^{\infty} e^{i\alpha x} f(\beta) e^{-i\alpha\beta}\, d\beta\, d\alpha$, used to solve boundary value problems on infinite and semi-infinite domains, was stated. In Section 30.3, this theorem was the basis for the definition of the exponential form of the Fourier integral transform $F(\alpha) = \int_{-\infty}^{\infty} f(\beta) e^{-i\alpha\beta}\, d\beta$ and the inversion formula $f(x) = \frac{1}{2\pi} \int_{-\infty}^{\infty} F(\alpha) e^{i\alpha x}\, d\alpha$. In Section 30.3, $F(\alpha) = \frac{2}{1+\alpha^2}$, the Fourier transform of the function $f(x) = e^{-|x|}$, was computed by direct integration and the inversion integral was evaluated, with some difficulty, by direct integration. Alternatively, since the inversion integral is computed along the real line (where α is a real variable), it is generally possible to "close the contour" and evaluate the integral by the Cauchy residue theorem, as per Section 36.1.

EXAMPLE 36.18   By using the Cauchy residue theorem, we can invert $F(\alpha) = \frac{2}{1+\alpha^2}$, the Fourier transform of $f(x) = e^{-|x|}$. Now the Cauchy residue theorem states that

$$\int_C f(z)\, dz = 2\pi i \sum_{n=1}^{k} \text{Res}(f(z), z_n)$$

so the Fourier inversion integral becomes

$$f(x) = \frac{1}{2\pi} \int_{-\infty}^{\infty} F(\alpha)e^{i\alpha x}\, d\alpha = \frac{2\pi i}{2\pi} \sum_{n=1}^{k} \text{Res}(F(\alpha)e^{i\alpha x}, \alpha_n) = i \sum_{n=1}^{k} \text{Res}(F(\alpha)e^{i\alpha x}, \alpha_n)$$

(36.5)

There are two simple poles of $F(\alpha)$, one in the upper half-plane at $\alpha = i$ and one in the lower half-plane at $\alpha = -i$.

As in Figure 36.1(*a*), let C_1 be the positively oriented closed contour consisting of the real axis and C_R, the upper half of the circle $|\alpha| = Re^{i\theta}$. As in Figure 36.1(*b*), let C_2 be the positively oriented closed contour consisting of the real axis and C_r, the lower half of the same circle. On C_1, the positive orientation means a counterclockwise traverse of the semicircle and a passage from left to right along the real line. However, on C_2, the positive orientation (and consequent counterclockwise passage around the contour) means the real axis is traversed from right to left, against the natural direction of integration. This will induce a minus sign in the contour integral along C_2.

On either contour, the Cauchy residue theorem applies if we can show the contribution of the integral along the semicircles goes to zero as $R \to \infty$. In particular, along $\alpha = Re^{i\theta}$ we have $|e^{i\alpha x}| = |e^{ixRe^{i\theta}}| = e^{-xR\sin\theta}$. The decay of this magnitude for large R requires $x\sin\theta > 0$. When $x > 0$, this condition is satisfied when $0 < \theta < \pi$ because then, $\sin\theta > 0$. When $x < 0$, this condition is satisfied when $\pi < \theta < 2\pi$ because then, $\sin\theta < 0$. Thus, when $x > 0$, use contour C_1, which contains the one simple pole of $F(\alpha)$ at $\alpha = i$. When $x < 0$, use contour $-C_2$, which contains the one simple pole of $F(\alpha)$ at $\alpha = -i$. (The minus sign on contour C_2 appears because on it, the integration along the real axis is from right to left.) Therefore, the residues at each pole are

$$r_1 = \text{Res}(F(\alpha)e^{i\alpha x}, i) = -ie^{-x} \quad \text{and} \quad r_2 = \text{Res}(F(\alpha)e^{i\alpha x}, -i) = ie^x$$

When $x > 0$, the inversion integral is i times the residue r_1; and when $x < 0$, the inversion integral is i times $-r_2$. Hence, the value of the inversion integral is e^x when $x < 0$ and e^{-x} when $x > 0$. When $x = 0$, the integral on the left in (36.5) is 1, as we see in Exercise A1. Hence, the inversion integral, computed by the Cauchy residue theorem, reconstructs $f(x) = e^{-|x|}$.

The contour integration technique used in this example requires the contribution to the integral along either semicircle to go to zero as $R \to \infty$. Since $|F(\alpha)| \le \left|\frac{1}{1+\alpha^2}\right| \le \frac{1}{|\alpha|^2}$, we can use the discussion in Section 36.1 to conclude that the contributions from C_R or C_r tend to zero. In this example, $m = x > 0$ in the upper half-plane but $m = x < 0$ in the lower half-plane. On C_2, the positive orientation forces integration along the real axis to be from right to left, so the value of the contour integral is $-2\pi i$ times the residue at $\alpha = -i$. ❖

EXAMPLE 36.19   The Fourier transform of

$$f(x) = \begin{cases} 0 & x < 0 \\ xe^{-x} & x > 0 \end{cases} = xe^{-x}H(x)$$

(36.6)

where $H(x)$ is the Heaviside function, is $F(\alpha) = \frac{1}{(1+i\alpha)^2}$ and has a pole of order 2 at $\alpha = i$. Since $|F(\alpha)| = \left|\frac{1}{(1+i\alpha)^2}\right| \le \frac{1}{|\alpha|^2}$, the Fourier inversion integral can be evaluated by the Cauchy residue theorem according to the equation

$$f(x) = \frac{1}{2\pi} \int_{-\infty}^{\infty} F(\alpha)e^{i\alpha x}\, d\alpha = \frac{2\pi i}{2\pi}\text{Res}(F(\alpha)e^{i\alpha x}, i) = i\,\text{Res}(F(\alpha)e^{i\alpha x}, i)$$

On the contour C_1 from Example 36.18, the integration is counterclockwise, encloses the pole at $\alpha = i$, and has the residue $r_1 = \text{Res}(F(\alpha)e^{i\alpha x}, i) = -ixe^{-x}$. The contour C_2

from Example 36.18 encloses no poles of $F(\alpha)$, so the sum of residues inside C_2 is zero. Thus, if $x < 0$ so that contour C_2 is used, $f(x) = 0$. If $x > 0$ so that contour C_1 is used, $f(x) = xe^{-x}$. Hence, we have (36.6).   ❖

EXERCISES 36.3–Part A

A1. Show that when $x = 0$, the Fourier inversion integral in (36.5) evaluates to 1 in Example 36.18.

A2. Obtain the trigonometric form of the Fourier series for $f(x) = |x|$, $-1 \le x \le 1$.

A3. Obtain the exponential form of the Fourier series for the function in Exercise A2.

A4. Show the series in Exercises A2 and 3 are actually the same.

A5. Use the Cauchy residue theorem to evaluate the Fourier inversion integral for $F(\alpha) = \frac{1}{2+i\alpha}$.

EXERCISES 36.3–Part B

For the function $f(x)$ specified in each of Exercises B1–10, obtain the exponential form of the Fourier series and show that it is equivalent to the trigonometric form. Compare a graph of a partial sum of the exponential form with a graph of the function itself.

B1. Exercise B19, Section 10.1   **B2.** Exercise B20, Section 10.1

B3. Exercise B21, Section 10.1   **B4.** Exercise B22, Section 10.1

B5. Exercise B23, Section 10.1   **B6.** Exercise B24, Section 10.1

B7. Exercise B25, Section 10.1   **B8.** Exercise B26, Section 10.1

B9. Exercise B27, Section 10.1   **B10.** Exercise B28, Section 10.1

In Exercises B11–25, use the Cauchy residue theorem to evaluate the inversion integral for each of the following Fourier transforms. Where the parameter b appears, impose the restriction $b > 0$.

B11. $F(\alpha) = \dfrac{1}{b + i\alpha}$ **B12.** $F(\alpha) = \dfrac{1}{b - i\alpha}$

B13. $F(\alpha) = \dfrac{1}{(b + i\alpha)^2}$ **B14.** $F(\alpha) = \dfrac{1}{(b - i\alpha)^2}$

B15. $F(\alpha) = \dfrac{e^{-i\alpha}}{1 + \alpha^2}$ **B16.** $F(\alpha) = \dfrac{\alpha}{(b^2 + \alpha^2)^2}$

B17. $F(\alpha) = \dfrac{(b^2 - \alpha^2)}{(b^2 + \alpha^2)}$ **B18.** $F(\alpha) = e^{-b|\alpha|}$

B19. $F(\alpha) = \alpha e^{-b|\alpha|}$ **B20.** $F(\alpha) = e^{-b\alpha^2}$

B21. $F(\alpha) = \alpha e^{-\alpha^2}$ **B22.** $F(\alpha) = \dfrac{\sin b\alpha}{\alpha}$

B23. $F(\alpha) = \dfrac{b^2}{\alpha(b^2 - \alpha^2)} \sin \dfrac{\pi \alpha}{b}$ **B24.** $F(\alpha) = \dfrac{1 - \cos b\alpha}{\alpha}$

B25. $F(\alpha) = \dfrac{4}{1 + \alpha^4}$

36.4   The Root Locus

Introduction

In the study and design of feedback control systems, engineers make use of the *root-locus* diagram, which shows the variation of eigenvalues as a function of a system parameter. The eigenvalues are actually found as poles of an appropriate transfer function, a function of the complex variable s arising from the Laplace transform of the differential equations of the system. For each value of the system parameter, the poles are plotted in the complex plane. The resulting locus of points in the complex plane is called the *root locus*.

This section will give some mathematical insight into the construction of a root-locus diagram. We begin with a study of the "locus of roots" and reserve the phrase "root locus" specifically for the case pertinent to controls engineering. (For a more complete engineering perspective, see [70] or [31].)

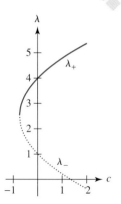

FIGURE 36.7   Example 36.20: The eigenvalues as a function of the parameter c (solid curve is λ_+, dotted curve is λ_-)

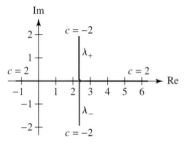

FIGURE 36.8   Example 36.20: In the complex plane, the locus of roots determined by $\lambda_\pm$ (curve in black is λ_+, while curve in color is λ_-)

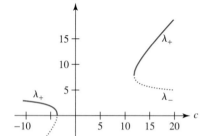

FIGURE 36.9   Example 36.21: The eigenvalues as a function of the parameter c (solid curve is λ_+, dotted curve is λ_-)

EXAMPLE 36.20   In Chapter 12, the mixing-tank problem led to $\mathbf{x}' = A\mathbf{x}$, the first-order linear system of ordinary differential equations with constant coefficients. The solution is given by $e^{At}\mathbf{x}_0$, where $\mathbf{x}_0 = \mathbf{x}(0)$ and e^{At} is the fundamental matrix. The eigenvalues of A determine stability of the equilibrium point at the origin.

Suppose the matrix A contains the parameter c, making the system performance dependent on the value of c. For example, the eigenvalues $\lambda_\pm$ of the matrix A given in (36.7) are graphed as functions of c in Figure 36.7

$$A = \begin{bmatrix} 1 & c \\ 3 & 4 \end{bmatrix} \Rightarrow \lambda_\pm = \frac{5}{2} \pm \frac{1}{2}\sqrt{9 + 12c} \qquad (36.7)$$

The solid curve is λ_+, the dotted, λ_-. For $c < -\frac{3}{4}$, the eigenvalues are complex with positive real part, so the equilibrium point at the origin is unstable. For $-\frac{3}{4} \le c < \frac{4}{3}$, there are two positive real eigenvalues, the origin is a node (out) and is again unstable. For $c > \frac{4}{3}$, one eigenvalue is positive and the other is negative, so the origin is a saddle, again unstable.

We have just seen that when c causes both eigenvalues to fall into the left half of the complex plane, the origin is an asymptotically stable equilibrium point but that when c causes at least one eigenvalue to fall into the right half of the complex plane, the origin is an unstable equilibrium point. A plot of the eigenvalues as points in the complex plane is an important tool in the design of feedback control systems that also can be modeled with systems of linear differential equations. In the complex s-plane, the curve traced by the eigenvalues as they vary with an appropriate parameter of the system is called a *root locus*. Figure 36.8 shows, for this example, the locus of roots as c varies in the interval $[-2, 4]$. (Our parameter c is not one that would be found in a typical problem in controls engineering.)

The initial and terminal points on each locus are indicated on the graph. The curve in black is the locus of λ_+, whereas the curve in color is the locus of λ_-. When $c = -2$, both eigenvalues are complex. In fact, they are complex conjugates. As c increases toward $c = -\frac{3}{4}$, the two eigenvalues approach the real axis. For $-\frac{3}{4} < c < \frac{4}{3}$, the eigenvalues remain along the positive real axis. For $c > \frac{4}{3}$, λ_+ remains on the positive real axis, but λ_- is now found on the negative real axis.   ❖

EXAMPLE 36.21   Figure 36.9 contains a graph of $\lambda_\pm$, the eigenvalues of the matrix A given in

$$A = \begin{bmatrix} c & -5 \\ 3 & 4 \end{bmatrix} \Rightarrow \lambda_\pm = \frac{1}{2}c + 2 \pm \frac{1}{2}\sqrt{c^2 - 8c - 44} \qquad (36.8)$$

The solid curves are branches of λ_+, the dotted, of λ_-. There are definitely values of c for which the eigenvalues are of opposite sign. There are also values of c for which the eigenvalues are complex. It is not yet clear if there are values of c for which both eigenvalues remain positive. Thus, we compute $\lim_{c \to \infty} \lambda_- = 4$ to determine if, for large c, the eigenvalue λ_- remains positive. Hence, for c large enough, both eigenvalues are positive. From $c^2 - 8c - 44$, the discriminant of the characteristic polynomial

$$\lambda^2 - (c + 4)\lambda + 4c + 15$$

we find $c = 4 \pm 2\sqrt{15}$ as the values for transition between real and complex eigenvalues.

Figure 36.10 shows the locus of roots for $-5 \le c \le 15$. For c in the range $-5 \le c \le 4 - 2\sqrt{15}$, λ_+ (in black) is along the positive real axis but λ_- (in color) is along the negative real axis. For c in the range $4 - 2\sqrt{15} \le c \le 4 + 2\sqrt{15}$, the eigenvalues are complex conjugates. The locus of roots consists of the upper and lower halves of a circle with radius

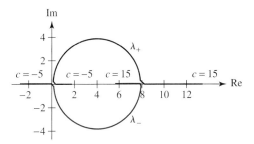

FIGURE 36.10 Example 36.21: In the complex plane, the locus of roots determined by $\lambda_{\pm}$ (curve in black is λ_+, while curve in color is λ_-)

4 and center at $(4, 0)$. For $c > 4 + 2\sqrt{15}$, both eigenvalues are on the positive real axis. We have already seen that λ_- tends toward the point $(4, 0)$. The locus of roots indicates that as $c \to \infty$, the eigenvalue λ_+ also tends to infinity along the positive real axis. ❖

Transfer Functions and Block Diagrams

Recall that for a differential equation of the form $y'' + 2y' + 10y = f(t)$, with the inert initial conditions $y(0) = y'(0) = 0$, the Laplace transform leads to

$$s^2 Y + 2sY + 10Y = F \quad \text{and} \quad Y(s) = \frac{F(s)}{s^2 + 2s + 10}$$

The function $f(t)$ is an "input" to the system, providing the system with a driving force or excitation function. The solution $y(t)$ is the system "output," so the function

$$T(s) = \frac{1}{s^2 + 2s + 10}$$

is the *transfer function* for the differential equation. Since $\frac{Y}{F} = T$, the transfer function is characterized as the ratio of Laplace transforms of the system output over the system input. (See Section 6.11 for a determination of the transfer function by means of the Dirac delta function.) Written as $Y(s) = T(s)F(s)$, the transfer function $T(s)$ determines how the action of the input, $f(t)$, will be *transferred* into the output, $y(t)$.

The engineering approach to feedback control systems is couched in the language of transfer functions and block diagrams. To illustrate this approach, consider the first-order linear system on the left in Table 36.2 The system parameter K is often called a *gain* in the controls literature. The Laplace transform of each equation, in concert with the inert initial conditions $x(0) = y(0) = 0$, leads to the equations in the center of Table 36.2. Solving the first of these equations for X and the second for Y gives the equations on the right in Table 36.2, where

$$G_1 = \frac{2}{s-1} \qquad G_2 = \frac{1}{s-4} \qquad G_3 = \frac{3}{s-4}$$

are transfer functions.

This form of the differential equations is modeled by the block diagram in Figure 36.11. The circle is a junction node, the squares are "blocks" representing the transfer functions, and the secret to reading the diagram is to realize that F and X represent a system *input* and *output*, respectively. The loop containing the blocks representing G_1 and G_2 is the

---

$x' = x + 2y$
$y' = -Kx + 4y + 3f(t)$

$sX = X + 2Y$
$sY = -KX + 4Y + 3F$

$X = G_1 Y$
$Y = -KG_2 X + G_3 F$

TABLE 36.2 Example of a linear feedback system

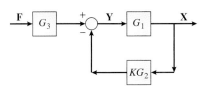

FIGURE 36.11 Block diagram for the linear feedback system whose equations appear in Table 36.2

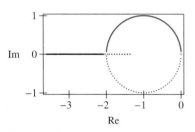

FIGURE 36.12   Root locus for Example 36.22

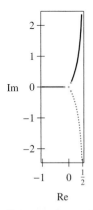

FIGURE 36.13   Root locus for Example 36.23

feedback loop. The surprise is that inhomogeneous first-order systems arising from mixing-tank problems and studied in Chapter 12 can be interpreted as feedback systems. Hence, feedback is inherently found in the coupling of the differential equations and not necessarily in an array of sensors and amplifiers.

Finally, solve for $X(s)$ and $Y(s)$ in terms of the transfer functions, obtaining

$$X = \frac{G_3 G_1 F}{1 + K G_1 G_2} \quad \text{and} \quad Y = \frac{G_3 F}{1 + K G_1 G_2}$$

The transfer function connecting the input $f(t)$ to the output $x(t)$ is then

$$T(s) = \frac{G_3 G_1}{1 + K G_1 G_2} = \frac{\frac{1}{s-1}}{1 + K \frac{2}{s-1}\frac{1}{s-4}}$$

The poles of $T(s)$ are values of s for which the denominator of $T(s)$ vanishes. The equation whose roots are the poles of $T(s)$ is

$$0 = 1 + K G_1 G_2 = 1 + \frac{2K}{(s-1)(s-4)} = \frac{(s-1)(s-4)+2K}{(s-1)(s-4)}$$

The numerator of the fraction on the right leads to the equation

$$s^2 - 5s + 4 + 2K = 0$$

which is the characteristic equation for the system matrix A given on the left in

$$A = \begin{bmatrix} 1 & 2 \\ -K & 4 \end{bmatrix} \Rightarrow \lambda_{\pm} = \frac{5}{2} \pm \frac{1}{2}\sqrt{9 - 8K} \qquad (36.9)$$

The denominator of $T(s)$ therefore vanishes for $s = \lambda_{\pm}$, the eigenvalues of A, which are given on the right in (36.9). The poles of $T(s)$ are precisely the eigenvalues of A. Thus, the stability of the system is determined by the location of the poles of the transfer function because these poles are the eigenvalues of the system matrix.

EXAMPLE 36.22   Suppose the denominator of a transfer function for a feedback system is $1 + kh(s)$, where $h(s) = \frac{s+1}{s^2}$. Figure 36.12 shows the root locus, drawn by solving the equation $1 + \frac{k(s+1)}{s^2} = 0$ for a number of values of k and graphing, in the complex plane, the resulting solution points. The values of k corresponding to the four segments of the locus can be determined by solving $1 + kh(s) = 0$ for $s = s_{\pm}(k) = -\frac{1}{2}k \pm \frac{1}{2}\sqrt{k^2 - 4k}$. (See Exercises A1–5.)

Another approach to studying the dependence of the characteristic roots on the gain k is to solve the equation $1 + \frac{k(s+1)}{s^2} = 0$ for $k = k(s) = -\frac{s^2}{s+1}$ rather than $s = s(k)$. But in the complex plane, s is the complex variable $s = x + iy$, so

$$k = k(x + iy) = -\frac{(x+iy)^2}{1+x+iy} = -\frac{x^3+y^2x+x^2-y^2}{x^2+2x+1+y^2} - i\frac{y(x^2+y^2+2x)}{x^2+2x+1+y^2} \qquad (36.10)$$

The equation $\mathcal{I}(k) = 0$ implicitly defines a curve $y = y(x)$, which, in this example, is $(x + 1)^2 + y^2 = 1$, the equation of a circle. ❖

EXAMPLE 36.23   For $h(s) = \frac{s+1}{s^3}$, the root locus, that is, the curve $s = s(k)$ defined implicitly by the equation $1 + kh(s) = 0$, is seen in Figure 36.13. The characteristic equation is the cubic $s^3 + ks + k = 0$, for which an exact, but tedious, solution exists. The root locus in Figure 36.13 was drawn numerically, by plotting as discrete points and for various values of k, solutions of the characteristic equation. (See Exercises A6 and 7.) ❖

EXERCISES 36.4–Part A

A1. In Example 36.22, solve the equation $1 + kh(s) = 0$ for $s_\pm(k)$.

A2. In Example 36.22, show that $\lim_{k\to\infty} s_-(k) = -\infty$, thereby tracing, in Figure 36.12, the solid black line segment to the left.

A3. In Example 36.22, show that $\lim_{k\to\infty} s_+(k) = -1$, thereby tracing, in Figure 36.12, the dotted black line segment toward the right, inside the circle.

A4. In Example 36.22, show that the upper half of the circle, the solid portion in color, is traced by $s_-(k)$, $0 \le k \le 4$.

A5. In Example 36.22, show that the lower half of the circle, the dotted portion in color, is traced by $s_+(k)$, $0 \le k \le 4$.

A6. In Example 36.23, the equation $s^3 + ks + k = 0$ has three branches, as shown in Figure 36.13. Obtain numerical evidence that the branch drawn in color is traced between $s = 0$ and $s = -1$ as k varies from 0 to infinity.

A7. Obtain numerical evidence that in Figure 36.13, the two branches drawn in black (both solid and dotted curves) are traced as k varies from 0 to infinity. Provide evidence that these branches are asymptotic to the vertical line $s = \frac{1}{2}$.

EXERCISES 36.4–Part B

B1. In (36.10), $\mathcal{I}(k) = 0$ and $\mathcal{R}(k) = k$ determine $x = x(k)$ and $y = y(k)$. For example, $\mathcal{I}(k) = 0$ can be solved for $y = y(x)$, and that substituted into $\mathcal{R}(k) = k$ to obtain $x = x(k)$, from which $y = y(x) = y(x(k)) = y(k)$. Use this strategy to obtain the locus of roots for the equation $s^2 + k(1 + s) = 0$.

For the matrices A in Exercises B2–9:

    (a) Obtain the eigenvalues $\lambda_\pm$ as a function of the parameter c.

    (b) Plot the eigenvalues as functions of c.

    (c) From the graph in part (b), analyze the stability of the linear system $\mathbf{x}' = A\mathbf{x}$.

    (d) Plot, in the complex s-plane, the locus of roots determined by the eigenvalues $\lambda_\pm$.

B2. $\begin{bmatrix} c & -3 \\ -7 & 1 \end{bmatrix}$ **B3.** $\begin{bmatrix} 4 & c \\ -5 & 12 \end{bmatrix}$ **B4.** $\begin{bmatrix} -1 & 11 \\ c & -3 \end{bmatrix}$

B5. $\begin{bmatrix} 11 & -6 \\ 12 & c \end{bmatrix}$ **B6.** $\begin{bmatrix} c & 5 \\ 4 & 6 \end{bmatrix}$ **B7.** $\begin{bmatrix} -12 & c \\ 1 & 1 \end{bmatrix}$

B8. $\begin{bmatrix} 4 & -6 \\ c & 11 \end{bmatrix}$ **B9.** $\begin{bmatrix} 12 & 6 \\ 7 & c \end{bmatrix}$

In each of Exercises B10–19, given the rational functions $G_k(s)$, $k = 1, 2, 3$:

    (a) Form the transfer function $T(s) = \frac{G_3 G_1}{1 + K G_1 G_2}$.

    (b) Find all poles of $T(s)$.

    (c) Plot a root-locus diagram.

    (d) Based on the root locus, analyze the stability of the control system associated with $T(s)$.

B10. $G_1 = \dfrac{1}{6s + 6}$, $G_2 = \dfrac{7}{8s + 1}$, $G_3 = \dfrac{1}{8s + 1}$

B11. $G_1 = \dfrac{1}{8s - 5}$, $G_2 = \dfrac{5}{4s + 9}$, $G_3 = \dfrac{1}{4s + 9}$

B12. $G_1 = \dfrac{1}{6s - 8}$, $G_2 = \dfrac{s + 9}{2s^2 + 9s - 8}$, $G_3 = \dfrac{1}{2s^2 + 9s - 8}$

B13. $G_1 = \dfrac{1}{5s + 1}$, $G_2 = \dfrac{5}{3s^2 + 4s}$, $G_3 = \dfrac{1}{3s^2 + 4s}$

B14. $G_1 = \dfrac{1}{s + 3}$, $G_2 = \dfrac{6s - 5}{8s^2 - 5}$, $G_3 = \dfrac{1}{8s^2 - 5}$

B15. $G_1 = \dfrac{1}{4s - 8}$, $G_2 = \dfrac{4s + 6}{8s^2 + 7s + 9}$, $G_3 = \dfrac{1}{8s^2 + 7s + 9}$

B16. $G_1 = \dfrac{4s + 1}{6s^2 + s + 6}$, $G_2 = \dfrac{1}{6s + 7}$, $G_3 = \dfrac{3}{6s + 7}$

B17. $G_1 = \dfrac{4s - 6}{9s^2 - 4s - 8}$, $G_2 = \dfrac{1}{9s - 7}$, $G_3 = \dfrac{5}{9s - 7}$

B18. $G_1 = \dfrac{1}{9s + 2}$, $G_2 = \dfrac{s^2 + 4s + 5}{9s^3 - 6s^2 + 9s - 2}$, $G_3 = \dfrac{1}{9s^3 - 6s^2 + 9s - 2}$

B19. $G_1 = \dfrac{1}{3s - 4}$, $G_2 = \dfrac{2s^2 - 5s + 6}{8s^3 + 7s^2 - 8s + 8}$, $G_3 = \dfrac{1}{8s^3 + 7s^2 - 8s + 8}$

36.5 The Nyquist Stability Criterion

The Winding Number

The circle $z = e^{i\theta}$, $0 \le \theta \le 2\pi$, encircles, or wraps around the origin, once. The closed contour described by $z = e^{i\theta}$, $0 \le \theta \le 4\pi$, is a circle traced twice and, therefore, encircles or wraps around the origin twice. In general, if C is a closed contour, the number of times

C wraps around the point $z = \zeta$ is called the *winding number* of C with respect to ζ and is given by

$$v(C, \zeta) = \frac{1}{2\pi i} \int_C \frac{dz}{z - \zeta} \qquad (36.11)$$

The integral in (36.11) reflects such a geometric property of the contour C because the antiderivative is a branch of the logarithm. The integral in (36.11) actually measures the change in the argument of the logarithm. If C wraps around ζ once, the argument of the logarithm will increase (or decrease) by 2π. Hence, the number of times C wraps around ζ is reflected by the change in the argument of the logarithm generated by the antiderivative for (36.11).

EXAMPLE 36.24   Let C be the circle $z(t) = e^{it}, 0 \le t \le 2\pi$, and take $\zeta = 0$. The winding number is then $v(C, 0) = \frac{1}{2\pi i} \int_0^{2\pi} i \, dt = 1$ since C encloses $z = 0$ once. This would be the result for any other z inside C. For example, the winding number about $z = \frac{1+i}{3}$ is

$$v\left(C, \frac{1+i}{2}\right) = \frac{1}{2\pi i} \int_0^{2\pi} \frac{i e^{it} \, dt}{e^{it} - \frac{1}{2}(1+i)} = 1$$

The integral defining the winding number is easily computed by Cauchy's residue theorem, since we have

$$v(C, \zeta) = \frac{1}{2\pi i} 2\pi i \, \mathrm{Res}\left(\frac{1}{z - \zeta}, \zeta\right) = 1$$

whenever C is a simple closed curve enclosing ζ just once.   ❖

EXAMPLE 36.25   Let C be the limaçon defined in polar coordinates (r, t) by $r(t) = 1 + 3\cos t$, a distinguishing feature of which is the way it loops over itself, as seen in Figure 36.14. We write C as $z(t) = r(t)e^{it}$, and compute the winding number about $z = 1$, a point inside the "inner" loop, as

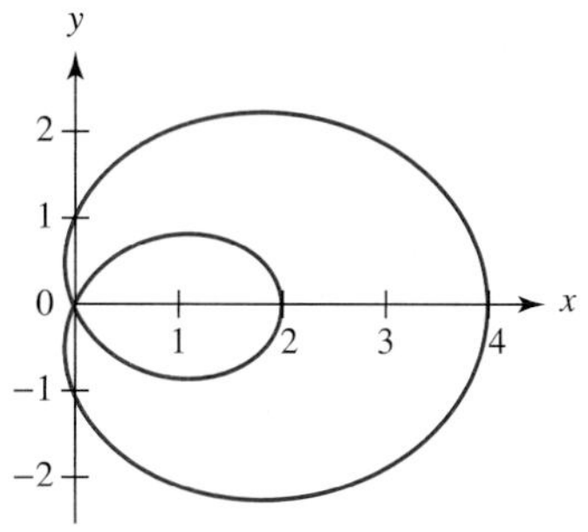

$$v(C, 1) = \frac{1}{2\pi i} \int_0^{2\pi} \frac{i(1 + 3e^{it})e^{it} \, dt}{(1 + 3\cos t)e^{it} - 1} = 2$$

indicating that the limaçon wraps twice around $z = 1$. The winding number about $\zeta = 3$, a point inside the limaçon but not inside the inner loop is

$$v(C, 3) = \frac{1}{2\pi i} \int_0^{2\pi} \frac{i(1 + 3e^{it})e^{it} \, dt}{(1 + 3\cos t)e^{it} - 3} = 1$$

indicating that the limaçon wraps around $z = 1$ but a single time.   ❖

FIGURE 36.14   The limaçon of Example 36.25

EXAMPLE 36.26   Let C be the closed contour in Figure 36.15, consisting of four semicircles taken from the circles

$$z_1(t) = 1 + 3e^{it} \qquad z_2(t) = 2 + 2e^{it} \qquad z_3(t) = 1 + e^{it} \qquad z_4(t) = 2e^{it}$$

Inside C there are points that are enclosed twice by the contour. The winding number about $z = 1$ is

$$v(C, 1) = \frac{1}{2\pi i}\left[\int_{-\pi}^{0} i \, dt + \int_0^{\pi} \frac{2ie^{it} \, dt}{1 + 2e^{it}} + \int_{-\pi}^{0} i \, dt + \int_0^{\pi} \frac{2ie^{it} \, dt}{2e^{it} - 1}\right] = 2$$

indicating that the contour C wraps around $z = 1$ twice. The point $z = 3$ is inside C but not inside the "inner" loop. Thus, the winding number about $z = 3$ is

$$v(C, 3) = \frac{1}{2\pi i} \left[\int_{-\pi}^{0} \frac{3i e^{it}\, dt}{3 e^{it} - 2} + \int_{0}^{\pi} \frac{2i e^{it}\, dt}{2 e^{it} - 1} + \int_{-\pi}^{0} \frac{i e^{it}\, dt}{e^{it} - 2} + \int_{0}^{\pi} \frac{2i e^{it}\, dt}{2 e^{it} - 3} \right] = 1$$

indicating that the contour C wraps around $z = 3$ but once.   ❖

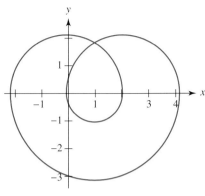

Figure 36.15   The closed contour of Example 36.26

The following is called "the principle of the argument."

The Principle of the Argument

1. D is a simply connected domain.
2. $G(z)$ is analytic in D, except, perhaps for a finite number of poles.
3. C is a simple closed positively-oriented contour in D.
4. $G(z)$ has neither poles nor zeros on C.
5. $G(z)$ has N zeros and P poles inside C with N reflecting the multiplicity of each zero and P reflecting the order of each pole.

$$\Longrightarrow \frac{1}{2\pi i} \int_{C} \frac{G'(z)}{G(z)}\, dz = N - P$$

A proof of this result can be found in [69], or in any similar text. The following example contains the key ideas from which the proof is typically constructed.

 **Example 36.27**   The points $z_1 = 2 + 3i$, $z_2 = 4 - i$, $z_3 = -5 + 2i$ are a zero and two simple poles, respectively, for the function $G = \frac{z - 2 - 3i}{(z - 4 + i)(z + 5 - 2i)}$. Then

$$\frac{G'}{G} = -\frac{z^2 - 4z - 6iz + 13 - 14i}{(z - 2 - 3i)(z - 4 + i)(z + 5 - 2i)}$$

has simple poles at all three points that lie within C, the circle $|z| = 7$. The Cauchy residue theorem then gives

$$\frac{1}{2\pi i} \int_{C} \frac{G'(z)}{G(z)}\, dz = \frac{2\pi i}{2\pi i} \sum_{k=1}^{3} \mathrm{Res}\left(\frac{G'}{G}, z_k \right) = (1 - 1 - 1) = N - P = (1 - 2) = -1$$

thereby verifying the principle of the argument in this case.   ❖

The following combines the constructs of the winding number and the principle of the argument.

The Principle of the Argument—Extended

1. The positively oriented, simple closed contour C is described by $z = f(t)$, $\alpha \le t \le \beta$.
2. $H(z)$ is analytic on C; inside C, $H(z)$ may have a finite number of poles.
3. $H(z) \neq \zeta_0$ on C.
4. Γ, in the plane of $\zeta = u + iv$, is the image of C under H; that is, Γ is given by $\zeta = F(t) = H(f(t))$, $\alpha \le t \le \beta$.
5. Inside C, the function $H(z) - \zeta_0$ has N zeros and P poles, counted with respect to multiplicity and order.
6. $v(\Gamma, \zeta_0)$ is the winding number of Γ about the point ζ_0.

$$\Longrightarrow v(\Gamma, \zeta_0) = N - P$$

In essence, this principle says that $N - P$ for the function $H(z)$ inside C is the winding number for Γ, the image of C under the mapping induced by $H(z)$. To find out something about the number of poles and zeros $H(z)$ has inside C, obtain the winding number for Γ, the image of C under the mapping of $w = H(z)$. If the winding number of Γ can be obtained geometrically, without resort to integration, then the principle gives a geometric tool for counting poles and zeros of $H(z)$.

If $H(z)$ is a polynomial, it has no poles in the finite complex plane. Furthermore, if the winding number $\nu(\Gamma, \zeta_0)$ were known, then this extension of the principle of the argument yields N, the number of zeros $H(z) - \zeta_0$ has inside C.

The proof of this principle is straightforward. Starting with the definition of the winding number, we have

$$\nu(\Gamma, \zeta_0) = \frac{1}{2\pi i} \int_\Gamma \frac{d\zeta}{\zeta - \zeta_0} = \frac{1}{2\pi i} \int_\alpha^\beta \frac{F'(t)\, dt}{F(t) - \zeta_0}$$

$$= \frac{1}{2\pi i} \int_\alpha^\beta \frac{H'(f(t)) f'(t)\, dt}{H(f(t)) - \zeta_0} = \frac{1}{2\pi i} \int_C \frac{H'(z)\, dz}{H(z) - \zeta_0}$$

and the last integral is just $N - P$, a result that follows by setting $G(z) = H(z) - \zeta_0$ in the principle of the argument.

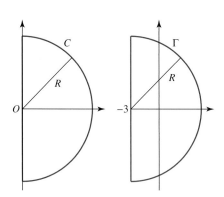

FIGURE 36.16 The closed contours C and Γ of Example 36.28

EXAMPLE 36.28 Let C be the contour consisting of the right half of the circle $|z| = R$ and that part of the imaginary axis satisfying $-iR \leq z \leq iR$. This is the D-shaped curve shown on the left in Figure 36.16. If $H(z) = z - 3$, the equation $H(z) = 0$ has one solution, $z = 3$, in the right half-plane. Thus, $H(z)$ has one zero and no poles in the right half-plane. The image of C under the map $H(z)$ is the contour Γ, shown on the right in Figure 36.16. Both C and Γ are traced in the counterclockwise (positive) sense. The contour Γ encircles the origin once, so the winding number with respect to $\zeta = 0$ is $\nu(\Gamma, 0) = 1$. Hence, $H(z)$ has exactly one zero in the right half-plane. We verify this by computing

$$\nu(\Gamma, 0) = \frac{1}{2\pi i} \int_C \frac{H'(z)\, dz}{H(z) - \zeta_0} = \frac{1}{2\pi i} \int_C \frac{H'(z)\, dz}{H(z)} = \frac{1}{2\pi i} \int_C \frac{dz}{z - 3} = 1 \qquad \diamondsuit$$

EXAMPLE 36.29 Let C be the contour defined in Example 36.28, and let $H(z) = (z - 2)(z - 3)$, a function with two zeros, $z = 2$ and $z = 3$, both in the right half-plane. The image of C under the mapping induced by $H(z)$ is Γ, as shown in Figure 36.17. The contour Γ encircles $\zeta_0 = 0$ twice, so the winding number is $\nu(\Gamma, 0) = 2$ and the function $H(z) - \zeta_0 = H(z)$ has two zeros in the right half-plane. Since

$$\frac{H'(z)}{H(z)} = \frac{2z - 5}{(z - 2)(z - 3)} = \frac{1}{z - 2} + \frac{1}{z - 3}$$

the winding number is $\nu(\Gamma, 0) = \frac{2\pi i}{2\pi i}[\operatorname{Res}(\frac{H'}{H}, 2) + \operatorname{Res}(\frac{H'}{H}, 3)] = 2$. $\qquad \diamondsuit$

EXAMPLE 36.30 With C again as in Example 36.28, let $H(z) = \frac{1}{10}z^3 + z^2 + 9z + 4$, a function with the three zeros -0.468, $-4.77 \pm 7.93i$, all in the left half-plane. There are no zeros in the right half-plane. Consequently, Γ, the image of C under the mapping induced by $H(z)$, should not encircle $\zeta_0 = 0$. Figure 36.18 shows Γ when $R = 12$. Careful observation shows the large loop (solid curve) is traced in the counterclockwise sense, whereas the small loop (dotted curve) is traced in the clockwise sense. Thus, there is no *net* encirclement of $\zeta_0 = 0$ by Γ and the winding number is correctly determined to be $\nu(\Gamma, 0) = 0$. $\qquad \diamondsuit$

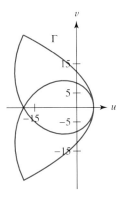

FIGURE 36.17   Example 36.29: Under $H = (z - 2)(z - 3)$, the image of the contour C in Figure 36.16

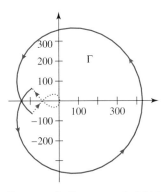

FIGURE 36.18   Example 36.30: Under $H = \frac{1}{10}z^3 + z^2 + 9z + 4$, the image Γ of the contour C in Figure 36.16 (solid portion of Γ is traced counterclockwise, whereas the dotted portion is traced clockwise)

The Nyquist Stability Criterion

In 1932, Harry Nyquist (1889–1976), an engineer at Bell Telephone Laboratories, articulated the use of the extended principle of the argument for determining where the poles of a transfer function in a linear feedback system are located. This usage has since been called the *Nyquist stability criterion*.

The transfer function of a feedback system (see Section 36.4) arises from a Laplace transform of the system. The poles of the transfer function are actually the eigenvalues or characteristic roots of the differential equation governing the system. If any of these poles (or eigenvalues) have positive real part, that is, if any pole is located in the right half-plane, the solution may contain exponential terms that grow without bound as time increases. Thus, determining whether or not there are poles of the transfer function in the right half-plane is tantamount to determining the stability of the feedback system.

If the transfer function of the feedback system is a completely reduced rational function of the form $T(z) = \frac{A(z)}{B(z)}$, the poles of $T(z)$ are the zeros of $B(z)$. Determining the location of the zeros of $B(z)$ is equivalent to determining the poles of $T(z)$; and the poles of $T(z)$ are the eigenvalues, or characteristic roots, of the system. The solution of the differential equation underlying the feedback system will contain multiples of exponential terms whose exponents contain the eigenvalues. If any of the eigenvalues have positive real parts (because they fall in the right half-plane), the solution might grow unbounded as time increases. Thus, poles of $T(z)$ falling in the right half-plane means the feedback system could be unstable.

Engineering texts on feedback control (e.g., [70] and [31]) contain additional detail on the use of the Nyquist stability criterion. In addition, we even find complex variables texts such as [99] discussing the notion of feedback systems and the Nyquist stability criterion.

EXERCISES 36.5–Part A

A1. Let C be a square of side 2, centered at the origin. Compute the winding number $v(C, 0) = 1$.

A2. Let C be the circle $z = 2e^{it}$, $0 \le t \le 2\pi$; and let $G(z) = \frac{z^2}{z-1}$. Determine $N - P$, where N and P are, respectively, the number of zeros and poles $G(z)$ has inside C.

A3. Evaluate $\frac{1}{2\pi i} \int_C \frac{G'(z)}{G(z)}\, dz$, thereby verifying the principle of the argument for the contour C and function $G(z)$ in Exercise A2.

A4. The polynomial $G(z) = (z - 1)(z - 2)$ has exactly two zeros and no poles in the right half-plane. Verify $N = 2$ using the principle of the argument.

A5. Repeat Exercise A4 for $G(z) = (z - 1)(z - 2)^2$.

EXERCISES 36.5–Part B

B1. Let C be the circle $z = 3 + 3e^{it}$, $0 \le t \le 2\pi$, and $G(z)$ be the polynomial in Exercise A5. The principle of the argument gives $v(C, 0) = 3 = N$. The extended principle of the argument gives $v(C, 0) = \frac{1}{2\pi i} \int_C \frac{G'(z)}{G(z)}\, dz = v(\Gamma, 0)$, where Γ is the image of C under $G(z)$. Obtain a graph of Γ. Does it wrap around $\zeta_0 = 0$ three times?

For the rational functions $G(z)$ in Exercises B2–11:

(a) Determine its zeros.

(b) Determine its poles.

(c) Evaluate (even numerically) the contour integral
$v = \frac{1}{2\pi i} \int_C \frac{G'(z)}{G(z)}\, dz$, where C is the "D-shaped" path

consisting of $z_1(t) = it$, $-R \le t \le R$, and $z_2(t) = Re^{it}$, $-\frac{\pi}{2} \le t \le \frac{\pi}{2}$, traversed in a counterclockwise sense with R large enough so that C encloses all zeros and poles in the right half-plane.

(d) Show that $\nu = N - P$, where N is the number of zeros of $G(z)$ inside C and P is the number of poles of $G(z)$ inside C.

(e) Sketch Γ, the image of C under the mapping $G(z)$. Does Γ wrap N times around $\zeta_0 = 0$?

B2. $G(z) = \dfrac{z^2 + 3z - 4}{6z^3 - 8z^2 - 11z - 12}$ **B3.** $G(z) = \dfrac{7z^2 - 4z + 1}{4z^3 - 1}$

B4. $G(z) = \dfrac{11z^2 - 12z - 10}{z^3 + 3z^2 + 7z - 1}$ **B5.** $G(z) = \dfrac{2z^2 - 5}{12z^3 + 7z^2 - 9z + 3}$

B6. $G(z) = \dfrac{4z^3 + 6z^2 + 3z - 5}{2z^4 - 9z^3 - 8z^2 - 10z + 7}$

B7. $G(z) = \dfrac{9z^3 - 2z^2 + 11z - 7}{9z^4 - 4z^3 + 6z^2 - 12z + 3}$

B8. $G(z) = \dfrac{5z^3 + 3z^2 + 7z - 9}{6z^4 + 5z^3 - 8z^2 + 5z - 3}$

B9. $G(z) = \dfrac{6z^3 - 12z^2 - 5z - 10}{3z^4 - 6z^3 - 7z^2 + 4z - 10}$

B10. $G(z) = \dfrac{7z^3 - 5z^2 + z + 5}{7z^4 + 11z^3 - 2z^2 - 2}$

B11. $G(z) = \dfrac{2z^3 + z^2 - 4z + 5}{7z^4 + 6z^3 + 4z^2 + 10z + 4}$

For the polynomials $G(s)$ in each of Exercises B12–21:

(a) Interpret $G(s)$ as the denominator of a transfer function $T(s)$, and determine the poles of $T(s)$ by finding the zeros of $G(s)$.

(b) Use the Principle of the Argument to verify the number of poles of $T(s)$ in the right half-plane. Thus, evaluate (even numerically) the contour integral $\nu = \frac{1}{2\pi i} \int_C \frac{G'(z)}{G(z)} \, dz$, where C is the "D-shaped" path consisting of $z_1(t) = it$, $-R \le t \le R$, and $z_2(t) = Re^{it}$, $-\frac{\pi}{2} \le t \le \frac{\pi}{2}$, traversed in a counterclockwise sense with R large enough so that C encloses all zeros and poles in the right half-plane.

(c) Sketch Γ, the image of C under the mapping induced by $G(z)$. Does Γ wrap the appropriate number of times around $\zeta_0 = 0$?

B12. $G(s) = 7s^2 - s - 1$ **B13.** $G(s) = 3s^2 - 4s + 6$

B14. $G(s) = 10s^2 + s + 3$ **B15.** $G(s) = 2s^3 - 11s^2 + 7s + 9$

B16. $G(s) = 10s^3 - 3s^2 + 2s + 11$

B17. $G(s) = 9s^3 - 2s^2 - 12s + 6$

B18. $G(s) = 5s^3 + s^2 - 5s - 3$

B19. $G(s) = 7s^4 + 6s^3 + 4s^2 + 10s + 4$

B20. $G(s) = 5s^4 + 7s^3 - 9s^2 + 12s + 8$

B21. $G(s) = 10s^4 + 7s^2 - 3s - 4$

36.6 Conformal Mapping

Complex Functions as Mappings

The complex function $w = f(z)$ defines a mapping of the z-plane to the w-plane. A point $z = x + iy$ in the z-plane is mapped by $f(z)$ to the point $w = u(x, y) + iv(x, y)$ in the w-plane. We have already seen how, for example, $f(z) = e^z$ maps a horizontal strip of width 2π onto the whole w-plane except for $w = 0$. In particular, we are interested in how a complex function treats the angle between two intersecting curves.

EXAMPLE 36.31   The curves $s_1(t) = e^{it}$ and $s_2(t) = 1 + e^{it}$, $0 \le t \le 2\pi$, intersect twice, since they are the intersecting circles shown in Figure 36.19. The uppermost point of intersection is $P = s_1(\frac{\pi}{3}) = \frac{1}{2}(1 + i\sqrt{3})$, found by equating real and imaginary parts of $s_1(t_1)$ and $s_2(t_2)$. At P, vectors tangent to $s_1(t)$ and $s_2(t)$ are $\mathbf{v}_1 = \frac{1}{2}(-\sqrt{3}\mathbf{i} + \mathbf{j})$ and $\mathbf{v}_2 = -\frac{1}{2}(\sqrt{3}\mathbf{i} + \mathbf{j})$ and the angle between them is $\theta = \frac{\pi}{3}$.

Now consider the mapping defined by $w = f(z) = z^2 = u + iv = x^2 - y^2 + 2xyi$. Figure 36.20 shows, in the w-plane, the images of the circles $s_1(t)$ and $s_2(t)$. The image of P is $P' = \frac{1}{2}(-1 + i\sqrt{3})$, and vectors tangent to the images at P' in the w-plane are $\mathbf{v}_1' = -(\sqrt{3}\mathbf{i} + \mathbf{j})$ and $\mathbf{v}_2' = -2\mathbf{j}$. The angle between $\mathbf{v}_1'$ and $\mathbf{v}_2'$ is again $\frac{\pi}{3}$, as is seen in Figure 36.20. The angle between curves intersecting in the z-plane is the same as the angle between the images under the map induced by $w = f(z) = z^2$. ❖

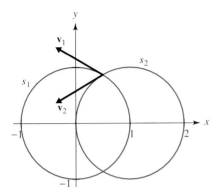

FIGURE 36.19   Intersecting circles in Example 36.31

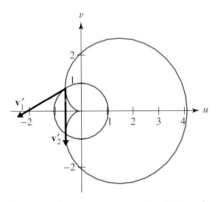

FIGURE 36.20   Image, under $f(z) = z^2$, of the circles in Figure 36.19

DEFINITION 36.1

The mapping of the z-plane to the w-plane defined by the function $w = f(z)$ is said to be *conformal* at z_0 if the angle between any two curves intersecting at z_0 is the same as the angle between the w-plane images of the two curves. A mapping that is conformal at every point in a domain is said to be conformal in that domain.

THEOREM 36.4   THE CONFORMAL MAPPING THEOREM

An analytic function $f(z)$ is conformal at every point z_0 for which $f'(z_0) \neq 0$.

PROOF   A constructive proof along the lines of Example 36.31 is straightforward. In the z-plane, let the two curves

$$z_1(t) = x_1(t) + iy_1(t) \quad \text{and} \quad z_2(t) = x_2(t) + iy_2(t)$$

intersect at $z_0 = z_1(t_1) = z_2(t_2)$. At z_0, construct tangent vectors

$$\mathbf{v}_1 = x_1'\mathbf{i} + y_1'\mathbf{j} \quad \text{and} \quad \mathbf{v}_2 = x_2'\mathbf{i} + y_2'\mathbf{j}$$

respectively, so that the angle $\theta = \arccos(\frac{\mathbf{v}_1 \cdot \mathbf{v}_2}{\|\mathbf{v}_1\|\|\mathbf{v}_2\|})$ is measured between them. The images of the curves under $w = f(z)$ are $w_1(t) = f(z_1(t))$ and $w_2(t) = f(z_2(t))$, and at $w_0 = f(z_0)$ vectors tangent to the images are $\mathbf{V}_1 = f'(z_0)\mathbf{v}_1$ and $\mathbf{V}_2 = f'(z_0)\mathbf{v}_2$, respectively, by the chain rule. Provided $f'(z_0) \neq 0$, the angle between $\mathbf{V}_1$ and $\mathbf{V}_2$ is then $\arccos(\frac{\mathbf{V}_1 \cdot \mathbf{V}_2}{\|\mathbf{V}_1\|\|\mathbf{V}_2\|}) = \arccos(\frac{(f')^2\mathbf{v}_1 \cdot \mathbf{v}_2}{(f')^2\|\mathbf{v}_1\|\|\mathbf{v}_2\|}) = \theta$.

A Dictionary of Conformal Maps

It is typical in complex variables texts to build intuition about conformal maps by studying a short "Dictionary of Conformal Maps." The following collection of examples is given in that spirit. (There is an actual "dictionary" of conformal maps [50] in print!)

EXAMPLE 36.32   The function $w = f(z) = z^2$, studied in Section 35.2, is analytic and has a nonvanishing derivative at every point except $z = 0$. The mapping of the z-plane to the w-plane induced by this function is conformal at every point except $z = 0$. The accompanying Maple worksheet contains several animations showing the images, in the w-plane, of regions in the z-plane being mapped dynamically by this function.   ❖

EXAMPLE 36.33   The mapping defined by the function $f(z) = \frac{1}{z} = u + iv = \frac{x}{x^2+y^2} + i\frac{y}{x^2+y^2}$ is called an *inversion* because it maps the interior of the unit circle to the exterior of the circle and the exterior of the unit circle to the interior, as suggested by Figure 36.21.

In addition, Table 36.3 summarizes the effect of $f(z)$ on circles and lines in the z-plane. Lines through the origin and circles not through the origin map to "the same things," that is, to the *same* geometric object with the *same* property. Lines not through the origin and circles through the origin map to "double opposites," that is, to the *other* geometric object with the *other* property. Alternatively, arbitrary lines map to objects through the origin and arbitrary circles map to objects not through the origin.

z-plane

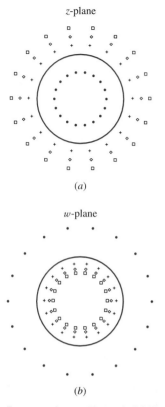

(a)

w-plane

(b)

FIGURE 36.21   Example 36.33: The inversion effected by $f(z) = \frac{1}{z}$

	Line	Circle
Through origin	Line through origin	Line not through origin
Not through origin	Circle through origin	Circle not through origin

TABLE 36.3   Example 36.33: Effect of the mapping $f(z) = \frac{1}{z}$ on lines and circles

Lines through Origin   Lines through the origin in the z-plane map to lines through the origin in the w-plane, as can be seen by obtaining the equations of the inverse mapping from the w-plane to the z-plane. Solve the equations $u = u(x, y)$, $v = v(x, y)$ for $x = x(u, v)$, $y = y(u, v)$, obtaining $x = \frac{u}{u^2+v^2}$, $y = -\frac{v}{u^2+v^2}$, so that the line $y = mx$ becomes $-\frac{v}{u^2+v^2} = \frac{mu}{u^2+v^2}$, or $v = -mu$ in the w-plane.

Lines not through Origin   Lines not through the origin in the z-plane map to circles through the origin in the w-plane, as can be seen by writing the equation of a line not through the origin as $ax + by + c = 0$, where $c \neq 0$. The equation of this line's image in the w-plane is then $\frac{au}{u^2+v^2} - \frac{bv}{u^2+v^2} + c = 0$, from which $au - bv + cv^2 + cu^2 = 0$ follows. Division by c, valid since $c \neq 0$, gives $\frac{a}{c}u - \frac{b}{c}v + v^2 + u^2 = 0$, which we recognize as the equation of a circle through the origin in the w-plane.

Circles through Origin   A circle through the origin in the z-plane maps to a line not through the origin in the w-plane, as can be seen by writing the equation of a circle through the origin of the z-plane as $x^2 + ax + y^2 + by = 0$, provided that a and b are not both zero. The equation of the w-plane image is then

$$\frac{u^2}{(u^2+v^2)^2} + \frac{au}{u^2+v^2} + \frac{v^2}{(u^2+v^2)^2} - \frac{bv}{u^2+v^2} = 0$$

which leads to $-bv + 1 + au = 0$, recognized as the equation of a line not through the origin in the w-plane.

Circles not through Origin   Circles not through the origin in the z-plane map to circles not through the origin in the w-plane, as can be seen by writing the equation of a circle not through the origin as $x^2 + y^2 + ax + by + c = 0$, where $c \neq 0$. The equation for the image in the w-plane is then

$$\frac{u^2}{(u^2+v^2)^2} + \frac{v^2}{(u^2+v^2)^2} + \frac{au}{u^2+v^2} - \frac{bv}{u^2+v^2} + c = 0$$

from which follows $cv^2 - bv + au + 1 + cu^2 = 0$, the equation of a circle not through the origin in the w-plane. In fact, dividing by $c \neq 0$ and completing the square gives

$$\left(u + \frac{a}{2c}\right)^2 + \left(v - \frac{b}{2c}\right)^2 + \frac{4c - a^2 - b^2}{4c^2} = 0$$

Evaluating at $(u, v) = (0, 0)$ in the left-hand side gives $\frac{1}{c} \neq 0$, so the image circle does not go through the origin in the w-plane.

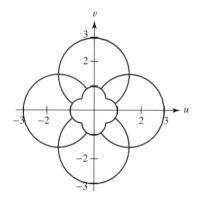

FIGURE 36.22   Example 36.33: Image, in the w-plane, of the grid lines in the z-plane

Grid Lines   Grid lines not through the origin in the z-plane will map to circles through the origin in the w-plane, as shown in Figure 36.22. Arcs crossing the u-axis are images of lines $x = $ constant, while arcs crossing the v-axis are images of lines $y = $ constant. Only those parts of the circles are drawn in the w-plane that correspond to the line *segments* used in the z-plane. The images cross at right angles in conformity with the conformal property of the map. (See Exercises A5 and 6.) ❖

EXAMPLE 36.34 The inversion $f(z) = \frac{1}{z}$ in Example 36.33 is a special case of the *linear fractional transformation* (also called the *Moebius* or *bilinear* transformation), defined by $f(z) = \frac{az+b}{cz+d}$, where $ad - bc \ne 0$. The condition on the coefficients a, b, c, d prevents $f(z)$ from reducing to a constant because it prevents the numerator from being proportional to the denominator. Indeed, if $ad - bc = 0$, then, for example, $a = \frac{bc}{d}$ and $f(z) = \frac{b}{d}$. The inverse of the mapping $w = f(z)$ is $z = -\frac{wd-b}{wc-a}$, again a linear fractional transformation (LFT). With respect to lines and circles, the LFT behaves much like the inversion of Example 36.33, except that the origin is replaced by the pole, $z = \frac{d}{c}$.

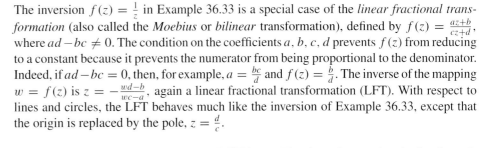

Upper Half-Plane to the Upper Half-Plane If a, b, c, d are real and $ad - bc > 0$, the LFT $f(z) = \frac{az+b}{cz+d}$ maps the upper half of the z-plane to the upper half of the w-plane. Indeed, if $y > 0$, then

$$\mathcal{I}(w) = \mathcal{I}(f(x + iy)) = \frac{y(ad - bc)}{c^2\left(x + \frac{d}{c}\right)^2 + c^2 y^2}$$

which, in light of the assumptions on y and $ad - bc$, is positive.

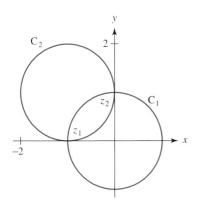

FIGURE 36.23   Mapping of a crescent in Example 36.34

Upper Half-Plane to Interior of Circle If $\mathcal{I}(z_0) > 0$, that is, if z_0 is in the upper half of the z-plane and α is real, then the LFT $w = f(z) = \rho e^{i\alpha} \frac{z-z_0}{z-\bar{z}_0}$ maps the upper half of the z-plane to the interior of the circle $|w| = \rho$. Indeed, the real axis is a line not through the pole $z = \bar{z}_0$, so it maps to a circle through the pole because a calculation shows $|w| = |f(x)| = \rho$. Since $z = z_0$ will map to $w = 0$, the upper half-plane maps to the *interior* of the circle $|w| = \rho$.

Interior of Circle to Interior of Circle If α is real, and $|z_0| < \rho$, then the LFT

$$w = f(z) = \rho^2 e^{i\alpha} \frac{z - z_0}{\bar{z}_0 z - \rho^2}$$

FIGURE 36.24   A crescent mapped by $f(z) = \frac{z+1}{z-i}$ in Example 36.34

maps the interior of the circle $|z| = \rho$ to the interior of the circle $|w| = \rho$. Indeed, a calculation shows $|w| = |f(\rho e^{i\theta})| = \rho$.

Crescent to Angle The LFT $w = f(z) = \rho e^{i\alpha} \frac{z-z_1}{z-z_2}$, where ρ is positive, α is real, and z_k, $k = 1, 2$, are the points of intersection of two circles, maps the crescent-shaped region between the circles C_k, $k = 1, 2$, to the wedge-shaped region between the two rays L_k, $k = 1, 2$, shown in Figure 36.23. Circle C_k maps to line L_k, with β, the angle between the intersecting arcs, being preserved in the w-plane.

As an example take $f(z) = \frac{z+1}{z-i}$ so that $z_1 = -1$ and $z_2 = i$ are the intersections of the circles $x^2 + y^2 = 1$ and $(x + 1)^2 + (y - 1)^2 = 1$ shown in Figure 36.24. At $z_1 = -1$, the angle between C_1 and C_2 is $\frac{\pi}{2}$. The images of the circles C_1 and C_2 are shown in Figure 36.25. Since

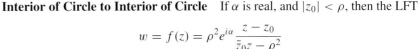

$$\frac{\mathcal{I}(f(e^{it}))}{\mathcal{R}(f(e^{it}))} = 1 \quad \text{and} \quad \frac{\mathcal{I}(f(i - 1 + e^{it}))}{\mathcal{R}(f(i - 1 + e^{it}))} = -1$$

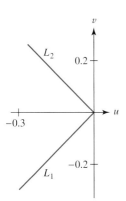

FIGURE 36.25   Under $f(z) = \frac{z+1}{z-i}$, the image of the crescent in Figure 36.24

the images are the lines $v = u$ and $v = -u$, respectively. Clearly, the angle between L_1 and L_2 is $\frac{\pi}{2}$. ❖

EXERCISES 36.6–Part A

A1. Let s_k and $\mathbf{v}_k$, $k = 1, 2$, be as in Example 36.31. If $f(z) = z^3$, obtain $\mathbf{v}'_k$, $k = 1, 2$, and show the angle between these vectors is still $\frac{\pi}{3}$.

A2. Find the image of the line $y = x$ under the mapping $f(z) = \frac{1}{z}$.

A3. Verify that under the mapping $f(z) = \frac{1}{z}$, the image of the line $x + y + 1 = 0$ is the circle $u^2 + u + v^2 - v = 0$.

A4. Verify that under the mapping $f(z) = \frac{1}{z}$, the image of the circle $z = 1 + e^{it}$, $0 \leq t \leq 2\pi$, is the vertical line $u = \frac{1}{2}$.

A5. Verify that under the mapping $f(z) = \frac{1}{z}$, the image of the line $x = c$ is the circle $(u - \frac{1}{2c})^2 + v^2 = \frac{1}{4c^2}$.

A6. Verify that under the mapping $f(z) = \frac{1}{z}$, the image of the line $y = c$ is the circle $u^2 + (v + \frac{1}{2c})^2 = \frac{1}{4c^2}$.

EXERCISES 36.6–Part B

B1. Let $\zeta = \frac{az+b}{cz+d}$ and $w = \frac{A\zeta+B}{C\zeta+D}$ be two bilinear maps with the composition $w(\zeta(z)) = \frac{Pz+Q}{Rz+S}$. Show that
$$\begin{bmatrix} P & Q \\ R & S \end{bmatrix} = \begin{bmatrix} A & B \\ C & D \end{bmatrix} \begin{bmatrix} a & b \\ c & d \end{bmatrix},$$ the product of two matrices of coefficients.

B2. Find the (acute) angle of intersection of the curves $y_1 = x^2$ and $y_2 = \frac{1}{x}$.

In Exercises B3–12, let y_k, $k = 1, 2$, be the curves given in Exercise B2. For the given function $f(z)$,

    (a) obtain Y_1 and Y_2, the images of y_1 and y_2 under the mapping $w = f(z)$.

    (b) obtain the angle of intersection of Y_1 and Y_2, showing it is the same as the angle between y_1 and y_2, as computed in Exercise B2.

B3. $f(z) = z^3$ **B4.** $f(z) = e^z$ **B5.** $f(z) = \dfrac{2z+3}{3z-5}$

B6. $f(z) = \dfrac{z-1}{z^2+z+1}$ **B7.** $f(z) = \sin z$ **B8.** $f(z) = \cos z$

B9. $f(z) = \tan z$ **B10.** $f(z) = \sinh z$

B11. $f(z) = \cosh z$ **B12.** $f(z) = \tanh z$

B13. Given the linear fractional transformation $w = \frac{z-1}{z+1}$, where $w = u + iv$,

    (a) find all fixed points, that is, all points for which $w(z) = z$.

    (b) show that the line $y = 0$ maps to the line $v = 0$.

    (c) show that the points satisfying $|z + 2| \leq 1$ map to the region determined by $u \geq 2$.

    (d) show that the region $y \leq 2(x + 1)$ maps to the region $v \leq 2(1 - u)$.

    (e) show that the region $x \leq 0$ maps to the region $|w| \geq 1$.

    (f) show that the region $|z - 1| \leq 1$ maps to the region $|w + \frac{1}{3}| \leq \frac{2}{3}$.

B14. Given the linear fractional transformation $w = \frac{2z+1}{z+i}$, where $w = u + iv$,

    (a) find all fixed points, that is, all points for which $w(z) = z$.

    (b) show that the region $|z| \leq 1$ maps to the region $v \leq \frac{3}{2} - 2u$.

    (c) show that the region $y \geq x - 1$ maps to the region $v \geq 3u - 6$.

    (d) show that the region $|z| \leq 2$ maps to the region $|w - \frac{1}{3}(8 + i)| \geq \frac{2}{3}\sqrt{5}$.

    (e) show that the region $y \leq x$ maps to the region $|w - \frac{1}{2}(1 + i)| \geq \sqrt{\frac{5}{2}}$.

B15. Given the linear fractional transformation $w = \frac{z}{z+4}$, where $w = u + iv$,

    (a) find all fixed points, that is, all points for which $w(z) = z$.

    (b) show that the region $y \geq 0$ maps to the region $v \geq 0$.

    (c) show that the region $x \leq 0$ maps to the region $|w - \frac{1}{2}| \geq \frac{1}{2}$.

    (d) show that the region $|z| \leq 4$ maps to the region $u \leq \frac{1}{2}$.

    (e) show that the region $x \geq -2$ maps to the region $|w| \leq 1$.

    (f) show that the line $x = -4$ maps to the line $u = 1$.

B16. Given the linear fractional transformation $w = 4\frac{z-1}{z-4}$, where $w = u + iv$,

    (a) find all fixed points, that is, all points for which $w(z) = z$.

    (b) show that the region $|z| \leq 2$ maps to the region $|w| \leq 2$.

    (c) show that the half-plane $x \leq 4$ maps to the half-plane $u \leq 4$.

    (d) show that the half-plane $x \geq 0$ maps to the region $|w - \frac{5}{2}| \geq \frac{3}{2}$.

A linear fractional transformation taking z_k to w_k, $k = 1, 2, 3$, can be found by writing $f(z) = \frac{az+b}{cz+d}$, using substitution to form three equations in the four unknown parameters a, b, c, d, and solving. Should any of the z_k or w_k be ∞, the point at infinity, use a finite point such as r, taking the limit as $r \to \infty$ after solving the resulting equations. In Exercises B17–21, find the linear fractional transformation taking the given z_k to the given w_k.

B17. $z_1 = 2, z_2 = -1, z_3 = 0 \to w_1 = 1, w_2 = 3, w_3 = -5$

B18. $z_1 = i, z_2 = 3, z_3 = -2 \to w_1 = 0, w_2 = -i, w_3 = 1$

B19. $z_1 = 5, z_2 = -1, z_3 = 1 \to w_1 = 2, w_2 = -1, w_3 = 4$

B20. $z_1 = \infty, z_2 = 1 + i, z_3 = 2 - 3i \to w_1 = 0, w_2 = i, w_3 = 1 - i$

B21. $z_1 = 1, z_2 = \infty, z_3 = 0 \to w_1 = \infty, w_2 = 3 + 4i, w_3 = i$

B22. Find a linear fractional transformation that maps the region $|z| \leq 1$ to the region $\mathcal{R}(w) \geq 0$, the right half-plane. *Hint:* A linear fractional transformation will map a circle through the pole to a line. Let $z = 1$ be the pole, and let the points $z = -1$ and i map to $w = 0$ and i, respectively.

B23. Find a linear fractional transformation that maps the region $\mathcal{I}(z) \leq 0$, the lower half-plane, to the region $|w - 3| \leq 2$. *Hint:* First, find some LFT that maps the line $y = 0$ to the circle $|w| = 2$. For example, let $z = 0, \infty, -1$ map to $w = 2i, -2, -2i$, respectively, thereby determining the LFT. Then, translate in the w-plane.

B24. Show that the linear fractional transformation $w = \frac{z-3}{3z-1}$ maps the region $A = \{|z| \geq 1 \cap |z - \frac{9}{2}| \geq \frac{5}{2}\}$ to the region $B = \{\frac{1}{5} \leq |w| \leq 1\}$. The region A is the complement of the interior of two circles, and the region B is an annulus centered at $w = 0$.

B25. If $w = f(z) = 2\sqrt{z+1} + \text{Log}(\frac{\sqrt{z+1}-1}{\sqrt{z+1}+1})$, where $w = u + iv$,

    (a) show that $f(z)$ takes the upper half-plane $y \geq 0$ to the region $\{0 \leq v \leq \pi, u < 0\} \cup \{u \geq 0, v \geq 0\}$.

    (b) find $z_k, k = 1, \ldots$, where $f(z_k) = w_k$, and $w_1 = -5$, $w_2 = -5 + \pi i, w_3 = \pi i, w_4 = 10i, w_5 = 5$.

36.7 The Joukowski Map

The Forward Map

The *Joukowski map*

$$w = f(z) = z + \frac{1}{z}$$

is of special interest in the application of complex variables to plane fluid flows. The real and imaginary parts of $f(z) = u + iv$ are

$$u(x, y) = x + \frac{x}{x^2 + y^2} \quad \text{and} \quad v(x, y) = y - \frac{y}{x^2 + y^2}$$

respectively. The w-plane image of $z = e^{i\theta}$, the unit circle in the z-plane, is $w = 2\cos\theta$. As θ varies through any range of 2π, w varies between -2 and 2. Hence, the unit circle in the z-plane maps to the segment $-2 \leq u \leq 2$ in the w-plane. The unit circle in the z-plane becomes a slit in the w-plane. Figure 36.26 shows what the map does to the rest of the z-plane.

The interior of the unit circle in the z-plane maps onto the w-plane. The upper half of the unit disk maps onto the lower half of the w-plane, while the lower half of the unit disk maps onto the upper half of the w-plane. In addition, the upper half of the z-plane exterior to the unit circle maps onto the upper half of the w-plane, and the lower half of the z-plane exterior to the unit circle maps onto the lower half of the w-plane. Thus, the Joukowski map is not one-to-one, and the inverse mapping is multivalued.

The Joukowski Airfoil

The Joukowski map takes circles that contain $z = 1$ in their interiors and pass through $z = -1$, to "airfoils" in the w-plane. Figure 36.27 shows, on the left, the circle through $z = -1$, with center $(\frac{1}{5}, \frac{1}{5})$ and radius $\frac{\sqrt{17}}{5}$. On the right in Figure 36.27 is the image of this circle under the Joukowski map. A more general version of the Joukowski airfoil is generated by the map $f(z) = z + \frac{k^2}{z}$, where $k > 0$. In this case, circles passing through $z = -k$ and containing $z = k$ in their interiors are mapped to Joukowski airfoils. If the circle passes through $z = k$ and contains $z = -k$ in its interior, the airfoil will face in the opposite direction.

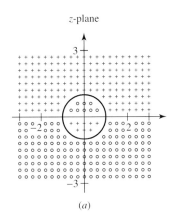

z-plane

(a)

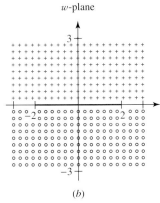

w-plane

(b)

FIGURE 36.26 The effect of the Joukowski map applied to the z-plane

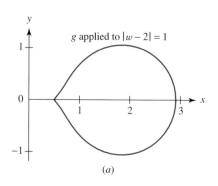

g applied to |w − 2| = 1

(a)

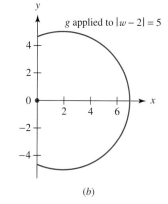

g applied to |w − 2| = 5

(b)

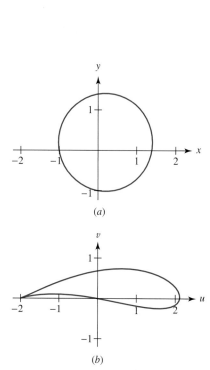

(a)

(b)

FIGURE 36.27   The Joukowski airfoil generated when the Joukowski map is applied to the circle $z = \frac{1}{5}[(1+i) + \sqrt{17}e^{i\theta}]$

FIGURE 36.28   Under $g(w)$, inverse image of $|w - 2| = 1$ in (a) and $|w - 2| = 5$ in (b), when the principal square root is used. Discontinuity is not restricted to $-2 \le u \le 2$, $v = 0$.

The Inverse Map

The inverse of the Joukowski map is obtained by solving the equation $w = f(z) = z + \frac{1}{z}$ for z. Since this map is two-to-one, we obtain $z = \frac{1}{2}(w \pm \sqrt{w^2 - 4})$ and define the inverse by

$$g(w) = \tfrac{1}{2}(w + \sqrt{w^2 - 4})$$

with the square root defined so that its discontinuity is along the line segment $-2 \le u \le 2$, $v = 0$, where $w = u + iv$. The principal square root does not satisfy this constraint, as shown in Figure 36.28 where, under $z = g(w)$, the inverse images of the circles $|w-2| = 1$ and $|w - 2| = 5$ can be seen. The image of the smaller circle (which cuts through the segment $-2 \le u \le 2$, $v = 0$) is on the left and is continuous. The image of the larger circle (which does not pass between $u = \pm 2$, $v = 0$) is on the right and is discontinuous!

The appropriate definition of the square root is

$$\sqrt{w^2 - 4} = \sqrt{|w^2 - 4|}e^{i/2[\text{Arg}(w-2)+\text{Arg}(w+2)]}$$

where Arg w is the principal argument, so that $-\pi < \text{Arg } w \le \pi$. Figure 36.29, with $\alpha = \text{Arg}(w + 2)$ and $\beta = \text{Arg}(w - 2)$, interprets this definition geometrically. Finally, Figure 36.30 shows in the z-plane, the pre-image of the w-plane under this branch of the inverse Joukowski map. It is consistent with our findings for the (forward) Joukowski map, $w = f(z) = z + \frac{1}{z}$.

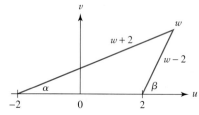

FIGURE 36.29   Geometric basis for the square root used to define the inverse of the Joukowski map

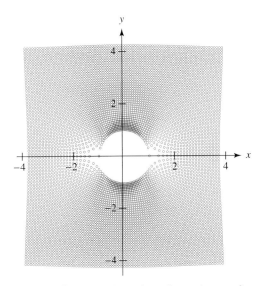

FIGURE 36.30 In the z-plane, the pre-image of the w-plane under the inverse of the Joukowski map

EXERCISES 36.7

In Exercises 1–18, $f(x) = z + \frac{4}{z}$.

1. Show that C_1, the circle $z_1(t) = -2 + 2i\sqrt{5} + 6e^{it}$, $-\pi \leq t \leq \pi$, passes through $z_b = 2$ and contains $z = -2$ in its interior.

2. Since C_1 passes through $z_b = 2$, it intersects C_2, the circle $z_2(\theta) = 2e^{i\theta}$, $-\pi \leq \theta \leq \pi$, at least once. Find z_a, the second point where C_1 and C_2 intersect.

3. Find those values of t_a and t_b (not θ) for which the two circles C_1 and C_2 intersect at z_a and z_b, respectively.

4. Show that for $t_a < t < t_b$, the corresponding arc on C_1 lies inside C_2.

5. Let $w = f(z) = z + \frac{4}{z}$ define a Joukowski map, where $w = u + iv$. Show that $f(z)$ takes circle C_1 to the Joukowski airfoil J.

6. Show that $B = f(z_b)$ is the cusp on the Joukowski airfoil J.

7. Show that $A = f(z_a)$ is the intersection of J with u-axis in the w-plane.

8. Show that $f(z)$ maps the arc in Exercise 4 to that portion of J lying between A and B.

9. For $f(z)$ as defined in Exercise 5, show that $z = \frac{1}{2}(w \pm \sqrt{w^2 - 16})$.

10. Define $g_+(w) = \frac{1}{2}(w + \sqrt{|w^2 - 16|}e^{i/2[\text{Arg}(w-4)+\text{Arg}(w+4)]})$ and show that it is discontinuous across the branch cut $-4 \leq u \leq 4$, $v = 0$.

11. Show that $g_+(w)$ maps the w-plane back to that part of the z-plane defined by $\{|z| > 2\} \cup \{|z| = 2, v \geq 0\}$, which is therefore outside the circle C_2 except for the portion of the circumference lying in the upper half-plane.

12. Let J' be that portion of the Joukowski airfoil J complimentary to the arc BA. Show that $g_+(w)$ maps J' to that portion of circle C_1 that is exterior to circle C_2. For example, show that $g_+(f(z_1(t)))$ for t in $[t_b, \pi] \cup [-\pi, t_a]$ is the appropriate portion of C_1.

13. Define $g_-(w) = \frac{1}{2}(w - \sqrt{|w^2 - 16|}e^{i/2[\text{Arg}(w-4)+\text{Arg}(w+4)]})$ and show that $g_-(w)$ maps arc AB on the airfoil J to that portion of circle C_1 lying inside C_2. For example, show that $g_-(f(z_1(t)))$ for t in $[t_a, t_b]$ is the appropriate portion of C_1.

14. Let ζ_k, $k = 1, 2$, denote the upper and lower intersections of circle C_1 with the imaginary axis in the z-plane. Find the points $w_k = f(\zeta_k)$, $k = 1, 2$, on the airfoil J. For $k = 1, 2$, find t_k, the values of t for which $z_1(t_k) = \zeta_k$.

15. Let ζ_3 denote the leftmost intersection of circle C_1 with the real axis in the z-plane. Find the point $w_3 = f(\zeta_3)$ on the airfoil J. Find t_3, the value of t for which $z_1(t_3) = \zeta_3$.

16. For $k = 1, 2, 3$, use $g_\pm(w)$, as appropriate, to map w_k back to ζ_k.

17. Find w_4 and w_5, the upper and lower points, respectively, where the airfoil J intersects the imaginary axis in the w-plane. Find w_6, the leftmost intersection of J with the real axis in the w-plane.

18. For $k = 4, 5, 6$, use $g_\pm(w)$, as appropriate, to determine ζ_k, the pre-image of w_k.

19. Repeat Exercises 1–18 for C_1, the circle given by $z_1(t) = -1 + 3i + 5e^{it}$, $-\pi \leq t \leq \pi$, and for C_2, the circle given by $z_2(\theta) = 3e^{i\theta}$, $-\pi \leq \theta \leq \pi$. Make all appropriate changes, taking, for example, $z_b = 3$ and $f(z) = z + \frac{9}{z}$.

36.8   Solving the Dirichlet Problem by Conformal Mapping

Introduction

The Dirichlet problem on an irregular planar region in the z-plane can be solved by mapping the region conformally to a "nicer" region in the w-plane where the transformed Dirichlet problem can be more easily solved. Suppose the mapping is given by $w = f(z) = u(x, y) + iv(x, y)$, and suppose that the solution found in the w-plane is the harmonic function $\phi(u, v)$. Then, $\psi(x, y) = \phi(u(x, y), v(x, y))$ is harmonic and, thus, solves the Dirichlet problem in the z-plane. Indeed, by direct calculation, we find

$$\psi_{xx} = \phi_{uu}u_x^2 + \phi_u u_{xx} + \phi_{vv}v_x^2 + \phi_v v_{xx} + 2\phi_{uv}u_x v_x$$
$$\psi_{yy} = \phi_{uu}u_y^2 + \phi_u u_{yy} + \phi_{vv}v_y^2 + \phi_v v_{yy} + 2\phi_{uv}u_y v_y$$

(36.12)

so that

$$\nabla^2\psi = \phi_u \nabla^2 u + \phi_v \nabla^2 v + \phi_{uu}\left(u_x^2 + u_y^2\right) + \phi_{vv}\left(v_x^2 + v_y^2\right) + 2\phi_{uv}(u_x v_x + u_y v_y)$$
$$= \left(v_x^2 + v_y^2\right)\nabla^2\phi$$

(36.13)

where $\nabla^2 u = \nabla^2 v = 0$ because the real and imaginary parts of an analytic function are harmonic, $u_x^2 + u_y^2 = v_x^2 + v_y^2$, and $u_x v_x + u_y v_y = 0$ by the Cauchy–Riemann equations. Since ϕ is harmonic, $\nabla^2\psi = 0$.

EXAMPLE 36.35   We will solve the Dirichlet problem consisting of Laplace's equation on the first quadrant and the boundary conditions $s(x, 0) = a$ and $s(0, y) = b$, where a and b are constants. Figure 36.31 summarizes the problem.

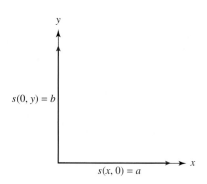

FIGURE 36.31   The BVP of Example 36.35

SOLUTION BY FOURIER SINE TRANSFORM   Taken with respect to x, the Fourier sine transform (see Section 30.7) converts $s_{xx}(x, y) + s_{yy}(x, y) = 0$, Laplace's equation, to

$$\alpha\sqrt{\frac{2}{\pi}}s(0, y) - \alpha^2 S(\alpha, y) + S_{yy}(\alpha, y) = 0$$

Applying the boundary condition $s(0, y) = b$ gives

$$\alpha\sqrt{\frac{2}{\pi}}b - \alpha^2 S(\alpha, y) + S_{yy}(\alpha, y) = 0$$

This, along with $S(\alpha, 0) = \sqrt{\frac{2}{\pi}}\frac{a}{\alpha}$, the transform of $s(x, 0) = a$, and the condition that $P(y)$ remain bounded as $y \to \infty$, is solved as an ordinary differential equation in $S(\alpha, y) = P(y)$, yielding

$$S(\alpha, y) = \frac{1}{\alpha}\sqrt{\frac{2}{\pi}}(b + (a - b)e^{-\alpha y})$$

with inverse

$$s(x, y) = b + \frac{2}{\pi}(a - b)\arctan\frac{x}{y} = a - \frac{2}{\pi}(a - b)\arctan\frac{y}{x}$$

(36.14)

The second form of the solution arises from the first by an application of the trig identity $\arctan\frac{1}{u} = \frac{\pi}{2} - \arctan u$.

SOLUTION BY INSPECTION   The real and imaginary parts of the analytic function $f(z) = u(x, y) + iv(x, y)$ are each harmonic functions. The imaginary part of the analytic function $f(z) = \operatorname{Log} z$ is $v = \operatorname{Arg} z = \arctan\frac{y}{x}$. Since $v(x, 0) = 0$ and $v(0, y) = \frac{\pi}{2}$, we can immediately write the solution to the given Dirichlet problem as $\frac{2}{\pi}(b-a)\arctan\frac{y}{x}+a$, which is equivalent to the solution in (36.14).

SOLUTION BY CONFORMAL MAPPING   The following solution, obtained by mapping the first quadrant to the upper half-plane, is more complicated than the solution obtained by inspection. However, it illustrates the key points in solving the Dirichlet problem by conformal mapping, so it is a fitting introduction to the ensuing examples.

The mapping $w = f(z) = z^2$ conformally maps the first quadrant to the upper half-plane. The positive x-axis maps to the positive x-axis, whereas the "positive part" of the imaginary axis maps to the "negative part" of the real axis. Along the horizontal axis, the boundary condition $s(x, 0) = a$ is imposed. Along the vertical axis, the condition $s(0, y) = b$ is imposed. Along the positive portion of the u-axis, the boundary condition $\sigma(u, 0) = a$ is imposed on the transformed harmonic function $\sigma(u, v)$. On the negative portion of the u-axis, the condition $\sigma(0, v) = b$ is imposed.

By inspection, $\sigma(u, v) = \frac{1}{\pi}(b-a)\mathrm{Arg}\,w + a$, so

$$s(x, y) = \frac{1}{\pi}(b-a)\mathrm{Arg}\,z^2 + a = a + \frac{1}{\pi}(b-a)\arctan\frac{2xy}{x^2-y^2}$$

Getting this solution to match that in (36.14) is a stiff exercise in trigonometric transformations. The difference of the solutions, $\frac{1}{\pi}(a-b)(\arctan\frac{2xy}{x^2-y^2} - 2\arctan\frac{y}{x})$, should vanish. The relevant factor is of the form $A - 2B$, where

$$A = \arctan\frac{2xy}{x^2-y^2} \Rightarrow \tan A = \frac{2xy}{x^2-y^2} \quad \text{and} \quad B = \arctan\frac{y}{x} \Rightarrow \tan B = \frac{y}{x}$$

Replacing $\tan A$ and $\tan B$ with their equivalents in the numerator of

$$\tan(A-2B) = \frac{\tan A \tan^2 B + 2\tan B - \tan A}{\tan^2 B - 2\tan A \tan B - 1} \tag{36.15}$$

and simplifying yield zero.   ❖

EXAMPLE 36.36

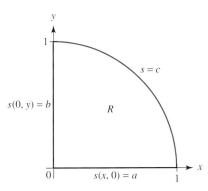

FIGURE 36.32   The BVP of Example 36.36

If the region R is that quarter of the unit disk lying in the first quadrant, we next solve the Dirichlet problem summarized by Figure 36.32 wherein a, b, c are constants.

The conformal map $\lambda = g(z) = z^2$ will transform the quarter-disk to the half-disk, and the subsequent conformal map $\zeta = h(\lambda) = \left(\frac{1+\lambda}{1-\lambda}\right)^2$ will take the half-disk to the upper half-plane. Alternatively, the composition

$$w = f(z) = h(g(z)) = \left(\frac{1+z^2}{1-z^2}\right)^2$$

maps the quarter-disk to the upper half-plane, directly. In either event, the solution in the upper half of the ζ-plane is then found by inspection.

Since the boundaries of the quarter-disk map to the u-axis and the boundary values are constants, the solution of the transformed Dirichlet problem is available by inspection. As in Example 36.35, write

$$W(w) = c + \frac{a-c}{\pi}\mathrm{Arg}\,w + \frac{b-c}{\pi}(\mathrm{Arg}(w-1) - \mathrm{Arg}\,w)$$

The solution in the z-plane is then

$$W(w(z)) = W(f(z)) = T(z) = c + \frac{a-b}{\pi}\mathrm{Arg}\left(\frac{1+z^2}{1-z^2}\right)^2 + \frac{b-c}{\pi}\mathrm{Arg}\left(\frac{2z}{1-z^2}\right)^2$$

$$\tag{36.16}$$

Write $z = re^{it}$, $0 \le t \le \frac{\pi}{2}$, and set $a = 100$, $b = -50$, $c = 25$ to obtain

$$H(r, t) = \frac{25}{\pi}(6 \arctan(\alpha_1, \alpha_2) - 3 \arctan(\alpha_1, r^2 \alpha_3) + \pi)$$

where

$$\alpha_1 = 4r^2(1 - r^4)\sin 2t \qquad \alpha_2 = 1 - 4r^4 + r^8 + 2r^4 \cos 4t \qquad \alpha_3 = 4(r^4 - 1)\cos^2 t - 8r^2$$

and $\arctan(\alpha, \beta)$ is the two-argument arctangent function. Figure 36.33 contains a graph of the level curves of $H(r, t)$ (in black), and the flow lines (in color), solutions of $\mathbf{R}'(p) = -\nabla H$, with $\mathbf{R} = x(p)\mathbf{i} + y(p)\mathbf{j}$. ❖

EXAMPLE 36.37

We will find the steady-state temperatures between two concentric circles if on the *inner* circle of radius r_i the temperature is held constant at ϕ_i and if on the *outer* circle of radius r_o the temperature is held constant at ϕ_o.

SOLUTION BY SEPARATION OF VARIABLES   In Section 29.1, we used the technique of separation of variables to solve the Dirichlet problem on the disk. In that solution, we worked in polar coordinates and allowed the boundary data to vary with the angle θ. The data in our present problem is independent of θ, so the solution will consist of just the (constant) eigensolution corresponding to the eigenvalue $\lambda = 0$. All other eigenfunctions contain θ and cannot appear in the solution to the present problem.

Substitute $\Phi(r, \theta) = R(r)$ into Laplace's equation in polar coordinates to obtain the ordinary differential equation $R'' + \frac{1}{r}R' = 0$, the solution of which is $R(r) = a + b \ln r$. After applying the boundary conditions, we then have

$$\Phi(r) = \phi_i + \frac{(\phi_o - \phi_i)(\ln r - \ln r_i)}{\ln r_o - \ln r_i} \tag{36.17}$$

SOLUTION BY CONFORMAL MAPPING   The function

$$w = \log z = \ln|z| + i \operatorname{Arg} z + 2k\pi i$$

which we write as $w = u + iv$ so that $u = \ln|z|$ and $v = \operatorname{Arg} z + 2k\pi$, maps the annulus to the infinite vertical strip $\ln r_i \le u \le \ln r_o$, $-\infty < v < \infty$. Indeed, since the inner circle is given by $z = r_i e^{i\theta}$ and the outer circle by $z = r_o e^{i\theta}$, with $-\pi < \theta \le \pi$, we have for the images of these circles, the vertical lines $\ln r_i + i(\theta + 2k\pi)$ and $\ln r_o + i(\theta + 2k\pi)$.

Now $H = c + du$ is a function that varies linearly from ϕ_i to ϕ_o as u varies from $\ln r_i$ to $\ln r_o$, where solving the equations $c + d \ln r_i = \phi_i$, $c + d \ln r_o = \phi_o$ for c and d lead to (36.17) if we set $u = \ln r$. ❖

EXAMPLE 36.38

We will find the steady-state temperatures exterior to the two circles $|z| = 1$ and $\left|z - \frac{9}{2}\right| = \frac{5}{2}$ if the boundary of the first circle is maintained at a temperature of ϕ_o and the boundary of the second at ϕ_i. The region exterior to the two circles is pictured in Figure 36.34.

The linear fractional transformation $w = f(z) = \frac{z-3}{3z-1}$ maps the region exterior to the two circles to an annulus centered at the origin. Computation shows $|f(e^{i\theta})| = 1$ and $\left|f(\frac{1}{2}(9 + 5e^{i\theta}))\right| = \frac{1}{5}$, so the smaller circle becomes the outer boundary of the annulus and the larger circle becomes the inner boundary.

The solution to the Dirichlet problem on the annulus was found in Example 36.37. Hence, we write $W(w) = \phi_i + (\phi_o - \phi_i)\ln(5|w|)/\ln 5$, so

$$\Phi(z) = \phi_i + \frac{(\phi_o - \phi_i)\ln\left(5\left|\frac{z-3}{3z-1}\right|\right)}{\ln 5} \tag{36.18}$$

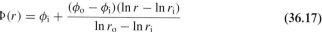

FIGURE 36.33   Level curves (in black) and flow lines (in color) for the solution of the Dirichlet problem of Example 36.36

is the solution to the original Dirichlet problem. In terms of x and y, we have

$$\Phi(x, y) = \phi_o + \frac{\phi_o - \phi_i}{2 \ln 5} \ln\left(\frac{x^2 - 6x + 9 + y^2}{9x^2 - 6x + 1 + 9y^2} \right) \qquad (36.19)$$

If $\phi_i = 100$ and $\phi_o = 25$, the solution becomes

$$\Phi_1(x, y) = 25 - \frac{75}{2 \ln 5} \ln\left(\frac{x^2 - 6x + 9 + y^2}{9x^2 - 6x + 1 + 9y^2} \right)$$

Figure 36.35 shows the isotherms in black and the flow lines in color for the solution Φ_1.

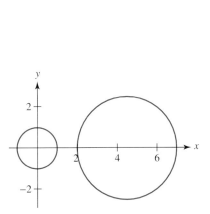

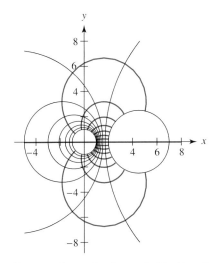

FIGURE 36.34 The region for Example 36.38 is exterior to both these circles

FIGURE 36.35 Isotherms (in black) and flow lines (in color) for Example 36.38

EXERCISES 36.8–Part A

A1. Verify the calculations in (36.12) and (36.13).

A2. Verify the trig formula in (36.15).

A3. Verify the calculations in (36.16).

A4. Show that (36.19) is a consequence of (36.18).

A5. If $z = x + iy$, show that $u(x, y) = \alpha + \frac{\beta - \alpha}{\theta} \operatorname{Arg} z$ is the solution of the Dirichlet problem for the sector $0 \le \operatorname{Arg} z \le \theta < \pi$, where $u = \alpha$ along $\operatorname{Arg} z = 0$ and $u = \beta$ along $\operatorname{Arg} z = \theta$.

EXERCISES 36.8–Part B

B1. If $z = x + iy$:

(a) Show that $u(x, y) = \alpha + \frac{\beta - \alpha}{\pi} \operatorname{Arg}(z - x_0)$ is the solution of the Dirichlet problem for the upper half-plane, where

$$u(x, 0) = \begin{cases} \alpha & x > x_0 \\ \beta & x < x_0 \end{cases}.$$

(b) Show that $\operatorname{Arg}(z - x_0) = \arctan(y, x - x_0)$.

(c) Let $\alpha = 3$, $\beta = -2$, and $x_0 = 1$. Obtain a graph of the solution surface for the resulting solution $u(x, y)$.

(d) Obtain the level curves (equipotentials or isotherms) for the solution in part (c).

(e) Obtain the flow lines for the solution in part (c).

B2. If $z = x + iy$:

   (a) Show that $u(x, y) = \alpha_0 + \frac{\alpha_1 - \alpha_0}{\pi} \operatorname{Arg}(z - x_0)$
     $+ \frac{\alpha_2 - \alpha_1}{\pi} \operatorname{Arg}(z - x_1)$ is the solution of the Dirichlet problem
     for the upper half-plane, where $u(x, 0) = \begin{cases} \alpha_0 & x > x_0 \\ \alpha_1 & x_1 < x < x_0 \\ \alpha_2 & x < x_1 \end{cases}$.

   (b) Let $\alpha_0 = 5$, $\alpha_1 = -3$, $\alpha_2 = 1$, $x_0 = 4$, $x_1 = -1$, and obtain a graph of the solution surface for the resulting solution $u(x, y)$.

   (c) Obtain the level curves (equipotentials or isotherms) for the solution in part (b).

   (d) Obtain the flow lines for the solution in part (b).

   (e) Generalize to $u(x, y) = \alpha_0 + \frac{1}{\pi} \sum_{k=0}^{n-1} (\alpha_{k+1} - \alpha_k) \operatorname{Arg}(z - x_k)$.

B3. If $z = x + iy$, solve the Dirichlet problem on the first quadrant, with $u(0, y) = \alpha_2$ and $u(x, 0) = \begin{cases} \alpha_1 & 0 < x < x_0 \\ \alpha_0 & x > x_0 \end{cases}$. Thus:

   (a) Use $w = z^2$ to map the first quadrant to the upper half of the w-plane, and apply the results of Exercise B2.

   (b) Let $\alpha_0 = -1$, $\alpha_1 = 2$, $\alpha_2 = 1$, and obtain the level curves of the resulting solution $\Psi(x, y)$.

   (c) Let $\alpha_0 = 3$, $\alpha_1 = -2$, $\alpha_2 = -1$, and obtain the level curves of the resulting solution $\Psi(x, y)$.

   (d) Let $\alpha_0 = 2$, $\alpha_1 = 0$, $\alpha_2 = 1$, and obtain the level curves of the resulting solution $\Psi(x, y)$.

B4. Let R be the lens-shaped region bounded by the circular arcs $\zeta_1(s) = i + e^{is}$, $-\frac{\pi}{2} \leq s \leq 0$, and $\zeta_2(t) = 1 + e^{it}$, $\frac{\pi}{2} \leq t \leq \pi$, which intersect at $z_1 = 0$ and $z_2 = 1 + i$. On R, solve the Dirichlet problem for $\Psi(x, y)$ if $\Psi(\zeta_1) = \alpha_1$ and $\Psi(\zeta_2) = \alpha_2$ on the boundary of R.

   (a) Sketch the region R, and verify the intersection points.

   (b) If $w = u + iv$, find a linear fractional transformation of the form $w = f(z) = e^{i\theta} \frac{z}{z - z_2}$ that maps R to the first quadrant, with the arc $\zeta_1(s)$ mapping to the nonnegative u-axis and the arc $\zeta_2(t)$ mapping to the nonnegative v-axis. (See Section 36.6.)

   (c) Use Exercise A5 to obtain $\psi(u, v)$, the solution to the transformed Dirichlet problem.

   (d) Obtain $\Psi(x, y)$.

   (e) If $\alpha_1 = 1$ and $\alpha_2 = 0$, obtain the level curves of the resulting solution $\Psi(x, y)$.

B5. If $\Psi(\zeta_1) = \begin{cases} a_1 & -\frac{\pi}{2} < s < -\frac{\pi}{3} \\ a_2 & -\frac{\pi}{3} < s < 0 \end{cases}$ with ζ_1 as in Exercise B4.

   (a) Use Exercise B3 to solve the Dirichlet problem on the region R from Exercise B4.

   (b) If, in addition, $a_1 = 2$, $a_2 = 3$, $\alpha_2 = -1$, obtain the level curves of the resulting solution $\Psi(x, y)$.

   (c) Superimpose on the level curves found in part (b), the flow lines of the solution $\Psi(x, y)$.

B6. Repeat Exercise B5 for
$$\Psi(\zeta_1) = \begin{cases} a_1 & -\frac{\pi}{2} < s < -\frac{\pi}{4} \\ a_2 & -\frac{\pi}{4} < s < 0 \end{cases}, \ a_1 = -1, a_2 = 2, \alpha_2 = 1$$

B7. Repeat Exercise B5 for
$$\Psi(\zeta_1) = \begin{cases} a_1 & -\frac{\pi}{2} < s < -\frac{\pi}{6} \\ a_2 & -\frac{\pi}{6} < s < 0 \end{cases}, \ a_1 = 0, a_2 = 5, \alpha_2 = -3$$

B8. Repeat Exercise B5 for
$$\Psi(\zeta_1) = \begin{cases} a_1 & -\frac{\pi}{2} < s < -\frac{\pi}{12} \\ a_2 & -\frac{\pi}{12} < s < 0 \end{cases}, \ a_1 = 3, a_2 = -2, \alpha_2 = 1$$

Let the circle $C_2 = \{|z - a| = r\}$ lie completely inside the circle $C_1 = \{|z - A| = R\}$, where, obviously, $R > r$. If z_1 and z_2 satisfy the equations $z_2 = \frac{R^2}{z_1 - A} + A$ and $z_2 = \frac{r^2}{z_1 - a} + a$, then the linear fractional transformation $w = f(z) = \frac{z - z_1}{z - z_2}$ maps Q, the region between C_1 and C_2, to an annulus centered at $w = 0$. The points z_1 and z_2 are on a line that simultaneously passes through both A and a, and $f(z)$ maps z_1 to $w = 0$ and z_2 to $w = \infty$. If z_1 is outside C_1, then C_1 will become the inner boundary of the annulus, else it will become the outer boundary. For the set of parameters in each of Exercises B9–13:

   (a) Sketch C_1 and C_2.

   (b) Determine z_1, z_2 and the LFT $f(z)$ that maps the region Q to an annulus, the image of C_1 becoming its outer boundary.

   (c) Find $R' = |f(C_1)|$, the outer radius of the annulus.

   (d) Find $r' = |f(C_2)|$, the inner radius of the annulus.

   (e) Solve the Dirichlet problem on Q, where $u(x, y)$ assumes the value $u_1 = 1$ on C_1 and $u_2 = 2$ on C_2.

   (f) Obtain a graph of the level curves of $u(x, y)$.

   (g) Obtain a graph of the flow lines for $u(x, y)$.

   B9. $R = 5, r = 2, A = -1, a = 1$

   B10. $R = 5, r = 2, A = 3, a = 1$

   B11. $R = 6, r = 2, A = -1, a = 1$

   B12. $R = 5, r = 3, A = 0, a = 1$

   B13. $R = 7, r = 1, A = -2, a = 2$

Let the circles $C_1 = \{|z - A| = R\}$ and $C_2 = \{|z - a| = r\}$ be completely disjoint, and let z_1 and z_2 satisfy the equations $z_2 = \frac{R^2}{z_1 - A} + A$ and $z_2 = \frac{r^2}{z_1 - a} + a$. Then, the linear fractional transformation $w = f(z) = \frac{z - z_1}{z - z_2}$ maps Q, the region exterior to both C_1 and C_2, to an annulus centered at $w = 0$. The points z_1 and z_2 are on a line that simultaneously passes through both A and a, and $f(z)$ maps z_1 to $w = 0$ and z_2 to $w = \infty$. If z_1 is inside C_1, then C_1 will become the inner boundary of the annulus, else it will become the outer boundary. For the set of parameters in each of Exercises B14–18:

   (a) Sketch C_1 and C_2.

   (b) Determine z_1, z_2 and the LFT $f(z)$ that maps the region Q to an annulus, the image of C_1 becoming its inner boundary.

   (c) Find $R' = |f(C_2)|$, the outer radius of the annulus.

   (d) Find $r' = |f(C_1)|$, the inner radius of the annulus.

(e) Solve the Dirichlet problem on Q, where $u(x, y)$ assumes the value $u_1 = -3$ on C_1 and $u_2 = 5$ on C_2.

(f) Obtain a graph of the level curves of $u(x, y)$.

(g) Obtain a graph of the flow lines for $u(x, y)$.

B14. $R = 5, r = 2, A = -1, a = 8$

B15. $R = 3, r = 1, A = -2, a = 3$

B16. $R = 3, r = 1, A = -2, a = 4$

B17. $R = 2, r = 3, A = -2, a = 5$

B18. $R = 1, r = 3, A = -2, a = 6$

36.9   Planar Fluid Flow

Preliminaries

The concepts and techniques of complex function theory prove efficacious in describing the planar flow of fluids. We therefore begin this discussion with a rephrasing of key concepts from Unit Four.

Let $\mathbf{V} = p(x, y)\mathbf{i} + q(x, y)\mathbf{j}$ be a continuous vector field in the plane. If $\mathbf{V} = \nabla\phi(x, y)$, that is, if $\mathbf{V}$ is the gradient of the scalar function $\phi(x, y)$, we say that $\mathbf{V}$ is *conservative*. Contrary to the usage of Unit Four, we do not take $\mathbf{V}$ as the negative of the gradient of Φ.

In the theory of fluid flows, $\mathbf{V}$ represents the velocity vector for an *inviscid, incompressible* steady-state planar fluid flow. An inviscid fluid experiences no friction, experiences no viscosity, and does not adhere to boundaries. An inviscid fluid would therefore slide along a boundary. An incompressible fluid has constant density.

A field for which the curl vanishes is said to be *irrotational*. If $\mathbf{V}$ has continuous partial derivatives, then being conservative is equivalent to being irrotational. (In Section 20.4 we saw that the curl of the gradient vanishes identically. In Section 21.5 we saw that irrotational fields of requisite smoothness were necessarily conservative.)

If the scalar potential $\phi(x, y)$ satisfies Laplace's equation, that is, if $\phi_{xx} + \phi_{yy} = 0$, then it is called *harmonic*.

A field for which the divergence vanishes is said to be *solenoidal*. In Section 21.5 we saw that a conservative field is solenoidal exactly when the scalar potential $\phi(x, y)$ is harmonic. (See Figure 21.13.)

The *circulation* c around a closed contour C is defined for the field $\mathbf{V}$, as in Section 19.1, by the line integral of the tangential component of $\mathbf{V}$ around C. Thus, $c = \int_C \mathbf{V} \cdot d\mathbf{r}$ and the line integral would be called work if $\mathbf{V}$ were a force field. If the circulation vanishes for every closed contour C, then $\mathbf{V}$ is said to be *circulation-free*. If $\mathbf{V}$ is sufficiently smooth and irrotational on a simply connected domain, then it is conservative and the line integral of the tangential component around every closed path vanishes. (See Section 21.5.) Thus, irrotational flows will be circulation-free if the domain of the field is simply connected.

The *flux* F of $\mathbf{V}$ across the closed contour C is defined, as in Section 19.2, by the line integral of the normal component of $\mathbf{V}$ around C. Thus, flux is given by the line integral $F = \int_C \mathbf{V} \cdot \mathbf{N} \, ds$, where $\mathbf{N}$ is a unit outward normal along C and s, the parameter along C, is arc length. If the flux through every closed contour C is zero, then $\mathbf{V}$ is said to be *source-free*. In Section 19.2 the divergence of the field $\mathbf{V}$ was shown to represent the flux of $\mathbf{V}$ through C, per unit area within C, in the limit as C was shrunk to a point. Hence, if a field is solenoidal, it will be source-free, provided the domain on which the field is solenoidal is simply connected.

If $\mathbf{V}$ is a conservative velocity field with continuous first partial derivatives, it is necessarily irrotational. (See Sections 20.4 and 21.5.) On a simply connected domain, $\mathbf{V}$ will be circulation free. On a multiply-connected domain, $\mathbf{V}$ could exhibit nonzero circulation for contours that cannot be shrunk to a point in the domain. If $\phi(x, y)$, the scalar potential for $\mathbf{V}$, is harmonic, then $\mathbf{V}$ is solenoidal, or divergence-free, as well. (See Section 21.5.) If $\phi(x, y)$ is harmonic on a simply connected domain, then $\mathbf{V}$ is source-free. If $\phi(x, y)$ is harmonic on a multiply-connected domain, then $\mathbf{V}$ could be locally divergence-free but still not be source-free.

Let $\mathbf{V}$ be conservative so that $\mathbf{V} = \nabla\phi(x, y)$. The level curves (contours) of $\phi(x, y)$ are defined by $\phi(x, y) = $ constant. Orthogonal to these "equipotentials" are the flow lines, the integral curves of the gradient field. These flow lines are the solution of the equations $\frac{dx}{dt} = \phi_x, \frac{dy}{dt} = \phi_y$. If these solutions are given implicitly as $y = y(x)$ by the function $\psi(x, y)$, then the level curves $\psi(x, y) = $ constant are the flow lines for the field $\mathbf{V}$. If $\psi(x, y)$ is also harmonic, then it is called the *harmonic conjugate* of $\phi(x, y)$.

If the harmonic function $\phi(x, y)$ is the real part of an analytic function $F(z)$, then $\psi(x, y)$, the imaginary part, is the harmonic conjugate and the level curves of each harmonic function are mutually orthogonal. If $\mathbf{V} = \nabla\phi$, with ϕ harmonic, then there is an analytic function $F(z) = \phi + i\psi$ for which ψ is the harmonic conjugate of ϕ and for which the level curves are the flow lines of the field $\mathbf{V}$. The function $F(z)$ is called the *complex potential* for the field $\mathbf{V}$, and the function $\psi(x, y)$ is called the *stream function* because its level curves (contours) are the flow lines for the fluid. In this context, the contours of the stream function are called *streamlines*. The velocity vector $\mathbf{V}$ is tangent to the streamlines, so fluid does not cross a streamline. Therefore, fluid boundaries must fall along the streamlines of the flow.

The following theorem from [69] gives necessary and sufficient conditions for the existence of a complex potential for a planar flow.

THEOREM 36.5

The functions $p(x, y)$ and $q(x, y)$, continuous on a domain D, define $\mathbf{V} = p(x, y)\mathbf{i} + q(x, y)\mathbf{j}$ as a source-free and circulation-free flow in D if and only if there exists in D, a single-valued analytic function $F(z) = \phi(x, y) + i\psi(x, y)$ for which $F'(z) = p - iq$.

If the analytic function $F(z) = \phi + i\psi$ exists and $F' = p - iq$, then by the definition of a complex derivative $F' = \phi_x + i\psi_x$ and by the Cauchy–Riemann equations $F' = \phi_x - i\phi_y$. Consequently, $p = \phi_x$ and $q = \phi_y$ and $\mathbf{V} = \overline{F'(z)}$. The derivative $F'(z)$ is called the *complex velocity*, even though $\mathbf{V}$, the velocity vector, is obtained from the *conjugate* of this derivative.

For a closed contour C, the circulation and flux integrals can be computed as the real and imaginary parts, respectively, of the integral

$$Q = \int_C F'(z)\, dz = \int_C (p - iq)(dx + i\, dy) = \int_C p\, dx + q\, dy + i \int_C p\, dy - q\, dx$$

The real part is recognized as the circulation integral, and the imaginary part as the flux integral. (See Sections 19.1 and 19.2.)

Finally, the magnitude of the velocity vector is called the *speed* and is denoted by $v = |\overline{F'(z)}| = \|\mathbf{V}\|$.

All these relationships are illustrated in the following examples of inviscid incompressible planar flows.

Elementary Inviscid Incompressible Planar Flows

The following examples illustrate elementary inviscid incompressible planar flows. In Section 36.10, more complicated flows will be obtained by applying conformal maps to these elementary flows.

EXAMPLE 36.39   The flow defined by the complex potential $F(z) = Az$, where $A > 0$, exhibits parallel streamlines. To see this, obtain the real and imaginary parts, $\phi = Ax$ and $\psi = Ay$, respectively, so the velocity vector is $\mathbf{V} = \nabla\phi = A\mathbf{i}$. Figure 36.36 shows the equipotentials (in color) and streamlines (in black) for the case $A = 2$. A single arrow shows the direction of the flow.

The complex velocity is then $F'(z) = p - iq = A$. The *conjugate* of the complex velocity is $V = A$, and the speed is $v = A$, computed both from $|F'(z)|$ and $\|\mathbf{V}\|$. Let the contour C be the circle $z(t) = re^{it}, 0 \le t \le 2\pi$, so

$$Q = \int_C F'(z)\,dz = Ar \int_0^{2\pi} ie^{it}\,dt = 0$$

and both the circulation and flux for C are zero.

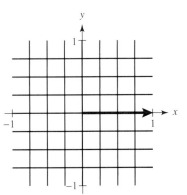

FIGURE 36.36   Streamlines (in black) and equipotentials (in color) for the flow in Example 36.39

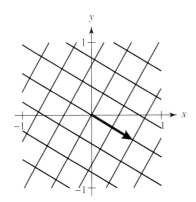

FIGURE 36.37   Streamlines (in black) and equipotentials (in color) for the flow in Example 36.40

EXAMPLE 36.40   The flow defined by the complex potential $F(z) = Ae^{i\lambda}z$, where $A > 0$, exhibits parallel but slanted streamlines. To see this, obtain the real and imaginary parts, $\phi = A(x\cos\lambda - y\sin\lambda)$ and $\psi = A(x\sin\lambda + y\cos\lambda)$, respectively; so the velocity vector is $\mathbf{V} = \nabla\phi = A(\cos\lambda\mathbf{i} - \sin\lambda\mathbf{j})$. Figure 36.37 shows the equipotentials and streamlines for the case $\lambda = \frac{\pi}{6}$ and $A = 2$. The streamlines are the black lines and the equipotentials are the color lines. A single arrow shows the direction of the flow. (Notice that in the complex potential $F(z)$ the factor $e^{i\lambda}$ appears but the streamlines make an angle of $-\lambda$ with the horizontal.)

The complex velocity is $F'(z) = Ae^{i\lambda}$, its conjugate is $V = Ae^{-i\lambda}$, and the speed is $v = |F'(z)| = \|\mathbf{V}\| = A$. If C is the circle $z(t) = re^{it}, 0 \le t \le 2\pi$, then

$$Q = \int_C F'(z)\,dz = Ae^{i\lambda}r \int_0^{2\pi} ie^{it}\,dt = 0$$

and both the circulation and flux for C are zero.

EXAMPLE 36.41   The flow defined by the complex potential $F(z) = Az^2$, where $A > 0$, models a flow perpendicular to a wall. To see this, obtain $\phi = A(x^2 - y^2)$ and $\psi = 2Axy$, the real and imaginary parts, respectively, of $F(z)$. The velocity vector is $\mathbf{V} = \nabla\phi = 2A(x\mathbf{i} - y\mathbf{j})$. Figure 36.38 shows the equipotentials (in color) and streamlines (in black) for the case $A = 2$. Arrows in the direction of $\mathbf{V}$ are included by graphing the gradient field for ϕ.

The complex velocity is $F'(z) = 2Az$; the speed is $v = |\overline{F'(z)}| = \|\mathbf{V}\| = 2A\sqrt{x^2 + y^2}$. If C is the circle $z(t) = 2i + re^{it}$, $0 \le t \le 2\pi$, then

$$Q = \int_C F'(z)\,dz = 2Ar \int_0^{2\pi} i(2i + re^{it})e^{it}\,dt = 0$$

and both the circulation and flux for C are zero.   ❖

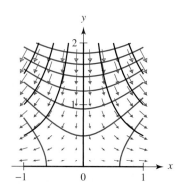

FIGURE 36.38   Streamlines (in black) and equipotentials (in color) for the flow in Example 36.41

EXAMPLE 36.42   The flow defined by the complex potential $F(z) = A\log(z - a)$, where $A > 0$, models flow emanating from a source located at $z = a$. To see this, use the principle logarithm and obtain

$$\phi = \tfrac{1}{2}A\ln((x - a)^2 + y^2) \quad\text{and}\quad \psi = A\arctan(y, x - a)$$

the real and imaginary parts, respectively, of $F(z)$. The velocity vector is then

$$\mathbf{V} = \nabla\phi = A\left(\frac{x - a}{(x - a)^2 + y^2}\mathbf{i} + \frac{y}{(x - a)^2 + y^2}\mathbf{j}\right)$$

Figure 36.39 shows the equipotentials and streamlines for the case $A = 2, a = 1$. The streamlines, which appear to be radial lines emanating from $z = a$, are in black; and the equipotentials, which are concentric circles centered at $z = a$, are in color. A single arrow shows the flow is radially outward.

The complex velocity is $F'(z) = \frac{A}{z-a}$, and the speed is

$$v = |\overline{F'(z)}| = \|\mathbf{V}\| = \frac{A}{\sqrt{(x - a)^2 + y^2}}$$

Far from $z = a$, the limiting value of $\mathbf{V} = \overline{F'(z)}$ is clearly 0. If C_1 is the circle $z(t) = 3 + e^{it}$, $0 \le t \le 2\pi$, so that $a = 1$ is not enclosed by it, then

$$Q = \int_{C_1} F'(z)\,dz = iA \int_0^{2\pi} \frac{e^{it}\,dt}{2 + e^{it}} = 0$$

suggesting that if the contour C does not enclose $z = a$, then the circulation and flux, around and through C, respectively, vanish. If C_2 is the circle $z(t) = 1 + e^{it}$, $0 \le t \le 2\pi$, so that $a = 1$ is now enclosed by it, then

$$Q = \int_{C_2} F'(z)\,dz = iA \int_0^{2\pi} dt = 2\pi iA$$

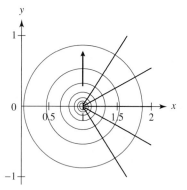

FIGURE 36.39   Streamlines (in black) and equipotentials (in color) for the flow in Example 36.42

suggesting there is a flux of $2\pi A$ through contours that enclose $z = a$. Such a flow is said to be circulation-free but with a flux of $2\pi A$ through contours enclosing $z = a$. For A positive, the flow is said to have a *source* of strength $2\pi A$ at the point $z = a$; while for A is negative, the flow is said to have a *sink* at $z = a$.   ❖

EXAMPLE 36.43   The flow defined by the complex potential $F(z) = -iB\log(z - a)$, where $B > 0$, models a flow with a point-source of circulation at $z = a$. To see this, obtain

$$\phi = B\arctan(y, x - a) \quad\text{and}\quad \psi = -\tfrac{1}{2}B\ln((x - a)^2 + y^2)$$

the real and imaginary parts, respectively, of $F(z)$. The velocity vector is

$$\mathbf{V} = \nabla\phi = B\left(\frac{-y}{(x-a)^2 + y^2}\mathbf{i} + \frac{x-a}{(x-a)^2 + y^2}\mathbf{j}\right)$$

Figure 36.40 shows the equipotentials and streamlines for the case $B = 1, a = 1$. The black curves are streamlines, concentric circles centered at $z = a$, and the lines in color are equipotentials. Arrows in the direction of $\mathbf{V}$ show the flow is in a counterclockwise direction around the circles.

The complex velocity is $F'(z) = -\frac{iB}{z-a}$, and the speed is

$$v = |\overline{F'(z)}| = \|\mathbf{V}\| = \frac{B}{\sqrt{(x-a)^2 + y^2}}$$

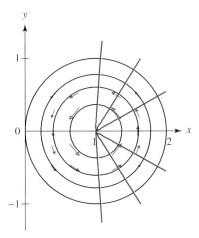

Far from $z = a$, the limiting value of $V = \overline{F'(z)}$ is clearly 0. If C_1 is the circle $z(t) = 3 + e^{it}, 0 \le t \le 2\pi$, so it does not enclose $a = 1$, we have

$$Q = \int_{C_1} F'(z)\,dz = B\int_0^{2\pi} \frac{e^{it}\,dt}{2 + e^{it}} = 0$$

suggesting that if the contour C does not enclose $z = a$, then the circulation and flux, around and through C, respectively, vanish. If C_2 is the circle $z(t) = 1 + e^{it}, 0 \le t \le 2\pi$, so that it does enclose $a = 1$, we have

$$Q = \int_{C_2} F'(z)\,dz = B\int_0^{2\pi} dt = 2\pi B$$

suggesting the circulation around a contour that encloses $z = a$ is $c = 2\pi B$. Thus, this flow is flux-free but exhibits a circulation of $2\pi B$ around a contour enclosing $z = a$ and is said to have a *vortex* of strength $2\pi B$ at the point $z = a$.   ❖

FIGURE 36.40   Streamlines (in black) and equipotentials (in color) for the flow in Example 36.43

EXAMPLE 36.44   The flow defined by the complex potential

$$F(z) = \frac{m}{2\pi}\log\left(\frac{z-a}{z}\right) = \frac{m}{2\pi}\log(z-a) - \frac{m}{2\pi}\log z$$

where $m > 0$, has a source of strength m at $z = a$ and a sink of strength m at $z = 0$. To see this, obtain

$$\phi = \frac{m}{4\pi}\ln\left(\frac{(x-a)^2 + y^2}{x^2 + y^2}\right) \quad \text{and} \quad \psi = \frac{m}{2\pi}\arctan\left(\frac{ay}{x^2 + y^2}, \frac{x^2 - ax + y^2}{x^2 + y^2}\right)$$

the real and imaginary parts, respectively, of the complex potential $F(z)$. The velocity vector is

$$\mathbf{V} = \nabla\phi = \frac{am}{2\pi(x^2 + y^2)((x-a)^2 + y^2)}\left((x^2 - ax - y^2)\mathbf{i} + y(2x-a)\mathbf{j}\right)$$

Figure 36.41 shows the equipotentials and streamlines for the case $m = 1, a = 1$. The streamlines are the black lines, and the equipotentials are the lines in color. Arrows in the direction of $\mathbf{V}$ are included by graphing the gradient field for ϕ.

The complex velocity and the speed are given, respectively, by

$$F'(z) = \frac{am}{2\pi z(z-a)} \quad \text{and} \quad v = \frac{am}{2\pi\sqrt{x^2 + y^2}\sqrt{(x-a)^2 + y^2}}$$

Far from $z = a$, the limiting value of $V = \overline{F'(z)}$ is 0. If a C_1 does not enclose $z = 0$ or $z = a$, then the circulation and flux, around and through C_1, respectively, vanish. If a

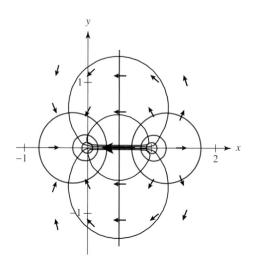

FIGURE 36.41   Streamlines (in black) and equipotentials (in color) for the flow in Example 36.44

contour C_2 encloses $z = a$ but not $z = 0$, then the circulation around C_2 vanishes but the flux through C_2 is m. If a contour C_3 encloses $z = 0$ but not $z = a$, then the circulation around C_3 vanishes but the flux through C_3 is $-m$. Thus, $z = a$ is a source of strength m and $z = 0$ is a sink of strength m. For example, take $a = 1$ and on $C_k, k = 1, \ldots, 3$, take $0 \le t \le 2\pi$, obtaining the results summarized in Table 36.4.   ❖

C_1	$z(t) = 3 + e^{it}$	$Q = \displaystyle\int_{C_2} F'(z)\,dz = \dfrac{m}{2\pi}\displaystyle\int_0^{2\pi}\dfrac{ie^{it}\,dt}{(+e^{it})(2+e^{it})} = 0$
C_2	$z(t) = 1 + \tfrac{1}{2}e^{it}$	$Q = \displaystyle\int_{C_2} F'(z)\,dz = \dfrac{m}{2\pi}\displaystyle\int_0^{2\pi}\dfrac{i\,dt}{1+\tfrac{1}{2}e^{it}} = im$
C_3	$z(t) = \tfrac{1}{2}e^{it}$	$Q = \displaystyle\int_{C_2} F'(z)\,dz = \dfrac{m}{2\pi}\displaystyle\int_0^{2\pi}\dfrac{i\,dt}{\tfrac{1}{2}e^{it}-1} = -im$

TABLE 36.4   Circulation and flux for the contours in Examle 36.44

EXAMPLE 36.45   The flow defined by the complex potential $F(z) = -\sigma e^{i\lambda}/2\pi z$ exhibits, at $z = 0$, a *doublet of strength* σ, with doublet-axis along the ray through the origin defined by $e^{i\lambda}$. To see this, obtain

$$\phi = -\frac{\sigma}{2\pi}\frac{x\cos\lambda + y\sin\lambda}{x^2 + y^2} \quad \text{and} \quad \psi = -\frac{\sigma}{2\pi}\frac{x\sin\lambda - y\cos\lambda}{x^2 + y^2}$$

the real and imaginary parts, respectively, of $F(z)$. The velocity vector is

$$\begin{aligned}\mathbf{V} &= \nabla\phi \\ &= \frac{\sigma}{2\pi(x^2+y^2)^2}\left[((x^2-y^2)\cos\lambda + 2xy\sin\lambda)\mathbf{i} + ((y^2-x^2)\sin\lambda + 2xy\cos\lambda)\mathbf{j}\right]\end{aligned}$$

Figure 36.42 shows the equipotentials and streamlines for the case $\sigma = 1, \lambda = \frac{\pi}{3}$. The streamlines are the black lines, and the equipotentials are the lines in color. Arrows in the direction of $\mathbf{V}$ are included by graphing the gradient field for ϕ.

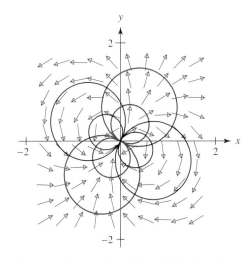

FIGURE 36.42   Streamlines (in black) and equipotentials (in color) for the flow in Example 36.45

The complex velocity is $F'(z) = \sigma e^{i\lambda}/2\pi z^2$, and the speed is

$$v = |\overline{F'(z)}| = \|\mathbf{V}\| = \frac{\sigma}{2\pi(x^2 + y^2)}$$

Far from $z = a$, the limiting value of $V = \overline{F'(z)}$ is 0. For any contour C not through $z = 0$, $Q = \int_C F'(z)\,dz = 0$. For example, if C is the contour $z(t) = 3 + e^{it}, 0 \le t \le 2\pi$, we have $Q = \int_0^{2\pi} ie^{it}\,dt/(3 + e^{it})^2 = 0$. If C is the contour $z(t) = e^{it}, 0 \le t \le 2\pi$, then $Q = \int_0^{2\pi} ie^{-it}\,dt = 0$. Thus, the doublet is locally circulation-free and source-free. In fact, Q can be evaluated directly by residues, noting that $F(z)$ is a multiple of $\frac{1}{z}$ and so $F'(z)$ is a multiple of $\frac{1}{z^2}$. Thus, the residue at $z = 0$ is zero, and $Q = 0$ is the correct result.

In fact, the doublet is obtained from the complex potential in Example 36.44 by allowing the source and sink to coalesce. Thus, the doublet is free of both circulation and flux because the source and sink cancel each other out. The exact manner in which the source and sink are allowed to approach each other is left to the exercises.   ❖

EXERCISES 36.9

Harmonic functions attain their maximum and minimum on the boundary. More precisely, let R be a closed bounded region consisting of a domain D (a connected open set) and a boundary B. If u is harmonic in D and continuous on B, then on B there are two points z_1 and z_2, not necessarily unique, for which $u(z_1) \le u(z) \le u(z_2)$ holds for all z in R. Let R be the disk $|z - a| \le 1$, where $a = 2(1 + i)$. In Exercises 1–5, obtain the harmonic functions $\phi = \mathcal{R}(F)$ and $\psi = \mathcal{I}(F)$ and show each attains its maximum and minimum on the boundary of R. (Graph each as a surface, find extreme points in the interior of R, and then find extreme points on the boundary of R by using the Lagrange multiplier method (Section 18.4) or by parametrizing the boundary, thereby making u on the boundary a function of a single parameter.)

1. $F(z) = \arcsin z$ **2.** $F(z) = \arccos z$

3. $F(z) = \operatorname{arcsinh} z$ **4.** $F(z) = \operatorname{arccosh} z$ **5.** $F(z) = \dfrac{1}{z}$

For the flow determined by the complex potential $F(z)$ in Exercises 6–10:

(a) Obtain the potential function $\phi(x, y)$ and the stream function $\psi(x, y)$.

(b) Obtain a graph of the equipotentials.

(c) Obtain a graph of the streamlines.

(d) Graph the arrows of the gradient field $\nabla\phi$, and compare to the graph of the streamlines.

(e) Demonstrate that $V = \overline{F'(z)}$, the velocity, is equivalent to $\mathbf{V} = \nabla\phi$.

(f) Compute the speed $v = |F'(z)|$ and graph it as a surface defined on the xy-plane.

(g) Compute $Q = \int_C F'(z)\,dz$, where C is the generic circle $z(t) = a + be^{it}$, $-\pi \le t \le \pi$, and show that $\mathcal{R}(Q) = \int_C \nabla\phi \cdot d\mathbf{r}$ and $\mathcal{I}(Q) = \int \phi_x\,dy - \phi_y\,dx$.

(h) Obtain the streamlines from $\phi(x, y)$ by integrating the gradient field $\nabla\phi$. Eliminate the parameter along the streamline by writing the single differential equation $\frac{dy}{dx} = \phi_y/\phi_x$ and solving it to obtain a first integral $f(x, y) = $ constant. Show that the level curves of $f(x, y)$ are equivalent to the streamlines determined by $\psi(x, y) = $ constant.

6. $F(z) = \cos z$ **7.** $F(z) = \sinh z$

8. $F(z) = \frac{z+1}{z-1}$ **9.** $F(z) = \arctan z$ **10.** $F(z) = \cosh z$

For each complex potential given in Exercises 11–15:

(a) Obtain the harmonic function $\phi(x, y) = \mathcal{R}(F)$.

(b) Show that $\psi(x, y) = \phi(u(x, y), v(x, y))$ is harmonic, if $f(z) = z^2 = u + iv$.

(c) Show that $\psi(x, y) = \phi(u(x, y), v(x, y))$ is harmonic, if $f(z) = z + \frac{1}{z} = u + iv$.

11. $F(z) = \tanh z$ **12.** $F(z) = \operatorname{arctanh} z$

13. $F(z) = \frac{z+1}{z+2}$ **14.** $F(z) = \frac{2z+i}{3z-1}$ **15.** $F(z) = \sin z$

36.10   Conformal Mapping of Elementary Flows

Conformal Maps of Fluid Flows

Conformal mappings are often used in conjunction with complex potentials when solving planar flow problems. To find a flow over a complicated region in the z-plane, a conformal map is typically used to transform the region in the z-plane to a simpler region in the w-plane. The transformed flow problem is solved in the w-plane and the solution mapped back to the z-plane. Sometimes, however, simple flow in the z-plane is conformally mapped to a complicated flow in the w-plane. Clearly, this is just a role-reversal between the z- and w-planes, but when the conformal map is not easily inverted, this seems to be the prevailing usage.

In either event, the technique is viable because a conformal map of a complex potential is again a complex potential. Thus, under a conformal transformation, streamlines in the z-plane map to streamlines in the w-plane. To be precise, let $w = f(z)$ be analytic with $f'(z) \ne 0$ in a domain D. Then (see, e.g., [69] or [53]), $f(z)$ maps D conformally and also has a unique inverse for which $z = g(w)$. Let $z_k = x_k + iy_k$, $k = 1, 2$, be two points on a streamline of $\psi(x, y)$ so that $\psi(x_k, y_k) = c$, $k = 1, 2$. Let $w_k = f(z_k) = u_k + iv_k$, $k = 1, 2$, be the images, in the w-plane, of the two points in the z-plane so that $x_k = x_k(u_k, v_k)$ and $y_k = y_k(u_k, v_k)$ for $k = 1, 2$. Then, for $k = 1, 2$,

$$\psi(x_k, y_k) = \psi(x_k(u_k, v_k), y_k(u_k, v_k)) = \Psi(u_k, v_k) = c$$

Two points on a streamline in the z-plane map to two points on a streamline in the w-plane. The stream functions $\psi(x, y)$ and $\Psi(u, v)$ have the same constant values on corresponding streamlines. The correspondence is a consequence of the existence of an inverse map, itself a consequence of conformality.

Given the principle that conformal transformations map streamlines to streamlines, we next depict how a flow behaves under a conformal transformation. Let D_z be a domain in the z-plane mapped to the domain D_w in the w-plane by the conformal transformation $w = f(z) = u(x, y) + iv(x, y)$. Thus, $z = x + iy$ and $w = u + iv$. Let $G(w) = \Phi(u, v) + i\Psi(u, v)$ be a complex potential for a flow in the w-plane. Then the complex potential defined by

$$\begin{aligned} F(z) &= G(w(z)) = G(f(z)) \\ &= \Phi(u(x, y), v(x, y)) + i\Psi(u(x, y), v(x, y)) \\ &= \phi(x, y) + i\psi(x, y) \end{aligned}$$

is the complex potential for the pre-image, in D_z, of the flow in D_w.

In most cases, the domain D_z is the complicated one, and D_w is simpler. The flow problem is solved in the w-plane by finding $G(w)$, and the solution is brought back to the z-plane by writing $F(z) = G(f(z))$. However, there are some important cases where the domain D_z is the simpler one, and a flow $F(z)$ is mapped by $w = f(z)$ to D_w, the more complicated domain. This is generally the approach taken when the inverse map from the complicated domain to the simpler domain is algebraically difficult.

In the rest of this section, we give examples of these ideas of flow mapping by conformal transformations.

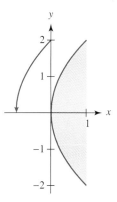

FIGURE 36.43   Flow around the parabola $y^2 = 4x$

EXAMPLE 36.46   **Flow around a parabola**   This example of flow around a parabola is adapted from [34]. The parabola $y^2 = 4x$, along with a representation of the flow, is shown in Figure 36.43. The mapping $w = f(z) = \sqrt{z-1}$ will transform the parabola $y^2 = 4x$ into the horizontal line $v = 1$, provided the discontinuity in the square root function occurs along the positive real axis. (See Exercise A1.) The principal square root has its branch cut along the negative real axis, but an appropriate branch of $\sqrt{z-1}$ will satisfy $0 \le \theta < 2\pi$, where θ is the angle made by the segment $z - 1$ and the positive real axis. (If $\text{Arg}(z - 1) \ge 0$, then $\theta = \text{Arg}(z - 1)$; if $\text{Arg}(z - 1) < 0$, then $\theta = \text{Arg}(z - 1) + 2\pi$. Define the square root by $\sqrt{z-1} = \sqrt{|z-1|}e^{i\theta/2}$. See Exercises B1 and B2.)

The complex potential for the flow in the w-plane is $G(w) = Aw$. Therefore, the complex potential for the flow in the z-plane is $F(z) = G(w(z)) = G(f(z)) = A\sqrt{z-1}$, where A is determined by the condition that the complex velocity at the vertex of the parabola should be $\mathbf{V} = -i$. Thus, the condition $\overline{F'(0)} = -i$ gives $A = -2$, and the desired complex potential in the z-plane is then $F(z) = -2\sqrt{z-1}$. Writing $F(z)$ as $\phi(x, y) + i\psi(x, y)$, the streamlines are the level curves of $\psi(x, y) = \mathcal{I}(F(x+iy))$. The streamlines are shown in Figure 36.44. (Also, see Exercise B3.)   ❖

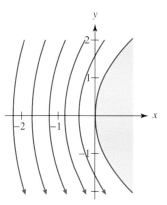

FIGURE 36.44   Streamlines for flow around the parabola $y^2 = 4x$

EXAMPLE 36.47   **Flow against a parabola**   This example of flow directed *at* the vertex of the parabola $y^2 = 4x$ is adapted from [34]. If the square root is defined as in Exercise B2, the function $f(z) = \sqrt{z-1}$ maps the parabola to the line $v = 1$. The negative real axis is mapped to $v \ge 1$ on the positive imaginary axis. Thus, $f(z)$ maps the z-plane to the w-plane as shown in Figure 36.45.

Since the flow in the z-plane is along the negative real axis and against the vertex of the parabola, the flow in the w-plane will be downward and against the "wall" along $v = 1$. The solution for flow against a wall is given in Example 36.41, Section 36.9, provided we shift the point $w = i$ in the w-plane to the origin.

Therefore, define the mapping $\zeta = h(w) = w - i$, with the complex potential in the ζ-plane being given by $H(\zeta) = A\zeta^2$. Then, the complex potential in the w-plane is

$$G(w) = H(h(w)) = G = A(w - I)^2$$

Finally, $F(z)$, the complex potential for the flow in the z-plane, is given by

$$F(z) = G(w(z)) = G(f(z)) = A(\sqrt{z-1} - I)^2$$

The constant A is determined by the condition that as $z \to \infty$, the velocity tends toward, say, 1. Thus, the condition $\lim_{z \to \infty} \overline{F'(z)} = 1$ leads to $A = 1$. Writing the complex potential $F(z)$ as $F(x + iy) = \phi(x, y) + i\psi(x, y)$, we have, in the upper half-plane,

$$\phi = A(x - 2 + \sqrt{\lambda - 2x + 2}) \quad \text{and} \quad \psi = \frac{A}{2}\sqrt{\lambda + 2x - 2}(\sqrt{\lambda - 2x + 2} - 2)$$

where $\lambda = 2\sqrt{(x-1)^2 + y^2}$. In the lower half-plane, $\phi(x, y)$ is the same but $\psi(x, y)$ is replaced with its negative. Figure 36.46 shows the streamlines and the velocity field (gradients of $\phi(x, y)$) for the case $A = 1$.   ❖

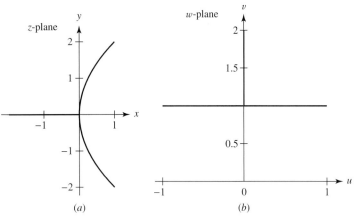

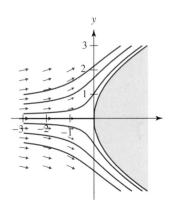

FIGURE 36.45   Mapping the z-plane by $w = \sqrt{z-1}$

FIGURE 36.46   Streamlines for flow against the parabola $y^2 = 4x$

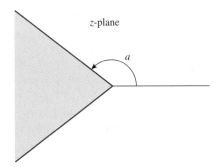

FIGURE 36.47   The wedge in Example 36.48

EXAMPLE 36.48   **Flow past a wedge**   To determine the flow of a fluid moving from left to right past the symmetric wedge shown in Figure 36.47, note that the mapping $w = f(z) = z^{\pi/a}$ will take the edges of the wedge to the negative real axis in the w-plane. In fact, the lines $z = re^{\pm ia}$ both map to the negative real axis since $f(re^{\pm ia}) = -r^{\pi/a}$.

The complex potential for the flow in the w-plane is then $G(w) = Aw$, so the complex potential for the flow in the z-plane is

$$F(z) = G(w(z)) = G(f(z)) = Az^{\pi/a}$$

Write $F(z)$ as $F(x + iy) = \phi(x, y) + i\psi(x, y)$, where

$$\phi = A(x^2 + y^2)^{\pi/2a} \cos\left(\frac{\pi}{a}\arctan(y, x)\right)$$

$$\psi = A(x^2 + y^2)^{\pi/2a} \sin\left(\frac{\pi}{a}\arctan(y, x)\right)$$

where the two-argument arctangent function takes values in $[-\pi, \pi]$. For the case $A = 1$, $a = \frac{3\pi}{4}$, a graph of the streamlines and velocity vector field is given in Figure 36.48.

For flow from left to right *against* a wedge, note that replacing z with $-z$ will rotate the wedge through an angle of π. Thus, write $F_1(z) = A(-z)^{\pi/a}$ and obtain the velocity field in Cartesian coordinates as the real and imaginary parts of $V = \overline{F_1'(x + iy)}$. In particular, if $a = \frac{3\pi}{4}$, we find

$$V = \frac{4A}{3\left(x^2 + y^2\right)^{1/3}}\left[(x\cos\lambda + y\sin\lambda)\mathbf{i} + (y\cos\lambda - x\sin\lambda)\mathbf{j}\right]$$

where $\lambda = \frac{4}{3}\arctan(-y, -x)$. In particular, at $z = -1$, the velocity vector should be proportional to the unit vector $\mathbf{i}$. At this point, the velocity vector is actually $-\frac{4}{3}A\mathbf{i}$, so we will set $A = -1$. The streamlines and the velocity vectors can be seen in Figure 36.49.   ❖

EXAMPLE 36.49   **Flow into a closed channel**   The flow into a closed channel can be found by mapping the channel, given in the z-plane as the upper half-plane region $-\frac{\pi}{2} \leq \mathcal{R}(z) \leq \frac{\pi}{2}$, to the w-plane by the function $w = f(z) = \sin z$. The boundaries of the channel map to the real axis. Hence, the complex potential in the w-plane is given by the function $G(w) = Aw$, so the complex potential for the flow into the closed channel in the z-plane is given by

$$F(z) = G(w(z)) = G(f(z)) = A\sin z$$

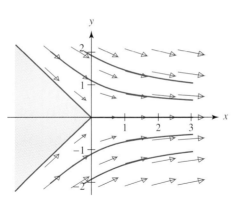

FIGURE 36.48   The streamlines and velocity vector field for flow *past* a wedge in Example 36.48

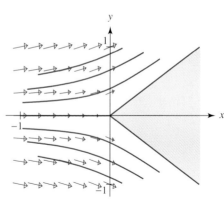

FIGURE 36.49   The streamlines and velocity vector field for flow *against* a wedge in Example 36.48

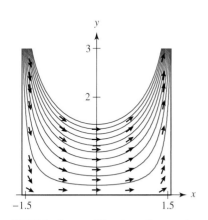

FIGURE 36.50   The streamlines and velocity vector field for flow into a closed channel in Example 36.49

the real and imaginary parts of which are

$$\phi = A \sin x \cosh y \quad \text{and} \quad \psi = A \cos x \sinh y$$

respectively. The velocity field can be obtained as

$$\nabla\phi = A(\cos x \cosh y \mathbf{i} + \sin x \sinh y \mathbf{j})$$

or as $\overline{F'(x + iy)}$. A sketch of the flow when $A = 1$ is given in Figure 36.50.   ❖

EXAMPLE 36.50   **Flow through a slit**   The flow through a slit can be determined by mapping the z-plane to the w-plane with the function $w = f(z) = \arcsin z$. This function maps $x \leq -1, y = 0$ on the negative x-axis to the vertical line $u = -\frac{\pi}{2}$ in the w-plane and maps $x \geq 1, y = 0$ on the positive x-axis to the vertical line $u = \frac{\pi}{2}$ in the w-plane. Consequently, if the slit is coincident with $-1 \leq x \leq 1$ on the x-axis, its image under $f(z) = \arcsin z$ will be the strip $-\frac{\pi}{2} \leq u \leq \frac{\pi}{2}, -\infty < v < \infty$, in the w-plane.

The complex potential for the flow in the w-plane is $G(w) = -iAw$, $A > 0$, consistent with an upward flow through the channel, since $V = \overline{G'(w)} = iA$. The complex potential for the flow in the z-plane is then

$$F(z) = G(w(z)) = G(f(z)) = -iA \arcsin z$$

for which the real and imaginary parts, respectively, are

$$\phi = A \operatorname{csgn}(y - ix) \ln(\alpha + \beta + \sqrt{(\alpha + \beta)^2 - 1}) \quad \text{and} \quad \psi = A \arcsin(\beta - \alpha)$$

where $\alpha = \frac{1}{2}\sqrt{(x + 1)^2 + y^2}, \beta = \frac{1}{2}\sqrt{(x - 1)^2 + y^2}$, and

$$\operatorname{csgn} z = \begin{cases} 1 & \mathcal{R}(z) > 0 \text{ or } \mathcal{R}(z) = 0 \text{ and } \mathcal{I}(z) > 0 \\ -1 & \mathcal{R}(z) < 0 \text{ or } \mathcal{R}(z) = 0 \text{ and } \mathcal{I}(z) < 0 \end{cases}$$

For the case $A = 1$, Figure 36.51 shows the streamlines as determined by $\psi(x, y)$ and the velocity field as determined by $\phi(x, y)$.   ❖

EXAMPLE 36.51   **Flow past a cylinder**   The circle $|z| = \sigma$ is the cross-section of a cylinder whose axis is perpendicular the z-plane. The complex potential for flow past this cylinder when $V_\infty = Ae^{i\lambda}$, $A > 0$, can be found by mapping the z-plane to the ζ-plane to eliminate the slant to the far-field velocity and then mapping the ζ-plane to the w-plane by the Joukowski map to

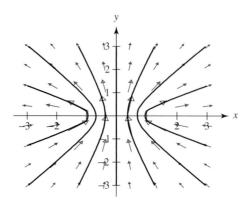

FIGURE 36.51   The streamlines and velocity vector field for flow through a slit in Example 36.50

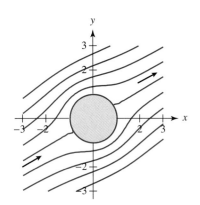

FIGURE 36.52   The streamlines for flow past a cylinder in Example 36.51

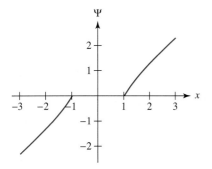

FIGURE 36.53   Example 36.51: Graph of $\psi(0, y)$, the stream function along the y-axis

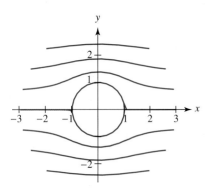

FIGURE 36.54   Example 36.51: Streamlines when $\lambda = 0$ and $A = \sigma = 1$

flatten the cylinder to a line segment. The flow in the w-plane is brought back to the z-plane through the ζ-plane.

In the z-plane, first rotate the flow by the angle $-\lambda$ so that the velocity vector, far from the origin, will be just A. This is accomplished by the conformal map $\zeta = f(z) = ze^{-i\lambda}$. The circle $|z| = \sigma$ maps to the circle $|\zeta| = \sigma$. Map the ζ-plane to the w-plane with the conformal map $w = g(\zeta) = \zeta + \frac{\sigma^2}{\zeta}$, which takes the circle $|\zeta| = \sigma$ to the slit $-2\sigma \le u \le 2\sigma, v = 0$, in the w-plane. This slight generalization of the Joukowski map appeared in Section 36.7. The rest of the ζ-plane is mapped two-to-one onto the w-plane.

The complex potential for flow past the slit in the w-plane is then $G(w) = Aw$, so the complex potential in the ζ-plane is

$$G_1(\zeta) = G(w(\zeta)) = G(g(\zeta))$$

and the complex potential in the z-plane is

$$F(z) = G(g(f(z))) = A\left(e^{-i\lambda}z + \frac{\sigma^2 e^{i\lambda}}{z}\right)$$

Since $V = \overline{F'(z)} = Ae^{-i\lambda} - \sigma^2 e^{i\lambda}/z^2$, we have $V_\infty = Ae^{i\lambda}$. Determined by $F'(z) = 0$, there are two stagnation points, $\pm\sigma e^{i\lambda}$, both on the circle $|z| = \sigma$, and both diametrically opposite each other. Writing $F(z)$ as $F(x + iy) = \phi(x, y) + i\psi(x, y)$, the real and imaginary parts are, respectively,

$$\phi = A(x\cos\lambda + y\sin\lambda)\left(1 + \frac{\sigma^2}{x^2 + y^2}\right) \quad \text{and} \quad \psi = (y\cos\lambda - x\sin\lambda)\left(1 - \frac{\sigma^2}{x^2 + y^2}\right)$$

On the circle $|z| = \sigma$, the stream function $\psi(x, y)$ has the value zero. Figure 36.52 shows the flow for the case $A = 1, \lambda = \frac{\pi}{6}, \sigma = 1$. The streamlines are the level curves of $\psi(x, y)$, the stream function, and the two representative velocity vectors superimposed on the flow are obtained from $\nabla\phi(x, y)$.

Each streamline crossing the y-axis outside the circle $|z| = 1$ is a contour along which the stream function $\psi(x, y)$ maintains a constant but different value. The variation of the stream function along the y-axis is shown Figure 36.53.

When $\lambda = 0$, the stream function simplifies to $\psi_h = Ay(1 - \frac{\sigma^2}{x^2 + y^2})$. Then, for the case $A = 1, \sigma = 1$, this further simplifies. If polar coordinates are introduced, this simplified stream function becomes $\psi_{hp} = \frac{r^2 - 1}{r}\sin\theta$, and Figure 36.54 shows the resulting flow. ❖

EXAMPLE 36.52

Circulant flow past a cylinder   If flow past the disk $|z - a| = \sigma$ has nonzero circulation c and $V_\infty = Ae^{i\lambda}$, its complex potential (as found, e.g., in [69]), is

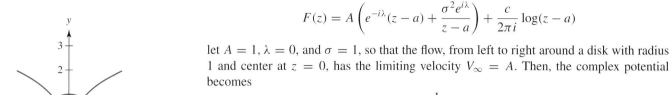

$$F(z) = A\left(e^{-i\lambda}(z - a) + \frac{\sigma^2 e^{i\lambda}}{z - a}\right) + \frac{c}{2\pi i}\log(z - a)$$

let $A = 1$, $\lambda = 0$, and $\sigma = 1$, so that the flow, from left to right around a disk with radius 1 and center at $z = 0$, has the limiting velocity $V_\infty = A$. Then, the complex potential becomes

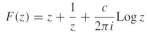

$$F(z) = z + \frac{1}{z} + \frac{c}{2\pi i}\operatorname{Log} z$$

where we have switched to the principal logarithm. For any contour C enclosing the disk $|z| = 1$ in its interior, the flow has circulation c. For example, if C is the circle $z(t) = 2e^{it}$, $0 \le t \le 2\pi$, then

$$Q = \int_C F'(z)\,dz = \frac{i}{2\pi}\int_0^{2\pi}(4\pi e^{it} - \pi e^{-it} - ic)\,dt = c$$

On the other hand, the circulation around a contour that does not contain the disk $|z| = 1$ in its interior will be zero. For example, around the circle $z(t) = 3 + e^{it}$, $0 \le t \le 2\pi$, we find

$$Q = \int_C F'(z)\,dz = i\int_0^{2\pi}\left(1 + \frac{c}{2\pi i}\frac{1}{3 + e^{it}} - \frac{1}{(3 + e^{it})^2}\right)e^{it}\,dt = 0$$

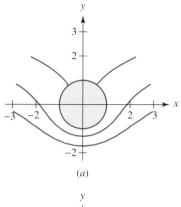

Writing $F(z)$ in polar coordinates as $F(re^{(i\theta)}) = \phi(r, \theta) + i\psi(r, \theta)$, we find the real and imaginary parts to be, respectively,

$$\phi = \left(r + \frac{1}{r}\right)\cos\theta + \frac{c}{2\pi}\arctan(r\sin\theta, r\cos\theta) \quad \text{and} \quad \psi = \left(r - \frac{1}{r}\right)\sin\theta - \frac{c}{2\pi}\ln r$$

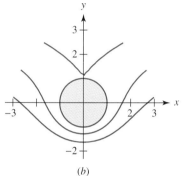

Figure 36.55 shows the flow for two values of c. On the left, $c = 10$, while on the right, $c = 12.6$; thus suggesting the stagnation points on the disk move as c is increased, even to the extent that the stagnation points move off the boundary of the disk!

As a function of the circulation c, the location of the stagnation points is determined by solving the equation $F'(z) = 0$, giving the two solutions $s_\pm = \frac{i}{4\pi}(c \pm \sqrt{c^2 - 16\pi^2})$. On the boundary of the disk, the stagnation points are distinct except when $c = \pm 4\pi$. (For a disk of radius σ, the values of c for which the two stagnation points coalesce are found to be $\pm 4\pi\sigma$, as dealt with in the exercises.) In our example, for $c > 4\pi$, the stagnation point at s_+ is actually on the positive imaginary axis and above the disk. If, say $c = 5\pi$, then the stagnation point is at $s_+ = 2i$.

The variation of the stream function $\psi(r, \theta)$ on contours above the disk can be seen in the graph of $\psi(r, \frac{\pi}{2})$ with $c = 5\pi$, shown in Figure 36.56. The vertical axis represents, when the stagnation point is at $z = 2i$, values of the stream function on streamlines crossing the y-axis above the disk $|z| = 1$. On the surface of the disk, the stream function has the value zero. Then, the values of ψ become negative, are negative at the stagnation point, and don't become positive until well above the stagnation point.

At the stagnation point above the disk, $\psi = \psi_{\text{crit}} = \frac{1}{2}(3 - 5\ln 2) \stackrel{\circ}{=} -0.23$. Set $c = 5\pi$ and solve $\psi(r, \theta) = \psi_{\text{crit}}$ for $\theta = \theta(r) = \arcsin\frac{r(3 + 5\ln(r/2))}{2(r^2 - 1)}$ as one branch of the arcsine function and $\pi - \theta(r)$ as the other. Figure 36.57 shows the streamlines through the stagnation point above the disk. The curve in color is the branch defined by $\theta(r)$; the black, by $\pi - \theta(r)$.

❖

FIGURE 36.55   Example 36.52: (a) streamlines for $c = 10$ and (b) for $c = 12.6$

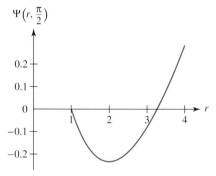

FIGURE 36.56   Example 36.52: Graph of $\psi(r, \frac{\pi}{2})$, the stream function on the y-axis above the cylinder

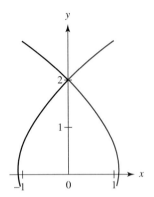

FIGURE 36.57 Example 36.52: Streamlines through the stagnation point when $c = 5\pi$ (branch defined by $\theta(r)$ is in color, while the branch defined by $\pi - \theta(r)$ is in black)

EXAMPLE 36.53

Flow past a line segment   In the plane of $z = x + iy$, the complex potential $F(z)$ for a fluid flowing past the line segment $x = 0$, $-1 \le y \le 1$, and having the limiting velocity $V_\infty = Ae^{i\lambda}$ is found by a series of two mappings. First, the z-plane is mapped to the ζ-plane by the function $\zeta = f(z) = -2iz$, which rotates the segment on the imaginary z-axis to the real axis in the ζ-plane. In addition, the segment is stretched so that it lies between $\zeta = \pm 2$. Then, the ζ-plane is mapped to the w-plane by the function $w = g(\zeta) = \frac{1}{2}(\zeta + \sqrt{\zeta^2 - 4})$, which takes the segment between $\zeta = \pm 2$ to the circle $|w| = 1$. The observant reader will recognize this as the inverse of the Joukowski map.

Circulation-free flow past this disk in the w-plane has the complex potential $G(w) = B(e^{-i\tau}w + e^{i\tau}/w)$. In the z-plane, the complex potential $F(z)$ is given by the composition

$$F(z) = G(w(\zeta(z))) = G(g(f(z)))$$

$$= B\left(e^{-i\tau}(-iz + \sqrt{-z^2 - 1}) + \frac{e^{i\tau}}{-iz + \sqrt{-z^2 - 1}}\right)$$

$$= 2B(\sqrt{z^2 + 1}\sin\tau - iz\cos\tau)$$

The parameters $B = \frac{A}{2}$ and $\tau = \frac{\pi}{2} - \lambda$ are determined by the condition

$$Ae^{i\lambda} = V_\infty = \lim_{z\to\infty} \overline{F'(z)} = 2B(\sin\tau + i\cos\tau)$$

Hence, we have $F(z) = A(\sqrt{z^2 - 1}\cos\lambda - iz\sin\lambda)$.

At this point, we have typically written $F(z)$ as $F(x + iy) = \phi(x, y) + i\psi(x, y)$ and obtained, by direct computation, the real and imaginary parts of $F(z)$. However, the presence of the square root would force us to determine a proper branch cut, and in this instance, the principal square root turns out to be inappropriate. Instead of struggling with the definition of this branch cut, we employ an alternate device that works here because the radical can be isolated.

Set $A = 1$ and $F(z) = \phi + i\psi$, isolate the radical, square both sides, replace z by $x + iy$, and equate real and imaginary parts to obtain the two equations

$$1 + x^2 - y^2 = -\frac{1}{\cos^2\lambda}\left[(\psi + (x + y)\sin\lambda - \phi)(\psi + (x - y)\sin\lambda + \phi)\right]$$

$$2xy = \frac{2}{\cos^2\lambda}\left[(\psi + x\sin\lambda)(\phi - y\sin\lambda)\right]$$

which can be solved for $\phi = \phi(x, y, \lambda)$ and $\psi = \psi(x, y, \lambda)$. The resulting expressions are large and cumbersome, but for $\lambda = 0$ the relevant solutions are

$$\psi(x, y, 0) = \frac{1}{\sqrt{2}}\sqrt{\alpha + \sqrt{\beta_+\beta_-}} \quad \text{and} \quad \phi(x, y, 0) = \frac{xy}{\psi(x, y, 0)}$$

where $\alpha = y^2 - x^2 - 1$, $\beta_\pm = x^2 + (y \pm 1)^2$. In the upper half-plane, the complex potential is $F(z) = \phi + i\psi$, and in the lower half-plane it is $-F(z)$. Figure 36.58 shows the streamlines for $\lambda = \frac{1}{2}$. The accompanying Maple worksheet contains animations of the flow field and of the velocity vectors as λ varies from 0 to 1. ❖

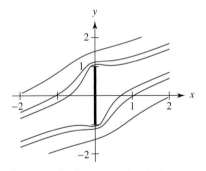

FIGURE 36.58 Example 36.53: Streamlines when $\lambda = \frac{1}{2}$

EXAMPLE 36.54

Flow down a step   Flow with complex potential $F(z) = Az$ in the z-plane is mapped by the function $w = f(z) = \frac{1}{\pi}(\sqrt{z^2 - 1} + \text{arccosh}\,z)$ to flow down a step in the w-plane. This is an example where the simpler flow in the z-plane is mapped to a more complicated flow in the w-plane. Because of the algebraic difficulty of writing an expression for the inverse map from the w-plane back to the z-plane, the flow is studied in the w-plane.

In $f(z)$, use the definition

$$\sqrt{z^2 - 1} = \sqrt{|z^2 - 1|} e^{i/2[\text{Arg}(z+1) + \text{Arg}(z-1)]}$$

so that the square root is discontinuous (has its branch cut) along $-1 \le x \le 1$, $y = 0$ on the real axis. In addition, this root will map the half-ray $x < -1$, $y = 0$ to the half-ray $u < 0$, $v = 1$ in the w-plane. (The principal square root would map the same half-ray to the half-ray $u > 0$, $v = 1$ in the w-plane. The accompanying Maple worksheet contains an extended discussion of the behavior of the square root function in this example.) Figure 36.59 shows the image, in the w-plane, of the real axis in the z-plane. The real axis becomes the "step down" in the w-plane.

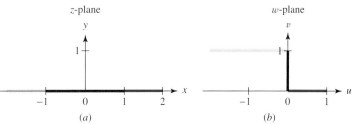

FIGURE 36.59   The mapping $f(z)$ in Example 36.54

Analytically, the complex potential for flow in the w-plane is $G(w) = F(z(w))$, where $F(z) = Az$. However, given the complexity of the mapping $w = f(z)$, it is difficult to obtain an expression for $z = z(w)$. If such an expression were available, the complex potential in the w-plane would then be $G(w) = Az(w)$. Figure 36.60 shows the streamlines in the w-plane, images of the lines $y = $ constant in the z-plane.   ❖

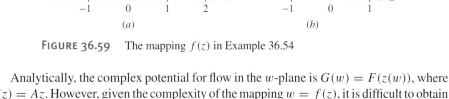

FIGURE 36.60   Streamlines in Example 36.54

EXAMPLE 36.55   **Flow over the Joukowski airfoil**   Let C be the circle $|z - a| = \sigma$, passing though $z = k$ and containing $z = -k$ in its interior. The Joukowski map $w = z + \frac{k^2}{z}$ transforms C to an airfoil. Hence, flow past circle C in the z-plane is carried by the Joukowski map to flow past the airfoil in the w-plane. The complex potential

$$F(z) = A\left(e^{-i\lambda}(z - a) + \frac{\sigma^2 e^{i\lambda}}{z - a}\right) + \frac{c}{2\pi i} \log(z - a)$$

mapped to the w-plane becomes the complex potential $G(w) = F(z(w))$. Thus, the Joukowski map must be inverted to provide $z = z(w)$, which is then substituted into $F(z)$. The flow around the airfoil is then studied in the w-plane and is not mapped back to the z-plane.

For example, on the left in Figure 36.61 is $z(t) = a + \sigma e^{it}$, the circle C through $z = k = 1$, with $a = -\frac{1}{9}(3 - 4i\sqrt{3})$ and $\sigma = \frac{8}{9}\sqrt{3}$. On the right is the image of the circle under $w = z + \frac{1}{z}$, the Joukowski map, with $k = 1$.

The complex potential for circulant flow past the circle C, with limiting velocity $V_\infty = A = 1$, is

$$F(z) = \left(z - a + \frac{\sigma^2}{z - a}\right) + \frac{c}{2\pi i} \text{Log}(z - a)$$

where now, we use the principal logarithm and the definitions of a and σ given above. The streamlines for the flow past the airfoil in the w-plane are the images of the streamlines for the flow past C in the z-plane. To get the streamlines in the z-plane, write $F(z)$ as $F(x+iy) = \phi(x,y) + i\psi(x,y)$, where

$$\phi = \beta_1 + \frac{64}{27}\frac{\beta_1}{\beta_1^2 + \beta_2^2} + \frac{c}{2\pi}\arctan(\beta_2, \beta_1)$$

$$\psi = \beta_2 - \frac{64}{27}\frac{\beta_2}{\beta_1^2 + \beta_2^2} - \frac{c}{4\pi}\ln(\beta_1^2 + \beta_2^2)$$

with $\beta_1 = x + \frac{1}{3}$, $\beta_2 = y - \frac{4}{9}\sqrt{3}$. Figure 36.62 shows the streamlines, the level curves of ψ, when $c = 1$.

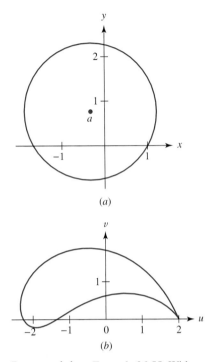

(a)

(b)

FIGURE 36.61   Example 36.55: With $k = 1$, the Joukowski map takes the cylinder in (a) to the airfoil in (b)

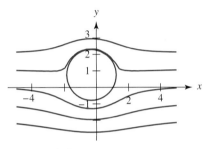

FIGURE 36.62   Example 36.55: For $c = 1$, streamlines in the z-plane

Write the Joukowski map $w = z + \frac{1}{z}$ as $w(x+iy) = u + iv$, where $u = x + \frac{x}{x^2+y^2}$ and $v = y - \frac{y}{x^2+y^2}$. Transform each point in Figure 36.62 by the function $T : (x, y) \to (u(x, y), v(x, y))$, thereby obtaining Figure 36.63, showing the image in the w-plane of the flow in the z-plane.   ❖

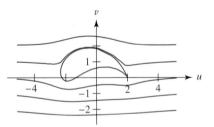

FIGURE 36.63   Example 36.55: For $c = 1$, streamlines in the w-plane

EXERCISES 36.10–Part A

A1. Define $w = f(z) = \sqrt{z-1} = u + iv$ using the principal square root. Show that $f(z)$ is continuous across the positive real axis. Then show that with *this* definition, $f(z)$ transforms $y^2 = 4x$ to $v = \pm 1$. *Hint:* Write $z = x + iy$, and obtain $(x - 1) + iy =$

$u^2 - v^2 + 2uvi$. Equate real and imaginary parts, set $x = \frac{y^2}{4}$, and eliminate u.

A2. In Example 36.47, with $A = 1$, evaluate $\nabla\phi$ at $(x, y) = (-1, -1)$, $(-1, 0)$, and $(-1, 1)$.

A3. The two-argument arctangent function appears in Example 36.48. Verify that $\frac{\partial}{\partial x}\arctan(y, x) = \frac{\partial}{\partial x}\arctan\frac{y}{x}$ and that $\frac{\partial}{\partial y}\arctan(y, x) = \frac{\partial}{\partial y}\arctan\frac{y}{x}$.

A4. In Example 36.48, let $a = \frac{3\pi}{4}$ and evaluate $\nabla\phi$ at $(x, y) = (1, 1)$ when $A = 1$.

A5. In Example 36.49, evaluate $\nabla\phi$ at $(x, y) = (\frac{\pi}{4}, \frac{1}{2})$ when $A = 1$.

A6. In Example 36.51, let $\sigma = 1$, $A = 1$, $\lambda = \frac{\pi}{6}$ and evaluate $\nabla\phi$ at $(x, y) = (0, \frac{3}{2})$.

A7. In Example 36.53, evaluate $\psi(0, 1.2, 0)$ and $\psi(0, -1.2, 0)$ when $A = 1$.

A8. In Example 36.55, let $c = 1$, $A = 1$ and evaluate $\psi(0, \frac{5}{2})$.

EXERCISES 36.10–Part B

B1. Define $s(z) = \sqrt{|z|}e^{i\operatorname{Arg}(z)/2}$, where $-\pi < \operatorname{Arg} z \le \pi$. Provide numerical, graphical, or analytical evidence, showing that $s(z)$ is just $\sqrt{z}$, the principal square root.

B2. Define $S(z) = \sqrt{|z|}e^{i\arg(z)/2}$, where $0 \le \arg z < 2\pi$.

   (a) Show that $S(z)$ is discontinuous across the positive real axis.

   (b) Show that $\arg z = \begin{cases} \operatorname{Arg} z & \mathcal{I}(z) \ge 0 \\ 2\pi + \operatorname{Arg} z & \mathcal{I}(z) < 0 \end{cases}$ and that

   $S(z) = \begin{cases} s(z) & \mathcal{I}(z) \ge 0 \\ -s(z) & \mathcal{I}(z) < 0 \end{cases}$, where $s(z)$ is as defined in Exercise B1.

   (c) With this definition of the square root, show $\sqrt{z-1}$ maps $y^2 = 4x$ to $v = 1$, not $v = \pm 1$.

B3. Verify that the streamlines shown in Figure 36.44 can be found as the contours of the function $-2\mathcal{I}(F(z-1))$. Thus, by distinguishing between the upper and lower half-planes, the ordinary square root can be used to define this flow.

B4. As a function of y, graph $\psi(-2, y)$, the stream function whose streamlines are shown in Figure 36.46.

B5. Obtain and analyze the flow against the parabola $y^2 = x$. Graph the streamlines, the equipotentials, and the velocity vector field.

B6. Obtain and analyze the flow against the wedge $-\frac{\pi}{6} \le \operatorname{Arg} z \le \frac{\pi}{6}$. Graph the streamlines, the equipotentials, and the velocity vector field.

B7. Obtain and analyze the complex potential for flow into the closed channel $-3 \le x \le 3$, $y \ge 0$. Graph the streamlines, the equipotentials, and the velocity vector field.

B8. Obtain and analyze the complex potential for flow through the slit $-2 \le x \le 2$, $y = 0$. Graph the streamlines, the equipotentials, and the velocity vector field.

B9. Define $\sigma(z) = \sqrt{|z^2 - 8|}e^{i[\operatorname{Arg}(z - 2\sqrt{2}) + \operatorname{Arg}(z + 2\sqrt{2})]/2}$, where $-\pi < \operatorname{Arg} z \le \pi$.

   (a) Show that $\sigma(z)$ is discontinuous on the real axis for $-2\sqrt{2} \le \mathcal{R}(z) \le 2\sqrt{2}$.

   (b) Show that $\sigma(z) = \begin{cases} s(z^2 - 8) & \mathcal{R}(z) > 0 \\ -s(z^2 - 8) & \mathcal{R}(z) < 0 \end{cases}$, where $s(z)$ is as defined in Exercise B1.

   (c) Show that $w = f(z) = \frac{1}{4}(z + \sigma(z))$ maps the exterior of the ellipse $x^2 + 9y^2 = 9$ to the exterior of the unit circle in the w-plane.

   (d) Show that $z = g(w) = 2w + \frac{1}{w}$ is the inverse of $f(z)$ and that $g(w)$ maps the exterior of the unit circle in the w-plane to the exterior of the ellipse $x^2 + 9y^2 = 9$ in the z-plane.

   (e) Since $G(w) = A[e^{-i\lambda}w + e^{i\lambda}/w]$ is the complex potential for flow around a circle in the w-plane, obtain $F(z) = G(f(z))$ as the complex potential for flow around the ellipse $x^2 + 9y^2 = 9$ in the z-plane.

   (f) If $A = 1$, obtain a graph of the streamlines (contours of $\psi(x, y) = \mathcal{I}(F(x + iy))$) for each of $\lambda = 0, -\frac{\pi}{24}, -\frac{\pi}{12}, -\frac{\pi}{6}$.

   (g) Show that the ellipse $x^2 + 9y^2 = 9$ is the streamline for which $\psi(x, y) = 0$.

B10. Obtain the complex potential for circulant flow past the ellipse in Exercise B9. Obtain graphs of the streamlines for the case $A = 1$ and $\lambda = 0$, at various values of c, the circulation constant. Determine the location of any stagnation points. How do the stagnation points vary with c? Can the stagnation points be made to move off the ellipse?

B11. Define $r(z) = \sqrt{|z^2 + 1|}e^{i[\arg(z - i) + \arg(z + i)]/2}$, where $-\frac{\pi}{2} < \arg(z \pm i) \le \frac{3\pi}{2}$.

   (a) Show that $r(z)$ is continuous except on the imaginary axis between $\pm i$ where it has a branch cut.

   (b) Show that $\arg(z - i) = \begin{cases} \operatorname{Arg}(z - i) - 2\pi & x < 0, y < 1 \\ \operatorname{Arg}(z - i) & \text{else} \end{cases}$.

   (c) Show that $\arg(z + i) = \begin{cases} \operatorname{Arg}(z + i) - 2\pi & x < 0, y < -1 \\ \operatorname{Arg}(z + i) & \text{else} \end{cases}$.

   (d) Show that

   $r(z) = \begin{cases} \sqrt{|z^2 + 1|}e^{i[\operatorname{Arg}(z - i) + \operatorname{Arg}(z + i) - 2\pi]/2} & x < 0, |y| < 1 \\ \sqrt{|z^2 + 1|}e^{i[\operatorname{Arg}(z - i) + \operatorname{Arg}(z + i)]/2} & \text{else} \end{cases}$.

   (e) Show that $r(z) = \begin{cases} s(z^2 + 1) & \mathcal{R}(z) > 0 \\ -s(z^2 + 1) & \mathcal{R}(z) < 0 \end{cases}$, where $s(z)$ is as defined in Exercise B1.

B12. The complex potential for flow past the finite segment between $\pm i$ on the imaginary axis was shown to be $F(z) = \cos\lambda\sqrt{z^2 + 1} - iz\sin\lambda$. For $\lambda = \frac{\pi}{12}, \frac{\pi}{6}$, and $\frac{\pi}{4}$, obtain a graph of the streamlines as the contours of $\mathcal{I}(F(x + iy))$, where $r(z)$ is as given in Exercise B11(e).

B13. In the upper half-plane, the flow whose complex potential is $F(z) = z$ is mapped to the w-plane by the function given in Exercise 25, Section 36.6. Graph, in the w-plane, the images of the streamlines $y = $ constant in the z-plane. Interpret the flow in the w-plane.

B14. In the upper half-plane, the flow whose complex potential is $F(z) = z$ is mapped to the w-plane by the function given in Exercise 26, Section 36.6. Graph, in the w-plane, the images of the streamlines $y = $ constant in the z-plane. Interpret the flow in the w-plane.

B15. In the upper half-plane, the flow whose complex potential is $F(z) = z$ is mapped to the w-plane by the function given in Exercise 28, Section 36.6. Graph, in the w-plane, the images of the streamlines $y = $ constant in the z-plane. Interpret the flow in the w-plane.

B16. In the first quadrant of the z-plane, the flow whose complex potential is $F(z) = z^2$ is mapped to the w-plane by the function given in Exercise 27, Section 36.6. Graph, in the w-plane, the images of the streamlines in the z-plane. Interpret the flow in the w-plane.

Chapter Review

1. For what class of functions $F(\alpha, \beta)$ will the substitution $z = e^{i\theta}$ transform the integral $\int_0^{2\pi} F(\cos\theta, \sin\theta)\, d\theta$ to one for which the residue theorem applies? Show how to make the transformation.

2. Using the technique detailed in Question 1, evaluate $\int_0^{2\pi} \frac{d\theta}{2+\cos\theta+\sin\theta}$.

3. What conditions on $f(x)$ will allow the integral $\int_{-\infty}^{\infty} f(x)\, dx$ to be evaluated by contour integration? Describe the process by which the contour is closed and the integral evaluated.

4. Evaluate $\int_{-\infty}^{\infty} \frac{x^2-1}{(x^2+1)(x^2+4)}\, dx$ by the technique detailed in Question 3.

5. What conditions on $f(x)$ will allow the integral $\int_{-\infty}^{\infty} e^{imx} f(x)\, dx$ to be evaluated by contour integration? Describe the process by which the contour is closed and the integral evaluated.

6. Evaluate $\int_{-\infty}^{\infty} \frac{xe^{-2x}\, dx}{x^2+1}$ by the technique detailed in Question 5.

7. Use residues and an indented contour to obtain the Cauchy principal value of $\int_{-\infty}^{\infty} \frac{x\, dx}{(x^2+1)(x-3)}$.

8. Obtain the Cauchy principal value of $\int_0^{\infty} \frac{x^{1/3}\, dx}{(x^2+1)(x-1)}$.

9. Invert the Laplace transform $F(s) = \frac{1}{s^2+1}$ by evaluating the Bromwich inversion formula via residues.

10. Obtain the exponential form of the Fourier series for $f(t) = 1 - t^2$, $-1 \le t \le 1$. Show it is equivalent to the trigonometric form.

11. Obtain the roots of $x^2 - (c+5)x + 10 = 0$, and plot them as a function of the parameter c. Then plot the locus of roots in the complex plane.

12. How many times does the contour C, given by $z(t) = (2 + 5\sin t)e^{it}$, $0 \le t \le 2\pi$, wrap around the point $2i$? Verify by computing the winding number $v(C, 2i)$.

13. State the *principle of the argument*. Illustrate it with the function $G(z) = \frac{z}{z-1}$ and a circle with radius 2 and center at the origin.

14. What is the extended principle of the argument? Explain how it can be used to determine the number of zeros a polynomial has within a

contour C. Illustrate with the polynomial $z^2 + z + 1$ and the contour C, which is the unit circle centered at the origin.

15. What is the *conformal property* of an analytic function?

16. The linear fractional transformation $w = \frac{az+b}{cz+d}$ is completely specified by how it maps three distinct points. Obtain the LFT that maps $z = 0, 1, i$ to $w = i, 0, 1$.

17. Verify that the LFT $f(z) = \frac{2z+1}{z-1}$, with pole at $z = 1$, has as its inverse, another LFT with a pole at $w = 2$.

18. Verify that the LFT in Question 17 maps:

  (a) $z(t) = t + (t-1)i$, a line through the pole $z = 1$, to a line through the pole $w = 2$.

  (b) $z(t) = t + i$, a line not through the pole $z = 1$, to a circle through the pole $w = 2$.

  (c) $z(t) = e^{it}$, a circle through the pole $z = 1$, to a line not through the pole $w = 2$.

  (d) $z(t) = 2e^{it}$, a circle not through the pole $z = 1$, to a circle not through the pole $w = 2$.

19. State the Joukowski map and describe what it does to the z-plane.

20. Describe how to create airfoils with the Joukowski map.

21. Describe how to obtain the inverse of the Joukowski map.

22. Describe how the mapping $w = z^2$ is used to solve the Dirichlet problem on the first quadrant of the xy-plane when the prescribed boundary values are $s(x, 0) = a$ along the x-axis and $s(0, y) = b$ along the y-axis.

23. For a plane fluid flow, what is the *complex potential*? What is the *complex velocity*? How does the complex potential capture the *velocity field* and the *streamlines* (which are the flow lines of the velocity field)?

24. Why must a streamline for an incompressible and frictionless flow fall on a boundary?

25. How are the *flux* and *circulation* for a plane flow computed from the complex potential?

26. Describe the flow whose complex potential is:

   (a) $F(z) = Ae^{i\lambda}z$, $A > 0$.   **(b)** $F(z) = Az^2$, $A > 0$.

   (c) $F(z) = A\log(z - a)$, $A > 0$.

   (d) $F(z) = -iB\log(z - a)$, $B > 0$.

27. Describe how the mapping $w = \sqrt{z - 1}$ is used to solve for the flow around the parabola $y^2 = 4x$. Show how the mapping takes the parabola to a simpler geometry in the w-plane where the flow is easily obtained. Then show how mapping the flow back to the z-plane provides the solution to the original problem.

28. Describe how the mapping $w = \sqrt{z - 1} - i$ maps flow against the parabola $y^2 = 4x$ to a simpler flow in the w-plane where the complex potential is more easily obtained. Then show how mapping the flow back to the z-plane gives the desired solution to the original flow problem.

29. Describe how the mapping $w = \sin z$ is used to solve for the flow into the closed channel $-\frac{\pi}{2} \leq \mathcal{R}(z) \leq \frac{\pi}{2}$.

30. Describe how the problem of planar flow past a cylinder is solved by conformal mapping to a simpler geometry.

31. Describe how the problem of planar flow down a step is solved by mapping a simpler flow to the desired "complicated" flow. The technique in this case is therefore not the same as the technique used in Questions 27–30.

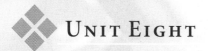

Numerical Methods

As it stands, Unit Eight is by far the longest unit in the text. However, the attention devoted to numeric computations is by no means unwarranted. Problems solvable "by hand" and even by closed-form analytic techniques are a minuscule portion of the spectrum of engineering and scientific analysis. By far, the greatest portion of modern applications requires numeric computation of some form. For as much attention as we have given numeric methods, we have barely scratched the surface.

The calculators and computers with which we typically execute numerical methods are fixed-precision machines. Their architecture limits the number of digits with which each number can be expressed. This is a source of roundoff error in all numeric work. Thus, we must begin with a discussion of accuracy so that we can determine if roundoff has invalidated a computation. We must also examine the way roundoff error occurs and what steps can be taken to prevent or mitigate it.

Numeric algorithms are judged by their ability to provide accurate results with a minimum of computational effort. Viable iterative methods generate a sequence of values that converge to the correct result. The rate at which this convergence occurs is an important characteristic of all such methods.

Up to this point in the text, we have freely made use of packaged software for numerically solving algebraic or differential equations and evaluating integrals. In this unit we discuss methods for solving single algebraic equations and systems, both linear and nonlinear, stressing how to select, use, and assess such solvers. A generation ago, use of a numerical method required the ability to write computer programs. Today, the typical user of a numerical solver will invoke a professionally written package and will not write a computer program. However, today's consumer of numerical solvers will still need to understand how such solvers work, what their strengths and weaknesses might be, and how to interpret their results.

We then take up the study of methods for passing smooth curves through discrete data points. Closely related to this problem of interpolation is our discussion of techniques for approximating continuous functions. In a sense, if discrete data points are selected along the graph of an unknown function, the approximation problem becomes the interpolation problem. This is the problem of approximating discrete data. We can pass smooth curves through each of the data points, either a single curve or a piecewise composite of arcs that form a "spline." Alternatively, we can try to represent the data points with a least-squares curve, one that passes as near to as many of the data points as possible.

There are two chapters on numeric calculus, both differentiation and integration. Our discussion of numeric differentiation introduces the very important Richardson extrapolation algorithm that takes two approximations, one less accurate, one more accurate, and from them, extrapolates a "best" approximation.

Numeric integration techniques such as the Trapezoid rule and Simpson's rule are generally staples of the elementary calculus course. Hence, we review these equispaced methods and embed them in the general family of Newton–Cotes quadrature formulas. We describe how an adaptive Simpson's rule would be constructed, and we discuss the Gauss–Legendre quadrature formulas that are not equispaced. Error is minimized by varying the nodes at which the integrand is sampled!

We conclude the unit with a chapter devoted to the numeric computation of eigenvalues. None of the methods used for this task begin with the characteristic polynomial. Early in the unit the problem of computing the zeros of a polynomial is shown to be inherently unstable. Modern methods for computing eigenvalues are iterative and generally involve results from the matrix algebra we studied in Unit Six.

Chapter 37

Equations in One Variable —Preliminaries

I N T R O D U C T I O N   We begin our study of numerical mathematics with a discussion of *accuracy* and *error.* When any algorithm yields a numeric answer, the shrewd practitioner immediately asks "how accurate is this answer?" It is not reasonable to compute to 20 decimal places an answer that cannot possibly be accurate to more than three. Consequently, in a world where access to software tools is so prevalent, the user must bring to computation the ability to understand and interpret the outcomes of numeric calculations.

A first step in this understanding and interpretation is distinguishing between a numeric answer being correct to a specified number of *significant digits* or being correct to a specified number of *decimal places.* Typically, computation is done on hardware devices that can compute with only a fixed number of digits. This limitation generates *roundoff errors,* errors attributable to the *finite precision* of the computing devices being used. Users of computer algebra systems that, in software, emulate machines of arbitrary precision cannot ignore the issue of roundoff error by deciding that the cure is simply to use more digits. No matter the precision with which a computation is executed, the potential for roundoff always exists and must be understood and dealt with.

We use root-finding algorithms to discuss the *rate of convergence* of an iterative process to its limit. It is not enough that a method eventually provides an answer. It must do so rapidly. Not only can it be expensive to perform excessive calculation, but the additional arithmetic can make the accumulation of roundoff error more likely. Hence, the notion of the rate of convergence is the focus of a discussion in which the terms *linear, superlinear,* and *quadratic* convergence appear.

37.1   Accuracy and Errors

Introduction

Although the chapter on numerical methods comes late in this text, we have not eschewed using numerical computations because we have relied on computer algebra systems to supply easily accessed numerical routines. Thus, we have solved differential equations, both ordinary and partial, computed eigenvalues and eigenvectors, and solved both linear and nonlinear algebraic equations by invoking numeric algorithms built into modern software packages.

This book is predicated on the conviction that the typical student will not write code for implementing numerical methods. The prevalence of readily available high-quality code suggests that writing and debugging code for simple tasks like finding roots is akin to computing square roots "by hand." However, we recognize the value of having the student understand the underlying algorithms and of having the student gain an awareness of the potential problems that routinely arise during numeric computations.

EXAMPLE 37.1 Lest the student falsely believe that access to a numeric solver permits thinking to cease, consider the following example tailored after a similar one found in [98] as reported in [2]. Start with the polynomial $f(x) = \prod_{k=1}^{20}(x - k)$, whose roots are exactly the integers $1, 2, \ldots, 20$. In expanded form, the polynomial is

$$
\begin{aligned}
f(x) = {} & x^{20} - 210x^{19} + 20615x^{18} - 1256850x^{17} + 53327946x^{16} - 1672280820x^{15} \\
& + 40171771630x^{14} - 756111184500x^{13} + 11310276995381x^{12} \\
& - 135585182899530x^{11} + 1307535010540395x^{10} - 10142299865511450x^{9} \\
& + 63030812099294896x^{8} - 311333643161390640x^{7} + 1206647803780373360x^{6} \\
& - 3599979517947607200x^{5} + 8037811822645051776x^{4} - 12870931245150988800x^{3} \\
& + 13803759753640704000x^{2} - 8752948036761600000x + 2432902008176640000
\end{aligned}
$$

Subtract $2^{-23} = 0.1192092896 \times 10^{-6}$ from -210, the coefficient of x^{19}, a small perturbation, indeed. The zeros of the perturbed polynomial, computed to 30 digits and rounded to 25 places, are listed in Table 37.1.

0.9999999999999999999999990	2.00000000000000000009762004
2.99999999999805232975829	4.000000000261023189249615
4.9999999275515378879647781	6.0000069439522985742482479
6.99969723393588907378813	8.00726760345344220917021
8.91725024848811866085120	
10.0952661450674392149903 ± 0.6435009039623016783251460i	
11.793633881014642158676 14 ± 1.652329728287250339820714i	
13.99235813719123104650426 ± 2.5188300697937639492740 17i	
16.7307374661039805550417 0 ± 2.8126248944615912593629 97i	
19.50243940059550307499439 ± 1.940330346820607238200480i	
20.84690810162158266985602	

TABLE 37.1   Zeros of the perturbed polynomial in Example 37.1

A number of complex zeros have appeared, some of which are at a considerable distance from a real integer, as can be seen from Figure 37.1, a plot, in the complex plane, of the zeros listed in Table 37.1. We have accurately computed the zeros of the perturbed polynomial that is very near the original polynomial, but the zeros of each are considerably separated.

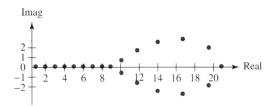

FIGURE 37.1   Zeros of the perturbed polynomial in Example 37.1

We must conclude that the roots of a polynomial are extremely sensitive to small changes in the coefficients. This sensitivity is not caused by the solution technique but is inherent in the polynomial.

The perturbation in a coefficient could easily have been introduced by the very act of entering the polynomial into a computing device that stores the coefficients with finite precision in binary arithmetic. Alternatively, there could have been roundoff in calculations leading to the stored polynomial. In either event, the associated solver then accurately finds the zeros of the perturbed polynomial. This is but one example of difficulties that can occur when computing numerically with finite-precision devices. Later in this section we will examine numeric errors that can enter calculations performed on such devices and conclude that some basic knowledge of numeric computation is essential for intelligent use of all numeric software. ❖

Rounding to k Decimal Places

To round x, a positive decimal number, to k decimal places, chop $x + 0.5 \times 10^{-k}$ after the kth decimal digit. To round a negative decimal number, round its absolute value and then restore the sign. Table 37.2 shows the effect of rounding the numbers 123.0065492 and -123.00654982 to k decimal digits.

k	123.0065492	-123.00654982
0	123	-123
1	123.0	-123.0
2	123.01	-123.01
3	123.007	-123.007
4	123.0065	-123.0065
5	123.00655	-123.00655
6	123.006549	-123.006550
7	123.0065492	-123.0065498
8	123.0065492	-123.00654982

TABLE 37.2   Rounding numbers to k decimal places

Accurate to k Decimal Places

We say that X approximates x to k decimal places provided $|x - X| < \frac{1}{2} \times 10^{-k}$. As a consequence of this definition, if both x and X are rounded to k decimal places, then the kth decimals in the rounded versions differ by no more than one unit.

For example, the two numbers $x = 12.34$ and $X = 12.387$ differ by $|x - X| = 0.047 < 0.5 \times 10^{-1} = 0.05$ so X approximates x to 1 decimal place. Rounding x and X to the $k = 1$ decimal place, we obtain $x_r = 12.3$ and $X_r = 12.4$, respectively, and x_r and X_r differ in the first decimal place by no more than one unit. When X approximates x to k decimal places, we sometimes say that these two numbers *agree* to k decimal places. However, this does not mean that both numbers are necessarily the *same* when rounded to k decimal places.

Significant Figures

The leading (or most) significant figure in a decimal number is the leftmost nonzero digit. The least significant figure is the rightmost digit. All digits in between are considered

k Significant Digits	X = 0.0026450
1	0.003
2	0.0026
3	0.00265
4	0.002645
5	0.0026450

TABLE 37.3   Significant figures

significant. Thus, zeros to the left of the first significant figure are not considered significant digits. For example, in the number $X = 0.0026450$, the first significant digit is 2 and the rightmost zero is the fifth significant digit. The result of rounding X to k significant digits is shown in Table 37.3.

Accurate to k Significant Figures

If $|x - X| < \frac{1}{2} \times 10^{-k}|x|$, or equivalently, if

$$x - \tfrac{1}{2} \times 10^{-k}|x| < X < x + \tfrac{1}{2} \times 10^{-k}|x|$$

then the floating-point number X approximates x to k significant figures. Thus, $X = x \pm \sigma$, where $\sigma < \frac{1}{2} \times 10^{-k}|x|$; so if x and X are both rounded to k significant figures, their kth significant digits will differ by at most 1. Table 37.4 illustrates this for $x = \pi = 3.141592654$. As a function of k, the number of significant digits, the error band on either side of x is then $d(k) = 0.5 \times 10^{-k}|x|$. The number of significant figures is given in the first column. The second column contains the intervals $[x - d(k), x + d(k)]$; numbers interior to these intervals approximate π to k significant digits. In the third column, the bounds on that interval are rounded to k significant digits, and finally, in the last column, π is rounded to k significant digits.

k	$[x - d(k), x + d(k)]$	Interval Rounded	x Rounded
1	[2.984513021, 3.298672287]	[3.0, 3.0]	3.0
2	[3.125884691, 3.157300617]	[3.1, 3.2]	3.1
3	[3.140021858, 3.143163450]	[3.14, 3.14]	3.14
4	[3.141435574, 3.141749734]	[3.141, 3.142]	3.142
5	[3.141576946, 3.141608362]	[3.1416, 3.1416]	3.1416
6	[3.141591083, 3.141594225]	[3.14159, 3.14159]	3.14159
7	[3.141592497, 3.141592811]	[3.141592, 3.141593]	3.141593
8	[3.141592638, 3.141592670]	[3.1415926, 3.1415927]	3.1415927

TABLE 37.4   Approximating π to k significant figures

For example, the numbers $X = 3.1415699864$ and π, rounded to 5 significant figures, are both 3.1416, but rounded to 6 significant figures, are respectively 3.14157 and 3.14159. Thus, X approximates π to 5 significant figures but not to 6. In fact, X lies in the interval [3.141576946, 3.141608362], an interval inside of which all members approximate π to 5 significant figures.

Machine Epsilon

The typical computer or calculator is a *fixed-precision* device. The number of digits the device can manipulate is fixed by its hardware configuration. Hence, numeric computing must take into account the limitations imposed by this architecture. One of these limitations is the existence of ε_M, the *machine epsilon*, the smallest positive number the device can add to 1 while recognizing the sum is different than 1.

To determine ε_M computationally, find the smallest positive ε for which $1 + \varepsilon \neq 1$. For example, if a computing device gives 1.000000001 for $1 + 10^{-9}$ but 1 for $1 + 10^{-10}$, we would conclude that $10^{-10} < \varepsilon_M \leq 10^{-9}$ and we would call the device a *10 significant-digit device*.

EXAMPLE 37.2   A more dramatic example of the effects of finite precision on a calculation is the following computation of a derivative by discrete arithmetic. Evaluation of the difference quotient $\frac{f(x+h)-f(x)}{h}$ for decreasing values of h can produce unexpected results on a finite-precision device. Table 37.5 lists the results of this calculation for the function $f(x) = e^x$, at $x = 1$, with $h = 10^{-k}$. The second column shows the results on a 10 significant-digit device, and the third column, on a 15 significant-digit device.

k	**10 digits**	**15 digits**
1	2.85884196	2.8588419548738
2	2.7319187	2.731918655787
3	2.719642	2.71964142253
4	2.71842	2.7184177470
5	2.7183	2.718295419
6	2.719	2.71828318
7	2.72	2.7182819
8	2.8	2.718281
9	3.0	2.71828
10	0	2.7182
11		2.718
12		2.718
13		2.7
14		2.0
15		0

TABLE 37.5   Example 37.2: Loss of significant figures when evaluating a difference quotient on a finite-precision computing device

As h gets smaller, the accuracy of the calculation in the second column first increases and then decreases dramatically until the value computed for $e = f'(1)$ is completely corrupted. In the third column, again, as h decreases, the approximation to $e = f'(1)$ at first improves and then worsens until it becomes completely erroneous. With an increase in the number of digits available, there is a delay in the onset of catastrophic error. However, no matter how many digits are used, at some point, as long as the calculation is performed in fixed-precision arithmetic, catastrophic error can occur.   ❖

Error Propagation

Numeric computing on fixed-precision devices is subject to *roundoff errors* attributable to the limited number of digits such machines provide. These errors are listed in Table 37.6.

The existence of a machine epsilon is the source of negligible addition. The example of numeric differentiation in the previous section illustrates negligible addition in $f(1 + h)$, subtractive cancellation in $f(1 + h) - f(x)$, and error magnification in the division by h.

To examine each of these errors in detail, define the numbers $p = 654.9581$, $q = 655.1389$, $r = 0.014897$, $s = 128761$ and $P = 655.0$, $Q = 655.1$, $R = 0.01490$, $S = 1.288 \times 10^5$, their representations when rounded to four significant digits. Table 37.7 summarizes five exact arithmetic calculations, the answers to which have been rounded to four significant digits. Table 37.8 shows those same calculations performed on a device that carries just four significant digits.

Error	Description
Negligible addition	Adding (or subtracting) two numbers of sufficiently different magnitudes that the result rounds to the larger number
Creeping roundoff	The accumulation of error arising from repeated rounding to k significant digits
Error magnification	An erroneous number is multiplied (divided) by a large (small) magnitude
Subtractive cancellation	Subtracting two nearly equal numbers whose difference resides in significant digits beyond the capacity of the computing device to record

TABLE 37.6 Types of roundoff errors possible on a finite-precision computing device

Exact Arithmetic	Rounded to 4 Significant Digits
$p - r = 654.9581 - 0.014897 = 654.943203$	654.9
$q + s = 655.1389 + 128761 = 129416.1389$	1.294×10^5
$qs = (655.1389)(128761) = 84356339.90$	8.436×10^7
$\frac{p}{r} = \frac{654.9581}{0.014897} = 43965.77163$	4.397×10^4
$p - q = 654.9581 - 655.1389 = -0.1808$	-0.1808

TABLE 37.7 Exact arithmetic rounded to 4 significant digits

Computing with 4 Significant Digits
$P - R = 655.0 - 0.01490 = 655.0$
$Q + S = 655.1 + 1.288 \times 10^5 = 1.295 \times 10^5$
$QS = (655.1)(1.288 \times 10^5) = 8.438 \times 10^7$
$\frac{P}{R} = \frac{655.0}{0.01490} = 4.396 \times 10^4$
$P - Q = 655.0 - 655.1 = -0.1$

TABLE 37.8 The calculations in Table 37.7 performed on a device carrying 4 significant digits

NEGLIGIBLE ADDITION   Comparing $p - r$ (rounded) to $P - R$, we see that roundoff error has crept into the fourth significant digit. Comparing $q + s$ (rounded) to $Q + S$, we see that again there is a difference in the fourth significant digit.

ERROR MAGNIFICATION   Comparing qs (rounded) to QS, we find the two answers would differ, in absolute terms, by the substantial error of -20000. Comparing $\frac{p}{r}$ (rounded) to $\frac{P}{R}$, we again find a difference in the fourth significant digit.

CREEPING ROUNDOFF   In each of the calculations $p - r$, qs, $\frac{p}{r}$, and $q + s$, the result of working in four significant digits as opposed to working "exactly" and then rounding is a loss of precision in the fourth significant digit. Each of these calculations illustrates *creeping roundoff*, the gradual loss of precision as repeated rounding errors accumulate.

SUBTRACTIVE CANCELLATION   Comparing $p - q = -0.1808$ to $P - Q = -0.1$, we again find significant error introduced by working in fixed-precision arithmetic.

Minimizing Roundoff Errors

Even implementing familiar "exact" formulas on fixed-precision devices can lead to errors. For example, the quadratic formula gives $-5000 \pm \sqrt{24999999}$ as the exact solutions to the equation $x^2 + 10000x + 1 = 0$. To 10 significant digits, these solutions are -0.000100 and -9999.999900, but to 7 significant digits, they are 0 and -10000.00. Thus, the familiar exact formula for the roots of the quadratic equation might not be enough to guarantee accurate results on a fixed precision device. The problem is the subtractive cancellation, which occurs in the first root.

For use with fixed-precision arithmetic, the familiar quadratic formula can be transformed (by rationalizing the numerators) into the more accurate

$$r_1 = -\operatorname{signum}(b)\frac{2c}{|b| + \sqrt{b^2 - 4ac}} \qquad r_2 = -\operatorname{signum}(b)\frac{|b| + \sqrt{b^2 - 4ac}}{2a} \qquad \textbf{(37.1)}$$

if the roots are real. In 7 significant digits, the roots of the given quadratic equation become, with these new formulas, -0.000100 and -10000.00, respectively. Thus, even on a 7 significant-figure device, the modified form of the quadratic formula produces correct results, whereas the "usual" form does not.

EXERCISES 37.1–Part A

A1. Express sin 1 as a decimal number rounded to 6 decimal digits.

A2. Express tan 2 as a decimal number rounded to 6 significant digits.

A3. Find the interval containing all numbers that approximate e to 7 decimal digits.

A4. Find the interval containing all numbers that approximate e to 7 significant digits.

A5. Find the interval containing all numbers that approximate 2 to 5 decimal digits, and then find the interval containing all numbers that approximate 2000 to 5 decimal digits.

A6. Find the interval containing all numbers that approximate 2 to 5 significant digits, and then find the interval containing all numbers that approximate 2000 to 5 significant digits.

EXERCISES 37.1–Part B

B1. Find the machine epsilon for at least one computing device to which you have access.

In Exercises B2–11:

   (a) Round the given number to 2, 3, 4, 5, and 6 decimal places.

   (b) Round the given number to 3, 4, 5, 6, and 7 significant figures.

B2. 252.7813620   **B3.** 89.65556978   **B4.** −1393.357143

B5. 0.24808182873   **B6.** −614.9285714   **B7.** −20.88039867

B8. 6.292929293   **B9.** −5.262836186

B10. 2097.558824   **B11.** −0.06886228

In the pairs (x, y) given in each of Exercises B12–23, y is an approximation to x, which is assumed to be exact.

   (a) Determine the highest number of significant figures to which y approximates x.

   (b) Determine the highest number of decimal digits to which y approximates x.

B12. (68.73243953, 68.73237080)

B13. (172.0217518, 171.8497300)

B14. (6719.398767, 6786.592755)

B15. $(0.05850096615, 0.05850681625)$

B16. $(0.7410299638, 0.7410298897)$

B17. $(0.002418045892, 0.002418070072)$

B18. $(-190.376626, -190.376816)$

B19. $(9777.216670, 9777.216572)$

B20. $(-0.0002147347557, -0.0002147347578)$

B21. $(-0.429583981, -0.429626940)$

B22. $(-19.6505527, -19.6503562)$

B23. $(0.0006524736264, 0.0006524736916)$

For the quadratic equations given in Exercises B24–33, find k, the smallest number of digits for which a k significant-figure device, using the quadratic formula in the form $\frac{-b \pm \sqrt{b^2 - 4ac}}{2a}$, will compute *both* roots correct to 7 significant figures. Find the smallest k for which (37.1) gives

both roots correct to 7 significant figures. Use appropriate software or a graph to determine the "correct" roots.

B24. $45x^2 + 351120470x + 8 = 0$

B25. $37x^2 - 90x + 0.127127845 \times 10^{-7} = 0$

B26. $80x^2 + 134266483x - 23 = 0$

B27. $88x^2 - 77x + 0.502150031 \times 10^{-5} = 0$

B28. $40x^2 - 542501919x - 67 = 0$

B29. $66x^2 + 53x - 0.750235777 \times 10^{-6} = 0$

B30. $81x^2 - 623851705x + 63 = 0$

B31. $92x^2 - 9x - 0.453825812 \times 10^{-8} = 0$

B32. $65x^2 + 202356318x + 6 = 0$

B33. $66x^2 + 15x - 0.67945509 \times 10^{-9} = 0$

37.2 Rate of Convergence

Linear Convergence of an Iterative Sequence

Consider the sequence $\{x_0, x_1, \ldots\}$ generated by the iteration $x_{k+1} = g(x_k)$. For example, if $g(x) = 1 + \frac{x}{2}$ and x_0, the "starting value," is taken as $\frac{9}{10}$, then the first few members of the resulting sequence are listed in the second column of Table 37.9. The third column contains the increments $\Delta x_k = x_{k+1} - x_k$, and the fourth column contains the ratios $\Delta x_{k+1}/\Delta x_k$.

k	x_k	$\Delta x_k = x_{k+1} - x_k$	$\dfrac{\Delta x_{k+1}}{\Delta x_k}$
0	$\frac{9}{10} = 0.9000000000$	$\frac{11}{20}$	$\frac{1}{2}$
1	$\frac{29}{20} = 1.450000000$	$\frac{11}{40}$	$\frac{1}{2}$
2	$\frac{69}{40} = 1.725000000$	$\frac{11}{80}$	$\frac{1}{2}$
3	$\frac{149}{80} = 1.862500000$	$\frac{11}{160}$	$\frac{1}{2}$
4	$\frac{309}{160} = 1.931250000$	$\frac{11}{320}$	$\frac{1}{2}$
5	$\frac{629}{320} = 1.965625000$	$\frac{11}{640}$	$\frac{1}{2}$
6	$\frac{1269}{640} = 1.982812500$	$\frac{11}{1280}$	$\frac{1}{2}$
7	$\frac{2549}{1280} = 1.991406250$	$\frac{11}{2560}$	$\frac{1}{2}$
8	$\frac{5109}{2560} = 1.995703125$	$\frac{11}{5120}$	$\frac{1}{2}$
9	$\frac{10,229}{5120} = 1.997851563$	$\frac{11}{10,240}$	$\frac{1}{2}$
10	$\frac{20,469}{10,240} = 1.998925781$	$\frac{11}{20480}$	
11	$\frac{40,949}{20,480} = 1.999462891$		

TABLE 37.9 Linear convergence of the iterates $x_{k+1} = 1 + \dfrac{x_k}{2}$

Although it would be easy to argue that the sequence $\{x_k\}$ seems to be converging to the fixed point $x = 2$, the solution of the equation $x = g(x)$, we are presently more interested in quantifying the *rate* at which the sequence converges.

If the sequence $\{x_k\}$ converges, the increments Δx_k in the third column must shrink to zero. (The converse is *not* true. If $x_k = \sum_{n=1}^{k} \frac{1}{n}$, the kth partial sum of the divergent harmonic series, then $x_{k+1} - x_k = \frac{1}{k+1}$; so the increments Δx_k go to zero, but x_k, the members of the sequence, diverge to infinity.)

The sequence $\{x_k\}$ is said to *converge linearly* provided the ratio of increments $\Delta x_{k+1}/\Delta x_k$ tends to a constant C_L, where $0 < |C_L| < 1$. For this sequence, the ratios of successive increments, found in the fourth column of Table 37.9, are all exactly $C_L = \frac{1}{2}$ and the convergence of the sequence to $x = 2$ is linear. In fact, we call this convergence *exactly* linear since $\Delta x_{k+1} = C_L \Delta x_k$ for all $k > 0$.

The sequence $\{x_k\}$ is said to *diverge linearly* if the ratio of increments $\Delta x_{k+1}/\Delta x_k$ tends to a constant C_L where $|C_L| > 1$.

The following theorem, whose proof can be found in [60], explains the behavior of linearly convergent or divergent iteration sequences.

THEOREM 37.1   LINEAR CONVERGENCE THEOREM

1. $X = g(X)$, so $x = X$ is a fixed point of the iteration $x_{k+1} = g(x_k)$.
2. $g'(x)$ is continuous in a neighborhood of the fixed point X.
3. $g'(X) \neq 0$

$\Longrightarrow$

1. $\{x_k\}$ converges to X linearly, with $C_L = g'(X)$ if $0 < |g'(X)| < 1$.
2. $\{x_k\}$ diverges linearly, with $C_L = g'(X)$ if $|g'(X)| > 1$.
3. $\{x_k\}$ converges or diverges *slowly* if $g'(X) = \pm 1$.

In the preceding example, we found $C_L = \frac{1}{2}$, which is the value of $g'(2)$.

GENERAL STRUCTURE OF A LINEARLY CONVERGING SEQUENCE   It is informative to determine a pattern for a sequence $\{s_k\}$ that is *exactly* linearly convergent. For example, if $s_0 = 0$ and $s_1 = r$, the general term in a sequence that converges exactly linearly with convergence constant C is

$$s_k = r \sum_{n=0}^{k-2} C^n = r \frac{C^{k-1} - 1}{C - 1} \tag{37.2}$$

(See Exercise A1.) The increments are then $\Delta s_k = s_{k+1} - s_k = rC^k$; so clearly, the ratios of successive increments are precisely the constant C. Appropriate conditions on C and r would then guarantee *convergence* rather than divergence.

Quadratic Convergence of an Iterative Sequence

Consider a sequence $\{x_0, x_1, \dots\}$ generated by the iteration $x_{k+1} = g(x_k)$, where $g(x) = \frac{2x^2 + 3}{4x + 1}$, a function with fixed points $x = -\frac{3}{2}, 1$. Table 37.10 lists, in column 2, the first few members of the sequence for which $x_0 = 2$. Apparently, the sequence converges very rapidly to the fixed point $x = 1$. Column 3 lists the corresponding increments, and column

k	x_k	$\Delta x_k = x_{k+1} - x_k$	$\dfrac{\Delta x_{k+1}}{(\Delta x_k)^2}$
0	2.0	-0.7777777778	-0.3396226415
1	1.222222222	-0.2054507338	-0.3947041787
2	1.016771488	-0.01666046496	-0.3999644756
3	1.000111024	-0.0001110185766	-0.3999999984
4	1.000000005	$-0.4930049719 \times 10^{-8}$	
5	1.000000000		

TABLE 37.10 Quadratic convergence of the iterates $x_{k+1} = \dfrac{2x_k^2 + 3}{4x_k + 1}$

4 lists the ratios $\Delta x_{k+1}/(\Delta x_k)^2$. The rapid convergence seen in this example is quantified by the following definition.

The sequence $\{x_k\}$ is said to *converge quadratically* provided the ratio of increments $\Delta x_{k+1}/(\Delta x_k)^2$ tends to a constant $C_Q \neq 0, \pm\infty$. If $\Delta x_{k+1} = C_Q(\Delta x_k)^2$ for all $k > 0$, we say the sequence is *exactly quadratically convergent*.

Quadratic convergence is much faster than linear convergence because each successive increment is proportional to the *square* of the preceding one. In this example, it appears that the ratios $\Delta x_{k+1}/(\Delta x_k)^2$ are tending toward the constant $C_Q = -\frac{2}{5}$, confirming the quadratic convergence of the sequence $\{x_k\}$ to the fixed point $x = 1$.

The following theorem whose proof can be found in [60], explains the behavior of the quadratically convergent iteration sequence.

THEOREM 37.2   QUADRATIC CONVERGENCE THEOREM

1. $X = g(X)$, so $x = X$ is a fixed point of the iteration $x_{k+1} = g(x_k)$.

2. $g''(x)$ is continuous in a neighborhood of the fixed point X.

3. $g'(X) = 0$

$\Longrightarrow$ $\{x_k\}$ converges to X quadratically, with $C_Q = -\frac{1}{2}\, g''(X)$.

In the preceding example, we found $C_Q = -\frac{2}{5}$, which is the value of $g''(x) = \frac{100}{(4x+1)^3}$ evaluated at $x = 1$.

GENERAL STRUCTURE OF A QUADRATICALLY CONVERGING SEQUENCE   It is informative to determine a pattern for a sequence $\{s_k\}$ that is *exactly* quadratically convergent. For example, Table 37.11 lists s_k, $\Delta s_k = s_{k+1} - s_k$, and the ratios $\Delta s_{k+1}/(\Delta s_k)^2$ for a sequence whose convergence is exactly quadratic, with convergence constant C, and with starting values $s_0 = 0$ and $s_1 = r$.

Aitken's Acceleration Formula

Aitken's formula

$$y_k = s_k - \frac{(s_k - s_{k-1})^2}{s_k - 2s_{k-1} + s_{k-2}}$$

k	s_k	$\Delta s_k = s_{k+1} - s_k$	$\dfrac{\Delta s_{k+1}}{(\Delta s_k)^2}$
0	0	r	C
1	r	$r^2 C$	C
2	$r^2 C + r$	$r^4 C^3$	C
3	$r^4 C^3 + r^2 C + r$	$r^8 C^7$	C
4	$r^8 C^7 + r^4 C^3 + r^2 C + r$	$r^{16} C^{15}$	C
5	$r^{16} C^{15} + r^8 C^7 + r^4 C^3 + r^2 C + r$	$r^{32} C^{31}$	
6	$r^{32} C^{31} + r^{16} C^{15} + r^8 C^7 + r^4 C^3 + r^2 C + r$		

TABLE 37.11   The structure of a sequence which is exactly quadratically convergent

applied to three consecutive members of the exactly linearly convergent sequence $s_k = r \sum_{n=0}^{k-2} C^n = (rC^{k-1} - 1)/(C - 1)$, where $|C| < 1$, yields $y_k = \frac{r}{1-C}$, the exact value of the fixed point of the sequence. (See Exercise A2.) Alternatively, if the Aitken formula is applied to $s_n, n = k - 2, k - 1, k$, three successive iterates of a sequence that is linearly convergent but not exactly linearly convergent, then y_k will be closer to the fixed point X than s_k. Steffensen's implementation of Aitken acceleration, to be studied in Section 38.3, can make a linearly divergent iteration converge and can make some linearly converging sequences converge quadratically.

Superlinear Convergence of an Iterative Sequence

Superlinear convergence is the middle ground between linear convergence and quadratic convergence. A sequence $\{x_k\}$ converging superlinearly would have its increments $\Delta x_k = x_{k+1} - x_k$ obey $|\Delta x_{k+1}|/|\Delta x_k|^p \to C_S$, where $1 < p < 2$. Superlinear convergence is more rapid than linear but not as rapid as quadratic.

EXAMPLE 37.3   For an example of a sequence that converges superlinearly, consider the sequence $\{x_k\}$ generated by the iteration $x_{k+1} = g(x_k, x_{k-1})$, where $g(x, y) = \frac{2xy+3}{2x+2y+1}$. This iteration function requires two starting values, which we take as $x_0 = 5$ and $x_1 = 4$. Then, the next iterate would be $x_2 = g(x_1, x_0)$. The fixed points, $x = -\frac{3}{2}, 1$, are the solutions of the equation $x = g(x, x)$. For this sequence, Table 37.12 lists x_k, $\Delta x_k = x_{k+1} - x_k$, and the ratios

k	x_k	$\Delta x_k = x_{k+1} - x_k$	$\dfrac{\lvert\Delta x_{k+1}\rvert}{\lvert\Delta x_k\rvert^p}$	k	x_k	$\Delta x_k = x_{k+1} - x_k$	$\dfrac{\lvert\Delta x_{k+1}\rvert}{\lvert\Delta x_k\rvert^p}$
0	5	1	1.737	7	1.000019452607	$0.1943915535387 \times 10^{-4}$	0.5644
1	4	1.736842105263	0.2877	8	$1 + 0.1345 \times 10^{-7}$	$0.1345156307500 \times 10^{-7}$	0.5697
2	2.263157894737	0.7028466106901	0.7016	9	$1 + 0.1047 \times 10^{-12}$	$0.1046671870034 \times 10^{-12}$	0.5655
3	1.560311284047	0.3966090364565	0.6042	10	$1 + 0.5632 \times 10^{-21}$	$0.5631792892315 \times 10^{-21}$	0.5676
4	1.163702247590	0.1352519276245	0.6800	11	$1 + 0.2358 \times 10^{-34}$	$0.2357855691984 \times 10^{-34}$	
5	1.028450319966	0.0267203357964	0.6006	12	1.0		
6	1.001729984169	0.0017105315622	0.5826				

TABLE 37.12   The superlinear convergence of the sequence in Example 37.3

$|\Delta x_{k+1}|/|\Delta x_k|^p$, where $p = \frac{1+\sqrt{5}}{2} = 1.618033989$. The computations were actually done with 50 digits in Maple. The empirical evidence supports the claim that the convergence to $x = 1$ is superlinear with $p = \frac{1+\sqrt{5}}{2}$ and C_S approximately 0.6.   ❖

GENERAL STRUCTURE OF A SEQUENCE CONVERGING SUPERLINEARLY   If the sequence

$$s_k = \sum_{n=0}^{k-1} r^{(p^n)} C^{(p^n-1)/(p-1)} \qquad k \ge 1$$

converges, the convergence will be superlinear with order p and positive, finite convergence constant C. This structure is derived in the exercises.

EXERCISES 37.2–Part A

A1. Establish (37.2) by recognizing the sum as a geometric sum.

A2. Show that the application of Aitken's formula to three successive iterates of the sequence $s_k = r(C^{k-1} - 1)/(C - 1)$ yields $y_k = \frac{r}{1-C}$.

A3. Using Table 37.11 with $r = \frac{1}{2}$, $s_0 = 0$, and $C = \frac{3}{5}$, find $s_k, k = 1, 2, 3$, the next three terms of the corresponding sequence that converges exactly quadratically.

A4. Find all fixed points of $x_{k+1} = g(x_k)$ if $g(x) = x^2 + 2x - 20$. Will the sequence $\{x_k\}$ converge to any of these fixed points?

A5. Find all fixed points of $x_{k+1} = g(x_k)$ if $g(x) = \frac{x^2+20}{2x+1}$. Will the sequence $\{x_k\}$ converge to any of these fixed points?

EXERCISES 37.2–Part B

B1. Find all fixed points of $x_{k+1} = g(x_k, x_{k-1})$ if $g(x, y) = \frac{xy+20}{x+y+1}$. Generate the first few members of the sequence $\{x_k\}$ if $x_0 = 0$ and $x_1 = 1$.

In Exercises B2–11:

   (a) Show empirically that the iteration $x_{k+1} = g(x_k)$, starting from the given x_0, converges linearly.

   (b) Empirically determine the convergence constant C_L.

   (c) By solving the equation $x = g(x)$, determine the fixed point $\bar{x}$ to which the iteration in part (a) converges.

   (d) Compare C_L with $g'(\bar{x})$.

B2. $g(x) = 6x^3 + x^2 - 7x - 4, x_0 = -\frac{3}{4}$

B3. $g(x) = 3x^3 + x^2 - 4x + 2, x_0 = \frac{1}{2}$

B4. $g(x) = 7x^3 + 2x^2 - 2x - 1, x_0 = -\frac{3}{5}$

B5. $g(x) = -4x^3 - 8x^2 - 2x + 1, x_0 = -1.3$

B6. $g(x) = -3x^3 + 5x^2 - 3x + 1, x_0 = 0.9$

B7. $g(x) = 3x^4 - 2x^3 - 2x^2 + 2, x_0 = \frac{1}{2}$

B8. $g(x) = x^4 - 2x^3 - 5x^2 - 3x - 1, x_0 = -\frac{1}{2}$

B9. $g(x) = 3x^4 - 4x^3 - 4x^2 + 4x + 2, x_0 = -\frac{7}{10}$

B10. $g(x) = 4x^4 + 4x^3 - 2x^2 - 3x - 1, x_0 = -\frac{1}{2}$

B11. $g(x) = 3x^4 + 2x^3 - 2x^2 - 2x - 1, x_0 = -0.9$

In Exercises 12–21:

   (a) Show empirically that the iteration $x_{k+1} = g(x_k)$, starting from the given x_0, converges quadratically.

   (b) Empirically determine the convergence constant C_Q.

   (c) By solving the equation $x = g(x)$, determine the fixed point $\bar{x}$ to which the iteration in part (a) converges.

   (d) Show that $g'(\bar{x}) = 0$.

   (e) Compare C_Q with $-\frac{1}{2}g''(\bar{x})$.

B12. $g(x) = \dfrac{2x^3 + 3x^2 - 5}{3x^2 + 6x - 2}, x_0 = -4$

B13. $g(x) = \dfrac{6x^3 + 3x^2 + 1}{9x^2 + 6x + 4}, x_0 = \dfrac{1}{2}$

B14. $g(x) = \dfrac{8x^3 + 4x^2 - 3}{12x^2 + 8x + 1}, x_0 = -1$

B15. $g(x) = \dfrac{6x^3 + 5x^2 + 1}{x(9x + 10)}, x_0 = -2$

B16. $g(x) = \dfrac{10x^3 - 3x^2 - 4}{15x^2 - 6x + 2}, x_0 = -1$

B17. $g(x) = \dfrac{3x^4 + 4x^3 + 5x^2 + 1}{4x^3 + 6x^2 + 10x - 1}, x_0 = -1$

B18. $g(x) = \dfrac{3x^4 + 10x^3 + 3x^2 + 4}{4x^3 + 15x^2 + 6x - 1}, x_0 = -4$

B19. $g(x) = \dfrac{12x^4 - 8x^3 - x^2 + 2}{16x^3 - 12x^2 - 2x + 3}$, $x_0 = -1$

B20. $g(x) = \dfrac{12x^4 + 10x^3 + x^2 + 4}{16x^3 + 15x^2 + 2x + 2}$, $x_0 = -1$

B21. $g(x) = \dfrac{6x^4 + 8x^3 - 5x^2 + 4}{8x^3 + 12x^2 - 10x - 5}$, $x_0 = -2$

In Exercises B22–31:

    (a) Use the given x_0 with $x_1 = 1.02x_0$ to obtain the iterates $x_{k+1} = g(x_k, x_{k-1})$.

    (b) Show empirically that the convergence is superlinear with parameter $p = \frac{1+\sqrt{5}}{2} = 1.618033989$.

    (c) Solve the equation $x = g(x, x)$ for the fixed point(s) of the iteration.

B22. $g(x, y) = \dfrac{5x^2y - 3xy + 5xy^2 + 1}{5x^2 - 3x + 5xy - 3 - 3y + 5y^2}$, $x_0 = 2$

B23. $g(x, y) = \dfrac{x^2y - 4xy + xy^2 + 4}{x^2 - 4x + xy + 4 - 4y + y^2}$, $x_0 = 2$

B24. $g(x, y) = \dfrac{3x^2y + 5xy + 3xy^2 + 5}{3x^2 + 5x + 3xy - 5 + 5y + 3y^2}$, $x_0 = -1$

B25. $g(x, y) = \dfrac{5x^2y - xy + 5xy^2 - 2}{5x^2 - x + 5xy - 5 - y + 5y^2}$, $x_0 = -2$

B26. $g(x, y) = \dfrac{2xy + 1}{2x + 1 + 2y}$, $x_0 = -2$

B27. $g(x, y) = \dfrac{5xy + 4}{5x - 2 + 5y}$, $x_0 = -2$

B28. $g(x, y) = \dfrac{x^2y + xy^2 + 1}{x^2 + xy + y^2}$, $x_0 = 2$

B29. $g(x, y) = \dfrac{x^2y + xy^2 + 2}{x^2 + xy + y^2}$, $x_0 = 2$

B30. $g(x, y) = \dfrac{yx^3 + y^2x^2 + xy^3 + 2}{x^3 + yx^2 + y^2x + y^3}$, $x_0 = -2$

B31. $g(x, y) = \dfrac{xy + 1}{x + 1 + y}$, $x_0 = -2$

For the sequences $x_k, k = 0, 1, \ldots$, given in Exercises B32 and 33:

    (a) Show the convergence is linear.

    (b) Determine the limit of the sequence.

    (c) Determine the convergence constant C_L.

    (d) Apply Aitken's formula to $x_k, k = 5, 6, 7$, thereby obtaining the limit from part (b).

B32. $x_k = 3 - \dfrac{1}{4}\dfrac{1}{3^{k-1}}$ **B33.** $x_k = 2 - \dfrac{2}{9}\dfrac{1}{4^{k-1}}$

B34. If the sequence $\{s_k\}$ converges quadratically, show that
$$s_k = \sum\nolimits_{n=0}^{k-1} r^{(2^n)} C^{(2^n - 1)}.$$

B35. If the sequence $\{s_k\}$ converges superlinearly with order p, show that $s_k = \sum\nolimits_{n=0}^{k-1} r^{(p^n)} C^{(p^n - 1)/(p-1)}$.

In Exercises B36–40:

    (a) Show that $x = a$ is a fixed point of the iteration $x_{k+1} = g(x_k)$.

    (b) Find all values of a for which the convergence of the iteration is linear.

    (c) Find all values of a for which the convergence of the iteration is quadratic.

B36. $g(x) = x^3 + (8 - a)x^2 - (19 + 8a)x + 20a$

B37. $g(x) = x^4 - x^3a - 75x^2 + (251 + 75a)x - 250a$

B38. $g(x) = x^3 - (8 + a)x^2 + (8a - 19)x + 20a$

B39. $g(x) = x^3 - (5 + a)x^2 + (5a - 83)x + 84a$

B40. $g(x) = x^4 - (12 + a)x^3 + 12x^2a + 257x - 256a$

Chapter Review

1. What does it mean to say that x is a five decimal place approximation to π?

2. What does it mean to say that x approximates π to five significant figures?

3. What is the *machine epsilon* of a computing device? How can it be determined experimentally?

4. What is a *fixed-precision* computing device? What is a 10 significant-figure computing device? How can the precision of a computing device be determined experimentally?

5. Name and describe four different types of roundoff error.

6. The truncation error of a numeric algorithm would be the inherent mathematical error made by the approximations used in the algorithm. How would this differ from roundoff error?

7. What does it mean for an iterative numerical technique to converge (a) linearly; (b) quadratically; and (c) superlinearly?

8. Under what conditions will the fixed-point iteration $x_{k+1} = g(x_k)$ converge linearly? Under what conditions will it converge quadratically?

9. What is Aitken's acceleration formula? Where might it be used, and what effect might it be expected to have?

Chapter 38

Equations in One Variable — Methods

I N T R O D U C T I O N   Chapter 38 examines five methods for finding roots of equations (or zeros of functions). Two methods (*fixed-point iteration* and *bisection*) converge linearly, which is slow convergence. One method (*Newton–Raphson*) generally converges quadratically, which is rapid convergence. Two methods (*secant* and *Muller*) converge superlinearly, which is faster than linear but not as fast as quadratic. Other methods appear in the exercises.

We begin by describing *fixed-point iteration,* a method that *converges linearly*. This is not a production method. It converges far too slowly and is too difficult to arrange so that convergence is assured. It is discussed so that the shortcomings of linear convergence will be fully understood.

Next, we describe the *bisection* algorithm, which, if it starts by bracketing a root, is guaranteed to converge. However, it converges at a rate that is, on average, linear, with convergence constant $\frac{1}{2}$. Unfortunately, the bisection method does not generalize to higher dimensions.

The *Newton–Raphson* method is generally *quadratically convergent* if started near enough to a root. Although the method can be deceived by pathological examples, it is generally regarded as a viable and robust solver. The method's need for derivatives can sometimes be a difficulty, but in an era of computer algebra systems that can provide symbolic expressions for even complicated derivatives, this objection is not as formidable as it once was.

The *secant method* is a popular alternative to Newton's method because it replaces the derivative with a difference quotient. However, the secant method *converges superlinearly* and not quadratically, so it is slightly slower than Newton's method.

Newton's method can actually be used in the complex plane, where, however, it must usually be started very near to a complex root. *Muller's method* is an alternative because its iterates can move from the real line to the complex plane and back to the real line. It will generally converge to some root for almost any starting point. Again, in comparison to the quadratic convergence of Newton's method, Muller's method *converges superlinearly,* a slightly slower rate.

38.1   Fixed-Point Iteration

Finding Roots by Fixed-Point Iteration

The method we will call "fixed-point iteration" is called *functional iteration* in [49], *fixed-point iteration* in [16], [42], [38], and [32], and *repeated substitution* in [60] and [85]. In addition, a case could be made for the terms *Picard iteration* and *linear iteration*.

To find (by fixed-point iteration) a root of the equation $f(x) = 0$, write the equation in the form $x = g(x)$ and apply the iteration $x_{k+1} = g(x_k)$. The starting value x_0 will matter, and so will the form chosen for $g(x)$. According to Theorem 37.1, if $x = X$ is a fixed point of the iteration, $g'(x)$ is continuous in a neighborhood of X, and $0 < |g'(X)| < 1$, then the sequence $\{x_k\}$ will converge linearly to X provided x_0 is sufficiently near X.

The practical difficulties of writing the equation $f(x) = 0$ in terms of a function $g(x)$ for which the inequality $0 < |g'(X)| < 1$ holds are substantial. Even if that hurdle is cleared, the resulting linear convergence is usually slow enough to warrant using a more powerful technique. Fixed-point iteration is included here to show how iteration works and to see how slow linear convergence actually is.

EXAMPLE 38.1   Use fixed-point iteration to find the roots of the equation $f(x) = 0$, where

$$f(x) = (x - 1)(x - 2)(x - 3) = x^3 - 6x^2 + 11x - 6 \qquad (38.1)$$

Solving for the x^3 term, we get $x^3 = 6x^2 - 11x + 6$, from which follows the iteration

$$x = g(x) = (6x^2 - 11x + 6)^{1/3}$$

The roots of $f(x) = 0$ are obviously $x = 1, 2, 3$, so we obtain $g'(1) = \frac{1}{3}$, $g'(2) = \frac{13}{12}$, and $g'(3) = \frac{25}{27}$. Fixed-point iteration can be expected to converge for x_0 near $x = 1$ and $x = 3$ but will diverge for x_0 near $x = 2$. Table 38.1 summarizes iterations starting from $x_0 = 5, 0$, and 2.0001. The iteration starting from $x_0 = 5$ converges very slowly to $x = 3$ because $C_L = g'(3)$ is so close to 1. The iteration starting from $x_0 = 0$ converges linearly to $x = 1$, and the iteration starting from $x_0 = 2.0001$ diverges slowly but surely.

k	x_k	x_k	x_k
0	5	0	2.0001
1	4.657009508	1.817120593	2.000108332
2	4.395093028	1.799098003	2.000117359
3	4.189904570	1.779020177	2.000127138
4	4.025726919	1.756601438	2.000137731
5	3.892025504	1.731508797	2.000149207
10	3.486190313	1.550110656	2.000222621
15	3.290498975	1.231037079	2.000332147
20	3.182355198	1.007146103	2.000495531
25	3.117832357	1.000031235	2.000739231

TABLE 38.1   Fixed-point iterates for Example 38.1

The function $g(x)$ is not the only way $f(x) = 0$ can be rearranged. For example, we could solve for the x^2 term, obtaining $x^2 = \frac{1}{6}x^3 + \frac{11}{6}x - 1$, from which the iteration

$$x = g_1(x) = \frac{1}{6}\sqrt{6x^3 + 66x - 36}$$

results. Again checking the derivative at each of the three fixed points, we find $g_1'(1) = \frac{7}{6}$, $g_1'(2) = \frac{23}{24}$, and $g_1'(3) = \frac{19}{18}$. This time, we are assured of convergence for x_0 near $x = 2$ but divergence for x_0 near $x = 1$ or $x = 3$. But again, the convergence to the fixed point at $x = 2$ will be slow because the derivative is nearly 1 at that point. Indeed, starting with $x_0 = 2.2$ yields a slowly converging sequence for which $x_1 = 2.192715212$ and $x_{25} = 2.075054707$.

Finally, we can solve the equation $f(x) = 0$ for the term in just x, obtaining the iteration

$$x = g_2(x) = \frac{1}{11}(6 + 6x^2 - x^3)$$

Again evaluating the derivative at the fixed points, we find $g_2'(1) = \frac{9}{11}$, $g_2'(2) = \frac{12}{11}$, and $g_2'(3) = \frac{9}{11}$, so we would have slow convergence to $x = 1$ or $x = 3$ and divergence for x_0 near $x = 2$. ❖

Graphics

Geometrically, fixed-point iteration is explained in terms of a graph of the line $y = x$ and the curve that is the graph of $g(x)$. Clearly, where these two graphs intersect is a fixed point. Figure 38.1, generally called a "web diagram," shows graphs of $y(x) = x$, $g(x) = 1 + \frac{x}{2}$, and the iterates $\{\frac{1}{10}, \frac{21}{20}, \frac{61}{40}, \frac{141}{80}, \frac{301}{160}\}$ generated by $x = g(x)$. The vertical lines represent the progress of the iteration, starting from $x_0 = \frac{1}{10}$. The stair-step pattern arises from making an output value, which is a y-coordinate, into an input value, which is an x-coordinate, by reflecting across the line $y = x$. The accompanying Maple worksheet contains an animation that represents this dynamically.

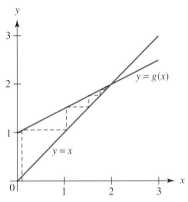

FIGURE 38.1 The progress of fixed-point iteration when $g(x) = 1 + \frac{x}{2}$

EXERCISES 38.1

1. If $f(x)$ is as given in (38.1), investigate the fixed-point iteration determined by writing $x = f(x) + x = g(x)$.

2. If $f(x)$ is as given in (38.1), for what values of a will the fixed-point iteration determined by $x = \frac{1}{a}(f(x) + ax) = g(x)$ converge when x_0 is near 2?

For the cubics given in each of Exercises 3–12:

(a) Determine all real zeros.

(b) Arrange the equation $f(x) = 0$ as the fixed-point iteration $x_{k+1} = g_1(x_k)$ by isolating x in the linear term in $f(x)$. On the same set of axes, graph the line $y = x$ and the curve $y = g_1(x)$. Carry out fixed-point iteration from an appropriate starting point x_0 and show it either converges to a real zero r found in part (a) or diverges. In either case, show that the outcome is consistent with the value of $g_1'(r)$. If the iteration converges, check Δx_k to see how many zeros appear just after the decimal point. A hallmark of linear convergence is the appearance of one more zeros at each step.

(c) Repeat part (b) for $x_{k+1} = g_2(x_k)$, a branch determined by isolating x in the quadratic term in $f(x)$.

(d) Repeat part (b) for $x_{k+1} = g_3(x_k)$, a real branch determined by isolating x in the cubic term in $f(x)$.

3. $f(x) = 60x^3 + 95x^2 + 95x - 18$

4. $f(x) = 46x^3 - 20x^2 + 52x + 31$

5. $f(x) = 38x^3 - 23x^2 - 14x - 58$

6. $f(x) = 46x^3 - 12x^2 - 81x - 63$

7. $f(x) = 54x^3 - 71x^2 + 71x + 17$

8. $f(x) = 85x^3 + 36x^2 + 35x - 11$

9. $f(x) = 90x^3 + 31x^2 + 47x - 50$

10. $f(x) = 48x^3 + 36x^2 + 73x - 40$

11. $f(x) = 6x^3 - 65x^2 + 52x + 99$

12. $f(x) = 39x^3 - 49x^2 + 76x + 15$

In each of Exercises 13–20:

(a) Determine r_1 and r_2, $(r_1 < r_2)$, the fixed points of the iteration $x_{k+1} = g(x_k)$.

(b) Determine all values of p for which the iteration converges linearly to r_1.

(c) Determine all values of p for which the iteration converges linearly to r_2.

(d) Determine all values of p for which the iteration converges quadratically to r_1.

(e) Determine all values of p for which the iteration converges quadratically to r_2.

(f) Determine initial values x_0 for which fixed-point iteration is likely to converge to r_1 when $p = 16$.

(g) Determine initial values x_0 for which fixed-point iteration is likely to converge to r_2 when $p = -11$.

13. $g(x) = \dfrac{x^2 + px - 6}{p - 1}, p \neq 1$ **14.** $g(x) = \dfrac{x^2 + px - 3}{p - 2}, p \neq 2$

15. $g(x) = \dfrac{x^2 + px - 28}{p + 3}, p \neq -3$

16. $g(x) = \dfrac{x^2 + px - 32}{p + 4}, p \neq -4$

17. $g(x) = \dfrac{x^2 + px - 84}{p - 5}, p \neq 5$ **18.** $g(x) = \dfrac{x^2 + px - 33}{p - 8}, p \neq 8$

19. $g(x) = \dfrac{x^2 + px - 55}{p - 6}, p \neq 6$ **20.** $g(x) = \dfrac{x^2 + px - 54}{p - 3}, p \neq 3$

38.2   The Bisection Method

Bisection—the Concept

If a continuous function $f(x)$ changes sign between $x = a$ and $x = b$, then there is at least one point c, $a < c < b$, for which $f(c) = 0$. For example, Figure 38.2 shows

$$f(x) = \arctan x - \sqrt{2x - 5} \tag{38.2}$$

has a root in the interval $[3, 4]$.

To turn this basic idea into an algorithm, evaluate $f(x)$ at the midpoint of the interval and determine the half-interval in which the sign-change resides. Keep the interval, now half the original, in which the root resides. Continue until the root has been bracketed with sufficient accuracy.

In Figure 38.2, $m = \frac{3+4}{2} = \frac{7}{2}$, the midpoint of the interval $[3, 4]$, is marked with an arrow. The sign-change resides in the left half-interval, namely, in $[3, \frac{7}{2}]$. Hence, the subinterval $[3, \frac{7}{2}]$ is retained and the process repeats. All that is needed to flesh out an algorithm is an efficient method for determining which half-interval contains the sign-change. This is done by examining the sign of the products $f(a)f(m)$. If this is negative, then the function has opposite signs at these two endpoints, so the sign-change is in the left half-interval $[a, m]$. If the product is positive, then the sign-change in $[a, b]$ has to be in the right half-interval $[m, b]$.

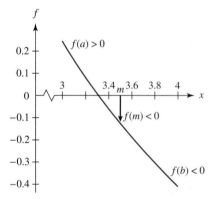

FIGURE 38.2   The bisection method applied to $f(x) = \arctan x - \sqrt{2x - 5}$ which has a zero in the interval $[3, 4]$

Code for a Bisection Algorithm

The following steps constitute the bisection algorithm.

1. Test that $f(a)f(b) < 0$.

2. Obtain $m = \frac{b+a}{2}$.

3. If $f(a)f(m) < 0$, then $b = m$. Return to step 2.

4. If $f(a)f(m) > 0$, then $a = m$. Return to step 2.

5. If $f(a)f(m) = 0$, then stop since m is a root of $f(x) = 0$.

EXAMPLE 38.2   Table 38.2 contains the output of the bisection algorithm applied to the function $f(x)$ given in (38.2). The first column records the number of interval-halvings; the second gives the interval in which the root is located; the third is the midpoint of the interval just halved; the fourth is the function value, or residual, at the midpoint; and the fifth is the length of the interval in which the root is located. Clearly, the original interval is reduced by a factor of

k	$[\alpha, \beta]$	m	$f(m)$	$\lvert\beta - \alpha\rvert$
1	[3.000000000, 3.500000000]	3.500000000	−0.121716894	$\frac{1}{2}$
2	[3.250000000, 3.500000000]	3.250000000	0.047552524	$\frac{1}{4}$
3	[3.250000000, 3.375000000]	3.375000000	−0.040134776	$\frac{1}{8}$
4	[3.312500000, 3.375000000]	3.312500000	0.002854552	$\frac{1}{16}$
5	[3.312500000, 3.343750000]	3.343750000	−0.018840941	$\frac{1}{32}$
6	[3.312500000, 3.328125000]	3.328125000	−0.008044899	$\frac{1}{64}$
7	[3.312500000, 3.320312500]	3.320312500	−0.002608295	$\frac{1}{128}$
8	[3.316406250, 3.320312500]	3.316406250	0.000119823	$\frac{1}{256}$
9	[3.316406250, 3.318359375]	3.318359375	−0.001245058	$\frac{1}{512}$
10	[3.316406250, 3.317382813]	3.317382813	−0.000562824	$\frac{1}{1024}$
11	[3.316406250, 3.316894532]	3.316894532	−0.000221552	$\frac{1}{2048}$
12	[3.316406250, 3.316650391]	3.316650391	−0.000050878	$\frac{1}{4096}$

TABLE 38.2   Results of the bisection method used in Example 38.2

$1/2^k$ after k steps. Thus, the ultimate resolution of the method is the computing device's ability to distinguish two endpoints separated by a distance of $\lvert b - a\rvert/2^k$. When this distance becomes smaller than the machine epsilon, no further refinement of the root can take place. ❖

EXERCISES 38.2–Part A

A1. After n iterations of the bisection method, x_n, the last midpoint computed, becomes the endpoint of the last interval kept and the latest approximation to the bracketed root. Hence, the maximum error in x_n is the length of the last subinterval, or $\lvert b - a\rvert/2^n$. If x_n must be determined with an error no greater than ε, how large must n be?

A2. Starting with the interval $[a, b] = [0, 3]$, obtain the first three subintervals when the bisection method is applied to the function $f(x) = x - 1$.

A3. The bisection method samples the function $f(x)$ at the midpoint of the current interval $[\alpha, \beta]$. The method of *false-position* is an alternate bracketing method wherein $f(x)$ is sampled at ξ, the x-intercept of the line-segment connecting the points $(\alpha, f(\alpha))$ and $(\beta, f(\beta))$. Show that $\xi = \frac{\beta f(\alpha) - \alpha f(\beta)}{f(\alpha) - f(\beta)}$.

A4. Starting with the interval $[a, b] = [0, 3]$, obtain the first three subintervals when the method of false-position is applied to function $f(x) = x - 1$.

EXERCISES 38.2–Part B

B1. In the language of your choice, implement the bisection algorithm. Test your code by finding the zeros of $f(x) = x^2 - 3x + 2$.

B2. In the language of your choice, implement the method of false-position as defined in Exercise A2. Test your code by finding the zeros of $f(x) = x^2 - 3x + 2$.

For the functions $g(x)$ given in each of Exercises B3–12:

  (a) Use a graph to locate all real zeros.

  (b) Using the bisection method and the results of Exercise A1, compute the value of each real zero to a tolerance of 10^{-7}.

  (c) Use the method of false-position to compute each real zero, correct to 7 significant figures.

B3. $g(x) = 2x^2 - 6x - 3$ **B4.** $g(x) = 5x^3 + 2x^2 - 3x + 3$
B5. $g(x) = 3x^4 - 3x^3 + 5x^2 + 3x - 1$

B6. $g(x) = \dfrac{3x^3 - 7x^2 - 6x + 7}{7x^3 + 6x^2 - 3x + 6}$ **B7.** $g(x) = \dfrac{6x^3 - 9x^2 - 4x - 9}{8x^2 + 9x - 7}$ **B10.** $g(x) = \dfrac{3x^3 - 9x^2 - 8x - 1}{9x^2 - 4x + 1}$

B8. $g(x) = \dfrac{x^3 - 7x^2 + x + 7}{2x^4 + 9x^3 - 2x^2 + x + 1}$ **B11.** $g(x) = \dfrac{7x^3 - 9x^2 + 4x + 6}{9x^4 - x^3 + 9x + 5}$

B9. $g(x) = \dfrac{4x^3 - x^2 - 2x - 9}{7x^3 + 5x^2 + 2x + 1}$ **B12.** $g(x) = \dfrac{5x^3 - x^2 - 4x + 5}{8x^4 + 7x^3 + 9x^2 + 4x - 8}$

38.3 Newton–Raphson Iteration

Newton's Method—the Concept

Newton–Raphson iteration finds zeros of a function $f(x)$ by finding, instead, zeros of approximating tangent lines. The sequence $\{x_k\}$ generated by this method generally converges quadratically to a root of the equation $f(x) = 0$. The closer the initial point, x_0, is to the root, the greater the likelihood, in general, the sequence will converge to that root.

For example, consider the function $f(x) = \arctan x - \sqrt{2x - 5}$ that has a zero in the interval $[3, 4]$. The equation of the tangent line at $x_0 = 2.6$ is

$$y = 6.235134095 - 2.107201999x$$

found as the first-degree Taylor polynomial at x_0. The x-intercept for this tangent line is $x_1 = 2.958963639$, which is closer to the root than x_0, as seen in Figure 38.3.

The equation of the tangent line constructed at x_1 is

$$y = 3.071906228 - 0.9412418573x$$

and its x-intercept is $x_2 = 3.263673629$. The equation of the tangent line constructed at x_2 is $y = 2.398324602 - 0.7233284742x$, and its x-intercept is $x_3 = 3.315678406$. It turns out that x_3 is within 0.000899215 of the root, which is actually at $x = 3.316577621$.

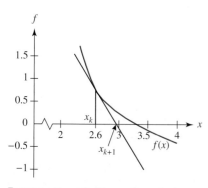

FIGURE 38.3   In Newton's method, a root of $f(x) = 0$ is approximated by the x-intercept of a tangent line

Toward a Newton Algorithm

From our discussion, we see that the process of forming Newton iterates consists of the following steps.

1. Let x_k be the last Newton iterate found.

2. At x_k, construct the tangent line $y = f'(x_k)(x - x_k) + f(x_k)$.

3. Solve for the x-intercept on this tangent line, obtaining $x = x_k - f(x_k)/f'(x_k)$.

4. Declare this x-intercept to be the next Newton iterate so that $x_{k+1} = x_k - f(x_k)/f'(x_k)$.

Thus, fixed-point iteration with $x = g(x) = x - \frac{f(x)}{f'(x)}$ generates the Newton iterates.

EXAMPLE 38.3   The second column of Table 38.3 gives the Newton iterates, starting with $x_0 = 2.6$, for the function $f(x) = \arctan x - \sqrt{2x - 5}$. The third column gives the increments $\Delta x_k = x_{k+1} - x_k$, and the fourth column gives the ratios $\Delta x_{k+1}/(\Delta x_k)^2$. The entries in the fourth column are consistent with the claim that the ratios $\Delta x_{k+1}/(\Delta x_k)^2$ converge to a constant C_Q approximately 0.3097. Theorem 37.2 indicates this constant should be $-\frac{1}{2}g''(X) = 0.3097038499$, where X is the fixed point and $g(x) = x - \frac{f(x)}{f'(x)}$. Also, we note the doubling of the number of zeros after the decimal point in Δx_k. This is a hallmark of quadratic convergence.

k	x_k	$\Delta x_k = x_{k+1} - x_k$	$\dfrac{\Delta x_{k+1}}{(\Delta x_k)^2}$
0	2.6	0.35896	2.364752940
1	2.9589636391	0.30471	0.5601055206
2	3.2636736290	0.52005×10^{-1}	0.3323962178
3	3.3165773705	0.89896×10^{-3}	0.3100656806
4	3.3165773705	0.25058×10^{-6}	0.3097039502
5	3.3165776210	0.19446×10^{-13}	
6	3.3165776210		

TABLE 38.3 Newton iteration applied to $f(x) = \arctan x - \sqrt{2x - 5}$ in Example 38.3

A computation shows that $g'(X) = 0$ for the Newton iteration function

$$g(x) = x - \frac{\arctan x - \sqrt{2x - 5}}{\frac{1}{x^2+1} - \frac{1}{\sqrt{2x-5}}}$$

Thus, if $g'(X) \neq 0$ for the fixed-point iteration $x = g(x)$, the convergence to the fixed point X is linear; whereas if $g'(X) = 0$ but $g''(X) \neq 0$, then the convergence is quadratic. ❖

For the general Newton iteration function $G(x) = x - \frac{F(x)}{F'(x)}$, we find $G'(x) = \frac{F(x)F''(x)}{(F'(x))^2}$, so $G'(x) = 0$ at a zero of $F(x)$ provided $F'(x) \neq 0$ at that zero. Moreover,

$$G''(x) = \frac{F''(x)}{F'(x)} + F(x) \left(\frac{F'''(x)}{(F'(x))^2} - 2 \frac{(F''(x))^2}{(F'(x))^3} \right)$$

so $G''(X) = \frac{F''(X)}{F'(X)}$ at X, a zero of $F(x)$. Thus, the quadratic convergence constant $C_Q = -\frac{1}{2} G''(X)$ must also equal $-\frac{1}{2} \frac{F''(X)}{F'(X)}$.

Sufficient Conditions for Convergence

If $f(r) = 0$ but $f'(r) \neq 0$; and if $f(x)$, $f'(x)$, $f''(x)$ are continuous in a neighborhood of $x = r$, then Newton iteration, which is actually fixed-point iteration with the iteration function $g(x) = x - \frac{f(x)}{f'(x)}$, converges to r if x_0 is sufficiently close to r.

Multiple Roots

DEFINITION 38.1

A function $f(x)$ is said to have a zero of order m at $x = r$ if $f, f', f'', \ldots, f^{(m-1)}$ all vanish at r but $f^{(m)}(r) \neq 0$.

Thus, if the function and the first $m-1$ derivatives vanish at $x = r$ but the mth derivative does not, then the function has a zero of order m at $x = r$. Equivalently, a function with a zero of order m at $x = r$ can be written in the form $f(x) = (x - r)^m Q(x)$, where $Q(r) \neq 0$, an obvious representation based on a Taylor expansion at $x = r$.

EXAMPLE 38.4

The function $f(x) = \ln(\cos x) + \frac{1}{2} \sin x^2$ has a zero at $x = 0$, as seen from its graph in Figure 38.4. The flatness of the curve in the neighborhood of $x = 0$ suggests the zero is of higher order than $m = 1$. To determine the order of the zero, obtain the Taylor expansion $-\frac{1}{12}x^4 - \frac{19}{180}x^6 - \frac{17}{2520}x^8 + \cdots$, which obviously factors to $(x - 0)^4 Q(x)$, where $Q(x) = -\frac{1}{12} - \frac{19}{180}x^2 - \frac{17}{2520}x^4 + \cdots$, and $Q(0) \neq 0$. Alternatively, compute $f(0) = f'(0) = f''(0) = f'''(0) = 0$ and $f^{(4)}(0) = -2$. The first nonzero derivative at $x = 0$ is the fourth, again indicating that the zero at $x = 0$ is of the fourth order. Hence, $m = 4$ at $x = 0$. ❖

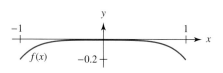

FIGURE 38.4   The function $f(x)$ in Example 38.4

LINEAR CONVERGENCE AT MULTIPLE ROOTS   At a zero of order m, Newton's method converges linearly, with convergence constant $C_L = \frac{m-1}{m}$. A function of the form $F(x) = (x - r)^m Q(x)$, with $Q(r) \neq 0$, has a zero of order m at $x = r$. For the associated Newton iteration function $G(x) = x - \frac{F(x)}{F'(x)}$, we find by direct computation that $G'(r) = \frac{m-1}{m} \neq 0$ whenever $m > 1$. Consequently, the convergence is linear, not quadratic, and $C_L = \frac{m-1}{m}$.

EXAMPLE 38.5

The Newton iteration function for $f(x) = \ln(\cos x) + \frac{1}{2} \sin x^2$ is

$$g(x) = x - \frac{\ln(\cos x) + \frac{1}{2} \sin x^2}{x \cos x^2 - \tan x}$$

Column 2 of Table 38.4 lists the Newton iterates x_k, starting with $x_0 = 1$. Column 3 lists the increments $\Delta x_k = x_{k+1} - x_k$, and column 4 lists the ratios $\Delta x_{k+1}/\Delta x_k$. Motivated by Theorem 37.2, we compute $g'(0) = \frac{3}{4}$, consistent with the claim $C_L = \frac{m-1}{m} = \frac{4-1}{3} = \frac{3}{4}$ and with the empirical evidence in the fourth column of Table 38.4. ❖

k	x_k	$\Delta x_k = x_{k+1} - x_k$	$\dfrac{\Delta x_{k+1}}{\Delta x_k}$
0	1	−0.191613	0.8515147856
1	0.808387	−0.163162	0.8384593354
2	0.645225	−0.136804	0.8247462283
3	0.508421	−0.112829	0.8085081933
4	0.395592	−0.091223	0.7920919548
5	0.304369	−0.072257	0.7780616363
6	0.232112	−0.056220	0.7676206188
7	0.175891	−0.043156	0.7606068775
8	0.132735	−0.032825	0.7562119694
9	0.099910	−0.024822	
10	0.075088		

TABLE 38.4   Newton iteration applied to $f(x) = \ln(\cos x) + \frac{1}{2} \sin x^2$ in Example 38.5

MODIFIED NEWTON ITERATION   For a function of the form $F(x) = (x - r)^m Q(x)$, with $Q(r) \neq 0$, which has a zero of order m at $x = r$, the modified Newton iteration function

$$G_m(x) = x - \frac{m F(x)}{F'(x)}$$

generates a fixed-point iteration that converges at least quadratically. Direct computation leads to $G_m'(r) = 0$ and $G_m''(r) = \frac{2 Q'(r)}{m Q(r)}$. Since the first derivative vanishes at $x = r$, the modified Newton iteration converges more rapidly than linearly. If $Q'(r) \neq 0$, the convergence is quadratic, with $C_Q = -\frac{1}{2}(\frac{2}{m})\frac{Q'(r)}{Q(r)}$. If $Q'(r) = 0$, then the convergence is even more rapid than quadratic. In fact, if $Q'(r) = 0$, then $G_m'''(r) = \frac{6Q''(r)}{mQ(r)}$, which means that for cubic convergence of the modified Newton iteration function, the convergence constant would be $C_C = \frac{1}{3!}(\frac{6}{m})\frac{Q''(r)}{Q(r)}$.

EXAMPLE 38.6   The modified Newton iteration function for $f(x) = \ln(\cos x) + \frac{1}{2}\sin x^2$ is

$$G_m(x) = x - 4\frac{\ln(\cos x) + \frac{1}{2}\sin x^2}{x \cos x^2 - \tan x}$$

Column 2 of Table 38.5 lists the Newton iterates x_k, starting with $x_0 = 1$. Column 3 lists the increments $\Delta x_k = x_{k+1} - x_k$, and column 4 lists the ratios $\Delta x_{k+1}/(\Delta x_k)^3$. Using $Q(x) = \frac{f(x)}{x^4}$, we find $C_C = \frac{Q''(0)}{4Q(0)} = \frac{19}{30} = 0.6\overline{33}$, in agreement with the ratios in the fourth column.   ❖

k	x_k	$\Delta x_k = x_{k+1} - x_k$	$\dfrac{\Delta x_{k+1}}{(\Delta x_k)^3}$
0	1	-0.76645341	0.5023607709
1	0.2335465946	-0.22618979	0.6357048088
2	$0.7356806958 \times 10^{-2}$	$-0.73565548 \times 10^{-2}$	0.6333377124
3	$0.2521498847 \times 10^{-6}$	$-0.25214988 \times 10^{-6}$	
4	$0.1015333389 \times 10^{-19}$		

TABLE 38.5   Modified Newton iteration applied to $f(x) = \ln(\cos x) + \frac{1}{2}\sin x^2$ in Example 38.6

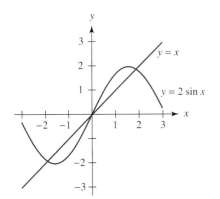

FIGURE 38.5   The three solutions of $x = 2\sin x$ in Example 38.7

STEFFENSEN'S METHOD   As interesting as the modified Newton iteration might be, the reader should be skeptical of its practical value. In order to determine the order of a zero, one must know the location of the zero, in which case, why recompute it?

As a more practical alternative in the face of linear convergence of Newton's method at an apparent multiple root, consider applying Aitken's acceleration process to a Newton iteration when it appears that convergence is being retarded by an apparent multiple root. The following implementation of this acceleration technique is attributed to Steffensen.

When x_{k-2}, x_{k-1}, x_k, three successive Newton iterates, appear to be converging linearly, use Aitken's formula

$$x_{k+1} = x_k - \frac{(x_k - x_{k-1})^2}{x_k - 2x_{k-1} + x_{k-2}} \tag{38.3}$$

to generate x_{k+1}, the next iterate in the sequence. From x_{k+1}, compute x_{k+2} and x_{k+3}, two additional Newton iterates, and again apply Aitken acceleration to the triple $(x_{k+1}, x_{k+2}, x_{k+3})$. Continue in this manner until convergence. (The reader is cautioned that Steffensen's algorithm is something entirely different in [49].)

EXAMPLE 38.7   The fixed-point iteration $x = g(x) = 2\sin x$ should converge linearly to $r = 1.895494267$, one of the three solutions of the equation $x = 2\sin x$. (See Figure 38.5.) The convergence is linear since $g'(r) = -0.6380450482 \neq 0$.

Column 2 of Table 38.6 lists x_k, the Steffensen iterates computed via (38.3). Column 3 lists $\Delta x_k = x_{k+1} - x_k$, the increments between the Steffensen iterates; and column 4 lists

k	x_k	$\Delta x_k = x_{k+1} - x_k$	$\dfrac{\Delta x_{k+1}}{(\Delta x_k)^2}$
0	1	1.2324295	−0.2619674388
1	2.232429	−0.3978977	0.3770358299
2	1.834532	0.596933×10^{-1}	0.3560330600
3	1.894225	0.126865×10^{-2}	0.3689291231
4	1.895494	0.593780×10^{-6}	0.3691627331
5	1.895494	0.130156×10^{-12}	0.3691628421
6	1.895494	0.625398×10^{-26}	
7	1.895494		

TABLE 38.6 Fixed-point iteration accelerated by Steffensen's method, converges quadratically in Example 38.7

the ratios $\Delta x_{k+1}/(\Delta x_k)^2$, whose relative constancy argues for the quadratic convergence of the iteration. ❖

Newton Iteration in the Complex Plane

Newton's method will converge to complex roots if the iteration can be moved off the real line. This can be done by starting the iteration at a complex number or by having the iteration itself generate a complex value. Thus, a Newton iteration can leave the real line to converge to a point in the complex plane or meander through the complex plane to converge to a point on the real line.

In the days before the availability of electronic computing devices, working with complex numbers was tedious. In the early days of computing when the only realistic alternative was the language FORTRAN, manipulations with complex numbers became feasible but not easy. In modern computer algebra systems, complex numbers are routinely handled and a Newton iteration that swings into the complex plane does not require a change of data type as was required in a FORTRAN program.

Lest there be premature celebration, we point out that the essential difficulty with using Newton's method in the complex plane is locating the roots accurately enough so that viable starting values can be obtained. Because there is "so much extra room" in the complex plane, iterations that do not start sufficiently near to a root might converge to a different root or to the desired root more slowly than expected.

EXAMPLE 38.8 An example of a Newton iteration that leaves the real line of its own accord, converging to a fixed point in the complex plane, is afforded by the equation $f(x) = x - \sqrt{x-5} = 0$. Of course, this example is just a disguised form of the quadratic equation $x^2 - x + 5 = 0$, which has the complex roots $\frac{1 \pm i\sqrt{19}}{2} = 0.5 \pm 2.179449472i$. Starting from $x_0 = 10$, the Newton iterates are then 0, $0.4761904763 + 2.129588551i$, $0.4999847542 + 2.179417993i$, and $0.5000000001 + 2.179449472i$. As predicted, Newton iteration has left the real line and converged to a fixed point in the complex plane. ❖

EXAMPLE 38.9 The equation $f(x) = \sqrt{x} - \frac{x}{3} = 0$ is a disguised form of the quadratic equation $x^2 - 9x = 0$, whose roots are the real numbers 0 and 9. The Newton iterates, starting from the real number $x_0 = -5$, are listed in Table 38.7. The iteration, starting from the real line, has wandered into the complex plane, only to converge to the (real) fixed point $x = 9$. ❖

$x_0 = -5$	$x_4 = 8.952612441 - 0.087681943i$
$x_1 = 1.551724139 + 2.313173770i$	$x_5 = 8.999844224 + 0.00023038420i$
$x_2 = 4.309230249 - 3.113854023i$	$x_6 = 9.000000000 - 0.19938 \times 10^{-8}i$
$x_3 = 7.917951296 + 1.367009384i$	$x_7 = 9.000000000$

TABLE 38.7 In Example 38.9, Newton iteration passes through the complex plane

Anomalies

In [24], two functions are given for which Newton's method exhibits pathological behavior. For the first, there is no zero, but Newton's method will claim there is. For the second, there is a zero that Newton's method will never find. Cycling, where the iterates repeat cyclically, is another possibility with Newton's method. We give examples of each of these behaviors.

EXAMPLE 38.10

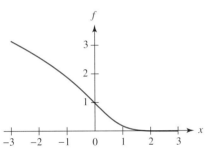

FIGURE 38.6 The function in Example 38.10 has no real zeros

Taken from [24], the function $f(x) = \dfrac{e^{-x/2}(x+\sqrt{x^2+1})}{\sqrt{x+\sqrt{x^2+1}}}$, part of whose graph is shown in Figure 38.6, has no real zeros. The behavior of the function as $|x| \to \infty$ is determined by the limits $\lim_{x\to-\infty} f(x) = \infty$ and $\lim_{x\to\infty} f(x) = 0$.

Newton's method applied to this function will falsely claim $f(x) = 0$ has a root $x = r > 0$. Column 2 of Table 38.8 lists x_k, the Newton iterates starting from $x_0 = 0$; and column 3 lists the increments $\Delta x_k = x_{k+1} - x_k$. Although the iterates in column 2 do not seem to be converging, the increments Δx_k appear to be decreasing, and any test based on the shrinking of the increments will suggest there is a root at some sufficiently large x. In fact, if computed in exact arithmetic, the iterates are $x_k = \sqrt{k}$, so the sequence of iterates does not converge but the increments $\Delta x_k = \sqrt{k+1} - \sqrt{k}$ tend to zero. The limit $\lim_{k\to\infty} \sqrt{k+1} - \sqrt{k}$, an example of the indeterminate form $\infty - \infty$, is seen to be zero if the two radicals are expanded for large k, giving $\Delta x_k = O(x^{-1/2})$.

Thus, if Newton's method is implemented to stop when successive increments become sufficiently small, it is possible for the iteration to declare there is a root where in reality there is none. For this example, even the residual $f(\sqrt{k})$ will be deceptively small for k sufficiently large because the right "tail" of $f(x)$ tends to zero. ❖

k	x_k	$\Delta x_k = x_{k+1} - x_k$
0	0	1.0
1	1.0	0.414213562
2	1.414213562	0.317837245
3	1.732050807	0.267949192
4	1.999999999	0.236067978
5	2.236067977	0.213421766
6	2.449489743	0.196261568
7	2.645751311	0.182675815
8	2.828427126	0.171572875
9	3.000000001	0.162277660
10	3.162277661	0.154347130
11	3.316624791	

TABLE 38.8 Newton iteration applied to the function in Example 38.10

EXAMPLE 38.11   Also taken from [24], the function $f(x) = \sqrt[3]{x}e^{-x^2}$, graphed in Figure 38.7, has a single zero, $x = 0$, to which Newton's method cannot converge. As in Example 38.10, $f(x)$ tends to zero as $|x|$ gets large.

The Newton iteration function is

$$g(x) = x - \frac{f(x)}{f'(x)} = \frac{2x(3x^2 + 1)}{6x^2 - 1}$$

for which $g'(0) = -2$. Hence, Newton iteration for this function cannot converge. Worse yet, any stopping rule based on the increments $\Delta x_k = x_{k+1} - x_k$ will declare there is a root at some $|x| > 0$ because as $|x_k|$ gets large, the increments get small. Indeed, if at some point in the iteration we have $x_k = a$, then the increment $x_{k+1} - x_k$ is $g(a) - a$; that difference is plotted as a function of a in Figure 38.8. Since the iteration is known to diverge, a will eventually get large; so the corresponding increments will become vanishingly small. Thus, since $|x_k|$ must eventually get large, the limiting value of the increments will be zero. ❖

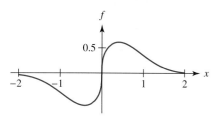

FIGURE 38.7   The function $f(x) = \sqrt[3]{x}e^{-x^2}$ in Example 38.11

EXAMPLE 38.12   The function $f(x) = \sqrt{x^2 - 2x - 1}$ has two real zeros, $1 \pm \sqrt{2}$. The Newton iteration function

$$g(x) = x - \frac{f(x)}{f'(x)} = \frac{x + 1}{x - 1}$$

has the property that $g(g(x)) = x$. Thus, no matter where the iteration is started, the second iterate will be the initial one; hence, Newton's method will simply cycle between x_0, x_1, and $x_2 = x_0$. The sufficient condition for convergence of a fixed-point iteration is $|g'(r)| < 1$, but we find $g'(1 \pm \sqrt{2}) = -1$. At neither zero of $f(x)$ is the sufficient condition for convergence satisfied. ❖

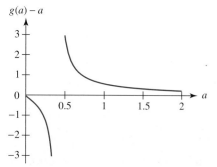

FIGURE 38.8   The function $g(a) - a$ plotted as a function of a in Example 38.11

EXERCISES 38.3–Part A

A1. Obtain the Newton iteration function $g(x) = x - \frac{f(x)}{f'(x)}$ when $f(x) = x^2 - 6x + 8$.

A2. For $g(x)$ in Exercise A1, show that $g'(2) = g'(4) = 0$. Determine C_Q for fixed-point iteration with $g(x)$.

A3. If $x_0 = 2.5$, obtain the next three Newton iterates for $g(x)$ in Exercise A1.

A4. Obtain the Newton iteration function $g(x) = x - \frac{f(x)}{f'(x)}$ when $f(x) = x^2(x - 1)$.

A5. For $g(x)$ in Exercise A4, show that $g'(0) = \frac{1}{2}$. Explain why this is so.

EXERCISES 38.3–Part B

B1. Show that $x = 0$ is a triple zero of $f(x) = \tan x - xe^{-x^2}$ by
  (a) showing $f(0) = f'(0) = f''(0) = 0$ but $f'''(0) \neq 0$.
  (b) using a Maclaurin series to write $f(x) = x^3 Q(x)$, where $Q(0) \neq 0$.

B2. For $f(x)$ in Exercise B1, form the modified Newton iteration function $G_3 = x - \frac{3f(x)}{f'(x)}$. Show that $G_3'(0) = 0$ and that $G_3''(0) = \frac{2Q'(0)}{3Q(0)}$.

In Exercises B3–14:

(a) Obtain the Newton iteration function $g(x) = x - \frac{f(x)}{f'(x)}$.

(b) Use Newton's method to compute each of the real zeros of $f(x)$.

(c) Show empirically that the convergence is quadratic, and estimate C_Q, the quadratic convergence constant.

(d) For each real zero r, show analytically that $g'(r) = 0$.

(e) For each real zero r, verify that $g''(r) = \frac{f''(r)}{f'(r)}$ and that
$$C_Q = -\frac{g''(r)}{2}.$$

B3. $f(x) = 6x^3 + 8x^2 + 10x + 1$ **B4.** $f(x) = 9x^3 - 3x^2 - 7x + 1$

B5. $f(x) = 4x^3 - x^2 - 5x + 12$ **B6.** $f(x) = 11x^3 + 5x^2 + 4x + 6$

B7. $f(x) = 7x^4 + x^3 - 7x^2 + x + 1$

B8. $f(x) = 2x^4 - 3x^3 + 3x^2 - x - 3$

B9. $f(x) = 9x^4 + 4x^3 + 6x^2 - 3x - 2$

B10. $f(x) = 3x^4 - 5x^3 - 5x^2 - 5x + 1$

B11. $f(x) = \dfrac{97 + 149x - 24x^2}{84x^2 - 94x - 57}$ **B12.** $f(x) = \dfrac{120x^2 + 67x - 61}{25x^2 + 37x - 46}$

B13. $f(x) = \dfrac{246x^2 + 194x - 39}{74x^2 + 73x - 64}$ **B14.** $f(x) = \dfrac{83x^2 - 17x - 81}{69x^2 + 9x - 58}$

In Exercises B15–24, let $g(x) = x - \frac{f(x)}{f'(x)}$ be the Newton iteration function and $g_m(x) = x - m\frac{f(x)}{f'(x)}$ be the modified Newton iteration function.

(a) Verify that $x = r$ is a zero of $f(x)$, find m, the order of that zero, and determine $Q(x)$, the companion factor of $(x - r)^m$.

(b) Show empirically that Newton's method converges linearly at the multiple zero $x = r$, and estimate C_L, the convergence constant.

(c) Show that $C_L = \frac{m-1}{m}$.

(d) Show that $g'(r) = C_L$, in agreement with part (c).

(e) Apply Steffensen's method, verifying empirically that the convergence to the multiple zero is now quadratic.

(f) Show that fixed-point iteration with $g_m(x)$ converges quadratically to the multiple zero, $x = r$.

(g) Empirically estimate C_Q, the quadratic convergence constant for $g_m(x)$.

(h) Show that the estimate in part (g) is consistent with $C_Q = -\frac{1}{m}\frac{Q'(r)}{Q(r)}$, provided $Q'(r) \neq 0$.

B15. $f(x) = 4x^6 - 16x^5 + 30x^4 - 40x^3 + 40x^2 - 24x + 6, r = 1$

B16. $f(x) = -4x^4 + 24x^3 + 4x^2 - 160x + 64, r = 4$

B17. $f(x) = 4x^5 - 91x^4 + 735x^3 - 2401x^2 + 2401x, r = 7$

B18. $f(x) = -4x^4 + 16x^3 - 19x^2 + 12x - 12, r = 2$

B19. $f(x) = -2x^4 + 31x^3 - 147x^2 + 245x - 343, r = 7$

B20. $f(x) = \dfrac{40x^3 + 624x^2 + 1512x - 7776}{87x^2 + 82x - 79}, r = -9$

B21. $f(x) = \dfrac{48x^3 - 293x^2 + 40x + 1456}{48x^2 + 21x - 1}, r = 4$

B22. $f(x) = \dfrac{38x^4 - 468x^3 + 1752x^2 - 1420x - 1950}{46x^2 + 46x - 71}, r = 5$

B23. $f(x) = \dfrac{64x^4 - 1066x^3 + 4696x^2 - 1536x - 4608}{20x^2 + 75x + 15}, r = 8$

B24. $f(x) = \dfrac{7x^5 + 6x^4 - 111x^3 - 38x^2 + 732x - 744}{19x^2 + 54x + 18}, r = 2$

In Exercises B25–34, use Newton iteration to determine all complex zeros in the complex plane.

B25. $f(x) = 4x^4 - 5x^3 + 5x^2 - 3x + 1$

B26. $f(x) = 7x^4 - 5x^3 - 9x^2 - 6x + 3$

B27. $f(x) = 5x^4 - x^3 - 3x^2 + 5x - 8$

B28. $f(x) = 4x^4 + 2x^3 - x^2 + 12x + 3$

B29. $f(x) = 8x^4 - 5x^3 + 6x^2 - 12x - 9$

B30. $f(x) = 3x^4 + 4x^3 - 6x^2 - 7x - 10$

B31. $f(x) = 2x^4 - 9x^3 + 2x^2 + 9x + 10$

B32. $f(x) = 7x^4 - 7x^3 + 4x^2 - 10x - 2$

B33. $f(x) = 5x^4 + 7x^3 - x^2 - 5x - 7$

B34. $f(x) = 2x^4 + 2x^3 + 2x - 11$

38.4 The Secant Method

Slope Methods

Newton's method, which is fixed-point iteration $x_{k+1} = g(x_k)$, with $g(x) = x - \frac{f(x)}{f'(x)}$ as the iteration function, is an example of a *slope method*. Such methods approximate the root of $f(x) = 0$ with the x-intercept of a line of appropriate slope. In Newton's method, the slope is that of the tangent line at x_k. In the secant method, the approximating x-intercept is obtained from a secant line connecting $(x_k, f(x_k))$ and $(x_{k-1}, f(x_{k-1}))$.

The Secant Algorithm

The secant method is the fixed-point iteration defined by

$$x_{k+1} = x_k - \frac{f(x_k)(x_k - x_{k-1})}{f(x_k) - f(x_{k-1})} \tag{38.4}$$

which is of the form $x_{k+1} = g(x_k, x_{k-1})$, and hence computes a new member of the iteration sequence $\{x_k\}$ from *two* previous members. Thus, to start the iteration, *two* initial guesses must be made. However, it is not uncommon to take $x_1 = (1 \pm \varepsilon)x_0$, where $\varepsilon = 0.01$. This represents a 1% perturbation in x_0, and the sign rarely matters.

The convergence rate for the secant method is better than linear but not quite quadratic. In fact, it is *superlinear*, with $p = \frac{1+\sqrt{5}}{2} = 1.618034$, so increments between iterates tend to obey $|\Delta x_{k+1}|/|\Delta x_k|^p \to C_S$, where C_S is a nonzero constant.

The iteration formula (38.4) can be obtained from Newton's method by approximating the derivative $f'(x_k)$ with the difference quotient $(f(x_k) - f(x_{k-1}))/(x_k - x_{k-1})$. Alternatively, let x_{k-1} and x_k be two successive approximations to the root r. The secant line through the points $(x_{k-1}, f(x_{k-1}))$ and $(x_k, f(x_k))$ has the equation

$$y = \frac{f(x_k) - f(x_{k-1})}{x_k - x_{k-1}}(x - x_k) + f(x_k) \tag{38.5}$$

Setting $y = 0$ and solving for x gives the x-intercept x_{k+1}, whose formula is precisely (38.4). (See Exercise A1.)

EXAMPLE 38.13   The equation $f(x) = \arctan x - \sqrt{2x - 5} = 0$ has a root in the interval $[3, 4]$, as seen in Section 38.3. The secant-iteration function for this equation is

$$x_{k+1} = x_k - \frac{(\arctan x_k - \sqrt{2x_k - 5})(x_k - x_{k-1})}{\arctan x_k - \sqrt{2x_k - 5} - \arctan x_{k-1} + \sqrt{2x_{k-1} - 5}}$$

The secant iterates x_k, starting with $x_0 = 2.6$ and $x_1 = (1 + 0.02)x_0$, are listed in column 2 of Table 38.9. Column 3 lists the increments $\Delta x_k = x_{k+1} - x_k$, and column 4 lists the ratios $|\Delta x_{k+1}|/|\Delta x_k|^p$, with $p = \frac{1+\sqrt{5}}{2}$. The eventual constancy of these ratios supports the claim that the secant method converges superlinearly, with parameter $p = \frac{1+\sqrt{5}}{2}$. (The

k	x_k	$\Delta x_k = x_{k+1} - x_k$	$\dfrac{|\Delta x_{k+1}|}{|\Delta x_k|^p}$
0	2.6	0.052	41.98
1	2.652	0.35116101	1.139
2	3.0031610085	0.20952293	1.154
3	3.2126839339	$0.92046650 \times 10^{-1}$	0.543
4	3.3047305839	$0.11449900 \times 10^{-1}$	0.547
5	3.3161804843	$0.39567358 \times 10^{-3}$	0.469
6	3.3165761578	$0.14630393 \times 10^{-5}$	0.497
7	3.3165776209	$0.17999261 \times 10^{-9}$	0.477
8	3.3165776210	$0.81566365 \times 10^{-16}$	0.489
9	3.3165776210	$0.45468705 \times 10^{-26}$	0.482
10	3.3165776210	$0.1148604 \times 10^{-42}$	
11	3.3165776210	0.1×10^{-48}	

TABLE 38.9   Secant method iteration in Example 38.13

computations were done in extended precision—at 50 digits—with a computer algebra system.) ❖

Multiple Roots

Like Newton's method, when the secant method converges to a root of multiplicity $m > 1$, it converges linearly. For example, as we saw in Section 38.3, the equation $f(x) = \ln(\cos x) + \frac{1}{2}\sin x^2 = 0$ has a root of multiplicity $m = 4$ at $x = 0$. Column 2 of Table 38.10 lists the secant iterates x_k, starting with $x_0 = 1$ and $x_1 = (1 + 0.02)x_0$. Column 3 lists the increments $\Delta x_k = x_{k+1} - x_k$, and column 4 lists the ratios $\Delta x_{k+1}/\Delta x_k$. The eventual constancy of these ratios supports the claim that the secant method converges linearly, with C_L approximately 0.819. (The computations were done in extended precision.)

k	x_k	$\Delta x_k = x_{k+1} - x_k$	$\dfrac{\Delta x_{k+1}}{\Delta x_k}$
0	1	0.02	−10.1583921
1	1.02	−0.203168	0.4750245
2	0.816832	−0.096510	1.1752695
3	0.720322	−0.113425	0.7979081
4	0.606898	−0.090503	0.9113336
⋮	⋮	⋮	⋮
27	0.712579×10^{-2}	-0.128849×10^{-2}	0.8191889
28	0.583730×10^{-2}	-0.105552×10^{-2}	0.8191835
29	0.478178×10^{-2}	-0.864663×10^{-3}	0.8191799
30	0.391711×10^{-2}	-0.708314×10^{-3}	
31	0.320880×10^{-2}		

TABLE 38.10 Secant method applied to $f(x) = \ln(\cos x) + \frac{1}{2}\sin x^2$ which has a zero of order $m = 4$ at $x = 0$

EXERCISES 38.4–Part A

A1. Obtain the x-intercept for (38.5); show that if it is taken as x_{k+1}, the formula for it is precisely (38.4).

A2. Let $x_0 = 0$, $x_1 = 1$, and $f(x) = x^2 - 6x + 8$. Obtain x_k, $k = 2, 3, 4$, the next three iterates generated by the secant method.

EXERCISES 38.4–Part B

B1. It is shown in [49] that at a simple zero $x = r$, the convergence constant in the secant method is $C_S = \left|\frac{f''(r)}{2f'(r)}\right|^{p-1}$, where $p - 1 = \frac{1+\sqrt{5}}{2} - 1 = \frac{\sqrt{5}-1}{2} \doteq 0.618033989$. Show that the data in Table 38.9 are consistent with this claim.

In Exercises B2–7, use the secant method to find the real zeros of the given $f(x)$. Show that the convergence is superlinear, with $p = \frac{1+\sqrt{5}}{2} \doteq 1.618033989$. Show that C_S is as described in Exercise B1.

B2. $f(x)$ in Exercise B3, Section 38.3

B3. $f(x)$ in Exercise B4, Section 38.3
B4. $f(x)$ in Exercise B7, Section 38.3
B5. $f(x)$ in Exercise B8, Section 38.3
B6. $f(x)$ in Exercise B11, Section 38.3
B7. $f(x)$ in Exercise B12, Section 38.3

For the functions given in each of Exercises B8–13:

(a) Show that the secant method converges linearly at the given multiple zero.

(b) Apply Steffensen's algorithm (from Section 38.3) to the linearly converging sequence in part (a) as follows. Let $x_0 = r + 1$, $x_1 = r + 0.9$ for the secant method, and apply acceleration, just once, to x_{10}, x_{11}, x_{12}. Compare the accelerated value to x_{12}.

B8. Exercise B15, Section 38.3   **B9.** Exercise B16, Section 38.3

B10. Exercise B17, Section 38.3   **B11.** Exercise B20, Section 38.3

B12. Exercise B22, Section 38.3   **B13.** Exercise B24, Section 38.3

For the functions $f(x)$ given in Exercises B14–19:

(a) Write $g(x) = x - \frac{f(x)}{f'(x)} - \frac{f''(x)}{2f'(x)} \cdot \left(\frac{f(x)}{f'(x)}\right)^2$ as the iteration function.

(b) Use the iteration $x_{k+1} = g(x_k)$ to find the real zeros of $f(x)$.

(c) Show analytically that $g'(r) = g''(r) = 0$ at $x = r$, a simple zero of $f(x)$. Hence, the rate of convergence for the iteration is cubic.

B14. $f(x)$ in Exercise B5, Section 38.3

B15. $f(x)$ in Exercise B6, Section 38.3

B16. $f(x)$ in Exercise B9, Section 38.3

B17. $f(x)$ in Exercise B10, Section 38.3

B18. $f(x)$ in Exercise B13, Section 38.3

B19. $f(x)$ in Exercise B14, Section 38.3

In Exercises B20–23, apply the iteration $x_{k+1} = g(x_k)$, where $g(x) = x - \frac{f(x)}{f'(x)} - \frac{f''(x)}{2f'(x)} \cdot \left(\frac{f(x)}{f'(x)}\right)^2$. Each function has a multiple zero. What is the rate of convergence of this iteration at a multiple zero? Obtain analytical and/or empirical evidence to support your answer.

B20. Exercise B18, Section 38.3   **B21.** Exercise B19, Section 38.3

B22. Exercise B21, Section 38.3   **B23.** Exercise B23, Section 38.3

38.5 Muller's Method

Description

Newton's method replaces $f(x)$ with the line tangent to $f(x)$ at $x = x_k$ and then takes x_{k+1} as the x-intercept of this tangent line. The secant method replaces $f(x)$ with a (secant) line through the two points $(x_{k-1}, f(x_{k-1}))$ and $(x_k, f(x_k))$ and then takes x_{k+1} as the x-intercept on this line.

In Muller's method, $f(x)$ is replaced by a quadratic polynomial (a parabola) through the three points $(x_{k-2}, f(x_{k-2}))$, $(x_{k-1}, f(x_{k-1}))$, and $(x_k, f(x_k))$ and then x_{k+1} is taken as the "x-intercept" (there are two) that is closest to x_k. We use the phrase "x-intercept" carefully because the zeros of a quadratic may be complex numbers, in which case the geometric notion of "crossing an axis" might not apply.

EXAMPLE 38.14   The function $f(x) = \arctan x - \sqrt{2x - 5}$ has a real zero in the interval $[3, 4]$ but evaluates to a complex number if $x < \frac{5}{2}$. This will allow us to explore the behavior of Muller's method as its iterates enter the complex plane. Start Muller's iteration with the three points $(x_k, f(x_k))$, $k = 1, \ldots, 3$, given by $(1, f(1))$, $(2, f(2))$, $(3, f(3))$ and interpolated by the quadratic

$$y = (-0.58993 + 0.13397i)x^2 + (2.0915 + 0.33013i)x - 0.71621 - 2.1962i$$

whose two zeros are $0.14334 + 1.2201i$ and $3.1073 + 0.077755i$, at distances 3.1 and 0.13, respectively, from $(3, f(3))$. Selecting the root corresponding to the smaller of these two increments yields the next Muller iterate, $x_4 = 3.1073 + 0.077755i$.

The three points $(x_k, f(x_k))$, $k = 2, \ldots, 4$, are interpolated by the quadratic

$$y = (-0.056216 - 0.87243i)x^2 + (-0.57703 + 5.3621i)x + 2.4861 - 8.2346i$$

whose two zeros are $2.8194 + 1.0794i$ and $3.2590 - 0.026353i$, at distances 1.04 and 0.18, respectively, from $(x_4, f(x_4))$. Selecting the root corresponding to the smaller of these two increments yields the next Muller iterate, $x_5 = 3.2590 - 0.026353i$.

$x_6 = 3.318162834 + 0.147 \times 10^{-4}i$
$x_7 = 3.316571952 - 0.159 \times 10^{-6}i$
$x_8 = 3.316577620 - 0.799 \times 10^{-10}i$
$x_9 = 3.316577623 - 0.262 \times 10^{-11}i$

TABLE 38.11   Muller's method in Example 38.14

The three points $(x_k, f(x_k))$, $k = 3, \ldots, 5$, are interpolated by a quadratic, and the iteration continues, yielding the additional Muller iterates listed in Table 38.11.

The ninth Muller iterate gives a reasonable approximation to the known root $r = 3.316577621$, but computing $x_{10} = 3.316572477 - 0.53 \times 10^{-4}i$ yields a less accurate result caused by roundoff error in the application of the quadratic formula. A strategy for implementing the quadratic formula, similar to that described in Section 37.1, is used in the Muller algorithm developed next. ❖

Implementation

Each cycle of the Muller algorithm interpolates three points labeled (x_k, f_k), $k = 1, \ldots, 3$, where $f_k = f(x_k)$. The interpolating polynomial is written in the form $y = AZ^2 + BZ + C$, where $Z = x - x_3$, $C = f_3$, and A and B are given in Table 38.12. The zeros of the interpolating quadratic are

$$Z_\pm = \frac{-B \pm \sqrt{B^2 - 4AC}}{2A} = -\frac{2C}{B \pm \sqrt{B^2 - 4AC}}$$

and the zero closest to x_3 is $x_3 + \min\{|Z_\pm|\}$. Hence, the correct choice for Z is $-\frac{2C}{\sigma}$, where $\sigma = \max\{|B \pm \sqrt{B^2 - 4AC}|\}$. The complete algorithm appears in a procedure written in the accompanying Maple worksheet.

$$\Delta x_1 = x_2 - x_1 \qquad \Delta f_1 = f_2 - f_1 \qquad A = \frac{\frac{\Delta f_2}{\Delta x_2} - \frac{\Delta f_1}{\Delta x_1}}{\Delta x_1 + \Delta x_2}$$

$$\Delta x_2 = x_3 - x_2 \qquad \Delta f_2 = f_3 - f_2 \qquad B = A\Delta x_2 + \frac{\Delta f_2}{\Delta x_2}$$

TABLE 38.12   Implementing Muller's method

EXAMPLE 38.15 If the Muller algorithm in the accompanying Maple worksheet is applied to the function $f(x) = \arctan x - \sqrt{2x - 5}$, with starting values $x_k = k$, $k = 1, \ldots, 3$, the iterates in Table 38.13 are obtained. The greater care used in implementing the quadratic formula results in improved accuracy. ❖

$x_1 = 1$
$x_2 = 2$
$x_3 = 3$
$x_4 = 3.107324776 + 0.0777546i$
$x_5 = 3.258957576 - 0.0263527i$

$x_6 = 3.318162834 + 0.147429 \times 10^{-4}i$
$x_7 = 3.316571953 - 0.159366 \times 10^{-6}i$
$x_8 = 3.316577621 - 0.561067 \times 10^{-10}i$
$x_9 = 3.316577621 - 0.153784 \times 10^{-14}i$

TABLE 38.13   Muller iterates in Example 38.15

Comments

Like Newton's method and the secant method, Muller's method converges linearly at a repeated root. In general, though, Muller's method converges superlinearly, with parameter

p equal to the positive root of $x^3 - x^2 - x - 1 = 0$. (See, e.g., [4].) This value is usually listed in the literature as 1.840, but it is actually 1.839286755.

First published in 1956, Muller's method has found favor because it typically converges to *some* root of a given polynomial, no matter what the starting values. Moreover, it will find both real and complex zeros of transcendental functions.

EXERCISES 38.5

1. Use Muller's method to find the positive root of $x^3 - x^2 - x - 1 = 0$.

In Exercises 2–5:

(a) Use Muller's method to find the real and complex zeros of the given functions.

(b) For each zero computed in part (a), show empirically (where possible) that the convergence is superlinear, with $p = 1.839286755$.

(c) For each zero $x = r$ computed in part (a), show (where possible) that the convergence constant C_M is consistent with the claim $C_M = \left| \frac{f'''(r)}{6f'(r)} \right|^{(p-1)/2}$, made in [4], where p is given in part (b).

2. $f(x)$ in Exercise B25, Section 38.3

3. $f(x)$ in Exercise B28, Section 38.3

4. $f(x)$ in Exercise B31, Section 38.3

5. $f(x)$ in Exercise B34, Section 38.3

In Exercises 6–11, investigate the convergence of Muller's method for the indicated multiple zeros.

6. Exercise B15, Section 38.3   **7.** Exercise B16, Section 38.3

8. Exercise B17, Section 38.3   **9.** Exercise B20, Section 38.3

10. Exercise B22, Section 38.3   **11.** Exercise B24, Section 38.3

Chapter Review

1. Given an equation of the form $f(x) = 0$, describe how fixed-point iteration might be used to obtain a root. Discuss the strong and weak points of this technique. Discuss the convergence properties. Give examples of convergent and divergent fixed-point iterations.

2. Demonstrate the first few steps of fixed-point iteration used to solve the equation $x^3 + x - 1 = 0$.

3. Describe the bisection algorithm. What are its strong and weak points? What is its average rate of convergence?

4. Demonstrate the first few steps of the bisection method applied to the equation $x^3 + x - 1 = 0$.

5. State the Newton iteration algorithm. Under what conditions does the method converge quadratically? Under what conditions will it converge linearly?

6. Illustrate the first few steps of Newton iteration for obtaining a root of the equation $x^3 + x - 1 = 0$.

7. What does it mean to say that for the equation $f(x) = 0$, $x = r$ is a root of multiplicity m?

8. Discuss the convergence of Newton's method at a root of multiplicity m.

9. Describe Steffensen's algorithm. When can it be used, and what effect can it be expected to have?

10. Describe three ways in which Newton iteration can fail.

11. State the algorithm for the secant method. What is the anticipated rate of convergence of this method? What are the advantages of the method? Discuss the convergence at a root of multiplicity m.

12. Illustrate the first few steps of the secant method when applied to finding a root of the equation $x^3 + x - 1 = 0$.

13. Describe Muller's method. What are its advantages? What are the rates of convergence at isolated roots and multiple roots?

Chapter 39

Systems of Equations

INTRODUCTION Numeric solvers for linear systems of algebraic equations are at the heart of many computational techniques. The reader should not think of solving such systems as merely an end in itself. The principal tool for solving linear equations is *Gaussian elimination,* first seen in Section 12.5. Other tools include the *LU decomposition* of Section 34.1, but that factorization is, in fact, accomplished by the arithmetic of Gaussian elimination. These earlier sections did not consider the effects of roundoff error generated by computing on a finite-precision device. Mostly, the computations were done in exact arithmetic via the symbolic capacity of a computer algebra system.

Chapter 39 revisits *Gaussian elimination* and examines it from the perspective of accuracy in a climate susceptible to roundoff error. We demonstrate how, on a finite-precision machine, dramatic roundoff errors can occur. *Pivoting strategies* for decreasing roundoff error are considered in the exercises. The *operation counts* for Gaussian elimination are obtained in the section. For large systems, perhaps on the order of five million equations in five million unknowns, the total number of arithmetic operations Gaussian elimination consumes becomes a significant component of the solution strategy. The special case of the *tridiagonal system* is examined as a contrast to the general case. However, the extensive theory that tailors different strategies for different types of matrices are beyond the scope of this text.

We then discuss *condition numbers,* a device for quantizing the inherent tendency for a linear system to propagate roundoff error during numeric solution. Matrices with large condition numbers may generate solutions with almost no accurate significant digits, and such matrices can be found in small problems arising from a commonplace task such as least-squares fitting of a linear model to data. The condition number is, unfortunately, expensive to compute, so in practice, variants (see the exercises) are used to estimate the likelihood of catastrophic roundoff error.

If $\mathbf{x}_0$ is an approximate solution of the system $A\mathbf{x} = \mathbf{y}$ and if $\mathbf{r} = A\mathbf{x}_0$ is the *residual vector, iterative improvement* is a method for obtaining from $\mathbf{r}$ a better approximation to the actual solution. Packaged system solvers designed for finite-precision machines actually use this technique for obtaining the maximum accuracy with the resources available.

We also describe the *method of Jacobi,* an *iterative technique* that is the generalization of fixed-point iteration. Hence, the convergence is linear; but for problems so large that the full set of equations cannot be fit into memory at the same time, the Jacobi method is a realistic alternative. We also examine *Gauss–Seidel iteration,* a variant of Jacobi's method in which newly computed values are used immediately in the on-going calculations. Generally, Gauss–Seidel iteration converges more quickly than Jacobi's method, but the convergence is still linear.

The original *relaxation* technique, a pencil-and-paper iteration, was devised prior to World War II in an era when digital computers weren't yet even a dream. The technique is included as a pointer to the more modern *successive overrelaxation technique* (SOR), which can be loosely compared to Gauss–Seidel iteration with a damping factor called the *relaxation parameter*. If the relaxation parameter can be optimally chosen, the convergence properties of the iteration are greatly enhanced.

The chapter concludes with a discussion of iterative techniques for solving nonlinear systems of algebraic equations. *Fixed-point iteration* applied to such a system, converging linearly, is nothing more than the Jacobi method for nonlinear equations. The obvious variant is a fixed-point method that uses the immediate-update strategy of Gauss–Seidel iteration. *Newton iteration,* the generalization of Newton's method for a single equation in a single unknown is treated here, even though it is also an iterative algorithm. However, its quadratic convergence makes it a useful strategy, and some of the devices that make the method more efficient are discussed.

39.1   Gaussian Arithmetic

Gauss Elimination

REVIEW   In Section 12.5 we saw how to apply Gaussian elimination to solve the system of algebraic equations $A\mathbf{x} = \mathbf{y}$ by row-reducing $[A, \mathbf{y}]$, the augmented matrix, to an upper triangular form and using back substitution to determine $\mathbf{x}$, the vector of unknowns. In the exact arithmetic of Section 12.5, we would compute the solution of the system $A\mathbf{x} = \mathbf{y}$, where A, $\mathbf{y}$, and $[A, \mathbf{y}]$ are

$$
A = \begin{bmatrix} 0 & 2 & -4 \\ \frac{7}{10,000} & 4 & 8 \\ 9 & 7 & 6 \end{bmatrix}
\quad
\mathbf{y} = \begin{bmatrix} 4 \\ 6 \\ 1 \end{bmatrix}
\quad
[A, \mathbf{y}] = \begin{bmatrix} 0 & 2 & -4 & 4 \\ \frac{7}{10,000} & 4 & 8 & 6 \\ 9 & 7 & 6 & 1 \end{bmatrix}
$$

by row reduction of the augmented matrix to upper triangular form and back substitution, giving

$$
\begin{bmatrix} 9 & 7 & 6 & 1 \\ 0 & 2 & -4 & 4 \\ 0 & 0 & \frac{71,993}{4500} & -\frac{179,909}{90,000} \end{bmatrix}
\quad
\mathbf{x} = \begin{bmatrix} -\frac{84,000}{71,993} \\ \frac{1,259,951}{719,930} \\ -\frac{179,909}{1,439,860} \end{bmatrix}
\overset{\circ}{=} \begin{bmatrix} -1.166780104 \\ 1.750102093 \\ -0.1249489534 \end{bmatrix}
$$

where the exact solution x has also been expressed in floating-point form.

The astute reader will have noticed that because $A_{11} = 0$, Gaussian elimination cannot begin for this matrix until row 1 is interchanged with either row 2 or row 3. Since the upper triangular matrix produced by Gaussian elimination will have its first row unchanged and that row is the original third row, it is clear that the first and third rows were switched at the outset.

NUMERIC SOLUTION   We next examine the process of Gaussian elimination in floating-point arithmetic on a fixed-precision device. We simulate work on a five-digit device. Upon interchanging rows 1 and 2, converting the augmented matrix to floating-point form, and using row 1 to create a zero in the (3, 1)-position, we get

$$By = \begin{bmatrix} 0.00070000 & 4 & 8 & 6 \\ 0 & 2 & -4 & 4 \\ 9 & 7 & 6 & 1 \end{bmatrix} \rightarrow \begin{bmatrix} 0.00070000 & 4 & 8 & 6 \\ 0 & 2 & -4 & 4 \\ 0.0001 & -51,421 & -102,850 & -77,141 \end{bmatrix} = By1$$

Surprisingly, we do not obtain a zero in the (3, 1)-position because roundoff error is greatly exacerbated by our restriction to five significant digits. Indeed, to five digits, the multiplier $-\frac{By_{31}}{By_{11}}$ is $-12,857$, so that multiplication of row 1 by this factor yields -8.9999, and not -9, to be added to the (3, 1)-element. That is why we obtained 0.0001 instead of 0 for the new (3, 1)-entry.

In the typical implementation of Gaussian elimination on a fixed-precision device, this entry would not be calculated. A zero would simply be understood to exist at this location. However, we can see the kind of roundoff error that enters a calculation performed on a fixed-precision device.

Finally, row 2 is used to create a zero at the (3, 2)-location by adding to row 3, $-\frac{By_{32}}{By_{22}} = 25,711$ times row 2, giving the left-hand matrix in

$$\begin{bmatrix} 0.00070000 & 4 & 8 & 6 \\ 0 & 2 & -4 & 4 \\ 0.0001 & 1 & -205,690 & 25,699 \end{bmatrix} \rightarrow \begin{bmatrix} 0.00070000 & 4 & 8 & 6 \\ 0 & 2 & -4 & 4 \\ 0 & 0 & -205,690 & 25,699 \end{bmatrix}$$

(39.1)

Again, it is disconcerting to see a nonzero entry at the (3, 2)-position in the matrix on the left in (39.1); but when 25,711 multiplies the 2 in the (2, 2)-location, we get 51,422, not 51,421 as expected. Hence, we have another example of roundoff error being introduced because of the fixed-precision of the computing device. Remembering, however, that the typical Gaussian elimination code would simply write a zero in this location, we simply transform the upper triangular matrix to the right-hand matrix in (39.1). This is the matrix Gaussian elimination on a five-digit device would produce. If back substitution is now used to compute the solution $\mathbf{x}$, we will obtain the right-hand vector in

$$x_{\text{exact}} = \begin{bmatrix} -1.1668 \\ 1.7501 \\ -0.12495 \end{bmatrix} \qquad \begin{bmatrix} -1.2571 \\ 1.7501 \\ -0.12494 \end{bmatrix} = x_{\text{Gaussian}}$$

Finally, we consider the effect of interchanging not rows 1 and 2 but rows 1 and 3. The augmented matrix and its upper-triangularization are then

$$\begin{bmatrix} 9 & 7 & 6 & 1 \\ 0.00070000 & 4 & 8 & 6 \\ 0 & 2 & -4 & 4 \end{bmatrix} \rightarrow \begin{bmatrix} 9 & 7 & 6 & 1 \\ 0 & 3.9995 & 7.9995 & 5.9999 \\ 0 & 0 & -8.0002 & 0.9997 \end{bmatrix}$$

(39.2)

Back substitution now yields the right-hand vector in

$$\begin{bmatrix} -1.1668 \\ 1.7501 \\ -0.12495 \end{bmatrix} \qquad \begin{bmatrix} -1.2857 \\ 1.7501 \\ -0.12494 \end{bmatrix} \qquad \begin{bmatrix} -1.1668 \\ 1.7501 \\ -0.12496 \end{bmatrix}$$

with the left-hand vector being the exact solution and the middle vector being the solution arising from the interchange of rows 1 and 2. To quantize the difference in the solutions, we compute the Euclidean norm of the differences as measured against the exact solution, obtaining 0.11892 for the first solution by Gaussian elimination and 0.0000229 for the second. Interchanging the first and third rows produces less error than interchanging the first and second rows. This reduction in error is a result of dividing by 9 rather than by the much smaller 0.0007. Avoiding divisions by a small number generally leads to a reduction of roundoff error.

Hence, rows are interchanged, or pivoted, not only to avoid division by zero, but also to avoid division by small numbers. The strategy of interchanging rows to bring the largest remaining column element to the (k, k)-location when creating zeros in column k is called *partial pivoting* and is one of the techniques used in numeric code for Gaussian elimination to reduce roundoff error.

In the next section we will examine in more detail systems for which roundoff-reduction techniques are essential. Surprisingly, some linear systems of very modest size can generate debilitating roundoff error even on computing devices carrying 16 or more significant digits.

Counting Operations

Another important aspect of numeric Gaussian elimination is the total number of arithmetic operations that must be performed to solve $A\mathbf{x} = \mathbf{y}$, a linear system of n equations in n unknowns. Knowing how many operations are needed and having an idea of how fast each operation can be performed on a given computer enable us to determine the feasibility of solving large systems that arise in practice.

Some texts simply total all four kinds of arithmetic operations, and some distinguish between them. Table 39.1, essentially a reproduction of Table 3-1 on page 135 of [60], lists the number of operations it takes to row-reduce A to an upper triangular matrix U ($A \rightarrow U$), the number of operations it takes to row-reduce the column $\mathbf{y}$ in the augmented matrix $[A, \mathbf{y}]$ ($\mathbf{y} \rightarrow \mathbf{w}$), and the number of operations it takes to complete back substitution.

	$N(\pm)$	$N(*)$	$N(/)$	**Row Totals**
$A \rightarrow U$	$\dfrac{n(n-1)(2n-1)}{6}$	$\dfrac{n(n-1)(2n-1)}{6}$	$\dfrac{n(n-1)}{2}$	$\dfrac{n(n-1)(4n+1)}{6}$
$\mathbf{y} \rightarrow \mathbf{w}$	$\dfrac{n(n-1)}{2}$	$\dfrac{n(n-1)}{2}$	0	$n(n-1)$
Back substitution	$\dfrac{n(n-1)}{2}$	$\dfrac{n(n-1)}{2}$	n	n^2
Column totals	$\dfrac{n(n-1)(2n+5)}{6}$	$\dfrac{n(n-1)(2n+5)}{6}$	$\dfrac{n(n+1)}{2}$	$\dfrac{n(4n^2+9n-7)}{6}$

TABLE 39.1   Count of arithmetic operations needed for solving $A\mathbf{x} = \mathbf{y}$ by Gaussian elimination and back substitution

In the lower right-hand corner is found the total number of arithmetic operations needed to solve an $n \times n$ system by Gaussian elimination. This total does not include the effort expended making comparisons for the sake of pivoting (interchanging rows to reduce the chance of roundoff error). For large n, the number of operations grows as the dominant term, that is, as $\frac{2}{3}n^3$. Hence, it is common to say that the number of operations is $O(n^3)$, that is, order of n^3. This should be compared to the number of operations performed on the vector **y** and the number of operations needed to execute back substitution. Both of these totals are $O(n^2)$, so most computational energy goes into reducing the matrix A to upper triangular form.

Table 39.2 lists the total number of arithmetic operations needed to upper-triangularize and solve by back substitution $n \times n$ systems of various sizes. For a modest system with $n = 1000$, it takes nearly 700 billion arithmetic operations to compute a solution. For n equal to one million, a small system when designing an aircraft, the total is so large that a computer executing one billion operations per second would take more than $\frac{666,668,166,665,500,000}{(10^9)(3600)(24)(365)} =$ 21.140 years to produce a solution. Thus we see one of the reasons for the development of supercomputers and parallel-processing machines in which Gaussian arithmetic is performed simultaneously rather than sequentially.

n	Total Number of Operations
3	28
4	62
10	805
10^2	681,550
10^3	668,165,500
10^6	666,668,166,665,500,000

TABLE 39.2 Total number of arithmetic operations needed to solve various $n \times n$ systems by Gaussian arithmetic

DERIVING THE OPERATION COUNTS   Deriving the results in Table 39.1 requires

$$\sum_{k=1}^{n-1} k = \frac{1}{2}n(n-1) \quad \text{and} \quad \sum_{k=1}^{n-1} k^2 = \frac{1}{6}n(n-1)(2n-1)$$

formulas for the sum of consecutive integers and their squares, respectively.

Our approach is to consider the case $n = 4$, look for patterns, and generalize. To make the case $n = 4$ more concrete, consider the augmented matrix

$$[A, \mathbf{y}] = \begin{bmatrix} x & x & x & x & y_1 \\ t & x & x & x & y_2 \\ t & t & x & x & y_3 \\ t & t & t & x & y_4 \end{bmatrix} \tag{39.3}$$

in which the numbers at the locations marked with the letter t will be converted to zeros during the upper triangularization as A is transformed to U. Divisions take place only when the multipliers are formed. Column 1 requires the formation and use of 3 multipliers; column 2, 2 multipliers; and column 3, 1 multiplier. The number of divisions required for upper-triangularizing A is

$$(n-1) + (n-2) + \cdots + 1 = \sum_{k=1}^{n-1} k = \frac{n(n-1)}{2}$$

As A is upper-triangularized to U, additions are performed when a row is added to a row below it. The tally of such additions is given in Table 39.3, from which we conclude that the total number of additions needed to upper-triangularize A is

$$(n-1)^2 + (n-2)^2 + \cdots + 1^2 = \sum_{k=1}^{n-1} k^2 = \frac{n(n-1)(2n-1)}{6}$$

As A is upper-triangularized to U, multiplications are performed when a multiplier multiplies the entries in a row being used to create zeros in a row beneath it. When row 1 makes a zero in rows 2, 3, or 4, there are 3 multiplications. When row 2 makes a zero in rows 3 or 4, there are 2 multiplications. When row 3 makes a zero in row 4, there is 1

	Number of Additions Per Row		
Row 1 makes 0's in Column 1 $\Rightarrow$	3 in row 2	3 in row 3	3 in row 4
Row 2 makes 0's in Column 2 $\Rightarrow$		2 in row 3	2 in row 4
Row 3 makes 0's in Column 3 $\Rightarrow$			1 in row 4

TABLE 39.3   Number of arithmetic operations needed for upper-triangularizing a 4×4 matrix

multiplication. Thus, the total number of multiplications needed to upper-triangularize A is

$$(n-1)^2 + (n-2)^2 + \cdots + 1^2 = \sum_{k=1}^{n-1} k^2 = \frac{n(n-1)(2n-1)}{6}$$

To help visualize the Gaussian elimination steps applied to the column vector $\mathbf{y}$, look at the schematic of the augmented matrix $[A, \mathbf{y}]$ in (39.3). No divisions are visited upon an element in the fifth column. However, each time a row is used to create a zero, that row is multiplied by the multiplier, and the row in which the zero is created has an addition made in its fifth column. Thus, when row 1 is used to make zeros in rows 2, 3 and 4, there are 3 multiplies and 3 adds. When row 2 is used to make zeros in rows 3 and 4, there are 2 multiplies and 2 adds. When row 3 is used to make zeros in row 4, there is 1 multiply and 1 add. Hence, there are an additional

$$(n-1) + (n-2) + \cdots + 1 = \sum_{k=1}^{n-1} k = \frac{n(n-1)}{2}$$

multiplications and an equal number of additional additions required to transform the vector $\mathbf{y}$ into the vector $\mathbf{w}$.

To count the operations required for back substitution, consider the schematic of the augmented matrix $[U, \mathbf{w}]$ appearing in

$$[U, \mathbf{w}] = \begin{bmatrix} u_{11} & u_{12} & u_{13} & u_{14} & w_1 \\ 0 & u_{22} & u_{23} & u_{24} & w_2 \\ 0 & 0 & u_{33} & u_{34} & w_3 \\ 0 & 0 & 0 & u_{44} & w_4 \end{bmatrix} \tag{39.4}$$

Solving for w_4 requires 1 division, 0 multiplications, and 0 additions. Solving for w_3 requires 1 division, 1 multiplication, and 1 addition. Solving for w_2 requires 1 division, 2 multiplications, and 2 additions. Solving for w_1 requires 1 division, 3 multiplications, and 3 additions. Consequently, during back substitution there will be an additional n divisions,

$$1 + 2 + \cdots + (n-1) = \sum_{k=1}^{n-1} k = \frac{n(n-2)}{2}$$

multiplications, and an equal number of additions. The total number of operations of all types will then be the sum

$$2\sum_{k=1}^{n-1} k^2 + 5\sum_{k=1}^{n-1} k + n = \frac{1}{6}n(4n^2 + 9n - 7)$$

The Tridiagonal System

DEFINITION 39.1

The $n \times n$ system of equations $A\mathbf{x} = \mathbf{y}$ is called *tridiagonal* if the only nonzero elements in A appear on the main diagonal, the first diagonal above the main diagonal, and the first diagonal below the main diagonal. Such a matrix is called *banded,* and the tridiagonal matrix has three contiguous bands, as suggested by the following schematic.

$$
\begin{bmatrix}
d_1 & u_1 & 0 & 0 & 0 & 0 & . & . & 0 \\
l_2 & d_2 & u_2 & 0 & 0 & 0 & . & . & 0 \\
0 & l_3 & d_3 & u_3 & 0 & 0 & . & . & 0 \\
0 & 0 & l_4 & d_4 & u_4 & 0 & . & . & 0 \\
0 & 0 & 0 & . & . & . & . & . & 0 \\
0 & 0 & 0 & 0 & . & . & . & . & 0 \\
0 & 0 & 0 & 0 & 0 & . & . & . & 0 \\
0 & 0 & 0 & 0 & 0 & 0 & l_{n-1} & d_{n-1} & u_{n-1} \\
0 & 0 & 0 & 0 & 0 & 0 & 0 & l_n & d_n
\end{bmatrix}
$$

We designate the n elements on the main diagonal by $d_k, k = 1, \ldots, n$; the $n-1$ elements on the diagonal above the main diagonal by $u_k, k = 1, \ldots, n-1$; and the $n-1$ elements on the diagonal below the main diagonal by $l_k, k = 2, \ldots, n$. The symbols l_1 and u_n are not used.

Such systems are of interest both because they are readily solvable and because they frequently occur in such applications as implicit schemes for solving boundary value problems by finite-difference techniques. We mention here, and verify momentarily, that such a system can be solved (reduced to upper triangular form and solved with back substitution) with $8n - 7$ total arithmetic operations. Hence, there is an O(n) rather than an O(n^3) solution scheme for the tridiagonal system. For example, if n is a modest 100,000, then $8n - 7$ is approximately 8×10^5; whereas an unthinking Gaussian elimination solution would involve 666,681,666,550,000 (approximately 7×10^{14}) arithmetic operations, a factor of about a billion more.

OPERATION COUNTS We will next show that Table 39.4 tallies the numbers of operations needed to solve an $n \times n$ tridiagonal system of linear equations. As before, the table lists the number of operations needed to row-reduce A to an upper triangular matrix U ($A \to U$), the number of operations needed to row-reduce the column $\mathbf{y}$ in the augmented matrix $[A, \mathbf{y}]$ ($\mathbf{y} \to \mathbf{Y}$), and the number of operations needed to complete the solution by back substitution.

The following schematic of $[A, \mathbf{y}]$, the augmented matrix for a 6×6 tridiagonal system $A\mathbf{x} = \mathbf{y}$, will be useful in generalizing the operations count of a Gaussian elimination process

	$N(\pm)$	$N(*)$	$N(/)$	**Row Totals**
$A \to U$	$n-1$	$n-1$	$n-1$	$3(n-1)$
$\mathbf{y} \to \mathbf{Y}$	$n-1$	$n-1$	0	$2(n-1)$
Back substitution	$n-1$	$n-1$	n	$3n-2$
Column totals	$3(n-1)$	$3(n-1)$	$2n-1$	$8n-7$

TABLE 39.4   Count of operations needed for solving a tridiagonal system by Gaussian elimination and back substitution

that is specifically tailored to the sparse structure of A.

$$[A, \mathbf{y}] = \begin{bmatrix} d_1 & u_1 & 0 & 0 & 0 & 0 & y_1 \\ l_2 & d_2 & u_2 & 0 & 0 & 0 & y_2 \\ 0 & l_3 & d_3 & u_3 & 0 & 0 & y_3 \\ 0 & 0 & l_4 & d_4 & u_4 & 0 & y_4 \\ 0 & 0 & 0 & l_5 & d_5 & u_5 & y_5 \\ 0 & 0 & 0 & 0 & l_6 & d_6 & y_6 \end{bmatrix}$$

To create a zero in the $(k, k-1)$-position, $k = 2, \ldots, 6$, compute

$$l_k = -\frac{l_k}{d_{k-1}} \qquad d_k = d_k + l_k u_{k-1} \qquad y_k = y_k + l_k y_{k-1}$$

Notice that l_k, d_k, and y_k are all over-written by new values, thus eliminating the need for additional storage. However, this notation should not be allowed to obscure the change in values of $d_2, \ldots, d_6$ and of $y_2, \ldots, y_6$. In the representation of the upper-triangularized matrix these calculations have produced, we will use $D_2, \ldots, D_6$ and $Y_2, \ldots, Y_6$ to denote these new values.

In general, reduction of the tridiagonal A to the upper triangular U will require $n-1$ additions and an equal number of multiplications and divisions. Conversion of the vector $\mathbf{y}$ to a new vector $\mathbf{Y}$ will entail $n-1$ additions and an equal number of multiplications.

For the case $n = 6$, the augmented matrix $[A, \mathbf{y}]$ will now be in the form

$$[U, \mathbf{Y}] = \begin{bmatrix} d_1 & u_1 & 0 & 0 & 0 & 0 & y_1 \\ 0 & D_2 & u_2 & 0 & 0 & 0 & Y_2 \\ 0 & 0 & D_3 & u_3 & 0 & 0 & Y_3 \\ 0 & 0 & 0 & D_4 & u_4 & 0 & Y_4 \\ 0 & 0 & 0 & 0 & D_5 & u_5 & Y_5 \\ 0 & 0 & 0 & 0 & 0 & D_6 & Y_6 \end{bmatrix}$$

Back substitution is accomplished via the computations

$$Y_6 = \frac{Y_6}{D_6} \qquad Y_k = \frac{Y_k - u_k Y_{k+1}}{D_k} \qquad k = 5, 4, 3, 2 \qquad y_1 = \frac{y_1 - u_1 Y_2}{d_1}$$

Again, the updated values of entries in the rightmost column are stored "in place," eliminating the need to allocate additional memory. The desired solution $\mathbf{x}$ is then the rightmost column of the augmented matrix after back substitution has been performed.

In general, back substitution requires $n-1$ additions, an equal number of multiplications, and n divisions. Hence, the general tridiagonal system can be solved with just $8n-7$ arithmetic operations.

For $k = 1$ to n execute
If $d_k = 0$ then QUIT else if $k < n$ then

$$l_{k+1} = -\frac{l_{k+1}}{d_k}$$

$$d_{k+1} = k_{k+1} + l_{k+1} u_k$$

$$y_{k+1} = y_{k+1} + l_{k+1} y_k$$

Next k

$$y_n = \frac{y_n}{d_n}$$

For $k = n-1$ down to 1 in steps of -1, execute

$$y_k = \frac{y_k - u_k y_{k+1}}{d_k}$$

Next k

PSEUDOCODE A numeric algorithm for solving $A\mathbf{x} = \mathbf{y}$, an $n \times n$ tridiagonal system of linear equations, could be programmed in a language such as FORTRAN or C according to the pseudocode in the margin on page 980. The syntax unique to each programming language would be used to implement constructs such as for-loops and if-statements. Also, each program would have to be given a mechanism for initializing the variables with the data in the matrix A and the vector $\mathbf{y}$.

There must be a test for $d_k = 0$ since it is possible for one of the elements on the main diagonal to become zero during the calculation. If this happens, the algorithm fails. This does not mean there is no solution, because the matrix A need not be singular. It simply means that the standard intervention of a row-interchange can't be used here since that would then destroy the tridiagonal structure of the matrix.

EXERCISES 39.1

1. If the lower right entry in Table 39.1 is $F(n)$, the total number of arithmetic operations needed to solve a linear system:

(a) Graph $F(n)$ and $\frac{2}{3}n^3$, $n \leq 100$, on the same set of axes.

(b) For $n = 10^5$ and $n = 10^6$, obtain the percent error made when using $\frac{2}{3}n^3$ instead of $F(n)$.

When upper-triangularizing the $n \times n$ matrix $A = \{a_{ij}\}$ by Gaussian elimination, the kth row is used to create zeros in the kth column, below the main diagonal. If $a_{kk} = 0$ and the kth row is interchanged with the first row below it for which $a_{jk} \neq 0$, $j > k$, then the strategy of *naive pivoting* has been applied. For the vector $\mathbf{y} = [1, 2, 3, 4]^T$ and the matrices A in Exercises 2–7, apply Gaussian elimination with naive pivoting and back substitution to the augmented matrix $[A, \mathbf{y}]$, work in four significant figures, and solve $A\mathbf{x} = \mathbf{y}$. Compare the solution to an exact solution, and explain any differences that may be observed.

2. $\begin{bmatrix} 43 & -62 & 54 & -30 \\ 77 & 66 & -5 & 31 \\ 44 & -4 & 40 & -31 \\ 36 & 90 & 48 & -71 \end{bmatrix}$ **3.** $\begin{bmatrix} 15 & 18 & 33 & 25 \\ 60 & 72 & 72 & -23 \\ 63 & -38 & -39 & 79 \\ -95 & -51 & 79 & 29 \end{bmatrix}$

4. $\begin{bmatrix} 96 & 84 & -25 & 17 \\ -48 & -42 & 11 & 29 \\ -82 & -6 & -89 & -10 \\ -25 & -41 & 88 & 97 \end{bmatrix}$ **5.** $\begin{bmatrix} -18 & -27 & 94 & -15 \\ 12 & 18 & 41 & -97 \\ -43 & 16 & 14 & 58 \\ -95 & 37 & -4 & 50 \end{bmatrix}$

6. $\begin{bmatrix} -77 & 66 & -64 & -51 \\ 63 & -54 & 97 & -66 \\ 84 & 10 & -21 & -17 \\ -57 & -82 & 45 & 17 \end{bmatrix}$ **7.** $\begin{bmatrix} -28 & -32 & 0 & 89 \\ -77 & -88 & -77 & 30 \\ 91 & -12 & -85 & -68 \\ -74 & 13 & -85 & 69 \end{bmatrix}$

Gaussian elimination is implemented with the strategy of *partial pivoting* if the row used to create zeros in the kth column is the one for which $|a_{ik}| = \max_{j \geq k}\{|a_{jk}|\}$. Dividing by the largest coefficient available in a column tends to reduce roundoff error. For the matrices A in Exercises 8–13 and the vector $\mathbf{y} = [-2, 3, 1, -5]^T$, apply Gaussian elimination with partial pivoting to the augmented matrix $[A, \mathbf{y}]$, work in four significant figures, and solve $A\mathbf{x} = \mathbf{y}$. Compare the solution to an exact solution, and explain any differences that might be observed.

8. $\begin{bmatrix} 43 & -62 & 77 & 66 \\ 54 & -57 & 99 & -61 \\ -50 & -12 & -18 & 31 \\ -26 & -62 & 13 & -47 \end{bmatrix}$ **9.** $\begin{bmatrix} -91 & -47 & -61 & 41 \\ -58 & -90 & 53 & -17 \\ 94 & 83 & -86 & 23 \\ -84 & 19 & -50 & 88 \end{bmatrix}$

10. $\begin{bmatrix} -53 & 85 & 49 & 78 \\ 17 & 72 & -99 & -85 \\ -86 & 30 & 80 & 72 \\ 66 & -29 & -91 & -53 \end{bmatrix}$ **11.** $\begin{bmatrix} -19 & -47 & 68 & -72 \\ -87 & 79 & 43 & -66 \\ -53 & -61 & -23 & -37 \\ 31 & -34 & -42 & 88 \end{bmatrix}$

12. $\begin{bmatrix} -76 & -65 & 25 & 28 \\ -61 & -60 & 19 & 29 \\ -66 & -32 & 78 & 39 \\ 94 & 68 & -17 & -98 \end{bmatrix}$ **13.** $\begin{bmatrix} -36 & 40 & 22 & 53 \\ -88 & -43 & -73 & 25 \\ 41 & -59 & 62 & -55 \\ 25 & 39 & 40 & 61 \end{bmatrix}$

If $\mathbf{y} = [3, 2, 1, 4, 5, 6]^T$, use the tridiagonal algorithm to solve $A\mathbf{x} = \mathbf{y}$ for each of the matrices A in Exercises 14 and 15.

14. $\begin{bmatrix} -5 & 7 & 0 & 0 & 0 & 0 \\ -1 & -1 & -2 & 0 & 0 & 0 \\ 0 & 1 & 10 & 4 & 0 & 0 \\ 0 & 0 & -6 & 6 & -9 & 0 \\ 0 & 0 & 0 & 8 & -5 & -2 \\ 0 & 0 & 0 & 0 & -4 & 4 \end{bmatrix}$

15. $\begin{bmatrix} 68 & -65 & 0 & 0 & 0 & 0 \\ 8 & 20 & 93 & 0 & 0 & 0 \\ 0 & -80 & -5 & 23 & 0 & 0 \\ 0 & 0 & -24 & -63 & -36 & 0 \\ 0 & 0 & 0 & 37 & 60 & 95 \\ 0 & 0 & 0 & 0 & 52 & 31 \end{bmatrix}$

16. Apply the tridiagonal algorithm to the augmented matrix
$$[A, \mathbf{y}] = \begin{bmatrix} 1 & 1 & 0 & 0 & 6 \\ 2 & 2 & 2 & 0 & 7 \\ 0 & 3 & 3 & 3 & 8 \\ 0 & 0 & 4 & 4 & 9 \end{bmatrix}.$$ Show that A is not singular but that the algorithm fails because a diagonal entry becomes zero. What is the solution of the system?

17. If $\mathbf{y} = [1, 2, 3, 4, 5, 6]^T$ and $A = \begin{bmatrix} -3 & 10 & -1 & 0 & 0 & 0 \\ 10 & -3 & 7 & 6 & 0 & 0 \\ -2 & -9 & -2 & -9 & 6 & 0 \\ 0 & 8 & 8 & -1 & -1 & 1 \\ 0 & 0 & -9 & -9 & 3 & 4 \\ 0 & 0 & 0 & -10 & -4 & 7 \end{bmatrix}$, apply Gaussian elimination and back substitution to the augmented matrix $[A, \mathbf{y}]$, noting where adding zero or multiplying by zero occurs.

18. Generalize the tridiagonal algorithm for a matrix with five contiguous bands.

19. Obtain the total number of arithmetic operations needed to execute the algorithm developed in Exercise 18. If $n = 10^5$, compare the number of operations needed to solve a five-banded system with the algorithm of Exercise 18 to the number of operations needed to solve the same system with ordinary Gaussian elimination and back substitution.

39.2 Condition Numbers

EXAMPLE 39.1

This example explores the dependence of $\mathbf{x}$, the solution to the linear system $A\mathbf{x} = \mathbf{y}$, on a small parameter in the matrix A. We assume c, the small parameter, is roundoff error in one entry of A introduced by the act of entering A into a finite-precision computing device. This gives the matrix $A(c)$ in Table 39.5, which also gives $\mathbf{x}_0$, the solution to $A(0)\mathbf{x} = \mathbf{y}$, and $\mathbf{x}_c$, the solution to the perturbed system $A(c)\mathbf{x} = \mathbf{y}$. The determinant of $A(c)$ is $1 + 91c$, and the determinant of $A(0)$ is 1. As long as $c \neq -\frac{1}{91}$, $A(c)$ is not singular.

The ∞-norm of the difference $\Delta \mathbf{x} = \mathbf{x}_c - \mathbf{x}_0$ is given by $\|\Delta \mathbf{x}\|_\infty = 7462 \left| \frac{c}{1+91c} \right|$, the largest of the absolute values of the components of the vector. Figure 39.1 shows a graph of this magnitude, from which we can see that large changes in the solution are possible with just small changes in the perturbation parameter c. Alternatively, the variability in the solution is given by the ratio

$$\frac{\|\Delta \mathbf{x}\|_\infty}{\|\mathbf{x}_c\|_\infty} = \frac{7462 \left| \frac{c}{1+91c} \right|}{\max \left\{ \left| \frac{c-9}{1+91c} \right|, \frac{82}{|1+91c|} \right\}}$$

a measure of the relative change in the perturbed solution, graphed in Figure 39.2. Again, we see that small changes in c can cause large relative changes in the solution. ❖

$$A(c) = \begin{bmatrix} 1+c & 9 \\ 10 & 91 \end{bmatrix}$$

$$\mathbf{y} = \begin{bmatrix} 1 \\ 1 \end{bmatrix}$$

$$\mathbf{x}_0 = \begin{bmatrix} 82 \\ -9 \end{bmatrix}$$

$$\mathbf{x}_c = \begin{bmatrix} \dfrac{82}{1+91c} \\ \dfrac{c-9}{1+91c} \end{bmatrix}$$

TABLE 39.5   The system $A\mathbf{x} = \mathbf{y}$, and its solution, in Example 39.5

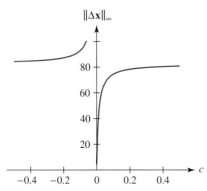

FIGURE 39.1   A plot of $|\Delta\mathbf{x}|_\infty$ vs. c in Example 39.1

Our measurement of the changes in the solution were made with exact arithmetic. The existence of these changes is not attributable to the computational process but is inherent in the matrix A itself. This property of the matrix is measured by its *condition number*, explored next.

The Condition Number

The *condition number* of a matrix A is defined as $\text{cond}(A) = \|A\| \|A^{-1}\|$, where the condition number depends on the specific norm used on the right.

Using either the ∞-norm or the 1-norm, $\text{cond}(A(0)) = 10{,}100$, where $A(0)$ refers to the matrix defined in Table 39.5. Using the 2-norm, $\text{cond}(A(0)) = 8462.999883$. Note that the matrix $A(0)$, whose determinant is 1, has such a large condition number. Matrices with large condition numbers are said to be *ill-conditioned* and have the potential to multiply the relative error in the matrix by a large number, establishing a large upper bound for the relative error in the solution of the system $A\mathbf{x} = \mathbf{y}$.

For the system in Example 39.1, we have already examined the relative error in the solution $\mathbf{x}_c$. Let us now obtain the relative error in $A(0)$. Thus, we compute $\Delta A(0) = A(c) - A(0) = \begin{bmatrix} c & 0 \\ 0 & 0 \end{bmatrix}$, and using the ∞-norm, we find $\frac{\|\Delta A(0)\|}{\|A(0)\|} = \frac{|c|}{101}$ as the relative error in $A(0)$. Figure 39.3 contains a graph (solid lines) of the relative error in $\mathbf{x}_c$, along with a graph (dotted lines) of $\text{cond}(A(0))$ times the relative error in $A(0)$. The inequality governing the placement of the solid and dotted lines is derived in the next subsection.

We next look at the singular value decomposition of the matrix $A(0)$. In Section 34.5, we saw that $A = U\Sigma V^T$, where U and V are orthogonal matrices and Σ is a diagonal matrix

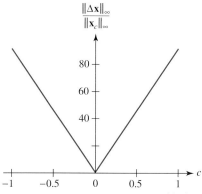

FIGURE 39.2 A plot of the ratio $\dfrac{|\Delta\mathbf{x}|_\infty}{|\mathbf{x}_c|_\infty}$ vs. c in Example 39.1

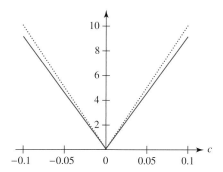

FIGURE 39.3 Example 39.1: Relative error in $\mathbf{x}_c$ (solid), and $\mathrm{cond}(A(0))$ times the relative error in $A(0)$ (dotted)

of singular values σ_j, $j = 1, \ldots, k$, where k is the rank of A. These singular values are the square roots of the eigenvalues of the matrix $A^{\mathrm{T}}A$. If A is an $r \times c$ matrix, then U is $r \times r$, while V is $c \times c$. Denoting the columns of U by $\mathbf{u}_i$ and the columns of V by $\mathbf{v}_i$, we also have, from Section 34.5, the outer product expansion of A, namely, $A = \sigma_1 \mathbf{u}_1 \mathbf{v}_1^{\mathrm{T}} + \cdots + \sigma_k \mathbf{u}_k \mathbf{v}_k^{\mathrm{T}}$.

The singular values for $A(0)$ in Example 39.1 are exactly $\frac{\sqrt{8465 \pm \sqrt{8461}}}{2}$ and approximately 91.99456442, 0.01087020. The small singular value is more indicative of ill-conditioning than the value of the determinant. This has a theoretical basis. In the outer product expansion of A, σ_k is the smallest singular value. If the term that contains it were deleted, the remaining sum in the expansion would be the rank $k - 1$ approximation to A having smallest 2-norm. Thus, if A has a small singular value, then it is "close" to a singular matrix.

In addition, the 2-norm of A is equal to the largest singular value and the 2-norm condition number of A is the ratio of the largest and smallest singular values. Thus, a large singular value can raise the condition number, but a very small singular value can raise the condition number higher.

The connections between the singular values and both the norm of A and its condition number are detailed in the spectral radius theorem, which can be found in texts such as [60] or [64].

Derivation of the Bound on Relative Error

We next relate the relative change in the solution $\mathbf{x}$ to a relative change in the matrix A, establishing the bound

$$\frac{\|\Delta\mathbf{x}\|}{\|\mathbf{x}_0 + \Delta\mathbf{x}\|} \le \mathrm{cond}\, A \frac{\|\Delta A\|}{\|A\|} \tag{39.5}$$

To this end, let the linear system $A\mathbf{x} = \mathbf{y}$ have the solution $\mathbf{x}_0$, let ΔA be a small change in A, and let $B = A + \Delta A$ be a perturbation of A. Finally, let the perturbed system $B\mathbf{x} = \mathbf{y}$ have solution $\mathbf{x}_c = \mathbf{x}_0 + \Delta\mathbf{x}$. Then $B\mathbf{x}_c = \mathbf{y}$ or

$$(A + \Delta A)(\mathbf{x}_0 + \Delta\mathbf{x}) = \mathbf{y} = A\mathbf{x}_0$$

Removing the parentheses, we get

$$A\mathbf{x}_0 + A\Delta\mathbf{x} + \Delta A(\mathbf{x}_0 + \Delta\mathbf{x}) = \mathbf{y} = A\mathbf{x}_0$$

or $A\Delta\mathbf{x} + \Delta A(\mathbf{x}_0 + \Delta\mathbf{x}) = 0$. Solve for $A\Delta\mathbf{x}$, obtaining $A\Delta\mathbf{x} = -\Delta A(\mathbf{x}_0 + \Delta\mathbf{x})$; then multiply both sides by A^{-1}, the inverse of A, which gives $\Delta\mathbf{x} = -A^{-1}\Delta A(\mathbf{x}_0 + \Delta\mathbf{x})$. Compute the norm of both sides, obtaining

$$\|\Delta\mathbf{x}\| = \|A^{-1}\Delta A(\mathbf{x}_0 + \Delta\mathbf{x})\|$$

From Section 33.5 we have, as general properties of matrix norms, $\|M\mathbf{x}\| \le \|M\|\|\mathbf{x}\|$ and $\|MN\| \le \|M\|\|N\|$. Applying these estimates, we then get

$$\|\Delta\mathbf{x}\| \le \|A^{-1}\|\|\Delta A\|\|\mathbf{x}_0 + \Delta\mathbf{x}\|$$

Divide both sides by $\|\mathbf{x}_0 + \Delta\mathbf{x}\|$ to get

$$\frac{\|\Delta\mathbf{x}\|}{\|\mathbf{x}_0 + \Delta\mathbf{x}\|} \le \|A^{-1}\|\|\Delta A\|$$

On the right, multiply and divide by the norm of A, obtaining, finally,

$$\frac{\|\Delta\mathbf{x}\|}{\|\mathbf{x}_0 + \Delta\mathbf{x}\|} \le \{\|A^{-1}\|\|A\|\}\frac{\|\Delta A\|}{\|A\|}$$

The factor $\|A^{-1}\|\|A\|$ is called the *condition number* of A and, in the upper bound for the relative error in the solution, is the factor by which the relative error in A might be multiplied.

EXAMPLE 39.2   The exact solution $\mathbf{x}_e$ for the system $A\mathbf{x} = \mathbf{y}$, is given in (39.6), along with A and $\mathbf{y}$.

$$
A = \begin{bmatrix} 9 & -2 & 9 \\ -4 & 3 & 8 \\ 8 & 8 & 2 \end{bmatrix} \qquad \mathbf{y} = \begin{bmatrix} 1 \\ 1 \\ 1 \end{bmatrix} \qquad \mathbf{x}_e = \begin{bmatrix} \frac{5}{234} \\ \frac{1}{13} \\ \frac{25}{234} \end{bmatrix} \tag{39.6}
$$

The matrix A has the relatively small condition number 1.884725008. To simulate the introduction of modest error into the matrix A, convert the augmented matrix $[A, \mathbf{y}]$ to the floating-point form and row-reduce it to upper triangular form, using partial pivoting and just three significant digits. The result, the solution obtained by back substitution, and the exact solution to three significant digits are then

$$
[A, \mathbf{y}] \to \begin{bmatrix} 9.00 & -2.00 & 9.00 & 1.00 \\ 0 & 9.78 & -6.00 & 0.111 \\ 0 & 0 & 13.3 & 1.42 \end{bmatrix} \qquad \mathbf{x} = \begin{bmatrix} 0.0212 \\ 0.0770 \\ 0.107 \end{bmatrix} \qquad \mathbf{x}_e = \begin{bmatrix} 0.0214 \\ 0.0769 \\ 0.107 \end{bmatrix}
$$

Rounding to three significant digits puts error in the third significant digit of the entries of A, but using the 2-norm, cond(A) is less than 2. The error gets multiplied at worst by a factor of 2, so the first two significant digits in the solution should still be accurate. In fact, the computed solution agrees with the exact solution to two significant digits.   ❖

EXAMPLE 39.3   By writing A as an outer product expansion, we will construct a matrix with the prescribed singular values 25, 10, and $\frac{1}{1000}$. To do this, we need the column vectors $\mathbf{u}_j$ and $\mathbf{v}_j$ in the orthogonal matrices U and V, respectively, which we get by orthonormalizing the two random matrices U_1 and V_1 in (39.7).

$$
U_1 = \begin{bmatrix} 0 & -2 & 3 \\ 3 & -1 & -3 \\ 1 & 0 & -4 \end{bmatrix} \qquad V_1 = \begin{bmatrix} -1 & 2 & 5 \\ 0 & 2 & 1 \\ 3 & -1 & -4 \end{bmatrix} \tag{39.7}
$$

The matrices U and V are then

$$
U = \begin{bmatrix} 0 & -\frac{2\sqrt{10}}{\sqrt{41}} & -\frac{1}{\sqrt{41}} \\ \frac{3}{\sqrt{10}} & -\frac{1}{\sqrt{410}} & \frac{2}{\sqrt{41}} \\ \frac{1}{\sqrt{10}} & \frac{3}{\sqrt{410}} & -\frac{6}{\sqrt{41}} \end{bmatrix} \qquad V = \begin{bmatrix} -\frac{1}{\sqrt{10}} & \frac{3}{\sqrt{26}} & \frac{6}{\sqrt{65}} \\ 0 & \frac{2\sqrt{2}}{\sqrt{13}} & -\frac{\sqrt{5}}{\sqrt{13}} \\ \frac{3}{\sqrt{10}} & \frac{1}{\sqrt{26}} & \frac{2}{\sqrt{65}} \end{bmatrix}
$$

The matrix $A = 25\mathbf{u}_1\mathbf{v}_1^{\mathrm{T}} + 10\mathbf{u}_2\mathbf{v}_2^{\mathrm{T}} + \frac{1}{1000}\mathbf{u}_3\mathbf{v}_3^{\mathrm{T}}$ is then

$$
A = \begin{bmatrix} -\frac{150{,}003}{500}\sigma & -\frac{79{,}999}{200}\sigma & -\frac{50{,}001}{500}\sigma \\ -\frac{15}{2} - \frac{3747}{250}\sigma & -\frac{2001}{100}\sigma & \frac{45}{2} - \frac{1249}{250}\sigma \\ -\frac{5}{2} + \frac{11{,}241}{250}\sigma & \frac{6003}{100}\sigma & \frac{15}{2} + \frac{3747}{250}\sigma \end{bmatrix}
$$

where $\sigma = \frac{1}{\sqrt{2665}}$. The determinant of A is $\frac{1}{4}$, not particularly small. However, the condition number of A is large, since it is 25,000, the ratio of 25 and $\frac{1}{1000}$. Finally, the 2-norm of A, the largest singular value, is 25.

We have constructed a matrix A with a small singular value. It has modest determinant and 2-norm but a large condition number. How long are the vectors $\mathbf{r}_j$ that make up the rows of A? The 2-norms of these row vectors whose components define the equations in the system $A\mathbf{x} = \mathbf{y}$ are then found to be 24.69, 9.57, and 4.87, respectively.

In Section 12.4 we interpreted the determinant of a 3×3 matrix as the volume of the parallelepiped spanned by the row vectors in the matrix. For the matrix A that we have

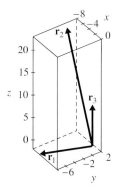

FIGURE 39.4   The row vectors in the matrix of Example 39.3

constructed, that volume is $\det(A) = \frac{1}{4}$, and the row vectors spanning the parallelepiped with that volume are not particularly small. However, they are nearly coplanar, an observation consistent with A's small singular value. To show that the rows of A are nearly coplanar, we obtain the length of the orthogonal projection of each row, in turn, on the plane spanned by the other two rows.

The length of the projection of $\mathbf{r}_3$ onto the span of $\mathbf{r}_1$ and $\mathbf{r}_2$ is 1.1×10^{-3}, and the length of the projection of $\mathbf{r}_2$ onto the span of $\mathbf{r}_1$ and $\mathbf{r}_3$ is 3.2×10^{-3}, and the length of the projection of $\mathbf{r}_1$ onto the span of $\mathbf{r}_2$ and $\mathbf{r}_3$ is 6.4×10^{-3}. That the orthogonal projections are so small is evidence that the three vectors are nearly coplanar. Figure 39.4 contains a graph of these vectors.

The exact solution of the system $A\mathbf{x} = \mathbf{y}$, where $\mathbf{y}$ is the vector given in (39.6), and its floating-point equivalent are

$$\mathbf{x}_e = \begin{bmatrix} -\frac{2}{125} - \frac{23.079}{2050}\sqrt{2665} \\ \frac{9614}{1025}\sqrt{2665} \\ \frac{6}{125} - \frac{7693}{2050}\sqrt{2665} \end{bmatrix} \stackrel{\circ}{=} \begin{bmatrix} -581.21 \\ 484.22 \\ -193.68 \end{bmatrix}$$

The augmented matrix $[A, \mathbf{y}]$, rounded to five significant figures, and its upper triangular form obtained on a simulated five significant-digit device are

$$[A, \mathbf{y}] = \begin{bmatrix} -5.8114 & -7.7484 & -1.9372 & 1 \\ -7.7903 & -0.38762 & -0.38762 & 1 \\ -1.6290 & 1.1629 & 7.7903 & 1 \end{bmatrix} \rightarrow \begin{bmatrix} -7.7903 & -0.38762 & -0.38762 & 1 \\ 0 & 7.4592 & -18.649 & 0.25402 \\ 0 & 0 & -0.0045 & 0.83325 \end{bmatrix}$$

Back substitution then gives, with errors in the second significant digits, the solution

$$\mathbf{x}_5 = \begin{bmatrix} -555.67 \\ 462.90 \\ -185.17 \end{bmatrix}$$

To understand the role of the condition number in predicting this loss of accuracy, measure the relative error in the computed solution by $\frac{\|\mathbf{x}_e - \mathbf{x}_5\|_\infty}{\|\mathbf{x}_5\|_\infty} = 0.0459$. This suggests there could be errors in the second significant digit, which is what is actually observed.

The increment in the matrix A, ascribed to rounding, is

$$\Delta A = \begin{bmatrix} 0.7560 & -10.8408 & -6.4147 \\ 3.2115 & -0.68681 & -22.2628 \\ 0.36549 & 6.0604 & 3.2115 \end{bmatrix} \times 10^{-5}$$

and the condition number times the relative error in A is then $25{,}000\frac{\|\Delta A\|_\infty}{\|A\|_\infty} = 0.214$. The condition number is on the order of 10^4, so errors in the fifth significant digit can be magnified to become errors in the first. We observed errors in the second significant digit.   ❖

Another Bound on Relative Error

We next relate the relative change in the solution $\mathbf{x}$ to a relative change in the vector $\mathbf{y}$, establishing the bound

$$\frac{\|\Delta\mathbf{x}\|}{\|\mathbf{x}_0\|} \le \text{cond}(A)\frac{\|\Delta\mathbf{y}\|}{\|\mathbf{y}\|} \tag{39.8}$$

To this end, let the linear system $A\mathbf{x} = \mathbf{y}$ have the solution $\mathbf{x}_0$ and let $\Delta\mathbf{y}$ be a small change in $\mathbf{y}$ so that $\mathbf{Y} = \mathbf{y} + \Delta\mathbf{y}$ is a perturbation of $\mathbf{y}$. Finally, let the perturbed system $A\mathbf{x} = \mathbf{Y}$ have solution $\mathbf{x}_p = \mathbf{x}_0 + \Delta\mathbf{x}$. Then $A\mathbf{x}_p = \mathbf{Y}$, or

$$A(\mathbf{x}_0 + \Delta\mathbf{x}) = \mathbf{Y} = \mathbf{y} + \Delta\mathbf{y}$$

Removing the parentheses and replacing $\mathbf{y}$ with $A\mathbf{x}_0$, we get

$$A\mathbf{x}_0 + A\Delta\mathbf{x} = \mathbf{Y} = A\mathbf{x}_0 + \Delta\mathbf{y}$$

or $A\Delta\mathbf{x} = \Delta\mathbf{y}$. Solving for $\Delta\mathbf{x}$, we obtain $\Delta\mathbf{x} = A^{-1}\Delta\mathbf{y}$. Next, compute the norm of both sides, obtaining $\|\Delta\mathbf{x}\| = \|A^{-1}\Delta\mathbf{y}\|$. Again from Section 33.5 we invoke the matrix norm property $\|M\mathbf{x}\| \le \|M\|\|\mathbf{x}\|$, obtaining

$$\|\Delta\mathbf{x}\| \le \|A^{-1}\|\|\Delta\mathbf{y}\|$$

Divide both sides by $\|\mathbf{y}\|$ to get

$$\frac{\|\Delta\mathbf{x}\|}{\|\mathbf{y}\|} \le \|A^{-1}\|\frac{\|\Delta\mathbf{y}\|}{\|\mathbf{y}\|} \tag{39.9}$$

From $A\mathbf{x}_0 = \mathbf{y}$, we obtain the estimate

$$\|\mathbf{y}\| = \|A\mathbf{x}_0\| \le \|A\|\|\mathbf{x}_0\|$$

In the denominator of the fraction on the left in (39.9), replace $\|\mathbf{y}\|$ with a larger term, namely, with $\|A\|\|\mathbf{x}_0\|$, thereby strengthening the inequality. This gives

$$\frac{\|\Delta\mathbf{x}\|}{\|A\|\|\mathbf{x}_0\|} \le \|A^{-1}\|\frac{\|\Delta\mathbf{y}\|}{\|\mathbf{y}\|}$$

Now, multiply through by $\|A\|$, giving

$$\frac{\|\Delta\mathbf{x}\|}{\|\mathbf{x}_0\|} \le \{\|A\|\|A^{-1}\|\}\frac{\|\Delta\mathbf{y}\|}{\|\mathbf{y}\|}$$

On the left we have the relative error in $\mathbf{x}_0$, and on the right we have $\mathrm{cond}(A)$ times the relative error in $\mathbf{y}$. Again we see the role played by $\mathrm{cond}(A)$ is estimating the propagation of error when solving a linear system.

EXERCISES 39.2–Part A

A1. Let $A = A(0)$ and $\mathbf{y}$ be defined as in Example 39.1. Let $\Delta A = \begin{bmatrix} 0 & c \\ 0 & 0 \end{bmatrix}$, and define $B = A + \Delta A$. Obtain $\mathbf{x}_0$, the solution of $A\mathbf{x} = \mathbf{y}$, and $\mathbf{x}_c$, the solution of $B\mathbf{x} = \mathbf{y}$.

A2. Obtain $\Delta\mathbf{x} = \mathbf{x}_c - \mathbf{x}_0$ and graph $\|\Delta\mathbf{x}\|_\infty$ vs. c.

A3. Evaluate $\|\Delta x\|_\infty/\|\mathbf{x}_c\|_\infty$ at $c = 0.5$.

A4. Graph $\mathrm{cond}\,(A)\frac{\|\Delta A\|_\infty}{\|A\|_\infty}$ vs. c.

A5. At $c = 0.5$, show (39.5) holds for the quantities computed in Exercises A1–4.

EXERCISES 39.2–Part B

B1. Let $A = A(0)$ and $\mathbf{y}$ be as defined in Example 39.1. Define $\Delta\mathbf{y} = \begin{bmatrix} c \\ 0 \end{bmatrix}$ and $\mathbf{w} = \mathbf{y} + \Delta\mathbf{y}$, and obtain $\mathbf{x}_c$, the solution of $A\mathbf{x} = \mathbf{w}$. Obtain $\Delta\mathbf{x} = \mathbf{x}_c - \mathbf{x}_0$, where $\mathbf{x}_0$ is the solution of $A\mathbf{x} = \mathbf{y}$.

B2. If $\Delta\mathbf{x}, \mathbf{x}_0, \Delta\mathbf{y}$, and $\mathbf{y}$ are as determined in Exercise 1, graph $\frac{\|\Delta x\|_\infty}{\|x_0\|_\infty}$ and $\mathrm{cond}\,(A)\frac{\|\Delta y\|_\infty}{\|y\|_\infty}$ as functions of c. Does the graph support the validity of (39.8)?

If $\mathbf{y} = [1, 2, 3]^T$, $C = \begin{bmatrix} 0 & 0 & a \\ 0 & 0 & 0 \\ 0 & 0 & 0 \end{bmatrix}$, and the matrices A, and B are as given in each of Exercises B3–6,

(a) obtain $\mathbf{x}$, the solution of $A\mathbf{x} = \mathbf{y}$.

(b) obtain $\mathbf{x}_1$, the solution of $(A + C)\mathbf{x} = \mathbf{y}$.

(c) obtain $\Delta\mathbf{x} = \mathbf{x}_1 - \mathbf{x}$, and compute $\alpha = \frac{\|\Delta x\|}{\|x_1\|}$, $\mathrm{cond}\,A$, and $\beta = \frac{\|C\|}{\|A\|}$.

(d) on the same set of axes, plot α and the product $\beta\,\mathrm{cond}\,A$ for $0 \le a \le 1$.

(e) obtain $\mathbf{z}$, the solution of $B\mathbf{z} = \mathbf{y}$.

(f) obtain $\mathbf{z}_1$, the solution of $(B + C)\mathbf{z} = \mathbf{y}$.

(g) obtain $\Delta\mathbf{z} = \mathbf{z}_1 - \mathbf{z}$, and compute $\mu = \frac{\|\Delta z\|}{\|z_1\|}$, $\mathrm{cond}\,B$, and $\nu = \frac{\|C\|}{\|B\|}$.

(h) on the same set of axes, plot μ and the product ν cond B for $0 \le a \le 1$.

(i) observe that cond A is large but cond B is modest, that det A is modest but det B is not modest, and that both graphs are consistent with the inequality (39.5).

(j) use the 2-norm to compute cond A, and show it is the ratio of the largest singular value of A to the smallest. Consequently, the large condition number of A indicates that A is close to a singular matrix even though the determinant of A is not "small."

B3. $A = \begin{bmatrix} 1 & -11 & 8 \\ -6 & -1 & -4 \\ 11 & 10 & 2 \end{bmatrix} \quad B = \begin{bmatrix} -10 & 11 & -5 \\ -2 & -2 & -2 \\ -10 & -5 & 6 \end{bmatrix}$

B4. $A = \begin{bmatrix} 4 & -1 & 6 \\ -4 & -9 & -7 \\ -3 & 6 & -4 \end{bmatrix} \quad B = \begin{bmatrix} -5 & 10 & -9 \\ 5 & -10 & -2 \\ -12 & -7 & -6 \end{bmatrix}$

B5. $A = \begin{bmatrix} -8 & -11 & 8 \\ -10 & -1 & 9 \\ 7 & 10 & -7 \end{bmatrix} \quad B = \begin{bmatrix} -3 & -3 & -7 \\ -12 & -10 & -6 \\ -4 & -9 & 0 \end{bmatrix}$

B6. $A = \begin{bmatrix} -1 & 0 & -1 \\ -11 & -5 & -10 \\ 10 & -7 & 11 \end{bmatrix} \quad B = \begin{bmatrix} -11 & 9 & 10 \\ -9 & 8 & 4 \\ 11 & -2 & 7 \end{bmatrix}$

If $\mathbf{y} = [-2, 3, 5]^T$, $\Delta\mathbf{y} = [0, 0, a]^T$, and the matrices A and B are as given in each of Exercises B7–10,

(a) obtain $\mathbf{x}$, the solution of $A\mathbf{x} = \mathbf{y}$, and $\mathbf{x}_1$, the solution of $A\mathbf{x} = \mathbf{y} + \Delta\mathbf{y}$.

(b) obtain $\mathbf{z}$, the solution of $B\mathbf{z} = \mathbf{y}$, and $\mathbf{z}_1$, the solution of $B\mathbf{z} = \mathbf{y} + \Delta\mathbf{y}$.

(c) obtain $\Delta\mathbf{x} = \mathbf{x}_1 - \mathbf{x}$, and $\Delta\mathbf{z} = \mathbf{z}_1 - \mathbf{z}$.

(d) obtain $\alpha = \frac{\|\Delta\mathbf{x}\|}{\|\mathbf{x}\|}$, $\beta = \frac{\|\Delta\mathbf{z}\|}{\|\mathbf{z}\|}$, and $\lambda = \frac{\|\Delta\mathbf{y}\|}{\|\mathbf{y}\|}$.

(e) obtain cond A, cond B, det A, and det B.

(f) on the same set of axes, plot α and λ cond A for $a \ge 0$.

(g) on the same set of axes, plot β and λ cond B for $a \ge 0$.

(h) observe that cond A is large but cond B is modest, that det A is modest but det B is not, and that both graphs are consistent with inequality (39.8).

(i) use the 2-norm to compute cond A, and show it is the ratio of the largest singular value of A to the smallest. Consequently, the large condition number of A indicates that A is close to a singular matrix even though the determinant of A is not "small."

B7. $A = \begin{bmatrix} 8 & -9 & -5 \\ 9 & 12 & 4 \\ 1 & -2 & -1 \end{bmatrix} \quad B = \begin{bmatrix} -3 & -3 & 3 \\ -8 & -3 & 7 \\ -9 & -1 & 6 \end{bmatrix}$

B8. $A = \begin{bmatrix} -4 & 8 & -9 \\ -9 & -3 & 4 \\ 2 & -7 & 8 \end{bmatrix} \quad B = \begin{bmatrix} 7 & 5 & -4 \\ 8 & 0 & -12 \\ -10 & -8 & -11 \end{bmatrix}$

B9. $A = \begin{bmatrix} 8 & 2 & 1 \\ 4 & 7 & 6 \\ -10 & -11 & -9 \end{bmatrix} \quad B = \begin{bmatrix} -7 & 9 & -1 \\ -10 & 0 & 4 \\ -2 & 8 & -9 \end{bmatrix}$

B10. $A = \begin{bmatrix} -7 & -12 & -9 \\ -5 & 10 & 11 \\ 2 & -10 & -10 \end{bmatrix} \quad B = \begin{bmatrix} 7 & 3 & -6 \\ 5 & 3 & -4 \\ -11 & 1 & 10 \end{bmatrix}$

If $\mathbf{y} = [-7, 4, 2]^T$ and the matrix A is as given in each of Exercises B11–14,

(a) obtain $\mathbf{x}_0$, the exact solution of $A\mathbf{x} = \mathbf{y}$.

(b) let B be the matrix obtained if A is rounded to four significant digits, and obtain $\alpha = \frac{\|B - A\|}{\|A\|}$.

(c) let $\mathbf{z}$ be the solution of $B\mathbf{x} = \mathbf{y}$ computed by Gaussian elimination and back substitution performed in four significant-digit arithmetic.

(d) let $\Delta\mathbf{x} = \mathbf{z} - \mathbf{x}_0$, and compute $\beta = \frac{\|\Delta\mathbf{x}\|}{\|\mathbf{z}\|}$ and cond A.

(e) estimate the accuracy in $\mathbf{z}$ using (39.5) and the product α cond A.

(f) compare the estimated accuracy in part (e) with β computed in part (d). How useful is (39.5) in estimating the accuracy of a solution computed in fixed precision arithmetic?

B11. $A = \frac{1}{123,310} \begin{bmatrix} -543 & 1010 & -2480 \\ -417 & 435 & 3205 \\ 1720 & -20 & 1270 \end{bmatrix}$

B12. $A = \frac{1}{158,996} \begin{bmatrix} 2247 & 4590 & -4025 \\ -550 & 4608 & -2482 \\ -2823 & 814 & -887 \end{bmatrix}$

B13. $A = \frac{1}{11,138} \begin{bmatrix} 1082 & -2454 & -172 \\ -1290 & 2164 & 483 \\ 1344 & -2410 & -296 \end{bmatrix}$

B14. $A = \frac{1}{82,144} \begin{bmatrix} -5130 & 2603 & 1093 \\ 6576 & -2424 & -8 \\ 5798 & -1725 & -819 \end{bmatrix}$

In [16] we find the approximation cond $A \approx \frac{\|\mathbf{u}\|}{\|\mathbf{z}\|} 10^k$, where k is the number of significant digits with which the approximate solution $\mathbf{z}$ has been computed for the system $A\mathbf{x} = \mathbf{y}$, $\mathbf{u}$ is the solution of $A\mathbf{x} = \mathbf{r}$, with $\mathbf{r} = A\mathbf{z}$ being the residual vector and the norm is the ∞-norm. For $\mathbf{y} = [1, 2, 3, 4]^T$ and each matrix A in Exercises B15–18:

(a) Let B be the result of rounding A to five significant digits, and let $\mathbf{z}$ be the solution of $B\mathbf{x} = \mathbf{y}$ computed by Gaussian elimination and back substitution in five-digit arithmetic.

(b) Compute the residual $\mathbf{r} = B\mathbf{z} - \mathbf{y}$ in 10-digit arithmetic.

(c) In five-digit arithmetic, compute $\mathbf{u}$, the solution of $B\mathbf{x} = \mathbf{r}$.

(d) Compare cond A with $\frac{\|\mathbf{u}\|}{\|\mathbf{z}\|} 10^5$.

B15. $A = \frac{1}{12,057,292} \begin{bmatrix} -9128 & 1280 & -157,020 & 218,512 \\ 352,822 & 185,647 & -1,343,713 & 1,059,206 \\ 114,386 & 118,693 & -383,555 & 405,526 \\ 448,330 & 98,283 & -1,365,133 & 1,065,978 \end{bmatrix}$

B16. $A = \frac{1}{43,666,854} \begin{bmatrix} -281,727 & 214,182 & 527,493 & -241,751 \\ 51,252 & 425,562 & -618,612 & -243,230 \\ -581,037 & 95,778 & 122,121 & -342,973 \\ -532,194 & 125,604 & 217,596 & 138,356 \end{bmatrix}$

B17. $A = \frac{1}{18,962,039}\begin{bmatrix} -75,420 & 399,548 & 21,607 & -41,271 \\ 282,466 & -276,517 & 169,993 & -70,199 \\ 221,763 & -49,468 & -23,557 & -66,458 \\ -400,578 & 559,296 & -245,774 & -314,239 \end{bmatrix}$

B18. $A = \frac{1}{9,661,790}\begin{bmatrix} -100,915 & -284,840 & 94,525 & 181,335 \\ 152,985 & 154,160 & -133,245 & -83,895 \\ 16,908 & 57,394 & -126,132 & -193,622 \\ -64,490 & 211,950 & 62,800 & 161,360 \end{bmatrix}$

39.3 Iterative Improvement

The Residual Vector

If $\mathbf{x}_0$ is an approximate solution of the linear system $A\mathbf{x} = \mathbf{y}$, then $\mathbf{r} = \mathbf{y} - A\mathbf{x}_0$ is called the *residual vector* for $\mathbf{x}_0$. Of course, if $\mathbf{x}_0$ were the exact solution, then $\mathbf{r}$ would be $\mathbf{0}$, the zero vector.

EXAMPLE 39.4   A symmetric $n \times n$ matrix whose entries are $a_{ij} = \frac{1}{i+j-1}$ is called a *Hilbert matrix*. Such matrices have large condition numbers and inverses containing only integers. As such, they provide useful examples and counterexamples. The 5×5 Hilbert matrix and its inverse are

$$A = \begin{bmatrix} 1 & \frac{1}{2} & \frac{1}{3} & \frac{1}{4} & \frac{1}{5} \\ \frac{1}{2} & \frac{1}{3} & \frac{1}{4} & \frac{1}{5} & \frac{1}{6} \\ \frac{1}{3} & \frac{1}{4} & \frac{1}{5} & \frac{1}{6} & \frac{1}{7} \\ \frac{1}{4} & \frac{1}{5} & \frac{1}{6} & \frac{1}{7} & \frac{1}{8} \\ \frac{1}{5} & \frac{1}{6} & \frac{1}{7} & \frac{1}{8} & \frac{1}{9} \end{bmatrix} \quad A^{-1} = \begin{bmatrix} 25 & -300 & 1050 & -1400 & 630 \\ -300 & 4800 & -18,900 & 26,880 & -12,600 \\ 1050 & -18,900 & 79,380 & -117,600 & 56,700 \\ -1400 & 26,880 & -117,600 & 179,200 & -88,200 \\ 630 & -12,600 & 56,700 & -88,200 & 44,100 \end{bmatrix}$$

The condition number of A is the large number 943,656. Thus, Hilbert matrices are extremely ill-conditioned, allowing us to test for stability and roundoff when solving linear systems.

The vector $\mathbf{y}$ and $\mathbf{x}_e$, the exact solution to the system $A\mathbf{x} = \mathbf{y}$, are

$$\mathbf{y} = \begin{bmatrix} 1 \\ 1 \\ 1 \\ 1 \\ 1 \end{bmatrix} \quad \mathbf{x}_e = \begin{bmatrix} 5 \\ -120 \\ 630 \\ -1120 \\ 630 \end{bmatrix}$$

Clearly, the residual vector for the exact solution is $\mathbf{r} = \mathbf{0}$. Suppose, however, that we are working numerically on a five-digit device. Let A_5 be A expressed to five significant figures, that is

$$A_5 = \begin{bmatrix} 1.0000 & 0.50000 & 0.33333 & 0.25000 & 0.20000 \\ 0.50000 & 0.33333 & 0.25000 & 0.20000 & 0.16667 \\ 0.33333 & 0.25000 & 0.20000 & 0.16667 & 0.14286 \\ 0.25000 & 0.20000 & 0.16667 & 0.14286 & 0.12500 \\ 0.20000 & 0.16667 & 0.14286 & 0.12500 & 0.11111 \end{bmatrix}$$

and let $\mathbf{x}_0$ be the approximate solution found by back substitution with the upper triangular form of the augmented matrix $[A_5, \mathbf{y}]$. Let the residual for $\mathbf{x}_0$ be $\mathbf{r}_0$, computed in double precision, which, here, is 10 digits. As can be seen in (39.10), the residual appears small, but $\|\mathbf{x}_e - \mathbf{x}_0\|_\infty = 108.4$, so the ∞-norm of the difference between the exact and approximate solutions is large. For ill-conditioned systems, the residual vector is a poor reflection of the accuracy of the solution. The residual vector can be small, while the solution generating it is far from accurate.

$$\mathbf{x}_0 = \begin{bmatrix} 4.2800 \\ -104.94 \\ 561.00 \\ -1011.6 \\ 575.47 \end{bmatrix} \quad \mathbf{r}_0 = \begin{bmatrix} -0.0040000 \\ -0.00166669 \\ -0.00166669 \\ -0.00146422 \\ 0.00003173 \end{bmatrix} \tag{39.10}$$

Surprisingly, however, even for ill-conditioned systems, the residual can be used to improve the accuracy of an inaccurate solution, as we shall now show. ❖

Iterative Improvement

The process of refining an inaccurate solution using the residual vector is called *iterative improvement* in [60] and [38]; *iterative refinement* in [16], [49] and [19]; and *residual correction* in [60]. The steps of one cycle of the iterative improvement algorithm are as follows.

1. Let $\mathbf{x}_k$ be a solution of the linear system $A\mathbf{x} = \mathbf{y}$.
2. In double precision, compute $\mathbf{r}_k = \mathbf{y} - A\mathbf{x}_k$, the residual vector for $\mathbf{x}_k$.
3. Solve the system $A\mathbf{e}_k = \mathbf{r}_k$ for $\mathbf{e}_k$, the error vector.
4. Define $\mathbf{x}_{k+1} = \mathbf{x}_k + \mathbf{e}_k$, the improved solution.

Before discussing the algorithm, we illustrate with a continuation of the example just given.

EXAMPLE 39.5 In (39.10) we list $\mathbf{x}_0$, the five-digit solution of $A\mathbf{x} = \mathbf{y}$, and $\mathbf{r}_0$, the residual computed in double precision. The system $A\mathbf{e}_0 = \mathbf{r}_0$ must be solved for $\mathbf{e}_0$ with the same Gaussian arithmetic as was used for obtaining $\mathbf{x}_0$. Thus, using A_5, Gaussian elimination, and back substitution we obtain $\mathbf{e}_0$ and the first improved estimate, $\mathbf{x}_1 = \mathbf{x}_0 + \mathbf{e}_0$. Since $\|\mathbf{x}_1 - \mathbf{x}_e\| = 7.9$ is approximately $\frac{1}{13}\|\mathbf{x}_0 - \mathbf{x}_e\|$, we conclude that an improvement has been made.

$$\mathbf{e}_0 = \begin{bmatrix} 0.67100 \\ -13.993 \\ 64.016 \\ -100.48 \\ 50.514 \end{bmatrix} \quad \mathbf{x}_1 = \begin{bmatrix} 4.9510 \\ -118.93 \\ 625.02 \\ -1112.1 \\ 625.98 \end{bmatrix}$$

Repeating the cycle of computing the residual $\mathbf{r}_k$ for $\mathbf{x}_k$, the error vector $\mathbf{e}_k$ as the solution of $A\mathbf{e}_k = \mathbf{r}_k$, and the improved $\mathbf{x}_{k+1} = \mathbf{x}_k + \mathbf{e}_k$, we obtain the following vectors $\mathbf{x}_2, \mathbf{x}_3$, and $\mathbf{x}_4$. For these vectors, the norms $\|\mathbf{x}_k - \mathbf{x}_e\|$ turn out to be 0.6, 0.03, and 0; so iterative improvement, at least for this example, is effective.

$$\mathbf{x}_2 = \begin{bmatrix} 4.9963 \\ -119.92 \\ 629.63 \\ -1119.4 \\ 629.70 \end{bmatrix} \quad \mathbf{x}_3 = \begin{bmatrix} 4.9997 \\ -119.99 \\ 629.97 \\ -1120.0 \\ 629.98 \end{bmatrix} \quad \mathbf{x}_4 = \begin{bmatrix} 5.0000 \\ -120.00 \\ 630.00 \\ -1120.0 \\ 630.00 \end{bmatrix} \quad ❖$$

EXAMPLE 39.6 We give one more example of the iterative improvement scheme, illustrating a more efficient technique for obtaining $\mathbf{e}_k$, the solutions of the systems $A\mathbf{e}_k = \mathbf{r}_k$. In Section 34.1 we saw the LU factorization of a matrix A. In fact, if $A = L_1U$ is the Doolittle decomposition, then the system $A\mathbf{x} = \mathbf{y}$ becomes $L_1 U\mathbf{x} = \mathbf{y}$ and setting $U\mathbf{x} = \mathbf{c}$ means the system $L_1\mathbf{c} = \mathbf{y}$ can be solved for $\mathbf{c}$ by forward substitution and $U\mathbf{x} = \mathbf{c}$ can be solved for $\mathbf{x}$ by back substitution. The labor of Gaussian elimination is expended once when factoring A to $L_1 U$, and any number of systems involving A can be solved by simply performing a forward substitution and a back substitution pair. Since the labor of factoring A is $O(n^3)$ but that of either a forward substitution or back substitution is $O(n^2)$, the computational savings are worth the effort of learning how to use factorization techniques. (See Section 39.1 for a discussion of operation counts.)

Using a strategy from Section 39.2, we construct a 3×3 matrix A with prescribed singular values and large condition number. Let the unit vectors $\mathbf{u}_k, k = 1, \ldots, 3$, be the columns of the orthogonal matrix Q in (39.11); and obtain A as the outer product $A = \sigma_1 \mathbf{u}_1 \mathbf{u}_1^{\mathrm{T}} + \sigma_2 \mathbf{u}_2 \mathbf{u}_2^{\mathrm{T}} + \sigma_3 \mathbf{u}_3 \mathbf{u}_3^{\mathrm{T}}$, where $\sigma_k, k = 1, \ldots, 3$, are the singular values 25, 10, and $\frac{1}{1000}$.

$$
Q = \begin{bmatrix} -\dfrac{4}{\sqrt{53}} & -\dfrac{234}{\sqrt{308,089}} & \dfrac{55}{\sqrt{5813}} \\[2mm] \dfrac{1}{\sqrt{53}} & \dfrac{498}{\sqrt{308,089}} & -\dfrac{32}{\sqrt{5813}} \\[2mm] \dfrac{6}{\sqrt{53}} & -\dfrac{73}{\sqrt{308,089}} & \dfrac{42}{\sqrt{5813}} \end{bmatrix} \quad A = \begin{bmatrix} \dfrac{114,916,813}{12,323,560} & \dfrac{14,598,168}{7,702,225} & -\dfrac{331,685,757}{30,808,900} \\[2mm] \dfrac{14,598,168}{7,702,225} & \dfrac{328,177,409}{38,511,125} & \dfrac{154,427,346}{38,511,125} \\[2mm] -\dfrac{331,685,757}{30,808,900} & \dfrac{154,427,346}{38,511,125} & \dfrac{1,321,270,873}{77,022,250} \end{bmatrix}
$$

$$\tag{39.11}$$

Even though the matrix Q contains radicals, A contains just rational numbers because the singular value decomposition of the symmetric A is $U \Sigma U^{\mathrm{T}}$. The matrix A has the modest determinant $\frac{1}{4}$ but the large (2-norm) condition number 25,000. Thus, A is nearly singular and, with a large condition number, is ill-conditioned. Yet, it does not have an exceptionally small determinant!

In five-digit arithmetic, the Doolittle decomposition of A (Section 34.1) is $EL_1 U$, where E is a permutation matrix ($E^{\mathrm{T}} = E^{-1}$) recording row interchanges caused by partial pivoting. Hence, A is approximated by

$$
EL_1 U = \begin{bmatrix} 0 & 0 & 1 \\ 0 & 1 & 0 \\ 1 & 0 & 0 \end{bmatrix} \begin{bmatrix} 1 & 0 & 0 \\ -0.17604 & 1 & 0 \\ -0.86615 & 0.58179 & 1 \end{bmatrix} \begin{bmatrix} -10.766 & 4.0099 & 17.154 \\ 0 & 9.2275 & 7.0298 \\ 0 & 0 & 0.0021265 \end{bmatrix}
$$

$$
= \begin{bmatrix} 9.3250 & 1.8953 & -10.766 \\ 1.8953 & 8.5216 & 4.0100 \\ -10.766 & 4.0099 & 17.154 \end{bmatrix}
$$

where roundoff causes the $(2, 3)$-entry in $EL_1 U$ to differ from the $(2, 3)$-entry. The system $A\mathbf{x} = \mathbf{y}$ becomes $EL_1 U\mathbf{x} = \mathbf{y}$. To allow forward substitution with the lower triangular L_1, multiply this last equation by E^{T}, the inverse of E, to obtain $L_1 U\mathbf{x} = E^{\mathrm{T}}\mathbf{y} = \mathbf{b}$. Then, as usual, set $U\mathbf{x} = \mathbf{c}$, and solve $L_1\mathbf{c} = \mathbf{b}$ for $\mathbf{c}$ by forward substitution and $U\mathbf{x} = \mathbf{c}$ for $\mathbf{x}$ by back substitution.

If $\mathbf{y}$ is the leftmost vector in (39.12), then $\mathbf{x}_e$, the exact solution to the system $A\mathbf{x} = \mathbf{y}$, is the second vector in (39.12). To the right of the arrow it is approximated to five significant digits. Lastly, the vector $\mathbf{b} = E^{\mathrm{T}}\mathbf{y}$ is given on the right in (39.12).

$$
\mathbf{y} = \begin{bmatrix} 1 \\ 2 \\ 3 \end{bmatrix} \quad \mathbf{x}_e = \begin{bmatrix} \dfrac{8,526,850,633}{7,702,225} \\[2mm] -\dfrac{4,958,902,987}{7,702,225} \\[2mm] \dfrac{13,023,744,981}{15,404,450} \end{bmatrix} \rightarrow \begin{bmatrix} 1107.1 \\ -643.83 \\ 845.45 \end{bmatrix} \quad \mathbf{b} = \begin{bmatrix} 3 \\ 2 \\ 1 \end{bmatrix} \tag{39.12}
$$

The solution $\mathbf{x}_0$, computed from the $L_1 U$ decomposition with five-digit arithmetic, and the residual $\mathbf{r}_0$, computed in double precision, are given in (39.13). For this solution we also find $\|\mathbf{x}_0 - \mathbf{x}_e\| \doteq 203$, so $\mathbf{x}_0$ contains considerable error. The correction vector $\mathbf{e}_0$, the solution of $A\mathbf{e}_0 = \mathbf{r}_0$, is found from $L_1 U \mathbf{e}_0 = E^{\mathrm{T}}\mathbf{r}_0$ and is the rightmost vector in (39.13).

$$\mathbf{x}_0 = \begin{bmatrix} 1310.1 \\ -762.02 \\ 1000.6 \end{bmatrix} \quad \mathbf{r}_0 = \begin{bmatrix} 0.99547 \\ 0.245892 \\ -1.62732 \end{bmatrix} \quad \mathbf{e}_0 = \begin{bmatrix} -240.29 \\ 139.88 \\ -183.60 \end{bmatrix} \quad \textbf{(39.13)}$$

The improved solution $\mathbf{x}_1 = \mathbf{x}_0 + \mathbf{e}_0$, along with $\mathbf{x}_6$, the fifth improvement generated by the process of iterative refinement, are given in (39.14). Since $\|\mathbf{x}_1 - \mathbf{x}_e\| = 37.3$ and $\|\mathbf{x}_6 - \mathbf{x}_e\| = 0.01$, we conclude that iterative improvement was successful once again.

$$\mathbf{x}_1 = \begin{bmatrix} 1069.8 \\ -622.14 \\ 817.00 \end{bmatrix} \quad \mathbf{x}_6 = \begin{bmatrix} 1107.1 \\ -643.83 \\ 845.46 \end{bmatrix} \quad \textbf{(39.14)}$$

Justification

We conclude with a few words on why iterative improvement works.

Just as $\mathbf{x}_0$ is an approximate solution of the system $A\mathbf{x} = \mathbf{y}$, so also is $\mathbf{e}_k$ an approximate solution of $A\mathbf{e}_k = \mathbf{r}_k$. Consequently, we write

$$\mathbf{e}_k \doteq A^{-1}\mathbf{r}_k = A^{-1}(\mathbf{y} - A\mathbf{x}_k) = A^{-1}\mathbf{y} - A^{-1}A\mathbf{x}_k = \mathbf{x}_e - \mathbf{x}_k$$

Thus, the correction $\mathbf{e}_k$ is approximately the difference between the exact solution $\mathbf{x}_e$ and the approximate solution $\mathbf{x}_k$, so $\mathbf{x}_e$ is then approximately $\mathbf{x}_k + \mathbf{e}_k$.

EXERCISES 39.3

1. The $n \times n$ Hilbert matrix $H_n = \{\frac{1}{i+j-1}\}$, $1 \le i, j \le n$, notoriously ill-conditioned even though its inverse is nice enough to contain only integers, is a useful example against which to test most matrix algorithms. Let $\mathbf{y} = [1, 2, 3, 4, 5]^{\mathrm{T}}$, and let $A = H_5$.

(a) Obtain $\mathbf{x}_e$, the exact solution of $A\mathbf{x} = \mathbf{y}$.

(b) Let B be the result of rounding A to five significant figures. Using five-digit arithmetic, compute $\mathbf{x}_1$, the solution of $B\mathbf{x} = \mathbf{y}$. (Use an $L_1 U$ decomposition of B and forward/back substitutions.) Compute $\|\mathbf{x}_e - \mathbf{x}_1\|$ as a measure of the accuracy of $\mathbf{x}_1$.

(c) On the basis of cond A and the work of Section 39.2, estimate the accuracy in $\mathbf{x}_1$.

(d) Apply iterative improvement to $\mathbf{x}_1$, being sure to compute each residual $\mathbf{r}_k$ in 10-digit arithmetic. Compute $\|\mathbf{x}_e - \mathbf{x}_k\|$ for each iteratively improved solution $\mathbf{x}_k$, $k = 2, \dots$. How accurately can the solution be computed using iterative improvement?

The matrices A in Exercises 2–5 are well-conditioned, thus reducing the need for iterative improvement. For each such matrix and the vector $\mathbf{y} = [-7, 4, 2]^{\mathrm{T}}$:

(a) Obtain $\mathbf{x}_e$, the exact solution of $A\mathbf{x} = \mathbf{y}$.

(b) Let B be the result of rounding A to five significant figures. Using five-digit arithmetic, compute $\mathbf{x}_1$, the solution of $B\mathbf{x} = \mathbf{y}$. (Use an $L_1 U$ decomposition of B and forward/back

substitutions.) Compute $\|\mathbf{x}_e - \mathbf{x}_1\|$ as a measure of the accuracy of $\mathbf{x}_1$.

(c) Apply iterative improvement to $\mathbf{x}_1$, being sure to compute each residual $\mathbf{r}_k$ in 10-digit arithmetic. Compute $\|\mathbf{x}_e - \mathbf{x}_k\|$ for each iteratively improved solution $\mathbf{x}_k$, $k = 2, \dots$. How useful is iterative improvement?

2. A in Exercise B11, Section 39.2

3. A in Exercise B12, Section 39.2

4. A in Exercise B13, Section 39.2

5. A in Exercise B14, Section 39.2

For the matrices A in Exercises 6–9, and the vector $\mathbf{y} = [1, 2, 3, 4]^{\mathrm{T}}$:

(a) Obtain $\mathbf{x}_e$, the exact solution of $A\mathbf{x} = \mathbf{y}$.

(b) Let B be the result of rounding A to five significant figures. Using five-digit arithmetic, compute $\mathbf{x}_1$, the solution of $B\mathbf{x} = \mathbf{y}$. (Use an $L_1 U$ decomposition of B and forward/back substitutions.) Compute $\|\mathbf{x}_e - \mathbf{x}_1\|$ as a measure of the accuracy of $\mathbf{x}_1$.

(c) Apply iterative improvement to $\mathbf{x}_1$, being sure to compute each residual $\mathbf{r}_k$ in 10-digit arithmetic. Compute $\|\mathbf{x}_e - \mathbf{x}_k\|$ for each iteratively improved solution $\mathbf{x}_k$, $k = 2, \dots$. How useful is iterative improvement?

6. A in Exercise B15, Section 39.2

7. A in Exercise B16, Section 39.2

8. A in Exercise B17, Section 39.2

9. A in Exercise B18, Section 39.2

10. The process of iterative improvement is often "strong enough" to use in the following way. Suppose the matrix $A(\varepsilon)$ depends on the small parameter ε so that $\mathbf{x}$, the solution of the system $A(\varepsilon)\mathbf{x} = \mathbf{y}$, will likewise depend on ε. Let $A_0 = A(0)$ and $A_1 = A(\varepsilon_1)$ for a fixed ε_1. Let $\mathbf{z}$ be the solution of $A_0\mathbf{x} = \mathbf{y}$ and iteratively refine $\mathbf{z}$ using A_1 to compute the residuals $\mathbf{r}_k$ but still using A_0 to compute the error vectors $\mathbf{e}_k$ via the equations $A_0\mathbf{e}_k = \mathbf{r}_k$. The iteratively

refined solutions converge to $\mathbf{X}$, the solution of $A_1\mathbf{x} = \mathbf{y}$. For example, if $\mathbf{y} = [1, 2, 3]^{\mathrm{T}}$ and

$$A(\varepsilon) = \begin{bmatrix} -78\epsilon^2 + 62\epsilon + 11 & 88\epsilon^2 + \epsilon + 30 & 81\epsilon^2 - 5\epsilon - 28 \\ 4\epsilon^2 - 11\epsilon + 10 & 57\epsilon^2 - 82\epsilon - 48 & -11\epsilon^2 + 38\epsilon - 7 \\ 58\epsilon^2 - 94\epsilon - 68 & 14\epsilon^2 - 35\epsilon - 14 & -9\epsilon^2 - 51\epsilon - 73 \end{bmatrix}:$$

(a) Obtain $A_1 = A(0.1)$ and $\mathbf{X}$, the exact solution of $A_1\mathbf{x} = \mathbf{y}$.

(b) Obtain $A_0 = A(0)$ and $\mathbf{z}$, the solution of $A_0\mathbf{x} = \mathbf{y}$.

(c) Perform iterative improvement on $\mathbf{z}$, being sure to compute the residual vectors $\mathbf{r}_k$ with A_1, but the error vectors $\mathbf{e}_k$ with A_0. For each improvement $\mathbf{x}_k$, compute $\|\mathbf{X} - \mathbf{x}_k\|$, and show that these norms get smaller and that $\mathbf{x}_k$ converges to $\mathbf{X}$.

39.4 The Method of Jacobi

Fixed-Point Iteration for a Linear System

FORMULATING AN ITERATION FUNCTION   If the ith equation in the $n \times n$ linear system $A\mathbf{x} = \mathbf{y}$ is solved for the jth variable in such a way that each variable appears exactly once on a left-hand side of the rearranged equations and the equations reordered so the indices on the variables on the left-hand sides of the equations are in sequence, the equations can be written as

$$x_j = x_j + \frac{1}{a_{ij}}\left(y_i - \sum_{k=1}^{n} a_{ik}x_k\right) \qquad j = 1, \ldots, n$$

Defining the right-hand side of each equation as a function $g_j(\mathbf{x})$ and the vector of functions $[g_1(\mathbf{x}), \ldots, g_n(\mathbf{x})]^{\mathrm{T}}$ as the vector-valued function $\mathbf{g}(\mathbf{x})$, the rearranged equations become $\mathbf{x} = \mathbf{g}(\mathbf{x})$, an equation that suggests a vector version of fixed-point iteration. The fixed-point iteration just described is called *Jacobi's method,* or the *method of simultaneous displacements* [60], [38].

At each iteration in the method of Jacobi, $x_{k+1} = g(x_k)$, so a new value for *every* variable is computed, in sequence, from the values generated in the previous iteration. The method converges slowly, at a linear rate, as measured by $\|x_k - x\|$, where $x = g(x)$ is a fixed point of the iteration. The method is amenable to implementation on a parallel-processor machine where variables could be updated simultaneously on dedicated processors.

EXAMPLE 39.7   To clarify the process of putting the linear system $A\mathbf{x} = \mathbf{y}$ into the form $\mathbf{x} = \mathbf{g}(\mathbf{x})$, consider the 4×4 system of linear equations defined by the matrix A and vectors $\mathbf{x}$ and $\mathbf{y}$ in

$$A = \begin{bmatrix} -6 & 11 & 3 & 10 \\ 10 & -3 & -2 & -8 \\ -9 & -5 & -5 & -12 \\ -1 & -4 & 12 & -3 \end{bmatrix} \qquad \mathbf{x} = \begin{bmatrix} x_1 \\ x_2 \\ x_3 \\ x_4 \end{bmatrix} \qquad \mathbf{y} = \begin{bmatrix} 7 \\ -9 \\ 8 \\ -2 \end{bmatrix} \qquad \textbf{(39.15)}$$

If the ith equation in $A\mathbf{x} = \mathbf{y}$ is solved for x_i, that is, if the first equation is solved for x_1, the second, for x_2, the third, for x_3, and the fourth, for x_4, the equations in $\mathbf{x} = g(\mathbf{x})$ would be

$$x_1 = -\tfrac{7}{6} + \tfrac{11}{6}x_2 + \tfrac{1}{2}x_3 + \tfrac{5}{3}x_4 \qquad x_3 = -\tfrac{8}{5} - \tfrac{9}{5}x_1 - x_2 - \tfrac{12}{5}x_4$$

$$x_2 = 3 + \tfrac{10}{3}x_1 - \tfrac{2}{3}x_3 - \tfrac{8}{3}x_4 \qquad x_4 = \tfrac{2}{3} - \tfrac{1}{3}x_1 - \tfrac{4}{3}x_2 + 4x_3$$

A more convenient form of these equations is given by

$$x_j = x_j + \frac{1}{a_{ij}} \left(y_i - \sum_{k=1}^{n} a_{ik} x_k \right) \qquad j = 1, \ldots, n$$

the advantage being that on the right, the sum $\sum_{k=1}^{n} a_{ik} x_k$ runs through all n variables in each equation. The alternative is the need to program each sum to skip one variable in each equation. ❖

Three Definitions

The variable x_k is the *strictly dominant variable* of the equation $\sum_{j=1}^{n} a_{ij} x_j = y_i$ if $|a_{ik}| > \sum_{j=1, j \neq k}^{n} |a_{ij}|$. Thus, the absolute value of the coefficient of x_k is strictly greater than the sum of the absolute values of all the *other* coefficients in the equation.

The matrix A is a *strictly dominant* matrix if each equation in the linear system $A\mathbf{x} = \mathbf{y}$ has a strictly dominant variable, each in a different column. Thus, every row of A contains a *dominant entry*, an entry whose absolute value is greater than the sum of all the other entries in that row, and no two rows have a dominant entry in the same column.

The matrix A is *strictly diagonally dominant* if it is strictly dominant, and the dominant entry in each row is a_{kk}, the diagonal element. In essence, the rows of a *strictly dominant* matrix can be rearranged to form a *strictly diagonally dominant* matrix.

EXAMPLE 39.8 The matrix A_1 is strictly diagonally dominant because of the inequalities to its right.

$$A_1 = \begin{bmatrix} 15 & 1 & -8 & -4 \\ -11 & 38 & 12 & 11 \\ 2 & -5 & -13 & 4 \\ -10 & 3 & 12 & -30 \end{bmatrix} \quad \begin{array}{l} |a_{11}| = 15 > |1| + |-8| + |-4| = 13 \\ |a_{22}| = 38 > |-11| + |12| + |11| = 34 \\ |a_{33}| = 13 > |2| + |-5| + |4| = 11 \\ |a_{44}| = 30 > |-10| + |3| + |12| = 25 \end{array}$$

(39.16)

The matrix A_2 is not strictly diagonally dominant because of the inequalities to its right.

$$A_2 = \begin{bmatrix} 0 & 0 & 1 & 9 \\ -6 & -1 & 4 & -6 \\ 6 & -7 & -7 & 0 \\ -11 & 6 & -8 & -1 \end{bmatrix} \quad \begin{array}{l} |a_{11}| = 0 < |0| + |1| + |9| = 10 \\ |a_{22}| = 1 < |-6| + |4| + |-6| = 16 \\ |a_{33}| = 7 < |6| + |-7| + |0| = 13 \\ |a_{44}| = 1 < |-11| + |6| + |-8| = 25 \end{array}$$

The matrix A_3 is not strictly diagonally dominant because of the inequalities to its right.

$$A_3 = \begin{bmatrix} -12 & -29 & 11 & -3 \\ -12 & -9 & 4 & 27 \\ -20 & -3 & 7 & 5 \\ -1 & 6 & 24 & 12 \end{bmatrix} \quad \begin{array}{l} |a_{11}| = 12 < |-29| + |11| + |-3| = 43 \\ |a_{22}| = 0 < |-12| + |4| + |27| = 43 \\ |a_{33}| = 7 < |-20| + |-3| + |5| = 28 \\ |a_{44}| = 12 < |-1| + |6| + |24| = 31 \end{array}$$

In A_3, however, $a_{12} = -29$ is a dominant element in row 1, $a_{24} = 27$ is a dominant element in row 2, $a_{31} = -20$ is a dominant element in row 3, and $a_{43} = 24$ is a dominant element in row 4. Thus, A_3 is strictly dominant, but not strictly diagonally dominant. Moreover, if row 1 becomes row 2, row 2 becomes row 4, row 3 becomes row 1, and row 4 becomes row 3, A_3 will be the strictly diagonally dominant matrix B where the strictly dominant elements in each row are now on the main diagonal.

$$B = \begin{bmatrix} -20 & -3 & 7 & 5 \\ -12 & -29 & 11 & -3 \\ -1 & 6 & 24 & 12 \\ -12 & -9 & 4 & 27 \end{bmatrix}$$

❖

Analysis

If the Jacobi method converges, it does so linearly and slowly. However, if A is a strictly diagonally dominant matrix, then Jacobi iteration converges for any starting vector $\mathbf{x}_0$, providing each equation is solved for its dominant (diagonal) variable. (See [16] for a proof.)

In the event the linear system $A\mathbf{x} = \mathbf{y}$ does not have a strictly dominant matrix A, the equations should be rearranged to make the system matrix as close as possible to a diagonally dominant matrix. Thus, as many equations as possible should be solved for their dominant variables.

EXAMPLE 39.9   The matrix A_1 in (39.16) has been shown to be strictly diagonally dominant. The vector $\mathbf{y}$ and the exact solution $\mathbf{x}_e$ of the corresponding linear system $A\mathbf{x} = \mathbf{y}$ are given in (39.17).

$$\mathbf{y} = \begin{bmatrix} 1 \\ 2 \\ 3 \\ 4 \end{bmatrix} \qquad \mathbf{x}_e = \begin{bmatrix} -\frac{40{,}951}{196{,}183} \\ \frac{34{,}356}{196{,}183} \\ -\frac{77{,}063}{196{,}183} \\ -\frac{3069}{15{,}091} \end{bmatrix} \overset{\circ}{=} \begin{bmatrix} -0.2087387796 \\ 0.1751222073 \\ -0.3928118135 \\ -0.2033662448 \end{bmatrix} \tag{39.17}$$

Solving for the ith variable in the ith equation, we therefore obtain a convergent fixed-point iteration of the form $\mathbf{x} = \mathbf{g}(\mathbf{x})$, whose equations are

$$x_1 = \frac{1}{15} - \frac{1}{15}x_2 + \frac{8}{15}x_3 + \frac{4}{15}x_4 \qquad x_3 = -\frac{3}{13} + \frac{2}{13}x_1 - \frac{5}{13}x_2 + \frac{4}{13}x_4$$

$$x_2 = \frac{1}{19} + \frac{11}{38}x_1 - \frac{6}{19}x_3 - \frac{11}{38}x_4 \qquad x_4 = -\frac{2}{15} - \frac{1}{3}x_1 + \frac{1}{10}x_2 + \frac{2}{5}x_3$$

Table 39.6 lists some of the first 20 iterates of the iteration $\mathbf{x}_{k+1} = \mathbf{g}(\mathbf{x}_k)$, starting with the initial vector $\mathbf{x}_0 = [10, 10, 10, 10]^T$. The rightmost column of Table 39.6 lists $\|\mathbf{x}_k - \mathbf{x}_e\|$,

k	x_1	x_2	x_3	x_4	$\|\mathbf{x}_k - \mathbf{x}_e\|_\infty$
0	10.0	10.0	10.0	10.0	10.3928
1	7.400000000	−3.105263158	0.538461538	1.533333334	7.6087
2	0.9697525866	1.580836707	2.573819163	−2.695141700	2.9666
3	0.6152766536	0.3007370314	−1.518865013	0.7310271402	1.1260
4	−0.5685032383	0.4987664423	−0.0268479453	−0.9158978532	0.7125
5	−0.2251427612	0.1616714764	−0.7918792385	0.0953052122	0.3991
⋮	⋮	⋮	⋮	⋮	⋮
16	−0.2088100049	0.1751546057	−0.3926759303	−0.2035262613	0.16×10^{-3}
17	−0.2087111395	0.1751049996	−0.3928844679	−0.2032849099	0.81×10^{-4}
18	−0.2087546922	0.1751296076	−0.3927759168	−0.2034062407	0.40×10^{-4}
19	−0.2087307936	0.1751178430	−0.3928294143	−0.2033458419	0.20×10^{-4}
20	−0.2087424350	0.1751241711	−0.3928026285	−0.2033763835	0.10×10^{-4}

TABLE 39.6   Jacobi iterates in Example 39.9

the norm of the difference between the kth iterate and the exact solution. It should be clear from this data that the iteration converges to the fixed point, but converges slowly. In fact, the convergence is linear, as an inspection of the norms in the rightmost column suggests. When the number of zeros before the first significant figure does not double at each iteration, the convergence is not quadratic. Since it takes several iterations to add another zero before the first significant digit, the convergence appears to be linear. ❖

EXERCISES 39.4

1. Obtain the exact solution of the linear system $2x + 3y = 9$, $5x + 4y = 9$. Then, solve each equation for its dominant variable, and obtain $\mathbf{x}_k$, $k = 1, 2, 3$, by Jacobi iteration starting from $\mathbf{x}_0 = \mathbf{0}$.

In Exercises 2–11:

(a) Rearrange the rows of the given matrix C so that it becomes a strictly diagonally dominant matrix A.

(b) Let $\mathbf{y} = [1, \ldots, n]^T$, where $n = 4$ or 5, as appropriate, and obtain $\mathbf{x}_e$, the exact solution of $A\mathbf{x} = \mathbf{y}$.

(c) Using $x_i^{(k)} = (y_i - \sum_{j=1, j \neq i}^{n} a_{ij} x_j^{(k-1)})/a_{ii}$ and taking $\mathbf{x}_0 = \mathbf{0}$ as the initial vector, implement Jacobi iteration in terms of the individual equations. Monitor convergence by computing $\sigma_k = \|\mathbf{x}_e - \mathbf{x}_k\|$, and note the number of iterations required for this measure to be less that 10^{-5}.

(d) If $A = \{a_{ij}\}$, let $D = \text{diag}(a_{11}, \ldots, a_{nn})$ be a diagonal matrix whose entries are the (main) diagonal elements in A. Let $B = D - A$, then obtain $T = D^{-1}B$ and $\mathbf{v} = D^{-1}\mathbf{y}$. The matrix form of Jacobi iteration for $A\mathbf{x} = \mathbf{y}$ is then $\mathbf{x}_{k+1} = T\mathbf{x}_k + \mathbf{v}$.

(e) Obtain ρ, the maximum magnitude of the eigenvalues of T.

(f) Implement the matrix form of Jacobi iteration, and compare σ_k from part (c) with $\rho^k \|\mathbf{x}_e - \mathbf{x}_0\|$. These numbers should be approximately the same.

(g) Using the ∞-norm, verify empirically that the bound $\|\mathbf{x}_e - \mathbf{x}_k\| \leq \|T\|^k \|\mathbf{x}_1 - \mathbf{x}_0\|/(1 - \|T\|)$ is valid.

2. $\begin{bmatrix} 1 & 8 & 8 & 25 \\ -22 & 6 & 6 & -7 \\ -6 & 0 & 11 & -2 \\ -9 & 21 & 3 & -7 \end{bmatrix}$ **3.** $\begin{bmatrix} -9 & -3 & -19 & -6 \\ 7 & 15 & -5 & -1 \\ 19 & -3 & -8 & 6 \\ -6 & 4 & -1 & -18 \end{bmatrix}$

4. $\begin{bmatrix} -2 & 1 & 10 & 1 \\ -11 & 1 & -7 & -2 \\ 4 & 7 & -1 & 0 \\ -9 & 0 & -7 & 19 \end{bmatrix}$ **5.** $\begin{bmatrix} -5 & 6 & 2 & -22 \\ 17 & 7 & -1 & 1 \\ 7 & -3 & 25 & -7 \\ -2 & -19 & 8 & -4 \end{bmatrix}$

6. $\begin{bmatrix} 9 & -4 & 9 & 26 \\ 8 & -24 & 5 & 8 \\ 27 & -7 & 9 & 7 \\ -3 & -7 & 18 & 3 \end{bmatrix}$ **7.** $\begin{bmatrix} 9 & 4 & -21 & 7 \\ -7 & 28 & 9 & -6 \\ 18 & 5 & 1 & 3 \\ -1 & 9 & -3 & 21 \end{bmatrix}$

8. $\begin{bmatrix} 10 & 5 & 17 & 86 & 49 \\ -19 & 77 & -13 & -21 & -23 \\ 39 & 33 & -11 & 33 & 119 \\ -47 & -18 & 121 & 5 & -42 \\ -145 & -42 & 16 & -45 & -35 \end{bmatrix}$

9. $\begin{bmatrix} -17 & 9 & 0 & -12 & 44 \\ -31 & 67 & 15 & -15 & 4 \\ 39 & -25 & -83 & 2 & 13 \\ 9 & -46 & -16 & -107 & -30 \\ 67 & 45 & -1 & 0 & 16 \end{bmatrix}$

10. $\begin{bmatrix} -6 & 46 & -93 & -14 & 22 \\ 9 & -34 & 10 & -100 & 44 \\ 9 & 34 & 18 & 16 & -82 \\ 41 & -90 & -14 & 2 & 31 \\ 149 & 49 & 26 & 33 & -35 \end{bmatrix}$

11. $\begin{bmatrix} -14 & 77 & 9 & 21 & -31 \\ -44 & 33 & -25 & 48 & 152 \\ 5 & 34 & -79 & -25 & 6 \\ 90 & -30 & -24 & -9 & 18 \\ -32 & -29 & 21 & 113 & -22 \end{bmatrix}$

For each of the matrices A given in Exercises 12–17:

(a) Show that $\rho < 1$, where ρ is the maximum magnitude of the eigenvalues of $T = D^{-1}(D - A)$, as defined for Exercises 2–11.

(b) Obtain $\mathbf{x}_e$, the exact solution of $A\mathbf{x} = \mathbf{y}$, where $\mathbf{y} = [2, -5, 3]^T$.

(c) Show that Jacobi iteration for the system in part (b) converges for $\mathbf{x}_0 = \mathbf{0}$. Actually, it is not necessary for A to be strictly diagonally dominant for Jacobi iteration to converge from any starting point. More generally, the iteration converges for any starting point if the condition in part (a) holds.

12. $\dfrac{1}{5}\begin{bmatrix} 20 & 4 & -3 \\ -2 & -5 & 6 \\ -3 & 1 & 15 \end{bmatrix}$ **13.** $\dfrac{1}{6}\begin{bmatrix} 18 & 5 & 5 \\ 5 & -12 & 9 \\ 0 & 1 & 42 \end{bmatrix}$

14. $\dfrac{1}{12}\begin{bmatrix} 24 & 8 & 0 \\ -7 & 36 & 9 \\ 3 & -5 & 48 \end{bmatrix}$ **15.** $\dfrac{1}{8}\begin{bmatrix} 8 & -9 & -3 \\ 3 & 16 & 2 \\ 4 & -1 & 24 \end{bmatrix}$

16. $\dfrac{1}{10}\begin{bmatrix} 30 & 2 & 9 \\ 0 & 40 & -6 \\ -3 & 9 & 50 \end{bmatrix}$ **17.** $\dfrac{1}{16}\begin{bmatrix} -48 & 7 & 11 \\ 10 & 80 & 10 \\ 9 & 2 & 32 \end{bmatrix}$

Gauss–Seidel Iteration

The Gauss–Seidel Method

In an attempt to speed up the convergence of the Jacobi fixed-point iteration $\mathbf{x} = \mathbf{g}(\mathbf{x})$ for solving the linear system $A\mathbf{x} = \mathbf{y}$, the Gauss–Seidel method uses newly generated values for the x_k's just as soon as they become available. Thus, the Jacobi method computes $\mathbf{x}_{m+1}$ from $\mathbf{g}(\mathbf{x}_m)$ by a recipe of the form

$$x_k^{m+1} = \frac{1}{a_{kk}}\left(y_k - \sum_{s \neq k} a_{ks}x_s^m\right) \qquad k = 1, \ldots, n$$

or even of the form

$$x_k^{m+1} = \frac{1}{a_{kk}}\left(y_k - \sum_{s=1}^{k-1} a_{ks}x_s^m - \sum_{s=k+1}^{n} a_{ks}x_s^m\right) \qquad k = 1, \ldots, n$$

where the kth component of the vector $\mathbf{x}_m$ is denoted by x_k^m.

The Gauss–Seidel method computes the components of $\mathbf{x}_{m+1}$ by a recipe of the form

$$x_k^{m+1} = \frac{1}{a_{kk}}\left(y_k - \sum_{s=1}^{k-1} a_{ks}x_s^{m+1} - \sum_{s=k+1}^{n} a_{ks}x_s^m\right) \qquad k = 1, \ldots, n \tag{39.18}$$

The difference in the two methods is subtle. In the first summation on the right-hand side in (39.18), newly generated values for $x_s, s = 1, \ldots, k-1$, denoted by x_s^{m+1}, are used as soon as available. Although this generally speeds up convergence, it makes the method less amenable to implementation on multiprocessor computers.

Because the method immediately uses updated information, it is sometimes called the *method of successive displacements.*

Analysis

CONVERGENCE   If Gauss–Seidel iteration converges, it converges linearly. However, it generally converges whenever Jacobi iteration converges and at a faster rate, but there are linear systems for which Jacobi iteration converges and Gauss–Seidel does not and linear systems for which Gauss–Seidel iteration converges and Jacobi does not. (See [94] for details.)

Like the Jacobi method, the Gauss–Seidel method converges from any starting vector when A is a strictly diagonally dominant matrix. Alternatively [9], if the real matrix A is symmetric and positive definite with strictly positive entries in the main diagonal, Gauss–Seidel iteration will also converge. (The real symmetric matrix A is positive definite if all its eigenvalues are positive, guaranteeing that $\mathbf{x}^T A\mathbf{x} > 0$ for all $\mathbf{x} \neq \mathbf{0}$. See Section 33.4.)

 **EXAMPLE 39.10**   The matrix A in (39.19) is clearly symmetric but not diagonally dominant since, for example, $|a_{22}| = 16 < |12| + |5| + |-8| = 25$. It is positive definite, however, since its eigenvalues are 0.71, 11.8, 34.4, and 41.1.

$$A = \begin{bmatrix} 32 & 12 & -4 & 9 \\ 12 & 16 & 5 & -8 \\ -4 & 5 & 24 & -11 \\ 9 & -8 & -11 & 16 \end{bmatrix} \tag{39.19}$$

❖

RATE OF CONVERGENCE   Write the matrix A in the form $A = L + D + U$, where L is zero on and above the main diagonal, D is zero off the main diagonal, and U is zero on and below the main diagonal. Then [38] shows that the rates of convergence for the Jacobi and Gauss–Seidel iterations are determined by the largest magnitude of the eigenvectors of the respective matrices

$$B_J = D^{-1}(-L - U) = D^{-1}(D - A) \quad \text{and} \quad B_{GS} = (-D - L)^{-1}U = (U - A)^{-1}U$$

In fact, if the largest magnitude of the eigenvalues of B_J or B_{GS} is less than one, the corresponding method converges, whether or not the matrix A is strictly diagonally dominant.

EXAMPLE 39.11   For the matrix A in (39.19), the matrices L, D, and U are

$$L = \begin{bmatrix} 0 & 0 & 0 & 0 \\ 12 & 0 & 0 & 0 \\ -4 & 5 & 0 & 0 \\ 9 & -8 & -11 & 0 \end{bmatrix} \quad D = \begin{bmatrix} 32 & 0 & 0 & 0 \\ 0 & 16 & 0 & 0 \\ 0 & 0 & 24 & 0 \\ 0 & 0 & 0 & 16 \end{bmatrix} \quad U = \begin{bmatrix} 0 & 12 & -4 & 9 \\ 0 & 0 & 5 & -8 \\ 0 & 0 & 0 & -11 \\ 0 & 0 & 0 & 0 \end{bmatrix}$$

and the matrices B_J and B_{GS} are

$$B_J = \begin{bmatrix} 0 & -\frac{3}{8} & \frac{1}{8} & -\frac{9}{32} \\ -\frac{3}{4} & 0 & -\frac{5}{16} & \frac{1}{2} \\ \frac{1}{6} & -\frac{5}{24} & 0 & \frac{11}{24} \\ -\frac{9}{16} & \frac{1}{2} & \frac{11}{16} & 0 \end{bmatrix} \quad B_{GS} = \begin{bmatrix} 0 & -\frac{3}{8} & \frac{1}{8} & -\frac{9}{32} \\ 0 & \frac{9}{32} & -\frac{13}{32} & \frac{91}{128} \\ 0 & -\frac{31}{256} & \frac{27}{256} & \frac{809}{3072} \\ 0 & \frac{1099}{4096} & -\frac{823}{4096} & \frac{34,147}{49,152} \end{bmatrix}$$

The magnitudes of their eigenvalues are [0.441, 0.507, 0.900, 0.962] and [0, 0.119, 0.274, 0.926], respectively.

The magnitudes of the eigenvalues of each matrix are no greater than one, so both methods will converge for this matrix, but the Gauss–Seidel method will converge slightly faster since 0.926, the maximum magnitude of its eigenvalues, is slightly smaller than 0.962, the maximum magnitude of the eigenvalues of the matrix B_J. ❖

Implementation

Solve the linear system $A\mathbf{x} = \mathbf{y}$ by Gauss–Seidel iteration, with A given by the symmetric positive definite matrix in (39.19), and $\mathbf{y} = [1, 2, 3, 4]^T$. The exact solution, $\mathbf{x}_e$, is given by

$$\mathbf{x}_e = \begin{bmatrix} -\frac{3506}{1687} \\ \frac{5434}{1687} \\ \frac{1222}{1687} \\ \frac{5951}{1687} \end{bmatrix} \overset{\circ}{=} \begin{bmatrix} -2.078245406 \\ 3.221102549 \\ 0.724362774 \\ 3.527563723 \end{bmatrix}$$

Rather than write two sums as in (39.18), it's easier to assign new values to the same symbols, namely, the x_k, $k = 1, \ldots, 4$, so that whenever an x_k is called as an input in the computation, its value will already have been updated. This strategy is then implemented by writing the general equation in the form

$$x_k = x_k + \frac{1}{a_{kk}} \left(y_k - \sum_{s=1}^{4} a_{ks} x_s \right) \qquad k = 1, \ldots, 4$$

where we solved the kth equation for the kth variable. Table 39.7 lists the components of $\mathbf{x}_k$

k	x_1	x_2	x_3	x_4	$\|\mathbf{x}_k - \mathbf{x}_e\|_\infty$
0	10.0	10.0	10.0	10.0	12.08
1	−5.2812500	5.9609375	2.5862630	7.9792277	4.45
2	−4.1249765	6.4001390	1.7612877	6.9812540	3.45
3	−4.1121189	6.1493138	1.3582813	6.5715421	3.04
4	−3.9532037	5.9512110	1.2382539	6.3005821	2.77
5	−3.8177111	5.7516201	1.1782275	6.0833039	2.56
⋮	⋮	⋮	⋮	⋮	⋮
16	−2.8253574	4.3083080	0.91540601	4.6227592	1.10
17	−2.7700908	4.2278833	0.90127381	4.5417434	1.01
18	−2.7189123	4.1534079	0.88818704	4.4667207	0.94
19	−2.6715198	4.0844418	0.87606834	4.3972477	0.87
20	−2.6276330	4.0205773	0.86484610	4.3329139	0.81

TABLE 39.7   Gauss–Seidel iterates in Example 39.11

and, in its rightmost column, $\|\mathbf{x}_k - \mathbf{x}_e\|$. The iteration, starting with $\mathbf{x}_0 = [10, 10, 10, 10]^T$, seems to be converging, but slowly, as determined by the magnitudes of the eigenvalues of the matrix B_{GS}.

EXERCISES 39.5

1. Let $A = L + D + U$, where L, D, U are described just after Example 39.10.

   (a) Show that $A\mathbf{x} = \mathbf{y}$ can be put into the form $\mathbf{x} = B_{GS}\mathbf{x} + \mathbf{v}$, where $\mathbf{v} = (L + D)^{-1}\mathbf{y}$. *Hint:* Isolate $(L + D)\mathbf{x}$ in $A\mathbf{x} = \mathbf{y}$.

   (b) For the case $n = 2$, use the equations $ax_1 + bx_2 = y_1$ and $cx_1 + dx_2 = y_2$ to show that $\mathbf{x} = B_{GS}\mathbf{x} + \mathbf{v}$ is actually $\mathbf{x}_{k+1} = B_{GS}\mathbf{x}_k + \mathbf{v}$ and, therefore, represents Gauss–Seidel iteration. *Hint:* The first equation in $\mathbf{x}_{k+1} = B_{GS}\mathbf{x}_k + \mathbf{v}$ is just $ax_1 + bx_2 = y_1$ solved for x_1, and the second is $cx_1 + dx_2 = y_2$ solved for x_2, after which x_1 is replaced by its equivalent from the first equation.

In Exercises 2–11, set $\mathbf{y} = [1, 2, 3, 4]^T$, and let A be the diagonally dominant form of the given matrix C and write $A = L + D + U$.

   (a) Use the matrix formulation $\mathbf{x}_{k+1} = B_{GS}\mathbf{x}_k + \mathbf{v}$, where $\mathbf{v} = (D + L)^{-1}\mathbf{y}$, to implement Gauss–Seidel iteration, starting at $\mathbf{x}_0 = \mathbf{0}$. Note the value of k for which $\|\mathbf{x}_e - \mathbf{x}_k\|_\infty < 10^{-5}$.

   (b) Using the ∞-norm, test the validity of the inequality $\|\mathbf{x}_e - \mathbf{x}_k\| \le \|B_{GS}\|^k \|\mathbf{x}_1 - \mathbf{x}_0\| / (1 - \|B_{GS}\|)$.

   (c) Using the ∞-norm, test the validity of the inequality $\|\mathbf{x}_e - \mathbf{x}_k\| \le \|B_{GS}\| \|\mathbf{x}_e - \mathbf{x}_0\|$.

   (d) Write the equations for Gauss–Seidel iteration in the form (39.18).

   (e) Make successive substitutions in the equations written in part

   (d) to show that the result is the set of equations implemented in matrix notation in part (a).

2. the matrix in Exercise 2, Section 39.4

3. the matrix in Exercise 3, Section 39.4

4. the matrix in Exercise 4, Section 39.4

5. the matrix in Exercise 5, Section 39.4

6. the matrix in Exercise 6, Section 39.4

7. the matrix in Exercise 7, Section 39.4

8. the matrix in Exercise 8, Section 39.4

9. the matrix in Exercise 9, Section 39.4

10. the matrix in Exercise 10, Section 39.4

11. the matrix in Exercise 11, Section 39.4

For the matrices A given in Exercises 12–17 and $\mathbf{y} = [-3, 7, 5]^T$:

   (a) Obtain $\mathbf{x}_e$, the exact solution of the system $A\mathbf{x} = \mathbf{y}$.

   (b) Obtain the matrix $B_{GS} = (U - A)^{-1}U$, where $A = L + D + U$ as in Exercise B1.

   (c) Show that the matrix A is positive definite.

   (d) Show that the eigenvalues of B_{GS} all have magnitude less than one.

(e) Starting from $\mathbf{x}_0 = \mathbf{0}$, implement the matrix form of Gauss–Seidel iteration and determine the number of iterations required for $\|\mathbf{x}_e - \mathbf{x}_k\|_\infty < 10^{-5}$.

(f) Write the Gauss–Seidel scheme in the form (39.18), and then make the appropriate substitutions to bring the equations into the matrix form implemented in part (e).

12. $\begin{bmatrix} 12 & 4 & 2 \\ 4 & 2 & 1 \\ 2 & 1 & 11 \end{bmatrix}$ **13.** $\begin{bmatrix} 11 & 6 & 2 \\ 6 & 10 & 2 \\ 2 & 2 & 4 \end{bmatrix}$ **14.** $\begin{bmatrix} 8 & 6 & 1 \\ 6 & 7 & 3 \\ 1 & 3 & 11 \end{bmatrix}$

15. $\begin{bmatrix} 3 & 1 & 1 \\ 1 & 7 & 4 \\ 1 & 4 & 9 \end{bmatrix}$ **16.** $\begin{bmatrix} 11 & 2 & 7 \\ 2 & 1 & 2 \\ 7 & 2 & 9 \end{bmatrix}$ **17.** $\begin{bmatrix} 12 & 4 & 2 \\ 4 & 5 & 6 \\ 2 & 6 & 10 \end{bmatrix}$

For the matrices A given in Exercises 18–23 and $\mathbf{y} = [6, -7, 1, 9]^T$:

(a) Obtain $\mathbf{x}_e$, the exact solution of the system $A\mathbf{x} = \mathbf{y}$.

(b) Obtain the matrix $B_{GS} = (U - A)^{-1}U$, where $A = L + D + U$.

(c) Show that the matrix A is positive definite.

(d) Show that the eigenvalues of B_{GS} all have magnitude less than one.

(e) Starting from $\mathbf{x}_0 = \mathbf{0}$, implement the matrix form of Gauss–Seidel iteration, and determine the number of iterations required for $\|\mathbf{x}_e - \mathbf{x}_k\|_\infty < 10^{-5}$.

(f) Write the Gauss–Seidel scheme in the form (39.18), and then make the appropriate substitutions to bring the equations into the matrix form implemented in part (e).

18. $\begin{bmatrix} 12 & 7 & 7 & 3 \\ 7 & 12 & 6 & 7 \\ 7 & 6 & 11 & 1 \\ 3 & 7 & 1 & 8 \end{bmatrix}$ **19.** $\begin{bmatrix} 10 & 3 & 7 & 6 \\ 3 & 8 & 2 & 2 \\ 7 & 2 & 6 & 4 \\ 6 & 2 & 4 & 6 \end{bmatrix}$

20. $\begin{bmatrix} 7 & 2 & 4 & 2 \\ 2 & 8 & 8 & 3 \\ 4 & 8 & 10 & 4 \\ 2 & 3 & 4 & 3 \end{bmatrix}$ **21.** $\begin{bmatrix} 10 & 8 & 4 & 5 \\ 8 & 9 & 5 & 3 \\ 4 & 5 & 9 & 6 \\ 5 & 3 & 6 & 7 \end{bmatrix}$

22. $\begin{bmatrix} 12 & 7 & 5 & 10 \\ 7 & 9 & 6 & 6 \\ 5 & 6 & 9 & 5 \\ 10 & 6 & 5 & 9 \end{bmatrix}$ **23.** $\begin{bmatrix} 12 & 7 & 4 & 6 \\ 7 & 10 & 8 & 6 \\ 4 & 8 & 9 & 3 \\ 6 & 6 & 3 & 6 \end{bmatrix}$

39.6 Relaxation and SOR

Relaxation—the Concept

The idea of *relaxation* has been attributed to the British engineer, Richard Southwell, [86], although some claim that Gauss was familiar with the concept. Conceived in the "by-hand" era before computers, the idea underlies the more modern method of *successive overrelaxation* (SOR). Our exposition will be in terms of the linear system $A\mathbf{x} = \mathbf{y}$, for which the strictly diagonally dominant matrix A, the vector $\mathbf{y}$, and the exact solution $\mathbf{x}_e$ are given in

$$A = \begin{bmatrix} -12 & -1 & 1 \\ 1 & 13 & 7 \\ 3 & 4 & 8 \end{bmatrix} \qquad \mathbf{y} = \begin{bmatrix} -11 \\ 48 \\ 35 \end{bmatrix} \qquad \mathbf{x}_e = \begin{bmatrix} 1 \\ 2 \\ 3 \end{bmatrix} \qquad (39.20)$$

The residual vector is $\mathbf{R}(\mathbf{x}) = A\mathbf{x} - \mathbf{y}$, where, of course, $\mathbf{R}(\mathbf{x}_e) = \mathbf{0}$. The vanishing of the kth component of $\mathbf{R}$ is equivalent to the satisfaction of the kth equation in the system $A\mathbf{x} = \mathbf{y}$. Solving each such equation for its dominant variable gives

$$x_1 = x_1(x_2, x_3) = \tfrac{1}{12}(11 - x_2 + x_3)$$
$$x_2 = x_2(x_1, x_3) = \tfrac{1}{13}(48 - x_1 - 7x_3)$$
$$x_3 = x_3(x_1, x_2) = \tfrac{1}{8}(35 - 3x_1 - 4x_2)$$

where the equations are interpreted as functions that determine the value of x_k that annihilates the kth component of $\mathbf{R}$.

If we start an iteration process at $\mathbf{x}_0 = \mathbf{0}$, the residual vector would initially be $\mathbf{R} = -\mathbf{y}$, in which the largest absolute value is 48, occurring in the second equation. If we could make the largest residual zero, that is, if we could *relax the largest residual to zero*, we should be led to a more accurate solution. Our strategy, then, is to keep $x_1 = x_3 = 0$ and adjust x_2 to make the second component of **R** become zero.

The value of x_2 that drives the second component of **R** to zero is $x_2(0, 0) = \frac{48}{13}$. If x_2 is set equal to this value while keeping $x_1 = x_3 = 0$, then the next member of the sequence of iterates is $\mathbf{x}_1 = [0, x_2(0, 0), 0]^T = [0, \frac{48}{13}, 0]^T$.

Since the vector **x** has changed to $\mathbf{x}_1$, the residual vector **R** will change to $\mathbf{R}(\mathbf{x}_1)$, whose largest magnitude is in the third component. Relax that third component to zero by computing $x_3(0, \frac{48}{13}) = \frac{263}{104}$, giving $\mathbf{x}_2 = [0, \frac{48}{13}, \frac{263}{104}]^T$.

The largest magnitude in $\mathbf{R}(\mathbf{x}_2)$ is the second, so relax that component to zero by computing $x_2(0, \frac{263}{104}) = 2.33$, giving $\mathbf{x}_3 = [0, 2.33, \frac{263}{104}]^T$. The largest magnitude in $\mathbf{R}(\mathbf{x}_3)$ is the first, so relax that component to zero by computing $x_1(2.33, \frac{263}{104}) = 0.933$, giving $\mathbf{x}_4 = [0.933, 2.33, \frac{263}{104}]^T$. The largest magnitude in $\mathbf{R}(\mathbf{x}_4)$ is the third, so relax that component to zero by computing $x_3(0.933, 2.33) = 2.86$, giving $\mathbf{x}_5 = [0.933, 2.33, 2.86]^T$. The largest magnitude in $\mathbf{R}(\mathbf{x}_5)$ is the second, so relax that component to zero by computing $x_2(0.933, 2.86) = 2.08$, giving $\mathbf{x}_6 = [0.933, 2.08, 2.86]^T$. The largest magnitude in $\mathbf{R}(\mathbf{x}_6)$ is the third, so relax that component to zero by computing $x_3(0.933, 2.08) = 2.98$, giving $\mathbf{x}_7 = [0.933, 2.08, 2.86]^T$. The largest magnitude in $\mathbf{R}(\mathbf{x}_7)$ is the second, so relax that component to zero by computing $x_2(0.933, 2.98) = 2.01$, giving $\mathbf{x}_8 = [0.933, 2.01, 2.86]^T$. All these calculations are summarized in Table 39.8, an unsophisticated summary of Southwell's more efficient tableau.

k	$\mathbf{R}(\mathbf{x}_k)$	x_i	$\mathbf{x}_{k+1}$	k	$\mathbf{R}(\mathbf{x}_k)$	x_i	$\mathbf{x}_{k+1}$
0	$\begin{bmatrix} 11 \\ -48 \\ -35 \end{bmatrix}$	$x_2(0,0) = \frac{48}{13}$	$\begin{bmatrix} 0 \\ \frac{48}{13} \\ 0 \end{bmatrix}$	4	$\begin{bmatrix} 0 \\ 0.933 \\ -2.65 \end{bmatrix}$	$x_3(0.933, 2.33) = 2.86$	$\begin{bmatrix} 0.933 \\ 2.33 \\ 2.86 \end{bmatrix}$
1	$\begin{bmatrix} 7.31 \\ 0 \\ -20.2 \end{bmatrix}$	$x_3(0, \frac{48}{13}) = \frac{263}{104}$	$\begin{bmatrix} 0 \\ \frac{48}{13} \\ \frac{263}{104} \end{bmatrix}$	5	$\begin{bmatrix} 0.331 \\ 3.25 \\ 0 \end{bmatrix}$	$x_2(0.933, 2.86) = 2.08$	$\begin{bmatrix} 0.933 \\ 2.08 \\ 2.86 \end{bmatrix}$
2	$\begin{bmatrix} 9.83 \\ 17.7 \\ 0 \end{bmatrix}$	$x_2(0, \frac{263}{104}) = 2.33$	$\begin{bmatrix} 0 \\ 2.33 \\ \frac{263}{104} \end{bmatrix}$	6	$\begin{bmatrix} 0.581 \\ 0 \\ -1.00 \end{bmatrix}$	$x_3(0.933, 2.08) = 2.98$	$\begin{bmatrix} 0.933 \\ 2.08 \\ 2.98 \end{bmatrix}$
3	$\begin{bmatrix} 11.2 \\ 0 \\ -5.45 \end{bmatrix}$	$x_1(2.33, \frac{263}{104}) = 0.933$	$\begin{bmatrix} 0.933 \\ 2.33 \\ \frac{263}{104} \end{bmatrix}$	7	$\begin{bmatrix} 0.706 \\ 0.875 \\ 0 \end{bmatrix}$	$x_2(0.933, 2.98) = 2.01$	$\begin{bmatrix} 0.933 \\ 2.01 \\ 2.98 \end{bmatrix}$

TABLE 39.8   Southwell's relaxation technique

Although it can become tedious, selecting the component with the largest magnitude in the residual vector is something that can be done "by-hand" for a system of modest size. Clearly, having to pick, repeatedly, the largest number from a list of 10,000 would be highly susceptible to error. Moreover, having a computer make the many comparisons required to select the largest magnitude is not efficient, and this version of the relaxation algorithm is not a viable method for solving large problems on a computer. However, the underlying concept does motivate the method of successive overrelaxation, the SOR technique, which is capable of being implemented on a computer.

Successive Overrelaxation

Southwell discovered that often, when driving a component of the residual to zero, the other components would move in the same direction. For example, the third component of the residual $\mathbf{R}(\mathbf{x}_1)$ is driven to zero by increasing it from -20 to 0. The first component of $\mathbf{R}(\mathbf{x}_2)$ is then driven upward from 7.31 to 9.83, and the second component is driven upward from 0 to 17.7.

This led Southwell to "overcompensate" or *overrelax*, pushing the residual, not to zero, but beyond zero. All that is needed is a more efficient way of implementing the calculations on a computer. Formulating the system $A\mathbf{x} = \mathbf{y}$ as if for performing Gauss–Seidel iteration is the basis for this efficiency.

Assuming that the kth equation has been solved for x_k, we write the linear system in the form

$$x_k^{m+1} = x_k^m + \frac{1}{a_{kk}} \left(y_k - \sum_{s=1}^{n} a_{ks} x_s^m \right)$$

as in the Jacobi method. Then, recognizing that the expression in parentheses is actually the kth component of the residual vector $\mathbf{R}$, we write $x_k^{m+1} = x_k^m + \mathbf{R}(\mathbf{x}_m)/a_{kk}$. By introducing the relaxation factor $\omega > 0$, this equation is modified to

$$x_k^{m+1} = x_k^m + \frac{\omega}{a_{kk}} \mathbf{R}(\mathbf{x}_m)$$

Finally, write the general equation in the form

$$x_k^{m+1} = (1 - \omega)x_k^m + \frac{\omega}{a_{kk}} \left(y_k - \sum_{s=1}^{k-1} a_{ks} x_s^{m+1} - \sum_{s=k+1}^{n} a_{ks} x_s^m \right)$$

which is now consonant with the Gauss–Seidel method, except for the introduction of ω, the relaxation factor. Clearly, if $\omega = 1$, the SOR technique is precisely the Gauss–Seidel method. If A is a symmetric positive definite matrix and $0 < \omega < 2$, then the SOR technique converges from any initial vector. (See [67] for a proof.)

Implementation

Implementing the SOR method for small systems is more conveniently done in matrix form. To accomplish this, write the linear system $A\mathbf{x} = \mathbf{y}$ in the form $\mathbf{y} - A\mathbf{x} = \mathbf{0}$ and multiply through by ω, giving $\mathbf{0} = \omega(\mathbf{y} - A\mathbf{x})$. Write $A = L + D + U$, where L is zero on and above the main diagonal, U is zero on and below the main diagonal, and D is zero off the main diagonal. In addition, add $D\mathbf{x}$ to each side of the equation, thereby obtaining

$$D\mathbf{x} = D\mathbf{x} - \omega(L + D + U)\mathbf{x} + \omega\mathbf{y}$$

Bring the term $-\omega L\mathbf{x}$ to the left, giving

$$(D + \omega L)\mathbf{x} = [(1 - \omega)D - \omega U]\mathbf{x} + \omega\mathbf{y}$$

Let the vector $\mathbf{x}$ on the left side be the updated one, namely, $\mathbf{x}_{m+1}$, while keeping the $\mathbf{x}$ on the right side as the "old" one, namely, $\mathbf{x}_m$. This gives

$$(D + \omega L)\mathbf{x}_{m+1} = [(1 - \omega)D - \omega U]\mathbf{x}_m + \omega\mathbf{y}$$

Solve for $\mathbf{x}_{m+1}$ to obtain

$$\mathbf{x}_{m+1} = (D + \omega L)^{-1}[(1 - \omega)D - \omega U]\mathbf{x}_m + \omega(D + \omega L)^{-1}\mathbf{y}$$

which has the fixed-point form

$$\mathbf{x}_{m+1} = B\mathbf{x}_m + \mathbf{C} = \mathbf{g}(\mathbf{x}_m)$$

where $B = (D + \omega L)^{-1}[(1 - \omega)D - \omega U]$ and $\mathbf{C} = \omega(D + \omega L)^{-1}\mathbf{y}$.

EXAMPLE 39.12   Consider the linear system $A\mathbf{x} = \mathbf{y}$ for which the symmetric, positive-definite matrix A, the vector $\mathbf{y}$, and the exact solution $\mathbf{x}_e$ are given in (39.21). The matrix A is obviously symmetric. Its eigenvalues are 0.36, 10.7, and 19.0, so it is also positive definite.

$$A = \begin{bmatrix} 8 & -5 & -7 \\ -5 & 14 & 1 \\ -7 & 1 & 8 \end{bmatrix} \qquad \mathbf{y} = \begin{bmatrix} -23 \\ 26 \\ 19 \end{bmatrix} \qquad \mathbf{x}_e = \begin{bmatrix} 1 \\ 2 \\ 3 \end{bmatrix} \qquad (39.21)$$

Decompose A as $L + D + U$, where L, D, and U are given by

$$L = \begin{bmatrix} 0 & 0 & 0 \\ -5 & 0 & 0 \\ -7 & 1 & 0 \end{bmatrix} \qquad D = \begin{bmatrix} 8 & 0 & 0 \\ 0 & 14 & 0 \\ 0 & 0 & 8 \end{bmatrix} \qquad U = \begin{bmatrix} 0 & -5 & -7 \\ 0 & 0 & 1 \\ 0 & 0 & 0 \end{bmatrix}$$

The matrix $B = (D + \omega L)^{-1}[(1 - \omega)D - \omega U]$ is

$$B = \begin{bmatrix} 1 - \omega & \frac{5}{8}\omega & \frac{7}{8}\omega \\ \frac{5}{14}\omega(1 - \omega) & \frac{25}{112}\omega^2 - \omega + 1 & \frac{5}{16}\omega^2 - \frac{1}{14}\omega \\ \frac{1}{112}\omega(5\omega - 98)(\omega - 1) & -\frac{25}{896}\omega^3 + \frac{43}{64}\omega^2 - \frac{1}{8}\omega & -\frac{5}{128}\omega^3 + \frac{347}{448}\omega^2 - \omega + 1 \end{bmatrix}$$

and the vector $\mathbf{C} = \omega(D + \omega L)^{-1}\mathbf{y}$ is

$$\mathbf{C} = \begin{bmatrix} -\frac{23}{8}\omega \\ -\omega(\frac{115}{112}\omega - \frac{13}{7}) \\ \omega(\frac{115}{896}\omega^2 - \frac{1231}{448}\omega + \frac{19}{8}) \end{bmatrix}$$

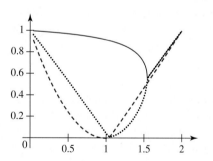

FIGURE 39.5   The eigenvalues of $B = (D + \omega L)^{-1}[(1 - \omega)D - \omega U]$ in Example 39.12

The eigenvalues of B determine the rate of convergence of the SOR iteration. Since we wish to pick ω so that the largest magnitude of the eigenvalues is as small as possible, we provide Figure 39.5, a graph of the three eigenvalues as a function of ω. The smallest maximum in the absolute values of the eigenvalues will occur at the lowest point for which a horizontal line lies above all three curves. From the graph, the optimal value of ω appears to be approximately 1.55.

For $\omega = 1.55$ the matrix B and the vector $\mathbf{C}$ become

$$\begin{bmatrix} -0.55 & 0.9687500000 & 1.356250000 \\ -0.3044642857 & -0.013727679 & 0.6400669643 \\ -0.6869475447 & 1.316526926 & 1.165401088 \end{bmatrix} \qquad \begin{bmatrix} -4.456250000 \\ 0.4117187500 \\ -2.442309569 \end{bmatrix}$$

respectively. Starting with the initial vector $[10, 10, 10]^T$, Table 39.9 lists the SOR iterates $\mathbf{x}_k, k = 1, \ldots, 10$, and $\|\mathbf{x}_k - \mathbf{x}_e\|_\infty$, the norms of the differences between the iterate and the exact solution.

If $\mathbf{x}_{10}$ were computed with $\omega = 1.75$, we would find $\|\mathbf{x}_{10} - \mathbf{x}_e\|_\infty = 1.03$, indicating that convergence with this value of ω is slower than with $\omega = 1.55$.   ❖

Equivalence of Forms

When $\omega = 1$, the SOR method reduces precisely to the Gauss–Seidel method, an equivalence we will now verify for the system in Example 39.12, whose Gauss–Seidel equations

k	x_1	x_2	x_3	$\|\mathbf{x}_k - \mathbf{x}_e\|_\infty$
0	10.0	10.0	10.0	9.00
1	13.29375000	3.630468746	15.50749512	12.51
2	12.78124436	6.240244067	11.27764306	11.78
3	9.85460544	3.653069007	10.13607283	8.85
4	7.40967639	3.848960551	7.41004743	6.41
5	5.24698535	2.845826190	6.17056897	4.25
6	3.78363634	2.724709863	4.89108132	2.78
7	2.73584174	2.352952246	4.24575613	1.74
8	2.07676629	2.264020314	3.72404447	1.08
9	1.64553353	2.131975680	3.45171011	0.66
10	1.38543983	2.090771093	3.25672532	0.39

TABLE 39.9   Example 39.12: SOR iterates with $\omega = 1.55$

are

$$x_1^{m+1} = \tfrac{5}{8}x_2^m + \tfrac{7}{8}x_3^m - \tfrac{23}{8}$$

$$x_2^{m+1} = \tfrac{5}{14}x_1^{m+1} - \tfrac{1}{14}x_3^m + \tfrac{13}{7}$$

$$x_3^{m+1} = \tfrac{7}{8}x_1^{m+1} - \tfrac{1}{8}x_2^{m+1} + \tfrac{19}{8}$$

When $\omega = 1$, the matrix B and the vector $\mathbf{C}$ will be

$$B = \begin{bmatrix} 0 & \frac{5}{8} & \frac{7}{8} \\ 0 & \frac{25}{112} & \frac{27}{112} \\ 0 & \frac{465}{896} & \frac{659}{896} \end{bmatrix} \quad \mathbf{C} = \begin{bmatrix} -\frac{23}{8} \\ \frac{93}{112} \\ -\frac{219}{896} \end{bmatrix}$$

Consequently, the SOR equations in the fixed-point iteration $\mathbf{x}_{m+1} = B\mathbf{x}_m + \mathbf{C}$ are

$$\begin{bmatrix} x_1^{m+1} \\ x_2^{m+1} \\ x_3^{m+1} \end{bmatrix} = \begin{bmatrix} \frac{5}{8}x_2^m + \frac{7}{8}x_3^m - \frac{23}{8} \\ \frac{25}{112}x_2^m + \frac{27}{112}x_3^m + \frac{93}{112} \\ \frac{465}{896}x_2^m + \frac{659}{896}x_3^m - \frac{219}{896} \end{bmatrix}$$

and we want to show these equations are indeed the Gauss–Seidel equations. Clearly, the first equation in each formulation is the same. In the second Gauss–Seidel equation, replace x_1^{m+1} from the first, obtaining $x_2^{m+1} = \frac{25}{112}x_2^m + \frac{27}{112}x_3^m + \frac{93}{112}$, the second equation in the SOR formulation. In the third Gauss–Seidel equation, replace x_2^{m+1} from the second equation. This introduces another x_1^{m+1}. Now replace *all* x_1^{m+1} from the first equation, and the result will be $x_3^{m+1} = \frac{465}{896}x_2^m + \frac{659}{896}x_3^m - \frac{219}{896}$, the third equation of the SOR formulation. Thus, we have verified that the matrix form of the SOR iteration actually uses the updated values $\mathbf{x}_{m+1}$ in the same way that Gauss–Seidel iteration does.

EXERCISES 39.6

1. Write $A = \begin{bmatrix} 9 & 0 \\ 2 & 5 \end{bmatrix}$ in the form $L + D + U$. Then:

(a) Obtain $B = (D + \omega L)^{-1}[(1 - \omega)D - \omega U]$ and its eigenvalues.

(b) As a function of ω, plot the magnitudes of the eigenvalues of B, and determine the value of ω for which the maximum magnitude is minimized.

For the matrices A given in each of Exercises 2–7 and $\mathbf{y} = [1, 2, 3]^T$:

(a) Obtain $\mathbf{x}_e$, the exact solution of the system $A\mathbf{x} = \mathbf{y}$.

(b) Implement Southwell's relaxation technique as summarized in Table 39.8.

2. $\begin{bmatrix} 14 & 7 & -9 \\ 5 & 15 & -8 \\ 3 & 5 & -9 \end{bmatrix}$ **3.** $\begin{bmatrix} 4 & 0 & 1 \\ -5 & 13 & -7 \\ 1 & -1 & 3 \end{bmatrix}$ **4.** $\begin{bmatrix} 8 & 3 & 3 \\ -9 & -13 & 3 \\ 9 & -9 & -25 \end{bmatrix}$

5. $\begin{bmatrix} 6 & 0 & 1 \\ 1 & -7 & 1 \\ -4 & 2 & 9 \end{bmatrix}$ **6.** $\begin{bmatrix} 12 & -7 & -3 \\ -7 & 8 & 0 \\ 8 & 5 & -14 \end{bmatrix}$ **7.** $\begin{bmatrix} 11 & -8 & 2 \\ 9 & -12 & 1 \\ 9 & 5 & 15 \end{bmatrix}$

For matrices A given in Exercises 8–13:

(a) Obtain the SOR matrix $B(\omega) = (D + \omega L)^{-1}[(1 - \omega)D - \omega U]$, where $A = L + D + U$.

(b) Obtain $\lambda_{\pm}(\omega)$, the two eigenvalues of $B(\omega)$.

(c) Plot $|\lambda_{\pm}(\omega)|$ for $0 < \omega < 2$, and from the graph, estimate $\hat{\omega}$, the value of ω for which $\max\{|\lambda_{\pm}(\omega)|\}$ is as small as possible. For several of the matrices, at least one of $|\lambda_{\pm}(\omega)|$ is greater than 1 for $0 < \omega < 2$, implying that there is no value of ω for which SOR would converge.

8. $\begin{bmatrix} -46 & -35 \\ 33 & 45 \end{bmatrix}$ **9.** $\begin{bmatrix} 81 & -26 \\ -54 & -60 \end{bmatrix}$ **10.** $\begin{bmatrix} -84 & 51 \\ 93 & -18 \end{bmatrix}$

11. $\begin{bmatrix} 63 & 78 \\ -70 & 49 \end{bmatrix}$ **12.** $\begin{bmatrix} -4 & -22 \\ -77 & -44 \end{bmatrix}$ **13.** $\begin{bmatrix} -94 & -89 \\ -81 & 41 \end{bmatrix}$

For the system $A\mathbf{x} = \mathbf{y}$ determined by $\mathbf{y} = [-3, 7, 5]^T$ and for the matrix A given in each of Exercises 14–19:

(a) Obtain the SOR matrix $B(\omega) = (D + \omega L)^{-1}[(1 - \omega)D - \omega U]$.

(b) Obtain $\lambda_i(\omega)$, $i = 1, 2, 3$, the eigenvalues of $B(\omega)$, and graph their magnitudes as functions of ω, $0 < \omega < 2$.

(c) From the graph in part (b), determine $\hat{\omega}$, the value of ω for which $\max\{|\lambda_1(\omega)|, |\lambda_2(\omega)|, |\lambda_3(\omega)|\}$ is as small as possible.

(d) Starting from $\mathbf{x}_0 = \mathbf{0}$, implement SOR iteration with $\omega = \hat{\omega}$ and determine the number of iterations required for $\|\mathbf{x}_e - \mathbf{x}_k\|_\infty < 10^{-5}$. Compare to the number of iterations

required by Gauss–Seidel iteration, as determined in part (e) of Exercises 12–17, Section 39.5.

14. Exercise 12, Section 39.5   **15.** Exercise 13, Section 39.5

16. Exercise 14, Section 39.5   **17.** Exercise 15, Section 39.5

18. Exercise 16, Section 39.5   **19.** Exercise 17, Section 39.5

For the system $A\mathbf{x} = \mathbf{y}$ determined by $\mathbf{y} = [6, -7, 1, 9]^T$ and for the matrix A given in each of Exercises 20–25:

(a) Obtain the SOR matrix $B(\omega) = (D + \omega L)^{-1}[(1 - \omega)D - \omega U]$.

(b) For each $\omega_k = \frac{k}{10}$, $k = 1, \ldots, 19$, obtain $\lambda_i(\omega_k)$, $i = 1, \ldots, 4$, the eigenvalues of $B(\omega_k)$ and determine $\sigma_k = \max\{|\lambda_i(\omega_k)|, i = 1, \ldots, 4\}$.

(c) From a graph of (σ_k, ω_k), determine $\hat{\omega}$, the value of ω that minimizes the maximum values of the eigenvalues of $B(\omega)$.

(d) Starting from $\mathbf{x}_0 = \mathbf{0}$, implement SOR iteration with $\omega = \hat{\omega}$ as found in part (c), determining the number of iterations required for $\|\mathbf{x}_e - \mathbf{x}_k\|_\infty < 10^{-5}$. Compare to the number of iterations required by Gauss–Seidel iteration, as determined in part (e) of Exercises 18–23, Section 39.5.

20. Exercise 18, Section 39.5   **21.** Exercise 19, Section 39.5

22. Exercise 20, Section 39.5   **23.** Exercise 21, Section 39.5

24. Exercise 22, Section 39.5   **25.** Exercise 23, Section 39.5

If A is a positive-definite tridiagonal matrix, then the optimal value of the relaxation parameter is $\hat{\omega} = \frac{2}{1+\sqrt{1-\rho^2}}$, where ρ is the maximum absolute value of the eigenvalues of $B_J = D^{-1}(D - A)$, the iteration matrix for Jacobi iteration. In each of Exercises 26–29, a positive-definite tridiagonal matrix A is determined by the main diagonal d, the (equal) subdiagonal l and superdiagonal u. For each such matrix A:

(a) Obtain B_J and its eigenvalues, and from them, determine ρ and $\hat{\omega}$.

(b) If $\mathbf{y} = [1, 2, 3, 4, 5, 6]^T$, obtain $\mathbf{x}_e$, the exact solution of the system $A\mathbf{x} = \mathbf{y}$.

(c) Starting at $\mathbf{x}_0 = \mathbf{0}$, implement the SOR iteration with $\omega = \hat{\omega}$, determining an $\mathbf{x}_k$ for which $\|\mathbf{x}_e - \mathbf{x}_k\|_\infty < 10^{-5}$.

26. $d = [16, 24, 24, 21, 3, 21]$ **27.** $d = [14, 22, 10, 25, 14, 8]$
$u = l = [15, 8, 13, 5, 2]$ $u = l = [8, 8, 6, 1, 5]$

28. $d = [20, 21, 22, 22, 15, 2]$ **29.** $d = [5, 21, 23, 24, 15, 4]$
$u = l = [10, 11, 14, 9, 1]$ $u = l = [2, 11, 4, 12, 2]$

39.7   Iterative Methods for Nonlinear Systems

Introduction

Numerically solving nonlinear systems of algebraic equations can be difficult, especially if one has no idea where possible solutions might lie. Iterative methods that converge when started near a solution often have to be started *very* near to a solution. Systems of two equations in two unknowns can be examined graphically for the location of solutions. Systems of three equations in three unknowns have solutions at the intersections of three surfaces in space, but the corresponding graph is often difficult to interpret.

Fixed-point iteration, in essence a nonlinear version of Jacobi's method, will converge slowly, if at all. Just as for linear systems, convergence can be accelerated by the Gauss–Seidel strategy of using new information immediately. And Newton's method, deferred to Section 39.8, converges quadratically when started close enough to a solution.

EXAMPLE 39.13   The nonlinear system

$$f(x, y) = 12 + 3x + 8y + 12xy + 9x^2 - 12y^2 = 0 \qquad \textbf{(39.22a)}$$

$$g(x, y) = -5 - 12x - 6y + 9xy + 2x^2 + 11y^2 = 0 \qquad \textbf{(39.22b)}$$

can be analyzed by examining a graph of the implicit functions $y(x)$ determined by each equation. Using a solid curve for the function determined by $f(x, y) = 0$ and a dotted curve for the function determined by $g(x, y) = 0$, we have Figure 39.6, on the basis of which we obtain the two solutions $(0.2475, -0.6904)$ and $(36.78, -18.36)$. (See Exercise 1.)

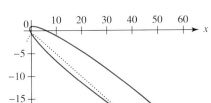

FIGURE 39.6   A graph of the implicit functions $y(x)$ determined by $f(x, y)$ and $g(x, y)$ in Example 39.13

SOLUTION BY FIXED-POINT ITERATION   To set up a fixed-point iteration, the analog of Jacobi iteration for linear systems, solve $f(x, y) = 0$ for x and $g(x, y) = 0$ for y, obtaining

$$x = -\tfrac{1}{6} - \tfrac{2}{3}y \pm \tfrac{1}{6}\sqrt{64y^2 - 24y - 47} \quad \text{and} \quad y = \tfrac{3}{11} - \tfrac{9}{22}x \pm \tfrac{1}{22}\sqrt{256 + 420x - 7x^2}$$
$$\textbf{(39.23)}$$

Choosing a branch that contains the solution is a challenge. We resort to "using the solution to find the solution" by evaluating each branch at the second fixed point, obtaining 36.78 and -12.63, respectively, for x and -11.19 and -18.36, respectively, for y. Hence, the desired fixed-point iteration $\mathbf{x} = \mathbf{g}(\mathbf{x})$ is defined by the vector-valued function

$$\mathbf{G}(x, y) = \begin{bmatrix} -\tfrac{1}{6} - \tfrac{2}{3}y + \tfrac{1}{6}\sqrt{64y^2 - 24y - 47} \\ \tfrac{3}{11} - \tfrac{9}{22}x - \tfrac{1}{22}\sqrt{256 + 420x - 7x^2} \end{bmatrix} \qquad \textbf{(39.24)}$$

The alternative to using the solution to pick the right branches is to investigate each possible combination of branches at the expense of greater complexity and tedium.

Starting from the point $(35, -15)$, Table 39.10 lists both $\mathbf{x}_k$ and $\|\mathbf{x}_k - \mathbf{x}_e\|$, where $\mathbf{x}_e$ corresponds to the point $(36.78, -18.36)$. The convergence is linear and, hence, slow. Fixed-point iteration should not ordinarily be the first solution technique tried when solving systems of nonlinear algebraic equations.

Use of new information as it is generated will usually accelerate convergence. Fixed-point iteration modified to implement this idea resembles Gauss–Seidel iteration for linear systems.

SOLUTION BY GAUSS–SEIDEL ITERATION   Gauss–Seidel iteration is implemented by writing the fixed-point iteration $\mathbf{x} = \mathbf{g}(\mathbf{x})$ in the form $x_{k+1} = G_1(x_k, y_k)$, $y_{k+1} = G_2(x_{k+1}, y_k)$, where G_1 and G_2 are the components of the vector-valued function $\mathbf{G}(\mathbf{x})$ defined in (39.24).

Table 39.11 records the result of a computation similar to that which led to Table 39.10. Comparing the two iterations, we find that the Gauss–Seidel technique speeds up convergence. For example, the norm of the difference between the fixed point and the fortieth Jacobi iterate is approximately 0.018, whereas for the Gauss–Seidel iterate it is 0.0002. ❖

k	x_k	y_k	$\|\mathbf{x}_k - \mathbf{x}_e\|$
0	35.0	−15.0	3.36397253
1	30.04952597	−17.67641645	6.73408283
2	35.40743061	−15.70067081	2.66330172
3	31.45236082	−17.83463779	5.33124798
4	35.72412590	−16.27041574	2.09355679
5	32.59297098	−17.95715073	4.19063782
⋮	⋮	⋮	⋮
36	36.76895194	−18.33275941	0.03121312
37	36.72113609	−18.35837669	0.06247271
38	36.77240896	−18.34011502	0.02385751
39	36.73585832	−18.35969661	0.04775048
40	36.77505077	−18.34573871	0.01823382

TABLE 39.10   Jacobi-style fixed-point iterates in Example 39.13

k	x_k	y_k	$\|\mathbf{x}_k - \mathbf{x}_e\|$
0	35.0	−15.0	3.36397253
1	30.04952597	−15.70067081	6.73408283
2	31.45236082	−16.27041574	5.33124798
3	32.59297098	−16.72794150	4.19063782
4	33.50886737	−17.09158938	3.27474143
5	34.23680267	−17.37821207	2.54680613
⋮	⋮	⋮	⋮
36	36.78296389	−18.36372626	0.6449×10^{-3}
37	36.78311609	−18.36378436	0.4927×10^{-3}
38	36.78323238	−18.36382875	0.3764×10^{-3}
39	36.78332123	−18.36386267	0.2876×10^{-3}
40	36.78338911	−18.36388859	0.2197×10^{-3}

TABLE 39.11   Example 39.13: Fixed-point iterates obtained using the Gauss–Seidel strategy

EXAMPLE 39.14   A system of three nonlinear equations in three unknowns is generally much more difficult to analyze and solve. Consider, for example, the functions

$$F_1(x, y, z) = 8z + 3z^2 + 2y^2 + xy - 4x^2 + 9xz - 4$$
$$F_2(x, y, z) = -z + 4z^2 - 4y^2 + 6xy - 6x^2 - 4yz + 39$$
$$F_3(x, y, z) = 9y - 2z - 3y^2 + 4xy - 8yz + 20$$

for which the equations $F_k(x, y, z) = 0, k = 1, \ldots, 3$, have the solution $P = (-1, 2, 1)$. The existence of multiple solutions is more difficult to determine in higher dimensions. There are actually three additional real solutions given to three significant figures by $(-3.05, -4.52, 0.785)$, $(-3.11, -1.84, -1.27)$, and $(1.95, -1.23, 0.675)$.

SOLUTION BY FIXED-POINT ITERATION   In an attempt to find the solution P by fixed-point iteration, solve $F_1(x, y, z) = 0$ for x, $F_2(x, y, z) = 0$ for y and $F_3(x, y, z) = 0$ for z, obtaining

$$x = \tfrac{1}{8}y + \tfrac{9}{8}z \pm \tfrac{1}{8}\sqrt{33y^2 + 18yz + 129z^2 + 128z - 64}$$
$$y = \tfrac{3}{4}x - \tfrac{1}{2}z \pm \tfrac{1}{4}\sqrt{20z^2 - 15x^2 - 12xz - 4z + 156}$$
$$z = \frac{1}{2}\frac{4xy - 3y^2 + 9y + 20}{4y + 1}$$

Again, we must select correct branches for both x and y. To this end, evaluate the four relevant branches at the known solution P, obtaining $\tfrac{15}{4}$, -1 and 2, $-\tfrac{9}{2}$, respectively. Hence, we define the fixed-point iteration $\mathbf{x} = \mathbf{g}(\mathbf{x})$ by

$$x = g_1(x, y, z) = \tfrac{1}{8}y + \tfrac{9}{8}z - \tfrac{1}{8}\sqrt{33y^2 + 18yz + 129z^2 + 128z - 64}$$

$$y = g_2(x, y, z) = \tfrac{3}{4}x - \tfrac{1}{2}z + \tfrac{1}{4}\sqrt{20z^2 - 15x^2 - 12xz - 4z + 156}$$

$$z = g_3(x, y, z) = \frac{1}{2}\frac{4xy - 3y^2 + 9y + 20}{4y + 1}$$

Starting at the point $(-\tfrac{1}{2}, 1, \tfrac{1}{2})$, Table 39.12 lists $\mathbf{x}_k$, the Jacobi iterates, and $\|\mathbf{x}_k - P\|$, the norm of the distance from $\mathbf{x}_k$ to the solution at P. The slow linear convergence can be accelerated with the Gauss–Seidel strategy, using new information as soon as it is available.

k	x_k	y_k	z_k	$\|\mathbf{x}_k - \mathbf{x}_e\|_\infty$
0	−0.5	1.0	0.5	1.0
1	−0.389605496	2.519936408	2.400000000	1.400000000
2	−1.500007043	2.619849438	0.889102762	0.619849438
3	−1.234647038	1.497478462	0.316595061	0.683404939
4	−0.517220669	1.858542861	1.384462811	0.482779331
5	−1.062720206	2.396354651	1.334999773	0.396354651
6	−1.219017372	1.948565677	0.668515225	0.331484775
7	−0.891672792	1.810822843	0.946357041	0.189177157
8	−0.915948167	2.101582428	1.213199967	0.213199967
9	−1.090452718	2.072952361	0.954917274	0.090452718
10	−1.016122105	1.916757919	0.899897417	0.100102583

TABLE 39.12   Jacobi-style fixed-point iterates in Example 39.14

SOLUTION BY GAUSS–SEIDEL ITERATION   Table 39.13 gives the results of a Gauss–Seidel iteration implemented with

$$x_{k+1} = g_1(x_k, y_k, z_k) \qquad y_{k+1} = g_2(x_{k+1}, y_k, z_k) \qquad z_{k+1} = g_3(x_{k+1}, y_{k+1}, z_k)$$

k	x_k	y_k	z_k	$\|\mathbf{x}_k - \mathbf{x}_e\|_\infty$
0	−0.5	1.0	0.5	1.0
1	−0.389605496	2.610778061	0.829312078	0.610778061
2	−1.224306938	1.790993560	1.085565949	0.224306938
3	−0.952199857	2.041084011	0.987393118	0.047800143
4	−1.012036733	1.989356559	1.003546239	0.012036733
5	−0.997027331	2.002606825	0.999150406	0.002972669
6	−1.000737743	1.999351630	1.000212520	0.000737743
7	−0.999817110	2.000160645	0.999947420	0.000182890
8	−1.000045351	1.999960160	1.000013044	0.000045351
9	−0.999988755	2.000009880	0.999996765	0.000011245
10	−1.000002789	1.999997551	1.000000801	0.28×10^{-5}

TABLE 39.13   Example 39.14: Fixed-point iterates obtained using the Gauss–Seidel strategy

in a manner that parallels the computation leading to Table 39.12. Comparing the two iterations, we find that the Gauss–Seidel technique speeds up convergence. For example, the norm of the difference between the fixed point and the tenth Jacobi iterate is approximately 0.10, whereas for the Gauss–Seidel iterate it is 0.0000028. ❖

EXERCISES 39.7

1. In Example 39.13, (39.23) defines the four functions $x_{\pm}(y)$ and $y_{\pm}(x)$. Were we not to know where to look for solutions to equations (39.22a) and (39.22b), we would have to examine four cases such as $x = x_{+}(y_{+}(x))$, $x = x_{+}(y_{-}(x))$, $x = x_{-}(y_{+}(x))$, and $x = x_{-}(y_{-}(x))$. Each of these equations can be solved with a method from Chapter 38. Use this approach to determine the solutions of the equations in Example 39.13.

In Exercises 2–11:

(a) Verify that $f(x, y) = g(x, y) = 0$ has a solution at the given $\hat{\mathbf{x}}$.

(b) Solve $f(x, y) = 0$ for the x appearing in its linear term and solve $g(x, y) = 0$ for the y appearing in its linear term, to obtain $\mathbf{x} = \mathbf{G}(\mathbf{x}) = [\hat{f}(x, y), \hat{g}(x, y)]^{\mathrm{T}}$.

(c) Show that $\|\nabla \hat{f}(\hat{\mathbf{x}})\|_1 < 1$ and that $\|\nabla \hat{g}(\hat{\mathbf{x}})\|_1 < 1$.

(d) Show that $J(\hat{\mathbf{x}}) = \begin{bmatrix} \hat{f}_x(\hat{\mathbf{x}}) & \hat{f}_y(\hat{\mathbf{x}}) \\ \hat{g}_x(\hat{\mathbf{x}}) & \hat{g}_y(\hat{\mathbf{x}}) \end{bmatrix}$, the Jacobian matrix for $\mathbf{G}(\mathbf{x})$, evaluated at $\mathbf{x} = \hat{\mathbf{x}}$, has no eigenvalue with magnitude greater than one.

(e) Implement the fixed-point iteration $\mathbf{x}_{k+1} = \mathbf{G}(\mathbf{x}_k)$, showing it converges to $\hat{\mathbf{x}}$ if $\mathbf{x}_0$ is close enough to $\hat{\mathbf{x}}$. (Try $\mathbf{x}_0 = \mathbf{0}$.) Determine the number of iterations required for $\|\hat{\mathbf{x}} - \mathbf{x}_k\|_\infty < 10^{-5}$.

(f) Show empirically that $\|\hat{\mathbf{x}} - \mathbf{x}_k\|_\infty \le \sigma^k \|\mathbf{x}_1 - \mathbf{x}_0\|_\infty / (1 - \sigma)$, where $\sigma = \|J(\hat{\mathbf{x}})\|_\infty$.

(g) Starting from the same $\mathbf{x}_0$ as in part (d), implement the Gauss–Seidel strategy $x_{k+1} = \hat{f}(x_k, y_k)$, $y_{k+1} = \hat{g}(x_{k+1}, y_k)$, determining the number of iterations required for $\|\hat{\mathbf{x}} - \mathbf{x}_k\|_\infty < 10^{-5}$. Compare the efficiency of the two iteration schemes.

(h) On the same set of axes, graph, in a neighborhood of the fixed point $\mathbf{x}$, the functions respectively determined by the equations $f(x, y) = 0$ and $g(x, y) = 0$. The curves should intersect at the given fixed point.

2. $f(x, y) = -8xy^2 + 9y^3 + 8x^2 + 4xy + 7y^2 - 50x + 4y + 56 \quad \hat{\mathbf{x}} = \begin{bmatrix} 2 \\ 1 \end{bmatrix}$
$g(x, y) = -6x^3 - 6x^2y - 9y^3 - 9x^2 - 4y^2 - 192y + 313$

3. $f(x, y) = 2x^3 - 9xy^2 + 9xy - 504x + y - 1527 \quad \hat{\mathbf{x}} = \begin{bmatrix} -2 \\ -5 \end{bmatrix}$
$g(x, y) = 4xy^2 - 9y^3 - 6x^2 + 8y^2 + x - 900y - 5599$

4. $f(x, y) = 3x^3 - 7x^2y + 7x^2 - 6xy - 40x + 56 \quad \hat{\mathbf{x}} = \begin{bmatrix} 1 \\ 2 \end{bmatrix}$
$g(x, y) = x^3 - 2xy^2 + 7xy - 4x - 11y + 19$

5. $f(x, y) = xy^2 + 7y^3 - 9xy - 494x - 8y - 468 \quad \hat{\mathbf{x}} = \begin{bmatrix} -2 \\ -4 \end{bmatrix}$
$g(x, y) = 5x^3 + 4xy^2 + 6y^3 + 9x^2 - 5x - 500y - 1494$

6. $f(x, y) = -6x^3y - 8xy^3 - 9x^3 - 3x^2 - 150x + 320 \quad \hat{\mathbf{x}} = \begin{bmatrix} 2 \\ -1 \end{bmatrix}$
$g(x, y) = 5x^3y + 6x^2y^2 + xy^3 + 4x^2y + 5x^2 - 50y - 36$

7. $f(x, y) = 9x^4 - 7x^3y + 3x^2y^2 - 8xy^2 - y^3 + 9y^2 - 2000x - 9772 \quad \hat{\mathbf{x}} = \begin{bmatrix} -4 \\ -2 \end{bmatrix}$
$g(x, y) = -7x^3y - 6x^3 + 3xy^2 - 9y^3 + 8xy - 999y - 1574$

8. $f(x, y) = 5x^3 - 5x^2y - 3y^3 + 5xy + y^2 - 448x + 406 \quad \hat{\mathbf{x}} = \begin{bmatrix} 2 \\ -5 \end{bmatrix}$
$g(x, y) = -2x^2 + 8xy - 7y^2 + 8x - 154y - 523$

9. $f(x, y) = -5x^2y - 5xy^2 + 8y^2 - 77x + 193 \quad \hat{\mathbf{x}} = \begin{bmatrix} 3 \\ -1 \end{bmatrix}$
$g(x, y) = 6x^2 + 2xy - y^2 + x - 50y - 100$

10. $f(x, y) = 9y^3 - 4x^2 - 7xy - 8y^2 - 113x + 9y - 170 \quad \hat{\mathbf{x}} = \begin{bmatrix} -2 \\ -1 \end{bmatrix}$
$g(x, y) = 9x^2 - 4xy + 4y^2 - 6x - 44y - 88$

11. $f(x, y) = -xy - 5y^2 - 41x + 5y + 15 \quad \hat{\mathbf{x}} = \begin{bmatrix} -1 \\ 4 \end{bmatrix}$
$g(x, y) = x^2 - 9xy - 9y^2 - 9x - 209y + 934$

In Exercises 12–18, $\mathbf{x} = [x, y, z]^{\mathrm{T}}$ and $\mathbf{F}(\mathbf{x}) = [f(\mathbf{x}), g(\mathbf{x}), h(\mathbf{x})]^{\mathrm{T}}$. In each:

(a) Verify that $\mathbf{F}(\hat{\mathbf{x}}) = \mathbf{0}$.

(b) By respectively solving $f(\mathbf{x})$, $g(\mathbf{x})$, and $h(\mathbf{x})$ for the x, y, z appearing in the linear term, obtain $\mathbf{x} = \mathbf{G}(\mathbf{x})$.

(c) Show that $\|\nabla f(\hat{\mathbf{x}})\|_1 < 1$, $\|\nabla g(\hat{\mathbf{x}})\|_1 < 1$, $\|\nabla h(\hat{\mathbf{x}})\|_1 < 1$.

(d) Show that $J(\hat{\mathbf{x}}) = \begin{bmatrix} f_x(\hat{\mathbf{x}}) & f_y(\hat{\mathbf{x}}) & f_z(\hat{\mathbf{x}}) \\ g_x(\hat{\mathbf{x}}) & g_y(\hat{\mathbf{x}}) & g_z(\hat{\mathbf{x}}) \\ h_x(\hat{\mathbf{x}}) & h_y(\hat{\mathbf{x}}) & h_z(\hat{\mathbf{x}}) \end{bmatrix}$, the Jacobian matrix for $\mathbf{G}(\mathbf{x})$, evaluated at $\mathbf{x} = \hat{\mathbf{x}}$, has no eigenvalue with magnitude greater than one.

(e) Implement the fixed-point iteration $\mathbf{x}_{k+1} = \mathbf{G}(\mathbf{x}_k)$, showing it converges to $\hat{\mathbf{x}}$ if $\mathbf{x}_0 = \mathbf{0}$. Determine the number of iterations required for $\|\hat{\mathbf{x}} - \mathbf{x}_k\|_\infty < 10^{-5}$.

(f) Show empirically that $\|\hat{\mathbf{x}} - \mathbf{x}_k\|_\infty \le \sigma^k \|\mathbf{x}_1 - \mathbf{x}_0\|_\infty / (1 - \sigma)$, where $\sigma = \|J(\hat{\mathbf{x}})\|_\infty$.

(g) Again starting from $\mathbf{x}_0 = \mathbf{0}$, implement the Gauss–Seidel strategy $x_{k+1} = f(x_k, y_k, z_k)$, $y_{k+1} = g(x_{k+1}, y_k, z_k)$, $z_{k+1} = h(x_{k+1}, y_{k+1}, z_k)$, determining the number of iterations required for $\|\hat{\mathbf{x}} - \mathbf{x}_k\| < 10^{-5}$. Compare the efficiency of the two iteration schemes.

12. $\mathbf{F}(\mathbf{x}) = \begin{bmatrix} -4x^2y + 2xyz + y^2z + yz^2 - 9yz - 500x + 1920 \\ 8y^4 - 7y^3z + y^2z^2 - 5y^2 - 1501y - 4z + 9578 \\ 9x^3y - 4x^2y^2 - 2x^2z^2 - 2x^2z - 6y^2 - 7z^2 - 3000z + 41833 \end{bmatrix}$

$\hat{\mathbf{x}} = [5, 5, 12]^{\mathrm{T}}$

13. $\mathbf{F}(\mathbf{x}) = \begin{bmatrix} 8x^2z^2 - 4y^2z^2 - 7yz^2 + 5z^3 + 3z^2 - 1504x + 5187 \\ -5x^3y - 3x^3z - xyz - 8y^2z + z^3 - 9x - 1000y + 4841 \\ -2xz^3 + 2xy^2 - 2xz^2 - 8xy + 6xz - 2yz - 200z - 714 \end{bmatrix}$

$\hat{\mathbf{x}} = [3, 5, -3]^T$

14. $\mathbf{F}(\mathbf{x}) = \begin{bmatrix} -9x^3z - 8xyz^2 + 4x^3 - 4z^2 - 1000x + 2508 \\ x^2yz + 2x^2z^2 - 7xy^3 + 4x - 300y - 5z + 623 \\ 2x^2z^2 + 3xyz^2 + 2xz^3 + 9x - 500z + 1194 \end{bmatrix}$

$\hat{\mathbf{x}} = [2, 2, 3]^T$

15. $\mathbf{F}(\mathbf{x}) = \begin{bmatrix} 8x^4 + 5x^2y^2 - 5x^2y + 2yz^2 - 8yz - 1500x - 5610 \\ -7x^4 - 4x^2yz - 3xy^3 - 5y^2z^2 - xz^2 + 9y^2z - 1000y + 3888 \\ -2x^2y^2 - 2x^2z^2 - 8xz^2 - 6x^2 + x - y - 400z - 1474 \end{bmatrix}$

$\hat{\mathbf{x}} = [-3, 3, -4]^T$

16. $\mathbf{F}(\mathbf{x}) = \begin{bmatrix} 6x^2y^2 + 7xyz^2 - 3y^4 + 9yz^3 + 2xy^2 + 6xy - 1200x - 3480 \\ -4z^4 - 7x^3 - 4y^2z - 5xy + 6x - 400y + 1437 \\ 9x^3y + 6xyz^2 + yz^3 - 3x^2 + 5xy - 4xz - 1500z + 2615 \end{bmatrix}$

$\hat{\mathbf{x}} = [-3, 4, 1,]^T$

17. $\mathbf{F}(\mathbf{x}) = \begin{bmatrix} -4x^4 + 9x^2yz - 3xy^3 - xyz^2 + 5yz - 1200x - 2756 \\ 4x^3z - 2x^2y^2 - 6x^2z + 5z^3 + 2xy - 700y + 1156 \\ -5y^2z^2 + 3z^4 - 8xz^2 + 3y^2z - 9y^2 - 1500z - 6748 \end{bmatrix}$

$\hat{\mathbf{x}} = [-3, 2, -4]^T$

18. $\mathbf{F}(\mathbf{x}) = \begin{bmatrix} 8xy^3 + 4y^4 + 9y^2z^2 - 7x^3 + xy^2 + 4yz - 2500x - 4286 \\ -xy^3 - yz^3 + 5xz^2 - 9z^3 - 7y^2 - 500y - 946 \\ 5x^3y + 3y^2z^2 - 4yz^3 + 8x^2 + 6xy - 1500z + 4759 \end{bmatrix}$

$\hat{\mathbf{x}} = [-1, -3, 4]^T$

39.8 Newton's Iteration for Nonlinear Systems

Generalization from the Single-Variable Case

We have already seen, in Section 38.3, the single-variable Newton iteration $x_{n+1} = x_n - f(x_n)/f'(x_n)$ for solving the scalar equation $f(x) = 0$. The natural generalization to the vector case $\mathbf{f}(\mathbf{x}) = \mathbf{0}$ is the iteration $\mathbf{x}_{n+1} = \mathbf{x}_n - J^{-1}(\mathbf{x}_n)\mathbf{f}(\mathbf{x}_n)$, where J is the Jacobian matrix whose entries are $\delta f_i / \delta x_j$, where f_i is the ith component of the vector $\mathbf{f}$. The Jacobian matrix plays the role of the derivative in the single-variable case. Since it is a matrix, the reciprocal of f' in the single-variable case generalizes to the matrix inverse.

EXAMPLE 39.15 As we saw in Section 39.7, the nonlinear system $f(x, y) = 0$, $g(x, y) = 0$, where

$$f(x, y) = 12 + 3x + 8y + 12xy + 9x^2 - 12y^2$$
$$g(x, y) = -5 - 12x - 6y + 9xy + 2x^2 + 11y^2$$

has the two real solutions $(0.247523687, -0.69040990)$ and $(36.783609, -18.3639725)$. The Jacobian matrix is

$$J = \begin{bmatrix} f_x & f_y \\ g_x & g_y \end{bmatrix} = \begin{bmatrix} 18x + 12y + 3 & 12x - 24y + 8 \\ 4x + 9y - 12 & 9x + 22y - 6 \end{bmatrix}$$

so the right-hand side of the iteration $\mathbf{x}_{n+1} = \mathbf{x}_n - J^{-1}(\mathbf{x}_n)\mathbf{f}(\mathbf{x}_n)$ is then the vector-valued function

$$\mathbf{g}(\mathbf{x}) = \begin{bmatrix} \dfrac{-144xy + 32 + 240xy^2 - 144y + 246x^2y - 168x + 57x^3 - 70x^2 - 16y^2}{78 + 31x - 366y + 492xy + 480y^2 + 114x^2} \\ 3\dfrac{57xy - 43 + 82xy^2 + 56y + 19x^2y + 46x + 80y^3 + 38x^2 - 37y^2}{78 + 31x - 366y + 492xy + 480y^2 + 114x^2} \end{bmatrix}$$

(39.25)

Newton's method consists of the fixed-point iteration $\mathbf{x} = \mathbf{g}(\mathbf{x})$. Starting at the point $(35, -15)$, Table 39.14 lists $\mathbf{x}_k$, the Newton iterates, and $\|\mathbf{x}_k - \mathbf{x}_e\|$, where $\mathbf{x}_e$ corresponds to the second real solution. The rapid (quadratic) convergence of the iteration is apparent from the doubling of the number of zeros after the decimal point in the value of the norm of the difference between each iterate and the fixed point, $\mathbf{x}_e$.

k	x_k	y_k	$\|\mathbf{x}_k - \mathbf{x}_e\|_\infty$
0	35.0	−15.0	3.364
1	50.99149362	−26.75906930	14.208
2	40.35974485	−20.42798275	3.576
3	37.11933214	−18.55449085	0.3357
4	36.78701010	−18.36587974	0.003401
5	36.78360934	−18.36397272	0.5443×10^{-6}

TABLE 39.14   Newton iterates in Example 39.15, computed at extended precision

Just as with the single-variable case, the derivative (Jacobian matrix) of the iteration function $\mathbf{g}(\mathbf{x})$, evaluated at $\mathbf{x} = \mathbf{x}_e$, is the zero matrix, as can be seen by a tedious but straightforward calculation.   ❖

Derivation

A derivation of Newton's method for the two-variable case is instructive. Let $\mathbf{x}_e = (x_e, y_e)$ satisfy $f(x_e, y_e) = g(x_e, y_e) = 0$. Use Taylor polynomials constructed at $\mathbf{x}_k$ to approximate $f(x, y)$ and $g(x, y)$ at $\mathbf{x}_{k+1}$. Assuming that $\mathbf{x}_{k+1}$ is close enough to $\mathbf{x}_e$ for $f(x_{k+1}, y_{k+1})$ and $g(x_{k+1}, y_{k+1})$ both to be essentially zero, we obtain, to a first approximation,

$$0 = f(x_{k+1}, y_{k+1}) = f(x_k, y_k) + f_x(x_k, y_k)(x_{k+1} - x_k) + f_y(x_k, y_k)(y_{k+1} - y_k)$$
$$0 = g(x_{k+1}, y_{k+1}) = g(x_k, y_k) + g_x(x_k, y_k)(x_{k+1} - x_k) + g_y(x_k, y_k)(y_{k+1} - y_k)$$

which can be written in the form

$$-\mathbf{f}(\mathbf{x}_k) = \begin{bmatrix} f_x & f_y \\ g_x & g_y \end{bmatrix}\begin{bmatrix} x_{k+1} - x_k \\ y_{k+1} - y_k \end{bmatrix} = J(\mathbf{x}_k)(\mathbf{x}_{k+1} - \mathbf{x}_k)$$

and from which $\mathbf{x}_{k+1} = \mathbf{x}_k - J^{-1}(\mathbf{x}_k)\mathbf{f}(\mathbf{x}_k)$ follows.

Practical Implementation

In actual practice, the Jacobian matrix J is *not* inverted in any realistic implementation of Newton's method. The iteration is based on the equation $\mathbf{f}(\mathbf{x}_k) = J(\mathbf{x}_k)(\mathbf{x}_k - \mathbf{x}_{k+1}) = J(\mathbf{x}_k)\mathbf{u}_k$. The *increment* $\mathbf{u}_k$ is computed by solving the linear system $J(\mathbf{x}_k)\mathbf{u}_k = \mathbf{f}(\mathbf{x}_k)$ with any of the methods of Sections 39.1–39.6. The next iterate, namely, $\mathbf{x}_{k+1}$, is then obtained as $\mathbf{x}_k - \mathbf{u}_k = \mathbf{x}_k - (\mathbf{x}_k - \mathbf{x}_{k+1}) = \mathbf{x}_{k+1}$.

EXAMPLE 39.16   We saw in Section 39.7 that the equations

$$F_1(x, y, z) = 8z + 3z^2 + 2y^2 + xy - 4x^2 + 9xz - 4 = 0$$
$$F_2(x, y, z) = -z + 4z^2 - 4y^2 + 6xy - 6x^2 - 4yz + 39 = 0$$
$$F_3(x, y, z) = 9y - 2z - 3y^2 + 4xy - 8yz + 20 = 0$$

have four real solutions, one of which is $(-1, 2, 1)$. Let the vector-valued function $\mathbf{F}(\mathbf{x})$ and its Jacobian matrix $J(\mathbf{x})$ be given, respectively, by

$$\mathbf{F}(\mathbf{x}) = \begin{bmatrix} F_1(x, y, z) \\ F_2(x, y, z) \\ F_3(x, y, z) \end{bmatrix} \qquad J(\mathbf{x}) = \begin{bmatrix} y - 8x + 9z & 4y + x & 8 + 6z + 9x \\ 6y - 12x & 6x - 8y - 4z & 8z - 4y - 1 \\ 4y & 9 - 6y + 4x - 8z & -2 - 8y \end{bmatrix}$$

Table 39.15 is obtained when Newton's method is started at $\mathbf{x}_0 = [10, 10, 10]^T$. During each cycle, $\mathbf{F}(\mathbf{x}_k)$ and $J(\mathbf{x}_k)$ are computed. The increment $\mathbf{u}_k$ is obtained as the solution of the linear system $J(\mathbf{x}_k)\mathbf{u}_k = \mathbf{f}(\mathbf{x}_k)$, and $\mathbf{x}_{k+1}$ is obtained as $\mathbf{x}_{k+1} = \mathbf{x}_k - \mathbf{u}_k$. The iteration converges rapidly from a considerable distance from the solution.   ❖

k	x_k	y_k	z_k	$\|\mathbf{x}_k - \mathbf{x}_e\|_\infty$
0	10.0	10.0	10.0	11.0
1	4.530201245	5.768865724	4.588308158	5.469798755
2	1.612863602	4.039272089	1.780733948	2.917337643
3	−0.662756139	3.093654054	0.538427972	2.275619741
4	−0.931874105	2.101094586	0.897097449	0.992559468
5	−0.999753527	1.998291828	0.997168715	0.102802758
6	−0.999999605	2.000000333	1.000002630	0.002833915

TABLE 39.15   Newton iterates in Example 39.16

EXERCISES 39.8

1. Use a computer algebra system to obtain $J_{\mathbf{g}}(\mathbf{x})$, the Jacobian matrix for the iteration function $\mathbf{g}(\mathbf{x})$ given in (39.25). Show that $J_{\mathbf{g}}(\mathbf{x}_e)$ is the zero matrix, where $\mathbf{x}_e$ is either of the two solutions of the system in Example 39.15.

For the 2 × 2 systems $\mathbf{F}(\mathbf{x}) = \mathbf{0}$ in Exercises 2–11:

(a) Obtain the Newton iteration function $\mathbf{G}(\mathbf{x}) = \mathbf{x} - J^{-1}(\mathbf{x})\mathbf{F}(\mathbf{x})$, where $J(\mathbf{x})$ is the Jacobian matrix for the vector-valued function $\mathbf{F}(\mathbf{x})$.

(b) Show that at the fixed point, the Jacobian matrix for $\mathbf{G}(\mathbf{x})$ is the zero matrix, thereby verifying that Newton's method for systems converges quadratically to a simple zero of $\mathbf{F}(\mathbf{x})$.

(c) Implement Newton iteration as the fixed-point iteration $\mathbf{x}_{k+1} = \mathbf{G}(\mathbf{x}_k)$, starting with $\mathbf{x}_0$ near enough to $\hat{\mathbf{x}}$ to ensure convergence to $\hat{\mathbf{x}}$ and noting the number of iterations required to obtain $\|\hat{\mathbf{x}} - \mathbf{x}_k\|_\infty < 10^{-5}$.

(d) Implement Newton iteration whereby, at each step of the iteration, $\mathbf{F}(\mathbf{x}_k) = J(\mathbf{x}_k)\mathbf{u}_k$ is solved for $\mathbf{u}_k$ and $\mathbf{x}_{k+1}$ is obtained as $\mathbf{x}_{k+1} = \mathbf{x}_k - \mathbf{u}_k$. Do not invert $J(\mathbf{x})$; rather, solve for $\mathbf{u}_k$ via a linear-system solver.

2. Exercise 2, Section 39.7   **3.** Exercise 3, Section 39.7

4. Exercise 4, Section 39.7   **5.** Exercise 5, Section 39.7

6. Exercise 6, Section 39.7   **7.** Exercise 7, Section 39.7

8. Exercise 8, Section 39.7   **9.** Exercise 9, Section 39.7

10. Exercise 10, Section 39.7   **11.** Exercise 11, Section 39.7

For the 3 × 3 systems $\mathbf{F}(\mathbf{x}) = \mathbf{0}$ in Exercises 12–18:

(a) Obtain the Newton iteration function $\mathbf{G}(\mathbf{x}) = \mathbf{x} - J^{-1}(\mathbf{x})\mathbf{F}(\mathbf{x})$, where $J(\mathbf{x})$ is the Jacobian matrix for the vector-valued function $\mathbf{F}(\mathbf{x})$.

(b) Implement Newton iteration as the fixed-point iteration $\mathbf{x}_{k+1} = \mathbf{G}(\mathbf{x}_k)$, starting with $\mathbf{x}_0$ near enough to $\hat{\mathbf{x}}$ to ensure convergence to $\hat{\mathbf{x}}$ and noting the number of iterations required to obtain $\|\hat{\mathbf{x}} - \mathbf{x}_k\|_\infty < 10^{-5}$.

(c) Show that $\|\hat{\mathbf{x}} - \mathbf{x}_k\|_\infty$ converges to zero quadratically, and that the Jacobian matrix $\mathbf{G}(\mathbf{x})$ is the zero matrix.

(d) Implement Newton iteration whereby, at each step of the iteration, $\mathbf{F}(\mathbf{x}_k) = J(\mathbf{x}_k)\mathbf{u}_k$ is solved for $\mathbf{u}_k$ and $\mathbf{x}_{k+1}$ is obtained as $\mathbf{x}_{k+1} = \mathbf{x}_k - \mathbf{u}_k$. Do not invert $J(\mathbf{x})$; rather, solve for $\mathbf{u}_k$ via a linear-system solver.

12. Exercise 12, Section 39.7   **13.** Exercise 13, Section 39.7

14. Exercise 14, Section 39.7   **15.** Exercise 15, Section 39.7

16. Exercise 16, Section 39.7   **17.** Exercise 17, Section 39.7

18. Exercise 18, Section 39.7

For each of the 3 × 3 systems $\mathbf{F}(\mathbf{x}) = \mathbf{0}$ in Exercises 19–28 where $\mathbf{x} = [x, y, z]^T$:

(a) Verify that $F(\hat{\mathbf{x}}) = 0$.

(b) Obtain the Newton iteration function $\mathbf{G}(\mathbf{x}) = \mathbf{x} - J^{-1}(\mathbf{x})\mathbf{F}(\mathbf{x})$, where $J(\mathbf{x})$ is the Jacobian matrix for the vector-valued function $\mathbf{F}(\mathbf{x})$.

(c) Show that the Jacobian matrix for $\mathbf{G}(\mathbf{x})$ is the zero matrix, thereby verifying that Newton's method for systems converges quadratically to a simple zero of $\mathbf{F}(\mathbf{x})$.

(d) Implement Newton iteration as the fixed-point iteration $\mathbf{x}_{k+1} = \mathbf{G}(\mathbf{x}_k)$, starting with $\mathbf{x}_0$ near enough to $\hat{\mathbf{x}}$ to ensure convergence to $\hat{\mathbf{x}}$ and noting the number of iterations required to obtain $\|\hat{\mathbf{x}} - \mathbf{x}_k\|_\infty < 10^{-5}$.

(e) Implement Newton iteration whereby, at each step of the
iteration, $\mathbf{F}(\mathbf{x}_k) = J(\mathbf{x}_k)\mathbf{u}_k$ is solved for $\mathbf{u}_k$ and $\mathbf{x}_{k+1}$ is obtained
as $\mathbf{x}_{k+1} = \mathbf{x}_k - \mathbf{u}_k$. Do not invert $J(\mathbf{x})$; rather, solve for $\mathbf{u}_k$ via a
linear-system solver.

19. $\mathbf{F}(\mathbf{x}) = \begin{bmatrix} 5z^3x - z^2x^2 - 9zy^3 + 4y^4 - 5z^3 - 1400 \\ -6z^3y - 9z^2x^2 + 3z^2xy + 3x^4 + 3x^3y + 9z^2y - 55353 \\ -7z^2y^2 - 4zy^3 + 2y^4 - 3z^2y + 9zx^2 - 7z + 11695 \end{bmatrix}$

$\hat{\mathbf{x}} = [-12, -5, -5]^{\mathrm{T}}$

20. $\mathbf{F}(\mathbf{x}) = \begin{bmatrix} -8z^3y + 4zx^3 + 6zxy^2 + z^3 + 7z^2x + 6zx + 6250 \\ -5zx^3 + 5zx^2y + 7z^2y + 9x^3 + 4xy + 6y - 5522 \\ -5zx^3 - xy^3 + 8zy^2 - 7y^3 + 3z^2 - 3164 \end{bmatrix}$

$\hat{\mathbf{x}} = [5, -3, -5]^{\mathrm{T}}$

21. $\mathbf{F}(\mathbf{x}) = \begin{bmatrix} 4z^2xy + 6z^2y^2 - 6x^2y^2 - 4z^2y + 9zxy + 4x + 122 \\ -8z^2y^2 - 5xy^3 - 5z^2y + 3x^3 + 8x^2 + 87 \\ -6zx^2y + 6z^3 - 9zy^2 - 7x^2y + 4y^3 - 7xy + 13 \end{bmatrix}$

$\hat{\mathbf{x}} = [-4, -1, -1]^{\mathrm{T}}$

22. $\mathbf{F}(\mathbf{x}) = \begin{bmatrix} 3z^2x + 3zxy - 4x^2y - 2xy^2 - 7z^2 + 8z + 140 \\ -3zx^2 - 8zxy - x^3 + 9zy + 9z + 22 \\ 3z^3 - 6x^2y + xy^2 - 6z^2 - 9y^2 + 8y + 276 \end{bmatrix}$

$\hat{\mathbf{x}} = [-4, 2, -2]^{\mathrm{T}}$

23. $\mathbf{F}(\mathbf{x}) = \begin{bmatrix} 2x^3 - 4x^2y + 7y^3 - 5zy - 4y^2 + 322 \\ 7z^2y + 3zx^2 + 3zxy - 5zy^2 - 5x^3 + 3z^2 - 635 \\ 3zx^2 - 2zx - 9zy + xy + x - 3y - 269 \end{bmatrix}$

$\hat{\mathbf{x}} = [-5, -4, 2]^{\mathrm{T}}$

24. $\mathbf{F}(\mathbf{x}) = \begin{bmatrix} -4xy^2 - 2y^3 - 9z^2 + 2zy - 2z + 72 \\ -5z^2y + 8zxy + 7zy^2 + 7zx - 5xy - 8y^2 - 370 \\ 4z^2x - 3xy^2 + 4y^3 - 2zy - 5x + 2y + 255 \end{bmatrix}$

$\hat{\mathbf{x}} = [-5, -2, 4]^{\mathrm{T}}$

25. $\mathbf{F}(\mathbf{x}) = \begin{bmatrix} -7z^2 + 9zy + 3xy - 2y^2 + 8z + 212 \\ -6z^2 + 8zy - 9x^2 + 5x + 8y + 268 \\ 2zx - 9zy - 7x^2 - 9xy - 7x - 3y - 161 \end{bmatrix}$

$\hat{\mathbf{x}} = [4, -5, 2]^{\mathrm{T}}$

26. $\mathbf{F}(\mathbf{x}) = \begin{bmatrix} -5z^2 + 5zx - 3x^2 - 3y^2 + 8y + 44 \\ -7z^2 - 6zx + 9z - 9y + 99 \\ -4z^2 - 2zy - 6y^2 - 3y + 33 \end{bmatrix}$

$\hat{\mathbf{x}} = [4, -1, 3]^{\mathrm{T}}$

27. $\mathbf{F}(\mathbf{x}) = \begin{bmatrix} 6z^2 + 7zy + 3x^2 + 7y^2 - 8x - 231 \\ -2zxy - 7x^2y - 3xy^2 + 3zx + 9y^2 - 432 \\ 8zx^3 + 4zy^3 - 4x^3y - 4y^4 + 4z^3 + 8x^2y \end{bmatrix}$

$\hat{\mathbf{x}} = [-3, -3, -3,]^{\mathrm{T}}$

28. $\mathbf{F}(\mathbf{x}) = \begin{bmatrix} -8z^2xy + x^4 - 8x^3y + 5x^2y^2 - 5z^2y - 9zy^2 - 597 \\ -z^2y - 9zx^2 - 8x^2y - 5y^3 - 5z^2 + 6xy - 32 \\ -3z^2 + x^2 + 7z + 7x + 8y + 84 \end{bmatrix}$

$\hat{\mathbf{x}} = [1, -2, -4]^{\mathrm{T}}$

29. Broyden's method, as described in [16], is an example of a
quasi-Newton algorithm wherein the need to invert the Jacobian
matrix (or the equivalent, thereof) is circumvented. The
convergence of Broyden's method is only superlinear, but that is an
acceptable trade-off for the great reduction in computational effort
effected. To solve $\mathbf{F}(\mathbf{x}) = \mathbf{0}$, starting with $\mathbf{x}_0$ and $\mathbf{x}_1$, implement the
following steps in a computer language of your choice.

(a) $A_0 = J(\mathbf{x}_0)$ and $A_0^{-1} = J^{-1}(\mathbf{x}_0)$ **(b)** $\mathbf{y}_k = \mathbf{F}(\mathbf{x}_k) - \mathbf{F}(\mathbf{x}_{k-1})$

(c) $\mathbf{s}_k = \mathbf{x}_k - \mathbf{x}_{k-1}$ **(d)** $A_k^{-1} = A_{k-1}^{-1} + \dfrac{\left(\mathbf{s}_k - A_{k-1}^{-1}\mathbf{y}_k\right)\mathbf{s}_k^{\mathrm{T}}A_{k-1}^{-1}}{\mathbf{s}_k^{\mathrm{T}}A_{k-1}^{-1}\mathbf{y}_k}$

(e) $\mathbf{x}_{k+1} = \mathbf{x}_k - A_k^{-1}\mathbf{F}(\mathbf{x}_k)$

In Exercises 30–38, use Broyden's method to solve the given system.
Try starting at the same $\mathbf{x}_0$ as before, and using $\mathbf{x}_1 = [1, 1]^{\mathrm{T}}$ seeing if
the method converges to $\hat{\mathbf{x}}$. Also, monitor the number of steps required
to achieve the accuracy entailed by $\|\hat{\mathbf{x}} - \mathbf{x}_k\|_\infty < 10^{-5}$, and compare to
the number of steps taken by Newton's method.

30. Exercise 2, Section 39.7   **31.** Exercise 3, Section 39.7

32. Exercise 4, Section 39.7   **33.** Exercise 12, Section 39.7

34. Exercise 13, Section 39.7   **35.** Exercise 14, Section 39.7

36. Exercise 19, Section 39.8   **37.** Exercise 20, Section 39.8

38. Exercise 21, Section 39.8

Chapter Review

1. If $A = \begin{bmatrix} 5 & 6 & 9 \\ -7 & 7 & 2 \\ 6 & -9 & 6 \end{bmatrix}$ and $\mathbf{y} = \begin{bmatrix} -58 \\ -38 \\ 3 \end{bmatrix}$, demonstrate the use of
Gaussian elimination to solve the system $A\mathbf{x} = \mathbf{y}$ by upper-
triangularizing the augmented matrix $[A, \mathbf{y}]$ and using back
substitution.

2. Describe the condition number for a matrix and show how it is used
to obtain an upper bound to the relative error in a numerically
computed solution to a system of linear equations.

3. What is *partial pivoting* in the context of Gaussian arithmetic for numerically solving systems of linear equations. What reasoning motivates this strategy?

4. What is the asymptotic estimate of the number of arithmetic operations Gaussian elimination and back substitution require in the solution of an $n \times n$ system of linear equations?

5. What is a *tridiagonal system,* and why is there a specialized algorithm for solving such a system?

6. What is *iterative improvement*? Describe how it is implemented. Justify why it works.

7. Describe the iterative method of Jacobi. Use the system in Question 1 to illustrate the first two iterations that start from $\mathbf{x}_0 = \mathbf{0}$.

8. Is the matrix A in Question 1 strictly diagonally dominant? If the method of Jacobi were implemented for the system in Question 1, would convergence be likely?

9. Give an example of a strictly diagonally dominant matrix.

10. Describe the Gauss–Seidel iterative method. Use the system in Question 1 to illustrate the first two iterations that start from $\mathbf{x}_0 = \mathbf{0}$. Is it likely that the iteration will converge? Why?

11. Demonstrate at least two steps in Southwell's relaxation solution of the system in Question 1.

12. Describe the SOR technique. Give its matrix form. Under what conditions on the iteration matrix can convergence be expected?

13. Demonstrate a Jacobi-style iteration for solving the nonlinear system $x^2 + y^2 = 1$, $x^2 - y^2 = \frac{1}{2}$.

14. Demonstrate how the system in Question 13 might be solved with an iterative technique patterned after the Gauss–Seidel algorithm.

15. Demonstrate how the system in Question 13 might be solved by Newton iteration. For efficiency, how should Newton iteration for systems be implemented?

Chapter *40*

Interpolation

INTRODUCTION   Passing a smooth curve through a set of points is called *interpolation,* the curve being said to *interpolate the points.* Chapter 40 examines several of the classic techniques for constructing an interpolating polynomial.

We begin with *naive interpolation* in which a set of algebraic equations, formed by substituting the data points into a polynomial template, is constructed and solved. Unfortunately, the resulting linear system generally has a large condition number, and the technique is not practical for anything but the smallest systems. Hence, the section goes on to consider *Lagrange interpolation* in which a set of basis polynomials, the Lagrange interpolating polynomials, are first constructed and then combined to form the desired interpolating polynomial. The section concludes with a look at the *polynomial wiggle problem* whereby a high-order interpolating polynomial, interpolating between equispaced nodes, can exhibit oscillations of a large magnitude between the nodes. Thus, although the polynomial passes through the given points, it is probably not a very good representation of the function that the data points are approximating.

We then discuss *Newton's method* of forming an interpolating polynomial from a *divided difference table.* The two techniques, that of Lagrange and that of Newton, yield the exact same polynomial, but in two different forms. The Newton polynomial is built up degree-by-degree, as new data points are added in from the divided difference table.

When interpolating a continuous function, if the nodes are spaced according to the zeros of the Chebyshev polynomials, a *minimax polynomial* results. This polynomial minimizes the maximum deviation, or wiggle, exhibited by the interpolating polynomial and the polynomial wiggle problem is greatly alleviated.

Spline interpolation is a variant of the interpolation problem. Instead of passing a single polynomial through all the points in a data set, many lower order polynomials are passed through contiguous subsets of the data points. Thus, the interpolation is accomplished by a piecewise polynomial function of consistently low degree. We therefore detail spline interpolation with linear, quadratic, and cubic polynomials, the first being the case where a collection of straight line segments connects consecutive points in a collection of points.

A Bezier curve is described parametrically by polynomials of equal degree. A cubic Bezier curve would be determined by four points, of which just the first and last are actually interpolated by the curve. The two "interior" points, called *control points,* merely influence the shape of the curve. The Bezier spline is a collection of contiguous Bezier curves, the advantage being that changing one control point only affects a single component Bezier curve. The whole spline does not have to be recalculated, which is quite the contrary to the case for the classic polynomial spline.

40.1   **Lagrange Interpolation**

The Need for Interpolation

Discrete data, typically obtained by experiment or sampling, can be difficult to work with in an analytic context. Thus, it is often useful to pass a smooth curve through the data points. The problem of *interpolation* is that of finding an appropriate smooth curve which passes through each of the points.

Naive Interpolation

Interpolating a set of points with the function $y = f(x)$ amounts to solving a set of equations. The simplest examples are afforded by polynomials, but any function with sufficient parameters could be used. To pass the parabola $y = f(x) = ax^2 + bx + c$ through the three data points $(1, 2)$, $(-4, 7)$, $(6, -13)$, solve the three equations on the left in (40.1)

$$
\begin{aligned}
a + b + c &= 2 \\
16a - 4b + c &= 7 \\
36a + 6b + c &= -13
\end{aligned}
\qquad
A = \begin{bmatrix} 1 & 1 & 1 \\ 16 & -4 & 1 \\ 36 & 6 & 1 \end{bmatrix}
\qquad \textbf{(40.1)}
$$

obtaining $y = \frac{1}{5}(19 - 8x - x^2)$ as the interpolating parabola. Clearly, the three equations determining the parameters a, b, c are $f(1) = 2$, $f(-4) = 7$, and $f(6) = -13$.

Ill-Conditioning

Systems of equations arising from interpolation problems are usually highly ill-conditioned. For example, the system matrix for the equations of the previous section, given on the right in (40.1), has a condition number (using the 1-norm) of 57.24. (See Section 39.2.) Were we to increase the size of the problem slightly, the condition number of the coefficient matrix would correspondingly increase.

For example, interpolating the five points

$$
(-63, 52) \quad (-57, 83) \quad (-78, -79) \quad (-12, -25) \quad (-45, 83),
$$

with the fourth-degree polynomial $y = a_0 + a_1 x + a_2 x^2 + a_3 x^3 + a_4 x^4$, generates five equations for which the system matrix A is given on the left in (40.2). This is a *Vandermonde matrix* (40.2), whose general form is given on the right in (40.2). The condition number of A is the exceptionally large 3.0×10^9. Working in exact arithmetic causes no problems, but should this system be solved by a numeric method from Chapter 39, the large condition number signals the possibility of severe roundoff error.

$$
A = \begin{bmatrix}
1 & -63 & 3969 & -250{,}047 & 15{,}752{,}961 \\
1 & -57 & 3249 & -185{,}193 & 10{,}556{,}001 \\
1 & -78 & 6084 & -474{,}552 & 37{,}015{,}056 \\
1 & -12 & 144 & -1728 & 20{,}736 \\
1 & -45 & 2025 & -91{,}125 & 4{,}100{,}625
\end{bmatrix}
=
\begin{bmatrix}
1 & x_0 & x_0^2 & \cdots & x_0^n \\
1 & x_1 & x_1^2 & \cdots & x_1^n \\
1 & x_2 & x_2^2 & \cdots & x_2^n \\
\vdots & \vdots & \vdots & \ddots & \vdots \\
1 & x_n & x_n^2 & \cdots & x_n^n
\end{bmatrix}
\qquad \textbf{(40.2)}
$$

Lagrange Interpolating Polynomials

Naive interpolation through the $n + 1$ points (x_k, y_k), $k = 0, \ldots, n$, can lead to ill-conditioned systems of equations. To avoid having to solve such systems, we can resort to the Lagrange basis of interpolating polynomials $L_k(x) = \prod_{i=0, i \neq k}^{n} (x - x_i)/(x_k - x_i)$, $k =$

$0, \ldots, n$, which are explicitly given by

$$L_k(x) = \frac{(x - x_0) \cdots (x - x_{k-1})(x - x_{k+1}) \cdots (x - x_n)}{(x_k - x_0) \cdots (x_k - x_{k-1})(x_k - x_{k+1}) \cdots (x_k - x_n)}$$

These polynomials then satisfy

$$L_k(x_j) = \begin{cases} 1 & j = k \\ 0 & j \neq k \end{cases}$$

so that the nth-degree polynomial interpolating the points (x_k, y_k), $k = 0, \ldots, n$, can be written as $f(x) = \sum_{k=0}^{n} y_k L_k(x)$.

Each Lagrange basis polynomial is of degree n since each numerator contains n linear factors. Corresponding to each factor $x - x_i$ in the numerator, the denominator contains the factor $x_k - x_i$. In the kth polynomial, the numerator skips the factor $x - x_k$ and the denominator skips the corresponding factor $x_k - x_k$.

EXAMPLE 40.1   If $n = 2$, the $n + 1 = 3$ Lagrange interpolating polynomials of degree $n = 2$ are

$$L_0 = \frac{(x - x_1)(x - x_2)}{(x_0 - x_1)(x_0 - x_2)} \qquad L_1 = \frac{(x - x_0)(x - x_2)}{(x_1 - x_0)(x_1 - x_2)} \qquad L_2 = \frac{(x - x_0)(x - x_1)}{(x_2 - x_0)(x_2 - x_1)} \qquad ❖$$

EXAMPLE 40.2   The Lagrange interpolating polynomials of degree $n = 2$ generated by the data points that give (40.1) are

$$L_0 = -\tfrac{1}{25}(x + 4)(x - 6) \qquad L_1 = \tfrac{1}{50}(x - 1)(x - 6) \qquad L_2 = \tfrac{1}{50}(x - 1)(x + 4) \qquad ❖$$

EXAMPLE 40.3   If $n = 3$, the $n + 1 = 4$ Lagrange interpolating polynomials of degree $n = 3$ are

$$L_0 = \frac{(x - x_1)(x - x_2)(x - x_3)}{(x_0 - x_1)(x_0 - x_2)(x_0 - x_3)} \qquad L_2 = \frac{(x - x_0)(x - x_1)(x - x_3)}{(x_2 - x_0)(x_2 - x_1)(x_2 - x_3)}$$

$$L_1 = \frac{(x - x_0)(x - x_2)(x - x_3)}{(x_1 - x_0)(x_1 - x_2)(x_1 - x_3)} \qquad L_3 = \frac{(x - x_0)(x - x_1)(x - x_2)}{(x_3 - x_0)(x_3 - x_1)(x_3 - x_2)} \qquad ❖$$

EXAMPLE 40.4   The Lagrange polynomials for the $n + 1 = 5$ points $(-25, -43)$, $(-22, -20)$, $(-24, 37)$, $(-69, 13)$, $(88, 88)$ are

$$L_0 = -\frac{(x + 22)(x + 24)(x + 69)(x - 88)}{14{,}916} \qquad L_3 = \frac{(x + 25)(x + 22)(x + 24)(x - 88)}{14{,}610{,}420}$$

$$L_1 = -\frac{(x + 25)(x + 24)(x + 69)(x - 88)}{31{,}020} \qquad L_4 = \frac{(x + 25)(x + 22)(x + 24)(x + 69)}{218{,}569{,}120}$$

$$L_2 = \frac{(x + 25)(x + 22)(x + 69)(x - 88)}{10{,}080}$$

The fourth-degree polynomial interpolating the given points is then

$$f(x) = \sum_{k=0}^{4} L_k(x) y_k = \tfrac{221{,}875{,}439}{30{,}818{,}245{,}920} x^4 + \tfrac{220{,}127{,}311}{1{,}100{,}651{,}640} x^3 - \tfrac{1{,}423{,}147{,}234{,}529}{30{,}818{,}245{,}920} x^2$$

$$- \tfrac{4{,}665{,}156{,}224{,}837}{2{,}201{,}303{,}280} x - \tfrac{278{,}419{,}407{,}759}{11{,}673{,}578} \qquad ❖$$

The Polynomial Wiggle Problem

In [60], we find the following warning about the "polynomial wiggle problem."

> "If data points P_k lie on a curve that is not polynomial-like, high-degree polynomial curves will oscillate between successive P_ks when forced to go ... through them; and once such oscillations start, increasing the degree further generally increases their maximum amplitude."

We give two examples of this phenomenon. First, consider the function $f(x) = \frac{1}{x}$, which is certainly not polynomial-like since it has both a horizontal and vertical asymptote. If we select the $n + 1 = 9$ points $(\frac{1}{k}, k)$, $k = 1, \ldots, 9$, whose nodes are equally spaced in the interval $[0, 1]$, the eighth-degree interpolating polynomial passing through them is

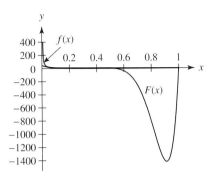

FIGURE 40.1 Graph of eighth-degree polynomial interpolating $(\frac{1}{k}, k)$, $k = 1, \ldots, 9$

$$F(x) = 362{,}880x^8 - 1{,}026{,}576x^7 + 1{,}172{,}700x^6 - 723{,}680x^5 + 269{,}325x^4$$
$$-63{,}273x^3 + 9450x^2 - 870x + 45$$

which is graphed in Figure 40.1. The extreme oscillation near the end of the interval should be a reminder about the dangers of using interpolating polynomials for nonconforming data.

A more famous example is due to the German mathematician Carl Runge. For the function $f(x) = \frac{1}{1+x^2}$, the eleven points $(k, f(k))$, $k = -5, \ldots, 5$, with nodes evenly spaced along the x-axis, are interpolated by the tenth-degree polynomial

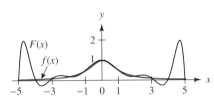

FIGURE 40.2 Graph of tenth-degree polynomial interpolating $(k, \frac{1}{1+k^2})$, $k = -5, \ldots, 5$

$$F(x) = -\frac{1}{44{,}200}x^{10} + \frac{7}{5525}x^8 - \frac{83}{3400}x^6 + \frac{2181}{11{,}050}x^4 - \frac{149}{221}x^2 + 1 \qquad \textbf{(40.3)}$$

Since $f(x)$ is not polynomial-like, having a horizontal asymptote, Figure 40.2, a graph of $f(x)$ and $F(x)$, shows dramatic oscillation in $F(x)$ near the ends of the interval.

EXERCISES 40.1–Part A

A1. If $n = 1$, write the $n + 1 = 2$ Lagrange interpolating polynomials of degree $n = 1$.

A2. Obtain L_0 and L_1, the Lagrange interpolating polynomials corresponding to the data points $(-2, 3)$ and $(4, 7)$.

A3. Show that the equation of the line connecting the two points in Exercise A2 is $y = 3L_0 + 7L_1$.

A4. Interpolate the $n + 1 = 3$ points $\{(-1, 1), (2, -3), (3, 5)\}$ via Lagrange interpolation.

A5. Write the Vandermonde matrix for the data points in Exercise A4.

EXERCISES 40.1–Part B

B1. Use appropriate software to verify that the condition number of the matrix A in (40.2) is indeed 3.0×10^9.

In Exercises B2–11:

   (a) Use naive interpolation (forming $n + 1$ equations in $n + 1$ unknowns) to obtain $p_n(x)$, the nth-degree polynomial that interpolates the points $(x_k, f(x_k))$, $k = 0, \ldots, n$.

   (b) Write the equations in part (a) in the form $A\mathbf{x} = \mathbf{y}$, and determine the condition number of the matrix A.

   (c) Obtain $P_n(x) = \sum_{k=0}^{n} L_k(x)f(x_k)$, where $\{L_k(x)\}$ are the Lagrange interpolating polynomials, and show that $P_n(x) = p_n(x)$.

   (d) Compare $p_n(x)$ to $f(x)$ by plotting both on the given domain.

   (e) Provided $f^{(n+1)}$ exists on the closed interval I containing c and the interpolated nodes $x_0, \ldots, x_n$, the error an interpolating polynomial makes at $x = c$ is given by $f(c) - p_n(c) = f^{(n+1)}(\xi)/(n + 1)! \prod_{k=0}^{n}(c - x_k)$, for some ξ in the open interval I. If $M = \max_{x \in I} |f^{(n+1)}(x)|$, then

$|f(c) - p_n(c)| \le \frac{M}{(n+1)!} \prod_{k=0}^{n} |c - x_k| = \varepsilon(c)$. Plot $E = |f(c) - p_n(c)|$ and $\varepsilon(c)$ on the same set of axes.

(f) A viable approach to estimating $f'(c)$ from the data points $(x_k, y_k), k = 0, \ldots, n$, is to interpolate the data points with $p_n(x)$ and approximate $f'(c)$ with $p_n'(c)$. Test the validity of this procedure by graphing $f'(x)$ and $p_n'(x)$ on the same set of axes.

B2. $f(x) = 10xe^{-x}, 0 \le x \le 7; x_k = k, k = 0, \ldots, 7 = n$

B3. $f(x) = \frac{x^2}{1 + x^2}, -3 \le x \le 3; x_k = -3 + k, k = 0, \ldots, 6 = n$

B4. $f(x) = x + \frac{9}{x} \cdot \frac{1}{2}, \frac{1}{2} \le x \le \frac{15}{2}; x_k = \frac{1}{2} + k, k = 0, \ldots, 7 = n$

B5. $f(x) = \arctan x, 0 \le x \le 7; x_k = k, k = 0, \ldots, 7 = n$

B6. $f(x) = e^{-x} \cos x, 0 \le x \le 5; x_k = k, k = 0, \ldots, 5 = n$

B7. $f(x) = \sqrt{1 - x^2}, 0 \le x \le 1; x_k = \frac{k}{5}, k = 0, \ldots, 5 = n$

B8. $f(x) = \frac{x}{1 + x^2}, 0 \le x \le 7; x_k = k, k = 0, \ldots, 7 = n$

B9. $f(x) = \frac{x}{1 + x}, 0 \le x \le 7; x_k = k, k = 0, \ldots, 7 = n$

B10. $f(x) = \frac{\cosh x}{1 + x^2}, 0 \le x \le \frac{7}{2}; x_k = \frac{k}{2}, k = 0, \ldots, 7 = n$

B11. $f(x) = e^{-x} \tan x, 0 \le x \le \frac{7}{5}; x_k = \frac{k}{5}, k = 0, \ldots, 7 = n$

For the nodes $x_k, k = 0, \ldots, n$, given in Exercises B12–14:

(a) Obtain A, the Vandermonde matrix as given in (40.2).

(b) Obtain the condition number of A.

(c) Show that $\det A = \prod_{0 \le j < k \le n} (x_k - x_j)$.

B12. 2, 3, 5, 6   **B13.** $-2, 0, 5, 9, 11$ **B14.** 1, 3, 4, 7, 10

The *inverse interpolation* problem consists in estimating, from the data points $(x_k, y_k), k = 0, \ldots, n$, where the y_k are monotone, the value of x for which $y = y^*$. There are two obvious approaches. First, interpolate the data points with $p_n(x)$ and solve the equation $p_n(x) = y^*$ for x. Second, reverse the roles of x and y, interpolate the points (y_k, x_k) with $P_n(y)$, and take $P_n(y^*)$ as the desired x. For the data sets and the prescribed value of y^* in each of Exercises B15–19:

(a) Plot the data points.

(b) Use the first method (involving $p_n(x)$) to determine the value of x for which $y = y^*$.

(c) Use the second method (involving $P_n(y)$) to determine the value of x for which $y = y^*$. Do you expect the two methods to give the same value of x? Why do the two methods give the same value?

B15. $\{(1, 0), (\frac{3}{2}, \frac{47}{52}), (2, 6), (\frac{5}{2}, \frac{767}{44}), (3, \frac{55}{2})\}; y^* = 12$

B16. $\{(\frac{1}{2}, -\frac{21}{92}), (1, \frac{4}{9}), (\frac{3}{2}, \frac{751}{820}), (2, \frac{141}{113}), (\frac{5}{2}, \frac{5171}{3372}), (3, \frac{317}{176})\}; y^* = 1$

B17. $\{(-3, \frac{271}{130}), (-\frac{5}{2}, \frac{1251}{607}), (-2, \frac{39}{20}), (-\frac{3}{2}, \frac{233}{149}), (-1, \frac{3}{8}), (-\frac{1}{2}, -\frac{57}{35})\}; y^* = 0$

B18. $\{(-3, -\frac{15}{224}), (-\frac{5}{2}, -\frac{544}{7625}), (-2, -\frac{19}{227}), (-\frac{3}{2}, -\frac{192}{1493}), (-1, -\frac{5}{16}), (-\frac{1}{2}, -\frac{160}{161}), (0, -\frac{9}{5})\}; y^* = -\frac{1}{2}$

B19. $\{(\frac{1}{4}, -\frac{392}{25}), (\frac{1}{2}, -\frac{41}{5}), (\frac{3}{4}, -\frac{1432}{479}), (1, -\frac{4}{3}), (\frac{5}{4}, -\frac{1624}{2323}), (\frac{3}{2}, -\frac{53}{130}), (\frac{7}{4}, -\frac{584}{2285}), (2, -\frac{14}{83})\}; y^* = -1$

Interpolation through prescribed values of the function and its derivatives is called *osculatory interpolation*. Some problems of osculatory interpolation have a unique solution, some have no solution, and some have an infinite number of solutions. For the sets of $n + 1$ conditions in Exercises B20–24, use an nth-degree polynomial and naive interpolation to write $n + 1$ equations in $n + 1$ unknowns. If there is one solution to these equations, the associated osculatory interpolation problem has a unique solution. If there are an infinite number of solutions to these equations, the polynomial $p_n(x)$ is not unique. If there is no solution to the equations, try a higher degree polynomial. If there is a solution to the new set of equations, the interpolating polynomial, $p_{n+1}(x)$, will not be unique.

B20. $f(1) = 2, f'(2) = 0, f(3) = -5$

B21. $f(0) = 2, f(1) = 8, f'(1) = -2, f(2) = 1$

B22. $f(1) = -3, f(2) = 1, f'(2) = 4, f'(3) = -1$

B23. $f(-1) = 5, f(0) = 3, f'(1) = 1, f''(2) = 0$

B24. $f'(0) = 1, f(1) = 2, f'(1) = 1, f(2) = 3$

B25. Consider the osculatory interpolation problem in which the cubic $p_3(x) = \sum_{k=0}^{3} a_k x^k$ is to satisfy the conditions $f(0) = b_1$, $f(1) = b_2, f'(c) = b_3, f(2) = b_4$. Write the equations for this interpolation in the form $A\mathbf{x} = \mathbf{b}$, where $\mathbf{x} = [a_0, \ldots, a_3]^T$ and $\mathbf{b} = [b_1, \ldots, b_4]^T$, and determine all values of c for which A is singular. If $\hat{c}$ is such a value, then prescribing $f'(\hat{c})$ leads to either no solution or an infinite number of solutions, depending on the values assigned to $b_1, \ldots, b_4$.

Osculatory interpolation is called *Hermite interpolation* if, whenever $f^{(k)}(\hat{x})$ is prescribed, so also are $f(\hat{x}), f'(\hat{x}), \ldots, f^{(k-1)}(\hat{x})$. Hermite interpolation *always* has a unique solution. For the sets of conditions in Exercises B26–30, use naive interpolation to obtain the solution of the associated Hermite interpolation problem and then graph the interpolating polynomial.

B26. $f(0) = 1, f(1) = 2, f'(1) = -1, f(2) = 5$

B27. $f(1) = -1, f(2) = 3, f'(2) = 0, f''(2) = 1$

B28. $f(1) = -2, f'(1) = -1, f(3) = 5, f'(3) = 0$

B29. $f(1) = 2, f'(1) = -2, f(3) = -1, f'(3) = 0, f''(3) = \frac{1}{2}, f'''(3) = -\frac{1}{4}, f(4) = 1$

B30. $f(-3) = \frac{5}{2}, f(-2) = \frac{2}{3}, f'(-2) = -1, f(0) = 5, f'(0) = -5, f(3) = \frac{1}{3}, f'(3) = \frac{4}{3}, f''(3) = 0$

40.2   Divided Differences

Basic Definitions and Notation

The nth-degree polynomial interpolating the $n+1$ data points $(x_0, y_0), \ldots, (x_n, y_n)$ can be written in the form

$$P_n(x) = f[x_0] + \sum_{k=1}^{n} f[x_0, x_1, \ldots, x_k](x - x_0) \cdots (x - x_{k-1}) \qquad \textbf{(40.4)}$$

where $f[x_k] = y_k$ and $f[x_0, x_1, \ldots, x_k]$ is the kth *divided difference* defined recursively by

$$f[x_0, x_1, \ldots, x_k] = \frac{f[x_1, x_2, \ldots, x_k] - f[x_0, x_1, \ldots, x_{k-1}]}{x_k - x_0}$$

provided $x_k, k = 0, \ldots, n$, are distinct. This form of the interpolating polynomial is called the *Newton interpolating polynomial.*

Divided differences can be computed by hand in a *divided difference table* constructed according to the pattern shown in Table 40.1.

x	y	First DDs	Second DDs	Third DDs
x_0	$f[x_0]$			
		$f[x_0, x_1] = \dfrac{f[x_1] - f[x_0]}{x_1 - x_0}$		
x_1	$f[x_1]$		$f[x_0, x_1, x_2] = \dfrac{f[x_1, x_2] - f[x_0, x_1]}{x_2 - x_0}$	
		$f[x_1, x_2] = \dfrac{f[x_2] - f[x_1]}{x_2 - x_1}$		$f[x_0, x_1, x_2, x_3] = \dfrac{f[x_1, x_2, x_3] - f[x_0, x_1, x_2]}{x_3 - x_0}$
x_2	$f[x_2]$		$f[x_1, x_2, x_3] = \dfrac{f[x_2, x_3] - f[x_1, x_2]}{x_3 - x_1}$	
		$f[x_2, x_3] = \dfrac{f[x_3] - f[x_2]}{x_3 - x_2}$		$f[x_1, x_2, x_3, x_4] = \dfrac{f[x_2, x_3, x_4] - f[x_1, x_2, x_3]}{x_4 - x_1}$
x_3	$f[x_3]$		$f[x_2, x_3, x_4] = \dfrac{f[x_3, x_4] - f[x_2, x_3]}{x_4 - x_2}$	
		$f[x_3, x_4] = \dfrac{f[x_4] - f[x_3]}{x_4 - x_3}$		$f[x_2, x_3, x_4, x_5] = \dfrac{f[x_3, x_4, x_5] - f[x_2, x_3, x_4]}{x_5 - x_2}$
x_4	$f[x_4]$		$f[x_3, x_4, x_5] = \dfrac{f[x_4, x_5] - f[x_3, x_4]}{x_5 - x_3}$	
		$f[x_4, x_5] = \dfrac{f[x_5] - f[x_4]}{x_5 - x_4}$		
x_5	$f[x_5]$			

TABLE 40.1   Divided differences $f[x_j, \ldots, x_{j+i}]$

Table 40.2 is a more compact, but perhaps less informative, representation of the information in Table 40.1. Table 40.2 more aptly reveals some of the latent patterns and introduces a simpler notation for divided differences.

x	y	**First DDs**	**Second DDs**	**Third DDs**	**Fourth DDs**	**Fifth DDs**
x_0	f_0					
		$f_{0,1}$				
x_1	f_1		$f_{0,1,2}$			
		$f_{1,2}$		$f_{0,1,2,3}$		
x_2	f_2		$f_{1,2,3}$		$f_{0,1,2,3,4}$	
		$f_{2,3}$		$f_{1,2,3,4}$		$f_{0,1,2,3,4,5}$
x_3	f_{15}		$f_{2,3,4}$		$f_{1,2,3,4,5}$	
		$f_{3,4}$		$f_{2,3,4,5}$		
x_4	f_4		$f_{3,4,5}$			
		$f_{4,5}$				
x_5	f_5					

TABLE 40.2   Divided differences expressed as $f_{j,\ldots,j+i}$

The entries in the jth column of Table 40.2 are formed according to the prescriptions in Table 40.3.

j	**Entries in Column j**
3	$f_{k-1,k} = \dfrac{f_k - f_{k-1}}{x_k - x_{k-1}}, k = 1, \ldots, n$
4	$f_{k-1,k,k+1} = \dfrac{f_{k,k+1} - f_{k-1,k}}{x_{k+1} - x_{k-1}}, k = 1, \ldots, n-1$
5	$f_{k-1,\ldots,k+2} = \dfrac{f_{k,k+1,k+2} - f_{k-1,k,k+1}}{x_{k+2} - x_{k-1}}, k = 1, \ldots, n-2$
6	$f_{k-1,\ldots,k+3} = \dfrac{f_{k,k+1,k+2,k+3} - f_{k-1,k,k+1,k+2}}{x_{k+3} - x_{k-1}}, k = 1, \ldots, n-3$
7	$f_{k-1,\ldots,k+4} = \dfrac{f_{k,k+1,k+2,k+3,k+4} - f_{k-1,k,k+1,k+2,k+3}}{x_{k+4} - x_{k-1}}, k = 1, \ldots, n-4$

TABLE 40.3   The jth column of a divided-difference table

EXAMPLE 40.5   From the divided difference table in Table 40.4, we obtain the fifth-degree interpolating polynomial

$$P_5(x) = -\frac{1993}{112,266,000}x^5 + \frac{671}{226,800}x^4 - \frac{12,127}{3,402,000}x^3 - \frac{5,156,159}{10,206,000}x^2 - \frac{4076}{14,175}x - \frac{22,811}{17,820}$$

formed according to (40.4). Thus, $P_5(x)$ is the simplification of

$$16 - 10(x+13) + \frac{41}{28}(x+13)(x+10) - \frac{229}{2772}(x+13)(x+10)(x+6)$$

$$+ \frac{14,083}{4,490,640}(x+13)(x+10)(x+6)(x-5)$$

$$- \frac{1993}{112,266,000}(x+13)(x+10)(x+6)(x-5)(x-14)$$

k	x_k	y_k	**First DDs**	**Second DDs**	**Third DDs**	**Fourth DDs**	**Fifth DDs**
0	-13	16					
			$f_{0,1}=-10$				
1	-10	-14		$f_{0,1,2}=\frac{41}{28}$			
			$f_{1,2}=\frac{1}{4}$		$f_{0,1,2,3}=-\frac{229}{2772}$		
2	-6	-13		$f_{1,2,3}=-\frac{1}{44}$		$f_{0,1,2,3,4}=\frac{14{,}083}{4{,}490{,}640}$	
			$f_{2,3}=-\frac{1}{11}$		$f_{1,2,3,4}=\frac{49}{23{,}760}$		$f_{0,1,2,3,4,5}=-\frac{1993}{112{,}266{,}000}$
3	5	-14		$f_{2,3,4}=\frac{53}{1980}$		$f_{1,2,3,4,5}=\frac{10{,}973}{4{,}158{,}000}$	
			$f_{3,4}=\frac{4}{9}$		$f_{2,3,4,5}=\frac{943}{13{,}860}$		
4	14	-10		$f_{3,4,5}=\frac{131}{90}$			
			$f_{4,5}=15$				
5	15	5					

TABLE 40.4   The divided differences for Example 40.5

the terms of which are obtained by traversing the "diagonal" from y_0 to $f_{0,1,2,3,4,5}$ in Table 40.4.   ❖

Utility of the Newton Form

The Newton form of the interpolating polynomial is useful because it allows one more point to be interpolated by simply adding a term to the existing interpolating polynomial. To see why this is so, we make several observations.

First, from (40.4) we see that $f[x_0, \ldots, x_n]$ is the coefficient of x^n in the interpolating polynomial $P_n(x)$. (This is explored empirically in the exercises.) Hence, the order of the nodes $x_0, \ldots, x_n$ is irrelevant; that is, $f[x_0, \ldots, x_n] = f[x_1, x_0, x_2, \ldots, x_n]$, etc. Finally, as is shown in [60], if $P(x)$ interpolates the nodes $x_k, \ldots, x_{k+m-1}$, then

$$P(x) + f[x_k, \ldots, x_{k+m}](x - x_k) \cdots (x - x_{k+m-1})$$

interpolates the nodes $x_k, \ldots, x_{k+m}$. If $P(x)$ interpolates the nodes $x_{k+1}, \ldots, x_{k+m}$, then

$$P(x) + f[x_k, \ldots, x_{k+m}](x - x_{k+1}) \cdots (x - x_{k+m})$$

interpolates the nodes $x_k, \ldots, x_{k+m}$. In other words, to add a node to the nodes already interpolated, add the node to the divided difference symbol and multiply by the factors corresponding to all the nodes previously interpolated. It is this flexibility of the Newton form of the interpolating polynomial, which makes it useful in the applications.

EXAMPLE 40.6   The polynomial of degree zero that interpolates the single node x_3 is y_3. The polynomial of degree one that interpolates the nodes x_3 and x_4 is

$$P_1 = y_3 + f[x_3, x_4](x - x_3) \tag{40.5}$$

The polynomial of degree two that interpolates the nodes x_3, x_4, and x_2 is

$$P_2 = P_1 + f[x_2, x_3, x_4](x - x_3)(x - x_4) \tag{40.6}$$

The polynomial of degree three that interpolates the nodes x_2, x_3, x_4, and x_1 is

$$P_3 = P_2 + f[x_1, x_2, x_3, x_4](x - x_2)(x - x_3)(x - x_4) \tag{40.7}$$

EXERCISES 40.2–Part A

A1. Obtain a divided difference table for the three points (3, 2), (7, 9), (5, 4), in that order. What is $f_{0,1,2}$?

A2. Obtain a divided difference table for the three points (7, 9), (3, 2), (5, 4), in that order. What is $f_{0,1,2}$?

A3. Use the data in Table 40.4 to obtain the polynomial given by (40.5). Confirm this polynomial actually interpolates the appropriate points.

A4. Use the data in Table 40.4 to obtain the polynomial given by (40.6). Confirm this polynomial actually interpolates the appropriate points.

A5. Use the data in Table 40.4 to obtain the polynomial given by (40.7). Confirm this polynomial actually interpolates the appropriate points.

EXERCISES 40.2–Part B

B1. If the data points in Table 40.4 are P_k, $k = 0, \ldots, 5$, obtain a divided difference table for the points P_2, P_4, P_0, P_5, P_1, P_3. The rightmost entry in the new table is $f_{2,4,0,5,1,3}$. Show this is the same as $f_{0,1,2,3,4,5}$ in Table 40.4.

For the sets of data points in Exercises B2–11:

    (a) Obtain a divided difference table.

    (b) From the divided difference table, construct the Newton interpolating polynomial (40.4) for the data.

    (c) Interpolate the given data with $\sum_{k=0}^n L_k(x) y_k$, where $L_k(x)$ is the kth Lagrange interpolating polynomial.

    (d) Show that both forms of the interpolating polynomial agree.

    (e) Graph the interpolating polynomial.

B2. $\{(-16, 14), (-13, 15), (-6, 4), (-3, -10), (-1, 10), (10, -7)\}$

B3. $\{(-19, 16), (9, 18), (10, 2), (14, 10), (16, -7)\}$

B4. $\{(-19, -10), (-15, -20), (-12, 20), (-11, 20), (2, -17), (3, 4), (5, -18)\}$

B5. $\{(-19, 9), (-7, -18), (9, -17), (13, -17)\}$

B6. $\{(-13, -7), (-1, 7), (3, -2), (6, 14), (8, 8), (10, -6)\}$

B7. $\{(-13, -8), (0, -15), (1, -12), (7, 17), (12, 12)\}$

B8. $\{(-14, -7), (-13, 6), (-5, 4), (-4, -16), (-2, -14), (0, 0), (10, -14), (18, 13)\}$

B9. $\{(-17, 3), (2, -2), (11, -18), (18, -12)\}$

B10. $\{(-18, -9), (-9, -16), (-1, 13), (4, -20), (16, 1)\}$

B11. $\{(-10, 1), (-7, -17), (3, 1), (12, 12), (15, 10), (17, -13)\}$

For the polynomials $p_n(x)$ in Exercises B12–17:

    (a) Obtain the divided difference table for the points $(k, p_n(k))$, $k = 0, \ldots, n+3$.

    (b) Which column of the divided difference table first becomes constant?

    (c) Verify that $p_n(x) = \sum_{k=0}^n p_n[x_0, x_1, \ldots, x_k] \prod_{i=0}^{k-1}(x - x_i)$.

B12. $p_2(x) = 2x^2 + 6x + 5$ **B13.** $p_3(x) = 8x^3 - 8x^2 - 7x - 9$

B14. $p_4(x) = 9x^4 + 3x^3 - 4x^2 + 6x - 5$

B15. $p_5(x) = 5x^5 - 5x^4 + 4x^3 + 2x^2 + 2x + 7$

B16. $p_6(x) = 9x^6 + 7x^5 + x^4 - x^3 - 2x^2 + 1$

B17. $p_7(x) = 6x^6 + 8x^5 + 8x^3 - 5x^2 - 5x + 3$

In Exercises B18–22, use the given function and data points to

    (a) verify that if $p_n(x)$ interpolates $f(x)$ at $(x_k, f(x_k))$, $k = 0, \ldots, n$, then $f(t) - p_n(t) = f[x_0, x_1, \ldots, x_n, t] \prod_{k=0}^n (t - x_k)$, provided t is distinct from the nodes x_k.

    (b) verify that in each case, there is a point ξ in $[x_0, x_n]$ for which $f[x_0, x_1, \ldots, x_n] = \frac{1}{n!} f^{(n)}(\xi)$. (In general, the statement is true provided $f(x)$ has n continuous derivatives in $[a, b]$ and the nodes x_k are distinct points in $[a, b]$.) Here, verify the claim by finding ξ.

B18. $f(x) = \dfrac{-5x^4 - 11x^3 + 9x^2 - 12x - 8}{-7x^3 - 3x^2 - 3x - 3}$; $x_k = \dfrac{k-1}{2}$, $k = 0, \ldots, 5 = n$; $t = \frac{3}{4}$

B19. $f(x) = \dfrac{11 + 8x^3 - 2x^2 - 4x}{-3x^4 - 6x^3 - 7x^2 - x - 1}$; $x_k = k - 1$, $k = 0, \ldots, 6 = n$; $t = 7$

B20. $f(x) = \dfrac{-7x^3 - 6x^2 - 12x - 4}{-5x^4 - 9x^3 - 3x^2 - 3x - 1}$; $x_k = k$, $k = 0, \ldots, 4 = n$; $t = c$

B21. $f(x) = \dfrac{-11x^2 + 6x + 2}{-9x^4 - 9 - 5x^2 - 9x}$; $x_k = k - 1$, $k = 0, \ldots, 3 = n$; $t = 5$

B22. $f(x) = \dfrac{7x^2 - 4x - 6}{-4x^3 - 2x^2 - 2x - 3}$; $x_k = k$, $k = 0, \ldots, 5 = n$; $t = c$

B23. Demonstrate that $f[x_0, x_1, \ldots, x_n] = (-1)^n \prod_{k=0}^n 1/x_k$ for $f(x) = \frac{1}{x}$.

40.3 Chebyshev Interpolation

Polynomial Wiggle Problem—Review

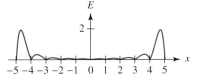

FIGURE 40.3 The magnitude of the error when $F(x) = \frac{1}{1+x^2}$ is interpolated by $F(x)$, a polynomial through $(k, f(k))$, $k = -5, \ldots, 5$, the nodes of which are equally spaced

In Section 40.1 we observed the *polynomial wiggle problem* wherein polynomial interpolation of data points from a curve that is not polynomial-like yields an interpolating function exhibiting dramatic oscillation between the interpolated points. We saw that for the function $f(x) = \frac{1}{1+x^2}$ and the 11 points $(k, f(k))$, $k = -5, \ldots, 5$, with equally spaced nodes, the interpolating polynomial $F(x)$ was given by (40.3). The magnitude of the error between $f(x)$ and $F(x)$, $E(x) = |F(x) - f(x)|$, is graphed in Figure 40.3. The maximum value of the error, $E_{max} = 1.91566$, located at $x_{max} = \pm 4.701093180$, is found by elementary calculus. The aim of this section is to show that with unequally spaced nodes, the maximum error of polynomial interpolation can be minimized. The agent of this reduced error is the set of Chebyshev polynomials, discussed next.

Chebyshev Polynomials

Table 40.5 lists $T_n(x), n = 0, \ldots, 7$, the first eight Chebyshev polynomials. In addition, the Chebyshev polynomials obey the recursive relationship

$$T_{n+1}(x) = 2xT_n(x) - T_{n-1}(x) \tag{40.8}$$

and are orthogonal, with weight $w(x) = \frac{1}{\sqrt{1-x^2}}$, on the interval $[-1, 1]$. Orthogonality of the Chebyshev polynomials is expressed by the integrals

$$\int_{-1}^{1} \frac{T_n(x)T_m(x)}{\sqrt{1-x^2}} \, dx = \begin{cases} 0 & m \neq n \\ \pi & m = n = 0 \\ \frac{\pi}{2} & m = n \neq 0 \end{cases} \tag{40.9}$$

$T_0 = 1$	$T_4 = 8x^4 - 8x^2 + 1$
$T_1 = x$	$T_5 = 16x^5 - 20x^3 + 5x$
$T_2 = 2x^2 - 1$	$T_6 = 32x^6 - 48x^4 + 18x^2 - 1$
$T_3 = 4x^3 - 3x$	$T_7 = 64x^7 - 112x^5 + 56x^3 - 7x$

TABLE 40.5 The first eight Chebyshev polynomials

Surprisingly, an alternate form of the Chebyshev polynomial is $T_n(x) = \cos(n \arccos x)$. To demonstrate this for a specific polynomial, set $\arccos x = t$ and expand the resulting $\cos nt$, obtaining, in the case $n = 7$,

$$64 \cos^7 t - 112 \cos^5 t + 56 \cos^3 t - 7 \cos t$$

Replacing each t with $\arccos x$ gives $64x^7 - 112x^5 + 56x^3 - 7x = T_7(x)$.

Chebyshev Interpolation

The following theorem from [60] describes Chebyshev interpolation.

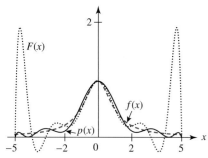

FIGURE 40.4   The Chebyshev interpolating polynomial $p(x)$ (solid), the equispaced interpolating polynomial $F(x)$ (dotted), and the function $f(x)$ (dashed)

THEOREM 40.1

1. $f^{(n)}(x)$, the nth derivative of $f(x)$, is continuous on the interval $[a, b]$.
2. $|f^{(n)}(x)| \le M_n$ on the interval $[a, b]$.
3. $\xi_1, \ldots, \xi_n$ are the n zeros of $T_n(x)$, the nth-degree Chebyshev polynomial.
4. In $[a, b]$, $x_k = a + \frac{1}{2}(b-a)(\xi_k + 1)$, $k = 1, \ldots, n$, are the images of $\xi_1, \ldots, \xi_n$ under the transformation $x = a + \frac{1}{2}(b-a)(\xi + 1)$, which takes $[-1, 1]$ onto $[a, b]$.
5. $p_{n-1}(x)$ is the polynomial of degree $n-1$ that interpolates the points $(x_1, f(x_1))$, $\ldots, (x_n, f(x_n))$.
6. c is in the interval $[a, b]$.

$$\Longrightarrow |f(c) - p_{n-1}(c)| \le \frac{M_n}{2^{2n-1}n!}(b-a)^n$$

Loosely speaking, this theorem says that to interpolate n points whose nodes lie in $[a, b]$, take the nodes as the images of the zeros of the Chebyshev polynomial $T_n(x)$ and the maximum error in the interpolating polynomial will as small as possible.

EXAMPLE 40.7   For the function $f(x) = \frac{1}{1+x^2}$, we will obtain the Chebyshev interpolating polynomial that interpolates 11 points in the interval $[-5, 5]$. Table 40.6 lists ξ_k, the 11 zeros of the Chebyshev polynomial $T_{11}(x)$, along with the points $(x_k, f(x_k))$, $k = 1, \ldots, 11$, where x_k is the image of ξ_k under the mapping $x = 5\xi$, which takes the interval $-1 \le \xi \le 1$ to $-5 \le x \le 5$.   ❖

ξ_k	x_k	$f(x_k)$
± 0.9898214419	± 4.949107210	0.0392254354
± 0.9096319954	± 4.548159977	0.0461132115
± 0.7557495744	± 3.778747872	0.0654495859
± 0.5406408175	± 2.703204088	0.1203758760
± 0.2817325568	± 1.408662784	0.3350834924
0	0	1.0

TABLE 40.6   The zeros of $T_{11}(x)$ and their images under the mapping $x = 5\xi$

The tenth-degree polynomial interpolating $(x_k, f(x_k))$, $k = 1, \ldots, 11$, is then

$$p = -0.478 \times 10^{-5}x^{10} + 0.333 \times 10^{-3}x^8 - 0.854 \times 10^{-2}x^6$$
$$+ 0.983 \times 10^{-1}x^4 - 0.499x^2 + 1$$

Figure 40.4 contains graphs of $p(x)$, the Chebyshev interpolating polynomial (solid), the equispaced interpolating polynomial (dotted), and the function $f(x)$ (dashed). Figure 40.5 shows a graph of $|p(x) - f(x)|$, the absolute value of the error made by the Chebyshev interpolating polynomial, revealing that for $x > 0$ the largest error is made somewhere in the interval $(0, 1)$. Elementary calculus gives the corresponding extreme point

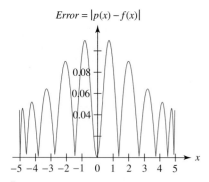

FIGURE 40.5   Graph of $|p(x) - f(x)|$, the absolute value of the error in the Chebyshev interpolating polynomial

as $(0.7757975, 0.1091535)$. The maximum error is considerably smaller than the error of nearly 2 in the equispaced interpolating polynomial.

EXERCISES 40.3

1. If $T_0(x) = 1$ and $T_1(x) = x$, use the recursion (40.8) to generate the Chebyshev polynomials $T_k(x), k = 2, \ldots, 5$.

In Exercises 2–6:

(a) Obtain $P_{n-1}(x)$, the polynomial of degree $n - 1$ that interpolates the n uniformly spaced nodes $x_k = a + (k - 1)h$, $k = 1, \ldots, n$, where $h = \frac{b-a}{n-1}$, and $[a, b]$ is the given interval.

(b) Obtain $p_{n-1}(x)$, the Chebyshev interpolating polynomial of degree $n - 1$.

(c) Determine M_n, a constant for which $|f^{(n)}(x)| \le M_n$ holds for x in $[a, b]$.

(d) Obtain $\varepsilon_1 = \max_x |f(x) - P_{n-1}(x)|$, $\varepsilon_2 = \max_x |f(x) - p_{n-1}(x)|$, and $\varepsilon_3 = M_n(b - a)^n/(2^{2n-1}n!)$.

(e) Verify that $\varepsilon_2 \le \varepsilon_3$ and that $\varepsilon_2 < \varepsilon_1$, thereby showing the error bound for the Chebyshev interpolating polynomial is valid.

2. $f(x) = \dfrac{-5x^4 - 11x^3 + 9x^2 - 12x - 8}{-7x^3 - 3x^2 - 3x - 3}$; $[-\frac{1}{2}, 2]$, $n = 6$

3. $f(x) = \dfrac{11 + 8x^3 - 2x^2 - 4x}{-3x^4 - 6x^3 - 7x^2 - x - 1}$; $[-1, 5]$, $n = 7$

4. $f(x) = \dfrac{-7x^3 - 6x^2 - 12x - 4}{-5x^4 - 9x^3 - 3x^2 - 3x - 1}$; $[0, 4]$, $n = 5$

5. $f(x) = \dfrac{-11x^2 + 6x + 2}{-9x^4 - 9 - 5x^2 - 9x}$; $[-1, 2]$, $n = 4$

6. $f(x) = \dfrac{7x^2 - 4x - 6}{-4x^3 - 2x^2 - 2x - 3}$; $[0, 5]$, $n = 6$

In Exercises 7–16:

(a) Obtain $P_{n-1}(x)$, the polynomial of degree $n - 1$ that interpolates the n uniformly spaced nodes $x_k = a + (k - 1)h$, $k = 1, \ldots, n$, where $h = \frac{b-a}{n-1}$. (Except for a renaming of the parameter n, these are the polynomials obtained in the exercises of Section 40.1.)

(b) Obtain $p_{n-1}(x)$, the Chebyshev interpolating polynomial of degree $n - 1$.

(c) Determine M_n, a constant for which $|f^{(n)}(x)| \le M_n$ holds for x in $[a, b]$.

(d) Obtain $\varepsilon_1 = \max_x |f(x) - P_{n-1}(x)|$, $\varepsilon_2 = \max_x |f(x) - p_{n-1}(x)|$, and $\varepsilon_3 = M_n(b - a)^n/(2^{2n-1}n!)$.

(e) Verify that $\varepsilon_2 \le \varepsilon_3$ and that $\varepsilon_2 < \varepsilon_1$, thereby showing the error bound for the Chebyshev interpolating polynomial is valid.

7. $f(x) = 10xe^{-x}, 0 \le x \le 7; n = 8$

8. $f(x) = \dfrac{x^2}{1 + x^2}, -3 \le x \le 3; n = 7$

9. $f(x) = x + \dfrac{9}{x}, \dfrac{1}{2} \le x \le \dfrac{15}{2}; n = 8$

10. $f(x) = \arctan x, 0 \le x \le 7; n = 8$

11. $f(x) = e^{-x}\cos x, 0 \le x \le 5; n = 6$

12. $f(x) = \sqrt{1 + x^2}, 0 \le x \le 1; n = 7$

13. $f(x) = \dfrac{x}{1 + x^2}, 0 \le x \le 7; n = 8$

14. $f(x) = \dfrac{x}{1 + x}, 0 \le x \le 7; n = 8$

15. $f(x) = \dfrac{\cosh x}{1 + x^2}, 0 \le x \le \dfrac{7}{2}; n = 8$

16. $f(x) = e^{-x}\tan x, 0 \le x \le \dfrac{7}{5}; n = 8$

17. The following construction sheds light on the meaning of "minimax." The error in polynomial interpolation is given by $f(x) - p_n(x) = f^{(n+1)}(\xi)/(n + 1)! \prod_{k=0}^{n}(x - x_k)$. In Chebyshev interpolation, the nodes are "moved" until the absolute value of the factor $\psi = \prod_{k=0}^{n}(x - x_k)$ is as small as possible. Consider the case $n = 1$ in which $\psi = (x - \alpha)(x - \beta)$ and the interval of interpolation is taken as $[-1, 1]$. Assume that $|\psi|$ is minimized when as much of the parabola ψ is above the x-axis as below. Thus, require $\psi(-1) = \psi(1)$ but $\psi(\hat{x}) = -\psi(1)$, where $\hat{x}$ locates the vertex of the parabola. These conditions determine α and β as $\pm\frac{1}{\sqrt{2}}$ and give $\psi = x^2 - \frac{1}{2} = \frac{1}{2}T_2(x)$, the Chebyshev polynomial of second degree.

(a) Verify the construction of ψ.

(b) Extend the construction just described to the case $n = 2$ in which $\psi = (x - \alpha)(x - \beta)(x - \gamma)$ is a cubic and the construction yields $\frac{1}{2^2}T_3(x)$, the Chebyshev polynomial of third degree. *Hint:* Require $\psi(-1) = -\psi(1)$, $\psi(x') = -\psi(x'')$, and $\psi(x') = \psi(1)$, where x' and x'' are the locations of the two horizontal tangents (extrema) for ψ.

18. From the representation $T_k(x) = \cos(k \arccos x)$, obtain the formula $x_n = \cos(\frac{2n+1}{2}\frac{\pi}{k})$, $n = 0, \ldots, k - 1$, for the zeros of $T_k(x)$. For $k = 2, \ldots, 8$, verify that this formula gives the correct zeros.

19. Let $x_0, \ldots, x_{k-1}$ be the k zeros of $T_k(x)$. For $k = 7$ and several values of i and j, verify

(a) $\sum_{n=0}^{k-1} T_i(x_n)T_j(x_n) = 0$ for $i \ne j$

(b) $\sum_{n=0}^{k-1} T_i(x_n)T_j(x_n) = \dfrac{k}{2}$ for $i = j \ne 0$

(c) $\sum_{n=0}^{k-1} T_0(x_n)T_0(x_n) = k$

40.4 Spline Interpolation

What Is a Spline?

Loosely speaking, a *spline* is a piecewise-smooth polynomial that interpolates the $n + 1$ points (x_k, y_k), $k = 0, \ldots, n$. More precisely, a spline of degree d interpolating $n+1$ points is a piecewise defined function consisting of n polynomial segments, one between each pair of consecutive points, each polynomial being of degree no more than d and joined at the interpolated points so as to form a function whose first $d - 1$ derivatives are all continuous.

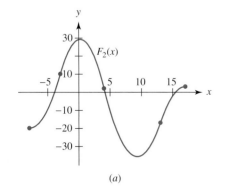

QUADRATIC SPLINE   A quadratic spline (degree $d = 2$) connecting the $n + 1 = 5$ points $(-8, -20)$, $(-3, 10)$, $(4, 2)$, $(13, -17)$, $(17, 3)$ is

$$F_2(x) = \begin{cases} \frac{1}{5}(284 + 96x + 6x^2) & -8 \leq x \leq -3 \\[4pt] \frac{1}{49}(1426 + 36x - 92x^2) & -3 \leq x \leq 4 \\[4pt] \frac{1}{567}(45{,}806 - 14{,}236x + 767x^2) & 4 \leq x \leq 13 \\[4pt] \frac{1}{252}(10{,}830x - 319x^2 - 91{,}163) & 13 \leq x \leq 17 \end{cases}$$

This spline is a piecewise function whose segments are quadratic polynomials. In Figure 40.6(a) we see a graph of the spline and the interpolated points, and in Figure 40.6(b) a graph of $F_2'(x)$, the first derivative. From the graph, we conclude that the first derivative is continuous.

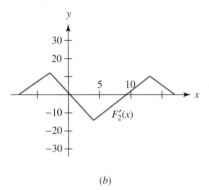

CUBIC SPLINE   A cubic spline (degree $d = 3$) joining the same $n + 1 = 5$ points is

$$F_3(x) = \begin{cases} \frac{9{,}094{,}052}{879{,}375} - \frac{9{,}106{,}354}{2{,}638{,}125}x - \frac{1{,}194{,}496}{879{,}375}x^2 - \frac{149{,}312}{2{,}638{,}125}x^3 & -8 \leq x \leq -3 \\[4pt] \frac{15{,}718{,}618}{1{,}231{,}125} - \frac{757{,}612}{738{,}675}x - \frac{676{,}646}{1{,}231{,}125}x^2 + \frac{122{,}846}{3{,}693{,}375}x^3 & -3 \leq x \leq 4 \\[4pt] \frac{59{,}221{,}586}{4{,}748{,}625} - \frac{1{,}271{,}612}{1{,}582{,}875}x - \frac{957{,}934}{1{,}582{,}875}x^2 + \frac{5141}{135{,}675}x^3 & 4 \leq x \leq 13 \\[4pt] \frac{25{,}675{,}649}{100{,}500} - \frac{120{,}049{,}619}{2{,}110{,}500}x + \frac{2{,}608{,}973}{703{,}500}x^2 - \frac{153{,}469}{2{,}110{,}500}x^3 & 13 \leq x \leq 17 \end{cases} \tag{40.10}$$

In Figure 40.7(a) we see a graph of the cubic spline $F_3(x)$, which seems to oscillate less than the quadratic spline $F_2(x)$ in Figure 40.6(a). In Figure 40.7(b) we see a graph of $F_3'(x)$, the first derivative of the cubic spline, and in Figure 40.7(c) we see a graph of $F_3''(x)$, the second derivative. From the graphs, we conclude that the first and second derivatives of $F_3(x)$ are continuous. Since $F_3''(-8) = 0 = F_3''(17)$, we have what is called a *natural* spline. Before the end of the section, this will be discussed further.

We next construct the cubic spline from first principles. To this end, start with the $n = 4$ cubic polynomials

$$g_1(x) = a_1 x^3 + b_1 x^2 + c_1 x + d_1 \qquad g_3(x) = a_3 x^3 + b_3 x^2 + c_3 x + d_3$$
$$g_2(x) = a_2 x^3 + b_2 x^2 + c_2 x + d_2 \qquad g_4(x) = a_4 x^3 + b_4 x^2 + x_4 c + d_4$$

We now have $n(d + 1) = 4 \times 4 = 16$ constants to determine. Prescribing the endpoint values for these four cubics gives the $2n = 8$ equations

$$\begin{aligned} -512a_1 + 64b_1 - 8c_1 + d_1 &= -20 & 64a_3 + 16b_3 + 4c_3 + d_3 &= 2 \\ -27a_1 + 9b_1 - 3c_1 + d_1 &= 10 & 2197a_3 + 169b_3 + 13c_3 + d_3 &= -17 \\ -27a_2 + 9b_2 - 3c_2 + d_2 &= 10 & 2197a_4 + 169b_4 + 13c_4 + d_4 &= -17 \\ 64a_2 + 16b_2 + 4c_2 + d_2 &= 2 & 4913a_4 + 289b_4 + 17c_4 + d_4 &= 3 \end{aligned} \tag{40.11}$$

FIGURE 40.6   The quadratic spline $F_2(x)$ and its first derivative

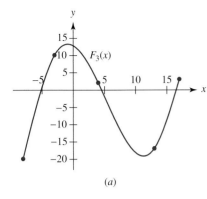

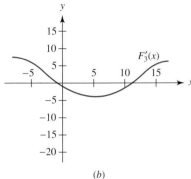

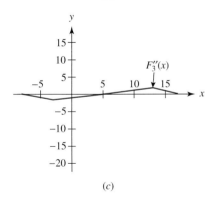

FIGURE 40.7 (*a*) the cubic spline $F_3(x)$; (*b*) its first derivative $F_3'(x)$; (*c*) the second derivative $F_3''(x)$

Continuity of the first derivative at each of the $(n+1) - 2 = 3$ interior points generates the three additional equations on the left in (40.12). Continuity of the second derivative at these same three points gives the three equations on the right in

$$27a_1 - 6b_1 + c_1 = 27a_2 - 6b_2 + c_2 \qquad -18a_1 + 2b_1 = -18a_2 + 2b_2$$
$$48a_2 + 8b_2 + c_2 = 48a_3 + 8b_3 + c_3 \qquad 24a_2 + 2b_2 = 24a_3 + 2b_3 \qquad \textbf{(40.12)}$$
$$507a_3 + 26b_3 + c_3 = 507a_4 + 26b_4 + c_4 \qquad 78a_3 + 2b_3 = 78a_4 + 2b_4$$

We have a total of $8 + 3 + 3 = 14$ equations in 16 unknowns. Using a computer algebra system, we can solve the equations (40.11) and (40.12) in terms of any two of the unknowns and write the cubic spline in terms of these two parameters. If we choose d_3 and d_4 as the arbitrary parameters, we are still at liberty to impose two additional conditions on the cubic spline.

There are four common endpoint conditions used to provide the additional two equations needed to specify the cubic spline completely. Prescribing the first derivatives at the endpoints yields the *clamped* spline. For example, if the derivatives at the endpoints are taken as zero, we have, in terms of the arbitrary parameters d_3 and d_4, the two additional equations

$$\frac{5,031,581}{194,922} - \frac{11,328}{3,398,759}d_4 - \frac{1,125,171}{799,708}d_3 = 0 \qquad \frac{18,847}{1989} - \frac{5152}{485,537}d_4 - \frac{1377}{28,561}d_3 = 0$$

Solving for d_3 and d_4, the resulting cubic spline is then

$$G_1(x) = \begin{cases} -\frac{2,443,933}{9,516,150}x^3 - \frac{1,842,913}{500,850}x^2 - \frac{6,500,744}{679,725}x + \frac{11,880,364}{1,586,025} & -8 \le x \le -3 \\[2mm] \frac{8171}{140,238}x^3 - \frac{2,248,345}{2,664,522}x^2 - \frac{1,407,530}{1,332,261}x + \frac{7,104,478}{444,087} & -3 \le x \le 4 \\[2mm] \frac{532,999}{10,277,442}x^3 - \frac{2,627,455}{3,425,814}x^2 - \frac{2,336,230}{1,712,907}x + \frac{12,045,022}{734,103} & 4 \le x \le 13 \\[2mm] -\frac{635,837}{2,030,112}x^3 + \frac{27,346,699}{2,030,112}x^2 - \frac{378,517,087}{2,030,112}x + \frac{1,661,551,985}{2,030,112} & 13 \le x \le 17 \end{cases}$$

which is graphed as the solid black curve in Figure 40.8. (In the accompanying Maple worksheet, the derivative at the left endpoint is assigned the value s and the resulting family of splines is animated to visualize how the endpoint condition influences the shape of the spline.)

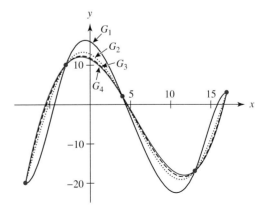

FIGURE 40.8 The four cubic splines G_1 (solid, in black), G_2 (dotted), G_3 (dashed), and G_4 (solid, in color)

At the endpoints, setting the second derivatives equal to zero gives the *free*, or *natural*, spline. This would correspond to prescribing the endpoint curvature to be zero. Doing that here gives the two additional equations

$$-\frac{671,974}{54,145} + \frac{73,089}{76,895}d_3 + \frac{2848}{1,307,215}d_4 = 0 \qquad \frac{101}{39} - \frac{1701}{57,122}d_3 - \frac{248}{28,561}d_4 = 0$$

and the cubic spline

$$G_2(x) = \begin{cases} \frac{9,094,052}{879,375} - \frac{9,106,354}{2,638,125}x - \frac{1,194,496}{879,375}x^2 - \frac{149,312}{2,638,125}x^3 & -8 \le x \le -3 \\[2mm] \frac{15,718,618}{1,231,125} - \frac{757,612}{738,675}x - \frac{676,646}{1,231,125}x^2 + \frac{122,846}{3,693,375}x^3 & -3 \le x \le 4 \\[2mm] \frac{59,221,586}{4,748,625} - \frac{1,271,612}{1,582,875}x - \frac{957,934}{1,582,875}x^2 + \frac{5141}{135,675}x^3 & 4 \le x \le 13 \\[2mm] \frac{25,675,649}{100,500} - \frac{120,049,619}{2,110,500}x + \frac{2,608,973}{703,500}x^2 - \frac{153,469}{2,110,500}x^3 & 13 \le x \le 17 \end{cases}$$ (40.13)

Comparing (40.10) with (40.13), we see $G_2(x) = F_3(x)$, so the cubic spline first examined is the natural spline with zero curvature at the endpoints. The spline $G_2(x)$ appears in Figure 40.8 as the dotted curve.

Equating the second derivatives at the first and second points will make the first cubic collapse to a quadratic. Similarly, equating the second derivatives at the next-to-last and last points will cause the last cubic to collapse to a quadratic. Thus, imposing the conditions

$$G''(x_0) = G''(x_1) \quad \text{and} \quad G''(x_n) = G''(x_{n+1})$$

gives the equations

$$-\frac{671,974}{54,145} + \frac{73,089}{76,895}d_3 + \frac{2848}{1,307,215}d_4 = \frac{33,959}{32,487} - \frac{85,023}{399,854}d_3 - \frac{1216}{3,398,759}d_4$$

$$\frac{339}{221} - \frac{729}{57,122}d_3 + \frac{704}{485,537}d_4 = \frac{101}{39} - \frac{1701}{57,122}d_3 - \frac{248}{28,561}d_4$$

and the cubic spline

$$G_3(x) = \begin{cases} -\frac{118,172}{168,147}x^2 - \frac{291,010}{168,147}x + \frac{623,996}{56,049} & -8 \le x \le -3 \\[2mm] \frac{10,334}{392,343}x^3 - \frac{548,186}{1,177,029}x^2 - \frac{1,200,016}{1,177,029}x + \frac{4,646,990}{392,343} & -3 \le x \le 4 \\[2mm] \frac{152,287}{4,539,969}x^3 - \frac{835,642}{1,513,323}x^2 - \frac{1,019,552}{1,513,323}x + \frac{51,679,010}{4,539,969} & 4 \le x \le 13 \\[2mm] \frac{127,121}{168,147}x^2 - \frac{990,965}{56,049}x + \frac{841,511}{9891} & 13 \le x \le 17 \end{cases}$$

As predicted, the first and last cubics in the spline have collapsed to quadratic functions. The spline $G_3(x)$ appears in Figure 40.8 as the dashed curve.

Finally, consider making the first and second cubics identical and the next-to-last and last cubics identical. These two conditions will generate the equations

$$g_1(x) = g_2(x) \quad \text{and} \quad g_{n-1}(x) = g_n(x)$$

In this example, these conditions become the equations

$$A(x^3 + 9x^2 + 27x + 27) = 0 \quad \text{and} \quad B(x^3 - 39x^2 + 507x - 2197) = 0$$

where $A = 6,163,516,528 - 562,406,733d_3 - 1,064,448d_4$ and $B = d_4 - d_3$. Hence, $A = B = 0$ determine d_3 and d_4, giving the cubic spline

$$G_4(x) = \begin{cases} \frac{156,757}{6,668,298}x^3 - \frac{1,435,963}{3,334,149}x^2 - \frac{6,786,827}{6,668,298}x + \frac{12,733,712}{1,111,383} & -8 \le x \le -3 \\[2mm] \frac{156,757}{6,668,298}x^3 - \frac{1,435,963}{3,334,149}x^2 - \frac{6,786,827}{6,668,298}x + \frac{12,733,712}{1,111,383} & -3 \le x \le 4 \\[2mm] \frac{210,839}{6,668,298}x^3 - \frac{1,760,455}{3,334,149}x^2 - \frac{4,190,891}{6,668,298}x + \frac{2,805,424}{256,473} & 4 \le x \le 13 \\[2mm] \frac{210,839}{6,668,298}x^3 - \frac{1,760,455}{3,334,149}x^2 - \frac{4,190,891}{6,668,298}x + \frac{2,805,424}{256,473} & 13 \le x \le 17 \end{cases}$$

The first two cubics are identical, and the last two cubics are identical. The spline $G_4(x)$ appears in Figure 40.8 as the solid curve in color.

Summary

A degree d polynomial spline that interpolates the $n + 1$ points (x_k, y_k), $k = 0, \ldots, n$, will be a piecewise function consisting of n polynomials of degree d, joined in such a way as to make a function whose first $d - 1$ derivatives are continuous. A polynomial of degree d contains $d + 1$ coefficients, so there are a total of $n(d + 1)$ parameters that must be determined to specify the spline. There are $2n$ interpolating conditions since each of the n polynomials must pass through two appropriate endpoints. Since each of the first $d - 1$ derivatives must be continuous at the $n - 1$ interior points, there are a total of $(d - 1)(n - 1)$ continuity conditions available. Thus, there are $d - 1$ conditions missing. Indeed, $(d + 1)n - \{2n + (d - 1)(n - 1)\} = d - 1$.

Numeric Algorithm for a Cubic Spline

Completely derived in [60], a formalism for numerically computing $G(x)$, a cubic spline that interpolates the $n + 1$ points (x_k, y_k), $k = 0, \ldots, n$, is stated next. The spline is described, not in terms of the $(d + 1)n = 4n$ coefficients of the individual cubics making up the spline, but rather, in terms of $\sigma_k = G''(x_k)$, $k = 0, \ldots, n$. Thus, the unknowns in the spline are the values of the *second derivative* of the spline at the nodes.

Cubic Spline Algorithm

1. Input x_k, $k = 0, \ldots, n$ and y_k, $k = 0, \ldots, n$.

2. Define $h_k = x_{k+1} - x_k$, $k = 0, \ldots, n - 1$.

3. Define $f[x_k, x_{k+1}] = (y_{k+1} - y_k)/(x_{k+1} - x_k)$, $k = 0, \ldots, n - 1$, the first divided differences as defined in Section 40.2.

4. To impose a condition on the left endpoint, select one equation from the following four types of endpoint conditions.

    (a) $2(h_0 + h_1)\sigma_1 + h_1\sigma_2 = 6(f[x_1, x_2] - f[x_0, x_1]) - h_0 G''(x_0)$

    (b) $(3h_0 + 2h_1)\sigma_1 + h_1\sigma_2 = 6(f[x_1, x_2] - f[x_0, x_1])$

    (c) $(h_0 + 2h_1)\sigma_1 + (h_1 - h_0)\sigma_2 = \dfrac{6h_1}{h_0 + h_1}(f[x_1, x_2] - f[x_0, x_1])$

    (d) $(\frac{3}{2}h_0 + 2h_1)\sigma_1 + h_1\sigma_2 = 3(2f[x_1, x_2] - 3f[x_0, x_1] + G'(x_0))$

    If (a) is chosen, input the value of $G''(x_0)$. If (d) is chosen, input the value of $G'(x_0)$.

5. At each of the $n - 3$ interior nodes x_k, $k = 2, \ldots, n - 2$, write the equation

    $$h_{k-1}\sigma_{k-1} + 2(h_{k-1} + h_k)\sigma_k + h_k\sigma_{k+1} = 6(f[x_k, x_{k+1}] - f[x_{k-1}, x_k])$$

6. To impose a condition on the right endpoint, select one equation from the following four types of right endpoint conditions.

    (a) $h_{n-2}\sigma_{n-2} + 2(h_{n-2} + h_{n-1})\sigma_{n-1} = 6(f[x_{n-1}, x_n] - f[x_{n-2}, x_{n-1}]) - h_{n-1}G''(x_n)$

    (b) $h_{n-2}\sigma_{n-2} + (2h_{n-2} + 3h_{n-1})\sigma_{n-1} = 6(f[x_{n-1}, x_n] - f[x_{n-2}, x_{n-1}])$

    (c) $(h_{n-2} - h_{n-1})\sigma_{n-2} + (2h_{n-2} + h_{n-1})\sigma_{n-1} = \dfrac{6h_{n-2}(f[x_{n-1}, x_n] - f[x_{n-2}, x_{n-1}])}{h_{n-2} + h_{n-1}}$

    (d) $h_{n-2}\sigma_{n-2} + (2h_{n-2} + \frac{3}{2}h_{n-1})\sigma_{n-1} = 3(3f[x_{n-1}, n_n] - 2f[x_{n-2}, x_{n-1}] - G'(x_n))$

    If (a) is chosen, input the value of $G''(x_n)$. If (d) is chosen, input the value of $G'(x_n)$. This is now a tridiagonal system of dimension $(n - 1) \times (n - 1)$ for σ_k, $k = 1, \ldots, n - 1$, a system that can be solved quickly with an O(n) algorithm.

7. After computing σ_k, $k = 1, \ldots, n-1$, compute σ_0 and σ_n from the following equations.

   (a) For type (a) endpoint conditions, use

   $$\sigma_0 = G''(x_0) \qquad \sigma_n = G''(x_n)$$

   (b) For type (b) endpoint conditions, use

   $$\sigma_0 = \sigma_1 \qquad \sigma_n = \sigma_{n-1}$$

   (c) For type (c) endpoint conditions, use

   $$\sigma_0 = \frac{(h_0 + h_1)\sigma_1 - h_0\sigma_2}{h_1} \qquad \sigma_n = \frac{(h_{n-2} + h_{n-1})\sigma_{n-1} - h_{n-1}\sigma_{n-2}}{h_{n-2}}$$

   (d) For type (d) endpoint conditions, use

   $$\sigma_0 = \frac{3}{h_0}(f[x_0, x_1] - G'(x_0)) - \frac{\sigma_1}{2}$$

   $$\sigma_n = \frac{3}{h_{n-1}}(G'(x_n) - f[x_{n-1}, x_n]) - \frac{\sigma_{n-1}}{2}$$

8. For $k = 0, \ldots, n-1$, compute the numbers A_k, B_k, C_k via the equations

   $$A_k = f[x_k, x_{k+1}] - \frac{h_k}{6}(\sigma_{k+1} + 2\sigma_k) \qquad B_k = \frac{\sigma_k}{2} \qquad C_k = \frac{\sigma_{k+1} - \sigma_k}{6h_k}$$

9. Define $z_k = x - x_k$ so that the individual cubics $g_k(x)$ making up the cubic spline are given by

   $$g_{k+1}(x) = y_k + z_k(A_k + z_k(B_k + z_kC_k)) \qquad x_k \le x \le x_{k+1} \qquad k = 0, \ldots, n-1$$

10. Comments:

   (a) Type (a) endpoint conditions are used when $G''(x_0)$ or $G''(x_n)$, the value of the second derivative at an endpoint, is specified. When the prescribed values are zero, these are the natural, or free, boundary conditions.

   (b) Type (b) endpoint conditions will cause $g_1(x)$ or $g_n(x)$ to collapse to quadratics.

   (c) Type (c) endpoint conditions will cause $g_1(x) = g_2(x)$ or $g_{n-1}(x) = g_n(x)$.

   (d) Type (d) endpoint conditions are used when $G'(x_0)$ or $G'(x_n)$, the value of the first derivative at an endpoint, is specified. These are the clamped boundary conditions.

EXAMPLE 40.8   Table 40.7 contains a list of $n + 1 = 9$ points (x_k, y_k), the eight differences $h_k = x_{k+1} - x_k$, and the first divided differences $f_{k,k+1} = f[x_k, x_{k+1}] = (y_{k+1} - y_k)/(x_{k+1} - x_k)$.

Table 40.8 contains the seven equations defining σ_k, $k = 1, \ldots, 7$. The first equation arises from the natural boundary condition $G''(x_0) = 0$ at the left endpoint. The next five equations arise from the interior-point equation

$$h_{k-1}\sigma_{k-1} + 2(h_{k-1} + h_k)\sigma_k + h_k\sigma_{k+1} = 6(f[x_k, x_{k+1}] - f[x_{k-1}, x_k])$$

The final equation arises from the natural boundary condition $G''(x_n) = 0$ at the right endpoint. Also included in Table 40.8 are the values of σ_k, $k = 1, \ldots, 7$, determined by these equations. The values of σ_0 and σ_8 were determined to be zero when we imposed the natural endpoint conditions.

Table 40.9 lists the values of the constants A_k, B_k, C_k, $k = 0, \ldots, n - 1 = 7$. The component cubics $g_k(x)$, $k = 1, \ldots, n = 8$, are obtained from $z_k = x - x_k$, and

$$g_{k+1}(x) = y_k + z_k(A_k + z_k(B_k + z_kC_k)) \qquad x_k \le x \le x_{k+1} \qquad k = 0, \ldots, n-1$$

k	(x_k, y_k)	$h_k = x_{k+1} - x_k$	$f_{k,k+1} = \dfrac{y_{k+1} - y_k}{h_k}$
0	$(-20, 3)$	$h_0 = 3$	$f_{0,1} = 0$
1	$(-17, 3)$	$h_1 = 9$	$f_{1,2} = -\frac{2}{3}$
2	$(-8, -3)$	$h_2 = 5$	$f_{2,3} = -\frac{13}{5}$
3	$(-3, -16)$	$h_3 = 5$	$f_{3,4} = \frac{11}{5}$
4	$(2, -5)$	$h_4 = 2$	$f_{4,5} = \frac{7}{2}$
5	$(4, 2)$	$h_5 = 6$	$f_{5,6} = -\frac{19}{6}$
6	$(10, -17)$	$h_6 = 3$	$f_{6,7} = \frac{22}{3}$
7	$(13, 5)$	$h_7 = 4$	$f_{7,8} = -\frac{9}{4}$
8	$(17, -4)$		

TABLE 40.7　Data points, h_k, and first divided-differences in Example 40.8

k	A_k	B_k	C_k
0	$-\dfrac{2,511,637}{52,294,920}$	0	$\dfrac{2,511,637}{470,654,280}$
1	$\dfrac{2,511,637}{26,147,460}$	$\dfrac{2,511,637}{52,294,920}$	$-\dfrac{20,830,429}{1,411,962,840}$
2	$-\dfrac{137,241,121}{52,294,920}$	$-\dfrac{763,283}{2,178,955}$	$\dfrac{92,868,289}{1,307,373,000}$
3	$-\dfrac{20,912,087}{26,147,460}$	$\dfrac{62,336,969}{87,158,200}$	$-\dfrac{30,137,909}{1,307,373,000}$
4	$\dfrac{241,783,913}{52,294,920}$	$\dfrac{1,609,953}{4,357,910}$	$\dfrac{19,478,113}{41,835,936}$
5	$\dfrac{13,444,981}{26,147,460}$	$-\dfrac{84,510,941}{34,863,280}$	$\dfrac{63,123,103}{209,179,680}$
6	$\dfrac{42,003,361}{10,458,984}$	$\dfrac{6,553,648}{2,178,955}$	$-\dfrac{33,153,709}{52,294,920}$
7	$\dfrac{129,295,987}{26,147,460}$	$-\dfrac{47,031,943}{17,431,640}$	$\dfrac{47,031,943}{209,179,680}$

The equations defining σ_k and their solution:

$$24\sigma_1 + 9\sigma_2 = -4 \qquad \sigma_1 = \frac{2,511,637}{26,147,460}$$

$$9\sigma_1 + 28\sigma_2 + 5\sigma_3 = -\frac{58}{5} \qquad \sigma_2 = -\frac{1,526,566}{2,178,955}$$

$$5\sigma_2 + 20\sigma_3 + 5\sigma_4 = \frac{144}{5} \qquad \sigma_3 = \frac{62,336,969}{43,579,100}$$

$$5\sigma_3 + 14\sigma_4 + 2\sigma_5 = \frac{39}{5} \qquad \sigma_4 = \frac{1,609,953}{2,178,955}$$

$$2\sigma_4 + 16\sigma_5 + 6\sigma_6 = -40 \qquad \sigma_5 = -\frac{84,510,941}{17,431,640}$$

$$6\sigma_5 + 18\sigma_6 + 3\sigma_7 = 63 \qquad \sigma_6 = \frac{13,107,296}{2,178,955}$$

$$3\sigma_6 + 14\sigma_7 = -\frac{115}{2} \qquad \sigma_7 = -\frac{47,031,943}{8,715,820}$$

TABLE 40.8　Example 40.8: The equations defining σ_k, $k = 1, \dots, 7$, and their solution

TABLE 40.9　The values of A_k, B_k, C_k, $k = 0, \dots, 7$, computed in Example 40.8

which appear in $G(x)$, given by

$$G(x) = \begin{cases} \frac{2,511,637}{470,654,280}x^3 + \frac{2,511,637}{7,844,238}x^2 + \frac{997,119,889}{156,884,760}x + \frac{1,052,648,209}{23,532,714} & -20 \le x \le -17 \\[4pt] -\frac{20,830,429}{1,411,962,840}x^3 - \frac{8,287,814}{11,766,357}x^2 - \frac{5,206,223,593}{470,654,280}x - \frac{1,905,000,572}{35,299,071} & -17 \le x \le -8 \\[4pt] \frac{92,868,289}{1,307,373,000}x^3 + \frac{73,786,214}{54,473,875}x^2 + \frac{7,072,166,663}{1,307,373,000}x - \frac{547,160,268}{54,473,875} & -8 \le x \le -3 \\[4pt] -\frac{30,137,909}{1,307,373,000}x^3 + \frac{110,635,559}{217,895,500}x^2 + \frac{3,750,999,317}{1,307,373,000}x - \frac{2,742,168,963}{217,895,500} & -3 \le x \le 2 \\[4pt] -\frac{19,478,113}{41,835,936}x^3 + \frac{110,270,189}{34,863,280}x^2 - \frac{63,832,763}{26,147,460}x - \frac{19,707,648}{2,178,955} & 2 \le x \le 4 \\[4pt] \frac{63,123,103}{209,179,680}x^3 - \frac{210,757,147}{34,863,280}x^2 + \frac{179,849,849}{5,229,492}x - \frac{25,343,352}{435,791} & 4 \le x \le 10 \\[4pt] -\frac{33,153,709}{52,294,920}x^3 + \frac{191,983,137}{8,715,820}x^2 - \frac{2,576,369,387}{10,458,984}x + \frac{4,589,328,251}{5,229,492} & 10 \le x \le 13 \\[4pt] \frac{47,031,943}{209,179,680}x^3 - \frac{799,543,031}{69,726,560}x^2 + \frac{39,553,529,213}{209,179,680}x - \frac{70,370,281,141}{69,726,560} & 13 \le x \le 17 \end{cases}$$

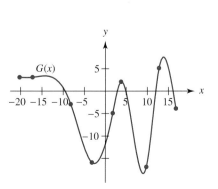

FIGURE 40.9　Graph of the cubic spline $G(x)$ computed in Example 40.8

A graph of the cubic spline and the data points is given in Figure 40.9

EXERCISES 40.4

1. Given the data points $\{(1, 5), (2, -3), (3, 7), (4, 2)\}$:

  (a) Graph the points (x_k, y_k), $k = 0, \ldots, n = 3$, and sketch a linear spline that interpolates them.

  (b) Obtain $F_0(x)$, the piecewise function that is an analytic representation of the linear spline sketched in part (a).

  (c) Obtain and graph $F_1(x)$, the natural quadratic spline for which $F_1'(x_0) = 0$.

  (d) Obtain and graph $F_2(x)$, the natural quadratic spline for which $F_2'(x_n) = 0$.

  (e) Obtain and graph $F_3(x)$, the clamped quadratic spline for which $F_3'(x_0) = 1$.

  (f) Obtain and graph $F_4(x)$, the clamped quadratic spline for which $F_4'(x_n) = 1$.

  (g) Obtain and graph $F_5(x)$, the quadratic spline for which $F_5'(x_0) = F_5'(x_1)$, noting that this condition causes $g_1(x)$, the first polynomial segment in the spline, to be linear.

  (h) Obtain and graph $F_6(x)$, the quadratic spline for which $F_6'(x_{n-1}) = F_6'(x_n)$, noting that this condition causes $g_n(x)$, the last polynomial segment in the spline, to be linear.

  (i) Obtain and graph $F_7(x)$, the quadratic spline for which the first and second segments are the same quadratic, that is, for which $g_1(x) = g_2(x)$.

  (j) Obtain and graph $F_8(x)$, the quadratic spline for which the next-to-last and last segments are the same quadratic, that is, for which $g_{n-1}(x) = g_n(x)$.

  (k) Obtain and graph $P_n(x)$, the nth-degree Newton (or Lagrange) polynomial that interpolates the given data.

In Exercises 2 and 3:

  (a) Graph the points (x_k, y_k), $k = 0, \ldots, n = 5$, and sketch a linear spline that interpolates them.

  (b) Obtain $F_0(x)$, the piecewise function that is an analytic representation of the linear spline sketched in part (a).

  (c) Obtain and graph $F_1(x)$, the natural quadratic spline for which $F_1'(x_0) = 0$.

  (d) Obtain and graph $F_2(x)$, the natural quadratic spline for which $F_2'(x_n) = 0$.

  (e) Obtain and graph $F_3(x)$, the clamped quadratic spline for which $F_3'(x_0) = 1$.

  (f) Obtain and graph $F_4(x)$, the clamped quadratic spline for which $F_4'(x_n) = 1$.

  (g) Obtain and graph $F_5(x)$, the quadratic spline for which $F_5'(x_0) = F_5'(x_1)$, noting that this condition causes $g_1(x)$, the first polynomial segment in the spline, to be linear.

  (h) Obtain and graph $F_6(x)$, the quadratic spline for which $F_6'(x_{n-1}) = F_6'(x_n)$, noting that this condition causes $g_n(x)$, the last polynomial segment in the spline, to be linear.

  (i) Obtain and graph $F_7(x)$, the quadratic spline for which the first and second segments are the same quadratic, that is, for which $g_1(x) = g_2(x)$.

  (j) Obtain and graph $F_8(x)$, the quadratic spline for which the next-to-last and last segments are the same quadratic, that is, for which $g_{n-1}(x) = g_n(x)$.

  (k) Obtain and graph $P_n(x)$, the nth-degree Newton (or Lagrange) polynomial that interpolates the given data.

2. $\{(-15, -5), (-9, 1), (-6, -17), (3, 18), (7, 17), (15, 8)\}$

3. $\{(-20, 8), (-16, -9), (-2, -5), (2, -3), (11, -16), (18, -14)\}$

In Exercises 4–6:

  (a) Obtain and graph $G_1(x)$, the natural cubic spline for which $G_1''(x_0) = G_1''(x_n) = 0$.

  (b) Obtain and graph $G_2(x)$, the cubic spline for which $G_2''(x_0) = 1$ and $G_2''(x_n) = -1$.

  (c) Obtain and graph $G_3(x)$, the cubic spline for which $G_3''(x_0) = G_3''(x_1)$ and $G_3''(x_{n-1}) = G_3''(x_n)$, noting that these conditions cause $g_1(x)$, the first segment polynomial in the spline, and $g_n(x)$, the last segment polynomial in the spline, to be quadratic.

  (d) Obtain and graph $G_4(x)$, the cubic spline for which $g_1(x) = g_2(x)$ and $g_{n-1}(x) = g_n(x)$, that is, for which the first and second cubics and next-to-last and last cubics are the same.

  (e) Obtain and graph $G_5(x)$, the clamped cubic spline for which $G_5'(x_0) = 0$ and $G_5'(x_n) = 1$.

  (f) Obtain and graph $G_6(x)$, the cubic spline for which $G_6'(x_0) = G_6''(x_0) = 0$.

  (g) Obtain and graph $G_7(x)$, the cubic spline for which $G_7''(x_0) = G_7''(x_1)$ and $g_{n-1}(x) = g_n(x)$.

4. $\{(-19, 2), (-8, 15), (-6, 5), (-5, 5), (-4, 7), (18, -9)\}$

5. $\{(-20, 4), (-18, -14), (-9, -16), (8, 15), (18, -4), (20, 15)\}$

6. $\{(-10, -2), (1, 16), (5, -10), (16, 17), (17, 9), (17, 19)\}$

For each function $f(x)$ and set of nodes x_k, $k = 0, \ldots, n$, prescribed in Exercises 7–10:

  (a) Obtain $G(x)$, the natural cubic spline $(G''(x_0) = G''(x_n) = 0)$ interpolating the points $(x_k, f(x_k))$, $k = 0, \ldots, n$.

  (b) Show that $\int_{x_0}^{x_n} [G''(x)]^2 \, dx \leq \int_{x_0}^{x_n} [f''(x)]^2 \, dx$.

7. Exercise B2, Section 40.1   **8.** Exercise B4, Section 40.1

9. Exercise B7, Section 40.1   **10.** Exercise B9, Section 40.1

In Exercises 11 and 12, use the following prescription to obtain the cubic *tension spline* $G_\tau(x)$, in which the parameter τ simulates "tension" in the spline. In each case, graph $G_\tau(x)$ for several values of τ in the range $\frac{1}{10} \le \tau \le 10$. (Try $\tau = 0.1, 1, 10$.) In fact, as $\tau \to 0$, $G_\tau(x)$ becomes a cubic spline, and as $\tau \to \infty$, $G_\tau(x)$ becomes a linear spline.

(a) For $k = 0, \ldots, n-1$, obtain $h_k = x_{k+1} - x_k$, $\alpha_k = 1/h_k - \tau/\sinh \tau h_k$, $\beta_k = \tau \coth \tau h_k - 1/h_k$, $\gamma_k = \tau^2(y_{k+1} - y_k)h_k$.

(b) Set $z_0 = z_n = 0$, and solve the equations $\alpha_{k-1}z_{k-1} + (\beta_{k-1} + \beta_k)z_k + \alpha_k z_{k+1} = \gamma_k - \gamma_{k-1}$, $k = 1, \ldots, n-1$, for z_k, $k = 1, \ldots, n-1$.

(c) On $[x_k, x_{k+1}]$, $G_\tau(x)$ is given by

$$G_\tau(x) = \frac{z_k \sinh \tau(x_{k+1} - x) + z_{k+1} \sinh \tau(x - x_k)}{\tau^2 \sinh \tau h_k}$$
$$+ \frac{1}{h_k}\left[\left(y_k - \frac{z_k}{\tau^2}\right)(x_{k+1} - x)\right.$$
$$+ \left.\left(y_{k+1} - \frac{z_{k+1}}{\tau^2}\right)(x - x_k)\right]$$

11. $\{(-16, -8), (-6, -20), (-2, 10), (5, -5), (8, 1), (20, -5)\}$

12. $\{(-18, 1), (-13, -14), (-6, 6), (-5, 12), (9, -18), (13, -4)\}$

In Exercises 13–15, obtain and graph $G(x)$, the cubic spline for which $G'''(x)$ is continuous at x_1 and x_{n-1}. This pair of continuity conditions is to hold in lieu of any of the options given in step (6) of the cubic spline numerical algorithm.

13. $\{(-19, 12), (-9, 10), (-8, 9), (-6, -15), (1, 9), (17, 10)\}$

14. $\{(-20, 6), (-17, 9), (-12, 11), (-11, -3), (-7, -1), (-2, 1)\}$

15. $\{(-15, 7), (-8, -19), (8, -15), (14, 12), (15, -4), (19, 16)\}$

In Exercises 16–20, the curve interpolating the given points (x_k, y_k), $k = 1, \ldots, n$, loops back on itself and, hence, must be represented parametrically. Let $t_k = k$; form the two sets of points (t_k, x_k), (t_k, y_k), $k = 1, \ldots, n$; and obtain two natural splines $G_x(t)$ and $G_y(t)$ interpolating the respective sets of points. Graph the resulting parametrically given curve $x = x(t)$ and $y = y(t)$.

16. $\{(-7, 20), (9, -7), (-18, -8), (20, -15), (8, 19), (12, 8)\}$

17. $\{(15, -4), (14, 16), (7, 5), (-19, 0), (-15, -20), (12, 20)\}$

18. $\{(4, 1), (-6, -15), (-2, -16), (-18, -5), (-15, 8), (-2, 2)\}$

19. $\{(-20, -13), (1, -20), (2, 15), (8, -7), (-11, 3), (-20, 6)\}$

20. $\{(9, -3), (-16, 11), (-4, -1), (-9, -3), (1, -10), (16, 9), (-12, 7)\}$

40.5 **Bezier Curves**

DEFINITION 40.1

Given the $n+1$ points $p_k = (x_k, y_k)$, $k = 0, \ldots, n$, and the corresponding vectors $\mathbf{P}_k = x_k\mathbf{i} + y_k\mathbf{j}$, the associated nth-degree *Bezier curve* is given by

$$\mathbf{R}(u) = \sum_{k=0}^{n} \binom{n}{k}(1-u)^{n-k}u^k\mathbf{P}_k \qquad (40.14)$$

where $\binom{n}{k} = \frac{n!}{k!(n-k)!}$ is the binomial coefficient "n choose k" and $0 \le u \le 1$.

EXAMPLE 40.9   The cubic Bezier curve associated with $n + 1 = 4$ points has the general form

$$\mathbf{R}(u) = 1(1-u)^3 u^0\mathbf{P}_0 + 3(1-u)^2 u^1\mathbf{P}_1 + 3(1-u)^1 u^2\mathbf{P}_2 + 1(1-u)^0 u^3\mathbf{P}_3$$

where $\binom{3}{0} = \binom{3}{3} = 1$, $\binom{3}{1} = \binom{3}{2} = 3$. The four points $(0, 0)$, $(-1, 2)$, $(7, 3)$, $(5, 0)$ determine the cubic Bezier curve

$$\mathbf{R}(u) = \begin{bmatrix} -3u + 27u^2 - 19u^3 \\ 6u - 3u^2 - 3u^3 \end{bmatrix}$$

which is graphed as the solid curve in Figure 40.10. The two intermediate points p_1 and p_2 are called *control points* and influence the shape of the Bezier curve. (The accompanying Maple worksheet contains an animation showing how the Bezier curve responds as these two control points move horizontally along two parallel lines.)

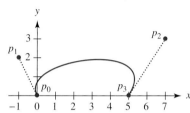

FIGURE 40.10   Example 40.9: The solid curve is the cubic Bezier curve, while the dotted lines are tangent at the endpoints

The dotted line segments connecting p_0 with p_1 and p_2 with p_3 appear to be tangent to the Bezier curve at the endpoints of the curve. Indeed, a computation shows $\mathbf{R}'(0) = 3(\mathbf{P}_1 - \mathbf{P}_0)$ and $\mathbf{R}'(1) = 3(\mathbf{P}_3 - \mathbf{P}_2)$, so lines connecting the first and second points and the third and fourth points must necessarily be tangent to the Bezier curve. In effect, the curve appears to be a flexible rod with "handles" attached at the ends. As the "handles" are manipulated, the rod flexes and assumes the shape of the Bezier curve. ❖

Constructing a Cubic Bezier Curve

The four points $(1, 5)$, $(4, 7)$, $(5, 12)$, $(9, 4)$ determine the cubic Bezier curve

$$\mathbf{R}(u) = \begin{bmatrix} 1 + 9u - 6u^2 + 5u^3 \\ 5 + 6u + 9u^2 - 16u^3 \end{bmatrix} \tag{40.15}$$

which, along with the four points, is graphed in Figure 40.11. To understand how the points p_1 and p_2 influence the shape of the Bezier curve, draw line segments connecting consecutive pairs of points p_k, p_{k+1}, $k = 0, 1, 2$. Then, using vector notation and for $k = 0, 1, 2$, let $\mathbf{Q}_{k+1}(u)$ describe the segment connecting p_k with p_{k+1}. In addition, for each fixed value of u in the interval $[0, 1]$, $\mathbf{Q}_k(u)$, $k = 1, 2, 3$, is a single point on the respective line segment. We can therefore connect the points $\mathbf{Q}_1(u)$ and $\mathbf{Q}_2(u)$ with the line segment $\mathbf{Q}_4(u)$ and the points $\mathbf{Q}_2(u)$ and $\mathbf{Q}_3(u)$ with the line segment $\mathbf{Q}_5(u)$. Finally, for each fixed value u in the interval $[0, 1]$, $\mathbf{Q}_4(u)$ and $\mathbf{Q}_5(u)$ are points that can be connected with the line segment $\mathbf{Q}_6(u)$.

As u varies from 0 to 1, the corresponding point on the line segment $\mathbf{Q}_6(u)$ traces the Bezier cubic, show in cyan, in Figure 40.12. The line segments $\mathbf{Q}_4(u)$ and $\mathbf{Q}_5(u)$ are dashed, and the line segment $\mathbf{Q}_6(u)$ is dotted. Analytic representations of the line segments $\mathbf{Q}_k(u) = \mathbf{P}_{k-1} + u(\mathbf{P}_k - \mathbf{P}_{k-1})$, $k = 1, 2, 3$, are

$$\mathbf{Q}_1 = \begin{bmatrix} 1 + 3u \\ 5 + 2u \end{bmatrix} \qquad \mathbf{Q}_2 = \begin{bmatrix} 4 + u \\ 7 + 5u \end{bmatrix} \qquad \mathbf{Q}_3 = \begin{bmatrix} 5 + 4u \\ 12 - 8u \end{bmatrix}$$

where, in each case, $0 \leq u \leq 1$. Analytic representations of the line segments $\mathbf{Q}_k(u) = \mathbf{Q}_{k-3} + u(\mathbf{Q}_{k-2} - \mathbf{Q}_{k-3})$, $k = 4, 5$, are

$$\mathbf{Q}_4 = \begin{bmatrix} 1 + 6u - 2u^2 \\ 5 + 4u + 3u^2 \end{bmatrix} \qquad \mathbf{Q}_5 = \begin{bmatrix} 4 + 2u + 3u^2 \\ 7 + 10u - 13u^2 \end{bmatrix}$$

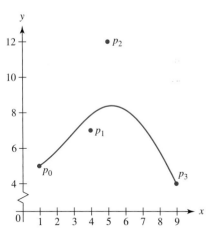

FIGURE 40.11 The Bezier curve determined by $(1, 5)$, $(4, 7)$, $(5, 12)$, and $(9, 4)$

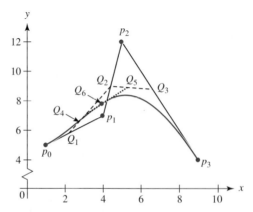

FIGURE 40.12 The cubic Bezier curve (in color) is the locus of point Q_6

Finally, the analytic representation of the line segment $\mathbf{Q}_6(u) = \mathbf{Q}_4 + u(\mathbf{Q}_5 - \mathbf{Q}_4)$ is

$$\mathbf{Q}_6 = \begin{bmatrix} 1 + 9u - 6u^2 + 5u^3 \\ 5 + 6u + 9u^2 - 16u^3 \end{bmatrix}$$

For each fixed value of u in the interval $[0, 1]$ there is a single point on the line segment $\mathbf{Q}_6(u)$. That point is a point on the Bezier cubic, as we can see by comparing $\mathbf{Q}_6(u)$ to the vector $\mathbf{R}$ (in (40.15)), which gave the Bezier curve by its definition.

Consequently, we can now imagine how the Bezier curve might have been discovered. Idly drawing four points on a scrap of paper, a doodler could have connected with solid lines, the consecutive pairs of points, then connected with dashed lines, the midpoints of the solid line segments, and connected with dotted lines, the midpoints of the dashed lines. If, say, the quarter points were similarly connected, and the one-third points likewise connected, etc., it is not hard to believe that the inspired doodler would have seen that the Bezier curve is the envelop of all the dotted lines. This speculation is animated and verified in the accompanying Maple worksheet.

The Cubic Bezier Spline

From Section 40.4, the natural cubic spline interpolating the $n + 1 = 4$ points $(0, 0)$, $(4, 7)$, $(7, 2)$, $(10, 5)$ is

$$Y = \begin{cases} \frac{1897}{636}x - \frac{49}{636}x^3 & 0 \le x \le 4 \\ -\frac{25{,}600}{1431} + \frac{31{,}291}{1908}x - \frac{1600}{477}x^2 + \frac{1159}{5724}x^3 & 4 \le x \le 7 \\ \frac{245{,}495}{2862} - \frac{53{,}479}{1908}x + \frac{2855}{954}x^2 - \frac{571}{5724}x^3 & 7 \le x \le 10 \end{cases}$$

with the graph shown in Figure 40.13. Both from this graph and the theory developed in Section 40.4, we know this spline has a continuous second derivative. However, if we wish to change the shape near one of the four interpolated points, we would have to recompute the complete spline. A small local change in the shape of the curve requires the recalculation of all the coefficients of all the component polynomials making up the spline.

Alternatively, we can interpolate the given points with three Bezier cubics by introducing control points such as $p_1 = (2, 2)$, $p_2 = (1, 6)$ between p_0 and p_3; introducing control points such as $p_4 = (5, 1)$, $p_5 = (8, -1)$ between p_3 and p_6; and introducing control points such as $p_7 = (8, 5)$, $p_8 = (12, -3)$ between p_6 and p_9. Then, three separate cubic Bezier curves can be constructed between p_0 and p_3; p_3 and p_6; and p_6 and p_9. In fact, these three Bezier curves are given by

$$\mathbf{R}_1 = \begin{bmatrix} 6u - 9u^2 + 7u^3 \\ 6u + 6u^2 - 5u^3 \end{bmatrix} \quad \mathbf{R}_2 = \begin{bmatrix} 4 + 3u + 6u^2 - 6u^3 \\ 7 - 18u + 12u^2 + u^3 \end{bmatrix} \quad \mathbf{R}_3 = \begin{bmatrix} 7 + 3u + 9u^2 - 9u^3 \\ 2 + 9u - 33u^2 + 27u^3 \end{bmatrix}$$

Each Bezier segment is independent from the others. This is clear even from an attempt to graph them. The parameter u ranges between 0 and 1 for each of the curves. The location of the graphs of these three segments is "built-into" the representation of each segment. Figure 40.14 shows these three Bezier curves, along with the control points, and the "handles" at the ends of the Bezier segments. The middle segment is dotted.

We have lost continuity of the derivative. The three Bezier curves form a continuous spline, but not necessarily a differentiable one. We have lost some degree of smoothness but have gained "local control." Thus, if control point p_2 is changed, only the first segment would be affected. The remaining portion of the interpolating spline would not have to be recalculated. We can see this analytically if we change p_2 to $(1, t)$ in an attempt to vary p_2 through a range of values. The influence of p_2 is felt only by $\mathbf{R}_1$, which, in terms of t,

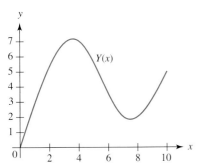

FIGURE 40.13 The natural cubic spline $Y(x)$ interpolating $(0, 0)$, $(4, 7)$, $(7, 2)$, and $(10, 5)$

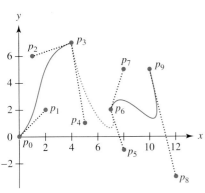

FIGURE 40.14 Cubic Bezier spline fitting the points interpolated in Figure 40.13

becomes

$$\mathbf{R}_1(u, t) = \begin{bmatrix} 6u - 9u^2 + 7u^3 \\ 6u - 12u^2 + 13u^3 + 3u^2t - 3u^3t \end{bmatrix}$$

Clearly, the Bezier spline allows local control of the shape of the interpolating function. Differentiability can be regained by choosing control points for which the slopes of contiguous "handles" are equal. (The accompanying Maple worksheet contains an animation showing the Bezier spline as t varies. It shows, as it should, that only $\mathbf{R}_1$ changes, and not $\mathbf{R}_2$ or $\mathbf{R}_3$.)

EXERCISES 40.5

1. Use (40.14) to obtain the quadratic Bezier curve determined by the points $p_0 = (1, 3)$, $p_1 = (5, 4)$, $p_2 = (7, -2)$. Sketch the curve. Show that the line segments connecting p_0 with p_1 and p_2 with p_1 are tangent to the Bezier curve at its endpoints.

2. Use (40.14) to obtain the cubic Bezier curve determined by the points $p_0 = (1, 3)$, $p_1 = (5, 4)$, $p_2 = (7, -2)$, $p_3 = (8, 8)$. Sketch the curve. Show that the line segments connecting p_0 with p_1 and p_2 with p_3 are tangent to the Bezier curve at its endpoints.

For the points p_k and the associated vectors $\mathbf{P}_k$, $k = 0, \ldots, 3$, in each of Exercises 3–7:

(a) Use (40.14) with $n = 3$ to obtain $\mathbf{R}(u) = [x(u), y(u)]^\mathrm{T}$, $0 \le u \le 1$, the cubic Bezier curve determined by the points.

(b) Show that the tangent vector $\mathbf{R}'(0)$ is proportional to the vector from $\mathbf{P}_0$ to $\mathbf{P}_1$.

(c) Show that the tangent vector $\mathbf{R}'(1)$ is proportional to the vector from $\mathbf{P}_3$ to $\mathbf{P}_2$.

(d) Carry out the construction corresponding to Figure 40.12 by obtaining the line segments $\mathbf{Q}_k(u) = \mathbf{P}_{k-1} + u(\mathbf{P}_k - \mathbf{P}_{k-1})$, $k = 1, 2, 3$; $\mathbf{Q}_k(u) = \mathbf{Q}_{k-3} + u(\mathbf{Q}_{k-2} - \mathbf{Q}_{k-3})$, $k = 4, 5$, and $\mathbf{Q}_6(u) = \mathbf{Q}_4 + u(\mathbf{Q}_5 - \mathbf{Q}_4)$. Show that $\mathbf{Q}_6(u) = \mathbf{R}(u)$.

(e) Graph the Bezier curve.

3. $(-16, 2)$, $(8, -9)$, $(11, -2)$, $(18, -20)$

4. $(-5, -3)$, $(-16, -14)$, $(-2, -16)$, $(20, 20)$

5. $(-6, 20)$, $(-8, -20)$, $(10, -5)$, $(1, -5)$

6. $(8, 3)$, $(-5, -6)$, $(-19, -4)$, $(2, 15)$

7. $(17, 1)$, $(16, 5)$, $(-10, 17)$, $(-2, 16)$

For the points p_k and the associated vectors $\mathbf{P}_k$, $k = 0, \ldots, 2$, in each of Exercises 8–12:

(a) Use 40.14 with $n = 2$ to obtain $\mathbf{R}(u) = [x(u), y(u)]^\mathrm{T}$, $0 \le u \le 1$, the quadratic Bezier curve determined by the points.

(b) Show that the tangent vector $\mathbf{R}'(0)$ is proportional to the vector from $\mathbf{P}_0$ to $\mathbf{P}_1$.

(c) Show that the tangent vector $\mathbf{R}'(1)$ is proportional to the vector from $\mathbf{P}_2$ to $\mathbf{P}_1$.

(d) Carry out the construction corresponding to Figure 40.12 by obtaining the line segments $\mathbf{Q}_k(u) = \mathbf{P}_{k-1} + u(\mathbf{P}_k - \mathbf{P}_{k-1})$, $k = 1, 2$, and $\mathbf{Q}_3(u) = \mathbf{Q}_1 + u(\mathbf{Q}_2 - \mathbf{Q}_1)$. Show that $\mathbf{Q}_3(u) = \mathbf{R}(u)$.

(e) Graph the Bezier curve.

8. $(-2, -18)$, $(1, -14)$, $(6, 12)$ **9.** $(-18, -4)$, $(-9, -6)$, $(-19, 17)$

10. $(-8, -6)$, $(12, 10)$, $(9, -15)$

11. $(9, 10)$, $(-11, -7)$, $(-12, -20)$

12. $(-2, -17)$, $(6, 9)$, $(11, -3)$

For the points p_k and the associated vectors $\mathbf{P}_k$, $k = 0, \ldots, 4$, in each of Exercises 13–17:

(a) Use (40.14) with $n = 4$ to obtain $\mathbf{R}(u) = [x(u), y(u)]^\mathrm{T}$, $0 \le u \le 1$, the quartic Bezier curve determined by the points.

(b) Show that the tangent vector $\mathbf{R}'(0)$ is proportional to the vector from $\mathbf{P}_0$ to $\mathbf{P}_1$.

(c) Show that the tangent vector $\mathbf{R}'(1)$ is proportional to the vector from $\mathbf{P}_4$ to $\mathbf{P}_3$.

(d) Carry out the construction corresponding to Figure 40.12 by obtaining the line segments $\mathbf{Q}_k(u) = \mathbf{P}_{k-1} + u(\mathbf{P}_k - \mathbf{P}_{k-1})$, $k = 1, \ldots, 4$; $\mathbf{Q}_k(u) = \mathbf{Q}_{k-4} + u(\mathbf{Q}_{k-3} - \mathbf{Q}_{k-4})$, $k = 5, 6, 7$; $\mathbf{Q}_k(u) = \mathbf{Q}_{k-3} + u(\mathbf{Q}_{k-2} - \mathbf{Q}_{k-3})$, $k = 8, 9$; and $\mathbf{Q}_{10}(u) = \mathbf{Q}_8 + u(\mathbf{Q}_9 - \mathbf{Q}_8)$. Show that $\mathbf{Q}_{10}(u) = \mathbf{R}(u)$.

(e) Graph the Bezier curve.

13. $(-18, 20)$, $(-8, -15)$, $(-7, 9)$, $(8, 12)$, $(20, -7)$

14. $(19, 8)$, $(15, 14)$, $(7, -19)$, $(-15, 12)$, $(-4, 16)$

15. $(5, 0)$, $(-20, 20)$, $(4, -6)$, $(-2, -18)$, $(-15, -2)$

16. $(1, -15)$, $(-16, -5)$, $(8, 2)$, $(-20, 1)$, $(2, 8)$

17. $(-8, -8)$, $(3, 4)$, $(8, -9)$, $(9, -4)$, $(14, -13)$

18. To determine if the control points $p_1 = (a, b)$ and $p_2 = (c, d)$ can be extracted from data, perform the following experiment.

    (a) Obtain $\mathbf{R}(u)$, the cubic Bezier curve through $p_0 = (0, 0)$ and $p_3 = (5, 0)$, controlled by p_1 and p_2.

    (b) Impose the conditions $\mathbf{R}(\frac{1}{3}) = [2, 3]^T$ and $\mathbf{R}(\frac{2}{3}) = [3, 7]^T$, and use the equations these conditions imply to calculate a, b, c, d, thereby determining p_1 and p_2 and the Bezier curve $R_1(u)$.

    (c) Now, impose conditions $\mathbf{R}(\frac{1}{4}) = [2, 3]^T$ and $\mathbf{R}(\frac{3}{4}) = [3, 7]^T$, and again use the equations these conditions imply to calculate

a, b, c, d, thereby determining $\hat{p}_1$ and $\hat{p}_2$ as the control points and the Bezier curve $\mathbf{R}_2(u)$.

    (d) Finally, impose the more proportional conditions $\mathbf{R}(\frac{2}{5}) = [2, 3]^T$ and $\mathbf{R}(\frac{3}{5}) = [3, 7]^T$, and again use the equations these conditions imply to calculate a, b, c, d, thereby determining $\tilde{p}_1$ and $\tilde{p}_2$ as the control points and the Bezier curve $\mathbf{R}_3(u)$.

    (e) Graph $\mathbf{R}_k(u)$, $k = 1, 2, 3$, on the same set of axes.

    (f) What does this experiment say about determining control points that make a Bezier curve assume a desired shape?

Chapter Review

1. Naive interpolation consists of passing a polynomial of degree n through $n + 1$ points by forming and solving a system of $n + 1$ equations in $n + 1$ unknowns. Demonstrate the process by interpolating a quadratic polynomial through the three points $(-1, 5), (1, 2), (3, 7)$. Why, in general, is this a numerically perilous technique?

2. Obtain the Lagrange polynomials for the three nodes in Question 1. Use them to construct the quadratic interpolating polynomial for the data.

3. Describe the "polynomial wiggle problem."

4. Form a divided difference table for the data in Question 1. Use the table to construct the quadratic polynomial that interpolates the data.

5. Illustrate how to approximate the function $f(x) = \frac{x+2}{x^2+2x+2}$, $-2 \leq x \leq 3$, with an interpolating quadratic based on three equispaced nodes.

6. Illustrate how to approximate the function in Question 5 with a quadratic Chebyshev interpolating polynomial. What is the advantage of Chebyshev interpolation?

7. Why isn't the data of Question 1 subjected to Chebyshev

interpolation? (Does Chebyshev interpolation make sense for a set of data points?)

8. What is a linear spline?

9. Construct a linear spline that interpolates the points $(0, 1), (2, -1), (3, 2), (4, 0)$.

10. What is a quadratic spline? How smooth is it? (How many derivatives will it have?) To what extent is a quadratic spline determined by the points through which it must pass?

11. Obtain the natural quadratic spline interpolating the data in Question 9. How many distinct quadratic polynomials will it take to interpolate these four data points?

12. What is a cubic spline? How smooth is it? (How many derivatives will it have?) To what extent is a cubic spline determined by the points through which it must pass?

13. Obtain the natural cubic spline interpolating the data in Question 9. How many distinct cubic polynomials will it take to interpolate these four data points?

14. Construct the cubic Bezier curve determined by the points in Question 9. Show that the tangent line at the first point passes through the second point, and show that the tangent line at the fourth point passes through the third point.

Chapter *41*

Approximation of Continuous Functions

I N T R O D U C T I O N   The interpolation problem of Chapter 40 is not unrelated to the problem of *approximating continuous functions,* the subject of Chapter 41. In the context of solving differential equations, the question of providing a solution often amounts to finding a useful representation of the solution. Sometimes an exact analytic formula can be given. Often, a numeric solution must be computed, and this can be given as a graph, as a table of discrete values, as a procedure that will yield the solution value at any given point, as a spline, as an interpolating polynomial, or as an approximating function or polynomial.

Chapter 41 discusses techniques for approximating a function. For purposes of development and exposition, the function being approximated is known. The approximating functions are generally polynomials but can be rational functions, the ratios of polynomials. The utility of an approximation is established only if there is a measure of how well the approximation succeeds.

Hence, the notion of norms again appears. The different ways to quantize how well an approximating function represents a given function are called *norms,* and the goal is to find approximations that have the smallest value for the norm of the difference between the function and its approximation. When the norm is the integral of the square of the difference between the function and its approximation and when the norm is minimized, the approximation is called the *least-squares approximation.* This is precisely the idea behind least-squares fits of linear models to data, but only here, the fit is between a function and its approximation.

We then discuss the *Padé approximation,* a *rational function* approximation to a given continuous function. The error made by the Padé is generally smaller than that made by a Taylor polynomial using the same number of coefficients, but like the Taylor approximation, the error is not uniform. In both cases, the error tends to increase at the ends of the interval over which the approximation is made.

The *Chebyshev approximation* is a partial sum of a Fourier–Chebyshev series, where the *Chebyshev polynomials,* orthogonal with weight function $w = \frac{1}{\sqrt{1-x^2}}$ on $[-1, 1]$, are the analog of sines and cosines in the trigonometric Fourier series. The error made by this approximation is more uniform than for either the Taylor or the Padé approximations and generally smaller.

The *Chebyshev–Padé approximation* is an approximation by a rational function in which powers of the variable have been replaced by Chebyshev polynomials of corresponding degree. The error made by this approximation is uniform and significantly smaller than that of any of the preceding approximations. Surprisingly, this is still not the best uniform approximation. That is given by the *minimax rational approximation,* computed by an iterative process that starts from a Chebyshev–Padé approximation. The minimax approximation minimizes the maximum deviation between the function and its approximation.

41.1 Least-Squares Approximation

Why Approximate?

Obtaining a floating-point value for $\sqrt{2}$, $\sin(1)$, $e^{1.7}$, $\ln 5.38$, or $J_0(2.94)$ seems to be a routine task in the presence of a modern scientific calculator or computer algebra system. But how are these computations done efficiently and accurately? How are functions evaluated on a computing device that knows only the four operations of arithmetic? And how can these evaluations be made with a minimum of the "expensive" (time-consuming) operations of multiplication and division?

In this chapter we will investigate some standard tools for approximating continuous functions.

Norms—the Yardsticks of Approximation

The number 178,563 is an approximation of π. It's not a very good approximation, but it's an approximation. If we had no measure of the "size" of each number, or of the "closeness" of the approximation, we would not have been able to say "it's not a very good approximation." So, there is no utility in making approximations if there is no corresponding measure of the validity of the approximation.

The "size" of a real or complex number x is simply $|x|$, its absolute value, which is an example of a *norm*. In particular, a norm is a function that yields a nonnegative real number $\|x\|$, is zero only when applied to the zero object, satisfies $\|\alpha x\| = |\alpha| \|x\|$ for every scalar α, and satisfies $\|x + y\| \leq \|x\| + \|y\|$, the *triangle inequality*. In effect, a norm is a function that has the essential properties of the absolute-value function on real or complex numbers.

The reader is reminded that Table 31.1 in Section 31.1 lists a variety of vector norms and that Section 33.5 describes related matrix norms. The notion of a norm is grounded in comparisons between real numbers and is extended to objects, like vectors and matrices, and, shortly, to functions.

A *metric* measures the length of the "gap" between two objects. For real or complex numbers we simply look at $|x - y|$, the absolute value (or norm) of the difference between the two numbers. Thus, a metric is induced by a norm. If there is a way of assigning a "size" to objects, then it is possible to assign a "size" to the difference between two objects by defining the induced metric $\|x - y\|$. For much of Chapter 39 we used the norm of the difference between vectors as a metric to monitor the convergence of iterative schemes for solving a linear system.

In this section we will consider two different ways of assigning a "size" to a function $f(x)$. Each way is a norm, and for each norm defined we have an associated metric or measure of the "distance" between two functions. The two norms we will consider in this section are the square-norm (Euclidean or 2-norm) and the infinity-norm (or the absolute-value norm).

THE SQUARE-NORM   The square-norm, already used in Section 10.6, measures the "size" of a function as the square root of the integral of its absolute value, squared. Thus, we have the definition $\| f(x) \|_2 = \sqrt{\int_a^b |f(x)|^2 \, dx}$. Consequently, the induced metric, the measure of the "distance" between the two functions $f(x)$ and $g(x)$, is the norm of the difference or

$$\| f(x) - g(x) \|_2 = \sqrt{\int_a^b |f(x) - g(x)|^2 \, dx}$$

Of course, we have assumed that both functions share the common domain $[a, b]$, and that domain is the interval over which the measurement of separation is to be made.

EXAMPLE 41.1   On the interval $[0, 2]$, the square-norm measure of the distance between the functions $f(x) = e^x$ and $g(x) = x$ is

$$\sqrt{\int_0^2 (e^x - x)^2 \, dx} = \sqrt{\tfrac{1}{2}e^4 - 2e^2 + \tfrac{1}{6}} = 3.561969889$$

Figure 41.1, a graph of the difference $f(x) - g(x)$ (solid) and the line $y = \| f(x) - g(x) \|_2$ (dotted), shows that $f(x)$ and $g(x)$ are separated by the distance $\| f(x) - g(x) \|_2$ only once in the interval $[0, 2]$. The value of the square-norm is an *average* of the separation between the functions.

Figure 41.2, containing graphs of $f(x)$ and $g(x)$, shows the distance between the functions is 1 at $x = 0$ and increases as x increases. The square-norm, like any norm for functions, uses a single number to express a property distributed over an interval.   ❖

INFINITY-NORM   The infinity-norm measures the "size" of the function $f(x)$ by the maximum of its absolute value, that is, by

$$\| f(x) \|_\infty = \max_{x \in [a,b]} \{ |f(x)| \}$$

Consequently, the "distance" between the functions $f(x)$ and $g(x)$ is the maximum of the absolute value of the difference between the functions. In particular, the infinity-norm measures the closeness of functions by

$$\| f(x) - g(x) \|_\infty = \max_{x \in [a,b]} \{ |f(x) - g(x)| \}$$

EXAMPLE 41.2   The infinity-norm measures the distance between the functions x and e^x as the maximum of $|e^x - x|$. One way to determine this maximum value is from the graph of $|e^x - x|$ in Figure 41.3, which shows that the maximum occurs at the endpoint $x = 2$, and so must be

$$\| x - e^x \|_\infty = |e^2 - 2| = 5.389056099$$

Measured this way, the distance between $f(x)$ and $g(x)$ seems larger than the distance measured by the square-norm.   ❖

Approximation by Taylor Polynomial

Throughout the earlier chapters of this text we used Taylor polynomials to approximate functions. For example, a third-degree Taylor polynomial constructed at $x = 1$, and which approximates the function $f(x) = e^x$ on the interval $[0, 2]$, is

$$p_3 = e + e(x - 1) + \tfrac{1}{2}e(x - 1)^2 + \tfrac{1}{6}e(x - 1)^3$$

Figure 41.4, a graph of $f(x) - p_3$, the error made by approximating $f(x)$ with p_3, reveals that the error is zero at $x = 1$, Lthe point at which the Taylor polynomial was constructed, but tends to be greater at the endpoints. As useful as Taylor polynomials are, the need for

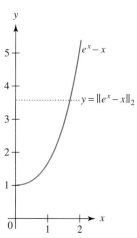

FIGURE 41.1   Example 41.1: The difference $e^x - x$ (solid) and the line $y = \| e^x - x \|_2$ (dotted)

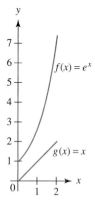

FIGURE 41.2   Example 41.1: Graphs of $f(x) = e^x$ and $g(x) = x$ showing how the distance between these functions increases from 1 as x varies from 0 to 2

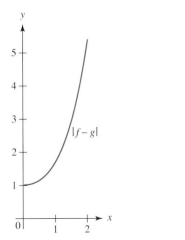

FIGURE 41.3 Graph of $|e^x - x|$ in Example 41.2

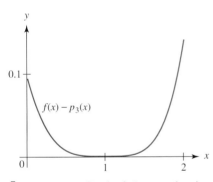

FIGURE 41.4 Graph of $e^x - p_3$, showing the error made when approximating e^x by p_3, its third-degree Taylor polynomial

more uniform approximations motivates the other approximation techniques of the next three sections.

The infinity-norm $\|e^x - p_3\|_\infty = 0.1403045564$ is just the maximum of $|e^x - p_3|$, the absolute value of the error. Figure 41.4 indicates the maximum error is located at the endpoint $x = 2$.

Best Square-Norm Approximations

If $f(x)$ is approximated by a function of the form $g(x) = \sum a_k \phi_k(x)$, then the square-norm applied to the difference $f(x) - g(x)$ will generate integrals containing terms like $\phi_i \phi_j$. If the functions ϕ_k are orthogonal, then these integrals will vanish, making the computation of the square-norm simpler. Hence, we distinguish between the cases of approximation by orthogonal and nonorthogonal functions.

NONORTHOGONAL CASE   On the interval $[0, 2]$, the best square-norm approximation to $f(x) = e^x$ by a quadratic polynomial from the family $g(x) = a + bx + cx^2$ is the one that minimizes the square-norm or, equivalently, its square. Thus, choose the parameters a, b, c so as to minimize

$$F(a, b, c) = \int_0^2 (e^x - a - bx - cx^2)^2 \, dx$$
$$= \tfrac{1}{2}e^4 + \tfrac{8}{3}b^2 - 2e^2 a + 8bc - 2e^2 b + \tfrac{32}{5}c^2 - 4e^2 c$$
$$+ 2a^2 + 4ab + \tfrac{16}{3}ac - \tfrac{1}{2} + 2a - 2b + 4c$$

via the ordinary techniques of the calculus. Thus, differentiating, setting the partial derivatives to zero, and dividing each equation by 2, we get the three *normal equations*

$$2a + 2b + \tfrac{8}{3}c - e^2 + 1 = 0$$
$$2a + \tfrac{8}{3}b + 4c - e^2 - 1 = 0$$
$$\tfrac{8}{3}a + 4b + \tfrac{32}{5}c - 2e^2 + 2 = 0$$

whose solution is $a = 3e^2 - 21$, $b = \tfrac{111}{2} - \tfrac{15}{2}e^2$, $c = -\tfrac{105}{4} + \tfrac{15}{4}e^2$. The optimal approximating quadratic is then

$$Y = 3e^2 - 21 + \left(\tfrac{111}{2} - \tfrac{15}{2}e^2\right)x + \left(-\tfrac{105}{4} + \tfrac{15}{4}e^2\right)x^2$$

giving, as the minimum value of the square-norm,

$$\|e^x - Y\|_2 = \tfrac{1}{2}\sqrt{144e^2 - 518 - 10e^4} = 0.1031721862$$

This example illustrates the nonorthogonal case. If the normal equations are written as the linear system $A\mathbf{x} = \mathbf{y}$, with the matrix A given by

$$A = \begin{bmatrix} 2 & 2 & \tfrac{8}{3} \\ 2 & \tfrac{8}{3} & 4 \\ \tfrac{8}{3} & 4 & \tfrac{32}{5} \end{bmatrix}$$

we see that A is not a diagonal matrix. This happens because the basis functions $1, x, x^2$, from which $g(x) = a + bx + cx^2$ was constructed, are not orthogonal on $[0, 2]$ under the inner product $\langle u, v \rangle = \int_0^2 u(x)v(x) \, dx$. Indeed, we find

$$\langle 1, x \rangle = \int_0^2 x \, dx = 2 \qquad \langle 1, x^2 \rangle = \int_0^2 x^2 \, dx = \tfrac{8}{3} \qquad \langle x, x^2 \rangle = \int_0^2 x^3 \, dx = 4$$

Were these functions mutually orthogonal, each of these three integrals would have evaluated to zero.

Although the matrix A has the modest determinant $\frac{32}{135} \doteq 0.237$, it has the large condition number 578.2, suggesting that the normal equations are typically ill-conditioned and become more so as the dimension of the system increases.

ORTHOGONAL CASE   To approximate, in the square-norm sense, $f(x) = e^x, 0 \le x \le 2$, with a member of the family

$$g(x) = a \sin \frac{\pi x}{2} + b \sin \pi x + c \sin \frac{3\pi x}{2}$$

minimize the square-norm or, equivalently, its square. Thus, choose the parameters a, b, c so as to minimize

$$F(a, b, c) = \int_0^2 \left(e^x - a \sin \frac{\pi x}{2} - b \sin \pi x - c \sin \frac{3\pi x}{2} \right)^2 dx$$

The normal equations, $\frac{1}{2}\frac{\partial F}{\partial a} = \frac{1}{2}\frac{\partial F}{\partial b} = \frac{1}{2}\frac{\partial F}{\partial c} = 0$, whose coefficient matrix is the diagonal matrix $A = I$, have the solution $a = 2\frac{\pi(1+e^2)}{4+\pi^2}, b = -\frac{(e^2-1)\pi}{1+\pi^2}, c = 6\frac{\pi(1+e^2)}{4+9\pi^2}$, so the optimal approximating function is then

$$Y = 2\frac{\pi(1+e^2)}{4+\pi^2} \sin \frac{\pi x}{2} - \frac{(e^2-1)\pi}{1+\pi^2} \sin \pi x + 6\frac{\pi(1+e^2)}{4+9\pi^2} \sin \frac{3\pi x}{2}$$

and the minimum value of the square-norm $\|e^x - Y\|_2$ is 2.458498213.

The coefficient matrix for the normal equations is a diagonal matrix because the basis functions $\phi_k = \sin \frac{k\pi x}{2}$, $k = 1, 2, 3$, are orthogonal on $[0, 2]$. Thus, we find

$$\langle \phi_n, \phi_m \rangle = \int_0^2 \sin \frac{n\pi x}{2} \sin \frac{m\pi x}{2} dx = 0 \qquad n \ne m \tag{41.1}$$

The matrix A has the modest determinant 1 and the equally modest condition number 1. Thus, the basis functions are orthogonal; the normal equations uncouple since only a single unknown appears in each; and, in this example, the system matrix A is no longer ill-conditioned.

EXERCISES 41.1

1. Verify (41.1).

For the functions $f(x)$ in each of Exercises 2–11:

(a) Obtain $\|f(x)\|_2$, the square-norm.

(b) On the given domain of the function, graph $f(x)$ and the horizontal line $y = \|f(x)\|_2$.

(c) Find all values of x, if any, for which $|f(x)| = \|f(x)\|_2$.

(d) Obtain $\|f(x)\|_\infty$, the infinity-norm.

(e) On the given domain, graph $f(x)$ and the horizontal lines $y = \pm\|f(x)\|_\infty$.

(f) On the given domain, find all values of x for which $|f(x)| = \|f(x)\|_\infty$.

2. $f(x) = \dfrac{x^2}{1+x^2}, -3 \le x \le 3$ **3.** $f(x) = e^{-x}\cos x, 0 \le x \le 2\pi$

4. $f(x) = 4x^3 - 7x^2 + 5x + 2, -2 \le x \le 2$

5. $f(x) = xe^{-x}, 0 \le x \le 5$ **6.** $f(x) = \sqrt{1-x^2}, -1 \le x \le 1$

7. $f(x) = e^{-x^2}, -2 \le x \le 2$

8. $f(x) = \dfrac{x^3 - 2x^2 - 5x + 7}{x^2 + x + 1}, 0 \le x \le 2$

9. $f(x) = \dfrac{100x}{7x^2 + 5x + 13}, 0 \le x \le 10$

10. $f(x) = e^{-x}\ln x, 1 \le x \le 5$ **11.** $f(x) = x(1-x), 0 \le x \le 1$

For each pair of functions $f(x)$ and $g(x)$ given in Exercises 12–21:

(a) Obtain $\|f(x) - g(x)\|_2$ and $\|f(x) - g(x)\|_\infty$.

(b) On the same set of axes, graph $f(x), g(x), f(x) - g(x)$, and the horizontal line $y = \|f(x) - g(x)\|_2$.

(c) Find all values of x, if any, for which $|f(x) - g(x)| = \|f(x) - g(x)\|_2$.

(d) On the same set of axes, graph $f(x) - g(x)$ and the horizontal lines $y = \pm \|f(x) - g(x)\|_\infty$.

(e) Find all values of x for which $|f(x) - g(x)| = \|f(x) - g(x)\|_\infty$.

12. $f(x) = \sin x, g(x) = \cos x, 0 \le x \le 2\pi$

13. $f(x) = \ln x, g(x) = e^{-x}, 1 \le x \le 3$

14. $f(x) = \cosh x, g(x) = \sinh x, 0 \le x \le 2$

15. $f(x) = x^2, g(x) = x^2 + \varepsilon, \varepsilon > 0; -1 \le x \le 1$

16. $f(x) = \dfrac{x}{1 + x^2}, g(x) = x - x^3, -2 \le x \le 2$

17. $f(x) = \cos x, g(x) = \dfrac{8}{3\pi}(\sin 2x + \frac{2}{5}\sin 4x), 0 \le x \le \pi$

18. $f(x) = \arctan x, g(x) = \frac{1}{4}\pi + \frac{5}{4}x - \frac{5}{6} - \frac{1}{2}x^2 + \frac{1}{12}x^3, 0 \le x \le 2$

19. $f(x) = x(\pi - x), g(x) = \dfrac{8}{\pi}(\sin x + \frac{1}{27}\sin 3x), 0 \le x \le \pi$

20. $f(x) = \sqrt{1 - x^2}, g(x) = 1 - \frac{1}{2}x^2 - \frac{1}{8}x^4 - \frac{1}{16}x^6, -1 \le x \le 1$

21. $f(x) = x \ln x, g(x) = -\frac{1}{3} - \frac{1}{2}x + x^2 - \frac{1}{6}x^3, \frac{1}{2} \le x \le \frac{5}{2}$

The functions $f(x)$ in Exercises 22–31 all have domain $[-1, 1]$. For each:

(a) Obtain the function $y_1 = \sum_{k=0}^{2} a_k x^k$ that minimizes $\|f(x) - y_1\|_2^2$.

(b) Obtain the function $y_2 = a_0 + a_1 \cos \pi x + a_2 \sin \pi x$ that minimizes $\|f(x) - y_2\|_2^2$.

(c) If $P_k(x), k = 0, \ldots, 2$, are the first three Legendre polynomials (see Section 10.7), obtain the function $y_3 = \sum_{k=0}^{2} A_k P_k(x)$ that minimizes $\|f(x) - y_3\|_2^2$.

(d) If $T_k(x), k = 0, \ldots, 2$, are the first three Chebyshev polynomials (see Section 40.3), obtain the function $y_4 = \sum_{k=0}^{2} C_k T_k(x)$ that minimizes $\|f(x) - y_4\|_2^2$.

(e) For which $y_k, k = 1, \ldots, 4$, is $\|f(x) - y_k\|_2$ smallest?

(f) On the same set of axes, graph $f(x)$ and $y_k, k = 1, \ldots, 4$.

22. $f(x) = \cos \dfrac{\pi}{2}x$ **23.** $f(x) = \sin \dfrac{\pi}{3}x$ **24.** $f(x) = x(1 - x)$

25. $f(x) = e^x$ **26.** $f(x) = xe^{-x}$ **27.** $f(x) = x^2 e^{-x}$

28. $f(x) = x^2(1 - x)$ **29.** $f(x) = x(1 - x)^2$

30. $f(x) = e^{-x} \cos \pi x$ **31.** $f(x) = e^x \sin \pi x$

For each function $f(x)$ and interval $a \le x \le b$ in Exercises 32–36:

(a) Obtain $\|f(x)\|_\infty$.

(b) For $k = 2, 3, \ldots$, obtain $\|f(x)\|_k = \left(\int_a^b |f(x)|^k dx\right)^{1/k}$, the k-norms of $f(x)$.

(c) Obtain $\lim_{k \to \infty} \|f(x)\|_k$, and compare to $\|f(x)\|_\infty$.

32. $f(x) = x^3, 0 \le x \le 1$ **33.** $f(x) = x^3, 0 \le x \le 10$

34. $f(x) = \frac{1}{2}, 0 \le x \le 10$ **35.** $f(x) = 2, 0 \le x \le 10$

36. $f(x) = \begin{cases} 1 & 0 \le x \le 9 \\ 0 & 9 < x < \infty \end{cases}$

The functions $f(x)$ in Exercises 37–46 all have domain $[-1, 1]$. For each:

(a) At $x = 0$, obtain $p_3(x)$, the third-degree Taylor (or Maclaurin) polynomial.

(b) Compare $\alpha = \|f(x) - p_3(x)\|_2$ with $\beta = \|f(x) - p_3(x)\|_\infty$.

(c) On the same set of axes, graph $|f(x) - p_3(x)|$ and the horizontal lines $y = \alpha, y = \beta$.

37. $f(x) = \sqrt{1 - x^2}$ **38.** $f(x) = x \arctan x$

39. $f(x) = \dfrac{x^2}{x^2 + x + 1}$ **40.** $f(x) = e^x \cosh x$

41. $f(x) = e^{x \cosh x}$ **42.** $f(x) = \dfrac{\cos x}{\cosh x}$

43. $f(x) = \cos(x) \cosh(x)$ **44.** $f(x) = \cos(x \cosh x)$

45. $f(x) = \ln \cosh x$ **46.** $f(x) = \sqrt{1 + 2\cos^2 x}$

When the weight $w(x)$ is a well-behaved positive function, an inner product between the real functions $f(x)$ and $g(x)$ is defined by $\langle f, g \rangle = \int_a^b w(x) f(x) g(x) dx$. Then the norm $\|f(x)\|_w$ is defined by $\sqrt{\langle f, f \rangle} = \sqrt{\int_a^b w(x) f^2(x) dx}$, and the metric $\|f - g\|_w$ is defined by $\sqrt{\langle f - g, f - g \rangle} = \sqrt{\int_a^b w(x)(f(x) - g(x))^2 dx}$. If $w(x) = e^{-x}$ and $[a, b] = [0, 1]$, obtain $\|f(x)\|_w, \|g(x)\|_w, \|f(x) - g(x)\|_w$, and a graph of $f(x) - g(x)$ for the functions $f(x)$ and $g(x)$ in Exercises 47–56.

47. $f(x) = \cos \pi x, g(x) = \sin \pi x$

48. $f(x) = 3x^2 - 5x + 7, g(x) = \arctan x$

49. $f(x) = \dfrac{x}{1 + 2x^2}, g(x) = \tan x$

50. $f(x) = \cosh x, g(x) = \sinh x$

51. $f(x) = x(1 - x), g(x) = \ln(1 + x)$

52. $f(x) = \dfrac{x^3 - x + 3}{x^2 + 2x + 5}, g(x) = \tanh x$

53. $f(x) = \operatorname{sech} x, g(x) = \sec x$

54. $f(x) = \sqrt{1 + 3x^2}, g(x) = \cosh \dfrac{4x}{3}$

55. $f(x) = J_1(7x), g(x) = \sin 2\pi x$

56. $f(x) = \operatorname{erf} 3x, g(x) = \dfrac{2}{\pi} \arctan 10x$

41.2 Padé Approximations

Rational-Function Approximations

A *rational function* is the ratio of two polynomials. As we saw in Section 41.1, approximating a function such as $f(x) = e^x$ with a Taylor polynomial yields an error that tends to increase at the endpoints of the approximating interval. In an attempt to approximate $f(x)$ with a more uniform error, we will therefore consider approximating with rational functions such as

$$r(x) = \frac{a_0 + a_1 x + a_2 x^2 + a_3 x^3}{1 + b_1 x + b_2 x^2 + b_3 x^3 + b_4 x^4 + b_5 x^5}$$

Without loss of generality, we illustrate the approximation based at $x = 0$. Were this not the case, the simple change of variables $z = x - x_0$ (accomplished by setting $x = z + x_0$) would restore the base $z = 0$ in place of the base $x = x_0$. Moreover, for $r(0)$ to be defined, the denominator must not vanish at $x = 0$. Hence, $b_0 \neq 0$, and we can divide by b_0 and take the lead coefficient in the denominator to be 1.

If $n = 3$ is the degree of the numerator and $m = 5$ is the degree of the denominator, the total number of parameters to determine is $N = n + m + 1 = 9$. The nine parameters $a_0, \ldots, a_3, b_1, \ldots, b_5$ are determined by the nine equations

$$f^{(k)}(0) = r^{(k)}(0) \qquad k = 0, \ldots, 8$$

where $f^{(k)}, k > 1$, denotes the kth derivative and $f^{(0)}$ is the function itself. The first four equations are

$$1 = a_0 \qquad 1 = a_1 - a_0 b_1 \qquad 1 = 2a_2 - 2a_1 b_1 + 2a_0 b_1^2 - 2a_0 b_2$$

$$1 = 6a_3 - 6a_2 b_1 + 6a_1 b_1^2 - 6a_1 b_2 - 6a_0 b_1^3 + 12a_0 b_1 b_2 - 6a_0 b_3$$

but the full set of nine fill more than a page. Surprisingly, these highly nonlinear equations have the simple and exact solution

$$R(x) = \frac{1 + \frac{3}{8} x + \frac{3}{56} x^2 + \frac{1}{336} x^3}{1 - \frac{5}{8} x + \frac{5}{28} x^2 - \frac{5}{168} x^3 + \frac{1}{336} x^4 - \frac{1}{6720} x^5}$$

The error made with this approximating rational function is then $e^x - R(x)$; and on the symmetric interval $[-1, 1]$, the infinity-norm of this error is 0.1721×10^{-6}, with the maximum error located at the endpoint, $x = 1$. Figure 41.5, a graph of the error, shows that again, the error increases at the endpoints of the approximating interval.

The approximating rational function $r(x)$ contained nine parameters. A Taylor polynomial approximating $f(x) = e^x$ and containing nine coefficients would be the eighth-degree polynomial $p_8 = \sum_{k=0}^{8} x^k / k!$. The infinity-norm of the error made when approximating $f(x)$ with this Taylor polynomial is $0.3058617775 \times 10^{-5}$, with this larger error again being located at an endpoint, $x = 1$.

Thus, the rational-function approximation seems to yield a smaller error with the same number of coefficients. Moreover, the rational function can be evaluated with fewer arithmetic operations, even if the Taylor polynomial is written in the nested (Horner) form

$$1 + \left(1 + \left(\tfrac{1}{2} + \left(\tfrac{1}{6} + \left(\tfrac{1}{24} + \left(\tfrac{1}{120} + \left(\tfrac{1}{720} + \left(\tfrac{1}{5040} + \tfrac{1}{40320} x\right) x\right) x\right) x\right) x\right) x\right) x\right) x$$

Starting inside the innermost parentheses, we count eight multiplications and eight additions. Alternatively, the rational function $R(x)$ can be more efficiently evaluated if it is written in continued-fraction form.

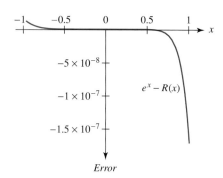

FIGURE 41.5 The error $e^x - R(x)$ when e^x is approximated by the rational function $R(x)$

Continued-Fraction Form of a Rational Function

By repeated long-division, a rational function such as that on the left in (41.2) can be expressed in the *continued-fraction* form achieved on the right.

$$\frac{x+2}{x^2+x+1} = \frac{1}{\left(\frac{x^2+x+1}{x+2}\right)} = \frac{1}{x+\frac{-x+1}{x+2}} = \frac{1}{x+\left[-1+\frac{3}{x+2}\right]} = \frac{1}{x-1+\frac{3}{x+2}} \quad (41.2)$$

Continued-Fraction Form of R(x)

If the rational function $R(x)$ is expressed in the continued-fraction form

$$R(x) = -\cfrac{20}{x^2-38x+758-\cfrac{10392}{x+10.44226328+\cfrac{21.92538896}{x+4.213319857+\cfrac{11.06363643}{x+3.344416863}}}}$$

it can be evaluated with four divisions and seven additions (or subtractions). Assuming a division is as expensive as a multiplication, the continued-fraction form of the rational function is therefore more efficient than the nested form of the polynomial.

Thus, the rational approximation makes a smaller error, and with less arithmetic, than an equivalent polynomial.

Padé Approximation

Texts such as [60] say that an approximating rational function with a numerator of degree n, denominator of degree m, and total number of parameters $N = n+m+1$, having $n = m$ when $n + m$ is even and $n = m + 1$ when $n + m$ is odd, is a *Padé approximation*. (The name Padé is pronounced with two syllables and sounds like *pad-DAY*.)

By way of contrast, [5], whom we will follow, defines the n, m Padé approximation to $f(x)$ by $[n/m] = P_n(x)/Q_m(x)$, where $P_n(x) = \sum_{k=0}^{n} a_k x^k$, a polynomial of degree at most n, and $Q_m(x) = \sum_{k=0}^{m} b_k x^k$, a polynomial of degree at most m, have no common factors, and typically the normalization $b_0 = 1$ is applied. (Also, [5] warns that some authors use the notation $[m/n]$ for $P_n(x)/Q_m(x)$.) Interestingly enough, [5] points out that Jacobi (1846) made the original discovery of Padé approximations, Frobenius (1881) studied the case where n and m differ by 1, and Padé (1892), in his thesis, studied the structure of the Padé table of approximations, depicted in Table 41.1.

[0/0]	[0/1]	[0/2]	[0/3]	[0/4]	$\cdots$
[1/0]	[1/1]	[1/2]	[1/3]	[1/4]	$\cdots$
[2/0]	[2/1]	[2/2]	[2/3]	[2/4]	$\cdots$
[3/0]	[3/1]	[3/2]	[3/3]	[3/4]	$\cdots$
[4/0]	[4/1]	[4/2]	[4/3]	[4/4]	$\cdots$
$\vdots$	$\vdots$	$\vdots$	$\vdots$	$\vdots$	$\ddots$

TABLE 41.1 Structure of the Padé table of approximations

Table 41.2 shows part of the Padé table for $f(x) = e^x$.

1	$\dfrac{1}{1-x}$	$\dfrac{1}{1-x+\frac{1}{2}x^2}$	$\dfrac{1}{1-x+\frac{1}{2}x^2-\frac{1}{6}x^3}$
$1+x$	$\dfrac{1+\frac{1}{2}x}{1-\frac{1}{2}x}$	$\dfrac{1+\frac{1}{3}x}{1-\frac{2}{3}x+\frac{1}{6}x^2}$	$\dfrac{1+\frac{1}{4}x}{1-\frac{3}{4}x+\frac{1}{4}x^2-\frac{1}{24}x^3}$
$1+x+\frac{1}{2}x^2$	$\dfrac{1+\frac{2}{3}x+\frac{1}{6}x^2}{1-\frac{1}{3}x}$	$\dfrac{1+\frac{1}{2}x+\frac{1}{12}x^2}{1-\frac{1}{2}x+\frac{1}{12}x^2}$	$\dfrac{1+\frac{2}{5}x+\frac{1}{20}x^2}{1-\frac{3}{5}x+\frac{3}{20}x^2-\frac{1}{60}x^3}$
$1+x+\frac{1}{2}x^2+\frac{1}{6}x^3$	$\dfrac{1+\frac{3}{4}x+\frac{1}{4}x^2+\frac{1}{24}x^3}{1-\frac{1}{4}x}$	$\dfrac{1+\frac{3}{5}x+\frac{3}{20}x^2+\frac{1}{60}x^3}{1-\frac{2}{5}x+\frac{1}{20}x^2}$	$\dfrac{1+\frac{1}{2}x+\frac{1}{10}x^2+\frac{1}{120}x^3}{1-\frac{1}{2}x+\frac{1}{10}x^2-\frac{1}{120}x^3}$

TABLE 41.2   Table of Padé approximations for $f(x) = e^x$

The leftmost column contains the Taylor polynomials of increasing degree. Corresponding to Table 41.2, Table 41.3 shows the maximum deviations from $f(x)$ on the interval $[-0.9, 0.9]$. The cubic Taylor polynomial containing four coefficients has a maximum error of 0.033, but the [2/1] Padé approximation, also with four coefficients, has the smaller maximum error of 0.019.

1.459603111	7.540396889	0.4794050914	0.1479588184
0.559603111	0.176760525	0.0296965691	0.0045658604
0.154603111	0.018968317	0.0021132326	0.0002194936
0.033103111	0.002171082	0.0001615241	0.0000120426

TABLE 41.3   For the interval $[-0.9, 0.9]$, the maximum deviation between $f(x) = e^x$ and its Padé approximations in Table 41.2

Numeric Algorithm

With a computer algebra system such as Maple, the Padé approximation can be obtained with the built-in **pade** function. Alternatively, the n, m Padé approximation can be constructed by solving the nonlinear equations

$$f^{(k)}(0) = r^{(k)}(0) \qquad k = 0, \ldots, n + m$$

for its coefficients. While there is an interesting structure to these equations, the following algorithm, based on coefficient matching, leads to a set of *linear* equations that can more easily be used for computing, perhaps numerically, the coefficients of $[n/m] = P_n(x)/Q_m(x)$, the n, m Padé approximation of $f(x)$, where $P_n(x) = \sum_{k=0}^{n} a_k x^k$ and $Q_m(x) = 1 + \sum_{k=1}^{m} b_k x^k$.

1. Expand $f(x)$ in a Taylor series of degree $n + m$ so that $f(x) = \sum_{k=0}^{n+m} c_k x^k$.
2. Require $\sum_{k=0}^{n+m} c_k x^k = P_n(x)/Q_m(x)$ to be an identity in x. Therefore:

   (a) Bring all terms to the left and add fractions, resulting in

   $$\frac{Q_m(x) \sum_{k=0}^{n+m} c_k x^k - P_n(x)}{Q_m(x)} = 0$$

   (b) Combine like-powers of x in the numerator, and equate to zero the coefficient of each such power.

(c) In the resulting set of equations, the first $n + m + 1$ are then solved for a_k, $k = 0, \ldots, n$, and b_k, $k = 1, \ldots, m$.

EXAMPLE 41.3 As an example of this algorithm, let $n = 3$ and $m = 5$, so that we have

$$P = a_0 + a_1 x + a_2 x^2 + a_3 x^3 \qquad Q = 1 + b_1 x + b_2 x^2 + b_3 x^3 + b_4 x^4 + b_5 x^5$$

$$F = c_0 + c_1 x + c_2 x^2 + c_3 x^3 + c_4 x^4 + c_5 x^5 + c_6 x^6 + c_7 x^7 + c_8 x^8$$

Then $F Q_m - P_n$, the numerator of $(F Q_m - P_n)/Q_m = 0$, becomes $\sum_{k=0}^{13} A_k x^k = 0$, giving the equations $A_k = 0$, $k = 0, \ldots, n + m = 8$, which are

$$-a_0 + c_0 = 0$$
$$-a_1 + c_1 + c_0 b_1 = 0$$
$$-a_2 + c_2 + c_1 b_1 + c_0 b_2 = 0$$
$$-a_3 + c_3 + c_2 b_1 + c_1 b_2 + c_0 b_3 = 0$$
$$c_4 + c_3 b_1 + c_2 b_2 + c_1 b_3 + c_0 b_4 = 0$$
$$c_5 + c_4 b_1 + c_3 b_2 + c_2 b_3 + c_1 b_4 + c_0 b_5 = 0$$
$$c_6 + c_5 b_1 + c_4 b_2 + c_3 b_3 + c_2 b_4 + c_1 b_5 = 0$$
$$c_7 + c_6 b_1 + c_5 b_2 + c_4 b_3 + c_3 b_4 + c_2 b_5 = 0$$
$$c_8 + c_7 b_1 + c_6 b_2 + c_5 b_3 + c_4 b_4 + c_3 b_5 = 0$$

Of these equations, the last $m = 5$ are independent of a_k, $k = 0, \ldots, n = 3$, so they can be solved for b_k, $k = 1, \ldots, m = 5$, and these values used in the first $m + 1 = 4$, thereby determining a_k, $k = 0, \ldots, n = 3$. In fact, the augmented matrix for the last $m = 5$ equations is

$$\begin{bmatrix} c_3 & c_2 & c_1 & c_0 & 0 & -c_4 \\ c_4 & c_3 & c_2 & c_1 & c_0 & -c_5 \\ c_5 & c_4 & c_3 & c_2 & c_1 & -c_6 \\ c_6 & c_5 & c_4 & c_3 & c_2 & -c_7 \\ c_7 & c_6 & c_5 & c_4 & c_3 & -c_8 \end{bmatrix} \qquad (41.3)$$

a matrix with sufficient structure to allow easy coding in a numeric programming language. (See Exercise B23.) Then, solving the first $n + 1 = 4$ equations for a_k, $k = 0, \ldots, n = 3$, we have

$$a_0 = c_0 \qquad a_1 = c_1 + c_0 b_1 \qquad a_2 = c_1 b_1 + c_2 + c_0 b_2 \qquad a_3 = c_3 + c_1 b_2 + c_0 b_3 + c_2 b_1$$

from which we can generalize the pattern $a_k = c_k + \sum_{n=1}^{k} b_n c_{k-n}$. ❖

EXERCISES 41.2–Part A

A1. If evaluating $x^2 = x \times x$ requires one multiplication, determine the total number of arithmetic operations (multiplications and additions) required to evaluate the polynomial $p(x) = 1 + 2x + 3x^2 + 4x^3$.

A2. Write the polynomial $p(x)$ (from Exercise A1) in nested (Horner) form, and count the number of arithmetic operations needed to evaluate it in that form.

A3. Determine the total number of arithmetic operations needed to evaluate the rational function $R(x) = \frac{1+x}{1+2x+5x^2}$.

A4. Write the denominator of $R(x)$ (from Exercise A3) in nested form, and again determine the number of arithmetic operations needed to evaluate $R(x)$.

A5. Verify that the continued-fraction form of $R(x)$ (from Exercise A3) is $R(x) = \frac{0.2}{x - 0.6 + 0.8/(x+1)}$. Count the number of arithmetic operations needed to evaluate $R(x)$ in this form.

EXERCISES 41.2–Part B

B1. Convert $R(x) = \frac{1+2x}{1+3x+x^2}$ to continued-fraction form.

The domain of the functions $f(x)$ in Exercises B2–11 is $[-1, 1]$. For each:

(a) Obtain [3/2], the 3, 2 Padé approximation, by writing $r(x) = \frac{P_3}{Q_2}$, $P_3 = \sum_{k=0}^{3} a_k x^k$, and $Q_2 = 1 + \sum_{k=1}^{2} b_k x^k$ and solving the equations $f^{(k)}(0) = r^{(k)}(0)$, $k = 0, \ldots, 3+2 = 5$, for the coefficients $a_0, \ldots, a_3$ and b_1, b_2.

(b) Let $F = p_5(x)$ be the fifth-degree Taylor polynomial constructed at $x = 0$, and obtain [3/2] by equating to zero the coefficients of x^k, $k = 0, \ldots, 3+2 = 5$, in the numerator of $\frac{FQ_2 - P_3}{Q_2}$.

(c) Obtain $\|f(x) - [3/2]\|_\infty$, and determine where in $[-1, 1]$ the maximum of $|f(x) - [3/2]|$ occurs.

(d) Count the number of operations needed for evaluation of [3/2].

(e) Write the numerator and denominator of [3/2] in Horner (nested) form, and count the number of operations needed for evaluation.

(f) Put [3/2] in continued-fraction form, and count the number of arithmetic operations needed for evaluation.

(g) Obtain $\|f(x) - p_5(x)\|_\infty$, and determine where in $[-1, 1]$ the maximum of $|f(x) - p_5(x)|$ occurs.

(h) Write $p_5(x)$ in Horner (nested) form, and count the number of arithmetic operations needed for evaluation.

(i) Graph $f(x)$ and [3/2] on the same set of axes.

B2. $f(x) = \cos\frac{\pi}{2}x$ **B3.** $f(x) = x(1-x)$ **B4.** $f(x) = x^2(1-x)$

B5. $f(x) = x(1-x)^2$ **B6.** $f(x) = xe^{-x}$ **B7.** $f(x) = \sqrt{1-x^2}$

B8. $f(x) = x \arctan x$ **B9.** $f(x) = \ln\cosh x$

B10. $f(x) = \frac{\cos x}{\cosh x}$ **B11.** $f(x) = \sqrt{1 + 2\cos^2 x}$

For each function $f(x)$ in Exercises B12–17, let $F(z) = f(z + x_0)$, where $x_0 = \frac{b+a}{2}$ is the midpoint of the interval $[a, b]$ over which $f(x)$ is defined. This centers $z = 0$ in the interval $[-h, h]$, where $h = \frac{b-a}{2}$.

(a) Obtain, for $F(z)$, the Padé table containing $[n/m]$, $0 \le n$, $m \le 3$.

(b) Obtain a table of the norms $\|F(z) - [n/m]\|_\infty$ corresponding to the table in part (a).

(c) Which Padé approximations in the table from part (a) have a smaller error than [3/0], the third-degree Taylor (or Maclaurin) polynomial for $F(z)$?

B12. $f(x)$ in Exercise 3, Section 41.1

B13. $f(x)$ in Exercise 5, Section 41.1

B14. $f(x)$ in Exercise 10, Section 41.1

B15. $f(x)$ in Exercise 13, Section 41.1

B16. $f(x)$ in Exercise 18, Section 41.1

B17. $f(x)$ in Exercise 21, Section 41.1

The text [38] claims a crude estimate of the error in $[n/m]$, the Padé approximation of $f(z)$, is given by $\frac{\beta z^\alpha}{Q_m}$, where βz^α is the first nonzero term past $b_n z^n$ in the Maclaurin expansion of $f(z)Q_m(z) - P_n(z)$. In the Maehly algorithm, $r = \frac{P_n}{Q_m}$, $F = P_{n+m}(z)$ is a Maclaurin expansion of $f(z)$, and $\frac{FQ_m - P_n}{Q_m} = 0$ leads to $[n/m]$. If $\hat{F}$, a Maclaurin polynomial for $f(z)$ of higher degree than F, is used in place of F, then $[n/m]$ is obtained from equating to zero the first $n + m + 1$ coefficients in $\hat{F}Q_m - P_n = 0$, and βz^α is the next nonvanishing term. In Exercises B18–22, test this claim for [4/4] with the given function $g(x)$. Since each of these functions is defined on an interval $[a, b]$, use the shift $f(z) = g(z + x_0)$, as in Exercises B12–17. Compute $\left\|\frac{\beta x^\alpha}{Q_4}\right\|_\infty$ and $\|f(z) - [4/4]\|_\infty$, then plot the magnitudes of the actual and estimated errors.

B18. $g(x) = e^{-x}$, $1 \le x \le 3$ **B19.** $g(x) = \sinh x$, $0 \le x \le 2$

B20. $g(x) = \ln(1+x)$, $1 \le x \le 5$ **B21.** $g(x) = \sin x$, $0 \le x \le 2$

B22. $g(x) = \arctan x$, $0 \le x \le 3$

B23. Generalize the pattern in (41.3) to $c_k + \sum_{n=1}^{m} b_n c_{k-n} = 0$, $k = n + 1, \ldots, n + m$, where $b_k = 0$ for $k > m$ and $c_k = 0$ for $k < 0$.

41.3 Chebyshev Approximation

Chebyshev Polynomials

In Section 40.3 we saw the Chebyshev polynomials $T_n(x)$ that satisfy (40.8) and the three-term recurrence relation $T_{n+1}(x) = 2xT_n(x) - T_{n-1}(x)$ and that are orthogonal, with weight function $w(x) = \frac{1}{\sqrt{1-x^2}}$, on $[-1, 1]$. Hence, they satisfy (40.9), the orthogonality conditions

$$\int_{-1}^{1} \frac{T_n(x)T_m(x)}{\sqrt{1-x^2}}\,dx = \begin{cases} 0 & m \neq n \\ \pi & m = m = 0 \\ \dfrac{\pi}{2} & m = n > 0 \end{cases}$$

The Fourier–Chebyshev Series

In Units Two, Three, Five, and Seven we have seen and/or used the Fourier series representation of the function $f(x)$ defined on the interval $[-L, L]$. Thus, we wrote $f(x) = \sum_{n=1}^{\infty} b_n \sin \frac{n\pi x}{L}$, where $b_n = \frac{2}{L} \int_0^L f(x) \sin \frac{n\pi x}{L}\, dx$.

In Units Two and Five we generalized the Fourier series to the Fourier–Legendre series for $f(x)$ defined on the interval $[-1, 1]$. Thus, we wrote $f(x) = \sum_{n=0}^{\infty} a_n P_n(x)$, where $P_n(x)$ is the nth Legendre polynomial and $a_n = \frac{2n+1}{2} \int_{-1}^{1} f(x) P_n(x)\, dx$.

Finally, we saw in Units Three and Five the Fourier–Bessel series for $f(x)$ defined on the interval $[0, \sigma]$. Thus, we wrote $f(x) = \sum_{n=0}^{\infty} A_n J_0(\lambda_n x)$, where $J_0(x)$ is the Bessel function of order zero, λ_n is the nth zero of $J_0(\sigma x)$, and

$$A_n = \frac{\int_0^{\sigma} x f(x) J_0(\lambda_n x)\, dx}{\int_0^{\sigma} x J_0^2(\lambda_n x)\, dx} = \frac{2 \int_0^{\sigma} x f(x) J_0(\lambda_n x)\, dx}{\sigma^2 J_1^2(\lambda_n \sigma)}$$

It should be no surprise, then, that the function $g(x)$, defined on the interval $[-1, 1]$, can be expanded in the Fourier–Chebyshev series $g(x) = \frac{A_0}{2} + \sum_{k=1}^{\infty} A_k T_n(x)$, where $A_k = \frac{2}{\pi} \int_{-1}^{1} [g(x) T_k(x) / \sqrt{1 - x^2}]\, dx$.

For the function $f(z)$ defined for z in the interval $[\alpha, \beta]$, the Fourier–Chebyshev series generalizes to

$$f(z) = \frac{\hat{a}_0}{2} + \sum_{k=1}^{\infty} \hat{a}_k T_k \left(\frac{2(z - \alpha)}{\beta - \alpha} - 1 \right)$$

$$\hat{a}_k = \frac{4}{\pi(\beta - \alpha)} \int_{\alpha}^{\beta} \frac{f(z) T_k \left(\frac{2(z-\alpha)}{\beta-\alpha} - 1 \right)}{\sqrt{1 - \left(\frac{2(z-\alpha)}{(\beta-\alpha)} - 1 \right)^2}}\, dz \tag{41.4}$$

Of course, this generalization is based on the linear transformation $z = \alpha + \frac{(\beta - \alpha)(x+1)}{2}$, which maps $-1 \le x \le 1$ to $\alpha \le z \le \beta$. (Equivalently, $x = \frac{2(z-\alpha)}{\beta-\alpha} - 1$ maps $\alpha \le z \le \beta$ to $-1 \le x \le 1$.)

$\hat{a}_0 = 5.429975096$
$\hat{a}_1 = 3.236988642$
$\hat{a}_2 = 1.113990242$
$\hat{a}_3 = 0.266347996$
$\hat{a}_4 = 0.04859825643$
$\hat{a}_5 = 0.007157295457$
$\hat{a}_6 = 0.0008829533638$
$\hat{a}_7 = 0.00009366852746$

TABLE 41.4 Fourier–Chebyshev coefficients for $f(z) = e^z$ in Example 41.4

EXAMPLE 41.4

To obtain the Fourier–Chebyshev series for $f(z) = e^z$ on the interval $-1 \le z \le 2$, write $x(z) = \frac{2(z-\alpha)}{\beta-\alpha} - 1 = \frac{2}{3}z - \frac{1}{3}$. The coefficients $\hat{a}_k = \frac{4}{3\pi} \int_{-1}^{2} e^z T_k(x(z))\, dz / \sqrt{1 - x(z)^2}$, $k = 0, \ldots, 7$, computed by numeric integration, are listed in Table 41.4.

The partial sum of the Fourier–Chebyshev series, $\frac{\hat{a}_0}{2} + \sum_{k=1}^{7} \hat{a}_k T_k(x(z))$, simplifies to

$$F_7 = 1 + 0.99998z + 0.49987z^2 + 0.16677z^3 + 4.1945 \times 10^{-2}z^4$$

$$+ 8.0994 \times 10^{-3}z^5 + 1.2525 \times 10^{-3}z^6 + 3.5086 \times 10^{-4}z^7$$

Figure 41.6 shows that the error $f(z) - F_7(z)$ is both smaller and more uniform than the error in the comparable Taylor polynomial approximation. In fact, the maximum value of this error, the infinity-norm of $f(z) - F_7(z)$, is $0.9496533650 \times 10^{-5}$, with the maximum value occurring at $z = 2$. ❖

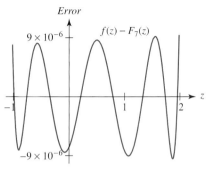

FIGURE 41.6 Example 41.4: The error $f(z) - F_7(z)$ made when approximating e^z with a Fourier–Chebyshev polynomial

A Comparison

The partial sum $F_7(z)$ contains the eight coefficients $\hat{a}_k$, $k = 0, \ldots, 7$. A Taylor polynomial with as many coefficients is $\sqrt{e} \sum_{k=0}^{7} \left(z - \frac{1}{2}\right)^k / k!$, where we have picked $z = \frac{1}{2}$, the midpoint of the interval $[-1, 2]$, as the point of expansion. The maximum error made by this polynomial is 0.001252930683, and it occurs at the endpoint $z = 2$.

[n/m]	Maximum Error
[0/7]	0.0180476732
[1/6]	0.0018070821
[2/5]	0.0004202679
[3/4]	0.0001744064
[4/3]	0.000119524870
[5/2]	0.000135075581
[6/1]	0.000271417184
[7/0]	0.001252930683

TABLE 41.5   Seven-coefficient Padé approximations of $f(z) = e^z$, and their maximum errors

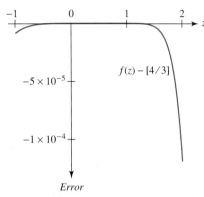

FIGURE 41.7   The error in the [4/3] Padé approximation of $f(z) = e^z$

All Padé approximations with seven coefficients and their maximum errors are listed in Table 41.5.

The smallest error is made by [4/3], the Padé approximation whose numerator is of degree $n = 4$ and whose denominator is of degree $m = 3$. Figure 41.7 contains a graph of the error in this approximation, showing the error is not uniform since it again increases greatly in the vicinity of the endpoint $z = 2$.

The maximum error of $0.9496533650 \times 10^{-5}$ in the eight-term Chebyshev approximation is considerably smaller than the errors in the Taylor or Padé approximations. In addition, the error in the Chebyshev approximation, shown in Figure 41.6, is far more uniform than the error in either the Taylor polynomial or the Padé approximation.

Chebyshev Economization

If $f(z)$ is approximated by a polynomial of degree n, we seek a new polynomial of lower degree that approximates $f(z)$ with about the same error. Since it has fewer coefficients, this new polynomial would then require less storage space on a computing device and fewer arithmetic operations for evaluation—hence, the name *economization*. The two equivalent algorithms we will present use Chebyshev polynomials—hence, the name *Chebyshev economization*.

On the interval $-1 \le z \le 2$, the fifth-degree Taylor polynomial $P_5(z) = \sum_{k=0}^{5} z^k/k!$ approximates $f(z) = e^z$ with a maximum error of 0.1223894323.

To find a polynomial of lower degree that approximates $f(z)$ with about the same error requires us to use Chebyshev polynomials that are defined on the interval $-1 \le x \le 1$. The change of variables

$$z = z(x) = \alpha + \frac{(\beta - \alpha)(x + 1)}{2} \longleftrightarrow x = x(z) = \frac{2(z - \alpha)}{\beta - \alpha} - 1 \quad \textbf{(41.5)}$$

linearly maps $\alpha \le z \le \beta$ to $-1 \le x \le 1$. With $\alpha = -1$ and $\beta = 2$ we have $z = \frac{1}{2} + \frac{3}{2}x$ and $x = \frac{2}{3}z - \frac{1}{3}$. Define $g_5(x)$ by

$$g_5(x) = P_5(z(x)) = \frac{6331}{3840} + \frac{633}{256}x + \frac{237}{128}x^2 + \frac{117}{128}x^3 + \frac{81}{256}x^4 + \frac{81}{1280}x^5$$

The Chebyshev economization algorithm in texts such as [16] and [38] is

$$g_{\mathrm{econ}}(x) = g_n(x) - \frac{A_n}{2^{n-1}} T_n(x)$$

where $g_{\mathrm{econ}}(x)$ is the economized polynomial and A_n is the coefficient of the highest power of x in $g_n(x)$. Here, $n = 5$, $A_n = \frac{81}{1280}$, and

$$g_{\mathrm{econ}}(x) = \frac{6331}{3840} + \frac{10{,}047}{4096}x + \frac{237}{128}x^2 + \frac{1017}{1024}x^3 + \frac{81}{256}x^4 \quad \textbf{(41.6)}$$

Already a polynomial of lower degree, $g_{\mathrm{econ}}(x)$ is now transformed to $p_{\mathrm{econ}}(z) = g_{\mathrm{econ}}(x(z))$ by the change of variables (41.5), giving

$$p_{\mathrm{econ}}(z) = \frac{61{,}681}{61{,}440} + \frac{6155}{6144}z + \frac{365}{768}z^2 + \frac{65}{384}z^3 + \frac{1}{16}z^4$$

The maximum error made on $-1 \le z \le 2$ when approximating $f(z)$ by $p_{\mathrm{econ}}(z)$ is then 0.1263445104, which is nearly the same error as that made when approximating $f(z)$ with $P_5(z)$.

The alternative algorithm, found in texts such as [60] and [41], again starts with $g_5(x) = P_5(z(x))$, but then expresses $g_5(x)$ in terms of Chebyshev polynomials by the substitutions in the rightmost column of Table 41.6.

Polynomial Form of $T_k(x)$	Chebyshev Expansion of x^k
$T_0(x) = 1$	$1 = T_0$
$T_1(x) = x$	$x = T_1$
$T_2(x) = 2x^2 - 1$	$x^2 = \frac{1}{2}T_2 + \frac{1}{2}$
$T_3(x) = 4x^3 - 3x$	$x^3 = \frac{1}{4}T_3 + \frac{3}{4}T_1$
$T_4(x) = 8x^4 - 8x^2 + 1$	$x^4 = \frac{1}{8}T_4 + \frac{1}{2}T_2 + \frac{3}{8}$
$T_5(x) = 16x^5 - 20x^3 + 5x$	$x^5 = \frac{1}{16}T_5 + \frac{5}{16}T_3 + \frac{5}{8}T_1$

TABLE 41.6   Chebyshev polynomials and the Chebyshev expansions of x^k

When each power of x in $g_5(x)$ is replaced by its equivalent in terms of the Chebyshev polynomials, the result is

$$G_C = \frac{82{,}733}{30{,}720}T_0 + \frac{6549}{2048}T_1 + \frac{555}{512}T_2 + \frac{1017}{4096}T_3 + \frac{81}{2048}T_4 + \frac{81}{20{,}480}T_5$$

which is economized by dropping the term in T_5, since this Chebyshev polynomial will contain the highest power of x. Deleting that term and replacing each symbol T_k by the corresponding Chebyshev polynomial from the leftmost column of Table 41.6 gives the economized polynomial $g_{\mathrm{econ}}(x)$ as stated in (41.6). As in the first algorithm, $p_{\mathrm{econ}}(z)$ is obtained as $g_{\mathrm{econ}}(x(z))$.

The theory of Chebyshev economization is best detailed in [16] and [41].

EXERCISES 41.3

1. If $f(x) = e^{-|x|}, -1 \le x \le 1$, obtain the Fourier series coefficients a_n and b_n up through index $n = 2$.

2. For the function in Exercise 1, obtain the Fourier–Legendre coefficients up through index $n = 2$.

3. For the function in Exercise 1, obtain the Fourier–Chebyshev coefficients up through index $n = 2$. Use numerical integration.

4. If $f(x) = e^{-x}, 0 \le x \le 1$, obtain the Fourier–Bessel coefficients up through index $n = 2$. Use numerical integration.

The functions $f(x)$ in Exercises 5–14 have domain $-1 \le x \le 1$. For each:

(a) Obtain $S_5(x) = \frac{A_0}{2}T_0 + \sum_{k=1}^{5} A_k T_k(x)$, a partial sum of the Fourier–Chebyshev expansion of $f(x)$.

(b) By replacing the symbols T_k with the Chebyshev polynomials $T_k(x)$, $k = 0, \ldots, 5$, obtain $S_5(x)$, a polynomial in x.

(c) Graph $f(x) - S_5(x)$, looking for uniformity in the error made by $S_5(x)$.

(d) Obtain $\| f(x) - S_5(x) \|_\infty$ and determine where in $[-1, 1]$ the error is this large.

(e) Compare $\| f(x) - S_5(x) \|_\infty$ to $\| f(x) - p_5(x) \|_\infty$, where $p_5(x)$ is the fifth-degree Taylor (or Maclaurin) polynomial at $x = 0$.

5. $f(x) = \cos\frac{\pi}{2}x$ **6.** $f(x) = x(1-x)$ **7.** $f(x) = x^2(1-x)$

8. $f(x) = x(1-x)^2$ **9.** $f(x) = xe^{-x}$ **10.** $f(x) = \sqrt{1-x^2}$

11. $f(x) = x\arctan x$ **12.** $f(x) = \ln\cosh x$

13. $f(x) = \dfrac{\cos x}{\cosh x}$ **14.** $f(x) = \sqrt{1 + 2\cos^2 x}$

In Exercises 15 and 16:

(a) Use (41.4) and (41.5) to obtain $s_5(z) = \frac{\hat{a}_0}{2} + \sum_{k=1}^{5} \hat{a}_k T_k(x(z))$, a partial sum of the Fourier–Chebyshev expansion of $g(z)$.

(b) Graph $g(z) - s_5(z)$, looking for uniformity in the error made by $s_5(z)$.

(c) Obtain $\| g(z) - s_5(z) \|_\infty$, and determine where in $[\alpha, \beta]$ the error is this large.

(d) Compare $\| g(z) - s_5(z) \|_\infty$ to $\| g(z) - p_5(z) \|_\infty$, where $p_5(z)$ is the fifth-degree Taylor polynomial constructed at $z = \frac{\alpha+\beta}{2}$, the midpoint of $[\alpha, \beta]$.

15. $g(z) = \dfrac{z^2}{1+z^2}, -3 \le z \le 3$ **16.** $g(z) = e^{-z}\cos z, 0 \le z \le 2\pi$

In Exercises 17–20, implement Chebyshev economization by executing the following steps.

(a) Apply (41.5) to $p_5(z)$, the fifth-degree Taylor polynomial constructed at $z = \frac{\alpha+\beta}{2}$, the midpoint of $[\alpha, \beta]$, to obtain $g_5(x)$.

(b) Obtain $g_{econ}(x) = g_5(x) - (A_n/2^{n-1} T_n(x))$, where n is the degree of $g_5(x)$.

(c) Use (41.5) to obtain $p_{econ}(z) = g_{econ}(x(z))$, the desired economized polynomial.

(d) Compare $\|g(z) - p_5(z)\|_\infty$ with $\|g(z) - p_{econ}(z)\|_\infty$, and graph $g(z), p_5(z)$, and $p_{econ}(z)$ on the same set of axes.

(e) Use Table 41.6 to replace each power of x in $g_5(x)$ with its equivalent in terms of the symbols T_k, and write the resulting expression in the form $\sum_{k=0}^{5} c_k T_k$.

(f) Drop the term containing T_5; then replace each remaining symbol T_k with the Chebyshev polynomial $T_k(x)$, thereby obtaining $g_{econ}(x)$, which should agree with the result in part (b).

17. $g(z) = 4z^3 - 7z^2 + 5z + 2, -2 \le z \le 2$

18. $g(z) = ze^{-z}, 0 \le z \le 5$ 19. $g(z) = \tanh z, -3 \le z \le 2$

20. $g(z) = e^{-z^2}, -2 \le z \le 2$

21. Obtain expressions comparable to (41.4) for a Fourier–Legendre series of a function $g(z)$ defined on the interval $[\alpha, \beta]$.

In Exercises 22 and 23, apply the results of Exercise 21 to each given function $g(z)$. In particular:

(a) Obtain $\sigma_5(z)$, a partial sum of the expansion developed in Exercise 21.

(b) Obtain $s_5(z) = \frac{\hat{a}_0}{2} + \sum_{k=1}^{5} \hat{a}_k T_k(x(z))$, a partial sum of the Fourier–Chebyshev expansion of $g(z)$.

(c) Obtain $\|g(z) - \sigma_5(z)\|_\infty$, and compare it to both $\|g(z) - s_5(z)\|_\infty$ and $\|g(z) - p_5(z)\|_\infty$, where $p_5(z)$ is the fifth-degree Taylor polynomial constructed at $z = \frac{\alpha+\beta}{2}$, the midpoint of $[\alpha, \beta]$.

22. $g(z) = \dfrac{z^3 - 2z^2 - 5z + 7}{z^2 + z + 1}, 0 \le z \le 2$

23. $g(z) = \dfrac{100z}{7z^2 + 5z + 13}, 0 \le z \le 10$

In Exercises 24 and 25, apply the formalism of Theorem 10.1, Section 10.1, and thereby

(a) obtain the partial sum $\psi = \frac{a_0}{2} + \sum_{k=1}^{3}(a_k \cos \frac{k\pi z}{L} + b_k \sin \frac{k\pi z}{L})$, where $\beta - \alpha = 2L$.

(b) obtain $s_5(z) = \frac{\hat{a}_0}{2} + \sum_{k=1}^{5} \hat{a}_k T_k(x(z))$, a partial sum of the Fourier–Chebyshev expansion of $g(z)$.

(c) obtain $\|g(z) - \psi\|_\infty$, and compare it to $\|g(z) - s_5(z)\|_\infty$, and $\|g(z) - p_5(z)\|_\infty$, where $p_5(z)$ is the fifth-degree Taylor polynomial constructed at $z = \frac{\alpha+\beta}{2}$, the midpoint of $[\alpha, \beta]$. Of course, ψ contains seven parameters, whereas the other approximations contain six.

24. $g(z) = e^{-z} \ln z, 1 \le z \le 5$ 25. $g(z) = z(1 - z), 0 \le z \le 1$

41.4 Chebyshev–Padé and Minimax Approximations

The Chebyshev–Padé Approximation

This section examines the Chebyshev–Padé approximation

$$\frac{\sum_{k=0}^{n} a_k T_k(z)}{1 + \sum_{k=1}^{m} b_k T_k(z)} \tag{41.7}$$

a rational function with small, uniform error, and which, in continued-fraction form, is economical to evaluate. Motivation for this approximation comes from the ideas seen in Sections 41.1–41.3.

On $-1 \le z \le 2$, the function $f(z) = e^z$ can be approximated via

1. the fifth-degree Taylor polynomial $P_5(z) = \sqrt{e} \sum_{k=0}^{5} \left(z - \frac{1}{2}\right)^k/k!$, with a nonuniform error whose maximum magnitude is less than 0.032925.

2. the Chebyshev series $S_5(z) = \frac{\hat{a}_0}{2} + \sum_{k=1}^{5} \hat{a}_k T_k(x(z))$, with a uniform error whose maximum magnitude is less than 0.98611×10^{-3}.

3. the 3, 2 Padé approximation $[3/2] = \frac{\sqrt{e}}{3} \frac{60+36(z-1/2)+9(z-1/2)^2+(z-1/2)^3}{24-8z+(z-1/2)^2}$, based at $z = \frac{1}{2}$, and having a nonuniform error whose maximum magnitude is less than 0.0100230.

Like P_5 and S_5, the 3, 2 Padé approximation has six coefficients but, its continued-fraction form,

$$9.067966991 + 0.5495737569z + \cfrac{83.53521107}{z - 6.657894737 + \cfrac{8.656509695}{z - 2.342105263}}$$

can be economically evaluated with just two divisions, one multiplication, and five additions.

Consequently, we seek a rational-function approximation to $f(z)$ that, like the Padé approximation, is economical to evaluate and, like the Chebyshev approximation, is very accurate. The desired approximation is the Chebyshev–Padé approximation (41.7). The 3, 2 Chebyshev–Padé approximation, obtained with an algorithm attributed to Clenshaw and Lord [36], is

$$CP_{3,2} = \frac{1.147328304 + 0.7309945667z + 0.2019501560z^2 + 0.02672316719z^3}{1.147393547 - 0.4165562786z + 0.04427970359z^2} \tag{41.8a}$$

$$= 10.23821925 + 0.6035082672z + \cfrac{97.18510371}{z - 6.944190547 + \cfrac{8.807502589}{z - 2.463195790}} \tag{41.8b}$$

and has an error with maximum magnitude no greater than 0.15734×10^{-3}, a bound that is roughly $\frac{1}{64}$ of that for the equivalent Padé approximation. In its continued-fraction form, it can be economically evaluated with just two divisions, one multiplication, and five additions. Finally, Figure 41.8 shows the uniformity of the error. Consequently, the Chebyshev–Padé approximation has a more uniform error that is significantly smaller than the error in other approximations and preserves the efficiency of the continued-fraction form.

A Chebyshev–Padé Algorithm

Texts such as [16], [60], [38], and [48] find the Chebyshev–Padé approximation of $f(z)$, $\alpha \le z \le \beta$, by imitating the technique for finding the Padé approximation. The difference $f(z) - P_n(z)/Q_m(z)$ is made to match its error term $\sum_{n+m+1}^{\infty} \alpha_k T_k(z)$ by matching the *numerator*, $Q_m(z)f(z) - P_n(z)$, with the error term. This is not the same as matching the *complete* difference with the error term, but it is simpler. Unfortunately, it results in an approximation that is not as good as the one obtained by the Clenshaw–Lord algorithm.

The algorithm that matches the numerator, $Q_m(z)f(z) - P_n(z)$, with the error term is attributed to Maehly by [36] and consists of the following steps.

1. Linearly map the interval $\alpha \le z \le \beta$ to $-1 \le x \le 1$ by the transformation $x(z) = \frac{2(z-\alpha)}{\beta-\alpha} - 1$ and its inverse $z(x) = \alpha + \frac{(\beta-\alpha)(x+1)}{2}$.
2. Obtain $g(x) = f(z(x))$.
3. Let $R = \frac{P}{Q}$, where $P = \sum_{k=0}^{n} a_k T_k$ and $Q = 1 + \sum_{k=1}^{m} b_k T_k$, treating T_k as a symbol. This step determines the parameters n and m.
4. Obtain the Chebyshev expansion $g(x) = \frac{A_0}{2} + \sum_{k=1}^{n+m} A_k T_k(x)$, writing the result with the symbols T_k not the Chebyshev polynomials $T_k(x)$. The coefficients A_k are determined by the integrals $A_k = \frac{2}{\pi} \int_{-1}^{1} [g(x)T_k(x)/\sqrt{1-x^2}]\,dx$.
5. Manipulate $R = g(x)$ to the form $\frac{gQ-P}{Q} = 0$ or, equivalently, $gQ - P = 0$.
6. The left side of $gQ - P = 0$ contains products of Chebyshev polynomials that are simplified with the identities $T_n T_m = \frac{1}{2}(T_{n+m} + T_{|n-m|})$, derived as part of Example 41.5. An equation of the form $\sum_{k=0} \phi_k T_k = 0$ results, where the ϕ_k are functions of the A_i, a_i, and b_i.

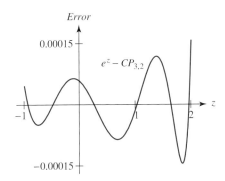

FIGURE 41.8 The error made when $f(z) = e^z$ is approximated by the 3, 2 Chebyshev–Padé rational function (41.8a)

7. Solve the $n + m + 1$ equations $\phi_k = 0$, $k = 0, \ldots, n + m$, for the $n + m + 1$ unknowns, $a_0, \ldots, a_n$ and $b_1, \ldots, b_m$.

EXAMPLE 41.5   This, the Maehly algorithm, is now illustrated for $f(z) = e^z$, $-1 \le z \le 2$. The interval $-1 \le z \le 2$ is mapped to $-1 \le x \le 1$ by the function $x(z) = \frac{2z-1}{3}$ and mapped back by the function $z(x) = \frac{3x+1}{2}$. Then, obtain $g(x) = f(z(x)) = e^{(3x+1)/2}$. Let $n = 3$ and $m = 2$ so that $P = \sum_{k=0}^{3} a_k T_k$ and $Q = 1 + \sum_{k=1}^{2} b_k T_k$. Expand $g(x)$ in the Chebyshev series $g(x) = \frac{A_0}{2} + \sum_{k=1}^{5} A_k T_k(x)$. The coefficients $A_k = \frac{2}{\pi} \int_{-1}^{1} [e^{(3x+1)/2} T_k(x)/\sqrt{1-x^2}] \, dx$ are listed in Table 41.7. Equate $R = \frac{P}{Q}$ with $g(x) = \frac{A_0}{2} + \sum_{k=1}^{5} A_k T_k$, and set the numerator to zero, obtaining

$$gQ - P = \left(\frac{A_0}{2} + \sum_{k=1}^{5} A_k T_k \right) \sum_{k=0}^{3} a_k T_k - (T_0 + b_1 T_1 + b_2 T_2) = 0$$

Expanding the left side leads to the surprisingly complex

$$\begin{aligned} 0 = &A_0 + A_0 b_1 T_1 + A_0 b_2 T_2 + 2A_1 T_1 + 2A_1 T_1^2 b_1 + 2A_1 b_2 T_2 T_1 + 2A_2 T_2 + 2A_2 b_1 T_1 T_2 \\ &+ 2A_2 T_2^2 b_2 + 2A_3 T_3 + 2A_3 b_1 T_1 T_3 + 2A_3 b_2 T_2 T_3 + 2A_4 T_4 + 2A_4 b_1 T_1 T_4 - 2a_3 T_3 \\ &+ 2A_4 b_2 T_2 T_4 + 2A_5 T_5 + 2A_5 b_1 T_1 T_5 + 2A_5 b_2 T_2 T_5 - 2a_0 - 2a_1 T_1 - 2a_2 T_2 \end{aligned}$$

(41.9)

The products of Chebyshev polynomials are simplified via the identity

$$T_n T_m = \tfrac{1}{2}(T_{n+m} + T_{|n-m|}),$$

which follows from writing $T_k(x)$ as $\cos(k \arccos x) = \cos k\theta$ and applying to the product $T_n T_m = \cos n\theta \cos m\theta$, the trig identity $\cos n\theta \cos m\theta = \frac{1}{2}(\cos(m+n)\theta + \cos(n-m)\theta)$. The absolute value on the difference $n - m$ can be accounted for by the evenness of $\cos \theta$ whereby $\cos(-\theta) = \cos(\theta)$.

Removing the products from (41.9) leads to $\sum_{k=0}^{7} \phi_k T_k = 0$, where the $n + m + 1 = 6$ equations $\phi_k = 0$ are linear and can be expressed in the form $A\mathbf{x} = \mathbf{y}$, where the matrix A and the vectors $\mathbf{x}$ and $\mathbf{y}$ are defined by

$$\begin{bmatrix} -2 & 0 & 0 & 0 & A_1 & A_2 \\ 0 & -2 & 0 & 0 & A_0 + A_2 & A_1 + A_3 \\ 0 & 0 & -2 & 0 & A_1 + A_3 & A_0 + A_4 \\ 0 & 0 & 0 & -2 & A_2 + A_4 & A_1 + A_5 \\ 0 & 0 & 0 & 0 & A_3 + A_5 & A_2 \\ 0 & 0 & 0 & 0 & A_4 & A_3 \end{bmatrix} \begin{bmatrix} a_0 \\ a_1 \\ a_2 \\ a_3 \\ b_1 \\ b_2 \end{bmatrix} = \begin{bmatrix} -A_0 \\ -2A_1 \\ -2A_2 \\ -2A_3 \\ -2A_4 \\ -2A_5 \end{bmatrix}$$

(41.10)

Although interesting, a symbolic solution of (41.10) is complicated. Hence, the floating-point form of the equations in (41.10), namely,

$$-a_0 + 1.618494321 b_1 + 0.5569951210 b_2 + 2.714987549 = 0$$
$$3.236988642 + 3.271982670 b_1 + 1.751668319 b_2 - a_1 = 0$$
$$2.739286677 b_2 + 1.113990242 - a_2 + 1.751668319 b_1 = 0$$
$$-a_3 + 0.5812942492 b_1 + 1.622072969 b_2 + 0.2663479964 = 0$$
$$0.5569951210 b_2 + 0.04859825644 + 0.1367526459 b_1 = 0$$
$$0.1331739982 b_2 + 0.02429912822 b_1 + 0.007157295460 = 0$$

$A_0 = 5.429975098$
$A_1 = 3.236988642$
$A_2 = 1.1139902420$
$A_3 = 0.2663479964$
$A_4 = 0.04859825644$
$A_5 = 0.00715729546$

TABLE 41.7   Fourier–Chebyshev coefficients for $g(x) = e^{(3x+1)/2}$, $-1 \le x \le 1$, in Example 41.5

is instead solved, yielding the rational function

$$R = \frac{1.879031459T_0 + 1.574036418T_1 + 0.3015685278T_2 + 0.02755554881T_3}{1 - 0.5313732123T_1 + 0.04321121568T_2}$$

$$= \frac{1.577462931 + 1.491369772x + 0.6031370556x^2 + 0.1102221952x^3}{0.9567887843 - 0.5313732123x + 0.08642243136x^2}$$

Finally, obtain $\rho(z) = R(x(z))$, where $\rho(z)$ is given by

$$\rho(z) = \frac{1.143272598 + 0.7506794224z + 0.2190732713z^2 + 0.03265842821z^3}{1.143515680 - 0.3926587777z + 0.03840996949z^2} \quad \textbf{(41.11)}$$

The maximum magnitude of the error made when using $\rho(z)$ to approximate $f(z) = e^z$ on $-1 \le z \le 2$ is $0.3755860225 \times 10^{-3}$.

The algorithm of Clenshaw and Lord yields the approximation (41.8a) for which the maximum magnitude of the error is $0.1573484994 \times 10^{-3}$, about half that of (41.11), because it uses a slightly different matching scheme than [16]. In fact, [48] suggests several alternatives, one of which is the collocation strategy $g(x_k) = R(x_k), k = 1, \ldots, n+m+1$, for $x_k, k = 1, \ldots, n+m+1$, in $-1 \le x \le 1$. Taking $x_k = -1 + \frac{2}{5}k, k = 0, \ldots, 5$, we obtain

$$\rho_c = \frac{1.148207901 + 0.7262159545z + 0.1980125881z^2 + 0.02548951481z^3}{1.148238067 - 0.4221292593z + 0.04569204506z^2}$$

for which the maximum magnitude of the error is $0.5527915183 \times 10^{-3}$. Other techniques are explored in the exercises.

Surprisingly, the rational approximation provided by the Clenshaw–Lord algorithm is not the best rational-function approximation. A better one is the *minimax* approximation, discussed next. ❖

The Minimax Approximation

The *minimax* rational function minimizes the maximum error in the approximation of $f(z)$. Typically computed by an iterative process attributed to Remes, the [3/2] minimax approximation for $f(z) = e^z$ on $[-1, 2]$ is given by Maple's **minimax** command as

$$\rho_{mm} = \frac{1.146452 + (0.7350169 + (0.2055243 + 0.02805070z)z)z}{1.146549 + (-0.4115607 + 0.04307749z)z}$$

where the calculations were performed at extended precision.

The maximum magnitude of the error made by this rational function is 0.841×10^{-4} and occurs at $x = 1.862$. Figure 41.9 shows the error to be uniform and small. Evaluation of the nested (Horner) form of the minimax approximation requires a total of five multiplications or divisions and five additions. On the other hand, in the continued-fraction form (41.12), evaluation requires just two divisions, a multiplication, and five additions.

$$10.9928 + 0.6511684z + \cfrac{104.7510}{z - 7.01510 + \cfrac{8.805248}{z - 2.538938}} \quad \textbf{(41.12)}$$

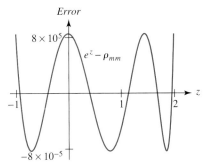

Error

8×10^{-5}

$e^z - \rho_{mm}$

-8×10^{-5}

FIGURE 41.9 The error made by the minimax approximation of $f(z) = e^z$

Additional information on the minimax approximation and the (second) algorithm of Remes can be found in [41], [75], and [72].

EXERCISES 41.4

1. Verify that the matching process in Maehly's algorithm for generating (41.7), the Chebyshev–Padé rational approximation of the function $f(x) = \frac{A_0}{2} + \sum_{k=1}^{n+m} A_k T_k(x)$, results in the system of equations $2a_0 = \sum_{k=1}^{m} A_k b_k$, $2a_j = \sum_{k=1}^{m} (A_{j+k} + A_{|j-k|}) b_k$, $k = 1, \ldots, n+m$, where $A_\alpha = 0$ if $\alpha > n + m$ and $a_\alpha = 0$ if $\alpha > n$.

Exercises 2–7, apply to any of the functions $f_k(x)$, $k = 1, \ldots, 10$, from the following list. Each function has the domain $-1 \le x \le 1$.

$$f_1(x) = \cos \frac{\pi}{2} x \qquad f_6(x) = \sqrt{1 - x^2}$$
$$f_2(x) = x(1 - x) \qquad f_7(x) = x \arctan x$$
$$f_3(x) = x^2(1 - x) \qquad f_8(x) = \ln \cosh x$$
$$f_4(x) = x(1 - x)^2 \qquad f_9(x) = \frac{\cos x}{\cosh x}$$
$$f_5(x) = xe^{-x} \qquad f_{10}(x) = \sqrt{1 + 2 \cos^2 x}$$

2. For a function $f_i(x)$ from the list provided:

(a) Use Maehly's algorithm to obtain $R_1(x)$ as a $(3, 2)$-Chebyshev–Padé rational approximation to $f_i(x)$.

(b) Graph $\varepsilon_1(x) = f_i(x) - R_1(x)$, the error of the approximation on the interval $[-1, 1]$.

(c) If $N = n + m = 3 + 2 = 5$, verify that $\varepsilon_1(x)$ has $N + 2 = 7$ local extrema at $x_0 = -1$, $x_{N+1} = 1$, and x_k, $k = 1, \ldots, N$, where $\varepsilon_1'(x_k) = 0$ for each x_k.

(d) If $k = 0, \ldots, N + 1$, obtain $|\varepsilon_1(x_k)|$, $m_1 = \min_k\{|\varepsilon_1(x_k)|\}$, and $M_1 = \max_k\{|\varepsilon_1(x_k)|\}$.

(e) Obtain $\|\varepsilon_1(x)\|_\infty$.

3. For the function $f_i(x)$ selected earlier:

(a) Determine $R_2(x)$, the $(3, 2)$-Chebyshev–Padé rational approximation to $f_i(x)$ obtained via collocation at ξ_k, $k = 1, \ldots, N + 1$, the $N + 1$ zeros of $T_{N+1}(x)$, where $N = n + m = 3 + 2 = 5$.

(b) Graph $\varepsilon_2(x) = f_i(x) - R_2(x)$, the error of the approximation on the interval $[-1, 1]$.

(c) Verify that $\varepsilon_2(x)$ has $N + 2 = 7$ extrema at $x_0 = -1$, $x_{N+1} = 1$, and x_k, $k = 1, \ldots, N$, where $\varepsilon_2'(x_k) = 0$ for each x_k.

(d) If $k = 0, \ldots, N + 1$, obtain $|\varepsilon_2(x_k)|$, $m_2 = \min_k\{|\varepsilon_2(x_k)|\}$, and $M_2 = \max_k\{|\varepsilon_2(x_k)|\}$.

(e) Obtain $\|\varepsilon_2(x)\|_\infty$.

4. For the function $f_i(x)$ selected earlier:

(a) Determine $R_3(x)$, the $(3, 2)$-Chebyshev–Padé rational approximation to $f_i(x)$ obtained via collocation at the $N + 1$ uniformly spaced nodes $\zeta_k = -1 + k\frac{1 - (-1)}{N}$, $k = 0, \ldots, N$, where $N = n + m = 3 + 2 = 5$.

(b) Graph $\varepsilon_3(x) = f_i(x) - R_3(x)$, the error of the approximation on the interval $[-1, 1]$.

(c) Verify that $\varepsilon_3(x)$ has $N + 2 = 7$ extrema at $x_0 = -1$, $x_{N+1} = 1$, and x_k, $k = 1, \ldots, N$, where $\varepsilon_3'(x_k) = 0$ for each x_k.

(d) If $k = 0, \ldots, N + 1$, obtain $|\varepsilon_3(x_k)|$, $m_3 = \min_k\{|\varepsilon_3(x_k)|\}$, and $M_3 = \max_k\{|\varepsilon_3(x_k)|\}$.

(e) Obtain $\|\varepsilon_3(x)\|_\infty$.

5. For the function $f_i(x)$ selected earlier:

(a) Determine $R_4(x)$, the $(3, 2)$-Chebyshev–Padé rational approximation to $f_i(x)$ obtained via a least-squares technique implemented by solving the $N + 1$ equations $\int_{-1}^{1} [(f_i(x)Q - P)]T_k(x)/\sqrt{1 - x^2}]\, dx = 0$, $k = 0, \ldots, N$, where $N = n + m = 3 + 2 = 5$.

(b) Graph $\varepsilon_4(x) = f_i(x) - R_4(x)$, the error of the approximation on the interval $[-1, 1]$.

(c) Verify that $\varepsilon_4(x)$ has $N + 2 = 7$ extrema at $x_0 = -1$, $x_{N+1} = 1$, and x_k, $k = 1, \ldots, N$, where $\varepsilon_4'(x_k) = 0$ for each x_k.

(d) If $k = 0, \ldots, N + 1$, obtain $|\varepsilon_4(x_k)|$, $m_4 = \min_k\{|\varepsilon_4(x_k)|\}$, and $M_4 = \max_k\{|\varepsilon_4(x_k)|\}$.

(e) Obtain $\|\varepsilon_4(x)\|_\infty$.

6. For the function $f_i(x)$ selected earlier:

(a) If possible, obtain $R_5(x)$, the $(3, 2)$-Chebyshev–Padé rational approximation to $f_i(x)$ given by Maple's **chebpade** command or the equivalent.

(b) Graph $\varepsilon_5(x) = f_i(x) - R_5(x)$, the error of the approximation on the interval $[-1, 1]$.

(c) Verify that $\varepsilon_5(x)$ has $N + 2 = 7$ extrema at $x_0 = -1$, $x_{N+1} = 1$, and x_k, $k = 1, \ldots, N$, where $\varepsilon_5'(x_k) = 0$ for each x_k.

(d) If $k = 0, \ldots, N + 1$, obtain $|\varepsilon_5(x_k)|$, $m_5 = \min_k\{|\varepsilon_5(x_k)|\}$, and $M_5 = \max_k\{|\varepsilon_5(x_k)|\}$.

(e) Obtain $\|\varepsilon_5(x)\|_\infty$.

7. For the function $f_i(x)$ selected earlier:

(a) If possible, obtain $R_6(x)$, the $(3, 2)$-minimax rational approximation to $f_i(x)$ given by Maple's **minimax** command or the equivalent.

(b) Graph $\varepsilon_6(x) = f_i(x) - R_6(x)$, the error of the approximation on the interval $[-1, 1]$.

(c) Verify that $\varepsilon_6(x)$ has $N + 2 = 7$ extrema at $x_0 = -1$, $x_{N+1} = 1$, and x_k, $k = 1, \ldots, N$, where $\varepsilon_6'(x_k) = 0$ for each x_k.

(d) If $k = 0, \ldots, N + 1$, obtain $|\varepsilon_6(x_k)|$, noting the relative constancy of these values.

(e) Obtain $\|\varepsilon_6(x)\|_\infty$.

8. To the extent possible, for the results obtained in Exercises 2–7, show that $m_k < \|\varepsilon_6(x)\|_\infty < M_k$, $k = 1, \ldots, 5$. Thus, the minimax approximation $R_6(x)$, if found, has a smaller error than any other rational approximation, but because it has a uniform error, its error is larger than the smallest extrema of another rational approximation.

For the functions $g(z)$ defined on the indicated intervals $[\alpha, \beta]$ in Exercises 9–18:

   (a) Obtain $R(z)$, a $(3, 2)$-Chebyshev–Padé rational approximation.

   (b) Graph $\varepsilon(z) = g(z) - R(z)$, the error of the approximation, and determine $\|g(z) - R(z)\|_\infty$.

   (c) If possible, compare $R(z)$ to the minimax approximation provided by Maple's **minimax** command.

9. $g(z) = \dfrac{z^2}{1 + z^2}, -3 \le z \le 3$ **10.** $g(z) = e^{-z}\cos z, 0 \le z \le 2\pi$

11. $g(z) = 4z^3 - 7z^2 + 5z + 2, -2 \le z \le 2$

12. $g(z) = ze^{-z}, 0 \le z \le 5$ **13.** $g(z) = \tanh z, -3 \le z \le 2$

14. $g(z) = e^{-z^2}, -2 \le z \le 2$

15. $g(z) = \dfrac{z^3 - 2z^2 - 5z + 7}{z^2 + z + 1}, 0 \le z \le 2$

16. $g(z) = \dfrac{100z}{7z^2 + 5z + 13}, 0 \le z \le 10$

17. $g(z) = e^{-z}\ln z, 1 \le z \le 5$ **18.** $g(z) = z(1 - z), 0 \le z \le 1$

Chapter Review

1. Obtain the least-squares approximation of $f(x) = x, 0 \le x \le 1$, using the functions $\sin k\pi x, k = 1, 2, 3$. How small is the square-norm of the difference between $f(x)$ and the approximation?

2. What is the infinity-norm of the difference between $f(x)$ and the least-squares approximation in Question 1?

3. Obtain the least-squares approximation of $f(x) = \sin x, 0 \le x \le \frac{\pi}{2}$, using the functions $1, x, x^2$. How small is the square-norm of the difference between $f(x)$ and the approximation?

4. What is the infinity-norm of the difference between $f(x)$ and the least-squares approximation in Question 3?

5. What is the advantage of performing a least-squares approximation with a set of *orthogonal* functions?

6. Obtain, at $x = 0$, the $[1/2]$ Padé approximation of $\sin x$.

7. Obtain the Chebyshev expansion of $f(x) = \frac{x+1}{x+2}, -1 \le x \le 2$. Sketch the calculations by which this series is determined.

8. Describe the difference between the Maehly algorithm for finding the Chebyshev–Padé approximation to $f(x)$ and the Clenshaw–Lord algorithm.

9. Describe the minimax rational approximation to $f(x)$.

Chapter *42*

Numeric Differentiation

INTRODUCTION   Chapter 42 is a short chapter, containing two essential ideas, namely, techniques for *numeric differentiation* and Richardson extrapolation. Because of the potential for subtractive cancellation in the difference quotient, numeric differentiation is inherently unstable on a finite precision computing device. A variety of *finite-difference formulas* for derivatives are obtained, and the notion of *polynomial exactness* appears. The degree of polynomial exactness refers to the degree of a polynomial for which a differentiation formula gives an exact answer. That and a method's order are the two ways of quantizing the accuracy of numeric differentiation.

Richardson extrapolation is a technique for generating more accurate numeric values from two values of lesser accuracy. This device is applied to numeric differentiation and is useful in that context because with it, the perils of subtractive cancellation are more likely to be avoided. Richardson extrapolation appears again in Chapter 43, where, in conjunction with the Trapezoid rule for numeric integration, it forms the method called *Romberg integration*. The theory that makes Richardson extrapolation viable also justifies error assessment in adaptive implementations of numeric integration and numeric methods for solving differential equations.

42.1 Basic Formulas

Naive Numeric Differentiating

In Section 25.4 we actually used the $O(h)$ *forward-difference* formula

$$f'(c) \overset{\circ}{=} \frac{f(c+h) - f(c)}{h} \tag{42.1}$$

to replace $u_t(x_n, t_m)$ with $(v_{n,m+1} - v_{n,m})/dt$ in the numeric solution of the heat equation on a finite rod. In Section 37.1 we used this formula to give an example of dramatic roundoff error so severe that subtractive cancellation could be made to invalidate numeric differentiation on an n-digit computing device, no matter how large n might be. In particular, our example was the computation of $f'(1) = e$ for the function $f(x) = e^x$. We wrote the difference quotient as $(e^{(1+h)} - e)/h$ and found for $h = 10^{-k}$, $k = 0, 1, \ldots, 10$, the results in Table 42.1.

As h decreases, the difference quotient begins to converge to e, but then, with further decrease in h, the difference quotient becomes wildly erroneous.

In this section we will study and derive more stable formulas for numeric differentiation. Formulas that obtain approximations of derivatives at $x = c$ using function values at and

h	$\dfrac{e^{(1+h)} - e}{h}$
1	4.670774271
10^{-1}	2.85884196
10^{-2}	2.7319187
10^{-3}	2.719642
10^{-4}	2.71842
10^{-5}	2.7183
10^{-6}	2.719
10^{-7}	2.72
10^{-8}	2.8
10^{-9}	3.0
10^{-10}	0

TABLE 42.1   At $x = 1$, and for $f(x) = e^x$, values of the difference quotient with $h = 10^{-k}$, $k = 0, \ldots, 10$

to the right of $x = c$ are called *forward-difference* formulas. Formulas that obtain such approximations using function values at and to the left of $x = c$ are called *backward-difference* formulas. Finally, formulas that obtain such approximations using function values at and symmetrically to the left and right of $x = c$ are called *central-difference* formulas.

Forward- and Backward-Difference Formulas

The naive formula (42.1) is an O(h) forward-difference expression for $f'(c)$ because the error it makes is proportional to h. Table 42.2 shows this for the calculation of $f'(1) = 3e^{3x}$, where $f(x) = e^{3x}$, by listing $h_k = 2^{-k}$, $k = 0, \ldots, 10$, the value of $f_k = (f(1 + h_k) - f(1))/h_k$, and the ratios $(f_k - f'(1))/h_k$. That these ratios tend to a constant is empirical evidence that the approximation is O(h).

h_k	$f_k = \dfrac{f(1 + h_k) - f(1)}{h_k}$	$\dfrac{f_k - f'(1)}{h_k}$
1	383.34	323.09
2^{-1}	139.86	159.21
2^{-2}	89.74	117.94
2^{-3}	73.11	102.83
2^{-4}	66.28	96.31
2^{-5}	63.17	93.28
2^{-6}	61.69	91.81
2^{-7}	60.97	91.10
2^{-8}	60.61	90.74
2^{-9}	60.43	90.56
2^{-10}	60.34	90.48

TABLE 42.2   For $f(x) = e^{3x}$, a demonstration that the truncation error for $f'(1) = \frac{f(1+h)-f(1)}{h}$ is O(h)

The formula approximating $f'(x)$ in Table 42.2 is an example of a 2-point formula. The following equispaced $(n + 1)$-point formulas for derivatives are O(h^2) forward-difference formulas when $h > 0$ and O(h^2) *backward-difference* formulas when $h < 0$.

$$f'(c) \doteq \frac{1}{2h}[-3f(c) + 4f(c + h) - f(c + 2h)] \tag{42.2a}$$

$$f''(c) \doteq \frac{1}{h^2}[2f(c) - 5f(c + h) + 4f(c + 2h) - f(c + 3h)] \tag{42.2b}$$

$$f'''(c) \doteq \frac{1}{2h^3}[-5f(c) + 18f(c + h) - 24f(c + 2h) + 14f(c + 3h) - 3f(c + 4h)] \tag{42.2c}$$

$$f^{(4)}(c) \doteq \frac{3f(c) - 14f(c + h) + 26f(c + 2h) - 24f(c + 3h) + 11f(c + 4h) - 2f(c + 5h)}{h^4}$$

$$\tag{42.2d}$$

Not only are these formulas of order O(h^2), they each have truncation errors of the form $Ch^2 + Dh^3 + \cdots$, as we will see. In addition, these formulas are exact for polynomials of degree 2, 3, 4, 5, respectively. Table 42.3 summarizes the results of using these formulas to compute $f^{(n-1)}(c)$ for the polynomials $p_n(x) = \sum_{k=0}^{n} a_k x^k$, $n = 2, \ldots, 5$, thereby verifying the polynomial exactness of each.

n	$p_n(x) = \sum_{k=0}^{n} a_k x^k$	$p_n^{(n-1)}(c) = f^{(n-1)}(c)$	**Exactness Degree**
2	$\sum_{k=0}^{2} a_k x^k$	$a_1 + 2a_2 c$	2
3	$\sum_{k=0}^{3} a_k x^k$	$2a_2 + 6a_3 c$	3
4	$\sum_{k=0}^{4} a_k x^k$	$6a_3 + 24a_4 c$	4
5	$\sum_{k=0}^{5} a_k x^k$	$24a_4 + 120a_5 c$	5

TABLE 42.3   Polynomial exactness of degree n for the $O(h^2)$ forward-difference formulas for $f^{(n-1)}(c)$

We conclude this discussion with derivations of these forward- and backward-difference differentiation formulas. We base the derivations on Taylor series expansions, thereby demonstrating the origin and nature of the error estimates.

DERIVATION OF FORMULA (42.1)   To obtain the $O(h)$ formula (42.1), solve the Taylor expansion $f(c + h) = \sum_{k=0}^{n} f^{(k)}(c)h^k/k! + O(h^{n+1})$, $n \geq 3$, for $f'(c)$, obtaining

$$f'(c) = \frac{f(c+h) - f(c)}{h} - \sum_{k=2}^{n} \frac{f^{(k)}(c)}{k!} h^{k-1} + O(h^n)$$

The first term is the formula under consideration, whereas the remaining terms give the truncation error in the form $Ch + Dh^2 + \cdots$. Thus, the formula is of order $O(h)$.

DERIVATION OF FORMULA (42.2a)   To derive formula (42.2a), write the Taylor expansions

$$f(c+h) = f(c) + f'(c)h + \frac{f''(c)}{2}h^2 + \frac{f'''(c)}{6}h^3 + \frac{f^{(4)}(c)}{24}h^4 + \cdots$$

$$f(c+2h) = f(c) + 2f'(c)h + 2f''(c)h^2 + \tfrac{4}{3}f'''(c)h^3 + \tfrac{2}{3}f^{(4)}h^4 + \cdots$$

and subtract the second equation from 4 times the first to eliminate the term in $f''(c)$. This gives

$$4f(c+h) - f(c+2h) = 3f(c) + 2f'(c)h - \tfrac{2}{3}f'''(c)h^3 - \tfrac{1}{2}f^{(4)}(c)h^4 + \cdots$$

to be solved for

$$f'(c) = \frac{1}{2h}[-3f(c) + 4f(c+h) - f(c+2h)] + \tfrac{1}{3}f'''(c)h^2 + \tfrac{1}{4}f^{(4)}(c)h^3 + \cdots$$

The first term on the right gives (42.2a), whereas the remaining terms indicate the truncation error is of the form $Ch^2 + Dh^3 + \cdots$. Thus, the formula is of order $O(h^2)$.

DERIVATION OF FORMULA (42.2b)   To derive formula (42.2b), write the Taylor expansions

$$f(c+h) = f(c) + f'(c)h + \frac{f''(c)}{2}h^2 + \frac{f'''(c)}{6}h^3 + \frac{f^{(4)}(c)}{24}h^4 + \frac{f^{(5)}(c)}{120}h^5 + \cdots$$

$$f(c+2h) = f(c) + 2f'(c)h + 2f''(c)h^2 + \tfrac{4}{3}f'''(c)h^3 + \tfrac{2}{3}f^{(4)}h^4 + \tfrac{4}{15}f^{(5)}(c)h^5 + \cdots$$

$$f(c+3h) = f(c) + 3f'(c)h + \tfrac{9}{2}f''(c)h^2 + \tfrac{9}{2}f'''(c)h^3$$
$$+ \tfrac{27}{8}f^{(4)}(c)h^4 + \tfrac{81}{40}f^{(5)}(c)h^5 + \cdots$$

To the first equation add $-\frac{4}{5}$ times the second and $\frac{1}{5}$ times the third, yielding

$$f(c+h) - \tfrac{4}{5}f(c+2h) + \tfrac{1}{5}f(c+3h) = \tfrac{2}{5}f(c) - \tfrac{1}{5}f''(c)h^2$$
$$+ \tfrac{11}{60}f^{(4)}(c)h^4 + \tfrac{1}{5}f^{(5)}(c)h^5 + \cdots$$

which, when solved for $f''(c)$, gives

$$f''(c) = \frac{1}{h^2}[2f(c) - 5f(c+h) + 4f(c+2h) - f(c+3h)]$$
$$+ \tfrac{11}{12}f^{(4)}(c)h^2 + f^{(5)}(c)h^3 + \cdots$$

The first term on the right gives (42.2b), whereas the remaining terms indicate that the truncation error is of the form $Ch^2 + Dh^3 + \cdots$. Thus, the formula is of order $\mathrm{O}(h^2)$.

DERIVATION OF FORMULA (42.2c)   To derive formula (42.2c), write the Taylor expansions $f(c+nh) = \sum_{k=0} f^{(k)}(c)(nh)^k/k!, n = 1, \ldots, 4$, and form the linear combination

$$f(c+h) - \tfrac{4}{3}f(c+2h) + \tfrac{7}{9}f(c+3h) - \tfrac{1}{6}f(c+4h)$$
$$= \tfrac{5}{18}f(c) + \tfrac{1}{9}f'''(c)h^3 - \tfrac{7}{36}f^{(5)}(c)h^5 - \tfrac{5}{18}f^{(6)}(c)h^6 + \cdots$$

Solving for $f'''(c)$ gives

$$f'''(c) = \frac{1}{2h^3}[-5f(c) + 18f(c+h) - 24f(c+2h) + 14f(c+3h) - 3f(c+4h)]$$
$$+ \tfrac{7}{4}f^{(5)}(c)h^2 + \tfrac{5}{2}f^{(6)}(c)h^3 + \cdots$$

The first term on the right gives (42.2c), whereas the remaining terms indicate that the truncation error is of the form $Ch^2 + Dh^3 + \cdots$. Thus, the formula is of order $\mathrm{O}(h^2)$.

DERIVATION OF FORMULA (42.2d)   To derive formula (42.2d), write the Taylor expansions $f(c+nh) = \sum_{k=0} f^{(k)}(c)(nh)^k/k!, n = 1, \ldots, 5$, and form the linear combination

$$f(c+h) - \tfrac{13}{7}f(c+2h) + \tfrac{12}{7}f(c+3h) - \tfrac{11}{14}f(c+4h) + \tfrac{1}{7}f(c+5h)$$
$$= \tfrac{3}{14}f(c) - \tfrac{1}{14}f^{(4)}(c)h^4 + \tfrac{17}{84}f^{(6)}(c)h^6 + \tfrac{5}{14}f^{(7)}(c)h^7 + \cdots$$

Solving for $f^{(4)}(c)$ gives

$$f^{(4)}(c) = \frac{1}{h^4}[3f(c) - 14f(c+h) + 26f(c+2h) - 24f(c+3h)$$
$$+ 11f(c+4h) - 2f(c+5h)] + \tfrac{17}{6}f^{(6)}(c)h^2 + 5f^{(7)}(c)h^3 + \cdots$$

The first term on the right gives (42.2d), whereas the remaining terms indicate that the truncation error is of the form $Ch^2 + Dh^3 + \cdots$. Thus, the formula is of order $\mathrm{O}(h^2)$.

Central-Difference Formulas

The following are $O(h^2)$ *central-difference* formulas with truncation errors of the form $Ch^2 + Dh^4 + Eh^6 + \cdots$. We have, in fact, used the first formula in Section 15.2 and the second in Sections 15.2, 24.6, 25.4, and 26.4. These are $(n + 1)$-point formulas: the first is a three-point formula even though $f(c)$ has a coefficient of zero! Finally, the formula for $f^{(n-1)}(c)$ is exact for polynomials of degree n. If, for $n = 2, \ldots, 5$, the formula for $f^{(n-1)}(c)$ is applied to the polynomial $p_n(x) = \sum_{k=0}^{n} a_k x^k$, the results listed in Table 42.3 are again obtained.

$$f'(c) \doteq \frac{1}{2h}[-f(c - h) + f(c + h)] \tag{42.3a}$$

$$f''(c) \doteq \frac{1}{h^2}[f(c - h) - 2f(c) + f(c + h)] \tag{42.3b}$$

$$f'''(c) \doteq \frac{1}{2h^3}[-f(c - 2h) + 2f(c - h) - 2f(c + h) + f(c + 2h)] \tag{42.3c}$$

$$f^{(4)}(c) \doteq \frac{1}{h^4}[f(c - 2h) - 4f(c - h) + 6f(c) - 4f(c + h) + f(c + 2h)] \tag{42.3d}$$

Table 42.4 contains empirical evidence that (42.3b), applied to $f(x) = e^{3x}$ at $c = 1$, is of order $O(h^2)$. Table 42.4 lists $h_k = 1/2^k$, $f_k = (e^{3(1-h_k)} - 2e^3 + e^{3(1+h_k)})/h_k^2$, and the ratios $(f_k - f''(1))/h_k^2$, $k = 0, \ldots, 7$. The constancy of the ratios supports the claim that (42.3b) is $O(h^2)$.

DERIVATION OF FORMULA (42.3a) To derive formula (42.3a), write the Taylor expansions $f(c \pm h) = \sum_{k=0}^{\infty} f^{(k)}(c)(\pm h)^k/k!$ and subtract the second from the first, obtaining

$$f(c + h) - f(c - h) = 2f'(c)h + \tfrac{1}{3}f'''(c)h^3 + \tfrac{1}{60}f^{(5)}(c)h^5 + \cdots$$

Solving for $f'(c)$ then gives

$$f'(c) = \frac{1}{2h}[-f(c - h) + f(c + h)] - \tfrac{1}{6}f'''(c)h^2 - \tfrac{1}{120}f^{(5)}(c)h^4 + \cdots$$

The first term on the right-hand side gives (42.3a), whereas the remaining terms indicate the truncation error is of the form $Ch^2 + Dh^4 + \cdots$. Thus, this formula is indeed of order $O(h^2)$.

DERIVATION OF FORMULA (42.3b) To derive formula (42.3b), add the two Taylor expansions $f(c \pm h) = \sum_{k=0}^{\infty} f^{(k)}(c)(\pm h)^k/k!$ to obtain

$$f(c - h) + f(c + h) = 2f(c) + f''(c)h^2 + \tfrac{1}{12}f^{(4)}(c)h^4 + \tfrac{1}{130}f^{(6)}(c)h^6 + \cdots$$

Solving for $f''(c)$ then gives

$$f''(c) = \frac{1}{h^2}[f(c - h) - 2f(c) + f(c + h)] - \tfrac{1}{12}f^{(4)}(c)h^2 - \tfrac{1}{360}f^{(6)}(c)h^4 + \cdots$$

The first term on the right-hand side establishes (42.3b), whereas the remaining terms indicate the truncation error is of the form $Ch^2 + Dh^4 + \cdots$. Thus, this formula is indeed of order $O(h^2)$.

DERIVATION OF FORMULA (42.3c) To derive (42.3c), write the four Taylor expansions $f(c \pm nh) = \sum_{k=0}^{\infty} f^{(k)}(c)(\pm nh)^k/k!$, $n = 1, 2$, and isolate $f'''(c)$ from the linear combi-

$$f_k = \frac{e^{3(1-h_k)} - 2e^3 + e^{3(1+h_k)}}{h_k^2}$$

h_k	f_k	$\dfrac{f_k - f''(1)}{h_k^2}$
1	364.258	183.488
2^{-1}	217.311	146.165
2^{-2}	189.404	138.145
2^{-3}	182.898	136.215
2^{-4}	181.300	135.737
2^{-5}	180.902	135.626
2^{-6}	180.803	135.937
2^{-7}	180.778	135.458

TABLE 42.4 For $f(x) = e^{3x}$, a demonstration that the truncation error for $f''(1) = \frac{1}{h}[f(1 - h) - 2f(1) + f(1 + h)]$ is $O(h^2)$

nation $f(c - 2h) - 2f(c - h) + 2f(c + h) - f(c + 2h)$, thereby obtaining

$$f'''(c) = \frac{1}{2h^3}[-f(c - 2h) + 2f(c - h) - 2f(c + h) + f(c + 2h)]$$

$$-\frac{f^{(5)}(c)}{4}h^2 - \frac{f^{(7)}(c)}{40}h^4 + \cdots$$

The first term on the right-hand side establishes (42.3c), whereas the remaining terms indicate the truncation error is of the form $Ch^2 + Dh^4 + \cdots$. Thus, this formula is indeed of order $O(h^2)$.

DERIVATION OF FORMULA (42.3d)   To derive (42.3d), write the four Taylor expansions $f(c \pm nh) = \sum_{k=0} f^{(k)}(c)(\pm nh)^k / k!, n = 1, 2$, and isolate $f^{(4)}(c)$ from the linear combination $f(c - 2h) - 4f(c - h) - 4f(c + h) + f(c + 2h)$, thereby obtaining

$$f^{(4)}(c) = \frac{1}{h^4}[f(c - 2h) - 4f(c - h) + 6f(c) - 4f(c + h) + f(c + 2h)]$$

$$-\frac{f^{(6)}(c)}{6}h^2 - \frac{f^{(8)}(c)}{80}h^4 + \cdots$$

The first term on the right-hand side establishes (42.3d), whereas the remaining terms indicate the truncation error is of the form $Ch^2 + Dh^4 + \cdots$. Thus, this formula is indeed of order $O(h^2)$.

Nonuniform Spacing

The forward-difference and central-difference formulas just studied, and derived from Taylor expansions, implement numerical differentiation for data with equally spaced nodes. Formulas for numerical differentiation of data with nonuniformly spaced nodes are derived from interpolating polynomials. Of course, equispaced nodes are a special case of this new theory, so we will be able to obtain all the previous formulas by this approach.

For example, to approximate $f'''(c)$ from the $n + 1 = 5$ points $(-5, y_0), (-2, y_1), (3, y_2), (6, y_3), (9, y_4)$ with nonuniformly spaced nodes, interpolate the points with the polynomial

$$P = \left(\frac{1}{2772}y_4 - \frac{1}{792}y_3 + \frac{1}{720}y_2 - \frac{1}{1320}y_1 + \frac{1}{3696}y_0\right)x^4$$

$$+ \left(\frac{5}{792}y_3 + \frac{13}{1320}y_1 - \frac{1}{231}y_0 - \frac{1}{90}y_2 - \frac{1}{1386}y_4\right)x^3$$

$$+ \left(\frac{3}{176}y_0 - \frac{5}{396}y_4 - \frac{41}{720}y_2 - \frac{3}{440}y_1 + \frac{47}{792}y_3\right)x^2$$

$$+ \left(-\frac{111}{440}y_1 + \frac{1}{77}y_4 - \frac{23}{264}y_3 + \frac{19}{60}y_2 + \frac{3}{308}y_0\right)x$$

$$+ \frac{27}{44}y_1 + \frac{3}{4}y_2 - \frac{15}{44}y_3 + \frac{5}{77}y_4 - \frac{27}{308}y_0$$

and differentiate the polynomial. Thus, we would have

$$P'''(c) = \left(-\frac{2}{77} + \frac{1}{154}c\right)y_0 + \left(\frac{13}{220} - \frac{1}{55}c\right)y_1 + \left(\frac{1}{30}c - \frac{1}{15}\right)y_2$$

$$+ \left(-\frac{1}{33}c + \frac{5}{132}\right)y_3 + \left(\frac{2}{231}c - \frac{1}{231}\right)y_4$$

as an approximation to the value of $f'''(c)$.

This result is of the form $f^{(k)}(c) \doteq \sum_{i=0}^{n} w_i(c)y_i$, whereby the kth derivative at $x = c$ is approximated by a weighted sum of the $n + 1$ function values $f(x_i) = y_i, i = 0, \ldots, n$.

The weights $w_i(c)$ are actually given by $w_i(c) = L_i^{(k)}(c)$, where

$$L_i(x) = \prod_{j=0, j\neq i}^{n} (x - x_j)/(x_i - x_j)$$

is the ith Lagrange interpolating polynomial.

To illustrate these relationships, use the five nodes x_i, $i = 0, \ldots, n = 4$, from the preceding discussion. The Lagrange interpolating polynomials for these nodes are

$$L_0 = \tfrac{1}{3696}(x + 2)(x - 3)(x - 6)(x - 9) \qquad L_3 = -\tfrac{1}{792}(x + 5)(x + 2)(x - 3)(x - 9)$$

$$L_1 = -\tfrac{1}{1320}(x + 5)(x - 3)(x - 6)(x - 9) \qquad L_4 = \tfrac{1}{2772}(x + 5)(x + 2)(x - 3)(x - 6)$$

$$L_2 = \tfrac{1}{720}(x + 5)(x + 2)(x - 6)(x - 9)$$

In terms of the Lagrange polynomials, the polynomial, interpolating the data points (x_i, y_i), $i = 0, \ldots, 4$, is $F(x) = \sum_{i=0}^{4} y_i L_i(x)$, which has to be the polynomial $P(x)$ obtained earlier. However, to find the weights $w_i(c)$, the representation in terms of Lagrange polynomials is preferable.

Next, compute $F'''(c)$ as an approximation to $f'''(c)$ and then express the result in the form $\sum_{i=0}^{4} w_i(c) y_i$, obtaining

$$\left(\frac{c - 4}{154}\right) y_0 + \left(\frac{13 - 4c}{220}\right) y_1 + \left(\frac{c - 2}{30}c\right) y_2 + \left(\frac{5 - 4c}{132}\right) y_3 + \left(\frac{2c - 1}{231}\right) y_4$$

Clearly, the coefficient of y_i is $w_i(c) = L_i^{(3)}(c)$ by linearity in the derivative of $F(x) = \sum_{i=0}^{4} y_i L_i(x)$.

We therefore have a procedure for numerically differentiating data with $n + 1$ nonuniformly spaced nodes. Simply interpolate the data with a polynomial of degree n and differentiate the polynomial. Formulas incorporating this idea are then $f^{(k)}(c) \doteq \sum_{i=0}^{n} w_i(c) y_i$, with $w_i(c) = L_i^{(k)}(c)$ and $L_i(x) = \prod_{j=0, j\neq i}^{n} (x - x_j)/(x_i - x_j)$. Details, proofs, and error-estimates can be found in texts such as [60] or [16]. (See Exercise 45.) We close, instead, with a demonstration that for equispaced nodes, the interpolating polynomial approach generates the same formulas as the Taylor expansion approach.

The Lagrange interpolating polynomials for the equispaced nodes $x_k = c + kh$, $k = 0, \ldots, 4$, are

$$L_0 = \frac{1}{24}\frac{(x - c - h)(x - c - 2h)(x - c - 3h)(x - c - 4h)}{h^4}$$

$$L_1 = -\frac{1}{6}\frac{(x - c)(x - c - 2h)(x - c - 3h)(x - c - 4h)}{h^4}$$

$$L_2 = \frac{1}{4}\frac{(x - c)(x - c - h)(x - c - 3h)(x - c - 4h)}{h^4}$$

$$L_3 = -\frac{1}{6}\frac{(x - c)(x - c - h)(x - c - 2h)(x - c - 4h)}{h^4}$$

$$L_4 = \frac{1}{24}\frac{(x - c)(x - c - h)(x - c - 2h)(x - c - 3h)}{h^4}$$

From the weights $w_i(c) = L_i^{(3)}(c)$, listed in Table 42.5, we have

$$f'''(c) \doteq \sum_{i=0}^{4} w_i(c) y_i = \frac{-5y_0 + 18y_1 - 24y_2 + 14y_3 - 3y_4}{2h^3}$$

which, except for the notation $y_i = f(c + ih)$, is identical to (42.2c), obtained earlier via Taylor expansions.

$$w_0 = -\frac{5}{2}\frac{1}{h^3}$$

$$w_1 = \frac{9}{h^3}$$

$$w_2 = -\frac{12}{h^3}$$

$$w_3 = \frac{7}{h^3}$$

$$w_4 = -\frac{3}{2}\frac{1}{h^3}$$

TABLE 42.5   The weights w_k, $k = 0, \ldots, 4$, in $f'''(c) = \sum_{i=0}^{4} w_i(c) y_i$

EXERCISES 42.1

1. Verify that (42.2a) is polynomial-exact of degree 2. See Table 42.3.

2. Verify that (42.2b) is polynomial-exact of degree 3. See Table 42.3.

3. Verify that (42.2c) is polynomial-exact of degree 4. See Table 42.3.

4. Verify that (42.2d) is polynomial-exact of degree 5. See Table 42.3.

5. Verify that (42.3a) is polynomial-exact of degree 2. See Table 42.3.

6. Verify that (42.3b) is polynomial-exact of degree 3. See Table 42.3.

7. Verify that (42.3c) is polynomial-exact of degree 4. See Table 42.3.

8. Verify that (42.3d) is polynomial-exact of degree 5. See Table 42.3.

The object of Exercises 9–18 is the recognition that the forward-difference formulas (42.2a)–(42.2d) are equivalent to the differentiation of an interpolating polynomial. For the nodes $x_k = -2+k$, $k = 0, 1, \ldots$, as appropriate, and for the given function $f(x)$:

(a) Apply each of (42.2a)–(42.2d) to approximate the relevant derivative at $c = -2$.

(b) Obtain $g_a(x), \ldots, g_d(x)$, the polynomial interpolating the points $(x_k, f(x_k))$ used for each differentiation formula in part (a).

(c) For each of $g_a(x), \ldots, g_d(x)$, evaluate the appropriate derivative at $x = c$, and compare to the result in part (a).

(d) To demonstrate empirically that each of (42.2a)–(42.2d) is $O(h^2)$, note that in these formulas the order of the derivatives varies from $n = 1$ to $n = 4$. Let $h_i = 2^{-i}$, $i = 0, 1, \ldots$, and let $g_i(x)$ be the polynomial interpolating the points $(c + kh_i, f(c + kh_i))$, $k = 0, \ldots, n + 1$. Evaluate $(f^{(n)}(c) - g_i^{(n)}(c))/h_i^2$ and show, for each $n = 1, \ldots, 4$, that these ratios become constant as i increases.

9. $f(x) = (x + 2)e^{-x-2}$ **10.** $f(x) = \sqrt{2 + 2x + x^2}$

11. $f(x) = \dfrac{9x^2 - 7x - 5}{4x^2 + 4x + 2}$ **12.** $f(x) = \dfrac{3x^3 + 5x^2 - 9x + 4}{5 + x}$

13. $f(x) = \dfrac{9x^2 + 5x - 6}{x - 7}$ **14.** $f(x) = x \sin x$

15. $f(x) = \sin(\cos x)$ **16.** $f(x) = \ln(x^2 + 5x + 2)$

17. $f(x) = \ln(\cos x + \cosh x)$ **18.** $f(x) = \sinh(\sin x)$

The object of Exercises 19–28 is the recognition that the central-difference formulas (42.3a)–(42.3d) are equivalent to the differentiation of an interpolating polynomial. For the nodes $x_k = -2, -2 \pm k$, $k = 1, 2, \ldots$, as appropriate, and for each of the following functions:

(a) Apply each of (42.3a)–(42.3d) to approximate the relevant derivative at $c = -2$.

(b) Obtain $g_a(x), \ldots, g_d(x)$, the polynomial interpolating the points $(x_k, f(x_k))$ used for each differentiation formula in part (a).

(c) For each of $g_a(x), \ldots, g_d(x)$, evaluate the appropriate derivative at $x = c$, and compare to the result in part (a).

(d) To demonstrate empirically that each of (42.3a)–(42.3d) is $O(h^2)$, note that in these formulas the order of the derivatives

varies from $n = 1$ to $n = 4$. Let $h_i = 2^{-i}$, $i = 0, 1, \ldots$, and let $g_i(x)$ be the polynomial interpolating the points $(c, f(c))$, $(c \pm kh_i, f(c \pm kh_i))$, $k = 1, \ldots, [\frac{n}{2}]$, where $[\cdot]$ is the greatest-integer function. Evaluate $(f^{(n)}(c) - g_i^{(n)}(c))/h_i^2$ and show, for each $n = 1, \ldots, 4$, that these ratios become constant as i increases.

19. $f(x) = xe^x$ **20.** $f(x) = 3 \sin x - 4 \cos x$

21. $f(x) = \dfrac{11x^2 - 12x - 10}{x^3 + 3x^2 + 7x - 1}$ **22.** $f(x) = \dfrac{x^3 + 3x^2 - 4x + 6}{8x^2 + 11x + 12}$

23. $f(x) = \dfrac{2x^2 - 9x - 8}{10x^2 - 7x + 1}$ **24.** $f(x) = 2e^x - \sin(\sin x)$

25. $f(x) = \sin x \tanh x$ **26.** $f(x) = \ln(x^2 + 2e^x)$

27. $f(x) = e^{-x^2} - \operatorname{sech} x$ **28.** $f(x) = \sqrt{1 + \ln(x + \cosh x)}$

29. This exercise shows symbolically that the forward-difference formulas can be derived by differentiating an appropriate interpolating polynomial.

(a) Let $g_n(x)$ interpolate the points $(c + kh, f(c + kh))$, $k = 0, \ldots, n + 1$, and show that $g_n^{(n)}(c)$, $n = 1, \ldots, 4$, are the right-hand sides in (42.2a)–(42.2d), respectively.

(b) Let $L_{n,k}(x)$ be the Lagrange interpolating polynomials for the nodes $x_k = c + kh$, $k = 0, \ldots, n + 1$, where $n = 1, \ldots, 4$ refer to the formulas for the derivatives $f^{(n)}(c)$. For each fixed $n = 1, \ldots, 4$, obtain the weights $w_{n,k} = L_{n,k}^{(n)}(c)$ and show that $\sum_{k=0}^{n+1} w_{n,k} f(x_k)$ are again the right-hand sides of (42.2a)–(42.2d).

30. This exercise shows symbolically that the central-difference formulas can be derived by differentiating an appropriate interpolating polynomial.

(a) Let $g_n(x)$ interpolate the points $(c, f(c))$, $(c \pm kh, f(c \pm kh))$, $k = 1, \ldots, [\frac{n}{2}]$, and show that $g_n^{(n)}(c)$, $n = 1, \ldots, 4$, are the right-hand sides in (42.3a)–(42.3d), respectively.

(b) Let $L_{n,k}(x)$ be the Lagrange interpolating polynomials for the nodes $x_k = c + kh$, $k = -[\frac{n}{2}], \ldots, [\frac{n}{2}]$, where $n = 1, \ldots, 4$ refer to the formulas for the derivatives $f^{(n)}(c)$. For each fixed $n = 1, \ldots, 4$, obtain the weights $w_{n,k} = L_{n,k}^{(n)}(c)$ and show that $\sum_{k=-[n/2]}^{[n/2]} w_{n,k} f(x_k)$ are again the right-hand sides of (42.3a)–(42.3d).

For the function $f(x)$, set of unequally spaced nodes x_k, $k = 0, \ldots, 4$, and value $x = c$ given in each of Exercises 31–40:

(a) Obtain the error $\varepsilon_1 = f''(c) - g_1''(c)$, where $g_1(x)$ is the polynomial that interpolates the points $P_k = (x_k, f(x_k))$, $k = 0, \ldots, n$.

(b) Obtain the error $\varepsilon_2 = f''(c) - g_2''(c)$, where $g_2(x)$ is a natural quadratic spline interpolating $\{P_k\}$.

(c) Obtain the error $\varepsilon_3 = f''(c) - g_3''(c)$, where $g_3(x)$ is a natural cubic spline interpolating $\{P_k\}$ (Section 40.4).

(d) Obtain the error $\varepsilon_4 = f''(c) - g_4''(c)$, where $g_4(x)$ is a natural quartic spline interpolating $\{P_k\}$ (Section 40.4).

(e) Obtain the error $\varepsilon_5 = f''(c) - g_5''(c)$, where $g_5(x)$ is the least-squares quadratic best fitting $\{P_k\}$ (Section 41.1).

(f) Obtain the error $\varepsilon_6 = f''(c) - g_6''(c)$, where $g_6(x)$ is the least-squares cubic best fitting $\{P_k\}$ (Section 41.1).

(g) Obtain the weights $w_k = L_k''(c)$, where $L_k(x), k = 0, \ldots, n$, are the Lagrange interpolating polynomials for the given nodes. Compute $\sum_{k=0}^{n} w_k f(x_k)$ and show it is the same as $g_1''(c)$.

31. $f(x) = e^{-x/5} \cos x$; $\{1, 1.7, 3.2, 3.3, 7.4\}, c = 4.6$

32. $f(x) = 5 \sin^2 x - 3 \cos 2x$; $\{3.7, 4.8, 6.0, 7.2, 8.7\}, c = 5.7$

33. $f(x) = \dfrac{3x^2 + 12x - 1}{7x^3 + 10 + 11x}$; $\{0.58, 2.5, 3.6, 5.2, 9.8\}, c = 6.7$

34. $f(x) = \dfrac{11x^3 - 9x^2 - 12x + 12}{11x^2 - 4x + 2}$; $\{0.68, 0.72, 1.4, 7.3, 7.8\}$, $c = 5.4$

35. $f(x) = \dfrac{4x^2 + 2x + 1}{10x^2 + 10x + 2}$; $\{2.6, 3.1, 6.5, 7.7, 8.2\}, c = 4.3$

36. $f(x) = \dfrac{1 + \sin x}{1 + e^{-x}}$; $\{5.2, 7.3, 7.5, 8.2, 8.9\}, c = 6.6$

37. $f(x) = \dfrac{x^2 - \sinh x}{\cos x + \cosh x}$; $\{3.8, 6.1, 8.7, 9.1, 9.9\}, c = 5.9$

38. $f(x) = \ln\left(1 + 2 \sin^2 \dfrac{x}{3}\right)$; $\{3.9, 6.0, 6.7, 8.0, 8.2\}, c = 5.2$

39. $f(x) = \dfrac{\sqrt{x+1} - \sqrt{x+2}}{\cosh \frac{x}{3} - \sin x}$; $\{1.1, 2.7, 2.8, 8.7, 9.6\}, c = 6.3$

40. $f(x) = e^{-\sqrt{\ln(x+1)}}$; $\{0.68, 2.2, 2.4, 5.1, 5.5\}, c = 3.7$

41. Rework Exercise 31 with the points (x_k, y_k), where $y_k = (1 + 0.01(-1)^k) f(x_k), k = 0, \ldots, 4$. Obtain the percent change in the errors $\varepsilon_n, n = 1, \ldots, 6$, because of the 1% distortion in function values.

42. Rework Exercise 31 with the points (x_k, y_k), where $y_k = (1 + 0.1(-1)^k) f(x_k), k = 0, \ldots, 4$. Obtain the percent change in the errors $\varepsilon_n, n = 1, \ldots, 6$, because of the 10% distortion in function values.

43. Develop the $O(h^4)$ central-difference formula $f'''(c) \overset{\circ}{=} \frac{1}{8h^3}(f_{-3} - 8f_{-2} + 13f_{-1} - 13f_1 + 8f_2 - f_3)$, where $f_k = f(c + kh), k = -3, \ldots, 3$,

(a) using the Taylor series, thereby showing the truncation error is of the form $Ch^4 + Dh^6 + \cdots$.

(b) by differentiating the polynomial that interpolates the nodes $x_k = c + kh, k = -3, \ldots, 3$.

(c) by finding the weights w_k in the approximation $f'''(c) \overset{\circ}{=} \sum_{k=-3}^{3} w_k f_k$.

44. Verify the polynomial exactness of the formula developed in Exercise 43.

45. Develop the $O(h^4)$ central-difference formula $f^{(4)}(c) \overset{\circ}{=} \frac{1}{6h^4}(-f_{-3} + 12f_{-2} - 39f_{-1} + 56f_0 - 39f_1 + 12f_2 - f_3)$, where $f_k = f(c + kh), k = -3, \ldots, 3$,

(a) using the Taylor series, thereby showing the truncation error is of the form $Ch^4 + Dh^6 + \cdots$.

(b) by differentiating the polynomial that interpolates the nodes $x_k = c + kh, k = -3, \ldots, 3$.

(c) by finding the weights w_k in the approximation $f^{(4)}(c) \overset{\circ}{=} \sum_{k=-3}^{3} w_k f_k$.

46. Verify the polynomial exactness of the formula developed in Exercise 45.

47. Let $L_k(x), k = 0, \ldots, n$, be the Lagrange interpolating polynomials for the nodes $x_k, k = 0, \ldots, n$. A theorem in [60] states that if c is a node x_k, then the truncation error in approximating $f'(c)$ with $p'(c) = \sum_{k=0}^{n} L_k'(c) f_k$, the derivative of the polynomial interpolating $(x_k, f_k), k = 0, \ldots, n$, is $((n+1)c^n - \sum_{k=0}^{n} L_k'(c)x_k^{n+1}) f^{(n+1)}(\xi)/(n+1)!$, for some ξ in the smallest closed interval containing all the nodes x_k. Verify this theorem for $f(x) = e^x$, with $x_k = k, k = 0, \ldots, 3$, and $c = 1$.

48. Show that the sum of the coefficients in formulas (42.2a)–(42.2d) sum to zero.

49. Show that the sum of the coefficients in formulas (42.3a)–(42.3d) sum to zero.

50. Show that, in general, if $f^{(k)}(c) \overset{\circ}{=} \sum_{i=0}^{n} w_i f_i$ is polynomial-exact of any positive degree, then $\sum_{i=0}^{n} w_i = 0$.

42.2 Richardson Extrapolation

Review

In Section 4.1, under the guise of estimating the per-step error when solving a differential equation, we established (4.2), a result that is essentially Richardson's extrapolation formula. Let us recast our earlier work by letting $E = y_{\text{exact}}$ be the exact value of a quantity being computed numerically with stepsize h in a function $f(h)$. We assume the truncation error in the approximation of E by $f(h)$ is of the form Ch^k, that is, the error is $O(h^k)$. We compute two estimates for E: $P = y_{\text{approx}}(h)$ at stepsize h and $Q = y_{\text{approx}}(\frac{h}{2})$ at stepsize $\frac{h}{2}$, giving the equations $E = P + Ch^k$ and $E = Q + C\left(\frac{h}{2}\right)^k$, from which we obtain

$$E - Q = \frac{Q - P}{2^k - 1} \qquad (42.4)$$

By dividing the difference between the more and less accurate approximations Q and P, respectively, by $2^k - 1$, we obtain an estimate of the error in Q, the more accurate approximation.

However, the Richardson extrapolation technique is based on the resulting expression $E = (2^k Q - P)/(2^k - 1)$ and its generalization. For the moment, we simply observe that the expression for E is just a weighted average of two approximations, with greater weight being given to the more accurate approximation. Because we started with two statements purporting to be exact representations of the exact value E, we ended up with a corresponding *equation* for E. However, the actual error estimate for the truncation error of $f(h)$ is more complex than simply Ch^k, so the value computed for E by $E = (2^k Q - P)/(2^k - 1)$ is not the exact value but merely a more accurate value. We have used two approximations to extrapolate to a more accurate value, but we typically don't obtain the exact value. However, we will be able to repeat the extrapolation process and thereby continue to obtain more accurate values from less accurate ones.

Richardson Extrapolation

Let E, the exact value of a quantity, be computed with stepsize h by the function $f(h)$ for which the truncation error is

$$\rho(h) = \sum_{k=0}^{m} c_k h^{s+k\Delta} + \phi(h)$$

where s and Δ are nonnegative integers and $\phi(h)$ goes to zero as h goes to zero. Two approximations to E using stepsizes h and $\frac{h}{r}$, where $r > 1$ and $E = f(h) + \rho(h)$, are

$$E = f(h) + c_0 h^s + c_1 h^{s+\Delta} + c_2 h^{s+2\Delta} + \cdots$$

$$E = f\left(\frac{h}{r}\right) + c_0 \left(\frac{h}{r}\right)^s + c_1 \left(\frac{h}{r}\right)^{s+\Delta} + c_2 \left(\frac{h}{r}\right)^{s+2\Delta} + \cdots$$

Assuming the equality of these two values for E and subtracting the first equation from r^s times the second, we solve for E, obtaining

$$E = \frac{r^s f\left(\frac{h}{r}\right) - f(h)}{r^s - 1} + \frac{h^{s+\Delta}(r^{-\Delta} - 1)c_1}{r^s - 1} + \frac{h^{s+2\Delta}(r^{-2\Delta} - 1)c_2}{r^s - 1} + \cdots$$

The truncation error's leading term, $c_0 h^s$, has been removed. The weighted average $F(h) = (r^s f\left(\frac{h}{r}\right) - f(h))/(r^s - 1)$ approximates E with a truncation error of the form $C_1 h^{s+\Delta} + C_2 h^{s+2\Delta} + \cdots$, where $C_k = c_k \lambda_k(r)$, $k = 1, \ldots$, for appropriate functions $\lambda_k(r)$.

The extrapolation process can be repeated with $F(h)$ now playing the role of $f(h)$ and $R(h) = \sum_{k=1}^{3} C_k h^{s+k\Delta}$ representing the truncation error. By the same strategy of computing E via $F(h)$ with the two stepsizes $\frac{h}{r^2}$ and $\frac{h}{r^3}$, we have

$$E = F\left(\frac{h}{r^2}\right) + C_1 \left(\frac{h}{r^2}\right)^{s+\Delta} + C_2 \left(\frac{h}{r^2}\right)^{s+2\Delta} + C_3 \left(\frac{h}{r^2}\right)^{s+3\Delta} + \cdots$$

$$E = F\left(\frac{h}{r^3}\right) + C_1 \left(\frac{h}{r^3}\right)^{s+\Delta} + C_2 \left(\frac{h}{r^3}\right)^{s+2\Delta} + C_3 \left(\frac{h}{r^3}\right)^{s+3\Delta} + \cdots$$

Subtracting the first equation from $r^{s+\Delta}$ times the second and solving for E, we have

$$E = A + \frac{h^{s+2\Delta}(r^{-2s-5\Delta} - r^{-2s-4\Delta})C_2}{r^{s+\Delta} - 1} + \frac{h^{s+3\Delta}(r^{-2s-8\Delta} - r^{-2s-6\Delta})C_3}{r^{s+\Delta} - 1}$$

where $A = (r^{s+\Delta}F(\frac{h}{r^3}) - F(\frac{h}{r^2}))/(r^{s+\Delta} - 1)$. Again, the truncation error's leading term, this time $C_1 h^{s+\Delta}$, has been eliminated, so the weighted average A approximates E with a truncation error of the form $D_2 h^{s+2\Delta} + D_3 h^{s+3\Delta} + \cdots$, where $D_k = C_k \sigma_k(r)$, $k = 2, \ldots$, and the $\sigma_k(r)$ are appropriate functions of r.

The extrapolation scheme these calculations support is summarized by the following Richardson extrapolation algorithm.

1. $E = f(h) + c_0 h^s + c_1 h^{s+\Delta} + c_2 h^{s+2\Delta} + c_3 h^{s+3\Delta} + \cdots$.
2.   Pick the factor $r > 1$ and an initial stepsize h.
3.   Compute the nonzero entries $A_{i,j}$ of the lower triangular array A, where both i and j start at 1, and

   (a) the first column of A is given by $A_{i,1} = f(h/r^{i-1})$.
   (b) the remaining nonzero entries of A are given by $A_{i,j} = (r^{s+(j-2)\Delta} A_{i,j-1} - A_{i-1,j-1})/(r^{s+(j-2)\Delta} - 1)$.
   (c) compute A by its rows, each row ending at its diagonal element. If two adjacent entries in a row differ by less than a prescribed tolerance, terminate the calculation. Alternatively, if successive diagonal elements differ by less than a prescribed tolerance, terminate the calculation and take the last diagonal element as the best approximation.

Table 42.6 contains a schematic illustrating the Richardson extrapolation table formed by this algorithm.

$A_{11} = f(h)$

$A_{21} = f\left(\frac{h}{r}\right)$ $A_{22} = \dfrac{r^s A_{21} - A_{11}}{r^s - 1}$

$A_{31} = f\left(\frac{h}{r^2}\right)$ $A_{32} = \dfrac{r^s A_{31} - A_{21}}{r^s - 1}$ $A_{33} = \dfrac{r^{s+\Delta} A_{32} - A_{22}}{r^{s+\Delta} - 1}$

$A_{41} = f\left(\frac{h}{r^3}\right)$ $A_{42} = \dfrac{r^s A_{41} - A_{31}}{r^s - 1}$ $A_{43} = \dfrac{r^{s+\Delta} A_{42} - A_{32}}{r^{s+\Delta} - 1}$ $A_{44} = \dfrac{r^{s+2\Delta} A_{43} - A_{33}}{r^{s+2\Delta} - 1}$

TABLE 42.6   Structure of the Richardson extrapolation table

There is a delicate balance between the benefits of rapid extrapolation (large r) and roundoff error (small h). See Tables 42.7–42.10 for a case study. A common value of r is $r = 2$, with h neither very small nor very large.

EXAMPLE 42.1   Table 42.7 contains the result of Richardson extrapolation applied to the forward-difference formula $g'(c) = \frac{g(c+h)-g(c)}{h}$, where $g(x) = e^x$ and $c = 1$. The truncation error, obtained in Section 42.1, is of the form $c_1 h + c_2 h^2 + c_3 h^3 + \cdots$ so that $s = 1$ and $\Delta = 1$. Thus, we have $f(h) = (e^{1+h} - e)/h$, along with $s = 1$, $\Delta = 1$, $r = 2$, and $h = 2$, so the entries in the first

i	$A_{i,1}$	$A_{i,2}$	$A_{i,3}$	$A_{i,4}$	$A_{i,5}$
1	8.68362755				
2	4.67077427	0.65792100			
3	3.52681448	2.38285470	2.957832597		
4	3.08824452	2.64967456	2.738614507	2.707297638	
5	2.89548017	2.70271582	2.720396241	2.717793632	2.718493365

TABLE 42.7   Richardson extrapolation applied to $g'(c) = \frac{g(c+h)-g(c)}{h}$ where $g(x) = e^x$, $c = 1$, $h = r = 2$

column of Table 42.7 are $A_{i,1} = f(2/2^{i-1})$, and, for $j > 1$, the entries in the jth column in Table 42.7 are computed with $A_{i,j} = (2^{2+(j-2)}A_{i,j-1} - A_{i-1,j-1})/(2^{2+(j-2)} - 1)$.

Table 42.8 lists $|e - A_{ij}|$, the absolute value of the error in each entry of Table 42.7. Indeed, accuracy increases across rows, with the most accurate value at the end of the last row.

i	$|e - A_{i,1}|$	$|e - A_{i,2}|$	$|e - A_{i,3}|$	$|e - A_{i,4}|$	$|e - A_{i,5}|$
1	5.9653457				
2	1.9524924	2.0603608			
3	0.8085327	0.3354271	0.239550769		
4	0.3699627	0.0686073	0.020332679	0.010984190	
5	0.1771983	0.0155660	0.002114413	0.000488196	0.000211537

TABLE 42.8   Absolute errors in the entries of Table 42.7

Table 42.9 is the Richardson table for $h = \frac{1}{100}$ and $r = 10$. For this contrasting experiment, we have taken h small and compensated with a large r.

i	$A_{i,1}$	$A_{i,2}$	$A_{i,3}$	$A_{i,4}$	$A_{i,5}$
1	2.7319187				
2	2.719642	2.7182779			
3	2.71842	2.7182842	2.7182843		
4	2.7183	2.7182867	2.7182867	2.7182867	
5	2.719	2.7190778	2.7190858	2.7190866	2.7190866

TABLE 42.9   Richardson extrapolation applied to $g'(c) = \frac{g(c+h)-g(c)}{h}$ where $g(x) = e^x$, $c = 1$, $h = \frac{1}{100}$, and $r = 10$

Table 42.10 lists $|e - A_{ij}|$, the absolute values of the errors in the entries of Table 42.9.

Now, because of roundoff error, the entries in the first column, values that are not extrapolated, decrease and then increase, as h/r^{i-1} decreases. Consequently, the error in the last entry in the last row now is greater than before. Hence, there is a delicate balance between the benefits of extrapolation and the perils of roundoff.   ❖

i	$\|e - A_{i,1}\|$	$\|e - A_{i,2}\|$	$\|e - A_{i,3}\|$	$\|e - A_{i,4}\|$	$\|e - A_{i,5}\|$
1	0.136×10^{-1}				
2	0.136×10^{-2}	0.391×10^{-5}			
3	0.138×10^{-3}	0.239×10^{-5}	0.246×10^{-5}		
4	0.182×10^{-4}	0.484×10^{-5}	0.486×10^{-5}	0.487×10^{-5}	
5	0.718×10^{-3}	0.796×10^{-3}	0.804×10^{-3}	0.805×10^{-3}	0.805×10^{-3}

TABLE 42.10   Absolute errors in the entries of Table 42.9

EXERCISES 42.2

1. Eliminate C and h from $E = P + Ch^k$ and $E = Q + C\left(\frac{h}{2}\right)^k$ to obtain $E = (2^k Q - P)/(2^k - 1)$ and then (42.4).

In Exercises 2–11:

(a) Apply Richardson extrapolation to (42.2a) to obtain, with error less than 10^{-7}, the value of $f'(-2)$, where $f(x)$ is the given function. Use $h = r = 2$ but be alert for the possibility of overwhelming roundoff errors. Compensate by computing with extended precision. Use the diagonal stopping rule whereby $f'(c) \doteq A_{k,k}$ when $|A_{k,k} - A_{k-1,k-1}| < 10^{-7}$, where $\{A_{i,j}\}$ are the entries of the Richardson table. Determine the actual error in $A_{k,k}$.

(b) Repeat part (a) for (42.2b) and $f''(-2)$. Take $h = \frac{4}{3}$ and $r = 2$.

(c) Repeat part (a) for (42.2c) and $f'''(-2)$. Take $h = 1$ and $r = 2$.

(d) Repeat part (a) for (42.2d) and $f^{(4)}(-2)$. Take $h = \frac{4}{5}$ and $r = 2$.

2. $f(x) = (x+2)e^{-x-2}$ **3.** $f(x) = \sqrt{2 + 2x + x^2}$

4. $f(x) = \dfrac{9x^2 - 7x - 5}{4x^2 + 4x + 2}$ **5.** $f(x) = \dfrac{3x^3 + 5x^2 - 9x + 4}{5 + x}$

6. $f(x) = \dfrac{9x^2 + 5x - 6}{x - 7}$ **7.** $f(x) = x \sin x$

8. $f(x) = \sin(\cos x)$ **9.** $f(x) = \ln(x^2 + 2x + 2)$

10. $f(x) = \ln(\cos x + \cosh x)$ **11.** $f(x) = \sinh(\sin x)$

In Exercises 12–21:

(a) Apply Richardson extrapolation to (42.3a) to obtain, with error less than 10^{-7}, the value of $f'(0)$, where $f(x)$ is the given function. Use $h = 1, r = 2$ but be alert for the possibility of overwhelming roundoff errors. Compensate by computing with extended precision. Use the diagonal stopping rule whereby $f'(c) \doteq A_{k,k}$ when $|A_{k,k} - A_{k-1,k-1}| < 10^{-7}$, where $\{A_{i,j}\}$ are the entries of the Richardson table. Determine the actual error in $A_{k,k}$.

(b) Repeat part (a) for (42.3b) and $f''(0)$. Take $h = 1, r = 2$.

(c) Repeat part (a) for (42.3c) and $f'''(0)$. Take $h = \frac{1}{2}, r = 2$.

(d) Repeat part (a) for (42.3d) and $f^{(4)}(0)$. Take $h = \frac{1}{2}, r = 2$.

12. $f(x) = xe^x$ **13.** $f(x) = 3 \sin x - 4 \cos x$

14. $f(x) = \dfrac{11x^2 - 12x - 10}{x^3 + 3x^2 + 7x - 1}$ **15.** $f(x) = \dfrac{x^3 + 3x^2 - 4x + 6}{8x^2 + 11x + 12}$

16. $f(x) = \dfrac{2x^2 - 9x - 8}{10x^2 - 7x + 2}$ **17.** $f(x) = 2e^x - \sin(\sin x)$

18. $f(x) = \sin x \tanh x$ **19.** $f(x) = \ln(x^2 + 2e^x)$

20. $f(x) = e^{-x^2} - \operatorname{sech} x$ **21.** $f(x) = \sqrt{1 + \ln(x + \cosh x)}$

In the following table, the numbers in the top row are nodes, x_k. The numbers in the first column are the values, at $x = 1$, of the derivatives of the functions $f_k(x), k = 1, \ldots, 5$, whose values at the nodes $x_k, k = 0, \ldots, 8$, are displayed in rows $2, \ldots, 6$, respectively. For the function f_k specified in each of Exercises 22–26, determine which of the following two alternatives yields the more accurate result.

(a) Richardson extrapolation applied to the differentiation formula $\frac{f(1+h)-f(1)}{h}$.

(b) Richardson extrapolation applied to (42.2a).

$f_k'(1)$	x	1	1.1	1.2	1.3	1.4	1.5	1.6	1.7	1.8
$\frac{28}{9}$	$f_1(x)$	$\frac{7}{3}$	$\frac{3547}{1356}$	$\frac{549}{193}$	$\frac{109}{36}$	$\frac{200}{63}$	$\frac{303}{92}$	$\frac{1109}{327}$	$\frac{10,297}{2964}$	$\frac{39}{11}$
$\frac{85}{169}$	$f_2(x)$	$\frac{4}{13}$	$\frac{499}{1380}$	$\frac{154}{365}$	$\frac{751}{1540}$	$\frac{226}{405}$	$\frac{43}{68}$	$\frac{316}{445}$	$\frac{1471}{1860}$	$\frac{424}{485}$
$\frac{9}{7}$	$f_3(x)$	$\frac{2}{7}$	$\frac{708}{1865}$	$\frac{43}{101}$	$\frac{1372}{3083}$	$\frac{216}{481}$	$\frac{4}{9}$	$\frac{311}{715}$	$\frac{2892}{6839}$	$\frac{138}{337}$
$\frac{34}{81}$	$f_4(x)$	$\frac{2}{9}$	$\frac{5752}{22,129}$	$\frac{977}{3369}$	$\frac{10,204}{32,523}$	$\frac{808}{2431}$	$\frac{128}{369}$	$\frac{347}{969}$	$\frac{23,236}{63,367}$	$\frac{1714}{4593}$
$-\frac{23}{121}$	$f_5(x)$	$\frac{4}{11}$	$\frac{430}{1243}$	$\frac{115}{348}$	$\frac{70}{221}$	$\frac{130}{427}$	$\frac{22}{75}$	$\frac{145}{512}$	$\frac{610}{2227}$	$\frac{160}{603}$

22. $f_1(x)$ **23.** $f_2(x)$ **24.** $f_3(x)$ **25.** $f_4(x)$ **26.** $f_5(x)$

27. In Section 42.1, the forward-difference formula $f'(c) = \frac{f(c+h)-f(c)}{h} + Ch + Dh^2 + \cdots$ was obtained.

(a) Use Richardson extrapolation to obtain an $O(h^3)$ forward-difference formula for $f'(c)$.

(b) Using $f(x) = e^x$ with $c = 1$, compute ratios of error over h^3 to verify empirically that your formula is actually $O(h^3)$.

28. In Section 42.1, the central-difference formula $f'(c) = \frac{f(c+h)-f(c-h)}{2h} + Ch^2 + Dh^4 + \cdots$ was obtained.

(a) Use Richardson extrapolation to obtain an $O(h^4)$ central-difference formula for $f'(c)$.

(b) Using $f(x) = e^x$ with $c = 1$, compute ratios of error over h^4 to verify empirically that your formula is actually $O(h^4)$.

Chapter Review

1. What does it mean to say that the forward-difference formula $\frac{1}{2h}[-3f(c) + 4f(c+h) - f(c+2h)]$ is an $O(h^2)$ approximation of $f'(c)$? What does it mean to say that this formula has polynomial exactness of degree 2?

2. At $x = 1$, apply the formula in Question 1 to $f(x) = e^{-x}$ and describe a calculation that would show empirically that the formula is indeed second order in h.

3. Use Taylor expansions to derive the formula in Question 1. Carry through enough terms so that the structure of the truncation error is evident.

4. Demonstrate how to use Lagrange interpolating polynomials for obtaining the formula in Question 1.

5. Describe the Richardson extrapolation algorithm.

6. Give a justification for the Richardson extrapolation algorithm.

7. What is the diagonal stopping rule for Richardson extrapolation?

8. For the function $f(x) = e^{-x}$, apply the Richardson extrapolation algorithm to the computation of $f'(1)$ with the forward-difference formula in Question 1. Start with $h = 1$, and let $r = 2$. Show at least one extrapolation step. In general, is it better to start with a small h or a large h? Why?

Chapter 43

Numeric Integration

INTRODUCTION   Chapter 43 is devoted to the ideas and methods of *numeric integration*. The student of elementary calculus has most likely seen such techniques as the *rectangular rule*, the *Trapezoid rule*, and *Simpson's rule*. It is even likely that the *error estimates* for these methods were seen as well. However, we begin with these three methods and the *midpoint rule*, with the rectangular rule being restricted to the *Left-end rule*, where heights of approximating rectangles are evaluated at the left end of the supporting subintervals.

The *recursive form* of the Trapezoid rule reuses previously computed function values when the stepsize is halved. This provides a considerable savings in computational effort when applying Richardson extrapolation to numeric integration based on the Trapezoid rule. The resulting method is called Romberg integration.

Gauss–Legendre quadratures evaluate the integrand at nodes that are not uniformly spaced. The evaluation nodes are the zeros of Legendre polynomials, and the methods so developed are *polynomial exact* to a degree that is more than double the number of function evaluations. These methods are often used in finite-element methods for the numeric solution of partial differential equations.

An *adaptive quadrature* based on Simpson's rule illustrates the concept of adaptivity. The basic adaptive idea is first seen in Section 4.8 where rkf45, the Runge–Kutta–Fehlberg algorithm for numerically solving differential equations, is described. The same process holds for numeric integration. From two approximate values for the integral, an estimate of the error in the more accurate value can be determined. If the estimated error is small enough, the more accurate value of the integral is accepted and the computation ends. If the estimated error is too large, the interval of integration is cut in half and the process repeated for each half. This continues until all of the original interval passes the error criterion.

A naive application of the adaptive strategy would not necessarily lessen the computation burden in numeric integration. In fact, it would probably increase it. Hence, a scheme that reuses enough of the previously computed function values is essential if adaptive quadrature is to be efficient.

Finally, we discuss the challenge of evaluating *iterated integrals* numerically. The techniques available are the one-dimensional ones from the earlier parts of the chapter. For example, when the outer integral in an iterated double integral requires the value of the inner integral, the inner integral must be evaluated numerically with one of the techniques from Sections 43.1–43.4. Thus, a numeric integration of the inner integral is performed for each function value required during the numeric integration of the outer integral. Coordinating the two integration processes with respect to data structures and computer code and with respect to error estimates is the challenge for the programmer.

43.1   Methods from Elementary Calculus

Newton–Cotes Integration Formulas

Some numeric integration methods that often appear in elementary calculus are listed in Tables 43.1 and 43.2. Table 43.1 shows the "one section" version for intervals $[\alpha, \beta]$ containing the minimum number of nodes to which each method applies. For the Midpoint rule, the notation $f_{k+1/2}$ designates the function value midway between the two nodes $\alpha = x_k$ and $\beta = x_{k+1}$.

Method	Integral	Value	Local Truncation Error
Left-end	$\displaystyle\int_{x_k}^{x_{k+1}} f(x)\,dx$	hf_k	$\dfrac{h^2}{2} f'(\xi)$
Trapezoid	$\displaystyle\int_{x_k}^{x_{k+1}} f(x)\,dx$	$\dfrac{h}{2}(f_k + f_{k+1})$	$-\dfrac{h^3}{12} f''(\xi)$
Midpoint	$\displaystyle\int_{x_k}^{x_{k+1}} f(x)\,dx$	$hf_{k+1/2}$	$\dfrac{h^3}{24} f''(\xi)$
Simpson-$\frac{1}{3}$	$\displaystyle\int_{x_{k-1}}^{x_{k+1}} f(x)\,dx$	$\dfrac{h}{3}(f_{k-1} + 4f_k + f_{k+1})$	$-\dfrac{h^5}{90} f^{(4)}(\xi)$

TABLE 43.1   Common Newton–Cotes quadratures for a single "section"

Rule	Approximation of $\displaystyle\int_a^b f(x)\,dx$	Global Truncation Error
Left-end	$h \displaystyle\sum_{k=0}^{n-1} f_k$	$\dfrac{h}{2} f'(\eta)(b-a)$
Trapezoid	$\dfrac{h}{2}\left(f_0 + f_n + 2\displaystyle\sum_{k=1}^{n-1} f_k\right)$	$-\dfrac{h^2}{12} f''(\eta)(b-a)$
Midpoint	$h \displaystyle\sum_{k=0}^{n-1} f_{k+1/2}$	$\dfrac{h^2}{24} f''(\eta)(b-a)$
Simpson-$\frac{1}{3}$	$\dfrac{h}{3}\left(f_0 + f_n + 4\displaystyle\sum_{k=1}^{n/2} f_{2k-1} + 2\displaystyle\sum_{k=1}^{n/2-1} f_{2k}\right)$	$-\dfrac{h^4}{180} f^{(4)}(\eta)(b-a)$

TABLE 43.2   Composite Newton–Cotes quadratures for the methods of Table 43.1

Table 43.2 lists the composite rules that apply in $[a, b]$, an interval consisting of at least two contiguous "sections." Obtained from Table 43.1, the rules in Table 43.2 appear in the form in which they are typically used.

Unifying Principles

The methods in Table 43.1 are obtained by integrating a polynomial that interpolates the $m + 1$ equispaced nodes $x_k, k = 0, \ldots, m$, in $[\alpha, \beta]$. The values of m for the methods in

Table 43.1 are $m = 0$ for the Left-end and Midpoint rules, $m = 1$ for the Trapezoid rule, and $m = 2$ for the Simpson-$\frac{1}{3}$ rule (henceforth called Simpson's rule). In each case, we obtain results of the form

$$\int_\alpha^\beta f(x)\,dx \doteq \sum_{k=0}^m w_k f(x_k) \tag{43.1}$$

where the weights w_k are computed by integrating $P_m(x)$, an mth-degree polynomial that interpolates the $m + 1$ points $(x_k, f(x_k)) = (x_k, f_k), k = 0, \ldots, m$. Indeed, if the interpolating polynomial is written in terms of the Lagrange polynomials $L_k(x) = \prod_{i=0, i\neq k}^m (x - x_i)/(x_k - x_i)$, the weights w_k are given by

$$w_k = \int_\alpha^\beta L_k(x)\,dx \tag{43.2}$$

and the truncation error is given by

$$\left[\frac{\beta^{m+2} - \alpha^{m+2}}{m+2} - \sum_{k=0}^m w_k x_k^{m+1}\right]\frac{f^{(m+1)}(\xi)}{(m+1)!} \tag{43.3}$$

with ξ in $[\alpha, \beta]$, provided $\prod_{k=0}^m (x - x_k)$ does not change sign in (α, β). The formulas in the "Value" column of Table 43.1 can all be obtained from (43.1) if the weights are given by (43.2). Except for Simpson's rule, the truncation errors can be determined by (43.3). The truncation error for Simpson's rule must be obtained by an alternate technique, as we will show. All such formulas are exact for polynomials whose degree is at least m. (See [60] for a proof of these claims.)

Quadrature formulas are called *closed* if they sample $f(x)$ at the endpoints α and β and *open* otherwise. The methods listed in Tables 43.1 and 43.2 are closed, except for the Midpoint rule, which is open since it does not sample the endpoints of the interval $[\alpha, \beta]$.

For integration on the interval $[a, b]$, three of the composite methods of Table 43.2 use the $n + 1$ equispaced nodes $x_k = a + kh, k = 0, \ldots, n$, where $h = \frac{b-a}{n}$. For the Midpoint rule, although the nodes are defined by $x_k = a + kh$, where $h = \frac{b-a}{n}$, the function evaluations are midway between the nodes—hence the notation $f_{k+1/2} = f(a + (k + \frac{1}{2})h), k = 0, \ldots, n-1$. For Simpson's rule, n must be even because the interval $[a, b]$ is a composite of "double-panels," pairs of contiguous subintervals $[x_{2k}, x_{2k+1}]$ and $[x_{2k+1}, x_{2k+2}], k = 0, \ldots, \frac{n}{2}$. In Simpson's (composite) rule, the first sum is over function values at nodes with odd indices, while the second sum is over nodes with even indices.

Comparing the truncation errors as listed in Tables 43.1 and 43.2 shows the order of the error in the composite case is always one less. This loss of order is explained in detail for the Left-end rule as part of its treatment in the paragraph on Derivations on page 1076.

In each case in Table 43.2, $a < \eta < b$ and $f_0 = a, f_n = b$. Since η is not known, the error estimates are used in a "worst case" sense wherein the absolute value of the actual error is bounded by the maximum of the absolute value of the estimate, obtained by estimating the maximum absolute value of the relevant derivative.

There may be a tendency in elementary calculus to prefer Simpson's rule over the Trapezoid rule because, as compared to the $O(h^2)$ error of the latter, the error of the former is $O(h^4)$, one order higher than expected. However, in Section 43.2 we will see *Romberg integration*, the effective integration technique generated by Richardson extrapolation of the Trapezoid rule.

EXAMPLE 43.1  With $F(x) = e^x$, we will use each of the methods in Table 43.2 to approximate, with an error no worse than 10^{-3}, the value of

$$\lambda = \int_0^1 F(x)\,dx = e - 1 \stackrel{\circ}{=} 1.718281828$$

LEFT-END RULE On the interval $[0, 1]$ the maximum value of $|F'(x)| = e^x$ is e. An n for which $\frac{b-a}{n}\frac{e}{2}(b-a) < 10^{-3}$ holds, where we have replaced h with $\frac{b-a}{n}$, is a solution of $\frac{e}{2n} = 10^{-3}$, or $n = 1359.140914$. Therefore, we take $n = 1360$ for a guarantee that the composite Left-end rule will compute λ with an error no worse than 10^{-3} and obtain 1.717650184 as an approximation to λ. The actual error made with this n is 0.000631644, somewhat smaller than the required tolerance. By trial and error, we can determine $n = 859$ is the smallest n for which the error bound is satisfied.

TRAPEZOID RULE On the interval $[0, 1]$ the maximum value of $|F''(x)| = e^x$ is e. An n for which $\left(\frac{b-a}{n}\right)^2\frac{e}{12}(b-a) < 10^{-3}$ holds, where we have again replaced h with $\frac{b-a}{n}$, is a solution of $\frac{e}{12n^2} = 10^{-3}$, or $n = 15.05069719$. Therefore, we take $n = 16$ for a guarantee that the composite Trapezoid rule will compute λ with an error no worse than 10^{-3} and obtain 1.718841128 as an approximation to λ. The actual error made with this n is 0.000559300, somewhat smaller than the required tolerance. By trial and error, we can determine $n = 12$ is the smallest n for which the error bound is satisfied.

MIDPOINT RULE On the interval $[0, 1]$ the maximum value of $|F''(x)| = e^x$ is e. An n for which $\left(\frac{b-a}{n}\right)^2\frac{e}{24}(b-a) < 10^{-3}$ holds, where we have again replaced h with $\frac{b-a}{n}$, is a solution of $\frac{e}{24n^2} = 10^{-3}$, or $n = 10.64245004$. Therefore, we take $n = 11$ for a guarantee that the composite Midpoint rule will compute λ with an error no worse than 10^{-3} and obtain 1.717690276 as an approximation to λ. The actual error made with this n is 0.000591552, somewhat smaller than the required tolerance. By trial and error, we can determine $n = 9$ is the smallest n for which the error bound is satisfied.

SIMPSON'S RULE Because the $(n + 1)$-point composite Simpson's rule will obtain λ with an error no worse than 10^{-3} for $n = 2$, too few points to experience the composite rule, we instead illustrate approximating λ with an error no worse than 10^{-5}.

On the interval $[0, 1]$ the maximum value of $|F^{(4)}(x)| = e^x$ is e. An n for which $\left(\frac{b-a}{n}\right)^4\frac{e}{180}(b-a) < 10^{-5}$ holds, where we have again replaced h with $\frac{b-a}{n}$, is a solution of $\frac{e}{180n^4} = 10^{-5}$, or $n = 6.233837751$. Therefore, we take $n = 8$ for a guarantee that the composite Simpson's rule will compute λ with an error no worse than 10^{-5} and obtain 1.718284155 as an approximation to λ. The actual error made with this n is 0.2327×10^{-5}, somewhat smaller than the required tolerance. By trial and error, we can determine $n = 6$ is the smallest n for which the error bound is satisfied. ❖

Polynomial Exactness

We next demonstrate polynomial-exactness for each of the four methods of Table 43.1. We therefore show exactness for the $(m + 1)$-point Newton–Cotes formulas, not the $(n + 1)$-point composite formulas because if the former are polynomial exact, the latter must also be exact.

LEFT-END RULE   If the zero-degree polynomial $p_0(x) = c$ is integrated on the interval $[a, a + h]$, the value of the integral $\int_a^{a+h} p_0(x)\, dx = ch$ is given exactly by the Left-end rule $h p_0(a) = hc$.

TRAPEZOID RULE   If the first-degree polynomial $p_1(x) = Ax + B$ is integrated over the interval $[a, a + h]$, the value of the integral

$$\int_a^{a+h} Ax + B\, dx = \tfrac{1}{2} h(Ah + 2Aa + 2B) \tag{43.4}$$

is precisely the value of $\frac{h}{2}[(Aa + B) + A(a + h) + B]$, as given by the Trapezoid rule.

MIDPOINT RULE   If the first-degree polynomial $p_1(x) = Ax + B$ is integrated over the interval $[a, a + h]$, the value of the integral is again given by (43.4). The Midpoint rule gives

$$h\left[A\left(\frac{a + (a + h)}{2}\right) + B\right] = \frac{1}{2}h(Ah + 2Aa + 2B)$$

Thus, the Midpoint rule is polynomial-exact of degree one.

SIMPSON'S RULE   If the third-degree polynomial $p_3(x) = c_0 + c_1 x + c_2 x^2 + c_3 x^3$ is integrated over the interval $[a, a + 2h]$, we obtain

$$\int_a^{a+2h} p_3(x)\, dx = 4c_3 h^4 + \left(\tfrac{8}{3} c_2 + 8 c_3 a\right) h^3$$
$$+ (4 c_2 a + 2 c_1 + 6 c_3 a^2) h^2 + 2(c_0 + c_1 a + c_3 a^3 + c_2 a^2) h$$

whereas the value given by Simpson's rule is $\frac{h}{3}(p_3(a) + p_3(a + 2h) + 4 p_3(a + h))$. A computation shows this is precisely $\int_a^{a+2h} p_3(x)\, dx$, so Simpson's rule is exact for third-degree polynomials, not just second-degree polynomials as would be expected from the value $m = 2$.

Derivations

Some Newton–Cotes integration formulas and their truncation errors are easily obtained from the general theory on integrating interpolating polynomials. Some are more easily obtained directly from Taylor series expansions. We illustrate both techniques in the ensuing derivations. In the exercises, we explore a different version of the theorem on integration of interpolating polynomials and again derive both the Newton–Cotes formulas and their truncation errors.

The single-section $(m+1)$-point formulas are used to form the companion $(n+1)$-point composite formulas. The way the error terms then combine to form a truncation-error term of order one less is detailed for the Left-end rule. The truncation errors for the other rules are obtained with similar reasoning.

LEFT-END RULE   Using Taylor's theorem to write $f(x)$ as $f(x) = f(x_k) + f'(\xi)(x - x_k)$ for some ξ in $[x, x_{k+1}]$, we have

$$\int_{x_k}^{x_{k+1}} f(x)\, dx = f(x_k)(x_{k+1} - x_k) + f'(\xi)\frac{(x_{k+1} - x_k)^2}{2} = h f_k + \frac{h^2}{2} f'(\xi)$$

establishing both the integration rule and the truncation error. (Since $x - x_k$ does not change sign in $[x_k, x_{k+1}]$, the second mean value theorem for integrals [91] justifies the integration. See Exercise B22.)

The composite Left-end rule would join n contiguous panels to form the interval $[a, b]$ containing the $n + 1$ nodes $x_k, k = 0, \ldots, n$. Clearly, we will have $\int_a^b f(x)\, dx = h \sum_{k=0}^{n-1} f_k + R_n$, where R_n is the truncation error. That R_n is $\frac{h}{2} f'(\eta)(b - a)$ follows from the sum $\frac{h^2}{2} \sum_{k=0}^{n-1} f'(\xi_k)$ *intuitively* by assuming that all the derivative values in the sum are nearly equal, so there is some one point η at which the derivative can be evaluated but once, and this value multiplied by n. If one h in h^2 is replaced by $h = \frac{b-a}{n}$, then the truncation error is $\frac{h}{2} \frac{b-a}{n} f'(\eta) n = \frac{h}{2}(b - a) f'(\eta)$.

A more rigorous argument for the equality $n f'(\eta) = \sum_{k=0}^{n-1} f'(\xi_k)$ rests on the Intermediate Value theorem, which says that a continuous function takes on every value between its maximum and minimum on a closed interval. In particular, write $\Psi = \sum_{k=0}^{n-1} f'(\xi_k)$ so that $n f'_{\min} \leq \Psi \leq n f'_{\max}$ and $f'_{\min} \leq \frac{\Psi}{n} \leq f'_{\max}$ are immediate. Consequently, $\frac{\Psi}{n}$ is a value that f' must assume in the interval $[a, b]$, provided f' is continuous on $[a, b]$, so there exists an η in $[a, b]$ for which $\frac{\Psi}{n} = f'(\eta)$, and hence, $\Psi = \sum_{k=0}^{n-1} f'(\xi_k) = n f'(\eta)$.

This same proof will suffice for each of the remaining derivations of truncation errors, provided the appropriate derivative is used in place of the first.

TRAPEZOID RULE If the two points (x_k, f_k) and (x_{k+1}, f_{k+1}) are interpolated by a first-degree polynomial and the polynomial integrated, we will have the Trapezoid rule. In fact, the interpolating polynomial is $P(x) = L_k(x) f_k + L_{k+1}(x) f_{k+1}$, where $L_k = (x - x_{k+1})/(x_k - x_{k+1})$ and $L_{k+1} = (x - x_k)/(x_{k+1} - x_k)$. The weights are $w_k = \frac{1}{2}(x_{k+1} - x_k)$ and $w_{k+1} = \frac{1}{2}(x_{k+1} - x_k)$, so the quadrature formula $\int_{x_k}^{x_{k+1}} f(x)\, dx = \sum_{k=0}^{1} w_k f_k$ becomes

$$\int_{x_k}^{x_{k+1}} f(x)\, dx = \frac{1}{2}(x_{k+1} - x_k)(f_k + f_{k+1}) = \frac{h}{2}(f_k + f_{k+1})$$

The truncation error of the single-section formula is given by

$$\left[\frac{\beta^{m+2} - \alpha^{m+2}}{m + 2} - \sum_{k=0}^{m} w_k x_k^{m+1} \right] \frac{f^{(m+1)}(\xi)}{(m + 1)!} \tag{43.5}$$

where $m = 1$, $\alpha = x_k$, and $\beta = x_{k+1}$, so $-\frac{1}{12}(x_{k+1} - x_k)^3$ is the multiplier of the second derivative. Since $h = x_{k+1} - x_k$, (43.5) is therefore $-\frac{h^3}{12} f''(\xi)$.

If there are n nodes in the interval $[a, b]$, then we must sum $\frac{h}{2}(f_k + f_{k+1})$ across the resulting n panels. In the leftmost panel, $k = 0$; but in the rightmost panel, $k = n - 1$. Hence, we must evaluate the sum

$$\sum_{k=0}^{n-1}(f_k + f_{k+1}) = \sum_{k=0}^{n-1} f_k + \sum_{k=0}^{n-1} f_{k+1}$$

The two sums on the right have most terms in common. The first sum on the right only has f_0 not in common with the second sum on the right, and the second sum on the right only has f_n not in common with the first sum on the right. Hence, all the common terms are doubled, and the result is $f_0 + f_n + 2 \sum_{k=1}^{n-1} f_k$, from which follows the composite Trapezoid rule.

The truncation error for the composite Trapezoid rule follows from $-\frac{h^3}{12} \sum_{k=0}^{n-1} f''(\xi_k)$ by an argument identical to that made for the truncation error of the composite Left-end rule.

MIDPOINT RULE If $x = c$ is the midpoint of the interval $I = [x_k, x_{k+1}] = [c - \frac{h}{2}, c + \frac{h}{2}]$, the single-section Midpoint rule can be derived by integrating the Taylor expansion

$f(x) = f(c) + f'(x - c) + \frac{1}{2}f''(\xi)(x - c)^2$ over I, yielding

$$\int_{c-h/2}^{c+h/2} f(x)\, dx = \int_{c-h/2}^{c+h/2} \left(f(c) + f'(x - c) + \frac{1}{2}f''(\xi)(x - c)^2 \right) dx$$

$$= hf(c) + \frac{h^3}{24}f''(\xi)$$

(43.6)

Since $f(c) = f(x_k + \frac{h}{2})$, (43.6) establishes both the Midpoint rule and its truncation error. (Again, since $(x - c)^2$ does not change sign in the interval of integration, the second mean value theorem for integrals justifies the integration.)

The composite Midpoint rule and expression for the corresponding truncation both follow from summing the single-section results over n sections.

SIMPSON'S RULE   Although Simpson's rule itself can be obtained by integrating an interpolating polynomial, the truncation error cannot be obtained this way because $\prod_{k=0}^{2}(x - x_k)$ changes sign in the interval $[x_k, x_{k+2}]$. Instead, we obtain both Simpson's rule and its truncation error directly, from Taylor series expansions.

Let $x_k = c$, so $x_{k-1} = c - h$ and $x_{k+1} = c + h$, letting us take the interval $[\alpha, \beta]$ as $[x_{k-1}, x_{k+1}] = [c - h, c + h]$. Replacing $f(x)$ with its Taylor expansion constructed at $x = c$ and integrating termwise, we have

$$\int_{x_{k-1}}^{x_{k+1}} f(x)\, dx = \int_{x_{k-1}}^{x_{k+1}} \sum_{k=0} \frac{f^{(k)}(c)}{k!}(x - c)^k\, dx = 2hf(c) + \frac{h^3}{3}f''(c) + \frac{h^5}{60}f^{(5)}(c) + \cdots$$

(43.7)

Replacing $f''(c)$ with the central-difference expression

$$f''(c) = \frac{1}{h^2}(f(c - h) - 2f(c) + f(c + h)) + \frac{h^2}{12}f^{(4)}(c) + \cdots$$

the right side in (43.7) becomes

$$\frac{h}{3}(f(c - h) + 4f(c) + f(c + h)) - \frac{h^5}{90}f^{(4)}(c) + \cdots$$

giving both the one-section integration formula and its truncation error.

EXERCISES 43.1–Part A

A1. Evaluate $\lambda = \int_0^1 \sin x\, dx$.

A2. Approximate λ by the Left-end rule with $n = 5$. Determine the actual error made. Use the truncation error formula from Table 43.2 to obtain an upper bound on the error.

A3. Approximate λ by the Trapezoid rule with $n = 5$. Determine the actual error made. Use the truncation error formula from Table 43.2 to obtain an upper bound on the error.

A4. Approximate λ by the Midpoint rule with $n = 5$. Determine the actual error made. Use the truncation error formula from Table 43.2 to obtain an upper bound on the error.

A5. Approximate λ by Simpson's rule with $n = 6$. Determine the actual error made. Use the truncation error formula from Table 43.2 to obtain an upper bound on the error.

EXERCISES 43.1–Part B

B1. Obtain Simpson's rule by evaluating $\int_{x_0}^{x_2} \sum_{k=0}^{2} L_k(x) f_k\, dx$, where $L_k(x)$ is the kth Lagrange interpolating polynomial for the nodes $x_k, k = 0, 1, 2$.

In Exercises B2–6, select one function $f_k(x)$ from the given list and use it for all five exercises.

$f_1(x) = (x + 2)e^{-x-2}$

$f_2(x) = \sqrt{2 + 2x + x^2}$

$f_3(x) = \dfrac{9x^2 - 7x - 5}{4x^2 + 4x + 2}$

$f_6(x) = x \sin x$

$f_7(x) = \sin(\cos x)$

$f_8(x) = \ln(x^2 + 2x + 2)$

$$f_4(x) = \frac{3x^3 + 5x^2 - 9x + 4}{5 + x} \qquad f_9(x) = \ln(\cos x + \cosh x)$$

$$f_5(x) = \frac{9x^2 + 5x - 6}{x - 7} \qquad f_{10}(x) = \sinh(\sin x)$$

B2. For the function $f_k(x)$ selected from the given list:

(a) Use the error estimate in Table 43.2 to determine an n for which the Left-end rule is guaranteed to approximate $\lambda = \int_{-2}^{3} f(x)\,dx$ with an error smaller than 10^{-2}.

(b) Evaluate λ either analytically or numerically (using appropriate technology).

(c) If $n < 5000$, implement the Left-end rule and determine the actual error made.

(d) Find the smallest value of n for which the Left-end rule actually approximates λ with less than the prescribed error.

B3. Repeat Exercise B2 for the Trapezoid rule.

B4. Repeat Exercise B2 for the Midpoint rule.

B5. Repeat Exercise B2 for Simpson's rule.

B6. For the function $f_k(x)$ selected from the given list and stepsizes $h_k = 2^{-k}$, show empirically that the error made when evaluating $\lambda = \int_{-2}^{3} f(x)\,dx$ with

(a) the Left-end rule is O(h).   (b) the Trapezoid rule is O(h^2).

(c) the Midpoint rule is O(h^2).   (d) Simpson's rule is O(h^4).

In Exercises B7–11, function values are given at h-spaced nodes for a function $f_k(x)$. The values of $\lambda_k = \int_{1}^{1.8} f_k(x)\,dx, k = 1, \ldots, 5$, are 2.478140004, 0.4550629667, 0.3361125107, 0.256845281, 0.246162838, respectively. Approximate the relevant λ_k by each of the following methods, and determine which method gives the most accurate results.

(a) The Left-end rule.   (b) The Trapezoid rule.

(c) The Midpoint rule.   (d) Simpson's rule.

(e) Integrating a linear spline that interpolates the given data points; compare the result with part (b).

(f) Integrating a natural quadratic spline that interpolates the given data points.

(g) Integrating a natural cubic spline that interpolates the given data points.

(h) Integrating a polynomial that interpolates the given data points.

B7. $f_1(x)$ from Exercise 22 in Section 42.2

B8. $f_2(x)$ from Exercise 23 in Section 42.2

B9. $f_3(x)$ from Exercise 24 in Section 42.2

B10. $f_4(x)$ from Exercise 25 in Section 42.2

B11. $f_5(x)$ from Exercise 26 in Section 42.2

In Exercises B12–21, functions $f_k(x), k = 1, \ldots, 10$, and nodes $x_0, \ldots, x_4$ are given, thus determining the points $P_n = (x_n, f_k(x_n)), n = 0, \ldots, 4$. For each:

(a) Evaluate $\lambda_k = \int_{x_0}^{x_4} f_k(x)\,dx$.

(b) Approximate λ_k by applying the Left-end rule to successive pairs of points.

(c) Approximate λ_k by applying the Trapezoid rule to successive pairs of points.

(d) Approximate λ_k by integrating a polynomial that interpolates $P_0, \ldots, P_4$.

(e) Approximate λ_k by integrating a linear spline that interpolates $P_0, \ldots, P_4$. Compare to part (c).

(f) Approximate λ_k by integrating a natural quadratic spline that interpolates $P_0, \ldots, P_4$.

(g) Approximate λ_k by integrating a natural cubic spline that interpolates $P_0, \ldots, P_4$.

(h) Approximate λ_k by integrating a natural quartic spline that interpolates $P_0, \ldots, P_4$.

(i) Approximate λ_k by applying the Left-end rule with uniformly spaced nodes and $n = 4$.

(j) Approximate λ_k by applying the Trapezoid rule with uniformly spaced nodes and $n = 4$.

(k) Approximate λ_k by applying the Midpoint rule with uniformly spaced nodes and $n = 4$.

(l) Approximate λ_k by applying Simpson's rule with uniformly spaced nodes and $n = 4$.

(m) Which method gives the most accurate answer?

B12. the data from Exercise 31, Section 42.1

B13. the data from Exercise 32, Section 42.1

B14. the data from Exercise 33, Section 42.1

B15. the data from Exercise 34, Section 42.1

B16. the data from Exercise 35, Section 42.1

B17. the data from Exercise 36, Section 42.1

B18. the data from Exercise 37, Section 42.1

B19. the data from Exercise 38, Section 42.1

B20. the data from Exercise 39, Section 42.1

B21. the data from Exercise 40, Section 42.1

B22. The second mean value theorem for integrals says that if $f(x)$ and $g(x)$ are continuous functions defined on $[a, b]$, an interval over which $g(x)$ does not change sign, then $\int_a^b f(x)g(x)\,dx = f(\xi)\int_a^b g(x)\,dx$ for some ξ satisfying $a \le \xi \le b$. (A proof can be found in [91].) In each of the following examples, show that the equation $\int_a^b f(x)g(x)\,dx = f(\xi)\int_a^b g(x)\,dx$ has no solution for ξ in $[a, b]$, a consequence of $g(x)$ changing sign in $[a, b]$. The

astute reader will notice that $f(\xi) = \int_a^b f(x)g(x)\,dx /$
$\int_a^b g(x)\,dx$ has no solution if $\int_a^b g(x)\,dx$ is close to zero, the
fraction is large, and $|f(x)|$ is small.

(a) $f(x) = 1 + x^3$, $g(x) = x^3$, $[-1, \frac{11}{10}]$

(b) $f(x) = xe^{-x}$, $g(x) = x^2 - 3x - 1$, $[0, 5]$

(c) $f(x) = \sinh x$, $g(x) = 2 \sin x + 4 \cos x$, $[0, 4]$

(d) $f(x) = x^2 + 2x + 2$, $g(x) = e^x - \frac{1}{10}e^{2x}$, $[0, 3]$

(e) $f(x) = \sqrt{1 + 2\cos^2 x}$, $g(x) = 7x^3 - 8x^2 + 3x + 5$, $[-1, \frac{1}{5}]$

43.2 **Recursive Trapezoid Rule and Romberg Integration**

Recursive Trapezoid Rule

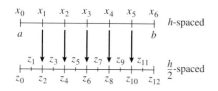

FIGURE 43.1   Schematic relating $T[h]$ to $T[\frac{h}{2}]$, where $T[h]$ is the result of approximating $\int_a^b f(x)\,dx$ with the Trapezoid rule when the stepsize is h

Let $T[h]$ be the value the Trapezoid rule gives for the integral $\int_a^b f(x)\,dx$ when used with stepsize h, so $T[\frac{h}{2}]$ is the value the Trapezoid rule gives when used with stepsize $\frac{h}{2}$. The schematic in Figure 43.1 helps visualize how some of the nodes used when computing $T[\frac{h}{2}]$ are nodes that were used when computing $T[h]$.

In particular, we observe that for $k = 1, \ldots, n-1$, the nodes z_{2k} coincide with the nodes x_k, and hence, we have $f(z_{2k}) = f(x_k)$, $k = 1, \ldots, n-1$. (In words, we might say that the interior nodes for $T[h]$ are the even interior nodes for $T[\frac{h}{2}]$.) Therefore, with stepsize $h = \frac{b-a}{n}$, we have $T[h] = h[(f_a + f_b)/2 + \sum_{k=1}^{n-1} f(x_k)]$, where $f_a = f(a)$ and $f_b = f(b)$. Then, with stepsize $\frac{h}{2}$, we have

$$T\left[\frac{h}{2}\right] = \frac{h}{2}\left(\frac{f_a + f_b}{2} + \sum_{k=1}^{2n-1} f(z_k)\right) = \frac{h}{2}\left(\frac{f_a + f_b}{2} + \sum_{k=1}^{n} f(z_{2k-1}) + \sum_{k=1}^{n-1} f(z_{2k})\right)$$

Since $f(z_{2k}) = f(x_k)$, $k = 1, \ldots, n-1$, we have $\sum_{k=1}^{n-1} f(z_{2k}) = \sum_{k=1}^{n-1} f(x_k)$ and, therefore, can write

$$T\left[\frac{h}{2}\right] = \frac{h}{2}\left(\frac{f_a + f_b}{2} + \sum_{k=1}^{n-1} f(x_k) + \sum_{k=1}^{n} f(z_{2k-1})\right)$$

$$= \frac{1}{2}\left(h\left\{\frac{f_a + f_b}{2} + \sum_{k=1}^{n-1} f(x_k)\right\} + h\sum_{k=1}^{n} f(z_{2k-1})\right)$$

$$= \frac{1}{2}\left(T[h] + h\sum_{k=1}^{n} f(z_{2k-1})\right)$$

This is the recursive form of the Trapezoid rule. The value of $T[\frac{h}{2}]$ can be obtained from the value of $T[h]$ by evaluating $f(x)$ at an additional $n-1$ nodes rather than at an additional $2n+1$ nodes. Hence, this recursive form saves about 50% of the computational energy needed to halve the stepsize in the application of the Trapezoid rule.

Romberg Integration

Romberg integration consists of applying Richardson extrapolation of Section 42.2 to the Trapezoid rule, with the recursive Trapezoid rule being used to economize function evaluations in the first column of a Richardson table. Since the first column of the resulting

Richardson table would contain $T[h_0/r^{i-1}]$, $i = 1, \ldots$, we take $r = 2$ so that we can compute this first column via the recursive form of the Trapezoid rule.

For an example, take the function $f(x) = xe^x/\sqrt{1+x^2}$ so that the definite integral $\Lambda = \int_0^1 f(x)\,dx$ has the approximate value $\lambda = 0.810208902342828$. With $s = 2$, $\Delta = 2$, $h = 1$, and $r = 2$, Table 43.3 gives the result of Richardson extrapolation applied to the values computed for Λ by means of the Trapezoid rule. The values of s and Δ are verified at the conclusion of this section.

0.9610577565					
0.8491941620	0.8119062973				
0.8200025601	0.8102720259	0.8101630740			
0.8126600517	0.8102125489	0.8102085834	0.8102093057		
0.8108218586	0.8102091273	0.8102088994	0.8102089043	0.8102089027	
0.8103621518	0.8102089159	0.8102089014	0.8102089013	0.8102089011	0.8102089010

TABLE 43.3 Romberg table for $\Lambda = \displaystyle\int_0^1 \frac{xe^x\,dx}{\sqrt{1+x^2}}$, with $r = s = 2, h = 1$

The ith row of column 1 in Table 43.3 contains the Trapezoidal approximations $T[1/2^{i-1}]$, $i = 1, \ldots$. Thus, the first invocation of the Trapezoid rule has $h = 1$ so that $x_0 = a$ and $x_n = b$, with $n = 1$. The second and successive columns contain the Richardson extrapolates built from the Trapezoidal values in column 1.

Table 43.4 contains $|\lambda - A_{ij}|$, the absolute value of the error in the i, j-entry in Table 43.3. The first column contains the errors in Trapezoidal approximations to Λ, approximations in which h is repeatedly halved for each iteration. Across each row we have the errors in successive Richardson extrapolates. Typically, the diagonal value at the end of the last row is accepted as the best approximation to the definite integral. Hence, a purely numeric implementation of Romberg integration would have as a stopping rule, the comparison of successive diagonal elements.

0.151					
0.390×10^{-1}	0.170×10^{-2}				
0.980×10^{-2}	0.631×10^{-4}	0.458×10^{-4}			
0.245×10^{-2}	0.365×10^{-5}	0.319×10^{-6}	0.403×10^{-6}		
0.613×10^{-3}	0.225×10^{-6}	0.294×10^{-8}	0.196×10^{-8}	0.357×10^{-9}	
0.153×10^{-3}	0.136×10^{-7}	0.943×10^{-9}	0.104×10^{-8}	0.124×10^{-8}	0.134×10^{-8}

TABLE 43.4 Absolute errors in the entries of Table 43.3

This table of errors shows the effect of roundoff in the extrapolation process. As we saw in Section 42.2, there is a balance between the stepsize h and the ratio r. Romberg integration based on the iterative Trapezoid rule must have $r = 2$, so we cannot manipulate the interplay between h and r to our advantage.

Numeric Implementation of Romberg Integration

The following steps sketch an implementation of Romberg integration that uses the iterated Trapezoid rule for generating the first column of the Richardson extrapolation table. Like Richardson extrapolation, the calculation proceeds by rows, with the diagonal element at the end of each row compared to the diagonal element of the preceding row. Computation is stopped by the diagonal stopping rule, which terminates the method when two successive diagonal elements are within a prescribed tolerance of each other. The complete Richardson table is printed upon termination of the calculations. In practice, just the diagonal element in the last row would be returned.

The initial stepsize is taken as $b - a$ and is halved for each succeeding Trapezoidal calculation, as required by the iterative Trapezoid rule. Hence $r = 2$, and since $s = \Delta = 2$, Richardson extrapolation proceeds with $r^{s+k\Delta} = 2^{2+2k} = 4^{k-1}$.

The output is formatted in an array, or matrix. Hence, there are no zero indices. Since the Romberg table is computed by rows, at the end of row i the row index remains i, but the column index becomes $j + 1$ outside the second loop. Thus, when comparing diagonal elements, the column index must be shifted back by one to compensate for the increment to j.

1. Input the parameters

    (a) $f(x)$, the function to be integrated;

    (b) a, b, the bounds on the interval of integration;

    (c) ε, the error tolerance in the stopping rule;

    (d) N, the maximum number of rows in the Richardson table.

2. Declare A to be a matrix so that the Richardson table is stored as a matrix. The upper left corner of the table corresponds to A_{11}.

3. Compute

    (a) $L = b - a$

    (b) $A_{11} = \dfrac{L}{2}(f(a) + f(b))$

4. For i from 2 to N, perform the following calculations.

    (a) $P = \dfrac{L}{2^{i-1}}$

    (b) $A_{i,1} = \frac{1}{2}A_{i-1,1} + P \displaystyle\sum_{k=1}^{2^{i-2}} f(a + (2k - 1)P)$

    (c) For j from 2 to i, perform the following calculations.

    (i) $A_{i,j} = \dfrac{4^{j-1}A_{i,j-1} - A_{i-1,j-1}}{4^{j-1} - 1}$

    (ii) If $|A_{i-1,j-2} - A_{i,j-1}| < \varepsilon$, return the matrix A, otherwise, continue.

If implemented with $\varepsilon = 10^{-8}$, these steps yield, except for several minute roundoff differences, the contents of Table 43.3. Note the second column of Table 43.3. The entries are exactly those that would be computed by the composite Simpson's rule because Richardson extrapolation of the Trapezoid rule yields Simpson's rule. Below, we will indeed verify the identity

$$S\left[\frac{h}{2}\right] = \frac{2^2 T\left[\frac{h}{2}\right] - T[h]}{2^2 - 1} = \frac{4T\left[\frac{h}{2}\right] - T[h]}{3}$$

where $S[h]$ is the value computed by the composite Simpson's rule using stepsize h.

Simpson's Rule as Richardson Extrapolation of the Trapezoid Rule

We next verify the claim that Richardson extrapolation applied to the Trapezoid rule yields Simpson's rule. As a consequence, we then have a proof that $s = \Delta = 2$ for Romberg integration. (The Trapezoidal rule is $O(h^2)$. One extrapolation of the Trapezoid rule yields Simpson's rule, for which we already know $\Delta = 2$. Hence, we are justified in carrying out Richardson extrapolation with $s = \Delta = 2$ when building the Romberg table.)

To prove that $\frac{1}{3}(4T[\frac{h}{2}] - T[h]) = S[\frac{h}{2}]$, write $T[\frac{h}{2}]$ in terms of the iterated Trapezoid rule, so that we have

$$\frac{1}{3}\left(4T\left[\frac{h}{2}\right] - T[h]\right) = \frac{1}{3}\left\{4\left(\frac{1}{2}\right)\left(T[h] + h\sum_{k=1}^{n} f(z_{2k-1})\right) - T[h]\right\}$$

$$= \frac{1}{3}\left\{2T[h] + 2h\sum_{k=1}^{n} f(z_{2k-1}) - T[h]\right\}$$

$$= \frac{1}{3}\left\{T[h] + 2h\sum_{k=1}^{n} f(z_{2k-1})\right\}$$

$$= \frac{1}{3}\left\{h\left(\frac{f_a + f_b}{2} + \sum_{k=1}^{n-1} f(x_k)\right) + 2h\sum_{k=1}^{n} f(z_{2k-1})\right\}$$

$$= \frac{\left(\frac{h}{2}\right)}{3}\left\{f_a + f_b + 2\sum_{k=1}^{n-1} f(x_k) + 4\sum_{k=1}^{n} f(z_{2k-1})\right\}$$

$$= \frac{\left(\frac{h}{2}\right)}{3}\left\{f_a + f_b + 2\sum_{k=1}^{n-1} f(z_{2k}) + 4\sum_{k=1}^{n} f(z_{2k-1})\right\}$$

$$= S\left[\frac{h}{2}\right]$$

where $x_k, z_{2k}, z_{2k-1}, k = 1, \ldots, n$, are defined according the schematic in Figure 43.1.

EXERCISES 43.2–Part A

A1. This exercise examines the computational savings that the *recursive* Trapezoid rule contributes to Romberg integration.

  (a) In closed form, obtain $N(i)$, the cumulative total for the number of function evaluations performed through i rows of a Romberg integration using the ordinary Trapezoid rule.

  (b) In closed form, obtain $\hat{N}(i)$, the cumulative total for the number of function evaluations performed through i rows of a Romberg integration using the recursive Trapezoid rule.

  (c) Compare, by graph or table of values, $N(i)$ with $\hat{N}(i)$, for $i = 1, \ldots, 10$.

A2. Where Simpson's rule is first considered in elementary calculus, some texts suggest successively doubling the number of points used in Simpson's rule and stopping when the digits no longer

change on the computing device being used. Show that this is an inefficient process by obtaining $\tilde{N}(i)$, the cumulative total for the number of function evaluations performed through i steps of this process if the process started with $n = 2$. Compare to the results in Exercise A1.

A3. Use the Trapezoid rule with $h = 1$ to approximate $\lambda = \int_0^1 \sin x \, dx$.

A4. Use the Trapezoid rule with $h = \frac{1}{2}$ to approximate λ in Exercise A3, and then perform a Richardson extrapolation based on the values computed in Exercises A3 and 4.

A5. Use Simpson's rule with $h = \frac{1}{2}$ to approximate λ in Exercise A3. Compare to the extrapolated value in Exercise A4. The results should agree.

EXERCISES 43.2–Part B

B1. Starting with $h = 1$, obtain the first three rows of the Romberg table that evaluates λ in Exercise A3.

For each function $f(x)$ in Exercises B2–11:

(a) Use Romberg integration with the diagonal element stopping-rule to approximate $\Lambda = \int_{-2}^{2} f(x)\,dx$ with an error less than 10^{-6}.

(b) Verify that the answer in part (a) indeed satisfies the prescribed tolerance requirement.

(c) Determine the actual number of function evaluations needed by the Romberg algorithm to obtain the answer in part (a).

(d) Use the error estimate for Simpson's rule (Table 43.2, Section 43.1) to estimate how many function evaluations it would take

Simpson's rule to obtain the same accuracy as required in part (a).

(e) Determine the actual number of function evaluations it would take Simpson's rule to approximate Λ to within the prescribed tolerance.

B2. $f(x) = (x + 2)e^{-x-2}$ **B3.** $f(x) = \sqrt{2 + 2x + x^2}$

B4. $f(x) = \dfrac{9x^2 - 7x - 5}{4x^2 + 4x + 2}$ **B5.** $f(x) = \dfrac{3x^3 + 5x^2 - 9x + 4}{5 + x}$

B6. $f(x) = \dfrac{9x^2 + 5x - 6}{x - 7}$ **B7.** $f(x) = x \sin x$

B8. $f(x) = \sin(\cos x)$ **B9.** $f(x) = \ln(x^2 + 2x + 2)$

B10. $f(x) = \ln(\cos x + \cosh x)$ **B11.** $f(x) = \sinh(\sin x)$

43.3   Gauss–Legendre Quadrature

Gauss–Legendre Quadrature by Polynomial Exactness

Gauss quadrature rules are $(n + 1)$-point open formulas of the form

$$\int_{\alpha}^{\beta} w(\xi) f(\xi)\,d\xi = \sum_{k=0}^{n} c_k f(\xi_k)$$

where $w(\xi)$ is a positive but fixed weight function; $\xi_k, k = 0, \ldots, n$, are $n + 1$ unevenly spaced nodes in the interval $[\alpha, \beta]$; and the nodal weights $c_k, k = 0, \ldots, n$, are appropriate constants. Such formulas are typically exact for polynomials of at least degree $2n + 1$.

When the weight function is $w(\xi) = 1$ and the interval $[\alpha, \beta]$ is $[-1, 1]$, the resulting formulas are called Gauss–Legendre quadratures. In fact, Table 43.5 lists several of the more common Gauss quadrature families.

$w(\xi)$	α	β	**Name**	$\displaystyle\int_{\alpha}^{\beta} w(\xi) f(\xi)\,d\xi$
1	-1	1	Gauss–Legendre	$\displaystyle\int_{-1}^{1} f(\xi)\,d\xi$
$\dfrac{1}{\sqrt{1 - \xi^2}}$	-1	1	Gauss–Chebyshev	$\displaystyle\int_{-1}^{1} \dfrac{f(\xi)}{\sqrt{1 - \xi^2}}\,d\xi$
$e^{-\xi}$	0	∞	Gauss–Laguerre	$\displaystyle\int_{0}^{\infty} e^{-\xi} f(\xi)\,d\xi$

TABLE 43.5   Gauss quadrature formulas

In this section we will explore Gauss–Legendre quadrature rules and leave the other methods for the exercises. We begin with the following examples.

CASE $n=1$ If $n = 1$, the associated Gauss–Legendre quadrature rule

$$\int_{-1}^{1} f(\xi)\,d\xi = c_0 f(\xi_0) + c_1 f(\xi_1)$$

contains four unknowns, namely, the two constants c_0 and c_1 and the two nodes ξ_0 and ξ_1. Requiring polynomial exactness of degree $2n + 1 = 2(1) + 1 = 3$ provides four conditions for determining the four unknowns since a polynomial of degree three contains the four basis elements $\{1, \xi, \xi^2, \xi^3\}$.

Imposing polynomial exactness for each of the polynomials ξ^k, $k = 0, \ldots, 3$, gives the equations and solution listed in Table 43.6, where the solution for which $\xi_0 < \xi_1$ holds has been selected. The resulting $n = 1$ Gauss–Legendre quadrature rule is then $\int_{-1}^{1} f(\xi)\,d\xi = f(-\frac{1}{\sqrt{3}}) + f(\frac{1}{\sqrt{3}})$.

$$\int_{-1}^{1} 1\,d\xi = 2 = c_0 + c_1 \qquad\qquad c_0 = 1$$

$$\int_{-1}^{1} \xi\,d\xi = 0 = c_0\xi_0 + c_1\xi_1 \qquad\qquad c_1 = 1$$

$$\int_{-1}^{1} \xi^2\,d\xi = \tfrac{2}{3} = c_0\xi_0^2 + c_1\xi_1^2 \qquad\qquad \xi_0 = -\tfrac{1}{\sqrt{3}}$$

$$\int_{-1}^{1} \xi^3\,d\xi = 0 = c_0\xi_0^3 + c_1\xi_1^3 \qquad\qquad \xi_1 = \tfrac{1}{\sqrt{3}}$$

TABLE 43.6   Nodes and weights for Gauss–Legendre quadrature when $n = 1$

EXAMPLE 43.2   For a test function, take $F(\xi) = \sqrt{3 - \xi}$ so that $\lambda_1 = \frac{4}{3}(4 - \sqrt{2})$, the value of the definite integral

$$\Lambda_1 = \int_{-1}^{1} \sqrt{3 - \xi}\,d\xi \tag{43.8}$$

is approximately 3.447715251. The $n = 1$ Gauss–Legendre quadrature formula gives

$$\sqrt{3 + \tfrac{1}{\sqrt{3}}} + \sqrt{3 - \tfrac{1}{\sqrt{3}}} \overset{\circ}{=} 3.447874791$$

with an error of 0.000159540.

By way of comparison, the Left-end rule, with $n = 2$ to force function evaluation at two nodes, yields the approximation $2 + \sqrt{3} \overset{\circ}{=} 3.732051$, with an error of 0.284335557; the Trapezoid rule with $n = 1$ and two function evaluations gives $2 + \sqrt{2} \overset{\circ}{=} 3.414213562$, with an error of 0.033501689; and the Midpoint rule, with $n = 2$ to force function evaluation at two nodes, yields the approximation $\frac{1}{\sqrt{2}}(\sqrt{5} + \sqrt{7}) \overset{\circ}{=} 3.451967523$, with error 0.004252272.

With the same number of function evaluations, Gauss–Legendre quadrature is significantly more accurate than any of the closed formulas familiar from elementary calculus. ❖

CASE $n=2$ If $n = 2$, the associated Gauss–Legendre quadrature rule

$$\int_{-1}^{1} f(\xi)\,d\xi = c_0 f(\xi_0) + c_1 f(\xi_1) + c_2 f(\xi_2)$$

contains six unknowns, namely, the three constants c_0, c_1, and c_2 and the three nodes ξ_0, ξ_1, and ξ_2. Requiring polynomial exactness of degree $2n + 1 = 2(2) + 1 = 5$ provides six conditions for determining the six unknowns since a polynomial of degree five contains the six basis elements $\{1, \xi, \xi^2, \xi^3, \xi^4, \xi^5\}$. Table 43.7 contains the six equations and their solution arising from imposing polynomial exactness for each of polynomials $\xi^k, k = 0, \ldots, 5$. The resulting $n = 2$ Gauss–Legendre quadrature rule is then

$$\int_{-1}^{1} f(\xi) \, d\xi = \tfrac{5}{9} f\left(-\sqrt{\tfrac{3}{5}}\right) + \tfrac{8}{9} f(0) + \tfrac{5}{9} f\left(\sqrt{\tfrac{3}{5}}\right) \tag{43.9}$$

$$\int_{-1}^{1} 1 \, d\xi = 2 = c_0 + c_1 + c_2 \qquad\qquad c_0 = \tfrac{5}{9}$$

$$\int_{-1}^{1} \xi \, d\xi = 0 = c_0\xi_0 + c_1\xi_1 + c_2\xi_2 \qquad\qquad c_1 = \tfrac{8}{9}$$

$$\int_{-1}^{1} \xi^2 \, d\xi = \tfrac{2}{3} = c_0\xi_0^2 + c_1\xi_1^2 + c_2\xi_2^2 \qquad\qquad c_2 = \tfrac{5}{9}$$

$$\int_{-1}^{1} \xi^3 \, d\xi = 0 = c_0\xi_0^3 + c_1\xi_1^3 + c_2\xi_2^3 \qquad\qquad \xi_0 = -\sqrt{\tfrac{3}{5}}$$

$$\int_{-1}^{1} \xi^4 \, d\xi = \tfrac{2}{5} = c_0\xi_0^4 + c_1\xi_1^4 + c_2\xi_2^4 \qquad\qquad \xi_1 = 0$$

$$\int_{-1}^{1} \xi^5 \, d\xi = 0 = c_0\xi_0^5 + c_1\xi_1^5 + c_2\xi_2^5 \qquad\qquad \xi_2 = \sqrt{\tfrac{3}{5}}$$

TABLE 43.7   Nodes and weights for Gauss–Legendre quadrature when $n = 2$

EXAMPLE 43.3   Evaluating (43.8) with (43.9) gives

$$\tfrac{5}{9}\sqrt{3 + \sqrt{\tfrac{3}{5}}} + \tfrac{8}{9}\sqrt{3} + \tfrac{5}{9}\sqrt{3 - \sqrt{\tfrac{3}{5}}} \doteq 3.447717777$$

with an error of 0.25258×10^{-5}. By way of comparison, the Left-end rule with $n = 3$ has four nodes but makes three function evaluations, yielding $\tfrac{2}{9}(6 + \sqrt{30} + 2\sqrt{6}) \doteq 3.639156680$, with an error of 0.191441429. The Trapezoid rule with $n = 2$ makes three function evaluations and yields $1 + \sqrt{3} + \tfrac{1}{\sqrt{2}} \doteq 3.439157589$, with an error of 0.008557662. The Midpoint rule with $n = 3$ has four nodes and three function evaluations, each midway between consecutive nodes, yielding $\tfrac{2}{9}(\sqrt{33} + 3\sqrt{3} + \sqrt{21}) \doteq 3.449620171$, with an error of 0.001904920. Finally, since $n = 2$ is an even integer, we can also apply Simpson's rule, obtaining $\tfrac{1}{3}(2 + \sqrt{2} + 4\sqrt{3}) \doteq 3.447472264$, with an error of 0.000242986.

Once again, with the same number of function evaluations, Gauss–Legendre quadrature is significantly more accurate than any of the closed methods familiar from elementary calculus.   ❖

CASE $n = 3$ If $n = 3$, the associated Gauss–Legendre quadrature rule

$$\int_{-1}^{1} f(\xi) \, d\xi = c_0 f(\xi_0) + c_1 f(\xi_1) + c_2 f(\xi_2) + c_3 f(\xi_3) \tag{43.10}$$

contains eight unknowns, namely, the four constants $c_k, k = 0, \ldots, 3$, and the four nodes $\xi_k, k = 0, \ldots, 3$. Requiring polynomial exactness of degree $2n + 1 = 2(3) + 1 = 7$ provides eight conditions for determining the eight unknowns since a polynomial of degree seven contains the eight basis elements $\xi^k, k = 0, \ldots, 7$. Table 43.8 contains the eight equations and their solution arising from imposing polynomial exactness for each of polynomials $\xi^k, k = 0, \ldots, 7$. Unfortunately, without the symmetry conditions $\xi_0 = -\xi_3, \xi_1 = -\xi_2, c_0 = c_3, c_1 = c_2$ and the ordering conditions $0 < \xi_2 < \xi_3$ the equations resist solution in exact arithmetic.

$$\int_{-1}^{1} 1\, d\xi = 2 = c_0 + c_1 + c_2 + c_3 \qquad\qquad c_0 = \tfrac{1}{2} - \tfrac{1}{36}\sqrt{30}$$

$$\int_{-1}^{1} \xi\, d\xi = 0 = c_0\xi_0 + c_1\xi_1 + c_2\xi_2 + c_3\xi_3 \qquad\qquad c_1 = \tfrac{1}{2} + \tfrac{1}{36}\sqrt{30}$$

$$\int_{-1}^{1} \xi^2\, d\xi = \tfrac{2}{3} = c_0\xi_0^2 + c_1\xi_1^2 + c_2\xi_2^2 + c_3\xi_3^2 \qquad\qquad c_2 = \tfrac{1}{2} + \tfrac{1}{36}\sqrt{30}$$

$$\int_{-1}^{1} \xi^3\, d\xi = 0 = c_0\xi_0^3 + c_1\xi_1^3 + c_2\xi_2^3 + c_3\xi_3^3 \qquad\qquad c_3 = \tfrac{1}{2} - \tfrac{1}{36}\sqrt{30}$$

$$\int_{-1}^{1} \xi^4\, d\xi = \tfrac{2}{5} = c_0\xi_0^4 + c_1\xi_1^4 + c_2\xi_2^4 + c_3\xi_3^4 \qquad\qquad \xi_0 = -\tfrac{1}{35}\sqrt{525 + 70\sqrt{30}}$$

$$\int_{-1}^{1} \xi^5\, d\xi = 0 = c_0\xi_0^5 + c_1\xi_1^5 + c_2\xi_2^5 + c_3\xi_3^5 \qquad\qquad \xi_1 = -\tfrac{1}{35}\sqrt{525 - 70\sqrt{30}}$$

$$\int_{-1}^{1} \xi^6\, d\xi = \tfrac{2}{7} = c_0\xi_0^6 + c_1\xi_1^6 + c_2\xi_2^6 + c_3\xi_3^6 \qquad\qquad \xi_2 = \tfrac{1}{35}\sqrt{525 - 70\sqrt{30}}$$

$$\int_{-1}^{1} \xi^7\, d\xi = 0 = c_0\xi_0^7 + c_1\xi_1^7 + c_2\xi_2^7 + c_3\xi_3^7 \qquad\qquad \xi_3 = \tfrac{1}{35}\sqrt{525 + 70\sqrt{30}}$$

TABLE 43.8   Nodes and weights for Gauss–Legendre quadrature when $n = 3$

This example calls attention to the difficulty of determining the nodes and nodal weights in exact form. Typically, these parameters are determined numerically in floating-point form and by more efficient techniques which we will illustrate for the case $n = 4$.

EXAMPLE 43.4   Evaluating (43.8) with (43.10) and the contents of Table 43.8 gives 3.447715300, with an error of 0.484×10^{-7}. By way of comparison, the Left-end rule with $n = 4$, so that there are four function evaluations, gives $1 + \tfrac{1}{4}(\sqrt{14} + 2\sqrt{3} + \sqrt{10}) \doteq 3.592009166$, with an error of 0.144293914. The Trapezoid rule with $n = 3$ implements four function evaluations, yielding $\tfrac{1}{9}(6 + 2\sqrt{30} + 4\sqrt{6} + 3\sqrt{2}) \doteq 3.443894535$, with an error of 0.003820716. The Midpoint rule with $n = 4$ implements four function evaluations and approximates Λ_1 as $\tfrac{1}{4}(\sqrt{15} + \sqrt{13} + \sqrt{11} + 3) \doteq 3.448789853$, with an error of 0.001074602.

Once again, with the same number of function evaluations, Gauss–Legendre quadrature is significantly more accurate than comparable closed methods familiar from elementary calculus. ❖

Gauss–Legendre Quadrature by Legendre Polynomials

The following theorem summarizes the creation of Gauss–Legendre quadrature formulas via the Legendre polynomials. (See [49], [16], or [60] for a proof.)

THEOREM 43.1

1. $\xi_k, k = 0, \ldots, n$, are the $n + 1$ zeros of $P_{n+1}(x)$, the Legendre polynomial of degree $n + 1$.

2. $L_k(x) = \prod_{i=0, i \neq k}^{n} (x - \xi_i)/(\xi_k - \xi_i)$, $k = 0, \ldots, n$, are the Lagrange interpolating polynomials for $\xi_k, k = 0, \ldots n$.

3. $c_k = \int_{-1}^{1} L_k(x)\, dx, k = 0, \ldots, n$

$\Longrightarrow$

1. The quadrature formula $\int_{-1}^{1} f(\xi)\, d\xi = \sum_{k=0}^{n} c_k f(\xi_k)$ is polynomial-exact of at least degree $2n + 1$.

2. The error of the approximation is given by

$$\frac{2^{2n+3}((n+2)!)^4 (2n+3)^2}{(n+2)^4 ((2n+3)!)^3} f^{(2n+2)}(\eta) \quad -1 < \eta < 1$$

We will now illustrate the construction of Gauss–Legendre quadrature formulas via this theorem. In the next part of the section, we will extend these Gauss–Legendre quadrature formulas to integrals of the form $\int_{a}^{b} g(x)\, dx$. In the final part of the section, we will examine the applicability of the error estimate given by this theorem and give an alternate form of the error estimate for integrals over the interval $[a, b]$.

CASE $n = 1$ Earlier we used polynomial exactness to construct the $n = 1$ Gauss–Legendre quadrature formula whose nodes and weights are given in Table 43.6. The zeros of the Legendre polynomial $P_2(x) = \frac{3}{2}x^2 - \frac{1}{2}$ are easily found to be $\pm\frac{1}{\sqrt{3}}$, in agreement with ξ_0 and ξ_1 of Table 43.6. The Lagrange interpolating polynomials are then $L_0 = \frac{1}{2} - \frac{1}{2}\sqrt{3}x$ and $L_1 = \frac{1}{2} + \frac{1}{2}\sqrt{3}x$, so the nodal weights are $c_0 = \int_{-1}^{1} L_0(x)\, dx = 1 = \int_{-1}^{1} L_1(x)\, dx = c_1$, again in agreement with Table 43.6.

CASE $n = 2$ Earlier we used polynomial exactness to construct the $n = 2$ Gauss–Legendre quadrature formula for which the nodes and weights are given in Table 43.7. The zeros of the Legendre polynomial $P_3(x) = \frac{5}{2}x^3 - \frac{3}{2}x$ are easily found to be $0, \pm\sqrt{\frac{3}{5}}$, in agreement with Table 43.7. For these nodes, the Lagrange interpolating polynomials $L_k(x)$ and the weights $c_k = \int_{-1}^{1} L_k(x)\, dx$ are

$$c_0 = \int_{-1}^{1} \frac{x}{6}\left(5x - \sqrt{15}\right) dx = \tfrac{5}{9}$$

$$c_1 = \int_{-1}^{1} \left(1 - \tfrac{5}{3}x^2\right) dx = \tfrac{8}{9}$$

$$c_2 = \int_{-1}^{1} \frac{x}{6}\left(5x + \sqrt{15}\right) dx = \tfrac{5}{9}$$

in agreement with Table 43.7.

Case $n = 3$ Earlier we used polynomial exactness to construct the $n = 3$ Gauss–Legendre quadrature formula for which the nodes and weights are given in Table 43.8. In agreement with Table 43.8, the zeros of the Legendre polynomial $P_4(x) = \frac{35}{8}x^4 - \frac{15}{4}x^2 + \frac{3}{8}$ are found to be $\pm\frac{1}{35}\sqrt{525 \pm 70\sqrt{30}}$ by noting that the equation to be solved is quadratic in x^2.

The Lagrange interpolating polynomials for these nodes are

$$L_0 = \xi_3\left[\frac{x^3}{48}\left(4 - \sqrt{30}\right) + \frac{x}{14}\left(\tfrac{23\sqrt{30}}{120} - 1\right)\right] + \frac{1}{48}\left[(7x^2 - 3)\sqrt{30} + 12\right]$$

$$L_1 = \xi_2\left[\frac{x^3}{48}\left(4 + \sqrt{30}\right) - \frac{x}{14}\left(\tfrac{23\sqrt{30}}{120} + 1\right)\right] + \frac{1}{48}\left[-(7x^2 - 3)\sqrt{30} + 12\right]$$

$$L_2 = \xi_1\left[\frac{x^3}{48}\left(4 + \sqrt{30}\right) - \frac{x}{14}\left(\tfrac{23\sqrt{30}}{120} + 1\right)\right] + \frac{1}{48}\left[-(7x^2 - 3)\sqrt{30} + 12\right]$$

$$L_3 = \xi_0\left[\frac{x^3}{48}\left(4 - \sqrt{30}\right) + \frac{x}{14}\left(\tfrac{23\sqrt{30}}{120} - 1\right)\right] + \frac{1}{48}\left[(7x^2 - 3)\sqrt{30} + 12\right]$$

and tedious calculation (or appropriate software) will show that $c_k = \int_{-1}^{1} L_k(x)\,dx$, $k = 0, \ldots, 3$, in agreement with Table 43.8.

Case $n = 4$ Finally, we will obtain the nodes for the $n = 4$ Gauss–Legendre quadrature formula by computing ξ_k, $k = 0, \ldots, 4$, the zeros of the Legendre polynomial $P_5(x) = \frac{63}{8}x^5 - \frac{35}{4}x^3 + \frac{15}{8}x$, as $0, \pm0.5384693101, \pm0.9061798459$. The Lagrange interpolating polynomials for these nodes are

$$L_0 = 1.146232575x^4 - 1.038692852x^3 - 0.332349219x^2 + 0.301168159x$$

$$L_1 = -3.246232572x^4 + 1.747996613x^3 + 2.665682549x^2 - 1.435388244x$$

$$L_2 = \tfrac{21}{5}x^4 - \tfrac{14}{3}x^2 + 1$$

$$L_3 = -3.246232572x^4 - 1.747996614x^3 + 2.665682548x^2 + 1.435388242x$$

$$L_4 = 1.146232570x^4 + 1.038692854x^3 - 0.3323492143x^2 - 0.3011681598x$$

and the nodal weights are

$$c_0 = \int_{-1}^{1} L_0(x)\,dx = 0.2369268858 = \int_{-1}^{1} L_4(x)\,dx = c_4$$

$$c_1 = \int_{-1}^{1} L_1(x)\,dx = 0.4786286687 = \int_{-1}^{1} L_3(x)\,dx = c_3$$

$$c_2 = \int_{-1}^{1} L_2(x)\,dx = \tfrac{128}{225} = 0.56\overline{8}$$

For the Gauss–Legendre quadrature formula $\int_{-1}^{1} f(\xi)\,d\xi = \sum_{k=0}^{4} c_k f(\xi_k)$, Table 43.9 contains the data that confirms the polynomial exactness of degree $2n + 1 = 2 \times 4 + 1 = 9$. After accounting for a modest amount of roundoff error from the numeric calculations, the table confirms the claim about polynomial exactness.

Integration on $[a, b]$

To extend the Gauss–Legendre quadrature rules to integrals of the form

$$\int_{a}^{b} g(x)\,dx \tag{43.11}$$

$\int_{-1}^{1} \xi^k \, d\xi$	$\sum_{i=0}^{4} c_i \xi^k$	$\int_{-1}^{1} \xi^k \, d\xi$	$\sum_{i=0}^{4} c_i \xi^k$
$\int_{-1}^{1} \xi^0 \, d\xi = 2$	1.999999999	$\int_{-1}^{1} \xi^5 \, d\xi = 0$	-0.4×10^{-9}
$\int_{-1}^{1} \xi^1 \, d\xi = 0$	0	$\int_{-1}^{1} \xi^6 \, d\xi = \frac{2}{7}$	0.2857142861
$\int_{-1}^{1} \xi^2 \, d\xi = \frac{2}{3}$	0.6666666665	$\int_{-1}^{1} \xi^7 \, d\xi = 0$	-0.3×10^{-9}
$\int_{-1}^{1} \xi^3 \, d\xi = 0$	-0.3×10^{-9}	$\int_{-1}^{1} \xi^8 \, d\xi = \frac{2}{9}$	0.2222222225
$\int_{-1}^{1} \xi^4 \, d\xi = \frac{2}{5}$	0.4	$\int_{-1}^{1} \xi^9 \, d\xi = 0$	-0.27×10^{-9}

TABLE 43.9   Confirmation of polynomial exactness of degree $2n + 1 = 2 \times 4 + 1 = 9$ for Gauss–Legendre quadrature when $n = 4$

we reason as follows. The formula

$$\int_{-1}^{1} f(\xi) \, d\xi = \sum_{k=0}^{n} w_k f(\xi_k)$$

is already available, but we want to evaluate (43.11), so linearly map the interval $a \leq x \leq b$ to the interval $-1 \leq \xi \leq 1$ by means of the transformation $\xi(x) = 2\frac{x-a}{b-a} - 1$, for which the inverse transformation is

$$x(\xi) = a + \frac{b-a}{2}(\xi + 1) \tag{43.12}$$

Define the transformed nodes

$$x_k = x(\xi_k) = a + \frac{b-a}{2}(\xi_k + 1) \qquad k = 0, \dots, n$$

and the function $f(\xi) = g(x(\xi))$ so that $f(\xi_k) = g(x(\xi_k)) = g(x_k), k = 0, \dots, n$. If we now start with the integral (43.11) and change of variables from x to ξ by means of (43.12), we obtain

$$\frac{b-a}{2} \int_{-1}^{1} g\left(a + \frac{b-a}{2}(\xi + 1)\right) d\xi$$

But $g(x(\xi))$ is $f(\xi)$, so we have

$$\int_{a}^{b} g(x) \, dx = \frac{b-a}{2} \int_{-1}^{1} g(x(\xi)) \, d\xi = \frac{b-a}{2} \int_{-1}^{1} f(\xi) \, d\xi$$

$$= \frac{b-a}{2} \sum_{k=0}^{n} c_k f(\xi_k) = \frac{b-a}{2} \sum_{k=0}^{n} c_k g(x_k)$$

a result illustrated in the following example.

EXAMPLE 43.5   The value of the definite integral $\Lambda_2 = \int_{-2}^{3} g(x) \, dx$, where $g(x) = \cosh\sqrt{1 + x + 2x^2}$, is approximately $\lambda_2 = 45.21287573$. The transformation (43.12) becomes $x = \frac{5}{2}\xi + \frac{1}{2}$. For the case $n = 4$, the nodes ξ_k, $k = 0, \ldots, 4$, the transformed nodes x_k, and the nodal weights c_k are listed in Table 43.10.

k	ξ_k	$x_k = \dfrac{5\xi + 1}{2}$	c_k
0	-0.9061798459	-1.765449615	0.2369268858
1	-0.5384693101	-0.8461732750	0.4786286687
2	0	$\frac{1}{2}$	0.5688888891
3	0.5384693101	1.846173275	0.4786286699
4	0.9061798459	2.765449615	0.2369268851

TABLE 43.10   Example 43.5: Nodes and weights for Gauss–Legendre quadrature applied to Λ_2 when $n = 4$

The $n = 4$ Gauss–Legendre quadrature formula

$$\int_a^b g(x)\,dx = \frac{b-a}{2} \sum_{k=0}^{4} c_k g(x_k)$$

will give 45.21154144, for which the absolute value of the error is 0.00133429. The $n = 4$ Gauss–Legendre quadrature formula evaluates $g(x)$ at $n + 1 = 5$ points, as does Simpson's rule with $n = 4$. The error made by Simpson's rule with this many function evaluations is 1.62638469, an error that is more than 1200 times as large.   ❖

Error Analysis

In this final part of the section, we will examine how the actual errors made in our examples of Gauss–Legendre quadrature compare with theoretical estimates. Table 43.11 summarizes the two forms of Gauss–Legendre quadrature we have just seen and error estimates for each.

Gauss–Legendre Quadrature	Error-Estimate Coefficient	Error Estimate	
$\displaystyle\int_{-1}^{1} f(\xi)\,d\xi = \sum_{k=0}^{n} c_k f(\xi_k)$	$\sigma_1(n) = \dfrac{2^{2n+3}((n+2)!)^4(2n+3)^2}{(n+2)^4((2n+3)!)^3}$	$E_1 = \sigma_1 f^{(2n+2)}(\eta)$	$-1 < \eta < 1$
$\displaystyle\int_a^b g(x)\,dx = \dfrac{b-a}{2} \sum_{k=0}^{n} c_k g(x_k)$	$\sigma_2(n) = \dfrac{\int_a^b \left[\prod_{k=0}^{n}(x - x_k)\right]^2 dx}{(2n+2)!}$	$E_2 = \sigma_2 g^{(2n+2)}(\eta)$	$a < \eta < b$

TABLE 43.11   Error-estimate formulas for Gauss–Legendre quadrature

In its third column, Table 43.12 lists the actual error found when approximating Λ_1 with the Gauss–Legendre quadrature of order n. The fourth column of Table 43.12 shows the common values for $\sigma_1(n)$ and $\sigma_2(n)$, the coefficients of the derivatives $F^{(2n+2)}(\eta)$ in the error estimates E_1 and E_2. For Λ_1, either of the estimates in Table 43.11 can be applied since here, $[a, b] = [-1, 1]$ and F and g can be identified with each other. The fifth column of Table 43.12 lists the maximum of the absolute value of $F^{(2n+2)}(\xi)$. In each case, the

n	$2n+2$	Actual Error	$\sigma_1(n) = \sigma_2(n)$	$\max\lvert F^{(2n+2)}\rvert$	Error Bound
1	4	0.160×10^{-3}	$\frac{1}{135}$	$\frac{15\sqrt{2}}{256}$	0.614×10^{-3}
2	6	0.253×10^{-5}	$\frac{1}{15,750}$	$\frac{945\sqrt{2}}{4096}$	0.208×10^{-4}
3	8	0.484×10^{-7}	$\frac{1}{3,472,875}$	$\frac{135,135\sqrt{2}}{65,536}$	0.840×10^{-6}

TABLE 43.12   Errors, and error bounds, for approximating Λ_1 by Gauss–Legendre quadratures of order $n = 1, 2, 3$

maximum occurs at the endpoint $\xi = 1$. The sixth column lists the estimate of error given by the theoretical error bounds E_1 or E_2 listed in Table 43.11. For each n, the actual error is smaller than the upper bound predicted by theory. However, the theoretical upper bounds are within an order of magnitude of the actual error, so theory proves useful in gauging errors in Gauss–Legendre quadratures.

     In Example 43.5, a fourth-order Gauss–Legendre quadrature approximated the definite integral Λ_2 as 45.21154144, with an error of 0.00133429. The nodes ξ_k and x_k, $k = 0, \ldots, 4$, are given in Table 43.10. Using the nodes ξ_k, $\sigma_2(4) = 0.808 \times 10^{-9}$, in agreement with the value of $\sigma_1(4)$. However, the error estimate E_2 must be used, so $\sigma_2(4)$ is evaluated with the nodes x_k, giving the value 0.1926×10^{-4}. The maximum of $\lvert g^{(10)}\rvert$ occurs at $x = 3$ and has the value 1631.77, so, as an upper bound for the error, theory predicts 0.0314321, which is slightly larger than the actual error.

EXERCISES 43.3

1. Use the 2-point ($n = 1$) Gauss–Legendre quadrature formula to approximate $\lambda = \int_0^1 \cos x\, dx$ and determine the actual error made. Compare to the error made when λ is approximated with the Left-end, Trapezoid, and Midpoint rules, each using exactly two function evaluations.

For Exercises 2–5, select a function $f_k(x)$ from the following list.

$f_1(x) = (x + 2)e^{-x-2}$ $f_6(x) = x \sin x$

$f_2(x) = \sqrt{2 + 2x + x^2}$ $f_7(x) = \sin(\cos x)$

$f_3(x) = \dfrac{9x^2 - 7x - 5}{4x^2 + 4x + 2}$ $f_8(x) = \ln(x^2 + 2x + 2)$

$f_4(x) = \dfrac{3x^3 + 5x^2 - 9x + 4}{5 + x}$ $f_9(x) = \ln(\cos x + \cosh x)$

$f_5(x) = \dfrac{9x^2 + 5x - 6}{x - 7}$ $f_{10}(x) = \sinh(\sin x)$

2. For the function selected from the given list:

  (a) Use 2-point ($n = 1$) Gauss–Legendre quadrature to approximate $\Lambda = \int_{-1}^{1} f_k(\xi)\, d\xi$.

  (b) Apply the appropriate error estimate from Table 43.11.

  (c) Compare the predicted error to the actual error.

3. Repeat Exercise 2 for 3-point ($n = 2$) Gauss–Legendre quadrature.

4. Repeat Exercise 2 for 4-point ($n = 3$) Gauss–Legendre quadrature.

5. Repeat Exercise 2 for 5-point ($n = 4$) Gauss–Legendre quadrature.

6. Obtain the nodes ξ_k and weights c_k for Gauss–Legendre quadrature for the case $n = 5$.

7. Repeat Exercise 6 for the case $n = 6$.

8. Repeat Exercise 6 for the case $n = 7$.

9. Repeat Exercise 6 for the case $n = 8$.

10. Repeat Exercise 6 for the case $n = 9$.

11. Repeat Exercise 6 for the case $n = 10$.

In Exercises 12–16:

  (a) Approximate $\Lambda = \int_{1}^{1.8} g(x)\, dx$ using $(n + 1)$-point Gauss–Legendre quadrature, with $n = 2, 3, 4$; in each case, determine the actual error made.

  (b) Approximate Λ using Simpson's rule with $n = 4$, and compare the actual error with part (a).

12. $g(x) = \dfrac{5x^3 + 8x^2 + 4x - 3}{8x^2 - 9x + 7}$

13. $g(x) = \dfrac{9x^2 - 9x + 4}{8x + 5}$

14. $g(x) = \dfrac{4x^2 + 7x - 9}{5x^3 + 7x^2 - 8x + 3}$

15. $g(x) = \dfrac{2x^3 + 9x^2 - 8x + 1}{9x^3 + 5x^2 + x + 3}$

16. $g(x) = \dfrac{3x + 1}{2x^2 + 8x}$

17. For the case $n = 1$, obtain the nodes ξ_k and the weights c_k for $(n + 1)$-point Gauss–Chebyshev quadrature $\int_{-1}^{1} \dfrac{f(\xi)}{\sqrt{1-\xi^2}} \, d\xi = \sum_{k=0}^{n} c_k f(\xi_k)$ using

   (a) polynomial exactness.

   (b) Chebyshev polynomials $T_k(\xi)$ for the nodes and Lagrange interpolating polynomials for the weights.

18. Repeat Exercise 17 for the case $n = 2$.

19. Repeat Exercise 17 for the case $n = 3$.

20. Approximate $\Lambda = \int_{-1}^{1} e^{\xi} \, d\xi / \sqrt{1 - \xi^2}$ with $(n + 1)$-point Gauss–Chebyshev quadrature, with $n = 1, 2, 3$. In each case, obtain the actual error made. How many function evaluations would it take to achieve the accuracy of the $n = 3$ Gauss–Chebyshev quadrature if Romberg integration were used?

21. To find a 2-point ($n = 1$) Gauss quadrature of the form $\int_{a}^{b} w(x) f(x) \, dx = \sum_{k=0}^{1} c_k f(x_k)$, where $w(x) = \sqrt{x}$, $[a, b] = [0, 4]$:

   (a) Using the inner product $\langle u, v \rangle = \int_{a}^{b} w(x) u(x) v(x) \, dx$, apply the Gram–Schmidt process to $\{1, x, x^2\}$ to obtain $p_0 = 1$, $p_1(x)$, and $p_2(x)$, polynomials orthogonal with respect to the given inner product.

   (b) Obtain the zeros of $p_2(x)$ as the nodes x_k.

   (c) Obtain $L_k(x)$, $k = 0, 1$, the Lagrange interpolating polynomials for the nodes x_0, x_1.

   (d) Obtain the weights $c_k = \int_{a}^{b} w(x) L_k(x) \, dx$.

22. Repeat Exercise 21 for $w(x) = e^x$, $[a, b] = [0, 2]$.

In Exercises 23–26:

   (a) Apply the Gauss quadrature of Exercise 21 to approximate the integral $\int_{0}^{4} \sqrt{x} f(x) \, dx$.

   (b) Determine the actual error made by this approximation.

23. $f(x) = xe^x$ **24.** $f(x) = 3 \sin x - 4 \cos x$

25. $f(x) = \dfrac{x^3 + 3x^2 - 4x + 6}{8x^2 + 11x + 12}$ **26.** $f(x) = 2e^x - \sin(\sin x)$

In Exercises 27–30:

   (a) Apply the Gauss quadrature of Exercise 22 to approximate the integral $\int_{0}^{2} e^x f(x) \, dx$.

   (b) Determine the actual error made by this approximation.

27. $f(x) = \sin x \tanh x$ **28.** $f(x) = \ln(x^2 + 2e^x)$

29. $f(x) = e^{-x^2} - \operatorname{sech} x$ **30.** $f(x) = \sqrt{1 + \ln(x + \cosh x)}$

43.4   Adaptive Quadrature

Motivation

Suppose Simpson's rule requires $n + 1$ points to evaluate the integral $\Lambda = \int_{a}^{b} f(x) \, dx$ to an accuracy of ε. It may well be that in one portion of the interval $[a, b]$ the function $f(x)$ varies more rapidly than in another, so that the number of function evaluations is actually being determined by only a small part of $[a, b]$. If so, it might be more efficient to integrate over each subsection of $[a, b]$, each time using just enough points appropriate for that subsection. Perhaps the total number of function evaluations can thereby be reduced.

By way of illustration, consider the function $f(x) = \cosh \sqrt{1 + x + 2x^2}$ so that

$$\Lambda = \int_{-2}^{3} f(x) \, dx \stackrel{\circ}{=} 45.21287573$$

Using Simpson's rule to approximate Λ with an error no greater than 10^{-4} takes 51 function evaluations because with $n = 48$, Simpson's rule gives an error of 1.08×10^{-4} and, with $n = 50$, it gives Λ to within 0.9×10^{-4}. There are 51 function evaluations because the nodes are indexed x_k, $k = 0, \ldots, 50$.

Suppose we split the interval $[-2, 3]$ exactly in half, forming the two subintervals $[-2, \frac{1}{2}]$ and $[\frac{1}{2}, 3]$, and then integrate separately over the two subintervals, permitting an error of at most $\frac{\varepsilon}{2} = 0.5 \times 10^{-4}$ on each subinterval. On each subinterval we have the integrals

$$\Lambda_1 = \int_{-2}^{1/2} f(x) \, dx \stackrel{\circ}{=} 6.594312496 \quad \text{and} \quad \Lambda_2 = \int_{1/2}^{3} f(x) \, dx \stackrel{\circ}{=} 38.61856323$$

It takes 19 ($n = 18$) function evaluations to approximate Λ_1 to within $\frac{\varepsilon}{2} = 0.5 \times 10^{-4}$ and 31 ($n = 15$) function evaluations for Λ_2. Hence, a total of $19 + 31 - 1 = 49$ function evaluations are needed with the interval-splitting technique. (Since the two subintervals are contiguous, the function need be computed at the midpoint of $[-2, 3]$ just once, not twice.) This is a slight savings over the original tally of 51, but the intent here is merely to provide a motivation for the interval splitting that constitutes the adaptive quadrature strategy.

Error-Test Mechanism

A key ingredient of an adaptive scheme is the error-test mechanism. In Section 4.1, in the context of numeric solutions of differential equations, we derived the error-test mechanism

$$ y_{\text{exact}} - y_{\text{approx}} \left(\frac{h}{2} \right) = \frac{y_{\text{approx}} \left(\frac{h}{2} \right) - y_{\text{approx}}(h)}{2^k - 1} $$

valid if the error in y_{approx} is $O(h^k)$. Thus, the error in the more accurate approximation (i.e., the one made with stepsize $\frac{h}{2}$) is measured by the difference between the more accurate $\frac{h}{2}$-approximation and the less-accurate h-approximation, all divided by $2^k - 1$. This very calculation was reviewed in Section 42.2 in the context of Richardson extrapolation.

Since Simpson's composite rule is $O(h^4)$, the factor in the denominator of the error-test estimate would be $2^4 - 1 = 15$.

We will show how the interval-splitting strategy is coupled with this error-test expression to produce an adaptive, or self-directed, numeric integration scheme.

Naive Implementation of Adaptive Quadrature

We next sketch a naive implementation of an adaptive quadrature scheme based on Simpson's rule. Since its purpose is merely to illustrate the strategy, no attempt is made to minimize the number of function evaluations. For the moment, we ignore the efficiencies that would accrue if we saved all function evaluations that get recomputed during the execution of our algorithm.

To evaluate the definite integral $\Lambda = \int_a^b f(x) \, dx$, we apply Simpson's rule with stepsize $h = b - a$ and again with stepsize $\frac{h}{2}$. The first computation is done by invoking Simpson's rule with $n = 2$, the second, with $n = 4$. As a consequence, we have two estimates of Λ, from which the error in the more accurate approximation can be determined by dividing the difference by 15. If the error is small enough, we accept the more accurate $\frac{h}{2}$-approximation and the computation is finished.

If the error is too large, the process is repeated on the interval $[a, \frac{a+b}{2}]$, with the interval $[\frac{a+b}{2}, b]$ set aside in a last-in, first-out *stack*, for processing later. If the numeric approximation for the left-hand subinterval is acceptable, its value is stored and the process repeated on the right-hand subinterval. The accuracy demanded of any subinterval is proportional to the ratio of the length of the subinterval to the length of the original interval. If the overall tolerance for Λ is ε, then the permitted error in any subinterval $[\alpha, \beta]$ is ε times the ratio of the length of $[\alpha, \beta]$ to the length of $[a, b]$.

Table 43.13 summarizes for

$$ \Lambda = \int_{-1}^{1} \sqrt{3 - x} \, dx \stackrel{\circ}{=} 3.447715250169 $$

an adaptive quadrature based on Simpson's rule, in which $\varepsilon = 10^{-8}$. The adaptive integration yielded the answer 3.447715248, whose error is $0.2 \times 10^{-8} < \varepsilon$.

k	Present Interval	Present Stack	P/F	k	Present Interval	Present Stack	P/F
1	$[-1, 1]$	[ ]	F	10	$[0, \frac{1}{2}]$	$[\frac{1}{2}, 1]$	F
2	$[-1, 0]$	$[0, 1]$	F	11	$[0, \frac{1}{4}]$	$[\frac{1}{2}, 1], [\frac{1}{4}, \frac{1}{2}]$	P
3	$[-1, -\frac{1}{2}]$	$[0, 1], [-\frac{1}{2}, 0]$	F	12	$[\frac{1}{4}, \frac{1}{2}]$	$[\frac{1}{2}, 1]$	P
4	$[-1, -\frac{3}{4}]$	$[0, 1], [-\frac{1}{2}, 0], [-\frac{3}{4}, -\frac{1}{2}]$	P	13	$[\frac{1}{2}, 1]$	[ ]	F
5	$[-\frac{3}{4}, -\frac{1}{2}]$	$[0, 1], [-\frac{1}{2}, 0]$	P	14	$[\frac{1}{2}, \frac{3}{4}]$	$[\frac{3}{4}, 1]$	P
6	$[-\frac{1}{2}, 0]$	$[0, 1]$	F	15	$[\frac{3}{4}, 1]$	[ ]	F
7	$[-\frac{1}{2}, -\frac{1}{4}]$	$[0, 1], [-\frac{1}{4}, 0]$	P	16	$[\frac{3}{4}, \frac{7}{8}]$	$[\frac{7}{8}, 1]$	P
8	$[-\frac{1}{4}, 0]$	$[0, 1]$	P	17	$[\frac{7}{8}, 1]$	[ ]	P
9	$[0, 1]$	[ ]	F				

TABLE 43.13   The dynamic stack of an adaptive Simpson's rule quadrature applied to $\Lambda = \int_{-1}^{1} \sqrt{3 - x}\, dx$

The left-hand column indexes the "cycles" executed by the code. The second column lists the "active interval," the interval over which integration is presently taking place. The third column lists the contents of the stack, the pile of subintervals over which integration has yet to take place. The fourth column lists the outcome of the error test applied to the integration on the "active interval." Where it lists *P (Pass)*, the integration was accurate enough, and no further splitting of that interval need take place. Where it lists *F (Fail)*, the integration was not accurate enough, and the "active integral" must be split in half, with the left-hand portion becoming the next "active interval" and the right-hand portion being added to the stack.

Taking the computation from the start, we begin with $[-1, 1]$ as the active, or present, interval. Simpson's rule is applied to this interval with $h = 2$ and $\frac{h}{2} = 1$. As seen from Table 43.13, this first integration is not accurate enough, so the interval is split into the subintervals $[-1, 0]$ and $[0, 1]$.

In cycle 2, the active interval is $[-1, 0]$ and the interval $[0, 1]$ is placed on the stack. Again, integration over the active interval is done twice, with $n = 2, 4$, respectively, in Simpson's rule. The outcome of the integrations is error-tested, and again, the integration over the active interval is not accurate enough. Hence, the active interval is split again, this time into the new active interval $[-1, -\frac{1}{2}]$ and the interval $[-\frac{1}{2}, 0]$, which is placed on the stack.

In cycle 3, the active integral is $[-1, -\frac{1}{2}]$ and integration over it is performed twice, with $n = 2, 4$, respectively, in Simpson's rule. The outcome of the two integrations is error-tested, and again, the error is too large. Once again, the active interval is split, this time into the new active interval $[-1, -\frac{3}{4}]$ and the interval $[-\frac{3}{4}, -\frac{1}{2}]$, which is placed on the stack.

In cycle 4, the active integral is $[-1, -\frac{3}{4}]$ and integration over it is performed twice, with $n = 2, 4$, respectively, in Simpson's rule. The outcome of the two integrations is passed to the error-test, and finally, the error is deemed small enough. The $n = 4$ value produced by Simpson's rule is accepted as the value of the integral over the active interval, and this number is added to an accumulator Σ that was initialized to 0 at the very start of the computation. No new subinterval is added to the stack. In fact, the last subinterval added to the stack at the end of cycle 3 is now removed from the stack and becomes the new active interval for cycle 5.

We leave it to the reader to continue this exposition, cycle-by-cycle. Instead, we turn our attention to some final aspects of the logic leading to Table 43.13.

1. Each time integration over a subinterval is deemed accurate enough for that value to be "kept," it is added to the accumulator Σ.

2. The computation ends when integration over the present interval is accepted as accurate *and* the stack is empty. Table 43.13 shows the stack is empty after cycles 8, 12, 14, and 16. However, the integrations in cycles 9, 13, and 15 do not pass the error test, so the integration does not terminate. Only after cycle 16 does the next integration pass the error test, and that is when the integration terminates.

3. When the integration process ends, the value accumulated in Σ is the desired approximation of Λ.

4. To program a "split" of the active interval $[\alpha, \beta]$, compute its midpoint $C = \frac{\alpha+\beta}{2}$ and form the right-hand subinterval $[C, \beta]$ before forming the left-hand subinterval $[\alpha, C]$. If β were overwritten by C first, the subinterval $[C, \beta]$ would be unavailable.

Minimizing Function Evaluations

It is highly inefficient to invoke a Simpson procedure twice for each cycle and to invoke Simpson's rule "fresh" for each subinterval. A little thought shows that repeated calls to Simpson's rule requires recalculating function values. If we can determine where these function values are being used, and can determine a way to store them, the number of function evaluations can be dramatically reduced.

In the example used to generate Table 43.13, for the two initial integrations, the active interval was $[-1, 1]$, so when Simpson's rule was applied with $n = 2$, $f(x)$ was evaluated at the three nodes $x = -1, 0, 1$. For that same interval, when Simpson's rule was applied with $n = 4$, $f(x)$ was evaluated at the five nodes $x = -1, -\frac{1}{2}, 0, \frac{1}{2}, 1$. However, only two of these nodes, namely, $x = \pm\frac{1}{2}$, are new nodes. Hence, if we had stored the values of $f(x)$ at the first three nodes initially, we would have needed to evaluate $f(x)$ at just two more nodes to complete the first cycle. The first cycle requires just five function evaluations.

The integrations in the first cycle were not accurate enough, so the initial interval was split into the two subintervals, $[-1, 0]$ and $[0, 1]$. Two new integrations were then performed in the new active interval, $[-1, 0]$. The $n = 2$ application of Simpson's rule required evaluating $f(x)$ at the three nodes $x = -1, -\frac{1}{2}, 0$. Had we stored all function values computed in cycle 1, we could have obtained this new Simpson's rule approximation without computing new function values.

The $n = 4$ application of Simpson's rule required evaluating $f(x)$ at the five nodes $x = -1, -\frac{3}{4}, -\frac{1}{2}, -\frac{1}{4}, 0$. However, only two of these five nodes—namely, $x = -\frac{3}{4}, -\frac{1}{4}$— are new. It takes just two new function evaluations to complete the integrations in cycle 2.

A similar analysis shows that after cycle 1 (which requires five function evaluations), all subsequent cycles can be completed with just two more function evaluations each. Thus, the 17 cycles we executed should have taken just $5 + 2 \times 16 = 37$ function evaluations. Had we used Simpson's rule with $n = 36$, we would have expended the equivalent computational energy and approximated Λ with an error of -0.3×10^{-8}, 50% more than the adaptive version's 0.2×10^{-8}.

However, for the function $f(x) = \frac{1}{x} + \frac{x^2}{1+x^2}$, the integral

$$\Lambda = \int_{1/10}^{5} f(x)\,dx \overset{\circ}{=} 7.538290891$$

would require 27 cycles, hence $5 + 2 \times 26 = 57$ points, to obtain Λ to within an error of only 10^{-4}. Direct invocation of Simpson's rule with $n = 56$ (and hence 57 function evaluations) yields an answer with an error of 0.008631295, considerably worse than the value obtained by the adaptive version of Simpson's rule. In fact, to obtain a comparable accuracy with a uniform Simpson's rule, we would need 205 function evaluations since the error with Simpson's rule and $n = 202$ is 0.00102 while the error made with $n = 204$ is 0.0000985.

EXERCISES 43.4

1. In the computing language of your choice, develop a procedure (or subroutine) that implements adaptive quadrature based on Simpson's rule. Have your procedure return the value of a definite integral and enough information from which to determine the number of function evaluations executed.

In Exercises 2–11:

(a) Use an adaptive quadrature program to approximate $\Lambda = \int_{-3}^{3} f(x)\,dx$ with an error no worse than 10^{-6}.

(b) Determine the number of function evaluations it took to obtain this approximation.

(c) Determine the accuracy that would be achieved using Simpson's rule with the same number of equispaced function evaluations.

(d) Determine the actual number of function evaluations that Simpson's rule would need to approximate Λ with an error no worse than 10^{-6}.

(e) Use the results of Exercise A1(b), Section 43.2, to determine the number of function evaluations Romberg integration, implemented with the recursive Trapezoid rule, needs for approximating Λ with an error no worse than 10^{-6}.

2. $f(x) = (x+2)e^{2(1-x)}$ **3.** $f(x) = \sqrt{2 + 2x + x^2}$

4. $f(x) = \dfrac{9x^2 - 7x - 5}{4x^2 + 4x + 2}$ **5.** $f(x) = \dfrac{3x^3 + 5x^2 - 9x + 4}{5 + x}$

6. $f(x) = \dfrac{9x^2 + 5x - 6}{x - 7}$ **7.** $f(x) = x \sin x$

8. $f(x) = \sin(\cos x)$ **9.** $f(x) = \ln(x^2 + 2x + 2)$

10. $f(x) = \ln(\cos x + \cosh x)$ **11.** $f(x) = \sinh(\sin x)$

In Exercises 12–15:

(a) Use the Trapezoid rule in an adaptive fashion to approximate, with an error no worse than 10^{-2}, each of the following integrals. Ignore concerns of efficiency; instead, concentrate on mastering the flow of the adaptive algorithm and the role of the error test. (If $T[h]$ is the output of the Trapezoid rule with stepsize h, then the error in $T[h]$ is approximately $\frac{1}{3}T[h] - T[2h]$.)

(b) Repeat the calculations with the Left-end rule where, if $L[h]$ is the output of the Left-end rule with stepsize h, then the error in $L[h]$ is approximately $L[h] - L[2h]$.

12. $\int_0^3 x^2 e^{-x}\,dx$ **13.** $\int_0^1 x^{15}\,dx$

14. $\int_0^5 e^{-x}\cos x\,dx$ **15.** $\int_0^\pi x e^x \sin x\,dx$

43.5 Iterated Integrals

The Double Integral

As an example of a multiple integral, the double integral is typically introduced in the multivariable calculus course by a discussion of volume under a surface. In particular, if $z = f(x, y)$ is a surface defined over a region R in the xy-plane, the volume of the region bounded by the surface and the region R is said to be given by

$$\lim_{M,N \to \infty} \sum_{m=0}^{M-1} \sum_{n=0}^{N-1} f(x_n, y_m)\,\Delta x_n\,\Delta y_m \tag{43.13}$$

a double limit of a double sum, provided appropriate care is taken with the definitions of the grid points (x_n, y_m) and the internodal distances Δx_n and Δy_m. The number determined by this expression is written as $\iint_R f(x, y)\,dA$ and is called the *double integral*.

EXAMPLE 43.6   We next demonstrate how we might implement the definition in (43.13). In Figure 43.2, the surface determined by $f(x, y) = x^2 y^2$ is graphed over the region R, the square defined by $-1 \le x, y \le 1$. Figure 43.3 contains a schematic of a rectangular grid constructed on R. Let there be N spaces in the x-direction and M spaces in the y-direction. This gives an x-spacing of $dx = \frac{b-a}{N} = \frac{1-(-1)}{N} = \frac{2}{N}$ and a y-spacing of $dy = \frac{d-c}{M} = \frac{1-(-1)}{M} = \frac{2}{M}$. Thus, the nodes in the grid have coordinates (x_n, y_m) with

$$x_n = -1 + n\frac{2}{N} \quad n = 0, 1, \ldots, N \qquad y_m = -1 + m\frac{2}{M} \quad m = 0, 1, \ldots, M$$

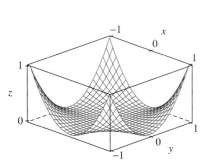

FIGURE 43.2   The surface determined by $f(x, y) = x^2 y^2$ in Example 43.6

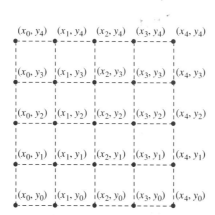

FIGURE 43.3   Example 43.6: A schematic of the grid over which $f(x, y) = x^2 y^2$ is to be evaluated

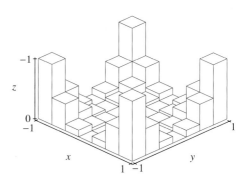

FIGURE 43.4   Example 43.6: A representation of the discretized volume bounded by the surface $f(x, y) = x^2 y^2$ and the plane $z = 0$

Figure 43.4 shows a collection of rectangular parallelepipeds (blocks) approximating the volume under the surface. At each node we evaluate the height of the surface and then construct a block of that height. For example, if $(1, 1)$ is a node, the height of the corresponding block would be $f(1, 1) = 1^2 \times 1^2 = 1$. We compute the volume of each such block and sum to get an approximation to the volume under the surface.

An approximation of the volume under the surface is

$$\sum_{m=0}^{M-1} \sum_{n=0}^{N-1} \left(-1 + n\frac{2}{N}\right)^2 \left(-1 + m\frac{2}{M}\right)^2 \frac{2}{N}\frac{2}{M}$$

and in the limit, as the number of blocks increases without bound and the area of their bases decreases, this double sum tends to the limiting value $\frac{4}{9}$. We therefore define the volume under this function and above the region R to be the number $\frac{4}{9}$.   ❖

The Iterated Double Integral

For continuous functions $f(x, y)$, the double integral can be evaluated by either of the iterated integrals

$$\int_a^b \int_{y_1(x)}^{y_2(x)} f(x, y)\, dy\, dx \quad \text{or} \quad \int_c^d \int_{x_1(y)}^{x_2(y)} f(x, y)\, dx\, dy$$

where each integral sign represents an integration with respect to one variable. For the function of Example 43.6, the two possible iterated double integrals and their evaluations

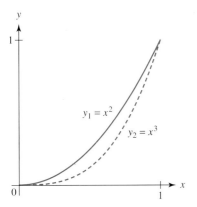

FIGURE 43.5   The region R bounded by $y_1 = x^2$ (solid) and $y_2 = x^3$ (dashed)

are

$$\int_{-1}^{1}\int_{-1}^{1} x^2 y^2 \, dy \, dx = \int_{-1}^{1} x^2 \left[\int_{-1}^{1} y^2 \, dy \right] dx = \int_{-1}^{1} x^2 \left(\frac{2}{3} \right) dx = \frac{4}{9}$$

$$\int_{-1}^{1}\int_{-1}^{1} x^2 y^2 \, dx \, dy = \int_{-1}^{1} y^2 \left[\int_{-1}^{1} x^2 \, dx \right] dy = \int_{-1}^{1} y^2 \left(\frac{2}{3} \right) dy = \frac{4}{9}$$

CURVED BOUNDARIES   The region R, bounded by the curves $y_1 = x^2$ and $y_2 = x^3$, and shown in Figure 43.5 with y_1 as the solid curve and y_2 as the dashed, has curved boundaries. The iterated integrals representing the volume bounded by this region and the function $f(x, y)$ are

$$\int_{0}^{1}\int_{x^3}^{x^2} x^2 y^2 \, dy \, dx = \int_{0}^{1} x^2 \left[\int_{x^3}^{x^2} y^2 \, dy \right] dx = \int_{0}^{1} x^2 \left(\frac{x^8 - x^{11}}{3} \right) dx = \frac{1}{108} \quad \textbf{(43.14a)}$$

$$\int_{0}^{1}\int_{\sqrt{y}}^{y^{1/3}} x^2 y^2 \, dx \, dy = \int_{0}^{1} y^2 \left[\int_{\sqrt{y}}^{y^{1/3}} x^2 \, dx \right] dy = \int_{0}^{1} y^2 \left(\frac{y - y^{3/2}}{3} \right) dy = \frac{1}{108} \quad \textbf{(43.14b)}$$

In evaluating the iterated double integral, the inner integral is computed first. For example, in evaluating (43.14a), the inner integral $\int_{x^3}^{x^2} y^2 \, dy$ determines a function $g(x) = \frac{1}{3}(x^8 - x^{11})$, which is then integrated in the remaining outer integral.

Numeric Evaluation of Iterated Integrals

Numeric evaluation of the iterated double integral takes its cue from the analytical evaluation just described. For example, in (43.14a), the inner integral determines $g(x)$. For each fixed value $x = \hat{x}$, the inner integral can be evaluated numerically, returning the value $g(\hat{x})$. In a numeric scheme for evaluating the outer integral, the integrand must be a function that returns the numeric value of g. A naive integration scheme would use a simple integrator such as Simpson's rule or the Trapezoid rule to evaluate $g(\hat{x}) = \int_{\hat{x}^3}^{\hat{x}^2} y^2 \, dy$.

For example, evaluating the outer integral with Simpson's rule and $n = 10$ requires that $g(x)$ be evaluated at 11 points. At each of these 11 points, a numeric integration must be performed. Suppose each of these integrations is again done with Simpson's rule and $n = 10$, so that each requires 11 function evaluations. The total number of function evaluations will then be $(n + 1)^2 = (10 + 1)^2 = 121$.

Were this scheme to be implemented for (43.14a), the integral would be approximated by 0.0091485179, with an error of 0.00011. Unfortunately, this naive strategy relinquishes all error control and is not viable for more demanding applications.

Error Control

The Romberg integration scheme in Section 43.2 can be modified to accept a function of *two* variables as integrand and returns, from the diagonal of the Romberg table, the single value that passes a prescribed tolerance test. Suppose this scheme is captured in a subroutine whose call is of the form **Inner-Romberg**$(f(\hat{x}, y), y_1(\hat{x}), y_2(\hat{x}), \varepsilon)$ and that returns the value of $g(\hat{x})$, computed to within a tolerance of ε.

Suppose another subroutine implements the Romberg integration scheme for a function of a single variable, again returning a single value satisfying a prescribed tolerance. Suppose, then, this integration is implemented with a subroutine whose call is of the form **Romberg**$(h(x), a, b, \varepsilon)$ and that returns the value of $\int_{a}^{b} h(x) \, dx$ to a tolerance of ε.

The iterated integral (43.14a) could therefore be evaluated with code that implemented the function calls implied by

$$\textbf{Romberg}(\textbf{Inner-Romberg}(f(\hat{x}, y), y_1(\hat{x}), y_2(\hat{x}), \varepsilon), a, b, \varepsilon)$$

A scheme such as this is implemented in the corresponding Maple worksheet. With $\varepsilon = 10^{-4}$, the scheme returns a value of 0.009259126807, with an error of 0.132452×10^{-6}, which is well within the prescribed tolerance of 10^{-4}.

Alternatively, an adaptive integration technique could be used for both the inner and outer integrals. In either event, provision has to be made for the integrand of the inner integral to be a function of two variables and the integrand of the outer to be a function of one variable.

EXERCISES 43.5

1. In the computing language of your choice, implement a Romberg-based numeric approximation of iterated double integrals. The code should have the ability to gage error and deal with regions having curved boundaries. (Maple code for this task will be found in the accompanying Maple worksheet.)

In Exercises 2–11, use the results of Exercise 1 to approximate the given integral numerically, with an error no worse than 10^{-4}. Where possible, determine the actual error made by the numeric integration.

2. $\displaystyle\int_0^\pi \int_0^{x^2} x \sin y \, dy \, dx$ 3. $\displaystyle\int_0^1 \int_0^{e^x} y \, dy \, dx$

4. $\displaystyle\int_1^2 \int_y^{y^2} x e^{xy} \, dx \, dy$ 5. $\displaystyle\int_0^2 \int_{\sqrt{1+x^2}}^{e^x} y^2 \sin x \, dy \, dx$

6. $\displaystyle\int_0^{2\pi} \int_{\sin(\cos x)}^{\cos(\sin x)} x y^2 \, dy \, dx$ 7. $\displaystyle\int_0^1 \int_{x^2}^{\sqrt{x}} x^2 y^3 \, dy \, dx$

8. $\displaystyle\int_0^1 \int_{x^4}^{\cosh x} y \sinh x \, dy \, dx$ 9. $\displaystyle\int_0^{2\pi} \int_0^{1-\cos\theta} r^3 \cos^2\theta \, dr \, d\theta$

10. $\displaystyle\int_0^\pi \int_1^{5+2\cos\theta} r\theta \, dr \, d\theta$ 11. $\displaystyle\int_0^{\pi/6} \int_0^{3\cos 3\theta} r^2\sqrt{1+\theta} \, dr \, d\theta$

Chapter Review

1. What is a Newton–Cotes formula for numeric integration?

2. What is an *open* formula? What is a *closed* formula?

3. State the Trapezoid rule and the composite Trapezoid rule. Show how the composite rule is obtained from the Newton–Cotes formula.

4. Detail how there is a loss of order in the truncation error for the composite Trapezoid rule.

5. Show how the expression for the truncation error of the composite Trapezoid rule can be used to estimate the number of function evaluations needed to obtain a desired accuracy.

6. What does it mean to say that the Trapezoid rule is polynomial exact of degree one?

7. State the algorithm for the recursive Trapezoid rule.

8. Suppose the composite Trapezoid rule, executed with $n = 20$, yields the value $T[h]$. Using the recursive Trapezoid rule, how many more function evaluations are required to obtain $T[\frac{h}{2}]$?

9. Describe the process of Romberg integration. Suppose the recursive Trapezoid rule were used to compute $T[h] = 5$ and $T[\frac{h}{2}] = 6$ in the first column of the Romberg table. What would be the extrapolated value on the diagonal in the second row? Suppose $T[\frac{h}{4}] = 6.2$ were the third entry in the first column of the Romberg table. What would be the extrapolated value on the diagonal in the third row?

10. Show how polynomial exactness of degree 3 is used to obtain the Gauss–Legendre quadrature formula for the case $n = 1$.

11. Obtain the Gauss–Legendre quadrature formula of Question 10 using an appropriate Legendre polynomial.

12. Show how the Gauss–Legendre quadrature formulas, derived for the interval $[-1, 1]$, are extended to the approximation of integrals on the interval $[a, b]$.

13. Describe the error-test mechanism that is at the heart of adaptive integration.

14. Describe the structure and functioning of an adaptive quadrature algorithm.

15. Describe the strategy for minimizing function evaluations in an adaptive Simpson's rule quadrature algorithm.

16. Describe how the iterated double integral $\int_a^b \int_{y_1(x)}^{y_2(x)} f(x, y) \, dy \, dx$ is evaluated numerically.

Chapter *44*

Approximation of Discrete Data

I N T R O D U C T I O N   Chapter 44 is related to Chapters 40 and 41. Chapter 40 treats the interpolation problem, that of passing a curve, generally a polynomial, through each point in a collection of points. In essence, Chapter 41 also treats passing a curve, either a polynomial or a rational function, through each point in a collection of points, the points being taken from the graph of a given continuous function. The emphasis in Chapter 41 is how well an approximating function approximates the given function.

In Chapter 44, a curve such as a polynomial is passed "through" a set of points, but this time, the curve need not intersect any of the points. A successful curve will minimize some measure of the difference between the points and the curve.

We begin with a discussion of the *least-squares regression line* whereby a linear function is made to fit a set of data points that may not all lie on any single straight line. This is the problem of "curve fitting," where a curve, this time a line, is fit to a set of data points. A linear function has two free parameters. If there are exactly two points in the data set, then fitting the function to the data becomes an interpolation problem. In general, then, there are more points than parameters in the fitting curve. Attempting to have each point lie on the fitting line leads to an overdetermined set of equations containing more equations than unknowns. Unless all the points were actually on one straight line, this set of equations has no solution. The least-squares solution is the one that minimizes the sum of the squares of the vertical deviations between the points and the fitting line.

In this context, a curve of the form $g(x) = \sum_{k=1}^{n} c_k \phi_k(x)$ is said to be a general linear model. The technique of least squares generalizes to fitting such models to data, the geometric aspects of which were seen in Section 33.6. The role of orthogonality in obtaining a least-squares fit is an essential insight here.

We conclude the chapter by considering the computationally challenging problem of the nonlinear least-squares fit. In particular, we compare two different linearization schemes with a nonlinear fit of the Michaelis–Menton model to a medical data set.

44.1   Least-Squares Regression Line

Introduction

In Section 33.6 we used the least-squares technique to fit a parabola to nine data points. Our intent was to motivate a study of the geometric structure of the least-squares problem. The present section uses the least-squares technique to fit a straight line to six data points, with an emphasis on the computational and visual aspects of the method. Often called "linear regression," the process of fitting a straight line to data can be found in calculus texts (as an example of multivariable optimization) and in science labs, as a tool of data analysis. The prevalence of calculators and computer programs that perform linear regression makes this topic readily accessible in today's classrooms.

◆◆ EXAMPLE 44.1   The six data points $(1, -22)$, $(3, -15)$, $(4, -8)$, $(7, 1)$, $(8, 4)$, $(9, 11)$, graphed as the black dots in Figure 44.1, do not appear to lie on a single straight line. If we attempt to fit these six points with the line $y = ax + b$, we will get the overdetermined set of six equations in two unknowns in Table 44.1.

SUM OF SQUARES   Figure 44.2 is the graph in Figure 44.1, along with a graph of the line $y = 3x - 20$ and the value of the sum of squares of the deviations of the points from this line. If the vertical black line segments are considered positive deviations, then the color segments would be negative deviations. If the totality of these deviations were to be added, some of the positive distances would cancel with some of the negative and the true "total deviation" of the points from the line would be smaller than it should be. If their absolute values were added, the "cancellation" problem would be avoided, but the sum of absolute values would not be a differentiable function. Instead, each deviation is *squared* so that the associated "error" is a positive number. The squares of all deviations between data points and the line are added and the sum is a differentiable function of a and b. The value of this sum, called the *sum of squares of deviations* or just *sum of squares,* appears in Figure 44.2 as the number in the oval inset. The figure was created with a software simulation that gives the user the opportunity to lower the value of the sum of squares by varying the slope and intercept. (See the software accompanying the text.)

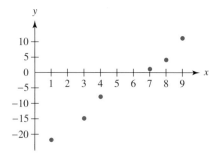

FIGURE 44.1   The six non-colinear data points of Example 44.1

$$a + b = -22$$
$$3a + b = -15$$
$$4a + b = -8$$
$$7a + b = 1$$
$$8a + b = 4$$
$$9a + b = 11$$

TABLE 44.1   Example 44.1: The equations arising when the line $y = ax + b$ is fit to the data points in Figure 44.1

MINIMIZING THE SUM OF SQUARES   The sum of squares of deviations

$$S = \sum_{k=1}^{6} [y_k - (ax_k + b)]^2 = 220a^2 + 58b - 78a + 64ab + 6b^2 + 911$$

is a function of the two parameters a and b. Figure 44.3, a graph of the surface defined by $z = S(a, b)$, shows there is a point (a, b) that minimizes the value of S.

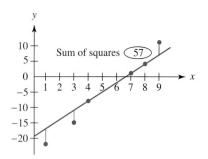

FIGURE 44.2   Example 44.1: The data points in Figure 44.1, the line $y = 3x - 20$, and the deviations of the data points from this line. Positive deviations are in black; negative, in color.

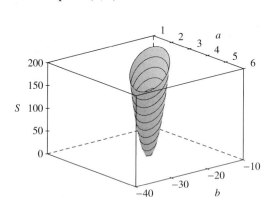

FIGURE 44.3   In Example 44.1, the surface $S(a, b) = \sum_{k=1}^{6} [y_k - (ax_k + b)]^2$

We can minimize $S(a, b)$ by differentiating with respect to a and b, setting the derivatives equal to zero, obtaining the equations

$$\frac{\partial S}{\partial a} = 440a + 64b - 78 = 0 \quad \text{and} \quad \frac{\partial S}{\partial b} = 64a + 12b + 58 = 0 \qquad \textbf{(44.1)}$$

The solution of (44.1) is $a = \frac{581}{148}$, $b = -\frac{1907}{74}$, so the least-squares line is

$$y = \frac{581}{148}x - \frac{1907}{74} \tag{44.2}$$

and the minimum sum of squares is $\frac{1563}{148}$. Figure 44.4 shows the least-squares line and the deviations, both positive and negative. A comparison with Figure 44.2 shows that S has been reduced by a factor of about 5, achieved by increasing the deviations for the third, fourth, and fifth points but decreasing them for the first, second and sixth.   ❖

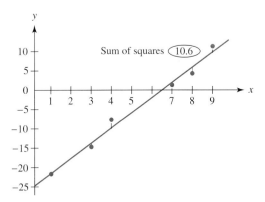

FIGURE 44.4   Example 44.1: The least-squares line, and the deviations when the sum of squares is minimized

Reduction to a Formula

From the general sum of squares of deviations $S = \sum_{k=1}^{N}[y_k - (ax_k + b)]^2$, we obtain the equations

$$-\frac{1}{2}\frac{\partial S}{\partial a} = \sum_{k=1}^{N}[y_k - (ax_k + b)]x_k = \sum_{k=1}^{N}x_k y_k - a\sum_{k=1}^{N}x_k^2 - b\sum_{k=1}^{N}y_k = 0$$

$$-\frac{1}{2}\frac{\partial S}{\partial b} = \sum_{k=1}^{N}[y_k - (ax_k + b)] = \sum_{k=1}^{N}y_k - a\sum_{k=1}^{N}x_k - b\sum_{k=1}^{N}1 = 0$$

Replacing $\sum_{k=1}^{N}1$ with N and solving for a and b give

$$a = \frac{\sum_{k=1}^{N}x_k \sum_{k=1}^{N}y_k - N\sum_{k=1}^{N}x_k y_k}{\left(\sum_{k=1}^{N}x_k\right)^2 - N\sum_{k=1}^{N}x_k^2}$$

$$b = \frac{\sum_{k=1}^{N}x_k \sum_{k=1}^{N}x_k y_k - \sum_{k=1}^{N}x_k^2 \sum_{k=1}^{N}y_k}{\left(\sum_{k=1}^{N}x_k\right)^2 - N\sum_{k=1}^{N}x_k^2} \tag{44.3}$$

To implement these formulas, sum x_k, x_k^2, y_k; sum the products $x_k y_k$; and square the sum of the x_k. In the late 1970s, the advent of affordable calculators that could obtain the first four sums from a single entry of the data impacted the teaching of statistics. Within a decade, however, even these calculators were obsolete as newer calculators appeared that solved for the parameters a and b, without the user having to manipulate the various sums. Simultaneously, spreadsheets and other computer software appeared and added the power of graphics to the task of regression and its analysis.

EXERCISES 44.1

1. Use the formulas in (44.3) to obtain the coefficients a and b in (44.2).

To the data set $\{(x_k, y_k)\}$, $k = 1, \ldots, N$, given in each of Exercises 2–11, fit the line $y = ax + b$.

(a) Write the equations $ax_k + b = y_k$, $k = 1, \ldots, N$, and show they are inconsistent.

(b) Form $S(a, b) = \sum_{k=1}^{N}(ax_k + b - y_k)^2$, the sum of squares of the deviations, and obtain its graph as a surface showing the existence of a minimum.

(c) Determine the critical point of $S(a, b)$ by the techniques of multivariable calculus.

(d) Corroborate the minimizing values of a and b using (44.3).

(e) Graph the data points and the least-squares regression line.

(f) Determine the minimum value of $S(a, b)$.

2. $\{(-9, -\frac{58}{5}), (-3, -\frac{88}{5}), (2, \frac{8}{5}), (4, 4), (10, \frac{224}{5}), (17, \frac{98}{5})\}$

3. $\{(-10, 27), (1, -\frac{1}{2}), (5, -\frac{63}{2}), (16, -38), (17, -\frac{243}{2}), (18, -\frac{81}{2})\}$

4. $\{(-20, -\frac{137}{2}), (-17, -58), (-12, -\frac{81}{2}), (-11, -111), (-7, -69),$ $(-2, -\frac{11}{2}), (9, 99), (10, \frac{73}{2})\}$

5. $\{(-20, -\frac{65}{2}), (-17, -84), (-7, -13), (-1, -12), (6, \frac{39}{2}), (9, 33),$ $(11, 42), (18, \frac{49}{2})\}$

6. $\{(-19, -\frac{167}{2}), (-15, -\frac{131}{2}), (-4, -16), (7, \frac{201}{2}), (12, 56),$ $(14, 195), (15, \frac{417}{2})\}$

7. $\{(-16, \frac{222}{5}), (-15, \frac{832}{5}), (-15, \frac{208}{5}), (-5, \frac{68}{5}), (-2, \frac{104}{5}),$ $(1, -\frac{16}{5}), (8, -\frac{114}{5})\}$

8. $\{(-19, -\frac{348}{5}), (5, \frac{336}{5}), (8, \frac{552}{5}), (12, 168), (16, \frac{282}{5})\}$

9. $\{(-9, -\frac{1037}{10}), (-4, -\frac{93}{10}), (4, \frac{289}{10}), (8, \frac{697}{10}), (9, \frac{141}{10})\}$

10. $\{(-12, -\frac{96}{5}), (-9, -\frac{833}{10}), (-4, -\frac{36}{5}), (1, \frac{3}{10}), (16, \frac{646}{5})\}$

11. $\{(-18, -27.3), (-17, -25.8), (-16, -137.7), (-14, -120.7),$ $(-10, -86.7), (3, 23.8), (7, 10.2), (9, 13.2), (11, 91.8)\}$

44.2   The General Linear Model

Introduction

Not all data can be meaningfully fit to a straight line. In Section 33.6, for example, we showed how to least-squares fit a parabola to a set of data. In general, when the least-squares technique is used to fit data with a curve of the form

$$g(x) = \sum_{k=1}^{n} c_k \phi_k(x)$$

we say we are building a *linear model* for the data. Although the curve $g(x)$ can be decidedly nonlinear, the parameters c_k, $k = 1, \ldots, n$, appear linearly in the model. The simplest case of the linear model is the one in which $\phi_k(x) = x^k$, making $g(x)$ just a polynomial of degree $n - 1$. In this section, we will show how to perform a least-squares fit by both a polynomial and more general functions with a linear structure.

 EXAMPLE 44.2   Table 44.2 lists 12 points on, or almost on, a curve $y(x)$ that was sketched on a sheet of graph paper. The curve that was originally sketched was intended to have a vertical asymptote at $x = 0$ and a horizontal asymptote for large x. Figure 44.5 contains a plot of these points and least-squares polynomials of degree 2, 3, 4, and 5, respectively, drawn as solid black, dashed, dotted, and solid color curves. The idea was to fit a general linear model to the data, and polynomials are the simplest functions to consider. Table 44.3 lists the least-squares polynomials.

| (1, 19) | (2, 14) | (3, 10) | (5, 6) | (9, 6) | (12, 9) |
| (13, 10) | (15, 12) | (18, 13) | (21, 12) | (25, $\frac{23}{2}$) | (28, 11) |

TABLE 44.2   The data for Example 44.2

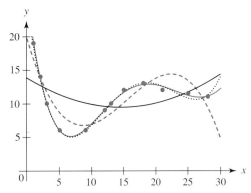

FIGURE 44.5 Example 44.2: The data points and least-squares polynomials of degree 2 (solid black), 3 (dashed), 4 (dotted), and 5 (solid color)

$f_2 = 13.893 - 0.60431x + 0.020706x^2$
$f_3 = 19.736 - 3.4584x + 0.27746x^2 - 0.005972x^3$
$f_4 = 25.447 - 7.3734x + 0.87488x^2 - 0.037836x^3 + 0.5481 \times 10^{-3}x^4$
$f_5 = 26.212 - 8.0756x + 1.0438x^2 - 0.053582x^3 + 0.1169 \times 10^{-2}x^4 - 0.8679 \times 10^{-5}x^5$

TABLE 44.3 The least-squares polynomials of Example 44.2

Not surprisingly, the quadratic polynomial (solid black) does not fit the data very well. The cubic polynomial (dashed) just begins responding to the data but does not provide as good a fit as do the polynomials of degree 4 and 5. The fit of these last two polynomials is remarkable since both capture the asymptotic behavior at $x = 0$; however, neither is asymptotic to a horizontal line past the last data point at $x = 28$. ❖

The Cubic Polynomial

We next fit the given data with a least-squares cubic constructed from first principles. Substituting each data point into the fitting function $g(x) = \sum_{k=0}^{3} a_k x^k$ gives the set of equations $A\mathbf{u} = \mathbf{y}$, where A, $\mathbf{u}$, and $\mathbf{y}$ are given in (44.4).

$$A = \begin{bmatrix} 1 & 1 & 1 & 1 \\ 1 & 2 & 4 & 8 \\ 1 & 3 & 9 & 27 \\ 1 & 5 & 25 & 125 \\ 1 & 9 & 81 & 729 \\ 1 & 12 & 144 & 1728 \\ 1 & 13 & 169 & 2197 \\ 1 & 15 & 225 & 3375 \\ 1 & 18 & 324 & 5832 \\ 1 & 21 & 441 & 9261 \\ 1 & 25 & 625 & 15{,}625 \\ 1 & 28 & 784 & 21{,}952 \end{bmatrix} \quad \mathbf{u} = \begin{bmatrix} a_0 \\ a_1 \\ a_2 \\ a_3 \end{bmatrix} \quad \mathbf{y} = \begin{bmatrix} 19 \\ 14 \\ 10 \\ 6 \\ 6 \\ 9 \\ 10 \\ 12 \\ 13 \\ 12 \\ \frac{23}{2} \\ 11 \end{bmatrix} \qquad \textbf{(44.4)}$$

As seen in (44.5), the fifth row in the reduced row-echelon form of the augmented matrix $[A, \mathbf{y}]$ represents the inconsistent equation $0 = 1$, so we know there is no solution to $A\mathbf{u} = \mathbf{y}$.

$$
\text{rref}([A, \mathbf{y}]) = \begin{bmatrix} 1 & 0 & 0 & 0 & 0 \\ 0 & 1 & 0 & 0 & 0 \\ 0 & 0 & 1 & 0 & 0 \\ 0 & 0 & 0 & 1 & 0 \\ 0 & 0 & 0 & 0 & 1 \\ 0 & 0 & 0 & 0 & 0 \\ \vdots & \vdots & \vdots & \vdots & \vdots \\ 0 & 0 & 0 & 0 & 0 \end{bmatrix}
\tag{44.5}
$$

A least-squares solution minimizes the sum of squares $S = \sum_{k=1}^{12} (g(x_k) - y_k)^2$ by solving the normal equations

$$
\frac{1}{2} \frac{\partial S}{\partial a_0} = 152a_1 + 2832a_2 + 60,860a_3 - \tfrac{267}{2} + 12a_0 = 0
$$

$$
\frac{1}{2} \frac{\partial S}{\partial a_1} = 1,411,944a_3 + 152a_0 + 2832a_1 + 60,860a_2 - \tfrac{3321}{2} = 0
$$

$$
\frac{1}{2} \frac{\partial S}{\partial a_2} = 1,411,944a_2 + 34,391,612a_3 + 2832a_0 + 60,860a_1 - \tfrac{63,605}{2} = 0
$$

$$
\frac{1}{2} \frac{\partial S}{\partial a_3} = 1,411,944a_1 + 34,391,612a_2 + 865,560,552a_3 + 60,860a_0 - \tfrac{1,383,309}{2} = 0
$$

whose solution is

$$
a_0 = \tfrac{21,833,515,650,511}{1,106,275,892,918} \qquad a_1 = -\tfrac{5,738,838,682,124}{1,659,413,839,377}
$$

$$
a_2 = \tfrac{1,227,806,681,771}{4,425,103,571,672} \qquad a_3 = -\tfrac{11,325,266,429}{1,896,472,959,288}
$$

The normal equations contain the same information as the original system $A\mathbf{u} = \mathbf{y}$. To see this, write the normal equations in the matrix form $M\mathbf{u} = \mathbf{b}$, where the matrix M and the vector $\mathbf{b}$ are

$$
M = \begin{bmatrix} 12 & 152 & 2832 & 60,860 \\ 152 & 2832 & 60,860 & 1,411,944 \\ 2832 & 60,860 & 1,411,944 & 34,391,612 \\ 60,860 & 1,411,944 & 34,391,612 & 865,560,552 \end{bmatrix} \qquad \mathbf{b} = \begin{bmatrix} \frac{267}{2} \\ \frac{3321}{2} \\ \frac{63,605}{2} \\ \frac{1,383,309}{2} \end{bmatrix}
\tag{44.6}
$$

A computation shows that $M = A^{\mathsf{T}}A$ and that $\mathbf{b} = A^{\mathsf{T}}\mathbf{y}$. Thus, the least-squares solution of the overdetermined system $A\mathbf{u} = \mathbf{y}$ is given by the solution of the determinate system $M\mathbf{u} = \mathbf{b}$.

We close this construction by noting the condition number of the matrix M is the very large 1.13×10^9. As elegant as the theory in these calculations might be, the ill-conditioning of the matrix M precludes its use in practical applications. We escaped the perils of round-off because we worked with exact arithmetic. When working in floating-point arithmetic, roundoff poses a significant difficulty in least-squares calculations. The cure lies in orthogonality, as we shall see in Section 44.3.

General Structure of the Linear Model

The general linear model fits the curve $g(x) = \sum_{s=1}^{n} c_s \phi_s(x)$ to the m points (x_k, y_k), $k = 1, \ldots, m$, by minimizing

$$S = \sum_{k=1}^{m} (g(x_k) - y_k)^2 = \sum_{k=1}^{m} \left(\sum_{s=1}^{n} c_s \phi_s(x_k) - y_k \right)^2$$

the sum of the squares of the deviations between the data points and the fitting curve $g(x)$. Differentiation with respect to each of the n parameters c_k, $k = 1, \ldots, n$, leads to n equations of the form

$$\frac{\partial S}{\partial c_r} = 2 \sum_{k=1}^{m} \left(\sum_{s=1}^{n} c_s \phi_s(x_k) - y_k \right) \phi_r(x_k) = 0$$

In each such equation, divide through by 2 and expand the outer sum to obtain

$$\sum_{k=1}^{m} \sum_{s=1}^{n} c_s \phi_s(x_k) \phi_r(x_k) - \sum_{k=1}^{m} \phi_r(x_k) y_k = 0$$

Then, move the rightmost sum to the right-hand side of the equation and interchange the order of summation on the left-hand side to obtain

$$\sum_{s=1}^{n} c_s \sum_{k=1}^{m} \phi_r(x_k) \phi_s(x_k) = \sum_{k=1}^{m} \phi_r(x_k) y_k$$

as the rth equation in a list of n such, where $r = 1, \ldots, n$. Shrewd observation recognizes this set of equations as the linear system

$$\begin{bmatrix} \sum_{k=1}^{m} \phi_1(x_k)\phi_1(x_k) & \sum_{k=1}^{m} \phi_1(x_k)\phi_2(x_k) & \cdots & \sum_{k=1}^{m} \phi_1(x_k)\phi_n(x_k) \\ \sum_{k=1}^{m} \phi_2(x_k)\phi_1(x_k) & \sum_{k=1}^{m} \phi_2(x_k)\phi_2(x_k) & \cdots & \sum_{k=1}^{m} \phi_2(x_k)\phi_n(x_k) \\ \vdots & \vdots & \cdots & \vdots \\ \sum_{k=1}^{m} \phi_r(x_k)\phi_1(x_k) & \sum_{k=1}^{m} \phi_r(x_k)\phi_2(x_k) & \cdots & \sum_{k=1}^{m} \phi_r(x_k)\phi_n(x_k) \\ \vdots & \vdots & \cdots & \vdots \\ \sum_{k=1}^{m} \phi_n(x_k)\phi_1(x_k) & \sum_{k=1}^{m} \phi_n(x_k)\phi_2(x_k) & \cdots & \sum_{k=1}^{m} \phi_n(x_k)\phi_n(x_k) \end{bmatrix} \begin{bmatrix} c_1 \\ c_2 \\ \vdots \\ c_r \\ \vdots \\ c_n \end{bmatrix} = \begin{bmatrix} \sum_{k=1}^{m} \phi_1(x_k)y_k \\ \sum_{k=1}^{m} \phi_2(x_k)y_k \\ \vdots \\ \sum_{k=1}^{m} \phi_r(x_k)y_k \\ \vdots \\ \sum_{k=1}^{m} \phi_n(x_k)y_k \end{bmatrix}$$

$$(44.7)$$

That this system is actually the set of normal equations $M\mathbf{u} = \mathbf{b}$ is established by extending the formalism as follows. For $r = 1, \ldots, n$, define the vectors

$$\mathbf{\Phi}_r = \begin{bmatrix} \phi_r(x_1) \\ \phi_r(x_2) \\ \vdots \\ \phi_r(x_m) \end{bmatrix} \qquad \mathbf{y} = \begin{bmatrix} y_1 \\ y_2 \\ \vdots \\ y_m \end{bmatrix} \qquad (44.8)$$

and on the space of functions continuous on a closed interval $[a, b]$ that contains x_k, $k =$

$1, \ldots, m$, define the *pseudo-inner product*

$$\langle \mathbf{\Phi}_\alpha, \mathbf{\Phi}_\beta \rangle = \sum_{k=1}^{m} \phi_\alpha(x_k) \phi_\beta(x_k)$$

An inner product, like the ordinary dot product, satisfies $\mathbf{u} \cdot \mathbf{u} = 0$ only when $\mathbf{u}$ is the zero vector. For a pseudo-inner product, $\langle \mathbf{u}, \mathbf{u} \rangle = 0$ can be satisfied by a nonzero vector $\mathbf{u}$. For this pseudo-inner product, any function $\psi(x)$ that interpolates the nodes $x_k, k = 1, \ldots, m$, vanishes at each node. Hence, $\psi(x)$ is not the zero function, but the pseudo-inner product of the vector $\mathbf{\Psi} = [\psi(x_1), \ldots, \psi(x_m)]^{\mathrm{T}}$ with itself must necessarily vanish.

The $m \times n$ matrix A in the system $A\mathbf{u} = \mathbf{y}$ that arises when we attempt to interpolate the m data points $(x_k, y_k), k = 1, \ldots, m$, with $g(x)$, is just $A = [\mathbf{\Phi}_1, \mathbf{\Phi}_2, \ldots, \mathbf{\Phi}_n]$, and $A\mathbf{u} = \mathbf{y}$ is

$$\begin{bmatrix} \phi_1(x_1) & \phi_2(x_1) & \cdots & \phi_n(x_1) \\ \phi_1(x_2) & \phi_2(x_2) & \cdots & \phi_n(x_2) \\ \vdots & \vdots & \cdots & \vdots \\ \phi_1(x_m) & \phi_2(x_m) & \cdots & \phi_n(x_m) \end{bmatrix} \begin{bmatrix} c_1 \\ c_2 \\ \vdots \\ c_n \end{bmatrix} = \begin{bmatrix} y_1 \\ y_2 \\ \vdots \\ y_m \end{bmatrix} \tag{44.9}$$

The product $A^{\mathrm{T}}A$ is then the matrix

$$\begin{bmatrix} \langle \mathbf{\Phi}_1, \mathbf{\Phi}_1 \rangle & \langle \mathbf{\Phi}_1, \mathbf{\Phi}_2 \rangle & \cdots & \langle \mathbf{\Phi}_1, \mathbf{\Phi}_n \rangle \\ \langle \mathbf{\Phi}_2, \mathbf{\Phi}_1 \rangle & \langle \mathbf{\Phi}_2, \mathbf{\Phi}_2 \rangle & \cdots & \langle \mathbf{\Phi}_2, \mathbf{\Phi}_n \rangle \\ \vdots & \vdots & \cdots & \vdots \\ \langle \mathbf{\Phi}_n, \mathbf{\Phi}_1 \rangle & \langle \mathbf{\Phi}_n, \mathbf{\Phi}_2 \rangle & \cdots & \langle \mathbf{\Phi}_n, \mathbf{\Phi}_n \rangle \end{bmatrix} \tag{44.10}$$

which is the same as the matrix in (44.7). Hence, we have shown that $A^{\mathrm{T}}A = M$. In addition, the product $A^{\mathrm{T}}\mathbf{y}$ is the vector

$$\begin{bmatrix} \langle \mathbf{\Phi}_1, \mathbf{y} \rangle \\ \langle \mathbf{\Phi}_2, \mathbf{y} \rangle \\ \vdots \\ \langle \mathbf{\Phi}_n, \mathbf{y} \rangle \end{bmatrix} \tag{44.11}$$

which is the same as the vector $\mathbf{b}$ on the right in (44.7). Thus, $A^{\mathrm{T}}\mathbf{y} = \mathbf{b}$, and multiplying $A\mathbf{u} = \mathbf{y}$ by A^{T} yields $M\mathbf{u} = \mathbf{b}$, the normal equations.

Finally, if we define the vector

$$\mathbf{G} = \begin{bmatrix} g(x_1) \\ g(x_2) \\ \vdots \\ g(x_m) \end{bmatrix}$$

then $A\mathbf{u} = \mathbf{G}$, that is,

$$A\mathbf{u} = \begin{bmatrix} c_1\phi_1(x_1) + c_2\phi_2(x_1) + \cdots + c_n\phi_n(x_1) \\ c_1\phi_1(x_2) + c_2\phi_2(x_2) + \cdots + c_n\phi_n(x_2) \\ \vdots \\ c_1\phi_1(x_m) + c_2\phi_2(x_m) + \cdots + c_n\phi_n(x_m) \end{bmatrix} = \mathbf{G} = \begin{bmatrix} g(x_1) \\ g(x_2) \\ \vdots \\ g(x_m) \end{bmatrix}$$

Consequently,

$$S = \sum_{k=1}^{m} (g(x_k) - y_k)^2 = \| A\mathbf{u} - \mathbf{y} \|_2^2 \tag{44.12}$$

Minimizing S, the sum of squares is equivalent to computing a vector projection.

EXAMPLE 44.3   For the data in Table 44.2, we have $g(x) = \sum_{k=0}^{3} a_k x^k$ so that the functions $\phi_k(x)$ are x^k, $k = 0, \ldots, 3$, respectively. The vectors $\mathbf{\Phi}_k$ and $\mathbf{y}$ are then

$$\mathbf{\Phi}_0 = \begin{bmatrix} 1 \\ 1 \\ 1 \\ 1 \\ 1 \\ 1 \\ 1 \\ 1 \\ 1 \\ 1 \\ 1 \\ 1 \end{bmatrix} \quad \mathbf{\Phi}_1 = \begin{bmatrix} 1 \\ 2 \\ 3 \\ 5 \\ 9 \\ 12 \\ 13 \\ 15 \\ 18 \\ 21 \\ 25 \\ 28 \end{bmatrix} \quad \mathbf{\Phi}_2 = \begin{bmatrix} 1 \\ 4 \\ 9 \\ 25 \\ 81 \\ 144 \\ 169 \\ 225 \\ 324 \\ 441 \\ 625 \\ 784 \end{bmatrix} \quad \mathbf{\Phi}_3 = \begin{bmatrix} 1 \\ 8 \\ 27 \\ 125 \\ 729 \\ 1728 \\ 2197 \\ 3375 \\ 5832 \\ 9261 \\ 15{,}625 \\ 21{,}952 \end{bmatrix} \quad \mathbf{y} = \begin{bmatrix} 19 \\ 14 \\ 10 \\ 6 \\ 6 \\ 9 \\ 10 \\ 12 \\ 13 \\ 12 \\ \frac{23}{2} \\ 11 \end{bmatrix}$$

The matrix $A = [\mathbf{\Phi}_0, \ldots, \mathbf{\Phi}_3]$ is the matrix in (44.4), and the product $A^{\mathsf{T}}A$ is the matrix M in (44.6). Alternatively, the elements of $M = \{m_{ij}\}$ are $m_{ij} = \langle \mathbf{\Phi}_i, \mathbf{\Phi}_j \rangle$. Finally, a calculation shows the equivalence of the two expressions for S in (44.12).   ❖

EXERCISES 44.2

1. Write the overdetermined system $\{3x = 5, 7x = 11\}$ in the matrix form $A\mathbf{u} = \mathbf{y}$. Obtain the normal equations $M\mathbf{u} = \mathbf{b}$ and the least-squares solution $\mathbf{u} = [x]$.

2. Write the overdetermined system $\{x + y = 1, x - 2y = 5,$ $2x - 3y = 7\}$ in the matrix form $A\mathbf{u} = \mathbf{y}$. Obtain the normal equations $M\mathbf{u} = \mathbf{b}$ and the least-squares solution $\mathbf{u} = [x, y]^{\mathsf{T}}$.

In Exercises 3–9, use the functions $\phi_k(x) = x^{k-1}$, $k = 1, 2, 3$, to fit a general linear model $g(x) = \sum_{k=1}^{3} c_k \phi_k$ to each of the following sets of data points $\{(x_k, y_k)\}$, $k = 1, \ldots, m$.

(a) Use calculus to minimize $S = \sum_{k=1}^{m}(g(x_k) - y_k)^2$, obtain the least-squares $g(x)$, compute $S_{\min}$, and draw a graph showing $g(x)$ and the data points.

(b) From the equations $g(x_k) = y_k$, form the matrix A and the vector $\mathbf{y}$, thereby expressing the equations as $A\mathbf{u} = \mathbf{y}$, where $\mathbf{u} = [c_1, c_2, c_3]^{\mathsf{T}}$. Show that the augmented matrix $[A, \mathbf{y}]$ has rank greater than 3, making the system inconsistent.

(c) Form the matrix $M = A^{\mathsf{T}}A$ and the vector $\mathbf{b} = A^{\mathsf{T}}\mathbf{y}$. Obtain the condition number of M, solve $M\mathbf{u} = \mathbf{b}$ for $\mathbf{u} = \hat{\mathbf{u}}$, and show that the solution is the same as in part (a).

(d) Form the matrix $P = A(A^{\mathsf{T}}A)^{-1}A^{\mathsf{T}}$ that projects vectors onto the column space of A, and compute $\mathbf{y}_A$, the component of $\mathbf{y}$ in the column space of A. Compute $A\hat{\mathbf{u}}$ and verify that $A\hat{\mathbf{u}} = \mathbf{y}_A$.

(e) Form the vectors $\mathbf{\Phi}_k = [\phi_k(x_1), \ldots, \phi_k(x_m)]^{\mathsf{T}}$, $k = 1, 2, 3$; compute $M_{ij} = \mathbf{\Phi}_i \cdot \mathbf{\Phi}_j$ and $b_k = \mathbf{\Phi}_k \cdot \mathbf{y}$. Compare to M and $\mathbf{b}$ in part (c).

(f) Compute $\|A\hat{\mathbf{u}} - \mathbf{y}\|_2^2$ and verify that it is $S_{\min}$.

3. $\{(-19, 1684), (-17, 1343), (-14, 2110), (-12, 659), (-9, 850),$ $(9, 951), (10, 1169), (17, 1425)\}$

4. $\{(-20, 6350), (-16, 4099), (-9, 1338), (-5, 435), (-3, 165),$ $(-2, 76), (2, 5), (8, 155), (11, 1712), (18, 4735)\}$

5. $\{(-18, 1518), (-16, 1206), (-14, 929), (-9, 1180), (-4, 86),$ $(1, -5), (4, 56), (15, 2873), (17, 3716)\}$

6. $\{(-19, 738), (-15, 462), (-8, 133), (-6, 223), (9, 448),$ $(10, 187), (12, 272), (17, 557)\}$

7. $\{(-17, 68), (-11, 72), (-4, -18), (-1, -8), (6, 106), (7, 40),$ $(11, 80), (12, 92), (18, 722), (20, 216)\}$

8. $\{(-18, 318), (-15, 193), (-6, -8), (-2, -14), (1, 15), (4, 72)\}$

9. $\{(-20, 411), (-13, 637), (-11, 109), (-7, 37), (3, 89), (15, 327)\}$

In Exercises 10–15, use the functions $\phi_k(x) = x^{k-1}$, $k = 1, \ldots, 4$, to fit a general linear model $g(x) = \sum_{k=1}^{4} c_k \phi_k$ to each of the following sets of data points $\{(x_k, y_k)\}$, $k = 1, \ldots, m$.

(a) Use calculus to minimize $S = \sum_{k=1}^{m}(g(x_k) - y_k)^2$, obtain the least-squares $g(x)$, compute $S_{\min}$, and draw a graph showing $g(x)$ and the data points.

(b) From the equations $g(x_k) = y_k$, form the matrix A and the vector $\mathbf{y}$, thereby expressing the equations as $A\mathbf{u} = \mathbf{y}$, where $\mathbf{u} = [c_1, \ldots, c_4]^{\mathsf{T}}$. Show that the augmented matrix $[A, \mathbf{y}]$ has rank greater than 4, making the system inconsistent.

(c) Form the matrix $M = A^{\mathsf{T}}A$ and the vector $\mathbf{b} = A^{\mathsf{T}}\mathbf{y}$. Obtain the condition number of M, solve $M\mathbf{u} = \mathbf{b}$ for $\hat{\mathbf{u}} = \hat{\mathbf{u}}$, and show that the solution is the same as in part (a).

(d) Form the matrix $P = A(A^{\mathsf T}A)^{-1}A^{\mathsf T}$ that projects vectors onto the column space of A, and compute $\mathbf{y}_A$, the component of $\mathbf{y}$ in the column space of A. Compute $A\hat{\mathbf{u}}$ and verify that $A\hat{\mathbf{u}} = \mathbf{y}_A$.

(e) Form the vectors $\boldsymbol\Phi_k = [\phi_k(x_1),\dots,\phi_k(x_m)]^{\mathsf T}$, $k = 1,\dots,4$; compute $M_{ij} = \boldsymbol\Phi_i \cdot \boldsymbol\Phi_j$ and $b_k = \boldsymbol\Phi_k \cdot \mathbf{y}$. Compare to M and $\mathbf{b}$ in part (c).

(f) Compute $\|A\hat{\mathbf{u}} - \mathbf{y}\|_2^2$ and verify that it is $S_{\min}$.

10. $\{(-9,-1972),(-5,-156),(6,1786),(7,723),(10,8742),$ $(11,2932),(17,11143),(18,13267),(20,18294)\}$

11. $\{(-13,-18798),(-11,-11541),(-10,-8747),(-8,-4591),$ $(5,901),(7,2514),(9,1355),(20,15417)\}$

12. $\{(-13,5338),(-9,2022),(-2,40),(-1,0),(4,88),(10,-440),$ $(15,-704),(18,-5624)\}$

13. $\{(-20,37720),(-18,27438),(-17,23085),(-16,4804),$ $(-1,-2),(1,-13),(3,-147),(7,-431),(8,-639)\}$

14. $\{(-18,8111),(-13,3205),(-10,1529),(-4,120),(7,-875),$ $(10,-3261),(11,-1134),(17,-4811),(19,-27525)\}$

15. $\{(-20,109394),(-19,23382),(-14,9164),(-11,4348),$ $(-10,3232),(-7,1052),(-5,1418),(1,-42),(13,-34064),$ $(20,-30236)\}$

In Exercises 16–21, use the functions $\phi_k(x) = x^{k-1}$, $k = 1,\dots,5$, to fit a general linear model $g(x) = \sum_{k=1}^{5} c_k\phi_k$ to each of the following sets of data points $\{(x_k,y_k)\}$, $k = 1,\dots,m$.

(a) Use calculus to minimize $S = \sum_{k=1}^{m}(g(x_k) - y_k)^2$, obtain the least-squares $g(x)$, compute $S_{\min}$, and draw a graph showing $g(x)$ and the data points.

(b) From the equations $g(x_k) = y_k$, form the matrix A and the vector $\mathbf{y}$, thereby expressing the equations as $A\mathbf{u} = \mathbf{y}$, where $\mathbf{u} = [c_1,\dots,c_5]^{\mathsf T}$. Show that the augmented matrix $[A,\mathbf{y}]$ has rank greater than 5, making the system inconsistent.

(c) Form the matrix $M = A^{\mathsf T}A$ and the vector $\mathbf{b} = A^{\mathsf T}\mathbf{y}$. Obtain the condition number of M, solve $M\mathbf{u} = \mathbf{b}$ for $\hat{\mathbf{u}} = \hat{\mathbf{u}}$, and show that the solution is the same as in part (a).

(d) Form the matrix $P = A(A^{\mathsf T}A)^{-1}A^{\mathsf T}$ that projects vectors onto the column space of A, and compute $\mathbf{y}_A$, the component of $\mathbf{y}$ in the column space of A. Compute $A\hat{\mathbf{u}}$ and verify that $A\hat{\mathbf{u}} = \mathbf{y}_A$.

(e) Form the vectors $\boldsymbol\Phi_k = [\phi_k(x_1),\dots,\phi_k(x_m)]^{\mathsf T}$, $k = 1,\dots,5$; compute $M_{ij} = \boldsymbol\Phi_i \cdot \boldsymbol\Phi_j$ and $b_k = \boldsymbol\Phi_k \cdot \mathbf{y}$. Compare to M and $\mathbf{b}$ in part (c).

(f) Compute $\|A\hat{\mathbf{u}} - \mathbf{y}\|_2^2$ and verify that it is $S_{\min}$.

16. $\{(-19,404626),(-18,325442),(-16,809838),$ $(-14,472648),(-13,350488),(-12,253717),(-8,12322),$ $(-1,31),(17,278122),(18,1394965)\}$

17. $\{(-13,86811),(-8,13017),(4,1939),(5,1311),(6,11536),$ $(7,22219),(11,146296),(14,394115),(17,871067),$ $(19,342491)\}$

18. $\{(-19,268923),(-12,11886),(-9,62191),(-4,747),$ $(-2,24),(4,-95),(5,-123),(10,1273),(12,3668)\}$

19. $\{(-20,392669),(-18,248569),(-16,148042),(-13,57946),$ $(-1,-10),(0,-1),(1,19),(14,160762),(20,155911)\}$

20. $\{(-16,263458),(-13,27463),(-8,3365),(-5,388),(-1,2),$ $(2,197),(7,16125),(14,55101),(15,287216)\}$

21. $\{(-20,476858),(-16,48037),(-14,111366),(1,2),(2,109),$ $(7,9693),(8,4011),(14,117619),(19,452771)\}$

In Exercises 22–26, use the functions $\phi_k(x) = e^{-(k-1)x}$, $k = 1,\dots,4$, to fit a general linear model $g(x) = \sum_{k=1}^{4} c_k\phi_k$ to each of the following sets of data points $\{(x_k,y_k)\}$, $k = 1,\dots,m$.

(a) Use calculus to minimize $S = \sum_{k=1}^{m}(g(x_k) - y_k)^2$, obtain the least-squares $g(x)$, compute $S_{\min}$, and draw a graph showing $g(x)$ and the data points.

(b) From the equations $g(x_k) = y_k$, form the matrix A and the vector $\mathbf{y}$, thereby expressing the equations as $A\mathbf{u} = \mathbf{y}$, where $\mathbf{u} = [c_1,\dots,c_4]^{\mathsf T}$. Show that the augmented matrix $[A,\mathbf{y}]$ has rank greater than 4, making the system inconsistent.

(c) Form the matrix $M = A^{\mathsf T}A$ and the vector $\mathbf{b} = A^{\mathsf T}\mathbf{y}$. Obtain the condition number of M, solve $M\mathbf{u} = \mathbf{b}$ for $\hat{\mathbf{u}} = \hat{\mathbf{u}}$, and show that the solution is the same as in part (a).

(d) Form the matrix $P = A(A^{\mathsf T}A)^{-1}A^{\mathsf T}$ that projects vectors onto the column space of A, and compute $\mathbf{y}_A$, the component of $\mathbf{y}$ in the column space of A. Compute $A\hat{\mathbf{u}}$ and verify that $A\hat{\mathbf{u}} = \mathbf{y}_A$.

(e) Form the vectors $\boldsymbol\Phi_k = [\phi_k(x_1),\dots,\phi_k(x_m)]^{\mathsf T}$, $k = 1,\dots,4$; compute $M_{ij} = \boldsymbol\Phi_i \cdot \boldsymbol\Phi_j$ and $b_k = \boldsymbol\Phi_k \cdot \mathbf{y}$. Compare to M and $\mathbf{b}$ in part (c).

(f) Compute $\|A\hat{\mathbf{u}} - \mathbf{y}\|_2^2$ and verify that it is $S_{\min}$.

22. $\{(0.060,2.50),(0.20,2.44),(0.54,2.15),(0.60,2.21),(0.64,2.10),$ $(0.70,2.21),(0.74,2.09),(0.76,2.00),(0.86,2.04),(0.94,2.05)\}$

23. $\{(0.040,2.27),(0.060,1.99),(0.16,2.01),(0.34,1.65),$ $(0.56,1.27),(0.78,1.00),(0.82,0.893),(0.88,0.834),(1,0.873)\}$

24. $\{(0.040,4.26),(0.10,4.01),(0.16,3.24),(0.28,2.65),(0.34,2.09),$ $(0.64,1.86),(0.68,1.59),(0.78,1.63),(0.96,1.06)\}$

25. $\{(0.14,-0.622),(0.16,-0.575),(0.18,-0.539),(0.30,-0.452),$ $(0.36,-0.423),(0.42,-0.401),(0.64,-0.394),(0.74,-0.335),$ $(0.98,-0.368)\}$

26. $\{(0.040,1.18),(0.16,0.934),(0.26,0.277),(0.34,0.0766),$ $(0.38,-0.000916),(0.74,-0.896),(0.92,-0.498),$ $(0.98,-0.550)\}$

In Exercises 27–31, use the functions $\phi_k(x) = x^{k-1}/(x^2 + k)$, $k = 1,\dots,4$, to fit a general linear model $g(x) = \sum_{k=1}^{4} c_k\phi_k$ to each of the following sets of data points $\{(x_k,y_k)\}$, $k = 1,\dots,m$.

(a) Use calculus to minimize $S = \sum_{k=1}^{m}(g(x_k) - y_k)^2$, obtain the least-squares $g(x)$, compute $S_{\min}$, and draw a graph showing $g(x)$ and the data points.

(b) From the equations $g(x_k) = y_k$, form the matrix A and the vector $\mathbf{y}$, thereby expressing the equations as $A\mathbf{u} = \mathbf{y}$, where $\mathbf{u} = [c_1,\dots,c_4]^{\mathsf T}$. Show that the augmented matrix $[A,\mathbf{y}]$ has rank greater than 4, making the system inconsistent.

(c) Form the matrix $M = A^TA$ and the vector $\mathbf{b} = A^T\mathbf{y}$. Obtain the condition number of M, solve $M\mathbf{u} = \mathbf{b}$ for $\mathbf{u} = \hat{\mathbf{u}}$, and show that the solution is the same as in part (a).

(d) Form the matrix $P = A(A^TA)^{-1}A^T$ that projects vectors onto the column space of A, and compute $\mathbf{y}_A$, the component of $\mathbf{y}$ in the column space of A. Compute $A\hat{\mathbf{u}}$ and verify that $A\hat{\mathbf{u}} = \mathbf{y}_A$.

(e) Form the vectors $\mathbf{\Phi}_k = [\phi_k(x_1), \ldots, \phi_k(x_m)]^T$, $k = 1, \ldots, 4$; compute $M_{ij} = \mathbf{\Phi}_i \cdot \mathbf{\Phi}_j$ and $b_k = \mathbf{\Phi}_k \cdot \mathbf{y}$. Compare to M and $\mathbf{b}$ in part (c).

(f) Compute $\|A\hat{\mathbf{u}} - \mathbf{y}\|_2^2$ and verify that it is $S_{\min}$.

27. $\{(0.24, 4.07), (0.30, 2.86), (0.34, 4.41), (0.44, 3.01), (0.48, 2.07), (0.70, 1.44), (0.78, 1.79), (1.06, 2.78), (1.18, 2.31)\}$

28. $\{(0.22, 22.0), (0.26, 8.19), (0.28, 14.7), (0.38, 6.09), (0.40, 3.59), (0.42, 5.62), (0.46, 2.04), (0.74, 1.05), (1.06, -0.443)\}$

29. $\{(0.28, 13.0), (0.60, 4.39), (0.62, 3.09), (0.82, 1.61), (1.04, 0.759), (1.06, 1.16), (1.20, 1.22), (1.42, 0.722), (1.48, 0.180), (1.56, 0.142)\}$

30. $\{(0.56, 10.6), (1.36, 1.84), (1.60, 2.77), (1.78, 2.02), (1.88, 3.35), (2.08, 2.38), (2.30, 2.19), (2.66, 3.34), (2.98, 2.50), (3.34, 3.12)\}$

31. $\{(0.58, 6.06), (0.82, 3.57), (1.04, 2.62), (1.60, 2.40), (1.74, 2.63), (2.08, 2.64), (3.30, 4.11), (3.38, 3.86), (3.84, 6.21), (3.92, 3.86)\}$

In Exercises 32–36, use the functions $\phi_k(x) = J_{k-1}(x)$, $k = 1, \ldots, 4$, where $J_\lambda(x)$ is the Bessel function of order λ, to fit a general linear model $g(x) = \sum_{k=1}^4 c_k\phi_k$ to each of the following sets of data points $\{(x_k, y_k)\}$, $k = 1, \ldots, m$.

(a) Use calculus to minimize $S = \sum_{k=1}^m (g(x_k) - y_k)^2$, obtain the least-squares $g(x)$, compute $S_{\min}$, and draw a graph showing $g(x)$ and the data points.

(b) From the equations $g(x_k) = y_k$, form the matrix A and the vector $\mathbf{y}$, thereby expressing the equations as $A\mathbf{u} = \mathbf{y}$, where $\mathbf{u} = [c_1, \ldots, c_4]^T$. Show that the augmented matrix $[A, \mathbf{y}]$ has rank greater than 4, making the system inconsistent.

(c) Form the matrix $M = A^TA$ and the vector $\mathbf{b} = A^T\mathbf{y}$. Obtain the condition number of M, solve $M\mathbf{u} = \mathbf{b}$ for $\mathbf{u} = \hat{\mathbf{u}}$, and show that the solution is the same as in part (a).

(d) Form the matrix $P = A(A^TA)^{-1}A^T$ that projects vectors onto the column space of A, and compute $\mathbf{y}_A$, the component of $\mathbf{y}$ in the column space of A. Compute $A\hat{\mathbf{u}}$ and verify that $A\hat{\mathbf{u}} = \mathbf{y}_A$.

(e) Form the vectors $\mathbf{\Phi}_k = [\phi_k(x_1), \ldots, \phi_k(x_m)]^T$, $k = 1, \ldots, 4$; compute $M_{ij} = \mathbf{\Phi}_i \cdot \mathbf{\Phi}_j$ and $b_k = \mathbf{\Phi}_k \cdot \mathbf{y}$. Compare to M and $\mathbf{b}$ in part (c).

(f) Compute $\|A\hat{\mathbf{u}} - \mathbf{y}\|_2^2$ and verify that it is $S_{\min}$.

32. $\{(1, 1.609), (2, 1.675), (3, -0.01184), (4, -0.7530), (5, -0.1353), (7, 1.041), (9, -0.3137)\}$

33. $\{(1, 1.315), (3, 1.191), (5, 0.3636), (6, 0.1528), (7, 0.07776), (8, -0.2773)\}$

34. $\{(0.20, 0.4249), (1.3, 1.703), (3.3, 0.9961), (5.6, -0.4003), (6.5, -0.5670), (6.9, -0.9468), (7, -0.4844), (7.6, -0.2618), (7.8, -0.3160), (9.4, 0.4108)\}$

35. $\{(0, 2.100), (1.2, 0.2304), (1.8, -0.06508), (4.6, 0.1420), (5.1, 0.6108), (6.3, 0.6364), (6.6, 0.4964), (8.2, -0.5039), (8.9, -0.6665), (9, -0.6642)\}$

36. $\{(0.50, -6.092), (1.2, -1.178), (2.5, 2.839), (3.3, 1.430), (3.5, 3.046), (4, 1.743), (4.5, -0.02484), (5.7, -1.567), (9.7, 1.413), (9.8, 1.354)\}$

In Exercises 37–41, use the functions $\phi_k(x) = \sin kx$, $k = 1, \ldots, 5$, to fit a general linear model $g(x) = \sum_{k=1}^5 c_k\phi_k$ to each of the following sets of data points $\{(x_k, y_k)\}$, $k = 1, \ldots, m$.

(a) Use calculus to minimize $S = \sum_{k=1}^m (g(x_k) - y_k)^2$, obtain the least-squares $g(x)$, compute $S_{\min}$, and draw a graph showing $g(x)$ and the data points.

(b) From the equations $g(x_k) = y_k$, form the matrix A and the vector $\mathbf{y}$, thereby expressing the equations as $A\mathbf{u} = \mathbf{y}$, where $\mathbf{u} = [c_1, \ldots, c_4]^T$. Show that the augmented matrix $[A, \mathbf{y}]$ has rank greater than 5, making the system inconsistent.

(c) Form the matrix $M = A^TA$ and the vector $\mathbf{b} = A^T\mathbf{y}$. Obtain the condition number of M, solve $M\mathbf{u} = \mathbf{b}$ for $\hat{\mathbf{u}} = \hat{\mathbf{u}}$, and show that the solution is the same as in part (a).

(d) Form the matrix $P = A(A^TA)^{-1}A^T$ that projects vectors onto the column space of A, and compute $\mathbf{y}_A$, the component of $\mathbf{y}$ in the column space of A. Compute $A\hat{\mathbf{u}}$ and verify that $A\hat{\mathbf{u}} = \mathbf{y}_A$.

(e) Form the vectors $\mathbf{\Phi}_k = [\phi_k(x_1), \ldots, \phi_k(x_m)]^T$, $k = 1, \ldots, 5$; compute $M_{ij} = \mathbf{\Phi}_i \cdot \mathbf{\Phi}_j$ and $b_k = \mathbf{\Phi}_k \cdot \mathbf{y}$. Compare to M and $\mathbf{b}$ in part (c).

(f) Compute $\|A\hat{\mathbf{u}} - \mathbf{y}\|_2^2$ and verify that it is $S_{\min}$.

37. $\{(1.9, 0.1312), (3.1, 0.1328), (4.1, -2.971), (5.9, -4.614), (6.4, 1.721), (6.6, 16.38), (7.7, 0.1390), (8, -0.5887), (9, 4.255), (9.4, 0.3169)\}$

38. $\{(2.2, 1.626), (2.5, 0.1581), (3.3, 0.1530), (3.9, -0.7341), (5.8, -0.2822), (6.9, 0.3997), (7.8, 2.399), (8, 0.8376), (8.3, 0.7503)\}$

39. $\{(1.2, -1.897), (1.5, -1.654), (2.2, -1.149), (5.5, 1.729), (6, 2.245), (6.2, 0.6745), (6.7, -1.070), (7.7, -5.223), (9.1, -2.225)\}$

40. $\{(-4.9, 4.213), (-4.5, 4.986), (-3.4, 1.310), (-2.4, -4.252), (-1.8, -4.992), (1.2, 3.510), (2.3, 2.996), (2.4, 4.252), (2.6, 3.558)\}$

41. $\{(-3.1, 0.03030), (-3, 0.2954), (-2.7, 0.5474), (-2.6, 0.4522), (-1.8, -2.951), (-0.40, -1.064), (0.90, 3.987), (2, 1.842), (2.1, 1.314), (3.1, -0.03030)\}$

The Role of Orthogonality

Motivation

In Section 44.2, we saw that in the general linear model of least-squares curve fitting, the function

$$g(x) = \sum_{s=1}^{n} c_s \phi_s(x)$$

is fitted to the m points $(x_k, y_k), k = 1, \ldots, m$, by minimizing

$$S = \sum_{k=1}^{m} (g(x_k) - y_k)^2 = \sum_{k=1}^{m} \left(\sum_{s=1}^{n} c_s \phi_s(x_k) - y_k \right)^2$$

the sum of squares of the deviations between the points and $g(x)$. The ordinary minimization techniques from multivariable calculus then give the normal equations in the form (44.7), that is, in the form $M\mathbf{u} = \mathbf{b}$. In Section 44.2, we also showed that defining the vectors $\mathbf{y}$ and $\mathbf{\Phi}_r, r = 1, \ldots, n$, as in (44.8), and the pseudo-inner product

$$\langle \mathbf{\Phi}_\alpha, \mathbf{\Phi}_\beta \rangle = \sum_{k=1}^{m} \phi_\alpha(x_k) \phi_\beta(x_k) \tag{44.13}$$

on the space of functions continuous on a closed interval $[a, b]$ that contains $x_k, k = 1, \ldots, m$, allowed us to write the normal equations in the form $M\mathbf{u} = \mathbf{b}$, where M is given by (44.10) and $\mathbf{b}$ is given by (44.11). Finally, we noted that, in general, the matrix M usually has a large condition number and, hence, tends to be ill-conditioned.

In this section we will consider the role of orthogonality in dealing with the ill-conditioning of the normal equations. We will look for polynomials $p_k(x), k = 1, \ldots, n$, for which the associated vectors

$$\mathbf{\Phi}_r = \begin{bmatrix} p_r(x_1) \\ p_r(x_2) \\ \vdots \\ p_r(x_n) \end{bmatrix} \qquad r = 1, \ldots, n$$

are orthogonal under the pseudo-inner product (44.13). If we can find such polynomials, the normal equations will reduce to the diagonal system

$$\begin{bmatrix} \langle \mathbf{\Phi}_1, \mathbf{\Phi}_1 \rangle & 0 & \cdots & 0 \\ 0 & \langle \mathbf{\Phi}_2, \mathbf{\Phi}_2 \rangle & \cdots & 0 \\ \vdots & \vdots & \ddots & \vdots \\ 0 & 0 & \cdots & \langle \mathbf{\Phi}_n, \mathbf{\Phi}_n \rangle \end{bmatrix} \mathbf{u} = \begin{bmatrix} \langle \mathbf{\Phi}_1, \mathbf{y} \rangle \\ \langle \mathbf{\Phi}_2, \mathbf{y} \rangle \\ \vdots \\ \langle \mathbf{\Phi}_n, \mathbf{y} \rangle \end{bmatrix}$$

from which the solution is immediately given by $c_k = \langle \mathbf{\Phi}_k, \mathbf{y} \rangle / \langle \mathbf{\Phi}_k, \mathbf{\Phi}_k \rangle, k = 1, \ldots, n$. Below, we detail an algorithm by means of which the desired set of orthogonal polynomials $p_k(x), k = 1, \ldots, n$, can be constructed.

An Algorithm for Generating Orthogonal Polynomials

Based on [60], the following algorithm generates a set of polynomials $p_k(x), k = 1, \ldots, n$, with the property that the normal equations are a diagonal system when fitting the m points $(x_k, y_k), k = 1, \ldots, m$, with the function $g(x) = \sum_{k=1}^{n} c_k p_k(x)$.

1. Define the vector $\mathbf{X} = [x_1, x_2, \ldots, x_m]^{\mathrm{T}}$.

2. Obtain $\lambda = \frac{1}{m}\sum_{k=1}^{m} x_k$, the average of the x_k's.

3. Define $p_0(x) = 1$ and $p_1(x) = x - \lambda$.

4. Define the vectors $\mathbf{P}_0 = \begin{bmatrix} p_0(x_1) \\ p_0(x_2) \\ \vdots \\ p_0(x_m) \end{bmatrix} = \begin{bmatrix} 1 \\ 1 \\ \vdots \\ 1 \end{bmatrix}$ and $\mathbf{P}_1 = \begin{bmatrix} p_1(x_1) \\ p_1(x_2) \\ \vdots \\ p_1(x_m) \end{bmatrix} = \begin{bmatrix} x_1 - \lambda \\ x_2 - \lambda \\ \vdots \\ x_m - \lambda \end{bmatrix} = \mathbf{X} - \lambda\mathbf{P}_0$.

5. Define the vectors $\mathbf{Q}_0 = \begin{bmatrix} x_1 p_0(x_1) \\ x_2 p_0(x_2) \\ \vdots \\ x_m p_0(x_m) \end{bmatrix} = \begin{bmatrix} x_1 \\ x_2 \\ \vdots \\ x_m \end{bmatrix} = \mathbf{X}$ and $\mathbf{Q}_1 = \begin{bmatrix} x_1 p_1(x_1) \\ x_2 p_1(x_2) \\ \vdots \\ x_m p_1(x_m) \end{bmatrix} = \mathbf{X} \boxtimes \mathbf{P}_1$,

   where we have use the notation $\mathbf{X} \boxtimes \mathbf{P}_1$ to indicate that corresponding components of the vectors $\mathbf{X}$ and $\mathbf{P}_1$ are multiplied. (In the computer program Matlab, this operation is implemented by the syntax $\mathbf{X}.*\mathbf{P}_1$. In Maple, the **zip** command is used.)

6. For $k = 2, 3, \ldots, n$, recursively compute $\alpha_k = \langle \mathbf{Q}_{k-1}, \mathbf{P}_{k-1} \rangle / \langle \mathbf{P}_{k-1}, \mathbf{P}_{k-1} \rangle$, $\beta_{k-1} = \langle \mathbf{Q}_{k-1}, \mathbf{P}_{k-2} \rangle / \langle \mathbf{P}_{k-2}, \mathbf{P}_{k-2} \rangle$, $p_k(x) = (x - \alpha_k)p_{k-1}(x) - \beta_{k-1}p_{k-2}(x)$, and

$$\mathbf{P}_k = \begin{bmatrix} p_k(x_1) \\ p_k(x_2) \\ \vdots \\ p_k(x_m) \end{bmatrix} \qquad \mathbf{Q}_k = \begin{bmatrix} x_1 p_k(x_1) \\ x_2 p_k(x_2) \\ \vdots \\ x_m p_k(x_m) \end{bmatrix} = \mathbf{X} \boxtimes \mathbf{P}_k$$

7. If the m points (x_k, y_k), $k = 1, \ldots, m$, are least-squares fitted with the function $g(x) = \sum_{k=1}^{n} c_k p_k(x)$, the normal equations $A^{\mathrm{T}}A\mathbf{u} = A^{\mathrm{T}}\mathbf{y}$ will be a diagonal system.

EXAMPLE 44.4 For the $m = 12$ points listed in Table 44.2, Section 44.2, $\lambda = \frac{1}{m}\sum_{k=1}^{m} x_k = \frac{38}{3}$. The vectors $\mathbf{X}$ and $\mathbf{y}$ are then

$$\mathbf{X} = [1, 2, 3, 5, 9, 12, 13, 15, 18, 21, 25, 28]^{\mathrm{T}}$$
$$\mathbf{y} = [19, 14, 10, 6, 6, 9, 10, 12, 13, 12, \tfrac{23}{2}, 11]^{\mathrm{T}}$$

The polynomials $p_0(x)$ and $p_1(x)$ are, respectively, 1 and $x - \frac{38}{3}$. The vectors $\mathbf{P}_0, \mathbf{P}_1, \mathbf{Q}_0$, and $\mathbf{Q}_1$ are

$$\mathbf{P}_0 = [1, 1, 1, 1, 1, 1, 1, 1, 1, 1, 1, 1]^{\mathrm{T}}$$
$$\mathbf{P}_1 = \tfrac{1}{3}[-35, -32, -29, -23, -11, -2, 1, 7, 16, 25, 37, 46]^{\mathrm{T}}$$
$$\mathbf{Q}_0 = [1, 2, 3, 5, 9, 12, 13, 15, 18, 21, 25, 28]^{\mathrm{T}} = \mathbf{X}$$
$$\mathbf{Q}_1 = [-\tfrac{35}{3}, -\tfrac{64}{3}, -29, -\tfrac{115}{3}, -33, -8, \tfrac{13}{3}, 35, 96, 175, \tfrac{925}{3}, \tfrac{1288}{3}]^{\mathrm{T}}$$

Table 44.4 lists $\alpha_k, \beta_{k-1}, p_k(x), k = 2, \ldots, 5$.

k	α_k	β_k	$p_k(x)$
2	14.89	75.56	$x^2 - 27.56x + 113.1$
3	15.44	60.57	$x^3 - 43.00x^2 + 477.9x - 978.5$
4	15.14	38.84	$x^4 - 58.13x^3 + 1090.0x^2 - 7143.0x + 1.042 \times 10^4$
5	13.36	52.1	$x^5 - 71.49x^4 + 1814.0x^3 - 1.946 \times 10^4 x^2 + 8.091 \times 10^4 x - 8.817 \times 10^4$

TABLE 44.4 Example 44.4: $\alpha_k, \beta_k, p_k(x), k = 2, \ldots, 5$

Respectively, the vectors $\mathbf{P}_k, k = 2, \dots, 5$, are

$$
\begin{bmatrix}
86.54 \\
61.98 \\
39.42 \\
0.2956 \\
-53.95 \\
-73.63 \\
-76.19 \\
-75.31 \\
-58.99 \\
-24.67 \\
49.09 \\
125.4
\end{bmatrix}
\begin{bmatrix}
-542.5 \\
-186.6 \\
95.35 \\
461.3 \\
569.2 \\
293.3 \\
165.3 \\
-108.6 \\
-474.3 \\
-642.0 \\
-277.5 \\
647.0
\end{bmatrix}
\begin{bmatrix}
4309.0 \\
44.14 \\
-2688.0 \\
-4688.0 \\
-1398.0 \\
1940.0 \\
2606.0 \\
2940.0 \\
933.5 \\
-2806.0 \\
-4643.0 \\
3451.0
\end{bmatrix}
\begin{bmatrix}
-2.497 \times 10^4 \\
9221.0 \\
2.287 \times 10^4 \\
1.514 \times 10^4 \\
-2.357 \times 10^4 \\
-1.791 \times 10^4 \\
-9542.0 \\
1.049 \times 10^4 \\
2.905 \times 10^4 \\
1.2 \times 10^4 \\
-3.961 \times 10^4 \\
1.683 \times 10^4
\end{bmatrix}
$$

and the vectors $\mathbf{Q}_k, k = 2, \dots, 5$, are

$$
\begin{bmatrix}
86.54 \\
124.0 \\
118.2 \\
1.478 \\
-485.5 \\
-883.5 \\
-990.4 \\
-1130.0 \\
-1062.0 \\
-518.1 \\
1227.0 \\
3511.0
\end{bmatrix}
\begin{bmatrix}
-542.5 \\
-373.2 \\
286.1 \\
2306.0 \\
5123.0 \\
3520.0 \\
2149.0 \\
-1628.0 \\
-8538.0 \\
-1.348 \times 10^4 \\
-6937.0 \\
1.812 \times 10^4
\end{bmatrix}
\begin{bmatrix}
4309.0 \\
88.27 \\
-8065.0 \\
-2.344 \times 10^4 \\
-1.258 \times 10^4 \\
2.327 \times 10^4 \\
3.388 \times 10^4 \\
4.41 \times 10^4 \\
1.68 \times 10^4 \\
-5.892 \times 10^4 \\
-1.161 \times 10^5 \\
9.664 \times 10^4
\end{bmatrix}
\begin{bmatrix}
-2.497 \times 10^4 \\
1.844 \times 10^4 \\
6.862 \times 10^4 \\
7.568 \times 10^4 \\
-2.121 \times 10^5 \\
-2.149 \times 10^5 \\
-1.24 \times 10^5 \\
1.574 \times 10^5 \\
5.229 \times 10^5 \\
2.521 \times 10^5 \\
-9.902 \times 10^5 \\
4.712 \times 10^5
\end{bmatrix}
$$

The matrix M is the diagonal matrix of inner products

$$
M = \begin{bmatrix}
12.0 & 0 & 0 & 0 & 0 & 0 \\
0 & 906.7 & 0 & 0 & 0 & 0 \\
0 & 0 & 5.492 \times 10^4 & 0 & 0 & 0 \\
0 & 0 & 0 & 2.133 \times 10^6 & 0 & 0 \\
0 & 0 & 0 & 0 & 1.111 \times 10^8 & 0 \\
0 & 0 & 0 & 0 & 0 & 5.378 \times 10^9
\end{bmatrix}
$$

Note that M still has the large condition number 0.45×10^9, but because the system is diagonal, there is little possibility of roundoff error invalidating the computation. (For a diagonal matrix, the condition numbers under the 1-norm, 2-norm, and ∞-norm are all the same. See Exercise 1.)

The vectors $\mathbf{b}$ and $\mathbf{u}$ in the normal equations $M\mathbf{u} = \mathbf{b}$ are then

$$
\mathbf{b} = \begin{bmatrix}
133.5 \\
-30.5 \\
1137.0 \\
-1.274 \times 10^4 \\
6.092 \times 10^4 \\
-4.668 \times 10^4
\end{bmatrix}
\qquad
\mathbf{u} = \begin{bmatrix}
11.13 \\
-3.364 \times 10^{-2} \\
2.071 \times 10^{-2} \\
-5.972 \times 10^{-3} \\
5.481 \times 10^{-4} \\
-8.679 \times 10^{-6}
\end{bmatrix}
$$

Three potential fitting functions $g_s(x) = \sum_{k=1}^{s} c_k p_k(x), s = 3, 4, 5$, are listed in Table 44.5. The small coefficient of x^5 in $g_5(x)$ would suggest little difference in the fits provided by $g_4(x)$ and $g_5(x)$. The sums of squares for each of these three functions are $S_3 = 34.42$, $S_4 = 1.035$, $S_5 = 0.630$, respectively. The small change in S between the quartic and quintic fitting functions is consistent with the observation about the coefficients.

$g_3(x) = 19.736 - 3.4584x + 0.27746x^2 - 0.0059718x^3$
$g_4(x) = 25.447 - 7.3734x + 0.87488x^2 - 0.037836x^3 + 0.00054813x^4$
$g_5(x) = 26.212 - 8.0756x + 1.0438x^2 - 0.053582x^3 + 0.0011686x^4 - 0.86792 \times 10^{-5}x^5$

TABLE 44.5   Example 44.4: Least-squares polynomials of degree 3, 4, and 5

Figure 44.6 shows the data points and graphs of all three fitting functions. The cubic function $g_3(x)$ (dashed) does not fit the data as well as the quartic $g_4(x)$ (dotted) or the quintic $g_5(x)$ (solid). As predicted, the data is fit nearly as well by $g_4(x)$ as by $g_5(x)$.   ❖

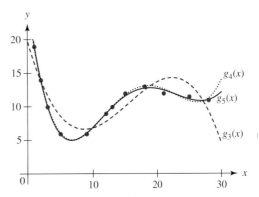

FIGURE 44.6   From Table 44.5, the least-squares polynomials $g_3(x)$ (dashed), $g_4(x)$ (dotted), and $g_5(x)$ (solid)

Index of Determination

For general linear models $g(x) = \sum_{k=1}^{n} c_k \phi_k(x)$ in which $\phi_1 = 1$, the *index of determination* [60] of the linear least-squares function $\tilde{g}$ is defined as

$$R(\tilde{g}) = \frac{\sum_{k=1}^{m}(\tilde{g}(x_k) - \hat{y})^2}{\sum_{k=1}^{m}(y_k - \bar{y})^2} = \frac{\mathbf{u} \cdot \mathbf{b} - m\hat{y}^2}{\sum_{k=1}^{m} y_k^2 - \hat{y}^2}$$

where $\hat{y} = \frac{1}{m}\sum_{k=1}^{m} y_k$ is the mean (or average) of the y_k's. If the number of data points exceeds the number of parameters by at least three and R is close to 1, then $\tilde{g}(x)$ fits the data well. In the fraction on the right, the symbol $\mathbf{u} \cdot \mathbf{b}$ represents the dot product between the vectors $\mathbf{u}$ and $\mathbf{b}$ pertinent to the model determined by $M\mathbf{u} = \mathbf{b}$.

(In the fraction on the left, the denominator is proportional to the *variance,* a measure of the dispersion of the y_k's about their mean $\hat{y}$. The numerator indicates how much of this dispersion is accounted for by $g(x)$.)

For the data on which the polynomials in Table 44.5 are based, $\hat{y} = \frac{89}{8}$, $m = 12$, $\sum_{k=1}^{m} y_k^2 = \frac{6481}{4}$, and the index of determination for each of the three models in Table 44.5 would be $R(g_3) = 0.745$, $R(g_4) = 0.9923$, $R(g_5) = 0.9953$. As already seen, both $g_4(x)$ and $g_5(x)$ fit the data well, with $g_5(x)$ fitting the data slightly better.

EXERCISES 44.3

1. Show that for a diagonal matrix D, the condition numbers based on the 1-norm, 2-norm, and ∞-norm are all the same and are, in fact, the ration of the largest diagonal element to the smallest.

For the data given in each of Exercises 2–8:

(a) Apply the orthogonal-polynomial algorithm to generate polynomials $p_k(x), k = 0, \ldots, 3$, orthogonal under the pseudo-inner product $\sum_{n=1}^{m} p_i(x_n)p_j(x_n)$.

(b) Using the formalism of Section 44.2, fit the data with the general linear model $g(x) = \sum_{k=0}^{3} c_k p_k(x)$, taking $\phi_k(x)$ as $p_k(x)$, so $\mathbf{\Phi}_k$ is $\mathbf{P}_k = [p_k(x_1), \ldots, p_3(x_m)]^T$ and $A = [\mathbf{P}_1, \ldots, \mathbf{P}_3]$.

(c) Show that $M = A^T A$ is a diagonal matrix with $M_{ij} = \mathbf{\Phi}_i \cdot \mathbf{\Phi}_j$, and obtain its condition number.

(d) Show that $b_k = \mathbf{\Phi}_k \cdot \mathbf{y}$ gives the vector $\mathbf{b} = A^T \mathbf{y}$.

(e) Show that $\mathbf{u} = [c_0, \ldots, c_3]^T$, the solution of $M\mathbf{u} = \mathbf{b}$, is given by $c_k = b_k/M_{kk} = (\mathbf{\Phi}_k \cdot \mathbf{y})/(\mathbf{\Phi}_k \cdot \mathbf{\Phi}_k)$.

(f) Graph the data points and the polynomials $g_s(x) = \sum_{k=0}^{s} c_k p_k(x), s = 1, 2, 3$.

(g) Show that $g_2(x)$ is the least-squares quadratic obtained in Section 44.2.

(h) For $s = 1, 2, 3$, obtain $S_s = \sum_{k=1}^{m}(g_s(x_k) - y_k)^2$ and $R(g_s)$, the indices of determination, and compare the fit of the polynomials $g_s(x)$.

2. the data from Exercise 3, Section 44.2

3. the data from Exercise 4, Section 44.2

4. the data from Exercise 5, Section 44.2

5. the data from Exercise 6, Section 44.2

6. the data from Exercise 7, Section 44.2

7. the data from Exercise 8, Section 44.2

8. the data from Exercise 9, Section 44.2

For the data given in each of Exercises 9–14:

(a) Apply the orthogonal-polynomial algorithm to generate polynomials $p_k(x), k = 0, \ldots, 4$, orthogonal under the pseudo-inner product $\sum_{n=1}^{m} p_i(x_n)p_j(x_n)$.

(b) Using the formalism of Section 44.2, fit the data with the general linear model $g(x) = \sum_{k=0}^{4} c_k p_k(x)$, taking $\phi_k(x)$ as $p_k(x)$, so $\mathbf{\Phi}_k$ is $\mathbf{P}_k = [p_k(x_1), \ldots, p_4(x_m)]^T$ and $A = [\mathbf{P}_1, \ldots, \mathbf{P}_4]$.

(c) Show that $M = A^T A$ is a diagonal matrix with $M_{ij} = \mathbf{\Phi}_i \cdot \mathbf{\Phi}_j$, and obtain its condition number.

(d) Show that $b_k = \mathbf{\Phi}_k \cdot \mathbf{y}$ gives the vector $\mathbf{b} = A^T \mathbf{y}$.

(e) Show that $\mathbf{u} = [c_0, \ldots, c_4]^T$, the solution of $M\mathbf{u} = \mathbf{b}$, is given by $c_k = b_k/M_{kk} = (\mathbf{\Phi}_k \cdot \mathbf{y})/(\mathbf{\Phi}_k \cdot \mathbf{\Phi}_k)$.

(f) Graph the data points and the polynomials $g_s(x) = \sum_{k=0}^{s} c_k p_k(x), s = 1, \ldots, 4$.

(g) Show that $g_3(x)$ is the least-squares cubic obtained in Section 44.2.

(h) For $s = 1, \ldots, 4$, obtain $S_s = \sum_{k=1}^{m}(g_s(x_k) - y_k)^2$ and $R(g_s)$, the indices of determination, and compare the fit of the polynomials $g_s(x)$.

9. the data from Exercise 10, Section 44.2

10. the data from Exercise 11, Section 44.2

11. the data from Exercise 12, Section 44.2

12. the data from Exercise 13, Section 44.2

13. the data from Exercise 14, Section 44.2

14. the data from Exercise 15, Section 44.2

For the data in each of Exercises 15–20:

(a) Apply the orthogonal-polynomial algorithm to generate polynomials $p_k(x), k = 0, \ldots, 5$, orthogonal under the pseudo-inner product $\sum_{n=1}^{m} p_i(x_n)p_j(x_n)$.

(b) Using the formalism of Section 44.2, fit the data with the general linear model $g(x) = \sum_{k=0}^{5} c_k p_k(x)$, taking $\phi_k(x)$ as $p_k(x)$, so $\mathbf{\Phi}_k$ is $\mathbf{P}_k = [p_k(x_1), \ldots, p_5(x_m)]^T$ and $A = [\mathbf{P}_1, \ldots, \mathbf{P}_5]$.

(c) Show that $M = A^T A$ is a diagonal matrix with $M_{ij} = \mathbf{\Phi}_i \cdot \mathbf{\Phi}_j$, and obtain its condition number.

(d) Show that $b_k = \mathbf{\Phi}_k \cdot \mathbf{y}$ gives the vector $\mathbf{b} = A^T \mathbf{y}$.

(e) Show that $\mathbf{u} = [c_0, \ldots, c_5]^T$, the solution of $M\mathbf{u} = \mathbf{b}$, is given by $c_k = b_k/M_{kk} = (\mathbf{\Phi}_k \cdot \mathbf{y})/(\mathbf{\Phi}_k \cdot \mathbf{\Phi}_k)$.

(f) Graph the data points and the polynomials $g_s(x) = \sum_{k=0}^{s} c_k p_k(x), s = 1, \ldots, 5$,

(g) Show that $g_4(x)$ is the least-squares quartic obtained in Section 44.2.

(h) For $s = 1, \ldots, 5$, obtain $S_s = \sum_{k=1}^{m}(g_s(x_k) - y_k)^2$ and $R(g_s)$, the indices of determination, and compare the fit of the polynomials $g_s(x)$.

15. the data from Exercise 16, Section 44.2

16. the data from Exercise 17, Section 44.2

17. the data from Exercise 18, Section 44.2

18. the data from Exercise 19, Section 44.2

19. the data from Exercise 20, Section 44.2

20. the data from Exercise 21, Section 44.2

For the data in each of Exercises 21–25:

(a) Apply the orthogonal-polynomial algorithm to generate polynomials $p_k(x), k = 0, \ldots, 5$, orthogonal under the pseudo-inner product $\sum_{n=1}^{m} p_i(x_n)p_j(x_n)$.

(b) Using the formalism of Section 44.2, fit the data with the general linear model $g(x) = \sum_{k=0}^{5} c_k p_k(x)$, taking $\phi_k(x)$ as $p_k(x)$, so $\mathbf{\Phi}_k$ is $\mathbf{P}_k = [p_k(x_1), \ldots, p_5(x_m)]^T$ and $A = [\mathbf{P}_1, \ldots, \mathbf{P}_5]$.

(c) Show that $M = A^TA$ is a diagonal matrix with $M_{ij} = \mathbf{\Phi}_i \cdot \mathbf{\Phi}_j$, and obtain its condition number.

(d) Show that $b_k = \mathbf{\Phi}_k \cdot \mathbf{y}$ gives the vector $\mathbf{b} = A^T\mathbf{y}$.

(e) Show that $\mathbf{u} = [c_0, \ldots, c_5]^T$, the solution of $M\mathbf{u} = \mathbf{b}$, is given by $c_k = b_k/M_{kk} = (\mathbf{\Phi}_k \cdot \mathbf{y})/(\mathbf{\Phi}_k \cdot \mathbf{\Phi}_k)$.

(f) Graph the data points and the polynomials $g_s(x) = \sum_{k=0}^{s} c_k p_k(x), s = 1, \ldots, 5$.

(g) Obtain the indices of determination $R(g_s)$, and compare to the index of determination $R(g)$ computed for the fitting-function computed in Section 44.2.

21. the data from Exercise 22, Section 44.2

22. the data from Exercise 23, Section 44.2

23. the data from Exercise 24, Section 44.2

24. the data from Exercise 25, Section 44.2

25. the data from Exercise 26, Section 44.2

For the data in each of Exercises 26–30:

(a) Apply the orthogonal-polynomial algorithm to generate polynomials $p_k(x), k = 0, \ldots, 5$, orthogonal under the pseudo-inner product $\sum_{n=1}^{m} p_i(x_n)p_j(x_n)$.

(b) Using the formalism of Section 44.2, fit the data with the general linear model $g(x) = \sum_{k=0}^{5} c_k p_k(x)$, taking $\phi_k(x)$ as $p_k(x)$, so $\mathbf{\Phi}_k$ is $\mathbf{P}_k = [p_k(x_1), \ldots, p_5(x_m)]^T$ and $A = [\mathbf{P}_1, \ldots, \mathbf{P}_5]$.

(c) Show that $M = A^TA$ is a diagonal matrix with $M_{ij} = \mathbf{\Phi}_i \cdot \mathbf{\Phi}_j$, and obtain its condition number.

(d) Show that $b_k = \mathbf{\Phi}_k \cdot \mathbf{y}$ gives the vector $\mathbf{b} = A^T\mathbf{y}$.

(e) Show that $\mathbf{u} = [c_0, \ldots, c_5]^T$, the solution of $M\mathbf{u} = \mathbf{b}$, is given by $c_k = b_k/M_{kk} = (\mathbf{\Phi}_k \cdot \mathbf{y}/(\mathbf{\Phi}_k \cdot \mathbf{\Phi}_k)$.

(f) Graph the data points and the polynomials $g_s(x) = \sum_{k=0}^{s} c_k p_k(x), s = 1, \ldots, 5$.

(g) Obtain the indices of determination $R(g_s)$, and compare to the index of determination $R(g)$ computed for the fitting-function computed in Section 44.2.

26. the data from Exercise 27, Section 44.2

27. the data from Exercise 28, Section 44.2

28. the data from Exercise 29, Section 44.2

29. the data from Exercise 30, Section 44.2

30. the data from Exercise 31, Section 44.2

For the data in each of Exercises 31–35:

(a) Apply the orthogonal-polynomial algorithm to generate polynomials $p_k(x), k = 0, \ldots, 5$, orthogonal under the pseudo-inner product $\sum_{n=1}^{m} p_i(x_n)p_j(x_n)$.

(b) Using the formalism of Section 44.2, fit the data with the general linear model $g(x) = \sum_{k=0}^{5} c_k p_k(x)$, taking $\phi_k(x)$ as $p_k(x)$, so $\mathbf{\Phi}_k$ is $\mathbf{P}_k = [p_k(x_1), \ldots, p_5(x_m)]^T$ and $A = [\mathbf{P}_1, \ldots, \mathbf{P}_5]$.

(c) Show that $M = A^TA$ is a diagonal matrix with $M_{ij} = \mathbf{\Phi}_i \cdot \mathbf{\Phi}_j$, and obtain its condition number.

(d) Show that $b_k = \mathbf{\Phi}_k \cdot \mathbf{y}$ gives the vector $\mathbf{b} = A^T\mathbf{y}$.

(e) Show that $\mathbf{u} = [c_0, \ldots, c_5]^T$, the solution of $M\mathbf{u} = \mathbf{b}$, is given by $c_k = b_k/M_{kk} = (\mathbf{\Phi}_k \cdot \mathbf{y})/(\mathbf{\Phi}_k \cdot \mathbf{\Phi}_k)$.

(f) Graph the data points and the polynomials $g_s(x) = \sum_{k=0}^{s} c_k p_k(x), s = 1, \ldots, 5$.

(g) Obtain the indices of determination $R(g_s)$, and compare to the index of determination $R(g)$ computed for the fitting-function computed in Section 44.2.

31. the data from Exercise 32, Section 44.2

32. the data from Exercise 33, Section 44.2

33. the data from Exercise 34, Section 44.2

34. the data from Exercise 35, Section 44.2

35. the data from Exercise 36, Section 44.2

For the data in each of Exercises 36–40:

(a) Apply the orthogonal-polynomial algorithm to generate polynomials $p_k(x), k = 0, \ldots, 5$, orthogonal under the pseudo-inner product $\sum_{n=1}^{m} p_i(x_n)p_j(x_n)$.

(b) Using the formalism of Section 44.2, fit the data with the general linear model $g(x) = \sum_{k=0}^{5} c_k p_k(x)$, taking $\phi_k(x)$ as $p_k(x)$, so $\mathbf{\Phi}_k$ is $\mathbf{P}_k = [p_k(x_1), \ldots, p_5(x_m)]^T$ and $A = [\mathbf{P}_1, \ldots, \mathbf{P}_5]$.

(c) Show that $M = A^TA$ is a diagonal matrix with $M_{ij} = \mathbf{\Phi}_i \cdot \mathbf{\Phi}_j$, and obtain its condition number.

(d) Show that $b_k = \mathbf{\Phi}_k \cdot \mathbf{y}$ gives the vector $\mathbf{b} = A^T\mathbf{y}$.

(e) Show that $\mathbf{u} = [c_0, \ldots, c_5]^T$, the solution of $M\mathbf{u} = \mathbf{b}$, is given by $c_k = b_k/M_{kk} = (\mathbf{\Phi}_k \cdot \mathbf{y})/(\mathbf{\Phi}_k \cdot \mathbf{\Phi}_k)$.

(f) Graph the data points and the polynomials $g_s(x) = \sum_{k=0}^{s} c_k p_k(x), s = 1, \ldots, 5$.

(g) Obtain the indices of determination $R(g_s)$, and compare to the index of determination $R(g)$ computed for the fitting-function computed in Section 44.2.

36. the data from Exercise 37, Section 44.2

37. the data from Exercise 38, Section 44.2

38. the data from Exercise 39, Section 44.2

39. the data from Exercise 40, Section 44.2

40. the data from Exercise 41, Section 44.2

Nonlinear Least Squares

Introduction

In Section 44.3 we developed an elegant formalism for least-squares fitting a general linear model to data. Unfortunately, there is no comparable formalism for fitting data with a function in which the parameters appear nonlinearly. For this problem, the two main options are linearizing the model or solving the nonlinear normal equations.

Table 44.6 lists several nonlinear models and some possible linearizations. The nonlinear models in L_3 and L_4 are the same but have two different linearizations. The linearization in L_3 follows from obvious algebra consequent to reciprocating each side of the nonlinear model, that is, from

$$\frac{1}{y} = \frac{b+x}{a} = \frac{b}{a} + \frac{1}{a}x$$

Index	Model	Linearization $w = A + Bz$	z	w	A	B	a	b
L_1	$y = ae^{bx}$	$\ln y = \ln a + bx$	x	$\ln y$	$\ln a$	b	e^A	B
L_2	$y = ax^b$	$\ln y = \ln a + b \ln x$	$\ln x$	$\ln y$	$\ln a$	b	e^A	B
L_3	$y = \dfrac{a}{b+x}$	$\dfrac{1}{y} = \dfrac{b}{a} + \dfrac{1}{a}x$	x	$\dfrac{1}{y}$	$\dfrac{b}{a}$	$\dfrac{1}{a}$	$\dfrac{1}{B}$	$\dfrac{A}{B}$
L_4	$y = \dfrac{a}{b+x}$	$y = \dfrac{a}{b} + \left(-\dfrac{1}{b}\right)xy$	xy	y	$\dfrac{a}{b}$	$-\dfrac{1}{b}$	$-\dfrac{A}{B}$	$-\dfrac{1}{B}$
L_5	$y = \dfrac{ax}{b+x}$	$\dfrac{1}{y} = \dfrac{1}{a} + \dfrac{b}{a}\dfrac{1}{x}$	$\dfrac{1}{x}$	$\dfrac{1}{y}$	$\dfrac{1}{a}$	$\dfrac{b}{a}$	$\dfrac{1}{A}$	$\dfrac{B}{A}$
L_6	$y = \dfrac{ax}{b+x}$	$y = a + (-b)\dfrac{y}{x}$	$\dfrac{y}{x}$	y	a	$-b$	A	$-B$

TABLE 44.6   Some nonlinear models and their linearizations

The linearization in L_4 appears in [60] and is obtained by multiplying each side of the nonlinear model by ay to get $a = by + xy$. Then, divide through by b to obtain $\frac{a}{b} = y + \frac{xy}{b}$. Solving for the isolated y on the right then gives $y = \frac{a}{b} - \frac{xy}{b}$.

The nonlinear models in L_5 and L_6, once again the same, are sometimes called the *Michaelis–Menton model*. To derive the linearization in L_5, write the reciprocal of each side of the nonlinear model, obtaining

$$\frac{1}{y} = \frac{b+x}{ax} = \frac{b}{a}\frac{1}{x} + \frac{1}{a} = \frac{1}{a} + \frac{b}{a}\frac{1}{x}$$

To obtain the linearization in L_6 that also appears in [60], multiply each side of the nonlinear model by $\frac{b+x}{x}$ to get $\frac{y}{x}(b + x) = a$. Then, distribute the quotient on the left, obtaining $b(\frac{y}{x}) + y = a$, from which follows $y = a + (-b)\frac{y}{x}$.

In the remainder of this section, we will illustrate the linearizations in L_5 and L_6 for the Michaelis–Menton model and demonstrate how to obtain and solve the nonlinear normal equations consequent to applying the least-squares technique to the nonlinear function $f(x) = \frac{ax}{b+x}$.

EXAMPLE 44.5 The data in Table 44.7 represent the 46 data points $(s_k, v_k), k = 1, \ldots, 46$, which pair blood-lactate concentrations s_k with voltages v_k measuring the steady-state saturation rates for an enzyme-catalyzed reaction in a pig weighing approximately 35 pounds. Figure 44.7 contains a graph of the data points. Theoretical considerations (the response v plateaus at high concentrations s) suggest the Michaelis–Menton model, so we will fit this data with the function $f(s) = \frac{as}{b+s}$.

k	s_k	v_k	k	s_k	v_k	k	s_k	v_k
1	0.50	0.484881	16	4.90	1.144826	31	9.54	1.860081
2	0.54	0.539144	17	5.05	1.258919	32	10.39	2.007124
3	0.67	0.614671	18	5.06	1.472855	33	10.64	1.845749
4	1.60	0.752003	19	5.43	1.466318	34	11.31	2.336491
5	1.88	0.851057	20	5.68	1.290979	35	11.50	2.088616
6	2.46	0.944543	21	5.75	1.592607	36	11.80	1.924875
7	2.80	1.070445	22	6.00	1.140909	37	12.18	1.836792
8	2.92	1.261938	23	6.29	1.410909	38	12.62	1.945948
9	3.15	1.167842	24	6.70	1.383805	39	13.89	2.170512
10	3.34	0.740801	25	7.04	1.756296	40	14.01	2.089855
11	4.10	0.993557	26	7.36	1.563731	41	14.10	2.174003
12	4.15	0.820775	27	8.28	1.694899	42	15.58	2.287603
13	4.18	1.062367	28	8.67	1.603329	43	15.75	2.379107
14	4.61	1.142383	29	9.14	2.052710	44	16.05	2.593504
15	4.75	1.430404	30	9.17	1.648573	45	16.28	2.422378
						46	17.95	2.497853

TABLE 44.7 The data set of Example 44.5

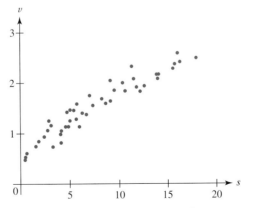

FIGURE 44.7 Plot of the data points in Example 44.5

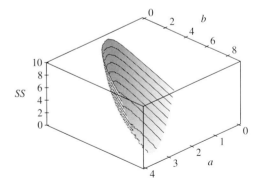

FIGURE 44.8 Example 44.5: The sum of squares of the deviations graphed as a function of the parameters a and b

FITTING THE MICHAELIS–MENTON FUNCTION TO THE DATA The Michaelis–Menton function $f(s) = \frac{as}{b+s}$ will be fitted to the given data using the linearizations in L_5 and L_6 in Table 44.6. The nonlinear least-squares fit will follow from the nonlinear normal equations obtained from the sum of squares $SS = \sum_{k=1}^{46} (f(s_k) - v_k)^2$. Figure 44.8, containing a graph of the sum of squares as a function of the parameters a and b, suggest there is but a single minimum for small values in the first quadrant of the ab-plane.

LINEARIZATION IN L_5   The linearization $\frac{1}{y} = \frac{1}{a} + \frac{b}{a}\frac{1}{x}$ requires the computation of the reciprocals $z_k = 1/s_k$ and $w_k = 1/v_k$. The least-squares regression line $w = A + Bz$ fitting the points (z_k, w_k) is

$$w = 0.5458858609 + 0.8061236141z$$

Since $a = \frac{1}{A}$ and $b = \frac{B}{A}$, we obtain $a_1 = 1.831884780$ and $b_1 = 1.476725579$ for the values of a and b computed by the linearization in L_5. With these values, the Michaelis–Menton model becomes

$$f_1(s) = \frac{1.831884780s}{1.476725579 + s}$$

The value of the sum of squares for this realization of the Michaelis–Menton model is then $SS_1 = 6.431917437$.

LINEARIZATION IN L_6   The linearization $y = a + (-b)\frac{y}{x}$ requires computation of the ratios $z_k = v_k/s_k$ consonant with the new independent variable $z = \frac{v}{s}$. The least-squares regression line $w = A + Bz$ is then

$$w = 2.078083532 - 1.921465149z$$

Since $a = A$ and $b = -B$, we obtain $a_2 = 2.078083532$ and $b_2 = 1.921465149$ for the parameters a and b as determined by the linearization in L_6. For the Michaelis–Menton model determined by this linearization, we obtain

$$f_2(s) = \frac{2.078083532\,s}{1.921465149 + s}$$

and find $SS_2 = 4.526899114$ for the value of the sum of squares of deviations from this version of the nonlinear model.

NONLINEAR FIT   The nonlinear least-squares fit is obtained by forming the nonlinear normal equations

$$\frac{1}{2}\frac{\partial SS}{\partial a} = \sum_{k=1}^{46} \frac{s_k}{b + s_k}\left(\frac{as_k}{b + s_k} - v_k\right) = 0$$

$$\frac{1}{2}\frac{\partial SS}{\partial b} = -\sum_{k=1}^{46} \frac{s_k}{(b + s_k)^2}\left(\frac{as_k}{b + s_k} - v_k\right) = 0$$

With 46 data points, these equations turn out to be strikingly cumbersome, in spite of the elegance of their symbolic representation. However, their solution gives $a_3 = 3.477708271$ and $b_3 = 8.224627001$, so the Michaelis–Menton function based on these parameters is

$$f_3(s) = \frac{3.477708271s}{8.224627001 + s}$$

The sum of squares of the deviations from this realization of the Michaelis–Menton model is then $SS_3 = 1.622578613$.

This problem illustrates a typical difficulty in nonlinear optimization. The minimum lies in a long, narrow "trough," making it difficult to locate and calculate. Indeed, the partial derivatives SS_a and SS_b at the critical point are the small numbers -0.513×10^{-8} and 0.69×10^{-9}, respectively, indicating the lack of sensitivity of the sum of squares to small changes in the parameters near the critical point. This is seen clearly in Figure 44.9, a contour plot for the surface defined by $SS(a, b)$. Contours corresponding to the four values $SS = 1.7, 1.8, 1.9, 2.0$, near $SS_3 = 1.622578613$ are the thin, elongated ovals depicted in the figure.

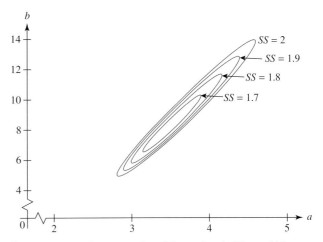

FIGURE 44.9   A contour plot of the surface in Figure 44.8

COMPARISONS   In this concluding part of the section we compare the three versions of the Michaelis–Menton model we have obtained. The sum of squares of the deviations for the three realizations of the Michaelis–Menton model were

$$SS_1 = 6.431917437 \qquad SS_2 = 4.526899114 \qquad SS_3 = 1.622578613$$

Without a frame of reference for comparing these numbers, the first two might not be alarming. However, a graph is more revealing. In addition to the data points, Figure 44.10 shows as dotted, the model determined by the linearization in L_5; as dashed, the model determined by the linearization in L_6; and as solid, the model determined by the nonlinear normal equations.

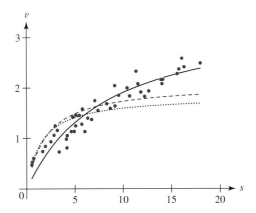

FIGURE 44.10   Example 44.5: Data points from Table 44.7, nonlinear model (solid), L_5-linearized model (dotted), and L_6-linearized model (dashed)

The curves determined by linearization methods are remarkable for their poor fit. Though they track each other closely, they are far from the data. Not every data set brings out the pitfalls of linearization as dramatically as this set. However, the outcome in this example should be kept in mind when fitting any nonlinear model to data. ❖

EXERCISES 44.4

1. Investigate the possibility of recovering $a = b = 1$ from the Michaelis–Menton model $y = \frac{x}{1+x}$ using the three (exact) data points $(1, \frac{1}{2})$, $(2, \frac{2}{3})$, $(3, \frac{3}{4})$.

In Exercises 2–4, fit $g(x) = \ln(1 + \sqrt{a + bx^2})$ to the given sets of data points $\{(x_k, y_k)\}$, $k = 1, \dots, m$.

(a) From the sum of squares $S = \sum_{k=1}^{m}(g(x_k) - y_k)^2$, form and solve the nonlinear normal equations $\frac{1}{2}\frac{\partial S}{\partial a} = 0$, $\frac{1}{2}\frac{\partial S}{\partial b} = 0$, thereby obtaining $g_1(x)$ as the fitting function.

(b) Show that $w = A + Bz$, where $z = x^2$, $w = (e^y - 1)^2$, $A = a$, $B = b$, is a linearization of $g(x)$.

(c) Fit $w = A + Bz$ to the data, and from this fit obtain $g_2(x)$ as the fitting function.

(d) Graph $g_1(x)$, $g_2(x)$, and the data points on the same set of axes.

(e) Determine $S_k = S(g_k)$, $k = 1, 2$; which technique gives the better fit?

2. $\{(1.465, 1.449), (2.198, 1.925), (2.726, 2.138), (4.294, 2.294), (4.537, 2.531), (5.254, 2.733), (6.441, 2.677), (6.765, 2.761), (8.129, 2.771), (8.455, 3.197)\}$

3. $\{(1.873, 1.403), (2.491, 1.532), (3.293, 2.013), (4.226, 1.783), (4.689, 2.354), (5.643, 2.401), (6.028, 2.261), (6.935, 2.319), (7.454, 2.363), (7.841, 2.587)\}$

4. $\{(1.10, 1.329), (1.14, 1.409), (1.16, 1.531), (1.58, 1.494), (1.61, 1.500), (2.34, 1.869), (2.46, 1.601), (3.01, 1.716), (3.64, 1.979), (4.58, 2.076)\}$

5. The data points given here were generated by adding random "noise" to values of the function $f(x) = \sqrt{x+2} + \sqrt{x+5}$. Fit the data with the function $g(x) = \sqrt{x+a} + \sqrt{x+b}$, and then graph $f(x)$, $g(x)$ and the data points.

$\{(0.910, 4.152), (1.02, 4.204), (1.24, 4.266), (1.25, 4.278), (1.75, 4.557), (3, 5.086), (3.48, 5.205), (3.87, 5.443), (4.56, 5.645), (4.88, 5.723)\}$

6. The data points given here were generated by adding random "noise" to values of the function $f(x) = \sqrt{x+3} + 2\sqrt{x+7}$. Fit the data with the function $g(x) = \sqrt{x+a} + 2\sqrt{x+b}$, and then graph $f(x)$, $g(x)$ and the data points. Compare to the results in Exercise 5. *Hint:* Sum of squares has more than one local minimum!

$\{(0.910, 7.556), (1.02, 7.618), (1.24, 7.851), (1.25, 7.759), (1.75, 8.151), (3, 8.805), (3.48, 8.926), (3.87, 9.254), (4.56, 9.456), (4.88, 9.806)\}$

7. The data points given here were generated by adding random "noise" to the values of the function $f(x) = 3e^{-2x}$. Fit the data with the function $g(x) = ae^{bx}$, obtaining $g_1(x)$, and then linearize according to L_1 in Table 44.6, obtaining $g_2(x)$. Graph the functions $f(x)$, $g_1(x)$, $g_2(x)$, and the given data. Compute $S(g_1)$ and $S(g_2)$.

$\{(0.27, 1.923), (0.72, 1.350), (1.25, 0.2955), (1.57, 0.1558), (1.78, 0.1109), (2.18, 0.01917), (3.93, 0.0004631), (4.36, 0.0002450), (4.79, 0.0003110)\}$

8. The data points given here were generated by randomly adding ± 2 to the values of the function $f(x) = 3e^{x/5}$. Fit the data with the function $g(x) = ae^{bx}$, obtaining $g_1(x)$, and then linearize according to L_1 in Table 44.6, obtaining $g_2(x)$. Graph the functions $f(x)$, $g_1(x)$, $g_2(x)$, and the given data. Compute $S(g_1)$ and $S(g_2)$.

$\{(0.23, 5.141), (1.81, 2.308), (1.91, 2.395), (2.52, 2.965), (2.66, 7.106), (3.39, 7.910), (4.17, 8.909), (4.18, 8.921), (4.35, 9.161), (4.61, 5.542)\}$

9. The data points given here were generated by adding random "noise" to the values of the function $f(x) = 5x^{7/4}$. Fit the data with the function $g(x) = ax^b$, obtaining $g_1(x)$, and then linearize according to L_2 in Table 44.6, obtaining $g_2(x)$. Graph the functions $f(x)$, $g_1(x)$, $g_2(x)$, and the given data. Compute $S(g_1)$ and $S(g_2)$.

$\{(0.79, 3.034), (1.13, 5.986), (1.58, 10.20), (2.51, 28.35), (2.52, 29.40), (2.82, 33.75), (4.10, 61.00), (4.25, 68.14), (4.32, 75.51)\}$

10. The data points given here were generated by adding random "noise" to the values of the function $f(x) = \frac{2}{5+x}$. Fit the data with the function $g(x) = \frac{a}{b+x}$, obtaining $g_1(x)$; linearize according to L_3 in Table 1, obtaining $g_2(x)$; and linearize according to L_4 in Table 44.6, obtaining $g_3(x)$. Graph the functions $f(x)$, $g_k(x)$, $k = 1, 2, 3$, and the given data. Compute $S(g_k)$, $k = 1, 2, 3$.

$\{(0.77, 0.3397), (1.48, 0.3117), (1.52, 0.3221), (2.21, 0.2968), (2.23, 0.2683), (2.49, 0.2617), (2.59, 0.2846), (2.74, 0.2351), (3.21, 0.2582), (3.25, 0.2521)\}$

For the data sets in Exercises 11 and 12, each of which was generated by adding random "noise" to the function $f(x) = \frac{3x}{7+x}$:

(a) Fit the data with the function $g(x) = \frac{ax}{b+x}$, obtaining $g_1(x)$.

(b) Linearize according to L_5 in Table 44.6, obtaining $g_2(x)$.

(c) Linearize according to L_6 in Table 44.6, obtaining $g_3(x)$.

(d) Graph the functions $f(x)$, $g_k(x)$, $k = 1, 2, 3$, and the given data.

(e) Compute $S(g_k)$, $k = 1, 2, 3$.

11. $\{(0.12, 0.05157), (1.34, 0.4386), (1.45, 0.5560),$
$(1.60, 0.5414), (2.52, 0.7226), (2.60, 0.8775), (3.34, 1.027),$
$(4.10, 1.141), (4.17, 1.042), (4.51, 1.234)\}$

12. $\{(0.13, 0.06564), (0.43, 0.2083), (1.10, 0.2037),$
$(1.25, 0.3636), (1.43, 0.2544), (1.50, 1.006), (1.97, 1.252),$
$(4.03, 2.083), (4.45, 1.749), (4.76, 0.4857)\}$

13. The data points given here were generated by randomly ± 0.25 to
the values of the function $f(x) = \frac{3x}{7+x}$. Fit the data with the
function $g(x) = \frac{ax}{b+x}$, obtaining $g_1(x)$ (a will lie in $[5, 6]$ and b will

lie in $[14, 15]$); linearize according to L_5 in Table 44.6, obtaining
$g_2(x)$; and linearize according to L_6 in Table 44.6, obtaining $g_3(x)$.
Graph the functions $f(x)$, $g_k(x)$, $k = 1, 2, 3$, and the given data.
Compute $S(g_k)$, $k = 1, 2, 3$. Comment on the efficacy of the
linearization process for data fitting.

$\{(0.07, 0.2797), (0.16, 0.3170), (0.86, 0.5782), (1.31, 0.2229),$
$(2.77, 1.101), (2.95, 0.6394), (3.01, 0.6521), (4.26, 1.385),$
$(4.72, 0.958), (4.94, 1.491)\}$

Chapter Review

1. Demonstrate the process of fitting the line $y = ax + b$ to the data
points $(0, 1)$, $(2, 0)$, $(3, 3)$. Obtain the sum of squares of the
deviations, and use the techniques of the calculus to minimize it.
Exhibit and solve the normal equations.

2. Rework Question 1 with matrix methods, starting with three
equations in two unknowns, leading to $A\mathbf{u} = \mathbf{y}$, and then
$A^{\mathsf{T}}A\mathbf{u} = A^{\mathsf{T}}\mathbf{y} = \mathbf{b}$.

3. Use the matrix methods of Question 2 to fit a quadratic to the data
points $(0, 1)$, $(2, 0)$, $(3, 3)$, $(4, 6)$. Show that this is an example of a

general linear model. Exhibit the vectors $\mathbf{\Phi}_k$, $k = 0, 1, 2$, the vector
$\mathbf{y}$, and the vector $\mathbf{b}$. Also, exhibit the matrices A and $A^{\mathsf{T}}A$.

4. Answer Question 3 by means of the algorithm for generating
orthogonal polynomials.

5. Describe the options available for fitting data with a nonlinear model
such as $y = \frac{ax}{b+x}$.

Chapter *45*

Numerical Calculation
of Eigenvalues

I N T R O D U C T I O N   The algorithm for the pencil-and-paper calculation of eigenvalues of a 2×2 or 3×3 matrix is not the algorithm used for numeric computation. With pencil and paper, the characteristic polynomial must be obtained and its zeros found. Solving the quadratic equation that results from a 2×2 matrix is possible with the quadratic formula. Solving the cubic that results from a 3×3 matrix is a significantly harder problem.

Chapter 45 describes several numeric methods for finding eigenvalues. None of these methods formulates or solves the characteristic equation. Essentially, this is because, as seen in Section 37.1, finding zeros of polynomials is inherently an unstable calculation on a finite-precision computing device. All numeric methods for finding eigenvalues are iterative.

This chapter contains the most difficult computational mathematics of the text. It is difficult mathematics because the numeric computation of eigenvalues is a difficult problem. Readers willing to rely on professionally packaged code for computing eigenvalues might easily treat the first four sections as optional since they merely detail the methods used in modern software. But those algorithms can be found already coded, and the reader need not recode them in order to execute them. With the right set of tools available, these first four sections can certainly be considered optional.

The various *power methods* converge linearly. Interesting as the underlying ideas might be, power methods are not generally found in production software. Except perhaps for large sparse matrices, there are better methods.

The *Householder reflection* is a computational tool of paramount importance in the more efficient algorithms found in effective modern software packages. Multiplying a vector **x** by its Householder reflection matrix yields a vector that is the geometric equivalent of a reflection across a plane or line and has zeros for all components below a designated one in **x**. In addition, the Householder matrix is both symmetric and orthogonal and has determinant -1.

A sequence of Householder reflections can be used to upper-triangularize a matrix A, thereby obtaining its QR decomposition without using the Gram–Schmidt process. At this point in the discussion, we simply have an alternative technique for obtaining the QR factorization of Section 34.3. The Gram–Schmidt process finds the factor Q, whereas the Householder reflection matrices lead to the factor R.

We are in possession of the QR algorithm from Section 34.4. This computational technique for finding eigenvalues requires repeated QR factorizations for which we now have two different processes. Efficient implementation of the QR algorithm is best performed on the *upper Hessenberg matrix,* a matrix that has only zeros below the diagonal, which is itself below the main diagonal. In other words, the upper Hessenberg matrix is one diagonal away from being an upper triangular matrix.

A sequence of Householder reflections can be used to put a matrix A into upper Hessenberg form. Moreover, this is accomplished with a similarity transform so that the resulting upper Hessenberg matrix has the same eigenvalues as A. Thus, the Householder reflections are used to precondition A, putting it into the form of an upper Hessenberg matrix with the same eigenvalues.

At this point there are two choices. The QR factorization of A can be completed using *annihilating Givens rotations,* or the *shifted QR-algorithm,* which converges rapidly for matrices in upper Hessenberg form. Thus, a matrix is put into upper Hessenberg form by Householder reflections and then factored into its QR form using Givens rotations. Alternatively, the shifted QR-algorithm can be applied to the upper Hessenberg form of A. When applied to such matrices, this algorithm converges very rapidly.

The chapter concludes with a discussion of the *generalized eigenvalue problem,* which appears, for example, when trying to obtain normal modes of vibration for coupled harmonic oscillators.

45.1 Power Methods

Introduction

The various *power methods* use iteration to compute eigenvalues and eigenvectors of the $n \times n$ matrix A. If the eigenvalues of A are arranged in order of decreasing magnitude, so that $|\lambda_1| \geq |\lambda_2| \geq \cdots \geq |\lambda_n|$, then any eigenvalues having largest magnitude will be called *dominant* and any having smallest magnitude, *least dominant.*

If $\mathbf{x}_0$ is an arbitrary initial vector, the iteration $\mathbf{x}_k = A^k \mathbf{x}_0, k = 1, 2, \ldots$, is the essence of the power method. Under the assumptions that $|\lambda_1| > |\lambda_2|$ and that A has n linearly independent eigenvectors $\mathbf{v}_1, \ldots, \mathbf{v}_n$, with $A\mathbf{v}_i = \lambda_i \mathbf{v}_i$, we can show how the power method leads to the dominant eigenvalue and its eigenvector.

First, write $\mathbf{x}_0$ in terms of the eigenvectors as

$$\mathbf{x}_0 = \alpha_1 \mathbf{v}_1 + \alpha_2 \mathbf{v}_2 + \cdots + \alpha_n \mathbf{v}_n$$

Then, since $A^k \mathbf{v}_i = \lambda_i^k \mathbf{v}_i$, we have for $\mathbf{x}_k = A^k \mathbf{x}_0$

$$\mathbf{x}_k = \alpha_1 \lambda_1^k \mathbf{v}_1 + \alpha_2 \lambda_2^k \mathbf{v}_2 + \cdots + \alpha_n \lambda_n^k \mathbf{v}_n = \lambda_1^k \left[\alpha_1 \mathbf{v}_1 + \alpha_2 \left(\frac{\lambda_2}{\lambda_1} \right)^k \mathbf{v}_2 + \cdots + \alpha_n \left(\frac{\lambda_n}{\lambda_1} \right)^k \mathbf{v}_n \right]$$

which, because of the strict dominance of λ_1, tends to $\alpha_1 \lambda_1^k \mathbf{v}_1$ as k increases.

Clearly, if we simply carry out the iteration $\mathbf{x}_k = A^k \mathbf{x}_0, k = 1, 2, \ldots$, the magnitude of $\mathbf{x}_k$ will grow if $|\lambda_1| > 1$ or will shrink to zero if $|\lambda_1| < 1$. (See the ensuing discussion of the *naive power method.*)

To extract the value of λ_1 from $\mathbf{x}_k$, some sort of scaling must be used at each step. Either divide $\mathbf{x}_k$ by the component with largest magnitude or consistently divide each iterate by the same component. Both options are explored in the discussion of the *scaled power method,* where we discover that the first technique converges slightly faster than the second.

Suppose we scale with the first component so that the power iteration consists of the two steps $\mathbf{y} = A\mathbf{x}_k$ and $\mathbf{x}_{k+1} = \frac{1}{y_1} \mathbf{y}$, where y_1, the first component of $\mathbf{y}$, is assumed to be nonzero. The first component of $\mathbf{x}_k$ is a 1, and the first component of $\mathbf{y}$ is approximately λ_1 since $\mathbf{x}_k$ is approximately proportional to the eigenvector $\mathbf{v}_1$. Then, $\mathbf{x}_k$ tends to that multiple of $\mathbf{v}_1$ in which the first component of the eigenvector is 1, and the values of y_1 tend to λ_1.

EXAMPLE 45.1   In (45.1), we have a matrix A and to its right, its eigenvalues.

$$A = \begin{bmatrix} 6 & 5 & -1 & 2 \\ -5 & -7 & 4 & -6 \\ 1 & 6 & 6 & 7 \\ 4 & 1 & 5 & -3 \end{bmatrix} \quad \begin{matrix} \lambda_1 = -10.08570923 \\ \lambda_2 = 9.458884520 \\ \lambda_3 = 2.950157561 \\ \lambda_4 = -0.323332851 \end{matrix} \quad (45.1)$$

Corresponding eigenvectors, scaled so that their first components are 1, are

$$\begin{bmatrix} 1.000000000 \\ -2.467783955 \\ 1.374495378 \\ -1.186147036 \end{bmatrix} \begin{bmatrix} 1.000000000 \\ 1.200899814 \\ 17.12862769 \\ 7.291506576 \end{bmatrix} \begin{bmatrix} 1.000000000 \\ -0.8267558725 \\ 0.02545954948 \\ 0.5546982303 \end{bmatrix} \begin{bmatrix} 1.000000000 \\ -1.698034948 \\ 0.1633114879 \\ 1.165076683 \end{bmatrix}$$

$$(45.2)$$

This matrix and its eigenpairs will be used in the following discussions.   ❖

Naive Power Method

If the power method is implemented without scaling, the magnitudes of the iterates $\mathbf{x}_k$ will either grow or shrink, unless, perchance, $|\lambda_1| = 1$. For the matrix A in Example 45.1, this is not the case, and the power iterates, starting with $\mathbf{x}_0 = [1, 1, 1, 1]^\mathrm{T}$, will include

$$\mathbf{x}_1 = \begin{bmatrix} 12 \\ -14 \\ 20 \\ 7 \end{bmatrix} \quad \mathbf{x}_5 = \begin{bmatrix} 42,412 \\ -77,816 \\ 168,910 \\ 10,591 \end{bmatrix} \quad \mathbf{x}_{10} = \begin{bmatrix} -3,094,896,548 \\ 9,601,408,696 \\ 4,178,333,965 \\ 8,208,895,693 \end{bmatrix}$$

Without some form of scaling, a naive power method just isn't useful.

Scaled Power Method

Next, we will execute the *scaled power method* in which the first component is always the scaling factor. The method then converges to $\mathbf{v}_1$.

For $k = 50, 100, 150,$ and 200, Table 45.1 contains the power iterates $\mathbf{x}_k$, along with the values of y_1, $|y_1 - \lambda_1|$, and $\|\mathbf{x}_k - \mathbf{v}_1\|$. The values of y_1 slowly converge to λ_1, the eigen-

	$k = 50$	$k = 100$	$k = 150$	$k = 200$
$\mathbf{x}_k$	$\begin{bmatrix} 1.000 \\ -2.510 \\ 1.194 \\ -1.283 \end{bmatrix}$	$\begin{bmatrix} 1.000 \\ -2.469 \\ 1.367 \\ -1.190 \end{bmatrix}$	$\begin{bmatrix} 1.000 \\ -2.468 \\ 1.374 \\ -1.186 \end{bmatrix}$	$\begin{bmatrix} 1.000 \\ -2.468 \\ 1.374 \\ -1.186 \end{bmatrix}$
y_1	-9.852557242	-10.07617438	-10.08532355	-10.08569364
$|y_1 - \lambda_1|$	0.233	0.953×10^{-2}	0.386×10^{-3}	0.156×10^{-4}
$\|\mathbf{x}_k - \mathbf{v}_1\|$	0.209	0.836×10^{-2}	0.338×10^{-3}	0.137×10^{-4}

TABLE 45.1   Scaled power iterates for the matrix of Example 45.1

value of largest magnitude, and the $\mathbf{x}_k$ slowly converge to $\mathbf{v}_1$, the corresponding eigenvector. Due note should be taken of the extremely large number of iterations required to obtain even modest accuracy. This is because the magnitude of λ_2 is close to the magnitude of λ_1. The ratio $\left|\frac{\lambda_2}{\lambda_1}\right| = 0.9378502101$ is raised to powers. The closer this ratio is to 1, the slower the rate of convergence for the scaled power method.

The speed of convergence can be increased slightly if, instead of dividing by the first component of $\mathbf{y}_k = A\mathbf{x}_{k-1}$, we divide by the component of largest magnitude. This means extra coding to determine which component to divide by in each iteration. As a consequence, it becomes difficult to compare $\mathbf{x}_k$ to $\mathbf{v}_1$ because we don't know beforehand the component with which to scale $\mathbf{v}_1$. With this modification, the $\mathbf{x}_{200}$ in (45.3) is obtained, along with the estimate $\lambda_1 \approx -10.08571680$, in error by 0.757×10^{-5}

$$\mathbf{x}_{200} = \begin{bmatrix} -0.4052214073 \\ 1.000000000 \\ -0.5569701743 \\ 0.4806547389 \end{bmatrix} \tag{45.3}$$

Inverse Power Method

If $(\lambda, \mathbf{x})$ is an eigenpair for the invertible matrix A, then $(\frac{1}{\lambda}, \mathbf{x})$ is an eigenpair for the matrix A^{-1}, the inverse of A. Indeed, by assumption we have $A\mathbf{x} = \lambda\mathbf{x}$ so that, upon multiplication by the inverse of A, we have $\mathbf{x} = \lambda A^{-1}\mathbf{x}$ or $\frac{1}{\lambda}\mathbf{x} = A^{-1}\mathbf{x}$. Thus, $\mathbf{x}$ is an eigenvector for A^{-1} and the eigenvalue is $\frac{1}{\lambda}$.

This observation clearly lets us compute the *least* dominant eigenvalue for A by computing the *most* dominant eigenvalue of the inverse, A^{-1}. This adaptation is called the *inverse power method* and is illustrated for the matrix A given in (45.1).

The inverse of A and its eigenvalues Λ_k are given in (45.4). The reciprocal relation of the eigenvalues of A^{-1} and A are noted to the right of A^{-1} in (45.4).

$$A^{-1} = \frac{1}{91}\begin{bmatrix} 327 & 242 & 39 & -175 \\ -503 & -387 & -65 & 287 \\ 46 & 41 & 13 & -21 \\ 345 & 262 & 52 & -203 \end{bmatrix} \qquad \begin{aligned} \Lambda_1 &= \frac{1}{\lambda_4} = -3.092788222 \\ \Lambda_2 &= \frac{1}{\lambda_3} = 0.3389649481 \\ \Lambda_3 &= \frac{1}{\lambda_2} = 0.1057207116 \\ \Lambda_4 &= \frac{1}{\lambda_1} = -0.0991501916 \end{aligned} \tag{45.4}$$

The eigenvectors of A^{-1}, scaled so that the first components are 1, and denoted by $\mathbf{V}_k$, are given in (45.5). Comparison with (45.2) shows $\mathbf{v}_k = \mathbf{V}_{5-k}, k = 1, \ldots, 4$, as expected.

$$\begin{bmatrix} 1.000000000 \\ -1.698034944 \\ 0.163311487 \\ 1.165076682 \end{bmatrix} \begin{bmatrix} 1.0000000000 \\ -0.8267558750 \\ 0.02545955023 \\ 0.5546982336 \end{bmatrix} \begin{bmatrix} 1.000000000 \\ 1.200899822 \\ 17.12862777 \\ 7.291506608 \end{bmatrix} \begin{bmatrix} 1.000000000 \\ -2.467783955 \\ 1.374495380 \\ -1.186147026 \end{bmatrix} \tag{45.5}$$

If we apply the scaled power method to the matrix A^{-1}, we should compute an approximation to the reciprocal of the least dominant eigenvalue of A, that is, to $\frac{1}{\lambda_4} = \Lambda_1 = -3.092788222$. After 200 iterations we compute, as an approximation to $\mathbf{V}_1 = \mathbf{v}_4$, the $\mathbf{x}_{200}$ shown in (45.6). Also shown are y_1, an approximation to Λ_1, the error $|y_1 - \Lambda_1|$, and $\|\mathbf{x}_{200} - \mathbf{V}_1\|$, the norm of the difference between the approximate eigenvector and the actual eigenvector. The apparent greater accuracy in the calculation of the least dominant eigenvalue stems from the comparatively smaller ratio $|\frac{\Lambda_2}{\Lambda_1}| = 0.1095984994$.

$$\mathbf{x}_{200} = \begin{bmatrix} 0.9999999999 \\ -1.698034944 \\ 0.1633114870 \\ 1.165076682 \end{bmatrix} \qquad \begin{array}{l} y_1 = -3.092788218 \\ |y_1 - \Lambda_1| = \quad 4 \times 10^{-9} \\ \|\mathbf{x}_{200} - \mathbf{V}_1\| = \quad 1 \times 10^{-10} \end{array} \qquad (45.6)$$

Efficiency in the Inverse Power Method

For a 4×4 matrix, computing the inverse with modern software tools is effortless. However, in more realistic problems, the inverse of a matrix is generally too expensive to compute. Instead, in the inverse power method, the equation $\mathbf{x}_{k+1} = A^{-1}\mathbf{x}_k$ is replaced by the equation $A\mathbf{x}_{k+1} = \mathbf{x}_k$ and $\mathbf{x}_{k+1}$ is obtained from an $L_1 U$ decomposition of A by forward and backward substitution. Thus, $L_1\mathbf{c} = \mathbf{x}_k$ is solved for $\mathbf{c}$ by forward substitution and $U\mathbf{x}_{k+1} = \mathbf{c}$ is solved for $\mathbf{x}_{k+1}$ by back substitution.

Shifted Inverse Power Method

The most and least dominant eigenvalues can be computed by the scaled and inverse power methods, respectively. To compute other eigenvalues, shift eigenvalues according to the following prescription.

The matrix A having eigenpair $(\lambda, \mathbf{x})$ is equivalent to the matrix $(A - \sigma I)$ having eigenpair $(\lambda - \sigma, \mathbf{x})$. This is verified by starting with $A\mathbf{x} = \lambda\mathbf{x}$ and subtracting $\sigma\mathbf{x}$ from each side to obtain $A\mathbf{x} - \sigma\mathbf{x} = \lambda\mathbf{x} - \sigma\mathbf{x}$ or $(A - \sigma I)\mathbf{x} = (\lambda - \sigma)\mathbf{x}$, which implies that $\lambda - \sigma$ is an eigenvalue of $A - \sigma I$, with eigenvector $\mathbf{x}$.

The utility of this observation hinges on finding σ near enough to λ_i so that the difference $\lambda_i - \sigma$ is now the least dominant eigenvalue of the matrix $A - \sigma I$. The inverse power method applied to $A - \sigma I$ will then find the eigenvector $\mathbf{x}$ and the eigenvalue $\omega_i = 1/(\lambda_i - \sigma)$ so that $\lambda_i = 1/\omega_i + \sigma$.

For the matrix A in (45.1), to compute $\lambda_2 = 9.458884520$, the eigenvalue with the second largest magnitude, set $\sigma = 7$ because the differences $\lambda_k - \sigma, k = 1, \dots, 4$, are then $-17.1, 2.5, -4.0$, and -7.3, respectively. Thus, $\lambda_2 - 7 = 2.5$ is the least dominant eigenvalue for the matrix $A - 7I$, and the inverse power method applied to this matrix will compute ω_2, an approximation to $\frac{1}{\lambda_2 - 7}$. The matrix $(A - 7I)^{-1}$, and $\mathbf{x}_{200}$, the resulting approximation to v_2, appear in (45.7). The corresponding estimate of ω_2 is 0.4066884797, so λ_2 is approximated as 9.458884502 with an error of just 1.8×10^{-8}.

$$(A - 7I)^{-1} = \begin{bmatrix} -\frac{216}{623} & -\frac{65}{623} & \frac{26}{623} & \frac{2}{89} \\ \frac{223}{1246} & -\frac{5}{1246} & \frac{1}{623} & \frac{7}{178} \\ \frac{59}{1246} & \frac{211}{1246} & \frac{207}{623} & \frac{25}{178} \\ -\frac{121}{1246} & \frac{53}{1246} & \frac{114}{623} & -\frac{3}{178} \end{bmatrix} \Rightarrow \mathbf{x}_{200} = \begin{bmatrix} 0.999999999 \\ 1.20089981 \\ 17.12862763 \\ 7.29150655 \end{bmatrix} \qquad (45.7)$$

Gerschgorin's Disk Theorem

Initial guesses for the shifts in the shifted inverse power method are usually taken as the diagonal elements in A. Motivation for this suggestion stems from the Gerschgorin Disk Theorem.

THEOREM 45.1

Each eigenvalue of the matrix A lies in at least one of the circular disks $D_k = \{z, |z - a_{kk}| \leq r_k\}$, where the radii of the disks are $r_k = \sum_{j=1}^n |a_{kj}|, j \neq k$. Moreover, if the union of any m disks intersects no other disk, then this union contains exactly m (possibly repeated) eigenvalues of A. Thus, the eigenvalues lie in the union of disks that have their centers at the diagonal elements of A and have radii that are the sums of the absolute values of the elements of the rows, exclusive of the diagonal elements. (See [60] for a proof.)

For the matrix A in (45.1), the centers and radii of the Gerschgorin disks are given in Table 45.2.

Centers	$(6, 0)$	$(-7, 0)$	$(6, 0)$	$(-3, 0)$
Radii	8	15	14	10

TABLE 45.2   Centers and radii of the Gerschgorin disks for the matrix in Example 45.1

A graph showing the disks and the actual eigenvalues (as black dots) is shown in Figure 45.1. The shift of $\sigma = 7$ is consistent with the information in Figure 45.1, but by no means is locating eigenvalues as mechanical as this one example might imply. The power method, and all its variants, is not really a modern "production" method for finding eigenvalues.

Our approach to numeric computation remains the same. The typical user will not write code to perform routine computations but, rather, will use professionally written and well-tested packages and libraries. For those who eventually find they must write their own code, the underlying ideas presented here will serve as guidelines.

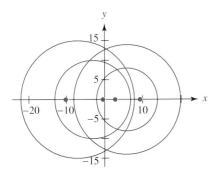

FIGURE 45.1   Eigenvalues and Gerschgorin disks for the matrix in Example 45.1

A Matrix with Repeated Eigenvalues

By design, the matrix A in (45.8) has the eigenvalue $\lambda = 3$ with algebraic multiplicity 4 and geometric multiplicity 2. Hence, A has just the two eigenvectors $\mathbf{v}_1$ and $\mathbf{v}_2$ given to the right of A in

$$A = \begin{bmatrix} \frac{10,623}{2650} & -\frac{2269}{2650} & -\frac{1608}{1325} & \frac{5207}{2650} \\ \frac{1199}{2650} & \frac{7303}{2650} & -\frac{754}{1325} & \frac{2641}{2650} \\ -\frac{1503}{2650} & -\frac{241}{2650} & \frac{5013}{1325} & -\frac{4177}{2650} \\ -\frac{889}{1325} & \frac{367}{1325} & \frac{1138}{1325} & \frac{1924}{1325} \end{bmatrix} \quad \mathbf{v}_1 = \begin{bmatrix} 23 \\ 3 \\ 17 \\ 0 \end{bmatrix} \quad \mathbf{v}_2 = \begin{bmatrix} -45 \\ -14 \\ 0 \\ 17 \end{bmatrix} \quad \textbf{(45.8)}$$

Applied to A with 500 iterations, the scaled power method yields 3.006 as an approximation to $\lambda = 3$ and $\mathbf{x}_{500}$, shown in (45.9), as the companion eigenvector. Because of the repeated eigenvalue, it took a large number of iterations to approximate the single eigenvalue $\lambda = 3$. The vector returned by the method is an approximation to a *linear combination* of the two eigenvectors of A. It is surprisingly difficult to demonstrate that $\mathbf{x}_{500}$ is close to being a linear combination of the eigenvectors $\mathbf{v}_1$ and $\mathbf{v}_2$.

$$\mathbf{x}_{500} = \begin{bmatrix} 1.000000000 \\ 0.705504797 \\ -1.588247513 \\ -1.183833025 \end{bmatrix} \quad \textbf{(45.9)}$$

EXERCISES 45.1

The matrices A_1 and A_2 are used in the exercises that follow.

$$A_1 = \frac{1}{300}\begin{bmatrix} 208 & -16 & 0 \\ 4 & 232 & -20 \\ -17 & -16 & 225 \end{bmatrix} \qquad A_2 = \frac{1}{186}\begin{bmatrix} 38 & 86 & -58 \\ 307 & -59 & -46 \\ 29 & 2 & -227 \end{bmatrix}$$

1. Show that for A_1, the maximum of the magnitudes of the eigenvalues is less than 1. Verify that the naive power method $A^k\mathbf{x}_0$ generates vectors of diminishing magnitudes.

2. Show that for A_2, the maximum of the magnitude of the eigenvalues is greater than 1. Verify that the naive power method $A^k\mathbf{x}_0$ generates vectors of increasing magnitudes.

There are two ways to implement the scaled power method. The vector $A^k\mathbf{x}_0$ can be consistently divided by, say, the first component, or by the component with largest magnitude. In Exercises 3 and 4, use the given matrix to quantize the difference in these two implementations. For example, the errors in the approximation of λ_1 can be compared after a fixed number of iterations, or the number of iterations it takes to reduce the errors to a fixed tolerance can be compared. Is the improvement with the second method significant?

3. Use matrix A_1.   **4.** Use matrix A_2.

For the matrices given in Exercises 5 and 6, apply the inverse power method to find the eigenvalue of smallest magnitude and its companion eigenvector. Implement the calculation in an "efficient" manner, solving the equation $A\mathbf{x}_{k+1} = \mathbf{x}_k$ for $\mathbf{x}_{k+1}$ using forward and back substitutions with an L_1U decomposition for A. (If $A = PL_1U$, where P is a permutation matrix so that $P^{-1} = P^T$, solve $L_1\mathbf{c} = P^T\mathbf{x}_k$ for $\mathbf{c}$ by forward substitution and $U\mathbf{x}_{k+1} = \mathbf{c}$ for $\mathbf{x}_{k+1}$ by back substitution.)

5. Use matrix A_1.   **6.** Use matrix A_2.

For the matrices given in Exercises 7 and 8, use the shifted inverse power method to find λ_2 and its companion eigenvector, where the eigenvalues λ_k have been ordered $|\lambda_1| > |\lambda_2| > |\lambda_3|$. (Because part of the "pleasure" of using this method is discovering appropriate shifts, we make no recommendations on the matter.)

7. Use matrix A_1 **8.** Use matrix A_2

For the matrices given in Exercises 9–12:

  (a) Obtain the eigenvalues (perhaps by using a software package).

  (b) Apply Gerschgorin's disk theorem, graphing the disks and eigenvalues.

  (c) Comment on the utility of the theorem to locate eigenvalues.

9. $\begin{bmatrix} 5 & 8 & 4 & -3 \\ 8 & -9 & 7 & 5 \\ 0 & 4 & 4 & 2 \\ 5 & -5 & 9 & -9 \end{bmatrix}$ **10.** $\begin{bmatrix} 6i & 3+6i & 8 & 2i \\ -5 & 5i & -4-3i & 1 \\ -5 & 6 & -1+8i & 2-4i \\ 4-2i & -7 & -9 & -7+3i \end{bmatrix}$

11. $\begin{bmatrix} -1-4i & -8i & -5+9i & -9 \\ 1 & -9-6i & -9 & -9+9i \\ 8+4i & 4 & 9+8i & -8 \\ -9 & -3-2i & 7 & 9+4i \end{bmatrix}$ **12.** $\begin{bmatrix} -5 & 5 & -7 & 1 \\ -1 & 3 & 8 & 3 \\ 3 & -9 & -6 & 3 \\ 9 & -9 & -5 & 6 \end{bmatrix}$

In the scaled power method, let μ_k be the kth approximation to λ_1, the eigenvalue of largest magnitude. For each of the matrices designated in Exercises 13 and 14, show that for k sufficiently large, $|\mu_{k+1} - \lambda_1|/|\mu_k - \lambda_1| \approx |\frac{\lambda_2}{\lambda_1}| < 1$, verifying the linear convergence of the method. *Caution:* If $\mathbf{x}_0$ is a linear combination of the eigenvectors corresponding to λ_1 and λ_2, then the limiting ratio turns out to be $|\frac{\lambda_3}{\lambda_1}|$. In Exercise 13, $\mathbf{x}_0 = [1, 1, 1]^T$, for example, will make this happen.)

13. Use matrix A_1.   **14.** Use matrix A_2.

An alternate implementation of the scaled power method begins with the unit vector $\mathbf{x}_0$, computes $\mathbf{v}_k = A\mathbf{x}_k$ and $\mu_k = \mathbf{x}_k \cdot \mathbf{v}_k$ as an approximation to λ_1, and then sets $\mathbf{x}_{k+1} = \mathbf{v}_k/\|\mathbf{v}_k\|_2$, the normalized version of $\mathbf{y}_k$. Apply this method to each matrix designated in Exercises 15 and 16, and show the convergence is still linear, with $|\mu_{k+1} - \lambda_1|/|\mu_k - \lambda_1| \overset{\circ}{=} |\frac{\lambda_2}{\lambda_1}| < 1$ holding for k large enough.

15. Use matrix A_1.   **16.** Use matrix A_2.

For A, a symmetric matrix, [16] shows the method in Exercises 15 and 16 satisfies $|\mu_{k+1} - \lambda_1|/|\mu_k - \lambda_1| \overset{\circ}{=} |\frac{\lambda_2}{\lambda_1}|^2 < 1$, so that the convergence, still linear, is more rapid than in the nonsymmetric case. For the given symmetric matrices in Exercises 17 and 18, apply the algorithm of Exercises 15 and 16 and verify the rate of convergence is as claimed.

17. Use the matrix in Exercise B30, Section 32.2.

18. Use the matrix in Exercise B33, Section 32.2.

19. Suppose the symmetric $n \times n$ matrix A has eigenpairs $(\lambda_k, \mathbf{v}_k), k = 1, \ldots, n$, with λ_1 as the eigenvalue of largest magnitude. Show that $B = A - \frac{\lambda_1}{\|\mathbf{v}_1\|^2}\mathbf{v}_1\mathbf{v}_1^T$ has eigenpairs $(0, \mathbf{v}_1), (\lambda_k, \mathbf{v}_k), k = 2, \ldots, n$. The matrix B is called the *Hotelling Deflation* of A.

For each of the matrices in Exercises 20 and 21, use Hotelling Deflation to find λ_2, the eigenvalue of second-largest magnitude. Unfortunately, roundoff error vitiates this approach to finding *all* eigenvalues of a matrix.

20. Use the matrix in Exercise 17.   **21.** Use the matrix in Exercise 18.

45.2 Householder Reflections

Introduction

If $r = \|\mathbf{x}\|_2 = \sqrt{\sum_{i=k}^{n} x_i^2}$ and signum $x = \pm 1, 0$, accordingly as x is positive, negative, or zero, then a *Householder reflection matrix* is an orthogonal matrix H that maps the vector

x to the vector **y** satisfying the conditions

$$y_i = x_i \quad i = 1, \dots, k-1 \qquad y_k = -r \operatorname{signum} x_k \qquad y_i = 0 \quad i = k+1, \dots, n$$

Thus, **y** agrees with **x** *above* the kth position, is zero *below* the kth position, and has $y_k = -r \operatorname{signum} x_k$ *at* the kth position. The vector **y** is displayed on the right in (45.10).

At the end of this section we will prove that if the vectors **u** and **y** are as defined by (45.10) and the matrix H by $H = I - 2\mathbf{u}\mathbf{u}^T$, then $H\mathbf{x} = \mathbf{y}$. We will also show that H is orthogonal, with $\det(H) = -1$, so it represents not just a pure rotation but a reflection (see Section 32.2). In fact, we will show the vector **y** is the reflection of **x** across the subspace orthogonal to the vector **u**.

$$\mathbf{u} = \frac{1}{\sqrt{2r(|x_k|+r)}} \begin{bmatrix} 0 \\ \vdots \\ 0 \\ \operatorname{signum}(x_k)(|x_k|+r) \\ x_{k+1} \\ \vdots \\ x_n \end{bmatrix} \qquad \mathbf{y} = \begin{bmatrix} x_1 \\ \vdots \\ x_{k-1} \\ -r \operatorname{signum} x_k \\ 0 \\ \vdots \\ 0 \end{bmatrix} \qquad (45.10)$$

In practice, Householder reflections are used to "zero out," or annihilate, entries of the vector **x**, creating zeros *below* the kth component. We will demonstrate two specific uses for these transformations. First, they provide an alternate path to the QR decomposition. In Section 34.3 we obtained the QR decomposition of a matrix A using the Gram–Schmidt process for finding Q. In Section 45.3 we will use Householder reflections to find R in the QR decomposition.

Second, Householder reflections can be used to precondition a matrix prior to application of the QR algorithm for finding eigenvalues. The goal of the preconditioning is the upper Hessenberg form of the matrix. Thus, in Section 45.4 we will see how to use Householder reflections to generate a similarity transform of A into upper Hessenberg form. (Similar matrices have the same eigenvalues.)

Articulation of the distinction between obtaining the QR decomposition by the Gram–Schmidt process or by Householder transformations follows from remarks in [89].

EXAMPLE 45.2   For the vector $\mathbf{x} = [3, -1, 6, -9, 6]^T$, define the eight parameters in Table 45.3.

k	1	2	3	4
σ_k	$\dfrac{\sqrt{2}\sqrt{3\sqrt{163}+163}}{308}$	$\dfrac{\sqrt{\sqrt{154}+154}}{\sqrt{308}}$	$\dfrac{\sqrt{6\sqrt{17}+51}}{\sqrt{102}}$	$\dfrac{\sqrt{3\sqrt{13}+13}}{\sqrt{26}}$
μ_k	$1 - \dfrac{3}{\sqrt{163}}$	$\sqrt{54}-1$	$\sqrt{17}-2$	$\sqrt{13}-3$

TABLE 45.3   The eight parameters defining the vectors $\mathbf{u}_k, k = 1, \dots, 4$, in Example 45.2

In terms of the parameters in Table 45.3, the vectors $\mathbf{u}_k, k = 1, \dots, 4$, for the Householder algorithm are given in (45.11). (See the software accompanying the text for a procedure that returns **u** when given a vector **x** and an integer k representing the component in

x below which annihilation is to occur.)

$$
\mathbf{u}_1 = \sigma_1 \begin{bmatrix} \frac{154}{\sqrt{163}} \\ -\mu_1 \\ 6\mu_1 \\ -9\mu_1 \\ 6\mu_1 \end{bmatrix}
\qquad
\mathbf{u}_2 = \sigma_2 \begin{bmatrix} 0 \\ -1 \\ \frac{2\mu_2}{51} \\ -\frac{2}{17} \\ \frac{2\mu_2}{51} \end{bmatrix}
\qquad
\mathbf{u}_3 = \sigma_3 \begin{bmatrix} 0 \\ 0 \\ 1 \\ -\frac{3\mu_3}{13} \\ \frac{2\mu_3}{13} \end{bmatrix}
\qquad
\mathbf{u}_4 = \sigma_4 \begin{bmatrix} 0 \\ 0 \\ 0 \\ -1 \\ \frac{\mu_4}{2} \end{bmatrix}
$$

$$(45.11)$$

The Householder reflection matrices $H_k = I - 2\mathbf{u}_k \mathbf{u}_k^{\mathrm{T}}, k = 1, \ldots, 4$, are given in (45.12)–(45.15). The reader can verify that $\det H_k = -1$ and that $H_k^{\mathrm{T}} = H_k^{-1}$ or, equivalently, that $H_k^{\mathrm{T}} H_k = I$, for $k = 1, \ldots, 4$. (These calculations appear in the software accompanying the text.)

$$
H_1 = \begin{bmatrix}
-3h_1 & h_1 & -6h_1 & 9h_1 & -6h_1 \\[2mm]
h_1 & \dfrac{3(51+h_1)}{154} & \dfrac{3(1-3h_1)}{77} & \dfrac{9(3h_1-1)}{154} & \dfrac{3(1-3h_1)}{77} \\[2mm]
-6h_1 & \dfrac{3(1-3h_1)}{77} & \dfrac{54h_1+59}{77} & \dfrac{27(1-3h_1)}{77} & \dfrac{18(3h_1-1)}{77} \\[2mm]
9h_1 & \dfrac{9(3h_1-1)}{154} & \dfrac{27(1-3h_1)}{77} & \dfrac{343h_1+73}{154} & \dfrac{27(1-3h_1)}{77} \\[2mm]
-6h_1 & \dfrac{3(1-3h_1)}{77} & \dfrac{18(3h_1-1)}{77} & \dfrac{27(1-3h_1)}{77} & \dfrac{54h_1+59}{77}
\end{bmatrix}
\qquad h_1 = \dfrac{1}{\sqrt{163}}
$$

$$(45.12)$$

$$
H_2 = \begin{bmatrix}
1 & 0 & 0 & 0 & 0 \\[2mm]
0 & -h_2 & 6h_2 & -9h_2 & 6h_2 \\[2mm]
0 & 6h_2 & \dfrac{4h_2+13}{17} & \dfrac{6(1-h_2)}{17} & \dfrac{4h_2-4}{17} \\[2mm]
0 & -9h_2 & \dfrac{6(1-h_2)}{17} & \dfrac{9h_2+8}{17} & \dfrac{6(1-h_2)}{17} \\[2mm]
0 & 6h_2 & \dfrac{4h_2-4}{17} & \dfrac{6(1-h_2)}{17} & \dfrac{4h_2+13}{17}
\end{bmatrix}
\qquad h_2 = \dfrac{1}{\sqrt{154}}
\qquad (45.13)
$$

$$
H_3 = \begin{bmatrix}
1 & 0 & 0 & 0 & 0 \\[2mm]
0 & 1 & 0 & 0 & 0 \\[2mm]
0 & 0 & -2h_3 & 3h_3 & -2h_3 \\[2mm]
0 & 0 & 3h_3 & \dfrac{2(9h_3+2)}{13} & \dfrac{6(1-2h_3)}{13} \\[2mm]
0 & 0 & -2h_3 & \dfrac{6(1-2h_3)}{13} & \dfrac{8h_3+9}{13}
\end{bmatrix}
\qquad h_3 = \dfrac{1}{\sqrt{17}}
\qquad (45.14)
$$

$$H_4 = \begin{bmatrix} 1 & 0 & 0 & 0 & 0 \\ 0 & 1 & 0 & 0 & 0 \\ 0 & 0 & 1 & 0 & 0 \\ 0 & 0 & 0 & -3h_4 & 2h_4 \\ 0 & 0 & 0 & 2h_4 & 3h_4 \end{bmatrix} \qquad h_4 = \frac{1}{\sqrt{13}} \qquad \text{(45.15)}$$

The vectors $\mathbf{y}_k$ obtained as the products $H_k\mathbf{x} = \mathbf{y}_k$, $k = 1, \ldots, 4$, are given in (45.16). Each Householder matrix H_k has been constructed to "zero out" the components in $\mathbf{x}$ *below* the kth. Thus, we see that $H_1\mathbf{x} = \mathbf{y}_1$, in which all components below the first are zero. We then have $H_2\mathbf{x} = \mathbf{y}_2$, in which all components of $\mathbf{x}$ below the second are zero. In $H_3\mathbf{x} = \mathbf{y}_3$, we see that all components of $\mathbf{x}$ below the third have been annihilated. Finally, in $H_4\mathbf{x} = \mathbf{y}_4$, we have all components below the fourth annihilated. Since there are no components of $\mathbf{x}$ below the fifth, there are no other Householder reflections we can inflict on this particular vector $\mathbf{x}$.

$$\mathbf{y}_1 = \begin{bmatrix} -\sqrt{163} \\ 0 \\ 0 \\ 0 \\ 0 \end{bmatrix} \quad \mathbf{y}_2 = \begin{bmatrix} 3 \\ \sqrt{154} \\ 0 \\ 0 \\ 0 \end{bmatrix} \quad \mathbf{y}_3 = \begin{bmatrix} 3 \\ -1 \\ -3\sqrt{17} \\ 0 \\ 0 \end{bmatrix} \quad \mathbf{y}_4 = \begin{bmatrix} 3 \\ -1 \\ 6 \\ -3\sqrt{13} \\ 0 \end{bmatrix} \quad \text{(45.16)}$$

In the remaining portion of the section, we illustrate the geometry that connects the vectors $\mathbf{x}$, $\mathbf{u}$, and $H\mathbf{x}$.

EXAMPLE 45.3   In this example, we construct a Householder reflection of the specific vector $\mathbf{x} = [2, 1]^{\mathsf{T}}$. In particular, we take $k = 1$ so that we "zero out" just the second component.

The general 2×2 matrix H is given by

$$H = \begin{bmatrix} h_1 & h_2 \\ h_3 & h_4 \end{bmatrix}$$

For H to be an orthogonal matrix, its transpose must be its inverse so that $H^{\mathsf{T}}H = I$. This generates the three conditions

$$h_1^2 + h_3^2 = 1 \qquad h_2^2 + h_4^2 = 1 \qquad h_1h_2 + h_3h_4 = 0$$

The requirement that H represent a reflection gives a fourth condition,

$$\det H = h_1h_4 - h_2h_3 = -1$$

There are two possible solutions to these four equations, leading to the matrices

$$H_1 = \begin{bmatrix} -h_4 & \sqrt{1-h_4^2} \\ \sqrt{1-h_4^2} & h_4 \end{bmatrix} \qquad H_2 = \begin{bmatrix} -h_4 & -\sqrt{1-h_4^2} \\ -\sqrt{1-h_4^2} & h_4 \end{bmatrix}$$

In agreement with (45.10), the parameter h_4 is determined by the condition that $H_j\mathbf{x} = \mathbf{y} = [-\sqrt{5}, 0]^{\mathsf{T}}$, $j = 1$ or 2. Using H_2 and equating corresponding components we have the equations

$$-\sqrt{1-h_4^2} - 2h_4 = -\sqrt{5} \quad \text{and} \quad -2\sqrt{1-h_4^2} + h_4 = 0$$

Generally, two equations in a single unknown might not have a solution if the two equations impose independent conditions on the unknown. However, there *is* a solution, namely, $h_4 = \frac{2}{\sqrt{5}}$. Surprisingly, the conditions imposed by $H_1\mathbf{x} = \mathbf{y}$ are not compatible and the equations

$$\sqrt{1 - h_4^2} - 2h_4 = -\sqrt{5} \quad \text{and} \quad 2\sqrt{1 - h_4^2} + h_4 = 0$$

have no solution.

Thus, the matrix H of the desired Householder reflection is given in (45.17), and we leave it to the reader to verify that $\det H = -1$, that $H^{\mathrm{T}} = H^{-1}$, and that $H\mathbf{x} = \mathbf{y} = [\sqrt{5}, 0]^{\mathrm{T}}$.

$$H = -\frac{1}{\sqrt{5}} \begin{bmatrix} 2 & 1 \\ 1 & -2 \end{bmatrix} \tag{45.17}$$

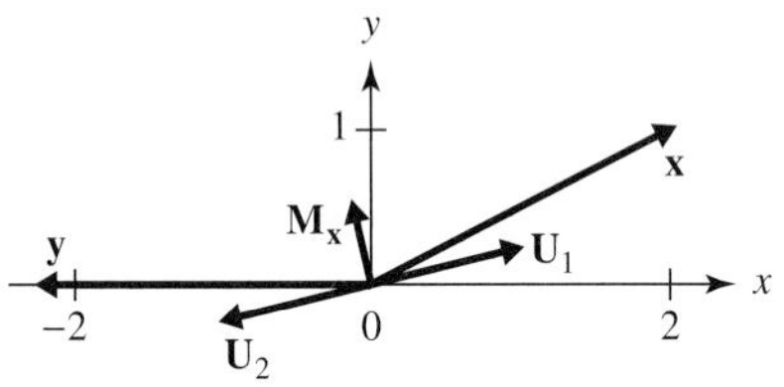

In Figure 45.2, the vectors $\mathbf{x}$ and $\mathbf{y}$ are drawn. A vector along the bisector of the angle between $\mathbf{x}$ and $\mathbf{y}$ is given by

$$\mathbf{M_x} = \tfrac{1}{2}(\mathbf{x} + \mathbf{y}) = \tfrac{1}{2}[2 - \sqrt{5}, 1]^{\mathrm{T}}$$

FIGURE 45.2   The vectors $\mathbf{x}, \mathbf{y}, \mathbf{M_x}, \mathbf{U}_1$ and $\mathbf{U}_2$ in Example 45.3

Unit vectors perpendicular to $\mathbf{M_x}$, namely, the vectors $\mathbf{U}_1$ and $\mathbf{U}_2$, are found by interchanging the components of $\mathbf{M_x}$, negating one component, and normalizing. Thus, $\mathbf{U}_1$ and $\mathbf{U}_2$ will point in opposite directions, and we have

$$\mathbf{U}_1 = \sqrt{\frac{5 - 2\sqrt{5}}{10}} \begin{bmatrix} \sqrt{5} + 2 \\ 1 \end{bmatrix} \qquad \mathbf{U}_2 = -\mathbf{U}_1$$

The vector component of $\mathbf{x}$ along $\mathbf{M_x}$ is given by the projection

$$\frac{\mathbf{x} \cdot \mathbf{M_x}}{\mathbf{M_x} \cdot \mathbf{M_x}} \mathbf{M_x} = \tfrac{1}{2}[2 - \sqrt{5}, 1]^{\mathrm{T}} = \mathbf{M_x}$$

The vector $\mathbf{y}$ is the sum of $\mathbf{M_x}$ and $\mathbf{U}_1$, while the vector $\mathbf{x}$ is the sum of $\mathbf{M_x}$ and $\mathbf{U}_2$.

The vector components of $\mathbf{x}$ along and perpendicular to $\mathbf{U}_1$ are, respectively,

$$\mathbf{x}_{\mathbf{U}_1} = (\mathbf{x} \cdot \mathbf{U}_1)\mathbf{U}_1 = \tfrac{1}{2}[2 + \sqrt{5}, 1]^{\mathrm{T}} \quad \text{and} \quad \mathbf{x}_{\perp \mathbf{U}_1} = \mathbf{x} - \mathbf{x}_{\mathbf{U}_1} = \tfrac{1}{2}[2 - \sqrt{5}, 1]^{\mathrm{T}} = \mathbf{M_x}$$

The component of $\mathbf{x}$ perpendicular to $\mathbf{U}_1$ is just $\mathbf{M_x}$. Hence, from $\mathbf{M_x} = \tfrac{1}{2}(\mathbf{x} + \mathbf{y})$, we have $\mathbf{y} = 2\mathbf{M_x} - \mathbf{x}$. Since $\mathbf{x}$ is the sum of its components along and perpendicular to $\mathbf{U}_1$, that is, since $\mathbf{x} = \mathbf{x}_{\mathbf{U}_1} + \mathbf{x}_{\perp \mathbf{U}_1}$, we have for $\mathbf{y}$

$$\mathbf{y} = 2\mathbf{M}_x - x = 2\mathbf{x}_{\perp \mathbf{U}_1} - (\mathbf{x}_{\mathbf{U}_1} + \mathbf{x}_{\perp \mathbf{U}_1}) = \mathbf{x}_{\perp \mathbf{U}_1} - \mathbf{x}_{\mathbf{U}_1}$$

Thus, $\mathbf{y}$ is the difference between the components of $\mathbf{x}$ perpendicular to and along $\mathbf{U}_1$. ❖

EXAMPLE 45.4   We next give a three-dimensional example of a reflection of the vector $\mathbf{x} = [2, 3, 4]^{\mathrm{T}}$ across the plane whose unit normal is $\mathbf{u} = [u_1, u_2, u_3]^{\mathrm{T}}$. For the reflection $\mathbf{y} = \mathbf{x}_{\perp \mathbf{u}} - \mathbf{x}_{\mathbf{u}}$, consider the case $k = 1$ so that we "zero out" the second and third components of $\mathbf{x}$. Using $\|\mathbf{u}\|^2 = 1$, the vectors $\mathbf{x}_{\mathbf{u}} = (\mathbf{x} \cdot \mathbf{u})\mathbf{u}$, $\mathbf{x}_{\perp \mathbf{u}} = \mathbf{x} - \mathbf{x}_{\mathbf{u}}$, and $\mathbf{y}$ become, respectively,

$$\begin{bmatrix} (2u_1 + 3u_2 + 4u_3)u_1 \\ (2u_1 + 3u_2 + 4u_3)u_2 \\ (2u_1 + 3u_2 + 4u_3)u_3 \end{bmatrix} \quad \begin{bmatrix} 2 - (2u_1 + 3u_2 + 4u_3)u_1 \\ 3 - (2u_1 + 3u_2 + 4u_3)u_2 \\ 4 - (2u_1 + 3u_2 + 4u_3)u_3 \end{bmatrix} \quad \begin{bmatrix} 2 - 2(2u_1 + 3u_2 + 4u_3)u_1 \\ 3 - 2(2u_1 + 3u_2 + 4u_3)u_2 \\ 4 - 2(2u_1 + 3u_2 + 4u_3)u_3 \end{bmatrix}$$

The reflection $\mathbf{y}$ must have zeros for its second and third components, and its length must be $\|\mathbf{x}\| = \sqrt{29}$. Thus, equating $\mathbf{y}$ with $[-\sqrt{29}, 0, 0]^\mathsf{T}$ gives three equations in u_k, $k = 1, 2, 3$, the solution of which leads to two vectors $\mathbf{u}$, namely,

$$\mathbf{U}_1 = \frac{\sqrt{1682 - 116\sqrt{29}}}{290} \begin{bmatrix} 2 + \sqrt{29} \\ 3 \\ 4 \end{bmatrix} \qquad \mathbf{U}_2 = -\mathbf{U}_1$$

Both unit vectors, simply pointing in opposite directions, generate the same reflection $\mathbf{y}$.

The projection of $\mathbf{x}$ onto the plane for which $\mathbf{U}_1$ (or $\mathbf{U}_2$) is the unit normal is

$$\mathbf{x} - \mathbf{x_u} = \mathbf{x_{\perp u}} = \frac{1}{2} \begin{bmatrix} 2 - \sqrt{29} \\ 3 \\ 4 \end{bmatrix} = \tfrac{1}{2}(\mathbf{x} + \mathbf{y}) = \mathbf{M}$$

The normal to the plane *containing* $\mathbf{x}$ and $\mathbf{y}$ can be found as the cross-product $\mathbf{N} = \mathbf{x} \times \mathbf{i} = [0, 4, -3]^\mathsf{T}$. The vector $\mathbf{N}$ must lie in the plane across which $\mathbf{x}$ is reflected, that is, the plane for which $\mathbf{U}_1$ (or $\mathbf{U}_2$) is the unit normal. This is confirmed by showing that $\mathbf{N}$ is orthogonal to either of the unit vectors and also to the vector $\mathbf{M}$. In fact, the orthogonality of $\mathbf{N}$ and $\mathbf{M}$ shows that $\mathbf{M}$ lies in the plane of reflection. Simple calculation shows $\mathbf{N} \cdot \mathbf{U}_1 = \mathbf{N} \cdot \mathbf{U}_2 = \mathbf{N} \cdot \mathbf{M} = 0$.

Figure 45.3 displays the vectors $\mathbf{x}$, $\mathbf{y}$, $\mathbf{M}$ and the two unit vectors $\mathbf{U}_1$ and $\mathbf{U}_2$. ❖

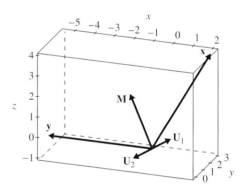

FIGURE 45.3   The vectors $\mathbf{x}$, $\mathbf{y}$, $\mathbf{M}$, $\mathbf{U}_1$ and $\mathbf{U}_2$ in Example 45.4

Obtaining H in Terms of $\mathbf{u}$

We have already seen the equation $\mathbf{y} = \mathbf{x_{\perp u}} - \mathbf{x_u}$ relating the vector $\mathbf{y}$ to the vectors $\mathbf{x}$ and $\mathbf{u}$. Since $\mathbf{x_{\perp u}} = \mathbf{x} - \mathbf{x_u}$, we have

$$\mathbf{y} = (\mathbf{x} - \mathbf{x_u}) - \mathbf{x_u} = \mathbf{x} - 2\mathbf{x_u}$$

Since $\mathbf{u}$ is a unit vector, the projection $\mathbf{x_u}$ is given by

$$\mathbf{x_u} = (\mathbf{x} \cdot \mathbf{u})\mathbf{u} = (\mathbf{x}^\mathsf{T}\mathbf{u})\mathbf{u} = \mathbf{u}(\mathbf{x}^\mathsf{T}\mathbf{u}) = \mathbf{u}\left(\mathbf{x}^\mathsf{T}\mathbf{u}\right)^\mathsf{T} = \mathbf{u}\mathbf{u}^\mathsf{T}\mathbf{x}$$

where we can go from the second to the third equality because $(\mathbf{x}^\mathsf{T}\mathbf{u})$ is a scalar and the transpose of a scalar is itself. Thus, the reflection becomes

$$\mathbf{y} = \mathbf{x} - 2\mathbf{u}\mathbf{u}^\mathsf{T}\mathbf{x} = (I - 2\mathbf{u}\mathbf{u}^\mathsf{T})\mathbf{x}$$

so the matrix H must be given by $H = I - 2\mathbf{u}\mathbf{u}^\mathsf{T}$.

We next show that if $\mathbf{u}$ is a unit vector, then $H = I - 2\mathbf{u}\mathbf{u}^\mathsf{T}$ is always an orthogonal matrix. Indeed, we need to show $H^\mathsf{T}H = I$, which we can do by remembering that the transpose operator distributes across sums. Thus, we write

$$H^\mathsf{T}H = \left(I - 2\mathbf{u}\mathbf{u}^\mathsf{T}\right)^\mathsf{T}\left(I - 2\mathbf{u}\mathbf{u}^\mathsf{T}\right) = (I - 2\mathbf{u}\mathbf{u}^\mathsf{T})(I - 2\mathbf{u}\mathbf{u}^\mathsf{T})$$

$$= I - 2\mathbf{u}\mathbf{u}^\mathsf{T} - 2\mathbf{u}\mathbf{u}^\mathsf{T} + 4\mathbf{u}\mathbf{u}^\mathsf{T}\mathbf{u}\mathbf{u}^\mathsf{T} = I - 4\mathbf{u}\mathbf{u}^\mathsf{T} + 4\mathbf{u}\mathbf{u}^\mathsf{T} = I$$

The product $\mathbf{u}\mathbf{u}^\mathsf{T}$ is a symmetric matrix since $\left(\mathbf{u}\mathbf{u}^\mathsf{T}\right)^\mathsf{T} = \left(\mathbf{u}^\mathsf{T}\right)^\mathsf{T}\mathbf{u}^\mathsf{T} = \mathbf{u}\mathbf{u}^\mathsf{T}$. Hence, we have the second equality. The fourth equality is true since $\mathbf{u}^\mathsf{T}\mathbf{u} = \mathbf{u} \cdot \mathbf{u} = 1$.

The proof that $H = I - 2\mathbf{u}\mathbf{u}^\mathsf{T}$ has determinant -1 is more subtle. The argument we present is from [64], in which a matrix A is constructed with $\mathbf{u}$ as the first column and the remaining columns taken as independent and orthogonal to $\mathbf{u}$ so $\det(A) \neq 0$. This can be done ultimately on the strength of the Gram–Schmidt orthogonalization process. Then, for such a matrix A, we establish

$$\det(HA) = \det H \det A = -\det A$$

The first equality is always true for the product of matrices. The second equality then implies that $\det H = -1$. To verify that vital second equality, consider the product

$$HA = (I - 2\mathbf{u}\mathbf{u}^\mathsf{T})A = A - 2\mathbf{u}\mathbf{u}^\mathsf{T}A$$

The product $\mathbf{u}^\mathsf{T}A$ is the row vector $[1, 0, \dots, 0]$, so $\mathbf{u}(\mathbf{u}^\mathsf{T}A) = [\mathbf{u}, \mathbf{0}, \dots, \mathbf{0}]$, and HA is the matrix A except that all the elements in the first column have been negated. Hence, $\det HA = -\det A$ and $\det H = -1$.

Thus, we have shown that for a unit vector $\mathbf{u}$, the matrix $H = I - 2\mathbf{u}\mathbf{u}^\mathsf{T}$ is both orthogonal and a reflection. Next, we will show how to determine $\mathbf{u}$ so that $H\mathbf{x} = \mathbf{y}$ for $k = 1, \dots, n - 1$.

Solving $H\mathbf{x} = \mathbf{y}$ for $\mathbf{u}$

Our construction of a unit vector $\mathbf{u}$ satisfying $(I - 2\mathbf{u}\mathbf{u}^\mathsf{T})\mathbf{x} = \mathbf{y}$ is motivated by [60] and is guided by Figure 45.4. The unit vector $\mathbf{u}$ is taken parallel to the vector $\mathbf{x} - \mathbf{y}$. The vector $\mathbf{x}$ is resolved into $\mathbf{x}_\mathbf{u}$ and $\mathbf{x}_{\perp\mathbf{u}}$, the vector components along and perpendicular to $\mathbf{u}$. The vector $\mathbf{x}$ is the sum of $\mathbf{x}_\mathbf{u}$ and $\mathbf{x}_{\perp\mathbf{u}}$, whereas the vector y is the sum of $\mathbf{x}_{\perp\mathbf{u}}$ and $-\mathbf{x}_\mathbf{u}$.

Analytically, we have $\mathbf{y} = \mathbf{x}_{\perp\mathbf{u}} - \mathbf{x}_\mathbf{u}$, a basis for the following fundamental claim. If $\|\mathbf{x}\| = \|\mathbf{y}\|$, taking $\mathbf{u} = \frac{\mathbf{x}-\mathbf{y}}{\|\mathbf{x}-\mathbf{y}\|}$ and defining the matrix $H = I - 2\mathbf{u}\mathbf{u}^\mathsf{T}$ means that $H\mathbf{x} = \mathbf{y}$. We offer three ways to see this.

First, since $\mathbf{u}$ is parallel to $\mathbf{x} - \mathbf{y}$, we see from Figure 45.4 that $\mathbf{y}$ is the sum of $\mathbf{x}_{\perp\mathbf{u}}$ and $-\mathbf{x}_\mathbf{u}$, and by earlier calculations, we know this leads to $H = I - 2\mathbf{u}\mathbf{u}^\mathsf{T}$.

Second, we provide a numeric example where $\|\mathbf{x}\| = \|\mathbf{y}\| = 1$ and $\mathbf{u} = \mathbf{x} - \mathbf{y}$.

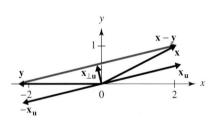

FIGURE 45.4 The vectors $\mathbf{x}$, $\mathbf{y}$, $\mathbf{x} - \mathbf{y}$, $\mathbf{x}_{\perp\mathbf{u}}$, $\mathbf{x}_\mathbf{u}$ and $-\mathbf{x}_\mathbf{u}$, where $\mathbf{u}$ is parallel to $\mathbf{x} - \mathbf{y}$

$$\mathbf{x} = \begin{bmatrix} 0.6119097028 \\ 0.6580915672 \\ 0.4387277114 \end{bmatrix} \qquad \mathbf{y} = \begin{bmatrix} -0.6592514624 \\ 0.4770898740 \\ -0.5811822103 \end{bmatrix} \qquad \mathbf{u} = \begin{bmatrix} 0.7752094520 \\ 0.1103827172 \\ 0.6219854991 \end{bmatrix}$$

The matrix $H = I - 2\mathbf{u}\mathbf{u}^\mathsf{T}$ is then given by (45.18), and a calculation shows $H\mathbf{x} = \mathbf{y}$.

$$H = \begin{bmatrix} -0.201899389 & -0.1711394514 & -0.9643380758 \\ -0.171139451 & 0.9756313115 & -0.1373128989 \\ -0.964338076 & -0.1373128989 & 0.2262680778 \end{bmatrix} \tag{45.18}$$

Third, we show directly that if $\|\mathbf{x}\| = \|\mathbf{y}\|$, $\mathbf{u} = \frac{\mathbf{x}-\mathbf{y}}{\|\mathbf{x}-\mathbf{y}\|}$, $H = I - 2\mathbf{u}\mathbf{u}^{\mathrm{T}}$, then $H\mathbf{x} = \mathbf{y}$. To do this, we need

$$\|\mathbf{x} - \mathbf{y}\|^2 = (\mathbf{x} - \mathbf{y}) \cdot (\mathbf{x} - \mathbf{y}) = \mathbf{x} \cdot \mathbf{x} - \mathbf{x} \cdot \mathbf{y} - \mathbf{y} \cdot \mathbf{x} + \mathbf{y} \cdot \mathbf{y}$$
$$= \|\mathbf{x}\|^2 + \|\mathbf{y}\|^2 - 2\mathbf{x} \cdot \mathbf{y} = 2\|\mathbf{x}\|^2 - 2\mathbf{x} \cdot \mathbf{y}$$

where the last equality is true in light of the assumption that $\|\mathbf{x}\| = \|\mathbf{y}\|$. Then, we have

$$H\mathbf{x} = \left(I - \frac{2(\mathbf{x} - \mathbf{y})(\mathbf{x} - \mathbf{y})^{\mathrm{T}}}{\|\mathbf{x} - \mathbf{y}\|^2} \right) \mathbf{x} = \frac{\mathbf{x}\|\mathbf{x} - \mathbf{y}\|^2 - 2(\mathbf{x} - \mathbf{y})(\mathbf{x} - \mathbf{y})^{\mathrm{T}}\mathbf{x}}{\|\mathbf{x} - \mathbf{y}\|^2}$$

for which, expanding the numerator and using $\mathbf{a}^{\mathrm{T}}\mathbf{b} = \mathbf{a} \cdot \mathbf{b}$, gives

$$\mathbf{x}\|\mathbf{x} - \mathbf{y}\|^2 - 2(\mathbf{x}\mathbf{x}^{\mathrm{T}} - \mathbf{x}\mathbf{y}^{\mathrm{T}} - \mathbf{y}\mathbf{x}^{\mathrm{T}} + \mathbf{y}\mathbf{y}^{\mathrm{T}})\mathbf{x}$$
$$= \mathbf{x}\|\mathbf{x} - \mathbf{y}\|^2 - 2(\mathbf{x}\mathbf{x}^{\mathrm{T}}\mathbf{x} - \mathbf{x}\mathbf{y}^{\mathrm{T}}\mathbf{x} - \mathbf{y}\mathbf{x}^{\mathrm{T}}\mathbf{x} + \mathbf{y}\mathbf{y}^{\mathrm{T}}\mathbf{x})$$
$$= \mathbf{x}\|\mathbf{x} - \mathbf{y}\|^2 - 2[\mathbf{x}(\mathbf{x} \cdot \mathbf{x}) - \mathbf{x}(\mathbf{y} \cdot \mathbf{x}) - \mathbf{y}(\mathbf{x} \cdot \mathbf{x}) + \mathbf{y}(\mathbf{y} \cdot \mathbf{x})]$$
$$= \mathbf{x}\|\mathbf{x} - \mathbf{y}\|^2 - 2[\mathbf{x}\|\mathbf{x}\|^2 - \mathbf{y}\|\mathbf{x}\|^2 - \mathbf{x}(\mathbf{x} \cdot \mathbf{y}) + \mathbf{y}(\mathbf{x} \cdot \mathbf{y})]$$
$$= \mathbf{x}[2\|\mathbf{x}\|^2 - 2(\mathbf{x} \cdot \mathbf{y})] - 2[\mathbf{x}\|\mathbf{x}\|^2 - \mathbf{y}\|\mathbf{x}\|^2 - \mathbf{x}(\mathbf{x} \cdot \mathbf{y}) + \mathbf{y}(\mathbf{x} \cdot \mathbf{y})]$$
$$= 2\mathbf{x}\|\mathbf{x}\|^2 - 2\mathbf{x}(\mathbf{x} \cdot \mathbf{y}) - 2\mathbf{x}\|\mathbf{x}\|^2 + 2\mathbf{y}\|\mathbf{x}\|^2 + 2\mathbf{x}(\mathbf{x} \cdot \mathbf{y}) - 2\mathbf{y}(\mathbf{x} \cdot \mathbf{y})$$
$$= \mathbf{y}[2\|\mathbf{x}\|^2 - 2(\mathbf{x} \cdot \mathbf{y})]$$
$$= \mathbf{y}\|\mathbf{x} - \mathbf{y}\|^2$$

and finally, $H\mathbf{x} = \frac{\mathbf{y}\|\mathbf{x}-\mathbf{y}\|^2}{\|\mathbf{x}-\mathbf{y}\|^2} = \mathbf{y}$.

All that remains is to describe $\mathbf{u}$ in terms of the vector $\mathbf{y}$, which is determined from $\mathbf{x}$ by "zeroing out" entries in $\mathbf{x}$. Start with the hypothesis $\|\mathbf{x}\| = \|\mathbf{y}\|$, stated as $\|\mathbf{x}\|^2 = \|\mathbf{y}\|^2$, to get

$$\sum_{i=1}^{n} x_i^2 = \sum_{i=1}^{k-1} x_i^2 + y_k^2 + \sum_{i=k+1}^{n} 0^2 \quad \text{and} \quad y_k^2 = \sum_{i=1}^{n} x_i^2 - \sum_{i=1}^{k-1} x_i^2 = \sum_{i=k}^{n} x_i^2 = r^2$$

Then define $r = \sqrt{\sum_{i=k}^{n} x_i^2}$ and $y_k = -\operatorname{signum}(x_k)r$, where $\operatorname{signum} x = \pm 1, 0$, accordingly as x is positive, negative, or zero. Thus, in the vector $\mathbf{x} - \mathbf{y}$, the kth component will be

$$x_k - y_k = x_k + \operatorname{signum}(x_k)r = \begin{cases} -(|x_k| + r) & x_k < 0 \\ |x_k| + r & x_k > 0 \end{cases} = \operatorname{signum}(x_k)(|x_k| + r)$$

where the difference has been expressed in a form that precludes subtractive cancellation generating roundoff error in this component. Next, obtain

$$\|\mathbf{x} - \mathbf{y}\|^2 = \sum_{i=1}^{k-1} 0^2 + (x_k - y_k)^2 + \sum_{i=k+1}^{n} x_i^2 = [\operatorname{signum}(x_k)(|x_k| + r)]^2 + \sum_{i=k+1}^{n} x_i^2$$
$$= x_k^2 + 2r|x_k| + r^2 + \sum_{i=k+1}^{n} x_i^2 = x_k^2 + 2r|x_k| + \sum_{i=k}^{n} x_i^2 + \sum_{i=k+1}^{n} x_i^2$$
$$= 2r|x_k| + \sum_{i=k}^{n} x_i^2 + \sum_{i=k}^{n} x_i^2 = 2r|x_k| + 2r^2 = 2r(|x_k| + r)$$

where $|x_k|^2 = x_k^2$ for x_k real. Consequently, the vectors $\mathbf{u}$ and $H\mathbf{x} = (I - 2\mathbf{u}\mathbf{u}^{\mathrm{T}})\mathbf{x} = \mathbf{y}$ are given by (45.10).

EXERCISES 45.2–Part A

A1. Let **u** be the normalized version of $2\mathbf{i} + 3\mathbf{j}$, and obtain the Householder reflection matrix $H = I - 2\mathbf{u}\mathbf{u}^{\mathrm{T}}$.

A2. Determine the equation of L_1, the line that passes through the origin and is orthogonal to **u** in Exercise A1.

A3. Sketch **u**, L_1, $\mathbf{x} = \mathbf{i} + \mathbf{j}$, and $\mathbf{y} = H\mathbf{x}$.

A4. Use (45.10) to obtain **u** for which $H = I - 2\mathbf{u}\mathbf{u}^{\mathrm{T}}$ is an annihilating Householder matrix for $\mathbf{x} = 3\mathbf{i} + 4\mathbf{j}$. Calculate $\mathbf{y} = H\mathbf{x}$.

A5. Let **x** and **y** be as in Exercise A4. Show that **u** is the normalized version of $\mathbf{x} - \mathbf{y}$.

EXERCISES 45.2–Part B

B1. Let **u** be the normalized version of $2\mathbf{i} - 3\mathbf{j} + \mathbf{k}$, and obtain the Householder reflection matrix $H = I - 2\mathbf{u}\mathbf{u}^{\mathrm{T}}$. Calculate $\mathbf{y} = H\mathbf{x}$, where $\mathbf{x} = \mathbf{i} + \mathbf{j} + \mathbf{k}$.

B2. Determine the equation of the plane that passes through the origin and is orthogonal to **u** in Exercise B1.

B3. Use (45.10) to obtain **u** for which $H = I - 2\mathbf{u}\mathbf{u}^{\mathrm{T}}$ is an annihilating Householder matrix for $\mathbf{x} = \mathbf{i} - 2\mathbf{j} + 7\mathbf{k}$, annihilating all components below the first. Calculate $\mathbf{y} = H\mathbf{x}$.

B4. Let **x** be as given in Exercise B3, and let $\mathbf{y} = -\|\mathbf{x}\|_2\mathbf{i}$. Show that normalizing $\mathbf{x} - \mathbf{y}$ yields the vector **u** obtained in Exercise B3.

In Exercises B5–8, construct Householder reflection matrices for every possible annihilation in the given vector **x**. Use (45.10) to obtain **u**, and from **u** obtain H. Verify that $H\mathbf{x}$ gives the vector **y** determined by (45.10).

B5. $\begin{bmatrix} 5 \\ 8 \\ 4 \\ -3 \end{bmatrix}$ **B6.** $\begin{bmatrix} 8 \\ -9 \\ 7 \\ 5 \end{bmatrix}$ **B7.** $\begin{bmatrix} -8 \\ -5 \\ 4 \\ 7 \\ -9 \end{bmatrix}$ **B8.** $\begin{bmatrix} 5 \\ 7 \\ -8 \\ 3 \\ 5 \end{bmatrix}$

For each vector **x** given in Exercises B9–11:

   (a) Obtain H_1, the Householder reflection matrix that maps **x** to $\mathbf{y}_1 = H_1\mathbf{x} = \alpha\mathbf{i}$ for some appropriate α.

   (b) Show $\|\mathbf{x}\| = \|\mathbf{y}_1\| = |\alpha|$.

   (c) Obtain H_2, the Householder reflection matrix that maps **x** to $\mathbf{y}_2 = H_2\mathbf{x} = x_1\mathbf{i} + \beta\mathbf{j}$ for some appropriate β.

   (d) Show $\|\mathbf{x}\| = \|\mathbf{y}_2\|$.   (e) Sketch **x**, $\mathbf{y}_1$, and $\mathbf{y}_2$.

B9. $\begin{bmatrix} -1 \\ -5 \\ 5 \end{bmatrix}$ **B10.** $\begin{bmatrix} -7 \\ 1 \\ -1 \end{bmatrix}$ **B11.** $\begin{bmatrix} -3 \\ 8 \\ 3 \end{bmatrix}$

In Exercises B12–14, vectors $\mathbf{v}_1$, $\mathbf{v}_2$, and **x** are given. For each such set of vectors:

   (a) From $\mathbf{v}_1 \times \mathbf{v}_2$, obtain **u**, a unit vector orthogonal to the plane spanned by $\mathbf{v}_1$ and $\mathbf{v}_2$.

   (b) Obtain the Householder reflection matrix $H = I - 2\mathbf{u}\mathbf{u}^{\mathrm{T}}$.

   (c) If $A = [\mathbf{v}_1, \mathbf{v}_2]$, obtain $P = A\left(A^{\mathrm{T}}A\right)^{-1}A^{\mathrm{T}}$, a matrix that projects onto the plane orthogonal to **u**.

   (d) Obtain $\mathbf{y} = H\mathbf{x}$, and show $\|\mathbf{x}\| = \|\mathbf{y}\|$.

   (e) Obtain $\mathbf{M} = \frac{1}{2}(\mathbf{x} + \mathbf{y})$ and show $\mathbf{M} = P\mathbf{x} = P\mathbf{y}$; interpret this equality geometrically.

   (f) Show $\mathbf{x} - P\mathbf{x} = -(\mathbf{y} - P\mathbf{y})$; interpret this equality geometrically.

B12. $\mathbf{v}_1 = \begin{bmatrix} 3 \\ -9 \\ -6 \end{bmatrix}$ $\mathbf{v}_2 = \begin{bmatrix} 3 \\ 9 \\ -9 \end{bmatrix}$ $\mathbf{x} = \begin{bmatrix} 1 \\ -7 \\ -1 \end{bmatrix}$

B13. $\mathbf{v}_1 = \begin{bmatrix} -1 \\ -4 \\ 2 \end{bmatrix}$ $\mathbf{v}_2 = \begin{bmatrix} 9 \\ -7 \\ -7 \end{bmatrix}$ $\mathbf{x} = \begin{bmatrix} -3 \\ -7 \\ -2 \end{bmatrix}$

B14. $\mathbf{v}_1 = \begin{bmatrix} -7 \\ 8 \\ 5 \end{bmatrix}$ $\mathbf{v}_2 = \begin{bmatrix} -3 \\ 6 \\ -8 \end{bmatrix}$ $\mathbf{x} = \begin{bmatrix} 2 \\ 9 \\ -8 \end{bmatrix}$

B15. Draw a sketch representing the geometry explored by Exercises B12–14.

For each pair of vectors $\{\tilde{\mathbf{x}}, \tilde{\mathbf{y}}\}$ given in Exercises B16–19:

   (a) Normalize each to **x** and **y**, respectively, thus ensuring $\|\mathbf{x}\| = \|\mathbf{y}\|$.

   (b) Set $\mathbf{u} = \frac{\mathbf{x}-\mathbf{y}}{\|\mathbf{x}-\mathbf{y}\|}$, and form the Householder matrix $H = I - 2\mathbf{u}\mathbf{u}^{\mathrm{T}}$.

   (c) Show that $H\mathbf{x} = \mathbf{y}$.

B16. $\tilde{\mathbf{x}} = [5, -1, -4]^{\mathrm{T}}, \tilde{\mathbf{y}} = [4, -8, -8]^{\mathrm{T}}$

B17. $\tilde{\mathbf{x}} = [7, 9, 4, 6]^{\mathrm{T}}, \tilde{\mathbf{y}} = [-3, 3, 2, 8]^{\mathrm{T}}$

B18. $\tilde{\mathbf{x}} = [-9, -4, -6, -9, -6]^{\mathrm{T}}, \tilde{\mathbf{y}} = [-9, -4, 8, 9, -7]^{\mathrm{T}}$

B19. $\tilde{\mathbf{x}} = [-1, -3, -9, -8, 9, -4]^{\mathrm{T}}, \tilde{\mathbf{y}} = [1, -9, 2, 9, -5, -2]^{\mathrm{T}}$

B20. If $\mathbf{x} = [a, b]^{\mathrm{T}}$ is a real vector and $H = \frac{1}{\sqrt{a^2+b^2}}\begin{bmatrix} a & b \\ b & -a \end{bmatrix}$:

   (a) Show that $H\mathbf{x} = [\sqrt{a^2 + b^2}, 0]^{\mathrm{T}}$.

   (b) Show that $\det H = -1$ and that $H^{\mathrm{T}} = H^{-1}$.

   (c) Determine the unit vector **u** from which H, a Householder reflection matrix, was constructed.

B21. Find a Householder matrix H so that $H\mathbf{i} = \mathbf{j}$ in the xy-plane.

For each pair of vectors **x** and **v** given in Exercises B22–26, obtain H, the Householder reflection matrix for which $\mathbf{y} = H\mathbf{x}$ is the reflection of **x** across the line whose direction is **v**. Also, obtain **y**, and sketch the vectors **x**, **v**, and **y**.

B22. $\mathbf{x} = \begin{bmatrix} 5 \\ -2 \end{bmatrix}$ $\mathbf{v} = \begin{bmatrix} 6 \\ 1 \end{bmatrix}$ **B23.** $\mathbf{x} = \begin{bmatrix} 4 \\ -6 \end{bmatrix}$ $\mathbf{v} = \begin{bmatrix} 8 \\ -9 \end{bmatrix}$

B24. $\mathbf{x} = \begin{bmatrix} 9 \\ -3 \end{bmatrix}$ $\mathbf{v} = \begin{bmatrix} 5 \\ 3 \end{bmatrix}$ **B25.** $\mathbf{x} = \begin{bmatrix} -2 \\ 7 \end{bmatrix}$ $\mathbf{v} = \begin{bmatrix} 1 \\ 3 \end{bmatrix}$

B26. $\mathbf{x} = \begin{bmatrix} 7 \\ 5 \end{bmatrix}$ $\mathbf{v} = \begin{bmatrix} 6 \\ -3 \end{bmatrix}$

For each given vector **x** given in Exercises B27–31, find a Householder reflection matrix that reflects **x** across the given plane. In each case, find the reflected vector **y**.

B27. $\mathbf{x} = [-7, 2, 3]^{\mathrm{T}}, 5x - 7y + z = 0$

B28. $\mathbf{x} = [-2, -1, -1]^{\mathrm{T}}, 3x + 8y + 3z = 0$

B29. $\mathbf{x} = [-3, 4, 1]^{\mathrm{T}}, 9x - 9y - 5z = 0$

B30. $\mathbf{x} = [-5, -8, 3]^{\mathrm{T}}, 2x + 9y - 7z = 0$

B31. $\mathbf{x} = [3, 7, 6]^{\mathrm{T}}, 2x + 9y - 8z = 0$

45.3 QR Decomposition via Householder Reflections

Introduction

Section 34.4 describes the use of the QR decomposition in the QR algorithm for computing eigenvalues of a matrix A. The QR decomposition as detailed in Section 34.3 applies the Gram–Schmidt orthogonalization process to orthonormalize the columns of A. This process yields the orthogonal factor Q (unitary, if A is complex), and from Q we obtain the upper triangular factor $R = Q^{\mathrm{T}}A$.

In this section we will see how the orthogonal Householder reflections can be used to transform A into R by successively "zeroing out" the columns of A. A sequence of reflection matrices H_k, $k = 1, \ldots, n-1$, are generated; and the product $H_n = H_{n-1}H_{n-2} \cdots H_1$ plays the role of Q^{T} since $H_n A = R$ and $A = H_n^{\mathrm{T}} R$.

As we saw in Section 45.2, Householder reflection matrices obtained as $H = I - 2\mathbf{u}\mathbf{u}^{\mathrm{T}}$ are necessarily symmetric. Hence, $A = H_1 H_2 \cdots H_{n-1} R$.

Implementation

The accompanying software suite contains an auxiliary procedure that returns H_k, the Householder reflection matrix that "zeros out," in the kth column of $A = \{a_{ij}\}$, the elements below a_{kk}, the diagonal element in column k. The basic technique is that described in Section 45.2. However, the $(n - k + 1)$-component vector $\mathbf{x}_k = [a_{kk}, a_{k+1,k}, \ldots, a_{nn}]^{\mathrm{T}}$ must be extracted from the kth column of A and normalized to form $\mathbf{u}_k$, which provides the $p \times p$ reflection matrix $h_p = I_p - 2\mathbf{u}_k\mathbf{u}_k^{\mathrm{T}}$, where $p = n - k + 1$. Then, the matrix h_p is inserted into H_k by the prescription

$$H_k = \begin{bmatrix} I_{k-1} & Z_1 \\ Z_1^{\mathrm{T}} & h_p \end{bmatrix}$$

where I_{k-1} is the $(k - 1) \times (k - 1)$ identity matrix and Z_1 is a $(k - 1) \times (n - k + 1)$ matrix of zeros.

EXAMPLE 45.5   Let A be the matrix.

$$A = \begin{bmatrix} 10 & 8 & 12 & -9 & -3 \\ -7 & -7 & 0 & 5 & -2 \\ -4 & 7 & -6 & -7 & -1 \\ 4 & 11 & -9 & -3 & 5 \\ 7 & -7 & 12 & -8 & -9 \end{bmatrix}$$

The Householder matrix H_1 is

$$H_1 = \begin{bmatrix} -0.6593804734 & 0.4615663314 & 0.2637521894 & -0.2637521894 & -0.4615663314 \\ 0.4615663314 & 0.8716126400 & -0.07336420573 & 0.07336420573 & 0.1283873600 \\ 0.2637521894 & -0.07336420573 & 0.9580775967 & 0.04192240327 & 0.07336420573 \\ -0.2637521894 & 0.07336420573 & 0.04192240327 & 0.9580775967 & -0.07336420573 \\ -0.4615663314 & 0.1283873600 & 0.07336420573 & -0.07336420573 & 0.8716126400 \end{bmatrix}$$

and $A_1 = H_1 A$ is then

$$A_1 = \begin{bmatrix} -15.16575089 & -6.330052544 & -12.66010509 & 10.87977781 & 3.626592604 \\ 0 & -3.014012526 & 6.859351679 & -0.5296758401 & -3.843225279 \\ 0 & 9.277707128 & -2.080370467 & -10.15981477 & -2.053271589 \\ 0 & 8.722292868 & -12.91962953 & 0.1598147668 & 6.053271589 \\ 0 & -10.98598747 & 5.140648321 & -2.470324160 & -7.156774721 \end{bmatrix}$$

As expected, the Householder reflection matrix H_1, when applied to A, has "zeroed out," in the first column of A, the entries below the main diagonal.

The Householder reflection matrix that annihilates, in the second column of A, all entries below the second, is

$$H_2 = \begin{bmatrix} 1 & 0 & 0 & 0 & 0 \\ 0 & -0.1764027560 & 0.5430014283 & 0.5104943948 & -0.6429828842 \\ 0 & 0.5430014283 & 0.7493625804 & -0.2356328937 & 0.2967866428 \\ 0 & 0.5104943948 & -0.2356328937 & 0.7784733793 & 0.2790193721 \\ 0 & -0.6429828842 & 0.2967866428 & 0.2790193721 & 0.6485667963 \end{bmatrix}$$

and $A_2 = H_2 A_2$ is

$$A_2 = \begin{bmatrix} -15.16575089 & -6.330052544 & -12.66010509 & 10.87977781 & 3.626592604 \\ 0 & 17.08597187 & -12.24040002 & -3.753396957 & 7.254870993 \\ 0 & 0 & 6.735651427 & -8.671816580 & -7.175906756 \\ 0 & 0 & -4.631382896 & 1.558733253 & 1.237305364 \\ 0 & 0 & -5.298645000 & -4.232303628 & -1.090921922 \end{bmatrix}$$

in which the appropriate number of zeros have appeared in the second column.

The Householder reflection matrix annihilating, in the third column of A, the elements below the main diagonal, is

$$H_3 = \begin{bmatrix} 1 & 0 & 0 & 0 & 0 \\ 0 & 1 & 0 & 0 & 0 \\ 0 & 0 & -0.6914478562 & 0.4754343079 & 0.5439320555 \\ 0 & 0 & 0.4754343079 & 0.8663643219 & -0.1528891118 \\ 0 & 0 & 0.5439320555 & -0.1528891118 & 0.8250835342 \end{bmatrix}$$

and $A_3 = H_3 A_2$ is

$$A_3 = \begin{bmatrix} -15.16575089 & -6.330052544 & -12.66010509 & 10.87977781 & 3.626592604 \\ 0 & 17.08597187 & -12.24040002 & -3.753396957 & 7.254870993 \\ 0 & 0 & -9.741372927 & 4.435098637 & 4.956635359 \\ 0 & 0 & 0 & -2.125375093 & -2.172924955 \\ 0 & 0 & 0 & -8.447196395 & -4.992477945 \end{bmatrix}$$

which has the requisite number of zeros in the third column.

Finally, the last Householder reflection matrix needed to upper-triangularize A is

$$H_4 = \begin{bmatrix} 1 & 0 & 0 & 0 & 0 \\ 0 & 1 & 0 & 0 & 0 \\ 0 & 0 & 1 & 0 & 0 \\ 0 & 0 & 0 & -0.2440022493 & -0.9697746658 \\ 0 & 0 & 0 & -0.9697746658 & 0.2440022493 \end{bmatrix}$$

and the upper triangular $A_4 = H_4 A_3$ is

$$A_4 = \begin{bmatrix} -15.16575089 & -6.330052544 & -12.66010509 & 10.87977781 & 3.626592604 \\ 0 & 17.08597187 & -12.24040002 & -3.753396957 & 7.254870993 \\ 0 & 0 & -9.741372927 & 4.435098637 & 4.956635359 \\ 0 & 0 & 0 & 8.710473364 & 5.371777208 \\ 0 & 0 & 0 & 0 & 0.889071724 \end{bmatrix}$$

The matrix A_4 is the upper triangular factor R in the QR decomposition of A. The factor Q is the transpose of $H_5 = H_4 \cdots H_1$, so $Q = H_1 \cdots H_4$, which we find is

$$Q = \begin{bmatrix} -0.6593804734 & 0.2239314794 & -0.6562919667 & 0.2210139795 & -0.1884297195 \\ 0.4615663314 & -0.2386905996 & -0.2999381978 & 0.0473700769 & -0.7986090863 \\ 0.2637521894 & 0.5074083748 & -0.3644262409 & -0.7288689812 & 0.0944195263 \\ -0.2637521894 & 0.5460874483 & 0.5804932902 & -0.0752309072 & -0.5381367225 \\ -0.4615663314 & -0.5806950406 & 0.0976677366 & -0.6418716504 & -0.1679630591 \end{bmatrix}$$

Notice that Q is not symmetric even though it is the product of symmetric factors. Symmetry is not preserved for products, but orthogonality is.

From [4] we learn that in a QR decomposition, once the signs on the main diagonal of R are fixed, the rest of the entries are unique.

We leave to the reader the task of verifying that the matrices $H_k, k = 1, \ldots, 4$, are orthogonal and have determinants of -1, thus showing they are Householder matrices. ❖

EXERCISES 45.3

1. If $A = \begin{bmatrix} 3 & 2 \\ 4 & 1 \end{bmatrix}$, obtain the Householder matrix $H = I - 2\mathbf{u}\mathbf{u}^\mathsf{T}$ that annihilates the second entry in the first column of A. Show that $HA = R$, an upper triangular matrix, so that $A = H^\mathsf{T} R = QR$ provides the QR decomposition of A.

For each $n \times n$ matrix A in Exercises 2–5:

(a) Obtain the Householder reflection matrices H_k, $k = 1, \ldots, n-1$, for which $H_n = \prod_{k=1}^{n-1} H_k = Q^\mathsf{T}$, where $R = Q^\mathsf{T} A$ and $A = QR$.

(b) Obtain the factors Q and R, and demonstrate that $A = QR$.

(c) Let $A = qr$ be a QR factorization of A obtained by the Gram–Schmidt process as in Section 34.3.

(d) If the factors Q and R differ from the factors q and r, reconcile the two decompositions by writing $(QD)(DR) = qr$, where D is a diagonal matrix of 1's and -1's so that $D^2 = I$.

2. $\begin{bmatrix} -9 & -4 & -1 \\ -2 & 1 & 7 \\ 5 & 2 & -1 \end{bmatrix}$
3. $\begin{bmatrix} -6 & 4 & 4 & 6 \\ 6 & -8 & 1 & 7 \\ -9 & 5 & -4 & 7 \\ 3 & -8 & -5 & 4 \end{bmatrix}$

4. $\begin{bmatrix} 9 & 5 & 6 & 7 & 3 \\ 4 & -9 & 2 & -1 & -2 \\ 5 & -7 & -5 & 8 & 3 \\ -1 & 7 & 3 & -5 & 6 \\ -4 & 5 & 2 & 6 & -7 \end{bmatrix}$

45.4 Upper Hessenberg Form, Givens Rotations, and the Shifted QR-Algorithm

Introduction

In Section 45.3 Householder reflection matrices were used to annihilate, in the kth column of a matrix A, all entries below the kth. A sequence of such Householder reflection matrices then upper-triangularized A, giving $R = H_{n-1} \cdots H_1 A$.

If, instead, Householder reflections were used to annihilate, in the kth column of $A = \{a_{ij}\}$, all components below $a_{k+1,k}$, then a succession of such H_k would transform A into *upper Hessenberg form*. Such a matrix has zeros for all entries below the subdiagonal. In other words, an upper Hessenberg matrix is "one diagonal away" from being an upper triangular matrix. In particular, the matrix $C = \{c_{ij}\}$ is in upper Hessenberg form if $c_{ij} = 0$ for $i \geq j + 2$.

EXAMPLE 45.6 A schematic for a 6×6 upper Hessenberg matrix appears in (45.19). The subdiagonal elements are denoted by $s_k, k = 1, \ldots, 5$. Should all the elements in that subdiagonal be zero, the matrix would be upper triangular.

$$\begin{bmatrix} a_{11} & a_{12} & a_{13} & a_{14} & a_{15} & a_{16} \\ s_1 & a_{22} & a_{23} & a_{24} & a_{25} & a_{26} \\ 0 & s_2 & a_{33} & a_{34} & a_{35} & a_{36} \\ 0 & 0 & s_3 & a_{44} & a_{45} & a_{46} \\ 0 & 0 & 0 & s_4 & a_{55} & a_{56} \\ 0 & 0 & 0 & 0 & s_5 & a_{56} \end{bmatrix} \qquad (45.19)$$

❖

EIGENVALUES FROM THE UPPER HESSENBERG FORM The QR algorithm finds the eigenvalues of a matrix A by the factorization $A = QR$, where the orthogonal factor Q can be obtained either by the Gram–Schmidt process or by a succession of Householder reflections. The eigenvalues of A are preserved neither in Q nor in R. They are preserved in $A_{k+1} = R_k Q_k$, which we showed in Section 34.4 to be a similarity transform on A. Thus, the matrices A_k converge to an upper triangular matrix in which the diagonal elements give the eigenvalues of A.

The upper Hessenberg form of the matrix A is obtained using each Householder reflection matrix to generate a similarity transformation on A, a transformation that therefore preserves the eigenvalues of A. The resulting upper Hessenberg form of A is then orthogonally similar to A and, therefore, contains the same eigenvalues as A.

Once A has been so transformed to an upper Hessenberg form containing the same eigenvalues, we have two alternatives, not necessarily mutually exclusive.

1. Use *Givens rotations*, defined after Example 45.7, to upper-triangularize the upper Hessenberg form of A, thereby completing the QR factorization of A.

2. Use the *shifted QR-algorithm*, which converges rapidly for an upper Hessenberg matrix. (See this section's penultimate subsection)

To explore either of these alternatives, it is useful to have tools for first obtaining the upper Hessenberg form of A.

Implementation

The software accompanying the text contains an auxiliary procedure for generating the Householder matrices that transform a matrix A to its upper Hessenberg form. It differs from the procedure used in Section 45.3 in that now, the Householder matrices must annihilate, in the kth column of A, all components below the *subdiagonal* instead of below the *diagonal*.

EXAMPLE 45.7   Transform the matrix

$$A = \begin{bmatrix} 9 & -2 & -7 & 4 & 6 \\ 6 & -9 & 2 & -1 & -1 \\ -4 & 8 & -8 & 9 & 3 \\ 3 & 4 & 8 & 4 & 7 \\ 8 & -8 & 0 & 7 & 4 \end{bmatrix} \qquad (45.20)$$

to upper Hessenberg form with a sequence of Householder matrices, the first of which is

$$H_1 = \begin{bmatrix} 1 & 0 & 0 & 0 & 0 \\ 0 & -0.5366563146 & 0.3577708764 & -0.2683281573 & -0.7155417528 \\ 0 & 0.3577708764 & 0.9167022588 & 0.06247330590 & 0.1665954824 \\ 0 & -0.2683281573 & 0.06247330590 & 0.9531450206 & -0.1249466118 \\ 0 & -0.7155417528 & 0.1665954824 & -0.1249466118 & 0.6668090352 \end{bmatrix}$$

and generates the similarity transform $A_1 = H_1^{-1} A H_1 = H_1^{\mathrm{T}} A H_1 = H_1 A H_1$, with A_1 given by

$$A_1 = \begin{bmatrix} 9 & -6.797646652 & -5.882991447 & 3.162243584 & 3.765982893 \\ -11.18033989 & -5.936000002 & -1.826287990 & -5.517373377 & -11.64212898 \\ 0 & -8.814767591 & -3.360352084 & 7.240157714 & -2.019844601 \\ 0 & -7.543234603 & 10.14397360 & 1.102099565 & -0.482535613 \\ 0 & -6.251670818 & -1.856577489 & 4.952645817 & -0.805747482 \end{bmatrix}$$

Both A and A_1 have the same eigenvalues, namely,

$$17.01106481, \ -16.77225262, \ -8.144827648, \ 7.397166441, \ 0.5088490206 \quad (45.21)$$

Continuing with the sequence of similarity transformations based on Householder reflections, we obtain

$$H_2 = \begin{bmatrix} 1 & 0 & 0 & 0 & 0 \\ 0 & 1 & 0 & 0 & 0 \\ 0 & 0 & -0.6688536171 & -0.5723712732 & -0.4743690173 \\ 0 & 0 & -0.5723712732 & 0.8036922646 & -0.1626956346 \\ 0 & 0 & -0.4743690173 & -0.1626956346 & 0.8651613525 \end{bmatrix}$$

and $A_2 = H_2 A_1 H_2$, with A_2 given by

$$A_2 = \begin{bmatrix} 9 & -6.797646652 & 0.338417119 & 5.296017035 & 5.534408497 \\ -11.18033989 & -5.936000002 & 9.902170637 & -1.494831960 & -8.308333048 \\ 0 & 13.17891892 & 5.315429519 & -5.081326641 & 4.281463723 \\ 0 & 0 & -5.041877096 & -9.352700813 & -3.483710161 \\ 0 & 0 & 0.5668673244 & 1.435661483 & 0.9732712911 \end{bmatrix}$$

Then, obtain

$$H_3 = \begin{bmatrix} 1 & 0 & 0 & 0 & 0 \\ 0 & 1 & 0 & 0 & 0 \\ 0 & 0 & 1 & 0 & 0 \\ 0 & 0 & 0 & -0.9937388429 & 0.1117278482 \\ 0 & 0 & 0 & 0.1117278482 & 0.9937388429 \end{bmatrix}$$

and $A_3 = H_3 A_2 H_3$, with A_3 given by

$$A_3 = \begin{bmatrix} 9 & -6.797646652 & 0.338417119 & -4.644510288 & 6.091469283 \\ -11.18033989 & -5.936000002 & 9.902170637 & 0.5572004084 & -8.423327628 \\ 0 & 13.17891892 & 5.315429519 & 5.527870386 & 3.686931114 \\ 0 & 0 & 5.073643977 & -8.996409178 & 4.604619294 \\ 0 & 0 & 0 & -0.3147523485 & 0.6169796558 \end{bmatrix}$$

$$\text{(45.22)}$$

The matrix A_3 is the desired upper Hessenberg form of A, and a computation shows that A_3 also has (45.21) for its eigenvalues.   ❖

Page 537 of [60] gives the details of how a product of the form HAH, where H is an annihilating Householder reflection, can be computed and stored in the memory allocated to A. Thus, reducing A to an upper Hessenberg form can be done using no more memory than it takes to store A itself.

Givens Rotations

Following [60], we define a *Givens rotation* as an $n \times n$ matrix $G_{ij} = \{g_{\alpha\beta}\}$ that coincides with I_n, the $n \times n$ identity matrix, except for the four entries

$$g_{jj} = c \quad g_{ji} = -s \quad g_{ij} = s \quad g_{ii} = c$$

where $c^2 + s^2 = 1$ and $1 \le j < i \le n$. The restriction $j < i$ means that there are $\frac{n(n-1)}{2}$ Givens rotations of dimension $n \times n$. This is perhaps best seen by considering the cases $n = 4$ and $n = 5$, where, with fixed c and s, the Givens rotations are

$$\begin{matrix} G_{21} & G_{31} & G_{41} \\ & G_{32} & G_{42} \\ & & G_{43} \end{matrix} \quad \text{and} \quad \begin{matrix} G_{21} & G_{31} & G_{41} & G_{51} \\ & G_{32} & G_{42} & G_{52} \\ & & G_{43} & G_{53} \\ & & & G_{54} \end{matrix}$$

In the first case, the number of Givens rotations is $3 + 2 + 1 = 6$, the sum of the first $n - 1$ integers. In the second case, the number of Givens rotations is $4 + 3 + 2 + 1 = 10$, again the sum of the first $n - 1$ integers. The sum of the first $n - 1$ integers is known to be $\sum_{k=1}^{n-1} k = \frac{1}{2}n(n-1)$.

The software accompanying the text contains an auxiliary procedure that creates the $n \times n$ Givens rotation G_{ij} with the special values c and s. For example, the 5×5 Givens rotations G_{42} and G_{41} are

$$G_{42} = \begin{bmatrix} 1 & 0 & 0 & 0 & 0 \\ 0 & c & 0 & -s & 0 \\ 0 & 0 & 1 & 0 & 0 \\ 0 & s & 0 & c & 0 \\ 0 & 0 & 0 & 0 & 1 \end{bmatrix} \quad G_{41} = \begin{bmatrix} c & 0 & 0 & -s & 0 \\ 0 & 1 & 0 & 0 & 0 \\ 0 & 0 & 1 & 0 & 0 \\ s & 0 & 0 & c & 0 \\ 0 & 0 & 0 & 0 & 1 \end{bmatrix}$$

In G_{ij}, the four special values $c, -s, s, c$ form the corners of a "square" whose lower left corner is g_{ij}.

A Givens rotation G_{ij} is orthogonal, the general proof of which we leave to the exercises. We illustrate the claim for G_{42} and G_{41}.

$$
G_{42}^{\mathrm{T}} G_{42} = \begin{bmatrix} 1 & 0 & 0 & 0 & 0 \\ 0 & c^2+s^2 & 0 & 0 & 0 \\ 0 & 0 & 1 & 0 & 0 \\ 0 & 0 & 0 & c^2+s^2 & 0 \\ 0 & 0 & 0 & 0 & 1 \end{bmatrix} \qquad G_{41}^{\mathrm{T}} G_{41} = \begin{bmatrix} c^2+s^2 & 0 & 0 & 0 & 0 \\ 0 & 1 & 0 & 0 & 0 \\ 0 & 0 & 1 & 0 & 0 \\ 0 & 0 & 0 & c^2+s^2 & 0 \\ 0 & 0 & 0 & 0 & 1 \end{bmatrix}
$$

In each case, if we recall that $c^2 + s^2 = 1$, we see that the products $G_{ij}^{\mathrm{T}} G_{ij}$ are just I_5. Hence, the transpose is the inverse and the matrices are orthogonal.

Multiplication of a vector $\mathbf{x}$ by the Givens rotation G_{ij} changes the jth component to $cx_j - sx_i$ and changes the ith component to $sx_j + cx_i$. Schematically, we have

$$
G_{ij}\mathbf{x} = [x_1 \cdots x_{j-1}, cx_j - sx_i, x_{j+1} \cdots x_{i-1}, sx_j + cx_i, x_{i+1} \cdots x_n]^{\mathrm{T}}
$$

The only components of $\mathbf{x}$ that change are the jth and the ith. This suggests we can construct a Givens rotation to annihilate the ith component of a vector, leaving all other components except the jth unchanged. We will call such a Givens rotation an *annihilating Givens rotation*.

Annihilating Givens Rotations

If the vector $\mathbf{x}$ satisfies $x_i \neq 0$, then the choices $c = x_j / \sqrt{x_j^2 + x_i^2}$ and $s = -x_i / \sqrt{x_j^2 + x_i^2}$ in the Givens rotation matrix G_{ij} implies that in the vector $G_{ij}\mathbf{x}$, the ith component will be 0 and the jth component will be $\sqrt{x_j^2 + x_i^2}$, that is, $(G_{ij}\mathbf{x})_i = 0$ and $(G_{ij}\mathbf{x})_j = \sqrt{x_j^2 + x_i^2}$.

EXAMPLE 45.8   The fourth component of the vector $\mathbf{x} = [-3, 4, 3, -1, 2]^{\mathrm{T}}$ can be annihilated with any of the Givens rotations listed in Table 45.4.

$$
\frac{1}{\sqrt{10}}\begin{bmatrix} -3 & 0 & 0 & -1 & 0 \\ 0 & 1 & 0 & 0 & 0 \\ 0 & 0 & 1 & 0 & 0 \\ 1 & 0 & 0 & -3 & 0 \\ 0 & 0 & 0 & 0 & 1 \end{bmatrix} \qquad \frac{1}{\sqrt{17}}\begin{bmatrix} 1 & 0 & 0 & 0 & 0 \\ 0 & 4 & 0 & -1 & 0 \\ 0 & 0 & 1 & 0 & 0 \\ 0 & 1 & 0 & 4 & 0 \\ 0 & 0 & 0 & 0 & 1 \end{bmatrix} \qquad \frac{1}{\sqrt{10}}\begin{bmatrix} 1 & 0 & 0 & 0 & 0 \\ 0 & 1 & 0 & 0 & 0 \\ 0 & 0 & 3 & -1 & 0 \\ 0 & 0 & 1 & 3 & 0 \\ 0 & 0 & 0 & 0 & 1 \end{bmatrix}
$$
$$
G_{41} \qquad\qquad\qquad\qquad G_{42} \qquad\qquad\qquad\qquad G_{43}
$$

TABLE 45.4   The Givens annihilating rotations G_{41}, G_{42}, and G_{43} in Example 45.8

Indeed, for the products $G_{41}\mathbf{x}$, $G_{42}\mathbf{x}$, and $G_{43}\mathbf{x}$, we find, respectively,

$$
G_{41}\mathbf{x} = \begin{bmatrix} \sqrt{10} \\ 4 \\ 3 \\ 0 \\ 2 \end{bmatrix} \qquad G_{42}\mathbf{x} = \begin{bmatrix} -3 \\ \sqrt{17} \\ 3 \\ 0 \\ 2 \end{bmatrix} \qquad G_{43}\mathbf{x} = \begin{bmatrix} -3 \\ 4 \\ \sqrt{10} \\ 0 \\ 2 \end{bmatrix} \qquad \diamondsuit
$$

EXAMPLE 45.9   A more geometrically accessible example consists of the vector $x = [1, 2, 3]^T$ and the two annihilating Givens rotations

$$G_{31} = \frac{1}{\sqrt{10}} \begin{bmatrix} 1 & 0 & 3 \\ 0 & 1 & 0 \\ -3 & 0 & 1 \end{bmatrix} \quad \text{and} \quad G_{32} = \frac{1}{\sqrt{13}} \begin{bmatrix} 1 & 0 & 0 \\ 0 & 2 & 3 \\ 0 & -3 & 2 \end{bmatrix}$$

which both annihilate the third component of **x**. Indeed, we find

$$\mathbf{V}_1 = G_{31}\mathbf{x} = \begin{bmatrix} \sqrt{10} \\ 2 \\ 0 \end{bmatrix} \quad \text{and} \quad \mathbf{V}_2 = G_{32}\mathbf{x} = \begin{bmatrix} 1 \\ \sqrt{13} \\ 0 \end{bmatrix}$$

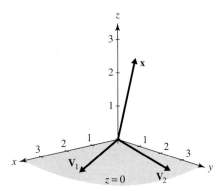

FIGURE 45.5   The vectors **x**, $\mathbf{V}_1 = G_{31}\mathbf{x}$, and $\mathbf{V}_2 = G_{32}\mathbf{x}$ in Example 45.9

as the two transformed versions of **x**. Figure 45.5 shows **x**, $\mathbf{V}_1$, and $\mathbf{V}_2$. The Givens rotation G_{31} rotates **x** so that it lies in the *xy*-plane. The Given rotation G_{32} rotates **x** so that it, too, lies in the *xy*-plane. Both $\mathbf{V}_1$ and $\mathbf{V}_2$ must lie in the *xy*-plane because the third component in each is zero. ❖

QR Factorization of an Upper Hessenberg Matrix via Givens Rotations

If the $n \times n$ matrix A is in upper Hessenberg form, then $M = G_{n,n-1}G_{n-1,n-2} \cdots G_{21}$, the product of $n - 1$ annihilating Givens rotations, will upper-triangularize A. Since $MA = R$ and M is orthogonal because it is the product of orthogonal Givens rotations, we have $A = M^T R$, so $M^T = Q$ and $A = QR$. With $n - 1$ annihilating Givens rotations, the upper Hessenberg matrix A is factored into Q and R.

We demonstrate this process with the upper Hessenberg matrix (45.22) derived from the matrix (45.20). Starting at the "top" of the subdiagonal, we annihilate the (2, 1)-entry in A_3 with the Givens rotation

$$G_{21} = \begin{bmatrix} 0.6270597126 & -0.7789711909 & 0 & 0 & 0 \\ 0.7789711909 & 0.6270597126 & 0 & 0 & 0 \\ 0 & 0 & 1 & 0 & 0 \\ 0 & 0 & 0 & 1 & 0 \\ 0 & 0 & 0 & 0 & 1 \end{bmatrix}$$

and then obtain $A_4 = G_{21}A_3$ as

$$A_4 = \begin{bmatrix} 14.35270009 & 0.361442635 & -7.501297913 & -3.346428352 & 10.38124453 \\ 0 & -9.017397363 & 6.472869460 & -3.268541782 & -0.536850320 \\ 0 & 13.17891892 & 5.315429519 & 5.527870386 & 3.686931114 \\ 0 & 0 & 5.073643977 & -8.996409178 & 4.604619294 \\ 0 & 0 & 0 & -0.3147523485 & 0.6169796558 \end{bmatrix}$$

Next, the (3, 2)-element in A_4 is annihilated by forming the Givens matrix

$$G_{32} = \begin{bmatrix} 1 & 0 & 0 & 0 & 0 \\ 0 & -0.5646942513 & 0.8253001895 & 0 & 0 \\ 0 & -0.8253001895 & -0.5646942513 & 0 & 0 \\ 0 & 0 & 0 & 1 & 0 \\ 0 & 0 & 0 & 0 & 1 \end{bmatrix}$$

and computing the product $A_5 = G_{32}A_4$ to get

$$\mathbf{A}_5 = \begin{bmatrix} 14.35270009 & 0.361442635 & -7.501297913 & -3.346428352 & 10.38124453 \\ 0 & 15.96863673 & 0.731632816 & 6.407879231 & 3.345981237 \\ 0 & 0 & -8.343652885 & -0.424028477 & -1.638926134 \\ 0 & 0 & 5.073643977 & -8.996409178 & 4.604619294 \\ 0 & 0 & 0 & -0.3147523485 & 0.6169796558 \end{bmatrix}$$

The $(4, 3)$-element in A_5 is annihilated by forming the Givens matrix

$$G_{43} = \begin{bmatrix} 1 & 0 & 0 & 0 & 0 \\ 0 & 1 & 0 & 0 & 0 \\ 0 & 0 & -0.8544304853 & 0.5195657280 & 0 \\ 0 & 0 & -0.5195657280 & -0.8544304853 & 0 \\ 0 & 0 & 0 & 0 & 1 \end{bmatrix}$$

and computing the product $A_6 = G_{43}A_5$ to get

$$A_6 = \begin{bmatrix} 14.35270009 & 0.361442635 & -7.501297913 & -3.346428352 & 10.38124453 \\ 0 & 15.96863673 & 0.731632816 & 6.407879231 & 3.345981237 \\ 0 & 0 & 9.765162911 & -4.311923027 & 3.792750828 \\ 0 & 0 & 0 & 7.907116924 & -3.082797248 \\ 0 & 0 & 0 & -0.3147523485 & 0.6169796558 \end{bmatrix}$$

Finally, the $(5, 4)$-element in A_6 is annihilated by forming the Givens matrix

$$G_{54} = \begin{bmatrix} 1 & 0 & 0 & 0 & 0 \\ 0 & 1 & 0 & 0 & 0 \\ 0 & 0 & 1 & 0 & 0 \\ 0 & 0 & 0 & 0.9992086732 & -0.03977470923 \\ 0 & 0 & 0 & 0.03977470923 & 0.9992086732 \end{bmatrix}$$

and computing the product $A_7 = G_{54}A_6$ to get

$$A_7 = \begin{bmatrix} 14.35270009 & 0.361442635 & -7.501297913 & -3.346428352 & 10.38124453 \\ 0 & 15.96863673 & 0.731632816 & 6.407879231 & 3.345981237 \\ 0 & 0 & 9.765162911 & -4.311923027 & 3.792750828 \\ 0 & 0 & 0 & 7.913378993 & -3.104897934 \\ 0 & 0 & 0 & 0 & 0.4938740591 \end{bmatrix}$$

The matrix A_7 is the factor R for the QR decomposition of the matrix A_3. The companion factor Q is found by transposing the product of the annihilating Givens rotations, that is, $Q = (G_{54}G_{43}G_{32}G_{21})^{\mathrm{T}}$, giving

$$Q = \begin{bmatrix} 0.6270597126 & -0.4398805534 & 0.5493006036 & 0.3337567304 & 0.01328559015 \\ -0.7789711909 & -0.3540970149 & 0.4421784562 & 0.2686689853 & 0.01069469377 \\ 0 & 0.8253001895 & 0.4824919832 & 0.2931636078 & 0.01166973183 \\ 0 & 0 & 0.5195657280 & -0.8537543516 & -0.03398472411 \\ 0 & 0 & 0 & -0.03977470923 & 0.9992086732 \end{bmatrix}$$

Reversing the factors Q and R in the QR algorithm gives another upper Hessenberg matrix that can be QR-factored again by $n - 1$ annihilating Givens rotations. The energy expended in bringing A into upper Hessenberg form is expended only once. The QR factorization of the upper Hessenberg matrix is accomplished with $n - 1$ annihilating Givens rotations.

Shifted QR

In Section 34.4 we saw the QR algorithm for finding the eigenvalues of a matrix A. The matrix A in (45.20) was specially chosen because its eigenvalues, listed in (45.21), have two pairs with nearly equal magnitudes. The first and second have nearly equal magnitudes, and the third and fourth also have nearly equal magnitudes.

The QR algorithm, like any power method, converges slowly in the presence of eigenvalues with nearly equal magnitudes. In fact, if we visit 50 iterations of the QR algorithm upon A, we obtain for $A_{51} = Q_{50}R_{50}$ the matrix

$$\begin{bmatrix} 15.17427984 & 8.398117280 & -2.584141572 & -2.082941449 & 3.447024830 \\ 6.987150649 & -14.93546771 & 0.5462226447 & -3.203958976 & 5.971972090 \\ 0 & 0 & -8.170707217 & -4.789143117 & 8.083392115 \\ 0 & 0 & 0.08412563135 & 7.423045996 & -6.147570340 \\ 0 & 0 & 0 & 0 & 0.5088490204 \end{bmatrix}$$

The QR algorithm is supposed to generate a sequence of matrices A_k that converge to an upper triangular matrix on the diagonal of which the eigenvalues are found. Even after 50 iterations we have elements as large as 6.987 on the subdiagonal. This is certainly not an efficient calculation.

Starting with the upper Hessenberg form A_3 in (45.22) fares no better, as the following recalculation of A_{51} shows.

$$\begin{bmatrix} 15.17427985 & -8.398117220 & 2.631951124 & 2.022193350 & 3.447024838 \\ -6.987150586 & -14.93546773 & 0.4714561593 & -3.215811154 & -5.971972064 \\ 0 & 0 & -8.271796584 & -4.423520230 & -7.938026296 \\ 0 & 0 & 0.4497485091 & 7.524135383 & 6.334160385 \\ 0 & 0 & 0 & 0 & 0.5088490192 \end{bmatrix}$$

There are still magnitudes as large as 6.987 on the subdiagonal.

A more effective strategy is the *shifted QR algorithm*, in which the QR decomposition is obtained for $A_k - z_k I_n = Q_k R_k$, not just A_k. The scalars z_k are the *shifts* and can be chosen as the lower right element on the main diagonal of A_k. Of course, the matrix I_n is the $n \times n$ identity matrix. The next iterate, A_{k+1}, is formed by reversing the factors and restoring the shift. Thus, we have the two steps

$$A_k - z_k I_n = Q_k R_k \qquad A_{k+1} = R_k Q_k + z_k I_n$$

The reader is referred to [4] for details, proofs of convergence, and an alternate strategy for selecting the shifts z_k. We merely demonstrate the increased rate of convergence for the problem at hand. Starting with A in its upper Hessenberg form A_3, we perform five iterations of the shifted QR algorithm, obtaining

$$\begin{bmatrix} 8.920147879 & -13.75868934 & 1.166510523 & 3.945434098 & -6.809074160 \\ -15.10104281 & -8.619714707 & 2.652155709 & -0.5944027224 & -2.159106872 \\ 0 & 0.3233152435 & 0.3775563492 & 10.46507754 & -9.975915415 \\ 0 & 0 & 5.688698425 & -1.186838537 & 0.3689963347 \\ 0 & 0 & 0 & 0 & 0.5088490192 \end{bmatrix}$$

The lower right element on the main diagonal of A_k is again nearly the eigenvalue of smallest magnitude. However, this took only 5 iterations, not 50. To obtain the remaining eigenvalues, "deflate" A_k by selecting the $(n-1) \times (n-1)$ submatrix

$$
\begin{bmatrix}
8.920147879 & -13.75868934 & 1.166510523 & 3.945434098 \\
-15.10104281 & -8.619714707 & 2.652155709 & -0.5944027224 \\
0 & 0.3233152435 & 0.3775563492 & 10.46507754 \\
0 & 0 & 5.688698425 & -1.186838537
\end{bmatrix}
$$

whose main diagonal coincides with the main diagonal of A_k but excludes the eigenvalue just computed.

Another five iterations of the shifted QR algorithm is applied to this submatrix. The shifts z_k are again taken as the lower right elements on the main diagonal, but on the main diagonal of the deflated matrix and its subsequent iterates. We thereby obtain

$$
\begin{bmatrix}
16.98307108 & -1.898878980 & 1.869823995 & -3.108769321 \\
-0.5214200038 & -16.78430654 & -2.511972534 & -0.2815075194 \\
0 & 0.4035024727 & 7.437214282 & -4.866455806 \\
0 & 0 & 0.5153017568 \times 10^{-6} & -8.144827814
\end{bmatrix}
$$

The lower right entry on the main diagonal of this iterate contains an estimate of the third eigenvalue, correct to five significant places. Another deflation is performed, resulting in the matrix

$$
\begin{bmatrix}
16.98307108 & -1.898878980 & 1.869823995 \\
-0.5214200038 & -16.78430654 & -2.511972534 \\
0 & 0.4035024727 & 7.437214282
\end{bmatrix}
$$

Again five iterations of the shifted QR algorithm are performed with the shifts being the lower right elements on the main diagonal of the new A_k's. We find

$$
\begin{bmatrix}
-7.259678460 & -14.50553086 & 3.447074461 \\
-15.91649747 & 7.498490645 & -0.1431097876 \\
0 & 0 & 7.397166630
\end{bmatrix}
$$

The lower right element on the main diagonal approximates the fourth eigenvalue 7.397166441 to five significant places. A final deflation gives the matrix

$$
\begin{bmatrix}
-7.259678460 & -14.50553086 \\
-15.91649747 & 7.498490645
\end{bmatrix}
$$

upon which we visit the final five iterations of the shifted QR algorithm. The shifts z_k are again taken as the lower right elements on the main diagonal of the depressed matrix and its iterates. We obtain

$$
\begin{bmatrix}
-16.77225263 & 1.410966615 \\
0 & 17.01106480
\end{bmatrix}
$$

The main diagonal of this final iterate contains approximations of the first two eigenvalues, 17.01106481 and -16.77225262. Both are accurate to seven significant places.

Summary

In practice, because of their potential inefficiency, power methods are not generally used for finding eigenvalues. However, in some special applications such as finding the eigenvalues of large sparse matrices, they can be useful. (See [4] for additional comments.)

Efficient codes for finding the eigenvalues of a matrix A generally begin with a reduction to upper Hessenberg form, a reduction performed by annihilating Householder reflections. The reduction is done "in place" by storing each of the matrices $A_{k+1} = H_k A_k H_k$ in the memory allocated to A itself.

Then, the shifted QR algorithm is applied to the upper Hessenberg form of A. Annihilating Givens rotations are used to perform the QR decomposition of the upper Hessenberg matrices generated by the QR process.

Writing and testing efficient code for finding eigenvalues would be a significant challenge for the neophyte. Fortunately, modern computer software packages provide such code in convenient wrappers.

EXERCISES 45.4

1. Determine if the product of upper Hessenberg matrices is an upper Hessenberg matrix.

2. Show that a general Givens rotation matrix G_{ij} is orthogonal.

For each vector **x** given in Exercises 3–5, obtain the annihilating Givens rotations G_{31}, G_{32}, G_{21}. Compute $\mathbf{u}_1 = G_{31}\mathbf{x}$, $\mathbf{u}_2 = G_{32}\mathbf{x}$, $\mathbf{u}_3 = G_{21}\mathbf{x}$. Sketch the vectors **x**, $\mathbf{u}_1$, $\mathbf{u}_2$, $\mathbf{u}_3$.

3. $\mathbf{x} = [-3, -7, -1]^\mathrm{T}$ **4.** $\mathbf{x} = [-9, 9, 1]^\mathrm{T}$ **5.** $\mathbf{x} = [3, 7, -5]^\mathrm{T}$

Exercises 6 and 7 use the following 5×5 matrices C_k, $k = 1, \ldots, 6$.

$$C_1 = \begin{bmatrix} 6 & 3 & 4 & -3 & -6 \\ 4 & 2 & 1 & 4 & 4 \\ 4 & -1 & -3 & 6 & 5 \\ 6 & 8 & 9 & 0 & 4 \\ -8 & -3 & 7 & -5 & 4 \end{bmatrix} \quad C_2 = \begin{bmatrix} -8 & 6 & 7 & 9 & 8 \\ 1 & -7 & -7 & -9 & 3 \\ 8 & -6 & 4 & 4 & 3 \\ 6 & -5 & -2 & -9 & -5 \\ 5 & 6 & 3 & 8 & 2 \end{bmatrix}$$

$$C_3 = \begin{bmatrix} 3 & 6 & 8 & 0 & 3 \\ 7 & 3 & -3 & 3 & 1 \\ 6 & 0 & -2 & -8 & 7 \\ -5 & 3 & -2 & 0 & 2 \\ 8 & 5 & 8 & 8 & -8 \end{bmatrix} \quad C_4 = \begin{bmatrix} 9 & 5 & 6 & 4 & 2 \\ -5 & -6 & -2 & -8 & 4 \\ 8 & 3 & 4 & 3 & -2 \\ 8 & -1 & -4 & 2 & -8 \\ 8 & 2 & 4 & -1 & 3 \end{bmatrix}$$

$$C_5 = \begin{bmatrix} -8 & 0 & 7 & 6 & 1 \\ 8 & -8 & -2 & -5 & 4 \\ 5 & 8 & -5 & -9 & 9 \\ 8 & -6 & 1 & 6 & 7 \\ -1 & 6 & 0 & 2 & 5 \end{bmatrix} \quad C_6 = \begin{bmatrix} 9 & 2 & 5 & 6 & 6 \\ 7 & -4 & 5 & -1 & -8 \\ 6 & 4 & 5 & 2 & -4 \\ 9 & 7 & 1 & 1 & -8 \\ 4 & -9 & 5 & -6 & -1 \end{bmatrix}$$

6. Fix a matrix $A = C_k$ from the given list, and for it

(a) obtain Householder transformation matrices H_1, H_2, H_3 and the matrix $H = \prod_{k=1}^{3} H_k$ for which $H^\mathrm{T} A H$ is both similar to A and in upper Hessenberg form.

(b) show that A and $B = H^\mathrm{T} A H$ have the same eigenvalues.

(c) verify that each H_k is both symmetric and orthogonal.

7. For the matrix $A = C_k$ selected in Exercise 6 and transformed to $B = HAH^\mathrm{T}$:

(a) Construct the annihilating Givens rotation matrices G_{21}, G_{32}, G_{43}, G_{54} that will upper-triangularize B.

(b) Obtain the upper triangular matrix $R = Q^\mathrm{T} B$, where $Q^\mathrm{T} = G_{54} G_{43} G_{32} G_{21}$.

(c) Show that the eigenvalues of R are not the same as the eigenvalues of A.

(d) Show that $B_1 = RQ$ is again upper Hessenberg and has the same eigenvalues as B.

(e) Construct the annihilating Givens rotation matrices G'_{21}, G'_{32}, G'_{43}, G'_{54} that will upper-triangularize B_1.

(f) Obtain the upper triangular matrix $R_1 = Q_1^\mathrm{T} B_1$, where $Q_1^\mathrm{T} = G'_{54} G'_{43} G'_{32} G'_{21}$.

(g) Show that the eigenvalues of R_1 are not the same as the eigenvalues of A.

(h) Show that $B_2 = R_1 Q_1$ is again upper Hessenberg and has the same eigenvalues as B_1.

(i) Describe the matrix $M = \lim_{k\to\infty} B_k$ if the sequence of matrices $\{B_k\}$ were formed by the construction started in this exercise.

For each of the symmetric matrices A given in Exercises 8–13:

(a) Obtain Householder transformation matrices H_1, H_2, H_3 and the matrix $H = \prod_{k=1}^{3} H_k$ for which $H^\mathrm{T} A H$ is both similar to A and in upper Hessenberg form.

(b) Show that A and $H^\mathrm{T} A H$ have the same eigenvalues.

(c) Verify that each H_k is orthogonal. Note that the upper Hessenberg form of a symmetric matrix is a tridiagonal matrix.

8. $\begin{bmatrix} 8 & 4 & -5 & 8 & 3 \\ 4 & -5 & 3 & 5 & -4 \\ -5 & 3 & -5 & -1 & -5 \\ 8 & 5 & -1 & 0 & -2 \\ 3 & -4 & -5 & -2 & -1 \end{bmatrix}$ **9.** $\begin{bmatrix} -9 & 5 & 4 & -9 & 4 \\ 5 & 8 & -3 & 7 & 4 \\ 4 & -3 & 8 & 5 & 2 \\ -9 & 7 & 5 & 0 & 5 \\ 4 & 4 & 2 & 5 & -5 \end{bmatrix}$

10. $\begin{bmatrix} 9 & -9 & 0 & 4 & 7 \\ -9 & 4 & -8 & 7 & -8 \\ 0 & -8 & -5 & -9 & 3 \\ 4 & 7 & -9 & 5 & 5 \\ 7 & -8 & 3 & 5 & -9 \end{bmatrix}$ **11.** $\begin{bmatrix} 4 & 2 & -5 & 1 & 3 \\ 2 & -1 & 5 & -1 & 3 \\ -5 & 5 & -7 & 3 & -9 \\ 1 & -1 & 3 & 8 & -6 \\ 3 & 3 & -9 & -6 & 3 \end{bmatrix}$ **12.** $\begin{bmatrix} 9 & -9 & 6 & -7 & 2 \\ -9 & -5 & 0 & -1 & 9 \\ 6 & 0 & 1 & -1 & -7 \\ -7 & -1 & -1 & -4 & -7 \\ 2 & 9 & -7 & -7 & -3 \end{bmatrix}$ **13.** $\begin{bmatrix} -7 & -2 & 8 & -6 & -8 \\ -2 & -7 & 5 & -8 & 1 \\ 8 & 5 & -3 & 2 & 9 \\ -6 & -8 & 2 & 9 & 5 \\ -8 & 1 & 9 & 5 & 1 \end{bmatrix}$

45.5 The Generalized Eigenvalue Problem

Introduction

In Section 12.17 we considered the mechanical system consisting of three springs attached to two masses and formulated the differential equation

$$MX'' + KX = 0$$

where M is a matrix whose entries are the masses and K is a matrix whose entries are related to the spring constants for the springs. The vector X contains the displacements of the masses from equilibrium, and primes on X represent differentiation with respect to time, t.

We then uncoupled the differential equations by multiplying through by the inverse of M to get

$$X'' + M^{-1}KX = 0$$

or

$$X'' + UX = 0$$

where $U = M^{-1}K$, and then diagonalized U, writing it as $U = PDP^{-1}$, to get

$$X'' + PDP^{-1}X = 0$$

Next, we multiplied through by the inverse of P to obtain

$$P^{-1}X'' + DP^{-1}X = 0$$

By defining $R = P^{-1}X$ we wrote the differential equation as $R'' + DR = 0$. Since D was a diagonal matrix, we uncoupled the equations and were led to the normal modes of vibration for the physical system.

Alternatively, we could have assumed a solution of the form $X = e^{\lambda t}Y$, in which case we would have faced a generalized eigenvalue problem of the form

$$\lambda^2 MY + KY = 0$$

In the present section, we will discuss the generalized eigenvalue problem of the form

$$Ax + \lambda Bx = 0 \qquad (45.23)$$

where A and B are $n \times n$ matrices and B is not singular.

EXAMPLE 45.10 By way of an example, consider the matrices

$$A = \begin{bmatrix} -4 & -6 \\ -7 & -5 \end{bmatrix} \quad \text{and} \quad B = \begin{bmatrix} -4 & -6 \\ 5 & 9 \end{bmatrix}$$

and write equation (45.23) as $(A - \lambda B)x = 0$, recognizing that the matrix $C = (A - \lambda B)$

must have a vanishing determinant if we want $\mathbf{x} \neq \mathbf{0}$. Thus, we obtain

$$\det C = \det \begin{bmatrix} -4 + 4\lambda & -6 + 6\lambda \\ -7 - 5\lambda & -5 - 9\lambda \end{bmatrix} = -22 + 28\lambda - 6\lambda^2 = 0$$

as the characteristic equation whose solutions are the eigenvalues 1 and $\frac{11}{3}$.

To find the corresponding solutions $\mathbf{x}$, substitute each eigenvalue, in turn, into the matrix C, and solve $C\mathbf{x} = \mathbf{0}$. Noting that any such x is in the null space of the respective matrix, we have the two generalized eigenvectors

$$\mathbf{x}_1 = \begin{bmatrix} -\frac{7}{6} \\ 1 \end{bmatrix} \quad \text{and} \quad \mathbf{x}_2 = \begin{bmatrix} -\frac{3}{2} \\ 1 \end{bmatrix}$$

Thus, the eigenpairs $(1, \mathbf{x}_1)$ and $(\frac{11}{3}, \mathbf{x}_2)$ satisfy (45.23). Moreover, these eigenpairs are actually the eigenpairs of the matrix

$$F = B^{-1}A = \begin{bmatrix} 13 & 14 \\ -8 & -\frac{25}{3} \end{bmatrix}$$

Our final observation is that the eigenpairs $(1, \mathbf{x}_1)$ and $(\frac{11}{3}, \mathbf{x}_2)$ are eigenpairs neither for A nor for B since the eigenpairs of A and B are, respectively,

$$\left(-11, \begin{bmatrix} 1 \\ \frac{7}{6} \end{bmatrix}\right), \left(2, \begin{bmatrix} 1 \\ -1 \end{bmatrix}\right) \quad \text{and} \quad \left(-1, \begin{bmatrix} -2 \\ 1 \end{bmatrix}\right), \left(6, \begin{bmatrix} 1 \\ -\frac{5}{3} \end{bmatrix}\right) \qquad ❖$$

In general, the eigenvectors of $F = B^{-1}A$ are not the eigenvectors of either A or B, as we have just seen. However, it is possible that they are, as the next example shows.

EXAMPLE 45.11   The matrices

$$A = \begin{bmatrix} -32 & -26 & 84 \\ 158 & 152 & -84 \\ 32 & 32 & -18 \end{bmatrix} \quad \text{and} \quad B = \begin{bmatrix} -2 & 1 & -7 \\ 9 & 6 & 7 \\ 2 & 2 & -2 \end{bmatrix}$$

each have the same set of eigenvectors, which are then the eigenvectors for $A\mathbf{x} = \lambda B\mathbf{x}$. Moreover, if $(\mu_k, \mathbf{x}_k)$ is an eigenpair for B and $(\lambda_k, \mathbf{x}_k)$ is an eigenpair for $A\mathbf{x} = \lambda B\mathbf{x}$, then $(\sigma_k, \mathbf{x}_k)$ is an eigenpair for A, with $\sigma_k = \lambda_k \mu_k$. This might seem like the only way for (45.23) to have a solution. Surprisingly, it is the exceptional way.

The eigenpairs for $A\mathbf{x} = \lambda B\mathbf{x}$ are most easily found from the matrix

$$F = B^{-1}A = \begin{bmatrix} 12 & 10 & -49 \\ 6 & 8 & 49 \\ 2 & 2 & 9 \end{bmatrix}$$

The eigenpairs of A, B, and F are then given in Table 45.5. All three matrices have the same eigenvectors, and the eigenvalues are related by $\sigma_k = \lambda_k \mu_k$.

Because A, B, and $F = B^{-1}A$ all share the same eigenvectors, it is possible to diagonalize A and B simultaneously with the transition matrix $S = [\mathbf{x}_1, \mathbf{x}_2, \mathbf{x}_3]$. The matrix S and the products $S^{-1}AS$ and $S^{-1}BS$ are given in (45.24), where the resulting diagonal matrices of eigenvalues can be seen. Thus, with the same transition matrix S, both A and B are similar to diagonal matrices.

$$S = \begin{bmatrix} 1 & 1 & 1 \\ -19 & -1 & -1 \\ -4 & -\frac{1}{7} & 0 \end{bmatrix} \quad S^{-1}AS = \begin{bmatrix} 126 & 0 & 0 \\ 0 & -18 & 0 \\ 0 & 0 & -6 \end{bmatrix} \quad S^{-1}BS = \begin{bmatrix} 7 & 0 & 0 \\ 0 & -2 & 0 \\ 0 & 0 & -3 \end{bmatrix}$$

$$\tag{45.24}$$

❖

Matrix	Eigenvalues	Eigenvectors
A	$126, -18, -6$	$\begin{bmatrix} 1 \\ -19 \\ -4 \end{bmatrix}, \begin{bmatrix} 1 \\ -1 \\ -\frac{1}{7} \end{bmatrix}, \begin{bmatrix} 1 \\ -1 \\ 0 \end{bmatrix}$
B	$7, -2, -3$	
F	$18, 9, 2$	

TABLE 45.5 The eigenvalues and eigenvectors of the matrices A, B, and $F = B^{-1}A$ in Example 45.11

In general, not every solution of the generalized eigenvalue problem leads to the simultaneous diagonalization of the two matrices A and B. We were able to diagonalize A and B simultaneously in Example 45.11 because of the special outcome that the eigenvectors of A and B were the same.

We will next examine conditions under which we are guaranteed simultaneous diagonalization of the two matrices A and B in the generalized eigenvalue problem $A\mathbf{x} = \lambda B\mathbf{x}$.

Symmetry

Recall that a real matrix B is positive definite if it is symmetric and has positive eigenvalues. (Positivity of the eigenvalues is equivalent to the condition that $\mathbf{x}^{\mathrm{T}} B\mathbf{x} > 0$ for all vectors $\mathbf{x} \neq \mathbf{0}$, as discussed in Section 33.4.) Therefore, a positive-definite matrix is invertible because a singular matrix must have at least one zero eigenvalue.

If, in the generalized eigenvalue problem $A\mathbf{x} = \lambda B\mathbf{x}$, A is symmetric and B is positive definite, then A and B can be simultaneously diagonalized by a matrix $S = [\mathbf{x}_1, \ldots, \mathbf{x}_n]$, where

$$A\mathbf{x}_k = \lambda_k B\mathbf{x}_k \qquad k = 1, \ldots, n$$
$$S^{\mathrm{T}} A S = \Lambda \qquad \text{and} \quad \Lambda = \begin{bmatrix} \lambda_1 & \cdots & 0 \\ \vdots & \ddots & \vdots \\ 0 & \cdots & \lambda_n \end{bmatrix}$$
$$S^{\mathrm{T}} B S = I$$

Solving the generalized eigenvalue problem for eigenpairs $(\lambda_k, \mathbf{x}_k)$ leads to the matrices S and Λ of the simultaneous diagonalization. However, neither diagonalization is a similarity transform, since S is not necessarily an orthogonal matrix. Recall that a similarity transformation $P^{-1} M P = D$ uses the inverse of P as one of the factors. If P is orthogonal, then the transpose is the inverse and the matrix M is said to be orthogonally similar to a diagonal matrix D. Since S will not necessarily be an orthogonal matrix, S^{T} will not necessarily be S^{-1}, so $S^{\mathrm{T}} A S = \Lambda$ and $S^{\mathrm{T}} B S = I$ will not necessarily be similarity transforms.

Because B is invertible, the eigenpairs $(\lambda_k, \mathbf{x}_k)$ can still be found from the matrix $F = B^{-1}A$. However, this approach vitiates the advantages of symmetry, since the inverse of a symmetric matrix is not necessarily symmetric.

An algorithm that preserves symmetry, finds S, and still simultaneously diagonalizes A and B will be given shortly. First, however, we give an example where A is symmetric, B is positive definite, and S is obtained from $B^{-1}A$.

In (45.25), the matrices A and B are clearly symmetric. The eigenvalues of B are $0.01, 0.05, 0.13$, so B is positive definite.

$$A = \frac{1}{1352} \begin{bmatrix} -41 & 107 & 135 \\ 107 & 67 & -105 \\ 135 & -105 & 215 \end{bmatrix} \qquad B = \frac{1}{2704} \begin{bmatrix} 347 & -7 & -29 \\ -7 & 43 & -15 \\ -29 & -15 & 131 \end{bmatrix} \qquad (45.25)$$

The matrix $F = B^{-1}A$ and its eigenpairs $(\lambda_k, \mathbf{p}_k)$ are given in

$$F = \frac{1}{26}\begin{bmatrix} 3 & 15 & 25 \\ 155 & 73 & -95 \\ 72 & -30 & 80 \end{bmatrix} \qquad \begin{array}{l} (-2, [-2, -1]^{\mathrm{T}}) \\[4pt] (3, [1, \frac{5}{2}, \frac{3}{2}]^{\mathrm{T}}) \\[4pt] (5, [0, -\frac{5}{3}, 1]^{\mathrm{T}}) \end{array} \tag{45.26}$$

The matrices $P = [\mathbf{p}_1, \mathbf{p}_2, \mathbf{p}_3]$, $\Lambda = \mathrm{diag}(\lambda_1, \lambda_2, \lambda_3)$, and $D = \mathrm{diag}(a, b, c)$ are given in

$$P = \begin{bmatrix} 1 & 1 & 0 \\ -2 & \frac{5}{2} & -\frac{5}{3} \\ -1 & \frac{3}{2} & 1 \end{bmatrix} \qquad \Lambda = \begin{bmatrix} -2 & 0 & 0 \\ 0 & 3 & 0 \\ 0 & 0 & 5 \end{bmatrix} \qquad D = \begin{bmatrix} a & 0 & 0 \\ 0 & b & 0 \\ 0 & 0 & c \end{bmatrix} \tag{45.27}$$

The matrix $S = [a\mathbf{p}_1, b\mathbf{p}_2, c\mathbf{p}_3] = PD$ must satisfy both $S^{\mathrm{T}}AS = \Lambda$ and $S^{\mathrm{T}}BS = I$. The columns of S are simply the *scaled* eigenvectors of $F = B^{-1}A$, where the scale factors $a = 2, b = 2, c = 3$ are determined by the equations

$$(PD)^{\mathrm{T}}B(PD) = \begin{bmatrix} \frac{1}{4}a^2 & 0 & 0 \\ 0 & \frac{1}{4}b^2 & 0 \\ 0 & 0 & \frac{1}{9}c^2 \end{bmatrix} = \begin{bmatrix} 1 & 0 & 0 \\ 0 & 1 & 0 \\ 0 & 0 & 1 \end{bmatrix} = I$$

The resulting matrix S, given on the left in (45.28), is not orthogonal, as verified by the product $S^{\mathrm{T}}S$, given on the right in (45.28). Computation shows that $S^{\mathrm{T}}AS = \Lambda$ and verifies that $S^{\mathrm{T}}BS = I$.

$$S = PD = \begin{bmatrix} 2 & 2 & 0 \\ -4 & 5 & -5 \\ -2 & 3 & 3 \end{bmatrix} \qquad S^{\mathrm{T}}S = \begin{bmatrix} 24 & -22 & 14 \\ -22 & 38 & -16 \\ 14 & -16 & 34 \end{bmatrix} \neq I \tag{45.28}$$

In the final part of this section, we find S (or its equivalent) with symmetry-preserving algorithms taken from [60], [27], and [64].

Symmetry-Preserving Algorithms

ALGORITHM 1   If A is symmetric and B is positive definite, the generalized eigenvalue problem (45.23) can be solved and the matrices A and B simultaneously diagonalized by a matrix S obeying both $S^{\mathrm{T}}AS = \Lambda$ and $S^{\mathrm{T}}BS = I$, via the following algorithm that preserves the symmetry in A and B.

1.   Use the Cholesky decomposition (Section 34.1) to factor B into the product LL^{T}, where L is a lower triangular matrix.
2.   Form the (symmetric) matrix $C = L^{-1}A\left(L^{-1}\right)^{\mathrm{T}}$.
3.   Obtain $\mathbf{v}_k, k = 1, \ldots, n$, the normalized eigenvectors of C, and $\lambda_1, \ldots, \lambda_n$, the corresponding eigenvalues.
4.   Obtain $\mathbf{x}_k = \left(L^{-1}\right)^{\mathrm{T}}\mathbf{v}_k, k = 1, \ldots, n$.
5.   The matrix S is given by $S = [\mathbf{x}_1, \ldots, \mathbf{x}_n]$, and the matrix Λ is the diagonal matrix with $\lambda_1, \ldots, \lambda_n$ on the diagonal.
6.   The eigenpairs $(\lambda_k, \mathbf{x}_k)$ satisfy the generalized eigenvalue problem (45.23).

EXAMPLE 45.12   Before explaining why these are the appropriate steps, we first illustrate them for the matrices A and B from (45.25). The factor L in the Cholesky decomposition $B = LL^T$ and the symmetric matrix $C = L^{-1}A\left(L^{-1}\right)^T$ are given in (45.29).

$$
L = \begin{bmatrix} \frac{\sqrt{347}}{52} & 0 & 0 \\ -\frac{7\sqrt{347}}{18044} & \sqrt{\frac{11}{694}} & 0 \\ -\frac{29\sqrt{347}}{18044} & -\frac{2\sqrt{2}}{\sqrt{3817}} & \frac{1}{\sqrt{22}} \end{bmatrix} \quad C = \begin{bmatrix} -\frac{82}{347} & \frac{1417}{347}\sqrt{\frac{2}{11}} & \frac{72\sqrt{2}}{\sqrt{3817}} \\ \frac{1417}{347}\sqrt{\frac{2}{11}} & \frac{12700}{3817} & -\frac{381\sqrt{347}}{3817} \\ \frac{72\sqrt{2}}{\sqrt{3817}} & -\frac{381\sqrt{347}}{3817} & \frac{32}{11} \end{bmatrix}
$$

$$(45.29)$$

The eigenvalues of C are $\lambda_1 = -2, \lambda_2 = 3, \lambda_3 = 5$. The corresponding eigenvectors, $\mathbf{v}_k, k = 1, 2, 3$, are on the left in Table 45.6. The vectors $\mathbf{x}_k = \left(L^{-1}\right)^T \mathbf{v}_k, k = 1, 2, 3$, are on the right in Table 45.6. The matrix $S = [\mathbf{x}_1, \mathbf{x}_2, \mathbf{x}_3]$ is then identical to the matrix S in (45.28). ❖

$\mathbf{v}_k, k = 1, 2, 3$			$\mathbf{x}_k, k = 1, 2, 3$		
$\begin{bmatrix} \frac{15}{\sqrt{347}} \\ -\frac{18\sqrt{2}}{\sqrt{3817}} \\ -\sqrt{\frac{2}{11}} \end{bmatrix}$	$\begin{bmatrix} \frac{11}{\sqrt{347}} \\ \frac{43}{\sqrt{7634}} \\ \frac{3}{\sqrt{22}} \end{bmatrix}$	$\begin{bmatrix} \frac{1}{\sqrt{347}} \\ \frac{67}{\sqrt{7634}} \\ -\frac{3}{\sqrt{22}} \end{bmatrix}$	$\begin{bmatrix} 2 \\ -4 \\ -2 \end{bmatrix}$	$\begin{bmatrix} 2 \\ 5 \\ 3 \end{bmatrix}$	$\begin{bmatrix} 0 \\ 5 \\ -3 \end{bmatrix}$

TABLE 45.6   The eigenvectors $\mathbf{v}_k$ and the vectors $\mathbf{x}_k = \left(L^{-1}\right)^T \mathbf{v}_k, k = 1, 2, 3$, in Example 45.12

Returning to our discussion of Algorithm 1, we note we have already verified that the matrix S simultaneously diagonalizes A and B and that its columns are the eigenvectors $\mathbf{x}_k$ for which the eigenpairs $(\lambda_k, \mathbf{x}_k)$ satisfy the generalized eigenvalue problem. All that remains is to justify the steps of the symmetry-preserving algorithm. After obtaining the Cholesky decomposition of B so that $B = LL^T$, the generalized eigenvalue problem (45.23) becomes $A\mathbf{x} = \lambda LL^T\mathbf{x}$, or $L^{-1}A\mathbf{x} = \lambda L^T\mathbf{x}$, upon left-multiplication by L^{-1}, the inverse of L. Between A and $\mathbf{x}$ on the left, insert $I = \left(L^T\right)^{-1} L^T$, obtaining

$$L^{-1}A\left(L^T\right)^{-1} L^T\mathbf{x} = \lambda L^T\mathbf{x}$$

Since the inverse of the transpose is the transpose of the inverse, that is, since $\left(L^T\right)^{-1} = \left(L^{-1}\right)^T$, we have

$$L^{-1}A\left(L^{-1}\right)^T \mathbf{x} = \lambda L^T\mathbf{x}$$

Defining the matrix $C = L^{-1}A\left(L^{-1}\right)^T$ and the vector $\mathbf{v} = L^T\mathbf{x}$ brings the generalized eigenvalue problem into the form $C\mathbf{v} = \lambda\mathbf{v}$, which is similar to the form $B^{-1}A\mathbf{x} = \lambda\mathbf{x}$, except that C is symmetric and $B^{-1}A$ generally is not.

The symmetry of C is established by showing $C^T = C$. Since C is of the form RAR^T, its transpose is $\left(R^T\right)^T A^T R^T$, which simplifies to $RAR^T = C$ because A is assumed symmetric.

If $\mathbf{v}_k, k = 1, \ldots, n$, are the normalized eigenvectors of C, $\lambda_k, k = 1, \ldots, n$, the corresponding eigenvalues, and $\mathbf{x}_k = \left(L^{-1}\right)^T \mathbf{v}_k$, then $S = [\mathbf{x}_1, \ldots, \mathbf{x}_n]$ will simultaneously diagonalize A and B by satisfying both $S^T A S = \Lambda$ and $S^T B S = I$. To see the second claim, note that the ij-entry in the product $S^T B S$ is given by

$$(S^T B S)_{ij} = \mathbf{x}_i^T B \mathbf{x}_j = \mathbf{v}_i^T L^{-1} B \left(L^{-1}\right)^T \mathbf{v}_j = \mathbf{v}_i^T L^{-1} L L^T \left(L^T\right)^{-1} \mathbf{v}_j$$

$$= \mathbf{v}_i^T \mathbf{v}_j = \delta_{ij} = \begin{cases} 1 & i = j \\ 0 & i \neq j \end{cases}$$

which indicates $S^T B S = I$.

The first equation is likewise verified with

$$(S^T A S)_{ij} = \mathbf{x}_i^T A \mathbf{x}_j = \mathbf{v}_i^T L^{-1} A \left(L^{-1}\right)^T \mathbf{v}_j = \mathbf{v}_i^T C \mathbf{v}_j = \mathbf{v}_i^T \lambda_j \mathbf{v}_j = \lambda_j \mathbf{v}_i^T \mathbf{v}_j = \lambda_j \delta_{ij}$$

so that $S^T A S = \Lambda$, a diagonal matrix whose main-diagonal entries are $\lambda_1, \ldots, \lambda_n$.

ALGORITHM 2   Detailed in [27] and sketched in [64], the following algorithm also preserves the symmetry in A and B but obtains S from a diagonalization of B. We still assume that B is positive definite and that the generalized eigenvalue problem is given by (45.23).

1.   For B, obtain the eigenpairs $(\sigma_k, \mathbf{u}_k)$, where the $\mathbf{u}_k$ are unit eigenvectors.

2.   Form $K = [\mathbf{u}_1, \ldots, \mathbf{u}_n]$ and the diagonal matrix P whose main-diagonal entries are $1/\sqrt{\sigma_k}, k = 1, \ldots, n$. Because B is symmetric and $\mathbf{u}_1, \ldots, \mathbf{u}_n$ are unit vectors, K is an orthogonal matrix.

3.   Form the matrix $M = (KP)^T A(KP)$ and obtain its eigenpairs $(\lambda_k, \mathbf{y}_k)$, where the $\mathbf{y}_k$ are unit eigenvectors.

4.   Form the matrix $W = [\mathbf{y}_1, \ldots, \mathbf{y}_n]$, an orthogonal matrix since M is symmetric and the $\mathbf{y}_k$ are unit vectors.

5.   The matrix $T = KPW$ satisfies both $T^T A T = \Lambda'$ and $T^T B T = I$, where Λ' is the diagonal matrix whose main-diagonal elements are $\lambda_1, \ldots, \lambda_n$. The columns of T, along with the λ_k, give the solutions to the generalized eigenvalue problem.

EXAMPLE 45.13   Using floating-point arithmetic, we illustrate Algorithm 2 by finding K, the orthogonal matrix whose columns are the normalized eigenvectors of B. The matrix K and the corresponding eigenvalues of B are given in

$$K = \begin{bmatrix} -0.0389430473 & -0.1247095323 & -0.9914287526 \\ -0.9838516582 & 0.1782489827 & 0.0162238945 \\ -0.1746978924 & -0.9760506302 & 0.1296372388 \end{bmatrix} \quad \begin{matrix} \sigma_1 = 0.0148148836 \\ \sigma_2 = 0.0480895058 \\ \sigma_3 = 0.1297731253 \end{matrix}$$

$$\text{(45.30)}$$

The diagonal matrix P containing $1/\sqrt{\sigma_k}, k = 1, 2, 3$, on the main diagonal, is

$$P = \begin{bmatrix} 8.215819287 & 0 & 0 \\ 0 & 4.560105000 & 0 \\ 0 & 0 & 2.775924296 \end{bmatrix}$$

The matrix $M = (KP)^T A(KP)$ is then

$$M = \begin{bmatrix} 2.261392409 & -1.451792172 & 2.246419733 \\ -1.451792172 & 4.167526127 & 0.716119721 \\ 2.246419734 & 0.716119721 & -0.428918542 \end{bmatrix}$$

The matrix W, whose columns are $\mathbf{y}_k$, the normalized eigenvectors of M, along with λ_k, the eigenvalues of M, are given in

$$W = \begin{bmatrix} -0.5349636425 & -0.6720252544 & -0.5120507367 \\ 0.8375677321 & -0.5013757467 & -0.2170314638 \\ -0.1108791932 & -0.5449811171 & 0.8310844643 \end{bmatrix} \quad \begin{matrix} \lambda_1 = 5 \\ \lambda_2 = 3 \\ \lambda_3 = -2 \end{matrix} \quad (45.31)$$

Finally, the matrix $T = KPW$ is given in (45.32). Comparing T to S in (45.28), we see that the first and third columns are interchanged and have opposite signs. Thus, calculation shows both that $T^{\mathrm{T}} A T = \Lambda' = \mathrm{diag}(5, 3, -2)$ and $T^{\mathrm{T}} B T = I$.

$$T = \begin{bmatrix} 0 & 2.0 & -2.0 \\ 5.0 & 5.0 & 4.0 \\ -3.0 & 3.0 & 2.0 \end{bmatrix} \quad (45.32)$$

❖

EXERCISES 45.5

The following pairs of matrices A_k and B_k, $k = 1, \ldots, 5$, are used in Exercises 1–3

$$A_1 = \begin{bmatrix} -5 & -2 & -9 \\ -2 & -1 & 5 \\ -9 & 5 & 8 \end{bmatrix} \quad B_1 = \begin{bmatrix} 7 & 1 & -2 \\ 1 & 4 & -3 \\ -2 & -3 & 7 \end{bmatrix}$$

$$A_2 = \begin{bmatrix} 4 & -3 & -9 \\ -3 & 8 & 7 \\ -9 & 7 & 5 \end{bmatrix} \quad B_2 = \begin{bmatrix} 7 & 1 & -2 \\ 1 & 4 & 1 \\ -2 & 1 & 3 \end{bmatrix}$$

$$A_3 = \begin{bmatrix} 0 & 4 & 2 \\ 4 & 4 & 5 \\ 2 & 5 & -5 \end{bmatrix} \quad B_3 = \begin{bmatrix} 9 & -2 & 5 \\ -2 & 4 & 1 \\ 5 & 1 & 8 \end{bmatrix}$$

$$A_4 = \begin{bmatrix} 9 & -9 & 0 \\ -9 & 4 & -8 \\ 0 & -8 & -5 \end{bmatrix} \quad B_4 = \begin{bmatrix} 8 & 6 & -2 \\ 6 & 9 & -2 \\ -2 & -2 & 1 \end{bmatrix}$$

$$A_5 = \begin{bmatrix} 4 & 7 & 5 \\ 7 & -9 & 7 \\ 5 & 7 & -8 \end{bmatrix} \quad B_5 = \begin{bmatrix} 2 & -1 & -2 \\ -1 & 6 & 5 \\ -2 & 5 & 9 \end{bmatrix}$$

1. Select a pair of matrices $A = A_k$, $B = B_k$ from the list.

   (a) Show B is positive definite.

   (b) Obtain $(\mu_k, \mathbf{z}_k)$, $k = 1, 2, 3$, the eigenpairs of $F = B^{-1}A$, and show that they are solutions of $A\mathbf{x} = \lambda B\mathbf{x}$.

   (c) Show that the eigenvalues in part (b) are the solutions of the characteristic equation $\det(A - \lambda B) = 0$.

2. For the pair of matrices selected in Exercise 1:

   (a) Apply Algorithm 1 for the simultaneous diagonalization of A and B, obtaining the vectors $\mathbf{x}_k$ and the matrices $S = [\mathbf{x}_1, \mathbf{x}_2, \mathbf{x}_3]$ and $\Lambda = \mathrm{diag}(\lambda_1, \lambda_2, \lambda_3)$.

   (b) Verify that $S^{\mathrm{T}} A S = \Lambda$ and that $S^{\mathrm{T}} B S = I$.

   (c) Verify that $(\lambda_k, \mathbf{x}_k)$, $k = 1, 2, 3$, are solutions of $A\mathbf{x} = \lambda B\mathbf{x}$.

   (d) Show that the sets $\{\mu_k\}$ and $\{\lambda_k\}$ contain the same members.

   (e) Show that if $\mu_k = \lambda_k$, then $\mathbf{z}_k$ is proportional to $\mathbf{x}_k$.

3. For the pair of matrices selected in Exercise 1:

   (a) Apply Algorithm 2 for the simultaneous diagonalization of A and B, obtaining the matrices K, P, M, W, and T, and λ_k, $k = 1, 2, 3$, the eigenvalues of M.

   (b) If $T = [\mathbf{t}_1, \mathbf{t}_2, \mathbf{t}_3]$, verify that $T^{\mathrm{T}} A T = \mathrm{diag}(\lambda_1, \lambda_2, \lambda_3)$ and that $T^{\mathrm{T}} B T = I$.

   (c) Show that the pairs $(\lambda_k, \mathbf{t}_k)$, $k = 1, 2, 3$, are solutions of $A\mathbf{x} = \lambda B\mathbf{x}$.

The following pairs of matrices A_k, B_k, $k = 1, \ldots, 5$, are used in Exercises 4 and 5.

$$A_1 = \begin{bmatrix} 5 & 7 \\ 7 & 3 \end{bmatrix} \quad B_1 = \begin{bmatrix} 6 & 3 \\ 3 & 5 \end{bmatrix} \quad A_2 = \begin{bmatrix} 9 & 2 \\ 2 & 3 \end{bmatrix} \quad B_2 = \begin{bmatrix} 2 & 1 \\ 1 & 3 \end{bmatrix}$$

$$A_3 = \begin{bmatrix} 7 & 6 \\ 6 & 8 \end{bmatrix} \quad B_3 = \begin{bmatrix} 2 & 1 \\ 1 & 7 \end{bmatrix} \quad A_4 = \begin{bmatrix} 2 & 5 \\ 5 & 9 \end{bmatrix} \quad B_4 = \begin{bmatrix} 5 & 4 \\ 4 & 6 \end{bmatrix}$$

$$A_5 = \begin{bmatrix} 8 & 2 \\ 2 & 1 \end{bmatrix} \quad B_5 = \begin{bmatrix} 7 & 5 \\ 5 & 4 \end{bmatrix}$$

4. Select a pair of matrices $A = A_k$, $B = B_k$ from the given list. Suppose the given pair of matrices A and B are the matrix coefficients of the vector differential equation $A\mathbf{x}'' + B\mathbf{x} = \mathbf{0}$ for which the initial condition is $\mathbf{x}(0) = [1, -2]^{\mathrm{T}}$, $\mathbf{x}'(0) = [2, -3]^{\mathrm{T}}$.

   (a) Multiply by A^{-1} to obtain $\mathbf{x}'' + U\mathbf{x} = \mathbf{0}$ write $U = PDP^{-1}$ (where D is a diagonal matrix), define $\mathbf{R} = P^{-1}\mathbf{x}$, and obtain $\mathbf{R}'' + D\mathbf{R} = \mathbf{0}$.

   (b) Solve the diagonalized system, and from this solution find the solution of the original initial value problem.

5. For the initial value problem chosen in Exercise 4:

   (a) Assume a solution of the form $\mathbf{x} = e^{\omega t} \mathbf{r}$, and obtain the algebraic equation $\omega^2 A\mathbf{r} + B\mathbf{r} = \mathbf{0}$.

   (b) Put this equation into the form $A\mathbf{r} = \lambda B\mathbf{r}$, and use either Algorithm 1 or 2 to obtain solutions $\mathbf{x}_k = e^{\omega_k t} \mathbf{r}_k$, $k = 1, 2$.

   (c) Use the solutions in part (b) to solve the given initial value problem.

Chapter Review

1. Demonstrate the use of the scaled power method for finding an eigenpair of the matrix $A = \begin{bmatrix} 1 & 2 \\ 3 & 4 \end{bmatrix}$.

2. Demonstrate the inverse power method for finding the other eigenpair of the matrix in Question 1. How should the calculations be implemented if computational efficiency is warranted?

3. Describe Gerschgorin's disk theorem.

4. Find the Householder reflection matrix that transforms $x = [1, 1, 1]^T$ to $y = [-\sqrt{3}, 0, 0]^T$.

5. Find the Householder reflection matrix that transforms
$$A = \begin{bmatrix} 1 & 2 & 3 \\ -2 & -1 & 0 \\ 3 & 1 & 1 \end{bmatrix} \text{ to upper Hessenberg form.}$$

6. Describe how Householder reflections are used to generate a QR decomposition of a matrix A.

7. Describe the Givens rotations. Give an example.

8. What is an annihilating Givens rotation?

9. Describe how annihilating Givens rotations are used to obtain the QR decomposition of a matrix in upper Hessenberg form.

10. Describe the shifted QR-algorithm for matrices in upper Hessenberg form. Where are the shifts obtained? Where do the eigenvalues appear?

Answers to Selected Exercises

CHAPTER 1

Exercises 1.2–Part A, Page 14

2. **(a)** linear **(b)** nonhomogeneous **(c)** first degree, first order **(d)** dependent variable y, independent variable t

5. **(a)** nonlinear **(b)** nonhomogeneous **(c)** first degree, first order **(d)** dependent variable y, independent variable x

13. **(a)** nonlinear **(b)** homogeneous **(c)** first degree, third order **(d)** dependent variable y, independent variable x

17. **(a)** boundary value **(b)** correct **(c)** nonlinear **(d)** second order   20. Upon substitution both sides equal $-\frac{1}{2}x + 3cx^3$; $c = \frac{3}{2}$.

Exercises 1.2–Part B, Page 15

2. Upon substitution both sides equal $\cos x$; $c = 2$.   3. Performing implicit differentiation on $x^2 y^3 - \frac{1}{4} y^4 = c$ gives $\frac{dy}{dx} = -2 \frac{xy}{3x^2 - y}$; $c = -\frac{1}{4}$.

7. Upon substitution both sides equal $13xe^{3/2x^2}$, moreover $y(0) = \frac{12}{3} = 4$.   10. Upon substitution both sides equal 0; $a = -\frac{2}{3}$, $b = -\frac{1}{3}$.

14. Upon substitution both sides equal 0; $a = 2$, $b = -4$.

Exercises 1.3–Part A, Page 17

1.

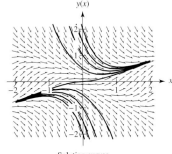

Solution curves

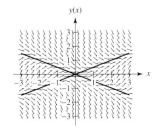

Nullclines

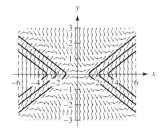

Isoclines

Exercises 1.3–Part B, Page 18

1.

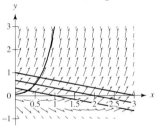

9. 0 is a source; 1 is a sink; 2 is a source.

Exercises 1.4–Part A, Page 19

1. Exact solution $\frac{1}{2}x^2 + 1$; $\phi_0 = 1$, $\phi_1 = \frac{1}{2}x^2 + 1$, $\phi_2 = \frac{1}{2}x^2 + 1, \ldots$　　　3. Exact solution $e^{1/2x^2}$; Maclaurin expansion $1 + \frac{1}{2}x^2 + \frac{1}{8}x^4 + \frac{1}{48}x^6 + O(x^8)$; $\phi_0 = 1$, $\phi_1 = \frac{1}{2}x^2 + 1$, $\phi_2 = \frac{1}{8}x^4 + \frac{1}{2}x^2 + 1$, $\phi_3 = \frac{1}{48}x^6 + \frac{1}{8}x^4 + \frac{1}{2}x^2 + 1$.

Exercises 1.4–Part B, Page 19

1. Upon substitution both sides equal $\frac{5}{3}e^{2x} - \frac{1}{2} - x$; $\phi_0 = 1$; $\phi_1 = \frac{1}{3}x^3 + 2x + 1$, $\phi_2 = \frac{1}{6}x^4 + \frac{1}{3}x^3 + 2x^2 + 2x + 1$, $\phi_3 = \frac{1}{15}x^5 + \frac{1}{6}x^4 + \frac{5}{3}x^3 + 2x^2 + 2x + 1$, $\phi_4 = \frac{1}{45}x^6 + \frac{1}{15}x^5 + \frac{5}{6}x^4 + \frac{5}{3}x^3 + 2x^2 + 2x + 1$; Taylor polynomial $1 + 2x + 2x^2 + \frac{5}{3}x^3 + \frac{5}{6}x^4 + O(x^5)$.

5. Upon substitution both sides equal $\cos x$; Taylor polynomial $1 - 2x + \frac{3}{2}x^2 - \frac{1}{3}x^3 - \frac{1}{8}x^4 + \frac{1}{12}x^5 + O(x^6)$; $\phi_0 = 1$; $\phi_1 = -x^2 - \cos x - 2x + 2$, $\phi_2 = \frac{1}{6}x^4 + \frac{4}{3}x^3 + 2\sin(x) - 3\cos(x) - 4x + 4$, $\phi_3 = -\frac{1}{90}x^6 - \frac{1}{5}x^5 - \frac{2}{3}x^4 + \frac{4}{3}x^3 + 10\sin x - 3\cos x - 12x + 4$, $\phi_4 = \frac{1}{2520}x^8 + \frac{4}{315}x^7 + \frac{1}{9}x^6 + \frac{2}{15}x^5 - \frac{2}{3}x^4 + 4x^3 + 8x^2 + 26\sin x + 13\cos x - 28x - 12$; Taylor polynomial for ϕ_4: $1 - 2x + \frac{3}{2}x^2 - \frac{1}{3}x^3 - \frac{1}{8}x^4 + \frac{7}{20}x^5 + O(x^6)$.

Exercises 1.5, Page 22

2. The determinant of the coefficient matrix equals zero. Infinitely many solutions $x = \frac{7}{3} + \frac{5}{3}t$, $y = t$.

3. The determinant of the coefficient matrix equals zero. No solutions, because adding two times the first equation to the second equation produces $0 = -7$.　　　7. The determinant of the coefficient matrix equals zero. Infinitely many solutions $x = 2 - 11t$, $y = t$, $z = 7t - 1$.

12. Substitution of $y = 0$ yields $0 = 0$. Substitution of $y = \left(\frac{3}{4}x\right)^{4/3}$ yields $\frac{1}{2}3^{1/3}2^{1/3}x^{1/3} = \frac{1}{2}3^{1/3}2^{1/3}x^{1/3}$. This does not contradict Picard's theorem because $\frac{\partial}{\partial y}y^{1/4} = \frac{1}{4}\frac{1}{y^{3/4}}$ is not continuous at 0.

14. Since both $f(x, y)$ and $f_y(x, y)$ are continuous in the entire plane, Picard's theorem guarantees existence and uniqueness on $|x| \le h$ where $h = \min\left\{\infty, \frac{b}{M}\right\} = \min\left\{\infty, \frac{b}{b^2+1}\right\} = \frac{b}{b^2+1}$. Since $\frac{b}{b^2+1}$ attains a maximum $\frac{1}{2}$, existence and uniqueness is guaranteed on $|x| \le \frac{1}{2}$. Observe that this result is based exclusively on the function $f(x, y) = 1 + y^2$ and not on the solution $\tan(x)$.

16. Picard's theorem does not guarantee existence and uniqueness, because $f_y(x, y) = \frac{1}{2}\frac{x}{\sqrt{xy}}$ is not continuous at $y = 0$; solutions $y = 0$ and $y = \frac{1}{9}x^3$.

CHAPTER 2

Exercises 2.1, Page 25

1. $k = \frac{1}{8}\ln 13$ years.　　　3. $11\frac{\ln 2}{\ln 3} = 6.9402$ years.　　　5. $5000\frac{\ln 100}{\ln 2} = 33{,}219$ years.　　　7. $k = \frac{1}{5}\ln 2 = 0.13863$ or 13.86% annually.

8. $\frac{dP}{dt} = \frac{k}{\sqrt{P}}$

Exercises 2.2–Part A, Page 27

1. $P(t) = \dfrac{Aa}{(a - A)e^{-kat} + A}$　　　2. $P_\infty = \frac{1}{2}\dfrac{ka + \sqrt{k^2a^2 - 4kh}}{k}$; $h_{max} = \frac{1}{4}ka^2$; new equilibrium level $\frac{1}{2}a$.

Exercises 2.2–Part B, Page 27

1. $a = \frac{230{,}736}{2309} = 99.929$, $P_0 = 10$, $k = \frac{2309}{115{,}368}\ln\frac{2209}{185} = 4.9634 \times 10^{-2}$; $P(t) = 1{,}153{,}680\dfrac{4{,}879{,}681^t}{11{,}545(4{,}879{,}681)^t + 103{,}823(34{,}225)^t}$

$P(\frac{3}{4}) = \frac{230{,}736}{2309 + 37\sqrt{185}} = 82.047$.　　　5. $y(t) = \dfrac{A}{A + (1 - A)e^{-t}}$, $A \in (0, 1)$; $t = \ln(\frac{1-A}{A})$.

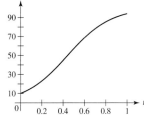

Exercises 2.3–Part A, Page 29

1. $\frac{dx}{dt} = rc - \frac{xr}{V}$　　　2. cV; c　Over time the contents of the tank becomes similar to the inflowing solution.

Exercises 2.3–Part B, Page 29

1. $\frac{dx}{dt} = 6 - \frac{3}{200}x$, $x(0) = 20$; $x(t) = 400 - 380e^{-3/200t}$

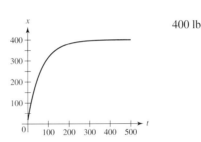

400 lb

5. $\frac{dx}{dt} = 6 - 2\frac{x}{100+t}$, $x(0) = 10$; $x(200) = \frac{5210}{9} = 578.89$ lb   **8.** $\frac{dx}{dt} = \frac{3}{2} - 5\frac{x}{300-2t}$, $x(0) = 30$; $\frac{1}{2}$ lb/gal

Exercises 2.4–Part B, Page 32

1. $U_0 = 1.5881°F$; $U_s = 29.888°F$, $k = -0.12846$

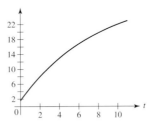

3. $U(t) = 1309.3 - 1203.2e^{0.0051154t}$

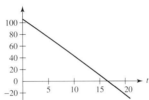

Observe from the graph that even though the ambient temperature is higher than the temperature of the object, the temperature of the object is decreasing rather than increasing.

4. Death could have occurred as much as 90.32 or as little as 81.30 minutes prior to 1.27 PM.

CHAPTER 3

Exercises 3.1–Part A, Page 36

1. If $x = x_0$ and $y = y_0$, the equation reduces to $0 = 0$.   **4.** $y(x) = -\frac{2}{3}\sqrt{3}\tanh(2(\arctan x + C)\sqrt{3})$ **5.** $x(t) = \dfrac{a + be^{k(-b+a)(t-C)}}{e^{k(-b+a)(t-C)} + 1}$

Exercises 3.1–Part B, Page 36

2. (a) $y(x) = -\frac{1}{\ln x - C}$ **(b)**

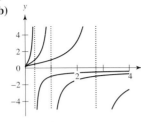

(c) $y(x) = -\frac{1}{\ln x + 1}$ **(d)**

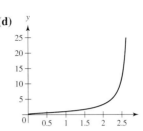

(e) $\displaystyle\int_1^y \frac{1}{t^2}\,dt = \int_1^x \frac{1}{t}\,dt \Rightarrow \frac{y-1}{y} = \ln(x) \Rightarrow y = -\frac{1}{\ln x - 1}$

Exercises 3.2–Part A, Page 38

1. $F(\lambda x, \lambda y) = (\lambda x)^k f(\frac{y}{x}) = \lambda^k F(x, y)$ **2.** $F(x, y) = x^k F(1, \frac{y}{x}) = x^k f(\frac{y}{x})$, with $f(u) = F(1, u)$

3. Substituting $F(x, y) = x^k f(\frac{y}{x})$ in $x\frac{\partial F}{\partial x} + y\frac{\partial F}{\partial y}$, followed by simplification, yields: $x\frac{\partial}{\partial x}x^k f(\frac{y}{x}) + y\frac{\partial}{\partial y}x^k f(\frac{y}{x}) = x^k k f(\frac{y}{x})$.

Exercises 3.2–Part B, Page 38

2. (a) $1 = \dfrac{Ce^{-4\arctan((x+2y)/x)}}{\sqrt{2xy + 2y^2 + x^2}}$ **(b)**

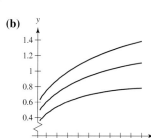

(c) $1 = \dfrac{\sqrt{5}e^{-4\arctan((x+2y)/x)+4\arctan(3)}}{\sqrt{2xy + 2y^2 + x^2}}$

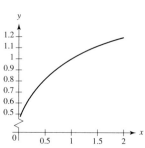

(d) $-\frac{1}{2}\ln(2\frac{y}{x} + 2\frac{y^2}{x^2} + 1) - 4\arctan(1 + 2\frac{y}{x}) = \ln x - \frac{1}{2}\ln 5 - 4\arctan 3$ To show equivalence with (c), exponentiate both sides and divide by the right hand side of the result.

6. $h = \frac{11}{16}, k = \frac{47}{32}; \frac{dw}{dt} = \frac{3t+2w}{7t-6w}; -\frac{3}{2}\ln 2 + \ln(16x - 11) - \frac{1}{2}\ln(-160yx - 454y + 103x + 298 + 192y^2 + 96x^2) - \frac{9}{47}\sqrt{47}\arctan(\frac{1}{47}\frac{(80x+227-192y)\sqrt{47}}{16x-11}) = -\ln 16 + \ln(16x - 11) + C$

Exercises 3.3–Part A, Page 41

3. $\frac{\partial M}{\partial y} = \frac{\partial N}{\partial x}$ yields $6xy^2 = 6xy^2$; $y = \frac{c}{x^{2/3}}$ **4.**

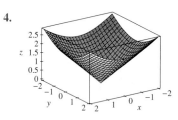 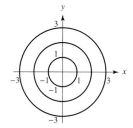

$y = \sqrt{(-x^2 + c^2)}, \; y = -\sqrt{(-x^2 + c^2)}; \; \dfrac{x\,dx}{\sqrt{x^2+y^2}} + \dfrac{y\,dy}{\sqrt{x^2+y^2}} = 0$

5. $-y\,dx + x\,dy = 0$ The equation is not exact; use separation of variables to obtain $y = Cx \Leftrightarrow f(x, y) = \frac{y}{x} = C$; $f_x\,dx + f_y\,dy = 0 \Leftrightarrow -\frac{y\,dx}{x^2} + \frac{dy}{x} = 0 \Leftrightarrow -y\,dx + x\,dy = 0$.

Exercises 3.3–Part B, Page 41

1. $-p\,dx + q\,dy = 0; \frac{\partial p}{\partial y} = -\frac{\partial q}{\partial x}$ **2.** $\frac{\partial p}{\partial y} = -\frac{\partial q}{\partial x} \Leftrightarrow 9 + 24x^2 = 9 + 24x^2; -9xy - 8x^3y + 9y^3 = c$

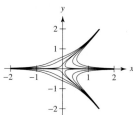

9. $\frac{\partial p}{\partial y} = -\frac{\partial q}{\partial x} \Leftrightarrow 24xy^2 + 15y^2 - 3x^2 - 60x^3y^2 = 24xy^2 + 15y^2 - 3x^2 - 60x^3y^2; -4x^2y^3 - 5xy^3 + x^3y + 5x^4y^3 = c$

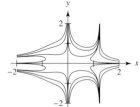

Exercises 3.4–Part A, Page 44

1. $U(t) = 1 + \frac{1}{5}\cos t + \frac{2}{5}\sin t + Ce^{-1/2t}$ Yes, the solution tends toward a periodic function with the same period 2π as U_s.

5. $u(x) = \frac{1}{x^2}; \frac{y}{x^2}\,dx - \frac{1}{x}\,dy = 0 \Rightarrow y = Cx$

Exercises 3.4–Part B, Page 45

1. $\frac{\partial}{\partial y}((2x + 3y)u(x, y)) = \frac{\partial}{\partial x}((5x + 7y)u(x, y)) \Leftrightarrow -2u(x, y) + (2x + 3y)\frac{\partial}{\partial y}u(x, y) - (5x + 7y)\frac{\partial}{\partial x}u(x, y) = 0$

6. $\frac{dy}{dx} + \frac{1}{2}\frac{y}{x} = \frac{8x^3 - 16x^2 + 2x - 1}{\sqrt{x}}$;

$y(x) = -\sqrt{x} + 2x^{7/2} - \frac{16}{3}x^{5/2} + x^{3/2} + \frac{C}{\sqrt{x}}$

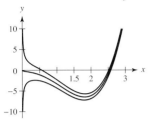

$C = -1, 0, 1$

13. $\frac{dy}{dx} + \frac{1}{2}\frac{y}{2-x} = \frac{10x^2 + 5x - 6}{x - 2}$;

$y(x) = -\frac{724}{3} + \frac{190}{3}x + \frac{20}{3}x^2 + C\sqrt{2 - x}$

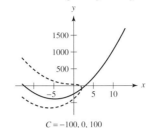

$C = -100, 0, 100$

17. $\frac{dx}{dt} = \frac{9}{2} - 4\frac{x}{500 - t}$, $x(0) = 75$; $\qquad \frac{3}{2}$

$x(t) = -\frac{27}{2,500,000,000}t^4 + \frac{27}{1,250,000}t^3 - \frac{81}{5000}t^2 + \frac{39}{10}t + 75$

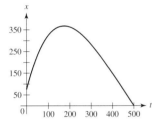

Exercises 3.5–Part A, Page 47

1. $ce^{-(P(x)+\alpha)} = ce^{-\alpha}e^{-P(x)} = Ce^{-P(x)}$ **3.** Rewrite the equation as $\frac{dy}{dx} = -py = f(x, y)$, then, since f and f_y are continuous in the entire xy-plane, the existence and uniqueness of the solution to the initial value problem $\frac{dy}{dx} + py = 0$, $y(x_0) = 0$ is guaranteed. Since $y = 0$ is a solution, it must be the only solution.   **5.** $x_h = Cx^2$, $x_p = -x^2 \cos x$

Exercises 3.5–Part B, Page 47

1. $\frac{dU}{dt} = k(U - 1 - e^{-t})$; $k = -1.205419580$;

$U(t) = 1 + 5.868085116e^{-t} - 3.868085116e^{-1.205419580t}$

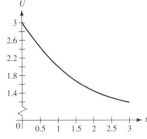

3. $\frac{dx}{dt} = 10 + 5te^{-t} - \frac{1}{20}x(t)$, $x(0) = 50$;

$x(t) = -\frac{2000}{361}e^{-t} - \frac{100}{19}te^{-t} + 200 - \frac{52,150}{361}e^{-1/20t}$

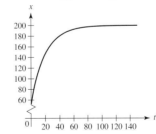

200 lb. If the incoming brine contains 2 lb of salt per gallon then the solution becomes $x(t) = 200 - 150e^{(-1/20t)}$. The first solution is always slightly larger than the second. See graph.

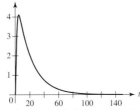

This makes sense, because in the second case the incoming brine contains less salt.

4. $\frac{dy}{dx} + 2\frac{y}{x} = -\frac{7x^3 - 6x^2}{x}$,

$y_h = \frac{C}{x^2}$, $y_p = -\frac{1}{10}x^2(14x - 15)$,

$y = \frac{C}{x^2} - \frac{1}{10}x^2(14x - 15)$

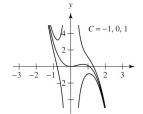

$C = -1, 0, 1$

9. $\frac{dy}{dx}y + 2\frac{\cos 2x}{\sin 2x}y = 16x + 3$;

$y_h = \frac{C}{\sin x \cos x}$; $y_p = -\frac{1}{2}\frac{16x\cos^2 x - 8\sin x\cos x - 8x + 3\cos^2 x}{\sin x \cos x}$;

$y = \frac{C}{\sin x \cos x} - \frac{1}{2}\frac{16x\cos^2 x - 8\sin x\cos x - 8x + 3\cos^2 x}{\sin x \cos x}$;

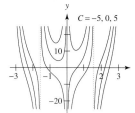

$C = -5, 0, 5$

19. $\frac{dy}{dx} - \frac{1}{2}\frac{y}{x} = -\frac{3x + 11}{x}$; $y_h = C\sqrt{x}$; $y_p = -6x + 22$; $y = C\sqrt{x} - 6x + 22$

26. $\frac{dy}{dx} - 2\frac{x}{x^2 - 4}y = -4x^2 + 7x + 5$; $y_h = C(x^2 - 4)$; $y_p = -\frac{1}{4}(x^2 - 4)(16x - 3\ln(x - 2) - 25\ln(x + 2))$;

$y = C(x^2 - 4) - \frac{1}{4}(x^2 - 4)(16x - 3\ln(x - 2) - 25\ln(x + 2))$

Exercises 3.6–Part A, Page 49

1. Substitute $z = y^{1-s}$ into $\frac{dz}{dt} + p(1 - s)z = r(1 - s)$ and simplify the resulting expression.

2. $\frac{dy}{dx} - 2\frac{y}{x} = (4x - 3)(4x + 3)y^2$; $\frac{dz}{dx} - 2\frac{z}{kx} = \frac{z^{k+1}(-9 + 16x^2)}{k}$ Choose $k = -1$; $\frac{dz}{dx} + 2\frac{z}{x} = 9 - 16x^2$.

5. $\frac{dy}{dx} + \frac{y}{x} = \frac{(x^2 + 5)y^{3/2}}{x^3}$; $\frac{dz}{dx} + \frac{z}{kx} = \frac{z^{(k+1)/2}(x^2 + 5)}{kx^3}$ Choose $k = -2$; $\frac{dz}{dx} - \frac{1}{2}\frac{z}{x} = -\frac{1}{2}\frac{x^2 + 5}{x^3}$.

Exercises 3.6–Part B, Page 49

1. $y(x) = \dfrac{1}{1 + e^{-x}C}$ **2.** It is not a Bernoulli equation.

3. $\frac{dy}{dx} - \frac{1}{2}\frac{y}{x} = -(2x^2 + 5x - 14)y^2$; $\frac{dz}{dx} + \frac{1}{2}\frac{z}{x} = 2x^2 + 5x - 14$; $y(x) = 21\dfrac{\sqrt{x}}{12x^{7/2} + 42x^{5/2} - 196x^{3/2} + 21C}$

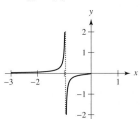

$C = -10i$

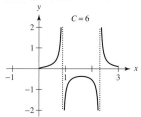

$C = 6$

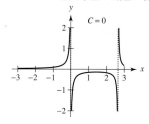

$C = 0$

7. $\frac{dy}{dx} + \frac{\sin x}{\cos x}y = -(4x^2 + 13x - 3)y^2$;

$\frac{dz}{dx} - \frac{\sin x}{\cos x}z = 4x^2 + 13x - 3$;

$y(x) = \dfrac{\cos x}{4x^2\sin x - 11\sin x + 8x\cos x + 13\cos x + 13x\sin x + C}$

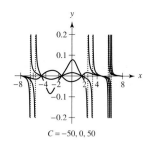

$C = -50, 0, 50$

CHAPTER 4

Exercises 4.1–Part A, Page 55

1. $y_1 = 1.20000$; $y_2 = 1.46500$ **3.** $\dfrac{y_{\text{approx}}\left(\frac{h}{2}\right) - y_{\text{approx}}(h)}{2^k - 1} = \dfrac{Ch^k - C\left(\frac{h}{2}\right)^k}{2^k - 1} = C\left(\dfrac{h}{2}\right)^k$

5. $y_1 := -1$, $y_2 := 1$, $y_3 := -1$, $y_4 := 1$, $y_5 := -1$; take $h = 0.0001$

Exercises 4.1–Part B, Page 55

2. **(a)** $y(t) = \frac{87{,}396}{9t^4 + 5t^2 - 5}$, $y(-6) = \frac{87{,}396}{11{,}839} = 7.3820424022299180674$ **(b)** $h = 0.1$, $y(-6) = 7.2181184243918785567$

(c)

s	$\left|\dfrac{y_{\text{exact}} - y_{\text{approx}}}{h}\right|$
1	1.3491576916414886760
2	1.5161943703554599664
3	1.6172854495413053544
4	1.6733790078835392128
5	1.7029993702267577888
6	1.7182297235447981888
7	1.7259535397271942528
8	1.7298430740664802816
9	1.7317948071854947840
10	1.7327724227253112832

(d)

s	$y_{\text{exact}} - y_{\text{approx}}(\tfrac{1}{2}h)$	$y_{\text{approx}}(\tfrac{1}{2}h) - y_{\text{approx}}(h)$	difference
2	0.3790485926	0.2955302532	-0.0835183394
3	0.2021606812	0.1768879114	-0.0252727698
4	0.1045861880	0.09757449320	-0.00701169480
5	0.05321873032	0.05136745767	-0.00185127265
6	0.02684733943	0.02637139089	-0.00047594854
7	0.01348401203	0.01336332740	-0.00012068463
8	0.006757199508	0.006726812521	-0.000030386987
9	0.003382411733	0.003374787775	$-0.7623958 \times 10^{-5}$
10	0.001692160569	0.001690251164	$-0.1909405 \times 10^{-5}$

(e)

s	y_{exact}	$2y_{\text{approx}}(\tfrac{1}{2}h) - y_{\text{approx}}(h)$
2	7.3820424022299180674	7.2985240628729324226
3	7.3820424022299180674	7.3567696324334567202
4	7.3820424022299180674	7.3750307074371388349
5	7.3820424022299180674	7.3801911295834669064
6	7.3820424022299180674	7.3815664536887293045
7	7.3820424022299180674	7.3819217176020681293
8	7.3820424022299180674	7.3820120152428923957
9	7.3820424022299180674	7.3820347782724219172
10	7.3820424022299180674	7.3820404928245668631

(f) $h = 0.1$, $y(-6) = 7.3820418889193688876$

(g)

s	$\left|\dfrac{y_{\text{exact}} - y_{\text{approx}}}{h^4}\right|$
1	0.0041870723765596704000
2	0.0047579289551821056000
3	0.0050688203689312256000
4	0.0052308689524555776000
5	0.0053135712715800576000
6	0.0053553451816189952000
7	0.0053763379482853376000
8	0.0053868635172634623999
9	0.0053920325104041983999
10	0.0053970627761012736000

(h)

s	$y_{\text{exact}} - y_{\text{approx}}(\tfrac{1}{2}h)$	$\tfrac{1}{15}y_{\text{approx}}(\tfrac{1}{2}h) - \tfrac{1}{15}y_{\text{approx}}(h)$
2	0.0000185856599811801	0.0000162070909035 8662
3	$0.12375049728836 \times 10^{-5}$	$0.115654366721976 \times 10^{-5}$
4	$0.798167259591 \times 10^{-7}$	$0.7717921646164 \times 10^{-7}$
5	$0.50674164501 \times 10^{-8}$	$0.498328730060 \times 10^{-8}$
6	$0.3192034472 \times 10^{-9}$	$0.31654753352 \times 10^{-9}$
7	$0.200284196 \times 10^{-10}$	$0.1994500184 \times 10^{-10}$
8	$0.12542269 \times 10^{-11}$	$0.125161285 \times 10^{-11}$
9	0.784644×10^{-13}	$0.7838417 \times 10^{-13}$
10	0.49086×10^{-14}	0.490372×10^{-14}

(i)

s	y_{exact}	$\tfrac{16}{15}y_{\text{approx}}(\tfrac{1}{2}h) - \tfrac{1}{15}y_{\text{approx}}(h)$
2	7.3820424022299180674	7.3820400236608404742
3	7.3820424022299180674	7.3820423212686124038
4	7.3820424022299180674	7.3820423995924085702
5	7.3820424022299180674	7.3820424021457889181
6	7.3820424022299180674	7.3820424022272621539
7	7.3820424022299180674	7.3820424022298346499
8	7.3820424022299180674	7.3820424022299154536
9	7.3820424022299180674	7.3820424022299179874
10	7.3820424022299180674	7.3820424022299180628

5. **(a)** $y(t) = \dfrac{\sqrt{3t + t\sqrt{9 + 84927744t^2}}}{2t^2}$; $y(9) = -5.33332127652920673349 0626$ **(b)** $h = 0.1$, $y(9) = -5.32583035176365611226238 8$

(c)

s	$\left\lvert \dfrac{y_{exact} - y_{approx}}{h} \right\rvert$
1	0.07843439248783619225341800
2	0.07619340831745341560100000
3	0.07512026002991237332042400
4	0.07459494246760486761774400
5	0.07433502919583199253724800
6	0.07420575058160208494681600
7	0.07414127974905152914995200
8	0.07410908632340039243673600
9	0.07409300009228084028979200
10	0.07408495959515706122035200

(d)

s	$y_{exact} - y_{approx}(\frac{1}{2}h)$	$y_{approx}(\frac{1}{2}h) - y_{approx}(h)$
2	−0.01904835207936335390025 0	−0.02016884416455474222645 9
3	−0.00939003250373904666505 3	−0.00965831957562430723519 7
4	−0.00466218390422530422610 9	−0.00472784859951374243894 4
5	−0.00232296966236974976678 9	−0.00233921424185555445932 0
6	−0.00115946485283753257729 4	−0.00116350480953221718949 5
7	−0.00057922874803946507148 4	−0.00058023610479806750581 0
8	−0.00028948861845078278295 6	−0.00028974012958868228852 8
9	−0.00014471289080523601619 1	−0.00014477572764554676676 5
10	−0.00007234859335464556759 8	−0.00007236429745059044859 3

(e)

s	y_{exact}	$2y_{approx}(\frac{1}{2}h) - y_{approx}(h)$
2	−5.33332127652920673349062 6	−5.33441768614398121816833
3	−5.33332127652920673349062 6	−5.33358956360109199406077 4
4	−5.33332127652920673349062 6	−5.33338694122449517170345 7
5	−5.33332127652920673349062 6	−5.33333752110869253818315 3
6	−5.33332127652920673349062 6	−5.33332531648590141810282 3
7	−5.33332127652920673349062 6	−5.33332228388596533592494 8
8	−5.33332127652920673349062 6	−5.33332152804034463299619 8
9	−5.33332127652920673349062 6	−5.33332133936604704424120 0
10	−5.33332127652920673349062 6	−5.33332129223330267837162 5

(f) $h = 0.1$, $y(9) = -5.333321276529177021179194$

(g)

s	$\left\lvert \dfrac{y_{exact} - y_{approx}}{h^4} \right\rvert$
1	$0.3058270233236000000000000 \times 10^{-9}$
2	$0.3004878372436480000000000 \times 10^{-9}$
3	$0.2976919640064000000000000 \times 10^{-9}$
4	$0.2962639328378880000000000 \times 10^{-9}$
5	$0.2955425999749120000000000 \times 10^{-9}$
6	$0.2951801334661120000000000 \times 10^{-9}$
7	$0.2949998486548480000000000 \times 10^{-9}$
8	$0.2948795696414720000000000 \times 10^{-9}$
9	$0.2954937499647999999999999 \times 10^{-9}$
10	$0.3067637441495040000000000 \times 10^{-9}$

(h)

s	$y_{exact} - y_{approx}(\frac{1}{2}h)$	$\frac{1}{15}y_{approx}(\frac{1}{2}h) - \frac{1}{15}y_{approx}(h)$
2	$-0.1173780614233 \times 10^{-11}$	$-0.11960272228995 \times 10^{-11}$
3	$-0.72678702150 \times 10^{-13}$	$-0.734067941388 \times 10^{-13}$
4	$-0.4520628858 \times 10^{-14}$	$-0.45438715528 \times 10^{-14}$
5	$-0.281851387 \times 10^{-15}$	$-0.2825851648 \times 10^{-15}$
6	$-0.17594107 \times 10^{-16}$	$-0.176171520 \times 10^{-16}$
7	$-0.1098955 \times 10^{-17}$	$-0.10996768 \times 10^{-17}$
8	-0.68657×10^{-19}	$-0.686865 \times 10^{-19}$
9	-0.4300×10^{-20}	-0.42905×10^{-20}
10	-0.279×10^{-21}	-0.2680×10^{-21}

(i)

s	y_{exact}	$\frac{16}{15}y_{approx}(\frac{1}{2}h) - \frac{1}{15}y_{approx}(h)$
2	−5.33332127652920673349062 6	−5.33332127652922898009929 4
3	−5.33332127652920673349062 6	−5.33332127652920746158261 6
4	−5.33332127652920673349062 6	−5.33332127652920675673332 3
5	−5.33332127652920673349062 6	−5.33332127652920673422440 6
6	−5.33332127652920673349062 6	−5.33332127652920673351367 3
7	−5.33332127652920673349062 6	−5.33332127652920673349135 0
8	−5.33332127652920673349062 6	−5.33332127652920673349065 7
9	−5.33332127652920673349062 6	−5.33332127652920673349061 9
10	−5.33332127652920673349062 6	−5.33332127652920673349061 7

13. **(a)** $y(t) = \frac{11}{5t^3} - \frac{3}{4t^2} + \frac{14}{3t} - \frac{43t^2}{60}$, $y(2) = -0.99583333333333333333$ **(b)** $h = 0.01$, $y(2) = -0.92155801415404984819$

(c)

$$
\begin{bmatrix}
s & \left|\frac{y_{\text{exact}} - y_{\text{approx}}}{h}\right| \\
1 & 5.4978395061728395062 \\
2 & 6.4780526817529732980 \\
3 & 6.9840075342886040498 \\
4 & 7.2288820525051161610 \\
5 & 7.3479611799361020467 \\
6 & 7.4065803031094071469 \\
7 & 7.4356582623578184755 \\
8 & 7.4501397923963020006 \\
9 & 7.4573662940130070682 \\
10 & 7.4609759940885130342
\end{bmatrix}
$$

(d)

$$
\begin{bmatrix}
s & y_{\text{exact}} - y_{\text{approx}}(\tfrac{1}{2}h) & y_{\text{approx}}(\tfrac{1}{2}h) - y_{\text{approx}}(h) \\
2 & -1.6195131704382433245 & -1.1294065826481764286 \\
3 & -0.8730009417860755062 & -0.7465122286521678182 \\
4 & -0.4518051282815697600 & -0.4211958135045057461 \\
5 & -0.2296237868730031889 & -0.2221813414085665711 \\
6 & -0.1157278172360844866 & -0.1138959696369187022 \\
7 & -0.0580910801746704568 & -0.0576367370614140298 \\
8 & -0.0291021085640480546 & -0.0289889716106224021 \\
9 & -0.0145651685429941544 & -0.0145369400210539002 \\
10 & -0.0072861093692270635 & -0.0072790591737670909
\end{bmatrix}
$$

(e)

$$
\begin{bmatrix}
s & y_{\text{exact}} & 2y_{\text{approx}}(\tfrac{1}{2}h) - y_{\text{approx}}(h) \\
2 & -0.99583333333333333333 & -0.5057267455432664375 \\
3 & -0.99583333333333333333 & -0.8693446201994256453 \\
4 & -0.99583333333333333333 & -0.9652240185562693193 \\
5 & -0.99583333333333333333 & -0.9883908878688967154 \\
6 & -0.99583333333333333333 & -0.9940014857341675489 \\
7 & -0.99583333333333333333 & -0.9953789902200769063 \\
8 & -0.99583333333333333333 & -0.9957201963799076808 \\
9 & -0.99583333333333333333 & -0.9958051048113930791 \\
10 & -0.99583333333333333333 & -0.9958262831378733607
\end{bmatrix}
$$

(f) $h = 0.1$, $y(2) = -0.99542551694636814077$

(g)

$$
\begin{bmatrix}
s & \left|\frac{y_{\text{exact}} - y_{\text{approx}}}{h^4}\right| \\
1 & 2.1116956552902748158 \\
2 & 3.4102782095350476723 \\
3 & 3.9857020261116057600 \\
4 & 4.1947422529651946291 \\
5 & 4.2684148561119857869 \\
6 & 4.2964545452679731610 \\
7 & 4.3081804966384369664 \\
8 & 4.3134612688397664255 \\
9 & 4.3159551017428923186 \\
10 & 4.3171635550443326669
\end{bmatrix}
$$

(h)

$$
\begin{bmatrix}
s & y_{\text{exact}} - y_{\text{approx}}(\tfrac{1}{2}h) & \tfrac{1}{15}y_{\text{approx}}(\tfrac{1}{2}h) - \tfrac{1}{15}y_{\text{approx}}(h) \\
2 & -0.01332139925599627997 & -0.008327305279976393068 \\
3 & -0.00097307178371865375 & -0.00082322183148517508 \\
4 & -0.00006400668720955192 & -0.00006060433976727345 \\
5 & -0.407067762004088 \times 10^{-5} & -0.3995733972634070 \times 10^{-5} \\
6 & -0.25608864696431 \times 10^{-6} & -0.254305931538438 \times 10^{-6} \\
7 & -0.1604922300815 \times 10^{-7} & -0.16002628263744 \times 10^{-7} \\
8 & -0.100430596360 \times 10^{-8} & -0.1002994469636 \times 10^{-8} \\
9 & -0.6280541277 \times 10^{-10} & -0.62766703389 \times 10^{-10} \\
10 & -0.392643738 \times 10^{-11} & -0.3925265026 \times 10^{-11}
\end{bmatrix}
$$

(i)

$$
\begin{bmatrix}
s & y_{\text{exact}} & \tfrac{16}{15}y_{\text{approx}}(\tfrac{1}{2}h) - \tfrac{1}{15}y_{\text{approx}}(h) \\
2 & -0.99583333333333333333 & -0.9908392393573134465 \\
3 & -0.99583333333333333333 & -0.99568348338109985474 \\
4 & -0.99583333333333333333 & -0.99582993098589105486 \\
5 & -0.99583333333333333333 & -0.99583325838968592651 \\
6 & -0.99583333333333333333 & -0.99583331550617907540 \\
7 & -0.99583333333333333333 & -0.99583333286738589000 \\
8 & -0.99583333333333333333 & -0.99583333333202183935 \\
9 & -0.99583333333333333333 & -0.99583333333329462402 \\
10 & -0.99583333333333333333 & -0.99583333333333216103
\end{bmatrix}
$$

16. (a) $y(t) = -3t^{5/2} - \frac{7}{2}t^{3/2} + \frac{15}{2}t^{-1/2}$, $y(2) = -21.566756826189699494$　　**(b)** $y(2) = -20.976346453651710192$

(c)

s	$\left\|\dfrac{y_{\text{exact}} - y_{\text{approx}}}{h}\right\|$
1	5.9429104555184901460
2	5.9140269696708665560
3	5.9054341697266924560
4	5.9023287584307443680
5	5.9010423623762397120
6	5.9004620181063693440
7	5.9001871085822877440
8	5.9000534140830499840
9	5.8999875000370273280
10	5.8999547754610462720

(d)

s	$y_{\text{exact}} - y_{\text{approx}}\left(\frac{1}{2}h\right)$	$y_{\text{approx}}\left(\frac{1}{2}h\right) - y_{\text{approx}}(h)$
2	−1.478506742417716639	−1.492948485341528434
3	−0.738179271215836557	−0.740327471201880082
4	−0.368895547401921523	−0.369283723813915034
5	−0.184407573824257491	−0.184487973577664032
6	−0.092194719032912021	−0.092212854791345470
7	−0.046095211785799123	−0.046099507247112898
8	−0.023047083648761914	−0.023048128137037209
9	−0.011523413086009819	−0.011523670562752095
10	−0.005761674585411178	−0.005761738500598641

(e)

s	y_{exact}	$2y_{\text{approx}}\left(\frac{1}{2}h\right) - y_{\text{approx}}(h)$
2	−21.566756826189699494	−21.581198569113511289
3	−21.566756826189699494	−21.568905026175743019
4	−21.566756826189699494	−21.567145002601693005
5	−21.566756826189699494	−21.566837225943106035
6	−21.566756826189699494	−21.566774961948132943
7	−21.566756826189699494	−21.566761121651013269
8	−21.566756826189699494	−21.566757870677974789
9	−21.566756826189699494	−21.566757083666441770
10	−21.566756826189699494	−21.566756890104886957

(f) $h = 0.1$, $y(2) = -21.566756553345677966$

(g)

s	$\left\|\dfrac{y_{\text{exact}} - y_{\text{approx}}}{h^4}\right\|$
1	0.0020922451322711360000
2	0.0024128146216243200000
3	0.0026682765423575040000
4	0.0028244575928975360000
5	0.0029096870265487360000
6	0.0029540350196449280000
7	0.0029766327216373760000
8	0.0029880516975001600000
9	0.0029940388819107839999
10	0.0029972686973173760000

(h)

s	$y_{\text{exact}} - y_{\text{approx}}\left(\frac{1}{2}h\right)$	$\frac{1}{15}y_{\text{approx}}\left(\frac{1}{2}h\right) - \frac{1}{15}y_{\text{approx}}(h)$
2	$-0.9425057115720 \times 10^{-5}$	$-0.80893509100817 \times 10^{-5}$
3	$-0.651434702724 \times 10^{-6}$	$-0.5849081608664 \times 10^{-6}$
4	$-0.43097802626 \times 10^{-7}$	$-0.405557933399 \times 10^{-7}$
5	$-0.2774893786 \times 10^{-8}$	$-0.26881939227 \times 10^{-8}$
6	$-0.176074208 \times 10^{-9}$	$-0.1732546385 \times 10^{-9}$
7	$-0.11088821 \times 10^{-10}$	$-0.109990258 \times 10^{-10}$
8	$-0.695710 \times 10^{-12}$	$-0.6928741 \times 10^{-12}$
9	-0.43569×10^{-13}	$-0.434760 \times 10^{-13}$
10	-0.2726×10^{-14}	-0.27229×10^{-14}

(i)

s	y_{exact}	$\frac{16}{15}y_{\text{approx}}\left(\frac{1}{2}h\right) - \frac{1}{15}y_{\text{approx}}(h)$
2	−21.566756826189699494	−21.566755490483493857
3	−21.566756826189699494	−21.566756759663157637
4	−21.566756826189699494	−21.566756823647690209
5	−21.566756826189699494	−21.566756826102999632
6	−21.566756826189699494	−21.566756826186879925
7	−21.566756826189699494	−21.566756826189609700
8	−21.566756826189699494	−21.566756826189666659
9	−21.566756826189699494	−21.566756826189699402
10	−21.566756826189699494	−21.566756826189699492

17. **(a)** $y(t) = \dfrac{(920{,}7197{,}824t^{34}+1{,}778{,}663{,}216t^{33}-9{,}763{,}030{,}079)^{1/4}}{187t^8}$, $y(2) = 2.3972670724285390400$ **(b)** $h = 0.1$, $y(2) = 2.5901906545021546936$

(c)

s	$\left\lvert\dfrac{y_{\text{exact}}-y_{\text{approx}}}{h}\right\rvert$
1	121.45868186108642696
2	8.9243694051861423632
3	5.1986347530016428632
4	0.0041047359376042144000
5	0.0041361325656049984000
6	0.0041521831962812160000
7	0.0041602731412752384000
8	0.0041643353827658240000
9	0.0041663711739617792000
10	0.0041673903102025728000

(d)

s	$y_{\text{exact}} - y_{\text{approx}}\left(\frac{1}{2}h\right)$	$y_{\text{approx}}\left(\frac{1}{2}h\right) - y_{\text{approx}}(h)$
2	2.2310923512965355908	58.498248579246677889
3	−0.6498293441252053579	2.8809216954217409487
4	−0.0002565459961002634	−0.6495727981291050945
5	−0.0001292541426751562	−0.0001272918534251072
6	−0.0000648778624418940	−0.0000643762802332622
7	−0.0000325021339162128	−0.0000323757285256812
8	−0.0000162669350889290	−0.0000162351988272838
9	$-0.81374436991441 \times 10^{-5}$	$-0.81294913897849 \times 10^{-5}$
10	$-0.40697170998072 \times 10^{-5}$	$-0.40677265993369 \times 10^{-5}$

(e)

s	y_{exact}	$2y_{\text{approx}}\left(\frac{1}{2}h\right) - y_{\text{approx}}(h)$
2	2.3972670724285390400	58.664423300378681338
3	2.3972670724285390400	5.9280181119754853466
4	2.3972670724285390400	1.7479508202955342089
5	2.3972670724285390400	2.3972690347177890890
6	2.3972670724285390400	2.3972675740107476718
7	2.3972670724285390400	2.3972671988339295716
8	2.3972670724285390400	2.3972671041648006852
9	2.3972670724285390400	2.3972670803808483992
10	2.3972670724285390400	2.3972670744190395103

(f) $h = 0.1$, $y(2) = -2.3974454619799884869$

(g)

s	$\left\lvert\dfrac{y_{\text{exact}}-y_{\text{approx}}}{h^4}\right\rvert$
1	211.74671270923663608
2	91799.540932228263140
3	19996.614987861805505
4	5.1993359095077208064
5	0.55387105436033351680
6	0.48714212083331235840
7	0.45618709401027215360
8	0.44126713557832171519
9	0.43394582781037117439
10	0.43032059366269255680

(h)

s	$y_{\text{exact}} - y_{\text{approx}}\left(\frac{1}{2}h\right)$	$\frac{1}{15}y_{\text{approx}}\left(\frac{1}{2}h\right) - \frac{1}{15}y_{\text{approx}}(h)$
2	358.59195676651665289	−23.023852481479290875
3	114.8819860810209486096	23.580664712366380285
4	− 0.0000793355699082599	0.3254710277272379130
5	$0.5282125991443 \times 10^{-6}$	$-0.532425216716028 \times 10^{-5}$
6	$0.290359330674 \times 10^{-7}$	$0.3327844440512 \times 10^{-7}$
7	$0.16994293556 \times 10^{-8}$	$0.182243358079 \times 10^{-8}$
8	$0.1027405112 \times 10^{-9}$	$0.10644592296 \times 10^{-9}$
9	$0.63147429 \times 10^{-11}$	$0.642838455 \times 10^{-11}$
10	$0.3913743 \times 10^{-12}$	$0.39489124 \times 10^{-12}$

(i)

s	y_{exact}	$\frac{16}{15}y_{\text{approx}}\left(\frac{1}{2}h\right) - \frac{1}{15}y_{\text{approx}}(h)$
2	2.3972670724285390400	−379.21854217556740474
3	2.3972670724285390400	21.095945703773970715
4	2.3972670724285390400	2.7228174357711710913
5	2.3972670724285390400	2.3972612199637727355
6	2.3972670724285390400	2.3972670766710503779
7	2.3972670724285390400	2.3972670725515432652
8	2.3972670724285390400	2.3972670724322444518
9	2.3972670724285390400	2.3972670724286526818
10	2.3972670724285390400	2.3972670724285425570

Exercises 4.2–Part A, Page 57

1. $f(t, y) = \lambda y \Rightarrow y_{k+1} = y_k + h\lambda y_k = (1 + h\lambda)y_k$ **4.** In this case $h = \frac{2}{|\lambda|}$ **5.** $y_k = y_0\left(1 + (-10)\left(\frac{1}{10}\right)\right)^k = y_0(0)^k = 0$

Exercises 4.2–Part B, Page 57

1.

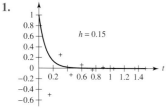
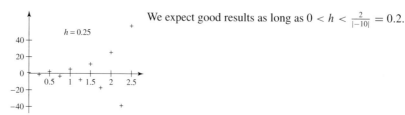

We expect good results as long as $0 < h < \frac{2}{|-10|} = 0.2$.

$$\begin{bmatrix} h = 0.1 \\ \begin{array}{ccc} x & y & y_{\text{exact}} \\ 0 & 2.0 & 2.0 \\ 1.0 & 5.987656250000000 & 6.000000306 \\ 2.0 & 14.98666763305664 & 15.00 \\ 3.0 & 27.98666666761041 & 28.00 \\ 4.0 & 44.98666666666759 & 45.00 \\ 5.0 & 65.98666666666666 & 66.00 \\ 6.0 & 90.98666666666668 & 91.00 \\ 7.0 & 119.9866666666667 & 120.00 \\ 8.0 & 152.9866666666667 & 153.00 \\ 9.0 & 189.9866666666667 & 190.00 \\ 10.0 & 230.9866666666666 & 231.00 \end{array} \end{bmatrix}$$

$$\begin{bmatrix} h = 0.2 \\ \begin{array}{ccc} x & y & y_{\text{exact}} \\ 0 & 2.0 & 2.0 \\ 1.0 & -26.88000000000001 & 6.000000306 \\ 2.0 & 1066.279999999997 & 15.00 \\ 3.0 & -33613.83999999973 & 28.00 \\ 4.0 & 0.1076582999999986 \times 10^7 & 45.00 \\ 5.0 & -0.3444915087999938 \times 10^8 & 66.00 \\ 6.0 & 0.1102375030279974 \times 10^{10} & 91.00 \\ 7.0 & -0.3527599793783899 \times 10^{11} & 120.00 \\ 8.0 & 0.1128831938002963 \times 10^{13} & 153.00 \\ 9.0 & -0.3612262201100780 \times 10^{14} & 190.00 \\ 10.0 & 0.1155923904358498 \times 10^{16} & 231.00 \end{array} \end{bmatrix}$$

$y_h = e^{-15t}$, $y_p = 2t^2 + 3t + 1$ Even though y_h stops contributing for positive values of t away from zero, the stepsize necessary for stability of the method is regulated by the inequality $|1 - 15h| < 1 \Rightarrow 0 < h < 0.133$.

9.

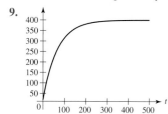
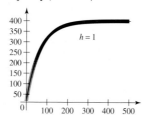

$|y_{\text{exact}} - y_{\text{approx}}| < 1.0550456$, on $t \in [0, 500]$

Exercises 4.3–Part A, Page 59

1. $\frac{1}{2} - \frac{1}{2}x + \frac{1}{4}x^2 - \frac{1}{12}x^3 + \frac{1}{48}x^4 - \frac{1}{240}x^5 + \frac{1}{1440}x^6 - \frac{1}{10,080}x^7 + \frac{1}{80,640}x^8 - \frac{1}{725,760}x^9 + \frac{1}{7,257,600}x^{10}$

3. $1 - x + \frac{1}{2}x^3 + \frac{1}{24}x^4 - \frac{11}{120}x^5 - \frac{1}{180}x^6 + \frac{67}{5040}x^7 + \frac{29}{40,320}x^8 - \frac{107}{72,576}x^9 - \frac{13}{181,440}x^{10}$

4. If the differential equation is written in the form $\frac{d^2y}{dx^2} = g(x, y(x), \frac{dy}{dx})$, then the function g (and its derivatives) cannot be evaluated at $x = 0$.

5. Substitute $y(x) = x$ in the differential equation. No, it does not contradict Exercise 4, because this solution cannot be generated using a Taylor series method.

Exercises 4.3–Part B, Page 59

1. (a) $y_{\text{approx}} = 1 - \frac{13}{2}t^2 + \frac{26}{3}t^3 - \frac{13}{8}t^4 - \frac{13}{3}t^5 + \frac{2587}{720}t^6 - \frac{299}{420}t^7 - \frac{19,279}{40,320}t^8 + \frac{221}{648}t^9 - \frac{27,157}{403,200}t^{10}$

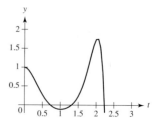

(b) $y_{\text{exact}} = 1 - \frac{13}{2}t^2 + \frac{26}{3}t^3 - \frac{13}{8}t^4 - \frac{13}{3}t^5 + \frac{2587}{720}t^6 - \frac{299}{420}t^7 - \frac{19,279}{40,320}t^8 + \frac{221}{648}t^9 - \frac{27,157}{403,200}t^{10}$ **(c)**

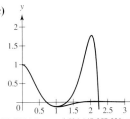

3. (a) $y_{\text{exact}}(t) = \frac{87,396}{5t^2+9t^4-5}$, $y(-6) = \frac{87,396}{11,839} = 7.382$ **(b)** $y_{\text{exact}} \approx \frac{1,438,704,761,959,840}{10,430,249,423,049} + \frac{164,126,041,874,144}{3,476,749,807,683}t + \frac{6,504,147,357,220}{1,158,916,602,561}t^2 + \frac{2,386,780,960,480}{10,430,249,423,049}t^3$
(c) $y_{k+1} = y_k(2,103,731,511,225h^2 + 596,695,240,120h^3 + 13,904,381,742,381h^2k^2 + 729h^{12}k^{12} - 61,236h^{11}k^{11} + 2,358,801h^{10}k^{10} -$
$764,700,592,880h^4k + 30,993,185,160h^7k^4 - 144,913,611,380h^6k^3 + 420,555,240h^9k^6 - 4,420,850,760h^8k^5 - 25,728,840h^{10}k^7 +$
$918,540h^{11}k^8 - 14,580h^{12}k^9 + 1,916,530,494,140h^4k^2 - 2,992,584,041,440h^3k + 7290h^{12}k^{10} + 6,591,467,421,280h^3k^2 +$
$181,402,946,475h^6k^4 - 727,685,295,540h^5k^3 + 3,686,249,520h^8k^6 - 31,023,654,480h^7k^5 - 300,487,320h^9k^7 + 16,082,145h^{10}k^8 -$
$510,300h^{11}k^9 + 435,759,932,580h^5k^2 - 2916h^{12}k^{11} + 224,532h^{11}k^{10} - 9,269,587,828,254h^2k + 801,038,197,190h^5k^4 -$
$2,813,578,332,440h^4k^3 + 22,758,873,984h^7k^6 - 159,719,537,698h^6k^5 - 2,317,577,040h^8k^7 + 165,286,170h^9k^8 - 7,862,670h^{10}k^9 -$
$17,784,269,852,454hk + 2,110,183,749,330h^4k^4 - 6,591,467,421,280h^3k^3 + 79,859,768,849h^6k^6 - 480,622,918,314h^5k^5 +$
$5,928,089,950,818h - 9,753,803,136h^7k^7 + 869,091,390h^8k^8 + 10,430,249,423,049 - 55,095,390h^9k^9)/(21,849 - 12,418hk +$
$2651h^2k^2 - 252h^3k^3 + 9h^4k^4)^3$, $h = 0.1$, $y_{\text{approx}}(-6) = 7.381908764$, error $= 0.000133638$ **(d)** $h = 0.05$, $y_{\text{approx}} = 7.382025117$, error
$= 0.000017285$

(e)

h	$\frac{y_{\text{approx}} - y_{\text{exact}}}{h^3}$
0.10000000000000000000	−0.13364495900473560000
0.05000000000000000000	−0.13825143678992480000
0.02500000000000000000	−0.14061383598020480000
0.01250000000000000000	−0.14181012677616640000
0.00625000000000000000	−0.14241208423505920000

(f)

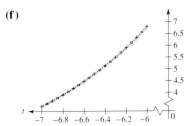

9. (a) $y_{\text{exact}}(t) = \frac{18}{5t^2+9t^4-5}$, $y(2) = \frac{6}{53} = 0.11321$ **(b)** $\frac{4160}{81} - \frac{7168}{81}t + \frac{3170}{81}t^2$ **(c)** $y_{k+1} = y_k(828hk + 3178h^2k^2 + 6076h^3k^3 +$
$6955h^4k^4 + 5076h^5k^5 + 2358h^6k^6 + 648h^7k^7 + 81h^8k^8 + 81 - 414h - 3178h^2k - 9114h^3k^2 - 13,910h^4k^3 - 12,690h^5k^4 - 7074h^6k^5 - 2268h^7k^6 -$
$324h^8k^7 + 1585h^2 + 7170h^3k + 14,925h^4k^2 + 17,820h^5k^3 + 12,555h^6k^4 + 4860h^7k^5 + 810h^8k^6)/(9 + 46hk + 59h^2k^2 + 36h^3k^3 + 9h^4k^4)^2$, $h =$
0.1, $y_{\text{approx}} = 0.1363495218$, error: $= 0.0231419746$ **(d)** $h = 0.05$, $y_{\text{approx}} = 0.1176571769$, error $= 0.0044496297$

(e)

h	$\frac{y_{\text{approx}} - y_{\text{exact}}}{h^2}$
0.10000000000000000000	2.3141974722643724420
0.05000000000000000000	1.7798519576705287440
0.02500000000000000000	1.5826089509250509120
0.01250000000000000000	1.4975483066273326720
0.00625000000000000000	1.4579903273736245760

(f)

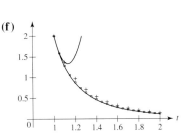

Exercises 4.4–Part A, Page 64

1. $y_1 = \frac{58}{25} = 2.32$; $y_2 = \frac{26,841}{10,000} = 2.6841$ **3.** $y_1 = \frac{3331}{1000} = 3.331$; $y_2 = \frac{44,825,267}{12,000,000} = 3.7354$ **5.** $y_1 = 0.9132946446$; $y_2 = 0.8504279079$

Exercises 4.4–Part B, Page 64

1. $y_{k+1} = y_k + \frac{h}{2}(ay_k + ay_{k+1}) \Rightarrow y_{k+1} = \frac{2+ha}{2-ha}y_k$ **3.** Substitute $y(t) = y_0e^{at}$ into the differential equation.

5. $y_{\text{exact}} = 2t^2 + 3t + 1 + e^{-15t}$

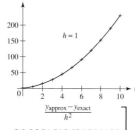

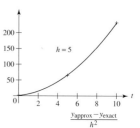

15.

$$\begin{bmatrix} h & \frac{y_{\text{approx}} - y_{\text{exact}}}{h^2} \\ 0.10000000000000000000 & 26.0051513621544100 \\ 0.05000000000000000000 & 28.1518639060855078 \\ 0.02500000000000000000 & 29.273279636088965 \\ 0.01250000000000000000 & 29.84450042689287 \\ 0.00625000000000000000 & 30.13249676338164 \\ 0.00312500000000000000 & 30.2770573421625 \\ 0.00156250000000000000 & 30.3494737261168 \\ 0.00078125000000000000 & 30.385715362301 \\ 0.00039062500000000000 & 30.40384446817 \end{bmatrix}$$

19.

$$\begin{bmatrix} h & \frac{y_{\text{approx}} - y_{\text{exact}}}{h^2} \\ 0.10000000000000000000 & -0.12619272285064639400 \\ 0.05000000000000000000 & -0.12536922283563788400 \\ 0.02500000000000000000 & -0.12516985999467780800 \\ 0.01250000000000000000 & -0.12512041101257881600 \\ 0.00625000000000000000 & -0.12510807301565388800 \\ 0.00312500000000000000 & -0.12510499002831564800 \\ 0.00156250000000000000 & -0.12510421937594777600 \\ 0.00078125000000000000 & -0.12510402671876505600 \\ 0.00039062500000000000 & -0.12510397855517900800 \end{bmatrix}$$

Exercises 4.5–Part A, Page 67

1. $y_0 = 0.3333333333$, $y_1 = 0.3460606761$, $y_2 = 0.3622411960$
$y_3 = 0.5882352941$, $y_4 = 0.6249876808$

3. $y_0 = 0.5000000000$, $y_1 = 0.5263157895$, $y_2 = 0.5555555556$,

5. $y = f_n - \dfrac{(f_{n-1} - f_n)(t - t_n)}{h}$, $y_{n+1} = y_n + \displaystyle\int_{t_n}^{t_n+h} f_n - \dfrac{(f_{n-1} - f_n)(t - t_n)}{h}\,dt = y_n + (\tfrac{3}{2}f_n - \tfrac{1}{2}f_{n-1})h$

Exercises 4.5–Part B, Page 67

1. $h = \frac{1}{50}$, $y_0 = 1$, $y_1 = 1.000394771$

2.

$$\begin{bmatrix} h & \frac{y_{\text{approx}} - y_{\text{exact}}}{h^2} \\ 0.50000000000000000000 & 0.22222222222222222200 \\ 0.25000000000000000000 & 0.25174603174603174560 \\ 0.12500000000000000000 & 0.28168234843003361280 \\ 0.06250000000000000000 & 0.29699798756829678080 \\ 0.03125000000000000000 & 0.30472101398986741760 \\ 0.01562500000000000000 & 0.30860253023530926080 \\ 0.00781250000000000000 & 0.31054912373753610240 \\ 0.00390625000000000000 & 0.31152400660968898560 \\ 0.00195312500000000000 & 0.31201186190673838080 \\ 0.00097656250000000000 & 0.31225589527425843200 \end{bmatrix}$$

5. tc:=[seq(t[n]-k*h, k=0..4)]; fc:=[seq(f[n-k], k=0..4)]; poly:=interp(tc, fc, s); method:=y[n+1]=y[n]+factor(int(poly, s=t[n]..t[n]+h));
$y_{n+1} = y_n + \frac{1}{720}h(1901 f_n + 251 f_{n-4} - 2774 f_{n-1} - 1274 f_{n-3} + 2616 f_{n-2})$

7. Substitute $y_k = r^k$ in the difference equation and note that $r^k \neq 0$ for $r \neq 0$.

8. $r^2 - (1 + \frac{3}{2}A)r + \frac{1}{2}A = 0$; $r_1 = \frac{1}{2} + \frac{3}{4}A + \frac{1}{4}\sqrt{4 + 4A + 9A^2}$, $r_2 = \frac{1}{2} + \frac{3}{4}A - \frac{1}{4}\sqrt{4 + 4A + 9A^2}$

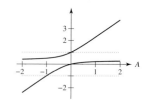

Exercises 4.6–Part A, Page 70

1. $y_4 = 0.4134585597$ **3.** $y_2 = 0.3627537798$ **5.** $y_4 = 0.6756782192$

Exercises 4.6–Part B, Page 71

2. T:=[seq(t[k]+n*h, n=-1..0)]; F:=[seq(f[k+n], n=-1..0)]; p[1]:=interp(T, F, s); predictor:=y[k+1]=y[k]+simplify(int(p[1],s=t[k]..t[k]+h));
Predictor: $y_{k+1} = y_k - \frac{1}{2}h(-3f_k + f_{k-1})$; T:=[seq(t[k]+n*h, n=0..1)]; F:=[seq(f[k+n], n=0..1)]; p[2]:=interp(T, F, s);
corrector:=y[k+1]=y[k]+simplify(int(p[2], s=t[k]..t[k]+h)); Corrector: $y_{k+1} = y_k + \frac{1}{2}h(f_{k+1} + f_k)$.

3.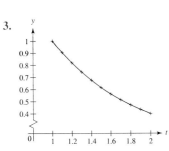

5. $$\begin{bmatrix} y_{\text{exact}} - y_{2,c} & \dfrac{C_c(y_{2,p} - y_{2,c})}{C_c - C_p} \\ -0.00001178725152467777 & -0.00003908006076437598545454 \end{bmatrix}$$

Exercises 4.7–Part A, Page 74

1. $y_4 = 0.4134523839$ **3.** Substitute $x_k = (-2)^k$ in the difference equation, and also observe that $(-2)^0 = 1$;
$[1, -2, 4, -8, 16, -32, 64, -128, 256, -512, 1024]$.

6. Solution: $\frac{1}{5}(-3)^k + \frac{9}{5}2^k$; $r_1 = -3$, $r_2 = 2$, $a = \frac{1}{5}$, $b = \frac{9}{5}$; $[2, 3, 9, 9, 45, 9, 261, -207, 1773, -3015, 13{,}653]$.

Exercises 4.7–Part B, Page 74

1.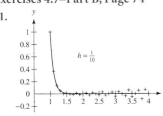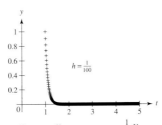

Be aware that eventually, even with $h = 0.01$, Milne's method will diverge. Just try to extend your computation to the interval $[1, 10]$.

2. Solve Section 4.1 problem B3; $$\begin{bmatrix} y_{\text{exact}} - y_4 & \frac{1}{29}y_{4,p} - \frac{1}{29}y_{4,c} \\ 0.24365563469 \times 10^{-8} & 0.214143797616 \times 10^{-8} \end{bmatrix}$$

7. **(a)** Solution: $(\frac{1}{2} - \frac{1}{6}i)(2 - 3i)^n + (\frac{1}{2} + \frac{1}{6}i)(2 + 3i)^n$ **(b)** and **(c)** $[1, 1, -9, -49, -79, 321, 2311, 5071, -9759, -104{,}959]$
(d) $r_1 = 2 - 3I$, $r_2 = 2 + 3I$, $a = \frac{1}{2} - \frac{1}{6}I$, $b = \frac{1}{2} + \frac{1}{6}I$.

Exercises 4.8, Page 76

2. Solve Section 4.1 problem B3;

t_k	$y(t_k)$	$y_{k,5}$	$\left\lvert y_{k,5} - y_{k,4} \right\rvert$	*Error*
6.1	-7.0723421515030	-7.0723421515022	0.685×10^{-10}	$-0.685818 \times 10^{-10}$
6.2	-7.1460202100500	-7.1460202100485	0.659×10^{-10}	$-0.659262 \times 10^{-10}$
6.3	-7.2209819764892	-7.2209819764873	0.633×10^{-10}	$-0.631642 \times 10^{-10}$
6.4	-7.2971760251442	-7.2971760251419	0.603×10^{-10}	$-0.603344 \times 10^{-10}$
6.5	-7.3745518811578	-7.3745518811548	0.574×10^{-10}	$-0.574713 \times 10^{-10}$
6.6	-7.4530601700514	-7.4530601700478	0.547×10^{-10}	$-0.546037 \times 10^{-10}$
6.7	-7.5326527414142	-7.5326527414101	0.517×10^{-10}	$-0.517578 \times 10^{-10}$
6.8	-7.6132827687735	-7.6132827687693	0.489×10^{-10}	$-0.489554 \times 10^{-10}$
6.9	-7.6949048277664	-7.6949048277620	0.464×10^{-10}	$-0.462136 \times 10^{-10}$
7.0	-7.7774749547400	-7.7774749547356	0.436×10^{-10}	$-0.435486 \times 10^{-10}$

CHAPTER 5

Exercises 5.1–Part A, Page 80

1. $\dfrac{d^2 y}{dt^2} + 2\dfrac{dy}{dt} + 10y = 0$ **2.** $\dfrac{d^2 y}{dt^2} + 3\dfrac{dy}{dt}\left\lvert\dfrac{dy}{dt}\right\rvert + 10y = 0$ **3.** Upon substitution and simplification both sides will equal zero.

5.

0.6308489604; 1.678046512 **6.** $m = 1$, $b = 2$, $k = 10$ It represents a damped motion.

7. $\phi = \arctan(\frac{1}{3}), \alpha = \sqrt{10}$ **8.** It provides the amplitude and the phase-shift.   **9.** $a = \frac{12}{17}\sqrt{17}, b = \frac{3}{17}\sqrt{17}$

10. Applying the initial conditions to the general solution given in Exercise 3 results in $A = -3$ and $-A + 3B = 0$.

Exercises 5.1–Part B, Page 80

1. $Ae^{-3t}\cos 2t + Be^{-3t}\sin 2t$ **2.** $A = 1, B = 1; y(t) = e^{-3t}\cos 2t + e^{-3t}\sin 2t$

3.
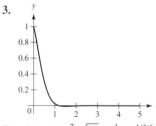
$\frac{3}{8}\pi = 1.1781$ **4.** $y(t) = Ae^{-4/3(6+\sqrt{30})t} + Be^{4/3(-6+\sqrt{30})t}$

5. $y(t) = (-\frac{7}{80}\sqrt{30} + \frac{1}{2})e^{-4/3(6+\sqrt{30})t} + \frac{1}{240}(21 + 4\sqrt{30})\sqrt{30}e^{4/3(-6+\sqrt{30})t}$

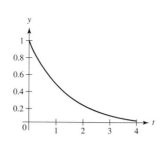

$t = 2.278913602$

Exercises 5.2–Part A, Page 84

1. **(a)** linear   **(b)** nonlinear   **(c)** nonlinear   **2.** **(a)** and **(c)** are nonhomogeneous   **(b)** is homogeneous

3. **(a)** 3   **(b)** 2   **(c)** 4   **4.** **(a)** $k = \frac{3}{11}$ **(b)** $k = \frac{36}{11}$ **5.** $\int_0^{3/4} \frac{36}{11}u\, du = \frac{81}{88} = 0.92045$ **6.** $k = \frac{48}{5}$

Exercises 5.2–Part B, Page 84

1. $\frac{1}{8}\frac{d^2y}{dt^2} + 6y = 0, y(0) = -\frac{13}{12}, \frac{dy}{dt}(0) = 0; y(t) = -\frac{5}{12}\cos(4\sqrt{3}t)$ **2.** -20 ft/s^2 **3.**

$\frac{5}{3}\sqrt{3}$ ft/s^2

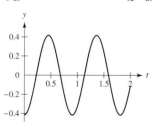

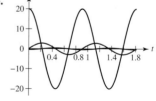

Exercises 5.3–Part A, Page 88

1. $y(t) = Ce^{-5/7t}$ **2.** $7\lambda + 5 = 0 \Rightarrow \lambda = -\frac{5}{7}$ **3.** Upon substitution both sides will equal zero.

7. **(a)** $\lambda^2 + 9\lambda + 20 = (\lambda + 5)(\lambda + 4) = 0$ **(b)** $-5, -4$ **(c)** $\{e^{-4t}, e^{-5t}\}$ **(d)** $y(t) = c_1e^{-4t} + c_2e^{-5t}$

Exercises 5.3–Part B, Page 88

1. $(7\lambda - 3)(2\lambda - 3)(5\lambda + 4) = 0; \lambda = \frac{3}{7}, \lambda = \frac{3}{2}, \lambda = -\frac{4}{5}; \{e^{3/2t}, e^{3/7t}, e^{-4/5t}\}; y(t) = c_1e^{3/2t} + c_2e^{3/7t} + c_3e^{-4/5t}$

2. **(a)** $(\lambda + 4)(\lambda + 3) = 0; -4, -3; \{e^{-3t}, e^{-4t}\}; c_1e^{-3t} + c_2e^{-4t}$ **(d)**

$t = 0.4444970568$

(b) $c_1 = -7, c_2 = 5, y(t) = -7e^{-3t} + 5e^{-4t}$

(c) $y(t) = -7e^{-3t} + 5e^{-4t}$

3. **(a)** $(\lambda + 6)(\lambda + 5) = 0; \lambda = -6, \lambda = -5; \{e^{-6t}, e^{-5t}\}; y(t) = c_1e^{-6t} + c_2e^{-5t}$ **(b)** $c_1 = -2, c_2 = 2, y(t) = -2e^{-6t} + 2e^{-5t}$

(c) $-2e^{-6t} + 2e^{-5t}$ **(d)**

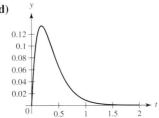

$t_{max} = \ln 6 - \ln 5 = \ln 1.2 \approx 0.18232$, $y_{max} = \frac{3125}{23,328} \approx 0.13395$

4. $b = 2\sqrt{15}$ **5.** $k = \frac{25}{12}$ **6.** $m = \frac{25}{28}$ **7.** $y(t) = 5e^{-t} - 4e^{-2t}$

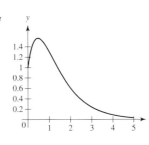

No, just look at this graph. The solution is first increasing before it starts to move down towards zero.

Exercises 5.4–Part A, Page 91

1. Upon substitution one obtains $0 = 0$, Wronskian: $-\frac{1}{t^6} \neq 0$ for $t > 0$

2. Upon substitution one obtains $0 = 0$, Wronskian: $-\frac{29}{t^6} \neq 0$ for $t > 0$ **3.** Upon substitution one obtains $0 = 0$, Wronskian: $-e^{-3t} \neq 0$, $\forall t$

4. Upon substitution one obtains $0 = 0$, Wronskian: $-43e^{-3t} \neq 0$, $\forall t$.   **5.** $\frac{d^2 y}{dt^2} + 2\frac{dy}{dt} - 15y = 0$ **6.** $\frac{d^3 y}{dt^3} - 3\frac{d^2 y}{dt^2} - 10\frac{dy}{dt} + 24y = 0$

Exercises 5.4–Part B, Page 91

1. False! Look at exercises A1 and 2.

2. (a) Set an arbitrary linear combination equal to zero, and differentiate this equation two times. This results in three equations:
$c_1 t e^t \sin t + c_2 t e^t \cos t + c_3 t^2 e^t = 0$, $c_1(e^t \sin t + t e^t \sin t + t e^t \cos t) + c_2(e^t \cos t + t e^t \cos t - t e^t \sin t) + c_3(2t e^t + t^2 e^t) = 0$,
$c_1(2e^t \sin t + 2e^t \cos t + 2t e^t \cos t) + c_2(2e^t \cos t - 2e^t \sin t - 2t e^t \sin t) + c_3(2e^t + 4t e^t + t^2 e^t) = 0 \Rightarrow c_2 = 0, c_1 = 0, c_3 = 0$.
(b) Wronskian: $-t^4 e^{3t}$ The set is linearly independent.

5. (a) $(\lambda + 2)(\lambda + 1) = 0$, $\lambda = -2$, $\lambda = -1$, $\{e^{-t}, e^{-2t}\}$ General solution: $c_1 e^{-t} + c_2 e^{-2t}$ **(b)** $-7e^{-t} + 4e^{-2t}$ **(c)** $-7e^{-t} + 4e^{-2t}$

(d)

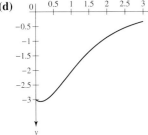

$t_{min} = \ln\frac{8}{7} \approx 0.133531393$, $y_{min} = -\frac{49}{16} = -3.062500000$

6. $y = \frac{1}{2}\sqrt{5}\cos(2t - \arctan\frac{1}{2})$

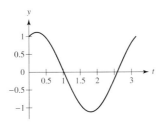

7. $ad - bc$ should not equal zero.

Exercises 5.5–Part A, Page 94

1. $\cos 3t = \frac{1}{2}e^{3it} + \frac{1}{2}e^{-3it}$, $\sin 3t = -\frac{1}{2}i(e^{3it} - e^{-3it})$; Wronskian: 3   **2.** Upon substitution and simplification both sides equal zero.

3. Upon substitution and simplification both sides equal zero.

4. $e^{3/8i\pi} = \frac{1}{2}\sqrt{2 - \sqrt{2}} + \frac{1}{2}i\sqrt{2 + \sqrt{2}}$; $\cos\frac{3\pi}{8} = \frac{1}{2}\sqrt{2 - \sqrt{2}}$ The second is the real part of the first.

5. $[1.0, 0.9950041653 - 0.09983341665i, 0.9800665778 - 0.1986693308i, 0.9553364891 - 0.2955202067i, 0.9210609940 - 0.3894183423i, 0.8775825619 - 0.4794255386i, 0.8253356149 - 0.5646424734i, 0.7648421873 - 0.6442176872i, 0.6967067093 - 0.7173560909i, 0.6216099683 - 0.7833269096i, 0.5403023059 - 0.8414709848i]$

6. $\sqrt{34}\cos(-2t - \arctan\frac{5}{3})$; amplitude: $\sqrt{34}$; period: π; frequency: $\frac{1}{\pi}$; angular frequency: 2   7. $-0.4161468366 = -0.4161468365$

8. (a) $z + \bar{z} = 4$; $z - \bar{z} = 6i$ 9. (a) $z_1 z_2 = 22 + 7i$ (b) $z_1 z_2 = -\sqrt{2} - 2\sqrt{3} + i(\sqrt{3} - 2\sqrt{2})$
   (c) $z_1 z_2 = \frac{1}{2}e^{-2}\cos 3 + \frac{1}{3}e^{-2}\sin 3 + i(-\frac{1}{3}e^{-2}\cos 3 + \frac{1}{2}e^{-2}\sin 3)$ 10. (a) $\frac{4}{41} + \frac{5}{41}i$ 11. (a) i (c) $\frac{22}{13} - \frac{7}{13}i$

Exercises 5.5–Part B, Page 95

1. (a) $e^{it} \approx 1 + it - \frac{1}{2}t^2 - \frac{1}{6}it^3 + \frac{1}{24}t^4 + \frac{1}{120}it^5 - \frac{1}{720}t^6 - \frac{1}{5040}it^7 + \frac{1}{40,320}t^8 + \frac{1}{362,880}it^9$
   (b) $e^{it} \approx 1 - \frac{1}{2}t^2 + \frac{1}{24}t^4 - \frac{1}{720}t^6 + \frac{1}{40,320}t^8 + i(t - \frac{1}{6}t^3 + \frac{1}{120}t^5 - \frac{1}{5040}t^7 + \frac{1}{362,880}t^9)$
   (c) $\cos t \approx 1 - \frac{1}{2}t^2 + \frac{1}{24}t^4 - \frac{1}{720}t^6 + \frac{1}{40,320}t^8$; $\sin t \approx t - \frac{1}{6}t^3 + \frac{1}{120}t^5 - \frac{1}{5040}t^7 + \frac{1}{362,880}t^9$
   (d) Compare the real and imaginary parts of (b) with the results found in (c).

2. (a) $y(t) = -2\cos 2t - \frac{1}{2}\sin 2t$ (b) $\{e^{2it}, e^{-2it}\}$ (c) $\{\frac{1}{2}e^{2it} + \frac{1}{2}e^{-2it}, \frac{1}{2i}(e^{2it} - e^{-2it})\}$ (d) $y(t) = c_1 \cos 2t + c_2 \sin 2t$
   (e) $-2\cos 2t - \frac{1}{2}\sin 2t$ (f) $\frac{1}{2}\sqrt{17}\cos(2t - \arctan\frac{1}{4} + \pi)$ See plot;

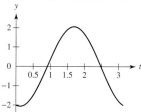

Exercises 5.6–Part A, Page 98

1. (a) $-1, -\frac{2}{3}$; overdamped   (b) $-\frac{5}{6} + \frac{1}{6}i\sqrt{23}, -\frac{5}{6} - \frac{1}{6}i\sqrt{23}$; underdamped; $T = \frac{12}{23}\pi\sqrt{23}$, $f = \frac{1}{12}\frac{\sqrt{23}}{\pi}$, $\omega = \frac{1}{6}\sqrt{23}$
   (c) $-1, -1$; critically damped.

2. (a) $-1 + 2i, -1 - 2i$; $\{e^{(-1-2i)t}, e^{(-1+2i)t}\}$; $\{e^{-t}\cos 2t, e^{-t}\sin 2t\}$; $c_1 e^{-t}\cos 2t + c_2 e^{-t}\sin 2t$
   (b) $-\frac{1}{2} + \frac{1}{3}i, -\frac{1}{2} - \frac{1}{3}i$; $\{e^{(-1/2+1/3i)t}, e^{(-1/2-1/3i)t}\}$; $\{e^{-1/2t}\cos\frac{1}{3}t, e^{-1/2t}\sin\frac{1}{3}t\}$; $c_1 e^{-1/2t}\cos\frac{1}{3}t + c_2 e^{-1/2t}\sin\frac{1}{3}t$

3. Yes! Let $b^2 = 4MK$.   4. $k = 2 + \frac{1}{18}\pi^2 = 2.5483$

5. m, b, and k can be determined up to a multiplicative constant; $m = \frac{1}{25}k, b = \frac{6}{25}k$, for any $k > 0$.

Exercises 5.6–Part B, Page 99

1. $m = 0.6176687004, m = 2.624609175$ 2. (a) $\lambda^2 + 2\lambda + 5 = 0$; $\lambda = -1 + 2i, \lambda = -1 - 2i$ (b) $\{e^{(-1+2i)t}, e^{(-1-2i)t}\}$,
   $\{\frac{1}{2}e^{(-1+2i)t} + \frac{1}{2}e^{(-1-2i)t}, \frac{1}{2i}(e^{(-1+2i)t} - e^{(-1-2i)t})\}$ (c) $\{e^{-t}\cos 2t, e^{-t}\sin 2t\}$ (d) $c_1 = 2, -c_1 + 2c_2 = 3$; $c_2 = \frac{5}{2}, c_1 = 2$
   (e) $e^{-t}(2\cos 2t + \frac{5}{2}\sin 2t)$ (f) $2e^{-t}\cos 2t + \frac{5}{2}e^{-t}\sin 2t$ (h)   (i) 3.466223946
   (g) $\frac{1}{2}\sqrt{41}e^{-t}\cos(2t - \arctan\frac{5}{4})$;
   $T = \pi, f = \frac{1}{\pi}, \omega = 2, \varphi = \arctan\frac{5}{4}$

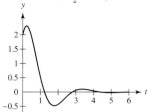

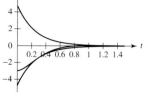

3. (a) $\lambda^2 + 8\lambda + 25 = 0$; $\lambda = -4 + 3i, \lambda = -4 - 3i$ (b) $\{e^{(-4+3i)t}, e^{(-4-3i)t}\}$; $\{\frac{1}{2i}(e^{(-4+3i)t} - e^{(-4-3i)t}), \frac{1}{2}e^{(-4+3i)t} + \frac{1}{2}e^{(-4-3i)t}\}$
   (c) $c_1 e^{-4t}\cos 3t + c_2 e^{-4t}\sin 3t$ (d) $c_1 = -3, -4c_1 + 3c_2 = 1$; $c_1 = -3, c_2 = -\frac{11}{3}$
   (e) $-3e^{-4t}\cos 3t - \frac{11}{3}e^{-4t}\sin 3t$ (f) $-3e^{-4t}\cos 3t - \frac{11}{3}e^{-4t}\sin 3t$ (h)   (i) 0.9645266633
   (g) $\frac{1}{3}\sqrt{202}e^{-4t}\cos(3t - \arctan\frac{11}{9} + \pi)$;
   $T = \frac{2}{3}\pi, f = \frac{3}{2}\frac{1}{\pi}, \omega = 3, \varphi = \arctan\frac{11}{9} - \pi$

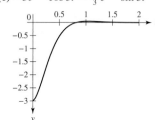

4. **(a)** $\lambda^2 + 8\lambda + 41 = 0$; $\lambda = -4 + 5i$, $\lambda = -4 - 5i$ **(b)** $\{e^{(-4+5i)t}, e^{(-4-5i)t}\}$; $\{\frac{1}{2}e^{(-4+5i)t} + \frac{1}{2}e^{(-4-5i)t}, \frac{1}{2i}(e^{(-4+5i)t} - e^{(-4-5i)t})\}$

(c) $c_1 e^{-4t} \cos 5t + c_2 e^{-4t} \sin 5t$ **(e)** $5e^{-4t} \cos 5t + \frac{16}{5}e^{-4t} \sin 5t$ **(f)** $5e^{-4t} \cos 5t + \frac{16}{5}e^{-4t} \sin 5t$

(d) $c_1 = 5$, $-4c_1 + 5c_2 = -4c_1 = 5$, $c_2 = \frac{16}{5}$

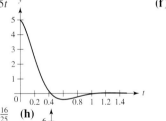

(g) $\frac{1}{5}\sqrt{881}e^{-4t} \cos(-5t + \arctan \frac{16}{25})$; $T = \frac{2}{5}\pi$, $f = \frac{5}{2}\frac{1}{\pi}$, $\omega = 5$, $\varphi = \arctan \frac{16}{25}$ **(h)**   **(i)** 1.020918998

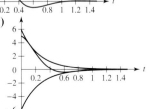

Exercises 5.7–Part A, Page 102

1. After three times integration one obtains: $\frac{dv}{dt} + v = 0$, $v = de^{-t}$, resulting in: $\frac{d^3u}{dt^3} = de^{-t}$, $u = a + bt + ct^2 + de^{-t}$ (where $-d$ has been replaced by d).   **2.** $\{1, t, t^2, e^{-t}\}$ **3.** The characteristic roots for the equation in u are: $0, 0, -4i, -4i$, resulting in the fundamental solution set $\{1, t, e^{-4it}, te^{-4it}\}$. Multiplication of this solution set by e^{2it} gives $\{e^{2it}, te^{2it}, e^{-2it}, te^{-2it}\}$, resulting in: $\{\cos 2t, \sin 2t, t \cos 2t, t \sin 2t\}$.   **4.** $\{1, t, e^t, te^t, t^2 e^t\}$; $c_1 + c_2 t + c_3 e^t + c_4 te^t + c_5 t^2 e^t$

5. $\{e^{-t} \cos t, e^{-t} \sin t, te^{-t} \cos t, te^{-t} \sin t\}$; $c_1 e^{-t} \cos(t) + c_2 e^{-t} \sin t + c_3 te^{-t} \cos t + c_4 te^{-t} \sin t$

6. $c_1 e^{-t} + c_2 te^{-t} + c_3 e^{-t} \cos 2t + c_4 e^{-t} \sin 2t$ **7.** $c_1 + c_2 t + c_3 \cos 3t + c_4 \sin 3t$

Exercises 5.7–Part B, Page 102

1. $y = e^{-3t} u$; $e^{-3t} \frac{d^2u}{dt^2} = 0$; $u = a + bt$; $y = e^{-3t}(a + bt)$; $w = e^{-6t}$

2. Upon substitution both sides equal zero; $y = x^3 u$; $x^4 \frac{du}{dx} + x^5 \frac{d^2u}{dx^2} = 0$; $u = a + b \ln x$; $y = x^3(a + b \ln x)$; $w = x^5$. The Wronskian is nonzero on any interval which does not contain the origin.

3. **(a)** $y(t) = -\frac{1}{5} \sin te^{-2t} + \frac{8}{5} \cos te^{-2t} + \frac{9}{5} \sin te^{-t} + \frac{7}{5} \cos te^{-t}$ **(b)**

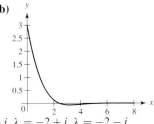

(c) $(\lambda^2 + 2\lambda + 2)(\lambda^2 + 4\lambda + 5) = 0$ **(d)** $\lambda = -1 + i$, $\lambda = -1 - i$, $\lambda = -2 + i$, $\lambda = -2 - i$

(e) $\{\cos te^{-t}, \sin te^{-t}, \cos te^{-2t}, \sin te^{-2t}\}$; $w = 5e^{-6t}$ **(f)** $c_1 \cos te^{-t} + c_2 \sin te^{-t} + c_3 \cos te^{-2t} + c_4 \sin te^{-2t}$

(g) $c_1 + c_3 = 3$, $-c_1 + c_2 - 2c_3 + c_4 = -3$, $-2c_2 + 3c_3 - 4c_4 = 2$, $2c_1 + 2c_2 - 2c_3 + 11c_4 = 1$; $c_2 = \frac{9}{5}$, $c_3 = \frac{8}{5}$, $c_4 = -\frac{1}{5}$, $c_1 = \frac{7}{5}$;

$y(t) = -\frac{1}{5} \sin te^{-2t} + \frac{8}{5} \cos te^{-2t} + \frac{9}{5} \sin te^{-t} + \frac{7}{5} \cos te^{-t}$

Exercises 5.8–Part A, Page 105

1. $T = 1.473537570$ s; $A = 1.76$ ft.   **2.** Using 62.5 lb/ft^3 as the density of water, one obtains $\rho = 97.65625005$.

3. $y(t) = -\frac{53}{75} \cos(\frac{10}{3}\sqrt{3}t)$ $T = \frac{1}{5}\pi\sqrt{3}$; $A = \frac{53}{75}$ **4.** $\omega = \sqrt{\frac{ab\rho g}{w}}$, $\frac{d^2y}{dt^2} + \omega^2 y = 0$

5. **(a)** fourth quadrant, (b) third quadrant, (c) second quadrant.

Exercises 5.8–Part B, Page 105

1. **(a)** $y(t) = \frac{1}{4}\cos(\frac{5}{3}\sqrt{10\pi}\,t) - \frac{\sqrt{10}}{50\sqrt{\pi}}\sin(\frac{5}{3}\sqrt{10\pi}\,t)$ **(b)**   **(c)** $T = \frac{3}{25}\sqrt{10\pi}$, $A = \frac{1}{100}\sqrt{625 + \frac{40}{\pi}}$

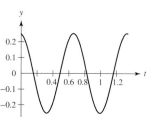

(d) $y(t) = \frac{1}{4}e^{-1/2bt}\cos(\frac{1}{6}\sqrt{1000\pi - 9b^2}\,t) + \frac{3}{4}\frac{e^{-1/2bt}\,b\sin(\frac{1}{6}\sqrt{1000\pi - 9b^2}\,t)}{\sqrt{1000\pi - 9b^2}} - 2\frac{e^{-1/2bt}\sin(\frac{1}{6}\sqrt{1000\pi - 9b^2}\,t)}{\sqrt{1000\pi - 9b^2}}$

(e) $\omega = \frac{1}{6}\sqrt{1000\pi - 9b^2}$ **(f)** $B := \frac{10}{3}\sqrt{10\pi}$

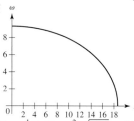

(g) $T := 12\frac{\pi}{\sqrt{1000\pi - 9b^2}}$; $[[b = \frac{1}{2}B, T = \frac{2}{25}\sqrt{30\pi} \approx 0.77665], [b = 0, T = \frac{3}{25}\sqrt{10\pi} \approx 0.67259]]$ **(h)**

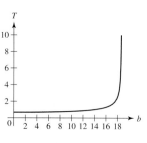

(i) $\omega = \frac{1}{6}\sqrt{1000\pi - 9b^2}$; $[[b = \frac{1}{2}B, \omega = \frac{5}{6}\sqrt{30\pi} \approx 8.09011], [b = 0, \omega = \frac{5}{3}\sqrt{10\pi} \approx 9.34165]]$

2. $\omega = \sqrt{\frac{ab\pi\rho g}{f}}$, $\frac{d^2 y}{dt^2} + \omega^2 y = 0$ 3. $\frac{7}{10}\sqrt{10}\cos(\sqrt{10}t + \arctan(\frac{3}{20}\sqrt{10}))$; $A = \frac{7}{10}\sqrt{10}$, $\omega = \sqrt{10}$,
$T = \frac{1}{5}\pi\sqrt{10}$, $F = \frac{1}{2}\frac{\sqrt{10}}{\pi}$, $\phi = -\arctan(\frac{3}{20}\sqrt{10})$

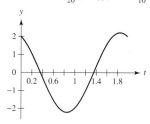

Exercises 5.9–Part A, Page 109

3. $y_p = 2$ 4. $y_p = \frac{1}{4}t$ 5. **(a)** $y_p = \frac{3}{2}$ **(b)** $y_p = \frac{1}{2}t - \frac{3}{4}$ **(c)** $y_p = te^{-t}$
6. The constant "a" merely generates a linear combination of y_1 and y_2; i.e., a solution of the homogeneous equation.

Exercises 5.9–Part B, Page 109

1. **(a)** $y_p := \frac{1}{15}x$ **(d)** $y_p = x^2\sin x - x^3\cos x$ **(e)** $y_p = x^5 e^x - 4x^4 e^x + 12x^3 e^x - 24x^2 e^x + 24xe^x$ **(f)** $y_p = -x^4\sin x + 2x^2\sin x - 2x^3\cos x$

2. **(a)** $y(t) = A\cos 2t + B\sin 2t - \frac{1}{4}\cos 2t - \frac{1}{2} - \frac{1}{4}\ln(-\cos t + \sin t)\sin 2t + \frac{1}{4}\ln(-\cos t - \sin t)\sin 2t$ **(b)** $y_h = A\cos 2t + B\sin 2t$,
$y_p = -\frac{1}{4}\cos 2t - \frac{1}{2} - \frac{1}{4}\ln(-\cos t + \sin t)\sin 2t + \frac{1}{4}\ln(-\cos t - \sin t)\sin 2t$ **(c)** $\cos 2t, \sin 2t$ **(d)** $W = 2$
(e) $y_p = \sin 2t\int\frac{1}{2}\cos 2t\tan^2 2t\,dt - \cos 2t\int\frac{1}{2}\sin 2t\tan^2 2t\,dt = -\frac{1}{2} + \frac{1}{4}\sin 2t\ln(\frac{1+\sin 2t}{\cos 2t})$ **(f)** Clearly the term $-\frac{\cos 2t}{4}$ in the particular
solution found in part (b) of this problem can be omitted. Observe that $-\frac{\cos 2t}{4}$ is a solution of the homogeneous equation.

4. **(a)** $y(t) = -\frac{1}{30}\cos t + \frac{1}{15}\sin t + Ae^{-3t}\cos 2t + Be^{-3t}\sin 2t$ **(b)** $y_h = Ae^{-3t}\cos 2t + Be^{-3t}\sin 2t$, $y_p = -\frac{1}{30}\cos t + \frac{1}{15}\sin t$
(c) $e^{-3t}\cos 2t, e^{-3t}\sin 2t$ **(d)** $W = 2e^{-6t}$ **(e)** $y_p = \frac{1}{2}e^{-3t}\sin 2t\int\cos 2t\sin t e^{3t}\,dt - \frac{1}{2}e^{-3t}\cos 2t\int\sin 2t\sin t e^{3t}\,dt = -\frac{1}{30}\cos t + \frac{1}{15}\sin t$
(f) In this case, the two versions of the particular solution totally agree.

6. **(a)** $y(x) = \frac{1}{6}\frac{1}{x^2} + \frac{A}{x^4} + \frac{B}{x^5}$ **(b)** $y_h = \frac{A}{x^4} + \frac{B}{x^5}$, $y_p = \frac{1}{6}\frac{1}{x^2}$ **(c)** $\frac{1}{x^4}, \frac{1}{x^5}$ **(d)** $W = -\frac{1}{x^{10}}$ **(e)** $y_p = -\frac{\int x^2\,dx - \int x\,dx\,x}{x^5} = \frac{1}{6}\frac{1}{x^2}$
(f) The two versions of the particular solution totally agree.

8. (a). $y(t) = -\frac{365}{193}e^{-3/4t}\cos(\frac{3}{4}\sqrt{3}t) + \frac{311}{579}e^{-3/4t}\sin(\frac{3}{4}\sqrt{3}t)\sqrt{3} + \frac{36}{193}\sin 2t - \frac{21}{193}\cos 2t$

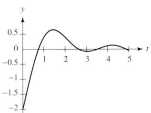

(b) $\lambda^2 + \frac{3}{2}\lambda + \frac{9}{4} = 0$; $\lambda = -\frac{3}{4} + \frac{3}{4}i\sqrt{3}$, $\lambda = -\frac{3}{4} - \frac{3}{4}i\sqrt{3}$; $e^{-3/4t}\cos(\frac{3}{4}\sqrt{3}t)$, $e^{-3/4t}\sin(\frac{3}{4}\sqrt{3}t)$; $y_h = c_1 e^{-3/4t}\cos(\frac{3}{4}\sqrt{3}t) + c_2 e^{-3/4t}\sin(\frac{3}{4}\sqrt{3}t)$;
$W = \frac{3}{4}e^{-3/2t}\sqrt{3}$; $y_p = \frac{1}{3}e^{-3/4t}\sin(\frac{3}{4}\sqrt{3}t)\sqrt{3}\int\cos(\frac{3}{4}\sqrt{3}t)\cos(2t)e^{3/4t}\,dt - \frac{\sqrt{3}}{3}e^{-3/4t}\cos(\frac{3}{4}\sqrt{3}t)\int\sin(\frac{3}{4}\sqrt{3}t)\cos(2t)e^{3/4t}\,dt =$
$-\frac{21}{193}\cos 2t + \frac{36}{193}\sin 2t$; $y_g = c_1 e^{-3/4t}\cos(\frac{3}{4}\sqrt{3}t) + c_2 e^{-3/4t}\sin(\frac{3}{4}\sqrt{3}t) - \frac{21}{193}\cos 2t + \frac{36}{193}\sin 2t$; $c_1 - \frac{21}{193} = -2$, $-\frac{3}{4}c_1 + \frac{72}{193} + \frac{3}{4}c_2\sqrt{3} = 3$;
$c_1 = -\frac{365}{193}$, $c_2 = \frac{311}{579}\sqrt{3}$; $y(t) = -\frac{365}{193}e^{-3/4t}\cos(\frac{3}{4}\sqrt{3}t) + \frac{311}{579}\sqrt{3}e^{-3/4t}\sin(\frac{3}{4}\sqrt{3}t) + \frac{36}{193}\sin 2t - \frac{21}{193}\cos 2t$

9. (a) $y(t) = -\frac{3}{40}\cos 2t - \frac{3}{20}\sin 2t - \frac{117}{40}e^{-2/5t}\cos(\frac{4}{5}t) + \frac{113}{80}e^{-2/5t}\sin(\frac{4}{5}t)$; (b) $y_p = -\frac{3}{40}\cos 2t - \frac{3}{20}\sin 2t$

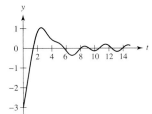

11. (a) $y_g = -\frac{1}{24}x + Ax^2 + Bx^5 + Cx^7$; $y_p = -\frac{1}{24}x$; x^2, x^5, x^7 (b) C_3: $2x^3\frac{du}{dx} + 20x^6\frac{dv}{dx} + 42x^8\frac{dw}{dx} = x$
(c) $\frac{du}{dx} = \frac{1}{15}\frac{1}{x^2}$, $\frac{dv}{dx} = -\frac{1}{6}\frac{1}{x^5}$, $\frac{dw}{dx} = \frac{1}{10}\frac{1}{x^7}$ (d) $u(x) = -\frac{1}{15}\frac{1}{x}$, $v(x) = \frac{1}{24}\frac{1}{x^4}$, $w(x) = -\frac{1}{60}\frac{1}{x^6}$ (e) $y_p = -\frac{1}{24}x$

Exercises 5.10–Part A, Page 116

1. (a) $y_p = c_1 te^{3t} + c_2\sin 5t + c_3\cos 5t$ (b) $y_p = c_1 + c_2\sin 2t + c_3\cos 2t$ (c) $y_p = c_1 te^{-t} + c_2\sin t + c_3\cos t$
(d) $y_p = c_1 e^t + c_2 e^{-3t}\sin t + c_2 e^{-3t}\cos t$ (e) $y_p = (c_1 + tc_2 + c_3 t^2)t^2 + (c_4 + c_5 t)te^{2t}$
(f) $y_p = c_1\sin 3t + c_2\cos 3t + c_3 t\sin 3t + c_4 t\cos 3t$ (g) $y_p = c_1 e^{4t}\sin t + c_2 e^{4t}\cos t + t(c_3 e^{4t}\sin t + c_4 e^{4t}\cos t)$
(h) $y_p = (c_1 + tc_2 + c_3 t^2)t$ (i) $y_p := c_1 e^{2t} + c_2 te^{2t} + c_3 t^2 e^{2t}$ (j) $y_p = c_1\cos t + c_2\sin t + (c_3\cos t + c_4\sin t)t$

2. $y_1 = \frac{1}{2}t - \frac{3}{4} - \frac{3}{10}\cos t + \frac{1}{10}\sin t + \frac{3}{2}e^{-t} - \frac{9}{20}e^{-2t}$; $y_2 = -\frac{3}{4} + \frac{1}{2}t + e^{-t} - \frac{1}{4}e^{-2t}$; $y_3 = -\frac{3}{10}\cos t + \frac{1}{10}\sin t + \frac{1}{2}e^{-t} - \frac{1}{5}e^{-2t}$; $y_1 = y_2 + y_3$

3. No, $y_1 \neq y_2 + y_3$ 4. $y_p = -\frac{1}{10}\cos 2t + \frac{1}{5}\sin 2t = -\frac{1}{10}\sqrt{5}\cos(2t + \arctan 2)$ 5. $\phi = \arctan(2\frac{\omega}{\sqrt{\omega^4+4}}, -\frac{\omega^2-2}{\sqrt{\omega^4+4}})$, $A = \frac{1}{\sqrt{\omega^4+4}}$

Exercises 5.10–Part B, Page 116

1. (a) $[-1, i, -i]$ (b) $\lambda^3 + \lambda^2 + \lambda + 1$ (c) $2y + 4\frac{dy}{dt} + 5\frac{d^2y}{dt^2} + 5\frac{d^3y}{dt^3} + 3\frac{d^4y}{dt^4} + \frac{d^5y}{dt^5} = 0$
(d) $\{-1, i, -i, -1+i, -1-i\} = \{-1+i, -1-i\} \bigcup \{-1, i, -i\}$
(e) $y(t) = c_1\sin t + c_2\cos t + c_3 e^{-t} + c_4 e^{-t}\sin t + c_5 e^{-t}\cos t$ Just let c_4 and c_5 equal zero.

2. (a) $-\frac{50}{13}te^{-2t} - \frac{447}{169}e^{-2t} - \frac{60}{169}\cos 3t - \frac{25}{169}\sin 3t$

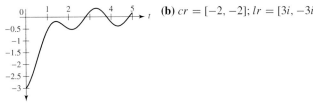

(b) $cr = [-2, -2]$; $lr = [3i, -3i]$

(c) $y_h = c_1 e^{-2t} + c_2 te^{-2t}$; $y_p = a_1\sin 3t + a_2\cos 3t$ (d) $a_1 = -\frac{25}{169}$, $a_2 = -\frac{60}{169}$; $y_p = -\frac{25}{169}\sin 3t - \frac{60}{169}\cos 3t$
(e) $y_g = c_1 e^{-2t} + c_2 te^{-2t} - \frac{25}{169}\sin 3t - \frac{60}{169}\cos 3t$; $c_1 - \frac{60}{169} = -3$, $-2c_1 + c_2 - \frac{75}{169} = 1$; $c_1 = -\frac{447}{169}$, $c_2 = -\frac{50}{13}$;
$y(t) = -\frac{50}{13}te^{-2t} - \frac{447}{169}e^{-2t} - \frac{60}{169}\cos 3t - \frac{25}{169}\sin 3t$ (f) $36y + 36\frac{dy}{dt} + 13\frac{d^2y}{dt^2} + 4\frac{d^3y}{dt^3} + \frac{d^4y}{dt^4} = 0$ (g) $\{3i, -3i, -2, -2\}$;
$y_g = c_1 e^{-2t} + c_2\sin 3t + c_3\cos 3t + c_4 te^{-2t}$ Choose c_1 and c_4 equal to zero.

Exercises 5.11–Part A, Page 121

1. (a) $\omega_N = \sqrt{\frac{7}{3}}$ (b) This system has no damping: $\omega_R = \omega_N = \sqrt{\frac{7}{3}}$ (c) $y(t) = \frac{1}{42}\sqrt{21}t\sin(\frac{1}{3}\sqrt{21}t) + c_1\cos(\frac{1}{3}\sqrt{21}t) + c_2\sin(\frac{1}{3}\sqrt{21}t)$

3. $\omega_N := \frac{2}{3}\sqrt{3}$; $\omega_R := \frac{1}{3}\sqrt{3}$; amplitude at resonance $\frac{1}{12}\sqrt{3}$

Exercises 5.11–Part B, Page 122

7.
$$\left[\frac{1}{\sqrt{-207\omega^2+64\omega^4+169}}, \frac{1}{\sqrt{-204\omega^2+64\omega^4+169}}, \frac{1}{\sqrt{-199\omega^2+64\omega^4+169}}, \frac{1}{\sqrt{-192\omega^2+64\omega^4+169}}, \frac{1}{\sqrt{-183\omega^2+64\omega^4+169}}, \frac{1}{\sqrt{-172\omega^2+64\omega^4+169}}, \frac{1}{\sqrt{-159\omega^2+64\omega^4+169}}, \frac{1}{\sqrt{-144\omega^2+64\omega^4+169}}\right]$$

8. $\omega_R = [\frac{3}{16}\sqrt{46}, \frac{1}{8}\sqrt{102}, \frac{1}{16}\sqrt{398}, \frac{1}{2}\sqrt{6}, \frac{1}{16}\sqrt{366}, \frac{1}{8}\sqrt{86}, \frac{1}{16}\sqrt{318}, \frac{3}{4}\sqrt{2}]$

9. $2\dfrac{m}{b\sqrt{-b^2+4mk}}$

10.

11. $\omega_N := \frac{1}{16}\sqrt{-b^2+416}$

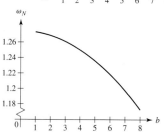

12. Clearly the ratio of the two frequencies is approximately one. **13.**

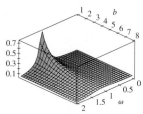

Exercises 5.12–Part A, Page 124

1. (a) $\cos(2\sqrt{2}\ln 2) - i\sin(2\sqrt{2}\ln 2)$ **2.** (a) $y(t) = c_1 t^{3/2}\cos(\frac{1}{2}\ln t) + c_2 t^{3/2}\sin(\frac{1}{2}\ln t)$ **3.** (c) $[t^3, t^3 \ln t]$

Exercises 5.12–Part B, Page 124

1. $\lambda^3 - \lambda^2 - 14\lambda + 24 = 0$; $\lambda = -4, 2, 3$; $y(t) = \frac{c_1}{t^4} + c_2 t^2 + c_3 t^3$ **2.** $\frac{d^3y}{dx^3} - \frac{d^2y}{dx^2} - 14\frac{dy}{dx} + 24y = 0$; $y(x) = c_1 e^{2x} + c_2 e^{3x} + c_3 e^{-4x}$; $y(\ln t) = c_1 t^2 + c_2 t^3 + \frac{c_3}{t^4}$ **3.** (c) $x^2\frac{d^2y}{dx^2} + (2-a)x\frac{dy}{dx} + by = 0$ **4.** (a) $y(t) = c_1 t + c_2 t^2$ (b) $\lambda^2 - 3\lambda + 2 = 0$; $\lambda = 1, 2$; $y(t) = c_1 t + c_2 t^2$ (c) $2y - 3\frac{dy}{dx} + \frac{d^2y}{dx^2} = 0$; $y(x) = c_1 e^x + c_2 e^{2x}$; $y(\ln t) = c_1 t + c_2 t^2$

12. (a) $y(t) = c_1 t^3\sin(2\ln t) + c_2 t^3\cos(2\ln t)$ (b) $\lambda^2 - 6\lambda + 13 = 0$; $\lambda = 3 + 2i, 3 - 2i$; $y(t) = c_1 t^3\cos(2\ln t) + c_2 t^3\sin(2\ln t)$ (c) $13y - 6\frac{dy}{dx} + \frac{d^2y}{dx^2} = 0$; $y(x) = c_1 e^{3x}\sin 2x + c_2 e^{3x}\cos 2x$; $y(\ln t) = c_1 t^3\sin(2\ln t) + c_2 t^3\cos(2\ln t)$

22. (a) $y(t) = \frac{c_1}{t^{5/3}} + \frac{c_2 \ln t}{t^{5/3}}$ (b) $9\lambda^2 + 30\lambda + 25 = 0$; $\lambda = -\frac{5}{3}, -\frac{5}{3}$ Let $y(t) = u(t)t^{-5/3}$; $t\frac{d^2u}{dt^2} + \frac{du}{dt} = 0$; $u(t) = c_1 + c_2 \ln t$; $y(t) = \frac{c_1 + c_2 \ln t}{t^{5/3}}$ (c) $25y + 30\frac{dy}{dx} + 9\frac{d^2y}{dx^2} = 0$; $y = c_1 e^{-5/3x} + c_2 x e^{-5/3x}$; $y(\ln t) = \frac{c_1 + c_2 \ln t}{t^{5/3}}$

Exercises 5.13–Part A, Page 128

1. (a) $y_g = c_1 e^{-2t} + c_2 e^{-t} + (-1 + \frac{1}{2}t)te^{-t}$; $y(t) = -e^{-2t} + e^{-t} + (-1 + \frac{1}{2}t)te^{-t}$ (b) $G(t, x) = \begin{cases} 0 & t < x \\ e^{x-t} - e^{2x-2t} & t > x \end{cases}$ (c) Subtraction of the results in (a) and (b) gives zero.   (d) Differentiate G with respect to t and compute the one-sided limits. The difference equals 1.

Exercises 5.13–Part B, Page 128

2. (a) $y(t) = -\frac{3}{10}t\cos t + \frac{17}{50}\cos t + \frac{1}{10}t\sin t + \frac{3}{25}\sin t - \frac{1}{2}e^{-t} + \frac{4}{25}e^{-2t}$ **(b)** $G(t, x) = \begin{cases} 0 & t < x \\ e^{x-t} - e^{2x-2t} & t > x \end{cases}$

(c) $y(t) = -\frac{3}{10}t\cos t + \frac{17}{50}\cos t + \frac{1}{10}t\sin t + \frac{3}{25}\sin t + (-\frac{1}{2}e^t + \frac{4}{25})e^{-2t}$

(d)
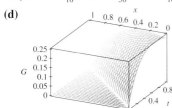
(e)
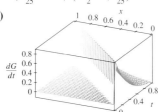
(f) Differentiate G with respect to t and compute the one-sided limits. The difference equals 1.

7. (a) $y(t) = -\frac{14}{291}\cos 3t + \frac{17}{291}\sin 3t + \frac{2}{3}e^{-3/2t} - \frac{60}{97}e^{-4/3t}$ **(d)**
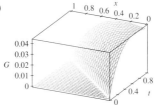
(e)
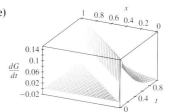

(b) $G(t, x) = \begin{cases} 0 & t < x \\ e^{4/3x-4/3t} - e^{3/2x-3/2t} & t > x \end{cases}$

(c) $y(t) = -\frac{14}{291}\cos 3t + \frac{17}{291}\sin 3t - \frac{60}{97}e^{-4/3t} + \frac{2}{3}e^{-3/2t}$

(f) Differentiate G with respect to t and compute the one-sided limits. The difference equals $\frac{1}{6}$.

9.   (a) $y(t) = \frac{3}{139}t^5 - \frac{3}{139}\frac{\cos(\frac{1}{5}\sqrt{19}\ln t)}{t^{1/5}} - \frac{78}{2641}\frac{\sqrt{19}\sin(\frac{1}{5}\sqrt{19}\ln t)}{t^{1/5}}$ **(b)** $G(t, x) = \begin{cases} 0 & t < x \\ \frac{1}{19}\frac{\sqrt{19}\sin(\frac{1}{5}\sqrt{19}\ln t - \frac{1}{5}\sqrt{19}\ln x)}{x^{4/5}t^{1/5}} & t > x \end{cases}$ **(c)** Integrating this G proves to

be a little too much for *Maple*. Instead, compute $y(t) = \int_1^t g(t, x)(3x^5)\,dx = \frac{3}{139}t^5 - \frac{78}{2641}\frac{\sin(\frac{1}{5}\sqrt{19}\ln t)\sqrt{19}}{t^{1/5}} - \frac{3}{139}\frac{\cos(\frac{1}{5}\sqrt{19}\ln t)}{t^{1/5}}$.

(d)
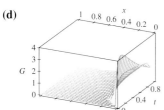
(e)
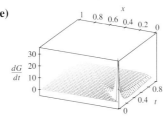
(f) Differentiate G with respect to t and compute the one-sided limits. The difference equals $\frac{1}{5}\frac{1}{x^2}$.

CHAPTER 6

Exercises 6.1–Part A, Page 133

1. No, because $\int_0^1 \frac{e^{-st}}{t^2}\,dt > \int_0^1 \frac{1-st}{t^2}\,dt$ and the latter integral diverges.   **4.** $\frac{3}{2}\frac{\sqrt{2}(s+5)}{s^2+25}$ **6.** $3\cos\sqrt{2}t$

Exercises 6.1–Part B, Page 134

1. (a) $\int_0^\infty te^{-st}\,dt = \frac{1}{s^2}, s > 0$ **(b)** $\int_0^\infty e^{2t}e^{-st}\,dt = \frac{1}{s-2}, s > 2$ **5. (a)** $\int_0^\infty 3t^2\sin(3t)e^{-st}\,dt = 54\frac{s^2-3}{(s^2+9)^3}$; $54\frac{s^2-3}{(s^2+9)^3}, s > 0$

12. (a) $f(t) = \begin{cases} 0 & t < 3 \\ 2(t-3) & 3 < t < 5 \\ 0 & t > 5 \end{cases}$ **(b)** $2\frac{(-e^{2s}+2s+1)e^{-5s}}{s^2}$

Exercises 6.2–Part A, Page 137

1. $3\frac{1}{s}\frac{1}{6} - \frac{\sqrt{3}-6}{s-\sqrt{3}} + \frac{1}{6}\frac{\sqrt{3}+6}{s+\sqrt{3}}$ **5. (c)** $-\frac{10s^3+11s^2+19s+22}{5s^4+8s^3+13s^2+16s+6}$ **6. (b)** $-4\cosh 2t + \frac{5}{2}\sinh 2t = -\frac{3}{4}e^{2t} - \frac{13}{4}e^{-2t}$

Exercises 6.2–Part B, Page 138

5. (a) $-\frac{6(\cos 1\sin 1)s - 18 - s^2 + s^2\cos^2 1}{s(s^2+36)}$ **(b)** Use $f(t) = \frac{1}{2} - \frac{1}{2}\cos 2\cos 6t - \frac{1}{2}\sin 2\sin 6t$ **(c)** $\int_0^\infty \sin(3t-1)^2e^{-st}\,dt = -\frac{1}{2}\frac{+s^2\cos 2+6s\sin 2-s^2-36}{s(s^2+36)}$
(This formula is equivalent to that given under (a).)   **7. (a)** $5\cos 2t - \frac{7}{2}\sin 2t$ **(b)** $5\frac{s}{s^2+4} - 7\frac{1}{s^2+4}$

12. (a) $F(s) = \frac{s^2-4s-5}{(s^2-4s+13)^2}$ **(b)** $f''(t) = 4e^{2t}\cos 3t - 6e^{2t}\sin 3t - 5te^{2t}\cos 3t - 12te^{2t}\sin 3t$ **(c)** $L[f''(t)] = \frac{4s^3-47s^2+104s-169}{(s^2-4s+13)^2}$

(d) $s^2F(s) - sf(0) - f'(0) = \frac{s^2(s^2-4s-5)}{(s^2-4s+13)^2} - 1$

14. **(a)** $Y(s) = -\frac{2s^3+18s+5s^2+39}{s^4+13s^2+4s^3+36s+36}$; $y(t) = -\frac{7}{13}te^{-2t} - \frac{314}{169}e^{-2t} - \frac{24}{169}\cos 3t - \frac{10}{169}\sin 3t$; $-\frac{7}{13}te^{-2t} - \frac{314}{169}e^{-2t} - \frac{24}{169}\cos 3t - \frac{10}{169}\sin 3t$

16. **(a)** $y(t) = \frac{5}{2}t^3 + 2e^{-3t}$ **(b)** $Y(s) = 15\frac{1}{s^4} + 2\frac{1}{s+3}$

Exercises 6.3–Part A, Page 140

1. **(b)** $g(t) = e^3 \cos(4t)$; $G(s) = \frac{e^3 s}{s^2+16}$; $F(s) = G(s+2) = \frac{e^3(s+2)}{(s+2)^2+16}$ 4. **(b)** $F(s) = \frac{s-3}{(s-3)^2+2}$; $f(t) = e^{3t}\cos(\sqrt{2}t)$

5. **(a)** $F(s) = \frac{5}{2}\frac{1}{s+3} + \frac{1}{2}\frac{1}{s+1}$; $f(t) = \frac{5}{2}e^{-3t} + \frac{1}{2}e^{-t}$ **(b)** $\frac{3s+4}{(s+2)^2-1}$; $G(s) = \frac{3s-2}{s^2-1}$; $g(t) = 3\cosh t - 2\sinh t = \frac{1}{2}e^t + \frac{5}{2}e^{-t}$;
   $f(t) = e^{-2t}g(t) = \frac{5}{2}e^{-3t} + \frac{1}{2}e^{-t}$

Exercises 6.3–Part B, Page 141

2. $Y(s) = -\frac{s^3+6s^2+9s-1}{s^4+8s^3+23s^2+30s+18}$; $y(t) = \frac{1}{5}te^{-3t} + \frac{4}{25}e^{-3t} - \frac{29}{25}e^{-t}\cos t + \frac{28}{25}e^{-t}\sin t$

4. **(a)** $g(t) = e^5 \cosh 2t$; $G(s) = \frac{e^5 s}{s^2-4}$; $F(s) = G(s+3) = \frac{e^5(s+3)}{(s+3)^2-4}$ **(b)** $\frac{e^5(s+3)}{(s+3)^2-4}$

9. **(a)** $F(s) = \frac{4s-3}{(s+4)^2-21}$; $G(s) = \frac{4s-19}{s^2-21}$; $g(t) = -\frac{19}{21}\sqrt{21}\sinh(\sqrt{21}t) + 4\cosh(\sqrt{21}t)$;
   $f(t) = e^{-4t}g(t) = -\frac{19}{21}e^{-4t}\sqrt{21}\sinh(\sqrt{21}t) + 4e^{-4t}\cosh(\sqrt{21}t)$

Exercises 6.4–Part A, Page 145

1. **(a)** $\frac{F(s+3)}{s+3}$ 2. **(d)** $L(\sin 3t) = 3\frac{1}{s^2+9} = G(s)$; $F(s) = -G'(s-2) = 6\frac{s-2}{((s-2)^2+9)^2}$

Exercises 6.4–Part B, Page 146

2. **(a)** $F(s) = \frac{s^2+6s+5}{(s^2+6s+13)^2s}$ **(b)** $g(x) = \cos 2x$; $G(s) = \frac{s}{s^2+4}$; $h(x) = xe^{-3x}g(x)$; $H(s) = -G'(s-3) = \frac{s^2-6s+5}{(s^2-6s+13)^2}$; $F(s) = \frac{H(s)}{s} = \frac{s^2-6s+5}{(s^2-6s+13)^2s}$

6. $f(t) = \cos 2t$; $F(s) = \frac{s}{s^2+4}$; $L[e^{-3t}\int_0^t \cos 2z\, dz] = \frac{1}{(s+3)^2+4}$ 17. **(a)** $f(t) = \frac{3}{4} - \frac{3}{4}e^{-4t}$ **(b)** $F(s) = 3\frac{1}{s(s+4)} = \frac{3}{4}\frac{1}{s} - \frac{3}{4}\frac{1}{s+4}$

23. **(a)** $f(t) = -\frac{e^{3/2t}\sin(\frac{1}{2}t)}{t}$ **(b)** $-\frac{d}{ds}\arctan(2s-3) = -2\frac{1}{1+(2s-3)^2}$

25. **(a)** $y(t) = -\frac{211}{82}e^{-5/6t}\cos(\frac{1}{6}\sqrt{23}t) - \frac{227}{1886}\sqrt{23}e^{-5/6t}\sin(\frac{1}{6}\sqrt{23}t) - \frac{14}{41}\sin 2t - \frac{35}{82}\cos 2t$ **(b)** $Y(s) = -\frac{9s^3-12s^2-36s-34}{3s^4+5s^3+16s^2+20s+16}$

Exercises 6.5–Part A, Page 150

1. **(a)** $g(t) = f(t-2)H(t-2)$ with $f(t) = (t+2)^2$; $F(s) = 2\frac{1+2s+2s^2}{s^3}$; $G(s) = F(s)e^{-2s} = 2\frac{(1+2s+2s^2)e^{-2s}}{s^3}$

2. **(a)** $F(s) = \frac{s}{s^2+1}$; $f(t) = \cos(t)$; $g(t) = \cos(t-1)H(t-1)$

Exercises 6.5–Part B, Page 150

7. **(a)**

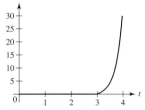

   (b) $f(t) = \begin{cases} 0 & t < \frac{1}{2} \\ e^{t+1} & t > \frac{1}{2} \end{cases}$; $\int_0^\infty f(t)e^{-st}\, dt = \frac{e^{3/2-1/2s}}{s-1}$

   (c) $f(t) = e^{(t+1)}H(t - \frac{1}{2})$

   (d) $F(s) = \frac{e^{3/2-1/2s}}{s-1}$ **(e)** $g(t) = e^{t+3/2}$; $G(s) = \frac{e^{3/2}}{s-1}$; $F(s) = G(s)e^{-1/2s} = \frac{e^{3/2-1/2s}}{s-1}$

11. **(a)** $f(t) = \frac{5}{4}H(t-3)\sinh(4t-12)$ **(b)** $G(s) = 5\frac{1}{s^2-16}$; $g(t) = \frac{5}{4}\sinh(4t)$; $f(t) = g(t-3)H(t-3)$ **(c)** $f(t) = \begin{cases} 0 & t < 3 \\ g(t-3) & t > 3 \end{cases}$;

Exercises 6.6–Part A, Page 153

1. $H(t) - H(t-1)$ 4. $tH(t) + (1-t)H(t-1)$; $\frac{1}{s^2} - \frac{e^{-s}}{s^2}$

5. $y(t) = (-\frac{1}{5}t + \frac{7}{25} - \frac{2}{25}e^{1-t}\cos(2t-2) + \frac{3}{50}e^{1-t}\sin(2t-2))H(t-1) + \frac{1}{5}t - \frac{2}{25} + \frac{2}{25}e^{-t}\cos(2t) - \frac{3}{50}e^{-t}\sin(2t)$

Exercises 6.6–Part B, Page 153

1. **(a)** $e^{-s}(L[t^2] + 2L[t] + L[1]) = e^{-s}(2\frac{1}{s^3} + 2\frac{1}{s^2} + \frac{1}{s})$ **(b)** $e^{-s}L[(t+1)^2] = e^{-s}(2\frac{1}{s^3} + 2\frac{1}{s^2} + \frac{1}{s})$

2. **(a)** $L[f(t)] = -6\frac{s+3}{((s+3)^2+16)^2} + 8\frac{(s+3)^3}{((s+3)^2+16)^3}$ **(b)** $L[f(t)] = \frac{d^2}{ds^2}L[\cos(4t)](s+3) = -6\frac{s+3}{((s+3)^2+16)^2} + 8\frac{(s+3)^3}{((s+3)^2+16)^3}$

4. **(a)** $L[f(t)] = \dfrac{e^{-s}(s^3\cos 1 + s\cos 1 - s^2\sin 1 - \sin 1 + s^2\cos 1 - \cos 1 - 2s\sin 1)}{s^4 + 2s^2 + 1}$

    (b) $L[f(t)] = -\dfrac{d}{ds}(e^{-s}L[\cos(t+1)]) = \dfrac{e^{-s}(s^3\cos 1 + s\cos 1 - s^2\sin 1 - \sin 1 + s^2\cos 1 - \cos 1 - 2s\sin 1)}{s^4 + 2s^2 + 1}$

9. **(a)** $f(t) = -\frac{1}{5}t + \frac{11}{25} - \frac{11}{5}e^{-5t}$ **(b)** $F(s) = -\frac{1}{5}\frac{1}{s^2} + \frac{11}{25}\frac{1}{s} - \frac{11}{25}\frac{1}{s+5}$ **(c)** $L[\int_0^t g(x)\,dx] = \frac{G(s)}{s}$; $g(x) = -\frac{1}{5} + \frac{11}{5}e^{-5x}$;

    $f(t) = \int_0^t g(x)\,dx = -\frac{1}{5}t + \frac{11}{25} - \frac{11}{25}e^{-5t}$

10. **(a)**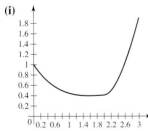

   (b) $F(s) = -\dfrac{e^{-3s}(2s-1)}{s^2}$

   (c) $g(t) = \begin{cases} 0 & t < 3 \\ t-5 & t > 3 \end{cases}$

   (d) $g(t) = t - 2$; $L[f(t)] = e^{-3s}L[g(t)] = -\dfrac{(2s-1)e^{-3s}}{s^2}$

   (e) $h(t) = t - 5$; $L[f(t)] = e^{-3s}L[h(t+3)] = -\dfrac{(2s-1)e^{-3s}}{s^2}$ **(f)** $\displaystyle\int_0^\infty (t-5)H(t-3)e^{-st}\,dt = -\dfrac{e^{-3s}(2s-1)}{s^2}$

14. **(a)**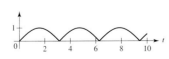

   (b) $\dfrac{s^2 + 4e^{-2s}s^2 + 4e^{-2s}s + 2e^{-2s}}{s^3}$

   (c) $\begin{cases} 1 & t < 2 \\ 1 + t^2 & t > 2 \end{cases}$

   (d) $g(t) = (t+2)^2$; $L[f(t)] = \dfrac{1}{s} + e^{-2s}L[g(t)] = \dfrac{s^2 + 4e^{-2s}s^2 + 4e^{-2s}s + 2e^{-2s}}{s^3}$

   (e) $h(t) = t^2$; $L[f(t)] = \dfrac{1}{s} + e^{-2s}L[h(t+2)] = \dfrac{s^2 + 4e^{-2s}s^2 + 4e^{-2s}s + 2e^{-2s}}{s^3}$

(f) $\displaystyle\int_0^\infty (1 + t^2H(t-2))e^{-st}\,dt = \dfrac{4e^{-2s}s^2 + 4se^{-2s} + 2e^{-2s} + s^2}{s^3}$ **(g)** $Y(s) = \dfrac{s^4 + s^3 + s^2 + 4e^{-2s}s^2 + 4e^{-2s}s + 2e^{-2s}}{s^3(s^2 + 2s + 2)}$;

$y(t) = \frac{1}{2}e^{-t}\cos t - \frac{1}{2}e^{-t}\sin t + \frac{1}{2} + \frac{1}{2}H(t-2) - \frac{1}{2}H(t-2)e^{-t+2}\cos(t-2) - \frac{3}{2}H(t-2)e^{-t+2}\sin(t-2) - H(t-2)t + \frac{1}{2}t^2H(t-2)$

(h) $y(t) = \begin{cases} \frac{1}{2}e^{-t}\cos t - \frac{1}{2}e^{-t}\sin t + \frac{1}{2} & t < 2 \\ \frac{1}{2}e^{-t}\cos t - \frac{1}{2}e^{-t}\sin t + 1 - \frac{1}{2}e^{2-t}\cos(t-2) - \frac{3}{2}e^{2-t}\sin(t-2) - t + \frac{1}{2}t^2 & t > 2 \end{cases}$

(i)

With the constant forcing function on $[0, 2]$, the damped oscillation approaches the steady state solution $y(t) = \frac{1}{2}$. For $t > 2$, the new forcing function $1 + t^2$ pushes the solution ever upward.

Exercises 6.7–Part A, Page 156

1. $\dfrac{e^{-\pi s} + 1}{(1 - e^{-\pi s})(s^2 + 1)}$

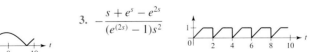

3. $-\dfrac{s + e^s - e^{2s}}{(e^{(2s)} - 1)s^2}$ (See plot.)

Exercises 6.7–Part B, Page 156

1. **(a)** $g(t) = \begin{cases} t & 0 \le t \le 1 \\ 1 & 1 \le t < 2 \end{cases}$

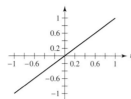

(b) $F(s) = \dfrac{(e^{-2s} + e^{-2s}s + s - 1)e^{s}}{(e^{-2s} - 1)s^2}$

3. $Y(s) = -\dfrac{e^{-s} - 1}{s(10 + s^2 + s^2 e^{-s} + 2s + 2se^{-s} + 10e^{-s})} \approx \dfrac{1 - 2e^{-s} + 2e^{-2s} - 2e^{-3s} + 2e^{-4s} - 2e^{-5s}}{s(s^2 + 2s + 10)}$

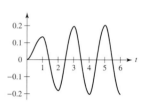

Exercises 6.8–Part A, Page 159

1. $\frac{1}{2}t^2$ 4. $\frac{1}{2}e^t + \frac{1}{2}\cos t - \frac{1}{2}\sin t - 1$

Exercises 6.8–Part B, Page 159

2. **(a)** $\int_0^t (t - x)^2 \cos 5x\, dx;$ $\int_0^t \cos(-5t + 5x)x^2\, dx$ **(b)** $-\frac{2}{125}\sin 5t + \frac{2}{25}t$ **(c)**

(d) $-\frac{2}{25}\frac{1}{s^2+25} + \frac{2}{25}\frac{1}{s^2}$

(e) $2\frac{1}{s^2(s^2+25)}$

(f) $-\frac{2}{125}\sin 5t + \frac{2}{25}t$

4. **(a)** $\frac{1}{54}t^2 e^{2/3t}$ **(b)** $g(t) = \frac{1}{3}e^{2/3t};$ $h(t) = \frac{1}{9}te^{2/3t};$ $\int_0^t \frac{1}{27}e^{2/3t-2/3x}xe^{2/3x}\, dx = \frac{1}{54}t^2 e^{2/3t}$

Exercises 6.9–Part A, Page 162

1. $H(s) = \dfrac{e^{-2s}}{s};$ $G(s) = \dfrac{s}{s^2 + 9};$ $H(t - 2)\cos 3t = \frac{1}{3}H(t - 2)\sin(3t - 6)$

2. (See plot.)

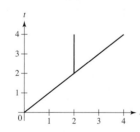

$\int_0^t H(x - 2)\cos(-3t + 3x)\, dx = \begin{cases} 0 & t < 2 \\ \int_2^t \cos(-3t + 3x)\, dx & t > 2 \end{cases} = \frac{1}{3}H(t - 2)\sin(3t - 6)$

Exercises 6.9–Part B, Page 162

1. $F(s) = \frac{1}{(s+1)^2+1};$ $G(s) = -\frac{1}{s^2-1} + 2\frac{s^2}{(s^2-1)^2};$ $\frac{1}{50}((5t - 4)e^t + 25te^{-t} + 4e^{-t}\cos t - 22e^{-t}\sin t)$

2. $\int_0^t e^{-x}(\sin x)(t - x)\cosh(t - x)\, dx = \frac{1}{50}((5t - 4)e^t + 25te^{-t} + 4e^{-t}\cos t - 22e^{-t}\sin t)$

3. **(a)** $F(s) = \dfrac{e^{-3s}}{s};$ $G(s) = \dfrac{s^2 - 1}{(s^2 + 1)^2};$ $H(t - 3)(t\sin(t - 3) - 3\sin(t - 3) + \cos(t - 3) - 1)$

   (b) $\int_0^t H(x - 3)(t - x)\cos(-t + x)\, dx = H(t - 3)(t\sin(t - 3) - 3\sin(t - 3) - 1 + \cos(t - 3))$

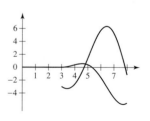

12. (a) $F_{1,3} = \dfrac{e^{-s}}{s} - \dfrac{e^{-3s}}{s}$; $F_{2,4} = \dfrac{e^{-2s}}{s} - \dfrac{e^{-4s}}{s}$; $H(t-3)t - 3H(t-3) - 2H(t-5)t + 10H(t-5) + H(t-7)t - 7H(t-7)$

(b) The integral is too difficult for *Maple*. Investigate the region on which the integrand is nonzero. It is bounded by the lines $x = 1$, $x = 3$, $t = x + 2$ and $t = x + 4$.

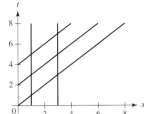

Assume $t \in (3, 5)$, then $\int_{1}^{t-2}(H(x-1) - H(x-3))(H(t-x-2) - H(t-x-4))\, dx = t - 3$ and if $t \in (5, 7)$, then

$\int_{t-4}^{3}(H(x-1) - H(x-3))(H(t-x-2) - H(t-x-4))\, dx = 7 - t$; $P_{1,3}(t) * P_{2,4}(t) = \begin{cases} 0 & t < 3 \\ t - 3 & 3 < t < 5 \\ 7 - t & 5 < t < 7 \\ 0 & t > 7 \end{cases}$

Exercises 6.10–Part A, Page 166

1. $y(t) = H(t-1)\sin(t-1)$; $y'(t) = \delta(t-1)\sin(t-1) + H(t-1)\cos(t-1) = \begin{cases} 0 & t < 1 \\ \cos(t-1) & t > 1 \end{cases}$

$\lim_{t \to 1^-} y'(t) = 0$; $\lim_{t \to 1^+} y'(t) = 1$

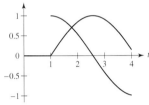

2. (a) $F = \dfrac{e^{-s}}{s^2 + 2s + 5}$; $y(t) = \frac{1}{2}H(t-1)e^{-t+1}\sin(2t-2)$

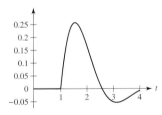

Exercises 6.10–Part B, Page 166

1. $f_a(t) = \begin{cases} \dfrac{t}{a^2} & \le t \le a \\ -\dfrac{t-2a}{a^2} & < t \le 2a \end{cases}$; $\displaystyle\int_0^\infty f_a(t)e^{-st}\, dt = \dfrac{e^{-2sa} - 2e^{-sa} + 1}{s^2 a^2}$; *(Hint:* Convert the function $f_a(t)$ to Heaviside form, before attempting the

integration); $\lim\limits_{a \to 0^-} \dfrac{e^{-2sa} - 2e^{-sa} + 1}{s^2 a^2} = 1$ **2.** $y(t) = \frac{1}{2}e^{-t}\sin 2t$ (for both)

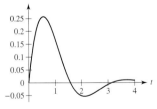

Exercises 6.11–Part A, Page 168

1. (a) $F(s) = -\dfrac{e^{-s} - 1}{s(s^2 + 2s + 5)}$; $y(t) = \frac{1}{5} - \frac{1}{5}e^{-t}\cos 2t - \frac{1}{10}e^{-t}\sin 2t - \frac{1}{5}H(t-1) + \frac{1}{5}H(t-1)e^{-t+1}\cos(2t-2) +$

$\frac{1}{10}H(t-1)e^{-t+1}\sin(2t-2)$ **(b)** $V(s) = \frac{1}{s^2+2s+5}$ **(c)** $v(t) = \frac{1}{2}e^{-t}\sin 2t$ **(d)** $g(t,x) = \frac{1}{2}e^{-t+x}\sin(2t-2x)$

(e) $\int_0^t \frac{1}{2}e^{-t+x}\sin(2t-2x)(1 - H(x-1))\, dx = \frac{1}{5} - \frac{1}{5}e^{-t}\cos 2t - \frac{1}{10}e^{-t}\sin 2t - \frac{1}{5}H(t-1) + \frac{1}{5}H(t-1)e^{-t+1}\cos(2t-2) +$

$\frac{1}{10}H(t-1)e^{-t+1}\sin(2t-2)$ **(f)** $U(s) = \dfrac{e^{-sx}H(x)}{s^2 + 2s + 5}$; $u(x) = \frac{1}{2}e^{-t+x}\sin(2t-2x)$

Exercises 6.11–Part B, Page 168

2. (a) $y(t) = -\frac{15}{26}\cos 3t - \frac{5}{13}\sin 3t + \frac{15}{26}e^{-t}\cos 2t + \frac{45}{52}e^{-t}\sin 2t$; $y(0) = 0$, $y'(0) = 0$

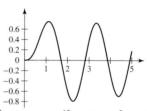

(b) $V(s) = \frac{1}{s^2+2s+5}$ **(c)** $v(t) = \frac{1}{2}e^{-t}\sin 2t$ **(d)** $\int_0^t v(t-x)f(x)\,dx = \frac{15}{26}e^{-t}\cos 2t + \frac{45}{52}e^{-t}\sin 2t - \frac{15}{26}\cos 3t - \frac{5}{13}\sin 3t$

(e) $e^{-t}\cos 2t$, $e^{-t}\sin 2t$; $g(t,x) = \frac{1}{2}e^{-t+x}\sin(2t-2x)$; $G(t,x) = \begin{cases} 0 & t < x \\ \frac{1}{2}e^{x-t}\sin(2t-2x) & t > x \end{cases}$ **(f)** $v(t-x) = -\frac{1}{2}e^{-t+x}\sin(-2t+2x)$

(g) $y(t) = \frac{1}{2}H(t-x)e^{-t+x}\sin(2t-2x)$ Of course, if $t > x$ this equals $v(t-x) = \frac{1}{2}e^{-t+x}\sin(2t-2x)$.

7. (a) $y(t) = -5 + t + 12e^{-1/3t} + \dfrac{(e^{3/2} + 4e^{1/2} - 12e^{1/6} + 6)e^{-1/3t}}{e^{(1/6)} - 1} - 6e^{-1/2t} + \dfrac{e^{-1/2t}(-e^{3/2} - 4e^{1/2} + 12e^{1/6} - 6)}{e^{1/6} - 1}$

(b) $y(t) = -5 + t - 4e^{-1/2t} - 2ae^{-1/2t} - 6be^{-1/2t} + 9e^{-1/3t} + 3ae^{-1/3t} + 6be^{-1/3t}$; $b = \frac{1}{6}\dfrac{(e^{1/6})^9 + 4(e^{1/6})^3 - 12e^{1/6} + 6}{e^{1/6} - 1}$, $a = 1$;

$y(t) = -(5e^{1/6} - 5 - te^{1/6} + t + 6e^{-1/3t} - e^{-1/3t+3/2} - 4e^{-1/3t+1/2} - 6e^{-1/2t+1/6} + e^{-1/2t+3/2} + \dfrac{4e^{-1/2t+1/2}}{e^{(1/6)} - 1}$;

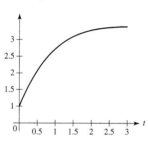

CHAPTER 7

Exercises 7.1–Part A, Page 174

1. $[0, \frac{1}{2}, 1, \frac{9}{8}, 1]$; $L := \lim_{k\to\infty}\dfrac{k^2}{2^k} = 0$ **5.** $[1.0, 1.187500000, 1.264404297, 1.299759667, 1.316757849]$; fixed points are $\frac{4}{3}$, and 4.

Exercises 7.1–Part B, Page 174

1. $g(x) = 1 + \frac{3}{16}x^2$ Observe that for $x \geq 5$, $x + \frac{1}{2} \leq g(x)$. This implies that $x_{n-1} + \frac{1}{2} \leq x_n$ and therefore $\lim_{n\to\infty}x_n = \infty$.

2. (a) $\left[2, \frac{6}{7}, \frac{12}{17}, \frac{20}{31}, \frac{30}{49}, \frac{42}{71}, \frac{56}{97}, \frac{72}{127}, \frac{90}{161}, \frac{110}{199}\right]$

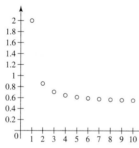

(b) $\lim_{k\to\infty}\dfrac{k^2+k}{2k^2-1} = \frac{1}{2}$

(c) 6; 501; 50001

(d) Take $N(\varepsilon) = \frac{3}{2\varepsilon}$.

6. (a) $\left[1, \frac{1}{2}, \frac{2}{9}, \frac{3}{32}, \frac{24}{625}, \frac{5}{324}, \frac{720}{117,649}, \frac{315}{131,072}, \frac{4480}{4,782,969}, \frac{567}{1,562,500}\right]$

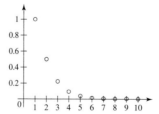

(b) $\lim_{k\to\infty}\dfrac{k!}{k^k} = 0$

(c) 4; 9; 14

(d) Take $N(\varepsilon) = \frac{1}{\varepsilon}$.

11. **(a)** $[-4, 2, -\frac{16}{9}, 2, -\frac{64}{25}, \frac{32}{9}, -\frac{256}{49}, 8, -\frac{1024}{81}, \frac{512}{25}]$ **(b)** $\lim\limits_{k \to \infty} \dfrac{(-1)^k 2^{k+1}}{k^2}$ does not exist.

(c) 9; 18; 25

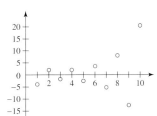

16. **(a)** [2.0, 5.062500000, 13.31829498, 35.52713679, 95.39621664, 257.0895515, 694.3931083, 1878.284768, 5085.770231, 13780.61234]

(b) $\lim\limits_{k \to \infty} (\frac{1}{k} + 1)^{k^2} = \infty$ **(c)** 3; 8; 12

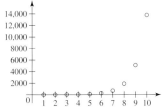

18. **(a)** $u_k = \dfrac{\left(-\frac{25}{13}\sqrt{13} - 7\right)\left(-\frac{2}{-3-\sqrt{13}}\right)^k}{-3 - \sqrt{13}} + \dfrac{\left(\frac{25}{13}\sqrt{13} - 7\right)\left(-\frac{2}{-3+\sqrt{13}}\right)^k}{-3 + \sqrt{13}}$ **(b)**

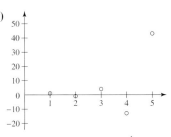

(c) The limit does not exist since $u_k = 2.109400392(0.3027756378)^k - 0.1094003939(-3.302775640)^k$ **(d)** $[4, -13, 43] = [4, -13, 43]$

23. **(a)** Linear, nonconstant coefficients   **(b)** $[2, -1, 3, -7, 24]$ **(c)**

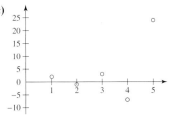

28. **(a)** $r = \frac{9}{5} - \frac{1}{5}\sqrt{6} \approx 1.310102051$ **(b)** [1.3, 1.310000000, 1.310100000, 1.310102010, 1.310102051, 1.310102051]

31. **(a)** 2.199999850   **(b)** [2.19, 2.191843290, 2.193343033, 2.194564511, 2.195560190, 2.196372362]

36. **(a)** $\left[2, \frac{5}{2}, \frac{25}{9}, \frac{425}{144}, \frac{221}{72}\right]$ **(c)** $\dfrac{\sinh \pi}{\pi}$ **(d)**

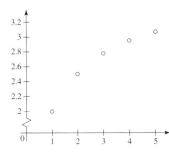
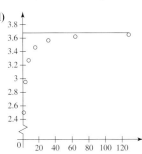

Exercises 7.2–Part A, Page 178

3. **(e)** $\dfrac{1}{2^n + 1} + \cdots + \dfrac{1}{2^{(n+1)}} > \dfrac{1}{2^{n+1}} + \cdots + \dfrac{1}{2^{n+1}} = \dfrac{2^n}{2^{n+1}} = \dfrac{1}{2}$ This implies that $\sum_{k=1}^{2^n} \frac{1}{k} \geqslant 1 + \sum_{m=0}^{n} \frac{1}{2} = 1 + \frac{n}{2}$

4. Evaluate the difference of the nth partial sum and ln(2) for $n = 1, 2, \ldots$ [0.3068528194, -0.1931471806, 0.1401861527, -0.1098138473, 0.0901861527, -0.0764805139, 0.0663766289, -0.0586233711, 0.0524877400, -0.0475122600]

Exercises 7.2–Part B, Page 178

2. Once the general term $\frac{1}{k}$ becomes smaller than the number of digits carried by a computer system, it will cease to contribute to the overall sum. The value returned will be the partial sum up to the point where $\frac{1}{k}$ adds "0" to the sum.

3. **(a)**

(b) $S = \left[2, 3, \frac{10}{3}, \frac{41}{12}, \frac{103}{30}, \frac{1237}{360}, \frac{433}{126}, \frac{69,281}{20160}, \frac{62,353}{18,144}, \frac{6,235,301}{1,814,400} \right]$ **(c)**

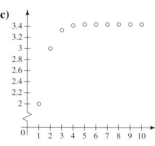

(d) This series converges (to $2e$).

6. **(a)**

(b) $S := [0, 0.3465735903, 0.7127776866, 1.059351277, 1.381238859, 1.679865437, 1.957852601, 2.217782794, 2.461918858, 2.692177367]$

(c)

(d) This series diverges (because the general term of this series is greater than the general term of the harmonic series, and the harmonic series diverges).

15. **(a)** $S := [0.04978706837, 0.05226582055, 0.05238923035, 0.05239537456, 0.05239568046, 0.05239569569, 0.05239569645, 0.05239569649, 0.05239569649, 0.05239569649]$ **(b)**

(c) $\lim_{n \to \infty} S_n = -\frac{e^{-3}}{e^{-3}-1}$

(d) $\lim_{k \to \infty} a_k = 0$

21. **(a)** $S := [0.8569292341, 0.8744215282, 0.8747785215, 0.8747858071, 0.8747859558, 0.8747859588, 0.8747859589, 0.8747859589, 0.8747859589, 0.8747859589]$ **(b)**

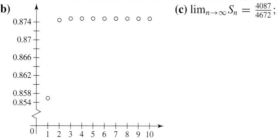

(c) $\lim_{n \to \infty} S_n = \frac{4087}{4672}$; $\lim_{k \to \infty} a_k = 0$

Exercises 7.3–Part A, Page 180

The ratio test: $\lim_{k\to\infty}\frac{1}{4}k\left(\frac{k+1}{k}\right)^k + \frac{1}{4}\left(\frac{k+1}{k}\right)^k = \lim_{k\to\infty}\frac{1}{4}(k+1)\left(1+\frac{1}{k}\right)^k = \infty$ The root test: $\lim_{k\to\infty}\frac{1}{4}k = \infty$

3. Use the ratio test: $\lim_{k\to\infty}\frac{2-\cos k}{k} = 0$; the series converges.

Exercises 7.3–Part B, Page 180

1. $f(x)$ is continuous, positive, and decreasing for $x \geq 1$ and $\int_1^\infty \frac{1}{x^2}\,dx = 1$.

4. (a)

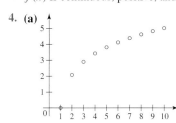

(b) The ratio test is inconclusive: $\lim_{k\to\infty}\frac{\ln^2 k}{\ln^2(k+1)} = 1$

(c) The root test is inconclusive: $\lim_{k\to\infty}\left(\frac{1}{\ln^2 k}\right)^{(1/k)} = 1$

(d) yes

(e) Observe that $\ln^2 k \leq k$ for all $k \geq 2$, and since the harmonic series diverges, so will this series.

10. (a) *Hint:* Use the **mul** command in your CAS to code the individual terms.   (b) The ratio test: $\rho = \lim_{k\to\infty}\frac{k+1}{2k+1} = \frac{1}{2}$

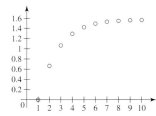

(c) The root test: $r = \lim_{k\to\infty}\left(\frac{\Gamma(k+1)\sqrt{\pi}}{\Gamma(k+\frac{1}{2})2^k}\right)^{(1/k)} = \frac{1}{2}$

(d) yes

(e) Does not apply, because the series diverges.

20. (a) $\left[\frac{1}{2}, \frac{1}{4}, \frac{9}{8}, \frac{1}{16}, d\frac{25}{32}, \frac{1}{64}, \frac{49}{128}, \frac{1}{256}, \frac{81}{512}, \frac{1}{1024}\right]$ (b) $\left[\frac{1}{2}, \frac{3}{4}, \frac{15}{8}, \frac{31}{16}, \frac{87}{32}, \frac{175}{64}, \frac{399}{128}, \frac{799}{256}, \frac{1679}{512}, \frac{3359}{1024}\right]$ (c) $\lim_{k\to\infty}\frac{a_{2k+1}}{a_{2k}} = \infty$; $\lim_{k\to\infty}\frac{a_{2k}}{a_{2k-1}} = 0$

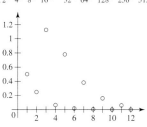

(d) $\lim_{k\to\infty} a_{2k}^{1/2k} = \frac{1}{2}$; $\lim_{k\to\infty} a_{2k-1}^{1/(2k-1)} = \frac{1}{2}$

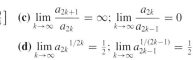

21. (a) Use the definition of σ_k to evaluate σ_1, σ_2, and σ_3.   (b) In the definition of σ_k, which contains a summation of the S_n, first write the individual S_n as a summation, then switch the order of the summations to obtain the desired result.   (c) 2   (d) $\frac{1}{2}$ (e) Investigate the sum of the partial sums and observe that $\frac{2}{3}\left(1-\frac{2}{k}\right) = \frac{2}{3}\frac{k-2}{k} \leq \sigma_k \leq \frac{2}{3}\frac{k+2}{k} = \frac{2}{3}\left(1+\frac{2}{k}\right)$. Taking the limit for $k \mapsto \infty$ of each part of this double inequality yields the desired result.   (f) Similar to (e); investigate the sum of the partial sums and observe that $\frac{3}{4}\left(1-\frac{3}{k}\right) = \frac{3}{4}\frac{k-3}{k} \leq \sigma_k \leq \frac{3}{4}\frac{k+3}{k} = \frac{3}{4}\left(1+\frac{3}{k}\right)$. Taking the limit for $k \mapsto \infty$ of each part of this double inequality yields the desired result

(g) $\sigma_k = \begin{cases} 0 & \text{if } k \text{ is even} \\ -\frac{k+1}{2k} & \text{if } k \text{ is odd} \end{cases}$ (h) *Maple* produces $\frac{1}{2}$.

Exercises 7.4–Part A, Page 184

1. (a) $\left[\frac{1}{3}, -\frac{1}{2}, \frac{1}{5}, -\frac{1}{4}, \frac{1}{7}, -\frac{1}{6}, \frac{1}{9}, -\frac{1}{8}, \frac{1}{11}, -\frac{1}{10}\right]$ (b) $\left[\frac{1}{3}, -\frac{1}{6}, \frac{1}{30}, -\frac{13}{60}, -\frac{31}{420}, -\frac{101}{420}, -\frac{163}{1260}, -\frac{641}{2520}, -\frac{4531}{27,720}, -\frac{7303}{27,720}\right]$

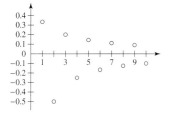

(c) Even though the series is alternating, $\{a_k\}$ (as in the Leibniz criterion) is not monotonically decreasing.   (d) Observe that if H_{2n+1} denotes the $(2n+1)$-st partial sum of the harmonic series; then $H_{2n+1} = 1 + \left(\sum_{k=1}^{n}\left(\frac{1}{2}\frac{1}{k} + \frac{1}{2k+1}\right)\right) = 1 + \left(\sum_{k=1}^{n}(|a_{2k-1}| + |a_{2k}|)\right)$. So the given series does not converge absolutely.   (e) $s_k = \sum_{m=1}^{k}(a_{2m-1} + a_{2m}) = \sum_{m=1}^{k}\left(\frac{1}{2m+1} - \frac{1}{2}\frac{1}{m}\right) = S_{2k+1} - 1$ (f) $[s_n - \ln(2) + 1]_{n=1}^{10} =$
$[0.6401861524, 0.1401861527, 0.3401861524, 0.0901861527, 0.2330432956, 0.0663766289, 0.1774877400, 0.0524877400,$
$0.1433968309, 0.0433968309]$ (g) $\sum_{m=1}^{\infty}\left(\frac{1}{2m+1} - \frac{1}{2}\frac{1}{m}\right) = \sum_{m=1}^{\infty}\left(-\frac{1}{2}\frac{1}{(2m+1)m}\right)$

3. (a) $\lim_{k\to\infty}\frac{k}{k+1} = 1$ (b) Even though the series is alternating, $\{a_k\}$ (as in the Leibniz criterion) is not monotonically decreasing

(c)
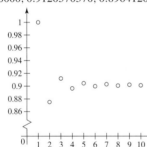

Observe that the sequence of even-indexed partial sums consists of negative terms and is monotone decreasing, while the sequence of odd-indexed partial sums consists of positive terms and is monotone increasing. The series must therefore diverge.

Exercises 7.4–Part B, Page 184

3. (a)

(b) $\frac{1}{(k+1)^3} \leq \frac{1}{k^3}$ $\lim_{k\to\infty}\frac{1}{k^3} = 0$

(c) $[1.0, 0.8750000000, 0.9120370370, 0.8964120370, 0.9044120370, 0.8997824074, 0.9026978593, 0.9007447343, 0.9021164764,$
$0.9011164764]$

(d) $n = 10$

(e) $\frac{14,420,574,181}{16,003,008,000} \approx= 0.9011164764$

(f) $hypergeom([1, 1, 1, 1], [2, 2, 2], -1) \approx 0.0004262010$

(g) According to the integral test, the series $\sum_{k=1}^{\infty}\frac{1}{k^3}$ will converge. Therefore the given series converges absolutely.

CHAPTER 8

Exercises 8.1–Part A, Page 189

2.   (See plot.)   4.   (See plot.)

Exercises 8.1–Part B, Page 189

1. $\lim_{k\to\infty}\int_0^1 f_k(x)\,dx = \lim_{k\to\infty}\frac{1}{k+1} = 0$; $\int_0^1 f(x)\,dx = 0$ 2. (a)

(b) $\left[\frac{1}{2}, 2\frac{\arctan 2}{\pi}, 2\frac{\arctan 3}{\pi}, 2\frac{\arctan 4}{\pi}, 2\frac{\arctan 5}{\pi}\right] \approx [0.5000000000, 0.7048327646, 0.7951672348, 0.8440417392, 0.8743340834]$

(c) 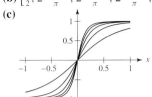   **5. (a)** 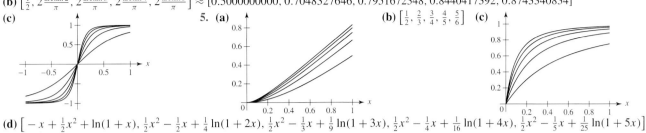   **(b)** $\left[\frac{1}{2}, \frac{2}{3}, \frac{3}{4}, \frac{4}{5}, \frac{5}{6}\right]$ **(c)**

(d) $\left[-x + \frac{1}{2}x^2 + \ln(1+x), \frac{1}{2}x^2 - \frac{1}{2}x + \frac{1}{4}\ln(1+2x), \frac{1}{2}x^2 - \frac{1}{3}x + \frac{1}{9}\ln(1+3x), \frac{1}{2}x^2 - \frac{1}{4}x + \frac{1}{16}\ln(1+4x), \frac{1}{2}x^2 - \frac{1}{5}x + \frac{1}{25}\ln(1+5x)\right]$

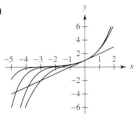

9. (a) $2\dfrac{\lim_{k\to\infty}\arctan(kx)}{\pi} = sgn(x)$ Observe that $x\,sgn(x) = |x|$.   **10. (a)**

(b) $\lim_{x\to 0}\dfrac{f_k(x) - f_k(0)}{x} = \lim_{x\to 0} 2\dfrac{\arctan(kx)}{\pi} = 0$

(c) $\lim_{x\to 0-}\dfrac{|x|}{x} = -1$; $\lim_{x\to 0+}\dfrac{|x|}{x} = 1$

(b) $[-1.0, -1.596071638, -2.180607124, -2.759002710, -3.333551485]$

Exercises 8.2–Part A, Page 190

1. $\lim_{k\to\infty} f_k(x) = \begin{cases} 1 & x = 0 \\ 0 & 0 < x \le \frac{\pi}{2} \end{cases}$ **2.** $\lim_{k\to\infty} f_k(x) = \begin{cases} 1 & x = 0 \le x < \frac{\pi}{2} \\ 0 & x = \frac{\pi}{2} \end{cases}$

Exercises 8.2–Part B, Page 191

1. $\lim_{k\to\infty}\int_0^1 f_k(x)\,dx = \ln 2$; $\int_0^1 f(x)\,dx = \ln 2$ **2. (a)** $f(x) = 0$; domain $[-10, 10]$ **(b)** $N(x, \varepsilon) = \left[\frac{|x|}{\varepsilon}\right]$, $[u]$ denotes the greatest integer function.

3. (a) $f(x) = 0$; domain $[0, 1]$ **(b)** $N(x, \varepsilon) = \left[\dfrac{\ln(\frac{x}{\varepsilon})}{x}\right]$ **5. (a)** $f(x) = 0$; domain $[0, 1]$ **(b)** $N(x, \varepsilon) = \begin{cases} 1 & x = 0 \\ \left[\frac{1}{x}\right] & 0 < x \le 1 \end{cases}$

11. (a) $f(x) = 0$; domain $[0, 1]$ **(b)** $N(x, \varepsilon) = \begin{cases} 1 & x = 0 \\ \left[\frac{1}{x}\right] & 0 < x \le 1 \end{cases}$ **(c)** $g(x) = piecewise\ (x = 0, \infty, 0 < x \le 1, 0)$ **(d)** No

(e) $F(x) = piecewise\ (x = 0, 0, 0 < x \le 1, 1)$ **(f)** No   **16. (a)** $f(x) = 0$; domain $[-1, 1]$ **(b)** $g(x) = 0$; domain $[-1, 1]$ **(c)** Yes

18. $f(x) = piecewise\ (x = 0, 0, 0 < x < 2\pi, -\frac{1}{2}x + \frac{1}{2}\pi, x = 2\pi, 0)$; $|f(0.5) - f_{2000}(0.5)| = 0.0003447398606$

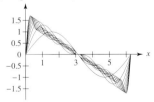

Exercises 8.3, Page 196

1. (a) $f(x) = 0$ **(b)** $f_k\left(\frac{1}{k}\right) = e^{-1}$ **(c)** $g(x) = \begin{cases} \infty & x = 0 \\ 0 & 0 < x \le 1 \end{cases}$ **(d)** The limit function is $h(x) = 0$. $\left|\int_0^x f_k(t)\,dt\right| < \varepsilon$, whenever $k >$

$N(\varepsilon) = \frac{1}{\varepsilon}$. *Hint:* Use the fact that $-1 \le (-kx - 1)e^{-kx} < 0$. It's fortuitous! The theorem does not apply, since the convergence of $f_k(x)$ is not uniform.

2. The limit function is $|x|$

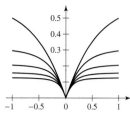

$N(\varepsilon) = \cot(\frac{1}{2}\varepsilon\pi)$

6. (a) The limit function is $f(x) = 0$; $f(\frac{1}{\sqrt{2k}}) = \sqrt{\frac{k}{2e}}$. **(b)** The limit function is $g(x) = \begin{cases} \infty & x = 0 \\ 0 & 0 < x \le 2 \end{cases}$. The convergence cannot be uniform since the limit function is not continuous, while the $g_k(x)$ are continuous. **(c)** The limit function is $h(x) = \begin{cases} 0 & x = 0 \\ \frac{1}{2} & 0 < x \le 2 \end{cases}$. The convergence cannot be uniform since the limit function is not continuous, while the $h_k(x)$ are continuous.

8. (a) The limit function is $f(x) = \begin{cases} 0 & x = 0 \\ 1 & 0 < x \le 1 \end{cases}$. Because the limit function is not continuous and the f_k are, the convergence cannot be uniform. **(b)** The limit function is $g(x) = \begin{cases} \infty & x = 0 \\ 0 & 0 < x \le 1 \end{cases}$. Because the limit function is not continuous and the derivatives of $f_k(x)$ are, the convergence cannot be uniform. **(c)** The limit function is $h(x) = x$; $N(\varepsilon) = \frac{1}{\varepsilon^2}$. *Hint:* Use the fact that $\ln(1+k) < \sqrt{k}$, whenever $1 \le k$. **(d)** Recall $f(x) = \begin{cases} 0 & x = 0 \\ 1 & 0 < x \le 1 \end{cases}$, and clearly $\int f(x)\,dx = x$, which in turn equals $\lim_{k\to\infty} \int f_k(x)\,dx = h(x) = x$. This is not a consequence of Theorem 2, because, $\{f_k(x)\}$ does not converge uniformly, the theorem does not apply.

Exercises 8.4–Part A, Page 199

1. The integral equals zero; $|f(x) - g(x)| \le |f(x^*) - g(x^*)| = \frac{2}{9}\sqrt{3}$ for $x^* = \frac{1}{2} \pm \frac{\sqrt{3}}{6}$.

3. *Hint:* To show that $f_k(x)$ converges in the mean to zero, use the u-substitution $u = x^{2k}$; observe that
$$\int_0^1 \frac{x^{2k}}{(1+x^{2k})^2}\,dx = \frac{1}{2k}\int_0^1 \frac{u^{1/2k}}{(1+u)^2}\,du < \frac{1}{2k}\int_0^1 \frac{1}{(1+u)^2}\,du; \quad f(x) = \begin{cases} 0 & 0 \le x < 1 \\ \frac{1}{2} & x = 1 \end{cases}.$$ The convergence is not uniform on [0, 1], because the sequence $f_k(x)$ converges pointwise to this discontinuous function.

Exercises 8.4–Part B, Page 199

1.

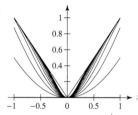

2. (a) $f(x) = 0$
(b) No, the convergence is not uniform. $f_k(\frac{1}{k}) = \frac{1}{2}$
(c) The limit function is again: $f(x) = 0$; $\lim_{k\to\infty}\int_0^\infty f_k(x)^2\,dx = \lim_{k\to\infty}\frac{1}{4}\frac{\pi}{k} = 0$
(d) D

11. (a) $f(x) = |x|$ **(b)** $\int_{-1}^1 (f_1(x) - |x|)^2\,dx = 0.2430698712$, $\int_{-1}^1 (f_2(x) - |x|)^2\,dx = 0.1078922519$, $\int_{-1}^1 (f_3(x) - |x|)^2\,dx = 0.05904471962$, $\int_{-1}^1 (f_4(x) - |x|)^2\,dx = 0.03691412747$, $\int_{-1}^1 (f_5(x) - |x|)^2\,dx = 0.02517651952$, $\int_{-1}^1 (f_6(x) - |x|)^2\,dx = 0.01824108593]$, $\int_{-1}^1 (f_7(x) - |x|)^2\,dx = 0.01381344015$, $\int_{-1}^1 (f_8(x) - |x|)^2\,dx = 0.01081847805$, $\int_{-1}^1 (f_9(x) - |x|)^2\,dx = 0.008699868768$, $\int_{-1}^1 (f_{10}(x) - |x|)^2\,dx = 0.007146820452$

13. $[x, (-\sqrt{2}+1)x^2 + \sqrt{2}x, (\sqrt{3} - \sqrt{6}+1)x^3 + (-2\sqrt{3}+\sqrt{6})x^2 + \sqrt{3}x, (3\sqrt{2} - 2 - 2\sqrt{3}+1)x^4 + (6+2\sqrt{3} - 6\sqrt{2})x^3 + (-6+3\sqrt{2})x^2 + 2x, (-\sqrt{5}+2\sqrt{15} - 2\sqrt{10}+1)x^5 + (-2\sqrt{5}+6\sqrt{10} - 4\sqrt{15})x^4 + (6\sqrt{5} - 6\sqrt{10}+2\sqrt{15})x^3 + (-4\sqrt{5}+2\sqrt{10})x^2 + \sqrt{5}x]$

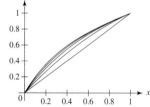

19. $[0.1600000000 - 0.02000000000z, -0.05200000000z^2 + 0.03200000000z + 0.4720000000, 0.01216000000z^3 - 0.07584000000z^2 - 0.02928000000z + 0.5420800000, 0.002488615385z^4 + 0.006025846150z^3 - 0.09077169231z^2 - 0.00376123077z + 0.5639532308, -0.001152000000z^5 + 0.007680000000z^4 + 0.01152000000z^3 - 0.1273600000z^2 - 0.00976000000z + 0.6223360000]$

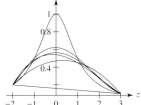

Exercises 8.5–Part A, Page 204

1. For all values of x. Use the Weierstrass M-test. $\dfrac{1 - x^{2k}}{k^3(1 + x^{2k})} \leq \dfrac{1}{k^3}$, and $\displaystyle\sum_{k=1}^{\infty} \dfrac{1}{k^3}$ converges.

3. Verify the conditions of Theorem 5. Let $f_k(x) = \dfrac{\sin kx}{k^3}$. Observe that (1) $f_k(x)$ is differentiable for all k and x, (2) $\sum_{k=1}^{\infty} f_k(0) = 0$, (3) $\left|\dfrac{\cos kx}{k^2}\right| \leq \dfrac{1}{k^2}$ and $\sum_{k=1}^{\infty} \dfrac{1}{k^2}$ converges, so $\sum_{k=1}^{\infty} \dfrac{\cos kx}{k^2}$ converges uniformly on $(-\infty, \infty)$.

Exercises 8.5–Part B, Page 205

1. $r < \dfrac{1}{LambertW(1)} \approx 1.7632$ *Hint:* Solve $|x \ln x| < 1$.

3. (a)

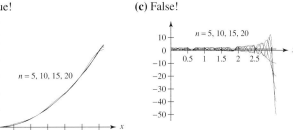

(b) $\left|\dfrac{(-1)^k x^{2k}}{(2k)!}\right| \leq \dfrac{\pi^{2k}}{(2k)!}$, and $\displaystyle\sum_{k=0}^{\infty} \dfrac{\pi^{2k}}{(2k)!}$ converges for all x according to the ratio test.

(c) $\displaystyle\sum_{k=0}^{\infty} \int_0^x \dfrac{(-1)^k t^{2k}}{(2k)!}\, dt = \sum_{k=0}^{\infty} \dfrac{x^{2k+1}(-1)^k}{(2k+1)(2k)!} = \sum_{k=0}^{\infty} \dfrac{(-1)^k x^{2k+1}}{(2k+1)!} = \sin x$ See Exercise B4.

(d) $\displaystyle\sum_{k=0}^{\infty} \dfrac{d}{dx} \dfrac{(-1)^k x^{2k}}{(2k)!} = \sum_{k=1}^{\infty} 2\dfrac{(-1)^k x^{2k} k}{x(2k)!} = \sum_{k=1}^{\infty} \dfrac{(-1)^k x^{2k-1}}{(2k-1)!} = \sum_{k=0}^{\infty} \dfrac{(-1)^{k+1} x^{2k+1}}{(2k+1)!} = -\sin x$

See Exercise B4

9. (a)

(b) True!

(c) False!

$(\sum f_k'(\pi) \to -\infty)$

$n = 5, 10, 15, 20$

$n = 5, 10, 15, 20$

$n = 5, 10, 15, 20$

20. (a) $\left[1, x, \dfrac{3}{2}x^2 - \dfrac{1}{2}, \dfrac{5}{2}x^3 - \dfrac{3}{2}x, \dfrac{35}{8}x^4 - \dfrac{15}{4}x^2 + \dfrac{3}{8}, \dfrac{63}{8}x^5 - \dfrac{35}{4}x^3 + \dfrac{15}{8}x, \dfrac{231}{16}x^6 - \dfrac{315}{16}x^4 + \dfrac{105}{16}x^2 - \dfrac{5}{16}, \dfrac{429}{16}x^7 - \dfrac{693}{16}x^5 + \dfrac{315}{16}x^3 - \dfrac{35}{16}x, \dfrac{6435}{128}x^8 - \dfrac{3003}{32}x^6 + \dfrac{3465}{64}x\right.$

(b) We verify the validity of this assertion by computing the associated dot products, $\int_{-1}^{1} P_k(x)P_j(x)\, dx$, the results of which are displayed in

the matrix: $\begin{bmatrix} 2 & 0 & 0 & 0 & 0 \\ 0 & \frac{2}{3} & 0 & 0 & 0 \\ 0 & 0 & \frac{2}{5} & 0 & 0 \\ 0 & 0 & 0 & \frac{2}{7} & 0 \\ 0 & 0 & 0 & 0 & \frac{2}{9} \end{bmatrix}$ (c) $[0, 1, 0, 0, 0, 0, 0, 0, 0, 0, 0]$; $[0, x, x, x, x, x, x, x, x, x, x]$

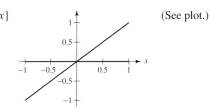

(See plot.)

(d) $[\frac{1}{3}, 0, \frac{2}{3}, 0, 0, 0, 0, 0, 0, 0, 0]$ $[\frac{1}{3}, \frac{1}{3}, x^2, x^2, x^2, x^2, x^2, x^2, x^2, x^2, x^2]$ (See plot.)

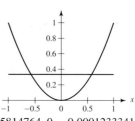

(e) $[0, 0, -1.519817754, 0, 0.5824466982, 0, -0.06611447688, 0, 0.003595814764, 0, -0.0001233341248]$

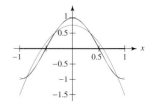

(f) $\left[0, -\frac{3}{2}, 0, \frac{7}{8}, 0, -\frac{11}{16}, 0, \frac{75}{128}, 0, -\frac{133}{256}, 0\right]$

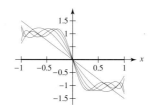

CHAPTER 9

Exercises 9.1–Part A, Page 209

1. $x - 2x^2$ **3.** $p_3(x) = 1 + 2x + 2x^2 + \frac{4}{3}x^3$; $p_3'(x) = 2 + 4x + 4x^2$

Exercises 9.1–Part B, Page 209

1. $\dfrac{1}{x+1} = 1 - x + \dfrac{x^2}{(1+x)^3} + \displaystyle\int_0^x \dfrac{6(xt - t^2/2)}{(1+t)^4}\, dt$; $\dfrac{1}{x+1} = \dfrac{1}{x+1}$

2. (a) $R_4(x) = \frac{1}{24}(\sin(\frac{2c-3}{3c+2})(\frac{2}{3c+2} - 3\frac{2c-3}{(3c+2)^2})^4 - 6\cos(\frac{2c-3}{3c+2})(\frac{2}{3c+2} - 3\frac{2c-3}{(3c+2)^2})^2(-\frac{12}{(3c+2)^2} + 18\frac{2c-3}{(3c+2)^3}) - 3\sin(\frac{2c-3}{3c+2})(-\frac{12}{(3c+2)^2} + 18\frac{2c-3}{(3c+2)^3})^2 - 4\sin(\frac{2c-3}{3c+2})(\frac{2}{3c+2} - 3\frac{2c-3}{(3c+2)^2})(\frac{108}{(3c+2)^3} - 162\frac{2c-3}{(3c+2)^4}) + \cos(\frac{2c-3}{3c+2})(-\frac{1296}{(3c+2)^4} + 1944\frac{2c-3}{(3c+2)^5}))(x-2)^4$ **(b)** $c = 2.178519784$

(c) $c = 2.323751209$ **(d)** $M := 0.979559308$ **(e)**   (See plot.)

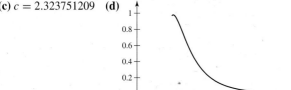

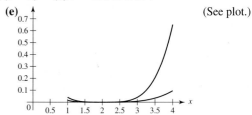

(f) $[1.465646879, 2.534353121]$ **(g)** $k = 12$

(h) $-0.2784055688 + 0.2015401512x - 0.07814958182(x-2.0)^2 + 0.02888468605(x-2.0)^3 - 0.01014516745(x-2.0)^4$

(See plots.)

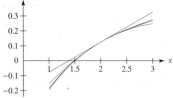

13. (a) 0.2452812035 **(b)** Use P_{51}; 0.2462610930

0.0509885424 **17. (a)** $x + \frac{1}{6}x^3$

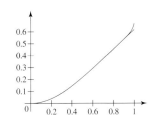

Exercises 9.2–Part A, Page 212

1. $\dfrac{1}{2+x} = \dfrac{1}{2}\dfrac{1}{1+\frac{x}{2}} = \dfrac{1}{2}\dfrac{1}{1-\left(-\frac{x}{2}\right)} = \dfrac{1}{2}\sum_{k=0}^{\infty}\left(-\dfrac{x}{2}\right)^k$

4. $\dfrac{1}{x+1} = \sum_{k=0}^{\infty}(-x)^k$; $\dfrac{1}{x+2} = \dfrac{1}{2}\sum_{k=0}^{\infty}\left(-\dfrac{1}{2}x\right)^k$; $\dfrac{1}{x+1} - \dfrac{1}{x+2} = \sum_{k=0}^{\infty}(-1)^k\left(1 - \dfrac{1}{2^{k+1}}\right)x^k$

Exercises 9.2–Part B, Page 212

1. $\frac{1}{x^2+2x+5} = \frac{1}{(x+1)^2+4} = \frac{1}{4}\sum_{k=0}^{\infty}(-1)^k\left(\frac{1}{2}x+\frac{1}{2}\right)^{2k}$

2. (a) $f(x) = \cos x$; $f^{2k-1}(0) = 0$; $f^{2k}(0) = (-1)^k$ **(b)** $0 \le \lim_{k\to\infty}|R_{k+1}| = \lim_{k\to\infty}\dfrac{|f^{2k+1}(c)|}{(2k+1)!}|x^{2k+1}| \le \lim_{k\to\infty}\dfrac{|x^{2k+1}|}{(2k+1)!} = 0$ The latter limit

equals zero, because it is the general term of a convergent series. **(c)** $\lim_{k\to\infty}\dfrac{a_{k+1}}{a_k} = \lim_{k\to\infty}\dfrac{1}{2}\dfrac{x^2}{(2k+1)(k+1)} = 0, \forall x$

7. (a) $\frac{5x-17}{4x^2+41x+45} = \sum_{k=0}^{\infty}(-1)^k\left(\frac{2}{9}9^{(-k)} - \frac{3}{5}4^k5^{(-k)}\right)x^k$ **(b)** $\lim_{k\to\infty}\dfrac{a_{k+1}}{a_k} = \dfrac{4}{5}x$; $R = \dfrac{5}{4}$ **(c)** $\frac{5x-17}{4x^2+41x+45} = \frac{2}{x+9} - \frac{3}{4x+5}$; $\dfrac{2}{x+9} = \sum_{k=0}^{\infty}2\dfrac{(-1)^kx^k}{9^{k+1}}$;

$-\dfrac{3}{4x+5} = \sum_{k=0}^{\infty}3\dfrac{(-1)^{k+1}4^kx^k}{5^{(k+1)}}$ **(d)** $\sum_{k=0}^{\infty}(-1)^k\left(\dfrac{2}{9^{k+1}} - 3\dfrac{4^k}{5^{k+1}}\right)x^k$

(e) $5x - 17 = 45a_0 + (45a_1 + 41a_0)x + \sum_{k=2}^{\infty}(4a_{k-2} + 41a_{k-1} + 45a_k)x^k$; $-17 = 45a_0 \Rightarrow a_0 = -\frac{17}{45}$; $5 = 45a_1 - \frac{697}{45} \Rightarrow a_1 = \frac{922}{2025}$;
$4a_{k-2} + 41a_{k-1} + 45a_k = 0, \forall k \ge 2 \Rightarrow a_k = \frac{2}{9}\left(-\frac{1}{9}\right)^k - \frac{3}{5}\left(-\frac{4}{5}\right)^k$

10). (a) $f(g(x)) = x$; $g(f(x)) = x$ for $-\frac{\pi}{2} \le x \le \frac{\pi}{2}$ **(b)** (See plot.)

(c) $f(x) = \sum_{k=0}^{\infty}\dfrac{(-1)^kx^{2k+1}}{(2k+1)!}$; $g(x) = \sum_{k=0}^{\infty}\dfrac{(2k)!4^{-k}x^{2k+1}}{(k!)^2(2k+1)}$; $a_0 + a_1x + a_2x^2 + a_3x^3 + a_4x^4 + a_5x^5 + a_6x^6 - \frac{1}{6}(a_0 + a_1x + a_2x^2 + a_3x^3 +$
$a_4x^4 + a_5x^5 + a_6x^6)^3 + \frac{1}{120}(a_0 + a_1x + a_2x^2 + a_3x^3 + a_4x^4 + a_5x^5 + a_6x^6)^5 = x \Rightarrow a_0 - \frac{1}{6}a_0^3 + \frac{1}{120}a_0^5 = 0$; $a_1 - \frac{1}{2}a_0^2a_1 +$
$\frac{1}{24}a_0^4a_1 - 1 = 0$; $a_2 - \frac{1}{2}a_1^2a_0 - \frac{1}{2}a_2a_0^2 + \frac{1}{12}a_1^2a_0^3 + \frac{1}{24}a_2a_0^4 = 0$; $a_3 - \frac{1}{6}a_1^3 - a_2a_0a_1 - \frac{1}{2}a_3a_0^2 + \frac{1}{6}a_2a_0^3a_1 + \frac{1}{24}a_3a_0^4 + \frac{1}{12}a_0^2a_1^3 = 0$;
$a_4 - \frac{1}{2}a_4a_0^2 - \frac{1}{2}a_1^2a_2 - a_3a_0a_1 - \frac{1}{2}a_0a_2^2 + \frac{1}{4}a_2a_0^2a_1^2 + \frac{1}{12}a_0^3a_2^2 + \frac{1}{6}a_3a_0^3a_1 + \frac{1}{24}a_0a_1^4 + \frac{1}{24}a_4a_0^4 = 0$; $\frac{1}{24}a_5a_0^4 - a_4a_0a_1 - \frac{1}{2}a_5a_0^2 - a_3a_0a_2 - \frac{1}{2}a_1a_2^2 + \frac{1}{120}a_1^5 - \frac{1}{2}a_1^2a_3 + \frac{1}{6}a_4a_0^3a_1 + \frac{1}{6}a_2a_0^3a_3 + \frac{1}{6}a_2a_0a_1^3 + \frac{1}{4}a_0^2a_1a_2^2 + \frac{1}{4}a_1^2a_0^2a_3 + a_5 = 0 \Rightarrow a_0 = 0$; $a_1 = 1$;
$a_2 = 0$; $a_3 = \frac{1}{6}$; $a_4 = 0$; $a_5 = \frac{3}{40}$ **13. (b)** $\frac{1}{2+x} = \frac{1}{4}\sum_{k=0}^{\infty}\left(\frac{1}{2} - \frac{1}{4}x\right)^k$

Exercises 9.3–Part A, Page 220

2. $\sin(x) \approx x - \frac{1}{6}x^3 + \frac{1}{120}x^5$; $\cos(x) \approx 1 - \frac{1}{2}x^2 + \frac{1}{24}x^4$; $\tan(x) \approx x + \frac{1}{3}x^3 + \frac{2}{15}x^5$; $\dfrac{x - \frac{1}{6}x^3 + \frac{1}{120}x^5}{1 - \frac{1}{2}x^2 + \frac{1}{24}x^4} \approx x + \frac{1}{3}x^3 + \frac{2}{15}x^5$

5. $J_0(x) = \sum_{k=0}^{\infty}\dfrac{(-1)^k4^{-k}x^{2k}}{(k!)^2}$; $\dfrac{d}{dx}J_0(x) = \sum_{k=1}^{\infty}2\dfrac{(-1)^k4^{-k}x^{2k}k}{(k!)^2x}$; $J_1(x) = \dfrac{1}{2}\sum_{k=0}^{\infty}\dfrac{(-1)^k4^{-k}x^{2k+1}}{(k!)^2(k+1)}$ *Hint:* Replace k by $m+1$ in the power series for $\frac{d}{dx}J_0(x)$.

Exercises 9.3–Part B, Page 220

1. $F(a) = \sqrt{a+4} + \frac{1}{2}a\ln(2 + \sqrt{a+4}) - \frac{1}{2}\sqrt{a+1} - \frac{1}{2}a\ln(1 + \sqrt{a+1})$; $\sqrt{a+x^2} \approx x + \frac{1}{2}\frac{a}{x} - \frac{1}{8}\frac{a^2}{x^3} + \frac{1}{16}\frac{a^3}{x^5} - \frac{5}{128}\frac{a^4}{x^7} + \frac{7}{256}\frac{a^5}{x^9}$;

$F(a) \approx \frac{3}{2} + \frac{1}{2}\ln(2)a - \frac{3}{64}a^2 + \frac{15}{1024}a^3 - \frac{105}{16,384}a^4 + \frac{1785}{524,288}a^5$

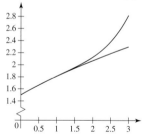

2. (a) $\displaystyle\sum_{k=0}^{\infty} \frac{(-1)^k x^{k+1}}{k+1}$

(b) $\displaystyle\lim_{k\to\infty} \frac{a_{k+1}}{a_k} = -x; \ R = 1$

(c) $\displaystyle\sum_{k=0}^{\infty} \frac{(-1)^k}{k+1}$ converges; $\displaystyle\sum_{k=0}^{\infty} - \frac{(-1)^{2k}}{k+1}$ diverges!

(d) $\displaystyle\sum_{k=0}^{\infty} \frac{(-1)^k}{k+1} = \ln 2 = \ln(1+1)$

(e)   (See plot.)

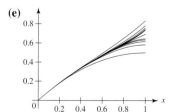

7. (a) $\displaystyle\sum_{k=0}^{\infty} a_k x^k = \sum_{k=0}^{\infty} \frac{(-1)^k x^{2k+1}}{(2k+1)!}; \ \sum_{k=0}^{\infty} b_k x^k = \sum_{k=0}^{\infty} \frac{x^k}{k!}$ **(b)** $\displaystyle\sum_{k=0}^{\infty} c_k x^k = \sum_{k=0}^{\infty} \frac{2^{1/2k}\sin(\frac{1}{4}k\pi)x^k}{k!}$ **(c)** $C_k = \displaystyle\sum_{n=0}^{k} a_n b_{k-n}; \{C_k\} = \{0, 1, 1, \frac{1}{3}, 0, -\frac{1}{30}\};$
$\{c_k\} = \{0, 1, 1, \frac{1}{3}, 0, -\frac{1}{30}\}$

17. (a) $\displaystyle\sum_{k=0}^{\infty} a_k x^k = \sum_{k=0}^{\infty} \frac{(-1)^k x^{2k+1}}{(2k+1)!}; \ \sum_{k=0}^{\infty} b_k x^k = \sum_{k=0}^{\infty} \frac{x^k}{k!}$ **(b)** $\displaystyle\sum_{k=0}^{\infty} d_k x^k = \sum_{k=0}^{\infty} \frac{2^{1/2k}\sin(\frac{3}{4}k\pi)x^k}{k!}$ **(c)** $d_0 = 0; \ d_1 + d_0 = 1; \ d_2 + d_1 + \frac{1}{2}d_0 = 0;$

$d_3 + d_2 + \frac{1}{2}d_1 + \frac{1}{6}d_0 = -\frac{1}{6}; \ d_4 + d_3 + \frac{1}{2}d_2 + \frac{1}{6}d_1 + \frac{1}{24}d_0 = 0; \ d_5 + d_4 + \frac{1}{2}d_3 + \frac{1}{6}d_2 + \frac{1}{24}d_1 + \frac{1}{120}d_0 = \frac{1}{120};$
$d_6 + d_5 + \frac{1}{2}d_4 + \frac{1}{6}d_3 + \frac{1}{24}d_2 + \frac{1}{120}d_1 + \frac{1}{720}d_0 = 0; \ \{d_k\} = \{0, 1, -1, \frac{1}{3}, 0, -\frac{1}{30}, \frac{1}{90}\}$ **(d)** $\sum_{k=0}^{6} s_k x^k = \sin 1 + (\cos 1)x +$
$(-\frac{1}{2}\sin 1 + \frac{1}{2}\cos 1)x^2 - \frac{1}{2}(\sin 1)x^3 + (-\frac{23}{24}\cos 1 - \frac{1}{4}\sin 1)x^4 + (-\frac{23}{120}\cos 1 - \frac{1}{24}\sin 1)x^5 + (\frac{11}{240}\sin 1 - \frac{37}{360}\cos 1)x^6$

(e) $f(g(x)) \approx \sum_{k=0}^{6} a_k \left(\sum_{m=0}^{6} b_m x^m \right)^k \approx \sum_{k=0}^{6} S_k x^k; \ f(g(x)) = \sum_{k=0}^{\infty} s_k x^k;$
$\{S_k\} = \{0.8414709848, 0.5403023059, -0.1505843394, -0.4207354920, -0.3229307250, -0.1386192282, -0.01696363818\};$
$\{s_k\} = \{0.8414709848, 0.5403023059, -0.1505843394, -0.4207354924, -0.3229307266, -0.1386192330, -0.01696365021\}$

34. (a) $f(x) \approx x - x^2 + \frac{5}{6}x^3 - \frac{5}{6}x^4 + \frac{101}{120}x^5; \ f'(x) \approx 1 - 2x + \frac{5}{2}x^2 - \frac{10}{3}x^3 + \frac{101}{24}x^4 - \frac{101}{20}x^5;$
$\frac{d}{dx}\left(x - x^2 + \frac{5}{6}x^3 - \frac{5}{6}x^4 + \frac{101}{120}x^5\right) = 1 - 2x + \frac{5}{2}x^2 - \frac{10}{3}x^3 + \frac{101}{24}x^4$

44. (b) $0.3862943611; \ x - 1 - \frac{1}{2}(x-1)^2 + \frac{1}{3}(x-1)^3 - \frac{1}{4}(x-1)^4 + \frac{1}{5}(x-1)^5 - \frac{1}{6}(x-1)^6 + \frac{1}{7}(x-1)^7 - \frac{1}{8}(x-1)^8 + \frac{1}{9}(x-1)^9; \ 0.3912698413$

CHAPTER 10

Exercises 10.1–Part A, Page 227

1. $a_0 = 2; \ a_n = 0, \forall n \geq 2; \ b_n = 2\frac{(-1)^n}{n\pi}; \ \lim_{n\to\infty} 0 = 0; \ \lim_{n\to\infty} 2\frac{(-1)^n}{n\pi} = 0$

5. $a_0 = 1 - \frac{\pi}{2}, \ a_n = -\frac{1+(-1)^n}{\pi n^2}; \ b_n = \frac{1+(-1)^n\pi - (-1)^n}{n\pi}$ We compute the overshoot of S_{50}. At 0: 8.882701730%; At π: 6.147235972 %.

Exercises 10.1–Part B, Page 228

1.

2. (a)

(b) $\displaystyle\frac{3}{4} + \sum_{n=1}^{\infty} \left(\frac{((-1)^n - 1)\cos(n\pi x)}{n^2\pi^2} - \frac{\sin(n\pi x)}{n\pi} \right)$

(c) $\displaystyle\lim_{n\to\infty} \frac{(-1)^n - 1}{n^2\pi^2} = 0; \ \lim_{n\to\infty} - \frac{1}{n\pi} = 0$

(d) 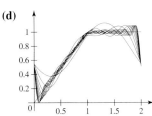 **(e)**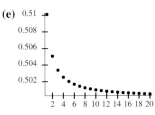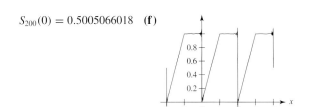

(g) We compute the overshoot of S_{100}. At 0: 8.9371663 %. **(h)** Using the partial sum of the first 500 terms: 1.333130894, compare this to the value of the integral 1.333333333.

18. **(a)** 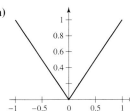 **(b)** $1 + \sum_{n=1}^{\infty} 2\frac{((-1)^n - 1)\cos(n\pi x)}{n^2 \pi^2}$ **(d)**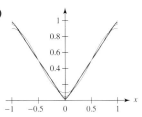

(c) $\lim_{n\to\infty} 2\frac{(-1)^n - 1}{n^2 \pi^2} = 0$; $\lim_{n\to\infty} 0 = 0$

(e) In this case there is no discontinuity. **(f)**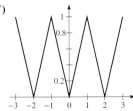

(g) In this case there is no discontinuity

(h) Using the partial sum of the first 500 terms: 0.6666666662, compare this to the value of the integral 0.6666666667.

Exercises 10.2, Page 231

1. Multiply the series together and integrate. Simplify the result using the orthogonality of the trigonometric functions involved.

2. **(a)** $F(x) = \begin{cases} \frac{1}{2}x^2 & 0 \le x < 1 \\ x - \frac{1}{2} & 1 \le x \le 2 \end{cases}$

(d)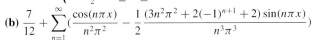

(b) $\frac{7}{12} + \sum_{n=1}^{\infty}\left(\frac{\cos(n\pi x)}{n^2\pi^2} - \frac{1}{2}\frac{(3n^2\pi^2 + 2(-1)^{n+1} + 2)\sin(n\pi x)}{n^3\pi^3}\right)$

(c) $\frac{3}{4}x + \sum_{n=1}^{\infty}\left(\frac{\sin(n\pi x)(-1)^n - \sin(n\pi x) + \cos(n\pi x)n\pi}{n^3\pi^3} - \frac{1}{n^2\pi^2}\right)$

(e) The result of termwise integration of the Fourier series of $f(x)$, will converge pointwise to $F(x)$, at all points of $[0, 2L]$ whether f is continuous at x or not. The Fourier series of $F(x)$, will converge pointwise to $F(x)$ on $(0, 2L)$, but convergence at 0 and $2L$ happens only if $F(0) = F(2L)$. **(f)** *Hints:* Observe that $\sum_{n=1}^{\infty}(-\frac{1}{n^2\pi^2}) = -\frac{1}{6}$, and expand $\frac{3x}{4} - \frac{1}{6}$ in a Fourier series. **(g)** Yes, Theorem 1 applies. The result of termwise integration of the Fourier series of $f(x)$, will converge pointwise to $F(x)$, at all points of $[0, 2L]$.

(h) $f'(x) = \begin{cases} 1 & 0 < x < 1 \\ 0 & 1 < x < 2 \end{cases}$ **(k)** 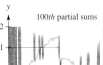 (See plot.) **(l)** Clearly, termwise differentiation is not valid in this case.

(i) $\frac{1}{2} + \sum_{n=1}^{\infty}\left(-\frac{((-1)^n - 1)\sin(n\pi x)}{n\pi}\right)$

(j) $\sum_{n=1}^{\infty}\left(-\frac{((-1)^n - 1)\sin(n\pi x)}{n\pi} - \cos(n\pi x)\right)$

(l) The evidence is negative. **(m)** Theorem 2 does not apply because $f(L)$ does not equal $f(0)$.

18. **(a)** $F(x) = \begin{cases} \frac{1}{2} - \frac{1}{2}x^2 & -1 \le x < 0 \\ \frac{1}{2} + \frac{1}{2}x^2 & 0 \le x \le 1 \end{cases}$ **(b)** $\frac{1}{2} + \sum_{n=1}^{\infty}\left(-\frac{(-1)^n}{n\pi} + 2\frac{1 + (-1)^n}{n^3\pi^3}\right)\sin(n\pi x)$ **(c)** $\frac{1}{2}x + \frac{1}{2} + \sum_{n=1}^{\infty} 2\frac{\sin(n\pi x)((-1)^n - 1)}{n^3\pi^3}$

(d)

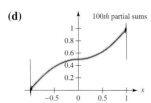

100*th* partial sums

(e) The result of termwise integration of the Fourier series of $f(x)$, will converge pointwise to $F(x)$, at all points of $[-L, L]$ whether f is continuous at x or not. The Fourier series of $F(x)$, will converge pointwise to $F(x)$ on $(-L, L)$, but convergence at $-L$ and L happens only if $F(-L) = F(L)$.
(f) *Hint:* Expand $\frac{x}{2} + \frac{1}{2}$ in a Fourier series.
(g) Yes, Theorem 1 applies. The result of termwise integration of the Fourier series of $f(x)$, will converge pointwise to $F(x)$, at all points of $[-L, L]$.

(h) $f'(x) = \begin{cases} -1 & -1 < x < 0 \\ 1 & 0 < x < 1 \end{cases}$

(k)

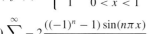

100*th* partial sums

(l) Clearly termwise differentiation is valid in this case. In fact, the series are identical.
(m) Yes, all the requirements of Theorem 2 are satisfied.

(i) $\displaystyle\sum_{n=1}^{\infty} -2\frac{((-1)^n - 1)\sin(n\pi x)}{n\pi}$

(j) $\displaystyle\sum_{n=1}^{\infty} -2\frac{((-1)^n - 1)\sin(n\pi x)}{n\pi}$

Exercises 10.3–Part A, Page 234

1. (a) $f(-x)g(-x) = (-f(x))(-g(x)) = f(x)g(x)$ **(b)** $f(-x)g(-x) = f(x)g(x)$ **(c)** $f(-x)g(-x) = (-f(x))g(x) = -f(x)g(x)$

3. (a) neither   **(b)** even   **(e)** odd

Exercises 10.3–Part B, Page 234

1. $\displaystyle\int_0^1 e^{-t^2}\,dt + \sum_{n=1}^{\infty}(2\int_0^1 e^{-t^2}\cos(n\pi t)\,dt)\cos(n\pi x)$

3. (a) $\displaystyle\sum_{n=1}^{\infty}2\frac{(2(-1)^{1+n} + 2 + n^2\pi^2)\sin(n\pi x)}{n^3\pi^3}$ **(b)** The function is odd.

(c)

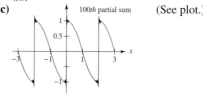

100*th* partial sum   (See plot.)

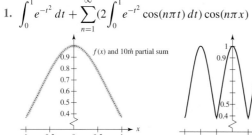

f(*x*) and 10*th* partial sum

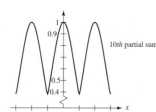

10*th* partial sum

10. The function is even.   **(c)** $b_n = \displaystyle\int_{-1}^{1}|x|\sin(n\pi x)\,dx = 0; a_0 = \int_{-1}^{1}|x|\,dx = 1 = \frac{2}{1}\int_0^1 |x|\,dx;$

$a_n = \displaystyle\int_{-1}^{1}|x|\cos(n\pi x)\,dx = 2\frac{(-1)^n - 1}{n^2\pi^2} = \frac{2}{1}\int_0^1 |x|\cos(n\pi x)\,dx, n \geq 1$

Exercises 10.4–Part A, Page 237

1. (a)

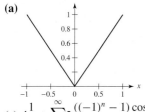

(b)

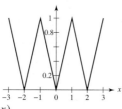

(c)

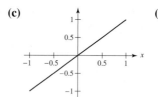

(d)

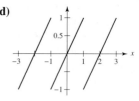

(e) $d\frac{1}{2} + \displaystyle\sum_{n=1}^{\infty}2\frac{((-1)^n - 1)\cos(n\pi x)}{n^2\pi^2}$ **(f)** (See plot.)

(g) $\displaystyle\sum_{n=1}^{\infty}2\frac{(-1)^{1+n}\sin(n\pi x)}{n\pi}$ **(h)** (See plot.)

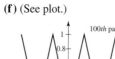
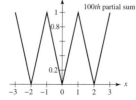

100*th* partial sum

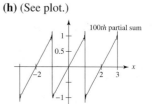

100*th* partial sum

Exercises 10.4–Part B, Page 237

2. (a) $g_{\text{odd}}(x) = \begin{cases} -f(-x) & -2 \le x < 0 \\ f(x) & 0 \le x \le 2 \end{cases}$ **(b)** $\displaystyle\sum_{n=1}^{\infty} -2\frac{(-2\sin(\frac{1}{2}n\pi)+(-1)^n n\pi)\sin(\frac{1}{2}n\pi x)}{n^2\pi^2}$ **(d)**

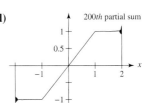

(c) See (b).

(e) $g_{\text{even}}(x) = \begin{cases} f(-x) & -2 \le x < 0 \\ f(x) & 0 \le x \le 2 \end{cases}$ **(f)** $\dfrac{3}{4} + \displaystyle\sum_{n=1}^{\infty} 4\frac{(-1+\cos(\frac{1}{2}n\pi))\cos(\frac{1}{2}n\pi x)}{n^2\pi^2}$ **(h)**

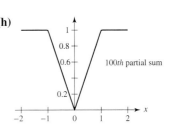

(g) See (f).

(i) Averaging the Fourier sine and cosine series of f will result in zero on the interval $[-2, 0)$, because $-f(-x) + f(-x) = 0$, and $f(x)$ on $[0, 2]$, because $(f(x) + f(x))/2 = f(x)$.

(j) $\dfrac{7}{8} + \displaystyle\sum_{n=1}^{\infty}\left(2\frac{(-1+\cos(\frac{1}{2}n\pi))\cos(\frac{1}{2}n\pi x)}{n^2\pi^2} + \frac{(-n\pi+2\sin(\frac{1}{2}n\pi))\sin(\frac{1}{2}n\pi x)}{n^2\pi^2}\right)$ **(k)**   (See plot.)

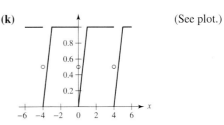

Exercises 10.5–Part A, Page 239

1. (a) $y_0(t) = \dfrac{1}{9}\dfrac{1-\cos 3t}{e^t}$ **(b)** $y_a(t) = \dfrac{1}{3}\dfrac{(a\sin 3t + 3e^{-at} - 3\cos 3t)e^{-t}}{a^2+9}$

(c) $\displaystyle\lim_{a\to 0}\frac{1}{3}\frac{ae^{-t}\sin 3t + 3e^{-(1+a)t} - 3e^{-t}\cos 3t}{a^2+9} = \frac{1}{9}e^{-t} - \frac{1}{9}e^{-t}\cos 3t$ **(d)** The mean square norm of the considered difference is given by:

$\sqrt{\int_0^\infty (e^{(-t)} - e^{(-(1+a)t)})^2\, dt} = \frac{1}{2}\sqrt{2}\sqrt{\frac{a^2}{(a+2)(1+a)}}$; $\lim_{a\to 0}\frac{1}{2}\sqrt{2}\sqrt{\frac{a^2}{(a+2)(1+a)}} = 0$

(e) $\sqrt{\int_0^\infty (\frac{1}{9}\frac{1-\cos 3t}{e^t} - \frac{1}{3}\frac{ae^{-t}\sin 3t + 3e^{-(1+a)t} - 3e^{-t}\cos 3t}{a^2+9})^2\, dt} = \frac{1}{260}\sqrt{130}\sqrt{\frac{(3a^2+15a+38)a^2}{(1+a)(a+2)(a^2+4a+13)}}$;

$\displaystyle\lim_{a\to 0}\frac{1}{260}\sqrt{130}\sqrt{\frac{(3a^2+15a+38)a^2}{(1+a)(a+2)(a^2+4a+13)}} = 0$

2. (a)

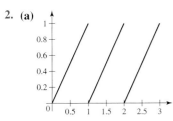

(b) $\dfrac{\int_0^1 te^{-st}\, dt}{1 - e^{(-s)}} = \dfrac{e^{-s}s + e^{-s} - 1}{(-1+e^{(-s)})s^2}$

Exercises 10.5–Part B, Page 240

1. $y_k(t) = \frac{1}{3}k(3k^2te^{-kt} + e^{-t}k^2\sin(3t) - 6e^{-t}k\cos(3t) + 6ke^{-kt} - 6kte^{-kt} - 2e^{-t}k\sin(3t) + 6e^{-t}\cos(3t) - 6e^{-kt} + 30te^{-kt} - 8e^{-t}\sin(3t))/$
$(10 - 2k + k^2)^2$; $\lim_{k\to\infty} y_k(t) = 0$

2. (a) $Y(s) = \dfrac{se^{-s} + e^{-s} - 1}{s^2(s^2 + 2s + 5)(e^{-s} - 1)}$ **(b)** $Y(s) \approx -\dfrac{1}{s^2(-s^2 - 2s - 5)} + \dfrac{\frac{1}{s}e^{-s}}{-s^2 - 2s - 5} + \dfrac{(e^{-s})^2}{s(-s^2 - 2s - 5)}$ **(c)** $y(t) \approx \frac{1}{5}t - \frac{2}{25} +$

$\frac{2}{25}e^{-t}\cos 2t - \frac{3}{50}e^{-t}\sin 2t + (-\frac{1}{5} + \frac{1}{5}e^{-t+1}\cos(2t-2) + \frac{1}{10}e^{-t+1}\sin(2t-2))H(t-1) + (-\frac{1}{5} + \frac{1}{5}e^{-t+2}\cos(2t-4) +$
$\frac{1}{10}e^{-t+2}\sin(2t-4))H(t-2)$

(d) We use the Fourier series: $\frac{1}{2} - \sum_{n=1}^{k} \frac{\sin(2n\pi t)}{n\pi}$. **(f)**

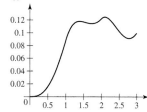

(e) The formulae are long and are best studied graphically. See (f).

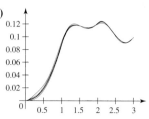

18. (a) $y(t) = (\frac{49}{25}\frac{e^2}{\sin 1} - \frac{29}{25}\frac{\cos 1}{\sin 1})e^{-2t}\sin t + \frac{29}{25}e^{-2t}\cos t + \frac{1}{5}t - \frac{4}{25}$ **(b)** See (a). **(c)** (See plot.)

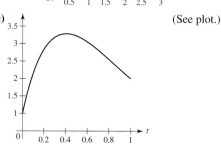

(d) $S_k(t) = \sum_{n=1}^{k} 2\dfrac{(-1)^{1+n}\sin(n\pi t)}{n\pi}$ **(e)**

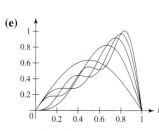

(f) The formulae are long and are best studied graphically. See (g). **(g)**

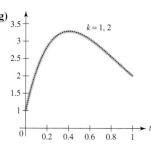

$k = 1, 2$

Exercises 10.6–Part A, Page 243

2. $s_1 = 2, s_2 = -\frac{4}{\pi}$ **5.** *Hint:* Use the identity: $\cos(mx)\cos(nx) = \frac{1}{2}\cos((m+n)x) + \frac{1}{2}\cos((m-n)x)$. **7.** $\int_0^\pi \sin x \sin 3x\, dx = 0$;

$\frac{df}{dx}(\frac{\pi}{4}) = \frac{1}{2}\sqrt{2}; \frac{dg}{dx}(\frac{\pi}{4}) = -\frac{3}{2}\sqrt{2}; \frac{df}{dx}(\frac{3\pi}{4}) = -\frac{1}{2}\sqrt{2}; \frac{dg}{dx}(\frac{3\pi}{4}) = \frac{3}{2}\sqrt{2}$

Exercises 10.6–Part B, Page 243

3. (a) $\{b_k t\} = \left[2, -1, \frac{2}{3}, -\frac{1}{2}, \frac{2}{5}\right]$

(b) $\left[-2\pi + \pi b_1 = 0,\ \pi + \pi b_2 = 0,\ -\frac{2}{3}\pi + \pi b_3 = 0,\ \frac{1}{2}\pi + \pi b_4 = 0,\ -\frac{2}{5}\pi + \pi b_5 = 0\right]$;

$\{b_k\} = [2, -1, \frac{2}{3}, -\frac{1}{2}, \frac{2}{5}]$

(c) $\left[\frac{1}{2}\pi b_1 = \pi, \frac{1}{2}\pi b_2 = -\frac{1}{2}\pi, \frac{1}{2}\pi b_3 = \frac{1}{3}\pi, \frac{1}{2}\pi b_4 = -\frac{1}{4}\pi, \frac{1}{2}\pi b_5 = \frac{1}{5}\pi\right]$

(d) $p_5 := 2\sin x - \sin 2x + \frac{2}{3}\sin 3x - \frac{1}{2}\sin 4x + \frac{2}{5}\sin 5x$

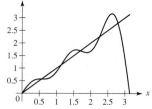

(e) $\int_0^\pi (f(x) - p_5(x))^2\, dx = -\frac{5269}{1800}\pi + \frac{1}{3}\pi^3$ **(f)** $\int_0^\pi p_5^2(x)\, dx = \frac{\pi}{2}\sum_{k=1}^5 b_k^2 = \frac{5269}{1800}\pi; \int_0^\pi f(x)^2\, dx = \frac{1}{3}\pi^3$

(g) $\{x_k\} = [0.1, 0.2, 1.0, 1.7, 2.7]$; $P_5 := 2.116589166\sin x - 1.124872737\sin 2x + 0.6446286627\sin 3x - 0.2645677405\sin 4x +$
$0.051516073\sin 5x$ **(h)** (See plot.) **(i)** $\int_0^\pi (f(x) - P_5(x))^2\, dx = 1.463720052$

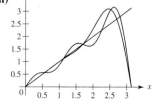

13. (a) $\{a_k\} = \left[\pi, -\frac{4}{\pi}, 0, -\frac{4}{9\pi}, 0, -\frac{4}{25\pi}\right]$ **(b)** $\frac{1}{2}a_0\pi - \frac{1}{2}\pi^2 = 0, a_1\pi + 4 = 0, a_2\pi = 0, a_3\pi + \frac{4}{9} = 0, a_4\pi = 0, \frac{4}{25} + a_5\pi = 0;$
$\{a_k\} = [\pi, -\frac{4}{\pi}, 0, -\frac{4}{9\pi}, 0, -\frac{4}{25\pi}]$ **(c)** $\frac{1}{2}\pi^2 = \frac{1}{2}A_0\pi, 0 = \frac{1}{2}A_2\pi, -\frac{2}{25} = \frac{1}{2}A_5\pi, -2 = \frac{1}{2}A_1\pi, -\frac{2}{9} = \frac{1}{2}A_3\pi, 0 = \frac{1}{2}A_4\pi$
(d) $p_5 = \frac{1}{2}\pi - 4\frac{\cos x}{\pi} - \frac{4}{9}\frac{\cos 3x}{\pi} - \frac{4}{25}\frac{\cos 5x}{\pi}$ **(e)** $\int_0^\pi (f(x) - p_5(x))^2\, dx = \frac{1}{12}\pi^3 - \frac{410,648}{50,625}\frac{1}{\pi}$

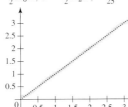

(f) $\int_0^\pi p_5^2\, dx = \frac{1}{202,500}\frac{50,625\pi^4 + 1,642,592}{\pi}; \frac{\pi}{4}a_0^2 + \frac{\pi}{2}\sum_{k=1}^5 a_k^2 = \frac{1}{202,500}\frac{50,625\pi^4 + 1,642,592}{\pi}; \int_0^\pi f(x)^2\, dx = \frac{1}{3}\pi^3$
(g) $\{x_k\} = [0.1, 0.2, 1.0, 1.7, 2.7, 3.0]; P_5 := 1.597746580 - 1.248896090\cos x + 0.04631135028\cos 2x - 0.08386740858\cos 3x -$
$0.08882569584\cos 4x - 0.1578683330\cos 5x$
(h)   **(i)** $\int_0^\pi (f(x) - P_5(x))^2\, dx = 0.04401576; \frac{1}{12}\pi^3 - \frac{410,648}{50,625}\frac{1}{\pi} = 0.001864922$

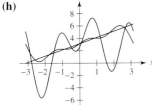

23. (a) $\{a_k\} = [2\pi, 0, 0, 0]; \{b_k\} = \left[2, -1, \frac{2}{3}\right]$ (See plot.)   **(e)** $\int_{-\pi}^\pi (f(x) - p_3)^2\, dx = \frac{2}{3}\pi^3 - \frac{49}{9}\pi$

(b) $-2\pi^2 + \pi a_0 = 0, 2a_1\pi = 0, 2a_2\pi = 0,$
    $2a_3\pi = 0, -4\pi + 2b_1\pi = 0,$
    $2\pi + 2b_2\pi = 0, -\frac{4}{3}\pi + 2b_3\pi = 0$

(c) See (b).

(d) $p_3 := \pi + 2\sin x - \sin 2x + \frac{2}{3}\sin 3x;$
(f) $\int_{-\pi}^\pi p_3^2(x)\, dx = 2\pi^3 + \frac{49}{9}\pi; \frac{\pi}{2}a_0^2 + \pi\sum_{k=1}^3 (a_k^2 + b_k^2) = 2\pi^3 + \frac{49}{9}\pi; \int_{-\pi}^\pi f(x)^2\, dx = \frac{8}{3}\pi^3$
(g) $\{x_k\} = [-2.8, -1.5, -0.5, 0.1, 1.2, 2.7, 3.01]; P_3 := 2.137654488 + 1.904594013\cos x + 1.709229628\sin x +$
$1.737964417\cos 2x - 1.122162776\sin 2x - 3.178577969\cos 3x + 2.011540815\sin 3x$
(h)   (See plot.)   **(i)** $\int_{-\pi}^\pi (f(x) - P_3(x))^2\, dx = 68.51998842; \int_{-\pi}^\pi (f(x) - p_3(x))^2\, dx = 3.56662446$

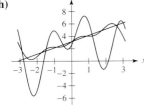

38. $\{v_k(x)\} = \{1, 1 - 2x, 1 - 6x + 6x^2, 1 - 12x + 30x^2 - 20x^3, 1 - 20x + 90x^2 - 140x^3 + 70x^4, 1 - 30x + 210x^2 - 560x^3 + 630x^4 - 252x^5\}$
39. $\{-0.00575321 + 0.2222222221 B_4 = 0, -0.055605469 + 0.2857142854 B_3 = 0, 0.1374170530 + 0.3999999996 B_2 =$
$0, 0.1205952214 + 0.6666666661 B_1 = 0, -0.6366197719 + 1.999999999 B_0 = 0, 0.00255930 + 0.1818181817 B_5 = 0\};$
$t_5(x) = 0.00030685770 - 0.007892638x + 3.151377003x^2 + 0.365738871x^3 - 7.055713355x^4 + 3.547189803x^5$

Exercises 10.7–Part A, Page 246

1. $\left[-\frac{1}{2}e^{-1} + \frac{1}{2}e, -3e^{-1}, -\frac{35}{2}e^{-1} + \frac{5}{2}e\right]$ **3.** For instance $\int_0^1 x\, dx = \frac{1}{2}$, so the dot product of x and 1 is non-zero;
$a = 75\ln 2 - 51, b = -408\ln 2 + 282, c = 390\ln 2 - 270$. The normal equations are now coupled.

Exercises 10.7–Part B, Page 246

1. $[0, -0.6079271018540266286, 0, -0.6530254515071300213, 0, 0.2930671186090945234, 0, -0.03390677275025224653, 0,$
$0.001850152855637060313, 0]; \sum_{k=0}^{10}(2\frac{s_k^2}{2k+1}) = 0.383993924849905317 95; \int_{-1}^1 x^2\cos(\pi x)^2\, dx = 0.3839939251545022190 6$

3. (a) $\{s_k\} = \left[0, \frac{3}{\pi}, 0, 7\frac{\pi^2 - 15}{\pi^3}, 0, 11\frac{\pi^4 - 105\pi^2 + 945}{\pi^5}\right]$

(b) See (a).

(c) See (a).

(d) $f_5 := 3\frac{x}{\pi} + 7\frac{(\pi^2 - 15)(\frac{5}{2}x^3 - \frac{3}{2}x)}{\pi^3} + 11\frac{(\pi^4 - 105\pi^2 + 945)(\frac{63}{8}x^5 - \frac{35}{4}x^3 + \frac{15}{8}x)}{\pi^5}$

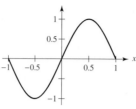

(e) 0.00003694684603

(f) $X := [-0.9, -0.3, -0.1, 0.13, 0.54, 0.8]$; $s_2 = 0.0129556010$, $s_3 = -1.127103955$, $s_0 = 0.00523873425$, $s_1 = 0.9637195018$, $s_5 = 0.2591528103$, $s_4 = 0.0029339603$; $F_5 := -0.000138831137 + 3.140286954x + 0.00843105037x^2 - 5.085346978x^3 + 0.01283607631x^4 + 2.040828381x^5$ **(g)**   (See plot.)   **(h)** 0.0007783701479

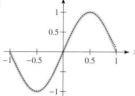

16. $[1, x, 2x^2 - 1, 4x^3 - 3x, 8x^4 - 8x^2 + 1, 16x^5 - 20x^3 + 5x]$ **18. (a)** $f_3(x) = -1.336037419x^3 + 0.2123519542x$
(b) 0.02973207946   **c)** $\{x_k\} = [-0.8, -0.1, 0.2, 0.7]$; $s_0 = -0.0736731188$, $s_1 = -1.250827998$, $s_2 = -0.0730184302$,
$s_3 = -0.7367344394$; $F_3(x) = -0.0006546886 + 0.959375320x - 0.1460368604x^2 - 2.946937758x^3$ **(d)** 0.6439695296
(e)   (See plot.)

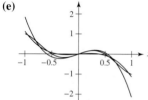

30. (a) $f_5(x) = -\frac{1}{192}x^4 + \frac{1}{8}x^3 - \frac{7}{8}x^2 + \frac{7}{4}x - \frac{1}{8}$ **(e)**

(b) 0.009375000000

(c) $F_5(x) = x - \frac{1}{6}x^3 + \frac{1}{120}x^5$

(d) 165.6500000

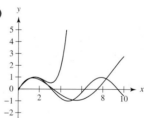

(See plot.)
Polynomials are not bounded on an infinite interval.

CHAPTER 11

Exercises 11.1–Part A, Page 251

1. Substitution into the differential equation and differentiating termwise yields $x^2y'' + (3x + 1)y' + y = \int_0^\infty -\frac{e^{-t}(t + xt^2 - 1 + xt)}{(1 + xt)^3} dt = 0$.

3. With $u = \frac{1}{(1+xt)^{k+1}}$ integration by parts yields $k! \int_0^\infty \frac{(-x)^k e^{-t}}{(1 + xt)^{(k+1)}} dt = k!(-x)^k + \int_0^\infty \frac{(k + 1)!(-x)^{k+1}e^{-t}}{(1 + xt)^{k+2}} dt$.

Exercises 11.1–Part B, Page 251

1. (a) $\left(\frac{2}{\sqrt{\pi}} \int_0^\infty e^{-t^2} dt\right)^2 = \left(\frac{2}{\sqrt{\pi}} \int_0^\infty e^{-s^2} ds\right)\left(\frac{2}{\sqrt{\pi}} \int_0^\infty e^{-t^2} dt\right) = \frac{4}{\pi} \int_0^\infty \int_0^\infty e^{-s^2-t^2} ds \, dt = \left(\frac{4}{\pi} \int_0^{\pi/2} 1 \, d\theta\right)\left(\int_0^\infty re^{-r^2} dr\right) = (2)(\frac{1}{2}) = 1$

(b) $\frac{2}{\sqrt{\pi}} \sum_{k=0}^\infty \frac{(-1)^k \int_0^x t^{2k} dt}{k!} = \frac{2}{\sqrt{\pi}} \sum_{k=0}^\infty \frac{(-1)^k x^{2k+1}}{k!(2k + 1)}$ **(c)**

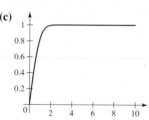

2. **(a)** $L\left[\dfrac{1}{\sqrt{1+t}}\right] = \dfrac{e^s\sqrt{\pi}\,\mathrm{erf}(\sqrt{s})}{\sqrt{s}} = -\dfrac{e^s\sqrt{\pi}(-1+\mathrm{erf}(\sqrt{s}))}{\sqrt{s}}$ **(b)** $\dfrac{1}{\sqrt{1+t}} = \displaystyle\sum_{k=0}^{\infty}\dfrac{(-1)^k(2k)!}{4^k(k!)^2}t^k$;

$L\left[\dfrac{1}{\sqrt{1+t}}\right] = \displaystyle\sum_{k=0}^{\infty}\dfrac{(-1)^k(2k)!}{4^k(k!)^2}L[t^k] = \sum_{k=0}^{\infty}\dfrac{(-1)^k(2k)!}{k!4^k s^{k+1}}$ **(c)** $\lim_{k\to\infty}\left|\dfrac{a_{k+1}}{a_k}\right| = \lim_{k\to\infty}\dfrac{1}{2}\dfrac{1+2k}{|s|}$ **(d)** In the integration by parts process, make sure you always integrate e^{-st}.   **(e)** $s = 1, n = 2, error = 0.2421278437$ **(f)** $s = 2, n = 3, error = 0.0463692298$; $s = 3, n = 4, error = 0.0115115812$; $s = 4, n = 5, error = 0.0031939938$; $s = 5, n = 6, error = 0.0009429715$

Exercises 11.2–Part A, Page 256

1. **(a)** $|x^2| \le 2|x|, |x| < 2$ **(b)** $|\sin x| \le 1, |x| < \infty$ **(c)** $|\sin x| \le |x|, |x| < \infty$ **(d)** $|x^2| \le |x|, |x| \le 1$ **(e)** $|e^{-x}| \le 1, x \ge 1$ **(f)** $|x| \le |x^2|, x \ge 1$ **(g)** $\lim_{x\to 0}\left|\dfrac{\cos x - 1}{x}\right| = 0$

2. **(a)** True! $\lim_{x\to\infty}e^{-x} = 0$ **(b)** True! $\lim_{x\to\infty}\dfrac{e^{-x}}{x} = 0$ **(c)** True! Observe that $\lim_{x\to 0}\dfrac{|\sinh x|}{|x|} = 1 \Rightarrow$ there exists an a such that $|\sinh x| \le 2|x|$ for all x with $|x| \le a$ **(d)** False! $\lim_{x\to 0}\dfrac{|\sinh x|}{|x|} = 1$ **(e)** True! $\left|\frac{1}{2}e^x - \frac{1}{2}e^{(-x)}\right| \le e^x$ for all x **(f)** False! $\lim_{x\to\infty}\left|\dfrac{\sinh x}{x}\right| = \infty$ **(g)** True! $\lim_{x\to\infty}\left|\dfrac{\ln x}{x}\right| = 0$ **(h)** False! $\lim_{x\to\infty}|\ln x| = \infty$

9. $\lim_{x\to a}\left|\dfrac{F(x)}{f(x)}\right| = 0$ and $\lim_{x\to a}\left|\dfrac{G(x)}{g(x)}\right| = 0 \Rightarrow \lim_{x\to a}\left|\dfrac{F(x)G(x)}{f(x)g(x)}\right| = \lim_{x\to a}\left(\left|\dfrac{F(x)}{f(x)}\right|\left|\dfrac{G(x)}{g(x)}\right|\right) = \left(\lim_{x\to a}\left|\dfrac{F(x)}{f(x)}\right|\right)\left(\lim_{x\to a}\left|\dfrac{G(x)}{g(x)}\right|\right) = 0$

Exercises 11.2–Part B, Page 256

2. **(a)** $\sin x = x - \frac{1}{6}x^3\cos\xi$ This implies that $|\sin x - x| \le \frac{1}{6}|x|^3$. (See plot.)

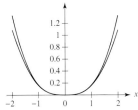

3. **(a)** $\lim_{x\to 0}\left|\dfrac{x^{1/2(k+1)}}{x^{1/2k}}\right| = \lim_{x\to 0}|x^{1/2}| = 0$ **(b)** $a_n = \dfrac{(-1)^{n+1}}{n}$ **(c)** According to the ratio test, the series converges for $0 \le x < 1$ and diverges for $x > 1$. Moreover, the series converges at $x = 1$.

4. $\lim_{x\to 0}\left|\dfrac{\ln(1+x^{k+1})}{\ln(1+x^k)}\right| = \dfrac{k+1}{k}\lim_{x\to 0}\left|\dfrac{x+x^{k+1}}{1+x^{k+1}}\right| = 0$ for $k \ge 1$. Moreover $\lim_{x\to 0}\ln(1+x) = 0$.   **5.** $\lim_{x\to 0}\left|\dfrac{\tan(x^{(1/2k+1/2)})}{\tan(x^{1/2k})}\right| = 0$

8. **(a)** $\lim_{x\to 0}\dfrac{\sin(x+\sqrt{x})}{\ln(1+x)}$ does not exist.   **(b)** $\sin(x+\sqrt{x}) = \tan(\sqrt{x}) + \tan(x) - \frac{1}{2}\tan(x^{3/2}) - \frac{1}{2}\tan(x^2) - \frac{5}{8}\tan(x^{5/2}) + O(\tan(x)^3)$

Exercises 11.3, Page 261

1. **(a)** $\dfrac{d}{ds}\displaystyle\sum_{m=0}^{\infty}\dfrac{(-1)^m(2m)!}{4^m m!s^{m+1}} = \sum_{m=0}^{\infty}\dfrac{(-1)^{m+1}(m+1)(2m)!}{4^m m!s^{m+2}} = -\dfrac{1}{s^2} + \dfrac{1}{s^3} - \dfrac{9}{4}\dfrac{1}{s^4} + \dfrac{15}{2}\dfrac{1}{s^5} - \dfrac{525}{16}\dfrac{1}{s^6} + O\left(\dfrac{1}{s^7}\right)$

(b) $\displaystyle\int_0^{\infty} -\dfrac{te^{-st}}{\sqrt{1+t}}\,dt = \dfrac{e^s\sqrt{\pi}\,\mathrm{erf}(\sqrt{s})}{\sqrt{s}} - \dfrac{1}{2}\dfrac{e^s\sqrt{\pi}\,\mathrm{erf}(\sqrt{s})}{s^{3/2}} - \dfrac{e^s e^{-s}}{s}$ **(c)** $-\dfrac{1}{s^2} + \dfrac{1}{s^3} - \dfrac{9}{4}\dfrac{1}{s^4} + \dfrac{15}{2}\dfrac{1}{s^5} - \dfrac{525}{16}\dfrac{1}{s^6} + O\left(\dfrac{1}{s^7}\right)$ **(d)** See (a) and (c).

5. **(a)** $F(x) = \frac{1}{x} - \frac{1}{x^2} + \frac{2}{x^4} - \frac{4}{x^5} + \frac{4}{x^6} - \frac{8}{x^8} + O(\frac{1}{x^9})$; $G(x) = \frac{1}{x} - \frac{1}{x^2} + \frac{1}{x^3} - \frac{1}{x^4} + \frac{1}{x^5} - \frac{1}{x^6} + \frac{1}{x^7} - \frac{1}{x^8} + O(\frac{1}{x^9})$
(b) $F(x)G(x) = f(x)g(x) = \frac{1}{x^2} - \frac{2}{x^3} + \frac{2}{x^4} - \frac{4}{x^6} + \frac{8}{x^7} - \frac{8}{x^8} + O(\frac{1}{x^9})$ **(c)** $[0, 0, 1, -2, 2, 0, -4, 8, -8]$
(d) $f(g(x)) = \frac{1}{2} - \frac{1}{4}\frac{1}{x^2} + \frac{3}{4}\frac{1}{x^3} - \frac{13}{8}\frac{1}{x^4} + \frac{3}{x^5} - \frac{79}{16}\frac{1}{x^6} + \frac{117}{16}\frac{1}{x^7} - \frac{307}{32}\frac{1}{x^8} + O(\frac{1}{x^9})$ The composition of the asymptotic series does not have a series development.   **(e)** $g(f(x)) = 1 - \frac{1}{x} + \frac{2}{x^2} - \frac{3}{x^3} + \frac{3}{x^4} - \frac{9}{x^6} + \frac{27}{x^7} - \frac{54}{x^8} + O(\frac{1}{x^9})$ The composition of the asymptotic series does not have a series development.   **(f)** $\frac{1}{f(x)} = x + 1 + \frac{1}{x} - \frac{1}{x^2} + \frac{1}{x^3} + O(\frac{1}{x^4})$ Observe that even though the expansions match, they are not an asymptotic series with respect to the original asymptotic basis.   **(g)** $\frac{1}{g(x)} = x + 1$; This is not an asymptotic series with respect to the original asymptotic basis.   **(h)** $\frac{f(x)}{g(x)} = 1 - \frac{1}{x^2} + \frac{2}{x^3} - \frac{2}{x^4} + \frac{4}{x^6} - \frac{8}{x^7} + \frac{8}{x^8} + O(\frac{1}{x^9})$ The series agree, and they are asymptotic expansions with respect to the original asymptotic basis.   **(i)** This particular function $f(x)$ is not sufficiently well behaved and the integral diverges.   **(j)** The integral of the first term diverges.   **(k)** This particular function $g(x)$ is not sufficiently well behaved and the integral diverges.

10. **(a)** $F(x) = \frac{1}{2} - \frac{1}{4}x^2 + \frac{1}{4}x^3 - \frac{1}{8}x^4 + \frac{1}{16}x^6 - \frac{1}{16}x^7 + \frac{1}{32}x^8 + O(x^9)$; $G(x) = 1 - x + x^2 - x^3 + x^4 - x^5 + x^6 - x^7 + x^8 + O(x^9)$
(b) $F(x)G(x) = f(x)g(x) = \frac{1}{2} - \frac{1}{2}x + \frac{1}{4}x^2 - \frac{1}{8}x^4 + \frac{1}{8}x^5 - \frac{1}{16}x^6 + \frac{1}{32}x^8 + O(x^9)$ **(c)** $[\frac{1}{2}, -\frac{1}{2}, \frac{1}{4}, 0, -\frac{1}{8}, \frac{1}{8}, -\frac{1}{16}, 0, \frac{1}{32}]$
(d) $f(g(x)) = \frac{2}{5} + \frac{3}{25}x - \frac{13}{125}x^2 + \frac{48}{3125}x^3 - \frac{158}{15,625}x^4 + \frac{468}{78,125}x^5 - \frac{1228}{390,625}x^6 + \frac{2688}{1,953,125}x^7 - \frac{3848}{...}x^8 + O(x^9)$; $F(G(x)) = \frac{13}{32} + \frac{1}{16}x + \frac{3}{16}x^2 - x^3 + \frac{51}{16}x^4 - \frac{67}{8}x^5 + O(x^6)$ Since $g(0) \ne 0$ the series do not match.   **(e)** $g(f(x)) = \frac{2}{3} + \frac{1}{9}x^2 - \frac{1}{9}x^3 + \frac{2}{27}x^4 - \frac{1}{27}x^5 + \frac{1}{81}x^6 - \frac{1}{243}x^8 + O(x^9)$; $G(F(x)) = \frac{171}{256} + \frac{27}{256}x^2 - \frac{27}{256}x^3 + \frac{21}{256}x^4 - \frac{15}{256}x^5 + O(x^6)$; Since $f(0) \ne 0$ the series do not match.

(f) $\frac{1}{f(x)} = \frac{1}{F(x)} = 2 + x^2 - x^3 + x^4 - x^5 + O(x^6)$ The expansions match, and are asymptotic series with respect to the original asymptotic basis.   **(g)** $\frac{1}{g(x)} = \frac{1}{G(x)} = 1 + x$ Again these are asymptotic series with respect to the original asymptotic basis.

(h) $\frac{f(x)}{g(x)} = \frac{F(x)}{G(x)} = \frac{1}{2} + \frac{1}{2}x - \frac{1}{4}x^2 + \frac{1}{8}x^4 - \frac{1}{8}x^5 + \frac{1}{16}x^6 - \frac{1}{32}x^8 + O(x^9)$ These series agree, and they are asymptotic expansions with respect to the original asymptotic basis.   **(i)** $f_1(x) = \frac{1}{2}x - \frac{1}{12}x^3 + \frac{1}{16}x^4 - \frac{1}{40}x^5 + \frac{1}{112}x^7 - \frac{1}{128}x^8 + O(x^9)$ **(j)** $\int_0^x F(t)\,dt = \frac{1}{2}x - \frac{1}{12}x^3 + \frac{1}{16}x^4 - \frac{1}{40}x^5 + \frac{1}{112}x^7 - \frac{1}{128}x^8 + O(x^9)$ The results agree.   **(k)** $\int_0^x g(t)\,dt = \int_0^x G(t)\,dt = x - \frac{1}{2}x^2 + \frac{1}{3}x^3 - \frac{1}{4}x^4 + \frac{1}{5}x^5 - \frac{1}{6}x^6 + \frac{1}{7}x^7 - \frac{1}{8}x^8 + O(x^9)$

CHAPTER 12

Exercises 12.1–Part A, Page 268

1. $-\frac{1}{10}x_\infty + \frac{1}{20}y_\infty = 0$, $x_\infty + y_\infty = 75$ These two equations determine x_∞ and y_∞.

4. (a) $x(t) = 3e^{7/2t}\cos(\frac{1}{2}\sqrt{119}t) - \frac{15}{119}e^{7/2t}\sqrt{119}\sin(\frac{1}{2}\sqrt{119}t)$, $y(t) = -\frac{31}{119}e^{7/2t}\sqrt{119}\sin(\frac{1}{2}\sqrt{119}t) - e^{7/2t}\cos(\frac{1}{2}\sqrt{119}t)$

Exercises 12.1–Part B, Page 269

1. $x(t) = (-8e^{-9t} + 9e^{-5t})a + (18e^{-9t} - 18e^{-5t})b$, $y(t) = (-4e^{-9t} + 4e^{-5t})a + (9e^{-9t} - 8e^{-5t})b$; $\lim_{t\to\infty}x(t) = 0$; $\lim_{t\to\infty}y(t) = 0$ The parametric curve approaches the origin, as $t \to \infty$.

2. (a) $\frac{dx}{dt} = -\frac{3}{200}x + \frac{1}{50}y$, $\frac{dy}{dt} = \frac{3}{200}x - \frac{1}{50}y$; $x(0) = 0$, $y(0) = 2250$ **(b)** Total amount of salt 2250 lb divided over 1750 gal, results in a limit concentration of $\frac{9}{7}$ lb/gal. So $x_\infty = \frac{9000}{7}$, $y_\infty = \frac{6750}{7}$ **(c)** $x_\infty = \frac{9000}{7}$, $y_\infty = \frac{6750}{7}$ **(d)** $x(t) = \frac{9000}{7} - \frac{9000}{7}e^{-7/200t}$, $y(t) = \frac{9000}{7}e^{-7/200t} + \frac{6750}{7}$

(h) $y = 2250 - x$

(i) $\frac{dy}{dx} = \frac{-\frac{3}{200}x - \frac{1}{50}y}{-\frac{3}{200}x + \frac{1}{50}y}$; $y(x) = 2250 - x$

Hint: Use the initial condition $y(0) = 2250$.

(e) 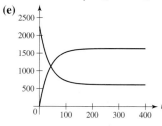   **(g)**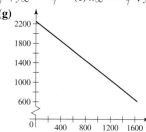

7. (a) $\frac{dx}{dt} = -\frac{1}{17}x + \frac{1}{44}y$, $\frac{dy}{dt} = \frac{1}{17}x - \frac{1}{44}y$; $x(0) = \frac{255}{4}$, $y(0) = 44$ **(b)** Concentration: $\frac{431}{1220}$ lb/gal. So $x_\infty = \frac{7327}{244}$, $y_\infty = \frac{4741}{61}$ **(c)** $x_\infty = \frac{7327}{244}$, $y_\infty = \frac{4741}{61}$ **(d)** $x(t) = \frac{7327}{244} + \frac{2057}{61}e^{-61/748t}$, $y(t) = \frac{4741}{61} - \frac{2057}{61}e^{-61/748t}$

(h) $y = -x + \frac{431}{4}$

(i) $\frac{dy}{dx} = \frac{\frac{1}{17}x - \frac{1}{44}y}{-\frac{1}{17}x + \frac{1}{44}y}$

(j) $y(x) = -x + \frac{431}{4}$

Hint: Use the initial condition $y(\frac{255}{4}) = 44$.

(e) 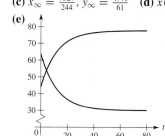   **(g)**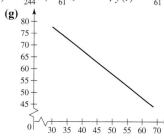

19. $r = -\frac{400}{39}\ln\frac{85}{144} = 5.4068$ **30. (a)** $x(t) = -0.7650300504e^{(2.817730064t)}\cos(12.64233945t) - 3.051954610e^{2.817730064t}\sin(12.64233945t) - 0.2349699496e^{(5.3645398728166t)}$, $y(t) = -3.975718508e^{2.817730064t}\cos(12.64233945t) - 5.920935674e^{2.817730064t}\sin(12.64233945t) + 5.975718508e^{(5.3645398728166t)}$, $z(t) = 3.254282568e^{2.817730064t}\cos(12.64233945t) + 3.553416603e^{2.817730064t}\sin(12.64233945t) - 2.254282568e^{(5.3645398728166t)}$

(b) 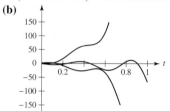   **(c)** 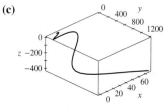   **(d)** It's equivalent, but the solution is huge.

Exercises 12.2–Part A, Page 272

1. Let x' and y' equal to zero and solve the resulting equations: $x_\infty = a\alpha$, $y_\infty = a\beta$.   **4. (a)** $\frac{dx}{dt} = \frac{3}{5} - \frac{3}{400}x + \frac{1}{125}y$, $\frac{dy}{dt} = \frac{3}{400}x - \frac{3}{125}y$; $x(0) = 140$, $y(0) = 50$ **(b)** In the steady state situation the entire system contains a 30% brine; $x_\infty = 120$, $y_\infty = \frac{75}{2}$ **(c)** Let x' and y' equal to zero and solve the resulting equations.

Exercises 12.2–Part B, Page 272

1. **(a)** $\frac{dx}{dt} = \frac{1}{2} - \frac{2}{225}x + \frac{3}{200}y$, $\frac{dy}{dt} = \frac{2}{225}x - \frac{1}{50}y$; $x(0) = \frac{153}{2}$, $y(0) = 28$ **(b)** $x_\infty = 225$, $y_\infty = 100$ **(c)** same as (b)   **(d)** $x(t) = 225 - \frac{297}{2}e^{-13/900t}\cosh(\frac{1}{900}\sqrt{133}t) - \frac{3429}{266}\sqrt{133}e^{-13/900t}\sin(\frac{1}{900}\sqrt{133}t)$, $y(t) = 100 - 72e^{-13/900t}\cosh(\frac{1}{900}\sqrt{133}t) - \frac{828}{133}\sqrt{133}e^{-13/900t}\sinh(\frac{1}{900}\sqrt{133}t)$

(e) 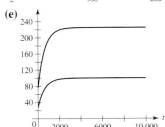   **(f)** $\lim_{t\to\infty}x(t) = 225$; $\lim_{t\to\infty}y(t) = 100$ **(g)**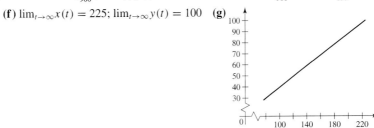

(h) $\frac{dy}{dx} = \frac{\frac{2}{225}x - \frac{1}{50}y}{\frac{1}{2} - \frac{2}{225}x + \frac{3}{200}y}$; $y(\frac{153}{2}) = 28$ **(i)** same as (d).

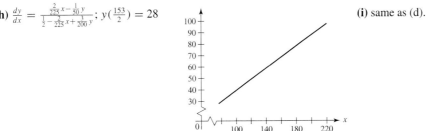

2. **(a)** $x(t) = 474e^{5t} - 117e^t - 364e^{-7t}$, $y(t) = -474e^{5t} - 78e^t + 560e^{-7t}$, $z(t) = 711e^{5t} - 39e^t - 672e^{-7t}$
(b) (See plot.)   **(c)**   **(d)** See (a).

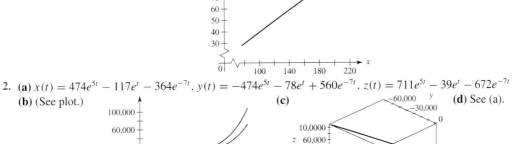

12. **(a)** $\frac{dx}{dt} = 4 - \frac{6}{625}x + \frac{1}{725}y$, $\frac{dy}{dt} = \frac{6}{625}x - \frac{6}{725}y$, $x(0) = 200$, $y(0) = \frac{203}{2}$ **(b)** and **(c)** $x_\infty = 500$, $y_\infty = 580$ **(d)**
$x(t) = 500 - 300e^{-162/18125t}\cosh(\frac{1}{18125}\sqrt{4494}t) - \frac{5575}{2996}e^{-162/18,125t}\sqrt{4494}\sinh(\frac{1}{18125}\sqrt{4494}t)$, $y(t) = 580 - \frac{957}{2}e^{-162/18125t}\cosh(\frac{1}{18,125}\sqrt{4494}t) - \frac{9657}{749}e^{-162/18,125t}\sqrt{4494}\sinh(\frac{1}{18,125}\sqrt{4494}t)$

(e) 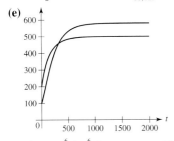   **(f)** $\lim_{t\to\infty}y(t) = 580$; $\lim_{t\to\infty}x(t) = 500$ **(g)**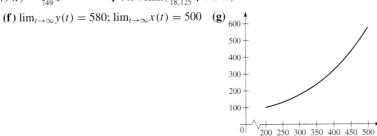

(h) $\frac{dy}{dx} = \frac{\frac{6}{625}x - \frac{6}{725}y}{4 - \frac{6}{625}x + \frac{1}{725}y}$; $y(200) = \frac{203}{2}$ **(i)** The answer is huge, but the solutions agree.

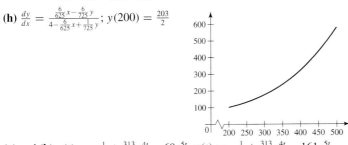

23. **(a)** and **(b)** $x(t) = -\frac{1}{4} + \frac{313}{4}e^{4t} - 69e^{5t}$, $y(t) = -\frac{1}{2} + \frac{313}{2}e^{4t} - 161e^{5t}$

Exercises 12.3–Part A, Page 276

2. $A\mathbf{x} = \mathbf{b}$; $A = \begin{bmatrix} 8 & -9 \\ 3 & -5 \end{bmatrix}$, $\mathbf{b} = \begin{bmatrix} 12 \\ 11 \end{bmatrix}$ **5.** $A = \begin{bmatrix} 1 & -1 \\ 1 & -1 \end{bmatrix}$, $B = \begin{bmatrix} 1 & 1 \\ 1 & 1 \end{bmatrix}$ **7.** $\mathbf{X}' = \begin{bmatrix} -10e^{-2t} \\ 8e^{-2t} \end{bmatrix}$, $A\mathbf{X} = \begin{bmatrix} -2e^{-2t} \\ 9e^{-2t} \end{bmatrix}$

Exercises 12.3–Part B, Page 277

1. **(b)** $B + 3C = \begin{bmatrix} 26 & 1 \\ 14 & -2 \end{bmatrix}$ **2.** **(d)** $AD = \begin{bmatrix} -\frac{11}{6} & -\frac{103}{28} \\ \frac{17}{3} & \frac{293}{28} \end{bmatrix}$, $DA = \begin{bmatrix} \frac{45}{14} & \frac{7}{2} \\ \frac{9}{2} & \frac{65}{12} \end{bmatrix}$ **3.** **(g)** $I + A + \frac{1}{2}A^2 + \frac{1}{6}A^3 = \begin{bmatrix} -\frac{41}{3} & -\frac{175}{2} \\ 35 & \frac{169}{3} \end{bmatrix}$

4. $\begin{bmatrix} x(t) \\ y(t) \end{bmatrix} = \begin{bmatrix} \frac{1}{89}\sqrt{89} - 1 \\ \frac{19}{178}\sqrt{89} + \frac{3}{2} \end{bmatrix} e^{(3-\sqrt{89})t/2} + \begin{bmatrix} -\frac{1}{89}\sqrt{89} - 1 \\ \frac{3}{2} - \frac{19}{178}\sqrt{89} \end{bmatrix} e^{(3+\sqrt{89})t/2}$ **16.** $A\mathbf{x} = \mathbf{b}$; $A = \begin{bmatrix} 2 & -3 \\ 4 & -5 \end{bmatrix}$, $\mathbf{b} = \begin{bmatrix} 7 \\ 9 \end{bmatrix}$

26. $A\mathbf{x} = \mathbf{b}$; $A = \begin{bmatrix} 2 & -3 & 5 \\ -1 & 4 & -8 \\ 12 & 7 & -13 \end{bmatrix}$, $\mathbf{b} = \begin{bmatrix} 7 \\ 9 \\ 17 \end{bmatrix}$ **36.** $A = \begin{bmatrix} -\frac{1}{25} & \frac{3}{200} \\ \frac{1}{25} & -\frac{3}{200} \end{bmatrix}$; $\mathbf{X}_0 = \begin{bmatrix} 0 \\ 20 \end{bmatrix}$

Exercises 12.4–Part A, Page 280

1. $|\mathbf{A} \times \mathbf{B}| = 29$ **3.** $x = -3$, $y = -4$

Exercises 12.4–Part B, Page 281

1. $\begin{bmatrix} 35 & 4 & -11 \\ -7 & 16 & 5 \\ 28 & 6 & 8 \end{bmatrix}$ **2.** $|\mathbf{A} \times \mathbf{B}| = 1$ **6.** $|(\mathbf{A} \times \mathbf{B}).\mathbf{C}| = 102$ **11.** $x = \frac{39}{17}$, $y = \frac{40}{17}$ **20.** $x = 1$, $y = -\frac{45}{2}$, $z = -\frac{25}{2}$

30. **(a)** $\det(AB) = \det(A)\det(B) = \det(BA) = 1023$ **(b)** $\det(ABC) = \det(A)\det(B)\det(C) = \det(BCA) = \det(ACB) = -12276$
(c) $\det(A - \lambda I) = 31 - 10\lambda + \lambda^2$ **(d)** $\det(A + B) = 125$, $\det(A) + \det(B) = 64$ **(e)** $\det(B - 2I) = 13$, $\det(B) - 2 = 31$

32. **(a)** and **(b)** $\det(A) = 10$ **(c)** $\begin{bmatrix} -10 & -10 & -20 & 10 \\ 34 & 30 & 67 & -25 \\ 8 & 10 & 19 & -5 \\ 14 & 10 & 22 & -10 \end{bmatrix}$

Exercises 12.5–Part A, Page 284

1. No, because the leftmost entries in each row are not necessarily ones.

2. False, because the leftmost nonzero entries in the rows need not be one.

4. $\begin{bmatrix} -3 & -3 & 3 & -8 & -3 \\ 7 & -9 & -1 & 6 & 7 \\ 5 & -4 & 8 & 0 & -12 \\ -10 & -8 & -11 & -7 & 9 \end{bmatrix}$; $\begin{bmatrix} -3 & -3 & 3 & -8 & -3 \\ 0 & -16 & 6 & -\frac{38}{3} & 0 \\ 0 & 0 & \frac{77}{8} & -\frac{149}{24} & -17 \\ 0 & 0 & 0 & \frac{1160}{231} & -\frac{1291}{77} \end{bmatrix}$; $\mathbf{x} = \begin{bmatrix} \frac{279}{58} \\ \frac{1361}{1160} \\ -\frac{4547}{1160} \\ -\frac{3873}{1160} \end{bmatrix}$

Exercises 12.5–Part B, Page 284

1. $\mathbf{x} = \begin{bmatrix} -51 \\ -91 \\ 14 \end{bmatrix}$ **5.** $\mathbf{x} = \begin{bmatrix} -\frac{14,589}{8014} \\ \frac{14,003}{8014} \\ \frac{6939}{4007} \end{bmatrix}$ **14.** $\mathbf{x} = \begin{bmatrix} -\frac{128}{5} \\ -19 \\ -\frac{219}{5} \\ 20 \end{bmatrix}$

18. **(a)** $-2x_2 + 11x_3 = -12$, $-10x_1 + 9x_2 - 4x_3 = -8$, $-4x_1 + x_2 + 10x_3 = -6$ **(b)** $\begin{bmatrix} 1 & -\frac{9}{10} & \frac{2}{5} & \frac{4}{5} \\ 0 & 1 & -\frac{11}{2} & 6 \\ 0 & 0 & 1 & -\frac{128}{27} \end{bmatrix}$

(c) $x_1 - \frac{9}{10}x_2 + \frac{2}{5}x_3 = \frac{4}{5}$, $x_2 - \frac{11}{2}x_3 = 6$, $x_3 = -\frac{128}{27}$ **(d)** $\mathbf{x} = \begin{bmatrix} -\frac{415}{27} \\ -\frac{542}{27} \\ -\frac{128}{27} \end{bmatrix}$ **(e)** $\begin{bmatrix} 1 & 0 & 0 & -\frac{415}{27} \\ 0 & 1 & 0 & -\frac{542}{27} \\ 0 & 0 & 1 & -\frac{128}{27} \end{bmatrix}$ **(f)** See **(d)**.

Exercises 12.6–Part A, Page 289

2. (a) $\operatorname{rank}(A) = 1$ **3. (a)** $\left\{ \begin{bmatrix} 1 \\ 0 \\ -\frac{6}{5} \end{bmatrix}, \begin{bmatrix} 0 \\ 1 \\ -\frac{9}{5} \end{bmatrix} \right\}$ **5. (a)** $\mathbf{x} = c_1 \begin{bmatrix} 1 \\ 0 \\ -\frac{6}{5} \end{bmatrix} + c_2 \begin{bmatrix} 0 \\ 1 \\ -\frac{9}{5} \end{bmatrix}$

Exercises 12.6–Part B, Page 289

1. (a) $\operatorname{rank}(A) = 2$; $\left\{ \begin{bmatrix} 1 \\ 4 \\ 3 \end{bmatrix} \right\}$ **2.** $\det(A) = 0$; $69 \begin{bmatrix} -10 \\ -12 \\ -45 \end{bmatrix} + 33 \begin{bmatrix} 26 \\ 13 \\ 96 \end{bmatrix} + 7 \begin{bmatrix} -24 \\ 57 \\ -9 \end{bmatrix} = \begin{bmatrix} 0 \\ 0 \\ 0 \end{bmatrix}$

11. We use the first matrix. **(a)** 0   **(b)** Does not apply   **(c)** $\mathbf{x} = \begin{bmatrix} t \\ -\frac{3}{2}t - \frac{3}{2}s \\ s \end{bmatrix}$ **(d)** $\left\{ \begin{bmatrix} 1 \\ -\frac{3}{2} \\ 0 \end{bmatrix}, \begin{bmatrix} 0 \\ -\frac{3}{2} \\ 1 \end{bmatrix} \right\}$ **(e)** 2

(f) $\begin{bmatrix} 1 & \frac{2}{3} & 1 \\ 0 & 0 & 0 \\ 0 & 0 & 0 \end{bmatrix}$; $c_1 \begin{bmatrix} 1 \\ -\frac{3}{2} \\ 0 \end{bmatrix} + c_2 \begin{bmatrix} 0 \\ -\frac{3}{2} \\ 1 \end{bmatrix}$ **(g)** $\left\{ \begin{bmatrix} 1 \\ -\frac{3}{2} \\ 0 \end{bmatrix}, \begin{bmatrix} 0 \\ -\frac{3}{2} \\ 1 \end{bmatrix} \right\}$ **(h)** $\operatorname{rank}(A) = 1$ **(i)** $1 + 2 = 3$.

15. (a) $\lambda_1 = -\frac{5}{2} + \frac{1}{2}\sqrt{149}$, $\lambda_2 = -\frac{5}{2} - \frac{1}{2}\sqrt{149}$ **(b)** $C_1 = \begin{bmatrix} \frac{17}{2} - \frac{1}{2}\sqrt{149} & 5 \\ -7 & -\frac{17}{2} - \frac{1}{2}\sqrt{149} \end{bmatrix}$; $C_2 = \begin{bmatrix} \frac{17}{2} + \frac{1}{2}\sqrt{149} & 5 \\ -7 & -\frac{17}{2} + \frac{1}{2}\sqrt{149} \end{bmatrix}$

(c) $N_1 = \left\{ \begin{bmatrix} 1 \\ -\frac{17}{10} + \frac{1}{10}\sqrt{149} \end{bmatrix} \right\}$; $N_2 = \left\{ \begin{bmatrix} 1 \\ -\frac{17}{10} - \frac{1}{10}\sqrt{149} \end{bmatrix} \right\}$

(d) $\mathbf{v}_1 = \begin{bmatrix} c_1 \\ c_1(-\frac{17}{10} + \frac{1}{10}\sqrt{149}) \end{bmatrix}$; $\mathbf{v}_2 = \begin{bmatrix} c_2 \\ c_2(-\frac{17}{10} - \frac{1}{10}\sqrt{149}) \end{bmatrix}$; $A\mathbf{v}_1 = \begin{bmatrix} -\frac{5}{2}c_1 + \frac{1}{2}c_1\sqrt{149} \\ \frac{117}{10}c_1 - \frac{11}{10}c_1\sqrt{149} \end{bmatrix}$; $A\mathbf{v}_2 = \begin{bmatrix} -\frac{5}{2}c_2 - \frac{1}{2}c_2\sqrt{149} \\ \frac{117}{10}c_2 + \frac{11}{10}c_2\sqrt{149} \end{bmatrix}$

20. (b) $N = \left\{ \begin{bmatrix} \frac{29}{9} \\ 1 \\ -\frac{10}{3} \end{bmatrix} \right\}$ **(c)** $\mathbf{X} = \begin{bmatrix} \frac{23}{3} + \frac{29}{9}t \\ t \\ -12 - \frac{10}{3}t \end{bmatrix}$ **(d)** $\mathbf{X}_p := \begin{bmatrix} \frac{23}{3} \\ 0 \\ -12 \end{bmatrix}$; $\mathbf{X}_h := \begin{bmatrix} \frac{29}{9}t \\ t \\ -\frac{10}{3}t \end{bmatrix}$

Exercises 12.7–Part A, Page 291

1. (a) $\det(A) = -2 \neq 0$ **(b)** $B_{1,1} + 2B_{2,1} = 1$, $B_{1,2} + 2B_{2,2} = 0$, $3B_{1,1} + 4B_{2,1} = 0$, $3B_{1,2} + 4B_{2,2} = 1 \Rightarrow$;

$B_{2,2} = -\frac{1}{2}$, $B_{2,1} = \frac{3}{2}$, $B_{1,2} = 1$, $B_{1,1} = -2$ **(c)** $\begin{bmatrix} 1 & 0 & -2 & 1 \\ 0 & 1 & \frac{3}{2} & -\frac{1}{2} \end{bmatrix}$ **4.** $AC = I \Rightarrow CA = I$, $B = IB = (CA)B = C(AB) = CI = C$

Exercises 12.7–Part B, Page 292

1. $\mathbf{X} = \begin{bmatrix} -\frac{14}{863} \\ -\frac{543}{863} \\ \frac{163}{863} \end{bmatrix}$; $\operatorname{rref}(A, \mathbf{Y}) = \begin{bmatrix} 1 & 0 & 0 & -\frac{14}{863} \\ 0 & 1 & 0 & -\frac{543}{863} \\ 0 & 0 & 1 & \frac{163}{863} \end{bmatrix}$ **2. (a)** $\det(A) = -765$ **(b)** $A^{-1} = \begin{bmatrix} -\frac{6}{85} & -\frac{7}{51} & \frac{4}{15} \\ -\frac{29}{255} & -\frac{14}{153} & \frac{11}{45} \\ \frac{1}{17} & \frac{1}{17} & 0 \end{bmatrix}$ **(c)** $AA^{-1} = A^{-1}A = I$

(d) Let $AB = I$, then $11B_{1,1} - 12B_{2,1} + 7B_{3,1} = 1$, $11B_{1,2} - 12B_{2,2} + 7B_{3,2} = 0$, $11B_{1,3} - 12B_{2,3} + 7B_{3,3} = 0$, $-11B_{1,1} + 12B_{2,1} + 10B_{3,1} = 0$, $-11B_{1,2} + 12B_{2,2} + 10B_{3,2} = 1$, $-11B_{1,3} + 12B_{2,3} + 10B_{3,3} = 0$, $B_{1,1} + 3B_{2,1} + 7B_{3,1} = 0$, $B_{1,2} + 3B_{2,2} + 7B_{3,2} = 0$, $B_{1,3} + 3B_{2,3} + 7B_{3,3} = 1 \Rightarrow B_{2,3} = \frac{11}{45}$, $B_{1,3} = \frac{4}{15}$, $B_{3,3} = 0$, $B_{3,2} = \frac{1}{17}$, $B_{2,2} = -\frac{14}{153}$, $B_{1,2} = -\frac{7}{51}$, $B_{3,1} = \frac{1}{17}$, $B_{2,1} = -\frac{29}{255}$, $B_{1,1} = -\frac{6}{85}$

(e) $\mathbf{X}_1 = \begin{bmatrix} -\frac{6}{85} \\ -\frac{29}{255} \\ \frac{1}{17} \end{bmatrix}$; $\mathbf{X}_2 = \begin{bmatrix} -\frac{7}{51} \\ -\frac{14}{153} \\ \frac{1}{17} \end{bmatrix}$; $\mathbf{X}_3 = \begin{bmatrix} \frac{4}{15} \\ \frac{11}{45} \\ 0 \end{bmatrix}$ **(f)** $\begin{bmatrix} 1 & 0 & 0 & -\frac{6}{85} & -\frac{7}{51} & \frac{4}{15} \\ 0 & 1 & 0 & -\frac{29}{255} & -\frac{14}{153} & \frac{11}{45} \\ 0 & 0 & 1 & \frac{1}{17} & \frac{1}{17} & 0 \end{bmatrix}$ **(g)** $\frac{\operatorname{adj}A}{\det(A)} = -\frac{1}{765} \begin{bmatrix} 54 & 105 & -204 \\ 87 & 70 & -187 \\ -45 & -45 & 0 \end{bmatrix}$

11. (a) and **(b)** $\left(-\frac{14}{51}, \frac{55}{51} \right)$

Exercises 12.8–Part A, Page 294

In section 12.7 we have introduced the transpose of a vector. From this point in the answer key, column vectors will be displayed as transposed row vectors; e.g., the vector $\mathbf{v} = \begin{bmatrix} 1 \\ 2 \end{bmatrix}$ will be displayed as $\mathbf{v} = [1, 2]^{\mathrm{T}}$.

1. $\Phi = \begin{bmatrix} -\frac{14+s}{72-17s+s^2} & \frac{6}{72-17s+s^2} \\ -\frac{5}{72-17s+s^2} & -\frac{3+s}{72-17s+s^2} \end{bmatrix}$ **2.** $\phi(t) = \begin{bmatrix} 6e^{8t} - 5e^{9t} & -6e^{8t} + 6e^{9t} \\ 5e^{8t} - 5e^{9t} & -5e^{8t} + 6e^{9t} \end{bmatrix}; \phi(0) = \begin{bmatrix} 1 & 0 \\ 0 & 1 \end{bmatrix}$

4. $\mathbf{v}' = \begin{bmatrix} 5 & 6 \\ -7 & -8 \end{bmatrix} \mathbf{v}; \phi = \begin{bmatrix} -6e^{-2t} + 7e^{-t} & -6e^{-2t} + 6e^{-t} \\ 7e^{-2t} - 7e^{-t} & 7e^{-2t} - 6e^{-t} \end{bmatrix}$

Exercises 12.8–Part B, Page 294

1. $\varphi' = \begin{bmatrix} 48e^{8t} - 45e^{9t} & -48e^{8t} + 54e^{9t} \\ 40e^{8t} - 45e^{9t} & -40e^{8t} + 54e^{9t} \end{bmatrix} = A\varphi$ **2. (a)** $\mathbf{v}(t) = [30e^{-5t} - 32e^t, -24e^t + 30e^{-5t}]^T$

(b)

(c) $\mathbf{v}(t) = [30, 30]^T e^{-5t} + [-32, -24]^T e^t$

(d) $A\mathbf{B}_1 = [-150, -150]^T = 5[-13, -30]^T = \lambda_1 \mathbf{B}_1; A\mathbf{B}_2 = [-32, -24]^T = 1[-32, -24]^T = \lambda_2 \mathbf{B}_2$

(e) $\phi(t) = \begin{bmatrix} -3e^{-5t} + 4e^t & 4e^{-5t} - 4e^t \\ -3e^{-5t} + 3e^t & 4e^{-5t} - 3e^t \end{bmatrix}$

(f) $\varphi(t)\mathbf{v}_0 = [30e^{-5t} - 32e^t, -24e^t + 30e^{-5t}]^T$

(g) $\mathbf{v}' = [-150e^{-5t} - 32e^t, -24e^t - 150e^{-5t}]^T = A\mathbf{v}, \mathbf{v}(0) = [-2, 6]^T$

(h) $\phi(0) = \begin{bmatrix} 1 & 0 \\ 0 & 1 \end{bmatrix}$ **(i)** $\mathbf{X}'_1 = [15e^{-5t} + 4e^t, 15e^{-5t} + 3e^t]^T = A\mathbf{X}_1, \mathbf{X}'_2 = [-20e^{-5t} - 4e^t, -20e^{-5t} - 3e^t]^T = A\mathbf{X}_2$

(j) $\phi' = \begin{bmatrix} 15e^{-5t} + 4e^t & -20e^{-5t} - 4e^t \\ 15e^{-5t} + 3e^t & -20e^{-5t} - 3e^t \end{bmatrix}$ **(k)** $\mathbf{X}_1(0) = \begin{bmatrix} 1 \\ 0 \end{bmatrix}, \mathbf{X}_2(0) = \begin{bmatrix} 0 \\ 1 \end{bmatrix}$

13. (a) $\mathbf{v}(t) = [-444e^{-2t} + 398e^{4t} + 35e^{9t}, 199e^{4t} - 222e^{-2t} + 35e^{9t}, -199e^{4t} + 333e^{-2t} - 140e^{9t}]^T$

(b)

(c) $\mathbf{v}(t) = [398, 199, -199]^T e^{4t} + [-444, -222, 333]^T e^{-2t} + [35, 35, -140]^T e^{9t}$

(d) $A\mathbf{B}_1 = [1592, 796, -796]^T = 4[398, 199, -199]^T = \lambda_1 \mathbf{B}_1,$

$A\mathbf{B}_2 = [888, 444, -666]^T = -2[-444, -222, 333]^T = \lambda_2 \mathbf{B}_2,$

$A\mathbf{B}_3 = [315, 315, -1260]^T = 9[35, 35, -140]^T = \lambda_3 \mathbf{B}_3$

(e) $\phi(t) = \begin{bmatrix} -10e^{4t} - e^{9t} + 12e^{-2t} & 26e^{4t} + 2e^{9t} - 28e^{-2t} & -4e^{-2t} + 4e^{4t} \\ -5e^{4t} - e^{9t} + 6e^{-2t} & 13e^{4t} + 2e^{9t} - 14e^{-2t} & -2e^{-2t} + 2e^{4t} \\ 5e^{4t} + 4e^{9t} - 9e^{-2t} & -13e^{4t} - 8e^{9t} + 21e^{-2t} & 3e^{-2t} - 2e^{4t} \end{bmatrix}$

(f) $\phi(t)\mathbf{v}_0 = [-444e^{-2t} + 398e^{4t} + 35e^{9t}, 199e^{4t} - 222e^{-2t} + 35e^{9t}, -199e^{4t} + 333e^{-2t} - 140e^{9t}]^T$

(g) $\mathbf{v}'(t) = [888e^{-2t} + 1592e^{4t} + 315e^{9t}, 796e^{4t} + 444e^{-2t} + 315e^{9t}, -796e^{4t} - 666e^{-2t} - 1260e^{9t}]^T; \mathbf{v}(0) = [-11, 12, -6]^T$

(h) $\phi(0) = \begin{bmatrix} 1 & 0 & 0 \\ 0 & 1 & 0 \\ 0 & 0 & 1 \end{bmatrix}$ **(i)** $\mathbf{X}'_1 = [-40e^{4t} - 9e^{9t} - 24e^{-2t}, -20e^{4t} - 9e^{9t} - 12e^{-2t}, 20e^{4t} + 36e^{9t} + 18e^{-2t}]^T = A\mathbf{X}_1, \mathbf{X}'_2 =$

$[104e^{4t} + 18e^{9t} + 56e^{-2t}, 52e^{4t} + 18e^{9t} + 28e^{-2t}, -52e^{4t} - 72e^{9t} - 42e^{-2t}]^T = A\mathbf{X}_2, \mathbf{X}'_3 = [8e^{-2t} + 16e^{4t}, 4e^{-2t} + 8e^{4t}, -6e^{-2t)} - 8e^{4t}]^T = A\mathbf{X}_3$

(j) $\varphi' = [\mathbf{X}'_1, \mathbf{X}'_2, \mathbf{X}'_3] = [A\mathbf{X}_1, A\mathbf{X}_2, A\mathbf{X}_3] = A[\mathbf{X}_1, \mathbf{X}_2, \mathbf{X}_3] = A\varphi$ **(k)** $\mathbf{X}_1(0) = [1, 0, 0]^T, \mathbf{X}_2(0) = [0, 1, 0]^T, \mathbf{X}_3(0) = [0, 0, 1]^T$

Exercises 12.9–Part A, Page 297

2. $Ae^{At} = A\sum_{k=0}^{\infty} \frac{(At)^k}{k!} = \sum_{k=0}^{\infty} \frac{A^{k+1}t^k}{k!} = \left(\sum_{k=0}^{\infty} \frac{A^k t^k}{k!}\right) A = e^{At} A$ **4.** $\phi(t) = \begin{bmatrix} 4e^{3t} - 3e^{5t} & -2e^{3t} + 2e^{5t} \\ 6e^{3t} - 6e^{5t} & -3e^{3t} + 4e^{5t} \end{bmatrix}$

Exercises 12.9–Part B, Page 297

1. $e^{At} \approx \begin{bmatrix} 1 + t - \frac{2}{3}t^3 - \frac{2}{3}t^4 - \frac{2}{5}t^5 & t + 2t^2 + 2t^3 + \frac{4}{3}t^4 + \frac{2}{3}t^5 \\ -t - 2t^2 - 2t^3 - \frac{4}{3}t^4 - \frac{2}{3}t^5 & 1 + 3t + 4t^2 + \frac{10}{3}t^3 + 2t^4 + \frac{14}{15}t^5 \end{bmatrix}; \sum_{k=0}^{4} \frac{(At)^k}{k!} = \begin{bmatrix} t - \frac{2}{3}t^3 - \frac{2}{3}t^4 + 1 & t + 2t^2 + 2t^3 + \frac{4}{3}t^4 \\ -t - 2t^2 - 2t^3 - \frac{4}{3}t^4 & 3t + 4t^2 + \frac{10}{3}t^3 + 2t^4 + 1 \end{bmatrix}$

2. (a) and (b) $\varphi(t) = \begin{bmatrix} -3e^{-5t} + 4e^t & -4e^t + 4e^{-5t} \\ 3e^t - 3e^{-5t} & 4e^{-5t} - 3e^t \end{bmatrix}$ **(c)** $\mathbf{X}_1 = [-3e^{-5t} + 4e^t, 3e^t - 3e^{-5t}]^T, \mathbf{X}_2 = [-4e^t + 4e^{-5t}, 4e^{-5t} - 3e^t]^T$

(d) $\sum_{k=0}^{4} \frac{(At)^k}{k!} = \begin{bmatrix} -\frac{71}{2}t^2 + \frac{379}{6}t^3 - \frac{1871}{24}t^4 + 1 & -24t + 48t^2 - 84t^3 + 104t^4 \\ 8t - 36t^2 + 63t^3 - 78t^4 & -23t + \frac{97}{2}t^2 - \frac{503}{6}t^3 + \frac{2497}{24}t^4 + 1 \end{bmatrix}$ **(e)** $Ae^{At} = \begin{bmatrix} 15e^{-5t} + 4e^t & -4e^t - 20e^{-5t} \\ 15e^{-5t} + 3e^t & -3e^t - 20e^{-5t} \end{bmatrix} = e^{At}A$

(f) $\phi'(t) = \begin{bmatrix} 15e^{-5t} + 4e^t & -4e^t - 20e^{-5t} \\ 15e^{-5t} + 3e^t & -3e^t - 20e^{-5t} \end{bmatrix} = Ae^{At}$ **(g)** $[\phi(t)]^{-1} = \begin{bmatrix} 4e^{-t} - 3e^{5t} & 4e^{5t} - 4e^{-t} \\ -3e^{5t} + 3e^{-t} & -3e^{-t} + 4e^{5t} \end{bmatrix} = \phi(-t)$

12. **(a) and (b)** $\varphi(t) = \begin{bmatrix} 12e^{-2t} - 10e^{4t} - e^{9t} & 2e^{9t} + 26e^{4t} - 28e^{-2t} & 4e^{4t} - 4e^{-2t} \\ -5e^{4t} + 6e^{-2t} - e^{9t} & -14e^{-2t} + 2e^{9t} + 13e^{4t} & 2e^{4t} - 2e^{-2t} \\ 5e^{4t} - 9e^{-2t} + 4e^{9t} & -8e^{9t} - 13e^{4t} + 21e^{-2t} & 3e^{-2t} - 2e^{4t} \end{bmatrix}$

(c) $\mathbf{X}_1 = [12e^{-2t} - 10e^{4t} - e^{9t}, -5e^{4t} + 6e^{-2t} - e^{9t}, 5e^{4t} - 9e^{-2t} + 4e^{9t}]^{\mathrm{T}}$,
$\mathbf{X}_2 = [2e^{9t} + 26e^{4t} - 28e^{-2t}, -14e^{-2t} + 2e^{9t} + 13e^{4t}, -8e^{9t} - 13e^{4t} + 21e^{-2t}]^{\mathrm{T}}; \mathbf{X}_3 = [4e^{4t} - 4e^{-2t}, 2e^{4t} - 2e^{-2t}, 3e^{-2t} - 2e^{4t}]^{\mathrm{T}}$

(d) $A e^{At} = \begin{bmatrix} -24e^{-2t} - 40e^{4t} - 9e^{9t} & 18e^{9t} + 104e^{4t} + 56e^{-2t} & 16e^{4t} + 8e^{-2t} \\ -12e^{-2t} - 20e^{4t} - 9e^{9t} & 18e^{9t} + 52e^{4t} + 28e^{-2t} & 8e^{4t} + 4e^{-2t} \\ 18e^{-2t} + 20e^{4t} + 36e^{9t} & -72e^{9t} - 52e^{4t} - 42e^{-2t} & -8e^{4t} - 6e^{-2t} \end{bmatrix} = e^{At} A$

(e) $\phi'(t) = \begin{bmatrix} e^{-2t} - 40e^{4t} - 9e^{9t} & 18e^{9t} + 104e^{4t} + 56e^{-2t} & 16e^{4t} + 8e^{-2t} \\ e^{-2t} - 20e^{4t} - 9e^{9t} & 18e^{9t} + 52e^{4t} + 28e^{-2t} & 8e^{4t} + 4e^{-2t} \\ e^{-2t} + 20e^{4t} + 36e^{9t} & -72e^{9t} - 52e^{4t} - 42e^{-2t} & -8e^{4t} - 6e^{-2t} \end{bmatrix} = A e^{At}$

(f) $[\phi(t)]^{-1} = \begin{bmatrix} -e^{-9t} - 10e^{-4t} + 12e^{2t} & 26e^{-4t} - 28e^{2t} + 2e^{-9t} & -4e^{2t} + 4e^{-4t} \\ -e^{-9t} - 5e^{-4t} + 6e^{2t} & 2e^{-9t} + 13e^{-4t} - 14e^{2t} & -2e^{2t} + 2e^{-4t} \\ -9e^{2t} + 4e^{-9t} + 5e^{-4t} & -13e^{-4t} - 8e^{-9t} + 21e^{2t} & -2e^{-4t} + 3e^{2t} \end{bmatrix} = \phi(-t)$

Exercises 12.10–Part A, Page 299

1. **(a)** $\mathbf{x}(t) = [(-6e^{5t} + 7e^{8t})\alpha + (-6e^{8t} + 6e^{5t})\beta, (7e^{8t} - 7e^{5t})\alpha + (7e^{5t} - 6e^{8t})\beta]^{\mathrm{T}}$ **(b)** $\mathbf{x}(t) = a[1, \frac{7}{6}]^{\mathrm{T}} e^{5t} + b[1, 1]^{\mathrm{T}} e^{8t}$
   (c) $A\mathbf{v}_1 = [5, \frac{35}{6}]^{\mathrm{T}} = \lambda_1 \mathbf{v}_1, A\mathbf{v}_2 = [8, 8]^{\mathrm{T}} = \lambda_2 \mathbf{v}_2$ **(d)** 0 in both cases, as expected
   (e) Bases for the nullspaces are given by $\{[1, \frac{7}{6}]^{\mathrm{T}}\}$ and $\{[1, 1]^{\mathrm{T}}\}$.

Exercises 12.10–Part B, Page 300

2.**(a)** $\mathbf{x}(t) = [(-3e^{-t} + 4e^{3t})\alpha + (12e^{3t} - 12e^{-t})\beta, (-e^{3t} + e^{-t})\alpha + (4e^{-t} - 3e^{3t})\beta]^{\mathrm{T}}$ **(g)**
   (b) $\mathbf{x}(t) = a[-4, 1]^{\mathrm{T}} e^{3t} + b[-3, 1]^{\mathrm{T}} e^{-t}$
   (c) $A\mathbf{v}_1 = [-12, 3]^{\mathrm{T}} = \lambda_1 \mathbf{v}_1, A\mathbf{v}_2 = [3, -1]^{\mathrm{T}} = \lambda_2 \mathbf{v}_2$
   (d) 0 in both cases
   (e) Bases for the nullspaces are given by $\{[-4, 1]^{\mathrm{T}}\}$ and $\{[-3, 1]^{\mathrm{T}}\}$.
   (f) See (b).

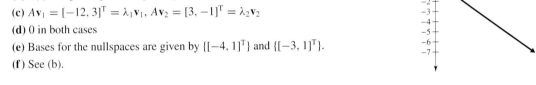

Exercises 12.11–Part A, Page 301

2. $\mathbf{x}(t) = -7e^{2t}[1, 1, -1]^{\mathrm{T}} - 9e^{3t}[1, 2, 2]^{\mathrm{T}} + 13e^{t}[1, 2, 1]^{\mathrm{T}}$

3. $\Psi(0) = \begin{bmatrix} 1 & 1 & 1 \\ 1 & 2 & 2 \\ -1 & 2 & 1 \end{bmatrix}; \Psi(t)\Psi^{-1}(0) = \begin{bmatrix} -4e^{t} + 2e^{2t} + 3e^{3t} & -2e^{3t} - e^{2t} + 3e^{t} & e^{3t} - e^{t} \\ 2e^{2t} - 8e^{t} + 6e^{3t} & 6e^{t} - 4e^{3t} - e^{2t} & 2e^{3t} - 2e^{t} \\ -2e^{2t} - 4e^{t} + 6e^{3t} & -4e^{3t} + e^{2t} + 3e^{t} & -e^{t} + 2e^{3t} \end{bmatrix}$;
   $\Psi(t)\Psi^{-1}(0)\mathbf{x}_0 = [13e^{t} - 7e^{2t} - 9e^{3t}, -7e^{2t} + 26e^{t} - 18e^{3t}, 7e^{2t} + 13e^{t} - 18e^{3t}]^{\mathrm{T}}$

Exercises 12.11–Part B, Page 301

2. **(a)** $\lambda_1 = -6, \mathbf{v}_1 = [-5, 1, -6]^{\mathrm{T}}, \lambda_2 = 2, \mathbf{v}_2 = [-13, 1, 66]^{\mathrm{T}}, \lambda_3 = -7, \mathbf{v}_3 = [-4, 1, -6]^{\mathrm{T}}$ **(b)** $A\mathbf{v}_1 = [30, -6, 36]^{\mathrm{T}} = \lambda_1\mathbf{v}_1$,
   $A\mathbf{v}_2 = [-26, 2, 132]^{\mathrm{T}} = \lambda_2\mathbf{v}_2, A\mathbf{v}_3 = [28, -7, 42]^{\mathrm{T}} = \lambda_3\mathbf{v}_3$ **(c)** $\mathbf{x}(t) = a_1 e^{-6t}[-5, 1, -6]^{\mathrm{T}} + a_2 e^{2t}[-13, 1, 66]^{\mathrm{T}} + a_3 e^{-7t}[-4, 1, -6]^{\mathrm{T}}$
   (d) $\{-5a_1 - 13a_2 - 4a_3 = -6, a_1 + a_2 + a_3 = 8, -6a_1 + 66a_2 - 6a_3 = 10\}; \{a_2 = \frac{29}{36}, a_3 = \frac{364}{9}, a_1 = -\frac{133}{4}\}$;
   $\mathbf{x}(t) = -\frac{133}{4}e^{-6t}[-5, 1, -6]^{\mathrm{T}} + \frac{29}{36}e^{2t}[-13, 1, 66]^{\mathrm{T}} + \frac{364}{9}e^{-7t}[-4, 1, -6]^{\mathrm{T}}$
   (e) $\phi(t) = \begin{bmatrix} -4e^{-7t} + 5e^{-6t} & \frac{95}{4}e^{-6t} - \frac{68}{3}e^{-7t} - \frac{13}{12}e^{2t} & -\frac{13}{72}e^{2t} + \frac{5}{8}e^{-6t} - \frac{4}{9}e^{-7t} \\ -e^{-6t} + e^{-7t} & \frac{17}{3}e^{-7t} + \frac{1}{12}e^{2t} - \frac{19}{4}e^{-6t} & \frac{1}{72}e^{2t} - \frac{1}{8}e^{-6t} + \frac{1}{9}e^{-7t} \\ 6e^{-6t} - 6e^{-7t} & \frac{11}{2}e^{2t} + \frac{57}{2}e^{-6t} - 34e^{-7t} & -\frac{2}{3}e^{-7t} + \frac{11}{12}e^{2t} + \frac{3}{4}e^{-6t} \end{bmatrix}$
   (f) $\mathbf{x}(t) = [-\frac{1456}{9}e^{-7t} + \frac{665}{4}e^{-6t} - \frac{377}{36}e^{2t}, -\frac{133}{4}e^{-6t} + \frac{364}{9}e^{-7t} + \frac{29}{36}e^{2t}, \frac{399}{2}e^{-6t} - \frac{728}{3}e^{-7t} + \frac{319}{6}e^{2t}]^{\mathrm{T}}$

Exercises 12.12–Part A, Page 305

1. $A\mathbf{v}_1 = [3, 3]^{\mathrm{T}} = 3\mathbf{v}_1; A\mathbf{v}_2 = [-3, 6]^{\mathrm{T}} = -3\mathbf{v}_2$ 3. $A - 3I = \begin{bmatrix} -2 & 2 \\ 4 & -4 \end{bmatrix}; A + 3I = \begin{bmatrix} 4 & 2 \\ 4 & 2 \end{bmatrix}$

Exercises 12.12–Part B, Page 305

2. **(a)** $\lambda_1 = -4, \mathbf{v}_1 = [-2, 1]^{\mathrm{T}}, \lambda_2 = 1, \mathbf{v}_2 = [-\frac{13}{4}, 1]^{\mathrm{T}}$ **(b)** $\det(A - \lambda I) = \lambda^2 + 3\lambda - 4 = 0; \lambda_1 = -4, \lambda_2 = 1$ **(c)** Making the rows
   proportional results in a homogeneous linear system which has non-trivial solutions if and only if $\det(A - \lambda I) = 0$.

(d) $\mathbf{v}_1 = [-2, 1]^\mathsf{T}$, $\mathbf{v}_2 = \left[-\frac{13}{4}, 1\right]^\mathsf{T}$ **(e)** $N_1 = \{[-2, 1]^\mathsf{T}\}$, $N_2 = \left\{\left[-\frac{13}{4}, 1\right]^\mathsf{T}\right\}$ **(f)** $1 + 1 = 2$. **(g)** $A\mathbf{v}_1 = [8, -4]^\mathsf{T} = \lambda_1 \mathbf{v}_1$,

$A\mathbf{v}_2 = \left[-\frac{13}{4}, 1\right]^\mathsf{T} = \lambda_2 \mathbf{v}_2$; rank$(A - \lambda_k I) = 1$ and dim(nul$(A - \lambda_k I)) = 1$, for both $k = 1, 2$ **(h)** $P := \begin{bmatrix} -2 & -\frac{13}{4} \\ 1 & 1 \end{bmatrix}$; $P^{-1}AP = \begin{bmatrix} -4 & 0 \\ 0 & 1 \end{bmatrix}$

(i) $\lambda_1 = -4$, $\theta_1 = -\arctan\left(\frac{1}{2}\right)$, $\mathbf{u}_1 = \left[\frac{2}{5}\sqrt{5}, -\frac{1}{5}\sqrt{5}\right]^\mathsf{T}$, $\lambda_2 = 1$, $\theta_2 = -\arctan\left(\frac{4}{13}\right)$, $\mathbf{u}_2 = \left[\frac{13}{185}\sqrt{185}, -\frac{4}{185}\sqrt{185}\right]^\mathsf{T}$

12. (a) $\lambda_1 = \lambda_2 = 7$, $\mathbf{v}_1 = \left[-\frac{8}{7}, 1\right]^\mathsf{T}$ **(b)** $\det(A - \lambda I) = \lambda^2 - 14\lambda + 49 = 0$; $\lambda = 7$ **(c)** A basis for the nullspace of $A - \lambda I$. is: $\left[-\frac{8}{7}, 1\right]^\mathsf{T}$
(d) $\left[-\frac{8}{7}, 1\right]^\mathsf{T}$ **(e)** rank$(A - 7I) = 1$, dim(nul$(A - 7I)) = 1$

22. (a) $\lambda_1 = -2$, $\mathbf{v}_1 = \left[1, -\frac{20}{11}, -\frac{12}{11}\right]^\mathsf{T}$, $\lambda_2 = 1$, $\mathbf{v}_2 = \left[-\frac{4}{3}, \frac{7}{3}, 1\right]^\mathsf{T}$, $\lambda_3 = 3$, $\mathbf{v}_3 = \left[-\frac{4}{3}, \frac{5}{3}, 1\right]^\mathsf{T}$ **(b)** $\det(A - \lambda I) = \lambda^3 - 2\lambda^2 - 5\lambda + 6 = 0$;
$\lambda = 1, 3, -2$ **(c)** rank$(A - \lambda_k I) = 2$ for each $k = 1, 2, 3$ **(d)** $\mathbf{v}_1 = \left[1, -\frac{20}{11}, -\frac{12}{11}\right]^\mathsf{T}$, $\mathbf{v}_2 = \left[-\frac{4}{3}, \frac{7}{3}, 1\right]^\mathsf{T}$, $\mathbf{v}_3 = \left[-\frac{4}{3}, \frac{7}{3}, 1\right]^\mathsf{T}$ **(e)** See (c).
(f) $P^{-1}AP = \begin{bmatrix} -2 & 0 & 0 \\ 0 & 1 & 0 \\ 0 & 0 & 3 \end{bmatrix}$

Exercises 12.13–Part A, Page 307

1. $x'' + x' - 12x = 0$; $x(t) = c_1 e^{-4t} + c_2 e^{3t}$; $\mathbf{x}(t) = [x(t), x'(t)] = [c_1 e^{-4t} + c_2 e^{3t}, c_1 e^{-4t} + 2c_2 e^{3t}]$ **4.** $M = \begin{bmatrix} 0 & 1 \\ 12 & -1 \end{bmatrix}$; $\lambda = -4, 3$

Exercises 12.13–Part B, Page 307

1. $\mathbf{v} = P\mathbf{x} \Rightarrow \mathbf{v}' = P\mathbf{x}'$, $\mathbf{v}' = M\mathbf{v} \Rightarrow P\mathbf{x}' = MP\mathbf{x} \Rightarrow \mathbf{x}' = P^{-1}MP\mathbf{x}$; $P = \begin{bmatrix} -\frac{5}{6}p_3 - \frac{7}{6}p_4 & \frac{7}{12}p_3 + \frac{11}{12}p_4 \\ p_3 & p_4 \end{bmatrix}$, for any real values of p_3 and p_4

4. (a) $\mathbf{x}(t) = c_1 e^t \left[-\frac{13}{4}, 1\right]^\mathsf{T} + c_2 e^{-4t}[-2, 1]^\mathsf{T}$ **(b)** $x'' + 3x' - 4x = 0$ **(c)** $x(t) = C_1 e^t + C_2 e^{-4t}$; $c_1 = -\frac{4}{13}C_1$, $c_2 = -\frac{1}{2}C_2$

14. $\mathbf{x}(t) = [y(t), y'(t)]^\mathsf{T}$; $A = \begin{bmatrix} 0 & 1 \\ -\frac{5}{3} & -\frac{2}{3} \end{bmatrix}$; $\mathbf{F}(t) = \left[0, -\frac{1}{3}\sin t\right]^\mathsf{T}$

Exercises 12.14–Part A, Page 310

2. $e^{3it}\mathbf{v}_1 = [\cos 3t + i\sin 3t, -2\cos 3t - 3\sin 3t + i(3\cos 3t - 2\sin 3t)]^\mathsf{T}$

5. $A - \lambda_2 I = \begin{bmatrix} 2 + 3i & 1 \\ -13 & -2 + 3i \end{bmatrix}$; rref$(A - \lambda_2 I) = \begin{bmatrix} 1 & \frac{2}{13} - \frac{3}{13}i \\ 0 & 0 \end{bmatrix}$

6. Complex roots of polynomials with real coefficients come in pairs. If a number r is a zero of a polynomial p with real coefficients, then its complex conjugate $\bar{r}$ is also a zero of p.

Exercises 12.14–Part B, Page 311

2. (a) $\lambda_1 = -3 + 3i$, $\mathbf{v}_1 = [i, 1]^\mathsf{T}$, $\lambda_2 = -3 - 3i$, $\mathbf{v}_2 = [-i, 1]^\mathsf{T}$ **(b)** $\phi(t) = \begin{bmatrix} e^{-3t}\cos 3t & -e^{-3t}\sin 3t \\ e^{-3t}\sin 3t & e^{-3t}\cos 3t \end{bmatrix}$;
$\mathbf{x}(t) = [-6e^{-3t}\cos 3t - 8e^{-3t}\sin 3t, -6e^{-3t}\sin 3t + 8e^{-3t}\cos 3t]^\mathsf{T}$
(c) $x'' + 6x' + 18x = 0$, $x(0) = -6$, $x'(0) = -6$; $x(t) = -6e^{-3t}\cos 3t - 8e^{-3t}\sin 3t$; $y = -\frac{1}{3}x' - x \Rightarrow y(t) = -6e^{-3t}\sin 3t + 8e^{-3t}\cos 3t$
(d) $[i, 1]^\mathsf{T}$ **(e)** $\mathbf{x}_1(t) = [-e^{-3t}\sin 3t, e^{-3t}\cos 3t]^\mathsf{T}$, $\mathbf{x}_2(t) = [e^{-3t}\cos 3t, e^{-3t}\sin 3t]^\mathsf{T}$ **(f)** $\mathbf{x}_1' = [3e^{-3t}\sin 3t - 3e^{-3t}\cos 3t,$
$-3e^{-3t}\cos 3t - 3e^{-3t}\sin 3t]^\mathsf{T} = A\mathbf{x}_1$, $\mathbf{x}_2' = [-3e^{-3t}\cos 3t - 3e^{-3t}\sin 3t, -3e^{-3t}\sin 3t + 3e^{-3t}\cos 3t]^\mathsf{T} = A\mathbf{x}_2$
(g) $\mathbf{x}(t) = a[-e^{-3t}\sin 3t, e^{-3t}\cos 3t]^\mathsf{T} + b[e^{-3t}\cos 3t, e^{-3t}\sin 3t]^\mathsf{T}$; $b = -6$, $a = 8$;
$\mathbf{x}(t) = [-8e^{-3t}\sin 3t - 6e^{-3t}\cos 3t, 8e^{-3t}\cos 3t - 6e^{-3t}\sin 3t]$ **(h)**

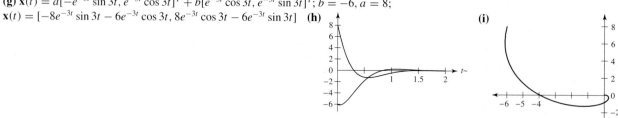

12. (a) $\mathbf{x}(t) = c_1 \left[e^{3t}\cos t, -e^{3t}\cos t, \frac{2}{5}e^{3t}\cos t - \frac{1}{5}e^{3t}\sin t\right]^\mathsf{T} + c_2 \left[e^{3t}\sin t, -e^{3t}\sin t, \frac{2}{5}e^{3t}\sin t + \frac{1}{5}e^{3t}\cos t\right]^\mathsf{T} + c_3 \left[-\frac{5}{4}e^{-t}, e^{-t}, \frac{3}{2}e^{-t}\right]^\mathsf{T}$
(b) $[-33e^{3t}\cos t + 326e^{3t}\sin t + 45e^{-t}, 33e^{3t}\cos t - 326e^{3t}\sin t - 36e^{-t}, 52e^{3t}\cos t + 137e^{3t}\sin t - 54e^{-t}]^\mathsf{T}$

$$\textbf{(c) } \phi = \begin{bmatrix} -4e^{3t}\cos t + 5e^{-t} + 38e^{3t}\sin t & -4e^{3t}\cos t + 5e^{-t} + 38e^{3t}\sin t & 5e^{3t}\sin t \\ 4e^{3t}\cos t - 4e^{-t} - 38e^{3t}\sin t & 5e^{3t}\cos t - 4e^{-t} - 40e^{3t}\sin t & -5e^{3t}\sin t \\ 6e^{3t}\cos t - 6e^{-t} + 16e^{3t}\sin t & 6e^{3t}\cos t - 6e^{-t} + 17e^{3t}\sin t & e^{3t}\cos t + 2e^{3t}\sin t \end{bmatrix};$$

$\mathbf{x}(t) = [-33e^{3t}\cos t + 326e^{3t}\sin t + 45e^{-t}, 33e^{3t}\cos t - 326e^{3t}\sin t - 36e^{-t}, 52e^{3t}\cos t + 137e^{3t}\sin t - 54e^{-t}]^{\mathrm{T}}$

Exercises 12.15–Part A, Page 313

3. $\mathbf{x} = a\mathbf{x}_1 + b\mathbf{x}_2 = ae^{2t}\mathbf{v} + b(t\mathbf{v} + [1,0]^{\mathrm{T}} + 3\mathbf{v})e^{2t} = (a+3b)e^{2t}\mathbf{v} + b(t\mathbf{v} + [1,0]^{\mathrm{T}})e^{2t}$, the coefficient a in (12.50) is now $a + 3b$.

5. (a) $u'' - 4u' + 4u = 0$ **(b)** $u(t) = c_1e^{2t} + c_2e^{2t}t$ **(c)** $\mathbf{x}(t) = [u(t), -\frac{1}{9}u'(t) - \frac{1}{9}u(t)]^{\mathrm{T}} = [c_1e^{2t} + c_2e^{2t}t, -\frac{1}{3}c_1e^{2t} - \frac{1}{3}c_2e^{2t}t - \frac{1}{9}c_2e^{2t}]^{\mathrm{T}}$
(d) $c_1 = b - 3a, c_2 = -3b$

Exercises 12.15–Part B, Page 313

1. (a) $(\lambda + 2)^3 = 0$ **(b)** $y(t) = c_1e^{-2t} + c_2te^{-2t} + c_3t^2e^{-2t}$ **(c)** $A = \begin{bmatrix} 0 & 1 & 0 \\ 0 & 0 & 1 \\ -8 & -12 & -6 \end{bmatrix}$ **(d)** $\lambda_1 = \lambda_2 = \lambda_3 = -2$

(e) $rref(A + 2I) = \begin{bmatrix} 1 & 0 & -\frac{1}{4} \\ 0 & 1 & \frac{1}{2} \\ 0 & 0 & 0 \end{bmatrix}$, $\dim(\mathrm{nul}(A+2I)) = n - \mathrm{rank}(A+2I) = 3 - 2 = 1$ **(f)** $[1, -2, 4]^{\mathrm{T}}$

(g) $\mathbf{x}(t) = [y, y', y'']^{\mathrm{T}} = [(c_1 + c_2t + c_3t^2)e^{-2t}, (-2c_3t^2 + (-2c_2 + 2c_3)t - 2c_1 + c_2)e^{-2t}, (4c_3t^2 + (4c_2 - 8c_3)t + 4c_1 - 4c_2 + 2c_3)e^{-2t}]^{\mathrm{T}}$

2. (a) $\phi = \begin{bmatrix} e^{7t} + 56te^{7t} & 64te^{7t} \\ -49te^{7t} & e^{7t} - 56te^{7t} \end{bmatrix}$; $\mathbf{x}(t) = [-3e^{7t} - 40te^{7t}, 35te^{7t} + 2e^{7t}]^{\mathrm{T}}$ **(b)** $\mathbf{x}(t) = a[-\frac{8}{7}e^{7t}, e^{7t}]^{\mathrm{T}} + b[e^{7t}(-\frac{8}{7}t - \frac{1}{49}), te^{7t}]^{\mathrm{T}}$

(c) $b = 35, a = 2$; $\mathbf{x}(t) = a\left[-\frac{8}{7}e^{7t}, e^{7t}\right]^{\mathrm{T}} + b\left[e^{7t}(-\frac{8}{7}t - \frac{1}{49}), te^{7t}\right]^{\mathrm{T}}$

13. (a) $\phi(t) = \begin{bmatrix} e^{-3t} + 4te^{-3t} + \frac{1}{2}t^2e^{-3t} & -4te^{-3t} - \frac{1}{2}t^2e^{-3t} & -te^{-3t} \\ 4te^{-3t} + \frac{1}{2}t^2e^{-3t} & e^{-3t} - 4te^{-3t} - \frac{1}{2}t^2e^{-3t} & -te^{-3t} \\ -te^{-3t} & te^{-3t} & e^{-3t} \end{bmatrix}$; $\mathbf{x}(t) = [-1 - 28t - 3t^2, -28t - 3t^2 + 5, 6t + 4]^{\mathrm{T}}e^{-3t}$

(b) Does not apply.   **(c)** Does not apply.
(d) $\mathbf{x}(t) = a[1,1,0]^{\mathrm{T}}e^{-3t} + b([0,0,-1]^{\mathrm{T}} + t[1,1,0]^{\mathrm{T}})e^{-3t} + c([1,0,4]^{\mathrm{T}} + t[0,0,-1]^{\mathrm{T}} + \frac{1}{2}t^2[1,1,0]^{\mathrm{T}})e^{-3t}$
(e) $a = 5, b = -28, c = -6$; $\mathbf{x}(t) = [(-1 - 28t - 3t^2)e^{-3t}, (-28t - 3t^2 + 5)e^{-3t}, (6t + 4)e^{-3t}]^{\mathrm{T}}$

27. (a) $\lambda = 2$ **(b)** $\mathbf{v}_1 = [4,1,0]^{\mathrm{T}}$; $\mathbf{v}_2 = [2,0,1]^{\mathrm{T}}$ **(c)** $\mathbf{x}(t) = a\mathbf{v}_1e^{2t} + b\mathbf{v}_2e^{2t} + c((\mathbf{v}_1 - \mathbf{v}_2)t + [1,0,0]^{\mathrm{T}})e^{2t}$

Exercises 12.16–Part A, Page 317

3. If $A = P^{-1}BP$ and $B = Q^{-1}CQ$, then $A = P^{-1}Q^{-1}CQP = (QP)^{-1}C(QP)$.

4. $P = \begin{bmatrix} 1 & 1 \\ -2 & -4 \end{bmatrix}$; $\mathbf{x}(t) = [e^{-t}c_1 + e^tc_2, -2e^{-t}c_1 - 4e^tc_2]^{\mathrm{T}}$

Exercises 12.16–Part B, Page 317

2. (a) $P = \begin{bmatrix} -4 & -3 \\ 1 & 1 \end{bmatrix}$; $D = \begin{bmatrix} 3 & 0 \\ 0 & -1 \end{bmatrix}$; $(-15\frac{d}{-ad+bc} - 4\frac{b}{-ad+bc})a + (-48\frac{d}{-ad+bc} - 13\frac{b}{-ad+bc})c = \alpha$,

$(-15\frac{d}{-ad+bc} - 4\frac{b}{-ad+bc})b + (-48\frac{d}{-ad+bc} - 13\frac{b}{-ad+bc})d = 0, (15\frac{c}{-ad+bc} + 4\frac{a}{-ad+bc})a + (48\frac{c}{-ad+bc} + 13\frac{a}{-ad+bc})c = 0,$

$(15\frac{c}{-ad+bc} + 4\frac{a}{-ad+bc})b + (48\frac{c}{-ad+bc} + 13\frac{a}{-ad+bc})d = \beta; a = -4c, b = -3d, c = c, d = d, \alpha = 3, \beta = -1; P = \begin{bmatrix} -3c & -4d \\ c & d \end{bmatrix}$;

$D = \begin{bmatrix} -1 & 0 \\ 0 & 3 \end{bmatrix}$ **(c)** $\mathbf{u}(t) = [e^{-t}C_2, e^{3t}C_1]^{\mathrm{T}}$ **(d)** $\mathbf{x}(t) = [-3e^{-t}C_2 - 4e^{3t}C_1, e^{-t}C_2 + e^{3t}C_1]^{\mathrm{T}}$ **(e)** $c_1 = -4C_1 - 3C_2, c_2 = C_2 + C_1$

8. (a) $P = \begin{bmatrix} -13 & -4 & -5 \\ 1 & 1 & 1 \\ 66 & -6 & -6 \end{bmatrix}$; $D = \begin{bmatrix} 2 & 0 & 0 \\ 0 & -7 & 0 \\ 0 & 0 & -6 \end{bmatrix}$
(b) $\mathbf{x}(t) = [-13e^{2t}C_1 - 4e^{-7t}C_3 - 5e^{-6t}C_2, e^{2t}C_1 + e^{-7t}C_3 + e^{-6t}C_2, 66e^{2t}C_1 - 6e^{-7t}C_3 - 6e^{-6t}C_2]^{\mathrm{T}}$
(c) $\mathbf{x}(t) = e^{At}\mathbf{c}$, where $\mathbf{c} = [-13C_1 - 5C_2 - 4C_3, C_1 + C_2 + C_3, 66C_1 - 6C_2 - 6C_3]^{\mathrm{T}}$, gives the solution in (b).

Exercises 12.17–Part A, Page 321

3. $M^{-1}K = \begin{bmatrix} \frac{k_1+k_3}{m_1} & -\frac{k_3}{m_1} \\ -\frac{k_3}{m_2} & \frac{k_2+k_3}{m_2} \end{bmatrix}$ **5.** $c_1 = \frac{1}{3}a_1 + \frac{2}{3}a_2, c_3 = -\frac{1}{3}a_1 + \frac{1}{3}a_2; c_2 = \frac{1}{3}b_1 + \frac{2}{3}b_2, c_4 = -\frac{1}{33}(b_1 - b_2)\sqrt{22}$

Exercises 12.17–Part B, Page 321

2. **(a)** $D = \begin{bmatrix} \frac{14}{3} + \frac{2}{3}\sqrt{31} & 0 \\ 0 & \frac{14}{3} - \frac{2}{3}\sqrt{31} \end{bmatrix}$ **(b)** $b_2 = -5\frac{\sqrt{31}b_1}{7\sqrt{31}+62}, a_2 = -5\frac{\sqrt{31}a_1}{7\sqrt{31}+62}; a_2 = -5\frac{\sqrt{31}a_1}{7\sqrt{31}-62}, b_2 = -5\frac{\sqrt{31}b_1}{7\sqrt{31}-62}$

   (c) Use the uncoupled system to obtain the desired result; $[-2.161349044\cos(2.894565524t) + 0.7953312575\sin(2.894565524t) + 1.161349044\cos(0.9771508366t) - 0.3092034807\sin(0.9771508366t), 0.5958880704\cos(2.894565524t) - 0.2192743508\sin(2.894565524t) + 1.404111930\cos(0.9771508366t) - 0.3738379071\sin(0.9771508366t)]^T$

   (d) 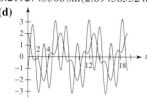   **(e)** 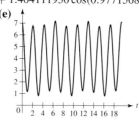   **(f)**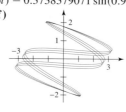

10. **(a)** $x_2'' - x_1 + 2x_2 = 0$, $x_1'' + 2x_1 - x_2 = \cos(t)$

    (b) $[x_1(t), x_2(t)]^T = \left[\frac{1}{4}\cos t - \frac{1}{4}\cos(\sqrt{3}t) + \frac{1}{4}\sin(t)t, -\frac{1}{4}\cos t + \frac{1}{4}\sin(t)t + \frac{1}{4}\cos(\sqrt{3}t)\right]^T$

    (c)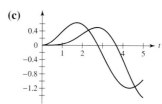

11. **(a)** $m_2 x_2'' - kx_1 + kx_2 = 0$, $m_1 x_1'' + kx_1 - kx_2 = 0$ **(b)** $x_1(t) = c_1 + c_2 t + c_3 e^{\alpha t} + c_4 e^{-\alpha t}$, $x_2(t) = c_1 + c_2 t - \frac{m_1}{m_2}c_3 e^{\alpha t} - \frac{m_1}{m_2}c_4 e^{-\alpha t}$, where $\alpha = \sqrt{-\frac{k(m_1+m_2)}{m_1 m_2}}$; The natural frequencies are 0 and $\sqrt{\frac{k(m_1+m_2)}{m_1 m_2}}$ **(c)** $u'' = 0$, $v'' + \frac{k(m_2+m_1)}{m_2 m_1}v = 0$; The normal modes are harmonic motions with angular frequency 0 and $\sqrt{\frac{k(m_1+m_2)}{m_1 m_2}}$.

12. Make the substitution $\mathbf{x} = P\mathbf{r}$ in $\mathbf{x}'' + A\mathbf{x}' + E\mathbf{x} = F$, resulting in $P\mathbf{r}'' + AP\mathbf{r}' + EP\mathbf{r} = G \Rightarrow \mathbf{r}'' + P^{-1}AP\mathbf{r}' + P^{-1}EP\mathbf{r} = P^{-1}G \Rightarrow \mathbf{r}'' + D_1\mathbf{r}' + D_2\mathbf{r} = P^{-1}G$

13. **(a)** $A := \begin{bmatrix} 9 & 2 \\ 2 & 9 \end{bmatrix}$; $E := \begin{bmatrix} 8 & 3 \\ 3 & 8 \end{bmatrix}$ **(b)** $P = \begin{bmatrix} p_1 & p_2 \\ -p_1 & p_2 \end{bmatrix}$ **(c)** $P = \begin{bmatrix} \frac{1}{2}\sqrt{2} & \frac{1}{2}\sqrt{2} \\ -\frac{1}{2}\sqrt{2} & \frac{1}{2}\sqrt{2} \end{bmatrix}$ **(d)** $v'' + 11v' + 11v = 0$, $u'' + 7u' + 5u = -\sqrt{2}\cos t$

    (e) $\mathbf{r}(t) = [u(t), v(t)]^T = [1.149933684c_1 e^{-0.8074175965t} - 0.1499336838c_1 e^{-6.192582405t} + 0.1856953382c_2 e^{-0.8074175965t} - 0.1856953382c_2 e^{-6.192582405t} - 0.08702852689\cos t - 0.1522999221\sin t, -0.1267831698c_3 e^{-9.887482195t} + 1.126783170c_3 e^{-1.112517807t} + 0.1139605764c_4 e^{-1.112517807t} - 0.1139605764c_4 e^{-9.887482195t}]^T$ **(f)** $\mathbf{x}(t) = [0.8131259057c_1 e^{-0.8074175965t} - 0.1060191245c_1 e^{-6.192582405t} + 0.1313064328c_2 e^{-0.8074175965t} - 0.1313064328c_2 e^{-6.192582405t} - 0.06153846151\cos t - 0.1076923077\sin t - 0.08964923908c_3 e^{-9.887482195t} + 0.7967560202c_3 e^{-1.112517807t} + 0.08058229634c_4 e^{-1.112517807t} - 0.08058229634c_4 e^{-9.887482195t}, -0.8131259057c_1 e^{-0.8074175965t} + 0.1060191245c_1 e^{-6.192582405t} - 0.1313064328c_2 e^{-0.8074175965t} + 0.1313064328c_2 e^{-6.192582405t} + 0.06153846151\cos t + 0.1076923077\sin t - 0.08964923908c_3 e^{-9.887482195t} + 0.7967560202c_3 e^{-1.112517807t} + 0.08058229634c_4 e^{-1.112517807t} - 0.08058229634c_4 e^{-9.887482195t}]^T$

Exercises 12.18–Part A, Page 323

2. $\phi^{-1}(t) = \begin{bmatrix} \frac{2}{3}e^{-3t} + \frac{1}{3}e^{3t} & -\frac{1}{3}e^{3t} + \frac{1}{3}e^{-3t} \\ -\frac{2}{3}e^{3t} + \frac{2}{3}e^{-3t} & \frac{1}{3}e^{-3t} + \frac{2}{3}e^{3t} \end{bmatrix}$

Exercises 12.18–Part B, Page 323

2. **(a)** $\mathbf{x}(t) = \left[\frac{1}{5}t + \frac{4}{25} - \frac{333}{50}e^{-5t} + \frac{1}{2}e^{-t}, \frac{4}{5}t - \frac{4}{25} + \frac{333}{50}e^{-5t} + \frac{3}{2}e^{-t}\right]^T$ **(b)** $\phi(t) = \begin{bmatrix} \frac{3}{4}e^{-5t} + \frac{1}{4}e^{-t} & \frac{1}{4}e^{-t} - \frac{1}{4}e^{-5t} \\ \frac{3}{4}e^{-t} - \frac{3}{4}e^{-5t} & \frac{1}{4}e^{-5t} + \frac{3}{4}e^{-t} \end{bmatrix}$

   (c) $\mathbf{x}_h(t) = \left[(\frac{3}{4}e^{-5t} + \frac{1}{4}e^{-t})c_1 + (\frac{1}{4}e^{-t} - \frac{1}{4}e^{-5t})c_2, (\frac{3}{4}e^{-t} - \frac{3}{4}e^{-5t})c_1 + (\frac{1}{4}e^{-5t} + \frac{3}{4}e^{-t})c_2\right]^T$

   (d) $\mathbf{x}_p(t) = \left[\frac{4}{25} + \frac{1}{5}t - \frac{4}{25}e^{(-5t)}, \frac{4}{5}t - \frac{4}{25} + \frac{4}{25}e^{(-5t)}\right]^T$ **(f)** 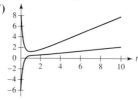   **(g)**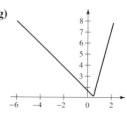

   (e) $c_1 = -6$, $c_2 = 8$

12. (a) $\mathbf{x}(t) = \left[-\frac{4}{25}\cos t + \frac{3}{25}\sin t - \frac{146}{25}e^{-3t}\cos 3t - \frac{197}{25}e^{-3t}\sin 3t, \ \frac{3}{25}\cos t + \frac{4}{25}\sin t + \frac{197}{25}e^{-3t}\cos 3t - \frac{146}{25}e^{-3t}\sin 3t \right]^{\mathrm{T}}$

(b) $\phi = \begin{bmatrix} e^{-3t}\cos 3t & -e^{-3t}\sin 3t \\ e^{-3t}\sin 3t & e^{-3t}\cos 3t \end{bmatrix}$ **(c)** $\mathbf{x}_h(t) = [e^{-3t}\cos(3t)c_1 - e^{-3t}\sin(3t)c_2, \ e^{-3t}\sin(3t)c_1 + e^{-3t}\cos(3t)c_2]^{\mathrm{T}}$

(d) $\mathbf{x}_p(t) = \left[\frac{3}{25}\sin t - \frac{4}{25}\cos t + \frac{3}{25}e^{-3t}\sin(3t) + \frac{4}{25}e^{(-3t)}\cos(3t), \ \frac{4}{25}\sin t + \frac{3}{25}\cos(t) + \frac{4}{25}e^{-3t}\sin(3t) - \frac{3}{25}e^{-3t}\cos(3t) \right]^{\mathrm{T}}$

(e) $c_1 = -6, c_2 = 8$ **(f)** **(g)**

22. (a) $\mathbf{x}(t) = \left[e^{-t} + \frac{211}{26}e^{-5t} - \frac{119}{10}e^{-3t} - \frac{14}{65}\cos t - \frac{8}{65}\sin t, \ \frac{633}{52}e^{-5t} - \frac{119}{10}e^{-3t} + \frac{3}{4}e^{-t} - \frac{3}{130}\cos t - \frac{11}{130}\sin t \right]^{\mathrm{T}}$

(b) $\phi = \begin{bmatrix} -2e^{-5t} + 3e^{-3t} & -2e^{-3t} + 2e^{-5t} \\ 3e^{-3t} - 3e^{-5t} & 3e^{-5t} - 2e^{-3t} \end{bmatrix}$ **(c)** $\mathbf{x}_h(t) = [(-2e^{-5t} + 3e^{-3t})c_1 + (-2e^{-3t} + 2e^{-5t})c_2, \ (3e^{-3t} - 3e^{-5t})c_1 + (3e^{-5t} - 2e^{-3t})c_2]^{\mathrm{T}}$

(d) $\mathbf{x}_p(t) = \left[-\frac{9}{10}e^{-3t} + e^{-t} - \frac{14}{65}\cos t - \frac{8}{65}\sin t + \frac{3}{26}e^{-5t}, \ \frac{3}{4}e^{(-t)} - \frac{3}{130}\cos t - \frac{11}{130}\sin t - \frac{9}{10}e^{-3t} + \frac{9}{52}e^{-5t} \right]^{\mathrm{T}}$ **(e)** $c_1 = -3, c_2 = 1$

(f) **(g)**

28. (a) $\mathbf{x}(t) = \left[-3t + \frac{127}{8} + \frac{595}{8}e^{-4t} - \frac{341}{4}e^{-2t}, \ t - \frac{21}{4} - \frac{119}{4}e^{-4t} + 31e^{-2t} \right]^{\mathrm{T}}$ **(b)** $\phi = \begin{bmatrix} -10e^{-4t} + 11e^{-2t} & \frac{55}{2}e^{-2t} - \frac{55}{2}e^{-4t} \\ -4e^{-2t} + 4e^{-4t} & 11e^{-4t} - 10e^{-2t} \end{bmatrix}$

(c) $\mathbf{x}_h(t) = \left[(-10e^{-4t} + 11e^{-2t})c_1 + (\frac{55}{2}e^{-2t} - \frac{55}{2}e^{-4t})c_2, \ (-4e^{-2t} + 4e^{-4t})c_1 + (11e^{-4t} - 10e^{-2t})c_2 \right]^{\mathrm{T}}$

(d) $\mathbf{x}_p(t) = \left[-\frac{121}{4}e^{(-2t)} - 3t + \frac{127}{8} + \frac{115}{8}e^{(-4t)}, \ t - \frac{21}{4} + 11e^{(-2t)} - \frac{23}{4}e^{(-4t)} \right]^{\mathrm{T}}$ **(e)** $c_1 = 5, c_2 = -4$

(f) **(g)**

34. (a) $\mathbf{x}(t) = \left[\frac{113}{49}t + \frac{576}{343} + \frac{2864}{49}te^{7t} + \frac{110}{343}e^{7t}, \ -\frac{16}{7}t - \frac{80}{49} - \frac{358}{7}te^{7t} + \frac{31}{49}e^{7t} \right]^{\mathrm{T}}$ **(b)** $\phi = \begin{bmatrix} e^{7t} + 56te^{7t} & 64te^{7t} \\ -49te^{7t} & e^{7t} - 56te^{7t} \end{bmatrix}$

(c) $\mathbf{x}_h(t) = [(e^{7t} + 56te^{7t})c_1 + 64te^{7t}c_2, \ -49te^{7t}c_1 + (e^{7t} - 56te^{7t})c_2]^{\mathrm{T}}$

(d) $\mathbf{x}_p(t) = \left[\frac{576}{343} + \frac{113}{49}t - \frac{576}{343}e^{7t} + \frac{512}{49}te^{7t}, \ -\frac{16}{7}t - \frac{64}{7}te^{7t} - \frac{80}{49} + \frac{80}{49}e^{7t} \right]^{\mathrm{T}}$

(e) $c_1 = 2, c_2 = -1$ **(f)** **(g)**

Exercises 12.19, Page 327

2. (a) $\lambda_{1,2} = 5 \pm \sqrt{37}$; Saddle **(b)** $\mathbf{x}_h(t) = c_1 e^{(5+\sqrt{37})t}\left[-\frac{4}{7} - \frac{1}{7}\sqrt{37}, 1 \right]^{\mathrm{T}} + c_2 e^{(5-\sqrt{37})t}\left[-\frac{4}{7} + \frac{1}{7}\sqrt{37}, 1 \right]^{\mathrm{T}}$

(c) $\mathbf{x}_1(t) = \left[(-\frac{11}{74}\sqrt{37} + \frac{1}{2})e^{(5+\sqrt{37})t}(-\frac{4}{7} - \frac{1}{7}\sqrt{37}) + \frac{1}{74}(11 + \sqrt{37})\sqrt{37}e^{(5-\sqrt{37})t}(-\frac{4}{7} + \frac{1}{7}\sqrt{37}), \ (-\frac{11}{74}\sqrt{37} + \frac{1}{2})e^{(5+\sqrt{37})t} + \frac{1}{74}(11 + \sqrt{37})\sqrt{37}e^{(5-\sqrt{37})t} \right]^{\mathrm{T}}$,

$\mathbf{x}_2(t) = \left[(\frac{3}{74}\sqrt{37} + \frac{1}{2})e^{(5+\sqrt{37})t}(-\frac{4}{7} - \frac{1}{7}\sqrt{37}) + \frac{1}{74}(-3 + \sqrt{37})\sqrt{37}e^{(5-\sqrt{37})t}(-\frac{4}{7} + \frac{1}{7}\sqrt{37}), \ (\frac{3}{74}\sqrt{37} + \frac{1}{2})e^{(5+\sqrt{37})t} + \frac{1}{74}(-3 + \sqrt{37})\sqrt{37}e^{(5-\sqrt{37})t} \right]^{\mathrm{T}}$,

$\mathbf{x}_3(t) = \left[(\frac{11}{74}\sqrt{37} - \frac{1}{2})e^{(5+\sqrt{37})t}(-\frac{4}{7} - \frac{1}{7}\sqrt{37}) - \frac{1}{74}(11 + \sqrt{37})\sqrt{37}e^{(5-\sqrt{37})t}(-\frac{4}{7} + \frac{1}{7}\sqrt{37}), \ (\frac{11}{74}\sqrt{37} - \frac{1}{2})e^{(5+\sqrt{37})t} - \frac{1}{74}(11 + \sqrt{37})\sqrt{37}e^{(5-\sqrt{37})t} \right]^{\mathrm{T}}$,

$\mathbf{x}_4(t) = \left[(-\frac{3}{74}\sqrt{37} - \frac{1}{2})e^{(5+\sqrt{37})t}(-\frac{4}{7} - \frac{1}{7}\sqrt{37}) - \frac{1}{74}(-3 + \sqrt{37})\sqrt{37}e^{(5-\sqrt{37})t}(-\frac{4}{7} + \frac{1}{7}\sqrt{37}), (-\frac{3}{74}\sqrt{37} - \frac{1}{2})e^{(5+\sqrt{37})t} - \frac{1}{74}(-3 + \sqrt{37})\sqrt{37}e^{(5-\sqrt{37})t})\right]^{\mathrm{T}}$

(d)
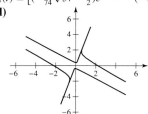

(e) $x' = 9x - 3y,\ y' = -7x + y$;

Check your directions, by looking at (f).

(f)
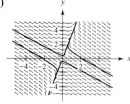

(g) Generate, and plot the separatrices

$e^{(5+\sqrt{37})t}\left[\frac{4}{7} + \frac{1}{7}\sqrt{37}, -1\right]^{\mathrm{T}},\ e^{(5+\sqrt{37})t}\left[-\frac{4}{7} - \frac{1}{7}\sqrt{37}, 1\right]^{\mathrm{T}},$

$e^{(5-\sqrt{37})t}\left[\frac{4}{7} - \frac{1}{7}\sqrt{37}, -1\right]^{\mathrm{T}},\ e^{(5-\sqrt{37})t}\left[-\frac{4}{7} + \frac{1}{7}\sqrt{37}, 1\right]^{\mathrm{T}}.$

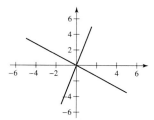

Exercises 12.20–Part A, Page 331

1. $\lambda_{1,2} = \pm 3i$; Center; Orbitally stable

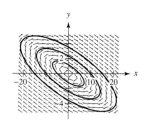

Exercises 12.20–Part B, Page 331

2. **(a)** $\lambda_{1,2} = \frac{3}{2} \pm \frac{1}{2}\sqrt{145}$; Saddle **(b)** Unstable **(c)** $\mathbf{x}_h(t) = c_1 e^{(3/2+1/2\sqrt{145})t}\left[\frac{15}{4} - \frac{1}{4}\sqrt{145}, 1\right]^{\mathrm{T}} + c_2 e^{(3/2-1/2\sqrt{145})t}\left[\frac{15}{4} + \frac{1}{4}\sqrt{145}, 1\right]^{\mathrm{T}}$

(d) $\mathbf{x}_1(t) = [0.7076136993e^{7.520797290t} + 0.2923863008e^{-4.520797290t}, 0.9567501392e^{7.520797290t} + 0.04324986086e^{-4.520797290t})]^{\mathrm{T}}$,

$\mathbf{x}_2(t) = [-0.7076136993e^{7.520797290t} - 0.2923863008e^{-4.520797290t}, -0.9567501392e^{7.520797290t} - 0.04324986086e^{-4.520797290t}]^{\mathrm{T}},\ \mathbf{x}_3(t) =$

$[-0.7076136993e^{7.520797290t} - 0.2923863008e^{-4.520797290t}, -0.9567501392e^{7.520797290t} - 0.04324986086e^{-4.520797290t}]^{\mathrm{T}},\ \mathbf{x}_4(t) =$

$[-0.9532958973e^{7.520797290t} + 1.953295897e^{-4.520797290t}, -1.288932059e^{7.520797290t} + 0.2889320586e^{-4.520797290t}]^{\mathrm{T}}$

(e)

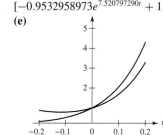
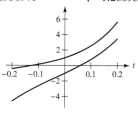
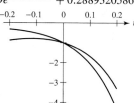
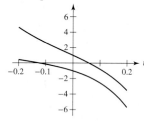

$[\lim_{t\to\infty}x_1(t), \lim_{t\to\infty}y_1(t)]^{\mathrm{T}} = [\infty, \infty]^{\mathrm{T}};\ [\lim_{t\to\infty}x_2(t), \lim_{t\to\infty}y_2(t)]^{\mathrm{T}} = [\infty, \infty]^{\mathrm{T}};\ [\lim_{t\to\infty}x_3(t), \lim_{t\to\infty}y_3(t)]^{\mathrm{T}} = [-\infty, -\infty]^{\mathrm{T}};$

$[\lim_{t\to\infty}x_4(t), \lim_{t\to\infty}y_4(t)]^{\mathrm{T}} = [-\infty, -\infty]^{\mathrm{T}}$ **(f)** See (h). **(h)**

(g) $x' = -6x + 10y,\ y' = -2x + 9y$;

Check your directions by looking at (h).

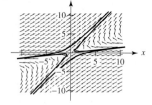

Exercises 12.21–Part A, Page 333

1. $(0, 0)$, $(-2, 1)$; The origin appears to be a saddle

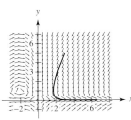

$(-2, 1)$ appears to be a center.   **4.** $x = 20$, $y = 14$
The y-population will vanish.

5. Integrate both sides of the differential equation from 0 to T, to obtain $0 = T - \frac{1}{2}\int_0^T y(t)\,dt$. Now solve for $\bar{y} = \frac{\int_0^T y(t)\,dt}{T}$.

Exercises 12.21–Part B, Page 333

1. $(0, 0)$, $(2, 0)$, $(0, 2)$, $(4, 4)$

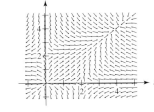

$(0, 0)$ Node out; $(2, 0)$ Saddle; $(0, 2)$ Saddle; $(4, 4)$ Node in.

2. **(a)** $(0, 0)$, $(3, 0)$, $(0, 3)$, $(18, 15)$ **(b)**

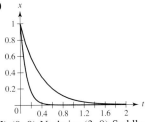

(c)

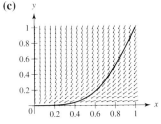

(d) $(0, 0)$ Node in; $(3, 0)$ Saddle; $(0, 3)$ Saddle; $(18, 15)$ Node out.

Exercises 12.22–Part A, Page 336

3. $\begin{bmatrix} 0 & 1 \\ -1 & 0 \end{bmatrix}$; $\lambda_{1,2} = \pm i$, so the origin is a center for the linearized system; $y(t) = c_1 \sin(-(1 + c_1)t + c_2)$, $x(t) = c_1 \cos(-(1 + c_1)t + c_2)$; These are clearly circles, so the origin is a center for the non-linear system.

Exercises 12.22–Part B, Page 336

1. **(a)** $(0, 0)$ **(e)** 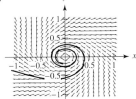   **(f)** It is an outward spiral.   **(h)** $\theta(t) = -t + c_1$; $r(t) = \frac{\pm 1}{\sqrt{c_2 - 2t}}$.

   (b) $A = \begin{bmatrix} 0 & 1 \\ -1 & 0 \end{bmatrix}$; $\lambda_{1,2} = \pm i$
   (d) Center

2. $(0, 0)$, $\lambda = -3, -9$, node in; $(3, 0)$, $\lambda = 3, -15$, saddle; $(0, 3)$, $\lambda = -6, 9$, saddle; $(18, 15)$, $\lambda = 58.37470931, 4.62529069$, node out

Exercises 12.23, Page 338

2. $x' = y$, $y' = -\frac{by}{mL} - \frac{g \sin x}{L}$; Equilibrium points are $[n\pi, 0]$. Use $b = \frac{1}{2}$, $L = 20$, $g = 32$, and $m = \frac{1}{2}$ to generate a plot. $(2n\pi, 0)$ represents a center, while $((2n - 1)\pi, 0)$ represents a saddle.

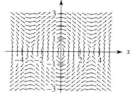

CHAPTER 13

Exercises 13.1, Page 341

1. **(a)** [[1, 3.141592654], [1.1, 3.422941139], [1.2, 3.719500953], [1.3, 4.026698028], [1.4, 4.341088334], [1.5, 4.660005718], [1.6, 4.981341922], [1.7, 5.303404449], [1.8, 5.624821973], [1.9, 5.944479696], [2.0, 6.261474045], [2.1, 6.575080133], [2.2, 6.884727770], [2.3, 7.189983290], [2.4, 7.490535351], [2.5, 7.786183465], [2.6, 8.076828413], [2.7, 8.362463932], [2.8, 8.643169275], [2.9, 8.919102336], [3.0, 9.190493146], [3.1, 9.457637600], [3.2, 9.720891309], [3.3, 9.980663530], [3.4, 10.23741113], [3.5, 10.49163258], [3.6, 10.74386194], [3.7, 10.99466290], [3.8, 11.24462287], [3.9, 11.49434707], [4.0, 11.74445278]]

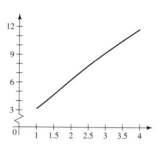

The results of the **rkn4** solution and Maple's solution are indistinguishable.

2. **(a)** $y(t) = \left(\frac{14003}{7225}\cos(\frac{1}{2}\sqrt{19}t) + \frac{1123}{137275}\sqrt{19}\sin(\frac{1}{2}\sqrt{19}t)\right)e^{-3/2t} + \left(\frac{447}{7225} - \frac{2}{85}t\right)\cos(3t) + \left(-\frac{354}{7225} + \frac{9}{85}t\right)\sin(3t)$

(b) -9.59243971344358, 0.347498327359458, 0.352966462263838, 0.354547353937737, 0.354650900447056

(c)

s	$\frac{y_{exact} - y_{approx}}{h^4}$
0	9.94709732136080
1	0.114548488924224
2	0.432933287266304
3	0.451600299970560
4	0.439580764798976

, clearly the method is fourth-order.

Exercises 13.2, Page 342

1.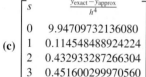

2.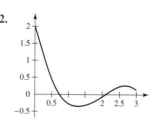

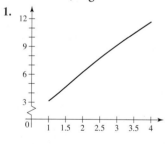

CHAPTER 14

Exercises 14.1–Part A, Page 348

1. $y_2 = -2f_2 = -\frac{2}{\sqrt{t}} + \sqrt{t} - \frac{1}{4}t^{3/2} + \frac{1}{24}t^{5/2} + O(t^{7/2})$; $y_1 = 4f_1 - 4y_2 = t^{3/2} - \frac{1}{6}t^{5/2} + \frac{1}{48}t^{7/2} - \frac{1}{480}t^{9/2} + O(t^{11/2})$

3. $2t^r r^2 + t^r r + 5t^{(r+1)}r + 7t^{(r+1)} - t^r = 0$ Setting the coefficient of t^r equal to zero yields: $2r^2 + r - 1 = 0$.

8. $12a_1 + 2 = 0$; $32a_2 + 4a_1 = 0$; $a_2 = \frac{1}{48}$, $a_1 = -\frac{1}{6}$

Exercises 14.1–Part B, Page 349

2. **(a)** $1 - 2t - \frac{5}{2}t^2 + 5t^3 - \frac{7}{12}t^4 - \frac{34}{15}t^5$ **(b)** $y_{exact} = \frac{3}{26}\cos(2t) + \frac{1}{13}\sin(2t) + \frac{23}{26}e^{-t}\cos(3t) - \frac{11}{26}e^{-t}\sin(3t) \approx 1 - 2t - \frac{5}{2}t^2 + 5t^3 - \frac{7}{12}t^4 - \frac{34}{15}t^5$

(c) same as (a) **(d)** same as (a) **(e)** *Hint:* Write $\cos(2t) = \sum_{m=0}^{\infty} \frac{\cos(\frac{1}{2}m\pi)2^m t^m}{m!}$; $(n^2 + 3n + 2)a_{n+2} + (2n + 2)a_{n+1} + 10a_n = \frac{\cos(\frac{1}{2}n\pi)2^n}{n!}$

(f) same as (e)

(g) same as (a)

(h) (See plot.)

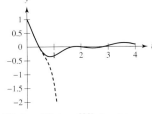

12. (a) $2 - t - \frac{29}{14}t^2 - \frac{635}{294}t^3 - \frac{859}{392}t^4 - \frac{3061}{1960}t^5$ **(b)** same as (a) **(c)** same as (a)
(d) $18a_{n-2} + (5n^2 + 6n + 6)a_n + (-9n^2 - 26n - 17)a_{n+1} + (7n^2 + 21n + 14)a_{n+2} = 0$ **(e)** same as (d) **(f)** same as (a)

27. (a) $y'' + (\frac{4}{3} - \frac{5}{6}\frac{1}{t})y' + (-\frac{7}{6}\frac{1}{t^2} + \frac{2}{3}\frac{1}{t})y = 0$ **(b)** $r^2 - \frac{11}{6}r - \frac{7}{6} = 0;\ \alpha = -\frac{1}{2},\ \beta = \frac{7}{3}$ **(c)** $-8na_n + (-6n^2 + 5n + 11)a_{1+n} = 0;$
$(-8n - \frac{68}{3})b_n + (-23 - 6n^2 - 29n)b_{1+n} = 0$ **(d)** $y_1(t) = \frac{1}{\sqrt{t}};\ y_2(t) = t^{7/3}(1 - \frac{68}{69}t + \frac{136}{261}t^2 - \frac{544}{2835}t^3 + \frac{544}{9963}t^4 - \frac{2176}{171.315}t^5)$
(e) $y(t) = \_C1t^{7/3}(1 - \frac{68}{69}t + \frac{136}{261}t^2 - \frac{544}{2835}t^3 + \frac{544}{9963}t^4 - \frac{2176}{171.315}t^5 + O(t^6)) + \frac{\_C2(1 + O(t^6))}{\sqrt{t}}$

42. (a) $y'' + (-\frac{3}{2} - \frac{2}{t})y' + (\frac{2}{t^2} + \frac{9}{2}\frac{1}{t})y = 0$ **(b)** $r^2 - 3r + 2 = 0;\ \alpha = 1,\ \beta = 2$
(c) $y(t) = \_C1t^2(1 - \frac{3}{4}t + O(t^6)) + \_C2(t\ln(t)(-3t + \frac{9}{4}t^2 + O(t^6)) + t(1 + \frac{3}{2}t - \frac{27}{4}t^2 + \frac{9}{16}t^3 + \frac{9}{128}t^4 + \frac{27}{2560}t^5 + O(t^6)))$
(d) $(-3n + 3)a_n + (2n^2 + 6n + 4)a_{1+n} = 0$ **(e)** $y_1(t) = t^2(1 - \frac{3}{4}t)$ **(f)** $(6 - 3n)b_n + (2n + 2n^2)b_{1+n} = 0$ **(g)** Does not apply
(h) $(6 - 3n)b_n + (2n + 2n^2)b_{1+n} = 0$; Choose $b_0 := 0,\ b_1 := 1;\ t(t - \frac{3}{4}t^2)$, which is essentially y_1

(i) $y_1(t) = \ln(t)t^2(1 - \frac{3}{4}t) - \frac{1}{3}t + (\frac{5}{2} - \frac{4}{3}d_2)t^2 + d_2t^3 - \frac{3}{16}t^4 - \frac{3}{128}t^5 - \frac{9}{2560}t^6$ Observe that $d_2(t^3 - \frac{4t^2}{3}) = -\frac{4d_2t^2(1 - \frac{3t}{4})}{3}$ is a multiple of the
first solution of this differential equation and can be safely omitted or included with any choice for d_2. It turns out that Maple makes the choice
$d_2 = \frac{9}{4}$ resulting in: $\ln(t)t^2(1 - \frac{3}{4}t) - \frac{1}{3}t - \frac{1}{2}t^2 + \frac{9}{4}t^3 - \frac{3}{16}t^4 - \frac{3}{128}t^5 - \frac{9}{2560}t^6$; Observe this is the answer obtained in (c), with $\_C1 = 0$ and
$\_C2 = -\frac{1}{3}$.

Exercises 14.2, Page 352

4. $y_2(t) = \dfrac{e^{-t}(1 - \frac{1}{t} + \frac{3}{2}\frac{1}{t^2} - \frac{3}{t^3} + O(\frac{1}{t^4}))}{t}$ **5. (a)** Observe that $f(t) = \frac{6t^2 - 8t - 9}{4t^2 + 3t + 8}$ and $f(\frac{1}{t}) = -\frac{6 + 8t + 9t^2}{4 + 3t + 8t^2}$ The zeros of the denominator of
$f(\frac{1}{t})$ are: $t = -\frac{3}{16} \pm \frac{1}{16}i\sqrt{119}$; So $f(\frac{1}{t})$ has a Mclaurin series with a radius of convergence $\frac{1}{2}\sqrt{2}$. A similar argument holds for $g(t)$.
(b) $f(t) = \frac{6t^2 - 8t - 9}{4t^2 + 3t + 8} = \frac{3}{2} - \frac{25}{8}\frac{1}{t} - \frac{93}{32}\frac{1}{t^2} + \frac{1079}{128}\frac{1}{t^3} + O(\frac{1}{t^4});\ g(t) = \frac{2t^2 + 9t - 1}{4t^2 + 3t + 8} = \frac{1}{2} + \frac{15}{8}\frac{1}{t} - \frac{85}{32}\frac{1}{t^2} - \frac{225}{128}\frac{1}{t^3} + O(\frac{1}{t^4});$
$a_0 = \frac{3}{2},\ b_0 = \frac{1}{2},\ a_1 = -\frac{25}{8},\ b_1 = \frac{15}{8}$ **(c)** $\{\lambda = -1,\ \alpha = 10\},\ \{\alpha = -\frac{55}{8},\ \lambda = -\frac{1}{2}\}$ **(d)** $y(t) = e^{-t}t^{10}(1 - \frac{118}{t} + \frac{5246}{t^2} - \frac{336.635}{3}\frac{1}{t^3});$
$y(t) = \dfrac{e^{-1/2t}(1 + \frac{4763}{32}\frac{1}{t} + \frac{28,468,561}{2048}\frac{1}{t^2} + \frac{208,343,062,235}{196,608}\frac{1}{t^3})}{t^{55/8}}$
(e) $p(x) = \frac{16x + 15x^2 + 16x^3 - 6}{x^2(4 + 3x + 8x^2)};\ q(x) = -\frac{2 - 9x + x^2}{x^4(4 + 3x + 8x^2)};$ Observe that $p(x)$ has a singularity of order 2 at the origin.

Exercises 14.3–Part A, Page 354

1. $[-3.179841638, 1.260128370, 1.919713268]; [-2.899262132, 1.449631066 - 0.2330036097i, 1.449631066 + 0.2330036097i]$

5. $x = 1 + a + 3a^2 + 12a^3 + 55a^4 + 273a^5 + O(a^6)$

Exercises 14.3–Part B, Page 354

1. (a) (See plot.) **(b)** $1 + a + 3a^2 + 12a^3 + 55a^4 + 273a^5$

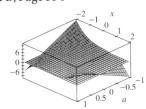

2. (a) $x = 0, 1, -1$ **(b)** **(c)** **(d)** $x_1 = 0 + a + a^3 + 3a^5$
$x_2 = 1 - \frac{1}{2}a - \frac{3}{8}a^2 - \frac{1}{2}a^3 - \frac{105}{128}a^4 - \frac{3}{2}a^5$
$x_3 = -1 - \frac{1}{2}a + \frac{3}{8}a^2 - \frac{1}{2}a^3 + \frac{105}{128}a^4 - \frac{3}{2}a^5$
(e) same as (d)

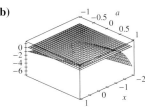

8. **(a)** $x = 1 + ae + a^2 e^2 + \frac{3}{2} a^3 e^3 + \frac{8}{3} a^4 e^4 + \frac{125}{24} a^5 e^5$ **(b)**

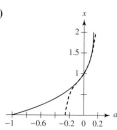

Exercises 14.4–Part B, Page 357

1. **(a)** $y(t) = \dfrac{1}{2} \dfrac{(1 + \sqrt{1 - 4a})e^{1/2(-1+\sqrt{1-4a})t}}{\sqrt{1 - 4a}} + \dfrac{1}{2} \dfrac{(-1 + \sqrt{1 - 4a})e^{-1/2(1+\sqrt{1-4a})t}}{\sqrt{1 - 4a}}$

    (b) $1 + (-t + 1 - e^{-t})a + (-2t + \frac{1}{2}t^2 + 3 - te^{-t} - 3e^{-t})a^2 + (-6t + \frac{3}{2}t^2 - \frac{1}{6}t^3 + 10 - e^{-t}(t + \frac{1}{2}t^2) - 10e^{-t} - 3e^{-t}t)a^3 + O(a^4)$

    (c) same as (b)   **(d)** (See plot.)   **(e)** (See plot.)

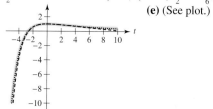 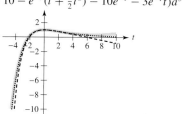

Exercises 14.5, Page 362

5. **(a)** $T = -\frac{193,347}{1,048,576} \pi a^5 A^{10} + \frac{30,345}{131,072} \pi a^4 A^8 - \frac{315}{1024} \pi a^3 A^6 + \frac{57}{128} \pi a^2 A^4 - \frac{3}{4} \pi a A^2 + 2\pi$ **(b)** (See 14.39.)

7. $u_1(\tau) = (\omega_1 A \tau - \frac{3}{8} A^3 \tau) \sin \tau - \frac{1}{32} A^3 \cos \tau + \frac{1}{32} A^3 \cos(3\tau)$ 10. **(a)** $x_0(t) = A \cos t$, $x_1(t) = -\frac{1}{6} A^2 \cos(2t) + \frac{1}{2} A^2 - \frac{1}{3} A^2 \cos t$,

    $x_2(t) = \frac{1}{48} A^3 \cos(3t) + \frac{29}{144} A^3 \cos t + \frac{5}{12} A^3 t \sin t + \frac{1}{9} A^3 \cos(2t) - \frac{1}{3} A^3$; Clearly $x_2(t)$ contains secular terms.

    (b) $\omega = 1 - \frac{5}{12} A^2 a^2 + \frac{5}{18} A^3 a^3$; $U(\tau) = A \cos \tau + (-\frac{1}{6} A^2 \cos(2\tau) + \frac{1}{2} A^2 - \frac{1}{3} A^2 \cos \tau)a + (\frac{1}{48} A^3 \cos(3\tau) + \frac{29}{144} A^3 \cos \tau + \frac{1}{9} A^3 \cos(2\tau) - \frac{1}{3} A^3)a^2 + (-\frac{2}{9} A^4 \cos(2\tau) + \frac{25}{48} A^4 - \frac{1}{432} A^4 \cos(4\tau) - \frac{119}{432} A^4 \cos \tau - \frac{1}{48} A^4 \cos(3\tau))a^3$ **(c)** Use $\tau = \omega t$ with ω as in (b).

    (d) $x(t)$ Yes, the dependence is clearly visible.   **(e)** $x(t)$

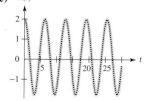 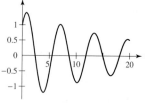

Exercises 14.6, Page 369

3. **(a)** $y(t) = \dfrac{e^{-1/2at} a \cos(\frac{1}{2}\sqrt{4 - a^2}t)}{a - 2} - 2 \dfrac{e^{-1/2at} \cos(\frac{1}{2}\sqrt{4 - a^2}t)}{a - 2} - \dfrac{e^{-1/2at}\sqrt{4 - a^2} \sin(\frac{1}{2}\sqrt{4 - a^2}t)}{a - 2}$

    (b) $y(t) = e^{-1/2at} A_0 \sin(t + \phi_0)$

    (c) With $a = 0.1$ and the initial conditions: $1.450000000 e^{(-0.05000000000t)} \sin(t + 38.46012460)$

6. **(a)** $u' = v$, $v' = -u - au^2$, $u(0) = 1$, $v(0) = 0$; equilibrium points $(0, 0)$ and $(-\frac{1}{a}, 0)$ **(b)** $(0, 0)$, It is a center in the linearized system and can be either a center or a spiral in the non-linear system; $(-\frac{1}{a}, 0)$ is a saddle in the linearized system and will be a saddle in the non-linear system.   **(c)** $y_p(t) = \cos t + (\frac{1}{6} \cos(2t) - \frac{1}{2} + \frac{1}{3} \cos t)a + (\frac{1}{48} \cos(3t) + \frac{29}{144} \cos t + \frac{5}{12} t \sin t + \frac{1}{9} \cos(2t) - \frac{1}{3})a^2$

    (d) $y_L(\tau) = \cos \tau + (-\frac{1}{2} + \frac{1}{6} \cos(2\tau) + \frac{1}{3} \cos \tau)a + (\frac{1}{48} \cos(3\tau) + \frac{29}{144} \cos \tau + \frac{1}{9} \cos(2\tau) - \frac{1}{3})a^2 + (\frac{2}{9} \cos(2\tau) - \frac{25}{48} + \frac{1}{432} \cos(4\tau) + \frac{1}{48} \cos(3\tau) + \frac{119}{432} \cos \tau)a^3$; $y_L(t) = \cos(t(1 - \frac{5}{12} a^2 - \frac{5}{18} a^3)) + (-\frac{1}{2} + \frac{1}{6} \cos(2t(1 - \frac{5}{12} a^2 - \frac{5}{18} a^3)) + \frac{1}{3} \cos(t(1 - \frac{5}{12} a^2 - \frac{5}{18} a^3)))a + (\frac{1}{48} \cos(3t(1 - \frac{5}{12} a^2 - \frac{5}{18} a^3)) + \frac{29}{144} \cos(t(1 - \frac{5}{12} a^2 - \frac{5}{18} a^3)) + \frac{1}{9} \cos(2t(1 - \frac{5}{12} a^2 - \frac{5}{18} a^3)) - \frac{1}{3})a^2 + (\frac{2}{9} \cos(2t(1 - \frac{5}{12} a^2 - \frac{5}{18} a^3)) - \frac{25}{48} + \frac{1}{432} \cos(4t(1 - \frac{5}{12} a^2 - \frac{5}{18} a^3)) + \frac{1}{48} \cos(3t(1 - \frac{5}{12} a^2 - \frac{5}{18} a^3)) + \frac{119}{432} \cos(t(1 - \frac{5}{12} a^2 - \frac{5}{18} a^3)))a^3$ **(e)** $A_0 \sin(t + \phi_0)$ After implementation of the initial conditions: $y_{KB}(t) = \sin(t + \frac{\pi}{2}) = \cos t$

(f)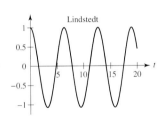

All solutions are very close

(g) Poincare   Lindstedt   Krylov-Bugoliubov

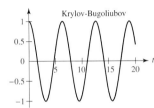

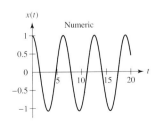

(h)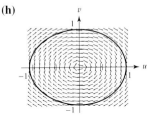

CHAPTER 15

Exercises 15.1–Part A, Page 374

5. $G(0,t) = t(H(t)-1) = 0$ for t on $[0,1]$; $G(1,t) = 0$; $y(x) = \int_0^1 G(x,t)t^2\, dt = -\frac{1}{12}x + \frac{1}{12}x^4$

6. $u(x) = x$ The temperature varies linearly inside the rod.   7. $u(x) = \frac{1}{12}x^4 - \frac{1}{6}x^3 + \frac{1}{12}x$

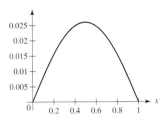

Exercises 15.1–Part B, Page 374

1. **(a)** $y_1(x) = e^{-x}$, $y_1(0) + y_1'(0) = 0$ **(b)** $y_2(x) = 3e^{-x+1} - 2e^{-2x+2}$, $y_2(1) - y_2'(1) = 0$ **(c)** $W(x) = 2e^{-3x+2}$
   (e) $y(x) = \frac{1}{2}x - \frac{3}{4} + \frac{3}{8}e^{-x-1} - \frac{1}{4}e^{-2x} + \frac{3}{8}e^{-x+1}$

2. **(a)** $y(x) = \dfrac{(-1-e^{3/2})e^{-1+x}}{-1+e^{1/2}} + \dfrac{(e+1)e^{-1+3/2x}}{-1+e^{1/2}}$ **(b)**

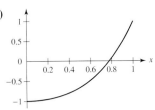

   (c) $y_g(x) = c_1 e^x + c_2 e^{3/2x}$
   (d) $-1 = c_1 + c_2$, $1 = c_1 e + c_2 e^{3/2}$;

   $$c_2 = \frac{(1+e)}{e(e^{1/2}-1)}, c_1 = -\frac{(1+e^{3/2})}{e(e^{1/2}-1)}$$

   (e) $y(x) = \dfrac{(-1-e^{3/2})e^{-1+x}}{-1+e^{1/2}} + \dfrac{(e+1)e^{-1+3/2x}}{-1+e^{1/2}}$

12. **(a)** $y_g(x) = c_1 e^{7/8x} + c_2 e^{9/2x}$ **(b)** $\left(\frac{103}{8} + \frac{119}{8}e^{7/8}\right)c_1 + (\frac{91}{2} + \frac{95}{2}e^{9/2})c_2 = 8$, $(\frac{27}{8} + \frac{43}{4}e^{7/8})c_1 + (\frac{43}{2} + 18e^{9/2})c_2 = -4$;

   $$c_1 = -\frac{5344e^{9/2} - 5664}{-1972 - 1143e^{9/2} + 2709e^{7/8} + 3886e^{43/8}}, c_2 = \frac{1256 + 2328e^{7/8}}{-1972 - 1143e^{9/2} + 2709e^{7/8} + 3886e^{43/8}}$$

   (c) $y(x) = \dfrac{(-5344e^{9/2} - 5664)e^{7/8x}}{-1972 - 1143e^{9/2} + 2709e^{7/8} + 3886e^{43/8}} + \dfrac{(1256 + 2328e^{7/8})e^{9/2x}}{-1972 - 1143e^{9/2} + 2709e^{7/8} + 3886e^{43/8}}$

   (d)

   (e) $y(0) = -0.6477974718$, $y(1) = -0.7448917049$ **(f)** Maple did not succeed.

22. (a) $y(x) = \dfrac{53}{81} + \dfrac{1}{9}x + \dfrac{(-\frac{53}{81}e^{9/8} + \frac{62}{81})e^{-1/5+1/5x}}{-1+e^{37/40}} + \dfrac{(-\frac{62}{81} + \frac{53}{81}e^{1/5})e^{-1/5+9/8x}}{-1+e^{37/40}}$

(b)

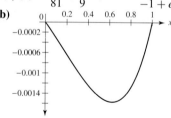

(c) $y_g(x) = \frac{53}{81} + \frac{1}{9}x + c_1 e^{1/5x} + c_2 e^{9/8x}$

(d) $\frac{53}{81} + c_1 + c_2 = 0$, $\frac{62}{81} + c_1 e^{1/5} + c_2 e^{9/8} = 0$; $c_1 = \dfrac{(-\frac{53}{81}e^{9/8} + \frac{62}{81})e^{-1/5}}{-1+e^{37/40}}$, $c_2 = \dfrac{(-\frac{62}{81} + \frac{53}{81}e^{1/5})e^{-1/5}}{-1+e^{37/40}}$

(e) same as (a)

(f) $G(x, t) = \begin{cases} -\dfrac{1}{37}\dfrac{(-e^{1/5t} + e^{9/8t})(e^{-1/5+1/5x} - e^{-9/8+9/8x})}{-e^{53/40t-9/8} + e^{53/40t-1/5}} & 0 \le x < t \\[4mm] -\dfrac{1}{37}\dfrac{(-e^{1/5x} + e^{9/8x})(e^{-1/5+1/5t} - e^{-9/8+9/8t})}{-e^{53/40t-9/8} + e^{53/40t-1/5}} & t < x \le 1 \end{cases}$

(g) $y(x) = -\frac{1}{81}(53 + 9x - 53e^{-37/40} - 9xe^{-37/40} - 53e^{1/5x} + 53e^{9/8x-37/40} + 62e^{-9/8+1/5x} - 62e^{-9/8+9/8x})$

32. (a) $y_h(x) = 0$ **(b)** $y(x) = \frac{1}{26} + \frac{25}{26}e^{-x}\cos(5x) - \frac{27}{26}e^{1/2\pi - x}\sin(5x)$ **(c)** $y_g(x) = \frac{1}{26} + c_1 e^{-x}\cos 5x + c_2 e^{-x}\sin 5x$

(d) $\frac{1}{26} + c_1 = 1$, $\frac{1}{26} + c_2 e^{(-1/2\pi)} = -1$ **(e)** $c_1 = \frac{25}{26}$, $c_2 = -\frac{27}{26}\frac{1}{e^{-1/2\pi}}$; $\frac{1}{26} + \frac{25}{26}e^{-x}\cos 5x - \frac{27}{26}\sin(5x)e^{1/2\pi - x}$

(f)

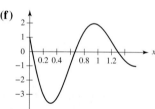

(g) $G(x, t) = \begin{cases} -\frac{1}{5}\sin(5x)\cos(5t)e^{-x+t} & 0 \le x < t \\ -\frac{1}{5}\sin(5t)\cos(5x)e^{-x+t} & t < x \le 1 \end{cases}$; $y(x) = \int_0^1 G(x, t)0.1\,dt$

Exercises 15.2–Part A, Page 378

1. $y(x) = -0.02352941176\cos x + 0.1058823529\sin x + 0.02352941176e^{-x}\cos(3x) + 17.95594223e^{-x}\sin(3x)$

Exercises 15.2–Part B, Page 378

1. $y(x) = (2a+b)e^{-x} - (a+b)e^{-2x}$, where $a = -0.1411201300$, $b = 0.9712219318$

2. (a) $y(x) = \frac{1}{30}xe^{-2x} + \frac{19}{900}e^{-2x} + \dfrac{\left(-\frac{1819}{900}e^{4/3} + \frac{49}{900}e^{-2} - 1\right)e^x}{-e + e^{4/3}} + \dfrac{\left(-\frac{49}{900}e^{-2} + \frac{1819}{900}e + 1\right)e^{4/3x}}{-e + e^{4/3}}$

(b)

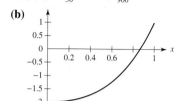

(c) $y'(0) = -0.019379023$

(d) $G(x, t) = \begin{cases} -\dfrac{(-e^x + e^{4/3x})(e^{t-1} - e^{-4/3+4/3t})}{-e^{7/3t-4/3} + e^{7/3t-1}} & 0 \le x < t \\[4mm] -\dfrac{(e^t - e^{4/3t})(-e^{-1+x} + e^{-4/3+4/3x})}{-e^{7/3t-4/3} + e^{7/3t-1}} & t < x \le 1 \end{cases}$

(e) $y(x) = -\frac{1}{900}(19e^{-1} + 30xe^{-1} - 19e^{-4/3} - 30xe^{-4/3} - 1819e^{3x-1} + 1819e^{10/3x-4/3} + 49e^{3x-13/3} - 49e^{10/3x-13/3} - 900e^{-7/3+3x} + 900e^{-7/3+10/3x})e^{-2x+1}/(e^{-1/3} - 1)$ **(f)** $y(x) = \frac{1}{30}xe^{-2x} + \frac{19}{900}e^{-2x} - 8.052974063e^x + 6.031862952e^{4/3x}$

(g) $[[0, -2.], [\frac{1}{10}, -1.988697965], [\frac{1}{5}, -1.944083449], [\frac{3}{10}, -1.857833410], [\frac{2}{5}, -1.720136278], [\frac{1}{2}, -1.519423766], [\frac{3}{5}, -1.242064045], [\frac{7}{10}, -0.8720094058], [\frac{4}{5}, -0.3903906907], [\frac{9}{10}, 0.2249502508], [1, 1.]]$ **(h)**

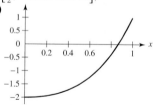

12. (a) $y(x) = \frac{1}{21} \dfrac{(3e^{1/3} - 1 - 41e^{3/2x} + 6e^{-7/6+3/2x} + 123e^{1/3+7/6x} - 6e^{-7/6+7/6x})e^{-x}}{3e^{1/3} - 1}$

(b)

[graph: curve rising from 2 to ~2.05 over x from 0 to 1, y-axis marks 2, 2.01, 2.02, 2.03, 2.04, 2.05; x-axis marks 0.2, 0.4, 0.6, 0.8, 1]

(c) $y'(0) = 0.082870995$

(d) $G(x,t) = \begin{cases} -3 \dfrac{e^{1/6x-1/6+1/6t}(e^{1/3x}-1)(e^{-1/3+1/3t}-3)}{-3e^{2/3t-1/6} + e^{2/3t-1/2}} & 0 \le x < t \\[2mm] -3 \dfrac{e^{1/6x-1/6+1/6t}(e^{1/3t}-1)(e^{-1/3+1/3x}-3)}{-3e^{2/3t-1/6} + e^{2/3t-1/2}} & t < x \le 1 \end{cases}$

(e) $y(x) = \frac{1}{21}(e^{-1/2} - 3e^{-1/6} + 41e^{3/2x-1/2} - 123e^{7/6x-1/6} - 6e^{3/2x-5/3} + 6e^{7/6x-5/3})e^{-x+1/6}/(-3 + e^{-1/3})$ **(f)** $\frac{1}{21}e^{-x} - 0.584e^{1/2x} + 2.537e^{1/6x}$

(g) $[[0, 2.], [\frac{1}{10}, 2.008147986], [\frac{1}{5}, 2.015906774], [\frac{3}{10}, 2.023168603], [\frac{2}{5}, 2.029825774], [\frac{1}{2}, 2.035770074], [\frac{3}{5}, 2.040892224],$
$[\frac{7}{10}, 2.045081350], [\frac{4}{5}, 2.048224471], [\frac{9}{10}, 2.050206003], [1, 2.050907264]]$ **(h)**

[graph: curve rising from 2 to ~2.05 over x from 0 to 1, y-axis marks 2, 2.01, 2.02, 2.03, 2.04, 2.05; x-axis marks 0.2, 0.4, 0.6, 0.8, 1]

27. (a) $y(x) = x\left(\frac{1}{2}\ln x + \frac{1}{2} - \frac{9}{16}\ln 3\right) + \frac{1}{x}\left(-\frac{3}{2} + \frac{9}{16}\ln 3\right)$ **(b)**

[graph: increasing curve passing through origin region, x-axis marks 1, 1.5, 2, 2.5, 3, y-axis marks -1, -0.5, 0, 0.5, 1]

(c) $y'(1) = 1.264061175$

(d) $y_g(x) = \frac{1}{2}\ln(x)x - \frac{1}{4}x + c_1 x + \frac{c_2}{x}$ **(e)** $-\frac{1}{4} + c_1 + c_2 = -1, \frac{3}{2}\ln 3 - \frac{3}{4} + 3c_1 + \frac{1}{3}c_2 = 1; c_1 = \frac{3}{4} - \frac{9}{16}\ln 3, c_2 = \frac{9}{16}\ln 3 - \frac{3}{2}$

(f) $y(x) = \frac{1}{2}\ln(x)x - \frac{1}{4}x + (\frac{3}{4} - \frac{9}{16}\ln 3)x + \frac{\frac{9}{16}\ln 3 - \frac{3}{2}}{x}$ **(g)** $G(x,t) = \begin{cases} \frac{1}{16}\dfrac{(x^2-1)(t^2-9)}{x} & 0 \le x < t \\[2mm] \frac{1}{16}\dfrac{(t^2-1)(x^2-9)}{x} & t < x \le 1 \end{cases}$

(h) $y(x) = -\frac{1}{16}\dfrac{8\ln(x)x^2 - 8x^2 + 24 + 9x^2\ln 3 - 9\ln 3}{x}$ **(i)** $[[1, -1.], [1.2, -0.7681114772], [1.4, -0.5608240471], [1.6, -0.3651906372],$
$[1.8, -0.1744139613], [2, 0.0152936615], [2.2, 0.2061271646], [2.4, 0.3993872182], [2.6, 0.5958490489], [2.8, 0.7959681174], [3, 1.]]$

(j) $\left(\frac{1}{2}x\ln x - 0.1179694045\right)x - \dfrac{0.8820305955}{x}$ **(k)**

[graph: increasing line through x-axis near x=2, x-axis marks 1, 1.5, 2, 2.5, 3, y-axis marks -1, -0.5, 0, 0.5, 1]

37. (a) $y(x) = \frac{4}{689}\cos 2x - \frac{25}{1378}\sin 2x + c_1 e^{7x} + c_2 e^{4/3x}$ **(b)** $\{\frac{191}{689} - 45c_1 - \frac{16}{3}c_2 - \frac{183}{689}\cos 2 - \frac{31}{689}\sin 2 + 47c_1 e^7 + \frac{22}{3}c_2 e^{4/3} = 4,$
$\frac{7}{689} + 15c_1 + \frac{28}{3}c_2 + \frac{107}{689}\cos 2 - \frac{76}{689}\sin 2 - 13c_1 e^7 + 4c_2 e^{4/3} = 2\}; c_1 = 0.0000259604620, c_2 = 0.1030061702$

(c) $y(x) = \frac{4}{689}\cos 2x - \frac{25}{1378}\sin 2x + 0.0000259604620 e^{7x} + 0.1030061702 e^{4/3x}$ **(d)** $y(0) = 0.1088376459, y'(0) = 0.1012388132$

(e) Maple fails to find the answer

[graph: curve rising from ~0.12 to ~0.4 over x from 0 to 1, y-axis marks 0.15, 0.2, 0.25, 0.3, 0.35, 0.4; x-axis marks 0.2, 0.4, 0.6, 0.8, 1]

(f) $[[0, 0.1103462696], [\frac{1}{10}, 0.1214549006], [\frac{1}{5}, 0.1346006006], [\frac{3}{10}, 0.1502529820],$
$[\frac{2}{5}, 0.1689416342], [\frac{1}{2}, 0.1912788802], [\frac{3}{5}, 0.2180092986], [\frac{7}{10}, 0.2501151570],$
$[\frac{4}{5}, 0.2890393433], [\frac{9}{10}, 0.3371563605], [1, 0.3987686582]]$

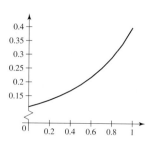

Exercises 15.3–Part A, Page 385

2. $\phi_0 = (-2 + e)x + 2$ 3. $R = 2b + 6cx - e^x$ 5. $y_2(x) = (-2 + e)x + 2 + (-108e + 294)x(x - 1) + (70e - 190)x(x^2 - 1)$

Exercises 15.3–Part B, Page 385

1. (a) $y_G(x) = 0.718281828x + 2. + 0.3703407888x(x - 1.) + 0.3147964947x(x^2 - 1.)$

3. (a) $y_e(x) = \dfrac{2425}{6912} + \dfrac{55}{288}x + \dfrac{1}{24}x^2 - \dfrac{1}{6912}\dfrac{(2425e^{8/5} + 2879)e^{3/5x}}{-e^{3/5} + e^{(8/5}} + \dfrac{1}{6912}\dfrac{(2879 + 2425e^{3/5})e^{8/5x}}{-e^{3/5} + e^{8/5}}$
(b) $R = 24c_2x^3 + (-165c_2 + 24c_1 - 1)x^2 + (-134c_1 + 126c_2 + 24)x + 105c_1 + 55c_2 - 55$
(c)

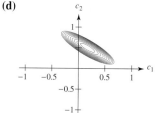

(d)

(e) $y_1(x) = x + 0.1714535005x(x - 1) + 0.5029923293x(x^2 - 1)$

(f) $\psi_1(x) = 105 - 110x + 24x(x - 1)$, $\psi_2(x) = 150x - 165x^2 + 55 + 24x(x^2 - 1)$ In either case the equations to be satisfied are:
$[\frac{33,163}{10}c_2 - \frac{22,043}{10} + \frac{46,913}{15}c_1 = 0, \frac{170,761}{35}c_2 + \frac{33,163}{10}c_1 - \frac{90,679}{30} = 0, c_1 = \frac{1,534,194,599}{8,948,167,223}, c_2 = \frac{13,502,578,411}{26,844,501,669}]$
(g) $y_2(x) = x + \frac{31078}{253,051}x(x - 1) + \frac{130,319}{253,051}x(x^2 - 1)$ (h) $y_3(x) = x + 0.1221067873x(x - 1.) + 0.5131101363x(x^2 - 1.)$
(i) $y_4(x) = x + 0.1167120326x(x - 1.) + 0.4997545775x(x^2 - 1.)$ (j) $y_5(x) = x + 0.1421180937x(x - 1) + 0.5009487037x(x^2 - 1)$
(k)

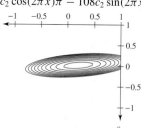

(l) 0.00003743513267, 0.000002775013713, 0.000002717524599,
0.00002709478662, 0.000002436542892
(m) $u = 0.2613605398$, $v = 0.6973911555$

9. (a) $y_e(x) = (-\dfrac{1}{40}x - \dfrac{3}{800} - \dfrac{1}{800}\dfrac{(-23e^{-1} - 800 + 3e^{-7/3})e^{10/9x}}{e^{(1/9)} - e^{(-7/3)}} - \dfrac{1}{800}\dfrac{(23e^{-1} - 3e^{1/9} + 800)e^{-4/3x}}{e^{(1/9)} - e^{(-7/3))}}e^{-x}$
(b) $R = (-7c_1 - 27c_1\pi^2)\sin(\pi x) + 120c_2\cos(2\pi x)\pi - 108c_2\sin(2\pi x)\pi^2 + 60 + 60c_1\cos(\pi x)\pi - xe^{-x} - 7x - 7c_2\sin(2\pi x)$
(c)

(d)

(e) $y_1(x) = x + 0.1971999664\sin(\pi x) + 0.0214251537\sin(2\pi x)$

(f) $\psi_1(x) = -27\sin(\pi x)\pi^2 + 60\cos(\pi x)\pi - 7\sin(\pi x)$; $\psi_2(x) = -108\sin(2\pi x)\pi^2 + 120\cos(2\pi x)\pi - 7\sin(2\pi x)$;
$\int_0^1 R\psi_k\,dx = 0 \Rightarrow [-9507.452925 + 55160.75689c_1 - 63955.03656c_2 = 0, -1242.363810 + 646636.8915c_2 - 63955.03650c_1 = 0]$,
$[c_1 = 0.1971999667, c_2 = 0.02142515383]$; $\frac{\partial S}{\partial c_k} = 0 \Rightarrow [110321.5138c_1 - 19014.90584 - 127910.0731c_2 = 0, 0.1293273784 \times 10^7$
$c_2 - 2484.72762 - 127910.0731c_1 = 0]$, $[c_1 = 0.1971999664, c_2 = 0.0214251537]$
(g) $y_2(x) = x + 0.2395646350\sin(\pi x) + 0.03789949370\sin(2\pi x)$ (h) $y_3(x) = x + 0.2525194691\sin(\pi x) + 0.02413669873\sin(2\pi x)$
(i) $y_4(x) = x + 0.2186372303\sin(\pi x) + 0.02348802950\sin(2\pi x)$ (j) $y_5(x) = x + 0.2403681672\sin(\pi x) + 0.03406883470\sin(2\pi x)$

(k)

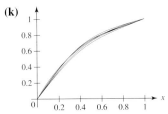

(l) 0.001114407803, 0.0001103894851, 0.0002258804084, 0.0003948232125, 0.0001027296791

(m) $u = 0.2722753697$, $v = 0.6626012022$

14. **(a)** $y_e(x) = 0.1040293916\cos(x)\sin(2.141592654x) + 0.1040293916\sin(x)\cos(2.141592654x) + 0.1943028548\cos(x)\cos(2.141592654x) - 0.1943028548\sin(x)\sin(2.141592654x) + 0.1249244088\sin(x)\cos(4.141592654x) - 0.1206534966\cos(x)\cos(4.141592654x) - 0.1249244088\cos(x)\sin(4.141592654x) - 0.1324728877\cos(x)e^{-4.x} + 1.207571974\sin(x)e^{4.-4.x} + 0.05882352941 - 0.1206534966\sin(x)\sin(4.141592654x)$ **(b)** $R = (17c_1 - c_1\pi^2 + 2)\sin(\pi x) - \frac{1}{4}\sin(\frac{1}{2}\pi x)\pi^2 + 16c_2\cos(2\pi x)\pi - 4c_2\sin(2\pi x)\pi^2 + 4\cos(\frac{1}{2}\pi x)\pi + 8c_1\cos(\pi x)\pi - 1 + 17\sin(\frac{1}{2}\pi x) + 17c_2\sin(2\pi x)$

(c)   **(d)**

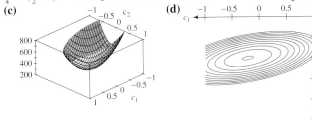

(e) $y_1(x) = \sin(\frac{1}{2}\pi x) - 0.2148730324\sin(\pi x) + 0.002492468535\sin(2\pi x)$ **(f)** $\int_0^1 R\psi_k\,dx = 0 \Rightarrow [341.2486116c_1 + 74.11231366 - 315.8273408c_2 = 0, 1515.948993c_2 - 315.8273408c_1 - 71.64123363 = 0]$, $[c_1 = -0.2148730322, c_2 = 0.002492468591]$; $\frac{\partial S}{\partial c_k} = 0 \Rightarrow [148.2246273 + 682.4972230c_1 - 631.6546816c_2 = 0, -143.2824672 + 3031.897986c_2 - 631.6546816c_1 = 0]$, $[c_1 = -0.2148730324, c_2 = 0.002492468535]$ **(g)** $y_2(x) = \sin(\frac{1}{2}\pi x) + 1.857450548\sin(\pi x) + 1.733129877\sin(2\pi x)$ **(h)** $y_3(x) = \sin(\frac{1}{2}\pi x) + 8.672870801\sin(\pi x) + 4.347237362\sin(2\pi x)$ **(i)** $y_4(x) = \sin(\frac{1}{2}\pi x) + 1.961001914\sin(\pi x) + 1.247392042\sin(2\pi x)$

(j) $y_5(x) = \sin\frac{1}{2}\pi x + 4.311186561\sin\pi x + 2.307418075\sin 2\pi x$ **(k)**

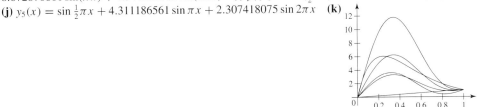

(l) 13.44952427, 3.725889204, 12.14315159, 3.874087058, 0.5505755259   **(m)** 0.3287755369; 0.7135107996

20. **(a)** $y_e(x) = 0.02958579882\cos x - 0.07100591716\sin x - 0.02958579882e^{1.500000000x} + 0.5048607115xe^{-1.500000000+1.500000000x}$

(b) $R = 32c_1 - 3c_2x - 12 - 42c_1x - 36c_2x^2 + 36c_2 + 9x + 9c_1x^2 + 9c_2x^3 - \cos x$

(c)   **(d)**   **(e)** $y_1(x) = x - 0.008532855333x(x-2) + 0.3375053608x(x^2 - 3)$

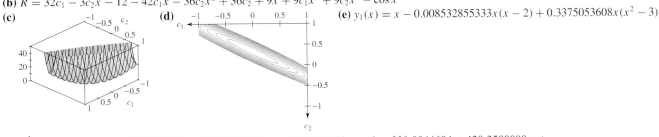

(f) $\int_0^1 R\psi_k\,dx = 0 \Rightarrow [-142.7947960 + 287.2000000c_1 + 430.3500000c_2 = 0, -230.8844684 + 430.3500000c_1 + 694.9714286c_2 = 0]$, $[c_1 = -0.008532849677, c_2 = 0.3375053572]$; $\frac{\partial S}{\partial c_k} = 0 \Rightarrow [-285.5895919 + 574.4000000c_1 + 860.7000000c_2 = 0, -461.7689369 + 860.7000000c_1 + 1389.942857c_2 = 0]$, $[c_1 = -0.008532855668, c_2 = 0.3375053611]$

(g) $y_2(x) = x - 0.1153918451x(x-2) + 0.3934737610x(x^2 - 3)$ **(h)** $y_3(x) = x - 0.09789258545x(x-2.) + 0.3924027147x(x^2 - 3.)$

(i) $y_4(x) = x - 0.0499673048x(x-2.) + 0.3476915661x(x^2 - 3.)$ **(j)** $y_5(x) = x - 0.09065911075x(x-2) + 0.3794035018x(x^2 - 3)$

(k)
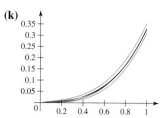

(l) 0.00001541336419, 0.000007488651317, 0.0001692199443, 0.0002172925221, 0.000004204434673   **(m)** $u = 0.2791420730$, $v = 0.7705196768$

28. (a) $y_e(x) = 1.566222222 - 1.026666667x + 0.2000000000x^2 + 3.396425278e^{0.7500000000 - 2.500000000x} - 4.136256271e^{0.7500000000 - 0.7500000000x}$

(b) $R = 26c_1 \sin(\pi x)\pi + (-15c_1 + 8c_1\pi^2)\cos(\pi x) + (-15c_2 + 32c_2\pi^2)\cos(2\pi x) + 52c_2 \sin(2\pi x)\pi + 20x + 26 + 15c_2 + 15c_1 - 3x^2$

(c)
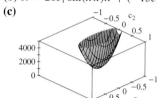

(d)

(e) $y_1(x) = x - 0.2967824228 + 0.2972858395\cos(\pi x) - 0.0005034167296\cos(2\pi x)$ **(f)** $\psi_1(x) = 8\cos(\pi x)\pi^2 + 26\sin(\pi x)\pi + 15 - 15\cos(\pi x)$, $\psi_2(x) = 32\cos(2\pi x)\pi^2 + 52\sin(2\pi x)\pi + 15 - 15\cos(2\pi x)$; $\int_0^1 R\psi_k\,dx = 0 \Rightarrow [7166.164678c_1 + 2130.286014 + 225.0000000c_2 = 0, 58817.24968c_2 + 37.2797265 + 225.0000000c_1 = 0, c_2 = 0.0005034167283, c_1 = -0.2972858395]$; $\frac{\partial S}{\partial c_k} = 0 \Rightarrow [450.0000000c_2 + 4260.572025 + 14332.32935c_1 = 0, 450.0000000c_1 + 117634.4993c_2 + 74.5594528 = 0, c_2 = 0.0005034167296, c_1 = -0.2972858395]$ **(g)** $y_2(x) = x - 33.10971754 + 20.56301239\cos(\pi x) + 12.54670515\cos(2\pi x)$

(h) $y_3(x) = x - 0.1674038060 + 0.3003342164\cos(\pi x) - 0.1329304104\cos(2\pi x)$ **(i)** $y_4(x) = x - 0.1958708135 + 0.2789497518\cos(\pi x) - 0.08307893833\cos(2\pi x)$ **(j)** $y_5(x) = x - 2.694300648 + 1.712663726\cos(\pi x) + 0.981636922\cos(2\pi x)$

(k)
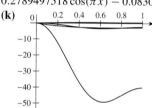

(l) 7.508674812, 1169.917443, 8.280309210, 8.113363516, 0.2766340718

(m) $u = 0.4139848220$; $v = 0.8439175331$

Exercises 15.4–Part A, Page 393

2. $b_1(x) = 3xH(x) - 6xH(x - \frac{1}{3}) + 3xH(x - \frac{2}{3}) + 2H(x - \frac{1}{3}) - 2H(x - \frac{2}{3})$, $b_2(x) = -H(x - \frac{1}{3}) + 4H(x - \frac{2}{3}) + 3xH(x - \frac{1}{3}) - 6xH(x - \frac{2}{3}) + 3xH(x - 1) - 3H(x - 1)$ **4.** $A = \begin{bmatrix} -\frac{34}{9} & \frac{23}{9} \\ \frac{41}{9} & -\frac{34}{9} \end{bmatrix}$

Exercises 15.4–Part B, Page 394

7. (a) $y(x) = \dfrac{21}{100} + \dfrac{1}{10}x + \dfrac{(\frac{169}{100} + \frac{121}{100}e^{5/8})e^{-5/8+2x}}{e^{11/8} - 1} + \dfrac{(-\frac{121}{100}e^2 - \frac{169}{100})e^{-5/8+5/8x}}{e^{11/8} - 1}$ **(b)** $Y'' - \frac{21}{8}Y' + \frac{5}{4}Y = -\frac{29}{8}x + \frac{73}{8}$

(c) $C = [-0.1404524267, 0.1267097136, -0.8653753166, -1.011004235, 0.2063583810, -0.3356858305]$ See the graph in (d).

(d)
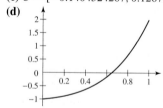

(e) $C = [-0.4895360087, -0.8403407602, -0.9701248947, -0.7534917225]$ See the graph in (g).

(f) $C = [-0.2563615107, -0.4856976182, -0.6807326824, -0.8324902876, -0.9299094158, -0.9593748457, -0.9041426571, -0.7436374746, -0.4525928723]$ See the graph in (g).

(g)

13. (a) $(\dfrac{1}{13} - \dfrac{66}{13}\cos(3x)e^{-2x} + \dfrac{1}{13}\dfrac{(-66\cos(3)e^{-3} - 66\sin(3)e^{-3} + \frac{1}{3}e^{-1} + \frac{26}{3})\sin(3x)e^{3-2x}}{-\sin(3) + \cos(3)})e^{-x}$ **(b)** $Y'' + 6Y' + 18Y = e^{-x} + 78 - 36x$

(c) $[-2.620297908, 0.7879018147, -0.7305695647, 1.674431980, 2.975569321, 3.262824375]$ See (d).

(d)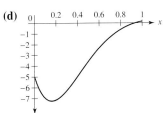

19. (a) $y(x) = \dfrac{1}{13}e^{-2x} + \dfrac{(-\frac{2}{13} - \frac{80}{39}e^{3/5} + 2e^6)e^{-3/5+9/2x}}{2e^{189/10} + 5} + \dfrac{(\frac{80}{39}e^{39/2} - \frac{5}{13} + 5e^6)e^{-3/5-9/5x}}{2e^{189/10} + 5}$ **(d)**

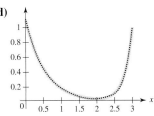

(b) $Y'' - \frac{27}{10}Y' - \frac{81}{10}Y = \frac{1}{10}e^{-2x} + \frac{513}{10} - \frac{81}{5}x; \; u(x) = -2x + 7$

(c) $C = [-6.086198464, -5.499030713, -4.486573561,$
     $-3.381211450, 0.533576468, -1.288960660]$ See the graph in (d).

25. (a) $y(x) = (2xe^x - 4e^x - \frac{3}{2}e^x \cos x + \frac{1}{2}(3\sin 1 - 2)e + \frac{1}{2}(-8e^{-1} + 3\sin 1 - 2)xe)e^{-x}$ **(d)**

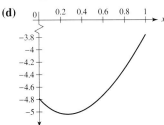

(b) $Y'' + 2Y'' + Y = -6x + 3\sin x - 1 - \frac{5}{2}x^2; \; u(x) = \frac{5}{2}x^2 - 2x$

(c) $[-4.812251130, -4.737266400,$
     $-4.587641202, -4.423532829, -4.293943981, -4.239363745]$ See the graph in (d).

CHAPTER 16

Exercises 16.1–Part A, Page 401

4. $\lambda_n = -\frac{1}{4}\left(\frac{\pi(1+2n)}{L}\right)^2; \; \phi_n = \cos\left(\frac{\pi(1+2n)}{2L}x\right)$ **5.** $L^* = 3\frac{d^2}{dx^2} - 4\frac{d}{dx} + 5; \; J(u,v) = 3(vu' - uv') + 4uv$

Exercises 16.1–Part B, Page 402

1. (a) $y(x) = c_1 e^{\sqrt{\lambda}x} + c_2 e^{-\sqrt{\lambda}x}$ **(b)** $c_1 + c_2 + 3c_1\sqrt{\lambda} - 3c_2\sqrt{\lambda} = 0; \; c_1 e^{\sqrt{\lambda}\pi} + c_2 e^{-\sqrt{\lambda}\pi} + 2c_1\sqrt{\lambda}e^{\sqrt{\lambda}\pi} - 2c_2\sqrt{\lambda}e^{-\sqrt{\lambda}\pi} = 0$

(c) $\begin{bmatrix} 1 + 3\sqrt{\lambda} & 1 - 3\sqrt{\lambda} \\ e^{(\sqrt{\lambda}\pi)} + 2\sqrt{\lambda}e^{\sqrt{\lambda}\pi} & e^{-\sqrt{\lambda}\pi} - 2\sqrt{\lambda}e^{-\sqrt{\lambda}\pi} \end{bmatrix} \begin{bmatrix} c_1 \\ c_2 \end{bmatrix} = \begin{bmatrix} 0 \\ 0 \end{bmatrix}; \; e^{-\sqrt{\lambda}\pi} + \sqrt{\lambda}e^{-\sqrt{\lambda}\pi} - 6\lambda e^{-\sqrt{\lambda}\pi} - e^{\sqrt{\lambda}\pi} + \sqrt{\lambda}e^{\sqrt{\lambda}\pi} + 6\lambda e^{\sqrt{\lambda}\pi} = 0$

(d) $\lambda = 0, \lambda = 0.09758526109$ **(e)** $-2\sin(x\pi) + 2x\cos(x\pi) - 12x^2 \sin(x\pi) = 0$

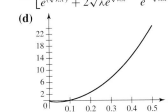
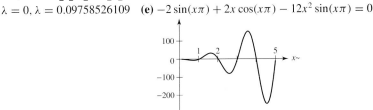

Yes, there are infinitely many purely imaginary solutions.

(f) $\phi_1 = \sin(1.043803405x) - 3.131410215\cos(1.043803405x); \; \phi_2 = \sin(2.025121443x) - 6.075364329\cos(2.025121443x);$
$\phi_3 = \sin(3.017249769x) - 9.051749307\cos(3.017249769x); \; \phi_4 = \sin(4.013076924x) - 12.03923077\cos(4.013076924x);$
$\phi_5 = \sin(5.010514411x) - 15.03154323\cos(5.010514411x); \; \phi_6 = \sin(6.008786212x) - 18.02635864\cos(6.008786212x)$

(g) $\int_0^\pi \phi_i\phi_j \, dx = \begin{bmatrix} 17.5 & 0 & 0 & 0 & 0 & 0 \\ 0 & 60.2 & 0 & 0 & 0 & 0 \\ 0 & 0 & 131. & 0 & 0 & 0 \\ 0 & 0 & 0 & 230. & 0 & 0 \\ 0 & 0 & 0 & 0 & 357. & 0 \\ 0 & 0 & 0 & 0 & 0 & 513. \end{bmatrix}$

(h)

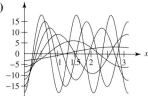

(i) $y(x) = c_1 x + c_2$; There is just the trivial solution $c_1 = 0$ and $c_2 = 0$.

2. (a) $\lambda_n = -\left(\frac{1+2n}{2}\right)^2$; $\phi_n = \cos(\frac{1}{2}(1+2n)x)$ **(b)** $\int_0^\pi \phi_i \phi_j \, dx =$ $\begin{bmatrix} \frac{1}{2}\pi & 0 & 0 & 0 \\ 0 & \frac{1}{2}\pi & 0 & 0 \\ 0 & 0 & \frac{1}{2}\pi & 0 \\ 0 & 0 & 0 & \frac{1}{2}\pi \end{bmatrix}$ The operator is self-adjoint and the eigenfunctions are orthogonal.

(c)

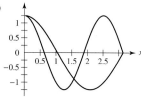

(d) Yes, the eigenspaces are one-dimensional.

12. (a) $\lambda_n = \frac{1}{4} + n^2$; $\phi_n = e^{-1/2x} \sin(nx)$ **(b)** $\int_0^\pi \phi_i \phi_j \, dx =$ $\begin{bmatrix} 0.3827144327 & 0.2086427837 & 0.06753784106 & 0.03209888979 \\ 0.2086427837 & 0.4502522738 & 0.2407416735 & 0.08274906653 \\ 0.06753784106 & 0.2407416735 & 0.4654634992 & 0.2503713404 \\ 0.03209888979 & 0.08274906653 & 0.2503713404 & 0.4710331479 \end{bmatrix}$

The operator is not self-adjoint, and the eigenfunctions are not orthogonal.   **(c)**

(d) Yes, the eigenspaces are one-dimensional.

24. (a) $L^* = x^2 \frac{d^2}{dx^2} + x \frac{d}{dx} - 6$; $J(u, v) = x^2(vu' - uv') + xuv$ **(b)** The operator is not self-adjoint.
   (c) $J(u, v)\big|_{x=0}^{x=\pi} = \pi^2(v(\pi)u'(\pi) - u(\pi)v'(\pi)) + \pi u(\pi)v(\pi)$ **(d)** 0   **(e)** $\pi u(\pi)v(\pi)$ **(f)** $\pi u(\pi)v(\pi)$ **(g)** $\pi u(\pi)v(\pi)$
   (h) $\pi u(\pi)v(\pi)$ **(i)** $\pi^2(v(0)u'(0) - u(0)v'(0)) + \pi u(0)v(0)$

Exercises 16.2–Part A, Page 406

2. $p(x) = \frac{1}{x}$, $q(x) = 1 - \frac{v^2}{x^2}$ **5.** $30, 360, \frac{15}{8}$

Exercises 16.2–Part B, Page 407

11. (b) $\mu_1 = 4.809651115$, $\mu_2 = 11.04015622$, $\mu_3 = 17.30745583$, $\mu_4 = 23.58306888$, $\mu_5 = 29.86183542$, $\mu_6 = 36.14212794$,
   $\mu_7 = 42.42327326$, $\mu_8 = 48.70494306$, $\mu_9 = 54.98695826$, $\mu_{10} = 61.26921294$

(c)

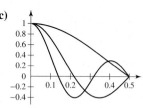

(d) $\int_0^{1/2} J_0(\mu_i r) J_0(\mu_j r) dr =$ $\begin{bmatrix} 0.03368926550 & 0 & 0 \\ 0 & 0.01447251732 & 0 \\ 0 & 0 & 0.009210793888 \end{bmatrix}$

(e) $A_1 = 0.4894450750$, $A_2 = 0.1179577229$,
$A_3 = 0.06720759094$, $A_4 = 0.03925210731$,
$A_5 = 0.02918459202$, $A_6 = 0.02093283345$,
$A_7 = 0.01711565354$, $A_8 = 0.01345662186$,
$A_9 = 0.01155372437$, $A_{10} = 0.009569030722$

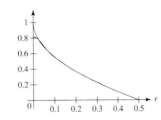

22. $y'' + \frac{y'}{x} + \frac{(x^2 - v^2)y}{x^2} = 0$

Exercises 16.3, Page 411

2.

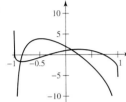

7. $c_0 = 0$, $yc_1 = 0.7377095900$, $c_2 = 0$, $c_3 = -0.03369954076$, $c_4 = 0$,
$c_5 = 0.0004341503171$, $c_6 = 0$, $c_7 = -0.2623664554 \times 10^{-5}$, $c_8 = 0$,
$c_9 = 0.9193460530 \times \times 10^{-8}$, $c_{10} = 0$; $\int_{-1}^{1} \sin^2 x \, dx - \sum_{k=0}^{10} c_k^2 = -0.1 \times 10^{-9}$

12. $c_0 = -0.3535533905$, $c_1 = 0.8164965810$, $c_2 = -0.1976423538$, $c_3 = -0.4677071734$,
$c_4 = 0.04419417381$, $c_5 = 0.2931509850$, $c_6 = -0.01991804498$, $c_7 = -0.2139541240$,
$c_8 = 0.01138857792$, $c_9 = 0.1685581954$, $c_{10} = -0.007383656395$, $c_{11} = -0.1390907516$,
$c_{12} = 0.005179004743$, $c_{13} = 0.1184079515$, $c_{14} = -0.003834846687$, $c_{15} = -0.1030868341$,
$c_{16} = 0.002954451223$, $c_{17} = 0.09127989485$, $c_{18} = -0.002346291566$, $c_{19} = -0.08190159100$, $c_{20} = 0.001908531303$;
$\int_{-1}^{1} f^2 \, dx - \sum_{k=0}^{20} c_k^2 = 0.062101493$

17. (a) $\Phi(z) = 1 - (-2 + z)^2$ **(b)**

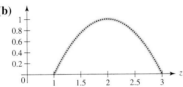

28. (a) $Q_0(x) = \operatorname{arctanh}(x)$, $Q_1(x) = x(\frac{1}{2}\ln(-\frac{1+x}{x-1}) - \frac{1}{x})$, $Q_2(x) = (\frac{3}{2}x^2 - \frac{1}{2})(\frac{1}{2}\ln(-\frac{1+x}{x-1}) - 3\frac{x}{3x^2-1})$,
$Q_3(x) = (\frac{5}{2}x^3 - \frac{3}{2}x)(\frac{1}{2}\ln(-\frac{1+x}{x-1}) - \frac{4}{9}\frac{1}{x} - \frac{25}{9}\frac{x}{5x^2-3})$

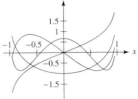

(b) Substitute the functions into the equation and simplify.   **(c)** $w(x) = \frac{C}{L_k(x)^2}$

Exercises 16.4, Page 414

1. (a) $\lambda_k = \frac{9}{4} + k^2\pi^2$, $\varphi_k(x) = e^{-3/2x}\sin(k\pi x)$

(b) $\{\mu_k\} = [11.94074806, 40.02724532, 83.77299040, 138.8958450, 200.0000000,$
$261.1041550, 316.2270096, 359.9727547, 388.0592519]$;
$\mathbf{u}_1 = [0.2350421184, 0.3843638658, 0.4548226193, 0.4596754192, 0.4155329789,$
$0.3397600920, 0.2484758729, 0.1552050363, 0.07015038021]^T$

(c)

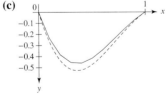

3. (a) $\{\lambda_k\} = [11.00687363, 39.26502026, 83.22320794, 138.6074491, 200.0000000,$
$261.3925509, 316.7767921, 360.7349797, 388.9931264]$;
$\mathbf{u}_1 = [0.1585935874, 0.2967633452, 0.3974907951, 0.4498736215, 0.4506144855,$
$0.4040480561, 0.3207975853, 0.2154398977, 0.1037340939]^T$, $\mathbf{u}_2 = [0.3070358890,$
$0.4886277961, 0.4750006069, 0.2810904526, -0.00402875055, -0.2604871836,$
$-0.3914216907, -0.3615873767, -0.2047124490]^T$

(b)

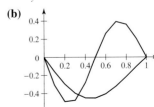

10. (a) $\{\lambda_k\} = [-0.2494863806, -2.208669595, -5.935257520, -11.06446576,$
$-17.09421169, -23.43426175, -29.46400768, -34.59321592, -38.31980385, -40.27898706];$
$\mathbf{u}_1 = [-0.4264014350, -0.4211517255, -0.4055318640, -0.3799264578, -0.3449660059,$
$-0.3015113434, -0.2506324721, -0.1935821975, -0.1317652873, -0.06670387898]^T$

(b)
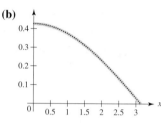

18. (a) $\{\lambda_k\} = [1.231051496, 4.073642324, 8.501081338, 14.07997996, 20.26423672,$
$26.44849348, 32.02739210, 36.45483112, 39.29742194];$
$\mathbf{u}_1 = [0.2402468162, 0.3900368246, 0.4582011971, 0.4597445082, 0.4125930182,$
$0.3349190487, 0.2431659819, 0.1507910303, 0.06766293567]^T$

(b)
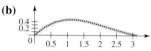

CHAPTER 17

Exercises 17.1–Part A, Page 420
2. $\mathbf{R}(p) = [3, -1, 2]^T + p[2, 8, -6]^T$ **3.** $y = -\frac{13}{2} + \frac{3}{2}x$

Exercises 17.1–Part B, Page 420
3. (a) $\mathbf{R}(p) = [p, 2p^2 - 3p + 1]^T$
(b) $[1, -3]^T, [1, 1]^T, [1, 5]^T$
(c) $\sqrt{10}, \sqrt{2}, \sqrt{26}$
(d) $\left[\frac{1}{10}\sqrt{10}, -\frac{3}{10}\sqrt{10}\right]^T, \left[\frac{1}{2}\sqrt{2}, \frac{1}{2}\sqrt{2}\right]^T, \left[\frac{1}{26}\sqrt{26}, \frac{5}{26}\sqrt{26}\right]^T$

(e)
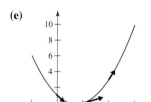

(f) $[1, 4p - 3]^T$ **(g)** $\sqrt{10 + 16p^2 - 24p}$

(h)
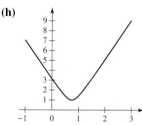

(i) $p_m = \frac{3}{4}, \left(\frac{3}{4}, -\frac{1}{8}\right)$

13. (a) $\left[\sqrt{2}, \frac{1}{10}\sqrt{90}\right]^T, \left[2, \frac{1}{10}\sqrt{10}\right]^T$
(b) $\frac{7}{60}\sqrt{10}, \frac{1}{20}\sqrt{385}$
(c) $\left[\frac{3}{7}\sqrt{5}, -\frac{2}{7}\right]^T, \left[\frac{1}{77}\sqrt{385}, -\frac{6}{77}\sqrt{154}\right]^T$

(d)
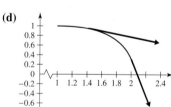

(e) $\mathbf{R}'(p) = \left[\frac{1}{2}\frac{1}{\sqrt{1+p}}, -\frac{p}{\sqrt{100-10p^2}}\right]^T$

(f) $\rho(p) = \frac{1}{10}\sqrt{-\frac{250-15p^2+10p^3}{(1+p)(-10+p^2)}}$

(g)
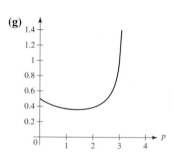

(h) $p_m = 1.400894313, (1.549481950, 0.8965207819)$

25. **(a)** $y - \frac{1}{2}\sqrt{2} = \frac{1}{2}\sqrt{2}(x - \frac{1}{4}\pi)$ **(e)** $3\sqrt{2} - 1 = 3\sqrt{2} - 1$

(b) $(\frac{1}{8}(24 - 4\sqrt{2} + \sqrt{2}\pi)\sqrt{2}, 3)$

(c) $\left[\frac{1}{8}(24 - 4\sqrt{2} + \sqrt{2}\pi)\sqrt{2} - \frac{1}{4}\pi, 3 - \frac{1}{2}\sqrt{2}\right]^{\mathrm{T}}$

(d) $\mathbf{R}'(\frac{\pi}{4}) = \left[1, \frac{1}{2}\sqrt{2}\right]^{\mathrm{T}}$

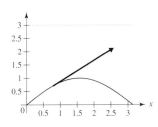

Exercises 17.2, Page 425

1. $\frac{5}{8}\sqrt{26} - \frac{1}{8}\ln(-5 + \sqrt{26}) + \frac{3}{8}\sqrt{10} - \frac{1}{8}\ln(-3 + \sqrt{10}) \approx 4.8891$

11. **(a)** 1.096434414 **(b)** 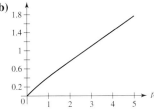   **(c)**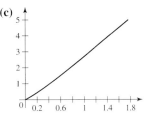

23. **(a)** 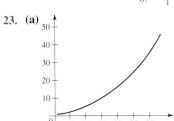   **(b)** $s(p) = 9p + \frac{1}{2}p^2$

(c) $\frac{ds}{dp} = 9 + p; \frac{dp}{ds} = \frac{1}{9+p}$

(d) $\left[-80 + 9\sqrt{81 + 2s} - \frac{1}{2}(-9 + \sqrt{81 + 2s})^2, 1 + 4(-9 + \sqrt{81 + 2s})^{3/2}\right]^{\mathrm{T}}$

(e) $\left[\frac{9}{\sqrt{81+2s}} - -\frac{9+\sqrt{81+2s}}{\sqrt{81+2s}}, 6\frac{\sqrt{-9+\sqrt{81+2s}}}{\sqrt{81+2s}}\right]^{\mathrm{T}}$

29. **(a)** $f(g(x)) = x = g(f(x))$ **(b)** $f'(x) = \cosh(x); g'(x) = \frac{1}{\sqrt{x^2+1}}$ **(c)** $\cosh(\mathrm{arcsinh}(x)) = \sqrt{x^2 + 1}$.

Exercises 17.3–Part A, Page 427

1. $\kappa = \frac{2}{25}\sqrt{5}$ **2.** $\kappa = \frac{2}{289}\sqrt{17}$

Exercises 17.3–Part B, Page 428

1. **(a)** $\mathbf{B_A} = \left[-\frac{8}{5}, \frac{16}{5}\right]^{\mathrm{T}}; \mathbf{B_{\perp A}} = \left[\frac{18}{5}, \frac{9}{5}\right]^{\mathrm{T}}$

2. **(a)** $\kappa(x) = 4\dfrac{e^{2x}}{(1 + 4e^{4x})^{3/2}}$ **(b)** 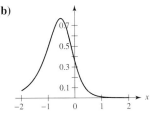   **(c)** $\kappa(0) = \frac{4}{25}\sqrt{5}$

(d) $\kappa = 4\dfrac{e^{2x}\sqrt{1 + 4e^{4x}}}{8e^{4x} + 16e^{8x} + 1}$

(e) $\mathbf{R}(p) = [p, e^{2p}]^{\mathrm{T}}$

(f) $v(p) = \sqrt{1 + 4e^{4p}}; \mathbf{T}(p) = \left[\frac{1}{\sqrt{1+4e^{4p}}}, 2\frac{e^{2p}}{\sqrt{1+4e^{4p}}}\right]^{\mathrm{T}}$

(g) $\kappa(0) = \frac{4}{25}\sqrt{5}$

7. **(a)** $\dfrac{2}{(8x^2 - 12x + 5)^{3/2}}$ **(b)** 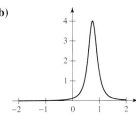   **(c)** $\kappa(0) = \frac{1}{25}\sqrt{10}$

(d) $x = \frac{3}{4}$: absolute maximum

(e) $\kappa = \dfrac{\sqrt{10 + 16x^2 - 24x}}{(5 + 8x^2 - 12x)^2}$

(f) $\mathbf{R}(p) = [p, 2p^2 - 3p + 1]^{\mathrm{T}}$

(g) $v(p) = \sqrt{10 + 16p^2 - 24p}; \mathbf{T}(p) = \left[\frac{1}{\sqrt{10+16p^2-24p}}, \frac{4p-3}{\sqrt{10+16p^2-24p}}\right]^{\mathrm{T}}$

(h) $\kappa(0) = \frac{1}{25}\sqrt{10}$

29. (a) $\mathbf{T}(s) = \left[\dfrac{9}{\sqrt{81+2s}}, -\dfrac{9+\sqrt{81+2s}}{\sqrt{81+2s}}, 6\dfrac{\sqrt{-9+\sqrt{81+2s}}}{\sqrt{81+2s}} \right]^{\mathrm{T}}$ **(b)** $\kappa = \dfrac{3}{\sqrt{-9+\sqrt{81+2s}}\,(81+2s)}$ **(c)** $\kappa(p) = \dfrac{3}{(p+9)^2\sqrt{p}}$ **(d)** $\kappa(s(p)) = \dfrac{3}{(81+18p+p^2)\sqrt{p}}$

35. (a) $(h, k) = \left(-\frac{5}{2}, \frac{9}{4}\right)$ **(b)**   **(c)** (See plot.)

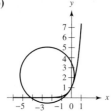   

(d) $(x + \frac{5}{2})^2 + (y - \frac{9}{4})^2 = \frac{125}{16}$, the point $(0, 1)$ is on the circle.   **(e)** Yes, both slopes are 2 at $(0, 1)$.   **(f)** Yes, the second derivatives agree and equal 4 at $(0, 1)$ **(g)** No, the third derivatives do not agree. The curve has a third derivative 8 at $(0, 1)$, while the circle of curvature has a third derivative $\frac{96}{5}$ at $(0, 1)$

43. (a) $x^3 - \frac{1}{4}y^2 = 0$ **(b)**

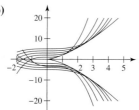

48. (a) $y = 3(2^{-4/3})x^{2/3} + \frac{1}{2}$ **(d)**

(b) $y - \mu^2 + \frac{1}{2}\frac{x-\mu}{\mu} = 0$

(c) $y = 3(2^{-4/3})x^{2/3} + \frac{1}{2}$

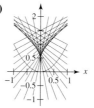

Exercises 17.4–Part A, Page 432

1. $\mathbf{N} = \left[-3\dfrac{p}{\sqrt{4+9p^2}}, 2\dfrac{1}{\sqrt{4+9p^2}}\right]^{\mathrm{T}}$ **2.** $\mathbf{T} = \left[-\frac{1}{2}\sqrt{2}\sin p, \frac{1}{2}\sqrt{2}\cos p, \frac{1}{2}\sqrt{2}\right]^{\mathrm{T}}$; $\mathbf{N} = [-\cos p, -\sin p, 0]^{\mathrm{T}}$

Exercises 17.4–Part B, Page 433

2. (a) See (e).   **(e)**   **(f)** $\kappa(1) = \sqrt{2}$

(b) $\mathbf{R} = [p, 2p^2 - 3p + 1]^{\mathrm{T}}$

(c) $\mathbf{T}(1) = \left[\frac{1}{2}\sqrt{2}, \frac{1}{2}\sqrt{2}\right]^{\mathrm{T}}$ **(g)** $\mathbf{T}'(p) = \left[-2\dfrac{4p-3}{(5+8p^2-12p)\sqrt{10+16p^2-24p}}, 2\dfrac{1}{(5+8p^2-12p)\sqrt{10+16p^2-24p}}\right]^{\mathrm{T}}$

(d) $\mathbf{N}(1) = \left[-\frac{1}{2}\sqrt{2}, \frac{1}{2}\sqrt{2}\right]^{\mathrm{T}}$ **(h)** $\mathbf{N}(1) = \left[-\frac{1}{2}\sqrt{2}, \frac{1}{2}\sqrt{2}\right]^{\mathrm{T}}$

12. (a) See (d).   **(d)**   **(e)** $\kappa = \frac{255}{3481}\sqrt{118}$ **(f)** $\mathbf{N}(1) = \left[-\frac{2}{59}\sqrt{59}, \frac{1}{59}\sqrt{3245}\right]^{\mathrm{T}}$

(b) $\mathbf{T}(1) = \left[\frac{1}{59}\sqrt{3245}, \frac{2}{59}\sqrt{59}\right]^{\mathrm{T}}$

(c) $\mathbf{N}(1) = \left[-\frac{2}{59}\sqrt{59}, \frac{1}{59}\sqrt{3245}\right]^{\mathrm{T}}$

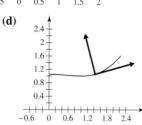

23. (a) $\mathbf{R}' = [6p, 2, 3p^2]^{\mathrm{T}}$; $\mathbf{R}'' = [6, 0, 6p]^{\mathrm{T}}$ **(b)** $\mathbf{T}(p) = \left[6\frac{p}{\sqrt{36p^2+4+9p^4}}, 2\frac{1}{\sqrt{36p^2+4+9p^4}}, 3\frac{p^2}{\sqrt{36p^2+4+9p^4}} \right]^{\mathrm{T}}$; $\mathbf{T}(1) = \left[\frac{6}{49}\sqrt{49}, \frac{2}{49}\sqrt{49}, \frac{3}{49}\sqrt{49} \right]^{\mathrm{T}}$

(c) $\kappa(1) = \frac{6}{343}\sqrt{17}$ **(d)** $\mathbf{T}'(p) = \left[-6\frac{9p^4-4}{(36p^2+4+9p^4)^{3/2}}, -36\frac{p(2+p^2)}{(36p^2+4+9p^4)^{3/2}}, 12\frac{p(9p^2+2)}{(36p^2+4+9p^4)^{3/2}} \right]^{\mathrm{T}}$; $\mathbf{T}'(1) = \left[-\frac{30}{343}, -\frac{108}{343}, \frac{132}{343} \right]^{\mathrm{T}}$

(e) $\mathbf{N}(1) = \left[-\frac{5}{119}\sqrt{17}, -\frac{18}{119}\sqrt{17}, \frac{22}{119}\sqrt{17} \right]^{\mathrm{T}}$ **(f)** $\mathbf{B}(1) = \left[\frac{2}{17}\sqrt{17}, -\frac{3}{17}\sqrt{17}, -\frac{2}{17}\sqrt{17} \right]^{\mathrm{T}}$

(g) $\mathbf{T}(p) = \left[6\frac{p}{\sqrt{36p^2+4+9p^4}}, 2\frac{1}{\sqrt{36p^2+4+9p^4}}, 3\frac{p^2}{\sqrt{36p^2+4+9p^4}} \right]^{\mathrm{T}}$; $\mathbf{B}(p) = \left[2\frac{p}{\sqrt{4p^2+9p^4+4}}, -3\frac{p^2}{\sqrt{4p^2+9p^4+4}}, -2\frac{1}{\sqrt{4p^2+9p^4+4}} \right]^{\mathrm{T}}$;

$\mathbf{N}(p) = \left[-\frac{9p^4-4}{\sqrt{4p^2+9p^4+4}\sqrt{36p^2+4+9p^4}}, -6\frac{p(2+p^2)}{\sqrt{4p^2+9p^4+4}\sqrt{36p^2+4+9p^4}}, 2\frac{p(9p^2+2)}{\sqrt{4p^2+9p^4+4}\sqrt{36p^2+4+9p^4}} \right]^{\mathrm{T}}$

30. (a), (b) and **(c)** $\tau = -2\frac{1}{4p^2+9p^4+4}$ **(d)**

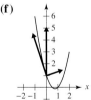

37. (a) $\frac{d\mathbf{T}}{ds} = \left[-6\frac{9p^4-4}{(36p^2+4+9p^4)^2}, -36\frac{p(2+p^2)}{(36p^2+4+9p^4)^2}, 12\frac{p(9p^2+2)}{(36p^2+4+9p^4)^2} \right]^{\mathrm{T}}$; $\frac{d\mathbf{N}}{ds} = \left[-8\frac{p(288p^2+1080p^4+80+648p^6+405p^8)}{(36p^2+4+9p^4)^2(4p^2+9p^4+4)^{3/2}}, 6-\frac{48p^2+112p^4+81p^{10}+1224p^6+486p^8-32}{(36p^2+4+9p^4)^2(4p^2+9p^4+4)^{3/2}}, \right.$

$\left. -2-\frac{432p^2-2448p^4+729p^{10}-504p^6+486p^8-32}{(36p^2+4+9p^4)^2(4p^2+9p^4+4)^{3/2}} \right]^{\mathrm{T}}$; $\frac{d\mathbf{B}}{ds} = \left[-2\frac{9p^4-4}{(4p^2+9p^4+4)^{3/2}\sqrt{36p^2+4+9p^4}}, -12\frac{p(2+p^2)}{(4p^2+9p^4+4)^{3/2}\sqrt{36p^2+4+9p^4}}, 4\frac{p(9p^2+2)}{(4p^2+9p^4+4)^{3/2}\sqrt{36p^2+4+9p^4}} \right]^{\mathrm{T}}$

Exercises 17.5–Part A, Page 436

1. Use $\mathbf{a} = \rho'\mathbf{T} + \kappa\rho^2\mathbf{N}$ and take the dot product with $\mathbf{T}$. Observe that the dot product of $\mathbf{T}$ and $\mathbf{N}$ will vanish.   **2.** $\kappa = \frac{1}{2}\frac{1}{(1+p^2)^{3/2}}$

Exercises 17.5–Part B, Page 436

2. (a)   **(b)** $\mathbf{R}''(p) = [0, 4]^{\mathrm{T}}$

(c) $\mathbf{T}(p) = \left[\frac{1}{\sqrt{10+16p^2-24p}}, \frac{4p-3}{\sqrt{10+16p^2-24p}} \right]^{\mathrm{T}}$; $\mathbf{N}(p) = \left[-\frac{4p-3}{\sqrt{10+16p^2-24p}}, \frac{1}{\sqrt{10+16p^2-24p}} \right]^{\mathrm{T}}$;

$\rho(p) = \sqrt{10+16p^2-24p}$;] $\rho'(p) = 4\frac{4p-3}{\sqrt{10+16p^2-24p}}$; $\kappa = \frac{\sqrt{2}}{(8p^2-12p+5)^{3/2}}$

(d) Yes, it is. $\rho'\mathbf{T} + \kappa\rho^2\mathbf{N} = [0, 4]^{\mathrm{T}} = \mathbf{R}''$

(e) $\mathbf{T}(0) = \left[\frac{1}{10}\sqrt{10}, -\frac{3}{10}\sqrt{10} \right]^{\mathrm{T}}$, $\mathbf{N}(0) = \left[\frac{3}{10}\sqrt{10}, \frac{1}{10}\sqrt{10} \right]^{\mathrm{T}}$, $\mathbf{R}''(0)_{\mathbf{T}(0)} = \left[-\frac{6}{5}, \frac{18}{5} \right]^{\mathrm{T}} = \rho'(0)\mathbf{T}(0)$, $\mathbf{R}''(0)_{\mathbf{N}(0)} = \left[\frac{6}{5}, \frac{2}{5} \right]^{\mathrm{T}} = \kappa(0)\rho(0)^2\mathbf{N}(0)$

(f)   **(g)** $\frac{d\rho}{dx} = 4\frac{4x-3}{\sqrt{10+16x^2-24x}}$; $\kappa\rho^2 = 2\frac{\sqrt{10+16x^2-24x}}{5+8x^2-12x}$

13. (a)   **(b)** $\mathbf{R}''(p) = \left[-\frac{1}{4}\frac{1}{(1+p)^{3/2}}, \frac{10}{(10+p^2)\sqrt{100+10p^2}} \right]^{\mathrm{T}}$

(c) $\mathbf{T}(p) = \left[\frac{\sqrt{5}\sqrt{10+p^2}}{\sqrt{50+7p^2+2p^3}}, \frac{\sqrt{1+p}\,p\sqrt{2}}{\sqrt{50+7p^2+2p^3}} \right]^{\mathrm{T}}$; $\mathbf{N}(p) = \left[-\frac{\sqrt{1+p}\,p\sqrt{2}}{\sqrt{50+7p^2+2p^3}}, \frac{\sqrt{5}\sqrt{10+p^2}}{\sqrt{50+7p^2+2p^3}} \right]^{\mathrm{T}}$;

$\rho(p) = \frac{1}{10}\frac{\sqrt{5}\sqrt{50+7p^2+2p^3}}{\sqrt{1+p}\sqrt{10+p^2}}$; $\rho'(p) = -\frac{1}{4}\frac{\sqrt{5}(p^4-8p^3+4p^2-8p+100)}{\sqrt{50+7p^2+2p^3}(1+p)^{3/2}(10+p^2)^{3/2}}$; $\kappa(p) = 5\frac{\sqrt{2}(p^3+30p+20)}{(50+7p^2+2p^3)^{3/2}}$

(d) Yes, it is. $\rho'\mathbf{T} + \kappa\rho^2\mathbf{N} = \left[-\frac{1}{4(1+p)^{3/2}}, \frac{\sqrt{10}}{(10+p^2)^{3/2}} \right]^{\mathrm{T}} = \mathbf{R}''$

(e) $\mathbf{R}''(1)_{\mathbf{T}(1)} = \left[-\frac{445}{10384}\sqrt{2}, -\frac{89}{57112}\sqrt{110} \right]^{\mathrm{T}} = \rho'(1)\mathbf{T}(1)$, $\mathbf{R}''(1)_{\mathbf{N}(1)} = \left[-\frac{51}{2596}\sqrt{2}, \frac{51}{5192}\sqrt{110} \right]^{\mathrm{T}} = \kappa(1)\rho(1)^2\mathbf{N}(1)$

(f)

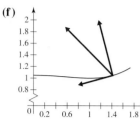

Note: the vectors in the accompanying figure have been scaled for visibility.

25. **(a)** $v(t_0) = \sqrt{29}$ **(b)** $\mathbf{T} = \left[\frac{3}{29}\sqrt{29}, -\frac{4}{29}\sqrt{29}, \frac{2}{29}\sqrt{29}\right]^{\mathrm{T}}$ **(c)** $v' = \frac{41}{29}\sqrt{29}$ **(d)** $\kappa = \frac{1}{841}\sqrt{117769}$ **(e)**
$\mathbf{B} = \left[-\frac{24}{4061}\sqrt{4061}, \frac{11}{4061}\sqrt{4061}, \frac{58}{4061}\sqrt{4061}\right]^{\mathrm{T}}$ **(f)** $\mathbf{N} = \left[\frac{254}{117769}\sqrt{117,769}, \frac{222}{117,769}\sqrt{117,769}, \frac{63}{117,769}\sqrt{117,769}\right]^{\mathrm{T}}$

Exercises 17.6, Page 439

1. 13333.33333 lb   **2.** 17997.99989 mi/hr   **6.** 5.401897895 revolutions per minute; 30.08000000 ft/s^2

8. $\mathbf{a} = [-64, 32]^{\mathrm{T}}$ **11.** 8.400000000 m/s   **14.** $\mathbf{F} = \left[-105,840\frac{\cos p}{14,161\cos^4 p + 5950\cos^2 p + 625}, -44100\frac{\sin p}{14161\cos^4 p + 5950\cos^2 p + 625}\right]^{\mathrm{T}}$

CHAPTER 18

Exercises 18.1–Part A, Page 442

2.

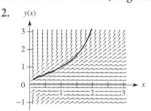

3.

Exercises 18.1–Part B, Page 443

2. (a) $(0, 0)$ **(b)**

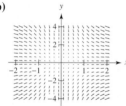

(c)

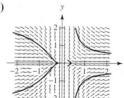

(d) $(0, 0)$ is unstable.

(e) The eigenvalues of the linearized system are -1 and 0; **(f)**
no conclusion can be drawn regarding the type of
critical point of the non-linear system.

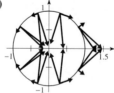

21.

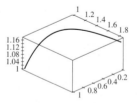

Exercises 18.2–Part A, Page 445

1. $D_{\mathbf{i}}f = \nabla f \cdot \mathbf{i} = f_x$, and $D_{\mathbf{j}}f = \nabla f \cdot \mathbf{j} = f_y$ **2.** $\frac{8}{13}\sqrt{13}$ **4.** $-\frac{58}{69}\sqrt{69}$

Exercises 18.2–Part B, Page 445

2. (a)

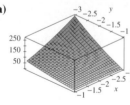

(b)

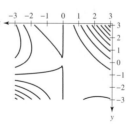

(c) $\nabla f = [-14xy - 6y^2 - 18x, -7x^2 - 12xy]$

(d) 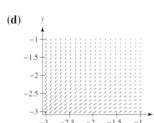   **(e)** 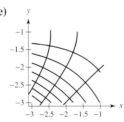   **(f)** Tail: $(-2, -2)$, Head: $(-46, -78)$ **(g)** $\frac{580}{73}\sqrt{73}$

12. $\frac{42{,}920}{1913}\sqrt{1913}$ **22.** $-\frac{15{,}248}{681}\sqrt{681}$ **32.** $\frac{3358}{1309}\sqrt{1309}$

44. (a) $-\frac{1}{4}\cos\theta + \frac{1}{2}\sin\theta$ **(b)** 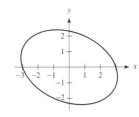   **(c)** $-\arctan 2 + \pi$ **(d)** $\left[-\frac{1}{5}\sqrt{5}, \frac{2}{5}\sqrt{5}\right]^{\mathsf{T}}$ **(e)** $\left[-\frac{1}{5}\sqrt{5}, \frac{2}{5}\sqrt{5}\right]^{\mathsf{T}}$

Exercises 18.3–Part A, Page 451

1. (a) $3x^2 + 2xy + 5y^2 = 27$

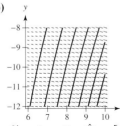

(b) $\mathbf{T} = \left[1, -\frac{5}{11}\right]^{\mathsf{T}}$
(c) $[10, 22]^{\mathsf{T}}$
(d) $(11, 24)$
(e) $2\sqrt{146}$ **(f)** $[-10, -22, 1]^{\mathsf{T}}$

Exercises 18.3–Part B, Page 451

2. (a) See (b).   **(b)**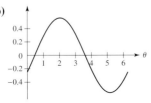

(c) Tail: $(8, -10)$; Head: $(1036, -266)$ **(g)**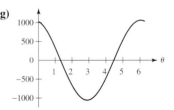
(d) $\mathbf{T} = \left[1, \frac{257}{64}\right]^{\mathsf{T}}$
(e) $\nabla f(8, -10) \cdot \mathbf{R}'(8, -10) = 0$
(f) $\delta(\theta) = 1028\cos\theta - 256\sin\theta$

(h) $\hat{\theta} = -\arctan\frac{64}{257} + 2\pi$ **(i)** $\mathbf{u}(\hat{\theta}) = \left[\frac{1}{70{,}145}\sqrt{4{,}633{,}007{,}105}, -\frac{64}{18{,}027{,}265}\sqrt{4{,}633{,}007{,}105}\right]^{\mathsf{T}}$
(j) $\frac{\nabla f(P)}{\|\nabla f(P)\|} = \left[\frac{257}{70{,}145}\sqrt{70{,}145}, -\frac{64}{70{,}145}\sqrt{70{,}145}\right]^{\mathsf{T}}$ **(k)** $\delta(\hat{\theta}) = \|\nabla f(P)\| = \frac{4}{257}\sqrt{4{,}633{,}007{,}105}$

14. (a) $[44, 280, -20]^{\mathsf{T}}$ **(f)**
(b) Tail: $(-11, -2, 3)$; Head: $(33, 278, -17)$
(c) $\mathbf{T}_x = [1, 0, \frac{11}{5}]$; $\mathbf{T}_y = [0, 1, 14]$
(d) $\nabla f \cdot \mathbf{T}_x = \nabla f \cdot \mathbf{T}_y = 0$
(e) $\delta(\theta, \varphi) = 44\cos(\theta)\sin(\phi) + 280\sin(\theta)\sin(\phi) - 20\cos(\phi)$

(g) $(\hat{\theta}, \hat{\phi}) = (\arctan\frac{70}{11}, \pi - \arctan\frac{1}{5}\sqrt{5021})$
(h) $\mathbf{u}(\hat{\theta}, \hat{\phi}) = \left[\frac{11}{174}\sqrt{6}, \frac{35}{87}\sqrt{6}, -\frac{5}{174}\sqrt{6}\right]^{\mathsf{T}}$
(i) $\frac{\nabla g(P)}{\|\nabla g(P)\|} = \left[\frac{11}{174}\sqrt{6}, \frac{35}{87}\sqrt{6}, -\frac{5}{174}\sqrt{6}\right]^{\mathsf{T}}$
(j) $\delta(\hat{\theta}, \hat{\varphi}) = \|\nabla g(P)\| = 116\sqrt{6}$

Exercises 18.4–Part A, Page 457

1. $\frac{17}{83}\sqrt{83}$

Exercises 18.4–Part B, Page 457

1. $\frac{30}{83}\sqrt{83}$; $\left(-\frac{7}{83}, -\frac{16}{83}, \frac{39}{83}\right)$; $[1, -2, 3]^{\mathsf{T}} - \left[-\frac{7}{83}, -\frac{16}{83}, \frac{39}{83}\right]^{\mathsf{T}} = \frac{30}{83}[3, -5, 7]^{\mathsf{T}}$

5. Min at $\left(\frac{1}{3}\sqrt{3}, \frac{1}{3}\sqrt{3}\right)$, $\left(-\frac{1}{3}\sqrt{3}, -\frac{1}{3}\sqrt{3}\right)$; Max at $(-1, 1)$, $(1, -1)$; (See plot.)   **9.** Max 6 at $(0, 2, 2)$, Min 2 at: $(0, -2, 2)$

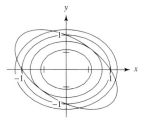

10. Max $\frac{49}{9}$ at $\left(\frac{4}{9}, \frac{1}{3}\sqrt{23}, \frac{2}{5}\right)$, $\left(\frac{4}{9}, -\frac{1}{3}\sqrt{23}, \frac{2}{5}\right)$, Min $-\frac{4}{3}\sqrt{13}$ at $\left(-\frac{8}{39}\sqrt{13}, 0, -\frac{12}{65}\sqrt{13}\right)$

Exercises 18.5–Part A, Page 460

1. $u(x) = \frac{1}{2}\frac{1}{x^2}$ **2. (a)** false   **(b)** true $u(x, y) = xe^{xy}y + c$ **(c)** true: $u(x, y) = \sin(xy)y + c$

3. $\frac{\partial}{\partial y}\left(\frac{xf\left(\sqrt{x^2+y^2}\right)}{\sqrt{x^2+y^2}}\right) = \frac{f'(\sqrt{x^2+y^2})yx}{x^2+y^2} - \frac{f(\sqrt{x^2+y^2})xy}{(x^2+y^2)^{3/2}} = \frac{\partial}{\partial x}\left(\frac{yf(\sqrt{x^2+y^2})}{\sqrt{x^2+y^2}}\right)$

Exercises 18.5–Part B, Page 460

2. (a)

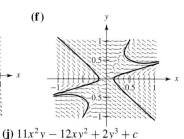

(b) (See Plot.)   **(c)** $[-22xy + 12y^2, -6y^2 - 11x^2 + 24xy]^{\mathrm{T}}$ **(d)** $(0, 0)$

(e)   **(f)**   **(g)** The case of an eigenvalue zero is not covered by this text.

(h) The system exhibits a saddle like behavior.

(i)   **(j)** $11x^2y - 12xy^2 + 2y^3 + c$

13. $u(x, y) = -11xy^2 - 12x^2y - 3y + c$ **28. (a)** $\mathbf{F} = [-5, 22yz - 8z, 11y^2 - 8y]^{\mathrm{T}}$ **(b)**

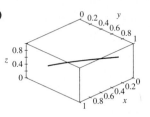

40. $u(x, y, z) = -4x^2z - 3xz^2 + 7yz^2 + c$

CHAPTER 19

Exercises 19.1–Part A, Page 466

1. 24; 48   **3.** $\frac{3008}{35}$ **5.** *Hint:* Write $x = x(t)$, $y = y(t)$, and $\mathbf{r}(t) = [x(t), y(t)]^{\mathrm{T}}$, then expand the integral on the left.

Exercises 19.1–Part B, Page 467

2. $-\dfrac{2,684,016,702}{5}$; $\dfrac{5,333,015,562,273}{2}$ **12.** $\dfrac{5,696,142,704,640}{7}$; $-\dfrac{66,795,668,098,653,968}{1155}$

22. (a)

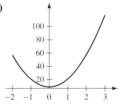

(b) Use a direction field.

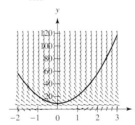

$\mathbf{F}(1, 21) = [-1910, 80710]^{\mathrm{T}}$; **(c)** $2,723,935,890$

$\mathbf{F_T}(1, 21) = \left[\dfrac{1,935,130}{577}, \dfrac{46,443,120}{577} \right]^{\mathrm{T}}$

32. (a)

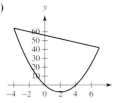

(b)

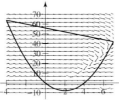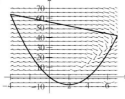

$\mathbf{F}(2, -8) = [648, -254]^{\mathrm{T}}$; $\mathbf{F_T}(2, -8) = [648, 0]^{\mathrm{T}}$ **(c)** $-\dfrac{244,129,358}{35}$

44. (a)

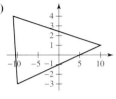

(b) (See plot.)

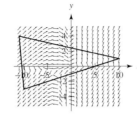

$\mathbf{F}(0, 17/7) = [4, 1]^{\mathrm{T}}$, $\mathbf{F}(0, -1) = [4, 1]^{\mathrm{T}}$; **(c)** -5592

$\mathbf{F_T}(0, 17/7) = \left[\dfrac{189}{50}, -\dfrac{27}{50} \right]^{\mathrm{T}}$,

$\mathbf{F_T}(0, -1) = \left[\dfrac{105}{26}, \dfrac{21}{26} \right]^{\mathrm{T}}$

Exercises 19.2–Part A, Page 470

1. $\dfrac{1}{6}$ **5.** -4π

Exercises 19.2–Part B, Page 471

1. (a)

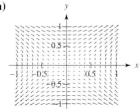

(b)

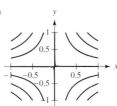

(c) $(0, 0)$ is a saddle. **(d)**

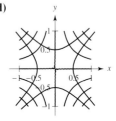

3. (a)

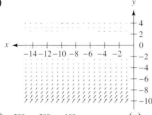

(b) $(0, 0)$, $\left(\dfrac{512}{1331}, \dfrac{8}{11} \right)$ Classification of the first point is not possible because linearization produces an eigenvalue zero. The second point is a saddle.

(c)

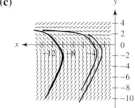

(d)

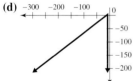

(e) $-\dfrac{41797}{2}\pi$

13. (a)

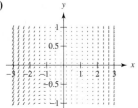

(b) $(-0.2071067810, 0.4571067810)$, $\lambda_1 = 2.594521170$, $\lambda_2 = -2.180307608$: saddle; $(1.207106781, -0.9571067810)$, $\lambda_{1,2} = -1.207106781 \pm 2.049328541i$: spiral sink.

(c)

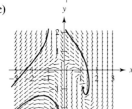

(d)

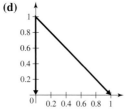

(e) 0

23. (a)

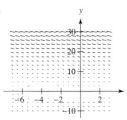

(b) and (c)

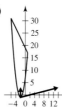

(d) $\frac{249,361}{15}$

33. (a)

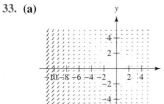

(b)

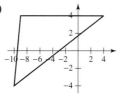

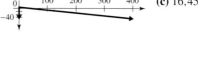

(c) 16,458

43. (a)

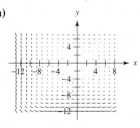

(b)

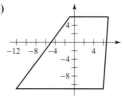

(c) $-\frac{57,970}{3}$

CHAPTER 20

Exercises 20.1–Part A, Page 476

1. (a) $\nabla \cdot \mathbf{F} = y^2 + x^2$ **(b)** $\nabla \cdot \mathbf{F} = y + z^2 + x^3$ **5.** $\nabla^2 u = 2y^3 z^4 + 6x^2 yz^4 + 12x^2 y^3 z^2$

Exercises 20.1–Part B , Page 476

4. $-\frac{14,036,560}{\sqrt{108,541}}$ **15. (a)**

(b)

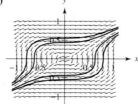

(c) 0   Yes, **F** is solenoidal.

26. (a) 58   **(b)** 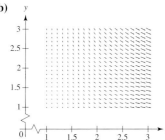   **(c)** 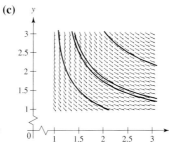   **(d)** $\frac{21}{4}\rho^4\pi + 58\rho^2\pi$ **(e)** 58   **(f)** $232\rho^2 + 28\rho^4$ **(g)** 58

38. (a) 100   **(b)** 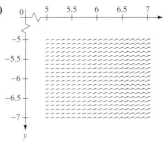   **(c)** 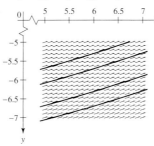   **(d)** $400\rho^2\pi$ **(e)** 100

48. $\frac{55{,}065}{121}$ **58.** $-112{,}959\mathbf{i} - 125{,}883\mathbf{j}$

Exercises 20.2–Part A, Page 479

1. (a) $(y\cos x - xe^y)\mathbf{k}$ **(b)** $(x - y)\mathbf{i} + (x - y)\mathbf{j}$ **4. (a)** irrotational   **(b)** irrotational

Exercises 20.2–Part B, Page 479

4. *Hint:* First compute the divergence of the cross product of **A** and **B**, then apply the condition that both fields are rotation free.

7. (a) **F** is neither solenoidal nor irrotational.   **(b)** 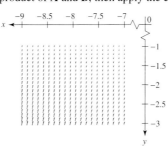   **(c)** 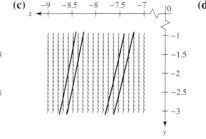   **(d)** $-77\mathbf{k}$

   (e) $\frac{3}{4}\rho^4\pi - 77\rho^2\pi$ **(f)** -77 **(g)** $-308\rho^2 + 4\rho^4$ **(h)** -77

18. (a) F is solenoidal, but not irrotational.   **(b)** 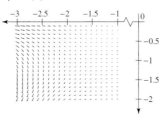   **(c)**   **(d)** $16\mathbf{k}$

   (e) $64\rho^2\pi - 272\rho^4\pi$ **(f)** 16 **27.** $-\frac{14{,}574}{\sqrt{6329}}$

Exercises 20.3, Page 483

1. The gradient of a vector, the divergence of a scalar, and the rotation of a scalar are all undefined.

2. (a) $\nabla(fg) = \left(-6\frac{x(-16z^2 + 8x^2z - 36xz + 15x^3)}{-4 + 3z}\right)\mathbf{i} + \left(6\frac{x^2(-64z + 24z^2 + 8x^2 - 48x + 9x^3)}{(-4 + 3z)^2}\right)\mathbf{k}$

   (b) $\nabla(\mathbf{u} \cdot \mathbf{v}) = (18xy^3 - 9y^4 - 6xz^2y - 9z^2y^2 + 108x^2zy - 216x^2 + 48xz + 32z^5 - 24xz^4 + 32yz^4 - 24xyz^3)\mathbf{i} + (27x^2y^2 - 36xy^3 - 3x^2z^2 - 18xz^2y + 36x^3z + 32xz^4 - 12x^2z^3)\mathbf{j} + (-6x^2zy - 18xzy^2 + 24x^2 + 36x^3y + 160xz^4 - 48x^2z^3 + 128xyz^3 - 36x^2yz^2)\mathbf{k}$

(c) $\nabla \cdot (f\mathbf{v}) = -108xz^2y + 54z^2y^2 - 243x^2zy + 162xzy^2 - 240x^2z^3 + 45x^3z^2 - 240xz^4$ **(d)** $\nabla \cdot (\mathbf{u} \times \mathbf{v}) =$
$-27xy^2 - 45x^2z^2 - 72xz^2y + 18x^2y^2 - 48xyz^3 + 72x^2zy - 108xzy^2 + 64xyz + 162xz^3 + 27x^2y - 16x^2z + 64xz^2 - 24z^4 + 4x^2y^3 - 32y^2z^2x$
(e) $\nabla \times (f\mathbf{v}) = (72x^2yz^2 - 96x^2z + 72x^3yz - 72x^3)\mathbf{i} + (-108x^2zy + 108xzy^2 - 81x^3y + 81x^2y^2 + 48z^5 + 108xz^4 - 81x^2z^3)\mathbf{j} +$
$(-48xyz^3 + 96xz^2 - 108x^2yz^2 + 216x^2z + 54x^2z^2 - 108xz^2y + 81x^3z - 162x^2zy)\mathbf{k}$
(f) $\nabla \times (\mathbf{u} \times \mathbf{v}) = (6x^2y^2z + 146x^2y - 8x^2z^3 - 81x^3 + 27x^2z - 54xyz - 72x^2zy + 72xzy^2 - 36x^2y^2 + 36xy^3 + 40xz^4 + 24y^2z^2x)\mathbf{i} +$
$(-146xy^2 - 54x^3z + 243x^2z^2 - 48x^2yz^2 - 8xyz^3 - 54xyz + 27zy^2 - 96xz^3 + 275x^2y + 64x^2z + 16xz^2 - 8y^3xz - 32x^2y^2z)\mathbf{j} + (72xz^2y - $
$36z^2y^2 + 72xzy^2 - 36y^3z - 8y^2z^3 + 6y^2z^2x - 8z^5 + 6xz^4 + 32x^2yz^2 - 32x^2z + 16x^2z^3)\mathbf{k}$

Exercises 20.4–Part A, Page 487

2. $\alpha = -1$ makes $\nabla \cdot \mathbf{v} = 0$ **3.** $2x\mathbf{i} + 2y\mathbf{j}$

Exercises 20.4–Part B, Page 487

2. (a) $\nabla \cdot \nabla f = \nabla^2 f = \frac{1}{98}$ **(c)** $(\nabla f) \times (\nabla g) = \nabla \times (f\nabla g) = -\nabla \times (g\nabla f) = \frac{93}{98}\mathbf{i} - \frac{45}{49}\mathbf{j} + \frac{29}{98}\mathbf{k}$
17. $u(x, y, z) = -6x - \frac{5}{2}x^2 + c(y, z)$ **24.** $\alpha = -10$

CHAPTER 21

Exercises 21.1–Part A, Page 491

2. $13\sqrt{6}$ **3. (a)** $d\sigma = \sqrt{1 + g_y{}^2 + g_z{}^2}\,dy\,dz$ **(b)** $d\sigma = \sqrt{1 + h_x{}^2 + h_z{}^2}\,dx\,dz$ **4.** $\frac{315}{2}\sqrt{155}$

Exercises 21.1–Part B, Page 491

2. (a) $\frac{4}{9}\pi\sqrt{155}$ **(b)** $\frac{109}{18}\sqrt{155}$ **(c)** $\frac{343}{54}\sqrt{155}$ **(d)** $\frac{6517}{648}\sqrt{155}$ **(e)** $\frac{5}{3}\pi\sqrt{62}$ **(f)** $\frac{9}{7}\pi\sqrt{155}$ **(g)** $5\pi\sqrt{155}$ **(h)** $\frac{27}{7}\sqrt{155}$ **(i)** $\frac{48}{5}\sqrt{155}$

6. 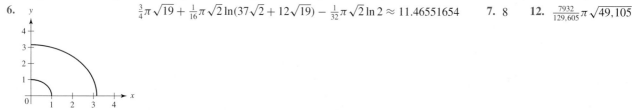 $\frac{3}{4}\pi\sqrt{19} + \frac{1}{16}\pi\sqrt{2}\ln(37\sqrt{2} + 12\sqrt{19}) - \frac{1}{32}\pi\sqrt{2}\ln 2 \approx 11.46551654$ **7.** 8   **12.** $\frac{7932}{129,605}\pi\sqrt{49,105}$

Exercises 21.2–Part A, Page 496

2. $\frac{25}{672}\pi\sqrt{5} - \frac{1}{3360}\pi$ **4.** $\frac{3}{2}\pi$

Exercises 21.2–Part B, Page 496

5. -1728π **15.** $\frac{1,892,625}{8}$ **25.** 0

Exercises 21.3–Part A, Page 501

1. $\iiint_V \nabla \cdot \mathbf{F}\,dv = \iint_S \mathbf{F} \cdot \mathbf{N}d\sigma = 0$.   **2.** $\iint_S [\nabla \times \mathbf{F}] \cdot \mathbf{N}d\sigma = \oint_C \mathbf{F} \cdot d\mathbf{r} = 0$
4. In section 20.3 it was proved that $\nabla \cdot (f\mathbf{F}) = f\nabla \cdot \mathbf{F} + \mathbf{F} \cdot \nabla f$, the divergence theorem then gives $\iint_S f\mathbf{F} \cdot \mathbf{N}d\sigma = \iiint_V \nabla \cdot (f\mathbf{F})\,dv$.

Exercises 21.3–Part B, Page 501

2. $\iiint_V \nabla \cdot \mathbf{F}\,dv = \iint_S \mathbf{F} \cdot \mathbf{N}d\sigma = 0$.   **9.** $\iiint_V \nabla \cdot \mathbf{F}\,dv = \iint_S \mathbf{F} \cdot \mathbf{N}d\sigma = \frac{64}{3}\pi$ **16.** $\iiint_V \nabla \cdot \mathbf{F}\,dv = \iint_S \mathbf{F} \cdot \mathbf{N}d\sigma = 4\pi$
23. $\iiint_V \nabla \cdot \mathbf{F}\,dv = \iint_S \mathbf{F} \cdot \mathbf{N}d\sigma = 0$ **30.** $\iint_S [\nabla \times \mathbf{F}] \cdot \mathbf{N}d\sigma = \oint_C \mathbf{F} \cdot d\mathbf{r} = 0$

Exercises 21.4–Part A, Page 505

3. (a) $\iint_R \nabla \cdot \mathbf{F}\,dx\,dy = \oint_C \mathbf{F} \cdot \mathbf{N}ds = \frac{35}{6}\sqrt{5}$ **(b)** $\iint_R [\nabla \times \mathbf{F}] \cdot \mathbf{k}\,dx\,dy = \oint_C \mathbf{F} \cdot d\mathbf{r} = -\frac{215}{84}\sqrt{5}$ **(c)** The area is $\frac{5}{6}\sqrt{5}$
4. (a) $\iint_R \nabla \cdot \mathbf{F}\,dx\,dy = \oint_C \mathbf{F} \cdot \mathbf{N}ds = 0$ **(b)** $\iint_R [\nabla \times \mathbf{F}] \cdot \mathbf{k}\,dx\,dy = \oint_C \mathbf{F} \cdot d\mathbf{r} = -144\pi$ **(c)** The area is 12π

Exercises 21.4–Part B, Page 505

11. (a) $\frac{8}{15}$ **(b)** $\iint_R \nabla \cdot \mathbf{F}\,dx\,dy = \oint_C \mathbf{F} \cdot \mathbf{N}ds = -\frac{128}{231}$, $\iint_R [\nabla \times \mathbf{F}] \cdot \mathbf{k}\,dx\,dy = \oint_C \mathbf{F} \cdot d\mathbf{r} = -\frac{64}{21}$
24. $\int_0^{2\pi} \int_0^1 (2r\cos t + 8)r^2 \sin t\,dr\,dt = 0$ **28.** $\int_0^{2\pi} \int_0^{1+\cos t} -(1 + r^3 \cos^3 t)r\,dr\,dt = -\frac{231}{64}\pi$

32. (a) 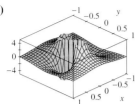 **(b)** $\lim_{x \to 0} -\frac{1}{x^2} = -\infty$; $\lim_{y \to 0} \frac{1}{y^2} = \infty$ **(f)** $\oint_{C_a'} \mathbf{F} \cdot \mathbf{T}\, ds = 0$; $\oint_{C_a'} \mathbf{F} \cdot \mathbf{N}\, ds = -2\pi$

Exercises 21.5–Part A, Page 512

7. $-\nabla u = (12xy + yz)\mathbf{i} + (6x^2 + xz)\mathbf{j} + xy\mathbf{k} = \mathbf{F}$ **8.** $-\nabla^2 u = -\dfrac{x(-\sin(\frac{y}{z})z^2 - \sin(\frac{y}{z})y^2 + 2\cos(\frac{y}{z})yz)}{z^4}$

Exercises 21.5–Part B, Page 512

3. (a) $\mathbf{F} = 6z\mathbf{i} + 15y^2\mathbf{j} + (-12z^2 + 6x)\mathbf{k}$ **(b)** $du = -6z\,dx - 15y^2\,dy + (12z^2 - 6x)\,dz$ **(c)** $\mathbf{F} \cdot d\mathbf{r} = 6z\,dx + 15y^2\,dy + (-12z^2 + 6x)\,dz$
(e) $u = 6ac - 5y^3 + 5b^3 + 4z^3 - 6xz - 4c^3$ **(f)** $v = 6ac - 5y^3 + 5b^3 + 4z^3 - 6xz - 4c^3$ **(g)** $u - v = 0$ **(h)** $6xz + 5y^3 - 4z^3$
(i) $6ca + 5b^3 - 4c^3$

12. (a) $\nabla \times \mathbf{F} = 0$ **(b)** $u = -5ab - 9ac + 5yx - 8bc + 8zy + 9zx - 9z + 9c$ **(c)** $v = -5ab - 9ac + 5yx - 8bc + 8zy + 9zx - 9z + 9c$
(d) $u - v = 0$ **(e)** $-5yx - 9zx - 8zy + 9z$ **(f)** $\nabla \cdot \mathbf{F} = 0$ **(g)** $\mathbf{A} = (-8xy - \frac{9}{2}x^2 + 9x + 8ay + \frac{9}{2}a^2 - 9a + 5yz + \frac{9}{2}z^2 - 5cy - \frac{9}{2}c^2)\mathbf{j} +$
$(8xz + \frac{5}{2}x^2 - 8az - \frac{5}{2}a^2)\mathbf{k}$ **(h)** $\mathbf{B} = (-4z^2 - 5xz + 4y^2 + 9xy - 9y)\mathbf{i} + (5yz + \frac{9}{2}z^2)\mathbf{j}$ **(i)** $\nabla \times (\mathbf{A} - \mathbf{B}) = 0$

22. $\nabla \times \mathbf{F} = 0$, $u = v = -2a^2c + 7b^2c + 7c^2b + 2x^2z - 7zy^2 - 7yz^2 + 2z^3 - 2c^3$ **37. (a)** $\nabla^2 u = 0$ **(b)** -666

40. $\iint_S \frac{\partial u}{\partial n}\, d\sigma = \iiint_V \nabla \cdot (\nabla u)\, dV = 0$

45. (a) $\nabla^2 u = 0$; $\iiint_R \|\nabla u\|^2\, dv = \iint_S u\frac{\partial u}{\partial n}\, d\sigma = \frac{1656}{35}\pi$ **(b)** $\iiint_R \|\nabla u\|^2\, dv = \iint_S u\frac{\partial u}{\partial n}\, d\sigma = \frac{384}{5}a^7 + 432a^5$
(c) $\iiint_R \|\nabla u\|^2\, dv = \iint_S u\frac{\partial u}{\partial n}\, d\sigma = \frac{2127}{20}\pi$

Exercises 21.6–Part A, Page 519

1. $\displaystyle\lim_{h \to 0} \frac{1}{8} \frac{(f(x, b+h, z) - f(y, b-h, z))\int_{c-h}^{c+h}\int_{a-h}^{a+h} 1\, dx\, dz}{h^3} = \lim_{h \to 0} \frac{1}{2} \frac{f(x, b+h, z) - f(y, b-h, z)}{h} = \frac{\partial f}{\partial y}(x, b, z)$

2. $\mathbf{F} \cdot \mathbf{N}\, d\sigma = \dfrac{(-2x^2 - y^2 + 3\rho^2)}{\sqrt{-x^2 - y^2 + \rho^2}}\, dx\, dy$

Exercises 21.6–Part B, Page 519

2. (a, b, c) $\nabla \cdot \mathbf{F} = 62$ **5.** (a, b, c) $\nabla \times \mathbf{F} = -50\mathbf{i} + 24\mathbf{j} - 5\mathbf{k}$ **8** (a, b, c) $\nabla\varphi = -540\mathbf{i} + 240\mathbf{j} + 72\mathbf{k}$ **12. (a)** $\frac{4}{5}\pi\mathbf{j}$ **(b)** $8\mathbf{j}$ **(c)** $\frac{3}{2}\pi\mathbf{j}$
14. (a) $\frac{24}{5}\pi\mathbf{k}$ **(b)** $48\mathbf{k}$ **(c)** $12\pi\mathbf{k}$ **17. (a)** 0 **(b)** 0 **(c)** 0 **20. (a)** 0 **(b)** $-28\mathbf{j}$ **(c)** $-7\pi\mathbf{j}$

CHAPTER 22

Exercises 22.1–Part A, Page 526

1. $(r, \theta) = (\sqrt{13}, -\arctan\frac{3}{2})$ 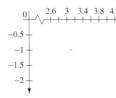 **2.** 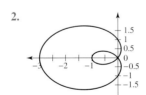   **4.** $\frac{1}{28}$

Exercises 22.1–Part B, Page 526

3. (a) 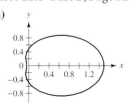   **(b)** $r = \frac{1}{2} + \cos\theta$ **(c)** 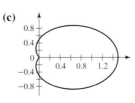   **(d)** (See plot.)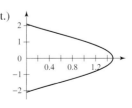

13. (a) 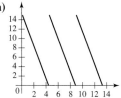 **(b)** (See plot.) 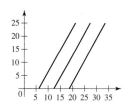 **(c)** $x = \frac{2}{31}v + \frac{7}{31}u,\ y = \frac{5}{31}u - \frac{3}{31}v$ **(d)**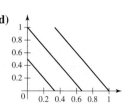

(e) 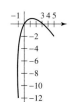 **(f)** $\left|\frac{\partial(x,y)}{\partial(u,v)}\right| = \frac{1}{31}$ 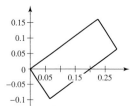 **(g)** 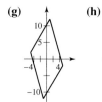 **(h)** 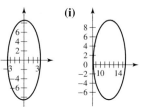 **(i)** **(j)**

(k) 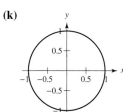 **(l)**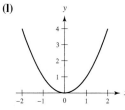

17. (a) $\left[-2\frac{\alpha v}{(\alpha^2+v^2)^2},\ \frac{\alpha^2-v^2}{(\alpha^2+v^2)^2}\right]$ **(b)** $\left[-\frac{u^2-\beta^2}{(u^2+\beta^2)^2},\ -2\frac{u\beta}{(u^2+\beta^2)^2}\right]$ **(c)** $\frac{1}{|u^2+v^2|^2}$ **(d)** $\frac{1}{|u^2+v^2|^2}$ **18. (a)** 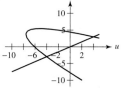 **(b) and (c)** $\frac{4}{3}$

20. (a) $-\frac{4}{3}\sqrt{2}+2\sqrt{3}-\frac{2}{3}$ **(b)** $y = v+2-2\sqrt{1-u+v},\ x = 1-\sqrt{1-u+v};\ \left|\frac{\partial(x,y)}{\partial(u,v)}\right| = -\frac{1}{2}\frac{\sqrt{1-u+v}}{-1+u-v}$ **(c)** area $= -\frac{4}{3}\sqrt{2}+2\sqrt{3}-\frac{2}{3}$

Exercises 22.2–Part A, Page 530

7. (22.7a) and (22.7b) are direct applications of the chain rule to $f_x = F_r r_x + F_\theta \theta_x$

Exercises 22.2–Part B, Page 530

2. (a) $\nabla f = 14xy\mathbf{i} + (7x^2+2)\mathbf{j}$ **(b)** $\nabla f = (21r^2\cos^2\theta\sin\theta + 2\sin\theta)\mathbf{e}_r + (-14r^2\cos\theta + 21r^2\cos^3\theta + 2\cos\theta)\mathbf{e}_\theta$
(c) $F = 7r^3\cos^2\theta\sin\theta + 2r\sin\theta$ **(d)** same as **(b)**.

7. (a) $\nabla\cdot\mathbf{F} = 18x + 18y$ **(b)** $\nabla\cdot\mathbf{F} = 18r(\cos\theta + \sin\theta)$ **(c)**
$\mathbf{G}(r,\theta) = (9r^2\cos^3\theta - 7r\cos\theta\sin\theta + 9r^2\sin\theta - 9r^2\sin\theta\cos^2\theta)\mathbf{e}_r + (-9r^2\sin\theta\cos^2\theta + 4r - 7r\cos^2\theta + 9r^2\cos\theta - 9r^2\cos^3\theta)\mathbf{e}_\theta$
(d) $\nabla\cdot\mathbf{G} = 18r(\cos\theta + \sin\theta)$

12. (a) $\nabla^2 f = 18y - 48xy$ **(b)** $\nabla^2 f = 18r\sin\theta - 48r^2\cos\theta\sin\theta$ **(c)** $F(r,\theta) = 9r^3\cos^2\theta\sin\theta - 8r^4\cos^3\theta\sin\theta$
(d) $\nabla^2 F = 18r\sin\theta - 48r^2\cos\theta\sin\theta$

17. (a) $x = \frac{2}{31}v + \frac{7}{31}u,\ y = -\frac{5}{31}u + \frac{3}{31}v$ **(b)** $\frac{\partial\mathbf{R}}{\partial u} = \frac{7}{31}\mathbf{i} - \frac{5}{31}\mathbf{j};\ \frac{\partial\mathbf{R}}{\partial v} = \frac{2}{31}\mathbf{i} + \frac{3}{31}\mathbf{j}$ **(c)** $\mathbf{e}_u = \frac{7}{74}\sqrt{74}\mathbf{i} - \frac{5}{74}\sqrt{74}\mathbf{j},\ \mathbf{e}_v = \frac{2}{13}\sqrt{13}\mathbf{i} + \frac{3}{13}\sqrt{13}\mathbf{j},$
$\mathbf{i} = \frac{3}{31}\sqrt{74}\mathbf{e}_u + \frac{5}{31}\sqrt{13}\mathbf{e}_v,\ \mathbf{j} = -\frac{2}{31}\sqrt{74}\mathbf{e}_u + \frac{7}{31}\sqrt{13}\mathbf{e}_v$ **(d)** $f_x = 3F_u + 5F_v;\ f_y = -2F_u + 7F_v$
(e) $\nabla f = \left(\frac{13}{31}\sqrt{74}F_u + \frac{1}{31}\sqrt{74}F_v\right)\mathbf{e}_u + \left(\frac{1}{31}\sqrt{13}F_u + \frac{74}{31}\sqrt{13}F_v\right)\mathbf{e}_v$ **(f)** $\nabla F = \left(\frac{6}{28,629,151}\sqrt{74}(2v+7u)(v-12u)(-5u+3v)^2\right)\mathbf{e}_u +$
$\left(\frac{1}{28,629,151}\sqrt{13}(2v+7u)(72v+97u)(-5u+3v)^2\right)\mathbf{e}_v$ **(g)** same as (f).

22. (a) $\nabla^2 F = (u^2 + v^2)^2(F_{uu} + F_{vv})$ **(b)** $\nabla^2 F = 6\frac{uv^2(v^2+2u^2)}{(u^2+v^2)^5}$ **(c)** $\nabla^2 f = 6xy^4 + 12x^3y^2$ **(d)** and **(e)** same as **(b)**.

25. (a) $\nabla \cdot \mathbf{F} = \frac{(f_u u^2 + f_u v^2 + fu + g_v u^2 + g_v v^2 + gv)}{(u^2+v^2)^{3/2}}$ **(b)** $\nabla \cdot \mathbf{F} = 7(u^2 - v^2)uv$ **(c)** $14xy$ **(d)** and **(e)** same as **(b)**.

Exercises 22.3–Part A, Page 537

1. $\dfrac{\frac{\partial}{\partial \rho}\rho^2 f(\rho,\theta,\phi)}{\rho^2} = 2\dfrac{f(\rho,\theta,\phi)}{\rho} + \dfrac{\partial}{\partial \rho}f(\rho,\theta,\phi)$; $\dfrac{\frac{\partial}{\partial\phi}g(\rho,\phi,\theta)\sin\phi}{\rho\sin\phi} = \dfrac{\frac{\partial}{\partial\phi}g(\rho,\phi,\theta)}{\rho} + \dfrac{g(\rho,\phi,\theta)\cos\phi}{\rho\sin\phi}$

3. $\left(\dfrac{\frac{\partial}{\partial\phi}h(\rho,\theta,\phi)}{\rho} + \dfrac{h(\rho,\theta,\phi)\cos\phi}{\rho\sin\phi} - \dfrac{\frac{\partial}{\partial\theta}g(\rho,\theta,\phi)}{\rho\sin\phi}\right)\mathbf{e}_\rho + \left(-\dfrac{h(\rho,\theta,\phi)}{\rho} - \dfrac{\partial}{\partial\rho}h(\rho,\theta,\phi) + \dfrac{\frac{\partial}{\partial\theta}f(\rho,\theta,\phi)}{\rho\sin\phi}\right)\mathbf{e}_\phi + \left(\dfrac{g(\rho,\theta,\phi)}{\rho} + \dfrac{\partial}{\partial\rho}g(\rho,\theta,\phi) - \dfrac{\frac{\partial}{\partial\phi}f(\rho,\theta,\phi)}{\rho}\right)\mathbf{e}_\phi$

Exercises 22.3–Part B, Page 538

5. (a) $\nabla^2 f = 200$ **(b)** $F = 2\rho^3\cos(\theta)\sin(\phi)\cos^2(\phi) - 8\rho^3\sin^3(\theta)\sin^3(\phi)$; $P = \left(3\sqrt{5}, -\arctan 2, -\arctan\frac{2}{5}\sqrt{5} + \pi\right)$ **(c)** $\nabla^2 F = 200$

   (d) $\nabla f = 50\mathbf{i} - 384\mathbf{j} - 40\mathbf{k}$ **(e)** $\nabla F = \frac{612}{5}\sqrt{5}\mathbf{e}_\rho - 246\mathbf{e}_\phi - \frac{284}{5}\sqrt{5}\mathbf{e}_\theta$ **(f)** $\nabla F = 50\mathbf{i} - 384\mathbf{j} - 40\mathbf{k}$

12. (a) $\nabla \cdot \mathbf{f} = 134$ **(b)** The expression is too long to display here.   **(c)** $P = \left(\sqrt{97}, \arctan\frac{5}{6} - \pi, \arctan(\frac{1}{582}\sqrt{97}\sqrt{5917}) - \pi\right)$

   (d) $\nabla \cdot \mathbf{F} = 134$ **(e)** $\nabla \times \mathbf{f} = -486\mathbf{i} + 41\mathbf{j}$ **(f)** $\nabla \times \mathbf{F} = -\frac{2711}{97}\sqrt{97}\mathbf{e}_\rho - \frac{16,266}{5917}\sqrt{5917}\mathbf{e}_\phi - \frac{2676}{61}\sqrt{61}\mathbf{e}_\theta$ **(g)** $\nabla \times \mathbf{F} = -486\mathbf{i} + 41\mathbf{j}$

22. (a) $\begin{bmatrix} \frac{x}{\sqrt{x^2+y^2+z^2}} & \frac{y}{\sqrt{x^2+y^2+z^2}} & \frac{z}{\sqrt{x^2+y^2+z^2}} \\ -\frac{y}{x^2+y^2} & \frac{x}{x^2+y^2} & 0 \\ \frac{zx}{(x^2+y^2+z^2)\sqrt{x^2+y^2}} & \frac{zy}{(x^2+y^2+z^2)\sqrt{x^2+y^2}} & -\frac{\sqrt{x^2+y^2}}{x^2+y^2+z^2} \end{bmatrix}$ **(b)** and **(c)** The expressions are too large to display here.

   (d) $F_{\rho\rho} + \frac{2}{\rho}F_\rho + \frac{1}{\rho^2}F_{\phi\phi} + \frac{\cos\phi}{\rho^2\sin\phi}F_\phi + \frac{1}{\rho^2\sin^2\phi}F_{\theta\theta}$

28. (a) $h_1 = 1$, $h_2 = u_1$, $h_3 = u_1\sin u_3$

29. (a) $\frac{\partial\mathbf{R}}{\partial u} = -\frac{u^2-v^2-w^2}{(u^2+v^2+w^2)^2}\mathbf{i} - 2\frac{vu}{(u^2+v^2+w^2)^2}\mathbf{j} - 2\frac{wu}{(u^2+v^2+w^2)^2}\mathbf{k}$; $\frac{\partial\mathbf{R}}{\partial v} = -2\frac{vu}{(u^2+v^2+w^2)^2}\mathbf{i} + \frac{u^2-v^2+w^2}{(u^2+v^2+w^2)^2}\mathbf{j} - 2\frac{wv}{(u^2+v^2+w^2)^2}\mathbf{k}$;

$\frac{\partial\mathbf{R}}{\partial w} = -2\frac{wu}{(u^2+v^2+w^2)^2}\mathbf{i} - 2\frac{wv}{(u^2+v^2+w^2)^2}\mathbf{j} + \frac{u^2+v^2-w^2}{(u^2+v^2+w^2)^2}\mathbf{k}$; $\begin{bmatrix} \frac{1}{(u^2+v^2+w^2)^2} & 0 & 0 \\ 0 & \frac{1}{(u^2+v^2+w^2)^2} & 0 \\ 0 & 0 & \frac{1}{(u^2+v^2+w^2)^2} \end{bmatrix}$ **(b)**

   (c) $h_1 = h_2 = h_3 = \frac{1}{u^2+v^2+w^2}$ **(d)** $\nabla\phi = ((u^2+v^2+w^2)\phi_u)\mathbf{e}_u + ((u^2+v^2+w^2)f_v)\mathbf{e}_v + \left((u^2+v^2+w^2)f_w\right)\mathbf{e}_w$

   (e) $\nabla \cdot \mathbf{A} = (u^2+v^2+w^2)\frac{\partial A_3}{\partial w} + (u^2+v^2+w^2)\frac{\partial A_1}{\partial u} + (u^2+v^2+w^2)\frac{\partial A_2}{\partial v} - 4A_3w - 4A_2v - 4A_1u$ **(f)** $\nabla \times \mathbf{A} =$

$\left((u^2+v^2+w^2)\frac{\partial A_3}{\partial v} + (-u^2-v^2-w^2)\frac{\partial A_2}{\partial w} - 2A_3v + 2A_2w\right)\mathbf{e}_u + \left((-u^2-v^2-w^2)\frac{\partial A_3}{\partial u} + (u^2+v^2+w^2)\frac{\partial A_1}{\partial w} - 2A_1w + 2A_3u\right)\mathbf{e}_v +$

$\left((u^2+v^2+w^2)\frac{\partial A_2}{\partial u} + (-u^2-v^2-w^2)\frac{\partial A_1}{\partial v} - 2A_2u + 2A_1v\right)\mathbf{e}_w$ **(g)** $\nabla^2\phi = (u^2+v^2+w^2)^2\phi_{uu} + (u^2+v^2+w^2)^2\phi_{vv} +$

$(u^2+v^2+w^2)^2\phi_{ww} - 2(u^2+v^2+w^2)u\phi_u - 2(u^2+v^2+w^2)v\phi_v - 2(u^2+v^2+w^2)w\phi_w$

CHAPTER 23

Exercises 23.1–Part A, Page 544

1. $\left(-\frac{1}{(x^2+y^2+1)^{3/2}} + \frac{3}{(x^2+y^2+9)^{3/2}}\right)d\sigma$ **2.** $\int_0^{2\pi}\int_0^1 \frac{(-\sqrt{r^2+9}r^2 - 9\sqrt{r^2+9} + 3\sqrt{r^2+1}r^2 + 3\sqrt{r^2+1})r}{(r^2+1)^{3/2}(r^2+9)^{3/2}}\,dr\,d\theta = \pi\sqrt{2} - \frac{3}{5}\pi\sqrt{10}$ **8.** $\int_0^{2\pi}\frac{2\sin t+1}{5+4\sin t}\,dt = 0$

Exercises 23.1–Part B, Page 544

3. (a) $\iint_S \mathbf{N}\cdot\frac{\mathbf{r}}{r^3}\,dr = \int_0^{2\pi}\int_0^{\sqrt{2}/2}\left(\frac{r^2+1}{(r^4-r^2+1)^{3/2}} - \frac{4}{(4r^2+1)^{3/2}}\right)r\,dr\,d\theta = 0$ **(b)** $\int_0^1\frac{2r(r^2+1)}{(-r^2+1+r^4)^{3/2}}\,dr = 4\pi$

   9. Implementing the condition gives a differential equation for $f(r)$, $0 = \nabla \cdot \mathbf{F} = 2f(r) + rf'(r)$, with solution $f(r) = \frac{c}{r^2}$.

Exercises 23.2–Part A, Page 548

1. $y_x = -\frac{f_x}{f_y}$; $y_z = -\frac{f_z}{f_y}$, $d\sigma = \frac{\|\nabla f\|}{|y|}\,dx\,dz$ **6.** 90

Exercises 23.2–Part B, Page 549

1. (a)

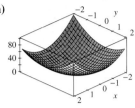

(b) Use the second derivative test.   **(c)**

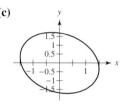

(d) $d\sigma = \sqrt{1 + 592x^2 + 336xy + 340y^2}\, dx\, dy$

(g)

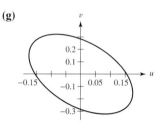

(e) $\int_{-3/2}^{3/2}\int_{-2/9x-1/9\sqrt{-104x^2+234}}^{-2/9x+1/9\sqrt{-104x^2+234}}\sqrt{1 + 592x^2 + 336xy + 340y^2}\, dy\, dx$

(f) $Z(u, v) = 1080u^2 + 552uv + 321v^2$

(h) $d\sigma = 51\sqrt{55{,}360u^2 + 32{,}224uv + 12{,}820v^2 + 1}\, du\, dv$

(i) $\int_{-1/17\sqrt{30}}^{1/17\sqrt{30}}\int_{-\frac{23}{90}v-1/180\sqrt{-7514v^2+780}}^{-\frac{23}{90}v+1/180\sqrt{-7514v^2+780}}51\sqrt{55{,}360u^2 + 32{,}224uv + 12{,}820v^2 + 1}\, du\, dv$

(j) Evaluate the Jacobian of the transformation and compare the integrands, the new integration region is given by (g). $|J| = 51$;
$\sqrt{1 + 592x^2 + 336xy + 340y^2} = \sqrt{55{,}360u^2 + 32{,}224uv + 12{,}820v^2 + 1}$

2. (a)

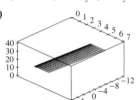

(b) $-341{,}879.380$ **(c)** $\frac{5904}{5}$ **6.** $\iint_S \mathbf{F}\cdot\mathbf{N}d\sigma = \frac{392{,}909{,}691}{77{,}268{,}297{,}715}$ **11.** $\iint_S \mathbf{F}\cdot\mathbf{N}d\sigma = \frac{2}{15}\pi$

Exercises 23.3, Page 552

1. Write the continuity equation as $\frac{\partial}{\partial x}(\rho u) = -(\frac{\partial}{\partial t}\rho)$ and integrate both sides with respect to x

3. (a) (See plot.)

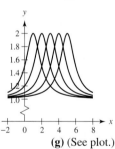

(b) $\frac{(1+(x-t)^2)f(t)}{(x-t)^2+2} + \frac{1}{(x-t)^2+2}$ **(d)**

(c) $\frac{\partial}{\partial x}u(t, t) = 0$

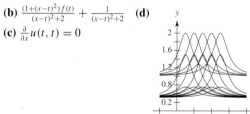

(e)

(f)

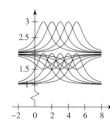

(g) (See plot.)   **(h)**

5. (a) $u(x, t) = \frac{f(t)}{\rho(x)}$ **(b)** $u(x, t) = -\frac{\frac{\partial}{\partial t}\rho(t)x + f(t)\rho(t)}{\rho(t)}$ **8.** In the continuity equation let $\mathbf{v} = \mathbf{E}$, and substitute $\mathbf{E} = -\nabla V$.

Exercises 23.4, Page 554

2. (a) $\iiint_V [f\nabla^2 g + (\nabla f)\cdot(\nabla g)]dv = \iint_S (f\nabla g)\cdot\mathbf{N}d\sigma = 6300\pi$ **(b)** $\iiint_V [f\nabla^2 g - g\nabla^2 f]dv = \iint_S [f\nabla g - g\nabla f]\cdot\mathbf{N}d\sigma = \frac{130{,}284}{25}\pi$

7. $(A\mathbf{x})\cdot\mathbf{y} = \mathbf{x}\cdot(A^T\mathbf{y}) = -992$

CHAPTER 24

Exercises 24.1–Part A, Page 563

1. $u(x,t) = \sum_{n=1}^{\infty} 4 \frac{(2(-1)^{1+n}-1)\sin(n\pi x)\cos(2n\pi t)}{n^3\pi^3}$ 3. $\frac{X''}{X} = -\lambda, X(0) = 0, X(\pi) = 0; \frac{1}{4}\frac{T''}{T} - \frac{3}{4} = -\lambda$

Exercises 24.1–Part B, Page 564

1. $\frac{\partial^2}{\partial t^2} f(x+ct) = \frac{\partial^2}{\partial x^2} c^2 f''(x+ct)$

3. **(a)** $2c^2 e^{(-(x+ct)^2)}(-1+2x^2+4xct+2c^2t^2) = 2c^2 e^{(-(x+ct)^2)}(-1+2x^2+4xct+2c^2t^2);$
   $2c^2 e^{(-(-x+ct)^2)}(-1+2x^2-4xct+2c^2t^2) = 2c^2 e^{(-(-x+ct)^2)}(-1+2x^2-4xct+2c^2t^2)$

   (b) 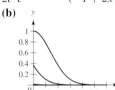   **(c)** 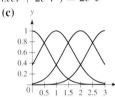   **(d)** 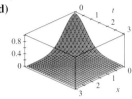   **(e)**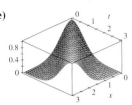

11. **(a)** $u_{tt} = -9\sin(2x)\cos(3t) = \frac{9}{4}u_{xx}; u(0,t) = 0, u(\pi,t) = 0, u(x,0) = \sin 2x, \frac{\partial u}{\partial t}(x,0) = 0$ **(b)** $f(x) = u(x,0) = \sin 2x$

   (c) 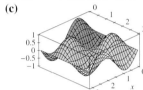   **(d)** 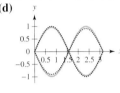   **(f)**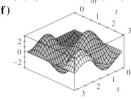

13. **(a)** $u(x,t) = \sum_{m=1}^{\infty} 4 \frac{\sin(\frac{1}{2}m\pi)\sin(mx)\cos(mt)}{\pi m^2}$ **(b)** 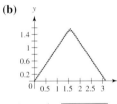   **(c)** 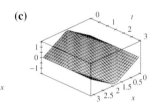   **(e)**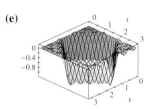

25. **(a)** $T_n(t) = e^{-1/2t}\cos(\frac{1}{2}\sqrt{-1+4n^2}t) + \frac{e^{-1/2t}\sin(\frac{1}{2}\sqrt{-1+4n^2}t)}{\sqrt{-1+4n^2}}$

   (b) $u(x,t) = \sum_{n=1}^{\infty} \frac{2}{5} \frac{((-1)^{1+n}+1)\left(e^{-1/2t}\cos(\frac{1}{2}\sqrt{-1+4n^2}t) + \frac{e^{-1/2t}\sin(\frac{1}{2}\sqrt{-1+4n^2}t)}{\sqrt{-1+4n^2}}\right)\sin nx}{\pi n^3}$

   (c) 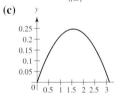   **(d)**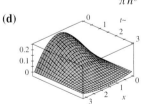

27. **(a)** $u(x,t) = \sum_{n=1}^{\infty} 4 \frac{\sin(\frac{1}{2}n\pi)\left(e^{-1/2t}\cos(\frac{1}{2}\sqrt{-1+4n^2}t) + \frac{e^{-1/2t}\sin(\frac{1}{2}\sqrt{-1+4n^2}t)}{\sqrt{-1+4n^2}}\right)\sin nx}{\pi n^2}$

   (b) 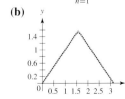   **(c)**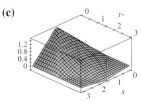

Exercises 24.2–Part A, Page 567

2. $u(x,t) = \sum_{n=1, n\neq 2}^{\infty} 8 \frac{((-1)^n - 1)\sin(\frac{1}{2}nx)\sin(nt)}{n^2\pi(n+2)(n-2)}$ **3.** Let $g(x) = \sin(\frac{\pi x}{L})$

Exercises 24.2–Part B, Page 567

2. **(a)** $u(x,t) = \sum_{n=1}^{\infty} -100(5\cos\frac{1}{5}n\pi - 5\cos\frac{2}{5}n\pi - n\pi\sin\frac{2}{5}n\pi + 5\cos\frac{3}{5}n\pi - n\pi\sin\frac{3}{5}n\pi - 5\cos(\frac{4}{5}n\pi))\sin(\frac{1}{5}n\pi x)\sin(\frac{1}{5}n\pi t)/(n^4\pi^4)$

(b) with 20 terms   **(c)**   **(e)**

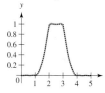 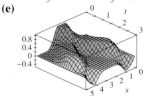

9. **(a)** $u(x,t) = \sum_{n=1}^{\infty} -12 \frac{(-2\sin(\frac{2}{3}n\pi) + \sin(\frac{1}{3}n\pi))e^{-1/2\pi t}\sin(\frac{1}{2}\sqrt{4n^2 - 1}\pi t)\sin(\frac{1}{3}n\pi x)}{n^2\pi^3\sqrt{4n^2 - 1}}$

(b)   **(c)**   **(e)**

 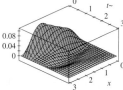

Exercises 24.3–Part A, Page 570

2. **(a)** $F(x) = 4\sin(x)x^2 + 2\sin^2(x)x + \frac{1}{3}\sin^3(x) - 4x^2 - 2x - \frac{1}{3}$
(b) $F'(x) = 4\cos(x)x^2 + 8\sin(x)x + 4\sin(x)x\cos(x) + 2\sin^2(x) + \sin^2(x)\cos(x) - 8x - 2$
(c) $F'(x) = 8\sin(x)x + 2\sin^2(x) - 8x - 2 + (2x + \sin(x))^2\cos(x)$

Exercises 24.3–Part B, Page 571

1. $u(x,t) = \frac{\cos(x-ct) - \cos(x+ct)}{2} = \frac{\sin x \sin ct}{c}$; In this case $g(y) = \sin y$ equals it's own odd periodic extension on $[0, \pi]$.

Exercises 24.4–Part A, Page 573

2. **(a)** Substitute the given expression into the differential equation and simplify.   **(b)** If f is twice differentiable, then substitution shows each term solves the PDE. The sum is a solution by linearity.

Exercises 24.4–Part B, Page 573

1. $u(x,t) = \sum_{n=1}^{\infty} b_n T_n(t) X_n(x)$; b_n denotes the n-th Fourier sine coefficient of f,   **(d)**   **(e)**
$X_n(x)$ is an eigenfunction of $\frac{X''}{X} = -\lambda$, $X(0) = 0$, $X(L) = 0$
and $T_n(t)$ is a corresponding solution of $T'' + 2\alpha T'(k + c^2\lambda)T = 0$.

(c) $u(x,t) = \sum_{n=1}^{\infty} -4 \frac{(20\cos(\frac{1}{10}n\pi) + \sin(\frac{1}{10}n\pi)n\pi - 20)\cos(\sqrt{\sqrt{2}+n^2}t)\sin(nx)}{\pi n^3}$

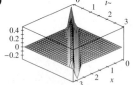 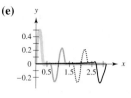

(f)

Exercises 24.5, Page 578

1. **(a)** $u_{tt} = c^2 u_{xx}$, $u_x(0,t) = 0$, $u_x(10,t) = 0$, $u(x,0) = \frac{1}{5}x$, $u_t(x,0) = 0$ **(b)** $u(x,t) = \sum_{n=0}^{\infty} a_n \cos(\frac{1}{10}n\pi ct)\cos(\frac{1}{10}n\pi x)$; a_n denotes the n-th Fourier cosine coefficient of f.

(c) 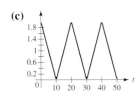 **(d)** 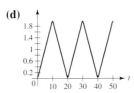 **(e)** 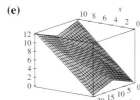 **(f)**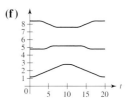

2. (a) $u_{tt} = c^2 u_{xx}$, $u(0, t) = 0$, $u_x(\pi, t) = 0$, $u(x, 0) = \frac{x}{\pi}$, $u_t(x, 0) = 0$ **(b)** $\sum_{n=0}^{\infty} b_n \cos(c(\frac{1}{2} + n)t) \sin((\frac{1}{2} + n)x)$; b_n denotes the n-th Fourier sine coefficient of f. **(c)** 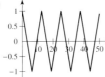 **(d)** 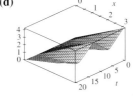 **(e)**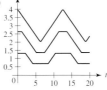

5. (a) $u(x, 0) = \left(\frac{x}{5}\right)^{3/2}$ **(b)** $u_{tt} = c^2 u_{xx}$, $u_x(0, t) = 0$, $u_x(5, t) = 0$, $u(x, 0) = \left(\frac{x}{5}\right)^{3/2}$, $u_t(x, 0) = 0$ **(b)** $\sum_{n=0}^{\infty} a_n \cos(\frac{1}{5} n\pi ct) \cos(\frac{1}{5} n\pi x)$; a_n denotes the n-th Fourier cosine coefficient of $u(x, 0)$.

(c) 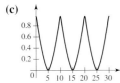 **(d)** 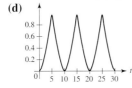 **(e)** 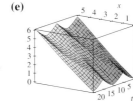 **(f)**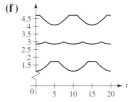

7. (a) $u_{tt} = c^2 u_{xx}$, $u_x(0, t) = 0$, $u_x(\pi, t) = 0$, $u(x, 0) = -\frac{x}{\pi}$, $u_t(x, 0) = 0$; $u(x, t) = \sum_{n=0}^{\infty} a_n \cos(nct) \cos(nx)$; a_n denotes the n-th Fourier cosine coefficient of $u(x, 0)$.

(b) 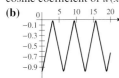 **(c)** 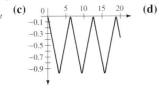 **(d)** 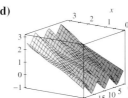 **(e)**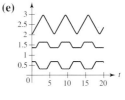

Exercises 24.6–Part A, Page 583

1. $u(x + h, t) + u(x - h, t) = 2u(x, t) + h^2 u_{xx}(x, t) \Rightarrow u_{xx}(x, t) = \frac{u(x-h,t) - 2u(x,t) + u(x+h,t)}{h^2}$.

Exercises 24.6–Part B, Page 583

1. 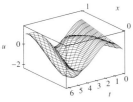 **3. (a)** 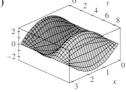 **(c)** 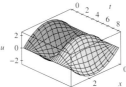 **(d)**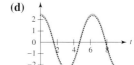

9. (b)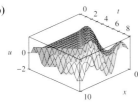

CHAPTER 25

Section 25.1, Exercises A, Page 590

1. $\frac{T'}{\kappa T} = \lambda \Rightarrow T' - \lambda \kappa T = 0$; $\frac{X''}{X} = \lambda \Rightarrow X'' - \lambda X = 0$. **4.** $0.8824969026, 0.6065306597, 0.1353352832$.

5. In case of non-homogeneous boundary conditions the method will fail because we don't know how to divide the non-zero boundary conditions among the infinitely many solutions.

Exercises 25.1–Part B, Page 590

1. 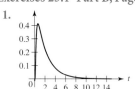   In the beginning the heat flow through the endpoint rapidly increases, but after a while the flow dies down as the temperature in the rod uniformly approaches zero.

3. **(a)** $u(x, t) = \sum_{n=1}^{\infty} 4 \dfrac{((-1)^{1+n} + 1)e^{-1/2n^2 t} \sin nx}{\pi n^3}$; **(b)** 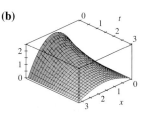 **(d)**

8. **(a)** 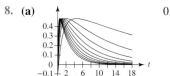 0.4840934508

9. **(a)** $u(x, t) = \sum_{n=1}^{\infty} 4 \dfrac{((-1)^{1+n} + 1)e^{t(h-n^2)} \sin nx}{\pi n^3}$ **(b)** 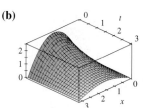 **(c)**

(d) 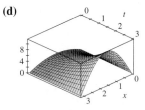   **(e)** For $h = \frac{1}{2}$ the exponents are all negative and the solution will converge to zero over time. When $h = 1$ there is a term which does not die off. When $h = \frac{3}{2}$ there even is a term which increases over time.

Exercises 25.2, Page 593

2. **(a)** $u(x, t) = \sum_{n=1}^{\infty} 108 \dfrac{(2(-1)^{1+n} - 1)e^{-4/27n^2\pi^2 t} \sin(\frac{1}{3}\pi nx)}{\pi^3 n^3}$; **(b)** 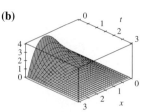 **(d)**

7. **(a)** $Q_0 = \frac{27}{4}M$

    (b) The expressions are too large to display here.

    (c) $\triangle Q_{total} = \frac{79,908,043,688}{121,550,625}\frac{M}{\pi^4}$

    (d) $M\int_0^3 U(x,0)\,dx = \frac{79,908,043,688}{121,550,625}\frac{M}{\pi^4}$

    (e) The expression is too large to display here.

(f)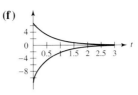

(g) $\frac{U_x(3,t)-U_x(0,t)}{\int_0^3 U_t(x,t)\,dx} = \frac{3}{4} = \frac{1}{\kappa}.$

12. **(a)** $u(x,t) = \sum_{n=1}^{\infty} a_n e^{-2n^2 t}\sin nx$ If $n \neq 2$ then $a_n = 6\frac{n((-1)^{1+n}+1)}{\pi(2+n)(-2+n)}, a_2 = 0$ **(b)**

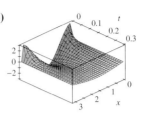

Exercises 25.3, Page 597

2. **(a)** $u_t = \frac{1}{2}u_{xx}, u_x(0,t)=0, u_x(\pi,t)=0, u(x,0)=x(\pi-x); u(x,t) = \frac{1}{6}\pi^2 + \sum_{n=1}^{\infty} -2\frac{((-1)^n+1)e^{-1/2n^2}t\cos nx}{n^2}$

    (b) 40 terms

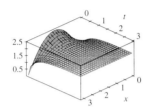

    (d) $\lim_{t\to\infty} u(x,t) = \frac{1}{6}\pi^2$ **(e)**

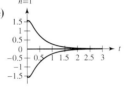

Exercises 25.4, Page 601

3. **(a)**

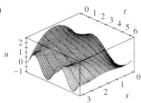

10. **(a)** $u(x,t) = e - 1 + \sum_{n=1}^{\infty} 2\frac{(e(-1)^n - 1)e^{-2n^2\pi^2 t}\cos \pi nx}{1+n^2\pi^2}$ **(b)**

    (c) exact: 1.656467786

                 numeric: 1.656465773

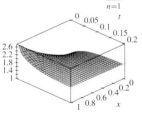

17.

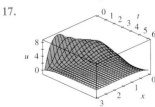
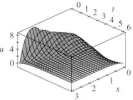

CHAPTER 26

Exercises 26.1–Part B, Page 606

2. **(a)** $u(x, y) = \sum_{n=1}^{\infty} 4 \dfrac{((-1)^{1+n} + 1)\sinh(n(1-y))\sin(nx)}{\pi n^3 \sinh(n)}$ **(c)**

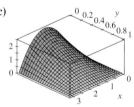   **(d)**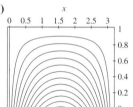

(b) $u(\frac{\pi}{2}, \frac{1}{2}) = 1.110562454$

(e) 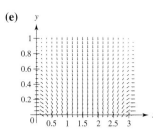   **(f)** 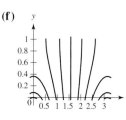   **(g)**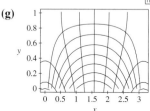

Exercises 26.2–Part A, Page 611

1. $-\dfrac{1}{2}\dfrac{\int_0^4 g(x)\sin\frac{1}{4}n\pi x\,dx}{\sinh n\pi} = -16 - \dfrac{\sin(\frac{1}{4}n\pi)n\pi - \sin(\frac{3}{4}n\pi)n\pi + 4\cos(\frac{1}{4}n\pi) - 4\cos(\frac{3}{4}n\pi)}{\sinh(n\pi)n^3\pi^3}$ **2.**

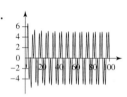

Exercises 26.2–Part B, Page 611

3. **(a)** $u(x, y) = \sum_{n=1}^{\infty} 12 \dfrac{(4(-1)^n - 4 + n^2\pi^2)\sinh(ny)\sin(nx)}{\pi n^5 \sinh(n)}$ **(c)**

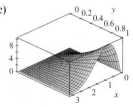   **(d)**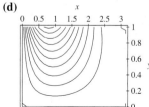

(b) $u(\frac{\pi}{2}, \frac{1}{2}) = 2.917603834$

(e) 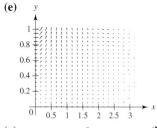   **(f)** 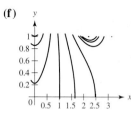   **(g)**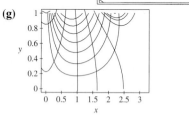

23. **(a)** 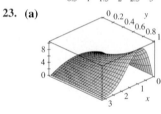   **(b)** 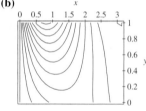   **(c)**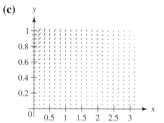

(d) 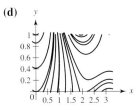   **(g)**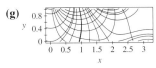

34. (a) 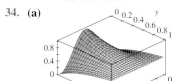   **(b)** 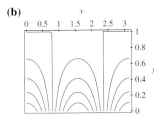   **(c)**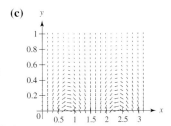

(d) 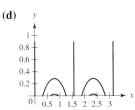   **(e)** 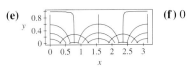   **(f)** 0

45. (a) 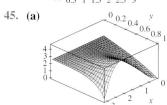   **(b)** 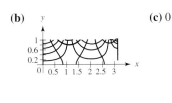   **(c)** 0

Exercises 26.3–Part A, Page 616

1. $A_n = -16 - \dfrac{4 + 4\cos\frac{1}{4}n\pi + n\pi\sin\frac{1}{4}n\pi + n\pi\sin\frac{3}{4}n\pi + 4\cos n\pi - 4\cos\frac{3}{4}n\pi}{\sinh(n\pi)n^3\pi^3}$

4. $u(x,y) = \sum_{n=1}^{\infty} A_n \sinh(\frac{n\pi(a-x)}{b})\sin(\frac{n\pi y}{b}) + B_n \sinh(\frac{n\pi x}{b})\sin(\frac{n\pi y}{b}))$; $A_n = 2\dfrac{\int_0^b f(y)\sin\frac{n\pi y}{b}\,dy}{b\sinh\frac{n\pi a}{b}}$; $B_n = 2\dfrac{\int_0^b F(y)\sin\frac{n\pi y}{b}\,dy}{b\sinh\frac{n\pi a}{b}}$

Exercises 26.3–Part B, Page 616

1. $u(x,y) = \sum_{n=1}^{\infty} A_n \cosh\dfrac{n\pi x}{b}\sin\dfrac{n\pi y}{b}$; $A_n = 2\dfrac{\int_0^b F(y)\sin\frac{n\pi y}{b}\,dy}{b\cosh\frac{n\pi a}{b}}$.

14. (a) $u(x,y) = \sum_{n=1}^{\infty} 4\dfrac{((-1)^{1+n}+1)\sinh(\pi n(-x+\pi))\sin\pi ny}{\pi^3 n^3 \sinh n\pi^2}$ **(c)** 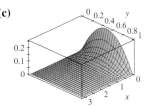   **(d)**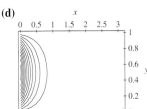

(b) $u(\frac{\pi}{2},\frac{1}{2}) = 0.001855494658$

(e) 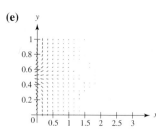   **(f)** 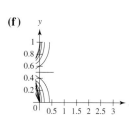   **(g)**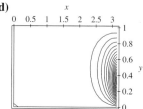

19. (a) $u(x, y) = \displaystyle\sum_{n=1}^{\infty} 4 \frac{((-1)^n + 2) \sinh n\pi x \sin \pi ny}{n^3 \pi^3 \sinh \pi^2 n}$ **(c)** 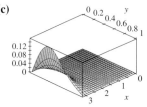   **(d)**

(b) $u(\frac{\pi}{2}, \frac{1}{2}) = 0.0009277473290$

(e) 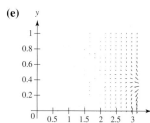   **(f)** 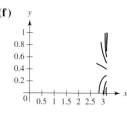   **(g)**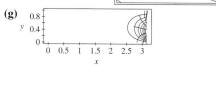

26. (a) 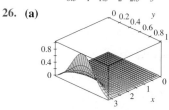   **(b)**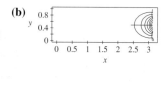

Exercises 26.4–Part A, Page 621

1. If $dx = dy$ then (26.14) becomes $v_{n-1,m} + v_{n+1,m} - 4v_{n,m} + v_{n,m+1} + v_{n,m-1} = 0$.

5. $\dfrac{v_{n-1,m} - 2v_{n,m} + v_{n+1,m}}{dx^2} + \dfrac{v_{n,m+1} - 2v_{n,m} + v_{n,m-1}}{dy^2} = 0, 1 \le n \le N - 1, 1 \le m \le M - 1$, and $-\dfrac{2v_{0,m} + 2v_{1,m}}{dx^2} +$
$\dfrac{v_{0,m+1} - 2v_{0,m} + v_{0,m-1}}{dy^2} = 0, 1 \le m \le M - 1$. The Dirichlet conditions along the other sides remain unchanged.

Exercises 26.4–Part B, Page 622

1. $\dfrac{v_{n-1,m} - 2v_{n,m} + v_{n+1,m}}{dx^2} + \dfrac{v_{n,m+1} - 2v_{n,m} + v_{n,m-1}}{dy^2} = 0, 1 \le n \le N - 1, 1 \le m \le M - 1$, and $\dfrac{v_{n-1,0} - 2v_{n,0} + v_{n+1,0}}{dx^2} +$
$\dfrac{2v_{n,1} - 2v_{n,0}}{dy^2} = 0, 1 \le n \le N - 1$, and $\dfrac{v_{n-1,M} - 2v_{n,M} + v_{n+1,M}}{dx^2} + \dfrac{-2v_{n,M} + 2v_{n,M-1}}{dy^2} = 0, 1 \le n \le N - 1$. The Dirichlet conditions along
the other sides remain unchanged.

4. $\dfrac{v_{n-1,m} - 2v_{n,m} + v_{n+1,m}}{dx^2} + \dfrac{v_{n,m+1} - 2v_{n,m} + v_{n,m-1}}{dy^2} = 0, 0 \le n \le N - 1, 0 \le m \le M - 1$, and $\dfrac{v_{n-1,0} - 2v_{n,0} + v_{n+1,0}}{dx^2} +$
$\dfrac{2v_{n,1} - 2v_{n,0}}{dy^2} = 0, 0 \le n \le N - 1$, and $-\dfrac{2v_{0,m} + 2v_{1,m}}{dx^2} + \dfrac{v_{0,m+1} - 2v_{0,m} + v_{0,m-1}}{dy^2} = 0, 0 \le m \le M - 1$. The Dirichlet conditions along the
other sides remain unchanged.

5. $\dfrac{v_{n-1,m} - 2v_{n,m} + v_{n+1,m}}{dx^2} + \dfrac{v_{n,m+1} - 2v_{n,m} + v_{n,m-1}}{dy^2} = 0, 0 \le n \le N - 1, 1 \le m \le M$, and $-\dfrac{2v_{0,m} + 2v_{1,m}}{dx^2} +$
$\dfrac{v_{0,m+1} - 2v_{0,m} + v_{0,m-1}}{dy^2} = 0, 1 \le m \le M$, and $\dfrac{v_{n-1,M} - 2v_{n,M} + v_{n+1,M}}{dx^2} + \dfrac{-2v_{n,M} + 2v_{n,M-1}}{dy^2} = 0, 0 \le n \le N - 1$. The Dirichlet
conditions along the other sides remain unchanged.

CHAPTER 27

Section 27.1, Page 626

3. **(a)** $u(x,t) = \dfrac{x}{\pi} + 4\dfrac{(\frac{1}{2}\pi - 1)^2 e^{-2t}\sin x}{\pi^2} + \sum_{n=2}^{\infty} 2\dfrac{(-1)^n e^{-2n^2 t}\sin nx}{\pi n}$ **(b)**

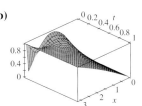

(c) 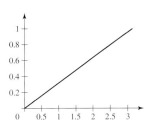 **(d)** $\dfrac{x}{\pi}$ **(f)**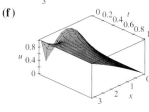

9. **(a)** $u(x,t) = -2x + 20 + \sum_{n=1}^{\infty} 240\dfrac{(-1)^{1+n} e^{-1/50 n^2 \pi^2 t}\sin\frac{1}{10}\pi nx}{\pi^3 n^3}$ **(b)**

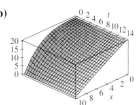

(c) 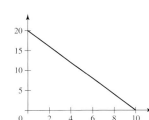 **(d)** $-2x + 20$ **(f)**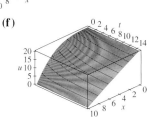

Section 27.2–Part A, Page 630

1. Because the eigenfunctions are orthogonal, the coefficients in the summation preceding (27.6) need to vanish. This produces (27.6).

2. Substitute (27.7) in (27.6) and simplify.

Section 27.2–Part B, Page 631

3. **(a)** $u(x,t) = \dfrac{x\cos t}{\pi} + \sum_{n=1}^{\infty}\left(2\dfrac{(-1)^n \cos t}{(4n^4+1)\pi n} - 4\dfrac{(-1)^n n\sin t}{(4n^4+1)\pi} - 2\dfrac{e^{-2n^2 t}(-1)^n}{(4n^4+1)\pi n}\right)\sin nx$

(b) 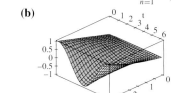 **(c)** 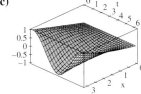 **(d)**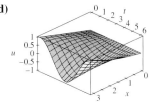

8. **(a)** $u(x,t) = \sin t - \dfrac{x\sin t}{\pi} + \sum_{n=1}^{\infty}\left(-4\dfrac{n\cos t}{(n^4+4)\pi} - 8\dfrac{\sin t}{(n^4+4)\pi n} + 4\dfrac{e^{-1/2n^2 t} n}{(n^4+4)\pi} - 2\dfrac{e^{-1/2n^2 t}(-1)^n}{\pi n}\right)\sin nx$

(b) 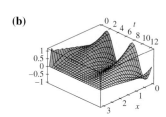   **(c)** 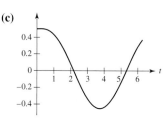   **(d)** 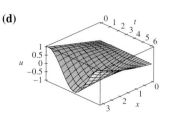   2. $x = \frac{3}{5}, \frac{6}{5}, \frac{9}{5}, \frac{12}{5}, 3, \frac{18}{5}$

CHAPTER 28

Exercises 28.1–Part A, Page 639

2. $x = \frac{3}{5}, \frac{6}{5}, \frac{9}{5}, \frac{12}{5}, 3, \frac{18}{5}$

Exercises 28.1–Part B, Page 639

2. **(a)** $x = 0, \frac{1}{2}\pi, \pi$ **(b)** $c = \frac{3}{2}$ **(c)** $T = \frac{2}{3}\pi$ 6.

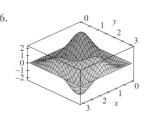

11. **(a)** $u(x, y, t) =$

$$\sum_{m=1}^{\infty} \sum_{n=1}^{\infty} (8(-1)^{1+n+m} + 2(-1)^{n+m}n^2\pi^2 + 8(-1)^m + n^2\pi^2(-1)^n + 4 + 4(-1)^{1+n}) \sin(nx)\sin(my)\cos\left(\frac{1}{2}\pi\sqrt{\frac{n^2}{\pi^2} + \frac{m^2}{\pi^2}}\,t\right)/(\pi m^3 n^5)$$

(b) 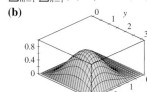   **(c)**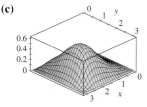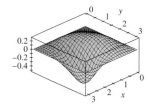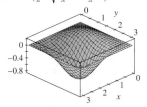

(d)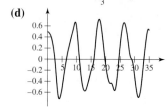

16. **(a)** $\sum_{m=1}^{\infty} \sum_{n=1}^{\infty} (16m(3e^{-\pi}(-1)^{1+m+n} + 3e^{-\pi}(-1)^m + e^{-\pi}(-1)^{m+n}m^2 - e^{-\pi}(-1)^m m^2 - (-1)^{m+n}e^{-\pi}\pi + e^{-\pi}(-1)^m\pi - (-1)^{m+n}e^{-\pi}m^2\pi + e^{-\pi}m^2(-1)^m\pi + 3(-1)^n - 3 - m^2(-1)^n + m^2 - (-1)^n\pi + \pi - \pi m^2(-1)^n + \pi m^2)\sin(nx)\sin(my)/((1+m^2)^3 n^3 \pi^2))$
(b) (30 terms in both x and y)   **(c)**

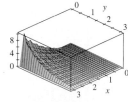   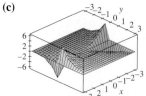

21. **(a)**

$$\frac{2}{27} + \sum_{m=1}^{\infty} \sum_{n=1}^{\infty} \left(\frac{4}{15} \frac{(4(-1)^{n+m}\pi^2 m^2 + 3(-1)^{1+n+m} + 3(-1)^{n+m})\cos(n\pi x)\cos(m\pi y)}{n^2\pi^6 m^4} - \frac{4}{15}\frac{(5(-1)^{1+n+m} - (-1)^{n+m})\cos(n\pi x)\sin(m\pi y)}{n^2\pi^5 m^3} \right) +$$
$$\sum_{n=1}^{\infty} \frac{2}{9}\frac{(-1)^n \cos(n\pi x)}{n^2\pi^2} + \sum_{m=1}^{\infty} \left(\frac{4}{45}\frac{(4m^2\pi^2(-1)^m + 3(-1)^{1+m} + 3(-1)^m)\cos(m\pi y)}{m^4\pi^4} + \frac{4}{45}\frac{(5(-1)^{(2+m)} + (-1)^m)\sin(m\pi y)}{m^3\pi^3} \right)$$

(b)

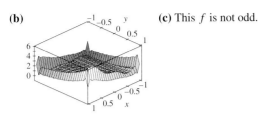

(c) This f is not odd.

Exercises 28.2–Part A, Page 643

2. $T'_{nm}(t) - \kappa(\lambda_n + \mu_m)T_{nm}(t) = 0 \Rightarrow T'_{nm}(t) - \kappa\pi^2\left(\frac{n^2}{a^2} + \frac{m^2}{b^2}\right)T_{nm}(t) = 0 \Rightarrow T_{nm}(t) = C_{nm}e^{-\kappa\pi^2((n^2/a^2)+(m^2/b^2))t}$

3. Using the orthogonality of the trigonometric functions and $\int_0^b \int_0^a \sin^2 \frac{n\pi x}{a} \sin^2 \frac{m\pi y}{b}\, dx\, dy = \frac{1}{4}ab$, one obtains that the d_{nm} in
$f(x, y) = \sum_{m=1}^{\infty}\sum_{n=1}^{\infty} d_{nm} \cos \frac{n\pi x}{a} \sin \frac{m\pi y}{b}$ satisfy $d_{nm} = \frac{4}{ab}\int_0^b \int_0^a f(x, y) \cos \frac{n\pi x}{a} \sin \frac{m\pi y}{b}\, dx\, dy$

Exercises 28.2–Part B, Page 643

2. (a) $u(x, y, t) = \sum_{m=1}^{\infty}\sum_{n=1}^{\infty} \frac{8}{5} \frac{(-(-1)^{n+m} + 2(-1)^{1+m} + (-1)^n + 2)\sin(nx)\sin(my)e^{-2\pi^2((n^2/\pi^2)+(m^2/\pi^2))t}}{\pi n^3 m^3}$

(b)

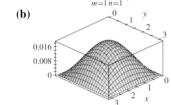 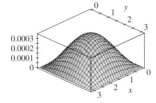 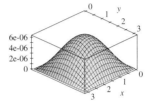

(c)

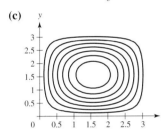 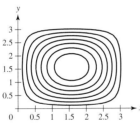 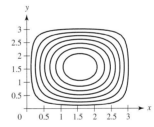

7. (a) $u(x, y, t) = \sum_{m=1}^{\infty}\sum_{n=1}^{\infty}\frac{4}{\pi^2}\left(\frac{1}{10}\frac{12(-1)^n(-1)^m - 12(-1)^m + 2\pi^2 n^2(-1)^m}{m^3 n^4} - \right.$

$\left. \frac{1}{5}\frac{-6 + 6(-1)^n + n^2\pi^2}{m^3 n^4}\right)\cos(nx)\sin(my)e^{-2\pi^2((n^2/\pi^2)+(m^2/\pi^2))t}\sum_{m=1}^{\infty}2\frac{\left(-\frac{1}{60}\frac{\pi^4(-1)^m}{m^3} + \frac{1}{60}\frac{\pi^4}{m^3}\right)\sin(my)e^{-2m^2 t}}{\pi^2}$

(b)

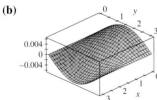 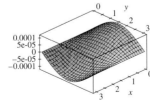 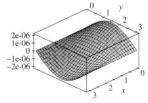

(d)

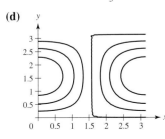 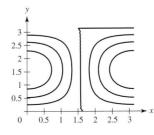 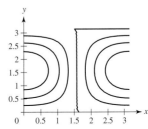

12. (a) $u(x, y, t) = \sum_{m=1}^{\infty} \sum_{n=1}^{\infty} \frac{4}{\pi^2} \left(-\frac{1}{5} \frac{\pi^2 n(-1)^n m(-1)^m + 2\pi^2 nm(-1)^m}{m^3 n^4} - \frac{1}{5} \frac{\pi(\pi n(-1)^n + 2\pi n)}{m^2 n^4} \right) \sin(nx) \cos(my) e^{-2\pi^2((n^2/\pi^2)+(m^2/\pi^2))t} +$

$\sum_{n=1}^{\infty} \frac{1}{15} \frac{\pi(\pi n(-1)^n + 2\pi n) \sin(nx) e^{-2n^2 t}}{n^4}$

(b)

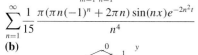 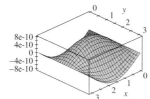 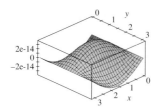

(d)

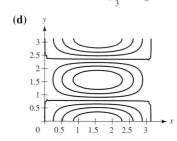 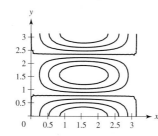 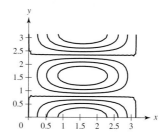

17. (a) $u(x, y, t) =$

$\sum_{m=0}^{\infty} \sum_{n=0}^{\infty} \frac{4}{\pi^2} \left(4 \frac{16\pi m(-1)^m(-1)^n + 8\pi(-1)^m(-1)^n}{(2m+1)^3(2n+1)^2} - 64 \frac{(-1)^n}{(2m+1)^3(2n+1)^2} \right) \sin(\frac{1}{2}(2n+1)x) \sin(\frac{1}{2}(2m+1)y) e^{-2(1/4(2n+1)^2 + 1/4(2m+1)^2)t}$

(b)

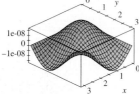 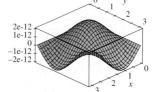

(d)

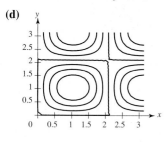 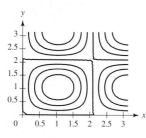 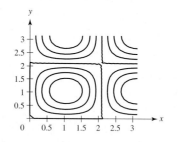

CHAPTER 29

Exercises 29.1–Part A, Page 648

3. With $\lambda = -\mu^2$, $\Theta(\theta) = \alpha_n \cos(\mu\theta) + \beta_n \sin(\mu\theta)$, $\Theta(-\pi) = \Theta(\pi) \Rightarrow -2\alpha \sin(\mu\pi) = 0$, $\Theta'(-\pi) = \Theta'(\pi) \Rightarrow 2\beta \sin(\mu\pi)\mu = 0$, so $\mu = n$, resulting in eigenfunctions $\alpha_n \cos n\theta + \beta_n \sin n\theta$ **4.** $R(r) = c_1 + c_2 \ln r$

Exercises 29.1–Part B, Page 648

2. (a) $u(r, \theta) = \frac{1}{2} r \sin\theta$ **(b)**

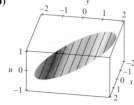   **(c)** 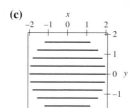   **(d)**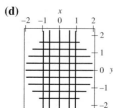

13. **(a)** $\dfrac{\int_{-\pi}^{\pi} \sin\theta\, d\theta}{\pi} = 0$ **(b)** $u(r,\theta) = r\sin\theta$ **(c)**   **(d) and (e)**

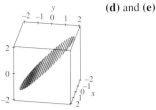

19. **(a)** $u(r,\theta) = \displaystyle\sum_{n=1}^{\infty} \frac{3}{2} \frac{16^{-n}((-1)^{1+n}+1)r^{4n}\sin 4n\theta}{\pi n^3}$ **(b)**

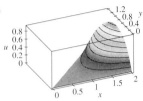

(c)

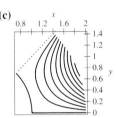

(d) Flow lines are perpendicular to level curves.

30. **(a)** $u(r,\theta) = \displaystyle\sum_{n=1}^{\infty} 24\, \frac{16^{-1+n}((-1)^{1+n}+1)(r^{4n}-(\frac{1}{r})^{4n})\sin 4n\theta}{\pi(256^n-1)n^3}$ **(b)**

(c)

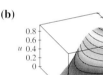
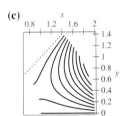

(d) Flow lines are perpendicular to level curves.

41. **(a)** $u(r,\theta) = \displaystyle\sum_{n=1}^{\infty} 24\, \frac{16^{-1+n}((-1)^{1+n}+1)(r^{4n}+(\frac{1}{r})^{4n})\sin 4n\theta}{\pi(256^n+1)n^3}$ **(b)**

(c)

(d) Flow lines are perpendicular to level curves.

Exercises 29.2, Page 654

4. **(a)** $u(r,z) = \displaystyle\sum_{k=1}^{\infty} A_k J_0(\lambda_k r)\cosh(\lambda_k z),\; A_k = \dfrac{2\int_0^{\sigma} r f(r) J_0(\lambda_k r)\, dr}{\sigma^2 \cosh(\lambda_k h) J_1^2(\lambda_k \sigma)}$ **(b)**

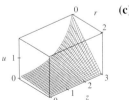

(c)

(d)

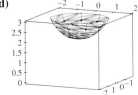
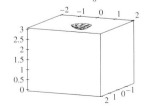

(e)

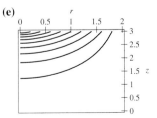

(f) Flow lines are perpendicular to level curves.

15. (a) $u(r, z) = \sum\limits_{k=1}^{\infty} A_k J_0(\lambda_k r) \sinh(\lambda_k z)$, $A_k = \dfrac{2 \int_0^\sigma r f(r) J_0(\lambda_k r)\, dr}{\sigma^2 \sinh(\lambda_k h) J_1^2(\lambda_k \sigma)}$ **(b)**

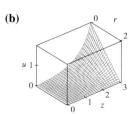
(c)
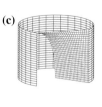

(d)
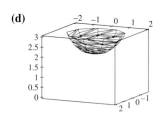 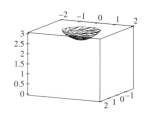 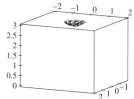
(e)
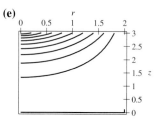

(f) Flow lines are perpendicular to level curves.

Exercises 29.3, Page 658

3. (a) $u(r, t) = \sum\limits_{k=1}^{\infty} A_k J_0(\lambda_k r) \cos(c\lambda_k t)$, $A_k = \dfrac{2 \int_0^\sigma r f(r) J_0(\lambda_k r)\, dr}{\sigma^2 J_1^2(\lambda_k \sigma)}$

(b)
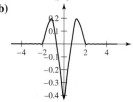 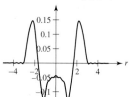 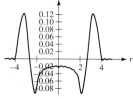

(d)
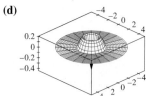 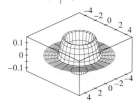 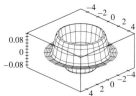

Exercises 29.4–Part A, Page 663

1. Make the substitution $z = \cos\phi$.   **5.** $R(\rho) = c_1 \rho^k + c_2 \rho^{-k-1}$

Exercises 29.4–Part B, Page 663

3. (a) $u(\rho, \phi) = \sum\limits_{k=0}^{\infty} A_k \rho^k P_k(\cos\phi)$, $A_k = \frac{2k+1}{2\sigma^k} \int_0^\pi f(\phi) P_k(\cos\phi) \sin\phi\, d\phi$

(b)
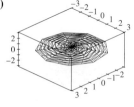 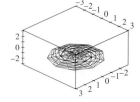 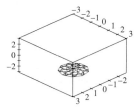

(c) 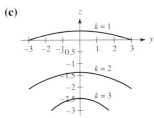 **(d)** 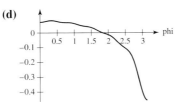 **(e)**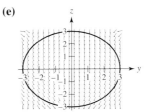

13. (a) $\sum_{k=0}^{\infty} A_k \rho^{-k-1} P_k(\cos\phi)$ **(b)**

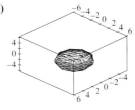 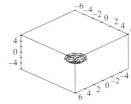 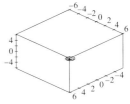

(c) 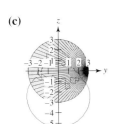 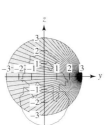 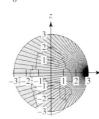 **(d)** 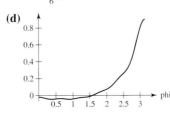 **(e)**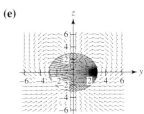

Exercises 29.5, Page 667

1. (b) $\alpha_k S^k + \beta_k S^{-k-1} = f_k$ **(c)** $a_k \sigma^k = \alpha_k \sigma^k + \beta_k \sigma^{-k-1}$ **(d)** $\kappa a_k k \sigma^{k-1} = \alpha_k k \sigma^{k-1} + \beta_k(-k-1)\sigma^{-k-2}$

(e) $\beta_k = \dfrac{S f_k k \sigma (\kappa-1)(S\sigma^2)^k}{\kappa k \sigma(\sigma^2)^k - \kappa k (S^2)^k S - k\sigma(\sigma^2)^k - k(S^2)^k S - (S^2)^k S}$; $a_k = \dfrac{f_k S^{(k+1)}(2k+1)}{-\kappa k \sigma(\sigma^2)^k + \kappa k(S^2)^k S + k\sigma(\sigma^2)^k + k(S^2)^k S + (S^2)^k S}$;

$\alpha_k = -\dfrac{f_k S^{(k+1)}(k+\kappa k+1)}{\kappa k \sigma(\sigma^2)^k - \kappa k(S^2)^k S - k\sigma(\sigma^2)^k - k(S^2)^k S - (S^2)^k S}$

(f) $U(\rho,\phi) = \displaystyle\sum_{k=0}^{\infty}\left(-\dfrac{f_k S^{k+1}(k+\kappa k+1)}{\kappa k \sigma(\sigma^2)^k - \kappa k(S^2)^k S - k\sigma(\sigma^2)^k - k(S^2)^k S - (S^2)^k S}\rho^k +\right.$

$\left.\dfrac{S f_k k \sigma (\kappa-1)(S\sigma^2)^k}{\kappa k \sigma(\sigma^2)^k - \kappa k(S^2)^k S - k\sigma(\sigma^2)^k - k(S^2)^k S - (S^2)^k S}\rho^{-k-1}\right) P_k(\cos\phi)$;

$u(\rho,\phi) = \displaystyle\sum_{k=0}^{\infty} \dfrac{f_k S^{k+1}(2k+1)}{-\kappa k \sigma(\sigma^2)^k + \kappa k(S^2)^k S + k\sigma(\sigma^2)^k + k(S^2)^k S + (S^2)^k S}\rho^k P_k(\cos\phi)$

2. (a) $U(\rho,\phi) = \sum_{k=0}^{\infty}(\alpha_k \rho^k + \beta_k \rho^{-k-1}) P_k(\cos\phi)$ **(c)** **(d)** Flow lines are perpendicular to level curves.

(b) Take eleven terms.

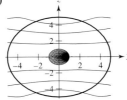

14. (b) $\alpha_k k S^{k-1} + \beta_k(-k-1)S^{-k-2} = f_k$ **(c)** $a_k \sigma^k = \alpha_k \sigma^k + \beta_k \sigma^{-k-1}$ **(d)** $\kappa a_k k \sigma^{k-1} = \alpha_k k \sigma^{k-1} + \beta_k(-k-1)\sigma^{-k-2}$

(e) $\alpha_k = \dfrac{f_k S^{k+2}(\kappa k+k+1)}{k(k(S^2)^k S\kappa + k(S^2)^k S + (S^2)^k S + \kappa k \sigma(\sigma^2)^k - k\sigma(\sigma^2)^k - (\sigma^2)^k \sigma + (\sigma^2)^k \sigma\kappa)}$;

$\beta_k = -\dfrac{f_k S^2 \sigma(\kappa-1)(\sigma^2 S)^k}{k(S^2)^k S\kappa + k(S^2)^k S + (S^2)^k S + \kappa k \sigma(\sigma^2)^k - k\sigma(\sigma^2)^k - (\sigma^2)^k \sigma + (\sigma^2)^k \sigma\kappa}$;

$a_k = \dfrac{f_k S^{k+2}(2k+1)}{k(k(S^2)^k S\kappa + k(S^2)^k S + (S^2)^k S + \kappa k \sigma(\sigma^2)^k - k\sigma(\sigma^2)^k - (\sigma^2)^k \sigma + (\sigma^2)^k \sigma\kappa)}$; Clearly a_0 and α_0 are undetermined. Moreover if $f_0 = 0$, then $\beta_0 = 0$ and the equation found in (c) dictates that $a_0 = \alpha_0$. Finally if $f_0 \neq 0$, then $\beta_0 \neq 0$ which contradicts the equation found in (d), so in this case there is no solution.

(f) $U(\rho, \phi) = \sum_{k=0}^{\infty} (\dfrac{f_k S^{k+2}(\kappa k + k + 1)}{k(k(S^2)^k S\kappa + k(S^2)^k S + (S^2)^k S + \kappa k\sigma(\sigma^2)^k - k\sigma(\sigma^2)^k - (\sigma^2)^k\sigma + (\sigma^2)^k\sigma\kappa)} \rho^k - $

$\dfrac{f_k S^2 \sigma(\kappa - 1)(\sigma^2 S)^k}{k(S^2)^k S\kappa + k(S^2)^k S + (S^2)^k S + \kappa k\sigma(\sigma^2)^k - k\sigma(\sigma^2)^k - (\sigma^2)^k\sigma + (\sigma^2)^k\sigma\kappa}) \rho^{-k-1} P_k(\cos\phi);$

$u(\rho, \phi) = \sum_{k=0}^{\infty} \dfrac{f_k S^{k+2}(2k + 1)}{k(k(S^2)^k S\kappa + k(S^2)^k S + (S^2)^k S + \kappa k\sigma(\sigma^2)^k - k\sigma(\sigma^2)^k - (\sigma^2)^k\sigma + (\sigma^2)^k\sigma\kappa)} \rho^k P_k(\cos\phi)$

15. (a) $u(\rho, \phi) = \sum_{k=0}^{\infty} a_k \rho^k P_k(\cos\phi)$; $U(\rho, \phi) = \sum_{k=0}^{\infty} (\alpha_k \rho^k + \beta_k \rho^{-k-1}) P_k(\cos\phi)$ **(b)** Take eleven terms.
(c) (See plot.)   **(d)** Flow lines are perpendicular to level curves.

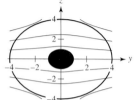

CHAPTER 30

Exercises 30.1–Part A, Page 674

2. (a) $L_x[f_x(x, y)] = \dfrac{as}{s^2+1}$ **(b)** $L_y[f_x(x, y)] = \dfrac{a\cos x}{s}$ **(c)** $L_x[f_y(x, y)] = -\dfrac{be^{-y}}{s}$ **(d)** $L_y[f_y(x, y)] = -\dfrac{b}{s+1}$ **(e)** $L_x[f_{xx}(x, y)] = -\dfrac{a}{s^2+1}$
   (f) $L_x[f_{yy}(x, y)] = \dfrac{be^{-y}}{s}$ **(g)** $L_y[f_{xx}(x, y)] = -\dfrac{a\sin x}{s}$ **(h)** $L_y[f_{yy}(x, y)] = \dfrac{b}{s+1}$

Exercises 30.1–Part B, Page 674

2. (a) $u(x, y) = (y - \frac{1}{6}x^2)^2 H(y - \frac{1}{6}x^2)$
(b)

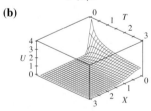

13. (a) $sU(x, s) = 2\dfrac{\partial^2}{\partial x^2} U(x, s)$

   (b) $U(x, s) = -10 \dfrac{e^{1/2\sqrt{2s}x} - e^{(p-1/2x)\sqrt{2s}}}{(s^2+1)(e^{p\sqrt{2s}} - 1)}$

   (c) $\lim_{p\to\infty} U(x, s) = \dfrac{10}{s^2+1} e^{-\sqrt{\frac{s}{2}}x}$

(d)

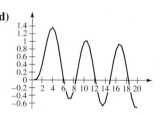

18. (a) $u(x, t) = \dfrac{5}{2} \dfrac{e^{-4\frac{x^2}{t}}}{\sqrt{t}\sqrt{\pi}}$

20. (a) $(H(t - \frac{x+c\pi}{c} + \pi) - H(t - \frac{x+c\pi}{c})) \sin(-t + \frac{x+c\pi}{c})$

(b)

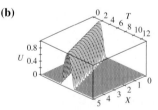

28. (a) $s(sU(x, s) - xe^{-x}) = \dfrac{\partial^2}{\partial x^2} U(x, s)$ **(b)** $U(x, s) = (s^3 e^{-x}x - sxe^{-x} - 2se^{-x} - \dfrac{s^5(s^2 p(e^{ps})^2 - p(e^{ps})^2 - 2(e^{ps})^2 + 2e^{p+ps})e^{sx}}{e^{p+ps}(s^4(e^{ps})^2 - 2s^2(e^{ps})^2 + (e^{ps})^2 - s^4 - 1 + 2s^2)} +$

$2\dfrac{s^3(s^2 p(e^{ps})^2 - p(e^{ps})^2 - 2(e^{ps})^2 + 2e^{p+ps})e^{sx}}{e^{p+ps}(s^4(e^{ps})^2 - 2s^2(e^{ps})^2 + (e^{ps})^2 - s^4 - 1 + 2s^2)} - \dfrac{s(s^2 p(e^{ps})^2 - p(e^{ps})^2 - 2(e^{ps})^2 + 2e^{p+ps})e^{sx}}{e^{p+ps}(s^4(e^{ps})^2 - 2s^2(e^{ps})^2 + (e^{ps})^2 - s^4 - 1 + 2s^2)} +$

$\dfrac{(s^2 p - p - 2 + 2e^{p+ps})s^5 e^{ps-p}e^{-sx}}{(s^4 - 2s^2 + 1)(-1 + (e^{ps})^2)} - 2\dfrac{(s^2 p - p - 2 + 2e^{(p+ps)})s^3 e^{(ps-p)}e^{(-sx)}}{(s^4 - 2s^2 + 1)(-1 + (e^{(ps)})^2)} + \dfrac{(s^2 p - p - 2 + 2e^{p+ps})s e^{ps-p}e^{-sx}}{(s^4 - 2s^2 + 1)(-1 + (e^{ps})^2)})/(s^4 - 2s^2 + 1)$

(c) $U(x, s) = \dfrac{s(s^2 e^{-x}x - xe^{-x} - 2e^{-x} + 2e^{-sx})}{s^4 - 2s^2 + 1}$ **(d)** $u(x, t) = \frac{1}{2}(x - t)e^{t-x} + \frac{1}{2}(x + t)e^{-x-t} + \frac{1}{2}((t - x)e^{t-x} + (-t + x)e^{-t+x})H(t - x)$

(e)

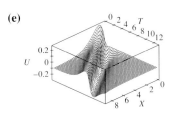

Exercises 30.2–Part A, Page 681

2. $\int_{-1}^{1} \cos(\alpha(\beta - x)) \, d\beta = -\frac{\sin(-\alpha + \alpha x) + \sin(\alpha + \alpha x)}{\alpha}$

Exercises 30.2–Part B, Page 681

2. (a)

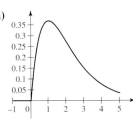

(b) $f(x) = \frac{1}{2} \frac{\int_{-\infty}^{\infty} \frac{e^{i\alpha x}}{(1 + i\alpha)^2} \, d\alpha}{\pi}$

(c) $f(x) = \frac{1}{\pi} \int_{-\infty}^{\infty} - \frac{(-1 + \alpha^2) \cos(\alpha x) - 2\alpha \sin(\alpha x)}{(1 + \alpha^2)^2} \, d\alpha$

(d) $A(\alpha) = -\frac{-1 + \alpha^2}{(1 + \alpha^2)^2}$; $B(\alpha) = 2 \frac{\alpha}{(1 + \alpha^2)^2}$

14. (a)

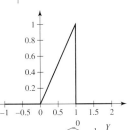

(b) $u(x, y) = \frac{1}{\pi} \int_0^{\infty} \int_{-\infty}^{\infty} f(\beta) \cos(\alpha(\beta - x)) e^{-\alpha y} \, d\beta \, d\alpha$

(c) $U(x, y) = \frac{1}{2} - \frac{2x \arctan(\frac{x-1}{y}) - y \ln(y^2 + x^2) + 2x \arctan(\frac{x}{y}) + y \ln(y^2 + 1 - 2x + x^2)}{\pi}$

(d)

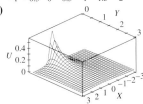

(e)

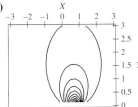

25. (a)

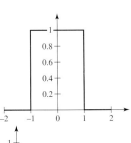

(b) $\int_{-\infty}^{\infty} f(\beta) \frac{\sinh(\alpha y)}{\sinh \alpha} \cos(\alpha(\beta - x)) \, d\beta = 2 \frac{\sinh(\alpha y) \sin(\alpha) \cos(\alpha x)}{\alpha \sinh(\alpha)}$

(d) $u(0, \frac{1}{2}) = 0.4725062710$ **(e)**

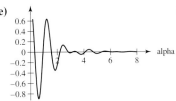 $u(5, \frac{1}{2}) = 0.1107956086 \times 10^{-5}$

37. (a)

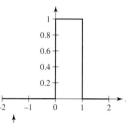

(b) $\int_{-\infty}^{\infty} f(\beta) \frac{\cosh(\alpha y)}{\cosh \alpha} \cos(\alpha(\beta - x))\, d\beta = \frac{\cosh(\alpha y)(\sin(\alpha)\cos(\alpha x) - \cos(\alpha)\sin(\alpha x) + \sin(\alpha x))}{\alpha \cosh(\alpha)}$

(d)

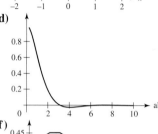

$u(0, \frac{1}{2}) = 0.4051093372$ **(e)**

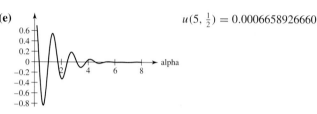

$u(5, \frac{1}{2}) = 0.0006658926660$

(f)

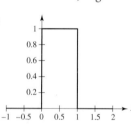

Exercises 30.3–Part B, Page 688

2. (a)

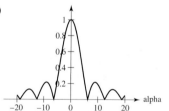

(b) $\mathcal{F}[f(x)] = -\dfrac{i(-e^{-i\alpha} + 1)}{\alpha}$

(c) $\mathcal{F}[f(x)] = -\dfrac{i(-e^{-i\alpha} + 1)}{\alpha}$

(d) $\frac{\pi}{2} \int_{-\infty}^{\infty} \mathcal{F}[f] e^{i\alpha x}\, d\alpha = \frac{1}{2} H(x) - \frac{1}{2} H(-x) - \frac{1}{2} H(x - 1) + \frac{1}{2} H(1 - x)$

(e) $\frac{1}{2} H(x) - \frac{1}{2} H(-x) - \frac{1}{2} H(x - 1) + \frac{1}{2} H(1 - x)$

(f)

(image at right)

12. (a) $\mathcal{F}[f(x)] = \dfrac{-1 + 2ie^{-2i\alpha}\alpha + e^{-2i\alpha}}{\alpha^2}$ $\mathcal{F}[f'(x)] = -\dfrac{i - 2\alpha e^{-2i\alpha} + ie^{-2i\alpha}}{\alpha}$

22. (a) $\mathcal{F}[f(x)] = \dfrac{(-\pi\alpha^2 + 2i\alpha + \pi)e^{-i\pi\alpha}}{\alpha^4 - 2\alpha^2 + 1} + 2\dfrac{i\alpha}{\alpha^4 - 2\alpha^2 + 1};$

$\mathcal{F}[xf(x)] = \dfrac{(-\pi^2 + 4i\pi\alpha^3 - 4i\pi\alpha + 2\pi^2\alpha^2 - \pi^2\alpha^4 + 2 + 6\alpha^2)e^{-i\pi\alpha}}{\alpha^6 - 3\alpha^4 + 3\alpha^2 - 1} + \dfrac{2 + 6\alpha^2}{\alpha^6 - 3\alpha^4 + 3\alpha^2 - 1}$

30. (a) $F(\alpha) = 2 - \dfrac{1 + e^{-i\pi\alpha}}{(\alpha - 2)(\alpha + 2)}$ **(b)** $g(x) = \frac{1}{2}(-\cos(2x) + 1)H(x) + \frac{1}{2}(-1 + \cos(2x))H(x - \pi)$ **(c)** $\mathcal{F}[g(x)] = -2\dfrac{i(-1 + e^{-i\pi\alpha})}{\alpha^3 - 4\alpha}$

Exercises 30.4–Part B, Page 693

2. (a) $u(x, t) = \frac{1}{4} H(x + 1 + t) - \frac{1}{4} H(-x - 1 - t) + \frac{1}{4} H(x + 1 - t) - \frac{1}{4} H(-x - 1 + t) - \frac{1}{4} H(x - 1 + t) + \frac{1}{4} H(-x + 1 - t) - \frac{1}{4} H(x - 1 - t) + \frac{1}{4} H(-x + 1 + t)$

(b) $u(x, t) = \frac{1}{2} H(x + 1 - t) - \frac{1}{2} H(x - 1 - t) + \frac{1}{2} H(x + 1 + t) - \frac{1}{2} H(x - 1 + t)$ **(c)**

(3D plot)

10. (a) $u(x, t) = \frac{1}{4}\sqrt{\pi}(\mathrm{erf}(x + t) + \mathrm{erf}(-x + t))$

(b) $u(x, t) = \frac{1}{4}\mathrm{erf}(x + t)\sqrt{\pi} + \frac{1}{4}\mathrm{erf}(-x + t)\sqrt{\pi}$

(c)

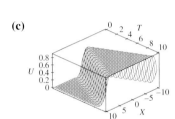

22. $u(x, t) = \dfrac{1}{\pi} \displaystyle\int_{-\infty}^{\infty} \dfrac{\cosh(\sqrt{-\alpha^2 + i\alpha t})\sin(\alpha)e^{i\alpha x}}{\alpha}\, d\alpha$

29. $u(x, t) = \dfrac{1}{2\pi} \displaystyle\int_{-\infty}^{\infty} (e^{-1/2t(1+\sqrt{1-4\alpha^2})}(4\alpha^2 + \sqrt{1-4\alpha^2} - 1) - (1 - 4\alpha^2 + \sqrt{1-4\alpha^2})e^{-1/2t(1-\sqrt{1-4\alpha^2})}\dfrac{e^{i\alpha x}\sin\alpha}{(-1+4\alpha^2)}\, d\alpha$

37. (a) $u(x, y) = \dfrac{1}{2} + \dfrac{1}{\pi}\arctan\dfrac{x}{y}$ **(b)**

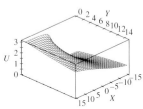

(c)

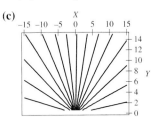

(d) The flow lines are perpendicular to the level curves.

Exercises 30.5, Page 695

2. (a) $u(x, t) = \dfrac{1}{2}x\,\mathrm{erf}\left(\dfrac{(\frac{1}{4}x+\frac{1}{4})\sqrt{2}}{\sqrt{t}}\right) + \sqrt{\dfrac{2}{\pi t}}\,te^{-1/8\frac{(x+1)^2}{t}} - \dfrac{1}{2}x\,\mathrm{erf}\left(\dfrac{(\frac{1}{4}x-\frac{1}{4})\sqrt{2}}{\sqrt{t}}\right) - \sqrt{\dfrac{2}{\pi t}}\,te^{-1/8\frac{(x-1)^2}{t}}$

(b)

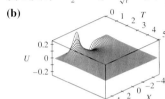

(c)

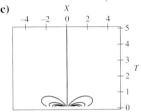

(d)

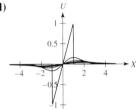

13. (a) $u(x, t) = (H(x + rt + 1) - H(x + rt - 1))\sin(\pi(x + rt))$

21. (a) $u(x, t) = \dfrac{1}{2}\,\mathrm{erf}\left(\dfrac{tr+1+x}{\sqrt{2t}}\right) - \dfrac{1}{2}\,\mathrm{erf}\left(\dfrac{tr-2+x}{\sqrt{2t}}\right)$

(b)

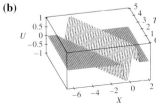

(c)

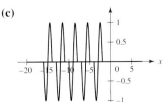

(b)

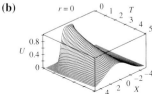
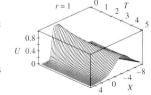

(c)

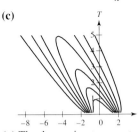

(d)

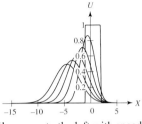

(e) $-t + \dfrac{1}{2}$ **(f)** $\mathrm{erf}\left(\dfrac{3}{4}\sqrt{\dfrac{2}{t}}\right)$

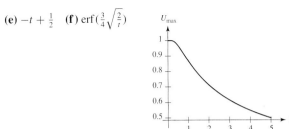

(g) The decreasing temperature profile moves to the left with speed r. **(h)** The decreasing temperature profile moves to the right with speed r.

32. (a)

(b) $\dfrac{e^{-(y^2+x^2/4\kappa t+1)}}{4\kappa t + 1}$ **(c)**

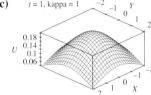

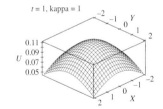

Exercises 30.6, Page 698

1. $u(x, y) = \int_{-\infty}^{\infty} \dfrac{1}{2} \dfrac{\sin(\frac{\pi y}{b}) f(x - \sigma)}{b(\cosh(\frac{\pi \sigma}{b}) + \cos(\frac{\pi y}{b}))} \, d\sigma$

2. (a) $u(x, y) = \dfrac{1}{\pi} \int_0^{\infty} \int_{-\infty}^{\infty} \dfrac{g(\beta) \sinh(\alpha(b - y)) \cos(\alpha(\beta - x))}{\sinh(b\alpha)} \, d\beta \, d\alpha$ (b) $u(x, y) = \int_{-\infty}^{\infty} \dfrac{1}{2} \dfrac{\sin(\frac{\pi(b-y)}{b}) g(x - \sigma)}{b(\cosh(\frac{\pi \sigma}{b}) + \cos(\frac{\pi(b-y)}{b}))} \, d\sigma$

   (c) The inverse Fourier transform will result in an absolutely integrable function over $(-\infty, \infty)$.

3. (a) $u(x, y) = \dfrac{1}{\pi} \int_0^{\infty} \int_{-\infty}^{\infty} \dfrac{f(\beta) \cosh(\alpha y) \cos(\alpha(\beta - x))}{\cosh(b\alpha)} \, d\beta \, d\alpha$ (b) $u(x, y) = \int_{-\infty}^{\infty} \dfrac{\cos(\frac{1}{2} \frac{\pi y}{b}) \cosh(\frac{1}{2} \frac{\pi \sigma}{b}) f(x - \sigma)}{b(\cosh(\frac{\pi \sigma}{b}) + \cos(\frac{\pi y}{b}))} \, d\sigma$

6. (a) $u(x, y) = \dfrac{1}{\pi} \int_{-\infty}^{\infty} \int_{-\infty}^{\infty} \dfrac{g(\beta) \sinh(\alpha(y - b)) \cos(\alpha(\beta - x))}{\cosh(b\alpha)\alpha} \, d\beta \, d\alpha$ (b) $u(x, y) = \dfrac{1}{2\pi} \int_{-\infty}^{\infty} \ln\left(\dfrac{\sin(\frac{1}{2} \frac{\pi(y-b)}{b}) + \cosh(\frac{1}{2} \frac{\pi \sigma}{b})}{\cosh(\frac{1}{2} \frac{\pi \sigma}{b}) - \sin(\frac{1}{2} \frac{\pi(y-b)}{b})}\right) g(x - \sigma) \, d\sigma$

   (c) The boundary $y = b$ is kept at temperature 0, while the local heat flux on the boundary $y = 0$ is prescribed by $u_y(x, 0) = g(x)$.

Exercises 30.7–Part A, Page 702   3. Integrate by parts twice.   5. $1 - H(x) = \begin{cases} 1 & x < 1 \\ 0 & x > 1 \end{cases}$ $H(1 - x) = \begin{cases} 0 & 1 - x < 0 \\ 1 & 1 - x > 0 \end{cases}$

Exercises 30.7–Part B, Page 702

2. (a) $\mathcal{F}_s\left[\frac{1}{x}\right] = \frac{1}{2}\sqrt{2\pi}$

17. (a)

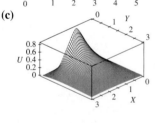

(b) $u(x, y) = 2(-\arctan(\frac{x-2}{y}) - \frac{1}{4}y \ln(y^2 + 4 - 4x + x^2) + \frac{1}{2}x \arctan(\frac{x-2}{y}) +$

$\frac{1}{4}y \ln(y^2 + 4 + 4x + x^2) + \arctan(\frac{x-1}{y}) + \arctan(\frac{1+x}{y}) - \arctan(\frac{2+x}{y}) +$

$\frac{1}{2}y \ln(y^2 + 1 - 2x + x^2) - x \arctan(\frac{x-1}{y}) - \frac{1}{2}y \ln(y^2 + 1 + 2x + x^2) +$

$x \arctan(\frac{1+x}{y}) - \frac{1}{2}x \arctan(\frac{2+x}{y}))/\pi$

(c)

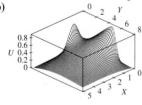

(d)

30. (a) $\frac{1}{\pi}((-6 - y) \arctan(\frac{6+y}{x}) + (-6 + y) \arctan(-\frac{6+y}{x}) + (6 + \frac{6}{5}y) \arctan(\frac{5+y}{x}) + (-\frac{6}{5}y + 6) \arctan(-\frac{5+y}{x}) + (2 + 2x) \arctan(\frac{x+1}{y}) + (2 -$

$2x) \arctan(\frac{x-1}{y}) + (-x - 2) \arctan(\frac{2+x}{y}) + (x - 2) \arctan(\frac{x-2}{y}) - y \ln(y^2 + 1 + 2x + x^2) + \frac{1}{2}y \ln(y^2 + 4 + 4x + x^2) + y \ln(y^2 + 1 - 2x + x^2) -$

$\frac{1}{2}x \ln(x^2 + 36 - 12y + y^2) - \frac{3}{5}x \ln(x^2 + 25 + 10y + y^2) + \frac{3}{5}x \ln(x^2 + 25 - 10y + y^2) - \frac{1}{2}y \ln(y^2 + 4 - 4x + x^2) + \frac{1}{2}x \ln(x^2 + 36 + 12y + y^2))$

(b)

(c)

Exercises 30.8, Page 707

3. (a) $\mathcal{F}_c[f(x)] = \frac{\sqrt{2}a}{\sqrt{\pi}(a^2 + \alpha^2)}$ (b) $\mathcal{F}_c[f(x)] = \frac{\sqrt{2}a}{\sqrt{\pi}(a^2 + \alpha^2)}$ (c) Except when $a = 0$, $f(x)$ is not an even function.   (d) $\mathcal{F}_c[f''(x)] = \frac{a^3 \sqrt{2}}{\sqrt{\pi}(a^2 + \alpha^2)}$

   (e) $\mathcal{F}_s[f''(x)] = \frac{\sqrt{2}a\alpha^2}{\sqrt{\pi}(a^2 + \alpha^2)}$ (f) $\mathcal{F}_c[f(\frac{x}{\lambda})] = \frac{\lambda\sqrt{2}a}{\sqrt{\pi}(a^2 + \alpha^2\lambda^2)}$; $\mathcal{F}_s[f(\frac{x}{\lambda})] = \frac{\lambda^2\sqrt{2}\alpha}{\sqrt{\pi}(a^2 + \alpha^2\lambda^2)}$

   (g) $\mathcal{F}_c[xf(x)] = -\frac{\sqrt{2}(-a^2 + \alpha^2)}{\sqrt{\pi}(a^2 + \alpha^2)^2}$; $\mathcal{F}_s[xf(x)] = 2\frac{\sqrt{2}a\alpha}{\sqrt{\pi}(a^2 + \alpha^2)^2}$

   (h) $\mathcal{F}_c[f(x)\cos(\omega x)] = \frac{\sqrt{2}a(a^2 + \alpha^2 + \omega^2)}{\sqrt{\pi}(a^2 + \alpha^2 + 2\alpha\omega + \omega^2)(a^2 + \alpha^2 - 2\alpha\omega + \omega^2)}$ $\mathcal{F}_s[f(x)\cos(\omega x)] = \frac{\sqrt{2}a(a^2 + \alpha^2 - \omega^2)}{\sqrt{\pi}(a^2 + \alpha^2 + 2\alpha\omega + \omega^2)(a^2 + \alpha^2 - 2\alpha\omega + \omega^2)}$

   (i) $\mathcal{F}_c[f(x)\sin(\omega x)] = -\frac{\sqrt{2}\omega(a^2 - \alpha^2 - \omega^2)}{\sqrt{\pi}(a^2 + \alpha^2 + 2\alpha\omega + \omega^2)(a^2 + \alpha^2 - 2\alpha\omega + \omega^2)}$ $\mathcal{F}_s[f(x)\sin(\omega x)] = 2\frac{\sqrt{2}a\alpha\omega}{\sqrt{\pi}(a^2 + \alpha^2 + 2\alpha\omega + \omega^2)(a^2 + \alpha^2 - 2\alpha\omega + \omega^2)}$

   (j) $\mathcal{F}_c[\int_x^{\infty} f(\sigma)d\sigma] = \frac{\sqrt{2}}{\sqrt{\pi}(a^2 + \alpha^2)}$ (k) $\mathcal{F}_c[f'(x)] = -\frac{\sqrt{2}a^2}{\sqrt{\pi}(a^2 + \alpha^2)}$

23. (a) $u(x, y) = -\frac{2}{\pi} \int_0^\infty \frac{\sin(\alpha) \sinh(\alpha(y-1)) \cos(\alpha x)}{\alpha \sinh(\alpha)} \, d\alpha$

(b)

(c)

(d) $\frac{2}{\pi} \int_{20}^\infty \frac{\sin(\alpha) \sinh(\frac{\alpha}{2}) \cos(2\alpha)}{\alpha \sinh(\alpha)} = -0.7387722634 \times 10^{-6}$

32. (a) $u(x, y) = \frac{2}{\pi} \int_0^\infty \frac{\sin(\alpha) \sinh(\alpha y) \cos(\alpha x)}{\alpha \sinh(\alpha)} \, d\alpha$

(b)

(c)

] **(d)** $\frac{2}{\pi} \int_{20}^\infty \frac{\sin(\alpha) \sinh(\frac{\alpha}{2}) \cos(2\alpha)}{\alpha \sinh(\alpha)} \approx -0.7387722634 \times 10^{-6}$

41. (a) $u(x, y) = \int_0^\infty 2 \frac{\cos(\alpha x) \sin(\alpha) \sinh(\alpha y)}{\pi \alpha \sinh(\alpha)} + \frac{\sqrt{2} \cos(\alpha x) \sinh(\alpha y)}{\sqrt{\pi} \alpha^2 \sinh(\alpha)} - \frac{\sqrt{2} \cos(\alpha x)}{\sqrt{\pi} \alpha^2} - \frac{\sqrt{2} \cos(\alpha x) \sinh(\alpha(y-1))}{\sqrt{\pi} \alpha^2 \sinh(\alpha)} \, d\alpha$

(b)

(c)

(d) $\int_{20}^\infty 2 \frac{\cos(2\alpha) \sin(\alpha) \sinh(\frac{1}{2}\alpha)}{\pi \alpha \sinh(\alpha)} + 2 \frac{\sqrt{2} \cos(2\alpha) \sinh(\frac{1}{2}\alpha)}{\sqrt{\pi} \alpha^2 \sinh(\alpha)} - \frac{\sqrt{2} \cos(2\alpha)}{\sqrt{\pi} \alpha^2} \, d\alpha \approx 0.0007724864554$

63. (a) $u(x, y) = \frac{\sqrt{2}}{\pi} \int_0^\infty \left(\frac{\sqrt{2}(\cos(\alpha) + \alpha \sin(\alpha) - 1) \cosh(\alpha y)}{\alpha \cosh(\alpha)} + \sqrt{2} \sin(\alpha) \left(\frac{\sinh(\alpha y)}{\alpha} - \frac{\sinh(\alpha) \cosh(\alpha y)}{\cosh(\alpha)\alpha} \right) \right) \frac{\cos(\alpha x)}{\alpha} \, d\alpha$

(b)

(c)

(d) $\frac{\sqrt{2}}{\pi} \int_{20}^\infty \left(\frac{\sqrt{2}(\cos(\alpha) + \alpha \sin(\alpha) - 1) \cosh(\frac{1}{2}\alpha)}{\alpha \cosh(\alpha)} + \sqrt{2} \sin(\alpha) \left(\frac{\sinh(\frac{1}{2}\alpha)}{\alpha} - \frac{\sinh(\alpha) \cosh(\frac{1}{2}\alpha)}{\cosh(\alpha)\alpha} \right) \right) \frac{\cos(2\alpha)}{\alpha} \, d\alpha \approx -0.6869479346 \times 10^{-6}$

CHAPTER 31

Exercises 31.1–Part A, Page 715

1. $\|\mathbf{P}\|_k = (2^k + 3^k + 4^k)^{\frac{1}{k}}$ **3.** $\cos^2 \alpha + \cos^2 \beta + \cos^2 \gamma = \frac{a^2+b^2+c^2}{|a|^2+|b|^2+|c|^2} = 1$ **4.** $\mathbf{u} = \left[\frac{1}{146}\sqrt{146}, \frac{4}{73}\sqrt{146}, -\frac{9}{146}\sqrt{146} \right]^T$

Exercises 31.1–Part B, Page 715

2. (a) $\|\mathbf{A}\|_1 = 11$; $\|\mathbf{A}\|_2 = \sqrt{69}$; $\|\mathbf{A}\|_3 = 521^{1/3}$; $\|\mathbf{A}\|_\infty = 8$ **(c)**   **(d)** $\|\mathbf{A} + \mathbf{B}\|^2 + \|\mathbf{A} - \mathbf{B}\|^2 = 352$

(b) $\|\mathbf{B}\|_1 = 17$; $\|\mathbf{B}\|_2 = \sqrt{107}$; $\|\mathbf{B}\|_3 = 713^{1/3}$; $\|\mathbf{B}\|_\infty = 7$ **(e)** $\mathbf{u} = \left[\frac{3}{143}\sqrt{286}, -\frac{15}{286}\sqrt{286}, \frac{5}{286}\sqrt{286} \right]^T$

(f) It is the half line starting at **A** in the direction of **B**

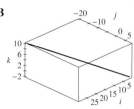

(g) $\{\alpha, \beta, \gamma\} = \left\{\frac{33}{20}, \frac{11}{20}, \frac{7}{4}\right\}$

(h) *Hint:* $\frac{1}{2}(\mathbf{A} + \mathbf{B}) = \mathbf{A} + \frac{1}{2}(\mathbf{B} - \mathbf{A})$

(i) $\frac{1}{2}(\mathbf{A} + \mathbf{B}) = \mathbf{A} + \frac{1}{2}(\mathbf{B} - \mathbf{A}); \frac{1}{3}(\mathbf{A} + \mathbf{B})$
$= \frac{1}{2}\mathbf{A} + \frac{1}{3}(\mathbf{B} - \frac{1}{2}\mathbf{A})$

(j) $\frac{1}{2}\mathbf{B} - \frac{1}{2}\mathbf{A} = \frac{1}{2}(\mathbf{B} - \mathbf{A})$

12. $\alpha = \frac{3705}{389}\sqrt{221}, \beta = \frac{1950}{389}\sqrt{685}$ **17.** $\mathbf{P}_1 - \mathbf{P}_4 = \mathbf{P}_2 - \mathbf{P}_3; \mathbf{P}_2 - \mathbf{P}_1 = \mathbf{P}_3 - \mathbf{P}_4$ **22.** See exercise (2i) and use $\mathbf{A} = \mathbf{P}_1 - \mathbf{P}_3, \mathbf{B} = \mathbf{P}_2 - \mathbf{P}_3$

Exercises 31.2–Part A, Page 719

2. $\langle x, y \rangle = (3x_1 + x_2)y_1 + (x_1 + 3x_2)y_2$

5. Observe $f(x_1, x_2) = \mathbf{x}^\mathsf{T} A\mathbf{x}$ has $(0, 0)$ as its only critical point, and $f_{xx}(0, 0) = 10$ is a minimum by the 2nd derivative test.

Exercises 31.2–Part B, Page 719

2. (a), (b) and **(c)** $\alpha = -\frac{22}{7}, \beta = \frac{31}{7}, \gamma = -6$ **8.** $\mathbf{A} \cdot \mathbf{B} = -19, \theta \approx 108°$

18. (a) and **(d)** $\mathbf{A} \cdot \mathbf{B} = -54$ **(b)** $(r\mathbf{A} + s\mathbf{B}) \cdot \mathbf{C} = 635$ **(c)** $\mathbf{A} \cdot (r\mathbf{B} + s\mathbf{C}) = -660$

23. (a) and **(d)** $\mathbf{A} \cdot \mathbf{B} = -43 - 70i$ **(b)** $(z_1\mathbf{A} + z_2\mathbf{B}) \cdot \mathbf{C} = 261 + 830i$ **(c)** $\mathbf{A} \cdot (z_1\mathbf{B} + z_2\mathbf{C}) = 1 - 942i$ **(d)** See (a).

28. (a) and **(b)** $\alpha = -\frac{3}{13}$

38. (a) and **(b)** $\alpha = -\frac{21,537}{17,258}, \beta = \frac{4643}{8629}$

(c)

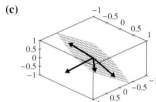

(d) Shortest vector $\mathbf{B} - \alpha\mathbf{A}$ occurs when $\mathbf{B} - \alpha\mathbf{A} \perp \mathbf{A}$

Exercises 31.3–Part A, Page 722

1. $\mathbf{A} \times (\mathbf{B} + \mathbf{C}) = [2, 2, -4]^\mathsf{T}$ **3.** $\mathbf{B} \times \mathbf{A} = [0, -2, 2]^\mathsf{T}$ **4.** $\mathbf{A} \times (\mathbf{B} \times \mathbf{C}) = [2, -2, 0]^\mathsf{T}$

Exercises 31.3–Part B, Page 722

2. $(\mathbf{A} \times \mathbf{B}) \times (\mathbf{C} \times \mathbf{D}) = [(a_3b_1 - a_1b_3)(c_1d_2 - c_2d_1) - (a_1b_2 - a_2b_1)(c_3d_1 - c_1d_3),$
$(a_1b_2 - a_2b_1)(c_2d_3 - c_3d_2) - (a_2b_3 - a_3b_2)(c_1d_2 - c_2d_1), (a_2b_3 - a_3b_2)(c_3d_1 - c_1d_3) - (a_3b_1 - a_1b_3)(c_2d_3 - c_3d_2)]^\mathsf{T}$

9. (a) and **(b)** $\mathbf{A} \times \mathbf{B} = 2\sqrt{6434}, \theta \approx 103°$ **19.** $\mathbf{C} = [264, -132, 143]^\mathsf{T}$

27. (a) $\mathbf{A} \times (\mathbf{B} \times \mathbf{C}) = [-1380, -633, -1226]^\mathsf{T}$ **(c)** $(\mathbf{A} \times \mathbf{B}) \times \mathbf{C} = [-1980, -73, -786]^\mathsf{T}$ **37.** area $= \frac{1}{2}\sqrt{12417}$

(b)

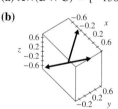

(d) $\mathbf{A} \cdot (\mathbf{B} \times \mathbf{C}) = 165$ **(e)**

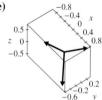

42. (a) $\mathbf{P}_2 - \mathbf{P}_1 = [-2, 5, 11]^\mathsf{T}; \mathbf{P}_3 - \mathbf{P}_4 = [-2, 5, 11]^\mathsf{T}$; Two opposite sides are parallel. **(b)** $\mathbf{P}_2 - \mathbf{P}_3 = [-3, -6, 13]^\mathsf{T};$
$\mathbf{P}_1 - \mathbf{P}_4 = [-3, -6, 13]^\mathsf{T}$; All opposite sides are parallel. **(c)** $\sqrt{17939}$ **(d)** $\sqrt{602}$; 11 **(e)** $(-\frac{5}{2}, -\frac{3}{2}, 0)$

50. (a) $\mathbf{T} = [45, -99, -117]^\mathsf{T}; \|\mathbf{T}\| = 27\sqrt{35}$ **(b)**

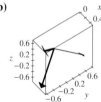

CHAPTER 32

Exercises 32.1 Page 726

1. **(a)** $\text{rref}\left(\begin{bmatrix} -7 & -3 \\ -6 & -3 \end{bmatrix}\right) = \begin{bmatrix} 1 & 0 \\ 0 & 1 \end{bmatrix}$ **(b)** and **(c)** $\mathbf{V} = [-10, 24]^{\mathrm{T}}$

11. **(a)** $\text{rref}\left(\begin{bmatrix} 2 & -3 & 6 \\ 6 & -9 & -3 \\ 8 & -2 & -2 \end{bmatrix}\right) = \begin{bmatrix} 1 & 0 & 0 \\ 0 & 1 & 0 \\ 0 & 0 & 1 \end{bmatrix}$ **(b)** and **(c)** $\mathbf{V} = \left[-\frac{81}{70}, \frac{23}{35}, -\frac{9}{7}\right]^{\mathrm{T}}$

21. **(a)** $\mathbf{Y} = [-84, -32]^{\mathrm{T}}$; $'\mathbf{Y} = \left[52, -\frac{280}{3}\right]^{\mathrm{T}}$ **(b)** $B = \begin{bmatrix} -70 & -27 \\ \frac{496}{3} & 65 \end{bmatrix}$; $B'\mathbf{V} = \left[52, -\frac{280}{3}\right]^{\mathrm{T}}$

31. **(a)** $\mathbf{Y} = [-180, 29, -18]^{\mathrm{T}}$; $'\mathbf{Y} = \left[-\frac{318}{35}, -\frac{26}{105}, -\frac{569}{21}\right]^{\mathrm{T}}$ **(b)** $B = \begin{bmatrix} \frac{509}{70} & \frac{573}{140} & \frac{183}{70} \\ \frac{43}{35} & \frac{1013}{210} & \frac{163}{105} \\ \frac{62}{7} & -\frac{419}{21} & \frac{61}{21} \end{bmatrix}$; $B'\mathbf{V} = \left[-\frac{318}{35}, \frac{-26}{105}, -\frac{569}{21}\right]^{\mathrm{T}}$

41. **(a)** $\mathbf{V} \cdot \mathbf{W} = 122$; $C = \begin{bmatrix} 85 & 39 \\ 39 & 18 \end{bmatrix}$ **(b)** $'\mathbf{W} = \left[-16, \frac{107}{3}\right]^{\mathrm{T}}$ **(c)** $'\mathbf{V}^{\mathrm{T}} C '\mathbf{W} = 122$

Exercises 32.2–Part A, Page 734

2. Compare the coefficients of x and of y.   3. $\arccos\left(\frac{\mathbf{U} \cdot \mathbf{X}}{\|\mathbf{U}\| \|\mathbf{X}\|}\right) = \frac{1}{8}\pi$

4. Both I and $\mathbf{uu}^{\mathrm{T}}$ are symmetric matrices, so H is symmetric. This implies $HH^{\mathrm{T}} = H^2 = I - 4\mathbf{uu}^{\mathrm{T}} + 4\mathbf{uu}^{\mathrm{T}}\mathbf{uu}^{\mathrm{T}} = I - 4\mathbf{u}(1 - \mathbf{u}^{\mathrm{T}}\mathbf{u})\mathbf{u}^{\mathrm{T}} = I$.

Exercises 32.2–Part B, Page 734

2. **(c)** and **(d)** See (a).   **(g)** j **(h)** $\|P\mathbf{v}\|^2 = \left|\frac{6}{37}\sqrt{37}a + \frac{1}{37}\sqrt{37}b\right|^2 + \left|-\frac{1}{37}\sqrt{37}a + \frac{6}{37}\sqrt{37}b\right|^2 = a^2 + b^2 = \|\mathbf{v}\|^2$

   (e) $\angle(\mathbf{x}, P\mathbf{x}) = 0$

   (f) $\|\mathbf{x}\| = \sqrt{101}$

12. **(c)** and **(d)** See   **(a)**   **(e)** $\left[\frac{5}{871}\sqrt{17,670} + \frac{1710}{871} + \frac{61}{871}\sqrt{155} + \frac{183}{871}\sqrt{114}, 1, \frac{1085}{871} + \frac{21}{871}\sqrt{17,670} + \frac{246}{871}\sqrt{114} + \frac{82}{871}\sqrt{155}\right]^{\mathrm{T}}$
   (f) $\left(\frac{5}{871}\sqrt{17,670} + \frac{1710}{871} + \frac{61}{871}\sqrt{155} + \frac{183}{871}\sqrt{114}\right)x + y + \left(\frac{1085}{871} + \frac{21}{871}\sqrt{17,670} + \frac{246}{871}\sqrt{114} + \frac{82}{871}\sqrt{155}\right)z = 0$
   (g) $\mathbf{v}_1 = \left[0, 1, -\frac{871}{1085 + 21\sqrt{17,670} + 246\sqrt{114} + 82\sqrt{155}}\right]^{\mathrm{T}}$; $\mathbf{v}_2 = \left[1, 1, -\frac{5\sqrt{17,670} + 2581 + 61\sqrt{155} + 183\sqrt{114}}{1085 + 21\sqrt{17,670} + 246\sqrt{114} + 82\sqrt{155}}\right]^{\mathrm{T}}$ **(h)** and **(i)** 0.8378750561
   (j) $\|\mathbf{v}_1\| = \|P\mathbf{v}_1\| = 1.006678308$ for both;   **(k)** $\|\mathbf{v}\| = \|P\mathbf{v}\| = \sqrt{a^2 + b^2 + c^2}$

20. $P = \begin{bmatrix} -\frac{2}{5}\sqrt{5} & \frac{1}{5}\sqrt{5} \\ \frac{1}{5}\sqrt{5} & \frac{2}{5}\sqrt{5} \end{bmatrix}$, $D = \begin{bmatrix} 6 & 0 \\ 0 & -14 \end{bmatrix}$ 30. $P = \begin{bmatrix} \frac{1}{6}\sqrt{2} & \frac{5}{38}\sqrt{38} & \frac{7}{57}\sqrt{19} \\ \frac{1}{6}\sqrt{2} & \frac{3}{38}\sqrt{38} & -\frac{11}{57}\sqrt{19} \\ -\frac{2}{3}\sqrt{2} & \frac{1}{19}\sqrt{38} & -\frac{1}{57}\sqrt{19} \end{bmatrix}$, $D = \begin{bmatrix} 3 & 0 & 0 \\ 0 & -7 & 0 \\ 0 & 0 & 12 \end{bmatrix}$

40. **(a)** $R = \begin{bmatrix} \frac{31}{60} & \frac{1}{12} - \frac{1}{10}\sqrt{10} & \frac{1}{30} + \frac{1}{4}\sqrt{10} \\ \frac{1}{12} + \frac{1}{10}\sqrt{10} & \frac{11}{12} & -\frac{1}{20}\sqrt{10} + \frac{1}{6} \\ \frac{1}{30} - \frac{1}{4}\sqrt{10} & \frac{1}{20}\sqrt{10} + \frac{1}{6} & \frac{17}{30} \end{bmatrix}$ **(c)** $x + 5y + 2z = 0$ **(d)** $\mathbf{u} = \mathbf{j} - \frac{5}{2}\mathbf{k}$,
   $R\mathbf{u} = -\frac{29}{40}\sqrt{10}\mathbf{i} + \left(\frac{1}{2} + \frac{1}{8}\sqrt{10}\right)\mathbf{j} + \left(-\frac{5}{4} + \frac{1}{20}\sqrt{10}\right)\mathbf{k}$, $\angle(\mathbf{u}, R\mathbf{u}) = \frac{1}{3}\pi$

Exercises 32.3, Page 741

2. **(a)**

   (b) $\mathbf{V} = (\cos p - p \sin p)\mathbf{i} + (\sin p + p \cos p)\mathbf{j}$
   (c) $'\mathbf{V} = \mathbf{e}_r + \mathbf{e}_\theta$
   (d) $\mathbf{i} = \cos(p)\mathbf{e}_r - \frac{1}{p}\sin(p)\mathbf{e}_\theta$, $\mathbf{j} = \sin(p)\mathbf{e}_r + \frac{1}{p}\cos(p)\mathbf{e}_\theta$
   (e) $\mathbf{e}_r = \cos(p)\mathbf{i} + \sin(p)\mathbf{j}$, $\mathbf{e}_\theta = -p\sin(p)\mathbf{i} + p\cos(p)\mathbf{j}$

9. (a)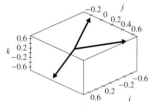

(b) $\mathbf{e}_u = -\frac{u^2-v^2}{(u^2+v^2)^2}\mathbf{i} + -2\frac{vu}{(u^2+v^2)^2}\mathbf{j}$; $\mathbf{e}_v = -2\frac{vu}{(u^2+v^2)^2}\mathbf{i} + \frac{u^2-v^2}{(u^2+v^2)^2}\mathbf{j}$ **(d)** $\mathbf{J}_\Phi = \begin{bmatrix} \frac{y^2-x^2}{(y^2+x^2)^2} & -2\frac{xy}{(y^2+x^2)^2} \\[2mm] -2\frac{xy}{(y^2+x^2)^2} & -\frac{y^2-x^2}{(y^2+x^2)^2} \end{bmatrix}$

(c) $\mathbf{J}_{\Phi^{-1}} = \begin{bmatrix} -\frac{u^2-v^2}{(u^2+v^2)^2} & -2\frac{vu}{(u^2+v^2)^2} \\[2mm] -2\frac{vu}{(u^2+v^2)^2} & \frac{u^2-v^2}{(u^2+v^2)^2} \end{bmatrix}$

16. (a) $\mathbf{V} = (\cos p - p\sin p)\mathbf{i} + (\sin p + p\cos p)\mathbf{j}$ **(b)** $'\mathbf{V} = (\frac{1}{2}(-2\sin(p)\sqrt{p^2(3\cos(p)^2+1)}\cos(p) + 3\sin(p)p\cos(p)^2$

$-2p\sqrt{p^2(3\cos(p)^2+1)} - 2\sqrt{p^2(3\cos(p)^2+1)}\cos(p)^2 p + 5p^2\cos(p) + 3p^2\cos(p)^3 + p\sin(p))/(\sqrt{2}\sqrt{p^2(3\cos(p)^2+1)} - 4p\cos(p)}$

$(3p\cos(p)^2 + p - 2\sqrt{p^2(3\cos(p)^2+1)}\cos(p)))\mathbf{e}_u + (-\frac{1}{2}(2\sqrt{p^2(3\cos(p)^2+1)}\cos(p)$

$-2\sqrt{p^2(3\cos(p)^2+1)}p\sin(p) - 3p\cos(p)^2 + 3p^2\cos(p)\sin(p) - p)/(\sqrt{2}\sqrt{p^2(3\cos(p)^2+1)} - 4p\cos(p)\sqrt{p^2(3\cos(p)^2+1)}))\mathbf{e}_v$

(c) $\mathbf{i}$ and $\mathbf{j}$ are the columns of $\mathbf{J}_\Phi$ **(d)** $\mathbf{e}_u$ and $\mathbf{e}_v$ are the columns of $\mathbf{J}_{\Phi^{-1}} = \mathbf{J}_\Phi^{-1}$

22. (a) $\mathbf{V} = (\cos p - p\sin p)\mathbf{i} + (\sin p + p\cos p)\mathbf{j}$ **(b)** $'\mathbf{V} = \left(\frac{1}{2}\frac{p\cos p+\sin p-p}{\sqrt{p}\sqrt{1-\cos p}}\right)\mathbf{e}_u + \left(\frac{1}{2} - \frac{\cos p+p\sin p+1}{\sqrt{p}\sqrt{1-\cos p}}\right)\mathbf{e}_v$ **(c)** $\mathbf{i}$ and $\mathbf{j}$ are the columns of $\mathbf{J}_\Phi$

(d) $\mathbf{e}_u$ and $\mathbf{e}_v$ are the columns of $\mathbf{J}_{\Phi^{-1}} = \mathbf{J}_\Phi^{-1}$

28. (a) $\mathbf{V} = (\cos p - p\sin p)\mathbf{i} + (\sin p + p\cos p)\mathbf{j}$ **(b)** $'\mathbf{V} = \left(-\frac{p\sin p+\cos p}{p^2}\right)\mathbf{e}_u + \left(-\frac{\sin p+p\cos p}{p^2}\right)\mathbf{e}_v$ **(c)** $\mathbf{i}$ and $\mathbf{j}$ are the columns of $\mathbf{J}_\Phi$

(d) $\mathbf{e}_u$ and $\mathbf{e}_v$ are the columns of $\mathbf{J}_{\Phi^{-1}} = \mathbf{J}_\Phi^{-1}$

40. (a) $\mathbf{V} = (\cos p - p\sin p)\mathbf{i} + (\sin p + p\cos p)\mathbf{j}$ **(b)** $'\mathbf{V} = \left(\frac{7}{31}\cos p - \frac{7}{31}p\sin p + \frac{2}{31}\sin p + \frac{2}{31}p\cos p\right)\mathbf{e}_u +$

$\left(\frac{5}{31}\cos p - \frac{5}{31}p\sin p - \frac{3}{31}\sin p - \frac{3}{31}p\cos p\right)\mathbf{e}_v$ **(c)** $\mathbf{i}$ and $\mathbf{j}$ are the columns of $\mathbf{J}_\Phi$ **(d)** $\mathbf{e}_u$ and $\mathbf{e}_v$ are the columns of $\mathbf{J}_{\Phi^{-1}} = \mathbf{J}_\Phi^{-1}$

48. (a)

(b) $\mathbf{V} = -\sin(p)\mathbf{i} + \cos(p)\mathbf{j} + \mathbf{k}$

(c) $'\mathbf{V} = \left(\frac{p}{\sqrt{p^2+1}}\right)\mathbf{e}_\rho + \mathbf{e}_\theta - \frac{1}{p^2+1}\mathbf{e}_\phi$

(d) $\mathbf{i}$, $\mathbf{j}$ and $\mathbf{k}$ are the columns of $\mathbf{J}_\Phi$, $\mathbf{i} = \frac{\cos(p)}{\sqrt{p^2+1}}\mathbf{e}_\rho - \sin(p)\mathbf{e}_\theta + \frac{p\cos(p)}{p^2+1}\mathbf{e}_\phi$,

$\mathbf{j} = \frac{\sin(p)}{\sqrt{p^2+1}}\mathbf{e}_\rho + \cos(p)\mathbf{e}_\theta + \frac{p\sin(p)}{p^2+1}\mathbf{e}_\phi$, $\mathbf{k} = \frac{p}{\sqrt{p^2+1}}\mathbf{e}_\rho - \frac{1}{p^2+1}\mathbf{e}_\phi$

(e) $\mathbf{e}_\rho, \mathbf{e}_\theta$ and $\mathbf{e}_\phi$ are the columns of $\mathbf{J}_{\Phi^{-1}} = \mathbf{J}_\Phi^{-1}$,

$\mathbf{e}_\rho = \frac{\cos(p)}{\sqrt{p^2+1}}\mathbf{i} + \frac{\sin(p)}{\sqrt{p^2+1}}\mathbf{j} + \frac{p}{\sqrt{p^2+1}}\mathbf{k}$, $\mathbf{e}_\theta = -\sin(p)\mathbf{i} + \cos(p)\mathbf{j}$, $\mathbf{e}_\phi = \cos(p)p\mathbf{i} + \sin(p)p\mathbf{j} - \mathbf{k}$

Exercises 32.4–Part A, Page 745

1. $\mathbf{e}^1 = \left[\frac{7}{31}, \frac{5}{31}\right]^T$; $\mathbf{e}^2 = \left[\frac{2}{31}, -\frac{3}{31}\right]^T$

Exercises 32.4–Part B, Page 745

2. $\mathbf{e}^1 = \left[-\frac{1}{86}, \frac{5}{43}\right]^T$; $\mathbf{e}^2 = \left[\frac{4}{43}, \frac{3}{43}\right]^T$

12. (a) $\mathbf{e}^1 = \left[-\frac{143}{14}, -\frac{11}{14}, -\frac{72}{7}\right]^T$; $\mathbf{e}^2 = \left[-\frac{59}{7}, -\frac{4}{7}, -\frac{60}{7}\right]^T$; $\mathbf{e}^3 = [-1, 0, -1]^T$ **(d)**

(c) $\mathbf{e}_1 \cdot (\mathbf{e}_2 \times \mathbf{e}_3) = -14$

22. (a) $\mathbf{J}_\Phi = \begin{bmatrix} \frac{y}{\sqrt{2}\sqrt{4x^2+y^2-4x}\sqrt{4x^2+y^2}} & -\frac{1}{2} - \frac{8x^2-y^2+4\sqrt{4x^2+y^2}x}{(\sqrt{4x^2+y^2}-2x)\sqrt{2}\sqrt{4x^2+y^2-4x}\sqrt{4x^2+y^2}} \\[3mm] -\frac{\sqrt{4x^2+y^2}-2x}{\sqrt{2}\sqrt{4x^2+y^2-4x}\sqrt{4x^2+y^2}} & \frac{1}{2}\frac{y}{\sqrt{2}\sqrt{4x^2+y^2-4x}\sqrt{4x^2+y^2}} \end{bmatrix}$; $\mathbf{J}_{\Phi^{-1}} = \begin{bmatrix} u & -v \\ 2v & 2u \end{bmatrix}$ **(b)** $\mathbf{e}_u = u\mathbf{i} + 2v\mathbf{j}$, $\mathbf{e}_v = -v\mathbf{i} + 2u\mathbf{j}$

(c)

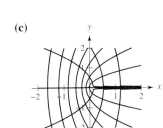

(d) $\left[\frac{u}{u^2+v^2}, \frac{1}{2}\frac{v}{u^2+v^2}\right]^{\mathrm{T}}; \left[-\frac{v}{u^2+v^2}, \frac{1}{2}\frac{u}{u^2+v^2}\right]^{\mathrm{T}}$ **(g)**

(e) $\mathbf{J}_\Phi = \begin{bmatrix} \frac{u}{u^2+v^2} & \frac{1}{2}\frac{v}{u^2+v^2} \\ -\frac{v}{u^2+v^2} & \frac{1}{2}\frac{u}{u^2+v^2} \end{bmatrix}$

(f) $\mathbf{J}_\Phi \mathbf{J}_{\Phi^{-1}} = I$

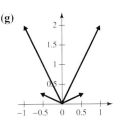

24. (a) $\mathbf{J}_{\Phi^{-1}} = \begin{bmatrix} -\frac{u^2-v^2}{(u^2+v^2)^2} & -2\frac{uv}{(u^2+v^2)^2} \\ -2\frac{uv}{(u^2+v^2)^2} & \frac{u^2-v^2}{(u^2+v^2)^2} \end{bmatrix}; \mathbf{e}_u = \left(\frac{\partial}{\partial u}\frac{u}{u^2+v^2}\right)\mathbf{i} + \left(\frac{\partial}{\partial u}\frac{v}{u^2+v^2}\right)\mathbf{j} = \left(-\frac{u^2-v^2}{(u^2+v^2)^2}\right)\mathbf{i} + \left(-2\frac{uv}{(u^2+v^2)^2}\right)\mathbf{j};$

$\mathbf{e}_v = \left(\frac{\partial}{\partial v}\frac{u}{u^2+v^2}\right)\mathbf{i} + \left(\frac{\partial}{\partial v}\frac{v}{u^2+v^2}\right)\mathbf{j} = \left(-2\frac{uv}{(u^2+v^2)^2}\right)\mathbf{i} + \left(\frac{u^2-v^2}{(u^2+v^2)^2}\right)\mathbf{j}$

(b)

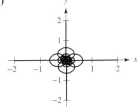

(c) $[-u^2+v^2, -2uv]^{\mathrm{T}}; [-2uv, u^2-v^2]^{\mathrm{T}}$ **(f)**

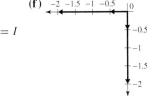

(d) $\mathbf{J}_\Phi = \begin{bmatrix} -u^2+v^2 & -2uv \\ -2uv & u^2-v^2 \end{bmatrix}$ **(e)** $\mathbf{J}_\Phi \mathbf{J}_{\Phi^{-1}} = I$

28. (a) $\mathbf{J}_\Phi \mathbf{J}_{\Phi^{-1}} = I$ **(b)**

Exercises 32.5, Page 750

1. $\mathbf{J}_{\Phi^{-1}} = \begin{bmatrix} \cos\theta & -r\sin\theta \\ \sin\theta & r\cos\theta \end{bmatrix}; \mathbf{J}_\Phi = \begin{bmatrix} \cos\theta & \sin\theta \\ -\frac{\sin\theta}{r} & \frac{\cos\theta}{r} \end{bmatrix};$ The rows of $\mathbf{J}_\Phi$ are the vectors $\mathbf{e}^r$ and $\mathbf{e}^\theta$ respectively.

3. (a) $\nabla f = (9y-14x)\mathbf{i} + 9x\mathbf{j}$ **(b)** $\nabla f = (9r\sin(2\theta) - 7r\cos(2\theta) - 7r)\mathbf{e}^r + (9r^2\cos(2\theta) + 7r^2\sin(2\theta))\mathbf{e}^\theta$
(c) Use the rows of $\mathbf{J}_{\Phi^{-1}}$, $\mathbf{i} = (\cos\theta)\mathbf{e}^r + (-r\sin\theta)\mathbf{e}^\theta$, $\mathbf{j} = (\sin\theta)\mathbf{e}^r + (r\cos\theta)\mathbf{e}^\theta$
(d) $\hat{f}(r,\theta) = 3 + 9r^2\sin(\theta)\cos(\theta) - 7r^2\cos^2(\theta), \frac{\partial\hat{f}}{\partial r} = 9r\sin(2\theta) - 7r\cos(2\theta) - 7r, \frac{\partial\hat{f}}{\partial\theta} = 9r^2\cos(2\theta) + 7r^2\sin(2\theta)$
(f) $\mathbf{e}^r = (\cos\theta)\mathbf{i} + (\sin\theta)\mathbf{j}, \mathbf{e}^\theta = (-\frac{\sin\theta}{r})\mathbf{i} + (\frac{\cos\theta}{r})\mathbf{j}, \nabla f: 9r\sin(2\theta) - 7r\cos(2\theta) - 7r, \frac{9r^2\cos(2\theta)+7r^2\sin(2\theta)}{r}$

14. (a) $\nabla f = (20xy + 27x^2)\mathbf{i} + (1 + 10x^2 + 36y^2)\mathbf{j}$ **(b)** $\nabla f = (25u^4v + 258u^2v^3 + \frac{27}{4}u^5 - \frac{27}{2}u^3v^2 + \frac{27}{4}uv^4 + 2v + 5v^5)\mathbf{e}^u +$
$(258u^3v^2 + 25uv^4 - \frac{27}{4}u^4v + \frac{27}{2}u^2v^3 - \frac{27}{4}v^5 + 2u + 5u^5)\mathbf{e}^v$ **(c)** Use the rows of $\mathbf{J}_{\Phi^{-1}}$; $\mathbf{i} = u\mathbf{e}^u - v\mathbf{e}^v, \mathbf{j} = 2v\mathbf{e}^u + 2u\mathbf{e}^v$
(d) $\hat{f}(u,v) = 6 + 2uv + 5u^5v + 86u^3v^3 + 5uv^5 + \frac{9}{8}u^6 - \frac{27}{8}u^4v^2 + \frac{27}{8}u^2v^4 - \frac{9}{8}v^6,$
$\frac{\partial\hat{f}}{\partial u} = 25u^4v + 258u^2v^3 + \frac{27}{4}u^5 - \frac{27}{2}u^3v^2 + \frac{27}{4}uv^4 + 2v + 5v^5, \frac{\partial\hat{f}}{\partial v} = 258u^3v^2 + 25uv^4 - \frac{27}{4}u^4v + \frac{27}{2}u^2v^3 - \frac{27}{4}v^5 + 2u + 5u^5$
(f) $\mathbf{e}^u = \left(\frac{u}{u^2+v^2}\right)\mathbf{i} + (\frac{1}{2}\frac{v}{u^2+v^2})\mathbf{j}; \mathbf{e}^v = (-\frac{v}{u^2+v^2})\mathbf{i} + (\frac{1}{2}\frac{u}{u^2+v^2})\mathbf{j}, \nabla f: \frac{1}{2}\frac{(25u^4v+258u^2v^3+\frac{27}{4}u^5-\frac{27}{2}u^3v^2+\frac{27}{4}uv^4+2v+5v^5)\sqrt{4u^2+v^2}}{u^2+v^2},$
$\frac{1}{2}\frac{(258u^3v^2+25uv^4-\frac{27}{4}u^4v+\frac{27}{2}u^2v^3-\frac{27}{4}v^5+2u+5u^5)\sqrt{4v^2+u^2}}{u^2+v^2}$

25. (a) $\nabla f = (y - 4x + 9x^2 + 3y^2)\mathbf{i} + (5 + x + 6xy + 18y^2)\mathbf{j}$ **(b)** $\nabla f = (\frac{3}{2}vu^2 - 2u^3 + 2uv^2 + \frac{9}{4}u^5 + \frac{3}{2}u^3v^2 - \frac{3}{4}uv^4 + 5v - \frac{1}{2}v^3 +$
$18u^2v^3)\mathbf{e}^u + (-\frac{3}{2}uv^2 + 2vu^2 - 2v^3 + \frac{3}{4}vu^4 - \frac{3}{2}u^2v^3 - \frac{9}{4}v^5 + 5u + \frac{1}{2}u^3 + 18u^3v^2)\mathbf{e}^v$
(c) Use the rows of $\mathbf{J}_{\Phi^{-1}}$; $\mathbf{i} = u\mathbf{e}^u - v\mathbf{e}^v, \mathbf{j} = v\mathbf{e}^u + u\mathbf{e}^v$ **(d)** $\hat{f}(u,v) = 5uv + \frac{1}{2}u^3v - \frac{1}{2}uv^3 - \frac{1}{4}u^4 + u^2v^2 - \frac{1}{2}v^4 + \frac{3}{8}u^6 + \frac{3}{8}u^4v^2 - \frac{3}{8}u^2v^4$
$-\frac{3}{8}v^6 + 6u^3v^3; \frac{\partial\hat{f}}{\partial u} = \frac{3}{2}vu^2 - 2u^3 + 2uv^2 + \frac{9}{4}u^5 + \frac{3}{2}u^3v^2 - \frac{3}{4}uv^4 + 5v - \frac{1}{2}v^3 + 18u^2v^3; \frac{\partial\hat{f}}{\partial v} = -\frac{3}{2}uv^2 + 2vu^2 - 2v^3 + \frac{3}{4}vu^4 - \frac{3}{2}u^2v^3 -$
$\frac{9}{4}v^5 + 5u + \frac{1}{2}u^3 + 18u^3v^2$ **(f)** $\mathbf{e}^u = \left(\frac{u}{u^2+v^2}\right)\mathbf{i} + (\frac{v}{u^2+v^2})\mathbf{j}, \mathbf{e}^v = (-\frac{v}{u^2+v^2})\mathbf{i} + (\frac{u}{u^2+v^2})\mathbf{j}, \nabla f: \frac{\frac{3}{2}vu^2-2u^3+2uv^2+\frac{9}{4}u^5+\frac{3}{2}u^3v^2-\frac{3}{4}uv^4+5v-\frac{1}{2}v^3+18u^2v^3}{\sqrt{u^2+v^2}},$
$-\frac{\frac{3}{2}uv^2+2vu^2-2v^3+\frac{3}{4}vu^4-\frac{3}{2}u^2v^3-\frac{9}{4}v^5+5u+\frac{1}{2}u^3+18u^3v^2}{\sqrt{u^2+v^2}}$

29. (a) $\nabla f = (7y + 27x^2 - 3y^2)\mathbf{i} + (7x + 14y - 6xy)\mathbf{j}$ **(b)** $\nabla f = \left(-\frac{21vu^4+7v^5-27u^4+42u^2v^2-3v^4-14v^3u^2-28u^3v^2-28uv^4}{(u^2+v^2)^4}\right)\mathbf{e}^u$

$+\left(-\frac{14u^3v^2+21uv^4+60u^3v-12uv^3-7u^5-14vu^4+14v^5}{(u^2+v^2)^4}\right)\mathbf{e}^v$ (c) Use the rows of $\mathbf{J}_{\Phi^{-1}}$; $\mathbf{i}=\left(-\frac{u^2+v^2}{(u^2+v^2)^2}\right)\mathbf{e}^u+\left(-2\frac{uv}{(u^2+v^2)^2}\right)\mathbf{e}^v$,

$\mathbf{j}=\left(-2\frac{uv}{(u^2+v^2)^2}\right)\mathbf{e}^u+\left(-\frac{u^2+v^2}{(u^2+v^2)^2}\right)\mathbf{e}^v$ (d) $\hat{f}(u,v)=\frac{7u^6+21u^4v^2+21u^2v^4+7v^6+7u^3v+7uv^3+7u^2v^2+7v^4+9u^3-3uv^2}{(u^2+v^2)^3}$,

$\frac{\partial\hat{f}}{\partial u}=-\frac{21vu^4+7v^5-27u^4+42u^2v^2-3v^4-14v^3u^2-28u^3v^2-28uv^4}{(u^2+v^2)^4}$, $\frac{\partial\hat{f}}{\partial v}=-\frac{14u^3v^2+21uv^4+60u^3v-12uv^3-7u^5-14vu^4+14v^5}{(u^2+v^2)^4}$ (f) $\mathbf{e}^u=(-u^2+v^2)\mathbf{i}+(-2uv)\mathbf{j}$,

$\mathbf{e}^v=(-2uv)\mathbf{i}+(u^2-v^2)\mathbf{j}$, $\nabla f: -\frac{21vu^4+7v^5-27u^4+42u^2v^2-3v^4-14v^3u^2-28u^3v^2-28uv^4}{(u^2+v^2)^3}$, $-\frac{14u^3v^2+21uv^4+60u^3v-12uv^3-7u^5-14vu^4+14v^5}{(u^2+v^2)^3}$

37. (a) $\nabla f=(y-14x+9y^2)\mathbf{i}+(11+x+18xy)\mathbf{j}$ **(b)** $\nabla f=(-620u^3-486uv^2+4320u^5-11232u^3v^2+3402uv^4+88u)\mathbf{e}^u$
$+(-486vu^2-360v^3-5616vu^4+6804u^2v^3+13122v^5-198v)\mathbf{e}^v$ **(c)** Use the rows of $\mathbf{J}_{\Phi^{-1}}$; $\mathbf{i}=10u\mathbf{e}^u+6v\mathbf{e}^v$, $\mathbf{j}=8u\mathbf{e}^u-18v\mathbf{e}^v$
(d) $\hat{f}(u,v)=44u^2-99v^2-155u^4-243u^2v^2-90v^4+720u^6-2808u^4v^2+1701u^2v^4+2187v^6$;
$\frac{\partial\hat{f}}{\partial u}=-620u^3-486uv^2+4320u^5-11232u^3v^2+3402uv^4+88u$; $\frac{\partial\hat{f}}{\partial v}=-486vu^2-360v^3-5616vu^4+6804u^2v^3+13122v^5-198v$
(f) $\mathbf{e}^u=\left(\frac{3}{38}\frac{1}{u}\right)\mathbf{i}+\left(\frac{1}{38}\frac{1}{u}\right)\mathbf{j}$, $\mathbf{e}^v=\left(\frac{2}{57}\frac{1}{v}\right)\mathbf{i}+\left(-\frac{5}{114}\frac{1}{v}\right)\mathbf{j}$, $\nabla f: \frac{1}{38}\frac{(-620u^3-486uv^2+4320u^5-11232u^3v^2+3402uv^4+88u)\sqrt{10}}{|u|}$,
$\frac{1}{114}\frac{(-486vu^2-360v^3-5616vu^4+6804u^2v^3+13122v^5-198v)\sqrt{41}}{|v|}$

45. (a) $\nabla f=(10-4y+24x-8xy)\mathbf{i}+(6-4x-4x^2)\mathbf{j}$ **(b)** $\nabla f=(60+96u+188v-540u^2+24uv+256v^2)\mathbf{e}^u$
$+(-22+188u+208v+12u^2+512uv+336v^2)\mathbf{e}^v$ **(c)** Use the rows of $\mathbf{J}_{\Phi^{-1}}$; $\mathbf{i}=3\mathbf{e}^u+2\mathbf{e}^v$, $\mathbf{j}=5\mathbf{e}^u-7\mathbf{e}^v$
(d) $\hat{f}(u,v)=60u-22v+48u^2+188uv+104v^2-180u^3+12u^2v+256uv^2+112v^3$; $\frac{\partial\hat{f}}{\partial u}=60+96u+188v-540u^2+24uv+256v^2$;
$\frac{\partial\hat{f}}{\partial v}=-22+188u+208v+12u^2+512uv+336v^2$ **(f)** $\mathbf{e}^u=\frac{7}{31}\mathbf{i}+\frac{2}{31}\mathbf{j}$, $\mathbf{e}^v=\frac{5}{31}\mathbf{i}-\frac{3}{31}\mathbf{j}$, $\nabla f:$
$\frac{1}{31}(60+96u+188v-540u^2+24uv+256v^2)\sqrt{53}$, $\frac{1}{31}(-22+188u+208v+12u^2+512uv+336v^2)\sqrt{34}$

CHAPTER 33

Exercises 33.1, Page 755

1. (a) $2A+3B=\begin{bmatrix}-45 & 19\\14 & 20\end{bmatrix}$, $A(5B-7C)=\begin{bmatrix}1020 & 28\\-1141 & 283\end{bmatrix}$, $ABC=\begin{bmatrix}156 & -108\\-1120 & 352\end{bmatrix}$ **(b)** $AB-BA=\begin{bmatrix}-26 & 129\\-184 & 26\end{bmatrix}$,
$AC-CA=\begin{bmatrix}-26 & -43\\31 & 26\end{bmatrix}$, $BC-CB=\begin{bmatrix}-13 & 43\\-135 & 13\end{bmatrix}$ **(c)** $I+A+A^2+A^3=\begin{bmatrix}-985 & 856\\1070 & -236\end{bmatrix}$ **(d)** $\det AC=1032$,
$\det(AB)^{\mathrm{T}}-\det A\det B=0$, $\det A^{-1}-\frac{1}{\det A}=0$ **(e)** $A^{-1}BA=\begin{bmatrix}\frac{250}{43} & \frac{7}{86}\\\frac{596}{43} & -\frac{465}{43}\end{bmatrix}$, $A^{\mathrm{T}}B=\begin{bmatrix}46 & 54\\-90 & 14\end{bmatrix}$ **(f)** $(AB)^{-1}-B^{-1}A^{-1}=\begin{bmatrix}0 & 0\\0 & 0\end{bmatrix}$,
$(AC)^{\mathrm{T}}-C^{\mathrm{T}}A^{\mathrm{T}}=\begin{bmatrix}0 & 0\\0 & 0\end{bmatrix}$ **(g)** $\mathrm{rref}(A)=\mathrm{rref}(B)=\mathrm{rref}(C)=\begin{bmatrix}1 & 0\\0 & 1\end{bmatrix}$

6. (a), (b), (c) and **(d)** $\mathbf{x}=\left[\frac{69}{5},-\frac{67}{5}\right]^{\mathrm{T}}$ **11. (a)** $\det A=30$; $A^{-1}=\begin{bmatrix}-\frac{1}{5} & -\frac{1}{10}\\-\frac{2}{15} & -\frac{7}{30}\end{bmatrix}$ **(b)** $\det A^{\mathrm{T}}=30$; $(A^{\mathrm{T}})^{-1}=\begin{bmatrix}-\frac{1}{5} & -\frac{2}{15}\\-\frac{1}{10} & -\frac{7}{30}\end{bmatrix}$ **(c)**
$\lambda_1=-3, \mathbf{e}_1=\left[1,\frac{4}{3}\right]^{\mathrm{T}}, \lambda_2=-10, \mathbf{e}_2=[-1,1]^{\mathrm{T}}$ **(e)** $e^{At}=\begin{bmatrix}\frac{4}{7}e^{-10t}+\frac{3}{7}e^{-3t} & \frac{3}{7}e^{-3t}-\frac{3}{7}e^{-10t}\\\frac{4}{7}e^{-3t}-\frac{4}{7}e^{-10t} & \frac{3}{7}e^{-10t}+\frac{4}{7}e^{-3t}\end{bmatrix}$

Exercises 33.2–Part A, Page 759

1. $\mathbf{B_A}=\left[\frac{34}{13},\frac{51}{13}\right]^{\mathrm{T}}$; $\mathbf{B_{\perp A}}=\left[-\frac{21}{13},\frac{14}{13}\right]^{\mathrm{T}}$ **5.** $P=P^2=\begin{bmatrix}\frac{2}{3} & \frac{1}{3} & -\frac{1}{3}\\\frac{1}{3} & \frac{2}{3} & \frac{1}{3}\\-\frac{1}{3} & \frac{1}{3} & \frac{2}{3}\end{bmatrix}$ **6.** $\mathbf{B}_{[A]}=[4,1,-3]^{\mathrm{T}}$; $\mathbf{B}_{\perp[A]}=[-2,2,-2]^{\mathrm{T}}$

Exercises 33.2–Part B, Page 759

2. (a) $\mathbf{B_A}=\left[\frac{172}{17},\frac{43}{17}\right]^{\mathrm{T}}$ **(c)** $P=P^2=\begin{bmatrix}\frac{16}{17} & \frac{4}{17}\\\frac{4}{17} & \frac{1}{17}\end{bmatrix}$ **(e)** $\alpha_{max}=-\frac{43}{34}$ **(f)**
(b) $\mathbf{B_{\perp A}}=\left[-\frac{36}{17},\frac{144}{17}\right]^{\mathrm{T}}$ **(d)** $PB=\left[\frac{172}{17},\frac{43}{17}\right]^{\mathrm{T}}$

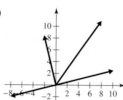

12. **(a)** $\operatorname{rref}(A) = \begin{bmatrix} 1 & 0 \\ 0 & 1 \\ 0 & 0 \end{bmatrix}$

(d) $\mathbf{B}_{[A]} = \left[-\frac{23}{18}, -\frac{14}{9}, \frac{73}{18} \right]^{\mathrm{T}}$ **(f)** $\alpha_2 = -\frac{17}{36}, \alpha_1 = -\frac{1}{36}$ **(g)**

(e) $\mathbf{B}_{\perp[A]} = \left[-\frac{121}{18}, -\frac{22}{9}, -\frac{55}{18} \right]^{\mathrm{T}}$

(b) $P = P^2 = \begin{bmatrix} \frac{41}{162} & -\frac{22}{81} & -\frac{55}{162} \\ -\frac{22}{81} & \frac{73}{81} & -\frac{10}{81} \\ -\frac{55}{162} & -\frac{10}{81} & \frac{137}{162} \end{bmatrix}$

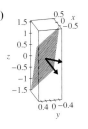

25. $P^2 = (A(A^{\mathrm{T}}A)^{-1}A^{\mathrm{T}})(A(A^{\mathrm{T}}A)^{-1}A^{\mathrm{T}}) = A(A^{\mathrm{T}}A)^{-1}(A^{\mathrm{T}}A)(A^{\mathrm{T}}A)^{-1}A^{\mathrm{T}} = A(A^{\mathrm{T}}A)^{-1}A^{\mathrm{T}} = P$

26. **(a)** $A(A^{\mathrm{T}}A)^{-1}A^{\mathrm{T}} = \begin{bmatrix} \frac{181}{230} & -\frac{63}{230} & -\frac{7}{23} \\ -\frac{63}{230} & \frac{149}{230} & -\frac{9}{23} \\ -\frac{7}{23} & -\frac{9}{23} & \frac{13}{23} \end{bmatrix}$ **(b)** $\mathbf{n} = [7, 9, 10]^{\mathrm{T}}, P = \mathbf{n}(\mathbf{n}^{\mathrm{T}}\mathbf{n})^{-1}\mathbf{n}^{\mathrm{T}} = \begin{bmatrix} \frac{49}{230} & \frac{63}{230} & \frac{7}{23} \\ \frac{63}{230} & \frac{81}{230} & \frac{9}{23} \\ \frac{7}{23} & \frac{9}{23} & \frac{10}{23} \end{bmatrix}$

Exercises 33.3, Page 762

1. $\mathbf{w}_1 = \left[\frac{5}{131}\sqrt{262}, -\frac{9}{262}\sqrt{262}, -\frac{9}{262}\sqrt{262} \right]^{\mathrm{T}}$, $\mathbf{w}_2 = \left[-\frac{1053}{1,827,581}\sqrt{13,951}\sqrt{131}, -\frac{716}{1,827,581}\sqrt{13,951}\sqrt{131}, -\frac{454}{1,827,581}\sqrt{13,951}\sqrt{131} \right]^{\mathrm{T}}$,
$\mathbf{w}_3 = \left[\frac{9}{13,951}\sqrt{2}\sqrt{13,951}, -\frac{107}{27,902}\sqrt{2}\sqrt{13,951}, \frac{127}{27,902}\sqrt{2}\sqrt{13,951} \right]^{\mathrm{T}}$

6. $\mathbf{w}_1 = \left[\frac{7}{115}\sqrt{115}, \frac{7}{115}\sqrt{115}, -\frac{1}{115}\sqrt{115}, \frac{4}{115}\sqrt{115} \right]^{\mathrm{T}}$, $\mathbf{w}_2 = \left[\frac{27}{575}\sqrt{23}, -\frac{42}{575}\sqrt{23}, -\frac{109}{575}\sqrt{23}, -\frac{1}{575}\sqrt{23} \right]^{\mathrm{T}}$,
$\mathbf{w}_3 = \left[\frac{2671}{2,756,450}\sqrt{110,258}, -\frac{2633}{1,378,225}\sqrt{110,258}, \frac{2643}{2,756,450}\sqrt{110,258}, \frac{2601}{1,378,225}\sqrt{110,258} \right]^{\mathrm{T}}$

10. $\mathbf{w}_1 = [-1, 1, -3]^{\mathrm{T}}, \mathbf{w}_2 = \left[\frac{57}{11}, -\frac{90}{11}, -\frac{49}{11} \right]^{\mathrm{T}}, \mathbf{w}_3 = \left[-\frac{8439}{1250}, -\frac{582}{125}, \frac{873}{1250} \right]^{\mathrm{T}}$

15. $\mathbf{w}_1 = [7, 11, -9, 4]^{\mathrm{T}}, \mathbf{w}_2 = \left[\frac{506}{267}, -\frac{845}{267}, -\frac{255}{89}, \frac{-283}{267} \right]^{\mathrm{T}}, \mathbf{w}_3 = \left[\frac{56,419}{6125}, \frac{144}{1225}, \frac{8053}{1225}, -\frac{10117}{6125} \right]^{\mathrm{T}}$

18. $\mathbf{w}_1 = [-0.8751130126, -0.07955572842, 0.4773343705]^{\mathrm{T}}, \mathbf{w}_2 = [-0.4082482906, -0.4082482906, -0.8164965812]^{\mathrm{T}}$,
$\mathbf{w}_3 = [0.2598279207, -0.9093977240, 0.3247849006]^{\mathrm{T}}$

23. $\mathbf{w}_1 = [0.2760262238, 0.8280786714, 0.4830458916, -0.06900655595]^{\mathrm{T}}$,
$\mathbf{w}_2 = [0.4083490719, -0.3539796488, 0.2591223573, -0.8005029966]^{\mathrm{T}}$, Be careful, the vectors are linearly dependent.

26. $1, x - \frac{1}{2}, x^2 + \frac{1}{6} - x, x^3 - \frac{1}{20} + \frac{3}{5}x - \frac{3}{2}x^2, x^4 + \frac{1}{70} - \frac{2}{7}x + \frac{9}{7}x^2 - 2x^3$

36. $\mathbf{w}_1 = [-6, 6, 8, 9]^{\mathrm{T}}, \mathbf{w}_2 = \left[\frac{20414}{3125}, \frac{26461}{3125}, -\frac{19927}{3125}, \frac{19379}{3125} \right]^{\mathrm{T}}, \mathbf{w}_3 = \left[-\frac{395,001}{334,519}, -\frac{2,370,674}{334,519}, -\frac{89,392}{334,519}, \frac{931,885}{334,519} \right]^{\mathrm{T}}$,
$\mathbf{w}_4 = \left[\frac{52,252,375}{17,131,476}, \frac{52,675}{8,565,738}, \frac{7,906,150}{4,282,869}, \frac{7,601,125}{17,131,476} \right]^{\mathrm{T}}$

Exercises 33.4–Part A, Page 767

1. $\left(\frac{47}{25}, -\frac{34}{25} \right)$ is a saddle.   2. $Q = \begin{bmatrix} 4 & 7 \\ 7 & 6 \end{bmatrix}, \lambda_{1,2} = 5 \pm 5\sqrt{2}$

Exercises 33.4–Part B, Page 767

2. **(a)** $\left(\frac{3}{8}, 2 \right)$ **(b)** $T = -48$, it is a saddle.   **(c)** $\lambda_1 = -2, \lambda_2 = 24$

12. **(a)** $\left(-\frac{1}{5}165^{1/3}, \frac{1}{60}165^{2/3} \right)$ 22. $Q = \begin{bmatrix} -54xy + 10 + 4y^2 & 21y^2 - 27x^2 + 8xy \\ 21y^2 - 27x^2 + 8xy & 42xy + 4x^2 \end{bmatrix}$
   (b) $\lambda_1 = -20.77714920, \lambda_2 = 34.74944822$, it is a saddle.

32. **(a)** $Q = \begin{bmatrix} 2 & -6 & 18 \\ -6 & 2 & -6 \\ 18 & -6 & 18 \end{bmatrix}$ **(b)** $\mathbf{x}^{\mathrm{T}}Q\mathbf{x} = x^2 - 6xy + 18xz + y^2 - 6yz + 9z^2$ **(c)** $P = \begin{bmatrix} 0 & \frac{1}{7}\sqrt{35} & \frac{1}{7}\sqrt{14} \\ \frac{3}{10}\sqrt{10} & \frac{1}{35}\sqrt{35} & -\frac{1}{14}\sqrt{14} \\ \frac{1}{10}\sqrt{10} & -\frac{3}{35}\sqrt{35} & \frac{3}{14}\sqrt{14} \end{bmatrix}$;

$D = \begin{bmatrix} 0 & 0 & 0 \\ 0 & -10 & 0 \\ 0 & 0 & 32 \end{bmatrix}$ **(d)** $u = \frac{3}{10}\sqrt{10}y + \frac{1}{10}\sqrt{10}z, v = \frac{1}{7}\sqrt{35}x + \frac{1}{35}\sqrt{35}y - \frac{3}{35}\sqrt{35}z, w = \frac{1}{7}\sqrt{14}x - \frac{1}{14}\sqrt{14}y + \frac{3}{14}\sqrt{14}z$

   (e) Q is indefinite.   **(g)** $\mathbf{x}_1 = \left[\frac{1}{7}\sqrt{35}, \frac{1}{35}\sqrt{35}, -\frac{3}{35}\sqrt{35} \right]^{\mathrm{T}}, \mathbf{x}_2 = \left[\frac{1}{7}\sqrt{14}, -\frac{1}{14}\sqrt{14}, \frac{3}{14}\sqrt{14} \right]^{\mathrm{T}}$

42. (a) $\mathbf{v}_{\max} = [2c, c, c]^{\mathrm{T}}$ **(b)** $\mathbf{v}_{\min} = [-c, c, c]^{\mathrm{T}}$

Exercises 33.5–Part A, Page 771

4. $\lambda_1 = 0.17825461$, $\lambda_2 = 20.92979274$, $\lambda_3 = 291.8919527$, $\sqrt{\lambda_1} = 0.4222020962$, $\sqrt{\lambda_2} = 4.574909042$, $\sqrt{\lambda_3} = 17.08484570$

Exercises 33.5–Part B, Page 771

5. $\|\mathbf{v}\|_1 = 20$; $\|\mathbf{v}\|_2 = 4\sqrt{13}$; $\|\mathbf{v}\|_3 = 2240^{1/3}$; $\|\mathbf{v}\|_\infty = 12$

15. (a) $\|A\|_1 = 16$; $\|A\|_2 = \frac{1}{2}\sqrt{373} + \frac{1}{2}\sqrt{29}$; $\|A\|_f = \sqrt{201}$; $\|A\|_\infty = 14$ **(b)** $\lambda_{1,2} = \frac{1}{2}\sqrt{373} \pm \frac{1}{2}\sqrt{29}$
   (c) $t_{\max} = \arctan(\frac{71}{76} - \frac{1}{76}\sqrt{10{,}817}) + \pi$

26. (a) 20

(b)

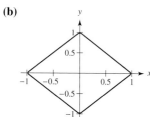

(d)

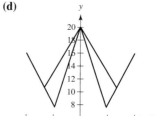

20

(c) $y = -|x| + 1 \Rightarrow \|A\mathbf{x}\|_1 = 5x - 10|x| + 10 + |11x + 10|x| - 10|$,

$y = |x| - 1 \Rightarrow \|A\mathbf{x}\|_1 = |5x + 10|x| - 10| + |11x - 10|x| + 10|$

29. (a) $\|A\|_1 = 9$ **(b)**

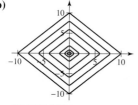

(c)

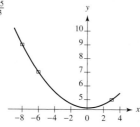

(d) $\max_{k=1\cdots12} \|A\mathbf{x}_k\| = 9$

37. (a) and **(b)** $\|A\|_4 = 12.33212576$

Exercises 33.6–Part B, Page 779

4. $y = \frac{7}{99}x^2 - \frac{1}{99}x + \frac{145}{33}$

9. $y = \frac{29}{570}x^3 + \frac{16}{95}x^2 - \frac{2453}{570}x - 8$

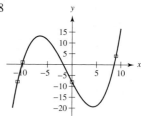

14. (a) $4a - 2b + c = 14$, $16a - 4b + c = 85$, $9a - 3b + c = 55$, $25a + 5b + c = 228$, $16a + 4b + c = 176$, $c = 23$,

$4a + 2b + c = 30$, $a + b + c = 45$, $a - b + c = 25$ **(b)** $A = \begin{bmatrix} 4 & -2 & 1 \\ 16 & -4 & 1 \\ 9 & -3 & 1 \\ 25 & 5 & 1 \\ 16 & 4 & 1 \\ 0 & 0 & 1 \\ 4 & 2 & 1 \\ 1 & 1 & 1 \\ 1 & -1 & 1 \end{bmatrix}$ **(c)** $\text{rref}([A, \mathbf{y}]) = \begin{bmatrix} 1 & 0 & 0 & 0 \\ 0 & 1 & 0 & 0 \\ 0 & 0 & 1 & 0 \\ 0 & 0 & 0 & 1 \\ 0 & 0 & 0 & 0 \\ 0 & 0 & 0 & 0 \\ 0 & 0 & 0 & 0 \\ 0 & 0 & 0 & 0 \\ 0 & 0 & 0 & 0 \end{bmatrix}$

(d) rank $A = 3$; rank$[A, \mathbf{y}] = 4$
(e) $S = -1362c - 2782b - 21234a + 1252a^2 + 196ab + 152ac + 76b^2 + 4bc + 9c^2 + 97{,}485$ **(f)** $a = \frac{34{,}345}{5082}$, $b = \frac{232{,}559}{25{,}410}$, $c = \frac{70{,}148}{4235}$
(g) $S_{\min} = \frac{3{,}126{,}802}{1815}$ **(h)** $a = \frac{34{,}345}{5082}$, $b = \frac{232{,}559}{25{,}410}$, $c = \frac{70{,}148}{4235}$

(i) $P =$
$$\begin{bmatrix}
\frac{67}{363} & \frac{159}{847} & \frac{23}{121} & -\frac{9}{121} & -\frac{5}{363} & \frac{127}{847} & \frac{71}{847} & \frac{307}{2541} & \frac{145}{847} \\[4pt]
\frac{159}{847} & \frac{18,569}{29,645} & \frac{1626}{4235} & \frac{458}{4235} & -\frac{81}{4235} & \frac{1959}{29,645} & -\frac{4003}{29,645} & -\frac{3666}{29,645} & \frac{1118}{29,645} \\[4pt]
\frac{23}{121} & \frac{1626}{4235} & \frac{169}{605} & -\frac{3}{605} & -\frac{14}{605} & \frac{244}{4235} & -\frac{57}{4235} & \frac{61}{4235} & \frac{492}{4235} \\[4pt]
-\frac{9}{121} & \frac{458}{4235} & -\frac{3}{605} & \frac{401}{605} & \frac{258}{605} & -\frac{348}{4235} & \frac{359}{4235} & \frac{87}{4235} & -\frac{424}{4235} \\[4pt]
-\frac{5}{363} & -\frac{81}{4235} & -\frac{14}{605} & \frac{258}{605} & \frac{587}{1815} & \frac{191}{4235} & \frac{667}{4235} & \frac{1202}{12705} & \frac{38}{4235} \\[4pt]
\frac{127}{847} & \frac{1959}{29,645} & \frac{244}{4235} & -\frac{348}{4235} & \frac{191}{4235} & \frac{7129}{29,645} & \frac{6093}{29,645} & \frac{7076}{29,645} & \frac{6252}{29,645} \\[4pt]
\frac{71}{847} & -\frac{4003}{29,645} & -\frac{57}{4235} & \frac{359}{4235} & \frac{667}{4235} & \frac{6093}{29,645} & \frac{6821}{29,645} & \frac{6817}{29,645} & \frac{4649}{29,645} \\[4pt]
\frac{307}{2541} & -\frac{3666}{29,645} & \frac{61}{4235} & \frac{87}{4235} & \frac{1202}{12,705} & \frac{7076}{29,645} & \frac{6817}{29,645} & \frac{22,247}{88,935} & \frac{5798}{29,645} \\[4pt]
\frac{145}{847} & \frac{1118}{29,645} & \frac{492}{4235} & -\frac{424}{4235} & \frac{38}{4235} & \frac{6252}{29,645} & \frac{4649}{29,645} & \frac{5798}{29,645} & \frac{6011}{29,645}
\end{bmatrix}$$

(j) $\mathrm{rref}([A,\,P\mathbf{y}]) =$
$$\begin{bmatrix}
1 & 0 & 0 & \frac{34,345}{5082} \\
0 & 1 & 0 & \frac{232,559}{25410} \\
0 & 0 & 1 & \frac{70,148}{4235} \\
0 & 0 & 0 & 0 \\
0 & 0 & 0 & 0 \\
0 & 0 & 0 & 0 \\
0 & 0 & 0 & 0 \\
0 & 0 & 0 & 0 \\
0 & 0 & 0 & 0
\end{bmatrix}$$

(k) $\mathbf{u} = \left[\frac{34,345}{5082},\ \frac{232,559}{25,410},\ \frac{70148}{4235}\right]^{\mathsf{T}}$ **(l)** $\mathbf{y}_{\perp[A]} = \left[-\frac{4099}{363},\ -\frac{13,067}{4235},\ \frac{3067}{605},\ -\frac{1984}{605},\ \frac{26,674}{1815},\ \frac{27,257}{4235},\ -\frac{135,101}{4235},\ \frac{159,139}{12,705},\ \frac{45,866}{4235}\right]^{\mathsf{T}}$ **(m)** $\|\mathbf{y}-P\mathbf{y}\|^2 = \frac{3,126,802}{1815}$

(n)

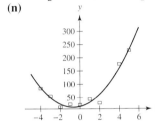

34. **(a)** rank $A = 5$ **(b)** $\{[0,1,0,0,0],\ [1,0,0,0,0],\ [0,0,0,0,1],\ [0,0,1,0,0],\ [0,0,0,1,0]\}$ **(c)** $\{\ \}$
(d) $\left\{\left[0,1,0,0,0,-\frac{6223}{22,444}\right],\ \left[1,0,0,0,0,\frac{9073}{22,444}\right],\ \left[0,0,1,0,0,\frac{18,187}{22,444}\right],\ \left[0,0,0,1,0,\frac{7082}{5611}\right],\ \left[0,0,0,0,1,\frac{19,923}{11,222}\right]\right\}$
(e) $\left\{\left[-\frac{9073}{6223},1,-\frac{18,187}{6223},\frac{-28,328}{6223},-\frac{39,846}{6223},\frac{22444}{6223}\right]\right\}$ **(f)** The row space and the column space both have dimension five. The null space has dimension zero, while the nullspace of the transpose matrix has dimension 1.   **(i)** This follows from the definition of the null space.
(j) This follows from the definition of the null space.

44. **(a)** $\mathrm{rref}([A,\mathbf{y}]) = \begin{bmatrix} 1&0&0&0&0 \\ 0&1&0&0&0 \\ 0&0&1&0&0 \\ 0&0&0&1&0 \\ 0&0&0&0&1 \end{bmatrix}$, (see row 5).   **(b)** $\mathbf{u} = \left[\frac{179,977,111}{276,424,169},\ \frac{297,856,765}{829,272,507},\ \frac{56,299,192}{118,467,501},\ \frac{10,111,149}{276,424,169}\right]^{\mathsf{T}}$

(c) $\mathrm{rref}([A^{\mathsf{T}}A,\,A^{\mathsf{T}}\mathbf{y}]) = \begin{bmatrix} 1&0&0&0&\frac{179,977,111}{276,424,169} \\[3pt] 0&1&0&0&\frac{297,856,765}{829,272,507} \\[3pt] 0&0&1&0&\frac{56,299,192}{118,467,501} \\[3pt] 0&0&0&1&\frac{10,111,149}{276,424,169} \end{bmatrix}$

CHAPTER 34

Exercises 34.1–Part A, Page 789

1. $U = \begin{bmatrix} 2 & -18 & 10 \\ 0 & 3 & 24 \\ 0 & 0 & -2 \end{bmatrix};\ L_1 = \begin{bmatrix} 1 & 0 & 0 \\ 4 & 1 & 0 \\ -3 & 8 & 1 \end{bmatrix}$ 2. $L_1 = \begin{bmatrix} 1 & 0 & 0 \\ 4 & 1 & 0 \\ -3 & 8 & 1 \end{bmatrix};\ D = \begin{bmatrix} 2 & 0 & 0 \\ 0 & 3 & 0 \\ 0 & 0 & -2 \end{bmatrix};\ U_1 = \begin{bmatrix} 1 & -9 & 5 \\ 0 & 1 & 8 \\ 0 & 0 & 1 \end{bmatrix}$

3. $L = \begin{bmatrix} 2 & 0 & 0 \\ 8 & 3 & 0 \\ -6 & 24 & -2 \end{bmatrix};\ U = \begin{bmatrix} 1 & -9 & 5 \\ 0 & 1 & 8 \\ 0 & 0 & 1 \end{bmatrix}$

Exercises 34.1–Part B, Page 789

2. **(a)** $U = \begin{bmatrix} -6 & 8 & 10 \\ 0 & -\frac{29}{3} & \frac{8}{3} \\ 0 & 0 & \frac{191}{29} \end{bmatrix};\ L_1 = \begin{bmatrix} 1 & 0 & 0 \\ -\frac{1}{6} & 1 & 0 \\ \frac{1}{3} & -\frac{10}{29} & 1 \end{bmatrix}$ **(b)** $L_1 = \begin{bmatrix} 1 & 0 & 0 \\ -\frac{1}{6} & 1 & 0 \\ \frac{1}{3} & -\frac{10}{29} & 1 \end{bmatrix};\ D = \begin{bmatrix} -6 & 0 & 0 \\ 0 & -\frac{29}{3} & 0 \\ 0 & 0 & \frac{191}{29} \end{bmatrix};\ U_1 = \begin{bmatrix} 1 & -\frac{4}{3} & -\frac{5}{3} \\ 0 & 1 & -\frac{8}{29} \\ 0 & 0 & 1 \end{bmatrix}$

(c) $L = \begin{bmatrix} -6 & 0 & 0 \\ 1 & -\frac{29}{3} & 0 \\ -2 & \frac{10}{3} & \frac{191}{29} \end{bmatrix}$; $U_1 = \begin{bmatrix} 1 & -\frac{4}{3} & -\frac{5}{3} \\ 0 & 1 & -\frac{8}{29} \\ 0 & 0 & 1 \end{bmatrix}$ **(d)** A is not positive definite.

12. (a) $L_1 = \begin{bmatrix} 1 & 0 & 0 \\ \frac{9}{7} & 1 & 0 \\ -1 & -\frac{28}{17} & 1 \end{bmatrix}$; $U = \begin{bmatrix} 7 & -11 & 0 \\ 0 & \frac{85}{7} & 11 \\ 0 & 0 & \frac{376}{17} \end{bmatrix}$ **(b)** $\mathbf{x}_1 = \left[\frac{91}{1880}, -\frac{113}{1880}, -\frac{19}{376}\right]^{\mathrm{T}}$; $\mathbf{x}_2 = \left[\frac{11}{470}, \frac{7}{470}, \frac{7}{94}\right]^{\mathrm{T}}$; $\mathbf{x}_3 = \left[-\frac{121}{1880}, -\frac{77}{1880}, \frac{17}{376}\right]^{\mathrm{T}}$

(e) $L = \begin{bmatrix} 7 & 0 & 0 \\ 9 & \frac{85}{7} & 0 \\ -7 & -20 & \frac{376}{17} \end{bmatrix}$; $U_1 = \begin{bmatrix} 1 & -\frac{11}{7} & 0 \\ 0 & 1 & \frac{77}{85} \\ 0 & 0 & 1 \end{bmatrix}$ **(f)** $T = \begin{bmatrix} 1 & 0 & 0 \\ -\frac{9}{7} & 1 & 0 \\ -\frac{19}{17} & \frac{28}{17} & 1 \end{bmatrix}$

22. (a) $L_1 = \begin{bmatrix} 1.0 & 0 & 0 \\ 0.2857142857 & 1.0 & 0 \\ -0.5714285714 & -0.03658536585 & 1.0 \end{bmatrix}$; $U = \begin{bmatrix} -7.0 & 1.0 & -5.0 \\ 0 & 11.71428571 & -0.5714285714 \\ 0 & 0 & -7.878048780 \end{bmatrix}$

(b) $L = \begin{bmatrix} -7.0 & 0 & 0 \\ -2.000000000 & 11.71428571 & 0 \\ 4.000000000 & -0.4285714284 & -7.878048780 \end{bmatrix}$; $U_1 = \begin{bmatrix} 0.9999999996 & -0.1428571428 & 0.7142857140 \\ 0 & 0.9999999999 & -0.04878048781 \\ 0 & 0 & 0.9999999998 \end{bmatrix}$

25. (a) $A = \begin{bmatrix} 5 & 45 & 15 \\ -20 & -180 & -60 \\ -45 & -405 & -143 \end{bmatrix}$ **(b)** $L = \begin{bmatrix} 1 & 0 & 0 \\ -4 & 1 & 0 \\ -9 & -2 & 1 \end{bmatrix}$; $U = \begin{bmatrix} 5 & 45 & 15 \\ 0 & 0 & 0 \\ 0 & 0 & -8 \end{bmatrix}$ **(c)** The procedure will fail because it hits a zero pivot.

Exercises 34.2–Part A, Page 792

1. $\det A = -2$; $\lambda_{1,2} = -2, 1$ **2.** $P = \begin{bmatrix} 1 & 1 \\ 2 & 3 \end{bmatrix}$ **3.** $P = \begin{bmatrix} 1 & 1 \\ 1 & \frac{3}{2} \end{bmatrix}$

Exercises 34.2–Part B, Page 793

5. (a) $P_A = \begin{bmatrix} 6 & 1 \\ 1 & 1 \end{bmatrix}$; $P_B = \begin{bmatrix} \frac{11}{4} & 4 \\ 1 & 1 \end{bmatrix}$ **(b)** $P = \begin{bmatrix} -4 & 17 \\ 0 & 1 \end{bmatrix}$ **15. (a)** $\mathbf{p}_1 = [1, 0, 0, 0]^{\mathrm{T}}$ **(b)** $\mathbf{p}_2 = [0, 1, 0, 0]^{\mathrm{T}}$; $\mathbf{p}_3 = [0, 0, 1, 0]^{\mathrm{T}}$; $\mathbf{p}_4 = [0, 0, 0, 1]^{\mathrm{T}}$; $A\mathbf{p}_2 = \mathbf{p}_1 + 2\mathbf{p}_2$, $A\mathbf{p}_3 = \mathbf{p}_2 + 2\mathbf{p}_3$, $A\mathbf{p}_4 = \mathbf{p}_3 + 2\mathbf{p}_4$

25. (a) $\lambda_{1,2,3} = 2, -3, -3$ **(b)** $J_A = \begin{bmatrix} 2 & 0 & 0 \\ 0 & -3 & 1 \\ 0 & 0 & -3 \end{bmatrix}$; $J_B = \begin{bmatrix} 2 & 0 & 0 \\ 0 & -3 & 0 \\ 0 & 0 & -3 \end{bmatrix}$

32. (a) $P = \begin{bmatrix} -6 & 1 \\ 9 & 0 \end{bmatrix}$; $J = \begin{bmatrix} -2 & 1 \\ 0 & -2 \end{bmatrix}$ **(b)** $e^{At} = \begin{bmatrix} e^{-2t} - 6te^{-2t} & -4te^{-2t} \\ 9te^{-2t} & e^{-2t} + 6te^{-2t} \end{bmatrix}$ **(c)** $e^{Jt} = \begin{bmatrix} e^{-2t} & te^{-2t} \\ 0 & e^{-2t} \end{bmatrix}$

Exercises 34.3, Page 796

7. (a) $\operatorname{rank}(A_1) = 2$; $\operatorname{rank}(A_2) = 3$ **(b)** $Q_1 = \begin{bmatrix} -\frac{26}{1739}\sqrt{3478} & -\frac{191}{6,415,171}\sqrt{12,830,342} \\ -\frac{7}{3478}\sqrt{3478} & \frac{283}{12,830,342}\sqrt{12,830,342} \\ -\frac{7}{3478}\sqrt{3478} & -\frac{3195}{12,830,342}\sqrt{12,830,342} \\ \frac{13}{1739}\sqrt{3478} & -\frac{774}{6,415,171}\sqrt{12,830,342} \end{bmatrix}$;

$R_1 = \begin{bmatrix} \sqrt{3478} & \frac{3795}{3478}\sqrt{3478} & \frac{931}{3478}\sqrt{3478} \\ 0 & \frac{57}{3478}\sqrt{12,830,342} & -\frac{63}{3478}\sqrt{12,830,342} \end{bmatrix}$; $Q_2 = \begin{bmatrix} -\frac{1}{90}\sqrt{30} & \frac{3131}{3,353,310}\sqrt{1,117,770} & -\frac{79577}{180,289,221,790}\sqrt{90,144,610,895} \\ \frac{11}{90}\sqrt{30} & \frac{389}{3,353,310}\sqrt{1,117,770} & \frac{423,203}{180,289,221,790}\sqrt{90,144,610,895} \\ \frac{1}{45}\sqrt{30} & -\frac{161}{1676655}\sqrt{1117770} & -\frac{215,739}{180,289,221,790}\sqrt{90,144,610,895} \\ \frac{2}{15}\sqrt{30} & -\frac{7}{558,885}\sqrt{1,117,770} & -\frac{358,611}{180,289,221,790}\sqrt{90,144,610,895} \end{bmatrix}$;

$R_2 = \begin{bmatrix} 3\sqrt{30} & \frac{161}{90}\sqrt{30} & \frac{2}{15}\sqrt{30} \\ 0 & \frac{1}{90}\sqrt{1,117,770} & -\frac{3652}{558,885}\sqrt{1,117,770} \\ 0 & 0 & \frac{2}{37,259}\sqrt{90,144,610,895} \end{bmatrix}$ **(d)** $\tilde{Q}_1 = \begin{bmatrix} -\frac{26}{1739}\sqrt{3478} & -\frac{191}{6,415,171}\sqrt{12,830,342} \\ -\frac{7}{3478}\sqrt{3478} & \frac{283}{12,830,342}\sqrt{12,830,342} \\ -\frac{7}{3478}\sqrt{3478} & -\frac{3195}{12,830,342}\sqrt{12,830,342} \\ \frac{13}{1739}\sqrt{3478} & -\frac{774}{6,415,171}\sqrt{12,830,342} \end{bmatrix}$;

$$\tilde{Q}_2 = \begin{bmatrix} -\frac{1}{90}\sqrt{30} & \frac{3131}{3,353,310}\sqrt{1,117,770} & -\frac{79577}{180,289,221,790}\sqrt{90,144,610,895} \\ \frac{11}{90}\sqrt{30} & \frac{389}{3,353,310}\sqrt{1,117,770} & \frac{423,203}{180,289,221,790}\sqrt{90,144,610,895} \\ \frac{1}{45}\sqrt{30} & -\frac{161}{1,676,655}\sqrt{1,117,770} & -\frac{215,739}{180,289,221,790}\sqrt{90,144,610,895} \\ \frac{2}{15}\sqrt{30} & -\frac{7}{558,885}\sqrt{1,117,770} & -\frac{358,611}{180,289,221,790}\sqrt{90,144,610,895} \end{bmatrix}$$

(e) $\tilde{R}_1 = \begin{bmatrix} \sqrt{3478} & \frac{3795}{3478}\sqrt{3478} & \frac{931}{3478}\sqrt{3478} \\ 0 & \frac{57}{3478}\sqrt{12,830,342} & -\frac{63}{3478}\sqrt{12,830,342} \end{bmatrix}$; $\tilde{R}_2 = \begin{bmatrix} 3\sqrt{30} & \frac{161}{90}\sqrt{30} & \frac{2}{15}\sqrt{30} \\ 0 & \frac{1}{90}\sqrt{1,117,770} & -\frac{3652}{558,885}\sqrt{1,117,770} \\ 0 & 0 & \frac{2}{37,259}\sqrt{90,144,610,895} \end{bmatrix}$

(f) R^{-1} exists only for the second matrix; $Q_2 Q_2^{\mathrm{T}} = \begin{bmatrix} \frac{9,636,411}{9,677,620} & -\frac{126,469}{9,677,620} & -\frac{580,783}{9,677,620} & \frac{209,293}{9,677,620} \\ -\frac{126,469}{9,677,620} & \frac{9,289,491}{9,677,620} & -\frac{1,782,403}{9,677,620} & \frac{642,313}{9,677,620} \\ -\frac{580,783}{9,677,620} & -\frac{1,782,403}{9,677,620} & \frac{1492,299}{9,677,620} & \frac{2,949,691}{9,677,620} \\ \frac{209,293}{9,677,620} & \frac{642,313}{9,677,620} & \frac{2,949,691}{9,677,620} & \frac{8,614,659}{9,677,620} \end{bmatrix}$

(g) $|R_{2,22}| = \frac{1}{90}\sqrt{1,117,770}$, $|R_{2,33}| = \frac{2}{37,259}\sqrt{90,144,610,895}$ **(h)** $B_1 = \begin{bmatrix} -52 & -63 \\ -7 & -3 \\ -7 & -60 \\ 26 & 3 \end{bmatrix}$; $P = Q_1 Q_1^{\mathrm{T}} = \begin{bmatrix} \frac{2910}{3689} & \frac{355}{3689} & \frac{737}{3689} & -\frac{1264}{3689} \\ \frac{355}{3689} & \frac{75}{3689} & -\frac{208}{3689} & -\frac{319}{3689} \\ \frac{737}{3689} & -\frac{208}{3689} & \frac{2987}{3689} & \frac{1229}{3689} \\ -\frac{1264}{3689} & -\frac{319}{3689} & \frac{1229}{3689} & \frac{1406}{3689} \end{bmatrix}$

(i) A_1 does not have full rank, while A_2 does.

17. (a) $M = \begin{bmatrix} 295.5119 & 43.8650 & 194.3208 \\ 43.8650 & 117.9937 & 19.7825 \\ 194.3208 & 19.7825 & 208.2692 \end{bmatrix}$; $C(M) = 11.71707814$ **(b)** $\mathbf{u} = [-0.7374763180, 0.3989617388, 0.1540718815]^{\mathrm{T}}$

(c) $Q = \begin{bmatrix} 0.5671750638 & 0.3989889761 & 0.2576692751 \\ 0.4799173617 & -0.7297048079 & -0.3058376261 \\ -0.4903882860 & -0.4251230460 & 0.07709365611 \\ 0.2175625373 & 0.1794610481 & -0.7586767171 \\ -0.4002219938 & 0.3088734665 & -0.5084644514 \end{bmatrix}$; $R = \begin{bmatrix} 17.19045956 & 2.551706069 & 11.30399099 \\ 0 & 10.55852718 & -0.8582600832 \\ 0 & 0 & 8.930418655 \end{bmatrix}$

(d) $\mathbf{u} = [-0.7374763178, 0.3989617386, 0.1540718814]^{\mathrm{T}}$ **(e)** $L = \begin{bmatrix} 0.05817180141 & 0 & 0 \\ -0.01405852693 & 0.09471017902 & 0 \\ -0.07498409845 & 0.009102145070 & 0.1119768331 \end{bmatrix}$

Exercises 34.4, Page 801

3. (a) $\lambda_{1,2,3} = 15.47690512, -9.857886436, -8.619018681$ **(b)** $r_1 = 0.6369417115$; $r_2 = 0.8743272442$ **(c)** After 100 iterations we find:
$\lambda_{1,2,3} = 15.47690515, -9.857888750, -8.619016363$ **(d)** The largest sub-diagonal element is $-0.2578878042 \times 10^{-6}$ which is in line with $r_2^{100} = 0.8743272442^{100} = 0.1470283656 \times 10^{-5}$.

13. (a) $H_{\mathbf{w}} = \begin{bmatrix} \frac{43}{67} & \frac{32}{67} & \frac{40}{67} & \frac{4}{67} \\ \frac{32}{67} & \frac{73}{201} & -\frac{160}{201} & -\frac{16}{201} \\ \frac{40}{67} & -\frac{160}{201} & \frac{1}{201} & -\frac{20}{201} \\ \frac{4}{67} & -\frac{16}{201} & -\frac{20}{201} & \frac{199}{201} \end{bmatrix}$ **(b)** $\mathbf{y} = \left[\frac{243}{67}, -\frac{302}{201}, -\frac{277}{201}, \frac{716}{201}\right]^{\mathrm{T}}$ **(c)** $\mathbf{x_w} = \left[-\frac{88}{67}, \frac{352}{201}, \frac{440}{201}, \frac{44}{201}\right]^{\mathrm{T}}$; $\mathbf{y_w} = \left[\frac{88}{67}, -\frac{352}{201}, -\frac{440}{201}, -\frac{44}{201}\right]^{\mathrm{T}}$

(d) $\mathbf{x}_{\perp\mathbf{w}} = \left[\frac{155}{67}, \frac{50}{201}, \frac{163}{201}, \frac{760}{201}\right]^{\mathrm{T}}$; $\mathbf{y}_{\perp\mathbf{w}} = \left[\frac{155}{67}, \frac{50}{201}, \frac{163}{201}, \frac{760}{201}\right]^{\mathrm{T}}$ **(e)** $H = \begin{bmatrix} \frac{43}{67} & \frac{32}{67} & \frac{40}{67} & \frac{4}{67} \\ \frac{32}{67} & \frac{73}{201} & -\frac{160}{201} & -\frac{16}{201} \\ \frac{40}{67} & -\frac{160}{201} & \frac{1}{201} & -\frac{20}{201} \\ \frac{4}{67} & -\frac{16}{201} & -\frac{20}{201} & \frac{199}{201} \end{bmatrix}$; $H^{\mathrm{T}}H = HH^{\mathrm{T}} = \begin{bmatrix} 1 & 0 & 0 & 0 \\ 0 & 1 & 0 & 0 \\ 0 & 0 & 1 & 0 \\ 0 & 0 & 0 & 1 \end{bmatrix}$

23. (a) $H_{\mathbf{w}} = \begin{bmatrix} -0.518321055 & -0.3455473703 & -0.08638684258 & 0.7774815832 \\ -0.3455473703 & 0.9213585396 & -0.01966036510 & 0.1769432859 \\ -0.08638684258 & -0.01966036510 & 0.9950849087 & 0.04423582147 \\ 0.7774815832 & 0.1769432859 & 0.04423582147 & 0.6018776068 \end{bmatrix}$ **(b)** $H_{\mathbf{w}}\mathbf{x} = [11.57583690, 0, 0, 0]^{\mathrm{T}}$;

$\|\mathbf{x}\| = 11.57583690$

28. (a) $A_1 = \begin{bmatrix} 18.78829424 & 4.896665923 & 1.437064995 & -4.896665925 \\ 0 & -2.842128040 & 12.32551077 & 1.899391110 \\ 0 & 2.842128040 & 2.674489237 & -14.89939111 \\ 0 & 9.157908023 & -4.106547944 & -3.524644813 \end{bmatrix}$

(b) $A_2 = \begin{bmatrix} 18.78829424 & 4.896665923 & 1.437064995 & -4.896665925 \\ 0 & 10.00113308 & -6.502945902 & -8.001359678 \\ 0 & 0 & 6.841101136 & -12.70842099 \\ 0 & 0 & 9.319113384 & 3.535101474 \end{bmatrix}$

(c) $A_3 = \begin{bmatrix} 18.78829424 & 4.896665923 & 1.437064995 & -4.896665925 \\ 0 & 10.00113308 & -6.502945902 & -8.001359678 \\ 0 & 0 & 11.56055963 & -4.670671972 \\ 0 & 0 & 0 & -12.33635804 \end{bmatrix}$

(d) $H = \begin{bmatrix} 0.2128985179 & 0.6386955549 & -0.6386955549 & -0.3725724069 \\ -0.7041695047 & 0.2872195798 & -0.2872195809 & 0.5823702724 \\ 0.2694397941 & 0.6876766018 & 0.6098384387 & 0.2874024479 \\ -0.6214673989 & 0.1915040280 & 0.3710390543 & -0.6628985592 \end{bmatrix}$;

$H^{\mathrm{T}}H = HH^{\mathrm{T}} = \begin{bmatrix} 1.000000001 & 0 & 0 & 0 \\ 0 & 1.000000000 & 0 & 0 \\ 0 \cdot & 0 & 0.9999999997 & 0 \\ 0 & 0 & 0 & 1.000000000 \end{bmatrix}$

(f) $H^{\mathrm{T}}A_3 = \begin{bmatrix} 3.999999998 & -6.000000009 & 8.000000009 & 11.00000001 \\ 12.00000002 & 6.000000000 & 6.999999996 & -11.00000001 \\ -12.00000002 & -6.000000011 & 8.000000008 & -1.999999991 \\ -7.000000007 & 3.999999987 & -1.000000003 & 4.000000006 \end{bmatrix}$

Exercises 34.5–Part A, Page 808

1. Multiply both sides of the stated equation by A. Then it becomes clear that $A\mathbf{x}$ is an eigenvector of AA^{T} with eigenvalue λ.

2. In (34.21) replace A with A^{T}.   **4.** $(\mathbf{u}_2\mathbf{u}_2{}^{\mathrm{T}})\,\mathbf{u}_2 = \mathbf{u}_2\,(\mathbf{u}_2{}^{\mathrm{T}}\mathbf{u}_2) = \mathbf{u}_2\,\|\mathbf{u}_2\|^2 = \mathbf{u}_2$

Exercises 34.5–Part B, Page 808

2. (a) $A^{\mathrm{T}}A = \begin{bmatrix} 1 & \frac{3}{5} \\ \frac{3}{5} & 1 \end{bmatrix}$; $\lambda_1 = \frac{8}{5}$, $\mathbf{v}_1 = [1, 1]^{\mathrm{T}}$, $\lambda_2 = \frac{2}{5}$, $\mathbf{v}_2 = [-1, 1]^{\mathrm{T}}$, $\sigma_1 = \frac{2}{5}\sqrt{10}$, $\sigma_2 = \frac{1}{5}\sqrt{10}$ **(b)** $\mathbf{u}_1 = \left[\frac{2}{5}\sqrt{5}, \frac{1}{5}\sqrt{5}\right]^{\mathrm{T}}$; $\mathbf{u}_2 = \left[\frac{1}{5}\sqrt{5}, -\frac{2}{5}\sqrt{5}\right]^{\mathrm{T}}$

(c) $U = \begin{bmatrix} \frac{2}{5}\sqrt{5} & \frac{1}{5}\sqrt{5} \\ \frac{1}{5}\sqrt{5} & -\frac{2}{5}\sqrt{5} \end{bmatrix}$; $V = \begin{bmatrix} \frac{1}{2}\sqrt{2} & -\frac{1}{2}\sqrt{2} \\ \frac{1}{2}\sqrt{2} & \frac{1}{2}\sqrt{2} \end{bmatrix}$; $\Sigma = \begin{bmatrix} \frac{2}{5}\sqrt{10} & 0 \\ 0 & \frac{1}{5}\sqrt{10} \end{bmatrix}$ **(d)** $U\Sigma V^{\mathrm{T}} = \begin{bmatrix} \frac{3}{5} & 1 \\ \frac{4}{5} & 0 \end{bmatrix}$

3. $W = \begin{bmatrix} \frac{1}{10}\sqrt{10} & \frac{3}{10}\sqrt{10} \\ \frac{3}{10}\sqrt{10} & -\frac{1}{10}\sqrt{10} \end{bmatrix}$; $\|A - W\|_F = \frac{1}{5}\sqrt{100 - 30\sqrt{10}}$; $WW^{\mathrm{T}} = \begin{bmatrix} 1 & 0 \\ 0 & 1 \end{bmatrix}$

5. $B = \begin{bmatrix} \frac{4}{5} & \frac{4}{5} \\ \frac{2}{5} & \frac{2}{5} \end{bmatrix}$; $\|A - B\|_2 = \|A - B\|_F = \frac{1}{5}\sqrt{10}$; $\|A - W\|_F = \frac{1}{5}\sqrt{100 - 30\sqrt{10}}$

6. (a) $k = 2$ **(b)** $U = \begin{bmatrix} -0.3417905767 & -0.8844032455 & -0.3178208631 \\ 0.1074948563 & -0.3727598508 & 0.9216805029 \\ -0.9336080856 & 0.2808576026 & 0.2224746042 \end{bmatrix}$; $V = \begin{bmatrix} -0.8110122690 & 0.09447627313 & 0.5773502692 \\ -0.4873249871 & -0.6551190912 & -0.5773502692 \\ -0.3236872819 & 0.7495953643 & -0.5773502692 \end{bmatrix}$;

$\Sigma = \begin{bmatrix} 9.033565692 & 0 & 0 \\ 0 & 6.032801247 & 0 \\ 0 & 0 & 0 \end{bmatrix}$ **(c)** $\sum_{i=1}^{1} \sigma_i\mathbf{u}_i\mathbf{v}_i^{\mathrm{T}} = \begin{bmatrix} 1.999999999 & 5.000000000 & -3.000000000 \\ -1.000000000 & 0.9999999995 & -2.000000000 \\ 7.000000000 & 3.000000000 & 3.999999999 \end{bmatrix}$

(d) $B = \begin{bmatrix} 2.504071447 & 1.504658601 & 0.9994128472 \\ -0.7875430710 & -0.4732227015 & -0.3143203694 \\ 6.839923362 & 4.110006336 & 2.729917026 \end{bmatrix}$; $\|A - B\|_2 = 6.032801247$

(e) $\lambda_{1,2,3} = 0, 81.60530911, 36.39469089$ **(g)** $V_1 = \begin{bmatrix} 19\frac{\frac{25}{19}+\frac{1}{19}\sqrt{511}}{\sqrt{2044+62\sqrt{511}}} & 19\frac{\frac{25}{19}-\frac{1}{19}\sqrt{511}}{\sqrt{2044-62\sqrt{511}}} \\ 19\frac{\frac{6}{19}+\frac{1}{19}\sqrt{511}}{\sqrt{2044+62\sqrt{511}}} & 19\frac{\frac{6}{19}-\frac{1}{19}\sqrt{511}}{\sqrt{2044-62\sqrt{511}}} \\ \frac{19}{\sqrt{2044+62\sqrt{511}}} & \frac{19}{\sqrt{2044-62\sqrt{511}}} \end{bmatrix}$; $V_2 = \begin{bmatrix} -\frac{1}{3}\sqrt{3} \\ \frac{1}{3}\sqrt{3} \\ \frac{1}{3}\sqrt{3} \end{bmatrix}$ **(h)** $\{[2, 5, -3]^T, [-1, 1, -2]^T\}$

(k) $U_1 = \begin{bmatrix} \frac{(\frac{23}{2}+\frac{7}{2}\sqrt{511})\sqrt{152278+5702\sqrt{511}}}{(59+\sqrt{511})(1022+31\sqrt{511})} & \frac{(\frac{23}{2}-\frac{7}{2}\sqrt{511})\sqrt{152278-5702\sqrt{511}}}{(1022-31\sqrt{511})(59-\sqrt{511})} \\ -\frac{57}{2}\frac{\sqrt{152278+5702\sqrt{511}}}{(59+\sqrt{511})(1022+31\sqrt{511})} & -\frac{57}{2}\frac{\sqrt{152278-5702\sqrt{511}}}{(1022-31\sqrt{511})(59-\sqrt{511})} \\ \frac{(\frac{269}{2}+5\sqrt{511})\sqrt{152278+5702\sqrt{511}}}{(59+\sqrt{511})(1022+31\sqrt{511})} & (\frac{269}{2}-5\sqrt{511})\sqrt{152278-5702\sqrt{511}} \end{bmatrix}$ **(l)** $U_2 = \begin{bmatrix} -\frac{1}{33}\sqrt{110} \\ \frac{29}{330}\sqrt{110} \\ \frac{7}{330}\sqrt{110} \end{bmatrix}$

Exercises 34.6–Part A, Page 813

1. If A is a square matrix with full rank, then Σ is an invertible diagonal matrix, and $\Sigma^{-1} = \Sigma^+$. The result follows.   2. $A^+ = \begin{bmatrix} 0 & \frac{5}{4} \\ 1 & -\frac{3}{4} \end{bmatrix}$

Exercises 34.6–Part B, Page 813

1. **(a)** This follows from the definition of Σ^+. **(b)** $\Sigma\Sigma^+$ is a block matrix with the identity matrix in the left top corner and zeros everywhere else. **(c)** $\Sigma^+\Sigma$ is a block matrix with the identity matrix in the left top corner and zeros everywhere else. **(d)** This follows directly from $A^+ = V\Sigma^+U^T$. **(e)** AA^+ is the projection matrix onto the column space of A. **(f)** A^+A is the projection matrix onto the row space of A.

2. **(a)** $\text{rank}(A) = 2$ **(b)** $A^+ = \begin{bmatrix} \frac{18,958}{468,837} & -\frac{6221}{468,837} & -\frac{14,185}{468,837} & -\frac{3763}{156,279} \\ -\frac{18,772}{468,837} & \frac{12,392}{468,837} & \frac{5044}{468,837} & \frac{2572}{156,279} \\ \frac{3299}{468,837} & -\frac{3976}{468,837} & \frac{1711}{468,837} & -\frac{119}{156,279} \end{bmatrix}$ **(c)** $(A^T)^+ = \begin{bmatrix} \frac{18,958}{468,837} & -\frac{18,772}{468,837} & \frac{3299}{468,837} \\ -\frac{6221}{468,837} & \frac{12,392}{468,837} & -\frac{3976}{468,837} \\ -\frac{14185}{468,837} & \frac{5044}{468,837} & \frac{1711}{468,837} \\ -\frac{3763}{156,279} & \frac{2572}{156,279} & -\frac{119}{156,279} \end{bmatrix}$

(e) $X^+ = \begin{bmatrix} -\frac{316,655}{468,837}, & \frac{269,924}{468,837}, & -\frac{34,849}{468,837} \end{bmatrix}^T$ **(g)** $U = \begin{bmatrix} 0.1657659827 & 0.7543546402 & -0.5523675100 & 0.3136253334 \\ -0.6373689962 & -0.3717920243 & -0.6734231328 & 0.04508589107 \\ 0.7180412379 & -0.3849560807 & -0.4880310949 & -0.3131313573 \\ 0.2251647892 & -0.3801691511 & 0.05671920769 & 0.8952961330 \end{bmatrix}$;

$\Sigma = \begin{bmatrix} 51.96519198 & 0 & 0 \\ 0 & 13.17644954 & 0 \\ 0 & 0 & 0 \\ 0 & 0 & 0 \end{bmatrix}$; $V = \begin{bmatrix} -0.6228720773 & 0.7410129476 & 0.2508588982 \\ -0.6263333277 & -0.6644783955 & 0.4076457096 \\ 0.4687610671 & 0.09678984145 & 0.8780061438 \end{bmatrix}$

(h) $V\Sigma^+U^T = \begin{bmatrix} 0.04043622836 & -0.01326900393 & -0.03025571786 & -0.02407873098 \\ -0.04003950199 & 0.02643136100 & 0.01075853655 & 0.01645774543 \\ 0.007036560681 & -0.008480559342 & 0.003649455995 & -0.000761458672 \end{bmatrix}$ **(i)** $Q = \begin{bmatrix} 2 & -\frac{15,652}{1143} \\ 17 & \frac{10,976}{1143} \\ -27 & \frac{364}{127} \\ -11 & \frac{6076}{1143} \end{bmatrix}$

(j) $P = \begin{bmatrix} \frac{275}{461} & -\frac{178}{461} & -\frac{79}{461} & -\frac{115}{461} \\ -\frac{178}{461} & \frac{251}{461} & -\frac{145}{461} & -\frac{1}{461} \\ -\frac{79}{461} & -\frac{145}{461} & \frac{306}{461} & \frac{142}{461} \\ -\frac{115}{461} & -\frac{1}{461} & \frac{142}{461} & \frac{90}{461} \end{bmatrix}$ **(k)** $P\mathbf{y} = \begin{bmatrix} -\frac{3979}{461}, & \frac{1625}{461}, & \frac{2516}{461}, & \frac{2192}{461} \end{bmatrix}^T$ **(l)** $\mathbf{v} = \begin{bmatrix} -\frac{6171}{5993} + \frac{8}{13}t_1, & t_1, & \frac{28}{13}t_1 - \frac{7877}{5993} \end{bmatrix}^T$ **(m)** $t_{min} = \frac{269,924}{468,837}$;

$X^+ = \begin{bmatrix} \frac{-316,655}{468,837}, & \frac{269,924}{468,837}, & -\frac{34,849}{468,837} \end{bmatrix}^T$ **(n)** $C = \begin{bmatrix} 1 & 0 \\ 0 & 1 \\ -\frac{2}{7} & -\frac{13}{28} \end{bmatrix}$ **(o)** $P' = \begin{bmatrix} \frac{953}{1017} & -\frac{104}{1017} & -\frac{224}{1017} \\ -\frac{104}{1017} & \frac{848}{1017} & -\frac{364}{1017} \\ -\frac{224}{1017} & -\frac{364}{1017} & \frac{233}{1017} \end{bmatrix}$ **(p)** $P'\mathbf{v} = \begin{bmatrix} -\frac{316,655}{468,837}, & \frac{269,924}{468,837}, & -\frac{34,849}{468,837} \end{bmatrix}^T$

(q) $\mathbf{V} = \begin{bmatrix} -\frac{2,159,392}{6,094,881} + \frac{8}{13}t_1, & -\frac{269,924}{468,837} + t_1, & -\frac{7,557,872}{6,094,881} + \frac{28}{13}t_1 \end{bmatrix}^T$ **(r)** $A\mathbf{V} = [0, 0, 0, 0]^T$

13. (b) $A = \begin{bmatrix} 1 & -30 & 18 & 6 & -30 \\ -16 & -143 & 18 & -9 & 42 \\ 46 & 73 & -140 & 35 & 74 \end{bmatrix}$ **(c)** $A^+ = \begin{bmatrix} \frac{792,449}{60,450,639} & -\frac{49,745}{60,450,639} & \frac{223,055}{60,450,639} \\ \frac{223,576}{20,150,213} & -\frac{222,623}{40,300,426} & -\frac{36,588}{20,150,213} \\ -\frac{1,894,849}{181,351,917} & \frac{289,453}{362,703,834} & \frac{2,347,733}{362,703,834} \\ \frac{2,991,022}{181,351,917} & -\frac{305,411}{362,703,834} & \frac{730,297}{181,351,917} \\ -\frac{4,492,340}{181,351,917} & \frac{879,452}{181,351,917} & -\frac{206,663}{181,351,917} \end{bmatrix}$ **(d)** $C = \begin{bmatrix} \frac{792,449}{60,450,639} & -\frac{49,745}{60,450,639} & \frac{223,055}{60,450,639} \\ \frac{223,576}{20,150,213} & \frac{222,623}{40,300,426} & \frac{36,588}{20,150,213} \\ -\frac{1,894,849}{181,351,917} & \frac{289,453}{362,703,834} & \frac{2,347,733}{362,703,834} \\ \frac{2,991,022}{181,351,917} & -\frac{305,411}{362,703,834} & \frac{730,297}{181,351,917} \\ -\frac{4,492,340}{181,351,917} & \frac{879,452}{181,351,917} & \frac{206,663}{181,351,917} \end{bmatrix}$

23. (a) $Q = \begin{bmatrix} -\frac{17}{314}\sqrt{314} & \frac{131}{219,486}\sqrt{219,486} & \frac{1}{699}\sqrt{699} \\ \frac{3}{314}\sqrt{314} & \frac{217}{219,486}\sqrt{219,486} & \frac{23}{699}\sqrt{699} \\ \frac{2}{157}\sqrt{314} & \frac{197}{109,743}\sqrt{219,486} & -\frac{13}{699}\sqrt{699} \end{bmatrix}$; $R = \begin{bmatrix} \sqrt{314} & \frac{114}{157}\sqrt{314} & -\frac{97}{314}\sqrt{314} & \frac{63}{314}\sqrt{314} \\ 0 & \frac{10}{157}\sqrt{219,486} & -\frac{3}{314}\sqrt{219,486} & -\frac{11}{314}\sqrt{219,486} \\ 0 & 0 & 0 & 0 \end{bmatrix}$

(b) Be careful. Because R does not have full row rank, the matrix RR^T is not invertible. Use R' and Q' instead;

$R' = \begin{bmatrix} \sqrt{314} & \frac{114}{157}\sqrt{314} & -\frac{97}{314}\sqrt{314} & \frac{63}{314}\sqrt{314} \\ 0 & \frac{10}{157}\sqrt{219,486} & -\frac{3}{314}\sqrt{219,486} & -\frac{11}{314}\sqrt{219,486} \end{bmatrix}$; $Q' = \begin{bmatrix} -\frac{17}{314}\sqrt{314} & \frac{131}{219,486}\sqrt{219,486} \\ \frac{3}{314}\sqrt{314} & \frac{217}{219,486}\sqrt{219,486} \\ \frac{2}{157}\sqrt{314} & \frac{197}{109,743}\sqrt{219,486} \end{bmatrix}$; $A^+ = \begin{bmatrix} -\frac{3632}{82249} & \frac{111}{82,249} & -\frac{83}{82,249} \\ \frac{722}{246,747} & \frac{2966}{246,747} & \frac{5192}{246,747} \\ \frac{1525}{164,498} & -\frac{341}{164,498} & \frac{243}{82,249} \\ -\frac{12,281}{493,494} & \frac{2863}{493,494} & -\frac{3005}{246,747} \end{bmatrix}$

33. If k denotes the rank of A, and U' denotes the matrix formed by the first k columns of U, then $AA^+ = (U \Sigma V^T)(V \Sigma^+ U^T) =$
$U \Sigma \Sigma^+ U^T = U \begin{bmatrix} I & 0 \\ 0 & 0 \end{bmatrix} U^T = U' I U'^T = U'(U'^T U')^{-1} U'^T = P$, where P is the projection matrix onto the column space of A.

CHAPTER 35

Exercises 35.1–Part A, Page 825

1. Use $\tan(u - v) = \frac{\tan u - \tan v}{1 + \tan u \tan v}$.   **2.** Substitute $A = \frac{3}{2}$ and $B = \frac{7}{4}$.   **4.** $\sqrt{41}\,\text{cis}\left(\arctan(\frac{4}{5})\right)$

Exercises 35.1–Part B, Page 825

2. (a) $|z| = 4\sqrt{5}$; $\text{Arg}(z) = \arctan(\frac{1}{2})$

(b) $\bar{z} = 8 - 4i$

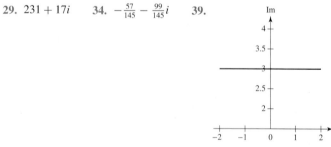

(c) $z + \bar{z} = 16$; $z - \bar{z} = 8i$
(d) $z\bar{z} = 80$
(e) $4\sqrt{5}\,\text{cis}\left(\arctan(\frac{1}{2})\right)$
(f) $\frac{z}{\bar{z}} = \frac{3}{5} + \frac{4}{5}i$; $\frac{\bar{z}}{z} = \frac{3}{5} - \frac{4}{5}i$

12. (a) $z_1 + z_2 = -5 + 12i = 13\,\text{cis}\left(-\arctan(\frac{12}{5}) + \pi\right)$; $z_1 - z_2 = 13 + 2i = \sqrt{173}\,\text{cis}\left(\arctan(\frac{2}{13})\right)$ **(b)** $z_1 z_2 = -71 - 43i$; $\frac{z_1}{z_2} = -\frac{1}{106} - \frac{83}{106}i$
(c) $\sqrt{6890}\,\text{cis}\left(\arctan(\frac{43}{71}) - \pi\right)$ **(d)** $\frac{1}{106}\sqrt{6890}\,\text{cis}(\arctan(83) - \pi)$ **(e)** $|z_1 - z_2|^2 = 173$ (f) $|z_1 + z_2| = 13 \le \sqrt{65} + \sqrt{106} = |z_1| + |z_2|$
(g) $\|z_1| - |z_2\| = -\sqrt{65} + \sqrt{106} \le \sqrt{173} = |z_1 - z_2|$

29. $231 + 17i$ **34.** $-\frac{57}{145} - \frac{99}{145}i$ **39.**

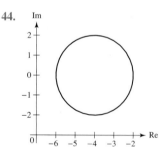

44.

Exercises 35.2–Part A, Page 828

2. $\tilde{w} = -2i = 2\,\text{cis}\left(-\frac{1}{2}\pi\right)$ **3.** $z_2 = -z_1 = 1 - i$

Exercises 35.2–Part B, Page 828

2. **(a)** $|z| = \sqrt{5}$; $\mathrm{Arg}(z) = \arctan 2 - \pi$ **(d)**

(e) $r_1 = 5^{1/4}\sin(\tfrac{1}{2}\arctan 2) - i5^{1/4}\cos(\tfrac{1}{2}\arctan 2)$;

$r_2 = -5^{1/4}\sin(\tfrac{1}{2}\arctan 2) + i5^{1/4}\cos(\tfrac{1}{2}\arctan 2)$

  (b) $\sigma = -3 + 4i = 5\,\mathrm{cis}(-\arctan(\tfrac{4}{3}) + \pi)$

(f)

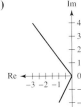

(g) $z^{3/2} = 5^{3/4}(-\sin(\tfrac{3}{2}\arctan 2) + i\cos(\tfrac{3}{2}\arctan 2)) = -3.330190677 - 0.3002831070i$;

$r_1^3 = (5^{1/4}\sin(\tfrac{1}{2}\arctan 2) - i5^{1/4}\cos(\tfrac{1}{2}\arctan 2))^3 = -3.330190674 - 0.3002831078i$;

$r_2^3 = (-5^{1/4}\sin(\tfrac{1}{2}\arctan 2) + i5^{1/4}\cos(\tfrac{1}{2}\arctan 2))^3 = 3.330190674 + 0.3002831078i$;

$\sqrt{z^3} = \tfrac{1}{2}\sqrt{22 + 10\sqrt{5}} + \tfrac{1}{2}i\sqrt{-22 + 10\sqrt{5}} = 3.330190677 + 0.3002831081i$

(h) $z^{5/2} = 5^{5/4}(\sin(\tfrac{5}{2}\arctan 2) - i\cos(\tfrac{5}{2}\arctan 2)) = 2.729624461 + 6.960664460i$;

$r_1^5 = (5^{1/4}\sin(\tfrac{1}{2}\arctan 2) - i5^{1/4}\cos(\tfrac{1}{2}\arctan 2))^5 = 2.729624454 + 6.960664453i$;

$r_2^5 = (-5^{1/4}\sin(\tfrac{1}{2}\arctan 2) + i5^{1/4}\cos(\tfrac{1}{2}\arctan 2))^5 = -2.729624454 - 6.960664453i$;

$\sqrt{z^5} = \tfrac{1}{2}\sqrt{-82 + 50\sqrt{5}} + \tfrac{1}{2}i\sqrt{82 + 50\sqrt{5}} = 2.729624466 + 6.960664460i$

12. **(a)** $\sqrt{z_1 z_2} = \sqrt{10}101^{1/4}\left(\dfrac{1}{\sqrt{1 + \frac{5\sqrt{101}+43}{5\sqrt{101}-43}}} - \dfrac{i\sqrt{5\sqrt{101}+43}}{\sqrt{5\sqrt{101}-43}\sqrt{1 + \frac{5\sqrt{101}+43}{5\sqrt{101}-43}}}\right) = 2.692466917 - 9.656571756i$

  (b) $\sqrt{z_1}\sqrt{z_2} = \tfrac{3}{2}\sqrt{-20 + 2\sqrt{101}} - \tfrac{1}{2}\sqrt{20 + 2\sqrt{101}} + i(\tfrac{1}{2}\sqrt{-20 + 2\sqrt{101}} + \tfrac{3}{2}\sqrt{20 + 2\sqrt{101}}) = -2.692466918 + 9.656571757i$

  (c) $\sqrt{\dfrac{z_1}{z_2}} = \dfrac{1}{101}\sqrt{1010}101^{1/4}\left(\dfrac{1}{\sqrt{1 + \frac{505\sqrt{101}+3737}{505\sqrt{101}-3737}}} - \dfrac{i\sqrt{505\sqrt{101}+3737}}{\sqrt{505\sqrt{101}-3737}\sqrt{1 + \frac{505\sqrt{101}+3737}{505\sqrt{101}-3737}}}\right) = 0.3621905044 - 0.9294381254i$

  (d) $\dfrac{\sqrt{z_1}}{\sqrt{z_2}} = \tfrac{3}{202}\sqrt{-20 + 2\sqrt{101}}\sqrt{101} + \tfrac{1}{202}\sqrt{20 + 2\sqrt{101}}\sqrt{101} + i(\tfrac{1}{202}\sqrt{-20 + 2\sqrt{101}}\sqrt{101} - \tfrac{3}{202}\sqrt{20 + 2\sqrt{101}}\sqrt{101}) = 0.3621905043 - 0.9294381252i$

Exercises 35.3–Part A, Page 831

2. $z^3 = x^3 - 3xy^2 + i(3x^2y - y^3)$ 3. $\{5^{1/3}\cos(\tfrac{1}{3}\arctan(\tfrac{3}{4})) + i5^{1/3}\sin(\tfrac{1}{3}\arctan(\tfrac{3}{4})), -5^{1/3}\sin(\tfrac{1}{3}\arctan(\tfrac{3}{4}) + \tfrac{1}{6}\pi) +$

$i5^{1/3}\cos(\tfrac{1}{3}\arctan(\tfrac{3}{4}) + \tfrac{1}{6}\pi), -5^{1/3}\cos(\tfrac{1}{3}\arctan(\tfrac{3}{4}) + \tfrac{1}{3}\pi) - i5^{1/3}\sin(\tfrac{1}{3}\arctan(\tfrac{3}{4}) + \tfrac{1}{3}\pi)\}$ 4. $w_1 = -2 + 2i = 2\sqrt{2}\,\mathrm{cis}(\tfrac{3}{4}\pi)$

Exercises 35.3–Part B, Page 831

2. **(a)** $|z| = \sqrt{34}$; $\theta = -\arctan(\tfrac{5}{3})$ **(b)** $z^3 = -198 - 10i = 34\sqrt{34}\,\mathrm{cis}\left(\arctan\left(\tfrac{5}{99}\right) - \pi\right)$ **(c)** $r = 34\sqrt{34}$ $w = -3\arctan(\tfrac{5}{3})$

  (d) $\{34^{1/6}(\cos(\tfrac{1}{3}\arctan(\tfrac{5}{3})) - i\sin(\tfrac{1}{3}\arctan(\tfrac{5}{3}))), 34^{1/6}(-\sin(-\tfrac{1}{3}\arctan(\tfrac{5}{3}) + \tfrac{1}{6}\pi) + i\cos(-\tfrac{1}{3}\arctan(\tfrac{5}{3}) + \tfrac{1}{6}\pi)),$

$34^{1/6}(-\cos(-\tfrac{1}{3}\arctan(\tfrac{5}{3}) + \tfrac{1}{3}\pi) - i\sin(-\tfrac{1}{3}\arctan(\tfrac{5}{3}) + \tfrac{1}{3}\pi))\}$

(e) $z^{2/3} = 2.504881547 - 2.054422804i$; $r_1^2 = 2.504881547 - 2.054422803i$;

$r_2^2 = -3.031623111 - 1.142079653i$; $r_3^2 = 0.5267415656 + 3.196502455i$;

$\sqrt[3]{z^2} = 2.504881548 - 2.054422803i$

(f) $z^{5/3} = -2.757469393 - 18.68767614i$; $r_1^5 = -2.757469387 - 18.68767614i$;

$r_2^5 = -14.80526760 + 11.73187659i$; $r_3^5 = 17.56273698 + 6.955799523i$;

$\sqrt[3]{z^5} = 17.56273697 + 6.955799549i$

23. Write the circle as $|z - z_0| = r$, then $r = \sqrt{|z_0|^2 - a}$.   25. *Hint:* $\cos(5\theta) = \mathcal{R}\,(\mathrm{cis}(5\theta))$.

31. Let $\{w_1, w_2, w_3\} = \{3.697293048 - 1.988523512i, -3.582990856 + 8.403739562i, -1.114302188 - 0.415216052i\}$ denote the three roots. Compute the midpoint m of $w_1 w_2$, then add $\frac{w_3 - m}{3}$. The result follows.

Exercises 35.4–Part A, Page 834

1. $z_1 = \sqrt{2}e^{1/4 i\pi}$ 4. $e^{\bar{z}} = e^{1-i} = e\cos 1 - ie\sin 1$ 5. $\pm\left(\tfrac{1}{2}2^{1/4}\sqrt{2 + \sqrt{2}} + \tfrac{1}{2}i2^{1/4}\sqrt{2 - \sqrt{2}}\right)$

Exercises 35.4–Part B, Page 834

2. **(d)** $e^{e^{z^2}} = e^{e^{-16}\cos 30}\cos(e^{-16}\sin 30) - ie^{e^{-16}\cos 30}\sin(e^{-16}\sin 30)$; $(e^{e^3\cos 5})^2\cos(e^3\sin 5)^2 - 2i(e^{e^3\cos 5})^2\cos(e^3\sin 5)\sin(e^3\sin 5) - (e^{e^3\cos 5})^2\sin(e^3\sin 5)^2$ **(e)** $z = \sqrt{34}e^{-i\arctan(5/3)}$ **(f)** $34^{1/4}\cos(\frac{1}{2}\arctan(\frac{5}{3})) - i34^{1/4}\sin(\frac{1}{2}\arctan(\frac{5}{3}))$;
$-34^{1/4}\cos(\frac{1}{2}\arctan(\frac{5}{3})) + i34^{1/4}\sin(\frac{1}{2}\arctan(\frac{5}{3}))$ **(g)** $34^{1/6}\cos(\frac{1}{3}\arctan(\frac{5}{3})) - i34^{1/6}\sin(\frac{1}{3}\arctan(\frac{5}{3}))$;
$-34^{1/6}\sin(-\frac{1}{3}\arctan(\frac{5}{3}) + \frac{1}{6}\pi) + i34^{1/6}\cos(-\frac{1}{3}\arctan(\frac{5}{3}) + \frac{1}{6}\pi)$; $-34^{1/6}\cos(-\frac{1}{3}\arctan(\frac{5}{3}) + \frac{1}{3}\pi) - i34^{1/6}\sin(-\frac{1}{3}\arctan(\frac{5}{3}) + \frac{1}{3}\pi)$

12. **(a)** $\frac{e^{z_1}}{e^{z_2}} = e^{-1}\cos 4 + ie^{-1}\sin 4$ **(b)** $\frac{z_1 z_2}{e^{z_1} + e^{z_2}} = -0.1995123695 + 0.01713768921i$
   (c) $z_1 e^{\bar{z}_2} + z_2 e^{\bar{z}_1} = 5e^4\cos 1 + 4e^5\cos 3 - e^5\sin 3 - 3e^4\sin 1 + i(-3e^4\cos 1 + e^5\cos 3 + 4e^5\sin 3 - 5e^4\sin 1)$ **(d)** $\frac{e^{z_1}\bar{z}_2}{e^{\bar{z}_1 z_2}} = \cos 34 + i\sin 34$.

22. **(a)** $xe^x\cos y - ye^x\sin y + i(ye^x\cos y + xe^x\sin y)$ **(b)**   **(c)**

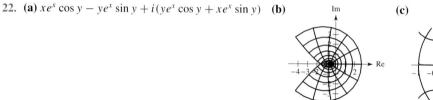

Exercises 35.5–Part A, Page 837

1. $\operatorname{Log} z_1 = \frac{1}{2}\ln 2 + \frac{1}{4}i\pi$; $\operatorname{Log} z_2 = \frac{1}{2}\ln 5 + i(-\arctan(\frac{1}{2}) + \pi)$; $\operatorname{Log}(z_1 z_2) = \frac{1}{2}\ln 10 + i(\arctan(\frac{1}{3}) - \pi)$; $\operatorname{Log}\frac{z_1}{z_3} = \frac{1}{2}\ln(\frac{2}{101}) + i(\arctan(\frac{11}{9}) - \pi)$
3. $\operatorname{Log} z_1 + \operatorname{Log} z_2 = \frac{1}{2}\ln 2 + \frac{1}{2}\ln 5 + i(\frac{5}{4}\pi - \arctan(\frac{1}{2})) = \operatorname{Log}(z_1 z_2) + 2\pi i$

Exercises 35.5–Part B, Page 837

2. **(a)** $\operatorname{Log} e^z = -2 + i(8 - 2\pi)$; $n_1 = -1$ **(b)** $N_1(z) = -1$ **(c)** $z_k = -2 + 8i + 2k\pi i$, $k = \pm 1, \pm 2, \pm 3$;
   $w = e^{z_k} = e^{-2}\cos 8 + ie^{-2}\sin 8$, $k = \pm 1, \pm 2, \pm 3$; $\log w = -2 + 8i + 2m\pi i$, $m \in \mathbb{Z}$

12. **(a)** $\operatorname{Log} z^w = -5\ln 3 - 5\ln 5 - 4\arctan(\frac{3}{4}) + 4\pi + i(4\ln 3 + 4\ln 5 - 5\arctan(\frac{3}{4}) - 3\pi) = w\operatorname{Log} z - 8\pi i$; $n_2 = -4$ **(b)** $N_2(z, w) = -4$

22. **(a)** $\operatorname{Log}(z_1 z_2) = \frac{1}{2}\ln 29 + \frac{1}{2}\ln 41 + i(-\arctan(\frac{17}{30}) + \pi) = \operatorname{Log} z_1 + \operatorname{Log} z_2 + 2\pi i$; $n_3 = 1$ **(b)** $N_3(z_1, z_2) = 1$

32. **(a)** $\operatorname{Log}\left(\frac{z_1}{z_2}\right) = \frac{1}{2}\ln 2 + \frac{1}{2}\ln 61 - \frac{1}{2}\ln 53 + i(-\arctan(\frac{15}{79}) + \pi) = \operatorname{Log} z_1 - \operatorname{Log} z_2 + 2\pi i$; $n_4 = 1$ **(b)** $N_4(z_1, z_2) = 1$

42. $z = 11\cos 12 - 11i\sin 12$; $\operatorname{Arg} z = -12 + 4\pi$

Exercises 35.6–Part A, Page 841

1. $2^i = e^{-2k\pi}\operatorname{cis}(\ln 2)$, for integer k.   2. $(2i)^i = e^{-1/2\pi - 2k\pi}\operatorname{cis}(\ln 2)$, for integer k.
5. $((-1 - i)^3)^{1/3} = \sqrt{2}\operatorname{cis}\left(\arctan(\sin(\frac{1}{12}\pi(-1 + 8k)), \cos(\frac{1}{12}\pi(-1 + 8k)))\right)$, for $k = 0, 1, 2$

Exercises 35.6–Part B, Page 841

2. **(a)** $e^{\operatorname{Log}(3-5i)} = 3 - 5i$ **(b)** $((3 - 5i)^{1/n})^n = 3 - 5i$, $n = 2, \cdots, 5$;
   $\{((3 - 5i)^n)^{1/n}\}_{n=2}^5 = \{3 - 5i, \sqrt{34}\sin(\frac{1}{3}\arctan(\frac{5}{99}) + \frac{1}{6}\pi) - i\sqrt{34}\cos(\frac{1}{3}\arctan(\frac{5}{99}) + \frac{1}{6}\pi), 1156^{1/4}\cos(-\frac{1}{4}\arctan(\frac{240}{161}) + \frac{1}{4}\pi) + i1156^{1/4}\sin(-\frac{1}{4}\arctan(\frac{240}{161}) + \frac{1}{4}\pi), 1156^{1/5}34^{1/10}\cos(\frac{1}{5}\arctan(\frac{1525}{717})) + i1156^{1/5}34^{1/10}\sin(\frac{1}{5}\arctan(\frac{1525}{717}))\}$

22. **(a)** $(zw)^a = 0.7971139227 \times 10^8 + 0.1238830966 \times 10^9 i$; $z^a w^a = 0.5191134384 + 0.8067777818i$ **(b)** and **(c)** $n_3 = 1$
   (d) $\left(\frac{z}{w}\right)^a = -16.10390854 - 64.53326170i$; $\frac{z^a}{w^a} = -16.10390853 - 64.53326169i$ **(e)** and **(f)** $n_4 = 0$

32. **(a)** $z^{ab} = -0.9771024323 \times 10^{14} + 0.1018864921 \times 10^{15} i$; $(z^a)^b = -636329.3735 + 663526.8281i$;
   $(z^b)^a = -0.004144039139 + 0.004321160161i$ **(b)** $n_2 = -1$ **(c)** $n_2(z, a) = -1$ **(d)** $n_2 = 1$ **(e)** $n_2(z, b) = 1$

Exercises 35.7–Part A, Page 846

4. $\sin\bar{z} = \sin(x)\cosh(y) - i\cos(x)\sinh(y) = \overline{\sin z}$ 7. $\cosh\bar{z} = \cosh(x)\cos(y) - i\sinh(x)\sin(y) = \overline{\cosh z}$

Exercises 35.7–Part B, Page 847

2. min: $\sin(1)$; max: $\sqrt{\cosh(1)^2 - 1}$ 18. **(a)**   **(b)**

22. $\sin(\frac{1}{2}\ln(13))\cosh(\arctan(\frac{3}{2})) + i\cos(\frac{1}{2}\ln(13))\sinh(\arctan(\frac{3}{2}))$

Exercises 35.8–Part A, Page 851

1. $r = e^{iz} \neq 0$, $w = \frac{1}{2}\left(r + \frac{1}{r}\right) \Leftrightarrow r^2 - 2wr + 1 = 0 \Leftrightarrow r = w + i\sqrt{1 - w^2} = e^{iz} \Leftrightarrow z = -i\log(w + i\sqrt{1 - w^2})$

3. $r = e^z \neq 0$, $w = \frac{1}{2}\left(r + \frac{1}{r}\right) \Leftrightarrow r^2 - 2wr + 1 = 0 \Leftrightarrow r = w + \sqrt{w^2 - 1} = e^z \Leftrightarrow z = \log(w + \sqrt{w^2 - 1})$

4. $w = \frac{r - \frac{1}{r}}{r + \frac{1}{r}} \Leftrightarrow r^2 = \frac{1+w}{1-w} \Leftrightarrow r = \sqrt{\frac{1+w}{1-w}} \Leftrightarrow z = \frac{1}{2}\log\left(\frac{1+w}{1-w}\right)$

Exercises 35.8–Part B, Page 852

2. In this problem Z denotes an integer, while B denotes a binary integer. **(a)** $z = \arcsin(1 - 3I) - 2\arcsin(1 - 3i)B + 2\pi Z + \pi B$
(b) $z = \arccos(1 - 3I) - 2\arccos(1 - 3i)B + 2\pi Z$ **(c)** $z = 1.461461854 - 0.3059438579i + \pi Z$
(d) $z = 1.824198702 - 1.233095218i + 2\pi i Z$ **(e)** $z = 1.864161544 - 1.263192677i + 2\pi i Z$
(f) $z = 0.09193119503 - 1.276795025i + \pi i Z$

10. **(a)** $\Re(\cot(x + iy)) = \frac{\sin x \cos x}{\sin^2 x + \sinh^2 y}$; $\Im(\cot(x + iy)) = -\frac{\sinh y \cosh y}{\sin^2 x + \sinh^2 y}$ **(b)** $|\cot(x + iy)| = \sqrt{\frac{\cosh^2 y - 1 + \cos^2 x}{\cosh^2 y - \cos^2 x}}$

16. $w = i\left(\frac{r + \frac{1}{r}}{r - \frac{1}{r}}\right) \Leftrightarrow r^2 = \frac{w-i}{w+i} \Leftrightarrow z = \frac{i}{2}\log\left(\frac{w-i}{w+i}\right)$ **23.** **(a)** $z = -0.00004872447884 - 1.000076624i$
(b) $z = 0.01293874902 - 0.003764914205i$ **(c)** $z = -0.003765833120 - 0.01293955699i$ **(d)**
$z = 0.9999896892 + 0.6685093531 \times 10^{-5}i$ **(e)** $z = 0.001406279660 + 0.004753891074i$ **(f)** $z = 0.001406233379 + 0.004753851458i$

Exercises 35.9–Part A, Page 856

1. If $f(z) = u(x, y) + iv(x, y)$ and $\overline{f(z)} = u(x, y) - iv(x, y)$ are both differentiable on the open connected set R, then the Cauchy–Riemann equations imply that u_x, u_y, v_x, and v_y all vanish throughout R. The desired result follows.

Exercises 35.9–Part B, Page 856

2. The region is open and connected. Verify the Cauchy–Riemann equations. $u = r^{(1/n)}\cos(\frac{\theta}{n})$; $v = r^{(\frac{1}{n})}\sin(\frac{\theta}{n})$; $\frac{\partial u}{\partial x} = \frac{r^{1/n}\cos(\theta/n)}{nr} = \frac{\partial v}{\partial y}$;
$\frac{\partial u}{\partial y} = \frac{r^{1/n}\sin(\theta/n)}{)}nr = -\frac{\partial v}{\partial x}$

7. $u = \frac{10x^2 - x - 21 + 10y^2}{25x^2 + 70x + 49 + 25y^2}$; $v = 29\frac{y}{25x^2 + 70x + 49 + 25y^2}$; $\frac{\partial u}{\partial x} = 29\frac{25x^2 + 70x + 49 - 25y^2}{(25x^2 + 70x + 49 + 25y^2)^2} = \frac{\partial v}{\partial y}$; $\frac{\partial u}{\partial y} = 290\frac{y(5x+7)}{(25x^2 + 70x + 49 + 25y^2)^2} = -\frac{\partial v}{\partial x}$

8. $u = \frac{\sin x \cos x}{\cos^2 x + \cosh^2 y - 1}$; $v = \frac{\sinh y \cosh y}{\cos^2 x + \cosh^2 y - 1}$; $\frac{\partial u}{\partial x} = \frac{2\cos^2 x \cosh^2 y - \cos^2 x - \cosh^2 y + 1}{\cos^4 x + 2\cos^2 x \cosh^2 y - 2\cos^2 x + \cosh^4 y - 2\cosh^2 y + 1} = \frac{\partial v}{\partial y}$;
$\frac{\partial u}{\partial y} = -2\frac{\sin x \cos x \sinh y \cosh y}{\cos^4 x + 2\cos^2 x \cosh^2 y - 2\cos^2 x + \cosh^4 y - 2\cosh^2 y + 1} = -\frac{\partial v}{\partial x}$

Exercises 35.10–Part A, Page 860

1. *Hint:* $|\sin(iy)| = |i\sinh y| = |\sinh y| = \sinh y$, when $y > 0$ **2.** $\nabla u \cdot \nabla v = u_x v_x + u_y v_y = v_y v_x - v_x v_y = 0$.

3. $\frac{\partial^2}{\partial x^2}(x^3 - 3xy^2) + \frac{\partial^2}{\partial y^2}(x^3 - 3xy^2) = 0$; $\frac{\partial^2}{\partial x^2}(3x^2 y - y^3) + \frac{\partial^2}{\partial y^2}(3x^2 y - y^3) = 0$

Exercises 35.10–Part B, Page 860

1. $v(x, y) = c \Rightarrow v_x x' + v_y y' = 0 \Rightarrow \frac{dy}{dx} = \frac{y'}{x'} = -\frac{v_x}{v_y} = \frac{u_y}{u_x}$

6. **(a)** $\frac{\partial^2}{\partial x^2}(\cos(x)\cosh(y)) + \frac{\partial^2}{\partial y^2}(\cos(x)\cosh(y)) = 0$; $\frac{\partial^2}{\partial x^2}(-\sin(x)\sinh(y)) + \frac{\partial^2}{\partial y^2}(-\sin(x)\sinh(y)) = 0$
(b)   **(c)** $c - \sin(x)\sinh(y) + \sin(x_0)\sinh(y_0)$ **(d)** $\cos(x)\cosh(y) + c$

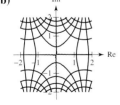

Exercises 35.11–Part B, Page 870

4. **(a)** $\int_C \frac{dz}{z^2+9} = \int_{C_{3i,1}} \frac{dz}{z^2+9} + \int_{C_{-3i,1}} \frac{dz}{z^2+9} = \frac{\pi}{3} - \frac{\pi}{3} = 0$ **(b)** See (a) **(c)** $\int_C \frac{dz}{z^2 + 9} = \int_0^{2\pi} \frac{15ie^{it}}{225e^{2it} + 9}dt = 0$.

19. $f''(2 + 3i) = \frac{2!}{2\pi i}\int_C \frac{f(z)}{(z - (2+3i))^3}dz = -\frac{821}{50653} - \frac{7299}{101306}i$, when C is the circle with radius 1 centered at $2 + 3i$.

24. $\int_{z_1}^{z_2} ze^z dz = -2e^{-1}\cos(3) - 3e^{-1}\sin(3) + 2e\sin(2) + i(-3e^{-1}\cos(3) + 2e^{-1}\sin(3) - 2e\cos(2))$

Exercises 35.12, Page 876

1. $f(z) = \frac{1}{3}\frac{1}{1-z} + \frac{1}{6}\frac{1}{1+\frac{z}{2}} = \frac{1}{3}\sum_{k=0}^{\infty} z^k + \frac{1}{6}\sum_{k=0}^{\infty}\left(-\frac{z}{2}\right)^k$ 2. $f(z) = \frac{1}{2} - \frac{1}{3}\frac{1}{1-\frac{z}{z}} - \frac{1}{6}\frac{1}{1+\frac{2}{z}} = \frac{1}{2} - \frac{1}{3}\sum_{k=0}^{\infty}\left(\frac{1}{z}\right)^k - \frac{1}{6}\sum_{k=0}^{\infty}\left(-\frac{2}{z}\right)^k$

3. **(a)** $\frac{1}{20}\sum_{k=0}^{\infty}\left(\frac{1+3i}{(-1+3i)^k} + \frac{1-3i}{(-1-3i)^k}\right)z^k$ **(b)** $\sqrt{10}$ **(c)** Application of the ratio test to the series $\sum_{k=0}^{\infty}\left(\frac{z}{-1+3i}\right)^k$ and $\sum_{k=0}^{\infty}\left(\frac{z}{-1-3i}\right)^k$ confirms $\sqrt{10}$ as the radius of convergence.   **13.** **(a)** Zeros: $(\frac{1}{14} + \frac{1}{14}\sqrt{29}, \frac{1}{14} - \frac{1}{14}\sqrt{29})$; Each has multiplicity one; Poles: $(3 + 9i, -5 + 6i)$; $3 + 9i$ has multiplicity one, $-5 + 6i$ has multiplicity two.

24. Series at $z = -5$: $\frac{5}{7}\frac{1}{z+5} + \sum_{k=0}^{\infty}\left(-\frac{1}{49}\frac{(z+5)^k}{14^k}\right)$; Series at $z = 9$: $\frac{2}{7}\frac{1}{z-9} + \sum_{k=0}^{\infty}\left(\frac{5}{98}\frac{(z-9)^k}{(-14)^k}\right)$; The radius of convergence is 14 for both series.

31. **(a)** In A: $\sum_{k=0}^{\infty}\left(-\frac{1}{84}\frac{(33^k + 4(-4)^k)z^k}{(-4)^k 3^k}\right)$; In B: $\frac{1}{7}\sum_{k=0}^{\infty}3^k\left(\frac{1}{z}\right)^{(k+1)} - \frac{1}{28}\sum_{k=0}^{\infty}(-1)^{(-k)}4^{(-k)}z^k$; In C: $-\frac{1}{12} + \sum_{k=0}^{\infty}\left(\frac{1}{84}\frac{(3(\frac{1}{3})^k + 4(-\frac{1}{4})^k)(\frac{1}{z})^k}{(-\frac{1}{4})^k(\frac{1}{3})^k}\right)$

Exercises 35.13–Part A, Page 880

2. **(a)** $\int_C \frac{7-z}{(z-1)(z+2)} = -2i\pi$ **(b)** $2\pi i(2 - 3) = -2i\pi$ **(c)** $2\pi i(\text{Res}(f(z), 1) + \text{Res}(f(z), -2)) = 2\pi i\left(\frac{6}{9} - \frac{9}{3}\right) = -2i\pi$

Exercises 35.13–Part B, Page 880

2. **(a)** $\int_C \frac{dz}{z^2+9} = 0$ **(b)** $\text{Res}(f(z), -3i) = \frac{1}{6}i$; $\text{Res}(f(z), 3i) = -\frac{1}{6}i$ 17. **(a)** $18 + 24(z - 1) + 14(z - 1)^2 + 3(z - 1)^3$ **(b)** 5; 5; 5; 3

27. $f''(2) = \frac{2!}{2\pi i}\int_C \frac{f(z)}{(z-2)^3}dz = 2!\,\text{Res}\left(\frac{f(z)}{(z-2)^3}, 2\right) = 0$

CHAPTER 36

Exercises 36.1–Part A, Page 891

1. $\int_C \frac{8(z^4-6z^2+1)}{2z^2(z^2+4z+1)} = 4\pi - \frac{4}{3}\pi\sqrt{3}$ 2. $2\pi i\left(\text{Res}\left(\frac{1+z^2}{4+z^2}, 1+i\right) + \text{Res}\left(\frac{1+z^2}{4+z^2}, -1+i\right)\right) = \frac{3}{4}\pi$.   4. $2\pi i\,\text{Res}\left(e^{2iz}\frac{1+z}{1+z^2}, i\right) = (1 + i)\pi e^{-2}$

Exercises 36.1–Part B, Page 891

2. $\int_C -\frac{2i}{az^2+2z+a}dz = 2\frac{\pi}{\sqrt{1-a^2}}$, with C the unit circle centered around the origin.   17. $2\pi i\,\text{Res}\left(\frac{(8z-9)}{(z^2-8z+25)(z+7+5i)}, 4 + 3i\right) = \left(\frac{89}{111} + \frac{16}{111}i\right)\pi$

27. $2\pi i\,\text{Res}\left(\frac{2z^2-z+4}{(z^2+12z+117)(z+i)}, -6 + 9i\right) = \left(-\frac{295}{204} + \frac{1075}{612}i\right)\pi e^{-45-30i}$

37. $2\pi i\left(\text{Res}\left(e^{9iz}\frac{9z^2-4z-6}{(z^2-4z+68)(z-6-9i)}, 6 + 9i\right) + \text{Res}\left(e^{9iz}\frac{9z^2-4z-6}{(z^2-4z+68)(z-6-9i)}, 2 + 8i\right)\right) = -\frac{135,654}{5185}\pi e^{-81}\sin(54) - \frac{60,768}{5185}\pi e^{-81}\cos(54) + \frac{789}{68}\pi e^{-72}\sin(18) + \frac{245}{17}\pi e^{-72}\cos(18)$

48. $2\pi i\,\text{Res}\left(\frac{5z^2-9z-1}{(z^2+14z+130)(z^2-8z+97)z}, -7 + 9i\right) + 2\pi i\,\text{Res}\left(\frac{5z^2-9z-1}{(z^2+14z+130)(z^2-8z+97)z}, -7 + 9i\right) + \pi i\,\text{Res}\left(\frac{5z^2-9z-1}{(z^2+14z+130)(z^2-8z+97)z}, 0\right) = -\frac{9233}{1,122,290}\pi$

57. $-\frac{1}{13,350}\pi\sqrt{389}^{5/6}\cos(\frac{1}{3}\arctan(\frac{5}{8})) - \frac{1}{4450}\pi\,895^{5/6}\sin(\frac{1}{3}\arctan(\frac{5}{8})) - \frac{1}{13,350}\pi\sqrt{389}^{5/6}\sin(\frac{1}{3}\arctan(\frac{5}{8})) + \frac{1}{4450}\pi\,895^{5/6}\cos(\frac{1}{3}\arctan(\frac{5}{8})) + \frac{1}{150}\pi 3^{1/6} = 0.02881987826$

Exercises 36.2–Part A, Page 895

1. $\frac{e^{(1+iy)t}}{1 + iy} = \left(\frac{\cos(ty)}{1 + y^2} + \frac{\sin(ty)y}{1 + y^2} + \frac{i(\sin(ty) - \cos(ty)y)}{1 + y^2}\right)e^t$ 2. The imaginary part of the integrand is an odd function of y

4. Direct substitution in the results of 1.0 and 2.0 yields the desired result.

Exercises 36.2–Part B, Page 895

2. $f(t) = \text{Res}\left(\frac{e^{st}}{(s+2)^2}, -2\right) = e^{-2t}t$ 12. $f(t) = \frac{1}{2} + \sum_{m=-\infty}^{\infty}\text{Res}\left(\frac{e^{st}}{s(1 + e^{2s})}, \frac{1}{2}i\pi + i\pi m\right) = \frac{1}{2} - \frac{2}{9}\frac{\sin(\frac{9}{2}\pi t)}{\pi} - \frac{2}{7}\frac{\sin(\frac{7}{2}\pi t)}{\pi} - \frac{2}{5}\frac{\sin(\frac{5}{2}\pi t)}{\pi} - \frac{2}{3}\frac{\sin(\frac{3}{2}\pi t)}{\pi} - 2\frac{\sin(\frac{1}{2}\pi t)}{\pi} + \cdots$

(b) $f(t) = H(t - 2) - H(t - 4) + H(t - 6) - H(t - 8) + H(t - 10)$

Exercises 36.3–Part A, Page 899

3. $\displaystyle\sum_{n=-\infty}^{-1} \frac{(-1+(-1)^n)e^{in\pi t}}{n^2\pi^2} + \frac{1}{2} + \sum_{n=1}^{\infty} \frac{(-1+(-1)^n)e^{in\pi t}}{n^2\pi^2}$

5. When $x > 0$, $f(x) = e^{-2x}$, and when $x < 0$, $f(x) = 0$, because $F(\alpha)$ has no poles in the lower half plane.

Exercises 36.3–Part B, Page 899

1. $\displaystyle\sum_{n=-\infty}^{-1}\left(2\frac{(-1)^n e^{in\pi x}}{n^2\pi^2}\right) + \frac{1}{3} + \sum_{n=1}^{\infty}\left(2\frac{(-1)^n e^{in\pi t}}{n^2\pi^2}\right) = \frac{1}{3} + \sum_{n=1}^{\infty}\left(4\frac{(-1)^n \cos(n\pi x)}{n^2\pi^2}\right)$

16. If $x > 0$ then $f(x) = i\,\mathrm{Res}\left(e^{i\alpha x}\dfrac{\alpha}{(b^2+a^2)^2}, bi\right) = \dfrac{1}{4}\dfrac{i\,e^{-bx}x}{b}$; When $x < 0$ then $f(x) = -i\,\mathrm{Res}\left(e^{i\alpha x}\dfrac{\alpha}{(b^2+a^2)^2}, -bi\right) = \dfrac{1}{4}\dfrac{i\,e^{bx}x}{b}$

Exercises 36.4–Part A, Page 903

1. $s_{\pm}(k) = -\frac{1}{2}k \pm \frac{1}{2}\sqrt{k^2 - 4k}$ 2. $\lim_{k\to\infty} -\frac{1}{2}k - \frac{1}{2}\sqrt{k^2 - 4k} = -\infty$ 3. $\lim_{k\to\infty} -\frac{1}{2}k + \frac{1}{2}\sqrt{k^2 - 4k} = -1$

6. $s(0) = 0$; $s(1500) = -0.9993346500$

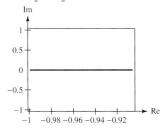

Exercises 36.4–Part B, Page 903

2. (a) $\lambda_{\pm} = \frac{1}{2}c + \frac{1}{2} \pm \frac{1}{2}\sqrt{c^2 - 2c + 85}$ (b)

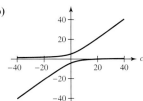

(c) The system is unstable for all values of c. If $c < 21$, the origin is a saddle. If $c > 21$ the origin is a source.

(d)

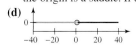

10. (a) $T(s) = \dfrac{1}{48s^2 + 54s + 6 + 7K}$ (b) $s_{\pm}(K) = -\dfrac{9}{16} \pm \dfrac{1}{48}\sqrt{441 - 336K}$ (c)

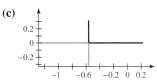

The system is stable for $K > -\frac{6}{7}$. If $K > \frac{21}{16}$, the poles are complex, but remain in the left half-plane.

Exercises 36.5–Part A, Page 907

1. $v(C, 0) = 1$ 2. $N = 2, P = 1$ 3. $\dfrac{1}{2\pi i}\int_0^{2\pi} \dfrac{2(\sin t + 3i(\cos t - 1))}{4\cos t - 5}\,dt = 1$

Exercises 36.5–Part B, Page 907

2. (a) zeros: $z = -4$, $z = 1$ (b) poles: $z = -0.5472305167 \pm 0.7241075608i$, $z = 2.427794367$ (c) Use $R = 3 \,; 0$
   (d) $N = 1, P = 1, v = 0$ (e)   Yes!, note $G(0) = \frac{1}{3}$.

12. (a) $s = \frac{1}{14} \pm \frac{1}{14}\sqrt{29}$ (b) Use $R = 2$; 0.9999999995   (c)   Yes, 1 time.

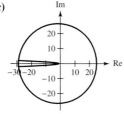

Exercises 36.6–Part A, Page 912

1. $\mathbf{v}'_1 = -3\mathbf{j}$; $\mathbf{v}'_2 = \frac{3}{2}\sqrt{3}\mathbf{i} - \frac{3}{2}\mathbf{j}$; $\theta = \frac{1}{3}\pi$ **2.** $w = \frac{1}{2t}(1 - i)$, i.e. the line $y = -x$

Exercises 36.6–Part B, Page 912

2. $\theta = \arccos(\frac{1}{10}\sqrt{10})$ **3.** (a) $Y_1 = t^3 - 3t^5 + i(3t^4 - t^6)$; $Y_2 = t^3 - \frac{3}{t} + i(3t - \frac{1}{t^3})$ (b) $\theta = \arccos(\frac{1}{10}\sqrt{10})$

13. (a) $z = \pm i$

(b) $Y_1 = \frac{t-1}{t+1}$

(c) The image of $z = -2 + e^{it}$ is given by $w = 2 - \frac{i\sin t}{-1+\cos t}$;
The picture shows the images of the circles with center -2 and radii $\frac{1}{2}$, 1, and 2 respectively.

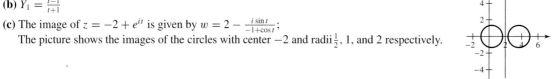

(d) The image of $z = t + 2(t+1)i$ is given by $w = \frac{t+\frac{3}{5}}{t+1} + \frac{4}{5}\frac{i}{t+1}$, i.e., $v = 2 - 2u$;
The picture shows the images of the lines $z = t + i(2t + 2 + A)$ for $A = -1, 0, 1$, respectively.

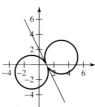

(e) The image of $z = A + ti$ is given by $w = -\frac{1+t^2}{1+t^2} + 2\frac{it}{1+t^2}$, i.e., $|w| = 1$;
The picture shows the images of the lines $y_3 = A + it$ for $A = -1, 0, 1$, and 2, respectively

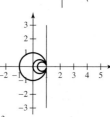

(f) The image of $z = 1 + e^{it}$ is given by $w = \frac{2\cos t+1}{5+4\cos t} + 2\frac{i\sin t}{5+4\cos t}$; which is equivalent to $|w + \frac{1}{3}| = \frac{2}{3}$;
The picture shows the images of the circles. with center 1 and radii $\frac{1}{4}, \frac{1}{3}, \frac{1}{2}$, 1 and $\frac{5}{4}$ respectively.

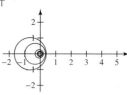

17. $f(z) = \frac{17z-10}{11z+2}$

Exercises 36.7, Page 915

1. $|z_1(0)| = 6$ and $|-2 - (-2 + 2\sqrt{5}i)| = 2\sqrt{5} < 6$ **5.**   **6.** $f(2) = 4$

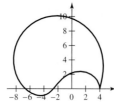

10. Plot the image of the circle with center 4 and radius 2.

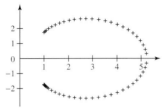

14. $\zeta_1 = i(2\sqrt{5}+4\sqrt{2}); \zeta_2 = i(2\sqrt{5}-4\sqrt{2}); w_1 = f(\zeta_1) = i(\frac{8}{3}\sqrt{5}+\frac{8}{3}\sqrt{2}); w_2 = f(\zeta_2) = i(\frac{8}{3}\sqrt{5}-\frac{8}{3}\sqrt{2}); t_1 = \arccos(\frac{1}{3}); t_2 = -\arccos(\frac{1}{3})$

Exercises 36.8–Part A, Page 919

2. Simplification of the difference of the two sides yields zero.

3. Observe that the function $W(w) = c + \frac{a-c}{\pi}Arg(w) + \frac{b-c}{\pi}(Arg(w-1) - Arg(w))$ has the property that $W(x) = c$ if $x \in (-2, 0)$, $W(x) = b$ if $x \in (0, 1)$, and $W(x) = a$ if $x \in (1, 2)$. Moreover the conformal mapping $w = f(z) = \left(\frac{1+z^2}{1-z^2}\right)^2$ maps the boundaries of the quarter unit disk in the first quadrant, exactly onto the given line segments. The solution in the z-plane can now readily be obtained.

Exercises 36.8–Part B, Page 919

1. (a) Observe that $Arg(s + ti)$ is harmonic, and therefore $u(x, y)$ is harmonic; Moreover $u(x, 0) = \alpha + \frac{\beta-\alpha}{\pi} 0 = \alpha$, when $x_0 < x$, while $u(x, 0) = \alpha + \frac{\beta-\alpha}{\pi}\pi = \beta$ when $x < x_0$ **(b)** $Arg(z - x_0) = Arg(x - x_0 + yi) = \arctan(y, x - x_0)$

(c) 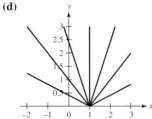 **(d)** 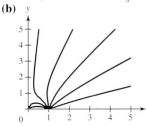 **(e)** The flow lines are orthogonal to the level curves.

3. (a) $u(x, y) = \alpha_0 + \frac{(\alpha_1-\alpha_0)\,Arg(z^2-x_0)}{\pi} + \frac{(\alpha_2-\alpha_1)\,Arg(z^2)}{\pi}$

(b) 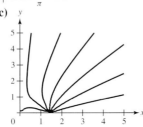 **(c)** 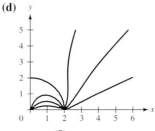 **(d)**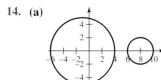

14. (a)  **(b)** $z_1 = 2; z_2 = \frac{22}{3}; f(z) = 3\frac{z-2}{3z-22}$

(c) $R' = 3$ **(d)** $r' = \frac{3}{5}$

(e) $u(x, y) = -3 + 8\dfrac{\ln(3\frac{\sqrt{x^2-4x+4+y^2}}{\sqrt{9x^2-132x+484+9y^2}})-\ln(\frac{3}{5})}{\ln(3)-\ln(\frac{3}{5})}$

(f) 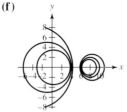 **(g)** The flow lines are orthogonal to the level curves.

Exercises 36.9, Page 927

1. $\varphi(x, y) = -\arcsin(-\frac{1}{2}\sqrt{x^2+2x+1+y^2}+\frac{1}{2}\sqrt{x^2-2x+1+y^2}); \psi(x, y) = \ln(\frac{1}{2}\sqrt{x^2+2x+1+y^2}+\frac{1}{2}\sqrt{x^2-2x+1+y^2}$
$+\sqrt{(\frac{1}{2}\sqrt{x^2+2x+1+y^2}+\frac{1}{2}\sqrt{x^2-2x+1+y^2})^2 - 1}); \min_R(\varphi) = \varphi(1.069110828, 2.365301723) = 0.3986118666;$
$\max_R(\varphi) = \varphi(2.462574282, 1.113419471) = 1.118655582; \min_R(\psi) = \psi(1.363725569, 1.228537203) = 1.300984786;$
$\max_R(\psi) = \psi(2.673820091, 2.738895449) = 2.036241351$

6. (a) $\phi(x, y) = \cos x \cosh y$; $\psi = -\sin x \sinh y$

(b)

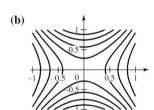

(c)

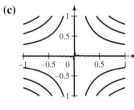

(d)

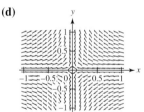

(e) $F = \phi + i\psi \Rightarrow \bar{F} = \phi - i\psi \Rightarrow V = \bar{F}' = \phi_x - i\psi_x = \phi_x + i\phi_y = \nabla\phi = -\sin x \cosh y\,\mathbf{i} + \cos x \sinh y\,\mathbf{j}$

(f) $v = \sqrt{|\sin x \cosh y|^2 + |\cos x \sinh y|^2}$

(g) All three integrals are zero.

(h) $\ln(\sinh y) + \ln(\sin x) = c_1$;
Combining the logarithms shows that this is
equivalent to $\psi(x, y) = -\sin x \sinh y = c_2$.

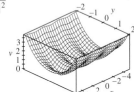

11. (a) $\phi(x, y) = \frac{\sinh x \cosh x}{\sinh^2 x + \cos^2 y}$ **(b)** $\psi(x, y) = \frac{\sinh(x^2 - y^2)\cosh(x^2 - y^2)}{\sinh^2(x^2 - y^2) + \cos^2(2xy)}$; $\nabla^2\psi = 0$ **(c)** $\psi(x, y) = \frac{\sinh\left(\frac{x(x^2+y^2+1)}{x^2+y^2}\right)\cosh\left(\frac{x(x^2+y^2+1)}{x^2+y^2}\right)}{\sinh^2\left(\frac{x(x^2+y^2+1)}{x^2+y^2}\right) + \cos^2\left(\frac{y(x^2+y^2-1)}{x^2+y^2}\right)}$; $\nabla^2\psi = 0$

Exercises 36.10–Part A, Page 936

1. The parabola maps onto the horizontal lines $v = 1$ and $v = -1$

4. $\nabla\phi(1, 1) = \left(\frac{1}{3}A2^{2/3} + \frac{1}{3}A2^{2/3}\sqrt{3}\right)\mathbf{i} + \left(\frac{1}{3}A2^{2/3} - \frac{1}{3}A2^{2/3}\sqrt{3}\right)\mathbf{j}$

6. $\nabla\phi\left(0, \frac{3}{2}\right) = \frac{13}{18}A\sqrt{3}\mathbf{i} + \frac{5}{18}A\mathbf{j}$

Exercises 36.10–Part B, Page 937

1. This is the definition of the principal value of the square root.
Plot the image of the unit circle centered in the origin.

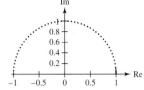

2. (a) Plot the image of the unit circle, centered at the origin. **(b)** The condition $\mathrm{Arg}(z) > 0$ is equivalent to $\Im(z) > 0$, while $\mathrm{Arg}(z) < 0$ is equivalent to $\Im(z) < 0$. In this particular branch of the square root, the branch cut lies along the positive Real axis. In the overlapping region ($0 < \Im(z)$) the functions coincide, outside of that region the relationship $\arg(z) = \mathrm{Arg}(z) + 2\pi$ introduced a negative sign.

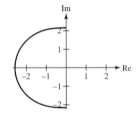

9. (a) Plot the images of the unit circles centered at $-2\sqrt{2}$, 0 and $2\sqrt{2}$.

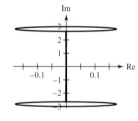

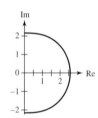

(b) Plot the image of right half of the unit circle centered at the origin under both transformations. The image under σ is dotted. Plot the image of left half of the unit circle center under both transformations. The image under σ is dotted.

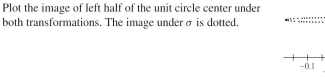
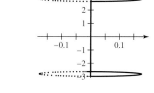

(c) Plot the image of the ellipse (thick) and of two exterior curves (thin).

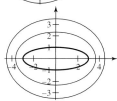

(d) Plot the image of the unit circle (thick) and of two exterior curves (thin).

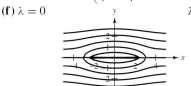

(e) $F(z) = A(e^{-i\lambda}\left(\dfrac{1}{4}z + \dfrac{1}{4}\sqrt{|z^2 - 8|}e^{(1/2i(\text{Arg}(z+2\sqrt{2})+\text{Arg}(z-2\sqrt{2})))}\right) + \dfrac{e^{i\lambda}}{\frac{1}{4}z + \frac{1}{4}\sqrt{|z^2-8|}e^{(1/2i(\text{Arg}(z+2\sqrt{2})+\text{Arg}(z-2\sqrt{2})))}})$

(f) $\lambda = 0$ $\lambda = -\dfrac{\pi}{24}$

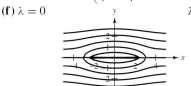
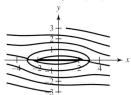

$\lambda = -\dfrac{\pi}{12}$ $\lambda = -\dfrac{\pi}{6}$

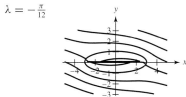
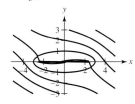

(g) Plot the contour $\psi(x, y) = 0$.

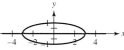

13. The graphic shows the images of a set of horizontal line segments in the upper half plane: $z_k = x + \dfrac{ki}{20}, k = 1, \ldots, 16, x \in [-8, 8]$. It is commensurate with the statement made in Exercise 25, Section 36.6, that this mapping takes the upper half plane to the union of the first quadrant and the strip $\{0 \le v \le \pi, u < 0\}$ in the second quadrant.

CHAPTER 37

Exercises 37.1–Part A, Page 948

1. $\sin 1 \approx .841471$ 2. $\tan 2 \approx -2.18504$ 4. $(2.718281692, 2.718281964)$

Exercises 37.1–Part B, Page 948

2. (a) {252.78, 252.781, 252.7814, 252.78136, 252.781362} **(b)** {253., 252.8, 252.78, 252.781, 252.7814}   **12. (a)** 5 **(b)** 3   **24.** 16; 1

Exercises 37.2–Part A, Page 953

3. $\left\{ \frac{1}{2}, \frac{13}{20}, \frac{1327}{2000} \right\}$ **5.** $x = -5, 4$; Yes, it will converge quadratically for both.

Exercises 37.2–Part B, Page 953

9. (a)

k	x_k	$\dfrac{\lvert x_{k+2} - x_{k+1} \rvert}{\lvert x_{k+1} - x_k \rvert}$
1	−0.6677	0.4266648038
2	−0.667113099	0.4367317599
3	−0.666862689	0.4410581372
4	−0.666753327	0.4429356277
5	−0.666705092	0.4438099696
6	−0.666683727	0.4441046193
7	−0.666674245	0.4440750416
8	−0.666670034	0.4443850267

(b) $C_L = 0.4443850267$ **(c)** $\bar{x} = -0.6666666667$ **(d)** $g'(\bar{x}) = 0.444444444$

12. (a)

k	x_k	$\dfrac{\lvert x_{k+2} - x_{k+1} \rvert}{\lvert x_{k+1} - x_k \rvert^2}$
1	−3.8636363636363636363636	0.4411014535799611956913476
2	−3.8552277278955589179451304	0.4402476087454627607488304
3	−3.8551965397489338068030530	0.4402767021015987052995771

(b) $C_Q = 0.4402767021$ **(c)** $\bar{x} = -3.855196539$ **(e)** $g''(\bar{x}) = 0.4402444065$

22. (a, b)

k	x_k	$\dfrac{\lvert x_{k+2} - x_{k+1} \rvert}{\lvert x_{k+1} - x_k \rvert^p}$
1	2.040	0.5045070474955341326191188
2	1.5443499392466585662211420	2.1705093318908211619535440
3	1.3822994704905489937773430	1.2319609224942530538949070
4	1.2680776793269240561436680	1.4751567433789816383011950
5	1.2312645502328277896644090	1.1954141699738184467715020
6	1.2242081719871345420865380	1.2004224356534277225046670

(c) 1.223809568

32. (a) $\displaystyle \lim_{k \to \infty} \frac{\lvert x_{k+2} - x_{k+1} \rvert}{\lvert x_{k+1} - x_k \rvert} = \lim_{k \to \infty} -\frac{\frac{1}{4}\frac{1}{3^{(k+1)}} + \frac{1}{4}\frac{1}{3^k}}{-\frac{1}{4}\frac{1}{3^k} + \frac{1}{4}\frac{1}{3^{(k-1)}}} = \frac{1}{3}$ **(b)** $\lim_{k \to \infty} x_k = 3$ **(c)** $C_L = \frac{1}{3}$ **(d)** $s_7 = 2.999656986$

36. (a) $g(a) = a$ **(b)** $(-10, -4 - \sqrt{34}) \cup (-4 + \sqrt{34}, 2)$ **(c)** $g'(a) = 0 \Leftrightarrow a = -4 \pm \sqrt{35}$

CHAPTER 38

Exercises 38.1, Page 957

2. $a > \frac{1}{2}$

3. (a) 0.1609394091   **(b)**

$g'(.1609394091) = -0.3709553320$; $\{[x_k, f(x_k)]\} = \{[0.2, 5.280], [0.1444210526, -2.11780795], [0.1667137679, 0.75620194],$
$[0.1587537475, -0.28407063], [0.1617439647, 0.10486621], [0.1606401099, -0.03896985], [0.1610503188, 0.01444649],$
$[0.1608982505, -0.00536031], [0.1609546749, 0.00198826], [0.1609337459, -0.00073758], [0.1609415098, 0.00027360]\}$

(c)

$g'(0.1609394091) = -3.259228175$; $\{[x_k, f(x_k)]\} = \{[0.2, 5.280], [0.1248156536i, -19.48000000 + 11.74081730i],$
$[0.4559006182 - 0.1355422055i, 47.48834917 - 29.53884613i], [0.1614869989 + 0.5800704493i, -41.67514463 + 63.91661435i]\}$

(d)

$g'(0.1609394091) = -26.93505922$; $\{[x_k, f(x_k)]\} = \{[.2, 5.280], [0.2154434689 + 0.3731590343i, -11.15188922 + 50.72509685i],$
$[0.8399775495 - 0.4395184361i, 116.8264101 - 162.6246159i], [0.9769813313 + 0.9728165398i, -34.88984663 + 384.8975739i]\}$

13. **(a)** $r_1 = -3, r_2 = 2$ **(b)** $\left(\frac{7}{2}, 6\right) \cup (6, \infty)$ **(c)** $(\infty, -4) \cup \left(-4, -\frac{3}{2}\right)$ **(d)** $p = 6$ **(e)** $p = -4$ **(f)** $x_0 \in \left(-\frac{31}{2}, -\frac{1}{2}\right)$
(g) $x_0 \in \left(-\frac{1}{2}, \frac{23}{2}\right)$

Exercises 38.2–Part A, Page 959

1. $n > \frac{\ln\left(\frac{|b-a|}{\varepsilon}\right)}{\ln 2}$; 2.0 $\{[0, 1.5], [0.75, 1.5], [0.75, 1.125]\}$ **3.** $0 - f(\alpha) = \frac{(f(\alpha)-f(\beta))(\xi-\alpha)}{\alpha-\beta} \Rightarrow \xi = \frac{f(\alpha)\beta-f(\beta)\alpha}{f(\alpha)-f(\beta)}$

Exercises 38.2–Part B, Page 959

3. (a)

(b) $a = -1$; $b = 0$; $n = 24$; $x = -0.4364916686$; $a = 3$; $b = 4$; $n = 24$; $x = 3.436491673$

(c) $a = -1$; $b = 0$; $n = 15$; $x = -0.4364916732$; $a = 3$; $b = 4$; $n = 15$; $x = 3.436491673$

Exercises 38.3–Part A, Page 966

1. $g(x) = x - \frac{x^2-6x+8}{2x-6} = \frac{x^2-8}{2(x-3)}$ **2.** $C_Q(2) = \frac{1}{2}$; $C_Q(4) = -\frac{1}{2}$ **3.** $X_1 = 1.750000000$; $X_2 = 1.975000000$; $X_3 = 1.999695122$

Exercises 38.3–Part B, Page 966

1. **(a)** $f'''(0) = 8$ **(b)** $Q(x) = \frac{4}{3} - \frac{11}{30}x^2 + \frac{139}{630}x^4 - \frac{449}{22.680}x^6 + O(x^7)$

3. **(a)** $g(x) = x - \frac{6x^3+8x^2+10x+1}{18x^2+16x+10}$ **(b)**

k	x_k	$\dfrac{x_{k+2} - x_{k+1}}{(x_{k+1} - x_k)^2}$
0	-0.2	-0.5831962496
1	-0.1031914894	-0.7210005817
2	-0.1086571391	-0.7132673601
3	-0.1086786778	-0.7132370297

(c) $C_Q = -0.713$

(d) This is an immediate consequence of the fact that $f'(r) \neq 0$ **(e)** $-\frac{1}{2}g''(r) = -\frac{1}{2}\frac{f''(r)}{f'(r)} = -0.7135$

15. **(a)** $m = 4$; $Q(x) = 4x^2 + 6$ **(b)**

k	x_k	$\dfrac{x_{k+2} - x_{k+1}}{x_{k+1} - x_k}$
0	2.0	0.8102350889
1	1.788461538	0.8027306605
2	1.617065654	0.7948228454
3	1.479480923	0.7870604556
4	1.370125435	0.7799080663
5	1.284056055	0.7736604583
6	1.216929852	0.7684313842
7	1.164996962	0.7641949627
8	1.125090100	0.7608434754
9	1.094593477	0.7582363356
10	1.071390320	0.7562316660
11	1.053796844	0.7547023753
12	1.040492100	0.7535419334
13	1.030450978	0.7526645338
14	1.022884572	0.7520027490
15	1.017189606	0.7515044199

; $C_L = 0.7515$ **(c)** $\frac{m-1}{m} = 0.7500$ **(d)** $g'(r) = 0.7500$

(e)

k	x_k	$\dfrac{x_{k+2} - x_{k+1}}{(x_{k+1} - x_k)^2}$
0	2.0	0.09084923758
1	0.8852603982	0.1448263128
2	0.9981536928	0.1499669537
3	0.9999994891	
4	1.000000000	

(f)

k	x_k	$\dfrac{x_{k+2} - x_{k+1}}{(x_{k+1} - x_k)^2}$
0	2.0	−0.2083448467
1	1.153846154	−0.2099527350
2	1.004676175	−0.2003718227
3	1.000004373	−0.2000003490
4	1.000000000	

(g) $C_Q = -0.2000003490$

(h) $-\dfrac{Q'(r)}{m\,Q(r)} = -0.2000000000$

25. $0.1049770865 \pm 0.7947096438i,\ 0.5200229135 \pm 0.3444279335i$

Exercises 38.4–Part A, Page 969

2. $x_2 = 1.600000000,\ x_3 = 1.882352941,\ x_4 = 1.981308409$

Exercises 38.4–Part B, Page 969

2. Establishing empirical evidence of superlinear convergence is not an easy task. The given table was created using extended precision arithmetic.

k	x_k	$\dfrac{\lvert x_{k+2} - x_{k+1}\rvert}{\lvert x_{k+1} - x_k\rvert^p}$
0	−0.2	1.564744673
1	0	0.2130629378
2	−0.1157407407	1.936385675
3	−0.1092364965	0.5112642622
4	−0.1086758746	1.073698717
5	−0.1086786793	0.6823007842
6	−0.1086786782	0.9033232874
7	−0.1086786782	0.7594937168
8	−0.1086786782	0.8454227020
9	−0.1086786782	0.7912333874
10	−0.1086786782	0.8242994759
11	−0.1086786782	0.8037040231
12	−0.1086786782	0.8163712079
13	−0.1086786782	0.8085191960

; $C_S = \left\lvert \dfrac{f''(r)}{2f'(r)} \right\rvert^{p-1} = 0.8115092723$

8. (a)

$$
\begin{bmatrix}
k & x_k & \dfrac{|x_{k+2} - x_{k+1}|}{|x_{k+1} - x_k|} \\
0 & 2.0 & 1.561321530 \\
1 & 1.9 & 0.7085725296 \\
2 & 1.743867847 & 0.9302856192 \\
3 & 1.633236892 & 0.8297439306 \\
4 & 1.530318506 & 0.8610706285 \\
5 & 1.444922600 & 0.8432283084 \\
6 & 1.371390693 & 0.8447320334 \\
7 & 1.309386508 & 0.8392136051 \\
8 & 1.257009587 & 0.8368628663 \\
9 & 1.213054162 & 0.8337978417 \\
10 & 1.176269499 & 0.8315135409 \\
11 & 1.145598526 & 0.8294120470 \\
12 & 1.120095197 & 0.8276842640 \\
13 & 1.098942429 & 0.8262093295 \\
14 & 1.081434615 & 0.8249839464 \\
15 & 1.066969496 & 0.8239609974
\end{bmatrix}
$$

(b) 0.9942313043

14. (a) $g(x) = \dfrac{960x^7 - 416x^6 - 420x^5 - 2763x^4 + 850x^3 + 2808x^2 - 2088x - 156}{(12x^2 - 2x - 5)^3}$

(b)

$$
\begin{bmatrix}
k & x_k & \dfrac{|x_{k+2} - x_{k+1}|}{|x_{k+1} - x_k|^3} \\
0 & -1.5 & 1.159587226 \\
1 & -1.632900000 & 0.7954714051 \\
2 & -1.635621939 & \\
3 & -1.635621955 & \\
4 & -1.635621955 &
\end{bmatrix}
$$

(c) $g'(r) = -\dfrac{1}{2}\dfrac{f(r)^2(f'(r)f'''(r) - 3f''(r)^2)}{f'(r)^4}$;

$g''(r) = -\dfrac{1}{2}f(r)(-6f'(r)^2 f''(r)^2 + 2f'(r)^3 f'''(r) + f'(r)^2 f''''(r)f(r) - 9f''(r)f'(r)f'''(r)f(r) + 12f''(r)^3 f(r))/f'(r)^5$; Clearly both expressions vanish at a point where $f(r) = 0$.

20.

$$
\begin{bmatrix}
k & x_k & \dfrac{|x_{k+2} - x_{k+1}|}{|x_{k+1} - x_k|} \\
0 & 1.5 & 0.2901525587 \\
1 & 1.845679012 & 0.3412700540 \\
2 & 1.945978662 & 0.3632028118 \\
3 & 1.980207929 & 0.3706922485 \\
4 & 1.992640095 & 0.3734005616 \\
5 & 1.997248603 & 0.3744024349 \\
6 & 1.998969422 & 0.3747762246 \\
7 & 1.999613701 & 0.3749161281 \\
8 & 1.999855161 & 0.3749685541
\end{bmatrix}
$$
; Convergence is linear; $g'(r) = 0.3749999990$.

Exercises 38.5, Page 972

2. (a)

$$
\begin{bmatrix}
k & x_k & \dfrac{|x_{k+2} - x_{k+1}|}{|x_{k+1} - x_k|^p} \\
0 & 0.05 - 0.8i & 7.326379935 \\
1 & 0.2 - 0.8i & 3.257712271 \\
2 & 0.1 - 1.0i & 0.05508457920 \\
3 & 0.1024453086 - 0.7927956524i & 7.445225594 \\
4 & 0.1049442648 - 0.7945375108i & 2.734020064 \\
5 & 0.1049767715 - 0.7947097645i & \\
6 & 0.1049770865 - 0.7947096439i &
\end{bmatrix}
$$
;

$$
\begin{bmatrix}
k & x_k & \dfrac{|x_{k+2} - x_{k+1}|}{|x_{k+1} - x_k|^p} \\
0 & 0.05 + 0.8i & 7.326379935 \\
1 & 0.2 + 0.8i & 3.257712271 \\
2 & 0.1 + 1.0i & 0.05508457920 \\
3 & 0.1024453086 + 0.7927956524i & 7.445225594 \\
4 & 0.1049442648 + 0.7945375108i & 2.734020064 \\
5 & 0.1049767715 + 0.7947097645i & \\
6 & 0.1049770865 + 0.7947096439i &
\end{bmatrix}
$$
;

$$
\begin{bmatrix}
k & x_k & \dfrac{|x_{k+2} - x_{k+1}|}{|x_{k+1} - x_k|^p} \\
0 & 0.5 - 0.3i & 13.16491894 \\
1 & 0.6 - 0.4i & 1.704545540 \\
2 & 0.3 - 0.2i & 0.03627708121 \\
3 & 0.5196059530 - 0.3411713905i & 9.417676218 \\
4 & 0.5200574229 - 0.3442061161i & 2.680060091 \\
5 & 0.5200224073 - 0.3444280602i & \\
6 & 0.5200229136 - 0.3444279332i &
\end{bmatrix}
\;;\;
\begin{bmatrix}
k & x_k & \dfrac{|x_{k+2} - x_{k+1}|}{|x_{k+1} - x_k|^p} \\
0 & 0.5 + 0.3i & 13.16491894 \\
1 & 0.6 + 0.4i & 1.704545540 \\
2 & 0.3 + 0.2i & 0.03627708121 \\
3 & 0.5196059530 + 0.3411713905i & 9.417676218 \\
4 & 0.5200574229 + 0.3442061161i & 2.680060091 \\
5 & 0.5200224073 + 0.3444280602i & \\
6 & 0.5200229136 + 0.3444279332i &
\end{bmatrix}
$$

(b) The claim of superlinear convergence cannot be verified for this particular problem.   **(c)** $C_M(0.520 \pm 0.344i) = 1.308006693$; $C_M(0.104 \pm 0.794i) = 1.180832955$

6.
$$
\begin{bmatrix}
k & x_k & \dfrac{|x_{k+2} - x_{k+1}|}{|x_{k+1} - x_k|} \\
0 & 0.8 & 0.2500000000 \\
1 & 1.2 & 1.299106920 \\
2 & 1.3 & 0.3709036054 \\
3 & 1.170089308 & 0.5449481914 \\
4 & 1.147828965 - 0.04273415660i & 1.308658941 \\
5 & 1.124215409 - 0.05421810285i & 0.7249531051 \\
6 & 1.090345143 - 0.06001484372i & 0.8376919628 \\
7 & 1.065762002 - 0.06404539927i & 0.8736739163 \\
8 & 1.044971644 - 0.06224634380i & 0.8197157050 \\
9 & 1.027369959 - 0.05749430616i & 0.8369417274 \\
10 & 1.013810847 - 0.05120923816i & 0.8364129837
\end{bmatrix}
$$
Convergence is linear.

CHAPTER 39

Exercises 39.1, Page 981

1. (a)

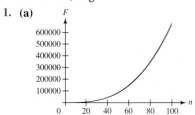

(b) $n = 10^5$: $0.00225\ \%$; $n = 10^6$: $0.00225\ \%$.

2. Numeric Solution: $\mathbf{x} = [0.2274, -0.09203, -0.4809, -0.3825]^T$; "Exact" Solution: $\mathbf{x} = [0.2261, -0.09119, -0.4774, -0.3800]^T$.

8. Numeric Solution: $\mathbf{x} = [-0.1500, 0.1949, 0.2203, -0.006542]^T$; "Exact" Solution: $\mathbf{x} = [-0.1501, 0.1947, 0.2203, -0.006549]^T$

14. $\mathbf{x} = [-0.01557632340, 0.4174454833, -1.200934580, 3.147975078, 2.454828660, 3.954828660]^T$.   **16.** $\mathbf{x} = \left[\frac{67}{12}, \frac{5}{12}, -\frac{5}{2}, \frac{19}{4} \right]^T$

Exercises 39.2–Part A, Page 986

1. (a) $\mathbf{x}_0 = [82, -9]^T$; $\mathbf{x}_c = \left[\frac{-82+c}{-1+10c}, \frac{9}{-1+10c} \right]^T$

2. $\Delta\mathbf{x} = \left[\frac{819c}{-1+10c}, \frac{90c}{-1+10c} \right]^T$

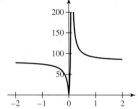

4.

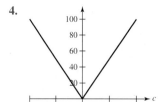

Exercises 39.2–Part B, Page 986

3. **(a)** $\mathbf{x} = [-199, 168, 256]^{\mathrm{T}}$ **(b)** $\mathbf{x}_1 = \left[-\frac{398+23a}{2+49a}, 8\frac{42+5a}{2+49a}, \frac{512}{2+49a}\right]^{\mathrm{T}}$ **(c)** $\triangle\mathbf{x} = \left[9728\frac{a}{2+49a}, -8192\frac{a}{2+49a}, -12544\frac{a}{2+49a}\right]^{\mathrm{T}}$;

$\alpha = 12544 \dfrac{\left|\frac{a}{2+49a}\right|}{\max(8\left|\frac{42+5a}{2+49a}\right|, \left|\frac{512}{2+49a}\right|, \left|\frac{398+23a}{2+49a}\right|)}$; cond $A = \frac{5681}{2}$; $\beta = \frac{1}{23}|a|$ **(d)**

(e) $\mathbf{z} = \left[-\frac{100}{311}, -\frac{109}{311}, \frac{-102}{311}\right]^{\mathrm{T}}$

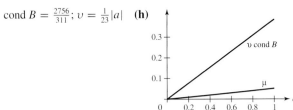

(f) $\mathbf{z}_1 = \left[2\frac{50+a}{-311+5a}, -\frac{-109+7a}{-311+5a}, \frac{102}{-311+5a}\right]^{\mathrm{T}}$ **(g)** $\triangle\mathbf{z} = \left[\frac{1122}{311}\frac{a}{-311+5a}, -\frac{1632}{311}\frac{a}{-311+5a}, \frac{510}{311}\frac{a}{-311+5a}\right]^{\mathrm{T}}$; $\mu = \frac{1632}{311}\dfrac{\left|\frac{a}{-311+5a}\right|}{\max(\frac{102}{|-311+5a|}, \left|\frac{-109+7a}{-311+5a}\right|, 2\left|\frac{50+a}{-311+5a}\right|)}$;

cond $B = \frac{2756}{311}$; $\upsilon = \frac{1}{23}|a|$ **(h)**

(i) det $A = -2$; det $B = 622$

(j) $\|A\|_2\|A^{-1}\|_2 = 1921.805761 = \frac{17.37094261}{0.009038864884} = \frac{\sigma_1}{\sigma_3}$

7. **(a)** $\mathbf{x} = [131, -420, 966]^{\mathrm{T}}$; $\mathbf{x}_1 = [131 + 24a, -420 - 77a, 966 + 177a]^{\mathrm{T}}$ **(b)** $\mathbf{z} = \left[\frac{1}{3}, 2, \frac{5}{3}\right]^{\mathrm{T}}$; $\mathbf{z}_1 = \left[\frac{1}{3} - \frac{4}{7}a, 2 - \frac{1}{7}a, \frac{5}{3} - \frac{5}{7}a\right]^{\mathrm{T}}$ **(c)**

$\triangle\mathbf{x} = [24a, -77a, 177a]^{\mathrm{T}}$; $\triangle\mathbf{z} = \left[-\frac{4}{7}a, -\frac{1}{7}a, -\frac{5}{7}a\right]^{\mathrm{T}}$ **(d)** $\alpha = \frac{59}{322}|a|$; $\beta = \frac{5}{14}|a|$; $\lambda = \frac{1}{5}|a|$ **(e)** cond $A = 5350$; cond $B = \frac{348}{7}$; det $A = 1$;

det $B = 21$ **(f)**   **(g)**   **(i)** $\|A\|_2\|A^{-1}\|_2 = 3277.188396 = \frac{16.59937332}{0.005065126357} = \frac{\sigma_1}{\sigma_3}$

11. **(a)** $\mathbf{x}_0 = [-10, -364, 202]^{\mathrm{T}}$ **(b)** $\alpha = 0.00008019459922$ **(c)** $\mathbf{z} = [-10.02065541, -364.0126093, 202.0141066]^{\mathrm{T}}$ **(d)**
$\triangle\mathbf{x} = [-0.02065541, -0.0126093, 0.0141066]^{\mathrm{T}}$; $\beta = 0.00005674366621$; cond $A = 6.613064634$ **(e)** $\|\triangle\mathbf{x}\| \leq 0.1930475599$
(f) $\|\triangle\mathbf{x}\| = 0.02065541$

15. **(a)** $\mathbf{z} = [326.71, 37.997, 243.70, 216.43]^{\mathrm{T}}$ **(b)** $\mathbf{r} = [0.000025357, 0.00024247, 0.000360857, 0.00042875]^{\mathrm{T}}$
(c) $\mathbf{u} = [0.043143, 0.016573, 0.025484, 0.017813]^{\mathrm{T}}$ **(d)** cond $A = 92.73061750$; $\frac{\|\mathbf{u}\|}{\|\mathbf{z}\|}10^5 = 13.20528909$

Exercises 39.3, Page 991

1. **(a)** $\mathbf{x}_e = [125, -2880, 14490, -24640, 13230]^{\mathrm{T}}$ **(b)** $\mathbf{x}_1 = [111.40, -2586.6, 13132., -22495., 12147.]^{\mathrm{T}}$; $\|\mathbf{x}_e - \mathbf{x}_1\| = 2145$ **(c)**
$\|\mathbf{x}_e - \mathbf{x}_1\| \leq \|\mathbf{x}_e\|$ cond $A \frac{\|A-B\|}{\|A\|} = 96881.6$ **(d)** $\mathbf{x}_1 = [124.00, -2858.6, 14390., -24482., 13150.]^{\mathrm{T}}$,
$\mathbf{x}_2 = [124.93, -2878.5, 14483., -24629., 13224.]^{\mathrm{T}}$, $\mathbf{x}_3 = [125.00, -2879.9, 14490., -24639., 13230.]^{\mathrm{T}}$,
$\mathbf{x}_4 = [125.00, -2879.9, 14490., -24640., 13230.]^{\mathrm{T}}$

2. **(a)** $\mathbf{x}_e = [-10, -364, 202]^{\mathrm{T}}$ **(b)** $\mathbf{x}_1 = [-9.9935, -363.99, 202.00]^{\mathrm{T}}$; $\|\mathbf{x}_e - \mathbf{x}_1\| = 0.01$ **(c)** $\mathbf{x}_2 = [-10.000, -364.00, 202.00]^{\mathrm{T}}$

6. **(a)** $\mathbf{x}_e = \left[\frac{865,110,701}{2,649,203}, \frac{100,712,501}{2,649,203}, \frac{645,360,867}{2,649,203}, \frac{573,200,259}{2,649,203}\right]^{\mathrm{T}} = [326.56, 38.016, 243.61, 216.37]^{\mathrm{T}}$ **(b)** $\mathbf{x}_1 = [326.71, 37.997, 243.70, 216.43]^{\mathrm{T}}$;
$\|\mathbf{x}_e - \mathbf{x}_1\| = 0.1549187$ **(c)** $\mathbf{x}_2 = [326.56, 38.016, 243.61, 216.37]^{\mathrm{T}}$

Exercises 39.4, Page 995

2. **(a)** $A = \begin{bmatrix} -22 & 6 & 6 & -7 \\ -9 & 21 & 3 & -7 \\ -6 & 0 & 11 & -2 \\ 1 & 8 & 8 & 25 \end{bmatrix}$ **(b)** $\mathbf{x}_e = \left[\frac{5221}{103,357}, \frac{8697}{103,357}, \frac{31657}{103,357}, \frac{3415}{103,357}\right]^{\mathrm{T}} = [0.050514, 0.084145, 0.30629, 0.033041]^{\mathrm{T}}$

(c) $\mathbf{x}_{14} = [0.05050667154, 0.08414344350, 0.3062806785, 0.03304675129]^{\mathrm{T}}$; $\sigma_{14} = 0.756552 \times 10^{-5}$

(d) $D = \begin{bmatrix} -22 & 0 & 0 & 0 \\ 0 & 21 & 0 & 0 \\ 0 & 0 & 11 & 0 \\ 0 & 0 & 0 & 25 \end{bmatrix}$; $T = \begin{bmatrix} 0 & \frac{3}{11} & \frac{3}{11} & -\frac{7}{22} \\ \frac{3}{7} & 0 & -\frac{1}{7} & \frac{1}{3} \\ \frac{6}{11} & 0 & 0 & \frac{2}{11} \\ -\frac{1}{25} & -\frac{8}{25} & -\frac{8}{25} & 0 \end{bmatrix}$; $\mathbf{v} = \left[-\frac{1}{22}, \frac{2}{21}, \frac{3}{11}, \frac{4}{25} \right]^{\mathrm{T}}$ **(e)** $\rho = 0.4951091354$

(f) $\mathbf{x}_{14} = [0.05050667159, 0.08414344344, 0.3062806785, 0.0330467513]^{\mathrm{T}}$; $\sigma_{14} = 0.756547 \times 10^{-5}$

12. **(a)** $\rho = 0.3813047054$ **(b)** $\mathbf{x}_e = \left[-\frac{100}{169}, \frac{985}{169}, \frac{250}{507} \right]^{\mathrm{T}} = [-0.59172, 5.8284, 0.4931]^{\mathrm{T}}$
    (c) $\mathbf{x}_{13} = [-0.5917140990, 5.828399731, 0.4930961944]^{\mathrm{T}}$, $\|\mathbf{x}_{13} - \mathbf{x}_e\| \le 10^{-5}$

Exercises 39.5, Page 998

2. **(a)** $\mathbf{x}_7 = [0.05051457853, 0.08414448153, 0.3062879762, 0.03304103036]^{\mathrm{T}}$; $\|\mathbf{x}_e - \mathbf{x}_7\| = 0.76262 \times 10^{-6}$

12. **(a)** $\mathbf{x}_e = \left[-\frac{17}{4}, \frac{167}{14}, \frac{1}{7} \right]^{\mathrm{T}}$ **(b)** $B_{GS} = \begin{bmatrix} 0 & -\frac{1}{3} & -\frac{1}{6} \\ 0 & \frac{2}{3} & -\frac{1}{6} \\ 0 & 0 & \frac{1}{22} \end{bmatrix}$ **(c)** for A: $\lambda = 0.5889664716, 9.685215241, 14.72581829$ **(d)** for B_{GS}:

    $\lambda = 0, 0.04545454545, 0.6666666667$ **(e)** $\mathbf{x}_{35} = [-4.249995918, 11.92856326, 0.1428571433]^{\mathrm{T}}$; $\|\mathbf{x}_e - \mathbf{x}_{35}\| = 0.817 \times 10^{-5}$

18. **(a)** $\mathbf{x}_e = \left[\frac{1214}{961}, -\frac{4555}{961}, \frac{1396}{961}, \frac{4437}{961} \right]^{\mathrm{T}} = [1.2633, -4.7399, 1.4527, 4.6171]^{\mathrm{T}}$

    (b) $B_{GS} = \begin{bmatrix} 0 & -\frac{7}{12} & -\frac{7}{12} \\ 0 & \frac{49}{144} & -\frac{23}{144} \\ 0 & \frac{49}{264} & \frac{11}{24} \end{bmatrix}$ **(c)** for A: $\lambda = 1.655908493, 4.393079379, 9.692847811, 27.25816432$ **(d)** for B_{GS}:

    $\lambda = 0, 0.4308202185, 0.4308202185$ **(e)** $\mathbf{x}_{31} = [1.263262222, -4.739845965, 1.452653012, 4.617060260]^{\mathrm{T}}$; $\|\mathbf{x}_e - \mathbf{x}_{31}\| = 0.8353 \times 10^{-5}$

Exercises 39.6, Page 1004

2. **(a)** $\mathbf{x}_e = \left[-\frac{178}{1003}, -\frac{24}{1003}, -\frac{407}{1003} \right]^{\mathrm{T}} = [-0.17747, -0.023928, -0.40578]^{\mathrm{T}}$
    (b) result: $\mathbf{x} = [-0.1774675972, -0.02392821535, -0.4057826520]^{\mathrm{T}}$

8. **(a)** $B(\omega) = \begin{bmatrix} 1 - \omega & -\frac{35}{46}\omega \\ \frac{11}{15}\omega(-1+\omega) & \frac{77}{138}\omega^2 + 1 - \omega \end{bmatrix}$ **(b)** $\lambda_{\pm} = -\omega + \frac{77}{276}\omega^2 + 1 \pm \frac{1}{276}\sqrt{42504\omega^2 - 42504\omega^3 + 5929\omega^4}$

    (c)

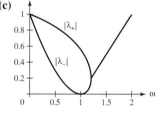

    $\hat{\omega} = 1.2$

14. **(a)** $B(\omega) = \begin{bmatrix} 1 - \omega & -\frac{1}{3}\omega & -\frac{1}{6}\omega \\ 2\omega(-1+\omega) & \frac{2}{3}\omega^2 + 1 - \omega & \frac{1}{3}\omega^2 - \frac{1}{2}\omega \\ -\frac{2}{11}\omega(-1+\omega)^2 & -\frac{2}{33}\omega^3 + \frac{5}{33}\omega^2 - \frac{1}{11}\omega & -\frac{1}{33}\omega^3 + \frac{5}{66}\omega^2 + 1 - \omega \end{bmatrix}$ **(b)**   **(c)** $\hat{\omega} = 1.25$

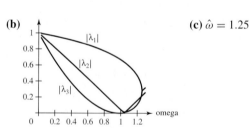

    (d) reward: 15 iterations were required, compared to 35 for Gauss–Seidel.

20. **(a)** $B(\omega) = \left[1 - \omega, -\frac{7}{12}\omega, -\frac{7}{12}\omega, -\frac{1}{4}\omega \right]$, $\left[\frac{7}{12}\omega(-1+\omega), \frac{49}{144}\omega^2 + 1 - \omega, \frac{49}{144}\omega^2 - \frac{1}{2}\omega, \frac{7}{48}\omega^2 - \frac{7}{12}\omega \right]$,
    $\left[-\frac{7}{22}\omega(\omega - 2)(-1+\omega), -\frac{49}{264}\omega^3 + \frac{11}{12}\omega^2 - \frac{6}{11}\omega, -\frac{49}{264}\omega^3 + \frac{85}{132}\omega^2 + 1 - \omega, -\frac{7}{88}\omega^3 + \frac{21}{44}\omega^2 - \frac{1}{11}\omega \right]$,
    $\left[\frac{1}{1056}\omega(42\omega^2 - 623\omega + 396)(-1+\omega), \frac{49}{2112}\omega^4 - \frac{475}{1152}\omega^3 + \frac{409}{352}\omega^2 - \frac{7}{8}\omega, \frac{49}{2112}\omega^4 - \frac{4793}{12672}\omega^3 + \frac{25}{32}\omega^2 - \frac{1}{8}\omega, \frac{7}{704}\omega^4 - \frac{791}{4224}\omega^3 + \frac{325}{528}\omega^2 + 1 - \omega \right]$

(c) $\hat{\omega} = 1.2$ **(d)** reward: 23 iterations were required, compared to 31 for Gauss–Seidel.

26. (a) $B_J = \begin{bmatrix} 0 & -\frac{15}{16} & 0 & 0 & 0 & 0 \\ -\frac{5}{8} & 0 & -\frac{1}{3} & 0 & 0 & 0 \\ 0 & -\frac{1}{3} & 0 & -\frac{13}{24} & 0 & 0 \\ 0 & 0 & -\frac{13}{21} & 0 & -\frac{5}{21} & 0 \\ 0 & 0 & 0 & -\frac{5}{3} & 0 & -\frac{2}{3} \\ 0 & 0 & 0 & 0 & -\frac{2}{21} & 0 \end{bmatrix}$; $\rho = 0.9617197192$; $\hat{\omega} = 1.569815585$

(b) $\mathbf{x}_e = \left[\frac{175,198}{136,033}, -\frac{177,809}{136,033}, \frac{238,939}{136,033}, \frac{-300,305}{136,033}, \frac{748,866}{136,033}, -\frac{32,454}{136,033}\right]^T = [1.2879, -1.3071, 1.7565, -2.2076, 5.505, -0.23857]^T$

(c) $\mathbf{x} = [1.287901658, -1.307096881, 1.756473991, -2.207585526, 5.505026400, -0.2385740700]^T$, 29 iterations are required for convergence of SOR. with $\omega = \hat{\omega}$

Exercises 39.7, Page 1008

2. (a) $f(2, 1) = 0$; $g(2, 1) = 0$ **(b)** $\mathbf{G}(x, y) = [-\frac{4}{25}xy^2 + \frac{9}{50}y^3 + \frac{4}{25}x^2 + \frac{2}{25}xy + \frac{7}{50}y^2 + \frac{2}{25}y + \frac{28}{25}, -\frac{1}{32}x^3 - \frac{1}{32}x^2y - \frac{3}{64}y^3 - \frac{3}{64}x^2 - \frac{1}{48}y^2 + \frac{313}{192}]^T$

(c) $\|\nabla f(\mathbf{x}_1)\|_1 = 0.9800000000$; $\|\nabla g(\mathbf{x}_1)\|_1 = 0.9947916667$ **(d)** $\lambda = 0.3415650255, 0.3415650255$ **(e)** with $\mathbf{x}_0 = [0, 0]^T$,

$\mathbf{x}_{13} = [2.000000248, 1.000002795]^T$, and $\|\hat{\mathbf{x}} - x_{13}\| = 0.2795 \times 10^{-5}$ **(f)** $\sigma = 0.9947916667$ **(g)** with $\mathbf{x}_0 = [0, 0]^T$,

$\mathbf{x}_{15} = [1.999998626, 1.000004501]^T$; and $\|\hat{\mathbf{x}} - x_{15}\| = 0.4501 \times 10^{-5}$ **(h)**

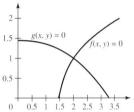

12. (a) $f(5, 5, 12) = 0$; $g(5, 5, 12) = 0$; $h(5, 5, 12) = 0$ **(b)** $\mathbf{G}(x, y, z) = [-\frac{1}{125}x^2y + \frac{1}{250}xyz + \frac{1}{500}y^2z + \frac{1}{500}yz^2 - \frac{9}{500}yz + \frac{96}{25},$

$\frac{8}{1501}y^4 - \frac{7}{1501}y^3z + \frac{1}{1501}y^2z^2 - \frac{5}{1501}y^2 - \frac{4}{1501}z + \frac{9578}{1501}, \frac{3}{1000}x^3y - \frac{1}{750}x^2y^2 - \frac{1}{1500}x^2z^2 - \frac{1}{1500}x^2z - \frac{1}{500}y^2 - \frac{7}{3000}z^2 + \frac{41,833}{3000}]^T$

(c) $\|\nabla f(\mathbf{x}_1)\|_1 = 0.8120000000$; $\|\nabla g(\mathbf{x}_1)\|_1 = 0.7921385743$; $\|\nabla h(\mathbf{x}_1)\|_1 = 0.7426666667$ **(d)**

$\lambda = 0.2804058191, 0.5185947642, 0.5185947642$ **(e)** $\mathbf{x}_{21} = [4.999999013, 5.000004533, 11.99999164]^T$ **(f)** $\sigma = 0.8120000000$

(g) $\mathbf{x}_{43} = [4.999991832, 5.000008324, 12.00000608]^T$

Exercises 39.8, Page 1011

2. (a) $\mathbf{G}(x, y) = [\frac{1}{2}(4830y + 1252x + 110x^2y - 3384y^3 + 84x^3 + 3056xy^2 + 228y^2x^3 + 144xy^4 - 192x^4y + 216y^3x^2 - 1164x^2 + 8635y^2 +$

$153y^2x^2 + 48x^3y + 28y^3x + 27y^4 + 12004 - 5776xy)/(171y^2x^2 + 108y^4 - 22y^3 + 1427y^2 - 12x^3 + 111xy^2 + 86xy - 1500x - 6x^2y -$

$184y + 222x^2 + 4800 - 144x^3y + 162y^3x), (-626y - 3008x + 7825 + 318x^2y + 442y^3 + 300x^3 + 31xy^2 - 24x^4 + 72y^5 - 96y^2x^3 +$

$108xy^4 + 114y^3x^2 - 279x^2 + 1352y^2 - 45y^2x^2 - 36x^3y + 60y^3x - 20y^4 - 336xy)/(171y^2x^2 + 108y^4 - 22y^3 + 1427y^2 - 12x^3 +$

$111xy^2 + 86xy - 1500x - 6x^2y - 184y + 222x^2 + 4800 - 144x^3y + 162y^3x)]^T$ **(b)** $J_\mathbf{G}(2, 1) = \begin{bmatrix} 0 & 0 \\ 0 & 0 \end{bmatrix}$ **(c)** with $\mathbf{x}_0 = [4, 0]^T$,

$\mathbf{x}_6 = [1.999994428, 1.000004464]^T$

12. (c) with $\mathbf{x}_0 = [4, 6, 10]^T$, $\mathbf{x}_4 = [5.000000002, 5.000000005, 11.99999999]^T$

19. (a) $\mathbf{F}(-12, -5, -5) = \mathbf{0}$ **(b)** $J_\mathbf{G}(-12, -5, -5) = \begin{bmatrix} 0 & 0 & 0 \\ 0 & 0 & 0 \\ 0 & 0 & 0 \end{bmatrix}$ **(c)** with $\mathbf{x}_0 = [-11, -4, -6]^T$,

$\mathbf{x}_4 = [-12.00000000, -5.000000001, -5.000000003]^T$ 30. with $\mathbf{x}_0 = [0, 0]^T$, $\mathbf{x}_7 = [2.000005569, 0.9999995278]^T$

CHAPTER 40

Exercises 40.1–Part A, Page 1017

1. $L_0(x) = \frac{x-x_1}{x_0-x_1}$; $L_1(x) = \frac{x-x_0}{x_1-x_0}$

2. $L_0(x) = -\frac{1}{6}x + \frac{2}{3}$; $L_1(x) = \frac{1}{6}x + \frac{1}{3}$

4. $f(x) = \frac{7}{3}x^2 - \frac{11}{3}x - 5$

Exercises 40.1–Part B, Page 1017

2. (a) $p_n(x) = 9.596726766x - 8.880334790x^2 + 3.746264282x^3 - 0.9021763333x^4 + 0.1279278366x^5 - 0.009939314523x^6 +$

$0.0003259661579x^7$ **(b)** $A = \begin{bmatrix} 1 & 0 & 0 & 0 & 0 & 0 & 0 & 0 \\ 1 & 1 & 1 & 1 & 1 & 1 & 1 & 1 \\ 1 & 2 & 4 & 8 & 16 & 32 & 64 & 128 \\ 1 & 3 & 9 & 27 & 81 & 243 & 729 & 2187 \\ 1 & 4 & 16 & 64 & 256 & 1024 & 4096 & 16384 \\ 1 & 5 & 25 & 125 & 625 & 3125 & 15,625 & 78,125 \\ 1 & 6 & 36 & 216 & 1296 & 7776 & 46,656 & 279,936 \\ 1 & 7 & 49 & 343 & 2401 & 16,807 & 117,649 & 823,543 \end{bmatrix}$; $\mathrm{cond}(A) = \frac{828,593,920}{9}$

(c) $P_n(x) = 10(\frac{1}{720}x^7 - \frac{3}{80}x^6 + \frac{59}{144}x^5 - \frac{37}{16}x^4 + \frac{319}{45}x^3 - \frac{223}{20}x^2 + 7x)e^{-1} + 20(-\frac{1}{240}x^7 + \frac{13}{120}x^6 - \frac{9}{8}x^5 + \frac{71}{12}x^4 - \frac{3929}{240}x^3 + \frac{879}{40}x^2 - \frac{21}{2}x)e^{-2} +$
$30(\frac{1}{144}x^7 - \frac{25}{144}x^6 + \frac{247}{144}x^5 - \frac{1219}{144}x^4 + \frac{389}{18}x^3 - \frac{949}{36}x^2 + \frac{35}{3}x)e^{-3} 40(-\frac{1}{144}x^7 + \frac{1}{6}x^6 - \frac{113}{72}x^5 + \frac{22}{3}x^4 - \frac{2545}{144}x^3 + \frac{41}{2}x^2 - \frac{35}{4}x)e^{-4} + 50(\frac{1}{240}x^7 -$
$\frac{23}{240}x^6 + \frac{69}{80}x^5 - \frac{185}{48}x^4 + \frac{134}{15}x^3 - \frac{201}{20}x^2 + \frac{21}{5}x)e^{-5} + 60(-\frac{1}{720}x^7 + \frac{11}{360}x^6 - \frac{19}{72}x^5 + \frac{41}{36}x^4 - \frac{1849}{720}x^3 + \frac{1019}{360}x^2 - \frac{7}{6}x)e^{-6} + 70(\frac{1}{5040}x^7 - \frac{1}{240}x^6 +$
$\frac{5}{144}x^5 - \frac{7}{48}x^4 + \frac{29}{90}x^3 - \frac{7}{20}x^2 + \frac{1}{7}x)e^{-7}$ **(d)**

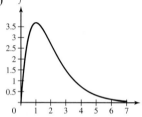

(e) $M = 80$; $\varepsilon(c) = \frac{1}{504}c(c-1)(c-2)(c-3)(c-4)(c-5)(c-6)(c-7)$

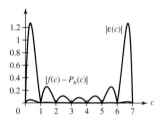

(f)

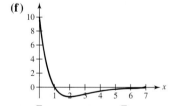

12. $A = \begin{bmatrix} 1 & 2 & 4 & 8 \\ 1 & 3 & 9 & 27 \\ 1 & 5 & 25 & 125 \\ 1 & 6 & 36 & 216 \end{bmatrix}$ **(b)** $\mathrm{cond}(A) = 6734$ **(c)** $\det(A) = 72$

17. (a)

(b) $x = -0.8943035503$ **(c)** $x = -132.1210881$; This is the polynomial wiggle at its best!

20. There is no quadratic osculating polynomial $p_2(x)$. The osculation cubic is: $p_3(x) = 3a - \frac{73}{2} + (-4a + 42)x + ax^2 - \frac{7}{2}x^3$, $\forall a \in \mathcal{R}$.

26. $p_3(x) = 1 + 6x - 8x^2 + 3x^3$

Exercises 40.2–Part A, Page 1022

1.
$$\begin{bmatrix} k & x & y & DD1 & DD2 \\ 0 & 3 & 2 & & \\ & & & \frac{7}{4} & \\ 1 & 7 & 9 & & \frac{3}{8} \\ & & & \frac{5}{2} & \\ 2 & 5 & 4 & & \end{bmatrix} ; f_{0,1,2} = \frac{3}{8}$$
3. $p_1(x) = -\frac{146}{9} + \frac{4}{9}x$

Exercises 40.2–Part B, Page 1022

2. (a)
$$\begin{bmatrix} k & x & y & DD1 & DD2 & DD3 & DD4 & DD5 \\ 0 & -16 & 14 & & & & & \\ & & & \frac{1}{3} & & & & \\ 1 & -13 & 15 & & -\frac{4}{21} & & & \\ & & & -\frac{11}{7} & & -\frac{5}{546} & & \\ 2 & -6 & 4 & & -\frac{13}{42} & & \frac{339}{18,200} & \\ & & & -\frac{14}{3} & & \frac{227}{840} & & -\frac{375,369}{239,439,200} \\ 3 & -3 & -10 & & \frac{44}{15} & & -\frac{40,767}{1,841,840} & \\ & & & 10 & & -\frac{8197}{34,320} & & \\ 4 & -1 & 10 & & -\frac{127}{143} & & & \\ & & & -\frac{17}{11} & & & & \\ 5 & 10 & -7 & & & & & \end{bmatrix}$$

(b) $P_5(x) = \frac{58}{3} + \frac{1}{3}x - \frac{4}{21}(x + 16)(x + 13) - \frac{5}{546}(x + 16)(x + 13)(x + 6) + \frac{339}{18,200}(x + 16)(x + 13)(x + 6)(x + 3) - \frac{375,369}{239,439,200}(x + 16)(x + 13)(x + 6)(x + 3)(x + 1)$ **(c)**

$P_5(x) = \frac{3,701,337}{115,115} + \frac{858,074,351}{32,650,800}x + \frac{968,174,619}{239,439,200}x^2 - \frac{89,357,399}{718,317,600}x^3 - \frac{783,039}{18,418,400}x^4 - \frac{375,369}{239,439,200}x^5$ **(e)**

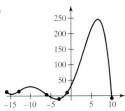

12. (a)
$$\begin{bmatrix} k & x & y & DD1 & DD2 & DD3 & DD4 & DD5 \\ 0 & 0 & 5 & & & & & \\ & & & 8 & & & & \\ 1 & 1 & 13 & & 2 & & & \\ & & & 12 & & 0 & & \\ 2 & 2 & 25 & & 2 & & 0 & \\ & & & 16 & & 0 & & 0 \\ 3 & 3 & 41 & & 2 & & 0 & \\ & & & 20 & & 0 & & \\ 4 & 4 & 61 & & 2 & & & \\ & & & 24 & & & & \\ 5 & 5 & 85 & & & & & \end{bmatrix}$$
(b) The fifth column of this tableau **(c)** $p_2(x) = 5 + 8x + 2x(x - 1)$.

18. (a) Clearly $f(t) - p_n(t) = 0$ for $t = x_k$, $k = 0, 1, \ldots, n$. This implies that $f(t) - p_n(t)$ equals some function of $x_0, \ldots, x_n, t$, times the product of the $t - x_k$. **(b)** From (a) we find that $f(t) - p_n(t)$, has at least $n + 1$ zeros on $[x_0, x_n]$, therefore the n-th derivative of this difference will have at least one zero on $[x_0, x_n]$, and noting the fact that the n-th derivative of $p_n(t)$ equals $n! f[x_0, \cdots, x_n]$ yields the desired result. $\xi = 0.09111223034$

Exercises 40.3, Page 1025

3. **(a)** $P_6(x) = 0.08174711402x^6 - 1.147898136x^5 + 5.765280127x^4 - 11.61652409x^3 + 4.166861647x^2 + 13.02831112x - 11.$ **(b)** $p_6(x) = 0.05540841278x^6 - 0.8014630580x^5 + 4.142973395x^4 - 8.560385190x^3 + 3.29952491x^2 + 8.797316165x - 6.961112122$ **(c)** $M_7 = 0.3809228067 \times 10^8$ **(d)** $\epsilon_1 = 8.564415162; \epsilon_2 = 4.187775308; \epsilon_3 = 258270.7646$

7. **(a)** $P_7(x) = 0.000325966158x^7 - 0.00993931455x^6 + 0.127927837x^5 - 0.902176337x^4 + 3.74626428x^3 - 8.880334797x^2 + 9.596726768x$ **(b)** $p_7(x) = 0.0003482288298x^7 - 0.01066417176x^6 + 0.1370140316x^5 - 0.9581051648x^4 + 3.922478274x^3 - 9.142597069x^2 + 9.727955517x + 0.0147311975$ **(c)** $M_8 = 80$ **(d)** $\epsilon_1 = 0.04852423200; \epsilon_2 = 0.01405198000; \epsilon_3 = 0.3490630255$

18. *Hint:* Solve the equation $k \arccos x = \left(n + \frac{1}{2}\right)\pi$ for x.

Exercises 40.4, Page 1032

2. **(a)**

(b) $F_0(x) = \begin{cases} 10 + x & -15 \le x \le -9 \\ -53 - 6x & -9 \le x \le -6 \\ \frac{19}{3} + \frac{35}{9}x & -6 \le x \le 3 \\ \frac{75}{4} - \frac{1}{4}x & 3 \le x \le 7 \\ \frac{199}{8} - \frac{9}{8}x & 7 \le x \le 15 \end{cases}$

(c) $F_1(x) = \begin{cases} \frac{65}{2} + 5x + \frac{1}{6}x^2 & -15 \le x \le -9 \\ -197 - 46x - \frac{8}{3}x^2 & -9 \le x \le -6 \\ -\frac{265}{9} + \frac{266}{27}x + \frac{161}{81}x^2 & -6 \le x \le 3 \\ -\frac{4651}{48} + \frac{3947}{72}x - \frac{793}{144}x^2 & 3 \le x \le 7 \\ \frac{58{,}081}{192} - \frac{17{,}077}{288}x + \frac{1523}{576}x^2 & 7 \le x \le 15 \end{cases}$

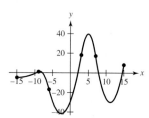

(d) $F_2(x) = \begin{cases} -\frac{685}{216}x^2 - \frac{676}{9}x - \frac{3345}{8} & -15 \le x \le -9 \\ \frac{433}{108}x^2 + \frac{1949}{36}x + \frac{327}{2} & -9 \le x \le -6 \\ -\frac{77}{324}x^2 + \frac{343}{108}x + \frac{191}{18} & -6 \le x \le 3 \\ -\frac{1}{2}x^2 + \frac{19}{4}x + \frac{33}{4} & 3 \le x \le 7 \\ \frac{9}{64}x^2 - \frac{135}{32}x + \frac{2537}{64} & 7 \le x \le 15 \end{cases}$

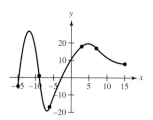

(e) $F_3(x) = \begin{cases} 10 + x & -15 \le x \le -9 \\ -\frac{7}{3}x^2 - 41x - 179 & -9 \le x \le -6 \\ \frac{152}{81}x^2 + \frac{257}{27}x - \frac{247}{9} & -6 \le x \le 3 \\ -\frac{757}{144}x^2 + \frac{3767}{72}x - \frac{4399}{48} & 3 \le x \le 7 \\ \frac{1451}{576}x^2 - \frac{16285}{288}x + \frac{55561}{192} & 7 \le x \le 15 \end{cases}$

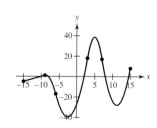

(f) $F_4(x) = \begin{cases} -\frac{649}{216}x^2 - \frac{640}{9}x - \frac{3165}{8} & -15 \le x \le -9 \\ \frac{397}{108}x^2 + \frac{1769}{36}x + \frac{291}{2} & -9 \le x \le -6 \\ -\frac{41}{324}x^2 + \frac{379}{108}x + \frac{155}{18} & -6 \le x \le 3 \\ -\frac{3}{4}x^2 + \frac{29}{4}x + 3 & 3 \le x \le 7 \\ \frac{17}{64}x^2 - \frac{223}{32}x + \frac{3377}{64} & 7 \le x \le 15 \end{cases}$

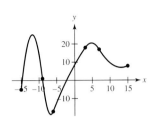

(g) $F_5(x) = \begin{cases} 10 + x & -15 \le x \le -9 \\ -\frac{7}{3}x^2 - 41x - 179 & -9 \le x \le -6 \\ \frac{152}{81}x^2 + \frac{257}{27}x - \frac{247}{9} & -6 \le x \le 3 \\ -\frac{757}{144}x^2 + \frac{3767}{72}x - \frac{4399}{48} & 3 \le x \le 7 \\ \frac{1451}{576}x^2 - \frac{16285}{288}x + \frac{55561}{192} & 7 \le x \le 15 \end{cases}$

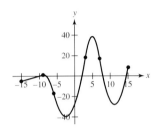

(h) $F_6(x) = \begin{cases} -\frac{1451}{432}x^2 - \frac{1433}{18}x - \frac{7095}{16} & -15 \le x \le -9 \\ \frac{947}{216}x^2 + \frac{4303}{72}x + \frac{735}{4} & -9 \le x \le -6 \\ -\frac{235}{648}x^2 + \frac{605}{216}x + \frac{463}{36} & -6 \le x \le 3 \\ -\frac{7}{32}x^2 + \frac{31}{16}x + \frac{453}{32} & 3 \le x \le 7 \\ -\frac{7}{32}x^2 + \frac{31}{16}x + \frac{453}{32} & 7 \le x \le 15 \end{cases}$

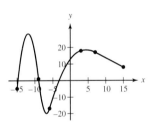

(i) $F_7(x) = \begin{cases} -\frac{7}{9}x^2 - \frac{53}{3}x - 95 & -15 \le x \le -9 \\ -\frac{7}{9}x^2 - \frac{53}{3}x - 95 & -9 \le x \le -6 \\ \frac{110}{81}x^2 + \frac{215}{27}x - \frac{163}{9} & -6 \le x \le 3 \\ -\frac{589}{144}x^2 + \frac{2927}{72}x - \frac{3223}{48} & 3 \le x \le 7 \\ \frac{1115}{576}x^2 - \frac{12589}{288}x + \frac{43801}{192} & 7 \le x \le 15 \end{cases}$

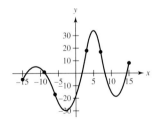

(j) $F_8(x) = \begin{cases} -\frac{1493}{432}x^2 - \frac{1475}{18}x - \frac{7305}{16} & -15 \le x \le -9 \\ \frac{989}{216}x^2 + \frac{4513}{72}x + \frac{777}{4} & -9 \le x \le -6 \\ -\frac{277}{648}x^2 + \frac{563}{216}x + \frac{505}{36} & -6 \le x \le 3 \\ -\frac{7}{96}x^2 + \frac{23}{48}x + \frac{551}{32} & 3 \le x \le 7 \\ -\frac{7}{96}x^2 + \frac{23}{48}x + \frac{551}{32} & 7 \le x \le 15 \end{cases}$

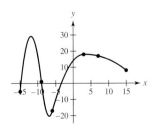

(k) $P_5(x) = \frac{60{,}871}{169{,}827{,}840}x^5 - \frac{1037}{8{,}491{,}392}x^4 - \frac{10{,}903}{101{,}088}x^3 + \frac{56911}{1{,}886{,}976}x^2 + \frac{41{,}234{,}507}{6{,}289{,}920}x + \frac{8951}{9984}$

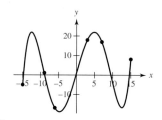

4. (a) $G_1(x) = \begin{cases} -\frac{16{,}486{,}267}{161{,}051} - \frac{20{,}592{,}806}{805{,}255}x - \frac{2{,}553{,}087}{1{,}610{,}510}x^2 - \frac{44{,}791}{1{,}610{,}510}x^3 & -19 \le x \le -8 \\[4pt] \frac{15{,}889{,}003}{73{,}205} + \frac{6{,}896{,}462}{73{,}205}x + \frac{35{,}637}{2662}x^2 + \frac{87{,}267}{146{,}410}x^3 & -8 \le x \le -6 \\[4pt] -\frac{577{,}057}{14{,}641} - \frac{2{,}490{,}682}{73{,}205}x - \frac{1{,}169{,}013}{146{,}410}x^2 - \frac{86{,}569}{146{,}410}x^3 & -6 \le x \le -5 \\[4pt] -\frac{154{,}857}{146{,}41} - \frac{1{,}224{,}082}{73{,}205}x - \frac{662{,}373}{146{,}410}x^2 - \frac{52{,}793}{146{,}410}x^3 & -5 \le x \le -4 \\[4pt] \frac{2{,}043{,}981}{161{,}051} + \frac{587{,}878}{805{,}255}x - \frac{259{,}713}{1{,}610{,}510}x^2 + \frac{9619}{3{,}221{,}020}x^3 & -4 \le x \le 18 \end{cases}$

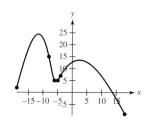

(b) $G_2(x) = \begin{cases} -\frac{119{,}761}{2{,}415{,}765}x^3 - \frac{3{,}745{,}663}{1{,}610{,}510}x^2 - \frac{159{,}484{,}519}{4{,}831{,}530}x - \frac{20{,}229{,}039}{161{,}051} & -19 \le x \le -8 \\[4pt] \frac{1{,}136{,}381}{1{,}756{,}920}x^3 + \frac{38{,}299}{2662}x^2 + \frac{44{,}228{,}443}{439{,}230}x + \frac{16{,}905{,}887}{73{,}205} & -8 \le x \le -6 \\[4pt] -\frac{568{,}661}{878{,}460}x^3 - \frac{2{,}608{,}219}{292{,}820}x^2 - \frac{8{,}580{,}769}{219{,}615}x - \frac{711{,}488}{14{,}641} & -6 \le x \le -5 \\[4pt] -\frac{222{,}257}{878{,}460}x^3 - \frac{876{,}199}{292{,}820}x^2 - \frac{2{,}085{,}694}{219{,}615}x + \frac{10{,}187}{14{,}641} & -5 \le x \le -4 \\[4pt] -\frac{159{,}239}{19{,}326{,}120}x^3 - \frac{177{,}359}{3{,}221{,}020}x^2 + \frac{5{,}439{,}856}{2{,}415{,}765}x + \frac{2{,}634{,}945}{161{,}051} & -4 \le x \le 18 \end{cases}$

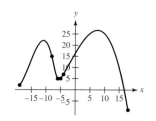

(c) $G_3(x) = \begin{cases} -\frac{66{,}351}{103{,}697}x^2 - \frac{1{,}668{,}926}{103{,}697}x - \frac{7{,}549{,}489}{103{,}697} & -19 \le x \le -8 \\[4pt] \frac{221{,}527}{414{,}788}x^3 + \frac{114{,}801}{9427}x^2 + \frac{8{,}964{,}370}{103{,}697}x + \frac{20{,}805{,}967}{103{,}697} & -8 \le x \le -6 \\[4pt] -\frac{115{,}613}{207{,}394}x^3 - \frac{1{,}549{,}155}{207{,}394}x^2 - \frac{3{,}259{,}961}{103697}x - \frac{3{,}642{,}695}{103{,}697} & -6 \le x \le -5 \\[4pt] -\frac{70{,}905}{207{,}394}x^3 - \frac{878{,}535}{207{,}394}x^2 - \frac{1{,}583{,}411}{103{,}697}x - \frac{848{,}445}{103{,}697} & -5 \le x \le -4 \\[4pt] -\frac{27{,}675}{207{,}394}x^2 + \frac{118{,}309}{103{,}697}x + \frac{1{,}420{,}515}{103{,}697} & -4 \le x \le 18 \end{cases}$

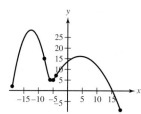

(d) $G_4(x) = \begin{cases} \frac{2{,}241{,}259}{13{,}458{,}588}x^3 + \frac{2{,}047{,}323}{407{,}836}x^2 + \frac{21{,}033{,}229}{517{,}638}x + \frac{116{,}697{,}407}{1{,}121{,}549} & -19 \le x \le -8 \\[4pt] \frac{2{,}241{,}259}{13{,}458{,}588}x^3 + \frac{2{,}047{,}323}{407{,}836}x^2 + \frac{21{,}033{,}229}{517{,}638}x + \frac{116{,}697{,}407}{1{,}121{,}549} & -8 \le x \le -6 \\[4pt] -\frac{5{,}399{,}015}{13{,}458{,}588}x^3 - \frac{23{,}321{,}091}{4{,}486{,}196}x^2 - \frac{139{,}142{,}819}{6{,}729{,}294}x - \frac{20{,}827{,}525}{1{,}121{,}549} & -6 \le x \le -5 \\[4pt] -\frac{525{,}743}{13{,}458{,}588}x^3 + \frac{1{,}045{,}269}{4{,}486{,}196}x^2 + \frac{43{,}604{,}881}{6{,}729{,}294}x + \frac{29{,}935{,}725}{1{,}121{,}549} & -5 \le x \le -4 \\[4pt] -\frac{525{,}743}{13{,}458{,}588}x^3 + \frac{1{,}045{,}269}{4{,}486{,}196}x^2 + \frac{43{,}604{,}881}{6{,}729{,}294}x + \frac{29{,}935{,}725}{1{,}121{,}549} & -4 \le x \le 18 \end{cases}$

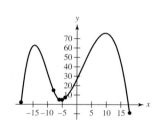

(e) $G_5(x) = \begin{cases} -\frac{79,430}{1,276,429}x^3 - \frac{3,516,643}{1,276,429}x^2 - \frac{6,801,392}{182,347}x - \frac{177,334,525}{1,276,429} & -19 \le x \le -8 \\ \frac{312,047}{464,156}x^3 + \frac{156,899}{10,549}x^2 + \frac{12,036,512}{116,039}x + \frac{27,517,801}{116,039} & -8 \le x \le -6 \\ -\frac{145,581}{232,078}x^3 - \frac{1,977,103}{232,078}x^2 - \frac{4,250,131}{116,039}x - \frac{5,055,485}{116,039} & -6 \le x \le -5 \\ -\frac{94,649}{232,078}x^3 - \frac{1,213,123}{232,078}x^2 - \frac{2,340,181}{116,039}x - \frac{1,872,235}{116,039} & -5 \le x \le -4 \\ \frac{23,889}{2,552,858}x^3 - \frac{564,017}{2,552,858}x^2 - \frac{181,319}{1,276,429}x + \frac{13,486,311}{1,276,429} & -4 \le x \le 18 \end{cases}$

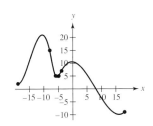

(f) $G_6(x) = \begin{cases} -\frac{156,941}{2,520,914}x^3 - \frac{3,474,222}{1,260,457}x^2 - \frac{94,073,769}{2,520,914}x - \frac{175,214,909}{1,260,457} & -19 \le x \le -8 \\ \frac{617,317}{916,696}x^3 + \frac{155,211}{10,417}x^2 + \frac{23,818,389}{229,174}x + \frac{27,232,105}{114,587} & -8 \le x \le -6 \\ -\frac{293,919}{458,348}x^3 - \frac{4,017,111}{458,348}x^2 - \frac{4,360,398}{114,587}x - \frac{5,307,080}{114,587} & -6 \le x \le -5 \\ -\frac{160,571}{458,348}x^3 - \frac{2,016,891}{458,348}x^2 - \frac{1,860,123}{114,587}x - \frac{1,139,955}{114,587} & -5 \le x \le -4 \\ \frac{30,013}{10,083,656}x^3 - \frac{810,351}{5,041,828}x^2 + \frac{914,097}{1,260,457}x + \frac{15,961,095}{1,260,457} & -4 \le x \le 18 \end{cases}$

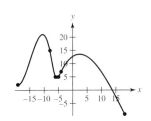

(g) $G_7(x) = \begin{cases} -\frac{75,850,843}{1,038,059} - \frac{666,027}{1,038,059}x^2 - \frac{16,755,932}{1,038,059}x & -19 \le x \le -8 \\ \frac{2,241,259}{4,152,236}x^3 + \frac{1,161,957}{94,369}x^2 + \frac{90,824,500}{1,038,059}x + \frac{211,030,309}{1,038,059} & -8 \le x \le -6 \\ -\frac{1,312,061}{2,076,118}x^3 - \frac{18,225,375}{2,076,118}x^2 - \frac{40,540,787}{1,038,059}x - \frac{51,700,265}{1,038,059} & -6 \le x \le -5 \\ -\frac{70,905}{2,076,118}x^3 + \frac{391,965}{2,076,118}x^2 + \frac{260,981}{45,133}x + \frac{25,871,985}{1,038,059} & -5 \le x \le -4 \\ -\frac{70905}{2,076,118}x^3 + \frac{391,965}{2,076,118}x^2 + \frac{260,981}{45,133}x + \frac{25,871,985}{1,038,059} & -4 \le x \le 18 \end{cases}$

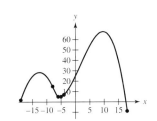

7. (a) $\begin{cases} 4.9006263x - 1.221831x^3 & 0 \le x \le 1 \\ -2.680108 + 12.94095x - 8.040325x^2 + 1.458276x^3 & 1 \le x \le 2 \\ 10.59727 - 6.975118x + 1.917709x^2 - 0.2013959x^3 & 2 \le x \le 3 \\ 4.068263 - 0.44610975x - 0.2586266x^2 + 0.04041917x^3 & 3 \le x \le 4 \\ 9.67138 - 4.648450x + 0.7919584x^2 - 0.04712958x^3 & 4 \le x \le 5 \\ 4.980575 - 1.833965x + 0.2290615x^2 - 0.009603116x^3 & 5 \le x \le 6 \\ 6.953093 - 2.820224x + 0.3934379x^2 - 0.01873514x^3 & 6 \le x \le 7 \end{cases}$

(b) $\int_0^7 [G''(x)]^2\,dx = 34.06104636 \le 124.9987319 = \int_0^7 [f''(x)]^2\,dx$

11. (a) $\tau = 0.1, 1, 10$ **(b)** if $\tau = 0.1$: $z_6 = 0, z_5 = -0.8300317786, z_4 = 2.758098187, z_3 = -4.008889185, z_2 = 2.544798588, z_1 = 0$; if $\tau = 1$: $z_6 = 0, z_5 = -0.8300317786, z_4 = 2.758098187, z_3 = -4.008889185, z_2 = 2.544798588, z_1 = 0$; if $\tau = 10$: $z_6 = 0, z_5 = -13.13717471, z_4 = 21.80901408, z_3 = -49.91184844, z_2 = 44.90981995, z_1 = 0$

(c)

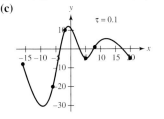
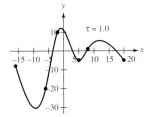
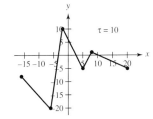

13. (a)
$$\begin{cases} -\frac{176,477,517}{373,114,280}x^3 - \frac{1,595,081,549}{93,278,570}x^2 - \frac{6,413,003,493}{33,919,480}x - \frac{30,375,236,378}{46,639,285} & -19 \le x \le -9 \\ -\frac{176,477,517}{373,114,280}x^3 - \frac{1,595,081,549}{93,278,570}x^2 - \frac{6,413,003,493}{33,919,480}x - \frac{30,375,236,378}{46,639,285} & -9 \le x \le -8 \\ \frac{307,694,681}{213,208,160}x^3 + \frac{5,389,155,407}{186,557,140}x^2 + \frac{9,532,293,951}{53,302,040}x + \frac{15,381,128,982}{46,639,285} & -8 \le x \le -6 \\ -\frac{76,035,599}{746,228,560}x^3 + \frac{803,215,943}{746,228,560}x^2 + \frac{2,232,920,301}{186,557,140}x - \frac{105,100,161}{26,651,020} & -6 \le x \le 1 \\ -\frac{76,035,599}{746,228,560}x^3 + \frac{803,215,943}{746,228,560}x^2 + \frac{2,232,920,301}{186,557,140}x - \frac{105,100,161}{26,651,020} & 1 \le x \le 17 \end{cases}$$

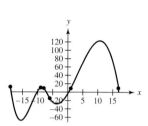

16. $G_x(t) = \begin{cases} -23 - \frac{3854}{209}t + \frac{10,797}{209}t^2 - \frac{3599}{209}t^3 & 1 \le t \le 2 \\ -\frac{105,663}{209} + \frac{147,430}{209}t - \frac{5895}{19}t^2 + \frac{9008}{209}t^3 & 2 \le t \le 3 \\ \frac{403,800}{209} - \frac{362,033}{209}t + \frac{104,976}{209}t^2 - \frac{519}{11}t^3 & 3 \le t \le 4 \\ -\frac{636,968}{209} + \frac{418,543}{209}t - \frac{90,168}{209}t^2 + \frac{6401}{209}t^3 & 4 \le t \le 5 \\ \frac{406,782}{209} - \frac{207,707}{209}t + \frac{35,082}{209}t^2 - \frac{1949}{209}t^3 & 5 \le x \le 6 \end{cases}$;

$G_y(t) = \begin{cases} 47 - \frac{2133}{209}t - \frac{5265}{209}t^2 + \frac{1755}{209}t^3 & 1 \le t \le 2 \\ \frac{50,591}{209} - \frac{63,285}{209}t + \frac{2301}{19}t^2 - \frac{3341}{209}t^3 & 2 \le t \le 3 \\ -\frac{172,483}{209} + \frac{159,789}{209}t - \frac{49,047}{209}t^2 + \frac{259}{11}t^3 & 3 \le t \le 4 \\ \frac{559,741}{209} - \frac{389,379}{209}t + \frac{88,245}{209}t^2 - \frac{6520}{209}t^3 & 4 \le t \le 5 \\ -\frac{653,384}{209} + \frac{338,496}{209}t - \frac{57,330}{209}t^2 + \frac{3185}{209}t^3 & 5 \le x \le 6 \end{cases}$

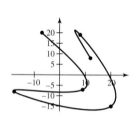

Exercises 40.5, Page 1036

2. $\mathbf{R}(u) = [1 + 12u - 6u^2 + u^3, 3 + 3u - 21u^2 + 23u^3]^{\mathrm{T}}$; $\mathbf{R}'(0) = [12, 3]^{\mathrm{T}}$, $\mathbf{P}_2 - \mathbf{P}_1 = [4, 1]^{\mathrm{T}}$; $\mathbf{R}'(1) = [3, 30]^{\mathrm{T}}$, $\mathbf{P}_4 - \mathbf{P}_3 = [1, 10]^{\mathrm{T}}$

3. (a) $\mathbf{R}(u) = [-16 + 72u - 63u^2 + 25u^3, 2 - 33u + 54u^2 - 43u^3]^{\mathrm{T}}$ **(b)** $\mathbf{R}'(0) = [72, -33]^{\mathrm{T}}$, $\mathbf{P}_2 - \mathbf{P}_1 = [24, -11]^{\mathrm{T}}$
(c) $\mathbf{R}'(1) = [21, -54]^{\mathrm{T}}$, $\mathbf{P}_4 - \mathbf{P}_3 = [7, -18]^{\mathrm{T}}$ **(d)** $\mathbf{Q}_1(u) = [-16 + 24u, 2 - 11u]^{\mathrm{T}}$; $\mathbf{Q}_1(u) = [8 + 3u, -9 + 7u]^{\mathrm{T}}$;
$\mathbf{Q}_3(u) = [11 + 7u, -2 - 18u]^{\mathrm{T}}$; $\mathbf{Q}_4(u) = [-16 + 48u - 21u^2, 2 - 22u + 18u^2]^{\mathrm{T}}$; $\mathbf{Q}_5(u) = [8 + 6u + 4u^2, -9 + 14u - 25u^2]^{\mathrm{T}}$;
$\mathbf{Q}_6(u) = [-16 + 72u - 63u^2 + 25u^3, 2 - 33u + 54u^2 - 43u^3]^{\mathrm{T}}$ **(e)**

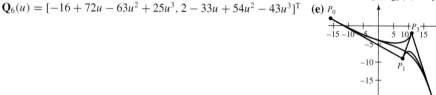

8. (a) $\mathbf{R}(u) = [-2 + 6u + 2u^2, -18 + 8u + 22u^2]^{\mathrm{T}}$ **(b)** $\mathbf{R}'(0) = [6, 8]^{\mathrm{T}}$, $\mathbf{P}_1 - \mathbf{P}_0 = [3, 4]^{\mathrm{T}}$ **(c)** $\mathbf{R}'(1) = [10, 52]^{\mathrm{T}}$, $\mathbf{P}_2 - \mathbf{P}_1 = [5, 26]^{\mathrm{T}}$
(d) $\mathbf{Q}_1(u) = [-2 + 3u, -18 + 4u]^{\mathrm{T}}$; $\mathbf{Q}_2(u) = [1 + 5u, -14 + 26u]^{\mathrm{T}}$; $\mathbf{Q}_3(u) = [-2 + 6u + 2u^2, -18 + 8u + 22u^2]^{\mathrm{T}}$
(e)

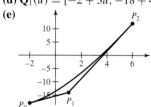

13. (a) $\mathbf{R}(u) = [-18 + 40u - 54u^2 + 92u^3 - 40u^4,\ 20 - 140u + 354u^2 - 320u^3 + 79u^4]^T$ **(b)** $\mathbf{R}'(0) = [40, -140]^T,\ \mathbf{P}_1 - \mathbf{P}_0 = [10, -35]^T$
   (c) $\mathbf{R}'(1) = [48, -76]^T,\ \mathbf{P}_4 - \mathbf{P}_3 = [12, -19]^T$ **(d)** $\mathbf{Q}_1(u) = [-18 + 10u,\ 20 - 35u]^T;\ \mathbf{Q}_2(u) = [-8 + u,\ -15 + 24u]^T;$
   $\mathbf{Q}_3(u) = [-7 + 15u,\ 9 + 3u]^T;\ \mathbf{Q}_4(u) = [8 + 12u,\ 12 - 19u]^T;\ \mathbf{Q}_5(u) = [-18 + 20u - 9u^2,\ 20 - 70u + 59u^2]^T;$
   $\mathbf{Q}_6(u) = [-8 + 2u + 14u^2,\ -15 + 48u - 21u^2]^T;\ \mathbf{Q}_7(u) = [-7 + 30u - 3u^2,\ 9 + 6u - 22u^2]^T;$
   $\mathbf{Q}_8(u) = [-18 + 30u - 27u^2 + 23u^3,\ 20 - 105u + 177u^2 - 80u^3]^T;\ \mathbf{Q}_9(u) = [-8 + 3u + 42u^2 - 17u^3,\ -15 + 72u - 63u^2 - u^3]^T;$
   $\mathbf{Q}_{10}(u) = [-18 + 40u - 54u^2 + 92u^3 - 40u^4,\ 20 - 140u + 354u^2 - 320u^3 + 79u^4]^T$ **(e)**

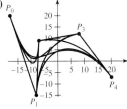

18. (a) $\mathbf{R}(u) = [3au + (-6a + 3c)u^2 + (5 + 3a - 3c)u^3,\ 3ub + (-6b + 3d)u^2 + (3b - 3d)u^3]^T$
   (b) $\mathbf{R}_1(u) = \left[\frac{19}{2}u - \frac{27}{2}u^2 + 9u^3,\ -\frac{9}{2}u + \frac{117}{2}u^2 - 54u^3\right]^T;\ p_1 = \left(\frac{19}{6}, -\frac{3}{2}\right);\ p_2 = \left(\frac{11}{6}, \frac{33}{2}\right)$
   (c) $\mathbf{R}_2(u) = \left[13u - 24u^2 + 16u^3,\ \frac{16}{3}u + \frac{112}{3}u^2 - \frac{128}{3}u^3\right]^T;\ \hat{p}_1 = \left(\frac{13}{3}, \frac{16}{9}\right);\ \hat{p}_2 = \left(\frac{2}{3}, 16\right)$ **(d)** $\mathbf{R}_3(u) = \left[5u, -\frac{125}{6}u + \frac{625}{6}u^2 - \frac{250}{3}u^3\right]^T;$
   $\tilde{p}_1 = \left(\frac{5}{3}, -\frac{125}{8}\right);\ \tilde{p}_2 = \left(\frac{10}{3}, \frac{125}{6}\right)$ **(e)**

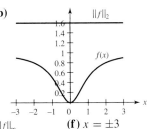

There are infinitely many control points, which make the Bezier curve go through one fixed set of four points.

CHAPTER 41

Exercises 41.1, Page 1042

2. (a) $\|f\|_2 = \frac{1}{10}\sqrt{630 - 300\arctan 3}$ **(b)**

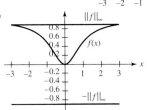

   (c) None

(d) $\|f\|_\infty = \frac{9}{10}$ **(e)**

   (f) $x = \pm 3$

12. (a) $\|f - g\|_2 = \sqrt{2\pi};\ \|f - g\|_\infty = \sqrt{2}$ **(b)**

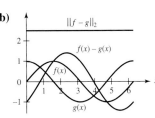

   (c) There are no real solutions to this equation.

(d)

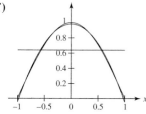

(e) $x = \frac{3}{4}\pi, \frac{7}{4}\pi$

22. (a) $y_1 = 3\frac{(5\pi^2 - 60)x^2}{\pi^3} + 3\frac{-\pi^2 + 20}{\pi^3}$ **(b)** $y_2 = \frac{2}{\pi}$ **(c)** $y_3 = 3\frac{(5\pi^2 - 60)x^2}{\pi^3} + 3 - \frac{\pi^2 + 20}{\pi^3}$ **(d)** $y_4 = 3\frac{(5\pi^2 - 60)x^2}{\pi^3} + 3 - \frac{\pi^2 + 20}{\pi^3}$
(e) $\|f - y_1\|_2 = 0.02441443924$; $\|f - y_2\|_2 = 0.4352361784$; $\|f - y_3\|_2 = 0.02441443924$; $\|f - y_4\|_2 = 0.02441443924$
(f)

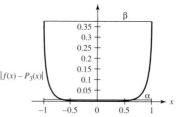

32. (a) $\|f\|_\infty = 1$ **(b)** $\|f\|_k = (1 + 3k)^{-\frac{1}{k}}$

37. (a) $p_3(x) = 1 - \frac{1}{2}x^2 - \frac{1}{8}x^4$ **(b)** $\alpha = 0.006046988$; $\beta = 0.3750000000$ **(c)**

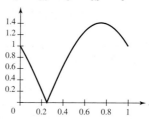

47. (a) $\|f\|_w = 0.5690943663$; $\|g\|_w = 0.5552046117$; $\|f - g\|_w = 0.7307536169$

Exercises 41.2–Part A, Page 1047

2. $1 + (2 + (3 + 4x)x)x$ Three multiplications and three additions.   **5.** 4 operations, 2 additions and 2 divisions.

Exercises 41.2–Part B, Page 1048

2. (a) $[3/2] = \frac{48 - 5\pi^2 x^2}{48 + \pi^2 x^2}$ **(c)** $\|f - [3/2]\|_\infty = 0.02329412857$; $-1.0, 1.0$ **(d)** Five multiplications/divisions and two additions/subtractions
(e) Five multiplications/divisions and two additions/subtractions   **(f)** Two multiplications/divisions and two additions/subtractions
(g) $\|f - p_5\|_\infty = 0.01996895776$; $x_{max} = -1, 1$ **(h)** $p_5(x) = 1. + (-1.233700551 + 0.2536695081x^2)x^2$ Four multiplications/divisions
and two additions/subtractions   **(i)**

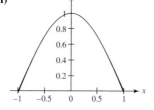

12. **(a)** $[-e^{-\pi}, -\frac{e^{-\pi}}{1+z}, -\frac{e^{-\pi}}{1+z+z^2}, -3\frac{e^{-\pi}}{3+3z+3z^2+2z^3}], [(-1+z)e^{-\pi}, (-1+z)e^{-\pi}, \frac{(-3+2z)e^{-\pi}}{3+z+z^2}, 3\frac{(-4+3z)e^{-\pi}}{12+3z+3z^2-z^3}], [(-1+z)e^{-\pi}, (-1+$

$z)e^{-\pi}, \frac{(-6+3z+z^2)e^{-\pi}}{6+3z+2z^2}, 3\frac{(-10-6z+9z^2)e^{-\pi}}{30+48z+21z^2+11z^3}], [(-1+z-\frac{1}{3}z^3)e^{-\pi}, \frac{1}{3}\frac{(-6+3z+3z^2-2z^3)e^{-\pi}}{2+z}, \frac{1}{3}\frac{(-60+30z+21z^2-11z^3)e^{-\pi}}{20+10z+3z^2}, \frac{1}{3}\frac{(-330+120z+138z^2-53z^3)e^{-\pi}}{110+70z+24z^2+5z^3}]$

(b)
$$\begin{bmatrix} 1.043213919 & 49.25793223 & 1.005591855 & 101.3704196 \\ 1.178974447 & 1.178974447 & 1.041237904 & 1.027542383 \\ 1.178974447 & 1.178974447 & 1.014714631 & 1.049735658 \\ 0.7323402113 & 162.7882892 & 0.6879756915 & 289.3653054 \end{bmatrix}$$

(c) $\|F - [3/0]\| = 0.7323402113$, only $[3/2]$ is smaller.

18. **(a)** $\|\frac{b_5 x^9}{Q_4}\|_\infty = 0.8941818118 \times 10^{-8}$; $\|G - [4/4]\|_\infty = 0.1496470170 \times 10^{-7}$

Exercises 41.3, Page 1051

1. $a_0 = 1.264241118$, $a_1 = 0.2516889099$, $a_2 = 0.03123247381$, $b_1 = 0$, $b_2 = 0$ **3.** $A_0 = 1.111645383$, $A_1 = 0$, $A_2 = -0.2419335080$

5. **(a)** $S_5(x) = 0.4720012157 - 0.4994032582T_2(x) + 0.02799207960T_4(x)$ **(b)** $S_5(x) = 0.9993965535 - 1.222743153x^2 + 0.2239366368x^4$
(c)   **(d)** $\|f - S_5\|_\infty = 0.00060344650000, \pm 0.8649407780$ **(e)** $\|f - p_5\|_\infty = 0.01996895776$

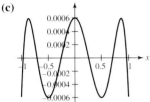

15. **(a)** $s_5(z) = 0.1845322548 + 0.2247310296z^2 - 0.01685760178z^4$ **(b)**

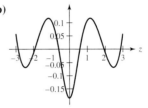

(c) $\|g - s_5\|_\infty = 0.1845322548\ 0$ **(d)** $\|g - p_5\|_\infty = 72.90000000$

18. **(a)** $g_5(x) = 0.2052124962 - 0.307818745x + 0.128257810x^2 + 0.106881508x^3 - 0.2004028289x^4 + 0.1670023572x^5$
(b) $g_{\mathrm{econ}}(x) = 0.2052124962 - 0.3600069816x + 0.128257810x^2 + 0.3156344545x^3 - 0.2004028289x^4$
(c) $p_{econ}(z) = 0.1774400044 + 0.4527968310z - 0.32337000430z^2 + 0.07150372929z^3 - 0.005130312420z^4$
(d) $\|g - p_5\|_\infty = 0.1670023574$; $\|g - p_{econ}\|_\infty = 0.1774400044$

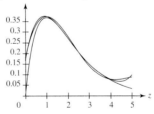

(e) $g_5(x) = 0.1941903404T_0(x) - 0.1232811407T_1(x) - 0.03607250950T_2(x) + 0.07890861363T_3(x) - 0.02505035361T_4(x) +$
$0.01043764733T_5(x)$ **(f)** $g_{econ}(x) = 2052124963 - 0.3600069816x + 0.1282578099x^2 + 0.3156344545x^3 - 0.2004028289x^4$

22. **(a)** $\sigma_5(z) = 7.031306465 - 12.75203101z + 7.462293256z^2 - 0.9806709512z^3 - 0.6200657562z^4 + 0.1867336843z^5$
(b) $s_5(z) = 7.014207256 - 12.58327593z + 7.024632549z^2 - 0.530864556z^3 - 0.8177881387z^4 + 0.2176148566z^5$
(c) $\|g - \sigma_5\|_\infty = 0.03130646500$; $\|g - s_5\|_\infty = 0.01420725600$; $\|g - p_5\|_\infty = 0.3703703700$

24. **(a)** $0.05184783285 - 0.03679426501\cos(\frac{\pi z}{2}) - 0.009437991726\sin(\frac{\pi z}{2}) + 0.01050425481\cos(\pi z) - 0.0009233791256\sin(\pi z) +$
$0.002933085745\cos(\frac{3\pi z}{2}) + 0.006327050379\sin(\frac{3\pi z}{2})$
(b) $s_5(z) = -0.8457568581 + 1.611422751z - 1.031009947z^2 + 0.3106980831z^3 - 0.04536006861z^4 + 0.002588433222z^5$
(c) $\|g - \psi\|_\infty = 0.02557853592$; $\|g - \sigma_5\|_\infty = 0.005597557199$; $\|g - s_5\|_\infty = 0.002582393612$; $\|g - p_5\|_\infty = 0.06177890469$

Exercises 41.4, Page 1056

2. **(a)** $R_1(x) = --\dfrac{0.8889310763 + 0.8898438804x^2}{0.8878978895 + 0.2242042210x^2}$ **(b)**

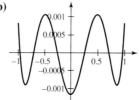

(c) $x_0 = -1$, $x_1 = -0.8624429066$, $x_2 = -0.4933522231$, $x_3 = 0$, $x_4 = 0.4933522231$, $x_5 = 0.8624429066$, $x_6 = 1$
(d) $\varepsilon_1(x_0) = 0.0008207914209$, $\varepsilon_1(x_1) = -0.0008931996$, $\varepsilon_1(x_2) = 0.0010632579$, $\varepsilon_1(x_3) = -0.001163632$, $\varepsilon_1(x_4) = 0.0010632579$, $\varepsilon_1(x_5) = -0.0008931996$, $\varepsilon_1(x_6) = 0.0008207914209$; $m_1 = 0.0008207914209$; $M_1 = 0.001163632$
(e) $\|\epsilon_1\|_\infty = 0.001163632454$

3. **(a)** $R_2(x) = \dfrac{0.8889511526 - 0.8898706254x^2}{0.8879247277 + 0.2241505446x^2}$ **(b)**

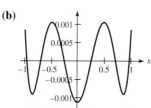

(c) $x_0 = -1$, $x_1 = -0.8620831573$, $x_2 = -0.4928180048$, $x_3 = 0$ $x_4 = 0.4928180048$, $x_5 = 0.8620831573$, $x_6 = 1$

(d) $\varepsilon_2(x_0) = 0.0008268078565$, $\varepsilon_2(x_1) = -0.0008960528$, $\varepsilon_2(x_2) = 0.0010592963$, $\varepsilon_2(x_3) = -0.001155982$, $\varepsilon_2(x_4) = 0.0010592963$, $\varepsilon_2(x_5) = -0.0008960528$, $\varepsilon_2(x_6) = 0.0008268078565$ $m_2 = 0.000826878565$; $M_2 = 0.001155982$
(e) $\|\epsilon_2\|_\infty = 0.001155981884$

5. **(a)** $R_4(x) = \dfrac{0.8890294897 - 0.8899739032x^2}{0.8880316574 + 0.2239366852x^2}$ **(b)**

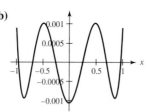

(c) $x_0 = -1$, $x_1 = -0.8606950817$, $x_2 = -0.4905955918$, $x_3 = 0$, $x_4 = 0.4905955918$, $x_5 = 0.8606950817$, $x_6 = 1$

(d) $\varepsilon_4(x_0) = 0.0008493166895$, $\varepsilon_4(x_1) = -0.0009083194$, $\varepsilon_4(x_2) = 0.0010445736$, $\varepsilon_4(x_3) = -0.001123645$, $\varepsilon_4(x_4) = 0.0010445736$, $\varepsilon_4(x_5) = -0.0009083194$, $\varepsilon_4(x_6) = 0.0008493166895$; $m_4 = 0.0008493166895$; $M_4 = 0.001123645$
(e) $\|\epsilon_4\|_\infty = 0.001123644964$

6. **(a)** $R_5(x) = \dfrac{0.8891893732 - 0.8901851387x^2}{0.8882489800 + 0.2235020401x^2}$ **(b)**

(c) $x_0 = -1$, $x_1 = -0.8578794742$, $x_2 = -0.4859952979$, $x_3 = 0$, $x_4 = 0.4859952979$, $x_5 = 0.8578794742$, $x_6 = 1$
(d) $\varepsilon_5(x_0) = 0.0008956729106$, $\varepsilon_5(x_1) = -0.0009336891$, $\varepsilon_5(x_2) = 0.0010150641$, $\varepsilon_5(x_3) = -0.001058705$, $\varepsilon_5(x_4) = 0.0010150641$, $\varepsilon_5(x_5) = -0.0009336891$, $\varepsilon_5(x_6) = 0.0008956729106$; $m_5 = 0.0008956729106$ $M_5 = 0.001058705$
(e) $\|\epsilon_5\|_\infty = 0.001058704509$

7. (a) $R_6(x) = \frac{0.8894307947 - 0.8905080542x^2}{0.8885695485 + 0.2228609030x^2}$ (b)

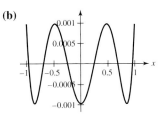

(c) $x_0 = -1$, $x_1 = -0.8536389477$, $x_2 = -0.4791788838$, $x_3 = 0$, $x_4 = 0.4791788838$, $x_5 = 0.8536389477$, $x_6 = 1$

(d) $\varepsilon_6(x_0) = 0.0009692547744$, $\varepsilon_6(x_1) = -0.0009703348$, $\varepsilon_6(x_2) = 0.0009703661$,
$\varepsilon_6(x_3) = -0.000969250$, $\varepsilon_6(x_4) = 0.0009703661$, $\varepsilon_6(x_5) = -0.0009703348$, $\varepsilon_6(x_6) = 0.0009692547744$; $m_6 = 0.000969250$;
$M_6 = 0.0009703661$ (e) $\|\epsilon_6\|_\infty = 0.0009703660256$

10. (a) $R(z) = -- \frac{0.1632281602 + 0.1654076615z - 0.04405923471z^2 + 0.003473589487z^3}{0.1650415865 - 0.02416230030z + 0.06152671580z^2}$ (b)

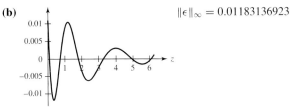

$\|\epsilon\|_\infty = 0.01183136923$

(c) $R_{\text{minimax}}(z) = \frac{0.2084131490 - 0.2114891686z + 0.05745879244z^2 - 0.004641841626z^3}{0.2091127136 - 0.0104223354z + 0.05563411229z^2}$; $\|\epsilon_{\text{minimax}}\|_\infty = 0.003345394873$

CHAPTER 42

Exercises 42.1, Page 1065

2. Subtract $f''(c)$ and expand the resulting expression in a Taylor series. Alternatively the formula can be tested with $f(x) = \sum_{k=0}^{3} a_k x^k$

9. (a) $f'(-2) \approx 0.6004235992$; $f''(-2) \approx -0.9060761451$; $f'''(-2) \approx 0.9985027767$; $f''''(-0.9590377328$

(b) $g_1(x) = -0.2325441580x^2 - 0.3297530328x + 0.2706705664$;
$g_2(x) = 0.0734979716x^3 - 0.0120502432x^2 - 0.1827570896x + 0.2706705664$;
$g_3(x) = -0.01548652632x^4 + 0.04252491898x^3 + 0.00343628304x^2 - 0.1517840371bx + 0.2706705664$;
$g_4(x) = 0.00244733791x^5 - 0.01548652632x^4 + 0.03028822945x^3 + 0.003436283018x^2 - 0.1419946853x + 0.2706705664$ (c) Yes,
(42.2a) through (42.2d) are the first, second, third and fourth derivatives of the appropriate interpolation polynomials.

(d)

i	$\dfrac{f'(c) - g_1'(c)}{h_i^2}$	$\dfrac{f''(c) - g_2''(c)}{h_i^2}$	$\dfrac{f'''(c) - g_3'''(c)}{h_i^2}$	$\dfrac{f''''(c) - g_4''''(c)}{h_i^2}$
0	0.3995764009	-1.093923855	2.001497225	-3.040962271
1	0.6192724870	-1.927934003	3.945743792	-6.618644814
2	0.7828654971	-2.633069064	5.787603182	-10.38232335
3	0.8836465857	-3.099604621	7.089250786	-13.21378594
4	0.9397189012	-3.369173271	7.868471642	-14.96758026
5	0.9693121134	-3.514236469	8.295560698	-15.94607067
6	0.9845164668	-3.589505324	8.519243497	-16.46321583
7	0.9922229850	-3.627846182	8.633722298	-16.72910245
8	0.9961026361	-3.647196059	8.691634351	-16.86391841
9	0.9980490984	-3.656916220	8.720760232	-16.93180019
10	0.9990239936	-3.661787651	8.735365849	-16.96586024
11	0.9995118578	-3.664226210	8.742679354	-16.98292014
12	0.9997558941	-3.665446201	8.746338784	-16.99145758
13	0.9998779384	-3.666056374	8.748169169	-16.99572816
14	0.9999389670	-3.666361506	8.749084529	-16.99786382
15	0.9999694830	-3.666514082	8.749542250	-16.99893234

19. (a) $f'(-2) \approx -0.1092591180$; $f''(-2) \approx 0.0241004865$; $f''' - 2) \approx 0.2551495139$; $f''''(-2) \approx 0.371676632$

(b) $g_1(x) = 0.0120502432x^2 - 0.061058145x - 0.4409878299$ $g_2(x) = 0.0120502432x^2 - 0.061058145x - 0.4409878299$;

$g_3(x) = 0.8403387008x + 0.623389862x^2 + 0.1664171294x^3 + 0.01548652632x^4$;

$g_4(x) = 0.8403387008x + 0.623389862x^2 + 0.1664171294x^3 + 0.01548652632x^4$ **(c)** Yes, (42.3a) through (42.3d) are the first, second, third and fourth derivatives of the appropriate interpolation polynomials.

(d)

i	$\dfrac{f'(c) - g_1'(c)}{h_i^2}$	$\dfrac{f''(c) - g_2''(c)}{h_i^2}$	$\dfrac{f'''(c) - g_3'''(c)}{h_i^2}$	$\dfrac{f''''(c) - g_4''''(c)}{h_i^2}$
0	-0.02607616520	-0.02410048667	-0.1198142306	-0.1010060642
1	-0.02341015829	-0.02293433862	-0.1058148872	-0.09279946308
2	-0.02276786702	-0.02265002086	-0.1025639855	-0.09086028949
3	-0.02260877867	-0.02257938608	-0.1017661144	-0.09038226688
4	-0.02256909892	-0.02256175508	-0.1015675644	-0.09026318046
5	-0.02255918475	-0.02255734906	-0.1015179841	-0.09023343499

29. (a) $g_1'(c) = -\frac{1}{2}\frac{3f(c) - 4f(c+h) + f(c+2h)}{h}$; $g_2''(c) = \frac{2f(c) - 5f(c+h) + 4f(c+2h) - f(c+3h)}{h^2}$; $g_3'''(c) = -\frac{1}{2}\frac{5f(c) - 18f(c+h) + 24f(c+2h) - 14f(c+3h) + 3f(c+4h)}{h^3}$;

$g_4''''(c) = \frac{3f(c) - 14f(c+h) + 26f(c+2h) - 24f(c+3h) + 11f(c+4h) - 2f(c+5h)}{h^4}$ **(b)** $\{-\frac{3}{2}\frac{1}{h}, \frac{2}{h}, -\frac{1}{2}\frac{1}{h}\}$; $\{\frac{2}{h^2}, -\frac{5}{h^2}, \frac{4}{h^2}, -\frac{1}{h^2}\}$; $\{-\frac{5}{2}\frac{1}{h^3}, \frac{9}{h^3}, -\frac{12}{h^3}, \frac{7}{h^3}, -\frac{3}{2}\frac{1}{h^3}\}$;

$\{\frac{3}{h^4}, -\frac{14}{h^4}, \frac{26}{h^4}, -\frac{24}{h^4}, \frac{11}{h^4}, -\frac{2}{h^4}\}$

31. (a) $\varepsilon_1 = -0.2513391738$ **(b)** $\varepsilon_2 = -0.3635833218$ **(c)** $\varepsilon_3 = 0.1132631118$ **(d)** $\varepsilon_4 = -0.0371483540$ **(e)** $\varepsilon_5 = -0.2379679108$

(f) $\varepsilon_6 = -0.1265706096x$ **(g)** $\{w_k\}_{k=0}^4 = \{-0.4630312868, 1.066769489, -0.872359964, 0.1763632267, 0.09225853610\}$;

$\sum_{k=0}^4 w_k f(x_k) = 0.1358443804 = g_1''(4.6)$

47. For this problem with $\xi = 1.488732157$, $[(n+1)c^n - \sum_{k=0}^n L_k'(c)x_k^{n+1}]\frac{f^{(n+1)}(\xi)}{(n+1)!} = f'(c) - p'(c) = 0.369289462$.

Exercises 42.2, Page 1070

2. (a) $A_{8,8} = 0.99999999999497311142$; error: $0.502688858 \times 10^{-11}$ **(b)** $A_{8,8} = -1.9999999999954249424$; error: $0.45750576 \times 10^{-11}$

(c) $A_{8,8} = 2.9999999999958641500$; error: $0.41358500 \times 10^{-11}$ **(d)** $A_{8,8} = -4.0000000000134955357$; error: $0.134955357 \times 10^{-10}$

12. (a) $A_{6,6} = 0.99999999999999179883$; error: 0.820117×10^{-14} **(b)** $A_{5,5} = 2.0000000000008615161$; error: $0.8615161 \times 10^{-12}$

(c) $A_{5,5} = 3.0000000000000161233$; error: 0.161233×10^{-13} **(d)** $A_{5,5} = 4.0000000000005467459$; error: $0.5467459 \times 10^{-12}$

22. (a) and (b) $f_1'(1) \approx 3.12156685$; error: 0.010455744

27. (a) $f'(c) = -\frac{21f(c) + 32f(c+\frac{1}{4}h) - 12f(c+\frac{1}{2}h) + f(c+h)}{3h} + O(h^3)$ **(b)**

k	h_k	$\dfrac{\lvert y_{approx}' - f'(1)\rvert}{h_k^3}$
0	1.0	0.02033267021
1	0.5000000000	0.01691523117
2	0.2500000000	0.01546367638
3	0.1250000000	0.01479357243
4	0.06250000000	0.01447149068
5	0.03125000000	0.01431358104
6	0.01562500000	0.01423539553
7	0.007812500000	0.01419649345
8	0.003906250000	0.01417708987
9	0.001953125000	0.01416739993
10	0.0009765625000	0.01416255790

CHAPTER 43

Exercises 43.1–Part A, Page 1078

2. $\lambda = 0.3740172474$; actual error: 0.0856804467; error bound: $< .1$

3. $\lambda = 0.4581643459$; actual error: 0.0015333482; error bound: < 0.003333333333

5. $\lambda = 0.4596996714$; actual error: 0.19773×10^{-5}; error bound: $< .8888888889 \times 10^{-5}$

Exercises 43.1–Part B, Page 1078

3. Using f_1: **(a)** $n = 46$ **(b)** $\lambda = 0.9595723180$ **(c)** error: 0.0010105135 **(d)** $n = 15$

5. Using f_1: **(a)** $n = 10$ **(b)** $\lambda = 0.9595723180$ **(c)** error: 0.0009965107 **(d)** $n = 6$

6. Using f_1: **(a)**

$$\begin{bmatrix} k & h & \dfrac{|L(h) - \lambda|}{h} \\ 0 & 1.0 & 0.09839854970 \\ 1 & 0.5000000000 & 0.05911647467 \\ 2 & 0.2500000000 & 0.03817446092 \\ 3 & 0.1250000000 & 0.02753411242 \\ 4 & 0.06250000000 & 0.02219255306 \\ 5 & 0.03125000000 & 0.01951909339 \\ 6 & 0.01562500000 & 0.01818202834 \\ 7 & 0.007812500000 & 0.01751345391 \\ 8 & 0.003906250000 & 0.01717916145 \\ 9 & 0.001953125000 & 0.01701201457 \end{bmatrix}$$

(b)

$$\begin{bmatrix} k & h & \dfrac{|T(h) - \lambda|}{h^2} \\ 0 & 1.0 & 0.08155368220 \\ 1 & 0.5000000000 & 0.08454321434 \\ 2 & 0.2500000000 & 0.08531837368 \\ 3 & 0.1250000000 & 0.08551395940 \\ 4 & 0.06250000000 & 0.08556296904 \\ 5 & 0.03125000000 & 0.08557522854 \\ 6 & 0.01562500000 & 0.08557829385 \end{bmatrix}$$

(c)

$$\begin{bmatrix} k & h & \dfrac{|M(h) - \lambda|}{h^2} \\ 0 & 1.0 & 0.03928207503 \\ 1 & 0.5000000000 & 0.04188402750 \\ 2 & 0.2500000000 & 0.04256139399 \\ 3 & 0.1250000000 & 0.04273247488 \\ 4 & 0.06250000000 & 0.04277535477 \\ 5 & 0.03125000000 & 0.04278608161 \\ 6 & 0.01562500000 & 0.04278876375 \end{bmatrix}$$

(d)

$$\begin{bmatrix} k & h & \dfrac{|S(h) - \lambda|}{h^4} \\ 1 & 0.5000000000 & 0.01594417140 \\ 2 & 0.2500000000 & 0.01653673264 \\ 3 & 0.1250000000 & 0.01668998098 \\ 4 & 0.06250000000 & 0.01672862265 \\ 5 & 0.03125000000 & 0.01673830383 \\ 6 & 0.01562500000 & 0.01674072543 \end{bmatrix}$$

7. **(a)** $\int_1^{1.8} f_1(x)\,dx \approx 2.415499275$ **(b)** $\int_1^{1.8} f_1(x)\,dx \approx 2.476105335$ **(c)** $\int_1^{1.8} f_1(x)\,dx \approx 2.482211868$ **(d)** $\int_1^{1.8} f_1(x)\,dx \approx 2.478140846$
(e) $\int_1^{1.8} f_1(x)\,dx \approx 2.476105336$ **(f)** $\int_1^{1.8} f_1(x)\,dx \approx 2.47814082$ **(g)** $\int_1^{1.8} f_1(x)\,dx \approx 2.477979391$ **(h)** $\int_1^{1.8} f_1(x)\,dx \approx 2.478139643$
(i) In this case, integration of the interpolation polynomial gives the best result.

12. **(a)** $\int_1^{7.4} e^{-\frac{x}{5}} \cos x\,dx = -0.3998575930$ **(b)** $\int_1^{7.4} e^{-\frac{x}{5}} \cos x\,dx \approx -0.9327099885$ **(c)** $\int_1^{7.4} e^{-\frac{x}{5}} \cos x\,dx \approx -0.8317929509$
(d) $\int_1^{7.4} e^{-\frac{x}{5}} \cos x\,dx \approx .4470812497$ **(e)** $\int_1^{7.4} e^{-\frac{x}{5}} \cos x\,dx \approx -0.8317929515$ **(f)** $\int_1^{7.4} e^{-\frac{x}{5}} \cos x\,dx \approx -1.723499224$
(g) $\int_1^{7.4} e^{-\frac{x}{5}} \cos x\,dx \approx -0.6948618837$ **(h)** $\int_1^{7.4} e^{-\frac{x}{5}} \cos x\,dx \approx -0.3695444806$ **(i)** $\int_1^{7.4} e^{-\frac{x}{5}} \cos x\,dx \approx -0.001805570816$
(j) $\int_1^{7.4} e^{-\frac{x}{5}} \cos x\,dx \approx -0.2758313418$ **(k)** $\int_1^{7.4} e^{-\frac{x}{5}} \cos x\,dx \approx -0.4397406968$ **(l)** $\int_1^{7.4} e^{-\frac{x}{5}} \cos x\,dx \approx -0.4311833903$ **(m)** Using the quartic spline gives the best result.

22. **(b)** On $[0, 5]$, $0 \le f(x) \le .367879441$; $\dfrac{\int_0^5 f(x)g(x)\,dx}{\int_0^5 g(x)\,dx} = 2.162178832 > f(\xi), \forall \xi \in [0, 5]$

Exercises 43.2–Part A, Page 1083

3. $T[1] = 0.4207354924$ 4. $T\left[\frac{1}{2}\right] = 0.4500805155$; Richardson extrapolation yields: 0.4598621900 5. $S\left[\frac{1}{2}\right] = 0.4598621899$

Exercises 43.2–Part B, Page 1084

2. **(a)** 0.9084218054 **(b)** $|\text{error}| \approx 0.2 \times 10^{-9}$ **(c)** 33 function evaluations **(d)** 71 function evaluations **(e)** 47 function evaluations.

Exercises 43.3, Page 1092

2. Using f_1: **(a)** $\Lambda_2 = 0.5387787308$ **(b)** $|\Lambda_2 - \Lambda| \le 0.008175098693$ **(c)** $|\Lambda_2 - \Lambda| = 0.0021681219$

5. Using f_1: **(a)** $\Lambda_4 = 0.5366106099$ **(b)** $|\Lambda_4 - \Lambda| \le 0.2674983948 \times 10^{-8}$ **(c)** $|\Lambda_4 - \Lambda| = 0.10 \times 10^{-8}$

6. $c_0 = 0.1713244915$, $c_1 = 0.3607615743$, $c_2 = 0.4679139382$, $c_3 = 0.4679139336$, $c_4 = 0.3607615737$, $c_5 = 0.1713244923$;
$\xi_0 = -0.9324695142$, $\xi_1 = -0.6612093865$, $\xi_2 = -0.2386191861$, $\xi_3 = 0.2386191861$, $\xi_4 = 0.6612093865$, $\xi_5 = 0.9324695142$

12. **(a)**

$$\begin{bmatrix} n & \Lambda_n & |\Lambda_n - \Lambda| \\ 2 & 2.478073693 & 0.000066311 \\ 3 & 2.478138895 & 0.1109 \times 10^{-5} \\ 4 & 2.478140069 & 0.65 \times 10^{-7} \end{bmatrix}$$

(b) $S_{10} = 2.478132120$, $|S_{10} - \Lambda| = 0.7884 \times 10^{-5}$

17. (a) $c_0 = \frac{1}{2}\pi$, $c_1 = \frac{1}{2}\pi$, $\xi_0 = -\frac{1}{2}\sqrt{2}$, $\xi_1 = \frac{1}{2}\sqrt{2}$ **(b)** $c_0 = \frac{1}{2}\pi$, $c_1 = \frac{1}{2}\pi$, $\xi_0 = -\frac{1}{2}\sqrt{2}$, $\xi_1 = \frac{1}{2}\sqrt{2}$

21. (a) $p_0 = 1$, $p_1 = x - \frac{12}{5}$, $p_2 = x^2 + \frac{80}{21} - \frac{40}{9}x$ **(b)** $x_0 = \frac{20}{9} - \frac{8}{63}\sqrt{70}$, $x_1 = \frac{20}{9} + \frac{8}{63}\sqrt{70}$ **(c)**
$L_0 = -\frac{9}{160}\sqrt{70}x + \frac{1}{8}\sqrt{70} + \frac{1}{2}$, $L_1 = \frac{9}{160}\sqrt{70}x - \frac{1}{8}\sqrt{70} + \frac{1}{2}$ **(d)** $c_0 = -\frac{4}{75}\sqrt{70} + \frac{8}{3}$, $c_1 = \frac{4}{75}\sqrt{70} + \frac{8}{3}$

23. (a) $\Lambda_{approx} = 281.2092536$ **(b)** $|\Lambda_{approx} - \Lambda| = 16.4604382$

Exercises 43.4, Page 1097

2. (a) $\Lambda_{approx} = -13.65046005$ **(b)** 225 function evaluations   **(c)** $|S_{224} - \Lambda| = 0.312 \times 10^{-5}$ **(d)** 301 function evaluations
(e) 257 function evaluations.

12. (a) $\Lambda_{approx} = 1.157142517$; 17 function evaluations   **(b)** $\Lambda_{approx} = 1.151738342$; 269 function evaluations.

Exercises 43.5, Page 1100

2. $\Lambda_{approx} = 5.149952789$; $|\Lambda_{approx} - \Lambda| = 0.23 \times 10^{-7}$

CHAPTER 44

Exercises 44.1, Page 1104

2. (a) $\{-9a + b = -\frac{58}{5}, -3a + b = -\frac{88}{5}, 2a + b = \frac{8}{5}, 4a + b = 4, 10a + b = \frac{224}{5}, 17a + b = \frac{98}{5}\}$; The first two equations imply
$a = -1$, $b = -\frac{103}{5}$, these values do not satisfy the third equation
(b) $S(a, b) = \frac{71352}{25} - \frac{9576}{5}a + 42ab + 499a^2 + 6b^2 - \frac{408}{5}b$

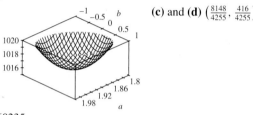

(c) and **(d)** $\left(\frac{8148}{4255}, \frac{416}{4255}\right)$

(e)

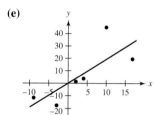

(f) $S_{min} = \frac{2,1623,064}{21,275} \approx 1016.360235$

Exercises 44.2, Page 1109

3. (a) $S_{min} = \frac{1,704,438,3651055}{17,006,139} \approx 0.1002248873 \times 10^7$; $g(x) = \frac{13,197,719,788}{17,006,139} - \frac{82,776,901}{68,024,556}x + \frac{173,922,145}{68,024,556}x^2$

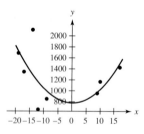

(b) $A = \begin{bmatrix} 1 & -19 & 361 \\ 1 & -17 & 289 \\ 1 & -14 & 196 \\ 1 & -12 & 144 \\ 1 & -9 & 81 \\ 1 & 9 & 81 \\ 1 & 10 & 100 \\ 1 & 17 & 289 \end{bmatrix}$; $\mathbf{y} = [1684, 1343, 2110, 659, 850, 951, 1169, 1425]^{\mathsf{T}}$; $\mathrm{rank}([A, \mathbf{y}]) = 4$

(c) $M = \begin{bmatrix} 8 & -35 & 1541 \\ -35 & 1541 & -10{,}331 \\ 1541 & -10{,}331 & 379{,}637 \end{bmatrix}$; $\mathbf{b} = [10{,}191, -55{,}451, 2{,}179{,}113]^{\mathrm{T}}$; $\mathrm{cond}(M) = \frac{437{,}671{,}652{,}186}{1{,}889{,}571} \approx 231{,}624.8779$, using the infinity

norm; $\hat{\mathbf{u}} = \left[\frac{13{,}197{,}719{,}788}{17{,}006{,}139}, -\frac{82{,}776{,}901}{68{,}024{,}556}, \frac{173{,}922{,}145}{68{,}024{,}556} \right]^{\mathrm{T}}$ **(d)** $\mathbf{y}_A = [\frac{9{,}762{,}461{,}218}{5{,}668{,}713}, \frac{52{,}230{,}793{,}187}{34{,}012{,}278}, \frac{44{,}019{,}248{,}093}{34{,}012{,}278}, \frac{19{,}707{,}247{,}711}{17{,}006{,}139}, \frac{33{,}811{,}782{,}503}{34{,}012{,}278},$

$\frac{16{,}533{,}395{,}197}{17{,}006{,}139}, \frac{34{,}677{,}662{,}321}{34{,}012{,}278}, \frac{8{,}470{,}597{,}645}{5{,}668{,}713}]^{\mathrm{T}}$ **(e)** $\Phi_1 = [1, 1, 1, 1, 1, 1, 1, 1]^{\mathrm{T}}$, $\Phi_2 = [-19, -17, -14, -12, -9, 9, 10, 17]^{\mathrm{T}}$,

$\Phi_3 = [361, 289, 196, 144, 81, 81, 100, 289]^{\mathrm{T}}$ **(f)** $\|A\hat{\mathbf{u}} - \mathbf{y}\|_2^2 = \frac{17{,}044{,}383{,}651{,}055}{17{,}006{,}139} = S_{\min}$

10. (a) $S_{\min} = \frac{53{,}257{,}289{,}786{,}210{,}631{,}643}{1{,}808{,}024{,}197{,}236} \approx 0.2945607137 \times 10^8$;

$g(x) = \frac{275{,}292{,}997{,}297{,}315}{452{,}006{,}049{,}309} + \frac{319{,}009{,}937{,}263{,}933}{1{,}808{,}024{,}197{,}236} x + \frac{9{,}830{,}956{,}137{,}955}{3{,}616{,}048{,}394{,}472} x^2 + \frac{1{,}865{,}959{,}211{,}801}{1{,}205{,}349{,}464{,}824} x^3$

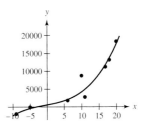

(b) $A = \begin{bmatrix} 1 & -9 & 81 & -729 \\ 1 & -5 & 25 & -125 \\ 1 & 6 & 36 & 216 \\ 1 & 7 & 49 & 343 \\ 1 & 10 & 100 & 1000 \\ 1 & 11 & 121 & 1331 \\ 1 & 17 & 289 & 4913 \\ 1 & 18 & 324 & 5832 \\ 1 & 20 & 400 & 8000 \end{bmatrix}$; $\mathbf{y} = [-1972, -156, 1786, 723, 8742, 2932, 11143, 13267, 18294]^{\mathrm{T}}$; $\mathrm{rank}([A, \mathbf{y}]) = 5$

(c) $M = \begin{bmatrix} 9 & 75 & 1425 & 20{,}781 \\ 75 & 1425 & 20{,}781 & 384{,}021 \\ 1425 & 20{,}781 & 384{,}021 & 6{,}732{,}885 \\ 20{,}781 & 384{,}021 & 6{,}732{,}885 & 125{,}632{,}725 \end{bmatrix}$; $\mathbf{b} = [54{,}759, 948{,}094, 16{,}001{,}498, 293{,}206{,}048]^{\mathrm{T}}$;

$\mathrm{cond}(M) = \frac{19{,}526{,}457{,}629{,}568{,}221{,}160}{150{,}668{,}683{,}103} \approx 0.1295986480 \times 10^9$, using the infinity norm;

$\hat{\mathbf{u}} = \left[\frac{275{,}292{,}997{,}297{,}315}{452{,}006{,}049{,}309}, \frac{319{,}009{,}937{,}263{,}933}{1{,}808{,}024{,}197{,}236}, \frac{9{,}830{,}956{,}137{,}955}{3{,}616{,}048{,}394{,}472}, \frac{1{,}865{,}959{,}211{,}801}{1{,}205{,}349{,}464{,}824} \right]^{\mathrm{T}}$ **(d)** $\mathbf{y}_A = [-\frac{3{,}412{,}190{,}120{,}703{,}353}{1{,}808{,}024{,}197{,}236}, \frac{-720{,}858{,}097{,}618{,}655}{1{,}808{,}024{,}197{,}236}, \frac{949{,}189{,}901{,}969{,}893}{452{,}006{,}049{,}309},$

$\frac{4{,}535{,}135{,}989{,}888{,}303}{1{,}808{,}024{,}197{,}236}, \frac{1{,}895{,}439{,}496{,}606{,}960}{452{,}006{,}049{,}309}, \frac{8{,}930{,}441{,}711{,}799{,}497}{1{,}808{,}024{,}197{,}236}, \frac{5{,}424{,}025{,}123{,}994{,}522}{452{,}006{,}049{,}309}, \frac{6{,}189{,}844{,}234{,}780{,}978}{452{,}006{,}049{,}309}, \frac{7{,}959{,}768{,}125{,}917{,}730}{452{,}006{,}049{,}309}]^{\mathrm{T}}$; $A\hat{\mathbf{u}} = \mathbf{y}_A$

(e) $\Phi_1 = [1, 1, 1, 1, 1, 1, 1, 1, 1]^{\mathrm{T}}$, $\Phi_2 = [-9, -5, 6, 7, 10, 11, 17, 18, 20]^{\mathrm{T}}$,

$\Phi_3 = [81, 25, 36, 49, 100, 121, 289, 324, 400]^{\mathrm{T}}$, $\Phi_4 = [-729, -125, 216, 343, 1000, 1331, 4913, 5832, 8000]^{\mathrm{T}}$

(f) $\|A\hat{\mathbf{u}} - \mathbf{y}\|_2^2 = \frac{53{,}257{,}289{,}786{,}210{,}631{,}643}{1{,}808{,}024{,}197{,}236} = S_{\min}$

22. (a) $S_{\min} = 0.02372863$; $g(x) = 2.462761245 - 3.181049767e^{-x} + 6.685919601e^{-2x} - 3.467315656e^{-3x}$

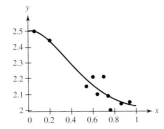

$$
\textbf{(b) } A = \begin{bmatrix}
1 & 0.9417645336 & 0.8869204367 & 0.8352702114 \\
1 & 0.8187307531 & 0.6703200460 & 0.5488116361 \\
1 & 0.5827482524 & 0.3395955256 & 0.1978986991 \\
1 & 0.5488116361 & 0.3011942119 & 0.1652988882 \\
1 & 0.5272924240 & 0.2780373005 & 0.1466069621 \\
1 & 0.4965853038 & 0.2465969639 & 0.1224564283 \\
1 & 0.4771139155 & 0.2276376884 & 0.1086091088 \\
1 & 0.4676664270 & 0.2187118870 & 0.1022842067 \\
1 & 0.4231620823 & 0.1790661479 & 0.07577400402 \\
1 & 0.3906278354 & 0.1525901058 & 0.05960594271
\end{bmatrix}
$$

$; y = [2.5, 2.44, 2.15, 2.21, 2.1, 2.21, 2.09, 2.0, 2.04, 2.05]^T;$

$\text{rank}([A, y]) = 5$ **(c)** $M = \begin{bmatrix}
10 & 5.674503163 & 3.500670315 & 2.362616088 \\
5.674503163 & 3.500670314 & 2.362616088 & 1.735116933 \\
3.500670315 & 2.362616088 & 1.735116934 & 1.366407964 \\
2.362616088 & 1.735116933 & 1.366407964 & 1.133399899
\end{bmatrix};$

$b = [21.79, 12.61920309, 7.968800415, 5.504904595]^T;$ $\text{cond}(M) = 0.1333848190 \times 10^7$, using the infinity norm. In contrast with problems 3 and 10, where were able to use exact arithmetic, this problem requires floating point computation. Therefore the large value of $\text{cond}(M)$ really does have an impact. Just look at the difference between the values of $A\hat{u}$ and y_A, computed in (d);
$\hat{u} = [2.462762777, -3.181057471, 6.685931975, -3.467321974]^T$ **(d)** $y_A = [2.497456311, 2.434357891, 2.191201035, 2.155523924,$
$2.134049930, 2.105362444, 2.088591386, 2.080926456, 2.049496667, 2.032140652]^T;$ $A\hat{u} = [2.500694620, 2.437140782, 2.193341170,$
$2.157580967, 2.136020209, 2.107231047, 2.090423343, 2.082729517, 2.051151089, 2.033687255]^T$ **(e)** $\Phi_1 = [1, 1, 1, 1, 1, 1, 1, 1, 1, 1]^T,$
$\Phi_2 = [0.9417645336, 0.8187307531, 0.5827482524, 0.5488116361, 0.5272924240, 0.4965853038, 0.4771139155, 0.4676664270,$
$0.4231620823, 0.3906278354]^T,$ $\Phi_3 = [0.8869204367, 0.6703200460, 0.3395955256, 0.3011942119, 0.2780373005, 0.2465969639,$
$0.2276376884, 0.2187118870, 0.1790661479, 0.1525901058]^T,$ $\Phi_4 = [0.8352702114, 0.5488116361, 0.1978986991, 0.1652988882,$
$0.1466069621, 0.1224564283, 0.1086091088, 0.1022842067, 0.07577400402, 0.05960594271]^T$ **(f)** $\|A\hat{u} - y\|_2^2 = 0.02372858747$

27. (a) $S_{\min} = 2.5730682$; $g(x) = \frac{6.828037253}{x^2+1} - 25.00205981\frac{x}{x^2+2} + 35.23458310\frac{x^2}{x^2+3} - 8.90058699\frac{x^3}{x^2+4}$

$$
\textbf{(b) } A = \begin{bmatrix}
0.9455370651 & 0.1166407465 & 0.01883830455 & 0.003406940063 \\
0.9174311927 & 0.1435406699 & 0.02912621359 & 0.006601466993 \\
0.8963786303 & 0.1607109094 & 0.03710360765 & 0.009550004860 \\
0.8378016086 & 0.2005835157 & 0.06062124248 & 0.02031285769 \\
0.8127438231 & 0.2152080344 & 0.07132243685 & 0.02614220877 \\
0.6711409396 & 0.2811244980 & 0.1404011461 & 0.07639198218 \\
0.6217358866 & 0.2990338905 & 0.1686065846 & 0.1029754362 \\
0.4708984743 & 0.3393520297 & 0.2724803570 & 0.2324568663 \\
0.4179903026 & 0.3478363401 & 0.3170020945 & 0.3046940138
\end{bmatrix}
$$

$; y = [4.07, 2.86, 4.41, 3.01, 2.07, 1.44, 1.79, 2.78, 2.31]^T;$

$\text{rank}([A, y]) = 5$

(c) $M = \begin{bmatrix}
5.135125981 & 1.408779685 & 0.6464200448 & 0.4082193012 \\
1.408779685 & 0.5511873996 & 0.3324707433 & 0.2497173577 \\
0.6464200448 & 0.3324707433 & 0.2342182558 & 0.1917231420 \\
0.4082193012 & 0.2497173577 & 0.1917231420 & 0.1645567286
\end{bmatrix};$

$b = [18.98338693, 5.330216808, 2.647460832, 1.834521782]^T;$ $\text{cond}(M) = 200905.7330$, using the infinity norm;
$\hat{u} = [6.828037253, -25.00205981, 35.23458310, -8.90058699]^T$
(d) $y_A = [4.173330465, 3.642932023, 3.324734710, 2.660721060, 2.449155880,$
$1.820972392, 1.793072047, 2.262645038, 2.615037475]^T;$ $A\hat{u} = [4.173339426, 3.642935009, 3.324732409, 2.660707387, 2.449136278,$
$1.820926192, 1.793013468, 2.262541804, 2.614909444]^T$ **(e)** $\Phi_1 = [0.9455370651, 0.9174311927, 0.8963786303, 0.8378016086,$
$0.8127438231, 0.6711409396, 0.6217358866, 0.4708984743, 0.4179903026]^T,$ $\Phi_2 = [0.1166407465, 0.1435406699, 0.1607109094,$

0.2005835157, 0.2152080344, 0.2811244980, 0.2990338905, 0.3393520297, 0.3478363401]^{T}, $\Phi_3 = [0.01883830455, 0.02912621359,$
0.03710360765, 0.06062124248, 0.07132243685, 0.1404011461, 0.1686065846, 0.2724803570, 0.3170020945]^{T}, $\Phi_4 = [0.003406940063,$
0.006601466993, 0.009550004860, 0.02031285769, 0.02614220877, 0.07639198218, 0.1029754362,
0.2324568663, 0.3046940138]^{T} **(f)** $\|A\hat{\mathbf{u}} - \mathbf{y}\|_2^2 = 2.573068254$

32. (a) $S_{\min} = 0.172225760$; $g(x) = -0.073033567 J_0(x) + 5.447793840 J_1(x) - 6.476566945 J_2(x) + 4.710220186 J_3(x)$

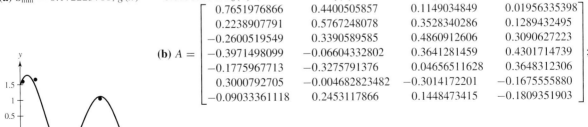

(b) $A = \begin{bmatrix} 0.7651976866 & 0.4400505857 & 0.1149034849 & 0.01956335398 \\ 0.2238907791 & 0.5767248078 & 0.3528340286 & 0.1289432495 \\ -0.2600519549 & 0.3390589585 & 0.4860912606 & 0.3090627223 \\ -0.3971498099 & -0.06604332802 & 0.3641281459 & 0.4301714739 \\ -0.1775967713 & -0.3275791376 & 0.04656511628 & 0.3648312306 \\ 0.3000792705 & -0.004682823482 & -0.3014172201 & -0.1675555880 \\ -0.09033361118 & 0.2453117866 & 0.1448473415 & -0.1809351903 \end{bmatrix};$

$\mathbf{y} = [1.609, 1.675, -0.01184, -0.7530, -0.1353, 1.041, -0.3137]^{\mathrm{T}}$; $\mathrm{rank}([A, \mathbf{y}]) = 5$

(c) $M = \begin{bmatrix} 0.9907579145 & 0.4385170863 & -0.2159056973 & -0.3061041245 \\ 0.4385170863 & 0.8130866132 & 0.4165073282 & -0.00375783635 \\ -0.2159056973 & 0.4165073282 & 0.6205700857 & 0.3958982602 \\ -0.3061041245 & -0.00375783635 & 0.3958982602 & 0.4914905947 \end{bmatrix};$

$\mathbf{b} = [2.273101973, 1.682263945, 0.1304186934, -0.2471487063]^{\mathrm{T}}$; $\mathrm{cond}(M) = 279.4615013$, using the infinity norm;
$\hat{\mathbf{u}} = [-0.073033580, 5.447793865, -6.476566979, 4.710220205]^{\mathrm{T}}$ **(d)** $\mathbf{y}_A = [1.689387371, 1.447724181, 0.1736666953,$
$-0.6628831287, -0.3547597547, 1.115498175, -0.4473526675]^{\mathrm{T}}$; $A\hat{\mathbf{u}} = [1.689387344, 1.447724205, 0.173666711,$
$-0.662883123, -0.354759749, 1.115498179, -0.4473526651]^{\mathrm{T}}$ **(e)** $\Phi_1 = [0.7651976866, 0.2238907791, -0.2600519549,$
$-0.3971498099, -0.1775967713, 0.3000792705, -0.09033361118]^{\mathrm{T}}$, $\Phi_2 = [0.4400505857, 0.5767248078, 0.3390589585,$
$-0.06604332802, -0.3275791376, -0.004682823482, 0.2453117866]^{\mathrm{T}}$, $\Phi_3 = [0.1149034849, 0.3528340286,$
$0.4860912606, 0.3641281459, 0.04656511628, -0.3014172201, 0.1448473415]^{\mathrm{T}}$, $\Phi_4 = [0.01956335398, 0.1289432495, 0.3090627223,$
$0.4301714739, 0.3648312306, -0.1675555880, -0.1809351903]^{\mathrm{T}}$ **(f)** $\|A\hat{\mathbf{u}} - \mathbf{y}\|_2^2 = 0.1722257984$

37. (a) $S_{\min} = 89.1933927$; $g(x) = 3.951066318 \sin x + 2.363710979 \sin(2x) + 5.391879107 \sin(3x) + 1.344585539 \sin(4x) +$
$0.887191007 \sin(5x)$

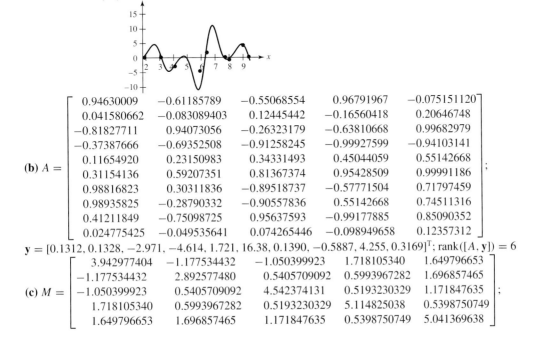

(b) $A = \begin{bmatrix} 0.94630009 & -0.61185789 & -0.55068554 & 0.96791967 & -0.075151120 \\ 0.041580662 & -0.083089403 & 0.12445442 & -0.16560418 & 0.20646748 \\ -0.81827711 & 0.94073056 & -0.26323179 & -0.63810668 & 0.99682979 \\ -0.37387666 & -0.69352508 & -0.91258245 & -0.99927599 & -0.94103141 \\ 0.11654920 & 0.23150983 & 0.34331493 & 0.45044059 & 0.55142668 \\ 0.31154136 & 0.59207351 & 0.81367374 & 0.95428509 & 0.99991186 \\ 0.98816823 & 0.30311836 & -0.89518737 & -0.57771504 & 0.71797459 \\ 0.98935825 & -0.28790332 & -0.90557836 & 0.55142668 & 0.74511316 \\ 0.41211849 & -0.75098725 & 0.95637593 & -0.99177885 & 0.85090352 \\ 0.024775425 & -0.049535641 & 0.074265446 & -0.098949658 & 0.12357312 \end{bmatrix};$

$\mathbf{y} = [0.1312, 0.1328, -2.971, -4.614, 1.721, 16.38, 0.1390, -0.5887, 4.255, 0.3169]^{\mathrm{T}}$; $\mathrm{rank}([A, \mathbf{y}]) = 6$

(c) $M = \begin{bmatrix} 3.942977404 & -1.177534432 & -1.050399923 & 1.718105340 & 1.649796653 \\ -1.177534432 & 2.892577480 & 0.5405709092 & 0.5993967282 & 1.696857465 \\ -1.050399923 & 0.5405709092 & 4.542374131 & 0.5193230329 & 1.171847635 \\ 1.718105340 & 0.5993967282 & 0.5193230329 & 5.114825038 & 0.5398750749 \\ 1.649796653 & 1.696857465 & 1.171847635 & 0.5398750749 & 5.041369638 \end{bmatrix};$

$\mathbf{b} = [10.90580910, 7.410770360, 23.35741273, 18.36156787, 22.04636341]^T$; $\text{cond}(M) = 34.9841491$, using the infinity norm;
$\hat{\mathbf{u}} = [3.951066318, 2.363710979, 5.391879106, 1.344585539, 0.887191007]^T$ **(d)** $\mathbf{y}_A = [0.5581867085, 0.5994389393,$
$-2.402376578, -10.21552524, 3.953705244, 9.187872436, -0.3457463422, -0.2517712027, 4.431236504, 0.3578183993]^T$;
$A\hat{\mathbf{u}} = [0.5581867138, 0.5994389406, -2.402376571, -10.21552525, 3.953705246, 9.187872427, -0.3457463434,$
$-0.2517711984, 4.431236503, 0.3578183997]^T$ **(e)** $\Phi_1 = [0.9463000877, 0.04158066243, -0.8182771111, -0.3738766648,$
$0.1165492049, 0.3115413635, 0.9881682339, 0.9893582466, 0.4121184852, 0.02477542545]^T$, $\Phi_2 = [-0.6118578909, -0.08308940282,$
$0.9407305567, -0.6935250848, 0.2315098251, 0.5920735147, 0.3031183567, -0.2879033167, -0.7509872468, -0.04953564088]^T$,
$\Phi_3 = [-0.5506855426, 0.1244544235, -0.2632317914, -0.9125824498, 0.3433149288, 0.8136737375, -0.8951873678,$
$-0.9055783620, 0.9563759284, 0.07426544558]^T$, $\Phi_4 = [0.9679196720, -0.1656041754, -0.6381066823,$
$-0.9992759921, 0.4504405943, 0.9542850945, -0.5777150444, 0.5514266812, -0.9917788534, -0.09894965755]^T$,
$\Phi_5 = [-0.07515112046, 0.2064674819, 0.9968297943, -0.9410314083, 0.5514266812,$
$0.9999118601, 0.7179745928, 0.7451131605, 0.8509035245, 0.1235731227]^T$ **(f)** $\|A\hat{\mathbf{u}} - \mathbf{y}\|_2^2 = 89.19339271$

Exercises 44.3, Page 1116

2. (a) $p_0 = 1$, $p_1 = x + \frac{35}{8}$, $p_2 = x^2 + \frac{9571}{3701}x - \frac{671{,}032}{3701}$, $p_3 = x^3 + \frac{15{,}825{,}263}{3{,}779{,}142}x^2 - \frac{869{,}741{,}981}{3{,}779{,}142}x - \frac{986{,}586{,}476}{1{,}889{,}571}$
   (b) $g(x) = \frac{14{,}142{,}303{,}900{,}329{,}972}{17967906496287} + \frac{261{,}992{,}150{,}985{,}283}{71{,}871{,}625{,}985{,}148}x + \frac{177{,}399{,}787{,}474{,}817}{71{,}871{,}625{,}985{,}148}x^2 - \frac{126{,}533{,}962{,}426}{5{,}989{,}302{,}165{,}429}x^3$

(c) $M = \begin{bmatrix} 8 & 0 & 0 & 0 \\ 0 & \frac{11{,}103}{8} & 0 & 0 \\ 0 & 0 & \frac{272{,}098{,}224}{3701} & 0 \\ 0 & 0 & 0 & \frac{7{,}985{,}736{,}220{,}572}{629{,}857} \end{bmatrix}$; $\text{cond}(M) = 0.1584831204 \times 10^7$, using the infinity norm

(f)

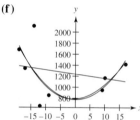

(h) $S_1 = 85{,}062.68468$, $S_2 = 56{,}5664.0009$,
$S_3 = 571{,}322.9385$; $R(g_1) = 0.05425217560$,
$R(g_2) = 0.3607751486$, $R(g_3) = 0.3643843655$

9. (a) $p_0 = 1.0$, $p_1 = x - 8.333333333$, $p_2 = x^2 - 11.13250000x - 65.56250000$, $p_3 = x^3 - 18.48637531x^2 - 57.75792690x + 1099.325481$,
$p_4 = x^4 - 21.34518066x^3 - 35.81046951x^2 + 1608.455004x - 1116.778547$
(b) $g(x) = 581.6316264 + 215.9267605x + 1.839598938x^2 + 1.024067238x^3 + 0.02454875791x^4$

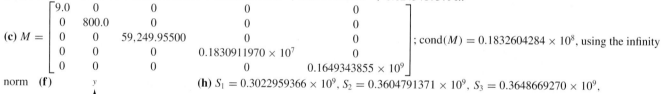

(c) $M = \begin{bmatrix} 9.0 & 0 & 0 & 0 & 0 \\ 0 & 800.0 & 0 & 0 & 0 \\ 0 & 0 & 59{,}249.95500 & 0 & 0 \\ 0 & 0 & 0 & 0.1830911970 \times 10^7 & 0 \\ 0 & 0 & 0 & 0 & 0.1649343855 \times 10^9 \end{bmatrix}$; $\text{cond}(M) = 0.1832604284 \times 10^8$, using the infinity

norm   **(f)**

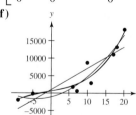

(h) $S_1 = 0.3022959366 \times 10^9$, $S_2 = 0.3604791371 \times 10^9$, $S_3 = 0.3648669270 \times 10^9$,
$S_4 = 0.3649663232 \times 10^9$; $R(g_1) = 0.7666201009$,
$R(g_2) = 0.9141722366$, $R(g_3) = 0.9252996370$, $R(g_4) = 0.9255517050$

21. (a) $p_0 = 1.0$, $p_1 = x - 0.6040000000$, $p_2 = x^2 - 0.9702624309x + 0.1517185083$,
$p_3 = x^3 - 1.589641494x^2 + 0.6883591996x - 0.05512226276$,
$p_4 = x^4 - 2.008119155x^3 + 1.311416846x^2 - 0.3022674903x + 0.01666919189$,
$p_5 = x^5 - 2.556531784x^4 + 2.344763204x^3 - 0.9134782460x^2 + 0.1356751963x - 0.005397054824$
(b) $g(x) = 2.301619111 + 5.311776274x - 38.83721438x^2 + 98.14035582x^3 - 106.3529509x^4 + 41.69593072x^5$
(c) M has diagonal elements $\{10.0, 0.6950400000, 0.04470465616, 0.001885276149, 0.0001280697219, 0.2092321138 \times 10^{-5}\}$;
$\text{cond}(M) = 0.4779381051 \times 10^7$, using the infinity norm

(f)

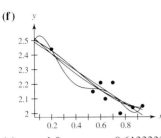

(g) $R(g_1) = 0.8861933609, R(g_2) = 0.9022653965,$
$R(g_3) = 0.9077519696, R(g_4) = 0.9077817541,$
$R(g_5) = 0.9219860725; R(g_{old}) = 0.9073432895$

26. (a) $p_0 = 1.0, p_1 = x - 0.6133333333, p_2 = x^2 - 1.407411936x + 0.3852570985,$
$p_3 = x^3 - 2.106253424x^2 + 1.307180183x - 0.2314309868,$
$p_4 = x^4 - 2.762682995x^3 + 2.638222075x^2 - 1.016929320x + 0.1320523070,$
$p_5 = x^5 - 3.583171673x^4 + 4.833531596x^3 - 3.031088660x^2 + 0.8730455673x - 0.009181385490$
(b) $g(x) = -5.692171218 + 93.20215800x - 312.3638260x^2 + 430.5308689x^3 - 257.5917505x^4 + 54.63279391x^5$ **(c)** M has diagonal
elements $\{9, 0.9160000000, 0.05645744689, 0.002911236122, 0.0002079804423, 0.8588300459 \times 10^{-5}\}$; cond$(M) = 0.1047937254 \times 10^7,$
using the infinity norm **(f)**

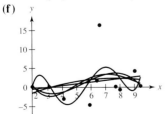

(g) $R(g_1) = 0.2895069893, R(g_2) = 0.6583444336,$
$R(g_3) = 0.6653860815, R(g_4) = 0.7663958207,$
$R(g_5) = 0.7696520847; R(g_{old}) = 0.6715832973$

31. (a) $p_0 = 1.0, p_1 = x - 4.428571430, p_2 = x^2 - 10.01197605x + 17.91017963,$
$p_3 = x^3 - 14.80916745x^2 + 60.30495793x - 60.96538818, p_4 = x^4 - 19.76605258x^3 + 129.5103099x^2 - 317.8199146x + 226.9400310,$
$p_5 = x^5 - 23.90130856x^4 + 207.7928358x^3 - 802.2101368x^2 + 1332.843380x - 727.8098744$ **(b)**
$g(x) = -6.435828256 + 15.18133567x - 9.126773495x^2 + 2.214854773x^3 - 0.2342331094x^4 + 0.008991397070x^5$ **(c)** M has diagonal
elements $\{7, 47.71428571, 268.8502995, 1129.706281\ 3903.317261, 5551.241286\}$; cond$(M) = 793.0344697,$ using the infinity norm.
(f)

(g) $R(g_1) = 0.2367150694, R(g_2) = 0.4235393459,$

$R(g_3) = 0.6700351677, R(g_4) = 0.9223170669, R(g_5) = 0.9999720954; R(g_{old}) = 0.9932521452$

36. (a) $p_0 = 1.0, p_1 = x - 6.210000000, p_2 = x^2 - 11.42161294x + 26.72721638,$
$p_3 = x^3 - 17.33819336x^2 + 89.62276748x - 129.0626011, p_4 = x^4 - 22.47877239x^3 + 174.9675033x^2 - 546.5605068x + 562.3309402,$
$p_5 = x^5 - 28.72129394x^4 + 311.7525247x^3 - 1577.435588x^2 + 3657.053809x - 3053.584691$ **(b)**
$g(x) = -162.9698314 + 203.4798526x - 91.51347118x^2 + 18.70200738x^3 - 1.763562908x^4 + 0.06233076738x^5$ **(c)** M has diagonal
elements $\{10, 56.36900000, 263.8824790, 998.4303302, 3533.644039, 7542.548394\}$; cond$(M) = 754.2548394,$ using the infinity norm
(f)

(g) $R(g_1) = 0.03865213281, R(g_2) = 0.05103911563,$
$R(g_3) = 0.1081762111, R(g_4) = 0.1166088092,$
$R(g_5) = 0.2150160939; R(g_{old}) = 0.7908463466$

Exercises 44.4, Page 1122

4. (a) $g_1(x) = \ln(1 + \sqrt{7.285076709 + 2.074543406x^2})$ **(b)** Solve the equation $y = \ln(1 + \sqrt{a + bx^2})$ for $a + bx^2$

(c) $g_2(x) = \ln(1 + \sqrt{8.046937627 + 2.007839527x^2})$ **(d)**

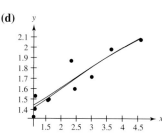

(e) $S(g_1) = 0.07887460831;$
$S(g_2) = 0.08155110762$

5. $g(x) = \sqrt{x + 4.537731834} + \sqrt{x + 2.346675573}$

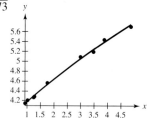

6. $g(x) = \sqrt{x + 10.17009985} + 2\sqrt{x + 3.617206074};$ Be careful, the sum of squares has more than one stationary point. Choose the one which results in the smallest residual

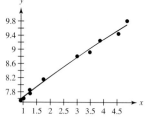

7. $g_1(x) = 3.074104719e^{-1.540386228x};$ $g_2(x) = 2.249689761e^{-1.680446565x}$

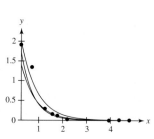

$S(g_1) = 0.1765379204;$
$S(g_2) = 0.7070043153$

10. $g_1(x) = \frac{2.194376417}{5.580351073+x};$ $g_2(x) = \frac{2.162895508}{5.490074715+x};$ $g_3(x) = \frac{2.433394416}{6.425524560+x}$

$S(g_1) = 0.001861354984;$
$S(g_2) = 0.001867671997;$
$S(g_3) = 0.001956725462$

CHAPTER 45

Exercises 45.1, Page 1130

3. Scaling the first component yields:

$$\begin{bmatrix} k & \lambda_k & \mathbf{v}_k \\ 50 & 0.8001055731 & [1.000000000, -2.001649366, 1.000000001] \\ 100 & 0.8000000120 & [0.9999999998, -2.000000188, 1.000000001] \\ 150 & 0.8000000002 & [1.000000000, -2.000000009, 1.000000004] \\ 200 & 0.8000000002 & [1.000000000, -2.000000009, 1.000000004] \end{bmatrix}$$; with errors:

$$
\begin{bmatrix}
k & E_{\lambda_k} & E_{\mathbf{v}_k} \\
50 & 0.0001055731 & 0.001649366000 \\
100 & 0.120 \times 10^{-7} & 0.1880027659 \times 10^{-6} \\
150 & 0.2 \times 10^{-9} & 0.9848857802 \times 10^{-8} \\
200 & 0.2 \times 10^{-9} & 0.9848857802 \times 10^{-8}
\end{bmatrix}
$$; $E_{\mathbf{v}_k}$ represents the 2-norm of the difference of the approximation and the exact

eigenvector, both normalized so that the first component of each equals one; Scaling the component with largest magnitude yields:

$$
\begin{bmatrix}
k & \lambda_k & \mathbf{v}_k \\
50 & 0.7999736327 & [0.9999999998, -2.001649361, \ 1.000000002]^{\mathrm{T}} \\
100 & 0.7999999972 & [1.000000000, \ -2.000000182, 1.000000003]^{\mathrm{T}} \\
150 & 0.8000000000 & [1.000000000, \ -2.000000002, 1.000000002]^{\mathrm{T}} \\
200 & 0.8000000000 & [1.000000000, \ -2.000000002, 1.000000002]^{\mathrm{T}}
\end{bmatrix}
$$; For reason of comparison, the printed vectors have again been

re-scaled, making their first component equal to one. Errors are given by: $$
\begin{bmatrix}
k & E_{\lambda_k} & E_{\mathbf{v}_k} \\
50 & 0.0000263673 & 0.001649361000 \\
100 & 0.28 \times 10^{-8} & 0.1820247236 \times 10^{-6} \\
150 & 0 & 0.2828427125 \times 10^{-8} \\
200 & 0 & 0.2828427125 \times 10^{-8}
\end{bmatrix}
$$; The second

method appears to be slightly faster than the first.

5. $\lambda = 0.6666666667$, $\mathbf{v} = [1.000000000, 0.5000000014, 1.000000003]^{\mathrm{T}}$

7. $\lambda = 0.7500000000$, $\mathbf{v} = [1.000000000, -1.062500000, -0.1718750002]^{\mathrm{T}}$; using a shift of $\frac{76}{100}$.

9. **(a)** $\{\lambda\} = \{10.95902794, -11.84940473 + 5.995958806i, -11.84940473 - 5.995958806i, 3.739781511\}$ **(b)**

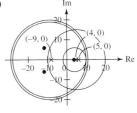

(c) The information provided by Gerschgorin's disk theorem is very crude.

13. $$
\begin{bmatrix}
k & \mu_k & \dfrac{|\mu_{k+1} - \lambda_1|}{|\mu_k - \lambda_1|} \\
25 & 0.9087174453 & 0.8332664404 \\
50 & 0.8090500620 & 0.9274811362 \\
75 & 0.8015897745 & 0.9355800601 \\
100 & 0.8003089803 & 0.9371175278
\end{bmatrix}
$$

15. $$
\begin{bmatrix}
k & \mu_k & \dfrac{|\mu_{k+1} - \lambda_1|}{|\mu_k - \lambda_1|} \\
25 & 0.8152690588 & 1.026822996 \\
50 & 0.8039290728 & 0.9361583523 \\
75 & 0.8007644761 & 0.9370536445 \\
100 & 0.8001513799 & 0.9374047532
\end{bmatrix}
$$

17. $$
\begin{bmatrix}
k & \mu_k & \dfrac{|\mu_{k+1} - \lambda_1|}{|\mu_k - \lambda_1|} \\
10 & 11.33267659 & 0.3634487291 \\
20 & 11.99998561 & 0.3402782776 \\
30 & 12.00000000 & 0.3402777778
\end{bmatrix}
$$

19. *Hint:* Eigenvectors of symmetric matrices, associated with distinct eigenvalues, are orthogonal.

20. $$
\begin{bmatrix}
k & \lambda_k \\
10 & -6.999588128 \\
20 & -6.999999914 \\
30 & -7.000000002
\end{bmatrix}
$$

Exercises 45.2–Part A, Page 1138

1. $H = \begin{bmatrix} \frac{5}{13} & -\frac{12}{13} \\ -\frac{12}{13} & -\frac{5}{13} \end{bmatrix}$ 3.

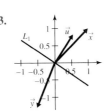

4. $\mathbf{u} = \left[\frac{2}{5}\sqrt{5}, \frac{1}{5}\sqrt{5}\right]^{\mathrm{T}}$; $\mathbf{y} = [-5, 0]^{\mathrm{T}}$

Exercises 45.2–Part B, Page 1138

1. $H = \begin{bmatrix} \frac{3}{7} & \frac{6}{7} & -\frac{2}{7} \\ \frac{6}{7} & -\frac{2}{7} & \frac{3}{7} \\ -\frac{2}{7} & \frac{3}{7} & \frac{6}{7} \end{bmatrix}$; $\mathbf{y} = [1, 1, 1]^{\mathrm{T}}$

3. $\mathbf{u} = [\frac{1}{78}\sqrt{195\sqrt{78}+3042}, -\frac{1}{53}\sqrt{10\sqrt{78}+156} + \frac{5}{2067}\sqrt{195\sqrt{78}+3042}, \frac{7}{106}\sqrt{10\sqrt{78}+156} - \frac{35}{4134}\sqrt{195\sqrt{78}+3042}]^{\mathrm{T}}$
$\approx [0.8849120051, -0.1279536300, 0.4478377046]^{\mathrm{T}}$; $\mathbf{y} = [-\sqrt{78}, 0, 0]^{\mathrm{T}}$

5. $H_1 = \begin{bmatrix} -\frac{5}{114}\sqrt{114} & -\frac{4}{57}\sqrt{114} & -\frac{2}{57}\sqrt{114} & \frac{1}{38}\sqrt{114} \\ -\frac{4}{57}\sqrt{114} & \frac{25}{89}+\frac{160}{5073}\sqrt{114} & \frac{80}{5073}\sqrt{114}-\frac{32}{89} & -\frac{20}{1691}\sqrt{114}+\frac{24}{89} \\ -\frac{2}{57}\sqrt{114} & \frac{80}{5073}\sqrt{114}-\frac{32}{89} & \frac{73}{89}+\frac{40}{5073}\sqrt{114} & -\frac{10}{1691}\sqrt{114}+\frac{12}{89} \\ \frac{1}{38}\sqrt{114} & -\frac{20}{1691}\sqrt{114}+\frac{24}{89} & -\frac{10}{1691}\sqrt{114}+\frac{12}{89} & \frac{80}{89}+\frac{15}{3382}\sqrt{114} \end{bmatrix}$; $\mathbf{y}_1 = [-\sqrt{114}, 0, 0, 0]^{\mathrm{T}}$;

$H_2 = \begin{bmatrix} 1 & 0 & 0 & 0 \\ 0 & -\frac{8}{89}\sqrt{89} & -\frac{4}{89}\sqrt{89} & \frac{3}{89}\sqrt{89} \\ 0 & -\frac{4}{89}\sqrt{89} & \frac{9}{25}+\frac{128}{2225}\sqrt{89} & -\frac{96}{2225}\sqrt{89}+\frac{12}{25} \\ 0 & \frac{3}{89}\sqrt{89} & -\frac{96}{2225}\sqrt{89}+\frac{12}{25} & \frac{16}{25}+\frac{72}{2225}\sqrt{89} \end{bmatrix}$; $\mathbf{y}_2 = [5, -\sqrt{89}, 0, 0]^{\mathrm{T}}$; $H_3 = \begin{bmatrix} 1 & 0 & 0 & 0 \\ 0 & 1 & 0 & 0 \\ 0 & 0 & -\frac{4}{5} & \frac{3}{5} \\ 0 & 0 & \frac{3}{5} & \frac{4}{5} \end{bmatrix}$; $\mathbf{y}_3 = [5, 8, -5, 0]^{\mathrm{T}}$

9. (a) $H_1 = \begin{bmatrix} -\frac{1}{51}\sqrt{51} & -\frac{5}{51}\sqrt{51} & \frac{5}{51}\sqrt{51} \\ -\frac{5}{51}\sqrt{51} & \frac{1}{102}\sqrt{51}+\frac{1}{2} & -\frac{1}{102}\sqrt{51}+\frac{1}{2} \\ \frac{5}{51}\sqrt{51} & -\frac{1}{102}\sqrt{51}+\frac{1}{2} & \frac{1}{102}\sqrt{51}+\frac{1}{2} \end{bmatrix}$ (b) $\|\mathbf{x}\| = \sqrt{51} = \|\mathbf{y}_1\| = |\alpha|$ (c) $H_2 = \begin{bmatrix} 1 & 0 & 0 \\ 0 & -\frac{1}{2}\sqrt{2} & \frac{1}{2}\sqrt{2} \\ 0 & \frac{1}{2}\sqrt{2} & \frac{1}{2}\sqrt{2} \end{bmatrix}$

(d) $\|\mathbf{x}\| = \sqrt{51} = \|\mathbf{y}_2\|$ (e)

12. (a) $\mathbf{u} = \left[\frac{15}{262}\sqrt{262}, \frac{1}{262}\sqrt{262}, \frac{3}{131}\sqrt{262}\right]^{\mathrm{T}}$ (b) $H = \begin{bmatrix} -\frac{94}{131} & -\frac{15}{131} & -\frac{90}{131} \\ -\frac{15}{131} & \frac{130}{131} & -\frac{6}{131} \\ -\frac{90}{131} & -\frac{6}{131} & \frac{95}{131} \end{bmatrix}$ (c) $P = \begin{bmatrix} \frac{37}{262} & -\frac{15}{262} & -\frac{45}{131} \\ -\frac{15}{262} & \frac{261}{262} & -\frac{3}{131} \\ -\frac{45}{131} & -\frac{3}{131} & \frac{113}{131} \end{bmatrix}$

(d) $\mathbf{y} = \left[\frac{101}{131}, -\frac{919}{131}, \frac{-143}{131}\right]^{\mathrm{T}}$; $\|\mathbf{x}\| = \sqrt{51} = \|\mathbf{y}\|$ (e) $M = \left[\frac{116}{131}, -\frac{918}{131}, \frac{-137}{131}\right]^{\mathrm{T}} = P\mathbf{x} = P\mathbf{y}$; Observe that $\mathbf{y}$ is the reflection of $\mathbf{x}$ in the plane V, through the origin and perpendicular to $\mathbf{u}$. Hence $\mathbf{x} + \mathbf{y}$ is in this plane and $\frac{\mathbf{x}+\mathbf{y}}{2}$ is exactly the projection of $\mathbf{x}$ (and of $\mathbf{y}$) on that plane.

(f) $\mathbf{x} - P\mathbf{x} = \left[\frac{15}{131}, \frac{1}{131}, \frac{6}{131}\right]^{\mathrm{T}}$; $\mathbf{y} - P\mathbf{y} = \left[-\frac{15}{131}, -\frac{1}{131}, -\frac{6}{131}\right]^{\mathrm{T}}$; Observe that $\mathbf{x} - P\mathbf{x}$ and $\mathbf{y} - P\mathbf{y}$ are the vectors pointing from the tips of $\mathbf{x}$ and $\mathbf{y}$ to the plane V mentioned in (e). These two vectors are parallel and have the same length, but obviously point in opposite directions.

16. (a) $\mathbf{x} = \left[\frac{5}{42}\sqrt{42}, -\frac{1}{42}\sqrt{42}, -\frac{2}{21}\sqrt{42}\right]^{\mathrm{T}}$, $\mathbf{y} = \left[\frac{1}{3}, -\frac{2}{3}, \frac{2}{3}\right]^{\mathrm{T}}$ (b) $H = \begin{bmatrix} \frac{100}{377}+\frac{1349}{15,834}\sqrt{42} & \frac{140}{377}-\frac{1429}{15,834}\sqrt{42} & \frac{90}{377}+\frac{362}{7917}\sqrt{42} \\ \frac{140}{377}-\frac{1429}{15,834}\sqrt{42} & \frac{196}{377}+\frac{563}{15,834}\sqrt{42} & \frac{126}{377}+\frac{356}{7917}\sqrt{42} \\ \frac{90}{377}+\frac{362}{7917}\sqrt{42} & \frac{126}{377}+\frac{356}{7917}\sqrt{42} & \frac{81}{377}-\frac{956}{7917}\sqrt{42} \end{bmatrix}$

(c) $H\mathbf{x} = \left[\frac{1}{3}, -\frac{2}{3}, \frac{2}{3}\right]^{\mathrm{T}} = \mathbf{y}$

20. (c) $\mathbf{u} = \left[\frac{a-\sqrt{a^2+b^2}}{\sqrt{2a^2-2a\sqrt{a^2+b^2}+2b^2}}, \frac{b}{\sqrt{2a^2-2a\sqrt{a^2+b^2}+2b^2}}\right]^{\mathrm{T}}$

22. $H = \begin{bmatrix} \frac{35}{37} & \frac{12}{37} \\ \frac{12}{37} & -\frac{35}{37} \end{bmatrix}$; $\mathbf{y} = \left[\frac{199}{37}, -\frac{10}{37}\right]^{\mathrm{T}}$

27. (a) $H = \begin{bmatrix} \frac{1}{3} & \frac{14}{15} & -\frac{2}{15} \\ \frac{14}{15} & -\frac{23}{75} & \frac{14}{75} \\ -\frac{2}{15} & \frac{14}{75} & \frac{73}{75} \end{bmatrix}$; $\mathbf{y} = \left[-\frac{13}{15}, -\frac{494}{75}, \frac{317}{75} \right]^{\mathrm{T}}$

Exercises 45.3, Page 1141

1. $H = \begin{bmatrix} -\frac{3}{5} & -\frac{4}{5} \\ -\frac{4}{5} & \frac{3}{5} \end{bmatrix}$; $HA = \begin{bmatrix} -5 & -2 \\ 0 & -1 \end{bmatrix}$

2. (a) $H_1 = \begin{bmatrix} -0.8581163303 & -0.1906925178 & 0.4767312946 \\ -0.1906925178 & 0.9804298387 & 0.04892540334 \\ 0.4767312946 & 0.04892540334 & 0.8776864917 \end{bmatrix}$; $H_2 = \begin{bmatrix} 1.0 & 0 & 0 \\ 0 & -0.9984499371 & 0.05565719242 \\ 0 & 0.05565719242 & 0.9984499371 \end{bmatrix}$

(b) $Q = \begin{bmatrix} -0.8581163303 & 0.2169304578 & 0.4653789209 \\ -0.1906925178 & -0.9761870602 & 0.1034175381 \\ 0.4767312946 & 0 & 0.8790490730 \end{bmatrix}$; $R = \begin{bmatrix} 10.48808848 & 4.195235392 & -0.9534625893 \\ 0 & -1.843908892 & -7.050239880 \\ 0 & 0 & -0.6205052271 \end{bmatrix}$;

$QR = \begin{bmatrix} -9 & -4 & -1 \\ -2 & 1 & 7 \\ 5 & 2 & -1 \end{bmatrix} = A$ **(c)** $q = \begin{bmatrix} -\frac{9}{110}\sqrt{110} & -\frac{2}{85}\sqrt{85} & -\frac{9}{374}\sqrt{374} \\ -\frac{1}{55}\sqrt{110} & \frac{9}{85}\sqrt{85} & -\frac{1}{187}\sqrt{374} \\ \frac{1}{22}\sqrt{110} & 0 & -\frac{1}{22}\sqrt{374} \end{bmatrix}$; $r = \begin{bmatrix} \sqrt{110} & \frac{2}{5}\sqrt{110} & -\frac{1}{11}\sqrt{110} \\ 0 & \frac{1}{5}\sqrt{85} & \frac{13}{17}\sqrt{85} \\ 0 & 0 & \frac{6}{187}\sqrt{374} \end{bmatrix}$

(d) $D = \begin{bmatrix} 1 & 0 & 0 \\ 0 & -1 & 0 \\ 0 & 0 & -1 \end{bmatrix}$

Exercises 45.4, Page 1150

1. No. A counter example is given by $A = \begin{bmatrix} 1 & 2 & 3 & 4 \\ 2 & 1 & 4 & 2 \\ 0 & 2 & 1 & 5 \\ 0 & 0 & 1 & 2 \end{bmatrix}$; $B = \begin{bmatrix} 2 & 3 & 9 & 1 \\ 8 & 1 & 3 & 1 \\ 0 & 1 & 4 & 1 \\ 0 & 0 & 1 & 2 \end{bmatrix}$; $AB = \begin{bmatrix} 18 & 8 & 31 & 14 \\ 12 & 11 & 39 & 11 \\ 16 & 3 & 15 & 13 \\ 0 & 1 & 6 & 5 \end{bmatrix}$

2. Let $G = G_{ij}$ denote an $n \times n$ Givens rotation matrix. Let $A = GG^{\mathrm{T}}$. Then observe that $A_{uv} = \sum_{k=1}^{n} G_{uk} G_{kv}^{\mathrm{T}} = \sum_{k=1}^{n} G_{uk} G_{vk} = \begin{cases} 1 & u = v \\ 0 & u \neq v \end{cases}$

3. (a) $G_{31} = \begin{bmatrix} -\frac{3}{10}\sqrt{10} & 0 & -\frac{1}{10}\sqrt{10} \\ 0 & 1 & 0 \\ \frac{1}{10}\sqrt{10} & 0 & -\frac{3}{10}\sqrt{10} \end{bmatrix}$, $\mathbf{u}_1 = [\sqrt{10}, -7, 0]^{\mathrm{T}}$; $G_{32} = \begin{bmatrix} 1 & 0 & 0 \\ 0 & -\frac{7}{10}\sqrt{2} & -\frac{1}{10}\sqrt{2} \\ 0 & \frac{1}{10}\sqrt{2} & -\frac{7}{10}\sqrt{2} \end{bmatrix}$, $\mathbf{u}_2 = [-3, 5\sqrt{2}, 0]^{\mathrm{T}}$;

$G_{21} = \begin{bmatrix} -\frac{3}{58}\sqrt{58} & -\frac{7}{58}\sqrt{58} & 0 \\ \frac{7}{58}\sqrt{58} & -\frac{3}{58}\sqrt{58} & 0 \\ 0 & 0 & 1 \end{bmatrix}$, $\mathbf{u}_3 = [\sqrt{58}, 0, -1]^{\mathrm{T}}$

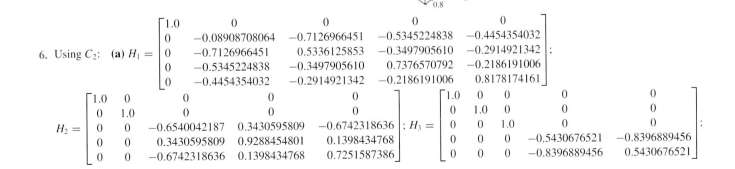

6. Using C_2:   **(a)** $H_1 = \begin{bmatrix} 1.0 & 0 & 0 & 0 & 0 \\ 0 & -0.08908708064 & -0.7126966451 & -0.5345224838 & -0.4454354032 \\ 0 & -0.7126966451 & 0.5336125853 & -0.3497905610 & -0.2914921342 \\ 0 & -0.5345224838 & -0.3497905610 & 0.7376570792 & -0.2186191006 \\ 0 & -0.4454354032 & -0.2914921342 & -0.2186191006 & 0.8178174161 \end{bmatrix}$;

$H_2 = \begin{bmatrix} 1.0 & 0 & 0 & 0 & 0 \\ 0 & 1.0 & 0 & 0 & 0 \\ 0 & 0 & -0.6540042187 & 0.3430595809 & -0.6742318636 \\ 0 & 0 & 0.3430595809 & 0.9288454801 & 0.1398434768 \\ 0 & 0 & -0.6742318636 & 0.1398434768 & 0.7251587386 \end{bmatrix}$; $H_3 = \begin{bmatrix} 1.0 & 0 & 0 & 0 & 0 \\ 0 & 1.0 & 0 & 0 & 0 \\ 0 & 0 & 1.0 & 0 & 0 \\ 0 & 0 & 0 & -0.5430676521 & -0.8396889456 \\ 0 & 0 & 0 & -0.8396889456 & 0.5430676521 \end{bmatrix}$;

$$H = \begin{bmatrix} 1.0 & 0 & 0 & 0 & 0 \\ 0 & -0.08908708064 & 0.5830602952 & 0.3667397166 & 0.7194485142 \\ 0 & -0.7126966451 & -0.2724506003 & 0.6198342985 & -0.1834110270 \\ 0 & -0.5345224838 & 0.6292247946 & -0.4418555928 & -0.3508924439 \\ 0 & -0.4454354032 & -0.4357608520 & -0.5348561096 & 0.5706388731 \end{bmatrix}$$

(b) $B = \begin{bmatrix} -8.0 & -13.89758458 & 3.768143904 & -1.716270822 & 4.439892886 \\ -11.22497216 & 2.047619048 & 3.303726311 & 0.9386230100 & 3.908137705 \\ 0 & 15.89766737 & -11.07725841 & 0.687631232 & -0.812570859 \\ 0 & 0 & -6.521095408 & -4.869458047 & -2.360178650 \\ 0 & 0 & 0 & -4.260727016 & 3.899097410 \end{bmatrix}$;

$\{\lambda\}_A = \{-20.92089615, -6.541861236, -6.182865448, 4.915959720, 10.72966312\}$;
$\{\lambda\}_B = \{-20.92089615, -6.541861193, -6.182865490, 4.915959720, 10.72966312\}$

7. Using C_2: **(a)** $G_{21} = \begin{bmatrix} -0.5803810002 & -0.8143450711 & 0 & 0 & 0 \\ 0.8143450711 & -0.5803810002 & 0 & 0 & 0 \\ 0 & 0 & 1 & 0 & 0 \\ 0 & 0 & 0 & 1 & 0 \\ 0 & 0 & 0 & 0 & 1 \end{bmatrix}$; $G_{32} = \begin{bmatrix} 1 & 0 & 0 & 0 & 0 \\ 0 & -0.6182739864 & 0.7859626438 & 0 & 0 \\ 0 & -0.7859626438 & -0.6182739864 & 0 & 0 \\ 0 & 0 & 0 & 1 & 0 \\ 0 & 0 & 0 & 0 & 1 \end{bmatrix}$;

$G_{43} = \begin{bmatrix} 1 & 0 & 0 & 0 & 0 \\ 0 & 1 & 0 & 0 & 0 \\ 0 & 0 & 0.6736496122 & -0.7390508777 & 0 \\ 0 & 0 & 0.7390508777 & 0.6736496122 & 0 \\ 0 & 0 & 0 & 0 & 1 \end{bmatrix}$; $G_{54} = \begin{bmatrix} 1 & 0 & 0 & 0 & 0 \\ 0 & 1 & 0 & 0 & 0 \\ 0 & 0 & 1 & 0 & 0 \\ 0 & 0 & 0 & -0.5009606455 & -0.8654700642 \\ 0 & 0 & 0 & 0.8654700642 & -0.5009606455 \end{bmatrix}$

(b) $R = \begin{bmatrix} 13.78404875 & 6.398425560 & -4.877332366 & 0.2317279545 & -5.759402151 \\ 0 & 20.22700124 & -9.418037056 & 1.741385160 & -1.471710248 \\ 0 & 0 & 80.823608230 & 4.340806276 & 1.369330974 \\ 0 & 0 & 0 & 4.923020672 & -2.371980719 \\ 0 & 0 & 0 & 0 & -3.685357569 \end{bmatrix}$

(c) $\{\lambda\}_A = \{-20.92089615, -6.541861236, -6.182865448, 4.915959720, 10.72966312\}$;
$\{\lambda\}_R = \{-3.685357569, 4.923020672, 8.823608230, 13.78404875, 20.22700124\}$, obvious from (b)

(d) $B_1 = \begin{bmatrix} -13.21052632 & -8.477533651 & -2.116886441 & 5.975698450 & 1.172974436 \\ -16.47175876 & -0.144079941 & 8.851206129 & -4.885853847 & 11.37867923 \\ 0 & 6.935026452 & -6.883109791 & -0.630230321 & -1.644611220 \\ 0 & 0 & -3.638362749 & 0.391496946 & 4.058506095 \\ 0 & 0 & 0 & 3.189566652 & 1.846219107 \end{bmatrix}$;

$\{\lambda\}_{B_1} = \{-20.92089616, -6.541861167, -6.182865514, 4.915959723, 10.72966311\}$ **(e)**

$G'_{21} = \begin{bmatrix} -0.6256510008 & -0.7801030862 & 0 & 0 & 0 \\ 0.7801030862 & -0.6256510008 & 0 & 0 & 0 \\ 0 & 0 & 1 & 0 & 0 \\ 0 & 0 & 0 & 1 & 0 \\ 0 & 0 & 0 & 0 & 1 \end{bmatrix}$; $G'_{32} = \begin{bmatrix} 1 & 0 & 0 & 0 & 0 \\ 0 & -0.6851487174 & 0.7284032090 & 0 & 0 \\ 0 & -0.7284032090 & -0.6851487174 & 0 & 0 \\ 0 & 0 & 0 & 1 & 0 \\ 0 & 0 & 0 & 0 & 1 \end{bmatrix}$;

$G'_{43} = \begin{bmatrix} 1 & 0 & 0 & 0 & 0 \\ 0 & 1 & 0 & 0 & 0 \\ 0 & 0 & 0.9392085973 & -0.3433470691 & 0 \\ 0 & 0 & 0.3433470691 & 0.9392085973 & 0 \\ 0 & 0 & 0 & 0 & 1 \end{bmatrix}$; $G'_{54} = \begin{bmatrix} 1 & 0 & 0 & 0 & 0 \\ 0 & 1 & 0 & 0 & 0 \\ 0 & 0 & 1 & 0 & 0 \\ 0 & 0 & 0 & -0.4053769815 & 0.9141496062 \\ 0 & 0 & 0 & -0.9141496062 & -0.4053769815 \end{bmatrix}$

(f) $R_1 = \begin{bmatrix} 21.11484885 & 5.416374620 & -5.580421098 & 0.072767949 & -9.610415414 \\ 0 & 9.520862024 & -0.088038509 & -5.747382267 & 3.052750692 \\ 0 & 0 & 10.59674910 & -5.009267707 & 3.909151580 \\ 0 & 0 & 0 & 3.489107942 & -0.643306112 \\ 0 & 0 & 0 & 0 & -6.005020624 \end{bmatrix}$

(g) $\{\lambda\}_{R_1} = \{-6.005020624, 3.489107942, 9.520862024, 10.59674910, 21.11484885\}$, obvious from

$$(f) \quad (h) \ B_2 = \begin{bmatrix} -17.43585687 & -13.02859631 & -5.384371677 & -8.018834123 & 5.624397677 \\ -7.427253848 & 4.017123083 & 6.105137465 & 4.366585851 & 2.316268777 \\ 0 & 7.718706049 & -5.099064865 & 6.491277307 & 4.994972848 \\ 0 & 0 & -1.197974986 & -1.916498469 & -2.734886931 \\ 0 & 0 & 0 & -5.489487239 & 2.434297134 \end{bmatrix};$$

$\{\lambda\}_{B_2} = \{-20.92089615, -6.541861127, -6.182865540, 4.915959719, 10.72966311\}$ **(i)** M will be an upper triangular matrix with the eigenvalues of A on the main diagonal.

8. (a) $H_1 = \begin{bmatrix} 1.0 & 0 & 0 & 0 & 0 \\ 0 & -0.3746343246 & 0.4682929058 & -0.7492686493 & -0.2809757435 \\ 0 & 0.4682929058 & 0.8404679400 & 0.2552512961 & 0.09571923603 \\ 0 & -0.7492686493 & 0.2552512961 & 0.5915979263 & -0.1531507776 \\ 0 & -0.2809757435 & 0.09571923603 & -0.1531507776 & 0.9425684584 \end{bmatrix};$

$H_2 = \begin{bmatrix} 1.0 & 0 & 0 & 0 & 0 \\ 0 & 1.0 & 0 & 0 & 0 \\ 0 & 0 & -0.4676446104 & -0.8154837466 & 0.3410202008 \\ 0 & 0 & -0.8154837466 & 0.5468836691 & 0.1894848583 \\ 0 & 0 & 0.3410202008 & 0.1894848583 & 0.9207609414 \end{bmatrix}; \ H_3 = \begin{bmatrix} 1.0 & 0 & 0 & 0 & 0 \\ 0 & 1.0 & 0 & 0 & 0 \\ 0 & 0 & 1.0 & 0 & 0 \\ 0 & 0 & 0 & -0.7356352068 & -0.6773779171 \\ 0 & 0 & 0 & -0.6773779171 & 0.7356352068 \end{bmatrix};$

$H = \begin{bmatrix} 1.0 & 0 & 0 & 0 & 0 \\ 0 & -0.3746343246 & 0.2962033473 & 0.7847706376 & 0.3950287916 \\ 0 & 0.4682929058 & -0.5685513926 & 0.1015533400 & 0.6686837726 \\ 0 & -0.7492686493 & -0.6540328952 & -0.1029066819 & -0.01573778747 \\ 0 & -0.2809757435 & 0.4015642701 & -0.6026874649 & 0.6297353320 \end{bmatrix}$

(b) $B = \begin{bmatrix} 8.0 & -10.67707825 & 0 & 0 & 0 \\ -10.67707825 & 0.2105263160 & 2.479649396 & 0 & 0 \\ 0 & 2.479649394 & -3.525513546 & -7.760053051 & 0 \\ 0 & 0 & -7.760053051 & 0.345112123 & 0.771445314 \\ 0 & 0 & 0 & 0.771445312 & -8.030124898 \end{bmatrix};$

$\{\lambda\}_A = \{-10.52836325, -8.037873696, -6.501073893, 6.462053498, 15.60525734\};$
$\{\lambda\}_B = \{-10.52836325, -8.037873698, -6.501073893, 6.462053497, 15.60525734\}$

Exercises 45.5, Page 1157

1. (a) We use A_1 and B_1: **(c)** $\mu_1 = -0.471245006$, $\mathbf{z}_1 = [-0.1394175539, -0.8954685321, 0.1615431331]^T$,
$\mu_2 = -1.819611306$, $\mathbf{z}_2 = [0.799311581, 0.05866334473, 0.4884726533]^T$,
$\mu_3 = 3.766266140$, $\mathbf{z}_3 = [-0.2408878726, 1.064822681, 0.9643300772]^T$; $A\mathbf{z}_k - \mu_k B\mathbf{z}_k = \mathbf{0}, k = 1, 2, 3$
(d) $\det(A - \lambda B) = 394 + 948\lambda + 180\lambda^2 - 122\lambda^3$, which has zeros: $\{\lambda\} = \{-1.819611301, -0.4712450048, 3.766266142\}$

2. (a) $S = \begin{bmatrix} -0.3761152742 & 0.06407780528 & 0.1008416593 \\ -0.02760397830 & 0.4115669599 & -0.4457612780 \\ -0.2298503236 & -0.07424695950 & -0.4036925725 \end{bmatrix}; \ \Lambda = \begin{bmatrix} -1.819611302 & 0 & 0 \\ 0 & -0.4712450048 & 0 \\ 0 & 0 & 3.766266146 \end{bmatrix}$

(b) $S^T A S = \begin{bmatrix} -1.819611301 & 0 & 0 \\ 0 & -0.4712450045 & 0 \\ 0 & 0 & 3.766266145 \end{bmatrix}; \ S^T B S = \begin{bmatrix} 0.9999999992 & 0 & 0 \\ 0 & 0.9999999995 & 0 \\ 0 & 0 & 1.000000001 \end{bmatrix}$

(c) $A\mathbf{x}_k - \lambda_k B\mathbf{x}_k = \mathbf{0}, k = 1, 2, 3$ **(d)** $\{\mu\} = \{-0.471245006, -1.819611306, 3.766266140\}$ **(e)** *Hint:* Normalize the first components and compare, or alternatively, normalize the vectors using the vector norm of your choice.

3. (a) $K = \begin{bmatrix} -0.04861051851 & 0.8303081063 & -0.5551805708 \\ -0.8410626694 & -0.3338386183 & -0.4256352451 \\ -0.5387491101 & 0.4462513033 & 0.7145692211 \end{bmatrix}; \ P = \begin{bmatrix} 0.6842058843 & 0 & 0 \\ 0 & 0.4255115464 & 0 \\ 0 & 0 & 0.3109724613 \end{bmatrix};$

$M = \begin{bmatrix} 2.574328238 & 0.7586308644 & -1.869330044 \\ 0.7586308638 & -1.632411766 & -0.02680883078 \\ -1.869330045 & -0.02680883118 & 0.5334933679 \end{bmatrix}; \ \{\lambda\} = \{-1.819611304, -0.471245006, 3.766266154\};$

$W = \begin{bmatrix} -0.2416399736 & -0.4520101111 & -0.8586599923 \\ 0.9533167718 & -0.2757279505 & -0.1231309468 \\ -0.1811001266 & -0.8483283308 & 0.4975357152 \end{bmatrix}; \ T = \begin{bmatrix} 0.3761152745 & 0.06407780548 & -0.1008416591 \\ 0.02760397864 & 0.4115669596 & 0.4457612779 \\ 0.2298503242 & -0.0742469599 & 0.4036925732 \end{bmatrix}$

(b) $T^{\mathsf{T}}AT = \begin{bmatrix} -1.819611304 & 0 & 0 \\ 0 & -0.4712450048 & 0 \\ 0 & 0 & 3.766266151 \end{bmatrix}$; $T^{\mathsf{T}}BT = \begin{bmatrix} 1.000000002 & 0 & 0 \\ 0 & 0.9999999996 & 0 \\ 0 & 0 & 1.000000002 \end{bmatrix}$

(c) $A\mathbf{y}_k - \lambda_k B\mathbf{y}_k = \mathbf{0}$, $k = 1, 2, 3$

4. Using A_1 and B_1: (a) $U = \begin{bmatrix} 0.08823529412 & 0.7647058824 \\ 0.7941176471 & -0.1176470588 \end{bmatrix}$; $D = \begin{bmatrix} -0.8007487068 & 0 \\ 0 & 0.7713369422 \end{bmatrix}$;

$P = \begin{bmatrix} -0.6522414459 & 0.7457784602 \\ 0.7582421209 & 0.6661940321 \end{bmatrix}$

(b) $\mathbf{R}(t) = [0.5000000000c_1 e^{0.8948456329t} + 0.5000000001c_1 e^{-0.8948456329t} + 0.5587555907c_2 e^{0.8948456329t} - 0.5587555907c_2 e^{-0.8948456329t}$, $c_3 \cos(0.8782579019t) + 1.138617709c_4 \sin(0.8782579019t)]^{\mathsf{T}}$; $\mathbf{x}(t) = [2.004650001 e^{0.8948456329t} - 0.5972753995 e^{-0.8948456329t} - 0.4073746010 \cos(0.8782579019t) - 0.3738327662 \sin(0.8782579019t), -2.330440786 e^{0.8948456329t} + 0.6943431278 e^{-0.8948456329t} - 0.3639023417 \cos(0.8782579019t) - 0.3339398644 \sin(0.8782579019t)]^{\mathsf{T}}$.

Bibliography

[1] Milton Abramowitz and Irene Stegun, eds., *Handbook of Mathematical Functions*, Dover Publ., New York, 1965.

[2] Forman S. Acton, *Numerical Methods that Work*, Harper & Row, New York, 1970.

[3] Tom M. Apostol, *Mathematical Analysis*, 2nd ed., Addison-Wesley Publ. Co., Reading, MA, 1974.

[4] Kendall E. Atkinson, *An Introduction to Numerical Analysis*, 2nd ed., John Wiley & Sons, New York, 1989.

[5] George A. Baker, Jr., *Essentials of Pade Approximants*, Academic Press, New York, 1975.

[6] Vernon D. Barger and Martin G. Olsson, *Classical Mechanics, A Modern Perspective*, 2nd ed., McGraw-Hill, New York, 1995.

[7] Eric B. Becker, Graham F. Carey, and J. Tinsley Odin, *Finite Elements, An Introduction*, Prentice-Hall, Englewood Cliffs, NJ, 1981.

[8] Richard Bellman, *Perturbation Techniques in Mathematics, Physics, and Engineering*, Holt, Rinehart and Winston, New York, 1966.

[9] Edward K. Blum, *Numerical Analysis and Computation: Theory and Practice*, Addison-Wesley Publ. Co., Reading, MA, 1972.

[10] Oskar Bolza, *Lectures on the Calculus of Variations*, G. E. Stetchert & Co., New York, 1931.

[11] Robert L. Borrelli and Courtney S. Coleman, *Differential Equations, A Modeling Approach*, Prentice-Hall, Englewood Cliffs, NJ, 1987.

[12] William E. Boyce and Richard C. DiPrima, *Elementary Differential Equations and Boundary Value Problems*, 4th ed., John Wiley & Sons, New York, 1986.

[13] Fred Brauer and John A. Nohel, *Ordinary Differential Equations: A First Course*, 2nd ed., W. A. Benjamin, New York, 1973.

[14] Fred Brauer and John A. Nohel, *Introduction to Differential Equations with Applications*, Harper & Row, New York, 1986.

[15] U. Brechtken-Manderscheid, *Introduction to the Calculus of Variations* (transl. by P. G. Engstrom), Chapman & Hall, London, 1991.

[16] Richard L. Burden and J. Douglas Faires, *Numerical Analysis*, 6th ed., Brooks/Cole, Pacific Grove, CA, 1997.

[17] Alan W. Bush, *Perturbation Methods for Engineers and Scientists*, CRC Press, Boca Raton, FL, 1992.

[18] Brice Carnahan, H. A. Luther, and James O. Wilkes, *Applied Numerical Methods*, John Wiley & Sons, New York, 1969.

[19] Steven C. Chapra and Raymond P. Canale, *Numerical Methods for Engineers*, 2nd ed., McGraw-Hill, New York, 1988.

[20] Tai L. Chow, *Classical Mechanics,* John Wiley & Sons, New York, 1995.

[21] Ruel V. Churchill, *Operational Mathematics,* 3rd ed., McGraw-Hill, New York, 1972.

[22] Harry F. Davis and Arthur David Snider, *Introduction to Vector Analysis,* 5th ed., Wm. C. Brown Publishers, Dubuque, Iowa, 1988.

[23] John W. Dettman, *Applied Complex Variables,* The Macmillan Company, New York, 1965.

[24] George C. Donovan, Arnold R. Miller, and Timothy J. Moreland, "Pathological Functions for Newton's Method," *The American Mathematical Monthly,* Vol. 100, No. 1, January, 1993.

[25] Henry E. Duckworth, *Electricity and Magnetism,* Holt, Rinehart and Winston, New York, 1960.

[26] Dominic G. B. Edelen, *Nonlocal Variations and Local Invariance of Fields,* American Elsevier Publ. Co., New York, 1969.

[27] Dominic G. B. Edelen and Anastasios D. Kydoniefs, *An Introduction to Linear Algebra for Science and Engineering,* 2nd ed., American Elsevier Publ. Co., New York, 1976.

[28] George M. Ewing, *Calculus of Variations with Applications,* Dover Publ., New York, 1985.

[29] Wade Ellis, Eugene Johnson, Ed Lodi, and Dan Schwalbe, *Maple V Flight Manual,* Brooks/Cole, Pacific Grove, CA, 1992.

[30] Marvin J. Forray, *Variational Calculus in Science and Engineering,* McGraw-Hill, New York, 1968.

[31] Gene F. Franklin, J. David Powell, and Abbas Emami-Naeini, *Feedback Control of Dynamic Systems,* Addison-Wesley Publ. Co., Reading, MA, 1986.

[32] Laurene V. Fausett, *Applied Numerical Analysis Using Matlab,* Prentice-Hall, Englewood Cliffs, NJ, 1999.

[33] Charles Fox, *An Introduction to the Calculus of Variations,* Dover Publ., New York, 1987.

[34] B. A. Fuchs and B. V. Shabat, *Functions of a Complex Variable and Some of Their Applications,* Volume 1, Addison-Wesley Publ. Co., Reading, MA, 1964.

[35] Watson Fulks, *Advanced Calculus,* John Wiley & Sons, New York, 1961.

[36] Keith O. Geddes, Block Structure in the Chebyshev-Padé Table, *SIAM J. Numer. Anal.,* Vol. 18, No. 5, October, 1981.

[37] I. M. Gelfand and S. V. Fomin, *Calculus of Variations* (transl. by Richard A. Silverman), Prentice-Hall, Englewood Cliffs, NJ, 1963.

[38] Curtis F. Gerald and Patrick O. Wheatley, *Applied Numerical Analysis,* 5th ed., Addison-Wesley Publ. Co., Reading, MA, 1994.

[39] Herbert Goldstein, *Classical Mechanics,* Addison-Wesley Publ. Co., Reading, MA, 1950.

[40] Martin Golubitsky and Michael Dellnitz, *Linear Algebra and Differential Equations Using Matlab,* Brooks/Cole, Pacific Grove, 1999.

[41] F. B. Hildebrand, *Introduction to Numerical Analysis,* 2nd ed., Dover Publ., New York, 1987.

[42] Joe D. Hoffman, *Numerical Methods for Engineers and Scientists,* McGraw-Hill, New York, 1992.

[43] Dio L. Holl, Clair G. Maple, and Bernard Vinograde, *Introduction to the Laplace Transform,* Appleton-Century-Crofts, Inc., New York, 1959.

[44] Roger A. Horn and Charles R. Johnson, *Matrix Analysis,* Cambridge Univ. Press, London, 1985.

[45] Ian Huntley and R. M. Johnson, *Linear and Nonlinear Differential Equations,* Halsted Press: a division of John Wiley & Sons, New York, 1983.

[46] Harold Jeffreys, *Asymptotic Approximations,* Oxford Univ. Press, London, 1962.

[47] Claes Johnson, *Numerical Solutions of Partial Differential Equations by the Finite Element Method,* Cambridge Univ. Press, London, 1987.

[48] Lee W. Johnson and R. Dean Riess, *Numerical Analysis,* Addison-Wesley Publ. Co., Reading, MA, 1977.

[49] David Kincaid and Ward Cheney, *Numerical Analysis: Mathematics of Scientific Computing*, 2nd ed., Brooks/Cole, Pacific Grove, CA, 1996.

[50] H. Kober, *Dictionary of Conformal Representations*, Dover Publ., New York, 1957.

[51] Erwin Kreyszig, *Advanced Engineering Mathematics*, 7th ed., John Wiley & Sons, New York, 1993.

[52] Steven J. Leon, *Linear Algebra with Applications*, 5th ed., Prentice-Hall, Englewood Cliffs, NJ, 1998.

[53] Norman Levinson and Raymond M. Redheffer, *Complex Variables*, Holden-Day, San Francisco, CA, 1970.

[54] Jonathan Lewin and Myrtle Lewin, *An Introduction to Mathematical Analysis*, Random House, New York, 1988.

[55] J. David Logan, *Applied Mathematics, A Contemporary Approach*, John Wiley & Sons, New York, 1987.

[56] Robert J. Lopez, *Maple via Calculus: A Tutorial Approach*, Birkhauser, Basel, 1994.

[57] Robert J. Lopez, A Separable Differential Equation: New Insights, *MapleTech*, Vol. 1, No. 1, 1994.

[58] Robert J. Lopez, Tips for Maple Instructors, *MapleTech*, Vol. 4, No. 3, 1997.

[59] R. J. Lopez, T. R. Chari, E. Moore, G. R. Peters, and A. Zielinski, Hydrodynamic Effects on Iceberg Gouging, *Cold Regions Science and Technology*, Vol. 4, 1981, pp. 55–61.

[60] Melvin J. Marin and Robert J. Lopez, *Numerical Analysis: A Practical Approach*, 3rd ed., Wadsworth, Belmont, CA, 1991.

[61] Jerrold E. Marsden and Michael J. Hoffman, *Basic Complex Analysis*, 2nd ed., W. H. Freeman and Co., San Francisco, CA, 1987.

[62] Ronald Elbert Mickens, *An Introduction to Nonlinear Oscillations*, Cambridge Univ. Press, London, 1981.

[63] Michael B. Monagan and Robert J. Lopez, Tips for Maple Users and Programmers, *MapleTech*, Vol. 3, No. 3, 1996.

[64] Ben Noble and James W. Daniel, *Applied Linear Algebra*, 3rd ed., Prentice-Hall, Englewood Cliffs, NJ, 1988.

[65] John M. H. Olmsted, *Advanced Calculus*, Prentice-Hall, Englewood Cliffs, NJ, 1961.

[66] Peter V. O'Neil, *Advanced Engineering Mathematics*, 4th ed., PWS-Kent Publishing Co., Boston, 1995.

[67] James M. Ortega, *Numerical Analysis, A Second Course*, Academic Press, New York, 1992.

[68] Richard Pavelle, ed., *Applications of Computer Algebra*, Kluwer Academic Publ., New York, 1985.

[69] Louis L. Pennisi, *Elements of Complex Variables*, Holt, Rinehart and Winston, New York, 1963.

[70] Charles L. Phillips and Royce D. Harbor, *Basic Feedback Control Systems*, alt. 2nd ed., Prentice-Hall, Englewood Cliffs, NJ, 1991.

[71] Enid R. Pinch, *Optimal Control and the Calculus of Variations*, Oxford Univ. Press, London, 1993.

[72] M. J. D. Powell, *Approximation Theory and Methods*, Cambridge Univ. Press, London, 1981.

[73] P. M. Prenter, *Splines and Variational Methods*, John Wiley & Sons, New York, 1975.

[74] Earl D. Railville, *Intermediate Differential Equations*, The Macmillan Company, New York, 1964.

[75] A. Ralston and P. Rabinowitz, *A First Course in Numerical Analysis*, 2nd ed., McGraw-Hill, New York, 1978.

[76] J. N. Reddy, *Energy and Variational Methods in Applied Mechanics*, John Wiley & Sons, New York, 1984.

[77] Hartley Rogers, Jr., *Multivariable Calculus with Vectors,* Prentice-Hall, Englewood Cliffs, NJ, 1999.

[78] Walter Rudin, *Principles of Mathematical Analysis,* 3rd ed., McGraw-Hill, New York, 1976.

[79] Hans Sagan, *Boundary and Eigenvalue Problems in Mathematical Physics,* John Wiley & Sons, New York, 1961.

[80] David A. Sanchez, Richard C. Allen, Jr., and Walter T. Kyner, *Differential Equations,* 2nd ed., Addison-Wesley Publ. Co., Reading, MA, 1988.

[81] Manuel Schwartz, Simon Green, and W. A. Rutledge, *Vector Analysis with Applications to Geometry and Physics,* Harper & Brothers, 1960.

[82] George F. Simmons, *Differential Equations with Applications and Historical Notes,* 2nd ed., McGraw-Hill, New York, 1991.

[83] Ian H. Sneddon, *The Use of Integral Transforms,* McGraw-Hill, New York, 1972.

[84] Murray R. Spiegel, *Schaum's Outline of Theory and Problems of Advanced Calculus,* McGraw-Hill, New York, 1963.

[85] Steven Schonefeld, *Numerical Analysis via Derive,* MathWare, 1994.

[86] Richard V. Southwell, *Relaxation Methods in Engineering Science,* Oxford Univ. Press, London, 1940.

[87] Ivar Stackgold, *Boundary Value Problems of Mathematical Physics, Volume I,* The Macmillan Company, New York, 1967.

[88] Gilbert Strang, *Linear Algebra and Its Applications,* 3rd ed., Harcourt Brace Jovanovich, San Diego, 1988.

[89] Gilbert Strang, *Introduction to Linear Algebra,* 2nd ed., Wellesley-Cambridge Press, Wellesley, MA, 1998.

[90] John C. Strikwerda, *Finite Difference Schemes and Partial Differential Equations,* Wadsworth & Brooks/Cole Advanced Books & Software, Pacific Grove, CA, 1989.

[91] Angus E. Taylor and W. Robert Mann, *Advanced Calculus,* 2nd ed., Xerox College Publishing, Lexington, MA, 1972.

[92] George B. Thomas, Jr., and Ross L. Finney, *Calculus and Analytic Geometry,* 8th ed., Addison-Wesley Publ. Co., Reading, MA, 1992.

[93] E. C. Titchmarsh, *The Theory of Functions,* 2nd ed., Oxford Univ. Press, London, 1939.

[94] Richard S. Varga, *Matrix Iterative Analysis,* Prentice-Hall, Englewood Cliffs, NJ, 1962.

[95] Paul Waltman, *A Second Course in Elementary Differential Equations,* Academic Press, New York, 1986.

[96] Robert Weinstock, *Calculus of Variations with Applications to Physics and Engineering,* McGraw-Hill, New York, 1952.

[97] David V. Widder, *Advanced Calculus,* 2nd ed., Dover Publ., New York, 1989.

[98] J. H. Wilkinson, The Evaluation of the Zeros of Ill-conditioned Polynomials, Part 1, *Numerische Mathematik,* Vol. 1, 1959, pp. 150–166.

[99] A. David Wunsch, *Complex Variables with Applications,* 2nd ed., Addison-Wesley Publ. Co., Reading, MA, 1994.

[100] Dennis G. Zill and Michael R. Cullen, *Advanced Engineering Mathematics,* PWS-Kent Publishing Co., Boston, 1992.

Index